TODAY'S TECHNICIAN™

CLASSROOM MANUAL

For Automotive Suspension & Steering Systems

SEVENTH EDITION

TODAY'S TECHNICIAN™

CLASSROOM MANUAL

For Automotive Suspension & Steering Systems

SEVENTH EDITION

Mark Schnubel

Australia • Brazil • Canada • Mexico • Singapore • United Kingdom • United States

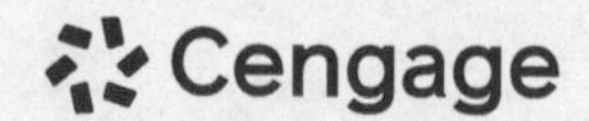

***Today's Technician: Automotive Suspension and Steering Systems*, Seventh Edition**
Mark Schnubel

SVP, GM Skills & Global Product Management: Jonathan Lau

Product Director: Matthew Seeley

Senior Product Manager: Katie McGuire

Product Assistant: Kimberly Klotz

Executive Director, Content Design: Marah Bellegarde

Learning Design Director: Juliet Steiner

Learning Designer: Mary Clyne

Vice President, Strategic Marketing Services: Jennifer Ann Baker

Marketing Director: Shawn Chamberland

Marketing Manager: Andrew Ouimet

Director, Content Delivery: Wendy Troeger

Manager, Content Deliver: Alexis Ferraro

Senior Content Manager: Meaghan Tomaso

Senior Digital Delivery Lead: Amanda Ryan

Senior Designer: Angela Sheehan

Service Provider/Compositor: SPi Global

Cover image(s): Photographicss/ShutterStock.com

Library of Congress Control Number: 2018960546

Book Only ISBN: 978-1-337-56734-3
Package ISBN: 978-1-337-56733-6

Cengage
200 Pier 4 Boulevard
Boston, MA 02210
USA

Cengage is a leading provider of customized learning solutions with employees residing in nearly 40 different countries and sales in more than 125 countries around the world. Find your local representative at: **www.cengage.com.**

To learn more about Cengage platforms and services, register or access your online learning solution, or purchase materials for your course, visit **www.cengage.com.**

Printed at CLDPC, USA, 05-24

CONTENTS

PREFACE

Thanks to the support the *Today's Technician* series has received from those who teach automotive technology, Cengage Learning, the leader in automotive-related textbooks, is able to live up to its promise to regularly provide new editions of texts of this series. We have listened and responded to our critics and our fans and present this new updated and revised seventh edition. By revising this series on a regular basis, we can respond to changes in the industry, changes in technology, changes in the accreditation process, and to the ever-changing needs of those who teach automotive technology.

The *Today's Technician* series features textbooks that cover all mechanical and electrical systems of automobiles and light trucks. Principally, the individual titles correspond to the areas for National Institute for Automotive Service Excellence (ASE) certification.

All titles in the *Today's Technician* series include remedial skills and theories common to all of the certification areas and advanced or specific subject areas that reflect the latest technological trends.

Today's Technician: Automotive Suspension & Steering Systems, 7e is designed to give students a chance to develop the same skills and gain the same knowledge that today's successful technician has. This edition also reflects the changes in the guidelines established by the ASE Education Foundation in 2017.

The purpose of the ASE Education Foundation is to evaluate technician training programs against standards developed by the automotive industry and recommend qualifying programs for accreditation by ASE. Programs can earn ASE accreditation upon the recommendation of the ASE Education Foundation. ASE Education Foundation's national standards reflect the skills that students must master. ASE Education Foundation accreditation ensures that the training programs meet or exceed industry-recognized, uniform standards of excellence.

The technician of today and for the future must know the underlying theory of all automotive systems and be able to service and maintain those systems. Dividing the material into two volumes, a Classroom Manual and a Shop Manual, provides the reader with the information needed to begin a successful career as an automotive technician without interrupting the learning process by mixing cognitive and performance learning objectives into one volume.

The design of Cengage Learning's *Today's Technician* series was based on features that are known to promote improved student learning. The design was further enhanced by a careful study of survey results, in which the respondents were asked to value particular features. Some of these features can be found in other textbooks, while others are unique to this series.

Each Classroom Manual contains the principles of operation for each system and subsystem. The Classroom Manual also discusses design variations in key components used by the different vehicle manufacturers, and considers emerging technologies that will be standard or optional features in the near future. This volume is organized to build upon basic facts and theories. Its primary objective is to help the reader gain an understanding of how each system and subsystem operates. This understanding is necessary to diagnose the complex automobiles of today and tomorrow. Although the basics contained in the Classroom Manual provide the knowledge needed for diagnostics, diagnostic procedures appear only in the Shop Manual. An understanding of the underlying theories is also a requirement for competence in the skill areas covered in the Shop Manual.

A spiral-bound Shop Manual delivers hands-on learning experiences with step-by-step instructions for diagnostic and repair procedures. Photo Sequences are used to illustrate some of the common service procedures. Other common procedures are listed and are accompanied with fine-line drawings and photos that let the reader visualize and conceptualize the finest details of the procedure. This volume explains the reasons for performing the procedures, as well as the circumstances when each particular service is appropriate.

The two volumes are designed to be used together and are arranged in corresponding chapters. Not only are the chapters in the volumes linked together, the contents of the chapters are also linked. The linked content is indicated by marginal callouts that refer the reader to the chapter and page where the same topic is addressed in the companion volume. This valuable feature saves users the time and trouble of searching the index or table of contents to locate supporting information in the other volume. Instructors will find this feature especially helpful when planning the presentation of material and when making reading assignments.

Both volumes contain clear and thoughtfully selected illustrations, many of which are original drawings or photos specially prepared for inclusion in this series. This means that the art is a vital part of each textbook and not merely inserted to increase the number of illustrations.

The page design of this series uses available margin space to deliver helpful information efficiently without interrupting the pedagogical lesson material. This information includes examples of concepts just introduced in the text, explanations or definitions of terms that are not defined in the text, examples of common trade jargon used to describe a part or an operation, and unique applications of the system or service described in the text. Many textbooks also include this information but insert it in the main body of text; this tends to interrupt the reader's thought process. By placing this information to the side of the main text, students can read through the text uninterrupted and refer to the additional information when it is best for them.

Jack Erjavec
Series Editor

HIGHLIGHTS OF THIS EDITION—CLASSROOM MANUAL

The text was updated to include the latest technology in suspension and steering systems. Updated and expanded coverage of basic electrical theory and hybrid electric vehicle safety, hybrid vehicle steering systems, driver assist systems, active steering systems, rear active steering (RAS), CAN bus networking, computer-controlled suspension systems, and adaptive cruise control systems. Expanded coverage of tires and wheels including tire plus sizing and rim offset considerations as well as expanded alignment theory coverage and sequencing of information. The text also includes the latest technology in driver assist systems including but not limited to vehicle stability control systems, traction control systems, active roll control, lane departure warning (LDW) systems, active cruise control, collision mitigation systems, telematics, and tire pressure monitoring systems (TPMS).

PREFACE

The first chapter explains the design and purpose of basic suspension and steering systems. This chapter provides students with the necessary basic understanding of suspension and steering systems. The other chapters in the book allow the student to build upon his or her understanding of these basic systems.

The second chapter explains all the basic theories required to understand the latest suspension and steering systems described in the other chapters. Students must understand these basic theories to comprehend the complex systems explained later in the text.

The other chapters in the book are designed to be stand alone to allow reordering of topics covered to fit individual program needs and explain all the current model systems and components such as wheel bearings, tires and wheels, shock absorbers and struts, front and rear suspension systems, computer-controlled suspension systems, steering columns and linkages, power steering pumps, steering gears and systems, updated information on four-wheel steering systems currently in production, frames, and four-wheel alignment. All of the art pieces have been replaced with color photos and color diagrams throughout the text to improve visual concepts of suspension and steering systems and components.

HIGHLIGHTS OF THIS EDITION—SHOP MANUAL

The chapters in the Shop Manual have been updated to explain the diagnostic and service procedures for the latest systems and components described in the Classroom Manual. Diagnostics is a very important part of an automotive technician's job. Therefore, proper diagnostic procedures are emphasized in the Shop Manual.

All Photo Sequences are in color with several new sequences added and existing sequences updated. These Photo Sequences illustrate the correct diagnostic or service procedure for a specific system or component. These Photo Sequences allow the students to visualize the diagnostic or service procedure. Visualization of these diagnostic and service procedures helps students to remember the procedures, and perform them more accurately and efficiently. The text covers the information required for the ASE test in Suspension and Steering Systems. New and updated Job Sheets have been created to meet current ASE Education Foundation tasks.

Chapter 1 explains the necessary safety precautions and procedures in an automotive repair shop. General shop safety and the required shop safety equipment are explained in the text. The text describes safety procedures when operating vehicles and various types of automotive service equipment. Correct procedures for handling hazardous waste materials are detailed in the text.

Chapter 2 describes suspension and steering diagnostic and service equipment and the use of service manuals. This chapter also explains employer and employee obligations and ASE certification requirements.

The other chapters in the text have been updated to explain the diagnostic and service procedures for the latest suspension and steering systems explained in the Classroom Manual. New job sheets related to the new systems and components have been added in the text. All of the art pieces have been replaced with color photos and color diagrams throughout the text to improve the student's visualization of diagnostic and service procedures.

Mark Schnubel

CLASSROOM MANUAL

Features of the Classroom Manual include the following:

Cognitive Objectives

These objectives outline the chapter's contents and identify what students should know and be able to do upon completion of the chapter. Each topic is divided into small units to promote easier understanding and learning.

Terms To Know List

A list of key terms appears immediately after the Objectives. Students will see these terms discussed in the chapter. Definitions can also be found in the Glossary at the end of the manual.

Cross-References to the Shop Manual

References to the appropriate page in the Shop Manual appear whenever necessary. Although the chapters of the two manuals are synchronized, material covered in other chapters of the Shop Manual may be fundamental to the topic discussed in the Classroom Manual.

Margin Notes

The most important terms to know are highlighted and defined in the margin. Common trade jargon also appears in the margin and gives some of the common terms used for components. This helps students understand and speak the language of the trade, especially when conversing with an experienced technician.

CHAPTER 1

SUSPENSION AND STEERING SYSTEMS

Upon completion and review of this chapter, you should be able to understand and describe:

- How strength and rigidity are designed into a unitized body.
- The advantages of reduced vehicle weight.
- The design of a short-and-long arm (SLA) front suspension system.
- How limited independent rear wheel movement is provided in a semi-independent rear suspension system.
- The advantage of an independent rear suspension system.
- The purposes of vehicle tires.
- The terms positive and negative offset as they relate to vehicle wheel rims.
- Three different loads that are applied to wheel bearings.
- The purposes of shock absorbers.
- The difference between a shock absorber and a strut.
- Two different types of computer-controlled shock absorbers.
- The advantages of computer-controlled suspension systems.
- Two types of steering linkages.
- How the rack is moved in a rack and pinion steering gear.
- How a power steering pump develops hydraulic pressure.
- The result of incorrect rear wheel toe.
- The front wheel caster.
- The results of excessive negative camber.

Terms to Know

Angular bearing loads	Negative offset	Thrust bearing loads
Carbon dioxide (CO_2)	Positive camber	Thrust line
Greenhouse gas	Positive caster	Toe-out
Jounce travel	Positive offset	Wheel alignment
Negative camber	Radial bearing load	Wheel offset
Negative caster	Rebound travel	Wheel shimmy

INTRODUCTION

The suspension system must provide proper steering control and ride quality. Performing these functions is extremely important to maintaining vehicle safety and customer satisfaction. For example, if the suspension system allows excessive vertical wheel oscillations, the driver may lose control of the steering when driving on an irregular road surface. This loss of steering control can result in vehicle collisions and personal inj... vertical wheel oscillations transfer undesirable vibration... passenger compartment, which causes custom...

The suspension system and ... provide normal ...

rotate or move the component. Bearings are precision-machined assemblies, which provide smooth operation and long life. When bearings are properly installed and maintained, bearing failure is rare.

AUTHOR'S NOTE Only 21 percent of the power developed by a vehicle engine actually gets to the drive wheels. Much of the engine energy is lost in overcoming friction and wind resistance. The U.S. Department of Energy (DOE) is working with a major bearing manufacturer to explore the use and advantage of roller bearings to reduce friction in vehicle engines. The DOE estimates that if all vehicle engines used roller-bearing, low-friction technology, 100 million barrels of oil could be saved in a year.

Shop Manual Chapter 4, page 148

BEARING LOADS

A thrust bearing load may be referred to as an axial load.

When a bearing load is applied in a vertical direction on a horizontal shaft, it is called a radial bearing load. If the vehicle weight is applied straight downward on a bearing, this weight is a radial load on the bearing. A thrust bearing load is applied in a horizontal direction (**Figure 4-1**). For example, while a vehicle is turning a corner, horizontal force is applied to the front wheel bearings. When an angular bearing load is applied, the angle of the applied load is somewhere between the horizontal and vertical positions. This can also be referred to as a combination load.

BALL BEARINGS

Ball bearings have round steel balls between the inner and outer races.

Front and rear wheel bearings may be ball bearings or roller bearings. Either type of bearing contains these basic parts:

1. Inner race, or cone
2. Rolling elements, balls, or rollers
3. Separator, also called a cage or retainer
4. Outer race, or cup

The inner race supports the inner side of the rolling elements in a bearing.

The rolling elements are the precision-machined balls or rollers between the inner and outer bearing races.

The inner race is an accurately machined component. The inner surface of the race is mounted on the shaft with a precision fit. The rolling elements are mounted on a very

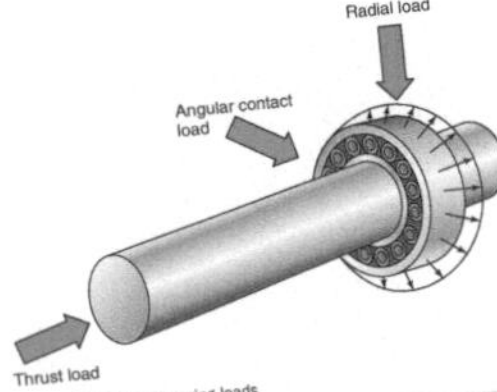

Figure 4-1 Types of bearing loads.

Author's Note

This feature includes simple explanations, stories, or examples of complex topics. These are included to help students understand difficult concepts.

114 Chapter 4

AUTHOR'S NOTE An understanding of wheel bearings and related service procedures is critical to maintain vehicle safety! If the technician's knowledge of wheel bearings and appropriate service procedures is inadequate, bearing failure may occur. Wheel bearing failure may cause a wheel to fly off a vehicle with disastrous results!

Shop Manual
Chapter 4, page 163

REAR-AXLE BEARINGS

On many rear-wheel-drive cars, the rear axles are supported by roller bearings mounted near the outer ends of the axle housing. The outer bearing race is pressed into the housing, and a machined surface on the axle contacts the in[illegible] roller surface. A seal is mounted in the axle housing on the outboard side [illegible] bearing is usually not seal[illegible] provides axle bea[illegible] of the [illegible] g (**Figure 4-22**). This type of axle [illegible] differential and rear axle housing [illegible] bricant leaks from the outer ends

[illegible] icles have sealed roller bearings [illegible] d on both sides, and an adapter [illegible] ring (**Figure 4-23**). The outer [illegible] ht press fit, and a seal is posi- [illegible] adapter ring. A retainer plate [illegible]. This retainer plate is bolted

[illegible] aring life. Bearing lubri- [illegible] urfaces from dirt and [illegible] anufacturing process, [illegible] *with grease.* [illegible] epacking at intervals [illegible] *ase specified by the*

Rear-axle bearings support the [illegible]

etainer ring
Oil seal collar
Oil seal

Steering Columns and Steering Linkage Mechanisms 223

Pitman Arm

The **pitman arm** connects the steering gear to the center link. This arm also supports the left side of the center link. Motion from the steering wheel and steering gear is transmitted to the pitman arm, and this arm transfers the movement to the steering linkage. This pitman arm movement forces the steering linkage to move to the right or left, and the linkage moves the front wheels in the desired direction. The pitman arm also positions the center link at the proper height to maintain the parallel relationship between the tie rods and the lower control arms.

Wear-type pitman arms have ball sockets and studs at the outer end, and this stud fits into the center link opening (**Figure 8-34**). The ball stud and socket are subject to wear, and pitman arm replacement is necessary if the ball stud is loose in the pitman arm. A non-wear pitman arm has a tapered opening in the outer end. A ball stud in the center link fits into this opening. The non-wear pitman arm only needs replacing if the arm is damaged or bent in a collision. The opening in the inner end of both types of pitman arms has serrations that fit over matching serrations on the steering gear shaft. A nut and lock washer retain the pitman arm to the steering gear shaft.

Idler Arm

An idler arm support is bolted to the frame or chassis on the opposite end of the center link from the pitman arm. The **idler arm** is connected from the support bracket to the center link. Two bolts retain the idler arm bracket to the frame or chassis. In some idler arms, a ball stud on the outer end of the arm fits into a tapered opening in the center link (**Figure 8-35**), whereas in other idler arms, a ball stud in the center link fits into a tapered opening in the idler arm.

A BIT OF HISTORY
For many years, rear-wheel-drive vehicles were equipped with parallelogram steering linkages. The smaller, lighter front-wheel-drive cars introduced in large numbers in the late 1970s and 1980s required a more compact, lighter steering gear. The rack and pinion steering gear is used almost exclusively on front-wheel-drive cars because of its reduced weight and space requirements.

A Bit of History

This feature gives the student a sense of the evolution of the automobile. This feature not only contains nice-to-know information, but also should spark some interest in the subject matter.

116 Chapter 4

Caution
If a bearing is operated without proper lubrication, bearing life will be very short.

lubricant is used for hypoid gears under normal conditions. The GL-5 lubricant is used in heavy-duty hypoid gears. Always use the vehicle manufacturer's specified differential gear oil.

The differential should be filled until the lubricant is level with the bottom of the filler plug opening in the differential housing. If the differential is overfilled, the bearings and seals may have excessive lubricant. Under this condition, the lubricant may leak past the seal. When the lubricant level is low in the differential, the lubricant may not be available in the axle housings. When this condition exists, the bearings do not receive enough lubrication and bearing life is shortened.

SUMMARY

- A bearing reduces friction, carries a load, and guides certain components such as pivots, shafts, and wheels.
- Radial bearing loads are applied in a vertical direction.
- Thrust bearing loads are applied in a horizontal direction.
- Angular bearing loads are applied at an angle between the vertical and horizontal.
- The inner bearing race is positioned at the center of the bearing and supports the rolling elements.
- The rolling elements in a bearing are positioned between the inner and outer races.
- The bearing separator keeps the rolling elements evenly spaced.
- The outer bearing race forms the outer ring on a bearing.
- A cylindrical ball bearing is designed primarily to withstand radial loads, but these bearings can handle a considerable thrust load.
- A snap ring can be mounted in a groove in the outer bearing race, and the snap ring retains the bearing in the housing.
- A bearing shield prevents dirt from entering the bearing, but it is not designed to keep lubricant in the bearing.
- Bearing seals keep lubricant in the bearing and prevent dirt from entering the bearing.
- Roller bearings are designed primarily to carry radial loads, but they can handle some thrust loads.
- Tapered roller bearings have excellent radial, thrust, and angular load-carrying capabilities.
- Needle roller bearings are very compact and are designed to carry radial loads. They will not carry thrust loads.
- Springless seals are used for wheel bearing seals in some wheel hubs.
- The garter spring provides additional force on the seal lip to compensate for lip wear, shaft movement, and bore eccentricity.
- Flutes on seal lips provide a pumping action to direct oil back into a housing.
- Bearing hub units are compact compared to bearings that are mounted in the wheel hub. This compactness makes bearing hub units suitable for front-wheel-drive cars.
- Some bearing hub units are bolted to the steering knuckle; other bearing hub units are pressed into the steering knuckle.
- Some steering knuckles contain two separate tapered roller bearings
- Rear-axle bearings are mounted between the drive axles and the housing on rear-wheel-drive cars.

Summary

Each chapter concludes with summary statements that contain the important topics of the chapter. These are designed to help the reader review the contents.

REVIEW QUESTIONS

Short Answer Essays

1. Define a radial bearing load.
2. Define a thrust bearing load, and give another term for this type of load.
3. Explain an angular bearing load.
4. Describe the main parts of a bearing, including the location and purpose of each part.
5. Describe the difference between a maximum-capacity ball bearing and an ordinary ball bearing.
6. Explain the design and purpose of bearing seals.
7. Explain the purpose of the sealer on the outside surface of a seal housing.
8. What does ASTM and NLGI stand for, and what is their function?

Review Questions

Short-answer essay, fill-in-the-blank, and multiple-choice questions follow each chapter. These questions are designed to accurately assess the student's competence in the stated objectives at the beginning of the chapter.

SHOP MANUAL

To stress the importance of safe work habits, the Shop Manual also dedicates one full chapter to safety. Other important features of this manual include the following:

Performance-Based Objectives

These objectives define the contents of the chapter and define what the student should have learned on completion of the chapter.

Terms To Know List

Terms in this list are also defined in the Glossary at the end of the manual.

Cautions and Warnings

Cautions appear throughout the text to alert the reader to potentially hazardous materials or unsafe conditions. Warnings advise the student of things that can go wrong if instructions are not followed or if an incorrect part or tool is used.

Margin Notes

The most important terms to know are highlighted and defined in the margin. Common trade jargon also appears in the margins and gives some of the common terms used for components. This feature helps students understand and speak the language of the trade, especially when conversing with an experienced technician.

Basic Tools Lists

Each chapter begins with a list of the basic tools needed to perform the tasks included in the chapter.

References to the Classroom Manual

References to the appropriate page in the Classroom Manual appear whenever necessary. Although the chapters of the two manuals are synchronized, material covered in other chapters of the Classroom Manual may be fundamental to the topic discussed in the Shop Manual.

Special Tools Lists

Whenever a special tool is required to complete a task, it is listed in the margin next to the procedure.

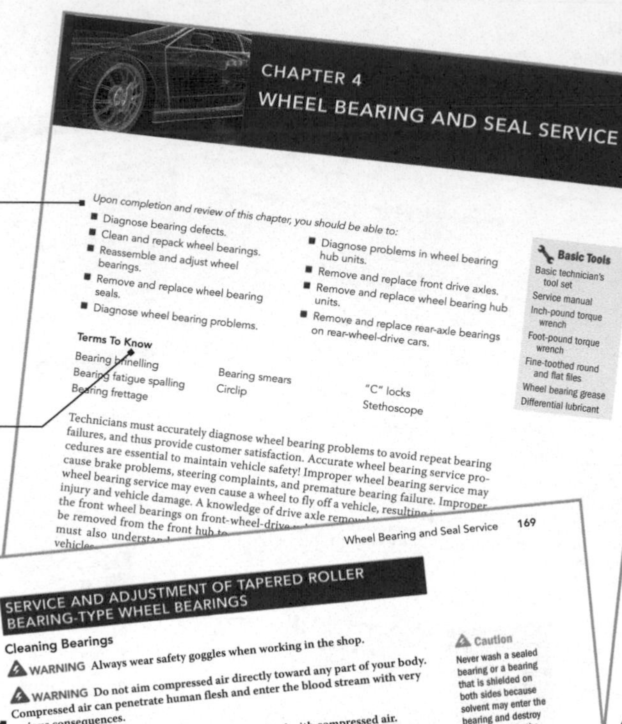
CHAPTER 4
WHEEL BEARING AND SEAL SERVICE

Upon completion and review of this chapter, you should be able to:
- Diagnose bearing defects.
- Clean and repack wheel bearings.
- Reassemble and adjust wheel bearings.
- Remove and replace wheel bearing seals.
- Diagnose wheel bearing problems.
- Diagnose problems in wheel bearing hub units.
- Remove and replace front drive axles.
- Remove and replace wheel bearing hub units.
- Remove and replace rear-axle bearings on rear-wheel-drive cars.

Basic Tools
Basic technician's tool set
Service manual
Inch-pound torque wrench
Foot-pound torque wrench
Fine-toothed round and flat files
Wheel bearing grease
Differential lubricant

Terms To Know
Bearing brinelling
Bearing fatigue spalling
Bearing frettage
Bearing smears
Circlip
"C" locks
Stethoscope

Technicians must accurately diagnose wheel bearing problems to avoid repeat bearing failures, and thus provide customer satisfaction. Accurate wheel bearing service procedures are essential to maintain vehicle safety! Improper wheel bearing service may cause brake problems, steering complaints, and premature bearing failure. Improper wheel bearing service may even cause a wheel to fly off a vehicle, resulting in injury and vehicle damage. A knowledge of drive axle removal ... the front wheel bearings on front-wheel-drive ... be removed from the front hub ... must also understand ... vehicles ...

Wheel Bearing and Seal Service 169

SERVICE AND ADJUSTMENT OF TAPERED ROLLER BEARING-TYPE WHEEL BEARINGS

Cleaning Bearings

WARNING Always wear safety goggles when working in the shop.

WARNING Do not aim compressed air directly toward any part of your body. Compressed air can penetrate human flesh and enter the blood stream with very serious consequences.

WARNING Do not spin the bearing at high speed with compressed air. Bearing damage or disintegration may result. Bearing disintegration may cause serious personal injury.

WARNING Never strike a bearing with a ball peen hammer. This action will damage the bearing and the bearing may shatter, causing severe personal injury.

Caution: Never wash a sealed bearing or a bearing that is shielded on both sides because solvent may enter the bearing and destroy the lubricant in the bearing, resulting in very short bearing life.

Two separate tapered roller bearings are used in the front wheel hubs of many rear-wheel-drive cars. The rear wheel hubs on some front-wheel-drive cars have the same type of bearings. Similar service and adjustment procedures apply to these tapered roller bearings.

These bearings should be cleaned, inspected, and packed with wheel bearing grease at the vehicle manufacturer's recommended service intervals. Pry the grease seal out of the inner hub opening with a seal removal tool, and discard the seal. This seal should always be replaced when the bearings are serviced. Do not attempt to wash sealed bearings or bearings that are shielded on both sides. If a bearing is sealed on one side, it may be washed in solvent and repacked with grease.

The outer bearing race on a tapered roller bearing may be called a bearing cup.

Bearings may be placed in a tray and lowered into a container of clean solvent. A brush may be used to remove old grease from the bearing (**Figure 4-3**). The bearings may be dried with compressed air after the cleaning operation. Be sure the shop air supply is free from moisture, which causes rust formation in the bearing. After all the old grease has been cleaned from the bearing, rinse the bearing in clean solvent and dry it thoroughly with compressed air.

Classroom Manual Chapter 4, page 98

When bearing cleaning is completed, bearings should be inspected for the defects illustrated in Figures 4-1 and 4-2. If any of these conditions are present on the bearing, replacement is necessary. Tapered roller bearings and their matching outer races must be replaced as a set. If the bearing installation is not done immediately, cover the bearings with a protective lubricant and wrap them in waterproof paper (**Figure 4-4**). Be sure to ...

Special Tool
Bearing driver

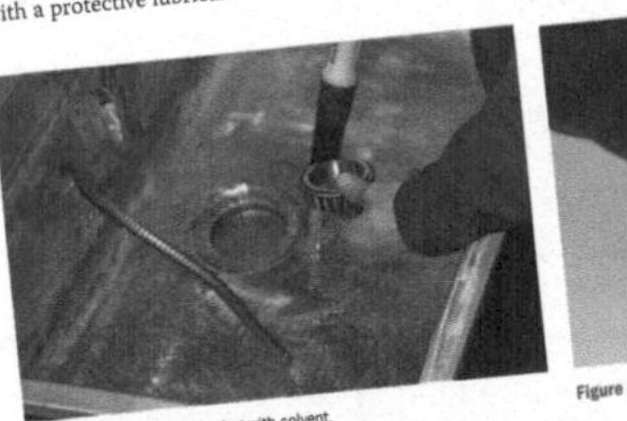
Figure 4-3 Cleaning a bearing with solvent.

Figure 4-4 Wrapping a bearing in waterproof paper.

"C" locks are split, circular metal rings that fit in rear axle grooves to retain the rear axles in the differential on some rear-wheel-drive vehicles.

... problems with a vehicle ... the transmission in drive. Keep away ... drive axles.

... noise may be diagnosed with the vehicle raised on a hoist. Be sure the safety mechanism is engaged after the vehicle is raised on the hoist. With the engine running and the transmission in drive, operate the vehicle at a moderate speed of 35–45 mph (56–72 km/h) and listen with a stethoscope placed on the rear axle housing directly over the axle bearings. If grinding or clicking noises are heard, the bearing must be replaced.

Many axle shafts in rear-wheel-drive cars have a roller bearing and seal at the outer end (**Figure 4-28**). These axle shafts are often retained in the differential with "C" locks that must be removed before the axles.

The rear-axle bearing removal and replacement procedure varies depending on the vehicle make and model year. Always follow the rear-axle bearing removal and replacement procedure in the manufacturer's service manual.

A typical rear axle shaft removal and replacement procedure on a rear-wheel-drive car with "C" lock axle retainers is as follows:

1. Loosen the rear wheel nuts and chalk-mark the rear wheel position in relation to the rear axle studs.
2. Raise the vehicle on a hoist and make sure the hoist safety mechanism is in place.
3. Remove the rear wheels and brake drums, or calipers and rotors.

214 Chapter 5

9. Connect the ohmmeter leads from the signal return terminal in the wiring harness side of the actuator connector to a chassis ground. The ohmmeter should indicate less than 10 ohms. If the ohmmeter reading is higher than specified, check the signal return wire and the programmed ride control (PRC) module.

SERVICE TIP Wiring harness colors vary depending on the vehicle make and model year. The wiring colors in the following steps are based on Ford vehicles. If you are working on a different make of vehicle, refer to the wire colors in the vehicle manufacturer's service information.

Service Tips

Whenever a shortcut or special procedure is appropriate, it is described in the text. Generally, these tips describe common procedures used by experienced technicians.

CASE STUDY

A customer complained about a squeaking noise in the rear suspension of a 2009 Dodge Caravan. The customer said the noise occurred during normal driving at lower speeds. The technician lifted the car on a hoist and made a check of all rear suspension bushings in shock absorbers, track bar, and trailing arms. The spring insulators and all suspension bolts were checked visually. All of these items were in good condition, and there was no evidence of a squeaking noise as the chassis was bounced gently. Because the exact source of chassis noise can sometimes be difficult to locate, the technician performed a visual check of bushings, insulators, and fasteners in the front suspension. No problems were found in the front suspension.

One of the first requirements for successful automotive diagnosis is to obtain as much information as possible from the customer. The customer with the squeaking rear suspension in a Dodge Caravan lived in a part of the country where cold temperatures occur in the winter. This complaint occurred in January. The technician questioned the customer further about the conditions when the squeaking suspension noise was heard, and the customer revealed that the noise occurred when the temperature was severely cold. The customer also indicated that the noise disappeared at warmer temperatures.

The technician informed the customer that the only way to find the exact cause of this annoying squeak was to leave the car on the lot at the shop all night and check it first thing in the morning when it was colder. The customer complied with this suggestion, and the technician drove the car into the shop immediately the next morning. The squeaking noise occurred as the car was driven across the parking lot. The technician lifted the car on a hoist and listened with a stethoscope at each rear suspension bushing as a coworker gently bounced the rear suspension. No squeaking noise was heard at any of the rear suspension bushings. However, when the stethoscope pickup was placed on the left rear shock absorber, the squeaking noise was loud and clear. The shock absorber was quickly removed, and the squeaking noise was gone when the rear chassis was bounced gently. Replacement of the left rear shock absorber corrected this complaint and made the customer happy.

Case Studies

Each chapter ends with a Case Study describing a particular vehicle problem and the logical steps a technician might use to solve the problem. These studies focus on system diagnosis skills and help students gain familiarity with the process.

ASE-STYLE REVIEW QUESTIONS

1. A slight oil film appears on the lower shock absorber oil chamber, and the shock absorber performs satisfactorily on a road test:
   Technician A says the shock absorber is satisfactory.
   Technician B says the shock absorber may contain excessive pressure.
   Who is correct?
   A. A only
   B. B only
   C. Both A and B
   D. Neither A nor B
2. While discussing a shock absorber and strut bounce test:
   Technician A says the shock absorber is satisfactory if the bumper makes two free upward bounces.
   Technician B says the bumper must be pushed downward with considerable force.
   Who is correct?
   A. A only
   B. B only
   C. Both A and B
   D. Neither A nor B

ASE-Style Review Questions

Each chapter contains ASE-style review questions that reflect the performance objectives listed at the beginning of the chapter. These questions can be used to review the chapter as well as to prepare for the ASE certification exam.

The rear coil springs ... cedure for spring removal from the fro...

INSTALLING STRUT CARTRIDGE, OFF-CAR

CUSTOMER CARE Check the cost of the strut cartridges versus the price of new struts. Give customers the best value for their repair dollar!

Customer Care

This feature highlights those little things a technician can do or say to enhance customer relations.

Many struts are a sealed unit, and thus rebuilding is impossible. However, some manufacturers supply a replacement cartridge that may be installed in the strut housing after the strut has been removed from the vehicle. Always follow the strut cartridge manufacturer's recommended installation procedure.

A strut cartridge contains the inner working part of the strut, which may be installed in the outer housing of the old strut.

Special Tool: Pipe cutter

The following is a typical off-car strut cartridge installation procedure:

1. Install a bolt and two nuts in the upper strut-to-knuckle mounting bolt hole. Place a nut on the inside and outside of the strut flange.
2. Clamp this bolt in a vise to hold the strut.
3. Locate the line groove near the top of the strut body, and use a pipe cutter installed in this groove to cut the top of the strut body.
4. After the cutting procedure, remove the strut piston assembly from the strut (**Figure 5-26**).
5. Remove the strut from the vise and dump the oil from the strut.
6. Place the special tool supplied by the vehicle manufacturer or cartridge manufacturer on top of the strut body. Strike the tool with a plastic hammer until the tool shoulder contacts the top of the strut body. This action removes burrs from the strut body and places a slight flare on the body.
7. Remove the tool from the strut body.
8. Install the required amount of oil in the strut, place the new cartridge in the strut body, and turn the cartridge until it settles into indentations in the bottom of the strut body.
9. Place the new nut over the cartridge.
10. Using a special tool supplied by the vehicle or cartridge manufacturer, tighten the nut to the specified torque.
11. Move the strut piston rod in and out several times to check for proper strut operation.

INSTALLING STRUT CARTRIDGE, ON-CAR

WARNING If a vehicle is hoisted or lifted in any way during an on-car strut cartridge replacement, the coil spring may fly off the strut, causing vehicle damage and personal injury.

On some vehicles, the front strut cartridge may be removed and replaced with the strut

Figure 7-2 Electronic stethoscope.

Riding Height Measurement

Regular inspection and proper main... tant to maintain vehicle safety. The ... tion. Other suspension components, ... they are worn. *Since incorrect riding ... measurement is critical*. Reduced r... causes rapid steering wheel return a... riding height is less than specified. Th... manufacturer's specified location, ... suspension system. On some vehicl... sured from the floor to the center of ... level floor or an alignment rack (**Fig**... cedure for measuring front and rear curb riding height.

PHOTO SEQUENCE 11
Typical Procedure for Measuring Front and Rear Riding Height

P11-1 Check the trunk for extra weight.

P11-2 Check the tires for normal inflation pressure.

P11-3 Park the car on a level shop floor or an alignment rack.

Photo Sequences

Many procedures are illustrated in detailed Photo Sequences. These photographs show the students what to expect when they perform particular procedures. They also familiarize students with a system or type of equipment that the school might not have.

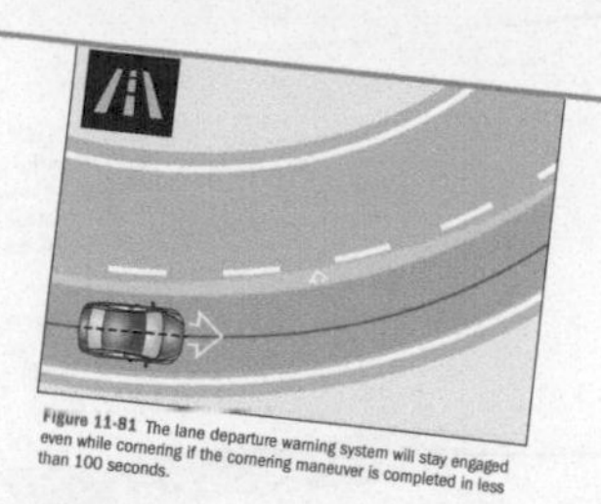

Figure 11-81 The lane departure warning system will stay engaged even while cornering if the cornering maneuver is completed in less than 100 seconds.

Some LDW systems have lane-departure prevention capabilities. If the vehicle is moving toward the lane markings on the left, the ABS and stability control computer lightly pulse the brakes on the right side of the vehicle. When the wheels on the right side slow down, the wheels on the left side turn faster than the wheels on the right side, and this action steers the car away from the highway markers on the left. Other LDW systems activate the electronic (electromechanical) steering on the vehicle to steer the vehicle away from the lane markers if a lane departure is starting to occur. The LDW system will engage the electromechanical power steering system to first cause a slight vibration of the steering wheel to warn driver to take corrective action to maintain lane position and then take corrective action to maintain lane position. If the driver exerts more than 2.21 ft.-lbs (3 Nm) of pressure on the steering wheel, the LDW system will disengage to allow a lane change without using turn signal. The LDW system will continue to function even while corning as long as the corning maneuver is completed in less than 100 seconds (**Figure 11-81**). The lane departure system can be adversely affect by weather and road surfaces. While the system is effective at detecting lane lines on asphalt, it can have issues detecting lane lines on concrete roads on bright sunny days. The system can also be affected by some wet surfaces due to reflection or snow-covered surfaces that cover lane lines. A dirty windshield that affects camera image will also cause lane departure system issues and will set a diagnostic trouble code "Lane Assist – no sensor visibility present".

AUTHOR'S NOTE Statistics compiled by one of the major car manufacturers indicate that 30 percent of all vehicle accidents in the United States involve rear-end collisions. In half of these collisions, the driver did not apply the brakes before collision occurred. The statistics also indicate that 75 percent of these rear-end collisions occur below 19 mph (30 km/h).

COLLISION MITIGATION SYSTEMS

Some vehicles are equipped with a collision mitigation system that uses radar and camera information to detect a vehicle in front. The collision mitigation computer is in constant communication with the PCM and ABS computers via the vehicle network. If the radar and camera signals indicate a vehicle in front is too close and a collision is imminent, the

Author's Note

This feature includes simple explanations, stories, or examples of complex topics. These are included to help students understand difficult concepts.

Rear Suspension Service 293

10. While diagnosing improper rear wheel tracking:
    A. Improper rear wheel tracking may be caused when both rear springs are sagged the same amount.
    B. A bent rear suspension tie rod does not affect rear wheel tracking.
    C. Improper rear wheel tracking may result in steering pull when driving straight ahead.
    D. Improper rear wheel tracking may cause front wheel shimmy.

ASE CHALLENGE QUESTIONS

1. A squeak in the rear suspension could be caused by all of the following EXCEPT:
   A. The suspension bushing.
   B. Weak spring leaves.
   C. Worn spring antifriction pads.
   D. A defective shock absorber.
2. A vehicle with a live axle coil spring rear suspension has become hard to steer with harsh ride quality. Which of the following could be the cause of this problem?
   A. Worn lateral link bushings
   B. Weak rear coil springs
   C. Bent rear shock rod
   D. Worn stabilizer bar bushings
3. A truck with a solid axle and parallel leaf spring rear suspension darts and acts erratic during turns. Which of the following is the most likely cause?
   A. Worn rear sway bar bushings
   B. Incorrect driveline angle
   C. Incorrect ride height
   D. Loose rear axle U-bolts
4. A car with independent rear suspension has excessive rear tire wear. An inspection of the rear tires shows they are worn on the inside edge and the tread is feathered.
   Technician A says the problem could be the tires are toeing out during acceleration.
   Technician B says worn control arm bushings could be the cause of the problem.
   Who is correct?
   A. A only
   B. B only
   C. Both A and B
   D. Neither A nor B
5. The steering on a front-wheel-drive car pulls to the right.
   Technician A says the strut on the right rear suspension assembly could be the problem.
   Technician B says worn bushings of the left rear lower arm assembly could be the problem.
   Who is correct?
   A. A only
   B. B only
   C. Both A and B
   D. Neither A nor B

ASE Challenge Questions

Each technical chapter ends with five ASE challenge questions. These are not more review questions; rather, they test the students' ability to apply general knowledge to the contents of the chapter.

ASE Practice Examination

A 50-question ASE practice exam, located in the Appendix, is included to test students on the content of the complete Shop Manual.

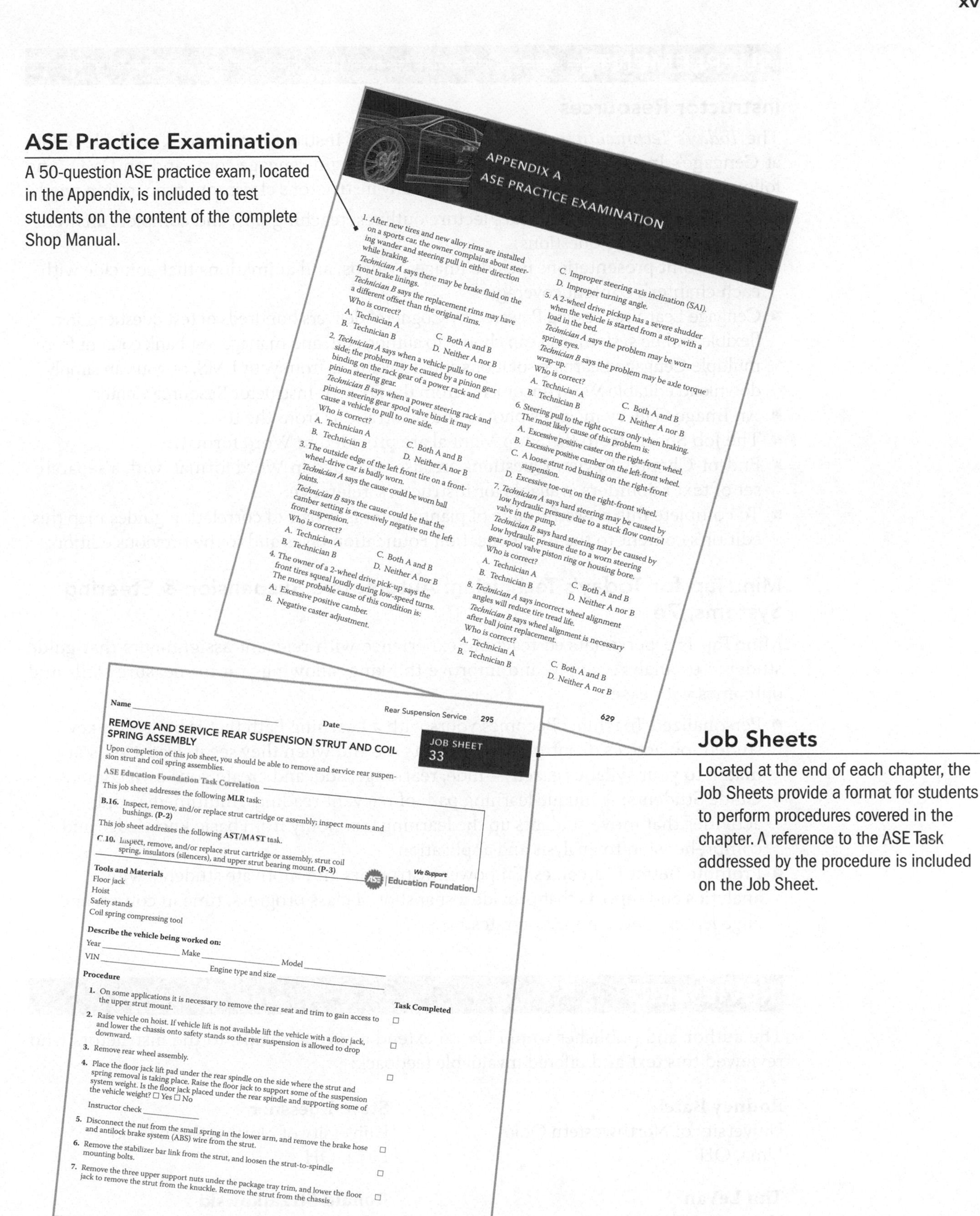

APPENDIX A
ASE PRACTICE EXAMINATION

1. After new tires and new alloy rims are installed on a sports car, the owner complains about steering wander and steering pull in either direction while braking.
Technician A says there may be brake fluid on the front brake linings.
Technician B says the replacement rims may have a different offset than the original rims.
Who is correct?
A. Technician A
B. Technician B
C. Both A and B
D. Neither A nor B

2. *Technician A* says when a vehicle pulls to one side; the problem may be caused by a pinion gear binding on the rack gear of a power rack and pinion steering gear.
Technician B says when a power steering rack and pinion steering gear spool valve binds it may cause a vehicle to pull to one side.
Who is correct?
A. Technician A
B. Technician B
C. Both A and B
D. Neither A nor B

3. The outside edge of the left front tire on a front-wheel-drive car is badly worn.
Technician A says the cause could be worn ball joints.
Technician B says the cause could be that the camber setting is excessively negative on the left front suspension.
Who is correct?
A. Technician A
B. Technician B
C. Both A and B
D. Neither A nor B

4. The owner of a 2-wheel drive pick-up says the front tires squeal loudly during low-speed turns. The most probable cause of this condition is:
A. Excessive positive camber.
B. Negative caster adjustment.
C. Improper steering axis inclination (SAI).
D. Improper turning angle.

5. A 2-wheel drive pickup has a severe shudder when the vehicle is started from a stop with a load in the bed.
Technician A says the problem may be worn spring eyes.
Technician B says the problem may be axle torque wrap-up.
Who is correct?
A. Technician A
B. Technician B
C. Both A and B
D. Neither A nor B

6. Steering pull to the right occurs only when braking. The most likely cause of this problem is:
A. Excessive positive caster on the right-front wheel.
B. Excessive positive camber on the left-front wheel.
C. A loose strut rod bushing on the right-front suspension.
D. Excessive toe-out on the right-front wheel.

7. *Technician A* says hard steering may be caused by low hydraulic pressure due to a stuck flow control valve in the pump.
Technician B says hard steering may be caused by low hydraulic pressure due to a worn steering gear spool valve piston ring or housing bore.
Who is correct?
A. Technician A
B. Technician B
C. Both A and B
D. Neither A nor B

8. *Technician A* says incorrect wheel alignment angles will reduce tire tread life.
Technician B says wheel alignment is necessary after ball joint replacement.
Who is correct?
A. Technician A
B. Technician B
C. Both A and B
D. Neither A nor B

629

Name ______________ Date ______________

Rear Suspension Service 295

REMOVE AND SERVICE REAR SUSPENSION STRUT AND COIL SPRING ASSEMBLY

JOB SHEET 33

Upon completion of this job sheet, you should be able to remove and service rear suspension strut and coil spring assemblies.

ASE Education Foundation Task Correlation

This job sheet addresses the following **MLR** task:

B.16. Inspect, remove, and/or replace strut cartridge or assembly; inspect mounts and bushings. **(P-2)**

This job sheet addresses the following **AST/MAST** task.

C.10. Inspect, remove, and/or replace strut cartridge or assembly, strut coil spring, insulators (silencers), and upper strut bearing mount. **(P-3)**

We Support ASE Education Foundation

Tools and Materials
Floor jack
Hoist
Safety stands
Coil spring compressing tool

Describe the vehicle being worked on:
Year ______ Make ______ Model ______
VIN ______ Engine type and size ______

Procedure — **Task Completed**

1. On some applications it is necessary to remove the rear seat and trim to gain access to the upper strut mount. ☐
2. Raise vehicle on hoist. If vehicle lift is not available lift the vehicle with a floor jack, and lower the chassis onto safety stands so the rear suspension is allowed to drop downward. ☐
3. Remove rear wheel assembly.
4. Place the floor jack lift pad under the rear spindle on the side where the strut and spring removal is taking place. Raise the floor jack to support some of the suspension system weight. Is the floor jack placed under the rear spindle and supporting some of the vehicle weight? ☐ Yes ☐ No ☐
Instructor check ______
5. Disconnect the nut from the small spring in the lower arm, and remove the brake hose and antilock brake system (ABS) wire from the strut. ☐
6. Remove the stabilizer bar link from the strut, and loosen the strut-to-spindle mounting bolts. ☐
7. Remove the three upper support nuts under the package tray trim, and lower the floor jack to remove the strut from the knuckle. Remove the strut from the chassis. ☐

Job Sheets

Located at the end of each chapter, the Job Sheets provide a format for students to perform procedures covered in the chapter. A reference to the ASE Task addressed by the procedure is included on the Job Sheet.

SUPPLEMENTS

Instructor Resources

The *Today's Technician* series offers a robust set of instructor resources, available online at Cengage's Instructor Resource Center (http://login.cengage.com) and on DVD. The following tools have been provided to meet any instructor's classroom preparation needs:

- An Instructor's Guide provides lecture outlines, teaching tips, and complete answers to end-of-chapter questions.
- PowerPoint presentations include images, videos, and animations that coincide with each chapter's content coverage.
- Cengage Learning Testing Powered by Cognero® delivers hundreds of test questions in a flexible, online system. You can choose to author, edit, and manage test bank content from multiple Cengage Learning solutions and deliver tests from your LMS, or you can simply download editable Word documents from the DVD or Instructor Resource Center.
- An Image Gallery includes photos and illustrations from the text.
- The Job Sheets from the Shop Manual are provided in Word format.
- End-of-Chapter Review Questions are also provided in Word format, with a separate set of text rejoinders available for instructors' reference.
- To complete this powerful suite of planning tools, a pair of correlation guides map this edition's content to the ASE Education Foundation tasks and to the previous edition.

MindTap for Today's Technician: Automotive Suspension & Steering Systems, 7e

MindTap is a personalized teaching experience with relevant assignments that guide students to analyze, apply, and improve thinking, allowing you to measure skills and outcomes with ease.

- Personalized Teaching: Becomes yours with a Learning Path that is built with key student objectives. Control what students see and when they see it. Use it as-is or match to your syllabus exactly—hide, rearrange, add, and create your own content.
- Guide Students: A unique learning path of relevant readings, multimedia, and activities that move students up the learning taxonomy from basic knowledge and comprehension to analysis and application.
- Promote Better Outcomes: Empower instructors and motivate students with analytics and reports that provide a snapshot of class progress, time in course, and engagement and completion rates.

REVIEWERS

The author and publisher would like to extend a special thanks to the instructors who reviewed this text and offered invaluable feedback:

Rodney Batch
University of Northwestern Ohio
Lima, OH

Tim LeVan
University of Northwestern Ohio
Lima, OH

Cory Peck
Vatterott College
St. Louis, MO

Steve Roessner
University of Northwestern Ohio
Lima, OH

Ronald Strzalkowski
Baker College Flint, MI

Claude F. Townsend III
Oakland Community College
Bloomfield Hills, MI

CHAPTER 1
SUSPENSION AND STEERING SYSTEMS

Upon completion and review of this chapter, you should be able to understand and describe:

- How strength and rigidity are designed into a unitized body.
- The advantages of reduced vehicle weight.
- The design of a short-and-long arm (SLA) front suspension system.
- How limited independent rear wheel movement is provided in a semi-independent rear suspension system.
- The advantage of an independent rear suspension system.
- The purposes of vehicle tires.
- The terms *positive* and *negative offset* as they relate to vehicle wheel rims.
- Three different loads that are applied to wheel bearings.
- The purposes of shock absorbers.
- The difference between a shock absorber and a strut.
- Two different types of computer-controlled shock absorbers.
- The advantages of computer-controlled suspension systems.
- Two types of steering linkages.
- How the rack is moved in a rack and pinion steering gear.
- How a power steering pump develops hydraulic pressure.
- The result of incorrect rear wheel toe.
- The front wheel caster.
- The results of excessive negative camber.

Terms To Know

Angular bearing loads
Carbon dioxide (CO_2)
Greenhouse gas
Jounce travel
Negative camber
Negative caster
Negative offset
Positive camber
Positive caster
Positive offset
Radial bearing load
Rebound travel
Thrust bearing loads
Thrust line
Toe-out
Wheel alignment
Wheel offset
Wheel shimmy

INTRODUCTION

The suspension system must provide proper steering control and ride quality. Performing these functions is extremely important to maintaining vehicle safety and customer satisfaction. For example, if the suspension system allows excessive vertical wheel oscillations, the driver may lose control of the steering when driving on an irregular road surface. This loss of steering control can result in vehicle collisions and personal injury. Excessive vertical wheel oscillations transfer undesirable vibrations from the wheel(s) to the passenger compartment, which causes customer dissatisfaction with the ride quality.

The suspension system and frame must also position the wheels and tires properly to provide normal tire life and proper steering control. If the suspension system does not position each wheel and tire properly, wheel alignment angles are incorrect and usually cause excessive tire tread wear. Improper wheel and tire position can also cause the steering to pull to one side. When the suspension system positions the wheels and tires properly, the steering should remain in the straight-ahead position if the car is driven straight ahead on a reasonably straight, smooth road surface. However, if the wheels and tires are not properly positioned, the steering can be erratic, and excessive steering effort is required to maintain the steering in the straight-ahead position.

The steering system is also extremely important to maintaining vehicle safety and reduce driver fatigue. For example, if a steering system component is suddenly disconnected, the driver may experience a complete loss of steering control, resulting in vehicle collision and personal injury. Loose steering system components can cause erratic steering, which causes the driver to continually turn the steering wheel in either direction to try and keep the vehicle moving straight ahead. This condition results in premature driver fatigue.

FRAMES AND UNITIZED BODIES

Some vehicles, such as rear-wheel-drive cars, sport utility vehicles (SUVs), and trucks, have a frame that is separate from the body (**Figure 1-1**). Other vehicles have a unitized body that combines the frame and body in one unit, eliminating the external frame (**Figure 1-2**). In a unitized body, the body design, rather than a heavy steel frame, provides strength and rigidity. All parts of a unitized body are load-carrying members, and these body parts are welded together to form a strong assembly.

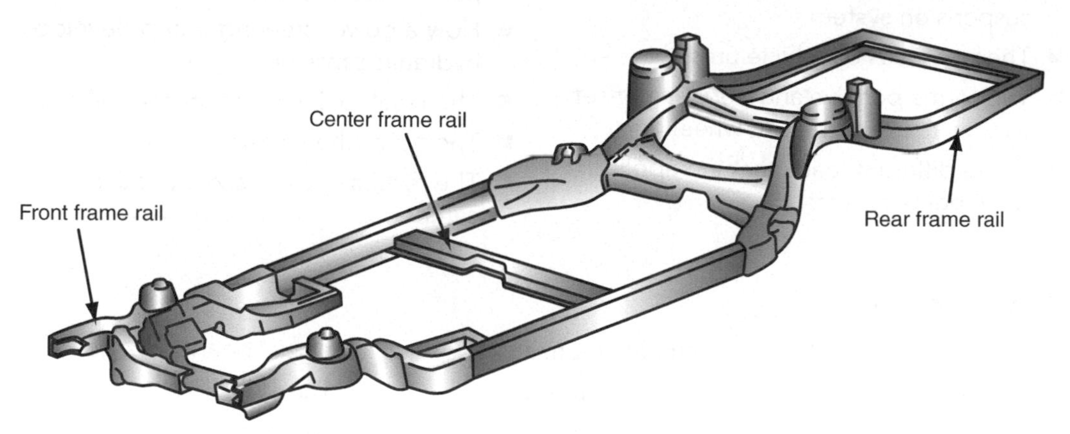

Figure 1-1 Vehicle frame.

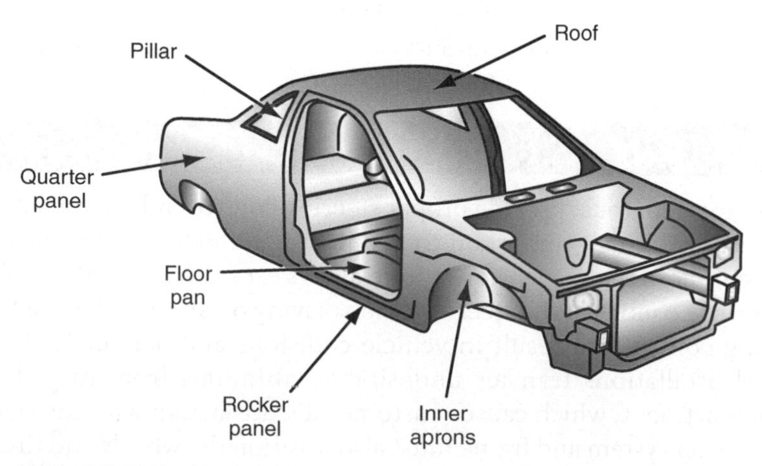

Figure 1-2 Unitized body design.

The frame or unitized body serves the following purposes:

1. Allows the vehicle to support its total weight, including the weight of the vehicle and cargo
2. Allows the vehicle to absorb stress when driving on rough road surfaces
3. Enables the vehicle to absorb torque from the engine and drivetrain
4. Provides attachment points for suspension and other components

Shop Manual
Chapter 10, page 443

The unitized body provides a steel box around the passenger compartment to protect passengers in a collision. In most unitized bodies, special steel panels are inserted in the doors to protect the vehicle occupants in a side collision. Some unitized body components are manufactured from high-strength or ultra high-strength steels. The unitized body design is the most popular design today. A steel cradle is mounted under the front of the unitized body to support the engine and transaxle (**Figure 1-3**). Rubber and steel mounts support the engine and transaxle on the cradle. Large rubber bushings are mounted between the cradle and the unitized body to help prevent engine vibration from reaching the passenger compartment. Some unitized bodies have a partial frame mounted under the rear of the vehicle to provide additional strength and facilitate the attachment of rear suspension components (**Figure 1-4**).

Vehicle weight plays a significant role in fuel consumption. One automotive design engineer states, "Fuel economy improvements are almost linear with weight reduction. A 30 percent reduction in vehicle weight provides approximately a 30 percent improvement in fuel economy." If a Toyota Prius weighs 3300 lb (1497 kg) and provides 50 miles per gallon (mpg), the same Prius would provide 55 mpg if it weighed 3000 lb (1360 kg).

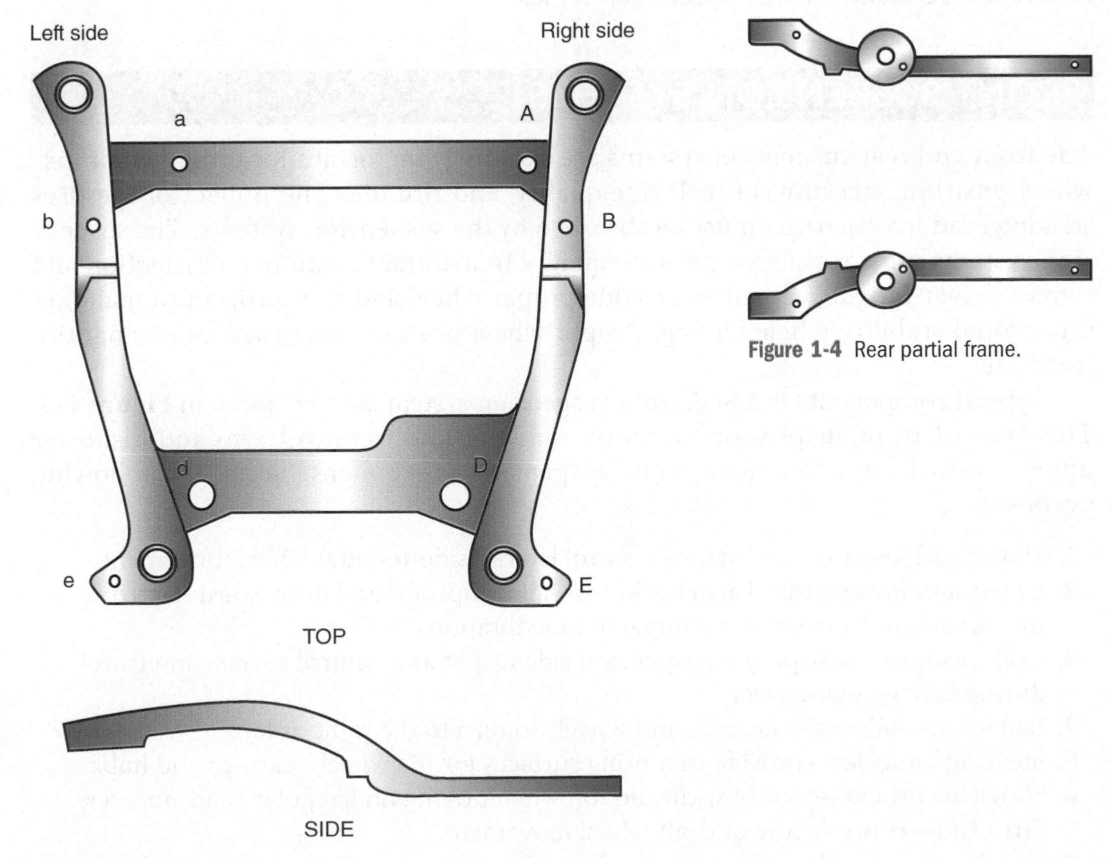

Figure 1-4 Rear partial frame.

Figure 1-3 Engine cradle.

Carbon dioxide is a by-product of the gasoline, diesel fuel, or ethanol combustion process.

A **greenhouse gas** is a gas that collects high in the earth's atmosphere and covers the earth like a blanket, resulting in more heat retained on the planet.

In the United States, it is estimated that automobiles contribute 1.5 billion tons of CO_2 to the atmosphere each year. Coal-burning power plants produce 2.5 billion tons of CO_2 each year.

Carbon dioxide (CO_2) emissions are a major concern for automotive manufacturers, because CO_2 is a **greenhouse gas** that contributes to global warming. Vehicle manufacturers are facing increasingly stringent CO_2 emissions standards. CO_2 emissions are proportional to fuel consumption. Reduced fuel consumption results in lower CO_2 emissions. Therefore, reducing vehicle weight results in less fuel consumption and lower CO_2 emissions. Reduced weight also contributes to improved vehicle performance.

When vehicle bodies, front and rear suspension systems, and steering systems are built from lighter-weight components, these items can make an important contribution to improved fuel economy and reduced CO_2 emissions. The components in these systems may be manufactured from high-strength steels, aluminum, magnesium, titanium, or carbon composites to reduce vehicle weight. The Corvette Z06 has a hydro-formed aluminum frame to reduce vehicle weight. Lightweight aluminum construction throughout the Jaguar XK results in a vehicle weight of 3671 lb (1665 kg), which is 450 lb (204 kg) lighter than its Mercedes-Benz SL500 competitor, which primarily uses steel construction. A new-model Ford Fiesta has been available in Europe since October 2008, and more recently in the United States and other countries. In this model, 55 percent of the unitized body structure is made from high-strength steels, making the body torsionally 10 percent stiffer than its predecessor and providing a very rigid safety cell surrounding the occupants. Ford says this model is 88 lb (44 kg) lighter than the previous model, and more fuel and CO_2 efficient.

It is estimated that for every gallon of gasoline burned, 28 lb (12 7 kg) of CO_2 is released into the atmosphere. Approximately 21 lb (9 53 kg) come directly from the tailpipe of a vehicle, and 7 lb (3 18 kg) come from the extraction, transportation, and refining of the gasoline.

High material and production costs have prevented the use of carbon composites in production vehicles. Carbon composites have been used in some exotic ultra high-performance cars or race cars, where cost was not a factor. Carbon composites provide reduced weight and improved crash energy absorption. New carbon composites and improved manufacturing processes may soon make the use of some carbon composite components a reality in high-production vehicles.

FRONT SUSPENSION SYSTEMS

When coil-type suspension springs are manufactured from titanium rather than steel, the weight saved per vehicle is 44–55 lbs (20–25 kg).

The front and rear suspension systems are extremely important for providing proper wheel position, steering control, ride quality, and tire life. The impact of the tires striking road irregularities must be absorbed by the suspension systems. The suspension systems must supply proper ride quality to maintain customer satisfaction and reduce driver fatigue, as well as provide proper wheel and tire position to maintain directional stability when driving. Proper wheel position also ensures normal tire tread life.

Typical components in a SLA front suspension system are illustrated in **Figure 1-5**. This type of front suspension system has a long lower control arm and a shorter upper control arm. The main front suspension components serve the following purposes:

The average car manufactured in the United States in 2018 contains 347 lb (157 kg) of aluminum. A decade ago, the average car contained 242 lb (110 kg) of aluminum, and in 1973, the average car contained 81 lb (37 kg) of aluminum.

1. Upper and lower control arms—control lateral (side-to-side) wheel movement.
2. Upper and lower control arm bushings—allow upward and downward control arm movement and absorb wheel impacts and vibrations.
3. Coil springs—allow proper suspension ride height and control suspension travel during driving maneuvers.
4. Ball joints—allow the knuckle and wheels to turn to the right or left.
5. Steering knuckles—provide mounting surfaces for the wheel bearings and hubs.
6. Shock absorbers—control spring action when driving on irregular road surfaces.
7. Strut rod—controls fore-and-aft wheel movement.
8. Stabilizer bar—reduces body sway when a front wheel strikes a road irregularity.

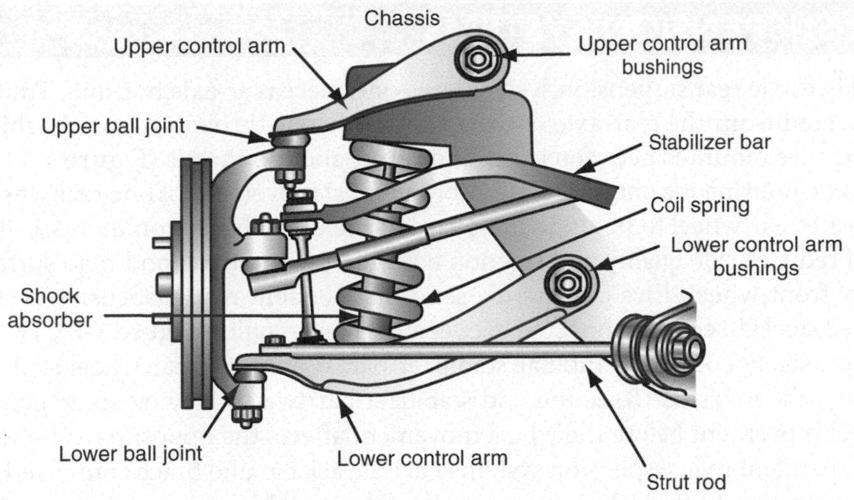

Shop Manual
Chapter 6, page 231

Figure 1-5 Typical short-and-long arm (SLA) front suspension system.

A MacPherson strut front suspension system has no upper control arm and ball joint; instead, a strut is connected from the top of the knuckle to an upper strut mount bolted to the reinforced strut tower in the unitized body (**Figure 1-6**). The strut supports the top of the knuckle and also performs the same function as the shock absorber in an SLA suspension system. The coil spring is mounted between a lower support on the strut and the upper strut mount. Insulators are mounted between the ends of the coil spring and the mounting locations. A bearing in the upper strut mount allows the strut and coil spring to rotate with the spindle when the front wheels are turned.

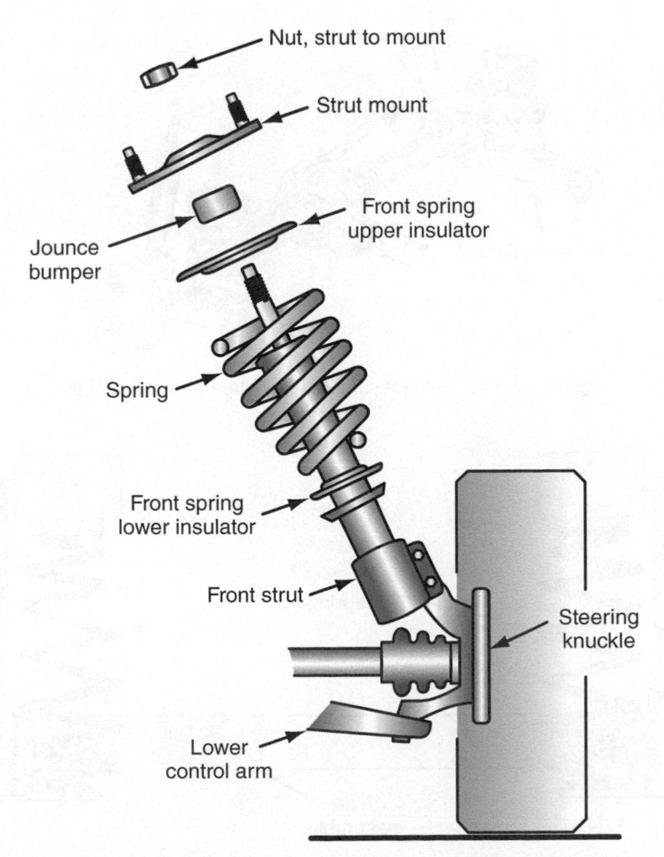

Figure 1-6 Typical MacPherson strut front suspension system.

REAR SUSPENSION SYSTEMS

A typical live axle rear suspension system has a one-piece rear-axle housing. Trailing arms are connected from the rear-axle housing to the chassis through rubber bushings. The coil springs are mounted between the trailing arms and the chassis (**Figure 1-7**). Because the rear-axle housing is a one-piece assembly, vertical movement of one rear wheel causes the opposite rear wheel to be tipped outward at the top. This action increases tire tread wear and reduces ride quality and traction between the tire tread and road surface.

Many front-wheel-drive cars have a semi-independent rear suspension system with an inverted steel U-section connected between the rear spindles (**Figure 1-8**). The inverted U-section usually contains a tubular stabilizer bar. When one rear wheel strikes a road irregularity, the inverted U-section and stabilizer bar twist, allowing some independent rear-wheel movement before the wheel movement affects the opposite rear wheel. Some semi-independent rear suspension systems have a track bar and brace connected from the inverted U-section to the chassis to reduce lateral rear-axle movement (**Figure 1-9**).

Many vehicles have an independent rear suspension system, wherein each rear wheel can move independently without affecting the position of the opposite rear wheel. This type of suspension system reduces rear tire wear and provides improved steering control. Independent rear suspension systems have a number of different configurations. A MacPherson strut independent rear suspension system has a strut and coil spring assembly connected from the top of the spindle through an upper strut mount to the chassis (**Figure 1-10**). No provision for strut rotation is required, because the rear wheels are not steered. Some independent rear suspension systems have a multilink design, wherein an adjustment link connected from the rear spindle to the chassis allows rear wheel position adjustment (**Figure 1-11**).

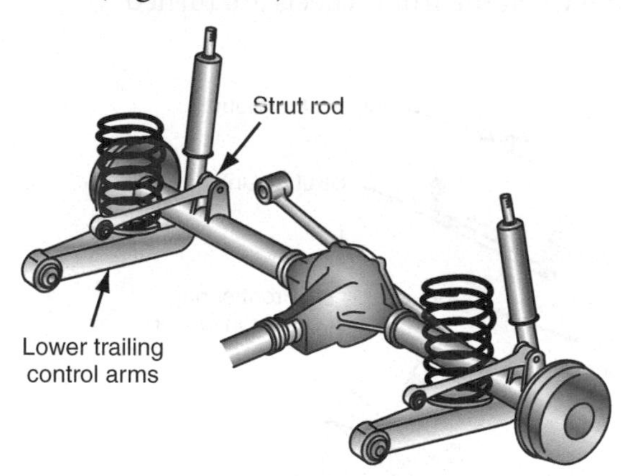

Figure 1-7 Live axle rear suspension system with coil springs.

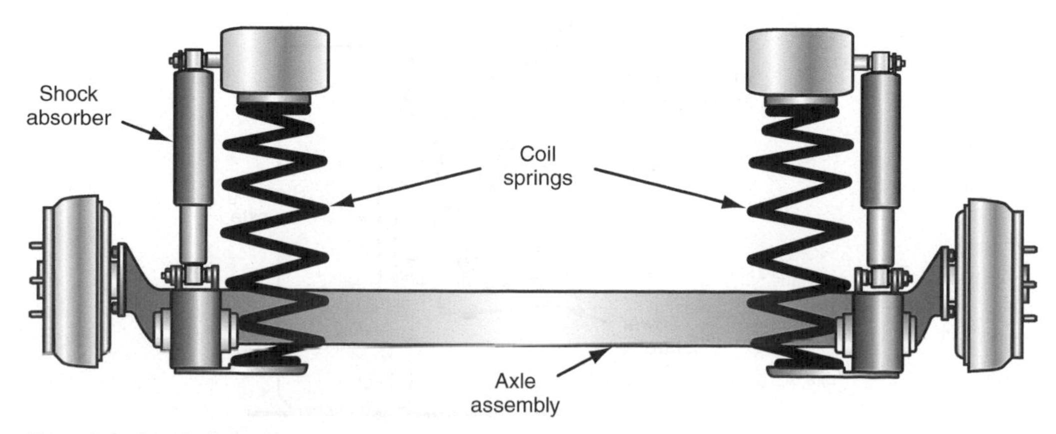

Figure 1-8 Semi-independent rear suspension system.

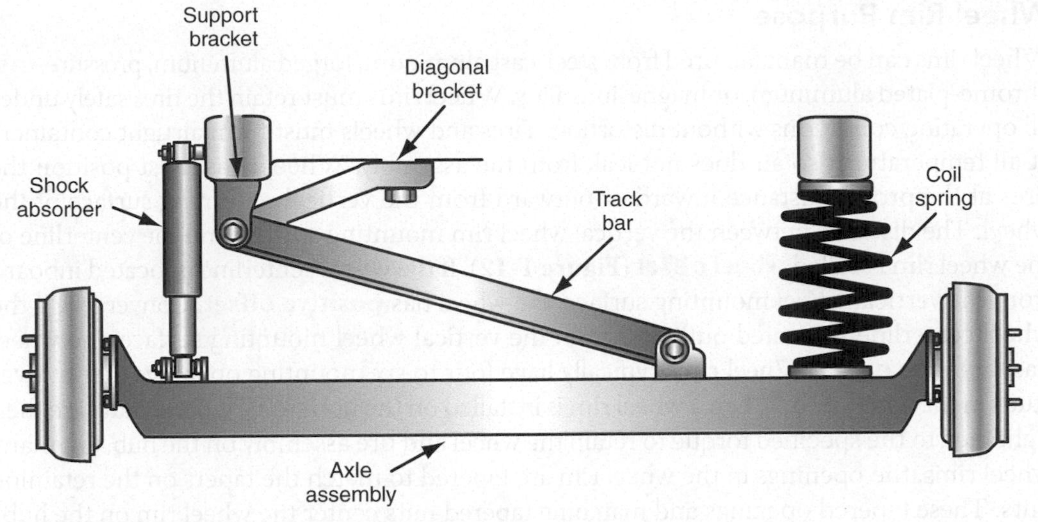

Figure 1-9 Semi-independent rear suspension system with track bar and brace.

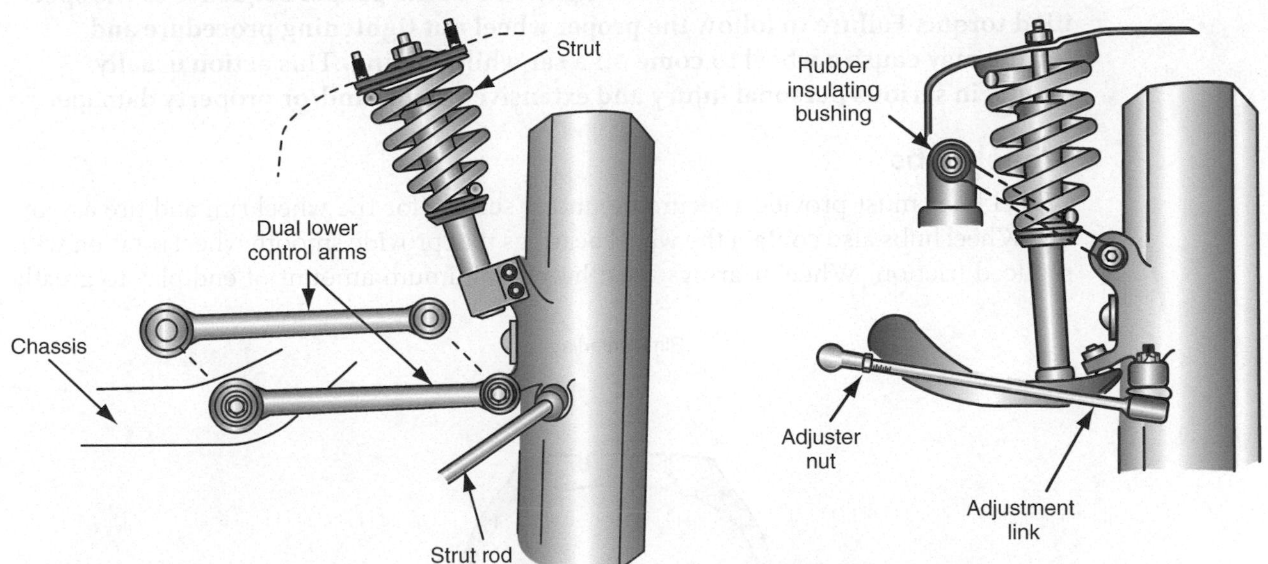

Figure 1-10 MacPherson strut independent rear suspension system.

Figure 1-11 Short-and-long arm independent rear suspension system.

TIRES, WHEELS, AND HUBS

Tire Purpose

Tires are extremely important because they play a large part in ensuring vehicle safety and ride quality. Tires are the only point of contact between the vehicle and the road surface. Vehicle tires provide these functions:

1. Tires must support the vehicle weight safely and firmly.
2. Tires must provide a comfortable ride.
3. Tires must supply adequate traction on various road surfaces to drive and steer the vehicle.
4. Tires must contribute to proper steering control and directional stability of the vehicle.
5. Tires must absorb high stresses when cornering, accelerating, and braking.
6. Tire treads must be designed to propel water off the tread and away from the tire when driving on wet highways. This action prevents water from lifting the tires off the road surface, which decreases tire traction.

Wheel Rim Purpose

Wheel rims can be manufactured from steel, cast aluminum, forged aluminum, pressure-cast chrome-plated aluminum, or magnesium alloy. Wheel rims must retain the tires safely under all operating conditions without distortion. Tires and wheels must form airtight containers at all temperatures so air does not leak from the assembly. Wheel rims must position the tires at the proper distance inward or outward from the vertical mounting surface of the wheel. The distance between the vertical wheel rim mounting surface and the centerline of the wheel rim is called **wheel offset** (**Figure 1-12**). If the wheel centerline is located inboard from the vertical wheel mounting surface, the wheel has **positive offset**. Conversely, if the wheel centerline is located outboard from the vertical wheel mounting surface, the wheel has **negative offset**. Wheel rims typically have four to six mounting openings that fit over studs in the wheel hub. When a wheel rim is installed on the hub studs, tapered nuts are then tightened to the specified torque to retain the wheel and tire assembly on the hub. On many wheel rims, the openings in the wheel rim are tapered to match the tapers on the retaining nuts. These tapered openings and matching tapered nuts center the wheel rim on the hub.

Shop Manual
Chapter 3, page 103

WARNING Wheel nuts must be tightened in the proper sequence to the specified torque. Failure to follow the proper wheel nut tightening procedure and torque may cause a wheel to come off a car while driving. This action usually results in serious personal injury and extensive vehicle and/or property damage.

Wheel Hubs

Wheel hubs must provide a secure mounting surface for the wheel rim and tire assembly. Wheel hubs also contain the wheel bearings that provide smooth wheel rotation with reduced friction. Wheel bearings must have a minimum amount of end play to greatly

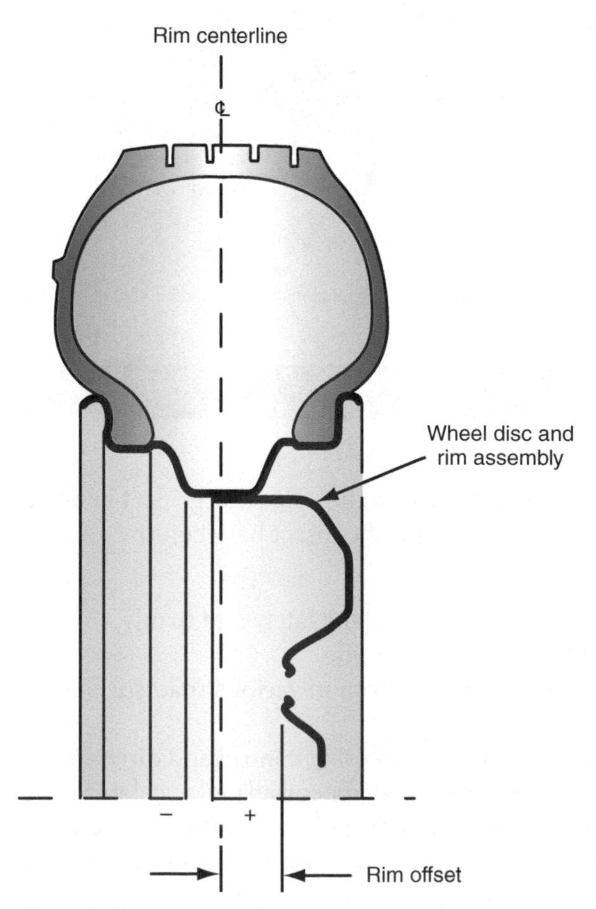

Figure 1-12 Wheel rim design.

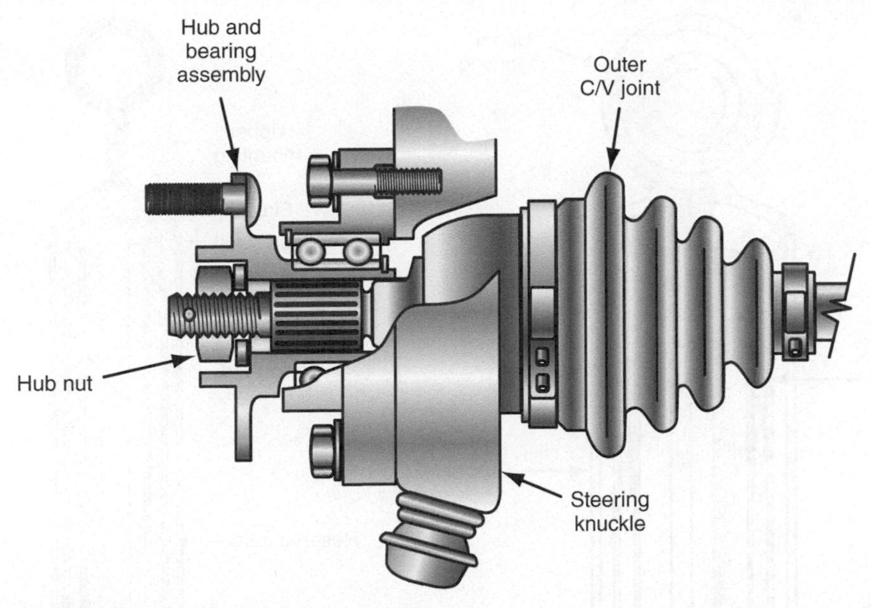

Figure 1-13 Wheel bearing and hub assembly.

reduce wheel lateral movement. The wheel hub and bearing assemblies must carry the load supplied by the vehicle weight, and these assemblies must also guide the wheel and tire assembly (**Figure 1-13**). The vehicle weight is supplied to the wheel hub and bearing assembly in a vertical direction. This type of bearing load is called a **radial bearing load**. When the vehicle turns a corner, the wheel hubs and bearings must carry **thrust bearing loads** supplied in a horizontal direction and **angular bearing loads** supplied in a direction between the horizontal and the vertical.

SHOCK ABSORBERS AND STRUTS

Each corner of the vehicle has a shock absorber or strut connecting the suspension system to the chassis. Shock absorbers control spring action and wheel oscillations to provide a comfortable ride. Controlling spring action and wheel oscillations also improve vehicle safety because the struts help to keep each tire tread in contact with the road surface. If the struts are worn out, excessive wheel oscillations when driving on irregular road surfaces can cause the driver to lose control of the vehicle. Struts also reduce body sway and lean while turning a corner. Struts reduce the tendency of the tire tread to lift off the road surface. This action improves tire tread life, traction, steering control, and directional stability.

Struts contain a sealed lower chamber filled with a special oil. Many shock absorbers have a nitrogen gas charge on top of the oil. This gas charge helps to prevent the shock absorber oil from foaming. A circular steel mount containing a rubber bushing is attached to the bottom end of the lower chamber, and this lower mounting is bolted to the suspension system. The upper strut housing is connected to a piston rod that extends into the lower chamber. A piston valve assembly is attached to the lower end of the piston rod (**Figure 1-14**). The upper strut mount is similar to the lower mounting, and the upper mount is bolted to the chassis.

Shop Manual
Chapter 5, page 196

When a wheel strikes a road irregularity, the wheel and suspension move upward, and the spring in the suspension system is compressed. This action forces the lower shock absorber chamber to move upward, and the oil must flow from below the shock absorber piston and valve to the area above the valve. Upward wheel movement is called **jounce travel**. The strut valves are designed to provide precise oil flow control, and thus control the speed of upward wheel movement.

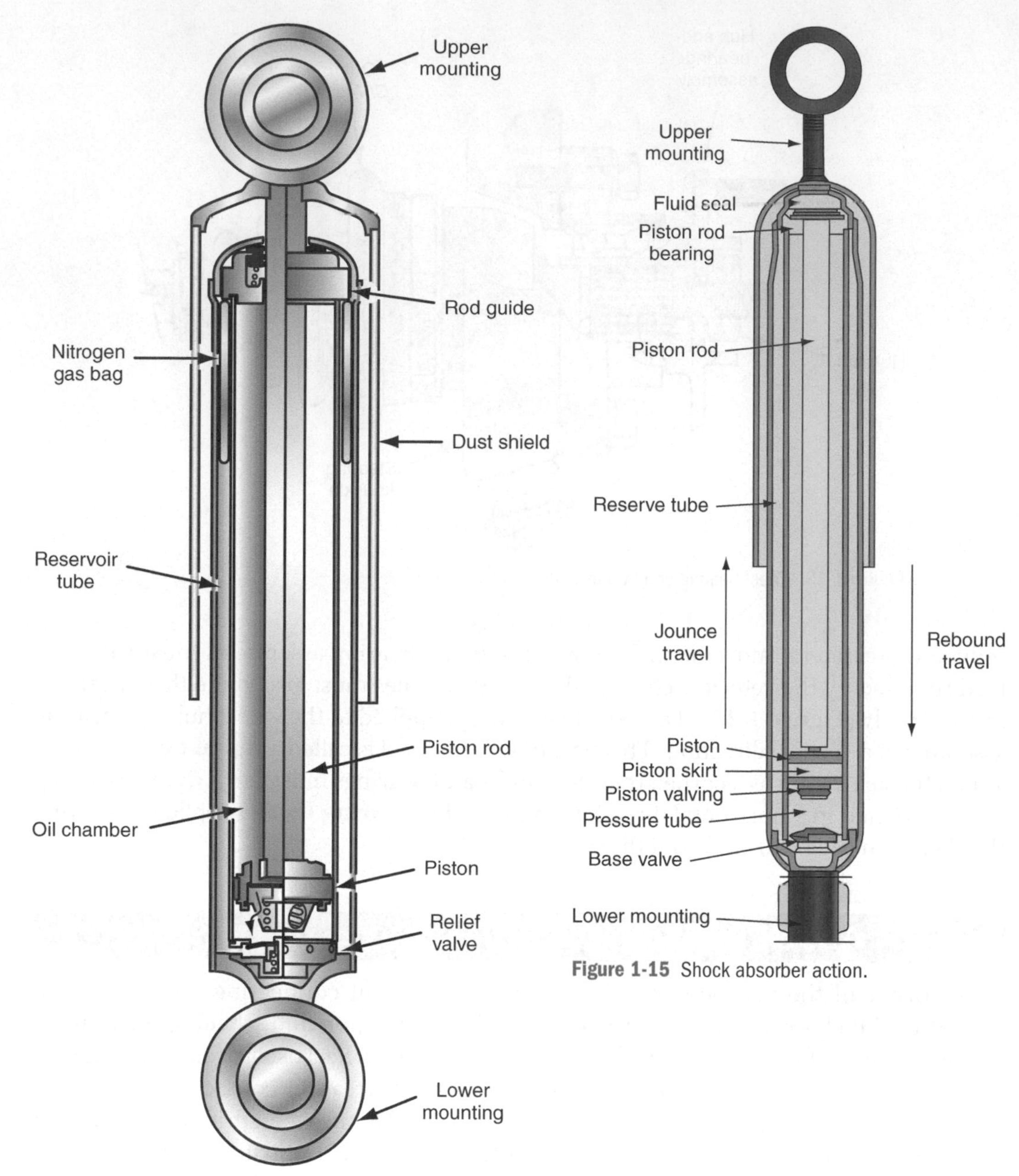

Figure 1-14 Shock absorber design.

Figure 1-15 Shock absorber action.

When a spring is compressed, it stores energy and then immediately expands with an equal amount of energy. When the spring expands, the tire and wheel assembly is forced downward. Under this condition, the lower strut chamber is forced downward, and oil must flow from above the shock absorber piston and valve to the area below the valve (**Figure 1-15**). Downward wheel movement is called **rebound travel**. The strut valves provide precise control of the oil flow, and this action controls spring action and wheel oscillations. Shock absorbers and valves are usually designed to provide more control during the rebound travel compared to the jounce travel.

Internal strut design is similar to shock absorber design, but struts also support the top of the steering knuckle. In most suspension systems, the lower end of the strut is attached to the top of the steering knuckle, and a special mount is connected between the upper end of the strut and the chassis (**Figure 1-16**). On front suspension systems, the upper strut mount must allow strut and spring rotation when the front wheels are turned to the right or the left. The upper strut mount isolates wheel and suspension vibrations from the chassis.

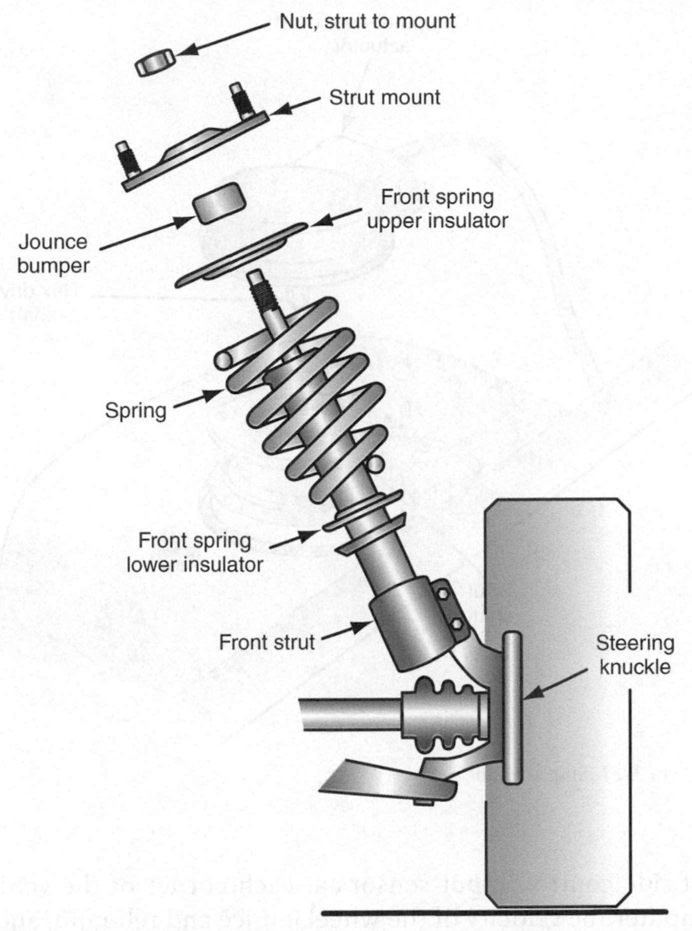

Figure 1-16 Front strut and spring assembly.

COMPUTER-CONTROLLED SUSPENSION SYSTEMS AND SHOCK ABSORBERS

Many vehicles are equipped with computer-controlled suspension systems that provide a soft, comfortable ride for normal highway driving, and then automatically and very quickly switch to a firm ride for hard cornering, braking, or fast acceleration. Computer-controlled suspension systems reduce body sway during hard cornering, and thus contribute to improved ride quality and vehicle safety. Some computer-controlled suspension systems are driver-adjustable with up to four suspension modes to allow the driver to tailor the ride quality to the driving style.

Some computer-controlled suspension systems have electronically actuated solenoids in each shock absorber or strut. These solenoids rotate the shock absorber or strut valves to adjust the valve openings and shock absorber control (**Figure 1-17**). Other shock absorbers or struts contain a magneto-rheological fluid that is a synthetic oil containing suspended iron particles. A computer-controlled electric winding is designed into the shock absorber housing. When there is no current flow through the winding, the iron particles are randomly dispersed in the oil. Under this condition, the oil consistency is thinner and the oil flows easily through the shock absorber valves to provide a softer ride. If the suspension computer supplies current flow to the shock absorber windings, the iron particles are aligned so the oil has a jelly-like consistency (**Figure 1-18**). This action instantly provides a much firmer ride. The computer can provide a large variation in current flow through the shock absorber windings and

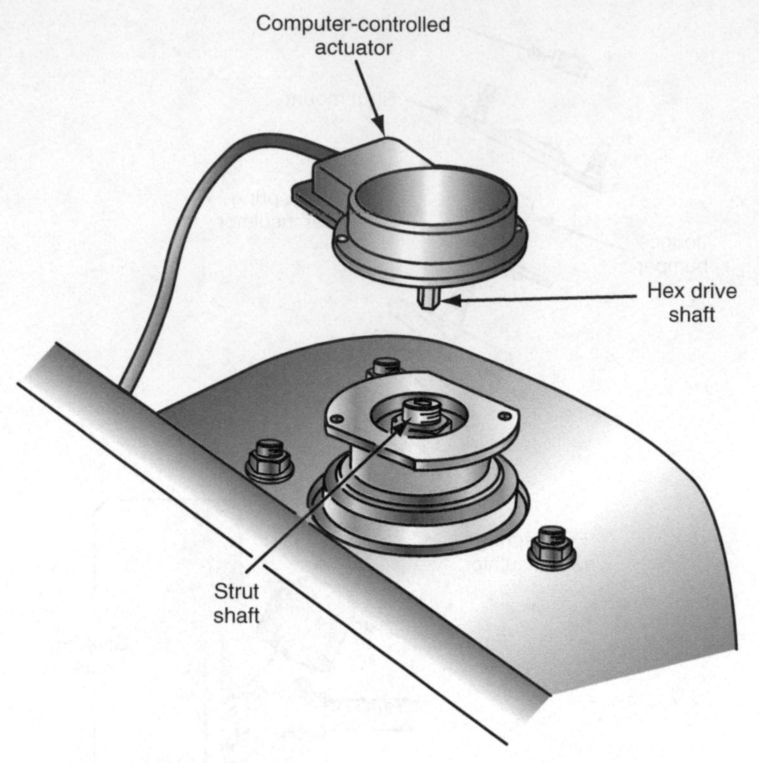

Figure 1-17 Strut actuator.

Shop Manual
Chapter 11, page 465

a wide range of ride control. Input sensors at each corner of the vehicle inform the suspension computer the velocity of the wheel jounce and rebound, and the computer uses these input signals to operate the shock absorber windings or actuators.

AUTHOR'S NOTE One of the advantages of computer-controlled shock absorbers and suspension systems is the speed at which modern computers can perform output functions. For example, a suspension computer can change the thickness of the magneto-rheological fluid in a shock absorber in about 1 millisecond (ms). When a wheel and tire strike a road irregularity and move upward, this fast computer action adjusts the thickness of the shock absorber fluid in relation to the wheel jounce velocity before the wheel moves downward in the rebound stroke and strikes the road surface.

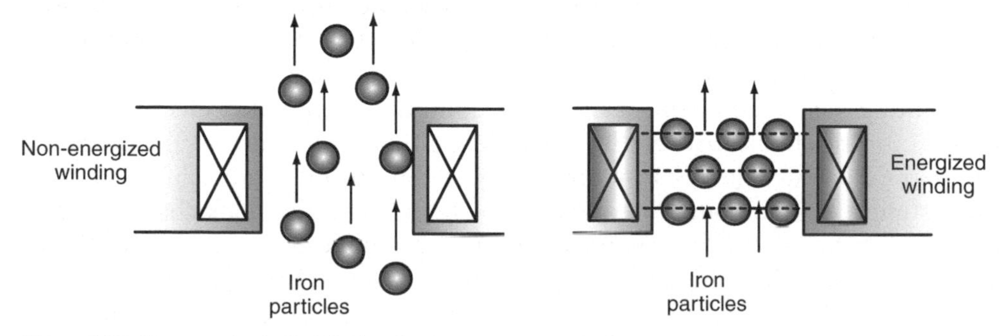

Figure 1-18 Magneto-rheological fluid action in a strut or shock absorber.

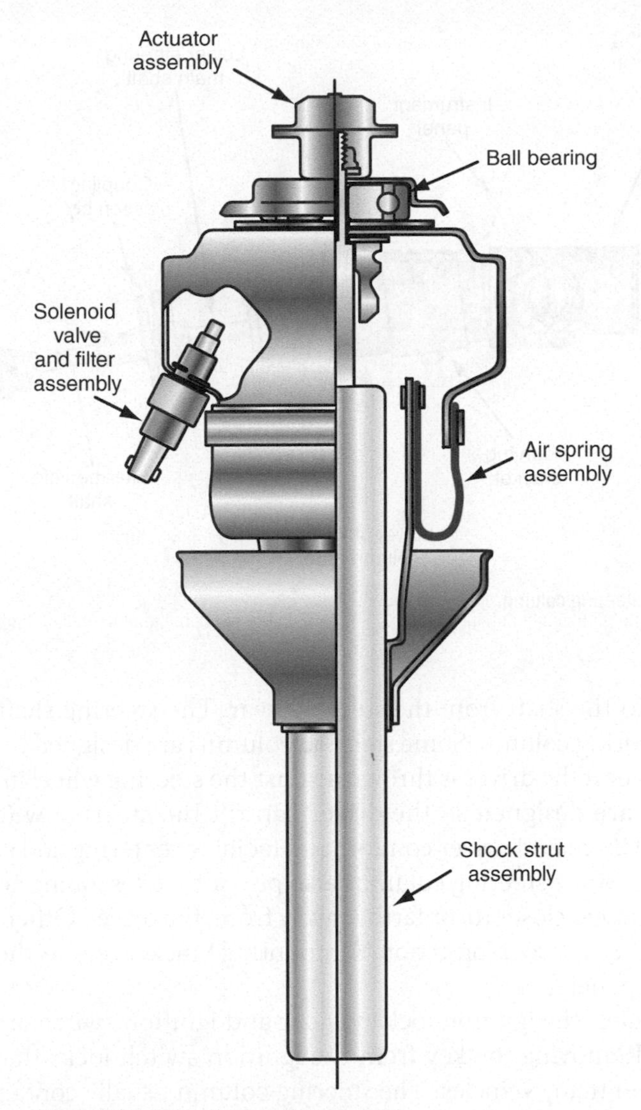

Figure 1-19 Air spring and strut assembly.

Some computer-controlled suspension systems have air springs in place of coil springs (**Figure 1-19**). Front and rear height sensors inform the suspension computer regarding the suspension height, and the computer operates an air compressor and air spring control valves to control the amount of air in the air springs, and thus control suspension height. Some air suspension systems also have computer-controlled shock absorbers or struts.

STEERING SYSTEMS

Steering systems are essential to vehicle safety, steering quality, and steering control. Steering system problems can cause the steering to pull to one side when driving straight ahead, as well as causing excessive steering effort, **wheel shimmy**, and/or excessive steering wheel free play. These problems all reduce vehicle safety and increase driver fatigue. Therefore, steering systems must be properly maintained.

Wheel shimmy may be defined as rapid inward and outward wheel and tire oscillations.

Steering Columns and Steering Linkage Mechanisms

The steering column connects the steering wheel to the steering gear. The steering wheel is connected to the steering shaft, and this shaft extends through the center of the steering column. The lower end of the steering shaft is connected through a universal joint or

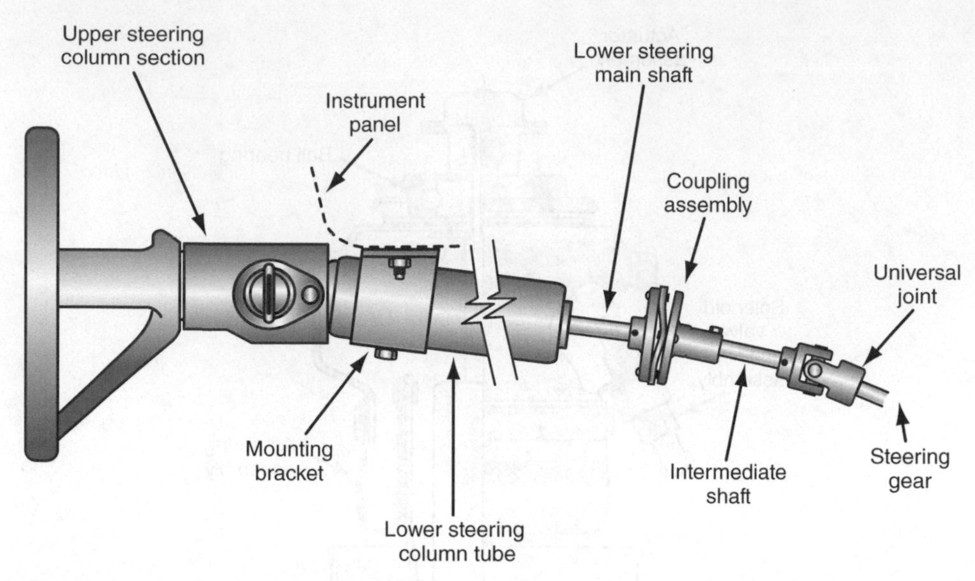

Figure 1-20 Tilt steering column.

flexible coupling to the shaft from the steering gear. The steering shaft is supported on bearings in the steering column. Some steering columns are designed to collapse or move away from the driver if the driver is thrown against the steering wheel in a collision. Some steering columns are designed so the driver can tilt the steering wheel downward or upward to provide increased driver comfort and facilitate entering and exiting the driver's seat (**Figure 1-20**). Some steering columns also provide a telescoping action so the steering wheel can be moved closer to, or farther away from, the driver. Other steering columns do not have any tilt or telescoping action. A mounting bracket retains the steering column to the instrument panel.

On most vehicles, the ignition lock cylinder and ignition switch are mounted in the steering column. Removing the key from the ignition switch locks the steering column and the gearshift on many vehicles. The steering column usually contains a combination signal light, wipe/wash, dimmer, and cruise control switch. This switch may be called a smart switch. The switch for the hazard warning lights is also mounted in the steering column. An air bag inflator module is mounted in the top of the steering wheel, and a clock-spring electrical connector under the steering wheel maintains electrical contact between the inflator module and the air bag electrical system.

Steering linkages connect the steering gear to the steering arms on the front wheels. In a parallelogram steering linkage, a pitman arm is connected from the steering gear to a center link (**Figure 1-21**). A pivoted idler arm bolted to the chassis supports the other end of the center link. Tie rods are connected from the center link to the steering arms attached to the front wheels. Pivoted ball studs are mounted in the inner ends of the tie rods, and outer tie rod ends are threaded into the tie rod adjusting sleeves. The outer tie rod ends contain pivoted ball studs, and these tapered studs fit into matching tapered openings in the outer ends of the steering arms. The pitman arm and idler arm position the center link so the tie rods are parallel to the lower control arms. This tie rod position is very important to maintain proper steering operation.

Caution

Regular inspection and maintenance of steering linkage components is very important for normal component life and maintaining vehicle safety.

Many vehicles have a rack and pinion steering linkage. In these linkages, the tie rods are connected through inner tie rod ends to the rack in the rack and pinion steering gear. Outer tie rod ends are connected from the tie rods to the steering arms (**Figure 1-22**). In this type of steering linkage, the steering gear mounting must position the tie rods so they are parallel to the lower control arms. The rack and pinion steering gear can be mounted on the cowl or the front cross member.

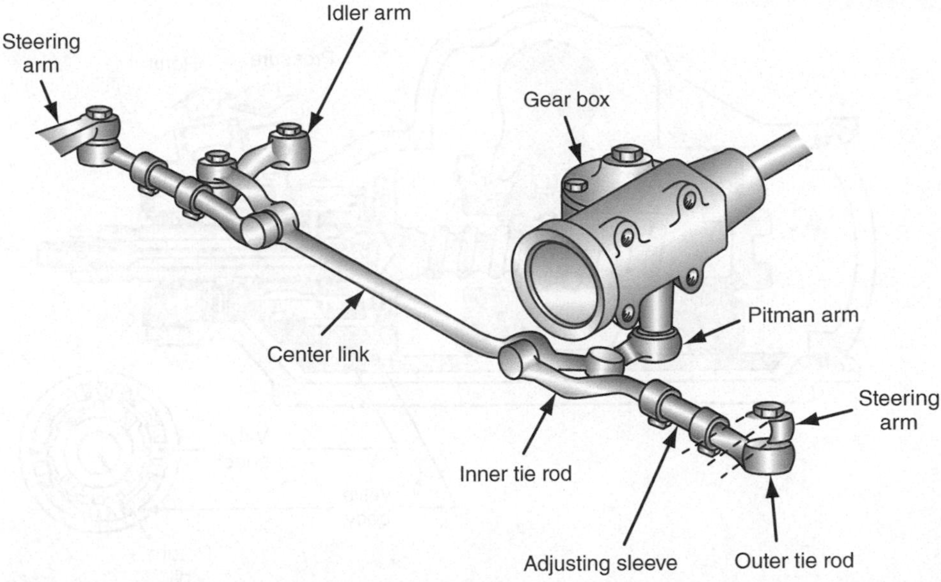

Figure 1-21 Parallelogram steering linkage.

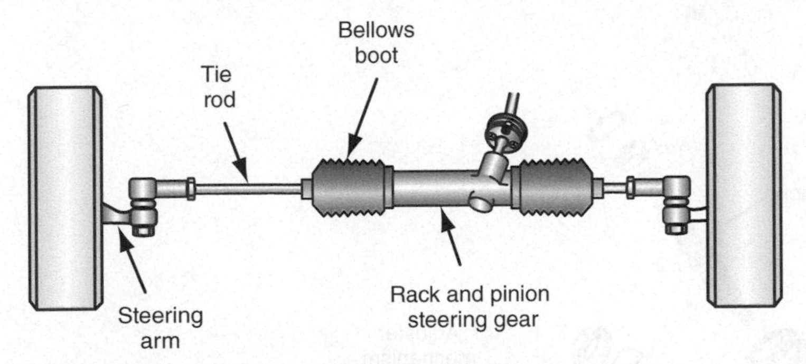

Figure 1-22 Rack and pinion steering gear and linkage.

Recirculating Ball Steering Gears

Some vehicles are equipped with a recirculating ball steering gear, wherein the steering shaft is attached to a worm gear in the steering gear. A ball nut with internal grooves is mounted over the worm gear. Ball bearings are mounted between the worm gear and ball nut grooves to reduce friction and provide reduced steering effort. Outer grooves on the ball nut are meshed with matching teeth on the sector shaft (**Figure 1-23**). The lower end of the sector shaft is splined to the pitman arm. When the steering wheel is turned, the ball nut moves upward or downward on the worm gear, which rotates the sector shaft to provide the desired steering action. Recirculating ball steering gears can be manual type with no hydraulic assist, or power type with hydraulic assist from the power steering pump.

Shop Manual
Chapter 14, page 611

Rack and Pinion Steering Gears

Rack and pinion steering gears and linkages are more compact than recirculating ball steering gears and parallelogram steering linkages. Therefore, rack and pinion steering gears are usually installed on smaller, front-wheel-drive vehicles. Rack and pinion steering gears transfer more road shock from the front wheels to the steering gear and steering wheel, because the tie rods are connected directly to the rack in the steering gear.

In a rack and pinion steering gear, a toothed rack is mounted on bushings in the rack housing. The rack teeth are meshed with teeth on a pinion gear mounted near one end of the gear. The pinion gear is mounted on bearings in the gear housing. The steering shaft from the steering column is attached to the upper end of the pinion gear (**Figure 1-24**).

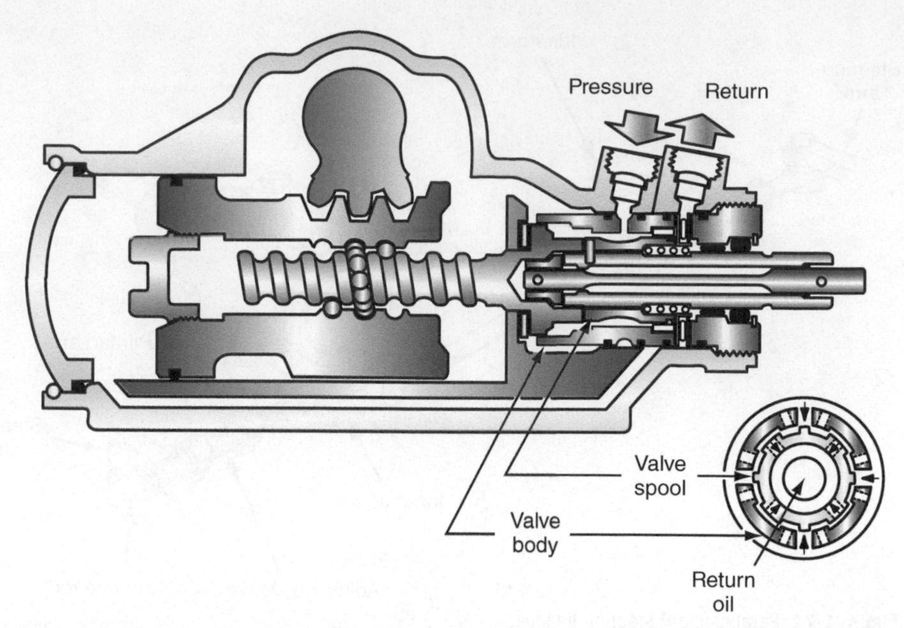

Figure 1-23 Power recirculating ball steering gear.

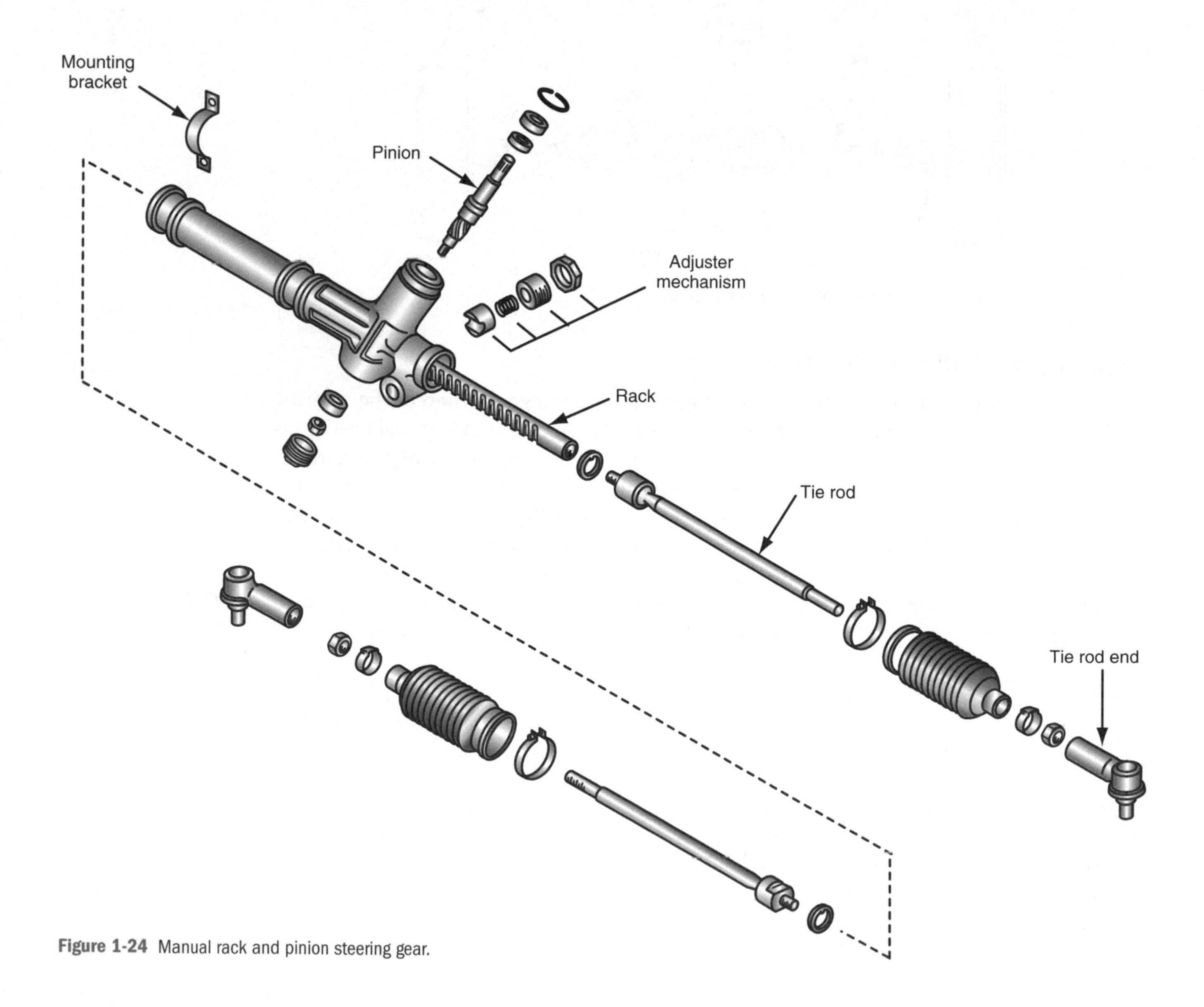

Figure 1-24 Manual rack and pinion steering gear.

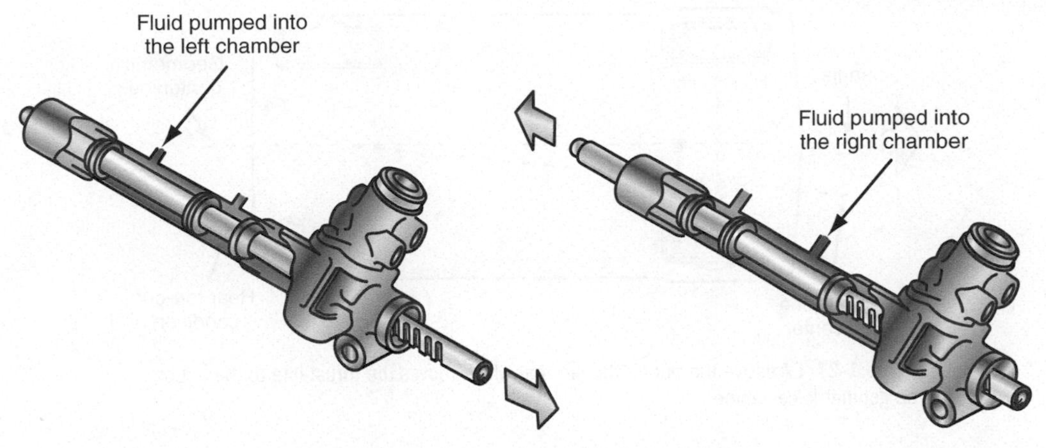

Figure 1-25 Power rack and pinion steering gear with sealed rack piston chambers.

When the steering wheel is turned, the rotation of the pinion gear moves the rack inward or outward to provide the desired steering action. Rack and pinion steering gears can be manual type or power assisted by fluid pressure from the power steering pump. Power rack and pinion steering gears have a piston near the center of the rack, and fluid pressure is supplied from the power steering pump to sealed chambers on either side of the rack piston to provide steering assistance (**Figure 1-25**).

WHEEL ALIGNMENT

Proper **wheel alignment** is extremely important for steering control, ride quality, and normal tire tread life. Improper wheel alignment may cause steering wander, steering pull to the right or left, or improper steering wheel return after turning a corner. Incorrect wheel alignment may contribute to harsh ride quality. Wheel alignment angles that are not within specifications may cause rapid tire tread wear.

Wheel alignment is the proper positioning of the tire and wheel assemblies on the vehicle to provide normal tire tread life, precise steering control, and satisfactory ride quality.

The **thrust line** is an imaginary line positioned at a 90-degree angle to the center of the rear-axle and extending forward.

Toe-out is a condition that occurs if the distance between the front edges of the front or rear tires is greater than the distance between the rear edges of the front or rear tires.

REAR WHEEL TRACKING

The rear wheels must track directly behind the front wheels to provide proper steering control. The front and rear wheels must be parallel to the vehicle centerline to provide proper rear wheel tracking. If the rear wheels are tracking directly behind the front wheels, the **thrust line** is positioned at the geometric centerline of the vehicle (**Figure 1-26**). If the left rear wheel has excessive **toe-out**, the thrust line is moved to the left of the geometric centerline (**Figure 1-27**). This improperly positioned thrust line causes the steering to pull to the right and also results in rapid tread wear on the left rear tire.

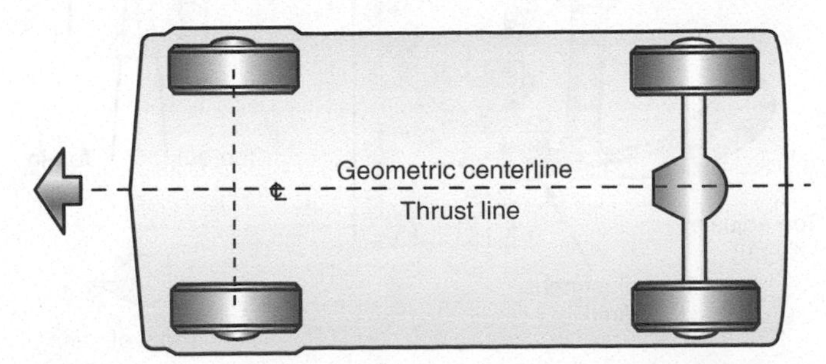

Figure 1-26 Front and rear wheels are parallel to the vehicle centerline.

Shop Manual
Chapter 9, page 375

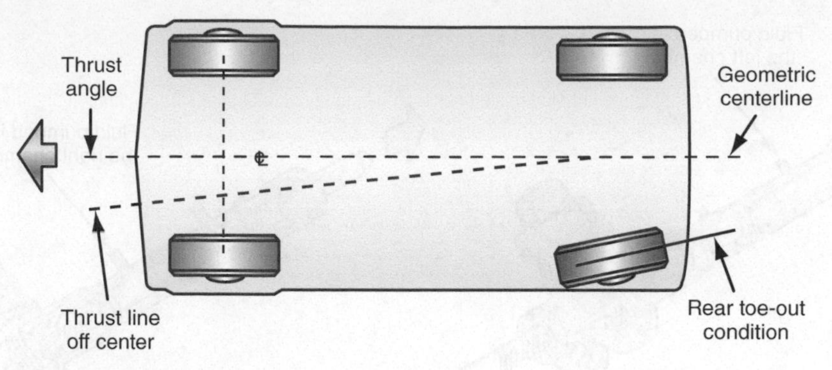

Figure 1-27 Excessive toe-out on the left rear wheel moves the thrust line to the left of the geometric centerline.

Rear Wheel Toe

A toe-out condition occurs when the distance between the front edges of the rear tires is greater than the distance between the rear edges of the rear tires. Toe-in occurs if the distance between the rear edges of the rear tires is greater than the distance between the front edges of the rear tires (**Figure 1-28**). On front-wheel-drive vehicles, driving forces tend to push the rear spindles backward. Therefore, these vehicles are usually designed with a zero toe-in or a slight toe-in. Improper rear wheel toe causes rapid tire tread wear because the wheel and tire assembly is being pushed sideways to a certain extent as the vehicle is driven. Steering pull to one side may also be a result of improper rear wheel toe.

Positive camber is present when the vertical centerline of the tire is tilted outward in relation to the true vertical centerline of the tire.

Negative camber occurs if the vertical centerline of the tire is tilted inward in relation to the true vertical centerline of the tire.

Rear Wheel Camber

Camber is the angle between the centerline of the wheel and tire in relation to the true vertical centerline of the wheel and tire viewed from the front. **Positive camber** occurs when the vertical centerline of the wheel and tire is tilted outward in relation to the true vertical centerline of the wheel and tire. **Negative camber** occurs if the vertical centerline of the wheel and tire is tilted inward in relation to the true vertical centerline of the wheel

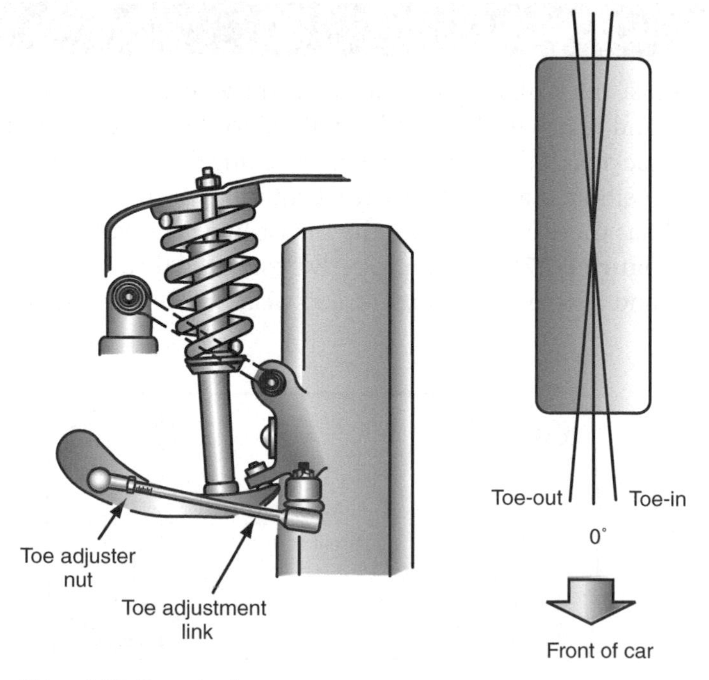

Figure 1-28 Rear wheel toe.

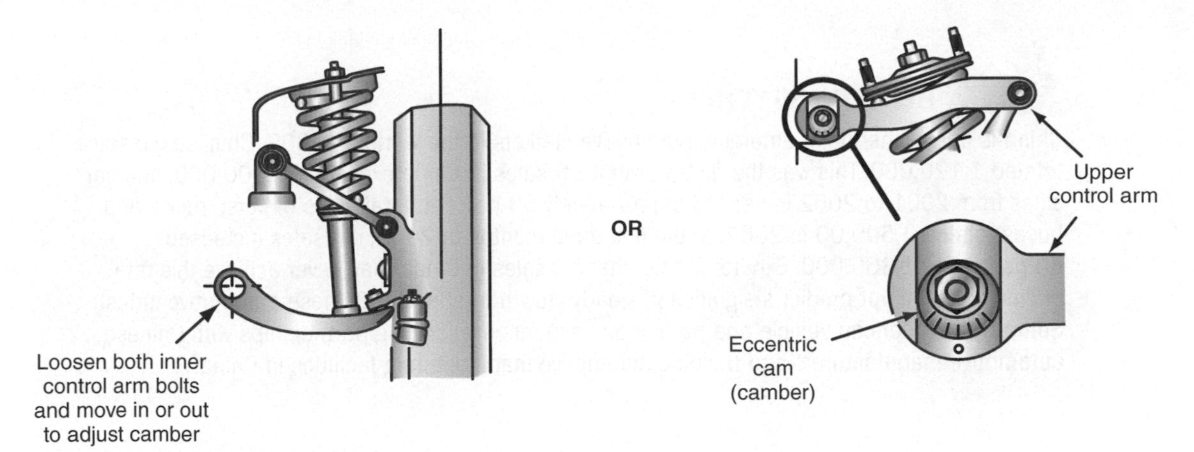

Figure 1-29 Rear wheel camber.

and tire (**Figure 1-29**). Excessive positive camber causes rapid wear on the outside edge of the tire tread, whereas excessive negative camber results in rapid wear on the inside edge of the tire tread. Because a tilted wheel and tire assembly always rolls in the direction of the tilt, improper camber angle on the front wheels may cause steering pull. Many front-wheel-drive vehicles have a slightly negative rear wheel camber that improves cornering stability.

Front Wheel Camber

Front wheel camber is the same as rear wheel camber except for the camber setting. Many front suspension systems are designed with a slightly positive camber.

Front Wheel Caster

Caster is the tilt of a line through the tire centerline (steering axis) in relation to the true vertical tire centerline viewed from the side. **Positive caster** occurs when the centerline of the tire is tilted rearward in relation to the true vertical centerline of the tire. **Negative caster** occurs when the centerline of the tire is tilted forward in relation to the centerline of the tire viewed from the side (**Figure 1-30**). Excessive negative caster causes the steering to wander when the vehicle is driven straight ahead. Excessive positive caster causes increased steering effort and rapid steering wheel return after turning a corner. Harsh ride quality is also a result of excessive positive caster because this condition causes the caster line to be aimed at road irregularities.

Positive caster is the angle of a line through the center of the tire that is tilted rearward in relation to the true vertical centerline of the tire viewed from the side.

Negative caster is the angle of a line through the center of the tire that is tilted forward in relation to the true vertical centerline of the tire viewed from the side.

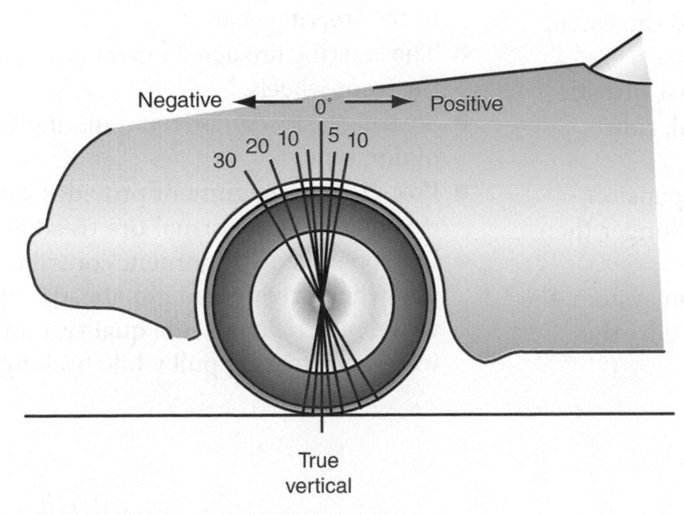

Figure 1-30 Positive and negative caster.

A BIT OF HISTORY

China is one of the major emerging automotive markets in the world. In 2002, Chinese car sales totaled 1,126,000. This was the first year that car sales in China exceeded 1,000,000, and car sales from 2001 to 2002 increased approximately 50 percent. Total sales of cars, trucks, and buses totaled 3,500,00 in 2002. In the first three months of 2003, car sales increased 40 percent to 1,360,000. Experts predict that car sales in China may never achieve this rate of growth again, but predict a significant, steady growth rate for the Chinese automotive industry. Some North American vehicle and parts manufacturers are forming partnerships with Chinese automotive manufacturers and building automotive manufacturing facilities in China.

Front Wheel Toe

Front wheel toe may be defined the same as rear wheel toe. However, the front wheel toe specification usually differs from the rear wheel toe specification. On a rear-wheel-drive car, driving forces tend to move the steering knuckles to a toe-out position. Therefore, these vehicles usually have a slight toe-in. Drive axle forces on a front-wheel-drive vehicle tend to move the front steering knuckles to a toe-in position, and this condition requires a slight toe-out specification.

SUMMARY

- The suspension system must provide proper steering control and ride quality.
- The steering system must maintain vehicle safety and reduce driver fatigue.
- All parts of a unitized body are load-carrying members, and these parts are welded together to form a strong assembly.
- Carbon dioxide (CO_2) is a by-product of the gasoline, diesel fuel, or ethanol combustion process.
- Greenhouse gases collect in the earth's upper atmosphere and form a blanket around the earth, which traps heat near the earth's surface.
- Front and rear suspension systems must provide proper wheel position, steering control, ride quality, and tire life.
- A short-and-long arm (SLA) front suspension system has a lower control arm that is longer than the upper control arm.
- In a MacPherson strut front suspension system, the top of the steering knuckle is supported by the lower end of the strut.
- Rear suspension systems can be live axle, semi-independent, or independent.
- Wheel rims are manufactured from steel, cast aluminum, forged aluminum, pressure-cast chrome-plated aluminum, or magnesium alloy.
- Wheel hubs contain the wheel bearings and support the load supplied by the vehicle weight.
- Bearing loads can be radial, thrust, or angular.
- Shock absorbers control spring action and wheel oscillations.
- The steering column connects the steering wheel to the steering gear.
- The steering linkage connects the steering gear to the front wheels.
- Steering gears can be recirculating ball or rack-and-pinion type.
- Proper wheel alignment provides steering control, ride quality, and normal tire tread life.
- Improper wheel alignment contributes to steering pull when driving straight ahead, improper steering wheel return, harsh ride quality, rapid tire tread wear, and steering pull while braking.

REVIEW QUESTIONS

Short Answer Essays

1. Explain how the engine and transaxle are supported in a front-wheel-drive vehicle with a unitized body.
2. Explain the purpose of coil springs in an SLA front suspension system.
3. Describe how the top of the steering knuckle is supported in a MacPherson strut front suspension system.
4. Explain the disadvantages of a live axle rear suspension system.
5. Explain the sources of CO_2 related to gasoline production and vehicle operation.
6. Describe the design of a wheel rim with positive offset.
7. Explain a radial bearing load.
8. Describe jounce wheel travel.
9. Explain the operation of a computer-controlled shock absorber that contains magneto-rheological fluid.
10. Explain why CO_2 is a harmful gas.

Fill-in-the-Blanks

1. In a rack and pinion steering gear, the inner ends of the tie rods are attached to the ends of the ______________.
2. The ______________ provides a steel box around the passenger compartment to provide passenger protection in a collision.
3. Tires are extremely important because they play a large part in providing ______________ and ______________.
4. Proper wheel alignment is extremely important to providing ______________ ______________, ______________ ______________, and normal tire ______________ ______________.
5. Excessive toe-out on a rear wheel may cause ______________ and rapid ______________.
6. The vehicle thrust line should be positioned at the ______________ of the vehicle.
7. If a wheel has negative camber, the tire centerline is tilted ______________ in relation to the true vertical tire centerline.
8. On a wheel rim the tapered openings match the tapers on the ______________.
9. Each corner of the vehicle has a ______________ or ______________ connected from the suspension system to the chassis.
10. Steering column connects the steering wheel to the ______________.

Multiple Choice

1. Unitized vehicle bodies have these special features and applications:
   A. Unitized bodies are typically used in large rear-wheel-drive vehicles.
   B. Unitized bodies have special steel panels in the hood and trunk lid to protect the vehicle occupants in a collision.
   C. Some members of a unitized body are load-carrying members.
   D. Unitized bodies have a steel box around the passenger compartment.
2. The purpose of the upper and lower control arms in an SLA front suspension system is to:
   A. Allow the steering knuckle to turn to the right or left.
   B. Control spring action on irregular road surfaces.
   C. Reduce body sway.
   D. Control lateral (side-to-side) wheel movement.
3. A semi-independent rear suspension system has:
   A. A rear-axle with an inverted steel U-section containing a tubular stabilizer bar.
   B. Upper and lower control arms.
   C. A one-piece rear-axle housing.
   D. A lower ball joint.

4. All of these statements about wheel rims are true EXCEPT:
   A. Wheel rims must position the tires at the proper distance inward or outward from the vertical wheel mounting surface.
   B. A wheel rim with a positive offset has the center of the tire positioned inboard from the vertical wheel mounting surface.
   C. A wheel rim with a neutral offset has the center of the tire positioned at the vertical center of the brake rotor.
   D. A wheel rim with a negative offset has the center of the tire positioned outboard from the vertical wheel mounting surface.
5. A thrust-type bearing load is applied in a:
   A. Horizontal direction.
   B. Vertical direction.
   C. Angular direction.
   D. Radial direction.
6. Shock absorbers control spring action and help to prevent:
   A. Improper wheel and tire position.
   B. Wheel oscillations.
   C. Excessive steering effort.
   D. Slow steering wheel return after a turn.
7. In a parallelogram steering linkage, the tie rods are parallel to the:
   A. Steering arms.
   B. Center link.
   C. Upper control arms.
   D. Lower control arms.
8. All of these statements about recirculating ball steering gears are true EXCEPT:
   A. Ball bearings are mounted between the worm gear and the ball nut.
   B. The lower end of the sector shaft is splined to the idler arm.
   C. As the steering wheel is turned, the ball nut moves upward or downward on the worm gear.
   D. The steering shaft is attached to the worm gear.
9. Excessive rear wheel toe-out on the left rear wheel causes:
   A. Steering pull to the right.
   B. Harsh ride quality.
   C. Reduced steering effort.
   D. Excessive wheel oscillations.
10. Proper wheel alignment is extremely important to provide all of the following EXCEPT:
    A. Steering control.
    B. Ride quality.
    C. Reduce body sway.
    D. Normal tire tread life.

CHAPTER 2
BASIC THEORIES

Upon completion and review of this chapter, you should be able to understand and describe:

- Newton's Laws of Motion.
- Work and force.
- Power.
- The most common types of energy and energy conversions.
- Inertia and momentum.
- Friction.
- Mass, weight, and volume.
- Static unbalance.
- Dynamic unbalance.
- The compressibility of gases and the noncompressibility of liquids.
- Atmospheric pressure and vacuum.
- Venturi operation.
- Voltage.
- Current.
- Resistance.
- Ohm's law.

Terms To Know

Ampere (A)	Horsepower	Torque
Atmospheric pressure	Inertia	Vacuum
Compressible	Mass	Voltage (V)
Dynamic balance	Momentum	Volume
Energy	Ohms (Ω)	Weight
Force	Ohm's law	Work
Friction	Static balance	

INTRODUCTION

An understanding of the basics is essential before you attempt a study of complex systems and components. Basic theories such as static balance, dynamic balance, and compressibility must be understood prior to a study of the components and systems in this book. If you have studied basic theories previously, the information in this chapter may be used as a review. A thorough study of this chapter will provide all the necessary background information you need before you study the suspension and steering systems in this book.

A BIT OF HISTORY

The automotive industry in the United States is a very large, dynamic industry. Total production of passenger cars, trucks, buses, and commercial vehicles has increased from 4192 in 1900 to 12,770,714 in 2000. Since 1920, the lowest vehicle production was 725,215 in 1945.

NEWTON'S LAWS OF MOTION

First Law

A body in motion remains in motion, and a body at rest remains at rest, unless some outside force acts on it. When a car is parked on a level street, it remains stationary unless it is driven or pushed. If the gas pedal is depressed with the engine running and the transmission in drive, the engine delivers power to the drive wheels and this force moves the car.

Second Law

A body's acceleration is directly proportional to the force applied to it, and the body moves in a straight line away from the force. For example, if the engine power supplied to the drive wheels increases, the vehicle accelerates faster.

Third Law

For every action, there is an equal and opposite reaction. A practical application of this law occurs when the wheel on a vehicle strikes a bump in the road surface. This action drives the wheel and suspension upward with a certain force, and a specific amount of energy is stored in the spring. After this action occurs, the spring forces the wheel and suspension downward with a force equal to the initial upward force.

WORK AND FORCE

Force is defined as energy applied to an object.

Work is defined as the result of applying a force.

When a **force** moves a certain mass a specific distance, **work** is done. When work is accomplished, the mass may be lifted or slid on a surface (**Figure 2-1**). Because force is measured in pounds and distance is measured in feet, the measurement for work is foot-pounds (ft.-lb). In the metric system, work is measured in Newton-meters (Nm). If a force moves a 3000-lb vehicle for 50 ft., 150,000 ft.-lb of work are produced. Mechanical force acts on an object to start, stop, or change the direction of the object. It is possible to apply force to an object and not move the object. For example, you may push with all your strength on a car stuck in a ditch and not move the car. Under this condition, no work is done. Work is only accomplished when an object is started, stopped, or redirected by mechanical force.

ENERGY

Energy may be defined as the ability to do work.

When **energy** is released to do work, it is called kinetic energy. This type of energy may also be referred to as energy in motion. Stored energy may be called potential energy. Energy is available in one of six forms:

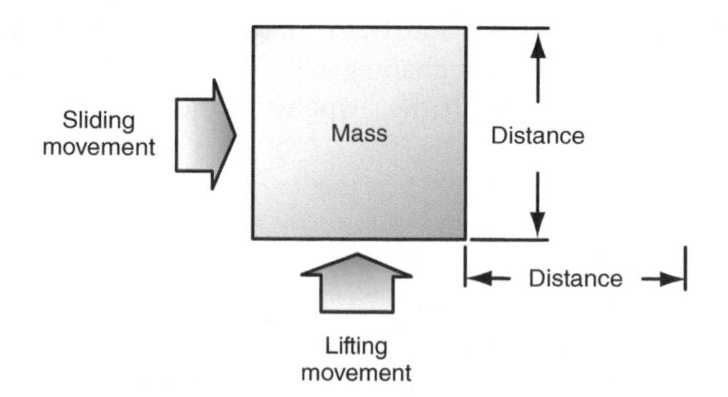

Figure 2-1 Work is accomplished when a mass is lifted or slid on a surface.

1. *Chemical energy* is contained in the molecules of different atoms. In the automobile, chemical energy is contained in the molecules of gasoline and also in the molecules of electrolyte in the battery.
2. *Electrical energy* is required to move electrons through an electric circuit. In the automobile, the battery is capable of producing electrical energy to start the vehicle, and the alternator produces electrical energy to power the electrical accessories and recharge the battery.
3. *Mechanical energy* is the ability to move objects. In the automobile, the battery supplies electrical energy to the starting motor, and this motor converts the electrical energy to mechanical energy to crank the engine. Because this energy is in motion, it may be called kinetic energy.
4. *Thermal energy* is energy produced by heat. When gasoline burns, thermal energy is released.
5. *Radiant energy* is light energy. In the automobile, radiant energy is produced by the lights.
6. *Nuclear energy* is the energy within atoms when they are split apart or combined. Nuclear power plants generate electricity with this principle. This type of energy is not used in the automobile.

ENERGY CONVERSION

Energy conversion occurs when one form of energy is changed to another form. Since energy is not always in the desired form, it must be converted to a usable form. Some of the most common automotive energy conversions are discussed in the following sections.

Chemical to Thermal Energy Conversion

Chemical energy in gasoline or diesel fuel is converted to thermal energy when the fuel burns in the engine cylinders.

Thermal to Mechanical Energy Conversion

Mechanical energy is required to rotate the drive wheels and move the vehicle. The piston and crankshaft in the engine and the drivetrain are designed to convert the thermal energy produced by the burning fuel into mechanical energy (**Figure 2-2**).

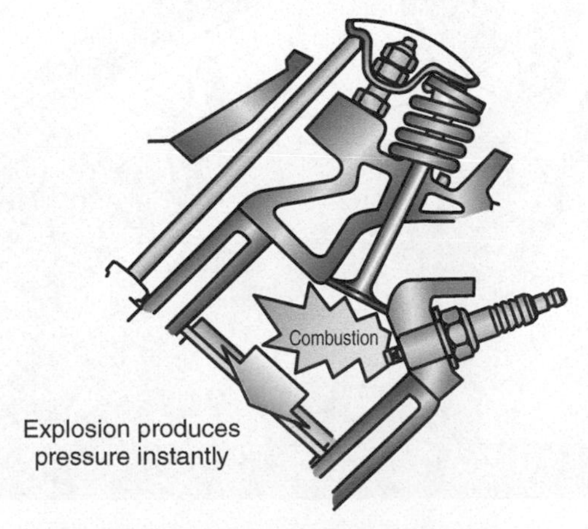
Combustion
Explosion produces pressure instantly

Figure 2-2 Thermal energy in the fuel is converted to mechanical energy in the engine cylinders. The piston, crankshaft, and drivetrain deliver this mechanical energy to the drive wheels.

Electrical to Mechanical Energy Conversion

The windshield wiper motor converts electrical energy from the battery or alternator to mechanical energy to drive the windshield wipers.

Mechanical to Electrical Energy Conversion

The alternator is driven by mechanical energy from the engine. The alternator converts this energy to electrical energy, which powers the electrical accessories on the vehicle and recharges the battery.

INERTIA

Inertia is defined as the tendency of an object at rest to remain at rest or the tendency of an object in motion to stay in motion.

The **inertia** of an object at rest is called static inertia, whereas dynamic inertia refers to the inertia of an object in motion. Inertia exists in liquids, solids, and gases. When you push and move a parked vehicle, you overcome the static inertia of the vehicle. If you catch a ball in motion, you overcome the dynamic inertia of the ball. Inertial forces are also referred to as apparent forces. Among the inertial forces are centrifugal and Coriolis forces. An example of inertial force can be observed if we place an object such as an apple on the passenger seat of a car while moving. To the driver the apple appears motionless. But when the brakes are applied suddenly with force, the apple travels forward in the direction of travel with great force and hits the dashboard, while the driver is restrained in the seat by the seat belt (**Figure 2-3**). From the driver's point of view, the apple's state of motion changes as it moves forward. It is the inertial force that causes the apple to move.

An example of centrifugal force can be observed when a driver makes a sharp and rapid turn into a tight corner. Again, if the driver has an apple sitting on the passenger seat as the vehicle is accelerated around a tight corner, the apple moves away from the center of rotation of the curve flying off the seat (**Figure 2-4**).

The Coriolis force is an inertial force caused when an object moves in relation to a rotating reference system. When the earth is the reference system the Coriolis force is caused by the rotation of the earth. This can be observed with pendulum swinging on a

Figure 2-3 As a driver slams on the brakes of a moving car, the apple that was sitting motionless on the passenger seat now flies forward due to inertial force.

Figure 2-4 As a driver makes a sharp cornering turn, the apple that was sitting motionless on the passenger seat moves away from the center of rotation of the curve, flying off the seat.

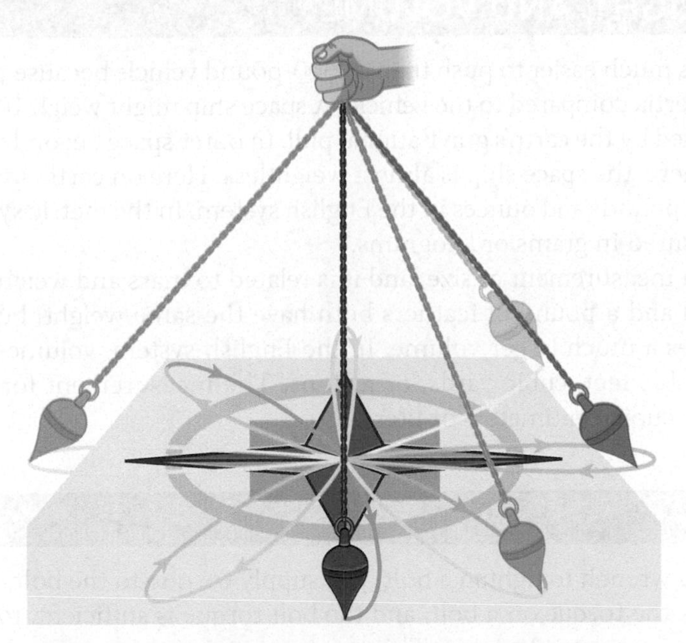

Figure 2-5 An example of the Coriolis force can be observed with a pendulum on a long wire. The pendulum will swing in overlapping elliptical rings. Over a 24-hour period of time, the plane of the oscillating pendulum will rotate 360 degrees.

long wire (**Figure 2-5**). The pendulum will swing in overlapping elliptical rings. Over a 24-hour period of time, the plane of the oscillating pendulum will rotate 360 degrees in a clockwise direction if it is located in the Northern Hemisphere. The Coriolis force is used in vehicle sensors to measure the yaw rate of the vehicle in motion and on some mass airflow sensors. Coriolis force plays an important role in ocean currents and the dynamics of weather.

MOMENTUM

When a force overcomes static inertia and moves an object, the object gains **momentum**.

Momentum is the product of an object's weight times its speed. Momentum is a type of mechanical energy. An object loses momentum if another force overcomes the dynamic inertia of the moving object.

FRICTION

Friction may be defined as the resistance to motion when the surface of one object is moved over the surface of another object.

Coefficient of drag (Cd) may also be called aerodynamic drag.

Friction may occur in solids, liquids, and gases. When a car is driven down the road, friction occurs between the air and the car's surface. This friction opposes the momentum, or mechanical energy, of the moving vehicle. Since friction creates heat, some of the mechanical energy from the vehicle's momentum is changed to heat energy in the air and body components. The mechanical energy from the engine must overcome the vehicle's inertia and the friction of the air striking the vehicle. Body design has a very dramatic effect on the amount of friction developed by the air striking the vehicle. The total resistance to motion caused by friction between a moving vehicle and the air is referred to as coefficient of drag (Cd). The study of Cd is not only very complicated but also very important. At 45 miles per hour (72 kilometers per hour), half of the engine's mechanical energy is used to overcome air friction, or resistance. Therefore, reducing a vehicle's Cd can be a very effective method of improving fuel economy and reducing CO_2 emissions.

MASS, WEIGHT, AND VOLUME

Mass is the measurement of an object's inertia.

Weight is the measurement of the earth's gravitational pull on the object.

Volume is the length, width, and height of a space occupied by an object.

A lawn mower is much easier to push than a 2500-pound vehicle because the lawn mower has very little inertia compared to the vehicle. A space ship might weigh 100 tons on earth where it is affected by the earth's gravitational pull. In outer space beyond the earth's gravity and atmosphere, the space ship is almost weightless. Here on earth, **mass** and **weight** are measured in pounds and ounces in the English system. In the metric system, mass and weight are measured in grams or kilograms.

Volume is a measurement of size, and it is related to mass and weight. For example, a pound of gold and a pound of feathers both have the same weight, but the pound of feathers occupies a much larger volume. In the English system, volume is measured in cubic inches, cubic feet, cubic yards, or gallons. The measurement for volume in the metric system is cubic centimeters or liters.

TORQUE

Torque is a force that does work with a twisting or turning force. However, movement does not have to occur.

When you pull a wrench to tighten a bolt, you supply **torque** to the bolt. If you pull on a wrench to check the torque on a bolt, and the bolt torque is sufficient, torque is applied to the bolt, but no movement occurs. This torque, or twisting force, is calculated by multiplying the force and the radius. For example, if you supply a 10-lb force on the end of a 2-ft. wrench to tighten a bolt, the torque is $10 \times 2 = 20$ ft.-lb (**Figure 2-6**). If the bolt turns during torque application, work is done. When a bolt does not rotate during torque application, no work is accomplished.

POWER

Horsepower is a measurement for the rate, or speed, at which work is done.

James Watt calculated that a horse could move 330 lb for 100 ft. in 1 minute (**Figure 2-7**). If you multiply 330 lb by 100 ft., the answer is 33,000 ft.-lb of work. Watt determined that one horse could do 33,000 ft.-lb of work in 1 minute. Thus, 1 **horsepower (HP)** is equal

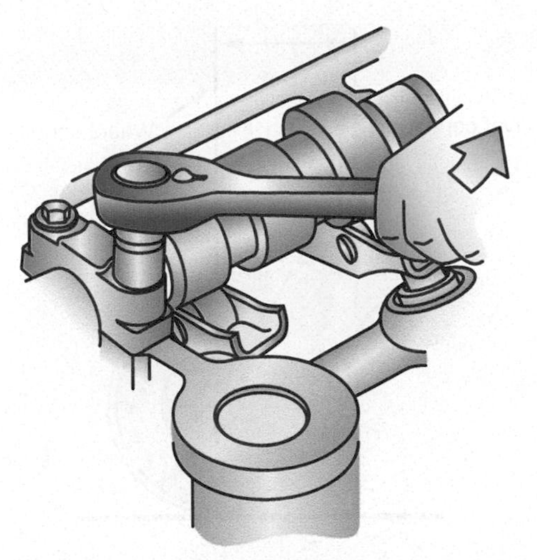

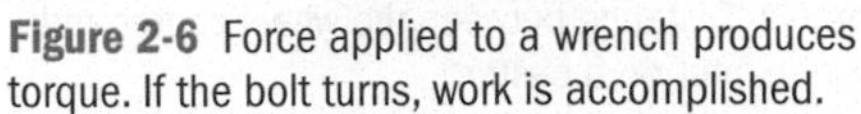

Figure 2-6 Force applied to a wrench produces torque. If the bolt turns, work is accomplished.

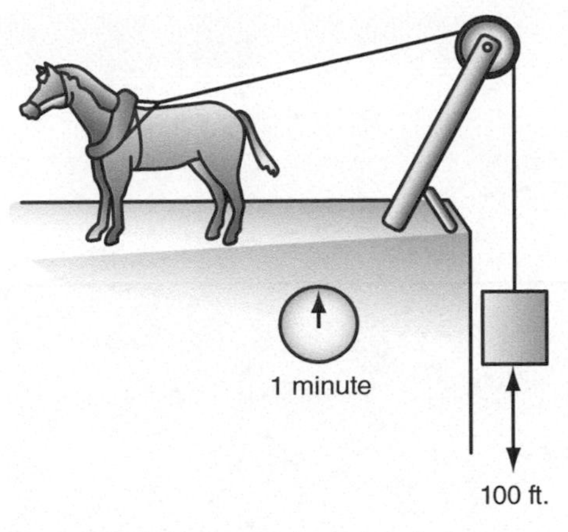

Figure 2-7 1 horsepower is produced when 330 lb are moved 100 ft. in 1 minute.

to 33,000 ft.-lb per minute, or 550 ft.-lb per second. 2 HP could do this same amount of work in one-half minute, or 4 HP could complete this work in one-quarter minute. If you push a 3000-lb (1360-kilogram) car for 11 ft. (3.3 meters) in one-quarter minute, you produce 4 HP. From this brief discussion about horsepower, we can understand that as power increases, speed also increases, or the time to do work decreases.

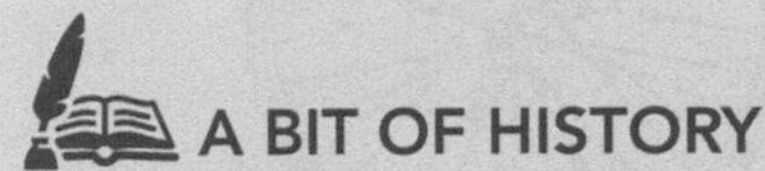

A BIT OF HISTORY

James Watt, of Scotland, is credited with being the first person to calculate power. He measured the amount of work that a horse could do in a specific time.

PRINCIPLES INVOLVING TIRES AND WHEELS IN MOTION

If you roll a cone-shaped piece of metal on a smooth surface, the cone does not move straight ahead. Instead, the cone moves toward the direction of the tilt on the cone. When you are riding a bicycle and want to make a left turn, if you tilt the bicycle to the left, it is much easier to complete the turn. The reason for this action is that a tilted, rolling wheel tends to move in the direction of the tilt. Similarly, if a tire and wheel on a vehicle are tilted, the tire and wheel tend to move in the direction of the tilt (**Figure 2-8**). This principle is used in front wheel alignment.

Shop Manual
Chapter 2, page 62

The casters on a piece of furniture are angled so the center of the caster wheel is some distance from the pivot center (**Figure 2-9**). When the furniture is moved, the casters turn on their pivots to bring the caster wheels into line with the pushing force on the furniture (**Figure 2-10**). This caster action causes the furniture to move easily in a straight line.

The weight of a bicycle rider is projected through the bicycle front fork to the road surface, and the tire pivots on the vertical centerline of the wheel when the handlebars are turned. Notice the centerline of the front fork is tilted rearward in relation to the vertical centerline of the wheel (**Figure 2-11**). Since the tire pivot point is behind the front fork centerline where the weight is projected against the road surface, the front wheel tends to return to the straight-ahead position after a turn. The wheel also tends to remain in the straight-ahead position as the bicycle is driven. This principle is applied in automotive front wheel alignment.

Tilted rolling tires and wheels always move in the direction of the tilt.

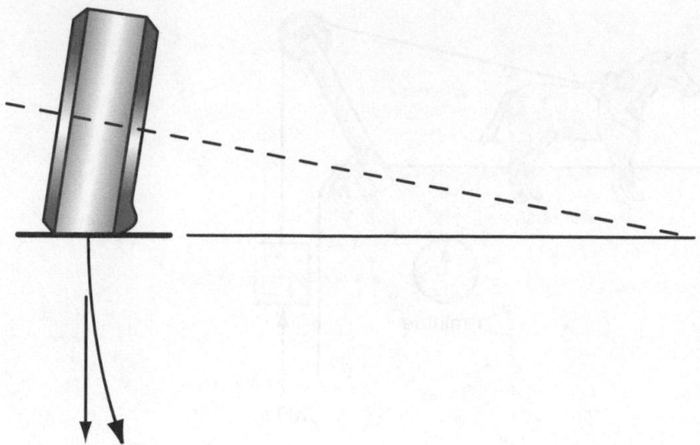

Figure 2-8 A tilted, rolling wheel tends to move in the direction of the tilt.

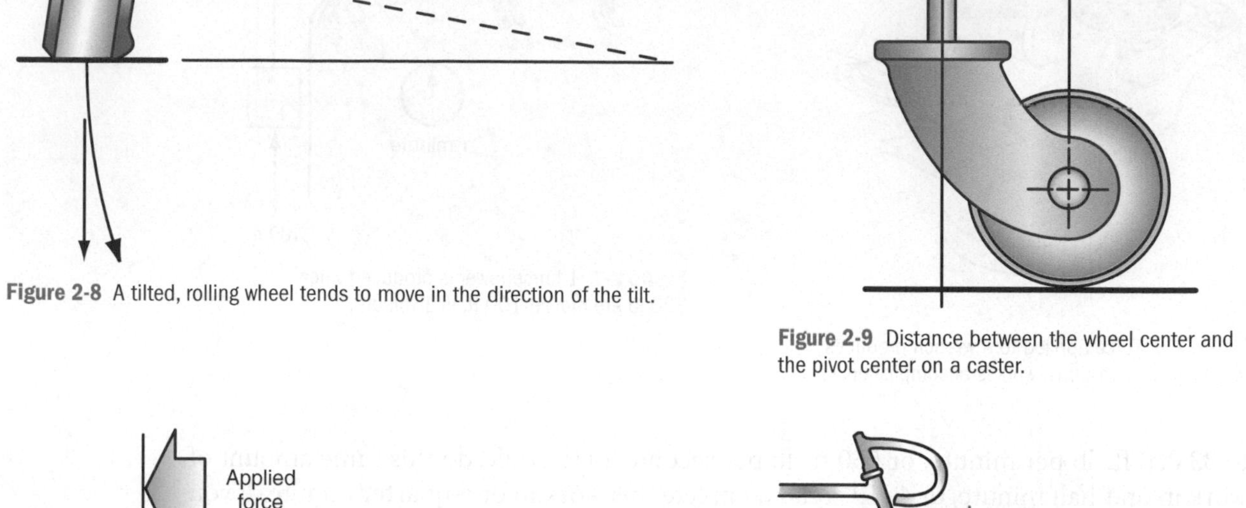

Figure 2-9 Distance between the wheel center and the pivot center on a caster.

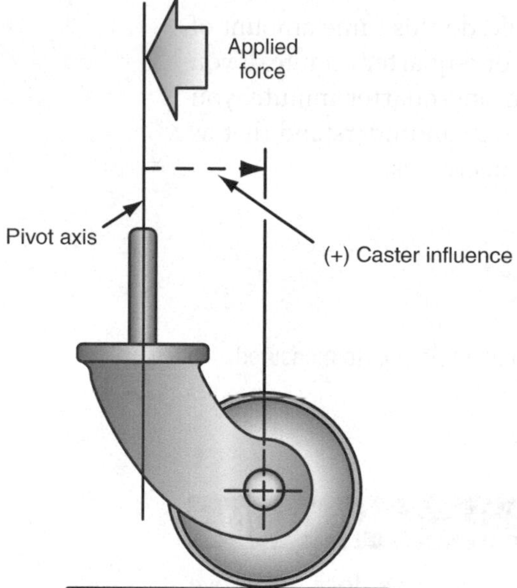

Figure 2-10 Caster aligned with the pushing force provides straight-ahead movement when furniture is pushed.

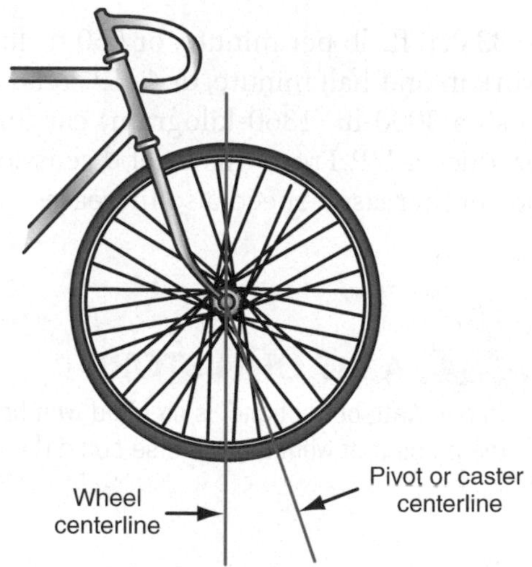

Figure 2-11 If a wheel centerline and pivot point are behind the front fork centerline pivot point, the wheel tends to return to the straight-ahead position after a turn. The wheel also tends to remain in the straight-ahead position as the bicycle is driven.

PRINCIPLES INVOLVING THE BALANCE OF WHEELS IN MOTION

Static balance refers to the equal distribution of weight around the center of a tire-and-wheel assembly viewed from the side.

When the weight of a tire and wheel assembly is distributed equally around the center of wheel rotation viewed from the side, the wheel and tire have proper **static balance**. Under this condition, the tire and wheel assembly has no tendency to rotate by itself, regardless of the wheel position. If the weight is not distributed equally around the center of wheel rotation, the wheel and tire are statically unbalanced (**Figure 2-12**). As the wheel and tire rotate, centrifugal force acts on this static unbalance and causes the wheel to "tramp" or hop. This wheel and tire action can often be seen while looking at tires on vehicles that may pass you on a highway.

When a ball is rotated on the end of a string, the ball and string form an angle with the axis of rotation. If the rotational speed is increased, the ball and string form a 90-degree

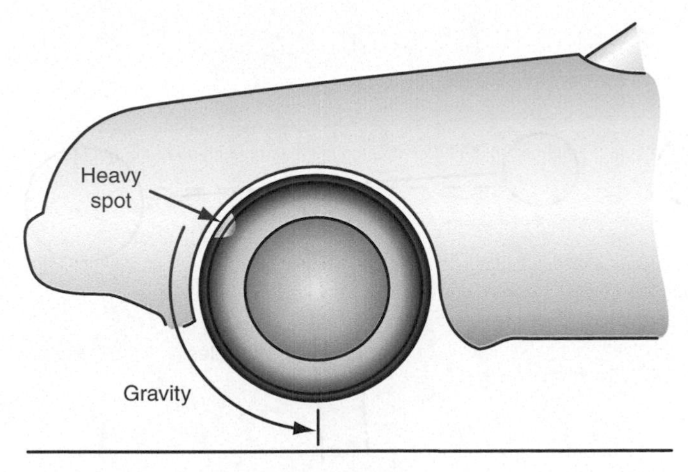

Figure 2-12 Static wheel unbalance caused by unequal weight distribution around the center of the wheel rotation.

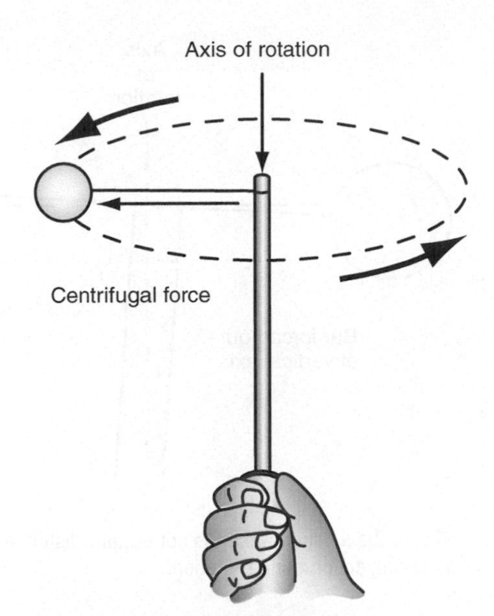

Figure 2-13 A weight tends to rotate at a 90-degree angle in relation to the axis of rotation.

angle with the axis of rotation (**Figure 2-13**). Any weight will always tend to rotate at a 90-degree angle to the axis of rotation.

If two balls are positioned on a metal bar so their weight is equally distributed on the centerline of the rotational axis of rotation, the path of rotation remains at a 90-degree angle to the centerline of the axis when the bar is rotated. Under this condition, the metal bar and the balls are in **dynamic balance** (**Figure 2-14**).

Dynamic balance refers to the equal distribution of weight on each side of a tire-and-wheel centerline viewed from the front or rear.

If weights are placed on a metal bar so their weights are not equally distributed in relation to the centerline of the rotational axis of rotation, the weights still tend to rotate at a 90-degree angle in relation to the rotational axis. This action forces the pivot out of its vertical axis (**Figure 2-15**). When the bar is rotated 180 degrees, the bar is forced out of its vertical axis in the opposite direction. If this condition is present, the bar has a wobbling action as it rotates. Under this condition, the bar is said to have dynamic unbalance, but static balance is maintained.

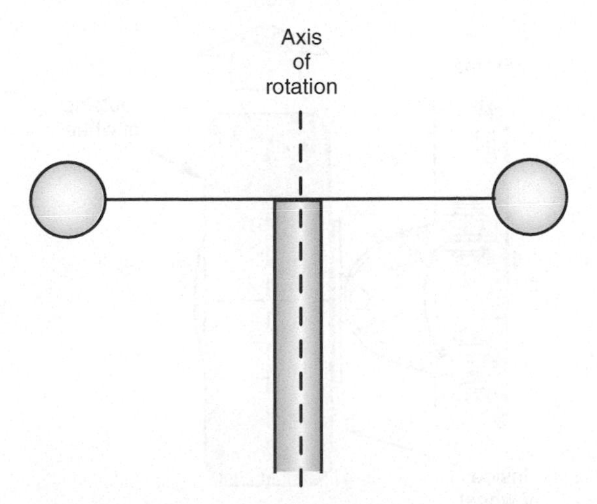

Figure 2-14 When weight on a metal bar is distributed equally on the centerline of the axis of rotation, the bar and balls remain in dynamic balance during rotation.

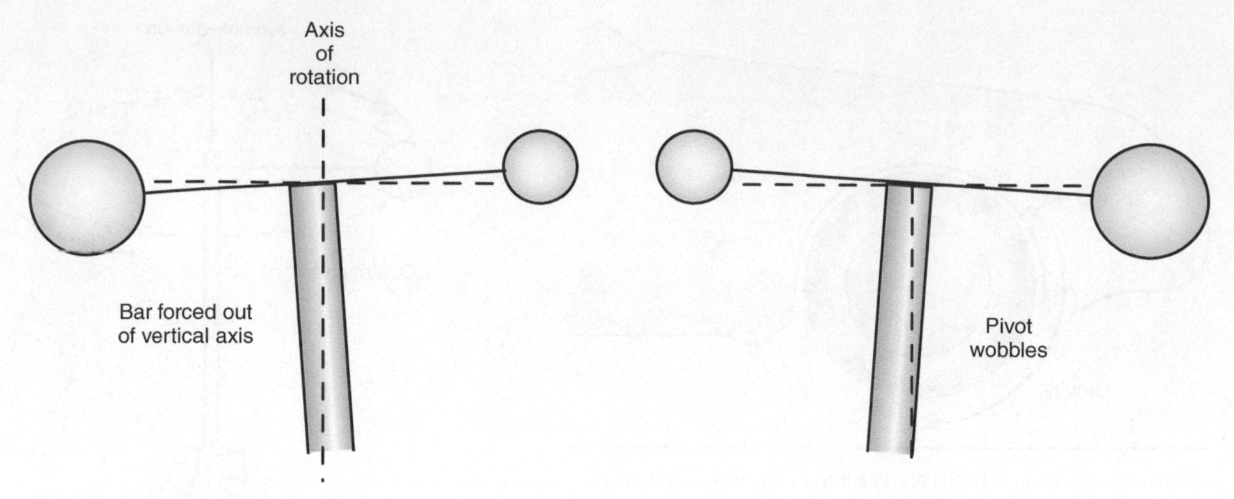

Figure 2-15 When weight is not equally distributed in relation to the centerline of rotational axis, dynamic unbalance causes a wobbling action during rotation.

Molecular energy may be defined as the kinetic energy available in atoms and molecules because of the constant electron movement within the atoms and molecules.

Similarly, when the weight on a tire and wheel is not distributed equally on both sides of the tire centerline viewed from the front, the tire and wheel are dynamically unbalanced. This condition produces a wobbling action as the tire and wheel rotate. The weight must be distributed equally in relation to the tire centerline to provide proper dynamic balance (**Figure 2-16**). These principles are used in automotive wheel balancing.

AUTHOR'S NOTE Improper static or dynamic wheel balance causes excessive tire tread wear, increased wear on suspension components, and driver fatigue.

PRINCIPLES INVOLVING LIQUIDS AND GASES

Molecular Energy

Remember that kinetic energy refers to energy in motion. Since electrons are constantly in motion around the nucleus in atoms or molecules, kinetic energy is present in all matter.

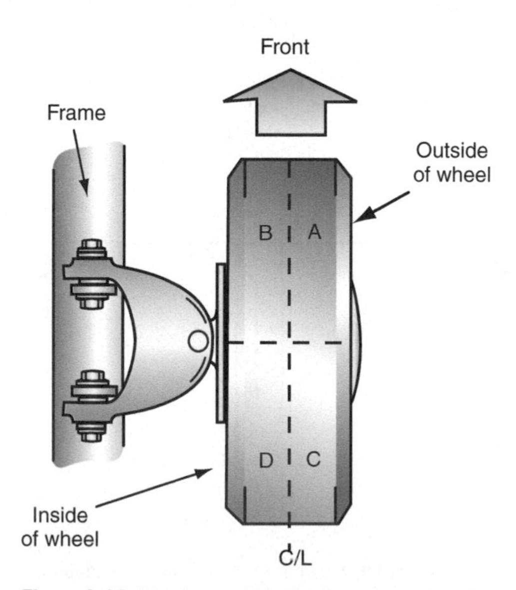

Figure 2-16 Weight must be distributed equally in relation to the tire-and-wheel centerline.

Kinetic energy in atoms and molecules increases as the temperature increases. A decrease in temperature reduces the kinetic energy. Molecules in solids move slowly compared to those in liquids or gases. Gas molecules move quickly compared to liquid molecules. Since gas molecules are in constant motion, they spread out to fill all the space available. At higher temperatures, gas molecules spread out more, whereas at lower temperatures, gas molecules move closer together.

Pressure may be defined as a force exerted on a given surface area.

Pressure and temperature are directly related. If you increase one, you also increase the other.

Pressure and volume are inversely proportional. If volume is decreased, pressure is increased.

Temperature

Temperature affects all liquids, solids, and gases. The volume of any matter increases as the temperature increases. Conversely, the volume decreases in relation to a reduction in temperature. When the gases in an engine cylinder are burned, the sudden temperature increase causes rapid gas expansion, which pushes the piston downward and causes engine rotation (**Figure 2-17**).

Pressure and Compressibility

Since liquids and gases are both substances that flow, they may be classified as fluids. If a nail punctures an automotive tire, the air escapes until the pressure in the tire is equal to atmospheric pressure outside the tire. When the tire is repaired and inflated, air pressure is forced into the tire. If the tire is inflated to 32 pounds per square inch (psi), or 220 kilopascals (kPa), this pressure is applied to every square inch on the inner tire surface. Pressure is always supplied equally to the entire surface of a container. Since air is a gas, the molecules have plenty of space between them. When the tire is inflated, the pressure in the tire increases, and the air molecules are squeezed closer together, or compressed. Under this condition, the air molecules cannot move as freely, but extra molecules of air can still be forced into the tire. Therefore, gases such as air are said to be **compressible**.

The air in the tire may be compared to a few balls on a billiard table without pockets. If a few more balls are placed on the table, the balls are closer together, but they can still move freely (**Figure 2-18**).

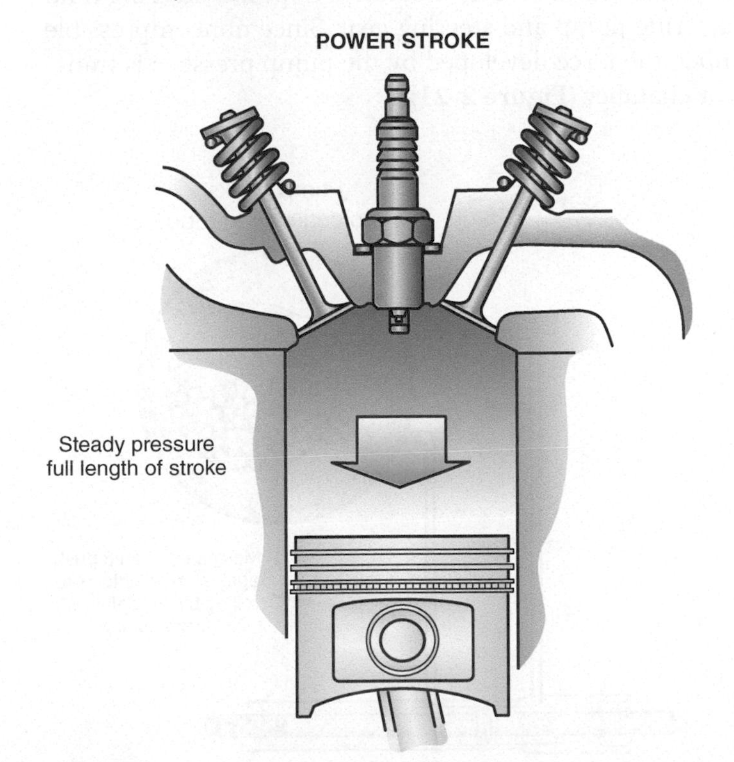
POWER STROKE
Steady pressure
full length of stroke

Figure 2-17 Hot, expanding gases push the piston downward and rotate the crankshaft.

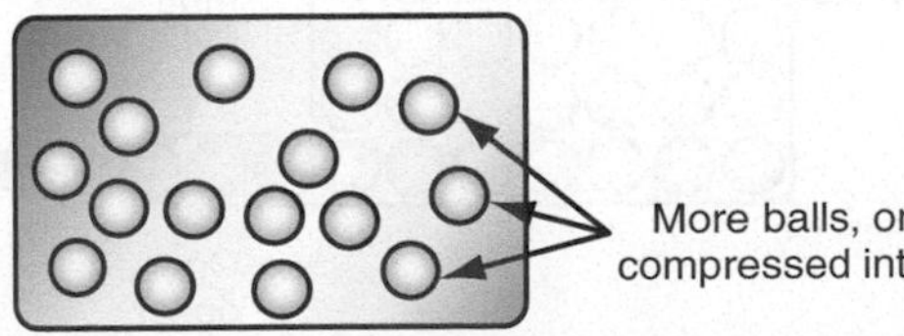
Balls are still free to move and bounce,
but with less space and more activity.
More balls, or molecules,
compressed into same space

Figure 2-18 Gases can be compressed much like more balls can be placed on a billiard table containing a few balls.

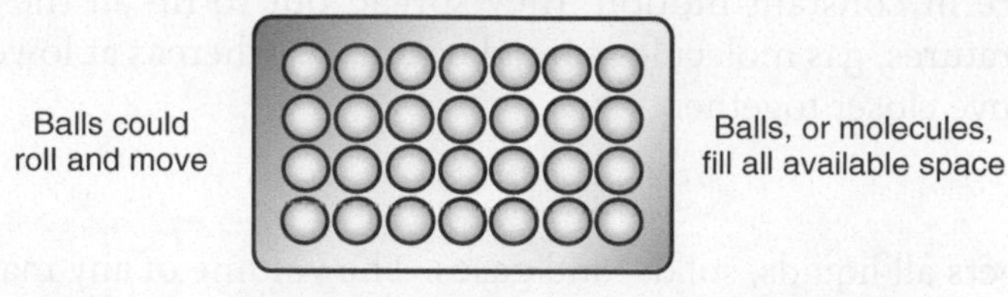

Figure 2-19 Liquids are noncompressible, just as more balls cannot be added to a billiard table with no pockets that is completely filled with balls.

If the vehicle is driven at high speed, friction between the road surface and the tires heats the tires and the air in the tires. When air temperature increases, the pressure in the tire also increases. Conversely, a temperature decrease reduces pressure.

If 100 cubic feet (2.8 cubic meters) of air is forced into a large truck tire, and the same amount of air is forced into a much smaller car tire, the pressure in the car tire is much greater.

Molecules in a liquid may be compared to a billiard table without pockets that is completely filled with balls. These balls can roll around, but no additional balls can be placed on the table because the balls cannot be compressed. Similarly, liquid molecules cannot be compressed (**Figure 2-19**).

Liquid Flow

If a tube is filled with billiard balls and the outlet is open, more balls may be added to the inlet. When each ball is moved into the inlet, a ball is forced from the outlet. If the outlet is closed, no more balls can be forced into the inlet (**Figure 2-20**).

The billiard balls in the tube may be compared to molecules of power steering fluid in the line between the power steering pump and steering gear. Since noncompressible fluid fills the line and gear chamber, the force developed by the pump pressure is transferred through the line to the gear chamber (**Figure 2-21**).

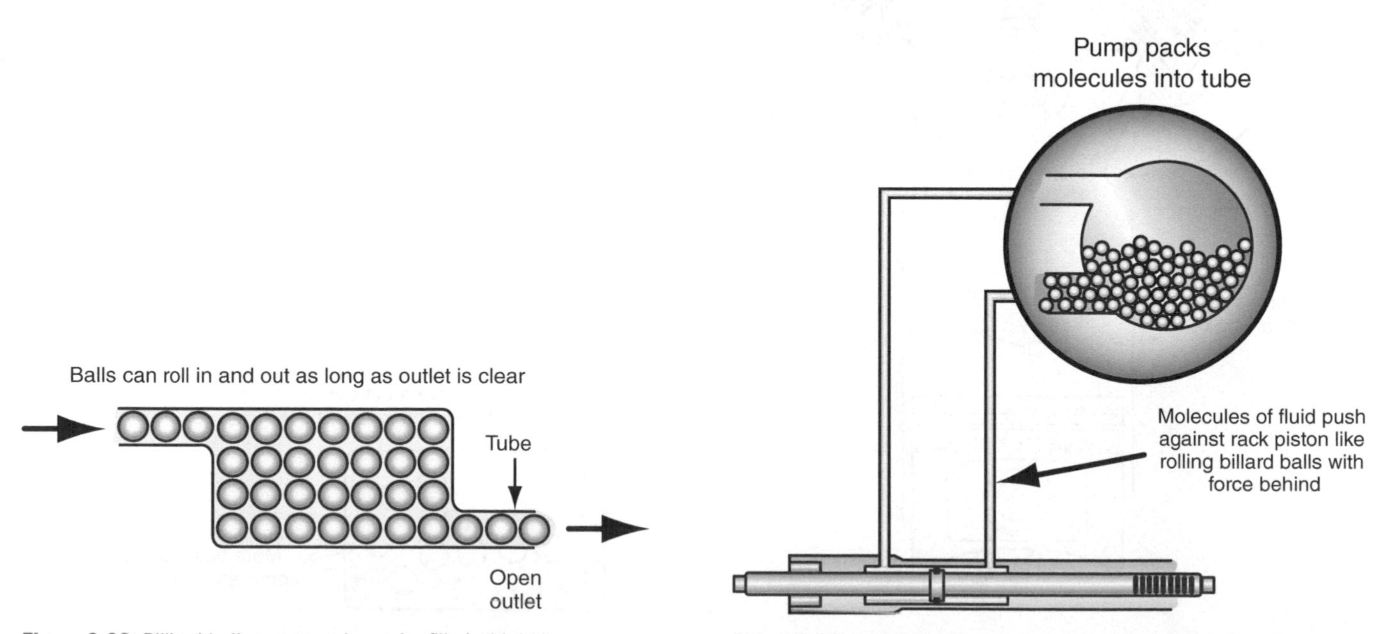

Figure 2-20 Billiard ball movement in a tube filled with balls compared to liquid flow.

Figure 2-21 Power steering pump pressure supplied to the steering gear chamber.

Hydraulic Fluid as a Flexible Machine

Hydraulic pressure has the same effect as a mechanical lever, because both of these items can multiply the input force to do more work. If a mechanical lever has a fulcrum at the center point and a 10-lb, or 4.5-kilogram (kg), weight on one end of the lever, 10 lb (4.5 kg) of weight are required on the other end of the lever to force the lever downward and raise the weight on the opposite end. If the lever is 5 ft. long and the fulcrum is placed 1 foot from the weight to be lifted, the lever has a 4 to 1 mechanical advantage. Under this condition, 2.5 lb (1.1 kg) of weight are required on the other end of the lever to lift the 10-lb (4.5-kg) weight (**Figure 2-22**). Therefore, a mechanical lever multiplies input force and makes it easier to do work.

When the power steering pump pressure supplied to the steering gear chamber is 1000 psi (7000 kPa), this pressure is applied to every square inch in the steering gear chamber (**Figure 2-23**). This pressure applied to the rack piston in the gear chamber acts like a mechanical lever and helps move the rack piston. Since the rack is connected through steering linkages and arms to the front wheels, the force on the rack piston helps the driver move the front wheels to the left or right during a turn (**Figure 2-24**).

Caution

Never completely block a pump outlet unless instructed to do so in the vehicle manufacturer's service manual. This action can result in extremely high pump pressure, damaged components, and personal injury.

An equal volume of hot air weighs less and exerts less pressure on the earth's surface than cold air.

Atmospheric Pressure

Since air is gaseous matter with mass and weight, it exerts pressure on the earth's surface. A 1-square-inch column of air extending from the earth's surface to the outer edge of the atmosphere weighs 14.7 psi at sea level. Therefore, **atmospheric pressure** is 14.7 psi at sea level (**Figure 2-25**).

Atmospheric pressure may be defined as the total weight of the earth's atmosphere.

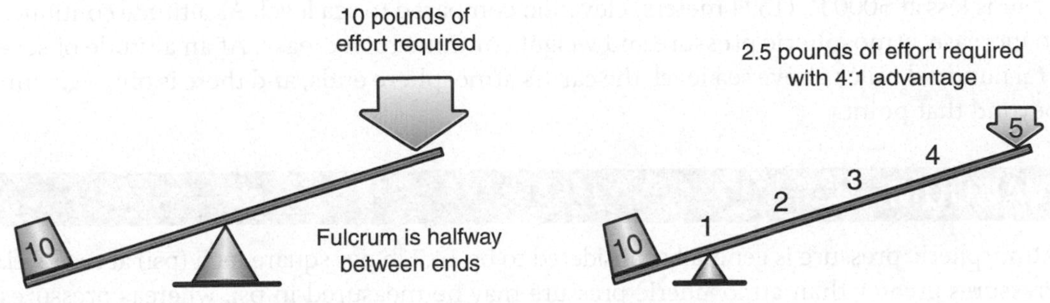

Figure 2-22 A mechanical lever multiplies input force and makes it easier to do work.

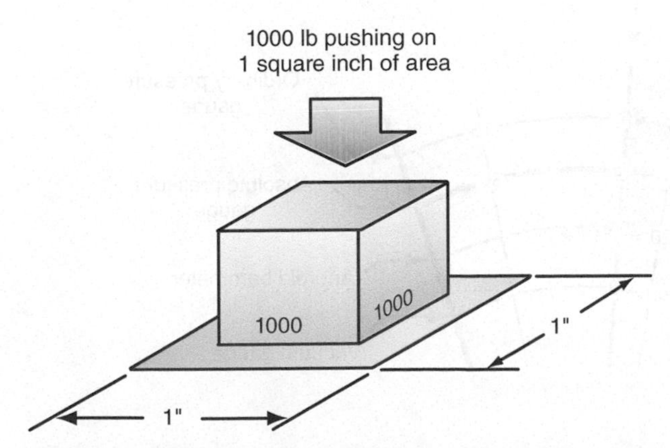

Figure 2-23 Hydraulic pressure is applied equally to every square inch in a container.

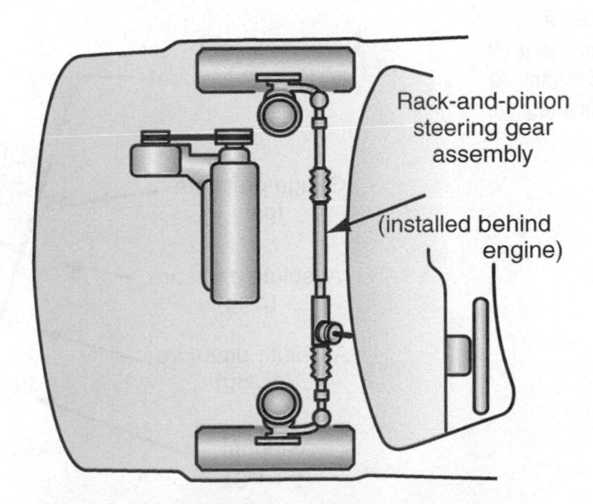

Figure 2-24 Steering gear, linkages, and arms connected to the front wheels.

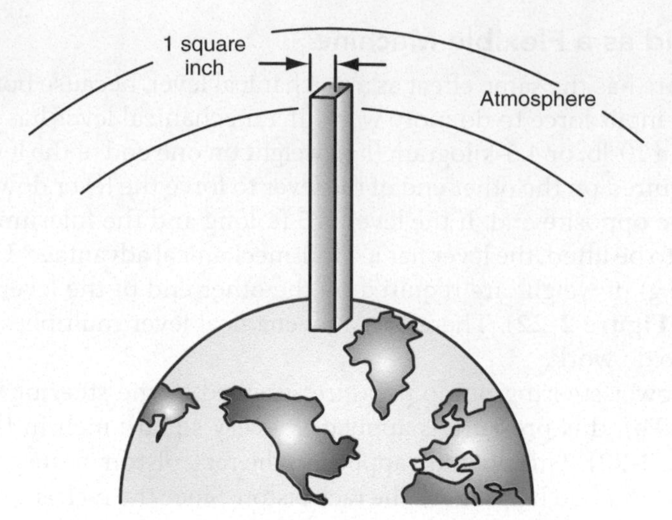

Figure 2-25 A column of air 1 in. square extending from the earth's surface at sea level to the outer edge of the atmosphere weighs 14.7 lb.

Atmospheric pressure decreases as altitude increases.

Atmospheric Pressure and Temperature

When air becomes hotter it expands, and this hotter air is lighter compared to an equal volume of cooler air. This hotter, lighter air exerts less pressure on the earth's surface compared to cooler air.

If the temperature decreases, air contracts and becomes heavier. Therefore, an equal volume of cooler air exerts more pressure on the earth's surface compared to hotter air.

Atmospheric Pressure and Altitude

When you climb above sea level, atmospheric pressure decreases. The weight of a column of air is less at 5000 ft. (1524 meters) elevation compared to sea level. As altitude continues to increase, atmospheric pressure and weight continue to decrease. At an altitude of several hundred miles above sea level, the earth's atmosphere ends, and there is only vacuum beyond that point.

VACUUM

Atmospheric pressure is generally considered to be 14.7 lb per square inch (psi) at sea level. Pressures greater than atmospheric pressure may be measured in psi, whereas pressures below atmospheric pressure are measured in **vacuum**, or psi absolute (psia) (**Figure 2-26**).

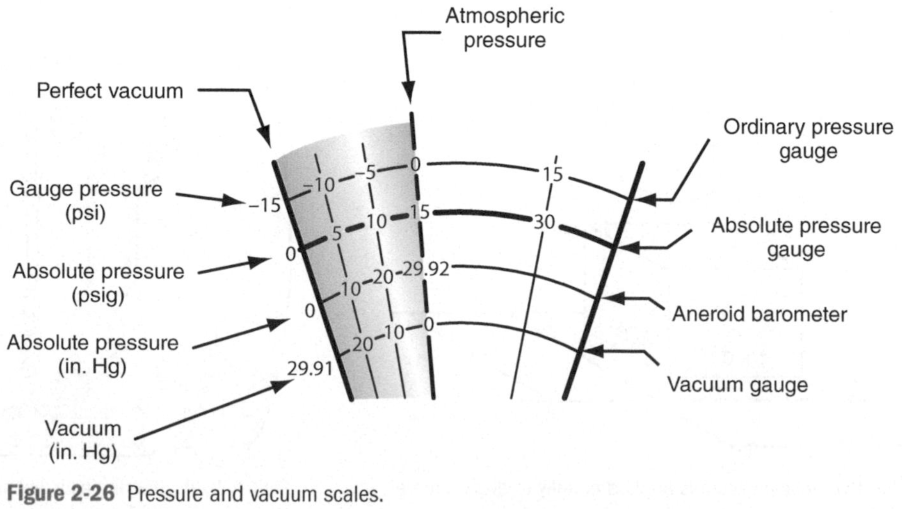

Vacuum may be defined as the absence of atmospheric pressure. A vacuum may be called a low pressure because it is a pressure less than atmospheric pressure.

Figure 2-26 Pressure and vacuum scales.

A conventional pressure gauge is used to measure pressures greater than atmospheric pressure. This type of pressure gauge indicates 0 psi at atmospheric pressure, and as the pressure increases it can read up to 15 psi. A conventional vacuum gauge indicates 0 inches of mercury (in. Hg) when atmospheric pressure is applied, and as the vacuum increases this gauge reads from 0 in. Hg to 29.91 in. Hg, and this may be considered a perfect vacuum. An absolute pressure gauge indicates absolute pressure in pounds per square inch gauge (psig). An absolute pressure gauge indicates 0 at a perfect vacuum, 15 psig at atmospheric pressure, and 30 psig at 15 psi on a conventional pressure gauge. An aneroid barometer reads in in. Hg absolute pressure, and thus it reads 0 in. Hg absolute at a perfect vacuum and 29.92 in. Hg absolute when atmospheric pressure is present.

Pressures above atmospheric pressure measured in psi are found in these automotive systems:

1. Fuel
2. Oil
3. Cooling
4. Power steering
5. Air conditioning
6. Turbocharger boost
7. Air springs
8. Gas-filled shock absorbers
9. Brake system during brake application

Pressures below atmospheric pressure measured in inches of vacuum are found in these automotive systems:

1. Manifold vacuum signal
2. Ported vacuum signal above the throttle
3. Carburetor venturi
4. Air-conditioning evacuation

Vacuum could be measured in psi, but in. Hg is most commonly used for this measurement. Let us assume that a manometer is partially filled with mercury, and atmospheric pressure is allowed to enter one end of the tube. If vacuum is supplied to the other end of the manometer, the mercury is forced downward by the atmospheric pressure. When this movement occurs, the mercury also moves upward on the side where the vacuum is supplied. If the mercury moves downward 10 in., or 25.4 centimeters (cm), where the atmospheric pressure is supplied, and upward 10 in. (25.4 cm) where the vacuum is supplied, 20 in. Hg is supplied to the manometer (**Figure 2-27**). The highest possible, or perfect, vacuum is approximately 29.9 in. Hg.

Liquids, solids, and gases tend to move from an area of high pressure to a low-pressure area.

Vacuum and atmospheric pressure are used in several automotive systems. For example, atmospheric pressure is present outside the compressor inlet on an electronic air suspension system. When the compressor is running, it creates a vacuum at the inlet and in the compressor cylinder. This pressure difference causes air to move from the atmosphere surrounding the inlet into the cylinder. The compressor develops high pressure at the discharge valve, and this pressure forces air into the air springs when the pressure is lower than at the compressor outlet (**Figure 2-28**).

Pumps use high and low pressure to move liquids or gases. For example, as a power steering pump rotates, it creates a high pressure at the pump outlet by pumping against a restriction, and a low pressure at the inlet by moving fluid through a restriction. This pressure difference causes fluid to flow through the power steering system.

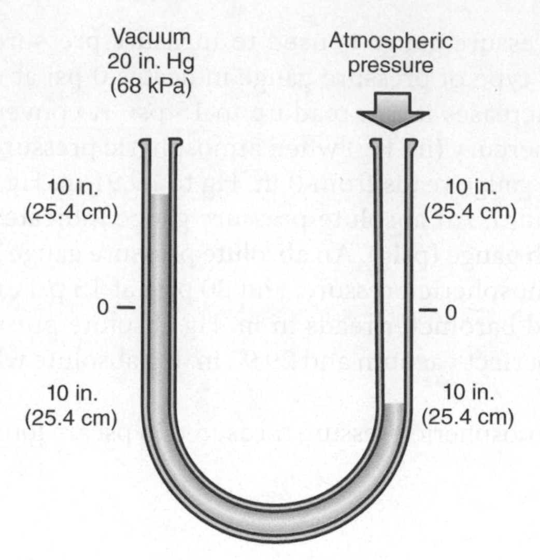

Figure 2-27 A vacuum of 20 in. Hg (68 kPa) supplied to mercury in a manometer.

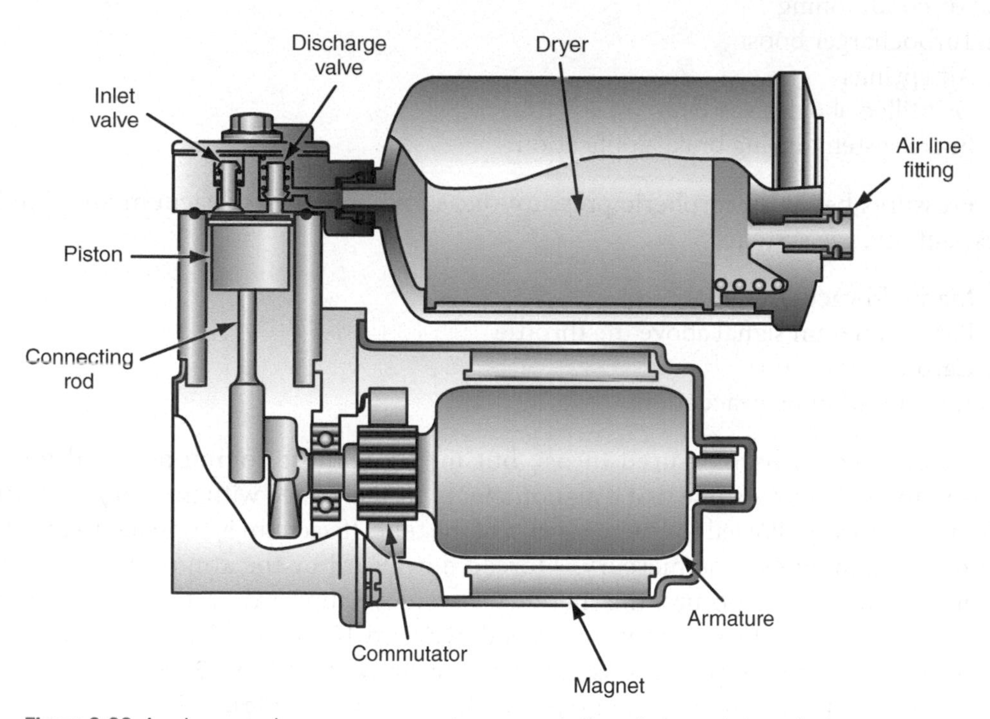

Figure 2-28 An air suspension compressor creates vacuum in the cylinder, and atmospheric pressure at the inlet forces air into the cylinder. The compressor provides high pressure at the discharge valve, which moves air into the air springs.

A venturi may be defined as a narrow area in a pipe through which a liquid or a gas is flowing.

VENTURI PRINCIPLE

If a gas or a liquid is flowing through a pipe and the pipe diameter is narrow in one place, the flow speeds up in the narrow area. This increase in speed causes a lower pressure in the narrow area, which may be defined as a venturi (**Figure 2-29**). Power steering pumps use a venturi principle to assist in the control of pump pressure.

Technicians must understand basic principles to comprehend the complex systems on modern vehicles. When basic principles are mastered, then technicians understand both the reason for, and the result of, these specific service procedures.

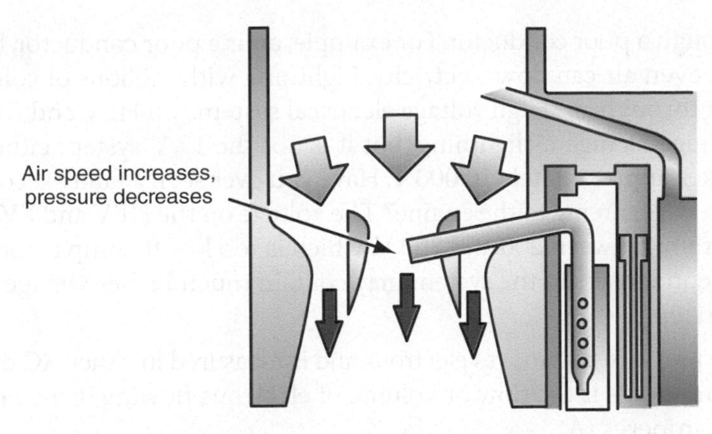
Air speed increases,
pressure decreases

Figure 2-29 A venturi increases airflow speed and produces a vacuum.

ELECTRICAL PRINCIPLES

We often think of suspension and steering components and systems as mechanical, but many subsystems are controlled or powered by electricity. Examples of these include the active steering system and the computer-controlled suspension system. Therefore, a basic understanding of some of the electrical principles, including voltage, amperage, and resistance, is required. The following section is meant to be a review or overview of these concepts. It is recommended that a technician receive in-depth electrical and electronic training as part of their overall education. Electronics is part of every major system on the automobile, and electrical failures have become routine complaints, though not always routine repairs.

The Basics: Voltage, Amperage, and Resistance

For many it is often easier to think of electricity in terms of water flowing in your home (hydraulic system principles) as there is a visible cause and effect. The flow of electricity is similar to the flow of water through your household plumbing. Where the water pressure is similar to electrical pressure or voltage, the water flow is similar to current flow in a conductor or amperage, and a restriction such as a kink in a hose is similar to the resistance in an electrical system. However, unlike water, electricity does not flow out the end of the wire and pour onto the ground if left open.

Voltage is a unit of measurement of electrical pressure.

Voltage (V) or electromotive force (EMF) is the electrical pressure and is measured in volts (V); it may be either direct current (DC) or alternating current (AC). In the automotive industry, especially on hybrid electric vehicle (HEV) and electric vehicle (EV) platforms, we deal with both. Current is the flow or rate of flow of electrons under pressure in a conductor between two points having a difference in potential and is measured in **amperes (A)** or amperage. A complete circuit is required for current to flow. A DC circuit requires a complete circuit or loop between positive (+) and negative (−) for current to flow. Resistance is the friction in an electrical circuit, which restricts the flow of electrons under pressure and is measured in **ohms (Ω)**. Electrical resistance is a load on the moving current that must be present to do any useful work. Resistance controls the amount of current flow in an electrical circuit. Electrical devices that use electricity to operate have a greater amount of resistance than a conductor (wire) and are considered loads in a circuit. A motor, light bulb, or solenoids are examples of electrical loads in a circuit; a load in a circuit is the electrical device that consumes electricity. Poor connections and corrosion are examples of unwanted electrical loads in a circuit. Resistance in an electrical circuit is measured in ohms (Ω). If there is resistance (Ω) in the conductor, electrons will not flow as readily.

Ohm is a unit of measurement of electrical resistance.

For all practical automotive purposes electricity only flows through a good conductor; in the majority of automotive wiring this is copper. It takes a high voltage (V) to flow current (A) through a poor conductor. For example, air is a poor conductor, but with a high enough voltage even air can flow electricity. Lightning with millions of volts is capable of flowing current through air! High voltage electrical systems on HEV and EV platforms do not contain voltage as high as lightning, but it is not the 12 V system either. To jump an inch of air, it takes approximately 10,000 V. Have you ever seen a faulty secondary ignition wire (spark plug wire) arcing to the engine? The voltage on the HEV and EV orange wiring harness is generally between 200 and 300 V, which is too low to jump through air, but the capacitors and condensers in the system may contain much higher voltage.

To summarize:

- Voltage is the pressure that moves electrons and is measured in either AC or DC volts (V).
- Current or amperage is the flow or volume of electrons flowing in a conductor and is measured in amperes (A).
- Electrical resistance is a load or opposition on the moving electrons (current) in a circuit and is measured in ohms (Ω).

Ohm's Law

Ohm's law defines the relationship between voltage, resistance, and current.

The amount of current flow is determined by the amount of resistance in the loop of the circuit. In a fixed voltage circuit, if the amount of resistance in the loop of the circuit is high, the current flow will be low and if the amount of resistance in the loop of the circuit is low, the current flow will be high. This inverse relationship is known as **Ohm's law** and is summarized by the mathematical equation in **Figure 2-30**. Ohm's law states that it requires 1 V of electrical pressure to move 1 A through 1 Ω of resistance. Mathematically,

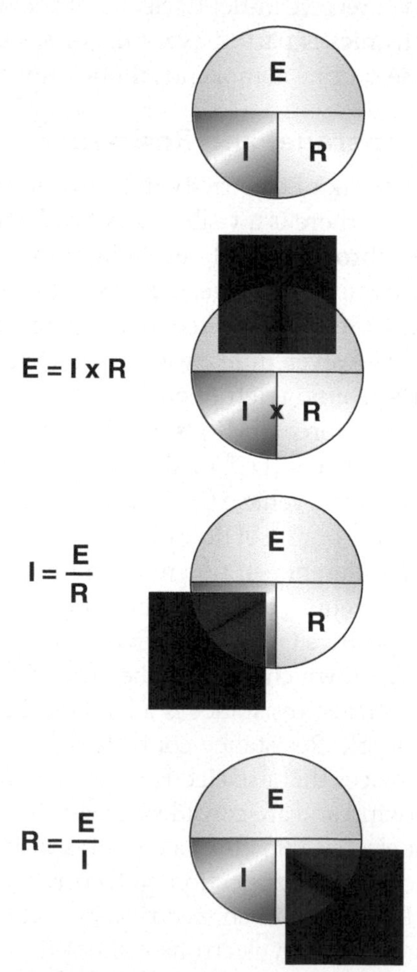

Figure 2-30 With Ohm's law, if you know two of the three electrical factors, you can calculate the third.

Ohm's law is expressed by the following equations, using the symbols E for voltage, I for current, and R for resistance:

To find voltage $E = I \times R$

To find current $I = E \div R$

To find resistance $R = E \div I$

For example, if a 12.6 V circuit has 2 Ω of resistance, you can use Ohm's law to determine the current flowing in the circuit as follows:

$I = E \div R$

$I = 12.6 \div 2$

$I = 6.3$ A

The current flowing in the circuit is 6.3 A.

These equations can be used to calculate the voltage, current, or resistance of a circuit. It is the understanding of this relationship that is most important to you as a technician. You are not generally crunching numbers; instead you are making measurements on a circuit with a digital multimeter (DMM) and trying to understand what this information means. Ohm's law can help to clear up what the meter is telling you for a given reading.

Types of Circuits and Using Digital Multimeters

There are three basic types of circuits that we will be dealing with, and a few ways to test these circuits for voltage, amperage, and resistance with a DMM. It is imperative that you understand these basic concepts and how to test a circuit with a DMM to avoid damage to the electrical circuit and components and/or damage to your expensive DMM. Remember Snap-On does not warranty misuse of equipment and service managers do not understand expensive mistakes! Take your time and think before you test. Remember the best place to begin your diagnosis is with the electrical diagram and system operation description contained in the vehicle service information database. As stated earlier this section is meant to be a review and not an in-depth training section, please refer to *Today's Technician Automotive Electricity & Electronics* for more information and training.

A series circuit is a circuit that provides a single path for current flow from the electrical source through all the circuit's components and back to the source (**Figure 2-31**).

Series Circuit laws:

- Current flow is the same at any point in the circuit (**Figure 2-32**).
- The sum of the individual voltage drops equals the source voltage (**Figure 2-33**).
- Total circuit resistance is the sum of the individual circuit resistances (**Figure 2-34**).

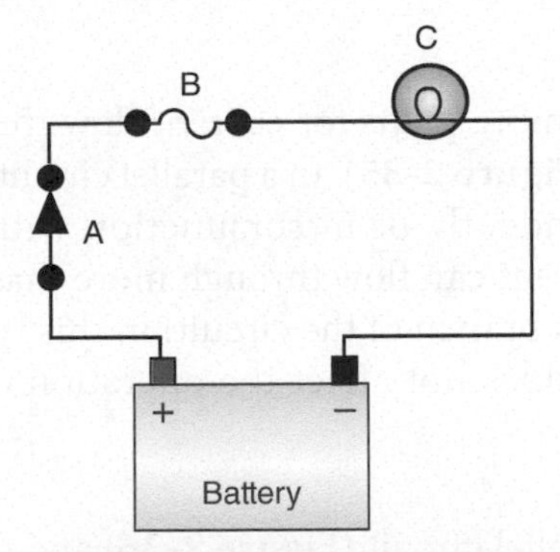

Figure 2-31 A simple series circuit including a switch (A), a fuse (B), and a lamp (C). For all article purposes the only load in the circuit is the lamp.

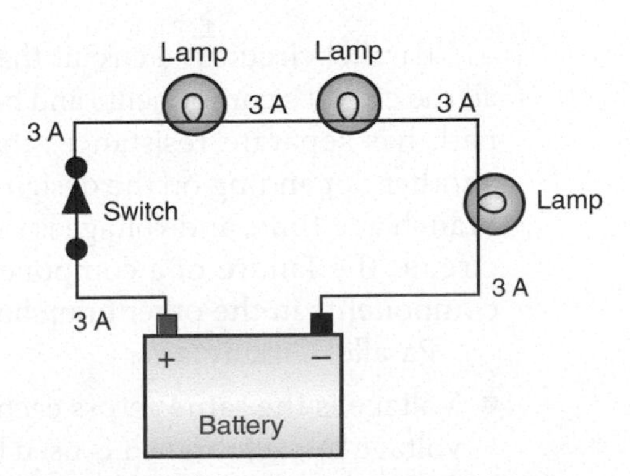

Figure 2-32 Current flow is the same at any point in the circuit.

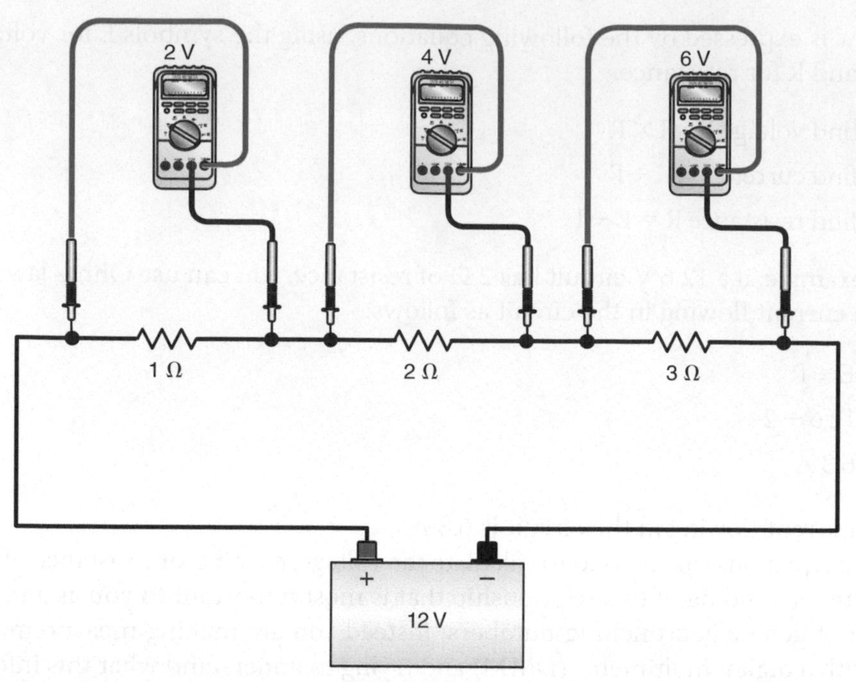

Figure 2-33 The sum of the individual voltage drops equals the source voltage (2V + 4V + 6V = 12V or source voltage).

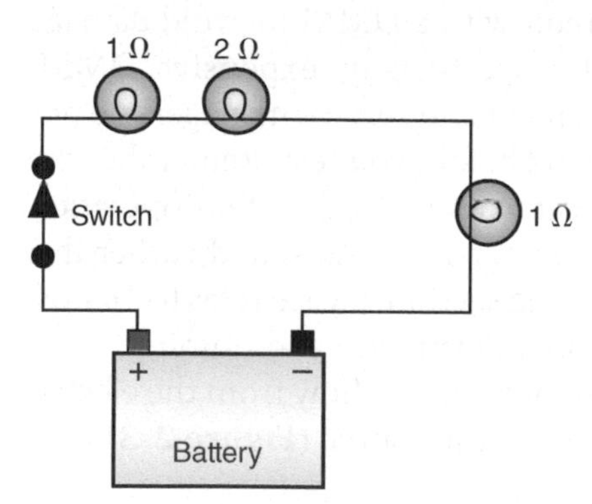

Figure 2-34 Total circuit resistance is the sum of the individual circuit resistances, 1Ω + 2Ω + 1Ω = 4Ω of total circuit resistance.

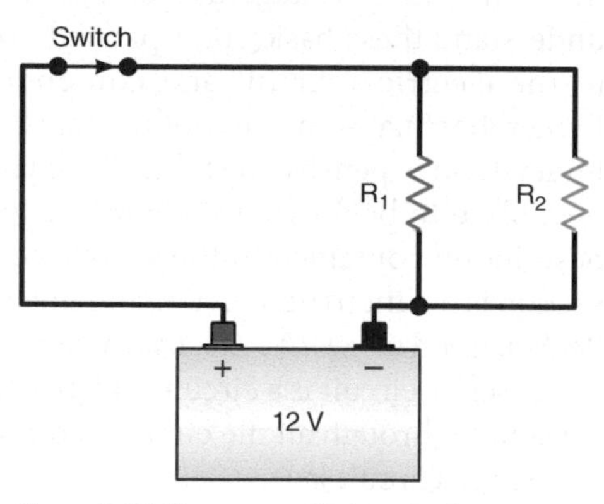

Figure 2-35 There are multiple paths for current to flow in a parallel circuit. If R_1 were to fail, R_2 would still work.

Parallel circuit is a circuit that provides two or more paths for current flow through all the circuit's components and back to the source (**Figure 2-35**). In a parallel circuit each path has separate resistances that operate independently or in conjunction with one another depending on the design of the circuit. Current can flow through more than one branch at a time, and voltage is the same across each branch of the circuit. In this type of circuit, the failure of a component in one branch does not affect the operation of the components in the other branches of the circuit.

Parallel Circuit laws:

- Voltage is the same across each branch of the parallel circuit (**Figure 2-36**) and the voltage in each branch is used by the load(s) in that branch. The voltage dropped across each parallel branch will be the same; but, if the branch contains more than

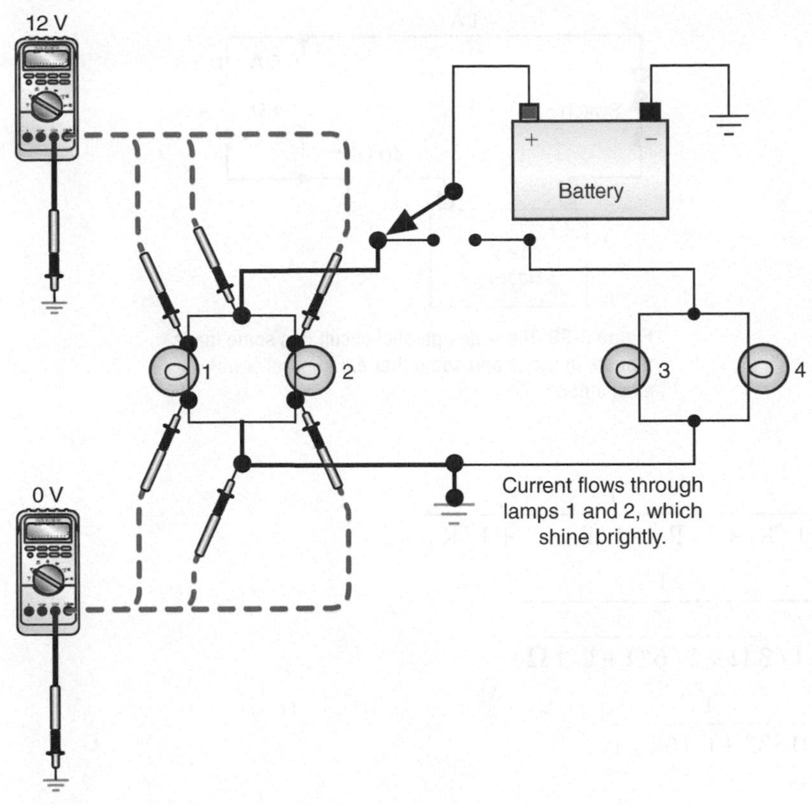

Figure 2-36 Voltage is the same across each branch of the parallel circuit, and the voltage in each branch is used by the load(s) in that branch.

one resistor, the voltage drop across each of them will depend on the resistance of each resistor in the branch.

- Total current in a parallel circuit is equal to the sum of the individual branch currents (**Figure 2-37**).
- Total resistance in a parallel circuit is always less than the smallest resistive branch (**Figure 2-38**).

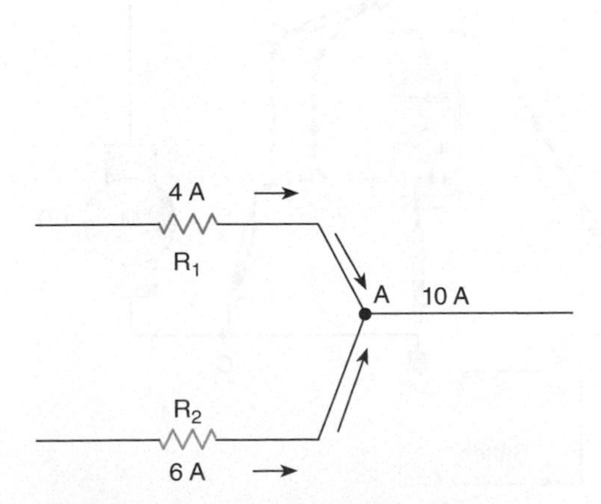

Figure 2-37 Total current in a parallel circuit is equal to the sum of the individual branch currents.

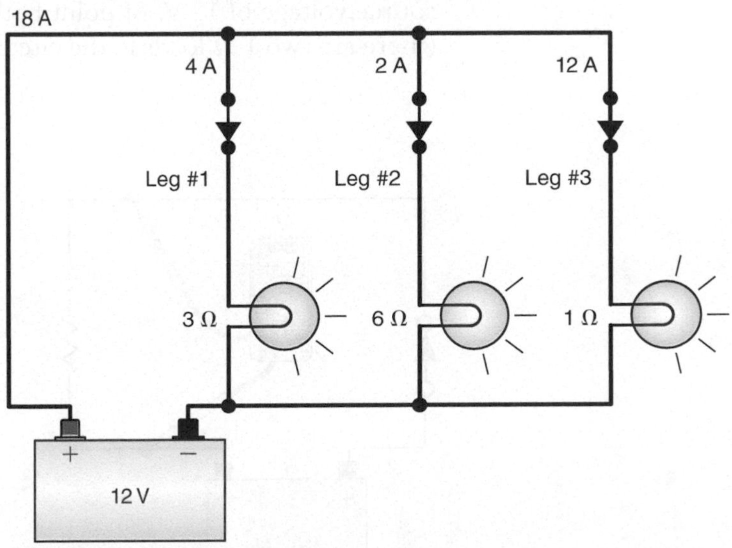

Figure 2-38 Total resistance in a parallel circuit is always less than the smallest resistive branch.

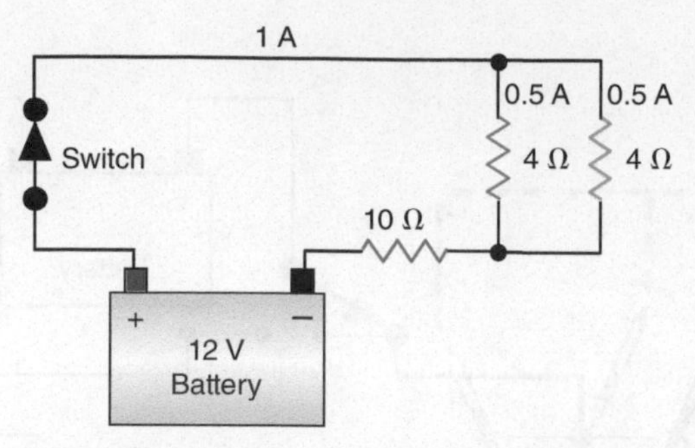

Figure 2-39 The series-parallel circuit has some loads that are in series and some that are in parallel with each other.

$$R_T = \frac{1}{1/R_1 + 1/R_2 + 1/R_3 + ... + 1/R_{10}}$$

$$R_T = \frac{1}{\frac{1}{1/3\,\Omega + 1/6\,\Omega + 1/1\,\Omega}}$$

$$R_T = \frac{1}{0.333 + 0.166 + 1}$$

$$R_T = \frac{1}{1.5}$$

$$R_T = 0.667\ \Omega$$

The series-parallel circuit has some loads that are in series and some that are in parallel with each other (**Figure 2-39**).

A voltmeter can be used to check for available voltage at the battery, terminals of any component or connectors. It can also be used to test voltage drops across electrical circuits, component loads, connectors, and switches. A voltmeter is connected in parallel with the circuit being tested (**Figure 2-40**). In **Figure 2-41** voltage is being tested in a closed 12 V series circuit with two loads. At test point A, the voltage should be the source voltage of 12 V. At point B, the 1 Ω resistor would have dropped half the voltage (there are two 1 Ω loads in the circuit) and the meter should read 6 V. At test point C, all

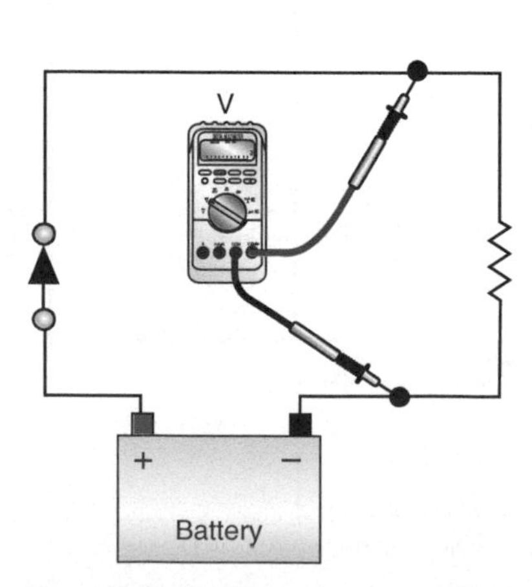

Figure 2-40 A voltmeter is connected in parallel with the circuit being tested.

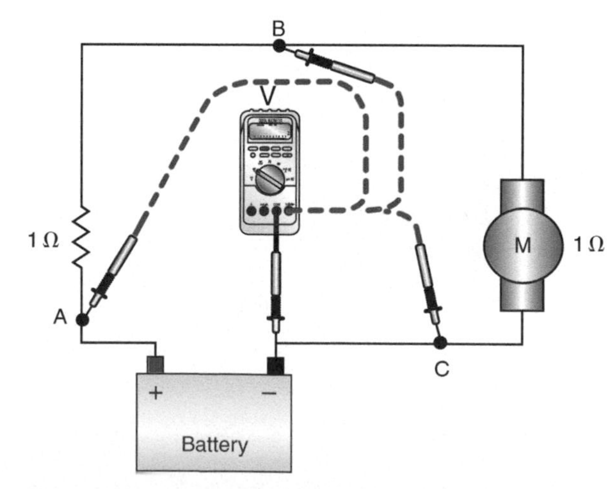

Figure 2-41 A voltmeter testing for voltage at various points in the series circuit.

the voltage should have been used up by the two loads in the circuit and the meter reading should be 0 V. These readings would indicate normal circuit operation.

One of the most useful tests to perform is the voltage drop test. In a circuit all of the voltage provided by the source power is used (dropped) by the circuit with nothing left over. The voltage is used by resistance in wiring, connectors, switches, and loads. This loss or use of voltage is called a voltage drop and is the amount of electrical energy converted into another form of energy. Voltage dropped in wiring, connectors, and switches is converted into heat energy. When a circuit or branch of a circuit has only one load (resistance) source voltage is dropped across that load (**Figure 2-42**). If there is more than one load in a circuit each load will use a portion of the voltage. The total of all voltage drops in a circuit should equal source voltage. All of the voltage must be used by the circuit. You should verify that the circuit is turned on and that source voltage is available at the load and that the load drops source voltage. If the component (load) that is in the circuit drops source voltage, then the component is faulty. In **Figure 2-43**, if both Lamp 1 and Lamp 2 are the same resistance value (size) they will both share the source voltage equally. Lamp 1 will use 6 V and Lamp 2 will use 6 V. If there is unwanted resistance in a circuit the load in the circuit will not drop all source voltage (**Figure 2-44**). There is some allowable voltage drop by circuit components other than the load, but it is generally limited to a maximum of:

0.2 V (200 mV) for wires and cables
0.3 V (300 mV) for a switch
0.1 V (100 mV) for a ground
0 V for a connection or connectors

Common circuit faults:

- Short circuits
- Short to ground
- Opens in a circuit
- High resistance (may be caused by corrosion or poor connections)
- Low voltage

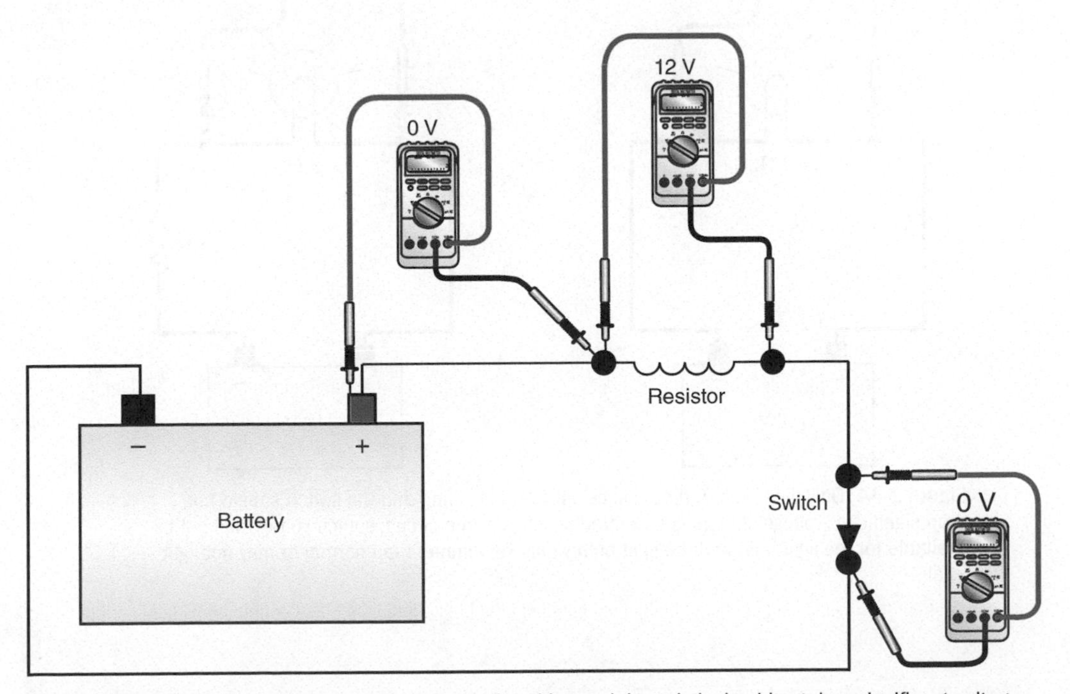

Figure 2-42 Only the load will drop source voltage; the wiring and the switch should not drop significant voltage.

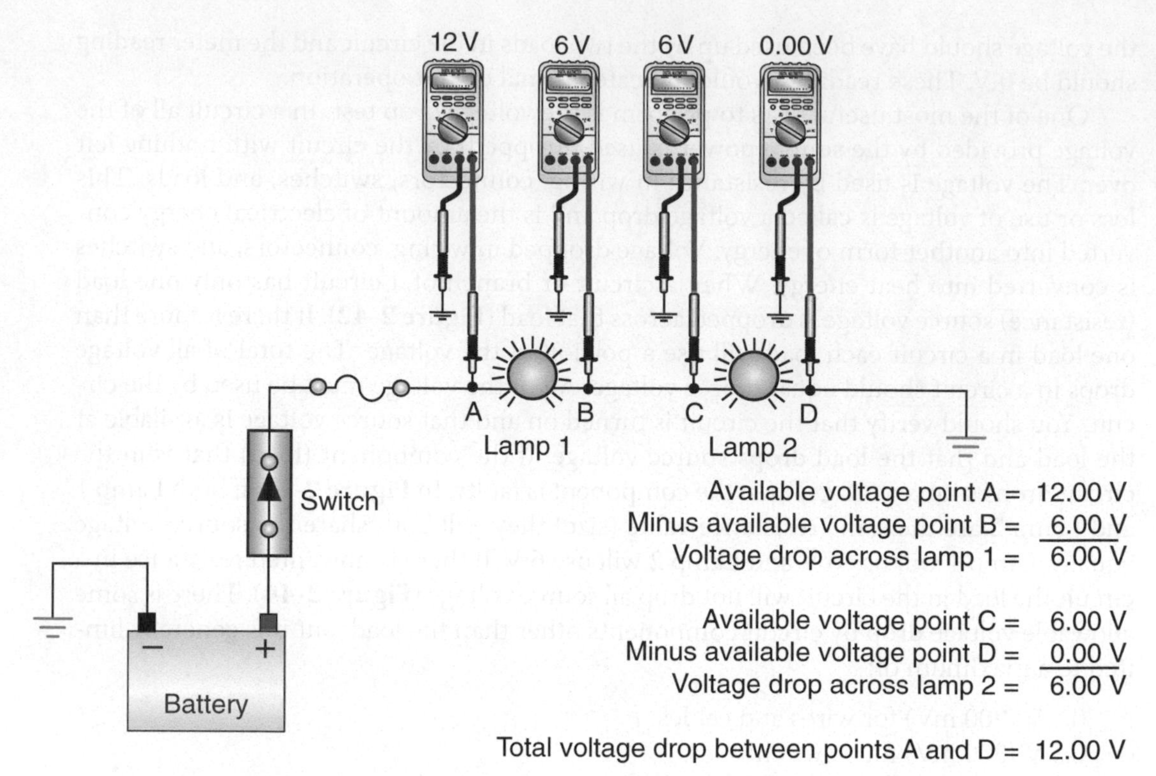

Figure 2-43 There should be source voltage before the first load. If there are two loads in a series circuit each load will drop a portion of the source voltage. There should be no voltage after the last load.

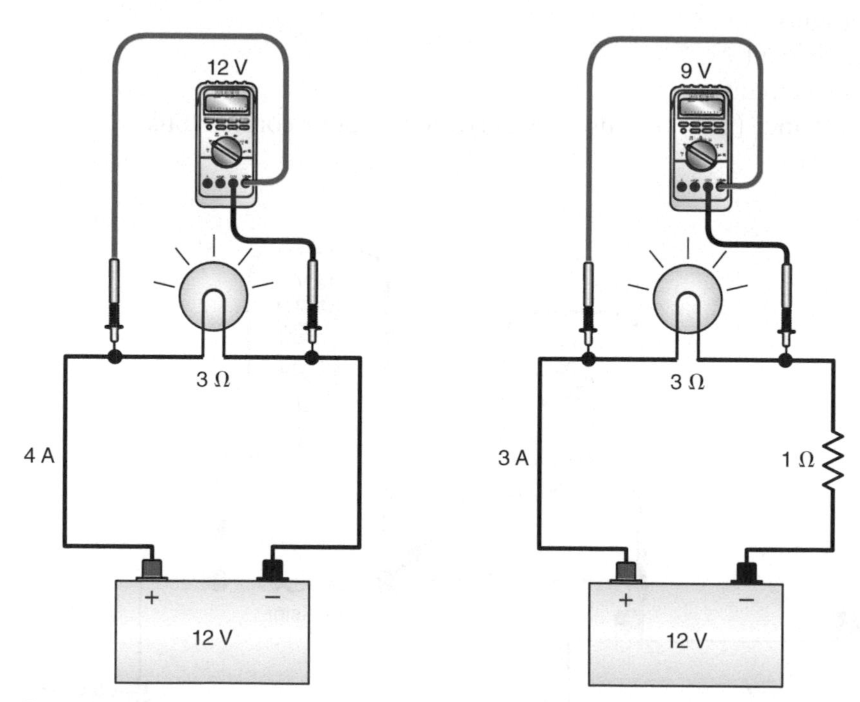

Figure 2-44 Only the load will drop source voltage; the wiring and the switch should not drop significant voltage. if there is unwanted resistance in a circuit, source voltage will not be available for the light bulb and the light bulb would be dimmer than normal or may not light at all.

TABLE 2-1 Circuit Test Chart

Type of Defect	Test Unit	Expected Results
Open	Ohmmeter	∞ infinite resistance between conductor ends
	Test light	No light after open
	Voltmeter	∅ volts at end of conductor after the open
Short to ground	Ohmmeter	∅ resistance to ground
	Test light	Lights if connected across fuse
	Voltmeter	Generally not used to test for ground
Short	Ohmmeter	Lower than specified resistance through load component ∅ resistance to adjacent conductor
	Test light	Light will illuminate on both conductors
	Voltmeter	A voltage will be read on both conductors
Excessive resistance	Ohmmeter	Higher than specified resistance through circuit
	Test light	Light illuminates dimly
	Voltmeter	Voltage will be read when connected in parallel over resistance

Table 2-1 shows common testing units and expected results when diagnosing the above common circuit faults. When using an ammeter to measure amperage, be sure to first turn power off in the circuit before connecting the DMM. The circuit must be opened (disconnected) from the load being tested. The DMM is placed in series in the circuit, recompleting the disconnected circuit (**Figure 2-45**). Always verify that the DMM is capable of handling the highest expected amperage in the circuit being tested. What size fuse protects the circuit being tested? Many DMMs are only internally protected to 10 A by their internal fuse and many automotive circuits can provide 20 or more amps. In these cases, an inductive amp probe is the best choice of tools to use to avoid DMM damage (**Figure 2-46**). The inductive probe eliminates the need to connect the DMM in series and is a safe noninvasive method of measuring amperage in a circuit.

When using an ohmmeter to measure resistance, the power from the circuit must be removed and circuit or component should be isolated (**Figure 2-47**). Ohmmeter leads are placed across or parallel with the component or circuit being tested.

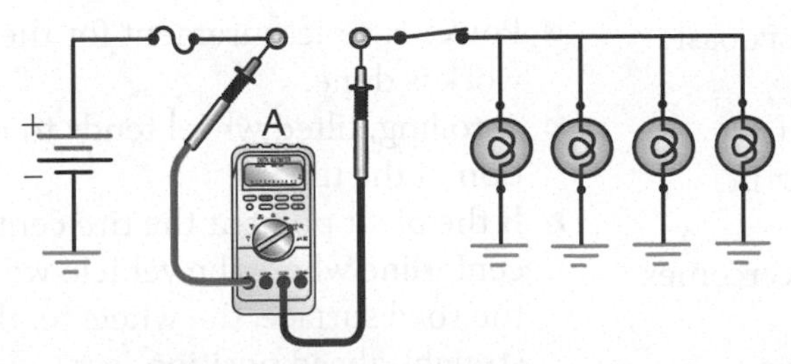

Figure 2-45 When testing amperage draw the circuit must be opened (disconnected) from the load being tested and the multimeter is placed in series in the circuit.

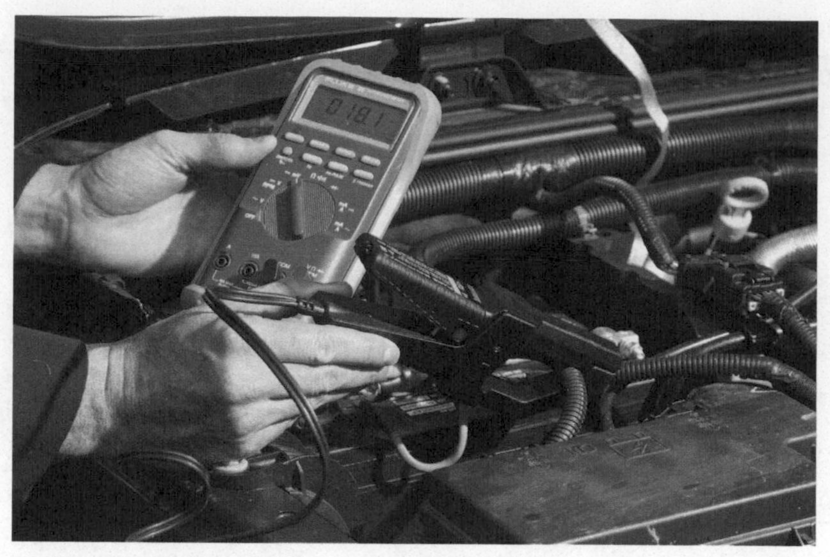

Figure 2-46 A multimeter with an inductive amp probe is a safe, noninvasive method of measuring amperage in a circuit.

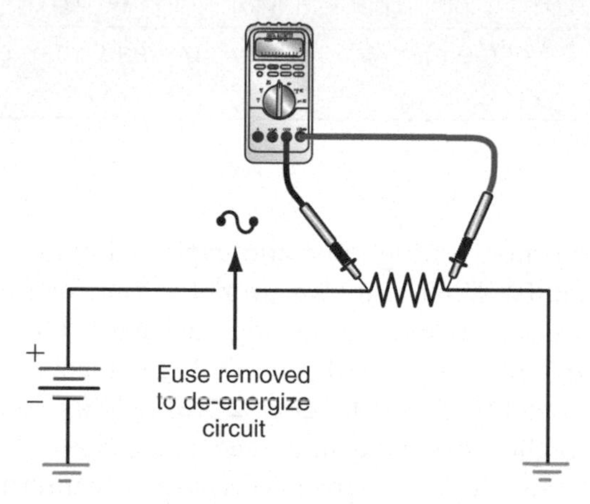

Figure 2-47 When measuring resistance with an ohmmeter, the meter is connected in parallel with power removed from the circuit and the circuit or component isolated.

SUMMARY

- Work is the result of applying a force.
- Force is measured in pounds and distance.
- Energy is the ability to do work; there are six basic types of energy.
- Inertia is the tendency of an object at rest to remain at rest, or the tendency of an object in motion to remain in motion.
- An object gains momentum when force overcomes static energy and moves the object.
- Friction is the resistance to motion when one object is moved over another object.
- Mass is a measurement of an object's inertia.
- Weight is a measurement of the earth's gravitational pull on an object.
- Volume is the length, width, and height of a space occupied by an object.
- Power is a measurement for the speed at which work is done.
- A rolling, tilted wheel tends to move in the direction of the tilt.
- If the pivot point at the tire centerline is behind the centerline where the vehicle weight is projected on the road surface, the wheel tends to remain in the straight-ahead position.
- To obtain static balance, weight must be distributed equally around the center of wheel rotation as viewed from the side.

- Weight must be distributed equally on both sides of the tire centerline viewed from the front of the tire to maintain proper dynamic balance.
- Torque is a twisting force that does work.
- Atmospheric pressure is the total weight of the earth's atmosphere.
- Vacuum is the absence of atmospheric pressure.
- Voltage (V) or electromotive force (EMF) is the electrical pressure and is measured in volts (V); it may be either direct current (DC) or alternating current (AC).
- Current is the flow or rate of flow of electrons under pressure in a conductor between two points having a difference in potential and is measured in amperes (A) or amperage.
- Resistance is defined as the opposition to current flow and is measured in ohms (Ω).
- Ohm's law defines the relationship between voltage, current, and resistance. It is the basic electrical law.

REVIEW QUESTIONS

Short Answer Essays

1. Describe Newton's first law of motion, and give an application of this law in automotive theory.
2. Explain Newton's second law of motion, and give an example of how this law is used in automotive theory.
3. Describe Newton's third law of motion, and give an example of how this law applies to an automotive suspension system.
4. Describe six different forms of energy.
5. Describe four different types of energy conversion.
6. Explain the difference between static and dynamic inertia.
7. Explain why a rotating, tilted wheel moves in the direction of the tilt.
8. Explain why the front wheel of a bicycle tends to remain in the straight-ahead position as the bicycle is driven.
9. Define static and dynamic balance.
10. Briefly define the term *current*.

Fill-in-the-Blanks

1. Improper static wheel balance causes wheel _______________ when driving the vehicle.
2. If the tire pivot point is behind the wheel-and-tire centerline where the vehicle weight is projected against the road surface, the wheel tends to remain in the _______________ position when driving the vehicle.
3. _______________ is a unit of measurement of electrical resistance.
4. _______________ is the flow or rate of flow of electrons under pressure in a conductor between two points having a difference in potential and is measured in _______________.
5. When one object is moved over another object, the resistance to motion is called _______________.
6. _______________ is the electrical pressure and it may be either direct current (DC) or alternating current (AC).
7. Torque is a force that does work with a/an _______________ action.
8. Power is a measurement for the rate at which _______________ is done.
9. To obtain proper dynamic balance, the weight must be distributed equally on both sides of a wheel-and-tire centerline viewed from the _______________ of the tire.
10. Vacuum is defined as the absence of _______________.

Multiple Choice

1. When an engine is running, the alternator converts:

   A. Thermal energy to mechanical energy.

   B. Electrical energy to mechanical energy.

   C. Chemical energy to electrical energy.

   D. Mechanical energy to electrical energy.

2. Work is accomplished during all of these conditions EXCEPT:
   A. When a mechanical force starts an object in motion.
   B. When a mechanical force is applied to an object, but the object does not move.
   C. When a mechanical force stops an object in motion.
   D. When a mechanical force redirects an object in motion.
3. All these statements about Newton's Laws of Motion are true EXCEPT:
   A. For every action there is an equal and opposite reaction.
   B. A body in motion remains in motion unless an outside force acts on it.
   C. A body's acceleration is directly proportional to the force applied to it.
   D. A body moves in an arc away from the force acting upon the object.
4. Torque is calculated by:
   A. Multiplying the force and the radius from the force to the object.
   B. Dividing the radius by the force.
   C. Adding the force and the distance of force movement.
   D. Subtracting the radius from the force to the object and the weight of the object.
5. When working with gases and liquids:
   A. Gases are not compressible.
   B. Liquids are not compressible.
   C. Temperature does not affect gas volume.
   D. Pressure is applied unevenly to the inside of a container surface.
6. When applying the principles of work and force:
   A. Work is accomplished when force is applied to an object that does not move.
   B. In the metric system the measurement for work is cubic centimeters.
   C. No work is accomplished when an object is stopped by mechanical force.
   D. If a 50-lb object is moved 10 ft., 500 ft.-lb of work are produced.
7. All of the following are true of voltage drops EXCEPT:
   A. Voltage drop can be measured with a voltmeter.
   B. All of the source voltage in a circuit must be dropped.
   C. Corrosion in a circuit does not cause a voltage drop.
   D. Voltage drop is the conversion of electrical energy into another form of energy.
8. A lever is 10 ft. (3 m) long and the fulcrum is positioned 1 ft. (0.304 m) from the end of the lever. A 5-lb (2.26-kg) weight is placed on the end of the lever nearest the fulcrum. The weight required on the opposite end of the lever to lift the 5-lb (2.26-kg) weight is:
   A. 0.368 lb (0.166 kg).
   B. 0.555 lb (0.251 kg).
   C. 0.714 lb (0.323 kg).
   D. 0.748 lb (0.339 kg).
9. A tire-and-wheel assembly that does not have proper dynamic balance has:
   A. Weight that is not distributed equally on both sides of the tire centerline.
   B. Worn tread on the inside edge of the tire.
   C. A tire with improperly positioned steel belts.
   D. Weight that is not distributed equally around the center of the tire and wheel.
10. When a car is driven on the road above 50 mph (80 km/h), the left front tire has a tramping action. The MOST likely cause of this condition is:
    A. Improper left front dynamic wheel balance.
    B. A bent left front wheel rim.
    C. Improper left front static wheel balance.
    D. Improper rim offset.

CHAPTER 3
TIRES AND WHEELS

Upon completion and review of this Chapter, you should be able to understand and describe:

- General tire function.
- Typical tire construction, and identify the purpose of each component in a tire.
- Three types of tire ply and belt designs.
- Tire ratings, and explain the meaning of each designation in the rating.
- The purpose of the tire performance criteria (TPC) rating.
- The difference between all-season tires and conventional tires.
- Two different types of tire load ratings.
- The Uniform Tire Quality Grading (UTQG) designations.
- The precautions to be observed when selecting replacement tires.
- Tire contact area, free tire diameter, and rolling tire diameter.
- The tire motion forces while a tire-and-wheel assembly is rotating on a vehicle.
- The importance of tire design quality as it relates to the tire motion forces.
- Wheel offset.
- Plus tire sizing.
- Wheel tramp, and explain how static unbalance causes wheel tramp.
- Wheel shimmy, and describe how dynamic unbalance results in wheel shimmy.
- Various types of tire pressure monitoring systems.
- The advantage of nitrogen tire inflation.
- Noise, vibration, and harshness analysis.

Terms To Know

All-season tires
Amplitude
Bead filler
Bead wire
Belt cover
Belted bias-ply tires
Bias-ply tires
Compact spare tires
Conicity
Cord plies
Dynamic balance
Electronic vibration analyzer (EVA)
Hertz
Hydroplaning
Liner
Load rating
Magnesium alloy wheels
Mud and snow tires
Natural vibration frequencies
Nitrogen tire inflation
Puncture sealing tires
Radial-ply tires
Replacement tires
Resonance
Road force variation
Run-flat tires
Sidewalls
Speed rating
Static balance
Synthetic rubber
Temperature rating
Tire belts
Tire chains
Tire contact area
Tire free diameter
Tire performance criteria (TPC)
Tire placard
Tire pressure
Tire rolling diameter
Tire treads
Tire valve
Traction ratings
Tread wear ratings
Uniform Tire Quality Grading (UTQG) System
Wheel rims
Wheel tramp

INTRODUCTION

Although tires are often taken for granted, they contribute greatly to the ride and steering quality of a vehicle. Tires also play a significant role in vehicle safety. Improper types of tires, incorrect inflation pressure, and worn-out tires create a safety hazard. When tires and wheels are out of balance, tire wear and driver fatigue are increased, which can create a driving hazard. Tires serve the following functions:

1. Tires cushion the vehicle ride to provide a comfortable ride for the occupant.
2. Tires must firmly support the vehicle weight.
3. Tires must develop traction to drive and steer the vehicle under a wide variety of road conditions. In other words, they must transmit traction and braking forces to the road.
4. Tires contribute to directional stability of the vehicle and must absorb all the stresses of accelerating, braking, and centrifugal force in turns.

TIRE DESIGN

Tires equipped on today's passenger cars and light trucks are of the radial ply design. Tire construction varies depending on the manufacturer and the type of tire. A typical modern tire contains these components (**Figure 3-1**):

The **bead wire** is a group of wire strands that retain the bead on the wheel rim.

The tire **liner** is made from synthetic gum rubber and seals the inside of the tire.

1. **Bead wire**
2. Bead filler
3. **Liner**, which functions as the air seal membrane
4. Sidewall with hard side compound
5. Rayon carcass plies
6. Steel belts
7. Jointless belt cover
8. Hard undertread compound
9. Hard high-grip tread compound

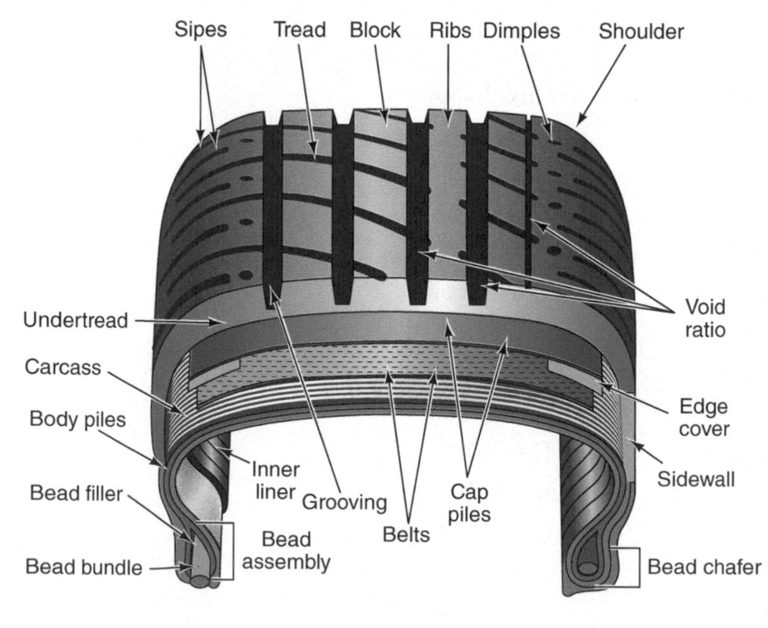

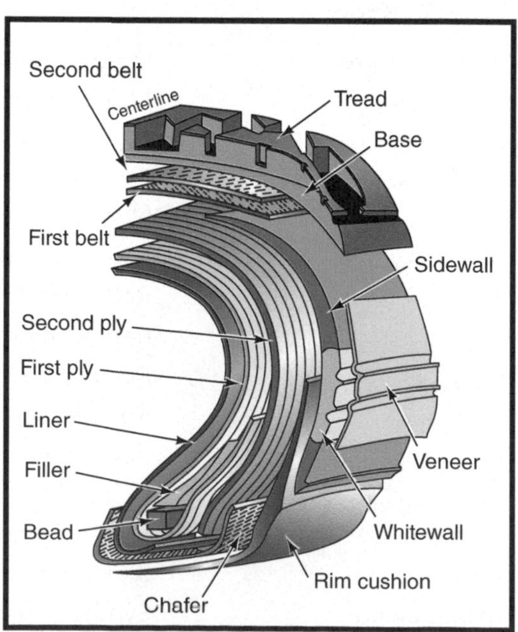

Figure 3-1 The typical tire design and construction.

The bead wire contains several turns of bronze-coated steel wire in a continuous loop. This bead wire is molded into the tire at the inner circumference and wrapped in the cord plies. The bead wire anchors the tire to the wheel. The **bead filler** above the bead reinforces the sidewall and acts as a rim extender.

Tire **sidewalls** are made from a blend of rubbers, which absorbs shocks and impacts from road irregularities, prevents damage to the plies, and also contains antioxidants and other chemicals that are gradually released to the surface of the sidewall during the life of the tire. These antioxidants help keep the sidewall from cracking and protect it from ultraviolet radiation and ozone attack. Since the sidewall must be flexible to provide ride quality, minimum thickness of this component is essential. Tire manufacturers have reduced sidewall thickness by 40 percent in recent years to reduce weight and heat buildup, improve ride quality, reduce rolling resistance, and improve fuel economy. Some off-road and performance tires are equipped with raised rubber that extends above the rim area and is referred to as a rim guard to offer added protection. A lettering and numbering arrangement for tire identification is located on the outside of the sidewall.

Shop Manual
Chapter 3, page 109

The **cord plies** surround both beads and extend around the inner surface of the tire to enable the tire to carry its load. The plies are molded into the sidewalls. Each ply is a layer of rubber with parallel cords imbedded in its body. The load capacity of a tire may be increased by adding more cords in each ply or by installing additional plies. The most common materials in tire plies are polyester, rayon, and nylon. Passenger car tires usually have two-cord plies, whereas the tires of many trucks and recreation vehicles are rated to carry the heavier loads of these vehicles and are equipped with a higher ply-rated tire. In general, tires with more plies have stiffer sidewalls, which provide less cushioning and reduced ride quality.

Steel is the most common material in **tire belts**, although other belt materials such as polyester have been used to some extent. Many tires contain two steel belts. The tire belts restrict ply movement and provide tread stability and resistance to deformation. This belt action provides longer tread wear and reduces heat buildup in the tire. Steel belts expand as wheel speed and tire temperature increase. Centrifugal force and belt expansion tend to tear the tire apart at high speeds and temperatures. Therefore, high-speed tires usually have a nylon jointless **belt cover**. This nylon belt cover contracts as it is heated and helps hold the tire together, providing longer tire life, improved stability, and better handling. Additionally, a tire may have a layer of DuPont Kevlar for added puncture proof protection under the tread layer.

Tire belts are positioned between the tread and the cord plies.

Tire treads are made from a blend of rubber compounds that are very resistant to abrasion wear. Spaces between the tire treads allow tire distortion on the road without scrubbing, which accelerates wear. Modern automotive tires contain two layers of tread materials. The first tread layer is designed to provide cool operation, low rolling resistance, and durability. The outer layer, or tread, is designed for long life and maximum traction. Tread rubber is a blend of many different synthetic and natural rubbers. Tire manufacturers may use up to thirty different types of **synthetic rubber** and eight types of natural rubber in their tires. Manufacturers blend these synthetic and natural rubbers in both tread layers to provide the desired traction and durability. Tire treads must provide traction between the tire and the road surface when the vehicle is accelerating, braking, and cornering. This traction must be maintained as much as possible on a wide variety of road surfaces. For example, on wet pavement, tire treads must be designed to drain off water between the tire and the road surface. This draining action is extremely important for maintaining adequate acceleration, braking, and directional control. Lines cut across the tread provide a wiping action, which helps dry the tire-road contact area. Most tire manufacturers add a PermaBlack compound to the rubber in their tires to maintain new-tire appearance throughout the life of the tire.

Tire treads are mounted between the sidewalls.

Synthetic rubber compounds are developed in laboratories, whereas natural rubber is found in nature.

A BIT OF HISTORY

In 1834, Charles Goodyear was a hardware merchant from Philadelphia with a great interest in a new substance imported from Brazil called rubber. When rubber was first imported to the United States, many entrepreneurs were interested in manufacturing products from rubber. However, these entrepreneurs soon discovered that rubber became bone-hard in cold weather, and then turned to a glue-like substance in very hot weather, and these characteristics ended the rubber manufacturing business at that time. Charles Goodyear persisted in experimenting with rubber! He experimented with mixing magnesia powder, nitric acid, and sulfur with rubber to reduce its stickiness. These substances improved the rubber quality, but it still was not perfected. Goodyear was often financially bankrupt and he had difficulty obtaining enough money to feed his children. He was even jailed on some occasions for failure to pay his bills, and ill health often plagued him.

By accident, Goodyear discovered that extreme heat changed the rubber to a weatherproof, gum-like material. He found that pressurized steam applied to rubber for four to six hours at 270°F (132°C) provided the most uniform rubber. Goodyear was now able to obtain financial backing to produce rubber. Goodyear was plagued with patent right problems and those trying to steal his rubber patents. However, he did succeed in discovering modern rubber! Neither Charles Goodyear nor his family were ever connected with the company named in his honor, today's Goodyear Tire and Rubber Company.

Today there is a cultivated rubber tree for every two human beings on Earth, and three million people are employed collecting the rubber from the trees in various countries. The United States imports approximately half of the rubber in the world and synthesizes an equal amount from petroleum. About 300,000 people in the United States are employed in the rubber industry, and they produce approximately $6 billion worth of products each year.

The synthetic gum rubber liner is bonded to the inner surface of the tire to seal the tire. All modern passenger car and light truck tires are the tubeless type. In these tires, the tire bead must provide an airtight seal on the rim, and both the tire and the wheel rim must be completely sealed. The inner liner of the tire provides an air tight seal.

Some heavy-duty truck tires and off-road equipment have inner tubes mounted inside the tire. On tube-type tires, the air is sealed in the inner tube, and the sealing qualities of the tire and wheel rim are not important.

Designing tires is a very complex engineering operation. The average all-season tire contains the components (by weight) listed in **Table 3-1**. Tire design varies depending on the operating conditions and the load capacity of the tire. A tire designed for improved steering and handling characteristics has a nylon bead reinforcement and a hard bead filler

TABLE 3-1 Tire Components by Weight

A typical P195/75-14 all-season tire contains:	
Synthetic rubber (30 types)	2.49 kg
Carbon Black (8 types)	2.27 kg
Natural Rubber (8 types)	2.04 kg
Chemicals, waxes, oils, pigments, etc. (40 types)	1.36 kg
Steel cord for belts	0.68 kg
Polyester and nylon	0.45 kg
Bead wire	0.23 kg
Total weight	9.42 kg

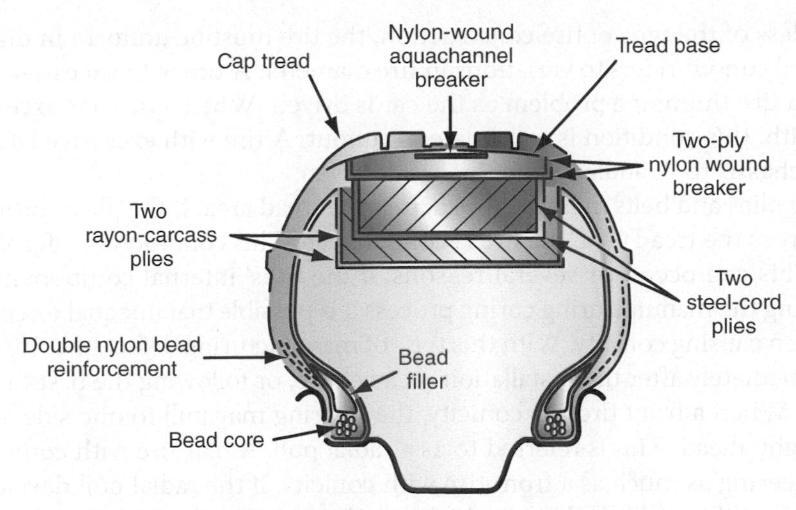

Figure 3-2 Tires designed for improved steering and handling with a nylon bead reinforcement and a hard bead filler with a slim tapered profile.

with a slim tapered profile (**Figure 3-2**). This type of tire is suitable for sports car operation because the design stiffens the tire and reduces tire deflection during high-speed cornering. However, this type of tire may provide slightly firmer ride quality.

TIRE PLY AND BELT DESIGN

The three basic tire construction designs that have been used are the bias, belted bias, and belted radial. In **bias-ply or belted bias-ply tires**, the cords crisscross each other. These cords are usually at an angle of 25° to 45° to the tire centerline. The belt-ply cord angle is usually 5° less than the cord angle in the tire casing. Two plies and two belts are most commonly used, but four plies and four belts may be used in some tires. Compared to a bias-ply tire, a belted bias-ply tire has greater tread rigidity. The belts reduce tread motion during road contact. This action provides extended tread life compared to a bias-ply tire.

Belted bias-ply tires have steel or polyester belts between the cord plies and the tread.

In **radial-ply tires**, the ply cords are arranged radially at a right angle to the tire centerline (**Figure 3-3**). Steel belts are most commonly used in radial tires, but other belt materials such as fiberglass, nylon, and rayon have also been used. The steel or fiberglass cords in the belts are crisscrossed at an angle of 10° to 30° in relation to the tire centerline. Many radial tires have two plies and two belts. Radial tires provide less rolling resistance, better steering characteristics, and longer tread life than bias-ply tires. As was stated earlier all modern vehicles are equipped with radial-ply tires.

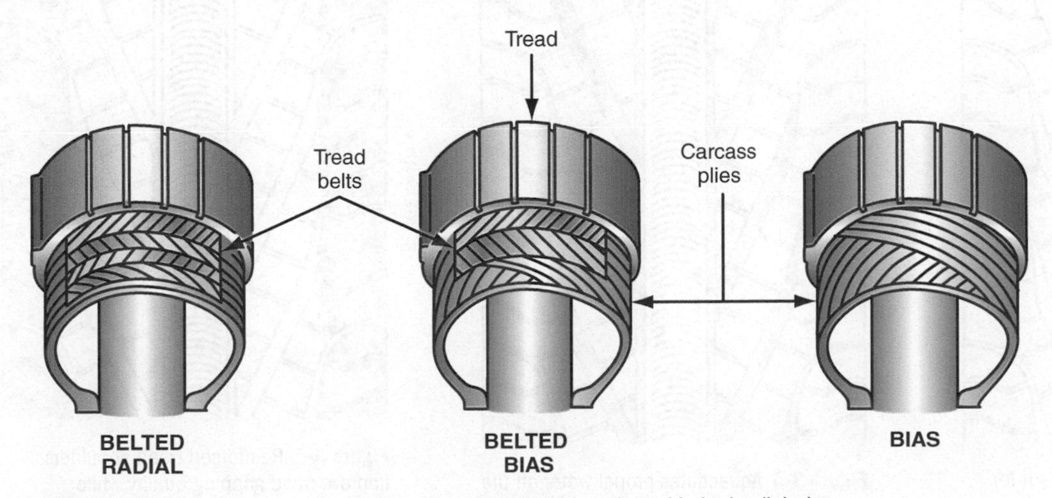

Figure 3-3 Three types of tire construction: bias-ply, belted bias-ply, and belted radial-ply.

Regardless of the type of tire construction, the tire must be uniform in diameter and width. Radial runout refers to variations in tire diameter. A tire with excessive radial runout causes a tire thumping problem as the car is driven. When a tire has excessive variations in width, this condition is called lateral runout. A tire with excessive lateral runout causes the chassis to "waddle" when the car is driven.

Tire **conicity** refers to a condition where the plies and/or belts are not level across the tire tread. When a tire has conicity, the plies and/or belts are somewhat cone shaped and may cause the vehicle to pull in one direction when driven.

The tire plies and belts must be level across the tread area. If the plies and/or belts are not level across the tread area, the tire is cone shaped. This condition is referred to as tire **conicity**. This can occur for several reasons; if the tires' internal components are misaligned during the manufacturing curing process it is possible that unequal forces internally could develop causing conicity. With this type of manufacturing defect, the tire pull will be present immediately after tire installation on a vehicle, or following the first time the tires are rotated. When a front tire has conicity, the steering may pull to one side as the car is driven straight ahead. This is referred to as a radial pull. A rear tire with conicity will not affect the steering as much as a front tire with conicity. If the radial pull develops later in the tires service life it is likely due to vehicle misalignment or driving conditions that have caused a slight angular wear pattern on the tire. This is when the tire tread wears faster on one side than the other (inside versus outside). If a tire pull is suspected, the front tires can be cross rotated. If the pull shifts direction or goes away, suspect a radial pull; further diagnosis will be required to determine which tire is affected. Often rotating the front tires to the rear tire position will eradicate this problem and even out the tire wear pattern. Additionally, the root cause of the initial improper wear pattern should be addressed, such as setting the vehicle alignment to manufacturer's preferred specifications.

TIRE TREAD DESIGN

Hydroplaning occurs when water on the pavement is allowed to remain between the pavement and the tire tread contact area. This action reduces friction between the tire tread and the road surface, and can contribute to a loss of steering control.

Vehicle tires have many different tire treads designed to provide the tire performance desired by the tire and vehicle manufacturers. A typical modern tire tread has these features:

1. An interlocking tread pattern for improved tread gripping quality on icy or slick roads (**Figure 3-4**).
2. Deeply carved aquachutes to propel water off the tread and away from the tire to reduce the possibility of **hydroplaning** when driving on wet pavement (**Figure 3-5**).
3. Reinforced tread shoulders to improve tread gripping quality when turning corners on dry pavement (**Figure 3-6**).

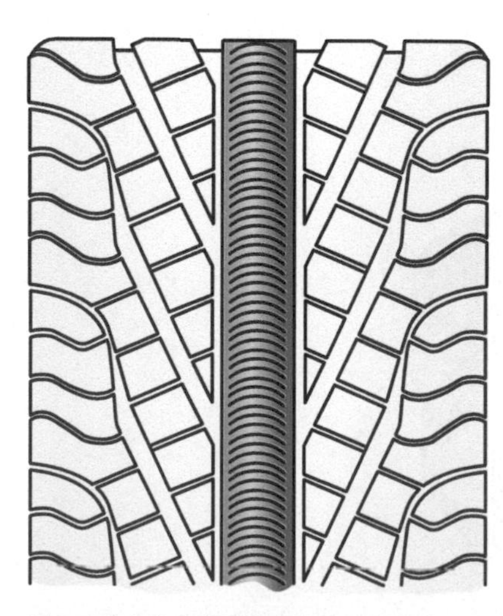

Figure 3-4 Interlocking tread pattern for improved traction on icy roads.

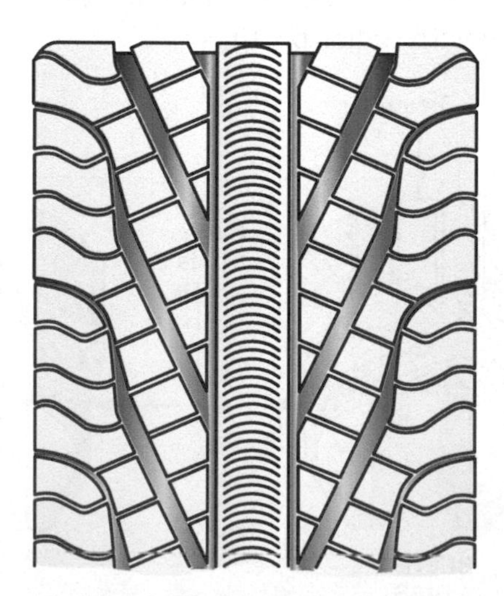

Figure 3-5 Aquachutes propel water off the tire tread.

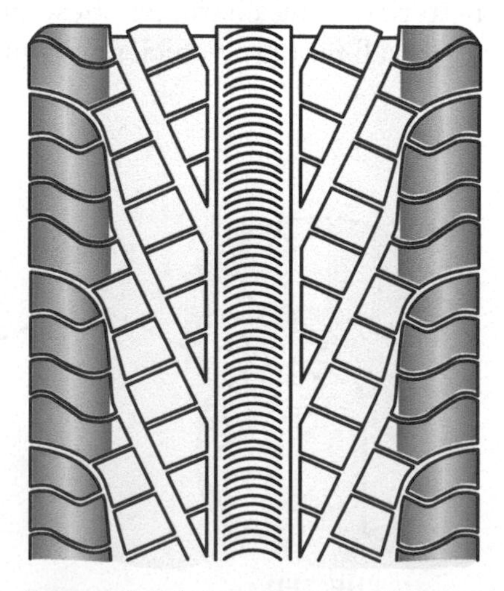

Figure 3-6 Reinforced tread shoulders improve tread gripping quality while cornering.

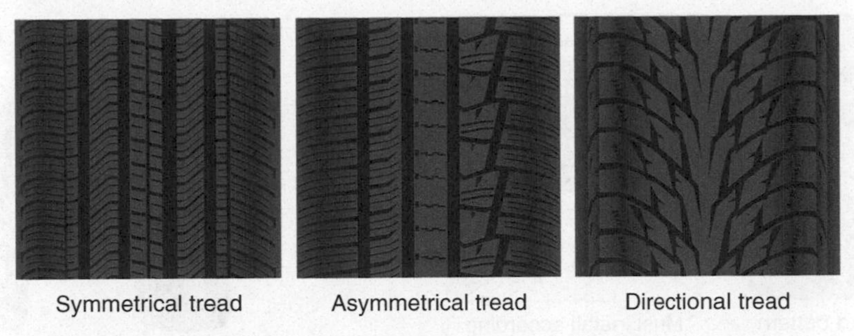

Figure 3-7 Tire tread pattern designs generally fall into symmetrical, asymmetrical, and directional designs.

Tire tread pattern designs generally fall into symmetrical, asymmetrical, and directional designs (**Figure 3-7**). Most tires manufactured have a nondirectional symmetrical tread design and can be mounted on a rim in either direction with no defined inside or outside preference on sidewall. Some tires feature an asymmetrical tire tread design that has a different tire tread pattern on the inside versus the outside tread surface area of the tire. An asymmetrical tire has a defined outside sidewall with markings identifying the inner side and outer side of the tire (**Figure 3-8**). If all four tires are the same size, the tires are rotated the same as symmetrical tires.

Directional tire tread designs have a unique pattern that is designed to rotate in a specific direction for optimum performance. Direction tread patterns can be either symmetrical or asymmetrical (**Figure 3-9**). A directional tire has a V-shaped tread pattern, and the tire has a defined sidewall marking indicating the direction of rotation with an arrow. When mounting four directional tires, two will be mounted with the arrow pointing in the clockwise direction and two will be mounted with the arrow pointing in the counterclockwise direction. When all four tires are the same size they can only be rotated front-rear and must remain on the original side of the vehicle (**Figure 3-10**). If the vehicle

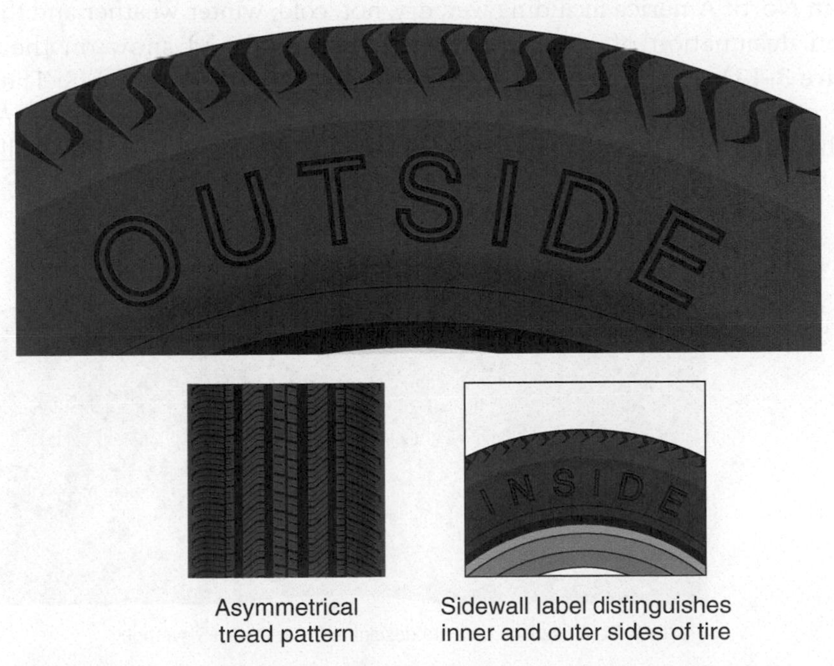

Figure 3-8 An asymmetrical tire has a defined sidewall marking indicating the outside or inside sidewall.

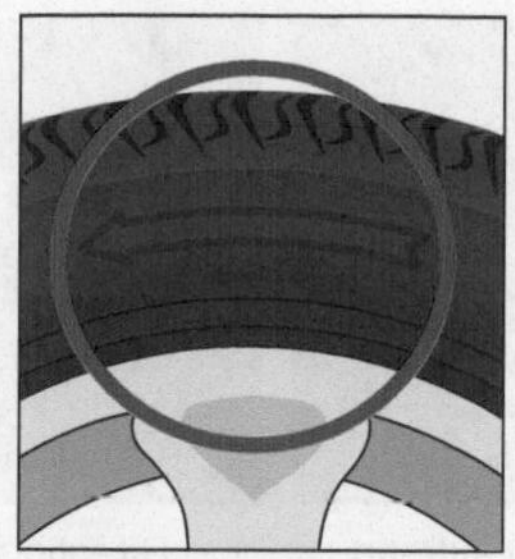

Figure 3-9 A directional tire has a V-shaped tread pattern and the tire has a defined sidewall marking indicating the direction of rotation with an arrow.

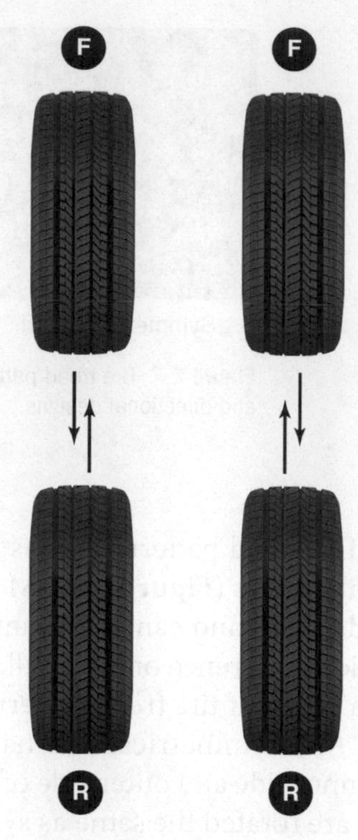

Figure 3-10 With directional tires, when all four tires are the same size they can only be rotated front-rear and must remain on the original side of the vehicle.

is equipped with directional tires but the front and rear tires are different sizes, then the tires are position specific and tire rotation is prohibited.

All-season tires are designed for all road conditions that a typical vehicle could experience in North America including wet, dry, hot, cold, winter weather and they have an all-season designation of M+S which stands for mud and snow on the tire side wall (**Figure 3-11**), variations of this symbol include M&S, M/S, and MS. These tires have a 37 percent higher average snow traction compared to non-all-season tires. All-season tires may have slightly improved performance in areas such as wet traction, rolling resistance,

Figure 3-11 All-season tires are designated with the M+S symbol.

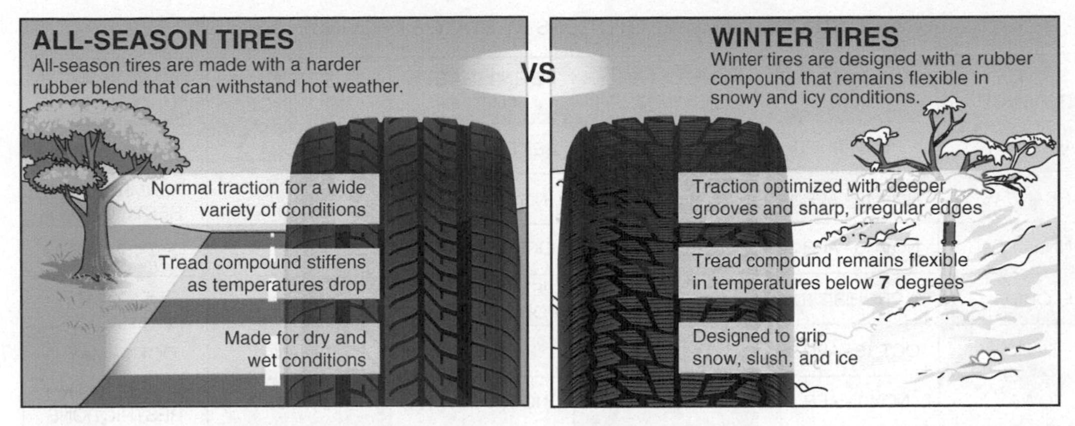

ALL-SEASON TIRES
All-season tires are made with a harder rubber blend that can withstand hot weather.
VS
WINTER TIRES
Winter tires are designed with a rubber compound that remains flexible in snowy and icy conditions.
Normal traction for a wide variety of conditions
Tread compound stiffens as temperatures drop
Made for dry and wet conditions
Traction optimized with deeper grooves and sharp, irregular edges
Tread compound remains flexible in temperatures below 7 degrees
Designed to grip snow, slush, and ice

Figure 3-12 All-season tires are generalist and while they perform adequately in most conditions, they do not provide the improved traction of a specialized winter tire for harsh snow and ice conditions.

and tread life. Improvements in tread design and tread compounds provide the superior performance in all-season tires. Though they are designated as an all-season tire, they do not perform as well as a winter/ice tire in extreme cold and harsh weather conditions (**Figure 3-12**).

Winter tires are designed to perform in extreme cold conditions, and remain flexible even at 7°F (−14°C) to provide improved traction performance. Winter tires carry the triple-peak mountain snowflake symbol, which is an indication that the tire meets minimum standards for harsh conditions traction (**Figure 3-13**). The Rubber Manufacturers Association sets the standards for approved tires that may carry the all-season and winter tire designation. Some winter tires are studdable, meaning they have a tread design that allows the installation of metal studs for enhanced ice and hard-packed snow performance (**Figure 3-14**). Studded tires provide improved traction on ice, but these tires are prohibited by law in many states because their use resulted in road surface damage. The states that do allow studded tires limit their use to certain winter months (**Figure 3-15**). A winter tire also often has sipes, which are small zigzag slits cut into the tread blocks to improve traction on slippery road surfaces. Some all-season tires are also equipped with sipes.

Figure 3-13 Winter tires are designated with M+S and have a triple peak mountain and snowflake symbol.

Figure 3-14 Winter tire tread design has a deep lug pattern, sipes for improved ice and slush traction and may also accept studs.

STUDDED TIRE RESTRICTIONS BY STATE & PROVINCE

STATE/ PROVINCE	PERMISSABLE DATES	STATE/ PROVINCE	PERMISSABLE DATES	STATE/ PROVINCE	PERMISSABLE DATES
ALABAMA	PROHIBITED	ALASKA	SEPT 30 - APR 15	ARIZONA	OCT 1 - MAY 1
ARKANSAS	NOV 15 - APR 15	CALIFORNIA	NOV 1 - APR 30	COLORADO	PERMITTED NO RESTRICTIONS
CONNECTICUT	NOV 15 - APR 30	DELAWARE	OCT 15 - APR 15	D.C.	OCT 15 - APR 15
FLORIDA	PROHIBITED	GEORGIA	PEMITTED ONLY FOR SNOW & ICE	HAWAII	PROHIBITED
IDAHO	OCT 15 - APR 30	ILLINOIS	PROHIBITED	INDIANA	OCT 1 - MAY 1
IOWA	NOV 1 - APR 1	KANSAS	NOV 1 - APR 15	KENTUCKY	PERMITTED NO RESTRICTIONS
LOUISIANA	PROHIBITED	MAINE	OCT 1 - MAY 1	MARYLAND	PROHIBITED
MASSACHUSETTS	NOV 2 - APR 30	MICHIGAN	PROHIBITED	MINNESOTA	PROHIBITED
MISSISSIPPI	PROHIBITED	MISSOURI	NOV 1 - MAR 31	MONTANA	OCT 1 - MAY 31
NEBRASKA	NOV 1 - APR 1	NEVADA	OCT 1 - APR 30	NEW HAMPSHIRE	NO RESTRICTIONS
NEW JERSEY	NOV 15 - APR 1	NEW MEXICO	NO RESTRICTIONS	NEW YORK	OCT 16 - APR 30
NORTH CAROLINA	NO RESTRICTIONS	NORTH DAKOTA	OCT 15 - APR 15	OHIO	NOV 1 - APR 15
OKLAHOMA	NOV 1 - APR 1	OREGON	NOV 1 - APR 1	PENNSYLVANIA	NOV 1 - APR 15
RHODE ISLAND	NOV 15 - APR 1	SOUTH CAROLINA	ALL YEAR. 1/16 Studs	SOUTH DAKOTA	OCT 1 - APR 30
TENNESSEE	OCT 1 - APR 15	TEXAS	PROHIBITED	UTAH	OCT 15 - MAR 31
VERMONT	NO RESTRICTIONS	VIRGINIA	OCT 15 - APR 15	WASHINGTON	NOV 1 - APR 1
WEST VIRGINIA	NOV 1 - APR 15	WISCONSIN	PROHIBITED	WYOMING	NO RESTRICTIONS
ALBERTA	NO RESTRICTIONS	BRITISH COLUMBIA	OCT 1 - APR 30	MANITOBA	OCT 1 - APR 30
NEW BRUNSWICK	OCT 16 - APR 30	NEWFOUNDLAND	NOV 1 - APR 30	N.W. TERRITORIES	NO RESTRICTIONS
NOVA SCOTIA	OCT 15 - APR 30	ONTARIO	PROHIBITED	P.E.I.	OCT 1 - MAY 31
QUEBEC	OCT 1 - MAY 1	SASKATCHEWAN	NO RESTRICTIONS	YUKON TERRITORY	NO RESTRICTIONS

Figure 3-15 Many states restrict the use of studded tires to winter months; some states prohibit their use all together.

One tire manufacturer installs 2 to 3 billion microscopic hollow shells or microbubbles in the tread material on one brand of their tires. These hollow shells are installed to 60 percent of the tread depth, and add rigidity to the tread material. When these hollow shells contact the road surface they break open, and the shell edges provide a gripping effect to improve traction. When driving on wet pavement, each time a hollow shell contacts the road surface and breaks open, a small amount of water is pulled into the hollow shell (**Figure 3-16**). Because this action is occurring at the millions of hollow shells in contact with the road surface, water is removed from between the tread and the road surface. This action reduces the possibility of hydroplaning. As the tire continues to rotate, water is expelled from the hollow shells when they move out of contact with the road surface.

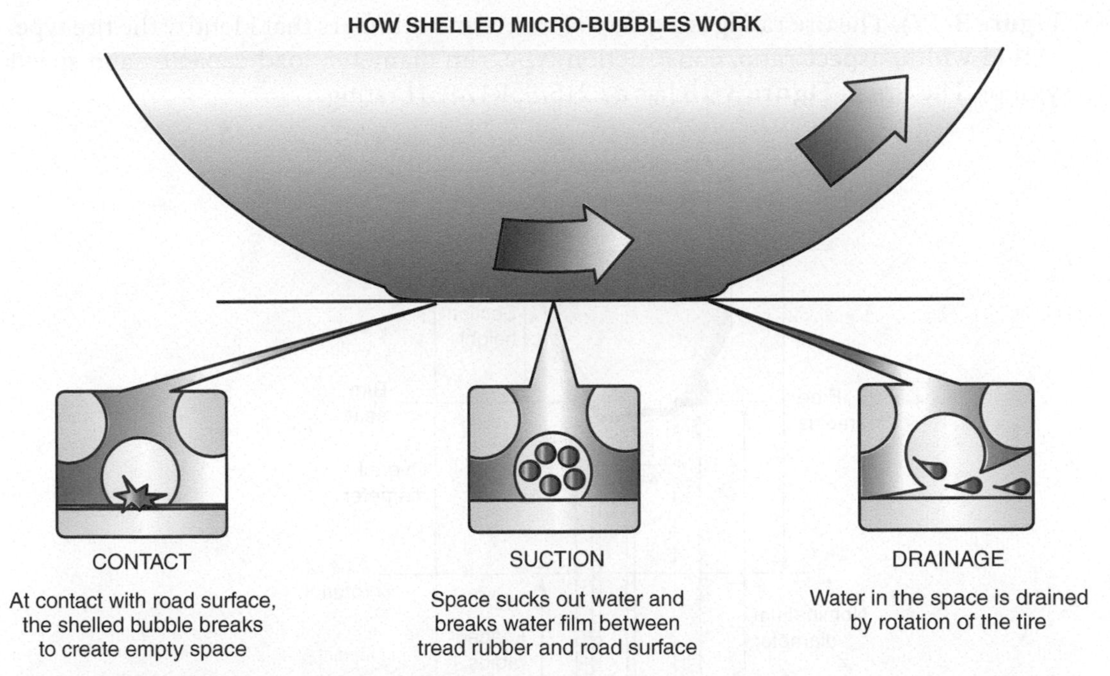
HOW SHELLED MICRO-BUBBLES WORK
CONTACT
At contact with road surface, the shelled bubble breaks to create empty space
SUCTION
Space sucks out water and breaks water film between tread rubber and road surface
DRAINAGE
Water in the space is drained by rotation of the tire

Figure 3-16 Advantages of micro-bubbles in the tire tread.

TIRE DEFECTS

Tire defects may be the result of manufacturing anomalies, road impact, or misuse. Manufacturing defects may be covered by the tire manufacturer's warranty. Possible defects include the following:

1. Separations and bulges, including tread or sidewall separation, or an open tread splice
2. Tread chunking, tearing, groove cracking, and shoulder cracking
3. Sidewall circumference fatigue, cracking, and weather cracking
4. Bead chafing, broken wires, or a pulled bead
5. Breaks in the sidewall or tread areas
6. Excessive radial or lateral runout
7. Conicity

If this inner liner is pinched or damaged, air can leak between the tire plies, forming a bubble on the sidewall or under the tread of the tire. This is often the result of tire impact damage such as hitting a pothole or curb, and is not a manufacturing defect. Inner liner damage can also result from an extreme underinflation condition that causes excessive heat buildup and flexing of the sidewall of the tire. It is critical that a tire suspected of being damaged or that has been driven underinflated be removed and inspected thoroughly before attempting to inflate or repair the assembly. A tire that is driven on while flat is almost always irreparable and must be replaced.

TIRE RATINGS AND SIDEWALL INFORMATION

A lot of essential information is molded into the sidewall of the average passenger car and light truck tire, including the tire rating. There are many dimensions related to tires and rims; not all this information is listed on the sidewall of the tire but is no less important

(**Figure 3-17**). The tire rating is a group of letters and numbers that identify the tire type, section width, aspect ratio, construction type, rim diameter, load capacity, and speed symbol. The tire in **Figure 3-18** has a P215/55R17 95H rating.

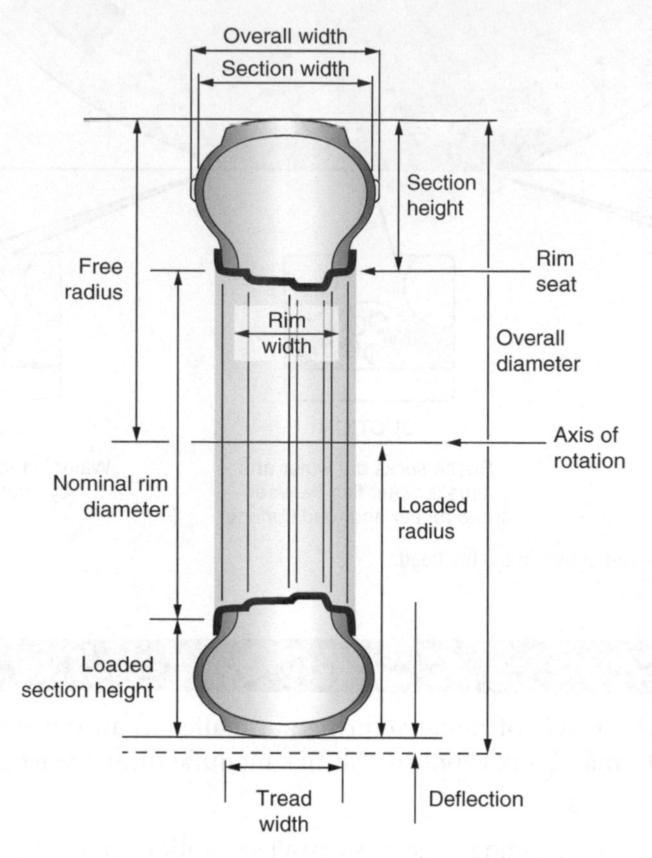

Figure 3-17 Various dimensions associated with tire and rim.

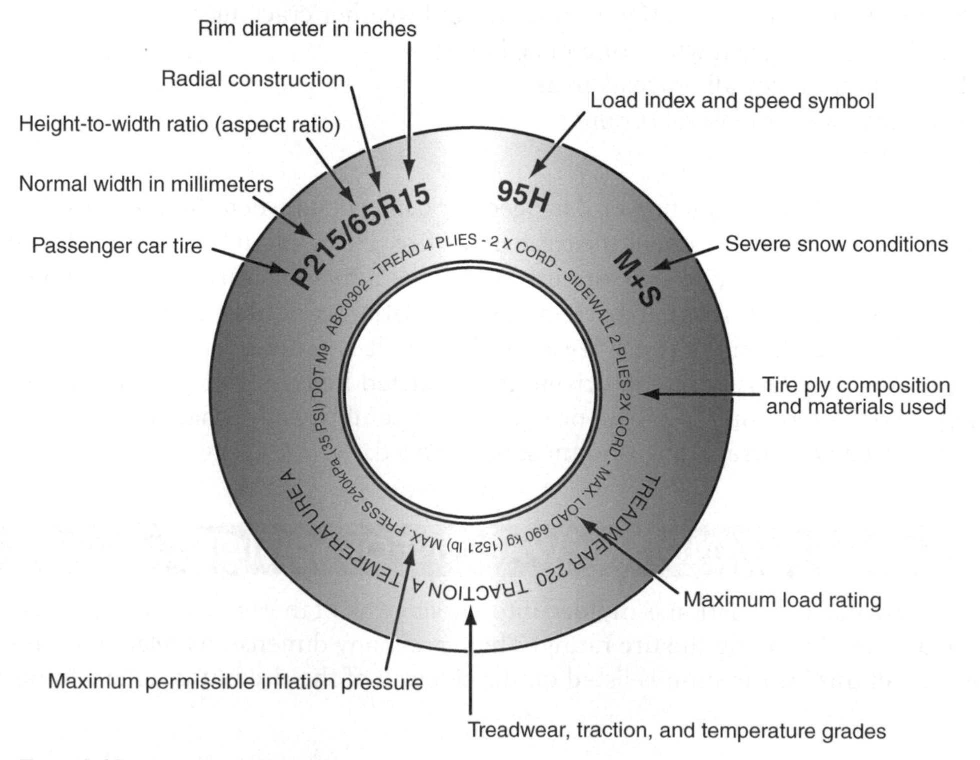

Figure 3-18 Tire sidewall information.

P indicates a passenger car tire
LT indicates a light truck tire
T indicates a temporary tire most commonly a compact spare tire

The complete DOT code must be stamped on one side of the tire sidewall. This code is a combination of letters and numbers and indicates the manufacturer plant code where the tire was manufactured, the batch code, and the week and year the tire was manufactured (**Figure 3-19**). It is not a unique number to a specific tire but rather an identification for a batch of tires produced. Beginning in 2000, the week and year the tire was manufactured are indicated by the last four digits of the DOT number (**Figure 3-20**). The first two digits indicate the week while the last two digits indicate the year of production. As an example, if the last four digits are 2514, then the tire was produced during the 25th week of 2014. On tires produced before 2000, the week and year the tire was manufactured are indicated by the last three digits of the DOT number (**Figure 3-21**). The first two digits indicate the week, while the last digit indicates the year of production. As an example, if the last three digits are 254, then the tire was produced during the 25th week of 1994.

Section Width, Aspect Ratio, and Ply Design

In the tire size P215/55R17, the number 215 is the section width of the tire in millimeters measured from sidewall to sidewall with the tire mounted on the recommended rim width. The section width, also referred to as the cross-section width, is the width of the tire from the widest point measured from one sidewall to the opposite sidewall.

The aspect ratio is a number that indicates the section height of the tire sidewall. In the tire size P215/55R17, the number 55 indicates the aspect ratio, which is the ratio of the height to the width. With a 55 aspect ratio, the tire's height is 55 percent of its width

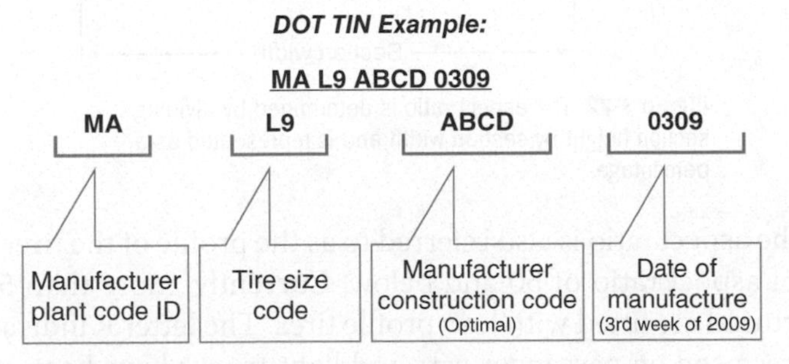

Figure 3-19 The tire DOT code contains batch information about where and when a tire was manufactured.

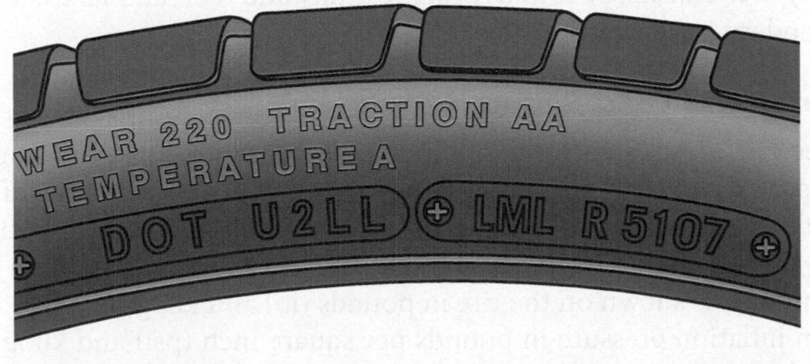

Figure 3-20 On tires produced beginning in 2000 the last four digits of the DOT code indicates among other things the week and year the tire was produced.

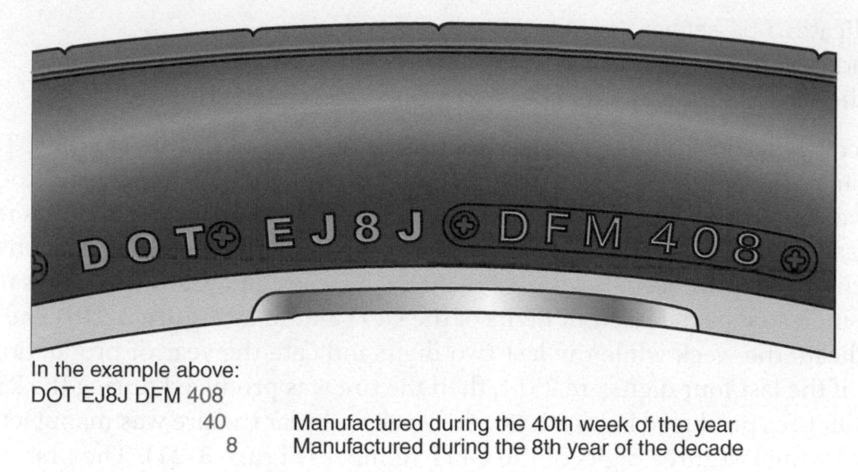

Figure 3-21 On tires produced before 2000, the last three digits of the DOT code indicates, among other things, the week and year the tire was produced.

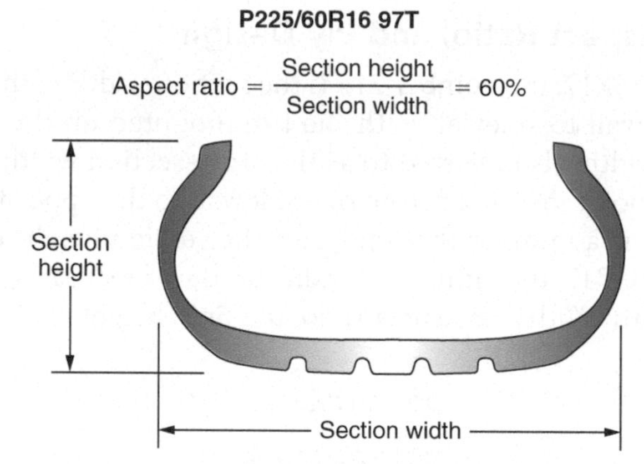

Figure 3-22 The aspect ratio is determined by dividing section height by section width and is represented as a percentage.

(**Figure 3-22**). The aspect ratio is also referred to as the profile of the tire. Low-profile tires are those with an aspect ratio of 50 and below. Currently, more than 50 percent of the vehicles manufactured are fitted with low-profile tires. The letter R indicates radial-ply tire construction. Tires used on passenger cars and light trucks have been of radial-ply construction since the late 1970s. If the tire construction was indicated by the letter B, the tire has a belted bias-ply construction. The letter D represents diagonal bias-ply tire design. Belted bias-ply tires can still be found on some utility and recreational trailers, but are not suitable for modern passenger vehicles.

Rim Diameter and Load Rating

In the tire size P215/55R17 95H, the number 17 is the rim diameter in inches. The load index is represented by the number 95. The tire in Figure 3-9 has load rating of 1521 pounds. The load index number correlates to the maximum rated load the tire can carry (**Figure 3-23**). A higher number represents the tire has a higher maximum load capacity. The maximum load is shown on the tire in pounds (lb) and kilograms (kg) together with the maximum inflation pressure in pounds per square inch (psi) and kilograms (kg). As an example, MAX LOAD 730 kg (1609 lb) and 240 kPa (35 psi) MAX PRESSURE COLD may be listed on the sidewall of the tire. This indicates the maximum load at a specified cold inflation pressure. Never exceed the maximum inflation pressure listed on the sidewall, or tire failure could result. Some tire manufacturers have used the letters B, C, D, or E to indicate the **load rating** most commonly on LT tires. The letter B indicates the lowest

The **load rating** indicates the tire's load-carrying capability.

Load Index	Maximum Load (lbs.)	Load Index	Maximum Load (lbs.)
74	827	88	1235
75	853	89	1279
76	882	90	1323
77	908	91	1356
78	937	92	1389
79	963	93	1433
80	992	94	1477
81	1019	95	1521
82	1047	96	1565
83	1074	97	1609
84	1102	98	1653
85	1135	99	1709
86	1168	100	1764
87	1201		

Figure 3-23 The load index number correlates to a number assigned to the maximum load the tire can carry.

load rating and the letter C represents a higher load rating. A tire with a D load rating is designed for light-duty trucks, and this tire will safely carry a load of 2623 pounds when inflated to 65 psi. The ply rating is not the actual number of sidewall plies used but rather an equivalent strength compared to earlier bias-ply tires. An E load range light truck tire may have the ply rating of a 10-ply tire but, in actuality, only have a 4-ply sidewall.

Aspect ratio is the percentage of a tire's height in relation to its width.

Speed Rating

In the tire size P215/55R17 95H, the letter H indicates the tire has a 130 miles per hour (mph) **speed rating**. Other speed ratings are as follows:

The **speed rating** indicates the tire's capability to withstand high speed.

Q – 99 mph
S – 112 mph
T – 118 mph
U – 124 mph
V – above 130 mph without service description
V – 149 mph with service description
Z – above 149 mph
W – 168 mph
Y – 186 mph

The aspect ratio may be referred to as the tire profile or series number.

Tire speed ratings do not suggest that vehicles can always be driven safely up to the maximum designated speed rating, because many different road and weather conditions may be encountered. Tire manufacturers do not endorse the operation of a vehicle in an unlawful or unsafe manner. Speed ratings are based on laboratory tests, and these ratings are not valid if tires are worn out, damaged, altered, underinflated, or overloaded. The condition of the vehicle may also affect high-speed operation.

Department of Transportation (DOT) Tire Grading

The United States Department of Transportation (DOT) and the National Highway Traffic Safety Administration (NHTSA) developed the **Uniform Tire Quality Grading (UTQG)** system to provide customers with standardized information related to a tire's tread wear, traction, and temperature capabilities to aid them with purchasing decisions. It is required that tire manufacturers grade most of their passenger car tires, except deep-treaded light truck tires, winter tires, temporary spare tires, and trailer tires. It should be noted that the DOT does not test the tires. It is left to the manufacturer to assign a grade based on their test results and those of independent testing companies they have hired.

Tread Wear Rating

The UTQG for **tread wear ratings** allows customers to compare tire tread life expectancies. The tread wear rating is based on the tire tread wear when tested under specific conditions on a regulated test track. Tire tread wear is monitored for a total of 7200 miles and then the data are extrapolated out for total tread life. A baseline tire has a tread wear rating of 100. A tire with a 150 tread wear rating would last 1.5 times as many miles on the test track as the baseline tire with a 100 tread wear rating. Tread wear ratings are valid only for comparison within a manufacturer's product line, and these ratings are not as valid for comparisons between tire manufacturers. This is one of the major flaws of the UTQG as it relates to tread wear. But you can expect a tire with a tread wear number of 600 to last longer than one rated at 500.

Traction Rating

The UTQG for **traction ratings** indicates the tire's straight line wet coefficient of traction and braking on both wet asphalt and wet concrete surfaces. The test does not evaluate the dry braking performance, wet cornering, dry cornering, or high-speed hydroplaning resistance. So while the information it provides is limited it does offer some ability to cross compare tires as long as the limits are understood.

To determine the traction rating, a properly inflated tire is placed on the axle of a skid trailer that is electronically monitored. Test conditions are carefully controlled to maintain test uniformity. A trailer speed of 40 mph is maintained over both wet concrete and wet asphalt. The brakes are momentarily applied on both surfaces while sensors gather information related to the tire's coefficient of friction, or braking g-force, as the tire slides, maintaining a constant speed of 40 mph. By design this test places less emphasis on tread design and more emphasis on the tread compound.

The results of the skid tests are averaged, and the traction rating is designated as AA, A, B, or C (**Figure 3-24**). An AA traction rating indicates the best traction, while a C rating indicates acceptable traction.

Temperature Rating

The UTQG for **temperature rating** indicates the tire's ability to generate or dissipate heat during tire operation. A tire that is unable to dissipate heat efficiently, or is unable to resist the destructive effects of internal heat buildup, will have a reduced ability to run at higher speeds.

Traction Grades	Asphalt g-Force	Concrete g-Force
AA	Above 0.54	0.41
A	Above 0.47	0.35
B	Above 0.38	0.26
C	Less than 0.38	0.26

Figure 3-24 The UTQG Traction Grade is determined by the tire's average coefficient of brake traction on both wet asphalt and wet concrete at a constant 40 mph in a straight line.

Temperature Grades	Speeds in mph
A	Over 115
B	Between 100 and 115
C	Between 85 and 100

Figure 3-25 The UTQG Temperature Grade indicates the tire's ability under a load to withstand heat buildup at various speeds.

To obtain the temperature rating, tires are tested on a laboratory test wheel under a load. The temperature rating indicates the tires' ability to operate at high speeds without failure. Extremely high temperatures may cause tire materials to degenerate and thus reduce tire life or cause tire failure. The temperature ratings indicate the tires' ability to withstand heat at various speeds and are A for best, B for intermediate, and C for acceptable (**Figure 3-25**). In laboratory testing a B-rated tire can withstand speeds between 100 and 115 mph without failure. The test is similar to those used to confirm a tire's speed rating, though it is not an exact test.

The UTQG is a good indicator of what can be expected from a particular tire. But it falls short of reflecting what actual real-world performance will be. It should be viewed as a useful tool for comparison only.

DOT regulations also require tire manufacturers to place the following information on tire sidewalls:

1. Size
2. Load range
3. Maximum load
4. Maximum pressure
5. Number of plies under the tread and in the sidewalls
6. Manufacturer's name
7. Tubeless or tube-type construction
8. Type of carcass construction
9. DOT approval number, including the manufacturer's code number and date of manufacture

Some tires have a **tire performance criteria (TPC)** number molded on the sidewall (**Figure 3-26**). This number represents that the tire meets the car manufacturer's performance standards for traction, endurance, dimensions, noise, handling, and rolling

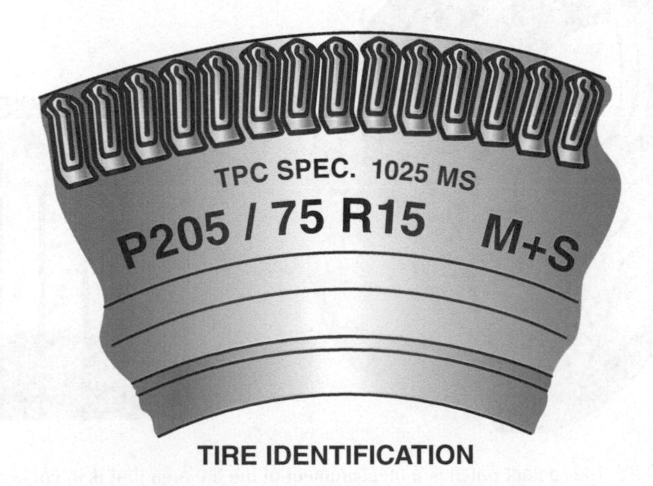

Figure 3-26 Tire performance criteria (TPC) number.

resistance. Most car manufacturers assign a different TPC number to each tire size. Replacement tires should have the same ratings and TPC number as the original tires.

TIRE CONTACT AREA

The **tire contact area**, or contact patch, refers to the area of the tire that is in contact with the road surface when the tire is supporting the vehicle weight. The **tire free diameter** is the distance of a horizontal line through the center of the spindle and wheel to the outer edges of the tread. The **tire rolling diameter** is the distance of a vertical straight line through the center of the spindle to the outer edges of the tread when the tire is supporting the vehicle weight. The rolling diameter is always less than the free diameter. The difference between the free diameter and the rolling diameter is referred to as deflection. Tire tread grooves take up excess rubber and prevent scrubbing as the tire deflects in the contact area. The rolling diameter, free diameter, and contact area are shown in **Figure 3-27**. The contact patch is a measurement of the tire area that is in contact with the road (**Figure 3-28**).

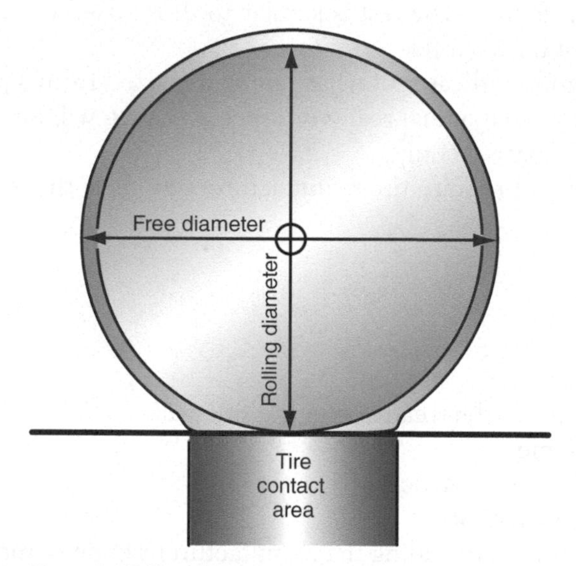

Figure 3-27 Tire rolling diameter, free diameter, and contact area.

Figure 3-28 The contact patch is a measurement of the tire area that is in contact with the road.

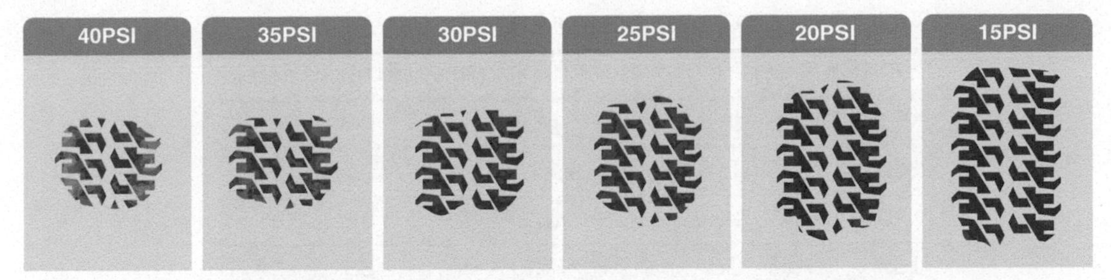

Figure 3-29 As air pressure is increased or decreased in a tire, the contact patch area will inversely change, decreasing as pressure is increased and increasing as pressure is decreased.

As air pressure is increased or decreased in a tire, the contact patch area will inversely change, decreasing as pressure is increased and increasing as pressure is decreased (**Figure 3-29**).

TIRE PLACARD AND INFLATION PRESSURE

The vehicle weight is supported by the correct air pressure exerted evenly against the entire interior tire surface, which produces tension in the tire carcass. Therefore, **tire pressure** is extremely important. Underinflation decreases the rolling diameter and increases the contact area, which results in excessive sidewall flexing and tread wear. Overinflation decreases the contact area, increases the rolling diameter, and stiffens the tire. This action results in excessive wear on the center of the tread. Tire pressure should be checked when the tires are cool. Since tire pressure normally increases at high tire temperatures, air pressure should not be released from hot tires. Excessive heat buildup in a tire may be caused by underinflation. This condition may lead to severe tire damage.

Tire pressure is the amount of air pressure in the tire.

AUTHOR'S NOTE There are many causes of excessive tire tread wear, such as improper wheel balance or alignment. However, it has been my experience that one of the most common causes of excessive tire tread wear is improper inflation pressure. Many vehicle owners or drivers do not check tire pressures regularly.

On many vehicles, the **tire placard** is permanently attached to the rear face of the driver's door. This placard provides tire information such as maximum vehicle load; tire size, including spare; and cold inflation pressure, including spare (**Figure 3-30**).

Tire pressure is carefully calculated by the vehicle manufacturer to provide satisfactory tread life, handling, ride, and load-carrying capacity. Most vehicle manufacturers recommend that tire pressures be checked cold once a month or prior to any extended trip. The manufacturer considers the tires to be cold after the vehicle has sat for three hours, or when the vehicle has been driven less than one mile. The tires should be inflated to the pressure indicated on the tire placard. Tire pressures may be listed in metric or English system values. Conversion charts provide pressures in either of these systems (**Table 3-2**). The tire pressure that is listed on the sidewall is the maximum pressure and weight load that the tire can safely handle and is not the recommended tire pressure (**Figure 3-31**). Vehicle manufacturer's recommended tire pressure is always less than the pressure stated on the tire sidewall.

Shop Manual
Chapter 3, page 114

TABLE 3-2 Tire Inflation Pressure Conversion Chart

Inflation Pressure Conversion Chart (Kilopascals to psi)			
kPa	**psi**	**kPa**	**psi**
140	20	215	31
145	21	220	32
155	22	230	33
160	23	235	34
165	24	240	35
170	25	250	36
180	26	275	40
185	27	310	45
190	28	345	50
200	29	380	55
205	30	415	60

Conversion: 6.9 kPa = 1 psi

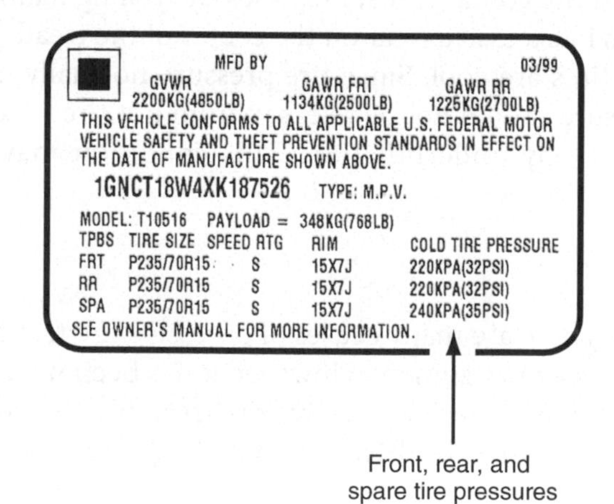

Figure 3-30 Tire placard.

Figure 3-31 The tire pressure that is listed on the sidewall is the maximum pressure and weight load that the tire can safely handle and is not the recommended tire pressure.

Nitrogen Tire Inflation

Some shops are presently equipped with **nitrogen tire inflation** equipment. When tires are inflated with compressed air, the air slowly passes through the tire walls and tread. Tires inflated with compressed air can lose as much as 12 psi in a 6-month period. When tire inflation pressure is lower than specified, tire tread wear is increased, while fuel economy and vehicle stability are decreased.

Nitrogen molecules are considerably larger than air molecules (**Figure 3-32**). Therefore, when a tire is inflated with nitrogen, the nitrogen passes through the tire walls and tread more slowly when compared with air. When tires are inflated with nitrogen, the tire inflation pressure remains more stable over a longer time period, which provides reduced tire tread wear, improved fuel economy, and increased vehicle stability. Another benefit from using nitrogen to inflate tires is reduced aluminum rim corrosion because the nitrogen does not contain any oxygen or moisture (water molecules). When tires are filled with air, the oxygen and water vapor in the air tends to react with the aluminum in the rims to cause corrosion.

SPECIALTY TIRES

Sport Utility Vehicle (SUV) and 4 × 4 Tires

SUV tires may be classified for use on pavement or off-road. SUV tires have greater load-carrying capacity compared to passenger car tires, because of the extra loads that may be carried in an SUV. A typical pavement SUV tire has these features:

1. A silica tread compound that provides low noise levels and exceptional wet braking capability.
2. An enhanced casing system and a stable tire contact area that supplies even tread wear and responsive handling quality.
3. Larger cables in the steel belts to provide increased strength and durability.
4. Full-depth, interlocking sections in the tread that supply excellent wet and snow traction.

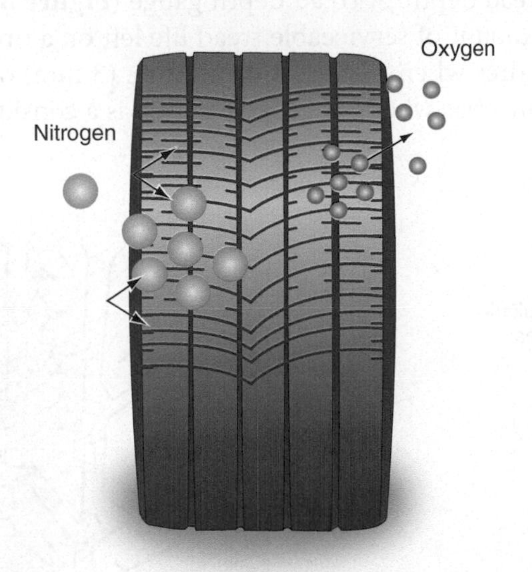

Figure 3-32 Oxygen and nitrogen leakage through tire walls and tread.

Compared to SUV tires for use on pavement, off-road SUV tires have stronger tread rubber to prevent cutting and chipping. Off-road SUV tires may also have thicker belt wire and more belt strands.

Tires designed for 4 × 4 vehicles may have some of the same features as SUV tires. A typical 4 × 4 tire has these additional features:

1. Two wide circumference grooves with a stepped profile to reduce hydroplaning.
2. Staggered shoulder blocks in the tread to improve lateral grip on slopes.
3. Gradual profile changes in the shoulder area of the tread to provide progressive breakaway during hard cornering.
4. Additional rubber at the base of the tread to reduce the possibility of tire damage.

Puncture sealing tires are available as an option on certain car lines, and some rubber companies sell these tires in the replacement tire market. These tires contain a special rubber sealing compound applied under the tread area during the manufacturing process. When a nail or other object up to 3/16 in. (4.76 mm) in diameter punctures the tread area, it picks up a coating of sealant. If the object is removed, the sealant sticks to the object and is pulled into the puncture. This sealant completely fills the puncture and forms a permanent seal to maintain tire inflation pressure (**Figure 3-33**). Puncture sealing tires usually have a special warranty. These tires can be serviced with conventional tire changing and balancing equipment. When repairing tires, the maximum repairable puncture size is ¼ in.

When snow tires are installed on a vehicle, these tires should be the same size and type as the other tires on the vehicle.

REPLACEMENT TIRES

Shop Manual
Chapter 3, page 110

Most tires have tread wear indicators built into the tread. When the tread wears down to 2/32 in. (1.5 mm) the minimum allowable tread depth, the wear indicators appear as bands across the tread (**Figure 3-34**). Car manufacturers typically recommend tire replacement when the wear indicators appear in one or more tread grooves at three locations around the tire. It is generally advisable that tires be replaced before they reach the minimum of 2/32 in. (1.5 mm) of tread depth. A tread depth gauge (**Figure 3-35**) should be used to determine the actual amount of serviceable tread life left on a tire. Technicians typically recommend replacing tires when there is only 4/32 in. (3 mm) of tread depth available. This is especially true in areas where wet road traction is a consideration. In areas where

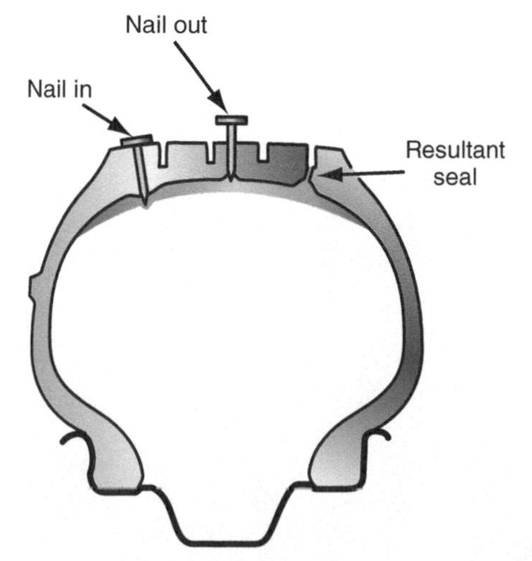

Figure 3-33 Puncture sealing tire.

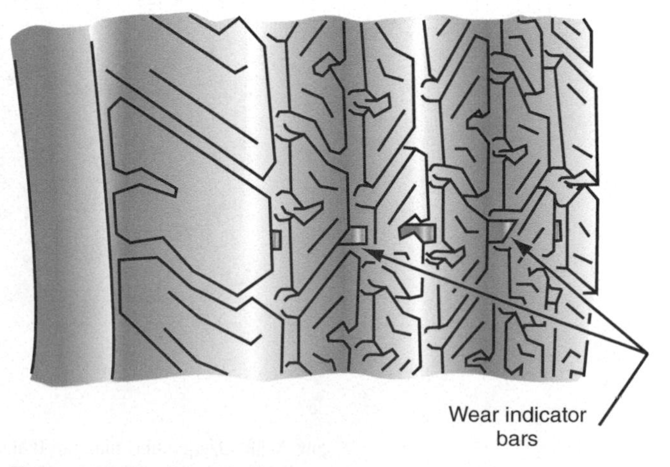

Figure 3-34 Tread wear indicators.

Figure 3-35 A tread depth gauge should be used for an accurate measurement of the serviceable tread available.

snow is a concern technicians often recommended all-season tire replacement when there is only 5/32 in. (4 mm) of tread depth available.

If **replacement tires** have a different size or construction type than the original tires, vehicle handling, ride quality, speedometer/odometer calibration, and antilock brake system (ABS) operation may be seriously affected. When replacement tires are a different size than the original tires, the vehicle ground clearance and tire-to-body clearance may be altered. Steering and braking quality may be seriously affected if different sizes or types of tires are installed on a vehicle. This does not include the compact spare tire, which is intended for temporary use. All vehicles manufactured after 2012 are equipped with ABS due to the requirement of stability control, and most vehicles manufactured since the early 1990s are equipped with ABS. When different-sized tires are installed on these vehicles, the ABS operation is abnormal, which may result in serious braking defects. When selecting replacement tires, the following precautions must be observed to maintain vehicle safety:

1. Replacement tires must be installed in pairs on the same axle. Never mix tire sizes or designs on the same axle. If it is necessary to replace only one tire, it should be paired with the tire having tread depth within 4/32 in. (3 mm) of the replacement tire.
2. The tire load rating must be adequate for the vehicle on which the tire is installed. Light-duty trucks, station wagons, and trailer-towing vehicles are examples of vehicles that require tires with higher load ratings than passenger car tires.
3. Snow tires should be the same type and size as the other tires on the vehicle.
4. A four-wheel-drive and all-wheel drive vehicle should have the same type and size of tires on all four wheels, and tread depth between tires should be within manufacturer specifications.
5. Do not install tires with a load rating less than the car manufacturer's recommended rating.
6. Replacement tire ratings should be equivalent to the original tire ratings in all rating designations.
7. When combining different tires front to rear on a vehicle, consult the car manufacturer's or tire manufacturer's recommendations.

Vehicles that are four-wheel-drive or all-wheel-drive, whether they are light trucks, SUVs, or passenger cars should have four matched tires. This includes tread depth. All tires should be within 2/32 in.–4/32 in. (1.5 mm–3 mm) of tread depth with one another depending on manufacturer's recommendations. The reason for this is that four-wheel-drive and all-wheel-drive vehicles are equipped with viscous couplings or differentials that are designed for momentary differences in wheel speeds. If all drive tires are not exactly the same size, the differentials and viscous couplings will be always forced to compensate for the different rolling diameters, resulting in unwanted heat buildup and potential wear to expensive drivetrain components. If a tire requires replacement because it was damaged, it may be necessary to replace all four tires due to too great a variation in tread depth created by installing one new tire. Refer to vehicle manufacturer's service information for specific details and allowable tread depth differences.

Tires should be rotated every 6000 miles as a general rule following manufacturer's recommendations for both frequency and rotation pattern (**Figure 3-36**). It is also important to torque the wheel lug nuts to specification following the recommended torque sequence or star pattern (**Figure 3-37**). Beginning in the 2008 model year, tire pressure monitoring systems (TPMS) became mandatory. Many vehicles with TPMS require that

Shop Manual
Chapter 3, page 102

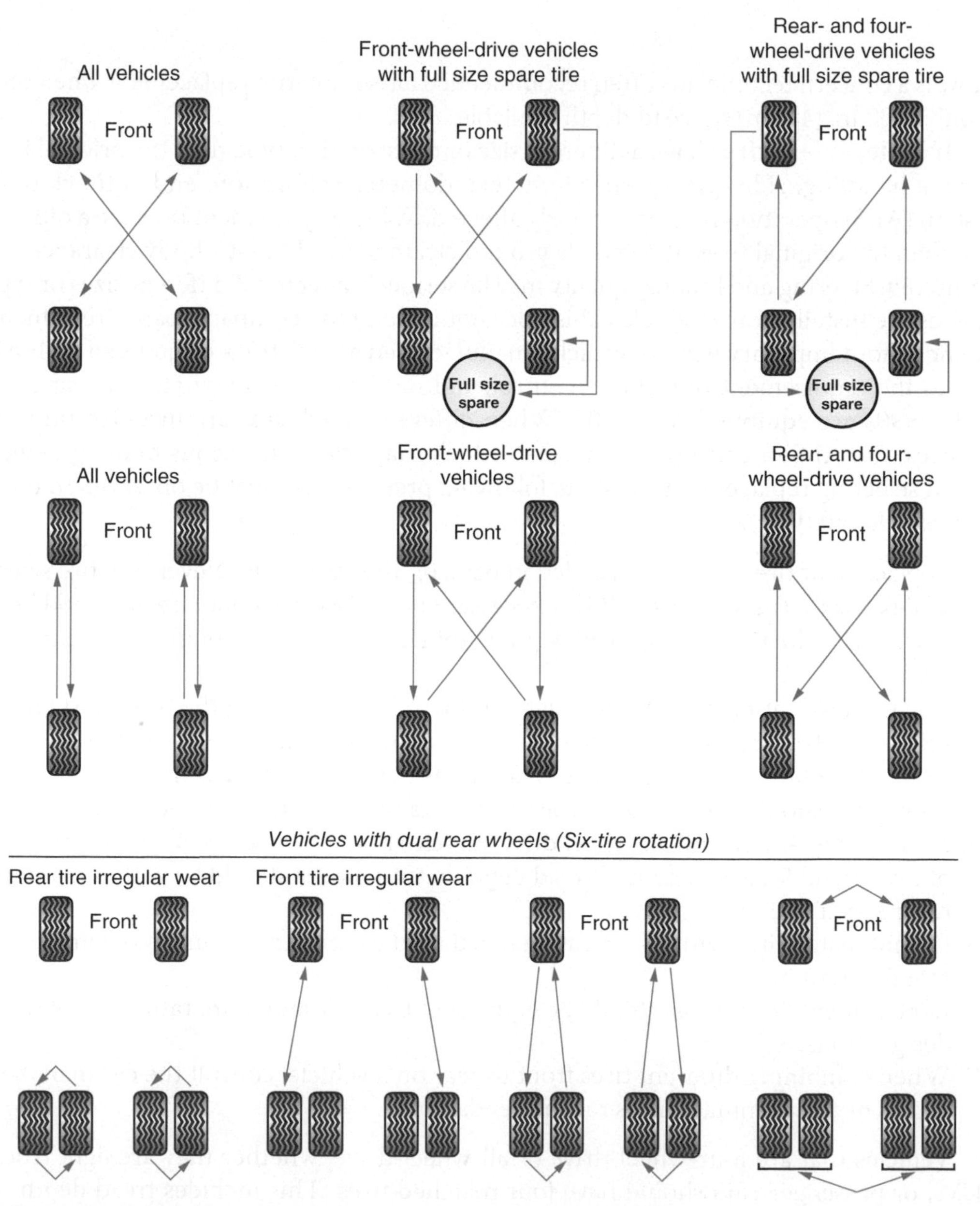

Figure 3-36 The typical tire rotation patterns when all tires are of the same size and type.

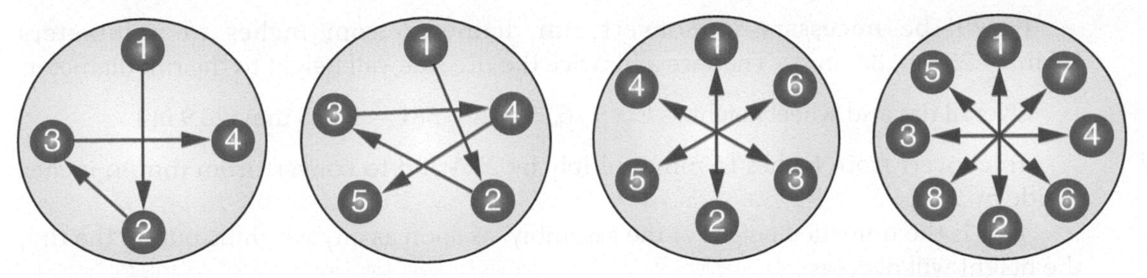

Figure 3-37 The typical rim torquing sequence.

the sensor position be relearned if any wheel and tire positions are changed (such as a tire rotation) since the last relearn procedure. Indirect TPMS systems that do not have a pressure sensor mount internally in the tire-and-wheel assembly do not require a relearn procedure. Consult the vehicle manufacturer's service information for exact procedures required following a tire rotation or replacement.

PLUS SIZING

Sometimes a customer does not like the appearance of their vehicle with the stock tire-and-wheel assembly and they would like to install a larger rim diameter to improve the looks and performance of the vehicle. This is referred to as plus sizing. It should be noted that a stock vehicle is designed to operate with a specific tire-and-wheel combination and deviating from allowed OEM combinations may adversely affect vehicle operation. Increasing a tire's contact patch may have a negative impact on wheel bearing loads and suspension component wear. It may also cause harsher ride quality due to less road isolation, which is caused by a narrower and stiffer tire side wall.

When the decision to plus size is made, you must remember that the overall height of the tire-and-wheel assembly must be maintained (**Figure 3-38**). By using a larger diameter wheel with a lower profile and a lower aspect number, it is possible to properly maintain the overall diameter of the tire and height of the assembly, keeping odometer and speedometer changes negligible. In addition, it is important to maintain the overall diameter of the tire to maintain the effectiveness of the antilock braking system, traction control, and vehicle stability system. It should be noted that when the decision is made to increase the rim diameter, it may also be necessary to change rim width and wheel offset. Refer to the section on rims for more detail regarding rim dimensions.

Now that we know what the tire size numbers mean, we can calculate the overall height of a tire-and-wheel assembly. We multiply the tire width by the aspect ratio to get the height of the tire:

$$\text{Tire sidewall height} = 235\text{ mm (width)} \times 0.75\text{ (aspect ratio percentage)}$$
$$= 176.25\text{ mm (64.94 in.)}$$

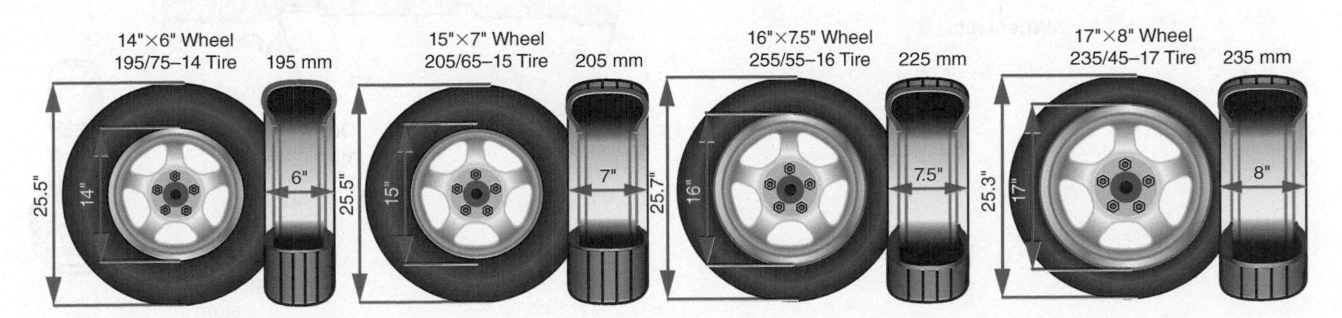

Figure 3-38 Plus sizing must maintain the overall height of the tire-and-wheel assembly.

It will be necessary to convert rim diameter from inches to millimeters (15 in. × 25.4 = 381mm). Then we add twice the tire sidewall height to the rim diameter.

Overall tire and wheel height = 2 × 176.25 mm + 381 = 733.5 mm (28.9 in.)

To convert from inches to mm multiply by 25.4 and to convert from mm to inches divide by 25.4.

This is the unloaded height of the assembly; as soon as any weight is put on the tire, the height will decrease.

Then use this information when comparing optional tire sizes. It should be within 0.25 of original assembly height.

TIRE VALVES

The **tire valve** allows air to flow into the tire, and it is also used to release air from the tire. The core in the center of the valve is spring loaded and allows air to flow inward while the tire is inflated (**Figure 3-39**). Once the tire is inflated, the valve core seals and prevents air flow out of the tire. The small pin on the outer end of the valve core may be pushed to unseat the valve core and release air from the tire. An airtight cap on the outer end of the valve keeps dirt out of the valve, and provides an extra seal against air leakage. A deep groove is cut around the inner end of the tire valve. When the rubber valve assembly is pulled into the wheel opening, this groove seals the valve in the opening. This type of valve stem is used on vehicles with indirect TPMS or on vehicles and trailers without TPMS. Steel valve stems are installed in some wheels. The lower end of the valve stem is threaded, and a nut retains the valve stem in the wheel (refer to Figure 3-39). Steel washers and sealing washers are located on the valve stem on the inside and outside of the wheel. The sealing washers are positioned next to the wheel.

On many vehicles equipped with TPMS, the rubber valve stem is mounted directly to the tire pressure sensor with a screw (**Figure 3-40**). The rubber valve stem and core can be served separately from the sensor, but care must be taken during both tire dismounting

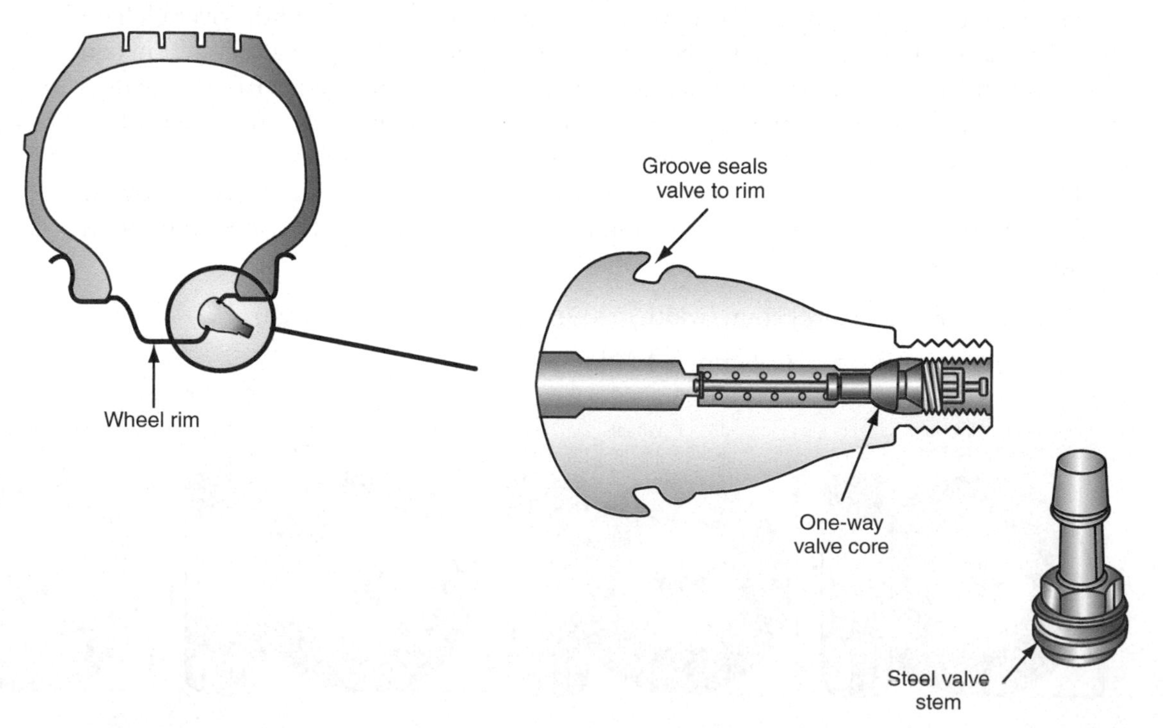

Figure 3-39 Valve stems and cores.

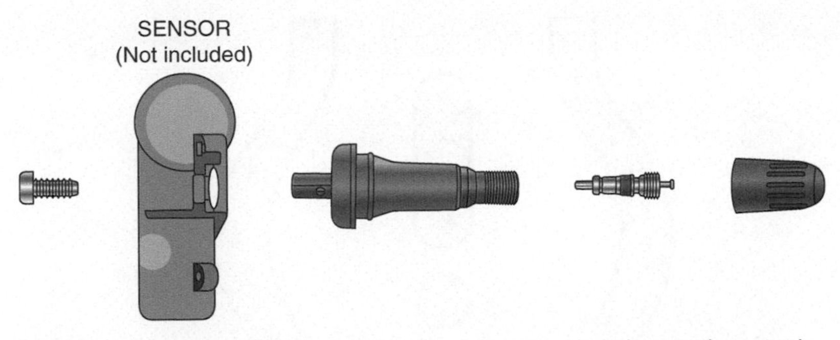

Figure 3-40 On some TPMSs, the rubber valve stem and core can be served separately from the tire pressure sensor.

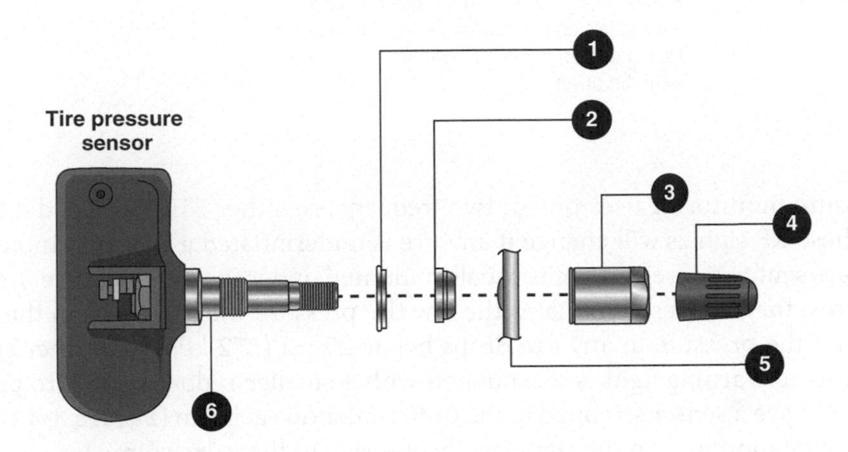

Figure 3-41 On some TPMSs, the metal valve stem is an integral part of the tire pressure sensor, and only the valve core and sensor-to-wheel seal can be serviced separately.

and mounting, as well as valve stem replacement, to avoid sensor damage. On some tire pressure monitoring systems, the metal valve stem is an integral part of the tire pressure sensor, and only the valve core and sensor to wheel seal can be serviced separately (**Figure 3-41**).

TIRE PRESSURE MONITORING SYSTEMS

The Transportation Recall Enhancement Accountability and Documentation Act passed by the U.S. government in October 2000 requires all new vehicles to have tire pressure monitoring systems no later than the 2008 model year. A TPMS illuminates a warning light in the instrument panel to inform the driver if one tire has low pressure (**Figure 3-42**).

Shop Manual
Chapter 3, page 114

Most direct TPMSs have a pressure sensor mounted to a rubber valve stem, or the sensor and valve stem are integrated into one assembly as was described earlier. Regardless of the sensor's mounting location, it transmits radio frequency (RF) signals. A vehicle with

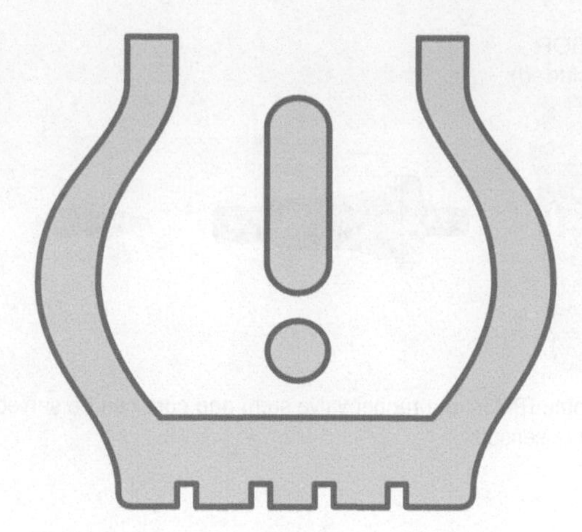

Figure 3-42 When the TPMS determines that a tire is low on air pressure, a warning image is illuminated on the instrument panel cluster to warn the driver.

direct pressure monitoring uses one of two frequencies, either 315 MHz and 433 MHz sensors. These RF signals will change if any tire is underinflated a specific amount. The RF signals are sent to a receiver that is usually mounted under the dash (**Figure 3-43**). On some systems, the receiver illuminates the low tire pressure warning light in the instrument panel if the pressure in any tire drops below 25 psi (172 kPa). On other systems, the tire pressure warning light is illuminated with a smaller reduction in tire pressure. Some TPMSs have a sensor strapped in the drop center on each rim (**Figure 3-44**). These sensors must be mounted on the rim directly opposite to the valve stem.

Some TPMSs provide continuous monitoring with the vehicle stopped and the ignition switch on. Other systems will not monitor tire pressure until the vehicle is moving above a specific speed, such as 25 mph (40 km/h). Some TPMSs have two warning lights: the low/flat tire warning light informs the driver regarding low tire pressure, and the service low tire pressure warning system (LTPWS) warning light informs the driver if there is a defect in the system (**Figure 3-45**). On some vehicles the low tire pressure receiver is connected to the data link connector (DLC) under the dash. A scan tool may be connected to the DLC to diagnose the TPMS. If a defect occurs in the system, a diagnostic trouble code (DTC) is stored in the receiver memory. A scan tool may be connected to the DLC to obtain the DTCs from the receiver.

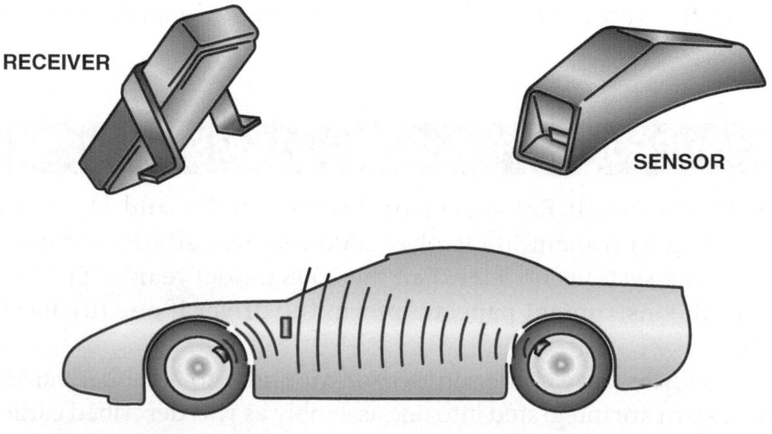

Figure 3-43 Receiver in a TPMS.

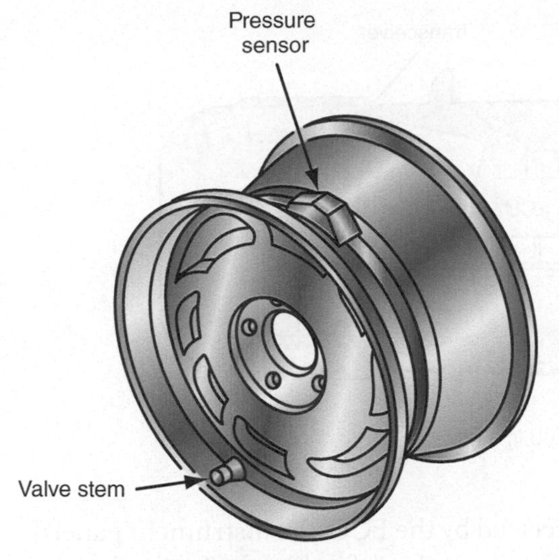

Figure 3-44 Tire pressure sensor.

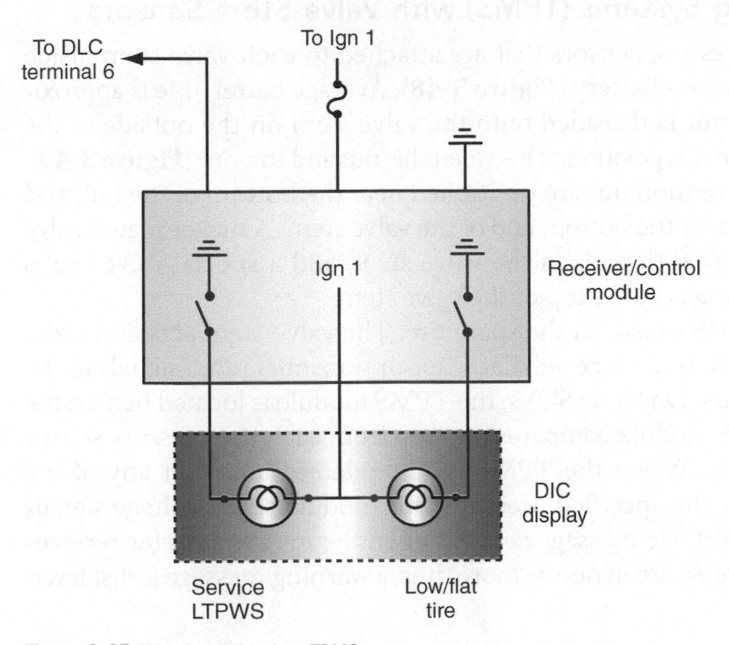

Figure 3-45 Warning lights in a TPMS.

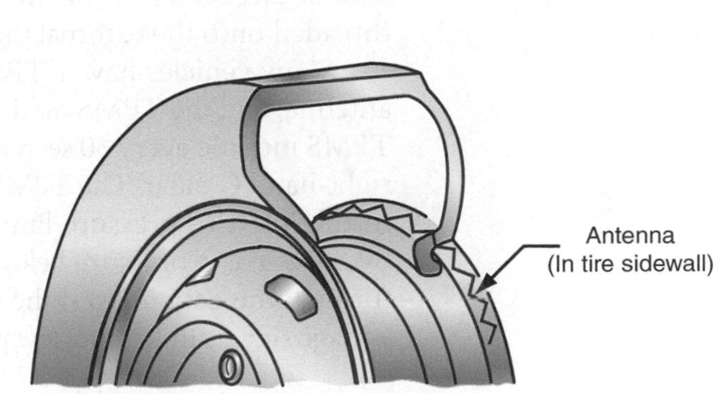

Figure 3-46 Antenna in a tire sidewall.

Other TPMSs use the wheel speed sensor signals on four-channel ABS to detect low tire pressure. A tire with low air pressure has a smaller diameter, and the wheel speed sensor generates a higher-frequency signal.

Some TPMSs have a miniature pressure sensor and computer chip about the size of a watch battery. This sensor and chip assembly is imbedded in the tire. The pressure sensor senses the actual tire pressure and sends voltage signals in relation to the tire pressure to a transceiver via a 360-degree circumferential antenna mounted in the tire sidewall close to the rim (**Figure 3-46**). A transceiver is mounted in each wheel well. The pressure sensor and computer chip do not require a battery, because they receive energy from the transceiver-generated field. Each transceiver is connected by data links to a central electronic control unit (ECU) (**Figure 3-47**). The transceivers relay voltage signals from the pressure sensors to the ECU. The computer chip calculates the recommended tire pressure based on data such as air temperature, tire pressure, and vehicle speed. Vehicle speed information is transmitted via data links from the powertrain control module (PCM) to

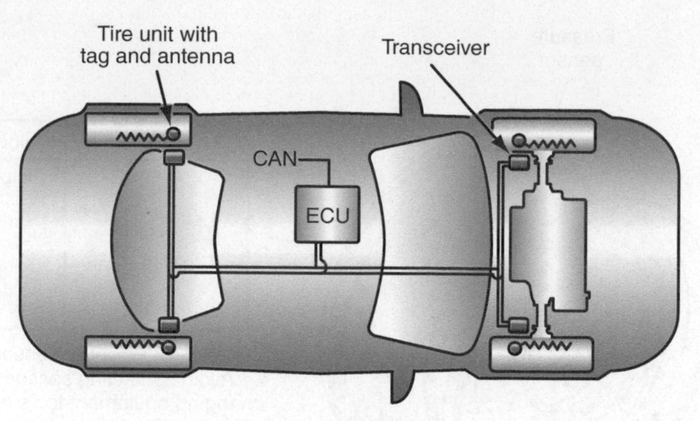

Figure 3-47 Transceivers and ECU in a TPMS.

the ECU. If a low tire pressure signal is received by the ECU, the instrument panel displays a warning message together with a calculated number of miles of safe travel.

Tire Pressure Monitoring Systems (TPMS) with Valve Stem Sensors

Most current TPMSs have pressure sensors that are attached to each valve stem inside the tire, and these sensors require a battery (**Figure 3-48**). Average battery life is approximately 10 years. A retention nut is threaded onto the valve stem on the outside of the wheel rim, and a sealing grommet is positioned between this nut and the rim (**Figure 3-49**). The internal threads on the retention nut are positioned near the bottom of the nut, and matching threads are located near the bottom end of the valve stem. A nickel-plated valve core is threaded into the internal threads in the valve stem, and a special valve cap is threaded onto the external threads at the top of the valve stem.

Many vehicles have a TPMS sensor in the spare tire. The valve stem acts as a radio antenna, and the TPMS module is the receiver. Each sensor transmits radio signals to the TPMS module every 60 seconds. On some SUVs, the TPMS module is located behind the right-hand C pillar. The TPMS module compares the data from each tire pressure sensor to the low tire pressure limits. When the TPMS module determines that any of the five tires has a pressure below the specified pressure, this module sends voltage signals through the data links to the vehicle message center. When the message center receives voltage signals indicating low pressure in one or more tires, a warning message is displayed

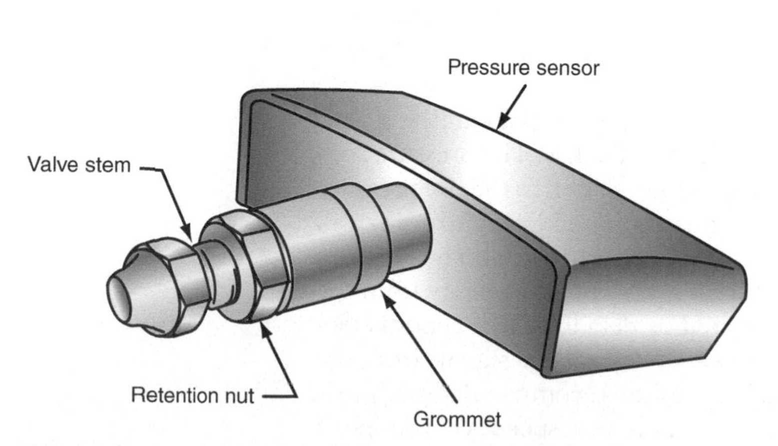

Figure 3-48 Tire pressure sensor mounted on the valve stem.

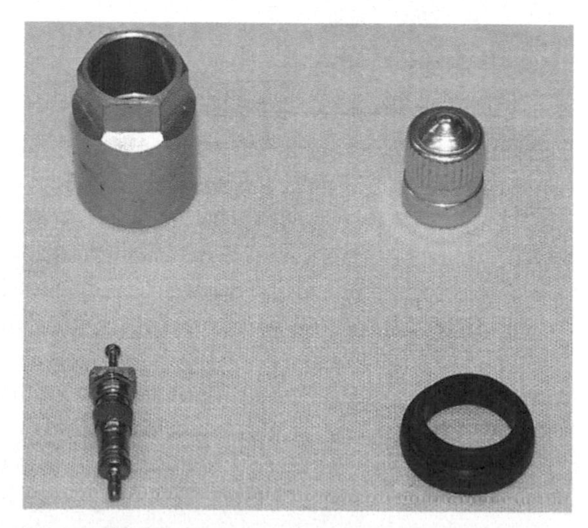

Figure 3-49 Tire pressure sensor retention nut, valve cap, valve core, and grommet.

in the message center, and the TPMS warning light is illuminated. The possible messages related to the TPMS are as follows:

1. WARNING TIRE VERY LOW—On some systems, this message is displayed if any tire has pressure less than 25 psi. An audible warning chime is heard when this message is displayed, and the TPMS warning light is illuminated.
2. CHECK TIRE PRESSURE—On some systems this message is displayed if any tire has pressure less than 30 psi. An audible warning chime is heard when this message is displayed, and the TPMS warning light is illuminated.
3. TIRE PRESSURE SENSOR FAULT—This message is displayed if one or more sensors are malfunctioning. If this message is displayed, the TPMS warning light flashes for 20 seconds.
4. TIRE PRESSURE MONITOR FAULT—This message is displayed if the TPMS module is defective or if all four sensors have failed. The TPMS warning light flashes for 20 seconds when this message is displayed.

Some TPMSs have the capability to sense a loss or an increase in tire pressure, and these systems also allow the driver to display the individual tire pressures and their locations on the driver information center (DIC) while the vehicle is being driven. This type of TPMS has conventional radio frequency–transmitting pressure sensors in each of the four valve stems. These sensors transmit signals to the antenna module, dash integration module (DIM), instrument panel cluster (IPC), and the DIC via the serial data circuit. The sensor's pressure accuracy from 14°F to 158°F (10°C to 70°C) is plus or minus 1 psi (7 kPa).

When the vehicle speed is less than 20 mph (32 km/h), the system remains in the stationary mode. In this mode, the sensors transmit data every 60 minutes to provide longer sensor battery life. When the vehicle speed is 20 mph (32 km/h) or greater, centrifugal force closes a roll switch in each sensor, and this action causes the sensors to enter the drive mode in which the sensors transmit signals every 60 seconds. The antenna module receives and translates each sensor signal into sensor presence, sensor mode, and tire pressure. The temperature and speed ratings of the TPMS system vary depending on the vehicle manufacturer. Always consult the vehicle manufacturer's specifications. The antenna module then transmits this information from each sensor to the DIC via the serial data circuit. The DIC displays an overhead view of the vehicle in which tire locations and pressures are displayed. If the TPMS senses a specific pressure gain or loss in any tire, a CHECK TIRE PRESSURE warning message is displayed on the DIC, and the low tire pressure warning indicator is illuminated in the IPC. Any defect in the TPMS system causes a SERVICE TIRE MONITOR warning message to be displayed on the DIC.

TPMS with Electronic Vehicle Information Center (EVIC) Display

Some TPMSs provide graphic tire pressure displays and warning messages on an EVIC display in the instrument panel. These systems may have four-tire or five-tire monitoring capabilities. A TPMS with five-tire monitoring capabilities also monitors the spare tire. These systems have valve stem sensors that broadcast tire pressure once per minute when the vehicle is moving at 25 mph (40 km/h) or faster. If the vehicle has a five-tire TPMS, the sensor in the spare tire transmits a signal every hour. Each valve stem sensor transmits a unique code so the EVIC module can determine the location of the sensor. If the wheels are rotated on the vehicle, or the spare tire and wheel are installed on the vehicle, the EVIC must be reprogrammed to recognize the new wheel and tire locations.

This type of TPMS provides a warning display in the EVIC if the tire pressure drops below a specific value or increases above a certain threshold. Typically, if the tire pressure in any tire decreases below 25 psi (172 kPa), the EVIC requests a chime warning, displays a LOW PRESSURE message, and indicates the location of the tire with low pressure. If the pressure in any tire exceeds 45 psi (310 kPa), the EVIC requests a chime warning and displays HIGH PRESSURE while indicating the location of the tire with high pressure. After a few seconds the EVIC reverts to a blinking display of the tire pressure. This blinking display continues for the entire ignition cycle or until an EVIC button is pressed. When an EVIC button is pressed, the blinking display returns after 60 seconds without the chime warning. If high or low pressure is detected in the spare tire on a five-tire system, SPARE HIGH PRESSURE or SPARE LOW PRESSURE appears on the EVIC display for 60 seconds during each ignition cycle.

WHEEL RIMS

Wheel rims are circular devices on which the tires are mounted. Wheel rims are bolted to the wheel hubs or axles.

Wheel rims may be manufactured from stamped or pressed steel discs that are riveted or welded together to form the circular rim, or may be cast or forged aluminum (**Figure 3-50**). Currently, less than 50 percent of the wheel rims installed by the original equipment manufacturers (OEMs) are stamped steel type.

A rim is equipped with a drop center, which allows tires to be more easily removed and installed (**Figure 3-51**). The closer the drop center is to the outside mounting flange the easier it is to remove and install the tire (**Figure 3-52**). Some performance rims and run-flat tire rims are equipped with a bead lock ridge, which aids in keeping the tire bead locked or secured to the rim (**Figure 3-53**). This is important especially during cornering maneuvers or in the case of a run-flat tire when it has lost all air pressure. These bead lock rims also commonly have a drop center that is located in the center of the rim farther from the outer mounting flange. Rims of this design make tire service much more challenging (**Figure 3-54**).

Shop Manual
Chapter 3, page 120

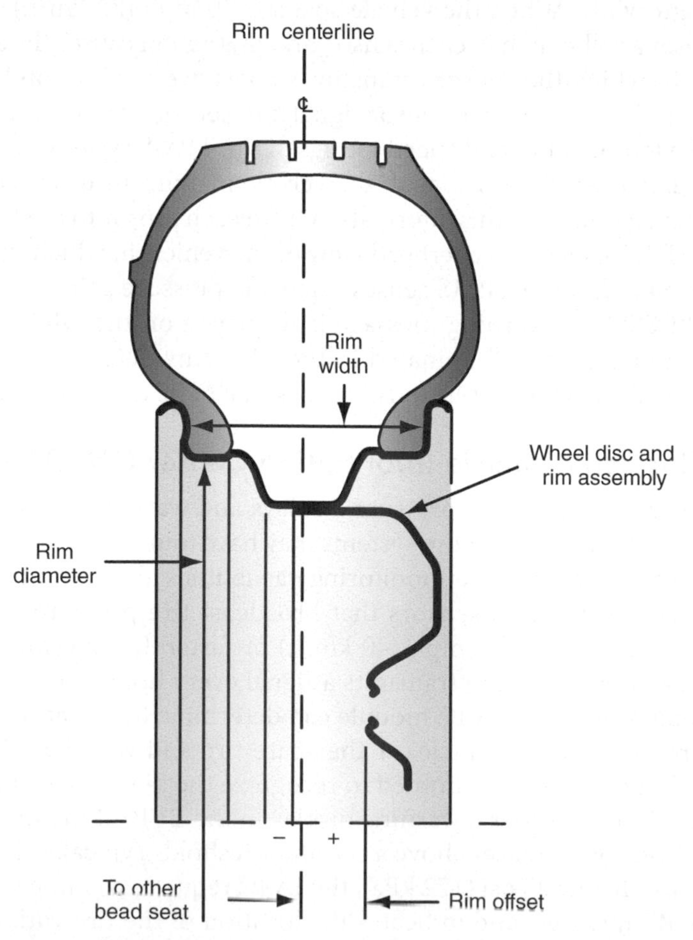

Figure 3-50 Wheel rim design.

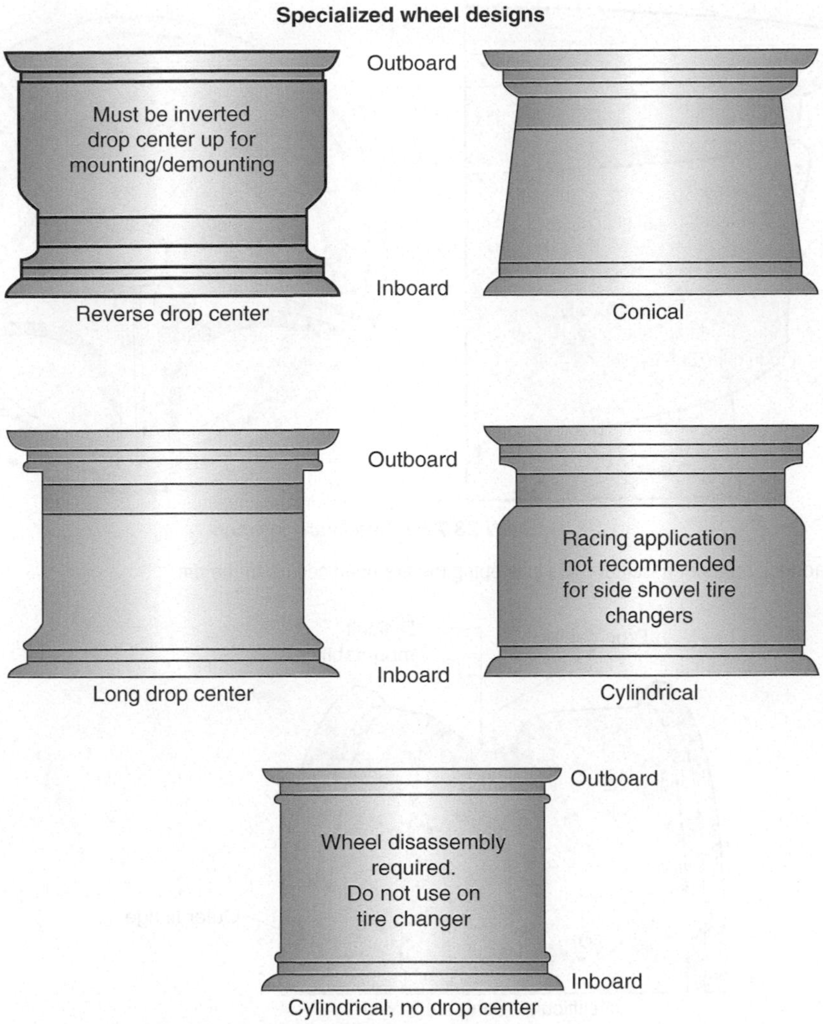

Specialized wheel designs

Outboard

Must be inverted drop center up for mounting/demounting

Inboard

Reverse drop center

Conical

Outboard

Racing application not recommended for side shovel tire changers

Inboard

Long drop center

Cylindrical

Outboard

Wheel disassembly required. Do not use on tire changer

Inboard

Cylindrical, no drop center

Figure 3-51 Various rim drop center designs.

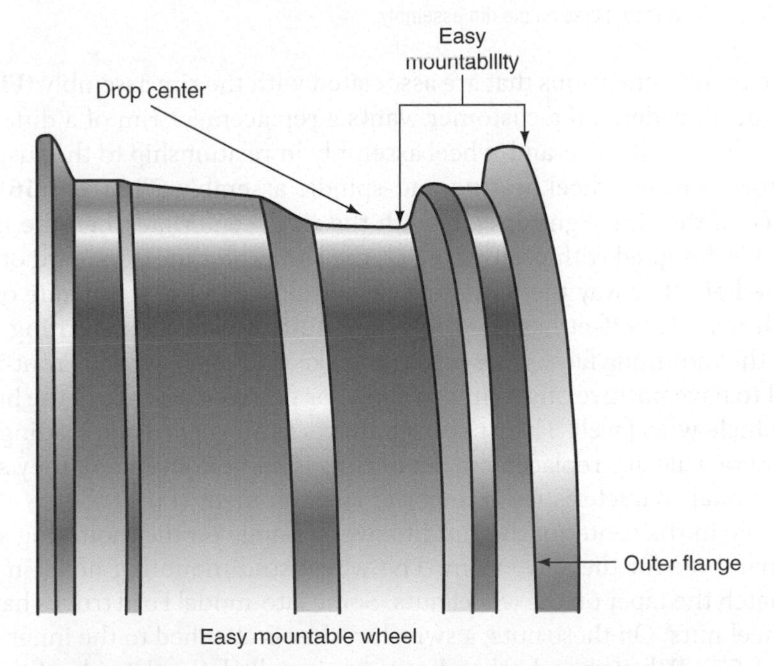

Figure 3-52 The closer the drop center is to the outer mounting flange, the easier it will be to dismount and mount tires on the rim assembly.

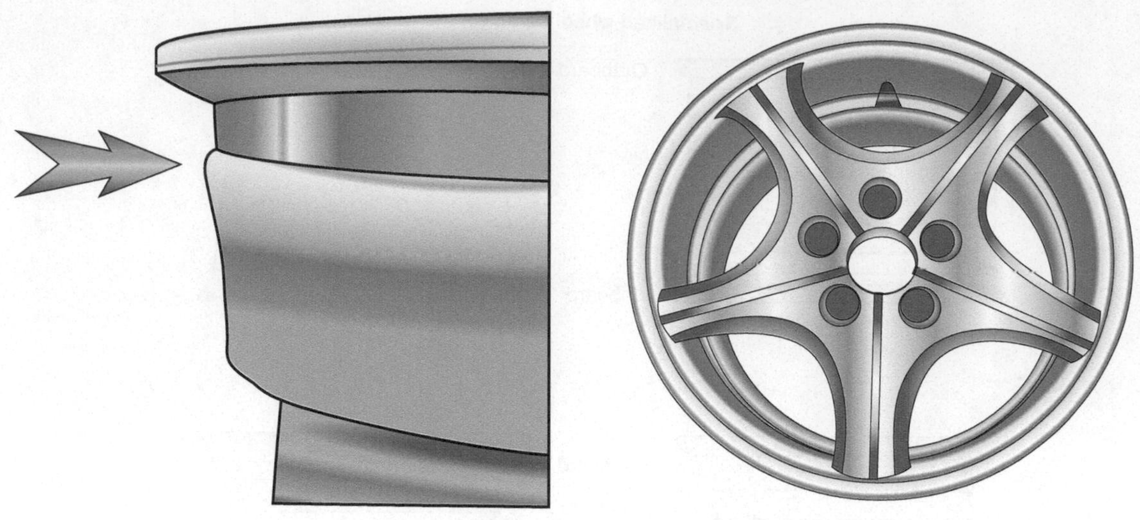

Figure 3-53 The bead lock on a rim assembly aids in keeping the tire bead secure to the rim.

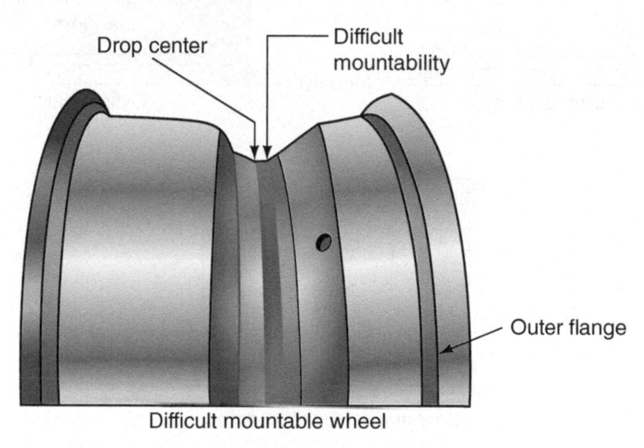

Figure 3-54 A rim with a drop center located in or close to the center of the assembly can make it very challenging to dismount and mount tires on the rim assembly.

There are many dimensions that are associated with the rim assembly (**Figure 3-55**), and all must be considered if a customer wants a replacement rim of a different design. The rim offset locates the tire-and-wheel assembly in relationship to the suspension and the lines of force on the wheel bearing-and-spindle assembly (**Figure 3-56**). When the mounting face of the rim aligns directly with the rim's centerline, then the rim has zero offset. If a rim is designed with positive offset, the rim centerline is inboard of the mounting face. Stated another way the mounting face is closer to the street side of the wheel. The rim with negative offset has the centerline outboard of the mounting face. Stated another way the mounting face is closer to the brake side of the wheel. Front-wheel-drive vehicles tend to have positive offset rims to allow for proper clearance of the hub assembly within the vehicle wheel well. The rim offset affects front suspension loading and operation. As a general rule, if a replacement set of rims is being considered, they should meet all the dimensional characteristics of the original equipment rim.

A large hole in the center of the rim fits over a flange on the mounting surface. The rim has a small hole for the valve stem. The wheel stud mounting holes in the rim are tapered to match the taper on the wheel nuts. Some late-model Ford trucks have non-self-centering wheel nuts. On these nuts, a swivel washer is attached to the inner side of each nut (**Figure 3-57**). When the wheel and nuts are installed, the flat side of the swivel nut fits against the wheel rim. The wheels are hub-centered, and an O-ring mounted on the

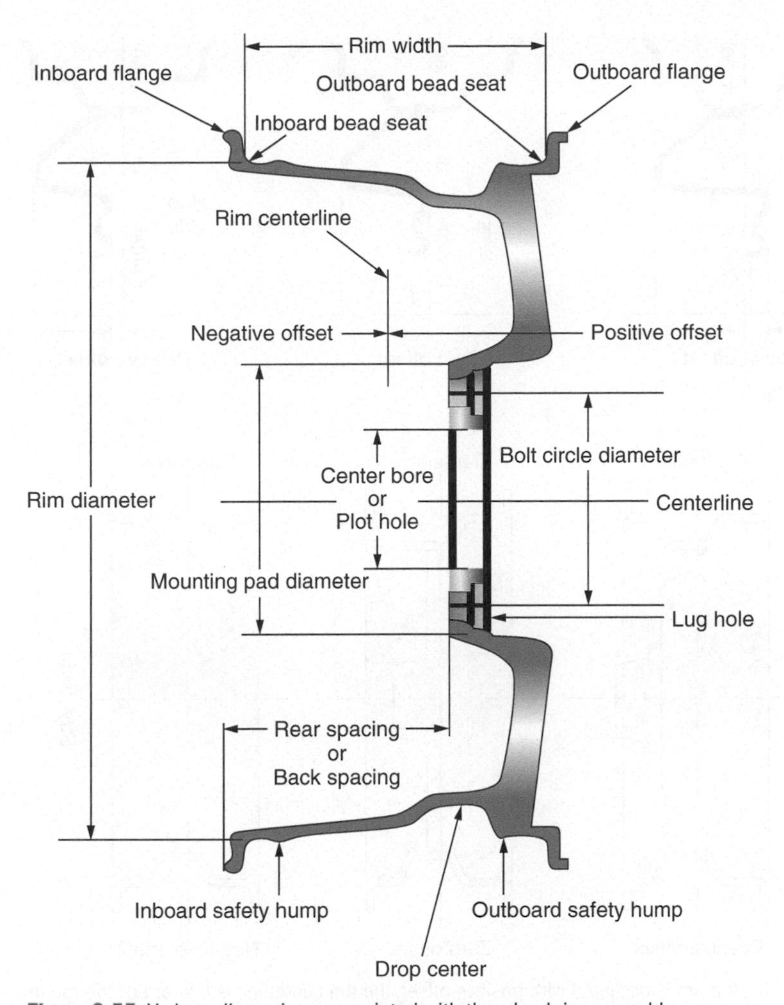

Figure 3-55 Various dimensions associated with the wheel rim assembly.

hub provides improved wheel centering (**Figure 3-58**). If this O-ring is missing or damaged, the wheel may not be properly centered, and this condition results in wheel vibrations. The O-ring also provides a seal to prevent corrosion.

Cars and light-duty trucks have four, five, six, or eight mounting stud openings and an equal number of matching studs in the hub or axle (**Figure 3-59**). The number of wheel studs depends on the vehicle weight and load. Heavier vehicles and/or increased vehicle load usually require more wheel mounting studs. These nuts usually have a taper on the inner side of the nuts, and this taper fits against a matching taper on the wheel stud openings (**Figure 3-60**). These tapers on the nuts and wheel openings center the wheel on the hub or axle; this is referred to as a stud piloted rim. Wheel stud openings are listed by the number of studs and the circle through the stud centers. For example, if wheel openings are listed as 5–4.5, there are five wheel studs and stud openings, and the centers of the wheel studs and wheel openings are on a 4.5-in. (11.4-cm) circle. If a wheel has an even number of studs, measure the distance between opposite stud centers to check the stud position and wheel circle. On a five-bolt wheel, measure the distance between two adjacent stud centers. The specified distances between adjacent wheel stud centers and the corresponding wheel circles are the following:

4.5-in. wheel circle	2.64 in. between adjacent stud centers
4.75-in. wheel circle	2.79 in. between adjacent stud centers
5.0-in. wheel circle	2.93 in. between adjacent stud centers
5.5-in. wheel circle	3.23 in. between adjacent stud centers

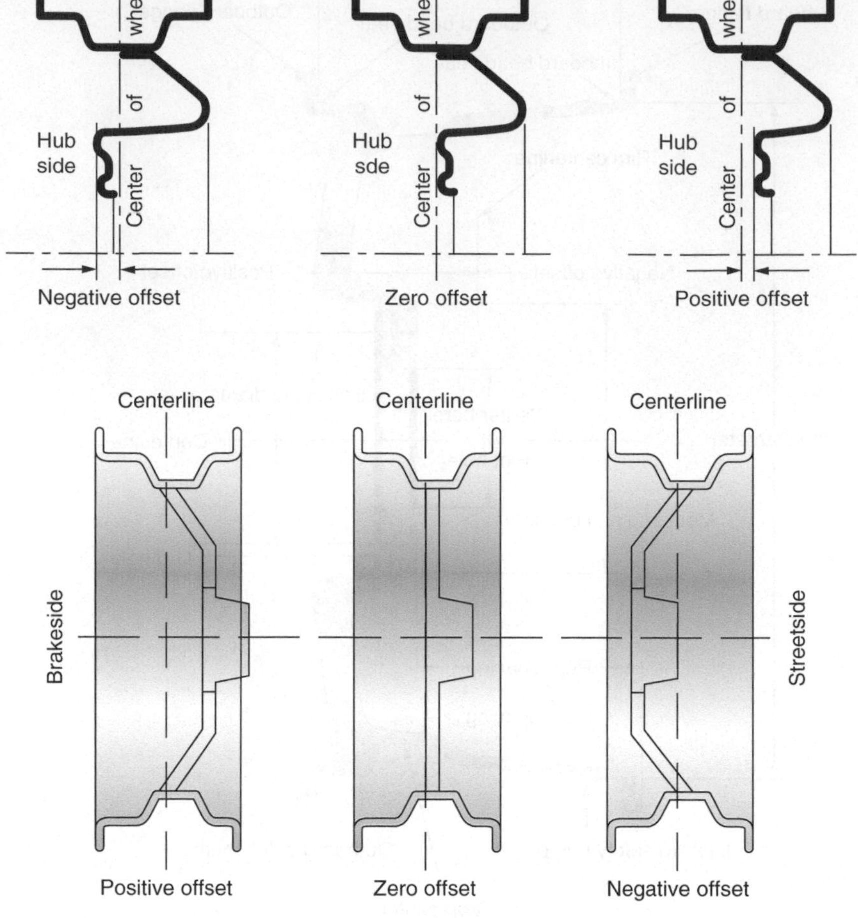

Figure 3-56 If a rim is designed with positive offset, the rim centerline is inboard of the mounting face. The rim with negative offset has the centerline outboard of the mounting face.

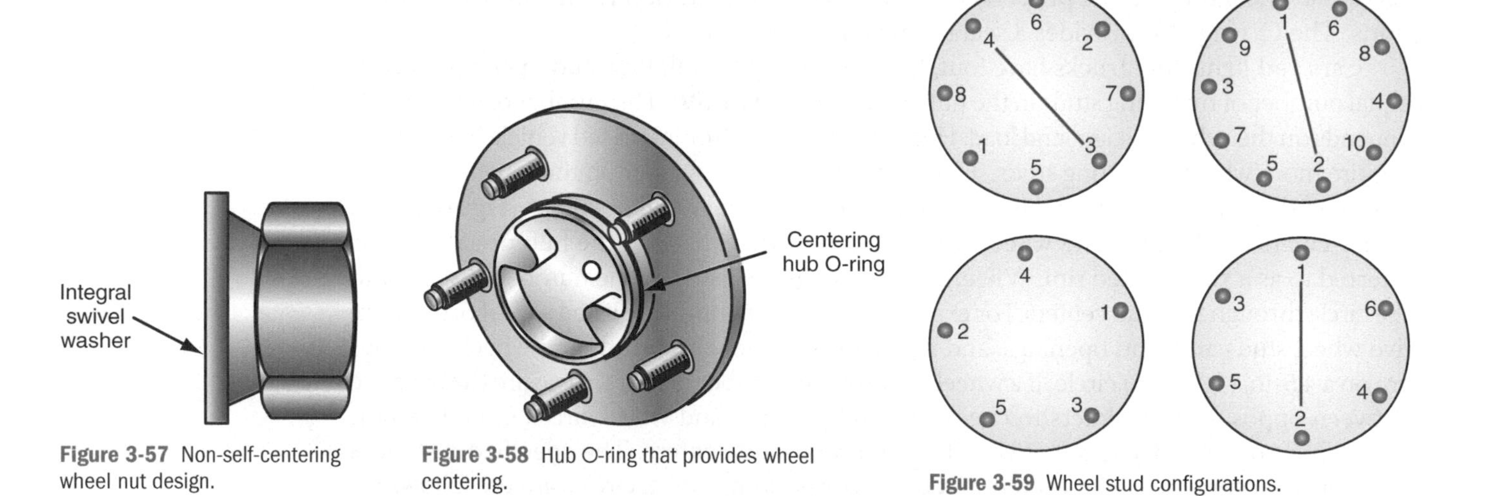

Figure 3-57 Non-self-centering wheel nut design.

Figure 3-58 Hub O-ring that provides wheel centering.

Figure 3-59 Wheel stud configurations.

Templates are available to measure the stud position on wheels that do not have an even number of studs.

The width of the wheel is measured between the vertical bead seats on the rim flanges. Rim diameter is the distance between the horizontal part of the bead seats measured through the rim center bottom. A drop center in the rim makes tire changing easier

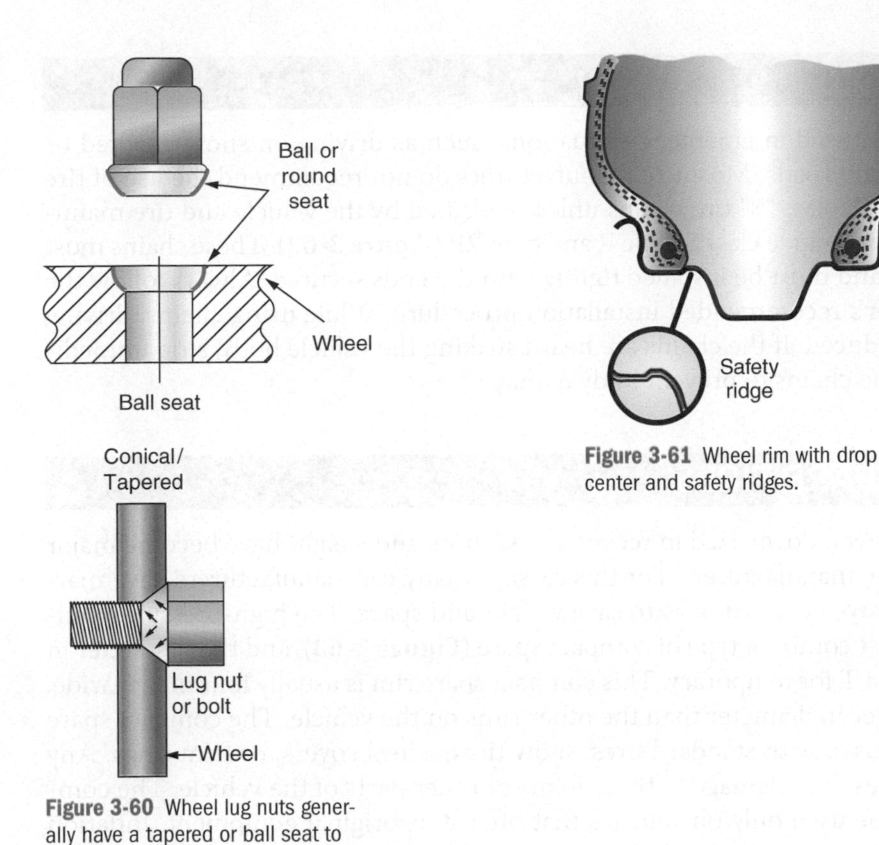

Figure 3-60 Wheel lug nuts generally have a tapered or ball seat to locate and secure the rim to the hub assembly.

Figure 3-61 Wheel rim with drop center and safety ridges.

(**Figure 3-61**). Rims have a safety ridge or bead lock ridge behind the tire bead locations, which help prevent the beads from moving into the drop center area if the tire blows out. If the tire blows out and a bead enters the drop center area, the tire may come off the wheel.

Cast aluminum wheels are commonly used on many vehicles by OEMs. These wheels may be polished, chromed, or painted. A cast aluminum wheel costs about twice as much as a stamped steel wheel.

Forged aluminum wheels are installed on some vehicles. They are more durable than cast aluminum wheels and have a bright finish. Forged aluminum wheels are used on heavy-load applications such as trucks. Some expensive sports cars are also equipped with forged aluminum wheels. A forged aluminum wheel is lighter than a steel wheel, but the cost of a forged aluminum wheel is considerably higher than a stamped steel or cast aluminum wheel.

Some vehicles are equipped with pressure-cast aluminum wheels that are designed to be chrome plated. Pressure-cast wheels may also be called squeeze-cast wheels. During the manufacture of a pressure-cast wheel, the molten metal is squeezed under pressure into the mold. This action squeezes all the air out of the metal and reduces the metal porosity, which provides an improved substrate for chrome plating. The chrome plating increases the cost of the pressure-cast aluminum wheel so it is considerably more expensive than other types of wheels.

Race cars may be equipped with **magnesium alloy wheels**, which are lighter than either steel or aluminum wheels.

Magnesium alloy wheels may be called "mag" wheels.

Replacement wheel rims must be the same as the original equipment wheels in load capacity, offset, width, diameter, and mounting configuration. An incorrect wheel can affect tire life, steering quality, wheel bearing life, vehicle ground clearance, tire clearance, and speedometer/odometer calibrations.

TIRE CHAINS

Tire chains are made from steel and wrap around the tire to improve traction.

Tire chains may be used in emergency situations, such as driving on snow-covered or ice-covered mountain roads. Most tire manufacturers do not recommend the use of tire chains. Use only SAE class "S" tire chains unless specified by the vehicle and tire manufacturer. Chain types include class S, type P, and type RP (**Figure 3-62**). These chains must be the proper size and must be installed tightly with the ends secured. Always follow the chain manufacturer's recommended installation procedure. While using chains, driving speed should be reduced. If the chains are heard striking the vehicle body, stop immediately and tighten the chains to prevent body damage.

COMPACT SPARE TIRES

Compact spare tires save onboard storage space, but are designed for very limited operation.

Because cars have been downsized in recent years, space and weight have become major concerns for vehicle manufacturers. For this reason, many car manufacturers have marketed cars with **compact spare tires** to save weight and space. The high-pressure mini-spare tire is the most common type of compact spare (**Figure 3-63**), and the first letter in the tire size will be a T for temporary. This compact spare rim is usually four inches wide, but is one inch larger in diameter than the other rims on the vehicle. The compact spare rim should not be used with standard tires, snow tires, wheel covers, or trim rings. Any of these uses may result in damage to these items or other parts of the vehicle. The compact spare should be used only on vehicles that offer it as original equipment. Inflation pressure in the compact spare should be maintained at 60 psi (415 kPa). The compact spare tire is designed for very temporary use until the conventional tire can be repaired or replaced. Limit driving speed to 50 mph (80 km/h) when the high-pressure mini-spare tire is installed on a vehicle.

The space-saver spare tire must be inflated with a special compressor. Battery voltage is supplied to the compressor from the cigarette lighter. This type of compact spare should be inflated to 35 psi (240 kPa). After the tire is inflated, be sure there are no folds in the sidewalls.

The lightweight-skin spare tire is a bias-ply tire with a reduced tread depth to provide an estimated 2000 mi. (3200 km) of tread life. Always inflate the lightweight-skin spare tire to the pressure specified on the tire placard.

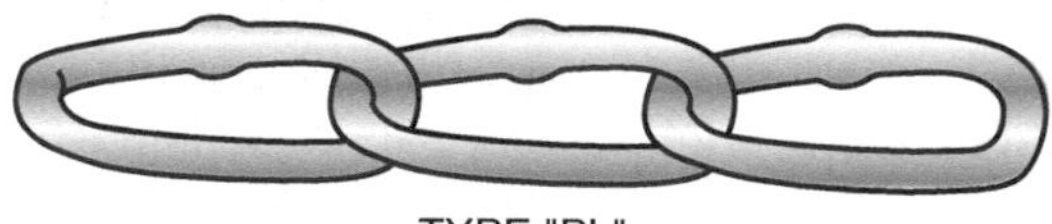

Figure 3-62 Types of tire chains.

Figure 3-63 Compact, high-pressure, mini-spare tire.

RUN-FLAT TIRES

Some tire manufacturers utilize **run-flat tires** on some of their vehicle platforms. Run-flat tires may be called extended mobility tires (EMT). These tires are standard equipment on the 1995 and late-model Corvette. Run-flat tires eliminate the need for a spare tire and a jack, saving both weight and space. Run-flat tires are designed with stiffer sidewalls that will partially support the vehicle weight without air pressure in the tire.

Run-flat tires are designed to operate at zero air pressure for limited mileage. Run-flat tires may be called self-supporting, extended mobility, or zero-pressure tires.

Run-flat tires must provide acceptable levels of inflated performance in the areas of comfort, ride, handling, and adequate deflated mobility. Run-flat tires share these same basic design objectives:

1. Minimize the difference between run-flat tires and conventional tires when the tires are inflated.
2. Enhance the handling and riding capabilities of run-flat tires when inflated.
3. Provide acceptable handling when run-flat tires have zero pressure on various vehicles.
4. Enhance low-pressure and zero-pressure bead retention on run-flat tires.
5. Provide sufficient zero-pressure durability so the vehicle can be driven a reasonable distance to a repair facility.

Run-Flat Tires with Sidewall Reinforcements

One of the most important requirements for run-flat tires is bead retention when running with low or zero pressure. Run-flat tires have improved beads to meet this requirement (**Figure 3-64**). Some run-flat tires have sidewall reinforcements that may be manufactured from flexible, low-hysteresis rubber, thermal-resistive materials, or metallic and/or textile tissues. Run-flat tires with a high aspect ratio require increased sidewall stiffness compared to run-flat tires with a low aspect ratio. Because of these sidewall reinforcements, run-flat tires are 20 to 40 percent stiffer compared to conventional tires. Therefore, run-flat tires may increase ride harshness and vertical firmness especially when the tire strikes a large road irregularity. This increase in ride harshness is not as noticeable when the tire contacts a smaller road irregularity. The increased stiffness in run-flat tires usually provides a small improvement in vehicle handling.

Shop Manual
Chapter 3, page 105

Run-flat tires with a low aspect ratio usually have the same or less rolling resistance compared to equivalent-size conventional tires. Because run-flat tires with a high aspect ratio have increased sidewall stiffness, these tires tend to have increased rolling resistance compared to conventional tires.

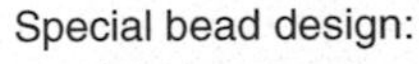
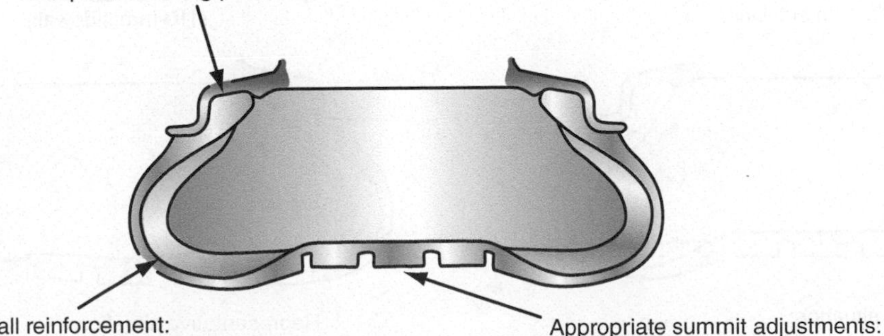
Special bead design:
* Enhanced retention after pressure loss
* Acceptable seating pressure
Sidewall reinforcement:
* Flexible low-hysteresis rubber
* Thermal-resistive material
* Metallic and/or textile tissues
Appropriate summit adjustments:
*Maintain inflated performance
(comfort & handling like std. tires)

Figure 3-64 Run-flat tire with sidewall reinforcement.

Caution

On a vehicle with PAX tires as original equipment, conventional tires and rims should not be substituted for the PAX tires and rims. This action will decrease ride quality and vehicle steering characteristics.

In moderate cornering and lane-change maneuvers, a run-flat tire with zero pressure provides a slight decrease in vehicle handling. The key word in the previous sentence is moderate. Run-flat tires are used with a tire pressure monitoring system, which has a warning light in the instrument panel to inform the driver when one tire has low pressure. Once the low tire pressure warning light is illuminated with the engine running, the driver should avoid high-speed driving and high-speed cornering, because run-flat tires provide reduced handling capabilities during high-speed cornering.

The zero-pressure durability of a run-flat tire varies depending on the vehicle weight, atmospheric temperature, and the tire aspect ratio. For example, a run-flat tire with a 40 aspect ratio supporting 880 lb (400 kg) provides 186 mi. (300 km) of driving at 55 mph (89 km/h) (**Figure 3-65**). A run-flat tire with a 60 aspect ratio supporting 1322 lb (600 kg) provides 50 mi. (80 km) of driving at 55 mph (89 km/h).

Run-Flat Tires with Support Ring

Michelin PAX run-flat tires were first introduced in 1998, but later discontinued in 2009. PAX run-flat tires were installed as original equipment on some Honda Odyssey, Toyota Sienna, Nissan Quest, and Acura RL models, as well as Audi A8 and A4 models in Europe. The letters "PAX" translate to the values of peace of mind, safety, and the future. The technology never gained traction on the world market, and has now become part of automotive history.

The run-flat tires in the PAX system have a flexible support ring mounted on a special rim to support the tire if deflation occurs. PAX tires do not have a stiffer sidewall, and so they provide excellent ride quality. When PAX tires are installed on a vehicle as original equipment, all the suspension system components such as springs and shock absorbers are engineered to go with the PAX tires.

Caution

Using conventional tire changers that do not have PAX capabilities on PAX run-flat tires and rims may damage the tire and rim. Tire deflection is the difference between the free diameter and the rolling diameter of the tire.

PAX run-flat tires have a vertical, mechanical locking bead system that "latches" the bead to the rim. The outer tire bead has a smaller diameter compared with the inside tire bead. Therefore, the outer edge of the rim has a smaller diameter compared with the inner edge (**Figure 3-66**). When mounted on the rim, the tire beads lock into vertical grooves in the outer edges of the rim. The outer edges of the inner and outer tire beads extend slightly outside the wheel rim edges. This design helps protect the rim. PAX tires may be driven for 125 mi. (201 km) at 55 mph (88 km/h) with zero air pressure. When mounting the tire on the rim, the tire is first positioned such that the tire beads are outside the rim. The outer (smaller-diameter) bead is then mounted onto the rim, followed by the inner (larger-diameter) bead. The disadvantages of this type of run-flat tire include the special rim that is required and the extra weight of the support ring. A TPMS is used with this type of run-flat tire.

Shop Manual
Chapter 3, page 110

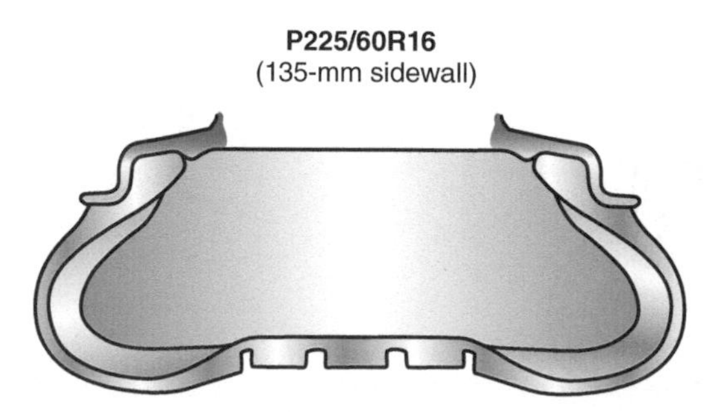

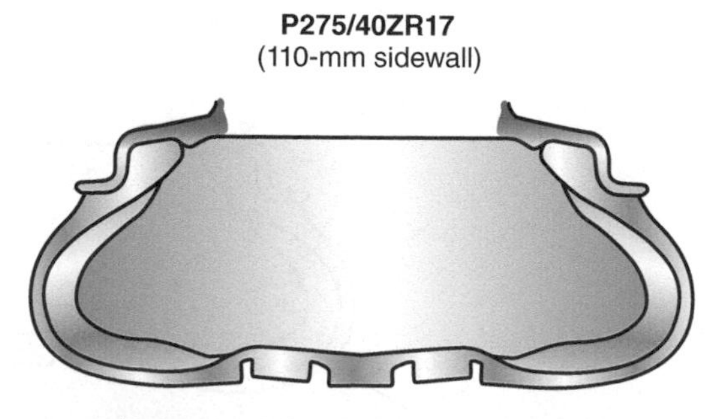

Figure 3-65 Run-flat tire zero-pressure durability.

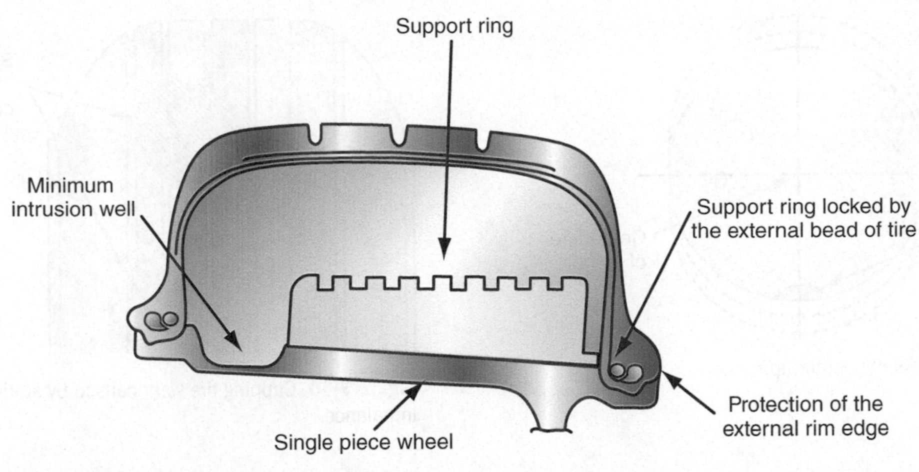

Figure 3-66 Run-flat tire with support ring.

STATIC WHEEL BALANCE THEORY

When a tire-and-wheel assembly has proper **static balance**, it has the weight equally distributed around its axis of rotation, and gravity will not force it to rotate from its rest position. If a vehicle is raised off the floor and a wheel is rotated in 120-degree intervals, a statically balanced wheel will remain stationary at each interval. When a wheel and tire are statically unbalanced, the tire has a heavy portion at one location. The force of gravity acting on this heavy portion will cause the wheel to rotate until the heavy portion is located near the bottom of the tire (**Figure 3-67**).

Static balance refers to the balance of a wheel in the stationary position.

Results of Static Unbalance

Centrifugal force may be defined as the force that tends to move a rotating mass away from its axis of rotation. As we have explained previously, a tire and wheel are subjected to very strong acceleration and deceleration forces when a vehicle is in motion. The heavy portion of a statically unbalanced wheel is influenced by centrifugal force. This influence attempts to move the heavy spot on a tangent line away from the wheel axis. This action tends to lift the wheel assembly off the road surface (**Figure 3-68**).

The wheel-lifting action caused by static unbalance may be referred to as **wheel tramp** (**Figure 3-69**). Wheel tramp action allows the tire to slip momentarily when it is lifted vertically. When the wheel and tire move downward as the heavy spot decelerates, the tire strikes the road surface with a pounding action. This repeated slipping and pounding action causes severe tire scuffing and cupping (**Figure 3-70**).

Wheel tramp may be defined as rapid upward and downward wheel and tire oscillations.

The vertical wheel motion from static unbalance is transferred to the suspension system and then absorbed by the chassis and body. This action causes rapid wear on

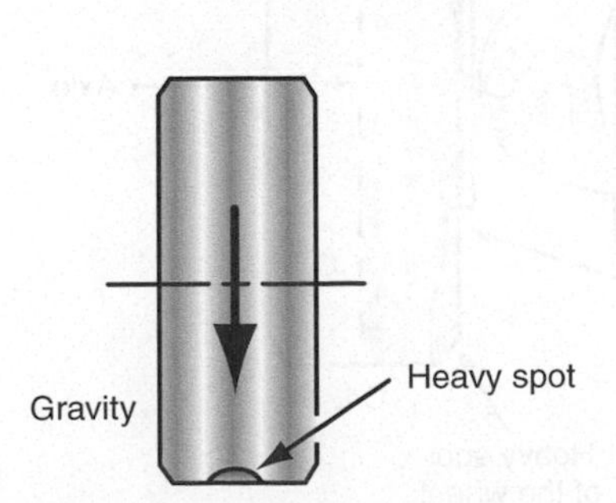

Figure 3-67 Static wheel imbalance.

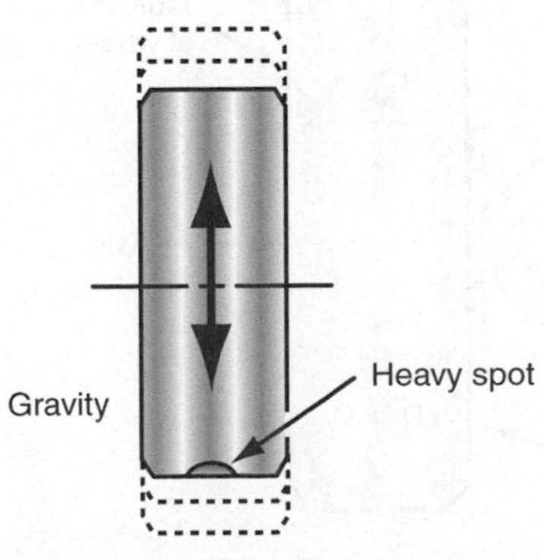

Figure 3-68 Effects of static imbalance.

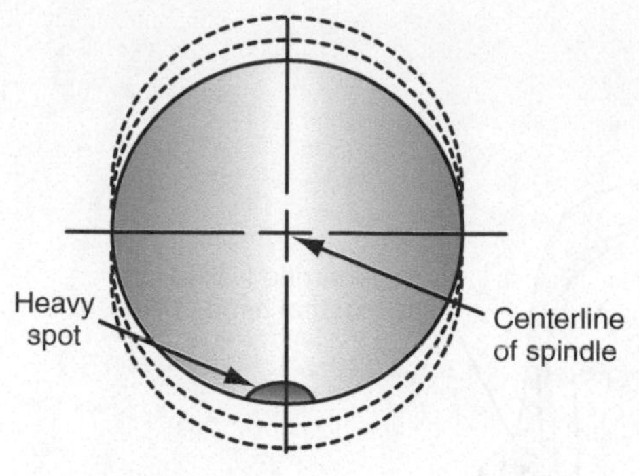

Figure 3-69 Wheel tramp.

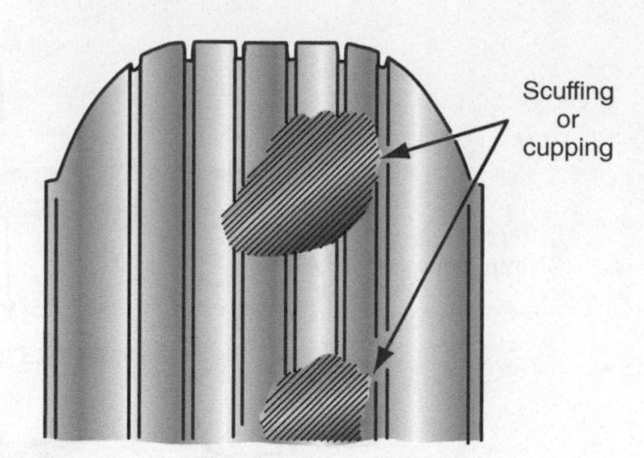

Figure 3-70 Cupping tire wear caused by static imbalance.

suspension and steering components. The wheel tramp action resulting from static unbalance is also transmitted to the passenger compartment, which causes passenger discomfort and driver fatigue.

When a vehicle is traveling at normal highway cruising speed, the average wheel speed is 850 revolutions per minute (rpm). A statically unbalanced tire-and-wheel assembly is an uncontrolled mass of weight in motion. When a vehicle is traveling at 60 mph (97 km/h) and a tire has 2 ounces (oz), or 57 grams (g), of static unbalance, the resultant pounding force is approximately 15 lb, or 6.8 kg, against the road surface.

DYNAMIC WHEEL BALANCE THEORY

Dynamic balance refers to the balance of a wheel in motion.

When a tire-and-wheel assembly has correct **dynamic balance**, the weight of the assembly is distributed equally on both sides of the wheel center viewed from the front. Dynamic wheel balance may be explained by dividing the tire into four sections (**Figure 3-71**). In Figure 3-71, if sections A and C have the same weight and sections B and D also have the same weight, the tire has proper dynamic balance. If a tire has dynamic unbalance, section D may have a heavy spot; thus, sections B and D have different weights (**Figure 3-72**).

Shop Manual
Chapter 3, page 125

From our discussion of dynamic balance, we can understand that a tire and wheel assembly may be in static balance, but have dynamic unbalance. Therefore, wheels must be in balance statically and dynamically. When a tire-and-wheel assembly is placed on a

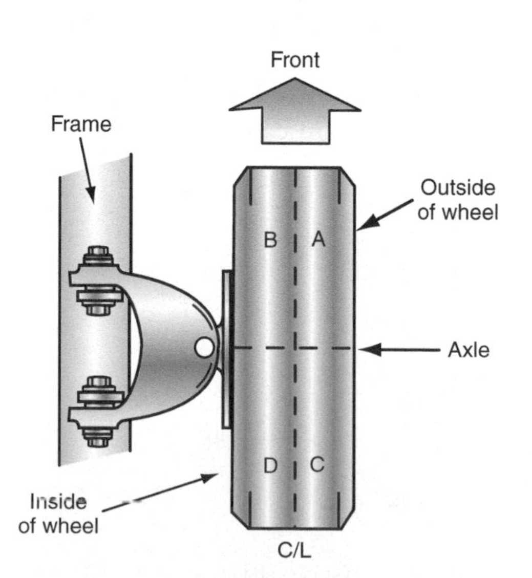

Figure 3-71 Dynamic wheel balance.

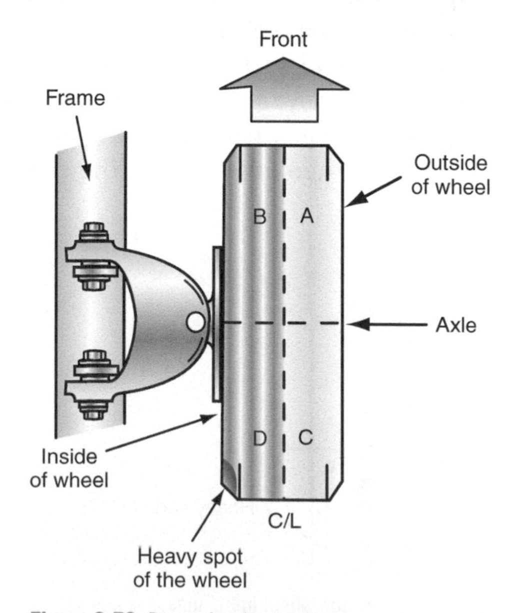

Figure 3-72 Dynamic wheel unbalance.

computer wheel balancer to check and correct for an imbalance condition, it is generally necessary to attach a wheel weight to the rim assembly to offset this condition. Wheel weights are generally either clip-on or stick-on weights. Clip-on wheel weights are attached to the outboard flange of the rim and come in an assortment of sizes and shapes depending on the shape of the rim flange (**Figure 3-73**).

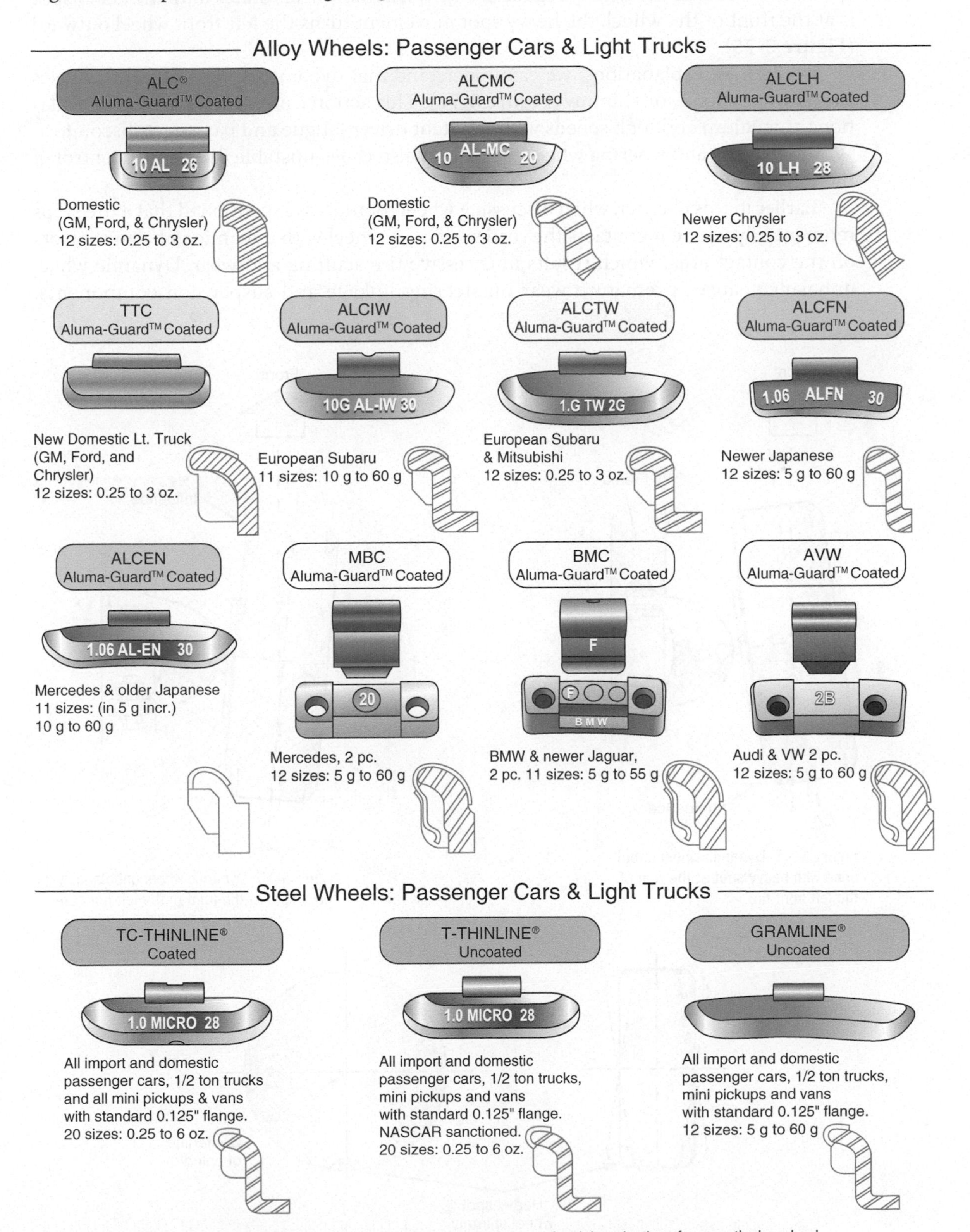

Figure 3-73 Wheels weights come in various dimensions depending on wheel rim assembly design.

Results of Dynamic Wheel Unbalance

When a dynamically unbalanced wheel is rotating, centrifugal force moves the heavy spot toward the tire centerline. The centerline of the heavy spot arc is at a 90° angle to the spindle. This action turns the true centerline of the left front wheel inward when the heavy spot is at the rear of the wheel (**Figure 3-74**). When the wheel rotates until the heavy spot is at the front of the wheel, the heavy spot movement turns the left front wheel outward (**Figure 3-75**).

From these explanations, we can understand that dynamic wheel unbalance causes lateral wheel shake, or shimmy (**Figure 3-76**). This action causes steering wheel oscillations at medium and high speeds with resultant driver fatigue and passenger discomfort. Wheel shimmy and steering wheel oscillations also cause unstable directional control of the vehicle.

Earlier in this chapter, when discussing wheel rotation, we mentioned that a tire stops momentarily where it contacts the road surface. A wheel with dynamic unbalance pivots on the contact area, which results in excessive tire scuffing and wear. Dynamic wheel unbalance causes premature wear on steering linkage and suspension components.

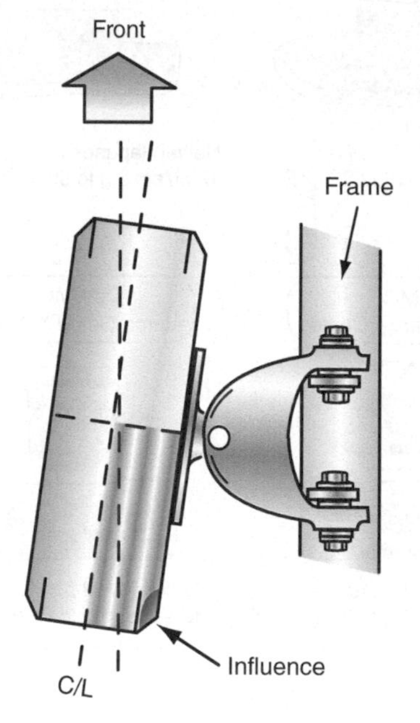

Figure 3-74 Dynamic wheel unbalance with heavy spot at the rear of the left front tire.

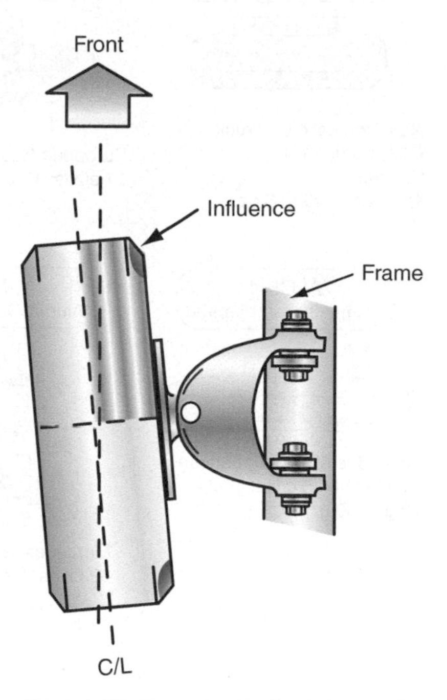

Figure 3-75 Dynamic wheel unbalance with heavy spot at the front of the left front tire.

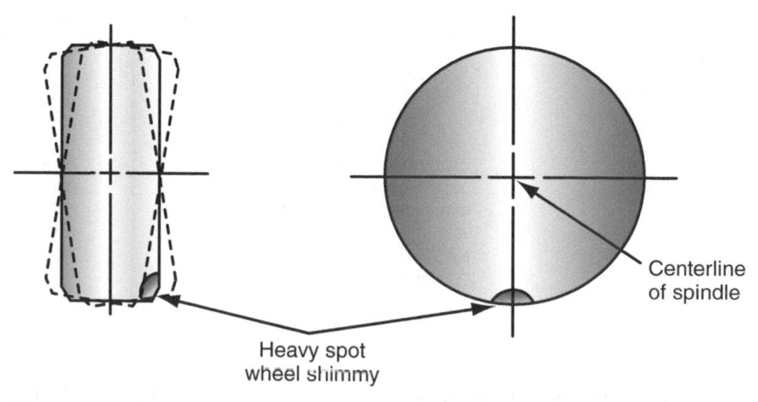

Figure 3-76 Dynamic wheel unbalance causes wheel shimmy.

Therefore, dynamic wheel balance is extremely important to provide normal tire life, reduce steering and suspension component wear, increase directional control, and decrease driver fatigue. The main purposes of proper wheel balance are as follows:

1. Maintains normal tire tread life.
2. Provides extended life of suspension and steering components.
3. Helps provide directional control of the vehicle.
4. Reduces driver fatigue.
5. Increases passenger comfort.
6. Helps maintain the life of body and chassis components.

TIRE MOTION FORCES

When a vehicle is in motion, wheel rotation subjects the tires to centrifugal force. The tires are also subjected to accelerating and decelerating forces because of their path of travel. If a vehicle is traveling at 55 mph, or 88.5 km/h, the part of the tire exactly fore and aft of the spindle is also traveling at 55 mph (88.5 km/h). At the exact top of the tire, the tire speed is 110 mph (177 km/h). The tire speed actually drops to zero at the exact bottom of the tire where the arc of deceleration ends and the arc of acceleration begins. Since the tires are subjected to strong acceleration and deceleration forces, the tire construction must be uniform. A soft spot in a tire will deflect farther than the surrounding area, and this area will be subjected to rapid wear as it strikes the road surface (**Figure 3-77**).

Tires must have equal stiffness in the sidewalls. If a tire does not have equal stiffness in all areas around the tire sidewalls, it may cause a vibration when driving. This vibration is called **road force variation** and produces a vibration similar to the vibration caused by improper wheel and tire balance. Imagine the tire as a collection of springs around the circumference of the tire. If the sidewall stiffness is not uniform a varied force is exerted, which creates a vibration. Hunter road force wheel balancers have a roller that is forced against the tire tread during the balance procedure, and this type of balancer detects road force variation (**Figure 3-78**).

Road force variation is a condition that occurs when tire sidewalls do not have equal stiffness around their complete area.

Shop Manual
Chapter 3, page 128

The radial force variation can be determined by examining the radial run out. Loaded radial runout of one thousands of an inch is equivalent to approximately one pound of

Figure 3-77 Think of a tire as sets of springs around its diameter. The variation in sidewall stiffness is referred to as road force variation and if excessive can create harmonic vibration and tire wear.

Figure 3-78 The roller on some wheel balancers senses road force variation.

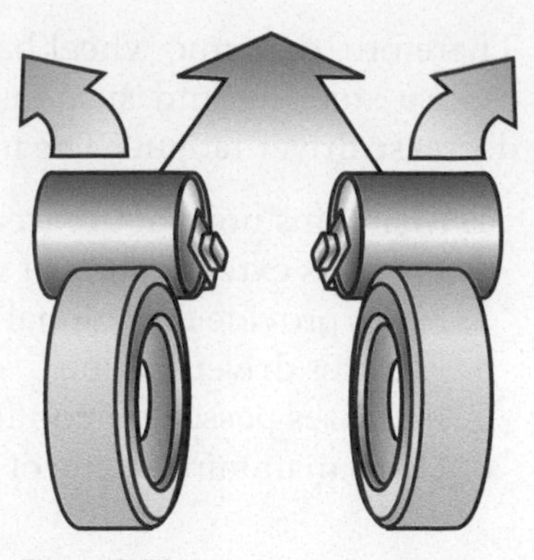
Figure 3-79 The roller on some wheel balancers senses concentricity.

radial force variation. As an example, at 50 mph a wheel imbalance of 1.5 oz (42 g) will result in the same amount of vibration as 0.30 in. (0.76 mm) of loaded radial runout or 30 lb of road force variation.

If excessive road force variation is detected, match mounting the rim and tire should be considered. In match mounting, the high point on the tire is matched with the low point on the rim runout or vice versa to decrease the rolling vibration in the wheel assembly.

Tires must be manufactured so the cords are concentric with the center of the tire. Tire conicity is a term used to describe a tire in which the cords are off-center in relation to the tire center. If the tire cords are not concentric with the tire center, the tire may cause a steering pull condition when driving. Modern wheel balancers use a roller pressed against the tire tread to detect conicity during the balance procedure (**Figure 3-79**).

NOISE, VIBRATION, AND HARSHNESS ANALYSIS

OEMs have made significant improvements in reducing vehicle noise, vibration, and harshness (NVH), because many customers prefer a quieter-running vehicle with reduced vibration and ride harshness. The suspension system plays a significant role in reducing NVH. Many of the suspension features described in this chapter are designed to reduce NVH. These features include track bars and braces, proper car riding height, hydraulic suspension system mounts, independent rear suspension systems, and large rubber insulating bushings on suspension system mountings.

However, the technician must understand that reducing NVH involves many mechanical and body or chassis components on the vehicle. For example, some vehicles have an optimized body structure, which is the major reason for reduced NVH. In this body design, lateral tie bars that connect the front longitudinal rails provide a stiffer front end. At the rear of the car, one-piece side rings with integral quarters eliminate rear pillar seams and provide a more precise door fit. The instrument panel and steering column are integrated solidly into the body structure by a cast magnesium beam. The door hinges are through bolted and thick spacer blocks are installed on these bolts to provide a very solid door attachment. Because the entire body is stronger and more rigid, the suspension can provide excellent ride quality and steering control without having to compensate for unwanted body flexing. In addition, appropriate body cavities like the dash panel are filled with expandable baffles to eliminate noise. The five-layer noise buffer in the dash panel contains these materials:

1. Fiberglass insulation mat
2. Viscoelastic energy-absorbing layer
3. Double steel panel
4. Single one-piece dash mat

A cast foam floor carpet system reduces noise transmitted through the floor pan and wheel wells. The door pillar and rocker panel cavities contain over 20 noise blockers.

Many engine refinements reduce vibration. For example, many engines now have a deep skirt block with the main bearings bolted through the sides of the block and vertically. Many engines now have cast oil pans and rocker covers rather than stamped steel components. The main bearing caps are contained in a one-piece casting on many engines to increase bottom end strength and reduce engine vibration. Some V-8 engines have a rubber intake manifold valley stuffer attached to the underside of the intake manifold to reduce vibration. Other vehicles have a slip yoke vibration damper mounted on the front of the drive shaft to reduce driveline vibration. Therefore, reducing NVH is a total vehicle concept.

Vibration Theory

Vibrations have these three elements:

1. Source—the cause of the vibration
2. Path—where the vibration travels through the vehicle
3. Responder—the component where the vibration is felt

For example, if the vehicle has an unbalanced tire, this is the vibration source. The vibration path is the steering and suspension system through which the vibration travels. The responder is the steering wheel, because this component is where the customer feels the vibration (**Figure 3-80**). When diagnosing vibration problems, locate and correct the source of the vibration. In the previous example of the unbalanced tire, installing a rigid brace from the steering column to the instrument panel and chassis may reduce the vibration experienced by the customer, but this does not solve the problem. To eliminate the problem, also diagnose the unbalanced tire condition and then balance the tire-and-wheel assembly.

Vibration may also produce noise. If a vehicle has a broken or improperly positioned tailpipe hanger that allows the tailpipe to contact the chassis, the customer may complain about a vibrating noise. The floor panel acts as a large speaker and amplifies the vibrating noise. The vibration path is through the exhaust system and chassis to the floor panel. The responder is the chassis and floor panel. In this case, after diagnosing the broken or improperly positioned tailpipe hanger, replace or reposition it to eliminate the vibration transfer path.

Clamp a yardstick to a table top with 18 in. (45 cm) hanging over the edge of the table (**Figure 3-81**). If the outer end of the yardstick is pulled upward or downward and then released, the end of the yardstick repeatedly vibrates up and down. Each vibration **cycle** begins

A **cycle** begins and ends at the same point.

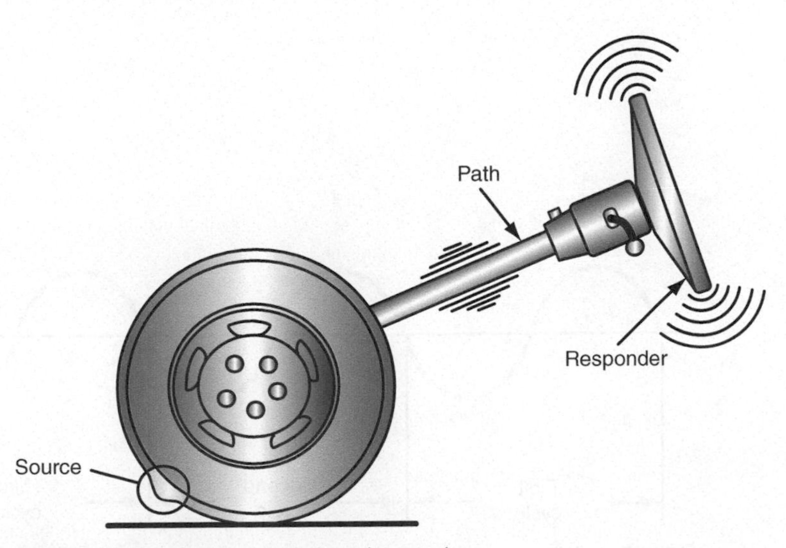

Figure 3-80 Vibration source, path, and responder.

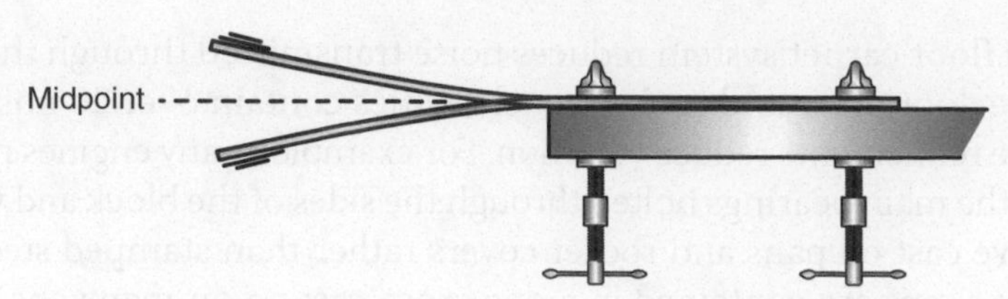

Figure 3-81 Theory of vibration cycles.

at the midpoint with the yardstick straight. From this point, the vibration cycle continues to the lowest point of travel and then moves up through the midpoint to the highest point of travel. The vibration cycle then returns to the midpoint where it begins over again.

The **frequency** of a vibration is the number of cycles per second or cycles per minute.

If the end of the vibrating yardstick completed 10 cycles per second, this is called the **frequency** of the vibration. To calculate the cycles per minute multiply the cycles per second by 60. In this example it is $10 \times 60 = 600$ cycles per minute.

If the yardstick is clamped to the table top with an 8-in. (20-cm) overhang, the end of the yardstick will vibrate much faster when the end of the yardstick is pulled upward or downward. Under this condition, the end of the yardstick may vibrate at a frequency of 30 cycles per second or 1800 cycles per minute.

Many vehicle vibrations are caused by an out-of-balance rotating component or engine firing pulses improperly isolated from the passenger compartment. Customers usually complain about vibrations that are felt in the steering wheel, instrument panel, frame or chassis, or the front or rear seat.

Vehicle vibrations may be tested with an **electronic vibration analyzer (EVA)** (**Figure 3-82**). The EVA has a vibration sensor mounted on the suspected vibration source. The EVA senses and records vibration cycles. A vibration cycle begins and ends at the same point and is continually repeated (**Figure 3-83**). Cycles per second are measured in **Hertz (Hz)**, and the Hz may be multiplied by 60 to obtain the cycles per minute. The **amplitude** of a vibration is the maximum value of the varying vibration. In **Figure 3-84**, 1 represents the maximum amplitude, 2 is the minimum amplitude, 3 is the zero-to-peak amplitude, and 4 indicates the peak-to-peak amplitude. The vibration amplitude may vary with the rotating speed of a component. For example, if a tire-and-wheel assembly is unbalanced, the amplitude of the resulting vibrations increases with wheel speed.

Natural vibration frequency is the frequency at which the object tends to vibrate.

All objects have **natural vibration frequencies**. The natural frequency of a front suspension system is 10 to 15 Hz. The suspension system design determines the natural frequency. The natural frequency of the suspension system is the same at all vehicle speeds.

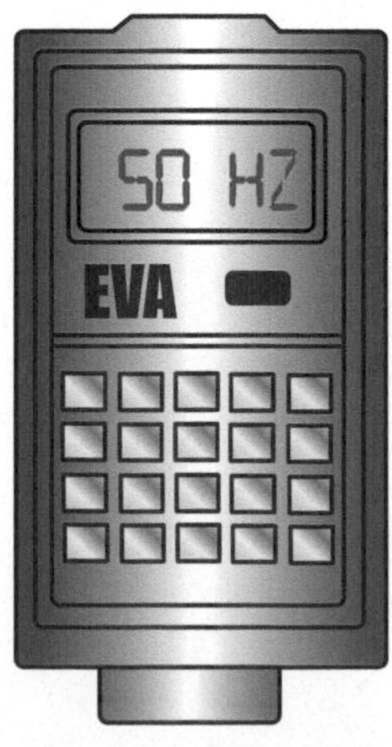

Figure 3-82 Electronic vibration analyzer (EVA).

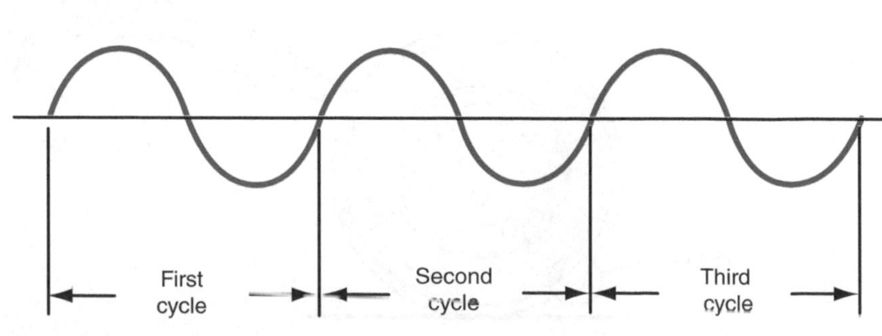

Figure 3-83 Vibration cycles.

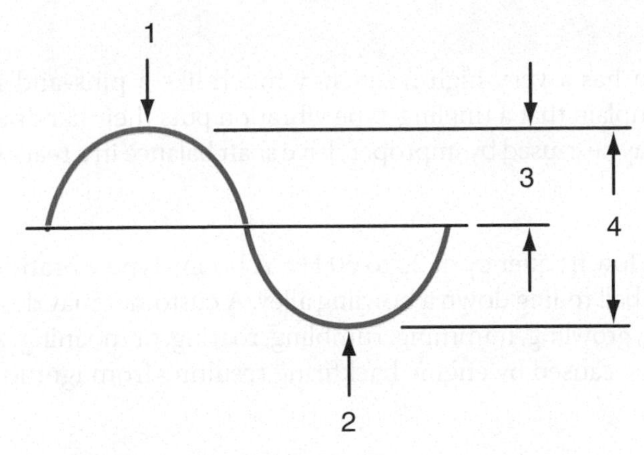

Figure 3-84 Amplitude of vibration cycles.

The vibration frequency caused by an unbalanced tire-and-wheel assembly increases with wheel speed. When the vibration frequency caused by the wheel balance problem intersects with the natural frequency of the suspension system, the suspension system begins to vibrate. This intersection point is called the **resonance**.

Resonance is the point where the frequency of a vibration from a defective component intersects with the natural frequency of the component or system where the component is located.

Vibration Classifications and Terminology

Vibrations that can be felt are:

1. Shake
2. Roughness
3. Buzz
4. Tingling

Vibrations that result in noise may be classified as:

1. Boom
2. Moan or drone
3. Howl
4. Whine

Shake

A shake is a vibration with a low frequency of 5 to 20 Hz, and is sometimes experienced in the steering wheel, seat, or console. Customers may refer to shake as shimmy, wobble, shudder, waddle, or hop. Most cases of a shake vibration are caused by brake rotors and drums or unbalanced tire-and-wheel assemblies. These defects cause a shake-type vibration that is vehicle-speed sensitive. The engine, clutch, or transmission may cause a shake-type vibration that is engine-speed sensitive.

Roughness

Roughness is a vibration with a higher frequency of 20 to 50 Hz. Holding a jigsaw to cut a piece of wood produces a roughness-type vibration. A roughness vibration may be caused by a defective wheel bearing. Prior to causing a roughness vibration, the defective wheel bearing would cause a howl.

Buzz

A buzz has a faster frequency of 50 to 100 Hz. Holding a vibrator-type electric razor produces a feeling similar to a buzz vibration. This type of vibration is usually felt in the vehicle floor or the seat. A buzz-type vibration is usually caused by defects in the exhaust system hangers, A/C compressor, or the engine.

Shop Manual
Chapter 3, page 133

Tingling

A tingling vibration has a very high frequency much like a pins-and-needles sensation. Customers may complain that a tingling-type vibration puts their hands and feet to sleep. A tingling vibration may be caused by improper drive shaft balance in a rear-wheel-drive vehicle.

Boom

A boom noise has a low frequency of 20 to 60 Hz. A boom-type vibration produces a noise similar to a bowling ball rolling down a bowling alley. A customer may describe a boom-type vibration as droning, growling, humming, rumbling, roaring, or moaning. A boom-type noise and vibration may be caused by engine backfiring resulting from ignition defects.

Moan or Drone

A moan or drone is a tone with a higher frequency of 60 to 120 Hz. A moan or drone produces a noise similar to a bumblebee in flight. Moan or drone may be caused by defects in the exhaust system or defective engine or transmission mounts.

Howl

A howl is a noise with frequency of 120 to 300 Hz, and this sound is much like the wind howling. A howling noise may be caused by worn differential bearings.

Whine

A whine is a high-pitched sound with a frequency of 300 to 500 Hz, and this sound is similar to a vacuum cleaner. A whine problem may be caused by meshing gears in the transmission of differential. When diagnosing vibration problems, the technician should match the vibration frequency to the rotating speed of a component, to help locate the component responsible for the vibration.

SUMMARY

- Tires are extremely important because they provide ride quality, support the vehicle's weight, provide traction for the drive wheels, and contribute to steering quality and directional stability.
- Tires have many design features and are carefully engineered to meet specific driving requirements.
- Tires may be bias-ply, belted bias-ply, or radial construction design.
- Tire ratings provide information regarding tire type, section width, aspect ratio, construction type, rim diameter, load capacity, and speed rating.
- Numeric tire ratings provide information about tire type, aspect ratio, construction type, and rim diameter.
- The tire performance criteria (TPC) number represents that the tire meets the car manufacturer's performance standards for traction, endurance, dimensions, noise, handling, and rolling resistance.
- The Uniform Tire Quality Grading (UTQG) designation includes tread wear, traction, and temperature ratings.
- Replacement tires must be the same type and size as the original tires to maintain vehicle safety.
- Replacement tires must have the same ratings as the original tires to maintain vehicle safety.
- The tire placard provides valuable information regarding the tires on the vehicle.
- Tires are subjected to severe acceleration and deceleration forces during normal operation.
- Replacement wheel rims must have the same width, diameter, offset, load capacity, and mounting configuration as the original rims to maintain vehicle safety.
- Static wheel unbalance causes wheel tramp and severe tire cupping.
- Dynamic wheel unbalance causes wheel shimmy, increased tire wear, unstable directional control, driver fatigue, and increased wear on suspension and steering components.

REVIEW QUESTIONS

Short Answer Essays

1. State general tire functions.
2. Interpret what each item of a tire size means, using P225/45R16 as your example.
3. What is the meaning of the service description 88H on the tire sidewall following the tire size?
4. Describe the structural difference between a bias-ply and radial-ply tire.
5. List and define the three major uniform tire quality grading system ratings.
6. What is the tire DOT code and how do you know the date the tire was manufactured?
7. Explain the difference between an A and C tire temperature rating.
8. Define tire contact area and tire deflection.
9. Explain the purpose of the wheel rim drop center and safety ridges.
10. Describe the results of static imbalance.

Fill-in-the-Blanks

1. To calculate the tire aspect ratio, the tire section height is divided by the _______________.
2. Tires equipped on today's passenger cars and light trucks are of the _______________ ply design.
3. In _______________ _______________, the ply cords are arranged radially at right angle to the tire centerline.
4. Replacement tires should be installed in pairs on the same _______________.
5. When a high-pressure mini-spare tire is installed on a vehicle, driving speed should not exceed _______________ mph.
6. Run-flat tires eliminate the need for a _______________ _______________ and a _______________.
7. Radial runout refers to variations in tire _______________.
8. Car manufacturers recommend that tire inflation pressures should be checked when the tires are _______________.
9. _______________ occurs when water on the pavement is allowed to remain between the pavement and the tire tread contact area.
10. _______________ _______________ vehicles tend to have positive offset rims to allow for proper clearance of the hub assembly within the vehicle wheel well.

Multiple Choice

1. All of the following statements about rim offset are correct EXCEPT:
   A. Zero offset, the rim centerline aligns with the mounting face of the rim.
   B. Positive offset, the rim centerline is inboard of the mounting face of the rim.
   C. Positive offset, the centerline is outboard of the mounting face of the rim.
   D. Negative offset, the centerline is outboard of the mounting face of the rim.
2. Tire pressure monitoring systems (TPMS) became required equipment on all vehicles sold in the United States beginning with what model year?
   A. 1996
   B. 2000
   C. 2006
   D. 2008
3. What is the maximum speed rating for a T rated tire?
   A. 112 mph
   B. 118 mph
   C. 124 mph
   D. 130 mph
4. Dynamic wheel unbalance can result in:
   A. Lateral wheel shimmy.
   B. Increased steering effort.
   C. Tire and wheel tramp.
   D. Normal tire tread life.
5. When a tire has a conicity problem:
   A. The tire has excessive lateral runout.
   B. The tire has too much radial runout.
   C. The tire is cone shaped and not level across the tread area.
   D. The tire has separation between the cord plies in the sidewall.

6. The Uniform Tire Quality Grading (UTQG) system provides consumers with useful information to help them purchase tires based on all of the following EXCEPT:
   A. Temperature rating.
   B. Traction rating.
   C. Speed rating.
   D. Tread wear rating.
7. Nitrogen tire inflation provides:
   A. Increased tire rolling resistance.
   B. Improved puncture resistance.
   C. More stable tire pressure.
   D. Improved ride quality.
8. The tread wear indicator bar becomes visible when tread depth is less than:
   A. 4/32 in.
   B. 2/16 in.
   C. 2/32 in.
   D. 3/16 in.
9. Static wheel unbalance causes:
   A. Cupped tire tread wear.
   B. Even wear on one edge of the tire tread.
   C. Even wear on the center of the tire tread.
   D. Feathered tire tread wear.
10. When installing replacement tires:
    A. Different tire sizes can be installed on the same axle.
    B. Snow tires can be a different type or size than the original tires on the vehicle.
    C. If the tire size is different than the original tire size, antilock brake system operation can be adversely affected.
    D. A four-wheel-drive vehicle can have different size tires on the front and rear wheels.

CHAPTER 4
WHEEL BEARINGS

Upon completion and review of this chapter, you should be able to understand and describe:

- The purposes of a bearing.
- Three different types of bearing loads.
- The basic parts in ball bearings or roller bearings, and describe the purpose of each part.
- The action between the balls and the race when a ball bearing is rotating.
- The purposes of bearing snap rings, shields, and seals.
- The load-carrying capabilities of ball bearings, roller bearings, tapered roller bearings, and needle roller bearings.
- The advantage of tapered roller bearings compared to other types of bearings.
- Seal design and purpose.
- The purpose of the garter spring behind a lip seal.
- The purpose of flutes on seal lips.
- Two different types of rear axle bearings in rear-wheel-drive cars, and give the seal location for each bearing type.
- Grease classifications.

Terms To Know

American Petroleum Institute (API)
Angular bearing load
Ball bearings
Bearing hub unit
Bearing seals
Bearing shields
Double-row ball bearings
Fluted lip seal
Inner race
Lithium-based grease
Needle roller bearing
Outer race
Radial bearing load
Rear-axle bearings
Roller bearing
Rolling elements
Separator
Single-row ball bearing
Snap ring
Society of Automotive Engineers (SAE)
Sodium-based grease
Springless seals
Spring-loaded seal
Tapered roller bearing
Thrust bearing load

INTRODUCTION

Many different types of bearings are used in the automobile. A bearing may be defined as a component that supports and guides one of these parts:

1. Pivot
2. Wheel
3. Rotating shaft
4. Oscillating shaft
5. Sliding shaft

While a bearing is supporting and guiding one of these components, the bearing is designed to reduce friction and support the load applied by the component and related assemblies. Since the bearing reduces friction, it also decreases the power required to rotate or move the component. Bearings are precision-machined assemblies, which provide smooth operation and long life. When bearings are properly installed and maintained, bearing failure is rare.

AUTHOR'S NOTE Only 21 percent of the power developed by a vehicle engine actually gets to the drive wheels. Much of the engine energy is lost in overcoming friction and wind resistance. The U.S. Department of Energy (DOE) is working with a major bearing manufacturer to explore the use and advantage of roller bearings to reduce friction in vehicle engines. The DOE estimates that if all vehicle engines used roller-bearing, low-friction technology, 100 million barrels of oil could be saved in a year.

Shop Manual
Chapter 4, page 166

BEARING LOADS

A **thrust bearing load** may be referred to as an axial load.

When a bearing load is applied in a vertical direction on a horizontal shaft, it is called a **radial bearing load**. If the vehicle weight is applied straight downward on a bearing, this weight is a radial load on the bearing. A **thrust bearing load** is applied in a horizontal direction (**Figure 4-1**). For example, while a vehicle is turning a corner, horizontal force is applied to the front wheel bearings. When an **angular bearing load** is applied, the angle of the applied load is somewhere between the horizontal and vertical positions. This can also be referred to as a combination load.

BALL BEARINGS

Ball bearings have round steel balls between the inner and outer races.

Front and rear wheel bearings may be **ball bearings** or roller bearings. Either type of bearing contains these basic parts:

1. Inner race, or cone
2. Rolling elements, balls, or rollers
3. Separator, also called a cage or retainer
4. Outer race, or cup

The **inner race** supports the inner side of the rolling elements in a bearing.

The **rolling elements** are the precision-machined balls or rollers between the inner and outer bearing races.

The **inner race** is an accurately machined component. The inner surface of the race is mounted on the shaft with a precision fit. The **rolling elements** are mounted on a very

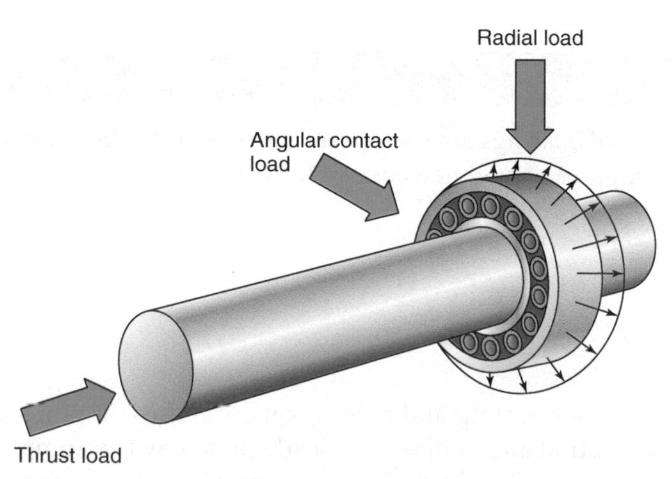

Figure 4-1 Types of bearing loads.

smoothly machined surface on the inner race. The surfaces of the rolling elements and the inner and outer races are case hardened to provide long bearing life. Positioned between the inner and outer races, the **separator** retains the rolling elements and keeps them evenly spaced. The rolling elements have precision-machined surfaces, and these elements are mounted between the inner and outer races. The **outer race** is the bearing's exterior ring. Both sides of this component have precision-machined surfaces. The outer surface of this race supports the bearing in the housing, and the inner surface is in contact with the rolling elements.

The **outer race** supports the outer side of the rolling elements and positions the bearing properly in the housing.

A **single-row ball bearing** has a crescent-shaped machined surface in the inner and outer races in which the balls are mounted (**Figure 4-2**). When a ball bearing is at rest, the load is distributed equally through the balls and races in the contact area. When one of the races and the balls begin to rotate, the bearing load causes the metal in the race to bulge out in front of the ball and flatten out behind the ball (**Figure 4-3**). This action creates a certain amount of friction within the bearing, and the same action is repeated for each ball, while the bearing is rotating. If metal-to-metal contact is allowed between the balls and the races, these components will experience very fast wear. Therefore, bearing lubrication is extremely important to eliminate metal-to-metal contact in the bearing and reduce wear.

A **single-row ball bearing** has one row of balls between the inner and outer races.

A ball bearing is designed primarily to handle radial loads. However, this type of bearing can also withstand a considerable amount of thrust load in either direction, even at high speeds. A maximum capacity ball bearing has extra balls for greater radial load-carrying capacity. Ball bearings are available in many different sizes for various applications (**Figure 4-4**).

Double-row ball bearings contain two rows of balls side by side. As in the single-row ball bearing, the balls in the double-row bearing are mounted in crescent-shaped grooves in the inner and outer races. The double-row ball bearing can support heavy radial loads and withstand thrust loads in either direction.

Ball Bearing Seals, Shields, and Snap Rings

For some applications, a ball bearing is held in place with a **snap ring**. A groove is cut around the outside surface of the outer race, and the snap ring is mounted in this groove. The snap ring may fit against a machined housing surface, or the outer circumference of the snap ring may be mounted in a groove in the housing. Ball bearings retained with a snap ring are not used on wheels because they are not designed to withstand high thrust loads encountered by wheel bearings.

A **snap ring** is made from spring steel and fits into a groove to retain a bearing on a shaft.

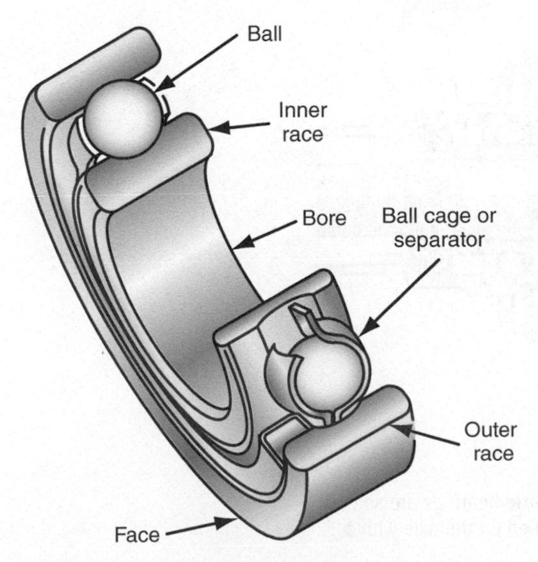

Figure 4-2 Parts of a ball bearing.

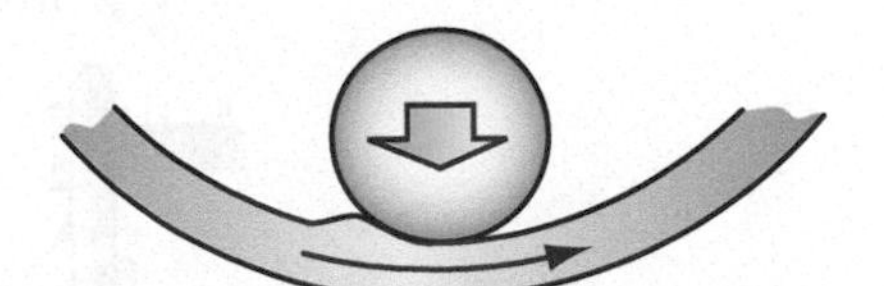

Figure 4-3 When a load is applied to a ball bearing, the metal in the race bulges out in front of the ball and flattens out behind the ball.

Figure 4-4 A ball bearing.

Bearing shields may be positioned between the inner and outer bearing races on the outside of the rolling elements.

A **bearing seal** may be mounted on the outside of the rolling elements. The seal is attached to the outer race, and the seal lip contacts the inner race.

Shop Manual
Chapter 4, page 172

Bearing shields cover the space between the two bearing races on one, or both, sides of the bearing. These shields are usually attached to the outer race, but space is left between the shield and the inner race. Bearing shields prevent dirt from entering the bearing, but excess lubrication can still flow through the bearing.

A **bearing seal** is a circular metal ring with a sealing lip on the inner edge. The seals are usually attached to the outer bearing race on each side of the bearing, and the lip surface contacts the inner race. The seal lip may have single, double, or triple lips made of synthetic or nonsynthetic rubber or elastomers. Lubricant is retained in the bearing by the seal, and the seal also keeps moisture, dirt, and contaminants out of the bearing. Some rear axle bearings on rear-wheel-drive cars are sealed on both sides and retained on the axle with a retainer ring (**Figure 4-5**).

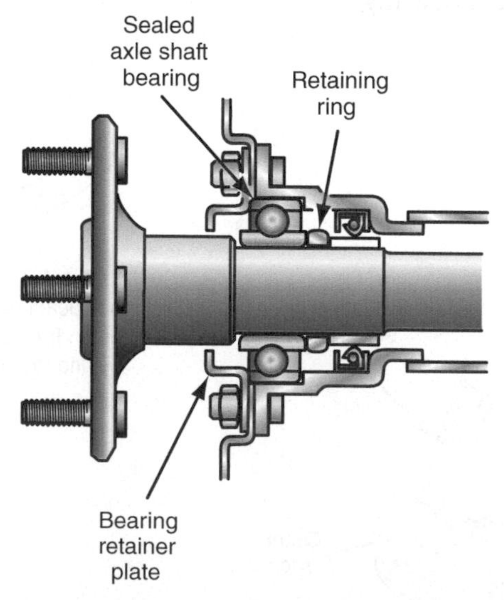

Figure 4-5 Some rear axle bearings are sealed on both sides and retained on the axle with a retainer ring.

ROLLER BEARINGS

Roller Bearing

A **roller bearing** contains precision-machined rollers that have the same diameter at both ends. These rollers are mounted in square-cut grooves in the outer and inner races (**Figure 4-6**). In the roller bearing, the races and rollers run parallel to one another. Roller bearings are designed primarily to carry radial loads, but they can withstand some thrust load. Since ball bearings do not withstand high thrust loads, they are usually not used in wheel bearings.

A **roller bearing** is sometimes called a nontapered roller bearing.

Tapered Roller Bearing

In a **tapered roller bearing**, the inner and outer races are cone-shaped. If imaginary lines extend through the inner and outer races, these lines taper and eventually meet at a point extended through the center of the bearing (**Figure 4-7**). The most important advantage of the tapered roller bearing compared to other bearings is an excellent capability to carry radial, thrust, and angular loads. In the tapered roller bearing, the rollers are mounted on cone-shaped precision surfaces in the outer and inner races. The bearing separator has an open space over each roller (**Figure 4-8**). Grooves cut in the side of the separator roller openings match the curvature of the roller. This design allows the rollers to rotate evenly

A **tapered roller bearing** has tapered rollers between the inner and outer races. Tapered surfaces on the races match the tapered roller surfaces.

Shop Manual
Chapter 4, page 167

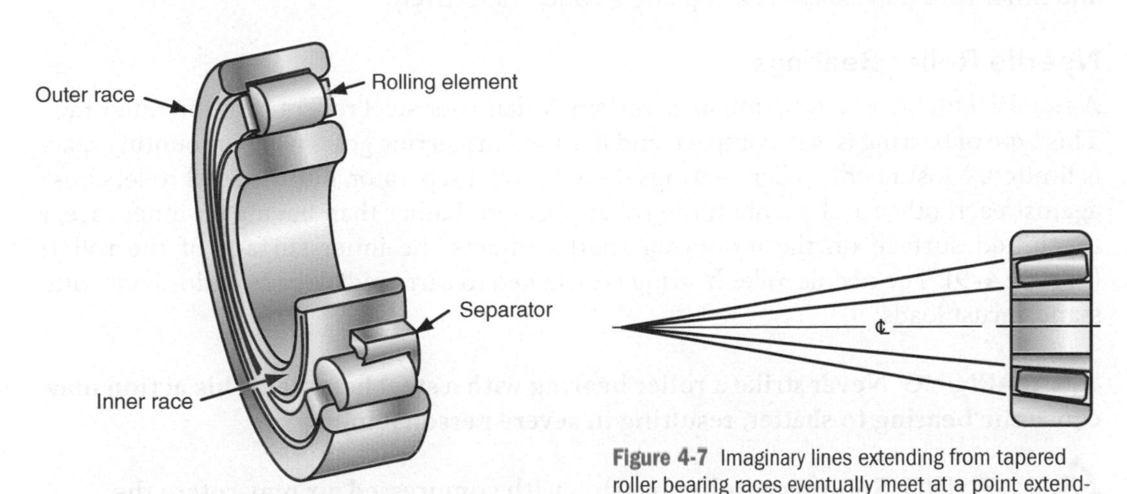

Figure 4-6 Parts of a roller bearing.

Figure 4-7 Imaginary lines extending from tapered roller bearing races eventually meet at a point extending from the bearing center.

Figure 4-8 Tapered roller bearings.

Figure 4-9 Needle roller bearing.

without interference between the rollers and the separator. Lubrication and proper end play adjustment are critical on tapered roller bearings. A tapered roller bearing outer and inner race may be called a cup and a cone, respectively.

Needle Roller Bearings

A **needle roller bearing** contains many small-diameter steel rollers in a thin outer race. This type of bearing is very compact, and it is used in steering gears where mounting space is limited. Most needle roller bearings do not have a separator, but the steel rollers push against each other and maintain the roller position. Rather than having an inner race, a machined surface on the mounting shaft contacts the inner surface of the rollers (**Figure 4-9**). The needle roller bearing is designed to carry radial loads; it does not withstand thrust loads.

WARNING Never strike a roller bearing with a steel hammer. This action may cause the bearing to shatter, resulting in severe personal injury.

WARNING Spinning a roller bearing with compressed air may rotate the bearing at extremely high speed and cause the bearing to disintegrate, resulting in serious personal injury.

SEALS

Seals are designed to keep lubricant in the bearing and prevent dirt particles and contaminants from entering the bearing. Static seals are used between two surfaces that do not move and dynamic seals are used between two surfaces where at least one surface moves. Wheel bearing seals are mounted in front and rear wheel hubs, and in rear axle housings on rear-wheel-drive cars. The metal seal case has a surface coating that resists corrosion and rust and acts as a bonding agent for the seal material. Seals have many different designs, including single lip, double lip, and fluted. A lip seal is usually encased in metal. The seal material is usually made of a synthetic rubber compound such as nitrile, silicon, polyacrylate, or a fluoroelastomer such as Viton. The actual seal material depends on the lubricant and contaminants that the seal encounters. All seals may be divided into two groups, springless and spring loaded. **Springless seals** are used in some front or rear wheel hubs, where they seal a heavy lubricant into the hub (**Figure 4-10**).

Springless seals do not have a garter spring behind the seal lip.

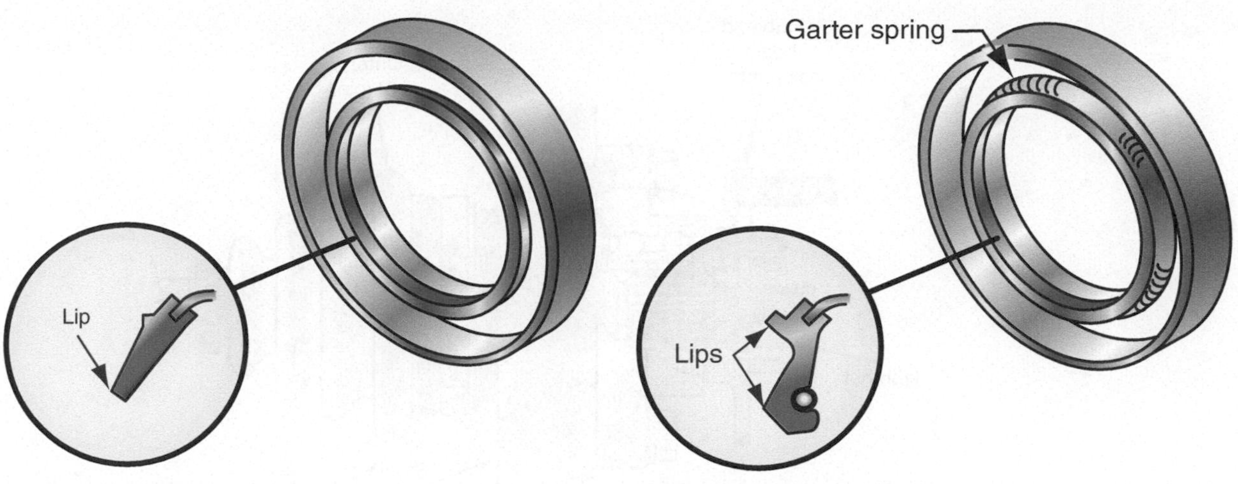

Figure 4-10 Springless seal.

Figure 4-11 Spring-loaded seal.

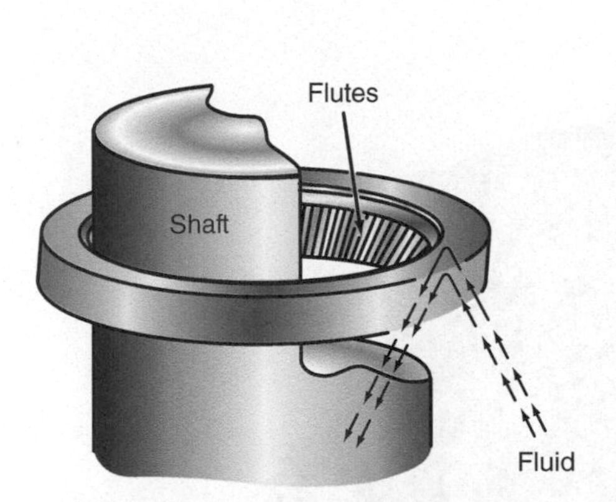

Figure 4-12 Fluted lip seal redirects oil back into the housing.

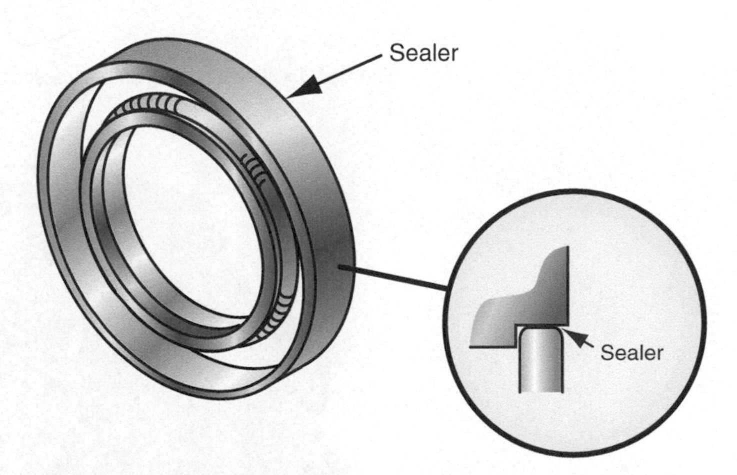

Figure 4-13 Sealer painted on the seal case prevents leaks between the case and the housing.

In a **spring-loaded seal**, the **garter spring** behind the seal provides additional force on the seal lip to compensate for lip wear, shaft movement, and bore eccentricity (**Figure 4-11**). A **fluted lip seal** may be used to direct oil back into a housing. This seal design provides a pumping action to redirect the oil back into the housing (**Figure 4-12**).

Some seals have a sealer painted on the outside surface of the metal seal housing. When the seal is installed, this sealer prevents leaks between the seal case and the housing (**Figure 4-13**). Seals should be replaced whenever it is necessary to remove one during a service procedure.

A **spring-loaded seal** has a garter spring behind the seal lip.

A **garter spring** is a circular coil spring mounted behind a seal lip.

A **fluted lip seal** has small ridges on the outer lip surface.

WHEEL BEARINGS

Some front-wheel-drive vehicles have front wheel bearing and hub assemblies that are bolted to the steering knuckles (**Figure 4-14**). The bearings are lubricated and sealed, and the complete bearing and hub assembly is replaced as a unit. The **bearing hub unit** is more compact than other types of wheel bearings mounted in the wheel hub. This type of bearing contains two rows of ball bearings with an angular contact angle of 32° (**Figure 4-15**). The inner bearing assembly bore is splined, and the inner ring extends to the outside to form a flange and spigot. The flange attached to the outer ring contains bolt holes, and bolts extend through these holes into the steering knuckle. This type of bearing attachment allows the bearing to become a structural member of the front suspension.

A **bearing hub unit** is a one-piece bearing and hub assembly that supports a front or rear wheel.

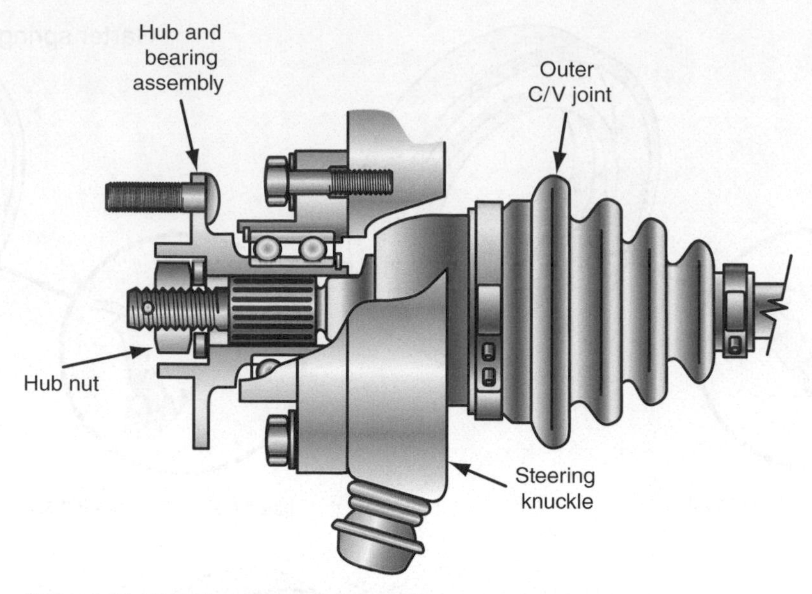

Figure 4-14 Wheel bearing and hub assembly.

Figure 4-15 Double-row, sealed wheel bearing hub unit.

Since the bearing outer ring is self-supporting, the main concern in knuckle design is fatigue strength rather than stiffness. The drive axle shaft transmits torque to the inner bearing race. This shaft is not designed to hold the bearing together. This type of wheel bearing is designed for mid-sized front-wheel-drive cars.

Shop Manual
Chapter 4, page 174

Each front drive axle has splines that fit into matching splines inside the bearing hubs (**Figure 4-16**). A hub nut secures the drive axle into the inner bearing race. Some wheel bearing hubs contain a wheel speed sensor for the antilock brake system (ABS).

Some front-wheel-drive vehicles have a sealed bearing unit that is pressed into the steering knuckle (**Figure 4-17**). The wheel hub is pressed into the inner bearing race, and the drive axle is splined into the hub. This type of bearing is designed for smaller front-wheel-drive cars. These bearings may contain two rows of ball bearings, or two tapered

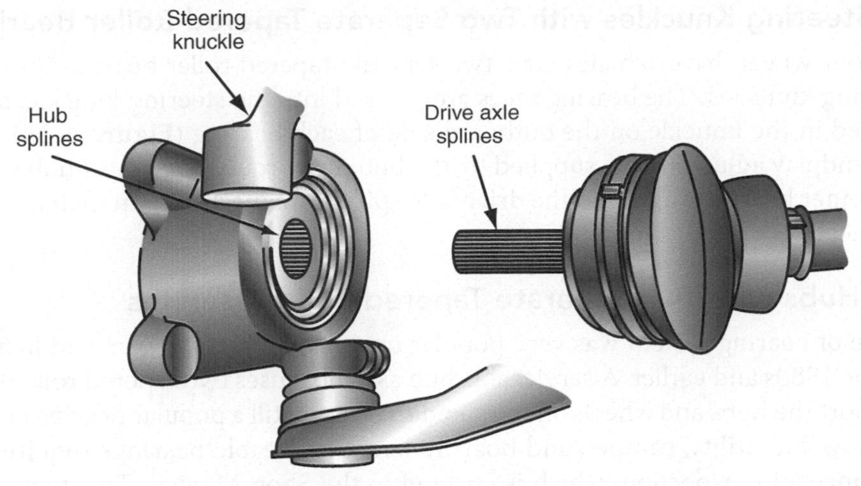

Figure 4-16 Front drive shaft installed in wheel bearing hub.

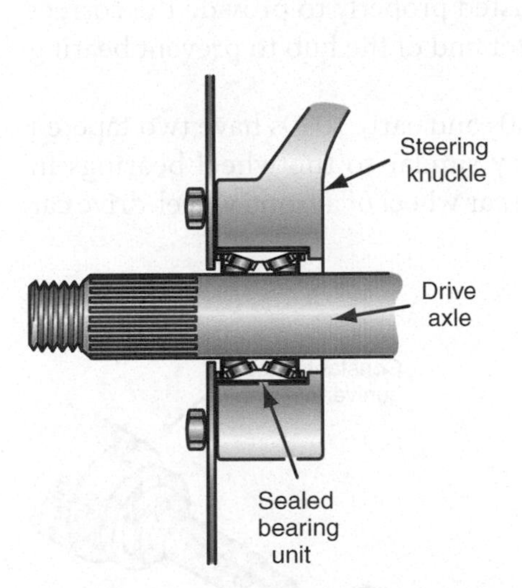

Figure 4-17 Steering knuckle with pressed-in bearing.

Figure 4-18 Cutaway view of front wheel bearing that is pressed into the steering knuckle.

roller bearings and a split inner race (**Figure 4-18**). The bearing containing two tapered roller bearings has more radial load capacity than the double-row ball bearing. However, the tapered roller bearing is more sensitive to misalignment. Both sides of the bearing are sealed, and a seal is positioned behind the bearing in the steering knuckle to keep contaminants out of the bearing area.

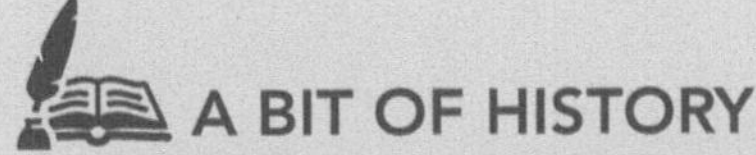

A BIT OF HISTORY

Nearly all rear-wheel-drive cars have two serviceable tapered roller bearings or ball bearings in the front wheel hubs. Most front-wheel-drive cars have sealed bearing hubs on the front wheels, and some have these units on the rear wheels. These sealed bearing hub units have largely replaced the tapered roller bearings or ball bearings in the front wheel hubs.

Front Steering Knuckles with Two Separate Tapered Roller Bearings

Other front-wheel-drive vehicles have two separate tapered roller bearings mounted in the steering knuckles. The bearing races are pressed into the steering knuckle, and seals are located in the knuckle on the outboard side of each bearing (**Figure 4-19**). Correct bearing endplay adjustment is supplied by the hub nut torque. The wheel hub is pressed into the inner bearing races, and the drive axle splines are meshed with matching splines in the wheel hub.

Wheel Hubs with Two Separate Tapered Roller Bearings

This type of bearing system was very popular on rear-wheel-drive cars and light trucks during the 1980s and earlier. A serviceable hub assembly uses two tapered roller bearings that support the hubs and wheels on the spindles. This is still a popular bearing configuration on towable utility, camper, and boat trailers. Serviceable bearings require routine maintenance and inspection, which is covered in the Shop Manual. This type of wheel bearing has the bearing races pressed into the hub. A grease seal is pressed into the inner end of the hub to prevent grease leaks and keep contaminants out of the bearings. The hub and bearing assemblies are retained on the spindle with a washer, adjusting nut, nut lock, and cotter pin. The adjusting nut must be adjusted properly to provide the correct bearing endplay. A grease cap is pressed into the outer end of the hub to prevent bearing contamination (**Figure 4-20**).

Some older front-wheel-drive cars from the 1980s and early 1990s have two tapered roller bearings in the rear wheel hubs that are very similar to the wheel bearings in **Figure 4-20**. The two tapered roller bearings in the rear wheel of a front-wheel-drive car are shown in **Figure 4-21**.

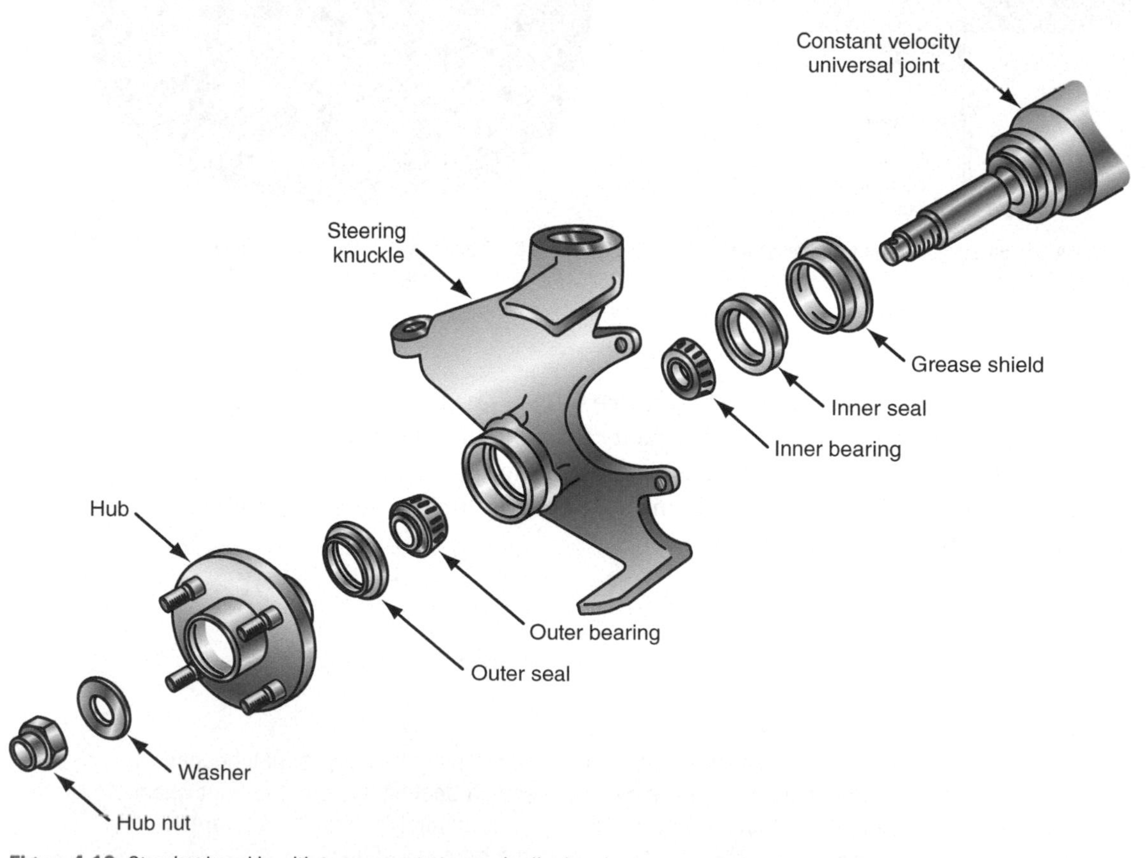

Figure 4-19 Steering knuckle with two separate tapered roller bearings.

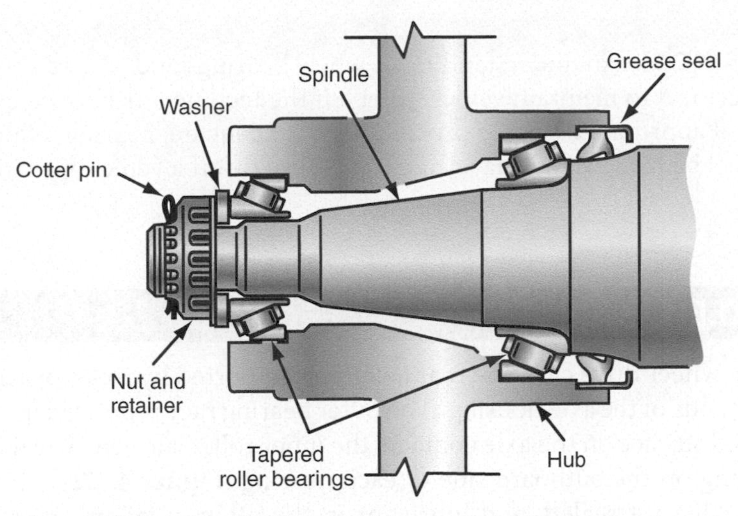

Figure 4-20 Front wheel bearing assembly, rear-wheel-drive car.

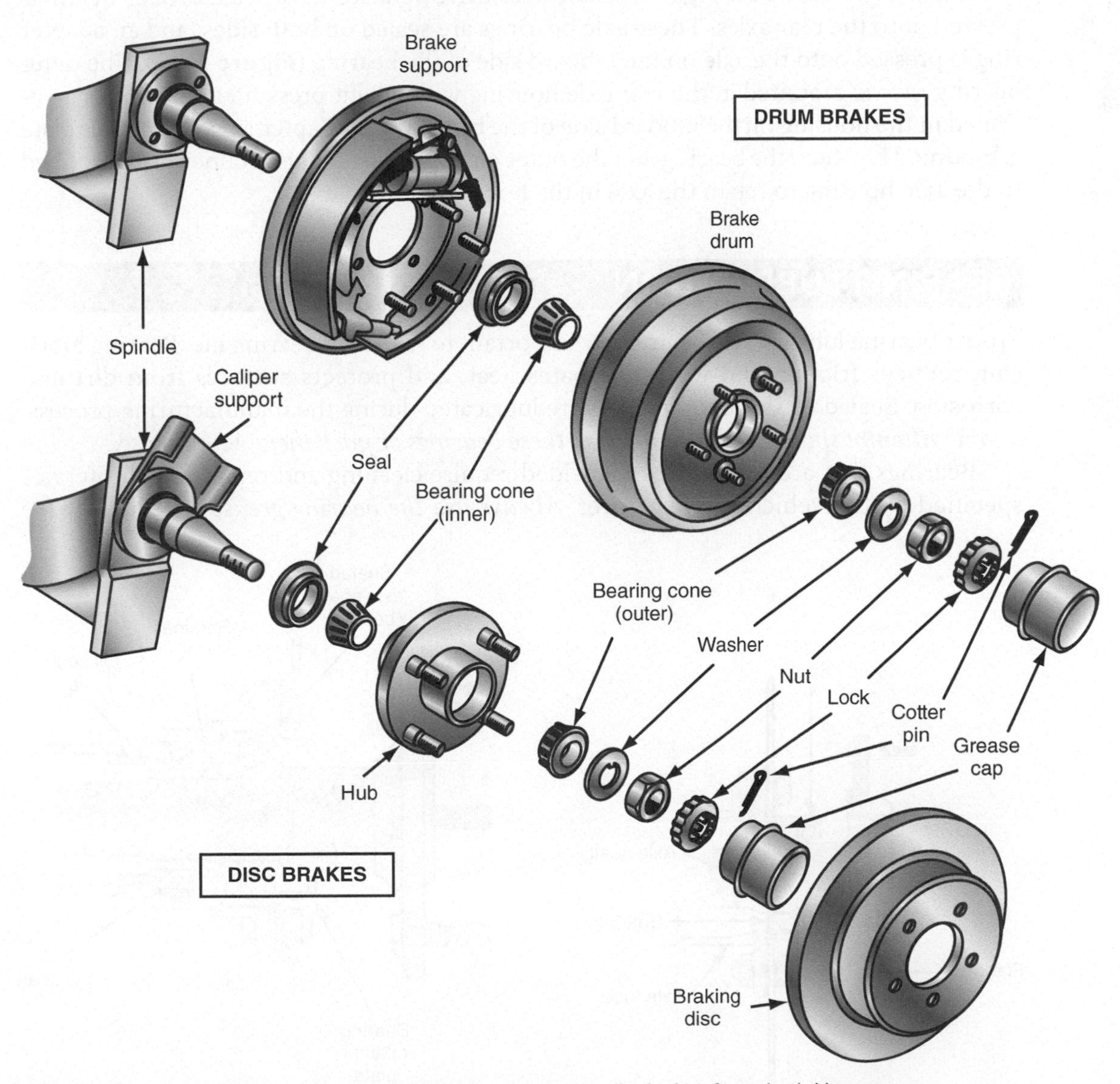

Figure 4-21 Rear wheel hub with tapered roller bearings for drum or disc brakes, front-wheel-drive car.

AUTHOR'S NOTE An understanding of wheel bearings and related service procedures is critical to maintain vehicle safety! If the technician's knowledge of wheel bearings and appropriate service procedures is inadequate, bearing failure may occur. Wheel bearing failure may cause a wheel to fly off a vehicle with disastrous results!

Shop Manual
Chapter 4, page 182

REAR-AXLE BEARINGS

On many rear-wheel-drive cars, the rear axles are supported by roller bearings mounted near the outer ends of the axle housing. The outer bearing race is pressed into the housing, and a machined surface on the axle contacts the inner roller surface. A seal is mounted in the axle housing on the outboard side of each bearing (**Figure 4-22**). This type of axle bearing is usually not sealed, and lubricant in the differential and rear axle housing provides axle bearing lubrication. The seals prevent lubricant leaks from the outer ends of the axle housing and keep dirt out of the bearings.

Rear-axle bearings support the rear axles in the rear axle housing on rear-wheel-drive vehicles.

Other **rear-axle bearings** on rear-wheel-drive vehicles have sealed roller bearings pressed onto the rear axles. These axle bearings are sealed on both sides, and an adapter ring is pressed onto the axle on the inboard side of the bearing (**Figure 4-23**). The outer bearing race is mounted in the rear axle housing with a light press fit, and a seal is positioned in the housing on the inboard side of the bearing and adapter ring. A retainer plate is mounted between the bearing and the outer end of the axle. This retainer plate is bolted to the axle housing to retain the axle in the housing.

Many wheel bearings require lithium-based or sodium-based grease.

A **lithium-based grease** has a specific amount of lithium mixed with the lubricant.

A **sodium-based grease** has a specific amount of sodium mixed with the lubricant.

BEARING LUBRICATION

Proper bearing lubrication is extremely important to maintain bearing life. Bearing lubricant reduces friction and wear, dissipates heat, and protects surfaces from dirt and corrosion. Sealed or shielded bearings are lubricated during the manufacturing process, *and no attempt should be made to wash these bearings or pack them with grease.*

Bearings that are not sealed or shielded require cleaning and repacking at intervals specified by the vehicle manufacturer. *Always use the bearing grease specified by the*

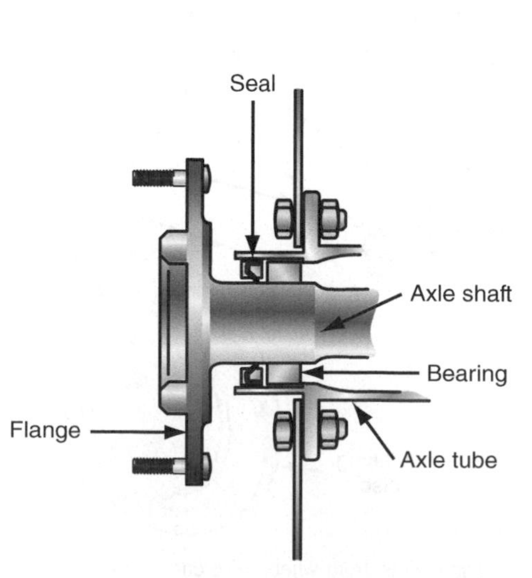

Figure 4-22 Rear-axle bearing, rear-wheel-drive car.

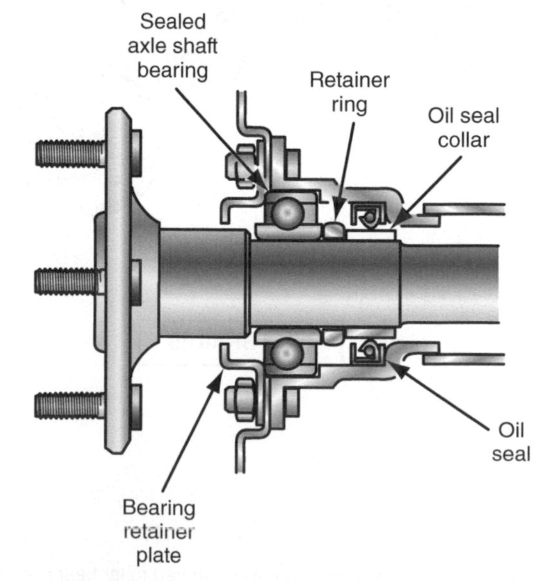

Figure 4-23 Rear-axle bearing and retainer.

vehicle manufacturer. Bearing lubricants may be classified as greases or oils. Grease is oil with a thickening agent added. Greases are named for the thickening agent:

- Aluminum
- Barium
- Calcium
- Lithium
- Sodium

ASTM International sets the standards for testing for grease. Classification ASTM D 4950 covers greases designed for the lubrication of wheel bearings and chassis components for passenger cars, trucks, and other vehicles. The National Lubricating Grease Institute (NLGI) classifies automotive greases into two main groups. The prefix designation for wheel bearing grease is G, and the prefix designation for chassis lubrication grease is L (**Figure 4-24**).

Service category L chassis grease:

- LA-Mild duty, non-critical applications with frequent relubrication intervals of less than 2000 mi. (3200 km). Performance offers oxidation resistance, shear stability, and corrosion and wear protection.
- LB-Mild to severe duty, high loads, and vibration with a usable temperature range of −40 − 248°*F*. Relubrication intervals of greater than 2000 mi. (3200 km). Performance offers oxidation resistance, shear stability, and corrosion and wear protection even under heavy loads and in the presence of water contamination.

Service category G wheel bearing grease:

- GA-Mild duty
- GB-Mild to moderate duty, with a usable temperature range of −40 − 248°*F* and occasional spikes to 320°*F*. Performance offers oxidation and evaporation resistance, shear stability, and corrosion and wear protection.
- GC-Mild to severe duty, with a usable temperature range of −40 − 320°*F* and occasional spikes to 392°*F*. Performance offers oxidation and evaporation resistance, shear stability, and corrosion and wear protection.

The most common grease designation used in automotive shops is GC and LB. A designation label GC/LB indicates that the grease is acceptable for bearings and chassis parts. Number 2 grease is the most common rating with the higher the number indicating the stiffer the consistency of the grease.

New bearings usually have a protective coating to prevent rust and corrosion. This coating should not be washed from the bearing. On rear-wheel-drive vehicles, where the rear-axle bearings are lubricated from oil in the differential housing, the type and level of oil in the housing are important.

Vehicle manufacturers usually recommend a **Society of Automotive Engineers (SAE)** No. 90 or SAE No. 140 hypoid gear oil in the differential. In very cold climates, the manufacturer may recommend an SAE No. 80 differential gear oil. The **American Petroleum Institute (API)** classifies gear lubricants as GL-1, GL-2, GL-3, GL-4, and GL-5. The GL-4

The **Society of Automotive Engineers (SAE)** is responsible for the establishment of many automotive standards.

The **American Petroleum Institute (API)** is responsible for establishing standards related to oils and lubricants.

NLGI SYMBOLS

Figure 4-24 The National Lubricating Grease Institute (NLGI) classifies automotive greases into two main groups, L and G.

lubricant is used for hypoid gears under normal conditions. The GL-5 lubricant is used in heavy-duty hypoid gears. Always use the vehicle manufacturer's specified differential gear oil.

The differential should be filled until the lubricant is level with the bottom of the filler plug opening in the differential housing. If the differential is overfilled, the bearings and seals may have excessive lubricant. Under this condition, the lubricant may leak past the seal. When the lubricant level is low in the differential, the lubricant may not be available in the axle housings. When this condition exists, the bearings do not receive enough lubrication and bearing life is shortened.

Caution

If a bearing is operated without proper lubrication, bearing life will be very short.

SUMMARY

- A bearing reduces friction, carries a load, and guides certain components such as pivots, shafts, and wheels.
- Radial bearing loads are applied in a vertical direction.
- Thrust bearing loads are applied in a horizontal direction.
- Angular bearing loads are applied at an angle between the vertical and the horizontal.
- The inner bearing race is positioned at the center of the bearing and supports the rolling elements.
- The rolling elements in a bearing are positioned between the inner and outer races.
- The bearing separator keeps the rolling elements evenly spaced.
- The outer bearing race forms the outer ring on a bearing.
- A cylindrical ball bearing is designed primarily to withstand radial loads, but these bearings can handle a considerable thrust load.
- A snap ring can be mounted in a groove in the outer bearing race, and the snap ring retains the bearing in the housing.
- A bearing shield prevents dirt from entering the bearing, but it is not designed to keep lubricant in the bearing.
- Bearing seals keep lubricant in the bearing and prevent dirt from entering the bearing.
- Roller bearings are designed primarily to carry radial loads, but they can handle some thrust loads.
- Tapered roller bearings have excellent radial, thrust, and angular load-carrying capabilities.
- Needle roller bearings are very compact and are designed to carry radial loads. They will not carry thrust loads.
- Springless seals are used for wheel bearing seals in some wheel hubs.
- The garter spring provides additional force on the seal lip to compensate for lip wear, shaft movement, and bore eccentricity.
- Flutes on seal lips provide a pumping action to direct oil back into a housing.
- Bearing hub units are compact compared to bearings that are mounted in the wheel hub. This compactness makes bearing hub units suitable for front-wheel-drive cars.
- Some bearing hub units are bolted to the steering knuckle; other bearing hub units are pressed into the steering knuckle.
- Some steering knuckles contain two separate tapered roller bearings
- Rear-axle bearings are mounted between the drive axles and the housing on rear-wheel-drive cars.

REVIEW QUESTIONS

Short Answer Essays

1. Define a radial bearing load.
2. Define a thrust bearing load, and give another term for this type of load.
3. Explain an angular bearing load.
4. Describe the main parts of a bearing, including the location and purpose of each part.
5. Describe the difference between a maximum-capacity ball bearing and an ordinary ball bearing.
6. Explain the design and purpose of bearing seals.
7. Explain the purpose of the sealer on the outside surface of a seal housing.
8. What does ASTM and NLGI stand for, and what is their function?

9. Explain how the proper bearing endplay adjustment is obtained when two tapered roller bearings are mounted in the steering knuckle.
10. Describe two different types of rear-axle bearings on rear-wheel-drive cars.

Fill-in-the-Blanks

1. A bearing can be described as a component that supports a _______________, _______________, or _______________.
2. A bearing is designed to support a load and _______________.
3. An angular bearing load is applied at an angle between the _______________ and the _______________.
4. A ball bearing is designed primarily to withstand _______________ loads.
5. Greases are named for the _______________ agent used.
6. Lubrication and proper_______________ adjustment are important on tapered roller bearings.
7. A needle roller bearing is not designed to carry _______________ loads.
8. A springless seal may be used to seal a _______________ lubricant into a hub.
9. When a vehicle is turning a corner, the front wheel bearings must carry a _______________ load.
10. When a rear-wheel-drive vehicle has cylindrical roller bearings mounted in the rear axle housing, the inner surface on the rollers contacts a machined surface on the _______________.

Multiple Choice

1. A tapered roller bearing has:
   A. Rollers that have the same diameter at both ends.
   B. Excellent ability to carry radial, thrust, and angular loads.
   C. Horizontal inner surfaces of the inner and outer races.
   D. A separator that allows the rollers to lightly contact each other.
2. A spring-loaded seal compensates for all of these conditions EXCEPT:
   A. Seal lip wear.
   B. Shaft movement.
   C. Bore eccentricity.
   D. Wear on the seal bore.
3. When two separate tapered roller bearings are located in a front steering knuckle on a front-wheel-drive car, the wheel bearing endplay adjustment is provided by:
   A. The bearing race position.
   B. Wheel nut torque.
   C. Hub nut torque.
   D. Wheel hub position.
4. Greases are named for the thickening agent. All of the following are common greases EXCEPT:
   A. Lithium
   B. Sodium
   C. Teflon
   D. Aluminum
5. All of the following statements are correct EXCEPT:
   A. Greases are named for their thickening agent.
   B. NLGI classifies gear lubricants.
   C. The prefix GC is the highest designation for automotive wheel bearing greases.
   D. The prefix LB is the highest designation for automotive wheel chassis greases.
6. When inspecting and servicing bearings:
   A. Some rear-wheel-drive vehicles have rear-axle bearings that are sealed on both sides.
   B. In a sealed bearing, the seal is usually attached to the inner race.
   C. Endplay adjustment is not critical on tapered roller bearings.
   D. A roller bearing can be removed by striking it with a steel hammer.
7. All of these statements about bearings are true EXCEPT:
   A. Tapered roller bearings have excellent capability to carry radial, thrust, and angular loads.
   B. Ball bearings are used in most wheel bearings.
   C. A roller bearing has rollers with the same diameter at each end.
   D. In a tapered roller bearing, the openings in the separator grooves match the curvature of the rollers.

8. While inspecting and servicing wheel hub seals:
   A. Many wheel hub seal lips are made from a plastic compound.
   B. Springless seals may be used in applications where they must seal a light fluid.
   C. A spring-loaded seal helps compensate for wheel hub bore eccentricity.
   D. The sealer on the outer surface of a seal housing improves seal and hub bore alignment.
9. When inspecting and servicing front wheel bearings on front-wheel-drive cars:
   A. Some one-piece front wheel bearing hubs are bolted to the front strut.
   B. In a one-piece front wheel bearing hub, the outer end of the drive axle is splined to the outer bearing race.
   C. When the steering knuckle contains two tapered roller bearings, the torque on the drive axle nut does not affect bearing endplay.
   D. Some front wheel bearings mounted in the steering knuckle contain two tapered roller bearings with a split inner race.
10. When inspecting and servicing rear wheel bearings on rear-wheel-drive cars:
    A. On some rear wheel bearings, a threaded adapter ring retains the bearing on the axle shaft.
    B. Roller-type rear-axle bearings may be lubricated from the oil supply in the differential.
    C. A GL-2 lubricant is recommended in hypoid-type differentials.
    D. An SAE 50 gear oil is recommended in many differentials.

CHAPTER 5
SHOCK ABSORBERS AND STRUTS

Upon completion and review of this chapter, you should be able to understand and describe:

- The three purposes of shock absorbers.
- How shock absorbers contribute to vehicle safety.
- Wheel jounce and rebound.
- Spring operation during wheel jounce and rebound.
- Shock absorber operation during wheel jounce.
- Shock absorber operation during wheel rebound.
- The advantages of nitrogen-gas-filled shock absorbers and struts.
- Shock absorber ratios.
- Travel-sensitive shock absorber operation.
- The operation of an adjustable shock absorber.
- The operation of load-leveling struts and shock absorbers.

Terms To Know

Active suspension systems
Adjustable struts
Gas-filled shock absorber
Heavy-duty shock absorbers
Jounce travel
Load-leveling shock absorbers
Magneto-rheological fluid
Rebound travel
Shock absorbers
Shock absorber ratios
Spring insulator
Struts
Travel-sensitive struts
Upper strut mount

INTRODUCTION

Two front **shock absorbers** or **struts** are connected from the front suspension to the chassis, and two rear shock absorbers or struts are attached between the rear suspension and the chassis. The internal strut design is very similar to shock absorber design, and struts perform the same control functions as shock absorbers. Shock absorbers have three main purposes:

Shock absorbers are devices used to control, or dampen, spring oscillations.

1. They control spring action and oscillations (bounce) to provide the desired ride quality and noise reduction.
2. They help prevent body sway (roll) and lean while cornering.
3. They reduce brake dive and acceleration squat, thereby reducing the tendency of a tire tread to lift off the road, which improves tire life, traction, and directional stability.

The kinetic energy of the suspension system spring action is converted to thermal energy as the shock absorber piston travels though the hydraulic fluid. This heat is then dissipated into the air flowing around the outer steel tube of the shock absorber. Because shock absorbers control spring action, spring oscillations, and chassis oscillations, they contribute to vehicle safety and passenger comfort. If the shock absorbers are worn out, excessive chassis oscillations may occur, particularly on rough road surfaces. These

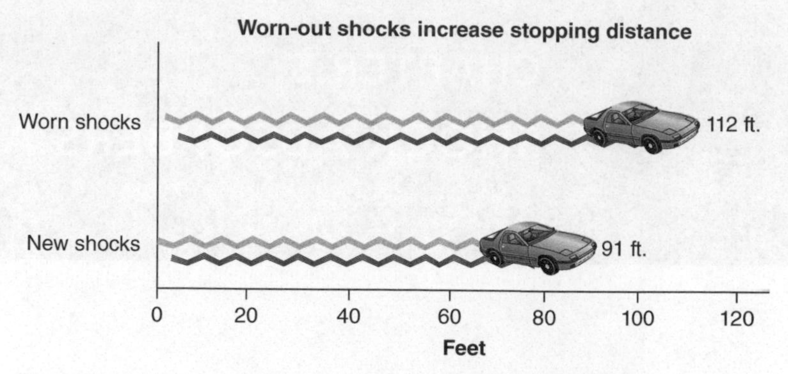

Figure 5-1 Worn-out shock absorbers can increase stopping distance.

excessive chassis oscillations may result in loss of steering control. Worn-out shock absorbers also cause excessive body lean and sway while cornering, which may cause the driver to lose control of the vehicle and can contribute to increased stopping distance (**Figure 5-1**). Shock absorbers are extremely important to providing longer tire life and improving vehicle handling, steering quality, and ride quality. A faulty shock absorber can cause tire cupping to occur. The two most common causes of tire cupping are a static imbalance condition of the wheel assembly and a weak hydraulic shock absorber.

Shock absorber design is matched to the deflection rate of the suspension spring to control the spring action. Several different types of shock absorbers are available for special service requirements.

SHOCK ABSORBER AND STRUT RATIOS

Most automotive shock absorbers and struts are a double-acting-type that controls spring action during jounce and rebound wheel movements. The piston and valves in many shock absorbers are designed to provide more extension control than compression control. An average shock absorber may have 70 percent of the total control on the extension cycle, and thus 30 percent of the total control is on the compression cycle. Shock absorbers usually have this type of design because they must control the heavier sprung body weight on the extension cycle. The lighter unsprung axle, wheel, and tire weight are controlled by the shock absorber on the compression cycle. A shock absorber with this type of design is referred to as a 70/30 type. **Shock absorber ratios** vary from 50/50 to 80/20.

Shock absorber ratios indicate the amount of control on the extension and compression cycles.

A shock absorber is mounted between the rear axle and the chassis on a front-wheel-drive car (**Figure 5-2**). Mounting bolts extend through hangers on the rear axle and

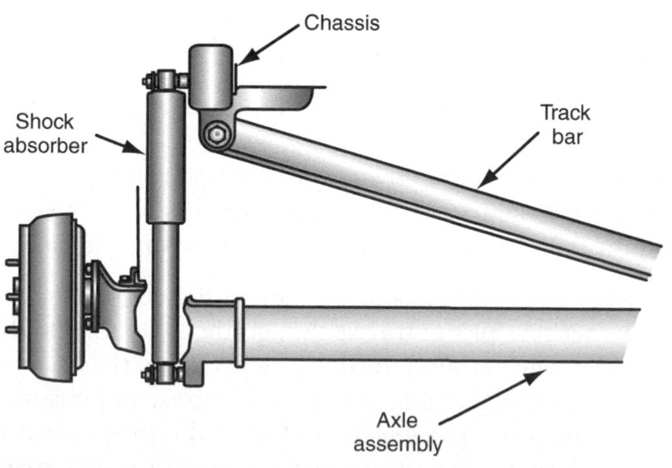

Figure 5-2 Rear shock absorber mounting on a front-wheel-drive car.

chassis. These bolts also pass through the isolating bushings on each of the shock absorbers. The isolating bushings are very important for preventing vibration and noise. Front shock absorbers may be mounted in a similar way between the lower control arms and the chassis.

TWIN-TUBE SHOCK ABSORBER DESIGN

The lower half of a shock absorber is a twin-tube steel unit filled with hydraulic oil and nitrogen gas (**Figure 5-3**). In some shock absorbers, the nitrogen gas is omitted. The outer tube is the reservoir tube, which stores excess hydraulic fluid and the inner tube is the pressure or working tube. A relief valve is located in the bottom of the unit to control fluid movement during compression, and a circular lower mounting is attached to the lower

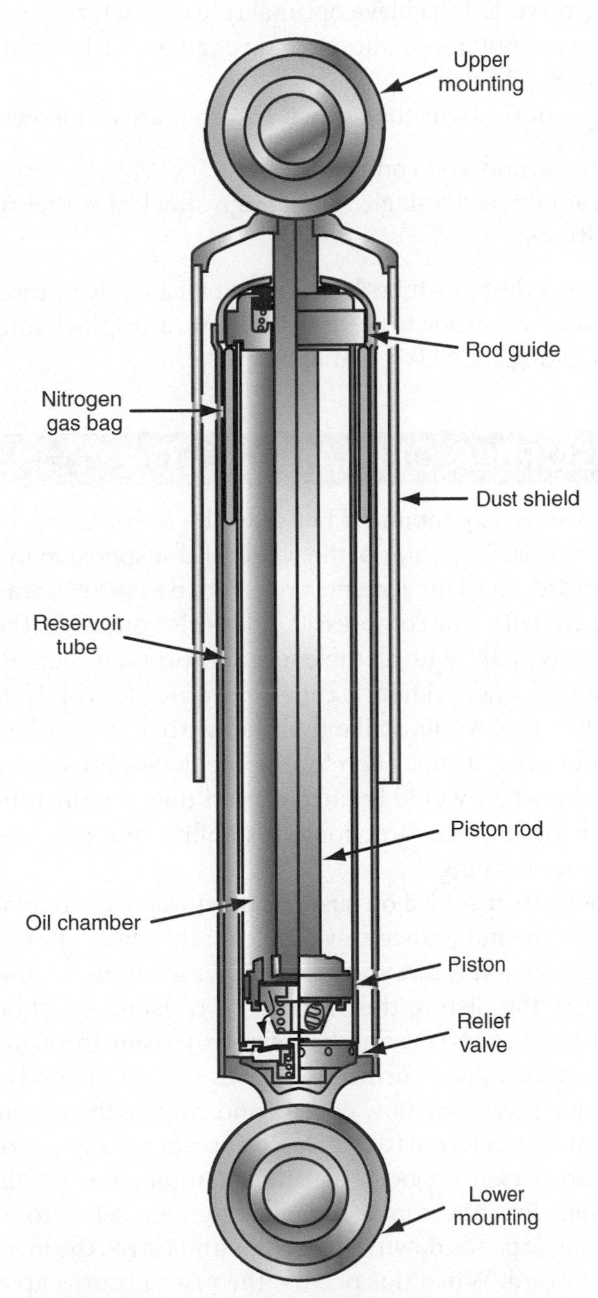

Figure 5-3 Shock absorber filled with hydraulic oil and nitrogen gas.

tube. This mounting contains a rubber isolating bushing, or grommet, to reduce the transmitted road noise and suspension vibration. The rubber bushings are flexible to allow for minimal movement as road forces are transmitted through the assembly. The upper mount is connected to the vehicle frame and the lower mount is typically connected to the lower control arm or axle tube of the suspension system. A piston and rod assembly is connected to the upper half of the shock absorber. This upper portion of the shock absorber has a dust shield that surrounds the lower twin-tube unit. The piston is precision fit in the inner cylinder of the lower unit. A piston rod guide and seal are located in the top of the lower unit. A circular upper mounting with a rubber bushing is attached to the top of the shock absorber. Shock absorber bore size refers to the diameter of the working piston. Generally, the larger the piston diameter, the higher the potential energy and damping control level of the shock absorber because of larger piston displacement and pressure area. As a general rule, the larger the piston area, the lower the internal operating temperature and pressure. To achieve optimal ride characteristics of stability and balance under a wide variety of driving conditions, ride engineers select piston size and valve rates for specific vehicle applications.

Shop Manual
Chapter 5, page 196

The advantages of the twin-tube shock absorber are as follows:

- Combines both comfort and control.
- Adjusts more rapidly than a single tube design shock absorber to changing weight and road conditions.

A disadvantage of the twin-tube design is that it can only be mounted in one direction. The twin-tube shock absorber is the most common original equipment manufacturer (OEM) design used on cars, SUVs, and light trucks.

SHOCK ABSORBER OPERATION

Shock absorbers are usually mounted between the lower control arms and the chassis. When a vehicle wheel strikes a bump, the wheel and suspension move upward in relation to the chassis. Upward wheel movement is referred to as **jounce travel**. This jounce action causes the spring to deflect or compress. Under this condition, the spring stores energy and springs back downward with all the energy absorbed when it deflected upward. This downward spring and wheel action is called **rebound travel**. If this spring action were not controlled, the wheel would strike the road with a strong downward force, and the wheel jounce would occur again. Therefore, some device must be installed to control the spring action, or the wheel would bounce up and down many times after it hit a bump, causing passenger discomfort, directional instability, and suspension component wear along with severe tire cupping.

Shock absorbers are installed on suspension systems to control spring action. When a wheel strikes a bump and jounce travel occurs, the shock absorber lower tube unit is forced upward. This action forces the piston downward in the lower tube unit. Because oil cannot leak past the piston, the amount of resistance a shock absorber develops depends on the piston size and the number of holes passing through it. The oil in the lower unit is forced through the piston orifices or valves to the upper oil chamber (**Figure 5-4**). These valves provide precise oil flow control and control the upward action of the wheel and suspension, which is referred to as a shock absorber compression stroke (**Figure 5-5**). Modern shock absorbers are velocity-sensitive damping devices that increase resistance the faster the suspension moves in effect adjusting themselves to road conditions.

When the spring expands downward in rebound travel, the lower shock absorber unit is also forced downward. When this occurs, the piston moves upward in the lower tube unit, and hydraulic oil is forced through the piston orifices or valves from the upper oil

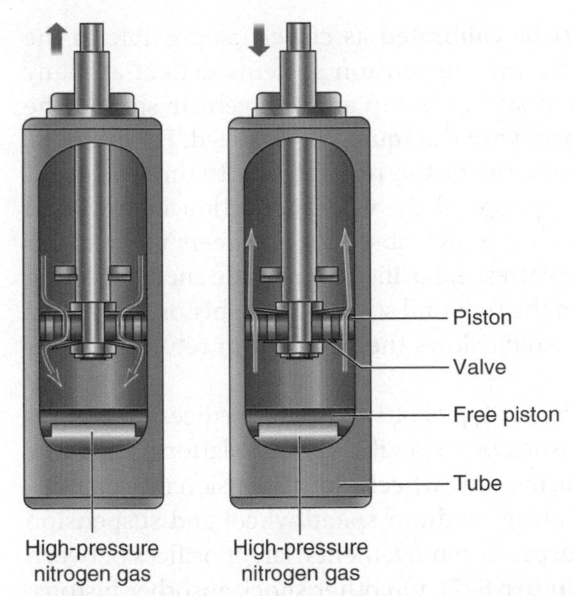

Figure 5-4 As the shock absorber is compressed (jounce) oil flows from the lower chamber to the upper chamber through piston passages and a one-way valve. As the shock absorber expands (rebound), oil flows from the upper chamber to the lower chamber through piston passages and the one-way valve. The ratio of fluid flow through the valved passages determines the shock absorbers rate or ratio.

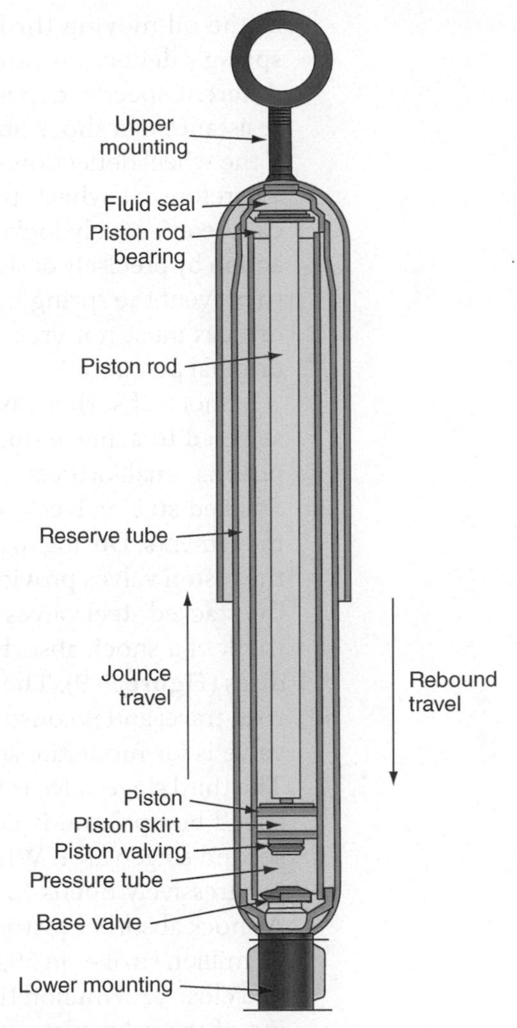

Figure 5-5 Shock absorber action.

Figure 5-6 Fluid flow during the jounce and rebound travel is restricted through valve passages in the working piston and base valve to achieve proper shock rate.

chamber to the lower oil chamber. Because the valves restrict oil flow with precise control, the downward suspension and wheel movement is controlled. Fluid flow during the jounce and rebound travel is restricted through valve passages in the working piston and base valve on a twin-tube design to achieve proper shock rate or ratio (**Figure 5-6**).

When the shock absorber piston moves, oil is forced through the piston. Because the piston valves and orifices resist the flow of oil, friction and heat are created. The resistance of the oil moving through the piston must be calibrated as closely as possible to the spring's deflection rate or strength. Wheels and suspension systems deflect at many different speeds, depending on the type and size of bump and the vehicle speed. The resistance of a shock absorber piston increases with the square of its speed. For example, if the wheel deflection speed increases 4 times, the piston resistance is 16 times as great. Therefore, if a wheel strikes a large bump at high speed, the wheel deflection and rebound can be effectively locked by the shock absorber. Shock absorber engineers prevent this action by precisely designing shock absorber valves and orifices to provide enough friction to prevent the spring from overextending on the rebound stroke. These piston valves and orifices must not create excessive friction, which slows the wheel from returning to its original position.

Shock absorber pistons have many different types of valves and orifices. Valving is selected to achieve optimal ride characteristics of a specific vehicle platform. In some pistons, small orifices control the oil flow during slow wheel and suspension movements. Stacked steel valves control the oil flow during medium speed wheel and suspension movements. During maximum wheel and suspension movements, larger orifices between the piston valves provide oil flow control (**Figure 5-7**). On other shock absorber pistons, the stacked steel valves alone provide oil flow control (**Figure 5-8**). As a simple example, think of a shock absorber with a multistage damping stack of three steel valve plates or discs (**Figure 5-9**). The first stage valve is for light bumps experienced during normal flat road travel and is constantly opening and closing as the vehicle is driven. The second stage valve is for moderate suspension movement that may occur during corning maneuvers. The third stage valve is for extreme suspension movement such as large bumps, pot holes, speed bumps, and it does not open as frequently for each mile driven as the first and second stage valve. When enough force is built up inside the shock absorber each valve progressively opens to allow fluid to bypass the piston seal offering variable damping. A shock absorber piston assembly may make as many as 1500 to 1900 strokes per mile or 75 million strokes in 50,000 mi. (80,467 km), which means that a first stage valve may open and close 37.5 million times in that same distance driven. In this design the constant flexing of the valve plate eventually causes metal fatigue of one or more of the valve plates

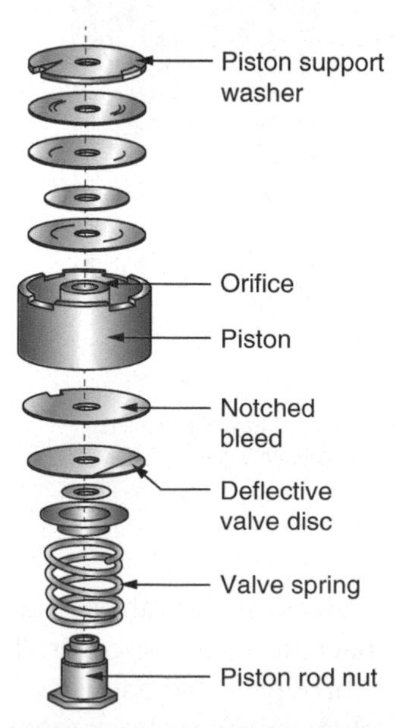

Figure 5-7 A combination of valve discs and orifice passages are used on some shocks to achieve proper shock rate.

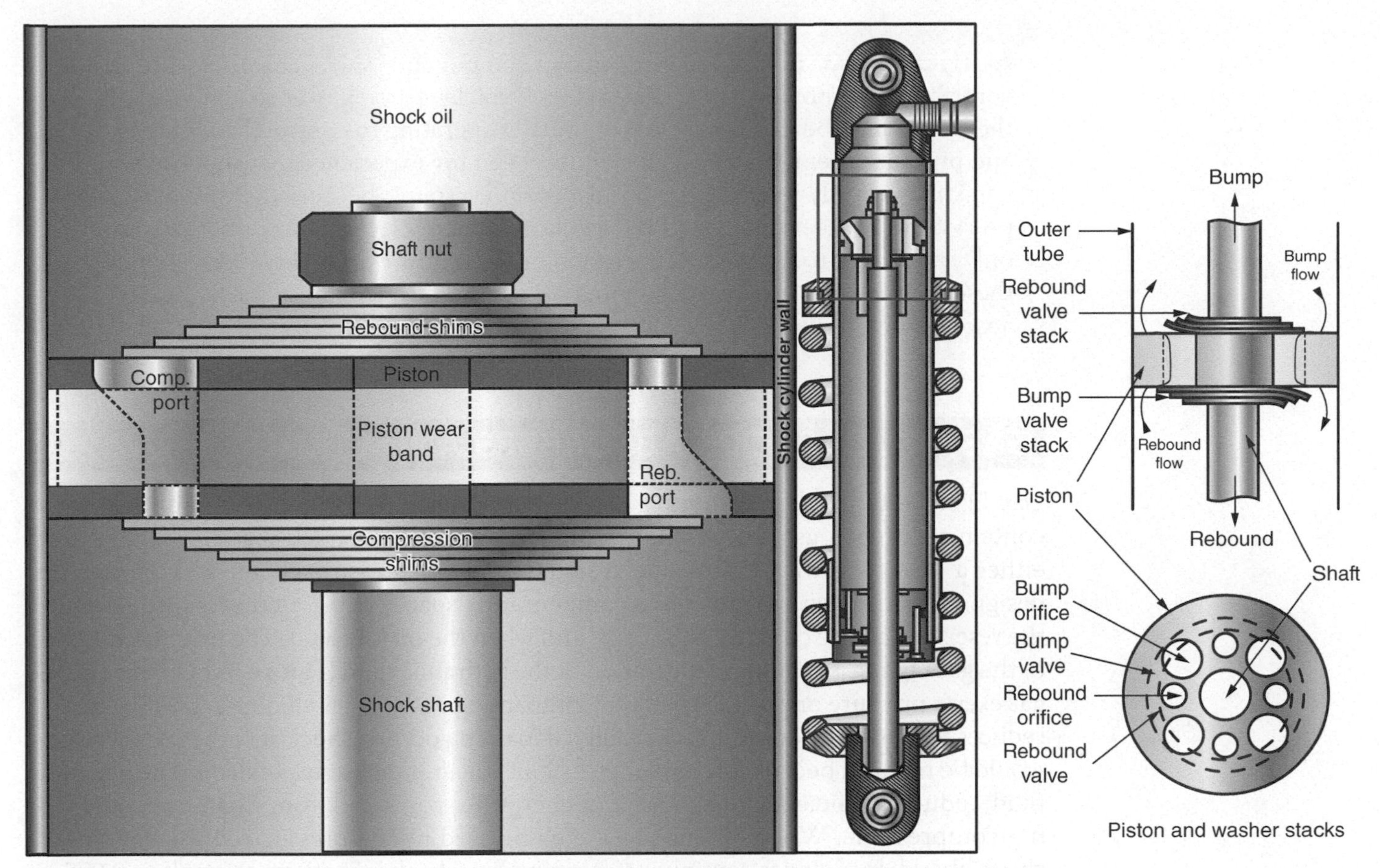

Figure 5-8 Stacked shims on the piston can be used to control fluid flow rates on both the rebound and compression cycles.

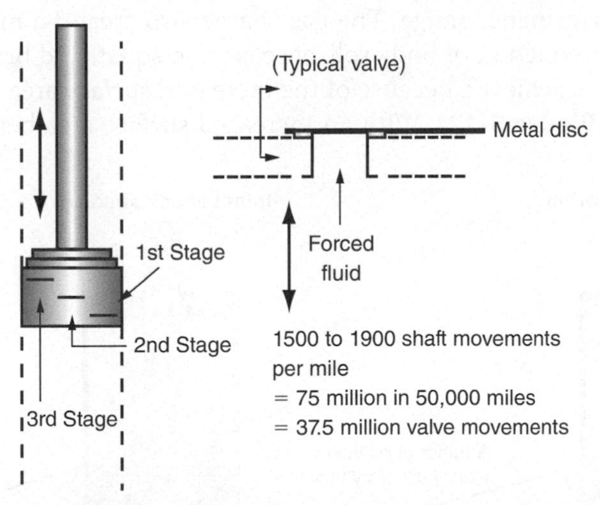

Figure 5-9 A simple stacked steel valve disc used to achieve proper shock damping rate.

effecting damping performance related to the pressure valve that has failed. Regardless of the piston orifice and valve design, the shock absorber must be precisely matched to absorb the spring's energy.

During fast upward wheel movement on the compression stroke, excessive pressure in the lower oil chamber forces the base valve open and thus allows oil to flow through this valve to the reservoir. A rebound rubber is located on top of the piston. If the wheel drops downward into a hole, the shock absorber may become fully extended. Under these conditions, the rebound rubber provides a cushioning action.

AUTHOR'S NOTE Of all the suspension components, shock absorbers and struts contribute the most to ride quality. As the vehicle is driven over road irregularities, the shock absorbers or struts are continually operating to control the spring action and provide acceptable ride quality. It has been my experience that shock absorbers and struts usually wear out first in suspension systems, because they are working every time a wheel strikes a road irregularity. Therefore, you must understand not only shock absorber and strut operation but also how ride quality is affected if these components are not functioning properly. Thus, you must know shock absorber and strut diagnosis and service procedures.

GAS-FILLED SHOCK ABSORBERS AND STRUTS

Gas-filled units are identified with a warning label. Many shock absorbers and struts today contain a nitrogen gas charge (**Figure 5-10**). A gas charge shock absorber may contain either a low-pressure in twin-tube designs or a high-pressure charge in a mono-tube design. The nitrogen gas provides a compensating space for the oil that is displaced into the reservoir on the compression stroke and when the oil is heated. The primary function of the gas charge is to minimize aeration of the hydraulic fluid (**Figure 5-11**). Because the gas exerts pressure on the oil, cavitation, or foaming of the oil, is eliminated, which in turn reduces fade or loss of damping capability. If foaming occurs, shock absorber performance would be reduced because of the compressibility of air bubbles suspended in the hydraulic fluid, reducing efficiency and transfer of energy. As you recall from Chapter 2, fluids are noncompressible. When oil bubbles are eliminated by the pressurized gas charge, the shock absorber provides continuous damping for wheel deflections as small as 0.078 in. (2.0 mm). By eliminating aeration the shock absorber is more responsive and predictable under a broader performance range. The gas charge also creates a mild boost in spring rate assisting in the reduction of body roll, acceleration squat, and brake dive. This mild boost in spring rate is achieved because of the increased surface area below the working piston than above (**Figure 5-12**). With an increased surface area below the piston as a

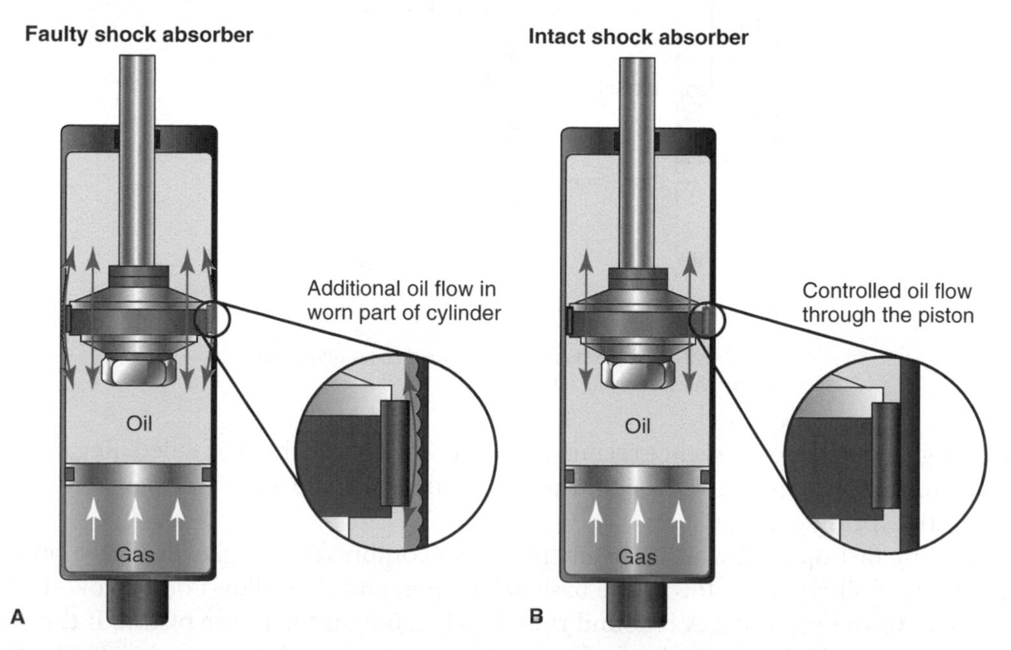

Figure 5-10 The nitrogen gas charge maintains pressure on the oil charge to eliminate aeration. As the cylinder bore and piston seal wear, oil will bypass the piston, reducing shock absorber effectiveness.

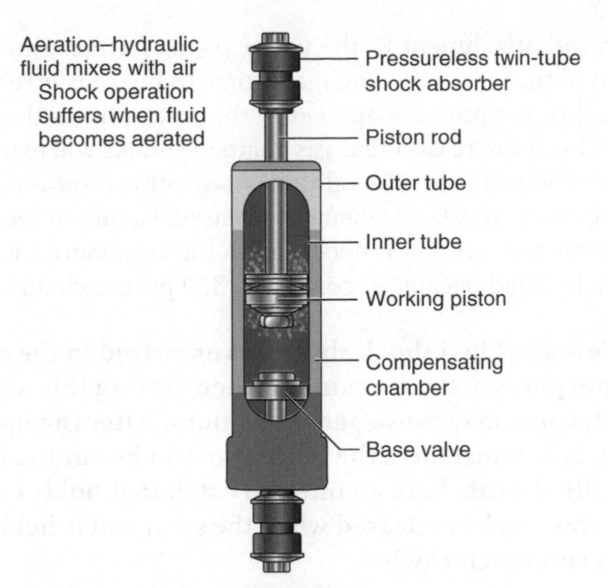

Figure 5-11 Shock absorbers that lose their gas charge or non-gas charged shocks are subject to aeration which decreases the absorbers effectiveness.

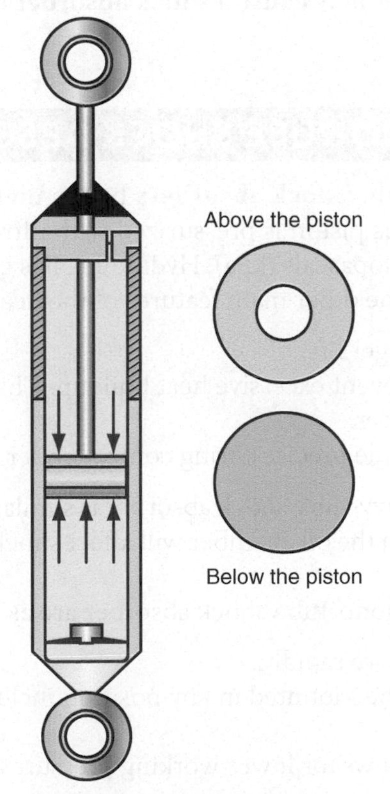

Figure 5-12 The surface area of the shock absorber working piston is greater at the bottom than it is at the top because of the loss of area due to the attachment area of the piston rod. This causes applied force to be greater below the piston than above the piston. Though, this does not take valving and orifices into account.

result of the piston rod attachment to the upper piston surface more pressurized fluid contacts the bottom of the piston causing more force to be transferred. The force applied to the piston is equal to the pressure applied to the piston times the surface area of the piston ($F = P \times A$). This is the reason that gas charged shocks will extend on their own. If a low-pressure (100–150 psi) **gas-filled shock absorber** is removed and compressed to its shortest length, it should reextend when it is released. Failure to reextend indicates that shock absorber or strut replacement is necessary. A high-pressure shock absorber should not be compressible by hand because it contains a 360 psi gas charge.

A **gas-filled shock absorber** contains a nitrogen gas charge to maintain pressure on the oil in the shock.

WARNING **New gas-filled shock absorbers are wired in the compressed position for shipping purposes. Exercise caution when cutting this wire strap because shock absorber extension may cause personal injury. After the upper shock absorber attaching bolt is installed, the wire strap can be cut to allow the unit to extend. Front gas-filled struts have an internal catch that holds them in the compressed position. This catch is released when the strut rod is held and the strut rotated 45 degrees counterclockwise.**

Caution

When drilling worn-out shock absorbers or struts to relieve the gas pressure prior to disposal, drill the shock absorber only at the vehicle manufacturer's specified location.

WARNING **Do not throw gas-filled shock absorbers or struts in the fire or apply excessive heat or flame to these units. These procedures may cause the unit to explode, resulting in personal injury.**

WARNING **Never apply heat to a shock absorber or strut chamber with an acetylene torch. This action may cause a shock absorber or strut explosion resulting in personal injury.**

HEAVY-DUTY MONO-TUBE SHOCK ABSORBER DESIGN

Some **heavy-duty mono-tube shock absorbers** have a dividing piston in the lower oil chamber. The area below this piston is pressurized with nitrogen gas to 360 pounds per square inch (psi), or 2.482 kilopascals (kPa). Hydraulic oil is contained in the oil chamber above the dividing piston. The other main features of the heavy-duty shock absorber are:

Heavy-duty mono-tube shock absorbers have several design features that provide improved durability compared with conventional shock absorbers.

1. High-quality seal for longer life.
2. Single tube design to prevent excessive heat buildup. This design may be called a mono-tube shock absorber.
3. Rising rate valve to provide precise spring control under all conditions.

The operation of the heavy-duty shock absorber is similar to that of the conventional type (**Figure 5-13**). A dent in the oil chamber will affect shock absorber operation on this type of design.

The advantages of the mono-tube shock absorber are as follows:

- Heat may be dissipated more rapidly.
- The shock assembly may be mounted in any position including upside down, reducing unsprung weight.
- Larger piston diameter allows for lower working pressures.

The disadvantages of the mono-tube shock absorber are as follows:

- Piston rod seal is subjected to internal damping pressure.
- Sufficient room must be provided by the suspension system layout to provide for its larger diameter.
- The single wall working chamber is susceptible to damage from rocks and other road debris. A dent will destroy the unit.
- Larger than a twin-tube design.

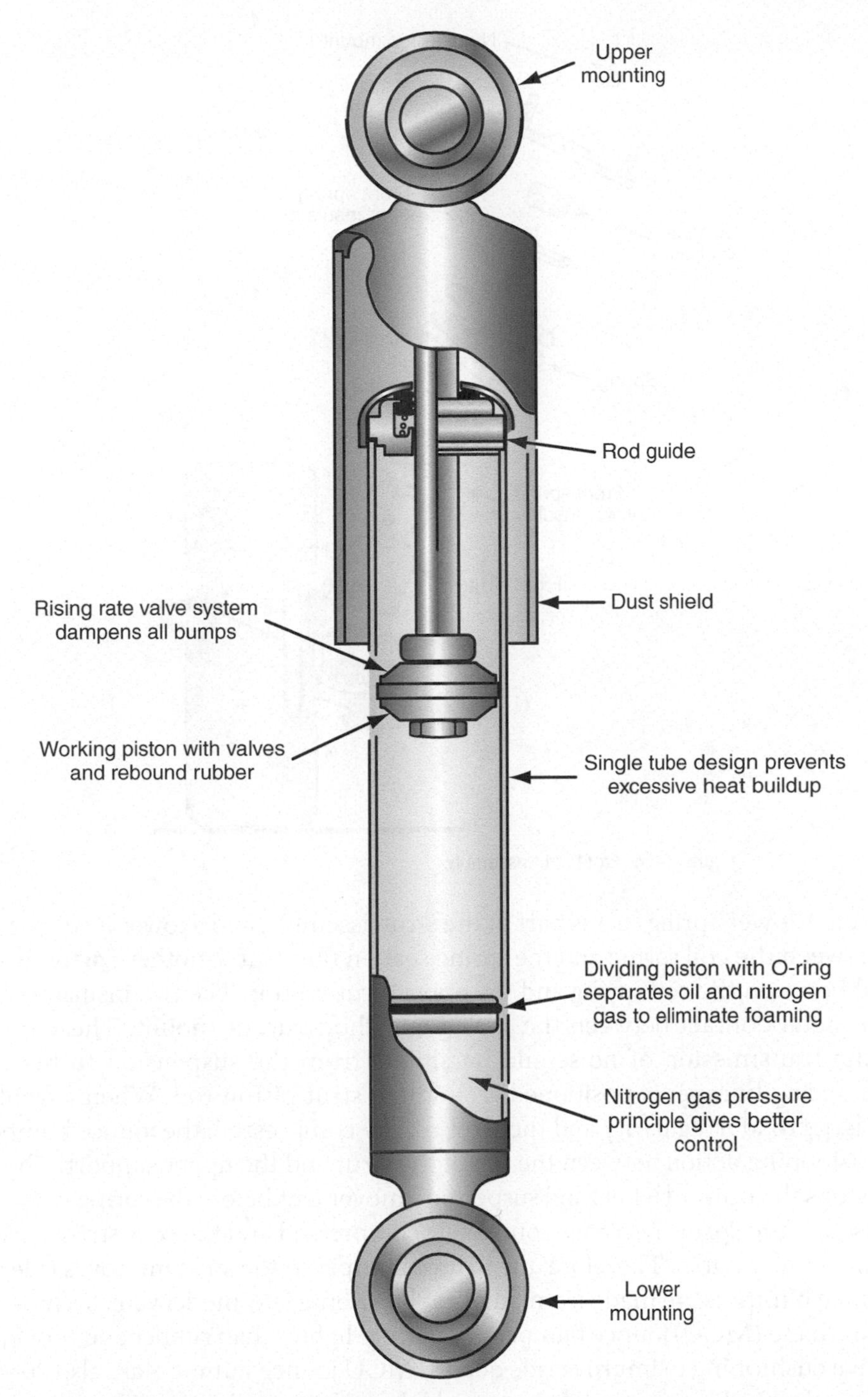
Upper
mounting
Rod guide
Dust shield
Rising rate valve system
dampens all bumps
Working piston with valves
and rebound rubber
Single tube design prevents
excessive heat buildup
Dividing piston with O-ring
separates oil and nitrogen
gas to eliminate foaming
Nitrogen gas pressure
principle gives better
control
Lower
mounting

Figure 5-13 Heavy-duty shock absorber.

STRUT DESIGN, FRONT SUSPENSION

A strut-type front suspension is used on most front-wheel-drive cars and some rear-wheel-drive cars. Internal strut hydraulic design is essentially the same as a shock absorber design, and the **struts** perform the same functions as shock absorbers damping spring oscillations. Some struts have a replaceable cartridge. In many strut-type suspension systems, the coil spring is mounted on the strut. The coil spring is largely responsible for proper curb riding height. A weak or broken coil spring reduces curb riding height and provides harsh riding. The lower end of the front suspension strut is bolted to the steering knuckle (**Figure 5-14**). An **upper strut mount** is attached to the strut, and this mount is bolted into the chassis

Struts are similar to shock absorbers, but struts are usually positioned between the knuckle and the chassis to provide knuckle support.

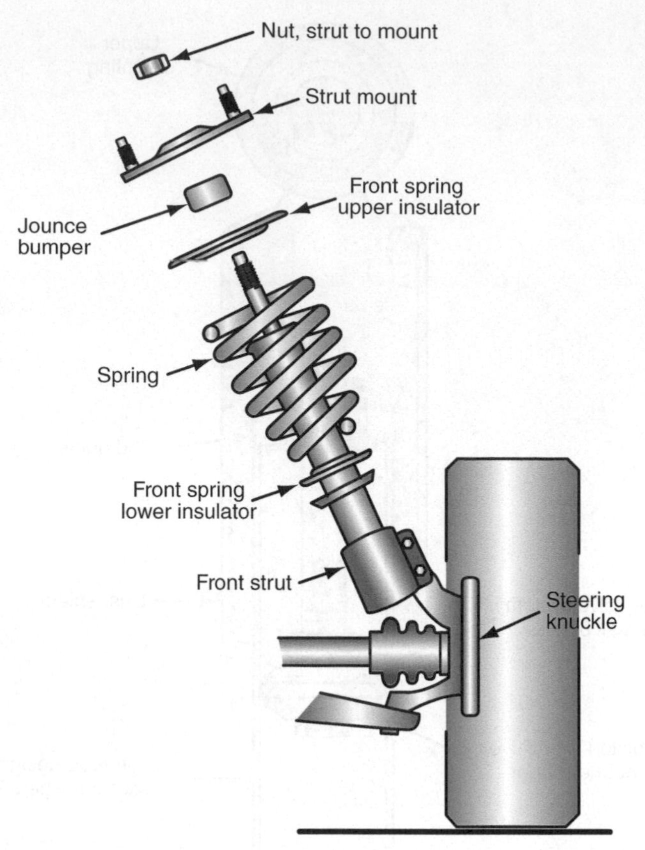

Figure 5-14 Front strut assembly.

strut tower. A lower spring seat is part of the strut assembly, and a lower insulator is positioned between the coil spring and the spring seat on the strut. Another **spring insulator** is located between the coil spring and the upper strut mount. The two insulators prevent metal-to-metal contact between the spring and the strut, or mount. These insulators reduce the transmission of noise and harshness from the suspension to the chassis. A rubber spring bumper is positioned around the strut piston rod. When a front wheel strikes a large road irregularity and the strut is fully compressed, the jounce bumper provides a cushioning action between the top of the strut and the upper support. The jounce bumper stops the upward wheel and suspension movement before the spring is completely compressed. If the spring becomes completely compressed and the coils strike each other, ride quality is very harsh. Therefore, the jounce bumper in the strut improves ride quality. Most jounce bumpers are made from butyl rubber. Some late-model vehicles have microcellular urethane (MCU) jounce bumpers, which are lighter than rubber and provide more progressive cushioning to improve ride quality. MCU jounce bumpers are also 20–40 percent lighter than rubber jounce bumpers, which reduces road noise transmission to the passenger compartment. In relation to temperature changes, MCU jounce bumpers remain more stable and provide improved ride quality regardless of the temperature. The upper strut mount contains a bearing, upper spring seat, and jounce bumper (**Figure 5-15**).

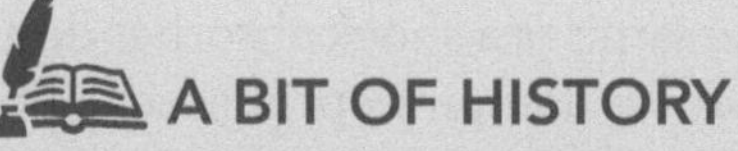

A BIT OF HISTORY

For many years, rear-wheel-drive cars were equipped with front and rear shock absorbers. Most front-wheel-drive cars are equipped with front struts, and some of these cars also have rear struts. Because massive numbers of front-wheel-drive cars have been introduced in the 1980s and 1990s, front and rear struts are now very common.

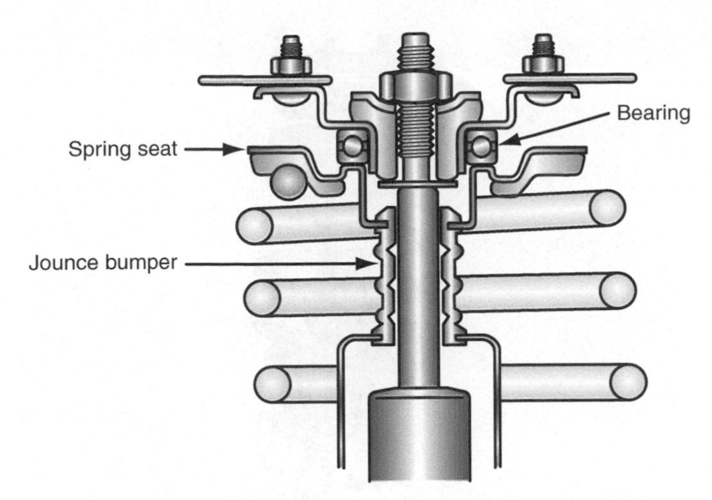

Figure 5-15 Upper strut mount.

When the front wheels are turned, the front strut and coil spring rotate with the steering knuckle. The strut-and-spring assembly rotates on the upper strut mount bearing.

Some cars have a multilink front suspension with an upper link connected from the chassis to the steering knuckle. The strut is connected from the upper link to the strut tower (**Figure 5-16**). A bearing is mounted between the upper link and the steering knuckle, and the wheel and knuckle turn on this bearing and the lower ball joint. Therefore, the coil spring and strut do not turn when the front wheels are turned, and a bearing in the upper strut mount is not required (**Figure 5-17**).

When strut replacement is necessary on many vehicles, complete strut assemblies are available that include a premium strut, the coil spring, upper strut mount and bearing plate (if equipped), upper spring seat, protective boot, and spring insulators in a fully-assembled unit ready for installation in the vehicle (**Figure 5-18**). These assemblies offer several benefits over strut servicing, which include greater safety (since there is no need to compress the coil spring), and ease of installation (since now special spring compressing tools are required). Monroe calls their assembly a Quick-Strut and Gabriel refers to theirs as ReadyMount. In addition to the previously mentioned benefits, these assemblies greatly

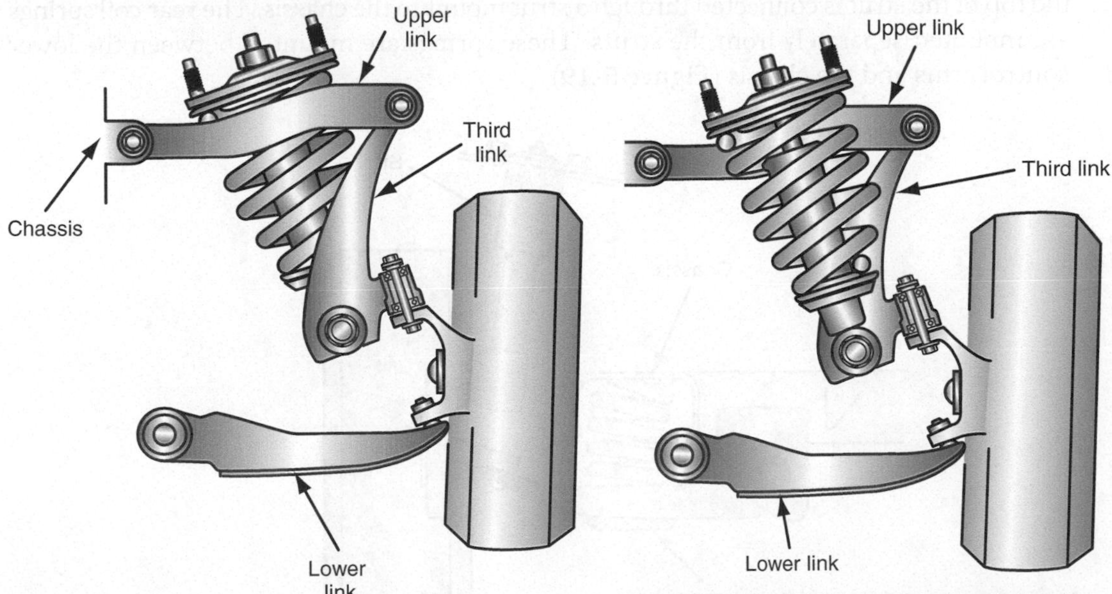

Figure 5-16 Multilink front suspension with strut connected between the upper link and the strut tower.

Figure 5-17 Multilink front suspension with knuckle and wheel pivots at the upper link bearing and lower ball joint, and a nonrotating upper strut mount.

Figure 5-18 Several aftermarket strut manufactures offer complete replacement strut service assemblies that are ready to be installed without the need for specialty tools.

reduce replacement time and are a cost-effective choice for consumers when mounts and springs show signs of wear. While the complete assembly is more expensive than a strut or cartridge alone, the labor savings achieved by eliminating the labor charge for disassembly and reassembly brings the cost to a comparable level with the added benefits of all-new complete assembly components.

Shop Manual
Chapter 5, page 200

SHOCK ABSORBER AND STRUT DESIGN, REAR SUSPENSION

In some rear suspension systems, the lower end of the strut is bolted to the spindle, and the top of the strut is connected through a strut mount to the chassis. The rear coil springs are mounted separately from the struts. These springs are mounted between the lower control arms and the chassis (**Figure 5-19**).

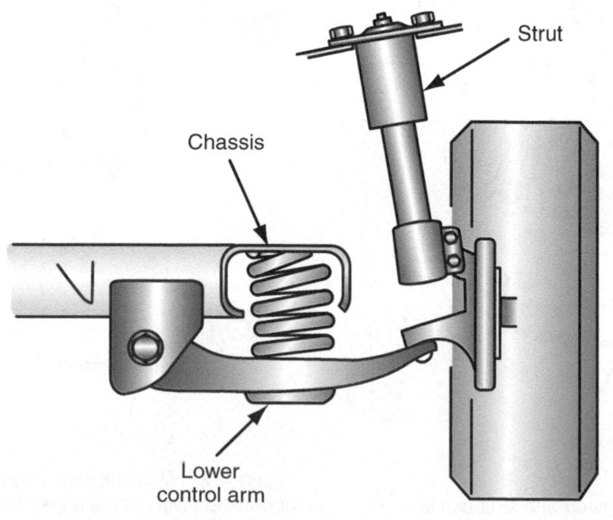

Figure 5-19 Rear suspension system with coil springs mounted separately from the struts.

In other rear suspension systems, the coil springs are mounted on the rear struts (**Figure 5-20**). An upper insulator is positioned between the top of the spring and the upper spring support, and a lower insulator is located between the bottom of the spring and the spring mount on the strut. A rubber spring bumper is positioned on the strut piston rod. If a rear wheel strikes a severe road irregularity and the strut is fully compressed, the spring bumper provides a cushioning action between the top of the strut and the upper support.

Shop Manual
Chapter 5, page 208

On some rear-wheel-drive light-duty trucks and sport utility vehicles (SUVs), the rear shock absorbers are slanted rearward and inward (**Figure 5-21**). On other light-duty trucks and SUVs, the rear shock absorbers are staggered on each side of the rear axle (**Figure 5-22**). Either of these shock absorber mountings improves ride quality and reduces noise, vibration, and harshness (NVH).

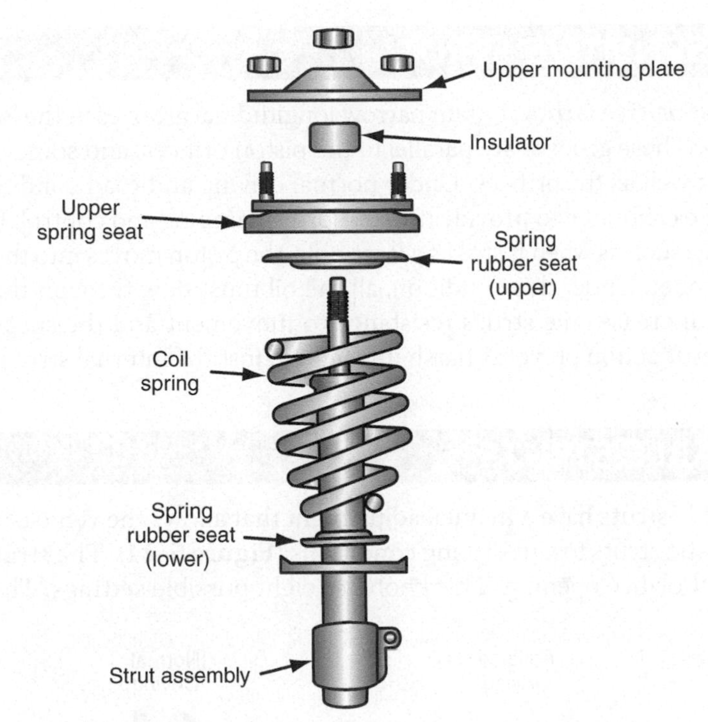

Figure 5-20 Rear suspension system with coil springs mounted on the struts.

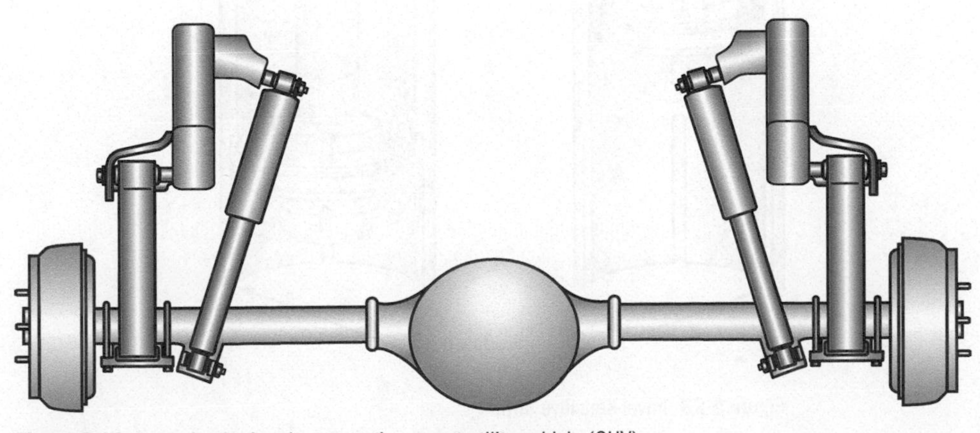

Figure 5-21 Rear shock absorber mounting, sport utility vehicle (SUV).

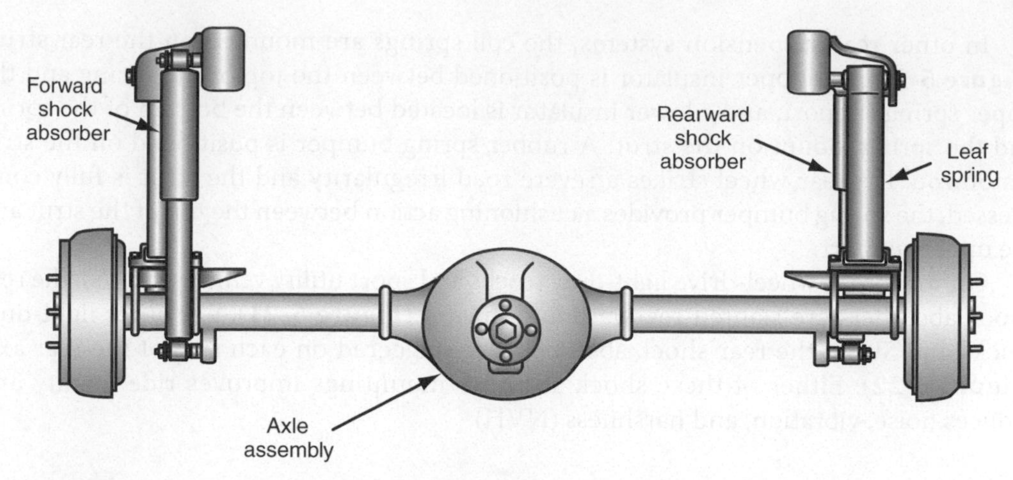

Figure 5-22 Shock absorbers staggered on each side of the rear axle.

TRAVEL-SENSITIVE STRUT

Travel-sensitive struts vary the amount of strut control in relation to strut travel.

Some **travel-sensitive struts** contain narrow longitudinal grooves in the lower oil chamber (**Figure 5-23**). These grooves are parallel to the piston orifices, and some oil flows through the grooves as well as the orifices. Under normal driving and road conditions, the orifices and grooves are calibrated to provide normal spring damping and control. If the front wheel drops suddenly, such as when it strikes a large hole, the piston moves into the narrow portion of the oil chamber. Under this condition, all the oil must flow through the piston orifices, which greatly increases the strut's resistance to movement and the suspension damping action. This strut action prevents harsh impacts against the internal strut rebound rubber.

ADJUSTABLE STRUTS

Some adjustable struts have a manual adjustment that allows the vehicle owner or technician to adjust the struts to suit driving conditions (**Figure 5-24**). The strut adjusting knob varies the strut orifice opening. This knob has eight possible settings. The factory setting

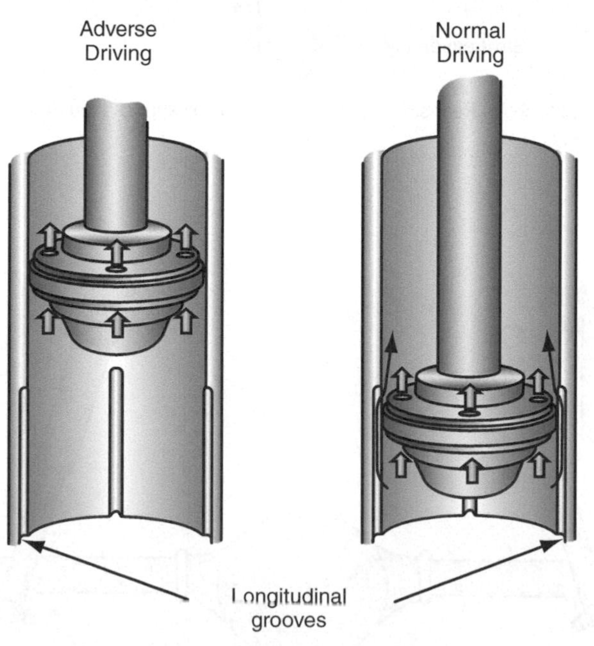

Figure 5-23 Travel-sensitive strut.

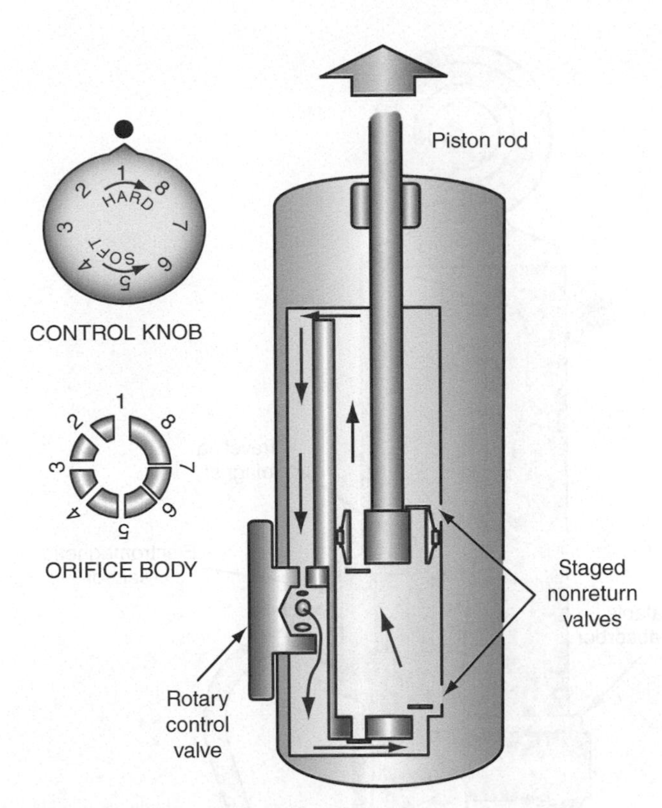

Figure 5-24 Manual adjustable strut.

is No. 3, which provides average suspension control. The No. 1 setting provides reduced spring control and the softest ride, whereas a No. 8 adjustment gives increased spring control and the hardest ride. The adjustment knob is usually accessible without raising the vehicle.

LOAD-LEVELING SHOCK ABSORBERS

Load-leveling rear shock absorbers or struts are used with an electronic height control system. An onboard air compressor pumps air into the rear shocks to raise the rear of the vehicle, and an electric solenoid releases air from the shocks to lower the rear chassis. An electromagnetic height sensor may be contained in the shock absorber, or an external sensor may be used (**Figure 5-25**). This sensor sends a signal to an electronic control module in relation to the rear suspension height. The module controls the air compressor and the exhaust solenoid to control air pressure in the shock absorbers. This action maintains a specific rear suspension trim height regardless of the load on the rear suspension. If a heavy package is placed in the trunk, the vehicle chassis is forced downward. However, the **load-leveling shock absorbers** extend to restore the original rear suspension height.

Load-leveling shock absorbers use air pressure supplied to the shock absorbers to maintain rear suspension height. Load-leveling shock absorbers may be called air shocks.

Aftermarket air shock absorbers are available. These shock absorbers contain an air valve connection. A shop air hose may be used to supply the desired pressure in these air shock absorbers.

Aftermarket spring-assisted shock absorbers are also available. These are conventional shock absorbers with a small coil spring mounted over them. Upper and lower spring seats are attached near the top and bottom of these shock absorbers to support and retain the spring. The coil springs on the shock absorbers help the springs in the suspension system support the vehicle weight.

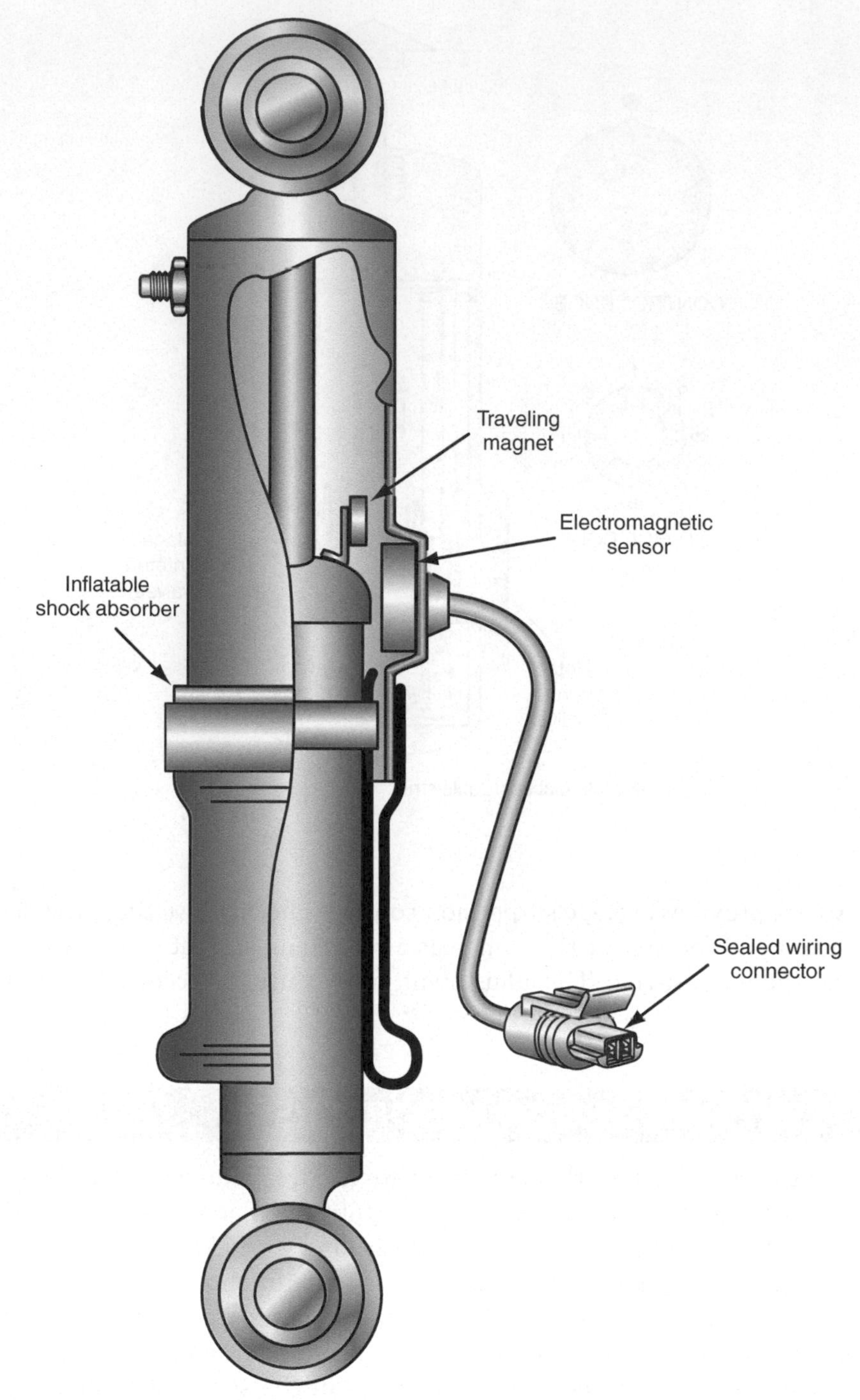

Figure 5-25 Load-leveling strut.

ELECTRONICALLY CONTROLLED SHOCK ABSORBERS AND STRUTS

Computer controlled suspension systems may be referred to as **active suspension systems.**

Many cars are equipped with computer-controlled **active suspension systems**. In these systems, a computer-controlled actuator is positioned in the top of each shock absorber or strut (**Figure 5-26**). The shock absorber or strut actuators rotate a shaft inside the piston rod, and this shaft is connected to the shock valve. Many of these systems have two modes: soft and firm. In the soft mode, the actuators position the shock absorber valves so there is less restriction to the movement of oil. When the computer changes the

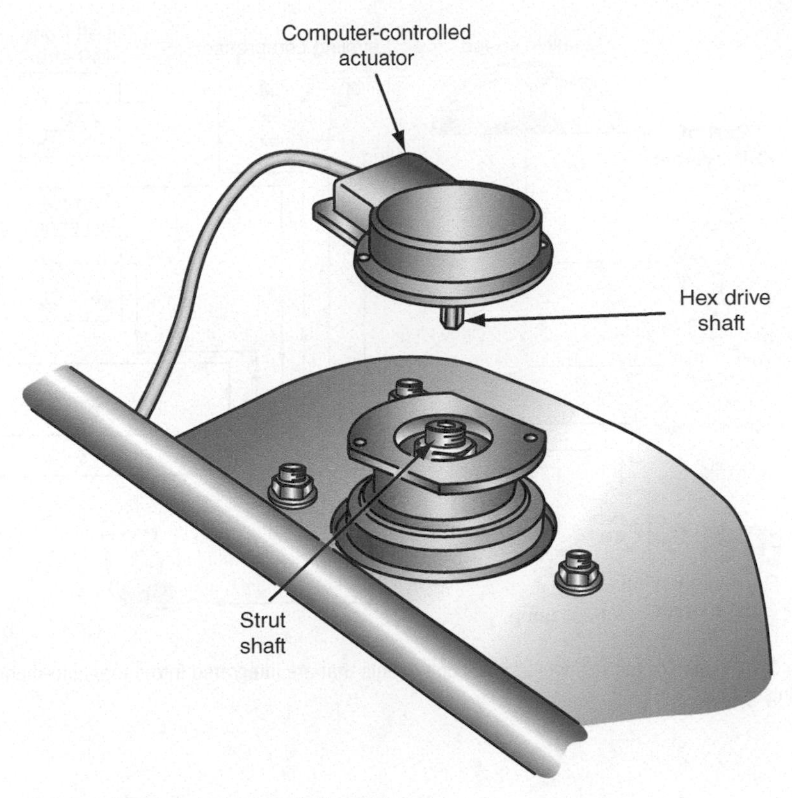

Figure 5-26 Computer-controlled strut actuator.

actuators to the firm mode, the actuators position the shock valves so they provide more restriction to oil movement, which provides a firmer ride.

Some electronically controlled shock absorbers and struts contain a synthetic **magneto-rheological fluid** that contains numerous small, suspended metal particles.

Magneto-rheological fluid A fluid that when subjected to a magnetic field will increase viscosity as the magnetic field is increased, to the point of becoming an elastic solid.

Delphi's trademarked name for the system is called MagneRide. There are no small moving parts or electromechanical valves to fail. The electro-rheological fluid viscosity changes when electrical current passes through the fluid itself, which contains magnetically soft particles of iron microspheres in a synthetic hydrocarbon-based fluid. Each shock absorber contains an electromagnetic coil winding inside the piston itself that is energized by the suspension computer (**Figure 5-27**). Only the fluid in the oil passage of the piston is involved in the viscosity change. The wires to the coil are routed through the piston rod to a connector at the top of the shock housing. If the shock absorber winding is not energized, the metal particles in the fluid align randomly in the fluid. Under this condition, the fluid has a mineral-oil-like consistency and the fluid moves easily through the shock absorber orifices. If the suspension computer energizes a shock absorber coil winding, the metal particles in the fluid are aligned into fibrous structures. When this occurs, the fluid has a jelly-like consistency for a firm ride (**Figure 5-28**). The computer can change the shock absorber damping characteristics almost instantaneously, in 1 millisecond, and can also supply a varying amount of current through the shock absorber windings to provide a wide variety of shock absorber damping characteristics. Depending on the amount of current supplied, the fluid's viscosity or resistance to flow can be varied from thinner than water to a near plastic or solid state or any consistency in-between. This makes the fluid infinitely adjustable in viscosity and achieves continuously variable real-time damping to suit almost every driving condition instantly.

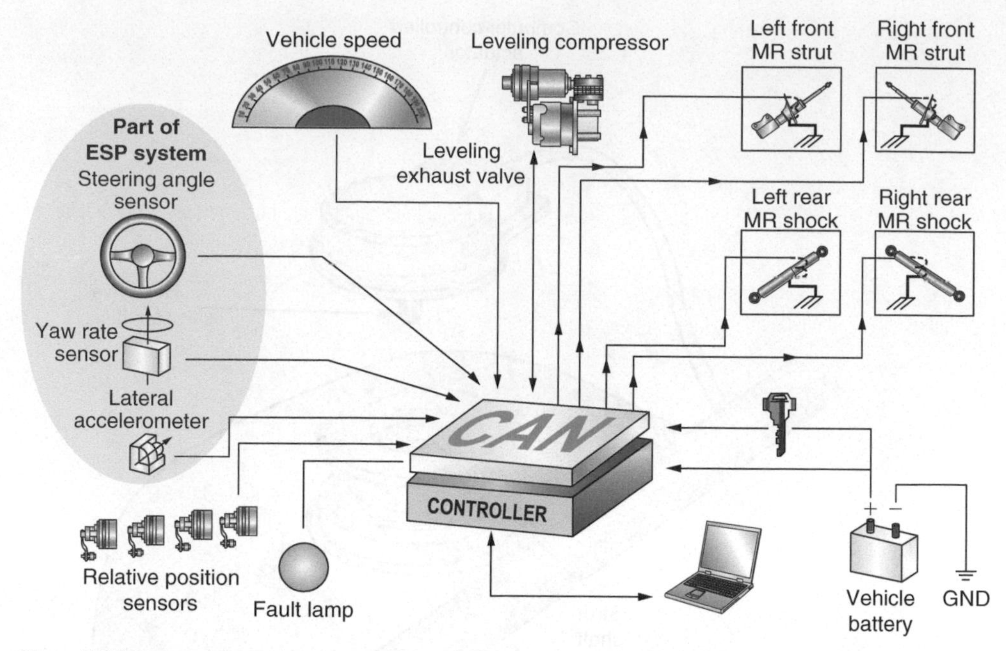

Figure 5-27 Some of the major electrical components that are integrated into a magneto-rheological damping system.

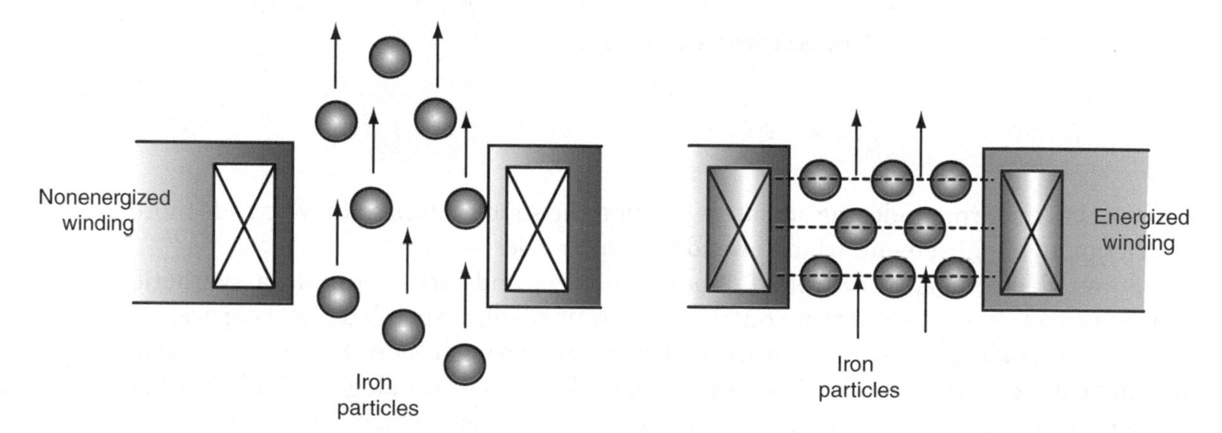

Figure 5-28 Magneto-rheological fluid action in a strut or a shock absorber.

The advantages of magneto-rheological-fluid-controlled shock absorber are as follows:

- Ability to smooth out the action of each tire.
- Reduced noise, bounce, and vibration giving a flatter ride by controlling body motions.
- Exceptional roll control during evasive steering maneuvers.
- Excellent handling by controlling weight transfer during lateral and longitudinal maneuvers. Enhanced road isolation by reducing high-frequency road noise through the dampers.
- When integrated with ABS and traction control, ensures maximum stability and balance on slippery surfaces and gravel.

SUMMARY

- Shock absorbers or struts play a very important role in ride quality, steering control, and tire life.
- Tire and wheel jounce travel occurs when a tire strikes a bump in the road surface and the tire and wheel move upward.
- Rebound tire and wheel travel occurs when the tire and wheel move downward after jounce travel.
- When a spring is deflected upward during jounce travel, it stores energy. The spring then expands downward in rebound travel with all the energy it stored during the jounce travel. If the spring action is not controlled, the energy in the spring during rebound travel drives the tire against the road surface with excessive force. This action drives the tire and wheel back upward in jounce travel and the wheel continues oscillating up and down.
- The shock absorbers control spring action and prevent excessive tire and wheel oscillations.
- During jounce travel, the piston moves downward in the lower shock absorber chamber; during rebound travel, this piston moves upward. Because the lower shock absorber chamber is sealed and filled with oil, this oil must flow past the piston during any piston movement.
- Valves and openings in the shock absorber piston provide precision control of the oil flow past the piston to control spring action. Shock absorber valves are matched to the amount of energy that may be stored in the spring.
- A nitrogen gas charge is located in the oil reservoir of many shock absorbers and struts to prevent oil cavitation or foaming, which provides more positive shock absorber action.
- Shock absorber ratio refers to the difference between the shock absorber control on the compression and extension cycle. Many shock absorbers provide more control on the extension cycle.
- Internal design is similar in shock absorbers and struts, but struts also support the coil spring.
- Most front struts are connected between the steering knuckle and the upper strut mount.
- Many rear struts are connected between the spindle and the upper support. A travel-sensitive shock absorber provides increased resistance to piston movement as the shock absorber is extended.
- Adjustable shock absorbers and struts have a manual adjustment that allows the technician or owner to adjust the strut orifice opening.
- Load-leveling shock absorbers have air pumped into the shock absorbers from an onboard compressor to maintain a specific rear suspension height regardless of the rear suspension load.
- Most front struts rotate on the upper strut mount bearing as the front wheels are turned.

REVIEW QUESTIONS

Short Answer Essays

1. Describe spring action during jounce and rebound travel.
2. Describe uncontrolled spring action without a shock absorber.
3. Explain shock absorber operation.
4. Describe the vehicle safety hazards created by worn-out shock absorbers.
5. Explain shock absorber ratio.
6. Explain the purpose of the nitrogen gas charge in shock absorbers and struts.
7. Explain the differences between heavy-duty mono-tube and conventional twin-tube shock absorbers.
8. Identify the purposes of an upper strut mount.
9. Explain the purpose of strut-and-spring insulators.
10. Describe the purpose of the rubber spring bumper on a strut piston rod.

Fill-in-the-Blanks

1. Shock absorbers control spring action and prevent excessive spring ________________.
2. Shock absorber design is matched to the ________________ rate of the spring.
3. Modern shock absorbers are ________________ ________________ damping devices that increase resistance the faster the suspension moves.

4. The nitrogen gas charge in a shock absorber prevents oil ______________ and ______________.
5. The single tube design in a heavy-duty shock absorber prevents excessive______________ ______________.
6. A shock absorber with a 70/30 ratio provides more control on the ______________ cycle.
7. The lower end of many front struts is bolted to the steering ______________.
8. When a higher number is selected on the adjustment knob of an adjustable strut, the strut provides a ______________ ride.
9. Load-leveling shock absorbers may have an internal ______________ ______________.
10. Travel-sensitive shock absorbers provide increased spring control when the shock absorber is ______________.

Multiple Choice

1. The main purpose of shock absorbers and struts is to:
   A. Control spring action.
   B. Prevent fore-and-aft wheel movement.
   C. Reduce lateral wheel movement.
   D. Prevent wheel shimmy.
2. All of these statements about shock absorber design and operation are true EXCEPT:
   A. The oil flow through the orifices and valves is matched to the spring's strength and deflection rate.
   B. A typical shock absorber may have 70 percent of the total control on the extension cycle.
   C. During jounce wheel travel the piston moves upward in the shock absorber lower tube unit.
   D. A nitrogen gas charge in a shock absorber prevents oil foaming.
3. Compared to a conventional shock absorber, a heavy-duty shock absorber has:
   A. Triple layers of steel in the lower tube unit.
   B. Higher viscosity oil.
   C. A larger diameter piston rod.
   D. A high-quality seal for longer life.
4. All of these statements about gas-filled shock absorbers are true EXCEPT:
   A. A hydrogen gas charge is used in most shock absorbers and struts.
   B. The gas charge creates a mild boost in spring rate.
   C. The primary function of the gas charge is to minimize aeration of the hydraulic fluid.
   D. A gas-filled shock absorber provides continuous damping action with very little wheel movement.
5. The magneto-rheological fluid used in some shock absorbers and struts contains:
   A. Transmission fluid.
   B. Small, suspended metal particles.
   C. Antifoaming agents.
   D. A dye for lead detection purposes.
6. During normal strut operation:
   A. Downward wheel movement is called wheel jounce.
   B. The strut prevents excessive wheel oscillations.
   C. When the wheel moves upward the strut piston also moves upward.
   D. The strut prevents wheel shimmy.
7. Travel-sensitive struts:
   A. Have an adjustment on the outside of the strut.
   B. Provide more resistance to oil movement when compressed.
   C. Provide more resistance to oil movement when driving on a smooth road surface.
   D. Bypass some of the oil past the piston when compressed.
8. Adjustable struts:
   A. May be adjusted with a control knob in the instrument panel.
   B. Provide five different settings.
   C. Provide softest ride with a No. 1 setting.
   D. Rotate the upper part of the strut to provide adjustable ride quality.

9. All of the following are benefits of replace faulty struts with a complete preassembled replacement strut assembly EXCEPT:
   A. Quicker, no need to disassemble existing strut.
   B. Safer, no exposure to dangers of compressing coil spring.
   C. Easier, reduce replacement time.
   D. Lower cost than strut assembly alone.

10. Electronically controlled shock absorbers and struts:
    A. Contain an electronically controlled actuator in the top of each strut.
    B. Usually have 12 different modes of strut operation from soft to very firm.
    C. Have an electronically controlled actuator that rotates the strut piston rod.
    D. Have an actuator that positions the strut valves to provide more oil restriction in the soft mode.

CHAPTER 6
FRONT SUSPENSION SYSTEMS

Upon completion and review of this chapter, you should be able to understand and describe:

- The causes of coil spring failure.
- The design and spring rate of a linear-rate coil spring.
- The difference between a regular-duty coil spring and a heavy-duty coil spring.
- The functions of three different types of coils in a variable-rate coil spring.
- Basic torsion bar action as a front wheel strikes a road irregularity.
- How friction and noise problems are reduced in a multiple-leaf spring.
- The advantages of a front suspension system with ball joints compared to earlier I-beam front suspension systems.
- Two types of load-carrying ball joints, and explain the location of the control arm in each type.
- How to recognize a worn ball joint from a visual inspection of the ball joint wear indicator.
- The mounting location and purpose of a stabilizer bar.
- The purpose of a strut rod.
- The two steering knuckle pivot points in a MacPherson strut front suspension system.
- The advantage of a short-and-long arm front suspension system compared with a suspension system with equal-length upper and lower control arms.
- Design of the twin I-beam front suspension.
- Vehicle ride height.
- The effect of sagged front springs on caster angle and directional stability.
- The effect of sagged front springs on camber angle.
- The advantages of hydraulic control arm bushings.
- The advantages of aluminum control arms.

Terms To Know

Compression loaded
Full-wire open-end springs
Gussets
Heavy-duty coil springs
Inactive coils
Linear-rate coil springs
Load-carrying ball joint
MacPherson strut front suspension systems
Mono-leaf springs
Multilink front suspension
Multiple-leaf springs
Non-load-carrying ball joint
Pigtail spring ends
Regular-duty coil springs
Short-and-long arm front suspension systems
Stabilizer bar
Taper-wire closed-end
Tension loaded
Transitional coils
Unsprung weight
Variable-rate coil springs
Wear indicators

INTRODUCTION

The front and rear suspension systems must perform several extremely important functions to maintain vehicle safety and owner satisfaction. The suspension system must provide steering control for the driver under all road conditions. Vehicle owners expect the suspension system to provide a comfortable ride. The suspension, together with the frame, must maintain proper vehicle tracking and directional stability. Another important purpose of the suspension system is to provide proper wheel alignment and minimize tire wear.

The impact of the front tires striking road irregularities must be absorbed and dissipated by the front suspension system. These impacts are distributed throughout the suspension, and this action isolates the vehicle passengers from road shock. The vehicle's ride characteristics are determined by the amount of impact energy that the suspension can absorb and by the rate at which the suspension dissipates these tire impacts. Ride characteristics are designed into the suspension system, and these characteristics are not adjustable. For example, the front suspension may be designed to provide a very soft ride in which all the tire impacts are absorbed and dissipated quickly by the suspension system. This type of suspension system provides a very comfortable ride, but it also allows excessive body lean during cornering, which reduces high-speed cornering and handling capabilities. Vehicles such as sports cars and sport utility vehicles (SUVs) are usually designed with a suspension system that provides a firm ride and absorbs and dissipates tire impacts more slowly. Although high-speed cornering and handling capabilities are improved, this type of suspension may transfer some road shock to the passenger compartment.

When the vehicle is driven over road irregularities, the front suspension system must allow the front wheels to move vertically while maintaining the tire's proper horizontal position in relation to the road surface. To provide this wheel action, the steering knuckle must be mounted between the upper and lower control arms on short-and-long arm (SLA) suspension systems or between the lower control arm and the lower end of the strut on MacPherson strut suspension systems. The upper and lower control arms must be pivoted on the inner ends; the steering knuckle pivots on the ball joints in SLA suspensions or on the lower ball joint and upper strut mount on MacPherson strut suspensions.

Springs absorb much of the shock from tire impacts with the road surface. When a front wheel strikes a road irregularity and moves upward, the coil spring compresses and absorbs energy during this movement. The coil spring immediately dissipates this energy as the spring moves back to its original state. Shock absorbers are installed in the front suspension system to dampen the oscillations of the coil springs.

A stabilizer bar is mounted on rubber insulating bushings and bolted to the chassis. The outer ends of the stabilizer bar are attached through links to the lower control arms. The stabilizer bar controls the amount of independent lower control arm movement. When one front wheel strikes a road irregularity and moves upward, the stabilizer bar transfers part of the lower control arm movement to the opposite lower control arm, which reduces and stabilizes body roll. Therefore, the stabilizer bar helps to define the suspension characteristics related to body roll.

All suspension components, including the frame and chassis, are designed or "tuned" to provide the ride and handling qualities that the manufacturer believes the average driver will desire. For example, the late-model Explorer Sport Trac frame has been stiffened 40 percent more than the Explorer frame. To achieve this increased stiffness, the frame side rails are thicker than the ones on the Explorer, and a new tubular crossmember has been added to the frame. **Gussets** have also been welded into the corners where the crossmembers meet the side rails. This frame design is matched to the torsion bar suspension to provide improved vehicle agility on and off the road.

Gussets are pieces of metal welded into the corner where two pieces of metal are attached together, which provide increased strength and stiffness.

SUSPENSION SYSTEM COMPONENTS

Coil Springs

The coil spring is the most commonly used spring for front and rear suspension systems. Coil springs are actually a coiled-spring steel bar. When a vehicle wheel strikes a road irregularity, the coil spring compresses to absorb shock, and then recoils back to its original installed height. Many coil springs contain a steel alloy that contains different types of steel mixed with other elements such as silicon or chromium. Coil springs may be manufactured by a cold or hot coiling process. The hot coiling process includes procedures for tempering and hardening the steel alloy. Coil springs are designed to carry heavy loads, but they must be light in weight. Many coil springs have a vinyl coating, which increases corrosion resistance and reduces noise.

Shop Manual
Chapter 6, page 230

Coil spring failures may be caused by these conditions:

1. Constant overloading
2. Continual jounce and rebound action
3. Metal fatigue
4. A crack or nick on the surface layer or coating

Coil springs do not have much ability to resist lateral movement. However, when coil springs are used on the drive wheels, the suspension usually has special bars to prevent lateral movement.

Linear-Rate Coil Springs. Coil springs are classified into two general categories: linear rate and variable rate. **Linear-rate coil springs** have equal spacing between the coils and one basic shape with a consistent wire diameter. When the load is increased on a linear-rate spring, the spring compresses and the coils twist or deflect. As the load is removed from the spring, the coils unwind, or flex, back to their original position. The spring rate is the load required to deflect the spring 1 in. Linear-rate coil springs have a constant spring rate, regardless of the load. For example, if 200 lb deflect the spring 1 in., 400 lb deflect the spring 2 in. The spring rate on linear springs is usually calculated between 20 and 60 percent of the total spring deflection.

Inactive coils are positioned at each end of a coil spring.

Transitional coils are positioned between the inactive coils and the active coils in the center of the spring.

The **active coils** are located in the center of a coil spring.

Variable-Rate Coil Springs. **Variable-rate coil springs** have a variety of wire sizes and shapes. The most common variable-rate coil springs have a consistent wire diameter with a cylindrical shape and unequally spaced coils (**Figure 6-1**).

The **inactive coils** at the end of the spring introduce force into the spring when the wheel strikes a road irregularity. When the **transitional coils** are compressed to their point of maximum load-carrying capacity, these coils become inactive. The **active coils** operate during the complete range of spring loading. When a stationary load is applied to

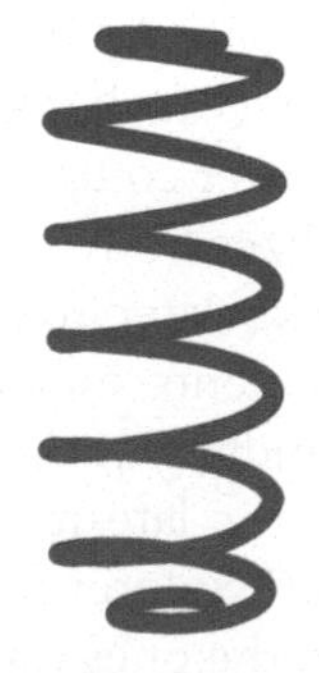

Figure 6-1 Variable-rate coil springs have consistent wire diameter and unequally spaced coils.

a variable-rate coil spring, the inactive coils theoretically support the load. If the load is increased, the transitional coils support the load until they reach maximum load-carrying capacity, and the active coils carry the remaining overload. This spring action provides automatic load adjustment while maintaining vehicle height.

Some variable-rate coil springs have a tapered wire, in which the active coils have the larger diameter and the inactive coils have the smaller diameter. Other variable-rate spring designs include truncated cone, double cone, and barrel shape. A variable-rate spring does not have a standard spring rate. This type of spring has an average spring rate based on the load at a predetermined spring deflection. *It is impossible to compare variable spring rates and linear spring rates because of this difference in spring rates.* Variable-rate coil springs usually have more load-carrying capacity than linear-rate coil springs in the same application.

Lightweight Coil Springs

A few sports cars are presently equipped with titanium coil springs. This type of coil spring reduces the weight on the front springs by 39 percent and 28 percent on the rear springs compared to steel coil springs. Decreasing the weight of the coil springs reduces the **unsprung weight** and improves ride control. The unsprung vehicle weight tends to force the wheel downward during rebound wheel travel. Higher unsprung weight drives the wheel downward with greater force and increases the impact force between the tire and the road surface, resulting in a harsh ride. To compensate for the high unsprung weight, the shock absorber damping rate must be increased. This suspension design reduces ride quality. When the unsprung weight is reduced, the wheel is forced downward with less force, reducing the shock absorber damping rate and improving ride quality.

Unsprung weight is the vehicle weight that is not supported by the coil springs, and sprung weight is the vehicle weight that is supported by the coil springs.

Heavy-Duty Coil Springs **Heavy-duty coil springs** are designed to carry 3 to 5 percent greater loads than regular-duty coil springs. The wire diameter may be up to 0.10 in. greater in a heavy-duty spring than in a regular-duty coil spring. This larger-diameter wire increases the load-carrying capacity of the spring. The free height of a heavy-duty coil spring is shorter than a regular-duty coil spring for the same application (**Figure 6-2**).

Selecting Replacement Springs When replacement coil springs are required, the technician must select the correct spring. The original part number is usually on a tag wrapped around one of the coils. However, this tag may have fallen off if the spring has been in service for very long. Some aftermarket suppliers stamp the part number on the end of the coil spring. If the original part number is available, the replacement springs may be ordered with the same part number. Most vehicle manufacturers recommend that both front or rear springs be replaced at the same time. The replacement springs must have the same type of ends as the springs in the vehicle. Coil spring ends may be square tapered, square nontapered, or tangential (**Figure 6-3**).

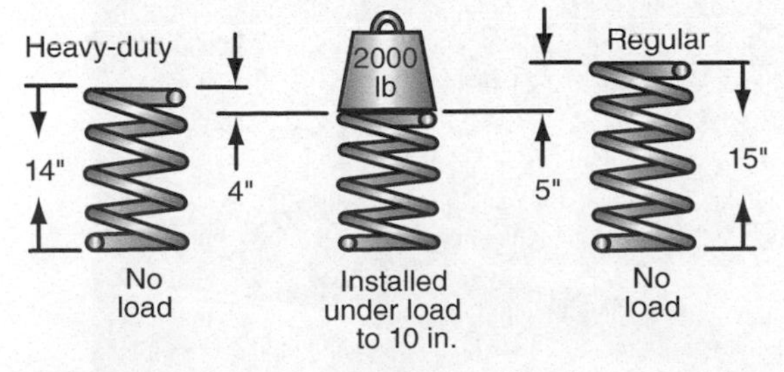

Figure 6-2 Comparison of heavy-duty and regular-duty coil springs.

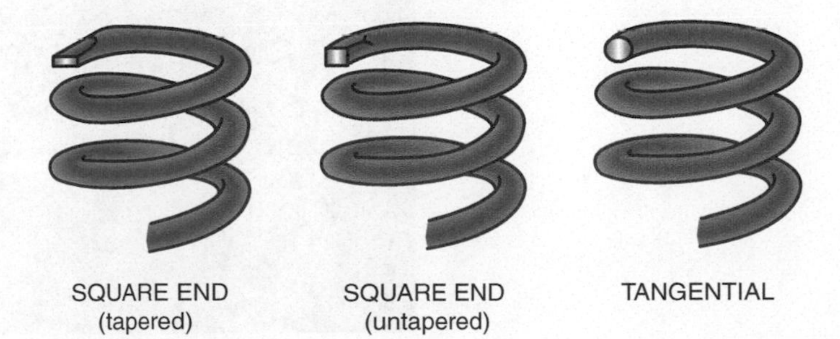

Figure 6-3 Types of coil spring ends.

If a wheel moves upward in relation to the chassis, this action is referred to as jounce travel. Downward wheel movement is called rebound travel.

Full-wire open-end springs have the ends cut straight off, and sometimes these ends are flattened, squared, or ground to a D-shape. **Taper-wire closed-end springs** are ground to a taper and wound to ensure squareness. **Pigtail spring ends** are wound to a smaller diameter. Springs are generally listed for front or rear suspensions.

Regular-duty coil springs are a close replacement for the original spring in the vehicle, and these springs may replace several different original equipment (OE) springs in the same vehicle. Linear-rate coil springs are usually found in regular-duty, heavy-duty, and sport suspension packages. Heavy-duty coil springs are required when the vehicle is carrying a continuous heavy load, such as trailer towing.

Variable-rate coil springs are generally used when automatic load leveling is required under increased loads. Variable-rate coil springs maintain the correct vehicle height under various loads and provide increased load-carrying capacity compared to heavy-duty coil springs. The technician must select the correct spring to meet the requirements of the vehicle and load conditions.

WARNING **A very large amount of energy is stored in a compressed coil spring. Always follow the spring service procedures in the vehicle manufacturer's service manual to avoid personal injury.**

WARNING **Never disconnect a suspension component that quickly releases the tension on a compressed coil spring. This action may cause personal injury.**

Torsion Bars

In some front suspension systems, torsion bars replace the coil springs. During wheel jounce, the torsion bar twists. During wheel rebound, the torsion bar unwinds back to its original position. A torsion bar may be thought of as a straight, flattened coil spring. One end of the heat-treated alloy steel torsion bar is attached to the vehicle frame, and the opposite end is connected to the lower control arm. A few older vintage vehicles had the end of the torsion bars connected to the upper control arm. Transversely mounted torsion bars were used in some front suspensions on older cars. Some light-duty trucks and SUVs are presently equipped with longitudinal torsion bars in the front suspension (**Figure 6-4**). On four-wheel-drive trucks, the use of torsion bars replaces coil springs, which allows more space for the front drive axles. The front end of the torsion bar is anchored in the lower control arm and the rear end of the torsion bars are anchored to a chassis

Figure 6-4 Longitudinal torsion bar.

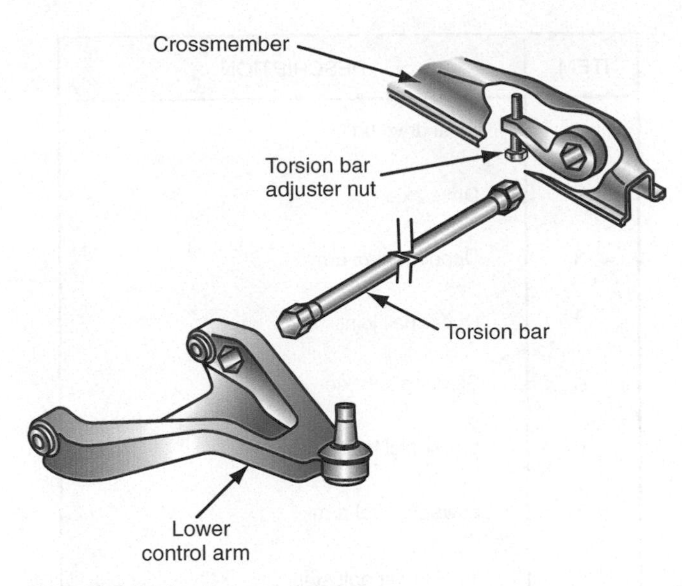

Figure 6-5 Torsion bar mounting.

crossmember (**Figure 6-5**). Since the twisting of the torsion bars supports the vehicle weight, the torsion bars replace the coil springs.

Because the lower control arm moves up and down as the wheel strikes road irregularities, this control arm action twists the torsion bar. The bar's natural resistance to twisting causes it to return to its original position. During the manufacturing process, torsion bars are prestressed to provide fatigue strength. These bars are directional. Torsion bars are marked right or left, and they must be installed on the appropriate side of the vehicle. Left and right on a vehicle are always viewed from the driver's seat.

A torsion bar is capable of storing a higher maximum energy compared with a loaded coil or a leaf spring. Shorter, thicker torsion bars have increased load-carrying capacity as compared to longer, thinner bars. Since torsion bars require less space than coil or leaf springs, they are usually found on front suspensions.

On a torsion bar front suspension system with upper and lower control arms (**Figure 6-6** and **Figure 6-7**) the front of the torsion bars has hex-shaped ends that are

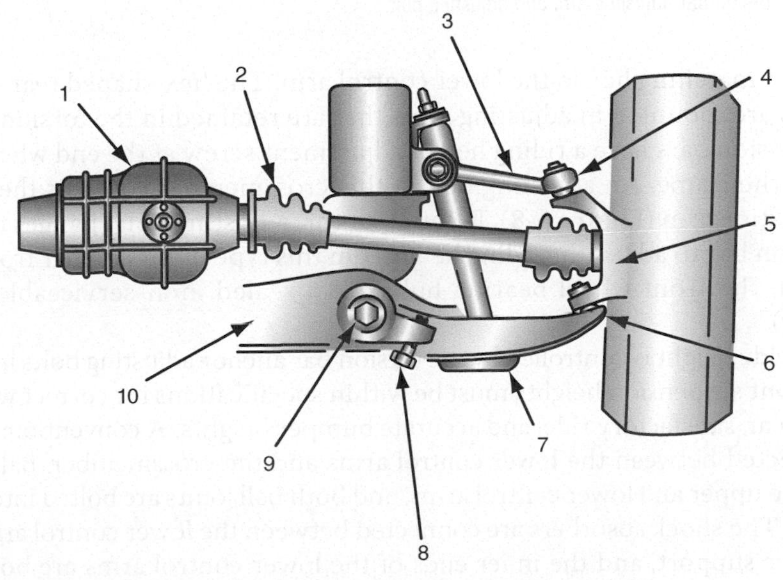

Figure 6-6 Four-wheel-drive light-duty truck torsion bar front suspension.

ITEM	DESCRIPTION
1	Final drive unit
2	Drive axle
3	Upper control arm
4	Upper ball joint
5	Steering knuckle
6	Lower ball joint
7	Lower control arm
8	Torsion bar adjuster
9	Torsion bar
10	Crossmember

Figure 6-7 Component identification light-duty four-wheel-drive truck torsion bar front suspension.

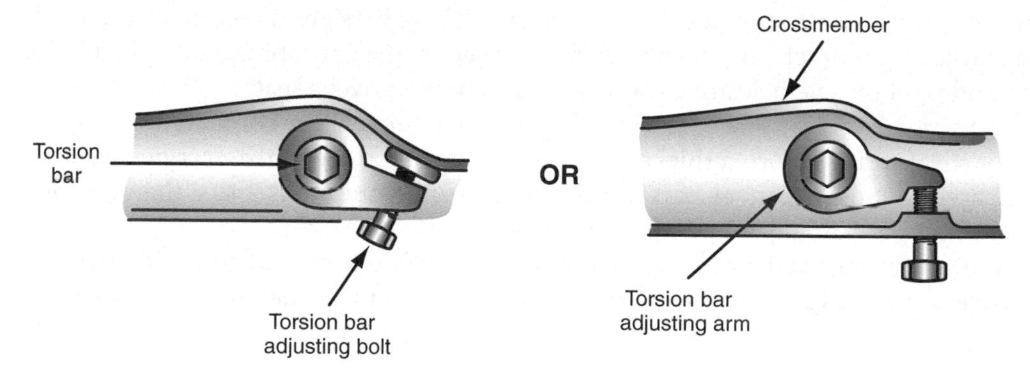

Figure 6-8 Torsion bar adjusting arm and adjusting bolt.

mounted in a matching hex in the lower control arm. The hex-shaped rear ends of the torsion bars are mounted in adjusting arms that are retained in the torsion bar crossmember. Torsion bars have a riding height adjustment screw at the end where they are attached to the frame. An adjusting bolt in this crossmember contacts the outer end of each adjusting arm (**Figure 6-8**). Rotation of this adjusting bolt changes the tension on the torsion bar to adjust the riding height. On this type of torsion bar front suspension system, the front wheel bearing hubs are a sealed, non-serviceable assembly (**Figure 6-9**).

Vehicle ride height is controlled by the torsion bar anchor adjusting bolts in the crossmember. Front suspension heights must be within specifications for correct wheel alignment, tire wear, satisfactory ride, and accurate bumper heights. A conventional stabilizer bar is connected between the lower control arms and the crossmember. Ball joints are located in the upper and lower control arms, and both ball joints are bolted into the steering knuckle. The shock absorbers are connected between the lower control arms and the crossmember support, and the inner ends of the lower control arms are bolted to the crossmember through an insulating bushing (**Figure 6-10**).

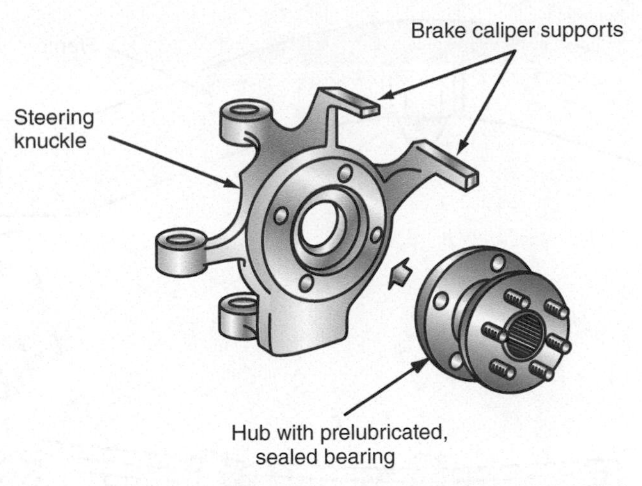

Figure 6-9 Four-wheel-drive torsion bar front suspension with sealed non-serviceable front wheel bearing hubs.

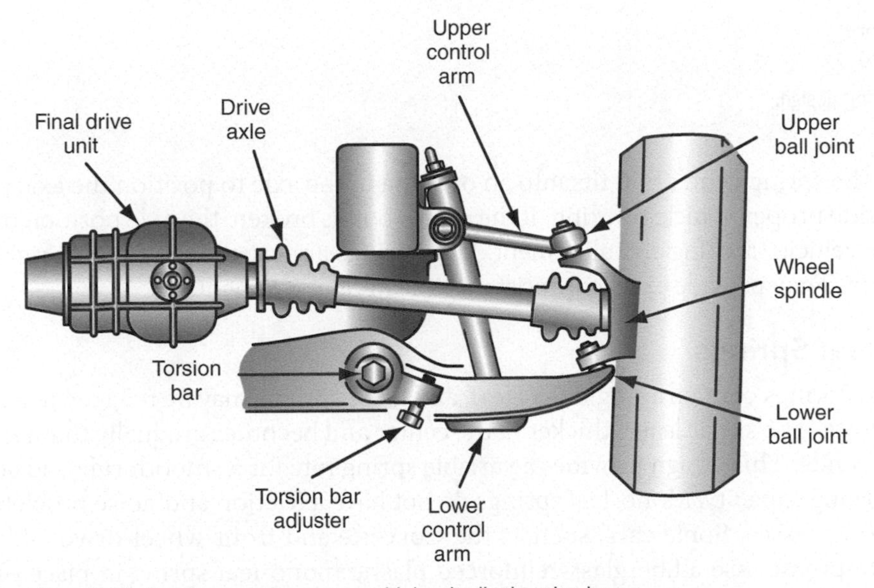

Figure 6-10 Front suspension system with longitudinal torsion bars.

Multiple-Leaf Springs

Leaf springs may be multiple-leaf or mono-leaf. **Multiple-leaf springs** have a series of flat steel leaves of varying lengths that are clamped together. A center bolt extends through all the leaves to maintain the leaf position in the spring. The upper leaf is called the main leaf, and this leaf has an eye on each end. An insulating bushing is pressed into each main leaf eye. The front bushing is attached to the frame, and the rear bushing is connected through a shackle to the frame. The shackle provides fore-and-aft movement as the spring compresses (**Figure 6-11**).

The main leaf is the longest leaf in the spring, and the other leaves get progressively shorter. Each spring leaf is curved in the manufacturing process. If this curve were doubled, it would form an ellipse. Therefore, leaf springs are referred to as semielliptical or quarter elliptical. Most leaf springs are semielliptical. The ellipse designation refers to how much of the ellipse the spring actually describes.

As a leaf spring compresses, it becomes progressively stiffer. When a leaf spring is compressed, the length of the leaves changes, and the leaves slide on each other. This sliding action could be a source of noise and friction. These noise and friction problems are reduced by interleaves, or spacers, made from zinc and plastic placed between the steel leaves. The

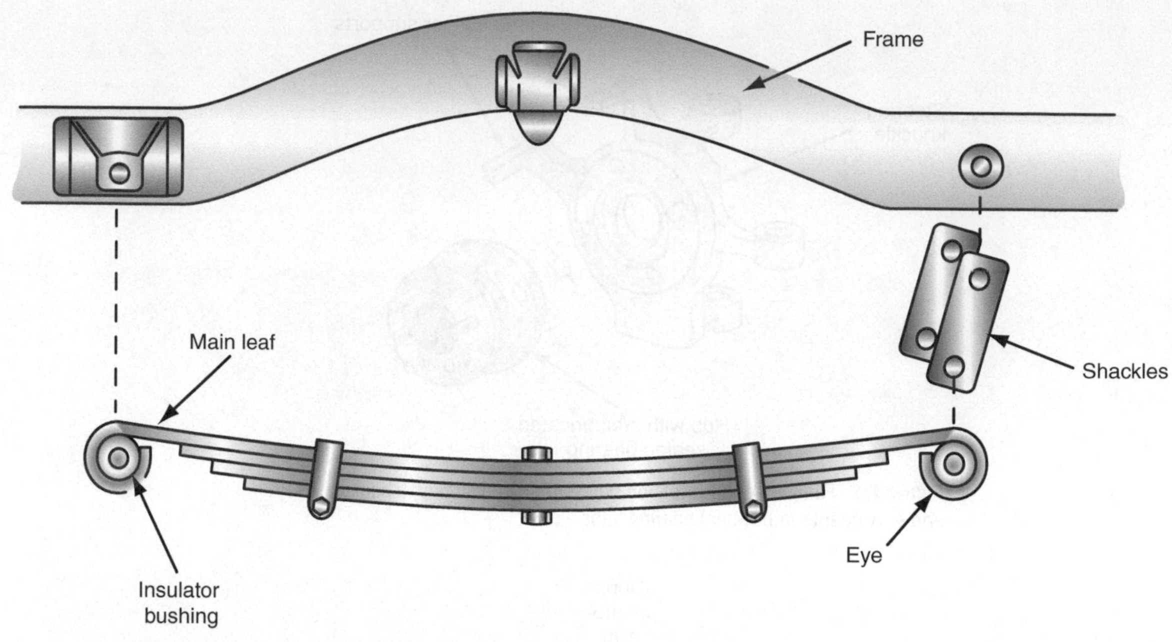

Figure 6-11 Leaf spring design.

head on the spring center bolt fits into an opening in the axle to position the axle properly and provide proper vehicle tracking. If the center bolt is broken, the axle position may shift and alter vehicle tracking and alignment. Leaf springs are usually mounted at right angles to the axle. They provide excellent resistance to lateral movement.

Mono-leaf Springs

Some leaf springs contain a single steel leaf, and these springs may be referred to as **mono-leaf springs**. The single leaf is thicker in the center and becomes gradually thinner toward the outer ends. This design provides a variable spring rate for a smooth ride and adequate load-carrying capacity. Mono-leaf springs do not have a friction and noise problem as the spring compresses. Some cars, such as the Corvette and front-wheel-drive Oldsmobile Cutlass Supreme, use a fiberglass-reinforced plastic mono-leaf spring in place of a steel spring to reduce weight (**Figure 6-12**). The mono-leaf spring may be mounted longitudinally or transversely, and this type of spring may be used in front or rear suspensions. Older model Chevrolet Astro vans have longitudinally mounted fiberglass mono-leaf rear springs.

Shop Manual
Chapter 6, page 213

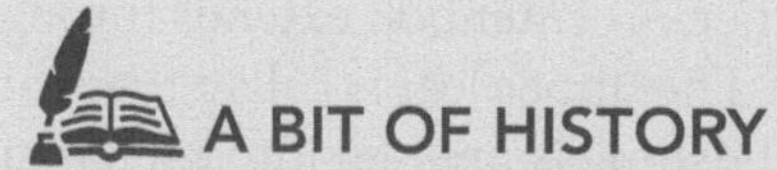

A BIT OF HISTORY

The transverse leaf-spring front and rear suspension is one of the oldest suspension systems in the automotive industry. It was used on Ford products from the early 1900s until 1948.

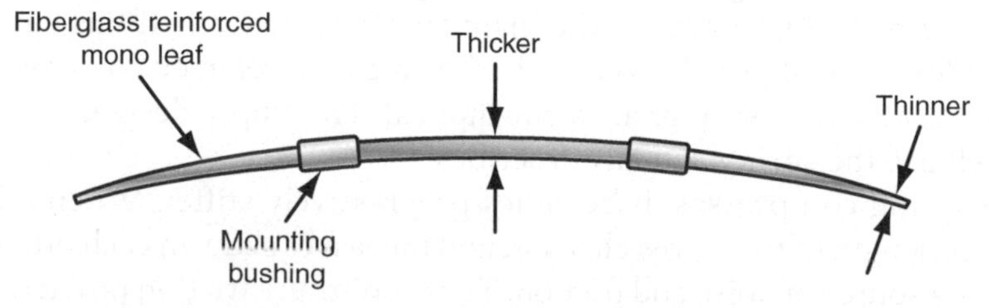

Figure 6-12 Transversely mounted fiberglass mono-leaf spring.

Ball Joints

The ball joints act as pivot points that allow the front wheels and spindles, or knuckles, to turn between the upper and lower control arms. Compared with the earlier I-beam and kingpin-type front suspension systems, ball-joint suspension systems are much simpler. A front suspension with ball joints reduces the number of load-carrying bearing surfaces. Compared with an I-beam front suspension with kingpins, a front suspension with ball joints has these advantages:

1. Reduced space requirements
2. Reduced unsprung weight
3. Easier alignment
4. More dependable steering control
5. Improved safety
6. Improved tire life
7. Reduced steering effort
8. Improved ride quality
9. Simplified service—no kingpin reaming or honing
10. Simplified lubrication

WARNING Worn ball joints may suddenly pull apart, resulting in loss of steering control and possibly a collision! Refer to the procedures in the Shop Manual Chapter 6 for measuring ball joint wear.

Load-Carrying Ball Joint

Ball joints may be grouped into two classifications, load carrying and non-load carrying. Ball joints may be manufactured with forged, stamped, cold-formed, or screw-machined housings. The coil spring is seated on the control arm to which the load-carrying ball joint is attached. For example, when the coil spring is mounted between the lower control arm and the chassis, the lower ball joint is a load-carrying joint (**Figure 6-13**). In a torsion bar suspension, the **load-carrying ball joint** is mounted on the control arm to which the torsion bar is attached. A load-carrying ball joint supports the vehicle weight.

Shop Manual
Chapter 6, page 230

In a load-carrying ball joint, the vehicle weight forces the ball stud into contact with the bearing surface in the joint. Load-carrying ball joints may be compression loaded or tension loaded. If the control arm is mounted above the lower end of the knuckle and rests on the knuckle, the ball joint is **compression loaded**. In this type of ball joint, the vehicle weight is pushing downward on the control arm. This weight is supported on the tire and

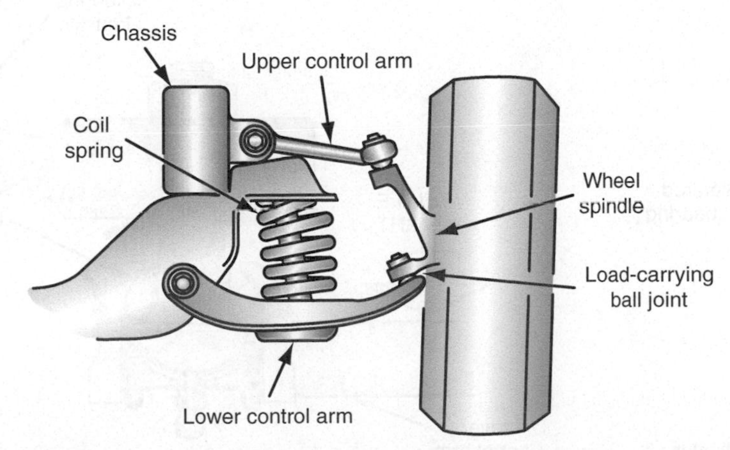

Figure 6-13 Load-carrying ball joint mounted on the control arm on which the spring is seated.

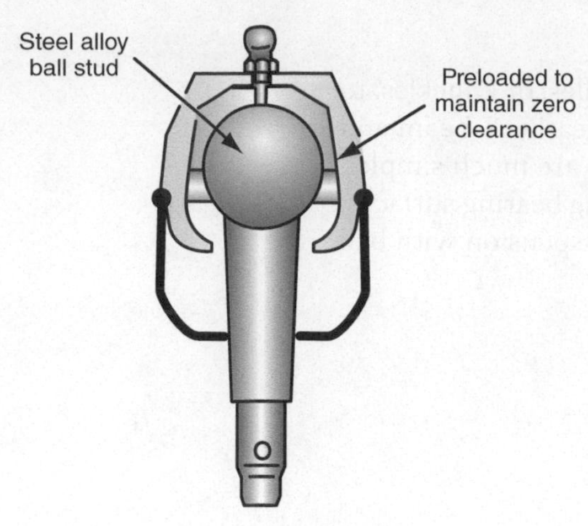

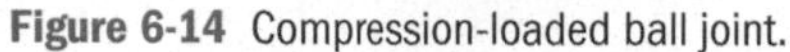
Figure 6-14 Compression-loaded ball joint.

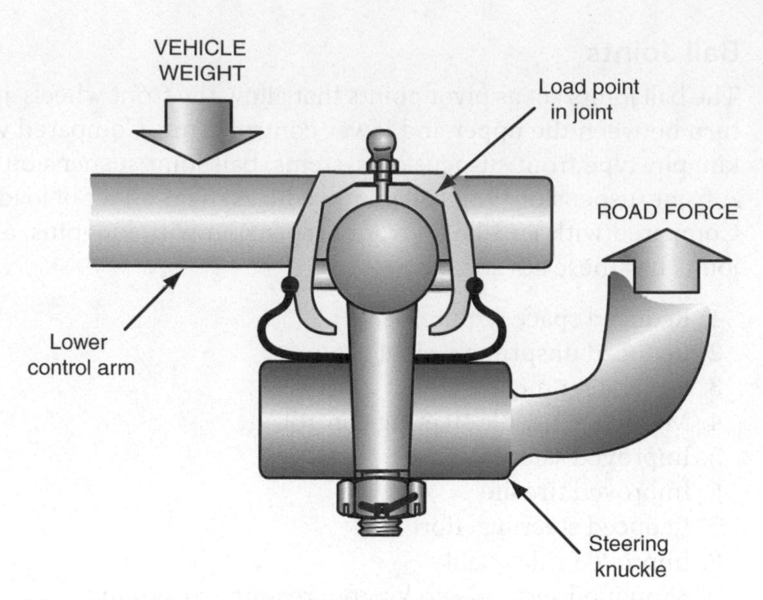

Figure 6-15 Compression-loaded ball joint mounting.

wheel, which are attached to the steering knuckle. Since the ball joint is mounted between the control arm and the steering knuckle, the vehicle weight squeezes the ball joint together (**Figure 6-14**). In this type of ball-joint mounting, the ball joint is mounted in the lower control arm and the ball joint stud faces downward (**Figure 6-15**).

When the lower control arm is positioned below the steering knuckle, the vehicle weight is pulling the ball joint away from the knuckle (**Figure 6-16**). This type of ball joint mounting is referred to as **tension loaded**. This type of ball joint is mounted in the lower control arm with the ball joint stud facing upward into the knuckle (**Figure 6-17**).

Since the load-carrying ball joint supports the vehicle weight, this ball joint wears faster compared with a non-load-carrying ball joint. Many load-carrying ball joints have built-in **wear indicators**. These ball joints have an indicator on the grease nipple surface that recedes into the housing as the joint wears. If the ball joint is in good condition, the grease nipple shoulder extends a specified distance out of the housing. If the grease nipple shoulder is even with or inside the ball joint housing, the ball joint is worn, and replacement is necessary (**Figure 6-18**).

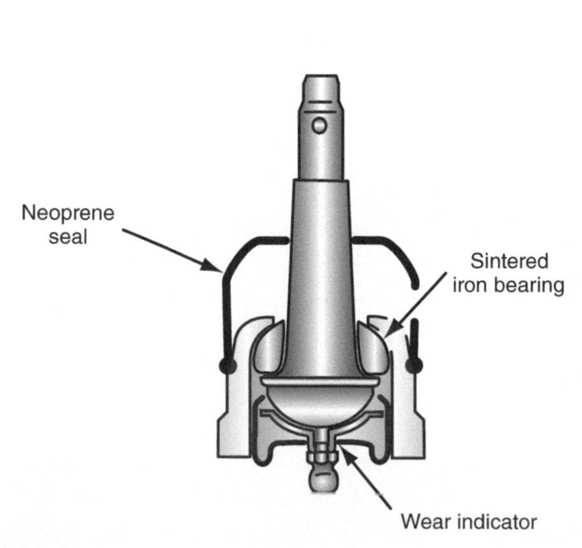

Figure 6-16 Tension-loaded ball joint.

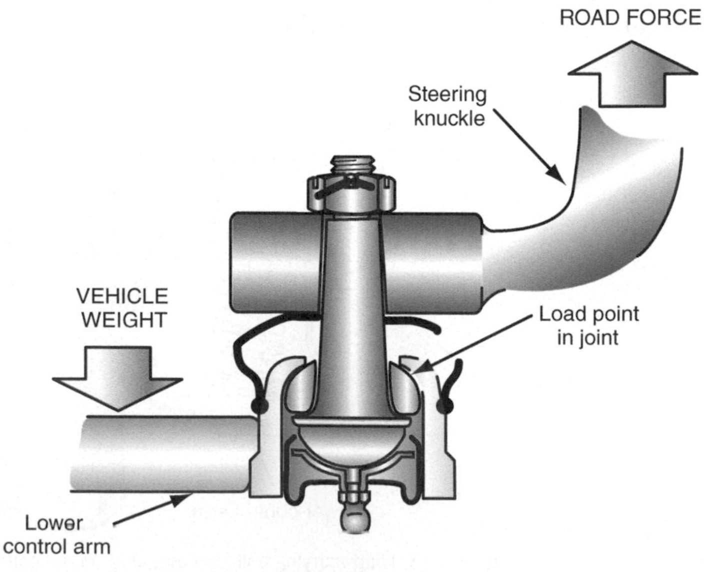

Figure 6-17 Tension-loaded ball joint mounting.

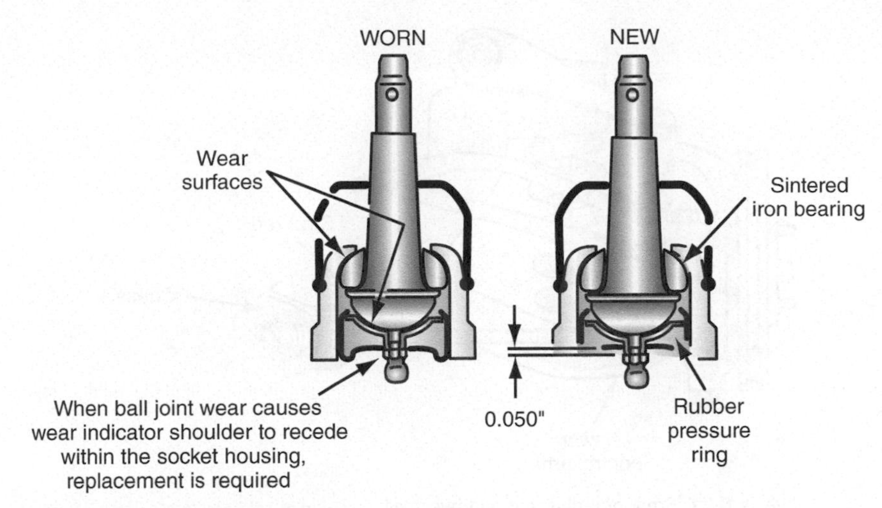

Figure 6-18 Ball joint wear indicator.

Most states have safety inspection procedures for testing ball joints and other safety-related components. These safety inspection procedures include specifications. Always follow the guidelines when performing safety inspections.

Non-Load-Carrying Ball Joint

A **non-load-carrying ball joint** may be referred to as a stabilizing or follower ball joint. A non-load-carrying ball joint is designed with a preload, which provides damping action (**Figure 6-19**). This ball joint preload provides improved steering quality and vehicle stability.

A **non-load-carrying ball joint** maintains knuckle position, but this ball joint does not support the vehicle weight.

Low-Friction Ball Joints

Low-friction ball joints are standard equipment on many vehicles. Low-friction ball joints provide precise low-friction movement of the ball socket in the ball joint. Compared to conventional ball joints, two-thirds of the internal friction is eliminated in a low-friction ball joint. The smooth ball socket movement in a low-friction ball joint provides improved steering performance, better steering wheel return, and longer ball joint life. Low-friction ball joints have a highly polished ball socket surface surrounded by a high-strength polymer bearing (**Figure 6-20**).

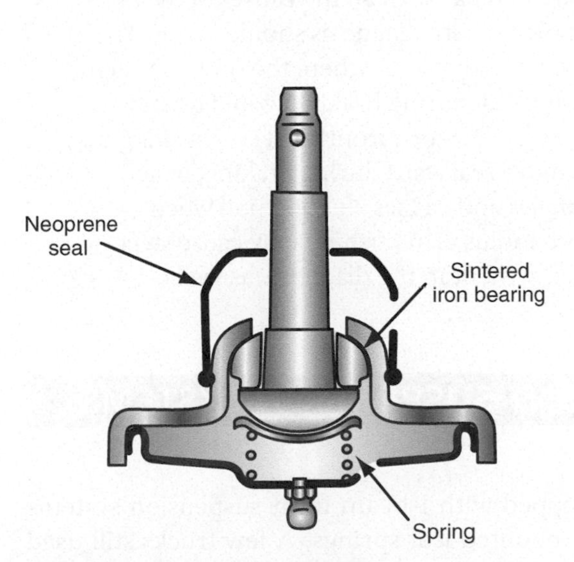

Figure 6-19 Non-load-carrying ball joint.

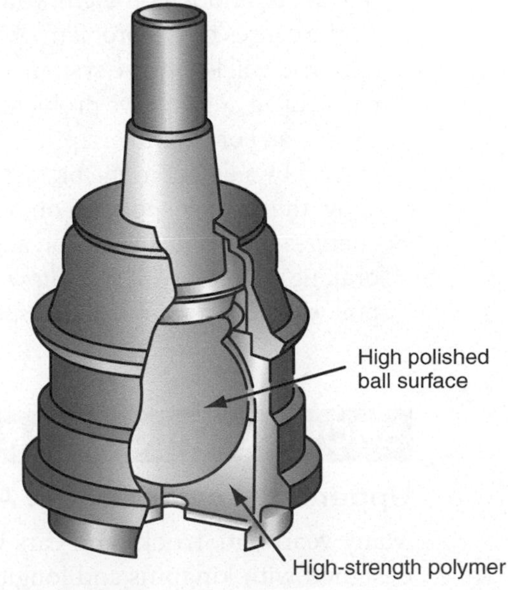

Figure 6-20 Low-friction ball joint.

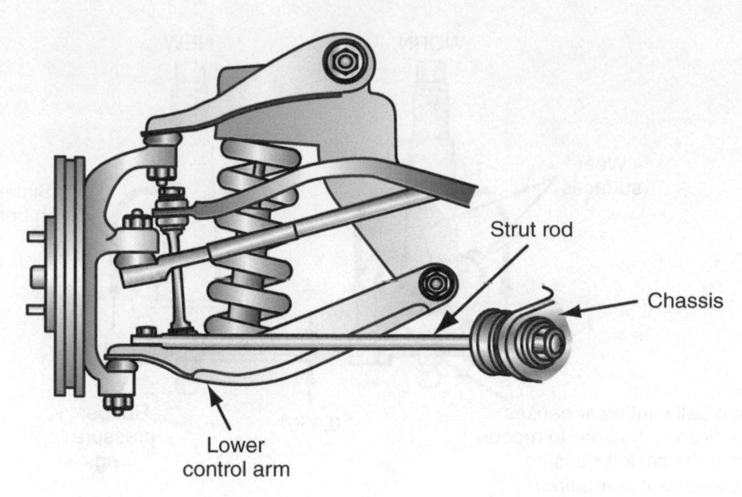

Figure 6-21 Strut, or radius, rod and bushings.

Strut Rod

On some front suspension systems, a strut rod is connected from the lower control arm to the chassis. The strut rod is bolted to the control arm, and a large rubber bushing surrounds the strut rod in the chassis opening. The outer end of the strut rod is threaded, and steel washers are positioned on each side of the strut rod bushing. Two nuts tighten the strut rod into the bushing (**Figure 6-21**). The strut rod prevents fore-and-aft movement of the lower control arm. In some suspension systems, the position of the strut rod nuts provides proper front wheel adjustment.

Shop Manual
Chapter 6, page 249

Some strut rods are presently manufactured from tubular steel to reduce strut rod and unsprung weight.

AUTHOR'S NOTE It has been my experience that the most important item when diagnosing automotive problems is your knowledge of automotive systems and components. You must understand the operation of automotive systems and the individual components within these systems. You must also be familiar with the problems that certain worn, defective, or misadjusted components can cause. When diagnosing an automotive problem, you must identify the exact problem, and then think of all the causes of the problem. When you are familiar with all the causes of the problem, you will know the system and component where diagnosis should begin. The cause of an automotive problem may not be in the system where the problem seems to appear. For example, a problem of steering pull during braking would usually be caused by a defect in the brake system. However, a worn front strut rod bushing may allow the front suspension on one side to move rearward during braking, which changes the front suspension alignment angles and causes steering pull when braking. Therefore, if you know that a worn radius arm bushing may cause steering pull when braking, you will inspect this component in the diagnosis as well.

SHORT-AND-LONG ARM FRONT SUSPENSION SYSTEMS

Upper and Lower Control Arms

Many years ago, trucks and cars were equipped with I-beam front suspension systems designed with kingpins and longitudinally mounted leaf springs. A few trucks still used this type of suspension until the late 1980s. The kingpins were retained in a vertical

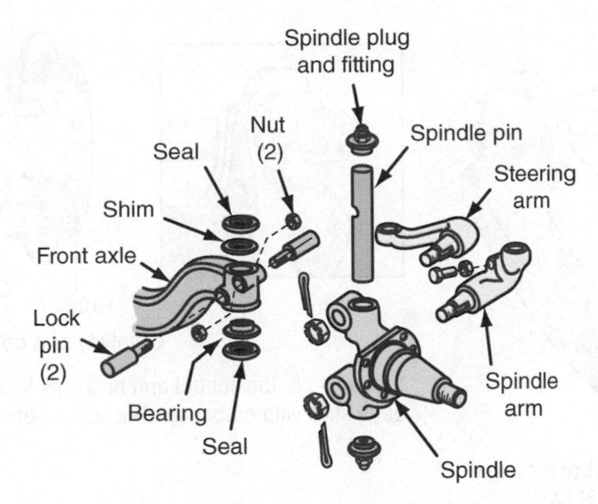

Figure 6-22 I-beam front suspension with kingpins.

Shop Manual
Chapter 6, page 231

opening in the ends of the axle with a lock bolt. Upper and lower bushings in the steering knuckle pivoted on the upper and lower ends of the kingpins (**Figure 6-22**). In this type of front suspension, if one front wheel moves upward or downward, some movement is transferred to the other end of the axle. This action tends to result in reduced ride quality, steering instability, and tire wear.

As automotive technology evolved, the control arm front suspension system replaced the I-beam front suspension system. This type of front suspension system has coil springs with upper and lower control arms. Since wheel jounce or rebound movement of one front wheel does not directly affect the opposite front wheel, the control arm suspension is an independent system. Many rear-wheel-drive cars have control arm front suspension systems. Early front suspension systems had equal-length upper and lower control arms. On these early suspension systems, the bottom of the tire moved in and out with wheel jounce and rebound travel. This action constantly changed the tire tread width and caused tire scuffing and wear problems (**Figure 6-23**).

An independent suspension system is one in which wheel jounce or rebound travel of one wheel does not directly affect the movement of the opposite wheel.

In later **short-and-long arm front suspension systems**, the upper control arm is shorter than the lower control arm. During wheel jounce and rebound travel in this suspension system, the upper control arm moves in a shorter arc than the lower control arm. This action moves the top of the tire in and out slightly, but the bottom of the tire remains in a more constant position (**Figure 6-24**). This short-and-long arm front suspension

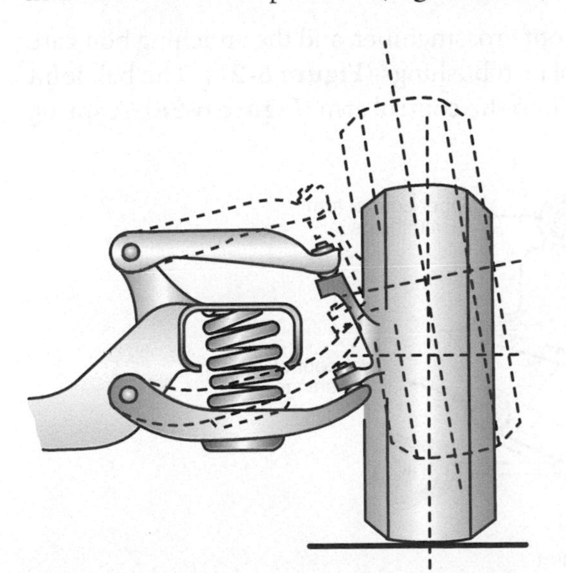

Figure 6-23 Early front suspension system with equal length upper and lower control arms.

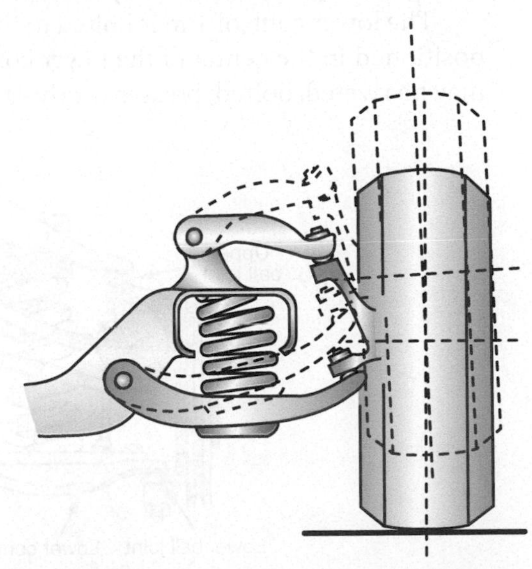

Figure 6-24 Short-and-long arm front suspension system.

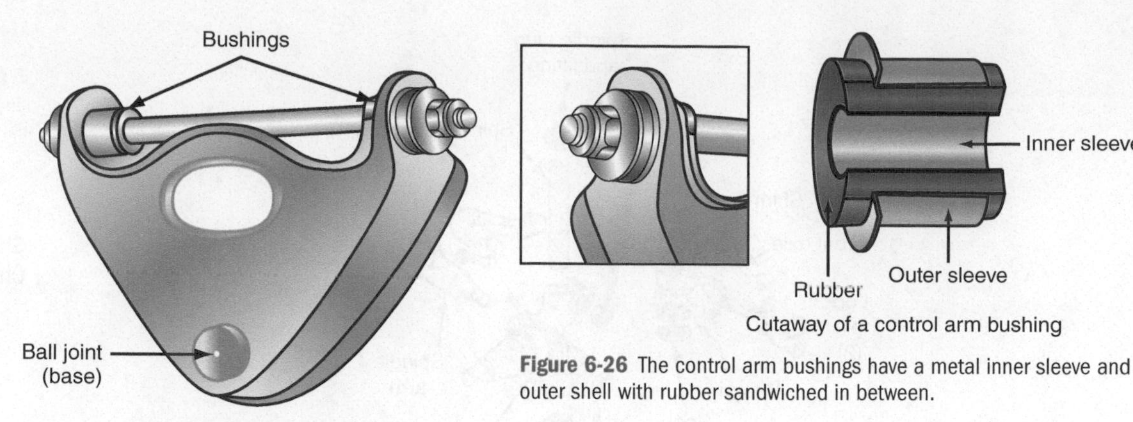

Figure 6-25 The inner end of the control arm contains large rubber insulating bushings, and the ball joint is attached to the outer end of the control arm.

Figure 6-26 The control arm bushings have a metal inner sleeve and outer shell with rubber sandwiched in between.

system provides reduced tire tread wear, improved ride quality, and better directional stability compared to I-beam suspension systems and suspension systems with equal-length upper and lower control arms.

The inner end of the control arm contains large rubber insulating bushings, and the ball joint is attached to the outer end of the control arm that attaches to the steering knuckle (**Figure 6-25**). A shaft or bolt passes through the eye of the bushing to connect it to the frame of the vehicle. The rubber bushings are used to separate and insulate parts while providing up-and-down pivot points for the suspension. The rubber cushions reduce noise and vibration from being transmitted to the vehicle body. The control arm bushings have a metal inner sleeve and outer shell with rubber sandwiched in between (**Figure 6-26**). The bushings are usually pressed into the control arms. The inner sleeve is bolted to the frame and remains stationary, whereas the outer shell of the bushing moves with the control arm. Thus, the control arm moves as the rubber is twisted and flexed allowing for up and down movement and some shock absorption. As was seen in other areas of the suspension system, bushings are found on shock absorbers, strut rods, leaf springs, and sway bars, among other places.

The lower control arm is bolted to the front crossmember, and the attaching bolts are positioned in the center of the lower control arm bushings (**Figure 6-27**). The ball joint may be riveted, bolted, pressed, or threaded into the control arm (**Figure 6-28**). A spring

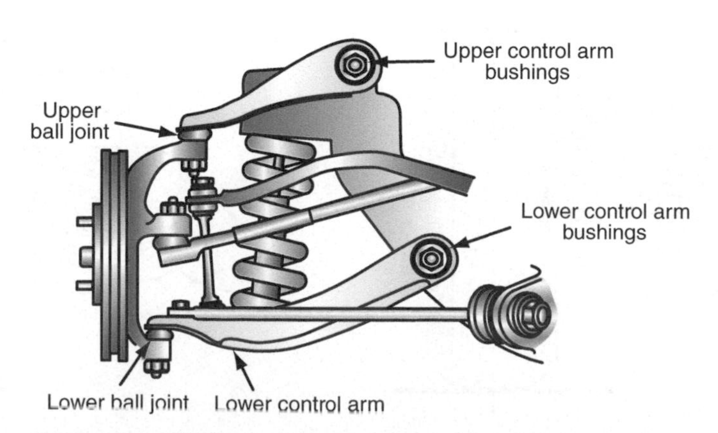

Figure 6-27 Ball joints and complete short-and-long arm front suspension system.

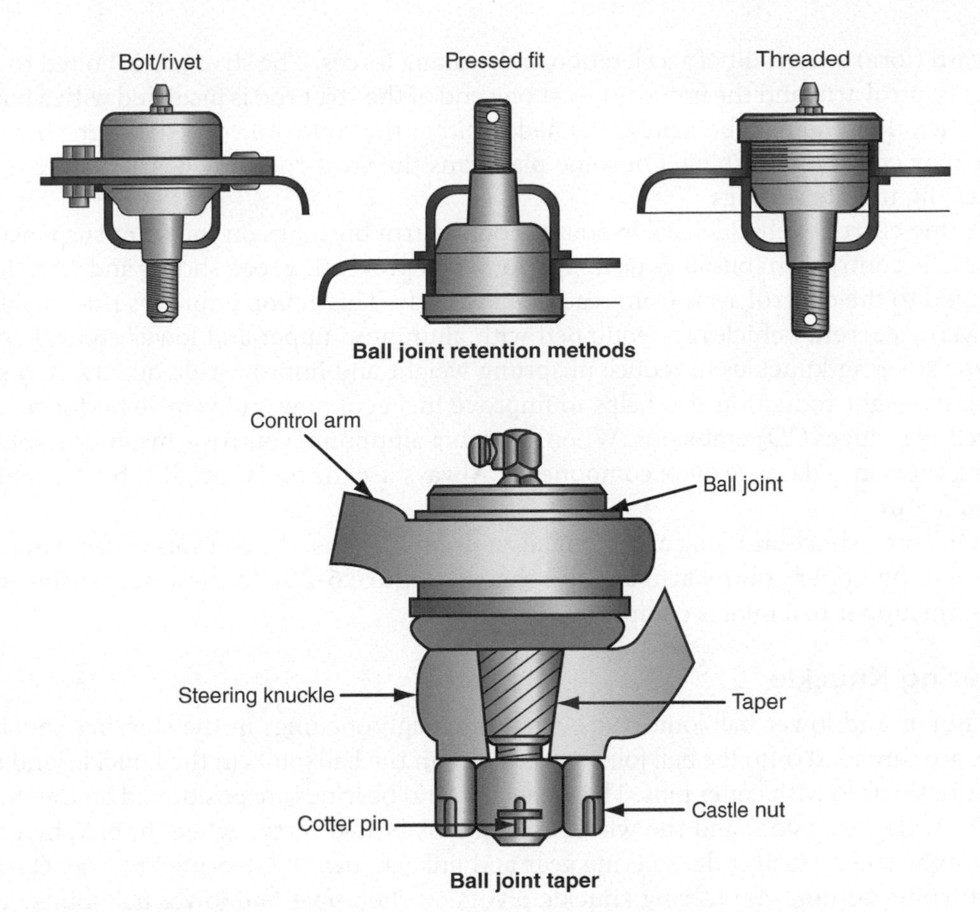

Figure 6-28 The ball joint may be riveted, bolted, pressed, or threaded into the control arm.

seat is located in the lower control arm. An upper control arm shaft is bolted to the frame, and rubber insulators are located between this shaft and the control arm.

On some suspension systems that utilize a lightweight lower control arm that has a single anchor point at the frame instead of two, a strut rod is used to provide the second anchor point (**Figure 6-29**). The strut rod provides stability to the control arm. This second anchor point is required to keep the control arm from moving backward (aft) or

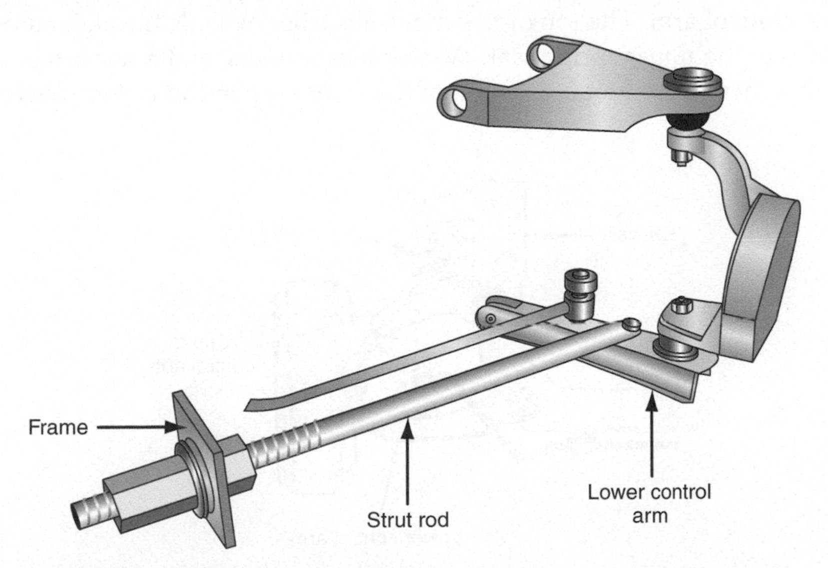

Figure 6-29 A strut rod is used with a control arm that uses only one anchor point at the frame to provide the second anchor point.

forward (fore) as a result of acceleration and braking forces. The strut rod is bolted to the lower control arm and the frame. At least one end of the strut rod is insulated with a bushing. The other end may be rigidly attached to either the control arm or the frame or both ends may contain a bushing. On some platforms the strut rod provides adjustment for wheel alignments settings.

Some current vehicles have hydraulic control arm bushings on the front suspension. Hydraulic control arm bushings do a superior job of preventing road shocks and vibrations supplied to the control arms from reaching the body. This action improves ride quality.

Many current vehicles are equipped with aluminum upper and lower control arms and/or steering knuckles to reduce unsprung weight and improve ride quality. Any significant weight reduction also helps to improve fuel economy and vehicle performance, as well as reduces CO_2 emissions. When servicing aluminum control arms, the use of the wrong tools may damage these components. Always use the tools specified by the vehicle manufacturer.

Shop Manual
Chapter 6, page 230

On some short-and-long arm front suspension systems, the coil spring is positioned between the upper control arm and the chassis (**Figure 6-30**). In these suspension systems, the upper ball joint is compression loaded.

Steering Knuckle

The upper and lower ball joint studs extend through openings in the steering knuckle. Nuts are threaded onto the ball joint studs to retain the ball joints in the knuckle, and the nuts are secured with cotter pins. The wheel hub and bearings are positioned on the steering knuckle extension, and the wheel assembly is bolted to the wheel hub. When the steering wheel is turned, the steering gear and linkage turn the steering knuckle. During this turning action, the steering knuckle pivots on the upper and lower ball joints. The upper and lower control arms must be positioned properly to provide correct tracking and wheelbase between the front and rear wheels. The control arm bushings must be in satisfactory condition to position the control arms properly.

Coil Spring and Shock Absorber

The coil spring is positioned between the lower control arm and the spring seat in the frame. A spring seat is located in the lower control arm, and an insulator is positioned between the top of the coil spring and the spring seat in the frame. The shock absorber is mounted in the center of the coil spring, and the lower shock absorber bushing is bolted to the lower control arm. The top of the shock absorber extends through an opening in the frame above the upper spring seat. Washers, grommets, and a nut retain the top of the shock absorber to the frame. Side roll of the front suspension is controlled by a steel

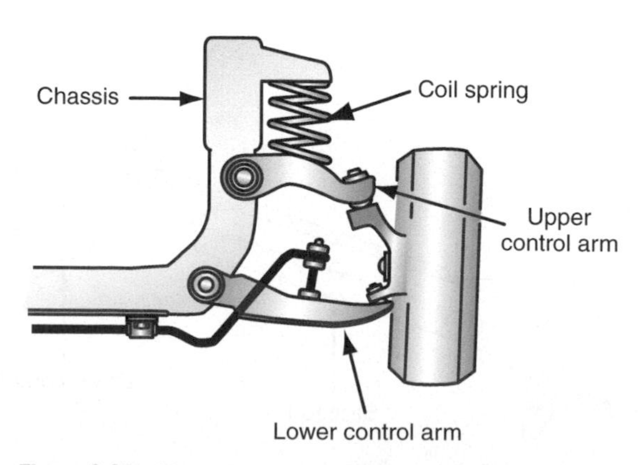

Figure 6-30 Short-and-long arm front suspension with the coil spring between the upper control arm and the chassis.

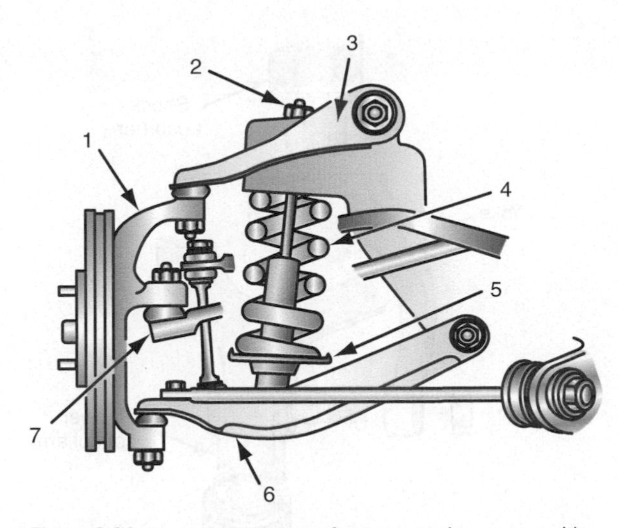

Figure 6-31 Short-and-long arm front suspension system with coil spring mounted on the shock absorber.

ITEM	DESCRIPTION
1	Steering knuckle
2	Nut, shock absorber upper mount
3	Upper control arm
4	Coil spring
5	Coil spring seat
6	Lower control arm
7	Tie rod end

Figure 6-32 Component identification in short-and-long arm suspension with the coil spring mounted on the shock absorber.

stabilizer bar, which is mounted to the lower control arms and the frame with rubber bushings.

On some later model short-and-long arm front suspension systems, the lower end of the coil spring is mounted on a seat attached to the shock absorber. A rubber bushing containing a metal bar is installed in the lower end of the shock absorber. This bushing is bolted to the lower control arm (**Figure 6-31** and **Figure 6-32**). The upper end of the coil spring is seated on an upper shock absorber mount. The rod in the center of the shock absorber extends through this mount, and a nut on the threaded end of the shock absorber rod retains the mount on the shock absorber. Bolts in the top of the upper shock absorber mount extend through the upper control arm.

The upper control arm is mounted high in the suspension system, and the upper end of the knuckle has a gooseneck shape. The lower control arm is made from stamped steel to reduce weight. The rear lower control arm bushing is mounted vertically and carries only fore-and-aft loads. This mounting allows the use of a softer rear bushing in the lower control arm. The horizontal front lower control arm bushing and the lower shock absorber mounting are aligned with the wheel center. This provides a direct path for lateral cornering loads. This design allows the use of a hard front lower control arm bushing.

When servicing this suspension system, a special spring compressing tool must be used to compress all the spring tension before loosening the nut on top of the shock absorber rod (**Figure 6-33** and **Figure 6-34**). After this nut is removed, the compressing tool is operated to gradually release the spring tension, and then the upper mount may be removed. (Refer to Chapter 5 in the Shop Manual for strut and coil spring service.)

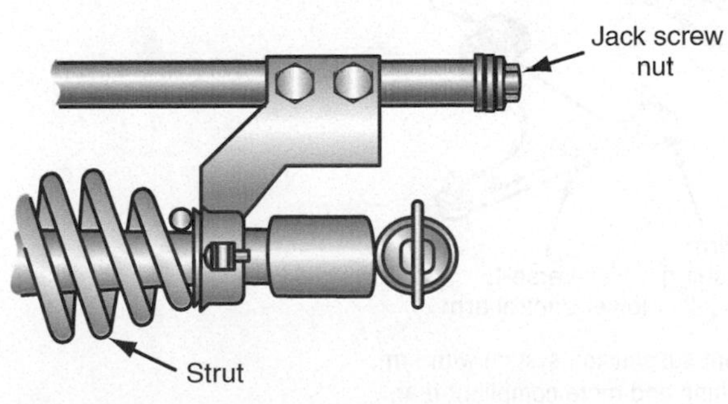

Figure 6-33 Lower end of spring compressing tool.

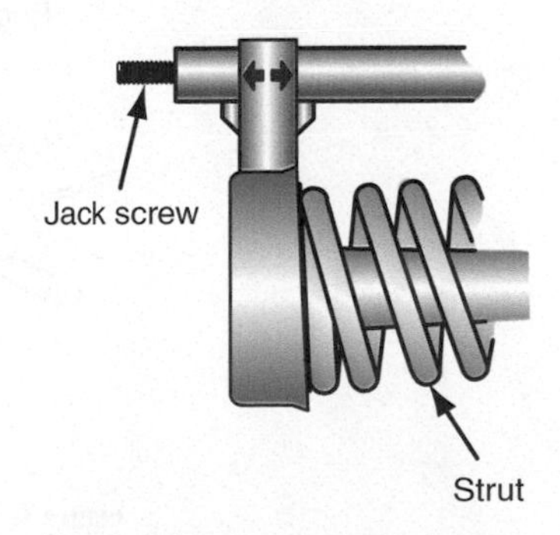

Figure 6-34 Upper end of spring compressing tool.

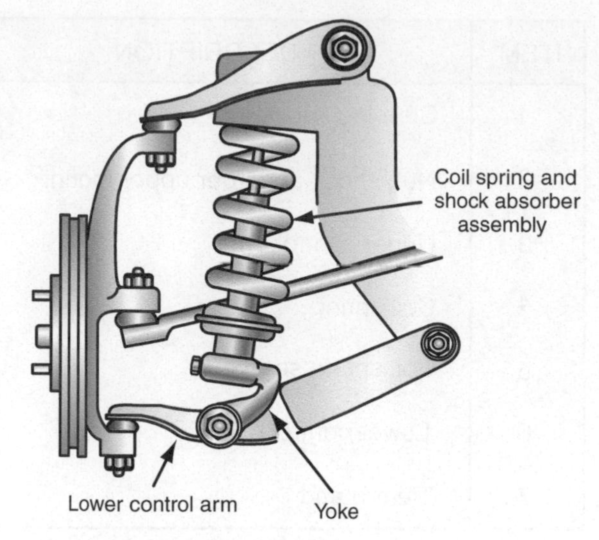

Figure 6-35 Short-and-long arm front suspension with yoke-type shock absorber mounting.

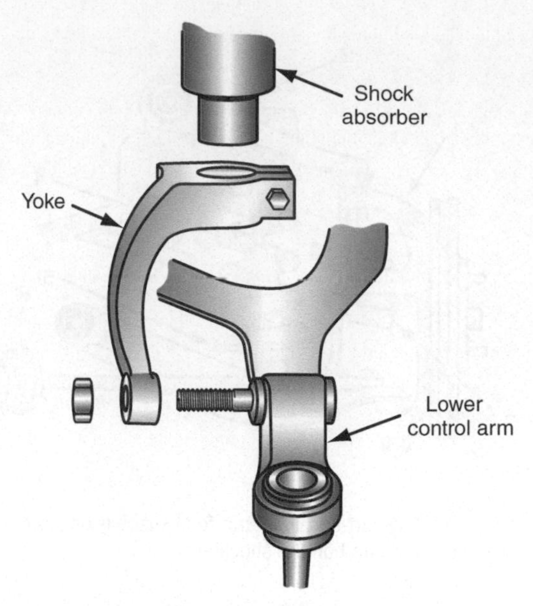

Figure 6-36 Shock absorber with yoke attachment to lower control arm.

Caution

Never loosen the nut on top of the shock absorber rod until a coil spring compressing tool is used to compress all the spring tension. Failure to follow this procedure may cause the spring tension to suddenly release, resulting in personal injury.

In the short-and-long arm front suspension system on the Trailblazer SUV, the coil spring is mounted over the shock absorber; the coil spring tension is applied against the upper mount and the lower spring seat on the shock absorber (**Figure 6-36**). A special yoke attaches to the lower end of the shock absorber, and the lower end of this yoke pivots on a bolt in the lower control arm (**Figure 6-37**). This type of suspension system provides a more positive shock absorber and spring mounting because the rubber bushing-type mounting between the shock absorber and the lower control arm is not required. The steering knuckle pivots on the upper and lower ball joints in the control arms. Therefore, the shock absorber and coil spring do not have to turn when the knuckle turns, which allow the use of a more rigid upper spring mount. This suspension design provides improved vehicle handling and ride characteristics.

Improved Designs in Short-and-Long Arm Front Suspension Systems

Some vehicles now have a reverse-L front suspension system. In these suspension systems, the rear attachment point on the lower control arm extends rearward and attaches to the engine cradle at a point farther toward the rear of the vehicle (Figure 6-37). The lower control arm has an L-shaped design. A firm bushing is installed at the location where the

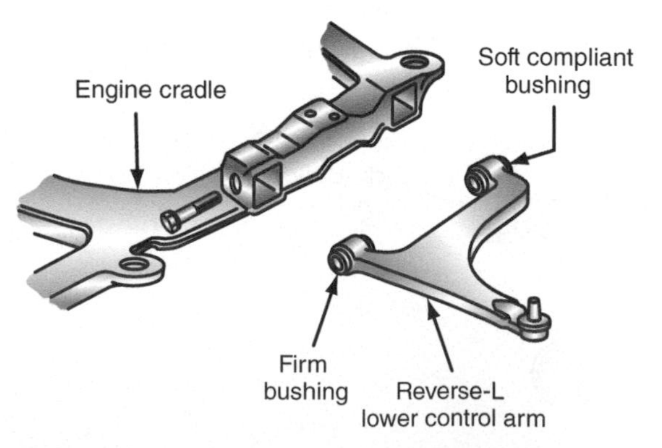

Figure 6-37 Reverse-L front suspension system with firm, front lower control arm bushing and more compliant rear, lower control arm bushing.

shorter forward leg of the lower control arm attaches to the cradle to control lateral control arm movement and quicken steering response. A more compliant bushing is installed between the longer rear leg and the rear attachment point to absorb the impact of longitudinal forces caused by road irregularities. This type of suspension separates the control of suspension loads into fore-and-aft control and side-to-side control to improve ride and steering quality.

Other modern short-and-long arm front suspension systems have high upper control arms that place the upper ball joints above the tires to provide suspension articulation that helps to keep the tire perpendicular to the road surface while cornering (**Figure 6-38**). This type of front suspension system has a lateral link and a tension strut to position the lower end of the knuckle and the lower ball joint rather than a lower control arm. The lateral link extends slightly forward from the lower ball joint to an attachment point on the chassis. Bushings are mounted in both ends of the lateral link. The tension strut extends rearward from the lower ball joint to the attachment location on the chassis. This type of front suspension also separates the control of suspension loads to improve ride quality and steering control. This type of front suspension may be called a four-link suspension system.

Sway Bar

The sway bar prevents excessive body roll or lean by resisting centrifugal forces. During a turn the inside of the suspension rises, and the outside of the suspension drops. The sway bar (stabilizer bar) is made of spring steel much like a torsion bar and is designed to absorb and transfer energy as it is twisted during suspension jounce and rebound. As the bar twists during a cornering maneuver the inside of the suspension is forced down and the outside of the suspension is forced up as energy is transmitted from one side to the other. In this manner, body roll is reduced, and the vehicle maintains a flatter profile with improved vehicle control and maneuverability during turns. If a sway bar were to fail or if one was not present, the inside tire of the vehicle would have less weight on it during a turn as the weight was transferred to the outside of the vehicle. This would reduce the ability of the driver to control the vehicle during the turn, thereby reducing the stability of the vehicle. A sway bar may be found on both the front

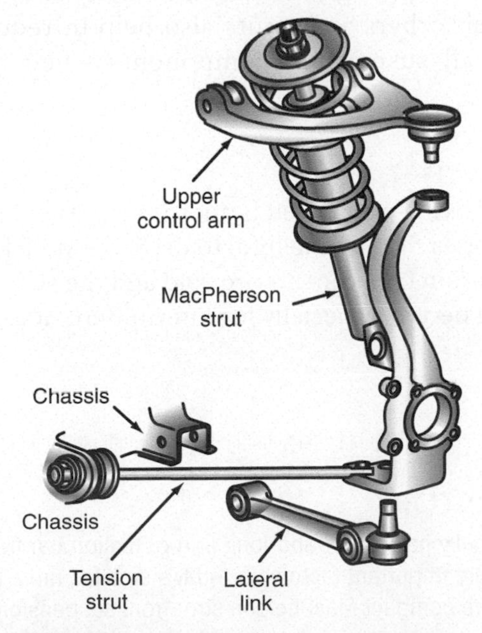

Figure 6-38 Front suspension system with lateral link and tension strut in place of lower control arm.

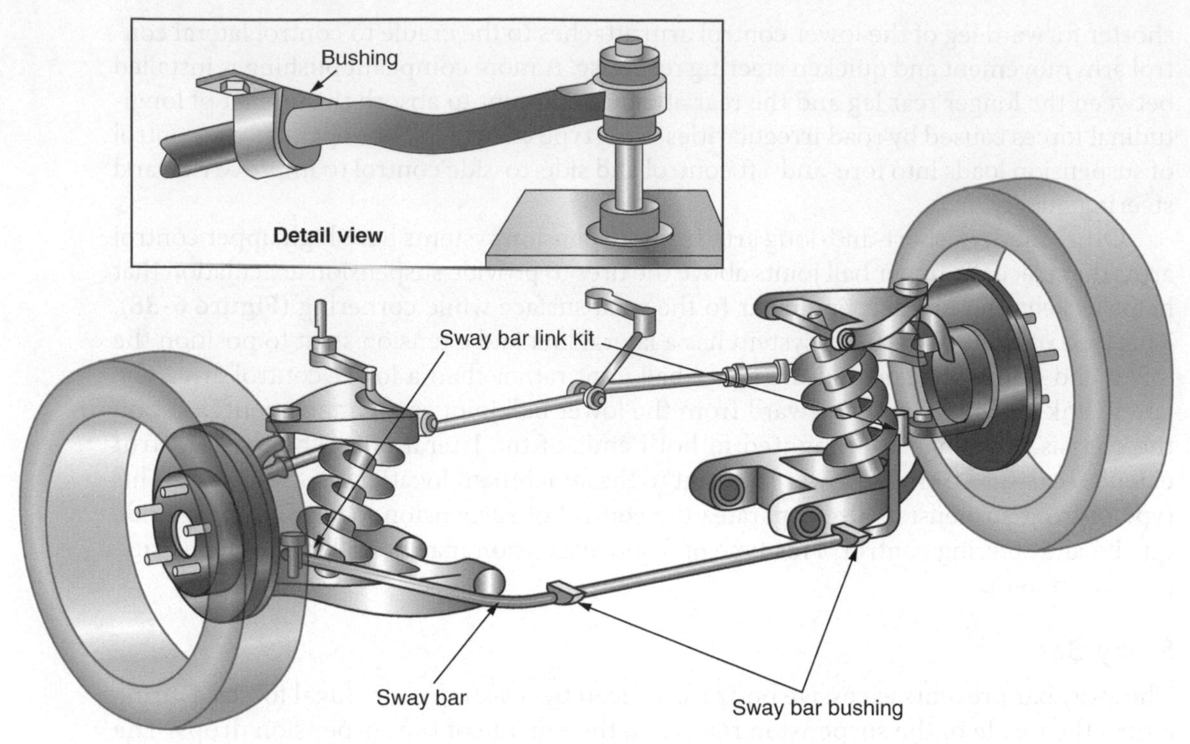

Figure 6-39 The sway bar (stabilizer bar) reduces body roll and is attached to the vehicle with mounting bushings and link pin bushings.

and rear suspension system of many vehicles. The front sway bar on a short-and-long arm suspension system connects both lower control arms to one another by sway bar links (**Figure 6-39**). Generally, each sway bar link has four bushings and a link pin and is a frequent area of wear and failure especially on higher-mileage vehicles. The sway bar is connected to the frame or unibody of the vehicle with one-piece sway bar mounting bushings and brackets. Some vehicles may exhibit a squeak or squawking noise over bumps, especially on cold mornings, as a result of faulty sway bar mounting bushings (**Figure 6-40**). Shock absorbers and struts also help in reducing body roll, so it is important to inspect all suspension components when excessive body role is suspected.

AUTHOR'S NOTE Driving over speed bumps is very useful to diagnose the presence of suspension noises. It is also helpful to drive the vehicle first thing in the morning when suspension components are cold and the suspension has settled from sitting for a long period, especially for intermittent noises.

A BIT OF HISTORY

Rear-wheel-drive cars usually have short-and-long arm or torsion bar front suspension systems. Since space and weight are important factors on today's smaller, more efficient front-wheel-drive cars, the lighter, more compact MacPherson strut front suspension is used on most of these cars.

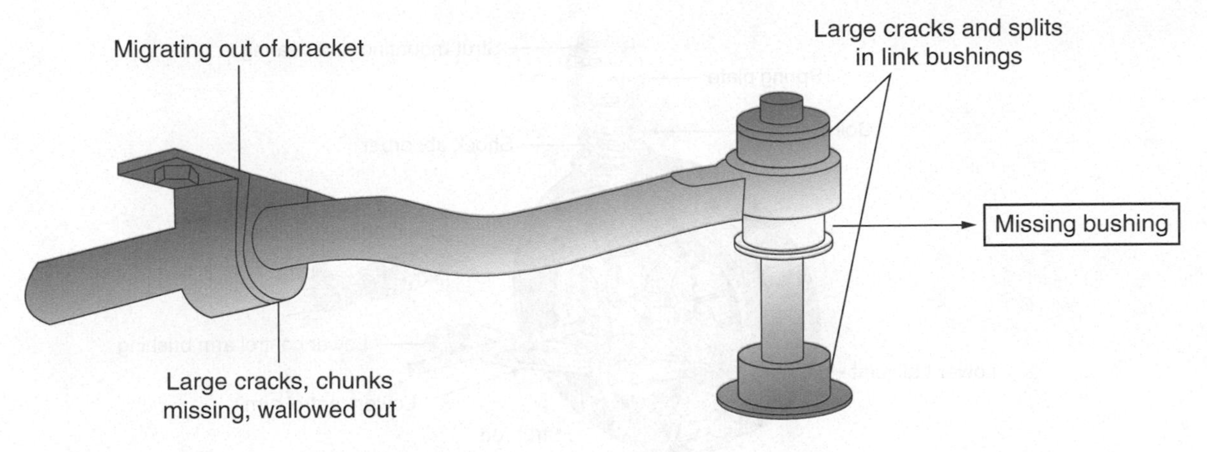

Figure 6-40 The sway bar (stabilizer bar) mounting bushings and link pin bushings are common areas of wear and damage on higher-mileage vehicles.

MACPHERSON STRUT FRONT SUSPENSION SYSTEM DESIGN

Lower Control Arms and Support

When smaller front-wheel-drive cars became popular, most of these cars had **MacPherson strut front suspension systems**. In these suspension systems, the lower end of the strut is bolted to the top of the steering knuckle, and the lower end of the knuckle is attached to the ball joint in the lower control arm (**Figure 6-41**). An upper strut mount connects the top of the strut to the chassis. An upper control arm is not required in this type of suspension system, because the strut supports the top of the steering knuckle. Since the upper control arm is not required in these suspension systems, they are more compact and therefore very suitable for smaller cars.

On some MacPherson strut front suspension systems, a steel support is positioned longitudinally on each side of the front suspension. These supports are bolted to the unitized body. The inner ends of the lower control arms contain large insulating bushings with a bolt opening in the bushing center. The control arm retaining bolts extend through the center of these bushings and openings in the support (**Figure 6-42**).

Road irregularities cause the tire and wheel to move up and down vertically, and the lower control arm bushings pivot on the mounting bolts during this movement. When the vehicle is driven over road irregularities, vibration and noise are applied to the tire and wheel. The control arm bushings help prevent the transfer of this noise and vibration to the support, the unitized body, and the passenger compartment. Proper location of the support and lower control arm is important to provide correct vehicle tracking. The supports also carry the engine and transaxle weight. Large rubber mounts are positioned between the supports and the engine and transaxle. These mounts absorb engine vibration.

A stabilizer bar may be called a sway bar.

Shop Manual
Chapter 6, page 232

Stabilizer Bar

The **stabilizer bar** is attached to the chassis and interconnects the lower control arms. Rubber insulating bushings are used at all stabilizer bar mounting positions. Some stabilizer bars are attached to the underside of the front crossmember, and the outer ends of the stabilizer bar are connected to the lower side of the front control arms (**Figure 6-43**). This type of torsion bar may be called direct contact, because the outer ends of the bar

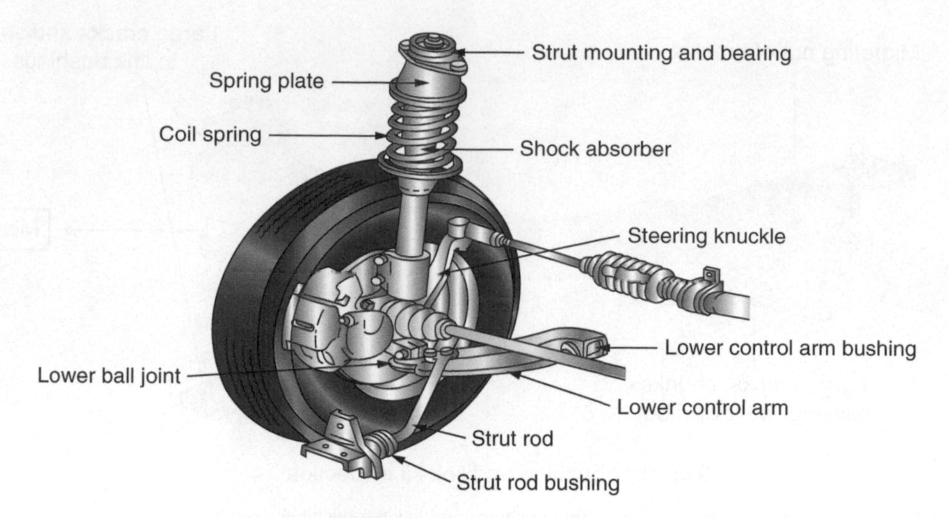

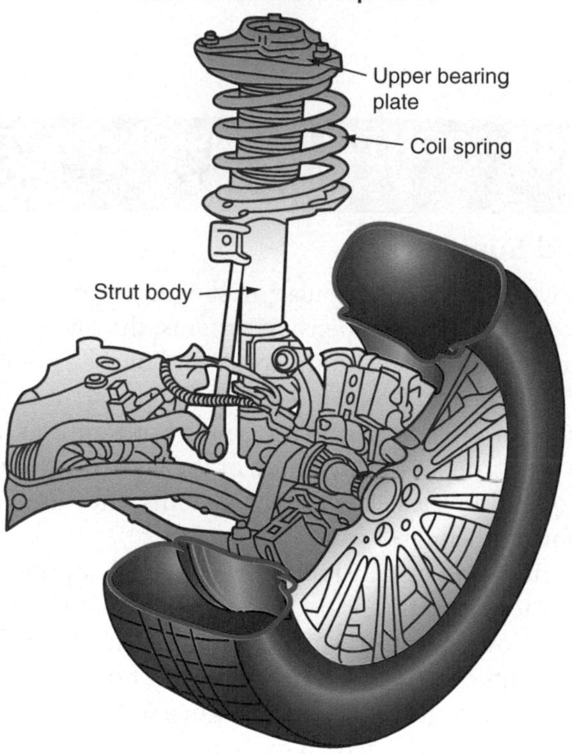

Figure 6-41 The MacPherson strut suspension system replaces the upper control arm and combines the shock absorber, spring, and upper pivot point into one assembly.

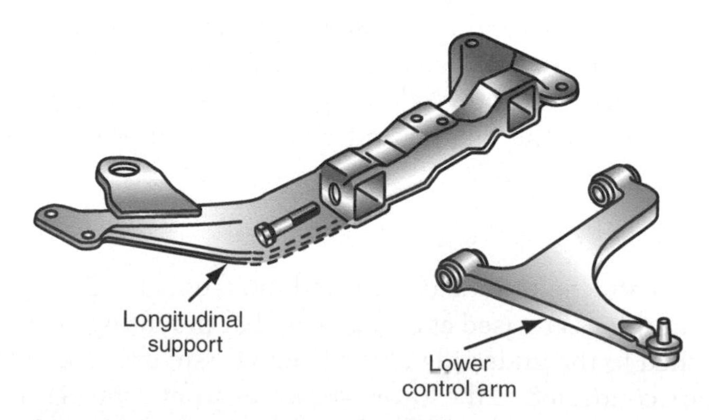

Figure 6-42 Lower control arm and support.

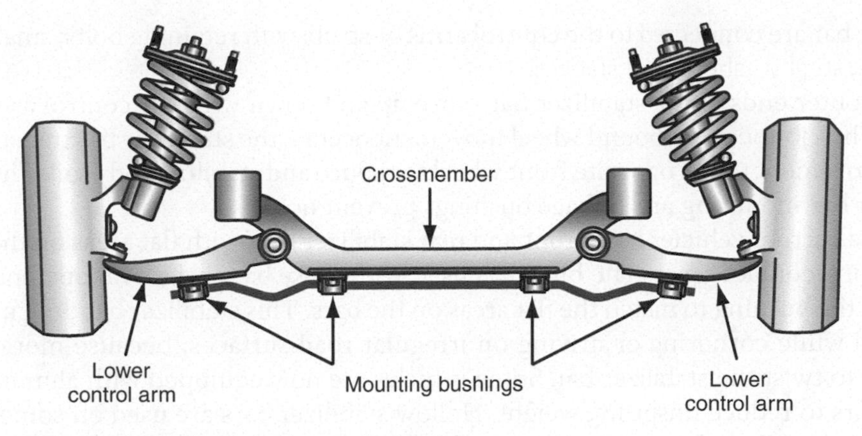

Figure 6-43 Stabilizer bar with mounting bushings.

are in direct contact with the lower control arms. Other stabilizer bars are attached to the upper side of the front longitudinal supports, and links are connected between the outer ends of the bar and the upper side of the front control arms (**Figure 6-44**). On some MacPherson strut front suspension systems, the stabilizer bars are attached to the upper side of the front subframe and the outer ends of the bar are linked to the front struts (**Figure 6-45**). Stabilizer bars with links between the outer ends of the bar and the suspension components may be called indirect contact. Large rubber bushings with steel mounting caps attach the stabilizer bar to the chassis. The linkages at the outer ends of the

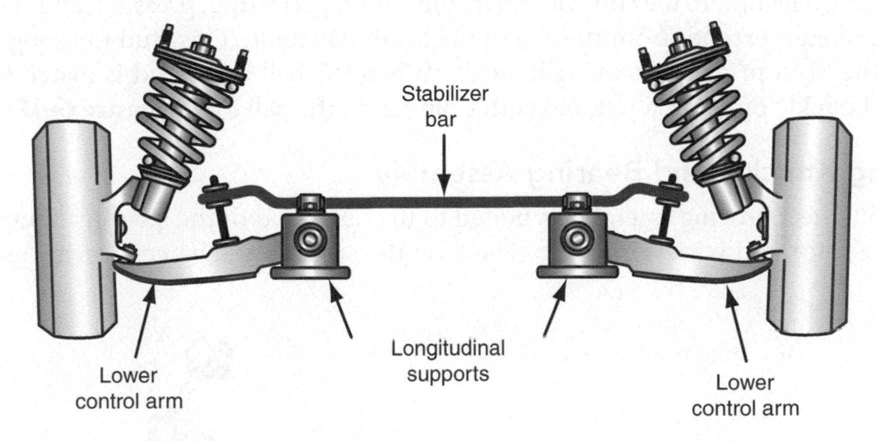

Figure 6-44 Stabilizer bar connected between the two lower control arms.

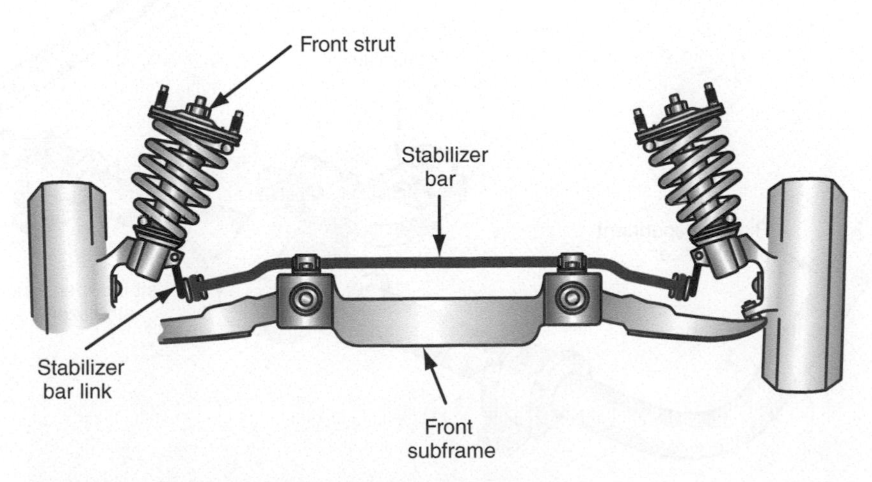

Figure 6-45 Stabilizer bar connected to the front struts.

stabilizer bar are connected to the control arms or struts with retaining bolts, small rubber bushings, steel washers, and sleeves.

The outer ends of the stabilizer bar move up and down with the control arm movement. When jounce or rebound wheel movement occurs, the stabilizer bar transmits part of this movement to the opposite front wheel to reduce and stabilize body roll. The rubber stabilizer bar mounting and linkage bushings prevent noise.

Some current vehicles have front and rear stabilizer bars with flat areas on the bars in the bushing contact areas. The bushings used with these bars have oval openings in the center of the bushing to match the flat areas on the bars. This stabilizer bar design reduces body roll while cornering or driving on irregular road surfaces, because more force is required to twist the stabilizer bar. Some vehicles are now equipped with aluminum stabilizer bars to reduce unsprung weight. Hollow stabilizer bars are used on some current vehicles for weight reduction.

An aftermarket electronically controlled stabilizer bar is available for off-road vehicle operation (**Figure 6-46**). In this type of stabilizer bar system, the driver may press a switch in the instrument panel, and the electronic control disconnects one side of the stabilizer bar from the opposite side, thus making this component ineffective. When the stabilizer bar is disconnected electronically, full vertical wheel travel is allowed on rough terrain. The driver can press the control switch again to return the stabilizer bar to normal operation for stabilized control of body roll during on-road operation.

Lower Ball Joint

The lower ball joint is attached to the outer end of the lower control arm. Methods used to attach the ball joint to the control arm include bolting, riveting, pressing, and threading. A threaded stud extends from the top of the lower ball joint. This stud fits snugly into a hole in the bottom of the steering knuckle. When the ball joint stud is installed in the steering knuckle opening, a nut and cotter pin retain the ball joint (**Figure 6-47**).

Steering Knuckle and Bearing Assembly

The front wheel bearing assembly is bolted to the outer end of the steering knuckle, and the brake rotor and wheel rim are retained on the studs in the wheel bearing assembly.

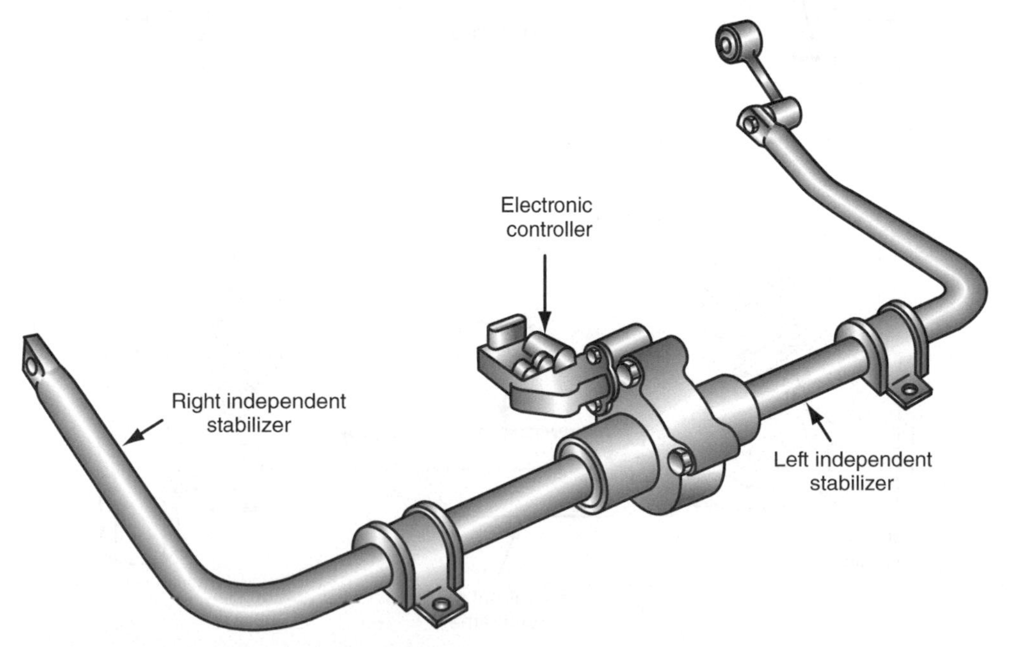

Figure 6-46 Electronically controlled stabilizer bar.

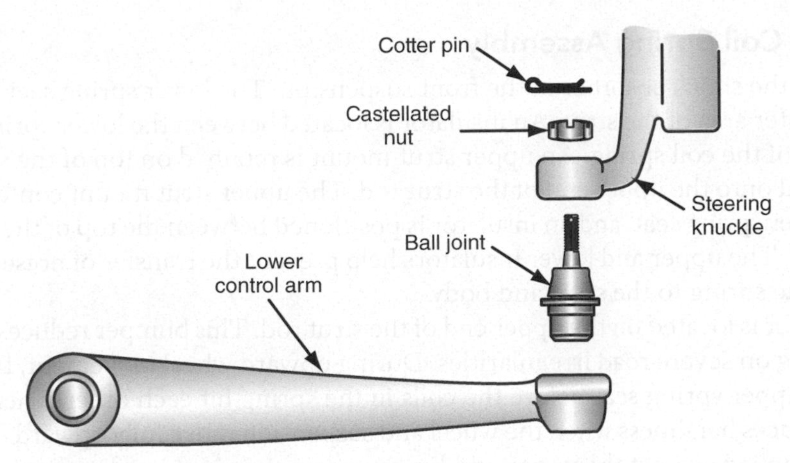

Figure 6-47 Lower ball joint.

This front wheel bearing assembly is a complete, non-serviceable, sealed unit. The front drive shaft is splined into the center of the wheel bearing hub. Thus, drive axle torque is applied to the front wheel. A tie rod end connects the steering linkage from the steering gear to the steering knuckle. The top end of the steering knuckle is bolted to the lower end of the strut (**Figure 6-48**). Many steering knuckles are manufactured from metal, but some current vehicles have aluminum steering knuckles to reduce vehicle weight and unsprung weight. Reducing unsprung weight improves ride quality and vehicle performance and reduces fuel consumption and CO_2 emissions.

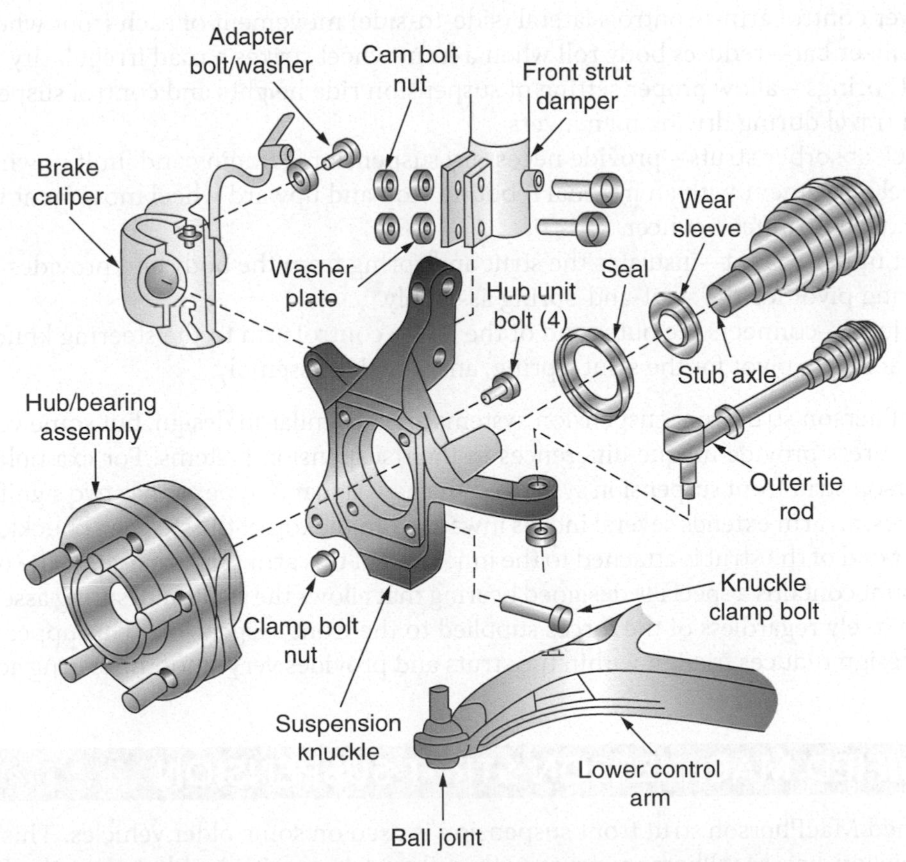

Figure 6-48 The front wheel bearing assembly is bolted to the outer end of the steering knuckle and the strut assembly is attached to the upper portion of the steering knuckle where the upper ball joint would have been located on a short-and-long-arm suspension system.

Strut and Coil Spring Assembly

The strut is the shock absorber in the front suspension. The lower spring seat is attached near the center area of the strut. An insulator is located between the lower spring seat and the bottom of the coil spring. An upper strut mount is retained on top of the strut with a nut threaded onto the upper end of the strut rod. The upper strut mount contains a bearing and upper spring seat, and an insulator is positioned between the top of the coil spring and the seat. The upper and lower insulators help prevent the transfer of noise and vibration from the spring to the strut and body.

A bumper is located on the upper end of the strut rod. This bumper reduces harshness while driving on severe road irregularities. During upward wheel movement, the bumper strikes the upper spring seat before the coils in the spring hit each other. Therefore, this bumper reduces harshness when the wheel and suspension move fully upward. The spring tension is applied against the upper and lower spring seats and insulators. However, the nut on top of the upper mount holds the spring in the compressed position between the upper and lower spring seats. When the steering wheel is turned, the steering linkage turns the steering knuckles to the right or left. During this front wheel turning action, the strut-and-spring assembly pivots on the lower ball joint and the upper strut mount bearing.

All the suspension-to-chassis mounting devices such as the lower control arm bushings and the upper strut mount must be positioned properly and be in satisfactory condition to provide correct vehicle tracking and the same wheelbase on both sides of the vehicle.

The purpose of the main components in a MacPherson strut front suspension system may be summarized as follows:

1. Lower control arm—controls lateral (side-to-side) movement of each front wheel.
2. Stabilizer bar—reduces body roll when a front wheel strikes a road irregularity.
3. Coil springs—allow proper setting of suspension ride heights and control suspension travel during driving maneuvers.
4. Shock absorber struts—provide necessary suspension damping and limit downward wheel movement with an internal rebound stop and upward wheel movement with an external jounce bouncer.
5. Strut upper mount—insulates the strut and spring from the body and provides a bearing pivot for the strut-and-spring assembly.
6. Ball joint—connects the outer end of the lower control arm to the steering knuckle and acts as a pivot for the strut, spring, and knuckle assembly.

MacPherson strut front suspension systems are all similar in design, but some vehicle manufacturers provide unique differences in their suspension systems. For example, the MacPherson strut front suspension system on the new Jaguar X-type car has two significant differences: an arm extends several inches inward from the top of the steering knuckle, and the lower end of the strut is attached to the inner end of this arm (**Figure 6-49**). The upper strut mount contains a specially designed bearing that allows the strut-and-spring assembly to rotate freely regardless of the forces supplied to the front suspension. This upper strut mount design reduces friction within the struts and provides very smooth steering action.

MODIFIED MACPHERSON STRUT SUSPENSION

A modified MacPherson strut front suspension is used on some older vehicles. This type of suspension has MacPherson struts with coil springs positioned between the lower control arms and the frame (**Figure 6-50**). The struts in these systems may be gas-filled or oil-filled.

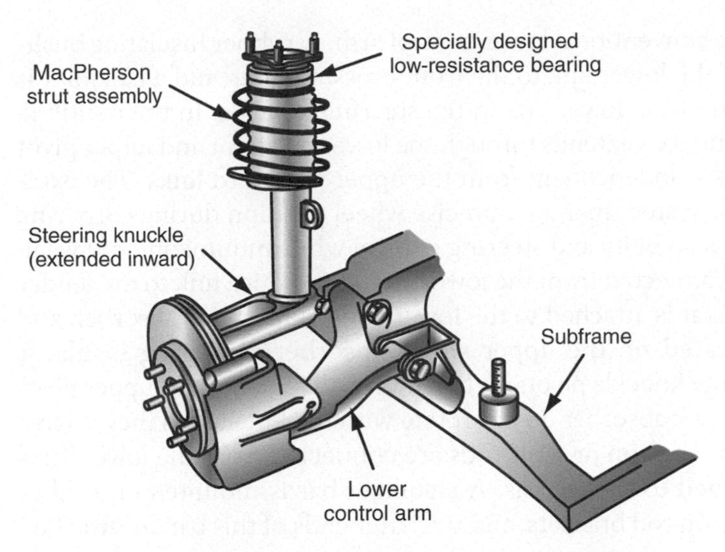

Figure 6-49 MacPherson strut front suspension system on Jaguar X-type car.

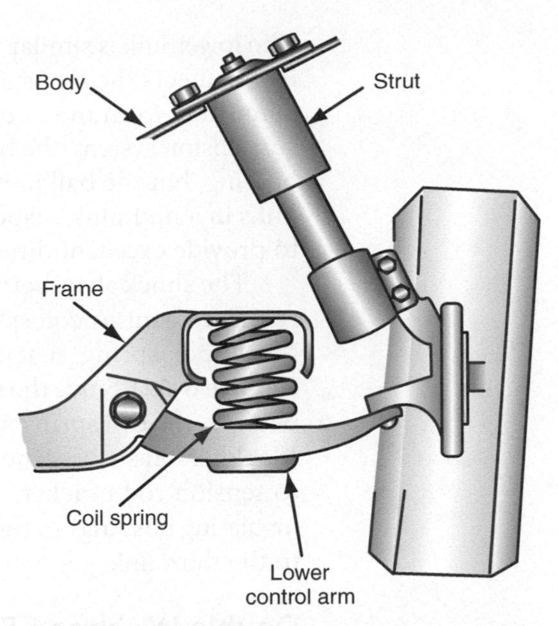

Figure 6-50 Modified MacPherson strut front suspension system.

HIGH-PERFORMANCE FRONT SUSPENSION SYSTEMS

Multilink Front Suspension System

High-performance suspension systems are usually installed on sports cars. In these cars, the driver expects improved steering quality, especially when driving and cornering at higher speeds.

In a **multilink front suspension**, a short upper link is attached to the chassis with a bracket, and the outer end of this upper link is connected to a third link. Large rubber insulating bushings are mounted in each end of the upper link. The lower end of the third link is connected through a heavy pivot bearing to the steering knuckle (**Figure 6-51**).

The **multilink front suspension** may be referred to as a double wishbone suspension because of the link design. It has upper and lower links, and a third link connects the upper link to the top of the knuckle through a bearing.

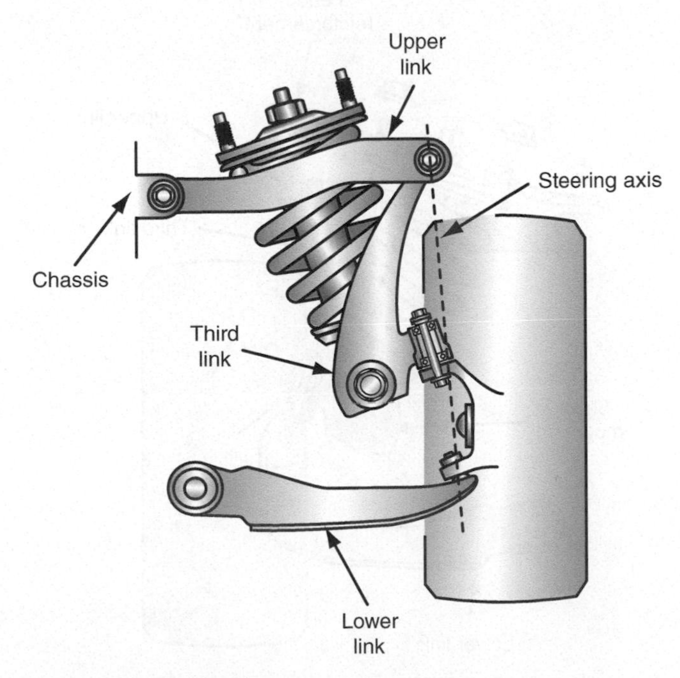

Figure 6-51 Multilink front suspension system.

The lower link is similar to the conventional lower control arm. A rubber insulating bushing connects the inner end of the lower link to the front crossmember, and a ball joint is connected from the outer end of the lower link to the steering knuckle. In the multilink suspension system, the ball joint axis extends through the lower ball joint and upper pivot bearing, but the ball joint axis is independent from the upper and third links. The extra links in a multilink suspension system maintain precise wheel position during cornering to provide excellent directional stability and steering control while minimizing tire wear.

The shock absorbers are connected from the lower end of the third link to the fender reinforcement. A coil spring seat is attached to the lower end of the shock absorber, and the upper spring seat is located on the upper shock absorber mounting insulator (**Figure 6-52**). Since the steering knuckle pivots on the lower ball joint and the upper pivot bearing, the coil spring and shock absorber do not rotate with the knuckle as they do in a MacPherson strut suspension. Tension or strut rods are connected from the lower links to tension rod brackets attached to the chassis. A stabilizer bar is mounted on rubber insulating bushings in the tension rod brackets, and the outer ends of this bar are attached to the third link.

Double Wishbone Front Suspension System

Double wishbone suspension systems provide increased suspension rigidity and maintain precise wheel position under all driving conditions to supply improved directional stability and steering control. In the double wishbone front suspension system, the upper and lower control arms are manufactured from lightweight, high-strength aluminum alloys designed for maximum strength and rigidity. These lighter control arms decrease the unsprung weight of the vehicle, which improves traction and ride quality. Since the upper and lower control arms have a wishbone shape, the term *double wishbone* is used for this type of suspension. Suspension rigidity is also increased by positioning the ball joints and steering knuckle inside the wheel profile (**Figure 6-53**). On each side of the car, the front suspension is attached to the chassis by a cast aluminum subframe. This design also reduces vehicle weight. The double wishbones are attached to the chassis at the most efficient locations to maintain precise wheel position and provide improved ride quality.

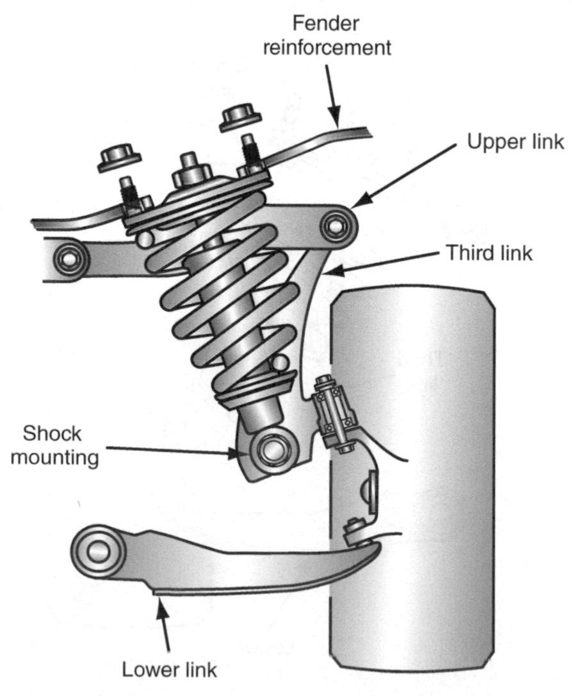

Figure 6-52 Complete multilink suspension system.

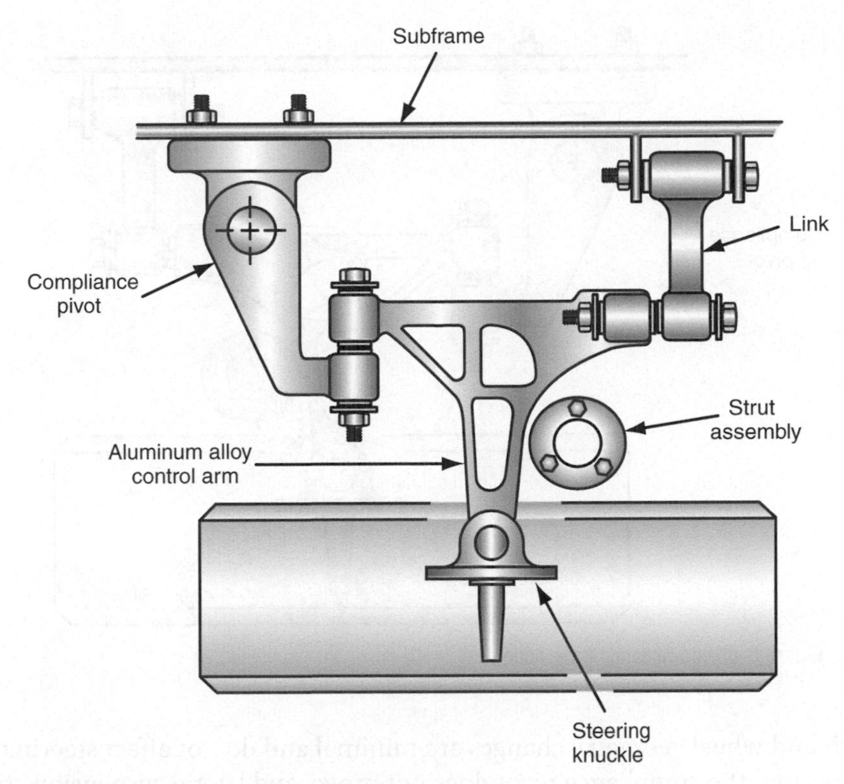

Figure 6-53 Double wishbone front suspension.

The front ends of the upper and lower control arms are attached to a compliance pivot assembly. Bushings are mounted in the front ends of the upper and lower control arms, and these bushings are bolted to the compliance pivot (**Figure 6-54**). When one of the front wheels is subjected to rearward force by hard braking or a road irregularity, the coil spring is compressed and the ride height is lowered. This rearward force on the front wheel twists the compliance pivot, allowing both control arms to pivot slightly. Under this condition, the upper and lower control arm movement allows the front wheel to move rearward a small amount, and this wheel movement absorbs energy to significantly improve ride quality (**Figure 6-55**). During this upper and lower control arm movement,

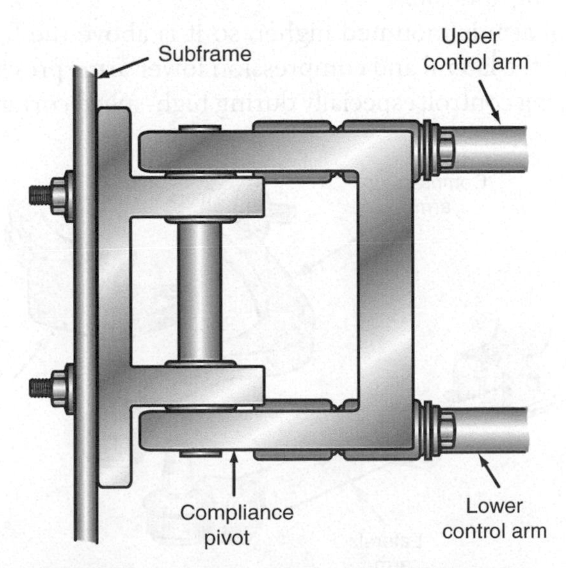

Figure 6-54 Compliance pivot on double wishbone front suspension.

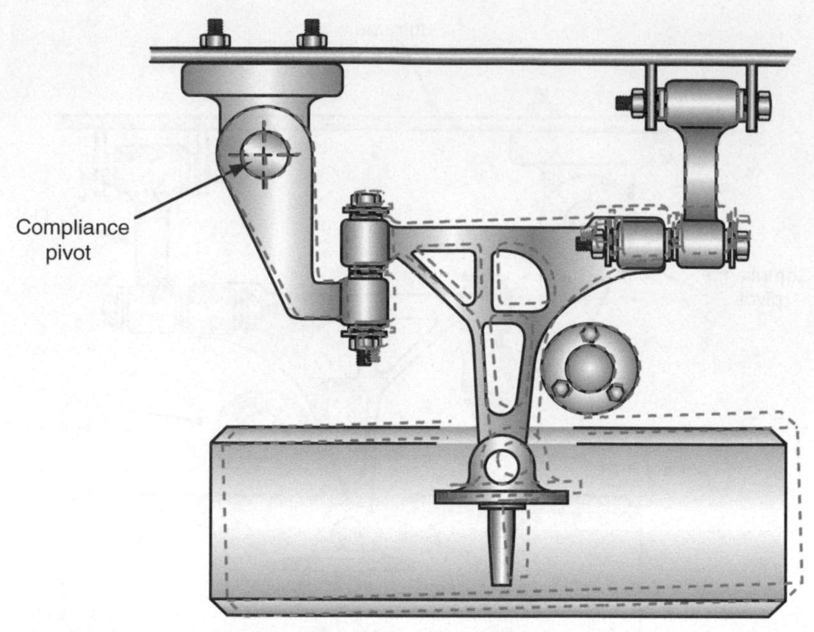

Figure 6-55 Compliance pivot action.

track width and wheel geometry changes are minimal and do not affect steering control. While cornering, the compliance pivot does not move, and lateral suspension stiffness is maintained to supply excellent steering control.

Multilink Front Suspension with Compression and Lateral Lower Arms

Some lateral multilink front suspensions have compression and lateral lower arms. The lateral arm prevents front wheel movement, and the compression arm prevents fore-and-aft front wheel movement. A rubber insulating bushing in the inner end of the lateral arm is bolted to the chassis. A second rubber insulating bushing near the outer end of the lateral arm is bolted to the damper fork. The upper end of the damper fork is bolted to the front strut (**Figure 6-56** and **Figure 6-57**). A ball joint in the outer end of the lateral arm is bolted into the steering knuckle. A rubber insulating bushing in the inner end of the compression arm is bolted to the chassis, and a ball joint in the outer end of this arm is bolted into the steering knuckle.

The upper control arm is mounted higher, so it is above the front tire. The higher upper control arm and the lateral and compression lower arms provide excellent suspension stability and steering control, especially during high-speed cornering or when driving

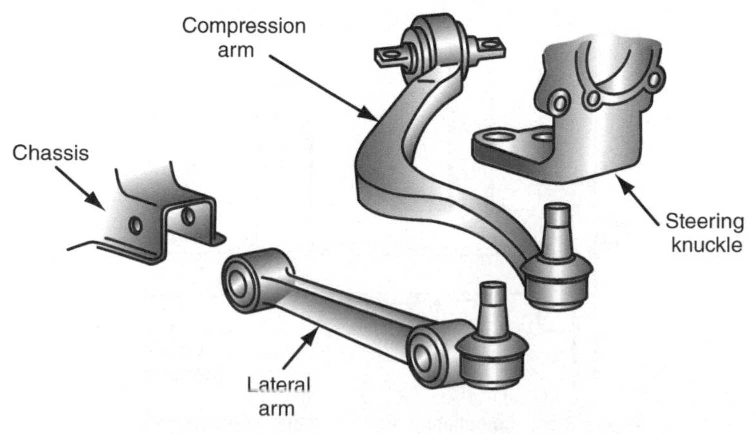

Figure 6-56 Multilink front suspension with lateral and compression lower arms.

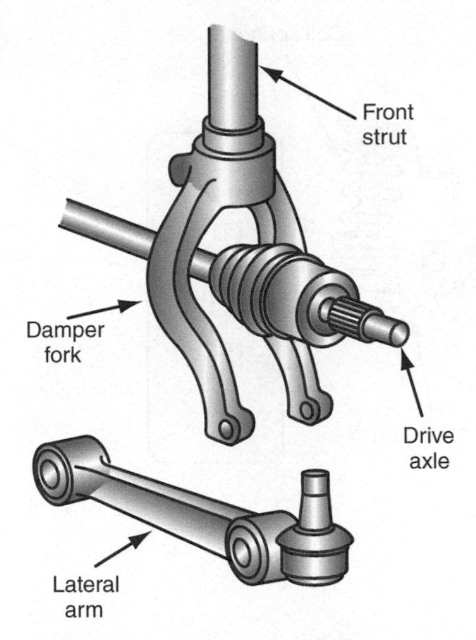

Figure 6-57 Multilink front suspension damper fork.

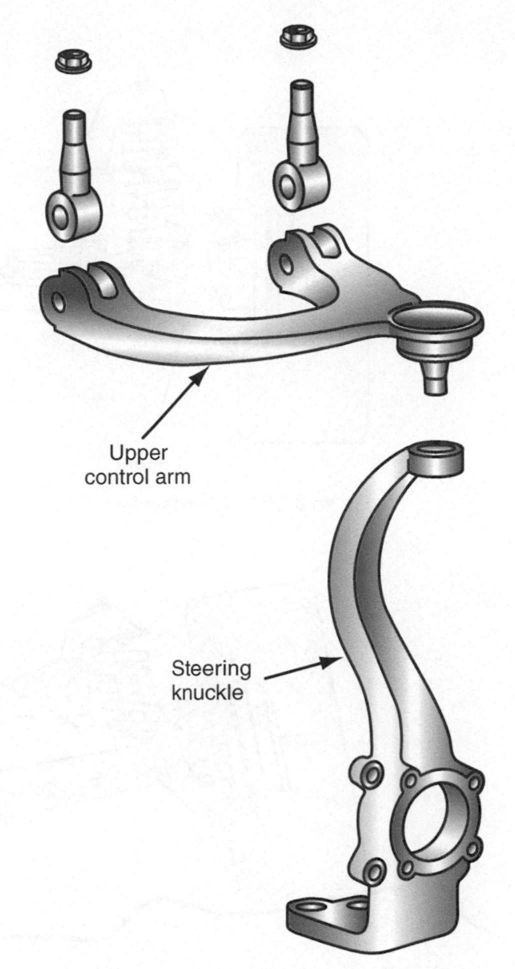

Figure 6-58 Multilink front suspension upper control arm.

on irregular road surfaces. A ball joint in the outer end of the upper control arm is attached to the top of the knuckle, and two shafts in the inner end of this arm are bolted into the strut tower (**Figure 6-58**). There are no provisions for camber or caster adjustments on this multilink front suspension.

Shop Manual
Chapter 6, page 231

Twin I-Beam Suspension Systems

Some Ford trucks and Ford two-wheel-drive Super Duty trucks are equipped with twin I-beam front suspension systems. In this type of suspension system, each front wheel is connected to a separate I-beam. The outer ends of the I-beam are connected to the spindles, and the inner ends of the beams are connected through a rubber pivot bushing to the chassis (**Figure 6-59**). Coil springs are positioned between the I-beams and the chassis to support the vehicle weight. Radius arms are connected rearward from each I-beam to the chassis to prevent longitudinal wheel movement. Since each front wheel can move independently in a twin I-beam suspension system, the problems associated with straight I-beam systems are greatly reduced.

In some older I-beam front suspension systems, kingpins were used to attach the I-beams to the spindles. In other I-beam suspension systems, ball joints connect the I-beams to the spindles.

Some Ford four-wheel-drive trucks used twin I-beam front suspension systems with a coil spring twin I-beam front suspension (**Figure 6-60**). The lower ends of the coil springs are seated on the twin I-beams, and the upper spring seat is positioned on the chassis. Heavy radius arms are bolted to the twin I-beams, and the other ends of the radius

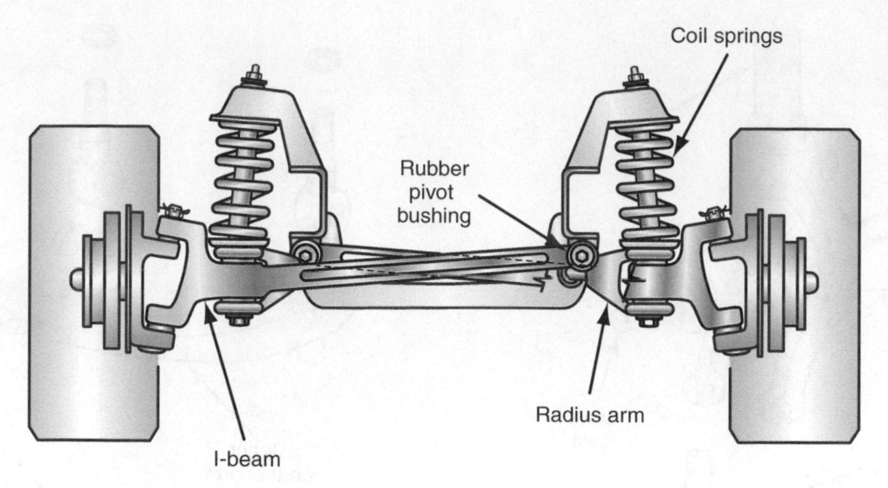

Figure 6-59 Twin I-beam front suspension.

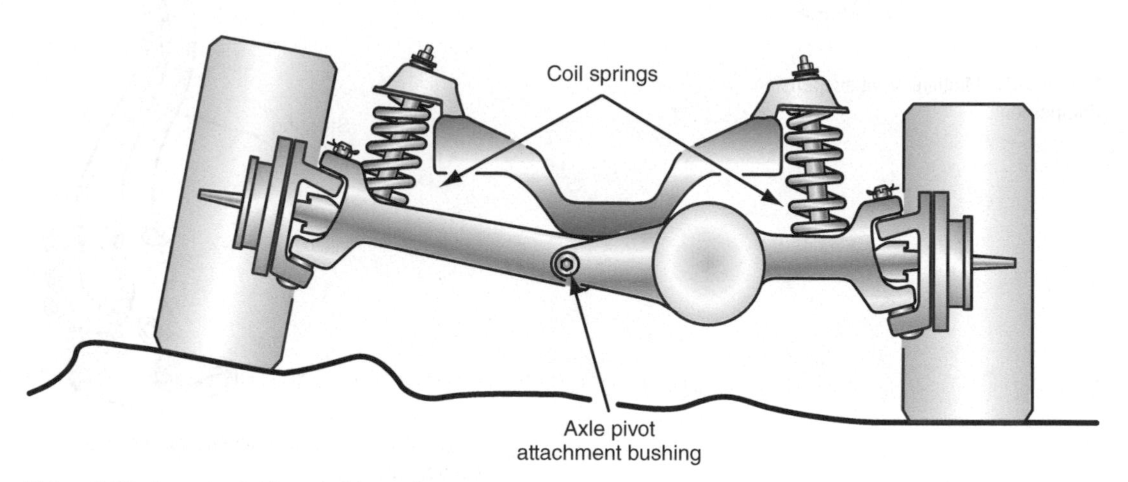

Figure 6-60 Four-wheel-drive twin I-beam front suspension with coil springs.

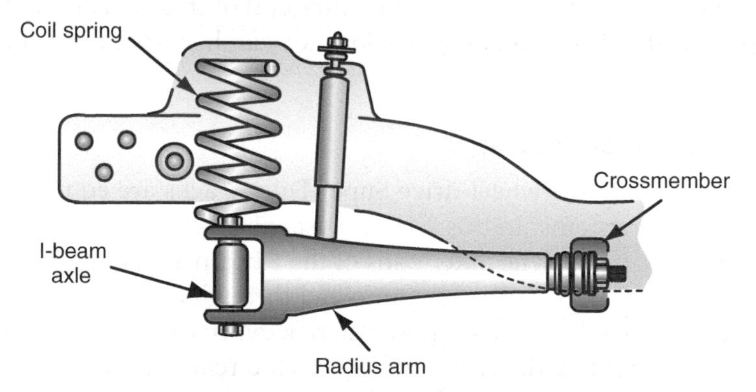

Figure 6-61 Radius arms on a twin I-beam front suspension system with coil springs.

arms are mounted in a bushing and frame bracket (**Figure 6-61**). The radius arms prevent axle movement. The bushings in the inner ends of the twin I-beams are mounted on pivots attached to the front crossmember. Since each front wheel can move independently, twin I-beam suspensions are independent suspension systems. Upper and lower ball joints are pressed in the ends of the steering knuckle. The studs on these ball joints extend through openings in the outer ends of the twin I-beams. The ball joint studs are

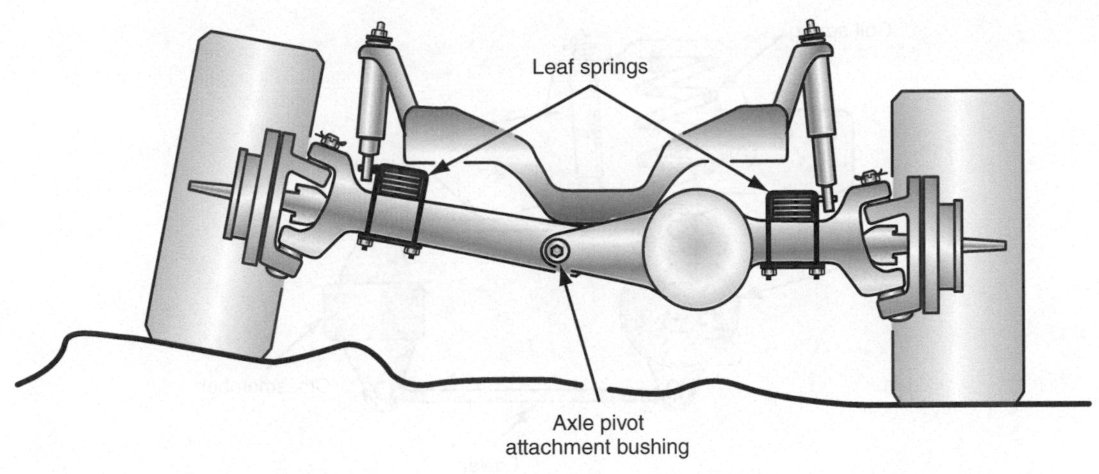

Figure 6-62 Four-wheel-drive twin I-beam front suspension system with leaf springs.

retained in the twin I-beams with nuts and cotter pins. Universal joints in the outer ends of the drive axles allow simultaneous wheel rotation and wheel turning to the right or left. The right-side drive axle also has an inner universal joint near the differential. Newer design Ford Super Duty four-wheel-drive trucks use a coil spring solid front axle assembly with heavy radius arms.

Older Ford F-250 and F-350 four-wheel-drive trucks used a leaf-spring twin I-beam front suspension system (**Figure 6-62**). U-bolts retain the leaf springs to the twin I-beams. A bushing in the front of the leaf-spring eye is bolted into a frame bracket. The bushing in the rear spring eye is connected to the frame through a conventional spring shackle (**Figure 6-63**). Since the leaf springs maintain the axle position, the radius arms are not required.

Some light-duty four-wheel-drive trucks have a straight front drive axle housing and a coil spring suspension system (**Figure 6-64**). Upper and lower radius rods control fore-and-aft drive axle movement, and a track bar controls lateral axle movement. Ball joints connect the steering knuckles to the outer ends of the drive axle housing.

Ford four-wheel drive Super Duty trucks use a similar design but with a single piece radius arm (**Figure 6-65**) to secure the front axle assembly with improved handling and stability.

Twin I-beam front suspension systems have recently been replaced with short-and-long arm or torsion bar suspension systems, because these systems are light weight and provide improved ride quality.

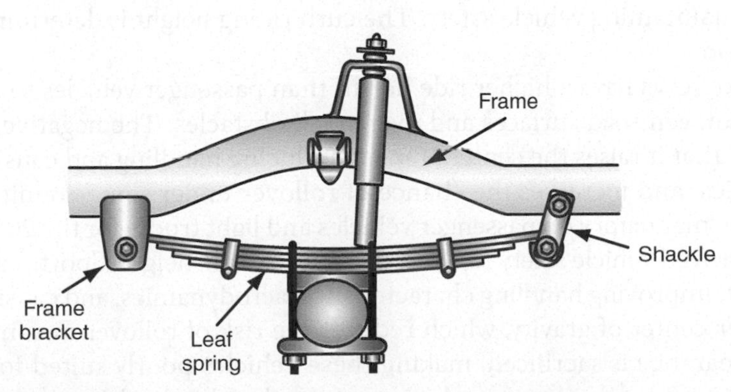

Figure 6-63 Leaf spring mounting, four-wheel-drive twin I-beam front suspension.

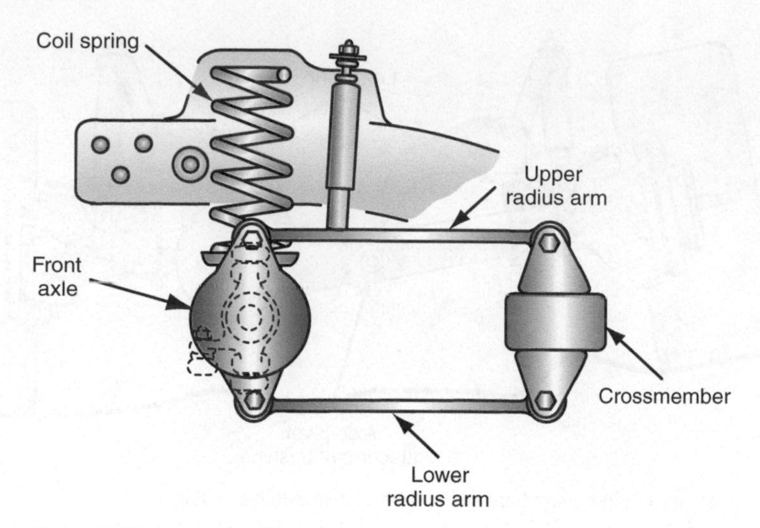

Figure 6-64 Four-wheel-drive front suspension with straight drive axle and coil springs.

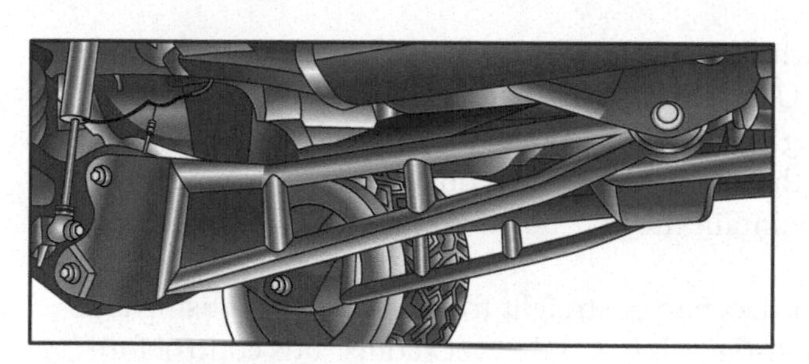

Figure 6-65 Ford Super Duty front radius arm.

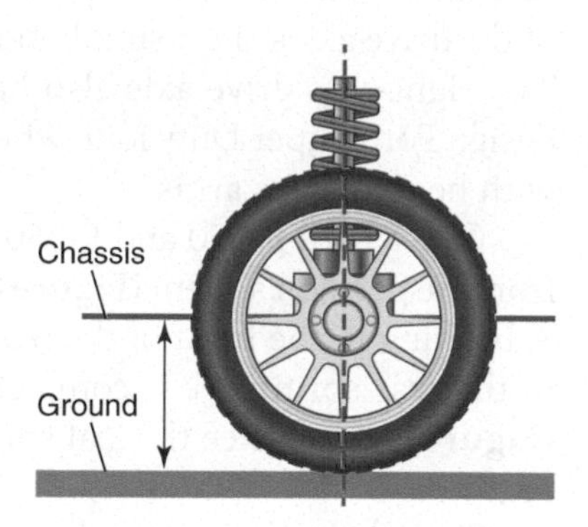

Figure 6-66 Ride height determines the amount of vehicle chassis ground clearance.

Ride Height

Vehicle **ride height** determines the amount of ground clearance from the bottom of the tire touching the ground and the lowest point of the vehicle's structure (**Figure 6-66**). Ride height determines the amount of vehicle clearance from the ground. Proper vehicle ride height is a compromise between practicality, handling, and stability without sacrificing safety. Regular inspection and proper maintenance of suspension systems are extremely important to maintaining vehicle safety. The curb riding height is determined mainly by spring condition.

Trucks and SUVs have a higher ride height than passenger vehicles to allow them to be driven on uneven road surfaces and over small obstacles. The negatives of increased ride height are that it raises the center of mass, reducing handling and causing a decrease in aerodynamics, and increases the chance of rollover under some conditions. Stability control became mandatory on passenger vehicles and light trucks for the 2012 model year, which has improved vehicle safety regardless of vehicle ride height. Sports cars have a very low ride height, improving handling characteristics, aerodynamics, and causing the vehicle to have a lower center of gravity, which reduces the risk of rollover in some conditions. But ground clearance is sacrificed, making these vehicles poorly suited for uneven and rough roads. Low ground clearance vehicles require that they be driven at very low speeds on very rough and uneven surfaces, and generally not at all on off-road conditions.

A vehicle's ride height should be maintained with manufacturer specifications. Most manufacturers allow up to 1 in. lower than specified ride height. If ride height is below manufacturer specifications it is often an indication of weak springs, but could also be caused by faulty suspension components, structural damage, or, on torsion bar suspensions, the need for height adjustment. Sagged springs cause insufficient curb riding height; therefore, the distance is reduced between the rebound bumper and its stop. This distance reduction causes the bumper to hit the stop frequently with resulting harsh ride quality.

Other suspension components such as worn control arm bushings will affect curb riding height. Since incorrect curb riding height affects most of the other suspension angles, this measurement is critical. Sagged springs change the normal operating arc of the lower ball joint. This action causes excessive lateral movement of the tire during wheel jounce and rebound with resulting tire wear (**Figure 6-67**).

Manufacturers specify vehicle ride height in their service information and the point at which ride height is measured which is critical (**Figure 6-68**). The curb riding height must be measured at the vehicle manufacturer's specified location, which varies depending on the type of suspension system. When the vehicle is on a level floor or an alignment rack, measure the curb riding height from the floor to the manufacturer's specified location on the chassis.

The measurement reference point may be on the fender, frame, body rocker panel, bumper, or suspension control arm. On some applications, such as light trucks, the measurement may be made between two points on the vehicle's suspension system.

Always verify where the measurement is supposed to be made (**Figure 6-69**). For most conventional ride height measurements, all that is required is a tape measure or ruler and possibly a level. Verify that the vehicle is equipped with the manufacturer-specified

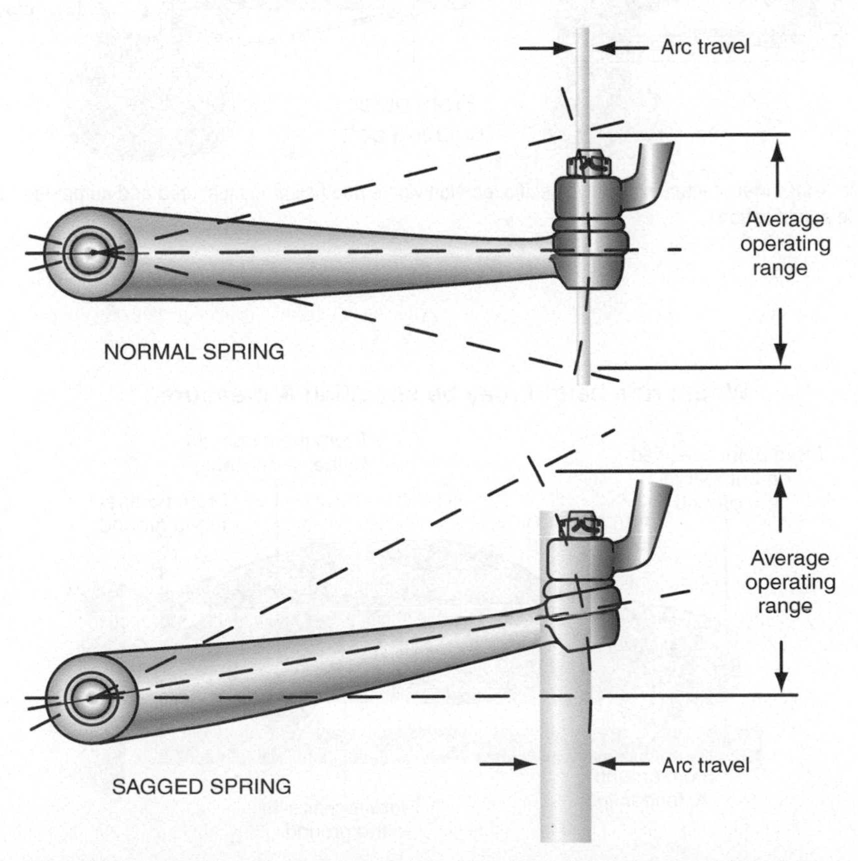

Figure 6-67 Sagged springs change the normal operating arc of the lower ball joint and cause excessive lateral tire movement during wheel jounce and rebound with resulting tire wear.

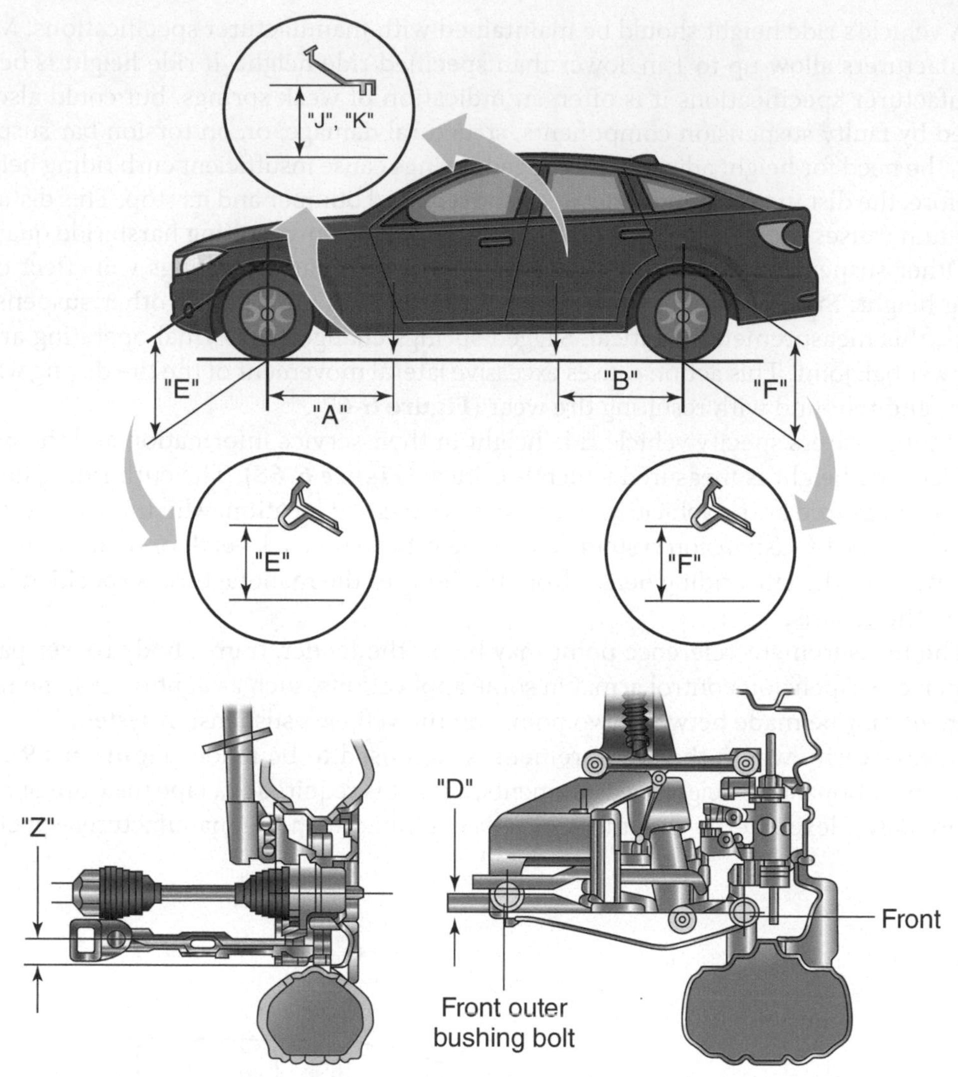

Figure 6-68 Manufacturers specify specific location where ride height is measured and compared to vehicle specifications.

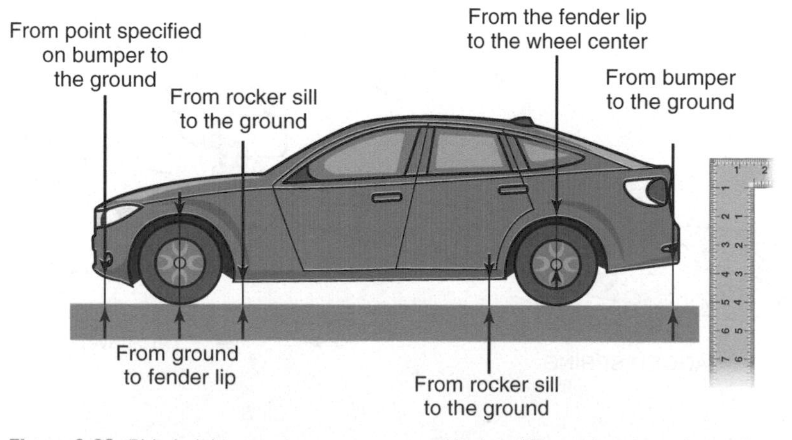

Figure 6-69 Ride height measurements are specified at different locations depending on the year, make, and model of vehicle.

tire and rim size. On vehicles where the tire and wheel assemblies have been replaced and that differ in overall assembly height (shorter or taller) than that specified by the manufacturer, ride height measurements will be affected. Shorter assemblies will decrease ride height, and taller assemblies will increase ride height if the measurement is being made from a vehicle reference point to the ground.

Caster angle is the tilt of an imaginary line through the center of the upper strut mount and ball joint in relation to the true vertical line through the center of the wheel and spindle as viewed from the side.

The vehicle's ride height should be measured at all four corners of the vehicle. Comparing side-to-side measurements on both the front and the rear of the vehicle will aid in identifying weak springs, frame damage, and body misalignment issues. A ride height difference of more than 1 in. could cause the vehicle to pull to one side.

Errors in ride height can cause caster measurements to be out of specifications. The caster angle is the number of degrees between the true vertical centerline of the tire and wheel, and an imaginary line through the center of the upper strut mount and lower ball joint (**Figure 6-70**). If the front of the vehicle is lower than specified, the caster measurement will be more negative (**Figure 6-71**). If the rear ride height is lower than specified, the caster measurement will be more positive (**Figure 6-72**).

If both the front and rear ride height are equally low or high, there will be no net change.

Incorrect vehicle ride height can cause camber measurements to be out of specifications (**Figure 6-73**); this is especially true if there is a side-to-side ride height difference in the front. Side-to-side ride height difference in the rear can affect rear camber readings on vehicles with independent rear suspension systems. The amount of camber change will depend on the specific vehicle geometry and the amount of ride height difference. If a spring is sagged on one side of the front suspension, the camber angle is negative on that

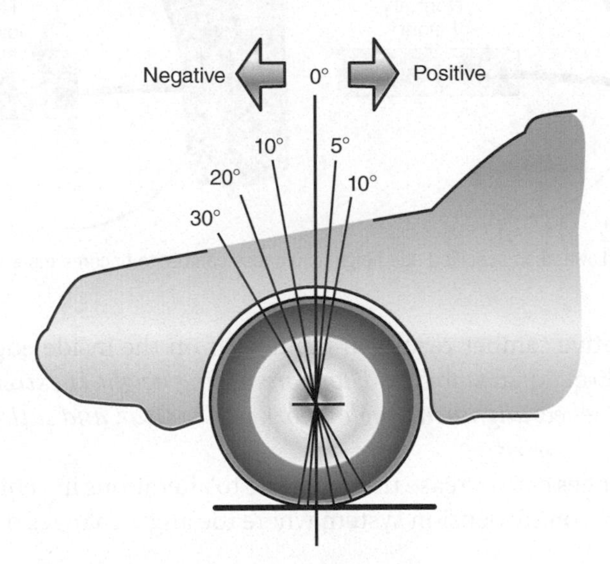

Figure 6-70 Positive and negative caster.

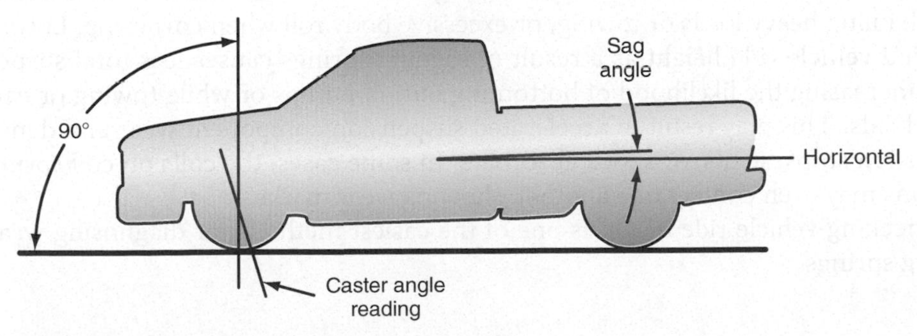

Figure 6-71 Effects of front spring sag on caster angle.

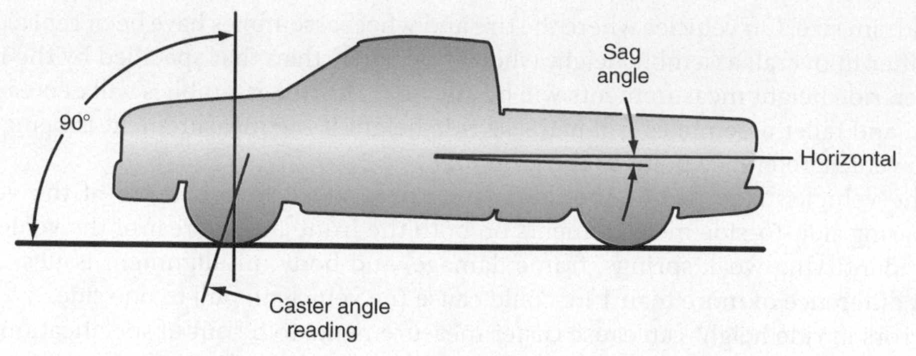

Figure 6-72 Effects of rear spring sag and incorrect curb riding height on caster angle.

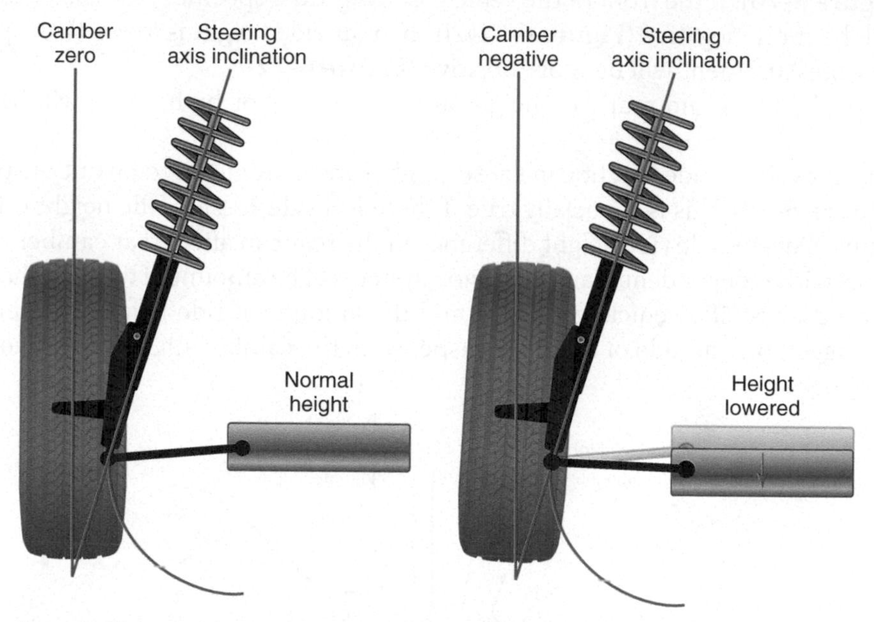

Figure 6-73 Lower than specified ride height can cause camber to become more negative.

side. Excessive negative camber results in rapid wear on the inside edge of the tire tread and a decrease in directional stability. *Therefore, riding height is extremely important in maintaining correct wheel alignment angles, normal tire wear, and satisfactory directional stability.*

Ride height changes can increase tire wear due to alterations in vehicle geometry. This is particularly evident on suspension system where toe angle changes occur with a change in ride height.

Vehicles with lower than specified ride height may generate customer complaints related to suspension bottoming, nose diving while braking, instability and wandering when hauling heavy loads or towing, or excessive body roll when cornering. Lower than specified vehicle ride height as a result of sagging springs causes less total suspension travel increasing the likelihood of bottoming out on bumps or while towing or carrying heavy loads. This may result in accelerated suspension component wear and damage to suspension stops, struts, or shock absorbers. In some cases, the coils on coil spring suspensions may even contact one another, showing wear marks.

Camber angle is the tilt of a line through the center of the wheel in relation to the vertical centerline viewed from the front of the wheel.

Checking vehicle ride height is one of the easiest methods for diagnosing weak and sagging springs.

SUMMARY

- Coil springs may be classified as linear rate or variable rate. A linear-rate coil spring has a constant spring rate regardless of spring load. Variable-rate coil springs have unequally spaced coils. The coils in these springs may be referred to as inactive, transitional, or active.
- Heavy-duty coil springs have larger-diameter wire than regular-duty coil springs. Most vehicle manufacturers recommend that springs be replaced in pairs. In a torsion bar suspension system, the torsion bars replace the coil springs. Torsion bars are directional, and they are marked right or left. In a leaf spring, zinc and plastic spacers between the leaves reduce noise and friction problems.
- Leaf springs may be mono-leaf or multiple-leaf design. Spring leaves may be manufactured from steel or fiberglass. Although many leaf springs are mounted longitudinally, some of these springs are mounted transversely.
- Ball joints may be classified as load carrying or non-load carrying. A load-carrying ball joint may be compression loaded or tension loaded. The load-carrying ball joint wears faster than the non-load-carrying ball joint.
- Many ball joints have wear indicators that provide visual ball joint wear inspection. A stabilizer bar reduces body roll or sway when one front wheel strikes a road irregularity. A strut rod prevents fore-and-aft lower control arm movement. In many MacPherson strut front suspension systems, the lower end of the strut is bolted to the steering knuckle, and the upper end of the strut is connected through an upper mount to the fender reinforcement.
- Torsion bars may be transversely or longitudinally mounted in a front suspension system. In a short-and-long arm suspension system, the coil springs may be mounted between the lower control arm and the frame or between the upper control arm and the chassis.
- In a twin I-beam suspension system, the spindles may be connected to the I-beams with kingpins or ball joints.
- Curb riding height is extremely important for maintaining normal tire wear and proper suspension alignment angles.

REVIEW QUESTIONS

Short Answer Essays

1. Explain the meaning of constant spring rate.
2. Describe the main difference in the design of heavy-duty coil springs compared with regular-duty coil springs, and compare the free diameter of each type of spring.
3. Explain the type of load condition that requires variable-rate coil springs, and describe the load-carrying advantage of these springs.
4. Describe the design of taper-wire closed-end springs.
5. Explain torsion bar action during wheel jounce and rebound.
6. Describe the action of a leaf spring as it compresses.
7. Explain the position of the lower control arm when the lower ball joint is compression loaded.
8. Describe the position of the wear indicator when a ball joint is worn.
9. Explain the basic purpose of the lower control arms.
10. Describe the effect of sagged front springs on caster angle and directional stability.

Fill-in-the-Blanks

1. Heavy-duty coil springs are designed to carry _______________ to _______________ percent greater loads than regular-duty coil springs.
2. A variable-rate coil spring provides automatic _______________ adjustment.
3. In a variable-rate spring, the _______________ coils operate during the complete range of spring loading.
4. The load-carrying ball joint is attached to the control arm on which the _______________ is seated.
5. If the lower control arm is mounted above the steering knuckle and rests on the knuckle, the lower ball joint is_______________.

6. When the grease nipple shoulder is extended from the ball joint housing, the ball joint is____________.
7. When one front wheel strikes a road irregularity, the stabilizer bar reduces______________.
8. The strut rod prevents ______________ and ______________ lower control arm movement.
9. When the front springs are sagged, the positive caster is______________.
10. If a front spring is sagged, the camber angle moves toward a ______________ position.

Multiple Choice

1. A 400-lb load deflects a linear-rate coil spring 1 in. An 800-lb load will deflect this coil spring:
   A. 1.5 in.
   B. 2 in.
   C. 3 in.
   D. 4 in.
2. Heavy-duty coil springs have:
   A. The same wire diameter as a regular-duty coil spring.
   B. A shorter free height compared to a regular-duty coil spring.
   C. A 10 percent increase in load-carrying capacity.
   D. A high aluminum content in the spring material.
3. A car's rear coil springs keep breaking. This problem is likely because of:
   A. Continual driving on rough road surfaces.
   B. Continual driving on curved roads.
   C. Excessive air pressure in the rear tires.
   D. Constant overloading of the rear suspension.
4. Aluminum control arms have all of these advantages EXCEPT:
   A. Reduce unsprung weight.
   B. Provide improved road feel.
   C. Provide improved ride quality.
   D. Contribute to improved fuel economy.
5. While discussing a torsion bar front suspension system:
   A. One end of the torsion bar is attached to the lower control arm.
   B. A suspension height adjustment is positioned on the end of the torsion bar connected to the control arm.
   C. Torsion bars may be used with coil springs in a front suspension system.
   D. Torsion bars eliminate the need for shock absorbers in the front suspension system.
6. While discussing leaf springs:
   A. A fiberglass mono-leaf spring is heavier than a steel mono-leaf spring.
   B. Some mono-leaf springs are mounted transversely.
   C. In a multiple-leaf spring, the lower leaf is called the main leaf.
   D. A multiple-leaf spring has the same stiffness regardless of how much it is compressed.
7. All these statements about ball joints are true EXCEPT:
   A. A load-carrying ball joint wears faster than a non-load-carrying ball joint.
   B. A non-load-carrying ball joint is designed with a preload.
   C. Wear indicators may be positioned in the side of the ball-joint housing.
   D. Low-friction ball joints have a highly polished ball socket surrounded by a high-strength polymer bearing.
8. While discussing MacPherson strut suspension systems:
   A. During a turn, the strut, spring, and knuckle rotate on the upper strut mount.
   B. The lower end of the coil spring is seated on the lower control arm.
   C. The strut is welded into the knuckle and these components are replaced as an assembly.
   D. The inner ends of the lower control arms are bolted to the strut.

9. While discussing short-and-long arm front suspension systems:
   A. The coil springs are positioned between the upper and lower control arms.
   B. Compared with a suspension system with equal length control arms, the short-and-long design reduces track width change and tire wear.
   C. The outer ends of the stabilizer bar may be attached to the steering knuckles.
   D. The coil spring is retained in the lower control arm with a clamp and bolt.

10. While discussing sagged springs:
    A. Sagged front springs cause harsh riding.
    B. Sagged front springs cause excessive positive caster on the front wheels and increased directional stability.
    C. Sagged rear springs cause excessive negative caster on the front wheels.
    D. Sagged rear springs cause excessive positive camber on the front wheels.

CHAPTER 7
REAR SUSPENSION SYSTEMS

Upon completion and review of this chapter, you should be able to understand and describe:

- A live axle rear suspension system.
- The advantages and disadvantages of a live axle leaf-spring rear suspension system.
- The movement of the rear axle housing during vehicle acceleration.
- How the differential torque is absorbed in a live axle coil spring rear suspension system.
- The purpose of a tracking bar in a live axle coil spring rear suspension system.
- The difference between a semi-independent and an independent rear suspension system.
- How individual rear wheel movement is provided in a semi-independent rear suspension system.
- The difference between a MacPherson strut and a modified MacPherson strut rear suspension.
- The advantage of attaching the differential housing to the chassis in an independent rear suspension system.
- How differential and suspension vibration, noise, and shock are insulated from the chassis in a multilink independent rear suspension system.
- How the top of the knuckle is supported in a multilink independent rear suspension system.
- The advantages of hydraulic suspension mounts.
- The effect of sagged rear springs on caster angle and steering.

Terms To Know

Asymmetrical leaf spring
Axle tramp
Axle windup
Braking and deceleration torque
Caster angle
Double wishbone rear suspension system
Five-link suspension
Independent rear suspension
Live axle rear suspension systems
Multilink independent rear suspension system
Rear suspension system
Rear wheel toe
Semielliptical springs
Semi-independent rear suspension
Solid axle beam
Sprung weight
Symmetrical leaf spring
Track bar
Unsprung weight

INTRODUCTION

A **rear suspension system** with two longitudinal leaf springs and a one-piece rear axle housing may be called a Hotchkiss drive.

The **rear suspension system** plays a very important part in ride quality and in the control of suspension and differential noise, vibration, and shock. Although the front wheels actually steer the vehicle, the rear suspension is also vital to steering control. The rear suspension must also provide adequate tire life and maintain tire traction on the road surface. Rear suspension systems described in this chapter include live axle, semi-independent, and independent. **Live axle rear suspension systems** are found on rear-wheel-drive trucks and vans, a few rear-wheel-drive cars, and some four-wheel-drive cars. Most front-wheel-drive vehicles have semi-independent or independent rear suspensions. Independent rear suspensions are also found on rear-wheel-drive cars and four-wheel-drive cars.

A **live axle rear suspension system** may be defined as one in which the differential axle housing, wheel bearings, and brakes act as a unit.

LIVE AXLE REAR SUSPENSION SYSTEMS

Leaf-Spring Rear Suspension

A leaf spring is mounted longitudinally on each side of the rear suspension on some rear-wheel-drive cars and trucks (**Figure 7-1**). These relatively flat springs provide excellent lateral stability and reduce side sway, which contribute to a well-controlled ride with very good handling characteristics. However, leaf-spring rear suspension systems have a lot of **unsprung weight**, and leaf springs require a considerable amount of space.

Unsprung weight refers to the weight that is not supported by the springs, which includes the weight of the suspension system.

A BIT OF HISTORY

The leaf-spring rear suspension was one of the first widely used rear suspension systems. Because this type of rear suspension system has weight and ride quality disadvantages, it has been replaced on many vehicles with independent or semi-independent rear suspension systems.

The **semielliptical springs** have steel leaves and zinc or plastic interleaves to reduce corrosion, friction, and noise. A large rubber bushing is installed in the front eye of the main spring leaf, and a bolt retains this bushing to the front spring hanger (**Figure 7-2**). The rear spring shackle is bolted to a rubber bushing in the rear main leaf eye, and the upper shackle bolt extends through a similar rubber bushing in the rear spring hanger (**Figure 7-3**). Some rear spring shackles contain threaded steel bushings or a slipper mount in which the end of the spring slides through the shackle. Shackle insulating bushings help prevent the transfer of noise and road shock from the suspension to the chassis and vehicle interior. When a rear wheel strikes a road irregularity, the spring is compressed and the spring length changes. The rear shackle provides fore-and-aft movement with variations in spring length.

Semielliptical springs have individual leaves stacked with the shortest leaf at the bottom and the longest leaf at the top.

Shop Manual
Chapter 7, page 288

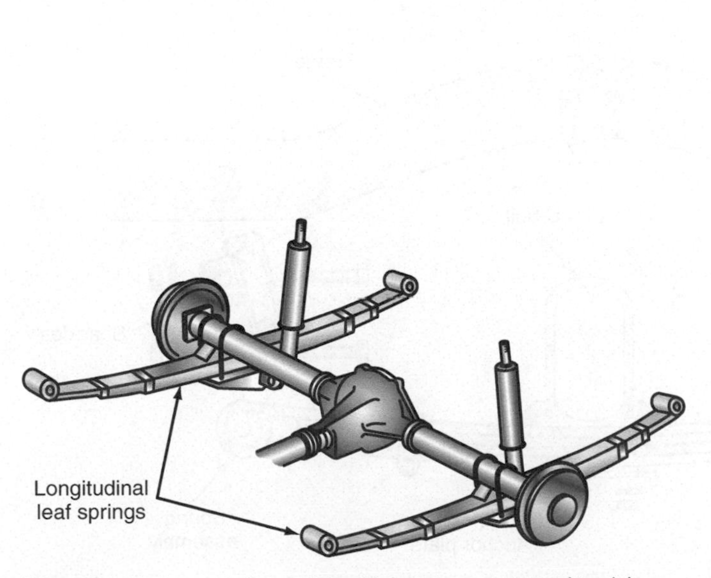

Figure 7-1 Rear suspension system with long torque arm and track bar.

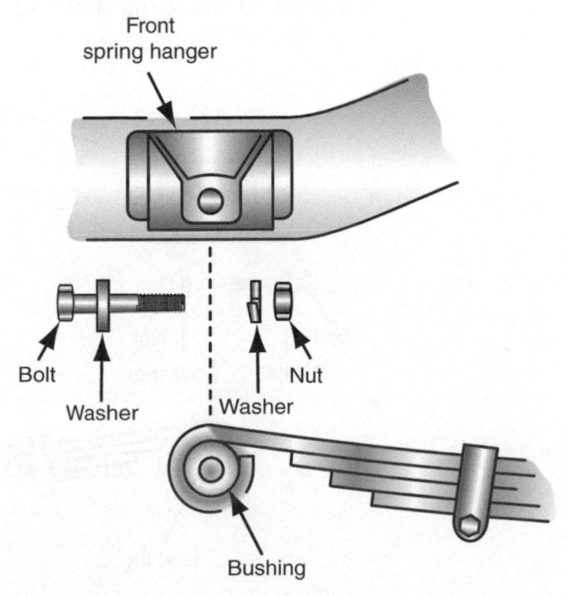

Figure 7-2 Rear leaf-spring eye bushing.

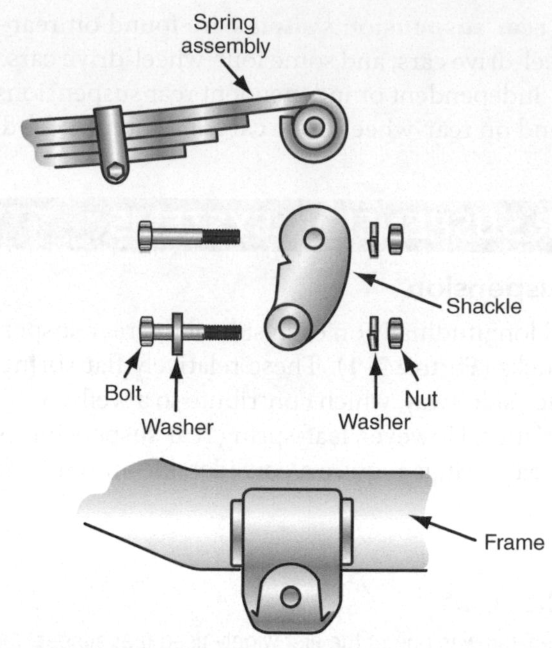

Figure 7-3 Rear leaf-spring shackle.

Because the differential axle housing is a one-piece unit, jounce and rebound travel of one rear wheel affects the position of the other rear wheel. This action increases tire wear and decreases ride quality and traction.

The differential axle housing is mounted above the springs, and a spring plate with an insulating clamp and U-bolts retains the springs to the rear axle housing (**Figure 7-4**). The shock absorbers are mounted between the spring plates and the frame.

Sprung weight refers to the weight carried by the springs, which includes the chassis and all components attached to the chassis.

The vehicle **sprung weight** is supported by the springs through the rear axle housing and wheels. When the vehicle accelerates, the rear wheels turn counterclockwise when viewed from the left vehicle side. One of Newton's laws of motion states that for every action there is an equal and opposite reaction. Therefore, when the wheels turn counterclockwise (when viewed from the left), the rear axle housing tries to rotate clockwise. This rear axle torque action is absorbed by the rear springs and the chassis moves downward (**Figure 7-5**). Engine torque supplied through the driveshaft to the differential tends to twist the differential housing and the springs. This twisting action may be referred to as

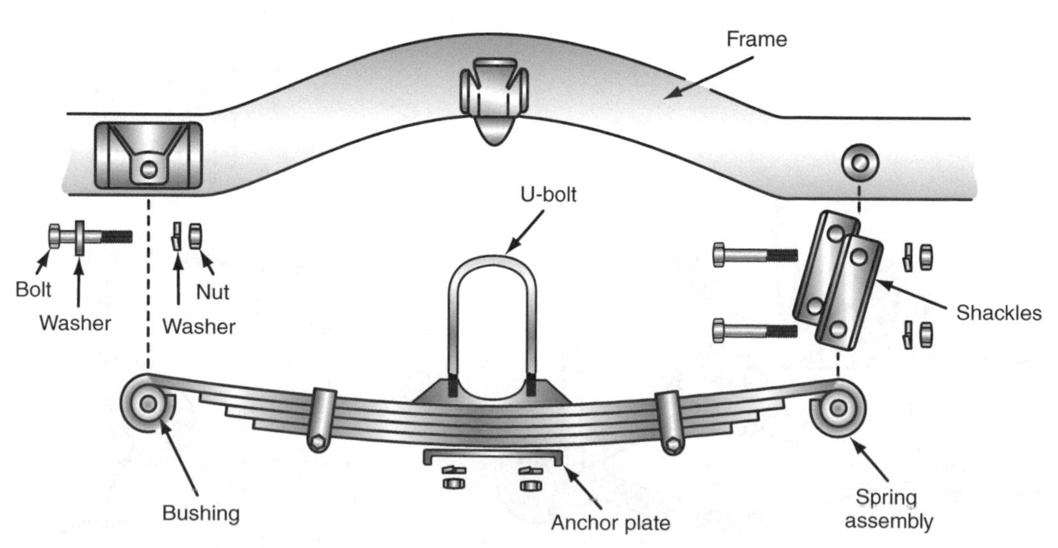

Figure 7-4 Individual leaf-spring suspension components.

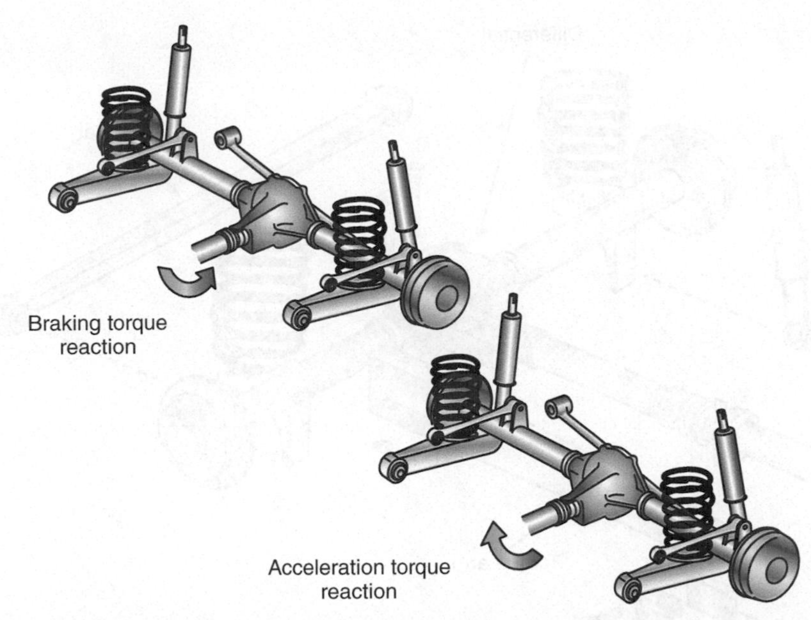

Figure 7-5 Rear axle torque action during acceleration and deceleration.

axle windup. Many leaf springs have a shorter distance from the center bolt to the front of the spring compared to the distance from the center bolt to the rear of the spring. This type of leaf spring is referred to as an **asymmetrical leaf spring**, and the shorter distance from the center bolt to the front of the spring resists axle windup. A **symmetrical leaf spring** has the same distance from the center bolt to the front and rear of the spring.

When braking and decelerating, the rear axle housing tries to turn counterclockwise. This rear axle torque action applied to the springs lifts the chassis. This action may be called **braking and deceleration torque**.

During hard acceleration, the entire power train twists in the opposite direction to engine crankshaft and drive shaft rotation. The engine and transmission mounts absorb this torque. However, the twisting action of the drive shaft and differential pinion shaft tends to lift the rear wheel on the passenger's side of the vehicle. Extremely hard acceleration may cause the rear wheel on the passenger's side to lift off the road surface. Once this rear wheel slips on the road surface, engine torque is reduced, and the leaf spring forces the wheel downward. When this rear tire contacts the road surface, engine torque increases and the cycle repeats. This repeated lifting of the differential housing is called **axle tramp**, and this action occurs on live axle rear suspension systems. Axle tramp is more noticeable on live axle leaf-spring rear suspension systems in which the springs have to absorb all the differential torque. For this reason, only engines with moderate horsepower were used with this type of rear suspension. Rear suspension and axle components such as spring mounts, shock absorbers, and wheel bearings may be damaged by axle

AUTHOR'S NOTE Leaf-spring rear suspension systems are still used on many light-duty trucks because of their load-carrying capability. However, today's design engineers have improved the ride quality of these suspension systems compared with past models. Ride quality in these leaf-spring suspension systems has been improved by installing longer leaf springs and using larger, improved rubber insulating bushings in the spring eye and shackle. Ride quality has also been improved by maximizing the shock absorber mounting location and matching the shock absorber design more closely to the leaf-spring jounce and rebound action. Optimizing the rear axle mounting position on the leaf springs also improves ride quality.

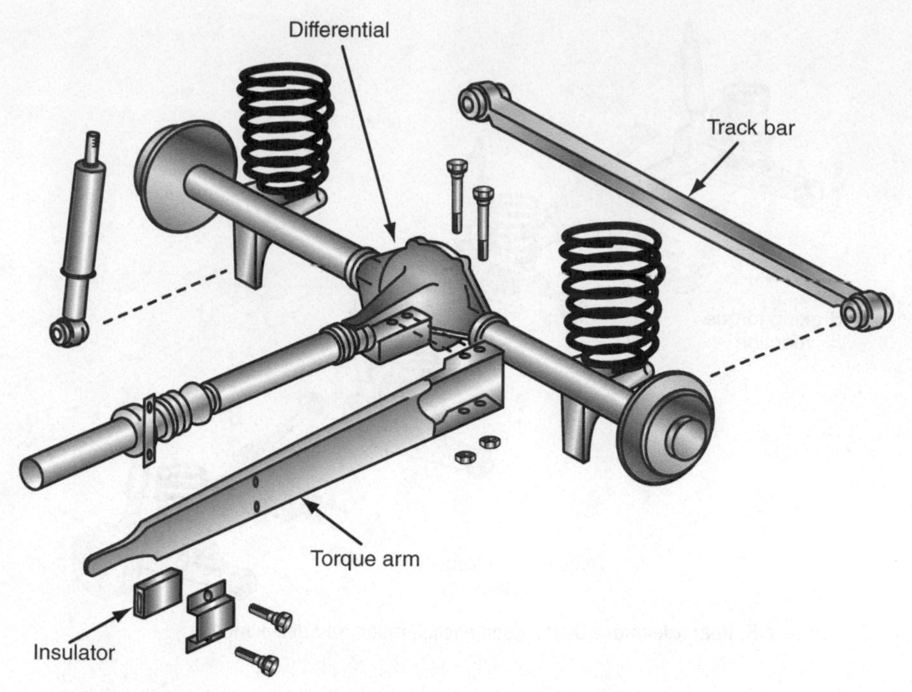

Figure 7-6 Rear suspension system with long torque arm and track bar.

tramp. Mounting one rear shock absorber in front of the rear axle and the other rear shock behind the rear axle helps reduce axle tramp.

In some cars with higher torque engines, a long torque arm is bolted to the rear axle housing (**Figure 7-6**). This torque arm helps prevent differential rotation during hard acceleration and braking. The front of this torque arm is mounted in a rubber insulator and bracket that is bolted to the back of the transmission housing. This long torque arm helps prevent differential rotation when high torque is delivered from the engine to the differential. The center bearing assembly on the drive shaft is bolted to the long torque arm. This rear suspension system has a track bar (tie rod) connected from the left side of the rear axle housing to the chassis to help prevent lateral rear axle movement. A track bar brace is connected from the chassis end of the track bar to the other side of the chassis to provide extra rigidity.

Coil Spring Rear Suspension

WARNING Compressed coil springs contain a large amount of energy. Never disconnect any suspension component that suddenly releases the coil spring tension. This action may result in personal injury and vehicle damage.

WARNING Always follow the vehicle manufacturer's recommended rear suspension service procedures in the service manual to avoid personal injury.

Some rear-wheel-drive cars have a coil spring rear suspension. Upper and lower suspension arms with insulating bushings are connected between the differential housing and the frame (**Figure 7-7**). The upper arms control lateral movement, and the lower trailing control arms absorb differential torque. In some rear suspension systems, the upper arms are replaced with strut rods. The front of the upper and lower arms contains large rubber bushings. When strut rods are used in place of the upper arms, both ends of these rods contain large rubber bushings to prevent noise and vibration transfer from the suspension to the chassis. The coil springs are usually mounted between the lower suspension arms and the frame, whereas the shock absorbers are mounted between the back of the suspension arms and the frame.

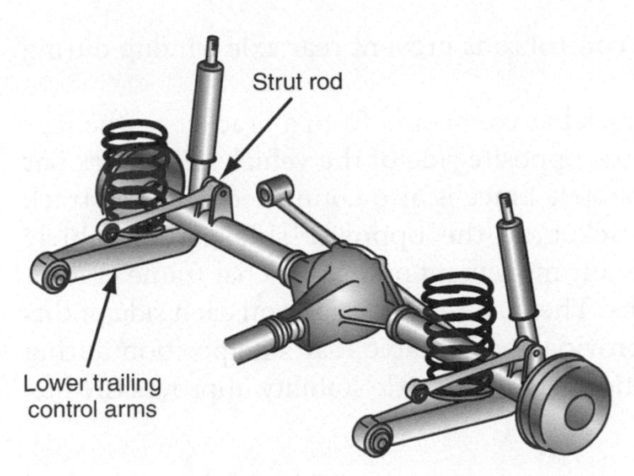

Figure 7-7 Live axle coil spring rear suspension.

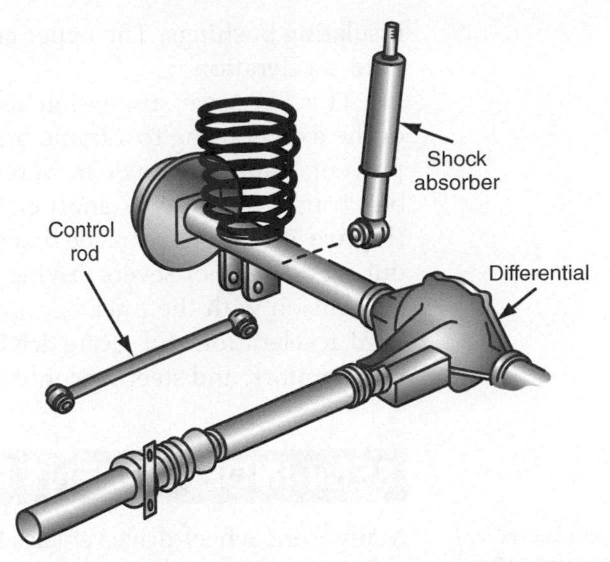

Figure 7-8 Rear suspension system with coil springs and shock absorbers mounted separately.

Some rear suspension systems have a **track bar** connected from one side of the differential housing to the chassis to prevent lateral chassis movement. Large rubber insulating bushings are positioned in each end of the track bar.

A **track bar** may be referred to as a Panhard rod or Watts rod.

Some late-model sport utility vehicles (SUVs) have a rear suspension system with coil springs and shock absorbers mounted separately from the springs (**Figure 7-8**). Notice that a control rod is mounted in rubber insulating bushings on the bottom of the rear axle housing, and the outer end of this bar is connected through links to the chassis. This rear suspension has an upper and a lower control rod on each side of the suspension. Each lower control rod is connected from a bracket on the lower side of the axle housing to a frame bracket. Rubber insulating bushings are located in both ends of the lower control rods. The frame bracket is located ahead of the rear axle. The upper control rod is connected from a bracket on top of the axle housing to a frame bracket near the lower control rod frame bracket (**Figure 7-9**). Both ends of the upper control rod contain rubber

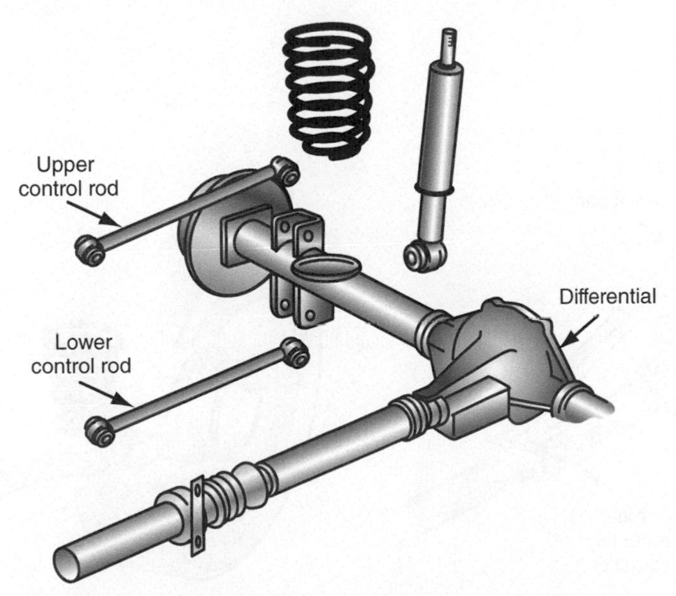

Figure 7-9 Rear suspension system with upper and lower control rods.

insulating bushings. The upper and lower control rods prevent rear axle windup during hard acceleration.

This SUV rear suspension also has a track bar connected from a bracket on the rear of the axle housing to a frame bracket on the opposite side of the vehicle. The track bar prevents lateral rear axle movement. A rear axle brace is also connected from the track bar frame bracket to another frame bracket on the opposite side of the vehicle (**Figure 7-10**). The rear axle brace prevents any movement of the track bar frame bracket during off-road or severe driving conditions. The dual rear axle rods on each side of this suspension with the track bar and brace provide a very stable rear axle position during hard acceleration and severe driving conditions. This rear axle stability improves tire life, ride comfort, and steering control.

SEMI-INDEPENDENT REAR SUSPENSION SYSTEMS

A **semi-independent rear suspension** allows some individual rear wheel movement when one rear wheel strikes a bump.

Shop Manual
Chapter 7, page 289

Many front-wheel-drive vehicles have a **semi-independent rear suspension** that has a **solid axle beam** connected between the rear trailing arms (**Figure 7-11**). A solid axle beam is usually a transverse inverted U-section channel connected between the rear wheels in a semi-independent rear suspension system. When one rear wheel strikes a bump, this beam twists to allow some independent wheel movement. Some of these rear axle beams are fabricated from a transverse inverted U-section channel.

In some rear suspension systems, the inverted U-section channel contains an integral tubular stabilizer bar. When one rear wheel strikes a road irregularity and the wheel moves upward, the inverted U-section channel twists, which allows some independent rear wheel movement. The trailing arms are connected to chassis brackets through rubber insulating bushings. In some semi-independent rear suspension systems, the coil springs are mounted on the rear struts, the lower spring seat is located on the strut, and the upper spring seat is positioned on the upper strut mount.

In other semi-independent rear suspension systems, the coil springs are mounted separately from the shock absorbers. Coil spring seats are located on the trailing arms, and the shock absorbers are connected from the trailing arms to the chassis. A crossmember connected between the trailing arms provides a twisting action and some independent rear wheel movement (**Figure 7-12**).

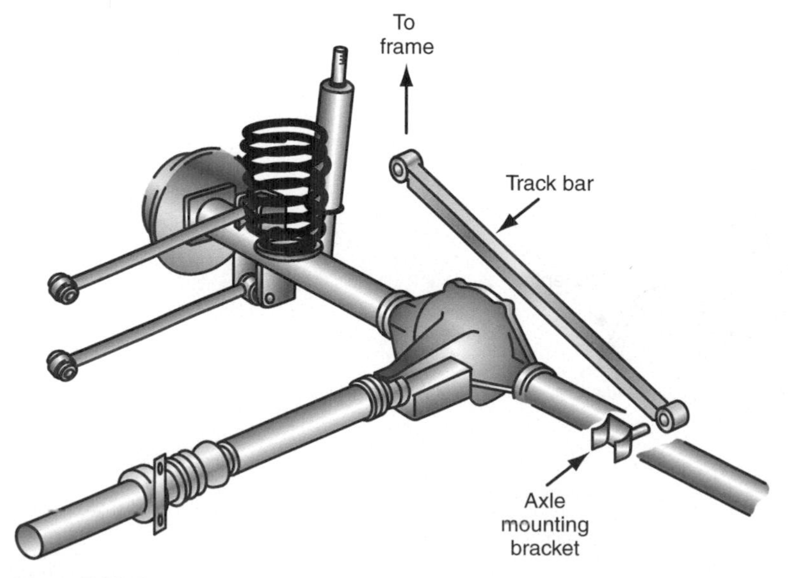

Figure 7-10 Track bar and brace.

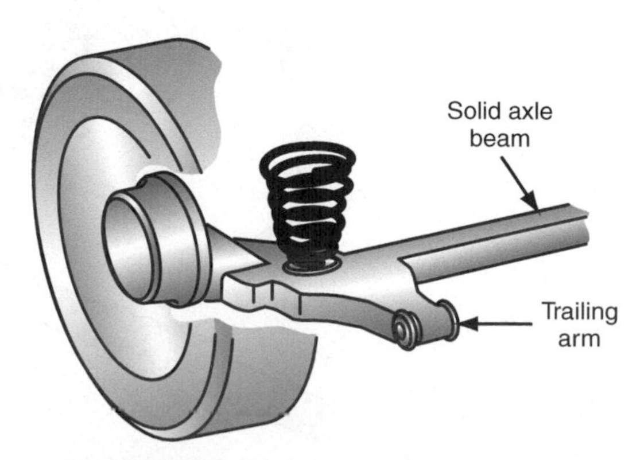

Figure 7-11 Semi-independent rear suspension system.

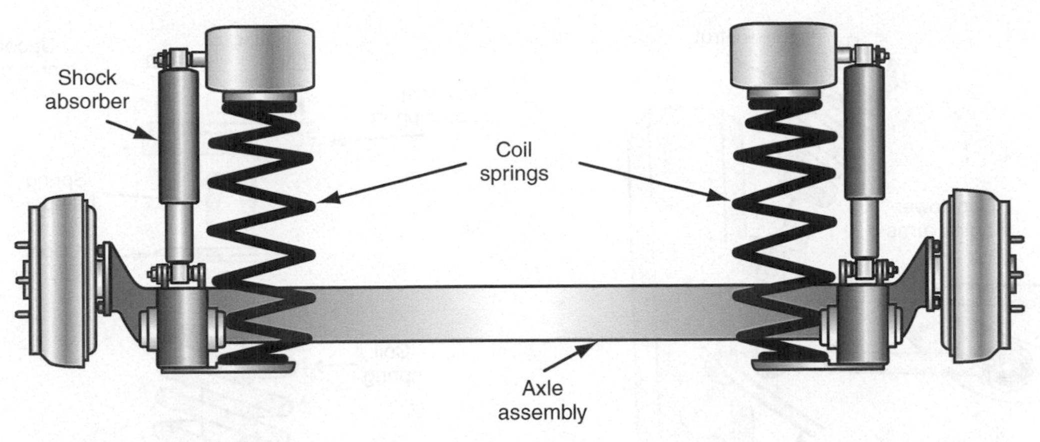

Figure 7-12 Semi-independent rear suspension system with coil springs and shock absorbers mounted separately.

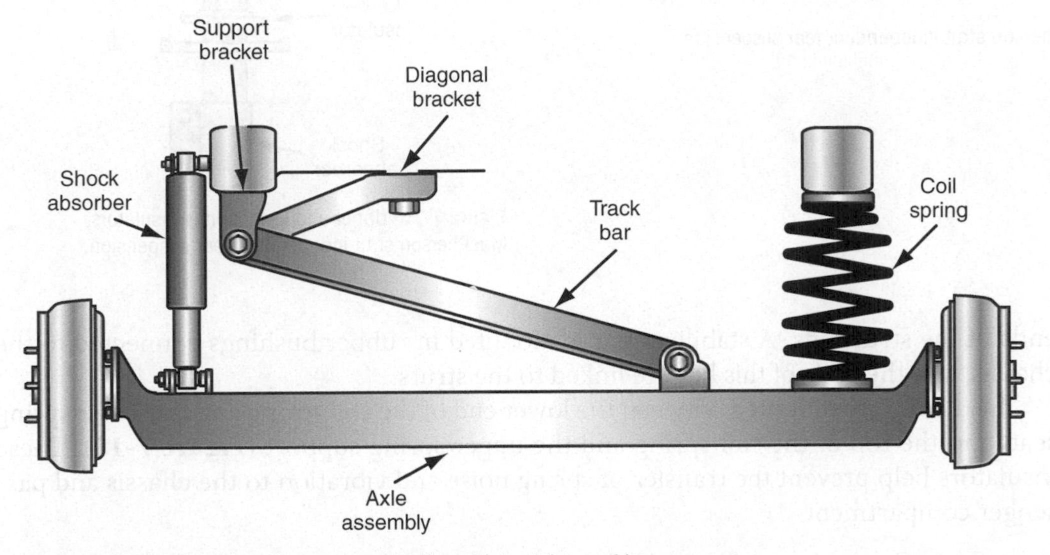

Figure 7-13 Semi-independent rear suspension with track bar and brace.

Some semi-independent rear suspension systems have a track bar connected from a rear axle bracket to a chassis bracket. In some applications, an extra brace is connected from this chassis bracket to the rear upper crossmember (**Figure 7-13**). The track bar and the brace prevent lateral rear axle movement.

INDEPENDENT REAR SUSPENSION SYSTEMS

MacPherson Strut Independent Rear Suspension System

In an **independent rear suspension** system, each rear wheel can move independently from the opposite rear wheel. Independent rear suspension systems may be found on front-wheel-drive and rear-wheel-drive vehicles. When rear wheel movement is independent, ride quality, tire life, steering control, and traction are improved. In a MacPherson strut rear suspension system, the coil springs are mounted on the rear struts. A lower spring seat is located on the strut, and the upper spring seat is positioned on the upper strut mount. This upper strut mount is bolted into the inner fender reinforcement. Dual lower control arms on each side of the suspension are connected from the chassis to the lower end of the spindle (**Figure 7-14**).

In an **independent rear suspension**, vertical movement of one rear wheel does not affect the opposite rear wheel.

Shop Manual
Chapter 7, page 290

The lower end of each strut is bolted to the spindle. Two strut rods are connected forward from the spindles to the chassis. Rubber insulating bushings are located in both

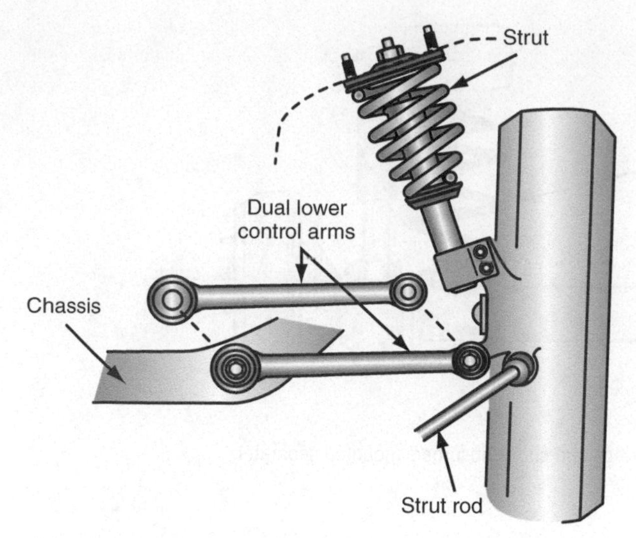

Figure 7-14 MacPherson strut independent rear suspension system.

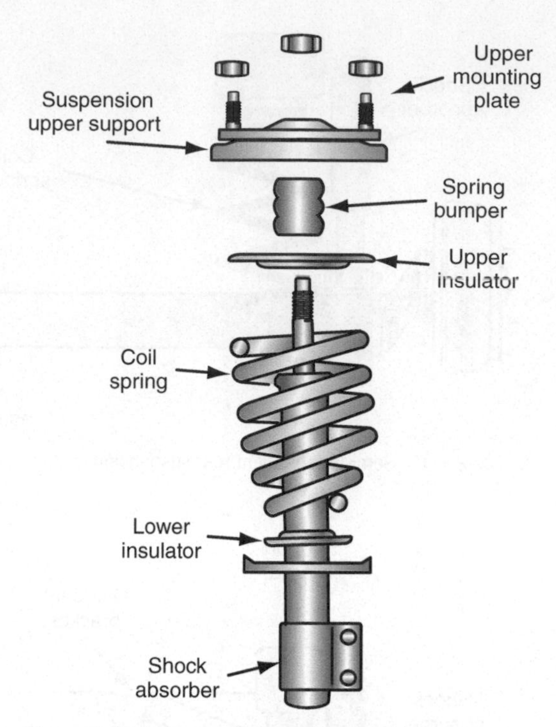

Figure 7-15 Upper and lower spring insulators, MacPherson strut independent rear suspension.

ends of the strut rods. A stabilizer bar is mounted in rubber bushings connected to the chassis, and the ends of this bar are linked to the struts.

Insulators are mounted between the lower end of the coil spring and the lower spring seat, and the top of the coil spring and the upper spring support (**Figure 7-15**). These insulators help prevent the transfer of spring noise and vibration to the chassis and passenger compartment.

Modified MacPherson Strut Independent Rear Suspension System

Some front-wheel-drive vehicles have a modified MacPherson strut independent rear suspension. Each side of the rear suspension has a shock strut, lower control arm, tie rod, forged spindle, and coil spring mounted between the lower control arm and the crossmember side rail (**Figure 7-16**).

The shock absorber strut has a rubber isolated top mount with a one-piece jounce bouncer dust shield. This top mount is attached to the body side panel and the lower end of the strut is bolted to the spindle. The stamped lower control arms are bolted to the crossmember and the spindle. A tie rod is connected from the spindle to the underbody. The purpose of each rear suspension component may be summarized as follows:

1. Stamped lower control arm—controls the lateral (side-to-side) wheel movement and contains the lower spring seat.
2. Tie rod—controls fore-and-aft wheel movement and positions the spindle properly.
3. Shock absorber strut—reacts to braking forces and provides suspension damping. A strut internal rebound stop provides rebound control, and an external jounce bumper supplies jounce control.
4. Coil spring—controls suspension travel, provides ride height control, and acts as a metal-to-metal jounce stop.
5. Forged spindle—supports the wheel bearings and attaches to the lower control arms, tie rod, brake assembly, and strut.

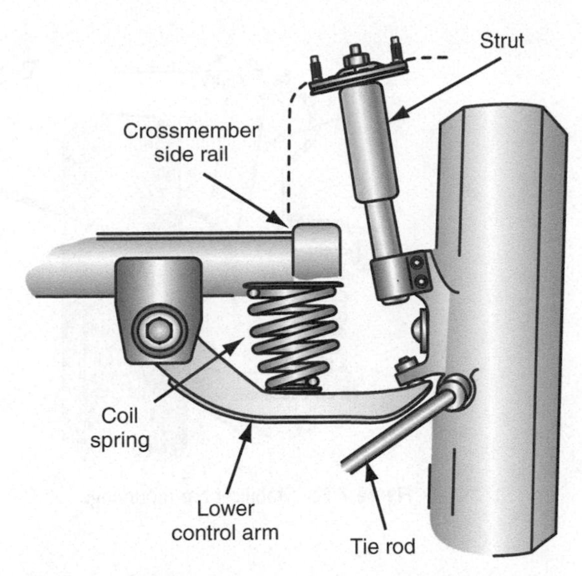

Figure 7-16 Modified MacPherson strut independent rear suspension.

6. Suspension bushings—insulate the chassis and passenger compartment from road noise and vibration.
7. Suspension fasteners—connect components such as the spindle and strut. These fasteners must always be replaced with equivalent quality parts, and each fastener must be tightened to the specified torque.

Independent Rear Suspension with Lower Control Arm and Ball Joint

Some front-wheel-drive cars have an independent rear suspension system with a ball joint pressed into the outer end of the lower control arm. The ball joint contains a conventional wear indicator. The upper end of the ball joint stud is bolted into the knuckle (**Figure 7-17**). The inner end of the lower control arm is connected to the chassis through two rubber insulating bushings (**Figure 7-18**).

The lower end of the strut is bolted to the knuckle, and the upper strut mount is bolted to the inner fender reinforcement (**Figure 7-19**). A stabilizer bar is mounted in bushings attached to the chassis, and the outer ends of this bar are linked to brackets connected to the strut (**Figure 7-20**).

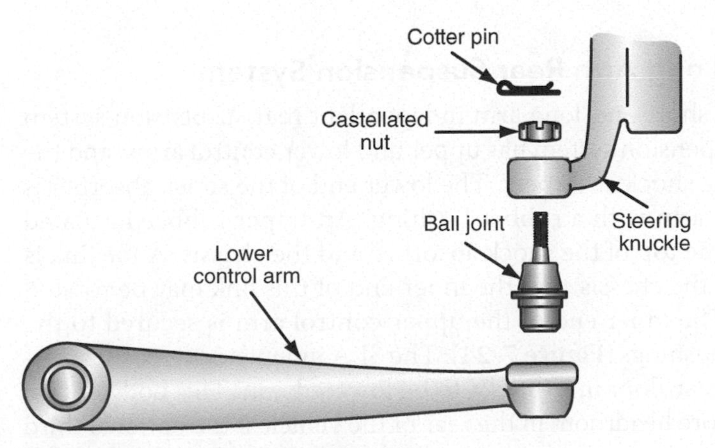

Figure 7-17 Lower control arm and ball joint.

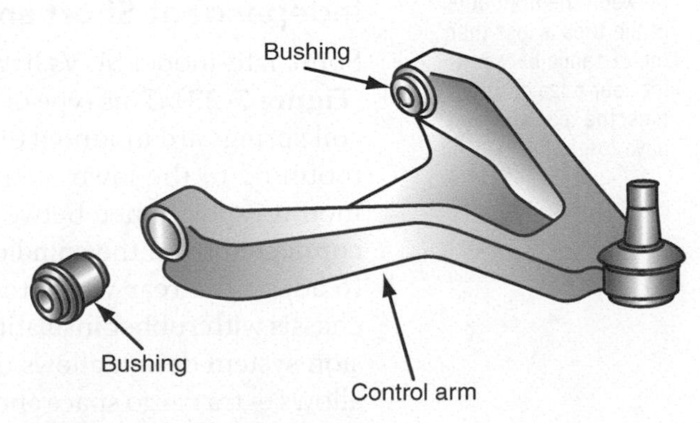

Figure 7-18 Lower control arm bushings.

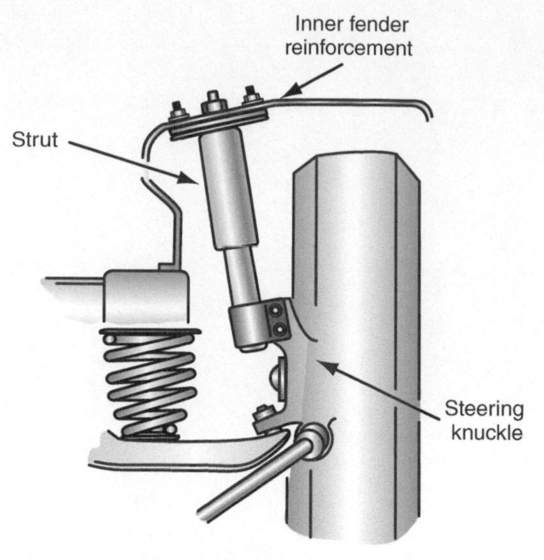

Figure 7-19 Upper and lower strut mounting.

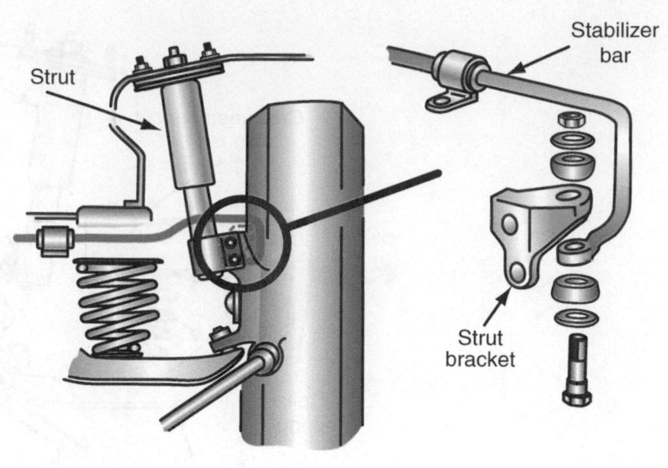

Figure 7-20 Stabilizer bar mounting.

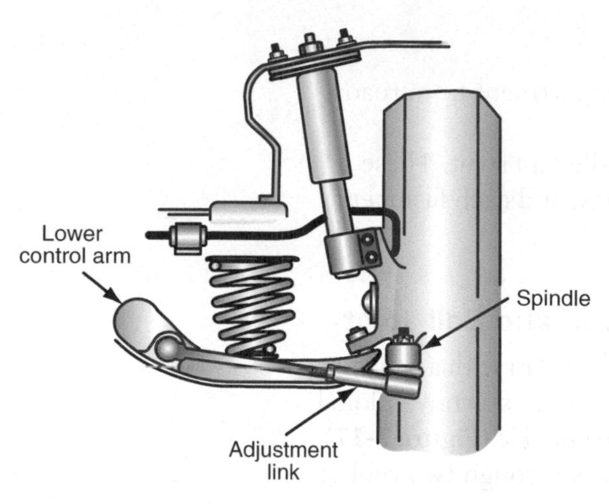

Figure 7-21 Suspension adjustment link.

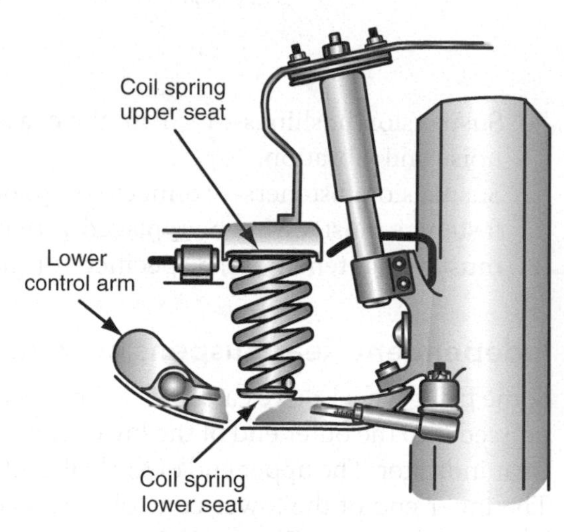

Figure 7-22 Coil spring mounting.

Rear wheel toe is the variation between the distance measured between the tires at the front edge and the rear edge. If the distance between the front edge of the tires is less than the distance between the rear edge of the tires, the rear wheels have toe-in.

A suspension adjustment link is connected from the lower control arm to the knuckle (**Figure 7-21**). This link provides a **rear wheel toe** adjustment. A coil spring seat is located in the lower control arm and the coil spring is mounted between this seat and an upper seat in the chassis (**Figure 7-22**). Upper and lower insulators are mounted between the coil spring and the seats.

Independent Short-and-Long Arm Rear Suspension System

Some late-model SUVs have a short-and-long arm independent rear suspension system (**Figure 7-23**). This type of suspension system has upper and lower control arms, and the coil springs are mounted on the shock absorbers. The lower end of the shock absorber is mounted to the lower control arm with a rubber bushing. An upper rubber-insulated mount is positioned between the top of the shock absorber and the chassis. A toe link is connected from the spindle to the chassis, and the inner end of this link may be rotated to adjust the rear wheel toe. The inner end of the upper control arm is secured to the chassis with rubber insulating bushings (**Figure 7-24**). The SLA independent rear suspension system design allows the rear floor in the SUV to be lowered 5 in. This body design allows extra cargo space and more headroom in the rear of the vehicle if the optional third seat is installed.

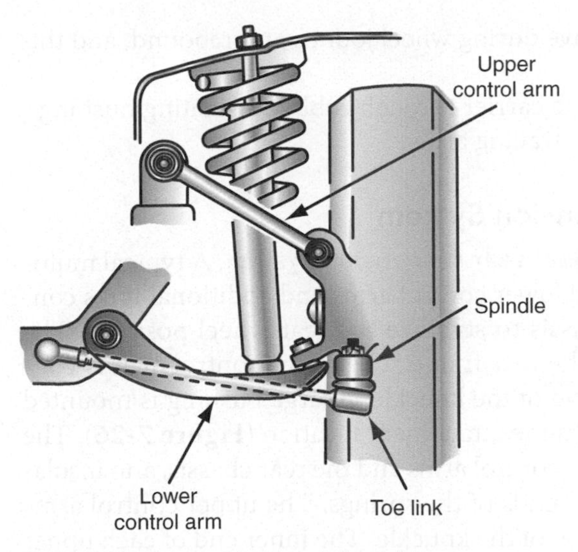

Figure 7-23 Short-and-long arm independent rear suspension system.

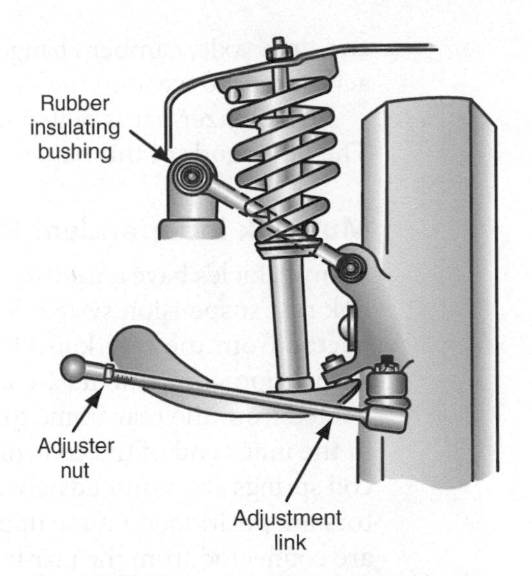

Figure 7-24 Upper control arm, short-and-long arm independent rear suspension system.

Independent Rear Suspension System with Rear Axle Carrier

Some import rear-wheel-drive cars have an independent rear suspension with a large rear axle carrier extending across the width of the chassis. A large extension on the rear axle carrier is bolted to the top of the differential, and the outer ends of this carrier are connected through heavy insulating bushings to chassis brackets (**Figure 7-25**).

A large final drive mount extends from the rear of the differential housing, and this mount is bolted to the chassis. When the differential is attached to the chassis, rather than being connected to the suspension, the unsprung weight is reduced. Two rubber insulating bushings connect the large trailing arms to the rear axle carrier. The rear axle carrier provides very stable differential and wheel position, which improves steering control, ride quality, and tire life. Lower coil spring seats are located on the trailing arms, and upper coil spring seats are positioned in the chassis.

Shock absorbers are connected from the back of the trailing arms to the chassis. The rear wheel bearings are mounted in the outer ends of the trailing arms. Drive axles with inner and outer drive joints are connected from the differential to the rear wheels. When the drive axles have inner and outer joints, rear wheel camber change is minimized during wheel jounce and rebound. On some early independent rear suspension systems in rear-wheel-drive cars, the drive axles had only inner joints. With this type of rear suspension

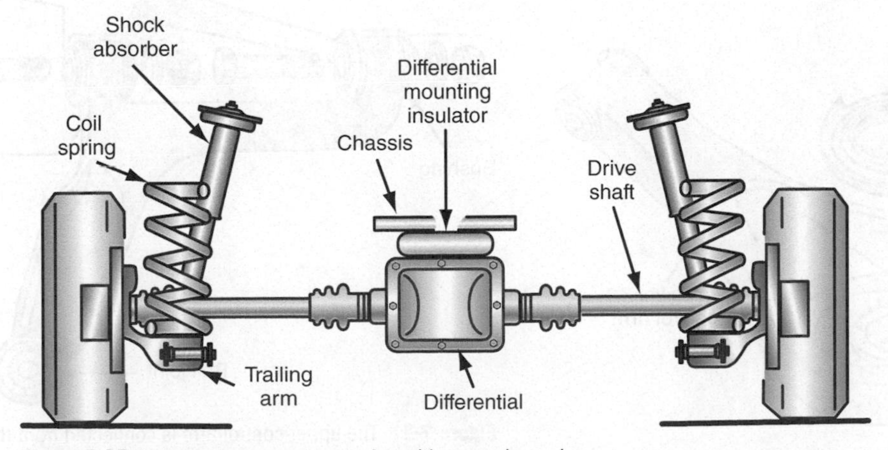

Figure 7-25 Independent rear suspension with rear axle carrier.

and drive axle, camber change was excessive during wheel jounce and rebound, and this action caused wear on the tire edges.

A stabilizer bar is bolted to the rear axle carrier through rubber insulating bushings. The outer ends of this bar are linked to the trailing arms.

Multilink Independent Rear Suspension System

Many vehicles have a **multilink independent rear suspension system**. A typical multilink rear suspension system has upper and lower control arms and additional links connected from the rear knuckles to the chassis to stabilize the rear wheel position. The suspension components are attached to the rear frame. The lower control arm is connected from the rear frame to the lower end of the knuckle. A large bushing is mounted in the inner end of the control arm at the frame attachment location (**Figure 7-26**). The coil springs are mounted between the lower control arms and the rear chassis, and insulators are positioned on the upper and lower ends of the springs. The upper control arms are connected from the rear frame to the top of the knuckle. The inner end of each upper control arm has two attachment points to frame, and large insulating bushings are installed at these attachment locations (**Figure 7-27**). A ball joint in the outer end of each upper control arm connects the control arm to the top end of the knuckle. Many current rear suspension systems have aluminum components such as upper and lower control arms. Aluminum components reduce weight and improve fuel mileage, which in turn reduces emissions.

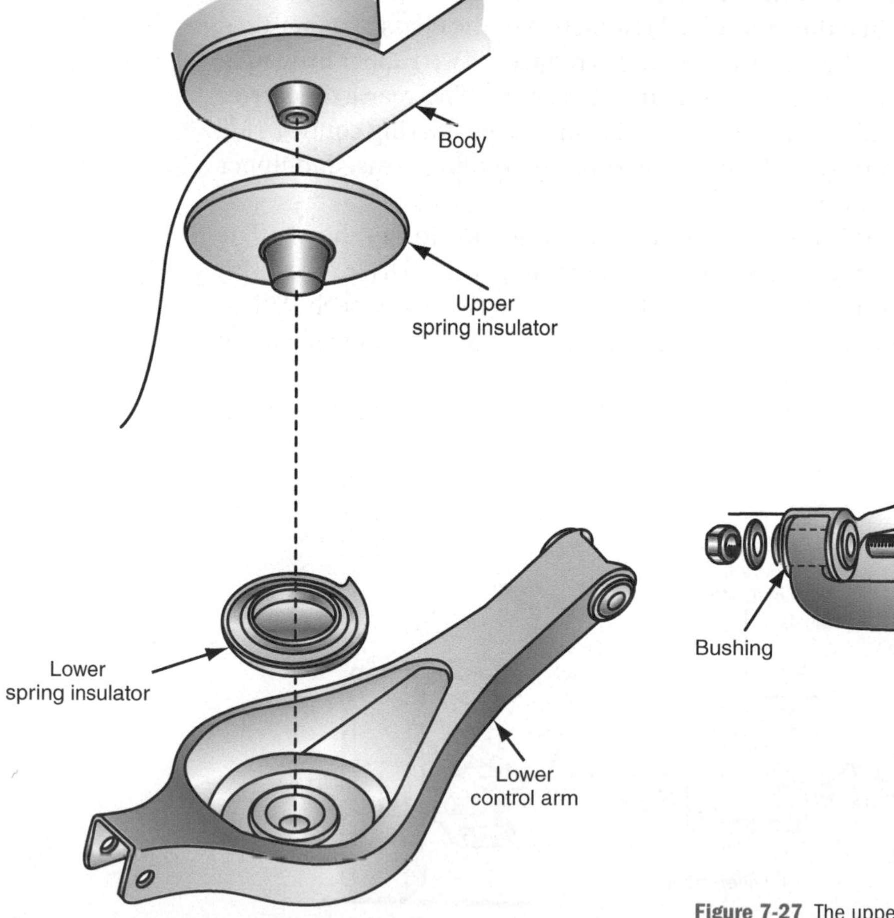

Figure 7-26 Lower control arm with upper and lower spring insulators.

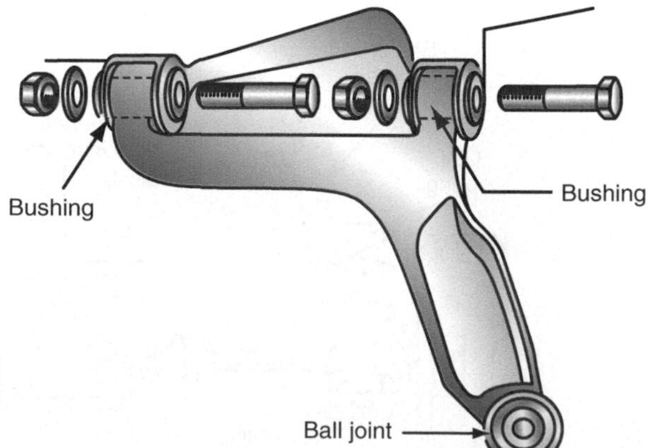

Figure 7-27 The upper control arm is connected from the frame to the top of the knuckle.

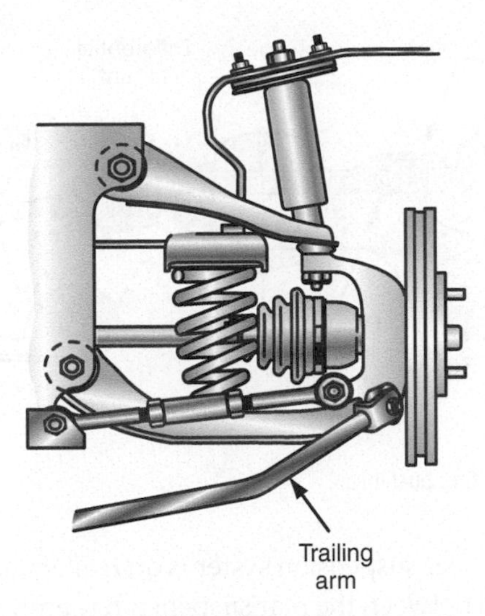

Figure 7-28 The trailing arm is connected from the lower end of the knuckle to the frame.

A trailing arm is connected from the lower end of the knuckle to the frame. Large insulating bushings are mounted at each trailing arm attachment location (**Figure 7-28**). The trailing arm prevents fore-and-aft wheel movement. An adjustment link is also connected from the lower end of the knuckle to the frame through insulating bushings (**Figure 7-29**). The adjustment link and lower control arm prevent lateral wheel movement, and this link also provides a method of rear wheel toe adjustment. This type of rear suspension may be called a **five-link suspension** because there are five attachments points between each knuckle and the frame. The lower control arm, trailing arm, and adjustment link have one attachment location, and the upper control arm has two attachment points. A five-link rear suspension provides a very stable rear wheel position, which improves steering control and rear tire tread life.

A five-link suspension has five attachment locations between each side of the suspension and the frame.

The shock absorbers are connected from the rear knuckles to the chassis (**Figure 7-30**). Two bolts retain the upper insulating shock absorber mount to the chassis. In this rear-wheel-drive car, the top of the differential is bolted to the upper part of the rear frame. Two large insulating bushings are mounted between the differential and the frame (**Figure 7-31**). Mounting the differential to the frame reduces the unsprung weight, which improves ride quality, because the unsprung weight forces the wheels downward during wheel rebound, which contributes to harsh ride quality.

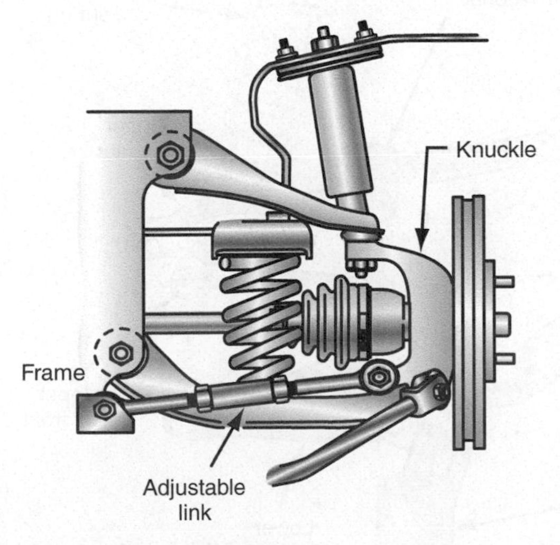

Figure 7-29 Adjustment link.

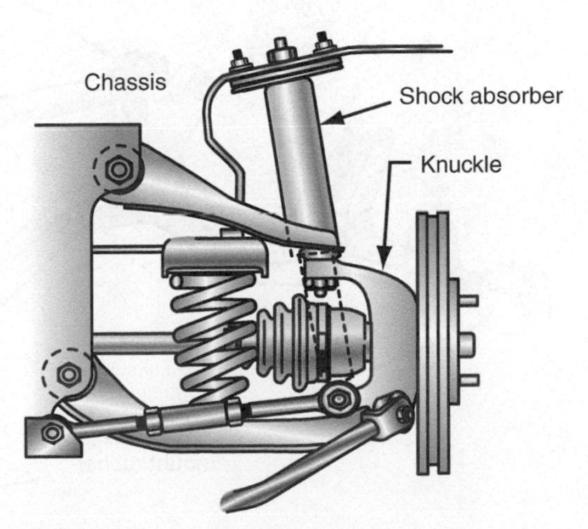

Figure 7-30 Shock absorber mounting.

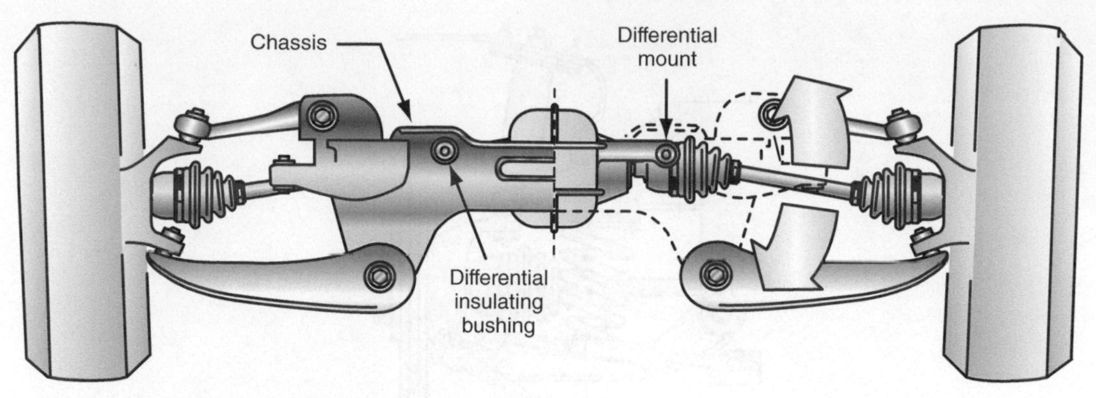

Figure 7-31 Differential to frame mounting bushings.

In some current multilink rear suspension systems on rear-wheel-drive vehicles, four large rubber mounts are positioned between the rear suspension member or frame and the chassis. These rubber mounts are designed such that they are soft in fore-and-aft movement but stiff in lateral movement. These rubber mounts reduce the transfer of road vibrations and shocks from the rear suspension system to the body. The stiff lateral movement in these bushings maintains lateral rear wheel position to reduce rear tire wear and improve vehicle stability.

Some multilink rear suspension systems have four hydraulic mounts mounted between the rear suspension member or frame and the chassis. The differential is bolted to the rear frame. These hydraulic mounts contain silicone oil that helps prevent noise, vibration, and shock transfer from the differential and suspension to the chassis and vehicle interior (**Figure 7-32**).

The rear suspension member is connected to the outer shell of the hydraulic mounts, and the inner shell is attached to the chassis. Silicone oil fills the area between the inner and outer shells in each mount (**Figure 7-33**). Noise, vibration, and shock are transferred from the differential and suspension to the rear suspension member, but the silicone oil in the hydraulic mounts prevents the transfer of these undesirable forces to the chassis and vehicle interior. These hydraulic mounts have superior noise and vibration dampening characteristics compared to rubber bushings.

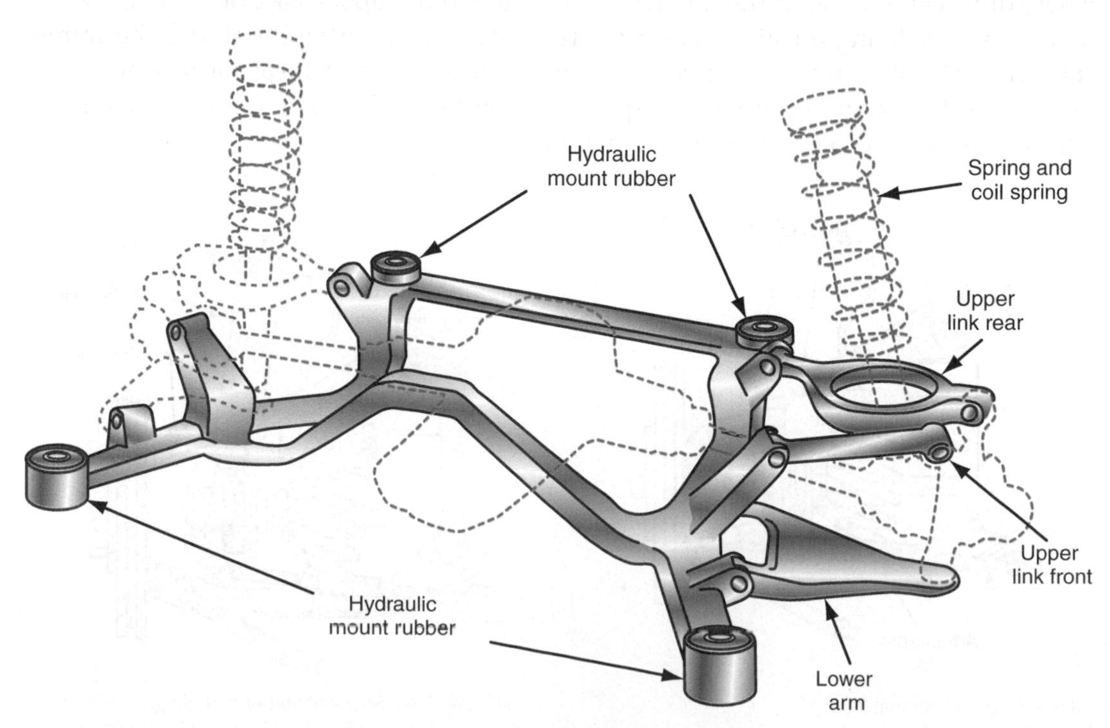

Figure 7-32 Multilink independent rear suspension system with hydraulic mounts.

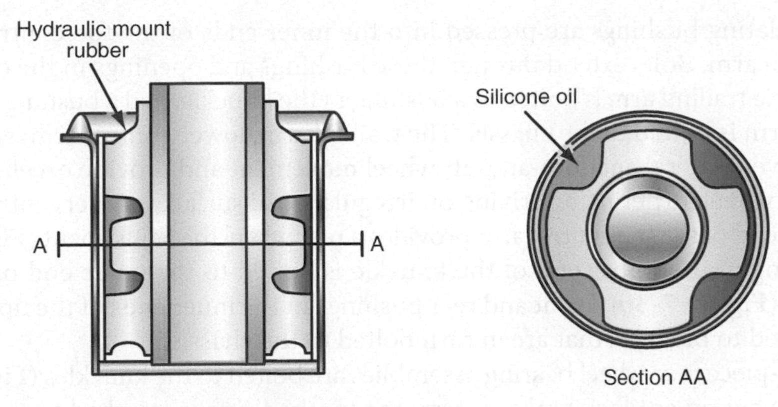

Figure 7-33 Hydraulic mount design.

Hydraulic bushing and mounts are gaining in popularity and may also be used as subframe mounts. Hydraulic bushings are tuned to the suspension system and are designed to work with the manufacturer tire and wheel combination to isolate noise frequencies and vibrations from the vehicle. Advancements in hydraulic bushings are in response to customer demands for quieter, better handling and smoother rides from today's vehicles. While inspecting the suspension system, these bushings should be inspected for leaks.

A rear upper link is connected from each side of the rear suspension member to the top of the knuckles. Both ends of this rear upper link contain rubber insulating bushings. The lower end of the shock absorber strut extends through a circular opening in the rear upper link. A front upper link is also connected from the rear suspension member to the knuckle. The top of the knuckle is supported by the front and rear upper links, rather than being supported by the shock absorber strut. The coil springs are mounted on the shock absorber struts, and the lower spring seat is attached to the strut. An upper spring seat is attached to the top of the shock absorber strut, and this upper seat is bolted into the inner fender reinforcement. The lower end of the shock absorber strut is connected to the back of the lower control arm. Lower control arms are connected from the rear suspension member to the lower end of the knuckles.

Some multilink rear suspension systems have a lower control arm and a toe control arm parallel to the lower control arm (**Figure 7-34**). A bolt in the outer end of the lower control arm extends through a bushing in the lower end of the steering knuckle. A ball joint in the outer end of the toe control arm is bolted into the lower end of the knuckle.

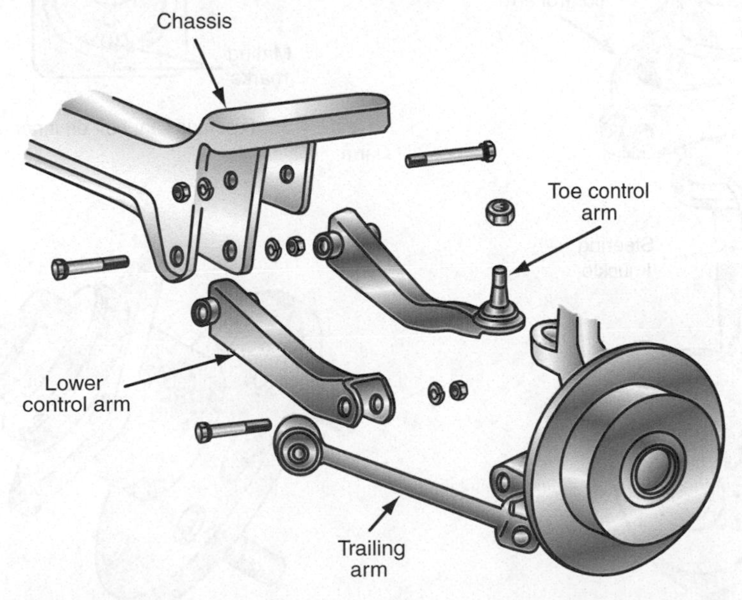

Figure 7-34 Multilink rear suspension with parallel lower control arm and toe control arm.

Rubber insulating bushings are pressed into the inner ends of the toe control arm and lower control arm. Bolts extend through these bushings and openings in the chassis. The rear end of the trailing arm is bolted to a bushing in the knuckle, and a bushing in the front end of this arm is bolted to the chassis. The trailing arm, lower control arm, and toe control arm prevent lateral and fore-and-aft wheel movement and provide excellent suspension rigidity when cornering or driving on irregular road surfaces. An eccentric cam bolt on the inner end of the toe control arm provides a rear wheel toe adjustment (**Figure 7-35**).

A bushing in the upper end of the knuckle is bolted to the outer end of the upper control arm (**Figure 7-36**). Front and rear bushings in the inner ends of the upper control arm are bolted to brackets that are in turn bolted to the chassis.

The one-piece rear wheel bearing assemblies are bolted to the knuckles (**Figure 7-37**). If the vehicle has an antilock brake system, the toothed ring is attached to the inner end of each rear wheel bearing and stub axle shaft, and a wheel speed sensor is mounted in the knuckle. The rear wheel bearings assemblies are non-serviceable.

Double Wishbone Rear Suspension

In the **double wishbone rear suspension system**, the upper and lower control arms are manufactured from lightweight, high-strength aluminum alloys designed for maximum strength and rigidity. These lighter control arms decrease the unsprung weight, which helps improve traction and ride quality. Suspension rigidity is also increased by positioning the ball joints and steering knuckle inside the wheel profile (**Figure 7-38**). Since the upper and lower control arms have a wishbone shape, the term *double wishbone* suspension system is used for this type of rear suspension. On each side of the car, the rear suspension is attached to the chassis by a cast aluminum subframe (**Figure 7-39**). This design also helps reduce vehicle weight and transmits suspension loads to the chassis at the most efficient locations.

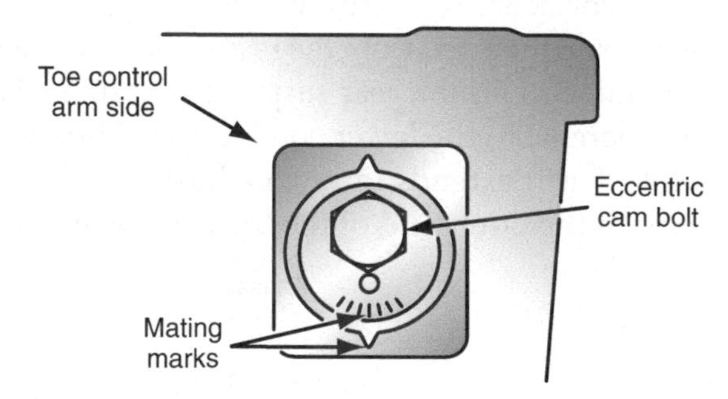

Figure 7-35 Eccentric cam bolt on inner end of toe control arm.

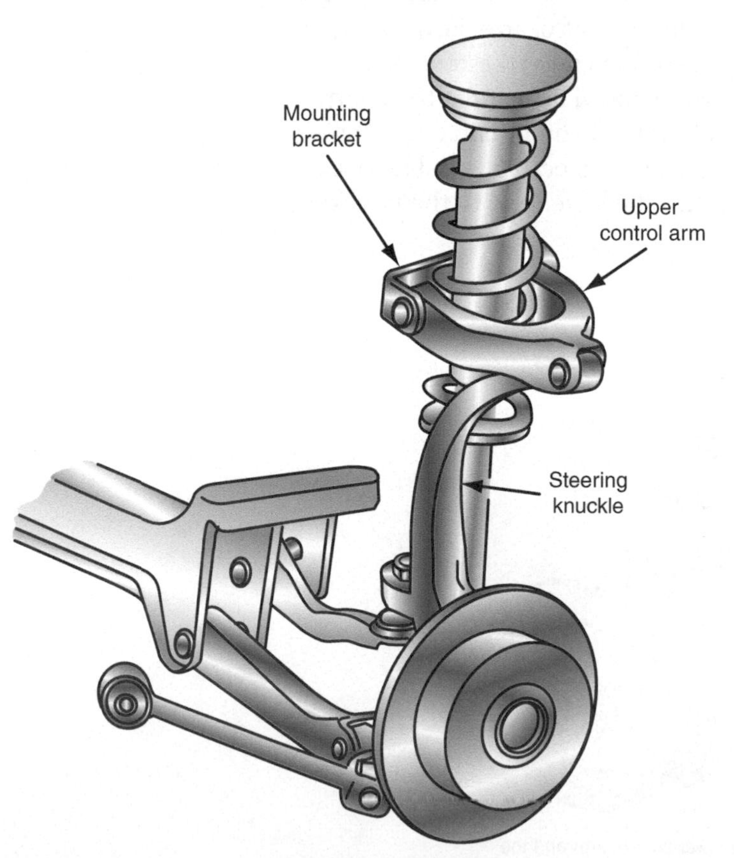

Figure 7-36 Upper control arm and strut in multilink rear suspension.

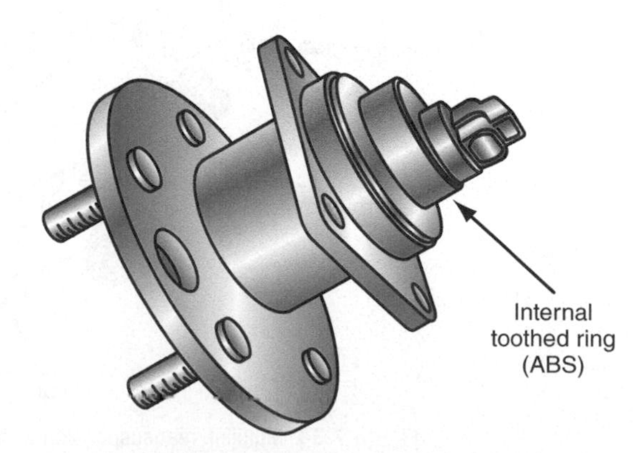

Figure 7-37 One-piece rear wheel bearing assembly.

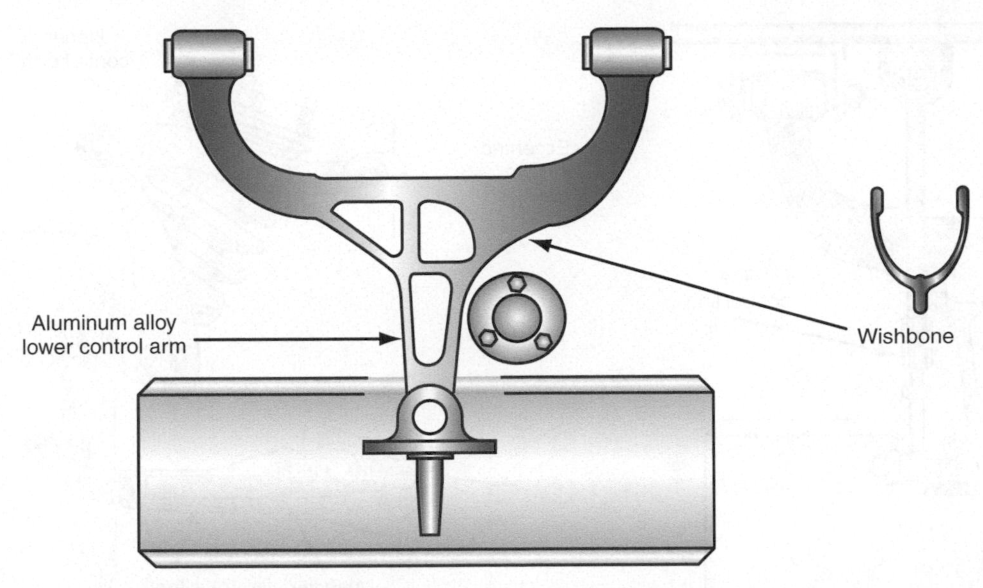

Figure 7-38 Double wishbone rear suspension.

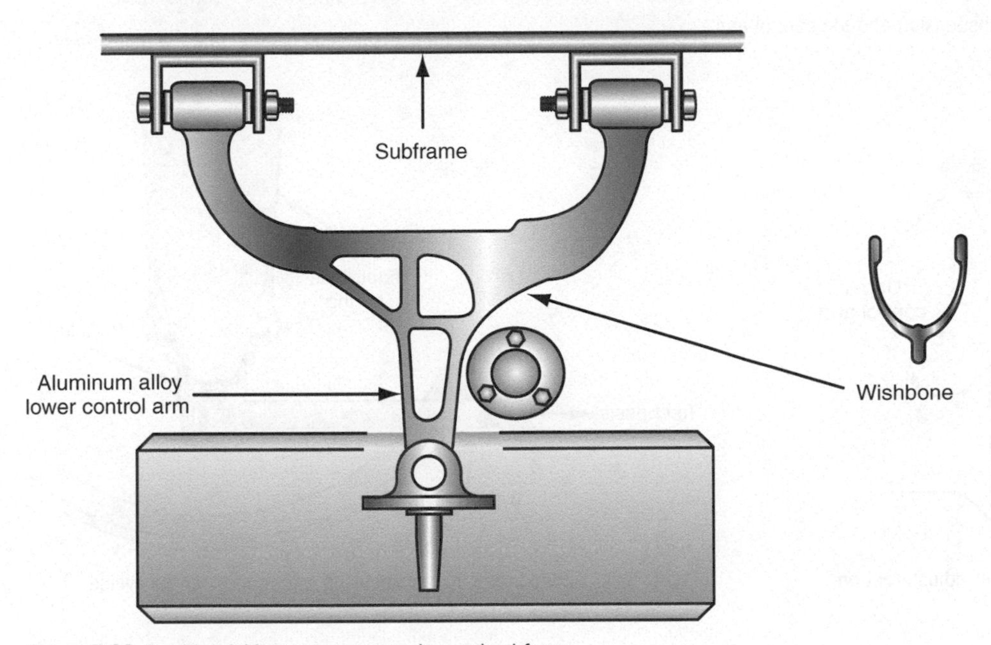

Figure 7-39 Double wishbone rear suspension and subframe.

Some independent rear suspension systems experience undesirable toe changes during wheel jounce and rebound. These toe changes cause vehicle instability during cornering and acceleration. In the double wishbone rear suspension system, the control arm design and the pivot locations on the rear toe control arm provide minimal change in toe-in during wheel jounce (**Figure 7-40**). This action results in extremely stable steering while cornering, accelerating, or driving on irregular road surfaces. The toe control arm may be lengthened or shortened to adjust rear wheel toe. An eccentric cam on the rear upper control arm bolt provides a rear wheel camber adjustment (**Figure 7-41** and **Figure 7-42**).

Independent Rear Suspension with Transverse Leaf Spring

Some cars have an independent rear suspension system with a transverse mono-leaf fiberglass spring. This type of spring is compact, lightweight, and corrosion free. Dual trailing arms are connected rearward from the chassis to the knuckle on each side of the suspension, and spindle support rods are attached from the center of the suspension to the bottom of the knuckle. Tie rods are connected from the rear of the knuckle to the center of the suspension (**Figure 7-43**).

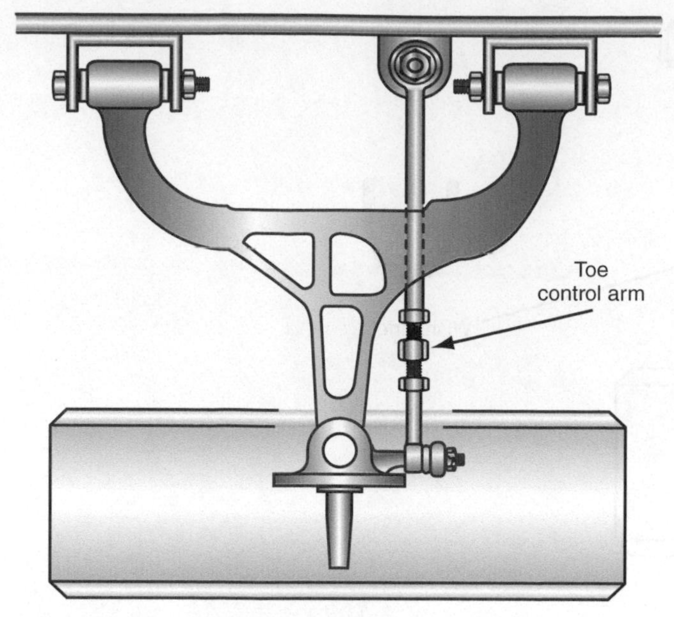

Figure 7-40 Double wishbone rear suspension and toe control arm.

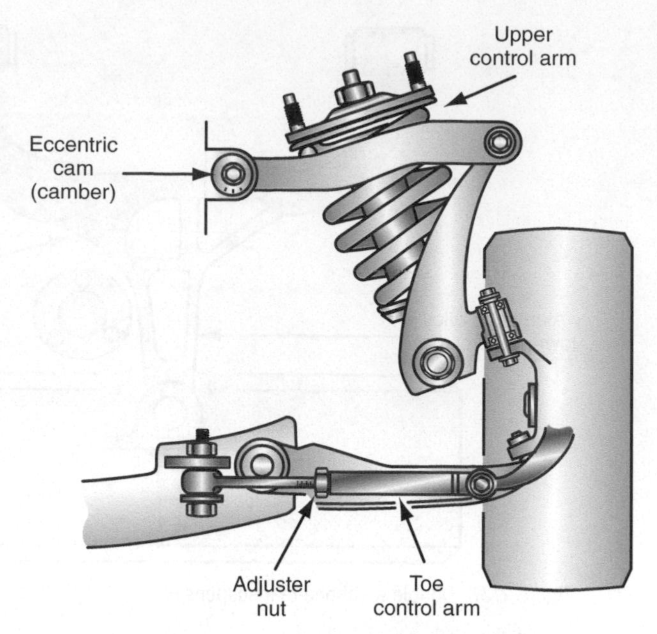

Figure 7-41 Rear suspension camber and toe adjustments.

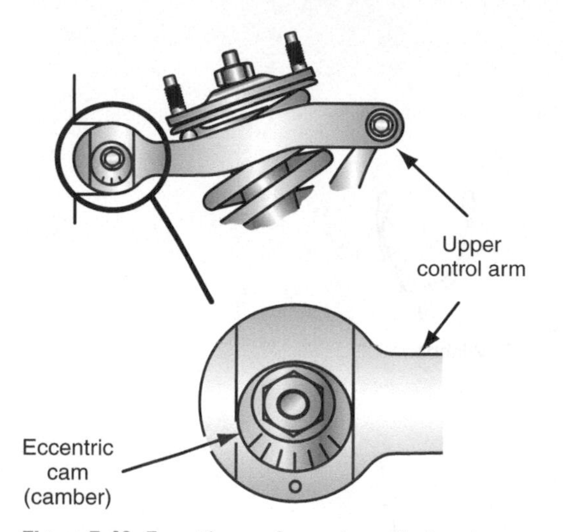

Figure 7-42 Eccentric cam for camber adjustment on rear upper control arm bolt.

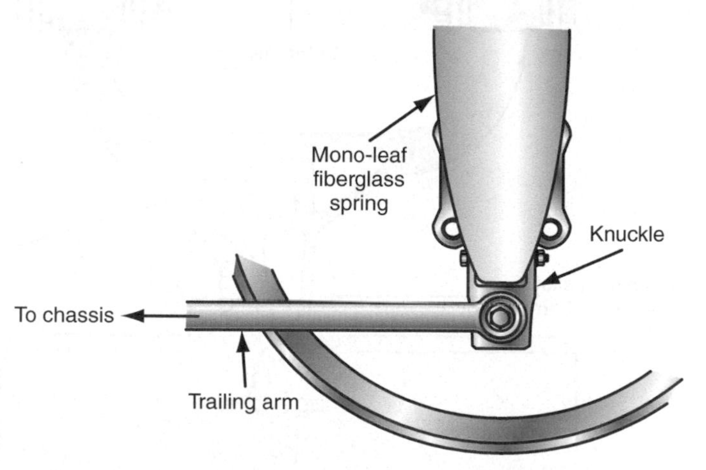

Figure 7-43 Independent rear suspension with mono-leaf transverse fiberglass spring.

Other independent rear suspension systems have a multiple-leaf transversely mounted rear spring. In these suspension systems, heavy control arms extend rearward from the chassis to the knuckles, and strut rods are connected from the bottom of the knuckles to the center of the suspension. The shock absorbers are connected from the lower end of the knuckles to the chassis. A suspension member connects the differential housing to the chassis. All suspension component mounting locations are insulated with rubber bushings.

CURB RIDING HEIGHT

Regular inspection and proper maintenance of suspension systems are extremely important to maintaining vehicle safety. *The curb riding height is determined mainly by spring condition.* Other suspension components such as control arm bushings will affect curb riding height if they are worn. Since incorrect curb riding height affects most of the other suspension angles, this measurement is critical. The curb riding height must be measured at the vehicle manufacturer's specified location, which varies depending on the type of suspension system. When the vehicle is on a level floor or an alignment rack, measure the curb riding height from the floor to the manufacturer's specified location (**Figure 7-44**).

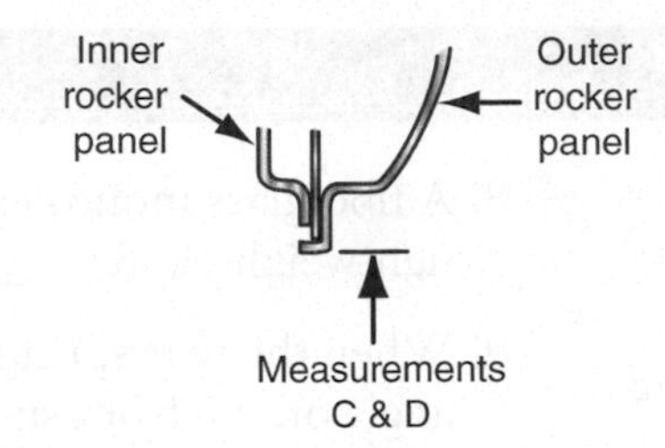

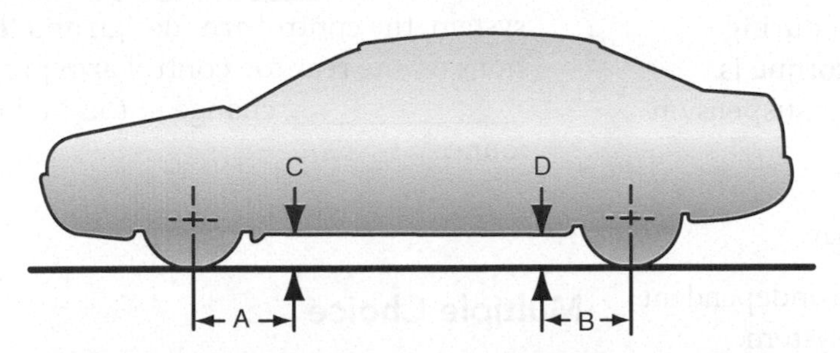

Figure 7-44 Front and rear curb ride height measurement locations.

SUMMARY

- A live axle rear suspension is one in which the differential housing, wheel bearings, and brakes act as a unit.
- A live axle rear suspension system provides good lateral stability, sway control, and steering characteristics, but this type of rear suspension causes increased tire wear with decreased ride quality and traction compared with other types of rear suspensions.
- Compared with other rear suspensions, the live axle leaf-spring suspension is more subject to axle tramp problems.
- In a live axle leaf-spring rear suspension system, the leaf springs absorb differential torque and provide lateral control.
- In a live axle coil spring rear suspension system, the lower control arms absorb differential torque and the upper arms control lateral movement.
- Live axle coil spring rear suspension systems may have a tracking bar to control lateral movement.
- A semi-independent rear suspension has a limited amount of individual rear wheel movement provided by a steel U-section channel or crossmember.
- In an independent rear suspension, each rear wheel can move individually without affecting the opposite rear wheel.
- Compared with a live axle rear suspension system, an independent rear suspension provides improved ride quality, steering control, tire life, and traction.
- In a MacPherson strut independent rear suspension, the coil springs are mounted on the struts.
- In a modified MacPherson strut independent rear suspension, the coil springs are mounted separately from the struts.
- In some independent rear suspension systems, the knuckle is positioned by a ball joint on the lower end and the strut on the upper end, and the coil spring is positioned between the lower control arm and the chassis.
- In a multilink independent rear suspension, the top of the knuckle is positioned by a rear upper link and a front upper link, and the lower end of the knuckle is positioned by a lower arm.
- Some rear-wheel-drive vehicles have an independent rear suspension with the differential connected through insulating mounts to the chassis.
- Some front-wheel-drive and rear-wheel-drive vehicles have an independent rear suspension with a mono-leaf or multiple-leaf transversely mounted leaf spring.

REVIEW QUESTIONS

Short Answer Essays

1. Describe a live axle rear suspension system.
2. Explain the disadvantages of a live axle leaf-spring rear suspension.
3. Describe the purpose of a leaf-spring shackle.
4. Explain the differential torque action during acceleration, and describe how this torque is absorbed in a live axle coil spring rear suspension.
5. Define axle tramp.
6. Describe the purpose of a tracking bar.
7. Explain the difference between a semi-independent and an independent rear suspension system.
8. Explain the advantages of an independent rear suspension compared with a live axle rear suspension.
9. Describe the advantage of mounting the differential to the chassis in a rear-wheel-drive car with an independent rear suspension.
10. Describe the components that position the upper end of the knuckle in a multilink rear suspension.

Fill-in-the-Blanks

1. A live axle leaf-spring rear suspension provides excellent ______________ stability.
2. During braking and deceleration, the front of the differential is twisted______________.
3. Axle tramp occurs during hard______________.
4. In many live axle coil spring rear suspensions, the differential torque is absorbed by the trailing lower______________ ______________.
5. In a modified MacPherson strut independent rear suspension, the coil spring seat is located on the______________ ______________.
6. In a multilink independent rear suspension, the rear suspension member supports the suspension and the______________.
7. In some multilink rear suspension systems, the rubber mounting bushings between the suspension member and the chassis are ______________ in the fore and aft direction and ______________ in the lateral direction.
8. A fiberglass mono-leaf rear spring is compact, lightweight, and ______________ free.
9. When the rear springs are sagged, the caster angle on the front suspension becomes more______________.
10. In the double ______________rear suspension system, the control arm design and the pivot locations on the rear toe control arm provide ______________ change in toe-in during wheel jounce.

Multiple Choice

1. Rear axle tramp is caused by:
   A. Irregular road surfaces.
   B. A bent rear control arm.
   C. Engine torque transmitted through the drive shaft.
   D. Improper rear wheel alignment.
2. All of these statements about a live axle leaf-spring rear suspension are true EXCEPT:
   A. During acceleration the front of the differential twists upward.
   B. The differential torque in a live axle rear suspension is applied to the springs.
   C. While decelerating and braking, the front of the differential twists downward.
   D. This type of rear suspension has a small amount of unsprung weight.
3. Rear axle tramp occurs during:
   A. Hard acceleration.
   B. Deceleration.
   C. High-speed driving.
   D. Cornering at high speed.
4. In a semi-independent rear suspension system, some individual wheel movement is provided by:
   A. The U-section channel and integral stabilizer bar.
   B. The struts.
   C. The trailing arms.
   D. The track bar and brace.

5. In a semi-independent rear suspension system:
   A. The track bar and brace absorb differential torque.
   B. The trailing arms prevent lateral wheel movement.
   C. The lower coil spring seats may be on the trailing arms.
   D. The lower end of the shock absorbers may be attached to the track bar.
6. Advancements in hydraulic bushings are in response to customer demands for:
   A. Quieter suspension.
   B. Lower ride height.
   C. Smoother ride quality.
   D. Improved performance.
7. All of these statements about modified MacPherson strut suspension systems are true EXCEPT:
   A. The lower control arms prevent lateral wheel movement.
   B. The tie rods control fore-and-aft wheel movement.
   C. The strut is bolted to the lower end of the spindle.
   D. The upper strut mount is retained on top of the strut.
8. In an independent rear suspension with a lower control arm and ball joint:
   A. The ball joint is pressed into the tie rod.
   B. The strut is bolted to the top of the knuckle.
   C. The lower coil spring seat is mounted on the strut.
   D. The suspension adjustment link is connected from the lower control arm to the chassis.
9. Sagged rear springs may cause:
   A. Slow steering wheel return after turning a corner.
   B. Decreased steering effort.
   C. Decreased positive caster on the front wheels.
   D. Harsh ride quality.
10. The adjustment link in a multilink rear suspension:
    A. Provides a rear wheel caster adjustment.
    B. Provides a rear wheel toe adjustment.
    C. Reduces fore-and-aft rear wheel movement.
    D. Absorbs engine torque transmitted through the drive shaft.

CHAPTER 8
STEERING COLUMNS AND STEERING LINKAGE MECHANISMS

Upon completion and review of this chapter, you should be able to understand and describe:

- How the steering column provides driver safety during a frontal collision.
- Two methods of steering column movement to protect the driver in a frontal collision.
- The purpose of a clock spring electrical connector.
- The mechanism that locks the steering wheel and gear shift when the ignition switch is in the lock position.
- A parallelogram steering linkage and explain the advantage of this type of linkage.
- The components in a parallelogram steering linkage.
- The purposes of the pitman arm and the idler arm.
- Two possible rack and pinion steering gear mountings.
- The rack and pinion steering linkage, and explain the advantages of this type of linkage.
- The design and operation of an active steering column.
- The design and operation of a driver protection module.

Terms To Know

Air bag deployment module
Center link
Clock spring electrical connector
Energy-absorbing lower bracket
Idler arm
Parallelogram steering linkage
Pitman arm
Pre-safe systems
Pyrotechnic
Rack and pinion
Steering linkage
Silencer
Spherical bearing
Spiral cable
Tie rod
Toe plate

INTRODUCTION

Steering columns play a significant part in steering control, safety, and driver convenience. The steering column connects the steering wheel to the steering gear. The column components must be in satisfactory condition to minimize free play and to provide adequate steering control. If steering column components such as universal joints are worn, column free play is excessive and steering control is reduced.

Most steering columns provide some method of column collapse during a collision. Some vehicles have plastic pins that shear off in the column jacket, gearshift tube, and steering shaft if the driver hits the steering wheel in a frontal collision. This shearing action of the plastic pins allows the column to collapse away from the driver. In other vehicles, the column-to-instrument panel mounting is designed to allow column movement if the driver hits the steering wheel in a collision. This action helps prevent driver injury.

All air bag-equipped vehicles have a driver's side air bag located in the upper side of the steering wheel, and most vehicles also have a passenger's side air bag mounted in the instrument panel. Some vehicles now have seat belt pretensioners that tighten the seat belts and hold the driver or passengers back against the seat if the vehicle is involved in a collision. Many vehicles now have side air bags mounted in the outer edge of the seat back, or an air bag curtain mounted just above the front and rear door openings. The side air bags have a separate module and sensors, and deploy only when the vehicle is involved in a side collision.

A recent development on some vehicles is the installation of **pre-safe systems** that react during the few milliseconds before a collision occurs to increase driver and passenger safety. The input signals to the pre-safe system include vehicle speed, braking torque, brake pedal application speed, wheel slip, vehicle acceleration around the vertical axis, spring compression and rebound travel, steering wheel rotational speed, and tire pressure. The pre-safe system recognizes and acts in three crucial situations:

1. Sideways skidding
2. Avoidance maneuvers
3. Emergency braking beyond ABS operation

The pre-safe system can differentiate between a drama and a crisis. For example, if the vehicle is driven over a patch of ice and the brakes are applied lightly without any vehicle skidding, the ABS system can operate to prevent wheel lockup, but the pre-safe system recognizes this condition as a drama and remains inoperative. The pre-safe system also remains inoperative during mild avoidance maneuvers.

When the inputs indicate an impending crisis driving situation, the pre-safe system performs these functions:

1. Pretensions the driver and front passenger seat belts so the driver and front passenger are optimally restrained in their seats during emergency braking or a skid prior to a collision. When a collision occurs, pyrotechnic seat belt tensioners provide increased seat belt tension. If a collision does not occur, the seat belts return to normal tension.
2. If the front passenger seat is moved too far forward, the pre-safe system moves the seat rearward before a collision occurs. If appropriate, the front passenger seat cushion and backrest are moved to a position that provides increased passenger protection if a collision occurs. For example, a steeply inclined seat backrest or a flat seat cushion may impair passenger restraint during a collision.
3. The electronically adjustable individual rear seat cushions are moved to the best possible angle to improve rear passenger restraint.
4. The sliding sunroof is closed to reduce the risk of occupant injury if a rollover occurs.

The brake assist system (BAS) also plays a crucial role in the pre-safe system operation. The BAS system operates beyond the ABS and electronic stability program (ESP) parameters. If the BAS detects an unstable driving condition such as severe vehicle skidding, it uses brake intervention and engine power reduction to correct the situation.

When the vehicle is involved in a frontal collision above a specific speed, the air bag sensor signals deploy the air bag or bags in a few milliseconds. As the driver is thrust forward during the frontal collision, the inflated air bag prevents the driver from striking the steering wheel, windshield, or instrument panel. The air bag deflates very quickly so it does not block the driver's vision in case the driver is still attempting to steer the vehicle. Since the driver's side air bag and some connecting components are mounted in the steering column, this column is a very important part of vehicle safety equipment.

Shop Manual
Chapter 8, page 310

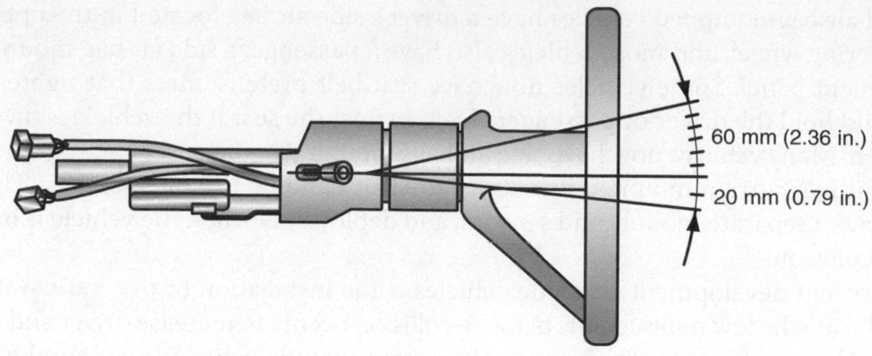

Figure 8-1 Tilt steering column.

Steering columns may be classified as non-tilting, tilting, and tilting or telescoping. Tilt steering columns facilitate driver entry to and exit from the front seat. These columns also allow the driver to position the steering wheel to suit individual comfort requirements (**Figure 8-1**). A tilting or telescoping steering column allows the driver to tilt and extend or retract the steering wheel. In this type of steering column, the driver has more steering wheel position choices.

NON-TILT STEERING COLUMN

Design

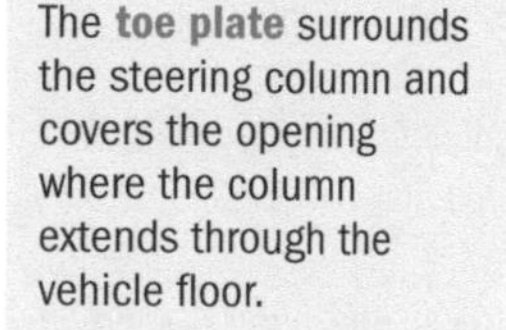

The **toe plate** surrounds the steering column and covers the opening where the column extends through the vehicle floor.

The **silencer** is mounted with the toe plate and helps to reduce the transmission of engine noise to the passenger compartment.

Many steering columns contain a two-piece steering shaft connected by two universal joints. A jacket and shroud surround the steering shaft. The upper shaft is supported by two bearings in the jacket. A **toe plate**, seal, and **silencer** surround the lower steering shaft and cover the opening where the shaft extends through the floor (**Figure 8-2**). The lower steering shaft is surrounded by a shield underneath the toe plate.

The lower universal joint couples the lower shaft to the stub shaft in the steering gear. In some steering columns, a flexible coupling is used in place of the lower universal joint (**Figure 8-3**).

Studs and nuts retain the steering column bracket to the instrument panel support bracket. The steering column is designed to protect the driver if the vehicle is involved in

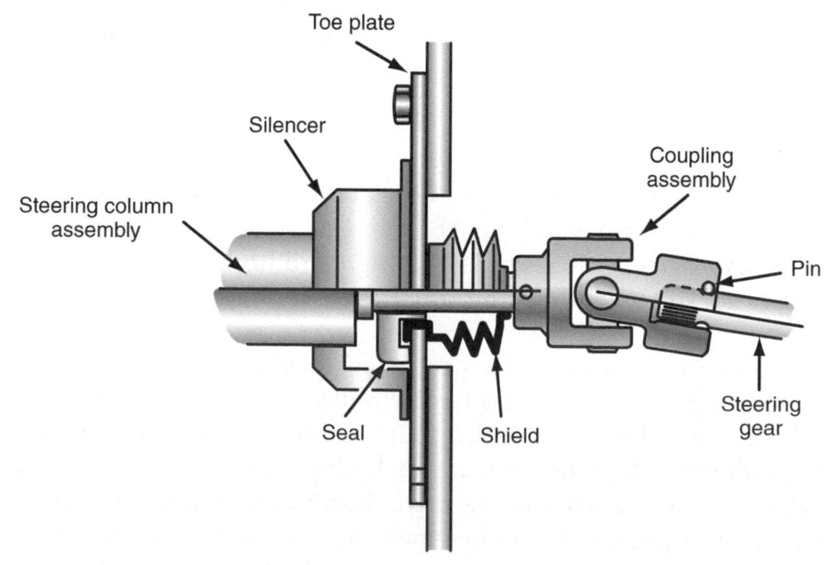

Figure 8-2 Toe plate and silencer.

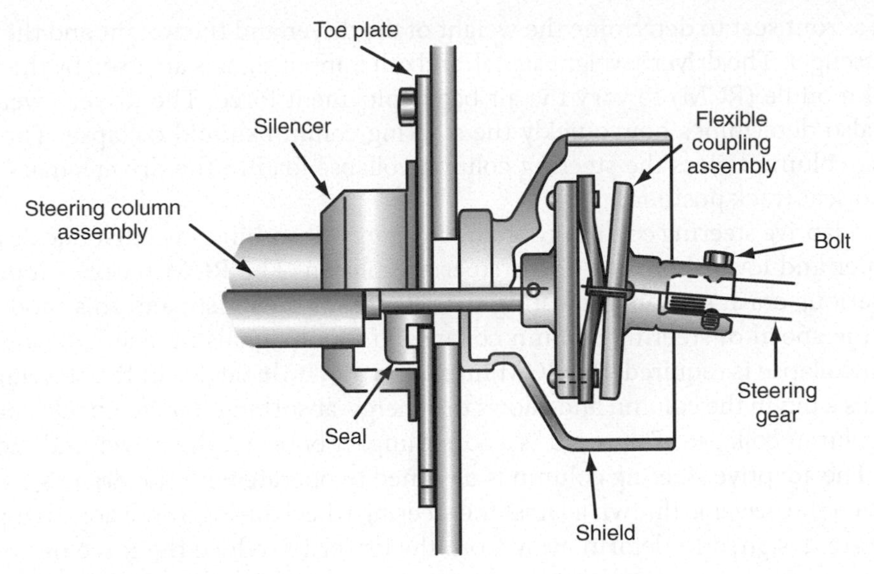

Figure 8-3 Flexible coupling.

a frontal collision. An **energy-absorbing lower bracket** and lower plastic adapter are used to connect the steering column to the instrument panel mounting bracket. This bracket allows the column to slide down if the driver is thrown forward into the wheel in a frontal collision. The mounting bracket is also designed to prevent rearward movement toward the driver in a collision.

Shop Manual
Chapter 8, page 315

In some steering columns, the outer column jacket is a two-piece unit retained with plastic pins (**Figure 8-4**). In this type of column, the lower steering shaft is a two-piece sliding unit retained with plastic pins (**Figure 8-5**). When the driver is thrown against the steering wheel in a frontal collision, the plastic pins shear off in the lower steering shaft and the outer column jacket. The shearing action of the plastic pins allows the steering column to collapse away from the driver, which reduces the impact as the driver hits the steering wheel.

A few current vehicles have steering columns with magnesium jackets or housings to reduce vehicle weight. Vehicle weight reduction improves fuel economy and reduces emissions.

An adaptive steering column has recently been introduced to the automotive market on some new models. The adaptive steering column collapses at two different speeds based on information received about the driver. Many vehicles have sensors in the lower

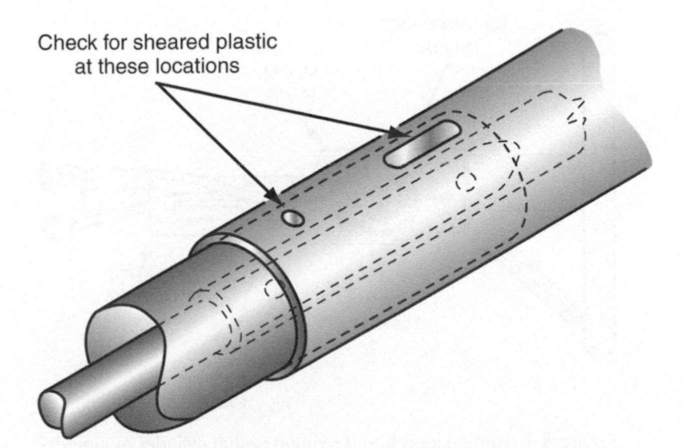

Figure 8-4 Injection plastic in collapsible outer steering column jacket.

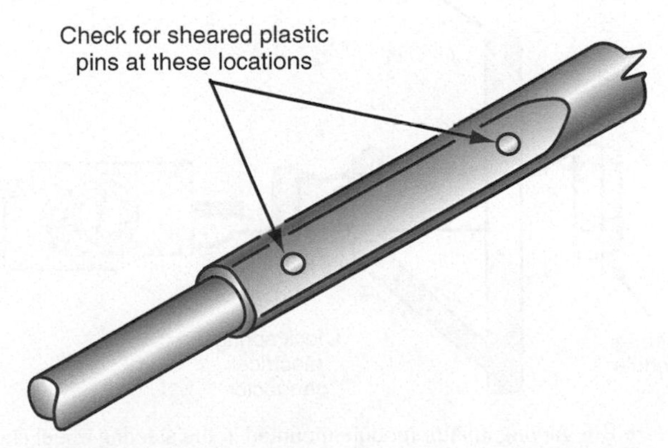

Figure 8-5 Injection plastic in collapsible lower steering shaft.

part of a front seat to determine the weight of the driver and the weight and the presence of a passenger. The driver's weight signal, and other input signals are used by the restraints control module (RCM) to vary the air bag deployment force. The driver's weight input signal also determines how quickly the steering column should collapse. The adaptive steering column tailors the steering column collapse load to the driver's mass, seat belt use, and seat track position.

An adaptive steering column contains an energy-absorbing steel that buckles between the upper and lower portions of the steering column. The RCM receives input signals from various crash sensors indicating the severity of the crash, and this module determines the speed of steering column collapse. If input signals indicate a faster steering column collapse is required, the RCM fires a **pyrotechnic** device in the steering column that pulls a pin in the column and allows the energy-absorbing steel to buckle and provide faster column collapse. The result is a softer impact between the driver and the steering wheel. The adaptive steering column is designed to operate with the driver's air bag.

A **pyrotechnic** device contains an explosive and an ignition source.

When the driver is thrown against the steering wheel during a collision, many steering wheels are designed to deform away from the driver to reduce the force on the driver's body.

WARNING Small amounts of sodium hydroxide are a by-product of an air bag deployment. Sodium hydroxide is a caustic chemical that causes skin irritation and eye damage. Always wear eye protection and gloves when servicing and handling a deployed air bag.

On many cars, the **air bag deployment module** is mounted in the top of the steering wheel (**Figure 8-6**). A **clock spring electrical connector**, or **spiral cable**, is mounted under the steering wheel. This component contains a ribbon-type conductor that maintains constant electrical contact between the air bag module and the air bag electrical system during steering wheel rotation.

An **air bag deployment module** contains the air bag and the inflator device.

The clock spring maintains positive electrical contact between the air bag module and the air bag electrical system regardless of steering wheel position. A clock spring electrical connector may be called a coil, **spiral cable**, or cable reel, depending on the manufacturer.

The steering wheel splines fit on matching splines on the top of the upper steering shaft, and a nut retains the wheel on the shaft. Most steering wheels and shafts have matching alignment marks that must be aligned when the steering wheel is installed.

An ignition switch cylinder is usually mounted in the upper right side of the column housing, and the ignition switch is bolted on the lower side of the housing (**Figure 8-7**). An operating rod connects the ignition switch cylinder to the ignition switch. Ignition switches are integral with the lock cylinder in some steering columns.

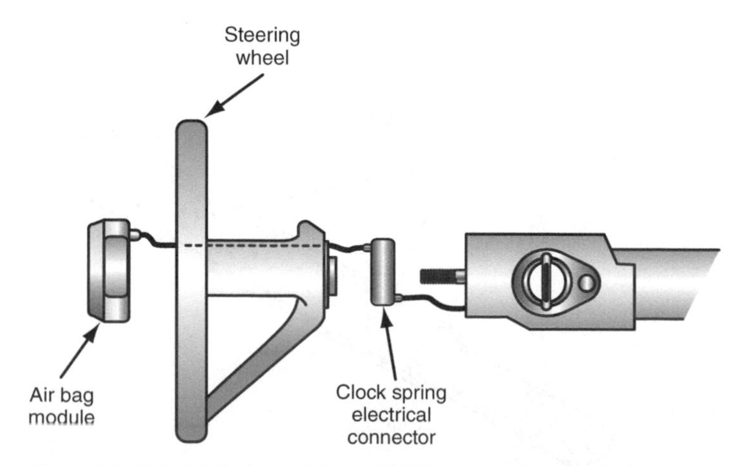

Figure 8-6 Air bag inflator module mounted in the steering wheel and clock spring electrical connector located under the steering wheel.

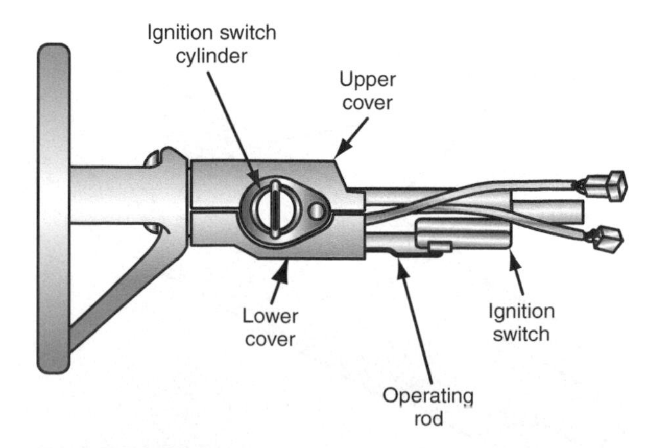

Figure 8-7 Ignition switch and ignition switch cylinder mounted in steering column.

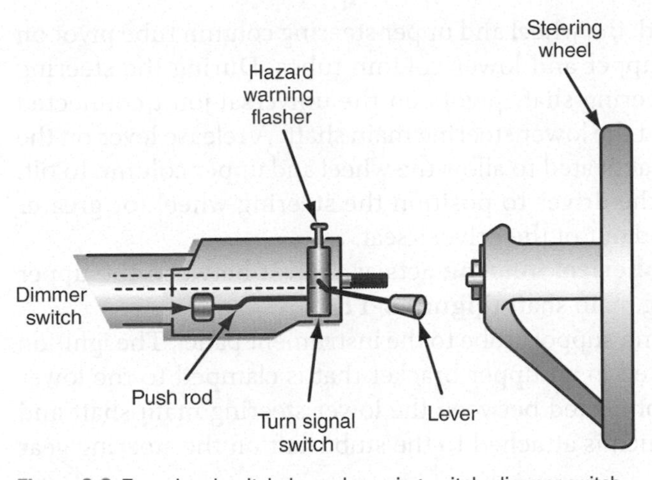

Figure 8-8 Turn signal switch, hazard warning switch, dimmer switch, and wipe or wash switch mounted in steering column.

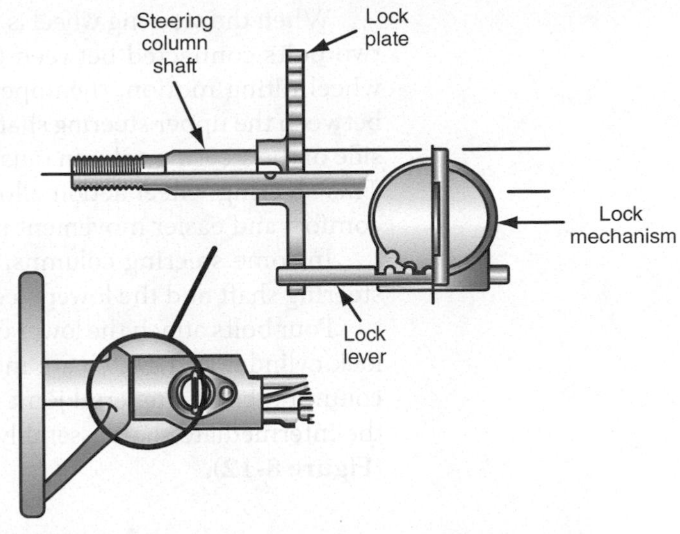

Figure 8-9 Upper steering column with locking plate and lever.

The turn signal switch and hazard warning switch are mounted on top of the steering column under the steering wheel. Lugs on the bottom of the steering wheel are used to cancel the signal lights after a turn is completed. On many vehicles, the signal light lever also operates the wipe or wash switch and the dimmer switch (**Figure 8-8**).

If the gear shift is mounted in the steering column, a tube extends from the gear shift housing to the shift lever at the lower end of the steering column. This shift lever is connected through a linkage to the transaxle or transmission shift lever. A lock plate is attached to the upper steering shaft, and a lever engages the slots in this plate to lock the steering wheel and gear shift when the gear shift is in Park and the ignition switch is in the Lock position (**Figure 8-9**).

The combination turn signal, hazard warning, wipe or wash, and dimmer switch are sometimes called a smart switch.

Multiaxial movement refers to movement in any direction around a pivot point.

Caution

When servicing a vehicle equipped with an air bag or bags, follow all service precautions in the vehicle manufacturer's service manual. Failure to follow these precautions may result in an expensive accidental air bag deployment.

Shop Manual
Chapter 8, page 317

TILT STEERING COLUMN

Design

Tilt steering columns have a short upper steering shaft connected to the steering wheel in the usual manner. This upper steering shaft is connected through a universal joint to the lower steering main shaft. An upper column tube surrounds the upper steering shaft, and the lower steering main shaft is supported on bushings in the lower column tube (**Figure 8-10**).

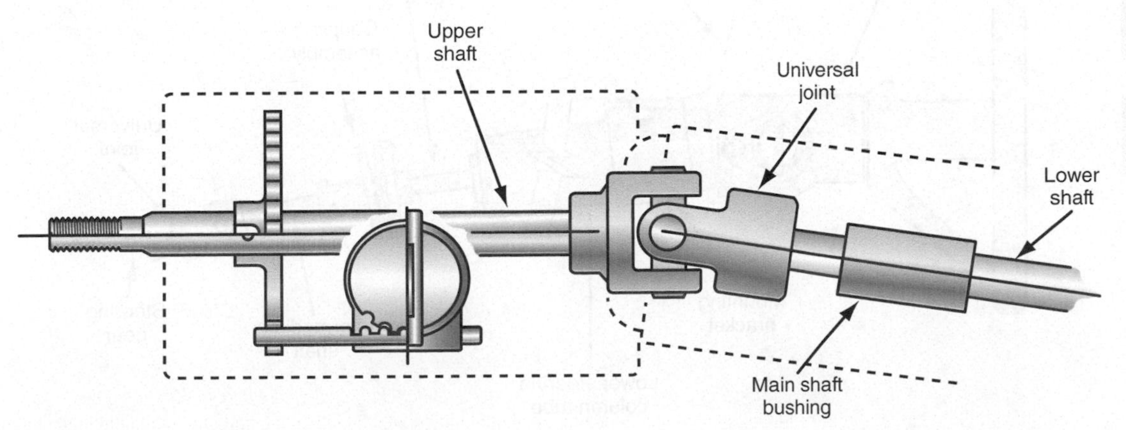

Figure 8-10 Tilt steering column components.

When the steering wheel is tilted, the wheel and upper steering column tube pivot on two bolts connected between the upper and lower column tubes. During the steering wheel tilting motion, the upper steering shaft pivots on the universal joint connected between the upper steering shaft and the lower steering main shaft. A release lever on the side of the steering column must be activated to allow the wheel and upper column to tilt. This steering wheel action allows the driver to position the steering wheel for greater comfort and easier movement in and out of the driver's seat.

In some steering columns, a **spherical bearing** acts as a pivot between the upper steering shaft and the lower steering main shaft (**Figure 8-11**).

Four bolts attach the lower column support tube to the instrument panel. The ignition lock cylinder and switch are mounted in an upper bracket that is clamped to the lower column tube. A universal joint is connected between the lower steering main shaft and the intermediate shaft assembly, which is attached to the stub shaft on the steering gear (**Figure 8-12**).

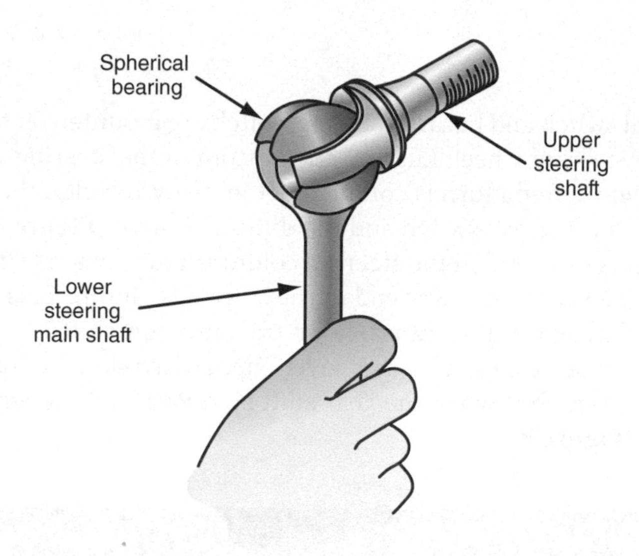

Figure 8-11 Spherical bearing acts as a pivot between the upper steering shaft and the lower steering main shaft in some tilt steering columns.

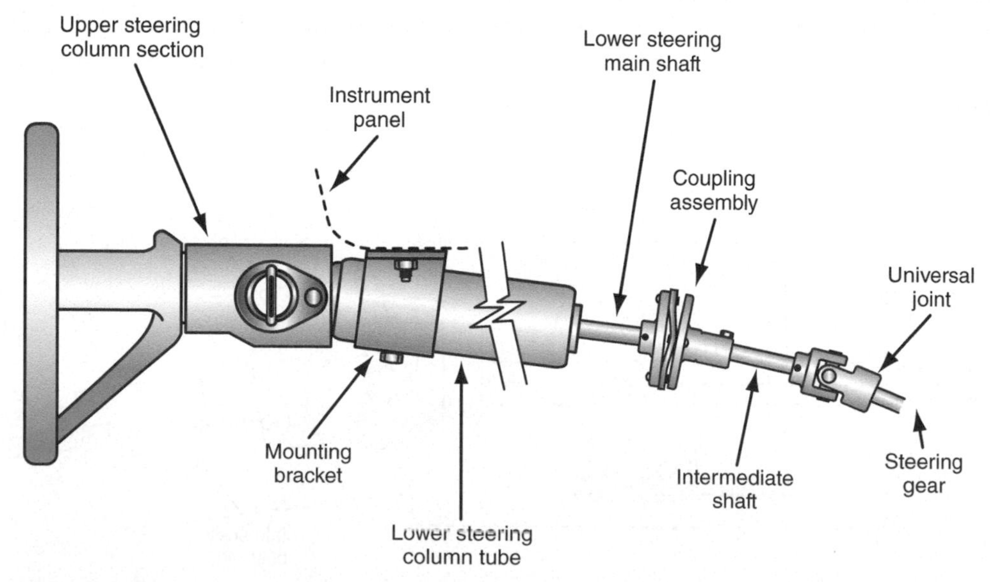

Figure 8-12 Tilt steering column with universal joint, intermediate shaft, combination switch, and steering wheel.

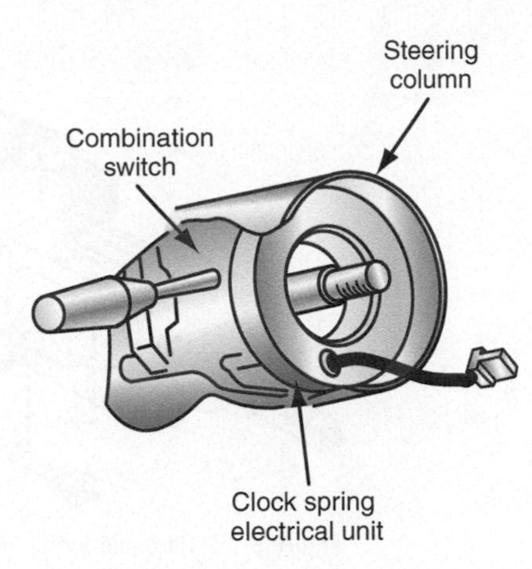

Figure 8-13 Clock spring electrical connector, or spiral cable, maintains electrical contact between the air bag inflator module and the air bag electrical system.

A steering wheel pad containing the air bag deployment module is mounted in the top of the steering wheel. Electrical contact between the air bag deployment module and the air bag electrical system is maintained by a clock spring electrical connector, or spiral cable, mounted directly under the steering wheel (**Figure 8-13**). A combination switch is mounted directly under the clock spring electrical connector. This switch contains the signal light switch, hazard light switch, dimmer switch, and wipe or wash switch. The upper side of the steering wheel also contains the horn switch contacts.

ELECTRONIC TILT AND TELESCOPING STEERING COLUMN

Some vehicles are equipped with an electronically controlled tilt and telescoping steering column. A driver-operated switch mounted in the steering column below the signal light lever controls the tilt and telescoping functions (**Figure 8-14**). A tilt and telescoping motor is mounted in the steering column power assembly. A potentiometer is mounted

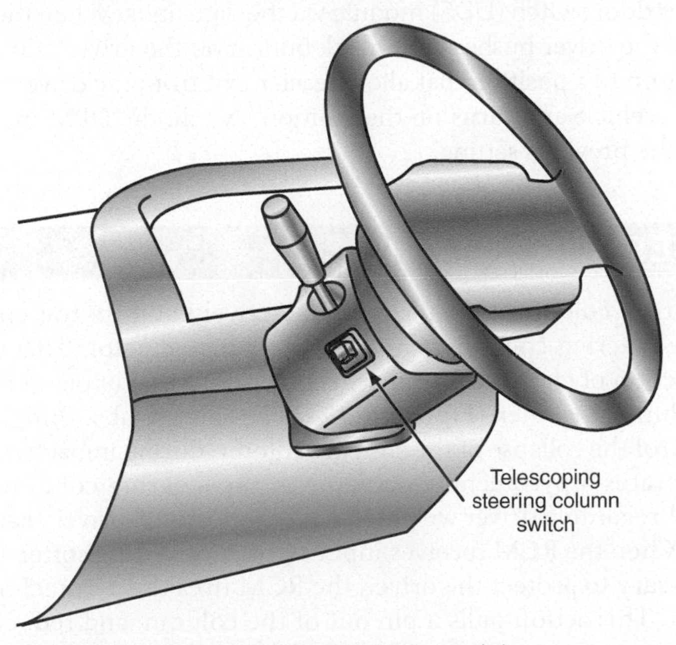

Figure 8-14 Tilt and telescoping steering column switch.

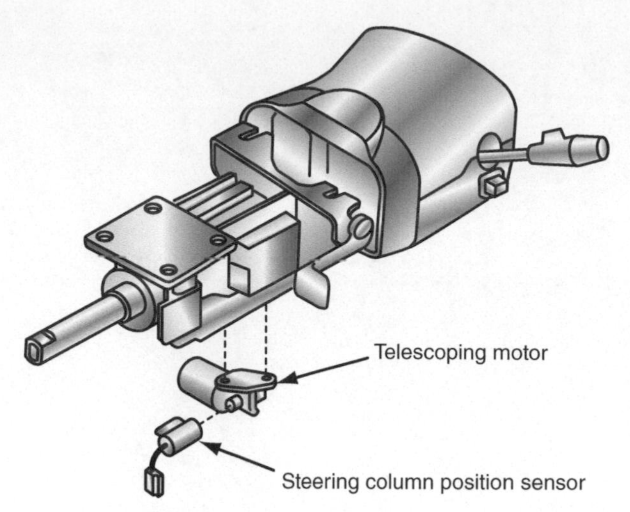

Figure 8-15 Tilt and telescoping motor and steering column position sensor.

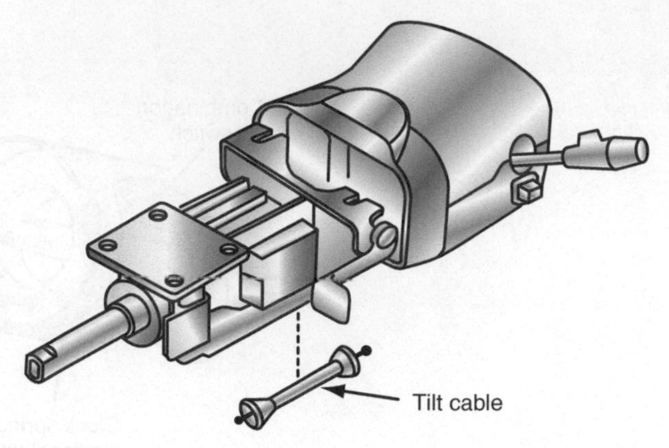

Figure 8-16 Tilt cable for tilt and telescoping steering column.

on the side of the tilt and telescoping motor, and this potentiometer is a steering column position sensor (**Figure 8-15**). This potentiometer sends a voltage signal to the driver position module (DPM) in relation to the motor and steering wheel position. The steering column position sensor signal is sent to the DPM, and the DPM uses this signal when storing and recalling steering column memory settings. A short tilt cable assembly is connected between the tilt and telescoping motor and the tilt mechanism in the steering column (**Figure 8-16**).

When the tilt and telescoping switch is operated by the driver, it sends voltage input signals to the DPM. If the driver operates the tilt and telescoping switch to request the telescoping function, the DPM operates the tilt and telescoping motor to move the steering wheel toward or away from the driver depending on the switch request. When the driver operates the tilt and telescoping switch to request the tilt function, the DPM operates the tilt and telescoping motor to tilt the wheel upward or downward depending on the switch request. The DPM is interconnected via data links to some of the other on-board computers. Therefore, the DPM can receive input signals from other computers such as the driver door switch (DDS) module via the data links. When the ignition switch is turned off and the driver pushes the unlock button on the driver's door, the DPM tilts the steering column to a position that allows easier exit from the driver's seat. When the driver enters the vehicle and turns on the ignition switch, the DPM moves the steering column back to the previous setting.

ACTIVE STEERING COLUMN

The active steering column in some vehicles actively adjusts the energy-absorbing capability of the steering column using a pyrotechnic actuator. This type of steering column has a section of energy-absorbing steel that holds the upper and lower halves of the steering column together (**Figure 8-17**). The energy-absorbing steel section is designed to control the collapse of the steering column during impacts.

A RCM operates a pyrotechnic actuator in the steering column. Input sensors inform the RCM regarding driver weight, front seat position, driver seat belt usage, and crash severity. When the RCM receives input signals indicating softer steering column collapse is necessary to protect the driver, the RCM fires the pyrotechnic device in the steering column. This action pulls a pin out of the column, and reduces the columns' resistance to collapse. The steering column collapse is designed to operate with the air

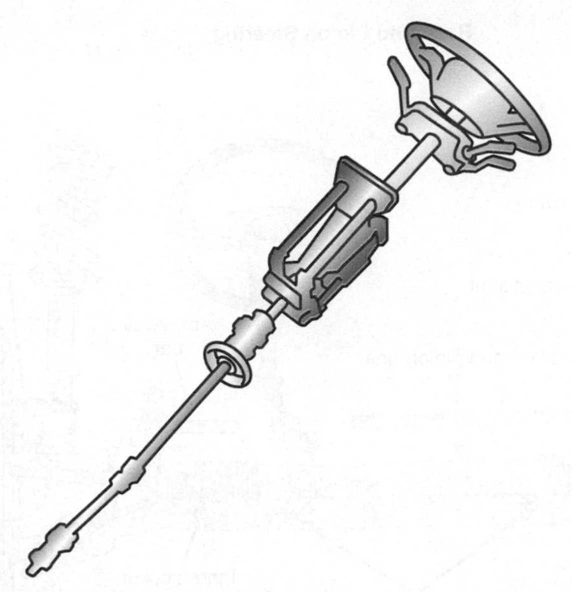

Figure 8-17 Active steering column.

bag level that is being activated. The ease of steering column collapse is different for a belted or unbelted driver. The active steering column helps the vehicle manufacturers to meet enhanced federal safety regulations, which require crash protection for 5th percentile, smaller female drivers, and 50th percentile, large male drivers with and without seat belts.

DRIVER PROTECTION MODULE

In the driver protection module, the steering column, knee bolster, and pedals are mounted in a module that allows these components to move away from the driver in a controlled manner during a vehicle crash (**Figure 8-18**). The driver protection module contains steel tubes supported by aluminum extrusions. This design allows the steering column, knee bolster, and pedals to move along the trajectory of the driver during a severe vehicle crash. This action helps to maintain air bag position. During a vehicle crash, the driver protection module movement may be controlled actively by a pyrotechnic device operated by an electronic module. The driver protection module will provide adequate crash protection for drivers from the 5th percentile to the 95th percentile. It is expected that this level of protection will be required by federal legislation in the future.

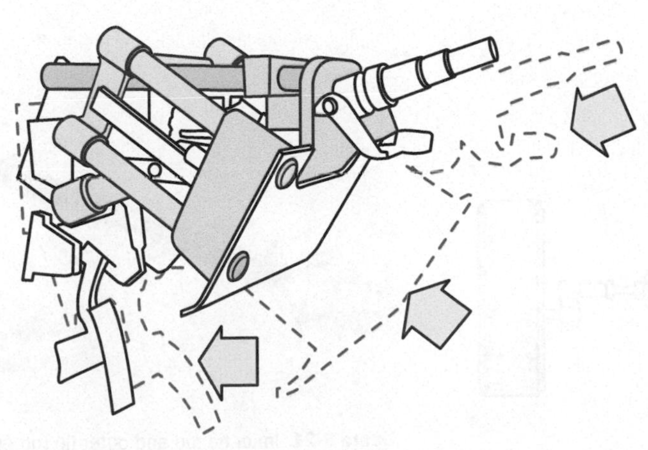

Figure 8-18 Driver protection module.

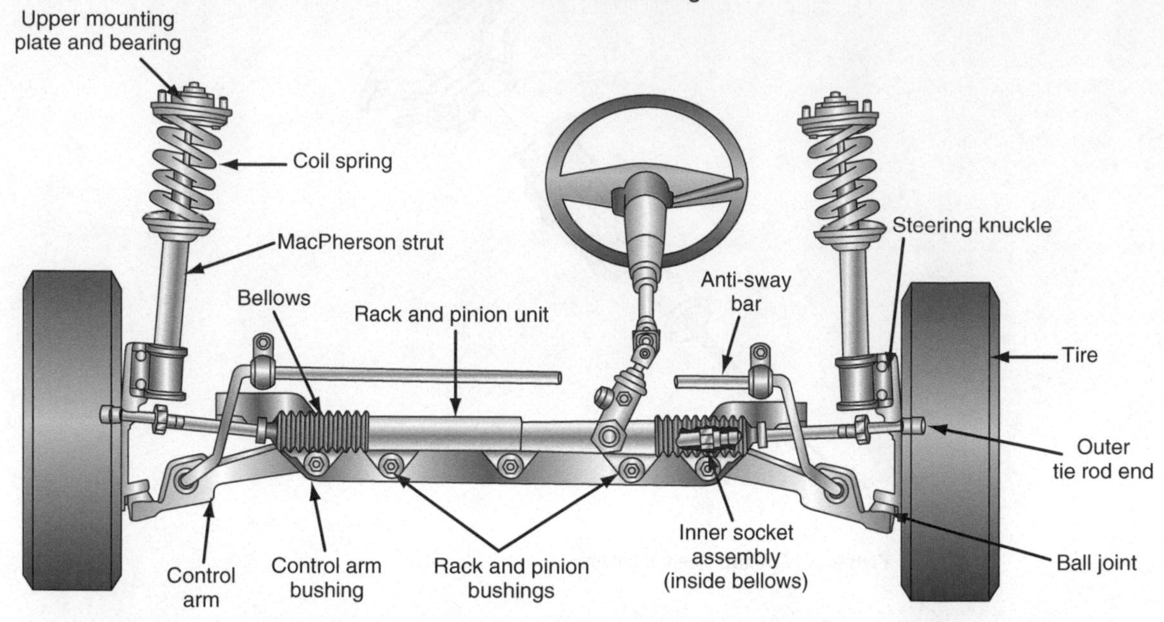

Figure 8-19 The rack and pinion steering linkage is used with rack and pinion steering gears.

STEERING LINKAGE MECHANISMS

Rack and Pinion Steering Linkage

The **rack and pinion steering linkage** is used with rack and pinion steering gears (**Figure 8-19**). In this type of steering gear, the rack is a rod with teeth on one side. This rack slides horizontally on bushings inside the gear housing. The rack teeth are meshed with teeth on a pinion gear, and this pinion gear is connected to the steering column. When the steering wheel is turned, the pinion rotation moves the rack sideways. Tie rods are connected directly from the ends of the rack to the steering arms. An inner tie rod end connects each tie rod to the rack, and bellows boots are clamped to the gear housing. The bellows boots keep dirt out of these joints (**Figure 8-20**). The inner tie rod end contains a spring-loaded ball socket. The outer tie rod ends connected to the steering arms are basically the same as those in parallelogram steering linkages (**Figure 8-21**). Some inner tie rod ends contain a bolt and a bushing. These tie rod ends are threaded onto the rack (**Figure 8-22**). Since the rack is connected directly to the tie rods, the rack replaces the center link in a parallelogram steering linkage.

Shop Manual
Chapter 8, page 327

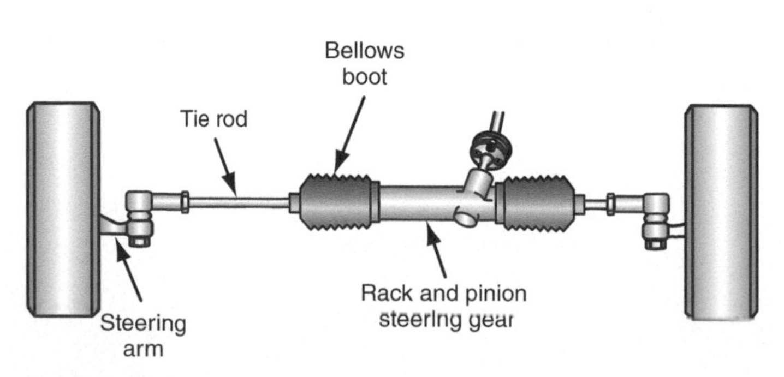

Figure 8-20 Rack and pinion steering gear.

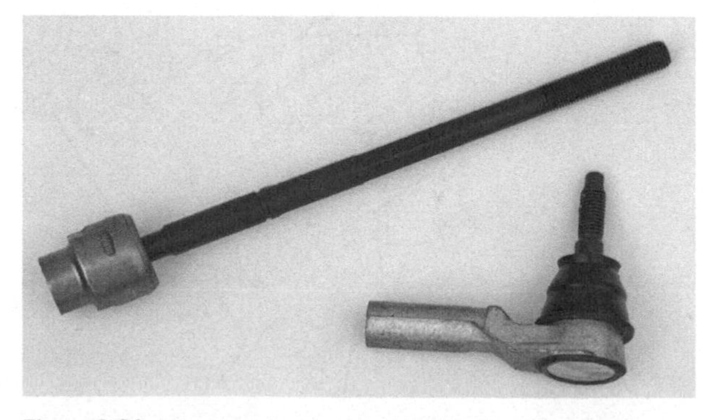

Figure 8-21 Inner tie rod and outer tie rod end, rack and pinion steering.

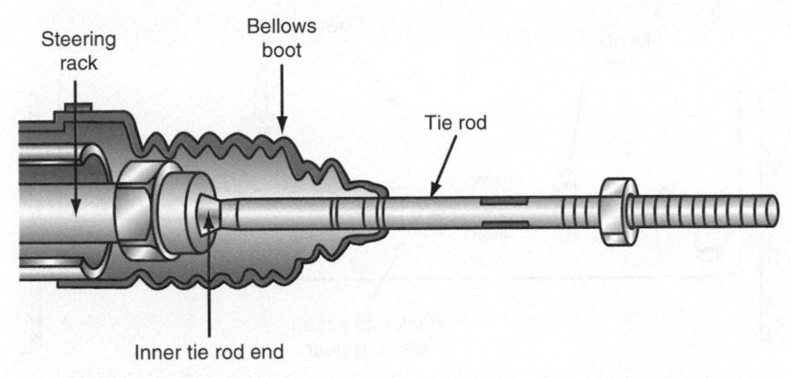

Figure 8-22 The inner tie rod connects directly to the rack and pinion steering gear and is protected by a bellows boot.

Some inner tie rod ends have a mirror-finished ball and a high-strength polymer bearing to ensure low torque, minimal friction, and extended life (**Figure 8-23**). A hardened alloy steel rod extends from the ball to the outer tie rod end and provides maximum strength and durability.

The rack and pinion steering gear may be mounted on the front subframe (**Figure 8-24**) or attached to the firewall behind the engine (**Figure 8-25**). Rubber

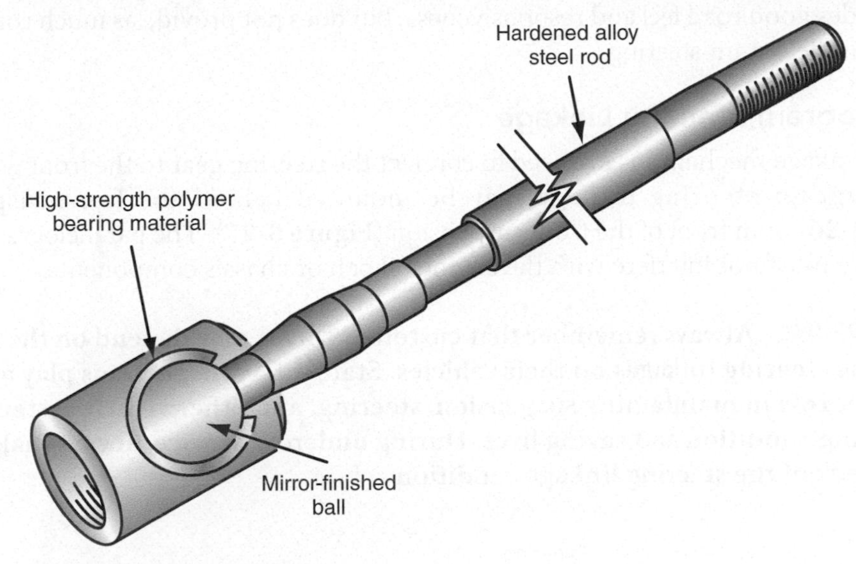

Figure 8-23 Inner tie rod design.

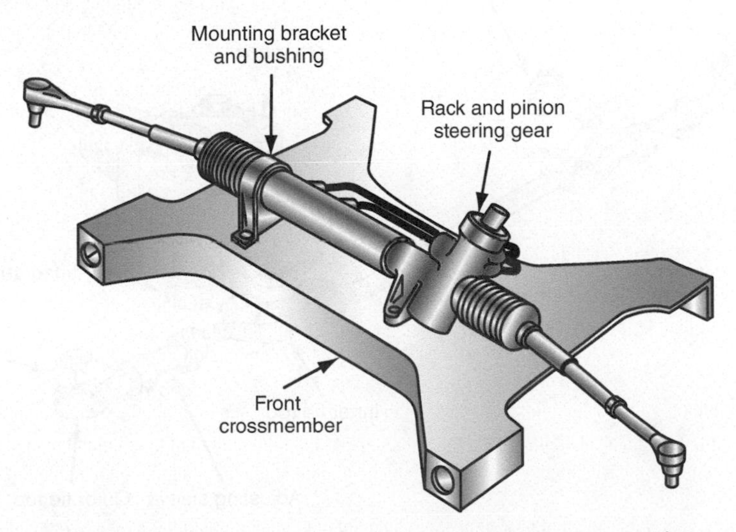

Figure 8-24 Rack and pinion steering gear mounting on front crossmember.

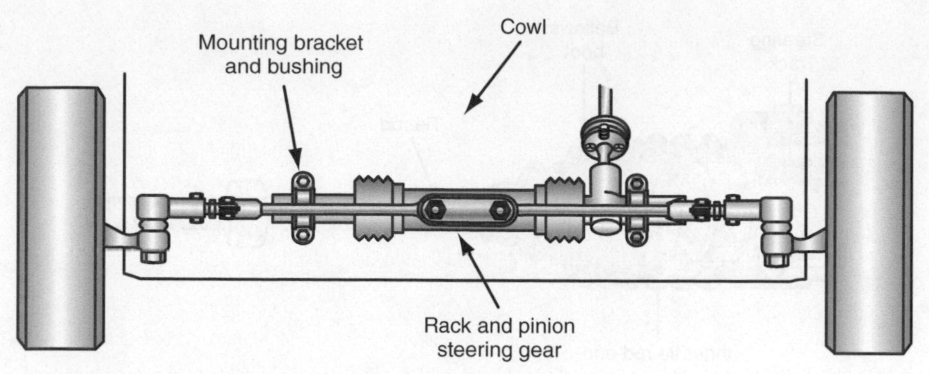

Figure 8-25 Rack and pinion steering gear mounted on cowl.

bushings insulate the steering gear, and these bushings are attached to the subframe or firewall. The rack and pinion steering gear is mounted at the proper height to position the tie rods and lower control arms parallel to each other. The number of friction points is reduced in a rack and pinion steering system, and this system is light and compact. Most of today's cars, sport utility vehicles (SUVs) and light trucks use rack and pinion steering. Since the rack and pinion is linked directly to the steering arms, this type of steering linkage provides good road feel and responsiveness, but does not provide as much road isolation as parallelogram steering.

Parallelogram Steering Linkage

A **parallelogram steering linkage** may be defined as one in which the tie rods are mounted parallel to the lower control arms.

Steering linkage mechanisms are used to connect the steering gear to the front wheels. A **parallelogram steering linkage** may be mounted behind the front suspension (**Figure 8-26**) or in front of the front suspension (**Figure 8-27**). The parallelogram steering linkage must not interfere with the engine oil pan or chassis components.

WARNING Always remember that customers' lives may depend on the condition of the steering linkages on their vehicles. State safety inspections play a very important role in maintaining suspension, steering, and other vehicle systems in safe driving condition and saving lives. During undercar service, always make a quick check of the steering linkage condition.

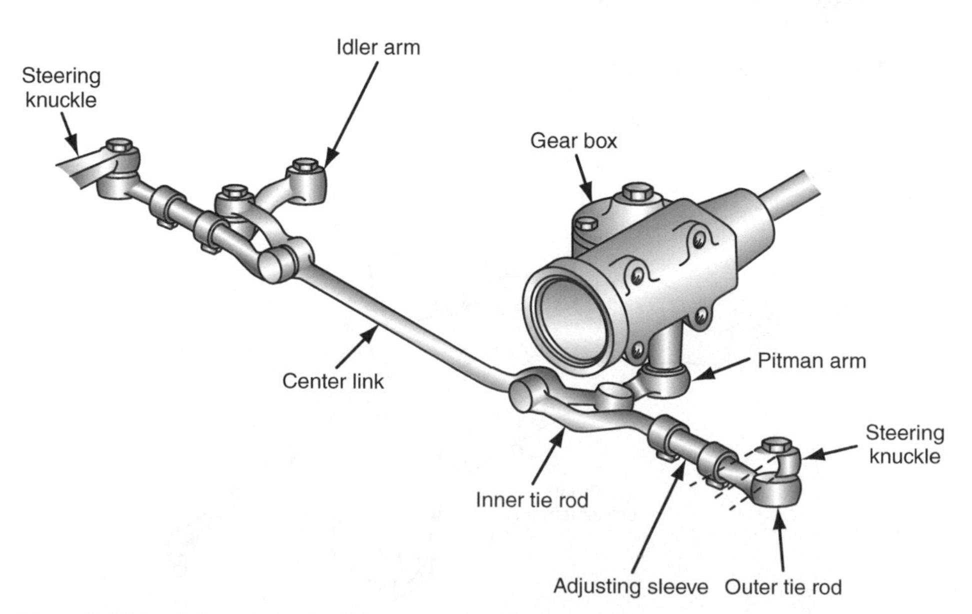

Figure 8-26 Parallelogram steering linkage behind the front suspension.

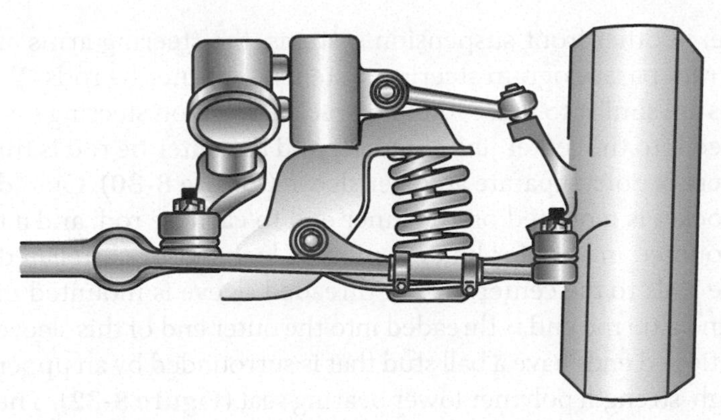

Figure 8-27 Parallelogram steering linkage in front of the front suspension.

Regardless of the parallelogram steering linkage mounting position, this type of steering linkage contains the same components. The main components in this steering linkage mechanism are:

1. Pitman arm
2. Center link
3. Idler arm assembly
4. Tie rods, inner
5. Tie rods, outer

Parallelogram steering linkages are found on independent front suspension systems (**Figure 8-28**). In a parallelogram steering linkage, the tie rods are connected parallel to the lower control arms. Road vibration and shock are transmitted from the tires and wheels to the steering linkage, and these forces tend to wear the linkages and cause steering looseness. If the steering linkage components are worn, steering control is reduced. Since loose steering linkage components cause intermittent toe changes, this problem increases tire wear. The wear points in a parallelogram steering linkage are the tie rod sockets and ends, idler arm, and center link end.

Tie Rods

The **tie rod** assemblies connect the center link to the steering arms, which are bolted to the front steering knuckles. In some front suspensions, the steering arms are part of the

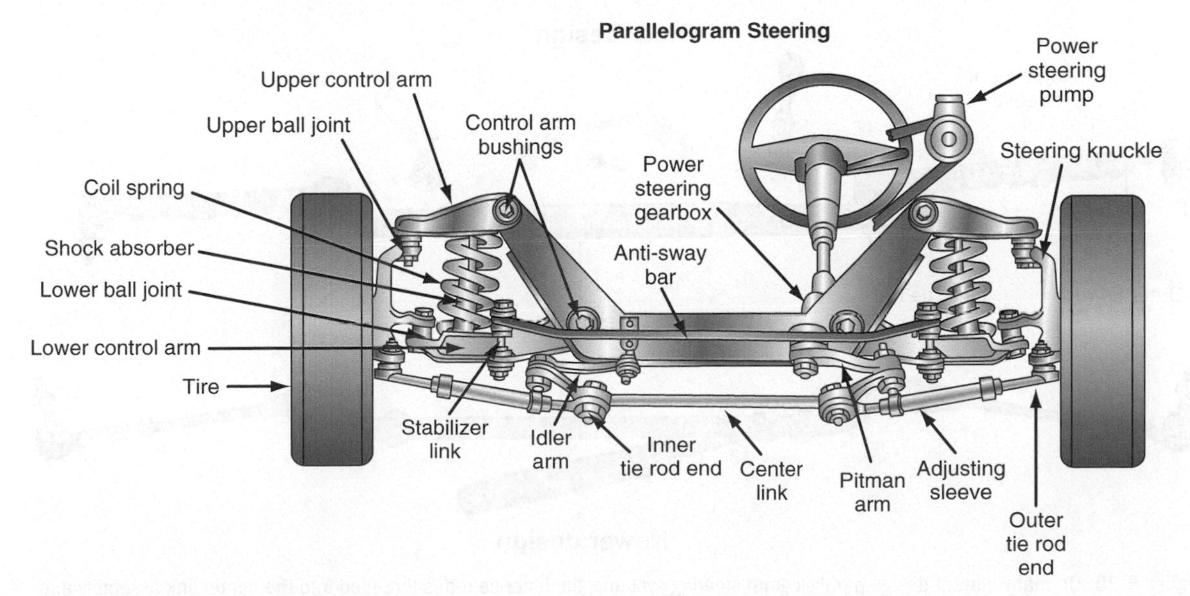

Figure 8-28 In a parallelogram steering linkage, the tie rods are connected parallel to the lower control arms.

steering knuckle; in other front suspension systems, the steering arms are bolted to the knuckle. On current parallelogram steering systems, the inner tie rods (**Figure 8-29**) and the outer tie rods are similar to those found on rack and pinion steering systems. The inner tie rod is threaded into the center link assembly and the outer tie rod is threaded onto the inner tie rod, there is not a separate adjuster sleeve (**Figure 8-30**). On older design inner tie rods, a ball socket is mounted on the inner end to each tie rod, and a tapered stud on this socket is mounted in a center link opening. A lock nut or castellated nut and cotter pin retain the tie rods to the center link. A threaded sleeve is mounted on the outer end of each tie rod, and a tie rod end is threaded into the outer end of this sleeve (**Figure 8-31**).

Some outer tie rod ends have a ball stud that is surrounded by an upper hardened steel bearing and a high-strength polymer lower bearing seat (**Figure 8-32**). The hardened steel

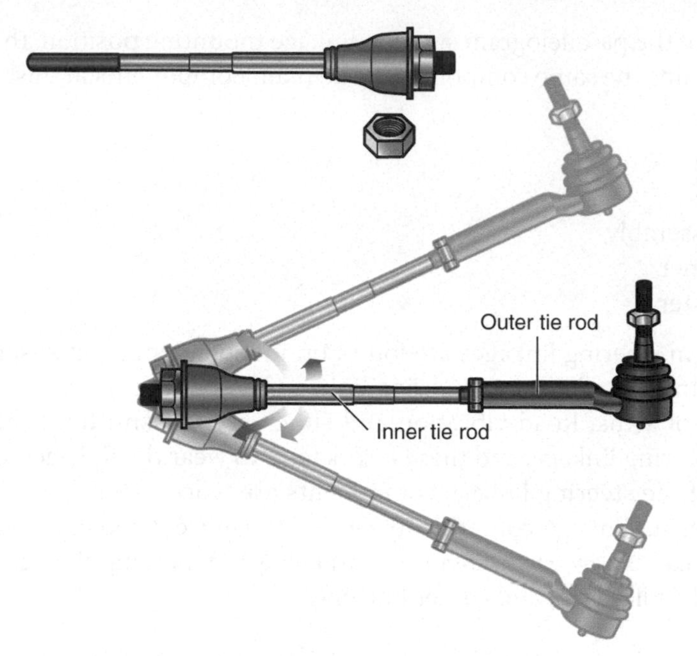

Figure 8-29 Inner tie rod design, which threads into the center link assembly, and outer tie rod threads onto inner tie rod, similar to a rack and pinion design.

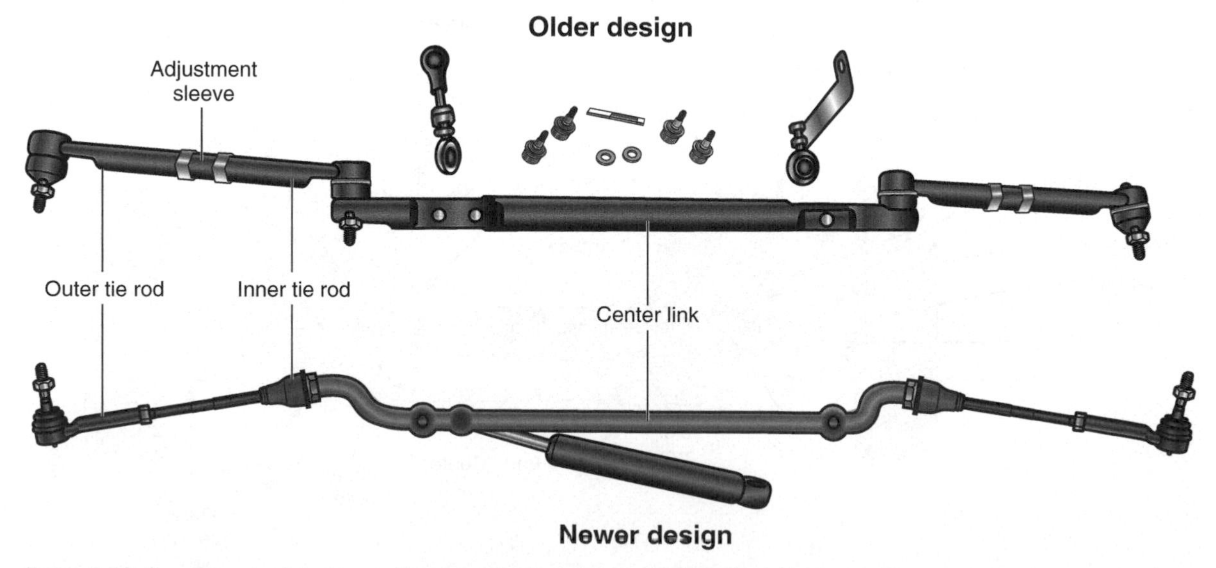

Figure 8-30 On many current design parallelogram steering systems, the inner tie rod is threaded into the center link assembly and the outer tie rod is threaded onto the inner tie rod; there is no separate adjuster sleeve.

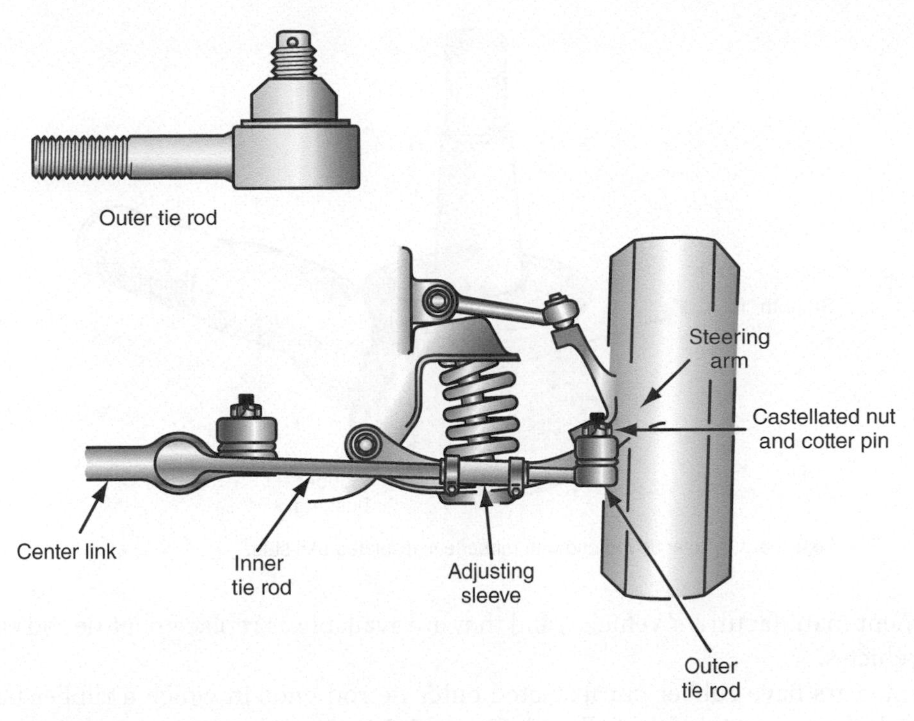

Figure 8-31 Tie rod design.

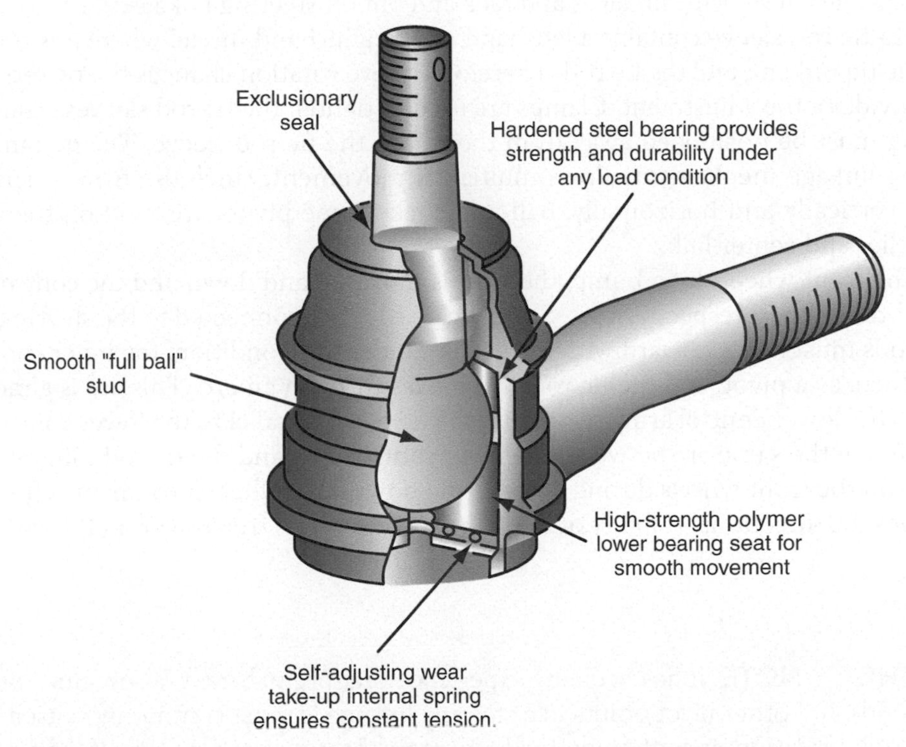

Figure 8-32 Outer tie rod end with hardened steel upper bearing and high-strength polymer lower bearing.

upper bearing provides strength and durability, and the polymer lower bearing seat provides smooth rotation of the ball stud in the tie rod end. An internal spring between the polymer lower bearing seat supplies self-adjusting action and constant tension on this seat. A seal in the upper part of the ball joint housing seals the ball stud to prevent contaminants from entering the tie rod end. These tie rod ends are installed on some original

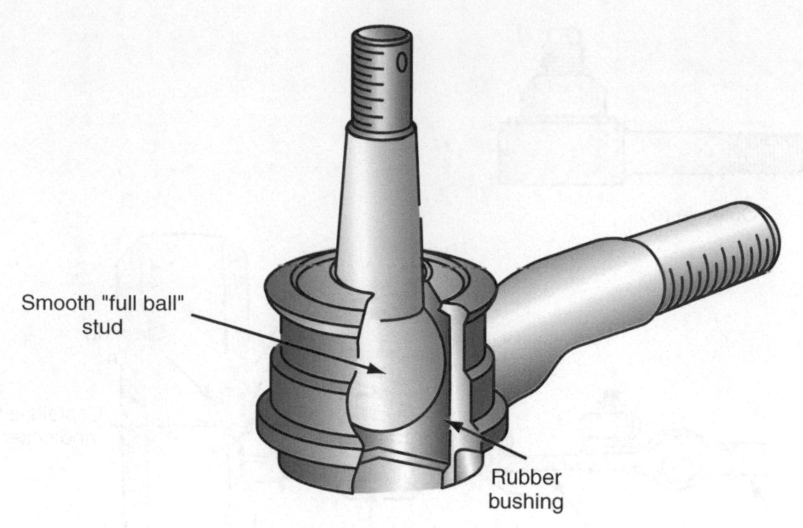

Figure 8-33 Outer tie rod end with rubber-encapsulated ball stud.

equipment manufacturers' vehicles, and they are available as replacement tie rod ends on most vehicles.

Some cars have rubber-encapsulated outer tie rod ends in which a rubber bushing surrounds the lower end of the ball stud (**Figure 8-33**). Special service procedures required on these tie rod ends are explained in the Shop Manual. Similar outer tie rod ends are used on parallelogram steering linkages and rack and pinion steering linkages.

Steering wheel free play is the amount of steering wheel movement before the front wheels start to move to the right or left.

Each tie rod sleeve contains a left-hand and a right-hand thread where it is threaded onto the tie rod end and the tie rod. Therefore, sleeve rotation changes the tie rod length and provides a toe adjustment. Clamps are used to tighten the tie rod sleeves. The clamp opening must be positioned away from the slot in the tie rod sleeve. The design of the steering linkage mechanism allows multiaxial movement, since the front suspension moves vertically and horizontally. Ball-and-socket-type pivots are used on the tie rod assemblies and center link.

If the front wheels hit a bump, the wheels move up and down and the control arms move through their respective arcs. Since the tie rods are connected to the steering arms, these rods must move upward with the wheel. Under this condition, the inner end of the tie rod acts as a pivot, and the tie rod also moves through an arc. This arc is almost the same as the lower control arm arc because the tie rod is parallel to the lower control arm. Maintaining the same arc between the lower control arm and the tie rod minimizes toe change on the front wheels during upward and downward wheel movement. This action improves the directional stability of the vehicle and reduces tread wear on the front tires.

AUTHOR'S NOTE It has been my experience that premature wear on outer tie rod ends and other pivot points in a steering linkage is most commonly caused by lack of lubrication or contamination that enters through broken seals. If the tie rod ends and other pivot points have grease fittings, lubrication at the manufacturer's recommended interval is important to maintain component life. However, you should not over-lubricate the steering linkage pivot points with a high-pressure grease gun, because this may rupture the seals. The steering linkage components are exposed to a large amount of water and dirt contamination. If the seals are leaking on any of the linkage pivot points, the pivots will soon be contaminated with moisture and dirt that acts as an abrasive to shorten component life.

Pitman Arm

The **pitman arm** connects the steering gear to the center link. This arm also supports the left side of the center link. Motion from the steering wheel and steering gear is transmitted to the pitman arm, and this arm transfers the movement to the steering linkage. This pitman arm movement forces the steering linkage to move to the right or left, and the linkage moves the front wheels in the desired direction. The pitman arm also positions the center link at the proper height to maintain the parallel relationship between the tie rods and the lower control arms.

Wear-type pitman arms have ball sockets and studs at the outer end, and this stud fits into the center link opening (**Figure 8-34**). The ball stud and socket are subject to wear, and pitman arm replacement is necessary if the ball stud is loose in the pitman arm. A non-wear pitman arm has a tapered opening in the outer end. A ball stud in the center link fits into this opening. The non-wear pitman arm only needs replacing if the arm is damaged or bent in a collision. The opening in the inner end of both types of pitman arms has serrations that fit over matching serrations on the steering gear shaft. A nut and lock washer retain the pitman arm to the steering gear shaft.

Idler Arm

An idler arm support is bolted to the frame or chassis on the opposite end of the center link from the pitman arm. The **idler arm** is connected from the support bracket to the center link. Two bolts retain the idler arm bracket to the frame or chassis. In some idler arms, a ball stud on the outer end of the arm fits into a tapered opening in the center link (**Figure 8-35**), whereas in other idler arms, a ball stud in the center link fits into a tapered opening in the idler arm.

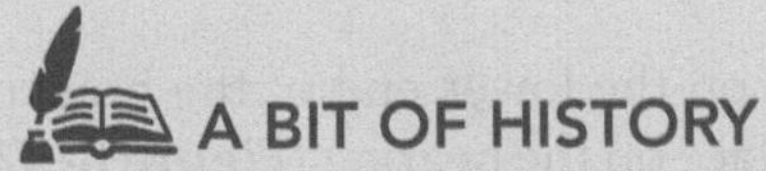

A BIT OF HISTORY

For many years, rear-wheel-drive vehicles were equipped with parallelogram steering linkages. The smaller, lighter front-wheel-drive cars introduced in large numbers in the late 1970s and 1980s required a more compact, lighter steering gear. The rack and pinion steering gear is used almost exclusively on front-wheel-drive cars because of its reduced weight and space requirements.

Figure 8-34 Pitman arm design.

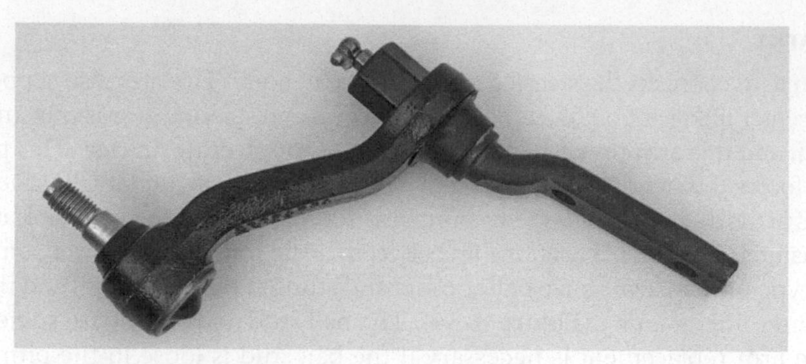

Figure 8-35 Idler arm design.

The idler arm supports the right side of the center link and helps maintain the parallel relationship between the tie rods and the lower control arms. The outer end of the idler arm is designed to swivel on the idler arm bracket, and this swivel is subject to wear. A worn idler arm swivel causes excessive vertical steering linkage movement and erratic toe. This action results in excessive steering wheel free play with reduced steering control and front tire wear.

Some idler arms contain an upper and lower gusher bearing that surrounds the bearing surface on the lower end of the bracket. A newly designed idler arm has a one-piece powdered metal gusher bearing that extends the full length of the friction surface on the lower end of the bracket (**Figure 8-36**). Compared with previous designs, this new design has 50 percent more bearing surface, which provides extended service life. A Belleville washer installed below the gusher bearing maintains a precision preload on the thrust washer, bracket, and gusher bearing.

Other idler arms have a conical machined surface on the lower end of the bracket. This conical surface is seated on a matching conical surface on the bearing (**Figure 8-37**). A coil spring between the lower end of the bracket and the housing maintains constant upward pressure on the bracket to maintain minimal endplay and constant turning torque as the bearing wears. Since this type of idler arm maintains minimal end play, more precise front suspension alignments are possible.

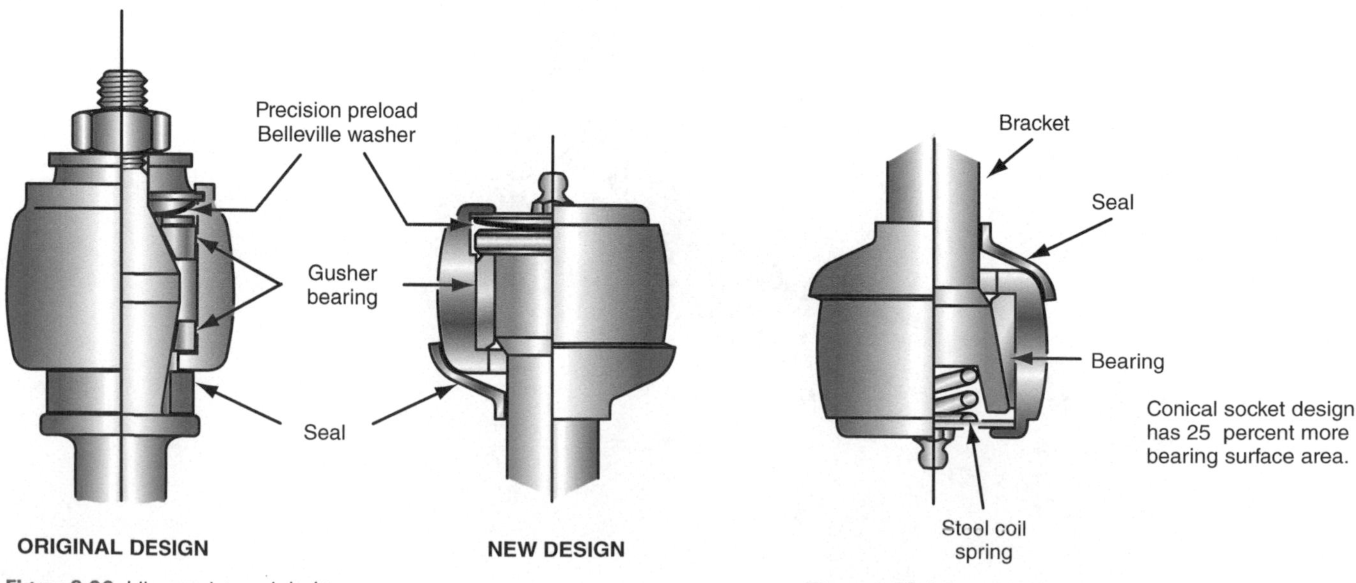

Figure 8-36 Idler arm internal design.

Figure 8-37 Idler arm with conical bearing.

Center Links

The **center link** controls the sideways steering linkage and wheel movement. The center link together with the pitman arm and idler arm provides the proper height for the tie rods, which is very important to minimizing toe change on road irregularities. Some center links have tapered openings in each end, and the studs on the pitman arm and idler arm fit into these openings. This type of center link may be called a taper end, or non-wear, link. Other wear-type center links have ball sockets in each end with tapered studs extending from the sockets (**Figure 8-38**). These tapered studs fit into openings in the pitman arm and idler arm, and they are retained with castellated nuts and cotter pins.

A **center link** may be referred to as a drag link, a steering link, or an intermediate link.

Haltenberger

A Haltenberger steering system is similar to the parallelogram and is used with a steering gear box. It was once popular on some light trucks, but over the last several decades this style steering system has primarily been used on just a few Jeep platforms. When the steering wheel is turned, the pitman arm will rotate. The pitman arm is connected to the drag link. The drag link is either connected to the tie rod assembly or directly to an arm on the left steering knuckle (**Figure 8-39**) depending on design. The drag link movement causes the tie rod assembly to move both the left and right steering arms thus turning the wheels. Some advantages to this system include its high strength, and the fact that it is less expensive than parallelogram linkage, as it has fewer parts. A major disadvantage is its less precise geometry.

STEERING DAMPER

A steering damper, or stabilizer, may be found on some parallelogram steering linkages. The steering damper is similar to a shock absorber. This component is connected from one of the steering links to the chassis or frame (**Figure 8-40**). When a front wheel strikes a road irregularity, a shock is transferred from the front wheel to the steering linkage, steering gear, and steering wheel. The steering damper helps absorb this road shock and

Figure 8-38 Center link design.

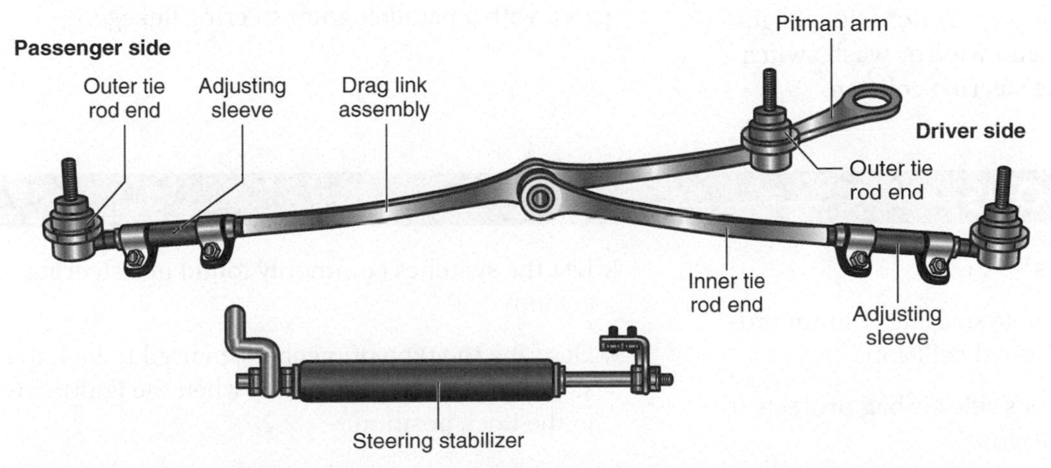

Figure 8-39 On a Haltenberger style steering system, the drag link is either connected to the tie rod assembly or directly to an arm on the left steering knuckle. The figure shows three variants of the design.

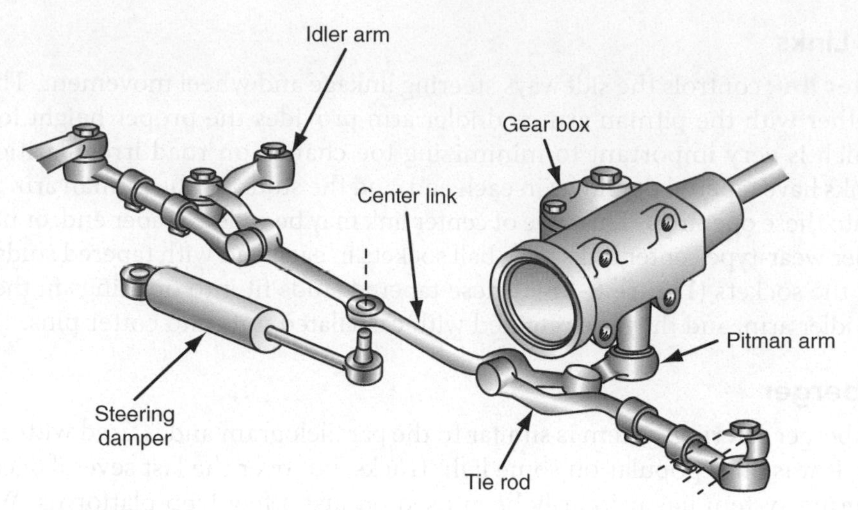

Figure 8-40 Steering gear damper and linkage.

prevent it from reaching the steering wheel. Heavy-duty steering dampers are available for severe road conditions such as those sometimes encountered by four-wheel-drive vehicles.

SUMMARY

- Steering columns help provide steering control, driver convenience, and driver safety. Many steering columns provide some method of energy absorption to protect the driver during a frontal collision.
- Steering wheels and columns now contain an air bag deployment module to protect the driver in a frontal collision.
- Tilt steering columns increase driver comfort and ease while driving or getting in or out of the driver's seat.
- A clock spring electrical connector supplies positive electrical contact between the air bag module in the steering wheel and the air bag electrical system.
- The ignition switch, dimmer switch, signal light switch, hazard switch, and wipe or wash switch may be mounted in the steering column.
- When the ignition switch is in the Lock position, a locking plate and lever in the upper steering column locks the steering wheel and the gear shift.
- In some tilt steering columns, the upper column housing pivots on two bolts, and the upper steering shaft pivots on a universal joint.
- In a parallelogram steering linkage, the tie rods are parallel to the lower control arms. The parallelogram steering linkage minimizes toe change as the control arms move up and down on road irregularities.
- A rack and pinion steering linkage has reduced friction points; it is lightweight and compact compared with a parallelogram steering linkage.

REVIEW QUESTIONS

Short Answer Essays

1. Explain how a collapsible steering column protects the driver in a frontal collision.
2. Explain how the driver's side air bag protects the driver in a frontal collision.
3. Describe the purpose of a clock spring.
4. List the switches commonly found in a steering column.
5. Describe the type of mechanism used to lock the steering wheel and gear shift when the ignition is in the Lock position.
6. Describe the pivot points in the upper shaft and upper column tube in a tilt steering wheel.

7. List the wear points in a parallelogram steering linkage.
8. List the five main components in a parallelogram steering linkage, and explain the purpose of each component.
9. Describe the basic design of a rack and pinion steering linkage.
10. Explain the advantages of a rack and pinion steering linkage compared with a parallelogram steering linkage.

Fill-in-the-Blanks

1. In some collapsible steering columns, _______________ in the outer column jacket and steering shaft shear off if the driver is thrown against the steering wheel in a frontal collision.
2. The driver's side air bag protects the driver if the vehicle is involved in a _______________ collision above a specific speed.
3. In an air bag system, a _______________ in the steering column maintains positive electrical contact between the air bag module and the air bag electrical system as the steering wheel is rotated.
4. In a parallelogram steering linkage, the center link connects the pitman arm to the _______________.
5. In a parallelogram steering linkage, the pitman arm and the _______________ position the center link and tie rods at the correct height.
6. In a wear-type pitman arm, a _______________ is positioned in the outer end of the pitman arm.
7. In a parallelogram steering linkage, the tie rods are parallel to the _______________.
8. The clamp opening in a tie rod sleeve must be positioned away from the _______________ in the tie rod sleeve.
9. In a rack and pinion steering system, the rack is connected directly to the _______________.
10. Compared with a parallelogram steering linkage, a rack and pinion steering linkage has a reduced number of _______________ points.

Multiple Choice

1. A typical air bag deployment time is:
   A. 1.5 minutes.
   B. 1 minute.
   C. 30 seconds.
   D. 40 milliseconds.
2. Many collapsible steering columns have:
   A. Plastic pins in the two-piece outer jacket.
   B. A collapsible bellows in the two-piece outer jacket.
   C. Steel pins in the two-piece lower steering shaft.
   D. A rubber spacer in the two-piece lower steering shaft.
3. The clock spring electrical connector:
   A. Maintains electrical contact between the air bag inflator module and the air bag electrical system.
   B. Is mounted above the steering wheel.
   C. Contains three spring-loaded copper contacts.
   D. Provides electrical contact between the signal light switch and the signal lights.
4. An active steering column has all of these components EXCEPT:
   A. A pyrotechnic actuator.
   B. A section of energy-absorbing steel.
   C. A telescoping cylinder.
   D. A pull-out pin that allows easier column collapse.
5. All of these statements about rack and pinion steering linkages are true EXCEPT:
   A. The tie rods are parallel to the lower control arms.
   B. The tie rod position depends on the steering gear mounting.
   C. The tie rods are connected to the pinion in the steering gear.
   D. The outer tie rod ends connect the tie rods to the steering arms.

6. All these statements about parallelogram steering linkages are true EXCEPT:
   A. Tie rod sleeves have the same type of thread in both ends of the sleeve.
   B. Loose steering linkages may cause excessive tire tread wear.
   C. Loose steering linkage causes excessive steering wheel free play.
   D. The pitman arm helps maintain the proper center link and tie rod height.
7. While discussing idler arms:
   A. Some idler arms contain a tapered roller bearing.
   B. The idler arm bracket is bolted to the upper control arm.
   C. A worn idler arm has no effect on front wheel toe.
   D. A partially seized idler arm bearing increases steering effort.
8. In a parallelogram steering linkage, the tie rods are parallel to the lower control arms to:
   A. Improve ride quality.
   B. Provide longer steering linkage life.
   C. Extend shock absorber and spring life.
   D. Reduce toe change during upward and downward front wheel movement.
9. A rack and pinion steering gear:
   A. Has tie rods that connect the rack directly to the steering arms.
   B. May be bolted to the vehicle frame.
   C. Has inner tie rod ends that are pressed onto the rack.
   D. Has more friction points compared with a parallelogram steering linkage.
10. While discussing steering linkages and dampers:
    A. In a rack and pinion steering gear, the rack positions the tie rods parallel to the lower control arms.
    B. A rack and pinion steering gear is used on most rear-wheel-drive cars.
    C. A steering damper is used on many front-wheel-drive vehicles.
    D. A defective steering damper may cause excessive steering effort.

CHAPTER 9

FOUR-WHEEL ALIGNMENT, PART 1 PRIMARY ANGLES

Upon completion and review of this chapter, you should be able to understand and describe:

- The effect of excessive positive camber on front suspension systems.
- Camber changes during front wheel jounce and rebound travel.
- The relationship between camber and vehicle directional stability.
- How front suspension camber and caster may be adjusted to compensate for road crown.
- The effects of positive and negative caster on directional control and steering effort.
- Positive and negative caster as they relate to ride quality.
- How higher- or lower-than-specified front or rear suspension height affects front suspension caster.
- Toe-in and toe-out on front suspension systems.
- The toe-in settings required on front-wheel-drive and rear-wheel-drive vehicles.
- Tire tread wear caused by excessive toe-in.
- The customer complaints that may arise from incorrect rear wheel toe and thrust line adjustments.
- Suspension and chassis defects that may cause improper rear wheel toe, thrust line, or camber.
- The variables that affect wheel alignment.
- Wheel alignment, and explain five reasons for performing a wheel alignment.
- Why four-wheel alignment is essential on front-wheel-drive unitized body cars.
- Thrust line and geometric centerline and describe the effect of an improper thrust line on vehicle steering.
- The result of improper rear wheel toe and its effect on the thrust line.
- The causes of improper rear wheel toe.
- The causes of rear axle offset.
- Rear axle sideset, setback, and dog tracking.
- Front and rear tire tread wear caused by inaccurate rear wheel toe and thrust line settings.
- Geometric centerline alignment and explain the problem with this type of alignment.
- Thrust line alignment and describe the shortcoming of this type of alignment.
- Total four-wheel alignment and explain the advantage of this type of alignment.
- The safety hazards created by incorrect wheel alignment or worn suspension and steering components.

Terms To Know

Bump steer
Directional stability
Four-wheel alignment
Geometric centerline
Geometric centerline alignment
High-frequency transmitter
Jounce
Memory steer
Motorist Assurance Program (MAP)
Negative camber

Negative caster	Road variables	Toe-out
Positive camber	Steering drift	Torque steer
Positive caster	Steering pull	Tracking
Rear axle offset	Steering wander	Wheel alignment
Rear axle sideset	Thrust angle	Wheel sensors
Rebound	Thrust line	Wheel setback
Receiver	Thrust line alignment	Wheel shimmy
Road crown	Toe-in	

INTRODUCTION

Automotive engineers design suspension and steering systems that provide satisfactory vehicle control with acceptable driver effort and road feel. The vehicle should have a tendency to go straight ahead without being steered. This tendency is called **directional stability**. A vehicle must have predictable directional control, which means the steering must provide a feeling that the vehicle will turn in the direction steered. The wheels must be reasonably easy to turn and tire wear should be minimized. These steering qualities and tire conditions are achieved when front and rear **wheel alignment** angles are within the vehicle manufacturer's specifications.

Wheel alignment may be defined as an adjustment and refitting of suspension parts to original specifications that ensure design performance.

The condition of suspension system components and wheel alignment is extremely important to maintaining driving safety and normal tire wear. Worn suspension components such as tie rod ends, ball joints, and control arms can suddenly fall apart and cause complete loss of steering. This disastrous event may result in not only some very expensive property damage, but also the loss of human life. When alignment angles are incorrect, an uncontrolled vehicle swerve or skid may occur during hard braking, resulting in a serious accident. Severe misalignment may reduce tire life to one-third of the normal expected tire life with correct alignment. After suspension components such as ball joints or control arms are replaced, wheel alignment is essential.

Shop Manual
Chapter 9, page 373

Technicians must be familiar with the symptoms that indicate incorrect rear wheel alignment. The rear wheel alignment procedures required to correct improper rear wheel alignment must be understood.

CAMBER FUNDAMENTALS

Camber

Camber is defined as the inward or outward tilt of the tire-and-wheel assembly as viewed from the front of the vehicle (**Figure 9-1**). A reference line begins at the base (ground) center of the tire and wheel and extends upward at a 90-degree angle to the road. Put another way, camber refers to the tilt of a line through the tire and wheel centerline in relation to the true vertical centerline of the tire and wheel. **Positive camber** is obtained when the top of the tire and wheel is tilted outward, away from the true vertical line of the wheel assembly. **Negative camber** occurs when the tire and wheel centerline tilts inward in relation to the wheel assembly true vertical centerline. The camber angle is measured and referred to the same for both the front and rear tires, meaning negative camber is the inward tilt of the tire at the top whether it is in the front or the rear of the vehicle.

Camber angle is displayed in degrees on an alignment computer. Camber is an angle that can affect both tire wear and directional stability, such as a pull or drift. If the side-to-side difference in front camber is greater than 0.5 to 0.75 degree, the vehicle may exhibit a pull or drift to the most positive side (**Figure 9-2**) as a general rule. This left

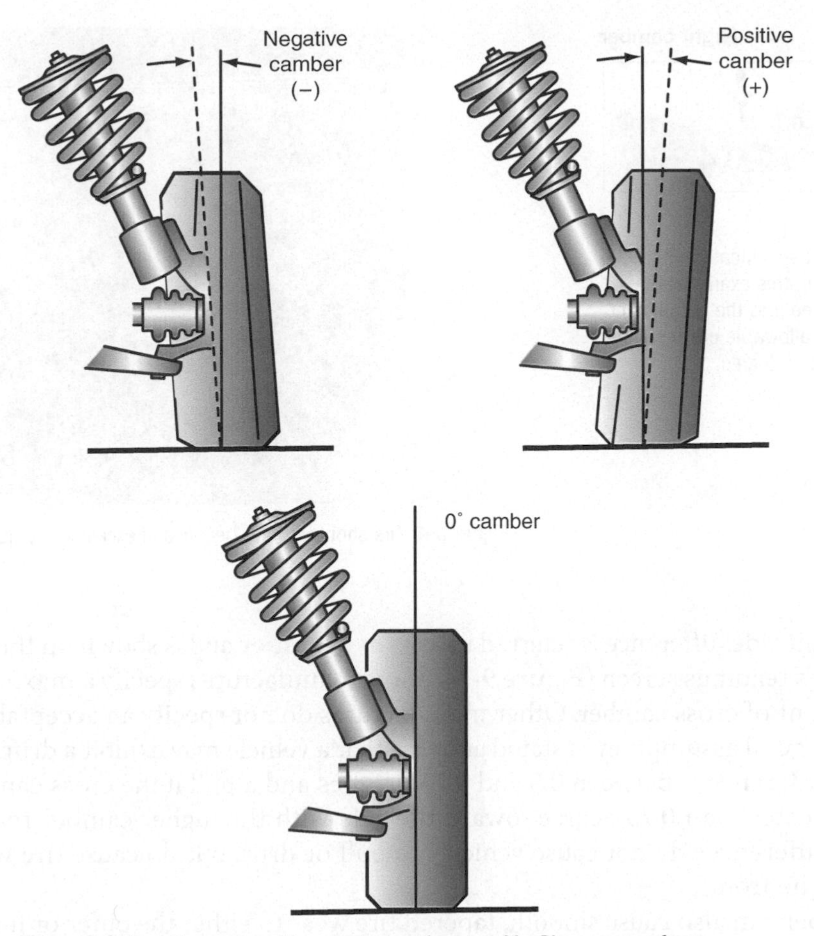

Figure 9-1 Negative and positive camber angles on a MacPherson strut front suspension system.

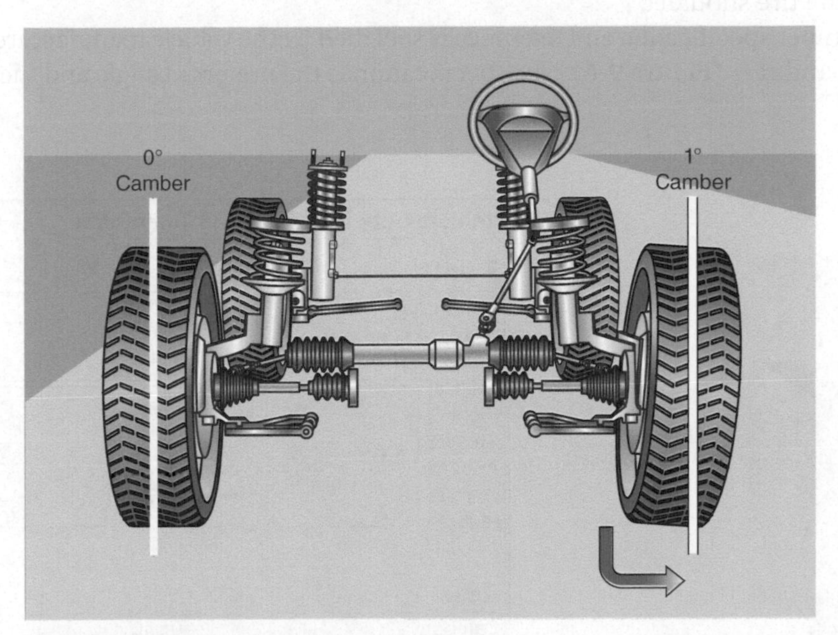

Figure 9-2 Cross camber is the side-to-side difference in camber measurements. This vehicle will pull right toward the most positive side.

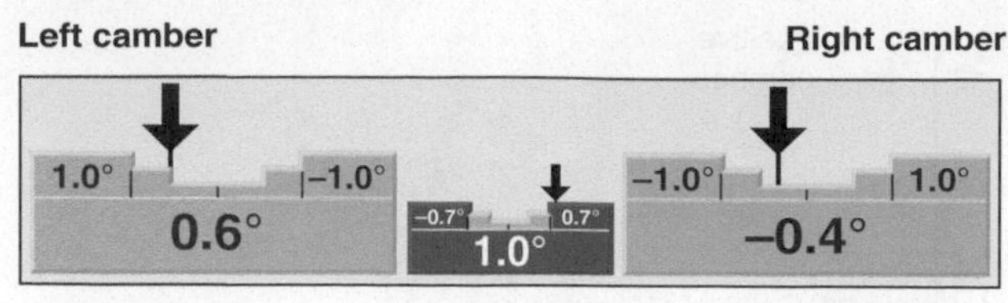

Figure 9-3 The left and right camber preferred specification is 0 degree and the tolerance is + or – 1.0 degree in this example. The actual left camber reading is positive 0.6 degree and the actual right camber reading is –0.4 degree. The maximum allowable cross camber is 0.7 degree with an actual cross camber of 1.0 degree.

Figure 9-4 Tire shoulder wear because of excessive camber.

side to right side difference is referred to as **cross camber** and is shown on the alignment computer's readings screen (**Figure 9-3**). Some manufactures specify a maximum allowable amount of cross camber. Other manufactures do not specify an acceptable amount and the general assumption as stated above is that a vehicle may exhibit a drift if the cross camber difference is between 0.5 and 0.75 degrees and a pull if the cross camber difference is greater than 0.75 degree toward the side with the higher camber reading. Rear camber differences do not cause vehicles to pull or drift, but do cause tire wear just as errors in the front.

Camber can also cause smooth, tapered tire wear to either the outer or inner edge of the tire tread around the entire tire's circumference, and is typically isolated to the tread shoulder area of the tire (**Figure 9-4**). Too much positive camber will cause outside edge wear to the tire shoulder (**Figure 9-5**). Too much negative camber will cause inside edge wear to the tire shoulder.

A camber specification and tolerance is specified by the vehicle manufacturer for both the front and rear (**Figure 9-6**). The specification is the preferred angle and the tolerance

Figure 9-5 Tire shoulder wear because of excessive positive camber.

Vehicle Specifications and Tolerances

Front	Spec.	Tol.
Left Camber	0 . 00°	1 . 00°
Right Camber	0 . 00°	1 . 00°
Cross Camber		0 . 70°
Left Ca		00°
Right C		00°
Cross C		70°
Total To		25°
Left SA		°
Right S		°

1.0° –1.0°

–0.0°

Rear		
Camber	–0 . 40°	0 . 75°
Total Toe	0 . 20°	0 . 30°
Thrust Angle		0 . 20°

Figure 9-6 Typical vehicle specification and tolerance data sheet with an alignment screen bar graph superimposed for camber.

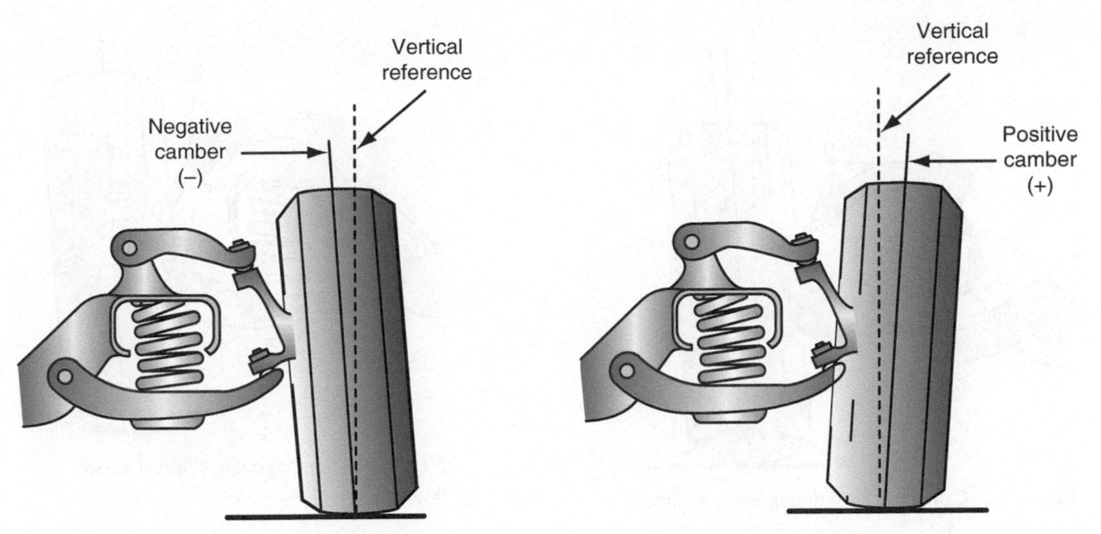

Figure 9-7 Negative and positive camber angles on a short-and-long arm front suspension system.

is the permitted amount that can be increased (+) or decreased (–) from the preferred angle. The tolerance is a means provided by the manufacturer to fine tune the suspension system to specific driving habits. It also allows for some change and wear that may naturally occur between alignments without a noticeable impact on directional stability or excessive tire wear.

During an alignment, camber is considered a live measurement because the computer is continually updating the sensor information on the computer measurement screen. Some alignment machines will allow you to adjust camber with the weight on or off the wheels using a program designed to allow the vehicle to be lifted. Hunter Engineering calls this program "Jack-Up Selected Axle."

The camber may be adjusted with the cam on the upper strut-to-steering knuckle bolt on some MacPherson strut front suspension systems. Negative and positive camber angles are the same on a short-and-long arm front suspension system as they are on a MacPherson strut suspension system (**Figure 9-7**).

Front cradle position is extremely important to provide proper front wheel camber on a front-wheel-drive vehicle. For example, if the vehicle is involved in a side collision and most of the collision force is supplied to the front wheel and chassis, the cradle may be pushed sideways. On the side of the vehicle impacted during the collision, the front suspension and cradle may be pushed inward, and the other side of the cradle may be forced outward. When the front suspension and cradle are moved inward, the camber on that side of the vehicle becomes more positive. When the cradle, suspension, and bottom of the wheel on the opposite side are moved outward, the camber is moved toward a negative position. Often this is noted as an equal and opposite difference in side-to-side camber.

DRIVING CONDITIONS AFFECTING CAMBER

Jounce and Rebound

Upward wheel movements are referred to as **jounce**, whereas downward wheel movements are termed **rebound**. On most modern suspension systems during wheel jounce, the top of the wheel moves outward and creates a more positive camber angle (**Figure 9-8**). During wheel rebound, it is desirable to have very little camber change to minimize tire tread wear. A short-and-long arm front suspension system will have a slightly negative camber during wheel rebound travel (**Figure 9-9**).

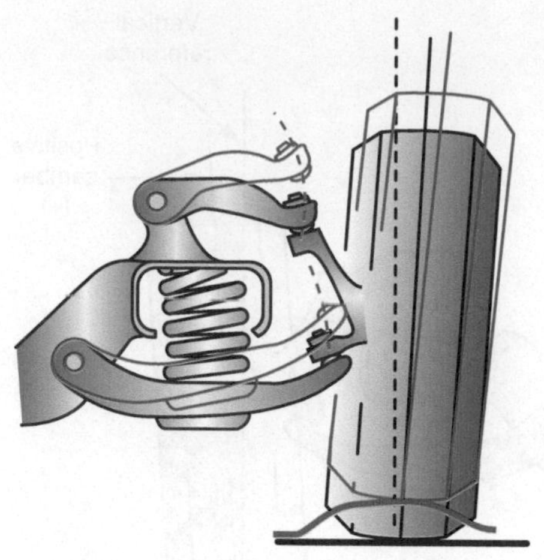

Figure 9-8 Camber change during wheel jounce.

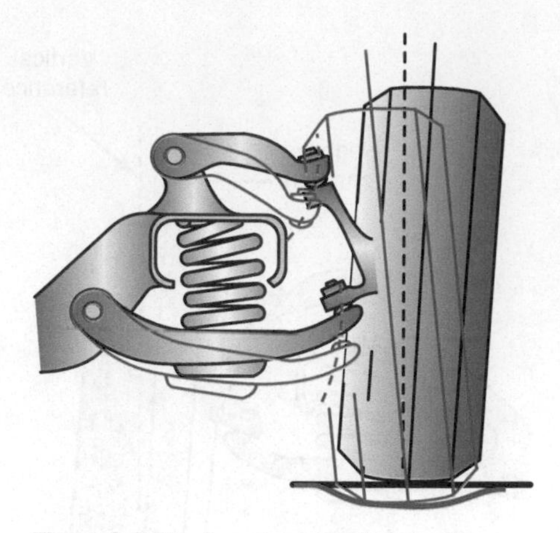

Figure 9-9 Camber change during wheel rebound.

Cornering Forces

During hard cornering at high speeds, centrifugal force attempts to move the vehicle sideways to the outside of the turn. Under this condition, more vehicle weight is transferred to the side of the vehicle on the outside of the curve. Therefore, the front suspension on the outside of the curve is forced downward, whereas the front suspension on the inside of the curve is lifted upward. When this action occurs, the camber on the inside wheel becomes less positive and the inside edge of the tire grips the road surface to help prevent sideways skidding (**Figure 9-10**).

Simultaneously, the camber on the outside wheel moves to a slightly more positive position. This action on the outside wheel causes the outside edge of the tire to grip the road surface, which helps prevent sideways skidding (**Figure 9-11**). Frequent, hard, high-speed cornering causes wear on the edges of the tire treads.

Tire Tread Wear

The camber angle may be referred to as one of the tire wear alignment angles. When the front wheels are adjusted to the manufacturer's specified camber setting, the front wheels will remain at, or very close to, the 0-degree camber position during average driving conditions. Therefore, maximum tire tread life and directional stability are maintained.

If a front wheel has excessive positive camber, the wheel is tilted outward and the vehicle weight is concentrated on the outside edge of the tire. Under this condition, the

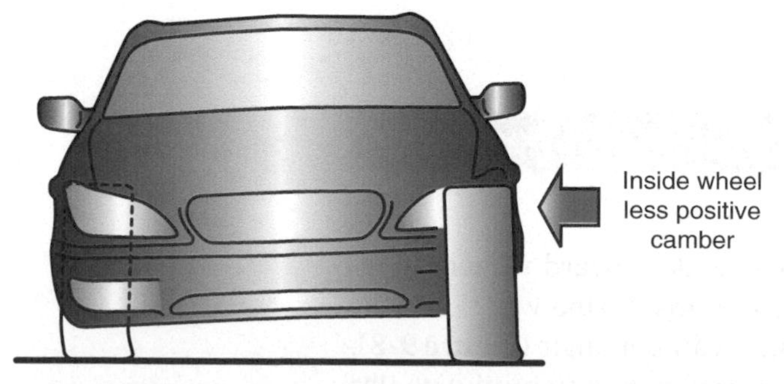

Figure 9-10 While cornering at high speed, the camber on the inside front wheel becomes less positive and the inside edge of the tire grips the road surface to resist lateral skidding.

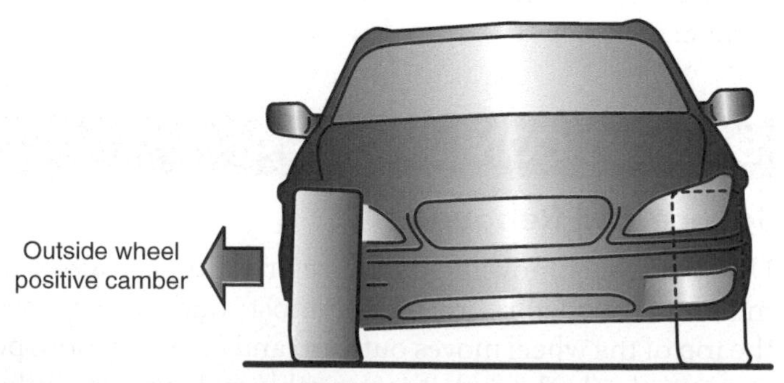

Figure 9-11 While cornering at high speed, the camber on the outside front wheel moves to a slightly more positive position and the outside edge of the tire grips the road surface to resist lateral kidding.

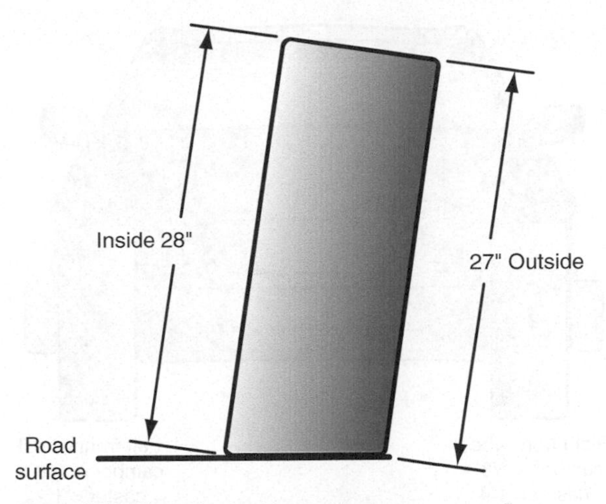

Figure 9-12 Excessive positive camber causes a smaller diameter on the outside edge of the tire compared to the inside edge of the tire.

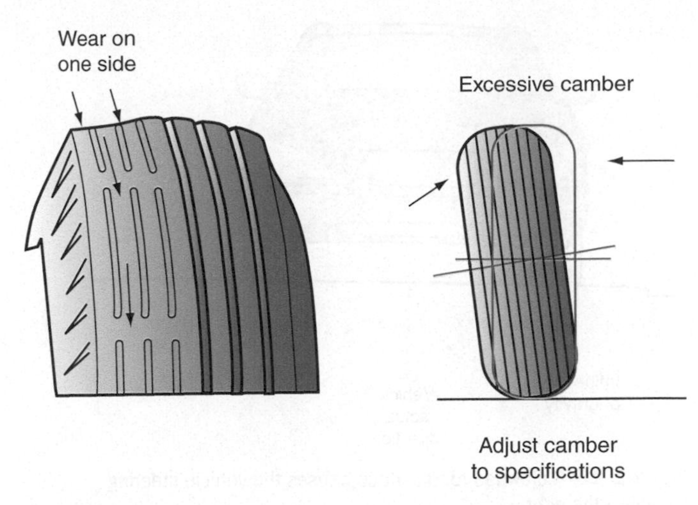

Figure 9-13 Tire tread wear caused by incorrect camber adjustment.

outside edge of the tire has a smaller diameter than the inside tire edge. Therefore, the outside tire edge has to complete more revolutions to travel the same distance as the inside tire edge. Because both edges are on the same tire, the outside edge must slip and scuff on the road surface as the tire and wheel revolve (**Figure 9-12**).

Shop Manual
Chapter 9, page 359

Excessive negative camber tilts the wheel inward and concentrates the vehicle weight on the inside edge of the tire. This condition causes wear and scuffing on the inside edge of the tire tread (**Figure 9-13**). Therefore, correct camber adjustment is extremely important to providing normal tire tread life.

WARNING Improper camber adjustment may result in rapid tire tread wear and steering pull, which is a vehicle safety hazard.

Road Crown

A wheel that is tilted tends to steer in the direction it is tilted (**Figure 9-14**). For example, a bicycle rider tilts the bicycle in the direction he or she wishes to turn, making the turning process easier.

When camber angles are equal on both front wheels, the camber steering forces are equal, and the vehicle tends to maintain a straight-line position. If the camber on the front wheels is significantly unequal, the vehicle will drift to the side with the greatest degree of positive camber.

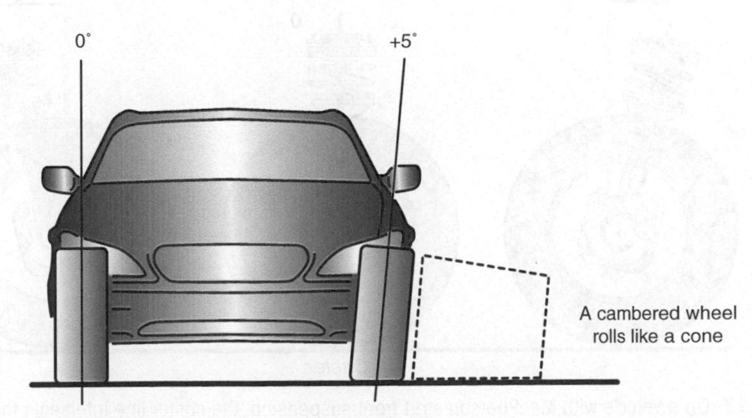

Figure 9-14 A wheel turns in the direction of the tilt.

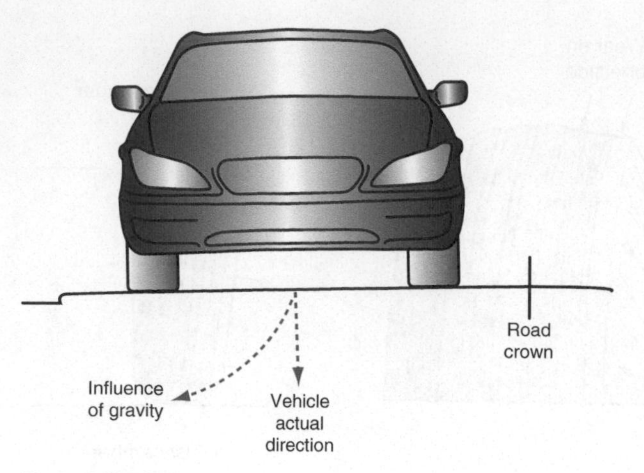

Figure 9-15 A crowned road surface causes the vehicle steering to pull to the right.

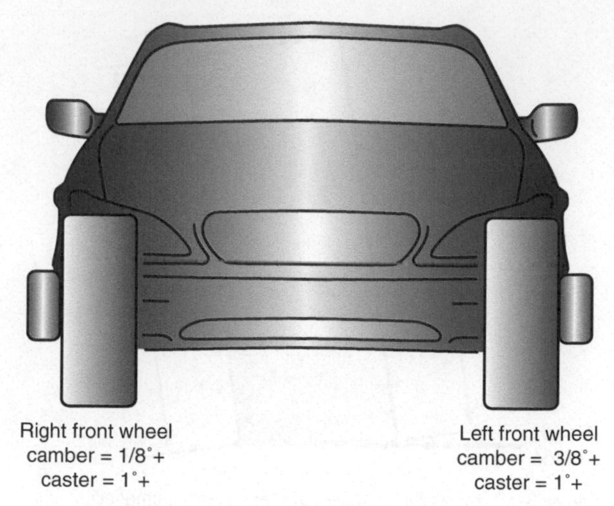

Figure 9-16 Camber setting to offset road crown.

Crowned highway design prevents water buildup on the driving surface. When a vehicle is driven on a crowned road, it is actually driven on a slight slope, which causes the vehicle steering to pull toward the right (**Figure 9-15**). Some car manufacturers use the pulling effect of camber to offset this pull to the right caused by road crown. In these vehicles, the left front wheel may have 0.25 to 0.5 degree more positive camber than the right front wheel.

When the camber on the front wheels is adjusted to the manufacturer's specifications, vehicle directional stability is maintained. If the camber adjustment on the left front wheel is used to offset road crown, the caster on both front wheels must be the same (**Figure 9-16**).

CASTER FUNDAMENTALS

Caster Definition

When viewed from the side, caster is the angle formed when a line is drawn through the upper and lower pivot points in reference to true vertical. On a MacPherson strut suspension system, the caster line intersects the center of the upper strut mount and lower ball joint (**Figure 9-17**). While on a short-and-long arm suspension system, the caster line intersects the center of the upper and lower ball joints (**Figure 9-18**). **Positive caster** occurs when the caster line is tilted backward toward the rear of the vehicle in relation to

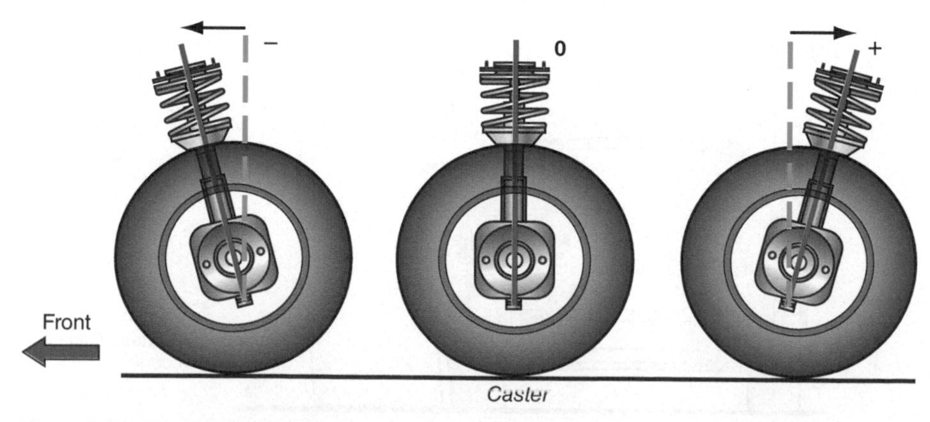

Figure 9-17 On a vehicle with MacPherson strut front suspension, the caster line intersects the upper strut mount and lower ball joint.

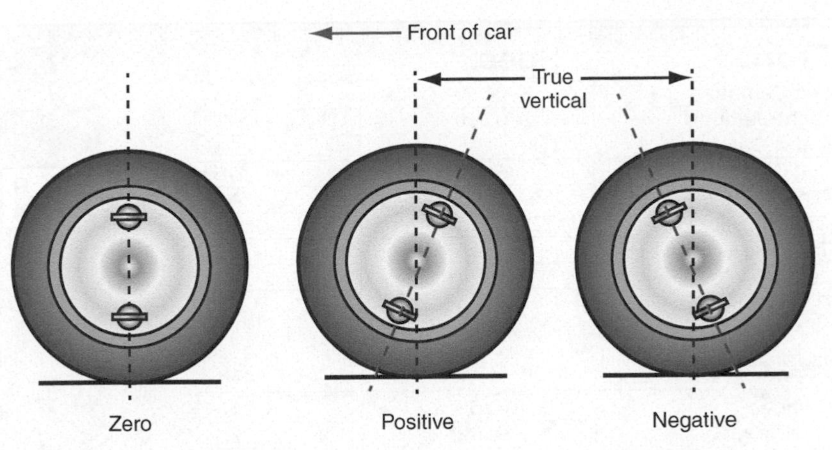

Figure 9-18 On a vehicle with upper and lower ball joints, the caster line intersects the upper and lower ball joints.

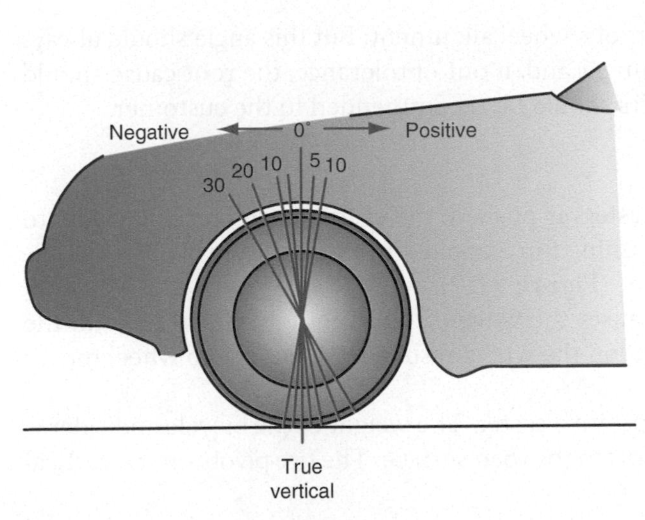

Figure 9-19 Positive and negative caster.

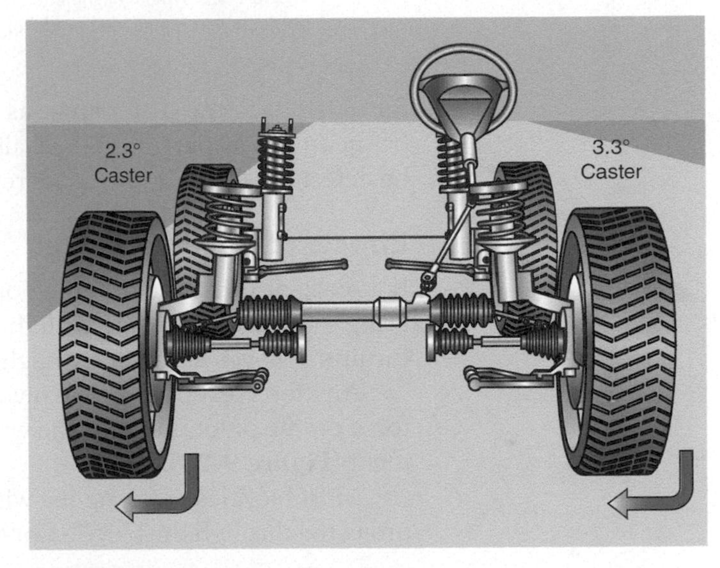

Figure 9-20 Cross caster is the side-to-side difference in caster and may cause a drift or pull to the side with the least positive caster.

the vertical centerline of the spindle and wheel viewed from the side (**Figure 9-19**). **Negative caster** occurs when the caster centerline is tilted toward the front of the vehicle in relation to the spindle and wheel vertical centerline.

Caster is not considered a direct tire wear angle, but is a directional stability angle and is used to improve cornering and steering wheel returnability. Caster is measured in degrees, and positive caster is specified on consumer production vehicles. Cross caster is the difference in caster between the right and left side of the vehicle (**Figure 9-20**). If one front wheel has more positive caster than the other front wheel, the steering pulls toward the side with the least amount of positive caster. In general, a vehicle may exhibit a drift if the cross caster difference is between 0.5 and 1.0 degrees and a pull if the cross caster difference is greater than 1.0 degree toward the side with the least positive caster. Rear caster is not measured.

The caster specification and tolerance are specified by the vehicle manufacturer for the front wheels (**Figure 9-21**). The specification is the preferred angle and the tolerance is the permitted amount that can be increased (+) or decreased (–) from the preferred angle. Like camber, the caster tolerance is a means provided by the manufacture to fine-tune the suspension system to specific driving habits. Many vehicles today do not allow

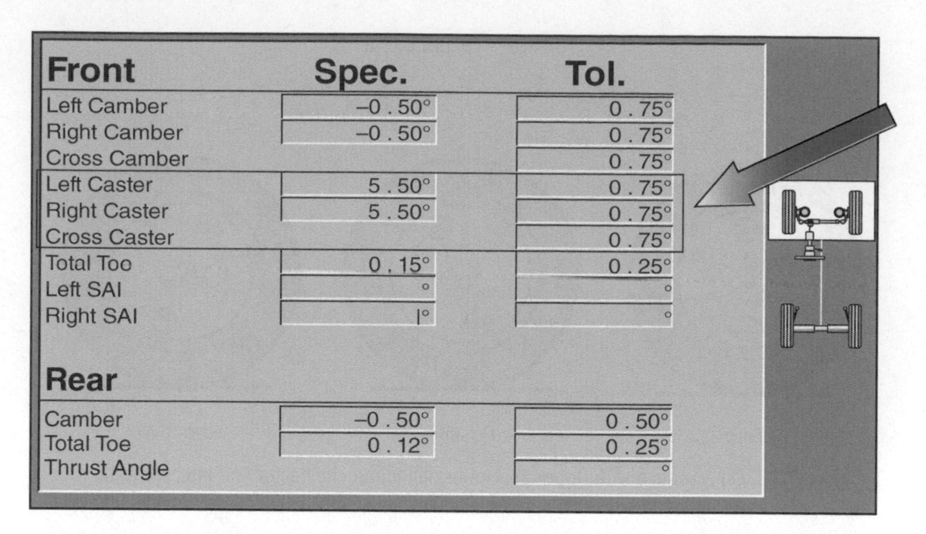

Front	Spec.	Tol.
Left Camber	–0 . 50°	0 . 75°
Right Camber	–0 . 50°	0 . 75°
Cross Camber		0 . 75°
Left Caster	5 . 50°	0 . 75°
Right Caster	5 . 50°	0 . 75°
Cross Caster		0 . 75°
Total Too	0 . 15°	0 . 25°
Left SAI	°	°
Right SAI	l°	°
Rear		
Camber	–0 . 50°	0 . 50°
Total Toe	0 . 12°	0 . 25°
Thrust Angle		°

Figure 9-21 Typical alignment specification sheet for a vehicle. Actual specifications vary between year, make, and model of vehicle.

for a change in the caster angle as part of a wheel alignment. But this angle should always be measured as part of a wheel alignment and, if out of tolerance, the root cause should be determined and appropriate repairs should be recommended to the customer.

Effects of Positive Caster

If a piece of furniture mounted on casters is pushed, the casters turn on their pivots to bring the wheels into line with the pushing force applied to the furniture. Therefore, the furniture moves easily in a straight line (**Figure 9-22**).

Any force exerted on the pivot causes the wheel to turn until it is lined up with the force on the pivot, because the weight on the wheel results in resistance to wheel movement (**Figure 9-23**).

Most bicycles are designed with positive caster. The weight of the bicycle and rider is projected through the bicycle front forks to the road surface. The tire pivots on the vertical

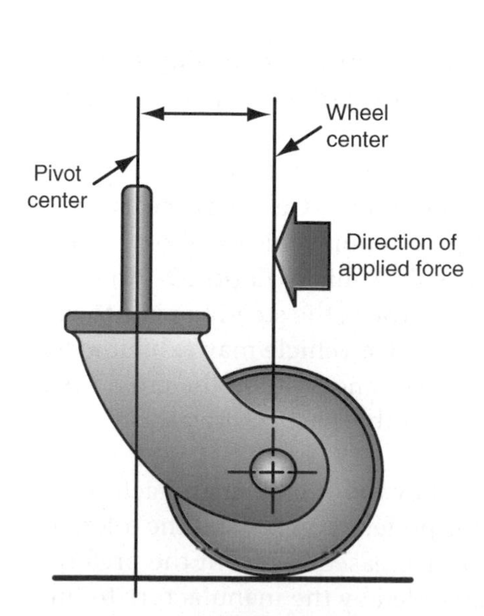

Figure 9-22 Caster action on furniture.

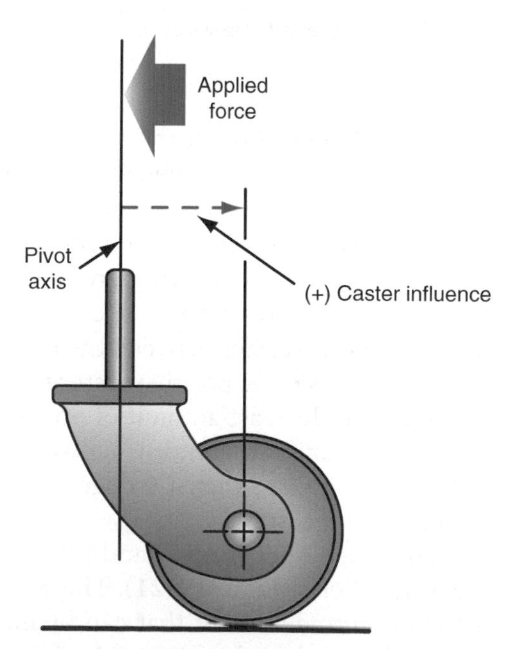

Figure 9-23 Furniture caster wheel aligned with the force on the pivot.

centerline of the spindle and wheel when the handlebars and front wheel are turned. Notice that the caster line through the center of the front forks is tilted rearward in relation to the vertical centerline of the spindle and wheel as viewed from the side (**Figure 9-24**).

Because the pivot point is behind the caster line where the bicycle weight is projected against the road surface, the front wheel tends to return to the straight-ahead position after a turn. The wheel also tends to remain in the straight-ahead position as the bicycle is driven. Therefore, the caster angle on a bicycle front wheel provides the same action as the caster angle on a piece of furniture.

Positive caster projects the vehicle weight ahead of the wheel centerline, whereas negative caster projects the vehicle weight behind the wheel centerline. Because positive caster causes a larger tire contact area behind the caster pivot point, this large contact area tends to follow the pivot point. This action tends to return the wheels to a straight-ahead position after a turn. It also helps maintain the straight-ahead position. Positive caster increases steering effort because the tendency of the tires to remain in the straight-ahead position must be overcome during a turn. The returning force to the straight-ahead position is proportional to the amount of positive caster. Positive caster helps maintain vehicle directional stability. Excessive positive caster is undesirable because it increases steering effort and creates a rapid steering wheel return.

Caster causes camber to change as the wheels are turned left or right. This effect is referred to as camber roll and is the effect that positive caster has on camber as the wheels are turned (**Figure 9-25**). If the caster angle were 0 degree, the front spindles would rotate horizontally in relation to the road surface. However, a positive caster angle causes the left front spindle to tilt toward the road surface during a left turn (**Figure 9-26**). Caster is the angle causing camber to increase or decrease when the wheels are turned, but camber is the angle causing the tire wear. This effect means that caster can be an indirect tire wear angle if the caster angle is excessively out of specifications. If caster is severely out of specifications, the tires may develop wear on both the inner and the outer

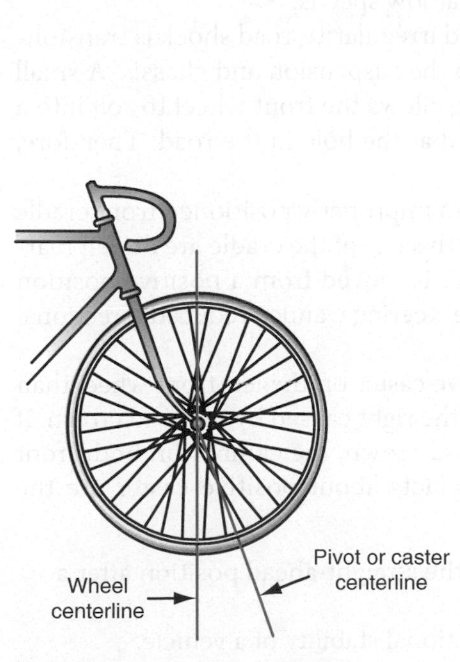

Figure 9-24 Caster line on a bicycle.

Figure 9-25 Camber roll is the effect that positive caster has on camber during a turn.

Figure 9-26 Left front spindle movement during a left turn.

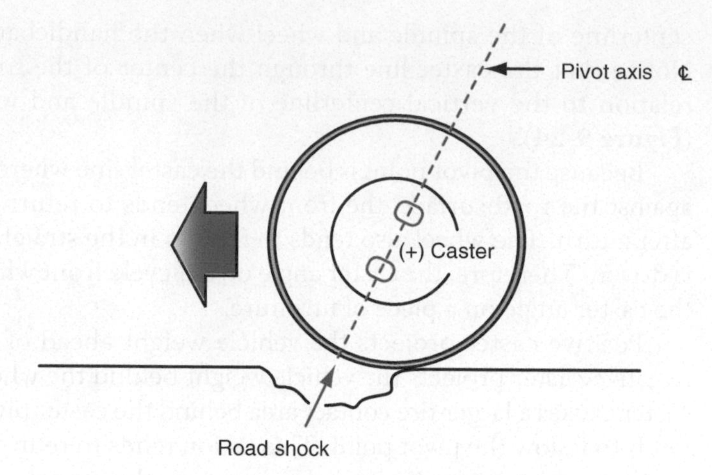

Figure 9-27 Harsh riding quality caused by excessive positive caster.

edges. Because the front wheels are the drive wheels on a front-wheel-drive vehicle, camber roll wear does occur on the front tires as part of the normal tire wear pattern on the front tires but the wear is minimal. Inner and outer edge wear becomes very noticeable on the front tires of a front-wheel-drive car if they are not rotated regularly. The severity of front tire wear because of camber roll is directly proportional to the amount of positive caster present.

This downward spindle movement tends to drive the tire into the road surface. Because this action cannot take place, the left side of the suspension and chassis is lifted. When the driver begins to return the wheel to the centered position, gravity forces the vehicle weight to its lowest position, which helps return the steering wheel to the straight-ahead position. Excessive positive caster increases the left front spindle downward tilt during a left turn, which increases the suspension and chassis lift. Therefore, excessive positive caster increases steering effort. The same action occurs at the right front spindle, but this spindle tilts downward during a right turn.

Harsh riding may be caused by excessive positive caster because the caster line is actually aimed at some road irregularities (**Figure 9-27**). Excessive positive caster may cause the front wheels to shimmy from side to side at low speeds.

When the caster line is aimed directly at the road irregularity, road shock is transmitted through the ball joint and upper strut mount to the suspension and chassis. A small degree of positive caster or a negative caster setting allows the front wheel to roll into a road depression without the caster line being aimed at the hole in the road. Therefore, this type of caster line improves ride quality.

Shop Manual
Chapter 9, page 387

Improper front wheel caster may be caused by an improperly positioned front cradle on a front-wheel-drive car. For example, if one or both sides of the cradle are driven rearward in a front-end collision, the front wheel caster is moved from a positive position toward a negative position. This condition may cause steering wander, reduced directional control, and slower steering wheel return.

Some car manufacturers recommend less positive caster on the left front wheel than on the right front wheel to offset the steering pull to the right caused by the road crown. If the caster adjustment is used to compensate for road crown, the camber on both front wheels should be the same. The most important facts about positive caster are the following:

1. Positive caster helps the front wheels return to the straight-ahead position after a turn.
2. Correct positive caster provides improved directional stability of a vehicle.

3. Excessive positive caster produces harsh riding quality.
4. Excessive positive caster promotes sideways front wheel shimmy.
5. The left front wheel may be adjusted with less positive caster than the right front wheel to compensate for road crown.

Rapid side-to-side wheel movement may be called **wheel shimmy**.

Effects of Negative Caster

Negative caster moves the centerline of the upper strut mount and lower ball joint behind the vertical centerline of the spindle and wheel at the road surface. If this condition is present, the friction of the tire causes the tire to pivot around the point where the centerline of the upper strut mount and ball joint meets the road surface. When this occurs, the wheel is pulled away from the straight-ahead position, which decreases directional stability.

Negative caster reduces steering effort. Because excessive positive caster increases road shock transmitted to the suspension and chassis, negative caster reduces this shock and improves ride quality. This improvement occurs because the front wheel rolls into a road depression without the caster line being aimed at the hole in the road. But, if caster is below manufacturers' specifications, poor steering wheel return may be experienced as well as a wondering feeling when driving straight ahead because of reduced directional stability that may result from a low or negative caster setting. Caster is never specified as a negative number. This is why we say that caster may pull to the side with the "least" positive caster because it is always specified to be a positive setting.

Effects of Suspension Height on Caster

When the rear springs become sagged or overloaded, the caster on the front wheels becomes more positive. This action explains why front wheel shimmy may occur when a trunk is severely overloaded.

If the rear suspension height is above the vehicle manufacturer's specification, the caster on the front wheels becomes less positive. Under this condition, the front wheel caster may change from positive to negative. This explains why a vehicle may have reduced directional stability and control when the rear suspension height is raised.

On many short-and-long arm front suspension systems, the inner front end of the upper control arm is higher than the inner rear side of this arm where it is attached to the frame. This design causes the front wheel caster to become more positive if the front suspension height is lowered. When the front suspension height is above the vehicle manufacturer's specification, the front wheel caster becomes less positive.

Most front-wheel-drive cars have a MacPherson strut front suspension system with a slightly positive caster setting. Therefore, the top strut mount is tilted rearward. If the front suspension height is lowered on this type of front suspension system, the front wheel caster becomes less positive. Conversely, if the suspension height is raised, positive caster increases. The most important facts about negative caster include the following:

1. Negative caster does not help return the front wheels to the straight-ahead position after a turn.
2. Negative caster contributes to directional instability and reduced directional control.
3. Negative caster does not contribute to front wheel shimmy.
4. Negative caster reduces road shock transmitted to the suspension and chassis.

SAFETY FACTORS AND CASTER

Directional Control

As explained earlier in this chapter, positive caster provides increased directional stability and control, whereas negative caster reduces directional stability. Therefore, front wheel caster must be adjusted to the manufacturer's specifications to maintain vehicle directional control and safe handling characteristics.

Suspension Height

We have already explained how incorrect front or rear suspension height results in changes in front wheel caster. Therefore, abnormal suspension heights may contribute to reduced directional control and unsafe steering characteristics.

WARNING Improper caster adjustment may result in decreased directional stability of the vehicle and reduced driving safety.

TOE DEFINITION

Total toe is the measurement difference between the distances between the fronts of the tires compared to the distance between the rears of the same tires (**Figure 9-28**). When the distance between the rear inside tire edges is greater than the distance between the front inside tire edges (**Figure 9-29**), the front wheels have a **toe-in** setting and is more commonly referred to as positive toe and is displayed on an alignment computer in degrees (**Figure 9-30**). **Toe-out** occurs when the distance between the inside front tire edges exceeds the distance between the inside rear tire edges and is more commonly referred to as negative toe and is displayed on an alignment computer as a negative number in degrees (**Figure 9-31**).

A total toe specification and tolerance is specified by the vehicle manufacturer for both the front and the rear (**Figure 9-32**). The specification is the preferred angle and the tolerance is the permitted amount that can be increased (+) or decreased (–) from the preferred angle. Manufacturers generally publish the toe specifications as total toe. On most vehicles individual right and left toe is adjustable independently by changing the length of right and left tie rod assemblies. The individual toe specification is one-half the total toe specification and may be measured in either degrees or inches. Individual toe is the difference between the front of one tire and a reference line and the rear of the tire and the same reference line. The reference line is either the centerline or the thrust line of the vehicle, depending on the type of alignment being performed (**Figure 9-33**). Because the front tires are the steering wheels, front individual toe will attempt to equalize while driving. If one front tire's toe specification is out of adjustment, tire wear will occur

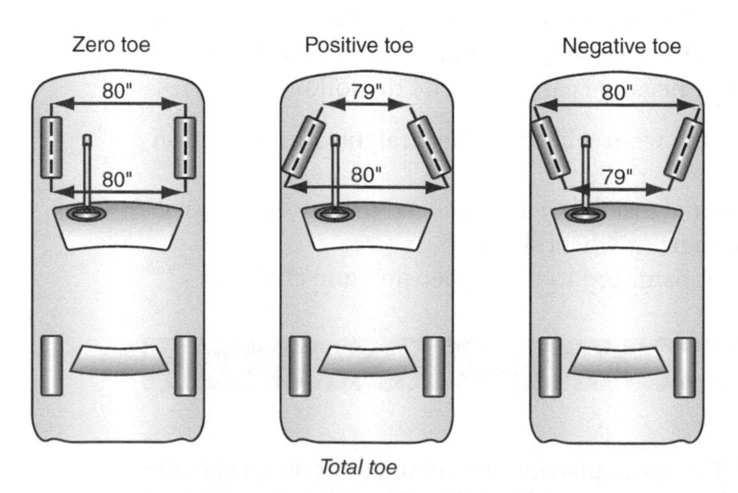

Figure 9-28 Total toe is the difference in distance measurements between the front measurement compared to the rear measurement.

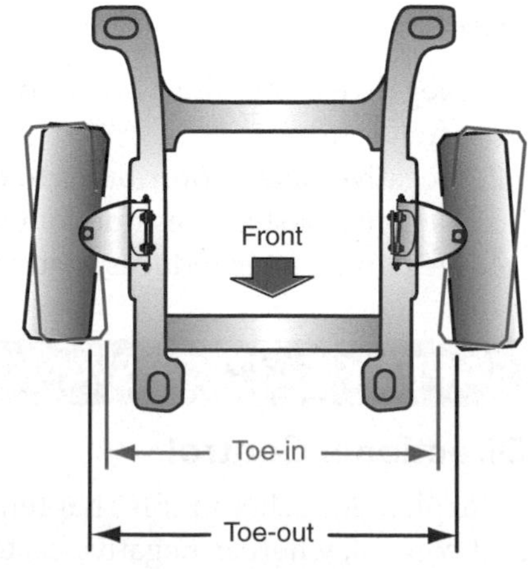

Figure 9-29 Toe-in and toe-out.

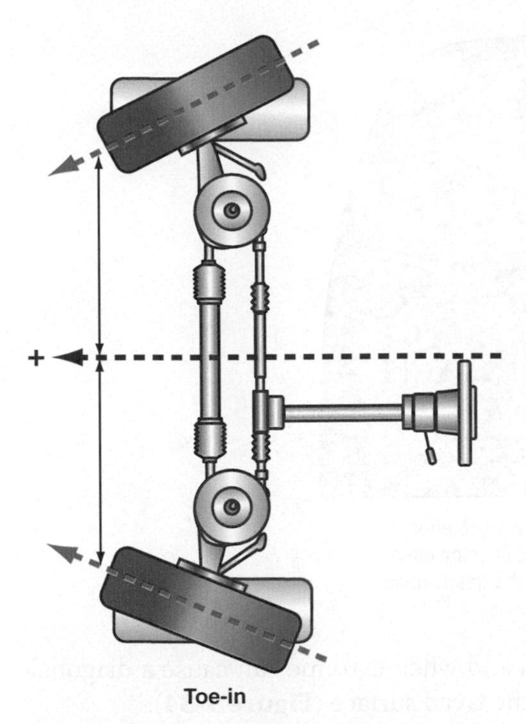

Figure 9-30 If distance between the rear inside tire edges is greater than the distance between the front inside tire edges, the front wheels have a positive (toe-in) setting.

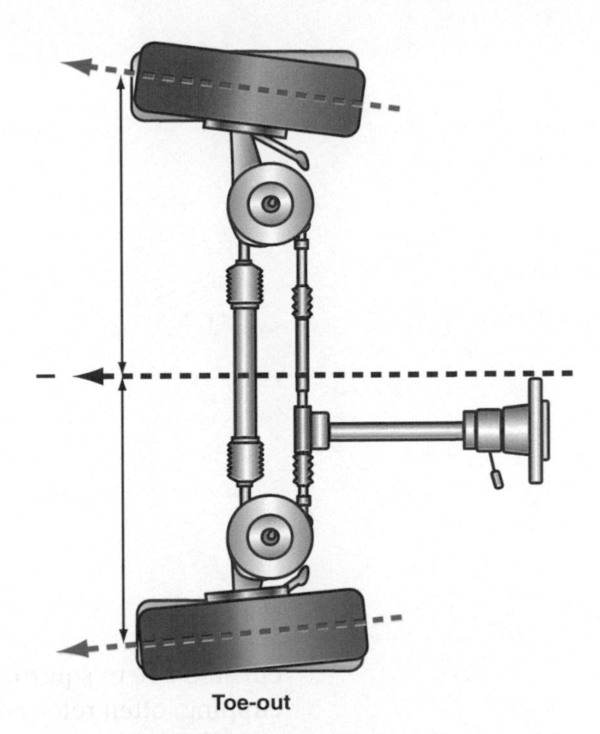

Figure 9-31 If distance between the inside front tire edges exceeds the distance between the inside rear tire edges, the front wheels have a negative (toe-out) setting.

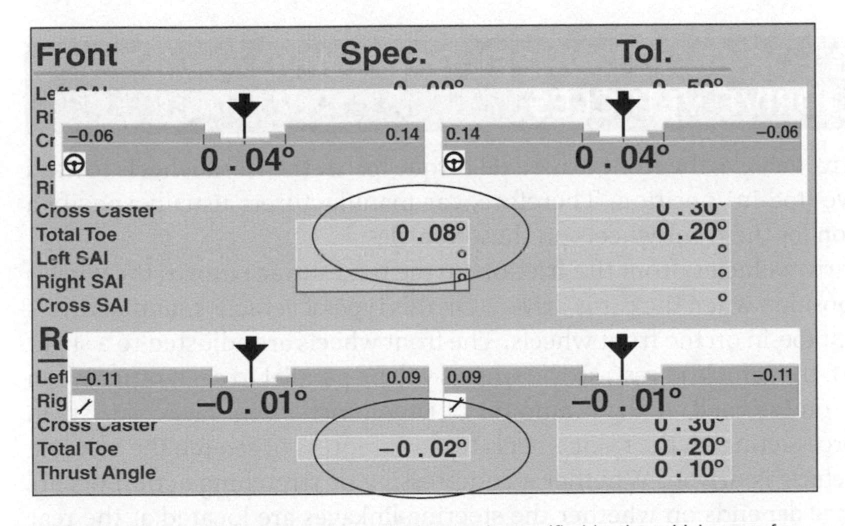

Figure 9-32 A total toe specification and tolerance is specified by the vehicle manufacturer for both the front and the rear. The individual toe specification is half (1/2) the total toe specification and may be measured in either degrees or inches.

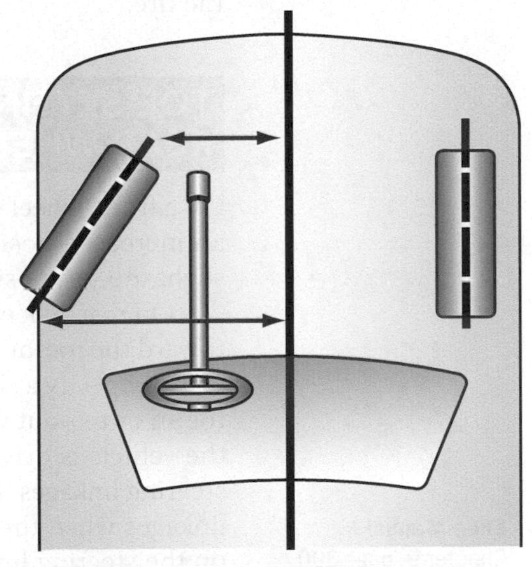

Figure 9-33 Individual toe is the difference between the front of one tire and a reference line and the rear of the tire and the same reference line.

on both front tires, and the steering wheel will not be centered when driving straight ahead. Changes in front individual toe can occur as a result of road impacts such as curbs, potholes, and severally rough roads or from loose or damaged components. Rear toe, unlike front toe, cannot equalize between the two rear tires because they are not steering wheels and the linkage is independent of one another. A rear toe error will cause a dog track condition, which is discussed in further detail in the thrust angle alignment section, and tire wear to the effected wheel. On the rear tires, a toe angle error will cause the

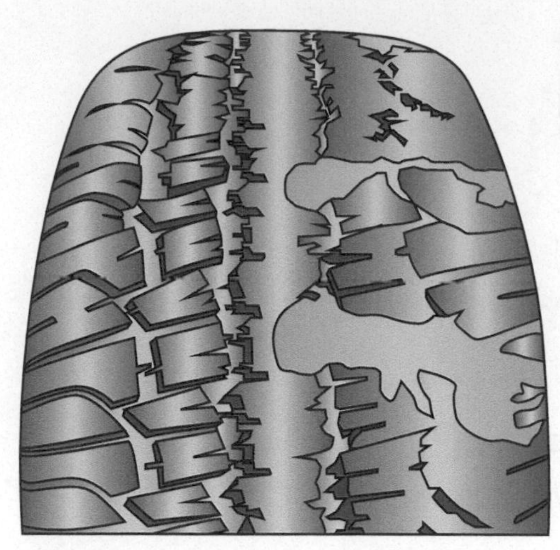

Figure 9-34 When rear tire's toe angle error is extreme, it can cause a diagonal cupping often referred to as diagonal wipe on the tread surface.

effected tire to squirm as it rolls down the road and when extreme can cause a diagonal cupping, often referred to as diagonal wipe on the tread surface (**Figure 9-34**).

Toe is considered a major tire wear angle and incorrect total toe will wear tires extremely quickly. Toe angles that are outside the allowed tolerance will cause tire wear. Positive toe will wear the outside edge of the tire and negative toe will wear the inside edge of the tire. The wear will begin on the shoulder of the tire and taper toward the center of the tire.

TOE SETTING FOR FRONT-WHEEL-DRIVE AND REAR-WHEEL-DRIVE VEHICLES

On a front-wheel-drive vehicle, the front drive axle torque forces the front wheels toward an increased positive (toe-in) position. Therefore, car manufacturers usually specify a slight toe-out position for the front wheels on these vehicles.

On rear-wheel-drive vehicles, front tire friction on the road surface moves the wheels toward the toe-out position when the car is driven. On this type of vehicle, manufacturers usually specify a slight toe-in on the front wheels. The front wheels are adjusted to a slight toe-in or toe-out with the vehicle at rest, so the wheels will be parallel to each other when the vehicle is driven on the road. A slight amount of lateral movement always exists in steering linkages. Forces acting on the front wheels try to compress or stretch the steering linkages when the vehicle is driven. Whether a compressing or stretching action occurs on the steering linkage depends on whether the steering linkages are located at the rear edge or front edge of the front wheels.

Shop Manual
Chapter 9, page 390

TOE ADJUSTMENT AND TIRE WEAR

When the front wheel toe is not adjusted to the manufacturer's specifications, front tire tread wear is excessive. If one of the tie rods is not parallel to the lower control arms, toe change on the front wheels is not equal during front wheel jounce and rebound. This action may result in bump steer when a front wheel strikes a road irregularity. While checking front wheel toe, improper toe change during suspension jounce and rebound should be checked to be sure the tie rods are parallel to the lower control arms. Technicians must understand front wheel toe and improper toe changes.

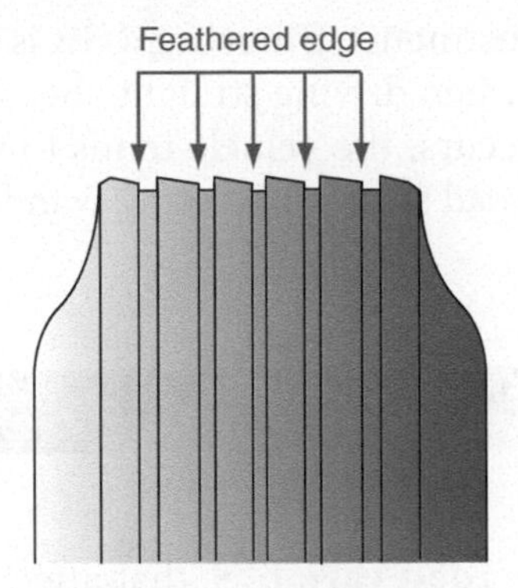

Figure 9-35 Feathered tire tread wear caused by incorrect toe adjustment.

Ford Motor Company has calculated that a toe-in error of 1/8 in. (3.17 mm) is equivalent to dragging the tires crosswise for 11 ft (3.3 m) for each mile the vehicle is driven. This crosswise movement causes severe feathered tire tread wear (**Figure 9-35**). Improper toe adjustment is the most common cause of rapid tire tread wear.

Excessive toe-out causes wear on the inside of the tire tread ribs and a sharp feathered edge on the outside of the tread ribs. If excessive toe-in is present, the tire tread wear is reversed.

 **Caution**

Improper toe adjustment causes rapid tire tread wear, which may result in tire failure.

Worn steering linkage components such as tie rod ends cause incorrect and erratic toe-in settings. If the front springs become weak, the front suspension height is lowered. When this occurs, the pitman arm and the idler arm move downward with the chassis. This action moves the tie rods to a more horizontal position that tends to push outward on the steering arms and increases front wheel toe-in. The toe-in change just described occurs when the steering linkage is located at the rear of the front wheels.

WARNING Worn steering components may suddenly become disconnected, causing complete loss of steering control, collision damage, and personal injury.

STEERING TERMINOLOGY

In the automotive service industry, certain terms are used for specific steering problems. Some of these problems are related to camber and caster, others are caused by various suspension or steering defects. Technicians must be familiar with both the steering terminology and the causes of the problems.

A front-wheel-drive vehicle with unequal-length drive axles produces some **torque steer** on hard acceleration. Some front-wheel-drive vehicles, especially those with higher horsepower, have equal-length front drive axles to reduce torque steer. On a front-wheel-drive vehicle, torque steer is aggravated by different tire tread designs on the front tires or uneven wear on the front tires.

Torque steer may be defined as the tendency of the steering to pull to one side during hard acceleration.

Bump steer is the tendency of the steering to veer suddenly in one direction when one or both of the front wheels strikes a bump.

When **memory steer** occurs, the vehicle doesn't want to steer straight ahead after a turn because the steering does not return to the straight-ahead position.

Bump steer occurs if the tie rods are not the same height, meaning one of the tie rods is not parallel to the lower control arm. This condition may be caused by a bent or worn idler arm or pitman arm on a parallelogram steering linkage. On a rack and pinion steering system, worn steering gear mounting bushings may cause this unparallel condition between one of the tie rods and the lower control arm.

Memory steer may be caused by a binding condition in the steering column or in the steering shaft universal joints. A binding upper strut mount may result in memory steer. Negative caster or reduced positive caster also causes memory steer.

Steering pull is the tendency of the steering to gradually pull to the right or left when the vehicle is driven straight ahead on a level road. Steering pull or drift may be caused by improper caster or camber adjustments. **Steering drift** is the tendency of the steering to slowly drift to the right or left when driving straight ahead on a level road.

Shop Manual
Chapter 9, page 354

When **steering wander** occurs, the vehicle tends to steer in either direction rather than straight ahead on a level road surface. Steering wander may be caused by improper caster adjustment.

WHEEL ALIGNMENT THEORY

Road Variables

Road variables are differences in road surface conditions, vehicle loads, and weather conditions.

Road crown refers to the high portion in the center of the road with a gradual slope to each side.

Vehicles are subjected to many **road variables** that affect wheel alignment. These variables must be counteracted by the suspension design and alignment, or steering would be very difficult. Some of the variables that affect wheel alignment and suspension design follow:

1. **Road crown** (the curvature of the road surface)
2. Bumps and holes
3. Natural crosswinds or crosswinds created by other vehicles
4. Heavy loads or unequal weight distribution
5. Road surface friction and conditions such as ice, snow, and water
6. Tire traction and pressure
7. Side forces while cornering
8. Drive axle forces in front-wheel-drive vehicles
9. Relationship between suspension parts, as the front wheels turn and move vertically when road bumps and holes are encountered

A desirable plan to reduce tire wear would be to place the front wheels and tires so they are perfectly vertical. The tires would then be flat on the road. However, if the wheels and tires are perfectly vertical, such variables as the driver entering the car, turning a corner, or adding weight to the luggage compartment would move the tire from its true vertical position. Therefore, tire wear and steering operation would be adversely affected.

Rather than allowing the variables to adversely affect tire wear and steering operation, the suspension and steering are designed with characteristics to provide directional stability, predictable directional control, and minimum tire wear. Wheel alignment angles are designed to provide these desired requirements despite road variables. Wheel alignment angles also control the **tracking** of the rear wheels in relation to the front wheels.

Vehicle **tracking** is the straightness of the rear wheels in relation to the front wheels.

A customer may enter your shop and request a wheel alignment for many reasons. The three most common reasons for customers requesting a wheel alignment are to straighten a steering wheel that is not centered (crooked), to prevent tire wear, or to address a vehicle handling issue such as a pull while driving straight ahead. In addition, a customer may request a wheel alignment for a steering wheel that is shaking. A shaking of the vehicle is not generally caused by a wheel alignment angle error. A tire or wheel balance or runout condition as well as bearing and axle faults are more likely causes of the steering wheel shake condition. Tire rotation and balancing are additional services that should be considered when a wheel alignment is performed, because customers are often expecting a smooth, vibration-free ride after an alignment is performed.

After an alignment is performed, the benefits your customer can expect are as follows:

- The steering wheel position is level and pointing straight down the road while driving the vehicle straight ahead.
- The steering wheel automatically returns to the straight-ahead position after making a right or left turning maneuver.

- The vehicle exhibits directional stability without pulling or wandering while driving straight ahead reducing driver fatigue.
- Improved tire tread life and fuel economy.
- Proper vehicle tracking whereby the front and rear wheels follow each other.
- Improved cornering performance, directional control, and feedback while maintaining optimum road isolation.

TYPES OF WHEEL ALIGNMENT

Geometric Centerline Alignment

Toe refers to the angle between the plane of a front or rear wheel and a reference line. Either the geometric centerline or the thrust line may be used as a toe reference line. When **geometric centerline alignment** is used as a reference for front wheel toe, the toe on each front wheel is adjusted to specifications using the geometric centerline as a reference (**Figure 9-36**). This type of wheel alignment has been used for many years. It may provide a satisfactory wheel alignment if the rear wheels are properly positioned and the thrust line is at the vehicle centerline. However, if the rear wheels are not positioned properly, the thrust line is not at the geometric centerline and steering problems will occur. Therefore, geometric centerline front wheel reference ignores rear wheel misalignment.

In a **geometric centerline alignment**, the front wheel toe is adjusted using the geometric centerline as a reference.

Thrust Line Alignment

If **thrust line alignment** is used, the thrust line created by the rear wheels is used as a reference for front wheel toe adjustment (**Figure 9-37**). When the front wheel toe is adjusted with a thrust line reference, and this line is not at the geometric centerline, neither the front nor rear wheels are parallel to the geometric centerline. Under this

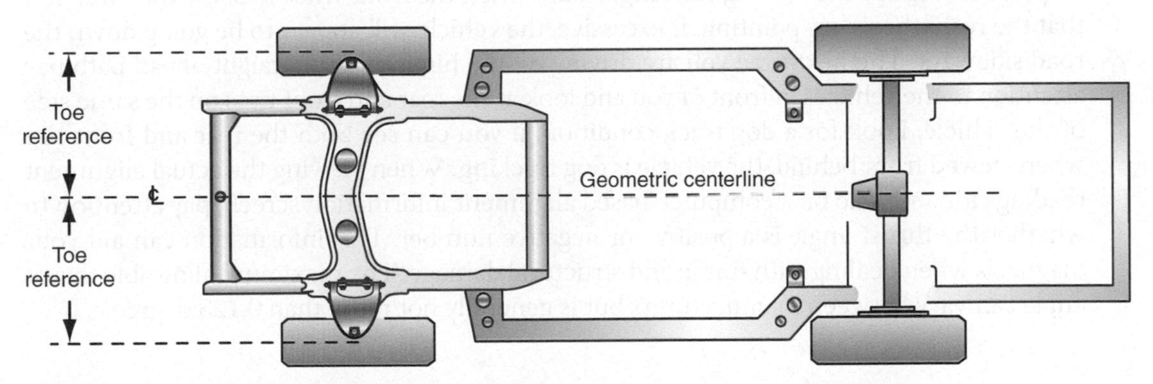

Figure 9-36 Front wheel alignment with the front wheel toe referenced to the geometric centerline.

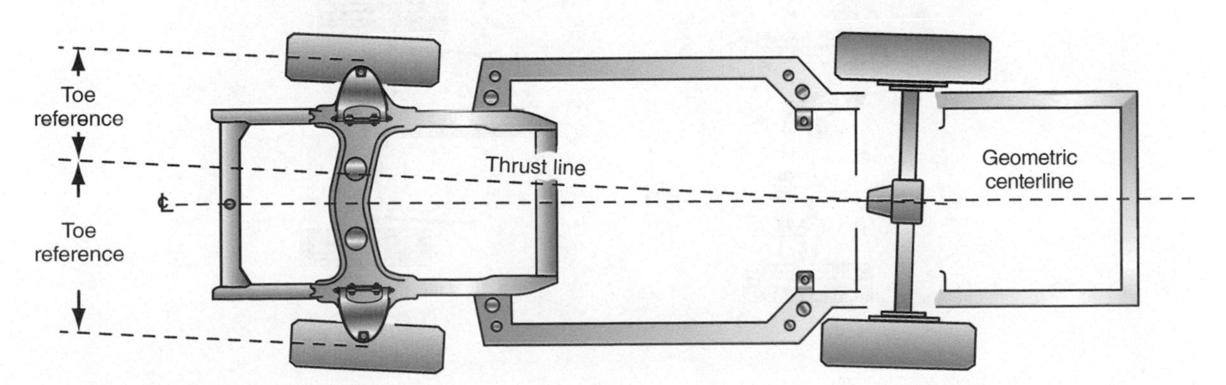

Figure 9-37 Thrust line alignment with the front wheel toe referenced to the thrust line.

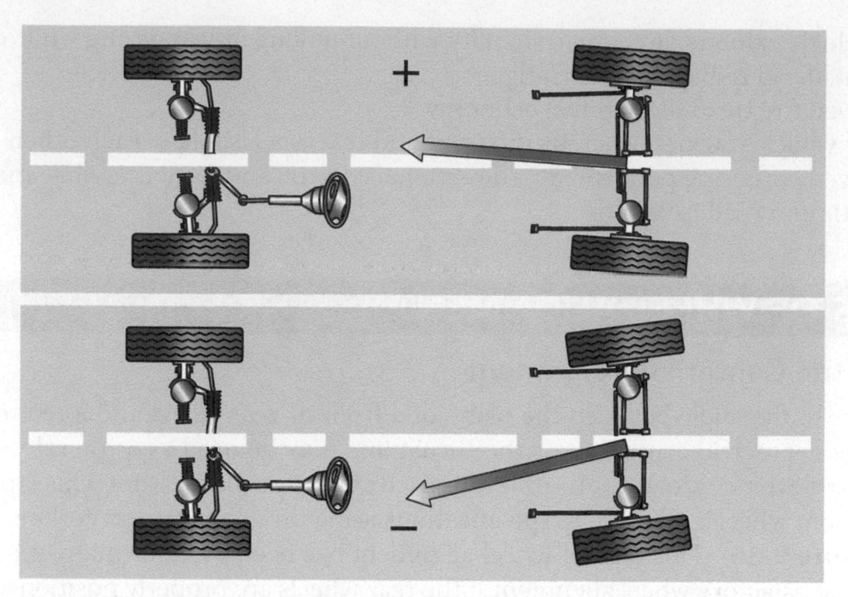

Figure 9-38 Positive and negative thrust line referenced to the vehicle centerline as viewed from above. A positive thrust line is to the right of the centerline and a negative thrust line is to the left of the centerline.

condition, none of the four wheels is facing straight ahead when the vehicle is driven straight ahead. This action results in excessive wear on all four tire treads. This type of alignment ensures a centered steering wheel when the vehicle is driven straight ahead.

A positive thrust line (angle) is to the right of the vehicle centerline when viewed from above and a negative thrust line (angle) is to the left of the vehicle centerline when viewed from above (**Figure 9-38**). If thrust angle error is excessive a visible dog track condition will be present (**Figure 9-39**). Dog tracking occurs when the front wheels follow the direction that the rear wheels are pointing. If excessive, the vehicle will appear to be going down the road sideways. The next time you are driving on the highway in a straight-ahead path pay attention to the vehicles in front of you and look at the rear and front tires on the same side of the vehicle. Look for a dog track condition; if you can see both the rear and front tire when viewed from behind, the vehicle is dog tracking. When viewing the actual alignment readings for a vehicle on a computer-based alignment information screen, pay attention to whether the thrust angle is a positive or negative number. This information can aid your diagnosis when dealing with frame and structural damage. The maximum allowable thrust angle can vary between manufacturers but is generally not more than 0.125 degree.

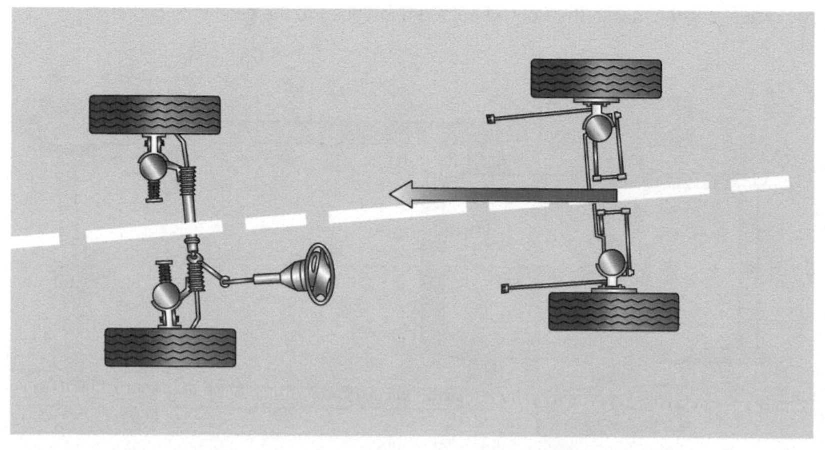

Figure 9-39 If a severe thrust angle condition exists a vehicle will dog track.

Four-Wheel Alignment

When a **four-wheel alignment** is completed, the thrust line position is measured in relation to the geometric centerline. Individual rear wheel toe-in is measured and adjusted to manufacturer's specifications. This adjustment moves the thrust line to the geometric centerline. Thrust angle is measured to be sure the thrust line is positioned at the geometric centerline. Front wheel toe is measured using the common geometric centerline and thrust line as a reference (**Figure 9-40**). When a four-wheel alignment is completed, all four wheels are parallel to the geometric centerline and the steering wheel is centered as the vehicle is driven straight ahead. Wheel aligners have different alignment capabilities. Modern computer wheel alignment systems have four-wheel alignment capabilities, and this type of wheel alignment must be performed on today's vehicles to ensure proper steering control and vehicle safety.

When performing a wheel alignment, the sequence that the angles are adjusted is critical, as a change in one angle may affect another angle. For example, a change in rear toe will change the position of the vehicles thrust angle, which is the reference line for front toe. So, a change in rear toe will also change the reading for front toe.

The sequence the wheel alignment angles must be adjusted in (if an angle adjustment is necessary) is:

1. Adjust rear camber (if a means of adjustment is available).
2. Adjust rear individual toe (if a means of adjustment is available).
3. Adjust front caster and camber (if a means of adjustment is available).
4. Adjust front individual toe.

The advantages of four-wheel alignment are the following:

1. *Improved fuel mileage.* After a four-wheel alignment, all four wheels are parallel, and this condition combined with proper tire inflation decreases rolling resistance, which improves fuel mileage.
2. *Longer tire life.* When all four wheels are aligned properly, tire tread wear is minimized.
3. *Improved vehicle handling.* When all four wheels are properly aligned and all steering and suspension components are in satisfactory condition, steering pulls, vibrations, and abnormal steering conditions are eliminated to ensure improved vehicle handling.
4. *Safer driving.* Proper alignment of all four wheels, plus inspection and replacement of all worn or defective steering and suspension components improves vehicle handling, and this reduces the possibility of a collision and provides safer driving.

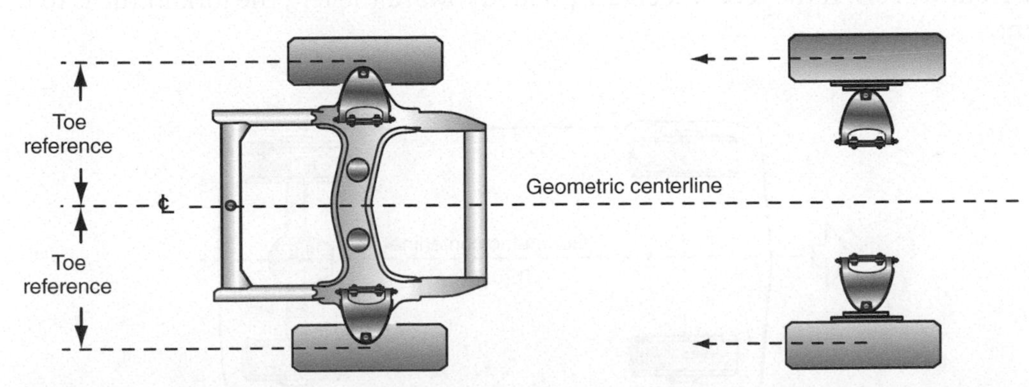

Figure 9-40 Four-wheel alignment with the thrust line adjusted so it is at the geometric centerline and the front wheel toe is referenced to the geometric centerline.

IMPORTANCE OF FOUR-WHEEL ALIGNMENT

Shop Manual
Chapter 9, page 360

Until the late 1970s, most vehicles in the United States were rear-wheel-drive. The majority of these vehicles had one-piece rear axle housings and frames. This type of vehicle design did not experience many rear wheel alignment problems and rear wheel alignment was usually not a concern in these years.

A unitized body may be called a unibody.

Beginning in the late 1970s, many domestic car makers began manufacturing front-wheel-drive cars with unitized bodies. The gasoline shortages in the 1970s and the introduction of federal corporate average fuel economy (CAFE) laws brought about a massive change to lighter weight, more fuel-efficient front-wheel-drive cars. A significant number of these cars had four-wheel independent suspension systems.

Cars with unitized bodies and independent or semi-independent rear suspension systems are more likely to experience rear wheel alignment problems compared with rear-wheel-drive vehicles with frames and one-piece rear axle housings. This is especially true after collision damage. Therefore, with the introduction of unitized bodies in massive numbers, four-wheel alignment became a necessity.

REAR WHEEL ALIGNMENT AND VEHICLE TRACKING PROBLEMS

Result of Proper Rear Wheel Alignment

The **thrust line** is an imaginary line at a 90-degree angle to the centerline of the rear wheels and projected forward.

The vehicle **geometric centerline** is an imaginary line through the exact center of the front and rear wheels.

A **rear axle offset** refers to a condition where the complete rear axle housing has rotated slightly, moving one rear wheel forward and the opposite rear wheel backward. Under this condition, the rear wheels are no longer parallel to the geometric centerline of the vehicle.

The driver uses the steering wheel to turn the front wheels and steer the vehicle in the desired direction. However, the rear wheels determine the direction of the vehicle to a large extent. When the **thrust line** is positioned at the **geometric centerline** and the front and rear wheels are parallel to the vehicle geometric centerline, the vehicle moves straight ahead with minimum guidance from the steering wheel (**Figure 9-41**).

Result of Improper Rear Wheel Alignment

Rear Axle Offset A rear axle offset may occur on a rear-wheel-drive vehicle with a one-piece rear axle housing or on a front-wheel-drive car with a trailing arm rear suspension. If the rear axle is offset, the thrust line is no longer at the vehicle centerline (**Figure 9-42**). The **thrust angle** is the angle between the geometric centerline and the thrust line.

With the rear axle offset problem shown in Figure 9-42, the thrust line is positioned to the left of the geometric centerline. Under this condition, the rear wheels steer the vehicle in a large clockwise circle. If the driver's hands are removed from the steering wheel, the steering pulls to the right. This steering action is similar to the rear wheels on a forklift. The rear wheels are used to steer the forklift because of the heavy load on the front wheels. If the rear wheels are pointed toward the left, the forklift turns to the right.

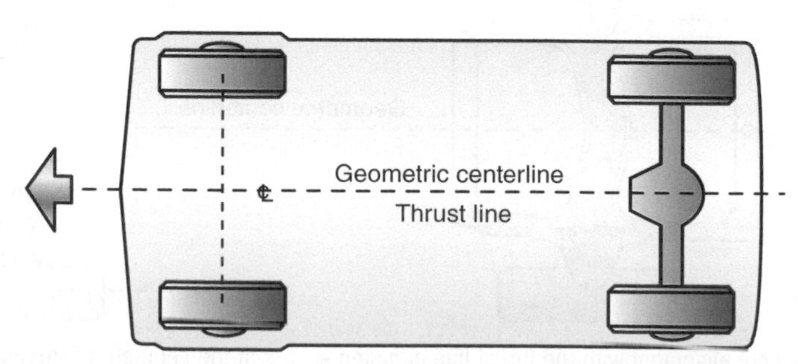

Figure 9-41 Front and rear wheels parallel to the vehicle centerline.

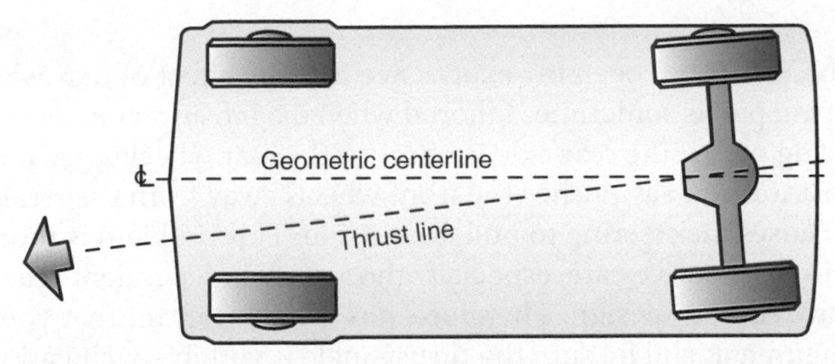

Figure 9-42 Rear axle offset and improperly positioned thrust line.

On a vehicle with this problem, the left front wheel toes out and the right front wheel toes in when the vehicle is driven straight ahead. This front wheel situation occurs because the front wheels try to compensate for the rear suspension defect. The front wheels are turned slightly to the left to compensate for the drift to the right caused by the rear suspension problem. Under this condition, both the front and rear tires may have feathered tire wear.

The ideal correction for this rear suspension problem is to reposition the rear suspension so the thrust line is at the vehicle centerline, then set the front wheel toe-in to the thrust line. The rear suspension problem that we have described also causes the steering wheel to be off-center when the vehicle is driven straight ahead.

Some of the causes of rear axle offset are:

1. A broken center bolt in rear leaf spring.
2. Worn shackles in rear leaf springs.
3. A bent frame.
4. A bent subframe or floor section, unitized body.
5. Worn trailing arm bushings.
6. Bent trailing arms.

Improper Rear Wheel Toe If the left rear wheel has excessive toe-out, the thrust line is moved to the left of the geometric centerline (**Figure 9-43**). This defect has basically the same effect on steering as rear axle offset. Improper toe on one rear wheel is most often encountered on vehicles with independent rear suspension.

Some of the causes of improper toe on one rear wheel are:

1. A bent one-piece rear axle.
2. A bent U section in trailing arm suspension.
3. A bent rear lower control arm.
4. A worn rear lower control arm bushing.
5. A bent rear spindle.
6. An improper rear toe adjustment.

Shop Manual
Chapter 9, page 375

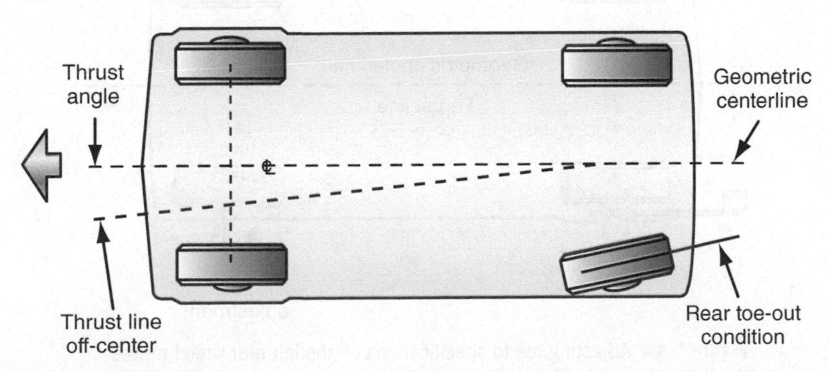

Figure 9-43 Excessive toe-out on the left rear wheel moves the thrust line to the left of the geometric centerline.

AUTHOR'S NOTE It has been my experience that the effect of rear wheel alignment on steering pull is sometimes ignored when diagnosing a customer complaint of steering pull. If the rear axle is offset or the rear wheel toe is incorrect, the rear wheel position actually pushes the front wheels away from the straight-ahead position and causes the steering to pull to one side. This problem is more likely to occur on front-wheel-drive cars, especially those with independent rear suspension and a unitized body design. Therefore, it is very important that you perform a four-wheel alignment, and be sure the thrust angle is within specifications before aligning the front suspension.

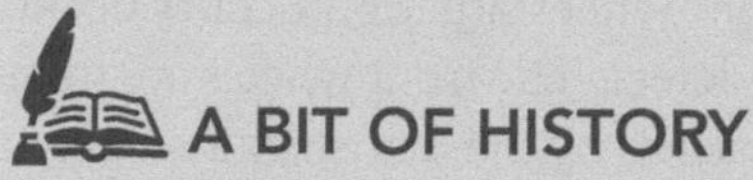

A BIT OF HISTORY

Early attempts at rear wheel alignment were slow and lacked precision. These attempts at rear wheel alignment included the use of a track bar and even backing the rear wheels of a car onto a front wheel aligner to align the rear wheels. To meet the need for fast, accurate front and rear wheel alignment, wheel alignment manufacturers designed computer wheel aligners. The technology in this equipment has greatly improved since the first models were introduced.

The necessary correction for this defect is to adjust the left rear wheel toe to the manufacturer's specifications. This adjustment moves the thrust line to the geometric centerline of the vehicle (**Figure 9-44**).

Rear Axle Sideset When **rear axle sideset** occurs, the rear wheels are parallel to each other, but the rear axle assembly has moved straight sideways so the geometric centerline is no longer at the proper vehicle centerline position (**Figure 9-45**). Because the rear wheels do not follow directly behind the front wheels when the vehicle is driven straight ahead, the vehicle dog tracks. Rear axle sideset does not affect steering pull as much as rear axle offset, but sideset will cause steering pull if it is severe.

Wheel Setback **Wheel setback** is a condition where one front or rear wheel is moved rearward in relation to the opposite wheel (**Figure 9-46**). This condition is usually caused by front-end collision damage, and thus it is most likely to be a front suspension problem. However, setback may occur on independent rear suspension systems. A severe setback condition causes steering pull.

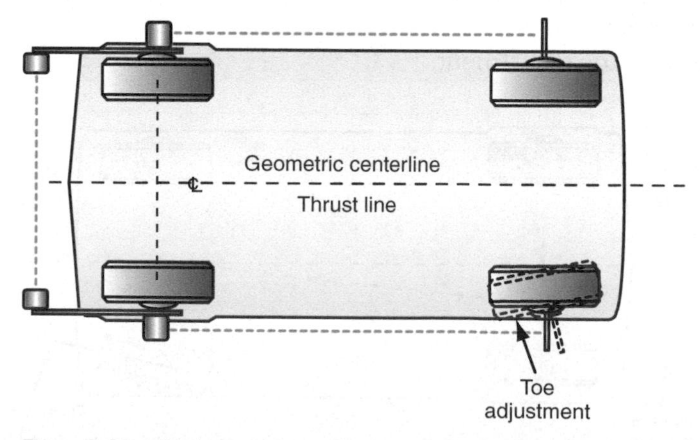

Figure 9-44 Adjusting toe to specifications on the left rear wheel moves the thrust line to the geometric centerline.

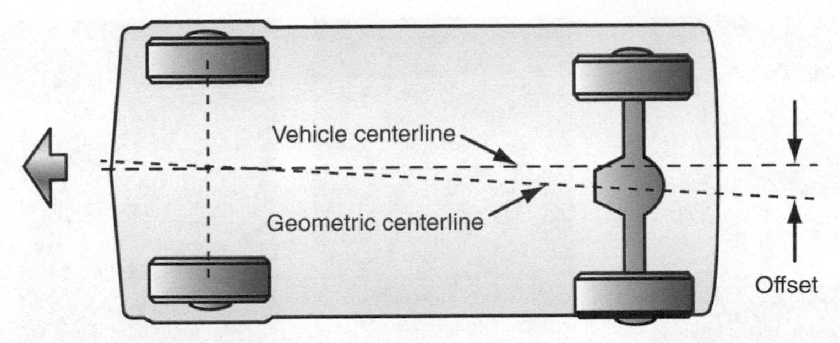

Figure 9-45 Rear axle offset occurs when the rear axle assembly has moved straight sideways, and the geometric centerline is shifted away from the proper vehicle centerline position.

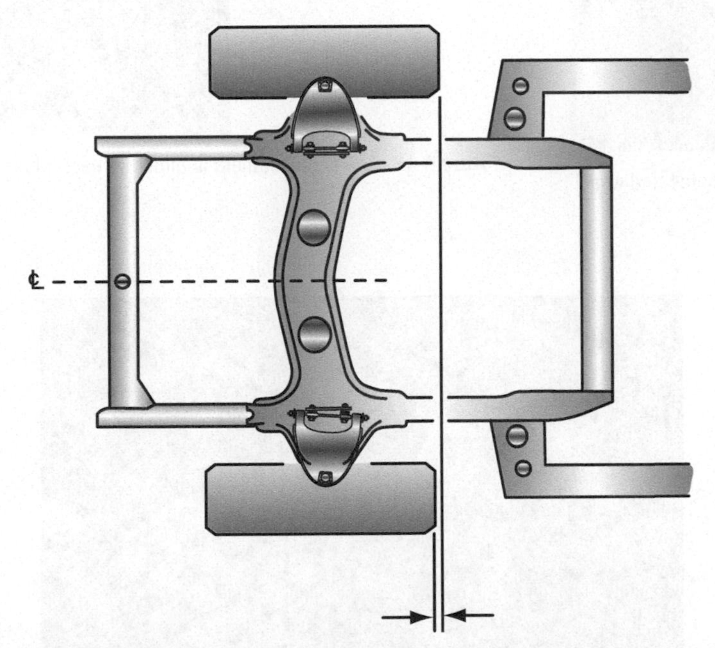

Figure 9-46 Front wheel setback occurs when one front wheel is moved rearward in relation to the opposite front wheel.

COMPUTER ALIGNMENT SYSTEMS

Computer Wheel Aligner Features

Some computer wheel aligners have four high-resolution digital cameras that measure wheel target position and orientation. The front and rear wheel alignment angles are sensed by the digital cameras and wheel targets and then displayed on the wheel alignment monitor. The vehicle is raised to a comfortable working height on the aligner lift, and two digital cameras are mounted in each end of a crossbar on a post in front of the vehicle (**Figure 9-47**). The post and crossbar height may be adjusted to match the vehicle height.

One digital camera is aimed at each wheel target (**Figure 9-48**). A target is mounted on each wheel with a self-centering adapter. The adapters are adjustable to fit rims up to 24 in. (60 cm) in diameter. The targets contain a polished aluminum faceplate that acts as a mirror (**Figure 9-49**). The targets do not contain any electronics, glass, or cables, and they do not require calibration. The targets are also lightweight and do not require any maintenance.

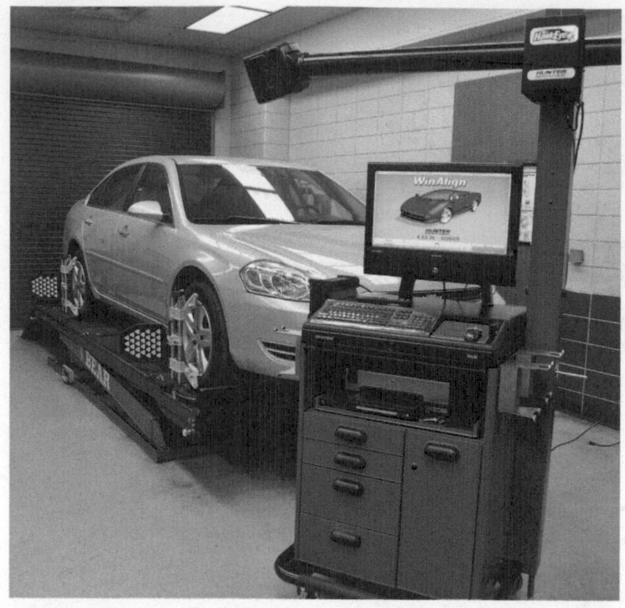

Figure 9-47 Computer wheel aligner with digital cameras and wheel targets.

Figure 9-48 One digital camera is aimed at each wheel target.

Figure 9-49 Each target contains a polished aluminum face plate with no electronics, glass, or cables.

This type of computer aligner is very simple to use. After the prealignment procedures and checks have been performed on the vehicle, and the vehicle is positioned at the desired height and position on the lift and turntables, the first step is to mount the wheel targets on the wheels. The second step is to roll the vehicle back a short distance until the onscreen indicators on the monitor turn green. The third step is to roll the vehicle forward and stop the vehicle so the front wheels are centered on the turntables. The fourth step is to observe all the front and rear camber and toe angles displayed on the monitor.

A remote control is wired into the monitor so the technician can perform the same functions when the monitor is not in the technician's view. A handheld remote control is used to measure the vehicle ride height (**Figure 9-50**). This remote is held against the

Figure 9-50 A handheld remote control transmits ride height measurements electronically to the console and monitor.

lower edge of the fender well to measure the ride height, and the remote transmits the ride height measurement electronically to the console and monitor.

On some computer wheel aligners, the **wheel sensors** contain a microprocessor and a **high-frequency transmitter** that acquire measurements and process data and then send these data to a **receiver** mounted on top of the wheel alignment monitor (**Figure 9-51** and **Figure 9-52**). This type of wheel sensor does not require any cables connected between the sensors and the computer wheel aligner. The data from this type of wheel sensor are virtually uninterruptible, even by solid objects. When these wheel sensors are stored on the computer wheel aligner, a "docking station" feature charges the batteries in the wheel sensors. The front wheel sensors have optical arms that project ahead of the tires to provide front toe readings. Some rear wheel sensors also have optical arms that project behind the rear tires (**Figure 9-53**). This type of rear wheel sensor measures rear wheel setback.

Wheel sensors are mounted on each wheel, and these sensors transmit wheel alignment data to the computer wheel aligner.

A **high-frequency transmitter** is contained in many wheel sensors. This transmitter sends data to a receiver on top of the monitor.

A **receiver** on the computer wheel aligner receives signals from the wheel sensors.

On some computer alignment systems, the reference signals between the wheel sensors are provided by light-emitting diodes (LEDs) or electronic signals. Cables are

Figure 9-51 Cordless wheel sensor containing a high-frequency transmitter.

Figure 9-52 Computer wheel aligner with data receiver on top of the monitor.

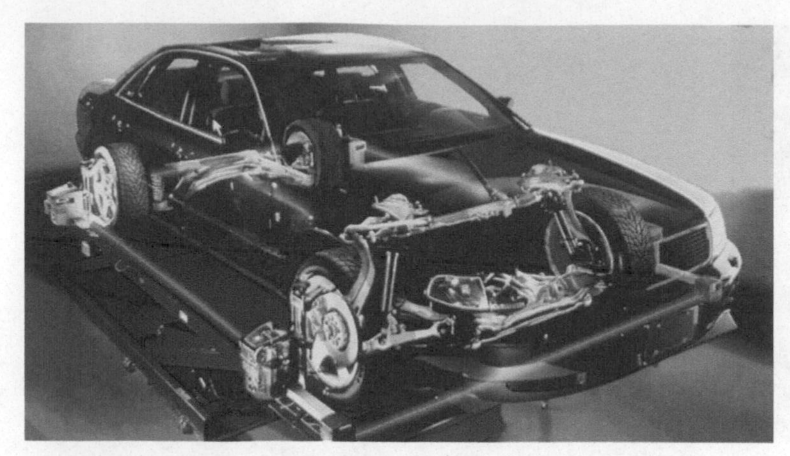

Figure 9-53 Front and rear wheel sensors with optical arms.

Figure 9-54 Front wheel sensor with optical arm and string connected to the rear wheel sensor.

connected from the wheel units to the computer aligner. Reference signals between the wheel units are provided by strings on some older model computer alignment systems (**Figure 9-54**). Most wheel sensors provide wheel runout compensation. Because the wheel unit is clamped to the rim, a bent rim will affect the alignment readings. On older computer wheel aligners, the technician had to perform a wheel runout check at each wheel. Newer wheel units use audible and visual prompts on the monitor screen to inform the technician if the wheel sensors require leveling or calibration. Wheel runout compensation may be completed by pressing a button on the wheel sensor. Wheel sensor leveling is done by adjusting the sensor level control.

Shop Manual
Chapter 9, page 363

Computer alignment systems have the capability to measure thrust angle and setback as well as other front and rear alignment angles. A typical computer wheel aligner has the following features:

1. Color monitor and computer (**Figure 9-55**).
2. Computer software program stored on a hard disc drive. This program contains vehicle four-wheel alignment procedures and diagnostic drawings.
3. Compact disc (CD) or digital video disk (DVD) drive. The vehicle specifications and alignment program may be contained on CDs or DVDs, and updates are possible by replacing the CDs or DVDs (**Figure 9-56**). Wheel alignment specifications include domestic and imported vehicles for the current year and a minimum of 10 previous years.

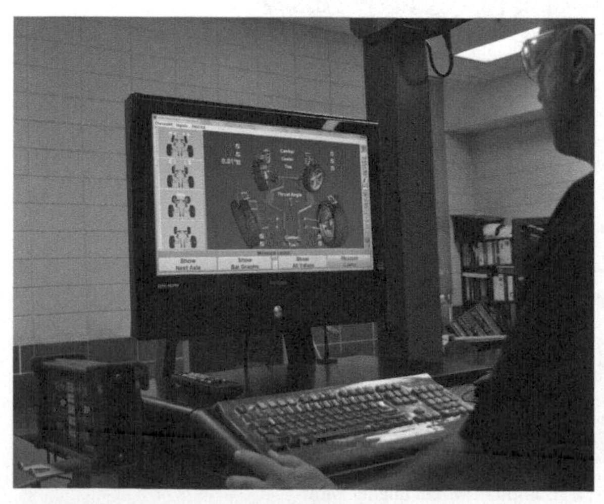

Figure 9-55 Computer wheel aligner.

Figure 9-56 CDs for wheel alignment program and specifications.

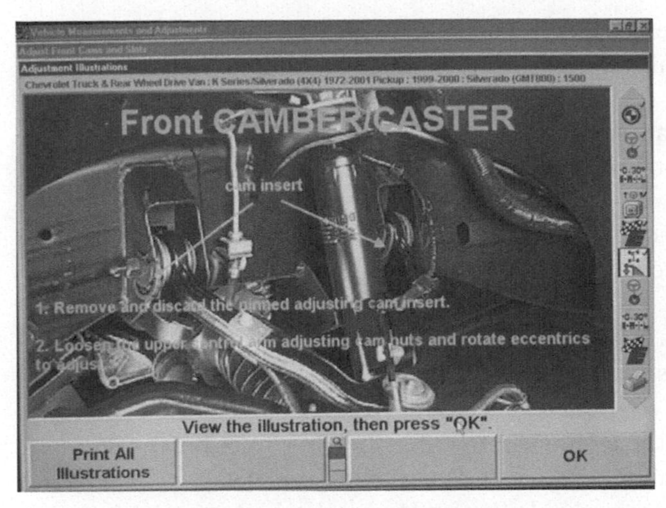

Figure 9-57 Point and click suspension inspection screen.

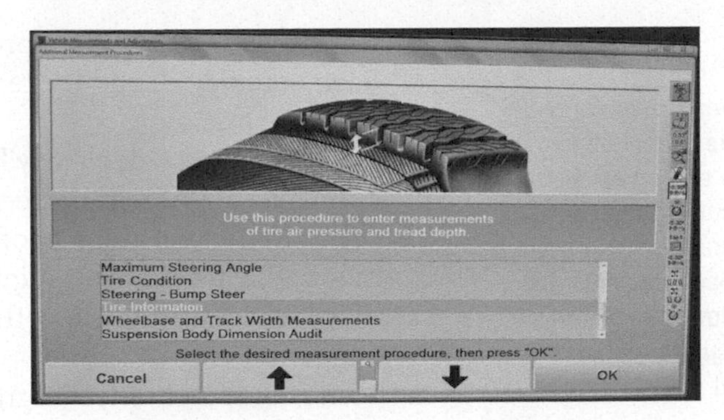

Figure 9-58 Tire inspection screen.

4. The software in some computer wheel aligners includes a CD or DVD image database containing 2100 digital photos of the adjustment and inspection points. In some computer aligners, a list of inspection points appears on the left of the screen. When the operator clicks on a listed inspection point, the matching photo appears, illustrating the selected inspection point (**Figure 9-57**). Special digital photos are available for inspection of the cradle and cradle-to-body alignment.
5. Live action inspection videos may be accessed instantly from the inspection screen. These inspection videos are part of the 142-video training library.
6. A tire inspection screen illustrates various types of tire tread wear and the related causes (**Figure 9-58**). This screen may be printed out and shown to the customer.
7. Measurement and adjustment screens illustrate suspension measurements. If a measurement is within specifications, the measurement is illustrated with a green bar and an arrow indicates the measurement. A red measurement bar indicates the adjustment is not within specifications (**Figure 9-59**). In the adjustment mode, the measurement remains on the screen, and as the necessary suspension adjustment is performed, the adjustment bar turns green when the arrow moves within specifications.
8. An alignment procedure bar is available in some computer aligners. This vertical bar contains an icon for each step in an alignment procedure (**Figure 9-60**). These

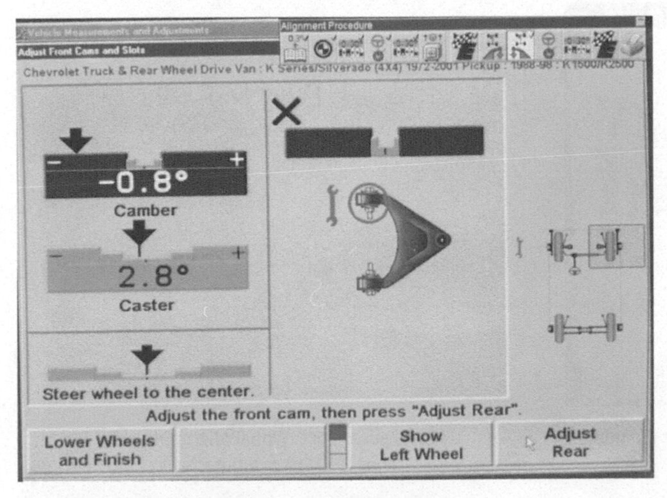

Figure 9-59 Wheel alignment adjustment screen.

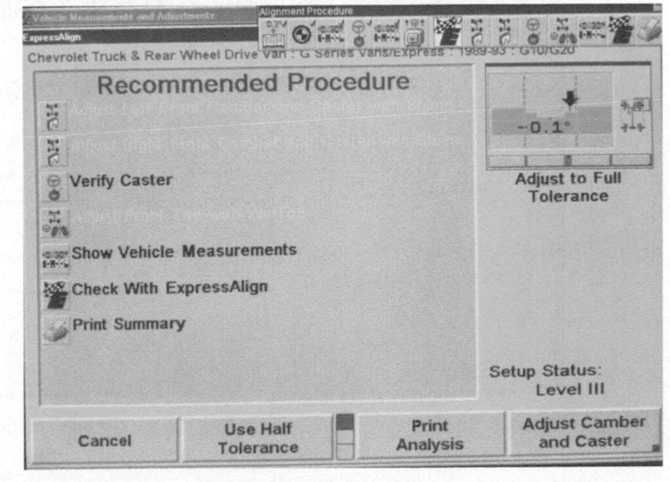

Figure 9-60 Wheel alignment procedure screen.

Caution

When using a computer wheel aligner, the equipment manufacturer's recommended alignment procedures must be followed to provide accurate readings and adjustments.

steps are arranged in the proper sequence. A check mark appears when a step has been completed, and the technician may select and click any step to move to that adjustment.

9. Correction kit videos may be accessed to illustrate the installation of aftermarket wheel alignment correction kits. Special tools videos are available in the computer wheel aligner software. These videos indicate the special wheel alignment adjustment tools required for the vehicle being aligned. Operation videos may be accessed to show the technician how to operate the computer wheel aligner.
10. Some computer wheel aligners have multilingual capabilities. These capabilities usually include three languages; however, up to 20 languages may be available from the equipment manufacturers.
11. Full-function keyboard allows the technician to enter such information as the customer's name and address and the make and year of the vehicle.
12. Set of four self-centering adjustable rim clamps and four-wheel sensors.
13. Full-function remote control that allows the technician to control the alignment program and see the results of suspension adjustments while working under the vehicle.
14. Color or laser printer provides detailed wheel alignment reports for the customer and specific alignment messages and diagrams for the technician.
15. In some computer wheel aligners, the **Motorist Assurance Program (MAP)** uniform inspection guidelines may be accessed for many steering, suspension, and brake components (**Figure 9-61**). The MAP program establishes uniform parts inspection guidelines to improve customer satisfaction with the automotive industry.
16. A variety of software programs are available with some computer wheel aligners. Much of the software may be extra-cost options. This software may include the following:
    - Wheel alignment software on CD or DVD: This software provides the technician with a very extensive vehicle information database as well as many patented adjustment and productivity features (**Figure 9-62**).
    - Specific computer software provides service bulletins and other information on domestic and import vehicles (**Figure 9-63** and **Figure 9-64**). Some computer software also provides labor estimates for automotive repairs.

The **Motorist Assurance Program (MAP)** improves automotive customer satisfaction by establishing uniform inspection guidelines.

Caution

Because wheel units contain sensitive electronic circuits, excessive shock, such as dropping them on the floor, can damage them.

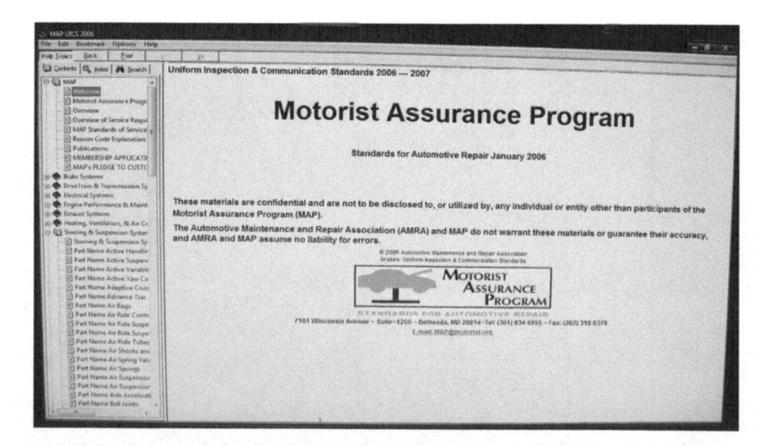

Figure 9-61 The Motorist Assurance Program (MAP) parts inspection guidelines may be accessed in computer wheel aligners.

Figure 9-62 Wheel Alignment software on CD or DVD for computer wheel aligners.

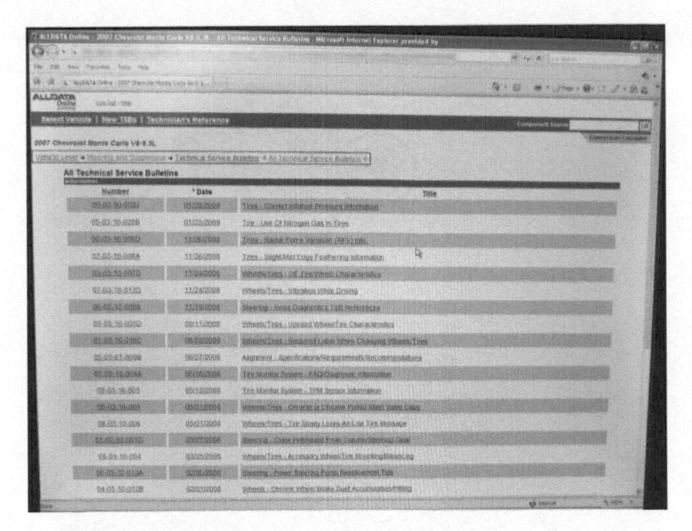

Figure 9-63 Some computer software provides service bulletin information for suspension and steering problems.

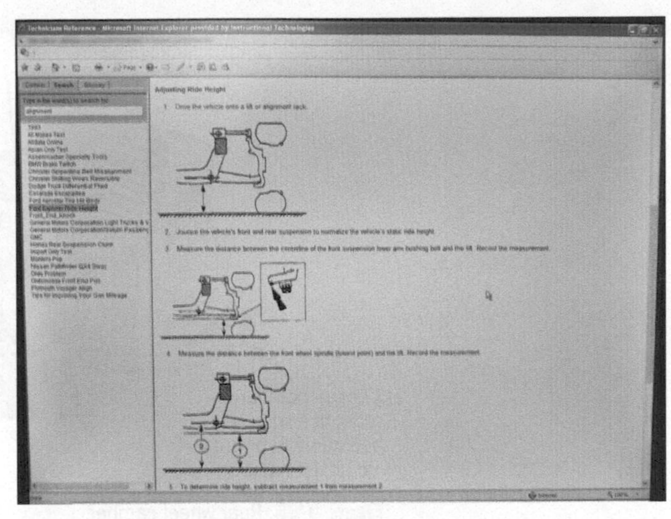

Figure 9-64 Specific computer software provides diagnostic and service information in some computer wheel aligners.

REAR WHEEL ALIGNMENT

Rear Wheel Alignment Diagnosis

Customer complaints that indicate rear wheel alignment problems are the following:

1. The front wheel toe is set correctly and the steering wheel is centered on the alignment rack, but the steering wheel is not centered when the vehicle is driven straight ahead.
2. The steering pulls to one side, but there are no worn suspension parts, defective tires, or improper front suspension alignment angles.
3. The vehicle is not overloaded and there are no worn suspension parts or improper front suspension alignment angles, but tire wear occurs.

Defects That Cause Incorrect Rear Wheel Alignment

These suspension or chassis defects cause incorrect rear wheel alignment:

1. Collision damage that results in a bent frame or distorted unitized body
2. A leaf-spring eye that is unwrapped or spread open
3. Leaf-spring shackles that are broken, bent, or worn
4. Broken leaf springs or leaf-spring center bolts
5. Worn rear upper or lower control arm bushings
6. Worn trailing arm bushings or dislocated trailing arm brackets
7. Bent components such as radius rods, control arms, struts, and rear axles

Rear Wheel Camber

Many front-wheel-drive vehicles have a slightly negative rear wheel camber that provides improved cornering stability. **Rear wheel camber** is basically the same as front wheel camber (**Figure 9-65**).

Rear wheel camber is the tilt of a line through the center of the rear tire and wheel in relation to the true vertical centerline of the tire and wheel.

During rear wheel jounce on a multilink rear suspension system, the camber becomes more positive, and the camber moves toward a negative position during wheel rebound. Because camber changes during wheel jounce, and rebound causes tire tread wear, the rear suspension is designed to minimize camber change during wheel jounce and rebound.

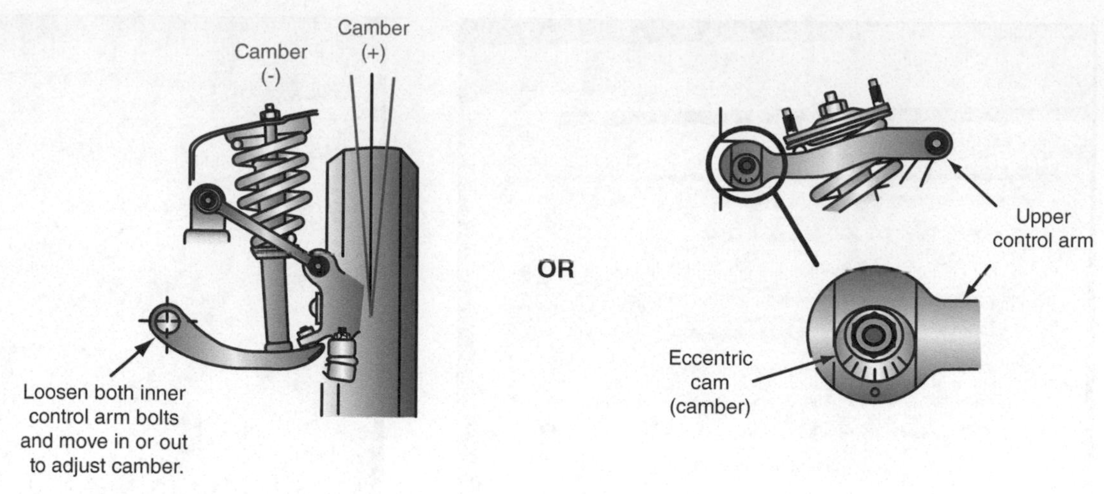

Figure 9-65 Rear wheel camber.

Shop Manual
Chapter 9, page 375

Rear wheel toe is the distance between the front edges of the rear tires in relation to the distance between the rear edges of the rear tires.

Rear Wheel Toe

On a front-wheel-drive vehicle, driving forces tend to push back the rear wheel spindles. Therefore, these rear wheels are designed with zero toe-in or a slight toe-in depending on the vehicle (**Figure 9-66**). Correct **rear wheel toe** is important to obtain normal tire life.

On a multilink rear suspension system with the toe adjustment link mounted in a level position, the toe moves toward a toe-in position during wheel jounce or rebound. These toe changes cause rear tire tread wear.

Steering Angle Sensor Reset

The last step of an alignment procedure that must be performed on some vehicle platforms is a reset procedure for the steering angle sensor (SAS) and other related sensors to update the vehicle geometry information stored in the vehicle's software system after an alignment (**Figure 9-67**). This procedure can be performed with an enhanced computer scan tool,

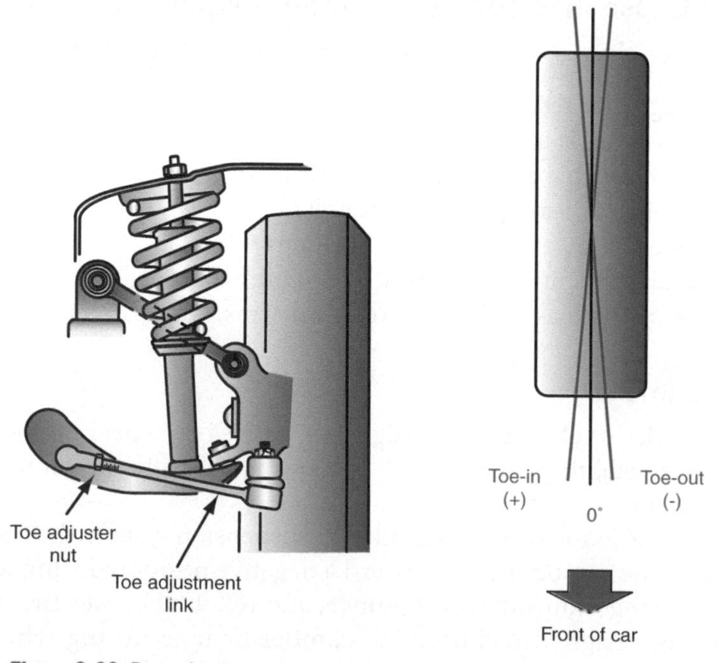

Figure 9-66 Rear wheel toe.

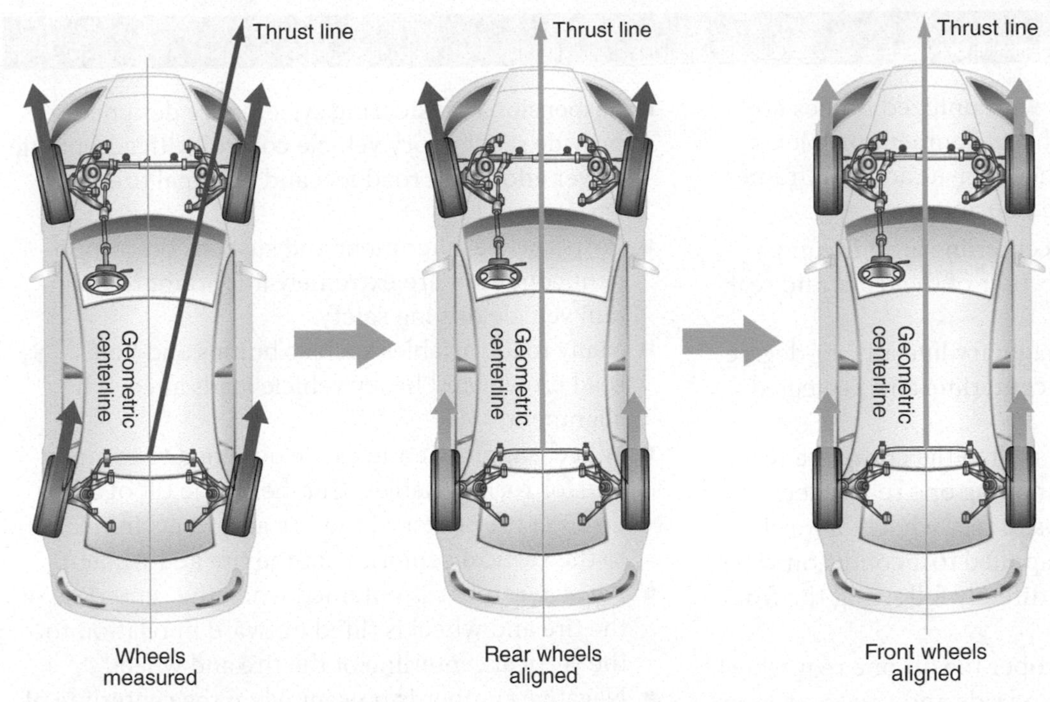

Figure 9-67 On some vehicles the last step after an alignment procedure is to update vehicle geometry in the vehicle's software system.

depending on model, or your alignment equipment may incorporate this feature. Hunter Engineering has an interface called "CodeLink" that works with the alignment program to relearn the SAS and other related sensor information at the end of the alignment procedure (**Figure 9-68**). Reset requirements vary between vehicle manufacturer year, make, and model of the platform. Always refer to manufacturer relearn service procedures. Other sensors that may be part of this reset procedure are the yaw rate sensor, torque angle sensor, deceleration sensor, and lane departure among others, depending on vehicle manufacturer model. Refer to specific vehicle information for specific steps and requirements.

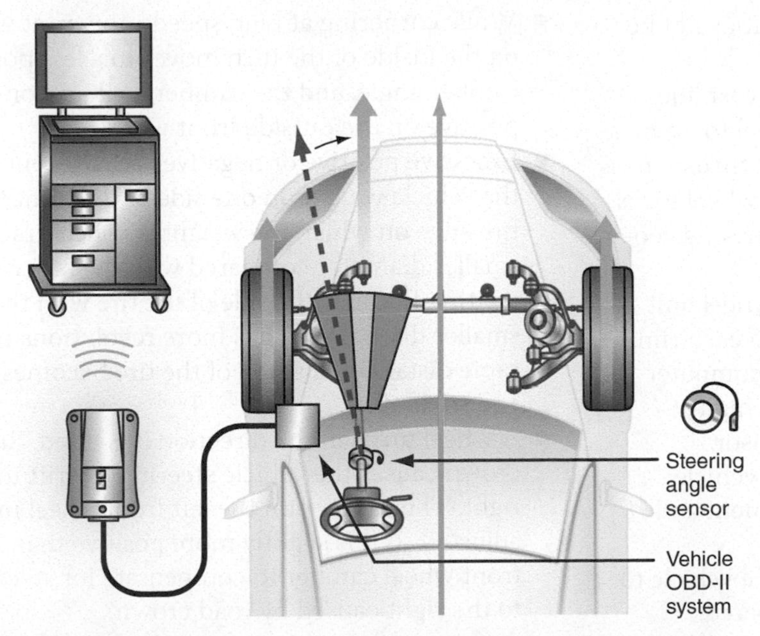

Figure 9-68 To update vehicle geometry in the vehicles software system a scan tool or interface module is connected to the vehicle's data link connector.

SUMMARY

- Front-wheel-drive cars with unitized bodies are more subject to rear wheel alignment problems compared with rear-wheel-drive cars with frames and one-piece rear axle housings.
- The vehicle geometric centerline is an imaginary line through the exact center of the front and rear wheels.
- The thrust line is an imaginary line at a 90-degree angle to the rear wheel centerline and projected forward.
- A rear axle offset occurs when the complete rear axle is rotated slightly, moving one rear wheel backward and the opposite rear wheel forward.
- Dog tracking is a term applied to a condition where the rear wheels are not directly following the front wheels.
- Rear axle offset or improper toe on one rear wheel causes steering pull to one side and tire tread wear.
- Rear axle sideset is a condition where the rear axle assembly has moved straight sideways and the geometric centerline is not positioned at the true vehicle centerline.
- Setback occurs when one front or rear wheel is moved backward in relation to the opposite wheel.
- In a geometric centerline front wheel alignment, the geometric centerline is used for a reference for front wheel toe. This type of alignment ignores thrust line position.
- In a thrust line front wheel alignment, the front wheel toe is checked using the thrust line as a reference, but the thrust line position may not be at the geometric centerline.
- In a total four-wheel alignment, the thrust line position is measured and the rear wheel toe is measured and adjusted as necessary so the thrust line is at the geometric centerline. The front wheel toe is measured using the common thrust line and geometric centerline as a reference.
- Computer alignment systems have a wheel unit mounted on a wheel clamp attached to each rim.
- Signals from the wheel sensors to the computer wheel aligner may be transmitted by a high-frequency transmitter in the wheel sensor.
- Computer alignment systems provide vehicle specifications plus diagrams of adjustment and inspection points.
- Directional stability is the tendency of a vehicle to travel straight ahead without being steered.
- Suspension and steering systems are designed to provide satisfactory vehicle control with acceptable driver effort and road feel and minimal tire tread wear.
- Proper wheel alignment and suspension component condition are extremely important to maintain vehicle driving safety.
- Many road variables such as bumps and holes, road crown, and heavy vehicle loads affect wheel alignment.
- Wheel alignment angles are designed to compensate for road variables. Camber is the tilt of a line through the center of the tire and wheel in relation to the vertical centerline of the tire and wheel.
- Positive camber is obtained when the centerline of the tire and wheel is tilted outward in relation to the vertical centerline of the tire and wheel.
- Negative camber is present when the centerline of the tire and wheel is tilted inward in relation to the vertical centerline of the tire and wheel.
- During wheel jounce, the top of the wheel moves outward and the camber becomes more positive.
- During wheel rebound, the top of the wheel moves inward and the camber becomes less positive or moves to a slightly negative position.
- While cornering at high speeds, centrifugal force attempts to move the vehicle to the outside of the turn. This force raises the front suspension on the inside of the turn while lowering the front suspension on the outside of the turn.
- While cornering at high speeds, the front wheel on the inside of the turn moves to a less positive camber angle, and the camber angle becomes more positive on the outside front wheel.
- Excessive positive or negative camber concentrates the vehicle weight on one side of the front tire. The tire edge on which the weight is concentrated has a smaller diameter compared with the other side of the tire. Because the side of the tire with the smaller diameter makes more revolutions to go the same distance, this side of the tire becomes worn and scuffed.
- A wheel turns in the direction it is tilted. Road crown causes the vehicle steering to drift to the right. The camber on the left front wheel may be adjusted so it is slightly more positive than the right front wheel camber to compensate for steering pull to the right caused by road crown.

- Caster is the tilt of a line that intersects the center of the lower ball joint and the center of the upper strut mount in relation to a vertical line through the center of the spindle and wheel as viewed from the side.
- Positive caster occurs when the centerline of the lower ball joint and upper strut mount is tilted rearward in relation to the centerline of the spindle and wheel, viewed from the side.
- Negative caster is obtained when the centerline of the lower ball joint and upper strut mount is tilted forward in relation to the centerline of the spindle and wheel.
- Positive caster increases directional stability, steering effort, and steering wheel returning force.
- Excessive positive caster results in harsh ride quality. Excessive positive caster may cause front wheel shimmy. Negative caster reduces directional stability and steering effort while improving ride quality.
- If the caster is different on the two front wheels, the steering pulls toward the side with the least positive caster.
- The caster adjustment on the left front wheel may be adjusted so it is less positive than the caster on the right front wheel to compensate for road crown.
- If the rear suspension height is lowered, the front wheel caster becomes more positive. When the front suspension height is lowered, the front wheel caster becomes more positive. Proper caster adjustment is very important to maintaining vehicle directional control and safety.
- Front wheel toe is the distance between the front edges of the tires compared with the distance between the rear edges of the tires.
- Toe-in is present when the distance between the front edges of the tires is less than the distance between the rear edges of the tires.
- Toe-out occurs when the distance between the front edges of the tires is greater than the distance between the rear edges of the tires.
- Most front-wheel-drive cars have a slight toe-out setting on the front wheels because drive axle forces tend to move the front wheels to a toe-in position.
- Most rear-wheel-drive cars have a slight toe-in setting because driving forces tend to move the front wheels to a toe-out position.
- The front wheel toe is adjusted with the vehicle at rest so the front wheels are straight ahead when the vehicle is driven.
- Improper toe adjustment results in feathered tire wear.

REVIEW QUESTIONS

Short Answer Essays

1. Explain why four-wheel alignment is essential on cars with semi-independent or independent rear suspension.
2. Explain why the rear wheel toe angle, thrust angle, and camber should be correct before adjusting the front suspension angles.
3. Describe the term total toe and what the difference is between a positive toe angle and a negative toe angle.
4. Explain what will happen if toe angles are outside the allowed tolerance.
5. Explain why excessive positive camber wears the outside edge of the tire tread.
6. Describe the type of tire tread wear caused by an improper toe setting.
7. Explain why positive caster provides increased directional stability.
8. Describe how positive caster provides increased steering wheel returning force.
9. Explain why positive caster causes harsh riding quality.
10. Define toe-in and toe-out.

Fill-in-the-Blanks

1. Excessive toe-out on the right rear wheel moves the thrust line to the ______________ of the geometric centerline.
2. If the thrust line is positioned to the left of the geometric centerline so the thrust angle is more than specified, the steering pulls to the ______________.

3. Directional stability refers to the tendency of a vehicle to travel straight ahead without being ______________.
4. When the front wheel camber is negative, the centerline of the tire and wheel is tilted ______________ in relation to the true vertical centerline of the tire and wheel.
5. Excessively ______________ camber on a front wheel assembly will cause the tire to wear on the inside edge.
6. Negative front wheel caster decreases ______________ and steering wheel ______________ ______________.
7. Raising the rear suspension height above the manufacturer's specification may change the front wheel caster from ______________ to ______________.
8. Front wheel shimmy may be caused by excessive ______________ caster.
9. Front wheel caster becomes more positive if the rear suspension height is ______________.
10. If the front suspension height is lowered on a MacPherson strut suspension, the front wheel caster becomes ______________ ______________.

Multiple Choice

1. While diagnosing front wheel toe problems:
   A. The front wheels are set to a straight-ahead position on most front-wheel-drive cars.
   B. Driving forces tend to move the front wheels toward a toe-in position on a rear-wheel-drive car.
   C. Improper toe adjustment may cause feathered wear on the front tire treads.
   D. Sagged front springs may increase the front wheel toe-out on a short-and-long arm suspension system.
2. "Feathering" type wear of a rear tire is likely caused by:
   A. Improper rear wheel camber alignment.
   B. Improper rear tire inflation.
   C. Improper rear wheel balance.
   D. Improper rear wheel toe alignment.
3. A front-wheel-drive vehicle with an independent rear suspension pulls to the right when driving straight ahead. All the front suspension alignment angles are within specifications. The most likely cause of this problem is:
   A. Excessive toe-out on the right rear wheel.
   B. Excessive negative camber on the right rear wheel.
   C. Excessive toe-out on the left rear wheel.
   D. Excessive positive camber on the left rear wheel.
4. All of these statements about the vehicle thrust line and steering pull are true EXCEPT:
   A. Excessive toe-out on the left rear wheel moves the thrust line to the left of the vehicle centerline.
   B. Excessive toe-in on the right rear wheel moves the thrust line to the right of the vehicle centerline.
   C. If the thrust line is moved to the left of the vehicle centerline, the steering tends to pull to the right.
   D. Excessive toe-out on the right rear wheel causes the steering to pull to the left.
5. While diagnosing suspension and wheel alignment problems:
   A. Road crown has no effect on vehicle steering or wheel alignment.
   B. Vehicle loads have no effect on wheel alignment.
   C. Steering angles are designed to reduce tire wear and provide directional control.
   D. Wheel alignment angles do not affect riding quality.
6. While diagnosing the vehicle geometric centerline and thrust line:
   A. The front and rear wheels should be parallel to the geometric centerline.
   B. The thrust angle is the difference between the front wheel camber and steering axis inclination (SAI) angles.
   C. If the thrust angle is more than specified, front wheel shimmy may occur.
   D. A bent front cradle may cause the thrust angle to be more than specified.

7. While driving straight ahead, a front-wheel-drive car pulls to the right. The most likely cause of this problem is:
   A. More positive camber on the left front wheel compared to the right front wheel.
   B. Sagged front springs and improper front wheel toe setting.
   C. Less positive caster on the right front wheel compared to the left front wheel.
   D. The SAI on the right front wheel is 1.5 degree more than the SAI on the left front wheel.
8. While adjusting front wheel camber and diagnosing camber-related problems:
   A. During front wheel jounce travel, the positive camber increases.
   B. Excessive positive camber on a front wheel causes premature wear on the inside edge of the tire tread.
   C. If the right front wheel camber is 1.5 degrees and the left front camber is 0.5 degree, the steering pulls to the left.
   D. Excessive positive camber on both front wheels may cause front wheel shimmy.
9. A driver complains of harsh riding and the suspension height is normal. The most likely cause of this problem is:
   A. Excessive negative camber on both front wheels.
   B. Excessive positive caster on both front wheels.
   C. The left front wheel has more positive caster than the right front wheel.
   D. Both front wheels have negative camber and negative caster.
10. Excessive positive caster may cause all of these problems EXCEPT:
    A. Front wheel shimmy.
    B. Harsh ride quality.
    C. Excessive steering effort.
    D. Slow steering wheel return.

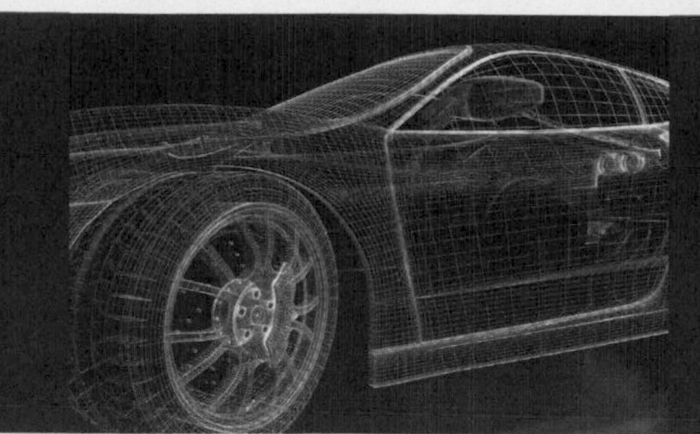

CHAPTER 10

FOUR-WHEEL ALIGNMENT, PART 2 DIAGNOSTIC ANGLES AND FRAME DAMAGE

Upon completion and review of this chapter, you should be able to understand and describe:

- The steering axis inclination (SAI) angle and the included angle.
- How SAI helps return the front wheels to the straight-ahead position.
- How SAI eliminates the need for excessive positive camber and caster.
- The effect that front suspension defects such as dislocated upper strut towers and bent struts or spindles have on SAI.
- Negative and positive scrub radius and the effect of each on steering quality.
- Setback on front suspension systems.
- How the front suspension system is designed to provide toe-out on turns.
- The purposes of a vehicle frame.
- Different types of frame construction.
- Different types of frame designs.
- Unitized body construction, and explain how this body design obtains its strength.
- Directional stability.
- Vehicle tracking, and explain how the four wheels on a vehicle must be positioned to obtain proper tracking.
- Wheelbase, and explain how the wheels on a vehicle must be positioned to provide correct wheelbase.
- Setback on a front wheel, and explain the effect that setback has on vehicle steering.
- Rear axle offset, and describe the effect of this problem on steering control.
- Rear axle sideset, and explain the effect of rear axle sideset on steering control.
- Side sway frame damage.
- Frame sag.
- Frame buckle.
- A diamond-frame condition.
- Frame twist.
- Aluminum space frame construction.

Terms To Know

Axle offset
Axle sideset
Axle thrust line
Channel frame
Complete box frame
Diamond-frame condition
Directional stability
Frame buckle
Frame sag
Frame twist
Geometric vehicle centerline
High-strength steels (HSS)
Hydro-forming
Included angle
King pin inclination (KPI)
Ladder frame design
Negative camber
Negative scrub radius
Perimeter-type frame
Positive scrub radius
Scrub radius
Side sway
Slip angle
Steering axis inclination (SAI)
Tracking
Tubular frame
Turning radius
Unitized body
Wheel setback
Wheelbase

INTRODUCTION

The frame and unitized body play an important role in vehicle geometry and alignment angles. It is important to understand the vehicle's structure elements and to be aware of how damage to the vehicle's structure elements can affect alignment angle. Many of the angles discussed in this chapter are not adjustable on their own but are used as diagnostic angles. These angles are often used in combination with one another as an aid in determining the root cause for the vehicle geometry being out of specification.

As an example, improper steering axis inclination (SAI) angles on either side of the front suspension may cause hazardous steering conditions while braking or accelerating. Therefore, technicians must be familiar with SAI and other related steering geometry if they are to diagnose routine drivability and handling issues.

Shop Manual
Chapter 10, page 431

STEERING AXIS INCLINATION DEFINITION

On front-wheel-drive vehicles with MacPherson strut front suspension systems, **steering axis inclination (SAI)** refers to the inward tilt of a line through the center of the top strut mount and the center of the lower ball joint in relation to the true vertical line through the center of the tire. These two lines are viewed from the front of the vehicle, and the SAI line always tilts inward in relation to the true vertical line (**Figure 10-1**).

Many rear-wheel-drive cars have a short-and-long arm front suspension system with unequal-length upper and lower control arms and a ball joint mounted in each control arm. The steering knuckle and spindle pivot on the ball joints as the wheels are turned. On this type of front suspension, the SAI line runs through the upper and lower ball joint centers. The **included angle** is the sum of the SAI angle and the camber angle (**Figure 10-2**).

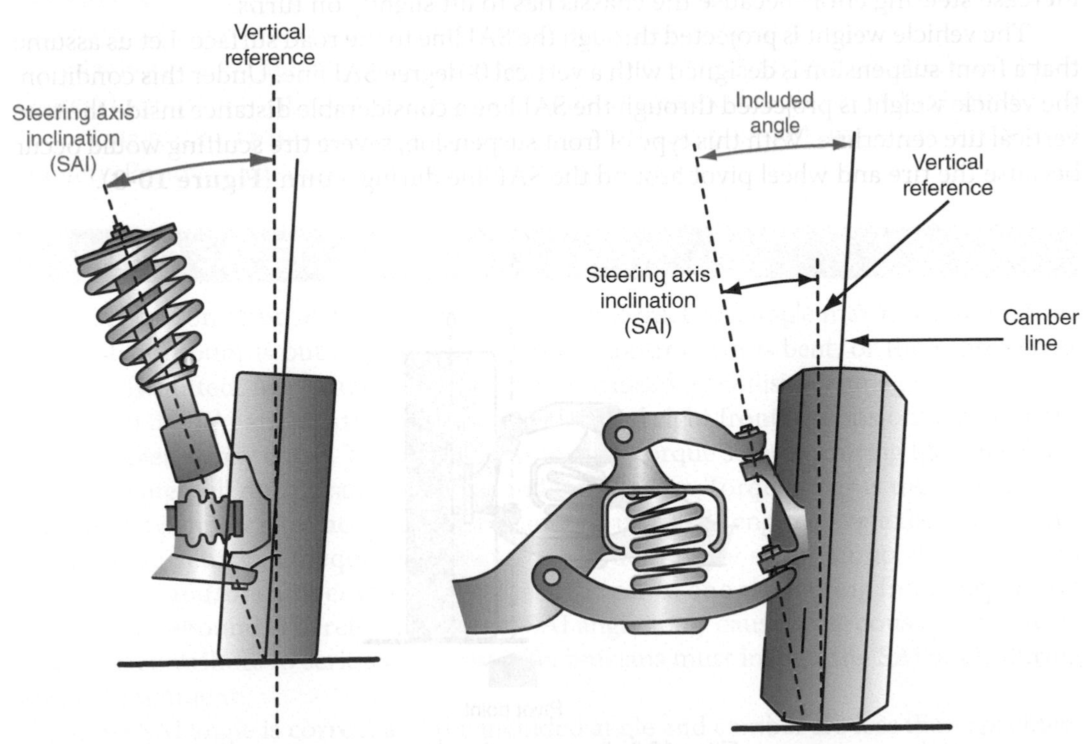

Figure 10-1 Steering axis inclination angle.

Figure 10-2 Steering axis inclination and included angle.

that requires inward or outward upper strut movement to adjust camber, the SAI angle changes with this movement. When an eccentric bolt between the strut and steering knuckle is used for camber adjustment, the SAI angle will not change if the camber is adjusted.

It is very important to remember that a camber adjustment on many front suspension systems also changes the SAI angle. For example, if the upper control arm on a front suspension system with upper and lower ball joints is shimmed outward to increase positive camber, the SAI angle will change with the camber angle. Therefore, the included angle remains the same. The camber may be adjusted to specification, but the SAI angle and included angle could be out of specification. This is especially true on MacPherson strut suspension systems, where collision damage may bend front struts or shift the upper strut towers. A service technician who inspects and adjusts the camber angle while ignoring the SAI, and included angles may be overlooking serious and dangerous front suspension defects.

SCRUB RADIUS

Scrub radius is the distance between the SAI line and the true vertical centerline of the tire at the road surface.

Scrub radius affects steering quality related to stability and returnability. However, scrub radius is not an alignment angle and it cannot be measured on conventional alignment equipment. Scrub radius is the distance from the point where the tire vertical line contacts the road to the location where the line through the upper strut center and the ball joint center meets the road surface. **Positive scrub radius** occurs when the line through the strut and ball joint meets the road surface inside the tire vertical centerline. A **negative scrub radius** is when the line through the strut and ball joint centers meets the road surface outside the tire and hub centerline (**Figure 10-5**).

Conventional short-and-long arm front suspension systems usually have positive scrub radius. Many front-wheel-drive vehicles have negative scrub radius.

When negative scrub radius is used in front-wheel-drive vehicles, straight-line braking is ensured and directional stability is maintained. As the vehicle moves forward, negative scrub radius tends to turn the front wheels inward. This action causes unequal forces applied to the steering to act inboard of the steering axis and pull the vehicle from any induced swerve. In a front-wheel-drive vehicle, a swerve to one side may be caused by one front wheel being on an ice patch while the other front wheel is on dry pavement. A failure of one-half the diagonal brake system, a sudden blowout of one front tire, or a grabbing brake on one front wheel will also cause a vehicle to swerve.

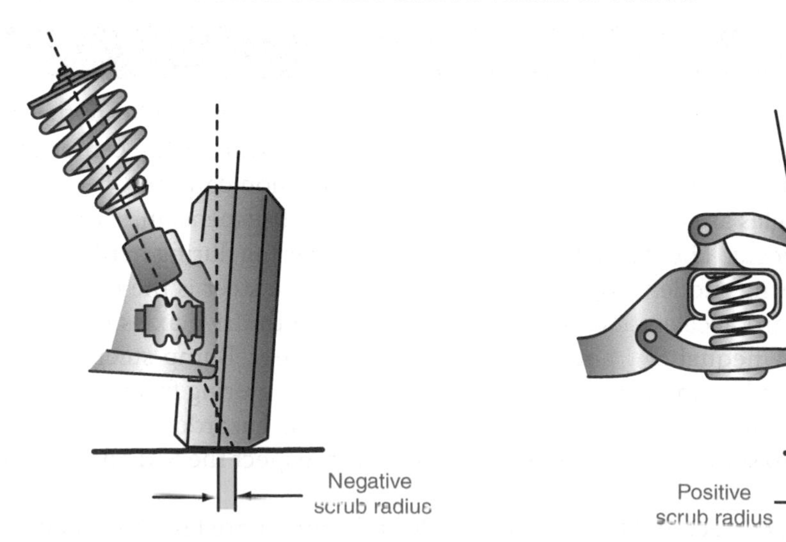

Figure 10-5 Scrub radius.

If a vehicle has positive scrub radius and the right front brake grabs, both the positive scrub radius and the grabbing brake tend to turn the right front wheel outward, and the vehicle pivots around the right front wheel. This action induces a swerve to the right.

When a right front brake grab occurs with negative scrub radius, the brake grab causes the right front wheel to turn outward, and the vehicle tends to pivot around the right front wheel. However, the negative scrub radius tends to turn the right front wheel inward. The two forces on the right front wheel cancel each other to maintain directional stability.

Scrub Radius and Safety

If front tires that are larger in diameter than specified by the car manufacturer are installed, directional control may be affected. Large tires raise the chassis farther from the road surface, which changes the scrub radius. The installation of larger front tires could change a positive scrub radius to a negative scrub radius.

Reversing the front rims so they are inside-out creates a significant scrub radius change and adversely affects directional control. This practice is not recommended by car manufacturers.

WARNING Installing larger tires, or different rims than the ones specified by the vehicle manufacturer, changes the scrub radius, which may result in reduced directional control, collision damage, and personal injury.

WHEEL SETBACK

Wheel setback is a condition in which one wheel is moved rearward in relation to the other wheel (**Figure 10-6**). Setback will not affect handling unless it is extreme. Collision damage may drive one front strut rearward and cause extreme setback. Setback can also occur on rear wheels, but it is more likely to occur on front wheels because of collision damage. Some computer four-wheel aligners have setback measuring capabilities.

TURNING RADIUS

Front and Rear Wheel Turning Action

When a vehicle turns a corner, the front and rear wheels must turn around a common center with respect to the **turning radius** (**Figure 10-7**). On most front suspension systems, the front wheels pivot independently at different distances from the center of the

Turning radius may be referred to as cornering angle or toe-out on turns.

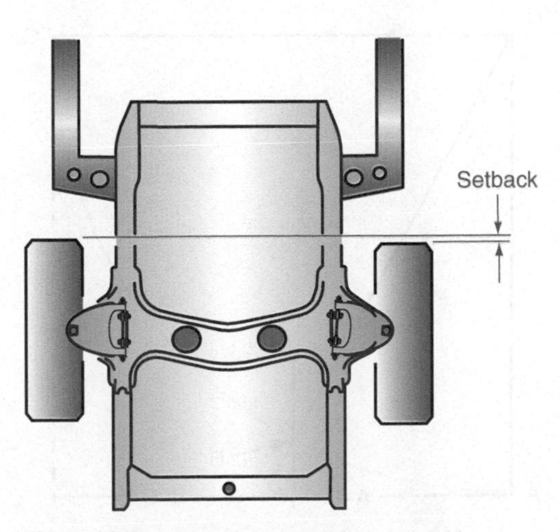

Figure 10-6 Wheel bearing and hub assembly.

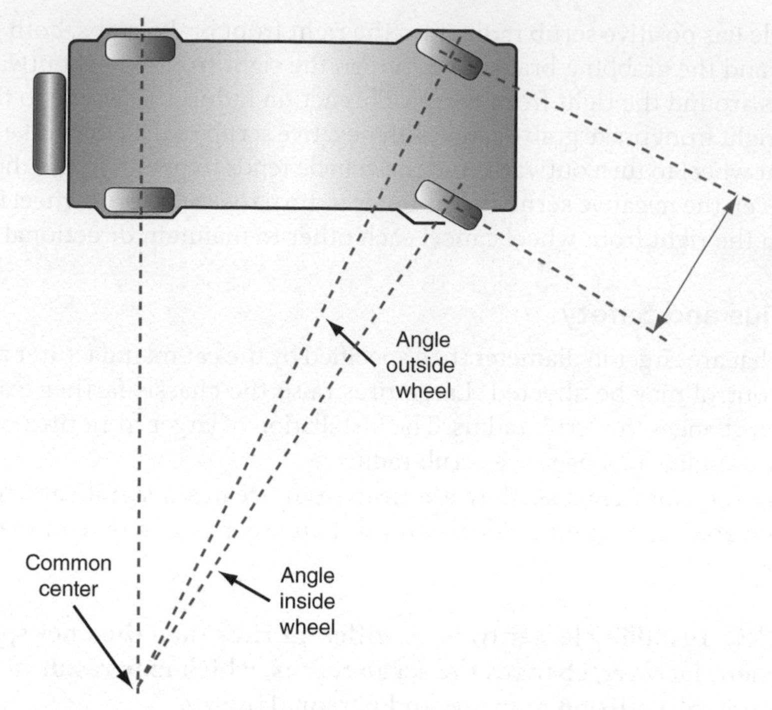

Figure 10-7 Front and rear wheels turning around a common center.

turn, and therefore the front wheels must turn at different angles. The inside front wheel must turn at a sharper angle than the outside wheel. This is because the inside wheel is actually ahead of the outside wheel. When this turning action occurs, both front wheels remain perpendicular to their turning radius, which prevents tire scuffing.

The turning angles of the front wheels are determined by steering arm design, and these angles are not adjustable. If the steering angles are not correct, the steering arms may be bent.

Steering Arm Design

An understanding of a lever moving in a circle is necessary before an explanation of steering arm design and operation. If a lever moves from point A to B, it pivots around point O and moves through a horizontal distance A to B (**Figure 10-8**).

Lever movement through arc B to C is much greater than the movement through arc A to B. However, during arc B to C, the lever moves through horizontal distance B to C and this distance is the same as horizontal distance A to B.

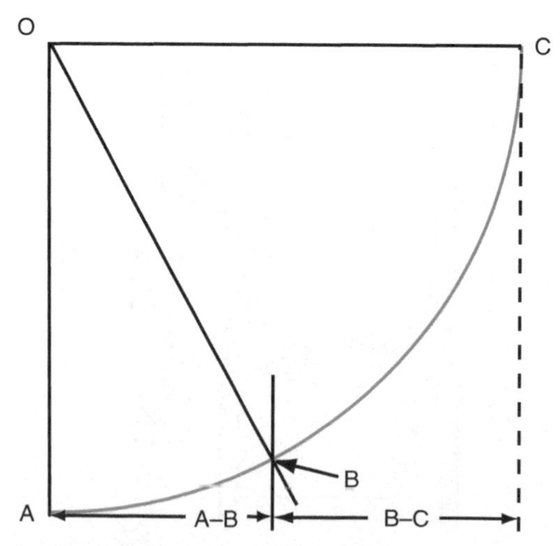

Figure 10-8 Lever movement in a circle.

The steering arms are connected from the tie rod ends to the steering knuckles. Steering arms and linkages maintain the front wheels parallel to each other when the vehicle is driven straight ahead. However, the steering arms are not parallel to each other. If the steering linkage is at the rear edge of the front wheels, the steering arms are closer together at the point where the tie rods connect than at their spindle pivot point (**Figure 10-9**). When the steering linkage is positioned at the front edge of the front wheels, the steering arms are closer together at their spindle pivot point than at the tie rod connecting point.

When the front wheels are turned on a vehicle with the steering linkage at the rear edge of the front wheels, the angle formed by the inside steering arm and linkage increases, whereas the angle of the outside steering arm and linkage decreases. The inside steering arm moves through the longer arc X, and the outside steering arm moves through shorter arc Z (**Figure 10-10**).

Therefore, the inside wheel turns at a sharper angle than the outside wheel. Because both steering arms are designed to have the same angle in the straight-ahead position, the inside front wheel always has a sharper angle regardless of the turning direction. The sharper inside wheel angle during a turn causes the inside wheel to toe out. Therefore, the term **toe-out on turns** is used for this steering action. If the front wheel turning angle is increased during a turn, the amount of toe-out on the inside wheel increases proportionally.

Toe-out on turns is the angle of the inside wheel in relation to the angle of the outside wheel during a turn.

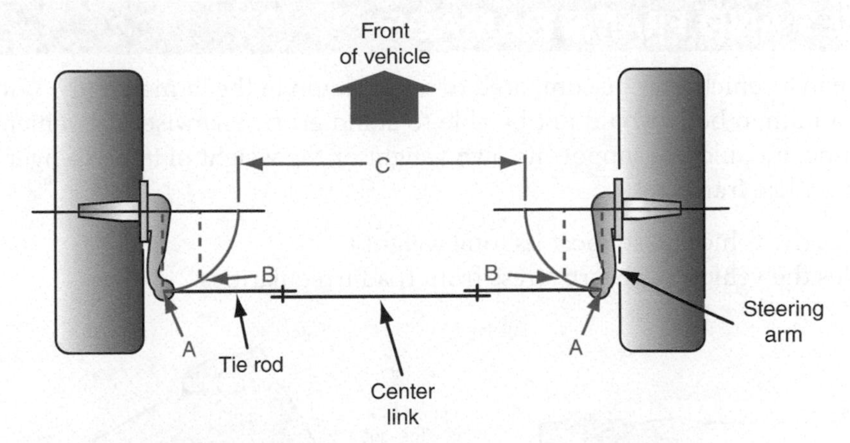

Figure 10-9 Steering arms are closer together at the point where the tie rods connect than at their spindle pivot points.

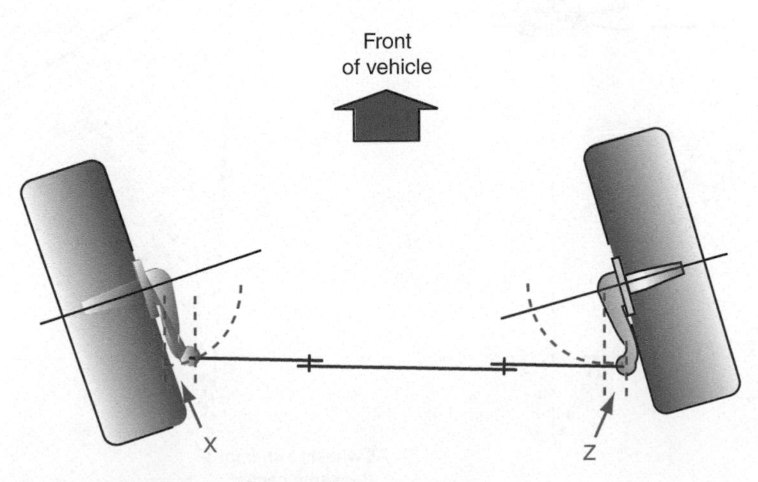

Figure 10-10 Steering arm operation while turning.

A BIT OF HISTORY

During a recent October Car Care Month, statistics were compiled from 65 inspection lanes in the United States. During these inspections, 29 percent of the cars failed the tire test. These figures indicate that almost one-third of the vehicles on the road have tire problems, and in many cases these problems create a safety hazard. These figures also tell us there is a tremendous business potential in tire sales, wheel balancing, and wheel alignment. Each time we perform even minor vehicle service, the tires should be inspected. This inspection may not only help us increase tire sales and the volume of wheel balancing and alignment service work, but in some cases, you may also save the customer's life!

Slip Angle

During a turn, centrifugal force causes all the tires to slip a certain amount. The amount of tire slip increases with speed and the sharpness of the turn. This tire slip action causes the actual center of the turn to be considerably ahead of the theoretical turn center (**Figure 10-11**).

The **slip angle** on different vehicles varies depending on such factors as vehicle weight and type of suspension.

Slip angle is the actual angle of the front wheels during a turn compared to the turning angle of the front wheels with the vehicle at rest.

FRAMES AND FRAME DAMAGE

The frame in a vehicle may be compared to the skeleton in the human body. Without the skeleton, a human body would not be able to stand erect. Likewise, if a vehicle did not have a frame, it could not support its own weight or the weight of its passenger or cargo load. The vehicle frame:

1. Enables the vehicle to support its total weight.
2. Enables the vehicle to absorb stress from road irregularities.

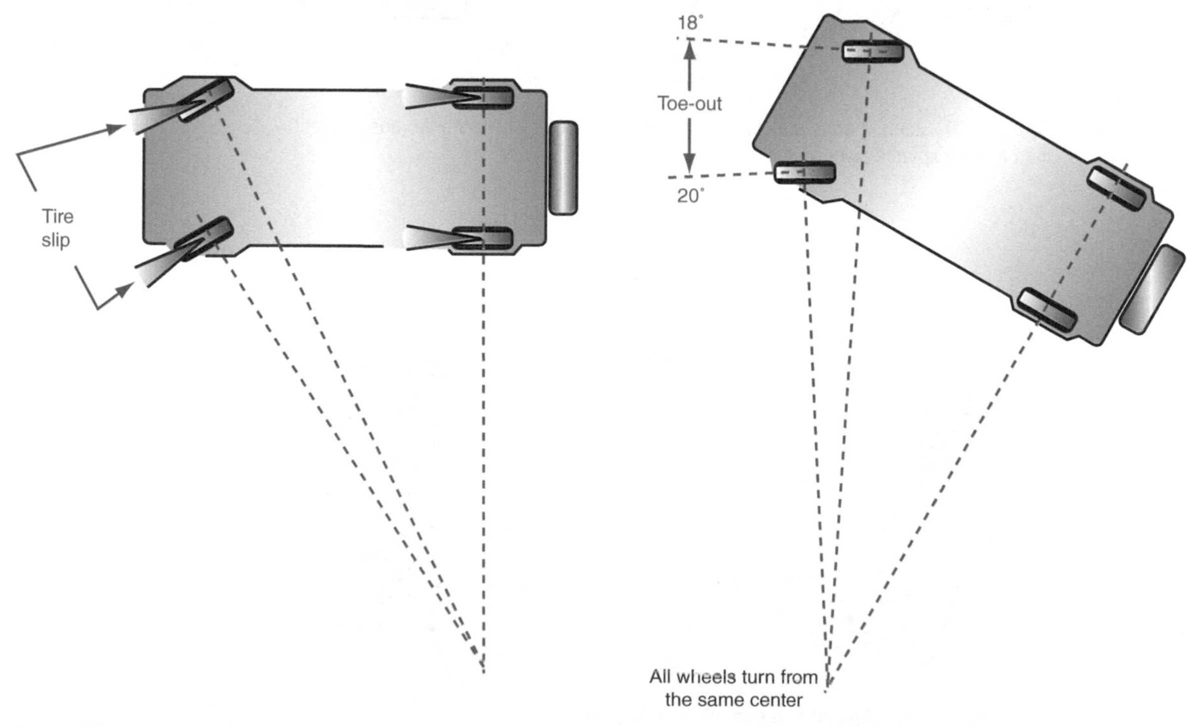

Figure 10-11 Slip angle during a turn.

3. Enables the vehicle to absorb torque from the engine and drive wheels.
4. Provides a main member for attachment of body and other components.

The frame, together with the front and rear suspension systems, must position the wheels properly to minimize tire tread wear and provide accurate steering control.

TYPES OF FRAMES AND FRAME CONSTRUCTION

Frame Construction

Vehicle frame construction may contain three types of steel members: **channel** (partial box) frame, **complete box frame**, or **tubular frame** (**Figure 10-12**). On modern vehicles, most frames include all three types.

Compared to a channel frame or a tubular frame, the complete box frame has increased torsional rigidity and improved crash performance. Therefore, in many frames the front section is a complete box design and the other part of the frame is a channel design. Some frames have X-shaped bracing at the rear of the frame for increased strength. Many vehicles now have frames or partial frames that are manufactured by a **hydro-forming** process. Hydro-formed frames are considerably more rigid than frames manufactured by heat-treating. Therefore, hydro-formed frames reduce the flexing of the frame and body components, which reduces squeaks and rattles to decrease noise, vibration, and harshness (NVH). Hydro-formed frames also provide more stable wheel position, which improves steering quality and reduces tire tread wear.

In a **channel frame**, each side of the frame is made from a U-shaped steel channel.

In a **complete box frame**, each side of the frame forms a metal box.

A **tubular frame** member is formed in an oval shape or circle.

Hydro-forming is the process of using extreme fluid pressure to shape metal.

Ladder-Type Frames

The **ladder frame design** is used on trucks and full-size vans. In this type of frame, the side rails have very little offset and are in a straight line between the front and rear wheels. The ladder-type frame has more crossmembers for increased load-carrying capacity and rigidity (**Figure 10-13**).

In a **ladder frame design**, crossmembers are mounted between the sides of the frame, and the crossmembers are similar to rungs on a ladder.

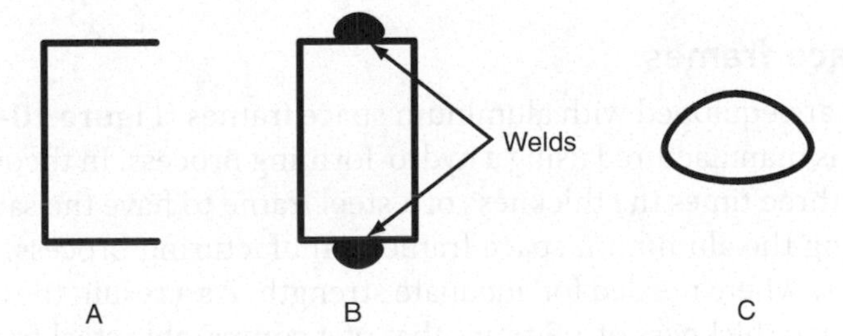

Figure 10-12 Types of frame construction: (A) channel or partial box, (B) complete box, (C) tubular crossmember.

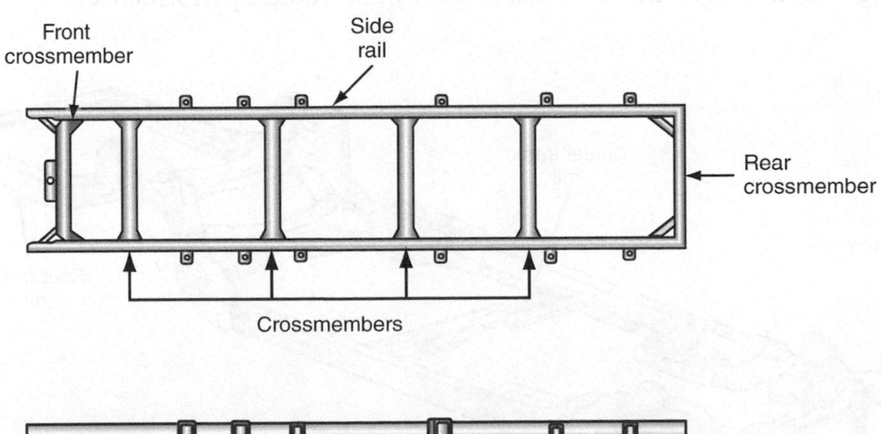

Figure 10-13 Ladder-type frame.

Perimeter-Type Frames

The **perimeter-type frame** is used in some rear-wheel-drive cars. This type of frame forms a border around the passenger compartment. The frame rails are stepped inward at the cowl area to provide increased strength, which supports the engine mounts and front suspension. This inward step at the cowl area also provides room for movement of the front wheels. Lateral support is provided by crossmembers welded between the frame rails near the front and rear of the frame. A transmission support member is welded or bolted between the frame rails at the back of the transmission. The rear frame kickup side rails support the rear suspension and the rear portion of the body weight. A front torque box is positioned just ahead of the transmission support member, and a rear torque box is located in front of the rear frame kickup side rails (**Figure 10-14**). These torque boxes are designed to absorb most of the impact during a side collision and thus reduce damage to other body components. The torque boxes also provide some protection for the vehicle occupants during a side collision.

Shop Manual
Chapter 10, page 440

The body components are bolted to the frame, but rubber insulating bushings are positioned between the body and frame mounting locations. These bushings help prevent the transfer of road noise and vibration from the suspension and frame to the body and vehicle interior. Rubber insulating mounts are positioned between the engine and transmission and the frame mounting positions. These engine and transmission mounts reduce the transfer of engine vibration to the frame, body, and vehicle interior. Many suspension components are also connected to the frame through rubber insulating bushings to reduce the transfer of harshness and vibration from the suspension to the frame while driving over road irregularities.

During the manufacturing process, most manufacturers apply a special coating to the frame to help prevent rust and corrosion. For example, many light-duty truck frames are coated with epoxy or wax. An epoxy-coated frame is black and very smooth, whereas a wax-coated frame is gray and sticky. If these coatings are scratched or damaged, frame rusting and corrosion may occur.

Aluminum Space frames

Some sports cars are equipped with aluminum space frames (**Figure 10-15**). The aluminum space frame is manufactured using a hydro-forming process. In theory, an aluminum frame must have three times the thickness of a steel frame to have the same strength as a steel frame. During the aluminum space frame manufacturing process, a manufacturer adds thickness only where needed for adequate strength. As a result, the aluminum space frame has an average thickness of 1.9 times that of a comparable steel frame. This aluminum frame weighs 285 lb (129 kg), which is 136 lb (62 kg) lighter than a steel frame. Reducing vehicle weight improves fuel economy and vehicle performance.

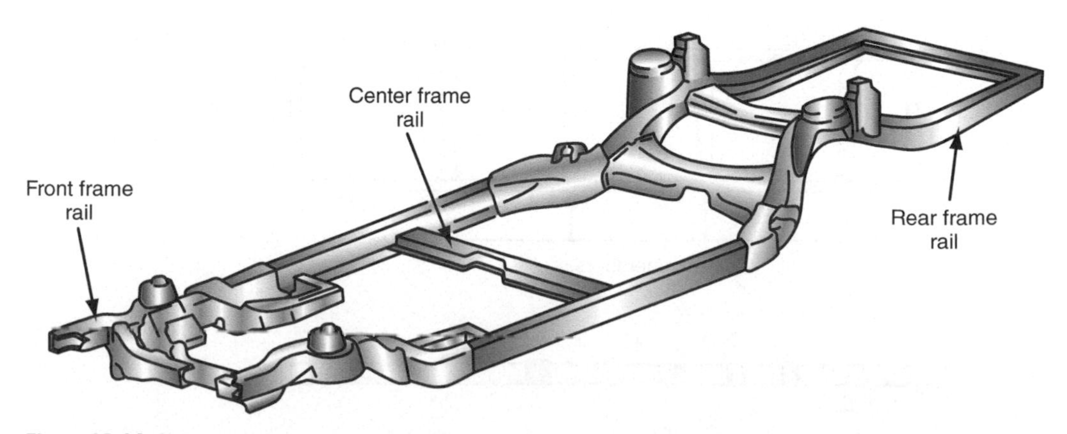

Figure 10-14 Perimeter-type frame components.

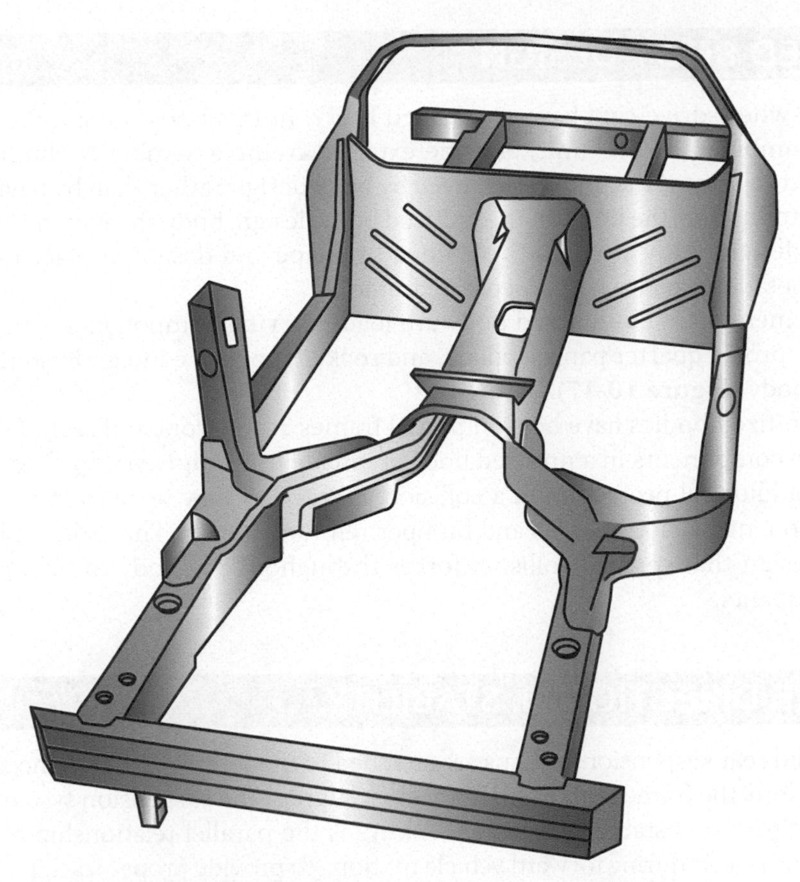

Figure 10-15 Aluminum space frame.

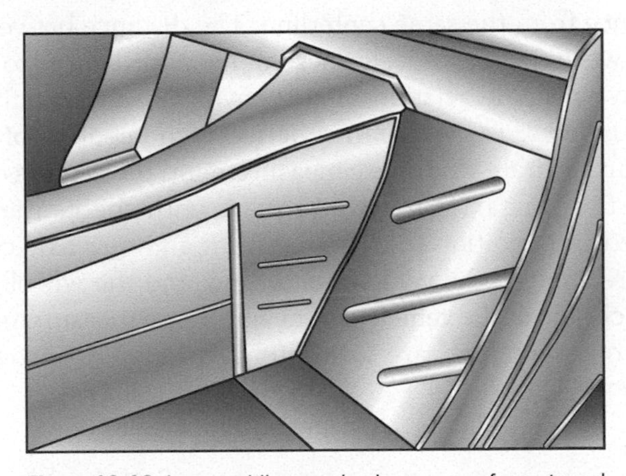

Figure 10-16 Laser welding on aluminum space frame tunnel.

Laser welding is used extensively on the tunnel in the aluminum space frame rather than spot welding (**Figure 10-16**). Laser welding adds stiffness and provides an improved seal compared to spot welding. Laser welding uses less heat per length of weld, which results in reduced aluminum distortion.

A BIT OF HISTORY

During the late 1940s, American Motors Corporation introduced the unitized body design in the United States. Some of these first-generation unitized bodies did not have partial frames. However, unitized bodies did not become popular until 1980 and 1981 with the introduction of General Motors X cars, Chrysler K cars, and other front-wheel-drive cars. These front-wheel-drive cars have partial frames and may be referred to as second-generation unitized bodies.

UNITIZED BODY DESIGN

Most front-wheel-drive cars have a **unitized body**. In this body design, the frame and body are combined as one unit, and the external frame assembly is eliminated. The strength and rigidity of the body is achieved by body design rather than by having a heavy steel frame to support the body. In the unitized body design, body sheet metal is fabricated into a box design. Body strength is obtained by shape and design in place of mass and weight of metal in vehicles with a separate frame.

All the members of a unitized body are load-carrying components. The floor pan, roof, inner aprons, quarter panels, pillars, and rocker panels are integrally joined to form a unitized body (**Figure 10-17**).

Most unitized bodies have bolt-on partial frames at the front and rear of the vehicle. Some of the components in a unitized body are made from **high-strength steels (HSS)** to provide additional protection in a collision. Ultra-HSS may be used in such unitized body components as door beams and bumper reinforcements. The unitized body has a complex design that spreads collision forces throughout the body to help protect the vehicle occupants.

VEHICLE DIRECTIONAL STABILITY

The front and rear suspension systems are attached to the frame, partial frame, or unitized body. Therefore, the frame or unitized body must support the suspension systems properly to provide directional stability. Vehicle **tracking** is the parallel relationship between the front and rear wheels during forward vehicle motion. To provide proper tracking, each front wheel must be at the same distance from the vehicle centerline, and each rear wheel must be at an equal distance from the same centerline. The distance between the front wheels and the distance between the rear wheels does not necessarily have to be the same, but all four wheels must have a parallel relationship to provide proper tracking (**Figure 10-18**).

Shop Manual
Chapter 10, page 435

The vehicle **wheelbase** is the distance between the centers of the front and rear wheels. To provide an accurate wheelbase measurement, the centers of the front spindles and the centers of the rear axles must be square with the centerline of the vehicle (**Figure 10-19**). For this condition to exist, the front and rear axle centers must be at a 90-degree angle in relation to the vehicle centerline, and the wheelbase measurements must be equal on each side of the vehicle. Equal wheelbase measurements on each side of the vehicle and proper tracking are absolutely essential to providing **directional stability**.

Directional stability refers to the tendency of a vehicle to remain in the straight-ahead position when driven straight ahead on a reasonably smooth, straight road.

The term dog tracking may be applied to axle offset or sideset, because in either of these conditions the rear wheels are not parallel to the front wheels. Similarly, many dogs run down the street with their rear ends out of line with their front ends.

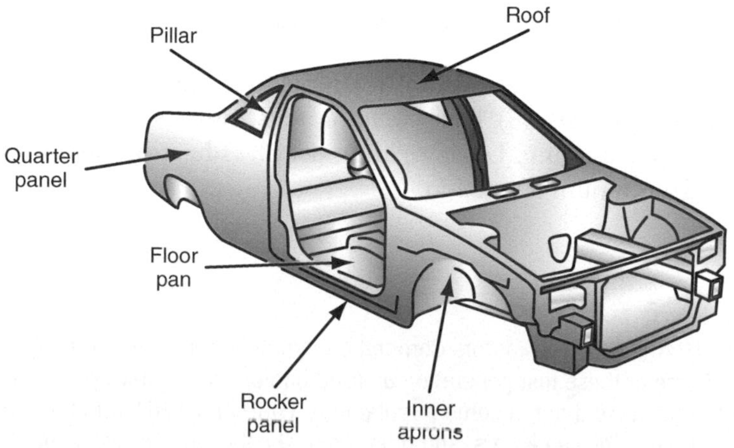

Pillar
Roof
Quarter panel
Floor pan
Rocker panel
Inner aprons

Figure 10-17 Unitized body design.

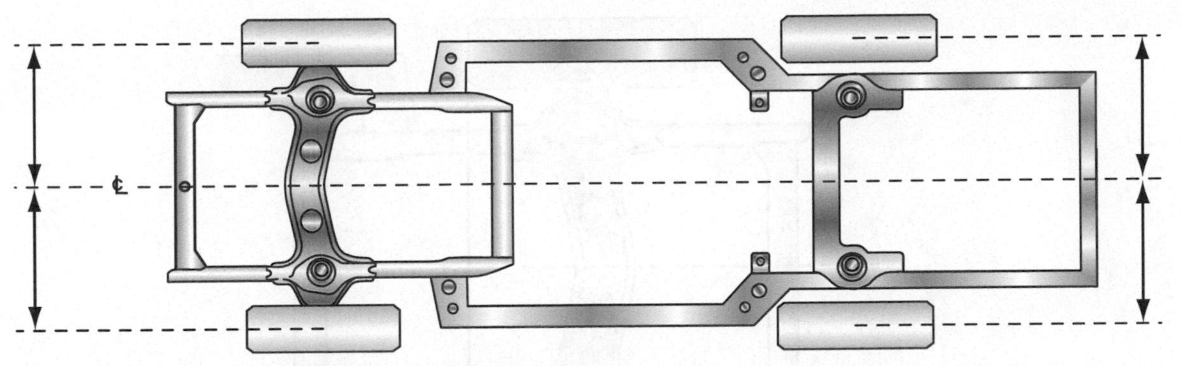

Figure 10-18 Front and rear wheels must be at the same distance from the vehicle centerline to provide proper tracking.

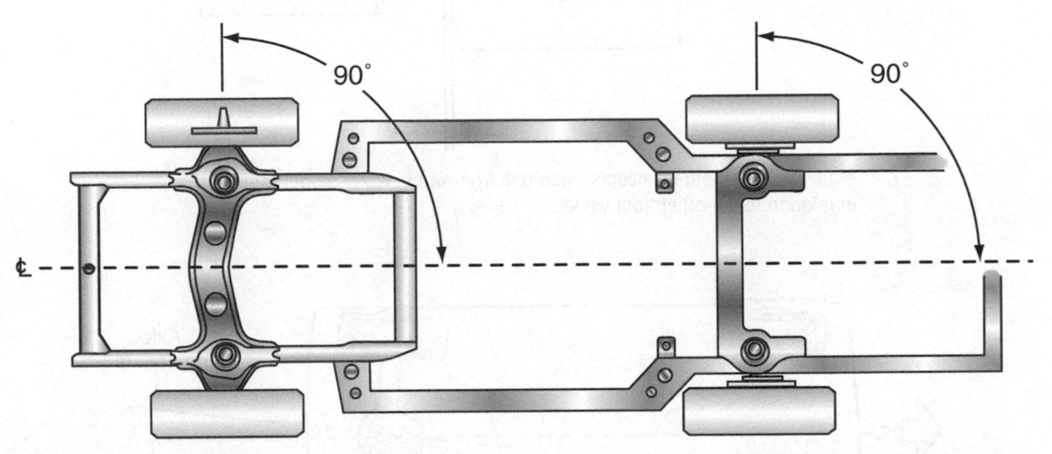

Figure 10-19 To provide accurate wheelbase measurement, the front spindle centerlines and the rear axle centers must be at a 90-degree angle to the vehicle centerline.

VEHICLE TRACKING

Wheel Setback

Front wheel setback occurs when the spindle on one front wheel is rearward in relation to the other front wheel. The centerline of each front wheel is still at a 90-degree angle to the vehicle centerline, but the front wheels no longer share the same centerline (**Figure 10-20**).

When left front wheel setback occurs, the vehicle has a tendency to steer to the left as it is driven straight ahead. Under this condition, the driver has to continually turn the steering wheel to the right to keep the vehicle moving straight ahead.

Axle Offset

If the rear axle is rotated, the **axle thrust line** is no longer at a 90-degree angle to the **geometric vehicle centerline** (**Figure 10-21**). This condition is referred to as **axle offset**. When the left side of the rear axle is rotated rearward, the steering pulls continually to the right. Under this condition, the driver has to turn the steering wheel to the left to keep the vehicle moving straight ahead.

The **axle thrust line** is a line extending forward from the center of the rear axle at a 90-degree angle.

The **geometric vehicle centerline** refers to the front-to-rear centerline of the vehicle body. The centerlines of the front and rear axles should be positioned on this geometric centerline.

Axle Sideset

When **axle sideset** occurs, the rear axle moves inward or outward, but the axle and vehicle centerlines remain at a 90-degree angle in relation to each other (**Figure 10-22**).

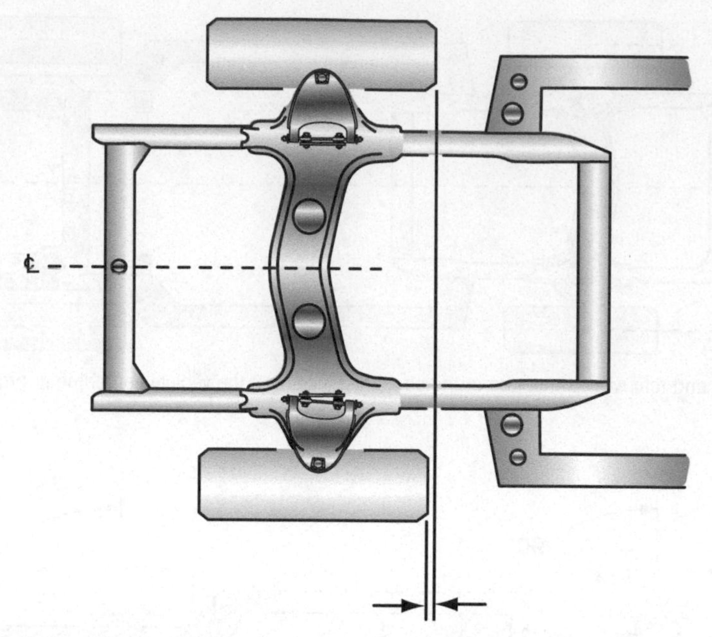

Figure 10-20 Setback occurs when one front wheel is positioned rearward in relation to the other front wheel.

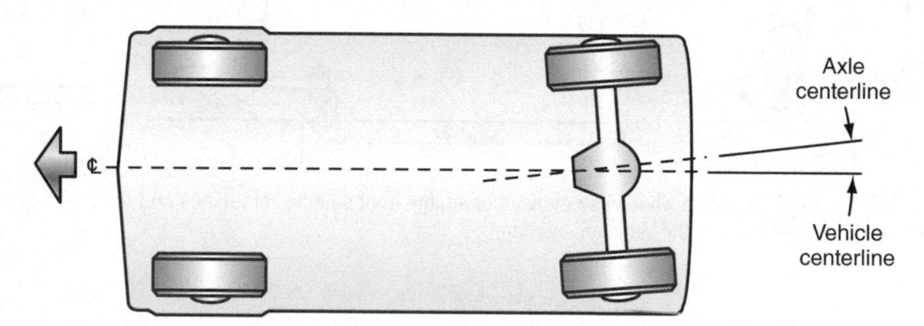

Figure 10-21 Rear axle offset occurs when the rear axle is rotated so the axle centerline and the vehicle centerline are no longer at a 90-degree angle.

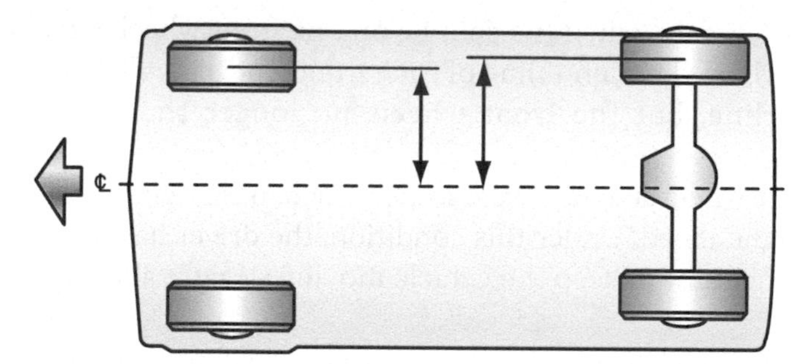

Figure 10-22 Axle sideset occurs when the rear axle moves inward or outward and the axle centerline remains at a 90-degree angle in relation to the vehicle centerline.

Under this condition, the front-to-rear axle thrust line is no longer at the geometric vehicle centerline. This condition also causes steering pull. The vehicle frame or unitized body and the front and rear suspension systems must have proper tracking and equal wheelbase measurements on each side of the vehicle to provide directional stability and steering control.

TYPES OF FRAME DAMAGE

Side Sway

Side sway on the front or rear of a vehicle is usually the result of the vehicle being involved in a collision that pushes the front or rear frame sideways (**Figure 10-23**). Under this condition, the wheelbase on one side of the vehicle is longer than the opposite side. This side sway condition causes the steering to pull to the side with the shorter wheelbase.

Side sway is a lateral misalignment that affects the frame or body centerline.

AUTHOR'S NOTE It has been my experience that frame damage is most commonly caused by abuse, and this problem is usually encountered on light-duty trucks or sport utility vehicles (SUVs). The frame damage may occur when the vehicle is overloaded and/or driven roughly on extremely uneven terrain. Another common cause of frame damage is from a vehicle collision. In this case, the frame damage was likely ignored or overlooked during the body repairs. Regardless of the cause, frame damage usually results in excessive tire tread wear and steering complaints.

Caution

When attaching any type of additional equipment to a vehicle frame, always try to use the holes that already exist in the frame. Do not drill holes in a vehicle frame except at locations specified by the vehicle manufacturer, and do not attach additional equipment to a vehicle frame unless this procedure is approved by the vehicle manufacturer.

Caution

Do not apply heat to a vehicle frame with an acetylene torch. When the metal is heated to a cherry red, the frame may be weakened.

Overall side sway occurs when the vehicle is hit directly on the side near the center in a collision. A vehicle frame is slightly V-shaped when it has overall side sway damage (**Figure 10-24**).

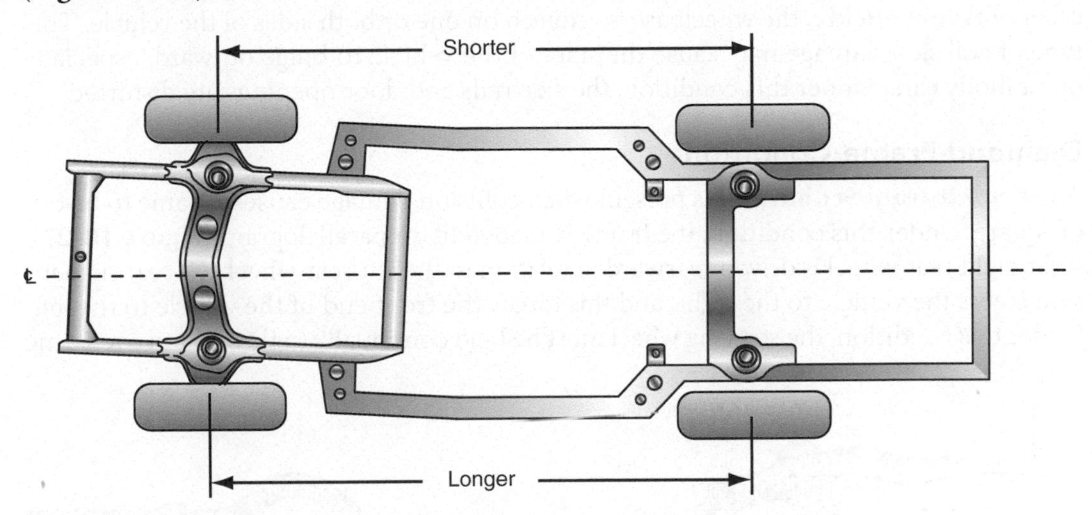

Figure 10-23 Side sway frame condition caused by collision damage.

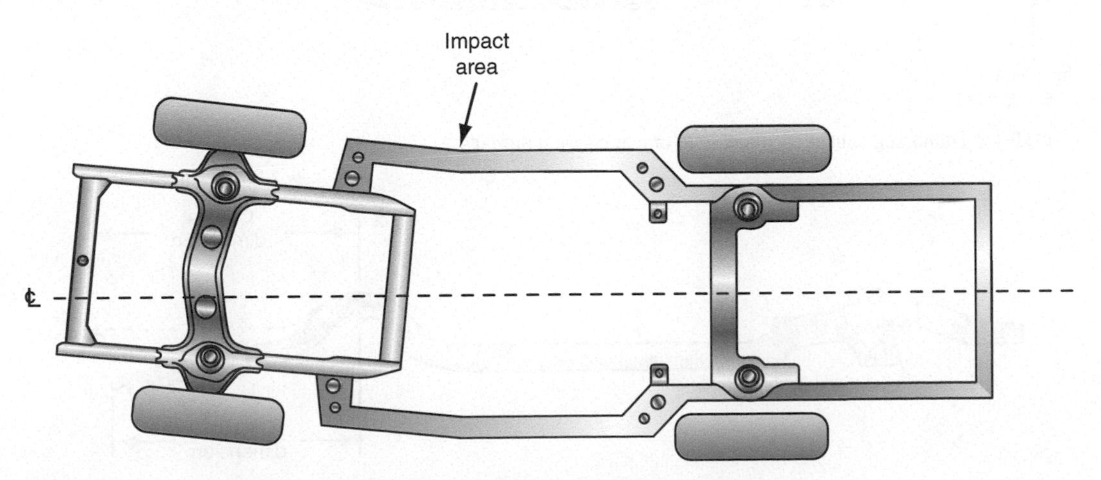

Figure 10-24 When a frame has overall side sway damage, it is slightly V-shaped.

WARNING Some types of frame damage cause steering pull when driving straight ahead, and this steering pull is increased during hard braking. Therefore, frame damage can create a safety hazard that leads to a collision involving personal injury and vehicle damage.

Sag

Frame sag usually occurs when the vehicle is involved in a direct front or rear collision. When this condition is present, the front and/or rear frame rails are moved upward in relation to the center of the frame (**Figure 10-25**). If one side of the vehicle sustained more collision force than the opposite side, the left and right wheel base measurements will also likely be different.

The front crossmember may receive sag damage in a collision. When this member is sagged, the upper control arms move closer together on a short-and-long arm suspension. If a MacPherson strut front suspension is sagged, the strut towers are moved closer together. In either type of front suspension, a sagged condition moves the top of the wheels inward to a **negative camber** position.

Negative camber occurs when the camber line through the center of the tire and wheel is tilted inward compared with the true vertical centerline of the tire and wheel.

Frame buckle is accordion-shaped damage on the front or rear of the frame, which causes the distance to be reduced between the cowl and front bumper or between the rear wheels and rear bumper. Frame buckle may be called frame crush or mash.

Frame Buckle

A buckle condition exists when the distance from the cowl to the front bumper is less than specified, or the measurement from the rear wheels to the rear bumper is less than specified (**Figure 10-26**). Frame buckle is caused by a direct front or rear collision. In many cases of **frame buckle**, the wheelbase is reduced on one or both sides of the vehicle. This type of collision damage may cause the sides of the vehicle to bulge outward, especially on unibody cars. Under this condition, the side rails and door openings are distorted.

Diamond-Frame Condition

A **diamond-frame condition** is present when collision damage causes a frame to be out of square. Under this condition, the frame is shaped like a parallelogram (**Figure 10-27**). If the right rear wheel is driven rearward in relation to the left rear wheel, the rear suspension steers the vehicle to the right, and this forces the front end of the vehicle to the left. Under this condition, the steering wheel must be held continually to the right to overcome

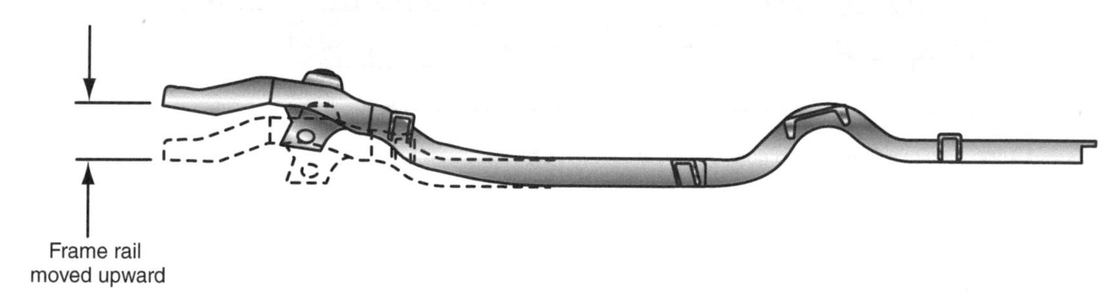

Figure 10-25 Frame sag caused by direct front or rear collision damage.

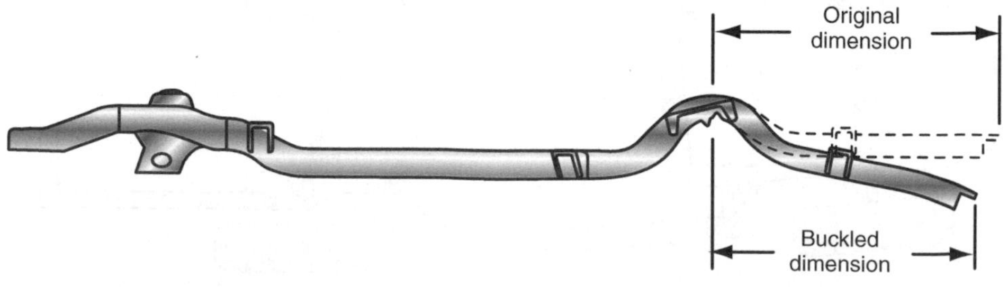

Figure 10-26 Rear frame buckle.

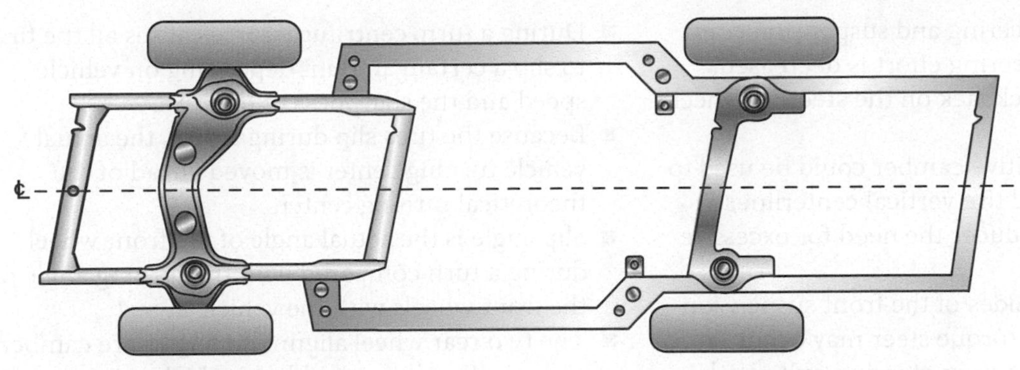

Figure 10-27 Diamond-frame condition.

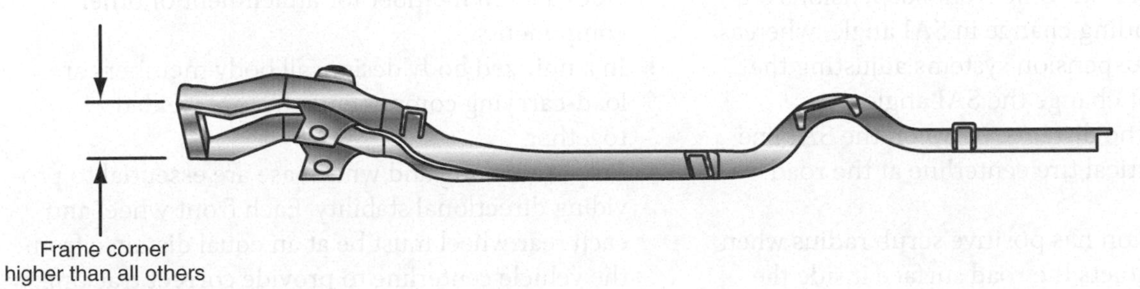

Figure 10-28 Frame twist.

the steering pull to the left. A diamond-frame condition usually occurs on vehicles with frames. Vehicles with unitized bodies seldom have this type of condition.

Frame Twist

A **frame twist** condition exists when one corner of the frame is higher than the other corners. When frame twist is present, the front or rear chassis does not sit level in relation to the road surface (**Figure 10-28**). Frame twist is usually caused by vehicle rollover.

SUMMARY

- Steering axis inclination (SAI) is the angle of a line through the center of the upper strut mount and lower ball joint in relation to the true vertical centerline of the tire viewed from the front of the vehicle.
- On short-and-long arm front suspension systems, the SAI line runs through the center of the upper and lower ball joints.
- The SAI line is always tilted toward the center of the vehicle. The included angle is the sum of the SAI and positive camber angle. If the camber is negative, the camber must be subtracted from the SAI angle to obtain the included angle.
- SAI causes the front spindles to move through an arc when the front wheels are steered to the right or left.
- Because the front spindles move through an arc, the chassis lifts as the front wheels are turned. This lifting action helps return the front wheels to the straight-ahead position after a turn.
- SAI also helps maintain the front wheels in the straight-ahead position. SAI reduces the need for excessive positive caster. SAI is not adjustable. If the SAI does not equal the manufacturer's specifications and the other steering angles are correct, some suspension component, such as a strut tower, is out of place.
- A suspension system with a 0-degree SAI line would have increased tire wear, greater steering effort, increased stress on suspension and steering components, and excessive road shock and kickback on the steering wheel.
- When the SAI line intersects the true vertical tire centerline at or near the road surface, tire life is

improved, stress on steering and suspension components is reduced, steering effort is decreased, and road shock and kickback on the steering wheel are minimized.

- Because excessive positive camber could be used to bring the SAI lines and tire vertical centerlines closer together, SAI reduces the need for excessive positive camber.
- If SAI angles on both sides of the front suspension are unequal, excessive torque steer may occur on hard acceleration, and severe steering pull may be present during hard braking.
- Adjusting camber on some front suspensions creates a corresponding change in SAI angle, whereas on other front suspension systems adjusting the camber does not change the SAI angle.
- Scrub radius is the distance between the SAI line and the true vertical tire centerline at the road surface.
- A front suspension has positive scrub radius when the SAI line contacts the road surface inside the true vertical tire centerline.
- A front suspension has negative scrub radius when the SAI line contacts the road surface outside the true vertical tire centerline.
- If the front tires are larger than the original tires specified by the vehicle manufacturer, a change occurs in scrub radius that may affect steering control.
- The installation of larger-than-specified front tires may change positive scrub radius to a negative scrub radius.
- Reversing the front rims so they are inside out results in a significant scrub radius change that adversely affects directional control.
- Wheel setback is a condition where one wheel is moved rearward in relation to the opposite wheel.
- Toe-out on turns is the turning angle of the wheel on the inside of the turn compared with the turning angle of the wheel on the outside of the turn.
- When the front wheel on the inside of a turn has turned 20 degrees outward, the front wheel on the outside of the turn may have turned 18 degrees.
- Turning radius is the amount of toe-out on turns. Toe-out on turns prevents tire scuffing. This angle is determined by the steering arm design.
- During a turn, centrifugal force causes all the tires to slip a certain amount depending on vehicle speed and the sharpness of the turn.
- Because the tires slip during a turn, the actual vehicle turning center is moved ahead of the theoretical turning center.
- Slip angle is the actual angle of the front wheels during a turn compared with the turning angle of the front wheels with the vehicle at rest.
- The two rear wheel alignment angles are camber and toe. The frame enables a vehicle to support its weight and absorb stress and torque. It also provides a main member for attachment of other components.
- In a unitized body design, all body members are load-carrying components that are welded together.
- Proper tracking and wheelbase are essential to providing directional stability. Each front wheel and each rear wheel must be at an equal distance from the vehicle centerline to provide correct tracking.
- To provide proper wheelbase, the centers of the front and rear suspensions must be at a 90-degree angle to the vehicle centerline.
- Front wheel setback, rear axle offset, and rear axle sideset cause the steering to pull to one side.
- Regardless of the type of front or rear suspension system, the suspension system and the frame must position the wheels properly to provide correct tracking and wheelbase.
- Frame side sway occurs when the front suspension is forced sideways in a collision, and one front wheel is forced rearward in relation to the opposite front wheel. Side sway may also occur on the rear suspension.
- Frame sag occurs when the front or rear frame rails are bent upward in relation to the center of the frame.
- Frame buckle is accordion-shaped damage on the front or rear of the frame that causes the distance to be reduced between the cowl and front bumper or between the rear wheels and rear bumper.
- A diamond-frame condition is present when one side of the frame is driven rearward in relation to the opposite side of the frame, and the front and rear wheels on one side of vehicle are rearward in relation to the wheels on the other side.
- Frame twist occurs when one corner of the frame is bent up higher than the other frame corners.

REVIEW QUESTIONS

Short Answer Essays

1. Define steering axis inclination (SAI).
2. Explain the included angle.
3. Explain how SAI helps to return the steering wheel to the center position after a turn.
4. Define negative scrub radius and positive scrub radius, including the type of suspension system on which each condition is used.
5. Describe front wheel setback.
6. Explain the necessary wheel position to provide proper tracking.
7. Describe the necessary axle position to provide correct wheelbase.
8. Explain the turning angle of each front wheel during a turn.
9. Explain the effects of front wheel setback.
10. Describe the effects of rear axle offset.

Fill-in-the-Blanks

1. A 3-degree difference in the SAI angle on each side of the front suspension may cause _______________ during hard braking.
2. A 3-degree difference in the SAI angle on each side of the front suspension may cause increased _______________ during hard acceleration.
3. When a front suspension has a positive scrub radius, the SAI line meets the road surface _______________ the true vertical centerline of the tire.
4. A negative scrub radius tends to turn the front wheels _______________when the car is driven.
5. Most rear-wheel-drive cars have a _______________ scrub radius.
6. Wheelbase is the distance between the front and rear wheel _______________.
7. Directional stability is the tendency of a vehicle to remain in the _______________ position when driven straight ahead on a level road.
8. During a turn, the _______________ front wheel turns at a sharper angle.
9. Tracking refers to the _______________ relationship between the front and rear wheels.
10. During a turn, centrifugal force causes all the tires to slip a certain amount, and the actual center of the turn is shifted _______________ of the theoretical center of the turn.

Multiple Choice

1. When measuring front wheel alignment angles, to calculate the included angle on the left front wheel when the camber on this wheel is positive:
   A. Add the camber to the toe setting.
   B. Add the camber to the SAI angle.
   C. Add the SAI to the caster angle.
   D. Subtract the SAI from the toe setting.
2. While diagnosing problems related to scrub radius:
   A. If the SAI line contacts the road surface inside the vertical tire and wheel centerline, the scrub radius is negative.
   B. Front-wheel-drive cars usually have a negative scrub radius and driving forces move the front wheels outward.
   C. If the SAI line contacts the road surface outside the tire and wheel vertical centerline, driving forces turn the front wheels outward.
   D. Larger-than-specified front tires may change the scrub radius from positive to negative.
3. While diagnosing a diamond-frame condition and frame twist:
   A. A diamond-frame condition causes the wheelbase to be unequal on the two sides of the vehicle.
   B. A diamond-frame condition does not affect steering pull and directional stability.
   C. Frame twist is usually caused when a vehicle is involved in a side collision.
   D. When frame twist occurs, the front or rear chassis does not sit level in relation to the road surface.
4. While diagnosing and adjusting turning radius:
   A. When a vehicle is making a left turn, the left front wheel turns at a sharper angle than the right front wheel.

B. Improper turning radius may be caused by a bent tie rod on a short-and-long arm suspension system.

C. Improper turning radius may be caused by an improperly positioned rack-and-pinion steering gear on a front-wheel-drive car.

D. Improper turning radius is adjusted by turning an eccentric strut-to-steering knuckle bolt.

5. All of these statements about unitized body design are true EXCEPT:

A. The frame and body are combined as one unit.

B. The external frame assembly is eliminated.

C. Some body members such as quarter panels do not contribute to body strength and rigidity.

D. Body strength is obtained by body shape and design.

6. A front-wheel-drive car has an improper toe-out on turns setting, and a visual check indicates all the steering linkage and suspension components are satisfactory. The most likely cause of this problem is:

A. A bent lower control arm.

B. A bent front strut.

C. A front strut tower that is out of position.

D. A bent steering arm.

7. All of these statements about SAI and front spindle movement are true EXCEPT:

A. When the steering wheel is turned, the front spindle movement is parallel to the road surface.

B. When the SAI angle is increased, the steering wheel returning force is increased.

C. The SAI angle tends to maintain the wheels in a straight-ahead position.

D. Greater SAI angle is necessary on front-wheel-drive cars to provide directional stability.

8. Front wheel setback occurs when one front wheel is:

A. Tilted inward from the true vertical position.

B. Moved rearward in relation to the opposite front wheel.

C. Tilted rearward from the true vertical position.

D. Inward from its original position.

9. A light-duty truck with a one-piece rear axle housing and a leaf-spring rear suspension has excessive toe-out on the left rear wheel and too much toe-in on the right rear wheel. The most likely cause of this problem is:

A. A broken center bolt in the left rear spring.

B. Both rear springs are sagged.

C. A bent rear axle housing.

D. Worn-out rubber bushings in the shock absorbers.

10. While discussing scrub radius:

A. Most front-wheel-drive cars have a positive scrub radius.

B. If the SAI line meets the road surface outside the tire vertical centerline, the scrub radius is positive.

C. Scrub radius is adjusted by shifting the upper strut tower on a MacPherson strut front suspension.

D. Incorrect scrub radius may be caused by larger-than-specified front tires.

CHAPTER 11
COMPUTER-CONTROLLED SUSPENSION SYSTEMS

Upon completion and review of this chapter, you should be able to understand and describe:

- The types of integrated computer networks used on vehicles.
- The operation of controller area network (CAN) system.
- The conditions that cause a programmed ride control (PRC) system to switch from the normal to the firm mode.
- How the firm ride condition is obtained in PRC struts.
- The major components in an electronic air suspension system.
- How air is forced into and exhausted from the air springs in an air suspension system.
- How an electronic air suspension system corrects low suspension trim height.
- The operation of an electronic air suspension system while driving the car with the doors closed and the brake pedal applied.
- The normal operation of the warning lamp in an electronic air suspension system.
- The three modes in the air suspension system on some modern four-wheel-drive sport utility vehicles (SUVs).
- How unnecessary rear suspension height corrections are prevented on irregular road surfaces with an air suspension system.
- The design of the struts and air springs in an automatic air suspension system.
- The design of an electronic rotary height sensor.
- Speed-leveling capabilities and the advantage of this function in a suspension control module.
- The operation of an automatic ride control (ARC) system in relation to transfer case modes.
- The inputs in an electronic suspension control (ESC) system.
- The advantages of an ESC system with magneto-rheological fluid in the shock absorbers compared to other computer-controlled suspension systems.
- The operation of the rear electronic level control system that is combined with the road sensing suspension system.
- The operation of the speed-sensitive steering system that is combined with the road sensing suspension system.
- The operation of a stability control system.
- The advantages of a traction control system.
- Various vehicle network systems.
- Active cruise control, lane departure warning, and collision-mitigation systems.
- The operation of active park assist system.

Terms To Know

Accelerometer
Actuator
Adaptive cruise control system
Air spring solenoid valve
Air springs
Antilock brake system (ABS)
Automatic level control (ALC)
Brake pressure modulator valve (BPMV)

- Brake pressure switch
- City safety system
- Controller Area Network (CAN)
- Cross-axis ball joints
- Damper solenoid valve
- Data bus network
- Electronic brake and traction control module (EBTCM)
- Electronic rotary height sensors
- Electronic suspension control (ESC) system
- Firm relay
- Hall element
- Height sensors
- Hold valve
- Lane departure warning (LDW) system
- Lateral accelerometer
- Lift/dive input
- Light-emitting diodes (LEDs)
- Lower vehicle command
- MagneRide system
- Network Photo diodes
- Programmed ride control (PRC) system
- Raise vehicle command
- Release valve
- Soft relay
- Speed-sensitive steering (SSS) system
- Stabilitrak®
- Steering sensor
- Steering wheel position sensor (SWPS)
- Throttle position sensor (TPS)
- Traction control system (TCS)
- Trim height
- Vehicle dynamic suspension (VDS)
- Vehicle speed sensor (VSS)
- Vehicle stability control system
- Vent valve
- Wheel position sensor
- Wheel speed sensors
- Yaw rate sensor

INTRODUCTION

We are all aware of the ever-accelerating electronics revolution that began in the early 1980s as computers were integrated into vehicle powertrain management. Most industries have felt the impact of this revolution, and the automotive industry is no exception. Computers have greatly influenced the way vehicles are designed and built. Most systems on the automobile, including the suspension system, have been impacted by the computer. Many drivers like a soft, comfortable ride while driving normally on the highway. However, many of these same drivers prefer a firm ride during hard cornering, severe braking, or fast acceleration. A firm ride under these driving conditions reduces body sway and front end dive or lift. Prior to the age of electronics, cars were designed to provide either a soft, comfortable ride or a firm ride. Drivers who wanted a firm ride selected a sports car with a suspension designed to supply the type of ride and handling characteristics they desired. Car buyers who wanted a softer ride purchased a family sedan with a suspension designed to provide a softer, more comfortable ride.

A BIT OF HISTORY

Suspension systems evolved slowly for many years. In the 1940s, many front suspensions were changed from I-beam to short-and-long arm. MacPherson strut suspensions replaced the short-and-long arm suspension systems on front-wheel-drive cars introduced in the late 1970s and 1980s. In the 2000s, we are experiencing an ever-expanding use of electronics technology in automotive suspension systems. Some computer-controlled suspension systems now have the capability to react to road or driving conditions in one millisecond to improve ride and handling quality. The use of electronics technology will continue to improve suspension systems so they provide better handling characteristics and safety with longer tire life.

Thanks to computer control, suspension system manufacturers can now provide a soft ride during normal highway driving, and then almost instantly switch to a firm ride during hard cornering, braking, fast acceleration, and high-speed driving. The computer-controlled suspension system allows the same car to meet the demands of both the driver who desires a soft ride, and the driver who wants a firm ride. Because computer-controlled suspension systems reduce body sway during hard cornering, these systems provide improved steering control.

Some computer-controlled suspension systems also supply a constant vehicle riding height regardless of the vehicle passenger or cargo load. This action maintains the vehicle's cosmetic appearance as the passenger and/or cargo load is changed. Maintaining a constant riding height also supplies more constant suspension alignment angles, which may provide improved steering control.

INTEGRATED ELECTRONIC SYSTEMS AND NETWORKS

Advantages of Integrated Electronic Systems and Networks

With the rapid advances in electronic technology, computer-controlled automotive systems have become integrated. Rather than having a separate computer for each electronic system, several of these systems may be controlled by one computer with control modules integrated. An example of this is the active park assist control module, which is integrated into the body control module (BCM). The active park assist system is an example of CAN data bus integration of various vehicle subsystems modules communicating with one another enabling a complex function. Vehicles without any integrated electronic systems may have many individual modules and computers. Because computers must have some protection from excessive temperature changes, extreme vibration, magnetic fields, voltage spikes, and oil contamination, it becomes difficult for engineers to find a suitable mounting place for this large number of computers. Integration of several electronic systems into one computer solves some of these computer mounting problems and reduces the length of wiring harness. The electronic suspension control (ESC) system explained in this chapter is another example of an integrated electronic system with suspension ride control, suspension level control, and speed-sensitive steering controlled by one computer.

Another method of reducing the number of wires on a vehicle is to interconnect many of the on-board computers with data links. A data link system may be referred to as a **network**. Some input sensor signals may be required by several computers. For example, on some vehicles the vehicle speed sensor (VSS) signal is required by the powertrain control module (PCM), suspension computer, transmission computer, cruise control module, and throttle control module. On many vehicles the VSS is hardwired to the PCM, and then the PCM relays the VSS signal to the other computers via the network.

In a **network** several computers are interconnected by data links.

Some vehicles use a front control module mounted near the front of the vehicle and a rear control module mounted near the rear of the vehicle. These vehicles also have a body computer module (BCM). The BCM, PCM, front and rear control modules, and other modules are interconnected by a network. The headlight switch may be hard-wired to the BCM. When the headlight switch is turned on, a voltage signal is sent to the BCM, and the BCM relays the appropriate LIGHTS ON message through the network to the front and rear control modules. These modules are hard-wired to the exterior lights. When the front and rear control modules receive a specific LIGHTS ON message, these modules turn on the appropriate front and rear lights. Connecting the BCM and the front and rear control modules via a network reduces the number of wires between the light switch and the front and rear lights. The headlight switch in no longer just a switch, but rather an input sensor to the BCM.

A BIT OF HISTORY

In 1996, the typical vehicle had six electronic control units (ECUs). In 2012, some luxury vehicles had up to 120 modules, 5 main networks, and 12 to 14 subnetworks depending on the on-board electronic equipment.

Some vehicles equipped with power windows and power door locks have a module in each door. These modules are hard-wired to the window motor and door lock controls in each door. These door modules are connected by a network to the BCM that is usually mounted under the dash. The window and door lock control switches are hard-wired to the BCM. When a WINDOW DOWN signal is sent from a window switch to the BCM, the BCM relays this signal through the network to the proper door module. When the door module receives the WINDOW DOWN signal, it supplies voltage to the window motor in the proper direction to roll the window down. Connecting the door modules to the BCM by a network greatly reduces the number of wires connected from the door switches into each door, and this design reduces wiring harness size and weight. Networks also reduce some of the problems associated with wiring harnesses. The previous examples are the type of networking that we see in a Controller Area Network (CAN) described in more detail later in the chapter.

Types of Networks

One type of network system introduced in the early 1980s is the Chrysler Collision Detection (CCD) network. The CCD system has a twisted pair of wires connected between the PCM, BCM, transmission control module (TCM), air bag control module (ACM), electromechanical instrument cluster (MIC), and vehicle theft security system (VTSS) module. On some models the CCD system is also connected to the data link connector (DLC), allowing the computers in the system to communicate with a scan tool connected to the DLC. The network system may be called a data bus. The CCD system operates at 2.5 V.

Another type of network system introduced in the 1980s is the universal asynchronous receive and transmit (UART) system. The UART data links are connected between various on-board computers and the DLC. The UART system operates at 5 V and transmits data at 8.2 kilobits per second. When sending data, the UART system toggles the voltage from 5 V to ground at a fixed bit pulse width. At rest, the UART network system has 5 V.

With the implementation of on-board diagnostic II (OBD-II) systems in 1996, improved communication was required between the PCM, other computers, and the scan tool. Class 2 networks were installed to meet this demand. Class 2 networks transmit data at 10.4 kilobits per second, and transmit data by toggling the voltage from 0 to 7 V. At rest this network system has 0 V. The programmable communication interface (PCI) networks were also introduced on some vehicles to increase the data communication requirements. The PCI network system is a single wire system. The PCI system operates between 0 V and 6 to 8 V. Communication on a network system is accomplished by sending a group of 0 and 1 signals (**Figure 11-1**). A long voltage pulse at a high voltage and a short pulse at a low voltage represent a 0 signal. Conversely, a short pulse at a high voltage and a long pulse at a low voltage indicate a 1 signal (**Figure 11-2**).

Today's vehicles are now equipped with CAN systems. A local area network (LAN) system is similar to the CAN system, and the LAN system is used on a significant number of vehicles. Some vehicles have a low-speed LAN system and a separate high-speed LAN system (**Figure 11-3**). The low-speed LAN system interconnects modules for

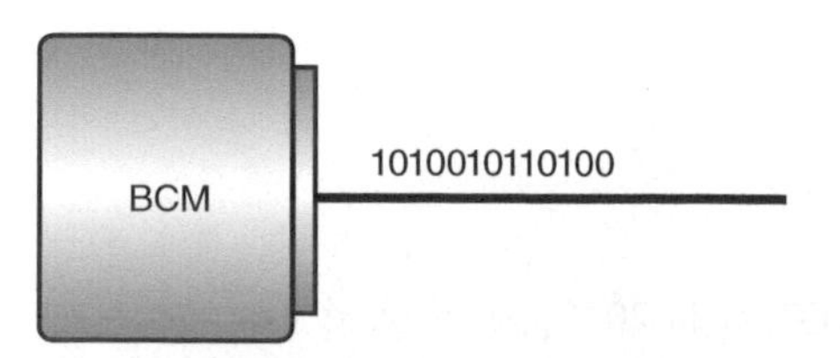

Figure 11-1 A data link system transmits data by using a group of 0 and 1 signals.

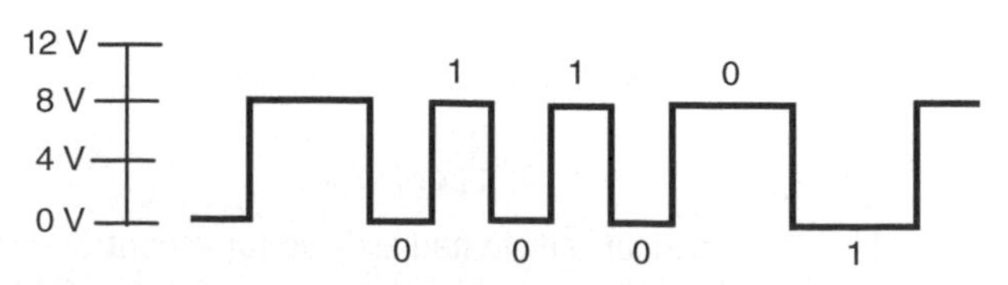

Figure 11-2 Voltage level and duration required for 0 and 1 signals.

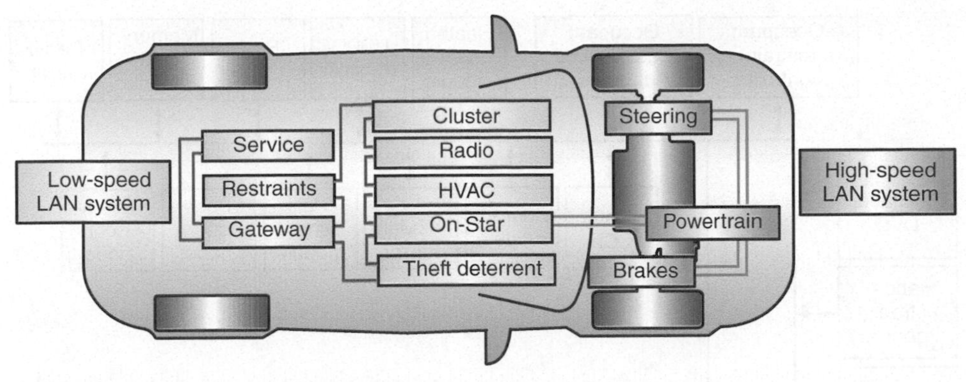

Figure 11-3 Low-speed LAN and high-speed LAN data link systems.

applications such as door locks, window motors, HVAC, and radio. The low-speed LAN system is a single-wire system. The high-speed LAN system interconnects modules such as the PCM, transmission, antilock brake modules, and suspension modules. The high-speed LAN system is a two-wire system that operates at 500 kilobits per second. The greatly increased data transmission speed capabilities of the LAN system enhance the communication between various computers in the system and the scan tool. Other high-speed data link systems on modern vehicles include FlexRay and local interconnect network (LIN). The FlexRay data link system transmits data at 10 megabits per second (Mbps). The Byteflight network used on some luxury cars has much in common with the FlexRay network. The network systems on a current SUV are shown in **Figure 11-4**. Some networks, such as the one illustrated in Figure 11-4, have a gateway module. The gateway module changes and directs the signals to go to the appropriate network within the complete network system. The gateway module is often combined within one of the other network computers.

Some networks, such as CAN, contain terminators located at both ends of the network. The terminators are usually positioned inside some of the network computers. Terminators provide electrical resistance to absorb data and prevent this data from being transmitted back into the network.

Most luxury vehicles presently have a media-oriented systems transport (MOST) network system in which the computers are interconnected by fiber-optic data links. The MOST system transmits data at 150 megabits per second. The data transmission rate depends on the model year of the vehicle and data link system. For example, early model MOST systems transmitted data at 22.6 Mbps and current MOST systems transmit data at 150 Mbps. This high data transmission speed is required on vehicles with navigation systems, CD changers, video systems, and satellite radios. The MOST system greatly reduces the number of wires in the wiring harnesses, but is more expensive compared to a network interconnected by wires.

Networks have collision resolution (CR) capabilities to prevent data collisions. The CR system varies depending on the network. In a single-wire CAN system, the system voltage is high when not transmitting data. When any computer wants to transmit data, it initiates a low voltage condition to begin transmission. As explained previously in this chapter, the data is a series of low and high voltages. The low voltage signal is the dominant voltage bit and the high voltage signal is the recessive voltage bit. The CR system uses the dominant and recessive voltage bits to determine transmission priority. If two computers transmit data at the same time, the CR system will recognize the computer with the most dominant bits and give priority to that computer. The low-priority computer stops its communication, and the high-priority computer continues data transmission. In a typical network, air bag, antilock brake, and suspension computers have high priority compared with audio computers.

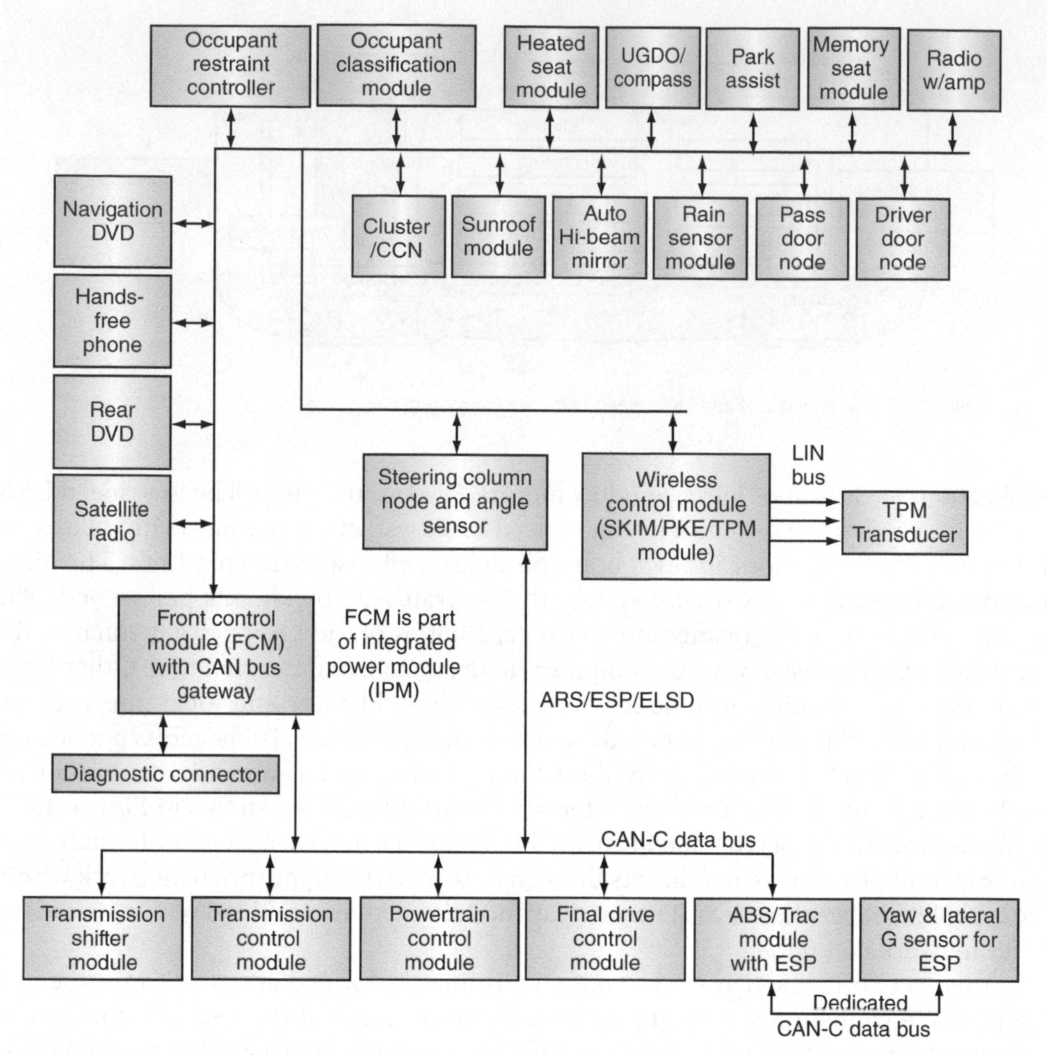

Figure 11-4 Data link systems on a current SUV.

DATA BUS NETWORK

Often information that is transmitted to and from a control module is in the form of a serial data stream. A serial data stream is a digitized code of ones and zeros that is known as a binary code. Each one or zero in a data stream is referred to as a binary digit or *bit* of information. A group of bits form a binary term or *byte*. The wire or wires over which serial data is transmitted is referred to as a **data bus network**. Often the data bus network is a twisted pair of wire (**Figure 11-5**). The twisted pair helps to eliminate the induction of electromagnetic-interference (EMF), which could disrupt or cloud the data signal. Any microprocessor that communicates on the data bus network is referred to as a node. A node may only be capable of transmitting (sending) information, or may have bi-directional capabilities allowing it to both send and receive data on the network. The two wires on a CAN bus are the Can High and the CAN Low. On the CAN network, the entire CAN High and CAN Low pins of all the nodes are connected together. Both the CAN High and CAN Low wires transmit the same data across the network but in what amounts to a mirror image of the information on the two lines. In other words, they are a check and balance for each other. In the event of a network or connection problem the system will see a difference in the data packet over the two lines and set a Network diagnostic trouble code known as a U-code.

Control modules (nodes) that may be multiplexed on a data bus network allowing them to share information and sensor data between one another. Examples of control

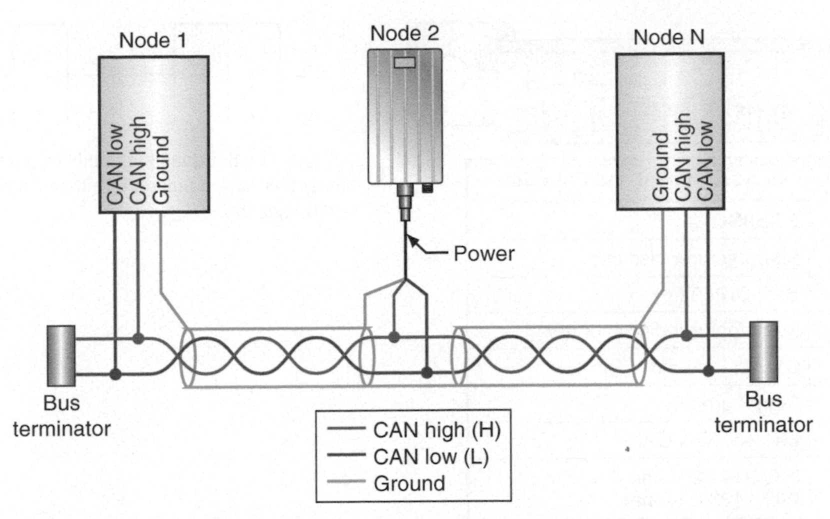

Figure 11-5 Twisted wire pair used to share information between the controllers such as Node 2 the Powertrain Control Module (PCM), Node 1 the Body Control Module (BCM), and Node N the Traction Control Module.

modules that may share data on a data bus network include the PCM, BCM, TCM, instrument panel cluster (IPC), and electronic brake control module (EBCM) to name a few (**Figure 11-6**). The data bus network eliminates the need to run hard wire from each sensor to each control module, instead the information is shared on the bus network. The

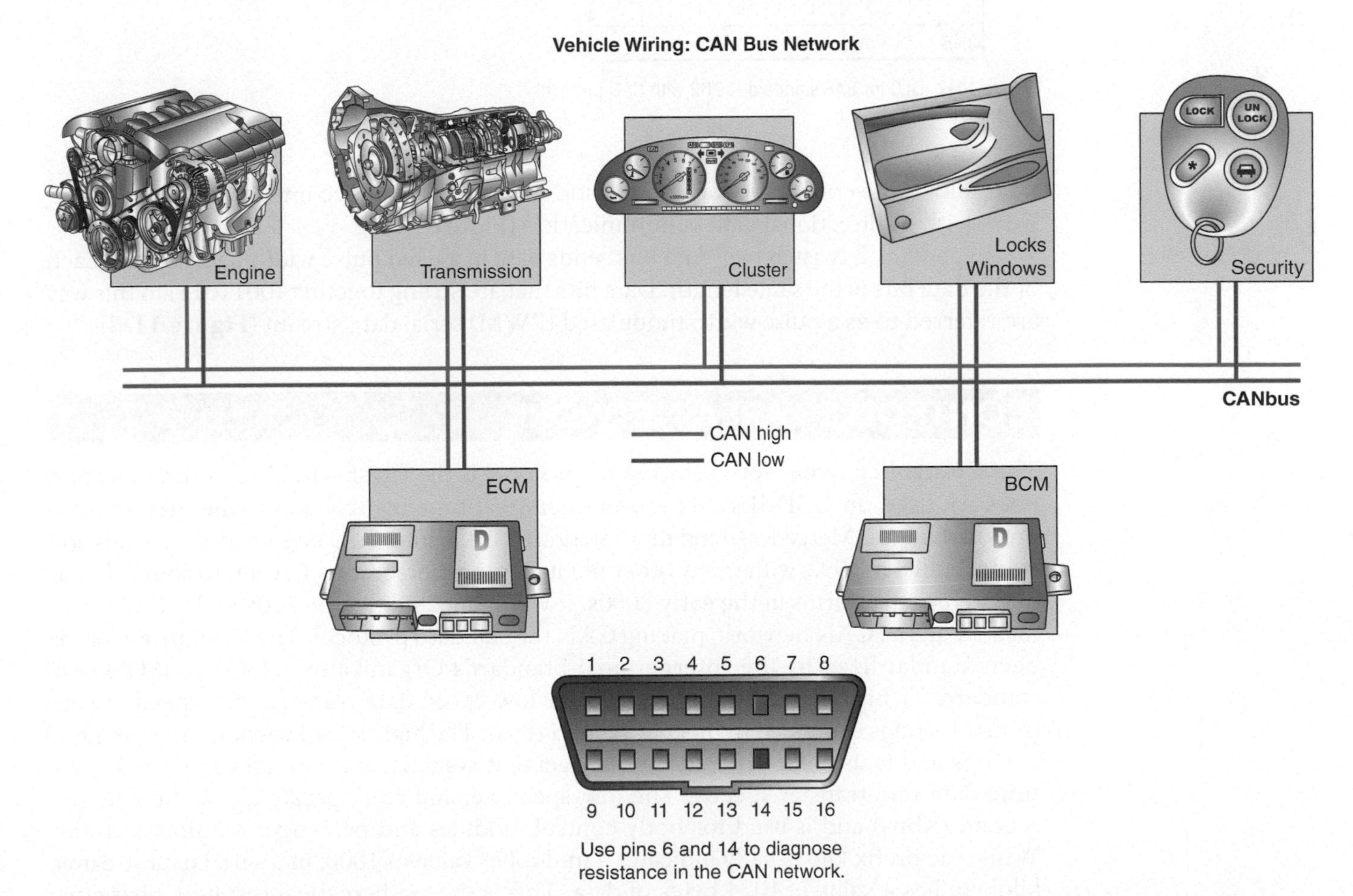

Figure 11-6 Data bus network used to share information between the controllers, such as the powertrain control module (PCM) and the transmission control module (TCM), as well as the DLC.

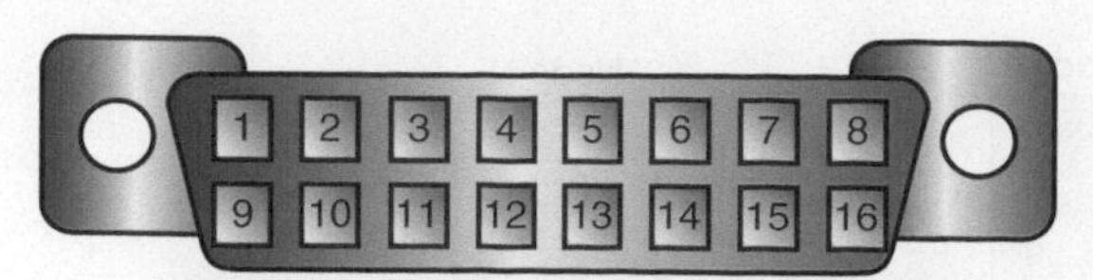

Terminal Assignment and Function	
PIN	**SAE/ISO**
1	Manufacturer discretionary
2	SAE J1850 (+)
3	Manufacturer discretionary
4	Chassis ground
5	Signal ground
6	ISO 15765-4 CAN-C (+)
7	ISO 9141-2 K-line ISO 1423-4 K-line
8	Manufacturer discretionary
9	Manufacturer discretionary
10	SAE J1850 (−)
11	Manufacturer discretionary
12	Manufacturer discretionary
13	Manufacturer discretionary
14	ISO 1565-4 CAN-C (−)
15	ISO 9141-2 L-line/ ISO 14230-4 L-line
16	Unswitched battery voltage

Figure 11-7 DLC for SAE standard J1962 with CAN protocol pin assignments.

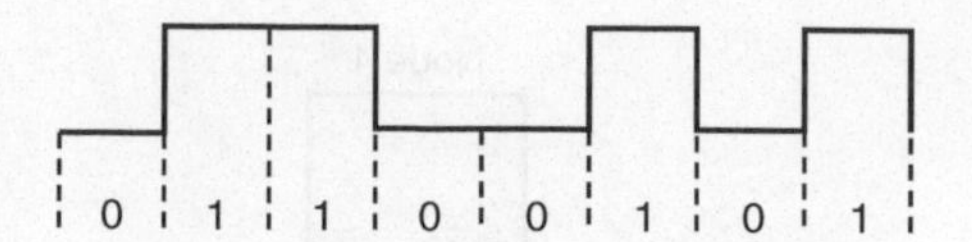

Figure 11-8 Equal length bits of data strung together form a pulse width modulated serial data stream.

DLC allows the connection of a diagnostic scan tool, which becomes a node on the network with bi-directional data communication (**Figure 11-7**).

In general, a twisted pair data bus sends data in a fixed pulse width data stream. Each of the data bits is the same length. Data bits that are strung together (0011011) in this way are referred to as a pulse width modulated (PWM) serial data stream (**Figure 11-8**).

CONTROLLER AREA NETWORK

The **Controller Area Network (CAN)** protocol is the latest serial bus communication network used on OBD-II systems and offers real time control and is the predominate protocol in use. Mercedes-Benz first integrated CAN into their engine and transmission control units in 1992, with many other manufacturers integrating CAN into some of their new vehicle platforms in the early 2000s. It was mandated that by 2008 all DLC communicated on the CAN network, making CAN the standard protocol. The CAN protocol has been standardized by the International Standards Organization (ISO) as ISO 11898 standard for high-speed and ISO 11519 for low-speed data transfer. The speed of data transmission is expressed in bits per second (bps). The high-speed version can operate at 1 Mbps and is used for powertrain management systems, and operates in virtually real time data rate transfer speeds. The low-speed version can operate at 125 kilobits per second (Kbps) and is used for body control modules and passenger comfort features. While the prefix kilo usually indicates a multiplier value of 1000, in a serial data stream a kilobyte has a value of 1024 bytes of data. This is the mathematical result of a base two numbering system (ones and zeros) carried to the tenth place. Additionally, a megabyte has a value of 1,048,576 bytes of data, which is one kilobyte (1024) squared. The CAN system has allowed for improved communication with on-board vehicle systems and is a true multiplexed network.

The Society of Automotive Engineers (SAE) has divided the speed of serial data transfer for automotive applications into three classes. Class A is the slowest transmission rate with speeds less than 10 kbps. Class A networks are used for low-priority data transmission; generally related to noncritical body control module functions such as memory seats. Class B networks are mid speed range networks with data transmission speeds between 10 kbps and 125 kbps; generally related to less critical devices such as HVAC, advanced lighting systems, and dash clusters. Class C networks have the fastest data transmission rate with speeds up to 1 bps. Class C networks are the most expensive to produce, and are used for "mission critical" data transmission that flow at "real-time" speeds. Examples of class C data include fuel control and ABS activation activity. The DLC is also connected to the class C network for improved on-board diagnostics.

CAN enables the use of enhanced diagnostics and more detailed DTCs. With CAN a scan tool is capable of communicating directly with sensors, independently of the PCM. The CAN protocol uses smart sensors. Each component contains its own control unit (microprocessor) called a "node". Each node on the network has the ability to communicate over a twisted pair of wires or a single wire called a data bus with all the other nodes on the network (bi-directional communication) without having to go through a central processing unit (**Figure 11-9**) unlike other multiplexed systems used in the past for data sharing. Every component on the network is independently capable of processing and communicating data over a common transmission line. Nodes transmit information (messages) with an identifier that prioritizes the message. The messages transmitted from a node are a package of data bits, which include a beginning of message signal, component identifier, message (sensor output signal), and an end of message signal. Since this is a bi-directional communication network, the control module receiving the data will send a signal back that the information was received. For this sophisticated communication protocol to function, the data transmission package must be a set size (number of bits) and format, and the information order must be consistent for all devices.

When multiple nodes need to send data simultaneously to the control module, the node will first see if the data bus is busy. The system uses collision detection similar to an Ethernet system. But unlike Ethernet, the CAN system can handle high data transmission rates. In essence the node is looking into traffic to see if a higher-priority node should be allowed to pass. Each CAN node on the network will have its own network unique identifier code, and nodes may be grouped based on function. The data message is then transmitted with its unique identifying code onto the network. Each node and control module on the network will perform an acceptance test of the transmission to determine if it is relative or not based on its identifier. Relative information is processed and non-relative information is ignored. Then the system segments transmissions based on the priority identifier (**Figure 11-10**) of the data package. The priority is determined by the unique number of the identifier, with lower number identifiers having higher priority. This guarantees higher-priority node identifier messages access to the network and lower-priority node messages will be automatically retransmitted in the next available bus cycle based on priority.

Since the CAN protocol technology allows for many nodes on one set of wiring, the overall vehicle wiring harness size is greatly reduced. A twisted pair wired network contains a CAN (+) and a CAN (-) wire. The CAN bus is a differential bus system where the data signal from the CAN (+) wire is a mirror image of the CAN (-) network wire (**Figure 11-11**). The combination of the twisted pair network wiring combined with the differential bus data eliminates the effect of EMF noise on the data transmission. Multiple networks on the vehicle can be linked together by gateways if necessary (**Figure 11-12**). Class C high-speed

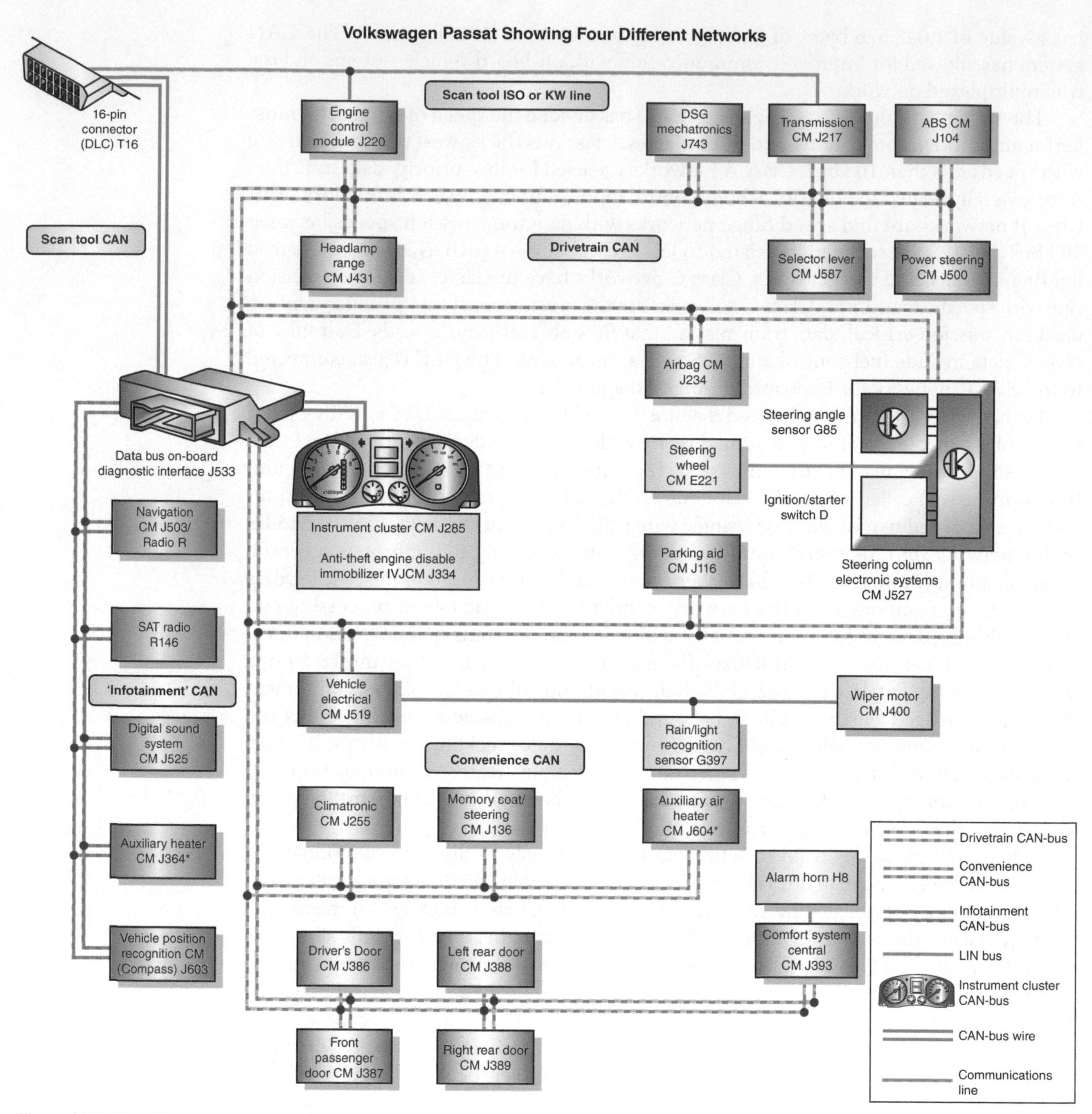

Figure 11-9 Four different communication networks on one vehicle.

data flows on one network, Class B mid-speed data flows on a second network, while Class A low-speed data flows on a third network. As an example, the intake air temperature (IAT) sensor will place its data on the network data bus allowing any control module on the network direct access to the information without the need for one control module (i.e., BCM) requesting the information from another control module (i.e., PCM). The BCM has direct access to information without having to request it from the PCM.

In 1996 the EPA specified that all vehicles be able to transmit generic scan tool data. However, proprietary data, any data other than P0 codes, and data steams were free to use any other protocol the manufacturer chose. The CAN PCM still transmits data to the DLC in SAE's generic scan tool protocol as specified by the EPA, for generic scan tool data

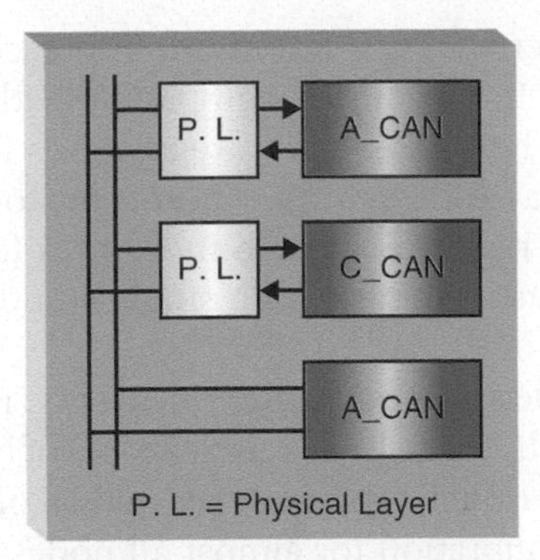

Figure 11-10 When multiple nodes are sending data simultaneously to the control module, the CAN system segments transmissions based on the priority identifier.

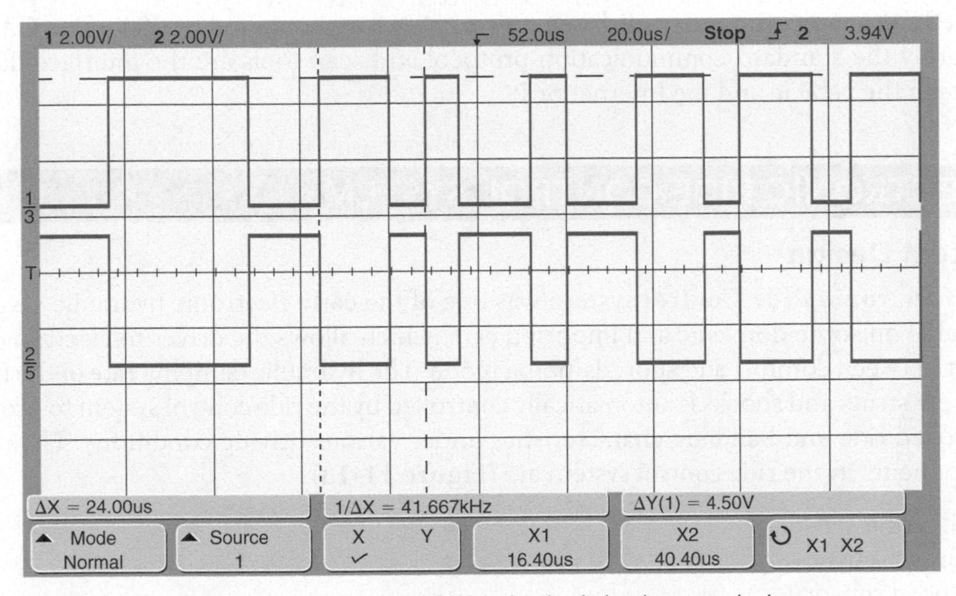

Figure 11-11 Waveform of the CAN+ and the CAN− twisted pair data bus network wires.

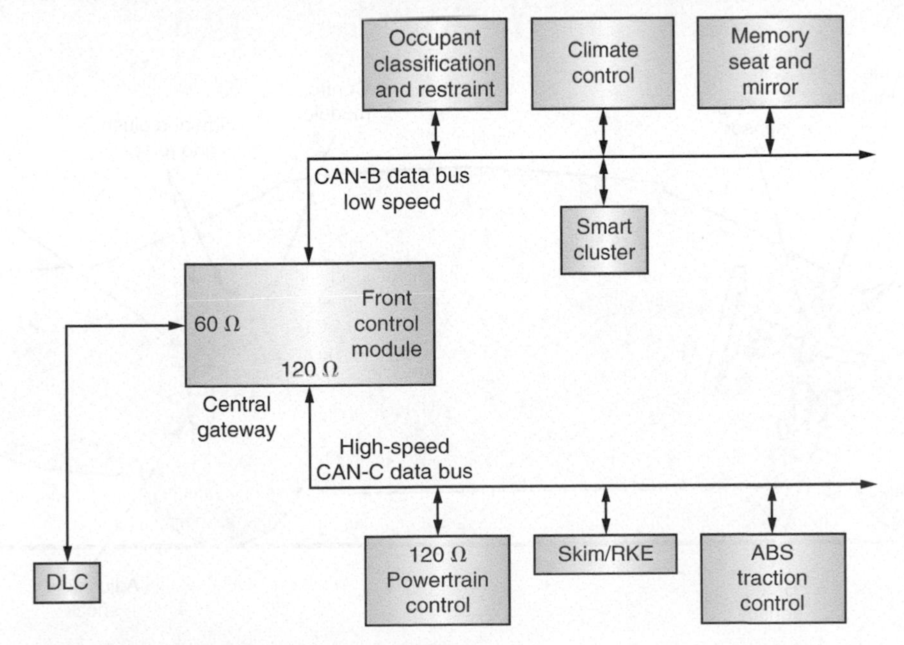

Figure 11-12 An example of a typical CAN bus network.

communication, such as generic DTC's. But in order to access all the functions available you will need to have a scan tool, which is compatible with CAN if that is the vehicle's network operation system. The EPA emission regulations for the 2008 model year have specified CAN as the new scan tool communications protocol for all vehicles sold in the United States, providing the repair technician more data for trouble shooting emission failures. With CAN the industry finally has a single standard for on-board diagnostic communication.

The CAN protocol still allows access to the typical DTC information and data streams, but with enhanced DTC detail. A scan tool is also capable of bi-directional communication directly with a smart sensor or actuator node as well as other control modules on the network. In addition, flash calibration for almost all nodes on the network will become common place. A smart sensor is capable of reporting the result of internal voltage drops, opens, grounds, and other self-test features. The network has the ability to take faulty sensors off line and can self-diagnose the difference between a faulty device or a circuit.

The EPA has required pass-thru flash programming using a standard PC connected to the Internet or using a data disk beginning in 2003 and required by 2008. The CAN is currently the standard communication protocol and scan tools are the interface device between the vehicle and the Internet or PC.

ELECTRONIC RIDE CONTROL SYSTEM

System Design

The **programmed ride control (PRC) system** adjusts shock absorber and strut damping.

The **Electronic Ride Control system** was one of the early electronic hydraulic systems available on some domestic and imported cars, which allows the driver to electronically select between comfort and sport damping mode. The hydraulic damping rate of the front and rear struts and shocks is automatically controlled by the ride control system to provide improved ride and handling characteristics under various driving conditions. The main components in the ride control system are (**Figure 11-13**):

1. Steering sensor
2. Brake sensor
3. Speed sensor

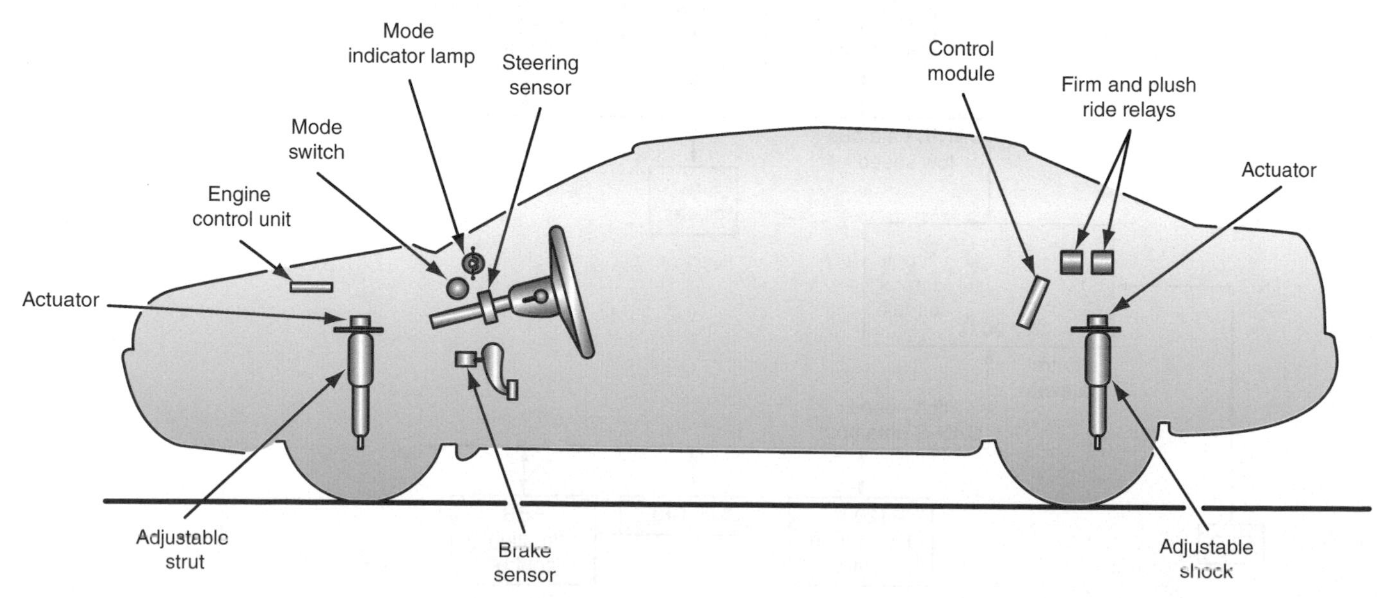

Figure 11-13 Programmed ride control (PRC) system components.

4. Struts and shocks with electric actuators
5. Control module
6. Powertrain control module (PCM)
7. Firm and plush shock relays
8. Mode select switch
9. Mode indicator light

Steering Sensor

The **steering sensor** is mounted on the steering column. This sensor contains a pair of **light-emitting diodes (LEDs)** and a matching pair of **photo diodes**. A slotted disc attached to the steering shaft rotates between the LEDs and photo diodes when the steering wheel is turned (**Figure 11-14**). This disc contains 20 slots spaced at 9-degree intervals. A signal is sent from the steering sensor to the control module in relation to the amount and speed of steering wheel rotation.

Brake Sensor

The brake sensor is a normally open (NO) switch mounted in the brake control valve assembly (**Figure 11-15**). When the brake fluid pressure reaches 400 pounds per square inch (psi) or 2758 kilopascals (kPa), the **brake pressure switch** closes and sends a signal to the control module.

Vehicle Speed Sensor

The vehicle speed sensor is usually mounted in the speedometer cable outlet of the transaxle or transmission (**Figure 11-16**). This sensor sends a vehicle speed signal to the control module. This signal is also used by the PCM.

Strut and Shock Actuators

An **actuator** is positioned in the top of each strut and shock (**Figure 11-17**). Each actuator contains a single pole armature, a pair of permanent magnets, and a position switch. When current is applied through the plush relay to the armature, the magnetic fields of the armature and the permanent magnets repel each other (**Figure 11-18**). This action causes clockwise armature rotation until the armature hits the internal stop. Under this condition, the leaf-spring switch is open in the position sensor circuit and no signal is returned to the ride control module.

An **actuator** in a strut or shock absorber varies damping action when activated and deactivated.

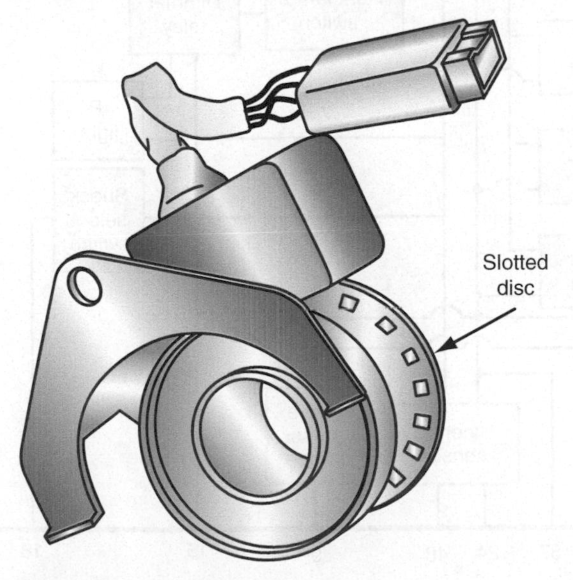

Figure 11-14 Steering sensor.

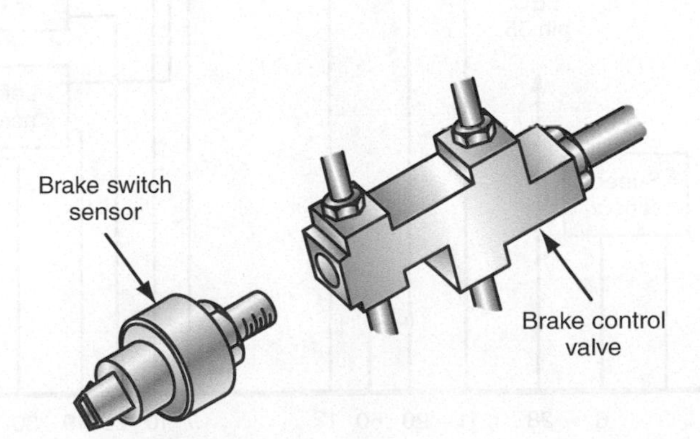

Figure 11-15 Brake switch.

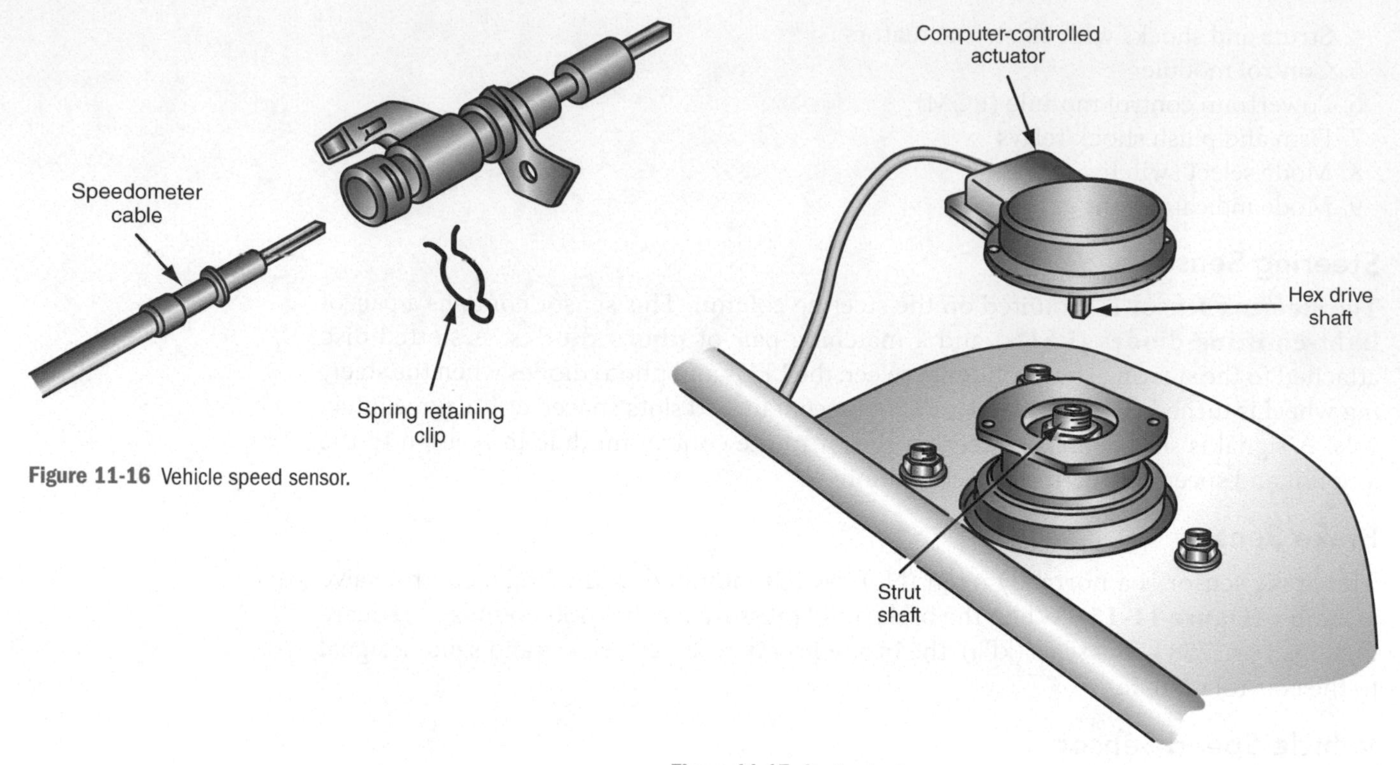

Figure 11-16 Vehicle speed sensor.

Figure 11-17 Strut actuator.

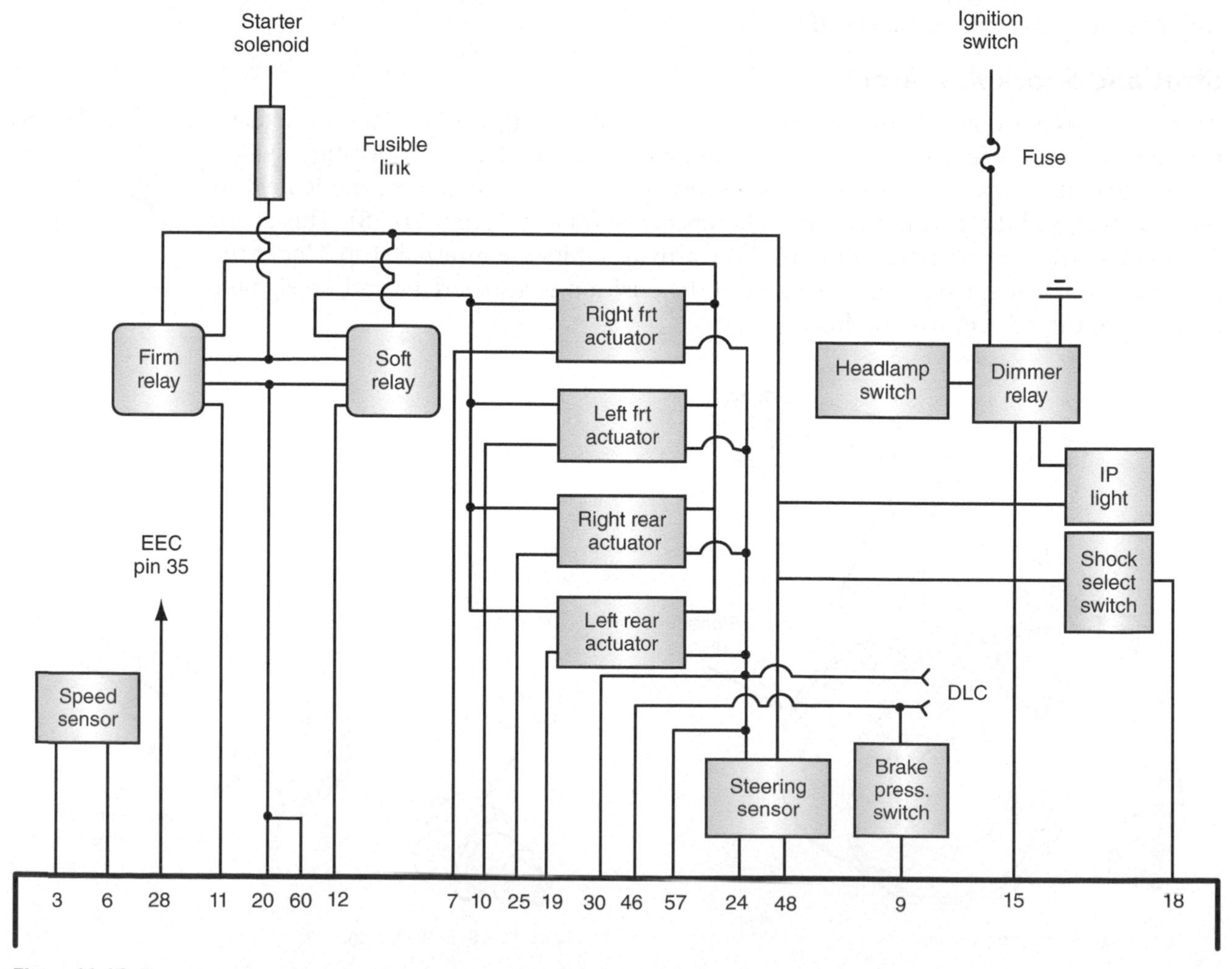

Figure 11-18 Firm relay, strut actuators, and other ride control system components.

If current is applied through the firm relay to the armature, there is an attraction between the magnetic fields of the armature and permanent magnets. This attraction causes counterclockwise armature rotation until the armature contacts the internal stop. The armature rotates an internal strut or shock valve to restrict oil movement and provide increased suspension damping. In the firm position, the leaf-spring switch closes and sends a feedback signal to the control module. The armature movement is 60 degrees and armature response time is 30 milliseconds (ms).

Operation

When the mode select switch is in the Auto position, the **firm relay** is de-energized and the **soft relay** is energized. Under this condition, current flows through the plush ride relay and the shock and strut actuators. This current is then routed to ground through the firm ride relay. If the vehicle is driven under normal speed and relatively straight-ahead conditions, this mode remains in operation.

The **firm relay** energizes the strut actuators in the firm mode.

The **soft relay** supplies voltage to the strut actuators in the soft mode.

The **throttle position sensor (TPS)** is usually a potentiometer connected to the throttle shaft. When the throttle is opened, a movable contact slides around a rotary variable resistor, and this contact movement changes the sensor voltage signal in relation to throttle opening. The TPS signal informs the PCM in the EECV system regarding the amount of throttle opening.

The following conditions cause the ride control system to switch from the auto mode to the firm mode:

1. Vehicle speed above 83 miles per hour (mph) or 133 kilometers per hour (km/h)
2. Engine acceleration at 90 percent throttle opening or 8 psi (55 kPa) turbo boost pressure
3. Lateral vehicle acceleration above 0.35 g
4. Brake pressure of 400 psi (2758 kPa) or more

The vehicle acceleration signal is sent from the **throttle position sensor (TPS)** to the PCM. This signal is relayed to the ride control module. Lateral vehicle acceleration is sensed by the steering sensor.

When the ride control module receives an input signal that requires firm ride control, the control module energizes the firm relay and de-energizes the soft relay. This action results in current flow from the firm ride relay through the shock and strut actuators, and the soft ride relay to ground. Therefore, shock and strut actuator current is reversed and the armature in each actuator moves the shock and strut valves to the Firm position.

The mode indicator light in the tachometer glows when the ride control system is in the Firm mode (**Figure 11-19**). During the first 80 seconds after the ignition switch is turned on, the PRC system does not respond to changes in vehicle direction. This action allows the ride control module to calculate the straight-ahead position.

If the mode select switch is placed in the Firm position, the system remains in the firm mode continually. In this mode, the mode indicator light remains on. In the firm mode, the shocks and struts provide approximately three times the damping action on the extension stroke compared to the normal mode.

Shop Manual
Chapter 11, page 467

Figure 11-19 Mode indicator light.

ELECTRONIC AIR SUSPENSION SYSTEM COMPONENTS

Air Springs

Air springs support the chassis weight in an air suspension system.

Caution

The ball joint studs in the upper ball joints do not have a press fit in the steering knuckle. When loosening these ball joint nuts, a hex holding feature on the ball joint stud prevents the stud from rotating when loosening the ball joint nuts. If the upper ball joint stud rotates in the aluminum knuckle, the knuckle opening may be damaged.

In an air suspension system, the **air springs** replace the coil springs in conventional suspension systems. These air springs have a composite rubber and plastic membrane that is clamped to a piston located in the lower end of the spring. An end cap is clamped to the top of the membrane and an air spring valve is positioned in the end cap. The air springs are inflated or deflated to provide a constant vehicle trim height. Front air springs are mounted between the control arms and the crossmember (**Figure 11-20**). The lower end of these air springs is retained in the control arm with a clip, and the upper end is positioned in a crossmember spring seat. The front shock absorbers are mounted separately from the air springs.

In some modern air suspension systems, the air springs are mounted and sealed on the shock absorbers (**Figure 11-21**). The lower end of the shock absorber is attached to the lower control arm through an insulating bushing, and the upper end of the shock absorber is attached to the chassis through an insulating mount. This type of air suspension system has aluminum front lower control arms and spindles, and forged steel upper control arms. Ball joints are mounted in the outer ends of the upper and lower control arms. The aluminum suspension components reduce the unsprung weight and improve ride quality. Reducing vehicle weight also improves fuel economy. The upper and lower ball joints are an integral part of the control arms, and these ball joints cannot be replaced separately.

Other vehicles have the air springs mounted over the front and rear struts, and these struts contain a solenoid actuator that varies the strut valve opening to control ride

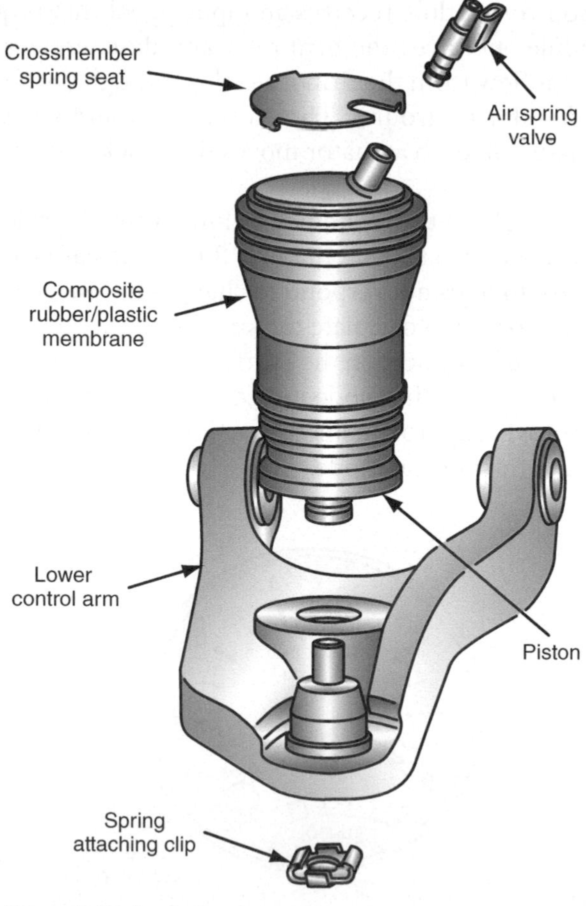

Figure 11-20 Front air spring.

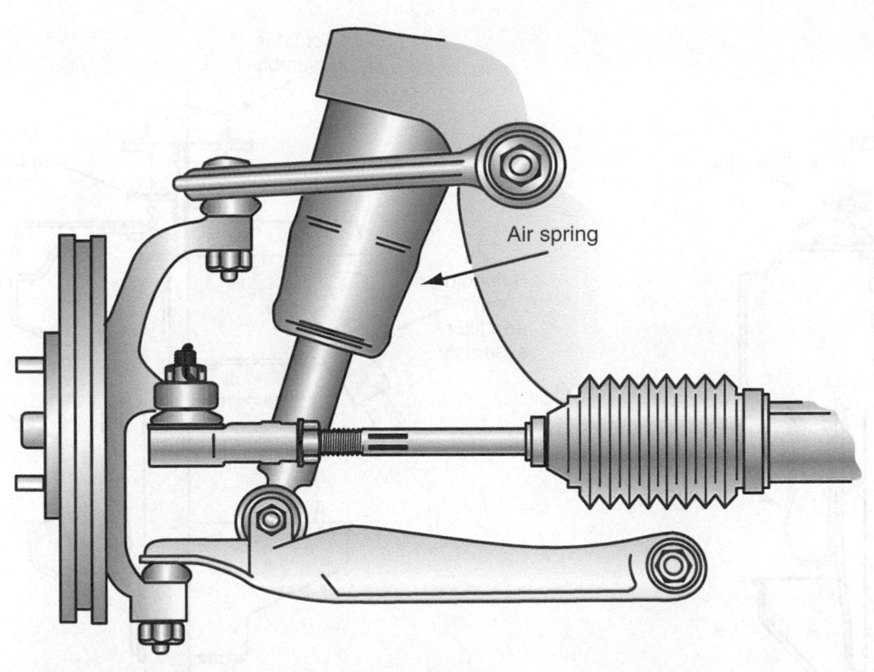

Figure 11-21 Air spring mounted on the shock absorber.

firmness (**Figure 11-22** and **Figure 11-23**). These strut actuators are similar to the ones explained previously in this chapter on the PRC system. Some of these air suspension systems that control ride firmness are not driver adjustable. The suspension module controls the strut firmness automatically in relation to the module inputs. This type of system may be called an automatic air suspension system. Some vehicles with solenoid actuators in the struts have up to four suspension modes that may be selected by the driver. One premium luxury car has these driver selectable suspension modes:

1. Comfort—provides a smooth luxurious ride.
2. Automatic—the suspension computer provides the best possible combination of comfort and handling based on speed, driver style, and road conditions.
3. Dynamic—stiffest, lowest, sportiest, most aerodynamic suspension mode.
4. Life—for rougher roads, steep approaches, and deep snow. This mode is used only for low-speed driving.

When the driver selects a suspension mode, the suspension module positions the strut actuators to provide the desired ride quality, and this module also adjusts the air spring pressure to provide the appropriate ride height.

This type of air suspension system dramatically reduces body roll and pitch that occurs during cornering and hard braking. Because this air suspension system lowers the ride height at higher speeds, it improves aerodynamic efficiency.

The rear air springs are similar to the front air springs and also have similar mountings. Some rear air springs are mounted between the rear suspension arms and the frame with the shock absorbers mounted separately from the air springs (**Figure 11-24**). Other rear air springs are mounted over the rear shock absorbers, and the shock absorbers are mounted between the lower control arms and the frame. Some modern air suspension systems on four-wheel-drive vehicles have rear knuckles with **cross-axis ball joints**. The lower cross-axis ball joint is a round insulating bushing mounted in the lower control arm, and a bolt attaches this bushing to the lower end of the knuckle. The upper cross-axis ball joint is a round insulating bushing mounted in the top of the knuckle, and a bolt attaches this bushing to the upper control arm (**Figure 11-25**). In this rear suspension system, the upper and lower control arms are made from aluminum. An adjustable toe link is

Cross-axis ball joints contain large insulating bushings in place of typical ball joints.

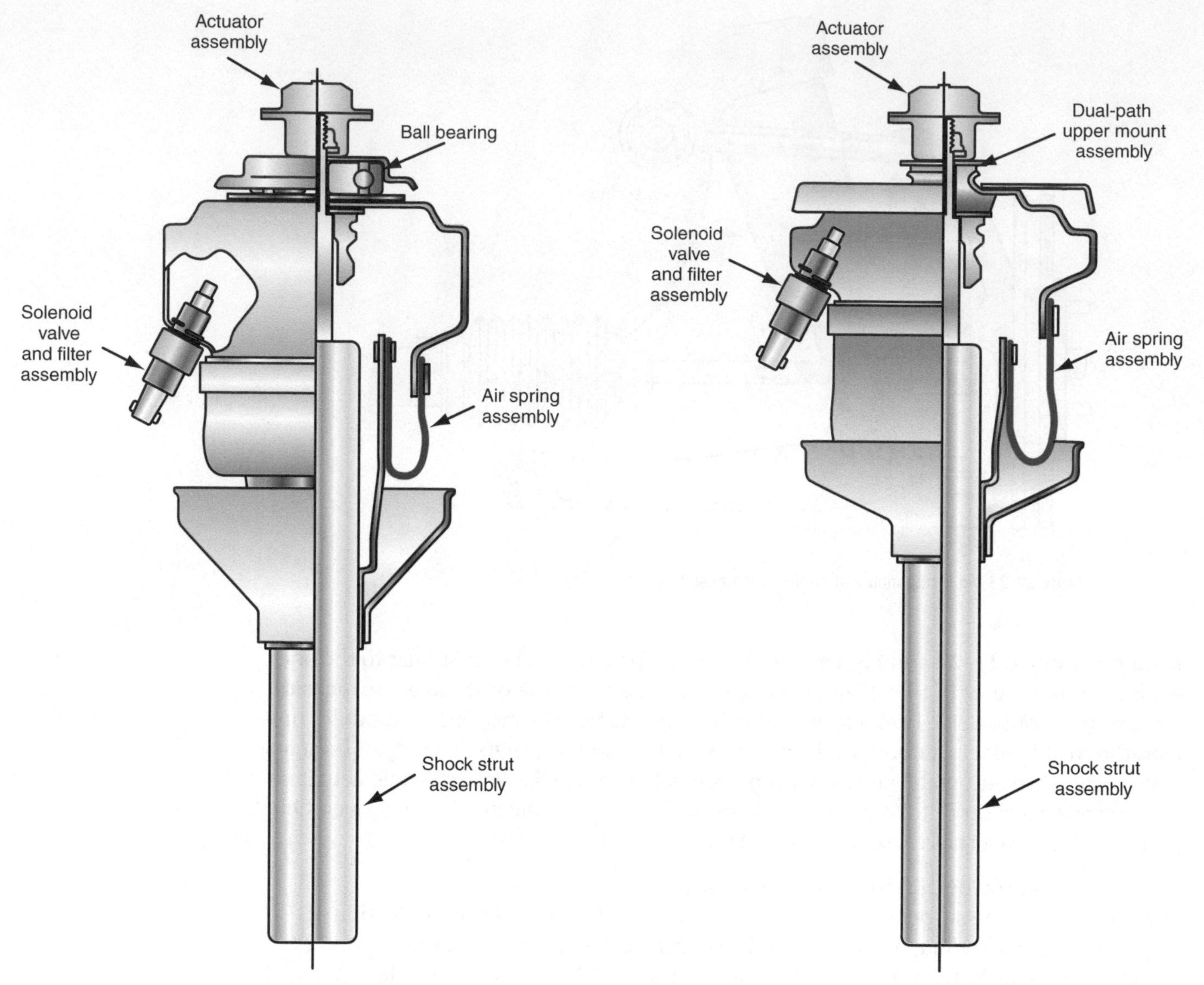

Figure 11-22 Front air spring and strut assembly.

Figure 11-23 Rear air spring and strut assembly.

connected from the knuckle to the frame to provide a rear toe adjustment. Rear wheel camber can be adjusted by installing a camber adjustment kit in place of the upper knuckle-to-control arm retaining bolt.

Air Spring Valves

An **air spring solenoid valve** allows air to flow into and out of an air spring.

An **air spring solenoid valve** is mounted in the top of each air spring (**Figure 11-26**). These valves are an electric solenoid-type valve that is normally closed. When the valve winding is energized, plunger movement opens the air passage to the air spring. Under this condition, air may enter or be exhausted from the air spring. Two O-ring seals are located on the end of the valves to seal them into the air spring cap. The valves are installed in the air spring cap with a two-stage rotating action similar to a radiator pressure cap.

Air Compressor

A single piston in the air compressor is moved up and down in the cylinder by a crankshaft and connecting rod (**Figure 11-27**). The armature is connected to the crankshaft, and therefore the rotating action of the armature moves the piston up and down. Armature rotation occurs when 12 V are supplied to the compressor input terminal. Intake and

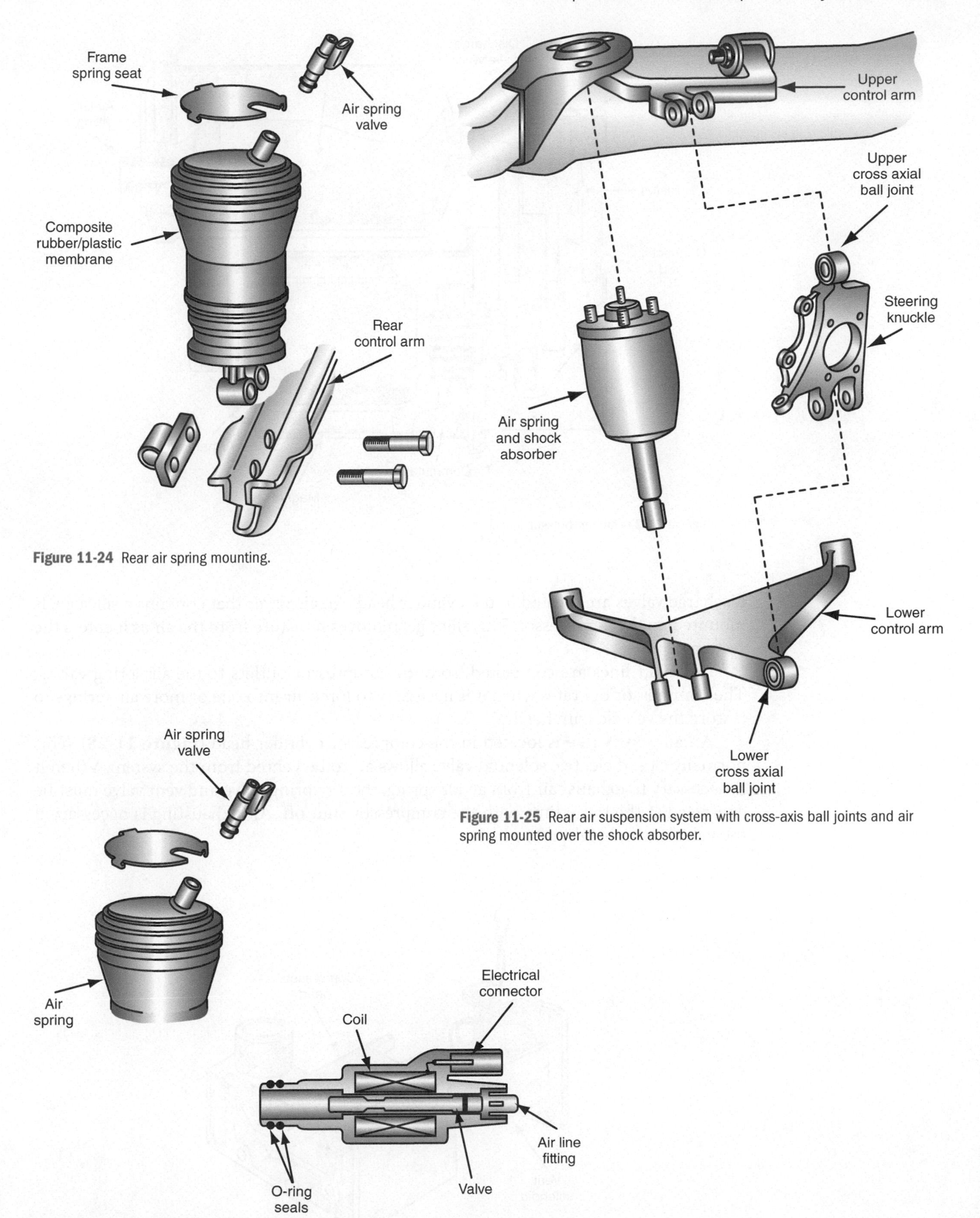

Figure 11-24 Rear air spring mounting.

Figure 11-25 Rear air suspension system with cross-axis ball joints and air spring mounted over the shock absorber.

Figure 11-26 Air spring valve.

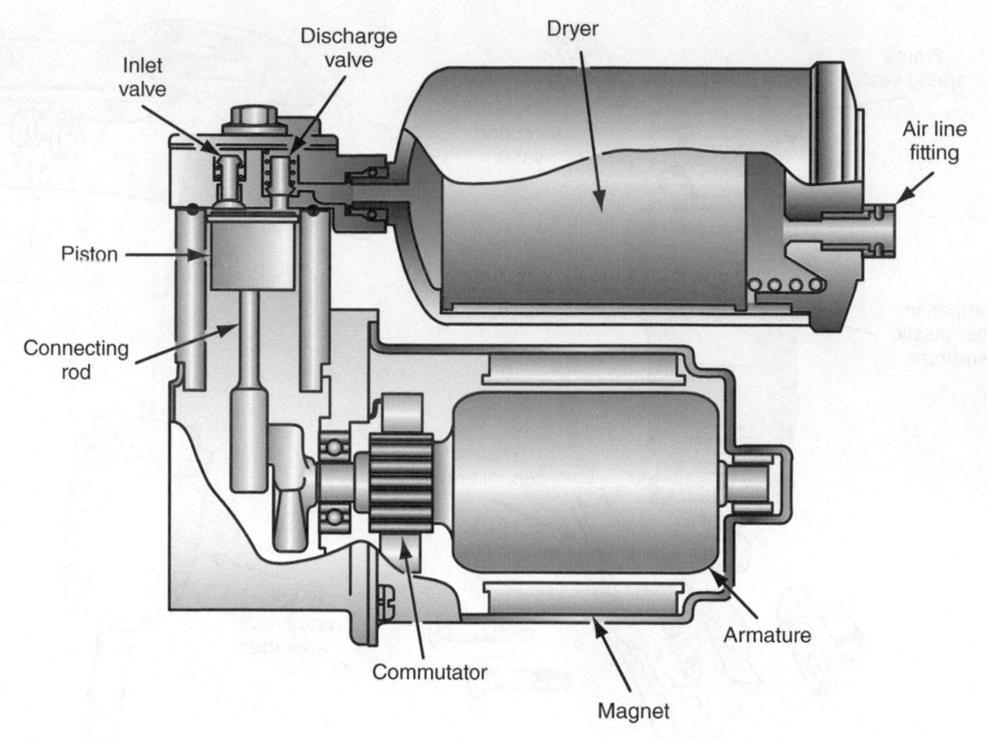

Figure 11-27 Air compressor.

discharge valves are located in the cylinder head. An air dryer that contains a silica gel is mounted on the compressor. This silica gel removes moisture from the air as it enters the system.

Nylon air lines are connected from the compressor outlets to the air spring valves. The compressor operates when it is necessary to force air into one or more air springs to restore the vehicle trim height.

An air **vent valve** is located in the compressor cylinder head (**Figure 11-28**). This normally closed electric solenoid valve allows air to be vented from the system. When it is necessary to exhaust air from an air spring, the air spring valve and vent valve must be energized at the same time with the compressor shut off. Air exhausting is necessary if the vehicle trim height is too high.

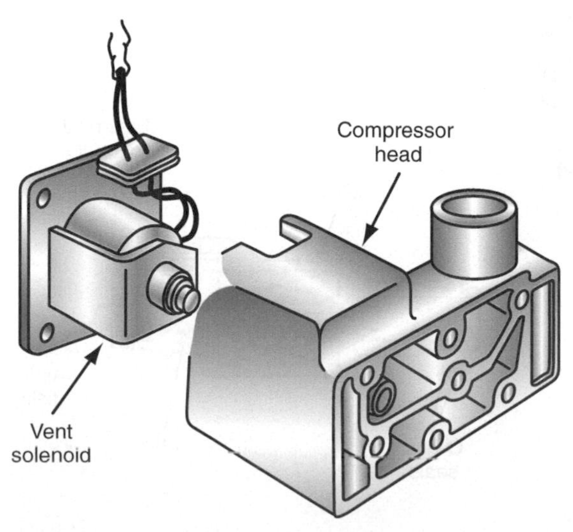

Figure 11-28 Air vent valve.

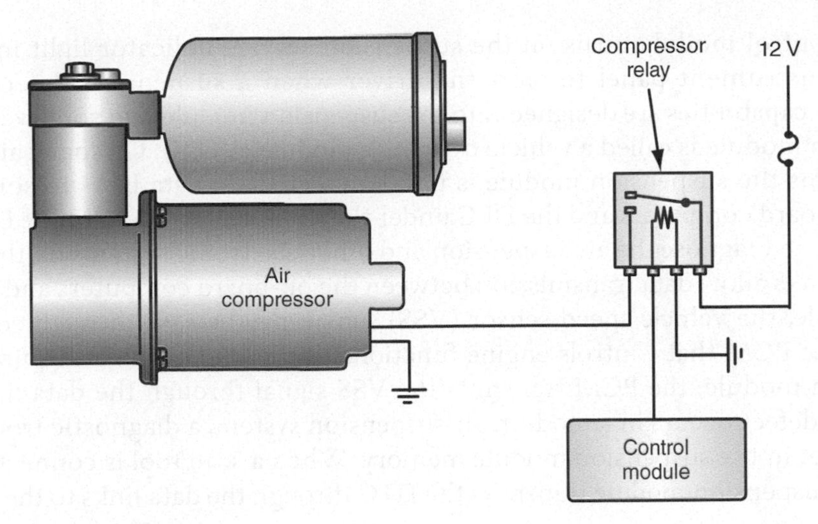

Figure 11-29 Compressor relay.

Compressor Relay

When the compressor relay is energized, it supplies 12 V through the relay contacts to the compressor input terminal (**Figure 11-29**). The relay contacts open the circuit to the compressor if the relay is de-energized. An electronic relay is used in some air suspension systems.

Control Module

The control module is a microprocessor that operates the compressor, vent valve, and air spring valves to control the amount of air in the air springs and maintain the trim height. The control module is located in the trunk (**Figure 11-30**) on some models. On other models, the module is mounted under the instrument panel above the parking brake.

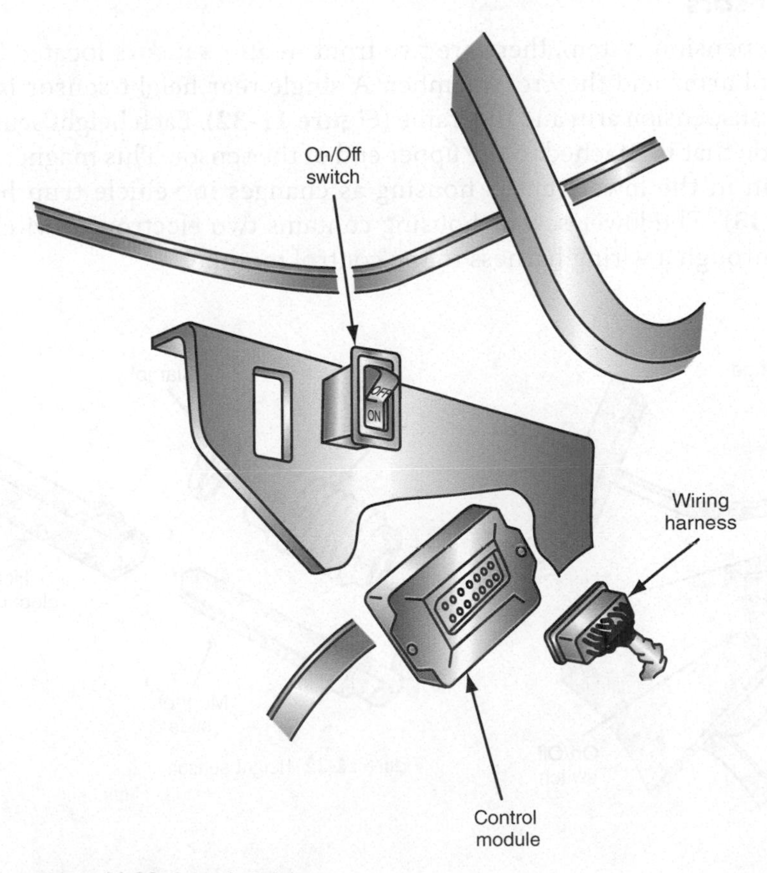

Figure 11-30 Control module.

The control module turns on the suspension service indicator light in the roof panel or instrument panel to alert the driver when a suspension defect occurs. Diagnostic capabilities are designed into the suspension module. On some vehicles the suspension module is called a vehicle dynamics module (VDM). On many air suspension systems the suspension module is interconnected via data links to some of the other on-board computers and the DLC under the dash. A scan tool may be connected to the DLC to diagnose the air suspension and other electronic systems on the vehicle. The data links allow data transmission between the on-board computers and the DLC. For example, the vehicle speed sensor (VSS) signal may be sent through connecting wires to the PCM that controls engine functions. If the VSS signal is required by the suspension module, the PCM transmits the VSS signal through the data links. If an electronic defect occurs in a modern air suspension system, a diagnostic trouble code (DTC) is set in the suspension module memory. When a scan tool is connected to the DLC, the suspension module transmits the DTC through the data links to the DLC and scan tool.

On/Off Switch

The On/Off switch opens and closes the 12 V supply circuit to the suspension module. This switch is located in the trunk (**Figure 11-31**). Depending on the vehicle make and model year, one or two panels in the trunk may have to be removed to access the On/Off switch. The On/Off switch must be turned off before the vehicle is hoisted, jacked, or towed. Certain air suspension service procedures may require this switch to be placed in the Off position.

WARNING If the vehicle is hoisted, jacked, or towed with the electronic air suspension switch in the On position, personal injury or vehicle damage may occur.

Height Sensors

Height sensors send an electric signal to the control module in relation to curb riding height.

In the air suspension system, there are two front **height sensors** located between the lower control arms and the crossmember. A single rear height sensor is positioned between the suspension arm and the frame (**Figure 11-32**). Each height sensor contains a magnet slide that is attached to the upper end of the sensor. This magnet slide moves up and down in the lower sensor housing as changes in vehicle trim height occur (**Figure 11-33**). The lower sensor housing contains two electronic switches that are connected through a wiring harness to the control module.

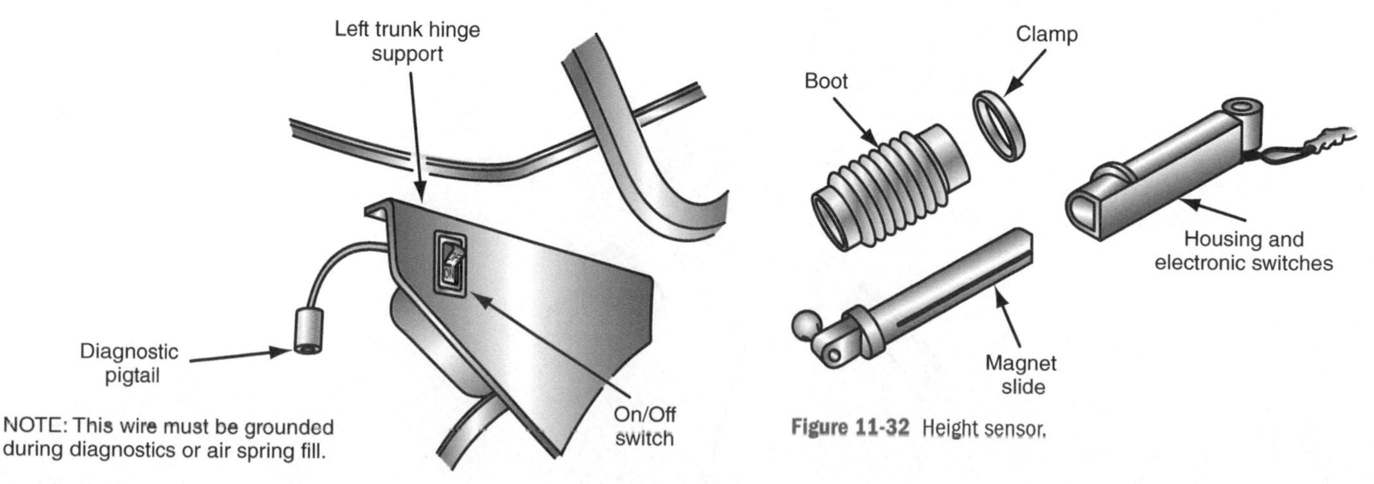

Figure 11-31 On/Off switch.

Figure 11-32 Height sensor.

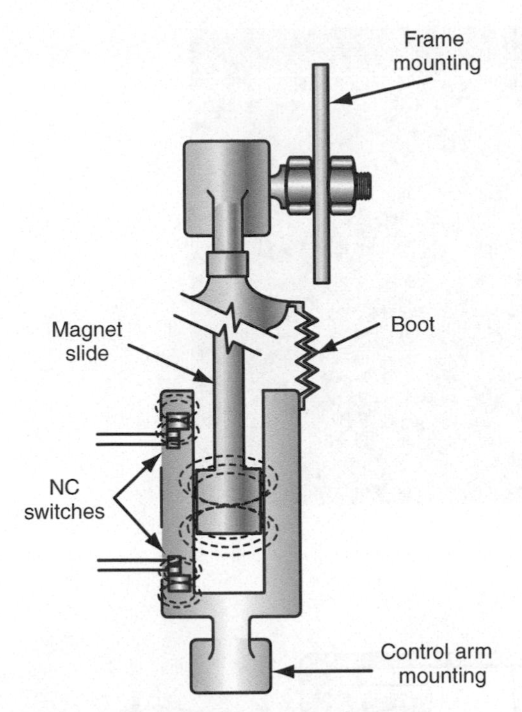

Figure 11-33 Rear height sensor mounting.

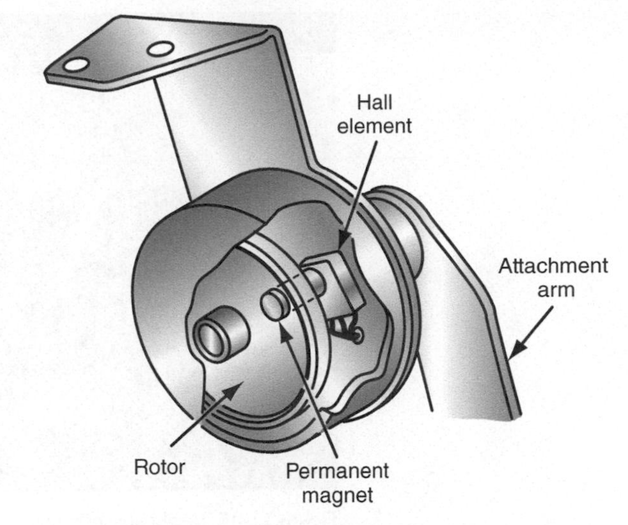

Figure 11-34 Electronic rotary height sensor.

When the vehicle is at **trim height**, the switches remain closed and the control module receives a trim height signal. If the magnet slide moves upward, the above trim switch opens and a **lower vehicle command** is sent from the height sensor to the module. When this signal is received by the module, it opens the appropriate air spring valve and the vent valve. This action exhausts air from the air spring and corrects the above trim height condition. Downward magnet slide movement closes the above trim switch and opens the below trim switch. If this action occurs, the height sensor sends a **raise vehicle command** to the module. When the control module receives this signal, it energizes the compressor relay and starts the compressor. The control module opens the appropriate air spring valve, and this action forces air into the air spring to correct the below trim height condition. The height sensors are serviced as a unit.

Some air suspension systems have **electronic rotary height sensors**. Each rotary height sensor contains a permanent magnet rotor and a **Hall element** (**Figure 11-34**). An arm on the height sensor is attached to the rotor. The height sensor body is mounted on the chassis and a linkage is connected from the sensor arm to the suspension (**Figure 11-35**). Suspension movement rotates the permanent magnet rotor and changes the voltage signal in the Hall element. The voltage signal from a rotary height sensor is proportional to trim height, above trim height, and below trim height.

Trim height refers to the distance between the chassis and the road surface measured at a specific location recommended by the vehicle manufacturer.

Electronic rotary height sensors have an internal rotating element and these sensors send voltage signals to the control module in relation to the curb riding height.

A **Hall element** is an electronic device that produces a voltage signal when the magnetic field approaches or moves away from the element.

Warning Lamp

When the control module senses a system defect, the module turns on the air suspension warning lamp in the roof console or instrument panel to inform the driver that a problem exists (**Figure 11-36**). If the air suspension system is working normally, the warning lamp will be on for one second when the ignition switch is turned from the Off to the Run position. After this time, the warning lamp should remain off. This lamp does not operate with the ignition switch in the start position. The warning lamp is used during the self-diagnostic procedure and the spring fill sequence.

On some vehicles, the air suspension warning light is replaced with a CHECK SUSPENSION message in the instrument panel message center. The suspension module provides a CHECK SUSPENSION message if an electrical defect occurs in the air

Caution

Never attempt to probe the electronic switches in slide-type height sensors. This action may damage the sensor.

Figure 11-35 Height sensor mounting.

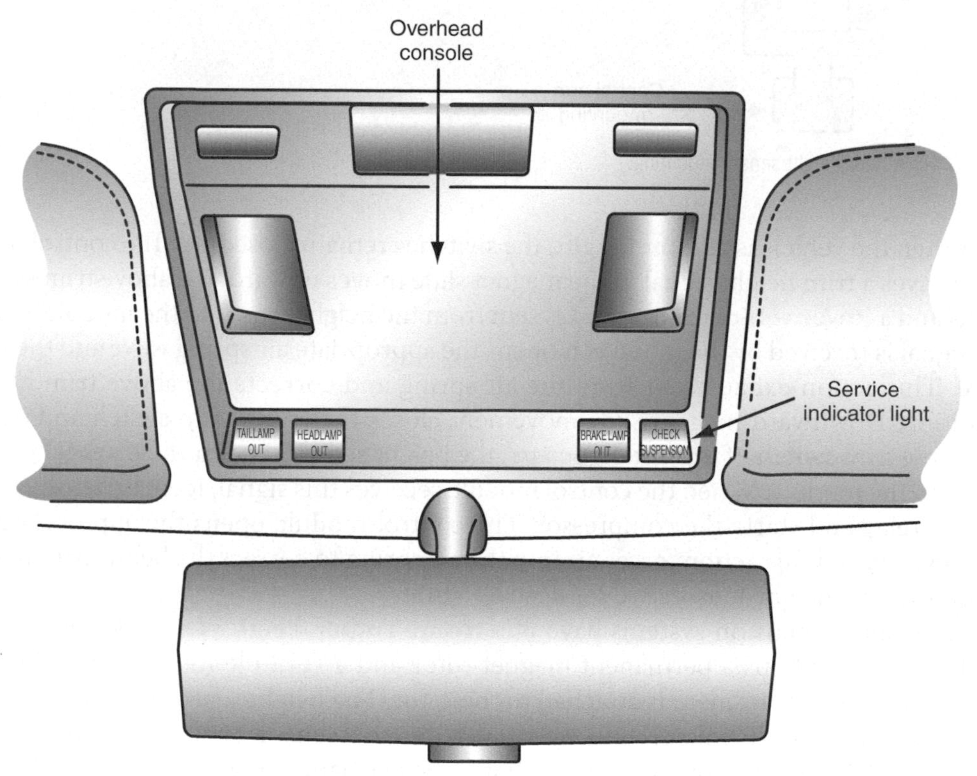

Figure 11-36 Air suspension warning light.

suspension system. An AIR SUSPENSION SWITCHED OFF message appears in the message center if the air suspension switch is in the Off position.

ELECTRONIC AIR SUSPENSION SYSTEM OPERATION

An electronically controlled suspension system may be called an active suspension system.

The operation of an air suspension system varies depending on the vehicle make and model year. The following section discusses typical air suspension operation.

Ignition Switch Off

The electronic air suspension system is fully operational for one hour after the ignition switch is turned from the Run to the Off position. During this time, lower vehicle commands are completed unless a height sensor was providing a high signal when the ignition switch was turned off. After a one-hour period, raise vehicle commands are acted upon and lower vehicle commands are ignored. The air compressor run time is limited to 15 seconds for rear springs and 30 seconds for front springs.

Ignition Switch in Run Position

When the ignition system has been in the Run position for less than 45 seconds, raise vehicle commands are completed immediately, but lower vehicle commands are ignored. If the ignition switch has been in the Run position for more than 45 seconds, the operation is as follows:

1. If a door is opened with the brake pedal released, raise vehicle commands are completed immediately, but lower vehicle commands are serviced after the door is closed. This action prevents an open door from catching on curbs or other objects.
2. If the doors are closed and the brake pedal is released, all commands are serviced by a 45-second averaging method to prevent excessive suspension height corrections on irregular road surfaces.
3. If the brake is applied and a door is open, raise vehicle commands are completed immediately, but lower vehicle commands are ignored.
4. When the doors are closed and the brake pedal is applied, all commands are ignored by the control module. If a command to raise the rear suspension is in progress under these conditions, this command will be completed. This action prevents correction of front end jounce while braking.

General Operation

When a height sensor sends a raise vehicle command to the control module and the other input signals are acceptable, the module grounds the compressor relay winding and starts the compressor. The module also opens the appropriate air spring valve to allow air flow into the air spring (**Figure 11-37**).

On non-computer-controlled suspension systems, the trim height may be called the curb riding height.

The rear air valve solenoids always operate together, but the front solenoids may be energized independently. When the correct chassis trim height is obtained, the control module opens the circuit from the compressor relay winding to ground and de-energizes the air spring valve. This action shuts off the compressor and traps the air in the air spring.

If a lower vehicle request is sent from a height sensor to the control module and the other input signals are acceptable, the control module opens the air vent valve and appropriate air spring valve. When this action occurs, air is released from the air spring until the correct trim height is obtained. The trim height signal from the height sensor to the module causes the module to close the air vent valve and the air spring valve.

Shop Manual
Chapter 11, page 468

Specific Control Module Operation

Commands are completed by the module in this order: rear up, front up, rear down, front down. When the ignition switch is in the Run position and a command cannot be completed within three minutes, the module turns on the air suspension warning lamp. This lamp remains on until the ignition switch is turned off. On some older models, all the control module memory is erased when the ignition switch is turned off. Therefore, the warning lamp may not indicate a defect immediately if the ignition switch is turned from the Off to the Run position. When a system defect causes the module to illuminate the warning lamp with the ignition in the Run position, other commands may be completed by the module. Commands from the front and rear height sensors are never completed simultaneously.

If an electrical defect occurs in a modern air suspension system, a DTC is set and retained in the suspension module memory. In these systems, DTCs are transmitted via the data links to the DLC under the instrument panel, and the DTCs may be displayed on a scan tool connected to the DLC.

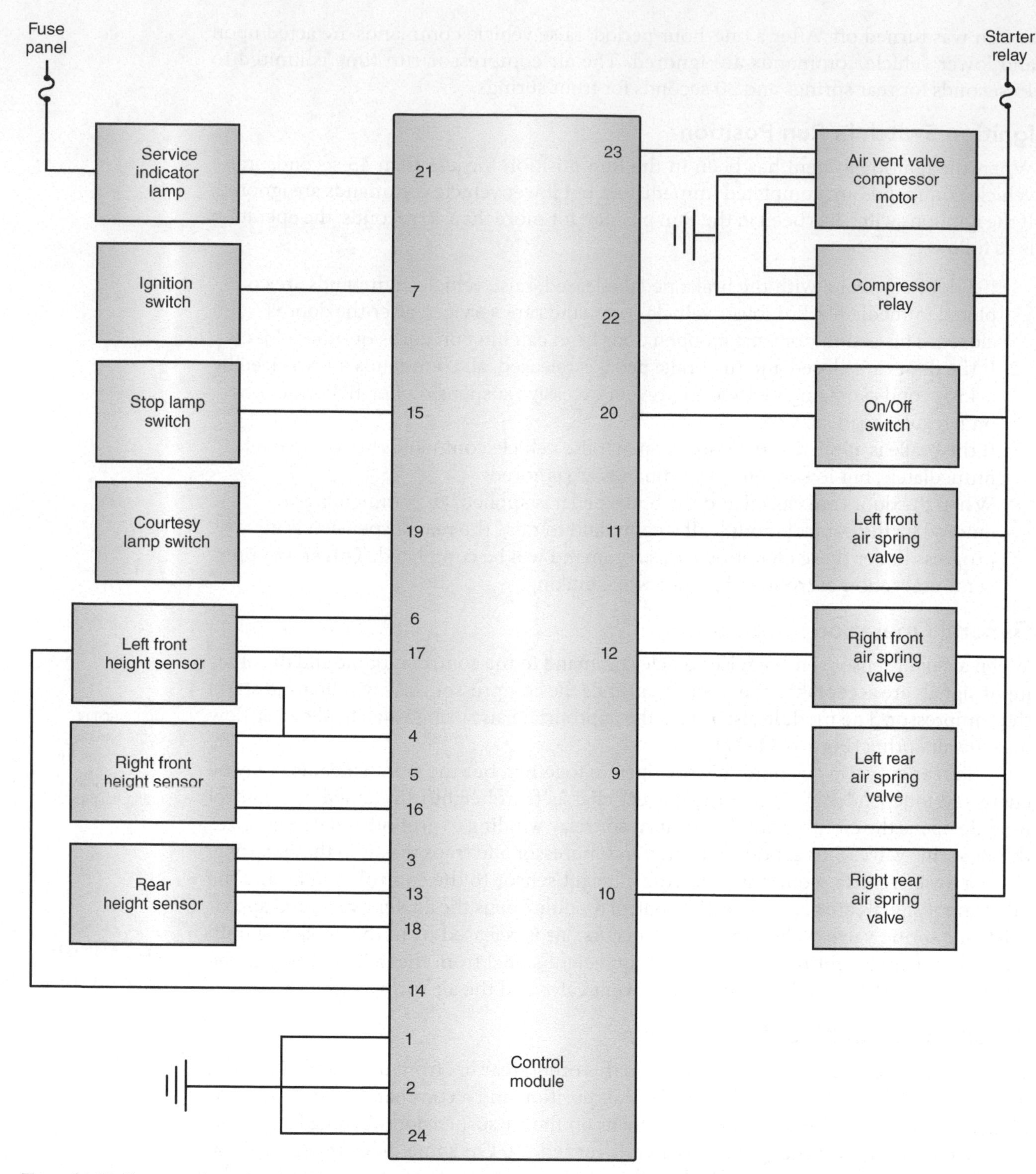

Figure 11-37 Air suspension wiring diagram.

AIR SUSPENSION SYSTEM DESIGN VARIATIONS

Rear Load-Leveling Air Suspension System

Some vehicles have air suspension on only the rear wheels, and these systems may be referred to as rear load-leveling air suspension systems. These air suspension systems have basically the same air springs, compressor and relay, On/Off switch, and rear height sensor as the air suspensions described previously. If additional weight is placed in the trunk, the rear height sensor signals the suspension control module to raise the rear suspension height to the specified trim height. When the extra weight is removed from the trunk, the suspension height is higher than specified, and the air suspension system lowers the rear trim height to the specified height.

Air Suspension System with Speed Leveling Capabilities

Some vehicles have an air suspension system on the front and rear wheels, which is similar to the air suspension systems described earlier in this chapter. However, these systems have an input signal from the vehicle speed sensor (VSS) to the suspension module. On some vehicles the signal from the VSS is sent to the powertrain control module (PCM) that controls engine functions. The PCM transmits the VSS signal through the data links from the PCM to the suspension module. When the VSS signal indicates the vehicle is traveling above 65 mph (105 km/h), the suspension module opens the vent valve and spring solenoid valves to lower the suspension height a specific amount. Under this condition, the vehicle is more dynamically efficient and this improves high-speed vehicle stability and fuel economy.

Special Air Suspension System Features on Four-Wheel-Drive Vehicles

Some four-wheel-drive vehicles with a conventional front torsion bar suspension and a leaf-spring rear suspension also have an air suspension system on all four wheels. These front and rear suspension systems maintain the specified ride height during two-wheel-drive operation. The air suspension system has the same components as described previously, except this air suspension system has front and rear air shock absorbers in place of air springs. An air solenoid is connected in the air line to each air shock absorber (**Figure 11-38**). The air shock absorbers also contain electrical solenoids that rotate the shock absorber valves and vary the ride firmness (**Figure 11-39**). This type of air suspension may be called an automatic ride control (ARC) system.

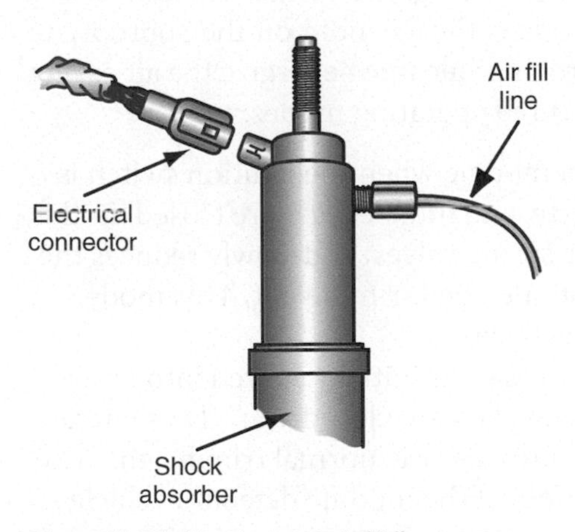

Figure 11-38 Electrical and air fill line connections to the shock absorber.

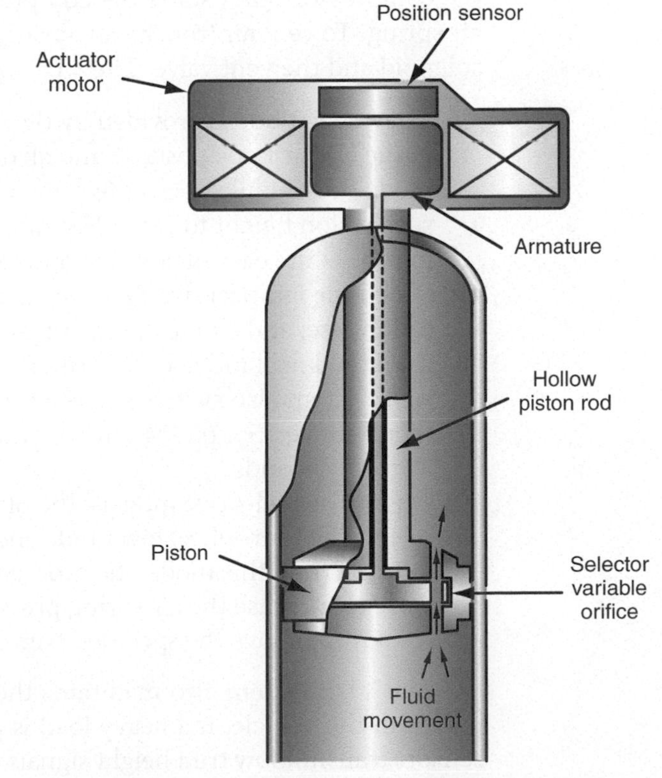

Figure 11-39 Strut actuator.

When driving on a smooth road surface in two-wheel drive, the suspension module positions the shock absorber valves to provide a soft ride. If the vehicle is driven on a rough road surface, the suspension's computer switches the shock absorber valves to the firm mode. When the driver selects four-wheel high mode, the suspension module starts the compressor and opens the air shock absorber valves, and increases the air shock absorber pressure until the suspension trim height is raised 1 in. (25.4 mm). If the vehicle speed exceeds 58 mph (93 km/h), the module returns the suspension height to the specified trim height. If the driver selects the four-wheel low mode, the suspension module operates the compressor and air shock absorber valves to increase the trim height 2 in. (50.8 mm). If the vehicle speed exceeds 30 mph (48 km/h in the four-wheel low mode), the module adjusts the suspension height to 1 in. (25.4 mm) above the specified trim height.

Shop Manual
Chapter 11, page 477

VEHICLE DYNAMIC SUSPENSION SYSTEM

Some SUVs are equipped with a **vehicle dynamic suspension (VDS)** system that is similar to the air suspension systems described previously in this chapter. The VDS system has these components:

1. Off/On service switch
2. Two front height sensors
3. One rear height sensor
4. Compressor with internal vent solenoid and air dryer
5. Control module
6. Air lines
7. Front and rear combined air springs and shock absorbers
8. Four air spring solenoids
9. Compressor relay

When increased air pressure is required in an air spring, the control module closes the compressor relay, starts the compressor, and opens the solenoid on the appropriate air spring. To vent air from an air spring, the control module must energize the air spring solenoid and the vent valve. The VDS system has three operating modes:

1. The kneel mode is provided by the suspension module when the ignition switch is in the Off or Lock position and all doors, liftgate, and liftgate glass are closed. In this mode the module opens the vent valve and air spring valves, and slowly reduces the suspension height to 1 in. (25.4 mm) below the specified trim height. This mode improves the ease of entering and exiting the vehicle.
2. When the ignition switch is On, and the transmission is initially shifted into Drive or Reverse, and all the doors, liftgate, and liftgate glass are closed, the VDS switches from the kneel mode to the trim mode, which provides the normal trim height. The VDS system also switches to the trim height mode if the module detects a vehicle speed above 15 mph (24 km/h). Transitions between modes require approximately 30 to 45 seconds.
3. On four-wheel-drive models, the off-road height mode is provided when the driver selects the four-wheel low mode and the vehicle speed is less than 25 mph (40 km/h). In this mode, the module starts the compressor and opens the air spring valve to increase the air spring pressure until the suspension height is 1 in. (25.4 mm) above the specified trim height.

The VDS system also maintains the specified trim height in relation to the weight placed in the vehicle. If a heavy load is placed in the rear of the vehicle, the rear height sensors transmit low trim height signals to the module, and the module opens the rear air spring solenoids and starts the compressor to restore the proper rear suspension trim height. The system stores front and rear trim height when a door or the rear liftgate is opened. The module maintains this suspension height even if weight is added to or removed from the vehicle. When all the doors, liftgate, and liftgate glass are closed, the system returns to normal operation. The VDS system makes limited height adjustments for 40 minutes after the ignition switch is turned Off.

ELECTRONIC SUSPENSION CONTROL (ESC) SYSTEM

General Description

The **electronic suspension control (ESC) system** design, which is favored by manufacturers, uses magneto-rheological fluid in the struts or shocks, which offers advanced damping technology. The magneto-rheological fluid offers rapid response to changing road conditions and driver requirements. Some manufactures have also integrated magneto-rheological fluid into engine mounts and other vehicle dampening mounts to improve driver comfort. The magneto-rheological system automatically controls damping forces in the front and rear struts or shock absorbers within milliseconds in response to various road and driving conditions. This system also offers improved performance of the stability control system on vehicles. The ESC system changes shock and strut damping forces in 1 to 12 milliseconds, whereas other suspension damping systems require a much longer time interval to change damping forces. It requires about 200 milliseconds to blink your eye. This gives us some idea how quickly the ESC system reacts. On some older models the ESC system may be called a continuously variable road sensing suspension (CVRSS).

The powertrain control module (PCM) is a computer that controls such output functions as electronic fuel injection, spark advance, emission devices, and transaxle shifting.

The ALC system is similar to the rear load-leveling suspension system explained previously in this chapter.

The ESC module receives inputs regarding vertical acceleration, wheel-to-body position, speed of wheel movement, vehicle speed, and lift/dive (**Figure 11-40**). The CVRSS module evaluates these inputs and controls a solenoid in each shock or strut to provide suspension damping control. The solenoids in the shocks and struts can react much faster compared with the strut actuators explained previously in some systems.

The ESC module also controls the speed-dependent steering system called MagnaSteer® and the automatic level control (ALC). This MagnaSteer® system is similar to the electronic variable orifice (EVO) steering explained in Chapter 13 under

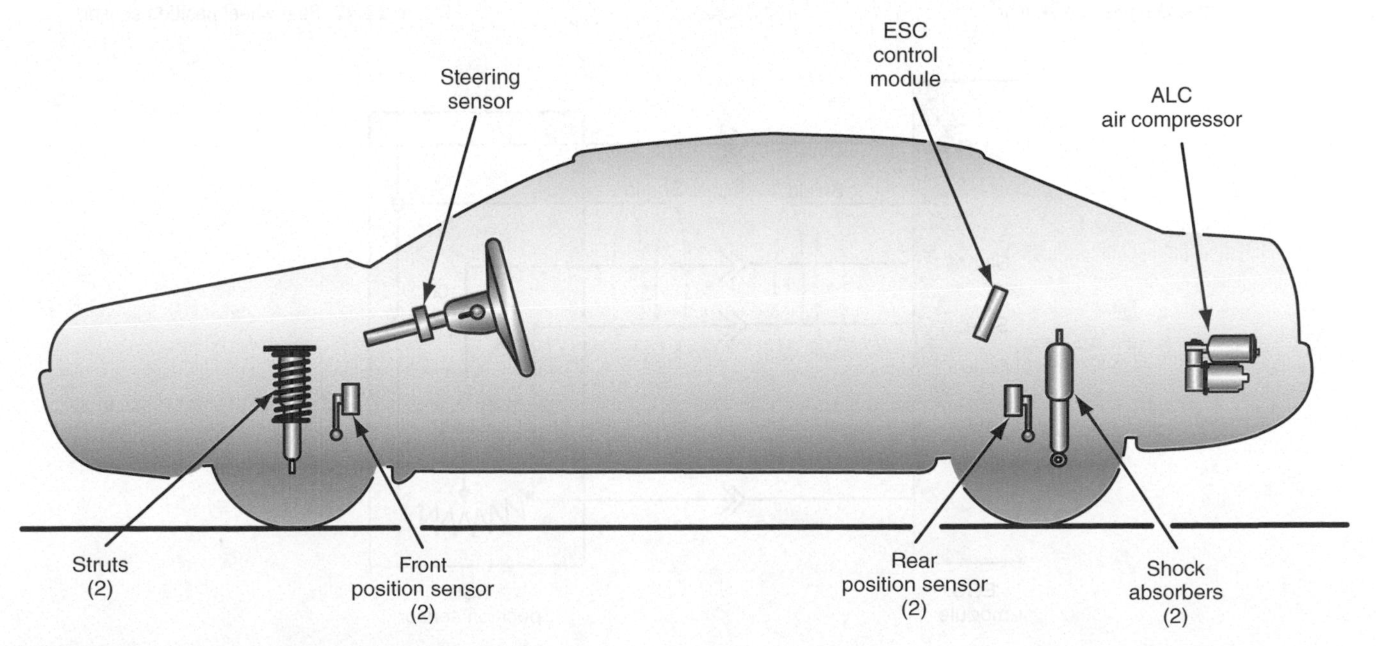

Figure 11-40 Electronic suspension control (ESC) system.

"Conventional and Electronic Rack and-Pinion Steering Gears." The ALC system controls the air pressure in the rear air shock absorbers to maintain the proper rear suspension height.

Inputs

Position Sensors A **wheel position sensor** is mounted at each corner of the vehicle between a control arm and the chassis (**Figure 11-41** and **Figure 11-42**). These sensor inputs provide analog voltage signals to the ESC module regarding relative wheel-to-body movement and the velocity of wheel movement (**Figure 11-43**). The rear position sensor inputs also provide rear suspension height information to the ESC module, and this

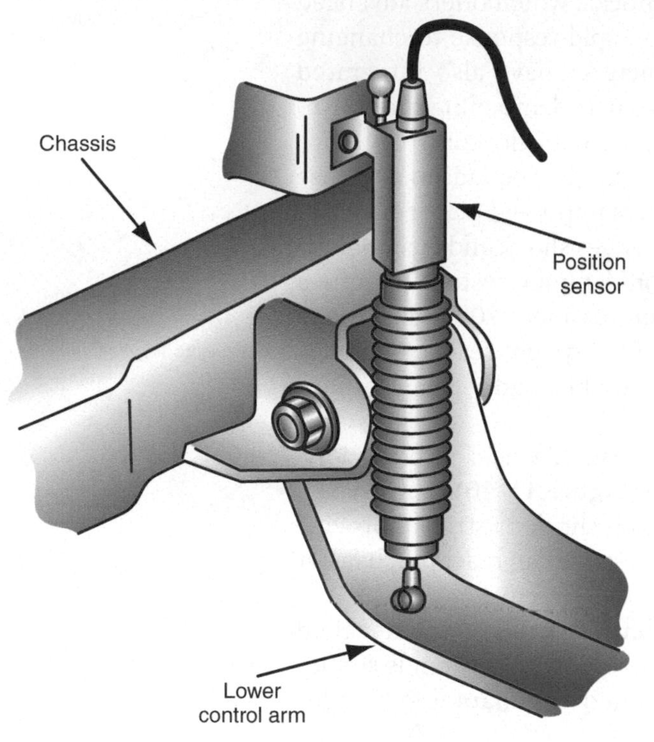

Figure 11-41 Front wheel position sensor.

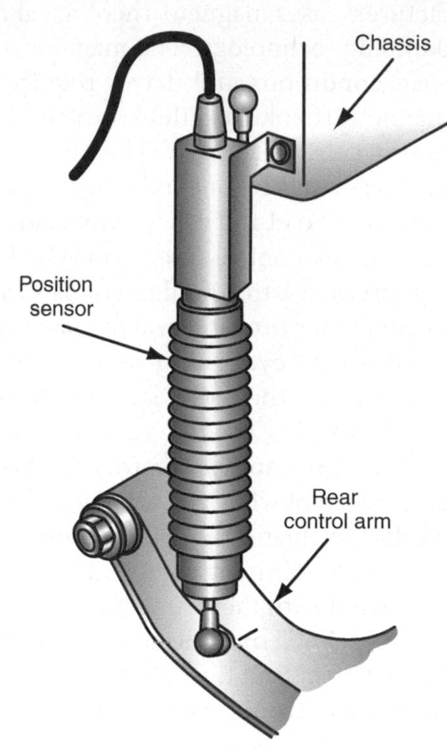

Figure 11-42 Rear wheel position sensor.

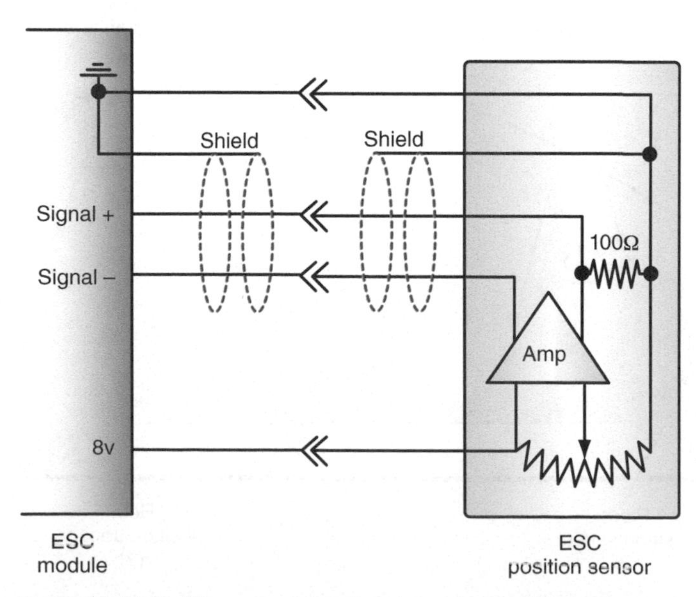

Figure 11-43 Position sensor internal design and wiring diagram.

information is used by the module to control the rear suspension trim height. All four position sensors have the same design. The wheel position sensors may be linear-type or rotary-type.

Accelerometer An **accelerometer** is mounted on each corner of the vehicle. These inputs send information to the ESC module in relation to vertical acceleration of the body. The front accelerometers are mounted on the strut towers (**Figure 11-44**), and the rear accelerometers are located on the rear chassis near the rear suspension support (**Figure 11-45**). All four accelerometers are similar in design, and they send analog voltage signals to the ESC module (**Figure 11-46**). On some later model vehicles, the four accelerometers are replaced by a single accelerometer under the driver's seat. On other late-model vehicles, the accelerometer(s) are eliminated.

Vehicle Speed Sensor The **vehicle speed sensor (VSS)** is mounted in the transaxle. This sensor sends a voltage signal to the PCM in relation to vehicle speed (**Figure 11-47**). The VSS signal is transmitted via data links from the PCM to the ESC module.

Lift/Dive Input The **lift/dive input** is sent from the PCM to the ESC module (**Figure 11-48**). Suspension lift information is obtained by the PCM from the throttle position, vehicle speed, and transaxle gear input signals. The PCM calculates suspension dive information from the rate of vehicle speed change when decelerating.

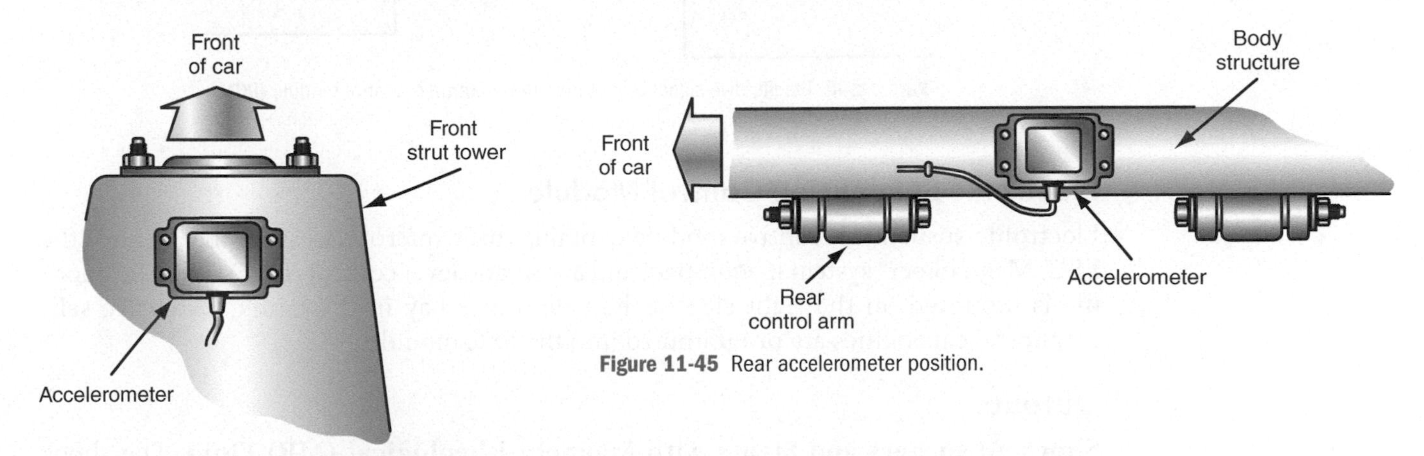

Figure 11-44 Front accelerometer mounting location.

Figure 11-45 Rear accelerometer position.

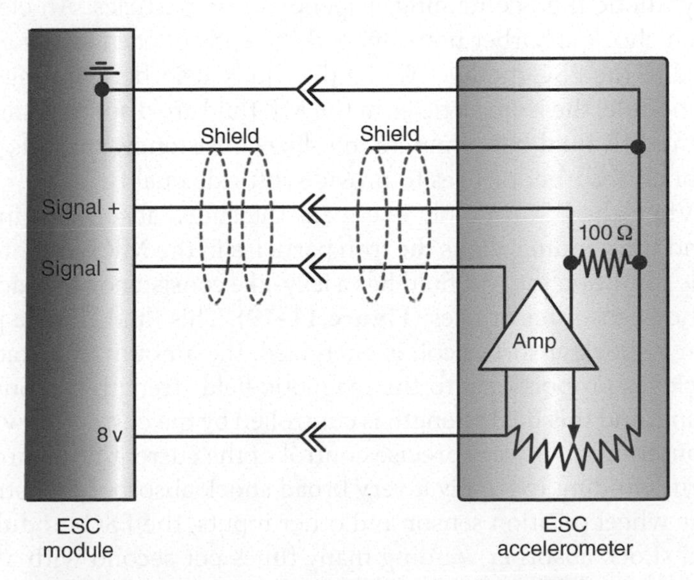

Figure 11-46 Accelerometer internal design and wiring diagram.

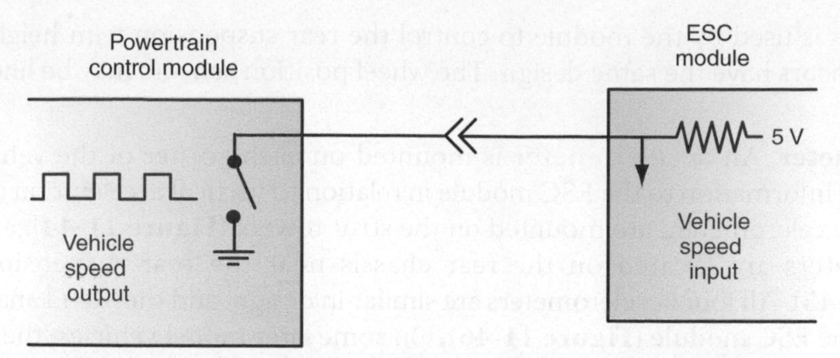

Figure 11-47 The vehicle speed sensor (VSS) signal is sent to the powertrain control module (PCM) and transmitted to the ESC module.

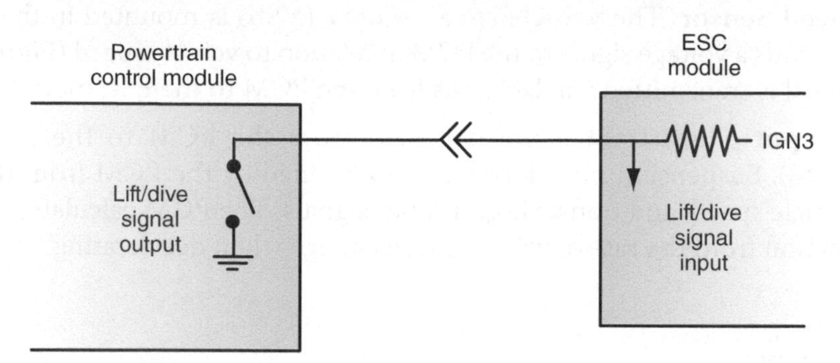

Figure 11-48 The lift/dive signal is sent from the powertrain control module (PCM) to the ESC module.

Electronic Suspension Control Module

Electronic suspension control module contains three microprocessors that control the ESC, MagnaSteer® system if equipped, and automatic level control (ELC). The ESC module is mounted on the right side of the electronics bay in the trunk. Extensive self-diagnostic capabilities are programmed into the ESC module.

Outputs

Shock Absorbers and Struts with Magneto-Rheological (MR) Fluid The shock absorbers or struts in the ESC system contain magneto-rheological (MR) fluid. This fluid is a synthetic fluid containing suspended iron particles. An electric winding is mounted in each shock absorber housing, and the ends of each shock absorber winding are connected to the ESC module. When the shock absorber windings are not energized by the ESC module, the iron particles in the MR fluid are dispersed randomly. Under this condition the MR fluid has a mineral oil–like consistency, and this fluid flows easily through the shock absorber orifices to provide soft ride quality.

When the ESC module energizes the shock absorber windings, the magnetic field around the winding aligns the iron particles in the MR fluid into thick fibrous structures. In this condition the MR fluid has a jelly-like consistency that does not flow easily through the shock absorber orifices (**Figure 11-49**). This fluid change provides firm ride quality. When a shock absorber coil is energized, the amount of attraction between the fibrous particles is proportional to the magnetic field strength surrounding the shock absorber winding, and this field strength is controlled by the current flow through the winding. The computer provides very precise control of the current flow through each shock absorber or strut winding to supply a very broad shock absorber and strut damping range. Based on the wheel position sensor and other inputs, the ESC module can energize each individual shock absorber winding many times per second with a **pulse width modulated (PWM)** signal. Therefore, the ESC system with magneto-rheological fluid in the shock

A **pulse width modulated** signal is a signal with a variable On time that may be used by a computer to control an output.

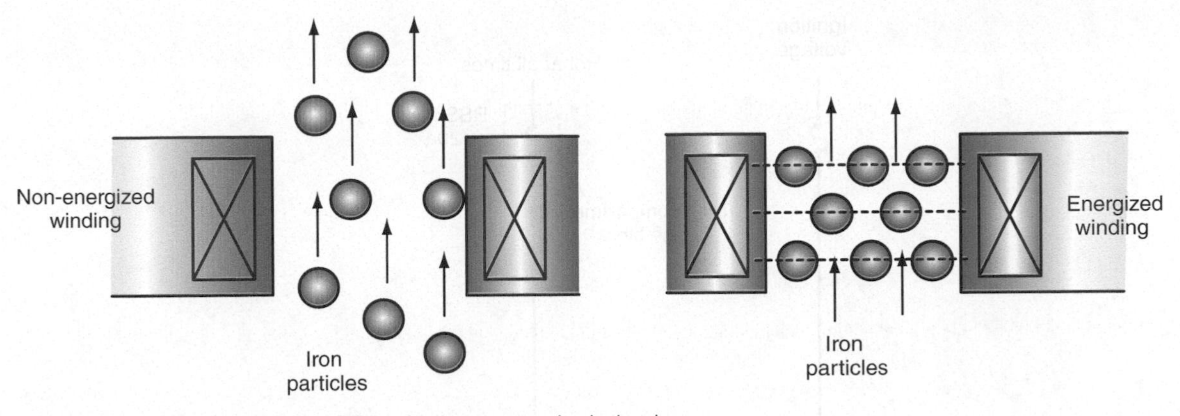

Figure 11-49 Magneto-rheological fluid action in a strut or shock absorber.

absorbers or struts provides an almost infinite variation in shock absorber damping. The ESC system can change the damping characteristics of the MR fluid in 1 millisecond. The ESC system may be called a **MagneRide system**.

The ESC system with MR fluid in the shock absorbers or struts provides greatly improved control of pitch and body roll motions, which supplies better road-holding capabilities, steering control, ride quality, and safety. Recent updates to the ESC system with MR fluid in the shock absorbers or struts include the following:

1. Improved computer software to provide a broader damping range and more precise control of the shock actuators.
2. Expanded sensor inputs including brake pressure, vehicle yaw rate, steering angle, and engine torque.
3. Improved computer networks to allow increased and faster data transmission.

These updates provide improved vehicle dynamics and ride quality without any increase in packaging or weight.

Damper Solenoid Valves On older designs and models that do not use magneto-rheological fluid, but instead use conventional hydraulic fluid, each strut or shock damper contains a solenoid that is controlled by the ESC module. Each **damper solenoid valve** provides a wide range of damping forces between soft and firm levels by increasing or decreasing fluid flow through or around the damper piston by decreasing or increasing the bypass passage size. While this system is responsive to changing road conditions and driving demands, it is not as responsive as the magneto-rheological fluid systems. Strut or shock absorber damping is controlled by the amount of current supplied to the damper solenoid in each strut or shock absorber. Battery voltage and ignition voltage are supplied through separate fuses to the ESC module (**Figure 11-50**). If the damper solenoids are not energized, the struts provide minimum damping force. When the damper solenoids are energized, the struts provide increased damping force for a firmer ride. The ESC module switches the voltage supplied to the damper solenoid in each strut on and off very quickly with a 2 kilohertz pulse width modulated action. If the ESC module keeps the damper solenoid in a strut energized longer on each cycle, current flow is increased through the strut damper solenoid. Under this condition, strut damping force is increased to provide a firmer ride. The ESC module provides precise, variable control of the current flow through each strut or shock damper solenoid to achieve a wide range of damping forces in the struts. The ESC system can change the shock absorber damping forces in 10 to 12 milliseconds.

Each damper solenoid is an integral part of the damper assembly and is not serviced separately. The ESC system operates automatically without any driver-controlled inputs.

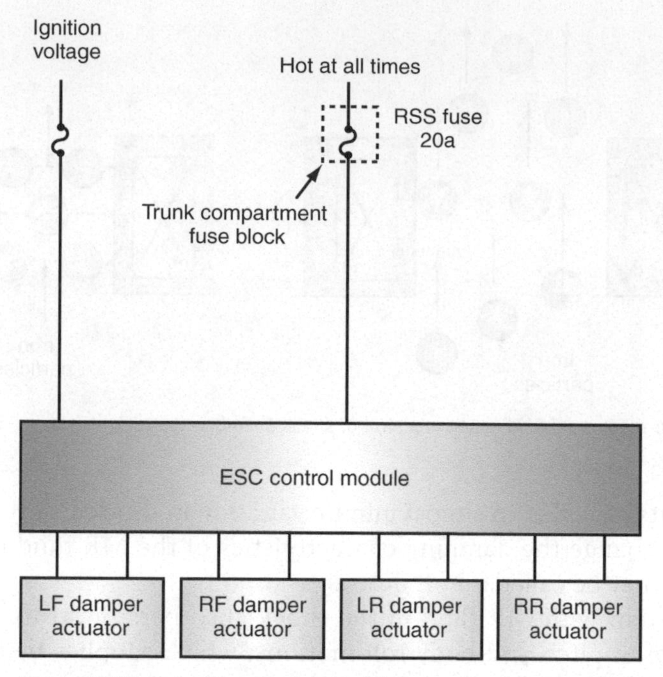

Figure 11-50 Strut damper solenoid circuit.

The fast reaction time of the ESC system provides excellent control over ride quality and body lift or dive, which provides improved vehicle stability and handling. Since the position sensors actually sense the velocity of upward and downward wheel movements and the damper solenoid reaction time is 11 to 12 milliseconds, the ESC module can react to these position sensor inputs very quickly. For example, if a road irregularity drives a wheel upward, the ESC module switches the damper solenoid to the firm mode before that wheel strikes the road again during the downward movement.

Rear Automatic Level Control

The **automatic level control (ALC)** system maintains the rear suspension trim height regardless of the rear suspension load. If a heavy object is placed in the trunk, the rear wheel position sensors send below trim height signals to the ESC module. When this signal is received, the ESC module grounds the ALC relay winding and closes the relay contacts that supply voltage to the compressor motor (**Figure 11-51**).

Once the compressor starts running, it supplies air through the nylon lines to the rear air shocks and raises the rear suspension height (**Figure 11-52**). When trim height signals are received from the rear wheel position sensors, the ESC module opens the compressor relay winding circuit and stops the compressor.

If a heavy object is removed from the trunk, the rear wheel position sensors send above trim height signals to the ESC module. Under this condition, the ESC module energizes the exhaust solenoid in the compressor assembly, and this action releases air from the rear air shocks. When the rear wheel position sensors send rear suspension trim height signals to the ESC module, this module shuts off the exhaust solenoid.

An independent ALC system is used on cars without the ESC system. In these systems, the computer is not required and a single suspension height sensor is used. This height sensor contains electronic circuits that control the compressor relay and the exhaust solenoid. This electronic circuit limits the compressor run time and the exhaust solenoid On time to seven minutes.

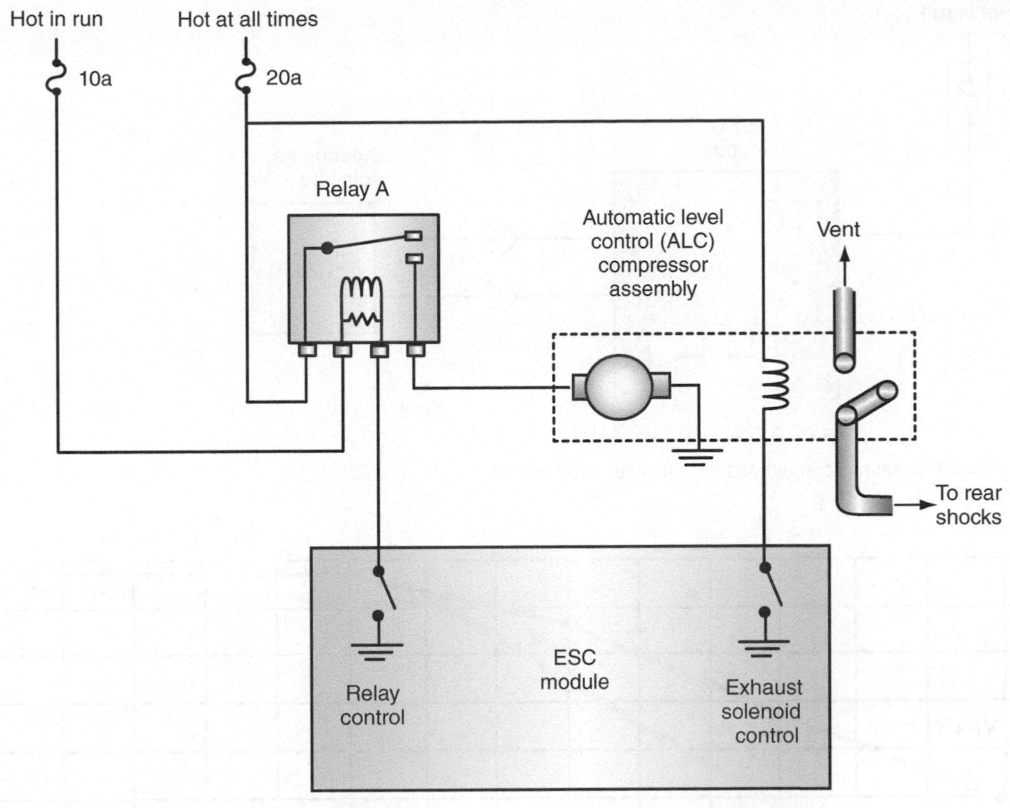

Figure 11-51 Rear automatic level control (ALC).

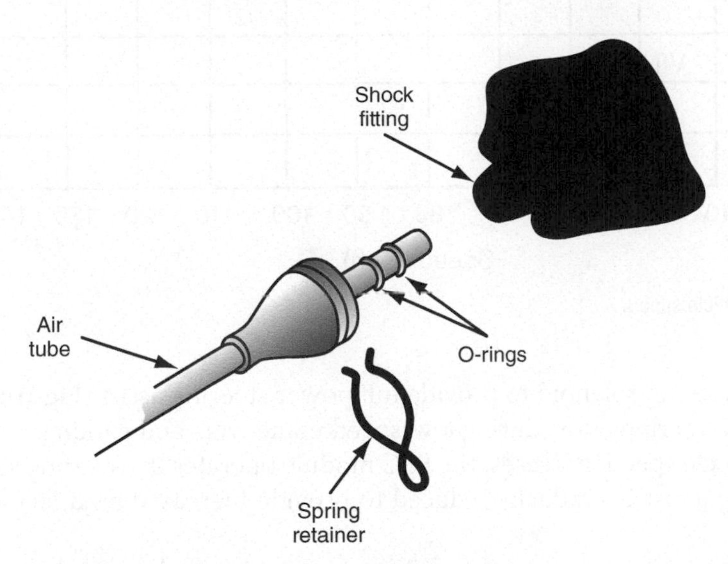

Figure 11-52 Nylon air line and rear shock air line fitting.

Speed-Sensitive Steering System

On some models the ESC module operates a solenoid in the **speed-sensitive steering (SSS) system** to control the power steering pump pressure in relation to vehicle speed (**Figure 11-53**). This action varies the power steering assist levels.

The ESC module varies the on time of the steering solenoid. This action may be referred to as pulse width modulation. When the solenoid is in the Off mode, the power steering pump supplies normal power assist. Below 10 mph (16 km/h), the computer

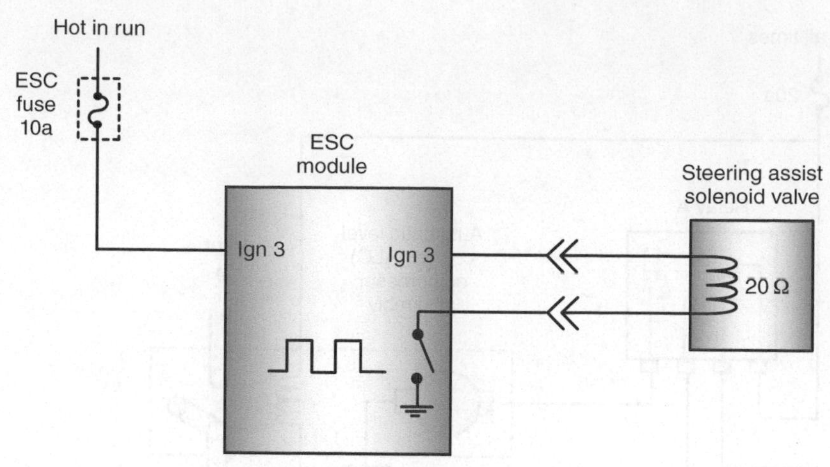

Figure 11-53 Steering solenoid and ESC module wiring diagram.

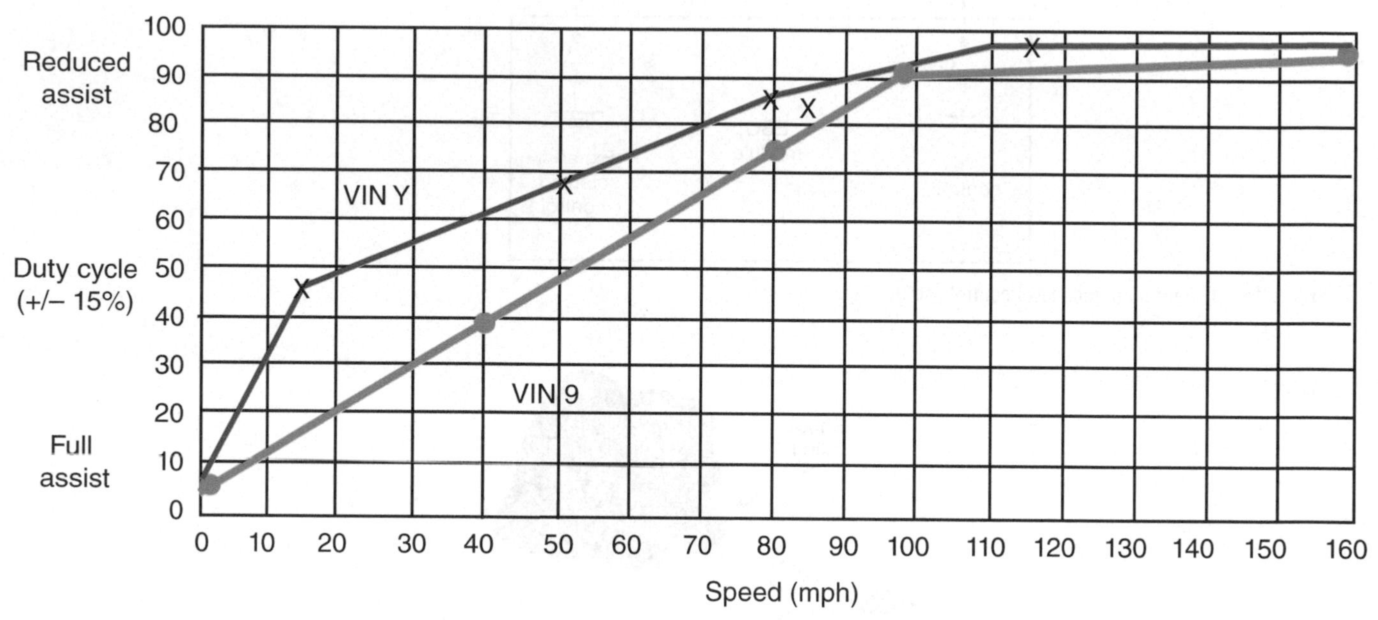

Figure 11-54 Power steering assist in relation to vehicle speed.

A **MagneRide** suspension system has magneto-rheological fluid in the shock absorbers.

operates the steering solenoid to provide full power steering assist (**Figure 11-54**). This action reduces steering effort during low-speed maneuvers and parking.

As the vehicle speed increases, the ESC module operates the steering solenoid so the power steering assist is gradually reduced to provide increased road feel and improved handling.

On later model cars, the speed-sensitive steering is called speed-dependent steering or MagnaSteer®. The module that controls the MagnaSteer® is contained in the electronic brake and traction control module (EBTCM).

VEHICLE STABILITY CONTROL

The National Highway Traffic Safety Administration (NHTSA) required that all new passenger vehicles sold in the United States to be equipped with electronic stability control (ESC) as of the 2012 model year. Some European models were equipped with ESC as early as 1995. A **vehicle stability control system** provides improved control if the vehicle begins to swerve sideways because of slippery road surfaces, excessive acceleration, or a combination of these two conditions. Therefore, a vehicle stability control system provides increased vehicle safety. Vehicle stability control systems have various brand names depending on the vehicle manufacturer. For example, on General Motors vehicles the vehicle stability control system is called **Stabilitrak®**. The Stabilitrak® system is available on many General Motors cars and some SUVs. The module that controls the Stabilitrak® system is combined with the **antilock brake system (ABS)** module and **traction control system (TCS)** module (**Figure 11-55**). This three-in-one module assembly is referred to as the **electronic brake and traction control module (EBTCM)**. The EBTCM is attached to the **brake pressure modulator valve (BPMV)**, and this assembly is mounted in the left front area in the engine compartment. A data link is connected between all the computers including the EBTCM and the ESC module (**Figure 11-56**). The combined EBTCM and ESC systems may be referred to as the integrated chassis control system 2 (ICCS2). Some sensors such as the **steering wheel position sensor (SWPS)** are hard-wired to both the EBTCM and the ESC module (**Figure 11-57**). The signals from other sensors may be sent to one of these modules and then transmitted to the other module on the data link. The data link also transmits data from these modules to the instrument panel cluster (IPC) during system diagnosis. This allows the IPC to display diagnostic information.

This book is concerned with suspension and steering systems. Because the stability control system operates in cooperation with the ABS and TCS systems, a brief description of these systems is necessary.

The **antilock brake system (ABS)** prevents wheel lockup during a brake application.

The **traction control system (TCS)** prevents drive wheel slippage.

The **electronic brake and traction control module (EBTCM)** controls ABS, TCS, and Stabilitrak® functions.

The **brake pressure modulator valve (BPMV)** controls brake fluid pressure to the wheel calipers or cylinders.

The **steering wheel position sensor** supplies a voltage signal in relation to the amount and speed of steering wheel rotation.

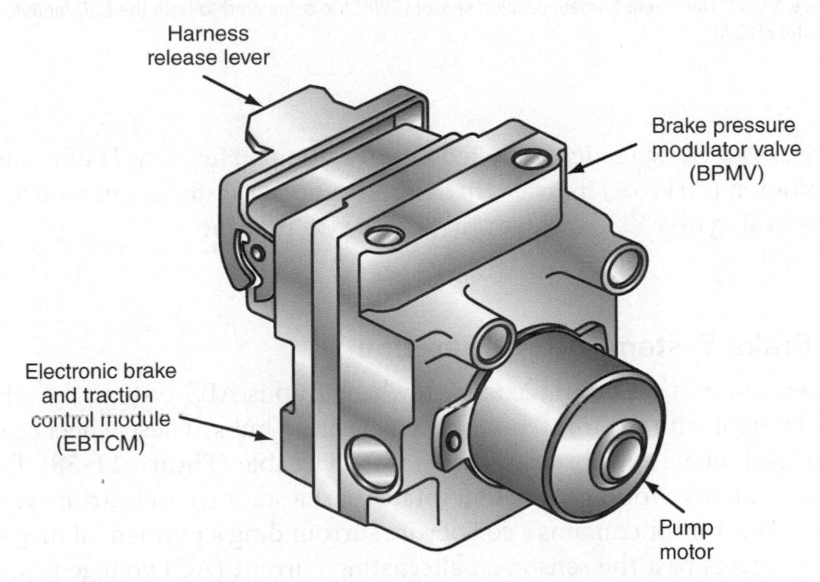

Figure 11-55 The electronic brake and traction control module (EBTCM) contains the antilock brake system (ABS), traction control system, and stability control modules.

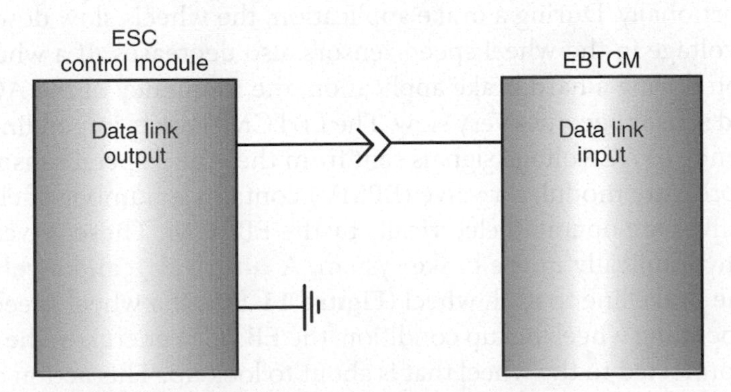

Figure 11-56 Data link between the EBTCM and ESC modules.

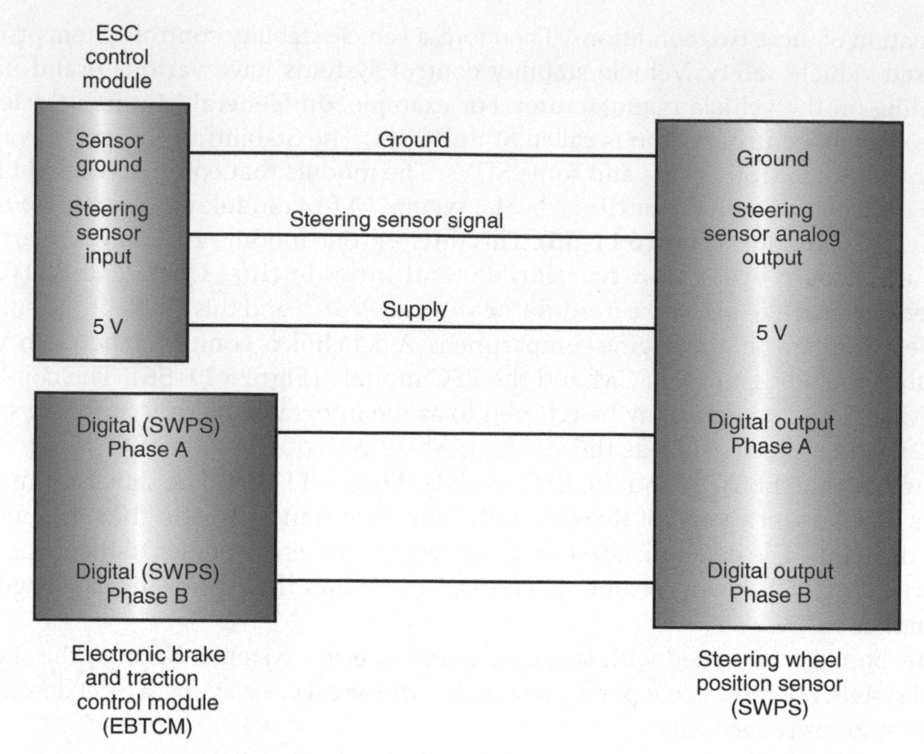

Figure 11-57 The steering wheel position sensor (SWPS) is connected to both the ESC module and the EBTCM.

AUTHOR'S NOTE Statistics compiled by the National Highway Traffic Safety Administration (NHTSA) indicate that stability control systems can reduce the incidence of single-vehicle accidents in SUVs by 63 percent.

Antilock Brake System (ABS) Operation

Wheel speed sensors are mounted at each wheel. In this ABS system, the wheel speed sensors are integral with the front or rear wheel bearing hubs. These wheel bearing hubs with the integral wheel speed sensors are non-serviceable (**Figure 11-58**). Each wheel speed sensor contains a toothed ring that rotates past a stationary electromagnetic wheel speed sensor. This sensor contains a coil of wire surrounding a permanent magnet. As the toothed ring rotates past the sensor, an alternating current (AC) voltage is produced in the sensor. This voltage signal from each wheel speed sensor is sent to the EBTCM. As wheel speed increases, the frequency of AC voltage produced by the wheel speed sensor increases proportionally. During a brake application, the wheels slow down, and the frequency of AC voltage in the wheel speed sensors also decreases. If a wheel is nearing a lockup condition during a hard brake application, the frequency of the AC voltage from that wheel speed sensor becomes very slow. The EBTCM detects impending wheel lockup from the frequency of AC voltage signals sent from the wheel speed sensors.

A **hold valve** opens and closes the fluid passage to each wheel caliper.

When energized, a **release valve** reduces pressure in a wheel caliper.

The brake pressure modulator valve (BPMV) contains a number of electrohydraulic valves. These valves are operated electrically by the EBTCM. These valves in the BPMV are connected hydraulically in the brake system. A **hold valve** and a **release valve** are connected in the brake line to each wheel (**Figure 11-59**). If a wheel speed sensor signal indicates an impending wheel lockup condition, the EBTCM energizes the normally open hold solenoid connected to the wheel that is about to lock up. This action closes the solenoid and isolates the wheel caliper from the master cylinder to prevent any further

Figure 11-58 Wheel speed sensor.

increase in brake pressure. If the wheel speed sensor signal still indicates an impending wheel lockup, the EBTCM keeps the hold solenoid closed and opens the normally closed release solenoid momentarily. This action reduces wheel caliper pressure to reduce brake application force and prevent wheel lockup. The EBTCM pulses the hold and the release solenoids on and off to supply maximum braking force without wheel lockup.

When the hold and the release valves are pulsated during a prolonged antilock brake function, the brake pedal fades downward as brake fluid flows from the release valves into the accumulators. To maintain brake pedal height during an antilock brake function, the EBTCM starts the pump in the BPMV at the beginning of an antilock function. When the pump motor is started, the pump supplies brake fluid pressure to the hold valves and wheel calipers. Pump motor pressure is also supplied back to the master cylinder. Under this condition, the driver may feel a firmer brake pedal and pedal pulsations and may hear the clicking action of the hold and the release solenoids.

ANTILOCK and BRAKE warning lights are mounted in the instrument panel. Both of these lights are illuminated for a few seconds after the engine starts. If the amber ANTILOCK light is on with the engine running, the EBTCM has detected an electrical fault in the ABS system. Under this condition, the EBTCM no longer provides an ABS function, but normal power-assisted brake operation is still available. When the red BRAKE warning light is illuminated with the engine running, the parking brake may be on, the master cylinder may be low on brake fluid, or there may be a fault in the ABS system.

Traction Control System (TCS) Operation

The EBTCM detects drive wheel spin by comparing the two drive wheel speed sensor signals. Wheel spin on both drive wheels is detected by comparing the wheel speed sensor signals on the drive wheels and non-drive wheels. If a wheel speed sensor signal informs the EBTCM that one or both drive wheels are spinning, the EBTCM enters the traction control mode. First, the EBTCM requests the PCM to reduce the amount of engine torque supplied to the drive wheels. The PCM reduces engine torque by retarding the ignition spark advance and shutting off the fuel injectors for a very short time. The PCM then sends a signal back to the EBTCM regarding the amount of torque delivered to the drive

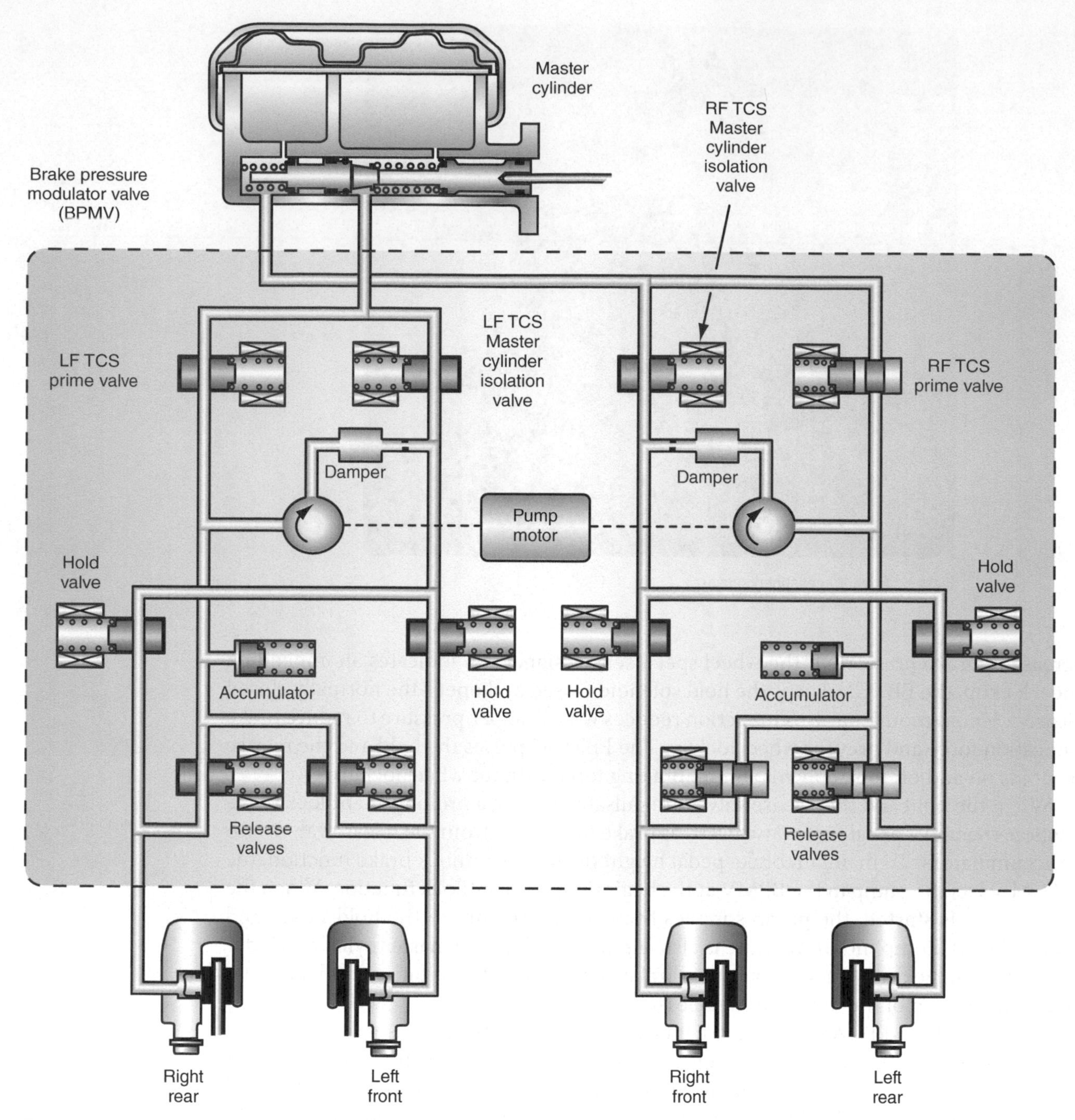

Figure 11-59 Brake pressure modulator valve (BPMV).

wheels. If one or both drive wheels continue to spin on the road surface, the EBTCM energizes the normally closed prime valve, closes the normally open isolation valve, and starts the pump in the BPMV. Under this condition, the prime valve opens and the pump begins to move brake fluid from the master cylinder through the prime valve to the pump inlet. The closed isolation valve prevents the pump pressure from being applied back to the master cylinder. Under this condition, the pump pressure is supplied through the normally open hold valve to the brake caliper on the spinning wheel. This action stops the wheel from spinning. If both drive wheels are spinning on the road surface, the EBTCM operates both prime valves and isolation valves to supply brake fluid pressure to both drive wheel brake calipers. During a TCS function, the EBTCM pulses the hold and the release solenoids on and off to control wheel caliper pressure. The EBTCM limits the traction control function to a short time period to prevent overheating brake components. During a TCS function, these messages may be displayed in the driver information center (DIC):

1. TRACTION ENGAGED is displayed after the TCS is in operation for three seconds.
2. TRACTION SUSPENDED is displayed if the EBTCM has discontinued the TCS function to prevent brake component overheating.
3. TRACTION OFF is displayed if the driver places the TCS switch on the instrument panel in the Off position.
4. TRACTION READY is displayed if the TCS switch is turned from Off to On.

During a TCS function, the EBTCM sends a signal through the data link to the PCM. When this signal is received, the PCM disables some of the fuel injectors to reduce engine torque. This action also helps to prevent drive wheel spin. The PCM disables the two injectors at the beginning of the firing order and in the center of the firing order. Depending on the speed of drive wheel spin, the PCM may disable every second injector in the firing order. The injectors are disabled for a very short time period. The TCS system improves drive wheel traction, and this system also prevents the tendency for the vehicle to swerve sideways when one drive wheel is spinning. Therefore, the TCS system increases vehicle safety.

Vehicle Stability Control

To provide stability control, the EBTCM uses two additional input signals from the **lateral accelerometer** and the **yaw rate sensor**. The lateral accelerometer is mounted under the front passenger's seat (**Figure 11-60**). The EBTCM sends a 5 V reference voltage to the lateral accelerometer. If the vehicle is driven straight ahead, the chassis has zero lateral acceleration. Under this condition, the lateral accelerometer provides a 2.5 V signal to the EBTCM. If the vehicle begins to swerve sideways because of slippery road conditions, high-speed cornering, or erratic driving, the lateral accelerometer signal to the EBTCM varies from 0.25 V to 4.75 V, depending on the direction and severity of the swerving action.

A **lateral accelerometer** supplies a voltage signal in relation to sideways movement of the chassis.

Yaw is erratic, side-to-side deviation from a course. The **yaw rate sensor** supplies a voltage signal in relation to rotational chassis speed during a sideways swerve.

The yaw rate sensor is mounted under the rear package shelf (**Figure 11-61**). Some yaw rate sensors contain a precision metal cylinder whose rim vibrates in elliptical shapes. The vibration and rotation of this metal cylinder is proportional to the rotational speed of the vehicle around the center of the cylinder. On some models, the lateral accelerometer and yaw rate sensors are combined in a single sensor. The sensor mounting location varies depending on the vehicle make and model year.

The EBTCM sends a 5 V reference voltage to the yaw rate sensor. If the vehicle chassis experiences zero yaw rate, the yaw rate sensor sends a 2.5 V signal to the EBTCM module.

Figure 11-60 Lateral accelerometer.

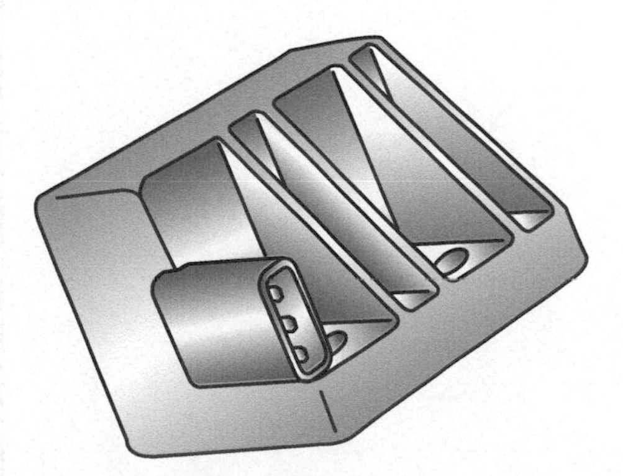

Figure 11-61 Yaw rate sensor.

If the vehicle begins to swerve sideways, the yaw rate sensor provides a 0.25V to 4.75V signal to the EBTCM, depending on the direction and severity of the swerve. The EBTCM also uses the wheel speed sensor signals for stability control. If the vehicle begins to swerve sideways, the EBTCM energizes the normally closed prime valve and closes the normally open isolation valve connected to the appropriate front wheel; then it starts the pump in the BPMV. Under this condition, the prime valve opens and the pump begins to move brake fluid from the master cylinder through the prime valve to the pump inlet. The closed isolation valve prevents the pump pressure from being applied back to the master cylinder. Under this condition, the pump pressure is supplied through the normally open hold valve to the brake caliper on the appropriate front wheel. Applying the brake on the front wheel pulls the vehicle out of the swerve and prevents the complete loss of steering control. If the EBTCM detects an electrical fault in the stability control system, STABILITY REDUCED is displayed in the DIC. If the EBTCM enters the stability control mode, STABILITY ENGAGED is indicated in the DIC.

In **Figure 11-62**, two vehicles driving side-by-side are negotiating a lane change to the left. The vehicle on the right has vehicle stability control, and the vehicle on the left

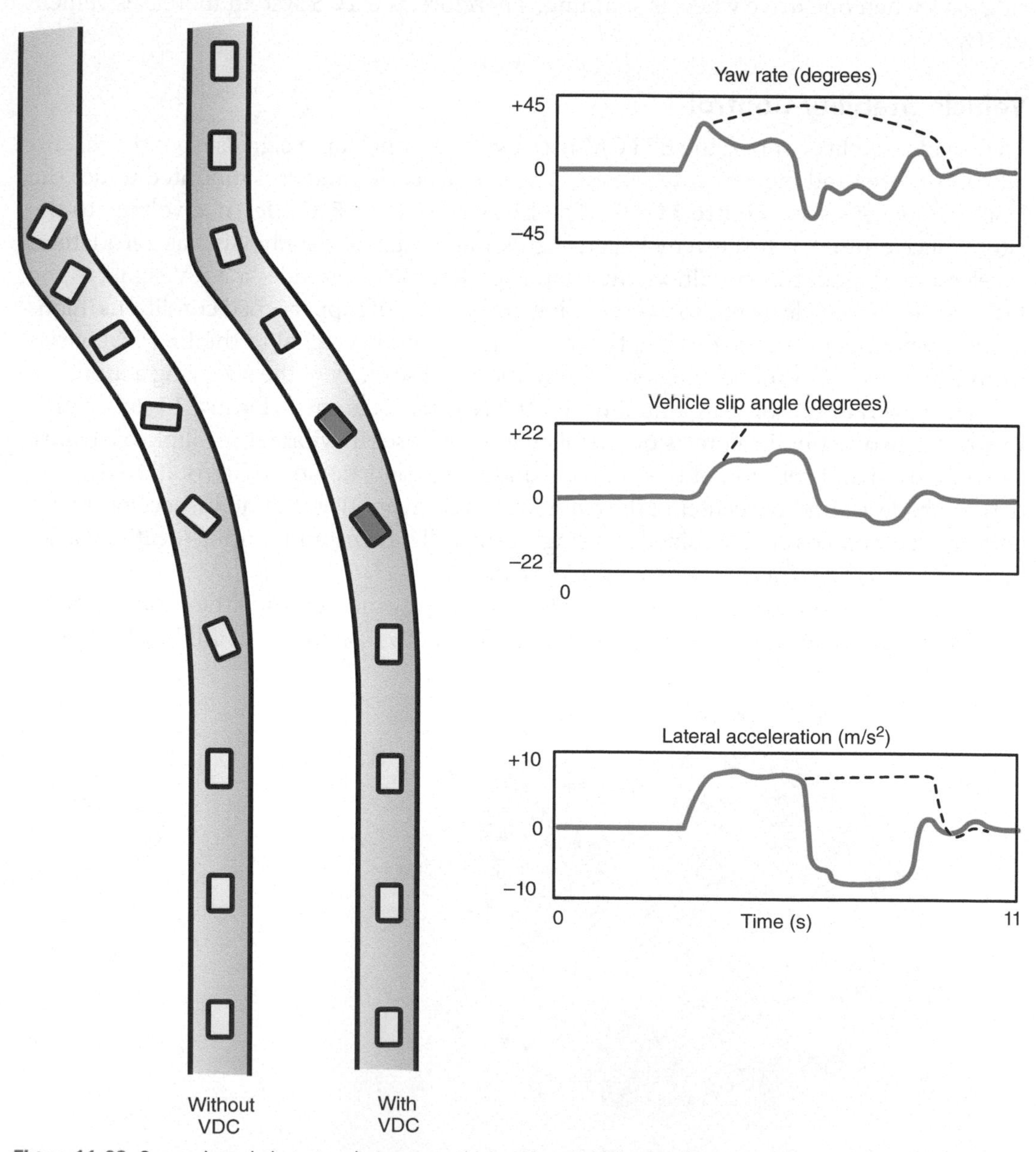

Figure 11-62 Comparison during a turn between a vehicle with a stability control system and a vehicle with no stability control system.

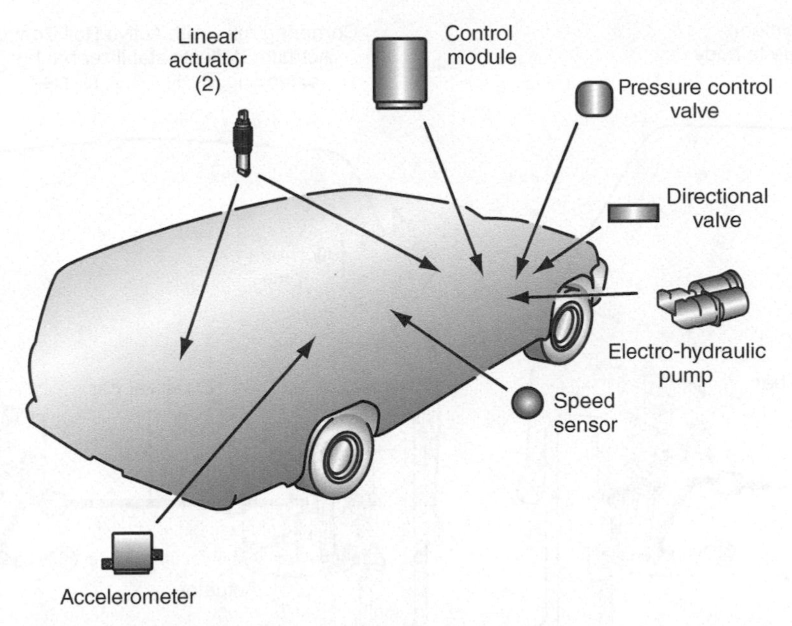

Figure 11-63 Active roll control system components.

does not have this system. As the vehicle on the left begins to turn, the rear of the vehicle begins to swing around. When the vehicle on the right starts to turn, the rear of the vehicle swerves slightly and the right front brake is applied by the vehicle stability control system to prevent this swerve. Further into the turn, the driver attempts to steer the car on the left, but this car enters into an uncontrolled swerve with loss of steering control. As the car on the right continues into the turn, the rear of the vehicle swerves slightly, but the vehicle stability control system again applies the right front brake momentarily to prevent this swerve. The car with the stability control system completes the turn while maintaining directional stability, but the vehicle without stability control goes into an uncontrolled swerve with complete loss of directional control. The vehicle stability control system improves vehicle safety! The other charts in **Figure 11-63** indicate that yaw rate, vehicle slip angle, and lateral acceleration are greatly reduced on a vehicle with a stability control system.

ACTIVE ROLL CONTROL SYSTEMS

Two independent automotive component manufacturers have developed active roll control systems in response to concerns about SUVs, which roll over more easily compared with cars because of their higher center of gravity. The active roll control system contains a control module, accelerometer, speed sensor, fluid reservoir, electrohydraulic pump, pressure control valve, directional control valve, and a hydraulic actuator in both the front and rear stabilizer bars (**Figure 11-63**). The accelerometer and speed sensor may be common to systems other than the active roll control. The electrohydraulic pump may also be used as the power steering pump. The active roll control system may be called a roll stability control (RSC) system or a dynamic handling system (DHS). Some active roll control systems have a gyro sensor that provides a voltage signal to the control module in relation to the vehicle roll speed and angle.

When the vehicle is driven straight ahead, the active roll control system does not supply any hydraulic pressure to the linear actuators in the stabilizer bars. Under this condition, both stabilizer bars move freely until the linear actuators are fully compressed. This action provides improved individual wheel bump performance and better ride quality. If the chassis begins to lean while cornering, the module operates the directional valve

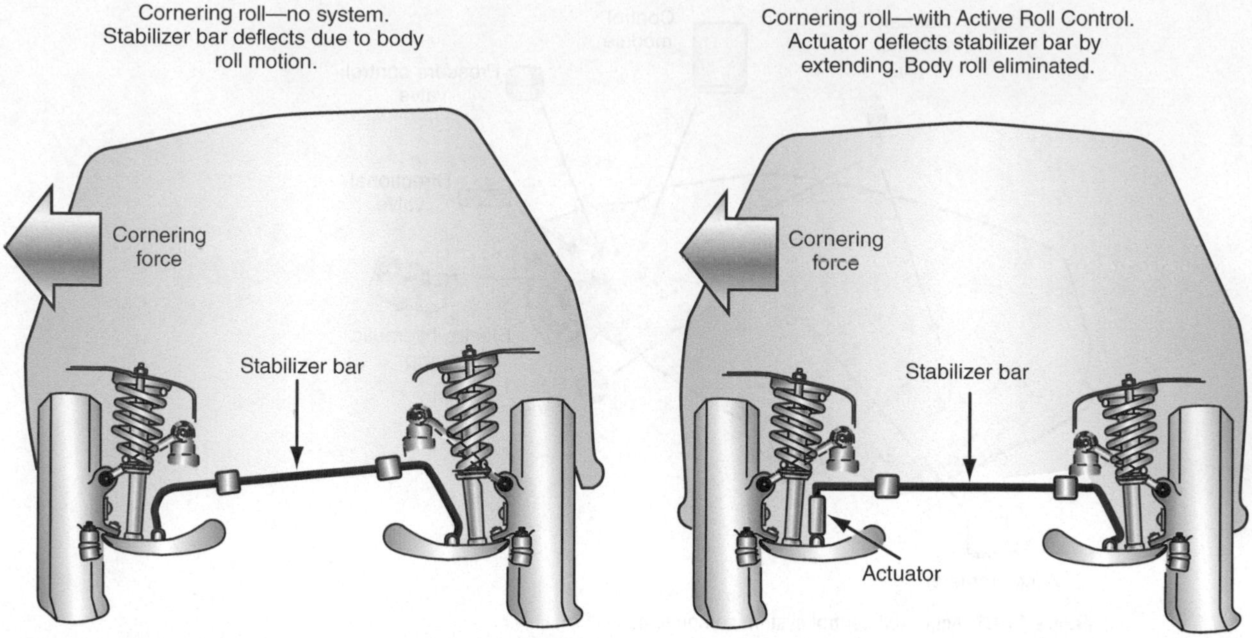

Cornering roll—no system. Stabilizer bar deflects due to body roll motion.

Cornering force

Stabilizer bar

Cornering roll—with Active Roll Control. Actuator deflects stabilizer bar by extending. Body roll eliminated.

Cornering force

Stabilizer bar

Actuator

Figure 11-64 Active roll control system operation while cornering.

so it supplies fluid pressure to the linear actuators in the stabilizer bars. This action stiffens the stabilizer bar and reduces body lean (**Figure 11-64**). The active roll control system increases safety by reducing body lean, which decreases the possibility of a vehicle rollover.

One vehicle manufacturer has chosen to reduce the yaw force on a new model by reducing the weight at the front and rear of the vehicle and shifting the weight toward the midpoint on the car. This was accomplished by manufacturing the hood, trunk lid, and bumper beams from aluminum. Engine weight was reduced by installing hollow camshafts, and using a lighter weight block and cylinder heads. These changes reduced the weight of the vehicle by 180 lb (80 kg), which reduced the yaw force and also improved fuel economy and performance.

ADAPTIVE CRUISE CONTROL (ACC) SYSTEMS

Adaptive cruise control (ACC) is available on some current vehicles. The ACC system has a long-range radar sensor mounted behind the front bumper. Radar stands for Radio Detection and Ranging. Radar is an electronic system developed for establishing the position of an object, and is based on the principle that electromagnetic waves are reflected by surfaces of an object and the waves returned are detected as an "echo" of the object. The ACC system measures the distance to the vehicle in front and the relative speed of that vehicle (**Figure 11-65**) using the radar. The radar emits a frequency modulated continuous wave at a 200 megahertz modulation and a carrier signal with a frequency of 76.5 gigahertz (**Figure 11-66**). The ACC sensor monitors the time it takes for the signal to be returned. The further the distance, the longer it takes for the signal to be returned (**Figure 11-67**). This radar sensor sends signals to the ACC computer. To determine the speed of the vehicle in front, a phenomenon known as the Doppler effect is used. An example of the Doppler effect can be demonstrated, as an emergency vehicle approaches, the siren sounds to be a constant high-pitch (high-frequency) noise. But as the emergency vehicle passes and moves farther away the tone now sounds lower (lower frequency). If the distance between the transmitter and the object increases, the frequency of the

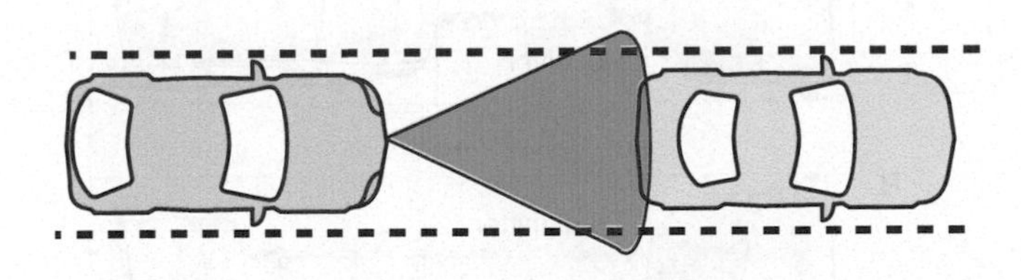

Figure 11-65 The radar sensor measures the distance to the vehicle in front and the relative speed of this vehicle.

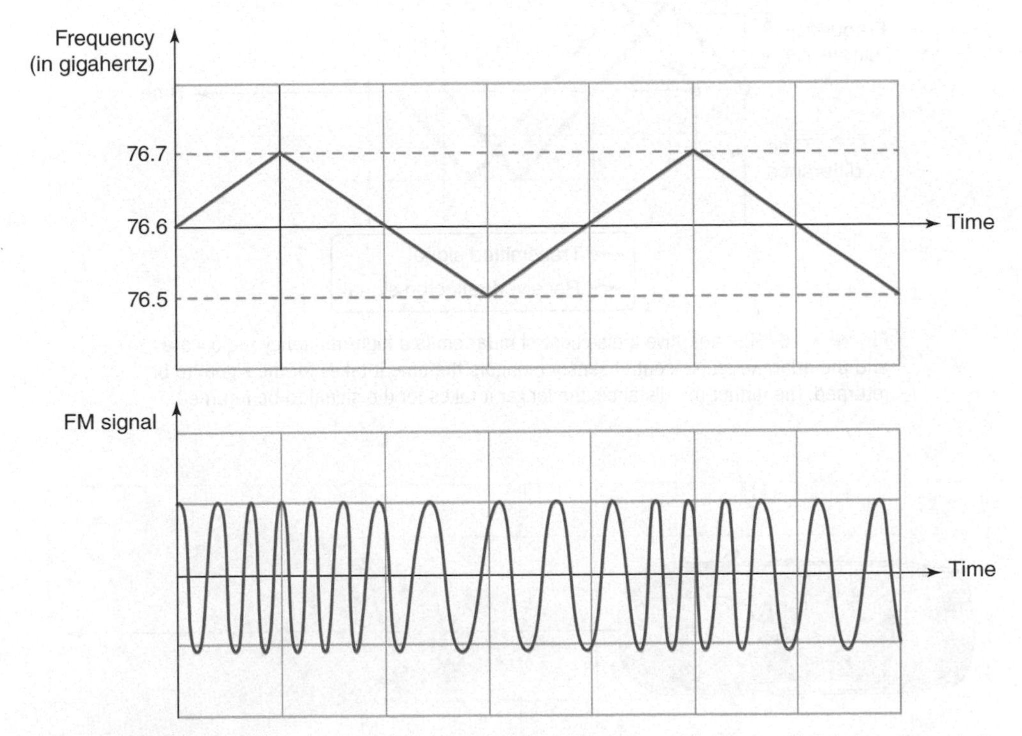

Figure 11-66 The adaptive cruise control radar emits a high-frequency radio wave modulated continuous at 200 megahertz and a carrier signal with a frequency of 76.5 gigahertz.

reflected wave decreases, and if the distance between the transmitter and the object decreases, the frequency of the reflected wave increases. The adaptive cruise control determines the vehicle in front has increased its speed by a decreasing frequency signal (ΔfD) received. The result is a differential frequency signal difference between the transmitted leading signal (Δf1) and the received trailing edge signal (Δf2). This signal difference is analyzed by the control module to determine vehicle in front's speed .The radar signal produced is a lobe-shaped signal pattern whereby the signal strength decreases as the distance from the transmitter is increased (**Figure 11-68**). The ACC system uses a three-beam radar to determine the angle the vehicle in front is traveling. The three radar receivers compare the ratio of the signal strengths (amplitude) returned to calculate angular information (**Figure 11-69**). The ACC computer uses the CAN network to remain in constant contact with the PCM and ABS computers. The ACC system can detect objects more than 330 ft. (100 m) ahead of the vehicle.

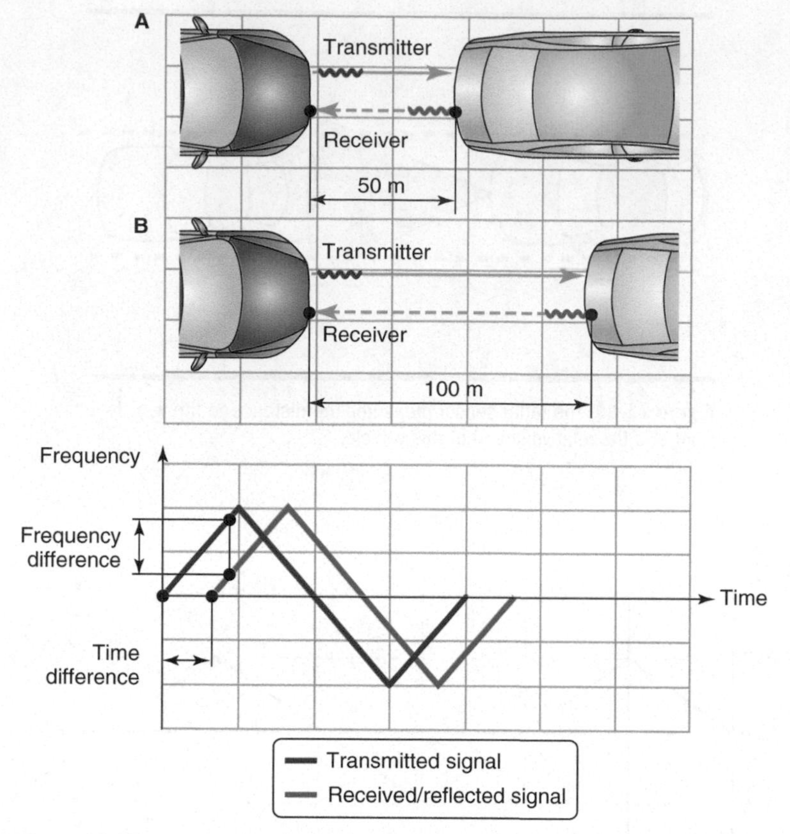

Figure 11-67 The adaptive cruise control radar emits a high-frequency radio wave and the adaptive cruise control sensor monitors the time it takes for the signal to be returned. The further the distance, the longer it takes for the signal to be returned.

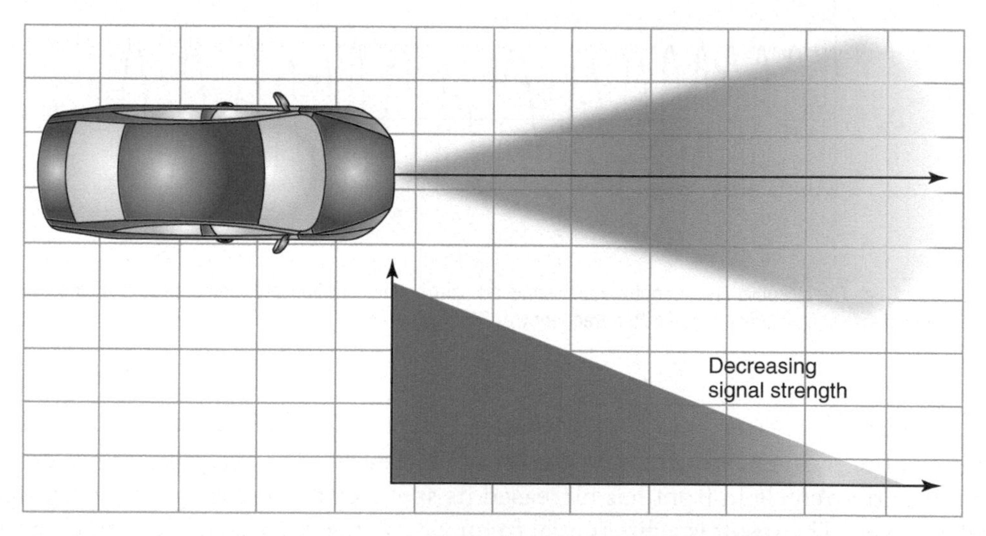

Figure 11-68 The adaptive cruise control radar signal strength decreases with increased distance from transmitter both in height and length.

A dash control allows the driver to set the distance between the vehicle and the vehicle ahead. A 1-second time delay behind the vehicle in front is used for sport driving mode and is best suited for stimulating driving and for slow-moving traffic. A 1.3- to 1.8-second delay behind the vehicle in front is used for standard driving mode, and is best suited for relaxed free-flowing traffic. A 2.3-second delay behind the vehicle in front is used for touring driving mode and is best suited for country roads or when hauling a trailer.

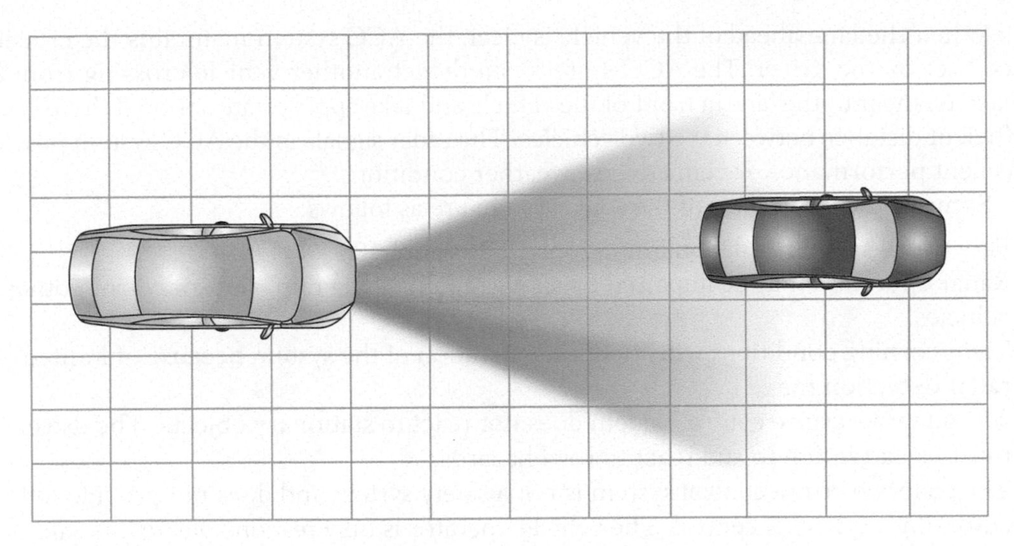

Figure 11-69 The adaptive cruise control uses a three-beam radar system in order to calculate angular information about the travel direction of the vehicle in front.

If the ACC system detects a slower-moving vehicle ahead and the distance to that vehicle becomes less than the driver-adjusted setting, the ACC system uses the throttle control and limited brake application, with a maximum possible braking action of 25 percent of full braking potential, to slow down the vehicle. When this action is taken, a small green car icon is illuminated in the head up display (HUD). If the distance to the vehicle in front is still decreasing after this action is taken, the icon in the HUD turns from green to yellow. When the ACC system senses that a collision is imminent, a large red car icon with yellow flashes is illuminated in the HUD and the ACC system activates a beeper.

The ACC system uses input information from the wheel speed sensors, steering angle and turning sensor to evaluate bends in the road. This information along with an average lane width programmed into the system is used to determine the fictitious lane the vehicle is traveling in and determine if there are vehicles around or in the travel lane (**Figure 11-70**).

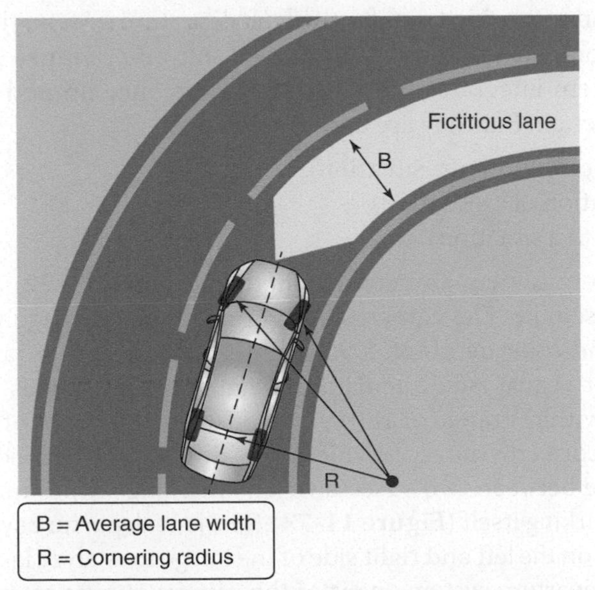

Figure 11-70 The adaptive cruise control system uses information from steering input sensors to determine bends in the road based on an average lane width, and monitors the road ahead.

When the lane ahead of the vehicle is clear, the ACC system maintains the cruising speed set by the driver. The ACC system can detect another vehicle crossing from an adjacent lane into the lane in front of the vehicle and take appropriate action if there is not sufficient distance between the two vehicles. The radar signals in the ACC system provide excellent performance even in adverse weather conditions.

Some of the limitations of the ACC system are as follows:

- Functional speed regulation limitations 18-124 mph (30-200 km/h)
- Radar operation may be impaired by snow, rain, freezing rain, and spray from other vehicles.
- Tight corning conditions will affect the operation of the system because of limited radar detection range.
- The adaptive cruise control system does not react to stationary objects. The driver must pay attention to and react to road hazards.
- The adaptive cruise control system is not a safety system and does not provide full autonomous driving control. The vehicle operator is fully responsible for the safe operation of the vehicle at all times. Regardless of whether advanced driver assistance systems are engaged.

LANE DEPARTURE WARNING (LDW) SYSTEMS

Some vehicles are equipped with **lane departure warning (LDW) systems**. The LDW system is just one example of driver assistance systems that have been integrated into vehicles and discussed in this chapter. A dash switch allows the driver to turn the LDW system on or off.

The introduction of electromechanical power steering systems that use an electric motor for power steering assist instead of hydraulic pressure and electronic throttle control has allowed for advancements in active and direct intervention with driving behavior. Advanced software systems have been developed to make use of existing vehicle sensors and actuators used already on other systems along with CAN networks to further increase driver safely and assist the driver with difficult driving maneuvers. In fact, because the LDW system is only available when bundled with other advanced driver control systems, the only additional hardware required are the LDW control unit with camera mounted in rearview mirror, and a warning lamp for the lane departure system; the rest of the system relies on existing components and the addition of software (**Figure 11-71**).

The LDW system integrates functions that were once limited only to humans or robots in science fiction. But they are here and include:

- Seeing (optical perception of a situation)
- Thinking (evaluation of a situation)
- Acting (reacting to a situation)

The LDW system camera and control module (**Figure 11-72**) are mounted in the rearview mirror assembly. The camera captures a gray scale image with a resolution of 640 x 480 pixels and color depth of 4096 gray scale, at a capture rate of 25 images per second allowing for virtual lane calculations even at higher speeds. The digitized road image is recorded within a range of 18–197 ft. (5.5–60 m) in front of the vehicle. For the system to function correctly, the recordable lane width must be between 8–15 ft. (2.45–4.6 m) and the distance between two successive lane marking stripes may not exceed twice the length of the marking itself (**Figure 11-73**). The lane departure system uses two trapezoidal image areas on the left and right side of the image to analyze lane markings (**Figure 11-74**). The lane departure system analyzes the camera images to create a virtual travel lane that the vehicle is traveling on (**Figure 11-75**). By looking for one or more gray scale values increasing on both the left and right lane line markings the system can calculate

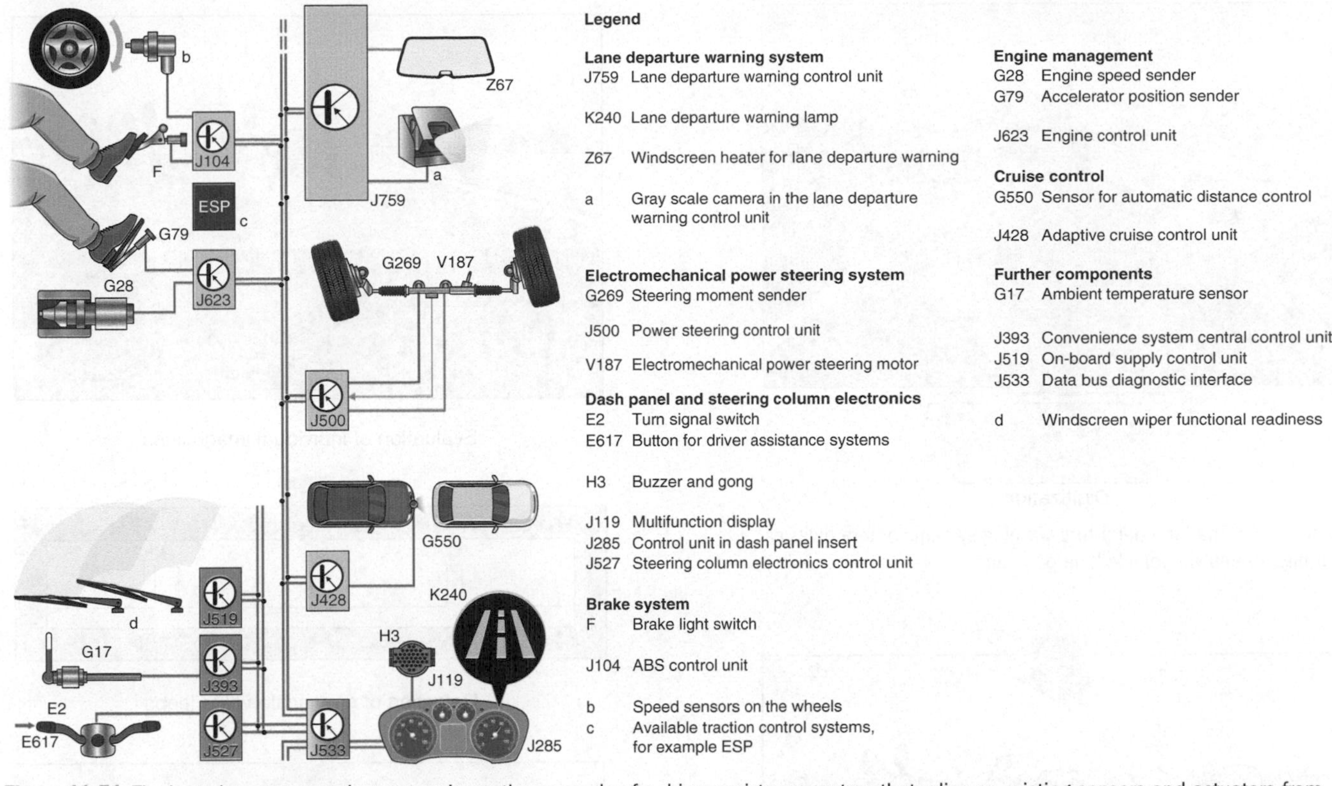

Figure 11-71 The lane departure warning system is another example of a driver assistance system that relies on existing sensors and actuators from other vehicle systems and advanced CAN network communication.

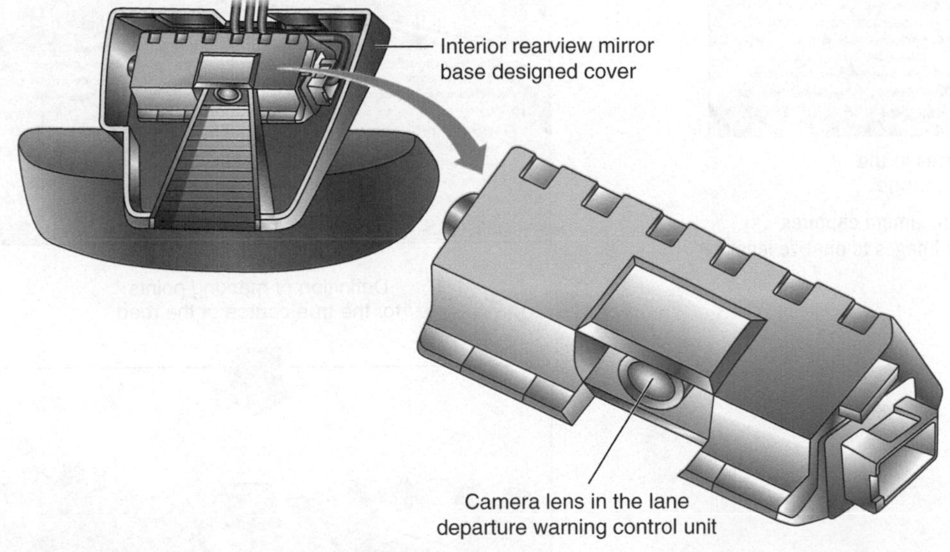

Figure 11-72 The lane departure warning system camera and control unit are mounted in the rearview mirror assembly.

vehicle position with precision. The system now calculates the true course of the road integrating in safety limits for vehicle position and the lateral orientation of the vehicle position within this virtual lane is determined (**Figure 11-76**). If the vehicle comes close to or crosses the virtual lane lines, steering correction is implemented. The system also has the ability to detect if the driver's hand is on the steering wheel. All roads have imperfections and when a driver grips a steering wheel these road imperfections cause the tires to move, which in turn applies torque to the steering linkage and steering gear. If the driver

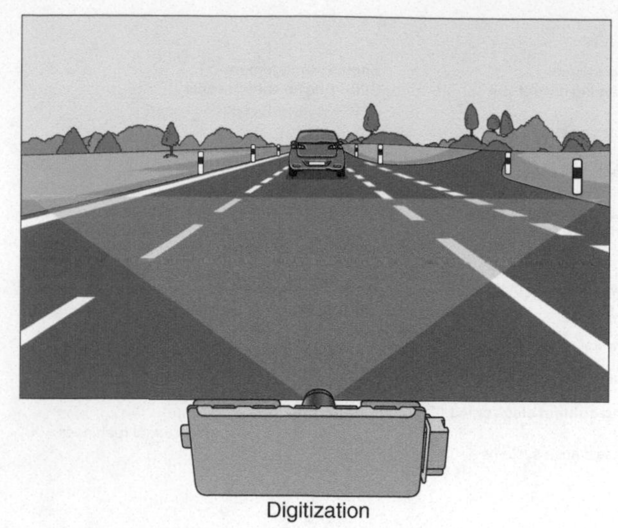

Figure 11-73 The lane departure warning system camera digitizes the image to analyze vehicle lane position.

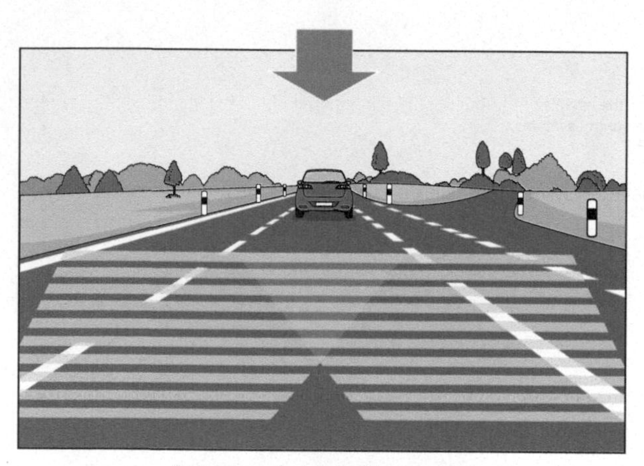

Selection of defined lines in the
trapezoidal detection range

Figure 11-74 The lane departure system camera captures 25 frames per second of two trapezoidal images to analyze lane markings and vehicle position.

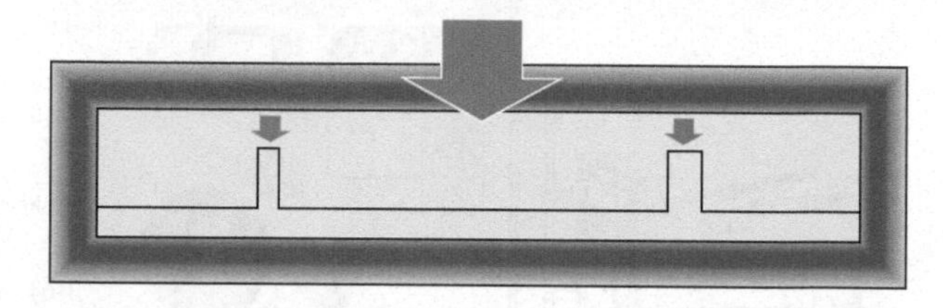

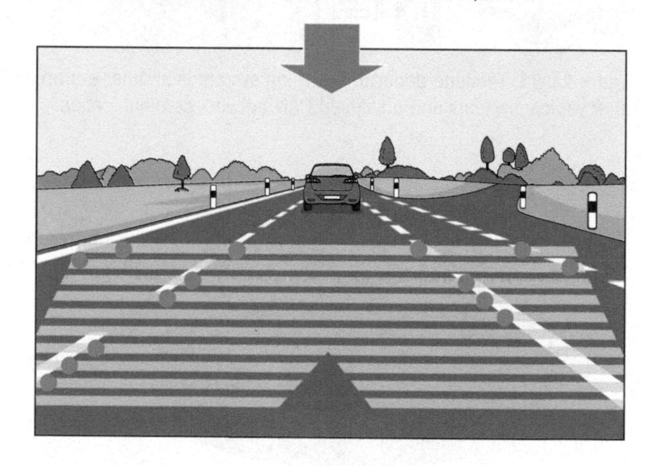

Definition of marking points
for the true course of the road

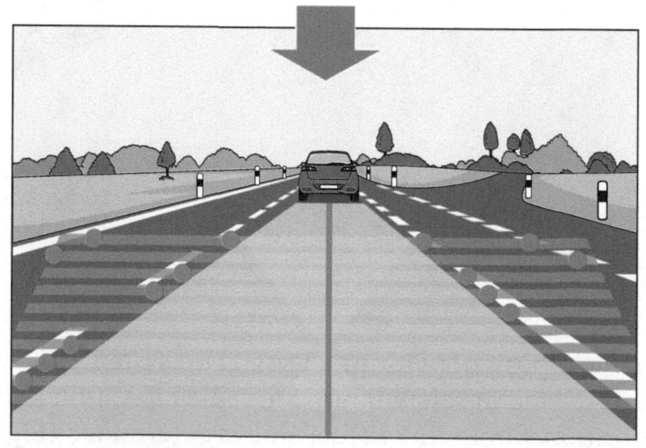

Figure 11-75 The lane departure system analyzes the camera images to create a virtual travel lane that the vehicle is traveling on.

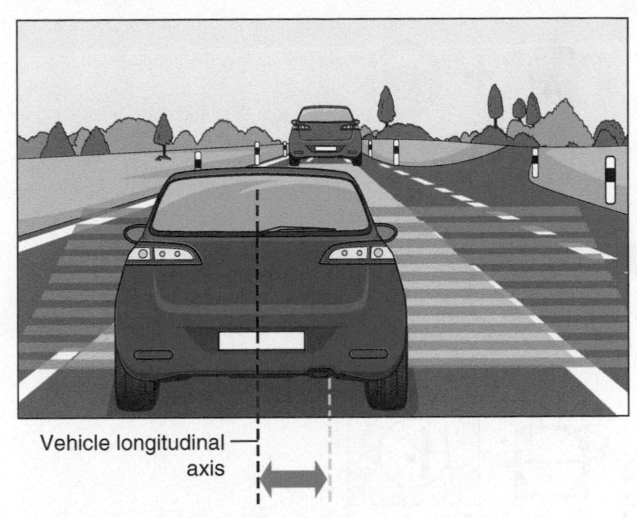

Figure 11-76 The lane departure system next calculates the virtual position of the vehicle with the travel lane.

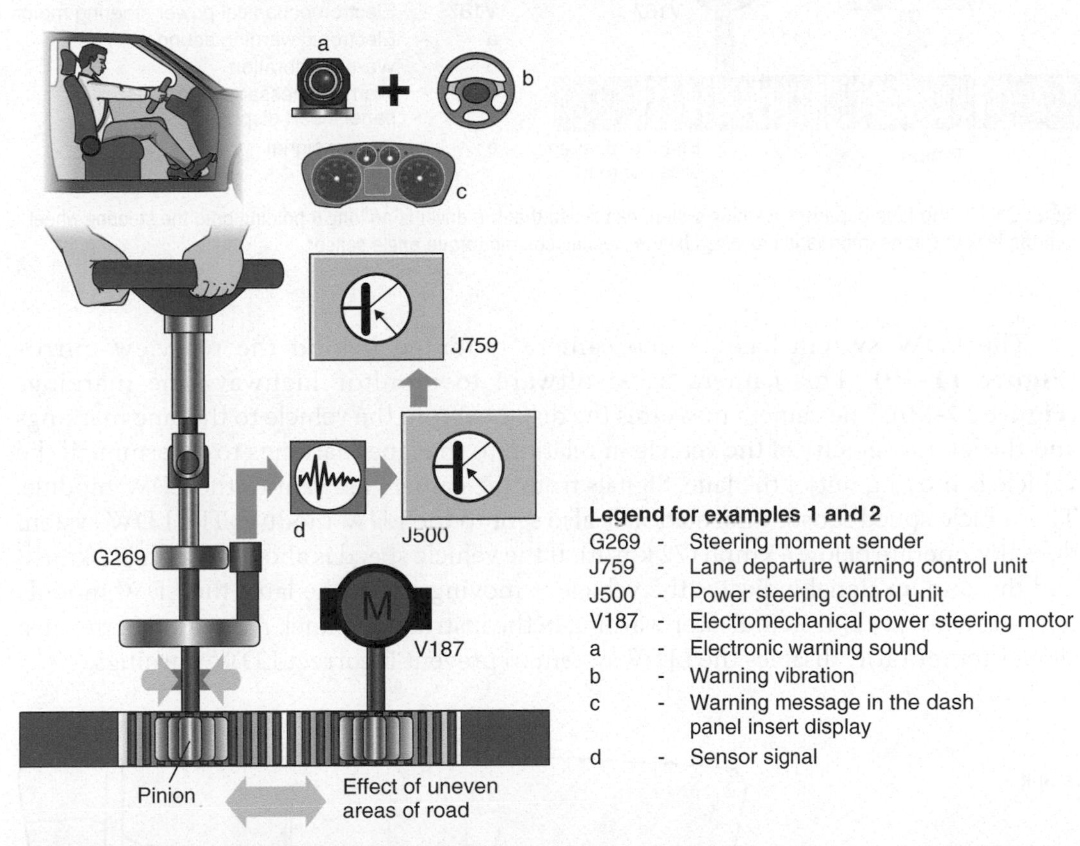

Figure 11-77 The lane departure warning system can sense that the driver is holding onto the steering wheel from torque information received by the steering column torque angle sensor.

is holding onto the steering wheel, the steering column torque angle sensor will detect this torque input (**Figure 11-77**). If the driver lets go of the steering wheel, the system senses the lack of torque resistance data received from the steering column torque angle sensor (**Figure 11-78**).

As advancements continue in the miniaturization of electronics and as computer artificial intelligence continues to improve, we will see more advanced driver assistance systems. One day the driver may be eliminated altogether.

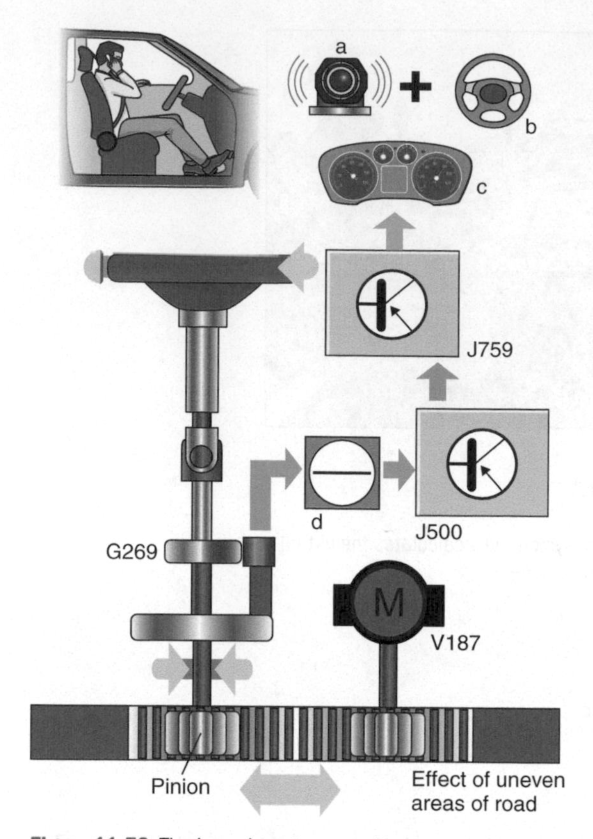

Legend for examples 1 and 2
G269 - Steering moment sender
J759 - Lane departure warning control unit
J500 - Power steering control unit
V187 - Electromechanical power steering motor
a - Electronic warning sound
b - Warning vibration
c - Warning message in the dash panel insert display
d - Sensor signal

Figure 11-78 The lane departure warning system can sense that the driver is no longer holding onto the steering wheel from the lack of torque information received by the steering column torque angle sensor.

The LDW system has a video camera mounted behind the rearview mirror (**Figure 11-79**). This camera uses software to monitor highway lane markings (**Figure 11-80**). The camera measures the distance from the vehicle to the lane markings and the lateral velocity of the vehicle in relation to the lane markings to determine if the vehicle is moving out of the lane. Signals from the camera are sent to the LDW module. The vehicle speed sensor information is also sent to the LDW module. The LDW system does not operate below 45 mph (72 km/h). If the vehicle speed is above 45 mph (72 km/h) and the camera signal indicates the vehicle is moving out of the lane, the LDW module operates a buzzer and an indicator warning in the instrument panel. Activation of the turn signals temporarily disables the LDW system to prevent incorrect LDW warnings.

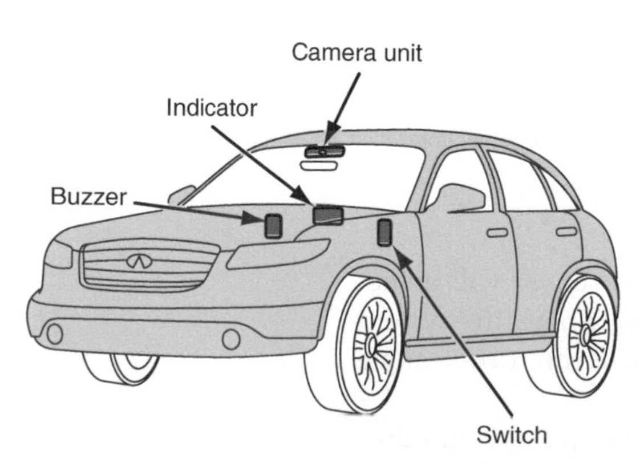

Figure 11-79 Components in the lane departure warning (LDW) system.

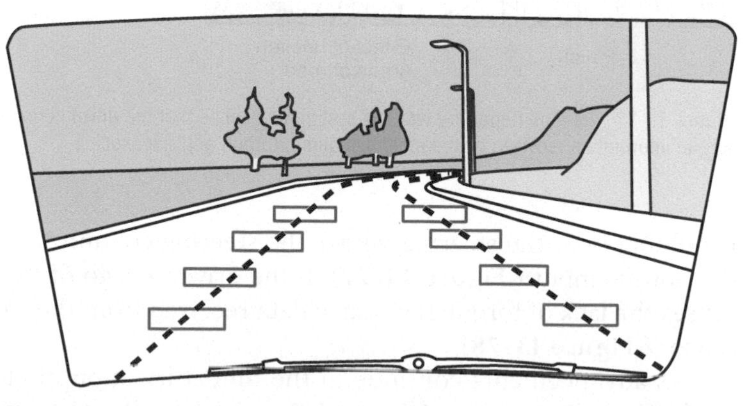

Figure 11-80 The LDW camera measures the distance from the vehicle to the highway lane markings.

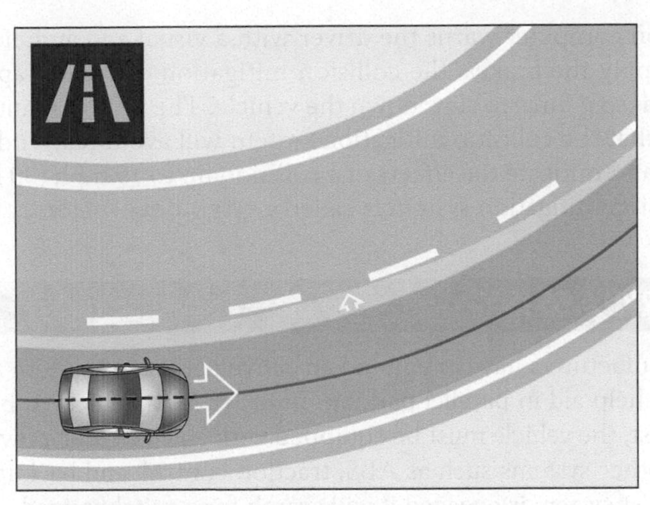

Figure 11-81 The lane departure warning system will stay engaged even while cornering if the cornering maneuver is completed in less than 100 seconds.

Some LDW systems have lane-departure prevention capabilities. If the vehicle is moving toward the lane markings on the left, the ABS and stability control computer lightly pulse the brakes on the right side of the vehicle. When the wheels on the right side slow down, the wheels on the left side turn faster than the wheels on the right side, and this action steers the car away from the highway markers on the left. Other LDW systems activate the electronic (electromechanical) steering on the vehicle to steer the vehicle away from the lane markers if a lane departure is starting to occur. The LDW system will engage the electromechanical power steering system to first cause a slight vibration of the steering wheel to warn driver to take corrective action to maintain lane position and then take corrective action to maintain lane position. If the driver exerts more than 2.21 ft.-lb (3 Nm) of pressure on the steering wheel, the LDW system will disengage to allow a lane change without using turn signal. The LDW system will continue to function even while corning as long as the corning maneuver is completed in less than 100 seconds (**Figure 11-81**). The lane departure system can be adversely affected by weather and road surfaces. While the system is effective at detecting lane lines on asphalt, it can have issues detecting lane lines on concrete roads on bright sunny days. The system can also be affected by some wet surfaces due to reflection or snow-covered surfaces that cover lane lines. A dirty windshield that affects camera image will also cause lane departure system issues and will set a diagnostic trouble code "Lane Assist – no sensor visibility present."

AUTHOR'S NOTE Statistics compiled by one of the major car manufacturers indicate that 30 percent of all vehicle accidents in the United States involve rear-end collisions. In half of these collisions, the driver did not apply the brakes before collision occurred. The statistics also indicate that 75 percent of these rear-end collisions occur below 19 mph (30 km/h).

COLLISION MITIGATION SYSTEMS

Some vehicles are equipped with a collision mitigation system that uses radar and camera information to detect a vehicle in front. The collision mitigation computer is in constant communication with the PCM and ABS computers via the vehicle network. If the radar and camera signals indicate a vehicle in front is too close and a collision is imminent, the collision mitigation computer warns the driver with a visual and audible warning. If the driver does not apply the brakes, the collision mitigation computer applies the brakes aggressively with a 0.5 g force to slow down the vehicle. The vehicle manufacturer's information indicates that the collision mitigation system will avoid rear-end collisions up to 9 mph (19 km/h) and mitigate the effects of a collision up to 19 mph (30 km/h). On some vehicles, the collision mitigation system is called a **city safety system**.

ACTIVE PARK ASSIST

Some vehicle manufacturers have integrated an active park assist system as a convenience to their drivers to help aid in parallel parking. In order for a vehicle to be equipped with active parking assist, the vehicle must be equipped with an electronic power steering assist motor as well as other systems such as ABS, traction control, and backup sensors. When the active park assist system is engaged it will search for a suitably sized space to park and display a graphic and a message on the display screen indicating that the system is searching for a space (**Figure 11-82**). The left or the right turn signal may be engaged to tell the system which side of the vehicle to search for a parking space. If the turn signal is not engaged, the system default is to search on the passenger side (**Figure 11-83**). Once the system locates a suitable space the display screen will notify the driver with an audible signal and message and direct the driver to pull forward slowly and stop when the next audible signal is heard and the screen displays a message to stop vehicle (**Figure 11-84**). When the vehicle is shifted into reverse, the system will re-verify that the parking spot is large enough to accept the vehicle. The drive controls the accelerator, gear shift, and the brakes. The active park assist system controls vehicle steering hands-free, steering the vehicle into the space. The system uses both audible and visual directions to direct driver's response (**Figure 11-85**). The active park assist system is designed to assist with parallel parking, but does not replace driver's attention and judgment. The active park assist system also has a park out feature that will automatically steer the vehicle out of the parking space.

Caution

The active park assist system sensors rely on ultrasonic waves. Sensors may not detect objects in heavy rain, objects with surfaces that absorb ultrasonic waves, or cause disruptive reflections.

Figure 11-82 When engaged the active park assist system will direct the driver with both audible and visual signals to detect a suitable space that will fit vehicle.

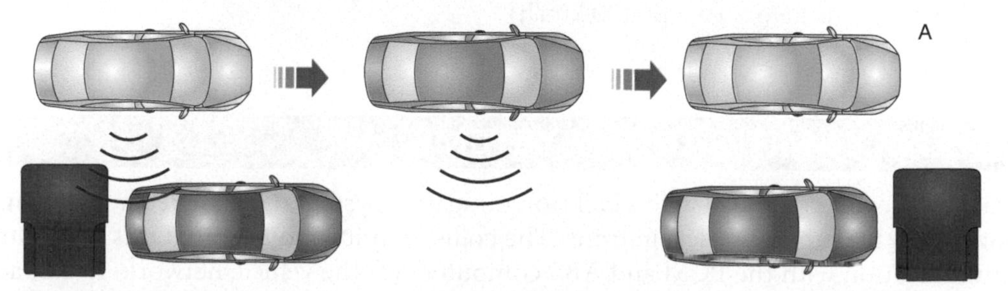

Figure 11-83 The active park assist system will by default search for a space on the passenger side.

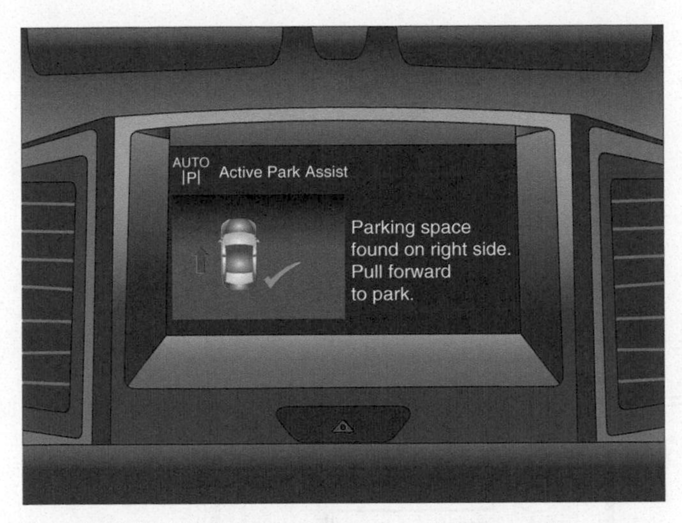

Figure 11-84 The active park assist system will detect a suitable space that will fit the vehicle and direct driver to pull forward.

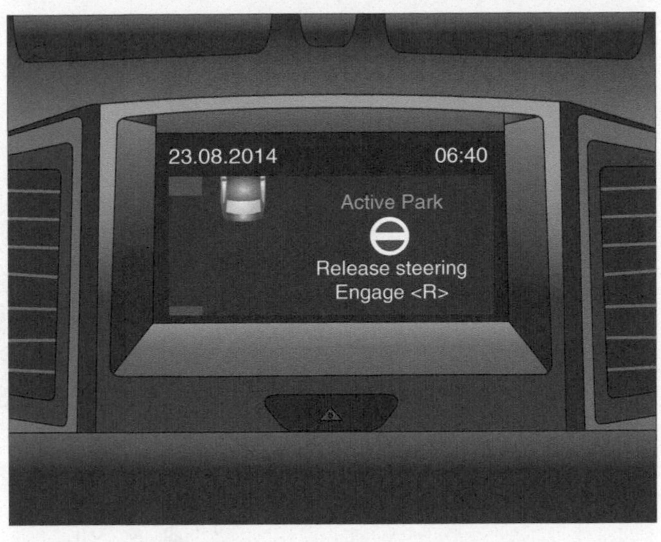

Figure 11-85 The active park assist system will direct the driver to place the vehicle in reverse and let go of the steering wheel.

The following conditions will deactivate the active park assist system:

- Holding on to the steering wheel
- Driving above 6 mph (10 km/h) during parking maneuver
- Driving above 0 mph (80 km/h) for 10 seconds during active parking spot search
- Turning off traction control
- Antilock brake activation or failure
- Traction control activation on a gravel surface or slippery road

The active park assist control module uses inputs from parking aid sensors, active park assist sensors, steering angle and torque sensor, as well as traction control module data and an advanced geometrical equation to analyze if vehicle will fit in a parking space and assist in parking vehicle. The active park assist system is an example of CAN data bus integration of various vehicle subsystems modules communicating with one another enabling a complex function (**Figure 11-86**). The subsystems that work with active park assist include the electromechanical power steering, powertrain control module, ABS, traction control, BCM, and steering column electronics. The active park assist sensors are narrow beam ultrasonic sonar sensors that continuously send out an inaudible ultrasonic sound waves at a consistent speed to detect the distance of other vehicles and obstacles around the vehicle and identify open parking spaces (**Figure 11-87**). The active park assist system ultrasonic sensors contain an emitter that sends out an ultrasonic sound wave beam and a piezoceramic receiver element to detect the return signal that has bounced off a detected object (**Figure 11-88**). The active park assist module calculates the echo time, the time it takes for the sound wave to return, to determine the distance of the object from the sensor (**Figure 11-89**). At sea level, atmospheric pressure of 14.7 psi (1 bar), sound with travel through 68°F (20°C) air at 1125 ft./sec (343 m/s). The speed at which sound travels depends on the density of the air, therefore the system uses information from the ambient air temperature sensor as a correction value in the distance calculations.

Fortunately, as with many advanced computer-controlled features on vehicles, a sophisticated self-diagnostic software system will set problem-specific diagnostic trouble codes to aid in pinpointing a system fault. These codes will be set as body (B) codes.

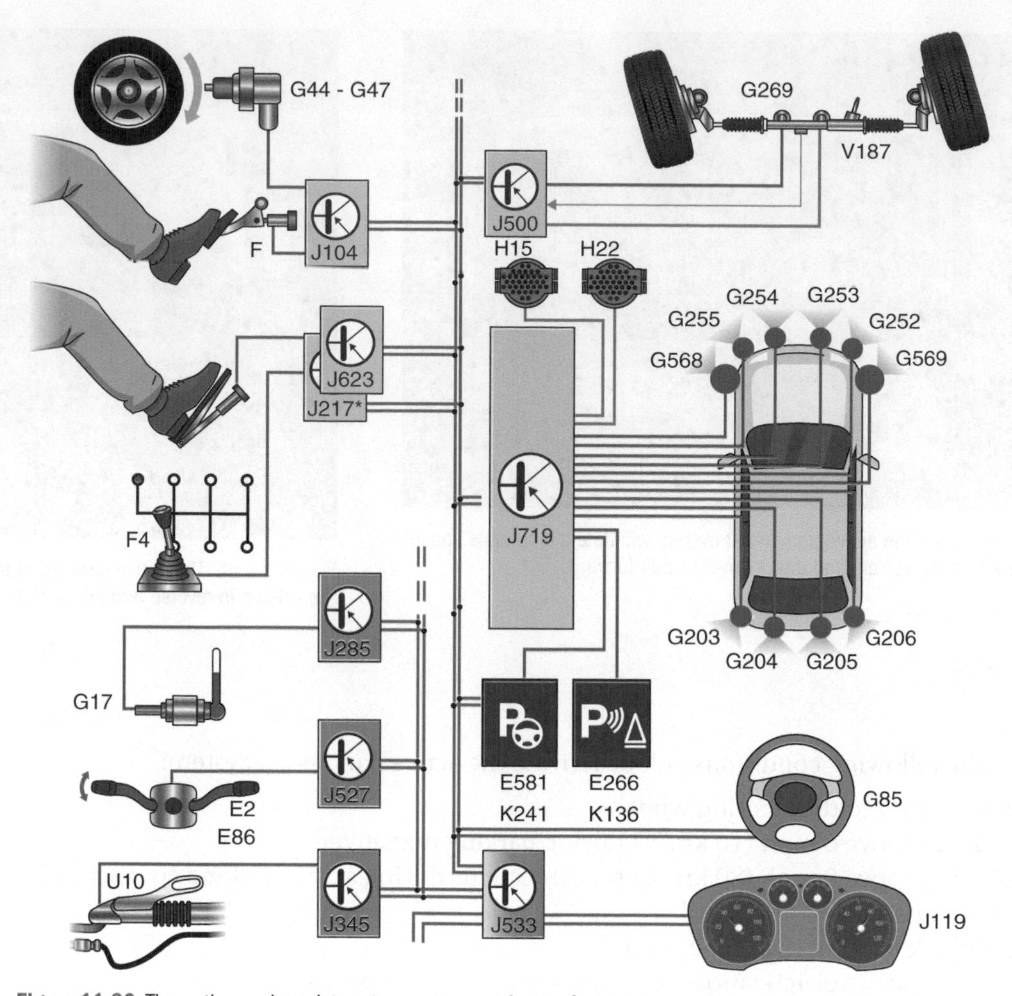

Figure 11-86 The active park assist system uses many inputs from various vehicle sensors.

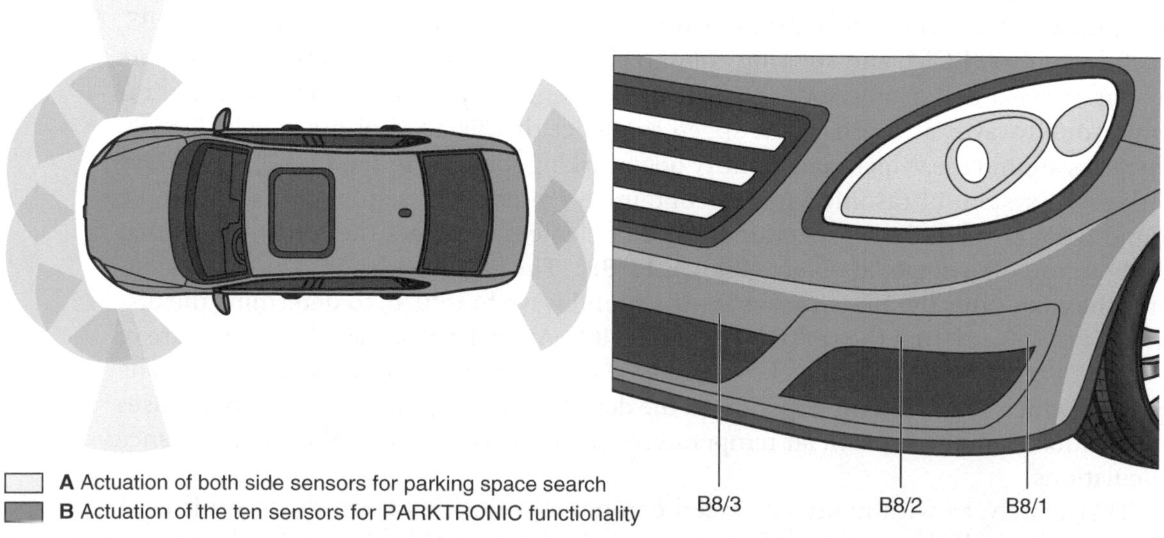

Figure 11-87 The active park assist system uses ultrasonic sensors that are always sending out a narrow ultrasonic beam to detect objects around the vehicle.

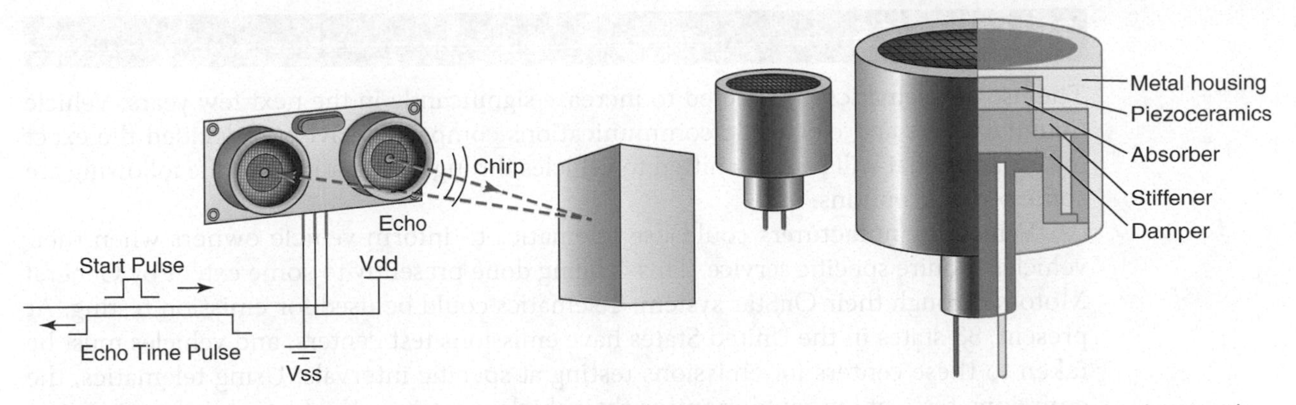

Figure 11-88 The active park assist system's ultrasonic sensors contain an emitter that sends out an ultrasonic beam and a piezoceramic receiver element; the active park assist module calculates the echo time to judge distance of objects.

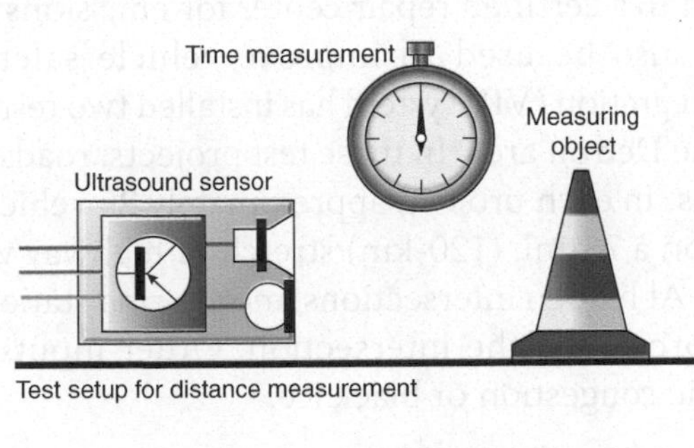

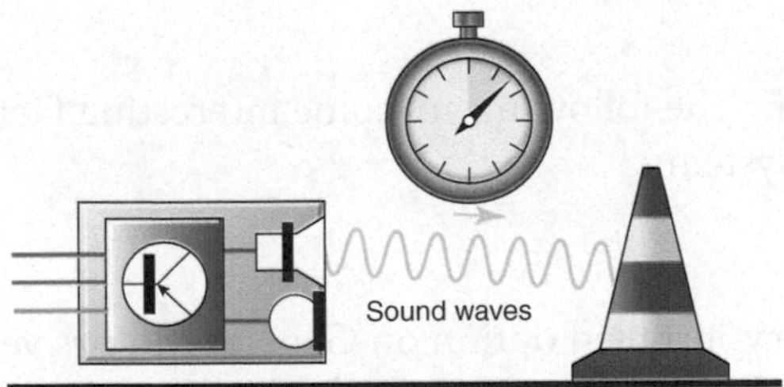

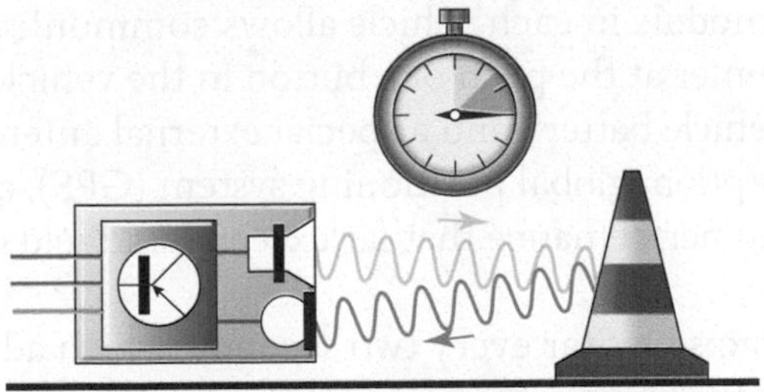

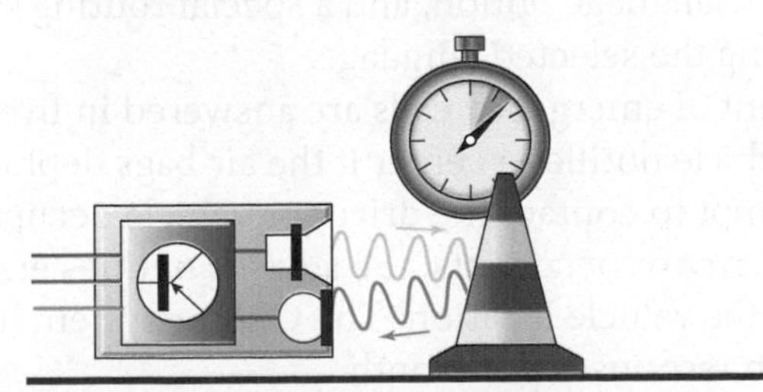

Figure 11-89 The active park assist system's ultrasonic sensors measure the amount of time it takes for the sound wave the sensor generated to be reflected off an object and return to the sensor to calculate the distance an object is from the vehicle.

TELEMATICS

The use of telematics is expected to increase significantly in the next few years. Vehicle manufacturers and electronic communications companies have not decided the exact information that will be transmitted to vehicles through telematics, but the following are some possible options.

Vehicle manufacturers could use telematics to inform vehicle owners when their vehicles require specific service. This is being done presently to some extent by General Motors through their OnStar system. Telematics could be used for emission testing. At present, 33 states in the United States have emissions test centers, and vehicles must be taken to these centers for emissions testing at specific intervals. Using telematics, the emissions test center could monitor the vehicle emissions levels, and if the emissions standards are exceeded, the test center could inform the vehicle owner or driver that the vehicle must be taken to a certified repair center for emissions service.

Telematics may also be used to improve vehicle safety. At present, a Vehicle Infrastructure and Integration (VII) system has installed two test projects, one in California and another one in the Detroit area. In these test projects, roadside beacons communicate with vehicle networks. In each project, approximately 25 vehicles supplied by car manufacturers are driven on a 75-mi. (120-km) stretch of highway where the communication beacons are installed. At hidden intersections, monitoring stations inform the test vehicles that vehicles are approaching the intersection. Other inputs inform the test vehicles regarding heavy traffic congestion or black ice.

Telematics may be defined as wireless communication between the telephone system and vehicle networks.

Interesting Fact

Each year in the United States, approximately 43,000 deaths are caused by traffic accidents. This is approximately the equivalent of a 747 aircraft crashing each week.

AUTHOR'S NOTE The following are some interesting facts about the OnStar communications system:

1. OnStar is a factory-installed option on General Motors vehicles. One-year OnStar service is included in the new GM vehicle purchase. There are approximately 7 million OnStar subscribers.
2. A sophisticated module in each vehicle allows communication between the vehicle and an OnStar center at the push of a button in the vehicle. This communication is enabled by the vehicle battery and a special external antenna on each vehicle that combine cell reception, global positioning system (GPS), and XM satellite radio to provide range and performance that far exceed handheld cell phones and GPS devices.
3. A customer contacts OnStar every two seconds, which adds up to about 5 million calls a month. At the beginning of an OnStar call, the driver must select the English, Spanish, or French language option, and a special routing system directs the call to an advisor speaking the selected language.
4. Ninety-five percent of emergency calls are answered in five seconds. The OnStar module on the vehicle notifies a center if the air bags deploy. An advisor calls the vehicle in an attempt to contact the driver or vehicle occupants. If the advisor cannot contact the driver or occupants, the advisor notifies emergency response personnel regarding the vehicle location. The OnStar system handles approximately 2200 vehicle crash responses per month.
5. The OnStar system processes about 600,000 requests for driving directions each month. OnStar subscribers purchase approximately 31 million minutes of OnStar time per month.
6. Current OnStar modules are connected to the CAN data network on the vehicle. This connection allows the OnStar module to monitor the powertrain, ABS,

traction control, and air bag systems. On current vehicles, the OnStar module can read about 1600 DTCs and monitor the engine oil life, emissions levels, and tire pressure. About 3 million OnStar subscribers have signed up for a monthly vehicle condition report that informs the subscriber regarding vehicle defects in any of the monitored systems. The monthly vehicle condition report also informs the subscriber regarding any recall repairs that have not been completed on their vehicle. Approximately, 61,000 calls are processed monthly for instant remote diagnosis.

7. If an OnStar subscriber is locked out of one's vehicle, he or she can phone an OnStar advisor and provide his or her OnStar number. The advisor can then unlock the doors on the subscriber's vehicle. If an OnStar advisor is informed that a vehicle is stolen, the advisor can use GPS to provide the vehicle location. Select 2009 General Motors vehicles have OnStar's Stolen Vehicle Slowdown service. If police are following a stolen vehicle, they can inform an OnStar advisor and request that the vehicle be stopped. The advisor will flash the vehicles external lights without the driver's awareness. When police confirm that the vehicle lights have flashed, the OnStar advisor will send a signal to the stolen vehicle that causes the electronic throttle control to remain continually in the idle position. All other vehicle functions such as steering and brakes operate normally. The OnStar subscriber cannot request that the vehicle be stopped. At present, OnStar receives about 700 stolen vehicle calls per month. It is estimated that there are 30,000 high-speed chases in the United States each year, and 25 percent of these end in injury; 300 deaths result from these chases.
8. OnStar has expanded beyond the United States and Canada and is now offered in China and Mexico.

SUMMARY

- In a programmed ride control (PRC) system, the steering sensor informs the control module regarding the amount and speed of steering wheel rotation.
- The PRC system switches from the normal to the firm mode during high-speed operation, braking, hard cornering, and fast acceleration.
- The struts and shock absorbers in some PRC systems provide three times as much damping action in the firm mode as in the normal mode.
- The accelerometer in a CCR system contains a mercury switch and this accelerometer sends a vehicle acceleration signal to the control module.
- In a CCR system, the accelerometer signal or the vehicle speed signal may inform the control module to switch from the normal to the firm mode.
- An electronic air suspension system maintains a constant vehicle trim height regardless of passenger or cargo load.
- To exhaust air from an air spring, the air spring solenoid valve and the vent valve in the compressor head must be energized.
- To force air into an air spring, the compressor must be running and the air spring solenoid valve must be energized.
- The air spring valves are retained in the air spring caps with a two-stage rotating action much like a radiator cap.
- An air spring valve must never be loosened until the air is exhausted from the spring. Voltage is supplied through the compressor relay points to the compressor motor. This relay winding is grounded by the control module to close the relay points.
- The On/Off switch in an electronic air suspension system supplies 12 V to the control module. This switch must be off before the car is hoisted, jacked, towed, or raised off the ground.
- If a car door is open, the control module does not respond to lower vehicle commands in an electronic air suspension system.
- When the brake pedal is applied and the doors are closed in an electronic air suspension system, the control module ignores all requests from the height sensors.
- In an electronic air suspension system, if the doors are closed and the brake pedal is released, all requests to the control module are serviced by a 45-second averaging method.
- If the control module in an electronic air suspension system cannot complete a request from a

height sensor in 3 minutes, the control module illuminates the suspension warning lamp.
- In an automatic air suspension system, the control module controls suspension height and strut damping automatically without any driver-controlled inputs.
- Rotary height sensors in automatic air suspension systems contain Hall elements. These sensors send voltage signals to the control module in relation to the amount and speed of wheel jounce and rebound.
- Some air suspension systems reduce trim height at speeds above 65 mph (105 km/h) to improve handling and fuel economy.
- The air suspension system on some four-wheel-drive vehicles increase suspension ride height when the driver selects four-wheel-drive high or four-wheel-drive low.
- The air suspension system on some four-wheel-drive vehicles have the capability to provide firmer shock absorber valving in relation to transfer case mode selection, vehicle speed, and operating conditions.
- The electronic suspension control system changes shock and strut damping forces in 10 to 12 milliseconds.
- In the electronic suspension control system, the module controls suspension damping, rear electronic level control, and speed-sensitive steering automatically without any driver-operated inputs.
- A vehicle stability control system applies one of the front brakes if the rear of the car begins to swerve out of control. This action maintains vehicle direction control.
- An adaptive cruise control (ACC) system applies the vehicle brakes lightly and warns the driver if the system senses inadequate distance to the vehicle in front.
- A lane departure warning (LDW) system warns the driver if the vehicle begins to leave the current lane. Some LDW systems apply the vehicle brakes on one side or steers the vehicle back into the current lane when lane departure begins to occur.
- A collision mitigation system warns the driver and applies the vehicle brakes aggressively if a rear-end collision is about to occur with the vehicle in front.

REVIEW QUESTIONS

Short Answer Essays

1. Describe the operation of the steering sensor in a programmed ride control (PRC) system.
2. Describe the purpose of the vehicle speed sensor signal in a programmed ride control (PRC) system.
3. Explain how air is forced into an air spring in a rear load-leveling air suspension system.
4. Describe the action taken by the control module if the control module in an electronic air suspension system receives a lower vehicle command from a rear suspension sensor with the doors closed, the brake pedal released, and the vehicle traveling at 60 mph (100 km/h).
5. Describe the action taken by the control module if the engine is running with a door open, and the control module receives a lower vehicle command from the height sensor in a rear load-leveling air suspension system.
6. List the conditions when the On/Off switch in an electronic air suspension system must be turned off.
7. Describe the conditions required for the control module to turn on the suspension warning lamp continually with the engine running in an electronic air suspension system.
8. Explain why the control module in an electronic air suspension system services all commands by a 45-second averaging method when the doors are closed and the brake pedal is released.
9. Explain why the suspension warning lamp in an electronic air suspension system may not indicate a defect immediately when the engine is started.
10. Explain how a vent solenoid is damaged by reversed battery polarity in a rear load-leveling air suspension system.

Fill-in-the-Blanks

1. The armature response time is ______________ milliseconds in a programmed ride control (PRC) system strut.
2. In a programmed ride control (PRC) system, if the car is accelerating with the throttle wide open, the PRC system is in the ______________ mode.

3. When the programmed ride control (PRC) mode switch is in the Auto position, the control module changes to the firm mode if lateral acceleration exceeds_______________.

4. In a programmed ride control (PRC) system, lateral vehicle acceleration is sensed from the _______________ sensor.

5. In a rear load-leveling air suspension system, the control module energizes the compressor relay when a _______________ command is received.

6. Two height sensors are mounted on the _______________ suspension in an electronic air suspension system.

7. In an electronic air suspension system two hours after the ignition switch is turned off, _______________ commands are completed, but _______________ commands are ignored.

8. In a rear load-leveling air suspension system, if the On/Off switch in the trunk is off, the system is_______________.

9. An electronic rotary height sensor contains a_______________ _______________.

10. In a continuously variable road sensing suspension system, the module senses vehicle lift and dive from some of the_______________ _______________ inputs.

Multiple Choice

1. While discussing a programmed ride control (PRC) system:
   A. The brake system pressure must be 300 psi (2068 kPa) before this mode change occurs.
   B. A PRC system switches from the auto mode to firm mode if the vehicle accelerates with 90 percent throttle opening.
   C. The PRC system switches to the firm mode if lateral acceleration exceeds 0.25 g.
   D. The mode indicator light in the tachometer is illuminated in the plush ride mode.

2. To increase the rear trim height in an electronic air suspension system, the control module:
   A. Starts the compressor and opens the vent valve.
   B. Starts the compressor and closes the rear air spring valves.
   C. Stops the compressor and opens the rear air spring valves.
   D. Starts the compressor and opens the rear air spring valves.

3. All these statements about air springs, shock absorbers, and struts are true EXCEPT:
   A. Air springs can be mounted separately from the shock absorbers.
   B. Air springs can be mounted over the front and rear struts.
   C. Some struts contain a solenoid actuator that controls strut firmness.
   D. Some air suspension systems have seven driver selectable operating modes.

4. The magneto-rheological fluid in the shock absorbers or struts in an ESC system contains:
   A. Automatic transmission fluid.
   B. Suspended iron particles.
   C. Power steering fluid.
   D. 5W-30 engine oil.

5. To sense the distance to the vehicle in front, an adaptive cruise control system uses:
   A. A video camera.
   B. A short-range radar signal.
   C. A digital camera.
   D. A long-range radar signal.

6. In an ESC system with magneto-rheological fluid in the shock absorbers or struts, the control module can vary the shock absorber firmness in:
   A. 1 millisecond.
   B. 5 milliseconds.
   C. 10 milliseconds.
   D. 152 milliseconds.

7. All of these statements about network systems are true EXCEPT:
   A. A network system can be a single-wire or dual-wire system.
   B. Some network systems operate between 0 V and 12 V.
   C. Some vehicles have two network systems.
   D. A fiber-optic network system has a very high data transmission rate.
8. In some traction control systems if the control module senses drive wheel spin, the first action taken by the control module is to:
   A. Retard the spark advance and shut off the fuel injectors.
   B. Apply the brakes on both non-drive wheels.
   C. Apply the brake on the spinning drive wheel.
   D. Apply the brake on the non-spinning drive wheel.
9. On a vehicle with a stability control system, if icy road conditions cause the vehicle to begin swerving sideways, the stability control module:
   A. Applies the brakes on both rear wheels.
   B. Applies the brakes on both front wheels.
   C. Applies the brake on one front wheel.
   D. Applies the brake on one front wheel and the opposite rear wheel.
10. All of these statements about a rear load-leveling suspension system are true EXCEPT:
    A. The On/Off switch is mounted in the vehicle trunk.
    B. The control module operates the compressor relay.
    C. If a door is open, the control module completes the lower suspension height commands.
    D. The rear suspension has one, non-serviceable suspension height sensor.

CHAPTER 12
POWER STEERING PUMPS

Upon completion and review of this chapter, you should be able to understand and describe:

- The difference between a conventional V-belt and a serpentine belt.
- The advantages of a serpentine belt compared with a conventional V-belt.
- The main components in a power-assisted rack and pinion steering system, and explain the steering gear mounting position.
- Two different types of power steering pump reservoirs.
- The difference between a hydro-boost power steering system and an integral power steering system.
- Three different types of power steering pump rotor designs.
- The power steering pump operation while driving with the front wheels straight ahead.
- The power steering pump operation while the vehicle is turning a corner.
- The power steering pump pressure relief operation, and explain when this operation occurs.
- The design and purpose of an electro-hydraulic power steering module.
- Electrohydraulic power steering (EHPS) systems.
- Hybrid electric vehicle (HEV) operation.

Terms To Know

Cam ring
Flow control valve
Hydro-boost power-assisted steering system
Integral fluid reservoir
Integral power-assisted steering system
Outlet fitting venturi
Parallel HEVs
Pressure relief ball
Rack and pinion power steering system
Regenerative braking
Remote fluid reservoir
Series HEVs
Vane-type power steering pumps
Venturi

INTRODUCTION

Power steering systems have contributed to reduced driver fatigue and made driving a more pleasant experience. Nearly all power steering systems at present use fluid pressure to assist the driver in turning the front wheels. Since driver effort required to turn the front wheels is reduced, driver fatigue is decreased. The advantages of power steering have been made available on many vehicles, and safety has been maintained in these systems.

There are several different types of power steering systems, including integral, rack and pinion, hydro-boost, and linkage type. In any of these systems, the power steering pump is the heart of the system because it supplies the necessary pressure to assist steering.

The power steering pump drive belt is a simple, but very important, component in the power steering system. A power steering pump in perfect condition will not produce the required pressure for steering assist if the drive belt is slipping. Various types of steering systems, drive belts, and pump designs are described in this chapter.

POWER STEERING PUMP DRIVE BELTS

Many power steering pumps are driven by a multi-ribbed belt that surrounds the crankshaft pulley and the power steering pump pulley. If a multi-ribbed belt is used to drive other components in addition to the power steering pump, such as the water pump, alternator, or air-conditioning compressor, it is referred to as a serpentine belt (**Figure 12-1**). The serpentine belt is much wider than a conventional V-belt, and the underside of the belt has a number of small ribbed grooves. Many serpentine belts have spring-loaded automatic belt tensioners that eliminate periodic belt tension adjustments. Since the serpentine belt may be used to drive all the belt-driven components, these components are placed on the same vertical plane, which saves a considerable amount of underhood space. The smooth back of the serpentine belt may also be used to drive one of the components. Regardless of the type of belt, the belt tension is critical. A power steering pump will never develop full pressure if the belt is slipping.

Serpentine belts may be constructed of either Neoprene or ethylene propylene diene M-class rubber (commonly called EPDM). The majority of belts are constructed of EPDM, but vehicles produced prior to 2001 or inexpensive aftermarket replacement belts may still be made of neoprene. Visually it is difficult to tell the two materials apart. Older neoprene belts were designed to last between 50,000–60,000 miles and often showed signs of wear by that time. EPDM belts are designed to last 80,000–100,000 miles and seldom show signs of outward visual wear unless there is a problem. The EPDM belt is more elastic than a standard neoprene belt and resists cracking even at higher mileage. A better indicator of when to replace EPDM belts is rib wear. Belts are designed to have clearance between the rib peaks and the pulley grooves. All belts are exposed to dirt, grit, rocks, road salt, and water. Over time, these contaminants cause the EPDM belt to gradually lose material on the ribbing similar to the way a tire wears out causing the belt to ride deeper in the grooves. A 5 to 10 percent loss of material is enough to cause belt slippage, overheating, and hydroplaning. Hydroplaning occurs when the belt ribs sit deeper in the pulley grooves

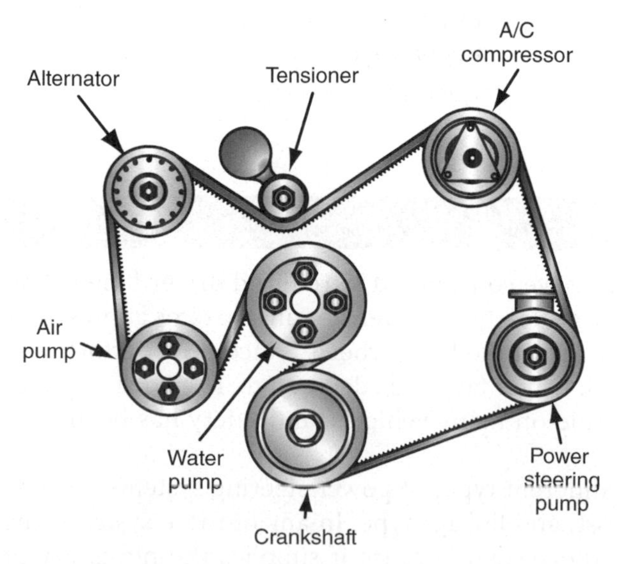

Figure 12-1 Serpentine belt.

due to wear and there is not enough room for water to escape. Instead, the water is trapped between the belt ribs and the pulley grooves lifting the belt away from the pulley and causing slippage. A slipping serpentine belt can cause check engine lights associated with misfire codes, air-conditioning compressor codes, reduced alternator output, reduced engine cooling, and poor air-conditioning performance to name a few. Belts should be checked for wear beginning at 50,000 miles. See Chapter 3 in the Shop Manual for further detail on diagnosing both Neoprene and EPDM belts, and how to measure for belt grove wear. It is wise to replace the drive belt when any pulley driven component is replaced.

Another change that has occurred in belts is the use of stretch fit belts that self tension on the drive pulleys. They do not require any mechanical adjustment and are used in limited applications, such as a single drive belt for air-conditioning compressor. As the name implies, they contain an elastomeric material and with the use of a special tool they are stretched on to the pulley system, there is no mechanical means to loosen or tighten the belt and as with the standard EPDM belt are designed to last 80,000–100,000 miles.

Some current vehicles are equipped with a stretchy belt that does not require a belt tensioner or belt adjustment. The stretchy belt has a similar appearance compared to a conventional serpentine belt. However, stretchy belts have tensile cords made from a polyamide material that is three times more elastic compared with the cords in a conventional serpentine belt. In a stretchy belt, the rubber layers around the cord layer have superior durability and improved adhesion to the cord layer to accommodate the stretching action. Eliminating the mechanical belt tensioner saves weight, space, and financial cost. The stretchy belt is usually installed only around two pulleys such as the air-conditioning compressor and crankshaft pulleys, and the remaining belt-driven components have a serpentine-belt drive.

If a stretchy belt requires replacement, the vehicle manufacturer recommends cutting the old belt to remove it. A special tool is used to lift and guide the new stretchy belt onto the pulley.

An earlier drive belt design was the V-belt. The sides of a V-belt are the friction surfaces that drive the power steering pump (**Figure 12-2**). If the sides of the belt are worn and the lower edge of the belt is contacting the bottom of the pulley, the belt will slip. The power steering pump pulley, crankshaft pulley, and any other pulleys driven by the V-belt *must be properly aligned*. If these pulleys are misaligned, excessive belt wear occurs.

WARNING **Always keep hands, tools, and equipment away from rotating belts and pulleys. If any of these items become entangled in rotating belts, personal injury and equipment damage may result.**

WARNING **Always keep long hair tied back while working in the automotive shop! If long hair is entangled in rotating belts and pulleys, personal injury will result.**

Shop Manual
Chapter 12, page 512

AUTHOR'S NOTE It has been my experience that the most common cause of power steering pump complaints is the pump drive belt. A loose, worn, or dry belt may cause a chirping noise at idle or squealing during acceleration. A loose or worn belt may cause a humming-type vibration noise. Intermittent or continual hard steering may be caused by a loose pump drive belt. A loose power steering pump belt may even cause a complaint of hard steering when driving on wet days, because the belt has more tendency to slip when it is wet. A loose belt may cause low power steering pump pressure when testing the pump. Therefore, when diagnosing power steering pump problems, you should always inspect the pump drive belt and measure the belt tension before any other diagnostic tests are performed.

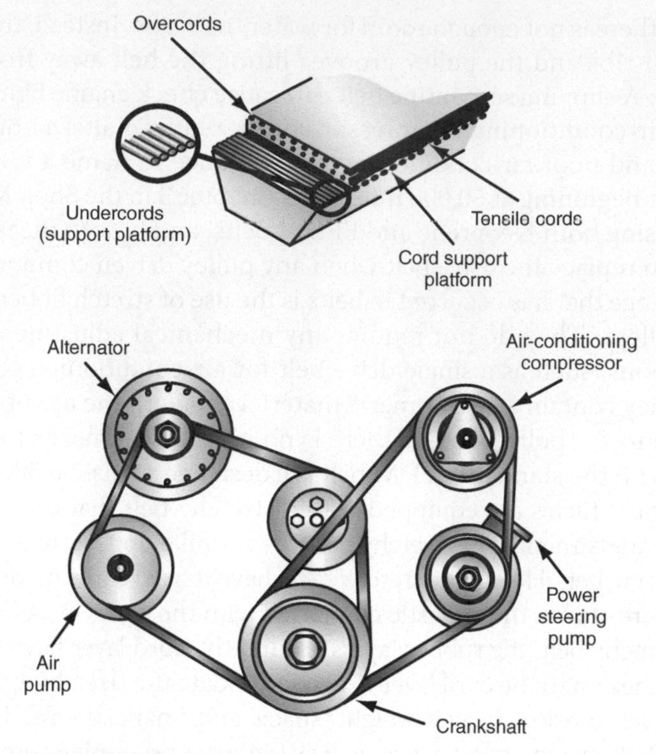

Figure 12-2 Conventional V-belt.

TYPES OF POWER-ASSISTED STEERING SYSTEMS

Rack and Pinion Steering System

The **rack and pinion power steering system** is used on most front-wheel-drive cars. In this steering system, the power steering pump is bolted to a bracket on the engine, and the pump is driven by a belt from the crankshaft. In most front-wheel-drive cars, the engine is mounted transversely, and the steering gear is mounted on the cowl behind the engine or on the crossmember below the engine (**Figure 12-3**).

In many power steering systems, an **integral fluid reservoir** is attached to the power steering pump. A dipstick is mounted in the reservoir for fluid level checking (**Figure 12-4**). A high-pressure hose and a low-pressure return hose are connected from the pump to the steering gear (**Figure 12-5**). The high-pressure hose is usually a steel-braided hose

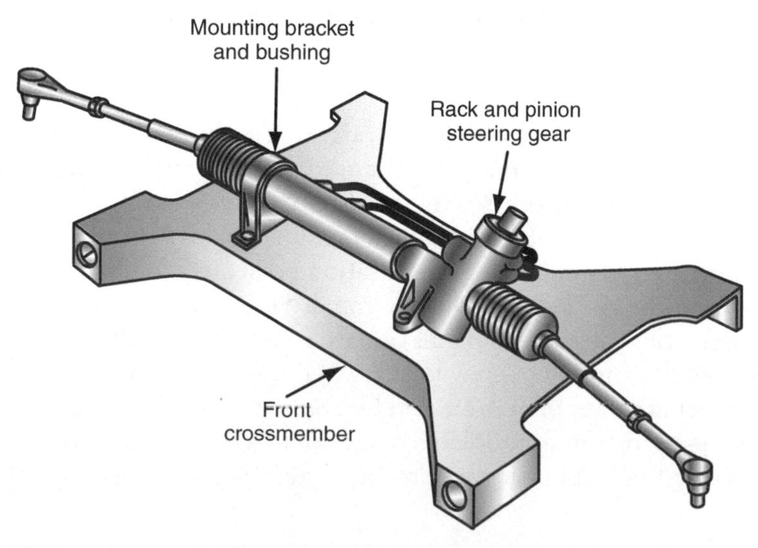

Figure 12-3 Power steering gear mounting.

Figure 12-4 Power steering reservoir and dipstick.

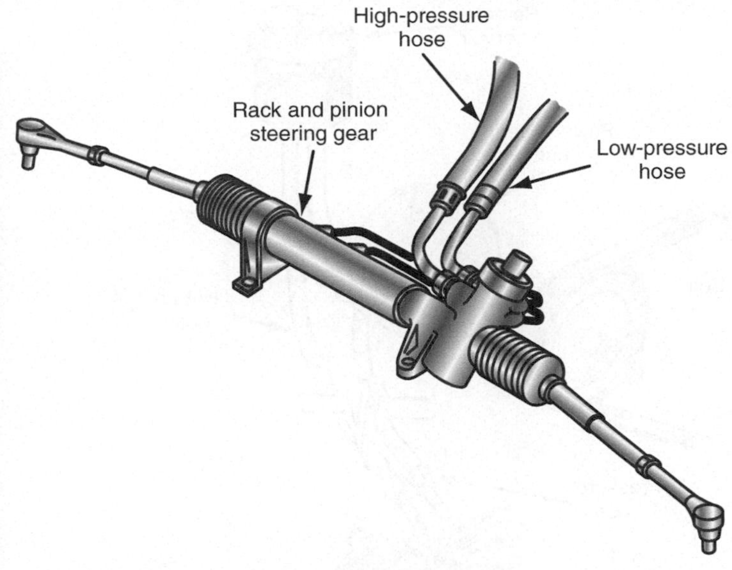

Figure 12-5 Power steering pump to gear hoses.

with appropriate fittings in each end. The return hose is a rubber-braided hose that is clamped to the fittings on the pump and gear.

Some power steering pumps have a **remote fluid reservoir**. This type of system is commonly used on small cars with limited underhood space, where it may be difficult to access an integral reservoir on the power steering pump. The remote reservoir is placed in a convenient position, and the return hose is connected from the steering gear to the reservoir. A second return hose is routed from the reservoir to the pump (**Figure 12-6**).

A **remote reservoir** is mounted externally from the power steering pump.

A BIT OF HISTORY

The number of independent automotive repair facilities and the number of individuals employed by these shops have increased steadily in the last decade. In 1995, there were 63,844 independent automotive mechanical repair facilities in the United States, and they employed 244,430 workers. In 2010, 723,400 technicians and mechanics were employed in the United States, with an expected growth of 17 percent (or 124,800 jobs) over the next 10 years.

Integral Power-Assisted Steering System

In the **integral power-assisted steering system**, the pump is bolted to a bracket on the engine, and the recirculating ball steering gear is mounted on the frame beside the engine. This type of steering system is used on many rear-wheel-drive cars and light-duty trucks. The pump is driven by a belt from the crankshaft and an integral reservoir is mounted on the pump. A high-pressure hose and a return hose are connected from the pump to the steering gear (**Figure 12-7**).

In an **integral power-assisted steering system**, the control valve and power cylinder are contained in the steering gear.

Hydro-Boost Power-Assisted Steering System

The **hydro-boost power-assisted steering system** is used on many light-duty trucks with diesel engines and on some cars. Since the hydro-boost system does not use manifold vacuum for brake assist, this system is suitable for diesel engines, which have reduced intake manifold vacuum compared with gasoline engines. The power steering on these systems is similar to an integral unit, but the power steering pump pressure is also applied to the hydro-boost unit in the brake master cylinder. Hydraulic lines are routed from the power steering pump to the steering gear, and another set of lines is connected from the pump to the master cylinder (**Figure 12-8**). In the hydro-boost system, the power steering

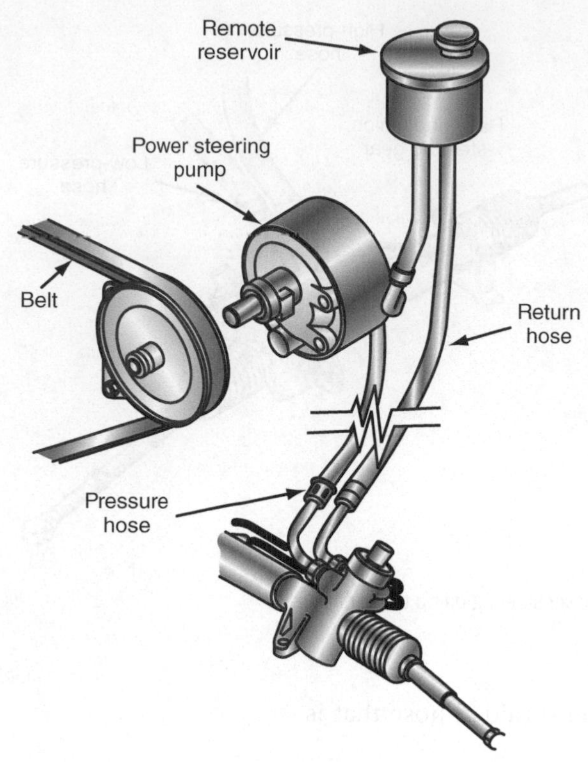

Figure 12-6 Remote reservoir on power-assisted rack and pinion steering system.

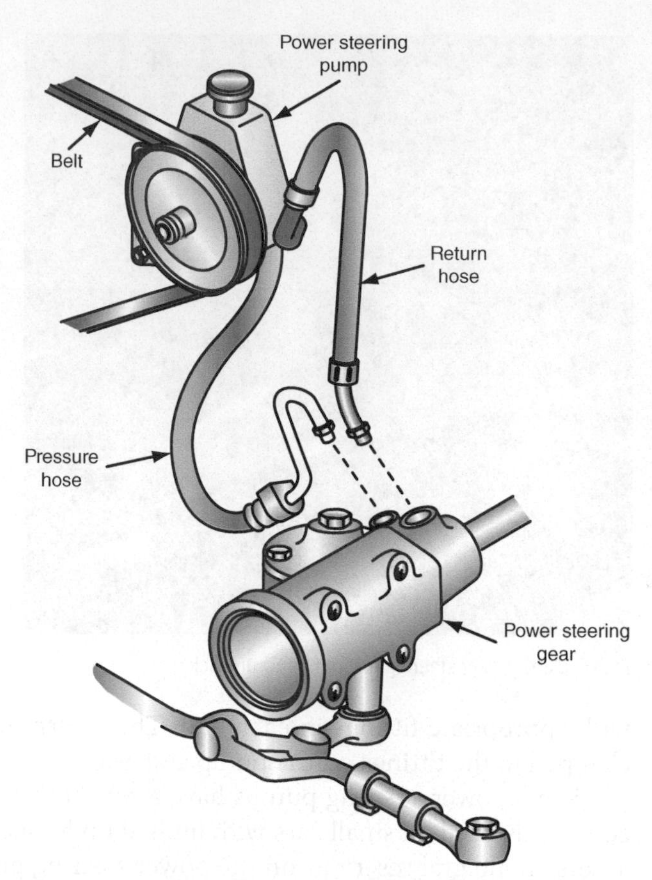

Figure 12-7 Integral power-assisted steering system.

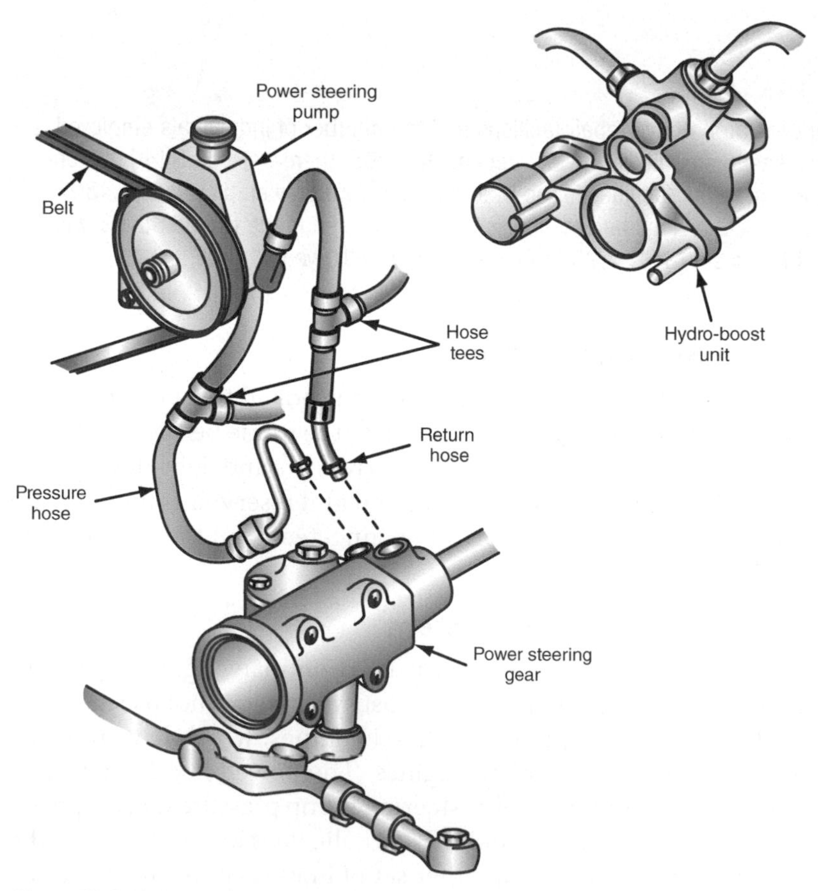

Figure 12-8 Hydro-boost power-assisted steering and brake system.

pump pressure applied to the master cylinder pistons acts as a brake booster to assist the driver in applying the brakes. The hydro-boost system does not have a conventional vacuum booster on the master cylinder.

POWER STEERING PUMP DESIGN

Various types of power steering pumps have been used by car manufacturers. Many **vane-type power steering pumps** have flat vanes that seal the pump rotor to the elliptical pump cam ring (**Figure 12-9**). Other vane-type power steering pumps have rollers to seal the rotor to the cam ring. In some pumps, inverted, U-shaped slippers are used for this purpose. The major differences in these pumps are in the rotor design and the method used to seal the pump rotor in the elliptical pump ring. The operating principles of all three types of pumps are similar.

A balanced pulley is pressed on the steering pump drive shaft. This pulley and shaft are belt-driven by the engine. A spring-loaded lip seal at the front of the pump housing prevents fluid leaks between the pump shaft and the housing. The oblong pump reservoir is made from steel or plastic. A large O-ring seals the front of the reservoir to the pump housing (**Figure 12-10**).

Shop Manual
Chapter 12, page 526

Smaller O-rings seal the bolt fittings on the back of the reservoir. The combination cap and dipstick keep the fluid reserve in the pump and vent the reservoir to the atmosphere. Some power steering pumps have a variable assist steering actuator in the back of

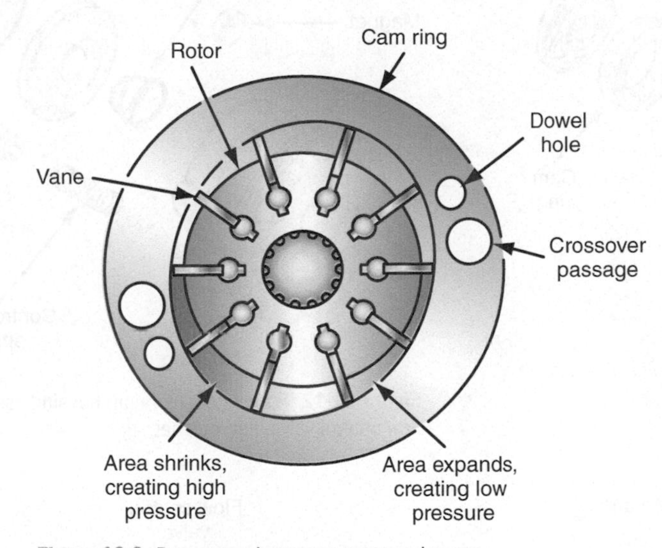

Figure 12-9 Power steering pump rotor and vanes.

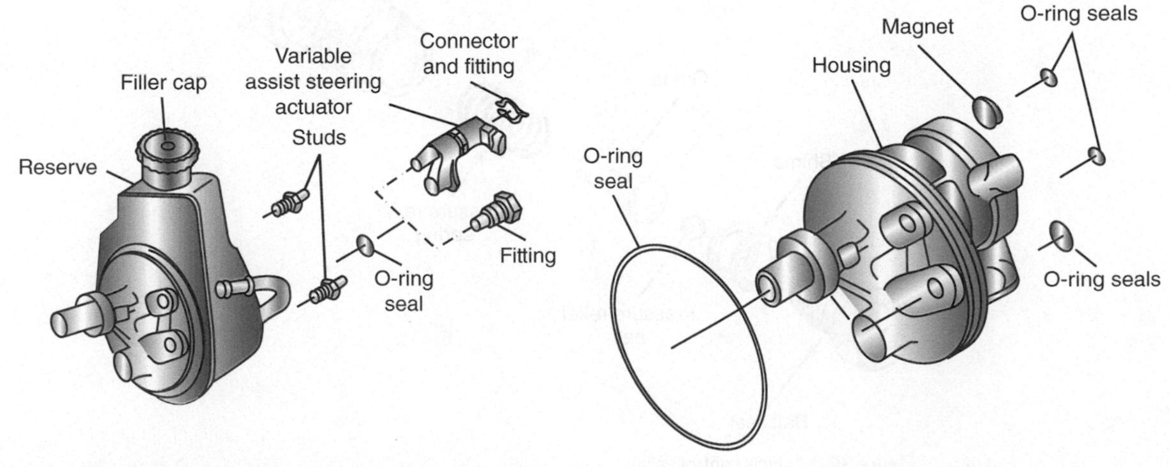

Figure 12-10 Power steering pump housing and reservoir.

the pump housing. The PCM operates this actuator to provide increased steering assist at low vehicle speeds.

The rotating components inside the pump housing include the shaft and rotor with the vanes mounted in the rotor slots. A seal between the output shaft and the housing prevents oil leaks around the shaft. As the pulley drives the pump shaft, the vanes rotate inside an elliptical opening in the **cam ring**. The cam ring remains in a fixed position inside the pump housing. A pressure plate is installed in the housing behind the cam ring (**Figure 12-11**).

A spring is positioned between the pressure plate and the end cover, and a retaining ring holds the end cover in the pump housing. The flow control valve is mounted in the pump housing, and a magnet is positioned on the pump housing to pick up metal filings rather than allowing them to circulate through the power steering system (**Figure 12-12**).

Caution

Avoid prying on the power steering pump reservoir, because this action can damage or puncture the reservoir.

The flow control valve is a precision-fit valve controlled by spring pressure and fluid pressure. Any dirt or roughness on the valve results in erratic pump pressure. The flow control valve contains a pressure relief ball (**Figure 12-13**). High-pressure fluid is forced past the control valve to the outlet fitting. A high-pressure hose connects the outlet fitting to the inlet fitting on the steering gear. A low-pressure hose returns the fluid from the steering gear to the inlet fitting in the pump reservoir.

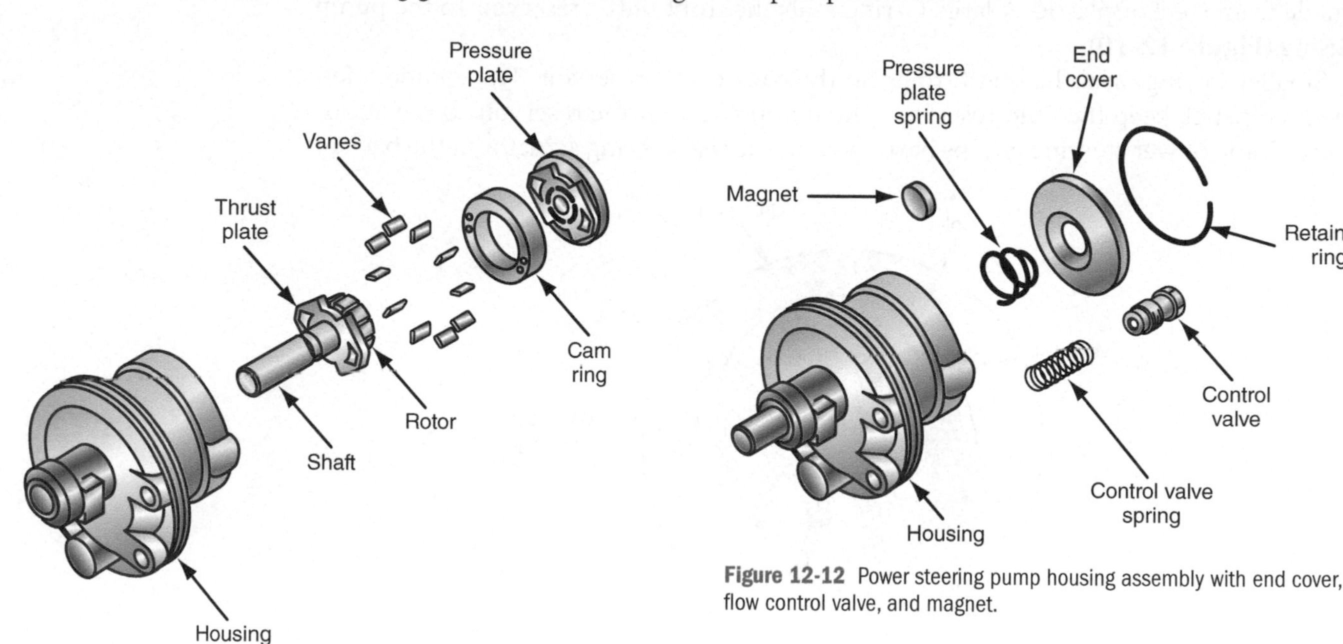

Figure 12-11 Power steering pump housing with shaft and rotor, vanes, cam ring, and pressure plate.

Figure 12-12 Power steering pump housing assembly with end cover, flow control valve, and magnet.

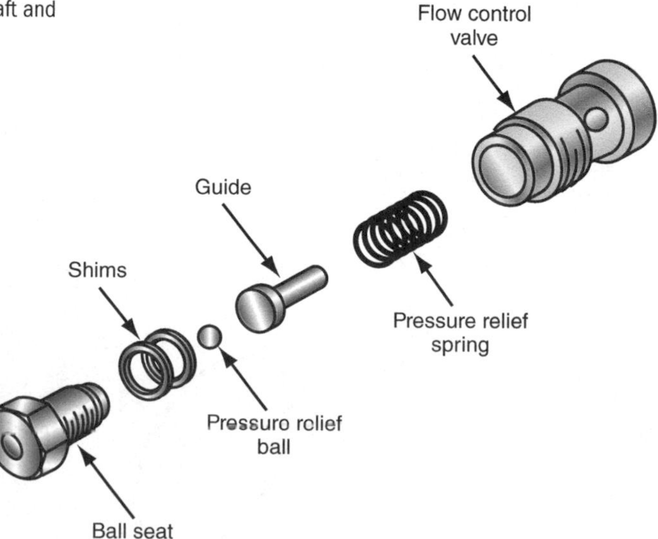

Figure 12-13 Flow control valve.

POWER STEERING PUMP OPERATION

Vane and Cam Action

As the belt rotates the rotor and vanes inside the cam ring, centrifugal force causes the vanes to slide out of the rotor slots. The vanes follow the elliptical surface of the cam. When the area between the vanes expands, a low-pressure area occurs between the vanes and fluid flows from the reservoir into the space between the vanes. As the vanes approach the higher portion of the cam, the area between the vanes shrinks and the fluid is pressurized.

High-pressure fluid is forced into two passages on the thrust plate. These passages reverse the fluid direction. Fluid is discharged through the cam crossover passages and pressure plate openings to the high-pressure cavity and **flow control valve**. The fluid is discharged from the flow control valve through the outlet fitting (**Figure 12-14**). Though most of the fluid is discharged from the pump, some fluid returns to the bottom of the vanes past the flow control valve and through the bypass passage.

The **flow control valve** controls the fluid from the pump to the steering gear.

A machined **venturi** orifice is located in the outlet fitting. Fluid passage applies pressure from the venturi area to the spring side of the flow control valve. When fluid flow in the venturi increases, the pressure in this area decreases.

A **venturi** is a narrow area in a pipe through which a liquid or gas is flowing. When liquid or gas flow increases in the venturi, the pressure drops proportionally. If liquid or gas flow decreases, the pressure in the venturi increases.

Power Steering Pump Operation at Idle Speed

At idle speed with the wheels straight ahead, the flow of fluid from the pump to the steering gear is relatively low. Under this condition, the steering gear valve is positioned so the fluid discharged from the pump is directed through the valve and the low-pressure hose to the pump vane inlet. With low fluid flow, the pressure is higher in the venturi and control orifice. This higher pressure is applied to the spring side of the flow control valve and helps the valve spring keep this flow control valve closed (**Figure 12-15**). Under this condition, pump pressure remains low, and all the fluid is discharged from the pump through the steering gear valve and back to the pump inlet.

Power Steering Pump Operation at Higher Engine Speeds

When engine speed is increased, the power steering pump delivers more fluid than the system requires. This increase in fluid flow creates a pressure decrease in the **outlet fitting venturi**. This pressure reduction is sensed at the spring side of the flow control valve, which allows the pump discharge pressure to force the flow control valve partially

The **outlet fitting venturi** controls the pressure supplied to the flow control valve.

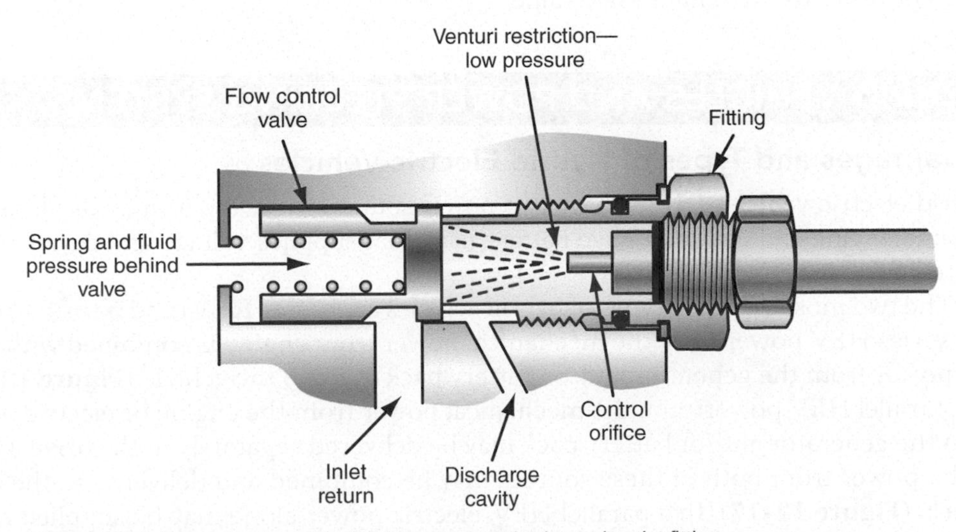

Figure 12-14 Power steering pump fluid flow to control valve and outlet fitting.

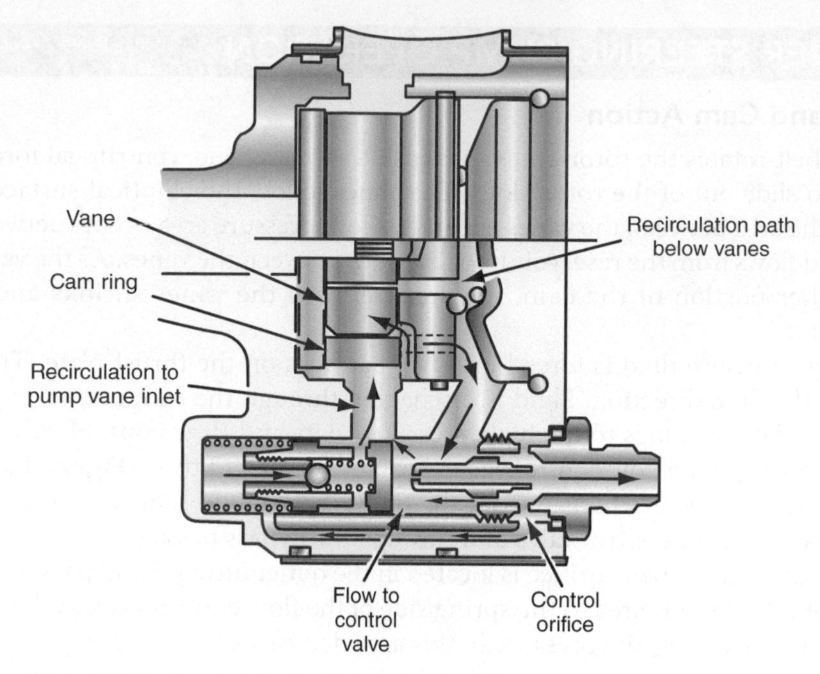

Figure 12-15 Flow control valve operation under various conditions.

open. Under this condition, the excess fluid from the pump is routed past the flow control valve to the pump inlet. If the steering wheel is turned, the fluid discharged from the pump rushes into the pressure chamber in the steering gear.

Power Steering Pump Operation During Pressure Relief

The **pressure relief ball** limits the maximum power steering pump pressure.

If the steering wheel is turned fully in either direction until the steering linkage contacts the steering stops, the rack piston is stopped. Under this condition, pump pressure could become extremely high and damage hoses or other components. When the rack piston stops, the flow from the pump decreases, but the high pump pressure is still directed through the steering gear valve to the rack piston. Since the flow of fluid through the venturi and control orifice is reduced, a higher pressure is present in this area. This extremely high pressure is also supplied to the spring end of the flow control valve. At a predetermined pressure, the **pressure relief ball** in the center of the flow control valve is unseated and the flow control valve moves to the wide-open position. This action allows some of the high pressure in the steering system to return to the pump inlet, which limits pump pressure to a maximum safe value.

Shop Manual
Chapter 12, page 523

HYBRID VEHICLES AND POWER STEERING SYSTEMS

Shop Manual
Chapter 12, page 531

Advantages and Types of Hybrid Electric Vehicles

Hybrid electric vehicles (HEVs) have two power sources: typically, a small displacement gasoline engine and a high-voltage battery pack that supplies voltage and current to an electric drive motor(s).

In a **series HEV**, electric power from the battery pack and/or generator and mechanical power from the engine are supplied together to the drive wheels.

The two most common types of hybrid vehicles are **series HEVs** and **parallel HEVs**. In a series HEV powertrain, the mechanical power from engine is combined with electric power from the generator and/or battery pack to drive the vehicle (**Figure 12-16**). In a parallel HEV powertrain, the mechanical power from the engine or electric power from the generator and/or battery pack may be delivered separately to the drive wheels or the power from both of these sources may be combined and delivered to the drive wheels (**Figure 12-17**). In a parallel HEV, electric power alone may be supplied to the

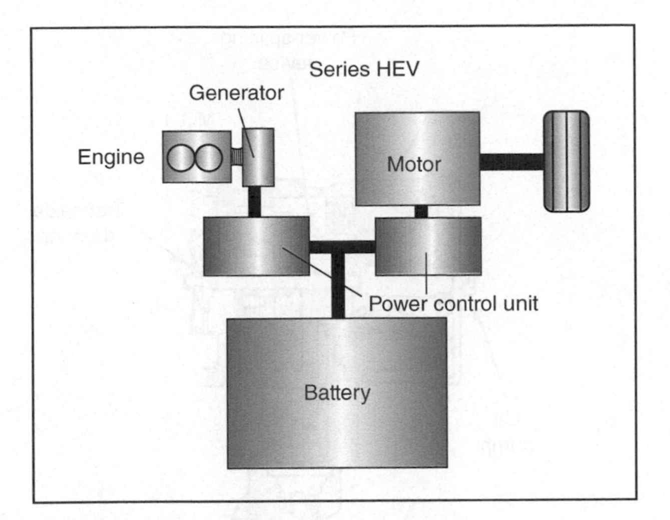

Figure 12-16 Series HEV.

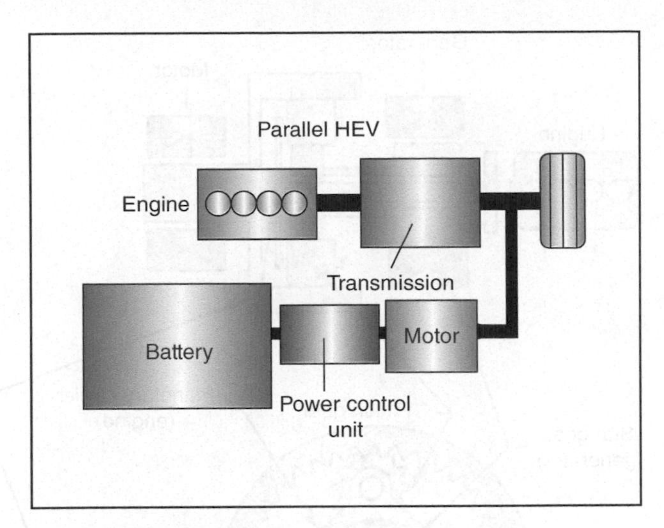

Figure 12-17 Parallel HEV.

drive wheels to drive the vehicle, whereas in a series HEV, electric power cannot drive the vehicle.

In a **parallel HEV**, electric power from the battery pack and/or generator and mechanical power from the engine may be supplied separately to the drive wheels or these two power sources may be combined.

The advantages of an HEV are increased fuel economy and reduced emissions. It is extremely important that all other systems and vehicle design are engineered to help achieve these objectives. Many HEVs have a smaller displacement engine compared with an equivalent-size non-hybrid vehicle. In a typical HEV, the smaller displacement engine coupled with the electric propulsion motor provides equivalent, or nearly equivalent, power compared with that provided in a non-hybrid vehicle. The smaller displacement engine in the HEV supplies improved fuel economy compared with that provided by the larger engine in the non-hybrid vehicle. Improved fuel economy is also achieved because the electric propulsion motor supplies some of the power to the drive wheels. In many HEVs, under specific operating conditions, the engine is shut off and power is supplied to the drive wheels only by the electric motor to provide additional fuel savings.

On many HEVs, the engine is stopped to conserve fuel and reduce emissions when the vehicle is standing still and the engine is warmed up and idling. Under this condition, the power steering must be active and ready to provide power steering assist if the driver turns the steering wheel. To meet this requirement, electrohydraulic power steering pumps are presently installed on some HEVs.

HYBRID POWERTRAIN COMPONENTS

Engine to Transaxle Coupling

In some HEVs, the engine and propulsion motors are coupled through a torque splitting planetary gear set to an electronically controlled continuously variable transaxle (CVT). The engine is coupled to the carrier in the planetary gear set (**Figure 12-18**). The propulsion motor is connected to the planetary ring gear, and the generator is attached to the planetary sun gear. The drive gear that transfers torque to the transaxle is attached to the electric motor drive shaft. This arrangement allows torque from the engine, the propulsion motor, or both the engine and the propulsion motor to be transferred to the transaxle and drive wheels. The sun gear is the smallest gear in the planetary gear set.

Propulsion Motor/Generator

The propulsion motor is an alternate current (AC) synchronous permanent magnet, liquid-cooled type. Some propulsion motors are rated at 40 hp (30 kW) at 940 to 2,000 rpm

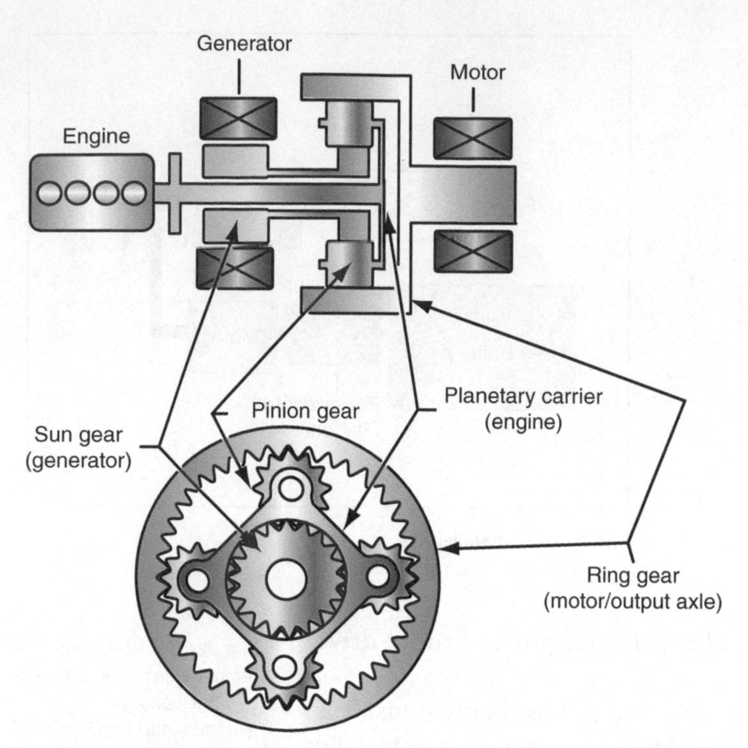

Figure 12-18 Planetary gear coupling between engine, electric motor, and transaxle.

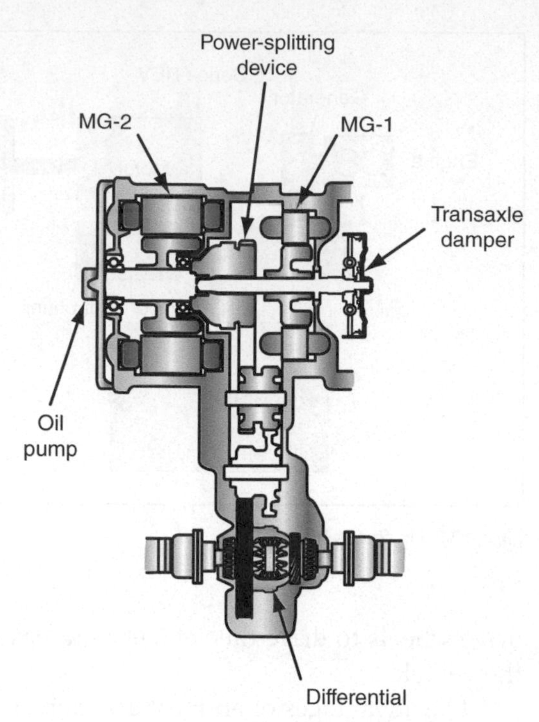

Figure 12-19 Electric drive motor (MG-2) and generator (MG-1).

and 225 ft.-lb (305 Nm) peak torque at 0 to 940 rpm. During deceleration and braking, the propulsion motor acts as a generator to recharge the batteries. This action is called **regenerative braking**. In **Figure 12-19**, the propulsion motor is identified as MG-2 and the generator is identified as MG-1. Some propulsion motors have higher torque and horsepower output depending on the motor design. Later model propulsion motors are capable of turning at speeds in excess of 12,000 rpm.

Regenerative braking occurs during vehicle deceleration when the drive motor becomes a generator and supplies current to recharge the batteries.

Generator/Starter

The generator is an AC synchronous type that produces voltage and current to recharge the batteries and power the electrical system. When starting the engine, current is supplied through the generator stator windings, and the generator cranks the engine. Under normal driving conditions, current from the generator is routed to the electric drive motor to increase the torque supplied to the drive wheels. In later model HEVs, generator design has been improved to allow higher generator rpm and improved generator output. In some current HEVs, the generator is capable of turning at 10,000 rpm, whereas some early model generators turned at 6,500 rpm.

Inverter

The inverter contains the power control unit that controls the voltage and current supplied by the batteries to the propulsion motor. The battery pack supplies high direct current (DC) voltage to the inverter and the inverter converts the DC voltage to three-phase AC voltage that is supplied to the propulsion motor. The inverter also controls the voltage and current supplied from the propulsion motor to the batteries during regenerative braking. Because the voltage and current supplied from the batteries to the propulsion motor may be very high, the inverter contains insulated gate bipolar transistors (IGBT) to control this circuit (**Figure 12-20**). This high voltage and current controlled by the inverter produces a considerable amount of heat, and the inverter is cooled by a dedicated cooling system to dissipate this heat. Heavy cables are connected between the batteries, inverter, and the propulsion motor. For easy identification, these cables and connectors have orange insulation.

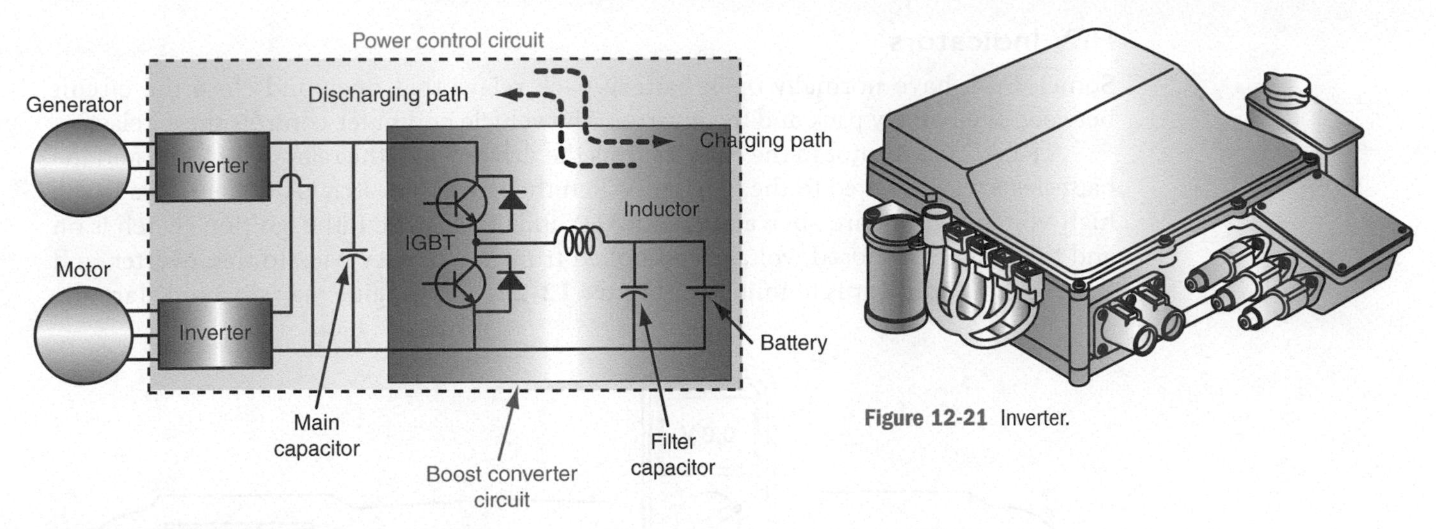

Figure 12-20 High-voltage power circuit in the inverter.

Figure 12-21 Inverter.

WARNING The high voltage in circuit between the batteries, inverter, and propulsion motor could electrocute a technician. It is very important to follow the vehicle manufacturer's recommended procedures when diagnosing and servicing HEVs.

WARNING Never touch, cut, pierce, or open any orange high-voltage cable, or high-voltage component. This action may cause severe electrical shock or electrocution.

The AC voltage and current supplied from the generator to the 12-V battery and electrical accessories is converted to DC voltage by the inverter and controlled in 14 V–15 V range. The inverter is illustrated with the cover removed in **Figure 12-21**.

Battery Pack and Related Cables

Some battery packs contain 240 cylindrical nickel-metal-hydride cells connected in series. Each cell has 1.2 V for a total of 288 V. Some HEVs have a higher voltage depending on the number of cells in the battery. The battery cells are connected in groups, with six cells in each group. The battery pack is installed behind the rear seat in some HEVs. The electrolyte is a gel in the nickel-metal-hydride battery pack; therefore, leakage is not a great concern. A cooling fan helps cool the battery pack. In the unlikely case of battery overcharging, a battery pack vent hose allows vapors from the battery pack to be vented outside the vehicle trunk.

Many battery packs have a long warranty period such as 10-year, 90,000 mi. (150,000 km). Positive and negative high-voltage cables are connected between the battery pack and the inverter. These cables are routed under the vehicle floor pan and they are completely insulated from the chassis to avoid any possibility of electrical shock when touching the chassis.

HEV components increase vehicle weight, which reduces vehicle performance and increases fuel consumption. Therefore, it is very important to reduce the weight of HEV components and also design other vehicle components with less weight.

Some HEVs are currently equipped with lithium-ion batteries, which are 40 percent lighter and take up 24 percent less space compared with a nickel-metal-hydride battery with the same power output.

AUTHOR'S NOTE One of the major Japanese vehicle manufacturers has just introduced a lithium-ion battery with a new internal design that has only one-half the volume but one and one-half times the power output compared with previous lithium-ion batteries.

HEV Indicators

Some HEVs have normally open battery-pack relays that open and close the circuit between the battery pack and the inverter. The vehicle computer controls these relays.

A READY indicator in the dash informs the driver when the relays are closed and the batteries are connected to the inverter. When the ignition switch is off, the relays in the high-voltage circuit are open and the READY indicator is off. If the ignition switch is on and the relays are closed, voltage is supplied from the battery pack to the inverter, and the READY indicator is illuminated (**Figure 12-22**). The engine may stop and start any

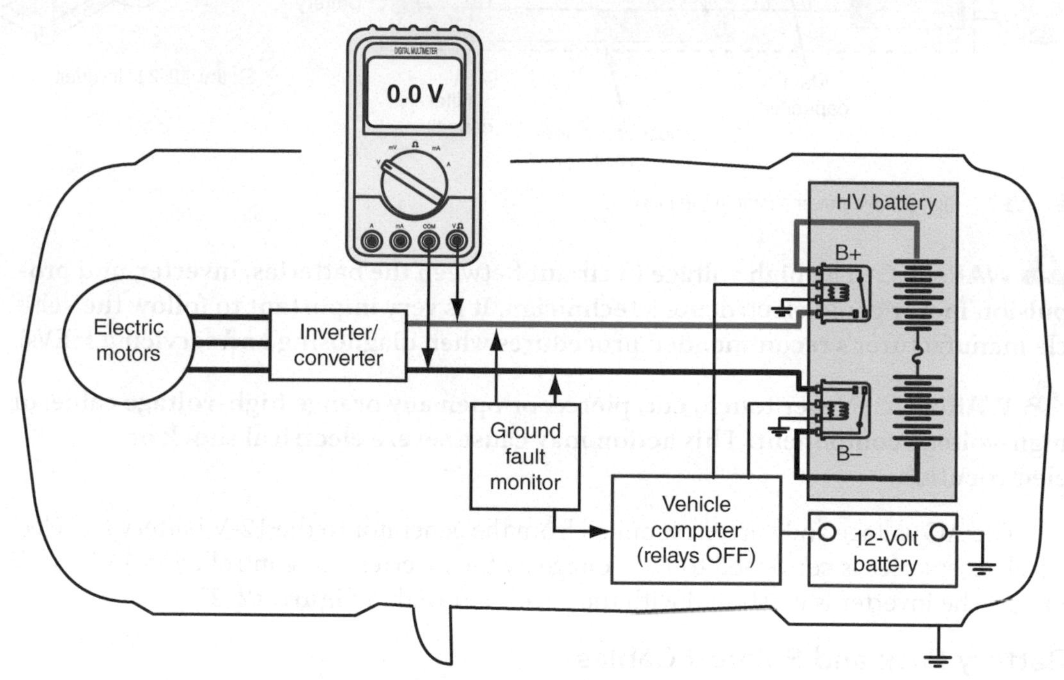

High-Voltage System—Vehicle Shut Off (READY -off)

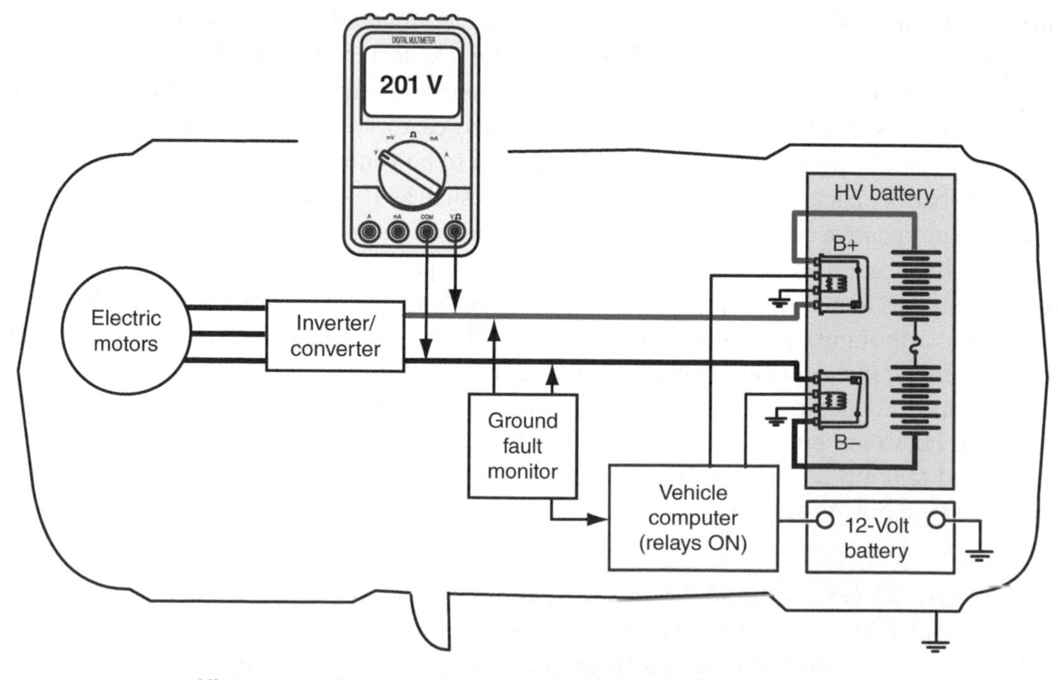

High-Voltage System—Vehicle On and Operational (READY -on)

Figure 12-22 High-voltage electrical system.

Figure 12-23 Master warning light.

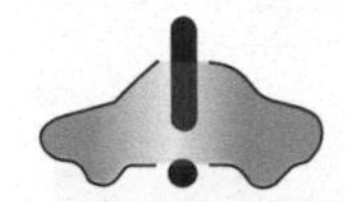

Figure 12-24 Hybrid warning light.

time while the READY indicator is illuminated. Never assume the electrical system is off just because the engine is stopped. The electrical system is off when the READY indicator is off. If the vehicle is involved in a collision, and the air bags are deployed, the battery pack relays automatically open and disconnect the voltage supplied from the battery pack to the inverter. When the engine is running, a ground fault monitor continuously monitors the high-voltage system for high voltage leakage to the metal chassis. If leakage occurs, the vehicle computer illuminates the master warning light in the instrument cluster (**Figure 12-23**) and the hybrid warning light in the liquid crystal display (LCD) (**Figure 12-24**). The conventional 12-V lead acid battery is located in the truck near the battery pack. The positive cable from the 12-V battery is routed to the engine compartment with the high-voltage cables from the battery pack. HEV systems and indicators vary depending on the vehicle make and model year.

HEV Operation

When the engine is at normal operating temperature, and the vehicle is starting off from a stop or operating at very low speed and light load, only electrical power from the battery pack is supplied to the propulsion motor to drive the vehicle (**Figure 12-25**).

During normal driving, mechanical power from the engine and electric power supplied from the generator to the propulsion motor are used to drive the vehicle (**Figure 12-26**).

If the engine is operating at wide throttle opening, mechanical power from the engine and electric power from the generator and battery pack are supplied to drive the vehicle (**Figure 12-27**).

During vehicle deceleration, the propulsion motor acts as a generator and supplies current to charge the batteries (**Figure 12-28**).

If the battery state of charge becomes low, more of the electric power from the generator is transmitted to the battery pack for recharging (**Figure 12-29**).

Power Steering and Hybrid Vehicles

Many HEVs have a stop/start function that allows the engine to be shut off when the vehicle has come to a complete stop, with the engine idling. When the ignition switch is on and the driver steps on the accelerator pedal, the vehicle starts off, and the engine may

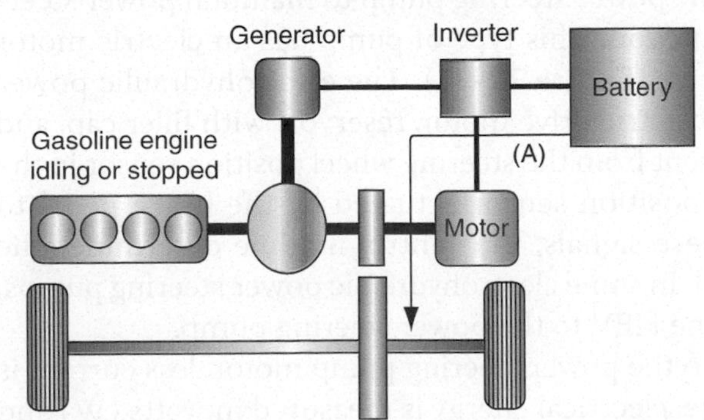

Figure 12-25 HEV starting off or operating at low speed and light load.

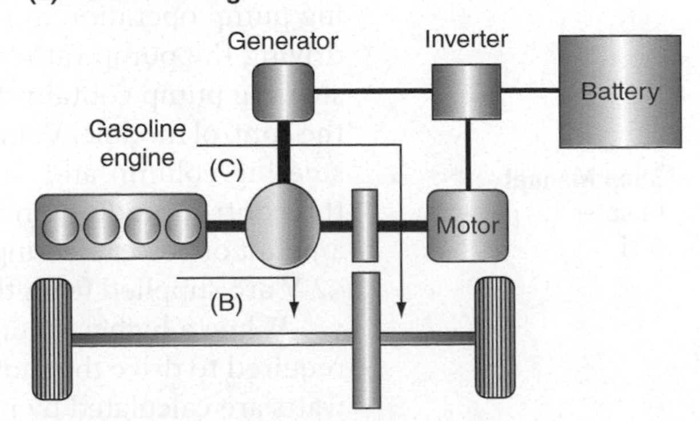

Figure 12-26 HEV operation during normal driving.

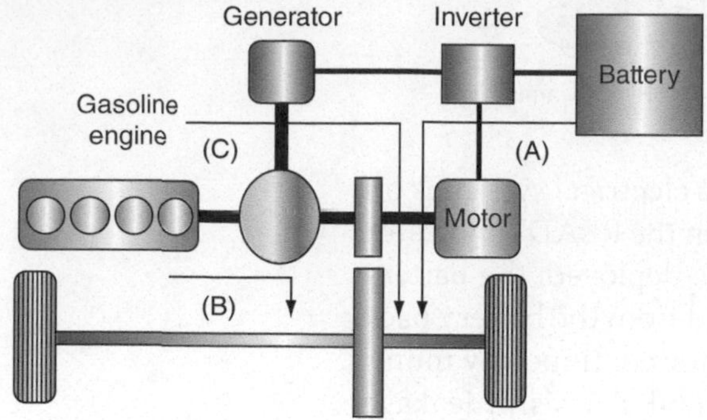

Figure 12-27 HEV operation at wide-open throttle.

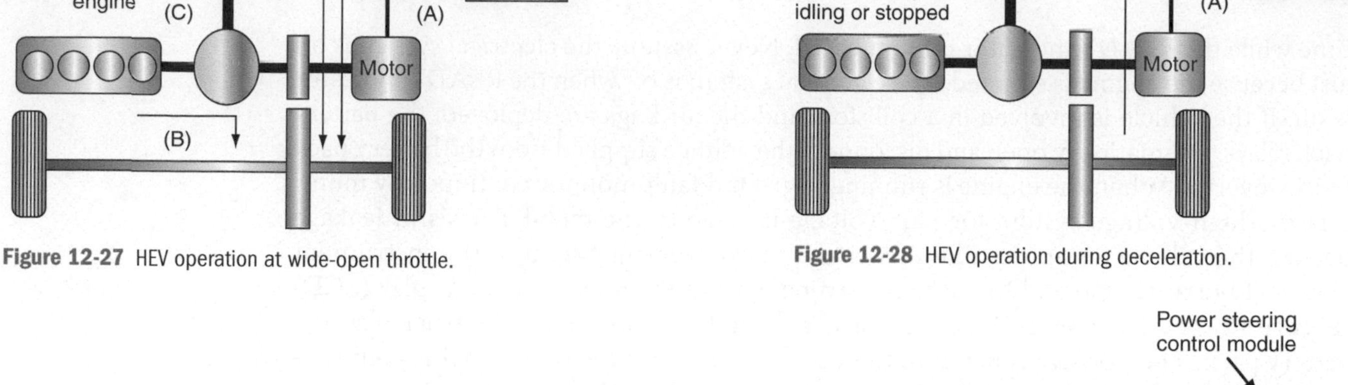

Figure 12-28 HEV operation during deceleration.

(5) Charging the batteries

Generator

Inverter

Battery

Gasoline engine

(C)

(D)

Motor

(B)

Figure 12-29 HEV operation with low battery state of charge.

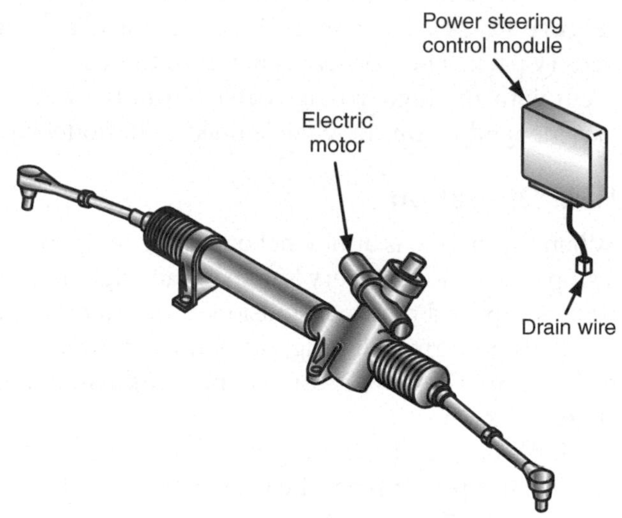

Figure 12-30 Electronic power steering (EPS).

restart at any time depending on throttle opening and other factors. The power steering must be operational even during the engine stop part of the stop/start function. To maintain power steering operation during the stop function, some vehicle manufacturers use an electronic power steering (EPS) gear. Some of these EPSs have an electric motor driving the steering gear pinion (**Figure 12-30**). A power steering control module supplies current to the EPS motor to supply the proper amount of steering assist. The EPS is fully operational when the ignition switch is turned on.

Other HEVs have an electrohydraulic power steering pump to maintain power steering pump operation during the stop function. This type of pump has an electric motor driving the pump rather than a belt drive (**Figure 12-31**). The electrohydraulic power steering pump contains the pump and electric drive motor, reservoir with filler cap, and the control module. Voltage signals are sent from the steering wheel position sensor in the steering column and the brake pedal position sensor actuated by the brake pedal to the control module. In response to these signals, the control module determines the amount of power steering assist required. In some electrohydraulic power steering pumps, 42 V are supplied from the inverter in the HEV to the power steering pump.

Shop Manual
Chapter 12, page 531

When a higher voltage is supplied to the power steering pump motor, less current is required to drive the motor. For example, electrical energy is measured in watts (W) and watts are calculated by multiplying the amperes by the volts. When 12 V are supplied to a headlight, 3 amperes may be required to illuminate the light. The electrical energy

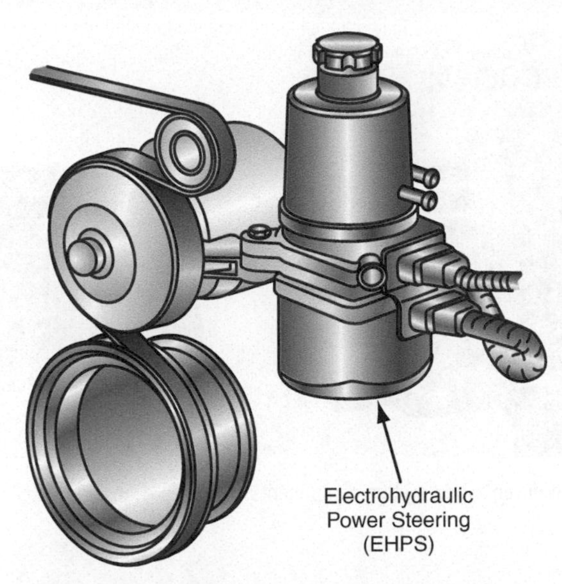

Figure 12-31 Electrohydraulic power steering (EHPS) module.

required to illuminate the light is 12 × 3 = 36 W. If 18 V are supplied to the headlight, only 2 amperes are required to illuminate the light. The advantage of increasing the voltage is that the same watts can be obtained with fewer amperes.

AUTHOR'S NOTE In 2008, one leading manufacturer's worldwide sales of HEVs surpassed 1.5 million units. The average silicon content on a non-hybrid vehicle consumes approximately 30 percent of an 8 in. (20.32 cm) silicon wafer. The silicon content of a hybrid vehicle requires the full 8 in. (20.32 cm) wafer. Approximately 50 percent of the cost of a hybrid vehicle is in the hardware and software. Toyota has taken 70 percent of the cost of the hybrid technology out of the hybrid Prius, improved its fuel economy, and increased the size of the chassis/body.

VARIOUS TYPES OF HEVs

Belt-Driven HEVs

Some HEVs have a belt-driven motor/generator with an improved drive belt and tensioner compared with a conventional generator (**Figure 12-32**). The latest belt-driven HEV system features a stop/start function, regenerative braking, intelligent battery charging, and 4 kW of energy are supplied from the battery pack to the motor/generator for up to 5 seconds during wide throttle, heavy load driving. Some belt-driven HEVs supply a modest electric-only propulsion below 8 mph (12.9 km/h). Lithium-ion batteries will soon be available in some belt-driven HEV systems.

Plug-In HEVs

Several major vehicle manufacturers are scheduled to introduce plug-in HEVs (PHEVs) in their showrooms during 2010. A PHEV may be driven only by an electric motor(s), but a small gasoline or diesel engine is installed in the vehicle for extra power or to drive a generator and recharge the batteries. A typical PHEV may be driven for 40 mi. (64 km) using only electric power. When the batteries reach a specific state of discharge, the engine starts and drives the generator to recharge the batteries.

Figure 12-32 Belt-driven hybrid system components.

Extended Range Electric Vehicle

An extended range electric vehicle (EREV) is similar to a PHEV. The EREV has a small gasoline or diesel engine that drives a generator to recharge the batteries. The engine only drives the generator; it cannot supply power to the drive wheels. The Chevrolet Volt is an EREV.

FUEL CELL VEHICLES

Fuel cell vehicles (FCVs) are not presently available in the automotive showrooms. However, many car manufacturers around the world are working on this technology. Several vehicle manufacturers have designed and produced FCVs, and a significant number of these vehicles have been released to various fleet owners and officials to gain field experience and determine how these vehicles perform in the real world.

An FCV has an electric-drive motor(s), and the fuel cell supplies electricity to this drive motor(s). The fuel cell must be supplied with hydrogen as a fuel, and the fuel cell produces electricity as the hydrogen and air flow through the cell. There are various types of fuel cells, but the most common type is the proton exchange membrane (PEM) fuel cell. The fuel cell actually contains a large group of cells. The only by-product from fuel cell operation is water vapor. A PEM fuel cell is compared to an internal combustion engine in **Figure 12-33**. The FCV requires a source of high voltage and current such as a battery pack and/or ultra-capacitor to supply extra current to the drive motor during operation at wide throttle openings. The disadvantages of FCVs are as follows:

1. The size, weight, and cost of the FCV components. Fuel cells are very expensive at present.
2. The lack and cost of hydrogen refueling facilities.
3. The cost and availability of FCVs.
4. Difficulty in storing hydrogen on-board vehicles; liquid hydrogen must be stored under pressure in special insulated tanks. It is difficult to store a sufficient amount of gaseous hydrogen on-board a vehicle.

The advantages of FCVs are as follows:

1. Reduced dependency on high-priced hydrocarbon fuels.
2. Hydrogen may be obtained from several different sources.
3. Greatly reduced tailpipe and evaporative emissions.

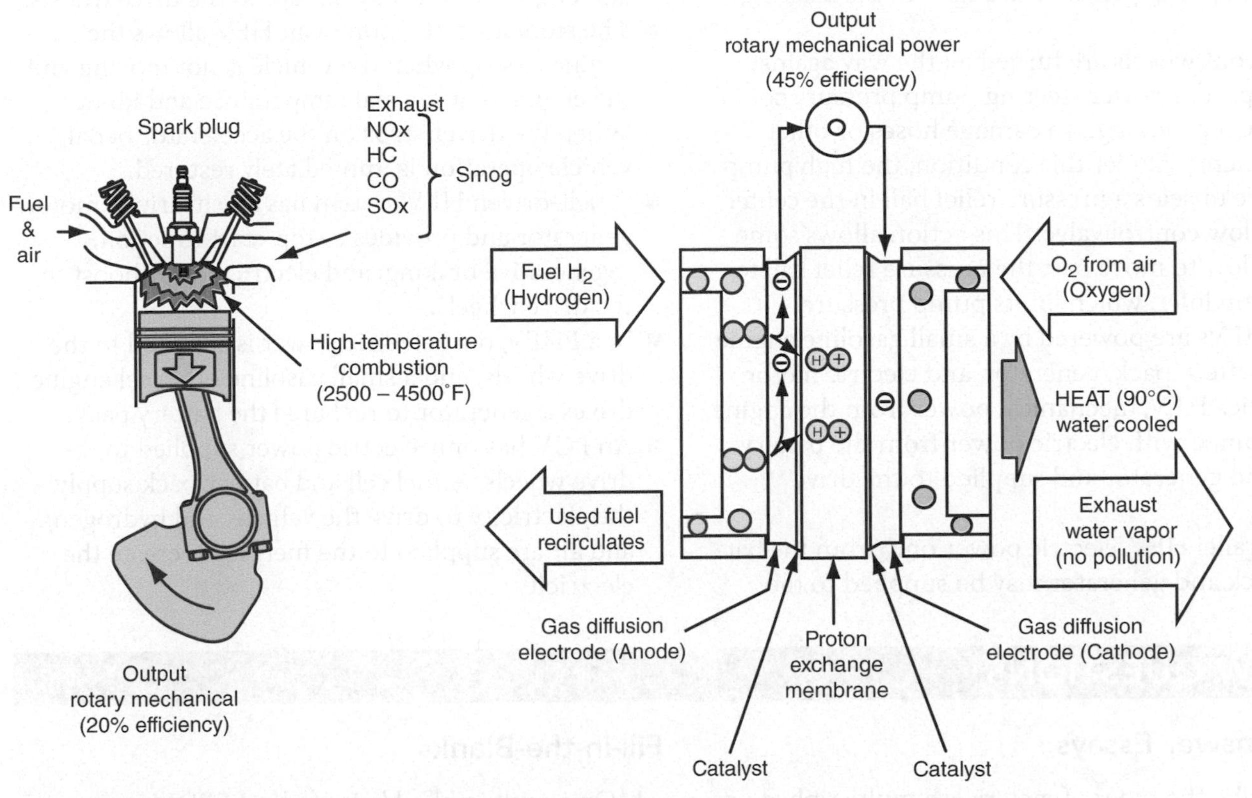

Figure 12-33 Comparison between an internal combustion engine and a fuel cell.

SUMMARY

- A multi-ribbed belt or a V-belt on older vehicles may be used to drive the power steering pump.
- A serpentine belt is a multi-ribbed belt that is used to drive all the belt-driven components. This allows these components to be on the same vertical plane.
- Proper belt tension is extremely important for adequate power steering pump operation. Many belts today are made of EPDM. A rack and pinion steering gear is usually mounted on the cowl or the front crossmember. Rack and pinion steering gears are used in many front-wheel-drive cars. Integral power steering gears are usually mounted on the vehicle frame. Integral power steering systems are found on many rear-wheel-drive cars and light-duty trucks. A power steering pump may have an integral or a remote fluid reservoir. In a hydro-boost power-assisted steering system, fluid pressure from the power steering pump is applied to the steering gear and the brake master cylinder. The power steering pump pressure acts as a booster to assist the driver in applying the brakes.
- A power steering pump may have a vane, roller, or slipper-type rotor assembly, but all three types of pumps operate on the same basic principle.
- The flow control valve in a power steering pump is moved toward the closed position by spring pressure and fluid pressure from the venturi in the pump outlet fitting.
- When a vehicle is driven at low speeds with the front wheels straight ahead, the power steering pump fluid flow is lower. Under this condition, the pressure in the outlet fitting venturi is higher. This higher pressure together with the spring tension keeps the flow control valve closed. In this valve position, a small amount of fluid is routed past the flow control valve to the pump inlet.
- If engine speed is increased, the pump delivers more fluid than the system requires. This increase in pump flow reduces pressure in the outlet fitting venturi. This pressure decrease is applied to the spring side of the flow control valve. This action allows the flow control valve to move toward the partially open position, and excessive pump flow is returned past the flow control valve to the pump inlet.
- If the driver turns the steering wheel, fluid rushes from the pump outlet into the steering gear pressure chamber. Under this condition, the flow

control valve moves toward the closed position to maintain pump pressure and flow to the steering gear.

- If the front wheels are turned all the way against the stops, the power steering pump pressure could become high enough to damage hoses or other components. Under this condition, the high pump pressure unseats a pressure relief ball in the center of the flow control valve. This action allows some pump flow to move past the pressure relief ball to the pump inlet, which limits pump pressure.
- Most HEVs are powered by a small gasoline engine and a battery pack, generator, and electric motor.
- In a series HEV, mechanical power from the engine is combined with electric power from the battery pack and generator and supplied to the drive wheels.
- In a parallel HEV, electric power only from the battery pack and generator may be supplied to the drive wheels or a combination of electric power and engine power may be sent to the drive wheels.
- The stop/start function in an HEV allows the engine to stop when the vehicle is not moving and the engine is at normal temperature and idling. When the driver steps on the accelerator pedal, vehicle operation is immediately restored.
- A belt-driven HEV system has a belt-driven motor/generator and provides a stop/start function, regenerative braking, and electric power boost to the drive wheels.
- In a PHEV, only electric power is supplied to the drive wheels, and a small gasoline or diesel engine drives a generator to recharge the battery pack.
- An FCV has only electric power supplied to the drive wheels. A fuel cell and battery pack supply the electricity to drive the vehicle, and hydrogen and air are supplied to the fuel cell to create the electricity.

REVIEW QUESTIONS

Short Answer Essays

1. Describe the types of materials a multi-ribbed or serpentine belt may be constructed of.
2. Explain two advantages of a serpentine belt.
3. List two possible locations for a power steering pump reservoir.
4. How is the power steering pump output effected as engine speed is increased but no additional power steering assist is required, such as in straight-ahead driving?
5. Describe how the brake boost pressure is obtained in a hydro-boost power steering system.
6. Explain how the fluid pressure is produced in a vane-type power steering pump.
7. List the forces that move the flow control valve toward the closed position in a power steering pump.
8. Describe the operation of a venturi.
9. Explain the operation of the flow control valve when the steering wheel is turned.
10. Describe the operation of the power steering pump when the steering wheel is rotated all the way to the right or left until the front wheels contact the stops.

Fill-in-the-Blanks

1. On a multi-ribbed belt made of EPDM a ______________ percent loss of material is enough to cause belt______________, ______________, and ______________.
2. Excessive belt wear occurs if pulleys are ______________.
3. Many serpentine belts have a spring-loaded ______________.
4. The master cylinder in a hydro-boost system does not have a conventional ______________.
5. The purpose of the vanes in the power steering pump rotor is to ______________ the rotor in the elliptical cam ring.
6. As the power steering pump shaft and rotor turn, ______________ ______________ causes the vanes to move outward against the cam ring.
7. A machined venturi is located in the power steering pump ______________ ______________.
8. When fluid movement through the venturi increases, the pressure in the venturi ______________.
9. With the engine running and the front wheels straight ahead, the power steering pump pressure is ______________.

10. The pressure relief ball in a power steering pump is forced open when the front wheels are turned against the ______________.

Multiple Choice

1. The most likely result of a power steering drive-belt that is worn is:
   A. Belt slipping.
   B. Belt breaking.
   C. Belt contamination.
   D. Wear on the power steering pump pulley.
2. All of these statements about a serpentine belt are true EXCEPT:
   A. A serpentine belt can be used to drive all the belt-driven components.
   B. The back of a serpentine belt can drive one component.
   C. A serpentine belt made of EPDM is susceptible to cracking.
   D. A serpentine belt can have an automatic belt tensioner.
3. A hydro-boost power steering system:
   A. Is used on many small front-wheel-drive vehicles.
   B. Uses pressure from the power steering pump to provide brake boost.
   C. Has a conventional vacuum brake booster.
   D. Has a very high-capacity power steering pump.
4. All of the following statements about multi-ribbed belts made of EPDM are true EXCEPT:
   A. EPDM belt is more elastic than a standard neoprene belt and resists cracking even at higher mileage.
   B. EPDM belts are designed to last 50,000–60,000 miles.
   C. Belts are designed to have clearance between the rib peaks and the pulley grooves.
   D. EPDM belt to gradually lose material on the ribbing similar to the way a tire wears.
5. All of these statements about integral-type power steering pump sealing are true EXCEPT:
   A. A large O-ring seal is positioned between the reservoir and the pump housing.
   B. A lip seal on the pump drive shaft prevents oil leaks around the shaft.
   C. Lip seals are mounted on bolt fittings on the back of the reservoir.
   D. The cap and dipstick provide a seal on the top of the reservoir neck.
6. A power steering system experiences repeated belt failures with normal power steering system operation. The most likely cause of this problem is:
   A. A misaligned power steering pump pulley.
   B. A loose power steering pump belt.
   C. Low fluid level in the power steering reservoir.
   D. A worn cam ring in the power steering pump.
7. During normal power steering pump operation:
   A. The flow control valve spring moves this valve toward the open position.
   B. The fluid pressure from the outlet fitting venturi moves the flow control valve toward the open position.
   C. Pump flow increases as the engine speed increases.
   D. The pressure relief valve opens when the engine speed is increased with the front wheels straight ahead.
8. All of these statements about power steering pump design and operation are true EXCEPT:
   A. An increase in fluid movement through the outlet fitting venturi causes a pressure decrease in the venturi.
   B. An increase in the flow control valve opening allows more fluid to return to the pump inlet.
   C. The rotor is attached to the pump shaft and must rotate with this shaft.
   D. Fluid pressure from the pump vanes moves the flow control valve toward the closed position.

9. All of the following statements about hybrid electric vehicles that have an electrohydraulic power steering pump are true EXCEPT:
   A. The power steering pump has an electric-driven motor.
   B. A control module determines the amount of power assist required.
   C. An electric motor drives the steering gear.
   D. Some are sullied 42 V from the HEV inverter.

10. A vehicle experiences repeated ruptures of the high-pressure power steering hose. The most likely cause of this problem is:
    A. A pressure relief ball sticking closed.
    B. A flow control valve sticking open.
    C. Worn pump vanes and cam ring.
    D. A partially restricted power steering return hose.

CHAPTER 13
RACK AND PINION STEERING GEARS AND FOUR-WHEEL STEERING

Upon completion and review of this chapter, you should be able to understand and describe:

- The advantages of a rack and pinion steering system compared to a recirculating ball steering gear and parallelogram steering linkage.
- How the tie rods are connected to the rack.
- The purpose of the rack bearing and adjuster plug.
- Two possible mounting positions for the rack and pinion steering gear.
- The fluid movement in a rack and pinion steering gear during a right turn.
- The fluid movement in a rack and pinion steering gear during a left turn.
- The operation of the spool valve and rotary valve.
- The operation of the power rack and pinion steering gear when hydraulic pressure is not available.
- The purpose of the breather tube.
- The main differences between a Saginaw and a TRW power rack and pinion steering gear.
- The advantages of an electronic variable orifice (EVO) steering system.
- The input sensors in an EVO steering system.
- The driving conditions when an EVO steering system provides increased power steering assistance.
- The operation of a rack-drive electronic power steering system during right and left turns.
- The operation of the steering shaft torque sensor in a column-drive electronic power steering system.
- The operation and advantages of active front steering (AFS).
- The advantages of four-wheel steering (4WS).
- How the rear steering rack is driven.
- The two inputs used by the control unit to operate the rear steering system.
- Negative phase and positive phase steering.
- The input sensors in an electronically controlled four-wheel steering system and give the location of each sensor.
- The type of signal produced by a wheel speed sensor.
- The steering action of the rear wheels in relation to vehicle speed in an electronically controlled 4WS system.
- The operation and advantages of rear active steering (RAS) and four-wheel active steering (4WAS).

Terms To Know

Analog voltage signal
Breather tube
Car access system (CAS)
Center take-off (CTO)
Column-drive EPS
Digital motor electronics (DME)
Electric locking unit (ELU)
Electronic control unit (ECU)
Electronic variable orifice (EVO) steering
Electronically controlled orifice (ECO) valve
Electronically controlled 4WS system
Encoded disc
End take-off (ETO)
Feel of the road
Four-wheel steering (4WS)

- Four-wheel active steering (4WAS)
- Helical gear teeth
- Lock-to-lock
- Magnasteer system
- Neutral phase steering
- Pinion
- Pinion angle sensor
- Pinion-drive EPS
- Positive phase steering
- Pulse width modulated (PWM) voltage signal
- Rack
- Rack-drive EPS
- Rack piston
- Rack seals
- Rear active steering (RAS)
- Rear steering actuator
- Rotary valve
- Safety and gateway module (SGM)
- Servotronic valve
- Sideslip
- Spool valve
- Spur gear teeth
- Steering angle
- Steering angle sensor
- Steering kickback
- Steering ratio
- Steering wheel rotation sensor
- Stop-to-stop
- Summation sensor
- Torsion bar
- Transistor
- Turning circle
- Variable effort steering (VES)
- Vehicle speed sensor (VSS)
- Yaw forces
- Yaw motion

INTRODUCTION

During the late 1970s and 1980s, the domestic automotive industry converted much of its production from larger rear-wheel-drive cars to smaller, lightweight, and more fuel-efficient front-wheel-drive cars. These front-wheel-drive cars required smaller, lighter components wherever possible. Manual and power rack and pinion steering gears are lighter and more compact than the recirculating ball steering gears and parallelogram steering linkages used on most rear-wheel-drive cars. Therefore, rack and pinion steering gears are ideally suited to these compact front-wheel-drive cars.

Steering systems have not escaped the electronics revolution. Many cars are presently equipped with electronic variable orifice (EVO) steering, which provides greater power assistance during low-speed cornering and parking for increased driver convenience. Some cars are now equipped with electronic power steering. In these systems, an electric motor in the steering gear provides steering assist.

MANUAL RACK AND PINION STEERING GEAR MAIN COMPONENTS

Shop Manual Chapter 13, page 556

Rack

The **rack** is a toothed bar that slides back and forth in a metal housing. The steering gear housing is mounted in a fixed position on the front crossmember or on the firewall. The rack takes the place of the idler and pitman arms in a parallelogram steering system and maintains the proper height of the tie rods so they are parallel to the lower control arms. The rack may be compared to the center link in a parallelogram steering linkage. Bushings support the rack in the steering gear housing. Sideways movement of the rack pulls or pushes the tie rods and steers the front wheels (**Figure 13-1**).

Spur gear teeth are cut so they are parallel to the gear rotational axis.

Helical gear teeth are curved to increase the amount of tooth contact between a pair of meshed gears. Helical gears tend to operate more quietly than spur gear teeth and provide increased strength compared with spur gear teeth.

Pinion

The **pinion** is a toothed shaft mounted in the steering gear housing so the pinion teeth are meshed with the rack teeth. The pinion may contain **spur gear teeth** or **helical gear teeth**. The upper end of the pinion shaft is connected to the steering shaft from the steering column. Therefore, steering wheel rotation moves the rack sideways to steer the front wheels. The pinion is supported on a ball bearing in the steering gear housing.

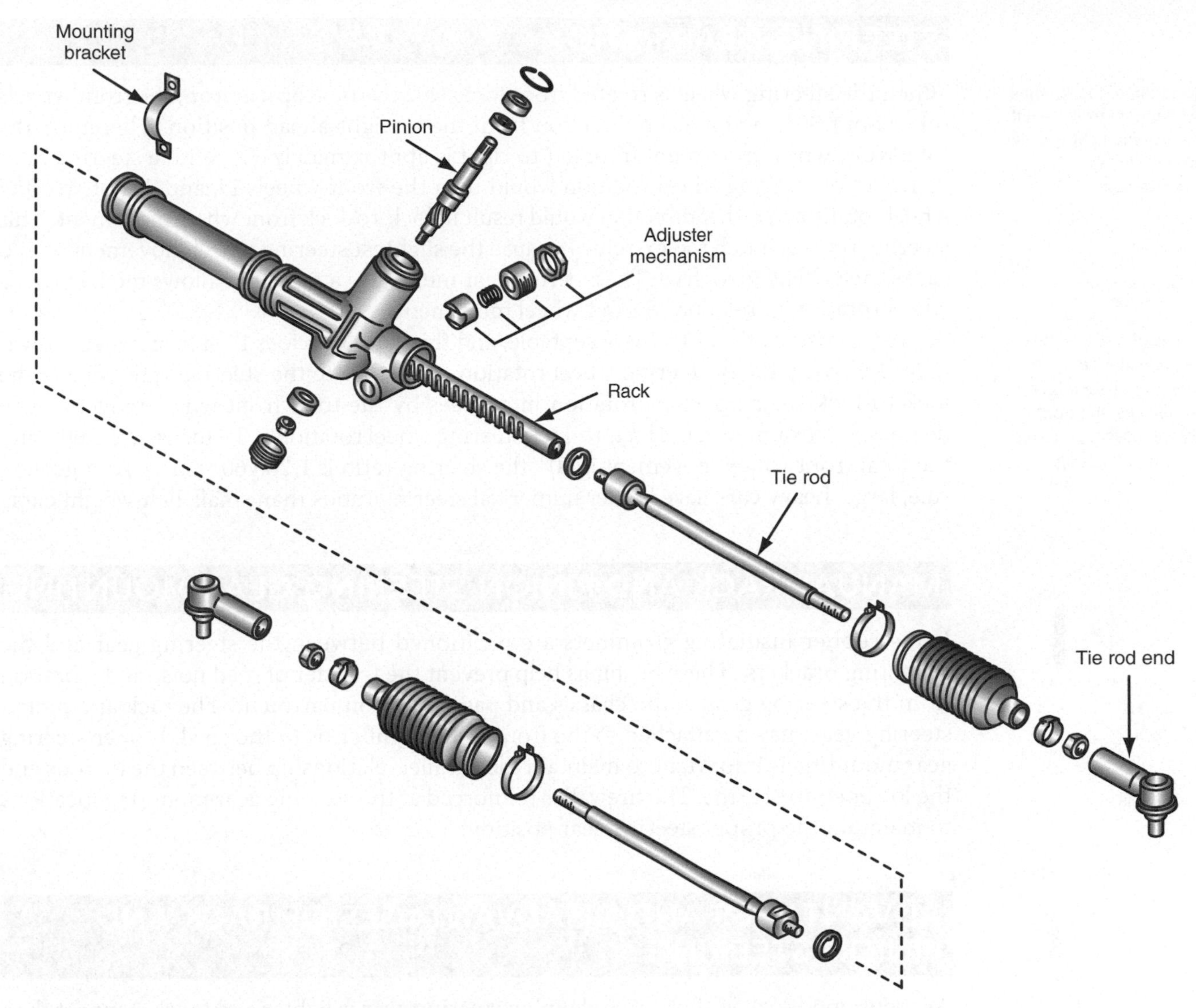

Figure 13-1 Manual rack and pinion steering gear components.

Tie Rods and Tie Rod Ends

The tie rods are similar to those used on parallelogram steering linkages. A spring-loaded ball socket on the inner end of the tie rod is threaded onto the rack. When these ball sockets are torqued to the vehicle manufacturer's specification, a preload is placed on the ball socket. A bellows boot is clamped to the housing and tie rod on each side of the steering gear, and these boots keep contaminants out of the ball socket and rack.

A tie rod end is threaded onto the outer end of each tie rod. These tie rod ends are similar to those used on parallelogram steering linkages. A jam nut locks the outer tie rod end to the tie rod.

Rack Adjustment

A rack bearing is positioned against the smooth side of the rack. A spring is located between the rack bearing and the rack adjuster plug that is threaded into the housing. This adjuster plug is retained with a locknut. The rack bearing adjustment sets the preload between the rack and pinion teeth, which affects **steering kickback**, harshness, and noise.

Steering kickback is the movement of the steering wheel caused by a front wheel striking a road irregularity.

STEERING GEAR RATIO

Rotation of the steering wheel from extreme left to extreme right is called **lock-to-lock** or **stop-to-stop**.

When the steering wheel is rotated from **lock-to-lock** or **stop-to-stop**, the front wheels turn about 30° each in each direction from the straight-ahead position. Therefore, the total front wheel movement from left to right is approximately 60°. With a steering ratio of 1:1, 1° of steering wheel rotation would turn the front wheels 1°, and 30° of steering wheel rotation in either direction would result in lock-to-lock front wheel movement. This steering ratio is much too extreme because the slightest steering wheel movement would cause the vehicle to swerve. The steering gear must have a ratio that allows more steering wheel rotation in relation to front wheel movement.

Steering ratio is the relationship between steering wheel rotation in degrees and front wheel movement to the right or left in degrees.

A **steering ratio** of 15:1 is acceptable, and this ratio provides 1° of front wheel movement for every 15° of steering wheel rotation. To calculate the steering ratio, divide the lock-to-lock steering wheel rotation in degrees by the total front wheel movement in degrees. For example, if the lock-to-lock steering wheel rotation is 3 5 turns, or 1,260°, and the total front wheel movement is 60°, the steering ratio is $1,260/60 = 21:1$. As a general rule, large, heavy cars have higher numerical steering ratios than small, lightweight cars.

MANUAL RACK AND PINION STEERING GEAR MOUNTING

Large rubber insulating grommets are positioned between the steering gear and the mounting brackets. These bushings help prevent the transfer of road noise and vibration from the steering gear to the chassis and passenger compartment. The rack and pinion steering gear may be attached to the front crossmember or to the cowl. Proper steering gear mounting is important to maintain the parallel relationship between the tie rods and the lower control arms. The firewall is reinforced at the steering gear mounting locations to maintain the proper steering gear position.

ADVANTAGES AND DISADVANTAGES OF RACK AND PINION STEERING

As mentioned earlier, the rack and pinion steering gear is lighter and more compact than a recirculating ball steering gear and parallelogram steering linkage. Therefore, the rack and pinion steering gear is most suitable for front-wheel-drive unibody vehicles.

Since there are fewer friction points in the rack and pinion steering than in the recirculating ball steering gear with a parallelogram steering linkage, the driver has a greater feeling of the road with rack and pinion steering gear. However, fewer friction points reduce the steering system's ability to isolate road noise and vibration. Therefore, drivers of a vehicle with a rack and pinion steering system may have more complaints of road noise and vibration transfer to the steering wheel and passenger compartment.

The **rack piston** is attached to the rack, and this piston moves horizontally in the steering gear chamber.

Rack seals are mounted at each end of the rack chamber and on the rack piston. The seals at each end of the rack chamber prevent leaks between the rack and the chamber.

POWER RACK AND PINION STEERING GEARS

Design and Operation

A power-assisted rack and pinion steering gear uses the same basic operating principles as a manual rack and pinion steering gear, but in the power-assisted steering gear, hydraulic fluid pressure from the power steering pump is used to reduce steering effort. A **rack piston** is integral with the rack, and this piston is located in a sealed chamber in the steering gear housing. Hydraulic fluid lines are connected to each end of this chamber, and **rack seals** are positioned in the housing at ends of the chamber. A seal is also located on the rack piston (**Figure 13-2**).

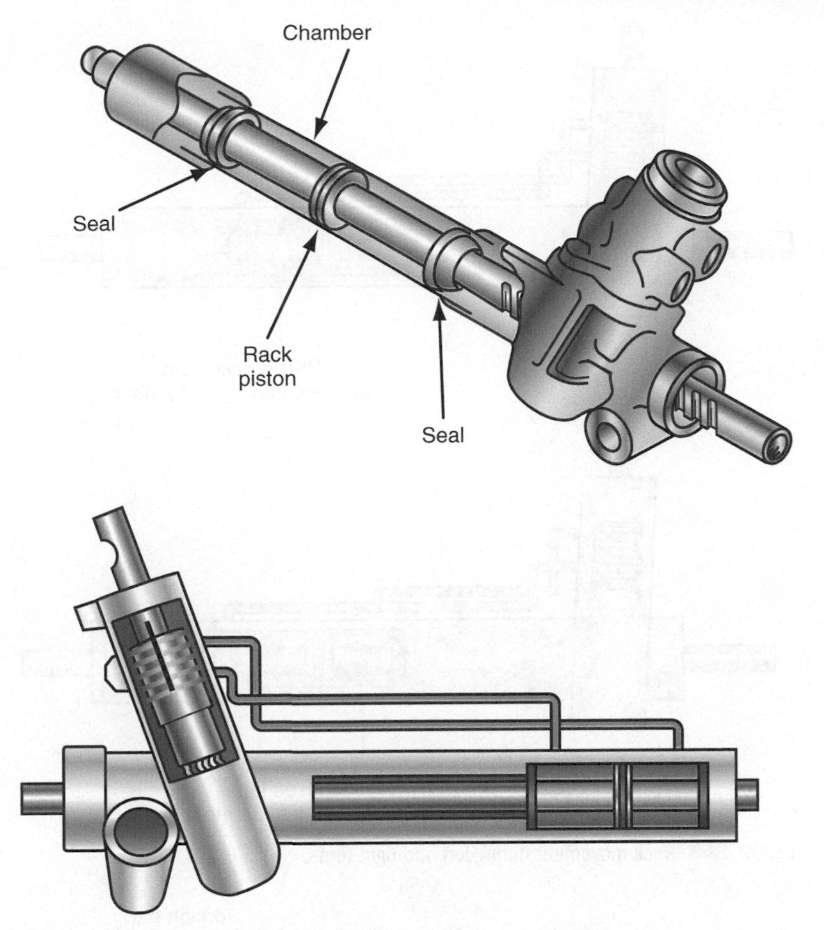

Figure 13-2 Hydraulic chamber in a power rack and pinion steering gear.

The following description assumes that the power steering rack is mounted behind the front wheel spindle, such as on the vehicle firewall or cradle assembly. When a driver is completing a left turn, fluid is pumped into the left side of the fluid chamber and exhausted from the right chamber area. This hydraulic pressure on the left side of the rack piston helps the pinion move the rack to the right (**Figure 13-3**, view A).

When a right turn is made, fluid is pumped into the right side of the fluid chamber, and fluid flows out of the left end of the chamber. Thus, hydraulic pressure is exerted on the right side of the rack piston, which assists the pinion gear in moving the rack to the left (**Figure 13-3**, view B). Since the steering gear is mounted behind the front wheels, rack movement to the left is necessary for a right turn, whereas rack movement to the right causes a left turn.

Rotary Valve and Spool Valve Operation

Fluid direction in the steering gear is controlled by a rotary valve attached to the pinion assembly (**Figure 13-4**). A stub shaft on the pinion assembly is connected to the steering shaft and wheel. The pinion is connected to the stub shaft through a **torsion bar** that twists when the steering wheel is rotated and springs back to the center position when the wheel is released. A rotary valve body contains an inner **spool valve** that is mounted over the torsion bar on the pinion assembly.

Torsion bar twisting during steering wheel rotation moves the spool valve in relation to the rotary valve.

When the front wheels are in the straight-ahead position, fluid flows from the pump through the high-pressure hose to the center **rotary valve** body passage. Fluid is then routed through the valve body to the low-pressure return hose and the pump reservoir (**Figure 13-5**).

The **rotary valve** is mounted over the spool valve on the steering gear pinion.

Many power steering systems contain a fluid cooler connected in the high-pressure hose between the pump and the steering gear. The fluid cooler is like a small radiator. Air

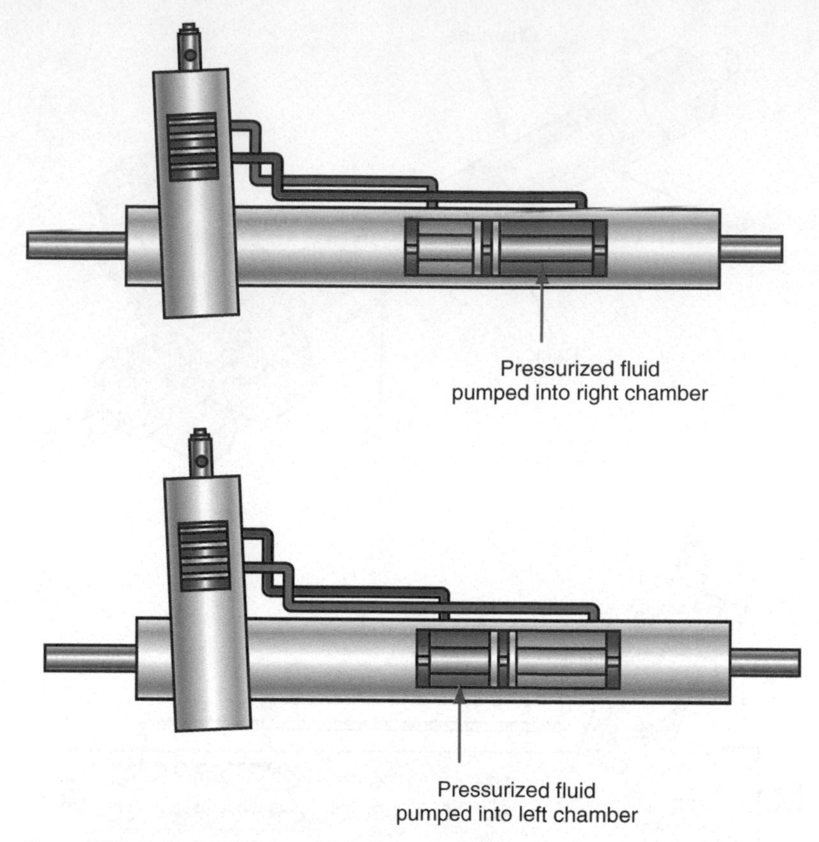

Figure 13-3 Rack movement during left and right turns.

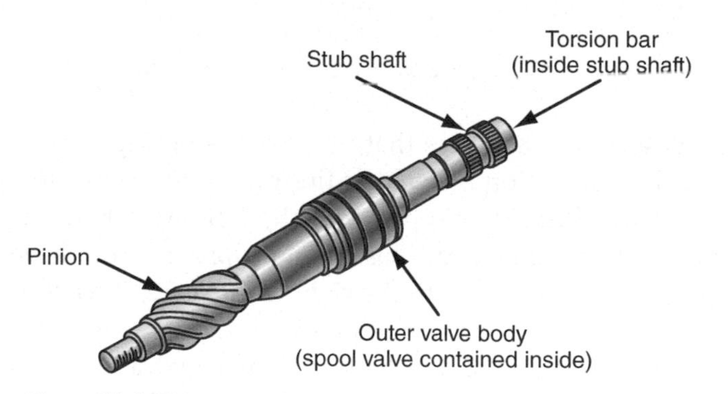

Figure 13-4 Pinion assembly for a power rack and pinion steering gear.

flows through the fins on the cooler and cools the fluid flowing through the internal cooler passages. Some power steering systems have a remote fluid reservoir connected in the low-pressure hose between the steering gear and the pump (**Figure 13-5**). Many power steering systems have a fluid filter, and this filter is often mounted in the remote reservoir.

Teflon rings or O-rings seal the rotary valve ring lands to the steering gear housing. A lot of force is required to turn the pinion and move the rack because of the vehicle weight on the front wheels. When the driver turns the wheel, the stub shaft is forced to turn. However, the pinion resists turning because it is in mesh with the rack, which is connected to the front wheels. This resistance of the pinion to rotation results in torsion bar twisting. During this twisting action, a pin on the torsion bar moves the spool valve with a circular motion inside the rotary valve. If the driver makes a left turn, the spool valve movement aligns the inlet center rotary valve passage with the outlet passage to the left side of the rack piston. Therefore, hydraulic fluid pressure applied to the left side of the rack piston assists the driver in moving the rack to the right.

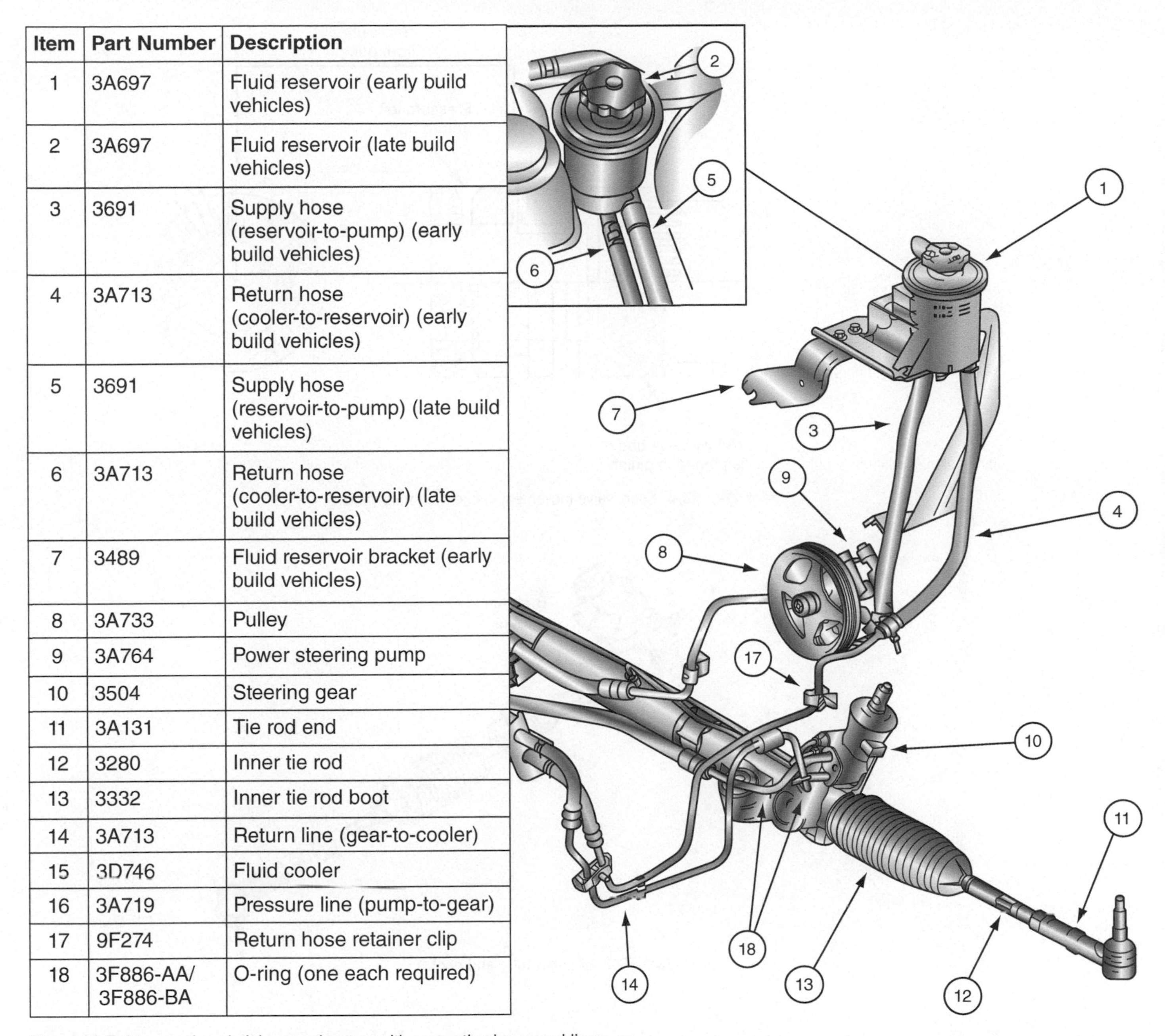

Item	Part Number	Description
1	3A697	Fluid reservoir (early build vehicles)
2	3A697	Fluid reservoir (late build vehicles)
3	3691	Supply hose (reservoir-to-pump) (early build vehicles)
4	3A713	Return hose (cooler-to-reservoir) (early build vehicles)
5	3691	Supply hose (reservoir-to-pump) (late build vehicles)
6	3A713	Return hose (cooler-to-reservoir) (late build vehicles)
7	3489	Fluid reservoir bracket (early build vehicles)
8	3A733	Pulley
9	3A764	Power steering pump
10	3504	Steering gear
11	3A131	Tie rod end
12	3280	Inner tie rod
13	3332	Inner tie rod boot
14	3A713	Return line (gear-to-cooler)
15	3D746	Fluid cooler
16	3A719	Pressure line (pump-to-gear)
17	9F274	Return hose retainer clip
18	3F886-AA/ 3F886-BA	O-ring (one each required)

Figure 13-5 Power rack and pinion steering gear with connecting hoses and lines.

When a right turn is made, twisting of the torsion bar moves the spool valve and aligns the center rotary valve passage with the outlet passage to the right side of the rack piston (**Figure 13-6**). Under this condition, hydraulic fluid pressure applied to the rack piston helps the driver move the rack to the left. The torsion bar provides a **feel of the road** to the driver.

When the steering wheel is released after a turn, the torsion bar centers the spool valve and power assistance stops. If hydraulic fluid pressure is not available from the pump, the power steering system operates like a manual system, but steering effort is higher. When the torsion bar is twisted to a designed limit, tangs on the stub shaft engage with drive tabs on the pinion. This action mechanically transfers motion from the steering wheel to the rack and front wheels. Since hydraulic pressure is not available on the rack piston, greater steering effort is required. If a front wheel raises going over a bump or drops into a hole, the tie rod pivots along with the wheel. However, the rack and tie rod still maintain the left-to-right wheel direction.

When the driver turns the steering wheel, the amount of feeling that he or she senses regarding front wheel turning is called **feel of the road**.

In power rack and pinion steering gears, a condition that causes excessive steering wheel turning effort when the vehicle is first started may be called "morning sickness."

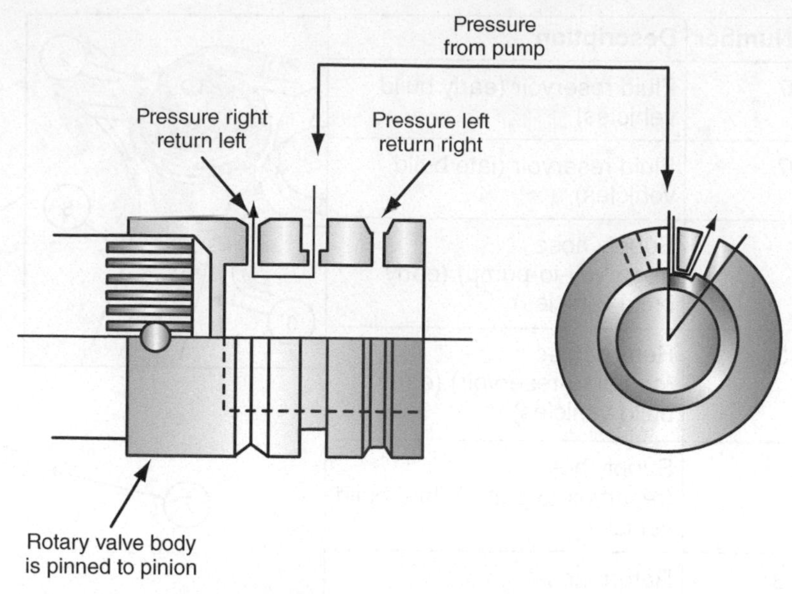

Figure 13-6 Spool valve movement inside the rotary valve.

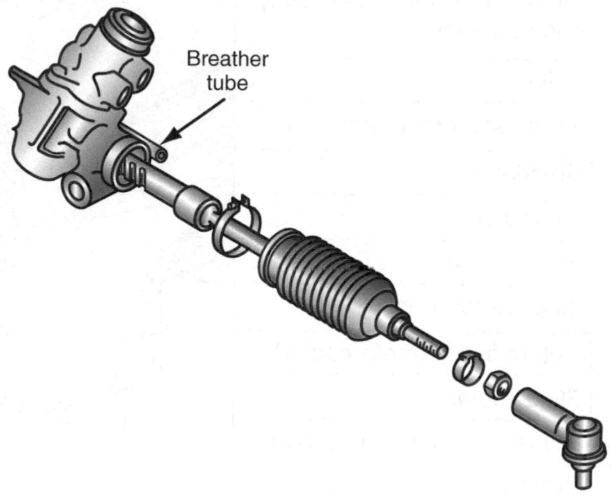

Figure 13-7 Breather tube and boot.

The rack boots are clamped to the housing and the rack. Since the boots are sealed and air cannot be moved through the housing, a **breather tube** is necessary to move air from one boot to the other when the steering wheel is turned (**Figure 13-7**). This air movement through the vent tube prevents pressure changes in the bellows boots during a turn.

AUTHOR'S NOTE It has been my experience that one of the most common problems with power rack and pinion steering gears is a condition that causes excessive steering wheel turning effort when the vehicle is first started in the morning. After the steering wheel is turned several times, the condition disappears. This problem is caused by grooves worn in the aluminum pinion housing by the seals on the control valve (**Figure 13-8**). When this condition is present, steering gear replacement is usually required. As an automotive technician, you will be required to diagnose and correct this problem.

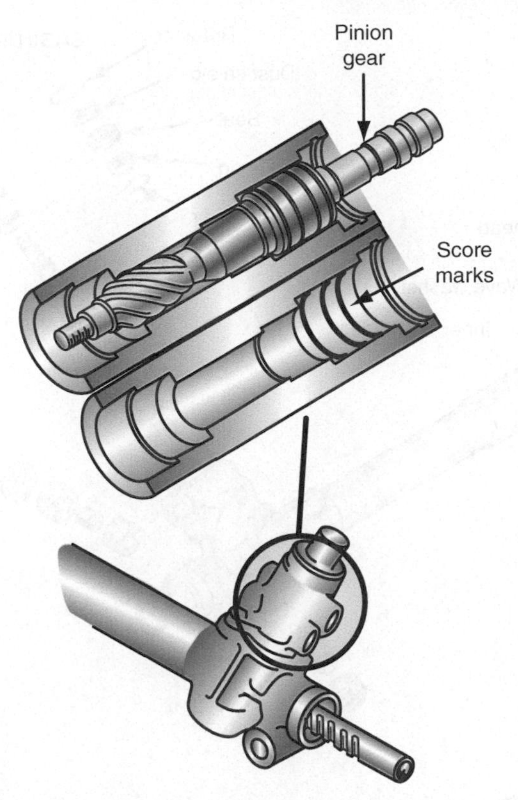

Figure 13-8 Score marks in a rack and pinion steering gear housing.

TYPES OF POWER RACK AND PINION STEERING GEARS

Power Rack and Pinion Steering Gear

Many vehicles have a rack and pinion steering gear manufactured by Saginaw (**Figure 13-9**). Some vehicles are equipped with a TRW rack and pinion steering gear. This type of steering gear is similar to the Saginaw gear except for the following differences:

1. Method of tie rod attachment
2. Bulkhead oil seal and retainer
3. Pinion upper and lower bearing hardware

On both the Saginaw and TRW power rack and pinion steering gears, the tie rods are connected to the ends of the rack. This type of steering gear may be referred to as **end take-off (ETO)**. On other power rack and pinion steering gears, the rack piston and cylinder are positioned on the right end of the rack and the tie rods are attached to a movable sleeve in the center of the gear (**Figure 13-10**). This type of steering gear may be called **center take-off (CTO)**.

The Toyota power rack and pinion steering gear has a removable control valve housing surrounding the control valve and pinion shaft (**Figure 13-11**). In this steering gear, claw washers are used to lock the inner tie rod ends to the rack. Apart from these minor differences, the Toyota power rack and pinion gear is similar to the Saginaw and TRW gears.

WARNING **When working on any power steering system, always wear protective gloves and use caution, because the system hoses, components, and fluid can be very hot if the system has been in operation for a period of time.**

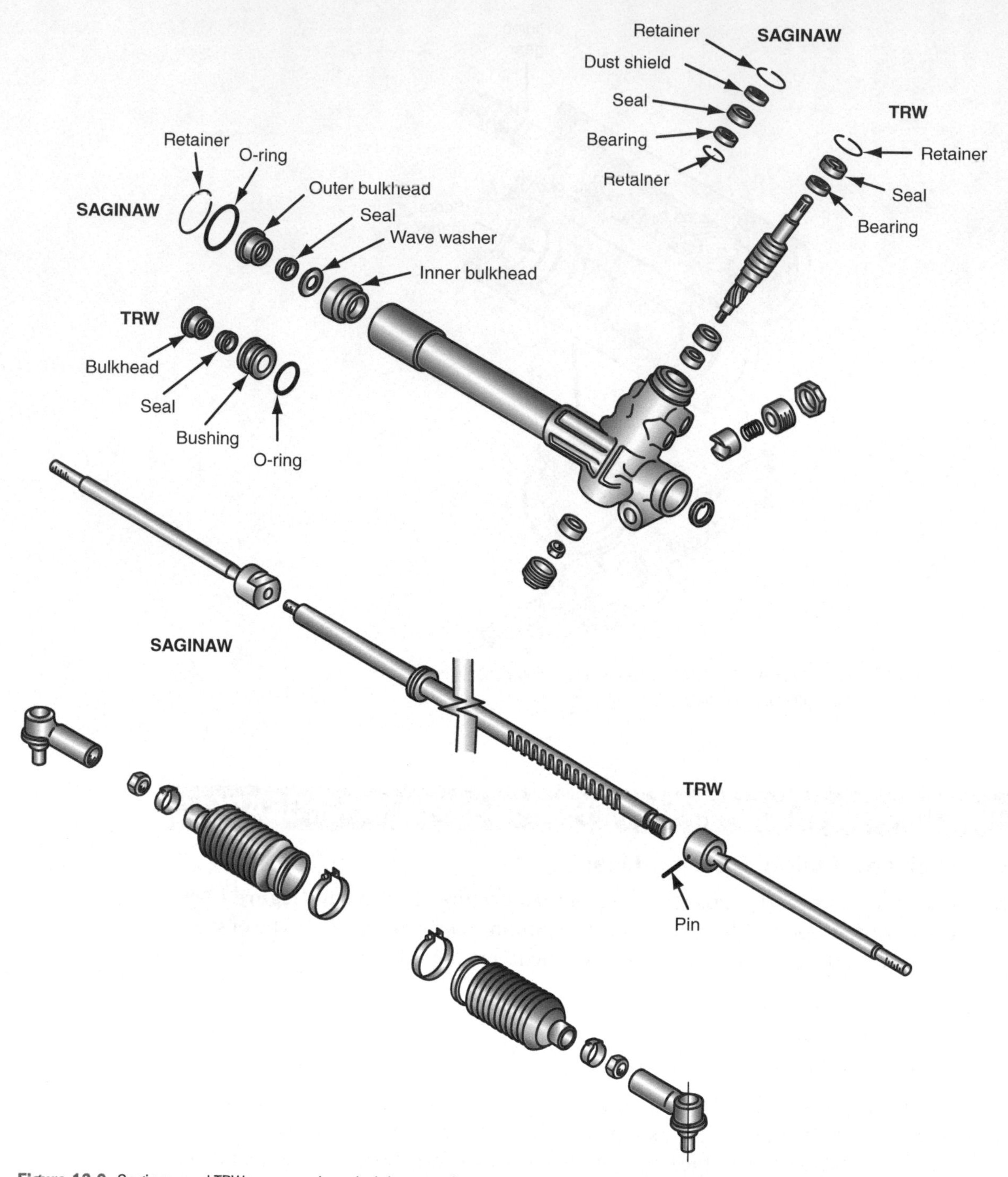

Figure 13-9 Saginaw and TRW power rack and pinion steering gear.

ELECTRONIC VARIABLE ORIFICE STEERING

Input Sensors

The **electronic variable orifice (EVO) steering** system is standard on many late-model vehicles. The EVO steering system provides high-power steering assistance during low-speed cornering and parking and normal power steering assistance at higher speeds for proper road feel. High-power steering assistance during low-speed cornering and parking increases driver convenience.

The **steering wheel rotation sensor** is mounted on the steering column, and a shutter disc attached to the steering shaft rotates through the sensor when the steering wheel

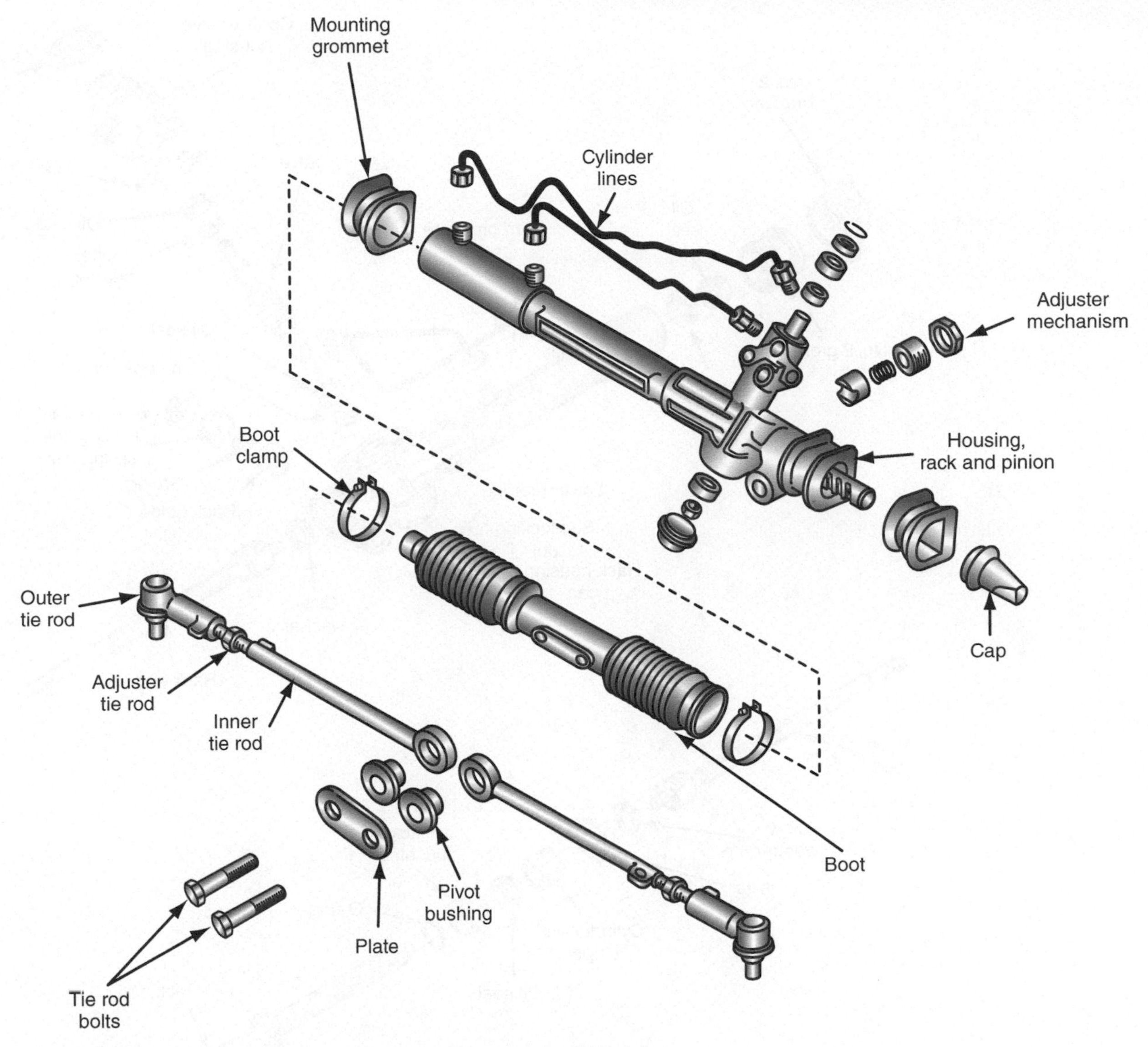

Figure 13-10 Power rack and pinion steering gear with center take-off tie rods.

is rotated. A row of slots is positioned near the outer edge of the shutter disc (**Figure 13-12**). When these slots rotate through the sensor, a steering wheel rotation speed signal is sent from the sensor to the control module.

The **vehicle speed sensor (VSS)** is mounted in the transaxle or transmission, and this sensor sends a signal to the control module in relation to vehicle speed (**Figure 13-13**). The VSS signal is also used for other purposes.

Never short across, or ground, any terminals in a computer-controlled system unless you are instructed to do so in the vehicle manufacturer's service manual. Such action may damage expensive electronic components.

Control Module

On some vehicles, the control module is mounted in the trunk (**Figure 13-14**). This module continually monitors the input signals from the VSS and the steering wheel rotation sensor (**Figure 13-15**). Some models have a combined EVO steering system and rear air suspension system. On these models, the control module operates the EVO steering system and the rear air suspension system. In the combined EVO steering and rear air suspension system, the inputs and outputs from both of these systems are connected to the same control module. (The rear air suspension system is explained in Chapter 11.) If the EVO system is used alone without the air suspension system, a different control module is required.

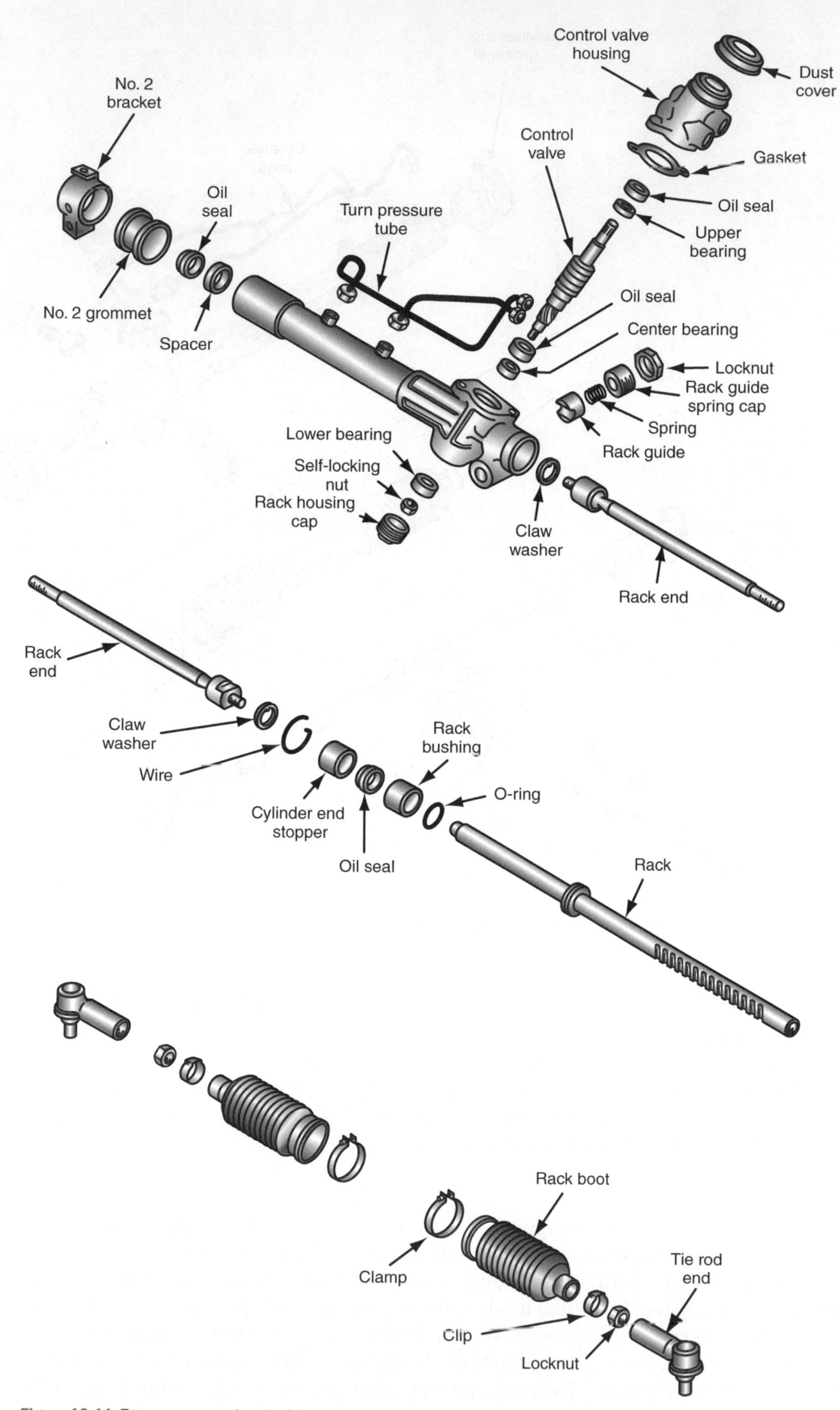

Figure 13-11 Toyota power rack and pinion steering gear.

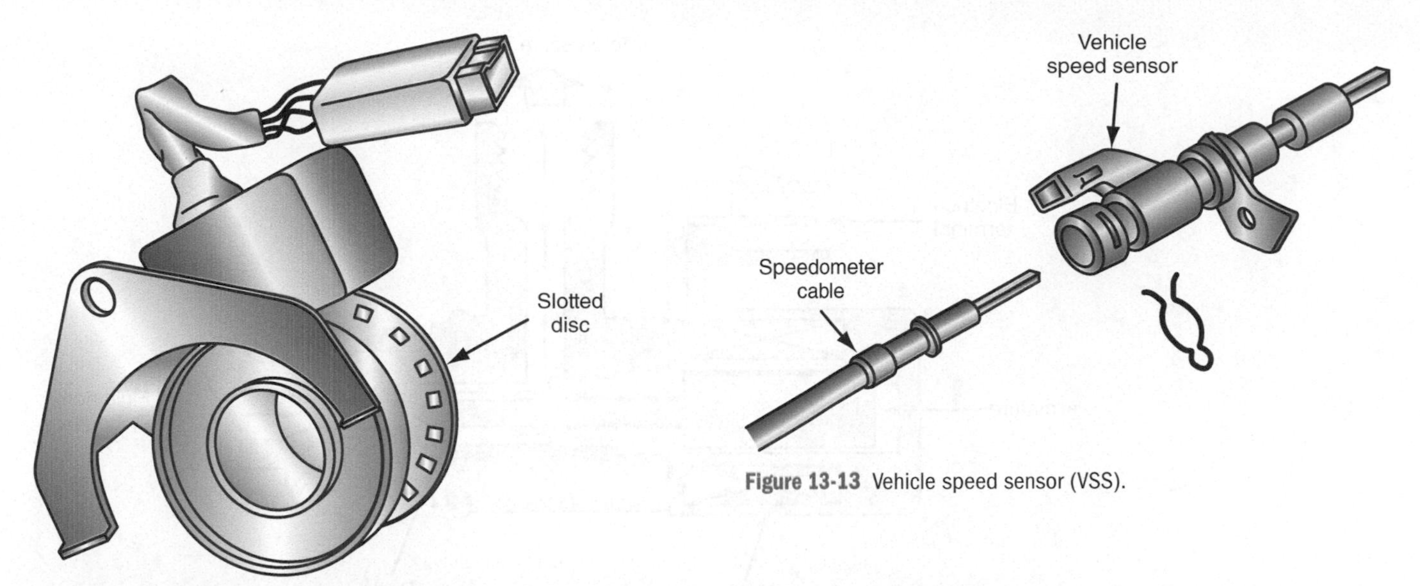

Figure 13-13 Vehicle speed sensor (VSS).

Figure 13-12 Steering wheel rotation sensor.

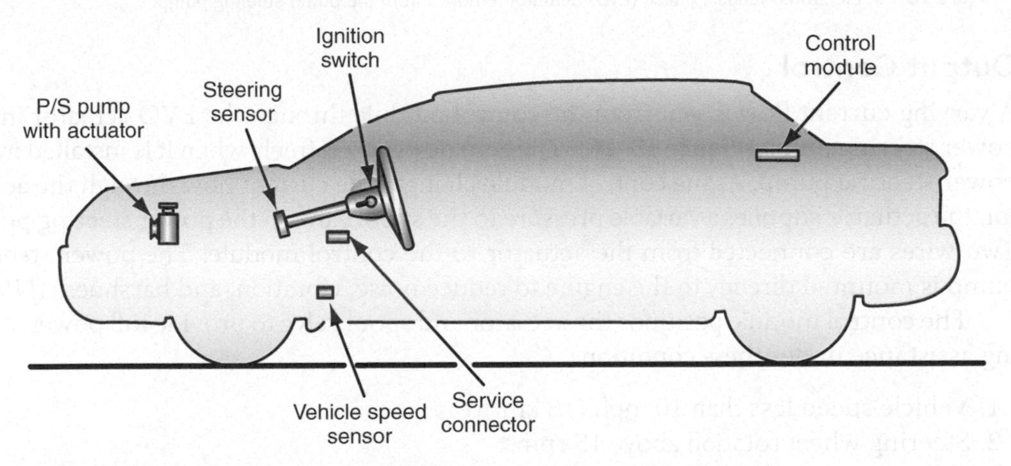

Figure 13-14 Control module and main components, electronic variable orifice steering.

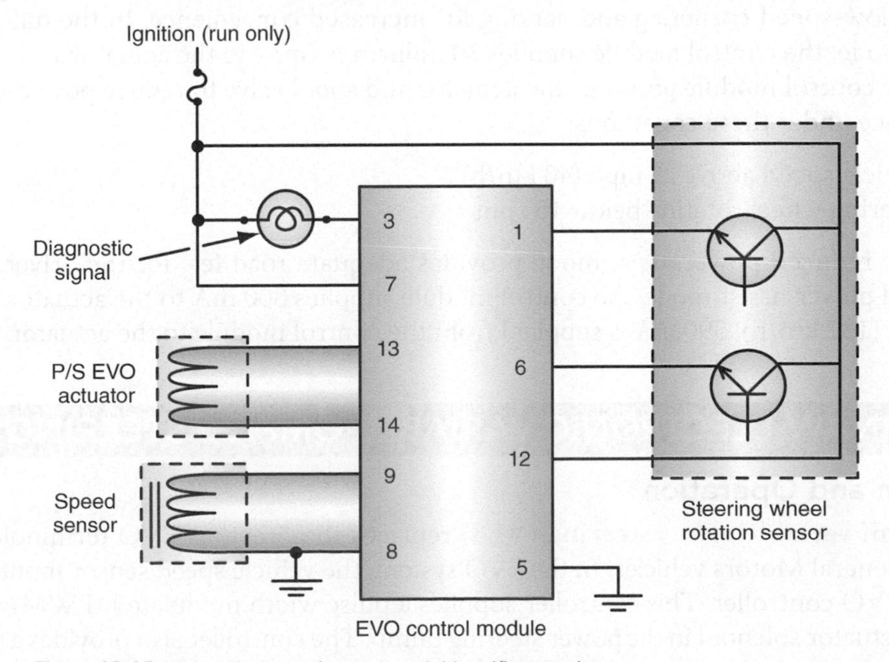

Figure 13-15 Wiring diagram, electronic variable orifice steering.

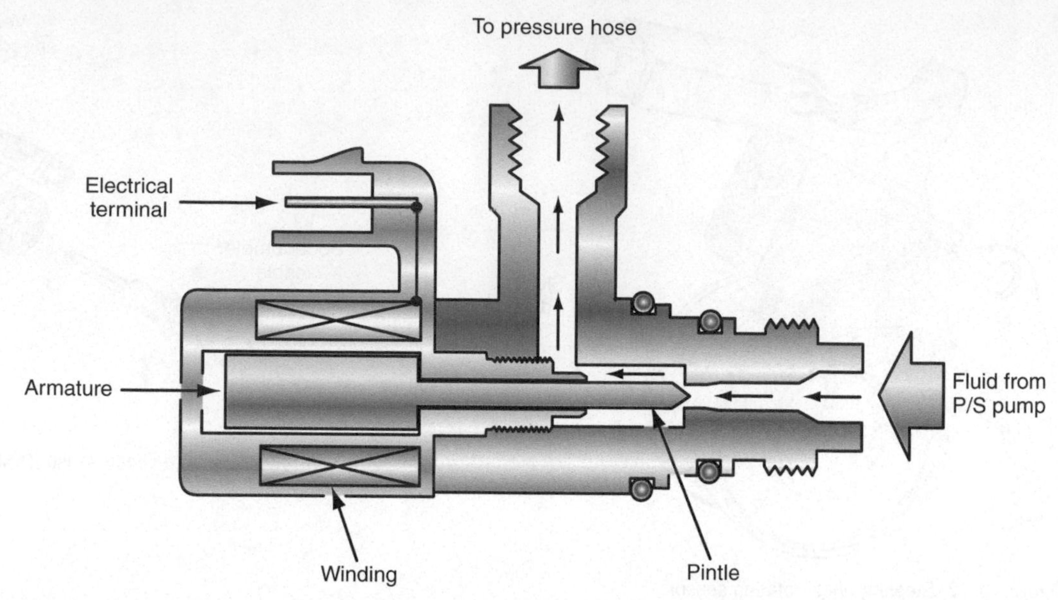

Figure 13-16 Electronic variable orifice (EVO) actuator removed from the power steering pump.

Output Control

A varying current flow is sent from the control module through the EVO actuator in the power steering pump (**Figure 13-16**). The actuator swivels freely when it is installed in the power steering pump. As the control module changes the current flow through the actuator, the actuator supplies a variable pressure to the spool valve in the power steering pump. Two wires are connected from the actuator to the control module. The power steering pump is mounted directly to the engine to reduce noise, vibration, and harshness (NVH).

The control module positions the actuator and spool valve to provide full power steering assistance under these conditions:

1. Vehicle speed less than 10 mph (16 km/h)
2. Steering wheel rotation above 15 rpm

The full power-assist mode reduces driver effort required to turn the steering wheel during low-speed cornering and parking for increased convenience. In the full power-assist mode, the control module supplies 30 milliamps (mA) to the actuator.

The control module positions the actuator and spool valve to reduce power steering assistance under these conditions:

1. Vehicle speed above 25 mph (40 km/h)
2. Steering wheel rotation below 15 rpm

The reduced power-assist mode provides adequate road feel for the driver. In the reduced power-assist mode, the control module supplies 300 mA to the actuator. Above 88 mph (132 km/h), 590 mA is supplied from the control module to the actuator.

SAGINAW ELECTRONIC VARIABLE ORIFICE STEERING

Design and Operation

The term **variable effort steering (VES)** replaces the previous EVO terminology on some General Motors vehicles. In the EVO system, the vehicle speed sensor input is sent to the EVO controller. This controller supplies a pulse width modulated (PWM) voltage to the actuator solenoid in the power steering pump. The controller also provides a ground connection for the actuator solenoid (**Figure 13-17**).

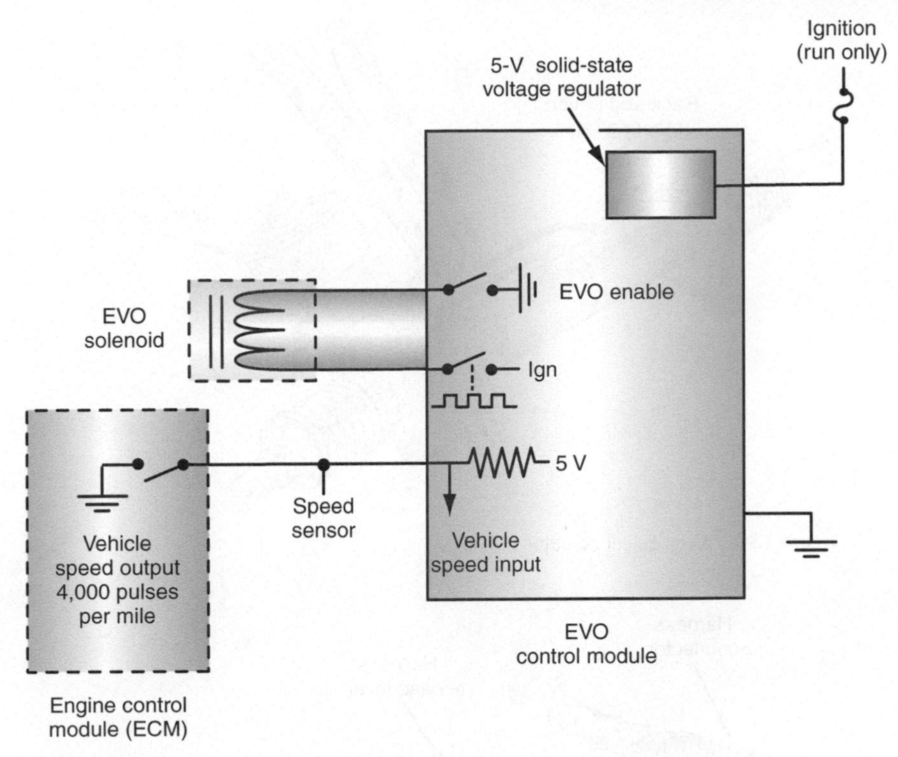

Figure 13-17 Electronic variable orifice (EVO) steering system.

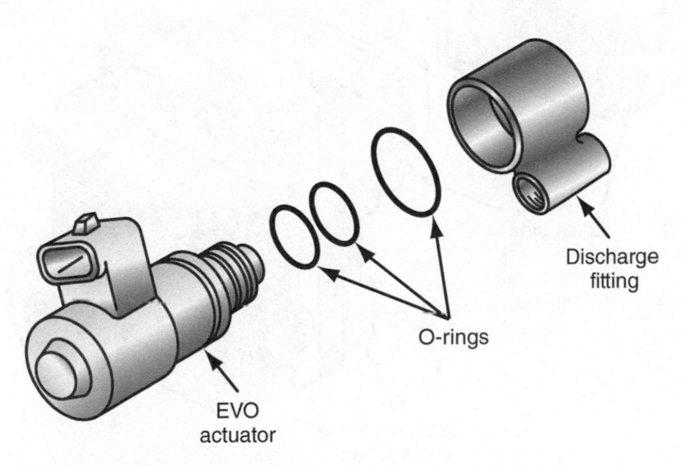

Figure 13-18 Actuator solenoid.

A **pulse width modulated (PWM) voltage signal** has a variable on time.

When the vehicle is operating at low speeds, the controller supplies a **pulse width modulated (PWM) voltage signal** to position the actuator solenoid plunger so the power steering pump pressure is higher (**Figure 13-18**). Under this condition, greater power assistance is provided for cornering or parking. If the vehicle is operating at higher speed, the controller changes the PWM signal to the actuator solenoid, and the solenoid plunger is positioned to reduce power steering pump pressure. This action reduces power steering assistance to provide improved road feel for the driver.

In a Magnasteer II VES system, as vehicle speed changes or lateral acceleration occurs, the amount of effort required to steer the vehicle can be varied dependent on the requirements. During low-speed operation, increased steering assist is provided for minimal steering effort to facilitate parking maneuvers and easy turning. At high speed, less assist is provided, which provides better road feel, firmer steering, and improved directional stability. If the system senses lateral acceleration, the steering effort will become firmer in an effort to reduce oversteering by the driver. The Electronic Brake Control Module

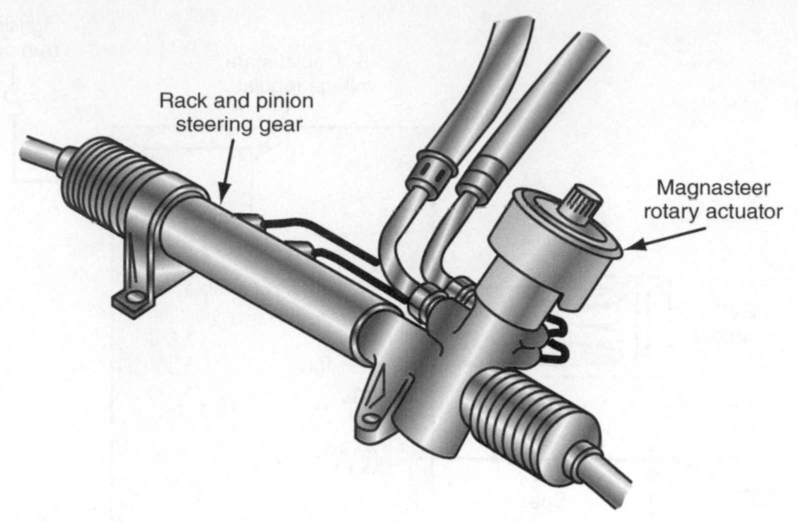

Figure 13-19 Magnasteer actuator solenoid.

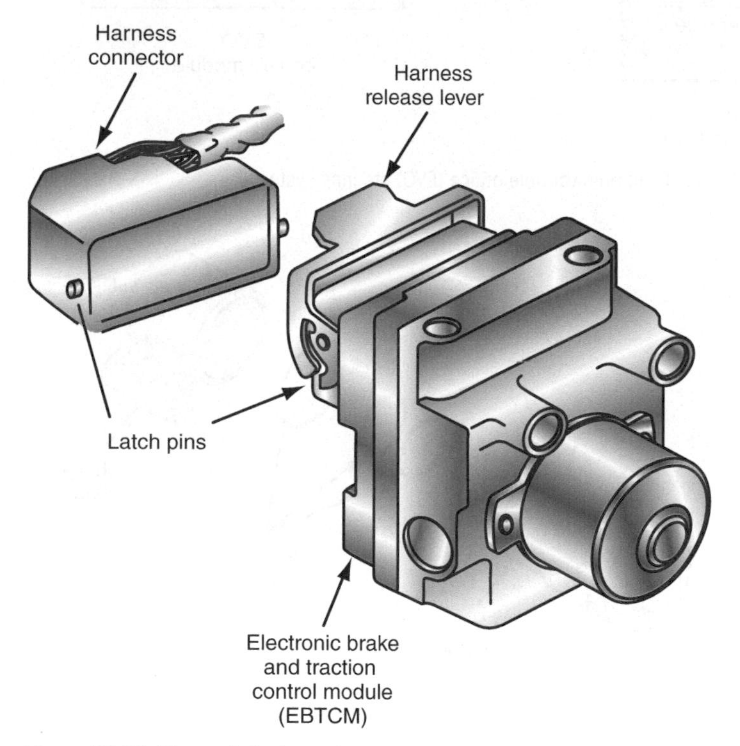

Figure 13-20 Electronic brake and traction control module (EBTCM).

The **Magnasteer system** uses an electromagnetic actuator, a multiple pole ring permanent magnet, a pole piece, and an electromagnetic coil in the steering gear to vary steering effort. The EBCM varies steering assist by regulating the amount of current flow to the electromagnet.

(EBCM) controls a bi-directional rotary magnetic actuator solenoid mounted in the steering gear (**Figure 13-19**). The computer that operates the **Magnasteer system** is combined with the electronic brake and traction control module (EBCM), and this module is usually mounted in the left-front corner of the engine compartment (**Figure 13-20**). Two wires are connected from the steering actuator to the EBCM (**Figure 13-21**).

The steering actuator solenoid contains a pole piece with 16 magnetic segments and a coil. A matching 16-segment permanent magnet is attached to the spool valve. As the steering wheel and the spool valve rotate, the 16 segments on the spool valve move into and out of alignment with the segments on the actuator pole piece. The EBCM can reverse the current flow through the steering actuator, and this action reverses the magnetic poles on the actuator segments. At low vehicle speeds, the EBCM supplies a negative current flow, 2–3 amperes, and command through the steering actuator so the magnetic poles on the actuator repel the

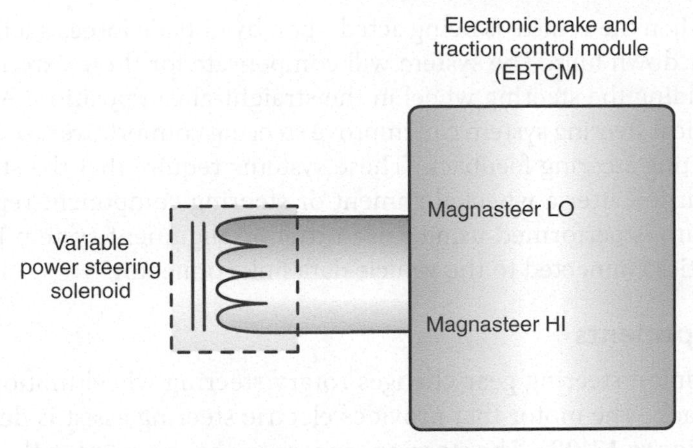

Figure 13-21 Wiring connections between the Magnasteer actuator solenoid and the EBTCM.

permanent magnet segments on the spool valve. This repelling action assists the driver to turn the steering wheel and reduces steering effort. At medium speeds of about 45 mph, no current is supplied, and steering is assisted by hydraulic pressure only. At higher speeds, the EBCM provides a positive current flow of up to 2–3 amperes through the steering actuator, and this action reverses the poles on the actuator segments. Under this condition, the actuator segments are attracted to the permanent magnet segments on the spool valve. This action increases steering effort to improve driver road feel. The EBCM also has the capability to vary the current flow through the steering actuator to provide variable steering effort in relation to vehicle speed. If the EBCM detects a malfunction in the circuitry or actuator the system will supply zero amps to the actuator and the steering will be assisted by hydraulic pressure only.

Shop Manual
Chapter 13, page 580

RACK-DRIVE ELECTRONIC POWER STEERING

In electro-mechanical power steering systems, there is no need for a hydraulic pump and associated plumbing. In addition to design and control benefits, eliminating hydraulic oil in the steering system is another important contribution to environmental protection. In addition, the system saves energy by reducing fuel consumption, saving space, reducing noise, and lowering emissions. Unlike engine-driven hydraulic power steering systems, which require a constant flow of hydraulic fluid, the electro-mechanical systems only draw energy when power assist is required.

Electronic power steering systems (EPS) may be classified as rack-drive, column-drive, or pinion-drive. All three types of EPS have a rack and pinion steering gear. In a **rack-drive EPS** the electric motor that provides steering assist is coupled to the rack in the steering gear. A **column-drive EPS** has an electric-assist motor coupled to the steering shaft in the steering column, and in a **pinion-drive EPS** the electric motor is coupled to the steering gear pinion. The EPS system is light and compact compared to rack and pinion steering gear with hydraulic steering assist, because the power steering pump and hoses are not required on the EPS system. Since the EPS does not require a power steering pump driven by engine power, this system minimizes engine power loss and reduces fuel consumption. The EPS system reduces steering kickback while providing a linear steering feel.

When steering assist is required, an electric motor is actuated based on steering wheel input response. Thus, the system provides power steering assistance based on driving conditions and driver commands. The electronic power steering system has an active return function to support steering wheel return back to the center position. The active return function produces balanced, accurate straight-line stability under various driving conditions. The straight-line stability function assists the driver in maintaining straight ahead control when the vehicle is being acted upon by outside forces such as cross winds or driving up or down hills. The system will compensate for these external forces when the driver is holding the steering wheel in the straight-ahead position. Additionally, the electro-mechanical steering system can improve steering comfort over rough uneven road surfaces by limiting steering feedback. These systems require that the steering position sensors be relearned after a wheel alignment or steering component replacement. The relearn procedure is performed using a scan tool or alignment system integrated with relearn capabilities connected to the vehicle data link connector.

System Components

The rack and pinion steering gear changes rotary steering wheel motion to transverse motion of the rack. The motor that provides electric steering assist is designed into the steering gear (**Figure 13-22**). The steering sensor may be mounted in the pinion shaft or can be integrated into the steering column. This sensor sends input voltage signals to the EPS control unit in relation to the direction, amount, and torque of the steering wheel rotation. Depending on road speed, steering force, steering angle, and steering wheel rotation speed, the control unit calculates the required amount of steering assist and energizes the electric motor to provide steering assist (**Figure 13-23**).

The transmission or transaxle is equipped with a vehicle speed sensor. This sensor sends voltage input signals to the transmission control module (TCM) in relation to output shaft rotational speed. These voltage signals are transmitted from the TCM to the electronic power steering (EPS) control unit. The TCM may be located in the rear of the vehicle (**Figure 13-24**) on some platforms; it may also be integrated into the powertrain control module (PCM).

The EPS control unit may be mounted on or near the steering gear (**Figure 13-25**). This control unit receives voltage input signals from the steering sensor, VSS, or differential speed sensor. When these signals are received, the EPS control unit calculates the proper amount and direction of steering assist. The EPS control unit then commands a power module in the EPS control unit to drive the electric motor in the steering gear and provide the proper direction and amount of steering assist. The EPS control unit also contains self-diagnostic capabilities.

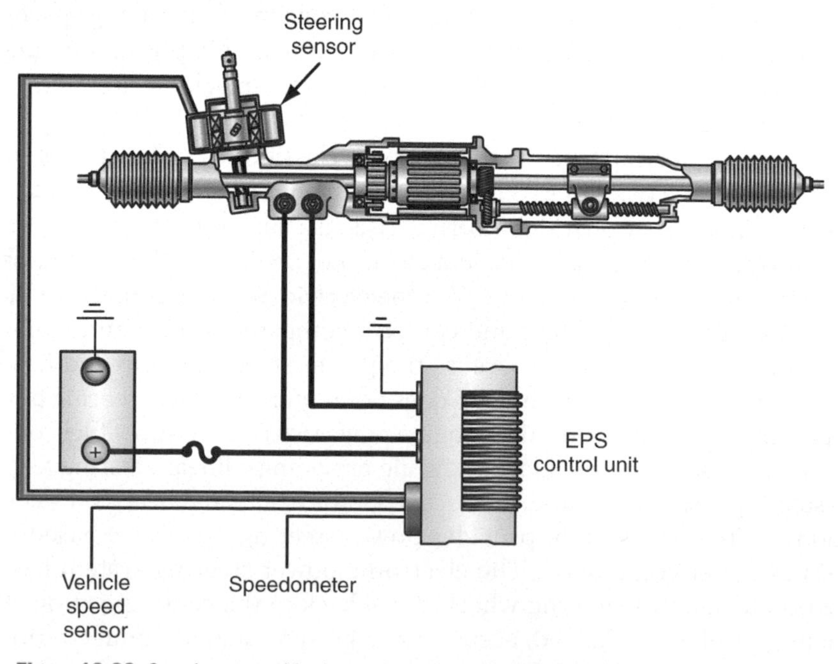

Figure 13-22 Steering gear with electronic assist and related system components.

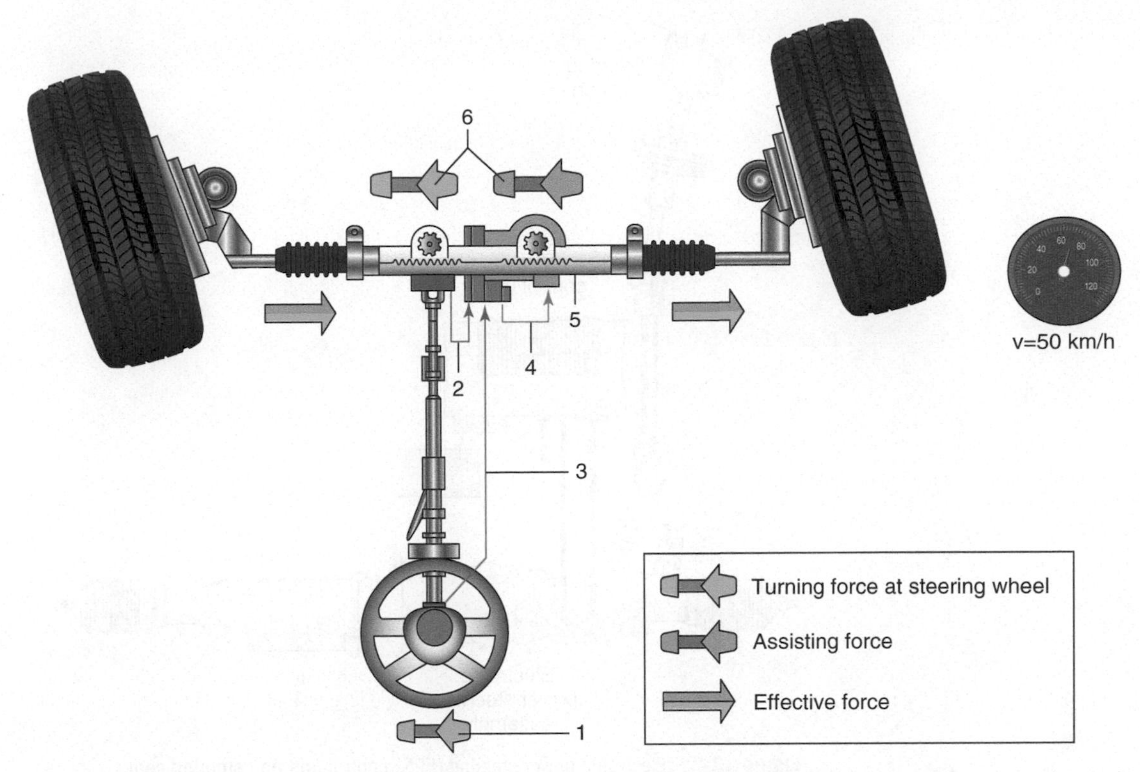

Figure 13-23 Depending on road speed, steering force, steering angle, and steering wheel rotation speed, the control unit calculates the required amount of steering assist and energizes the electric motor to provide steering assist.

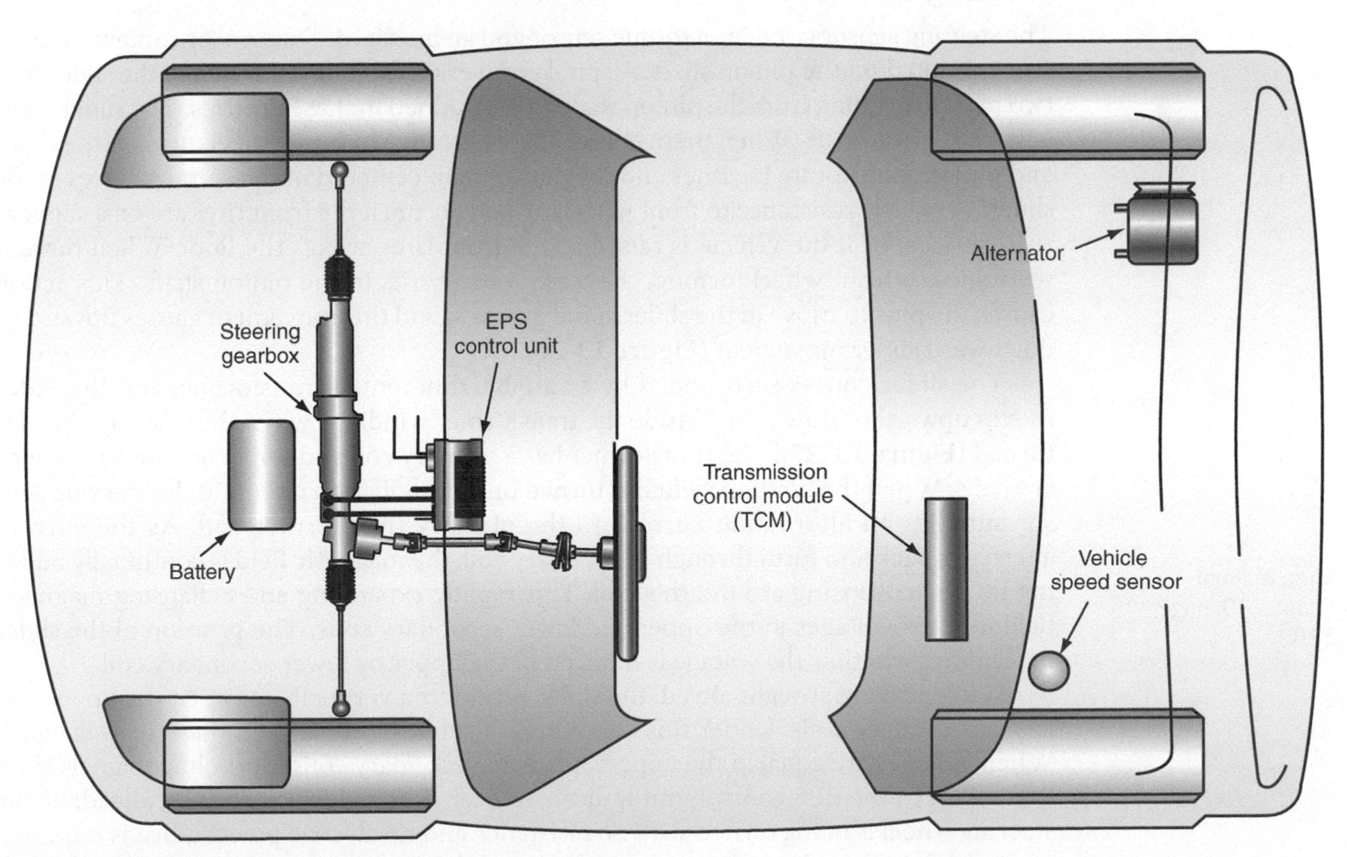

Figure 13-24 Electronic power steering (EPS) system component locations.

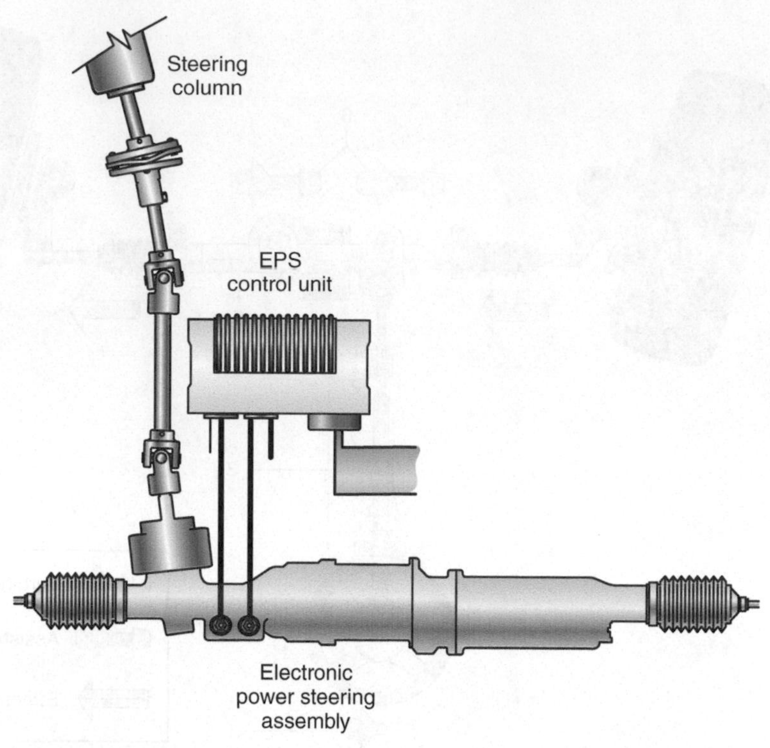

Figure 13-25 Electronic power steering (EPS) control unit and steering gear.

ELECTRONIC POWER STEERING SYSTEM OPERATION

Steering Sensor

The steering sensor contains a torque sensor and an interface. This sensor contains a slider core mounted on the pinion shaft. A spiral groove is located in each side of the slider, and two pins protruding from the pinion shaft are positioned in these grooves. The slider turns with the pinion shaft. When there is very little resistance to front wheel turning, the slider and pinion shaft rotate together and the pins remain centered in the spiral grooves in the slider. Very little resistance to front wheel turning occurs if the front tires are on a slippery road surface, or if the vehicle is raised so the front tires are off the floor. When there is resistance to front wheel turning, the torsion bar twists in the pinion shaft. This action causes the pins to move in the slider spiral grooves, and this movement causes upward or downward slider movement (**Figure 13-26**).

The slider core is surrounded by a variable differential transformer, and the slider moves upward or downward inside the transformer windings when the steering wheel is turned (**Figure 13-27**). The transformer has a primary coil and upper and lower secondary coils. When the ignition switch is turned on, an oscillation circuit in the steering sensor supplies an alternating current to the primary transformer coil. As the current alternates back and forth through the primary coil, the magnetic field is continually building up and collapsing around this coil. This rapidly expanding and collapsing magnetic field induces voltages in the upper and lower secondary coils. The position of the slider determines whether the voltage is induced in the upper or lower secondary coil.

Shop Manual
Chapter 13, page 585

While driving straight ahead, the slider is centered vertically between the upper and lower secondary coils. Under this condition, the voltage induced in these coils is equal. When voltages are equal in the upper and lower secondary coils, the voltage signals from these coils to the EPS control unit indicate the car is being driven straight ahead, or the steering wheel is being turned with no resistance and no electric power assist is required.

If the steering wheel is turned to the right, the slider moves upward. This slider position causes more induced voltage in the upper secondary coil and less induced voltage in

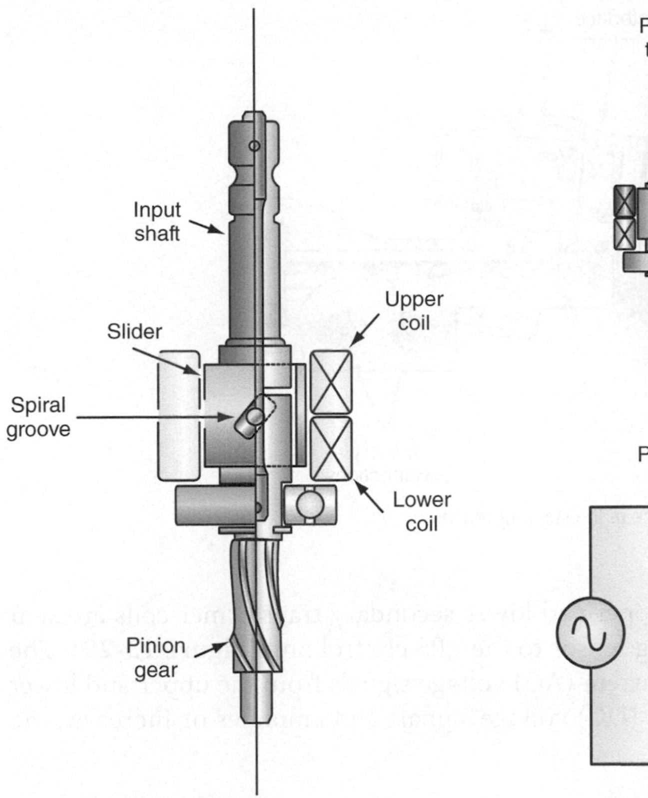

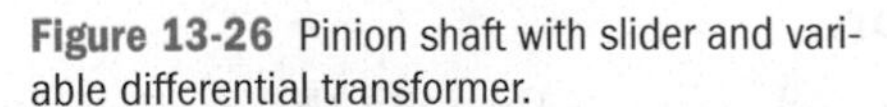
Figure 13-26 Pinion shaft with slider and variable differential transformer.

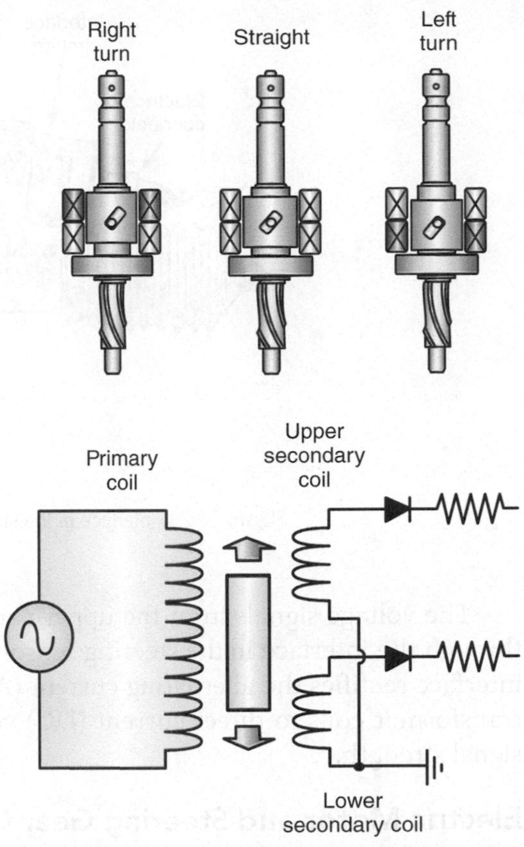

Figure 13-27 Slider movement when turning the steering wheel.

the lower secondary coil. When these voltage signals from the upper secondary coil are sent to the EPS control unit, this control unit supplies current to the electric motor on the rack so the motor rotates in the appropriate direction to provide the proper amount of steering assist to the right (**Figure 13-27**).

When the steering wheel is turned to the left, the slider moves downward. This slider position causes more induced voltage in the lower secondary coil and less induced voltage in the upper secondary coil. When these voltage signals from the lower secondary coil are sent to the EPS control unit, the control unit supplies current to the electric motor on the rack so the motor rotates in the appropriate direction to provide the proper amount of steering assist to the left (**Figure 13-28**).

Steering condition	Slider movement	Induction voltage on secondary coil
Steering to right (load steering)	Upward shift	Voltage on upper coil increases, and voltage on the lower decreases
Advancing straight ahead (no load steering)	Neutral	Voltage on the upper and lower coils are equal
Steering to left (load steering)	Downward shift	Voltage on lower coil increases, and voltage on the upper coil decreases

Figure 13-28 Summary of steering sensor and transformer operation.

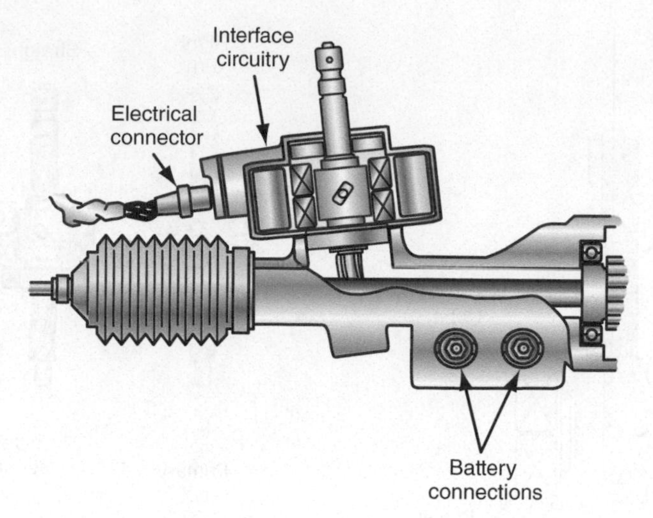

Figure 13-29 Interface in the steering sensor.

The voltage signals from the upper and lower secondary transformer coils are sent through the interface in the steering sensor to the EPS control unit (**Figure 13-29**). The interface rectifies the alternating current (AC) voltage signals from the upper and lower transformer coils to direct current (DC) voltage signals and amplifies or increases the signal strength.

Electric Motor and Steering Gear Operation

The armature in the electric motor is hollow, and the rack extends through the center of this armature. Ball bearings are mounted between the outer diameter of the armature and the steering gear housing to support the armature. Two spring-loaded brushes are mounted on opposite sides of the commutator on one end of the armature. A gear with helical teeth is mounted on the other end of the armature. The teeth on the armature gear are in constant mesh with a matching gear on the ball screw shaft (**Figure 13-30**). A recirculating ball screw nut is mounted on the ball screw shaft. Ball bearings are mounted in grooves in the ball screw shaft and recirculating ball screw nut. The recirculating ball screw is bolted to the steering gear rack. The ball screw shaft and recirculating ball screw nut are similar in design to the worm shaft and ball nut in a recirculating ball steering gear. A permanent magnet is mounted in the steering gear housing, and this magnet surrounds the armature core. There is a small clearance between the armature core and the permanent magnet.

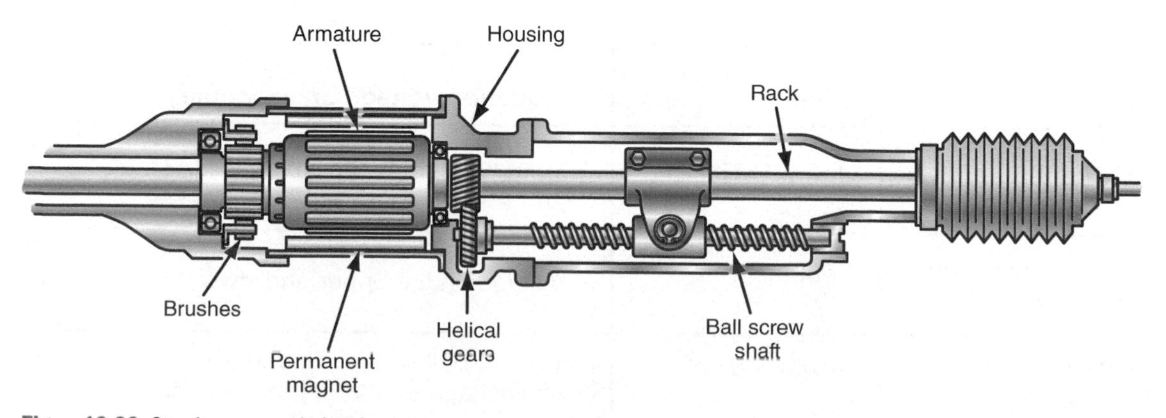

Figure 13-30 Steering gear and electric motor.

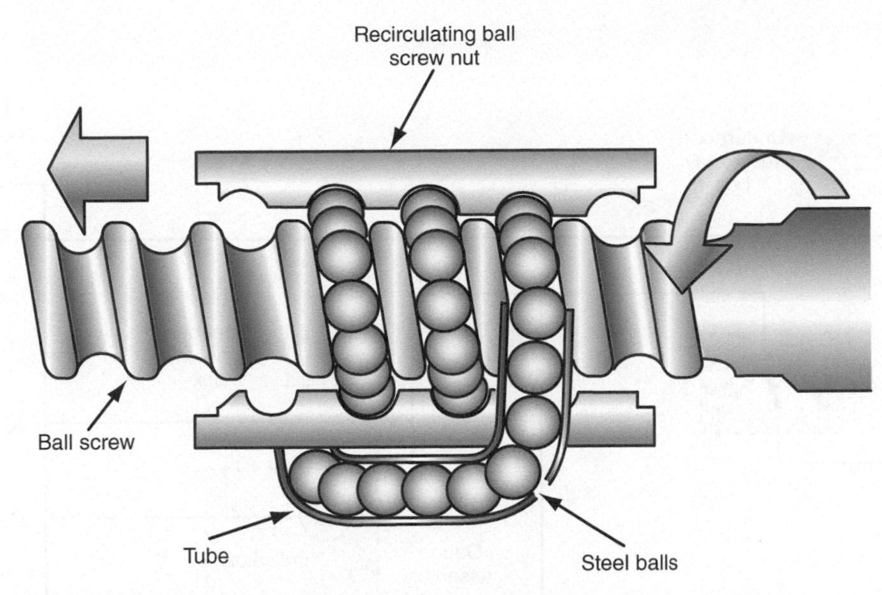

Figure 13-31 Ball screw shaft and recirculating ball screw nut.

When the steering wheel is turned, electric current is supplied from the power module in the EPS control unit through the brushes and armature windings to ground. This current flow creates strong magnetic fields around the armature windings. These magnetic fields around the armature windings react with the magnetic field of the permanent magnet and cause armature rotation in the proper direction to supply steering assist. When the armature rotates, the gear on the armature shaft drives the gear on the ball screw shaft. Ball screw shaft rotation moves the recirculating ball screw nut on the shaft. Since the recirculating ball screw nut is bolted to the steering gear rack, movement of this ball screw nut provides steering assist in the proper direction. The power module can reverse the armature rotation to provide steering assist in either direction by reversing the polarity of the brushes on the commutator. When the brush polarity is reversed, the current flow through the armature windings is reversed and this changes the direction of armature rotation.

The recirculating ball screw nut is designed so the ball bearings roll between the grooves in the ball screw shaft and the grooves in the recirculating ball screw nut. Ball bearings coming out of the recirculating ball screw nut move through a tube and re-enter the recirculating ball screw nut at the other end (**Figure 13-31**). The ball bearings in the grooves in the ball screw shaft and recirculating ball screw nut allow this nut to move on the shaft with very low friction.

Steering Gear Motor Current Control

The power module in the EPC control unit contains a driving circuit, current sensor, field effect transistor (FET) drive circuit, power relay, and fail-safe relay. When the ignition switch is turned on and the engine is cranked, the EPS control unit closes the power relay and fail-safe relay to make the EPS system operational. These relays actually close when the alternator begins producing voltage while the engine is cranking. Voltage signals are sent from the alternator and the ignition switch to the EPS control unit (**Figure 13-32** and **Figure 13-33**). The EPS system remains operational if the engine stalls and the ignition switch is on. If the engine stalls on a hydraulically assisted power steering system, power steering assist is lost and the steering wheel becomes very hard to turn. This action may result in a collision if the engine stalls while turning a corner. Since the EPS system is still operational if the engine stalls and the ignition switch is on, the EPS system reduces the possibility of a collision resulting from loss of power steering assist during an engine stall.

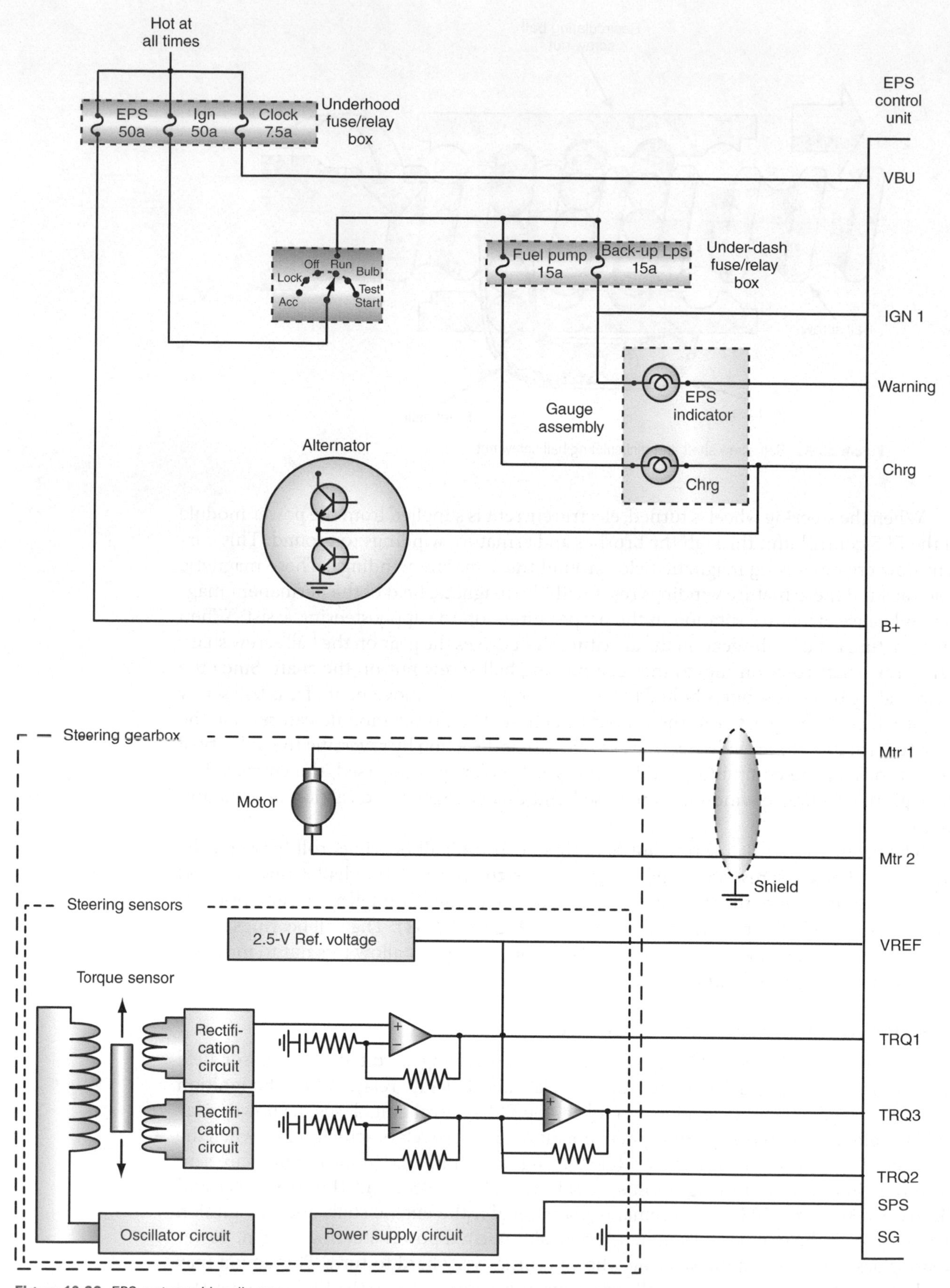

Figure 13-32 EPS system wiring diagram.

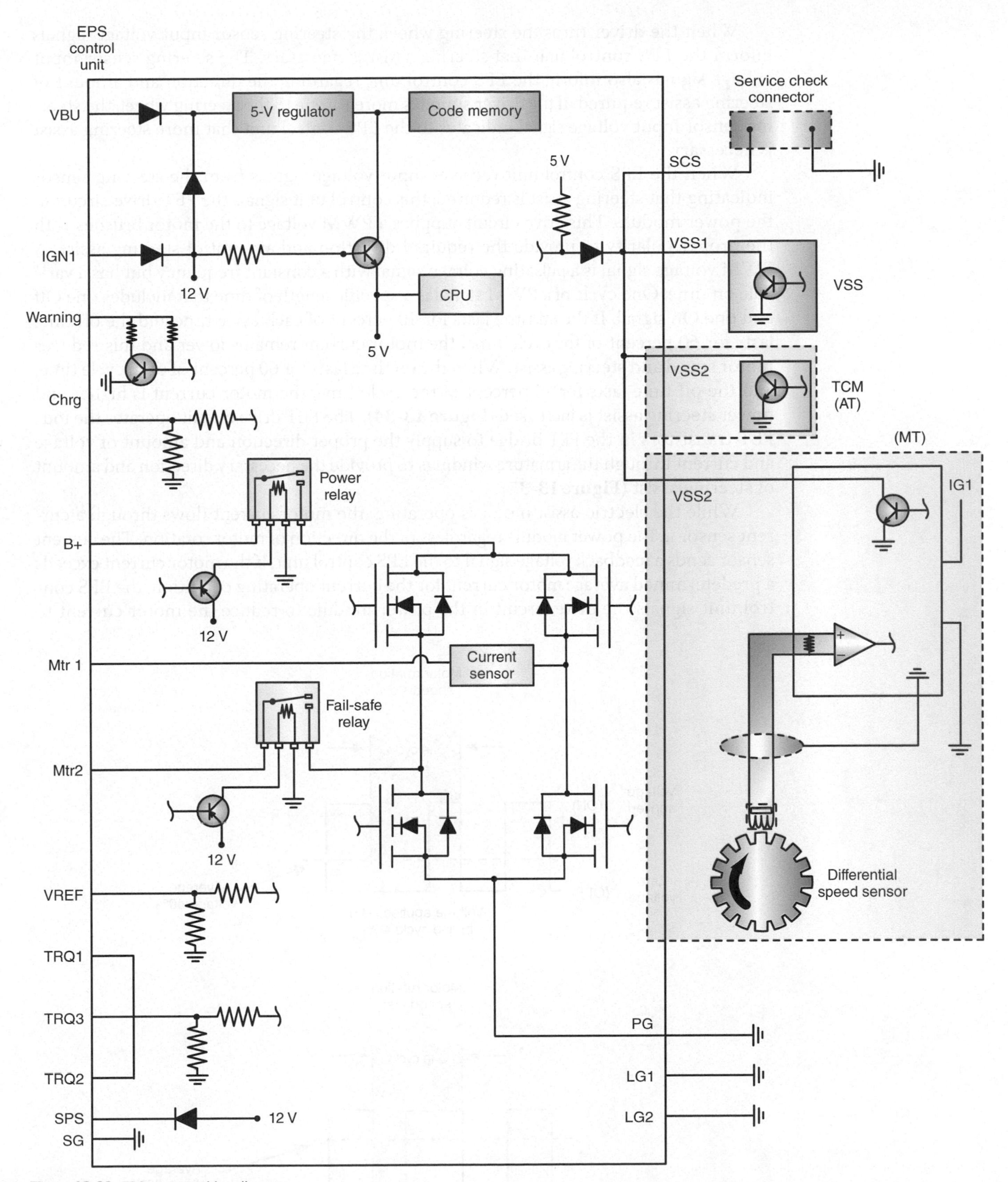

Figure 13-33 EPS system wiring diagram.

When the driver turns the steering wheel, the steering sensor input voltage signals inform the EPS control unit that steering assist is necessary. The steering sensor input voltage signals also inform the EPS control unit regarding the direction and amount of steering assist required. If the driver supplies more torque to the steering wheel, the steering sensor input voltage signal indicates to the EPS control unit that more steering assist is necessary.

When the EPS control unit receives input voltage signals from the steering sensor indicating that steering assist is required, this control unit signals the FET drive circuit in the power module. This drive circuit supplies a PWM voltage to the motor brushes with the proper polarity to provide the required direction and amount of steering assist. A PWM voltage signal is a pulsating voltage signal with a constant frequency but has a variable on time. One cycle of a PWM signal is a specific length of time that includes one Off and one On signal. If the on time lasts for 40 percent of each cycle time and the off time lasts for 60 percent of the cycle time, the motor current remains lower and this reduces motor speed and steering assist. When the on time lasts for 60 percent of each cycle time, and the off time lasts for 40 percent of the cycle time, the motor current is higher and power steering assist is increased (**Figure 13-34**). The FET drive circuit operates the four FET transistors in the FET bridge to supply the proper direction and amount of voltage and current through the armature windings to provide the necessary direction and amount of steering assist (**Figure 13-35**).

While the electric-assist motor is operating, the motor current flows through a current sensor in the power module regardless of the direction of motor rotation. The current sensor sends a feedback voltage signal to the EPS control unit. If the motor current exceeds a predetermined average motor current for the current operating condition, the EPS control unit signals the drive circuit in the power module to reduce the motor current to

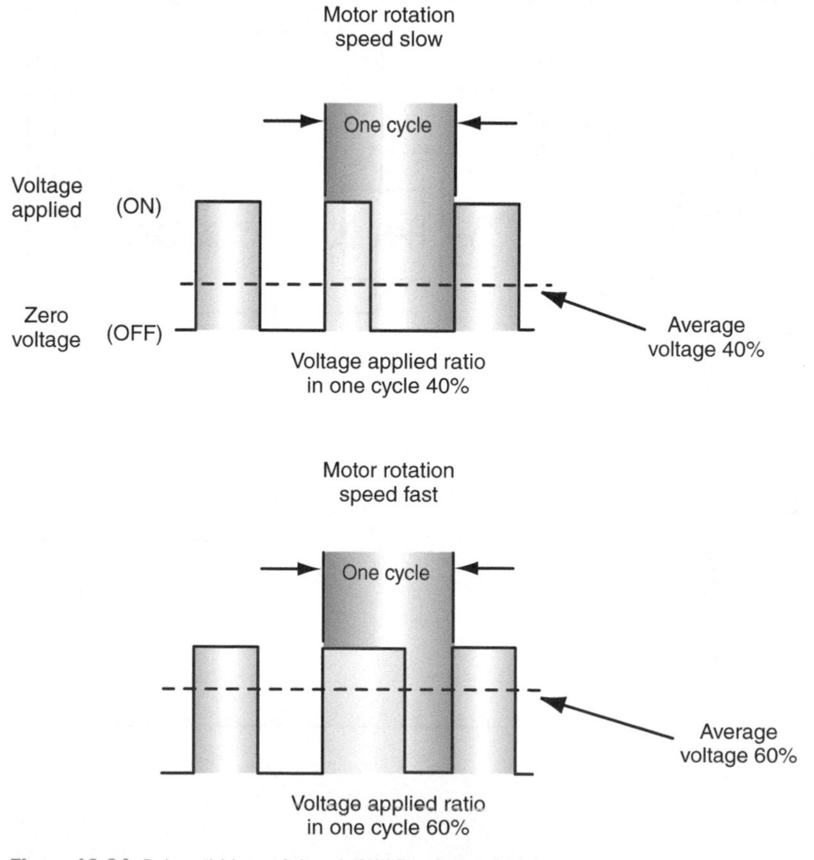

Figure 13-34 Pulse width modulated (PWM) voltage signal.

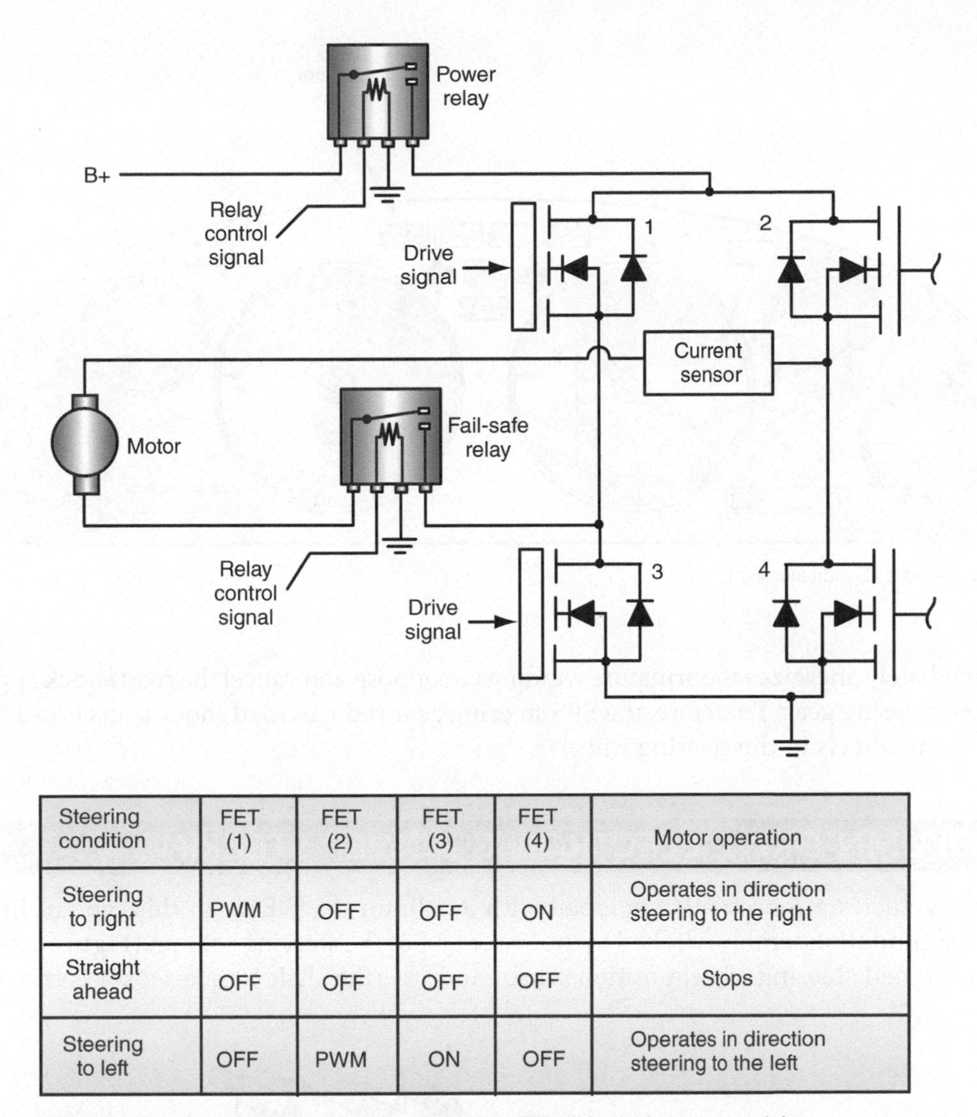

Steering condition	FET (1)	FET (2)	FET (3)	FET (4)	Motor operation
Steering to right	PWM	OFF	OFF	ON	Operates in direction steering to the right
Straight ahead	OFF	OFF	OFF	OFF	Stops
Steering to left	OFF	PWM	ON	OFF	Operates in direction steering to the left

Figure 13-35 Drive circuit and field effect transistor (FET) bridge in the power module.

prevent motor overheating. If the steering wheel is turned fully in one direction and held in this position, the motor current becomes much higher. Under this condition, the current sensor signals the EPS control unit, and this unit signals the drive circuit in the power module to reduce motor current to protect the motor.

If an electrical defect occurs in the EPS system, the EPS control unit illuminates an EPS indicator light in the instrument panel to inform the driver that a fault is present in the EPS system (**Figure 13-36**). Under this condition, the EPS unit opens the fail-safe and power relays to make the EPS system inoperative. When this action occurs, manual steering without any power assist is available.

If one of the front wheels strikes a large road irregularity, force from the front wheel is transferred to the steering gear rack. Rack movement attempts to move the ball screw nut on the ball screw shaft. This action tries to rotate the ball screw shaft and armature. A specific amount of force is required to move the ball screw nut and rotate the ball screw shaft and armature because the armature windings have to rotate through the magnetic field between the permanent magnets. This resistance to movement of the ball screw nut, ball screw shaft, and armature helps prevent road shock from being supplied through the steering gear to the steering wheel. When a very high road shock is transferred to the steering wheel, this road shock moves the pinion shaft and slider in the steering sensor. When this EPS control unit receives this steering sensor input signal, the EPS control unit

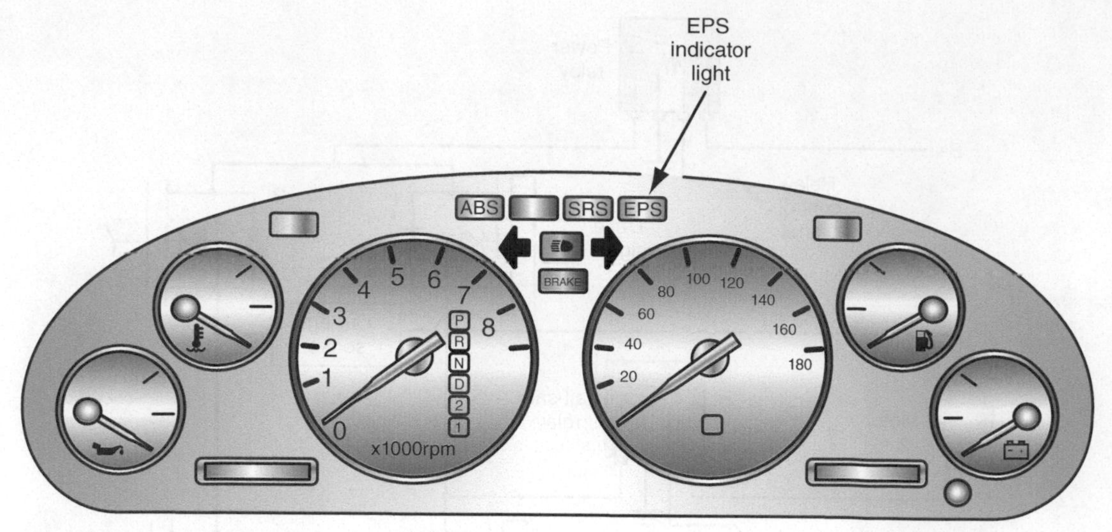

Figure 13-36 EPS indicator light.

immediately energizes the armature windings to oppose and cancel the road shock applied to the steering gear. Therefore, the EPS steering gear reduces road shock transferred from the front wheels to the steering wheel.

COLUMN-DRIVE ELECTRONIC POWER STEERING

Some vehicles are presently equipped with a column-drive EPS. In this type of EPS, a motor/module assembly is bolted to the lower end of the steering column (**Figure 13-37**). A combined steering wheel position sensor and steering shaft torque sensor is mounted

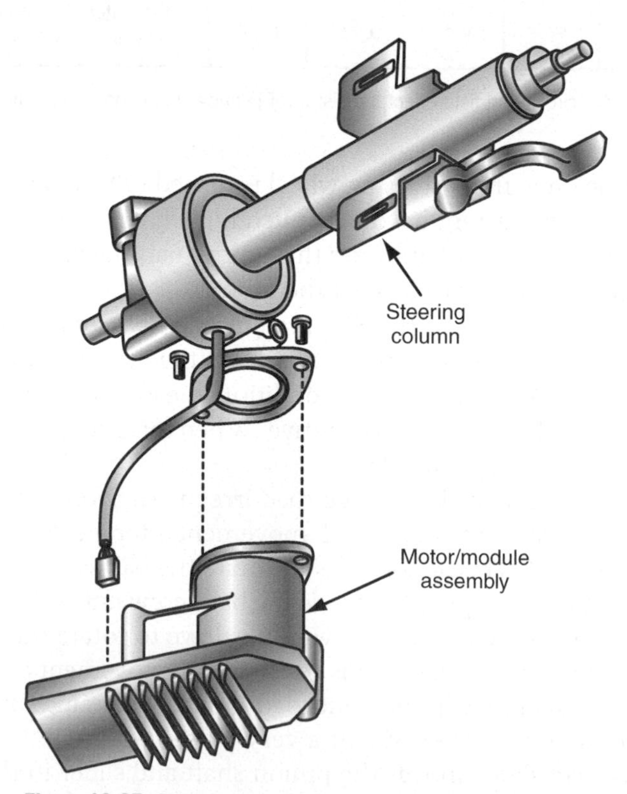

Figure 13-37 Column-drive electronic power steering system.

in the steering column at the motor/module attachment point. The module in the motor/module assembly is called the power steering control module (PSCM). Mounting the EPS motor/module assembly under the instrument panel provides underhood space for other components, subjects the assembly to less rigorous temperatures, and may provide better protection during a collision.

Input Sensors

The steering shaft torque sensor is the most important input used by the PSCM to supply the proper amount and direction of steering assist. The steering column contains a torque sensor input shaft connected from the steering wheel to the sensor, and an output shaft connected from the sensor to the steering shaft coupler. A torsion bar mounted inside the steering shaft torque sensor separates the input and output shafts. When the steering wheel is turned in either direction, the torque supplied from the steering wheel to the torsion bar causes this bar to twist. Increased torsion bar twisting results in higher-voltage signals from the steering shaft torque and position sensors. The steering shaft torque sensor is a dual sensor that provides two different voltage signals (**Figure 13-38**). During a right turn the voltage from sensor 1 increases and the voltage from sensor 2 decreases. The voltage signal range from each sensor is 0.25 V to 4.75 V. While completing a left turn, the voltage from sensor 1 decreases and the voltage from sensor 2 increases. The steering shaft torque sensor voltage signals inform the PSCM regarding the direction and amount of steering wheel torque.

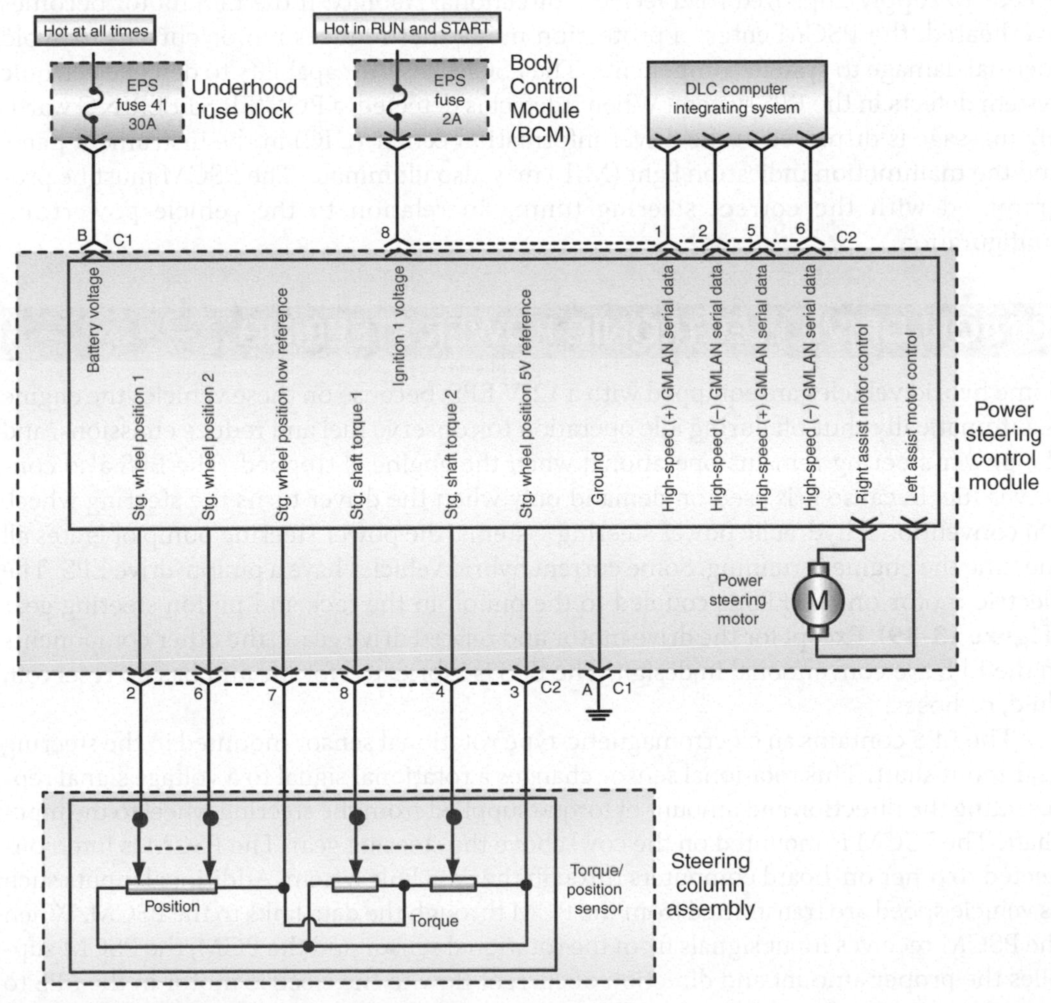

Figure 13-38 Wiring diagram for column-drive electronic power steering.

The steering wheel position sensor also contains dual voltage sensors that operate in the 0 V–5 V range. During steering wheel rotation, the voltage from these two sensors should remain within 2.5 V–2.8 V of each other. The steering wheel position sensor informs the PSCM regarding the position of the steering wheel. The combined steering shaft torque and position sensor assembly is serviced only as a unit.

The PSCM also receives a vehicle speed input signal from the PCM. This voltage signal is transmitted from the PCM to the PSCM through the interconnecting data links.

Power Steering Motor

The power steering motor is a 12-V brushless DC reversible motor rated at 65 amperes. This motor is coupled to the steering shaft through a worm gear and a reduction gear located in the steering column housing.

COLUMN-DRIVE ELECTRONIC POWER STEERING OPERATION

When the input signals indicate the driver is beginning to turn the steering wheel, the PSCM supplies the proper amount of current to the power steering motor to provide the necessary power steering assist. At low vehicle speeds, the PSCM provides more steering assist for easy steering wheel rotation during parking maneuvers. The PSCM reduces the motor current and provides reduced steering assist when the vehicle is driven at higher speeds to supply improved road feel and directional stability. If the EPS motor becomes overheated, the PSCM enters a protection mode that reduces motor current to avoid thermal damage to system components. The PSCM has the capability to detect electronic system defects in the EPS system. When a defect is detected, a POWER STEERING warning message is displayed in the driver information center (DIC) in the instrument panel and the malfunction indication light (MIL) may also illuminate. The PSCM must be programmed with the correct steering tuning in relation to the vehicle powertrain configuration.

PINION-DRIVE ELECTRONIC POWER STEERING

Some hybrid vehicles are equipped with a 12-V EPS, because on these vehicles the engine is automatically shut off during idle operation to conserve fuel and reduce emissions, and the power steering remains operational when the engine is stopped. The EPS also conserves fuel because it is used on demand only when the driver turns the steering wheel. On conventional hydraulic power steering systems, the power steering pump operates all the time the engine is running. Some current hybrid vehicles have a pinion-drive EPS. The electric motor on the EPS is coupled to the pinion in the rack and pinion steering gear (**Figure 13-39**). Except for the drive motor and related drive gears, the other components in the EPS are conventional in design. The EPS is lubricated for life and has no reservoir, fluid, or hoses.

The EPS contains an electromagnetic-type rotational sensor mounted in the steering gear input shaft. This rotational sensor changes a rotational signal to a voltage signal representing the direction and amount of torque supplied from the steering wheel to the input shaft. The PSCM is mounted on the cowl above the steering gear. The PSCM is interconnected to other on-board computers through the data link system. Additional inputs such as vehicle speed are transmitted from the PCM through the data links to the PSCM. When the PSCM receives input signals from the rotational sensor and the PCM, the PSCM supplies the proper amount and direction of current flow to the electric motor in the EPS to provide the required steering assist. A drain wire is connected from the PSCM to ground.

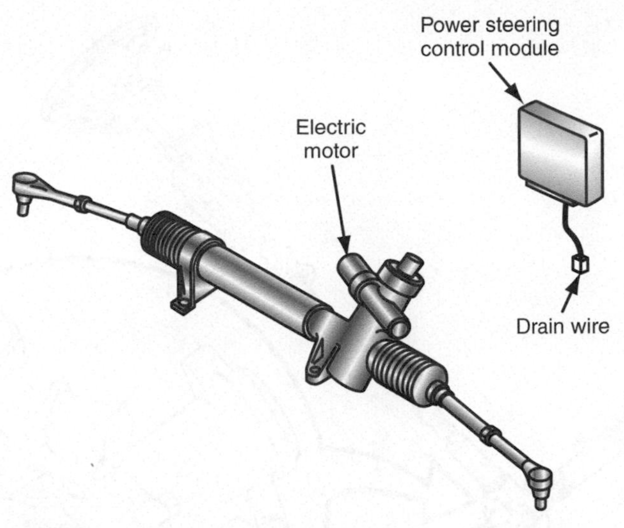

Figure 13-39 Pinion-drive electronic power steering (EPS) gear.

This wire prevents the EPS from emitting electromagnetic interference that could affect other electrical/electronic systems in the vehicle. If a defect occurs in the EPS electrical system, a SERVICE PWR STEERING message illuminates in the instrument panel.

Shop Manual
Chapter 13, page 587

ACTIVE STEERING SYSTEMS

Introduction

Active steering is available on a number of luxury vehicles. An active steering system contains many of the same components located in a conventional power rack and pinion steering system. The main difference in the active steering system is that the steering wheel and steering shaft are no longer connected directly to the pinion gear in the steering gear. In an active steering system, a dual planetary gear set is connected between the steering shaft and the pinion gear in the steering gear (**Figure 13-40**). The steering wheel and shaft are connected to one of the sun gears in a dual planetary gear set. The second sun gear in the planetary gear set is connected to the pinion gear in the steering gear. The two sun gears are connected by a set of planetary pinions. A brushless three-phase DC electric motor drives the ring gear on the planetary gear sets through a worm drive (Figure 13-41).

The **electronic control unit (ECU)** operates the electric motor to control the steering gear ratio and provide steering angle corrections to improve vehicle stability (**Figure 13-41**). In the active steering system, there is always a mechanical connection between the steering wheel and the pinion gear in the steering gear. The ECU may drive the electric motor and change the steering gear ratio or steering angle, but the mechanical connection between the steering wheel and the steering gear pinion always leaves the driver in control of the steering. The electric motor in the active steering system never provides turning forces to the front wheels.

ACTIVE STEERING SYSTEM COMPONENTS

DC Electric Motor and Motor Position Sensor

The electric drive motor has a wound stator and a permanent magnet rotor assembly. The ECU controls drive motor torque by a field-oriented control. The motor position sensor is mounted in the end of the electric motor, and this sensor informs the ECU regarding electric motor position.

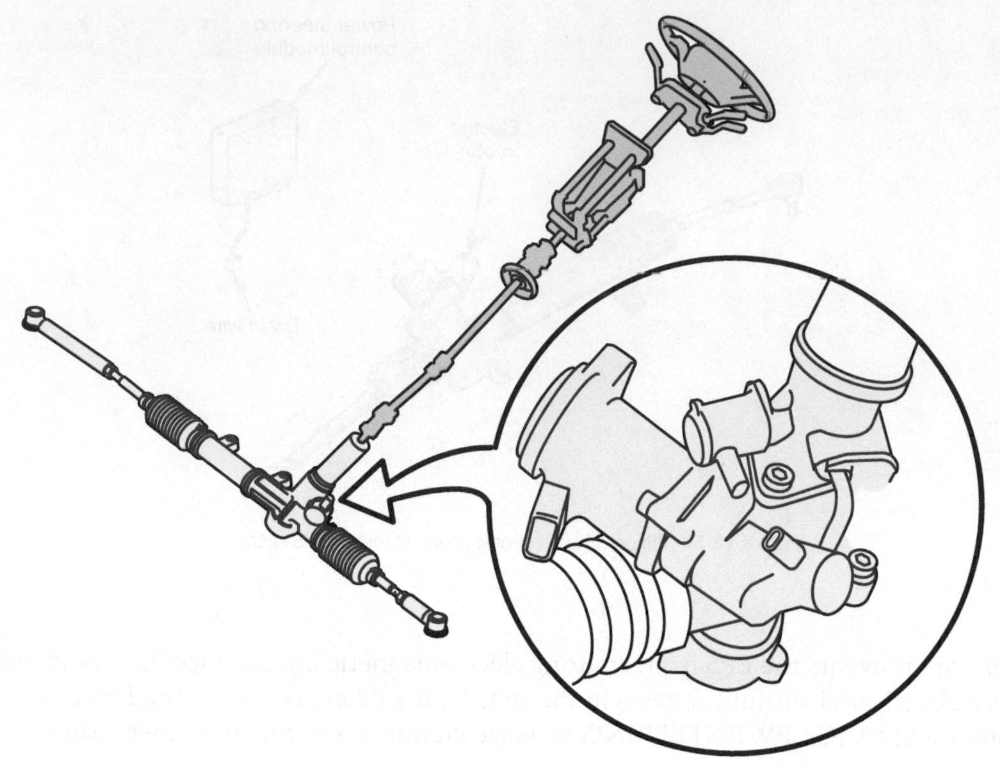

Figure 13-40 Active steering system.

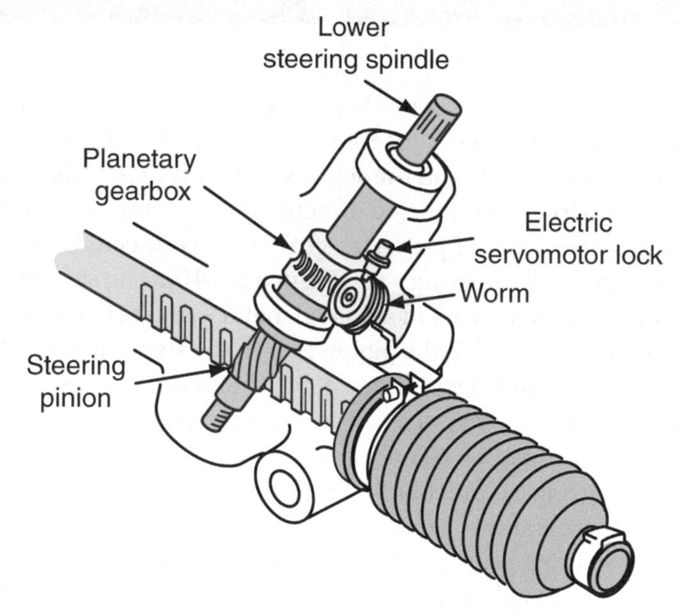

Figure 13-41 Active steering actuator.

Steering Angle Sensor

The **steering angle sensor** is mounted in the steering column switch assembly. The steering angle sensor is an optical sensor with no contacting parts, and this sensor is mounted to the circuit board near the top of the steering column (**Figure 13-43**). The main components in the steering angle sensor are an **encoded disc**, which is a slotted wheel, and an optical sensor. The encoded disc is attached to the steering shaft and rotates with this shaft. The encoded disc rotates within the optical sensor (**Figure 13-44**). The stationary part of

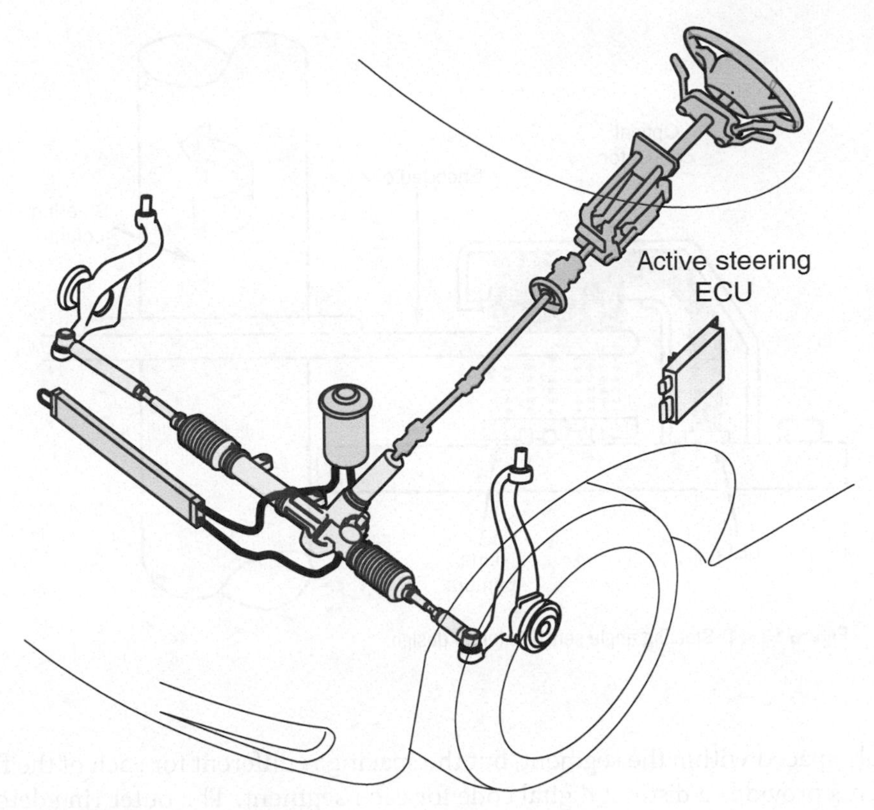

Figure 13-42 Active steering ECU.

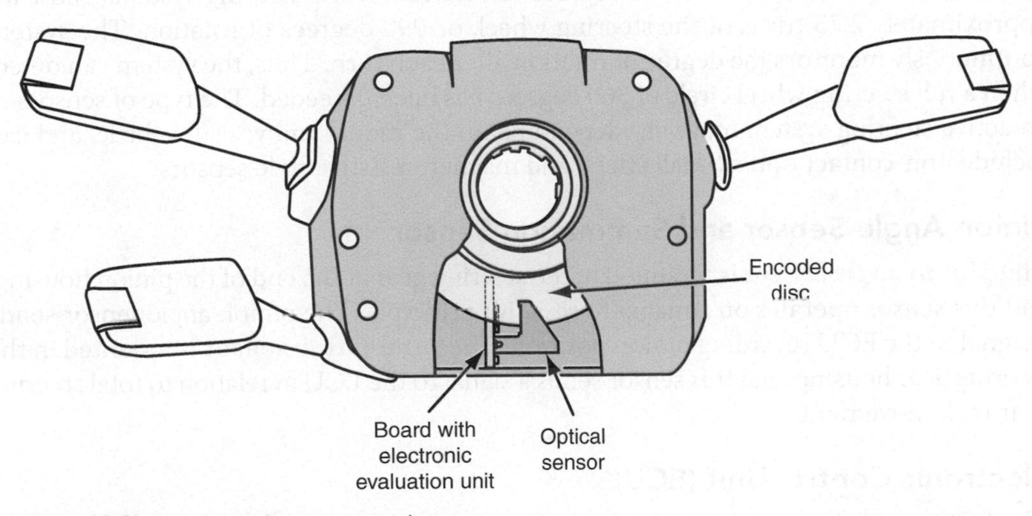

Figure 13-43 Steering angle sensor mounting.

the steering angle sensor contains a light-emitting diode (LED), fiber-optical conductor, and an optical sensor. An optical steering angle sensor uses an LED on one side of a slotted ring and a photocell on the opposite side. When the sensor is operating, light from the LED is projected onto the encoded disc through the optical conductor. As the steering wheel is turned, the slotted ring rotates causing light to alternately strike the photocell as the window (slot) is opened and then closed, in effect turning the photocell on and off, producing a digital square wave output (**Figure 13-45**). The optical sensor converts these light signals to voltage signals in relation to steering wheel rotation. The steering wheel sensor is comprised of two separate sensors, and the encoding disc is segmented into at least two slotted segments, an inner and an outer (**Figure 13-46**). The inner ring is separated into five distinct segments of 72 degrees each. Each of these five segments has slotted openings that

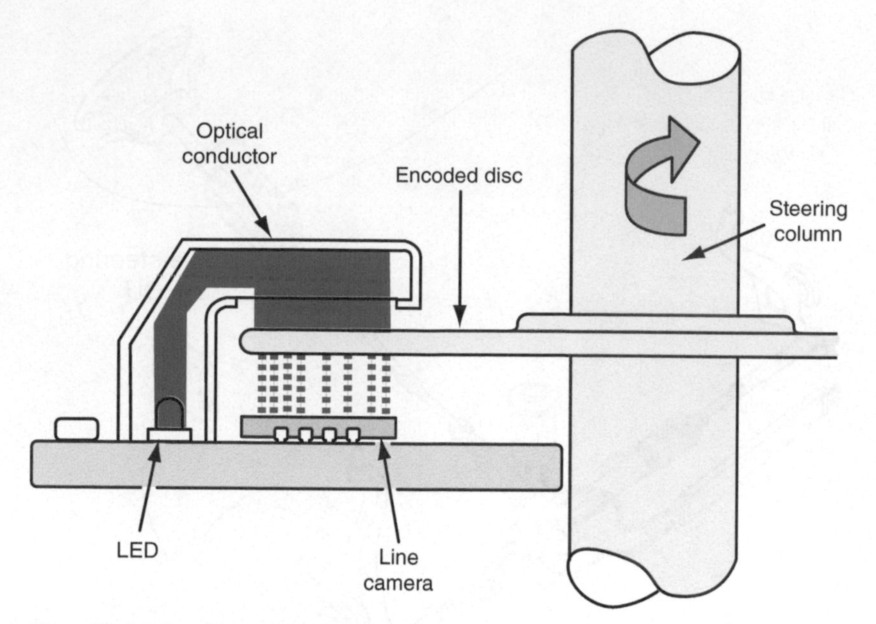

Figure 13-44 Steering angle sensor internal design.

are equally spaced within the segment, but the spacing is different for each of the five segments. This provides a distinct digital code for each segment. The outer ring determines the angle by using six photoelectric beam pairs of LEDs and photocells. The system can detect steering angles up to 1,044 degrees of rotation. Most steering systems allow for approximately 2.75 turns of the steering wheel, or 990 degrees of rotation. The system continuously monitors the degree of rotation after each turn. Thus, the system can detect when a full steering wheel circle of 360 degrees has been exceeded. The type of sensors in an active steering system may vary depending on the model and year of vehicle, and can include non-contact optical, Hall effect, and magnetoresistive style sensors.

Pinion Angle Sensor and Summation Sensor

The **pinion angle sensor** is mounted in the steering gear at the end of the pinion housing, and this sensor operates on a magneto resistive principle. The pinion angle sensor sends a signal to the ECU regarding pinion position. The **summation sensor** is mounted in the steering gear housing, and this sensor sends a signal to the ECU in relation to total steering gear rack movement.

Electronic Control Unit (ECU)

The ECU contains two microprocessors that control the electric motor, steering pump, servotronic valve, and electric locking unit. The ECU communicates with other ECUs and the active steering system input sensors via a CAN network.

Electric Locking Unit (ELU)

The **electric locking unit (ELU)** contains an electric solenoid and a lock pin. When the ELU solenoid is not energized by the ECU, the lock pin drops into one of the slots in the worm drive gear. This action locks the worm drive and drive motor (**Figure 13-47**). This locking action is taken by the ECU if a safety-related defect occurs in the active steering system. When the worm drive is locked, the driver maintains normal steering control. Under normal conditions, the ECU operates the ELU solenoid and pulls the lock pin away from the notches in the worm drive gear, and this allows normal active steering system operation.

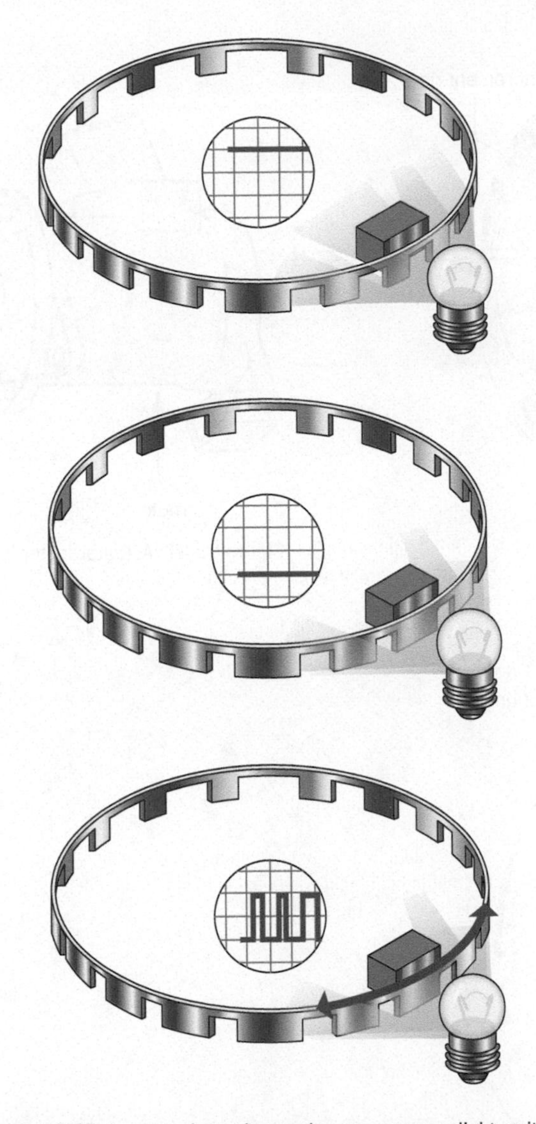

Figure 13-45 An optical steering angle sensor uses a light emitting diode on one side of a slotted ring and a photocell on the opposite side. As the steering wheel is turned, the slotted ring rotates causing light to alternately strike the photocell, in effect turning it on and off and producing a digital square wave output.

ACTIVE STEERING OPERATION

The actuator motor and planetary gear set have the capability to vary the steering gear ratio in relation to vehicle speed. The ECU receives voltage input signals from the system sensors, and in relation to these inputs, the ECU operates the actuator motor to vary the steering gear ratio. While cornering at low speeds, the steering gear ratio approximates 10:1. With this ratio, the driver does not have to rotate the steering wheel as much to obtain more front wheel turning action. When the vehicle is stationary, less than two steering wheel turns are required to turn the front wheels from lock-to-lock. This steering gear ratio allows the driver to apply the least amount of turning action to obtain a large amount of front wheel turning action. While parking and cornering at low speeds, this type of steering system requires less driver hand-to-hand action on the steering wheel, which increases safety and reduces driver fatigue.

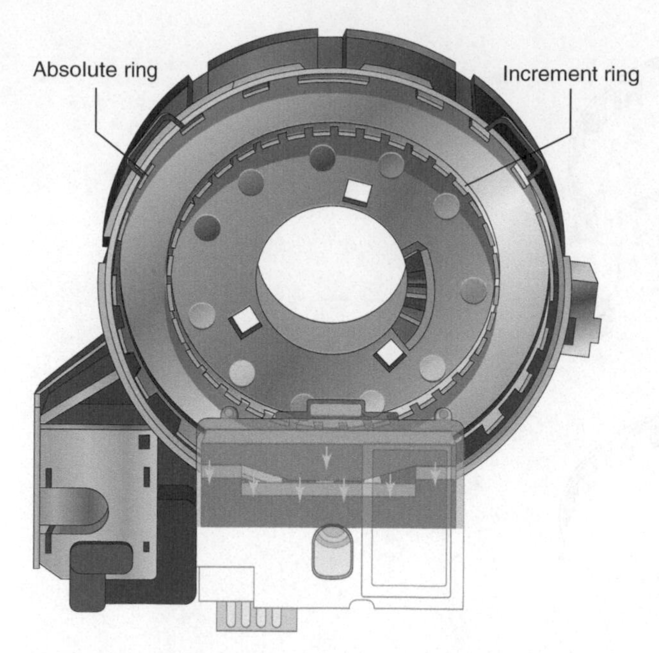

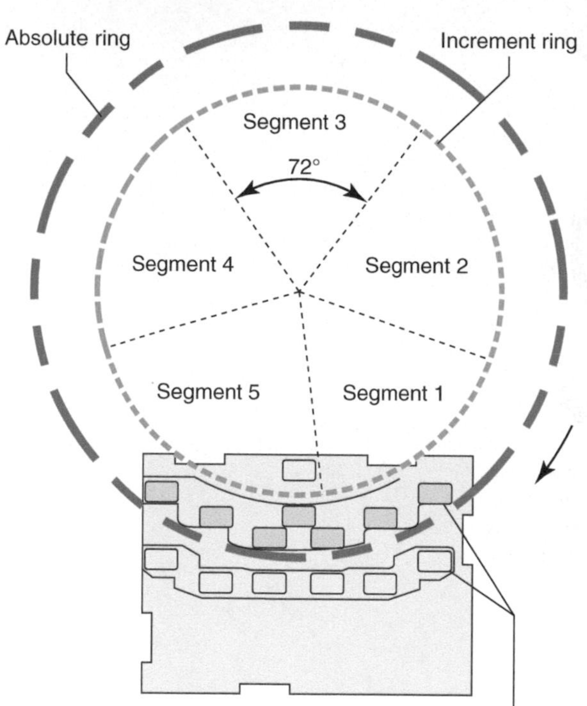

Figure 13-46 The steering wheel position sensor integrates two distinct sensors. The inner sensor has a ring separated into five distinct segments of 72 degrees each. The outer sensor ring determines the angle by using six photoelectric beam pairs of LEDs and photocells.

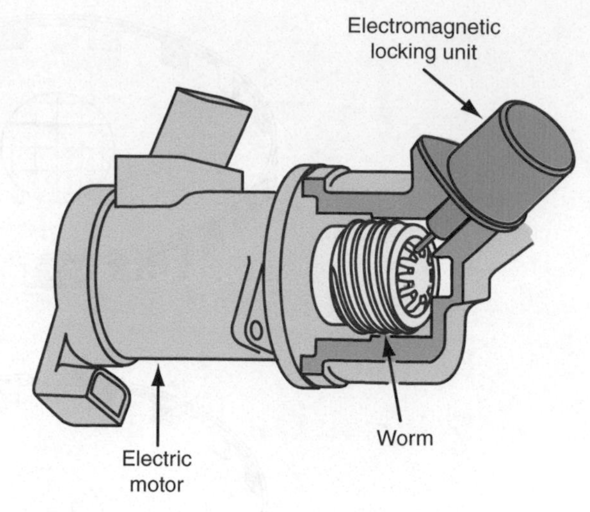

Figure 13-47 Actuator motor locking mechanism.

To reduce the steering ratio at low speeds, the actuator motor is driven in the same direction as the steering input, which decreases the steering ratio, and this action overdrives the steering input. This steering gear action may be compared to walking on an escalator. If you walk in the same direction as the escalator is moving, you multiply the total walking result.

As the vehicle speed increases, the ECU operates the actuator motor and planetary gear set to increase the steering ratio. At higher vehicle speeds, the steering gear ratio may be 20:1 and the increase in steering gear ratio dampens any sudden or excessive steering input by the driver. The increase in steering gear ratio under-drives the steering input. This action may be compared to walking against the movement of an escalator. The walking action of the person is canceled to some extent by the escalator movement, and the person has to walk more to cover the same distance. This steering gear action reduces the possibility of the driver causing the vehicle to go into a sideways swerve (**yaw motion**) by excessive steering wheel rotation. Any excessive steering wheel rotation by the driver is immediately counteracted by an increase in steering ratio by the active steering system. This active steering system action reduces the possibility of yaw vehicle motion. Many vehicles with active steering systems are also equipped with dynamic stability control (DSC), which greatly reduces the possibility of the vehicle swerving sideways out of control. On a vehicle equipped with active steering, the DSC system will not have to operate as frequently. If an electronic defect occurs in the active steering system, the system enters a fail silent mode in which the active steering system is inoperative and the driver has normal control of the steering.

Yaw motion is an erratic deviation from an intended course such as a sideways swerve when driving a vehicle.

A BIT OF HISTORY

Electronically controlled systems on vehicles have decreased emissions levels, increased fuel economy, and improved driver comfort and convenience. In the near future, electronic systems on vehicles may be interconnected with intelligent transportation systems (ITS) to speed up traffic flow and reduce accidents. Some traffic experts indicate that Americans lose 2 billion working hours each year because of traffic congestion and related delays. It is estimated these traffic delays cost American employers $40 billion per year. Therefore, we can appreciate that a system to increase traffic movement could save us billions of dollars each year.

Oakland County, Michigan, claims to be the world's largest operational test of an ITS system. This system uses 400 video cameras at 200 intersections to transmit information to a central computer that switches traffic light operation to match traffic flow. A similar system called Faster And Safer Travel Through Traffic Routing and Advanced Controls (FAST-TRAC) is installed in Oakland County, California. The ITS or FAST-TRAC systems may contain Advanced Traveler Information Systems (ATIS).

In these systems, on-board computers on vehicles receive information regarding driver information and route guidance from roadside beacons. The FAST-TRAC system in Oakland County, California, has decreased left-turn accidents by 69 percent and increased rush hour vehicle speeds by 19 percent. The U.S. Secretary of Transportation has indicated a commitment to installing ITS systems in 75 metropolitan areas in the United States with a goal of reducing travel time by 15 percent in these areas.

POWER STEERING SYSTEM

The electric-drive power steering pump and the active steering system are controlled by the same ECU, which contains two microprocessors. The power steering pump is a high-capacity vane-cell pump that is mounted immediately in front of the steering gear (**Figure 13-48**). The active steering system provides faster front wheel steering angles, and this action requires a higher-capacity power steering pump compared with conventional hydraulic-assisted power steering systems. The active steering ECU operates the **electronically controlled orifice (ECO) valve** in the power steering pump to control the fluid flow from the pump (**Figure 13-49**). The power steering pump supplies fluid

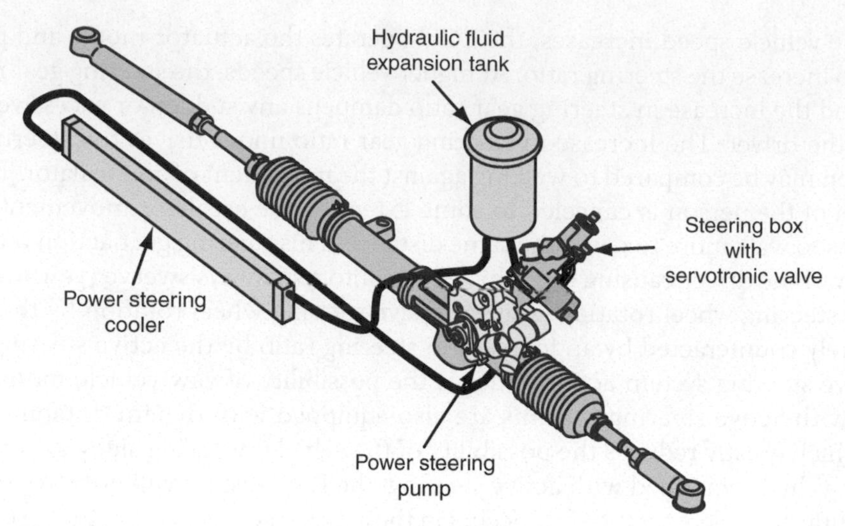

Figure 13-48 Electric-drive power steering system.

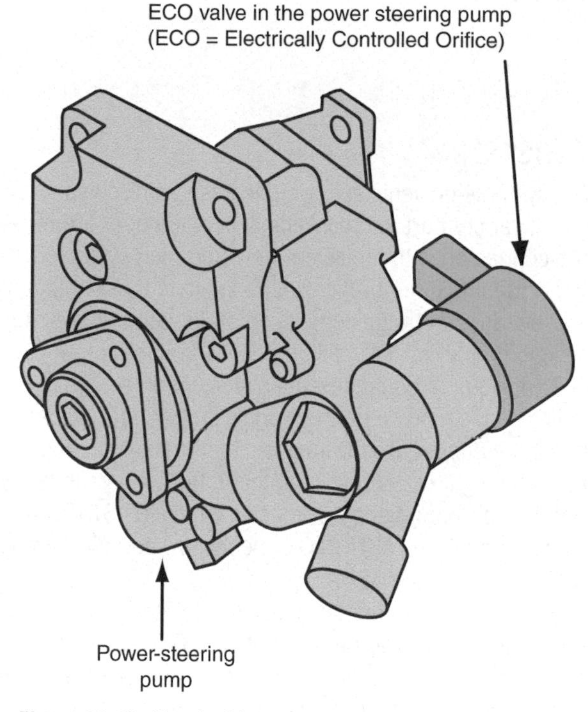

Figure 13-49 Electric-drive power steering pump.

pressure from the ECO valve in the pump to the **servotronic valve** in the steering gear. The active steering ECU operates the servotronic valve to increase power steering pump pressure and reduce steering effort and driver fatigue at low speeds (**Figure 13-50**). At higher vehicle speeds, the ECU operates the servotronic valve to reduce power steering pump pressure and increase steering effort to provide improved road feel and steering control. On some models, the active steering ECU directly operates the servotronic valve.

On other models, the software for the servotronic valve operation is stored in the active steering ECU, and this ECU sends signals through one of the CAN networks to the **safety and gateway module (SGM)**. On the basis of these input signals, the SGM operates the servotronic valve and the ECO valve.

The input signals required for servotronic valve operation are these:

1. Road speed signals from the wheel speed sensors to the DSC unit via one of the CAN networks to the active steering ECU.

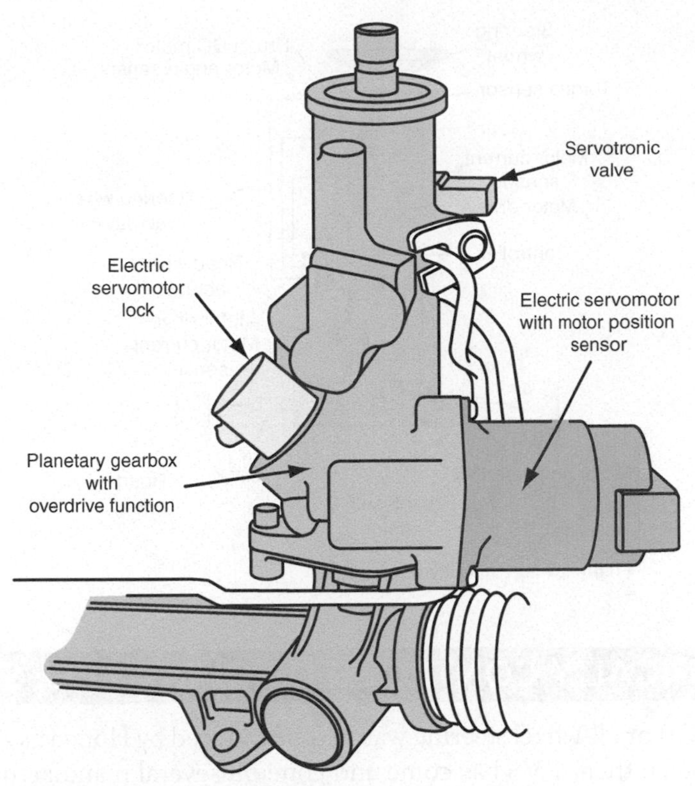

Figure 13-50 At low speeds the ECU operates the servotronic valve to increase power steering pump pressure.

2. Engine status signal from the **digital motor electronics (DME)** control unit via one of the CAN networks to the active steering ECU.
3. Terminal status from the **car access system (CAS)** via one of the CAN networks to the active steering ECU.

The servotronic valve is operated only when the engine is running. The servotronic valve enters a default mode if any of the input signals are incorrect, or if there is an electrical defect in the servotronic valve winding or connecting wires. In the default mode, the servotronic valve does not affect steering assist.

STEER-BY-WIRE SYSTEMS

The main difference between conventional steering systems and steer-by-wire systems is that steer-by-wire systems do not have any mechanical linkage between the steering wheel and the front wheels. In a steer-by-wire system, steering wheel input is supplied to torque sensor, gear, DC motor, and motor angle sensor (**Figure 13-51**). Input signals are sent to a controller from the motor angle sensor, torque sensor, and motor current sensor. On the basis of these input signals, the controller supplies output voltage signals to the motor drivers. The motor driver connected to the DC motor in the steering gear operates this motor to supply the desired steering angle.

Steer-by-wire systems are not available in automotive showrooms at present, but they may be available in the future. Steer-by-wire systems could have many possible advantages such as active steering control, improved vehicle maneuverability and stability, and steering system tuning to specific types of driving conditions. However, before steer-by-wire steering systems are commercialized, concerns regarding reliability and confirmation of advantages must be completely satisfied.

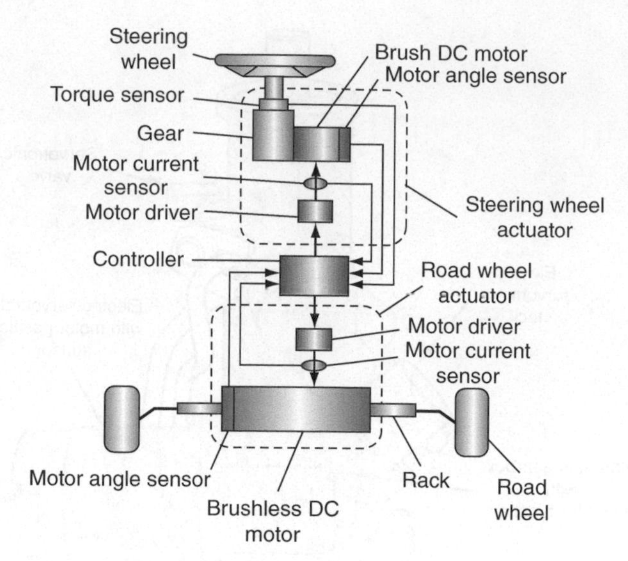

Figure 13-51 Steer-by-wire system.

ELECTRONICALLY CONTROLLED FOUR-WHEEL STEERING

Four-wheel (4WS) or all-wheel steering was first introduced by Honda as an option on the 1987 Prelude. Since then, 4WS has come and gone on several manufacturers' platforms, with one of the more notable and publicized being Quadra-steer offered on some General Motors trucks and full-size SUVs between 2002 and 2005. More recent entrants into 4WS were the 2016 Cadillac CT6 and the 2017 Audi Q7 platform, with other manufacturers—including Porsche, Infiniti, Lexus, Volvo, Renault, Ferrari, and Lamborghini—offering four-wheel steering on some models. The Audi Q7 offers a five-link independent suspension system with a rear-mounted electronic rear steering actuator assembly (**Figure 13-52**). Dynamic 4WS is designed to improve vehicle stability, comfort, and agility. While all the systems are somewhat similar with the utilization of a rear-mounted steering actuator assembly that resembles a rack and pinion assembly. Newer systems being introduced use the latest technology in electronic power assist, along with sophisticated computer programs that make full use of the CAN network system, communicating with other nodes on the network, including but not limited to active suspension system, ABS, stability control, and anti-rollover technology. While there are many benefits to 4WS, cost and system integration complexity are still major limiting factors that at least currently limit auto manufacturers desire to integrate 4WS. At least for now, 4WS will continue to be a novelty and not a mainstream feature.

In a **four-wheel steering (4WS)** system, all four wheels are turned to steer the vehicle.

If a car with conventional front wheel steering is parallel parked at a curb between two vehicles, this car may be driven from the parking space without hitting the car in front if the front wheels are turned all the way to the left (**Figure 13-53**, view A). When the same car is equipped with 4WS and the rear wheels steer in the same direction as the front wheels at low speed, the car will not steer out of the parking space without striking the vehicle in front (**Figure 13-53**, view B).

Steering angle is the number of degrees that the wheels turn when steered.

Turning circle is the diameter of a circle completed by a vehicle when the front wheels are turned fully to the right or left.

When the car in the same parking space has a 4WS system that steers the rear wheels in the opposite direction to the front wheels at low speed, the car steers out of the parking space with plenty of distance between the vehicle parked in front (**Figure 13-53**, view C). When the rear wheels steer in the opposite direction to the front wheels, the rear wheels steer toward the curb. This action causes the right rear tire to strike the curb if the car is parked close to the curb. Therefore, the maximum rear **steering angle** must be considerably less than the maximum front wheel steering angle to help prevent this problem. A car with 4WS has a smaller **turning circle**, or turning radius, compared to a vehicle with

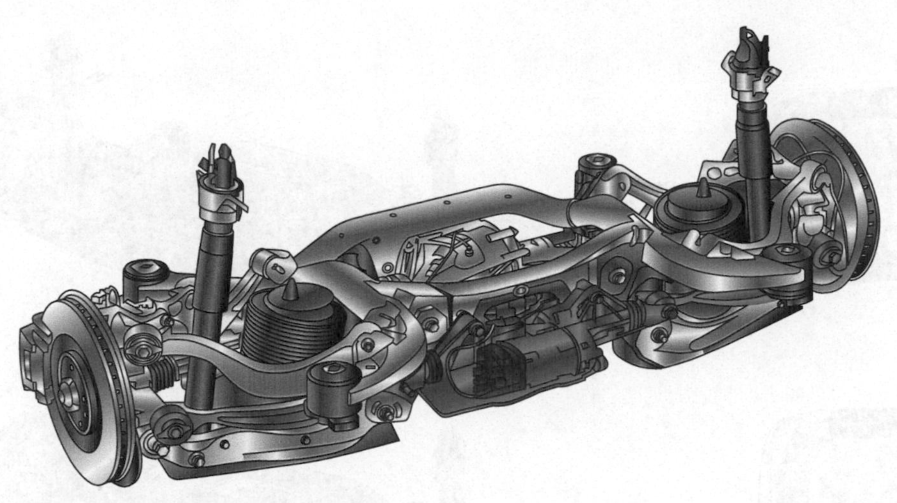

Figure 13-52 A five-link independent rear suspension assembly with rear steering actuator assembly that resembles a rack and pinion steering gear.

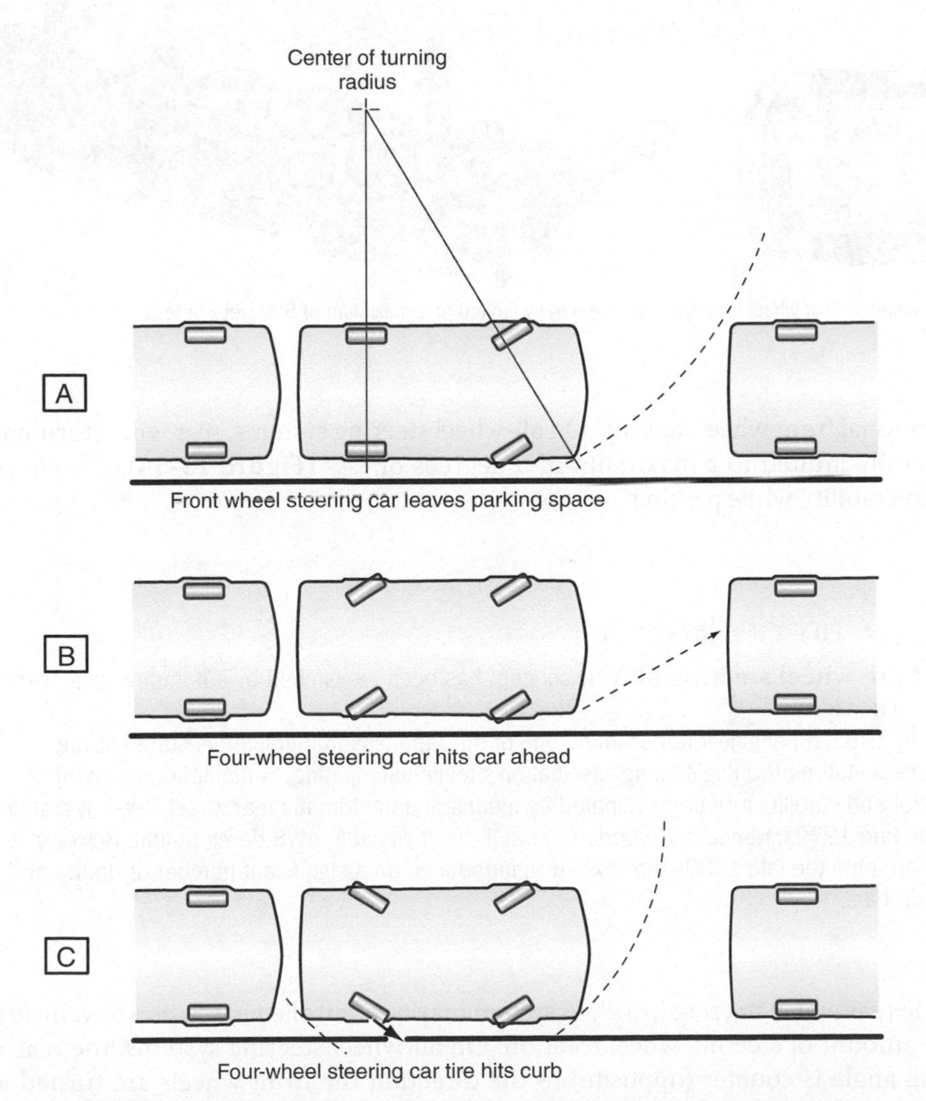

Center of turning radius
A
Front wheel steering car leaves parking space
B
Four-wheel steering car hits car ahead
C
Four-wheel steering car tire hits curb

Figure 13-53 Parallel parking with conventional front wheel steering and four-wheel steering.

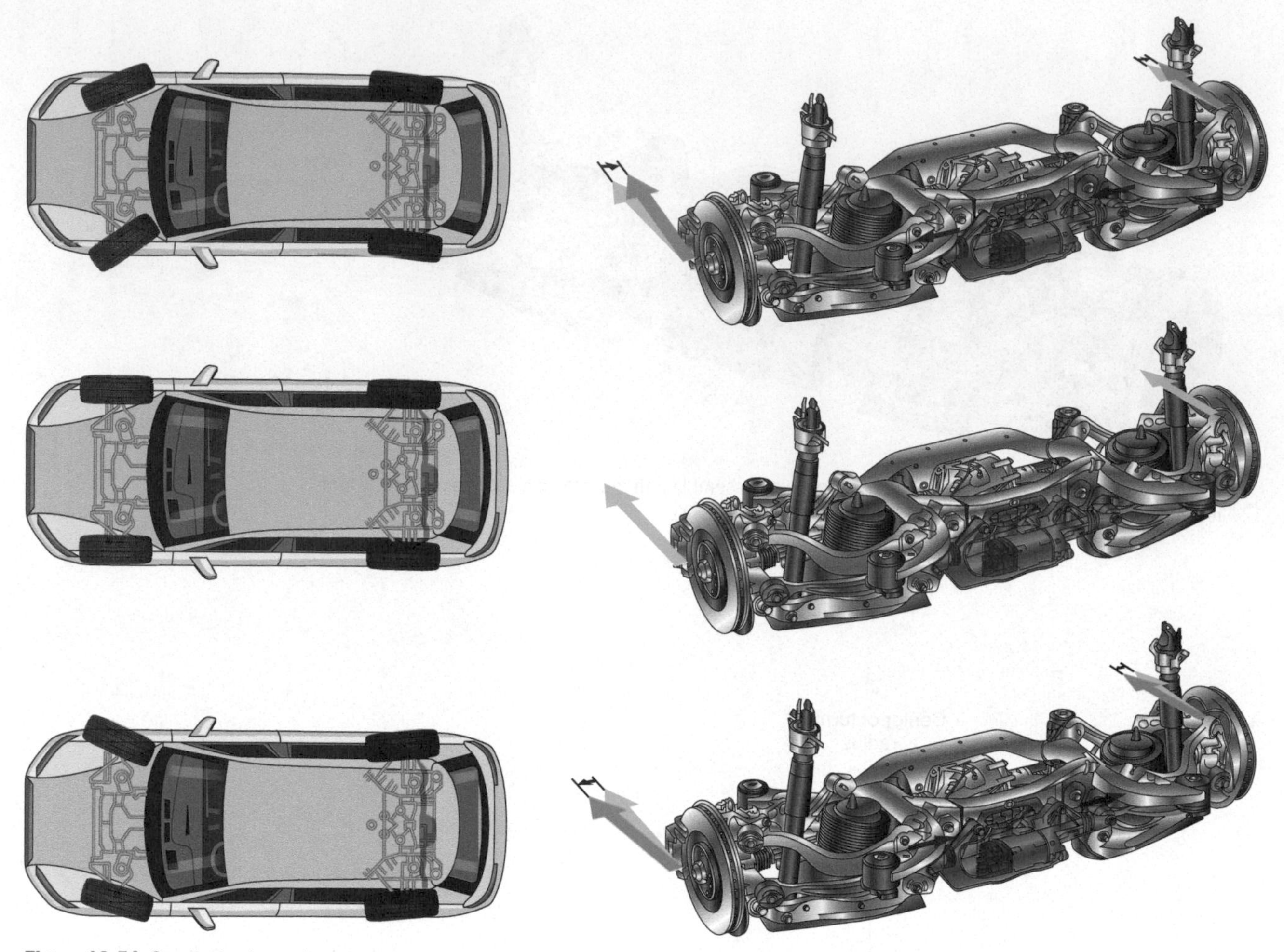

Figure 13-54 On all-wheel steering systems, rear wheel turning angle is generally limited to a maximum of 5 degrees or less.

conventional front wheel steering. On all-wheel steering systems, rear wheel turning angle is generally limited to a maximum of 5 degrees or less (**Figure 13-54**). This improves maneuverability while parking.

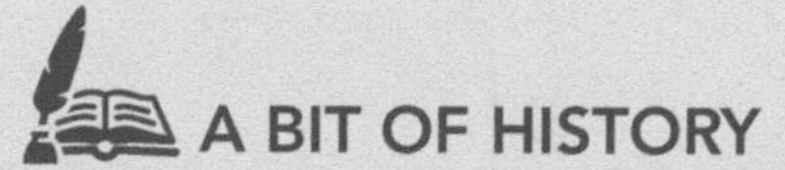

A BIT OF HISTORY

The **four-wheel steering (4WS)** concept has been researched by automotive engineers for several decades.

In 1962, an engineer representing one of the Japanese manufacturers stated at the Japanese Automotive Engineering Association's technical meeting, "A major improvement in control and stability may be anticipated by means of an automatic rear wheel steering system." In the late 1970s, Honda and Mazda were actively engaged in 4WS development. However, it was not until the late 1980s that 4WS was introduced on a significant number of Honda and Mazda cars.

The rear wheel steering in a 4WS system may be controlled in relation to vehicle speed or the amount of steering wheel rotation. On all-wheel steering systems, the rear wheel turning angle is counter (opposite) to the direction the front wheels are turned at low speeds below 30 mph (48 km/h), resulting in tighter turning radius (**Figure 13-55**).

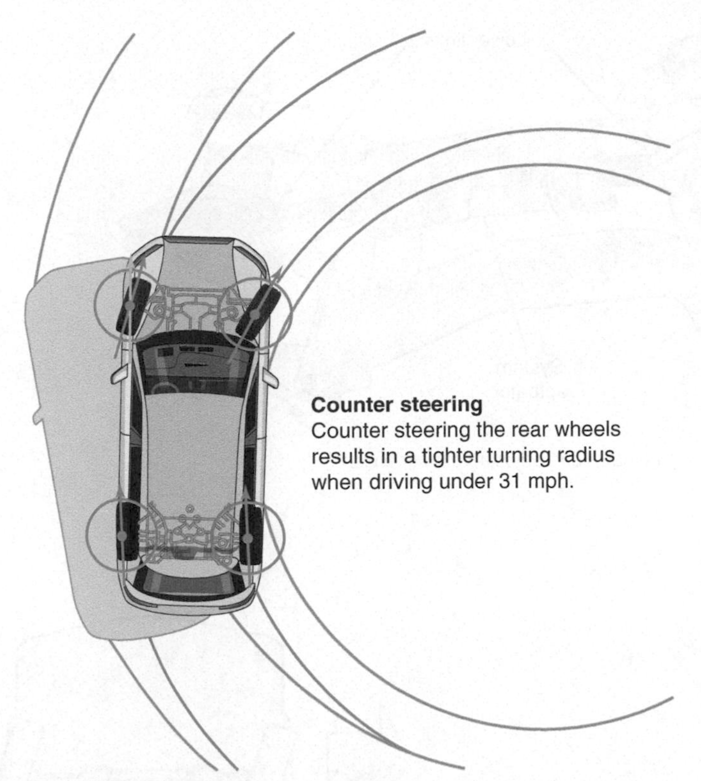

Counter steering
Counter steering the rear wheels results in a tighter turning radius when driving under 31 mph.

Figure 13-55 On all-wheel steering systems, the rear wheel turning angle is counter (opposite) to the direction the front wheels are turned at speeds below 30 mph, resulting in tighter all-wheel radius.

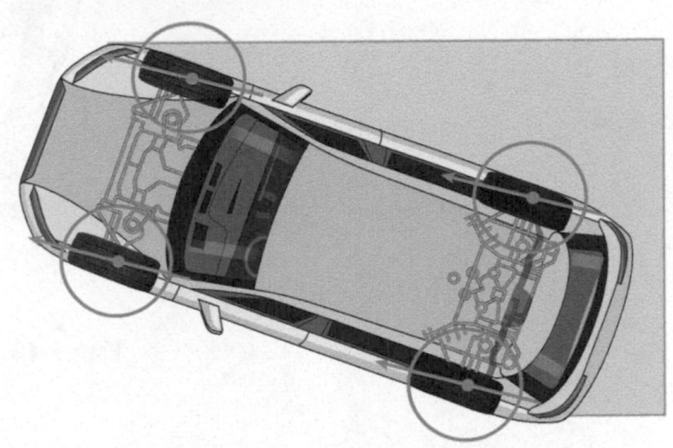

Parallel steering
When all four wheels turn the same direction, this results in precise lateral tracking at highway speeds.

Figure 13-56 On all wheel steering systems, the rear wheel turning angle is counter (opposite) to the direction the front wheels are turned at speeds below 30 mph resulting in tighter turning radius.

Allowing the rear wheels to steer in the opposite direction to the front wheels at low vehicle speeds has the following benefits:

- Smaller turning radius
- Improved agility
- Improved handling in low-speed traffic and urban driving
- Improved parking maneuverability

When the vehicle is operating at higher speeds or with a small amount of steering wheel rotation, the rear wheels are steered in the same direction as the front wheels (**Figure 13-56**). The advantage of steering the rear wheels in the same direction as the front wheels at highway speeds is precise lateral tracking; this is referred to as parallel steering. When a vehicle is cornering at higher speeds, centrifugal force tends to move the rear of the vehicle sideways. This action causes the rear tires to slip sideways on the road surface. This process may be called **sideslip**. The vehicle speed and the severity of the turn determine the amount of sideslip. If sideslip is excessive, the car may spin around, causing the driver to lose control. When the 4WS system steers the rear wheels in the same direction as the front wheels at higher speeds, sideslip is reduced and vehicle stability is improved. Allowing the rear wheels to steer in the same direction to the front wheels at higher vehicle speeds has these additional benefits:

- Improved vehicle stability
- Increased driving safety
- Improved driving dynamics

The higher-speed same-direction steering angle is considerably less than the low-speed opposite-direction steering angle. As was stated earlier, the maximum rear wheel steering angle is 5 degrees on most platforms during any steering maneuver, and even less on some platforms.

Shop Manual
Chapter 13, page 594

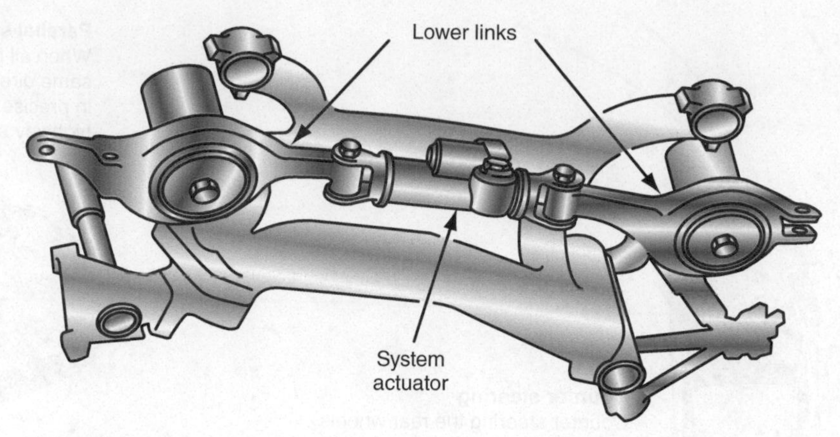
Lower links
System
actuator

Figure 13-57 Rear steering actuator and adjustable lower links.

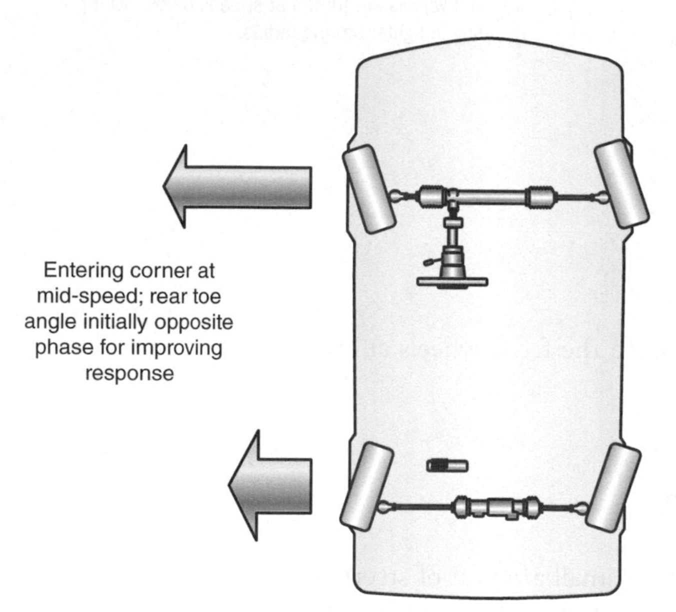
Entering corner at
mid-speed; rear toe
angle initially opposite
phase for improving
response

Figure 13-58 Rear wheel steering during negative phase operation.

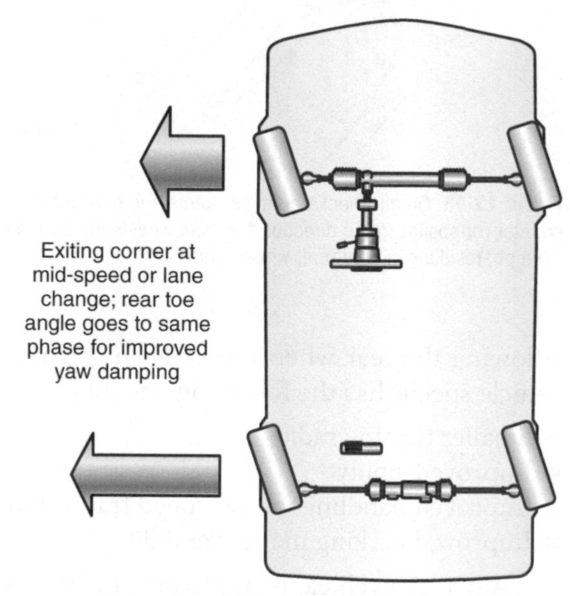
Exiting corner at
mid-speed or lane
change; rear toe
angle goes to same
phase for improved
yaw damping

Figure 13-59 Rear wheel steering during positive phase operation.

The four-wheel active steering (4WAS) improves steering and handling by adjusting the geometry of the rear suspension in relation to steering input and vehicle speed. The 4WAS control unit calculates the ultimate vehicle dynamics from the input signals from a group of sensors. In response to these inputs, the control unit operates the electric actuator to lengthen or shorten each lower rear suspension link (**Figure 13-57**).

When the vehicle begins turning a corner, driving around a curve, or making a lane change at mid-speed, the control unit operates the 4WAS actuator to initially turn the rear wheels in the negative phase mode in relation to the front wheels (**Figure 13-58**). This action provides improved turn-in response.

When the steering wheel rotation indicates the vehicle is exiting a corner or making a lane change at mid-speed, the control unit operates the 4WAS actuator to turn the rear wheels in the positive phase mode in relation to the front wheels (**Figure 13-59**). Under this condition yaw forces are reduced on the rear of the vehicle to provide improved vehicle stability.

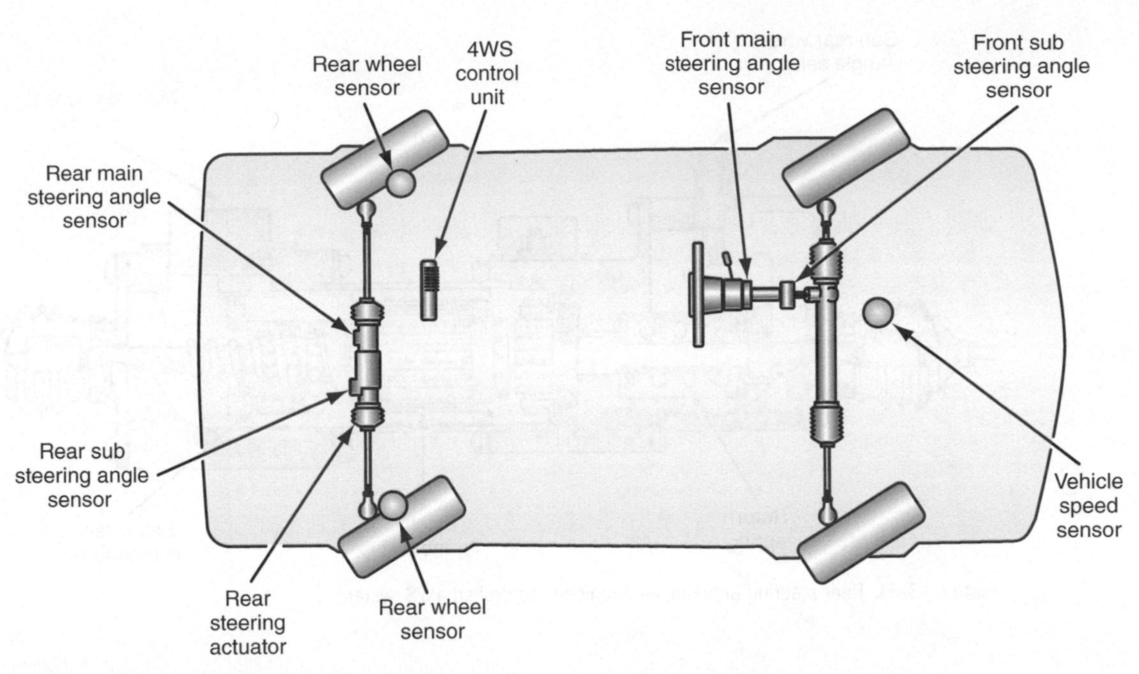

Figure 13-60 Electronically controlled 4WS with control unit mounted in the trunk.

On the electronically controlled 4WS system, there is no mechanical connection between the front steering gear and the **rear steering actuator**. The rear steering actuator is controlled by a 4WS control unit (**Figure 13-60**). The 4WS control unit in the electronically controlled system uses steering wheel rotational speed, vehicle speed, and front steering angle information to calculate and control the rear steering angle.

In an **electronically controlled 4WS system**, rear wheel steering is controlled electronically.

Input sensors information accessed by the electronically controlled 4WS system include, but are not limited to, the following:

1. Main rear wheel angle sensor.
2. Sub rear wheel angle sensor in the rear steering actuator.
3. Main steering wheel angle sensor in the steering column under the combination switch.
4. Sub front wheel steering angle sensor.
5. ABS wheel speed sensors.
6. Vehicle speed sensors.

Caution

Some electronic sensor wires have a special shield surrounding them to prevent electromagnetic interference (EMI) from other voltage sources. If this shield is damaged or removed, computer system operation may be adversely affected. Do not reroute sensor wires close to other voltage sources such as spark plug wires.

The rear steering actuator may be compared to an electric steering gear. This actuator contains an electric motor that drives a steering rack through a ball screw mechanism (**Figure 13-61**).Tie rods are connected from the steering actuator to the rear steering arms and spindles. A return spring inside the actuator moves the rear wheels to the straight-ahead position when the ignition switch is turned off or when a defect occurs in the 4WS system. Some manufacturers, such as Porsche, use individual rear steering actuators for both the left and right rear wheels (**Figure 13-62**).

When the engine is running, the 4WS control unit continually receives information from all the input sensors and associated control modules on the network. If the steering wheel is turned, the 4WS control unit analyzes information from the vehicle speed sensor, main steering wheel angle sensor, sub front wheel angle sensor, main rear wheel angle sensor, sub rear wheel angle sensor, and the rear wheel speed sensors, among others. The 4WS control unit calculates the proper rear wheel steering angle and then sends voltage command to the rear steering actuator(s) motor to provide this rear steering angle (**Figure 13-63**).

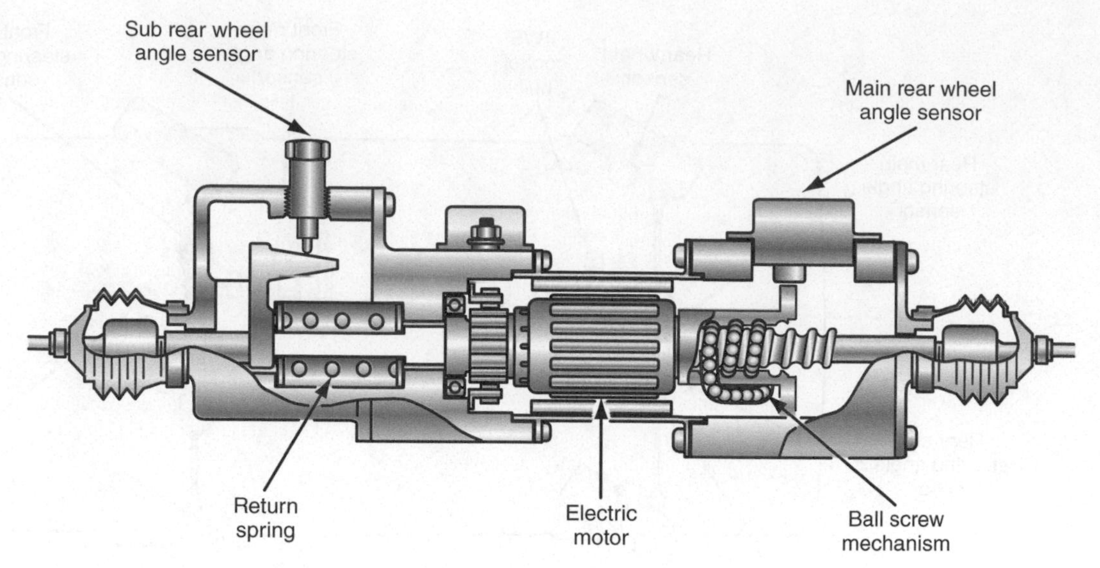

Figure 13-61 Rear steering actuator, electronically controlled 4WS system.

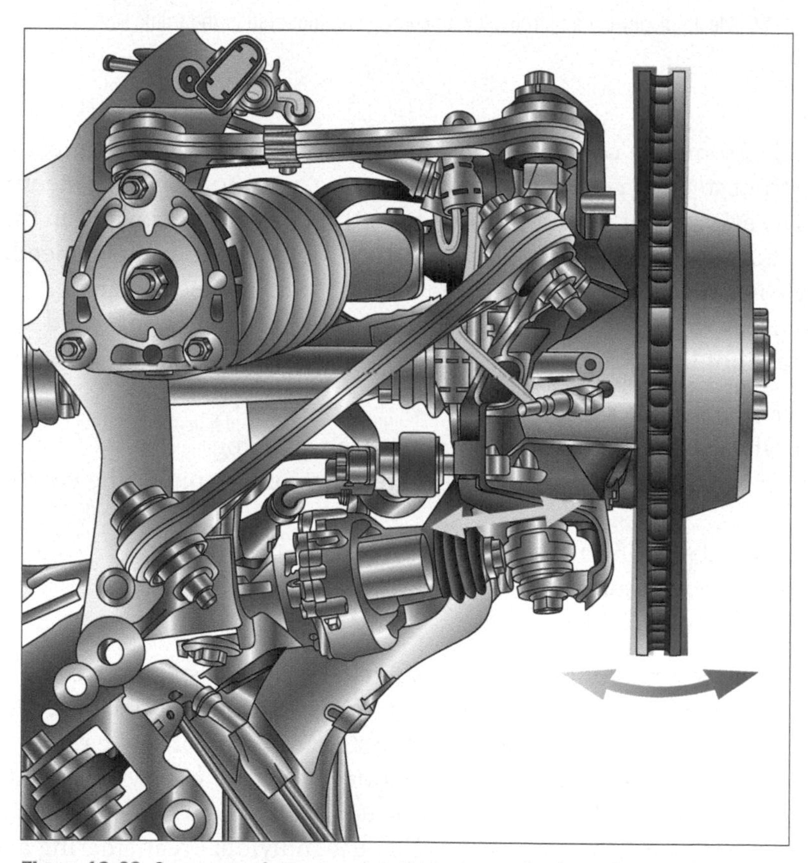

Figure 13-62 Some manufacturers use individual rear wheel steering actuators in unison to control rear wheel turning angle.

Transistors may be defined as automatic electronic relays that have no moving parts and are made from semiconductor materials.

Battery voltage is sent to the rear steering actuator motor through two heavy-duty output **transistors**. One of these transistors conducts current during a right turn, whereas the other transistor is activated during a left turn. The main rear wheel angle sensor and the sub rear wheel angle sensor send feedback signals to the 4WS control unit, indicating the proper rear steering angle has been supplied.

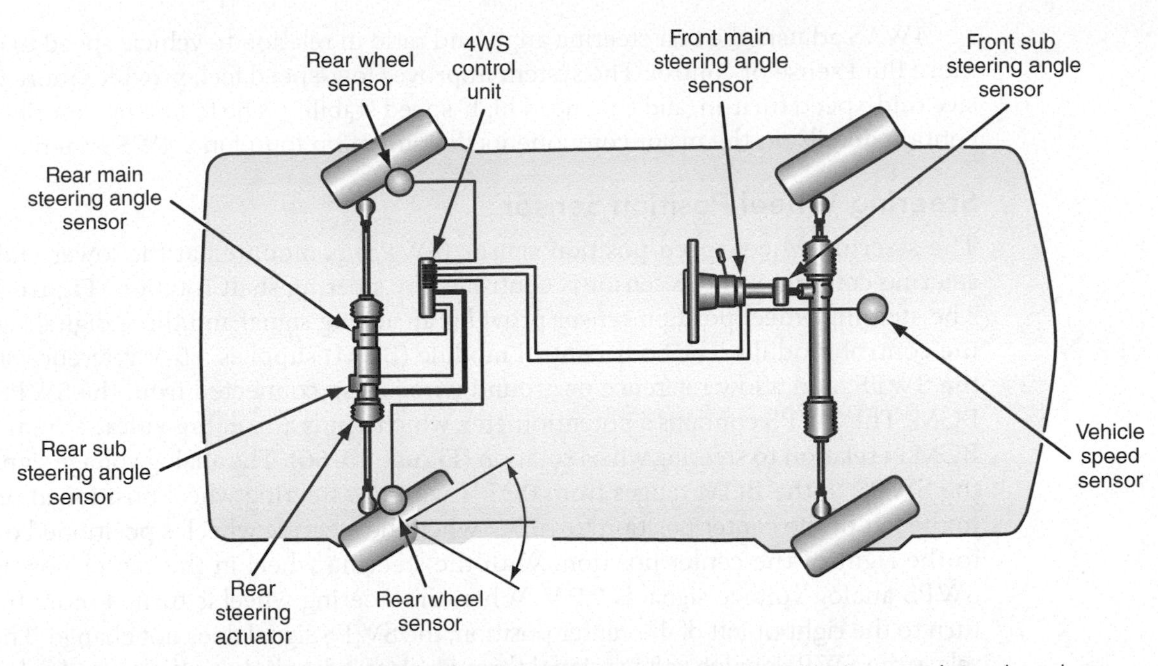

Figure 13-63 The 4WS control unit analyzes input sensor information, calculates the required rear steering angle, and operates the rear steering actuator motor to provide the proper rear steering angle.

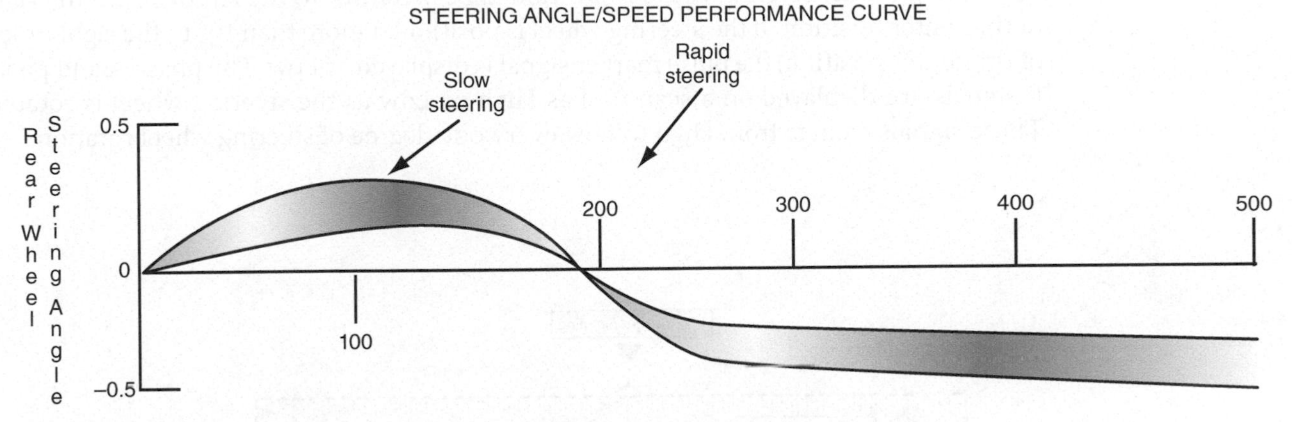

Figure 13-64 Rear steering angle in relation to vehicle speed and steering wheel rotation.

When the vehicle is operating at speeds less than 18 mph (29 km/h) to 30 mph (48 km/h), depending on the manufacturer, the rear wheels immediately begin to steer in the opposite direction to the front wheels if the steering wheel is turned (**Figure 13-64**). Maximum rear steering angle is generally 5° at 0 mph, but this maximum angle does vary depending on the manufacturer, from as little as 2.5° to as much as 12° (always refer to specific vehicle manufacturer service information for exact specifications). The rate of rear steering angle decreases in relation to vehicle speed.

When the vehicle speed increases above 30 mph (48 km/h), this speed varies depending on model year and manufacturer, the rear wheels steer in the same direction as the front wheels through the first 200° of steering wheel rotation. The rear steering angle reverts to the opposite phase if the steering wheel is rotated more than 200° in this vehicle speed range. When the vehicle speed is 60 mph (96 km/h) and the steering wheel rotation is 100°, the rear wheels steer about 1° in the same direction as the front wheels. If the steering wheel is rotated 500° slowly at this speed, the rear wheels are steered about 1° in the opposite direction to the front wheels.

4WAS adjusts the rear steering angle and ratio in relation to vehicle speed to create a more fluid sense of control. The system improves low-speed feel, provides more responsive mid-speed turn-in, and enhanced high-speed stability. The following description will contain details on the major components that are often found in a 4WS system.

Steering Wheel Position Sensor

The steering wheel speed/position sensor (SWPS) is mounted at the lower end of the steering column and this sensor is controlled by steering shaft rotation (**Figure 13-65**). The steering wheel position sensor provides an analog signal and three digital signals to the control module. The body control module (BCM) supplies a 5-V reference signal to the SWPS, and a low reference or ground wire is also connected from the SWPS to the BCM. The SWPS contains a potentiometer, which sends an **analog voltage signal** to the BCM in relation to steering wheel rotation (**Figure 13-66**). The analog voltage signal from the SWPS to the BCM ranges from 0.25 V with the steering wheel positioned one turn to the left of the center position to 4.75 V when the steering wheel is positioned one turn to the right of the center position. With the steering wheel in the center position, the SWPS analog voltage signal is 2.5 V. When the steering wheel is turned more than one turn to the right or left of the center position, the SWPS signal does not change. The BCM relays the SWPS analog voltage signal through class 2 data links to the rear wheel steering control module.

The SWPS sends digital signals through the phase A, phase B, and marker pulse circuits directly to the control module. The marker pulse digital signal is displayed on a scan tool as High if the steering wheel is positioned between 10° to the left or 10° to the right of the center position. If the steering wheel is positioned more than 10° to the right or left of the center position, the pulse marker signal is displayed as Low. The phase A and phase B signals are displayed on a scan tool as High or Low as the steering wheel is rotated. These signals change from High to Low every one degree of steering wheel rotation.

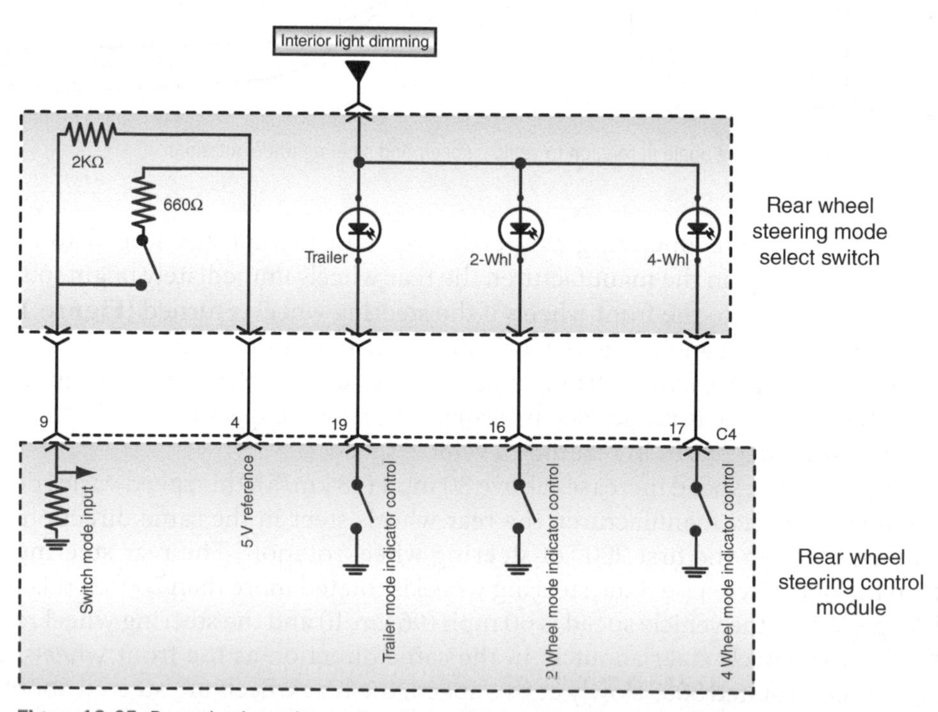

Figure 13-65 Rear wheel steering mode select switch inputs to the control module.

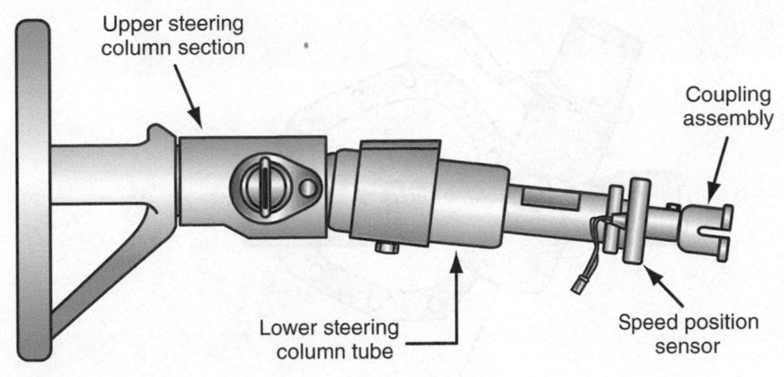

Figure 13-66 Steering wheel speed/position sensor.

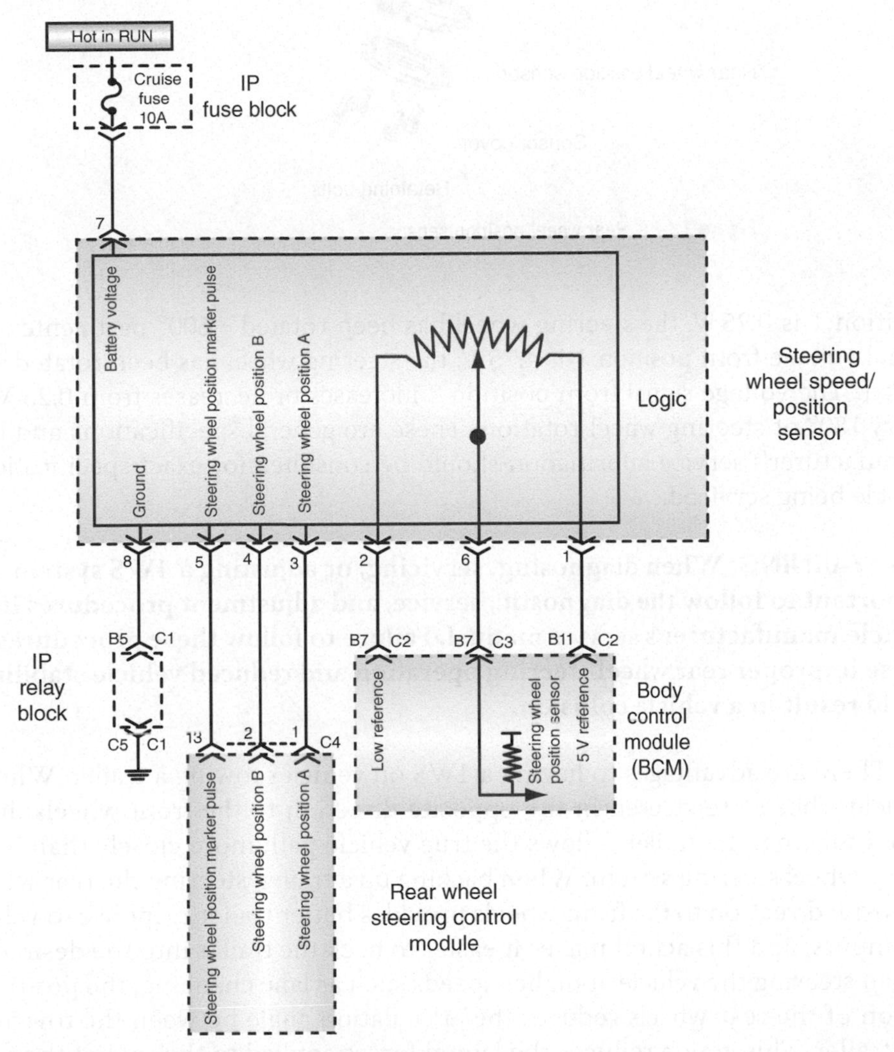

Figure 13-67 SWPS inputs to the control module.

Rear Wheel Position Sensor

The rear wheel position sensor is mounted on the lower side of the rear wheel steering gear (**Figure 13-67**). The rear wheel position sensor has two signal circuits connected to the rear wheel steering control module (**Figure 13-68**). The position 1 signal is a linear measurement of voltage per degree of rear steering position sensor rotation. For the position 1 input, the measurement in degrees is from −620° to the left to +620° to the right. The voltage signal from the position 1 input is 0.25 V to 4.75 V. If the signal voltage from

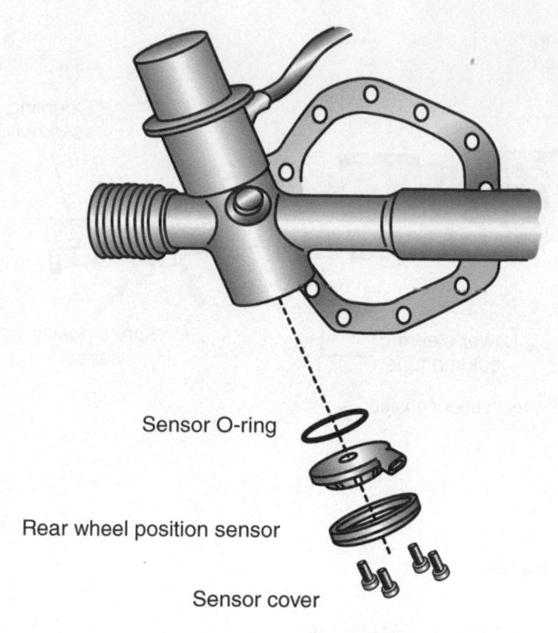

Figure 13-68 Rear wheel position sensor.

position 1 is 0.25 V, the steering wheel has been rotated −600° past center. When the signal voltage from position 1 is 4.75 V, the steering wheel has been rotated +600° past center. The voltage signal from position 2 increases or decreases from 0.25 V to 4.75 V every 180° of steering wheel rotation. These are general specifications and the vehicle manufacturer's service information should be consulted for exact specifications for the vehicle being serviced.

WARNING **When diagnosing, servicing, or adjusting a 4WS system, it is very important to follow the diagnostic, service, and adjustment procedures in the vehicle manufacturer's service manual. Failure to follow these procedures may cause improper rear wheel steering operation and reduced vehicle stability that could result in a vehicle collision.**

There are advantages to having a 4WS on vehicles towing a trailer. When the rear vehicle wheels are steered in the opposite direction to the front wheels during low-speed turning, the trailer follows the true vehicle path more closely than it does with a two-wheel steering system. When backing up a trailer, steering the rear wheels in the opposite direction to the front wheels provides better trailer response to vehicle steering inputs, and this action makes it easier to back the trailer into the desired position. When steering the vehicle at higher speeds such as lane changing, the positive steering action of the rear wheels reduces the articulation angle between the tow vehicle and the trailer. This action reduces the lateral forces applied to the rear of the tow vehicle by the trailer, which in turn reduces yaw velocity gain and improves trailer stability. A 4WS system reduces turning circle diameter from 45 ft. to 33.9 ft. (**Figure 13-69**) on some SUVs, and a similar proportional reduction in turning circles is experienced on small performance car platforms as well.

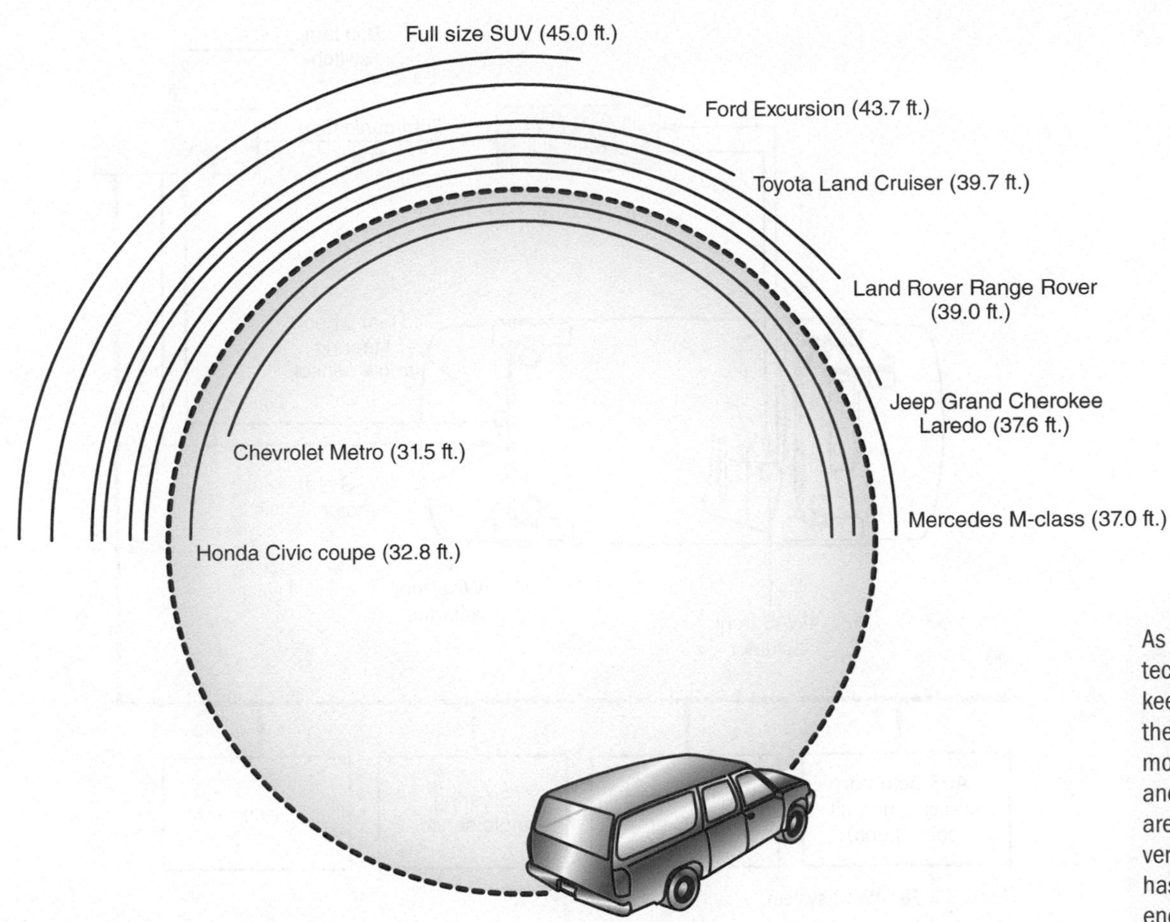

Figure 13-69 Turning circle diameter, four-wheel steering system.

As an automotive technician, you must keep up to date on the changes in automotive electronics, and these changes are occurring at a very rapid pace. It has been my experience that one of the best ways to keep up to date is to join professional technicians' organizations such as the National Automotive Service Task Force (NASTF), Automotive Service Association (ASA), or the International Automotive Technicians' Network (IATN). As a member of these organizations, you will be able to obtain information on the latest automotive electronics technology and the solutions to diagnostic problems related to automotive electronics.

FOUR-WHEEL ACTIVE STEERING (4WAS)

4WAS System Design

The 4WAS has a front ECU and a main (rear) ECU. The 4WAS system has a front steering actuator mounted coaxially in the front steering shaft (**Figure 13-70**). The front actuator contains a front wheel steering angle sensor, a front lock solenoid valve, a front motor, and a gear shaft (**Figure 13-71**). The front ECU operates the motor and gear shaft in the front actuator to change the steering ratio. The lock solenoid valve in the front steering actuator locks this actuator so the steering ratio cannot change if a defect occurs in the system.

The rear steering actuator body is attached to a chassis member, and the outer ends of the actuator shaft are linked to the rear wheels. The rear steering actuator contains a rod attached to the rear steering arms (item 1), a motor (item 4), motor shaft and drive gear (item 5), a driven gear (item 7), a housing assembly (item 3), and a rear wheel steering angle sensor (item 6) (**Figure 13-72**).

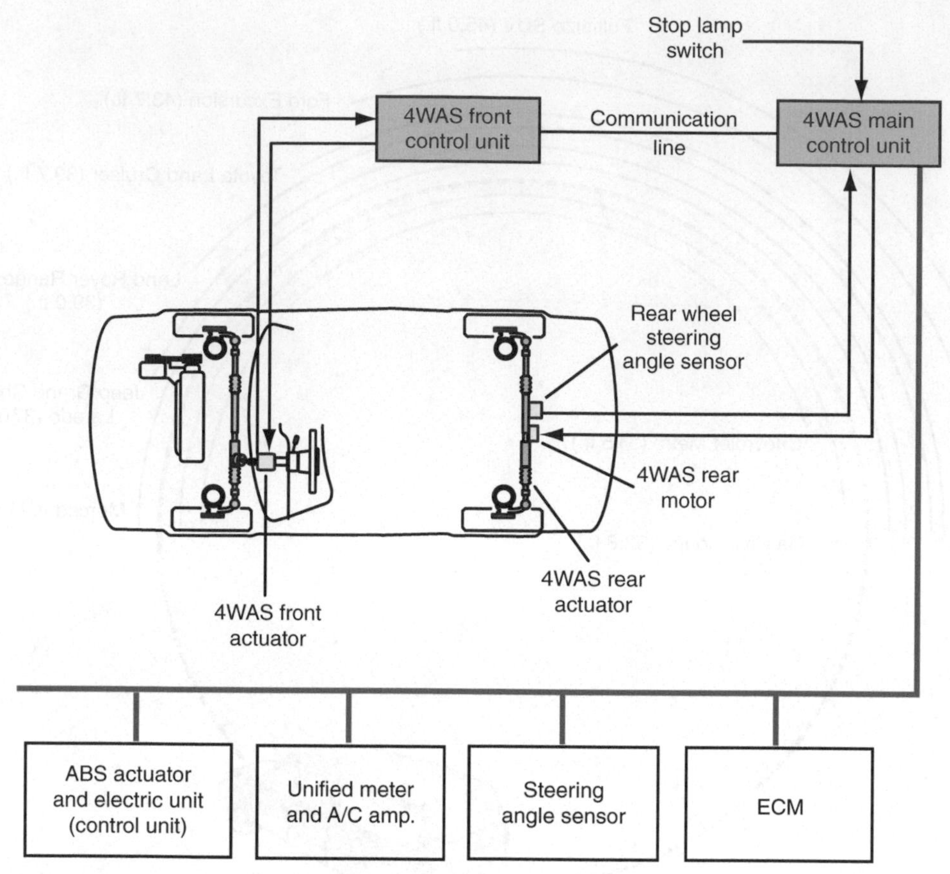

Figure 13-70 4WAS system.

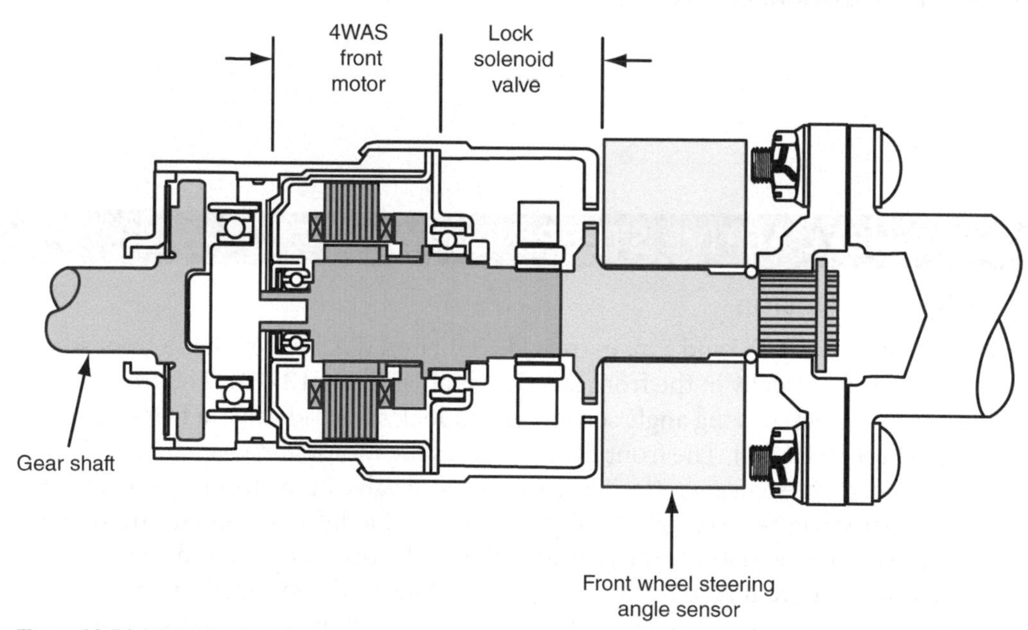

Figure 13-71 4WAS front actuator.

Four-wheel active steering (4WAS) may be defined as front and rear wheel steering in which a computer(s) can electronically change the front and rear steering angles.

Shop Manual
Chapter 13, page 533

The main ECU is mounted in the trunk, and controls the rear steering actuator, and the front ECU is mounted under the dash and controls the front steering actuator. The software in the ECUs contains a reference model that contains the desired dynamic steering characteristics. The ECUs operate the front and rear steering actuators to conform to the reference model in the software.

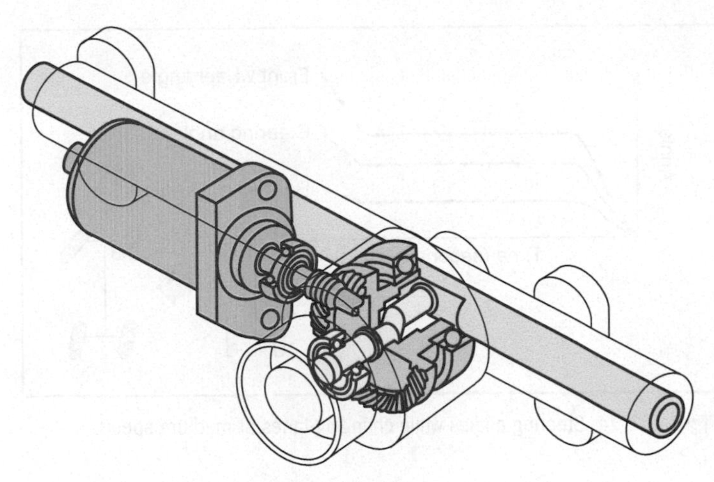

Figure 13-72 4WAS rear actuator.

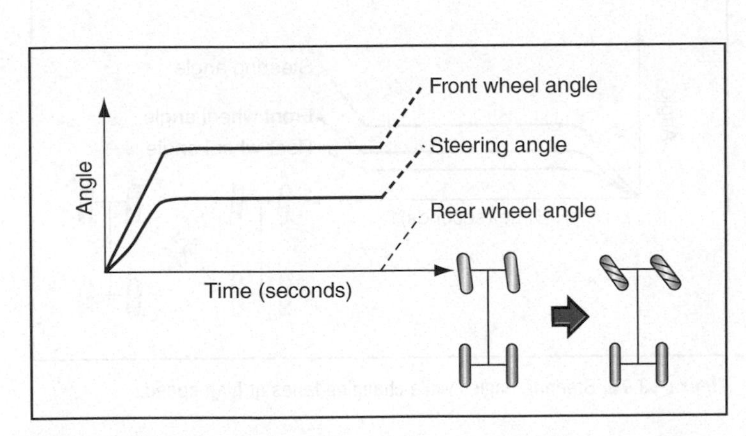

Figure 13-73 Steering angles while cornering at low speed.

The main control unit calculates the front and rear steering angles that will provide the best steering performance and vehicle stability based on the front and rear steering angle sensor inputs and the vehicle speed input. The vehicle speed input is sent from the antilock brake system (ABS) control unit through the CAN communication network to the main control unit. Engine speed signals are transmitted from the engine electronic control unit (ECU) via the CAN network to the main control unit. The 4WAS warning light is mounted in the unified meter in the instrument panel, and operated by the main control unit.

4WAS System Operation

When cornering at low speed the front control unit operates the front steering actuator to reduce the steering gear ratio. This action increases the front steering angle with less steering wheel movement and reduced driver steering effort (**Figure 13-73**).

When changing lanes in the medium-speed range, the front steering actuator can increase the front steering angle in relation to steering wheel rotation, and simultaneously the rear steering actuator steers the rear wheels a small amount in the same direction. When this action is taken, yaw motion, lateral force, and vehicle slide slip are reduced (**Figure 13-74**).

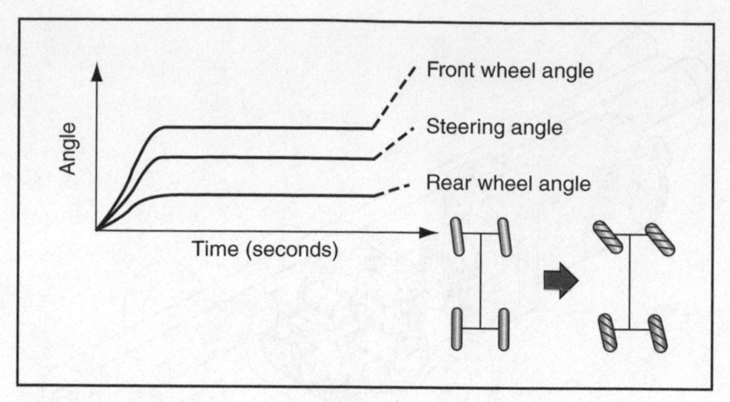

Figure 13-74 Steering angles while changing lanes at medium speed.

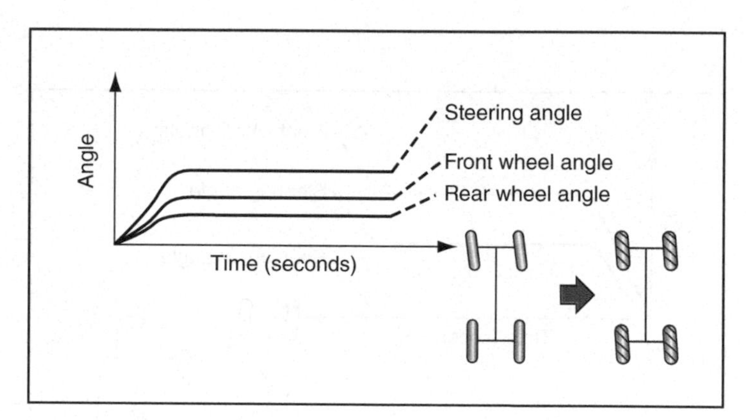

Figure 13-75 Steering angles while changing lanes at high speed.

When changing lanes at high speed, the front steering actuator increases the front steering ratio so the driver steering effort is increased, steering wheel movement is also increased, and steering angle is decreased in relation to steering wheel rotation. This action provides more road feel to the driver. Simultaneously, the rear steering actuator steers the rear wheels a few degrees in the same direction (**Figure 13-75**). This action provides improved steering response and vehicle stability.

On some Infiniti luxury car model packages, the 4WAS feature is an available option. This technology may or may not stay part of the luxury car segment. Often it depends more on whether the feature sells cars or not than it does on the viability of the technology. The 4WAS makes the steering and handling characteristics on this long-wheelbase car more nimble and agile compared to models without this technology.

SUMMARY

- Manual or power rack and pinion steering systems are lighter and more compact than recirculating ball steering gears and parallelogram steering linkages.
- The rack and pinion steering gear must be mounted so the rack maintains the tie rods in a parallel position in relation to the lower control arms.
- The rack takes the place of the idler arm and pitman arm in a parallelogram steering linkage.
- The pinion teeth may be spur or helical. The inner tie rods are connected to the rack through spring-loaded inner ball sockets. The rack bearing and adjuster plug maintains proper preload between the rack and pinion teeth.
- A rack and pinion steering gear may be mounted on the firewall or front crossmember. A rack and pinion steering system has fewer friction points than a recirculating ball steering gear and parallelogram steering linkage.
- Reducing the number of friction points in a steering system provides more road feel and reduces the steering system's ability to isolate road vibration and noise.
- Steering ratio is the relationship between steering wheel movement and front wheel movement to the right or left.
- During a left turn with a power rack and pinion steering gear, fluid pressure is directed to the left side of the rack piston, and fluid is released from the right side of this piston. This action moves the rack to the right to complete the left turn.
- During a right turn with a power rack and pinion steering gear, fluid pressure is directed to the right side of the rack piston, and fluid is released from the left side of this piston. Under this condition, the rack is forced to the left to help the driver complete the right turn.
- The fluid pressure in a power rack and pinion steering gear is directed to the appropriate side of the rack piston by the spool valve and rotary valve position.
- The spool valve position is controlled by torsion bar twisting during a right or left turn. If hydraulic pressure is not available in a power rack and pinion steering gear, stop tangs on the stub shaft and pinion provide steering action, but steering effort is much higher.
- The differences between Saginaw and TRW power rack and pinion steering gears are mainly in the method of tie rod attachment, bulkhead oil seal and retainer, and upper and lower pinion bearing hardware.
- The Toyota power rack and pinion steering gear has removable control valve housing and claw washers to lock the inner tie rods to the rack.
- The electronic variable orifice (EVO) steering system provides high-power steering assistance during low-speed cornering and parking and normal power assistance at higher speeds.
- The two inputs in the EVO system are the steering wheel rotation sensor and the vehicle speed sensor.
- The rear wheel steering system in a 4WS system is usually electronically controlled.
- Steering the rear wheels in the opposite direction to the front wheels at low speed provides a shorter turning circle, or radius, for easier maneuvering.
- Steering the rear wheels in the same direction as the front wheels at higher speeds reduces rear sideslip and improves vehicle stability while cornering or changing lanes.
- In an electronically controlled 4WS system, there is no mechanical connection between the front steering gear and the rear steering gear.
- In an electronically controlled 4WS system, the rear steering actuator contains an electric motor that is controlled by the 4WS control unit in response to various input sensor signals.
- In a 4WS system, the control unit uses inputs from the vehicle speed sensor (VSS) and the front steering position sensor to control the rear steering actuator.
- A higher output alternator is required on a 4WS system because of the higher current draw of the rear actuator.
- Negative phase steering occurs at lower speeds when the rear wheels are steered in the opposite direction as the front wheels.
- Positive phase steering occurs at higher speeds when the rear wheels are steered in the same direction as the front wheels.
- Steering the rear wheels in the same direction as the front wheels at higher speeds reduces the lateral forces applied to the rear of a vehicle, and this action reduces yaw velocity.
- When towing a trailer, steering the rear wheels in the same direction as the front wheels at higher speeds improves trailer stability.

REVIEW QUESTIONS

Short Answer Essays

1. Explain why the rack and pinion steering gear is ideally suited for unibody front-wheel-drive vehicles.
2. Explain the advantage of a 4WS system while parking a vehicle.
3. Explain the purpose of the front main steering wheel angle sensor in an electronically controlled 4WS system.
4. Explain the purpose of the bellows boots in a rack and pinion steering gear.
5. Describe two possible mounting positions for rack and pinion steering gears.
6. Explain why a rack and pinion steering gear provides improved feel of the road compared with a recirculating ball steering gear and parallelogram steering linkage.
7. Describe the fluid movement in the rack piston chamber during a right turn.
8. Explain how the spool valve is moved inside the rotary valve in a power rack and pinion steering gear.
9. Explain the difference between spur and helical pinion gear teeth and explain the advantage of the helical design.
10. Describe the driving conditions required for an electronic variable orifice (EVO) steering system to provide high-power steering assistance.

Fill-in-the-Blanks

1. The preload on the rack and pinion teeth affects steering harshness, noise, and_______________.
2. Compared to a recirculating ball steering gear and parallelogram steering linkage, a rack and pinion steering system has a reduced ability to isolate road noise and vibration because the rack and pinion system has fewer_______________.
3. During a left turn, the power steering pump forces fluid into the _______________ side of the rack chamber, and fluid is removed from the _______________ side of this chamber if the rack is mounted behind the front wheel spindle.
4. During a turn, fluid is directed to the appropriate side of the rack piston chamber by the _______________ valve and _______________ valve position.
5. When the torsion bar twists, it changes the position of the _______________ valve.
6. In an electronically controlled 4WS system, the electronic module operates a(n)_______________ _______________ to steer the rear wheels.
7. The front main steering angle sensor in an electronically controlled 4WS system is mounted in the_______________ _______________.
8. In an EVO steering system, the power steering assistance is decreased at higher speeds to provide improved_______________ _______________.
9. The input sensors in an EVO steering system are the vehicle speed sensor and the_______________ _______________ _______________ _______________.
10. During negative phase steering, the rear wheels are steered in the _______________ _______________ to the front wheels.

Multiple Choice

1. Compared to a recirculating ball steering gear, a rack and pinion steering gear provides:
   A. Improved capability to isolate road noise.
   B. Better capability to reduce steering harshness on road irregularities.
   C. Increased capability to absorb steering kickback.
   D. Increased road feel.
2. All of these statements about a 4WS system are true EXCEPT:
   A. A 4WS system reduces yaw forces on the vehicle at higher speeds.
   B. A 4WS system provides a smaller turning circle.
   C. A 4WS system reduces tire wear.
   D. A 4WS system improves vehicle maneuverability at low speeds.

3. All of these statements about a power rack and pinion steering gear mounted behind *the front wheel spindle are* true EXCEPT:
   A. During a left turn fluid pressure is directed to the left side of the rack piston.
   B. During a right turn fluid pressure is directed to the right side of the rack piston.
   C. Fluid is returned through the vent tube positioned in the bellows boot.
   D. The spool valve is moved by torsion bar twisting during a turn.
4. In a column-drive electric power steering system all of the following are true EXCEPT:
   A. A motor module assembly is bolted to the lower end of the steering column.
   B. The steering shaft torque sensor is the most important input used by the power steering control module.
   C. At low speeds, more steering assist is provided.
   D. The steering system may have a 20:1 steering ratio during low-speed cornering.
5. In a power rack and pinion steering gear the breather tube:
   A. Vents the right rack piston chamber during a left turn.
   B. Moves air from one boot to the other during a turn.
   C. Allows air into the rotary valve area while driving straight ahead.
   D. Prevents pressure buildup in both rack piston chambers when driving straight ahead.
6. The power steering pump belt breaks on a rack and pinion power steering system. Under this condition:
   A. The torsion bar may be broken when the front wheels are turned to the right or left.
   B. The steering operates like a manual steering system, and steering effort is higher.
   C. The spool valve no longer moves inside the rotary valve when the steering wheel is turned.
   D. Power steering fluid may be forced from the rack seals when the steering wheel is turned.
7. In electronic variable orifice (EVO) steering systems:
   A. Steering assistance is increased if the vehicle is cornering at high speeds.
   B. Steering assistance is increased if the vehicle is driven straight ahead at highway speeds.
   C. Full power steering assistance is provided if the vehicle speed is below 10 mph (16 km/h) and steering wheel rotation is above 15 rpm.
   D. The control unit increases steering assistance at higher speed to provide improved road feel.
8. If the lock-to-lock steering wheel rotation is 3.25 turns and the total front wheel movement is 78 degrees, the steering ratio is:
   A. 15.5:1
   B. 15:1
   C. 16:1
   D. 19.5:1
9. If an electrical defect occurs in a 4WS system, the:
   A. Rear wheels remain in the centered position.
   B. Positive phase rear wheel steering is canceled.
   C. Negative phase rear wheel steering is increased.
   D. Positive phase rear wheel steering angle is increased, and negative phase rear wheel steering is decreased.
10. All of the following are true about a Magnasteer power steering system EXCEPT:
    A. As vehicle speed changes or lateral acceleration occurs the amount of effort required to steer the vehicle can be varied depending on the requirements.
    B. At high speeds more assist is provided, which provides better road feel, firmer steering, and improved directional stability.
    C. If the system senses lateral acceleration the steering effort will become firmer in an effort to reduce oversteering by the driver.
    D. During low-speed operation increased steering assist is provided for minimal steering effort to facilitate parking maneuvers and easy turning.

CHAPTER 14
RECIRCULATING BALL STEERING GEARS

Upon completion and review of this chapter, you should be able to understand and describe:

- The purpose of a steering gear.
- The advantage of a recirculating ball steering gear compared with earlier worm and roller or cam and lever steering gears.
- The purpose of the worm shaft preload adjustment.
- How the sector shaft is rotated in a manual recirculating ball steering gear.
- The purpose of the interference fit between the sector shaft teeth and the recirculating ball teeth, and explain how this fit is obtained.
- The difference between constant ratio sector teeth and variable ratio sector teeth in a recirculating ball steering gear, and explain the advantage of the variable ratio sector teeth.
- Gear ratio.
- The term "faster" steering as it relates to steering gears.
- The power steering fluid movement in a power recirculating ball steering gear with the engine running and the front wheels straight ahead.
- The power steering fluid movement in a power recirculating ball steering gear during a left turn.
- The power steering fluid movement during a right turn.
- How kickback action is prevented in a power recirculating ball steering gear.

Terms To Know

Faster steering
Gear ratio
Gear tooth backlash
Interference fit
Kickback
Pitman shaft sector
Ross cam and lever steering gear
Rotary valve
Sector
Spool valve
Steering gear free play
Torsion bar
Vehicle wander
Worm and gear
Worm and roller steering gear
Worm and sector design
Worm shaft
Worm shaft bearing preload
Worm shaft end play

INTRODUCTION

In a **worm and gear** or **worm and sector design**, the sector teeth are in direct contact with the worm.

The purpose of the steering gear box is to provide a mechanical advantage that allows the driver to turn the front wheels with a reasonable amount of effort. In the early 1900s, steering gears were a **worm and gear** or **worm and sector design**. These steering gears gave the driver a mechanical advantage to turn the front wheels, but they created a lot of friction.

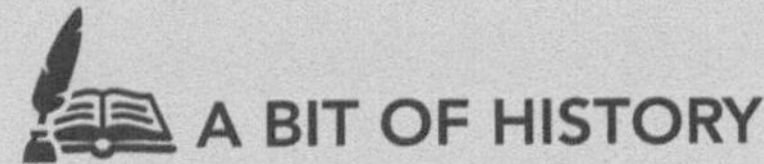

A BIT OF HISTORY

Steering gear design progressed from the crude, high-friction worm and gear of the early 1900s to the Ross cam and lever gear and Saginaw worm and roller gear of the 1920s. The Saginaw worm and roller steering gear was the forerunner of the modern, low-friction recirculating ball steering gear.

Steering **kickback** refers to a strong and sudden movement of the steering wheel in the opposite direction to which the steering wheel is turned. This kickback action tends to occur if a front wheel strikes a road irregularity during a turn.

The **Ross cam and lever steering gear** was introduced in 1923. The cam in this gear was a spiral groove machined into the end of the steering shaft. A pin on the pitman shaft was mounted in the spiral groove in the steering shaft. When the steering wheel and shaft were turned, the pin was forced to move, and this action rotated the pitman shaft. When a front wheel struck a road irregularity, this steering gear design prevented serious **kickback** on the steering wheel. However, this steering gear design still created a considerable amount of friction and required higher steering effort.

In the mid-1920s, Saginaw Steering Division of General Motors Corporation developed the **worm and roller steering gear**. In this steering gear, the **sector** became a roller, which greatly reduced friction and steering effort. The Saginaw worm and roller steering gear was the forerunner of the recirculating ball steering gear that has been widely used on rear-wheel-drive cars for many years.

A **sector** may be defined as a part of a gear.

The **worm shaft** is a spiral gear connected through the steering shaft to the steering wheel.

Worm shaft bearing preload is a condition where all end play is removed and there is a slight tension placed on the bearing.

MANUAL RECIRCULATING BALL STEERING GEARS

Design and Operation

In a recirculating ball steering gear, the steering wheel and shaft are connected to the **worm shaft**. Ball bearings support both ends of the worm shaft in the steering gear housing. A seal above the upper worm shaft bearing prevents oil leaks, and an adjusting plug is provided on the upper worm shaft bearing to adjust **worm shaft bearing preload**. Proper preloading of the worm shaft bearing is necessary to eliminate **worm shaft end play** and to prevent **steering gear free play** and **vehicle wander**. A ball nut is mounted over the worm shaft, and internal threads or grooves on the ball nut match the grooves on the worm shaft. Ball bearings run in ball nut and worm shaft grooves (**Figure 14-1**).

Worm shaft end play refers to the movement of this shaft between the bearings on which the shaft is mounted.

Steering gear free play refers to the amount of steering wheel rotation before the front wheels begin to turn right or left.

When the worm shaft is rotated by the steering wheel, the ball nut is moved up or down on the worm shaft. The gear teeth on the ball nut are meshed with matching gear teeth on the **pitman shaft sector**. Therefore, ball nut movement causes pitman shaft sector rotation. Since the pitman shaft sector is connected through the pitman arm and steering linkage to the front wheels, the front wheels are turned by the pitman shaft sector. The lower end of the pitman shaft sector is usually supported by a bushing or a needle bearing in the steering gear housing. A bushing in the side cover supports the upper end of this shaft.

Vehicle wander is the tendency of a vehicle to steer to the right or left as it is driven straight ahead.

When the front wheels are straight ahead, an **interference fit** exists between the sector shaft teeth and ball nut teeth. This interference fit eliminates **gear tooth backlash** when the front wheels are straight ahead and provides the driver with a positive feel of the road. Proper axial adjustment of the sector shaft is necessary to obtain the necessary interference fit between the sector shaft and worm shaft teeth. A sector shaft adjuster screw is threaded into the side cover to provide axial sector shaft adjustment (**Figure 14-2**).

The **pitman shaft sector** is meshed with the ball nut teeth.

Manual recirculating ball steering gears have sector gear teeth designed to provide a constant ratio, whereas power recirculating ball steering gears usually have sector gear teeth with a variable ratio (**Figure 14-3**). The sector gear teeth have equal lengths in a constant ratio steering gear, but the center sector gear tooth is longer compared with the other teeth in a variable ratio gear. The variable ratio steering gear varies the amount of

Gear tooth backlash refers to the movement between gear teeth that are meshed with each other.

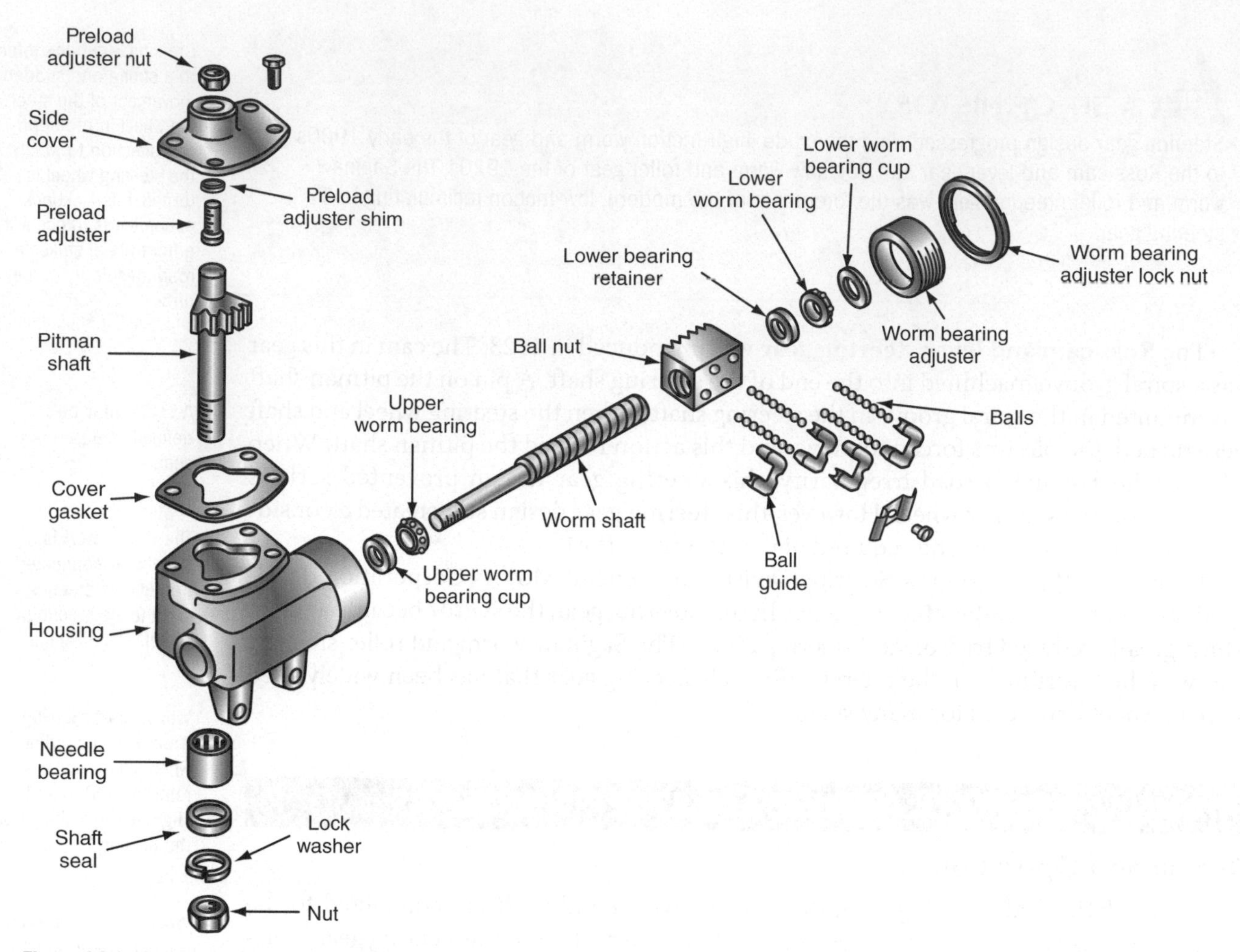

Figure 14-1 Manual recirculating ball steering gear design.

A steering gear with a lower numerical ratio may be called a **faster steering** gear compared with a steering gear with a higher numerical ratio.

Gear ratio refers to the relationship between the rotation of the drive and driven gears. If 13 turns of the drive gear are necessary to obtain one turn of the driven gear, the gear ratio is 13:1.

The **rotary valve** is mounted over the top of the spool valve in a steering gear.

mechanical advantage provided by the steering gear in relation to steering wheel position. This variable ratio provides **faster steering**. The steering **gear ratio** in a constant ratio manual steering gear is usually 15:1 or 16:1, whereas the average variable ratio steering gear ratio may be 13:1. When the same types of steering gears are compared, a higher numerical ratio provides reduced steering effort and increased steering wheel movement in relation to the amount of front wheel movement.

Many recirculating ball steering gears are bolted to the frame with hard steel bolts (**Figure 14-4**). These bolts must be tightened to the vehicle manufacturer's specified torque.

POWER RECIRCULATING BALL STEERING GEARS

Design and Operation

The ball nut and pitman shaft sector are similar in manual and power recirculating ball steering gears. In the power steering gear, a **torsion bar** is connected between the steering shaft and the worm shaft. Since the front wheels are resting on the road surface, they resist turning, and the parts attached to the worm shaft also resist turning. This turning resistance causes torsion bar twist when the wheels are turned, and this twist is limited to a predetermined amount. The worm shaft is connected to the **rotary valve** body, and the torsion bar pin also connects the torsion bar to the worm shaft. The upper end of the torsion bar is attached to the steering shaft and wheel. A stub shaft is mounted inside the

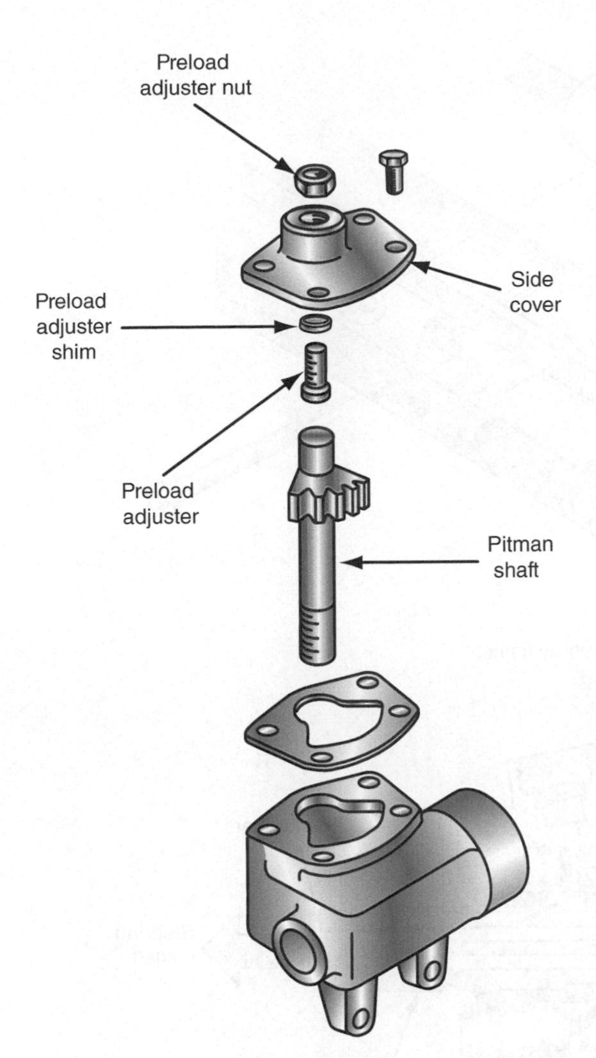

Figure 14-2 Sector shaft adjusting nut.

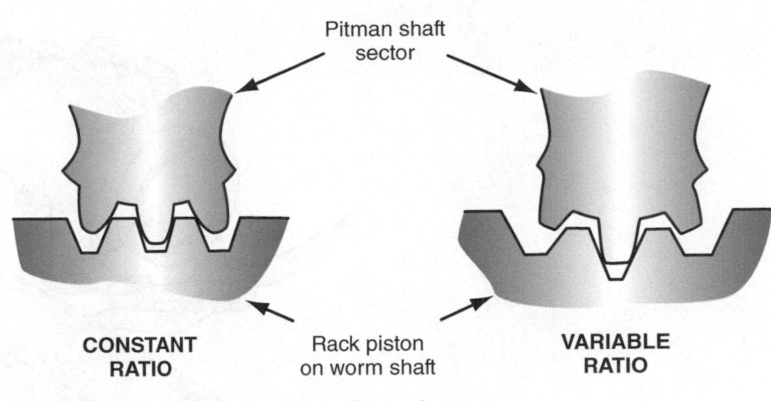

Figure 14-3 Constant and variable ratio steering gears.

The **spool valve** is mounted inside the rotary valve. The twisting action of the torsion bar moves the spool valve in relation to the rotary valve.

Shop Manual
Chapter 14, page 611

rotary valve and a pin connects the outer end of this shaft to the torsion bar. The pin on the inner end of the stub shaft is connected to the **spool valve** in the center of the rotary valve (**Figure 14-5**).

WARNING Many recirculating ball steering gears are mounted near the exhaust manifold, which can be extremely hot. Use caution and wear protective gloves when inspecting or servicing the steering gear to avoid burns to hands and arms.

When the car is driven with the front wheels straight ahead, oil flows from the power steering pump through the spool valve, rotary valve, and low-pressure return line to the pump inlet (**Figure 14-6**). In the straight-ahead steering gear position, oil pressure is equal on both sides of the recirculating ball piston, and the oil acts as a cushion that prevents road shocks from reaching the steering wheel.

If the driver makes a left turn, torsion bar twist moves the valve spool inside the rotary valve body so that oil flow is directed through the rotary valve to the left-turn holes in the spool valve (**Figure 14-7**). Since power steering fluid is directed from these left-turn holes to the upper side of the recirculating ball piston (**Figure 14-8**), this hydraulic pressure on the piston assists the driver in turning the wheels to the left.

When the driver makes a right turn, torsion bar twist moves the spool valve so that oil flows through the spool valve, rotary valve, and a passage in the housing to the pressure

Caution

Hard steel bolts can be used for steering gear mounting. Bolt hardness is indicated by the number of ribs on the bolt head. Harder bolts have five, six, or seven ribs on the bolt heads. Never substitute softer steel bolts in place of the original harder bolts, because these softer bolts may break, allowing the steering gear box to detach from the frame. This action results in a loss of steering control.

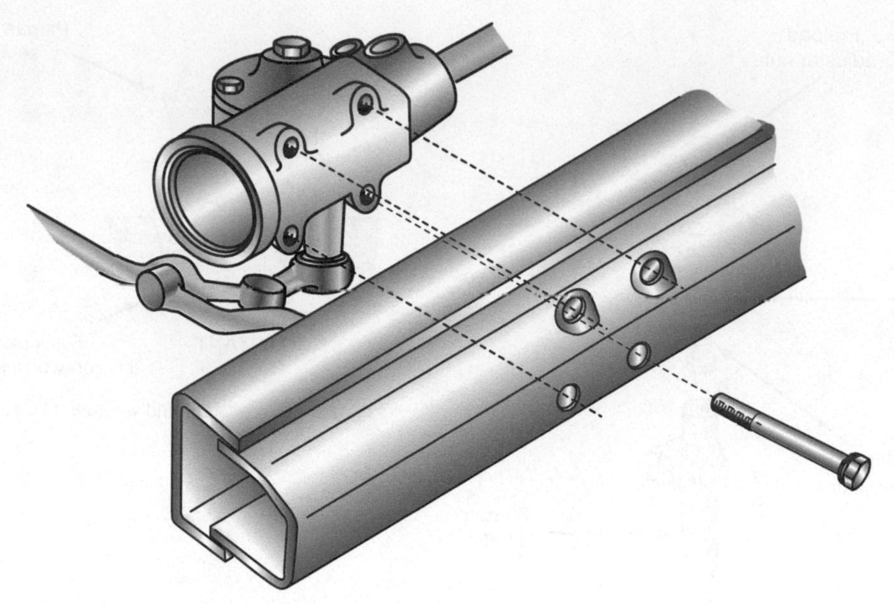

Figure 14-4 Steering gear mounting on the vehicle frame.

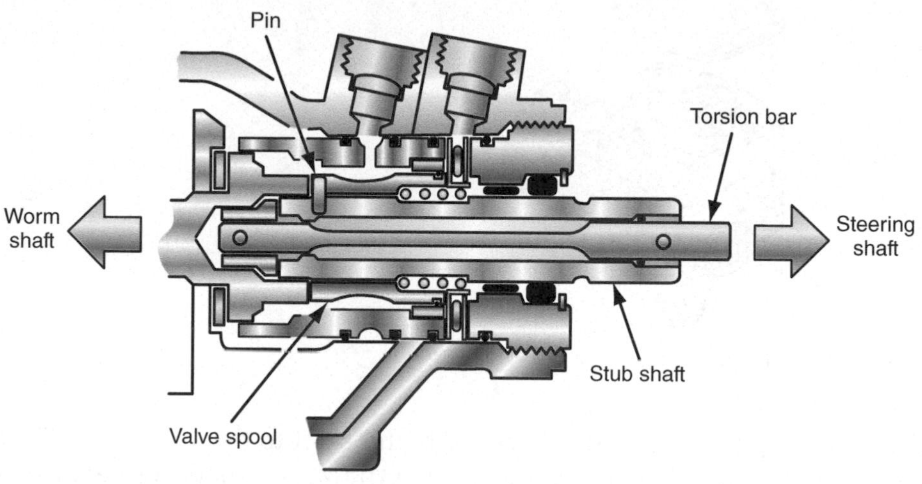

Figure 14-5 Torsion bar and stub shaft.

chamber at the lower end of the ball nut piston (**Figure 14-9**). During a right turn, hydraulic pressure applied to the lower end of the recirculating ball piston helps the driver turn the wheels.

During a turn, if a front wheel strikes a bump and the front wheels are driven in the direction opposite the turning direction, the recirculating ball piston tends to move against the hydraulic pressure and force oil back out the pressure inlet port. This action would create a kickback on the steering wheel, but a poppet valve in the pressure inlet fitting closes and prevents kickback action.

AUTHOR'S NOTE After road testing many vehicles to diagnose steering problems, it has been my experience that the rack and pinion steering gear and related linkage tends to transfer more road shock from the front wheels to the steering wheel compared with a recirculating ball steering gear. The design of the recirculating ball steering gear and the related steering linkage resists the transfer of road shock from the front wheels to the steering wheel. The reason for this resistance is the linkage design in which the steering arms are connected through the tie rods to the center link, and this link is connected through the pitman arm to the steering gear. In a rack and pinion steering gear, the tie rods are connected directly between the steering arms and the rack, and the rack is connected through the pinion to the steering wheel. Compared with a recirculating ball steering gear, a rack and pinion steering gear with its related linkage has fewer friction points and a more direct connection between the front wheels and the steering wheel. However, the rack and pinion steering gear does provide the driver with a greater feel of the road.

You need to be aware of the characteristics of these steering systems when diagnosing steering problems. When you understand the basic characteristics of these steering systems, you are able to immediately recognize normal and abnormal conditions, and this allows you to provide a fast, accurate diagnosis.

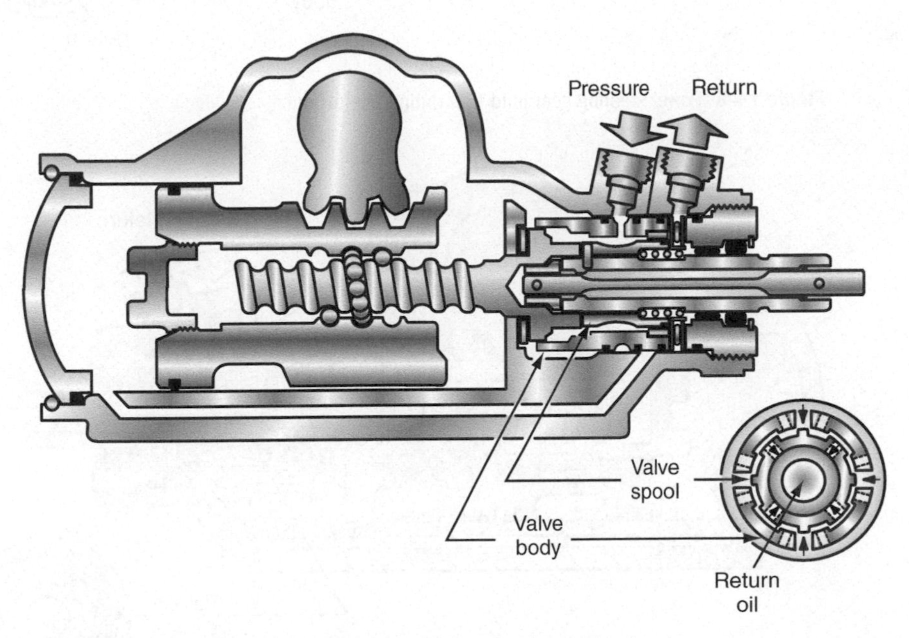

Figure 14-6 Power steering gear fluid flow with the wheels straight ahead.

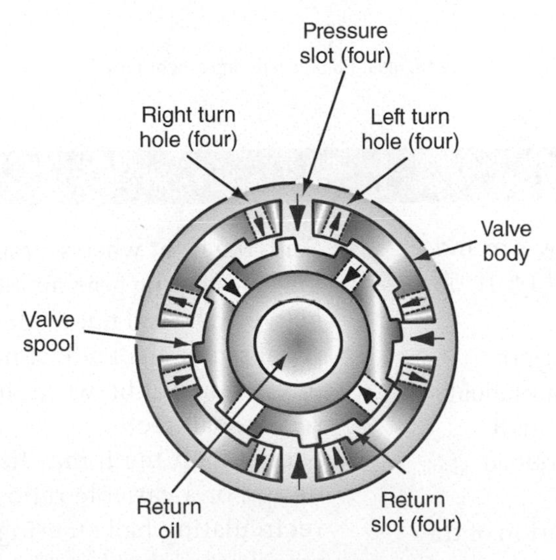

Figure 14-7 Spool valve position during a left turn.

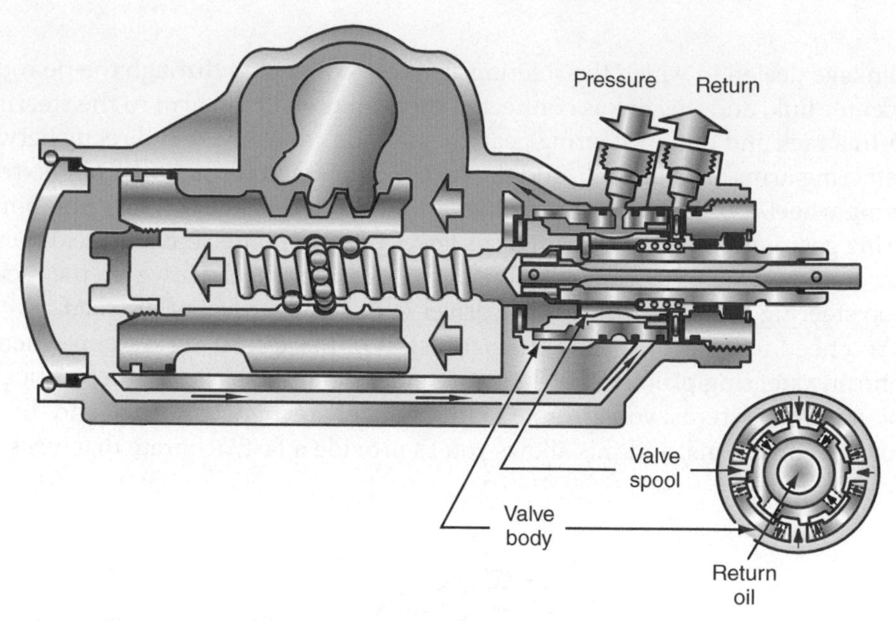

Figure 14-8 Power steering gear fluid flow during a left turn.

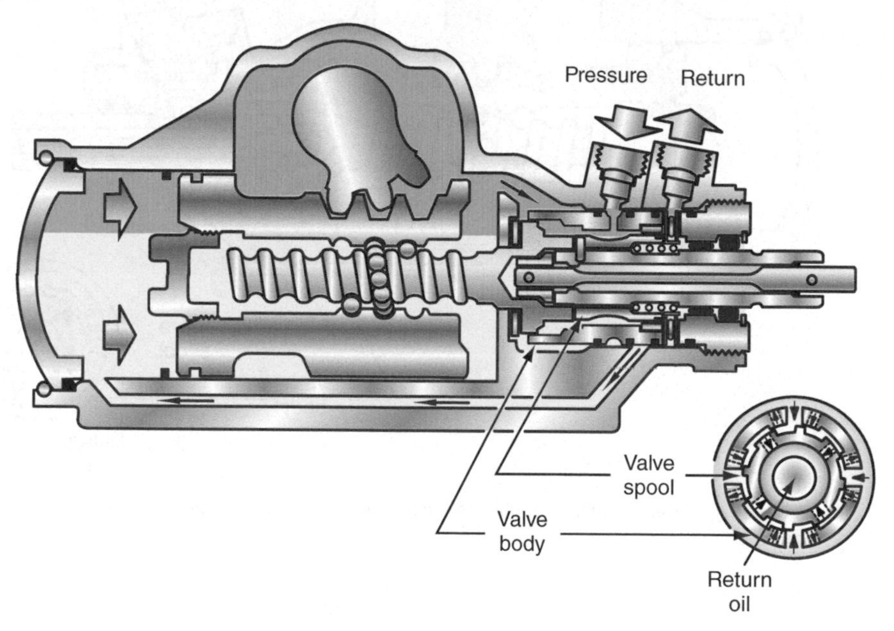

Figure 14-9 Power steering gear fluid flow during a right turn.

SUMMARY

- Early model steering gears such as the worm and gear or Ross cam and lever created a lot of friction and required higher steering effort.
- The modern recirculating ball steering gear reduces friction and steering effort. In a manual recirculating ball steering gear, a worm shaft adjusting plug provides a worm shaft preload adjustment.
- In a recirculating ball steering gear, rotation of the steering wheel causes the ball nut to move up and down on the worm shaft.
- With the front wheels straight ahead in a recirculating ball steering gear, an interference fit exists between the ball nut teeth and the sector shaft teeth.
- Axial sector shaft adjustment provides the proper interference fit between the ball nut teeth and the sector shaft teeth.
- Sector shaft teeth may have a constant ratio design or a variable ratio design. In a power recirculating ball steering gear with the front wheels straight ahead, equal pressure is applied to both sides of the recirculating ball piston, and

the fluid from the pump is directed through the spool valve and rotary valve to the return hose and pump inlet.

- In a power recirculating ball steering gear, the spool valve movement inside the rotary valve is controlled by torsion bar twist.
- When the front wheels are turned in a power recirculating ball steering gear, torsion bar twist moves the spool valve inside the rotary valve, and this valve movement directs the power steering fluid to the appropriate side of the recirculating ball piston to assist steering.

REVIEW QUESTIONS

Short Answer Essays

1. Describe the advantage of the manual recirculating ball steering gear compared with the previous cam and lever steering gear, and explain how this advantage is obtained.
2. Explain three purposes of the worm shaft preload adjustment.
3. Explain the purpose of the interference fit between the ball nut teeth and the sector shaft teeth in a manual recirculating ball steering gear, and describe how this interference fit is obtained.
4. Explain the difference in design between constant ratio and variable ratio sector shaft teeth.
5. Define the term faster steering.
6. Define steering gear ratio, and explain the effect of this ratio on steering effort with a manual recirculating ball steering gear.
7. Describe the flow of power steering fluid with the engine running and the front wheels straight ahead in a power recirculating ball steering gear.
8. Describe how torsion bar twist occurs in a recirculating ball steering gear.
9. Explain the purpose of torsion bar twist in a power recirculating ball steering gear.
10. Explain how kickbacks are prevented in a power recirculating ball steering gear.

Fill-in-the-Blanks

1. A manual recirculating ball steering gear provides reduced ______________ and ______________ compared with the earlier cam and lever manual steering gear.
2. If a bearing has preload, there is a slight ______________ on the bearing.
3. When the steering has free play, there is some ______________ movement before the front wheels start to turn.
4. The worm shaft end plug provides a worm shaft bearing ______________ adjustment.
5. In a manual recirculating ball steering gear, the interference fit between the ball nut and sector shaft teeth is obtained by proper ______________ sector shaft adjustment.
6. Compared with a constant ratio steering gear with a ratio of 13:1, a constant ratio steering gear with a ratio of 16:1 requires ______________ steering effort.
7. When a faster steering system is compared with a slower steering system, the faster system provides more front wheel movement with ______________ steering wheel rotation.
8. A variable ratio steering gear varies the amount of ______________ in relation to steering wheel position.
9. In a power recirculating ball steering gear with the front wheels straight ahead and the engine running, the fluid on each side of the recirculating ball piston helps prevent ______________ from reaching the steering wheel.
10. In a power recirculating ball steering gear during a turn, the ______________ movement directs the power steering fluid pressure to the appropriate side of the recirculating ball piston.

Multiple Choice

1. In a power recirculating ball steering gear, kickback is prevented by:
   A. The preload between the ball nut teeth and sector gear teeth.
   B. The bending of the torsion bar.
   C. A poppet valve in the pressure inlet fitting.
   D. The flexible coupling in the steering shaft.

2. When the worm shaft and bearings are properly installed and adjusted in a power recirculating ball steering gear:
   A. The worm shaft bearings should be preloaded.
   B. The worm shaft should have the specified endplay.
   C. The worm shaft endplay should be eliminated without any preload on the bearings.
   D. There must be an interference fit between the ball nut teeth and the worm shaft teeth.
3. All of these statements about recirculating ball steering gears are true EXCEPT:
   A. The steering gear provides a mechanical advantage.
   B. The steering gear reduces steering effort.
   C. Some early model steering gears had a worm and sector design.
   D. Ross cam and lever steering gears provided a very low internal friction.
4. A typical steering gear ratio in a manual steering gear is:
   A. 16:1.
   B. 18:1.
   C. 19:1.
   D. 22:1.
5. When comparing manual recirculating ball steering gears a higher numerical ratio provides:
   A. Increased steering effort.
   B. Increased steering wheel movement in relation to front wheel movement.
   C. Reduced kickback force on the steering wheel.
   D. Improved steering control.
6. In a manual recirculating ball steering gear:
   A. The interference fit between the sector shaft teeth and ball nut teeth is present when the front wheels are straight ahead.
   B. The proper interference fit between the sector shaft teeth and ball nut teeth is obtained by radial sector shaft movement.
   C. The interference fit between the sector shaft teeth and ball nut teeth becomes tighter when the front wheels are turned to the right or left.
   D. Tightening the worm shaft adjusting plug tightens the interference fit between the sector shaft teeth and ball nut teeth.
7. All of these statements about steering gear ratios are true EXCEPT:
   A. When a steering gear with a 15:1 ratio is compared with a steering gear with a 13:1 ratio, the gear with the 15:1 ratio requires more steering wheel rotation to turn the front wheels.
   B. When a steering gear with a 15:1 ratio is compared with a steering gear with a 13:1 ratio, the gear with the 15:1 ratio provides reduced steering effort.
   C. When a steering gear with a 15:1 ratio is compared with a steering gear with a 13:1 ratio, the gear with the 15:1 ratio may be called a faster steering gear.
   D. The 13:1 steering gear ratio may be used on a smaller, lighter car and the 15:1 steering gear ratio may be used on a larger, heavier vehicle.
8. In power recirculating ball steering gears:
   A. A variable ratio steering gear has sector shaft teeth with equal lengths.
   B. A variable ratio steering gear has a constant mechanical advantage regardless of steering wheel position.
   C. The torsion bar is connected between the steering shaft and the sector shaft.
   D. During a turn, power steering fluid is directed to the proper end of the ball nut piston by spool valve movement.
9. While driving with the front wheels straight ahead with a power recirculating ball steering gear:
   A. The torsion bar is twisted to move the spool valve inside the valve body.
   B. The fluid pressure is higher on the upper side of the ball nut piston compared with the lower side.
   C. The fluid pressure on each side of the ball nut piston cushions road shocks from reaching the steering wheel.
   D. The return passage from the steering gear valve body to the power steering pump is nearly closed.

10. When a vehicle with a power recirculating ball steering gear is making a right turn:
    A. Torsion bar twisting and spool valve movement has moved the spool valve and aligned the pressure ports with the right-turn ports.
    B. Fluid pressure is supplied to the upper side of the ball nut piston.
    C. Fluid pressure is exhausted from the lower side of the ball nut piston.
    D. Fluid pressure gradually leaks past the ball nut piston to reduce steering assist during long, gradual turns.

GLOSSARY

Note: **Terms are highlighted in color**, followed by **Spanish translation in bold**.

Accelerometer An input sensor that senses vehicle acceleration in a computer-controlled suspension system.

Acelerómetro Sensor de entrada que advierte la aceleración del vehículo en un sistema de suspensión controlado por computadora.

Active coils Located in the center area of a coil spring.

Muelles activos Ubicado en el centro de un muelle espiral.

Active suspension system A computer-controlled suspension system with double-acting solenoids at each wheel.

Sistema activo de suspensión Sistema de suspensión controlado por computadora con solenoides de doble acción montados en cada una de las ruedas.

Actuator An electronically operated solenoid in a strut or shock absorber that controls firmness.

Actuador Un solenoide operado electrónicamente en un tirante o un amortiguador que controla la firmeza.

Adaptive cruise control system Maintains a specific distance between the vehicle being driven and the vehicle being followed.

Sistema adaptable de marcha a velocidad de crucero Mantiene una distancia específica entre el vehículo que se conduce y el vehículo que se sigue.

Adjustable strut A strut with a manual-operated adjustment for strut firmness.

Tirante ajustable Un tirante de ajuste operado por mano que determina la firmeza del tirante.

Air bag deployment module The air bag and deployment canister assembly that is mounted in the steering wheel for the driver's side air bag or in the dash panel for the passenger's side air bag.

Unidad de despliegue del Airbag El conjunto del Airbag y elemento de despliegue montado en el volante de dirección para proteger al conductor, o en el tablero de instrumentos para proteger al pasajero.

Air spring An air-filled membrane that replaces the conventional coil springs in an air suspension system.

Muelle de aire Membrana llena de aire que reemplaza los muelles helicoidales convencionales en un sistema de suspensión de aire.

Air spring solenoid valve An electrically operated solenoid that allows air to flow in and out of an air spring.

Muelle de aire con válvula activada por solenoide Solenoide activado eléctricamente que permite el flujo libre de aire en un muelle de aire.

All-season tires Tires with special tread designed to improve traction on snow or ice, while providing acceptable noise levels on smooth road surfaces.

Neumáticos para toda época Neumáticos con una huella especial diseñada para mejorar la tracción en la nieve o en el hielo, a la misma vez que provee niveles aceptables de ruido cuando se conduce el vehículo en un camino cuya superficie es lisa.

American Petroleum Institute (API) An organization in charge of engine oil classifications and many other areas in the petroleum industry.

Instituto Americano del Petróleo (API) Organización que tiene a su cargo la clasificación del aceite de motor y varias otras áreas en la industria del petróleo.

Ampere (A) A unit of measurement of electrical current flow.

Amperio (A) El amperio es una unidad de medida del flujo de la corriente eléctrica.

Amplitude The extent of a vibratory movement.

Amplitud El magnitud de un movimiento de vibración.

Analog voltage signal A voltage signal that varies continuously within a specific range.

Señal de voltaje análogo Un señal que varía continuamente dentro de una banda específica.

Angular bearing load A load applied to a bearing in a direction between horizontal and vertical.

Carga del cojinete angular Una carga aplicada a un cojinete en una dirección entre horizontal y vertical.

Antilock brake system (ABS) A computer-controlled brake system that prevents wheel lockup during brake applications.

Sistema antiblocante de frenos Un sistema controlado por computadora que previene el bloqueo en las ruedas durante la aplicación de los frenos.

Asymmetrical An arrangement marked by irregularity such as different shapes, sizes, and positions.

Asimétrico Un arreglo caracterizado por la irregularidad tal como las diferencias de forma, tamaño y posición.

Asymmetrical leaf spring Has the same distance from the spring center bolt to each end of the spring.

Ballesta de hojas asimétricas Tiene la misma distancia desde el perno del centro del resorte a cada extremo del resorte.

Atmospheric pressure The pressure exerted on the earth by the atmosphere, which is 14.7 psi at sea level.

Presión atmosférica Presión que la atmósfera ejerce sobre la tierra. Dicha presión es 14.7 psi [peso sobre unidad de superficie] al nivel del mar.

Automatic level control (ALC) System maintains the rear suspension trim height regardless of the rear suspension load. If a heavy object is placed in the trunk, the rear wheel position sensors send trim height signals to the ESC module.

Sistema de Control de nivel automático (CNA) Mantiene la altura del centrado de la suspensión trasera sin tomar en cuenta la carga de la suspensión trasera. Si un objeto se coloca en la cajuela, los sensores de posición de la rueda trasera envían señales de altura por debajo del centrado al módulo de CSE.

Axle offset A condition in which the complete rear axle assembly has turned so one rear wheel has moved forward and the opposite rear wheel has moved rearward.

Descentralización del eje Condición que ocurre cuando todo el conjunto del eje trasero ha girado de manera que una de las ruedas traseras se ha movido hacia adelante y la opuesta se ha movido hacia atrás.

Axle sideset A condition in which the rear axle assembly has moved sideways from its original position.

Resbalamiento lateral del eje Condición que ocurre cuando el conjunto del eje trasero se ha movido lateralmente desde su posición original.

Axle thrustline Is a line extending forward from the center of the rear axle at a 90° angle.

Eje de directriz de presiones Es una línea que se extiende hacia enfrente desde el centro del eje trasero a un ángulo de 90°.

Axle tramp The repeated lifting of the differential housing during extremely hard acceleration.

Barreta del eje Levantamiento repetido del alojamiento del diferencial durante una aceleración sumamente rápida.

Axle windup The tendency of the rear axle housing to rotate in the opposite direction to the wheel-and-tire rotation during hard vehicle acceleration.

Bobinado del eje La tendencia de la carcasa del eje trasero de rotar en la dirección opuesta a la rotación de la rueda y el neumático durante la aceleración fuerte del vehículo.

Ball bearings Have round steel balls between the inner and outer races.

Cojinete de bolas Tiene bolas de acero entre las carreras interiores y exteriores.

Bead filler A piece of rubber positioned above the bead that reinforces the sidewall and acts as a rim extender.

Relleno de la pestaña de la llanta Pieza de caucho ubicada sobre la pestaña que sirve para reforzar la pared lateral y actúa como una extensión de la llanta.

Bead wire A group of circular wire strands molded into the inner circumference of the tire that anchors the tire on the rim.

Cable de pestaña Grupo de cordones circulares de cables moldeados en la circunferencia interior del neumático que sujetan el neumático a la llanta.

Bearing hub unit A complete wheel bearing and hub unit to which the wheel rim is bolted.

Conjunto de cubo y cojinete Conjunto total de cojinete y cubo de rueda al que se emperna la llanta de la rueda.

Bearing seals Circular metal rings with a sealing lip on the inner edge. Bearing seals are usually mounted on both sides of a bearing to prevent dirt and moisture from entering the bearing while sealing the lubricant in the bearing.

Juntas de estanqueidad del cojinete Anillos metálicos circulares con un reborde de estanqueidad en el borde interior. Normalmente las juntas de estanqueidad del cojinete se montan en ambos lados de un cojinete para evitar la entrada de suciedad y humedad mientras sellan el lubricante dentro del mismo.

Bearing shields Circular metal rings attached to the outer race that prevent dirt from entering the bearing but allow excess lubricant to flow out of the bearing.

Protectores para cojinetes Anillos metálicos circulares fijados a la arandela exterior que evitan la entrada de suciedad al cojinete, pero permiten la salida de exceso de lubricante del mismo.

Belt cover A nylon cover positioned over the belts in a tire that helps to hold the tire together at high speed, and provides longer tire life.

Cubierta de correa Cubierta de nilón ubicada sobre las correas de un neumático. Esta cubierta ayuda a conservar la solidez del neumático cuando se conduce el vehículo a gran velocidad, además de proporcionarle mayor durabilidad.

Belted bias-ply tire A type of tire construction with the cord plies wound at an angle to the center of the tire, and fiberglass or steel belts mounted under the tread area of the tire.

Neumático de correas con estrías diagonales Tipo de neumático fabricado con estrias arregladas a un ángulo con respecto al centro del neumático, y correas de fibra de vidrio o de acero montadas debajo de la huella del mismo.

Bias-ply tire A type of tire construction with the cord plies wound at an angle to the center of the tire and no belts surrounding the cords.

Neumático de estrías diagonales Un tipo de construcción de neumático con estrías arregladas en un ángulo con respecto al centro del neumático y sín correas envolviendo las estrías.

Brake-by-wire A computer-controlled brake system with no direct mechanical connection between the brake pedal and the master cylinder.

Frenado alámbrico Un sistema de frenos controlado por computadora sin una conexión mecánica directa entre el pedal de freno y el cilíndro maestro.

Brake pressure modulator valve (BPMV) A group of solenoid valves mounted in a single valve body that controls fluid pressure in the brake system during antilock brake, traction control, and vehicle stability control functions.

Válvula moderador de presión de frenado (BPMV) Un grupo de válvulas de solenoide montado en una sóla caja de válvula que controla la presión de los fluidos en el sistema de frenos durante las operaciones de freno antibloqueante, control de tracción, y control de la estabilidad del vehículo.

Brake pressure switch An electrical switch operated by brake fluid pressure.

Interruptor de presión del freno Interruptor eléctrico activado por la presión del disco de freno.

Braking and deceleration torque The torque applied to the rear axle assembly during deceleration and braking in a rear-wheel-drive vehicle.

Torsión de deceleración y frenaje La torsión impuesta en el ensamblado del eje trasero durante la deceleración y el frenado en un vehículo de tracción trasera.

Breather tube A small metal tube that allows air to flow between the bellows boots in a rack and pinion steering gear during a turn.

Tubo respiradero Pequeño tubo metálico que permite el flujo libre de aire entre las botas de fuelles en un mecanismo de dirección de cremallera y piñón durante un viraje.

Bump steer The tendency of the steering to veer suddenly in one direction when one or both front wheels strike a bump.

Cambio de dirección ocacionado por promontorios en el terreno Tendencia de la dirección a cambiar repentinamente de sentido cuando una o ambas ruedas delanteras golpea un promontorio.

Cam ring A metal ring with an elliptical-shaped inner surface on which the vanes make contact in a power steering pump.

Anillo de levas Un anillo metálico con una superficie interior de forma elíptica en la cual las aletas hacen contacto en una bomba de dirección hidraulica.

Car access system (CAS) A system that controls specific vehicle functions and operates in conjunction with the safety and gateway module on a vehicle with an active steering system.

Sistema de acceso al automóvil (CAS, en inglés) Un sistema que controla las funciones específicas del vehículo y que opera en conjunción con el módulo de la puerta de enlace y de seguridad de un vehículo con un sistema de dirección activo.

Carbon dioxide (CO_2) A greenhouse gas that is a by-product of gasoline or diesel fuel combustion.

Dióxido de carbono (CO_2) Un gas de efecto invernadero que es un subproducto de la combustión de la gasolina o del combustible diesel.

Caster angle The angle between an imaginary line through the center of the tire and wheel and another imaginary line through the centers of the lower ball joint and upper strut mount viewed from the side.

ángulo de inclinación El ángulo entre una línea imaginaria que pasa por el centro de un neumático y la rueda y otra línea que pasa por los centros de la articulación esférica inferior y la montadura superior del apoyadero visto de un lado.

Center link A long rod connected from the pitman arm to the idler arm in a steering linkage.

Biela motriz central Varilla larga conectada del brazo pitman al brazo auxiliar en un cuadrilátero de la dirección.

Center take-off (CTO) The tie rods are attached to a movable sleeve in the center of the gear.

Conexión central (Center take-off, CTO) Los brazos de dirección se conectan a una manga móvil en el centro del engranaje.

Channel frame A type of frame construction in which a steel channel is positioned on the top and bottom of the vertical frame web.

Armazón fabricado con ranuras Tipo de fabricación del armazón en la que una ranura de acero es colocada en las partes superior e inferior de la malla del armazón vertical.

City safety system On some vehicles, the collision mitigation system is called a city safety system.

Sistema de seguridad de la ciudad En algunos vehículos, el sistema de mitigación de choques se conoce como sistema de seguridad de la ciudad.

Clock spring electrical connector A conductive ribbon in a plastic case mounted on top of the steering column that maintains electrical contact between the air bag inflator module and the air bag electrical system.

Conector eléctrico de cuerda de reloj Cinta conductiva envuelta en una cubierta plástica montada en la parte superior de la columna de dirección, que mantiene contacto eléctrico entre la unidad infladora y el sistema eléctrico del Airbag.

Collision mitigation system Operates specific vehicle functions to help prevent a collision.

Sistema de alivio de colisiones Opera funciones especificas dei vehiculo para ayudar a evitar una colisión.

Column-drive EPS Has an electric assist motor coupled to the steering shaft in the steering column.

Transmisión de columna de GPE (guiado de propulsión electrónica) Tiene un motor de corriente eléctrica que acompaña al eje de dirección en la columna de la dirección.

Compact spare tire A spare tire and wheel that is much smaller than the other tires on the vehicle, and is designed for short-distance driving at low speed.

Neumático compacto de repuesto Conjunto de neumático y rueda de repuesto mucho más pequeño que los demás neumáticos del vehículo, diseñado para viajes de corta distancia a poca velocidad.

Complete box frame A type of vehicle frame shaped like a rectangular steel box.

Fabricación completa Tipo de armazón en forma de caja rectangular de acero.

Compressible A substance is compressible when an increase in pressure causes a decrease in volume.

Compresibile Una sustancia es compresible cuando un aumento en la presión produce una disminución en el volumen.

Compression loaded A suspension ball joint mounted so the vehicle weight is forcing the ball into the joint.

Cargada de presión Junta esférica de la suspensión montada de manera que el peso del vehículo presiona la bola hacia el interior de la junta.

Conicity Tire conicity refers to a condition where the plies and/or belts are not level across the tire tread. When a tire has conicity, the plies and/or belts are somewhat cone shaped and may cause the vehicle to pull in one direction when driven.

Conicidad La conicidad de un neumático se refiere al estado en el que las capas y/o las fajas no están niveladas en la trocha del neumático. Cuando un neumático presenta conicidad, las capas y/o las fajas tienen de alguna manera forma cónica y pueden hacer que el vehículo tire hacia una dirección cuando se conduce.

Controller Area Network (CAN) A high-speed serial bus communication network. The CAN protocol has been standardized by the International Standards Organization (ISO) as ISO 11898 standard for high-speed and ISO 11519 for low-speed data transfer.

Red de área de controlador (CAN) Una red de comunicaciones de bus serial de alta velocidad. El protocolo CAN ha sido estandarizado por la Organización Internacional de Estandarización (ISO) como la norma ISO 11898 para la transferencia de datos a alta velocidad y la norma ISO 11519 para la transferencia de datos a baja velocidad.

Cord plies Surround both tire beads and extend around the inner surface of the tire to enable the tire to carry its load.

Pliegues de cuerda Rodean ambas cejas de los neumáticos y se extienden alrededor de la superficie interor del neumático para permitir que el nuemático soporta su carga.

Cross-axis ball joints Contain large insulating bushings in place of typical ball joints.

Articulaciones de rótula a través del eje Contienen pasa tapas grandes en lugar de las rótulas esféricas comunes.

Cycle A series of events that repeat themselves regularly.

Ciclo Una serie de eventos que se repitan regularmente.

Damper solenoid valve An electronically operated hydraulic valve that controls shock absorber or strut damping.

Válvula amortiguador del solenoide Una válvula hidráulica operada electronicamente que controla el amortiguador o el amortiguamiento del apoyadero.

Data bus network Circuit over which serial data are transmitted.

Red de datos de bus Circuito por el cual se transmiten datos seriales.

Diamond-frame condition A frame that is diamond-shaped from collision damage.

Condición forma de diamante del armazón Armazón que ha adquirido la forma de un diamante a causa del impacto recibido durante una colisión.

Digital motor electronics (DME) A computer that controls specific vehicle functions and operates in conjunction with the safety and gateway module on a vehicle with an active steering system.

Sistema electrónico digital del motor (DME, en inglés) Una computadora que controla funciones específicas del vehículo y opera en conjunción con el módulo de la puerta de enlace y de seguridad de un vehículo con un sistema de dirección activo.

Digital voltage signal A voltage signal that is either high or low.

Señal digital de tensión Señal de tensión que es alta o baja.

Directional stability The tendency of the vehicle steering to remain in the straight-ahead position when driven straight ahead on a smooth, level road surface.

Estabilidad direccional Tendencia de la dirección del vehículo a permanecer en línea recta al ser así conducido en un camino cuya superficie es lisa y nivelada.

Double-row ball bearing A bearing with two adjacent rows of circular steel balls.

Cojinete de bolas de doble hilera Cojinete que contiene dos hileras adyacentes de bolas circulares de acero.

Double wishbone rear suspension system A multilink suspension system in which the links may be shaped like a wishbone.

Suspensión trasera de doble horquilla Sistema de suspensión de empalme múltiple en el que a las bielas motrices se les puede dar forma de espoleta.

Dynamic balance The balance of a wheel in motion.

Equilibrio dinámico Equilibrio de una rueda en movi-miento.

Electric locking unit (ELU) Locks the worm drive and drive motor in an active steering system if a safety-related defect occurs.

Unidad de bloqueo eléctrico (ELU, en inglés) Bloquea la transmisión por tornillo sinfín y el motor de impulsión de un sistema de dirección activo si se produce un defecto relacionado con la seguridad.

Electronic brake and traction control module (EBTCM) A module that controls antilock brake, traction control, and vehicle stability control functions.

Módulo electrónico de control de freno y tracción (EBTCM) Un módulo que controla las funciones de control del freno antibloqueo, control de tracción, y la estabilidad del vehículo.

Electronic control unit (ECU) A computer that receives input signals and controls output functions.

Unidad de control electrónico (ECU, en inglés) Una computadora que recibe señales de entrada y controla las funciones de salida.

Electronic rotary height sensor Has an internal rotating element and this sensor sends voltage signals to the control module in relation to the curb riding height.

Sensores electrónicos rotativos de altura Tienen un elemento rotativo interno, y estos sensores envían señales de voltaje al módulo de control con relación a la curva de la altura de rodaje.

Electronic suspension control (ESC) system Controls damping forces in the front struts and rear shock absorbers in response to various road and driving conditions.

Sistema de control de suspensión electrónica (CSE) Controla las fuerzas de amortiguamiento en las barras transversales frontales y los amortiguadores neumáticos traseros como respuesta a las variadas condiciones de manejo y del camino.

Electronic variable orifice (EVO) steering A computer-controlled power steering system in which the computer operates a solenoid to control power steering pump pressure in relation to vehicle speed.

Dirección de orificio variable electrónico Sistema de dirección hidráulica controlado por computadora en el que la computadora activa un solenoide para controlar la presión de la bomba de la dirección hidráulica de acuerdo a la velocidad del vehículo.

Electronic vibration analyzer (EVA) A tester that measures vibration amplitude.

Analizador electrónico de vibraciones (EVA) Un comprobador que mide el amplitud de las vibraciones.

Electronically controlled orifice (ECO) valve A computer-controlled solenoid valve in the power steering pump that controls fluid flow from the pump to the servotronic valve.

Válvula de orificio controlado electrónicamente (ECO, en inglés) Una válvula solenoide controlada por computadora que se encuentra en la bomba de dirección de potencia que controla el flujo de fluidos desde la bomba hasta la válvula servotrónica.

Electronically controlled 4WS system A four-wheel steering system in which the rear wheel steering is controlled electronically.

Dirección en las cuatro ruedas controlada electrónicamente Sistema de dirección en las cuatro ruedas en el que la dirección de la rueda trasera se controla electrónicamente.

Encoded disc Works with an optical steering angle sensor to supply a voltage signal in relation to steering wheel rotation.

Disco codificado Funciona con un sensor de ángulo de la dirección óptico para proporcionar una señal de voltaje en relación con la rotación del volante.

End take-off (ETO) The tie rods are connected to the ends of the rack.

Conexión en los extremos (End take-off, ETO) Los brazos de dirección se conectan a los extremos de la cremallera.

Energy The ability to do work.

Energía Capacidad para realizar un trabajo.

Energy-absorbing lower bracket A steering column bracket that protects the driver by allowing the column to move away from the driver when impacted by the driver in a collision.

Soporte absorbente de energía Un soporte de la columna de dirección que proteja al conductor permitiendo que la columna se aleja del conductor cuando éste la golpea en una colisión.

Faster steering A steering gear that turns the front wheels from lock-to-lock with less steering wheel rotation.

Dirección más rápida Mecanismo de dirección que gira las ruedas delanteras de un extremo al otro con menos movimiento del volante de dirección.

Road feel A feeling experienced by a driver during a turn when the driver has a positive feeling that the front wheels are turning in the intended direction.

Sensación del camino Sensación experimentada por un conductor durante un viraje cuando está completamente seguro de que las ruedas delanteras están girando en la dirección correcta.

Firm relay A relay in a computer-controlled suspension system that supplies voltage to the strut actuators and moves these actuators to the firm position.

Relé de fijación Relé en un sistema de suspensión controlado por computadora que le provee tensión a los accionadores de los montantes y los lleva a una posición fija.

Five-link suspension A five-link suspension has five attachment locations between each side of the suspension and the frame.

Suspensión de cinco enlaces Una suspensión de cinco enlaces tiene cinco lugares de conexión entre cada lado de la suspensión y el chasis.

Flow control valve A special valve that controls fluid movement in relation to system demands.

Válvula de control de flujo Válvula especial que controla el movimiento del fluido de acuerdo a las exigencias del sistema.

Fluted lip seal A seal lip with fine grooves that direct the lubricant back into the reservoir rather than leaking past the seal lip.

Junta de estanqueidad de reborde estriado Reborde de una junta de estanqueidad con ranuras finas que dirigen el lubricante hacia el tanque en vez de permitir que el mismo se escape del reborde.

Force Implies the exertion of strength, which may be physical or mechanical.

Fuerza Implica el esfuerzo excesivo físico o mecánico.

Four-wheel active steering (4WAS) A steering system in which the rear wheels are steered electronically in relation to the front wheel steering angle and vehicle speed.

Dirección activa con tracción en las cuatro ruedas (4WAS, en inglés) Un sistema de dirección en el que las ruedas traseras se direccionan electrónicamente en relación con el ángulo de dirección de las ruedas delanteras y la velocidad del vehículo.

Four-wheel alignment Measurement and adjustment of wheel alignment angles at the front and rear wheels.

Alineación de cuatro ruedas La medida y el ajuste de los ángulos de alineación de las ruedas en las ruedas delanteras y traseras.

Four-wheel steering (4WS) A steering system in which both the front and rear wheels turn right or left to steer the vehicle.

Dirección en las cuatro ruedas Sistema de dirección en el que tanto las ruedas traseras como las delanteras giran hacia la derecha o la izquierda para dirigir el vehículo.

Frame buckle A frame that is accordion-shaped to some extent from collision damage.

Encorvamiento del armazón Armazón que ha adquirido la forma de un acordeón a causa del impacto recibido durante una colisión.

Frame sag A frame that is bent downward in the center from collision damage.

Hundimiento del armazón Armazón torcido hacia abajo en el centro a causa del impacto recibido durante una colisión.

Frame twist A frame that is bent from collision damage so one corner is higher in relation to the opposite corner.

Torcedura del armazón Armazón torcido a causa del impacto recibido durante una colisión; como resultado de esta torcedura un ángulo será más alto con relación al ángulo opuesto.

Friction The resistance to motion when the surface of one object is moved over the surface of another object.

Fricción Resistencia al movimiento cuando la superficie de un objeto se mueve sobre la superficie de otro.

Full-wire open-end springs Coil spring ends that are cut straight off and sometimes flattened, squared, or ground to a D-shape.

Muelles de extremo abierto con cable cerrado Extremos de muelles helicoidales cortados en línea que a veces son aplanados, cuadrados o afilados en forma de D.

Garter spring A circular spring behind a seal lip.

Muelle jarretera Muelle circular ubicado detrás de un reborde de una junta de estanqueidad.

Gas-filled shock absorber Contains a gas charge and hydraulic fluid.

Amortiguadore lleno de gas Contienens una carga de gas y fluido hidráulico.

Gear ratio The relationship between the drive gear and the driven gear in a gear set.

Relación de engranaje La relación entre el engranaje de impulso y el engranaje mandado en un juego de engranajes.

Gear tooth backlash Movement between gear teeth that are meshed together.

Juego entre los dientes del engranaje Movimiento entre los dientes del engranaje que están endentados entre sí.

Geometric centerline The exact centerline of the vehicle chassis.

Línea central geométrica La línea central exacta del chasis del vehículo.

Geometric centerline alignment A wheel alignment procedure in which the front wheel toe is adjusted to specifications using the geometric centerline as a reference.

Alineación de la línea central geométrica Procedimiento de alineación de las ruedas en el que se ajusta el tope de la rueda delantera según las especificaciones utilizando la línea central geométrica como punto de referencia.

Geometric vehicle centerline An imaginary line through the exact center of the front and rear wheels.

Linea de eje central geométrica Línea imaginaria a través del centro exacto de las ruedas delanteras y traseras.

Greenhouse gas A gas that helps to form a blanket above the earth's atmosphere.

Gas de efecto invernadero Un gas que ayuda a formar un manto por encima de la atmósfera de la tierra.

Gussets Pieces of metal welded into a corner where two pieces of metal meet to increase component strength.

Esquinero Las piezas de metal soldadas en una esquina en donde se juntan dos piezas de metal para aumenter la fuerza del componente.

Hall element An electronic device that produces a voltage signal when the magnetic field approaches or moves away from the element.

Elemento Hall Es un dispositivo electrónico que produce una señal de voltaje cuando el campo magnético se acerca o se aleja del elemento.

Heavy-duty coil spring Compared with conventional coil springs, heavy-duty springs have larger wire diameter and 3 percent to 5 percent greater load-carrying capability.

Muelle helicoidal para servicio pesado Comparado con los muelles helicoidales convencionales, los muelles para servicio pesado tienen cables de diámetro más grande y una capacidad de carga de un 3% a un 5% mayor.

Heavy-duty shock absorbers Compared with conventional shock absorbers, heavy-duty shock absorbers have improved seals, a single tube to reduce heat, and a rising rate valve for precise spring control.

Amortiguadores para servicio pesado Comparados con los amortiguadores convencionales, los amortiguadores para servicio pesado tienen juntas de estanqueidad mejoradas, un tubo único para reducir el calor, y una válvula de vástago ascendente para un control preciso del muelle.

Height sensor Sends a signal to the suspension computer in relation to chassis height.

Sensor de altura Envía una señal a la computadora de suspensión referente a la altura del chasis.

Helical gear teeth Are positioned at an angle in relation to the gear centerline.

Dientes de engranaje helicoidales Ubicados a un ángulo con relación a la línea central del engranaje.

Hertz (Hz) A measurement for the speed at which an electronic signal or vibration cycles from high to low.

Hercio Una medición de la velocidad a la cual una señal electrónica o una vibración pasa de alta a baja.

High-frequency transmitter An electronic device that sends high-frequency voltage signals to a receiver. Some wheel sensors on computer-controlled wheel aligners send high-frequency signals to a receiver in the wheel aligner.

Transmisor de alta frecuencia Un dispositivo electróncio que manda las señales de alta frecuencia a un receptor. Algunos sensores de ruedas en los alineadores de ruedas controlados por computadora mandan los señales de alta frecuencia a un receptor en el alineador de ruedas.

High-strength steels (HSS) Steels that have above-average strength compared with other body components. High-strength steels are used in some unitized body parts.

Aceros de alta resistencia (HSS) Los aceros que tienen una resistencia superior a lo normal comparado a los otros componentes de la carrocería. Los aceros de alta resistencia se usan en algunos partes de un monocasco.

Hold valve A valve that prevents fluid return to the master cylinder during traction control or vehicle stability control functions.

Válvula de retención Una válvula que previene que los fluidos regresan al cilíndro maestro duante las funciones de control de tracción o control de estabilidad del vehículo.

Horsepower A measurement for the amount of power delivered by an internal combustion engine or an electric motor.

Caballo de fuerza Una medida de la cantidad de potencia entregada por un motor de combustión interna o por un motor eléctrico.

Hybrid vehicles Have two power sources that can supply torque to the drive wheels. In most applications the two power sources are a gasoline engine and an electric motor(s).

Vehículos híbridos Tienen dos fuentes de potencia que pueden proporcionar par motor a las ruedas de arranque. En la mayoría de las aplicaciones las dos fuentes de potencia son los motores de gasolina y eléctricos.

Hydroboost power-assisted steering system Uses fluid pressure from the power steering pump to supply brake power assistance.

Sistema de dirección asistido hidráulicamente Utiliza la presión del fluido de la bomba de la dirección hidráulica para reforzar la potencia del freno.

Hydro-forming Is the process of using extreme fluid pressure to shape metal.

Hidroformación Es el proceso de usar presión extrema de líquidos para darle forma a un metal.

Hydroplaning Occurs when water on the pavement is allowed to remain between the pavement and the tire tread contact area. This action reduces friction between the tire tread and the road surface, and can contribute to a loss of steering control.

Hidrodeslizamiento Sucede cuando se deja agua entre el pavimento y la superficie de rodadura. Esta acción reduce la fricción entre la superficie de rodadura y la superficie del camino, y puede contribuir a una pérdida de control en la dirección.

Idler arm A short, pivoted steering arm bolted to the vehicle frame and connected to the steering center link.

Brazo auxiliar Brazo de dirección corto y articulado, empernado al armazón del vehículo y conectado a la biela motriz central de dirección.

Inactive coils Inactive coils located at the top and bottom ends of a coil spring introduce force into the spring when a wheel strikes a road irregularity.

Bobinas inactivas Las bobinas inactivas localizadas en las extremidades superiores e inferiores de un resorte helicoidal introducen la fuerza en el resorte cuando una rueda choque contra una irregularidad en el camino.

Included angle The sum of the camber and steering axis inclination (SAI) angles.

Ángulo incluído La suma de los ángulos de la inclinación y la inclinación del eje de dirección (SAI).

Independent rear suspension A rear suspension system in which one rear wheel moves upward or downward without affecting the opposite rear wheel.

Suspensión trasera independiente Sistema de suspensión trasera en el que una rueda trasera se mueve de manera ascendente o descendente sin afectar la rueda trasera opuesta.

Inertia The tendency of an object to remain at rest, or the tendency of an object to stay in motion.

Inercia La tendencia de un objeto a permanecer inmóvil, o la tendencia de un objeto a continuar en movimiento.

Inner race The race in the center of a ball or tapered roller bearing that supports the balls or rollers.

Anillo interior El anillo en el centro de una bola o de un cojinete de rodillos cónicos que apoya las bolas o los rodillos.

Integral fluid reservoir A reservoir that is part of the power steering pump.

Tanque de líquido integral Tanque que forma parte de la bomba de la dirección hidráulica.

Integral power-assisted steering system A power steering system in which the power-assisted components are integral with the steering gear.

Sistema de dirección de astencia hidráulica integral Sistema de dirección hidráulica en el que los componentes asistidos forman parte integral del mecanismo de dirección.

Integral reservoir A reservoir that is part of the power steering pump.

Tanque de la bomba Tanque que forma parte de la bomba de la dirección hidráulica.

Interference fit A precision fit between two components that provides a specific amount of friction between the components.

Ajuste a interferencia Ajuste de precisión entre dos componentes que provee una cantidad específica de fricción entre los mismos.

Jounce Jounce is the action of the spring deflecting or compressing, storing energy.

Rebote Rebote es la acción del resorte al desviar o comprimir la energía de almacenamiento.

Jounce travel Upward wheel and suspension movement.

Sacudida Movimiento ascendente de la rueda y de la suspensión.

Kickback A force supplied to the steering wheel when one of the front wheels strikes a road irregularity.

Contragolpe Fuerza que se le suministra al volante de dirección cuando una de las ruedas delanteras golpea una irregularidad en el camino.

King pin inclination (KPI) The angle of a line through the center of the king pin in relation to the true vertical centerline of the tire viewed from the front of the vehicle.

Inclinación de la clavija maestra Ángulo de una línea a través del centro de la clavija maestra con relación a la línea central vertical real del neumático vista desde la parte frontal del vehículo.

Ladder frame design A frame with crossmembers between the side rails. These crossmembers resemble steps on a ladder.

Diseño de armazón de tipo escalera Armazón con travesaños entre las vigas laterales. Dichos travesaños se parecen a los peldaños de una escalera.

Lane departure warning (LDW) system Warns the driver if the vehicle drifts out of the lane in which it is driven.

Sistema de advertencia de salida del carril (LDW, en inglés) Advierte al conductor si el vehículo se aparta del carril en el que se está desplazando.

Lateral accelerometer Sends a voltage signal to the vehicle stability control computer in relation to lateral rear chassis movement.

Acelómetro lateral Manda una señal de voltaje a la computador de control de estabilidad del vehículo en relación con el movimiento lateral del trasero del chasis.

Lift/dive input A computer-controlled suspension system input signal regarding front end lift during acceleration and front end dive while braking.

Señal de entrada de ascenso y descenso Señal de entrada del sistema de suspensión controlado por computadora referente al ascenso del extremo delantero durante la aceleración y al descenso del extremo delantero durante el frenado.

Light-emitting diode (LED) A diode that emits light when current flows through the diode.

Diodo emisor de luz (LED) Diodo que emite luz cuando una corriente fluye a través del mismo.

Linear-rate coil spring A coil spring with equal spacing between the coils, one basic shape, and constant wire diameter. These springs have a constant deflection rate regardless of load. If 200 pounds compress the spring 1 inch, 400 pounds compress the spring 2 inches.

Muelle helicoidal de capacidad lineal Muelle helicoidal con separación igual entre los espirales, una forma básica, y un diámetro de cable constante. Estos muelles tienen una capacidad de desviación constante sin importar la carga. Si 200 libras comprimen el muelle una pulgada, 400 libras comprimen el muelle 2 pulgadas.

Liner A synthetic gum rubber layer molded to the inner surface of a tire for sealing purposes.

Revestimiento Capa de caucho sintético moldeada en la superficie interior de un neumático para sellarlo herméticamente.

Lithium-based grease A special lubricant that is used on the rack bearing in manual rack and pinion steering gears.

Grasa con base de litio Lubricante especial utilizado en el cojinete de la cremallera en mecanismos de dirección de cremallera y piñón manuales.

Live axle rear suspension system A live axle rear suspension system is one in which the differential housing, wheel bearings, and brakes act as a unit.

Sistemas de suspensión trasera del eje motriz La suspensión trasera del eje motriz es aquella en la cual la caja del diferencial, los rodamientos y los frenos actúan como una unidad.

Load-carrying ball joint A ball joint that supports the weight of the chassis.

Junta esférica con capacidad de carga Junta esférica que apoya el peso del chasis.

Load-leveling shock absorbers Shock absorbers to which air pressure is supplied to increase their load-carrying capability.

Amortiguadores con nivelación de carga Amortiguadores a los que se suministra presión de aire para aumentar su capacidad de carga.

Load rating A rating that indicates the load-carrying capability of a tire.

Clasificación de carga Clasificación que indica la capacidad de carga de un neumático.

Lock-to-lock A complete turn of the front wheels from full right to full left, or vice versa.

Vuelta completa de las ruedas Viraje completo de las ruedas delanteras desde la extrema derecha hasta la extrema izquierda, o viceversa.

Lower vehicle command A command, or signal, sent from a height sensor to the suspension computer that indicates the suspension height must be lowered.

Orden para bajar el vehículo Orden o señal enviada desde un sensor de altura a la computadora del sistema de suspensión que indica que debe bajarse la altura de la suspensión.

MacPherson strut front suspension system A suspension system in which the strut is connected from the steering knuckle to an upper strut mount, and the strut replaces the shock absorber.

Sistema de suspensión delantera de montante MacPherson Sistema de suspensión en el que el montante se conecta del muñón de dirección a un montaje del montante superior; el montante reemplaza el amortiguador.

Magnasteer system A power steering system that varies the steering effort in relation to vehicle speed.

Sistema Magnasteer Un sistema de dirección de potencia que varía el esfuerzo de dirección en relación con la velocidad del vehículo.

MagneRide system A suspension system with computer-controlled shock absorbers containing magneto-rheological fluid.

Sistema MagneRide Un sistema de suspensión con amortiguadores controlados por computadora que contienen fluido magnetoreológico.

Magnesium alloy wheels Wheels manufactured from magnesium mixed with other metals.

Ruedas de aleación de magnesio Ruedas fabricadas con magnesio mezclado con otros metales.

Magneto-rheological fluid (MR fluid) A synthetic fluid that contains numerous small, suspended metal particles. It can be found in some electronically controlled shock absorbers and struts.

Líquido magnetorreológico Líquido sintético que contiene numerosas partículas metálicas suspendidas y pequeñas. Puede encontrarse en algunos amortiguadores y barras transversales electrónicamente controlados.

Mass The measurement of an object's inertia.

Masa La medida de la inercia de un objeto.

Memory steer Occurs when the steering does not return to the straight-ahead position after a turn, and the steering attempts to continue turning in the original turn direction.

Dirección de memoria Condición que ocurre cuando la dirección no regresa a la posición de línea recta después de un viraje, y la dirección intenta continuar girando en el sentido original.

Momentum An object gains momentum when a force overcomes static inertia and moves the object.

Impulso Un objeto cobra impulso cuando una fuerza supera la inercia estática y mueve el objeto.

Mono leaf spring A leaf spring with a single leaf.

Muelle de lámina singular Muelle de lámina con una sola hoja.

Motorist Assurance Program (MAP) A program that establishes uniform parts inspection guidelines to improve customer satisfaction with the automotive industry.

Programa de aseguranza del conductor (MAP) Una programa que establece los requerimientos uniformes de la inspección de partes para mejorar la satisfacción del cliente con la industria automotríz.

Mud and snow tires Tires with a special tread that provides improved traction when driving in mud or snow.

Neumáticos para condiciones de lodo y nieve Los neumáticos que tienen una huella especial que provee tracción mejorada para conducir en el lodo o la nieve.

Multilink front suspension A suspension system in which the top of the knuckle is supported by two links connected to the chassis.

Suspensión delantera de bielas múltiples Un sistema de suspensión en el cual la parte superior del muñon se soporta por las dos bielas conectadas al chasis.

Multilink independent rear suspension system It has upper and lower links, and a third link connects the upper link to the top of the knuckle through a bearing.

Sistema de enlaces múltiples de suspensión trasera independiente Posee un enlace superior e inferior, y un tercer enlace que conecta el enlace superior a la parte superior del nudillo a través de un cojinete.

Multiple-leaf spring A leaf spring with more than one leaf.

Muelle de láminas múltiples Muelle de lámina con más de una hoja.

Natural vibration frequencies The frequency at which an object tends to vibrate.

Frecuencias de vibración naturales La frecuencia a la cual tiende a vibrar un objeto.

Needle roller bearing A bearing containing small circular rollers.

Cojinete de rodillos con agujas Cojinete que contiene rodillos circulares pequeños.

Negative camber Occurs when the camber line through the center of the tire is tilted inward in relation to the true vertical centerline of the tire viewed from the front of the vehicle.

Combadura negativa Condición que ocurre cuando la combadura a través del centro de un neumático se inclina hacia adentro con relación a la línea central vertical real del neumático vista desde la parte frontal del vehículo.

Negative caster The angle of a line through the center of the tire that is tilted forward in relation to the true vertical centerline of the tire viewed from the side.

Avance negativo del pivote de la rueda Es el ángulo de una línea a través del centro de la llanta que está inclinado hacia enfrente en relación con la verdadera línea central vertical de la rueda vista por un lado.

Negative offset A rim with negative offset has the centerline outboard of the mounting face.

Desviación negativa La llanta con desviación negativa tiene la línea central fuera de la borda de la superficie de montaje.

Negative-phase steering Occurs when the rear wheels are steered in the opposite direction to the front wheels.

Dirección de fase negativa Se aplica al modo cuando las ruedas traseras se mueven en dirección opuesta a las ruedas frontales.

Negative scrub radius Is present when the steering axis inclination line meets the road surface outside the true vertical centerline of the tire at the road surface.

Radio matorral negativo Condición que ocurre cuando la inclinación del pivote de dirección entra en contacto con la superficie del camino fuera de la línea central vertical real del neumático en la superficie del camino.

Network Data-transmitting wires that interconnect various computers on a vehicle.

Red Cables que transmiten datos que interconectan diversas computadoras de un vehículo.

Neutral phase steering Occurs when the rear wheels are centered in the straight-ahead position on a four-wheel steering system.

Direccional de fase neutral Sucede cuando las ruedas traseras se centran en una posición hacia enfrente.

Nitrogen tire inflation Tires inflated with nitrogen rather than air.

Inflado de neumáticos con nitrógeno Neumáticos inflados con nitrógeno en lugar de aire.

Non-load-carrying ball joint A ball joint that maintains suspension component location, but does not support the chassis weight.

Junta esférica sin capacidad de carga Junta esférica que mantiene la ubicación del componente de la suspensión, pero no apoya el peso del chasis.

Occupational Safety and Health Administration (OSHA) Regulates working conditions in the United States.

Dirección para Seguridad y Salud Industrial Rige las condiciones de trabajo en los Estados Unidos.

Ohms (Ω) The ohm is a unit of measurement of electrical resistance.

Ohmios (Ω) Ohmio es una unidad de medida para la resistencia eléctrica.

Ohm's law Ohm's law defines the relationship between voltage, resistance, and current.

Ley de Ohmio La Ley de Ohmio define la relación entre la tensión (o voltaje), la resistencia y la corriente.

Outer race The outer part of a bearing that supports the balls or tapered rollers.

Anillo exterior La parte exterior de un cojinete que apoya las bolas o los cojinetes de rodillo cónico.

Outlet fitting venturi A narrowing of the passage through the outlet fitting in a power steering pump.

Conexiones de salida venturi Estrechamiento del pasaje a través de la conexión de salida en la bomba de la dirección hidráulica.

Parallel HEVs A hybrid electric vehicle in which the drive wheels may be driven by the internal combustion engine, electric motor, or both.

HEV paralelos Un vehículo híbrido eléctrico en el cual las ruedas motrices se pueden impulsar mediante el motor de combustión interno, el motor eléctrico o ambos.

Parallelogram steering linkage A steering linkage in which the tie rods are parallel to the lower control arms.

Cuadrilátero de la dirección en paralelograma Cuadrilátero de la dirección en el que las barras de acoplamiento son paralelas a los brazos de mando inferiores.

Perimeter-type frame A vehicle frame in which the side rails are bent outward so these rails are near the side of the body.

Armazón de tipo perímetro Armazón del vehículo en el que las vigas laterales están dobladas hacia afuera para que queden cerca del lado de la carrocería.

Photo diode A diode that provides a voltage signal when light shines on the diode. The light source may be a light-emitting diode (LED).

Fotodiodo Diodo que provee una señal de tensión cuando es iluminado por la luz. La fuente de luz puede ser un diodo emisor de luz (LED).

Pigtail spring ends An end of a coil spring that is wound to a smaller diameter.

Extremos de muelle enrollados en forma de espiral Extremo de un muelle helicoidal devanado a un diámetro más pequeño.

Pinion The drive gear in rack and pinion steering gear.

Piñón Engranaje de mando en el mecanismo de dirección de cremallera y piñón.

Pinion angle sensor Provides a voltage signal in relation to steering pinion position.

Sensor de ángulo del piñón Proporciona una señal de voltaje de acuerdo con la posición del piñón de dirección.

Pinion-drive EPS The electric motor is coupled to the steering gear pinion.

GPE(guiado de propulsión electrónica) por tracción del piñón El motor eléctrico acompaña al piñón de la dirección.

Pitman arm A short steel arm connected from the steering gear sector shaft to the steering center link.

Brazo pitman Brazo corto de acero conectado del árbol del mecanismo de dirección a la biela motriz central de dirección.

Pitman shaft sector A shaft in a recirculating ball steering gear that is connected to the pitman shaft, and the gear teeth on the sector shaft are meshed with the worm gear.

Sector de eje pitman Un eje de un engranaje de bola recirculatoria que se conecta al eje pitman, y los dientes del engranaje del eje del sector se endentan con el engranaje sinfín.

Positive camber Occurs when the camber line through the centerline of the tire is tilted outward in relation to the true vertical tire centerline viewed from the front.

Combadura positiva Condición que ocurre cuando la combadura a través de la línea central del neumático se inclina hacia afuera con relación a la línea central vertical real vista desde la parte frontal.Is the angle of a line through the center of the tire that is tilted rearward in relation to the true vertical centerline of the tire viewed from the side.

Positive caster Is the angle of a line through the center of the tire that is tilted rearward in relation to the true vertical centerline of the tire viewed from the side.

Avance positivo del pivote de la rueda Es el ángulo de una línea a través del centro de la llanta que está inclinada hacia atrás en relación con la verdadera línea central vertical de la llanta vista por el lado.

Positive-phase steering Is applied to the mode when the rear wheels are steered in the same direction as the front wheels.

Guiado de fase positiva Se aplica cuando las ruedas traseras se guían en la misma dirección de las delanteras.

Positive offset Occurs when the wheel centerline is located inboard from the vertical wheel-mounting surface.

Desfasaje positivo Se produce cuando la línea central de las ruedas se encuentra dentro de la superficie vertical de montaje de las ruedas.

Positive scrub radius Occurs when the SAI line through the center of the lower ball joint and upper strut mount meets the road surface inside the true vertical centerline of the tire at the road surface as viewed from the front.

Radio matorral positivo Condición que ocurre cuando la línea SAI a través del centro de la junta esférica inferior y el montaje del montante superior entra en contacto con la superficie del camino dentro de la línea central vertical real del neumático en la superficie del camino vista desde la parte frontal.

Pre-safe systems React during the few milliseconds before a collision occurs to increase driver and passenger safety.

Presistema de protección Reacción durante los pocos milisegundos antes de que ocurra un choque para aumentar la seguridad del conductor y del pasajero.

Pressure relief ball A spring-loaded ball that opens at a specific pressure and limits pressure in a hydraulic system.

Bola de alivio de presión Bola con cierre automático que se abre a una presión específica y limita la presión en un sistema hidráulico.

Programmed ride control (PRC) system A computer-controlled suspension system in which the computer operates an actuator in each strut to control strut firmness.

Sistema de control programado del viaje Sistema de suspensión controlado por computadora en el que la computadora opera un accionador en cada uno de los montantes para controlar la firmeza de los mismos.

Pulse width modulated (PWM) voltage signal Is a signal with a variable on time that may be used by a computer to control an output.

Señal modulada de la anchura entre impulsos Es una señal con una variable de tiempo que puede usar una computadora para controlar una potencia de salida.

Pulse width modulation (PWM) A method of computer control in which the computer cycles an output on and off with a variable on time.

Modulación de duración de impulsos Método de control de computadoras en el que la computadora produce un ciclo de rendimiento a intérvalos; lo que produce un trabajo efectivo variable en cada ciclo.

Puncture sealing tires A tire with a special compound on the inner tire surface that seals punctures when the puncturing object is removed from the tire.

Neumáticos autoselladores Neumático con un compuesto especial en la superficie de la parte interior que sella pinchazos cuando se le remueve el objecto punzante al neumático.

Pyrotechnic Device that contains an explosive and an ignition source.

Pirotecnia Dispositivo que contiene un explosivo y una fuente de ignición.

Quadrasteer® A type of 4WS system used on some SUVs.

Quadrasteer® Un tipo de suspensión de cuatro ruedas utilizado en algunos modelos de SUV.

Rack A horizontal shaft in a rack and pinion steering gear containing teeth that are meshed with the pinion teeth.

Cremallera Eje horizontal en un mecanismo de dirección de cremallera y piñón que contiene los dientes que se engranan con los del piñón.

Rack and pinion power steering system A type of power steering in which the steering gear contains a pinion gear that is meshed with teeth on the horizontal rack.

Sistema de dirección hidráulica de cremallera y piñón Tipo de dirección hidráulica en la que el mecanismo de dirección contiene un piñón que se engrana con los dientes de la cremallera horizontal.

Rack and pinion steering linkage A rack and pinion steering gear with the tie rods connected from the rack ends or rack center to the steering arms.

Cuadrilátero de la dirección de cremallera y piñón Mecanismo de dirección de cremallera y piñón con las barras de acoplamiento conectadas de los extremos de la cremallera o del centro de la misma a los brazos de dirección.

Rack-drive EPS The electric motor that provides steering assist is coupled to the rack in the steering gear.

Cremallera de dirección GPE(guiado de propulsión electrónica) El motor eléctrico que proporciona ayuda de dirección acompaña a la cremallera en el mecanismo de dirección.

Rack piston A piston mounted on the rack in a power rack and pinion steering gear. Hydraulic pressure is supplied to this piston for power steering assistance.

Pistón de la cremallera Pistón montado en la cremallera en un mecanismo de dirección hidráulica de cremallera y piñón. A este pistón se le suministra presión hidráulica para reforzar la dirección hidráulica.

Rack seals Are positioned between the rack and the housing in a rack and pinion steering gear.

Juntas de estanqueidad de la cremallera Ubicadas entre la cremallera y el alojamiento en un mecanismo de dirección de cremallera y piñón.

Radial bearing load A load applied in a vertical direction.

Carga de marcación radial Carga aplicada en dirección vertical.

Radial-ply tires A tire in which the carcass plies are positioned at 90° in relation to the center of the tire, and steel or fiberglass belts are mounted under the tread.

Neumáticos de cordón radial Neumático fabricado con las estrias de armazón colocadas a un ángulo de 90° con relación al centro del neumático, y correas de acero o de fibra de vidrio montadas debajo de la huella.

Raise vehicle command A command, or signal, sent from a height sensor to the computer in a computer-controlled suspension system indicating the chassis height must be raised.

Orden para levantar el vehículo Orden o señal enviada desde un sensor de altura a la computadora en un sistema de suspensión controlado por computadora que indica que debe levantarse la altura del chasis.

Reactive Having the capability to cause a chemical reaction with another substance.

Reactivo Que tiene la capacidad de producir una reacción química con otra sustancia.

Real-time damping (RTD) system A term used to designate the road-sensing suspension system in on-board diagnostics.

Sistema de amortiguamiento en tiempo real Término utilizado para designar el sistema de suspensión con equipo sensor en pruebas de diagnóstico realizadas en el vehículo mismo.

Rear active steering (RAS) Has recently been introduced as standard equipment on the 2006 Infinity M series luxury car. The RAS makes the steering and handling characteristics on this long-wheelbase car more nimble and agile compared with models without this technology.

Dirección activa trasera (DAT) Últimamente se presentó como un equipo estándar en la serie del automóvil de lujo de la serie Infinity M del 2006. La DAT hace que la dirección y las características de manipulación de este automóvil de paso largo sean más ligeras y ágiles comparadas con los modelos que no poseen esta tecnología.

Rear axle bearing A bearing that supports the rear axle in the housing on a rear-wheel-drive vehicle.

Cojinete del eje trasero Cojinete que apoya el eje trasero en el alojamiento en un vehículo de tracción trasera.

Rear axle offset A condition in which the complete rear axle assembly has turned so one rear wheel has moved forward, and the opposite rear wheel has moved rearward.

Desviación del eje trasero Condición que ocurre cuando todo el conjunto del eje trasero ha girado de manera que una de las ruedas traseras se ha movido hacia adelante y la opuesta se ha movido hacia atrás.

Rear axle sideset A condition in which the rear axle assembly has moved sideways from its original position.

Resbalamiento lateral del eje trasero Condición que ocurre cuando el conjunto del eje trasero se ha movido lateralmente desde su posición original.

Rear steering actuator An assembly that controls rear wheel steering in some four-wheel steering systems.

Accionador de la dirección trasera Conjunto que controla la dirección de la rueda trasera en algunos sistemas de dirección en las cuatro ruedas.

Rear suspension system The rear suspension system plays a very important part in ride quality and in the control of suspension and differential noise, vibration, and shock.

Sistema de suspensión trasera El sistema de suspensión trasera desempeña un papel muy importante en la calidad de la conducción y en el control de la suspensión y el ruido, la vibración y el impacto en el diferencial.

Rear wheel camber The tilt of a line through the rear tire and wheel centerline in relation to the true vertical centerline of the rear tire and wheel.

Comba de las ruedas traseras El ángulo de una línea que atraviesa el eje mediano del neumático trasero y de la rueda en relación al eje mediano verdadero del neumático trasero y la rueda.

Rear wheel toe The distance between the front edges of the rear wheels in relation to the distance between the rear edges of the rear wheels.

Tope de la rueda trasera Distancia entre los bordes frontales de las ruedas traseras con relación a la distancia entre los bordes traseros de las ruedas traseras.

Rebound Rebound is the action of the spring expanding and releasing stored energy.

Rebote Rebote es la acción del resorte al expandir y liberar la energía de almacenamiento.

Rebound travel The downward spring and wheel action resulting from rebound. This downward spring and wheel action is called rebound travel.

Viaje del rebote Rebote es la acción del resorte al expandir y liberar la energía de almacenamiento. Esta acción descendente del resorte y la rueda se conoce como viaje del rebote.

Receiver An electronic device in some computer-controlled wheel aligners that receives high-frequency voltage signals from the wheel sensors.

Receptor Un dispositivo electrónico en algunos alineadores controlados por computadoras que recibe los señales de alta frequencia de los sensores de las ruedas.

Regenerative braking A system on a hybrid or electric vehicle that allows the drive wheels to turn the electric drive motor so it recharges the battery during vehicle deceleration.

Frenado regenerativo Un sistema en un vehículo híbrido o eléctrico que permite que las ruedas motrices giren el motor de impulsión eléctrica para que recargue la batería durante la desaceleración del vehículo.

Regular-duty coil spring A coil spring supplied to handle average loads to which the vehicle is subjected. This type of spring has smaller wire diameter compared with a heavy-duty spring.

Muelle helicoidal para servicio normal Muelle helicoidal provisto para sostener la carga normal a la que el vehículo está expuesto. Este tipo de muelle tiene un cable de un diámetro más pequeño comparado con un muelle para servicio pesado.

Release valve A valve that releases fluid pressure from a wheel caliper during antilock brake operation.

Válvula descargador Una válvula que descarga la presión de fluido del calíbre de la rueda durante la operación de frenado antibloqueante.

Remote fluid reservoir A reservoir mounted separately from the power steering pump.

Tanque de líquido remoto Un tanque montado por separado de la bomba de dirección hidráulica.

Remote reservoir A reservoir mounted separately from the power steering pump.

Depósito remoto Un depósito montado separadamente de la bomba de dirección hidráulica.

Replacement tires Tires purchased to replace the original tires that were supplied by the vehicle manufacturer.

Neumáticos de repuesto Neumáticos que se compran para reemplazar los neumáticos originales provistos por el fabricante del vehículo.

Resonance A reinforcement of sound in a vibrating body caused by sound waves from another body.

Resonancia Una fortificación del sonido en un cuerpo vibrante causada por las ondas sónicas de otro cuerpo.

Resource Conservation and Recovery Act (RCRA) States that hazardous material users are responsible for hazardous materials from the time they become a waste until the proper waste disposal is completed.

Ley de Conservación y Recuperación de Recursos Establece que los usuarios de materiales peligrosos se encarguen de estos materiales desde el momento en que se convierten en desperdicios hasta que se lleve a cabo la eliminación adecuada de los mismos.

Road crown The higher center of a road surface in relation to the edges of this surface.

Corona de camino Centro más alto de la superficie de un camino con relación a los bordes de la misma.

Road force variation Is a condition that occurs when tire sidewalls do not have equal stiffness around the complete area of the sidewalls.

Variación de la fuerza en carretera Es una condición que sucede cuando las paredes laterales de las llantas no tienen la rigidez equitativa alrededor del área completa de las paredes laterales.

Road variables Variables such as weight in the vehicle or road surface that affect wheel alignment while driving.

Variables del camino Condiciones variables, como por ejemplo el peso en el vehículo o la superficie del camino que afectan la alineación de la rueda durante un viaje.

Roller bearing A roller bearing contains precision machined rollers that have the same diameter at both ends. Designed primarily to carry radial loads, but they can withstand some thrust load.

Cilindro del cojinete El cilindro del cojinete contiene cilindros con torneado de precisión con el mismo diámetro en ambos extremos. Se diseñaron inicialmente para transportar cargas radiales, pero pueden soportar algunas cargas de empuje.

Rolling elements The balls or rollers and the separator in a bearing.

Elementos rodantes Las bolas o rodillos y el separador en un cojinete.

Ross cam and lever steering gear A type of steering gear used in the early 1900s that had a spiral groove in the lower end of the steering shaft meshed with a pin on the pitman shaft.

Mecanismo de dirección de leva y palanca Ross Tipo de mecanismo de dirección utilizado a principios de siglo que tenía una ranura espiral en el extremo inferior del eje de dirección engranado con un pasador en el árbol pitman.

Rotary valve A valve mounted with the spool valve in a power steering gear. The position of these two valves directs power steering fluid to the appropriate side of the rack or power piston.

Válvula rotativa Válvula montada con la válvula de carrete en un mecansimo de dirección hidráulica. La posición de estas dos válvulas dirige el fluido de la dirección hidráulica al lado apropiado de la cremallera o del pistón impulsor.

Run-flat tires Tires designed to operate safely without any air pressure for a specific distance.

Neumáticos de no presión Los neumáticos diseñados a operar sin peligro sín presión por distancias específicas.

Safety and gateway module (SGM) Operates the servotronic valve and the electronically controlled orifice valve in some power steering systems.

Módulo de puerta de enlace y seguridad (SGM, en inglés) Opera la válvula servotrónica y la válvula de orifi cio controlado electrónicamente en algunos sistemas de dirección de potencia.

Scrub radius The distance between the SAI line and the true vertical centerline of the tire at the road surface.

Radio matorral Distancia entre la línea SAI y la línea central vertical real del neumático en la superficie del camino.

Sector A shaft in a recirculating ball steering gear that is connected to the pitman shaft. The gear teeth on the sector shaft are meshed with the worm gear.

Sector Árbol en un mecanismo de dirección con bola recirculante que se conecta al árbol pitman. Los dientes del engranaje en el eje sector se endentan con el engranaje sinfín.

Semi-ellliptical springs Semi-elliptical springs have individual leaves stacked with the shortest leaf at the bottom and the longest leaf at the top.

Resortes semielípticos Los resortes semielípticos poseen láminas individuales apiladas con la lámina más corta en el fondo y la lámina más larga en la parte superior.

Semi-independent rear suspension A rear suspension system in which one rear wheel has a limited amount of movement without affecting the opposite rear wheel.

Suspensión trasera semi-independiente Sistema de suspensión trasera en el que una de las ruedas traseras tiene una cantidad limitada de movimiento sin afectar la rueda trasera opuesta.

Separator A component that prevents contact between two other parts.

Separador Componente que evita que otras dos piezas entren en contacto.

Series HEVs A hybrid electric vehicle in which the internal combustion engine and the electric drive motor both supply torque to the drive wheels.

HEV de serie Un vehículo híbrido eléctrico en el cual tanto el motor de combustión interna como el motor de impulsión eléctrica proporcionan potencia a las ruedas motrices.

Serpentine belt A ribbed V-belt drive system in which all the belt-driven components are on the same vertical plane.

Correa serpentina Sistema de transmisión con correa nervada en V en el que todos los componentes accionados por una correa se encuentran sobre el mismo plano vertical.

Servotronic valve A computer-controlled valve in an active steering system that controls power steering pump pressure in relation to the vehicle speed.

Válvula servotrónica Una válvula controlada por computadora en un sistema de dirección activo que controla la presión de la bomba de dirección de potencia de acuerdo con la velocidad del vehículo.

Shock absorber A hydraulic mechanism connected between the chassis and the suspension to control spring action.

Amortiguador Un mecanismo hidráulico conectado entre el chasis y la suspensión para controlar la acción del muelle.

Shock absorber ratio The amount of extension control compared with the amount of compression control.

Relación del amortiguador Cantidad de control de extensión comparado con la cantidad de control de compresión.

Short-and-long arm front suspension systems Suspension systems in which the upper control arm is shorter than the lower control arm.

Sistemas de suspensión de brazos largos y cortos Un sistema suspensión en el cual el brazo de control superior es más corto que el brazo de control inferior.

Sideslip The tendency of the rear wheels to slip sideways because of centrifugal force while a vehicle is cornering at high speed.

Patinaje Tendencia de las ruedas traseras a deslizarse lateralmente a causa de la fuerza centrífuga mientras un vehículo hace un viraje a gran velocidad.

Side sway Occurs when one side of a vehicle frame is bent inward.

Desviación Ocurre cuando un lado del bastidor del vehículo esta torcido hacia adentro.

Sidewall The area between the tread and the bead of a tire.

Pared lateral Área entre la huella y la pestaña de un neumático.

Silencer A component designed to reduce noise.

Silenciador Componente diseñado para disminuir el ruido.

Single-row ball bearing A bearing with a single row of balls.

Cojinete de bola de una sola fila Cojinete que contiene una sola fila de bolas.

Slip angle The actual angle of the front wheels during a turn compared with the turning angle of the front wheels with the vehicle at rest.

Ángulo de patinaje Ángulo real de las ruedas delanteras durante un viraje comparado con el ángulo de giro de las ruedas delanteras cuando se ha detenido la marcha del vehículo.

Snapring A circular steel ring with some tension designed to snap into a groove and retain a component.

Anillo de resorte Anillo circular de acero con un poco de tensión diseñado para ajustarse dentro de una ranura y sujetar un componente en su posición.

Society of Automotive Engineers (SAE) A society of professional engineers that provides many member and industry services, such as the development of industry standards and the communication of engineering information through publications and conferences.

Sociedad de Ingenieros de Automóviles (SAE) Sociedad de ingenieros profesionales que les provee muchos servicios a sus miembros y a la industria, como por ejemplo, el desarrollo de normas para la industria, y la comunicación de información sobre ingeniería mediante publicaciones y conferencias.

Sodium-based grease A special lubricant with a sodium base that may be required on some steering components.

Grasa a base de sodio Lubricante especial con una base de sodio que podrían necesitar algunos componentes de la dirección.

Soft relay A relay in a computer-controlled suspension system that supplies voltage to the strut actuators in the soft mode.

Relé blando Un relé en un sistema de suspensión controlado por computadora que suministra el voltaje a los accionadores de los montantes en un modo blando.

Solid axle beam A solid axle beam is usually a transverse inverted U-section channel connected between the rear wheels in a semi-independent rear suspension system.

Viga del eje sólido Una viga del eje sólido suele ser un canal transversal de sección U invertida, conectado entre las ruedas traseras en un sistema de suspensión trasera semi-independiente.

Speed rating A tire rating that indicates the maximum safe vehicle speed that a tire will withstand.

Clasificación de la velocidad Clasificación de un neumático que indica la velocidad máxima que podrá resistir un neumático.

Speed-sensitive steering (SSS) system A computer-controlled steering system that varies the steering effort in relation to vehicle speed.

Dirección sensible a la velocidad Sistema de dirección controlado por computadora que varía el esfuerzo necesario de la dirección de acuerdo a la velocidad del vehículo.

Spherical bearing A bearing shaped like a sphere and used in some tilt steering wheels.

Cojinete esférico Cojinete en forma de esfera utilizado en algunos volantes de dirección inclinables.

Spiral cable A conductive ribbon that is mounted in a plastic container on top of the steering column and maintains electrical contact between the air bag inflator module and the air bag electrical system. A spiral cable may be called a clock spring electrical connector.

Cable espiral Cinta conductiva montada en un recipiente plástico sobre la columna de dirección que mantiene el contacto eléctrico entre la unidad infladora y el sistema eléctrico del Airbag. El cable espiral se conoce también como conector eléctrico de cuerda de reloj.

Spool valve Positioned with the rotary valve in a power steering gear. These two valves direct power steering fluid to the appropriate side of the rack or power piston to provide steering assistance.

Válvula de carrete Ubicada con la válvula rotativa en un mecanismo de dirección hidráulica. Estas dos válvulas dirigen el fluido de la dirección hidráulica al lado apropiado de la cremallera o del pistón impulsor para reforzar la dirección.

Spring insulator Positioned between the ends of a coil spring and the spring mounting surfaces to reduce the transfer of noise and vibration from the spring to the chassis.

Aisladore de muelle Ubicados entre los extremos de un muelle helicoidal y las superficies para el montaje de muelles con el propósito de reducir la transferencia de ruido y la vibración del muelle al chasis.

Springless seal A seal lip with no spring behind the lip.

Junta de estanqueidad sin muelle Reborde de una junta de estanqueidad que no tiene un muelle detrás del mismo.

Spring-loaded seal A seal lip with a garter spring behind the lip to increase lip tension.

Junta de estanqueidad con cierre automático Reborde de una junta de estanqueidad con un muelle jarretera detrás del reborde para aumentar la tensión del mismo.

Sprung weight Is the vehicle weight that is supported by the coil springs.

Peso suspendido Es el peso del vehículo que soportan los muelles en espiral cilíndrica.

Spur gear teeth Gear teeth that are parallel to the centerline of the gear.

Dientes de engranaje rectos Dientes del engranaje que son paralelos a la línea central del engranaje.

Stabilitrak® A computer-controlled system that provides vehicle stability control by reducing sideways swerving.

Stabilitrak® Un sistema controlado por computadora que provee el control de estabilidad del vehículo asi disminuyendo viraje lateral.

Stabilizer bar A round steel bar, connected between the front or rear lower control arms, that reduces body sway.

Barra estabilizadora Una barra de acero circular conectada entre los brazos de control inferiores delanteros o traseros, que reducen el balanceo del cuerpo.

Static balance Refers to the balance of a tire and wheel at rest.

Equilibrio estático Se refiere al equilirio de un neumático y una rueda cuando se ha detenido la marcha del vehículo.

Steering angle The angle of the front wheel on the inside of a turn compared to the angle of the front wheel on the outside of the turn.

Ángulo de la dirección Ángulo de la rueda delantera en el interior de un viraje comparado con el ángulo de la rueda delantera en el exterior del viraje.

Steering angle sensor An optical-type sensor that supplies a voltage signal in relation to steering wheel rotation.

Sensor del ángulo de la dirección Un sensor de tipo óptico que proporciona una señal de voltaje de acuerdo con la rotación del volante

Steering axis inclination (SAI) The angle of a line through the center of the upper strut mount and lower ball joint in relation to the true vertical centerline of the tire viewed from the front of the vehicle.

Inclinación del pivote de dirección Ángulo de una línea a través del centro del montaje del montante superior y la junta esférica inferior con relación a la línea central vertical real del neumático vista desde la parte frontal del vehículo.

Steer-by-wire A computer-controlled steering system with no direct mechanical connection between the steering wheel and the steering gear.

Dirección alámbrica Un sistema de dirección controlado por computadora sín una conexión mecánica directa entre el volante de dirección y el aparato de dirección.

Steering drift The tendency of the steering to drift slowly to one side when the vehicle is driven straight ahead on a smooth, straight road surface.

Desviación de la dirección La tendencia de la dirección a desviarse poco a poco hacia un lado mientras que el vehículo se conduce en línea recta sobre un camino liso y nivelado.

Steering gear free play The amount of steering wheel movement before the front wheels begin to turn.

Juego en el aparato de dirección Cantidad de movimiento del volante de dirección antes de que las ruedas delanteras comiencen a girar.

Steering kickback Road force supplied back to the steering wheel.

Tensión de retroceso de la dirección Fuerza del camino que se vuelve a suministrar al volante. Sensor de posició.

Steering pull The tendency of the steering to pull constantly to one side or the other when driving the vehicle straight ahead on a smooth, level road surface.

Tiro en la dirección La tendencia de la dirección a tirar en una manera constante hacia un lado u otro mientras que el vehículo se conduce en línea recta sobre un camino liso y nivelado.

Steering ratio The number of degrees of steering wheel rotation in relation to the number of degrees of front wheel movement.

Relación de la dirección Número de grados de rotación del volante de dirección con relación al número de grados de movimiento de la rueda delantera.

Steering sensor A sensor that sends a voltage signal to the computer in relation to the amount and speed of steering wheel rotation in a programmed ride control (PRC) system.

Sensor de dirección Un sensor que manda una señal de voltaje a la computadora en relación a la cantidad y la velocidad de la rotación del volante de dirección en un sistema de viaje controlado (PRC).

Steering wander The tendency of the steering to pull to the right or left when the vehicle is driven straight ahead on a smooth road surface.

Desviación de la dirección Tendencia de la dirección a desviarse hacia la derecha o hacia la izquierda cuando se conduce el vehículo en línea recta en un camino cuya superficie es lisa.

Steering wheel angle sensor (SWAS) Provides a voltage signal in relation to steering wheel rotation.

Sensor del ángulo del volante (SWAS, en inglés) Proporciona una señal de voltaje de acuerdo con la rotación del volante.

Steering wheel position sensor (SWPS) A sensor that produces a voltage signal in relation to the amount and velocity of steering wheel rotation.

Sensor de posición de la dirección (SWPS) Un sensor que produce una señal de voltaje en relación a la cantidad y la velocidad de rotación del volante de dirección.

Steering wheel rotation sensor An input sensor in a computer-controlled suspension or steering system that sends a signal to the computer in relation to the amount and speed of steering wheel rotation.

Sensor de la rotación del volante de dirección Sensor de entrada en un sistema de suspensión controlado por computadora o en un sistema de dirección que le envía una señal a la computadora referente a la cantidad y a la velocidad de la rotación del volante de dirección.

Stop-to-stop Steering wheel rotation from extreme left to extreme right.

Parada a parada Rotación del volante de dirección desde la extrema izquierda hasta la extrema derecha.

Struts Components connected from the top of the steering knuckle to the upper strut mount that maintain the knuckle position and act as shock absorbers to control spring action.

Montantes Componentes conectados de la parte superior del muñón de dirección al montaje del montante superior que mantienen la posición del muñón y actúan como amortiguadores para controlar el movimiento de ascenso y descenso del muelle.

Summation sensor Provides a voltage signal in relation to steering rack position.

Sensor de suma Proporciona una señal de voltaje de acuerdo con la posición del soporte de la dirección.

Symmetrical leaf spring A symmetrical leaf spring has the same distance from the center bolt to the front and rear of the spring.

Hoja de muelle simétrica La hoja de muelle simétrica posee la misma distancia desde el centro del perno hasta la parte frontal y trasera del resorte.

Synthetic rubber A type of rubber developed in a laboratory.

Caucho sintético Tipo de caucho fabricado en un laboratorio.

Tapered roller bearing A bearing containing tapered roller bearings mounted between the inner and outer races.

Cojinete de rodillos cónicos Cojinete que contiene cojinetes de rodillos cónicos montados entre los anillos interiores y exteriores.

Taper-wire closed-end springs Coil spring ends that are tapered to a smaller diameter.

Muelles helicoidales de extremos cónicos Extremos de muelles helicoidales a los que se les da forma de cono para reducir su diámetro.

Telematics Using an in-vehicle phone for communication with the vehicle driver regarding vehicle diagnostics, emissions, or traffic- and safety-related concerns.

Telemática Uso de un teléfono incorporado al vehículo para comunicarse con el conductor del vehículo acerca de problemas de diagnóstico, emisiones, tránsito y seguridad.

Temperature rating A tire rating that indicates the ability of the tire to withstand heat.

Clasificación de la temperatura Clasificación de un neumático que indica la capacidad del neumático de resistir el calor.

Tension loaded A ball joint mounted so the vehicle weight tends to pull the ball out of the joint.

Cargada de tensión Junta esférica montada de manera que el peso del vehículo tiende a remover la bola de la junta.

Throttle position sensor (TPS) Sends an analog voltage signal to the engine computer in relation to throttle opening.

Sensor de posición del acelerador (TPS, en inglés) Envía una señal de voltaje analógica a la computadora del motor en relación con la apertura del acelerador.

Thrust angle The angle between the thrust line and the vertical centerline of the vehicle.

Ángulo de empuje El ángulo entre la línea de empuje y el eje mediano vertical del vehículo.

Thrust bearing load This type of load may be called an axial load.

Carga del cojinete de empuje Este tipo de carga puede llamarse una carga.

Thrust line An imaginary line positioned at a 90° angle to the center of the rear axle and extending forward.

Directriz de presiones Una línea imaginaria en posición de un ángulo de 90° con relación al centro del eje trasero y que se extiende hacia enfrente.

Thrust line alignment A wheel alignment in which the thrust line is used as a reference for front wheel toe adjustment.

Alineación de la línea de empuje Alineación de la rueda en la que se utiliza la línea de empuje como punto de referencia para el ajuste del tope de la rueda delantera.

Tie rod A rod connected from the steering arm to the rack or center link, depending on the type of steering linkage.

Barra de acoplamiento Varilla conectada del brazo de dirección a la cremallera o a la biela motriz central, dependiendo del tipo de cuadrilátero de la dirección.

Tire belts Belts that are placed under the tire tread to provide longer tread wear. Belts are usually made from steel or polyester.

Banda de refuerzo del neumático Las bandas que se colocan debajo de la banda de rodamiento para aumentar la durabilidad de la banda de rodamiento. Las bandas suelen ser fabricadas del acero o del poliester.

Tire chains May be placed over the tires to improve traction when driving on ice or snow.

Cadenas antideslizantes Pueden colocarse sobre los neumáticos para mejorar la tracción cuando se conduce el vehículo en el hielo o la nieve.

Tire contact area The part of the tire in contact with the road surface when the tire is supporting the vehicle weight.

Área de contacto del neumático Parte del neumático en contacto con la superficie del camino cuando el neumático apoya el peso del vehículo.

Tire free diameter The distance between the outer edges of the tread measured on a horizontal line through the center of the wheel.

Diámetro libre del neumático Distancia entre los bordes exteriores de la huella que se mide sobre una línea horizontal a través del centro de la rueda.

Tire performance criteria (TPC) Information molded on the tire sidewall indicating that the tire meets the manufacturer's performance standards for traction, endurance, dimensions, noise, handling, and rolling resistance.

Criterio sobre el rendimiento del neumático Información moldeada en la pared lateral del neumático que indica que el neumático cumple con las normas de rendimiento establecidas por el fabricante sobre la tracción, acción, dimensiones, ruido, movilización, y resistencia al rodaje.

Tire placard Often attached to the rear face of the driver's door, the tire placard provides information regarding maximum vehicle load, tire size, and cold inflation pressure.

Cartel del neumático Comúnmente fijado a la cara posterior de la puerta del conductor, el cartel del neumático provee información referente a la carga máxima del vehículo, el tamaño del neumático, y la presión fría de inflación.

Tire pressure The amount of air pressure contained in the tire to allow the tire to carry a load.

Presión del neumático La cantidad del presión de aire contenido en el neumático para permitir que el neumático soporta una carga.

Tire rolling diameter The distance between the outer edges of the tread measured on a vertical line through the center of the wheel when the tire is supporting the vehicle weight.

Diámetro del rodaje del neumático Distancia entre los bordes exteriores de la huella que se mide sobre una línea vertical a través del centro de la rueda cuando el neumático apoya el peso del vehículo.

Tire treads The part of the tire in contact with road surface.

Bandas de rodamiento del neumático La parte del neumático que se pone en contacto con la superficie del camino.

Tire valves Mounted in the wheel rim, the tire valves allow air to enter or be exhausted from the tire.

Válvulas del neumático Estas válvulas están montadas en la llanta de la rueda y permiten la entrada o la salida de aire desde el neumático.

Toe-in A condition that is present when the distance between the front edges of the front or rear wheels is less than the distance between the rear edges of the wheels.

Convergencia Condición que ocurre cuando la distancia entre los bordes frontales de las ruedas delanteras o traseras es menor que la distancia entre los bordes traseros de las mismas.

Toe-out A condition that is present when the distance between the front edges of the front or rear wheels is more than the distance between the rear edges of the wheels.

Divergencia Condición que ocurre cuando la distancia entre los bordes delanteros de las ruedas delanteras o traseras es mayor que la distancia entre los bordes traseros de las mismas.

Toe-out on turns The steering angle of the wheel on the inside of a turn compared with the steering angle of the wheel on the outside of the turn.

Divergencia durante un viraje El ángulo de la dirección de la rueda en el interior de un viraje comparado con el ángulo de la dirección de la rueda en el exterior del viraje.

Toe plate A metal plate surrounding the steering column and attached to the vehicle floor.

Placa de pie Una placa de metal que rodea la columna de dirección y conectada al piso del vehículo.

Torque A twisting force.

Par de torsión Fuerza de torcimiento.

Torque steer The tendency of the steering on a front-wheel-drive vehicle with unequal drive axles to pull to one side during hard acceleration.

Dirección de torsión Tendencia de la dirección en un vehículo de tracción delantera con ejes de mando desiguales a desviarse hacia un lado durante una aceleración rápida.

Torsion bar A steel bar connected from the chassis to the lower control arm. As the vehicle weight pushes the chassis downward, the torsion bar twists to support this weight. Torsion bars are used in place of coil springs.

Barra de torsión Barra de acero conectada del chasis al brazo de mando inferior. Mientras el peso del vehículo presiona el chasis hacia abajo, la barra de torsión se tuerce para apoyar este peso. Las barras de torsión se utilizan en lugar de los muelles helicoidales.

Track bar Some rear suspension systems have a track bar connected from one side of the differential housing to the chassis to prevent lateral chassis movement.

Barra transversal Algunos sistemas de suspensión trasera poseen una barra transversal conectada de un lado de la caja del diferencial al chasis, para evitar el movimiento lateral del chasis.

Tracking Refers to the position of the rear wheels in relation to the front wheels.

Encarrilamiento Se refiere a la posición de las ruedas traseras con relación a las ruedas delanteras.

Traction control system (TCS) A computer-controlled system that prevents drive wheel spinning.

Sistema de control de tracción (TCS) Un sistema controlado por computador que previene que giran las ruedas de propulsión.

Traction rating A tire rating indicating the traction capabilities of the tire.

Clasificación de tracción Clasificación de un neumático que indica las capacidades de tracción del mismo.

Transistors Fast-acting electronic switches with no moving parts.

Transistores Interuptores electrónicas de acción rápida sín partes móviles.

Transitional coils Coils located between the inactive and active coils in variable-rate coil springs.

Bobinas de transición Las bobinas localizadas entre las bobinas inactivas y activas en los resortes helicoidales de relación variable.

Travel-sensitive strut A strut with the capability to adjust its firmness in relation to the amount of piston travel inside the strut.

Montante sensible al movimiento Montante con la capacidad de ajustar su firmeza de acuerdo a la cantidad de movimiento del pistón dentro del montante.

Tread wear rating A tire rating indicating the wear capabilities of the tread that allow customers to compare tire life expectancy.

Clasificación del desgaste de un neumático Clasificación de un neumático que indica las capacidades de desgaste de la huella y le permite a los clientes comparar el índice de durabilidad del neumático.

Trim height The specified, or normal, suspension height in a computer-controlled suspension system.

Altura de la suspensión Altura especificada, o normal, de la suspensión en un sistema de suspensión controlado por computadora.

Tubular frame A frame member with a circular, or tubular, design.

Fabricación del armazón en forma tubular Pieza del armazón diseñada en forma circular o tubular.

Turning circle The turning angle of one front wheel in relation to the opposite front wheel during a turn.

Círculo de giro Ángulo de giro de una de las ruedas delanteras con relación a la rueda delantera opuesta durante un viraje.

Turning radius The turning angles of the front wheels around a common center point.

Diámetro del giro Los ángulos de giro de las ruedas frontales alrededor de un punto central común.

Uniform Tire Quality Grading (UTQG) System Information including tread wear, traction, and temperature ratings that is molded into the tire sidewall, and is required by the Department of Transportation (DOT).

Sistema de sobre la clasificación de la calidad uniforme de neumáticos Información moldeada en la pared lateral del neumático que incluye el desgaste de la huella, la tracción, y las clasificaciones de temperatura y que es requerida por el Ministerio de Transporte.

Unitized body A body design that does not have a frame because each body component is a supporting member.

Carrocería unificada Diseño de la carrocería que no tiene armazón porque cada uno de sus componentes es una pieza suplementaria.

Unsprung weight The vehicle weight that is not supported by the coil springs.

Peso no suspendido Es el peso del vehículo que no soportan los muelles en espiral cilíndrica.

Upper strut mount A mount connected between the strut and the strut tower. Front upper strut mounts contain a bearing that allows strut rotation.

Montaje de los montantes superiores Montaje conectado entre el montante y la torre del montante. Los montajes frontales y superiores contienen un cojinete que permite la rotación del montante.

Vacuum A pressure that is less than the atmospheric pressure.

Vacío Presión menor que la de la atmósfera.

Vane-type power steering pumps Pumps that have rotors with metal vanes that provide a seal between the rotor and the pump cam.

Bombas de dirección hidráulica tipo aletas Las bombas que tienen rotores con aletas de metal que proveen un sello entre el rotor y la leva de la bomba.

Variable effort steering (VES) A computer-controlled steering system that provides reduced steering effort at low vehicle speeds and increased steering effort at higher vehicle speeds.

Dirección de esfuerzo variable Sistema de dirección controlado por computadora que provee un menor esfuerzo de dirección cuando el vehículo viaja a velocidad baja y un mayor esfuerzo de dirección cuando el vehículo viaja a gran velocidad.

Variable-rate coil springs Rather than having a standard spring deflection rate, these springs have an average spring rate based on load at a predetermined deflection.

Muelles helicodal de capacidad variable En vez de tener una capacidad de desviación de muelle estándar, estos muelles tienen un valor promedio de elasticidad basado en la carga a una desviación predeterminada.

V-belt A drive belt with a V-shape.

Correa en V Correa de transmisión en forma de V.

Vehicle dynamic suspension (VDS) System found in some SUVs that is similar to air suspension systems.

Suspensión dinámica del vehículo (SDV) Sistema que se encuentra en algunas camionetas SUV que es similar a los sistemas de suspensión de aire.

Vehicle speed sensor (VSS) An input sensor that sends a voltage signal to the engine computer in relation to vehicle speed.

Sensor de la velocidad del vehículo Sensor de entrada que le envía una señal de tensión a la computadora del motor referente a la velocidad del vehículo.

Vehicle stability control system A computer-controlled system that prevents vehicle swerving, especially during hard acceleration on slippery road surfaces.

Sistema de control de estabilidad del vehículo Un sistema controlado por computadora que previene que se desvía el vehículo, especialmente durante una aceleración fuerte sobre un camino cuyo superficie es resbalosa.

Vehicle wander The tendency of the steering to pull to the left or the right when the vehicle is driven straight ahead on a smooth road surface.

Desviación de la marcha del vehículo Tendencia de la dirección a desviarse hacia la izquierda o hacia la derecha cuando se conduce el vehículo en línea recta en un camino cuya superficie es lisa.

Vent valve An electrically operated valve that vents air from an air spring in a computer-controlled suspension system.

Válvula de respiración Válvula activada eléctricamente que da salida al aire desde un muelle de aire en un sistema de suspensión controlado por computadora.

Venturi A narrow portion of an air passage.

Venturi Una porción estrecha de un pasaje de aire.

Vibration A rapid motion of particles, or a component that produces sound.

Vibración Un movimiento rápido de los partículos, o de un componente que produce un sonido.

Voltage (V) Voltage is a unit of measurement of electrical pressure.

Voltaje (V) El voltaje es una unidad de medida de la presión eléctrica.

Volume Volume is the length, width, and height of the space occupied by an object.

Volumen El volumen es la longitud, la anchura y la altura del espacio ocupado por un objeto.

Watts rod A rod connected from the chassis to the rear suspension to reduce body side sway, usually referred to as a tracking bar.

Barra wats Una barra conectada del chasis a la suspensión trasera para disminuir la oscilación lateral de la carrocería, suele referirse como una barra de tracción.

Wear indicators Rubber bars located near the bottom of tire treads. When the tread is worn to a specific depth, these bars become visible.

Indicadores del desgaste Las barras del caucho localizadas cerca la parte inferior de las bandas de rodamiento de los neumáticos. Cuando las bandas de rodamiento se han gastado a una profundidad específica, estas barras son visibles.

Weight The measurement of the earth's gravitational pull on an object.

Peso La medida de la atracción gravitacional de la tierra en un objeto.

Wheel alignment An adjustment and refitting of suspension parts to original specifications that ensures design performance.

Alineación de una rueda Puede definirse como un ajuste y una reparación de las piezas de la suspensión según las especificaciones originales, lo que asegura el rendimiento del diseño.

Wheelbase The distance between the center of the front and rear wheels.

Distancia entre ejes Distancia entre el centro de las ruedas delanteras y traseras.

Wheel offset The distance between the vertical wheel-mounting surface and the centerline of the wheel.

Desfasaje de rueda La distancia entre la superficie vertical de montaje de ruedas y la línea central de la rueda.

Wheel position sensor Is connected between each front and rear lower control arm and the chassis in a road-sensing suspension system. This sensor provides a computer input signal in relation to the amount and velocity of wheel movement.

Sensor para la posición de las ruedas Se conecta entre cada uno de los brazos de mando delantero y trasero y el chasis en un sistema de suspensión con equipo sensor. Este sensor envía una señal de entrada a la computadora referente a la cantidad y la velocidad del movimiento de la rueda.

Wheel rims Circular steel, aluminum, or magnesium components on which the tires are mounted. The wheel rim and tire assembly is bolted to the wheel hub.

Llantas de la rueda Componentes circulares de acero, aluminio o magnesio sobre los que se montan los neumáticos. El conjunto de la llanta de la rueda y el neumático se emperna al cubo de la rueda.

Wheel sensor An electronic unit attached to each wheel and connected to a four-wheel aligner.

Sensor de la rueda Una unedad electrónica prendida a cada rueda y conectada a un alineador de cuatro ruedas.

Wheel setback Occurs when one front or rear wheel is moved rearward in relation to the opposite front or rear wheel.

Retroceso de la rueda Ocurre cuando una rueda delantera o trasera se mueve hacia atrás con relación a la rueda delantera o trasera opuesta.

Wheel shimmy Rapid inward and outward oscillations of the front wheels.

Bailoteo de la rueda Oscilaciones rápidas hacia adentro y hacia afuera de las ruedas delanteras.

Wheel speed sensor Sends an AC voltage signal to the antilock brake system computer in relation to wheel speed.

Sensor de velocidad de la rueda Manda una señal de voltaje de corriente alterna a la computadora del sistema de freno antibloqueo en relación a la velocidad de la rueda.

Wheel tramp Rapid upward and downward wheel and tire movement.

Recorrido de la rueda Movimiento rápido ascendente y descendente de la rueda y el neumático.

Work The result of applying a force.

Trabajo Resultado de la aplicación de una fuerza.

Worm and gear or worm and sector design A steering gear developed in the early 1900s that required high steering effort because of internal friction.

Diseño del sinfín y engranaje o del sinfín y sector Un engranaje de dirección desarrollado en los principios de los años 1900 que requerían un gran esfuerzo en maniobro de dirección debido a la fricción interna.

Worm and roller steering gear A gear that contains a wormshaft and a roller-type sector.

Engranaje de dirección con sinfín y rodillo Un engranaje que contiene un eje de sinfín y un sector tipo rodillo.

Worm shaft The gear meshed with the pitman shaft sector in a recirculating ball steering gear.

Árbol del sinfín Engranaje endentado con el sector del árbol pitman en un mecanismo de dirección de bola recirculante.

Worm shaft bearing preload The amount of tension placed on the bearing by the adjustment procedure.

Carga previa del cojinete del árbol del sinfin Cantidad de tensión aplicada al cojinete por el procedimiento de ajuste.

Worm shaft end play The distance between the fully upward and fully downward worm shaft positions in a recirculating ball steering gear.

Holgadura del árbol del sinfín Distancia entre las posiciones completamente ascendente y descendente del árbol del sinfín en un mecanismo de dirección de bola recirculante.

Yaw forces Tend to cause the rear of the vehicle to swerve sideways.

Fuerzas de dirección tienden a causar que la parte posterior del vehículo viren bruscamente hacia los lados.

Yaw motion The tendency of the rear of a vehicle to swerve sideways during a turn.

Movimiento de derrape La tendencia de la parte posterior de un vehículo de desplazarse lateralmente durante un giro.

Yaw rate sensor Sends a voltage signal to the vehicle stability control computer in relation to sideways chassis movement.

Sensor de cantidad de desviación Manda una señal de voltaje a la computadora de control de estabilidad del vehículo en relación del movimiento lateral del chasis.

INDEX

T

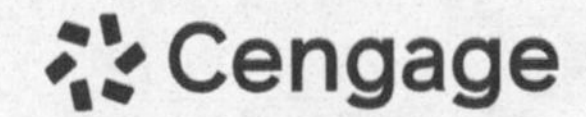

Today's Technician™: Classroom Manual for Automotive Electricity and Electronics, Eighth Edition

Barry Hollembeak

SVP, Product: Erin Joyner

VP, Product: Thais Alencar

Portfolio Product Director: Jason Fremder

Portfolio Product Manager: Emily Olsen

Learning Designer: Mary Clyne

Content Manager: Sangeetha Vijay, Lumina Datamatics Ltd.

Digital Project Manager: Elizabeth Cranston

VP, Product Marketing: Jason Sakos

Director, Product Marketing: Neena Bali

Content Acquisition Analyst: Erin McCullough

Production Service: Lumina Datamatics Ltd.

Designer: Felicia Bennett

Cover Image Source: IRINA SHI/Shutterstock.com.

Library of Congress Control Number: 2022921965

Classroom Manual ISBN: 978-0-357-76639-2

Package ISBN: 978-0-357-76638-5

Cengage
200 Pier 4 Boulevard
Boston, MA 02210
USA

Cengage is a leading provider of customized learning solutions. Our employees reside in nearly 40 different countries and serve digital learners in 165 countries around the world. Find your local representative at **www.cengage.com.**

To learn more about Cengage platforms and services, register or access your online learning solution, or purchase materials for your course, visit **www.cengage.com.**

Notice to the Reader

Publisher does not warrant or guarantee any of the products described herein or perform any independent analysis in connection with any of the product information contained herein. Publisher does not assume, and expressly disclaims, any obligation to obtain and include information other than that provided to it by the manufacturer. The reader is expressly warned to consider and adopt all safety precautions that might be indicated by the activities described herein and to avoid all potential hazards. By following the instructions contained herein, the reader willingly assumes all risks in connection with such instructions. The publisher makes no representations or warranties of any kind, including but not limited to, the warranties of fitness for particular purpose or merchantability, nor are any such representations implied with respect to the material set forth herein, and the publisher takes no responsibility with respect to such material. The publisher shall not be liable for any special, consequential, or exemplary damages resulting, in whole or part, from the readers' use of, or reliance upon, this material.

Printed Number: 6 Print Year: 2024
Printed in Mexico

TODAY'S TECHNICIAN™

CLASSROOM MANUAL

For Automotive Electricity and Electronics

EIGHTH EDITION

Barry Hollembeak

Cengage

Australia • Brazil • Canada • Mexico • Singapore • United Kingdom • United States

Shop Manual

To stress the importance of safe work habits, the Shop Manual dedicates one full chapter to safety. Other important features of this manual include the following:

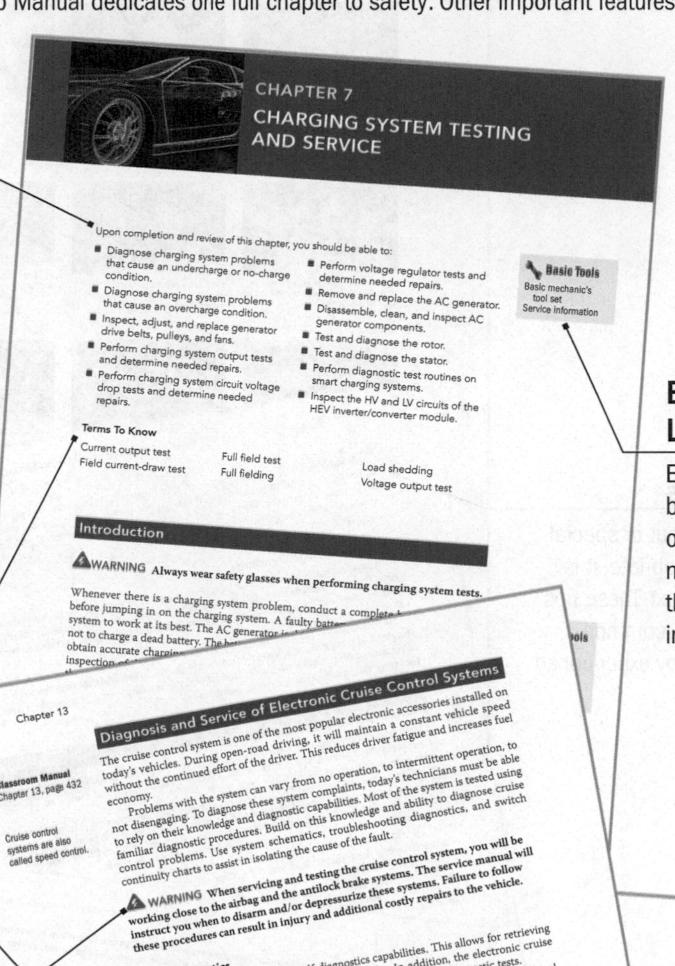

Performance-Based Objectives

These objectives define the contents of the chapter and define what the student should have learned upon completion of the chapter. These objectives also correspond with the list of required tasks for ASE certification. *Each ASE task is addressed*.

Although not designed to simply prepare someone for the ASE certification exams, the ASE task lists are one of the focuses of this textbook. These tasks are defined generically when the procedure is commonly followed and specifically when the procedure is unique for specific vehicle models. Procedures for imported- and domestic-model automobiles and light trucks are included.

Terms to Know List

Terms in this list are also defined in the Glossary at the end of the manual.

Special Tools List

Whenever a special tool is required to complete a task, it is listed in the margin next to the procedure.

Cautions and Warnings

Throughout the text, warnings alert the reader to potentially hazardous materials or unsafe conditions. Cautions advise the student of things that can go wrong if instructions are not followed or if an unacceptable part or tool is used.

Margin Notes

The most important Terms to Know are highlighted and defined in the margins. Common trade jargon also appears in the margins and gives some of the common terms used for components. This feature helps students understand and speak the language of the trade, especially when conversing with an experienced technician.

Basic Tools List

Each chapter begins with a list of the basic tools needed to perform the tasks included in the chapter.

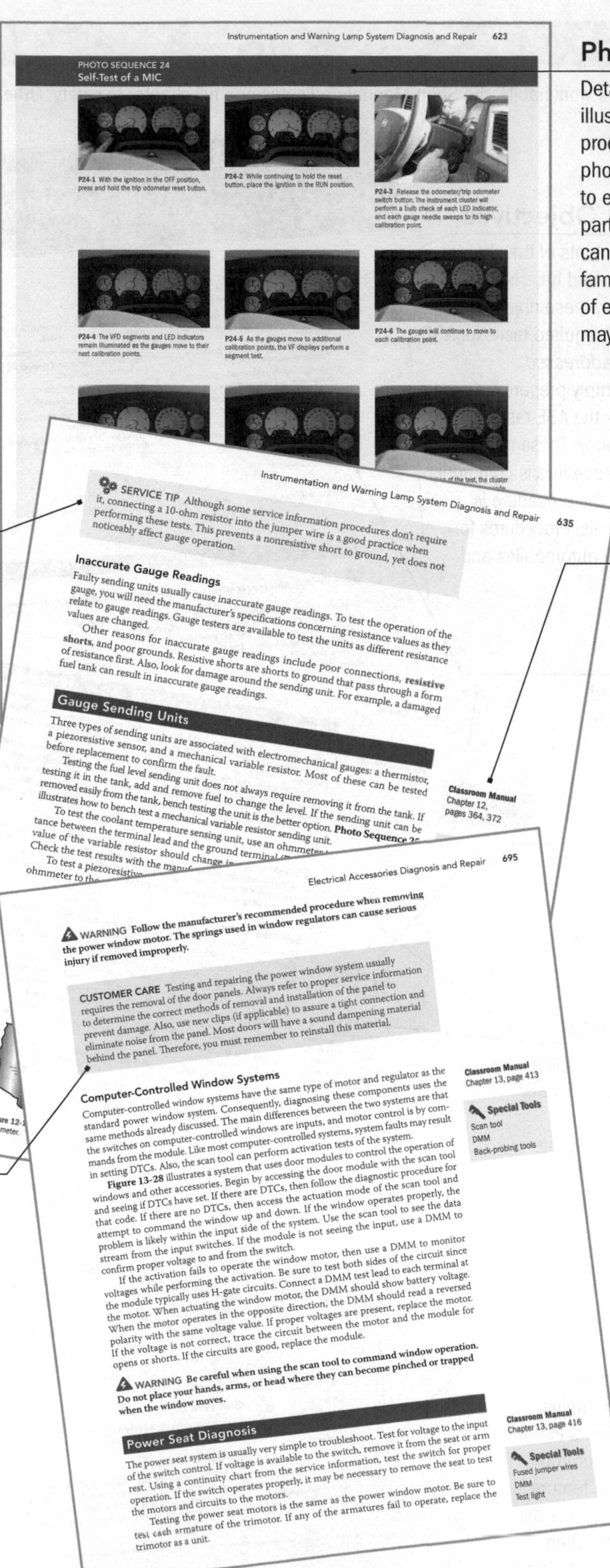

Photo Sequences

Detailed Photo Sequences illustrate many service procedures. These detailed photographs show students what to expect when they perform particular procedures. They can also provide the student a familiarity with a system or type of equipment, which the school may not have.

Cross-References to the Classroom Manual

Reference to the appropriate page in the Classroom Manual is given whenever necessary. Although the chapters of the two manuals are synchronized, material covered in other chapters of the Classroom Manual may be fundamental to the topic discussed in the Shop Manual.

Service Tips

Whenever a shortcut or special procedure is appropriate, it is described in the text. These tips generally describe common procedures used by experienced technicians.

Customer Care

This feature highlights those little things a technician can do or say to enhance customer relations.

Case Studies

Beginning with Chapter 3, each chapter ends with a case study describing a particular vehicle problem and the logical steps a technician might use to solve the problem.

ASE-Style Review Questions

Each chapter contains ASE-style review questions that reflect the performance-based objectives listed at the beginning of the chapter. These questions can be used to review the chapter as well as to prepare for the ASE certification exam.

ASE Challenge Questions

Each technical chapter ends with five ASE challenge questions. These are not mere review questions; rather, they test students' ability to apply general knowledge to the contents of the chapter. To answer these questions the student may need to do some research.

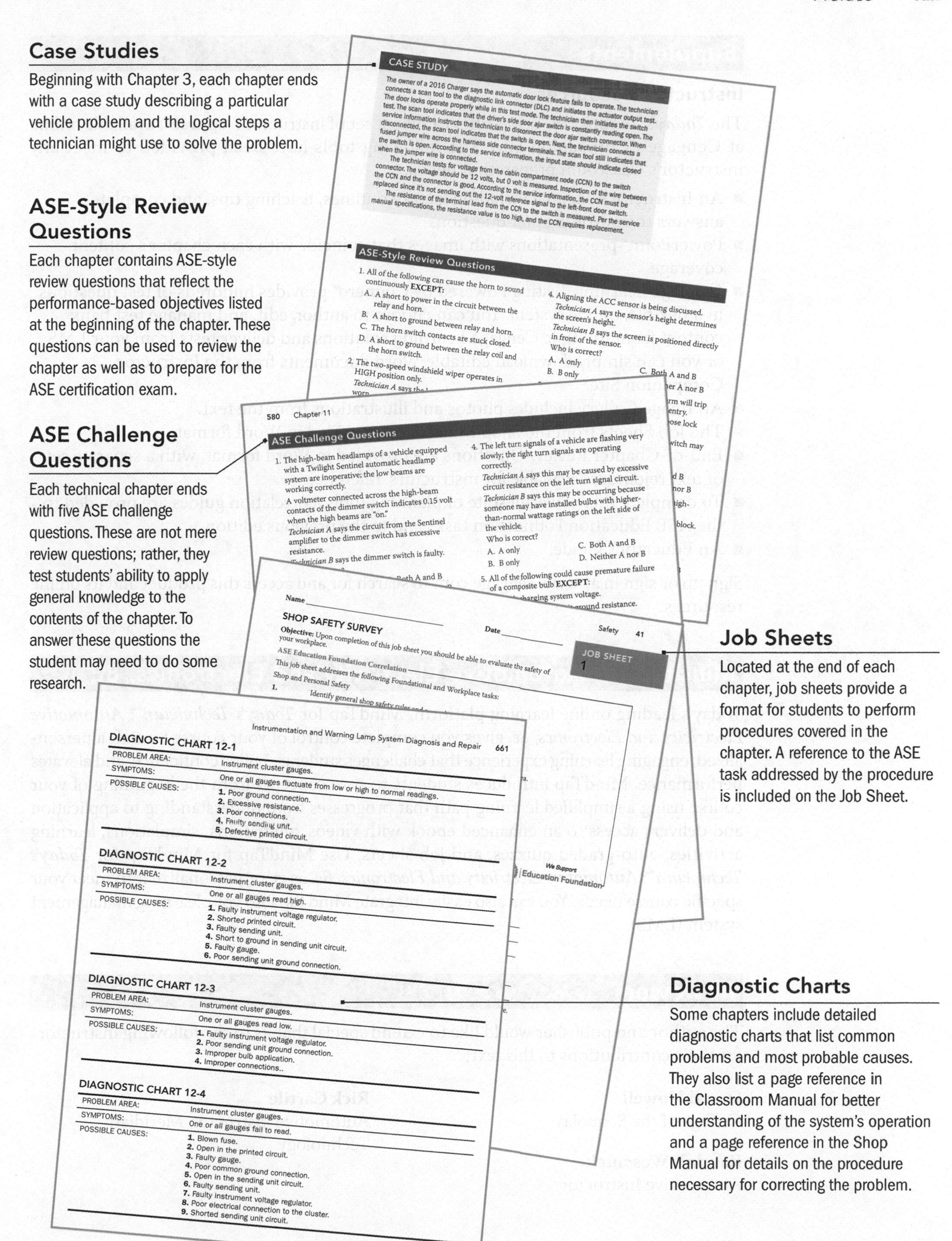

The owner of a 2016 Charger says the automatic door lock feature fails to operate. The technician connects a scan tool to the diagnostic link connector (DLC) and initiates the actuator output test. The door locks operate properly while in this test mode. The technician then initiates the switch test. The scan tool indicates that the driver's side door ajar switch is constantly reading open. The service information instructs the technician to disconnect the door ajar switch connector. When disconnected, the scan tool indicates that the switch is open. Next, the technician connects a fused jumper wire across the harness side connector terminals. The scan tool still indicates that the switch is open. According to the service information, the input state should indicate closed when the jumper wire is connected.

The technician tests for voltage from the cabin compartment node (CCN) to the switch connector. The voltage should be 12 volts, but 0 volt is measured. Inspection of the wire between the CCN and the connector is good. According to the service information, the CCN must be replaced since it's not sending out the 12-volt source signal to the left-front door switch.

The resistance of the terminal lead from the CCN to the switch is measured. Per the service manual specifications, the resistance value is too high, and the CCN requires replacement.

1. All of the following can cause the horn to sound continuously **EXCEPT:**
 A. A short to power in the circuit between the relay and horn.
 B. A short to ground between relay and horn.
 C. The horn switch contacts are stuck closed.
 D. A short to ground between the relay coil and the horn switch.

2. The two-speed windshield wiper operates in HIGH position only.
 Technician A says the l...
 worn...

4. Aligning the ACC sensor is being discussed.
 Technician A says the sensor's height determines the screen's height.
 Technician B says the screen is positioned directly in front of the sensor.
 Who is correct?
 A. A only
 B. B only
 C. Both A and B
 ...er A nor B

...rm will trip ...entry. ...ose lock ...witch may

...d B ...nor B ...high-...block.

CASE STUDY
ASE-Style Review Questions
580 Chapter 11
ASE Challenge Questions

1. The high-beam headlamps of a vehicle equipped with a Twilight Sentinel automatic headlamp system are inoperative; the low beams are working correctly.
 A voltmeter connected across the high-beam contacts of the dimmer switch indicates 0.15 volt when the high beams are "on."
 Technician A says the circuit from the Sentinel amplifier to the dimmer switch has excessive resistance.
 Technician B says the dimmer switch is faulty.
 ...Both A and B

4. The left turn signals of a vehicle are flashing very slowly; the right turn signals are operating correctly.
 Technician A says this may be caused by excessive circuit resistance on the left turn signal circuit.
 Technician B says this may be occurring because someone may have installed bulbs with higher-than-normal wattage ratings on the left side of the vehicle.
 Who is correct?
 A. A only
 B. B only
 C. Both A and B
 D. Neither A or B

5. All of the following could cause premature failure of a composite bulb **EXCEPT:**
 ...harging system voltage.
 ...ground resistance.

Name _____

SHOP SAFETY SURVEY

Objective: Upon completion of this job sheet you should be able to evaluate the safety of your workplace.

Date
Safety 41
JOB SHEET 1

ASE Education Foundation Correlation

This job sheet addresses the following Foundational and Workplace tasks:

Shop and Personal Safety

Identify general shop safety rules and...

We Support /Education Foundation

Instrumentation and Warning Lamp System Diagnosis and Repair 661
DIAGNOSTIC CHART 12-1
PROBLEM AREA: Instrument cluster gauges.
SYMPTOMS: One or all gauges fluctuate from low or high to normal readings.
POSSIBLE CAUSES: 1. Poor ground connection. 2. Excessive resistance. 3. Poor connections. 4. Faulty sending unit. 5. Defective printed circuit.
DIAGNOSTIC CHART 12-2
PROBLEM AREA: Instrument cluster gauges.
SYMPTOMS: One or all gauges read high.
POSSIBLE CAUSES: 1. Faulty instrument voltage regulator. 2. Shorted printed circuit. 3. Faulty sending unit. 4. Short to ground in sending unit circuit. 5. Faulty gauge. 6. Poor sending unit ground connection.
DIAGNOSTIC CHART 12-3
PROBLEM AREA: Instrument cluster gauges.
SYMPTOMS: One or all gauges read low.
POSSIBLE CAUSES: 1. Faulty instrument voltage regulator. 2. Poor sending unit ground connection. 3. Improper bulb application. 4. Improper connections..
DIAGNOSTIC CHART 12-4
PROBLEM AREA: Instrument cluster gauges.
SYMPTOMS: One or all gauges fail to read.
POSSIBLE CAUSES: 1. Blown fuse. 2. Open in the printed circuit. 3. Faulty gauge. 4. Poor common ground connection. 5. Open in the sending unit circuit. 6. Faulty sending unit. 7. Faulty instrument voltage regulator. 8. Poor electrical connection to the cluster. 9. Shorted sending unit circuit.

Job Sheets

Located at the end of each chapter, job sheets provide a format for students to perform procedures covered in the chapter. A reference to the ASE task addressed by the procedure is included on the Job Sheet.

Diagnostic Charts

Some chapters include detailed diagnostic charts that list common problems and most probable causes. They also list a page reference in the Classroom Manual for better understanding of the system's operation and a page reference in the Shop Manual for details on the procedure necessary for correcting the problem.

Supplements

Instructor Resources

The *Today's Technician*™ series offers a robust set of instructor resources, available online at Cengage's Companion Site. The following tools have been provided to meet any instructor's classroom preparation needs:

- An Instructor's Manual including lecture outlines, teaching tips, and complete answers to end-of-chapter questions.
- PowerPoint® presentations with images that coincide with each chapter's content coverage.
- Cengage Learning Testing Powered by Cognero® provides hundreds of test questions in a flexible, online system. You can choose to author, edit, and manage test bank content from multiple Cengage Learning solutions and deliver tests from your LMS, or you can simply download editable Word documents from the Instructor Companion Site.
- An Image Gallery includes photos and illustrations from the text.
- The Job Sheets from the Shop Manual are provided in Word format.
- End-of-Chapter Review Questions are provided in Word format, with a separate set of text rejoinders available for instructors' reference.
- To complete this powerful suite of planning tools, correlation guides are provided to the ASE Education Foundation tasks and to the previous edition.
- An Educator's Guide.

Sign up or sign in at www.cengage.com to search for and access this product and its online resources.

Mindtap for Automotive Electricity and Electronics, 8E

Today's leading online learning platform, MindTap for *Today's Technician*™: *Automotive Electricity and Electronics, 8e*, gives you complete control of your course to craft a personalized, engaging learning experience that challenges students, builds confidence, and elevates performance. MindTap introduces students to core concepts from the beginning of your course using a simplified learning path that progresses from understanding to application and delivers access to an enhanced ebook with videos, animations, simulations, learning activities, auto-graded quizzes, and job sheets. Use MindTap for MindTap for *Today's Technician*™: *Automotive Electricity and Electronics, 8e*, as is, or personalize it to meet your specific course needs. You can also easily integrate MindTap into your learning management system (LMS).

Reviewers

The author and publisher would like to extend special thanks to the following instructors for their contributions to this text:

Donal Howell
College of the Sequoias

Randall Wesenick
Automotive Instructor

Rick Cartile
Automotive Instructor, Meridian Technology

Derrek Keesling
Department Chair Automotive Technology College of Lake County

David Shields
Meridian Technology CenterMeridian Technology Center

Joseph Gumina
Professor of Automotive City College of San Francisco

Orlando Grijalva
Automotive Instructor Tarrant County College

Tony DeVillier
Associate Professor. Delgado Community College

Eric Gomez
Transportation Education Director Eastern New Mexico University Roswell

CHAPTER 1

INTRODUCTION TO AUTOMOTIVE ELECTRICAL AND ELECTRONIC SYSTEMS

Upon completion and review of this chapter, you should be able to:

- Explain the importance of learning automotive electrical systems.
- Explain the role of electrical systems in today's vehicles.
- Explain the interaction of the electrical systems.
- Describe the purpose of the starting system.
- Describe the purpose of the charging system.
- Describe the role of the computer in today's vehicles.
- Explain the purpose of vehicle communication networks.
- Describe the purpose of various electronic accessory systems.
- Explain the purpose of passive restraint systems.
- Explain what constitutes an alternate propulsion system.

Terms to Know

Air bag systems
Antitheft system
Automatic door locks (ADLs)
Bus
Charging system
Computer
Cruise control
Easy exit
Electric defoggers
Electric vehicle (EV)
Electrical accessories
Fuel cell
Hybrid electric vehicle (HEV)
Keyless entry system
Lighting system
Memory seat
Multiplexing
Network
Neutral safety switch
Passive restraint systems
Starting system
Vehicle instrumentation systems
Voltage regulator

Introduction

You are probably reading this book for one of two reasons. Either you are preparing yourself to enter the field of automotive service or you are expanding your skills to include automotive electrical systems. In either case, congratulations on selecting one of the most fast-paced segments of the automotive industry. Working with electrical systems can be challenging yet very rewarding; however, it can also be very frustrating at times.

For many people, learning electrical systems can be a struggle. It's my hope that I present the material to you in such a manner that you will not only understand electrical systems but also excel at it. There are many ways the theory of electricity can be explained, and many metaphors can be used. Some compare electricity to the flow of water, while others explain it in a purely scientific approach. Everyone learns differently. I am presenting electrical theory in a manner that I hope will be clear and concise. If you do not fully comprehend a concept, it's essential to discuss it with your instructor. Your instructor may be able to use a slightly different method of instruction to help you completely understand the concept. Electricity is somewhat abstract, so if you do have questions, be sure to ask your instructor.

Why Become an Electrical System Technician?

In the past, technicians could work their entire careers and almost avoid the vehicle's electrical systems altogether. They would specialize in engines, steering/suspension, or brakes. Today there's no system on the vehicle that is immune to the role of electrical circuits. Engine controls, electronic suspension systems, and antilock brakes are standard on today's vehicles. Even electrical systems that were once thought of as simple have evolved to computer controls. Headlights are now pulse-width modulated using high-side drivers and will automatically brighten and dim based on the light intensity of oncoming traffic. Today's vehicles are equipped with 20 or more computers, laser-guided cruise control, sonar park assist, infrared climate control, fiber optics, and radio frequency transponders and decoders. Simple systems have become more computer reliant. For example, the horn system on the 2021 Chrysler 300 uses three separate control modules and two bus networks to function. With the addition of tire pressure monitoring systems, even the tire and wheel assemblies have computers involved.

Today's technician must possess a comprehensive knowledge of electricity and electronics to succeed. The future will provide great opportunities for those technicians who have prepared themselves properly.

The Role of Electricity in the Automobile

In the past, electrical systems were basically stand-alone. For example, the ignition system was responsible only for supplying the voltage needed to fire the spark plugs. Ignition timing was controlled by vacuum and mechanical advance systems. Today there are very few electrical systems that are still independent and stand on their own.

Today, most manufacturers **network** their electrical systems together through computers. Networking means that systems can share information with each other. The result may be that a faulty component may cause several symptoms. Consider the following example. The wiper system can interact with the headlight system to turn on the headlights when the wipers are turned on. The wipers can interact with the vehicle speed sensor to provide for speed-sensitive wiper operation. The speed sensor may provide information to the antilock brake module. The antilock brake module can then share this information with the transmission control module, and the instrument cluster can receive vehicle speed information to operate the speedometer. If the vehicle speed sensor should fail, this could result in no antilock brake operation, and a warning light turned on in the dash. But it could also result in the speedometer not functioning, the transmission not shifting, and the wipers not operating properly.

Introduction to the Electrical Systems

This section aims to acquaint you with the electrical systems that are covered in this book. Here, we will define the purpose of these systems.

AUTHOR'S NOTE The discussion of the systems in this section of the chapter provides you with an understanding of their primary purpose. Some systems have secondary functions. These will be discussed in detail in later chapters.

The Starting System

The **starting system** is a combination of mechanical and electrical parts that work together to start the engine. The starting system is designed to change the electrical energy, which is being supplied by the battery, into mechanical energy. For this conversion to be accomplished, a starter is used. The basic starting system includes the following components (**Figure 1-1**):

1. Battery.
2. Cable and wires.
3. Ignition switch.
4. Starter solenoid or relay.
5. Starter motor.
6. Starter drive and flywheel ring gear.
7. Starter safety switch.

The starter motor (**Figure 1-2**) requires large amounts of current (up to 400 amps [A]) to generate the torque needed to turn the engine. The conductors used to carry this amount of current (battery cables) must be large enough to handle the current. These cables may be close to 1/2 inch (12.7 mm) in diameter. It would be impractical to place a conductor of this size into the wiring harness to the ignition switch. To control the high current, all starting systems contain some type of magnetic switch. There are two basic types of magnetic switches used: the solenoid and the relay.

The ignition switch is the power distribution point for most of the vehicle's primary electrical systems. The conventional ignition switch is spring loaded in the start position. This momentary contact automatically moves the contacts to the RUN position when the driver releases the key. All other ignition switch positions are detent positions.

Most of today's vehicles have replaced the conventional ignition switch with a module that provides switch statues and input data to other computers. For example, the ignition switch may broadcast that the ignition is in the RUN position as a "wake up" signal to the vehicle computers. This ignition module will broadcast to the powertrain control module (PCM) that the driver has placed the ignition into the START position, and the PCM activates the starter solenoid.

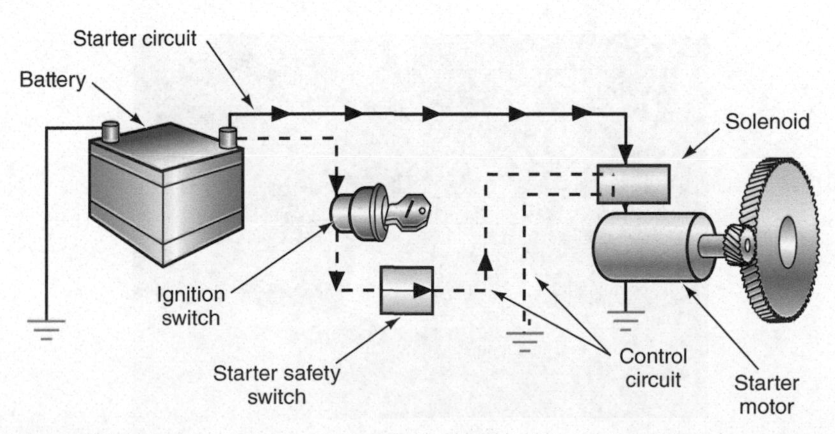

Figure 1-1 Major components of a conventional starting system.

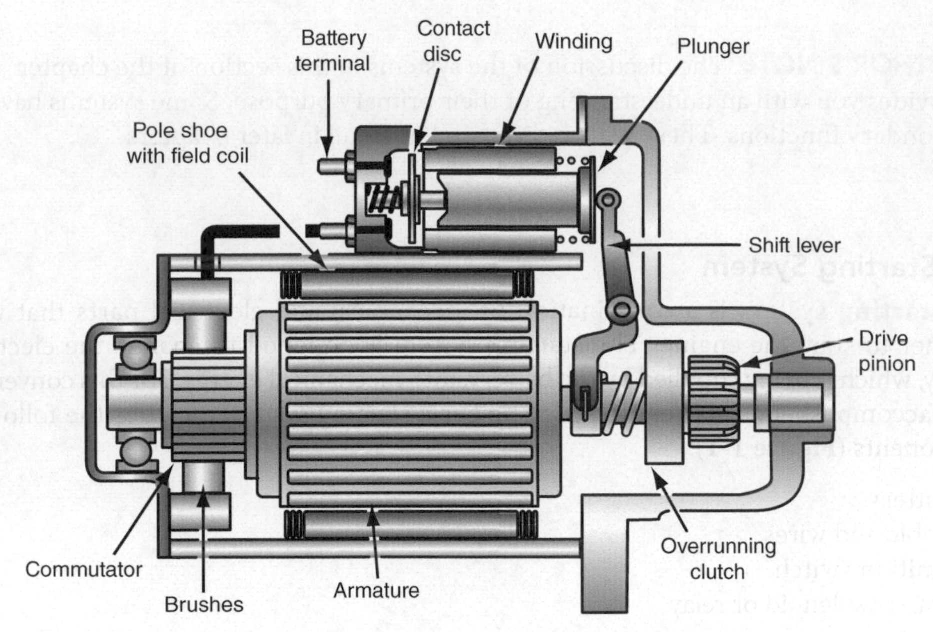

Figure 1-2 Starter motor.

The **neutral safety switch** is used on vehicles equipped with automatic transmissions. The switch is a safety feature that prevents the engine from cranking unless the transmission is in PARK or Neutral. The switch contacts open the starter control circuit when the transmission shift selector is in any position except PARK or NEUTRAL. Without this feature, the vehicle would lunge forward or backward once it's started, causing personal injury or property damage. The safety switch is connected into the starting system control circuit and is usually operated by the shift lever (**Figure 1-3**). When the transmission is in the PARK or NEUTRAL position, the switch allows current to flow to the starter circuit. If the transmission is in a gear position, the switch prevents current flow to the starter circuit. The safety switch function may also be incorporated within the transmission control module (TCM).

Most manual transmission equipped vehicles use a similar type of safety switch. The start/clutch interlock switch is usually operated by the movement of the clutch pedal (**Figure 1-4**).

The Charging System

The automotive storage battery cannot meet the demands of the various electrical systems for an extended period of time. Every vehicle must be equipped with a means of replacing the energy drawn from the battery. A **charging system** restores the electrical power to the battery

Figure 1-3 The neutral safety switch is usually attached to the transmission.

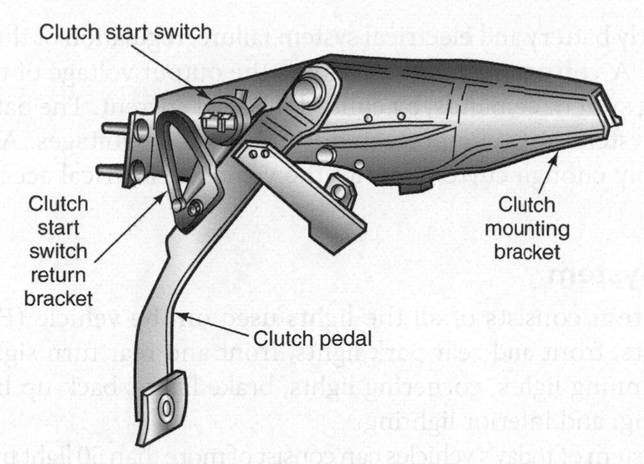

Figure 1-4 Most vehicles with a manual transmission use a clutch start switch.

that was used during engine starting. In addition, the charging system must be able to react quickly to high-load demands required of the electrical system. It's the vehicle's charging system that generates the current to operate all electrical accessories while the engine is running.

The purpose of the charging system is to convert the engine's mechanical energy into electrical energy to recharge the battery and run the electrical accessories. When the engine is first started, the battery supplies all the current required by the starting and ignition systems.

As illustrated in **Figure 1-5**, the entire charging system consists of the following components:

1. Battery.
2. Alternating current (AC) generator or direct current (DC) generator.
3. Drive belt.
4. Voltage regulator.
5. Charge indicator (lamp or gauge).
6. Ignition switch.
7. Cables and wiring harness.
8. Starter relay (some systems).
9. Fusible link (some systems).

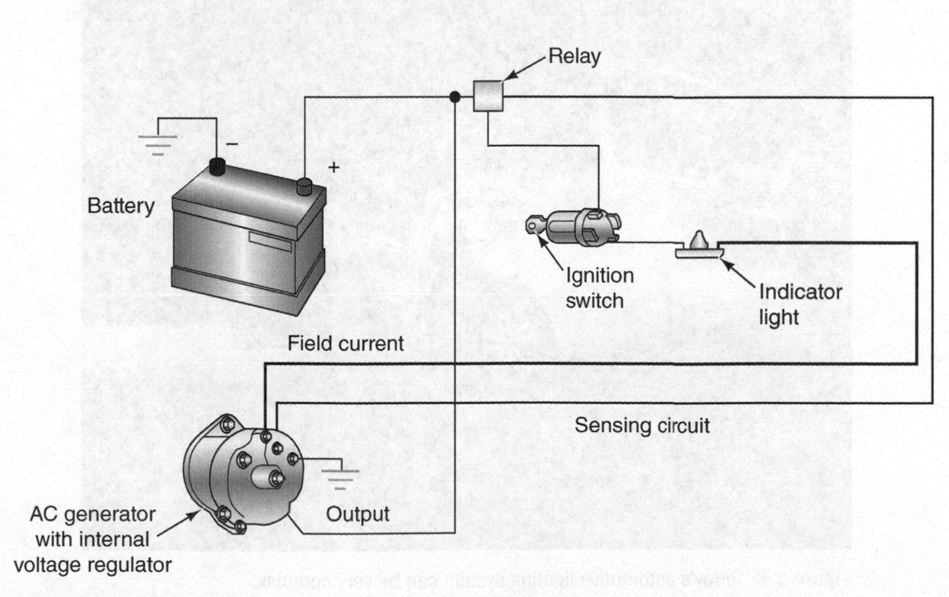

Figure 1-5 Components of a conventional charging system.

To prevent early battery and electrical system failure, regulation of the charging system is very important. A **voltage regulator** controls the output voltage of the AC generator, based on charging system demands, by controlling field current. The battery, and the rest of the electrical system, must be protected from excessive voltages. Also, the charging system must supply enough current to run the vehicle's electrical accessories when the engine is running.

The Lighting System

The **lighting system** consists of all the lights used on the vehicle (**Figure 1-6**). This includes headlights, front and rear park lights, front and rear turn signals, side marker lights, daytime running lights, cornering lights, brake lights, back-up lights, instrument cluster backlighting, and interior lighting.

The lighting system of today's vehicles can consist of more than 50 light bulbs and hundreds of feet of wiring. Incorporated within these circuits are circuit protectors, relays, switches, lamps, and connectors. In addition, more sophisticated lighting systems use computers and sensors. Since federal laws largely regulate the lighting circuits, the systems are similar among the various manufacturers. However, some variations exist in these circuits.

With the addition of computer controls in the automobile, manufacturers have incorporated several different lighting circuits or modified the existing ones. Some of the refinements made to the lighting system include automatic headlight washers, automatic headlight dimming, automatic on/off with time-delay headlights, and illuminated entry systems. Some of these systems use sophisticated body computer–controlled circuitry and fiber optics.

Some manufacturers have included such basic circuits as turn signals into their body computer to provide for pulse-width dimming in place of a flasher unit. The body computer can also control instrument panel lighting based on inputs that include if the side marker lights are on or off. By using the body computer to control many of the lighting circuits, the amount of wiring has been reduced. In addition, the use of computer control of these systems has provided a means of self-diagnosis in some applications.

Today, high-density discharge and light emitting diode (LED) headlamps are becoming increasingly popular options on many vehicles. These headlights provide improved lighting over conventional headlamps.

Figure 1-6 Today's automotive lighting system can be very complex.

Vehicle Instrumentation Systems

Vehicle instrumentation systems (**Figure 1-7**) monitor the various vehicle operating systems and provide information to the driver about their correct operation. Warning devices also provide information to the driver; however, they are usually associated with an audible signal. Some vehicles use a voice module to alert the driver to certain conditions.

Electrical Accessories

Electrical accessories provide for additional safety and comfort. There are many electrical accessories that can be installed into today's vehicles. These include safety accessories such as the horn, windshield wipers, and windshield washers. Comfort accessories include the blower motor, electric defoggers, power mirrors, power windows, power seats, and power door locks.

Horns. A horn is a device that produces an audible warning signal (**Figure 1-8**). Most automotive electrical horns operate by vibrating a diaphragm to produce a warning signal. The vibration of the diaphragm is repeated several times per second. As the diaphragm vibrates, it causes a column of air in the horn to vibrate. The vibration of the column of air produces sound.

The horn may also be a computer-generated sound. An electronic sound generator creates the sound. The driver can tune the tone of the horn to their preference.

Figure 1-7 The instrument panel displays various operating conditions.

Figure 1-8 Automotive horn.

Windshield Wipers. Windshield wipers are mechanical arms that sweep back and forth across the windshield to remove water, snow, or dirt (**Figure 1-9**). The motion of the wiper arm is transferred by a wiper motor that is connected to the wiper linkage. Most windshield wiper motors use permanent magnet or electromagnetic field motors.

Electric Defoggers. Electric defoggers heat the rear window to remove ice and/or condensation. Some vehicles use the same circuit to heat the outside mirrors. When electrical current flows through a resistance, heat is generated. Rear window defoggers use this principle of controlled resistance to heat the glass. The resistance is through a grid that is baked on inside of the glass (**Figure 1-10**). The system may incorporate a timer circuit that controls the relay.

Power Mirrors. Power mirrors are outside mirrors that are electrically positioned from the inside of the driver compartment. The electrically controlled mirror allows the driver to position the outside mirrors using a switch. The mirror assembly will use built-in motors.

Power Windows. Power windows are windows that are raised and lowered by the use of electrical motors. The motor used in the power window system is capable of operating in forward and reverse rotations. The power window system usually consists of the following components:

1. Master control switch.
2. Individual control switches.
3. Individual window drive motors.

Figure 1-9 Windshield wipers.

Figure 1-10 Rear window defogger grid.

Power Door Locks. Electric power door locks use either a solenoid or a motor to lock and unlock the door. Typically, power door lock systems provide for the locking or unlocking of all doors with a single button push. This system may also incorporate automatic door locking, as discussed later.

Computers

A **computer** is an electronic device that stores and processes data and can operate other devices (**Figure 1-11**). The use of computers on automobiles has expanded to include control and operation of several functions, including climate control, lighting circuits, cruise control, antilock braking, electronic suspension systems, and electronic shift transmissions. Some of these are functions of what is known as a body control module (BCM). Some body computer–controlled systems include direction lights, rear window defogger, illuminated entry, intermittent wipers, and other systems that were once thought of as basic.

A computer processes the physical conditions that represent information (data). The operation of the computer is divided into four basic functions:

1. Input.
2. Processing.
3. Storage.
4. Output.

Vehicle Communication Networks

All manufacturers now use a system of vehicle communications called **multiplexing (MUX)** to allow control modules to share information (**Figure 1-12**). Multiplexing provides the ability to use a single circuit to distribute and share data between several control modules throughout the vehicle. Because the data is transmitted through a single circuit, bulky wiring harnesses are eliminated.

Vehicle manufacturers will use multiplexing systems to enable different control modules to share information. A MUX wiring system uses **bus** data links that connect each module. The term *bus* refers to the transporting of data from one module to another. Each module can transmit and receive digital codes over the bus data links. The signal sent from a sensor can go to any one of the modules and can be shared by the other modules.

Figure 1-11 A control module is used to process data and operate different automotive systems.

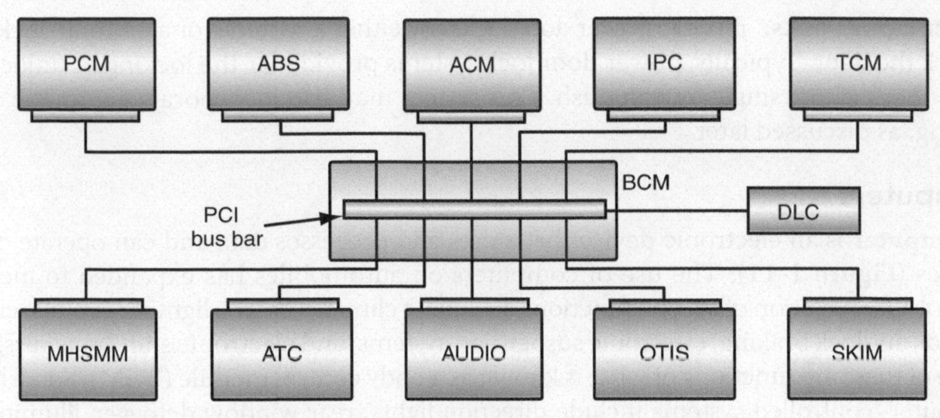

Figure 1-12 Automotive computers are networked together through multiplexing.

Electronic Accessory Systems

With the growing use of computers, most systems can be controlled electronically. This provides for improved monitoring of the systems for proper operation and the ability to detect if a fault occurs. The systems that are covered in this book include the following:

Cruise control systems are also referred to as *speed control.*

Electronic Cruise Control Systems. **Cruise control** is a system that allows the vehicle to maintain a preset speed with the driver's foot off the accelerator. Most cruise control systems are a combination of electrical and mechanical components.

Memory Seats. The **memory seat** feature allows the driver to program different seat positions that can be recalled at the push of a button. The memory seat feature is an addition to the basic power seat system. Most memory seat systems share the same basic operating principles, the difference being in programming methods and the number of positions that can be programmed. Most systems provide for two seat positions to be stored in memory.

An **easy exit** feature may be an additional function of the memory seat that provides for easier entrance and exit of the vehicle by moving the seat all the way back and down. Some systems also move the steering wheel up and to full retract.

Electronic Sunroofs. Some manufacturers have introduced electronic control of their electric sunroofs. These systems incorporate a pair of relay circuits and a timer function into the control module. Motor rotation is controlled by relays that are activated based on signals received from the slide, tilt, and limit switches.

Antitheft Systems. The **antitheft system** is a deterrent system designed to scare off would-be thieves by sounding alarms and/or disabling the ignition system. **Figure 1-13** illustrates many of the common components used in an antitheft system. These components include the following:

1. An electronic control module.
2. Door switches at all doors.
3. Trunk key cylinder switch.
4. Hood switch.
5. Starter inhibitor relay.
6. Horn relay.
7. Alarm.

In addition, many systems incorporate the exterior lights into the system. The lights are flashed if the system is activated.

Some systems use ultrasonic sensors to signal the control module if someone attempts to enter the vehicle through the door or window. The sensors can be placed to sense the parameter of the vehicle and sound the alarm if someone enters within the protected parameter distance.

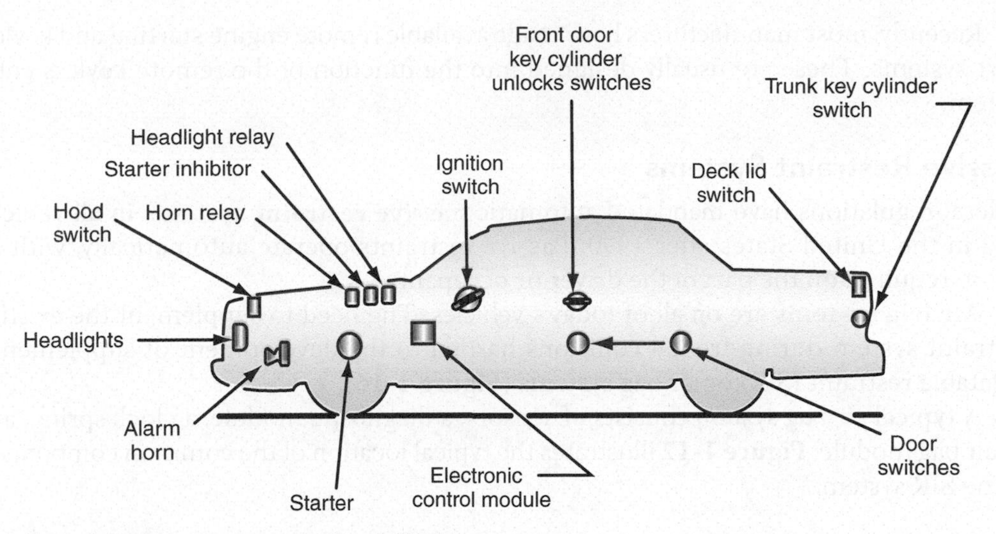

Figure 1-13 Typical components of an antitheft system.

Automatic Door Locks. **Automatic door locks (ADLs)** are passive systems that lock all doors when the required conditions are met. Many automobile manufacturers incorporate automatic door locks as an additional safety and convenience system. Most systems lock the doors when the gear selector is placed in DRIVE, the ignition switch is in RUN, and all doors are shut. Some systems will lock the doors when the gear shift selector is passed through the REVERSE position, while others do not lock the doors unless the vehicle is moving 15 mph or faster.

In addition, the system may unlock one or all doors when the vehicle is stopped and the driver's door is opened. The sequence of which the doors are unlocked may be a programmable feature that vehicle owners can set to their preference.

The system may use the body computer or a separate controller to control the door lock relays. The controller (or body computer) takes the place of the door lock switches for automatic operation.

Keyless Entry. The **keyless entry system** allows the driver to unlock the doors or the deck lid (trunk) from outside of the vehicle without using a key. The main components of the keyless entry system are the control module, a coded-button keypad located on the driver's door (**Figure 1-14**), and the door lock motors.

Some keyless entry systems can be operated remotely. Pressing a button on a hand-held transmitter will allow the operation of the system from distances of 25 to 50 feet (**Figure 1-15**).

Figure 1-14 Keyless entry system keypad.

Figure 1-15 Remote keyless entry system transponder.

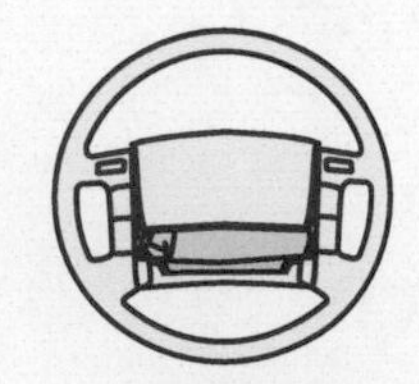
Stage 1

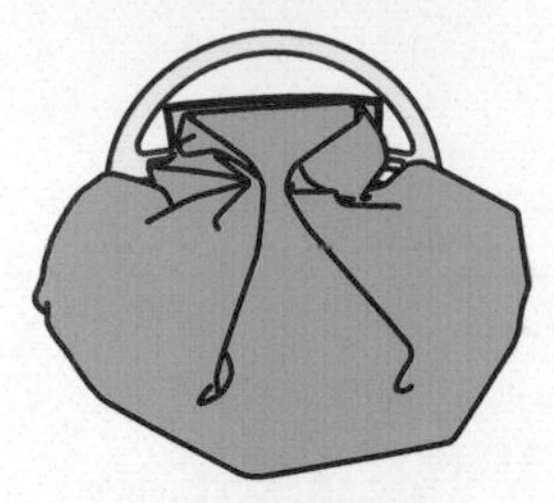
Stage 2

Stage 3

Figure 1-16 Air bag deployment sequence.

Recently, most manufacturers have made available remote engine starting and keyless start systems. These are usually designed into the function of the remote keyless entry system.

Passive Restraint Systems

Federal regulations have mandated automatic **passive restraint systems** in all vehicles sold in the United States after 1990. Passive restraints operate automatically, with no action required on the part of the driver or occupant.

Air bag systems are on all of today's vehicles. The need to supplement the existing restraint system during frontal collisions has led to the development of supplemental inflatable restraint (SIR) or air bag systems (**Figure 1-16**).

A typical air bag system consists of sensors, a diagnostic module, a clock spring, and an air bag module. **Figure 1-17** illustrates the typical location of the common components of the SIR system.

Alternate Propulsion Systems

Due to the increase in regulations concerning emissions and the public's desire to become less dependent on carbon fuel, many automotive manufacturers have developed alternative fuel or alternate power vehicles. Many automobile manufacturers have developed **electric vehicles (EVs)**. The primary advantage of an EV is a drastic reduction in noise and emission levels. General Motors introduced the EV1 electric car to the market in 1996. The original battery pack in this car contained twenty-six 12-volt (V) batteries that delivered electrical energy to a three-phase 102-kilowatt (kW) AC electric motor. The electric motor is used to drive the front wheels. The driving range is about 70 miles (113 km) of city driving or 90 miles (145 km) of highway driving.

With improvements in battery technology and more user-friendly charging methods, the 100% electric vehicle gained popularity in some market segments. The Nissan Leaf (**Figure 1-18**) is an example of a 100% electric, zero emissions vehicle. This vehicle uses an 80 kW synchronous electric motor and a single speed constant ratio transmission. The lithium-ion battery provides a driving range between 73 miles (117 km) and 109 miles (175 km) between charges.

EV battery limitation was a major stumbling block to most consumers. One method of improving the electric vehicle resulted in the addition of an on-board power generator that is assisted by an internal combustion engine, resulting in the **hybrid electric vehicle (HEV)**.

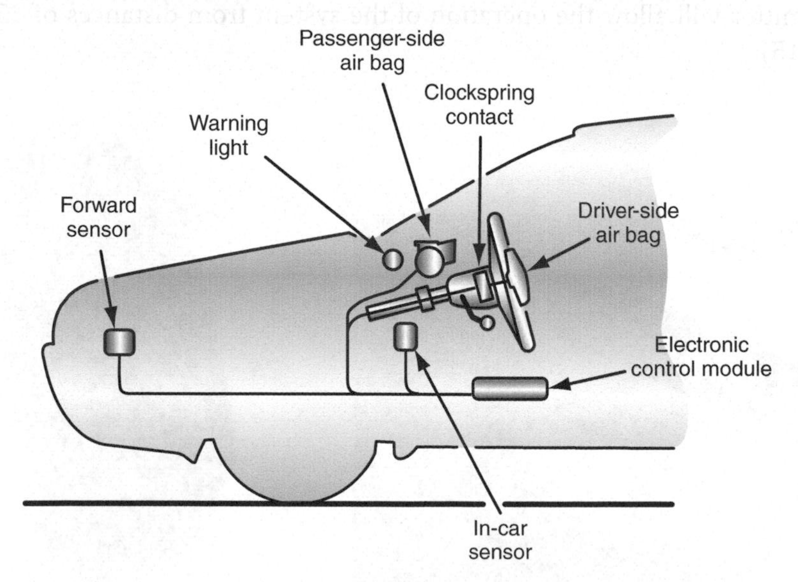

Figure 1-17 Typical location of components of the air bag system.

Figure 1-18 The Nissan Leaf is an all-electric vehicle.

Figure 1-19 HEV power system.

The HEV relies on power from the electric motor, the engine, or both (**Figure 1-19**). During normal driving conditions, the engine is the primary power source. When the vehicle moves from a stop and has a light load, the electric motor moves the vehicle. Power for the electric motor comes from stored energy in the battery pack. Engine power is also used to rotate a generator that recharges the storage batteries. The output from the generator may also be used to power the electric motor, which is run to provide additional power to the powertrain. A computer controls the operation of the electric motor depending on the power needs of the vehicle. During full throttle or heavy load operation, additional electricity from the battery is sent to the motor to increase the output of the powertrain.

The idea of **fuel cell**–powered vehicles has been researched for several years. The fuel cell vehicle combines the reach of conventional internal combustion engines with high efficiency, low fuel consumption, and minimal or no pollutant emission. At the same time, they are extremely quiet. Because they work with regenerative fuels such as hydrogen, they reduce the dependence on fossil fuels.

A fuel cell–powered vehicle (**Figure 1-20**) is basically an electric vehicle. As an electric vehicle, it uses an electric motor to supply torque to the drive wheels. The difference is that the fuel cell produces and supplies electric power to the electric motor instead of batteries. Many prototype fuel cell vehicles have been produced, with some placed in fleets in North America and Europe. Many vehicle manufacturers and independent laboratories are involved in fuel cell research and development programs.

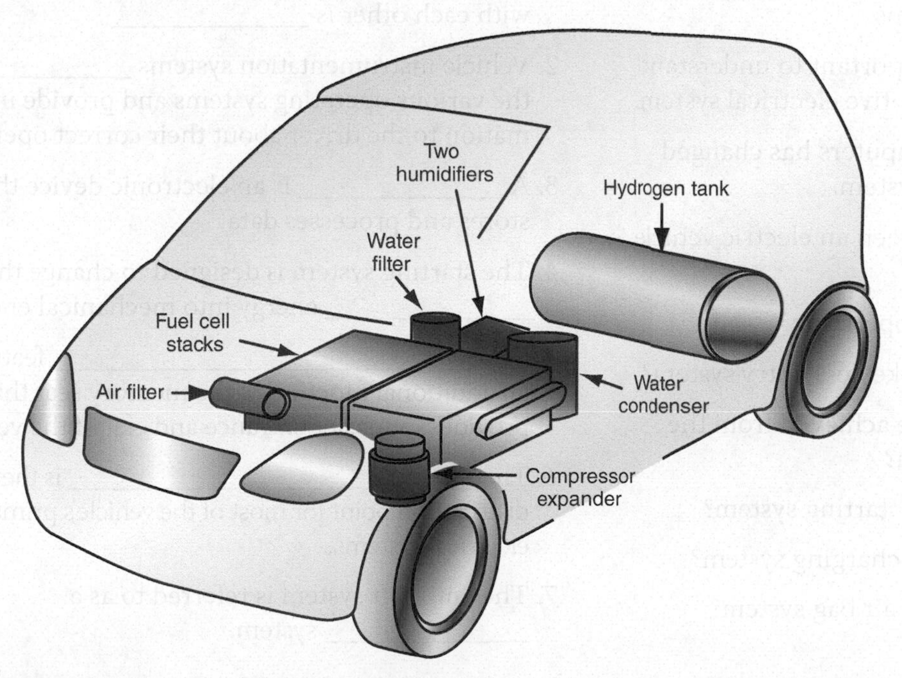

Figure 1-20 Fuel cell vehicle components.

Summary

- The starting system is a combination of mechanical and electrical parts that work together to start the engine.
- The charging system replaces the electrical power used by the battery and provides current to operate the electrical accessories while the engine is running.
- The lighting system consists of all the lights used on the vehicle.
- Vehicle instrumentation systems monitor the various vehicle operating systems and provide information to the driver.
- Electrical accessories provide additional safety and comfort.
- Many basic electrical accessory systems have electronic controls added to them to provide additional features and enhancement.
- Computers are electronic devices that gather, store, and process data.
- Most vehicles use a multiplexing system to share information between computer systems.
- The memory seat feature allows the driver to program different seat positions that can be recalled at the push of a button.
- Some manufacturers have introduced electronic control of their electric sunroofs. These systems incorporate a pair of relay circuits and a timer function into the control module.
- Antitheft systems are deterrent systems designed to scare off would-be thieves by sounding alarms and/or disabling the ignition system.
- Automatic door locks are a passive system used to lock all doors when the required conditions are met. Many automobile manufacturers incorporate the system as an additional safety and convenience feature.
- Passive restraints operate automatically, with no action required on the part of the driver or occupant.
- Electric vehicles powered by an electric motor run off a battery pack.
- The hybrid electric vehicle relies on power from the electric motor, the engine, or both.
- A fuel cell–powered vehicle is basically an electric vehicle, except that the fuel cell produces and supplies electric power to the electric motor instead of batteries.

Review Questions

Short-Answer Essays

1. Describe your level of comfort concerning automotive electrical systems.
2. Explain why you feel it's important to understand the operation of the automotive electrical system.
3. Explain how the use of computers has changed the automotive electrical system.
4. Explain the difference between an electric vehicle and a fuel cell vehicle.
5. Explain the basics of HEV operation.
6. What is the purpose of the keyless entry system?
7. What safety benefits can be achieved from the automatic door lock system?
8. What is the purpose of the starting system?
9. What is the purpose of the charging system?
10. What is the function of the air bag system?

Fill in the Blanks

1. The ability of computers to share information with each other is ______________.
2. Vehicle instrumentation systems ______________ the various operating systems and provide information to the driver about their correct operation.
3. A ______________ is an electronic device that stores and processes data.
4. The starting system is designed to change the ______________ energy into mechanical energy.
5. The ______________ ______________ feature is an additional function of the memory seat that provides for easier entrance and exit of the vehicle.
6. The ______________ ______________ is the power distribution point for most of the vehicle's primary electrical systems.
7. The antitheft system is referred to as a ______________ system.

8. The purpose of the charging system is to convert the ______________ energy of the engine into ______________ energy.
9. ______________ restraints operate automatically, with no action required on the part of the driver.
10. The ______________ ______________ ______________ uses an on-board power generator that is assisted by an internal combustion engine.

Multiple Choice

1. Electric vehicles power the motor by:
   A. A generator.
   B. A battery pack.
   C. An engine.
   D. None of these choices.
2. The charging system:
   A. Provides electrical energy to operate the electrical system while the engine is running.
   B. Restores the energy to the battery after starting the engine.
   C. Uses the principle of magnetic induction to generate electrical power.
   D. All of these choices.
3. The memory seat system:
   A. Operates separately of the power seat system.
   B. Requires the vehicle to be moving before the seat position can be recalled.
   C. Allows for the driver to program different seat positions that can be recalled at the push of a button.
   D. Can be equipped only on vehicles with manual position seats.
4. The following are true about the easy exit feature EXCEPT:
   A. It's an additional function of the memory seat.
   B. The driver's door is opened automatically.
   C. The seat is moved all the way back and down.
   D. The system may move the steering wheel up.
5. The following are components of the starting system EXCEPT:
   A. The flywheel ring gear.
   B. Neutral safety switch.
   C. Harmonic balancer.
   D. Battery.
6. Which of the following is the most correct statement?
   A. Automotive electrical systems are interlinked with each other.
   B. All automotive electrical systems function the same on every vehicle.
   C. Manufacturers are required by federal legislation to limit the number of computers used on today's vehicles.
   D. All of these choices.
7. Automotive horns operate on the principle of:
   A. Induced voltage.
   B. Depletion zone bonding.
   C. Frequency modulation.
   D. Vibrating a column of air.
8. The purpose of multiplexing is to:
   A. Increase circuit loads to a sensor.
   B. Prevent electromagnetic interference.
   C. Allow computers to share information.
   D. Prevent multiple system failures from occurring.
9. The following are true about the air bag system EXCEPT:
   A. It's an active system.
   B. It's a supplemental system.
   C. It's mandated by the federal government.
   D. Deployment is automatic.
10. Alternate propulsion systems include:
    A. Electric vehicles.
    B. Hybrid vehicles.
    C. Fuel cell vehicles.
    D. All of these choices.

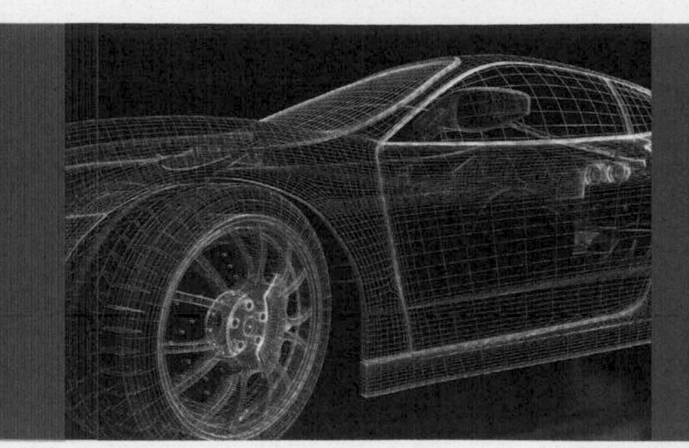

CHAPTER 2
BASIC THEORIES

Upon completion and review of this chapter, you should be able to:

- Explain the basics of electron flow.
- Describe the difference between insulators, conductors, and semiconductors.
- Define voltage, current, and resistance.
- Describe what voltage drop is.
- Define Ohm's law correctly.
- Use Ohm's law formulas to calculate the values of a circuit.
- Define Watt's law correctly.
- Use Watt's law formulas to calculate the values of a circuit.
- Explain the difference between AC and DC currents.
- Explain series, parallel, and series-parallel circuits.
- Apply electrical laws to series, parallel, and series-parallel circuits.
- Use Kirchoff's Voltage and Current Laws to determine circuit values.
- Explain the theory of electromagnetism.
- Explain the principles of induction.
- Describe how photovoltaics are used to generate electric current.
- Explain how electromagnetic interference can be suppressed.

Terms to Know

Alternating current (AC)
Ampere (A)
Atom
Balanced
Circuit
Closed circuit
Conductor
Continuity
Conventional theory
Current
Cycle
Direct current (DC)
Electromagnetic interference (EMI)
Electromagnetism
Electromotive force (EMF)
Electron
Electron theory
Equivalent series load
Ground
Induction
Insulator
Ion
Kirchhoff's current law
Kirchhoff's voltage law
Magnetic flux density
Mutual induction
Neutrons
Nucleus
Ohms
Ohm's law
Open circuit
Parallel circuit
Permeability
Photovoltaics (PV)
Power (P)
Protons
Reluctance
Resistance
Right-hand rule
Saturation
Self-induction
Semiconductors
Series-parallel circuit
Shell
Static electricity
Valence ring
Voltage
Voltage drop
Watt

Introduction

The electrical systems used in today's vehicles can be very complicated (**Figure 2-1**). However, by understanding the principles and laws that govern electrical circuits, technicians can simplify their job of diagnosing electrical problems. In this chapter, you will learn

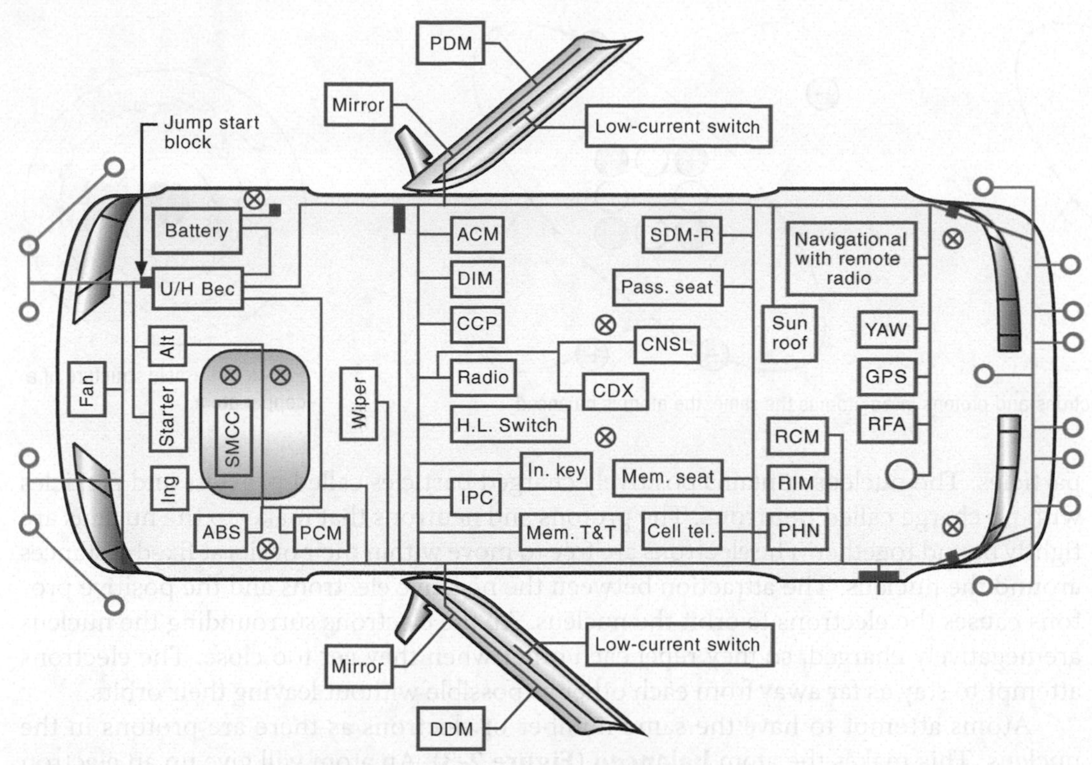

Figure 2-1 The electrical system of today's vehicles can be very complicated.

the laws that dictate electrical behavior, how circuits operate, the difference between types of circuits, and how to apply Ohm's law to each type of circuit. You will also learn the fundamental theories of semiconductor construction. Because magnetism and electricity are closely related, a study of electromagnetism and induction is included in this chapter.

Basics of Electron Flow

Because electricity is an energy form that cannot be seen, some technicians regard the vehicle's electrical system as being more complicated than it is. These technicians approach the vehicle's electrical system with some reluctance. Today's technician must understand that electrical behavior is confined to definite laws that produce predictable results and effects. To facilitate understanding the laws of electricity, a short study of atoms is presented.

Atomic Structure

An **atom** is the smallest part of a chemical element that still has all the characteristics of that element. An atom is constructed of a fixed arrangement of **electrons** in orbit around a **nucleus**, like planets orbiting the sun (**Figure 2-2**). Electrons are negatively charged

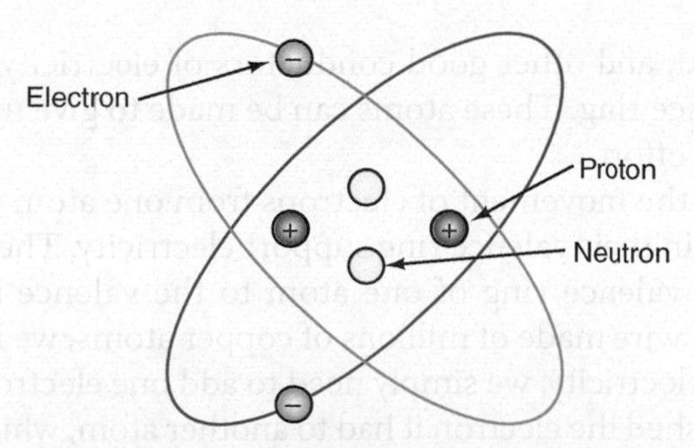

Figure 2-2 The composition of an atom.

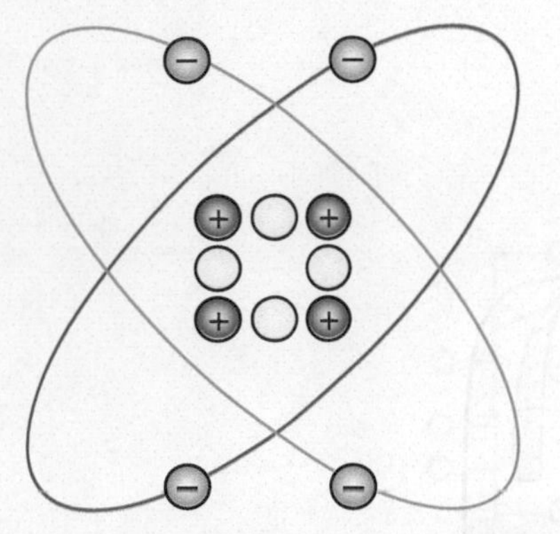

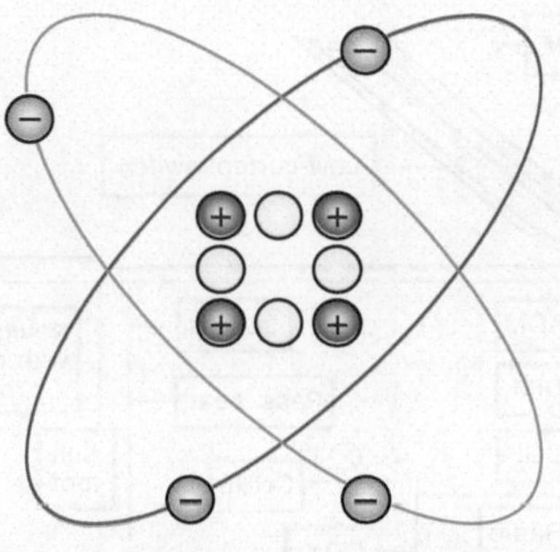

Figure 2-3 If the number of electrons and protons in an atom is the same, the atom is balanced.

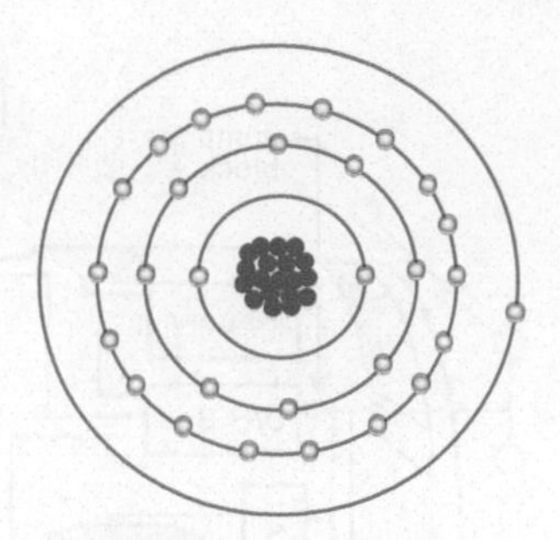

Figure 2-4 Basic structure of a copper atom.

particles. The nucleus contains positively charged particles called **protons** and particles with no charge called **neutrons**. The protons and neutrons that make up the nucleus are tightly bound together. The electrons are free to move within their orbits at fixed distances around the nucleus. The attraction between the negative electrons and the positive protons causes the electrons to orbit the nucleus. All the electrons surrounding the nucleus are negatively charged, so they repel each other when they get too close. The electrons attempt to stay as far away from each other as possible without leaving their orbits.

Like charges repel each other; unlike charges attract each other.

Atoms attempt to have the same number of electrons as there are protons in the nucleus. This makes the atom **balanced** (**Figure 2-3**). An atom will give up an electron to another atom or attract an electron from another atom to remain balanced. A specific number of electrons are in each of the electron orbit paths. The orbit closest to the nucleus has room for 2 electrons; the second orbit holds up to 8 electrons; the third holds up to 18; and the fourth and fifth hold up to 32 each. The number of orbits depends on the number of electrons the atom has. For example, a copper atom contains 29 electrons; 2 in the first orbit, 8 in the second, 18 in the third, and 1 in the fourth (**Figure 2-4**). The outer orbit, or **shell** as it's sometimes called, is referred to as the **valence ring**. In studying the laws of electricity, the only concern is with the electrons in the valence ring.

Since an atom seeks to be balanced, an atom that is missing electrons in its valence ring will attempt to gain other electrons from neighboring atoms. Also, if the atom has an excess number of electrons in its valence ring, it will try to pass them on to neighboring atoms.

Conductors and Insulators

To help explain why you need to know about these electrons and their orbits, let's continue to look at the atomic structure of copper. Copper is a metal and is the most commonly used **conductor** of electricity. A conductor is a substance that supports the flow of electricity through it. Since the copper atom has only 1 electron in the valance ring, for the valence ring to be filled, it would require 31 more electrons. Since there is only one electron, it is loosely tied to the atom and can be easily removed, making it a good conductor.

Copper, silver, gold, and other good conductors of electricity have only one or two electrons in their valence ring. These atoms can be made to give up the electrons in their valence ring with little effort.

Since electricity is the movement of electrons from one atom to another, atoms with one to three electrons in their valence ring support electricity. They allow the electron to easily move from the valence ring of one atom to the valence ring of another atom. Therefore, if we have a wire made of millions of copper atoms, we have a good conductor of electricity. To have electricity, we simply need to add one electron to one of the copper atoms. That atom will shed the electron it had to another atom, which will shed its original electron to another, and so on. As the electrons move from atom to atom, a force is released. This force is what we use to light lamps, run motors, and so on. We have electricity as long as we keep the electrons moving in the conductor.

Insulators are materials that don't allow electrons to flow through them easily. Insulators are atoms that have five to eight electrons in their valence ring. The electrons are held tightly around the atom's nucleus, and they can't be moved easily. Insulators are used to prevent electron flow or to contain it within a conductor. Insulating material covers the outside of most conductors to keep the moving electrons within the conductor.

A BIT OF HISTORY

The Greeks discovered electricity over 2,500 years ago. They noticed that when amber was rubbed with other materials, it was charged with an unknown force that had the power to attract objects such as dried leaves and feathers. The Greeks called amber "elektron." The word *electric* is derived from this word and means "to be like amber."

In summary, the number of electrons in the valence ring determines whether an atom is a good conductor or an insulator. Some atoms are neither good insulators nor good conductors; these are called **semiconductors**. In short:

1. Three or fewer electrons—conductor.
2. Five or more electrons—insulator.
3. Four electrons—semiconductor.

Electricity Defined

Electricity is the movement of electrons from atom to atom through a conductor (**Figure 2-5**). Electrons are attracted to protons. Since we have excess electrons on one end of the conductor, we have many electrons being attracted to the protons. This attraction acts to push the electrons toward the protons. This push is called electrical pressure or **electromotive force (EMF)**. The amount of electrical pressure is determined by the number of electrons that are attracted to the protons. The electrical pressure attempts to push an electron out of its orbit and toward the excess protons. If an electron is freed from its orbit, the atom acquires a positive charge because it now has one more proton than it has electrons. The unbalanced atom or **ion** attempts to return to its balanced state so it will attract electrons from the orbit of other balanced atoms. This starts a chain reaction as one atom captures an electron and another releases an electron. As this action continues to occur, electrons will flow through the conductor. A stream of free electrons forms, and an electrical current is started. This does not mean a single electron travels the length of the insulator; it means the overall effect is electrons moving in one direction. All this happens at almost the speed of light. The strength of the electron flow is dependent on the potential difference or voltage.

Random movement of electrons is not electric current; the electrons must move in the same direction.

It's often stated that the speed of light is 186,000 miles per second (299,000 kilometers per second). The actual speed of electricity depends on the composition of the wire and insulation and is slower than the speed of light due to the electrons bumping into each other and changing places.

The three elements of electricity are voltage, current, and resistance. How these three elements interrelate governs the behavior of electricity. Once the technician comprehends the laws governing electricity, understanding the function and operation of the various automotive electrical systems is an easier task. This knowledge will assist the technician in diagnosing and repairing automotive electrical systems.

An *E* can be used for the symbol to designate voltage (electromotive force). A *V* is also used as a symbol for voltage.

Voltage

Voltage can be defined as an electrical pressure (**Figure 2-6**) and is the electromotive force that causes the movement of the electrons in a conductor. In Figure 2-5, voltage is the force of attraction between the positive and the negative charges. An electrical

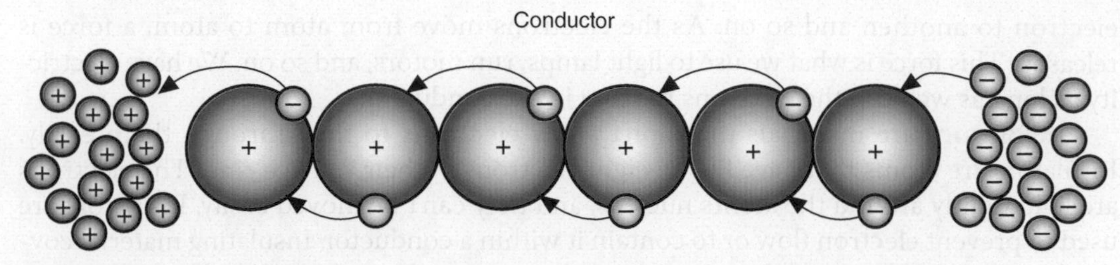

Figure 2-5 As electrons flow in one direction from one atom to another, an electrical current develops.

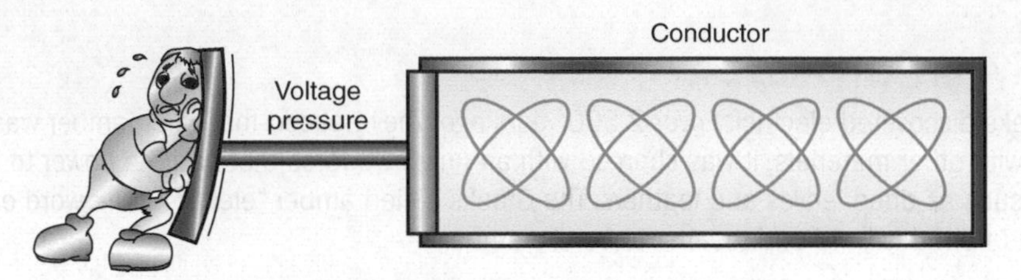

Figure 2-6 Voltage is the pressure that causes the electrons to move.

Shop Manual
Chapter 2,
pages 48, 54

pressure difference is created when there is a mass of electrons at one point in the circuit and a lack of electrons at another point in the circuit. In an automobile, the battery or generator applies the electrical pressure.

The amount of pressure applied to a circuit is stated in the number of volts. If a voltmeter is connected across the terminals of an automobile battery, it may indicate 12.6 volts. This means there is 12.6 volts of electrical pressure between the two battery terminals, and the voltmeter is reading that there is a difference in potential of 12.6 volts.

One volt (V) is the amount of pressure required to move one ampere of current through one ohm of resistance.

In a circuit that has current flowing, voltage will exist between any two points in that circuit (**Figure 2-7**). The only time voltage does not exist is when the potential drops to zero. Figure 2-7 shows that the voltage potential between points A and C and between points B and C is 12.6 volts. However, the pressure difference is zero between points A and B, and the voltmeter will indicate 0 volts.

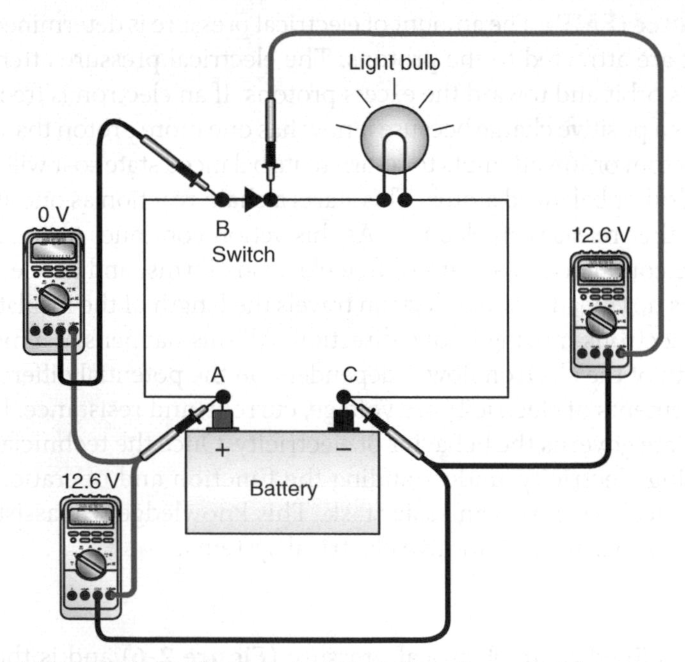

Figure 2-7 A simplified light circuit illustrating voltage potential.

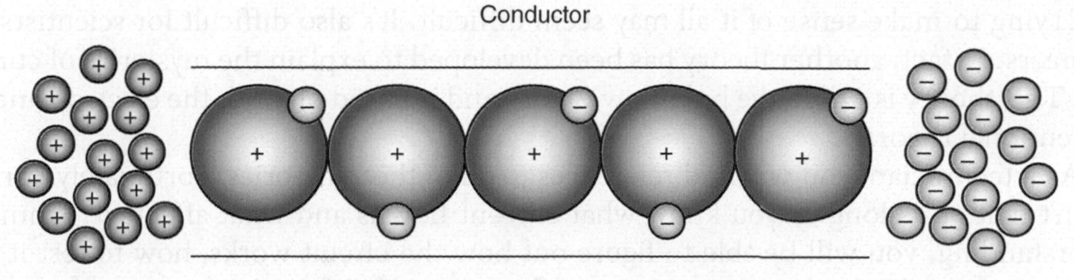

Figure 2-8 The rate of electron flow is called current and is measured in amperes.

Current

Current is defined as the *rate* (intensity) of electron flow (**Figure 2-8**) and is measured in amperes. Current is a measurement of the electrons passing any given point in the circuit in 1 second. Because the flow of electrons is so fast, it would be impossible to physically see electron flow. However, the rate of electron flow can be measured. Current will increase if pressure (voltage) is increased, provided circuit resistance remains constant.

Shop Manual
Chapter 2,
pages 48, 59

One **ampere (A)** represents the movement of 6.25 $\times$ 10^{18} electrons (or one coulomb) past one point in a conductor in 1 second.

An electrical current will continue to flow through a conductor as long as the electromotive force acts on the conductor's atoms and electrons. If a potential exists in the conductor, with a buildup of excess electrons at the end of the conductor farthest from the EMF and there is a lack of electrons at the EMF side, current will flow. The effect is called electron drift and accounts for the method in which electrons flow through a conductor.

An electrical current can be formed by the following forces: friction, chemical reaction, heat, pressure, magnetic induction, and photovoltaics (PV). Whenever electrons flow or drift in mass, an electrical current is formed. Six laws regulate this electrical behavior:

The symbol for current is "I," which stands for intensity. Also, "A" is used for amperage.

1. Like charges repel each other.
2. Unlike charges attract each other.
3. When an EMF acts on the conductor, a voltage difference is created in the conductor.
4. Electrons flow only when a voltage difference exists between the two points in a conductor.
5. Current tends to flow to ground in an electrical circuit as a return to source.
6. **Ground** is the baseline when measuring electrical circuits and is the point of lowest voltage. Also, it's the return path to the source for an electrical circuit. The ground circuit used in most automotive systems is through the vehicle chassis and/or engine block. In addition, ground allows voltage spikes to be directed away from the circuit by absorbing them.

So far, we have described current as the movement of electrons through a conductor. Electrons move because of a potential difference. This describes one of the common theories about current flow. The **electron theory** states that since electrons are negatively charged, current flows from the most negative to the most positive point within an electrical circuit. In other words, current flows from negative to positive. The electronic industry widely accepts this theory.

Another current flow theory is called the **conventional theory**. This theory describes current flow as being from positive to negative. The basic idea behind this theory is that although electrons move toward the protons, the energy or force released as the electrons move begins at the point where the first electron moved to the most positive charge. As electrons move in one direction, the released energy moves in the opposite direction. This theory is the oldest and serves as the basis for most automotive electrical diagrams.

Trying to make sense of it all may seem difficult. It's also difficult for scientists and engineers. In fact, another theory has been developed to explain the mysteries of current flow. This theory is called the hole-flow theory and is based on both the electron and the conventional theories.

As a technician, you will find references to all of these theories. Fortunately, it really doesn't matter as long as you know what current flow is and what affects it. From this understanding, you will be able to figure out how the circuit works, how to test it, and how to repair it. This text presents current flow as moving from positive to negative and electron flow as moving from negative to positive. Remember that current flow results from the movement of electrons, regardless of the theory.

Resistance

Shop Manual
Chapter 2, pages 48, 58

The **ohm** (Ω) is the unit of measurement for resistance of a conductor such that a constant current of 1 ampere in it produces a voltage of 1 volt between its ends.

The symbol for resistance is *R*.

The third component of electricity is **resistance**. Resistance is the opposition to current flow and is measured in **ohms** (Ω). In a circuit, resistance controls the amount of current. The size, type, length, and temperature of the material used as a conductor will determine its resistance. Devices that use electricity to operate (motors and lights) have a greater amount of resistance than the conductor.

A complete electrical **circuit** consists of the following: (1) a power source, (2) a load or resistance unit, and (3) conductors. Resistance (load) is required to change electrical energy to light, heat, or movement. There is resistance in any working device of a circuit, such as a lamp, motor, relay, coil, or other load component.

Five basic characteristics determine the amount of resistance in any part of a circuit:

1. The atomic structure of the material: The higher the number of electrons in the outer valence ring, the higher the material's resistance.
2. The length of the conductor: The longer the conductor, the higher the resistance.
3. The diameter of the conductor: The smaller the cross-sectional area of the conductor, the higher the resistance.
4. Temperature: Normally, an increase in the temperature of the conductor causes an increase in the resistance.
5. The physical condition of the conductor: If nicks or cuts damage the conductor, the resistance will increase because the conductor's diameter is decreased by these.

There may be unwanted resistance in a circuit. This could be in the form of a corroded connection or a broken conductor. In these instances, the resistance may cause the load component to operate at reduced efficiency or not to operate at all.

It does not matter if the resistance is from the load component or from unwanted resistance. Certain principles dictate its impact in the circuit:

1. Voltage always drops as current flows through the resistance.
2. An increase in resistance causes a decrease in current.
3. All resistances change the electrical energy into heat energy to some extent.

Voltage Drop Defined

Shop Manual
Chapter 2, page 54

Voltage drop occurs when current flows through a load component or resistance. Voltage drop is the amount of electrical energy that is converted as it pushes current flow through a resistance. Electricity is an energy. Energy cannot be created or destroyed, but it can be changed from one form to another. As electrical energy flows through a resistance, it's converted to some other form of energy, usually heat energy. The voltage drop over a resistance or load device indicates how much electrical energy is converted to another energy form. After a resistance, the voltage is lower than before the resistance.

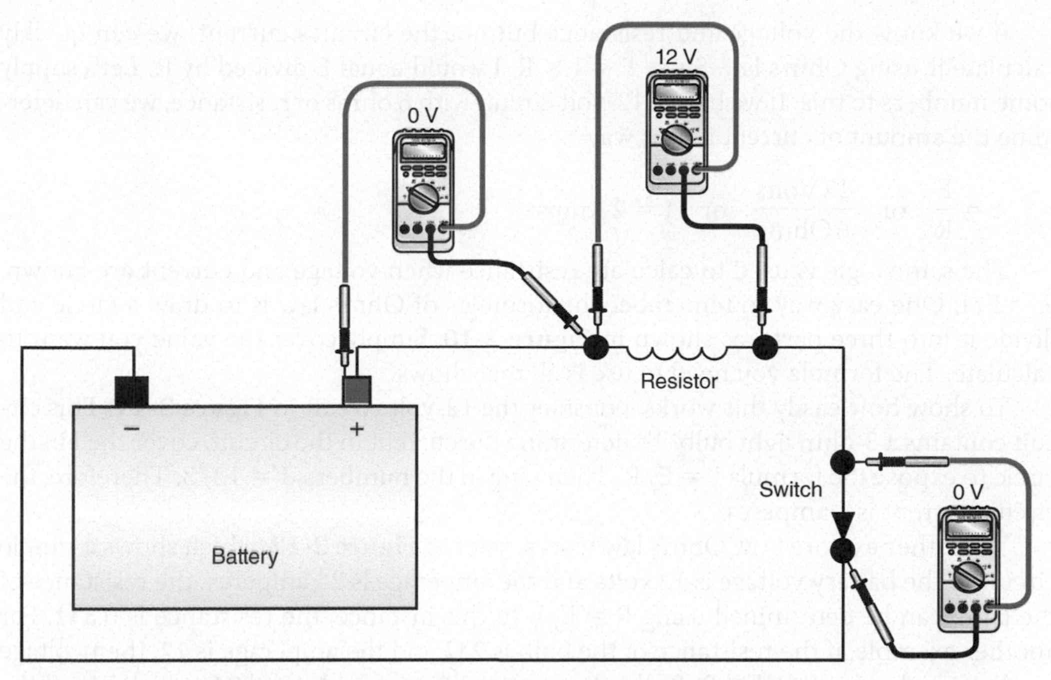

Figure 2-9 Using a voltmeter to measure voltage drop in different locations of a circuit.

Voltage drop can be measured with a voltmeter (**Figure 2-9**). With current flowing through a circuit, the voltmeter may be connected in parallel over the resistor, wire, or component to measure the voltage drop. The voltmeter indicates the amount of voltage potential between two points in the circuit. The voltmeter reading indicates the difference between the amount of voltage available to the resistor and the amount of voltage after the resistor.

There must be a voltage present for current to flow through a resistor. Kirchhoff's law basically states that the sum of the voltage drops in an electrical circuit will always equal source voltage. In other words, the circuit uses all of the source voltage.

Electrical Laws

Electricity is governed by well-defined laws. The most fundamental of these are Ohm's law and Watt's law. Today's technician must understand these laws in order to completely grasp electrical theory.

Ohm's Law

Understanding **Ohm's law** is the key to understanding how electrical circuits work. Ohm's law defines the relationship between current, voltage, and resistance. The law states that it takes 1 volt of electrical pressure to push 1 ampere of electrical current through 1 ohm of electrical resistance. This law can be expressed mathematically as follows:

$$1\,\text{Volt} = 1\,\text{Ampere} \times 1\,\text{Ohm}$$

This formula is most often expressed as follows: $E = I \times R$. E stands for electromotive force (electrical pressure or voltage), I stands for intensity (current or ampere), and R represents resistance. This formula is often used to find the amount of one electrical characteristic when the other two are known. For example, if we have 2 amps of current and 6 ohms of resistance in a circuit, we must have 12 volts of electrical pressure.

$$E = 2\,\text{Amps} \times 6\,\text{Ohms} \qquad E = 2 \times 6 \qquad E = 12\,\text{Volts}$$

If we know the voltage and resistance but not the circuit's current, we can quickly calculate it using Ohm's law. Since E = I × R, I would equal E divided by R. Let's supply some numbers to this. If we have a 12-volt circuit with 6 ohms of resistance, we can determine the amount of current in this way:

$$I = \frac{E}{R} \quad \text{or} \quad \frac{12\,\text{Volts}}{6\,\text{Ohms}} \quad \text{or} \quad I = 2\,\text{Amps}$$

The same logic is used to calculate resistance when voltage and current are known. R = E/I. One easy way to remember the formulas of Ohm's law is to draw a circle and divide it into three parts, as shown in **Figure 2-10**. Simply cover the value you want to calculate. The formula you need to use is all that shows.

The resistance of a lamp used in an automotive application will change when current passes through it because its temperature changes.

To show how easily this works, consider the 12-volt circuit in **Figure 2-11**. This circuit contains a 3-ohm light bulb. To determine the current in the circuit, cover the I in the circle to expose the formula I = E/R. Then plug in the numbers, I = 12/3. Therefore, the circuit current is 4 amperes.

To further explore how Ohm's law works, refer to **Figure 2-12**, which shows a simple circuit. If the battery voltage is 12 volts and the amperage is 24 amperes, the resistance of the lamp can be determined using R = E/I. In this instance, the resistance is 0.5 Ω. For another example, if the resistance of the bulb is 2 Ω and the amperage is 12, then voltage can be found using E = I × R. In this instance, voltage would equal 24 volts. Now, if the circuit had 12 volts and the resistance of the bulb was 2 Ω, then amperage could be determined by I = E/R. In this case, amperage would be 6 amperes.

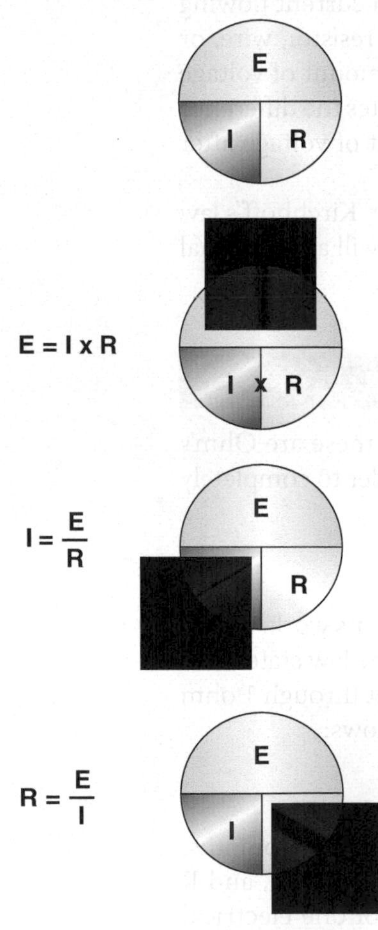

Figure 2-10 The mathematical formula for Ohm's law uses a circle to help understand the different formulas derived from it. To expose the formula to use, cover the unknown value.

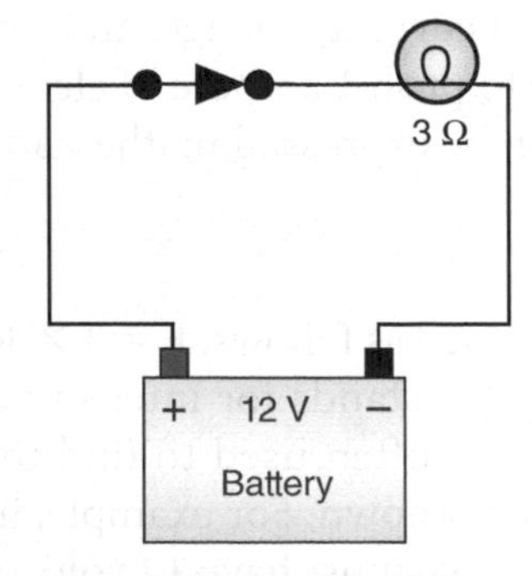

Figure 2-11 Simplified light circuit with 3 ohms of resistance in the lamp.

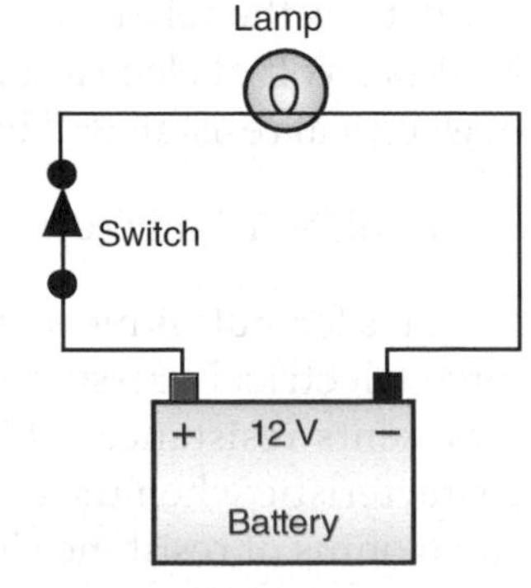

Figure 2-12 Simple lamp circuit to help use Ohm's law.

Ohm's law is the fundamental law of electricity. It states that the amount of current in an electric circuit is inversely proportional to the resistance and is directly proportional to the voltage in the circuit. For example, if the resistance decreases and the voltage remains constant, the amperage will increase. If the resistance stays the same and the voltage increases, the amperage will also increase.

Consider **Figure 2-13**; on the left side is a 12-volt circuit with a 3-ohm light bulb. This circuit will have 4 amps of current flowing through it. If a 1-ohm resistor is added to the same circuit (as shown to the right in Figure 2-13), total resistance is now 4 ohms. Because of the increased resistance, current dropped to 3 amps. Since the light bulb will have less current flowing through it, the light intensity will be less bright than before adding the additional resistance.

Another point to consider is voltage drop. Before adding the 1-ohm resistor, the light bulb dropped the source voltage (12 volts). With the additional resistance, the voltage drop of the light bulb decreased to 9 volts. The 1-ohm resistor dropped the remaining 3 volts. This can be proven by using Ohm's law. When the circuit current was 4 amps, the light bulb had 3 ohms of resistance. To find the voltage drop, we multiply the current by the resistance.

$E = I \times R$ or $E = 4 \times 3$ or $E = 12$ Volts

When the extra resistor was added to the circuit, the light bulb still had 3 ohms of resistance, but the current in the circuit decreased to 3 amps. Again, voltage drop can be determined by multiplying the current by the resistance.

$E = I \times R$ or $E = 3 \times 3$ or $E = 9$ Volts

The voltage drop of the additional resistor is calculated in the same way: $E = I \times R$ or $E = 3$ volts. The total voltage drop of the circuit is the same for both circuits. However, the voltage drop at the light bulb changed. This also would cause the light bulb to be dimmer. Ohm's law and its application will be discussed in greater detail later.

Watt's Law

Power (P) is the rate of doing electrical work. Power is expressed in watts. A **watt** is equal to 1 volt multiplied by 1 ampere. Another mathematical formula that expresses the relationship between voltage, current, and power is simply: $P = E \times I$ (**Figure 2-14**). Power measurements are measurements of the rate at which electricity is doing work.

Horsepower ratings can be converted to electrical power ratings using the conversion factor: 1 horsepower equals 746 watts.

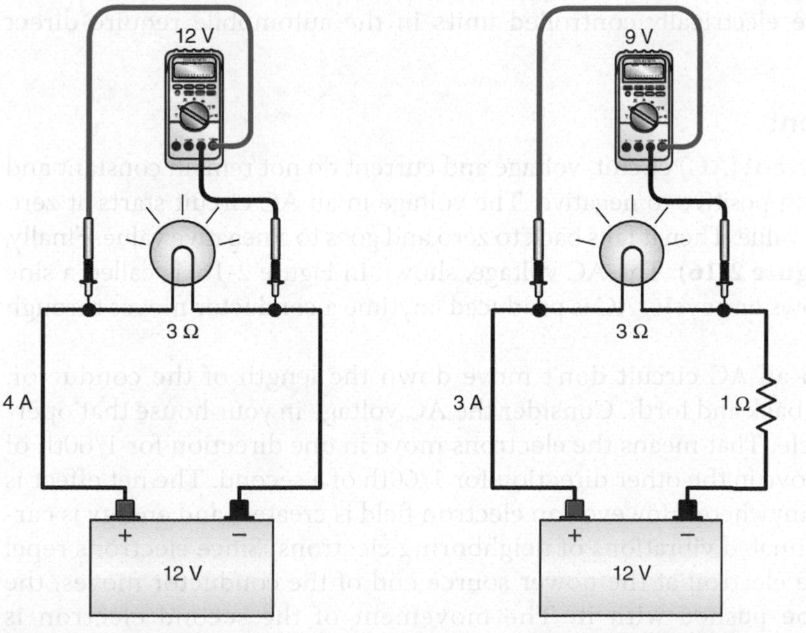

Figure 2-13 The light circuit in Figure 2-11 is shown with normal circuit values and with added resistance in series.

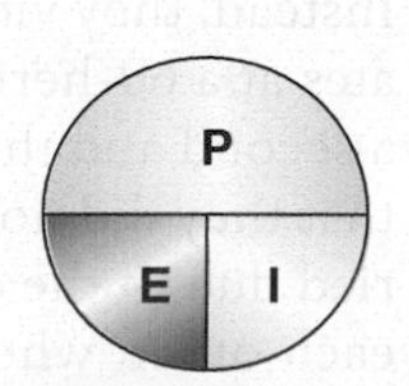

Figure 2-14 The mathematical formula for Watt's law.

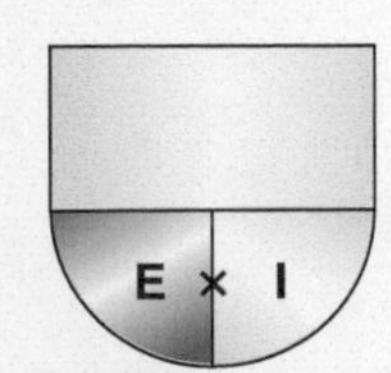

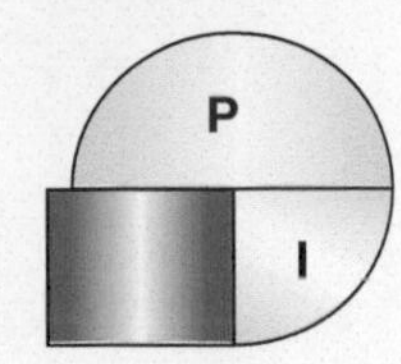

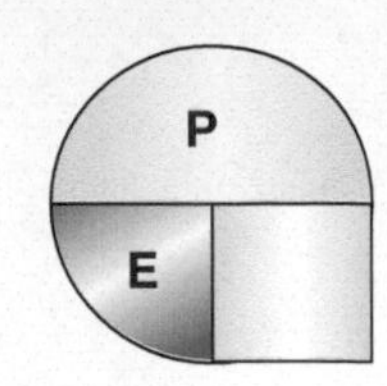

An excellent example to demonstrate the concept of electrical power is light bulbs. Household light bulbs are sold by wattage. A 100-watt bulb is brighter and uses more electricity than a 60-watt bulb.

Referring to Figure 2-13, the light bulb in the circuit on the left has a 12-volt drop at 4 amps. We can calculate the power the bulb uses by multiplying the voltage and the current.

$$P = E \times I \quad \text{or} \quad P = 12 \times 4 \quad \text{or} \quad P = 48\,\text{Watts}$$

The power output of the bulb is 48 watts. When the resistor was added to the circuit, the bulb dropped 9 volts at 3 amps. The power of the bulb is calculated in the same way as before.

$$P = E \times I \quad \text{or} \quad P = 9 \times 3 \quad \text{or} \quad P = 27\,\text{Watts}$$

This bulb produced 27 watts of power, a little more than half of the original. It would be almost half as bright. The key to understanding what happened is to remember the light bulb didn't change; the circuit changed.

Another example of using Watt's law is to determine the amperage if an additional accessory is added to the vehicle's electrical system. If the accessory is rated at 75 watts, the amperage draw would be:

$$I = P/E = 75/12 = 625\,\text{amps}$$

This tells the technician that this circuit will probably require a 10-amp-rated fuse.

Types of Current

There are two classifications of electrical current flow: direct current (DC) and alternating current (AC). The type of current flow is determined by the direction it flows and by the type of voltage that drives it.

Direct Current

Our study of electricity thus far has focused chiefly on **direct current (DC)**. In a DC circuit, electrons flow in one direction only. Voltage and current are constant if the switch is turned on or off (**Figure 2-15**). DC can be produced by a chemical reaction (such as in a battery). Most of the electrically controlled units in the automobile require direct current.

Alternating Current

In an **alternating current (AC)** circuit, voltage and current do not remain constant and change directions from positive to negative. The voltage in an AC circuit starts at zero and rises to a positive value. Then it falls back to zero and goes to a negative value. Finally, it returns to zero (**Figure 2-16**). The AC voltage, shown in Figure 2-16, is called a sine wave. Figure 2-16 shows one **cycle**. AC is produced anytime a conductor moves through a magnetic field.

The electrons in an AC circuit don't move down the length of the conductor. Instead, they vibrate back and forth. Consider the AC voltage in your house that operates at a 60-hertz cycle. That means the electrons move in one direction for 1/60th of a second and then move in the other direction for 1/60th of a second. The net effect is that they did not go anywhere. However, an electron field is created, and energy is carried due to the coordinated vibrations of neighboring electrons. Since electrons repel each other, when the electron at the power source end of the conductor moves, the next electron will be pushed with it. The movement of the second electron is

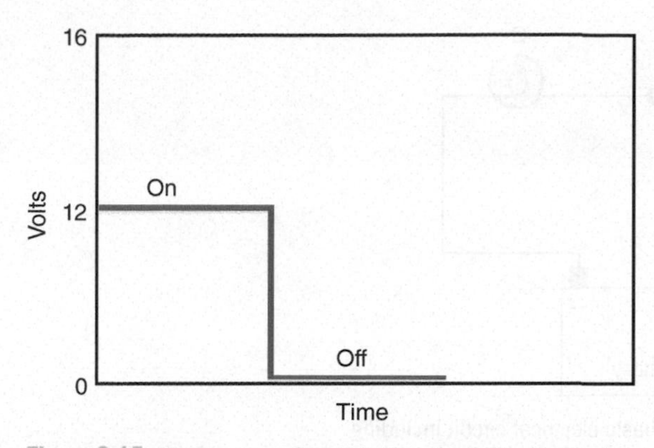

Figure 2-15 Direct current flow is in the same direction and remains constant on or off.

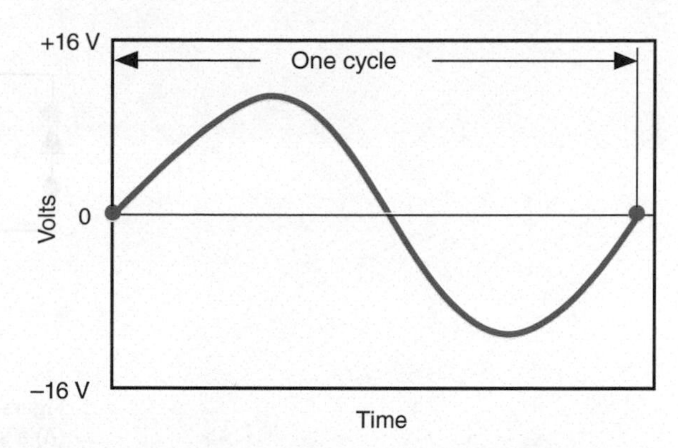

Figure 2-16 Alternating current does not remain constant.

transferred to the third electron. This continues until the end electron is reached. When the power source changes polarity and pulls the first electron back to its original position, the electron in front of it can also move back. All electrons will be pulled back to their original position. All electrons moving the same way as the electron at the beginning create ripples that travel through the conductor until reaching the load device. The ripples transfer electrical energy, and the end electron does the work without any electrons traveling very far.

The unrectified current produced within an AC generator is an example of alternating current found in the automobile. Circuits within the AC generator convert the alternating current to direct current.

The portion of the circuit from the positive side of the source to the load component is called the insulated side or "hot" side of the circuit. The portion of the circuit from the load component to the negative side of the source is called the ground side of the circuit.

Electrical Circuits

The electrical term **continuity** refers to the circuit being continuous. For current to flow, the electrons must have a continuous path from the source voltage to the load component and back to the source. A simple automotive circuit is made up of four parts:

1. Battery (power source).
2. Wires (conductors).
3. Load (light, motor, etc.).
4. A control device (switch).

The basic circuit (**Figure 2-17**) includes a switch to turn the circuit on and off, a protection device (fuse), and a load. When the switch is turned to the ON position, the circuit is referred to as a **closed circuit**. When the switch is in the OFF position, the circuit is referred to as an **open circuit**. In this instance, with the switch closed current flows from the battery's positive terminal through the light and returns to the battery's negative terminal. A complete circuit is formed when the switch is closed or turned on. The effect of opening and closing the switch to control electrical flow would be the same if the switch were installed on the ground side of the light.

The electrical term **closed circuit** means there are no breaks in the path, and current will flow. **Open circuit** means there will be no current flow since the path for electron flow is broken.

There are three different types of electrical circuits: (1) the series circuit, (2) the parallel circuit, and (3) the series-parallel circuit.

Series Circuit

A series **circuit** consists of one or more resistors (or loads) with only one path for current to flow. All of the current that comes from the positive side of the battery must pass through each resistor and then back to the negative side of the battery. If any of the components in the circuit fails, the entire circuit will not function.

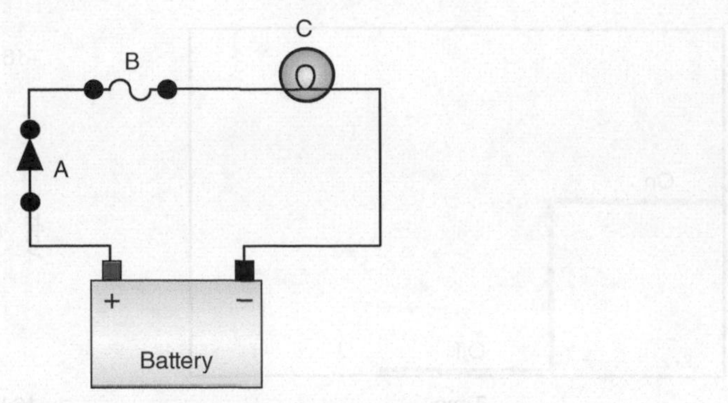

Figure 2-17 A basic electrical circuit including (A) a switch, (B) a fuse, and (C) a lamp.

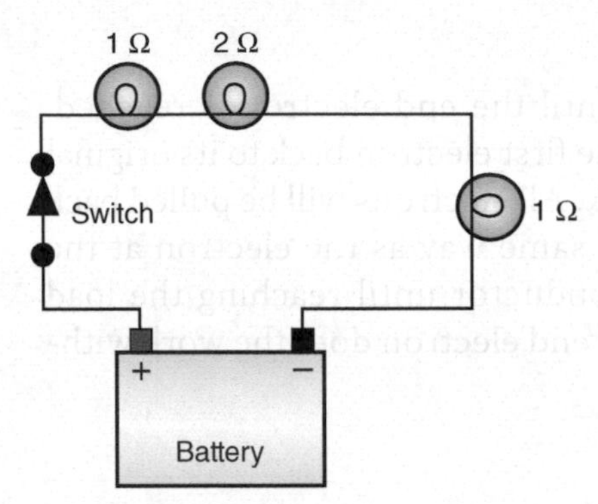

Figure 2-18 The total resistance in a series circuit is the sum of all resistances in the circuit.

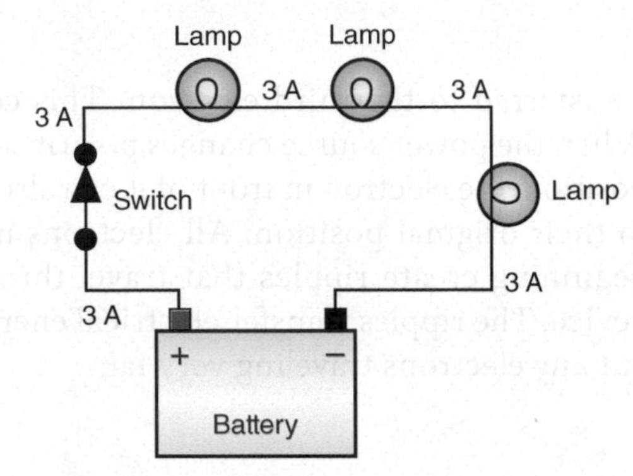

Figure 2-19 Regardless of where it's measured, amperage is the same at all points in a series circuit.

The total resistance of a series circuit is calculated by simply adding the resistances together. As an example, refer to **Figure 2-18**. Here is a series circuit with three light bulbs; 1 bulb has 2 ohms of resistance, and the other 2 have 1 ohm each. The total resistance of this circuit is 2 + 1 + 1 or 4 ohms.

The characteristics of a series circuit are as follows:

1. The total resistance is the sum of all resistances.
2. The current is the same at all points of the circuit (**Figure 2-19**).
3. The voltage drop across each resistance will be equal if the resistance values are the same (**Figure 2-20**).
4. The voltage drop across each resistance will be different if the resistance values are different (**Figure 2-21**).
5. The sum of all voltage drops equals the source voltage.

AUTHOR'S NOTE The symbols used for voltage, resistance, and current have changed over the years. In the past, the symbols were E for electromotive force (voltage), R for resistance, and I for intensity (current). Recently there has been a move to using the symbols V for voltage, R for resistance, and A for amps (current). As a result, Ohm's law can be expressed as $E = I \times R$, or $V = A \times R$. Since illustrations used to illustrate Ohm's law use V and A, the formula will be expressed as $V = A \times R$.

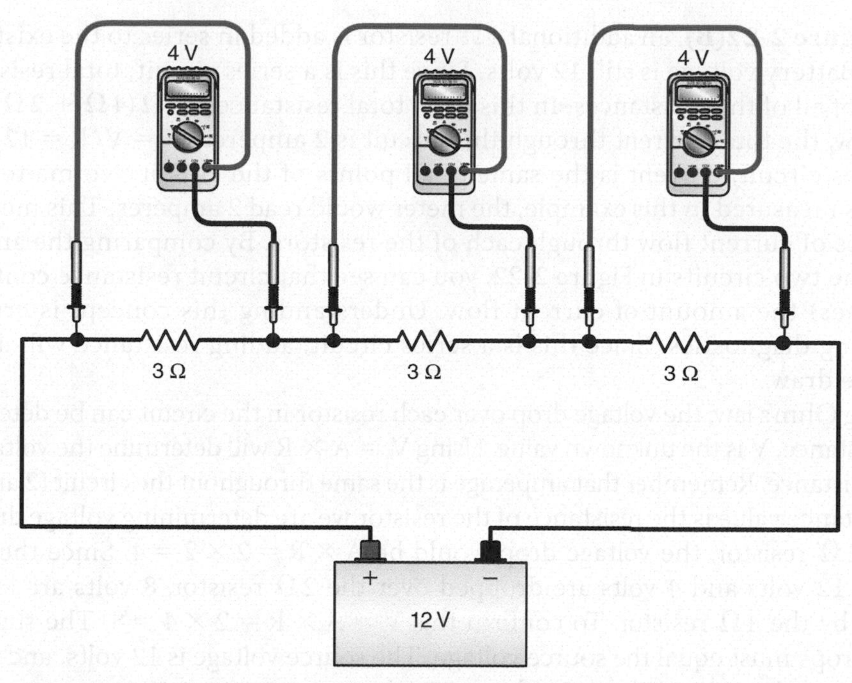

Figure 2-20 The voltage drop across each resistor in series will be the same if the resistance value of each is the same.

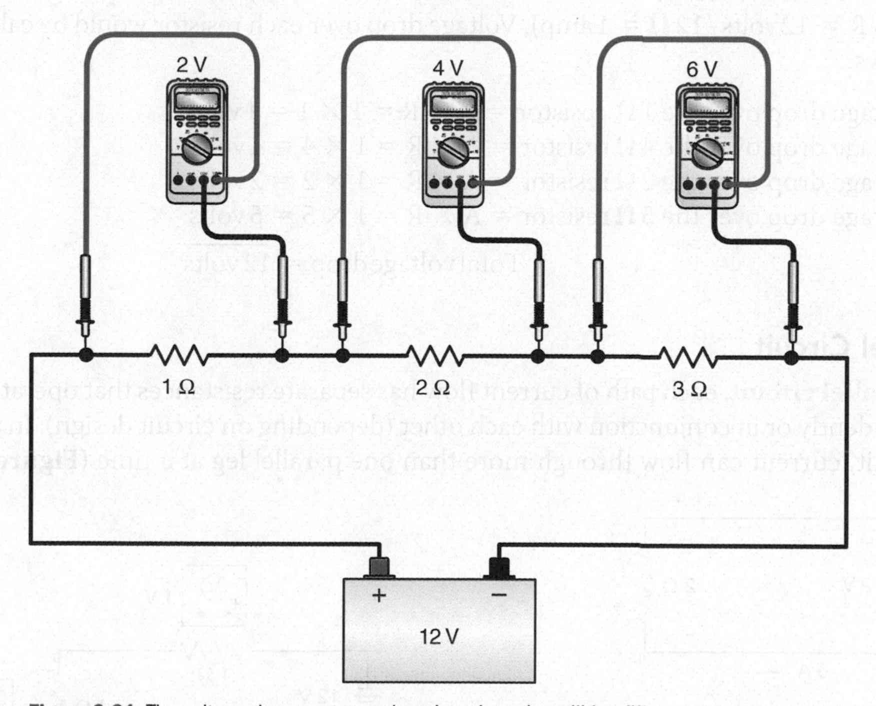

Figure 2-21 The voltage drop across each resistor in series will be different if the resistance value of each is different.

To illustrate the laws of the series circuit, refer to **Figure 2-22**. The illustration labeled (A) is a simple 12-volt series circuit with a 2 Ω resistor. Since the 2 Ω resistor is the only one in the circuit, all 12 volts are dropped across this resistor (V = A × R). Using Ohm's law, it can be determined that the current is 6 amperes (A = V/R).

In **Figure 2-22(B)**, an additional $4\,\Omega$ resistor is added in series to the existing $2\,\Omega$ resistor. Battery voltage is still 12 volts. Since this is a series circuit, total resistance is the sum of all of the resistances. In this case, total resistance is $6\,\Omega\,(4\,\Omega + 2\,\Omega)$. Using Ohm's law, the total current through this circuit is 2 amperes ($A = V/R = 12/6 = 2$). In a series circuit, current is the same at all points of the circuit. No matter where current is measured in this example, the meter would read 2 amperes. This means that 2 amperes of current flow through each of the resistors. By comparing the amperage flow of the two circuits in Figure 2-22, you can see that circuit resistance controls (or determines) the amount of current flow. Understanding this concept is critical to performing diagnostics. Since this is a series circuit, adding resistance will decrease amperage draw.

Using Ohm's law, the voltage drop over each resistor in the circuit can be determined. In this instance, V is the unknown value. Using $V = A \times R$ will determine the voltage drop over a resistance. Remember that amperage is the same throughout the circuit (2 amperes). The resistance value is the resistance of the resistor we are determining voltage drop over. For the $2\,\Omega$ resistor, the voltage drop would be $A \times R = 2 \times 2 = 4$. Since the battery provides 12 volts and 4 volts are dropped over the $2\,\Omega$ resistor, 8 volts are left to be dropped by the $4\,\Omega$ resistor. To confirm this $V = A \times R = 2 \times 4 = 8$. The sum of the voltage drops must equal the source voltage. The source voltage is 12 volts, and the sum of the voltage drops is 4 volts + 8 volts = 12 volts.

These calculations work for all series circuits, regardless of the number of resistors in the circuit. Refer to **Figure 2-23** for an example of a series circuit with four resistors. Total resistance is $12\,\Omega(1\,\Omega + 4\,\Omega + 2\,\Omega + 5\,\Omega)$. Total amperage is 1 amp ($A = V/R = 12\,\text{volts}/12\,\Omega = 1\,\text{amp}$). Voltage drop over each resistor would be calculated as follows:

1. Voltage drop over the $1\,\Omega$ resistor $= A \times R = 1 \times 1 = 1\,\text{volts}$
2. Voltage drop over the $4\,\Omega$ resistor $= A \times R = 1 \times 4 = 4\,\text{volts}$
3. Voltage drop over the $2\,\Omega$ resistor $= A \times R = 1 \times 2 = 2\,\text{volts}$
4. Voltage drop over the $5\,\Omega$ resistor $= A \times R = 1 \times 5 = 5\,\text{volts}$

$$\text{Total voltage drop} = \overline{12\,\text{volts}}$$

The legs of a **parallel circuit** are also called parallel branches or shunt circuits.

Parallel Circuit

In a **parallel circuit**, each path of current flow has separate resistances that operate either independently or in conjunction with each other (depending on circuit design). In a parallel circuit, current can flow through more than one parallel leg at a time (**Figure 2-24**).

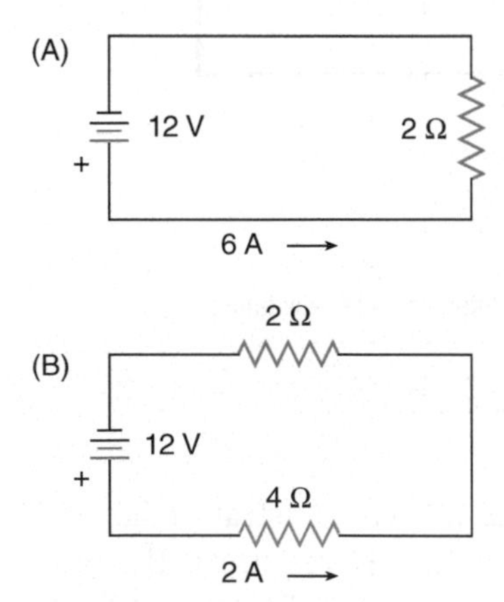

Figure 2-22 Circuit resistance controls, or determines, the amount of current flow.

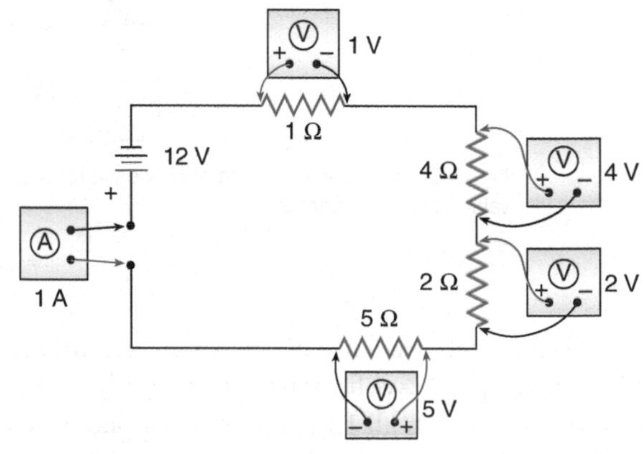

Figure 2-23 A series circuit used to demonstrate Ohm's law and voltage drop.

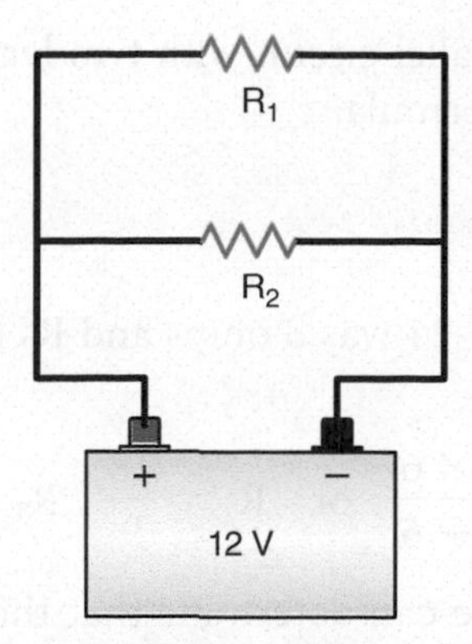

Figure 2-24 In a parallel circuit, there are multiple paths for current flow.

In this type of circuit, failure of a component in one parallel leg does not affect the components in the other legs of the circuit.

The characteristics of a parallel circuit are as follows:

1. The voltage applied to each parallel leg is the same.
2. The voltage dropped across each parallel leg will be the same; however, if the leg contains more than one resistor, the voltage drop across each of them will depend on the resistance of each resistor in that leg.
3. The total resistance of a parallel circuit will always be less than the resistance of any of its legs.
4. The current flow through the legs will be different if the resistances are different.
5. The sum of the current in each leg equals the total current of the parallel circuit.

Figuring total resistance is a bit more complicated for a parallel circuit than a series circuit. The total resistance in a parallel circuit is always less than the lowest individual resistance because the current has more than one path to flow. The method used to calculate total resistance depends on how many parallel branches are in the circuit, the resistance value of each branch, and personal preferences. Several methods of calculating total resistance are discussed. Choose the ones that work best for you.

If all resistances in the parallel circuit are equal, use the following formula to determine the total resistance:

$$R_T = \frac{\text{Value of one resistor}}{\text{Total number of branches}}$$

For example, if the parallel circuit, shown in Figure 2-24, had a 120 Ω resistor for and, total circuit resistance would be

$$R_T = 120\,\Omega/2 \text{ branches} = 60\,\Omega$$

Note that total resistance is less than any of the resistances of the branches. If a third branch were added in parallel that also had 120 Ω resistance, total circuit resistance would be

$$R_T = 120\,\Omega/3 \text{ branches} = 40\,\Omega$$

Note what happened when the third parallel branch was added. If more parallel resistors are added, more circuits are added, and the total resistance will decrease. With a decrease in total resistance, total amperage draw increases.

The total resistance of a parallel circuit with two legs or two paths for current flow can be calculated by using this formula:

$$R_T = \frac{R_1 \times R_2}{R_1 + R_2}$$

If the value of R_1 in Figure 2-24 was 3 ohms and R_2 had a value of 6 ohms, the total resistance can be found.

$$R_T = \frac{R_1 \times R_2}{R_1 + R_2} \quad \text{or} \quad R_T = \frac{3 \times 6}{3 + 6} \quad \text{or} \quad R_T = \frac{18}{9} \quad R_T = 2 \text{ ohms}$$

Based on this calculation, we can determine that the total circuit current is 6 amps (12 volts divided by 2 ohms). We can quickly determine other things about this circuit using Ohm's law and a basic understanding of electricity.

Each leg of the circuit has 12 volts applied to it; therefore, each leg must drop 12 volts. So, the voltage drop across R_1 is 12 volts, and the voltage drop across R_2 is also 12 volts. Using the voltage drops, we can quickly find the current that flows through each leg. Since R_1 has 3 ohms and drops 12 volts, the current through it must be 4 amps. R_2 has 6 ohms and drops 12 volts, and its current is 2 amps ($I = E/R$). The total current flow through the circuit is 4 + 2 or 6 amps. To calculate current in a parallel circuit, each shunt branch is treated as an individual circuit. To determine the branch current, simply divide the source voltage by the shunt branch resistance:

$I = E/R$

Referring to **Figure 2-25**, the total resistance of a circuit with more than two legs can be calculated with the following formula:

$$R_T = \frac{1}{\frac{1}{R_1} + \frac{1}{R_2} + \frac{1}{R_3} \cdots \frac{1}{R_n}}$$

Using Figure 2-25 and its resistance values, total resistance would be calculated by

$$R_T = \frac{1}{1/4 + 1/6 + 1/8}$$

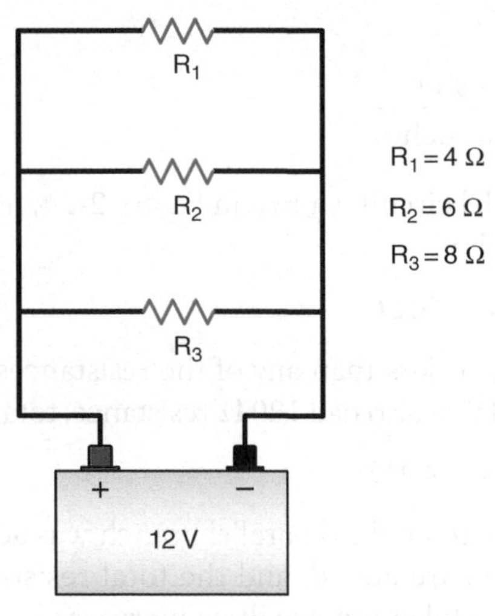

Figure 2-25 A parallel circuit with different resistances in each branch.

This means total resistance is equal to the reciprocal of the sum of $1/4 + 1/6 + 1/8$. The next step is to add $1/4 + 1/6 + 1/8$. To do this, the least common denominator must be found. In this case, the least common denominator is 24, so the formula now looks like this:

$$\frac{1}{6/24 + 4/24 + 3/24}$$

Now the fractions can be added together (remember to add only the numerator):

$$\frac{1}{13/24}$$

Since we are working with reciprocals, the formula now looks like this:

$$1 \times \frac{24}{13} = 1.85\,\Omega$$

Often it's much easier to calculate the total resistance of a parallel circuit by using total current. Begin by determining the current through each leg of the parallel circuit. The sum of all currents is the circuit's total current. Use basic Ohm's law to calculate the total resistance.

First, using the circuit illustrated in Figure 2-25, calculate the current through each branch:

1. Current through $R_1 = E/R = 12/4 = 3$ amperes
2. Current through $R_2 = E/R = 12/6 = 2$ amperes
3. Current through $R_3 = E/R = 12/8 = 1.5$ amperes

Add all of the current flow through the branches together to get the total current flow:

Total amperage $= 3 + 2 + 15 = 6.5$ amperes

Since this is a 12-volt system and total current is 6.5 amperes, total resistance is

$R_T = 12$ volts/6.5 amps $= 1.85\,\Omega$

This method can be mathematically expressed as follows:

$R_T = V_T/A_T$

Series-Parallel Circuits

The **series-parallel circuit** has some loads in series with each other, and some in parallel (**Figure 2-26**). To calculate the total resistance in this type of circuit, calculate the **equivalent series loads** of the parallel branches first. Next, calculate the series resistance and add it to the equivalent series load. For example, if the parallel portion of the circuit has two branches with $4\,\Omega$ resistance each and the series portion has a single load of $10\,\Omega$, use the following method to calculate the equivalent resistance of the parallel circuit:

The **equivalent series load**, or equivalent resistance, is the equivalent resistance of a parallel circuit plus the resistance in series and is equal to the equivalent resistance of a single load in series with the voltage source.

$$R_T = \frac{R_1 \times R_2}{R_1 + R_2} \quad \text{or} \quad \frac{4 \times 4}{4 + 4} \quad \text{or} \quad \frac{16}{8} \quad \text{or} \quad 2 \text{ ohms}$$

Then add this equivalent resistance to the actual series resistance to find the circuit's total resistance.

2 ohms + 10 ohms = 12 ohms

With the total resistance now known, total circuit current can be calculated. Because the source voltage is 12 volts, 12 is divided by 12 ohms.

$I = E/R$ or $I = 12/12$ or $I = 1$ amp

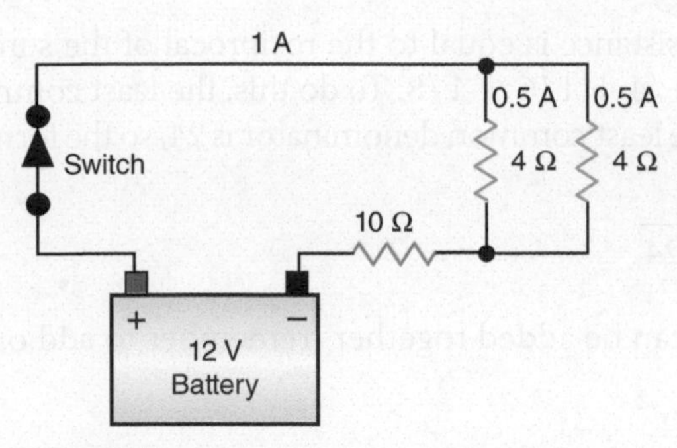

Figure 2-26 A series-parallel circuit with known resistance values.

The current flow through each parallel leg is calculated by using the resistance of each leg and voltage drop across that leg. To do this, you must first find the voltage drops. Since the circuit drops all 12 volts, we know that the parallel circuit drops some, and the rest is dropped by the resistor in series. We also know that the circuit current is 1 amp, the equivalent resistance value of the parallel circuit is 2 ohms, and the resistance of the series resistor is 10 ohms. Using Ohm's law, we can calculate the voltage drop of the parallel circuit:

$$E = I \times R \quad \text{or} \quad E = 1 \times 2 \quad \text{or} \quad E = 2\,\text{volts}$$

Two volts are dropped by the parallel circuit. This means 2 volts are dropped by each of the 4-ohm resistors. Using our voltage drop, we can calculate our current flow through each parallel leg:

$$I = E/R \quad \text{or} \quad I = 2/4 \quad \text{or} \quad I = 0.5\,\text{amps}$$

Since the resistance on each leg is the same, each leg has 0.5 amps through it. If we did this right, the sum of the amperages will equal the total current of the circuit. It does: $0.5 + 0.5 = 1$.

A slightly different series-parallel circuit is illustrated in **Figure 2-27**. In this circuit, a $2\,\Omega$ resistor is in series to a parallel circuit containing a $6\,\Omega$ and a $3\,\Omega$ resistor. To calculate total resistance, first find the resistance of the parallel portion of the circuit using $(6 \times 3)/(6 + 3) = 18/9 = 2$ ohms of resistance. Add this amount to the series resistance; $2 + 2 = 4$ ohms of total circuit resistance. Now total current can be calculated by $I = E/R = 12/4 = 3$ amperes. This means that 3 amperes are flowing through the series portion of the circuit. To figure how much amperage is in each parallel branch, the amount of applied voltage to each branch must be calculated. Since the series circuit has 3 amperes going through a $2\,\Omega$ resistor, the voltage drop over this resistor can be figured using $E = I \times R = 3 \times 2 = 6$. Since the source voltage is 12 volts, this means that 6 volts are

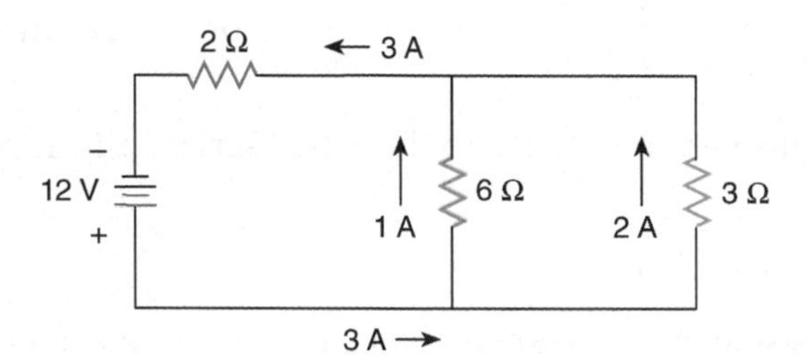

Figure 2-27 In a series-parallel circuit, the sum of the currents through the legs will equal the current through the series portion of the circuit.

applied to each of the resistors in parallel ($12 - 6 = 6$ volts) and that 6 volts are dropped over each of these resistors. Current through each branch can now be calculated:

Current through the $6\,\Omega$ branch is $I = E/R = 6/6 = 1$ ampere

Current through the $6\,\Omega$ branch is $I = E/R = 6/3 = 2$ ampere

The sum of the current flow through the parallel branches should equal total current flow ($1 + 2 = 3$ amperes).

Based on what was just covered, the characteristics of a series-parallel circuit can be summarized as follows:

1. Total resistance is the sum of the resistance value of the parallel portion and the series resistance.
2. Voltage drop over the parallel branch resistance is determined by the resistance value of the series resistor.
3. Total amperage is the sum of the current flow through each parallel branch.
4. The resistance in each parallel branch determines the amperage through each branch.

It's important to realize that the actual or measured values of current, voltage, and resistance may differ from the calculated values. The difference is caused by the effects of heat on the resistances. As the voltage pushes current through a resistor, the resistor heats up. The resistor changes the electrical energy into heat energy. This heat may cause the resistance to increase or decrease depending on the material it is made of. A good example of a resistance changing electrical energy into heat energy is a light bulb. A light bulb gives off light because the conductor inside the bulb heats up and glows when current flows through it.

Applying Ohm's Law

The primary importance of being able to apply Ohm's law is to predict what will happen if something else happens. Technicians use electrical meters to measure current, voltage, and resistance. When a measured value is not within specifications, you should be able to determine why. Ohm's law is used to do that.

Most automotive electrical systems are wired in parallel. Actually, the system is comprised of several series circuits wired in parallel. This allows each electrical component to work independently of the others. When one component is turned on or off, the operation of the other components should not be affected.

Figure 2-28 illustrates a 12-volt circuit with one 3-ohm light bulb. The switch controls the operation of the light bulb. When the switch is closed, current flows, and the bulb is lit. Four amps will flow through the circuit and the bulb.

$I = E/R$ or $I = 12/3$ or $I = 4$ amps

Figure 2-29 illustrates the same circuit with a 6-ohm light bulb added in parallel to the 3-ohm light bulb. With the switch for the new bulb closed, 2 amps will flow through that bulb. The 3-ohm is still receiving 12 volts and has 4 amps flowing through it. It will operate in the same way and with the same brightness as it did before we added the 6-ohm light bulb. The only thing that changed was circuit current, which is now $6\,(4 + 2)$ amps.

Leg #1 $I = E/R$ or $I = 12/3$ or $I = 4$ amps

Leg #2 $I = E/R$ or $I = 12/6$ or $I = 2$ amps

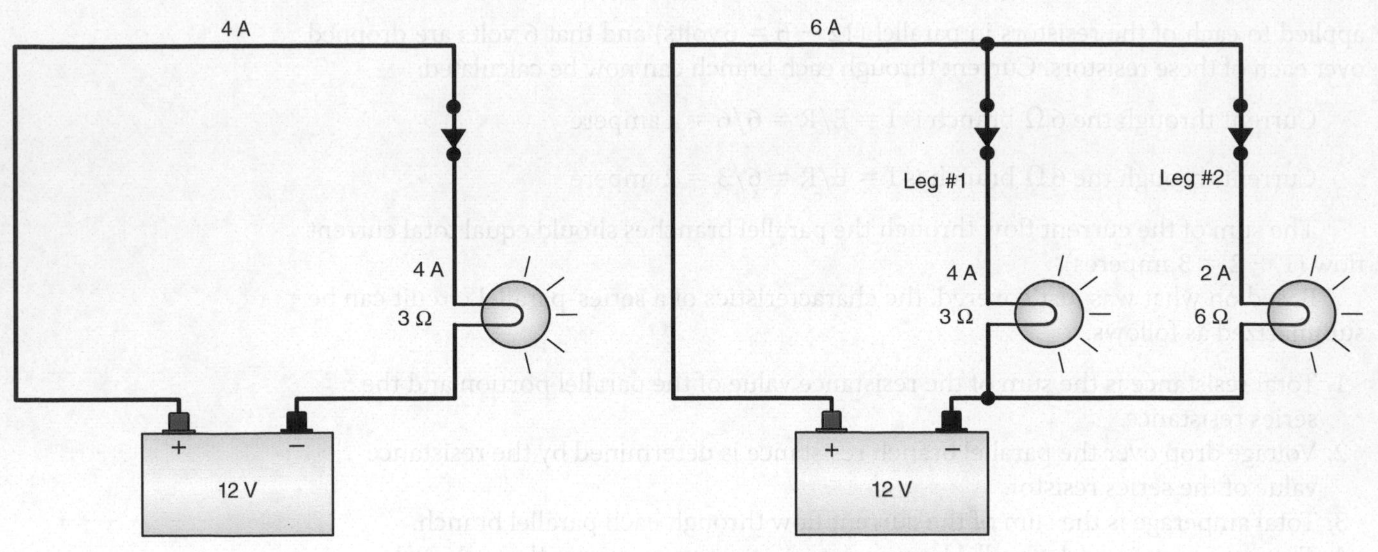

Figure 2-28 A simple light circuit.

Figure 2-29 Two light bulbs wired in parallel.

If the switch to the 3-ohm bulb is opened (**Figure 2-30**), the 6-ohm bulb works in the same way and with the same brightness as before we opened the switch. Two things happened in this case: the 3-ohm bulb is no longer lit, and the total circuit current dropped to 2 amps.

Figure 2-31 is the same circuit as Figure 2-30 except a parallel branch with a 1-ohm light bulb and a switch were added. With the switch for the new bulb closed, 12 amps will flow through that circuit. The other bulbs are working in the same way and with the same brightness as before. Again, total circuit resistance decreases, so the total circuit current increases. Total current is now 18 amps:

Leg #1 $I = E/R$ or $I = 12/3$ or $I = 4$ amps

Leg #2 $I = E/R$ or $I = 12/6$ or $I = 2$ amps

Leg #3 $I = E/R$ or $I = 12/1$ or $I = 12$ amps

Total current $= 4 + 2 + 12$ or 18 amps

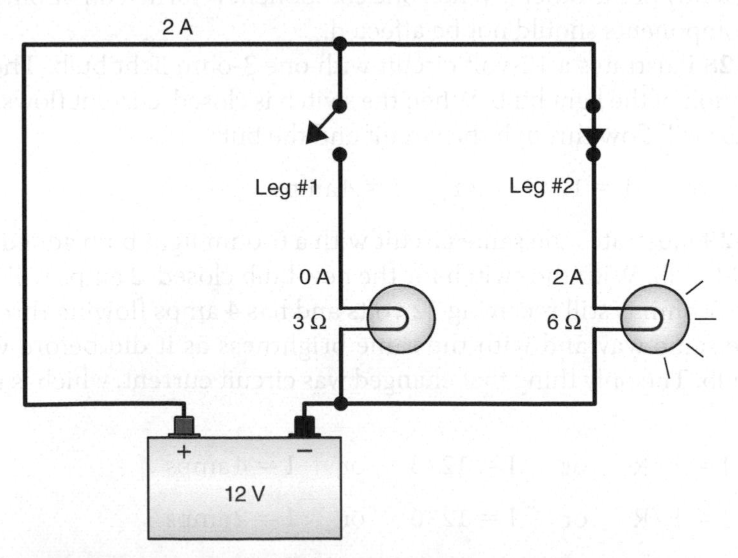

Figure 2-30 Two light bulbs wired in parallel; one is switched on, the other is switched off.

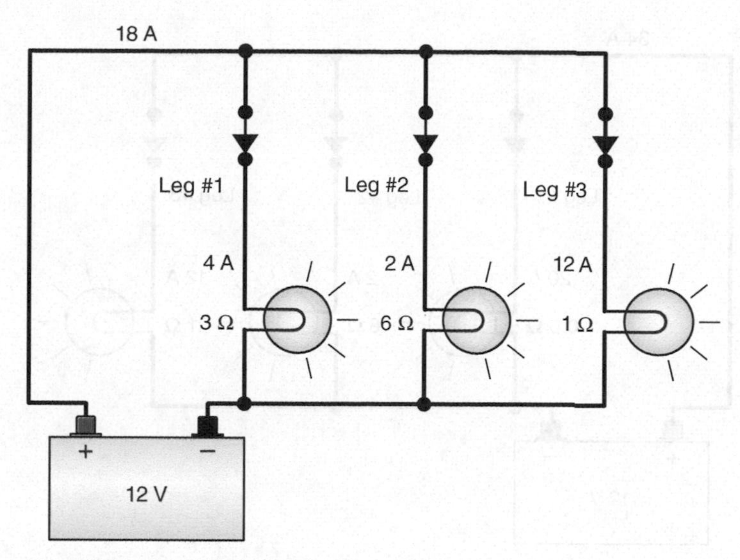

Figure 2-31 Three light bulbs wired in parallel.

When the switch for any of these bulbs is opened or closed, the only things that happen are the bulbs turn either off or on, and the total current through the circuit changes. Notice as we add more parallel legs, total circuit current goes up. There is a commonly used statement, "Current always takes the path of least resistance to ground." This statement is not totally correct. If this were a true statement, then parallel circuits would not work. However, as illustrated in the previous circuits, current flows to all bulbs regardless of the bulb's resistance. The resistances with lower values will draw higher currents, but all resistances will receive the current they allow. The statement is more accurate when expressed as, "Larger amounts of current will flow through lower resistances." Another way to state this would be, "The *majority* of the current flows the path of least resistance." This is very important to remember when diagnosing electrical problems.

From Ohm's law, we know that when resistance decreases, current increases. If we put a 0.6-ohm light bulb in place of the 3-ohm bulb (**Figure 2-32**), the other bulbs will work the same way and with the same intensity as before. However, 20 amps of current will flow through the 0.6-ohm bulb and raise the total circuit current to 34 amps. Lowering the resistance on the one leg of the parallel circuit increases the current throughout the circuit. This high current may damage the circuit or components. Current that is higher than the capability of the conductor to handle can cause the wires to burn. In this case, the wires that would burn are the wires that would carry the 34 amps or the 20 amps to the bulb, not the wires to the other bulbs.

Leg #1	$I = E/R$	or	$I = 12/0.6$	or	$I = 20$ amps
Leg #2	$I = E/R$	or	$I = 12/6$	or	$I = 2$ amps
Leg #3	$I = E/R$	or	$I = 12/1$	or	$I = 12$ amps

Total current $= 20 + 2 + 12$ or 34 amps

Let's see what happens when we add resistance to one of the parallel legs. An increase in resistance should cause a decrease in current. In **Figure 2-33**, a 1-ohm resistor was added after the 1-ohm light bulb. This resistor is in series with the light bulb, and the total resistance of that leg is now 2 ohms. The current through that leg is now 6 amps. Again, the other bulbs were not affected by the change. The only change to the whole circuit was in total circuit current, which now drops to 12 amps. The added resistance lowered the total circuit current and changed how the 1-ohm bulb works. This bulb will now drop only 6 volts. The added resistor will drop the remaining 6 volts. The 1-ohm bulb will be

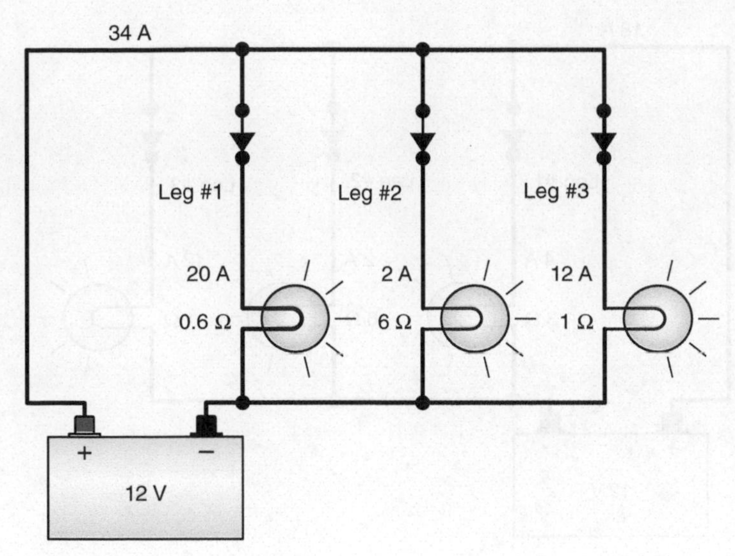

Figure 2-32 Parallel light circuit.

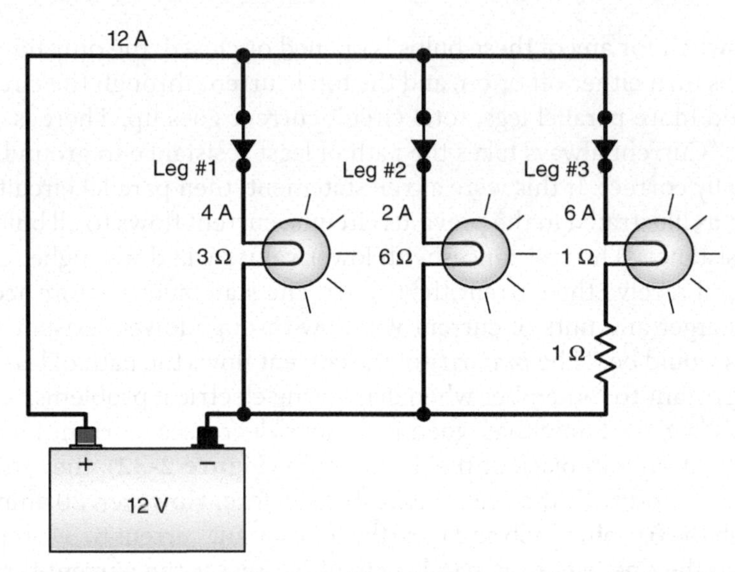

Figure 2-33 A parallel light circuit with one leg having a series resistance added.

much dimmer since its power rating dropped from 144 watts to 36 watts. Additional resistance causes the bulb to be dimmer. The bulb itself wasn't changed; only the resistance of that leg changed. The circuit, not the bulb, causes the dimness.

Leg #1 $I = E/R$ or $I = 12/3$ or $I = 4$ amps

Leg #2 $I = E/R$ or $I = 12/6$ or $I = 2$ amps

Leg #3 $I = E/R$ or $I = 12/1 + 1$ or $I = 12/2$ or $I = 6$ amps

Total current $= 4 + 2 + 6$ or 12 amps

Now let's see what happens when we add a resistance that is common to all of the parallel legs. In **Figure 2-34**, we added a 0.333-ohm resistor (0.333 was chosen to keep the math simple!) to the negative connection at the battery. This will cause the circuit's current to decrease; it will also change the operation of the bulbs in the circuit. The total resistance of the bulbs in parallel is 0.667 ohms.

$$R_T = \frac{1}{\frac{1}{3} + \frac{1}{6} + \frac{1}{1}} \text{ or } R_T = \frac{1}{0.333 + 0.167 + 1} \text{ or } R_T = \frac{1}{1.5} \text{ or } R_T = 0.667 \text{ ohms}$$

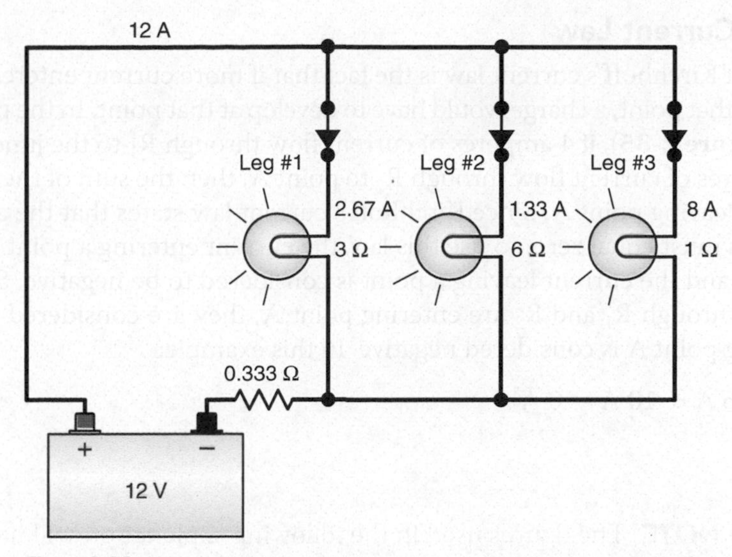

Figure 2-34 A parallel light circuit with a resistance in series to the whole circuit.

The circuit's total resistance is 1 ohm (0.667 + 0.333), which means the circuit current is now 12 amps. Because there will be a voltage drop across the 0.333-ohm resistor, each parallel leg will drop less than the source voltage. To find the amount of voltage dropped by the parallel circuit, we multiply the amperage by the resistance. Twelve amps multiplied by 0.667 equals 8. So, 8 volts will be dropped by the parallel circuit; the 0.333 resistor will drop the remaining 4 volts. The amount of current through each leg can be calculated by taking the voltage drop and dividing it by the resistance of the leg.

Leg #1 $I = E/R$ or $I = 8/3$ or $I = 2.667$ amps

Leg #2 $I = E/R$ or $I = 8/6$ or $I = 1.333$ amps

Leg #3 $I = E/R$ or $I = 8/1$ or $I = 8$ amps

Total circuit current = 2.6667 + 1.333 + 8 or 12 amps

The added resistance affected the operation of all the bulbs, because it was added to a point that was common to all of the bulbs. All of the bulbs would be dimmer and circuit current would be lower.

Kirchhoff's Laws

Ohm's law states that it takes one volt to push one ampere through one ohm of resistance. Ohm's law is the principal law of electricity and is used to determine electrical values. However, there are times when Ohm's law would be difficult to use in determining electrical values. This is true in circuits with no clearly defined series or parallel connections. Also, Ohm's law can be challenging to use if the circuit has more than one power source. In these instances, Kirchhoff's laws are used.

Kirchhoff stated two laws that described voltage and current relationships in an electric circuit. The first law, known as **Kirchhoff's voltage law**, states that the algebraic sum of the voltage sources and voltage drops in a closed circuit must equal zero. This law is actually the rule that states that the sum of the voltage drops in a series circuit must equal the source voltage.

The second law, known as **Kirchhoff's current law**, states that the algebraic sum of the currents entering and leaving a point must equal zero. This rule states that the total current flow in a parallel circuit will be the sum of the currents through all the circuit branches.

Kirchhoff's Current Law

The concept of Kirchhoff's current law is the fact that if more current entered a particular point than left that point, a charge would have to develop at that point. In the parallel circuit illustrated (**Figure 2-35**), if 4 amperes of current flow through R_1 to the junction at point A, and 6 amperes of current flow through R_2 to point A, then the sum of the two currents is 10 amperes leaving point A. Since Kirchhoff's current law states that the algebraic sum of the currents must equal zero, to use this law, the current entering a point is considered to be positive, and the current leaving a point is considered to be negative. Since the currents flowing through R_1 and R_2 are entering point A, they are considered positive. The current leaving point A is considered negative. In this example:

$$+4\text{ A} + 6\text{ A} - 10\text{ A} = 0\text{ A}$$

AUTHOR'S NOTE The figures used in the following explanation will have a source voltage of 120 volts. This amount is used so the example will be easier to understand and keep the values high enough to prevent using fractions.

Figure 2-36 illustrates a more complex series-parallel circuit. Ohm's law tells us that this circuit will have 2 amperes flowing through it since there is a total of 60 ohms of resistance. Since 2 amperes are flowing through R_1, 32 volts will be dropped, and the applied voltage at point B will be 88 volts. With 2 amperes flowing through R_3, the voltage drop over this resistor will be 40 volts. With 88 volts applied to point B and R_3 dropping 40 volts, that leaves 48 volts to be dropped over R_2 and the parallel circuit of R_4 through R_6. This means points B to E will have a current of 0.8 amperes, and the parallel branch will have 1.2 amperes.

Notice what occurs to the current at point B. Two amperes are flowing into point B from R_1. At this point, the current splits, with part flowing to R_2 and part to the circuit containing resistors R_4 through R_6. The current entering point B is considered positive, and the two currents leaving point B are considered negative. The equation would look like this:

$$+2\text{ A} - 0.8\text{ A} - 1.2\text{ A} = 0\text{ A}$$

At point E, there is 0.8 ampere of current entering from the circuit with resistor R_2 and 1.2 amperes of current entering from the circuit with resistors R_4 through R_6. Two amperes of current leaves point E and flows through R_3.

$$+0.8\text{ A} + 1.2\text{ A} - 2\text{ A} = 0\text{ A}$$

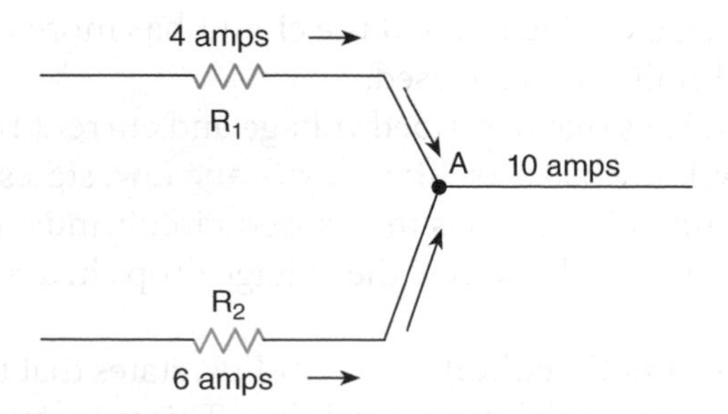

Figure 2-35 The sum of the currents entering and leaving a point in the circuit must equal zero.

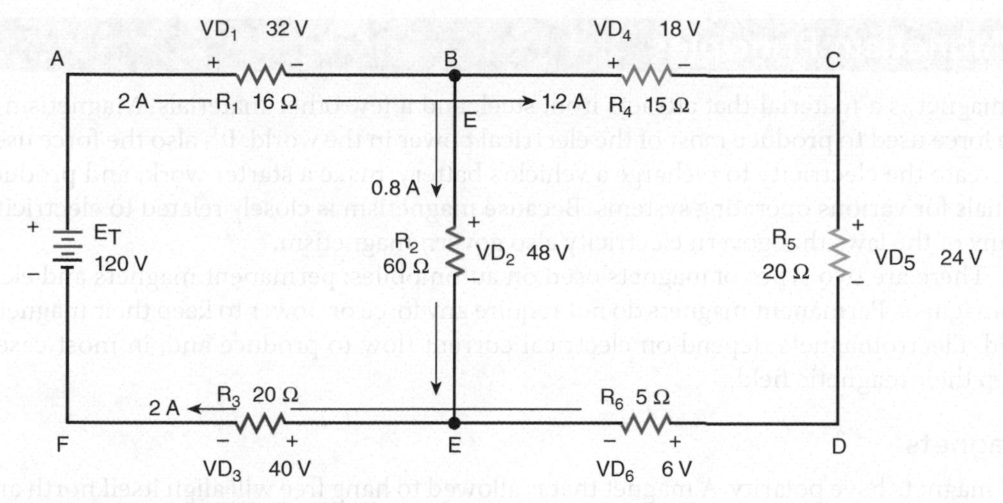

Figure 2-36 Illustration of the use of Kirchhoff's laws.

Kirchhoff's Voltage Law

Kirchhoff's voltage law states that the algebraic sum of the voltages around any closed loop must equal zero. To calculate the sum of the voltages, first establish which end of the resistive element is positive and which is negative. Referring to **Figure 2-36**, we will use the conventional theory of current flow being positive to negative. Therefore, the point at which current enters a resistor is marked positive, and the point where current leaves the resistor is marked negative.

AUTHOR'S NOTE It does not matter if you use the conventional theory of current flow or the electron theory; you only need to be consistent. The sum of voltage drops will be the same regardless of which current flow assumption you use.

The circuit illustrated in **Figure 2-36** has three separate closed loops. Closed-loop ACDF contains the voltage drops VD_1, VD_4, VD_5, VD_6, VD_3, and E_T. E_T is the source and must be included in the equation. The voltage drops for this loop are as follows:

$$+VD_1 + VD_4 + VD_5 + VD_6 + VD_3 - E_T = 0$$

$$+32\,V + 18\,V + 24\,V + 6\,V + 40\,V - 120\,V = 0\,V$$

The assumed direction of current flow determines each number's positive or negative sign. In this example, it's assumed that the current leaves point A and returns to point A. Current leaving point A enters R_1 at the positive side. Therefore, the voltage is considered to be positive (+ 32 V). The same is true for R_4, R_5, R_6, and R_3. However, the current enters the voltage source at the negative side, so ET is assumed to be negative.

Closed-loop ABEF contains voltage drops VD_1, VD_2, VD_3, and ET. Current will leave point A and return to point A through R_1, R_2, R_3, and the voltage source. The voltage drops are as follows:

$$+VD_1 + VD_2 + VD_3 - VD_T = 0$$

$$+32\,V + 48\,V + 40\,V - 120\,V = 0\,V$$

Closed-loop BCDE contains voltage drops VD_4, VD_5, VD_6, and VD_2. Current leaves point B and returns to point B, flowing through R_4, R_5, R_6, and R_2.

$$+VD_4 + VD_5 + VD_6 - VD_{26} = 0$$

$$+18\,V + 24\,V + 6\,V - 48\,V = 0\,V$$

Magnetism Principles

A magnet is a material that attracts iron, steel, and a few other materials. Magnetism is the force used to produce most of the electrical power in the world. It's also the force used to create the electricity to recharge a vehicle's battery, make a starter work, and produce signals for various operating systems. Because magnetism is closely related to electricity, many of the laws that govern electricity also govern magnetism.

There are two types of magnets used on automobiles: permanent magnets and electromagnets. Permanent magnets do not require any force or power to keep their magnetic field. Electromagnets depend on electrical current flow to produce and, in most cases, keep their magnetic field.

Magnets

All magnets have polarity. A magnet that is allowed to hang free will align itself north and south. The end facing north is called the north-seeking pole, and the end facing south is called the south-seeking pole. Like poles will repel each other, and unlike poles will attract each other. The magnetic attraction is the strongest at the poles. These principles are shown in **Figure 2-37**.

Magnetic flux density is a concentration of the lines of force (**Figure 2-38**). A strong magnet produces many lines of force, and a weak magnet produces fewer lines of force. Invisible lines of force leave the magnet at the north pole and enter again at the south pole. While inside the magnet, the lines of force travel from the south pole to the north pole (**Figure 2-39**).

The field of force (or magnetic field) is all the space outside the magnet that contains lines of magnetic force. Magnetic lines of force penetrate all substances; there is no known insulation against magnetic lines of force. The lines of force may be deflected only by other magnetic materials or by another magnetic field.

Electromagnetism

Electromagnetism uses the theory that whenever an electrical current flows through a conductor, a magnetic field is formed around the conductor (**Figure 2-40**). The number of lines of force and the strength of the magnetic field produced will directly proportional to the amount of current flow.

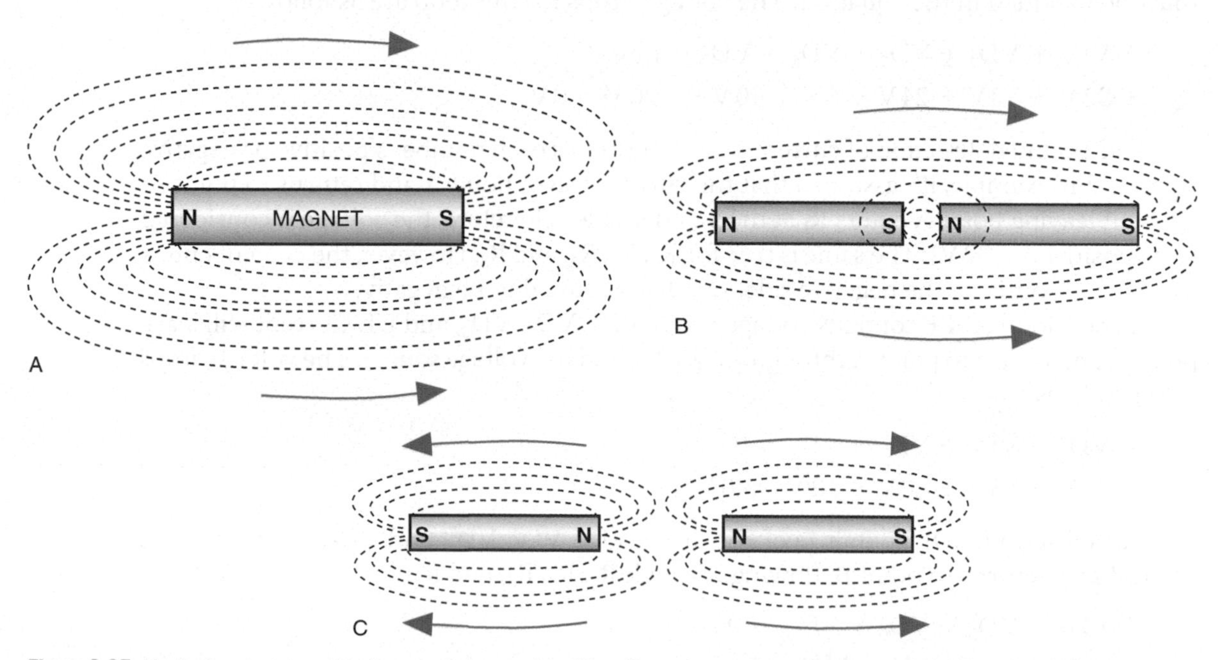

Figure 2-37 Magnetic principles: (A) all magnets have poles, (B) unlike poles attract each other, and (C) like poles repel.

Figure 2-38 Iron filings indicate the lines of magnetic flux.

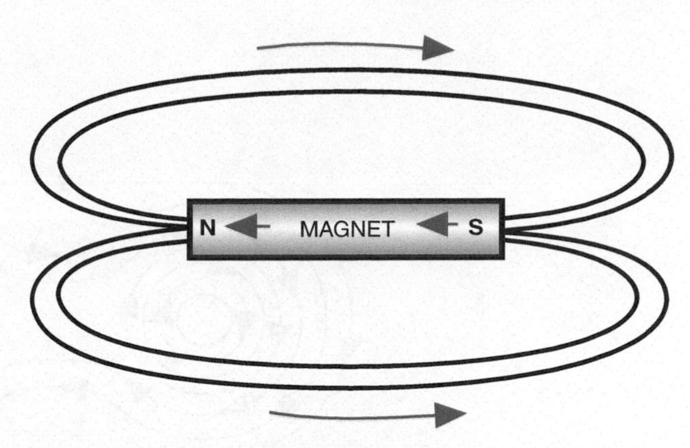

Figure 2-39 Lines of force through the magnet.

The **right-hand rule** exposes the direction of the lines of force. Using the conventional theory of current flow, the right hand is used to grasp the wire, with the thumb pointing in the direction of current flow. The fingers will point in the direction of the magnetic lines of force (**Figure 2-41**).

André Marie Ampère noted that current flowing in the same direction through two nearby wires would cause the wires to attract each other. Also, he observed that if current flow in one of the wires is reversed, the wires will repel each other. In addition, he found that if a wire is coiled with current flowing through the wire, the same magnetic field that surrounds a straight wire combines to form one larger magnetic field. This magnetic field has true north and south poles (**Figure 2-42**). Looping the wire doubles the flux density where the wire runs parallel to itself. **Figure 2-43** shows how these lines of force will join and add to each other.

Use the right-hand rule to determine which end of the coil is the north pole. Grasp the coil with your fingers pointing in the direction of current flow (+ to –), and the thumb will point toward the north pole (**Figure 2-44**).

As more loops are added, the fields from each loop will join and increase the flux density (**Figure 2-45**). To make the magnetic field even stronger, an iron core can be placed in the center of the coil (**Figure 2-46**). The soft iron core has high **permeability** and low **reluctance**, which provides an excellent conductor for the magnetic field to travel through the center of the wire coil.

Permeability is the term used to indicate the magnetic conductivity of a substance compared with the conductivity of air. The greater the permeability, the greater the magnetic conductivity, and the easier a substance can be magnetized or the more attracted it is to a magnet.

Reluctance is the term used to indicate a material's resistance to the passage of flux lines. Highly reluctant materials are not attracted to magnets.

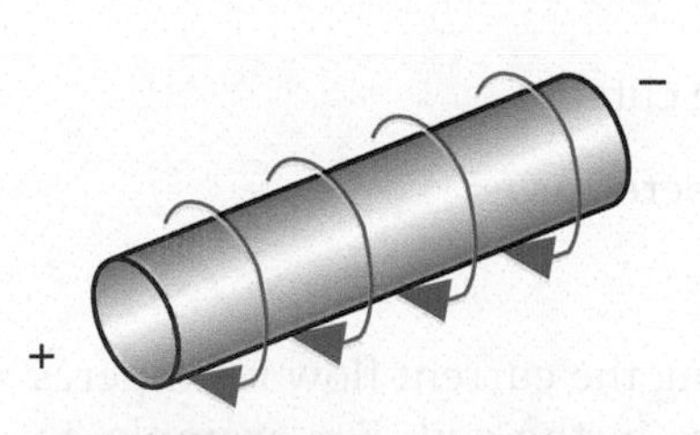

Figure 2-40 A magnetic field surrounds a conductor that has a current flowing through it.

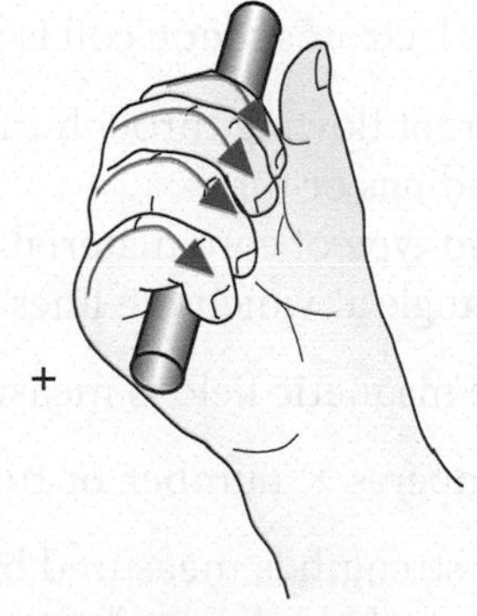

Figure 2-41 Right-hand rule to determine the direction of magnetic lines of force.

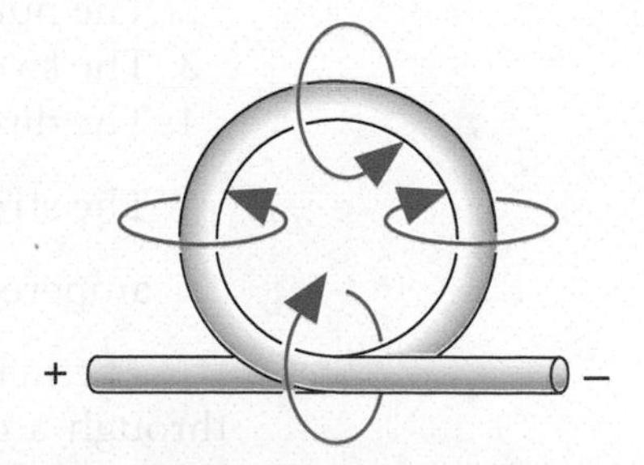

Figure 2-42 Looping the conductor increases the magnetic field.

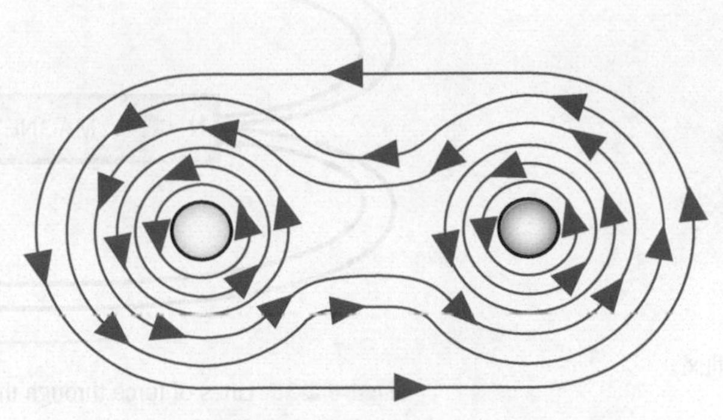

Figure 2-43 Lines of force join together and attract each other.

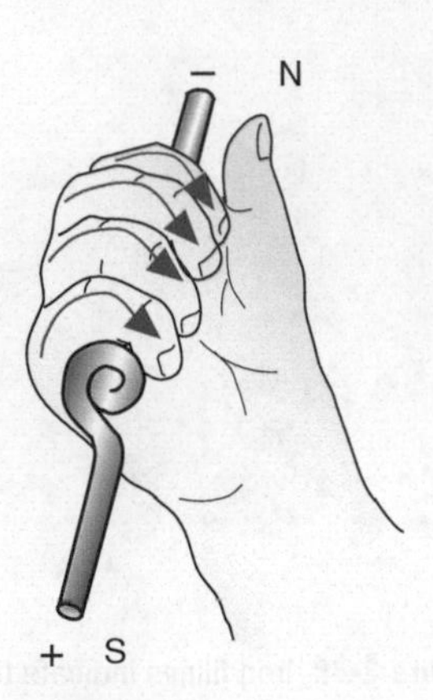

Figure 2-44 Right-hand rule to determine magnetic poles.

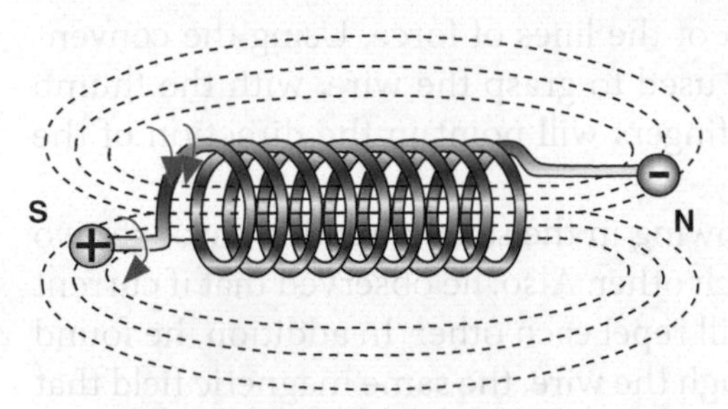

Figure 2-45 Adding more wire loops increases the magnetic flux density.

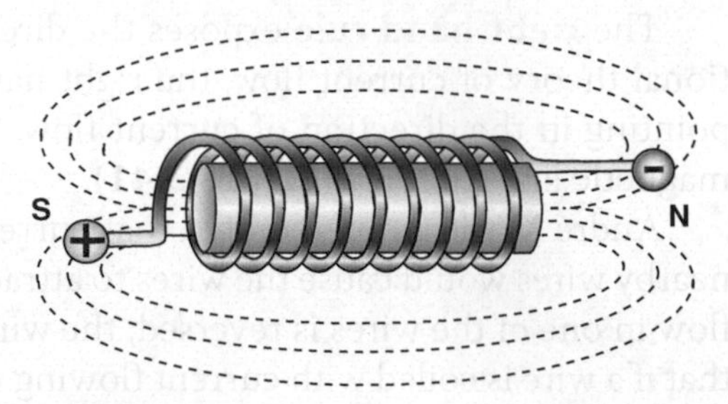

Figure 2-46 The addition of an iron core concentrates the flux density.

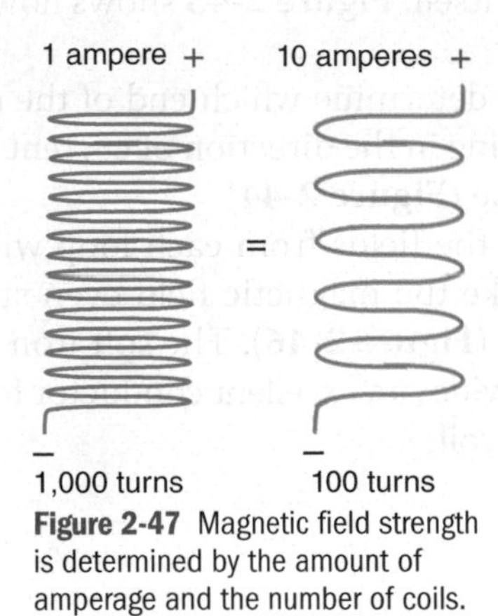

Figure 2-47 Magnetic field strength is determined by the amount of amperage and the number of coils.

The strength of an electromagnetic coil is affected by the following factors:

1. The amount of current flowing through the wire.
2. The number of windings or turns.
3. The size, length, and type of core material.
4. The direction and angle at which the lines of force are cut.

The strength of the magnetic field is measured in ampere-turns:

ampere-turns = amperes × number of turns

The magnetic field strength is measured by multiplying the current flow in amperes through a coil by the number of complete turns of wire in the coil. For example, in **Figure 2-47**, a 1,000-turn coil with 1 ampere of current would have a field strength of 1,000 ampere-turns. This coil would have the same field strength as a coil with 100 turns and 10 amperes of current.

Theory of Induction

Electricity can be produced by magnetic **induction**. Magnetic induction occurs when a conductor is moved through the magnetic lines of force (**Figure 2-48**) or when a magnetic field is moved across a conductor. A potential difference is set up between the ends of the conductor, and a voltage is induced. This voltage exists only when the magnetic field or the conductor is in motion.

The induced voltage can be increased by either increasing the speed at which the magnetic lines of force cut the conductor or by increasing the number of conductors that are cut. It's this principle that is behind the operation of all ignition systems, starter motors, and charging systems.

A common induction device is the ignition coil. As the current increases, the coil will reach a point of **saturation**. This is the point at which the magnetic strength eventually levels off and where current will no longer increase as it passes through the coil. The magnetic lines of force, which represent stored energy, will collapse when the applied voltage is removed. When the lines of force collapse, the magnetic energy is returned to the wire as electrical energy.

A desirable induction is called a **mutual induction.**

Mutual induction is used in ignition coils where a rapidly changing magnetic field in the primary windings creates a voltage in the secondary winding (**Figure 2-49**).

Self-induction is also referred to as counter EMF (CEMF) or as a voltage spike.

If voltage is induced in the wires of a coil when current is first connected or disconnected, it's called **self-induction**. The resulting current is in the opposite direction of the applied current and tends to reduce the magnetic force. Self-induction is governed by Lenz's law, which states that an induced current flows in a direction opposite the magnetic field that produces it.

Self-induction is generally not wanted in automotive circuits. For example, when a switch is opened, self-induction tends to continue to supply current in the same direction as the original current. This is because as the magnetic field collapses, it induces a voltage in the wire. According to Lenz's law, the voltage induced in a conductor tends to oppose a change in current flow. Self-induction can cause an electrical arc to occur across an opened switch. The arcing may momentarily bypass the switch and allow the circuit that was turned off to operate for a short period of time. The arcing will also burn the contacts of the switch.

Self-induction is commonly found in electrical components that contain a coil or an electric motor. A capacitor or clamping diode may be connected to the circuit to reduce the arc across the contacts. The capacitor will absorb the high-voltage arcs and prevent arcing across the contacts. Diodes are semiconductors that allow current flow in only one direction. A clamping diode can be connected in parallel to the coil and will prevent current flow from the self-induction coil to the switch. The capacitor and clamping diode will be discussed in Chapter 3.

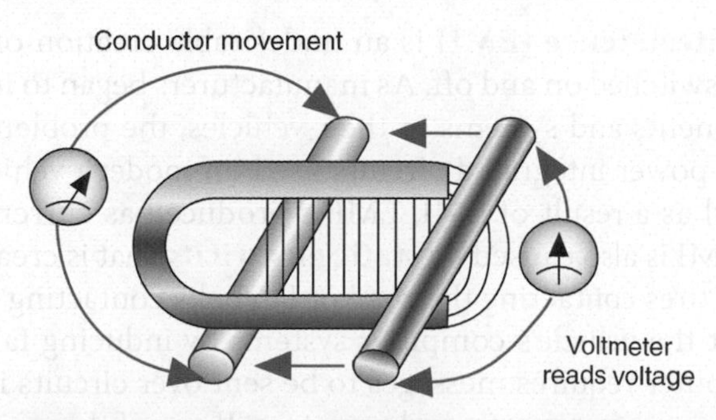

Figure 2-48 Moving a conductor through a magnetic field induces an electrical potential difference.

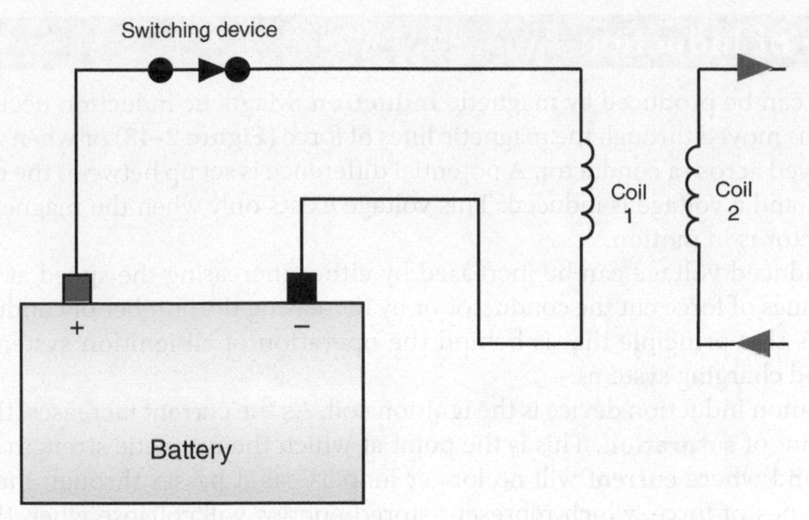

Figure 2-49 A mutual induction is used to create an electrical current in coil 2 if the current flow in coil 1 is turned off.

Magnetic induction is also the basis for a generator and many of the sensors on today's vehicles. In a generator, a magnetic field rotates inside a set of conductors. As the magnetic field crosses the wires, a voltage is induced. The amount of voltage induced by this action depends on the speed of the rotating field, the strength of the field, and the number of conductors the field cuts through. This principle will be discussed in greater detail in Chapter 7.

Magnetic sensors are used to measure speeds, such as engine, vehicle, and shaft speeds. These sensors typically use a permanent magnet. Rotational speed is determined by the passing of blades or teeth in and out of the magnetic field. As a tooth moves in and out of the magnetic field, the strength of the magnetic field is changed, and a voltage signal is induced. This signal is sent to a control device, where it's interpreted. This principle is discussed in greater detail in Chapter 10.

Photovoltaics

Photovoltaics (PV) is one of the forces that can generate electrical current by converting solar radiation through the use of solar cells. Solar cells make up solar panels that convert energy from the sun into direct current electricity. Some automotive manufacturers are looking to this energy source as a means to power some accessories and to recharge batteries.

EMI Suppression

Shop Manual
Chapter 2, page 61

Static electricity is electricity that is not in motion.

Electromagnetic interference (EMI) is an undesirable creation of electromagnetism whenever current is switched on and off. As manufacturers began to increase the number of electronic components and systems in their vehicles, the problem of EMI had to be controlled. The low-power integrated circuits used on modern vehicles are sensitive to the signals produced as a result of EMI. EMI is produced as current in a conductor is turned on and off. EMI is also caused by **static electricity** that is created by friction. The friction results from tires contacting the road or fan belts contacting the pulleys.

EMI can disrupt the vehicle's computer systems by inducing false messages to the computer. The computer requires messages to be sent over circuits in order to communicate with other computers, sensors, and actuators. If any of these signals are disrupted, the engine and/or accessories may turn off.

EMI can be suppressed by any one of the following methods:

1. Adding resistance to the conductors. This is usually done to high-voltage systems, such as the secondary circuit of the ignition system.
2. Connecting a capacitor in parallel and a choke coil in series with the circuit.
3. Shielding the conductor or load components with a metal or metal-impregnated plastic.
4. Increasing the number of paths to ground by using designated ground circuits. This provides a clear path to ground that is very low in resistance.
5. Adding a clamping diode in parallel to the component.
6. Adding an isolation diode in series to the component.

Summary

- An atom is constructed of a complex arrangement of electrons in orbit around a nucleus. If the number of electrons and protons is equal, the atom is balanced or neutral.
- A conductor allows electricity to flow through it easily.
- An insulator does not allow electricity to flow through it easily.
- Electricity is the movement of electrons from atom to atom. In order for the electrons to move in the same direction, an electromotive force (EMF) must be applied to the circuit.
- The electron theory defines electron flow as motion from negative to positive.
- The conventional theory of current flow states that current flows from a positive point to a less positive point.
- Voltage is an electrical pressure and is the difference between the positive and negative charges.
- Current is defined as the rate of electron flow and is measured in amperes. Amperage is the amount of electrons passing any given point in the circuit in 1 second.
- Resistance is defined as opposition to current flow and is measured in ohms (Ω).
- Ohm's law defines the relationship between current, voltage, and resistance. It's the fundamental law of electricity and states that the amount of current in an electric circuit is inversely proportional to the resistance of the circuit and is directly proportional to the voltage in the circuit.
- Wattage represents the measure of power (P) used in a circuit. Wattage is measured using Watt's law formula, which defines the relationship between amperage, voltage, and wattage.
- Direct current has a constant voltage and current that flows in one direction.
- In an alternating current circuit, voltage and current do not remain constant. AC changes direction from positive to negative and negative to positive.
- For current to flow, the electrons must have a complete path from the source voltage to the load component and back to the source.
- The series circuit provides a single path for current flow from the electrical source through all the circuit's components and back to the source.
- A parallel circuit provides two or more paths for current to flow.
- A series-parallel circuit is a combination of series and parallel circuits.
- The equivalent series load is the total resistance of a parallel circuit plus the resistance of the load in series with the voltage source.
- Resistance in any form causes a voltage drop in the circuit that reduces the electrical pressure available after the resistance.
- Kirchhoff's voltage law states that the algebraic sum of the voltage sources and voltage drops in a closed circuit must equal zero. This law is actually the rule that states that the sum of the voltage drops in a series circuit must equal the source voltage.
- Kirchhoff's current law states that the algebraic sum of the currents entering and leaving a point must equal zero. This rule states that the total flow of current in a parallel circuit will be the sum of the currents through all the circuit branches.

Review Questions

Short-Answer Essays

1. List and define the three elements of electricity.
2. Explain the basic principles of Ohm's law.
3. List and describe the three types of circuits.
4. Explain the principle of electromagnetism.
5. Describe the principle of induction.
6. Describe the basics of electron flow.
7. Define the two types of electrical current.
8. Describe the difference between insulators, conductors, and semiconductors.
9. Define "voltage drop."
10. What does the measurement of "watt" represent?

Fill in the Blanks

1. ______________ are negatively charged particles. The nucleus contains positively charged particles called ______________ and particles that have no charge called ______________.
2. A ______________ allows electricity to easily flow through it. An ______________ does not allow electricity to easily flow through it.
3. For the electrons to move in the same direction, there must be an ______________ applied.
4. The ______________ of current flow states that current flows from a positive point to a less positive point.
5. Resistance is defined as ______________ to current flow and is measured in ______________.
6. In a series circuit, the voltage drop across a resistance is determined by the ______________ value.
7. Kirchhoff's voltage law states that the ______________ of the ______________ ______________ in a series circuit will always ______________ the source voltage.
8. The ______________ of all the resistors in series is the total resistance of that series circuit.
9. ______________ is defined as an electrical pressure.
10. ______________ is defined as the rate of electron flow.

Multiple Choice

1. Which of the following methods can be used to form an electrical current?
   A. Magnetic induction.
   B. Chemical reaction.
   C. Heat.
   D. Both A and B.
   E. Neither A nor B.
2. In a parallel circuit:
   A. Total resistance is the sum of all of the resistances in the circuit.
   B. Total resistance is less than the lowest resistor.
   C. Amperage will decrease as more branches are added.
   D. All of these choices.
3. All of the following concerning voltage drop are true EXCEPT:
   A. All of the voltage from the source must be dropped before it returns to the source.
   B. Corrosion is not a contributor to voltage drop.
   C. Voltage drop is the conversion of electrical energy into another energy form.
   D. Voltage drop can be measured with a voltmeter.
4. All of the following concerning voltage are true EXCEPT:
   A. Voltage is the electrical pressure that causes electrons to move.
   B. Voltage will exist between any two points in a circuit unless the potential drops to zero.
   C. Voltage is A × R.
   D. In a series circuit, voltage is the same at all points in the circuit.

5. Wattage is:
   A. A measure of the total electrical work being performed per unit of time.
   B. Expressed as P = R × A.
   C. Both A and B.
   D. Neither A nor B.

6. Which answer is incorrect concerning a series circuit?
   A. Total resistance is the sum of all of the resistances in the circuit.
   B. Total resistance is less than the lowest resistor.
   C. Amperage will decrease as more resistance is added.
   D. All of these choices.

7. Which statement about electrical currents is correct?
   A. Alternating current can be stored in a battery.
   B. Alternating current is produced from a voltage and current that remain constant and flow in the same direction.
   C. Direct current is used for most electrical systems in the automobile.
   D. Direct current changes directions from positive to negative.

8. Induction:
   A. It is the magnetic process of producing a current flow in a wire without contact with the wire.
   B. Exists when the magnetic field or the conductor is in motion.
   C. Both A and B.
   D. Neither A nor B.

9. All of the following statements are true EXCEPT:
   A. If the resistance increases and the voltage remains constant, the amperage will increase.
   B. Ohm's law can be stated as A = V/R.
   C. If voltage is increased, amperage will increase.
   D. An open circuit does not allow current flow.

10. Which of the following statements is correct?
    A. An insulator is capable of supporting the flow of electricity through it.
    B. A conductor is not capable of supporting the flow of electricity.
    C. Both A and B.
    D. Neither A nor B.

CHAPTER 3

ELECTRICAL AND ELECTRONIC COMPONENTS

Upon completion and review of this chapter, you should be able to:

- Describe the common types of electrical components used in automotive circuits.
- Explain how each electrical components affect the electrical system.
- Explain the operation of switches.
- Explain the operation of relays and solenoids.
- Explain the function of fixed, stepped, and variable resistors.
- Explain the function of capacitors.
- Describe the basic operating principles of diodes and transistors.
- Explain the use of common diode types used in automotive applications.
- Describe the use of common transistor types.
- Explain the purpose of circuit protection devices.
- Describe the operation of the most common types of circuit protection devices used in automotive electrical circuits.
- Define circuit defects, including opens, shorts, grounds, and excessive resistance.
- Explain the effects that each type of circuit defect has on the operation of the electrical system.

Terms to Know

Anode
Avalanche diodes
Base
Bimetallic strip
Bipolar
Buzzer
Capacitance
Capacitor
Cathode
Circuit breaker
Clamping diode
Collector
Covalent bonding
Crystal
Darlington pair
Depletion-type FET
Dielectric
Diode
Electrostatic field
Emitter
Enhancement-type FET
Field-effect transistor (FET)
Fixed resistors
Forward-biased
Fuse
Fuse block
Fusible link
Ganged switch
Hole
Integrated circuit (IC)
ISO relays
Light-emitting diode (LED)
Load device
Maxi-fuse
Normally closed (NC) switch
Normally open (NO) switch
N-type material
Open
Overload
Peak reverse voltage (PRV)
Photodiode
Phototransistor
Pole
Positive plate
Positive temperature coefficient (PTC) thermistors
Potentiometer
Protection device
Reverse-biased
Rheostat
Short
Shorted circuit
Solenoid
Sound generator
Stepped resistor
Throw

Thyristor
Transistor
Turn-on voltage
Variable resistor
Wiper
Zener diode
Zener voltage

Introduction

This chapter introduces you to electrical and electronic components. These components include circuit protection devices, switches, relays, variable resistors, diodes, and different forms of transistors. Today's technician must comprehend the operation of these components and how they affect electrical system operation. With this knowledge, you will be able to accurately and quickly diagnose electrical failures.

To properly diagnose the electrical system's components and circuits, you must be able to use the test equipment designed for electrical system diagnosis. This chapter discusses the various types of test equipment used for diagnosing electrical systems. In addition, you will learn of the various types of defects that cause improper system operation. Also, you will learn how to use the appropriate equipment to locate the fault based on the symptoms.

Electrical Components

Electrical circuits require different components depending on the type of work they do and how they perform it. A light bulb may be wired directly to the battery, but it will remain on until the battery drains. Installing a switch will provide for control of the light circuit. However, if variable dimming of the light is required, a rheostat is also needed.

Automotive manufacturers may incorporate several electrical components into a circuit to achieve the desired results from the system. These components include switches, relays, solenoids, buzzers, various types of resistors, and capacitors.

Switches

A switch is the most common means of controlling electrical current flow to an accessory (**Figure 3-1**). A switch can control a circuit's on/off operation or direct current flow through various circuits. The contacts inside the switch assembly carry the current when they are closed. Opening the contacts stops the current flow.

Shop Manual
Chapter 3, page 121

A **normally open (NO) switch** will not allow current flow when it is in its rest position. The contacts are open until they are acted on by an outside force that closes them to complete the circuit. A **normally closed (NC) switch** will allow current flow when it is in its rest position. The contacts are closed until they are acted on by an outside force that opens them to stop current flow.

The simplest type of switch is the single-**pole**, single-**throw** (SPST) switch (**Figure 3-2**). This switch controls the on/off operation of a single circuit. The most common type of SPST switch design is the hinged pawl. The pawl acts as the contact and changes position as directed to open or close the circuit.

The term **pole** refers to the number of input circuits.

The term **throw** refers to the number of output circuits.

Some SPST switches are momentary contact switches. The horn switch on most vehicles is of this design. This switch usually has a spring that holds the contacts open until an outside force is applied and closes them. The switch contacts automatically open when the switch is released.

Some electrical systems may require a single-pole, double-throw switch (SPDT). The dimmer switch used in the headlight system is usually an SPDT switch. This switch has one input circuit with two output circuits. Depending on the position of the contacts, voltage is applied to the high-beam circuit or to the low-beam circuit (**Figure 3-3**).

Figure 3-1 Common types of switches used in the automotive electrical system.

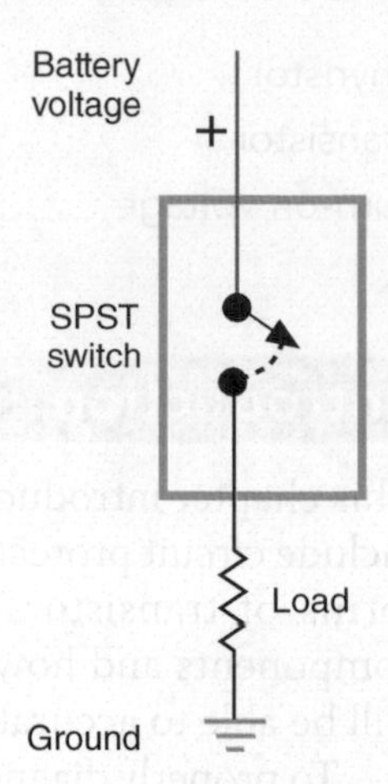

Figure 3-2 A simplified illustration of an SPST switch.

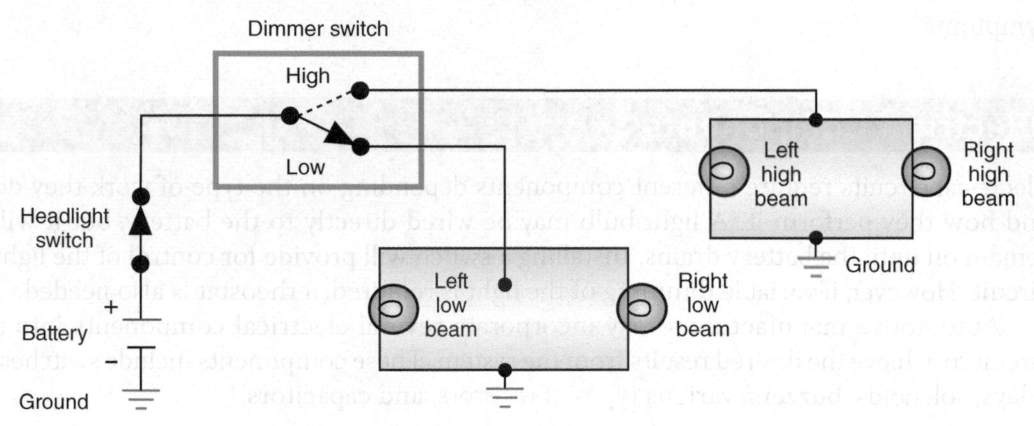

Figure 3-3 A simplified schematic of a headlight system using an SPDT dimmer switch.

One of the most complex switches is the **ganged switch**. Ignition switches are examples of this type of switch. In **Figure 3-4**, the five wipers are all ganged together and will move together. The starter relay terminal supplies battery voltage to the switch. Turning the switch into the START position moves all wipers to the "S" position. Wipers D and E will complete the circuit to ground to test the instrument panel warning lamps. Wiper B provides battery voltage to the ignition coil. Wiper C supplies battery voltage to the starter relay and the ignition module. There is no output from wiper A.

AUTHOR'S NOTE The dashed lines used in the switch symbol indicate that the wipers of the switch move together.

Once the engine starts, the wipers are moved to the RUN position. Wipers D and E are no longer in contact with any output terminals. Wiper A supplies battery voltage to the comfort controls and turn signals, wiper B supplies battery voltage to the ignition coil and other accessories, and wiper C supplies battery voltage to additional accessories. The jumper wire between terminals A and R of wiper C indicates that the listed accessories can operate with the ignition switch in the RUN or ACC position.

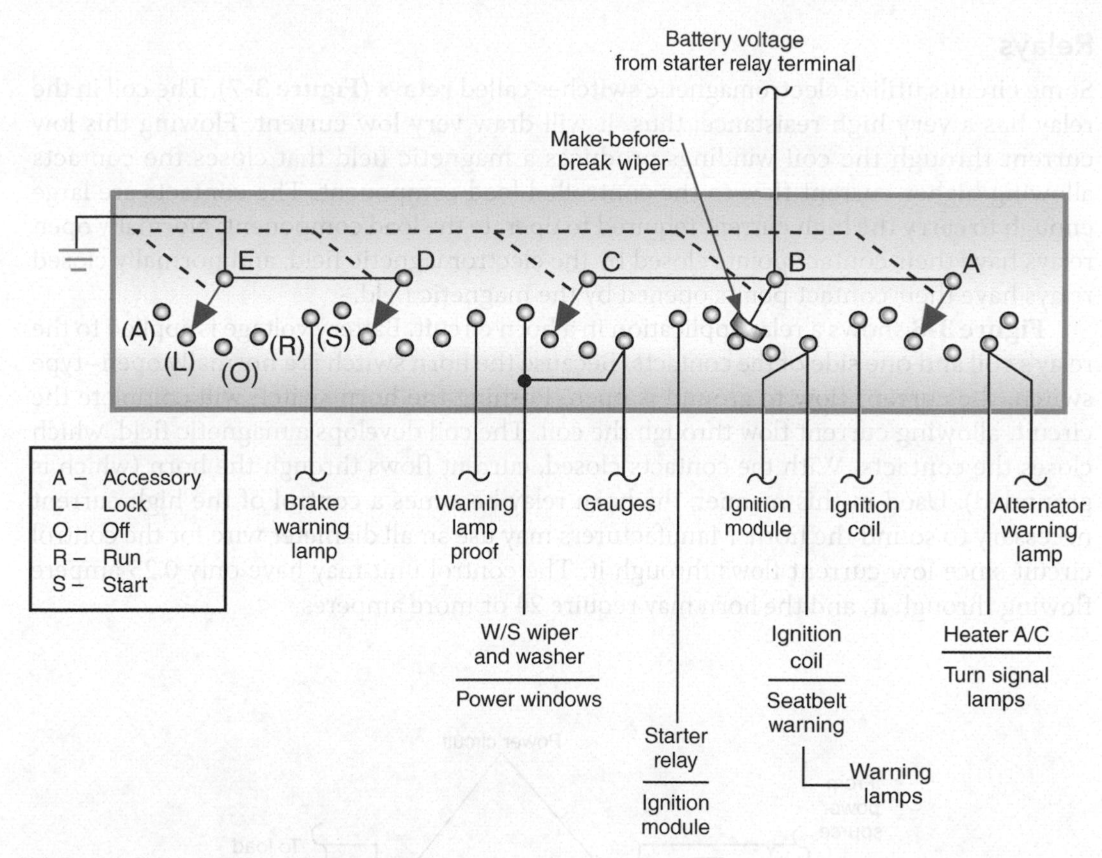

Figure 3-4 Illustration of an ignition switch.

Since mercury is an excellent conductor of electricity, many vehicle manufacturers use mercury switches to detect motion. This switch uses a partially filled capsule of mercury and has two electrical contacts located at one end. Placing the contacts above the mercury level makes the witch operate as a normally open switch (**Figure 3-5**). The switch completes the circuit when the rotation of the capsule causes the mercury to contact both electrical contacts (**Figure 3-6**). For example, if used as an underhood light switch, opening the hood tilts the capsule, and the mercury completes the circuit to illuminate the light. When the hood is shut, the capsule is tilted so the mercury does not complete the circuit.

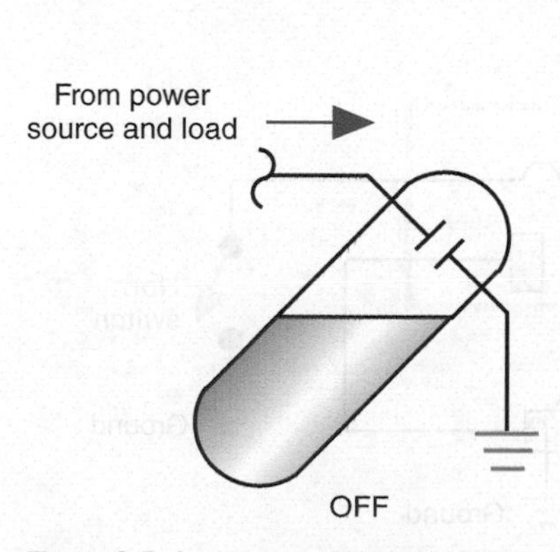

Figure 3-5 A mercury switch in the open position. The mercury is not covering the points.

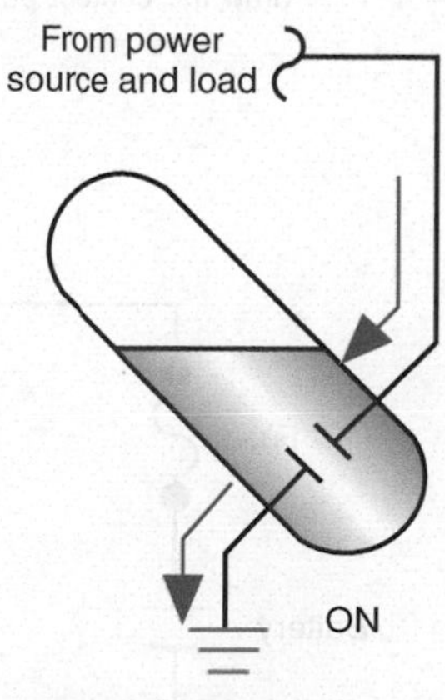

Figure 3-6 When the mercury switch is tilted, the mercury covers the points and closes the circuit.

Relays

Shop Manual
Chapter 3, page 123

Some circuits utilize electromagnetic switches called **relays** (**Figure 3-7**). The coil in the relay has a very high resistance; thus, it will draw very low current. Flowing this low current through the coil windings produces a magnetic field that closes the contacts allowing higher current flow to the controlled load component. The contacts are large enough to carry the high current required to operate the load component. Normally open relays have their contact points closed by the electromagnetic field, and normally closed relays have their contact points opened by the magnetic field.

The collapsing of the magnetic field around the coil produces voltage spikes. A clamping diode, or a resistor, installed in parallel to the relay's coil prevents the voltage spike from damaging sensitive electronic components. See page 69.

Figure 3-8 shows a relay application in a horn circuit. Battery voltage is applied to the relay's coil and one side of the contacts. Because the horn switch is a normally open–type switch, the current flow to ground is open. Pushing the horn switch will complete the circuit, allowing current flow through the coil. The coil develops a magnetic field, which closes the contacts. With the contacts closed, current flows through the horn (which is grounded). Used in this manner, the horn relay becomes a control of the high current necessary to sound the horn. Manufacturers may use small diameter wire for the control circuit since low current flows through it. The control unit may have only 0.25 ampere flowing through it, and the horn may require 24 or more amperes.

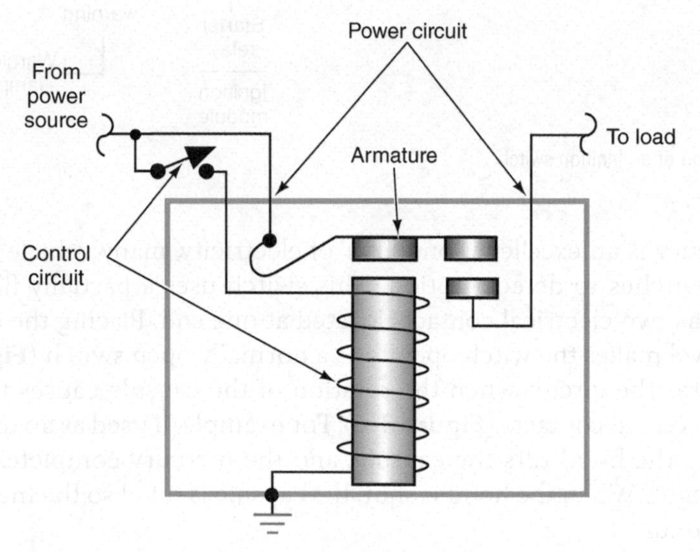

Figure 3-7 A relay uses electrical current to create a magnetic field to draw the contact point closed.

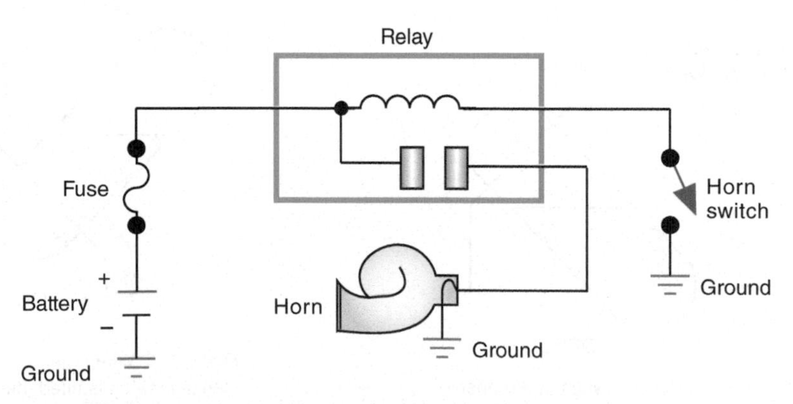

Figure 3-8 Using a relay in the horn circuit to reduce the required size of the conductors installed in the steering column.

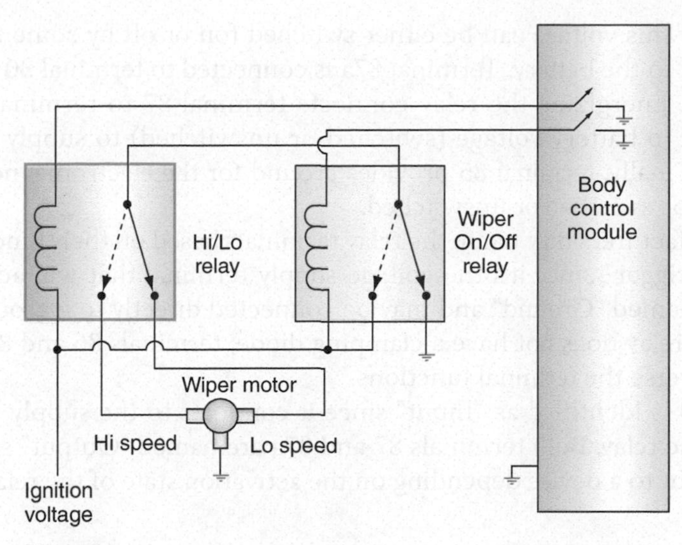

Figure 3-9 Using a relay as a diverter to control Hi/Lo wiper operation. The Hi/Lo relay diverts current to the different brushes of the wiper motor.

Relays also function as circuit diverters (**Figure 3-9**). In this example, the Hi/Lo wiper relay will direct current flow to either the high-speed brush or the low-speed brush of the wiper motor to control wiper speeds.

The coil side of the relay is rated based on its voltage. In the automotive industry, usually this means a rating of around 12 volts. The contact side of the relay is rated based on its current-carrying capabilities. Typically, automotive relays are rated to handle 20, 30, 35, or 40 amps. The relays can also be NO type or NC type. They may have different amperage requirements based on whether they are NC or NO relays. It's important to use the correct relay if you need to replace it.

ISO Relays. ISO relays conform to the specifications of the International Standards Organization (ISO) for common size and terminal patterns (**Figure 3-10**). Terminal identification is 30, 87a, 87, 86, and 85. In most applications, battery voltage connects

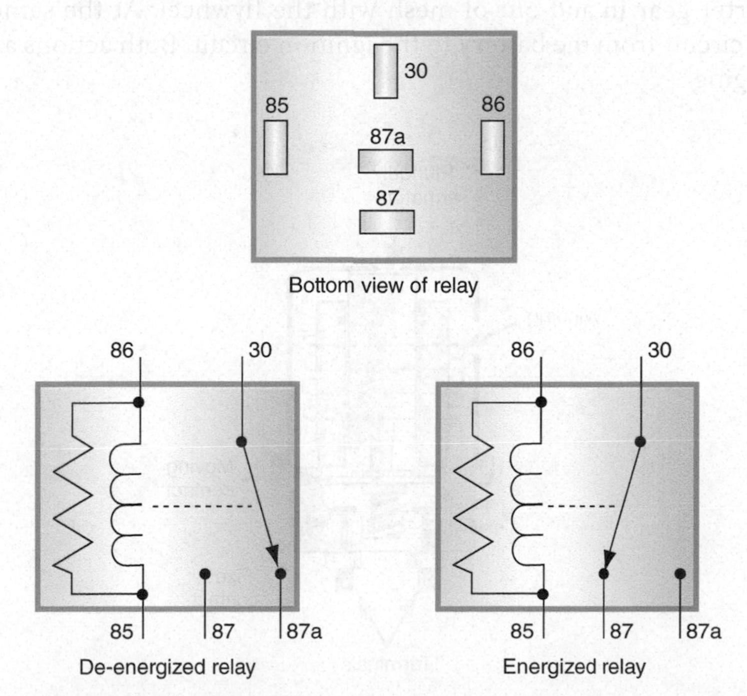

Figure 3-10 ISO relay terminal identification.

to Terminal 30. This voltage can be either switched (on or off by some type of switch) or come directly to the battery. Terminal 87a is connected to terminal 30 when the relay is de-energized. Energizing the relay connects terminal 87 to terminal 30. Terminal 86 is connected to battery voltage (switched or unswitched) to supply current to the electromagnet. Finally, terminal 85 provides ground for the electromagnet. Once again, the ground can be switched or unswitched.

Some manufacturers may name the relay terminals based on their function. Terminal 86 is named "Trigger" since it's the voltage supply terminal that will activate the coil. Terminal 85 is named "Ground" and may be connected directly to a ground location or switched. If the relay does not have a clamping diode, terminals 86 and 85 can be interchanged and reverse the terminal functions.

Terminal 30 is identified as "Input" since it connects to the supply voltage for the switch side of the relay. Both terminals 87 and 87a are named "Output" since either one can direct current to a device depending on the activation state of the relay.

Solenoids

A **solenoid** is an electromagnetic device that operates similarly to a relay; however, a solenoid uses a movable iron core. Solenoids can do mechanical work, such as switching electrical, vacuum, and liquid circuits. The iron core inside the solenoid coil is spring loaded (**Figure 3-11**). When current flows through the coil, the magnetic field created around the coil attracts the core and moves it into the coil. This example shows a solenoid that works as a high-amperage relay in that it uses low current to control high current. This would be a common type of starter motor solenoid.

A typical example of a solenoid designed to perform a mechanical function is an electric door lock or trunk release system that mechanically moves the locking mechanism. To do work, the core is attached to a mechanical linkage, which causes something to move. When current flows through the coil, it moves the core and ultimately moves the linkage. When current flow through the coil stops, the spring pushes the core back to its original position.

Solenoids may also switch a circuit on or off, in addition to causing a mechanical action. Such is the case with some starter motor–mounted solenoids. These devices move the starter gear in and out of mesh with the flywheel. At the same time, they complete the circuit from the battery to the ignition circuit. Both actions are necessary to start an engine.

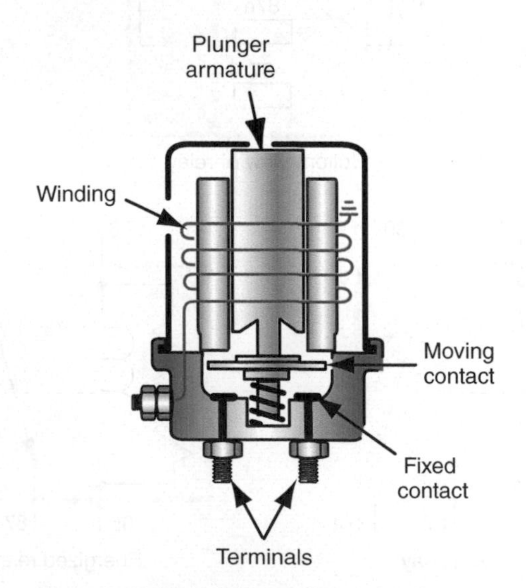

Figure 3-11 Typical electromechanical solenoid.

AUTHOR'S NOTE There are probably more solenoids on the vehicle than you may realize. For example, each fuel injector is really nothing more than a solenoid. As the magnetic field builds, it lifts the pintle off its seat, and fuel is allowed to flow past the pintle, through the orifices, and into the intake manifold.

Buzzers

A **buzzer**, or **sound generator**, is sometimes used to warn the driver of possible safety hazards by emitting an audio signal (such as when the seat belt is not buckled). A buzzer is similar in construction to a relay except for the internal wiring (**Figure 3-12**). The coil is supplied current through the normally closed contact points. Applying voltage to the buzzer flows current through the contact points to the coil. Energizing the coil attracts the contact arm to the magnetic field. Pulling the contact arm down opens the circuit, and the current flow to the coil stops, dissipating the magnetic field. The contact arm then closes again, and the circuit to the coil is closed. This opening and closing action occurs very rapidly. It's this movement that generates the vibrating signal and the buzzing sound.

Resistors

All circuits require resistance in order to operate. If the resistance performs a useful function, it's referred to as the **load device**. However, resistance can also control current flow and function as a sensing device for computer systems. A circuit may use several types of resistors. These include fixed resistors, stepped resistors, and variable resistors.

Fixed Resistors. Fixed resistors are usually made of carbon or oxidized metal (**Figure 3-13**). These resistors have a set resistance value to limit the amount of current flow in a circuit. Like any other electrical component, fixed resistors must be matched to the circuit. Resistors not only have different resistance values but are also rated based on their wattage. Color bands on the protective shell identify the resistance value (**Figure 3-14**). Usually, there are four or five color bands. When there are four bands, the first two are the digit bands, the third is the multiplier, and the fourth is the tolerance. On a resistor with five bands, the first three are digit bands.

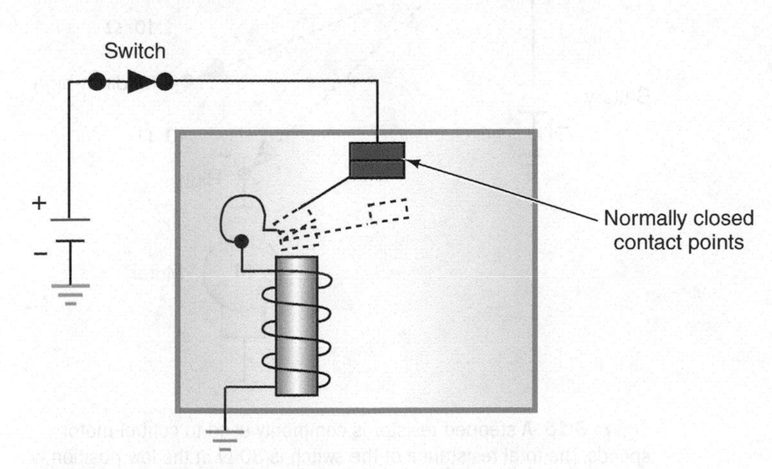

Figure 3-12 A buzzer reacts to the current flow to open and close rapidly, creating noise.

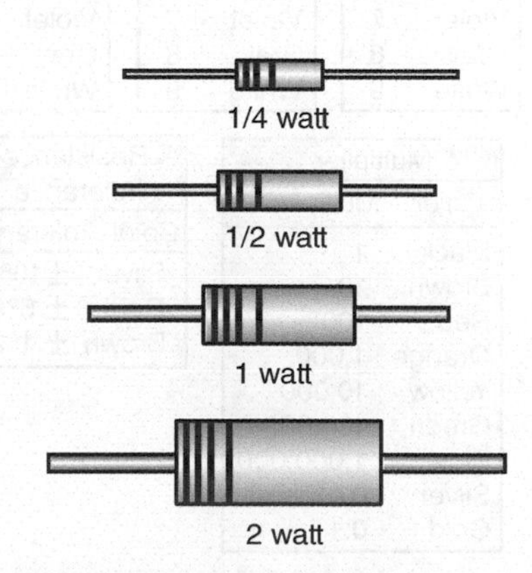

Figure 3-13 Fixed resistors.

For example, if the resistor has four color bands of yellow, black, brown, and gold, the resistance value is determined as follows:

The first color band (yellow) gives the first-digit value of 4.
The second color band (black) gives the second-digit value of 0.
The digit value is now 40. Multiply this by the value of the third band. In this case, brown has a value of 10, so the resistor should have 400 ohms of resistance (40 × 10 = 400).
The last band provides the tolerance range. Gold equals a tolerance range of ±5%.

Shop Manual
Chapter 3, page 126

Stepped Resistors. A **stepped resistor** is commonly used to control electrical motor speeds (**Figure 3-15**). A stepped resistor has two or more fixed resistor values. The stepped resistor can have an integral switch, or an external switch wired in series. By changing the position of the switch, resistance increases or decreases within the circuit. If the current flows through a low resistance, higher current flows to the motor and its speed increases. Placing the switch in the low-speed position adds more to the circuit. Less current flows to the motor, which causes the motor to operate at a reduced speed.

Computers can use stepped resistors to convert on/off digital signals into continuously variable analog signals.

Shop Manual
Chapter 3, page 126

Variable Resistors. **Variable resistors** provide for an infinite number of resistance values within a range. The most common types of variable resistors are rheostats and potentiometers. A **rheostat** is a two-terminal variable resistor used to regulate the

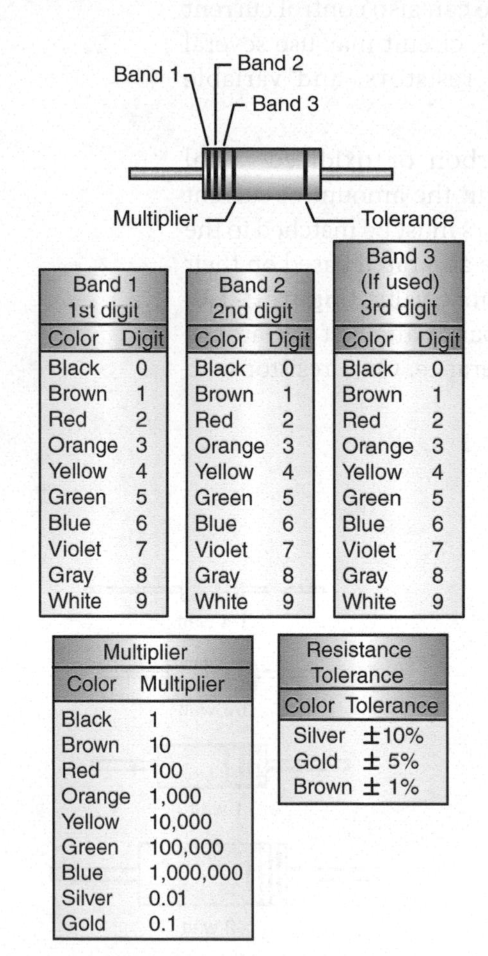

Band 1 1st digit Color	Digit
Black	0
Brown	1
Red	2
Orange	3
Yellow	4
Green	5
Blue	6
Violet	7
Gray	8
White	9

Band 2 2nd digit Color	Digit
Black	0
Brown	1
Red	2
Orange	3
Yellow	4
Green	5
Blue	6
Violet	7
Gray	8
White	9

Band 3 (If used) 3rd digit Color	Digit
Black	0
Brown	1
Red	2
Orange	3
Yellow	4
Green	5
Blue	6
Violet	7
Gray	8
White	9

Multiplier Color	Multiplier
Black	1
Brown	10
Red	100
Orange	1,000
Yellow	10,000
Green	100,000
Blue	1,000,000
Silver	0.01
Gold	0.1

Resistance Tolerance Color	Tolerance
Silver	±10%
Gold	± 5%
Brown	± 1%

Figure 3-14 Resistor color code chart.

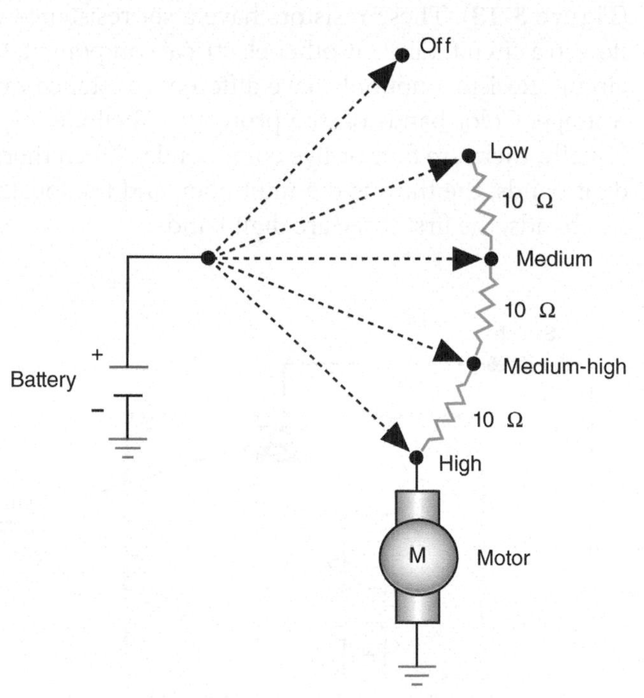

Figure 3-15 A stepped resistor is commonly used to control motor speeds. The total resistance of the switch is 30 Ω in the low position, 20 Ω in the medium position, 10 Ω in the medium-high position, and 0 Ω in the high position.

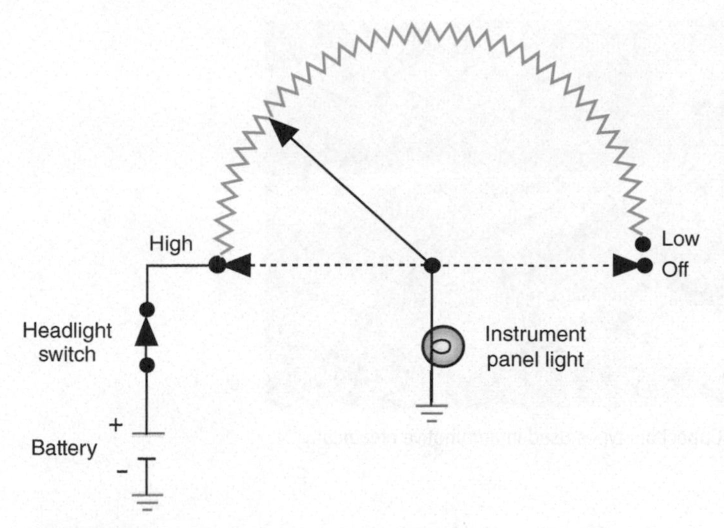

Figure 3-16 Using a rheostat to control the brightness of a lamp.

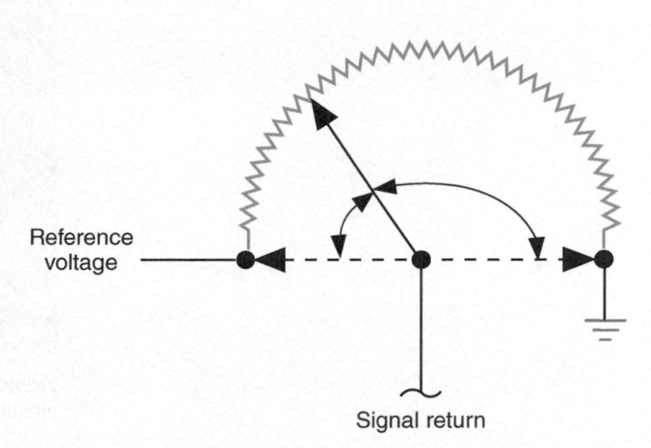

Figure 3-17 A potentiometer is used to send a signal voltage from the wiper.

strength of an electrical current. A rheostat has one terminal connected to the fixed end of a resistor and a second terminal connected to a movable contact called a **wiper** (**Figure 3-16**). By changing the position of the wiper on the resistor, the amount of resistance can be increased or decreased. The most common use of the rheostat is in the instrument panel lighting switch. Turning the switch knob changes the resistance and the intensity of the instrument lights.

A **potentiometer** is a three-wire variable resistor that acts as a voltage divider to produce a continuously variable output signal proportional to a mechanical position. When installing a potentiometer into a circuit (**Figure 3-17**):

- A terminal at one end of the wound resistor connects to a power source.
- The second wire connects to the opposite end of the wound resistor and is the ground return path.
- The third wire connects to the wiper contact. Moving the wiper over the resistor results in a variable voltage drop in the wiper's circuit.

Because current always flows through the wound resistor (and the amount of current does not change), the total voltage drop measured by the potentiometer is very stable. For this reason, the potentiometer is a common type of input sensor for the vehicle's on-board computers.

Capacitors

Some automotive electrical systems use a capacitor or condenser to store electrical charges (**Figure 3-18**). A capacitor uses the theory of **capacitance** to temporarily store electrical energy. Capacitance (C) is the ability of two conducting surfaces to store voltage. The two surfaces must be separated by an insulator.

Shop Manual
Chapter 3, page 128

A **capacitor** does not consume any power; however, it will store and release electrical energy. All voltage stored in the capacitor is returned to the circuit when the capacitor discharges. Because the capacitor stores voltage, it will also absorb voltage changes in the circuit. Providing for this storage of voltage controls damaging voltage spikes. Capacitors are also used to reduce radio noise. A capacitor blocks direct current. A small amount of current enters the capacitor and charges it.

A capacitor is made by wrapping two conductor strips around an insulating strip. The insulating strip, or **dielectric**, prevents the plates from coming in contact while keeping them very close to each other.

The insulator in a capacitor is called a **dielectric.** The dielectric can be made of ceramic, glass, paper, plastic, or even the air between the two plates.

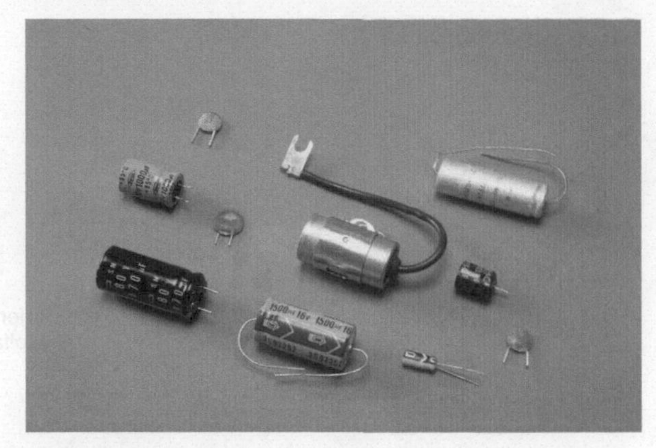

Figure 3-18 Capacitors types used in automotive electrical circuits.

The plate connected to the positive battery terminal is the **positive plate.**

An atom that has less electrons than protons is said to be a positive ion.

The field that is between the two oppositely charged plates is called the **electrostatic field.**

Capacitors operate on the principle that opposite charges attract and that there is a potential voltage between any two oppositely charged points. Typically, the capacitor is connected in parallel to the circuit's load device (**Figure 3-19**). When the switch is closed, the protons at the positive battery terminal will attract some of the electrons on one plate of the capacitor away from the area near the dielectric material. As a result, the atoms of the **positive plate** are unbalanced because there are more protons than electrons in the atom. This plate now has a positive charge because of the shortage of electrons (**Figure 3-20**). The positive charge of this plate will attract electrons on the other plate. The dielectric keeps the electrons on the negative plate from crossing over to the positive plate, resulting in the storage of electrons on the negative plate (**Figure 3-21**). The movement of electrons to the negative plate and away from the positive plate is an electrical current.

Current will flow through the capacitor until the voltage charges across the capacitor and across the battery equalize. Current flow through a capacitor is only the effect of the electron movement onto the negative plate and away from the positive plate. Electrons don't actually pass through the capacitor from one plate to another. The charges don't move through the **electrostatic field** and are stored on the plates as static electricity.

When the charges across the capacitor and battery equalize, there is no potential difference and no more current will flow through the capacitor (**Figure 3-22**). Current will now flow through the load components in the circuit (**Figure 3-23**).

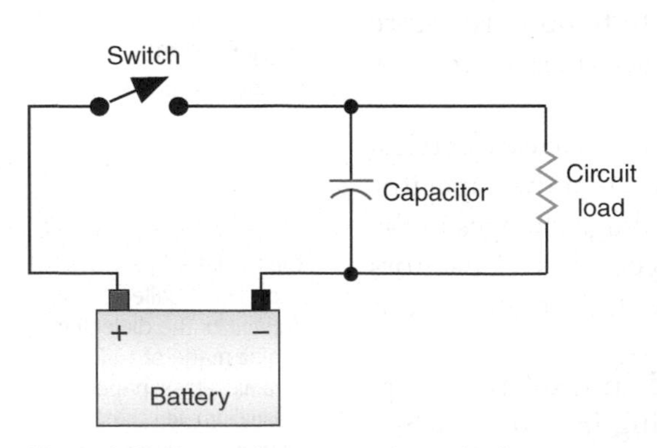

Figure 3-19 A capacitor connected to a circuit.

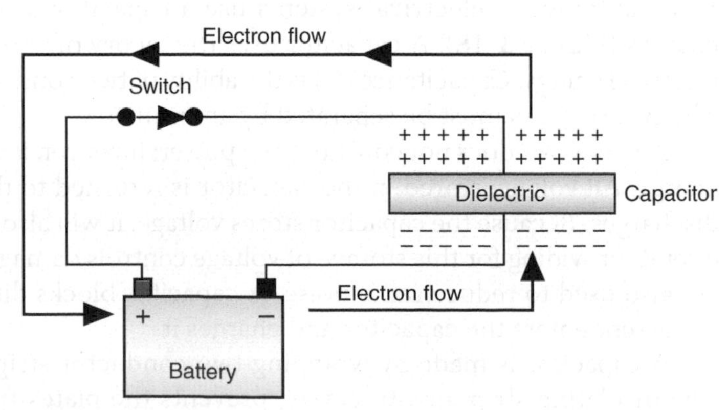

Figure 3-20 The positive plate sheds its electrons.

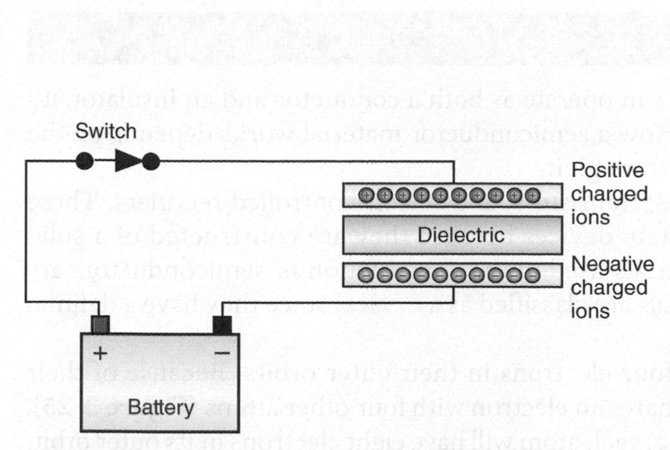

Figure 3-21 The electrons are stored on the negative plate.

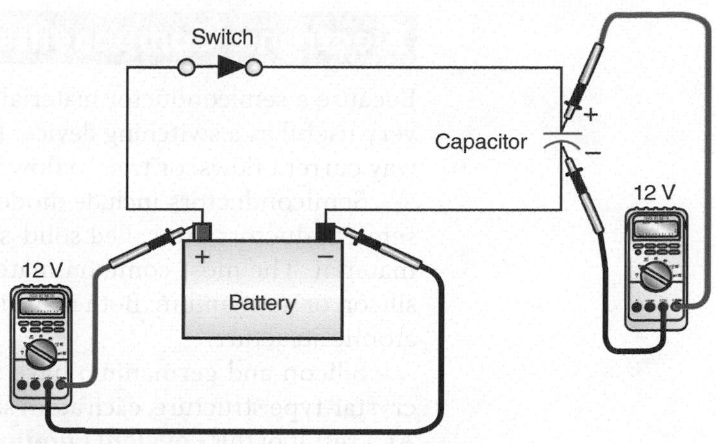

Figure 3-22 A capacitor when it's fully charged.

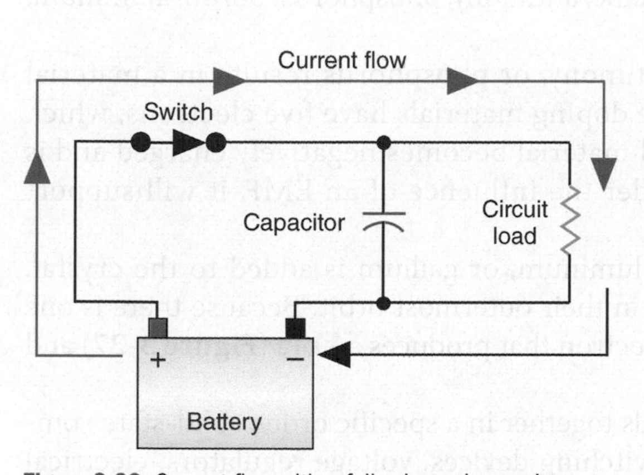

Figure 3-23 Current flow with a fully charged capacitor.

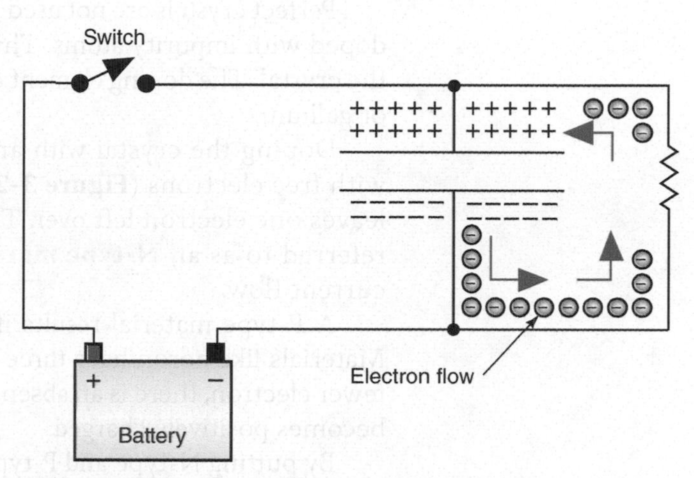

Figure 3-24 Current flow with the switch open and the capacitor discharging.

Opening the switch stops current flow from the battery through the resistor. However, the capacitor has a storage of electrons on its negative plate. Because the negative plate of the capacitor connects to the positive plate through the resistor, the capacitor acts as the source. The capacitor will discharge the electrons through the resistor until the positive plate and the negative plate atoms return to a balanced state (**Figure 3-24**).

If a high-voltage spike occurs in the circuit, the capacitor absorbs the additional voltage before it can damage the circuit components. A capacitor can also stop current flow quickly when a circuit is opened (such as in the ignition system). It can also store a high-voltage charge and then discharge it when a circuit needs the voltage (such as in some air bag systems).

Capacitors are rated in units called farads. A one-farad capacitor connected to a 1-volt source will store 6.28×10^{18} electrons. A farad is a large unit, thus most capacitors used in automotive applications are rated in picofarad (a trillionth of a farad) or microfarads (a millionth of a farad). In addition, the capacitor's voltage rating identifies the applied voltage limit before the dielectric breaks down. The voltage rating is related to the strength and thickness of the dielectric. The voltage rating increases with increasing dielectric strength and the thickness of the dielectric. The capacitance increases with the area of the plates and decreases with the thickness of the dielectric. The maximum voltage rating and capacitance determine the amount of energy a capacitor holds.

Electronic Components

Because a semiconductor material can operate as both a conductor and an insulator, it's very useful as a switching device. How a semiconductor material works depends on the way current flows, or tries to flow, through it.

Semiconductors include diodes, transistors, and silicon-controlled rectifiers. These semiconductors are called solid-state devices because they are constructed of a solid material. The most common materials used in the construction of semiconductors are silicon or germanium. Both materials are classified as a **crystal** since they have a definite atomic structure.

Silicon and germanium have four electrons in their outer orbits. Because of their crystal-type structure, each atom shares an electron with four other atoms (**Figure 3-25**). As a result of this **covalent bonding**, each atom will have eight electrons in its outer orbit. Filling all orbits means there are no free electrons; thus, the material (as a category of matter) falls somewhere between conductor and insulator.

Perfect crystals are not used in the manufacturing of semiconductors. The crystals are doped with impurity atoms. This doping adds a small percentage of another element to the crystal. The doping element can be arsenic, antimony, phosphorus, boron, aluminum, or gallium.

Doping the crystal with arsenic, antimony, or phosphorus results in a material with free electrons (**Figure 3-26**). These doping materials have five electrons, which leaves one electron left over. This doped material becomes negatively charged and is referred to as an **N-type material**. Under the influence of an EMF, it will support current flow.

A P-type material results if boron, aluminum, or gallium is added to the crystal. Materials like boron have three electrons in their outermost orbit. Because there is one fewer electron, there is an absence of an electron that produces a **hole** (**Figure 3-27**) and becomes positively charged.

By putting N-type and P-type materials together in a specific order, solid-state components are built that can be used for switching devices, voltage regulators, electrical control, and so on.

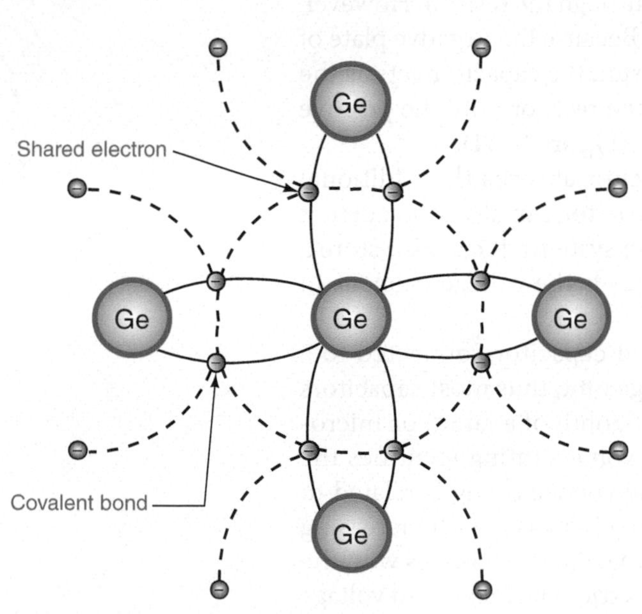

Figure 3-25 Crystal structure of germanium.

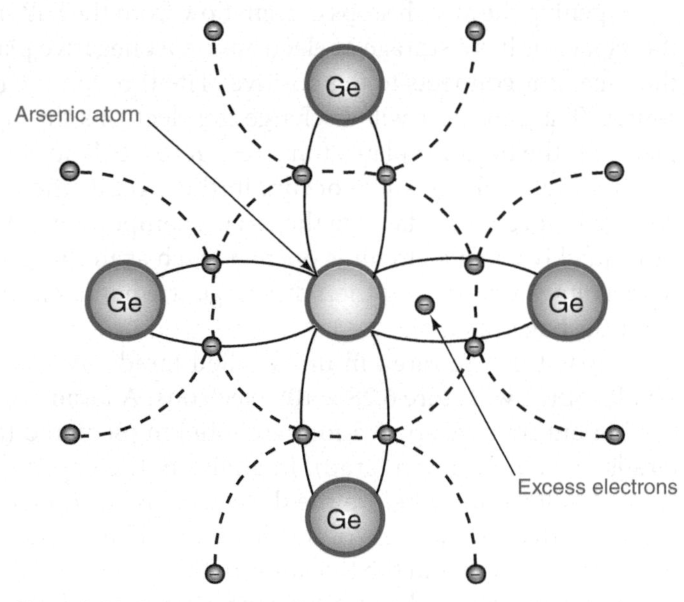

Figure 3-26 Germanium crystal doped with an arsenic atom to produce an N-type material.

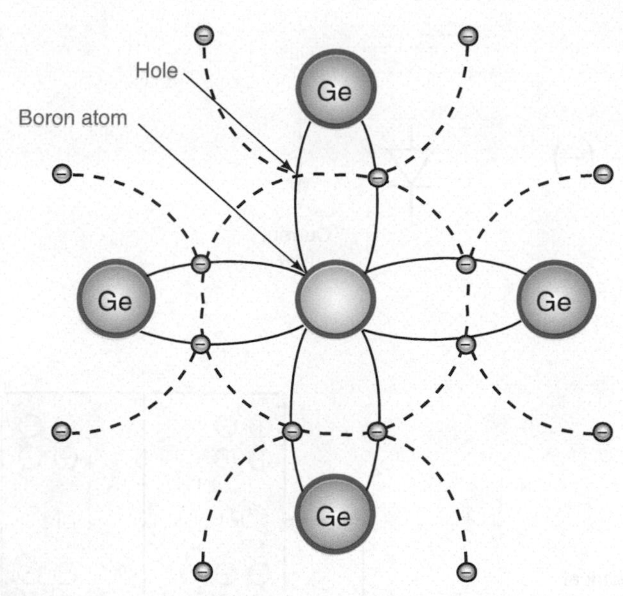

Figure 3-27 Germanium crystal doped with a boron atom to produce a P-type material.

Diodes

Shop Manual
Chapter 3, page 128

A **diode** is an electrical one-way check valve that will allow current to flow in one direction only. A diode is the simplest semiconductor device. Joining P-type semiconductor material with N-type material forms the diode. The N (negative) side of a diode is called the **cathode,** and the P (positive) side is the **anode** (**Figure 3-28**). The PN junction is the point where the cathode and anode join together. The outer shell of the diode will have a stripe painted around it. This stripe designates which end of the diode is the cathode.

During the production of a diode, the positive holes from the P region and the negative charges from the N region move toward the junction. This is because of the attraction of unlike charges to each other. Electrons from the N material will cross over and fill holes from the P material along the PN junction. This forms a depletion zone. When the charges cross over, the two halves are no longer balanced, and the diode builds up a network of internal charges. In the depletion zone, the material is an insulator since all of the holes are filled, and there are no free electrons. The internal characteristics change when the diode is incorporated within a circuit, and a voltage is applied. If the diode is **forward-biased**, current will flow (**Figure 3-29**). In this state, the depletion zone will become smaller as the electrons of the N region and the holes of the P region move across the junction. The negative region pushes electrons across the junction as the free electrons in the N region are repelled by the negative charge from the battery and are attracted to the P region of the diode. The repelling of like charges of the positive side of the battery and the P region of the diode results in the holes in the P region moving toward the N region. Since the depletion zone shrinks, the diode acts as a conductor when the diode is forward-biased.

Forward-biased means that a positive voltage is applied to the P-type material and a negative voltage to the N-type material.

If the diode is **reverse-biased**, there will be no current flow (**Figure 3-30**). The negative region will attract the positive holes away from the junction, and the positive region will attract electrons away from the junction. Since the electrons and holes are moving in the wrong direction, the depletion zone is enlarged, making the diode act as an insulator.

Reverse-biased means that the N-type material has a positive voltage applied, and the P-type material has a negative voltage applied.

When the diode is forward-biased, it will have a small voltage drop across it. The voltage drop is the **turn-on voltage**. On standard silicon diodes, this voltage is usually about 0.6 volt.

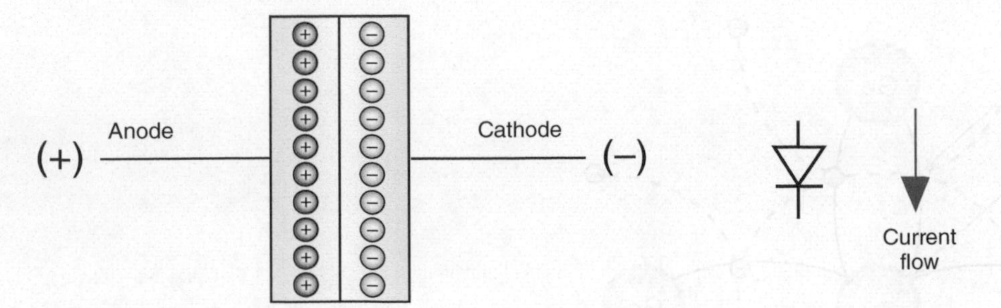

Figure 3-28 A diode and its symbol.

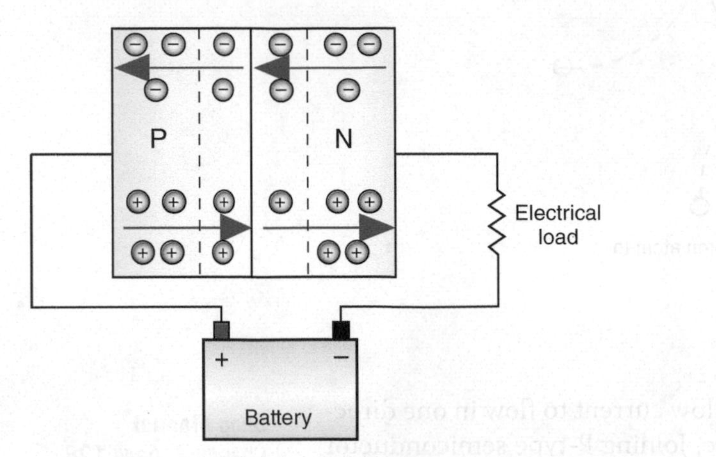

Figure 3-29 Forward-biased voltage causes current flow.

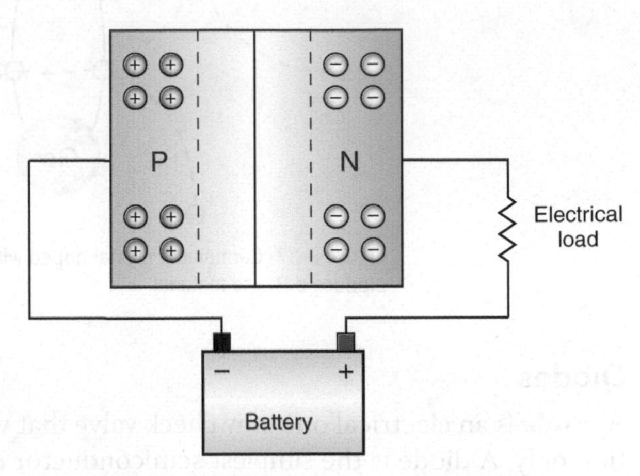

Figure 3-30 Reverse-biased voltage prevents current flow.

Diode Identification and Ratings. Since there are many different types of diodes and current-carrying capabilities, a diode identification system is used (**Figure 3-31**). The identification system uses a combination of numbers and letters to identify different types of semiconductor devices. This system also identifies transistors and many other special semiconductor devices.

The first two characters of the system identify the component. The first number indicates the number of junctions in the semiconductor device. Since this number is one less than the number of active elements, a 1 designates a diode, a 2 designates a transistor, and a 3 designates a tetrode (a four-element transistor). The letter "N" that follows the first number indicates that the device is a semiconductor.

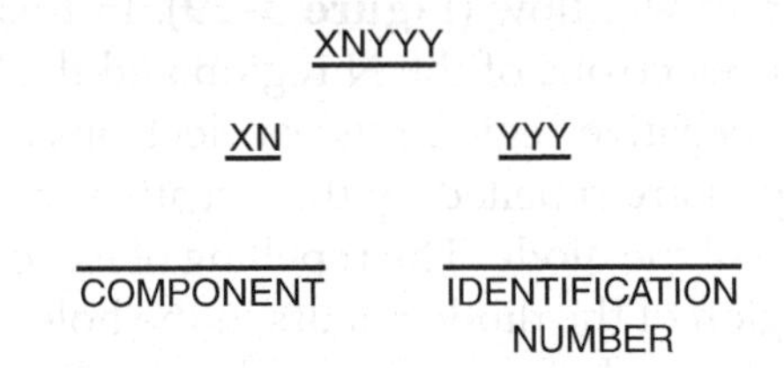

X - NUMBER OF SEMICONDUCTOR JUNCTIONS
N - A SEMICONDUCTOR
YYY - IDENTIFICATION NUMBER (ORDER OR REGISTRATION NUMBER)
ALSO INCLUDES SUFFIX LETTER (IF APPLICABLE) TO INDICATE

1. MATCHING DEVICES
2. REVERSE POLARITY
3. MODIFICATION

EXAMPLE - 1N345A (AN IMPROVED VERSION OF THE SEMICONDUCTOR DIODE TYPE 345)

Figure 3-31 Standard semiconductor identification markings.

The last series of characters that follow the "N" is a serialized identification number. This number may also contain a suffix letter after the third digit. Common suffix letters include "M" to describe matching pairs of separate semiconductor devices and "R" to indicate reverse polarity. In addition, suffix letters can indicate modified versions of the device. For example, a semiconductor diode designated as type 1N345A signifies a two-element diode (1) of semiconductor material (N) that is an improved version (A) of type 345.

To distinguish the anode from the cathode side of the diode, manufacturers generally code the cathode end of the diode. The identification can be a colored band or a dot, a "k," "+," "cath," or an unusual shape such as raised edge or taper. Some manufacturers use standard color code bands on the cathode side to not only identify the cathode end of the diode but also identify the diode by number.

Zener Diodes

Shop Manual
Chapter 3, page 129

As stated, if a diode is reverse-biased, it will not conduct current. However, by increasing the reverse voltage, a voltage level will be reached at which the diode will conduct in the reverse direction. This voltage level is referred to as **Zener voltage**. At the Zener voltage, the diode can conduct but will try to limit the voltage dropped across it to this Zener voltage. Reverse current can cause heat buildup that would destroy a simple PN-type diode. The point when the PN junction begins to break down is called the **peak reverse voltage (PRV)**. However, doping the diode with materials that withstand reverse current heat buildup protects the Zener diode.

A BIT OF HISTORY

The first diodes were vacuum tube devices (also known as thermionic valves). A vacuum surrounded arrangements of electrodes within a glass envelope (similar to light bulbs). Using this arrangement of a filament and plate, John Ambrose Fleming invented the diode in 1904. Current flow through the filament results in the generation of heat. The heat causes the emission of electrons into the vacuum, electrostatically drawing them to a positively charged outer metal plate (anode). Since the plate is not heated, the electrons will not return to the filament, even if the charge on the plate is made negative.

A **Zener diode** operates in reverse bias at the PRV region. When the PRV is reached, current flows in reverse bias. Limiting the voltage drop across the Zener to this voltage level prevents the voltage output from climbing any higher. This makes the Zener diode an excellent component for regulating voltage. For example, a Zener diode rated at 15 volts will not conduct in reverse bias when the voltage is below 15 volts. At 15 volts, it will conduct. The voltage across the load will not increase over 15 volts.

Figure 3-32 shows a simplified circuit with a Zener diode to provide a constant voltage level to the instrument gauge. In this example, the Zener diode is connected in series with the resistor and in parallel to the gauge. The Zener diode maintains a constant voltage drop. Since the total voltage drop in a series circuit must equal the amount of source voltage, the resistor drops any voltage greater than the Zener voltage. Although source voltage may vary (as a normal result of the charging system), causing different currents to flow through the resistor and Zener diode, the voltage dropped by the Zener diode remains the same. A gauge that requires the voltage to be limited to 7 volts requires a Zener diode rated at 7 volts.

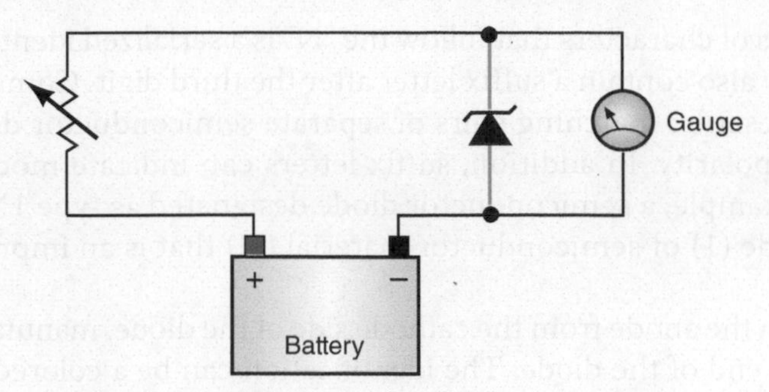

Figure 3-32 Simplified instrument gauge circuit using a Zener diode to maintain a constant voltage to the gauge. Note the symbol used for a Zener diode.

The Zener breaks down when system voltage reaches 7 volts. At this point, the Zener diode conducts reverse current, causing an additional voltage drop across the resistor. The amount of voltage to the instrument gauge will remain at 7 volts because the Zener diode makes the resistor drop the additional voltage to maintain this limit.

Here, you can see the difference between the standard and Zener diodes. When the Zener diode is reverse-biased, it holds the available voltage to a specific value.

Avalanche Diodes

Shop Manual
Chapter 3, page 130

Avalanche diodes are specialized diodes used as "relief valves" to protect electrical systems from excess voltages. Avalanche diodes break down at a well-defined reverse voltage without being destroyed. Like Zener diodes in operation, avalanche diodes conduct in the reverse direction when the reverse-bias voltage exceeds the breakdown voltage. However, the avalanche effect causes the breakdown when the reverse electric field moves across the PN junction and causes a wave of ionization (like an avalanche), leading to a large current.

Avalanche diodes are commonly used in automobile AC generators (alternators) to protect against voltage surges that can damage computer systems. Typically, these will go into avalanche at about 21 to 29 volts and conduct the current to ground instead of into the vehicle's electrical system. Voltages higher than the Zener voltage will result in the diode constantly conducting in the reverse-bias direction. This prevents the output voltage of the AC generator from going above the rating of the avalanche diode.

Light-Emitting Diodes

Shop Manual
Chapter 3, page 130

When the **light-emitting diode (LED)** is forward-biased, the holes and electrons combine, and current is allowed to flow through it (like a forward-biased standard diode). A side effect of this interaction is the generation of light. A lens built into the LED allows this light to be is seen (**Figure 3-33**). The light from an LED is not heat energy, as with other lights; it's electrical energy. Because of this, LEDs last longer than

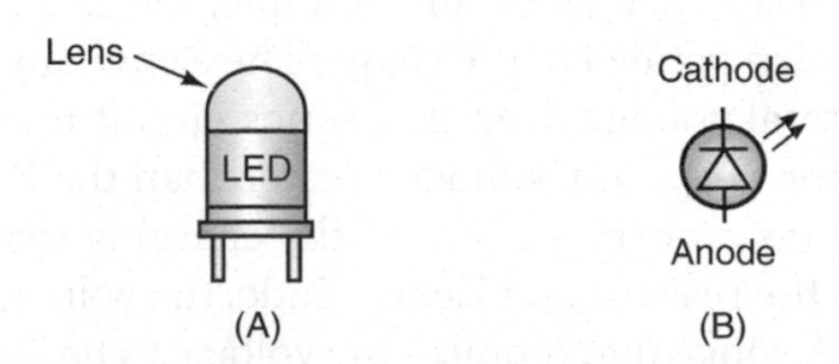

Figure 3-33 (A) A light-emitting diode uses a lens to emit the generated light. (B) Symbol for LED.

light bulbs. The material used to make the LED determines the color of the light emitted and the turn-on voltage.

Like standard silicon diodes, the LED has a constant turn-on voltage. However, this turn-on voltage is usually higher than that of standard diodes. The turn-on voltage defines the color of the light; 1.2 volts correspond to red, 2.4 volts to yellow.

Photodiodes

A **photodiode** also allows current to flow in one direction only. However, the direction of the current flow is opposite to that in a standard diode. Reverse current flow occurs only when the diode receives a specific amount of light. Automatic headlight systems often use this type of diode can be used in automatic headlight systems.

Clamping Diodes

Suddenly stopping the current flow through a coil (such as in a relay or solenoid) produces a voltage surge or spike. This surge results from the collapsing of the magnetic field around the coil. The movement of the field across the windings induces a very high-voltage spike, which can damage electronic components as it flows through the system. In some circuits, a capacitor can be used as a shock absorber to prevent component damage from this surge. Today's complex electronic systems may use a **clamping diode** to prevent the voltage spike. Installing a clamping diode in parallel with the coil provides a bypass for the electrons when the circuit is opened (**Figure 3-34**).

A **clamping diode** is nothing more than a standard diode; the term *clamping* refers to its function.

An example of using clamping diodes is on some air-conditioning compressor clutch circuits. Because the clutch operates by electromagnetism, opening the clutch coil circuit produces a voltage spike. If this voltage spike was left unchecked, it could damage the vehicle's on-board computers. The installation of the clamping diode in reverse bias prevents the voltage spike from reaching the computers.

Relays may also use a clamping diode. However, some use a resistor to dissipate the voltage spike. The two types of relays are not interchangeable.

Transistors

A **transistor** is a three-layer semiconductor used as a high-speed switching device. The word *transistor* is a combination of two words, *transfer* and *resist*. Like a switch, the transistor controls current flow in the circuit (**Figure 3-35**). It can allow a predetermined amount of current flow or resist this flow.

Shop Manual
Chapter 3, page 131

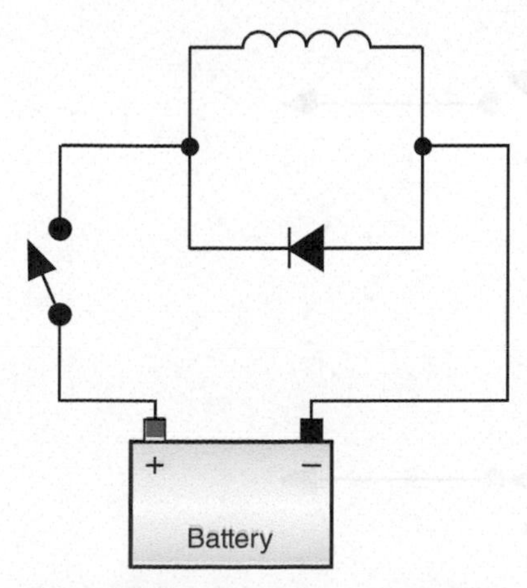

Figure 3-34 A clamping diode in parallel to a coil prevents voltage spikes when the switch is opened.

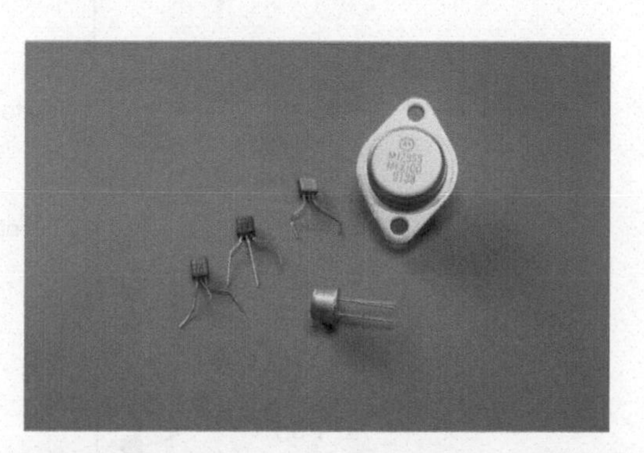

Figure 3-35 Examples of transistors used in automotive applications.

Transistors combine P-type and N-type materials in groups of three. The two possible combinations are NPN (**Figure 3-36**) and PNP (**Figure 3-37**).

The three layers of the transistor are referred to as the **emitter**, the **collector**, and the **base**. The emitter is the outside layer of the forward-biased diode that has the same polarity as the circuit side to which it's applied. The arrow on the transistor symbol refers to the emitter lead and points in the direction of positive current flow and to the N material. The collector is the outside layer of the reverse-biased diode. The base is the shared middle layer. Each layer has its own lead for connecting to different parts of the circuit. In effect, a transistor is two diodes that share a common center layer. When connecting a transistor to the circuit, the emitter-base junction will be forward-biased, and the collector-base junction will be reverse-biased.

In the NPN transistor, the emitter conducts current flow to the collector when the base is forward-biased. The transistor cannot conduct unless the voltage applied to the base leg exceeds the emitter voltage by approximately 0.7 volt. This means both the base and collector must be positive with respect to the emitter. The transistor acts as an opened switch with less than 0.7 volt applied to the base leg (compared to the voltage at the emitter). When the voltage difference is greater than 0.7 volt at the base (compared to the emitter voltage), the transistor acts as a closed switch (**Figure 3-38**).

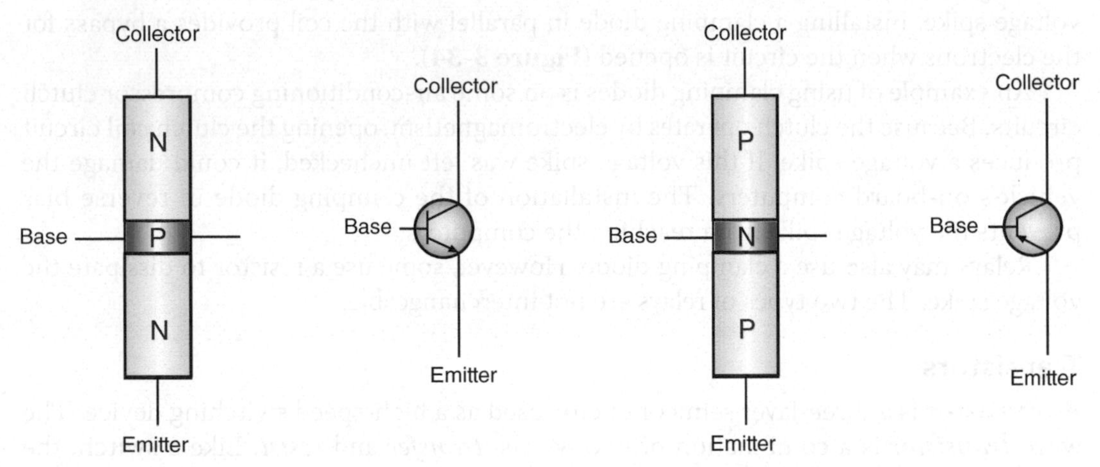

Figure 3-36 An NPN transistor and its symbol.

Figure 3-37 A PNP transistor and its symbol.

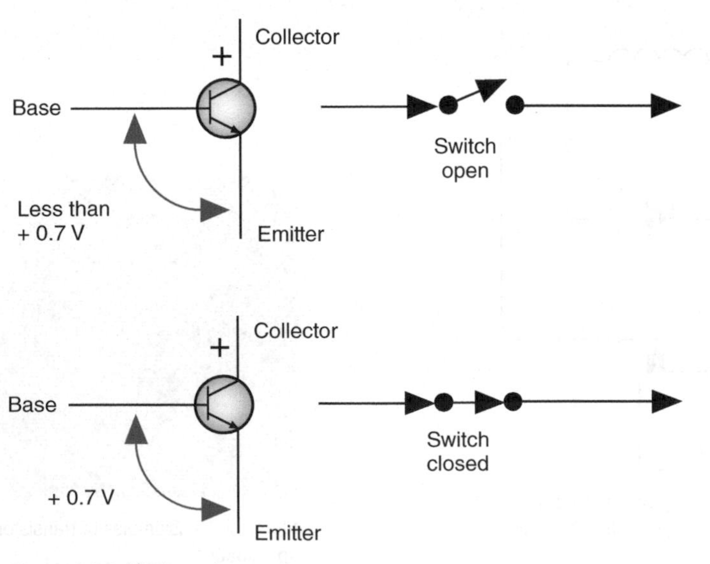

Figure 3-38 NPN transistor action.

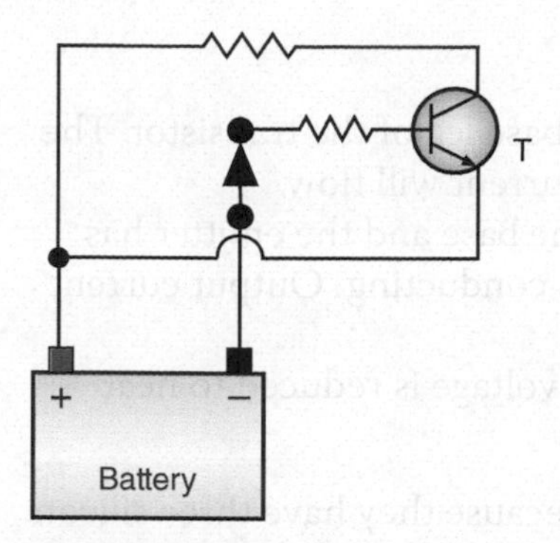

Figure 3-39 NPN transistor with reverse-biased voltage applied to the base. No current flow.

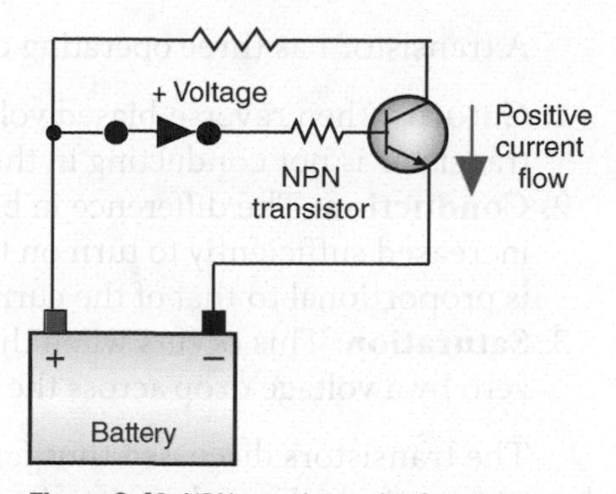

Figure 3-40 NPN transistor with forward-biased voltage applied to the base. Current flows.

An NPN transistor connected to a circuit normally has a reverse bias applied to the base-collector junction. If the emitter-base junction is also reverse-biased, no current will flow through the transistor (**Figure 3-39**). If the emitter-base junction is forward-biased (**Figure 3-40**), current flows from the emitter to the base. Because the base is a thin layer, and the collector has a positive voltage applied, electrons flow from the emitter to the collector.

In the PNP transistor, current will flow from the emitter to the collector when the base leg is forward-biased with a more negative voltage than that at the emitter (**Figure 3-41**). For current to flow through the emitter to the collector, both the base and the collector must be negative with respect to the emitter.

AUTHOR'S NOTE Current flow through transistors is always based on hole and/or electron flow.

Transistors function as high-speed electrical switches. It's also possible to control the amount of current flow through the collector because the output current is proportional to the amount of current through the base leg.

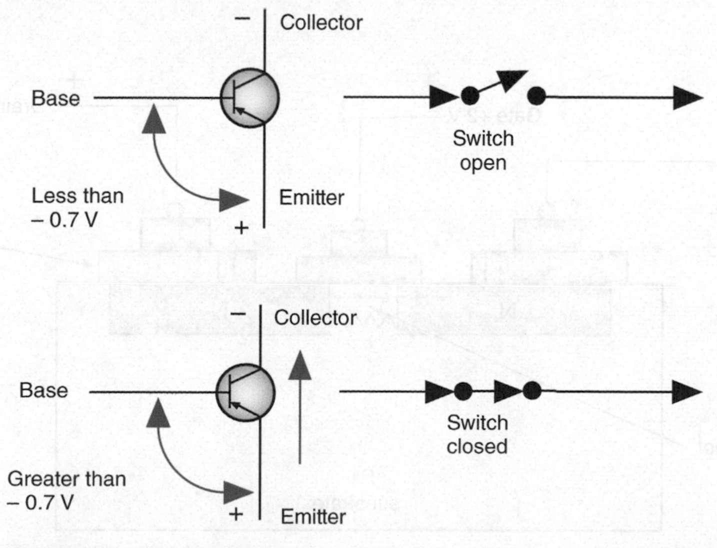

Figure 3-41 PNP transistor action.

A transistor has three operating conditions:

1. **Cutoff:** When reverse-biased voltage is applied to the base leg of the transistor. The transistor is not conducting in this condition, and no current will flow.
2. **Conduction:** The difference in bias voltage between the base and the emitter has increased sufficiently to turn on the transistor, and it is conducting. Output current is proportional to that of the current through the base.
3. **Saturation:** This occurs when the collector to emitter voltage is reduced to near zero by a voltage drop across the collector's resistor.

The transistors discussed thus far are called **bipolar** because they have three silicon layers, two of which are the same. Another type of transistor is the **field-effect transistor (FET)**. The FET's leads are the source, the drain, and the gate. The source supplies the electrons and is similar to the emitter in the bipolar transistor. The drain collects the current and is similar to the collector. The gate creates the electrostatic field that allows electron flow from the source to the drain (similar to the base).

While electrons are flowing from the source to the drain (electron theory), positive charges are flowing from the drain to the source (conventional theory).

The FET transistor does not require a constant bias voltage. A voltage needs to be applied to the gate terminal to get electron flow from the source to the drain. The source and the drain use the same type of doped material. They can be either N-type or P-type materials. A thin layer of either N-type or P-type material opposite the gate and drain separates the source and the drain.

Using **Figure 3-42**, if the source voltage has 0 volts applied to it and the drain has 6 volts applied to it, no current will flow between the two. However, applying a lower positive voltage to the gate forms a capacitive field between it and the channel. The voltage of the capacitive field attracts electrons from the source, and current will flow through the channel to the higher positive voltage of the drain.

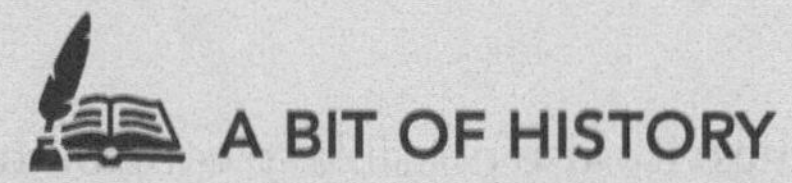

The development of the transistor was by a team of three American physicists: Walter Houser Brattain, John Bardeen, and William Bradford Shockley. They announced their achievement in 1948. In 1956, these physicists won the Nobel Prize in physics for this development.

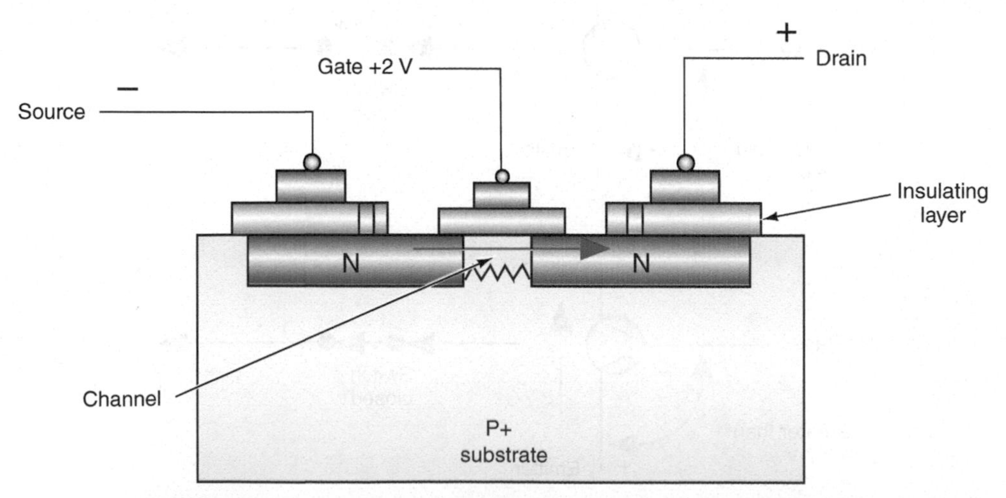

Figure 3-42 An FET uses a positive voltage to the gate terminal to create a capacitive field to allow electron flow.

Because the field effect improves current flow from the source to the drain, this type of FET is called an **enhancement-type FET**. This operation is like that of a NO switch. A **depletion-type FET** is like a NC switch, whereas the field effect cuts off current flow from the source to the drain.

Transistor Amplifiers

A transistor can also be used in an amplifier circuit to amplify the voltage. Amplifying the voltage is useful when using a very small voltage for sensing computer inputs but needing to boost that voltage to operate an accessory (**Figure 3-43**). The waveform showing the small signal voltage applied to the base leg may look like that shown in **Figure 3-44A**. The waveform showing the corresponding signal through the collector will be inverted (**Figure 3-44B**). Three things happen in an amplified circuit:

1. The amplified voltage at the collector is greater than that at the base.
2. The input current increases.
3. The pattern is inverted.

Shop Manual
Chapter 3, page 131

Some amplifier circuits use a **Darlington pair**, which is two transistors connected together. The first transistor in a Darlington pair is used as a preamplifier to produce a large current to operate the second transistor (**Figure 3-45**). The second transistor is

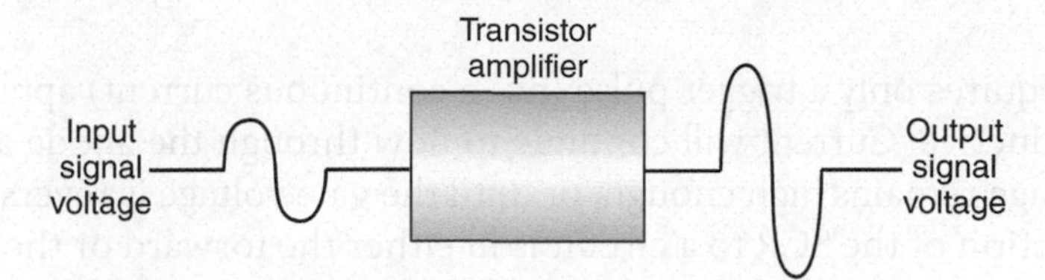

Figure 3-43 A simplified amplifier circuit.

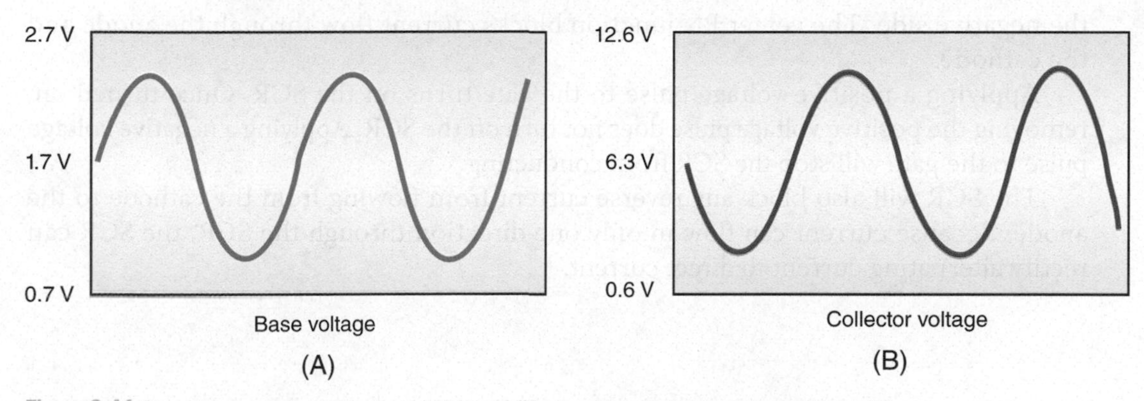

Figure 3-44 The voltage applied to the base (A) is amplified and inverted through the collector (B).

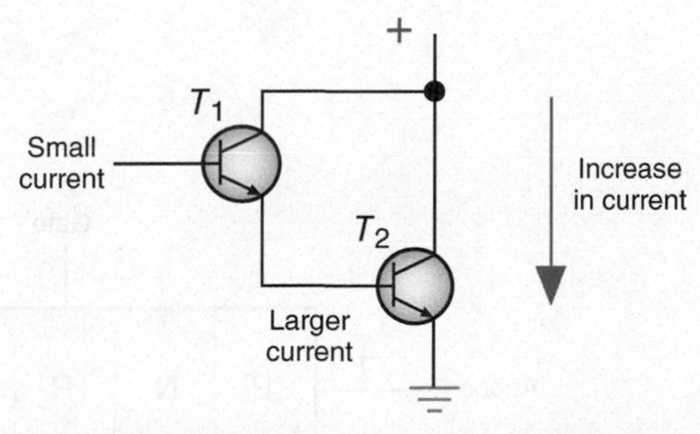

Figure 3-45 A Darlington pair used to amplify current. T1 acts as a preamplifier that creates a larger base current for T2, which is the final amplifier that creates a larger current.

isolated from the control circuit and is the final amplifier. The second transistor boosts the current to the amount required to operate the load component. Most electronic ignition system control modules use Darlington pairs.

Phototransistors

A **phototransistor** is a transistor that is sensitive to light. In a phototransistor, a small lens focuses incoming light onto the sensitive portion of the transistor (**Figure 3-46**). The light striking the transistor forms holes and free electrons, increasing the current flow through the transistor — the stronger the light intensity, the more current that will flow. Automatic headlight dimming circuits often use this type of phototransistor.

Thyristors

A **thyristor** is a semiconductor switching device composed of alternating N and P layers. It can be used to rectify current from AC to DC and to control power to light dimmers, motor speed controls, solid-state relays, and other applications where power control is needed.

The most common type of thyristor used in automotive applications is the silicon-controlled rectifier (SCR). Like the transistor, the SCR has three legs. However, it consists of four regions arranged PNPN (**Figure 3-47**). The three legs of the SCR are called the anode (or P-terminal), the cathode (or N-terminal), and the gate (one of the center regions).

The SCR requires only a trigger pulse (not a continuous current) applied to the gate to become conductive. Current will continue to flow through the anode and cathode as long as the voltage remains high enough, or until the gate voltage is reversed.

The connection of the SCR to a circuit is in either the forward or the reverse direction. Using Figure 3-47 of a forward-direction connection, note the P-type anode connection to the positive side of the circuit and the N-type cathode connection to the negative side. The center PN junction blocks current flow through the anode and the cathode.

Applying a positive voltage pulse to the gate turns on the SCR. Once turned on, removing the positive voltage pulse does not turn off the SCR. Applying a negative voltage pulse to the gate will stop the SCR from conducting.

The SCR will also block any reverse current from flowing from the cathode to the anode. Because current can flow in only one direction through the SCR, the SCR can rectify alternating current to direct current.

Figure 3-46 Phototransistor.

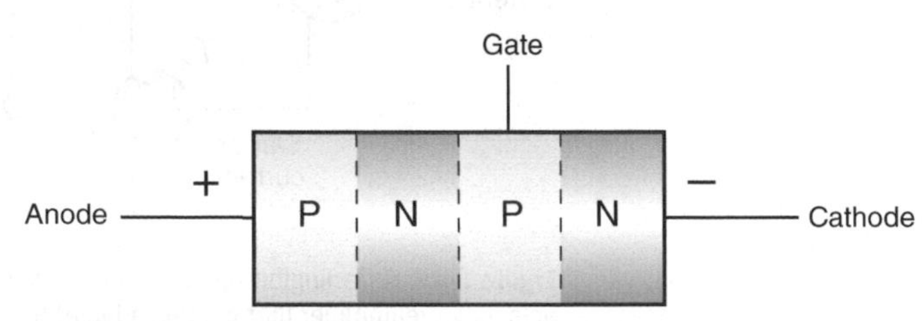

Figure 3-47 A forward-direction SCR.

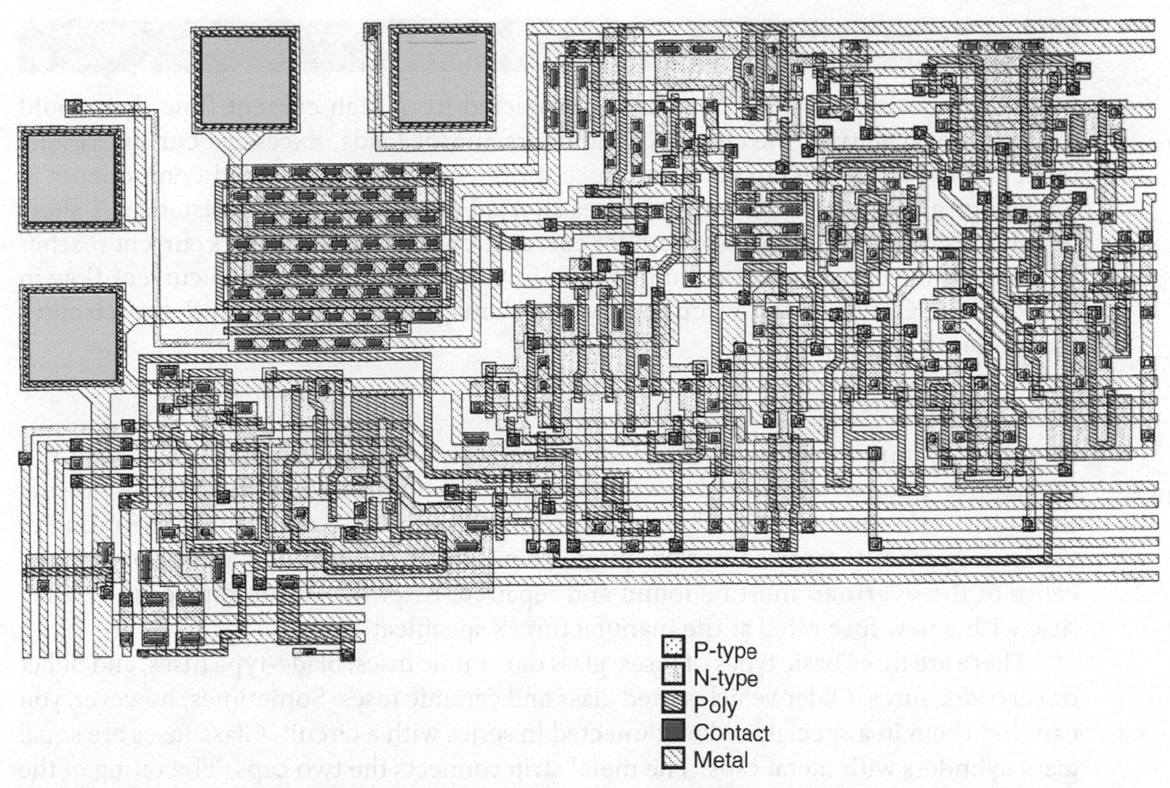

Figure 3-48 An enlarged illustration of an integrated circuit with thousands of transistors, diodes, resistors, and capacitors. The actual size can be less than 1/4 inch (6.35 mm) square.

Integrated Circuits

An **integrated circuit (IC)** is a complex circuit of thousands of transistors, diodes, resistors, capacitors, and other electronic devices formed onto a tiny silicon chip (**Figure 3-48**). A 1/4 inch (6.35 mm) square chip can have as many as 30,000 transistors placed on it.

A BIT OF HISTORY

Years ago, making an integrated circuit began with a large-scale drawing of the circuit. This drawing could have been room size. A mask for laying out the integrated circuit was made by taking photographs of the circuit drawing and reducing them to the actual size of the circuit. The mask was placed over the wafer to selectively expose the portion of the material to be etched away or the portion requiring selective deposition.

Making an IC chip is complex, and the entire process takes over 100 separate steps. Two of the most critical steps in their production involve lithography and photolithography. Lithography uses electron beams to make the mask that supplies the pattern of the circuit. Photolithography is the process of transferring the mask pattern onto the silicon wafer.

The mask is transferred to the wafer to identify areas that require etching or selective deposition. Etching uses chemical, physical, or both chemical and physical methods to remove unmasked areas of the thin film layer. The remaining substances form a chip pattern on the wafer. Since silicon is nonconductive, the wafer is doped with conductive P-type and N-type materials to form the conductive paths.

The small size of the integrated chip has made it possible for vehicle manufacturers to add several computer-controlled systems to the vehicle without taking up much space. Also, a single computer can perform several functions.

Circuit Protection Devices

Shop Manual
Chapter 3, page 116

Most automotive electrical circuits are protected from high current flow that would exceed the capacity of the circuit's conductors and/or loads. Excessive current results from a decrease in the circuit's resistance. Connecting too many load components in parallel or a component or wire becomes shorted decreases circuit resistance. A short is an undesirable, low-resistance path for current flow. When the circuit's current reaches a predetermined level, most circuit **protection devices** open and stop current flow in the circuit. Opening the circuit prevents damage to the wires and the circuit's components.

Fuses

Shop Manual
Chapter 3, page 116

Excess current flow in a circuit is called an **overload.**

The most common circuit protection device is the **fuse** (**Figure 3-49**). A fuse is a replaceable element that contains a metal strip that will melt when the current flowing through it exceeds its rating. The thickness of the metal strip determines the rating of the fuse. Most automotive fuses are rated from 3 to 30 amps. When a fuse "blows," the cause of the **overload** must be found and repaired. After the repair, replace the blown fuse with a new fuse rated at the manufacturer's specifications.

There are three basic types of fuses: glass or ceramic fuses, blade-type fuses, and bullet or cartridge fuses. Older vehicles used glass and ceramic fuses. Sometimes, however, you can find them in a special holder connected in series with a circuit. Glass fuses are small glass cylinders with metal caps. The metal strip connects the two caps. The rating of the fuse is usually marked on one of the caps.

Blade-type fuses are flat plastic units and are available in six different physical sizes (**Figure 3-50**). The plastic housing surrounds two male blade-type connectors. The metal strip connects these connectors inside the plastic housing. The rating of these fuses is on top of the plastic housing, and the plastic is color coded (**Figure 3-51**).

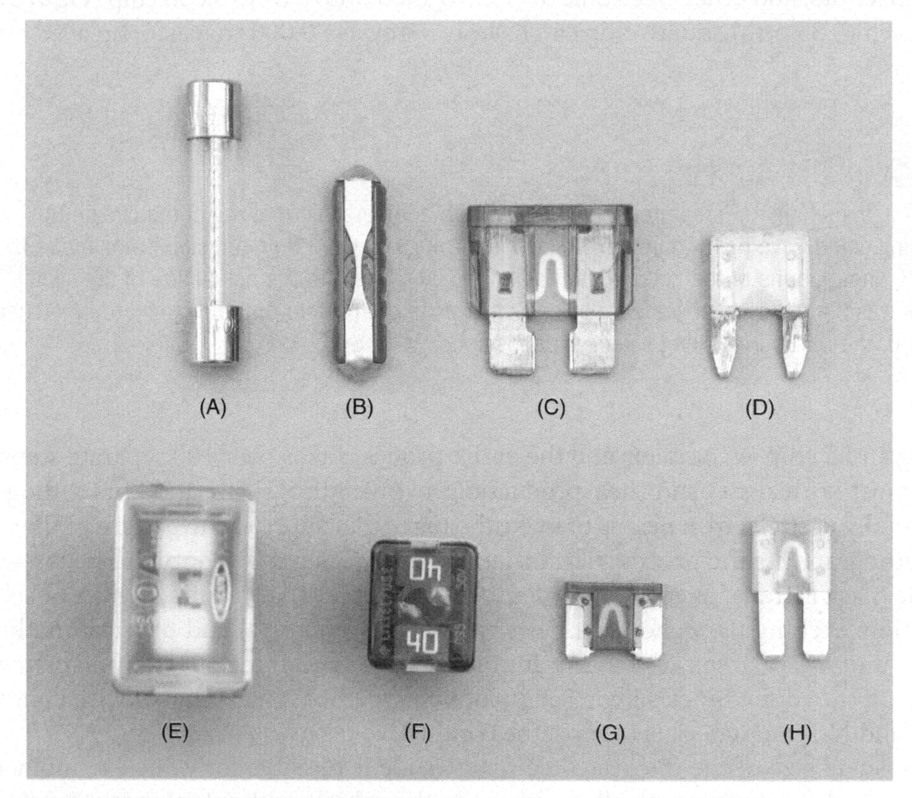

Figure 3-49 Common fuses (A) glass cartridge, (B) ceramic, (C) blade (auto-fuse), (D) mini, (E) maxi, (F) "F" type, (G) low-profile mini, and (H) MICRO.

Blade Size	Blade Group	Dimensions L × W × H	Common Ratings (Maximum Amperage)
Micro 2	APT, ATR	9.1 × 3.8 × 15.3 mm	5, 7.5, 10, 15, 20, 25, 30
Micro 3	ATL	14.4 × 4.2 × 18.1 mm	5, 7.5, 10, 15
Low-Profile Mini	APS, ATT	10.9 × 3.81 × 8.73 mm	2, 3, 4, 5, 7.5, 10, 15, 20, 25, 30
Mini	APM, ATM	10.9 × 3.6 × 16.3 mm	2, 3, 4, 5, 7.5, 10, 15, 20, 25, 30
Regular	APR, ATC (Closed), ATO (open), ATS[1]	19.1 × 5.1 × 18.5 mm	0.5, 1, 2, 3, 4, 5, 7.5, 10, 15, 20, 25, 30, 35, 40
Maxi	APX	29.2 × 8.5 × 34.3 mm	20, 25, 30, 35, 40, 50, 60, 70, 80, 100, 120

Figure 3-50 Fuse size and group classifications.

Auto-fuse

Current Rating	Color
3	Violet
5	Tan
7.5	Brown
10	Red
15	Blue
20	Yellow
25	Natural
30	Green

Maxi-fuse

Current Rating	Color
20	Yellow
30	Green
40	Amber
50	Red
60	Blue
70	Brown
80	Natural

Mini-fuse

Current Rating	Color
5	Tan
7.5	Brown
10	Red
15	Blue
20	Yellow
25	White
30	Green

Figure 3-51 Color coding for blade-type fuses. An auto-fuse is a standard blade-type fuse.

Figure 3-52 The MICRO 3 fuse protects two circuits.

Figure 3-53 Fuse boxes are typically located under the dash or in the engine compartment.

The low-profile mini, MICRO 2, and MICRO 3 fuses are the latest fuse types used in the automotive industry. All three of these fuse designs offer space savings. The low-profile mini fuse is a shorter version of the mini fuse and uses a different terminal design. The MICRO 3 fuse has three terminals and two fuse elements (**Figure 3-52**). The MICRO 3 fuse is essentially two fuses in one housing. The center terminal is common to both outputs. The MICRO 3 fuse offers additional circuit protection in less space.

Cartridge-type fuses are made of plastic or ceramic material. They have pointed ends, and the metal strip rounds from end to end. This type of fuse is much like a glass fuse, except the metal strip is not enclosed.

Usually, fuses are located in a central **fuse block** or power distribution box. However, fuses may also be found in relay boxes and electrical junction boxes. A common location for the fuse box is under the instrumental panel (**Figure 3-53**). Other locations for the fuse box include: in the glove box, in the engine compartment, behind kick panels, or various other places on the vehicle. Power distribution boxes are normally located in the engine

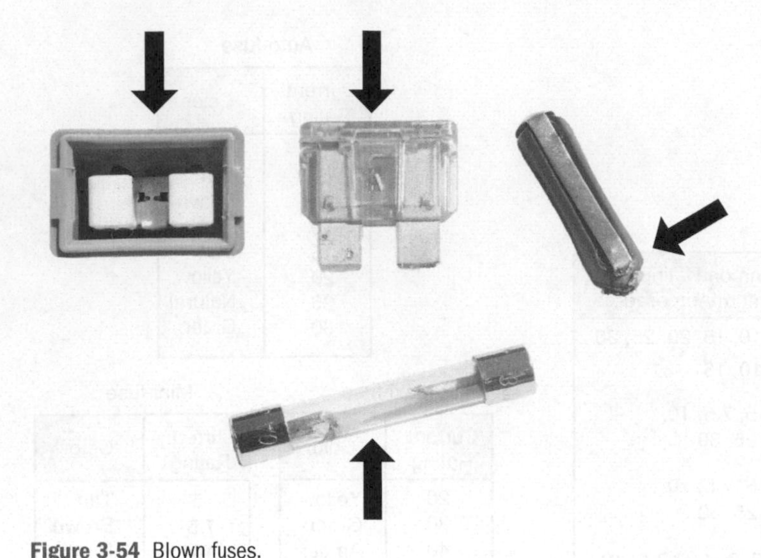
Figure 3-54 Blown fuses.

Figure 3-55 One fuse to protect the entire parallel circuit.

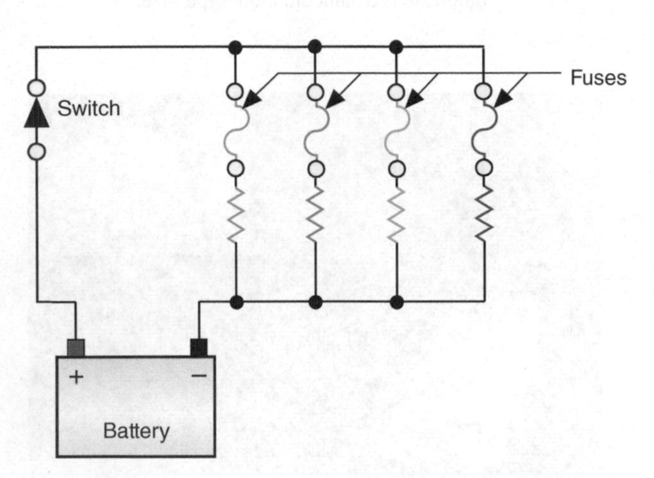

Figure 3-56 Fuses used to protect each branch of a parallel circuit.

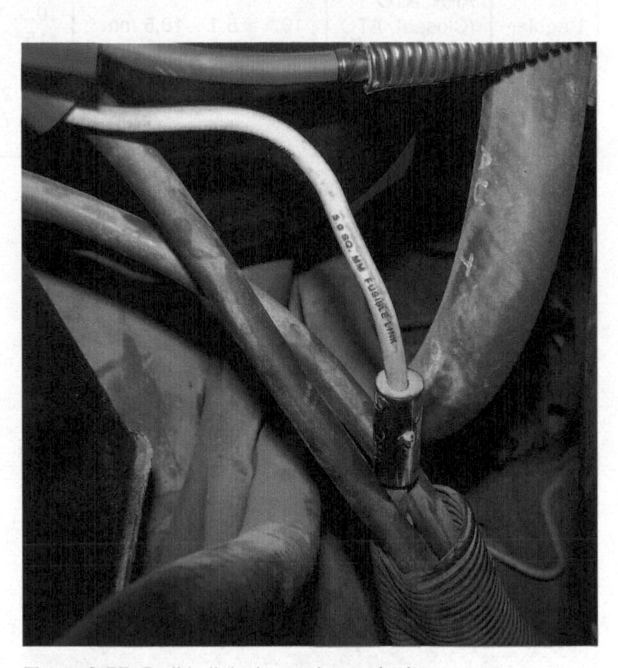
Figure 3-57 Fusible links located near the battery.

A "blown" fuse is identified by a burned-through metal wire in the capsule (**Figure 3-54**).

compartment and house fuses and relays. Fuse ratings, and the circuits they protect, may be identified on the inside of the fuse or power distribution box's cover. Of course, the vehicle owner's manual and the service information also have this information.

Typically, fuses are connected in series with the circuit and before all of the loads of the circuit (**Figure 3-55**). However, fuses may be placed before individual loads (**Figure 3-56**).

The correct fuse rating must be selected when adding accessories to the vehicle. Use the power formula to determine the correct fuse rating. The fuse selected should be rated slightly higher than the actual current draw to allow for current surges (5% to 10%).

Fusible Links

Shop Manual
Chapter 4, page 116

Fusible links are meltable conductor materials surrounded by a special heat-resistant insulation (**Figure 3-57**). When there is an overload in the circuit, the conductor link melts and opens the circuit. To properly test a fusible link, use an ohmmeter or continuity tester. A vehicle may have one or several fusible links to protect the main power wires before the circuit splits into individual circuits at the fuse box. Typical installation areas of the fusible links are near the battery or the starter solenoid. The size of the fusible link determines its current capacity. A fusible link is usually four wire sizes smaller (four numbers larger) than the wire size of the circuit it protects. The smaller the wire, the larger its number. A circuit that uses 14-gauge wire would require an 18-gauge fusible link for protection.

AUTHOR'S NOTE Some GM vehicles have the fusible link located at the main connection near the starter motor.

AUTHOR'S NOTE A "blown" fusible link is usually identified by bubbling of the insulator material around the link.

Maxi-Fuses

Many manufacturers use a **maxi-fuse** in place of fusible links (**Figure 3-58**). A maxi-fuse looks like a blade-type fuse, except it is larger and has a higher current capacity. Maxi-fuses allow the manufacturers to break down the electrical system into smaller circuits. If a fusible link burns out, many of the vehicle's electrical systems may be affected. By breaking down the electrical system into smaller circuits and installing maxi-fuses, the consequence of a circuit defect will not be as severe as it would have been with a fusible link. In place of a single fusible link, there may be many maxi-fuses depending on how the circuits are divided. Separating circuits in this manner can make your job of diagnosing a faulty circuit much easier.

Maxi-fuses are also called cartridge fuses.

Shop Manual Chapter 3, page 116

Maxi-fuses are less likely than a fusible link to cause an underhood fire when there is an overload in the circuit. If the fusible link burns in two, it is possible that the "hot" side of the link can contact the vehicle frame, and the wire can catch on fire.

Today many manufacturers are replacing maxi-fuses with "F"-type fuses. These are smaller versions of the maxi-fuses.

Circuit Breakers

A **circuit breaker** usually protects a circuit that is susceptible to an overload on a routine basis. A circuit breaker uses a **bimetallic strip** that reacts to excessive current (**Figure 3-59**). When an overload or circuit defect occurs that causes an excessive amount of current draw, the current flowing through the bimetallic strip causes it to heat. As the strip heats, it bends and opens the contacts. Once the contacts open,

Shop Manual Chapter 3, page 117

A **bimetallic strip** consists of two different types of metals. One strip will react more quickly to heat than the other, which causes the strip to flex in proportion to the amount of current flow.

Figure 3-58 Maxi-fuses have replaced the use of fusible links in many high current circuits.

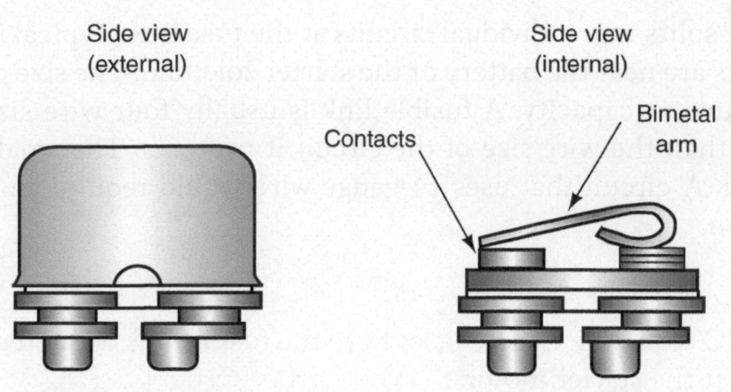

Figure 3-59 The circuit breaker uses a bimetallic strip that opens if current draw is excessive.

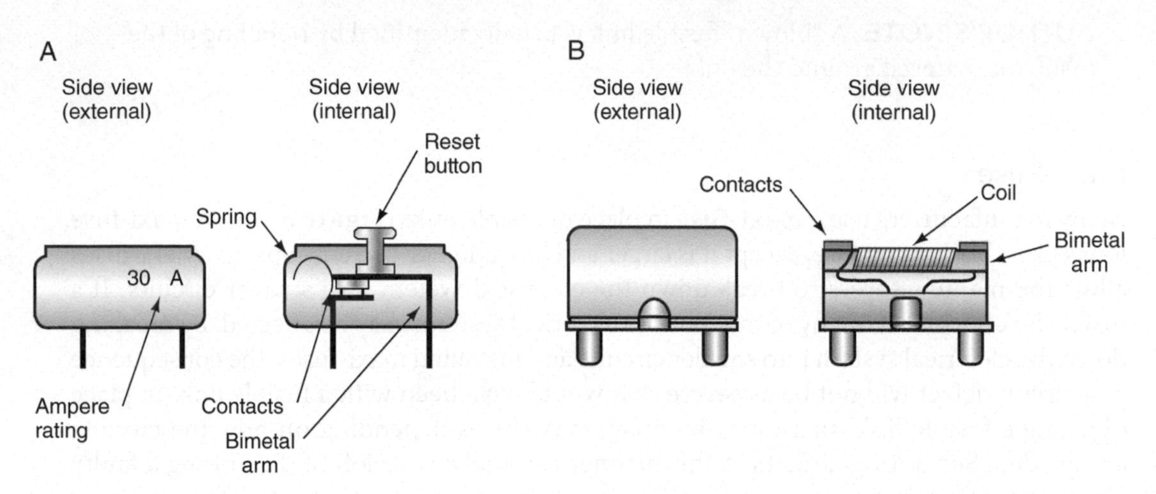

Figure 3-60 Non cycling circuit breakers. (A) Can be reset by pressing the button, while (B) requires being removed from the power to reset.

current can no longer flow. With no current flowing, the strip cools and closes again. If the cause of the excessive current is still in the circuit, the breaker will open again. The circuit breaker will continue to cycle open and close as long as the overload is in the circuit. This type of circuit breaker is self-resetting or "cycled." Some circuit breakers require manual resetting by pressing a button, while others must be removed from the power to reset (**Figure 3-60**).

An example of using a circuit breaker is in the power window circuit. Because the window is susceptible to jams due to ice buildup on the window, a current overload is possible. If this should occur, the circuit breaker will heat up and open the circuit before the window motor is damaged. If the operator continues to attempt to operate the power window, the circuit breaker will open and close until the cause of the jam is removed.

PTCs as Circuit Protection Devices

Shop Manual
Chapter 3, page 119

Automotive manufacturers must utilize means of providing reliable circuit protection, yet at the same time reduce vehicle weight and cost. Using fuses to protect multiple circuits can result in large, heavy, and complex wiring assemblies. Polymer, **positive temperature coefficient (PTC) thermistors** provide a lightweight alternative to protecting the circuit against shorts or overload conditions. A PTC thermistor increases in resistance as temperature increases. Because of its design, a PTC thermistor can trip (increase resistance to the point it becomes the load device in the circuit) during an over-current condition and reset after the fault is no longer present (**Figure 3-61**).

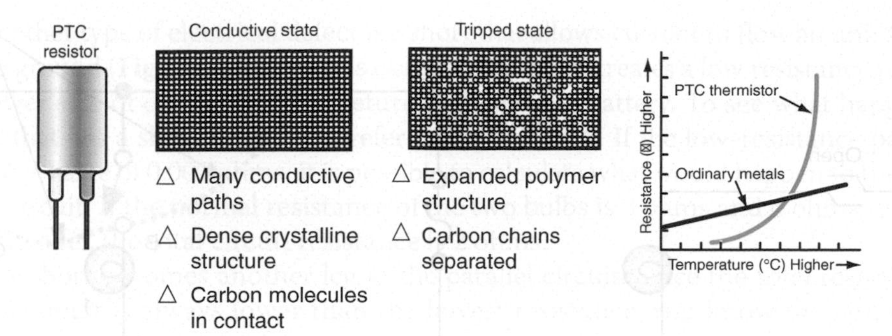

Figure 3-61 PTC operation.

Conductive polymers consist of specially formulated plastics and various conductive materials. At normal temperatures, the plastic materials form a crystalline structure. The structure provides a low-resistance conductive chain. The resistance is so low that it does not affect the operation of the circuit. However, if the current flow increases above the trip threshold, the additional heat causes the crystalline structure to change to an amorphous state. In this condition, the conductive paths separate, causing a rapid increase in the resistance of the PTC. The increased resistance reduces the current flow to a safe level.

Circuit Defects

All electrical defects fall into one of three classification types: open, short, or high resistance. Each one of these will cause a component to operate incorrectly or not at all. Understanding what each of these defects will do to a circuit is the key to the proper diagnosis of any electrical problem.

Open

An **open** is simply a break in the circuit (**Figure 3-62**). An open is caused by turning a switch off, a break in a wire, a burned-out light bulb, a disconnected wire or connector, or anything that opens the circuit. When a circuit is open, current does not flow, and the component doesn't work. Because there is no current flow, there are no voltage drops in the circuit. Source voltage is available everywhere in the circuit up to the point of the open. If the open is after a load device, source voltage is even available after the load.

Shop Manual
Chapter 3, page 106

Opens caused by a blown fuse will still cause the circuit not to operate, but the cause of the problem is the excessive current that blew the fuse. Nearly all other opens result from a break in the continuity of the circuit. These breaks can occur anywhere in the circuit.

Shorts

Short refers to a defect that allows the current to take a 'shortcut" instead of the intended path. **Shorted circuits** cause an increase in current flow that can burn wires or components.

Shop Manual
Chapter 3, page 108

The breaking down of a coil's insulation is an example of a short. The coil's windings are insulated from each other; however, if this insulation breaks down, copper-to-copper contact is made between the turns. Bypassing a part of the windings reduces the number of windings that the current will flow through — reduced current flow results in reduced coil effectiveness. Also, the increased current flow can generate excess heat.

resistance is now less than 0.001 ohms. Using Ohm's law, you can calculate the current flowing through the circuit.

A = V/R or A = 12/0.001 or A = 12,000 Amps

This amount of current would require a large wire. The 10-amp fuse would melt quickly when the short occurred, protecting the wires and light bulbs.

High Resistance

Shop Manual
Chapter 3, page 112

High-resistance problems occur when there is unwanted resistance in the circuit. The high resistance can come from a loose connection, corroded connection, corrosion in the wire, wrong size wire, and so on. Since the resistance becomes an additional load in the circuit, the effect is that the load component, with reduced voltage and current applied, operates with reduced efficiency. An example would be a taillight circuit with a load component (light bulb) rated at 50 watts. This bulb must draw 4.2 amperes at 12 volts (A 5 P/V) to be fully effective. This means the bulb must have a full 12 volts applied to it. However, if resistance is present at other points in the circuit, some of the 12 volts is dropped. With less voltage (and current) being available to the light bulb, the bulb will illuminate with less intensity.

Figure 3-67 illustrates a light circuit with unwanted resistance at the power feed for the bulb and at the negative battery terminal. When the circuit is operating correctly, the 2-ohm light bulb will have 6 amps of current flowing through it and drop 12 volts. The added resistance reduces the current to 3 amps, and the bulb drops only 6 volts. As a result, the bulb's illumination is very dim.

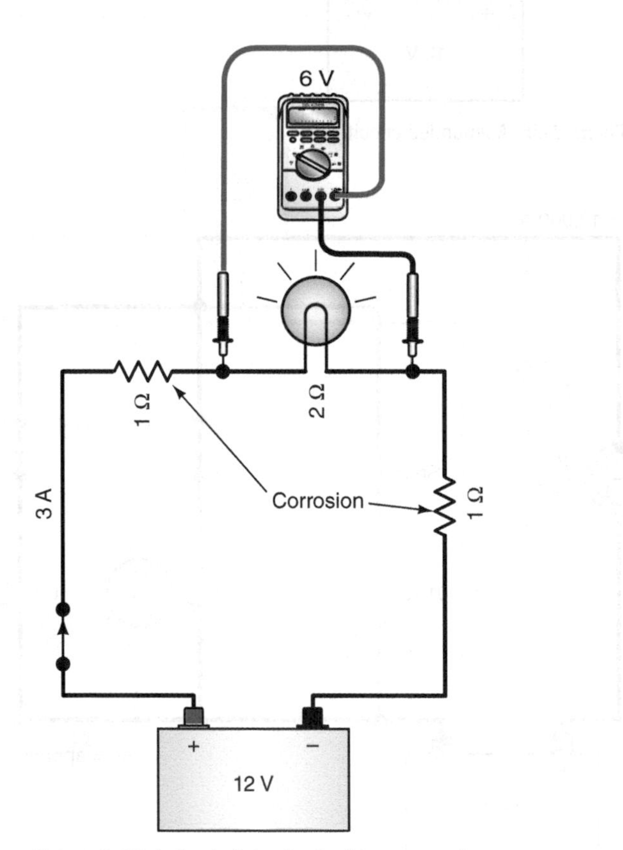

Figure 3-67 A simple light circuit with unwanted resistance.

Summary

- A switch can control the on/off operation of a circuit or direct the flow of current through various circuits.
- A normally open switch will not allow current flow when it is in its rest position. A normally closed switch will allow current flow when it is in its rest position.
- A relay is a device that uses a low current to control a high-current circuit.
- A buzzer is sometimes used to warn the driver of possible safety hazards by emitting an audio signal (such as when the seat belt is not buckled).
- A stepped resistor has two or more fixed resistor values. It is commonly used to control electric motor speeds.
- A variable resistor provides an infinite number of resistance values within a range. A rheostat is a two-terminal variable resistor used to regulate the strength of an electrical current. A potentiometer is a three-wire variable resistor that acts as a voltage divider to produce a continuously variable output signal proportional to a mechanical position.
- Capacitance is the ability of two conducting surfaces to store voltage.
- A diode is an electrical one-way check valve that will allow current to flow in one direction only.
- Forward bias refers to applying a positive voltage to the P-type material and a negative voltage to the N-type material. Reverse bias means that positive voltage is applied to the N-type material and negative voltage to the P-type material.
- A transistor is a three-layer semiconductor used as a very fast switching device.
- An integrated circuit is a complex circuit of thousands of transistors, diodes, resistors, capacitors, and other electronic devices formed onto a tiny silicon chip.
- The protection device interrupts current flow through the circuit it protects by creating an open (like turning off a switch) to prevent a complete circuit.
- Fuses are rated by amperage. Never install a larger rated fuse into a circuit than the one specified by the manufacturer. Doing so may damage or destroy the circuit.
- An open circuit is a circuit in which there is a break in continuity.
- A shorted circuit allows current to bypass part of the normal path.
- A short to ground allows current to return to ground before it has reached the intended load component.

Review Questions

Short-Answer Essays

1. Describe the use of three types of semiconductors.
2. What types of mechanical variable resistors are used on automobiles?
3. Define the meanings of opens, shorts, grounds, and excessive resistance.
4. Explain the effects that each type of circuit defect will have on the operation of the electrical system.
5. Explain the purpose of a circuit protection device.
6. Describe the most common types of circuit protection devices.
7. Describe the common types of electrical system (nonelectronic) components used and how they affect the electrical system.
8. Explain the basic concepts of capacitance.
9. Explain the difference between normally open (NO) and normally closed (NC) switches.
10. Explain the differences between forward-biasing and reverse-biasing a diode.

Fill in the Blanks

1. Never install a larger rated _______________ into a circuit than the one that was designed by the manufacturer.
2. A _______________ can control the on/off operation of a circuit or direct the flow of current through various circuits.
3. A normally _______________ switch will not allow current flow when it is in its rest position. A normally _______________ switch will allow current flow when it is in its rest position.

4. An ______________ is a complex circuit of many transistors, diodes, resistors, capacitors, and other electronic devices formed onto a tiny silicon chip.
5. When a ______________ voltage is applied to the P-type material of a diode and ______________ voltage is applied to the P-type material, the diode is reverse-biased. When a ______________ voltage is applied to the N-type material of a diode and ______________ voltage is applied to the P-type material, the diode is forward-biased.
6. A ______________ is used in electronic circuits as a very fast switching device.
7. A ______________ is an electrical one-way check valve that will allow current to flow in one direction only.
8. A ______________ is an electromechanical device that uses low current to control a high-current circuit.
9. A ______________ is a three-wire variable resistor that acts as a voltage divider. A ______________ is a two-terminal variable resistor used to regulate the strength of an electrical current.
10. The ______________ requires only a trigger pulse applied to the gate to become conductive.

Multiple Choice

1. All of the following are true concerning electrical shorts, EXCEPT:
   A. A short can add a parallel leg to the circuit, lowering the entire circuit's resistance.
   B. A short can result in a blown fuse.
   C. A short decreases amperage in the circuit.
   D. A short bypasses the circuit's intended path.
2. A "blown" fusible link is identified by:
   A. A burned-through metal wire in the capsule.
   B. Bubbling of the insulator material around the link.
   C. All of these choices.
   D. None of these choices.
3. All of the statements concerning circuit components are true, EXCEPT:
   A. A switch can control the on/off operation of a circuit.
   B. A switch can direct the flow of current through various circuits.
   C. A relay can be an SPDT-type switch.
   D. A potentiometer changes voltage drop due to the function of temperature.
4. Which of the following statements is correct?
   A. A Zener diode is an excellent component for regulating voltage.
   B. A reverse-biased diode lasts longer than a forward-biased diode.
   C. The switches of a transistor last longer than those of a relay.
   D. Thyristors require a constant current to the gate to be conductive.
5. The light-emitting diode (LED):
   A. Emits light when it is reverse-biased.
   B. Has a variable turn-on voltage.
   C. Has a light color defined by the materials used to construct the diode.
   D. Has a turn-on voltage that is usually less than standard diodes.
6. Which of the following is the correct statement?
   A. An open means there is continuity in the circuit.
   B. A short bypasses a portion of the circuit.
   C. High-amperage draw indicates an open circuit.
   D. High resistance in a circuit increases current flow.
7. Transistors:
   A. Can be used to control the on/off switching of a circuit.
   B. Can be used to amplify voltage.
   C. Control high current with low current.
   D. All of these choices.
8. All of the following are true, EXCEPT:
   A. Voltage drop can cause a lamp in a parallel circuit to burn brighter than normal.
   B. Excessive voltage drop may appear on either the insulated or the grounded return side of a circuit.
   C. Increased resistance in a circuit decreases current.
   D. A diode is an electrical one-way check valve.

9. Which statement is correct concerning diodes?
   A. Diodes are aligned to allow current flow in one direction only.
   B. Diodes can be used to rectify DC voltages into AC voltages.
   C. The stripe is on the anode side of the diode.
   D. Normal turn-on voltage of a standard diode is 1.5 volts.

10. A capacitor:
    A. Consumes electrical power.
    B. Stores electrical charges.
    C. Stores electrons on the positive plate.
    D. None of these choices.

CHAPTER 4
WIRING AND CIRCUIT DIAGRAMS

Upon completion and review of this chapter, you should be able to:

- Explain when single-stranded or multistranded wire should be used.
- Explain the use of resistive wires in a circuit.
- Describe the construction of spark plug wires.
- Explain how wire size is determined by the American Wire Gauge (AWG) and metric methods.
- Determine the correct wire gauge to be used in a circuit.
- Explain how temperature affects resistance and wire size selection.
- Explain the purpose and use of printed circuits.
- Explain the purpose of wiring diagrams.
- Identify the common wiring diagram electrical symbols that are used.
- Explain the purpose of the component locator.

Terms to Know

Common connection
Component locator
Electrical symbols
Gauge
Ground straps
Primary wiring
Printed circuit boards
Schematic
Secondary wiring
Splice
Stranded wire
Tracer
Wiring diagram
Wiring harness

Introduction

Trying to locate the cause of an electrical problem can be quite difficult if you do not have a good understanding of wiring systems and diagrams. Today's vehicles have a vast amount of electrical wiring that could stretch for half a mile or more (if laid end to end). Today's technician must be proficient at reading wiring diagrams in order to sort through this great maze of wires to isolate the root cause of the customer's issue.

In this chapter, you will learn how wiring harnesses (**Figure 4-1**) are made, how to read wiring diagrams, how to interpret wiring diagram symbols, and how terminals are used. It's also important to understand how to determine the correct type and size of wire to carry the anticipated amount of current. It's possible to cause an electrical problem by simply using the wrong size of wire. To perform repairs correctly, you must understand the influence that three factors have on a wire's resistance—length, diameter, and temperature.

Automotive Wiring

Shop Manual
Chapter 4, page 170

Primary wiring is the term used for conductors that carry low voltage. The insulation of primary wires is usually thin. **Secondary wiring** refers to wires used to carry high voltage, such as ignition spark plug wires. Secondary wires have much thicker insulation than primary wires.

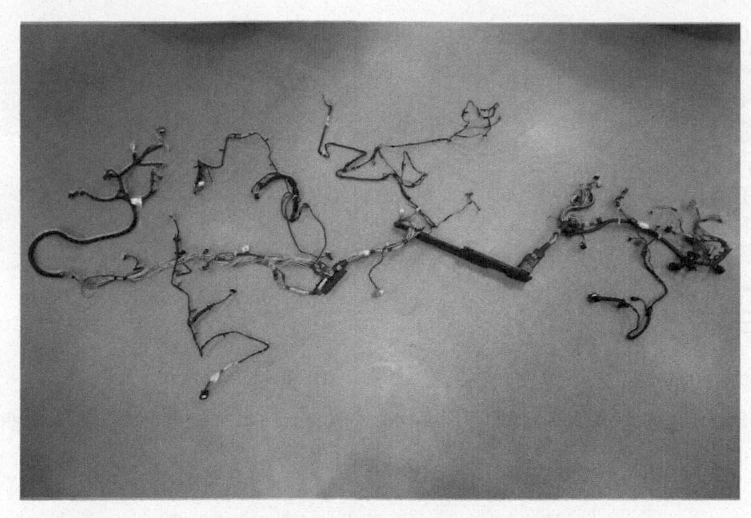

Figure 4-1 A small portion of the vehicle's wiring harness.

Most of the primary wiring conductors used in the automobile are made of several strands of copper wire wound together and covered with a polyvinyl chloride (PVC) insulation (**Figure 4-2**). Copper has low resistance and can be connected to easily by using crimping connectors or soldered connections. Other types of conductor materials used in automobiles include silver, gold, aluminum, and tin-plated brass.

AUTHOR'S NOTE Copper is used mainly because of its low cost and availability.

Stranded wire means the conductor is made of several individual wires that are wrapped together. Stranded wire is used because it's very flexible and has less resistance than solid wire of the same size. This is because electrons tend to flow on the outside surface of conductors. Since there is more surface area exposed in a stranded wire (each strand has its own surface), there is less resistance in the stranded wire than in the solid wire (**Figure 4-3**). The PVC insulation is used because it can withstand temperature extremes and corrosion. The insulation protects the wire from electrical shorts, corrosion, and other environmental factors. PVC insulation is also capable of withstanding battery acid, antifreeze, and gasoline.

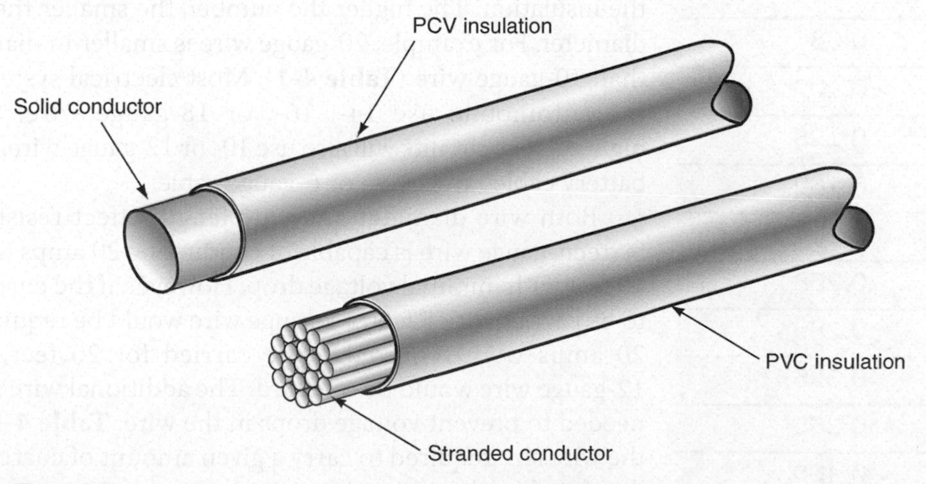

Figure 4-2 Comparison between solid and stranded primary wire.

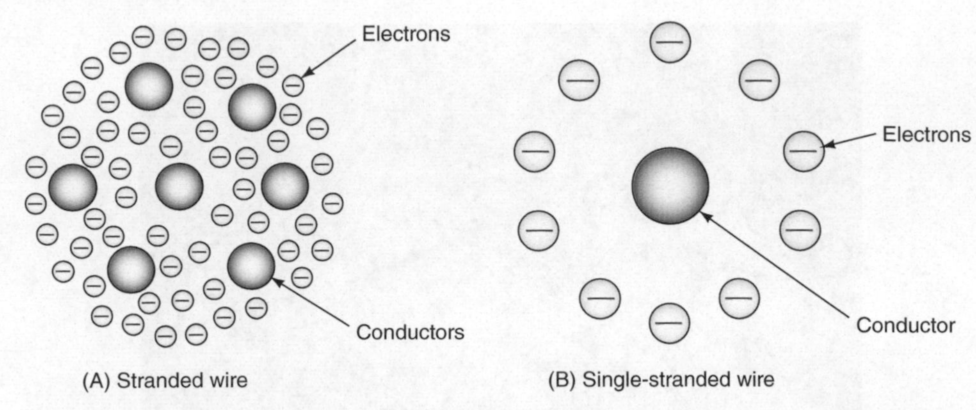

Figure 4-3 Stranded wire provides flexibility and more surface area for electron flow than a single-stranded solid wire.

AUTHOR'S NOTE General Motors has used single-stranded aluminum wire in limited applications where no flexing of the wire is expected. For example, it was used in some taillight circuits.

Wire Sizes

Consideration must be given for some margin of safety when selecting wire size. There are three major factors that determine the proper size of wire to be used:

1. The wire must have a large enough diameter, for the length required, to carry the necessary current for the load components in the circuit to operate properly.
2. The wire must be able to withstand the anticipated vibration.
3. The wire must be able to withstand the anticipated amount of heat exposure.

Wire size is based on the diameter of the conductor. The larger the diameter, the less the resistance. Two common standards are used to designate wire size: American Wire Gauge (AWG) and metric.

The AWG standard assigns a **gauge** number to the wire based on the diameter of the conductor not including the insulation. The higher the number, the smaller the wire diameter. For example, 20-gauge wire is smaller in diameter than 10-gauge wire (**Table 4-1**). Most electrical systems in the automobile use 14-, 16-, or 18-gauge wire. Some high-current circuits will also use 10- or 12-gauge wire. Most battery cables are 2-, 4-, or 6-gauge cable.

Both wire diameter and wire length affect resistance. Sixteen-gauge wire is capable of conducting 20 amps (A) for 10 feet with minimal voltage drop. However, if the current is to be carried for 15 feet, 14-gauge wire would be required. If 20 amps were required to be carried for 20 feet, then 12-gauge wire would be required. The additional wire size is needed to prevent voltage drops in the wire. **Table 4-2** lists the wire size required to carry a given amount of current for different lengths.

Table 4-1 Gauge and wire size chart.

American Wire Gauge Sizes	
Gauge Size	Conductor Diameter (inch)
20	0.032
18	0.040
16	0.051
14	0.064
12	0.081
10	0.102
8	0.128
6	0.162
4	0.204
2	0.258
1	0.289
0	0.325
2/0	0.365
4/0	0.460

Table 4-2 The distance the current must be carried is a factor in determining the correct wire gauge to use.

Total Approximate Circuit Amperes	Wire Gauge (for Length in Feet)								
12 V	**3**	**5**	**7**	**10**	**15**	**20**	**25**	**30**	**40**
1.0	18	18	18	18	18	18	18	18	18
1.5	18	18	18	18	18	18	18	18	18
2	18	18	18	18	18	18	18	18	18
3	18	18	18	18	18	18	18	18	18
4	18	18	18	18	18	18	18	16	16
5	18	18	18	18	18	18	18	16	16
6	18	18	18	18	18	18	16	16	16
7	18	18	18	18	18	18	16	16	14
8	18	18	18	18	18	16	16	16	14
10	18	18	18	18	16	16	16	14	12
11	18	18	18	18	16	16	14	14	12
12	18	18	18	18	16	16	14	14	12
15	18	18	18	18	14	14	12	12	12
18	18	18	16	16	14	14	12	12	10
20	18	18	16	16	14	12	10	10	10
22	18	18	16	16	12	12	10	10	10
24	18	18	16	16	12	12	10	10	10
30	18	16	16	14	10	10	10	10	10
40	18	16	14	12	10	10	8	8	6
50	16	14	12	12	10	10	8	8	6
100	12	12	10	10	6	6	4	4	4
150	10	10	8	8	4	4	2	2	2
200	10	8	8	6	4	4	2	2	1

Another factor that affects wire resistance is temperature. An increase in temperature creates a similar increase in resistance. A wire may have a known resistance of 0.03 ohm (Ω) per 10 feet at 70°F. When exposed to temperatures of 170°F, the resistance may increase to 0.04 ohm per 10 feet. Wires that are to be installed in areas that experience high temperatures, as in the engine compartment, must be of a size such that the increased resistance will not affect the operation of the load component. Also, the insulation of the wire must be capable of withstanding the high temperatures.

In the metric system, wire size is determined by the cross-sectional area of the wire. Metric wire size is expressed in square millimeters (mm^2). In this system, the smaller the number, the smaller the wire conductor. The approximate equivalent wire size of metric to AWG is shown in **Table 4-3**.

Ground Straps

Shop Manual
Chapter 4, page 183

Usually, there is not a direct metal-to-metal connection between the powertrain components and the vehicle chassis. Rubber mounts or bushings support the engine, transmission, and axle assemblies. The rubber acts as an insulator so any electrical components such as

Table 4-3 Approximate AWG to metric equivalents.

Metric Size (mm^2)	AWG (Gauge) Size	Ampere Capacity
0.5	20	4
0.8	18	6
1.0	16	8
2.0	14	15
3.0	12	20
5.0	10	30
8.0	8	40
13.0	6	50
19.0	4	60

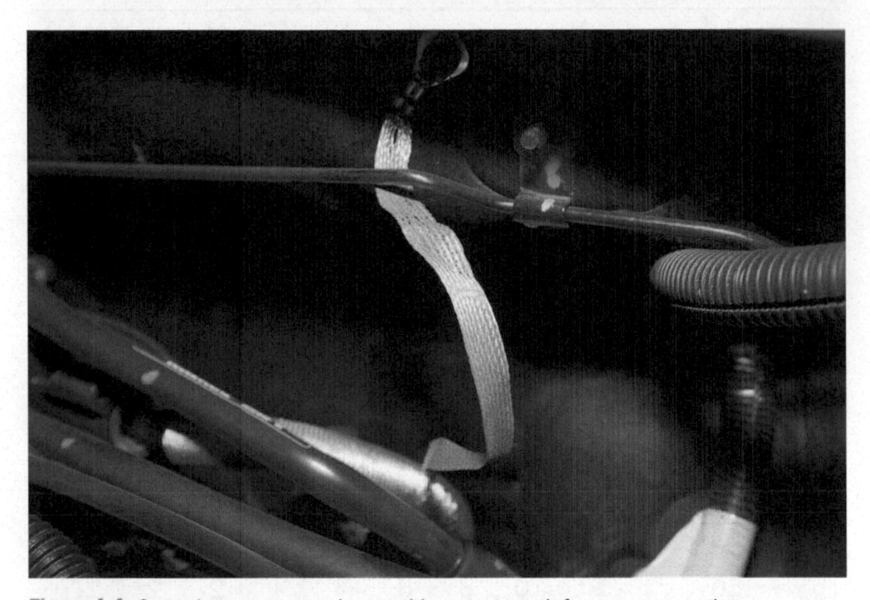

Figure 4-4 Ground straps are used to provide a return path for components that are insulated from the chassis.

actuators or sensors that are mounted to the powertrain components will not have a completed circuit back to the vehicle's battery. This is especially true if the negative battery cable is attached to the vehicle's chassis instead of the engine block. **Ground straps** between the powertrain components and the vehicle's chassis are used to complete the return path to the battery (**Figure 4-4**). In addition, ground straps suppress electromagnetic induction (EMI) and radiation by providing a low-resistance circuit ground path.

AUTHOR'S NOTE Ground straps are also referred to as bonding straps. Ground straps can be installed in various locations. Some of the most common locations are:

- Engine to bulkhead or fender.
- Across the engine mounts.
- Radio chassis to instrument panel frame.
- Air-conditioning evaporator valve to the bulkhead.

The ground strap can be a large gauge insulated-type cable or a braided strap. Even on vehicles with the battery negative cable attached to the engine block, ground straps are used to connect between the engine block and the vehicle chassis. The additional ground cable ensures a good, low-resistance ground path between the engine and the chassis.

Ground straps are also used to connect sheet metal parts such as the hood, fender panels, and the exhaust system even though there is no electrical circuit involved. In these cases, the strap is used to suppress EMI since the sheet metal could behave as a large capacitor. The air space between the sheet metal forms an electrostatic field and can interfere with any computer-controlled circuits that are routed near the sheet metal.

Terminals and Connectors

To perform the function of connecting the wires from the voltage source to the load component reliably, terminal connections are used. Today's vehicles can have as many as 500 separate circuit connections. The terminals used to make these connections must be able to perform with very low voltage drop. Terminals are constructed of either brass or steel. Steel terminals usually have a tin or lead coating. Some steel terminals have a gold plating to prevent corrosion. A loose or corroded connection can cause an unwanted voltage drop, resulting in poor operation of the load component. For example, a connector used in a light circuit that has as little as 10% voltage drop (1.2 volts [V]) may result in a 30% loss of lighting efficiency.

Shop Manual
Chapter 4,
pages 170, 180, 181

Terminals make the electrical connection at a component or connector and must be capable of withstanding the stress of normal vibration. Terminals can be either crimped or soldered to the conductor. **Figure 4-5** shows several different types of terminals used in the automotive electrical system. In addition, the following connectors are used on the automobile:

1. **Molded connector:** These connectors usually have one to four wires that are molded into a one-piece component (**Figure 4-6**). Although the male and female connector halves separate, the connector itself cannot be taken apart.
2. **Multiple-wire, hard-shell connector:** These connectors usually have a hard, plastic shell that holds the connecting terminals of separate wires (**Figure 4-7**). The wire terminals can be removed from the shell to be repaired.
3. **Bulkhead connectors:** These connectors are used when several wires must pass through the bulkhead (**Figure 4-8**).

Shop Manual
Chapter 4, page 181

Shop Manual
Chapter 4, page 181

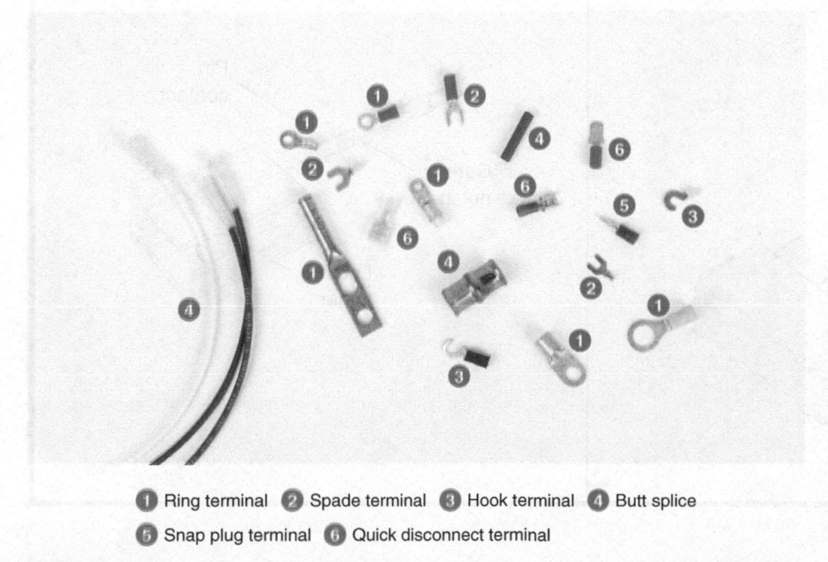

Figure 4-5 Examples of primary wire terminals and connectors used in automotive applications.

Figure 4-6 Molded connectors cannot be disassembled to replace damaged terminals or to test.

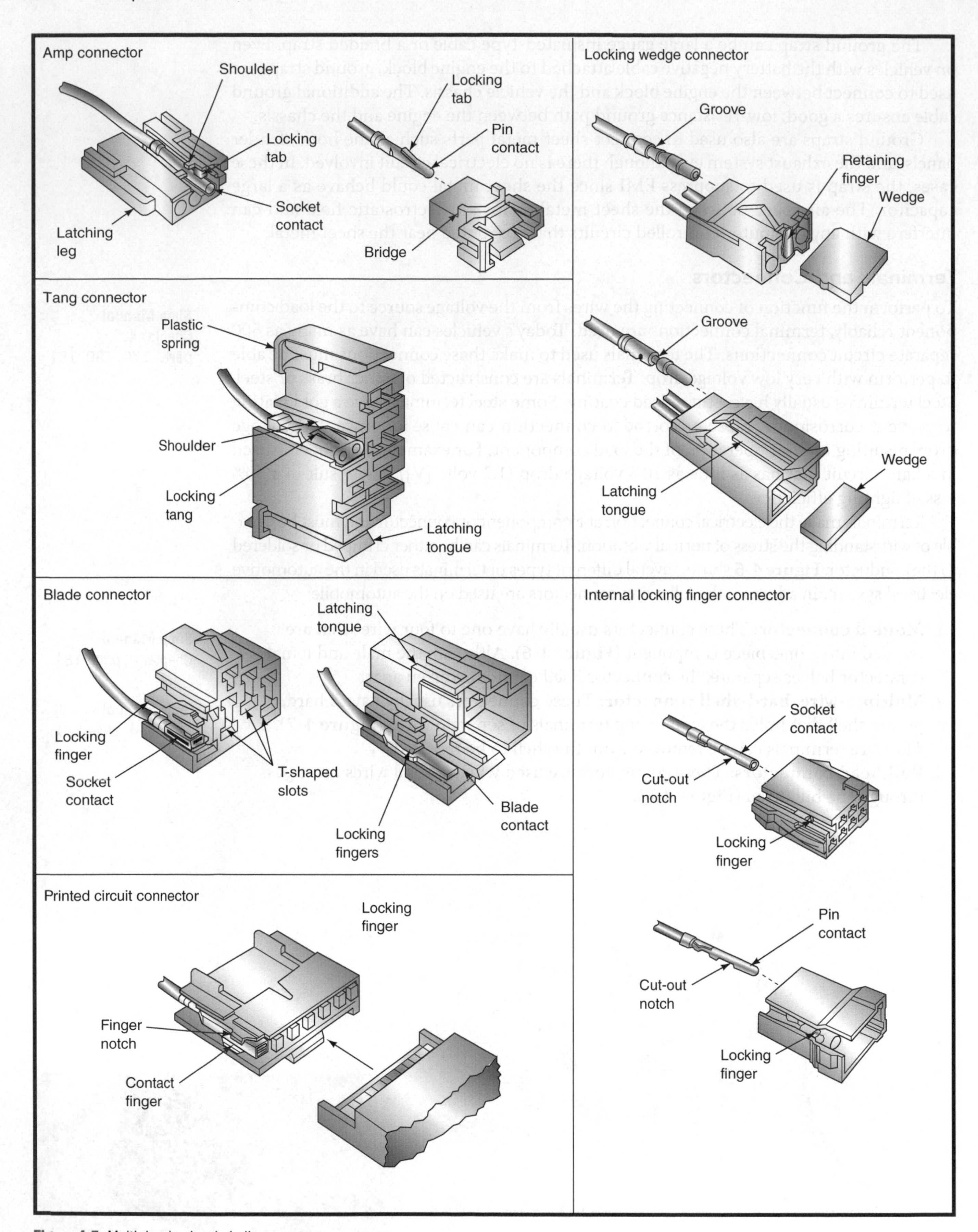

Figure 4-7 Multiple-wire, hard-shell connectors.

Figure 4-8 Bulkhead connector.

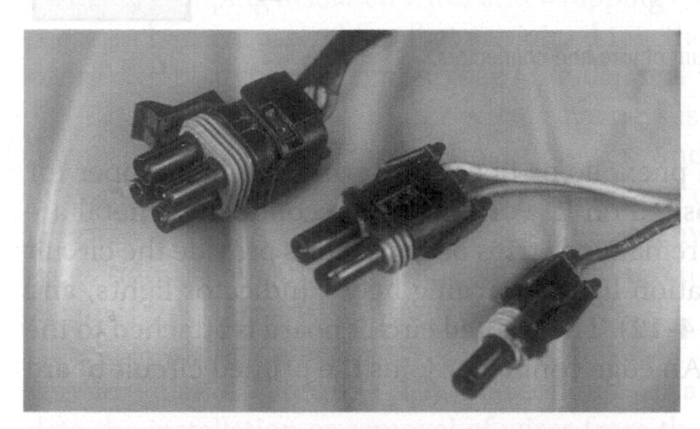

Figure 4-9 Weather-pack connector is used to prevent connector corrosion.

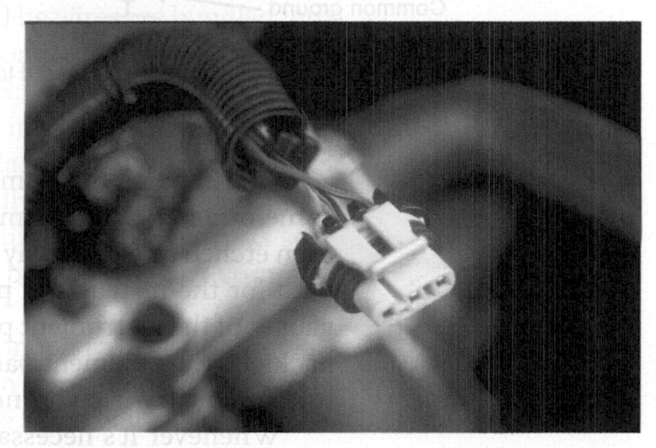

Figure 4-10 Metri-pack connector.

4. **Weather-pack connectors:** These connectors have rubber seals on the terminal ends and on the covers of the connector half (**Figure 4-9**). They are used on computer circuits to protect the circuit from corrosion, which may result in a voltage drop.
5. **Metri-pack connectors:** These are like the weather-pack connectors but do not have the seal on the cover half (**Figure 4-10**).
6. **Heat shrink–covered butt connectors:** Recommended for air bag applications by some manufacturers. Other manufacturers allow NO repairs to the circuitry, while still others require silver-soldered connections.

Shop Manual
Chapter 4, page 181

Shop Manual
Chapter 4, page 183

To reduce the number of connectors in the electrical system, a **common connection** can be used (**Figure 4-11**). Common connections are used to share a source of power or a common ground and are often called a **splice**. If there are several electrical components that are physically close to each other, a single common connection (splice) eliminates using a separate connector for each wire.

Printed Circuits

Printed circuit boards are used to simplify the wiring of the circuits they operate. Other uses of printed circuit boards include the inside of radios, computers, and some voltage regulators. Most instrument panels use printed circuit boards as circuit conductors.

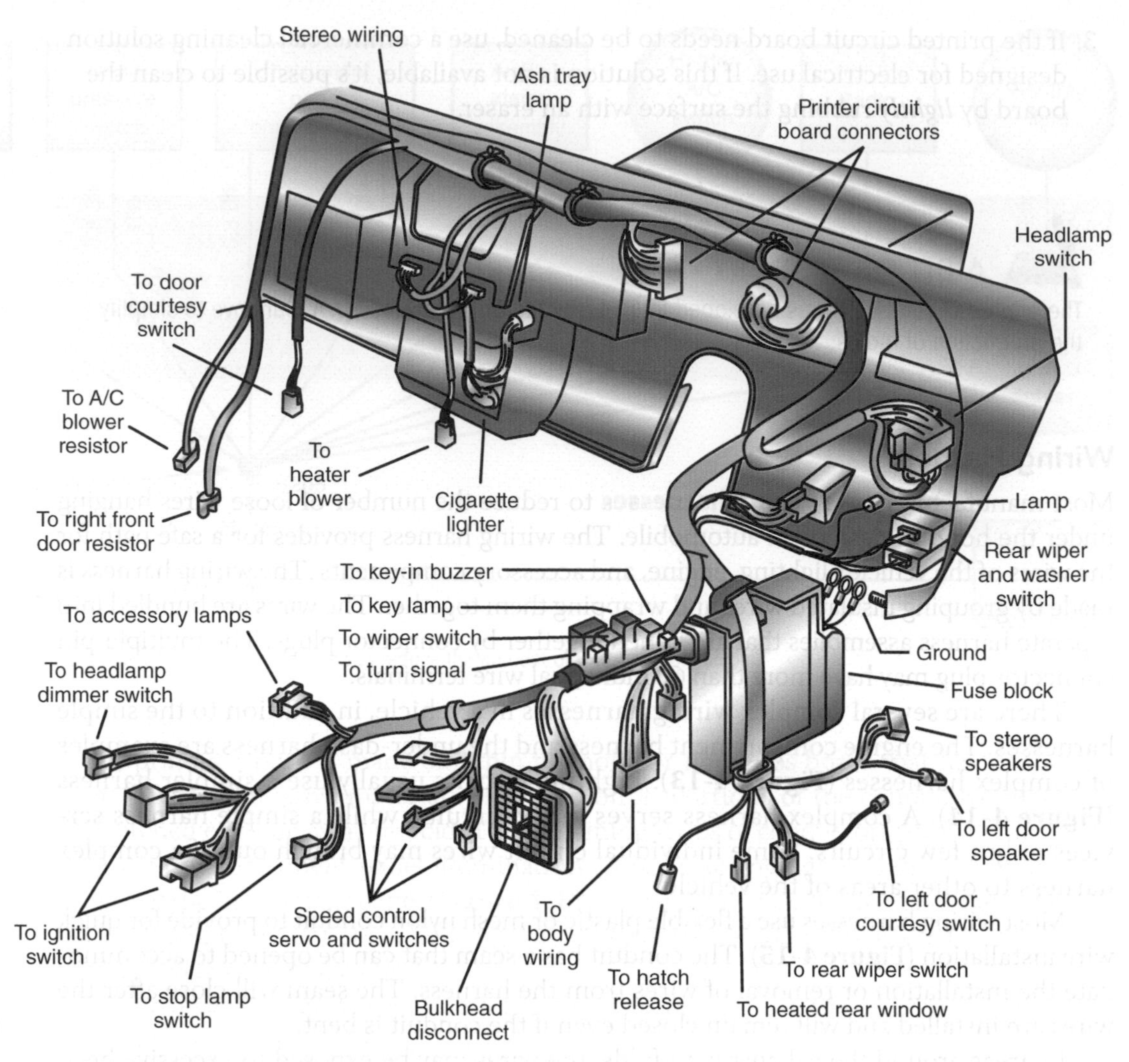

Figure 4-13 Complex wiring harness.

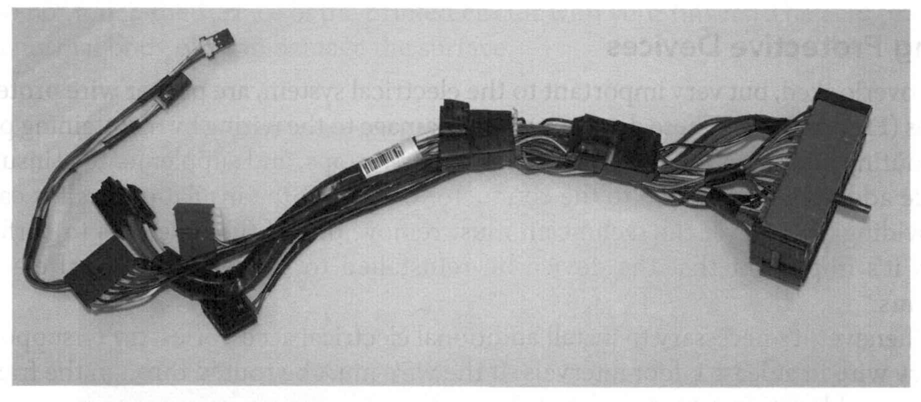

Figure 4-14 Simple wiring harness.

circuit on a vehicle. They also show where different circuits are interconnected, where they receive their power, where the ground is located, and the colors of the different wires. All of this information is critical to proper diagnosis of electrical problems. Some wiring diagrams also give additional information that helps you understand how a circuit

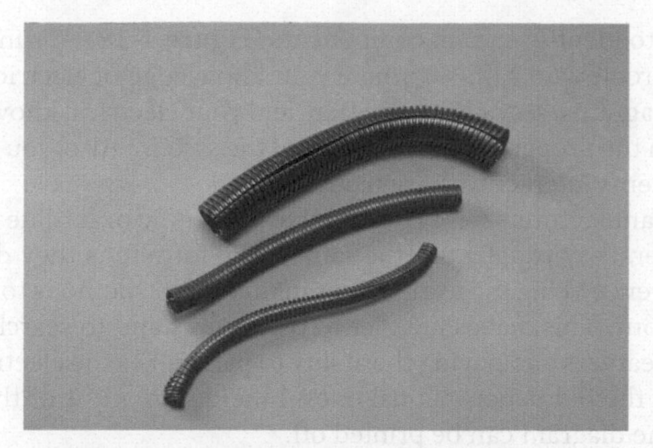

Figure 4-15 Flexible conduit used to make wiring harnesses.

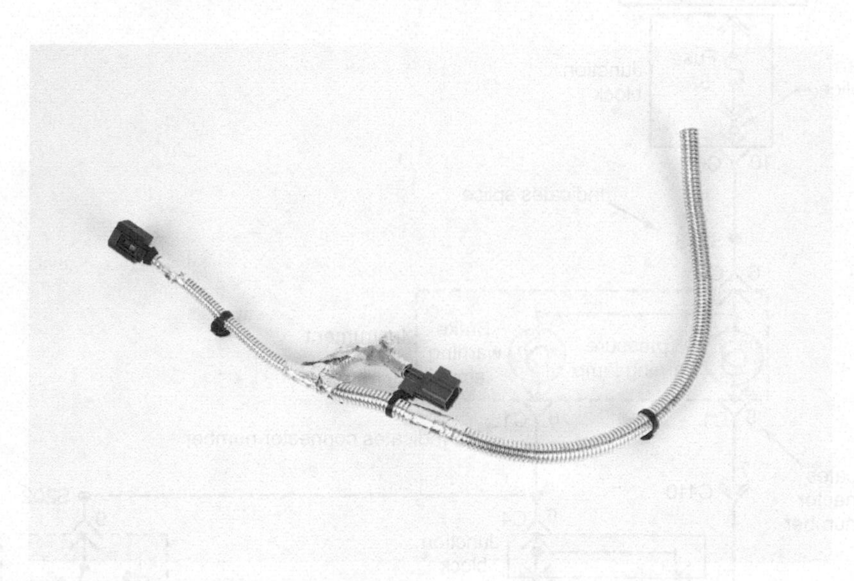

Figure 4-16 Special heat reflective conduit.

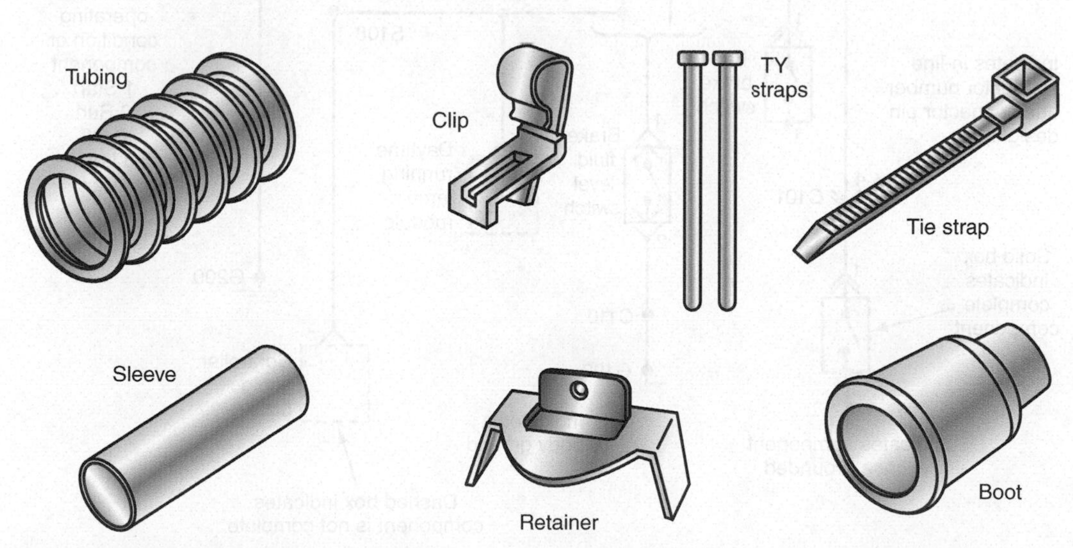

Figure 4-17 Common wire protection devices.

operates and how to identify certain components (**Figure 4-18**). Wiring diagrams do not explain how the circuit works; this is where your knowledge of electricity is required. By using the wiring diagrams, service information, and your electrical knowledge, you should be able to ascertain the proper operating design of the system. Also, you need to determine how electrical systems interact with each other.

Today most manufacturers have converted to electronic or on-line formats for wiring diagrams. However, they usually use the same basic format as they did for their paper manuals, the difference being the use of hyperlinks or other methods to navigate throughout the circuit layout. This makes it easier to trace wires and to search for components. Zoom in and out features and having the ability to pan makes the electronic format more user friendly then the old paper manual ones. In addition, most of the diagrams are in color. If needed, the diagram can be printed off.

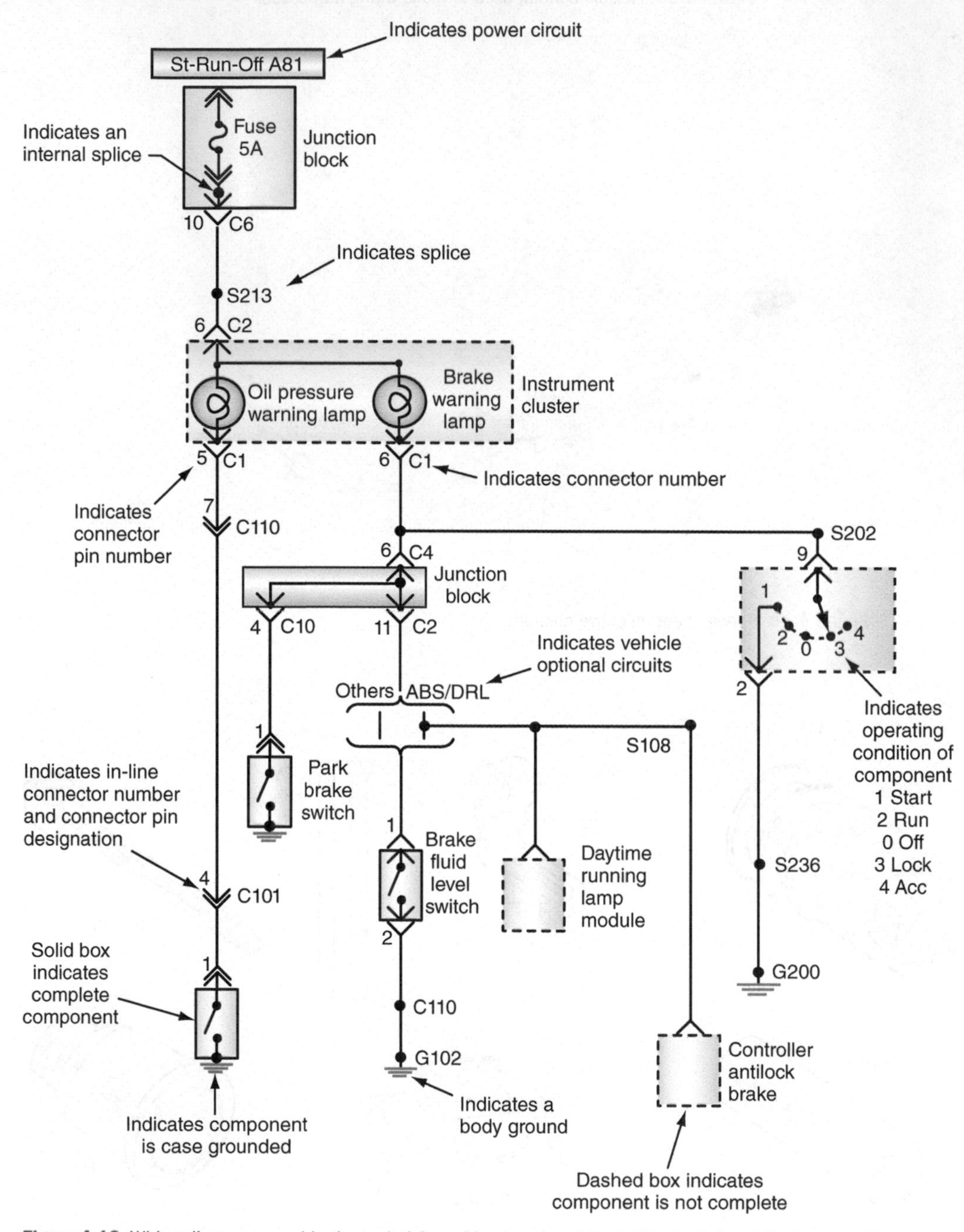

Figure 4-18 Wiring diagrams provide the technician with necessary information to accurately diagnose the electrical systems.

A wiring diagram can show the wiring of the entire vehicle or a single system (**Figure 4-19**). Wiring diagrams of the entire vehicle tend to look more complex and threatening than system diagrams. However, once you simplify the diagram to only those wires, connectors, and components that belong to an individual circuit, they become less complex and more valuable. A system wiring diagram is actually a portion of the total vehicle diagram. The system and all related circuitry is shown on just a few screens. System diagrams are often easier to use than vehicle diagrams simply because there is less information to sort through.

Remember that electrical circuits need a complete path for current to flow. A wiring diagram shows the insulated side of the circuit and the point of ground. Also, when lines (or wires) cross on a wiring diagram, this does not mean they connect. If wires are

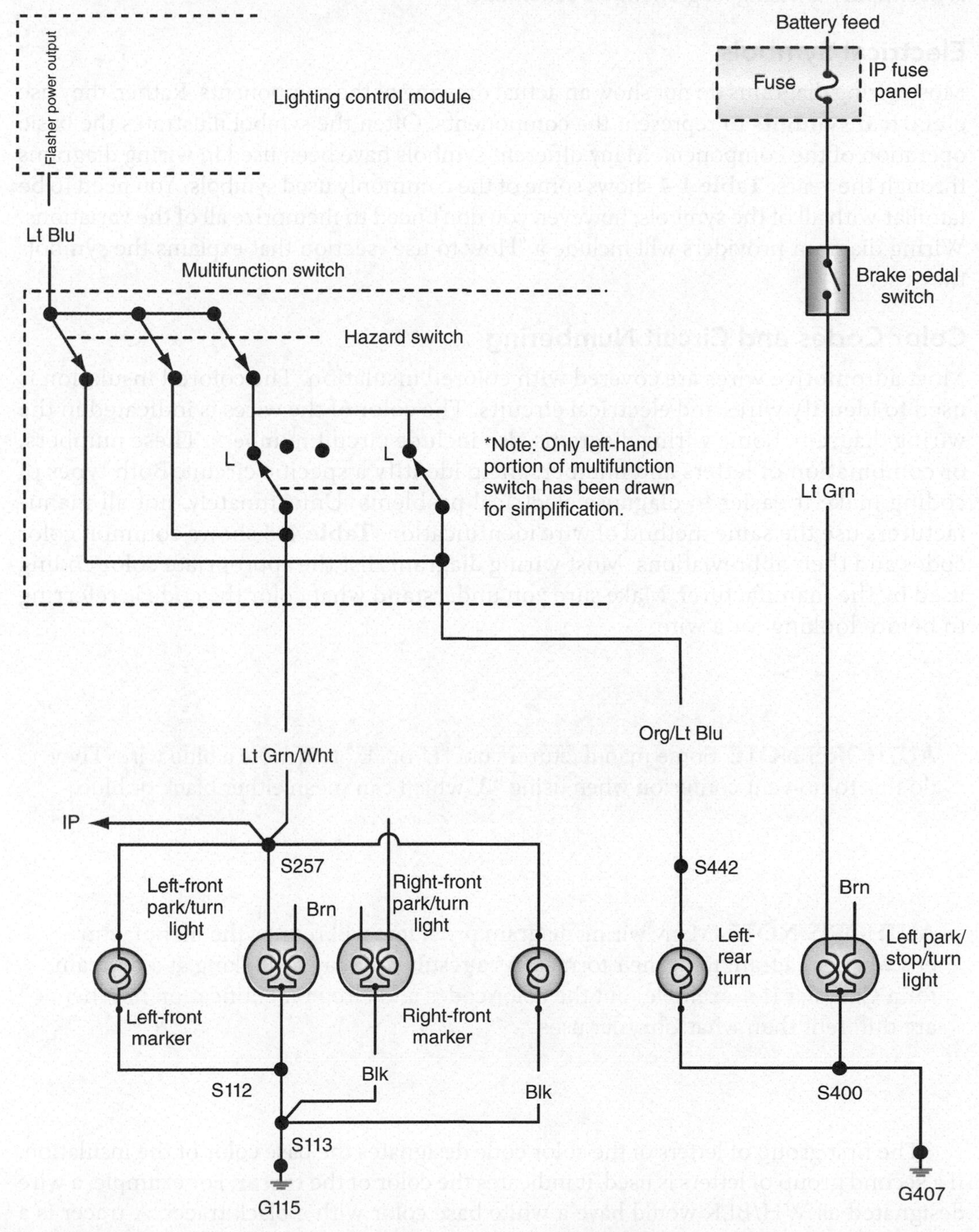

Figure 4-19 Wiring diagram illustrating only one specific system for easier reference.

connected, there will be a connector or a dot at the point where they cross. Most wiring diagrams do not show the location of the wires, connectors, or components in the vehicle. Some have location reference numbers displayed by the wires. After studying the wiring diagram, you will know what you are looking for. Then you move to the vehicle to find it.

In addition to entire vehicle and system-specific wiring diagrams, there are other diagrams that may be used to diagnose electricity problems. An electrical **schematic** shows how the circuit is connected. It does not show the colors of the wires or their routing. Schematics are what have been used so far in this book. They display a working model of the circuit. These are especially handy when trying to understand how a circuit works. Schematics are typically used to show the internal circuitry of a component or to simplify a wiring diagram. One of the troubleshooting techniques used by good electrical technicians is to simplify a wiring diagram into a schematic.

Electrical Symbols

Most wiring diagrams do not show an actual drawing of the components. Rather, they use **electrical symbols** to represent the components. Often the symbol illustrates the basic operation of the component. Many different symbols have been used in wiring diagrams through the years. **Table 4-4** shows some of the commonly used symbols. You need to be familiar with all of the symbols; however, you don't need to memorize all of the variations. Wiring diagram providers will include a "How to use" section that explains the symbols they use.

Color Codes and Circuit Numbering

Most automotive wires are covered with colored insulation. The colored insulation is used to identify wires and electrical circuits. The color of the wires is indicated in the wiring diagram. Some wiring diagrams also include circuit numbers. These numbers, or combination of letters and numbers, help identify a specific circuit. Both types of coding make it easier to diagnose electrical problems. Unfortunately, not all manufacturers use the same method of wire identification. **Table 4-5** shows common color codes and their abbreviations. Most wiring diagrams list the appropriate color coding used by the manufacturer. Make sure you understand what color the code is referring to before looking for a wire.

AUTHOR'S NOTE Some manufacturers use "L" or "U" to denote a blue wire. They do this to prevent confusion when using "B" which can mean either black or blue.

AUTHOR'S NOTE Many wiring diagram providers will redraw the manufacturer's wiring diagram to fit their format. As a result you may be looking at a diagram for a Chrysler (for example) but the color codes and circuit identification numbers are different than what Chrysler uses.

The first group of letters of the color code designates the base color of the insulation. If a second group of letters is used, it indicates the color of the **tracer**. For example, a wire designated as WH/BLK would have a white base color with a black tracer. A tracer is a thin or dashed line of a different color than the base color of the insulation.

Table 4-4 Common electrical and electronic symbols used in wiring diagrams.

COMPONENT	SYMBOL
Ammeter	
AND Gate	
Antenna	
Battery	
Capacitor	
Capacitor, Variable	
Cell	
Circuit Breaker/PTC device	
Clockspring	
Coaxial Cable	
Coil	
Crystal, Piezoelectric	
Diode	
Diode, Gunn	
Diode, Light-Emitting	
Diode, Photosensitive	
Diode, Photovoltaic	
Diode, Zener	
Dual Filament Lamp	
Female Contact	
Fuse	
Fusible link	
Gauge	
Ground, Chassis	
Ground, Earth	
Heating element	
Hot Bar	BATT AO
Inductor, Air-Core	
Inductor, Iron-Core	
In-Line Connectors	2 C123 2 C123
Integrated Circuit	
Inverter	
Jack, Coaxial	
Jack, Phono	
Lamp, Neon	
Male Contact	
Microphone	
Motor, One speed	
Motor, Reversible	
Motor, two Speed	
Multiple connectors	8 5 2 C123
Nand Gate	
Negative Voltage Connection	
NOR Gate	
Operational Amplifier	
OR Gate	
Outlet, Utility, 117-V	
Oxygen Sensor	
Page Reference	(BW-30-10)
Piezoelectric Cell	
Photocell, Tube	
Plug, Utility, 117-V	
Positive Voltage Connection	
Potentiometer	
Probe, Radio-Frequency	
Rectifier, Semiconductor	
Relay, DPDT	
Relay, DPST	
Relay, SPDT	
Relay, SPST	
Resistor	
Resonator	
Rheostat, Variable Resistor, Thermistor	
Shielding	-----

Table 4-4 *(continued)*

COMPONENT	SYMBOL
Signal Generator	
Single Filament Lamp	
Sliding Door Contact	
Solenoid	
Solenoid Valve	
Speaker	
Splice, External	S350
Splice, Internal	
Splice, Internal (Incompleted)	
Switch, Closed	
Switch, DPDT	
Switch, DPST	
Switch, Ganged	
Switch, Momentary-Contact	
Switch, Open	
Switch, Resistive Multiplex	
Switch, Rotary	
Switch, SPDT	
Switch, SPST	
Terminals	
Test Point	
Thermocouple	
Thyristor	
Tone Generator	
Transformer, Air-Core	
Transformer, Iron-Core	
Transformer, Tapped Primary	
Transformer, Tapped Secondary	
Transistor, Bipolar, npn	
Transistor, Bipolar, pnp	
Transistor, Field-Effect, N-Channel	
Transistor, Field-Effect, P-Channel	
Transistor, Metal-Oxide, Dual-Gate	
Transistor, Metal-Oxide, Single-Gate	
Transistor, Photosensitive	
Transistor, Unijunction	
Tube, Diode	
Tube, Pentode	
Tube, Photomultiplier	
Tube, Tetrode	
Tube, Triode	
Unspecified Component	
Voltmeter	V
Wattmeter	P
Wire Destination in Another Cell	
Wire Origin & Destination within Cell	
Wires	
Wires, Connected, Crossing	
Wires, Not Connected, Crossing	

Table 4-5 Common color codes used in automotive applications.

Color		Abbreviations	
Black	BLK	BK	B
Blue (Dark)	BLU DK	DB	DK BLU
Blue	BLU	B	L
Blue (Light)	BLU LT	LB	LT BLU
Brown	BRN	BR	BN
Glazed	GLZ	GL	
Gray	GR A	GR	G
Green (Dark)	GRN DK	DG	DK GRN
Green (Light)	GRN LT	LG	LT GRN
Maroon	MAR	M	
Natural	NAT	N	
Orange	ORN	O	ORG
Pink	PNK	PK	P
Purple	PPL	PR	
Red	RED	R	RD
Tan	TAN	T	TN
Violet	VLT	V	
White	WHT	W	WH
Yellow	YEL	Y	YL

Ford uses four methods of color coding its wires (**Figure 4-20**):

1. Solid color.
2. Base color with a stripe (tracer).
3. Base color with hash marks.
4. Base color with dots.

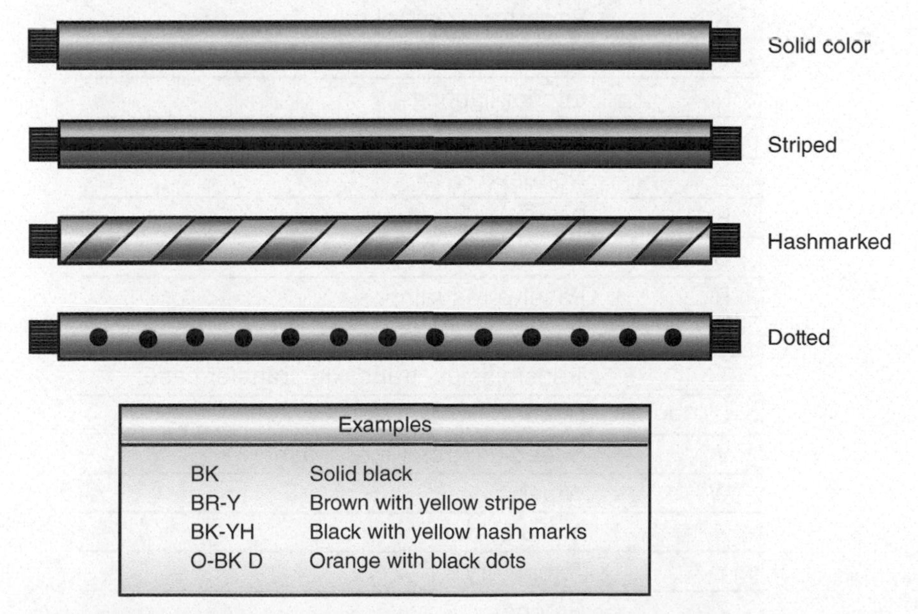

Figure 4-20 Four methods that Ford uses to color code circuit wires.

Chrysler uses a numbering method to designate the circuits on the wiring diagram (**Figure 4-21**). The circuit identification, wire gauge, and color of the wire are included in the wire number. A main circuit identification code that corresponds to the first letter in the wire number identifies the main circuits (**Table 4-6**).

General Motors uses numbers that include the wire gauge in metric millimeters, the wire color, circuit number, splice number, and ground identification (**Figure 4-22**). In this

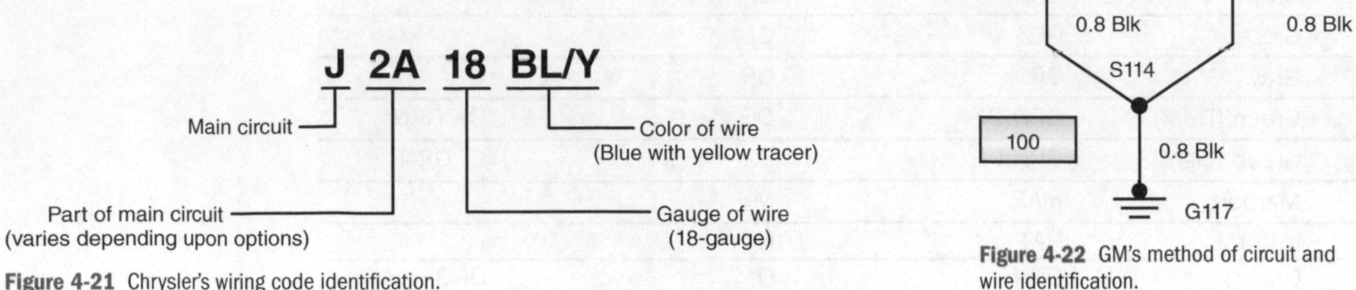

Figure 4-21 Chrysler's wiring code identification.

Figure 4-22 GM's method of circuit and wire identification.

Table 4-6 Chrysler's circuit identification codes.

Circuit Identification Code Chart	
Circuit	**Function**
A	Battery feed
B	Brake controls
C	Climate controls
D	Diagnostic circuits
E	Dimming illumination circuits
F	Fused circuits
G	Monitoring circuits (gauges)
H	Multiple
I	Not used
J	Open
K	Powertrain control module
L	Exterior lighting
M	Interior lighting
N	Multiple
O	Not used
P	Power option (battery feed)
Q	Power options (ignition feed)
R	Passive restraint
S	Suspension/steering
T	Transmission/transaxle/transfer case
U	Open
V	Speed control, wiper/washer
W	Wipers
X	Audio systems
Y	Temporary
Z	Grounds

example, the circuit is designated as 100, the wire size is 0.8 mm^2, the insulation color is black, the splice is numbered S114, and the ground is designated as G117.

Most manufacturers also number connectors, terminals, splices, and grounds for identification. The numbers correspond to their general location within the vehicle. The following is typical identification numbers:

100–199	Engine compartment forward of the dash panel
200–299	Instrument panel area
300–399	Passenger compartment
400–499	Deck area
500–599	Left front door
600–699	Right front door
700–799	Left rear door
800–899	Right rear door
900–999	Deck lid or hatch

Import manufacturers use different ways of illustrating their wiring diagrams. As an example of reading import wiring diagrams, consider the system that Toyota has used. One way Toyota differs from the methods we have discussed is that they rely heavily on ID numbers. When looking at a relay or junction block component, the ID number will be located in an oval next to the terminal identifiers (**Figure 4-23**). These ID numbers indicate which relay or junction block the component is located in.

Components and parts not part of the junction or relay block also use ID numbers made up of letters and numbers (**Figure 4-24**). Usually the letter matches the first letter of the component's name. In this example, the component ID number is "I1." The "I" is the first letter of idle air control valve. The ID number crosses over to the connector diagram and to the parts location table.

The ID numbers used for in-line connectors is different than that used for components. The connector ID number is located within the illustration of the connector (**Figure 4-25**). Also, notice the splice point identifier. The letter used for in-line connectors and splices is the same. The ID numbers will begin with "E" for Engine, "I" for Instrument cluster, and "B" for Body. These ID numbers correspond to how you use the location table.

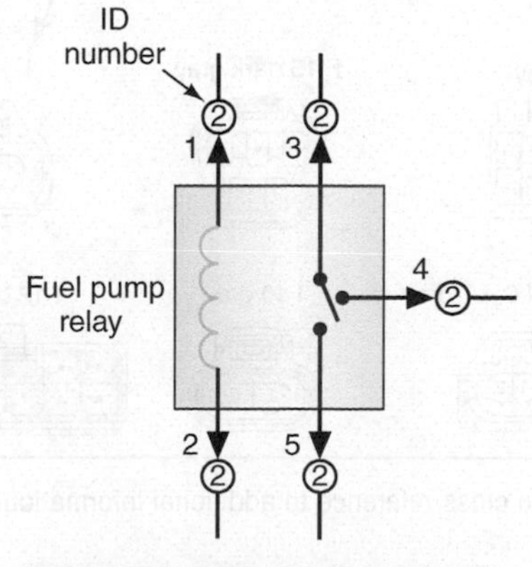

Figure 4-23 The ID number used with junction and relay block components identifies which block it's located in.

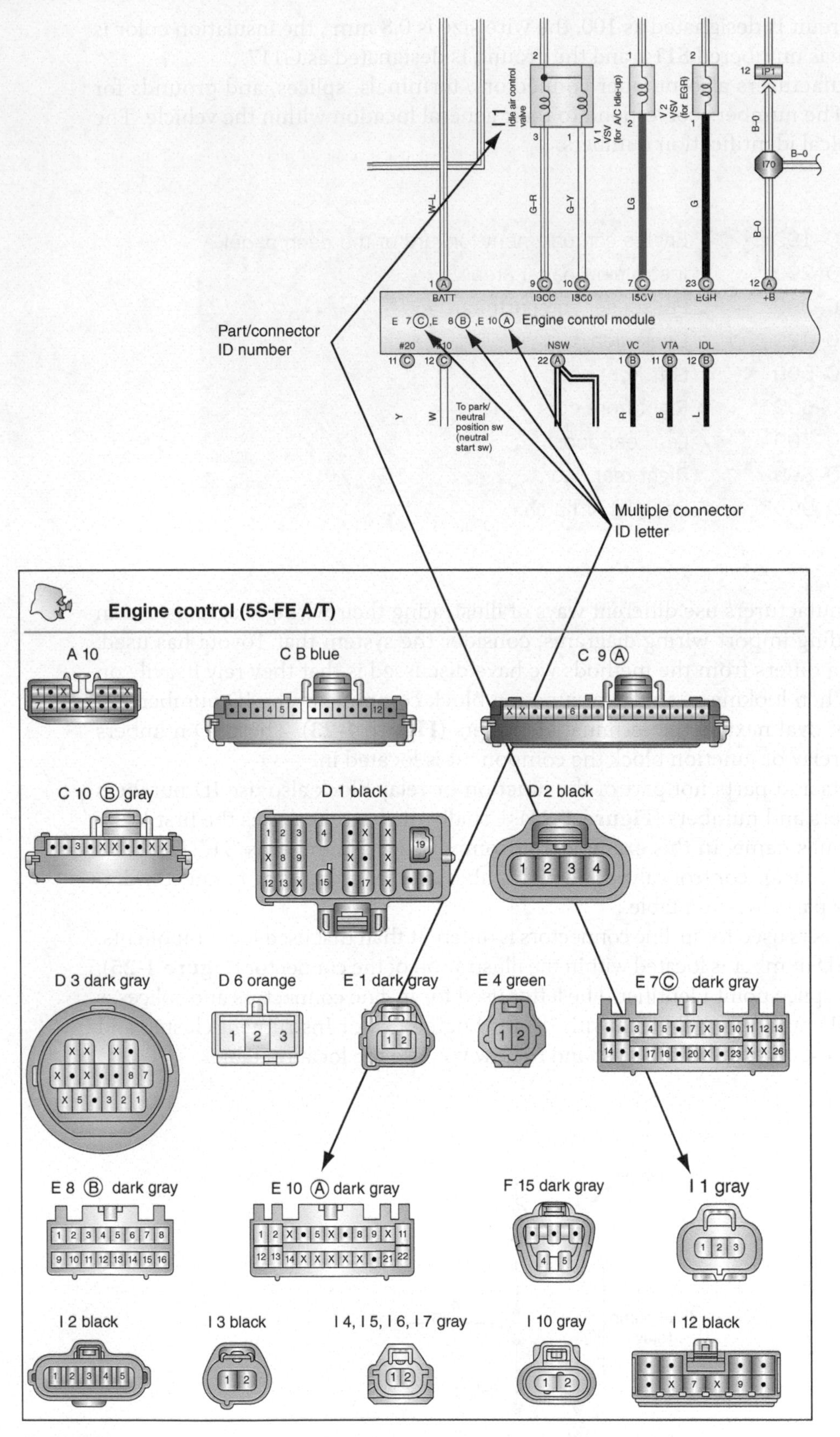

Figure 4-24 The component ID number is used to cross-reference to additional information.

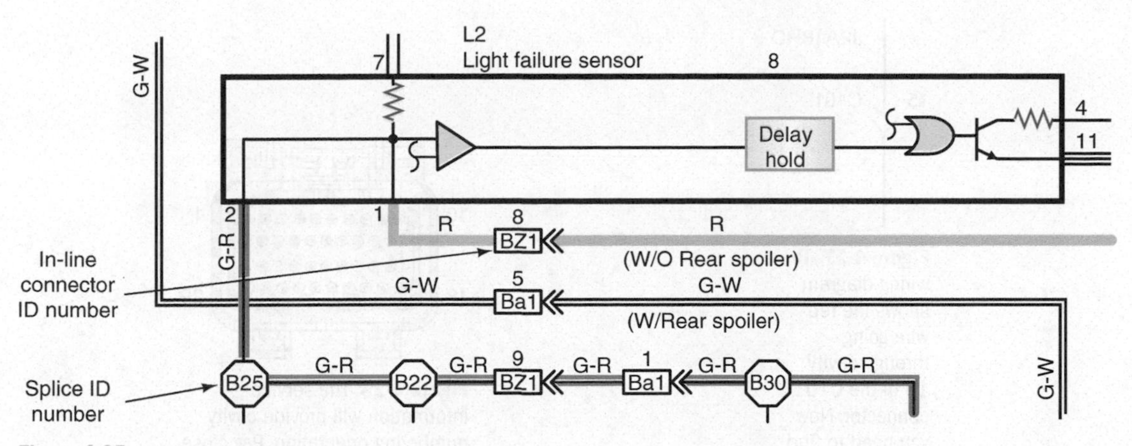

Figure 4-25 In-line connector and splice ID numbers.

An inverted triangle with a two-letter identifier are used to identify grounds (**Figure 4-26**). Like the in-line connector and splice identifiers, the first letter identifies which harness the ground is located in. Again, the ground ID is used to refer to the location chart to determine where it's located on the vehicle.

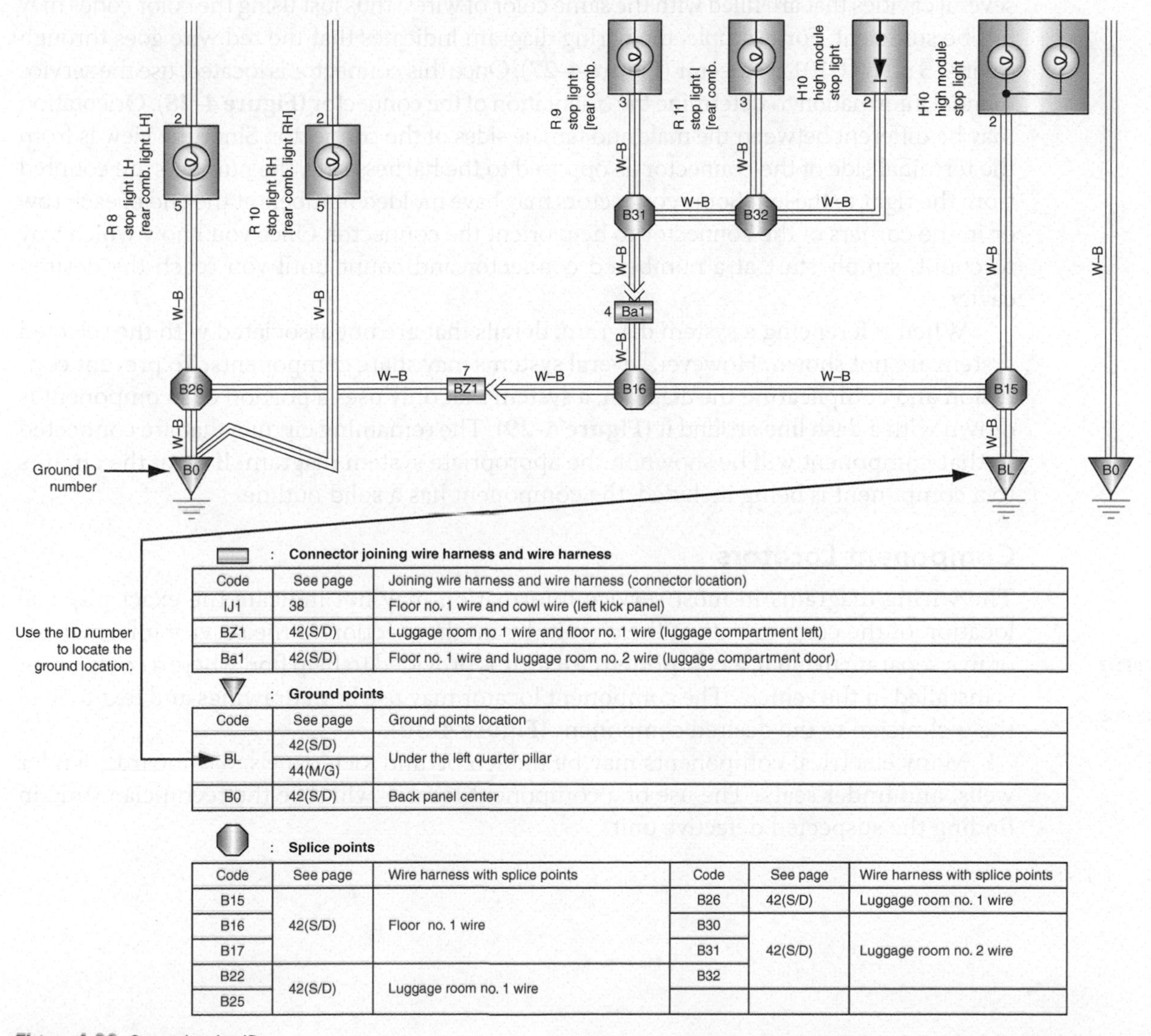

: **Connector joining wire harness and wire harness**

Code	See page	Joining wire harness and wire harness (connector location)
IJ1	38	Floor no. 1 wire and cowl wire (left kick panel)
BZ1	42(S/D)	Luggage room no 1 wire and floor no. 1 wire (luggage compartment left)
Ba1	42(S/D)	Floor no. 1 wire and luggage room no. 2 wire (luggage compartment door)

: **Ground points**

Code	See page	Ground points location
BL	42(S/D) 44(M/G)	Under the left quarter pillar
B0	42(S/D)	Back panel center

: **Splice points**

Code	See page	Wire harness with splice points	Code	See page	Wire harness with splice points
B15	42(S/D)	Floor no. 1 wire	B26	42(S/D)	Luggage room no. 1 wire
B16			B30	42(S/D)	Luggage room no. 2 wire
B17			B31		
B22	42(S/D)	Luggage room no. 1 wire	B32		
B25					

Figure 4-26 Ground point ID.

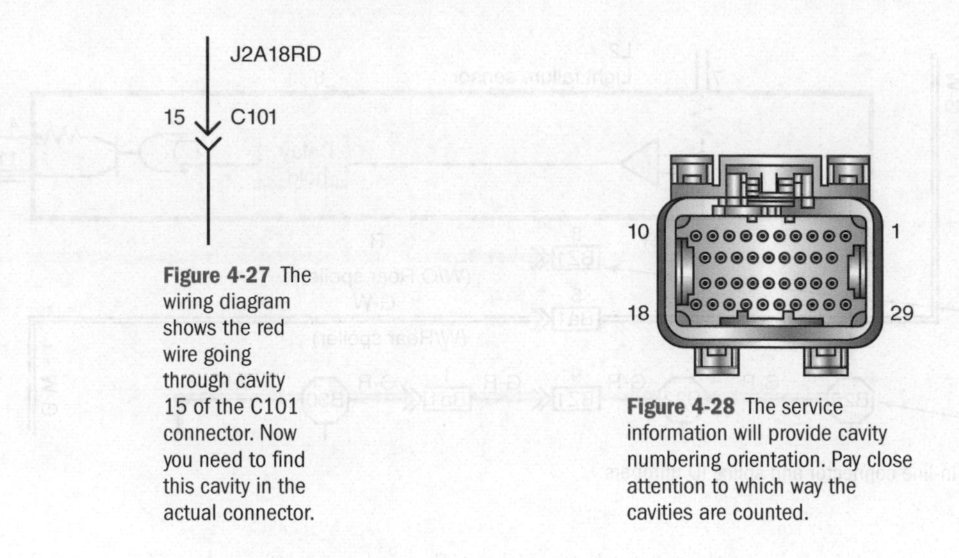

Figure 4-27 The wiring diagram shows the red wire going through cavity 15 of the C101 connector. Now you need to find this cavity in the actual connector.

Figure 4-28 The service information will provide cavity numbering orientation. Pay close attention to which way the cavities are counted.

When the wires pass through a bulk connector, you must properly identify the correct wire for the circuit you are diagnosing. Sometimes on larger bulk connectors, there may be several cavities that are filled with the same color of wires; thus just using the color codes may not be sufficient. For example, the wiring diagram indicates that the red wire goes through cavity 15 of the C101 connector (**Figure 4-27**). Once this connector is located, use the service manual information to determine the orientation of the connector (**Figure 4-28**). Orientation may be different between the male and female sides of the connector. Since the view is from the terminal side of the connector as opposed to the harness side, the numbers are counted from the right to the left. Some connectors may have molded numbers at the end of each row or in the corners of the connector to help orient the connector. Once you know which way to count, simply start at a numbered connector and count until you reach the desired cavity.

When referencing a system diagram, details that are not associated with the selected system are not shown. However, several systems may share components. To prevent confusion and complicating the diagram, a system that only uses a portion of a component is drawn with a dash line around it **(Figure 4-29)**. The remaining circuits that are connected to that component will be shown in the appropriate system diagram. If all of the circuits to a component is being included, the component has a solid outline.

Component Locators

Component locators are also called installation diagrams.

The wiring diagrams in most service information may not indicate the exact physical location of the components of the circuit. In another section of the service information, or in a separate manual, a **component locator** is provided to help find where a component is installed in the vehicle. The component locator may use both drawings and text to lead the technician to the desired component (**Figure 4-30**).

Many electrical components may be hidden behind kick panels, dashboards, fender wells, and under seats. The use of a component locator will save the technician time in finding the suspected defective unit.

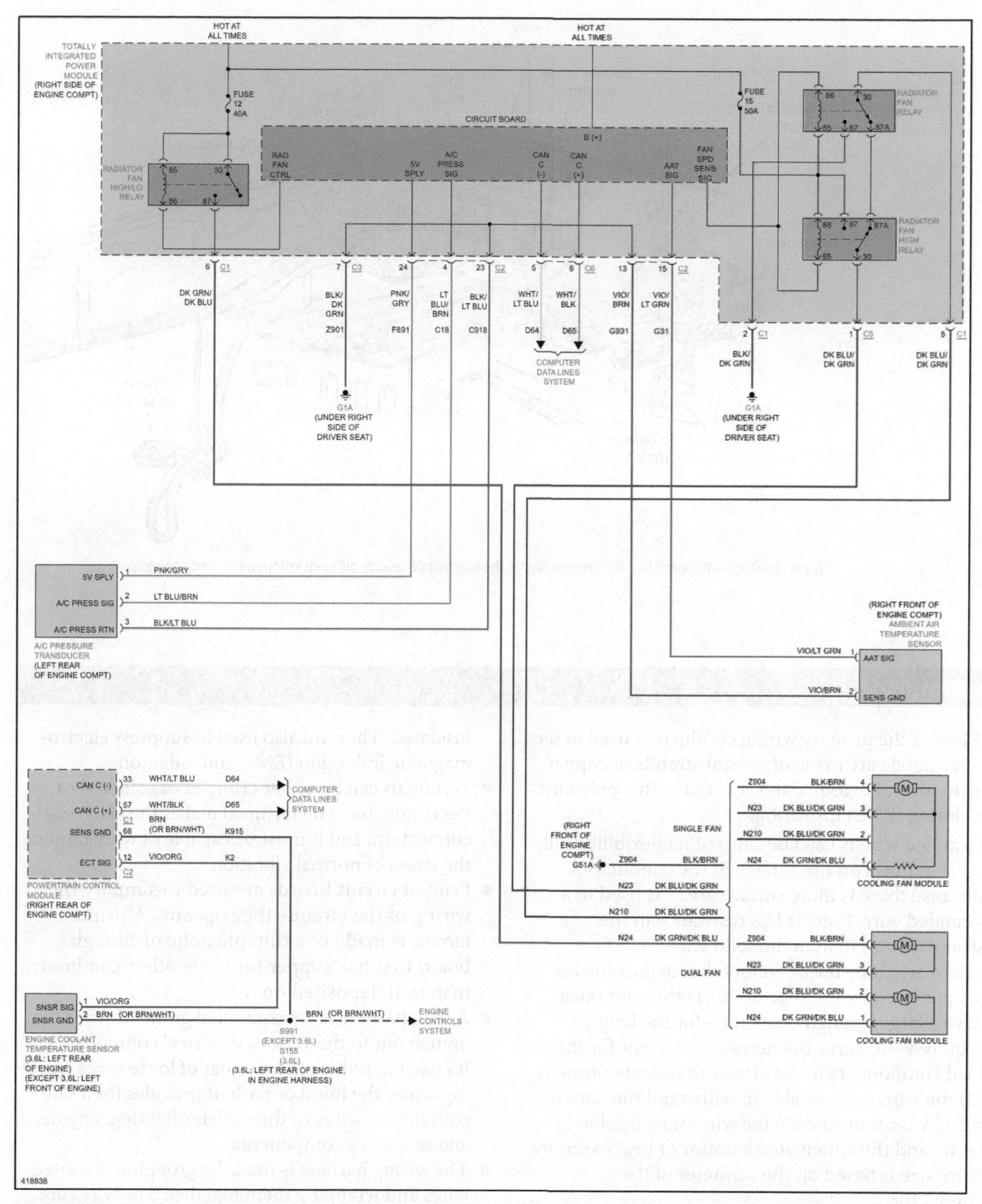

Figure 4-29 In this screenshot, a component drawn with a dashed line around it means only part of the component is being shown. A solid outline means the entire component is being shown. Courtesy of ProDemand.

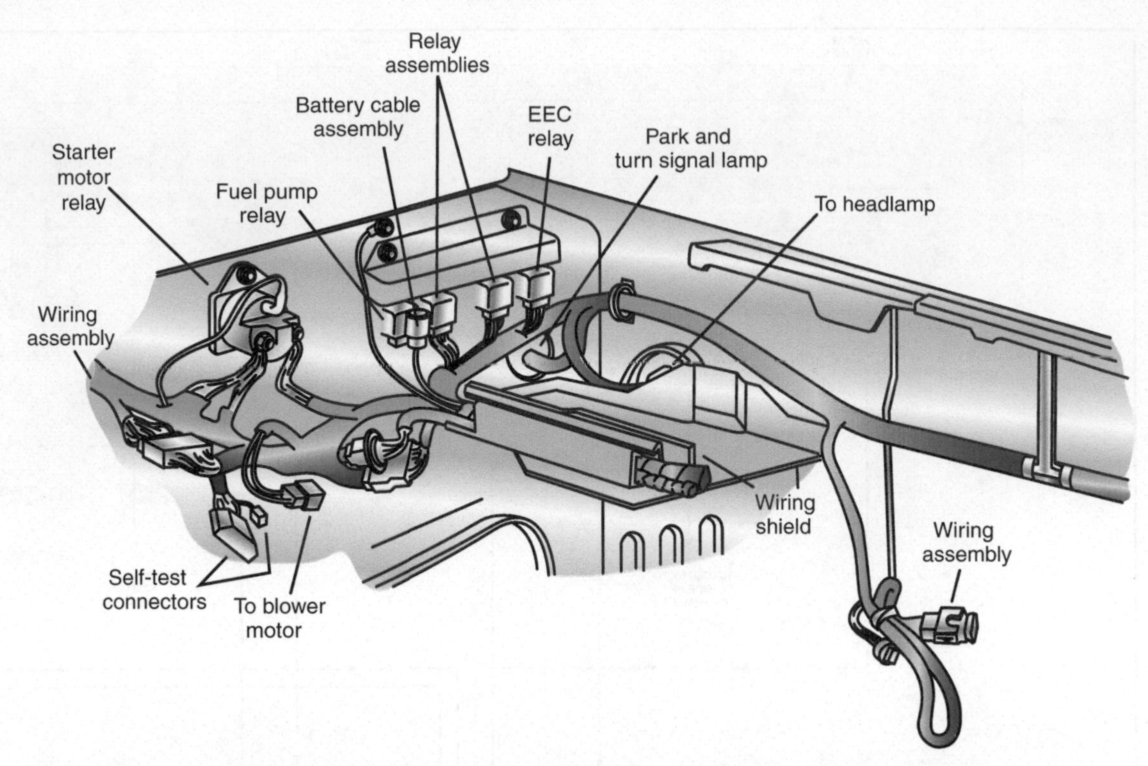

Figure 4-30 A component locator diagram shows the location of several different components and connectors.

Summary

- Most of the primary wiring conductors used in the automobile are made of several strands of copper wire wound together and covered with a polyvinyl chloride (PVC) insulation.
- Stranded wire is used because of its flexibility and current flows on the surface of the conductors. Because there is more surface area exposed in a stranded wire, there is less resistance in the stranded wire than in the solid wire.
- There are three major factors that determine the proper size of wire to be used: (1) the wire must have a large-enough diameter—for the length required—to carry the necessary current for the load components in the circuit to operate properly; (2) the wire must be able to withstand the anticipated vibration; and (3) the wire must be able to withstand the anticipated amount of heat exposure.
- Wire size is based on the diameter of the conductor.
- Factors that affect the resistance of the wire include the conductor material, wire diameter, wire length, and temperature.
- Ground straps are used to complete the return path to the battery between components that are insulated. They are also used to suppress electromagnetic induction (EMI) and radiation.
- Terminals can be either crimped or soldered to the conductor. The terminal makes the electrical connection, and it must be capable of withstanding the stress of normal vibration.
- Printed circuit boards are used to simplify the wiring of the circuits they operate. A printed circuit is made of a thin phenolic or fiberglass board that has copper (or some other conductive material) deposited on it.
- A wire harness is an assembled group of wires that branch out to the various electrical components. It's used to reduce the number of loose wires hanging under the hood or dash. It provides for a safe path for the wires of the vehicle's lighting, engine, and accessory components.
- The wiring harness is made by grouping insulated wires and wrapping them together. The wires are bundled into separate harness assemblies that are joined together by connector plugs.
- A wiring diagram shows a representation of actual electrical or electronic components and the wiring of the vehicle's electrical systems.

- The technician's greatest helpmate in locating electrical problems is the wiring diagram. Correct use of the wiring diagram will reduce the amount of time a technician needs to spend tracing the wires in the vehicle.
- In place of actual pictures, a variety of electrical symbols are used to represent the components in the wiring diagram.
- Color codes and circuit numbers are used to make tracing wires easier.
- In most color codes, the first group of letters designates the base color of the insulation. If a second group of letters is used, it indicates the color of the tracer.
- A component locator is used to determine the exact location of several of the electrical components.

Review Questions

Short-Answer Essays

1. Explain the purpose of wiring diagrams.
2. Explain how wire size is determined by the American Wire Gauge (AWG) and metric methods.
3. Explain the purpose and use of printed circuits.
4. Explain the purpose of the component locator.
5. Explain when single-stranded or multistranded wire should be used.
6. Explain how temperature affects resistance and wire size selection.
7. List the three major factors that determine the proper size of wire to be used.
8. List and describe the different types of terminal connectors used in the automotive electrical system.
9. What is the difference between a complex and a simple wiring harness?
10. Describe the methods the three domestic automobile manufacturers use for wiring code identification.

Fill in the Blanks

1. There is __________ resistance in the stranded wire than in the solid wire.
2. __________ complete the return path to the battery between components that are insulated.
3. Wire size is based on the __________ of the conductor.
4. In the AWG standard, the __________ the number, the smaller the wire diameter.
5. An increase in conductor temperature creates an __________ in resistance.
6. __________ connectors are used on computer circuits to protect the circuit from corrosion.
7. __________ are used to prevent damage to the wiring by maintaining proper wire routing and retention.
8. A wiring diagram is an electrical schematic that shows a __________ of actual electrical or electronic components (by use of symbols) and the __________ of the vehicle's electrical systems.
9. In most color codes, the first group of letters designates the __________ of the insulation. The second group of letters indicates the color of the __________.
10. A __________ is used to determine the exact location of several of the electrical components.

Multiple Choice

1. Automotive wiring is being discussed.
   Technician A says most primary wiring is made of several strands of copper wire wound together and covered with insulation.
   Technician B says the types of conductor materials used in automobiles include copper, silver, gold, aluminum, brass, and tin-plated brass.
   Who is correct?
   A. A only C. Both A and B
   B. B only D. Neither A nor B

2. Stranded wire use is being discussed.
   Technician A says there is less exposed surface area for electron flow in a stranded wire.
   Technician B says there is more resistance in the stranded wire than in the same gauge solid wire.
   Who is correct?
   A. A only
   B. B only
   C. Both A and B
   D. Neither A nor B

3. A Chrysler wiring diagram designation of M2 14 BK/YL identifies the main circuit as being:
   A. Climate control.
   B. Interior lighting.
   C. Wipers.
   D. None of these choices.

4. Ground straps are used to:
   A. Provide a return path to the battery between insulated components.
   B. Suppress electromagnetic induction.
   C. Both A and B.
   D. Neither A nor B.

5. The selection of the proper size of wire to be used is being discussed.
   Technician A says the wire must be large enough, for the length required, to carry the amount of current necessary for the load components in the circuit to operate properly.
   Technician B says temperature has little effect on resistance and it's not a factor in wire size selection.
   Who is correct?
   A. A only
   B. B only
   C. Both A and B
   D. Neither A nor B

6. Terminal connectors are being discussed.
   Technician A says good terminal connections will resist corrosion.
   Technician B says the terminals can be either crimped or soldered to the conductor.
   Who is correct?
   A. A only
   B. B only
   C. Both A and B
   D. Neither A nor B

7. Wire routing is being discussed.
   Technician A says to install additional electrical accessories that are necessary to support the primary wire in at least 10-foot intervals.
   Technician B says if the wire must be routed through the frame or body, use metal clips to protect the wire.
   Who is correct?
   A. A only
   B. B only
   C. Both A and B
   D. Neither A nor B

8. Printed circuit boards are being discussed.
   Technician A says printed circuit boards are used to simplify the wiring of the circuits they operate.
   Technician B says care must be taken not to touch the board with bare hands.
   Who is correct?
   A. A only
   B. B only
   C. Both A and B
   D. Neither A nor B

9. Wiring harnesses are being discussed.
   Technician A says a wire harness is an assembled group of wires that branches out to the various electrical components.
   Technician B says most underhood harnesses are simple harnesses.
   Who is correct?
   A. A only
   B. B only
   C. Both A and B
   D. Neither A nor B

10. Wiring diagrams are being discussed.
    Technician A says wiring diagrams give the exact location of the electrical components.
    Technician B says a wiring diagram will indicate what circuits are interconnected, where circuits receive their voltage source, and what color of wires are used in the circuit.
    Who is correct?
    A. A only
    B. B only
    C. Both A and B
    D. Neither A nor B

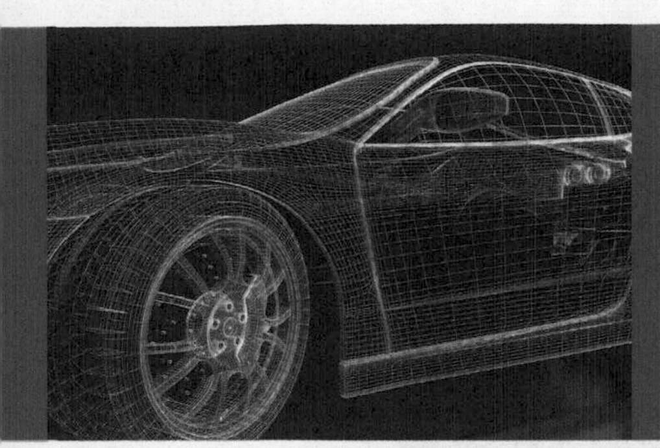

CHAPTER 5
AUTOMOTIVE BATTERIES

Upon completion and review of this chapter, you should be able to:

- Explain the purposes of a battery.
- Describe the construction of conventional, maintenance-free, hybrid, and recombination batteries.
- Explain the function of the main elements of a battery.
- Describe the chemical action that occurs to produce current in a battery.
- Describe the chemical reaction that occurs in a battery during cycling.
- Explain the differences, advantages, and disadvantages between battery types.
- Describe the different types of battery terminals used.
- Explain the meaning of common battery rating methods.
- Determine the correct battery to install into a vehicle.
- Explain the effects of temperature on battery performance.
- Describe the different loads or demands placed upon a battery during different operating conditions.
- Explain the primary reasons for battery failure.
- Define battery-related terms such as deep cycle, electrolyte solution, and gassing.
- Describe the function of EV and HEV high-voltage batteries.
- Describe the operation and purpose of ultra-capacitors.

Terms to Know

Absorbed glass mat (AGM) battery
Ampere-hour rating
Battery cables
Battery terminals
Cell element
Cold cranking amps (CCA)
Contactors
Cranking amps (CA)
Deep cycling
Electrochemical
Electrolyte
Energy density
Flooded
Gassing
Grid
Grid growth
Holddowns
Hybrid battery
Hydrometer
Maintenance-free battery
Material expanders
Memory effect
Radial grid
Recombination battery
Regenerative braking
Reserve capacity
Reserve-capacity rating
Specific gravity
Ultra-capacitors
Valve-regulated lead-acid (VRLA) battery

Introduction

Electrochemical refers to the chemical reaction of two dissimilar materials in a chemical solution that results in an electrical current. An automotive battery (**Figure 5-1**) is an electrochemical device capable of storing chemical energy for conversion to

Shop Manual
Chapter 5, page 218

Figure 5-1 Typical automotive 12-volt battery.

electrical energy. Connecting the battery to an external load, such as a starter motor, starts an energy conversion that results in an electrical current flowing through the circuit. The production of electrical energy in the battery is by the chemical reaction that occurs between two dissimilar plates immersed in an electrolyte solution. The automotive battery produces direct current (DC) electricity that flows in only one direction.

When discharging the battery (current flowing from the battery), the battery changes chemical energy into electrical energy. It's through this transformation that the battery releases stored energy. Charging the battery (current flowing through the battery from the charging system) reverses the process and converts electrical energy back into chemical energy. As a result, the battery can store energy until it's needed.

The automotive battery performs several important functions, including the following:

1. Powers the starting motor, ignition system, electronic fuel injection, and other electrical devices for the engine during cranking and starting.
2. Supplies all the electrical power for the vehicle accessories whenever the engine is not running or the charging system is not working.
3. Furnishes current for a limited time when electrical demands exceed charging system output.
4. Acts as a stabilizer of voltage for the entire automotive electrical system.
5. Stores energy for extended periods of time.

AUTHOR'S NOTE The battery does not store energy in electrical form. The battery stores energy in chemical form.

The largest demand placed on the battery occurs when it must supply current to operate the starter motor. The amperage requirements of a starter motor may be several hundred amperes. Ambient temperatures, engine size, and engine condition also affect this requirement.

After the engine starts, the vehicle's charging system works to recharge the battery and to provide the current to run the electrical systems. Alternating current (AC) generators can have a maximum output of 60 to 250 amps (A). This is usually enough to operate all of the vehicle's electrical systems and meet the demands of these systems. However, generator output is below its maximum rating under some conditions (such as the engine idling). If there are enough electrical accessories turned on during this time (heater, wipers, headlights, radio, etc.), the demand may exceed the AC generator output. During this time, the battery must supply the additional current.

Even with the ignition switch turned off, the battery must continue to provide electrical energy. Electrical loads that are present when the ignition switch is in the OFF position are called key-off or parasitic loads. Clocks, memory seats, engine computer memory, body computer memory, and electronic sound system memory are all examples of key-off loads. The total current draw of key-off loads is usually less than 50 mA.

Shop Manual
Chapter 5, pages 225, 233

If the vehicle's charging system fails; the battery must supply all of the current necessary to run the vehicle. The amount of time a battery can be discharged at a specific current rate until the voltage drops below a specified value is the battery's **reserve capacity**. Most batteries will supply a reserve capacity of 25 amps for approximately 120 minutes before discharging too low to keep the engine running.

The amount of electrical energy that a battery is capable of producing depends on the type, size, weight, active area of the plates, and the amount of sulfuric acid in the electrolyte solution.

The chemical molecular formula for sulfuric acid is H_2SO_4.

In this chapter, you will study the design and operation of different types of batteries currently used in automobiles. These include conventional batteries, maintenance-free batteries, hybrid batteries, recombination batteries, absorbed glass mat batteries, and valve-regulated batteries. In addition, there is a discussion of the high-voltage battery used in electric and electric hybrid vehicles.

Conventional Batteries

Seven basic components go into the construction of a conventional battery:

1. Positive plates.
2. Negative plates.
3. Separators.
4. Case.
5. Plate straps.
6. Electrolyte.
7. Terminals.

Test the condition of the battery when diagnosing any electrical problem. A poor performing battery affects the entire electrical system.

The difference between three-year and five-year batteries is the quantity of **material expanders** used to construct the plates and the number of plates used to build a cell. Material expanders are fillers used in place of the active materials. The use of expanders keep manufacturing costs low.

Lead dioxide is composed of small grains of particles. This gives the plate a high degree of porosity, allowing the electrolyte to penetrate the plate.

A plate, either positive or negative, starts with a **grid**. The grid is the frame structure with connector tabs at the top. The grid has horizontal and vertical grid bars that intersect at right angles (**Figure 5-2**). Generally, grids are made of lead alloys, usually antimony. The addition of about 5% to 6% antimony increases the grid's strength. A paste form of an active material made from ground lead oxide, acid, and material expanders is pressed into the grid. Applying a "forming charge" to the positive plate converts the lead oxide paste into lead dioxide. Applying a "forming charge" to the negative plate converts the paste into sponge lead.

Each **cell element** as alternately arranged negative and positive plates (**Figure 5-3**). Each cell element can consist of 9 to 13 plates. Separators made of microporous materials

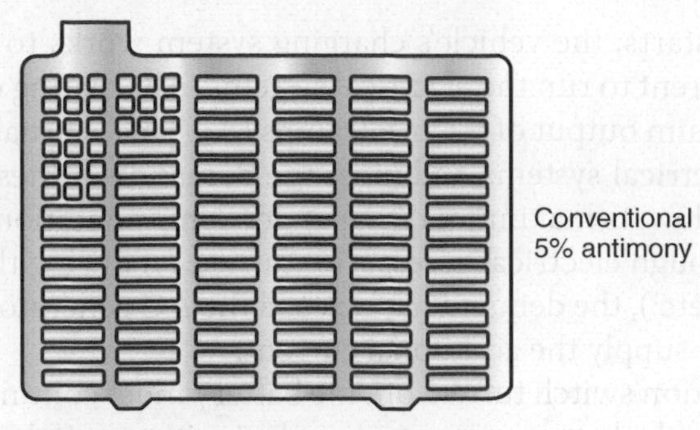

Figure 5-2 Conventional battery grid.

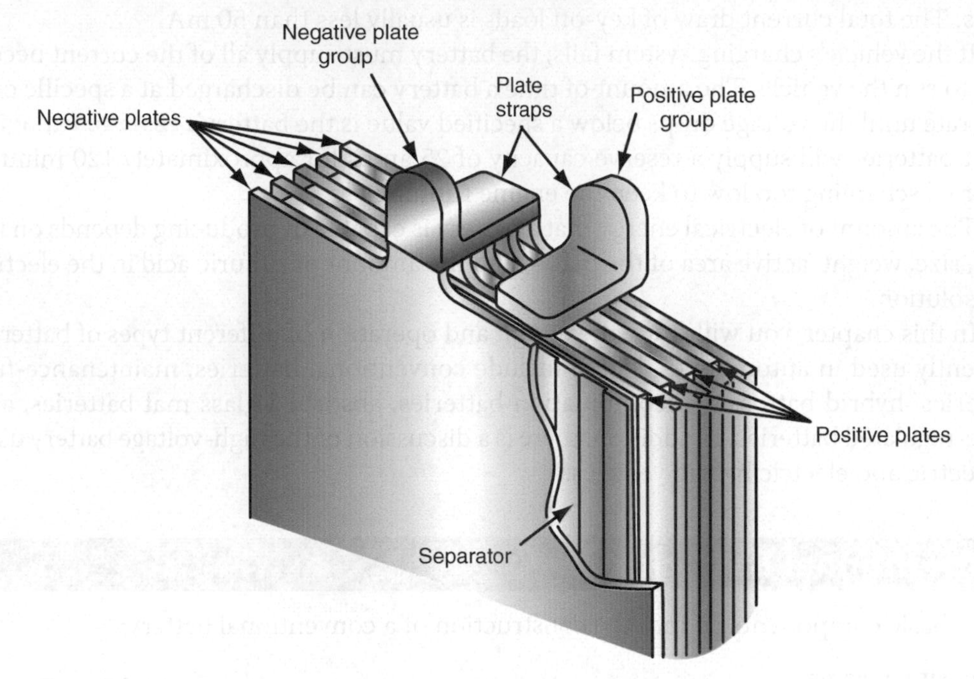

Figure 5-3 A battery cell consists of alternate positive and negative plates.

insulate the positive and negative plates from each other. The completion of the element's construction connects when all of the positive plates to each other and all of the negative plates to each other. The connection of the plates is by plate straps (**Figure 5-4**).

Usually, negative plate groups contain one more plate than positive plate groups to help equalize the chemical activity.

A typical 12-volt (V) automotive battery uses six cells connected in series with the positive side of a cell element connected to the negative side of the next cell element. (**Figure 5-5**). All six cells connect in this manner. By connecting the cells in series, the cell's current capacity and cell voltage remain the same. Each cell produces 2.1 volts; connecting them in series produces the 12.6 volts required by the automotive electrical system. The plate straps provide a positive cell connection and a negative cell connection. There are three cell connection types: over the partition, through the partition, or external (**Figure 5-6**).

Many batteries have envelope-type separators that retain active materials near the plates. The most common connection used to connect cell elements is through the partition. It provides the shortest path and the least resistance.

The cell elements are submerged in a cell case filled with an **electrolyte** solution of sulfuric acid diluted with water. The electrolyte solution used in automotive batteries consists of 64% water and 36% sulfuric acid by weight. Electrolyte is both conductive and reactive. Because the cells are completely submerged in liquid electrolyte, the lead-acid battery is also known as a **flooded battery.**

The battery case is made of polypropylene, hard rubber, and plastic base materials. The battery case must be capable of withstanding temperature extremes, vibration, and acid

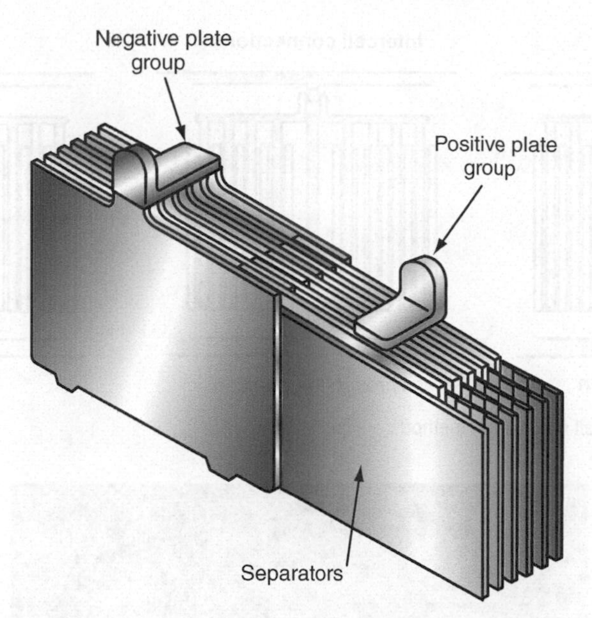

Figure 5-4 Construction of a battery element.

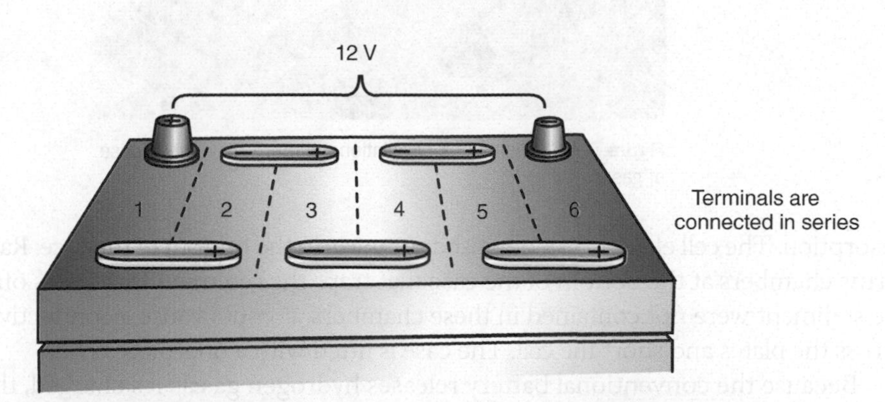

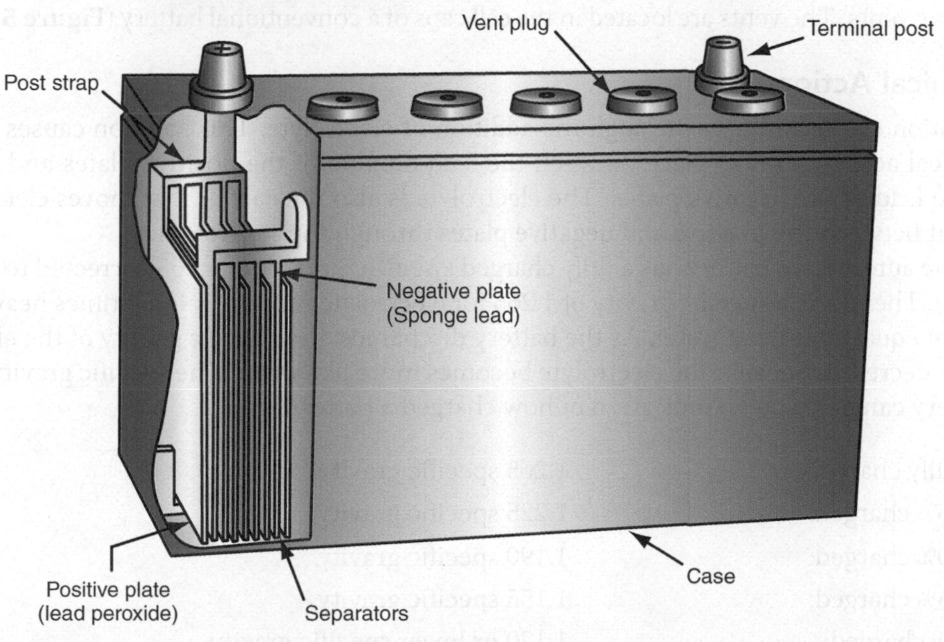

Figure 5-5 The 12-volt battery consists of six 2-volt cells connected in series.

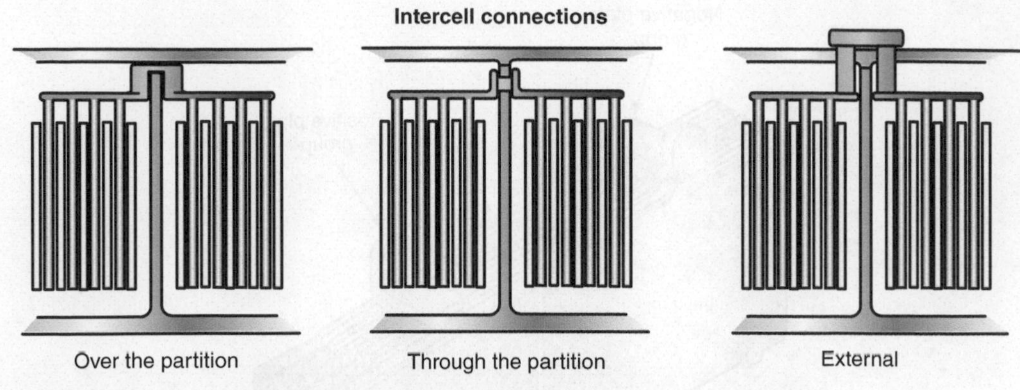

Figure 5-6 Three intercell connection methods.

Figure 5-7 The vents of a conventional battery allow the release of gases.

absorption. The cell elements sit on raised supports in the bottom of the case. Raising the cells forms chambers at the bottom of the case that traps the sediment that flakes off the plates. If the sediment were not contained in these chambers, it could cause a conductive connection across the plates and short the cell. The case is fitted with a one-piece cover.

Because the conventional battery releases hydrogen gas as it's charged, the case cover will have vents. The vents are located in the cell caps of a conventional battery (**Figure 5-7**).

Chemical Action

Shop Manual
Chapter 5, page 226

Activation of the battery is through the addition of electrolyte. This solution causes the chemical actions to take place between the lead dioxide of the positive plates and the sponge lead of the negative plates. The electrolyte is also the carrier that moves electric current between the positive and negative plates through the separators.

Shop Manual
Chapter 5, pages 223, 227

The automotive battery has a fully charged **specific gravity** of 1.265 corrected to 80° (27° C). Therefore, a specific gravity of 1.265 for electrolyte means it's 1.265 times heavier than an equal volume of water. As the battery discharges, the specific gravity of the electrolyte decreases because the electrolyte becomes more like water. The specific gravity of a battery can give you an indication of how charged a battery is.

Specific gravity is the weight of a given volume of a liquid divided by the weight of an equal volume of water. Water has a specific gravity of 1.000.

Fully charged:	1.265 specific gravity
75% charged:	1.225 specific gravity
50% charged:	1.190 specific gravity
25% charged:	1.155 specific gravity
Discharged:	1.120 or lower specific gravity

These specific gravity values may vary slightly according to the design of the battery. However, regardless of the design, the specific gravity of the electrolyte in all batteries will decrease as the battery discharges. Temperature of the electrolyte will also affect its specific gravity. All specific gravity specifications are based on a standard temperature of 80° (27° **C**). When the temperature is above that standard, the specific gravity is lower. When the temperature is below that standard, the specific gravity increases. Therefore, all specific gravity measurements must be corrected for temperature. A general rule to follow is to add 0.004 for every 10° F (5.5° C) above 80° F (27° C) and subtract 0.004 for every 10° F (5.5° C) below 80° F (27° C).

In operation, the battery is partially discharged and then recharged. This represents an actual reversing of the chemical action that takes place within the battery. The constant cycling of the charge and discharge modes slowly wears away the active materials on the cell plates. This action eventually causes the battery plates to sulfate. Battery replacement is necessary once the sulfation of the plates has reached the point of insufficient active plate area.

In the charged state, the positive plate material is essentially pure lead dioxide. The active material of the negative plates is spongy lead, PbO_2. The electrolyte is a solution of sulfuric acid H_2SO_4 and water. The voltage of the cell depends on the chemical difference between the active materials.

Figure 5-8 shows what happens to the plates and electrolyte during discharge. The electrolyte contains negatively charged sulfate ions (SO_4^{2-}) and positively charged hydrogen ions (H^+). Appling a load to the battery causes the sulfate ions to move to the negative plates and give up their negative charge. Any remaining sulfate combines with the sulfuric acid on the plates to form lead sulfate ($PbSO_4$). The lead sulfate becomes an electrical insulator. The accumulation of electrons creates an electric field that attracts hydrogen ions and repels sulfate ions, leading to a double-layer near the surface. The hydrogen ions screen the charged electrode from the solution, which limits further reactions unless a charge is allowed to flow out of electrode.

As the electrons become attracted to the positive plates, the oxygen (O_2) joins with the hydrogen ions from the electrolyte to form water (H_2O). The water dilutes the strength of the electrolyte and decreases the number of ions. Also, the lead sulfate formation on the positive and negative plates reduces the area of active material available to the ions.

The result of discharging is changing the of the positive plates and the PbO_2 of the negative plates into lead sulfate ($PbSO_4$). Because the plates are the same, the battery does not have an electrical potential.

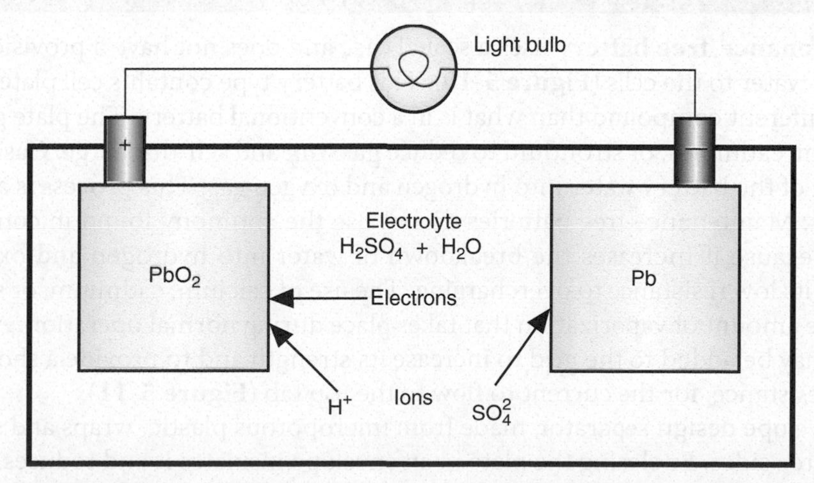

Figure 5-8 Chemical action that occurs inside of the battery during the discharge cycle.

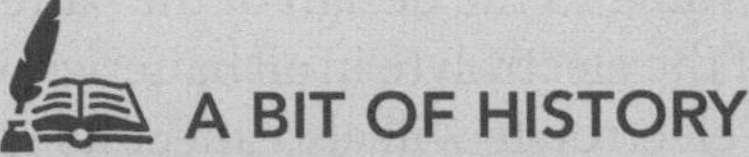

A BIT OF HISTORY

Lead acid batteries date back to 1859. Alexander Graham Bell used a primitive battery to make his first local call in 1876. After learning how to recharge lead acid batteries, they made their appearance in the automobile. These old-style batteries could not hold a charge very well and many mechanics believed that placing the battery on a concrete floor made them discharge faster. Although this fable has no truth to it, the idea has hung on for years.

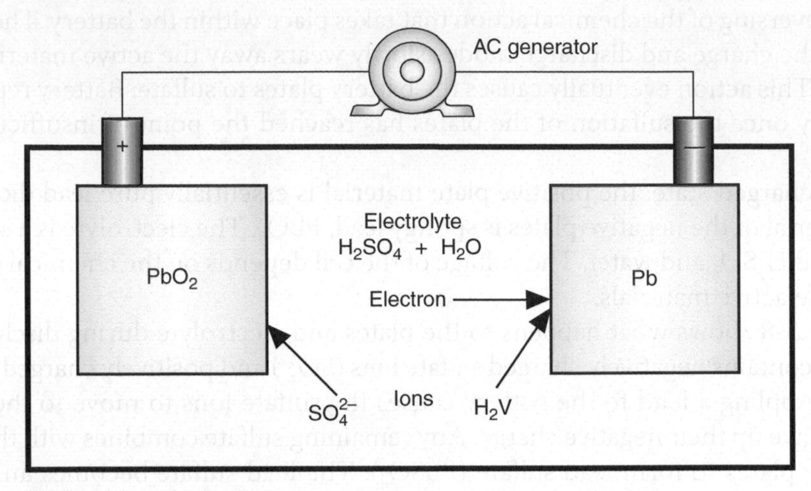

Figure 5-9 Chemical action inside of the battery during the charge cycle.

The charge cycle is exactly the opposite (**Figure 5-9**). The generator creates an excess of electrons at the negative plates, attracting the positive hydrogen ions to them. The hydrogen combines with the sulfate to revert it back to sulfuric acid (H_2SO_4) and lead. When most of the sulfate have been converted, the hydrogen rises from the negative plates. The water in the electrolyte splits into hydrogen and oxygen. The oxygen reacts with the lead sulfate on the positive plates to revert it back into lead dioxide. When the reaction is about complete, the oxygen bubbles rise from the positive plates. This puts the plates and the electrolyte back in their original form and the cell is charged.

Maintenance-Free Batteries

The **maintenance-free battery** uses a sealed case and does not have a provision for the addition of water to the cells (**Figure 5-10**). This battery type contains cell plates made of a slightly different compound than what is in a conventional battery. The plate grids contain calcium, cadmium, or strontium to reduce **gassing** and self-discharge. Gassing is the conversion of the battery water into hydrogen and oxygen gas. This process is also called electrolysis. Maintenance-free batteries do not use the antimony found in conventional batteries because it increases the breakdown of water into hydrogen and oxygen and because of its low resistance to overcharging. The use of calcium, cadmium, or strontium reduces the amount of vaporization that takes place during normal operation. Additional supports may be added to the grid to increase its strength and to provide a shorter path, with less resistance, for the current to flow to the top tab (**Figure 5-11**).

An envelope design separator, made from microporous plastic, wraps and seals each plate on three sides. Enclosing the plate in an envelope insulates it and reduces the shedding of the plate's active material.

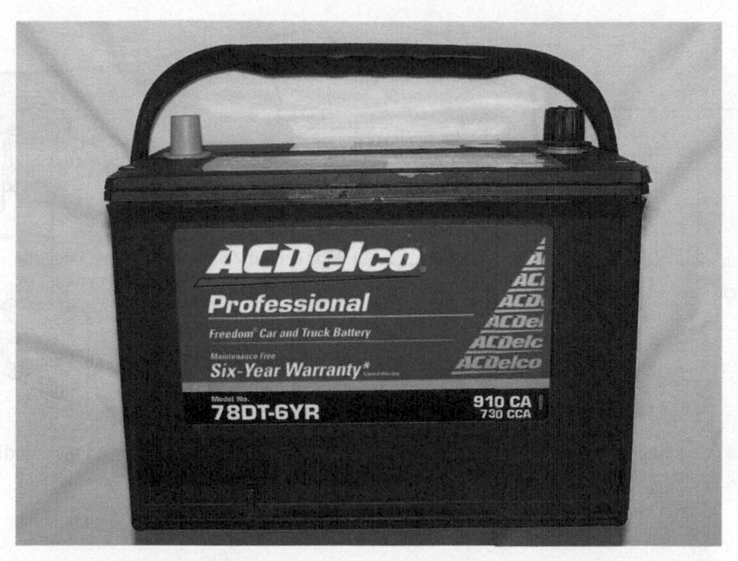

Figure 5-10 Maintenance-free batteries.

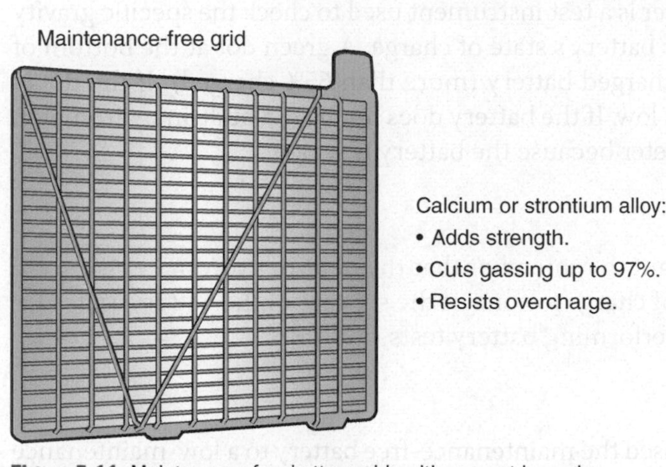

Figure 5-11 Maintenance-free battery grids with support bars give increased strength and faster electrical delivery.

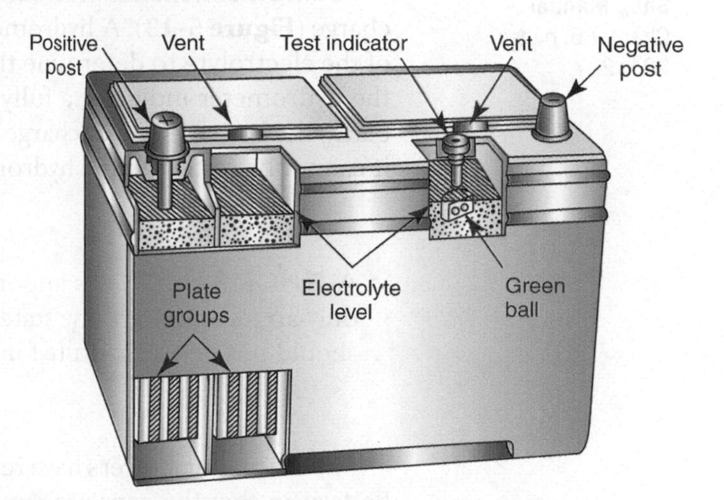

Figure 5-12 Construction of a maintenance-free battery.

AUTHOR'S NOTE Another term for the maintenance-free battery is sealed lead acid (SLA) battery.

Sealing the battery (except for a small vent) prevents the electrolyte and vapors from escaping (**Figure 5-12**). An expansion or condensation chamber allows the water to condense and drain back into the cells. Because the water cannot escape from the battery, it's not necessary to add water to the battery on a periodic basis. Containing the vapors also reduces the possibility of corrosion and discharge through the surface because of electrolyte on the surface of the battery. Vapors leave the case only when the pressure inside the battery is greater than atmospheric pressure.

AUTHOR'S NOTE If electrolyte and dirt accumulates on the top of the battery case, it may create a conductive connection between the positive and negative terminals, resulting in a constant discharge on the battery.

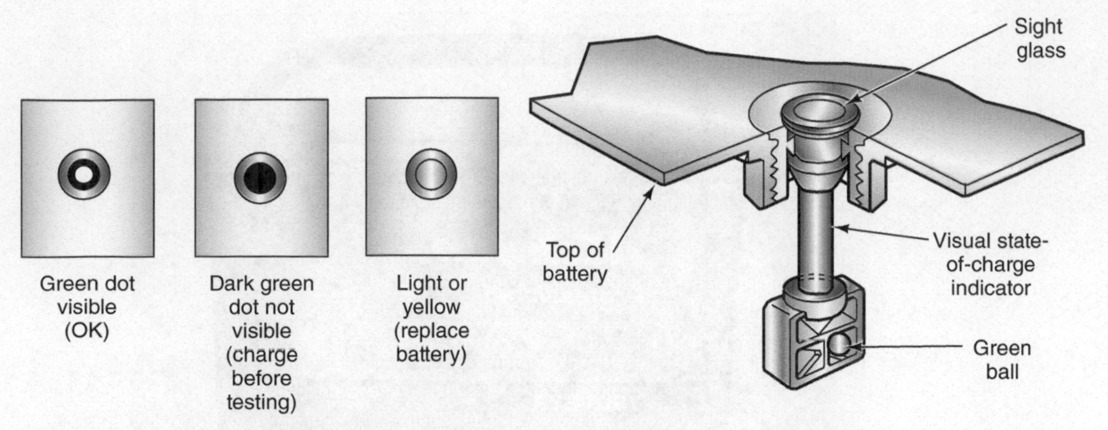

Figure 5-13 One cell of a maintenance-free battery has a built-in hydrometer, which gives indication of overall battery condition.

Shop Manual
Chapter 5, pages 221, 227

Some maintenance-free batteries have a built-in **hydrometer** to indicate the state of charge (**Figure 5-13**). A hydrometer is a test instrument used to check the specific gravity of the electrolyte to determine the battery's state of charge. A green dot at the bottom of the hydrometer indicates a fully charged battery (more than 65% charged). If the dot is black, the battery state of charge is low. If the battery does not have a built-in hydrometer, it cannot be tested with a hydrometer because the battery is sealed.

AUTHOR'S NOTE It's important to remember that the built-in hydrometer is only an indication of the state of charge for one of the six cells of the battery and should not be a substitute for performing battery tests.

Many manufacturers have revised the maintenance-free battery to a low-maintenance battery, in that the caps are removable for testing and electrolyte level checks. Also, the grid construction contains about 3.4% antimony. To decrease the distance and resistance of the path that current flows in the grid, and to increase its strength, the horizontal and vertical grid bars do not intersect at right angles (**Figure 5-14**).

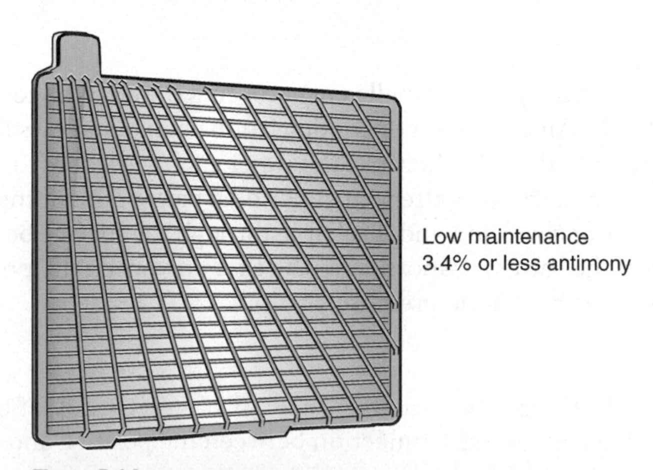

Figure 5-14 Low-maintenance battery grid with vertical grid bars intersecting at an angle.

The advantages of maintenance-free batteries over conventional batteries include:

1. A larger reserve of electrolyte above the plates.
2. Increased resistance to overcharging.
3. Longer shelf life (approximately 18 months).
4. Can be shipped with electrolyte installed, reducing the possibility of accidents and injury to technicians.
5. Higher cold cranking amps rating.

The major disadvantages of the maintenance-free battery include:

1. **Grid growth** when the battery is exposed to high temperatures.
2. Inability to withstand **deep cycling**.
3. Low reserve capacity.
4. Faster discharge by parasitic loads.
5. Shorter life expectancy.

Grid growth is a condition where the grid grows little metallic fingers that extend through the separators and short out the plates.

Deep cycling is to discharge the battery to a very low state of charge before recharging it.

Hybrid Batteries

AUTHOR'S NOTE The following discussion on hybrid batteries refers to a battery type and not to the batteries that are used in hybrid electric vehicles (HEVs).

The **hybrid battery** combines the advantages of the low-maintenance and maintenance-free batteries. The hybrid battery can withstand at least six deep cycles and still retain 100% of its original reserve capacity. The grid construction of the hybrid battery consists of approximately 2.75% antimony alloy on the positive plates and a calcium alloy on the negative plates. This allows the battery to withstand deep cycling while retaining reserve capacity for improved cranking performance. Also, the use of antimony alloys reduces grid growth and corrosion. The lead calcium has less gassing than conventional batteries.

Grid construction differs from other batteries in that the plates have a lug located near the center of the grid and the vertical bars use a **radial grid** pattern (**Figure 5-15**). This design provides a shorter path with less resistance for the current to follow to the lug This means the battery is capable of providing more current at a faster rate.

Making the separators from glass with a resin coating provides for low electrical resistance with high resistance to chemical contamination. This type of construction provides increases cranking performance and battery life.

Recombination Batteries

A recent variation of the automobile battery is the **recombination battery** (**Figure 5-16**). Another name for the recombination battery is "gel-cell battery." This battery does not use a liquid electrolyte. Instead, it uses separators that hold a gel-type material. The separators are placed between the grids and have very low electrical resistance. The spiral design provides a larger plate surface area than that in conventional batteries (**Figure 5-17**). In addition, the tight plate spacing results in decreased resistance. Because of this design, output voltage and current are higher than that in conventional batteries. The extra amount of available voltage (approximately 0.6 volt) assists in cold-weather starting. Also, gassing is virtually eliminated, and the battery can recharge faster.

Shop Manual
Chapter 5, page 224

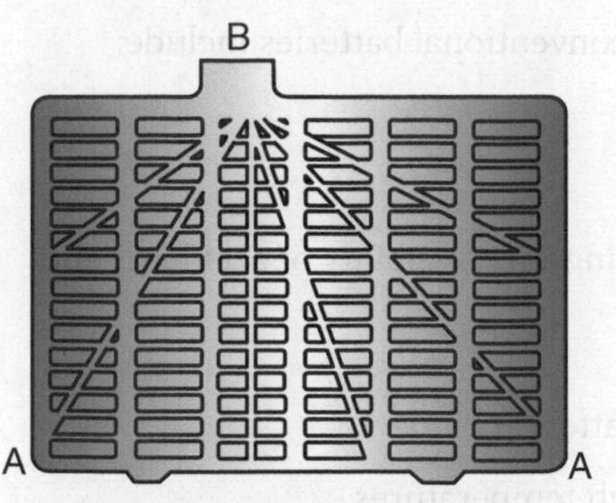

Figure 5-15 The hybrid battery grid construction allows for faster current delivery. Electrical energy at point "A" has a shorter distance to travel to get to the tab at point "B."

Figure 5-16 The recombination battery is spill proof.

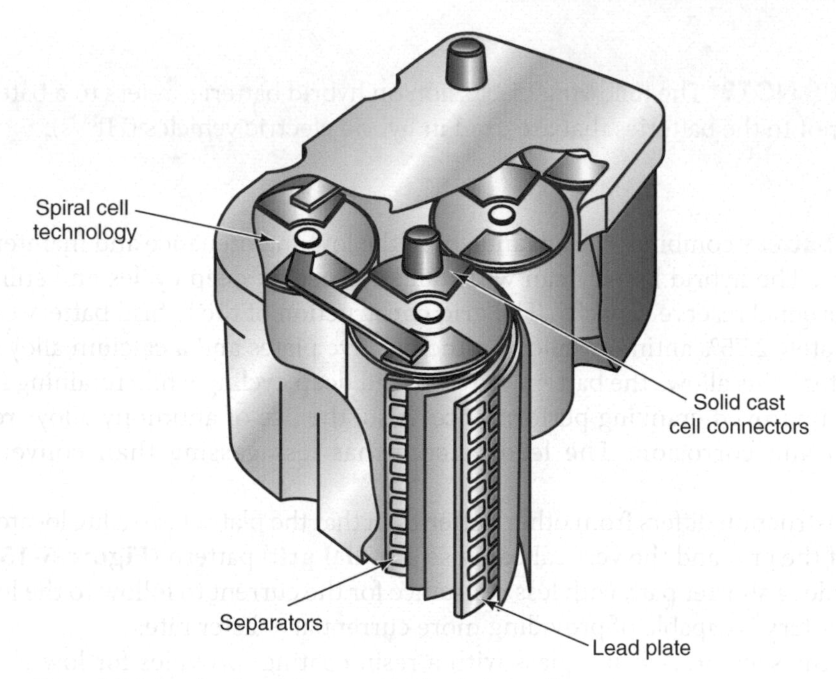

Figure 5-17 Construction of the recombination battery cells.

The following are some other safety features and advantages of the recombination battery:

1. Contains no liquid electrolyte that can spill from a cracked case.
2. Can be installed in any position, including upside down.
3. Is corrosion free.
4. Has very low maintenance because there is no electrolyte loss.
5. Can last as much as four times longer than conventional batteries.
6. Can withstand deep cycling without damage.
7. Can be rated over 800 cold cranking amperes.

Recombination batteries recombine the oxygen gas normally produced on the positive plates with the hydrogen given off by the negative plates. This recombination of oxygen and hydrogen produces water (H_2O) and replaces the moisture in the battery. The separators absorb the electrolyte solution of the recombination battery.

Special pressurized sealing vents traps the oxygen produced by the positive plates in the cell. The oxygen gases then travel to the negative plates through small fissures in the gelled electrolyte. There are between one and six one-way safety valves in the top of the battery. The safety valves are necessary for maintaining a positive pressure inside of the battery case. This positive pressure prevents oxygen from the atmosphere from entering the battery and causing corrosion. Also, the safety valves must release the excessive pressure produced due to overcharging of the battery.

Absorbed Glass Mat Batteries

A variation of the recombination battery is the **absorbed glass mat (AGM) battery**. Instead of using a gel, AGM batteries hold their electrolyte in a moistened fiberglass matting sandwiched between the battery's lead plates. The plates are made of high-purity lead and are tightly compressed into six cells. Separation of the plates is done by acid-permeated vitreous separators that act as sponges to absorb acid. Each cell is enclosed in its own cylinder within the battery case. This results in a sealed battery.

During normal discharging and charging of the battery, the hydrogen and oxygen sealed within the battery recombine to form water within the electrolyte. This process of recombining hydrogen and oxygen eliminates the need to add water to the battery.

AGM batteries are not easily damaged due to vibrations or impact. AGM batteries also have short recharging times and low internal resistance, which increases output.

HEVs and vehicles with Start/Stop capabilities also use 12-volt AGM batteries for operation of the vehicle accessories. Since most HEV's do not use the 12-volt battery to start the engine, these applications can use an AGM with very low energy capacity.

Valve-Regulated Batteries

All recombination batteries are valve-regulated batteries since they have one-way safety valves that control the internal pressure of the battery case. The valve will open to relieve any excessive pressure within the battery, but at all other times the valve closes and seals the battery.

A **valve-regulated lead-acid (VRLA) battery** is another variation of the recombination battery. Each cell contains a one-way check valve in the vent. In addition, the VRLA battery immobilizes the electrolyte and defuses the oxygen produced at the positive plates. Within the VRLA battery, the negative plate absorbs the oxygen produced on the positive plate. This causes a decrease in the amount of hydrogen produced at the negative plate. The small amount of hydrogen combines with the oxygen to produce water and returns to the electrolyte.

The VRLA uses a plate construction with a base of lead-tin-calcium alloy. The active material of one of the plates is porous lead dioxide, while the active material of the other plate is spongy lead. The electrolyte is sulfuric acid that absorbs into plate separators made of a glass-fiber fabric.

Battery Ratings

Shop Manual
Chapter 5, page 231

The Battery Council International (BCI) in conjunction with the Society of Automotive Engineers Battery have established battery capacity ratings. Battery cell voltage depends on the types of materials used in constructing the battery. Current capacity depends on several factors:

1. The size of the cell plates. The larger the surface area of the plates, the more chemical action that can occur. This means a greater current can be produced.

2. The weight of the positive and negative plate active materials.
3. The weight of the sulfuric acid in the electrolyte solution.

The battery's current capacity rating is an indication of its ability to deliver cranking power to the starter motor and of its ability to provide reserve power to the electrical system. The following explains the commonly used current capacity ratings.

Ampere-Hour Rating

The **ampere-hour rating** is the amount of steady current that a fully charged battery can supply for 20 hours at 80° F (26.7° C) without the terminal voltage falling below 10.5 volts. For example, if a battery can be discharged for 20 hours at a rate of 4.0 amps before its terminal voltage reads 10.5 volts, it would be rated at 80 ampere-hours.

Cold Cranking Amps

Cold cranking amps (CCA) is the most common method of rating automotive batteries. It's determined by the load, in amperes, that a battery is able to deliver for 30 seconds at 0° F (-17.7° C) without terminal voltage falling below 7.2 volts (1.2 volts per cell) for a 12-volt battery. The cold cranking rating is in total amperage and identified as 300 CCA, 400 CCA, 500 CCA, and so on. Some batteries are rated as high as 1,100 CCA.

Cranking Amps

Cranking Amps (CA) is an indication of the battery's ability to provide a cranking amperage at 32° F (0° C). This rating uses the same test procedure as the cold cranking rating or CCA discussed earlier, except it uses a higher temperature. To convert CA to CCA, divide the CA by 1.25. For example, a 650-CCA-rated battery is the same as 812 CA. It's important that you not misread the rating and mistake CCA as CA.

Reserve-Capacity Rating

The **reserve-capacity rating** is determined by the length of time, in minutes, that a fully charged battery can be discharged at 25 amps before battery voltage drops below 10.5 volts. This rating gives an indication of how long the vehicle can be driven, with the head-lights on, if the charging system should fail.

Battery Size Selection

Some of the aspects that determine the battery rating required for a vehicle include engine size, engine type, climatic conditions, and vehicle options. The requirement for electrical energy to crank the engine increases as the temperature decreases. Battery power drops drastically as temperatures drop below freezing (**Figure 5-18**). The engine also becomes harder to crank due to the tendency of oils to thicken when cold, which results in increased friction. For years the general rule was it takes 1 amp of cold cranking power per cubic inch of engine displacement. Therefore, a 200-cubic-inch displacement engine should

Temperature	% of Cranking Power
80°F (26.7°C)	100
32°F (0°C)	65
0°F (−17.8°C)	40

Figure 5-18 The effect temperature has on the cranking power of the battery.

have a battery of at least 200 CCA. To convert this into metric, it takes 1 amp of cold cranking power for every 16 cm^3 of engine displacement. A 1.6-liter engine should require at least a battery rated at 100 CCA. This rule does not apply to vehicles that have several electrical accessories. For this reason, the best method of determining the correct battery is to refer to the manufacturer's specifications.

It's permissible to install a battery with a higher capacity than that required by the manufacturer. However, the additional cost may not outweigh the benefits. Never install a battery rated below the manufacturer's recommendations. The selected battery should fit the battery holding fixture and the holddown must be able to be properly installed. It's important that the height of the battery not allow the terminals to short across the vehicle's hood when shut. BCI group numbers indicate the physical size and other features of the battery. This group number does not indicate the current capacity of the battery.

Battery Terminals

Shop Manual
Chapter 5, pages 221, 225

Battery terminals provide a means of connecting the battery plates to the vehicle's electrical system. All automotive batteries have two terminals. One terminal is a positive connection; the other is a negative connection. The battery terminals extend through the cover or the side of the battery case. The following are the most common types of battery terminals (**Figure 5-19**):

1. Post or top terminals: Positioned at the top of the battery, the positive post is typically larger than the negative post to prevent connecting the battery in reverse polarity. In addition, positive and negative symbols identify the posts.
2. Side terminals: Positioned in the side of the container near the top. The threaded terminals require a special bolt to connect the cables. Polarity identification is by positive and negative symbols.
3. L terminals: Used on specialty batteries and some imports. Polarity identification is by positive and negative symbols.

Battery Cables

Shop Manual
Chapter 5, page 221

Battery cables are high-current conductors that connect the battery to the vehicle's electrical system. Battery cables must be of a sufficient capacity to carry the current required to meet all electrical demands (**Figure 5-20**). Normal 12-volt cable size is usually 4 or 6 gauge. Various clamp and terminal designs assure a good electrical connection at each end of the cable. Connections must be clean and tight to prevent arcing, corrosion, and high-voltage resistance.

The positive cable is usually red (but not always), and the negative cable is usually black. The positive cable will fasten to the starter solenoid or relay. The negative cable

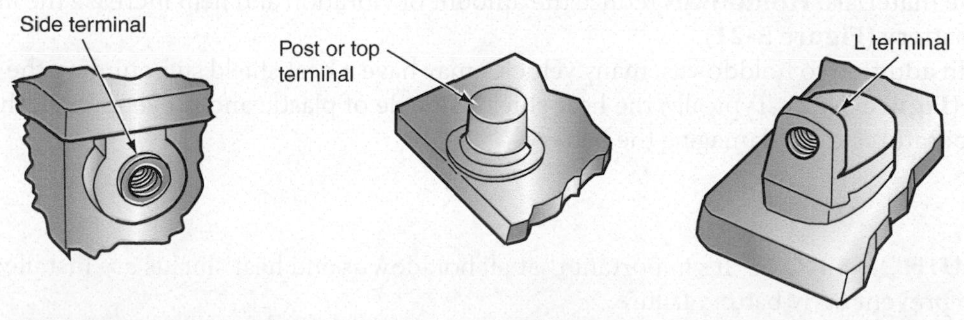

Figure 5-19 The most common types of automotive battery terminals.

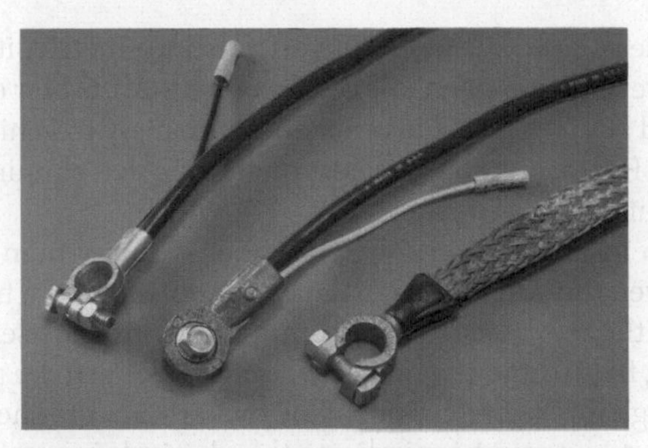

Figure 5-20 The battery cable carries the high current required to start the engine and supply the vehicle's electrical systems.

fastens to ground on the engine block or chassis. Some manufacturers use a negative cable with no insulation. Sometimes the negative battery cable may have a body grounding wire to help assure that the vehicle body is properly grounded.

A BIT OF HISTORY

Early automobiles had their storage battery mounted under the car. It wasn't until 1937 that the battery was located under the hood for better accessibility. Today, with the increased use of AGM batteries, manufacturers have "buried" the battery again. For example, to access the battery on some vehicles, you must remove the left front wheel and work through the wheel well. Also, some batteries are now located in the trunk area.

AUTHOR'S NOTE It's important to properly identify the positive and negative cables when servicing, charging, or jumping the battery. Do not rely on the color of the cable for this identification; use the markings on the battery case.

AUTHOR'S NOTE Pinch on battery cable clamps is a temporary repair only!

Battery Holddowns

Shop Manual
Chapter 5, pages 221, 235

All batteries must be secured in the vehicle to prevent damage and the possibility of shorting across the terminals if the battery tips. Normal vibrations cause the plates to shed their active materials. **Holddowns** reduce the amount of vibration and help increase the life of the battery (**Figure 5-21**).

In addition to holddowns, many vehicles may have a heat shield surrounding the battery (**Figure 5-22**). Typically, the heat shield is made of plastic and prevents underhood temperatures from damaging the battery.

AUTHOR'S NOTE It's important that all holddowns and heat shields are installed to prevent early battery failure.

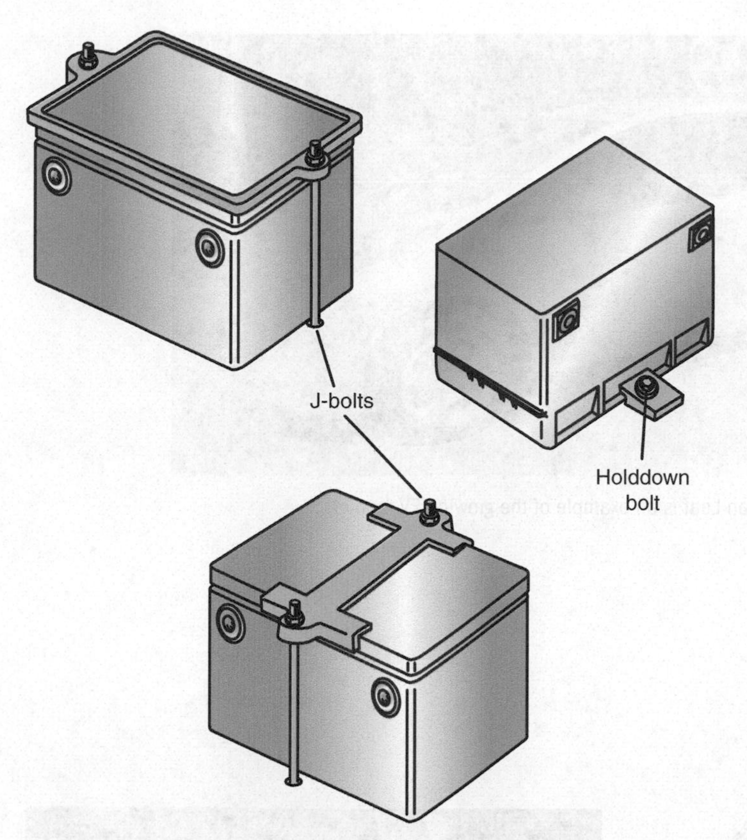

Figure 5-21 Different types of battery holddowns.

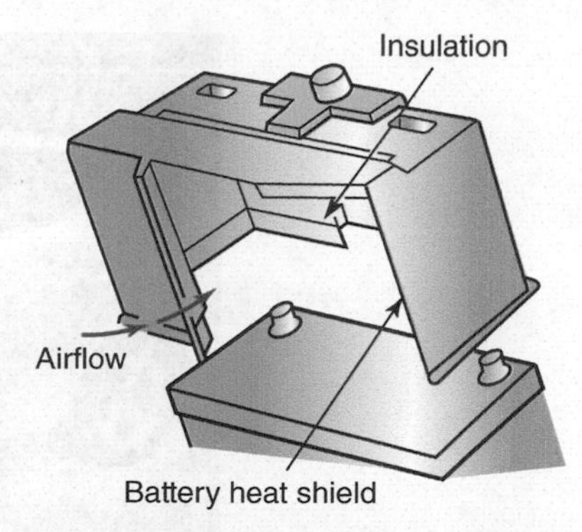

Figure 5-22 Some vehicles are equipped with a heat shield to protect the battery from excessive heat.

EV and HEV Batteries

Automotive manufacturers have been investigating the use of high-voltage (HV) batteries for several years. One of the first attempts was the use of 42-volt systems. Today, electric-drive vehicles require high-voltage batteries that can supply over 300 volts.

By connecting several lead-acid batteries in series, they can provide voltages high enough to power some electric vehicles (EVs). For example, the first-generation General Motors' EV used twenty-six 12-volt lead-acid batteries connected in series to provide 312 volts. The downside of this arrangement is that the battery pack weighed 1,310 pounds (595 kg). In addition, the travel distance between battery recharges was 55 to 95 miles (88 to 153 km). The next-generation EV used nickel-metal hydride (NiMH) batteries. These provided for a slightly longer traveling range between recharges. HEV technology has accelerated battery technology to the point that battery-powered EVs are becoming more practical (**Figure 5-23**).

Shop Manual
Chapter 5, page 240

The battery pack in an EV or HEV is typically made up of several cylindrical cells (**Figure 5-24**) or prismatic cells (**Figure 5-25**). There are several different technologies of HV batteries being used or in development.

Nickel–Cadmium (NiCad) Batteries

Nickel-cadmium (NiCad) cells have a role in EV and HEV because of several advantages that it has. These include being able to withstand many deep cycles, low cost of production, and long service life. Also, NiCad batteries perform very well when high-energy boosts are required.

NiCad batteries are also referred to as NiCd.

The negatives associated with the use of NiCad batteries include:

- Use toxic metals.
- Low **energy density.**

Figure 5-23 The Nissan Leaf is an example of the growing EV market.

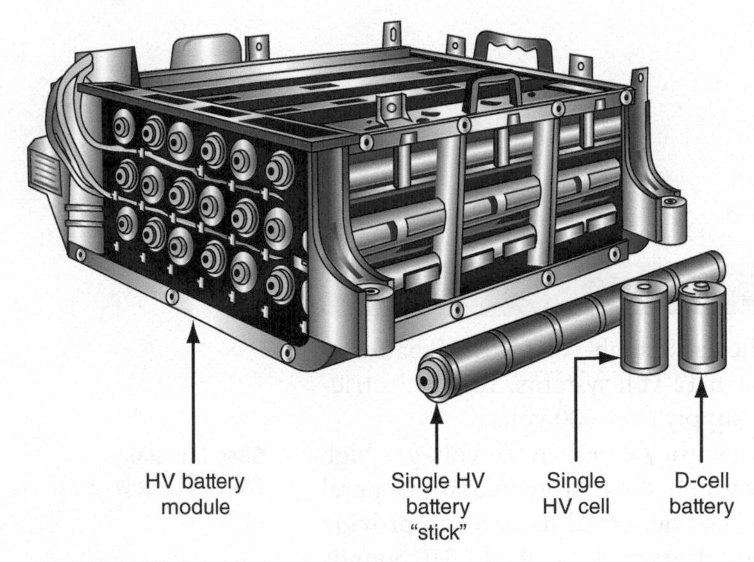

Figure 5-24 HV battery constructed of cylindrical cells.

Figure 5-25 Battery pack made of several prismatic cells.

- Require recharging if they are not used for a while.
- Suffer from the **memory effect**.

Energy density refers to the amount of energy available for a given amount of space.

Memory effect refers to the battery not being able to be fully recharged because it "remembers" its previous charge level. Recharging a battery before it's fully discharged results in a low battery charge. For example, consistently recharging a 50% discharged battery will eventually cause the battery to accept and hold only a 50% charge and not accept any higher charge.

The cathode (positive) electrode in a NiCad cell is a fiber mesh covered with nickel hydroxide **(Figure 5-26)**. The anode (negative) electrode is a fiber mesh covered with cadmium. The electrolyte is aqueous potassium hydroxide (KOH). The KOH is a conductor of ions and has little involvement in the chemical reaction process. During discharge, ions travel from the anode, through the KOH, and to the cathode. During charging, the opposite occurs. Each cell produces 1.2 volts.

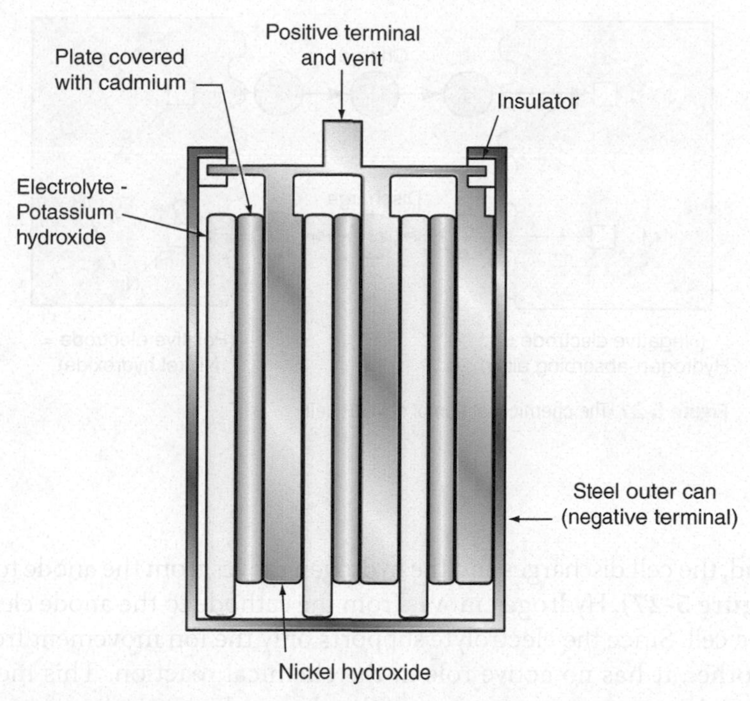

Figure 5-26 NiCad battery cell construction.

Nickel-Metal Hydride (NiMH) Batteries

NiMH batteries are very quickly replacing nickel–cadmium batteries. One reason for this is they are more environmentally friendly. NiMH batteries also have more capacity than NiCad batteries since they have a higher energy density. However, they have a lower current capacity when placed under a heavy load. Currently, the NiMH is the most common HV battery used in HEV.

The issue facing HEV manufacturers is that the NiMH battery has a relatively short service life due to the battery's subjection to several deep cycles of charging and discharging over its lifetime. In addition, NiMH cells generate heat during charging and require long charge times to prevent overheating.

The cathode electrode of the NiMH battery is a fiber mesh that contains nickel hydroxide. Hydrogen-absorbing metal alloys make up the anode electrode. The most commonly used alloys are compounds containing two to three of the following metals:

- Titanium
- Vanadium
- Zirconium
- Nickel
- Cobalt
- Manganese
- Aluminum.

The amount of hydrogen accumulated and stored by the alloy is far greater than the actual volume of the alloy.

A sheet of fine fibers saturated with an aqueous and alkaline electrolyte (KOH) separate the cathode electrode from the anode electrode. The cell components are placed in a metal housing, and then the unit is sealed. A safety vent allows high pressures to escape if needed.

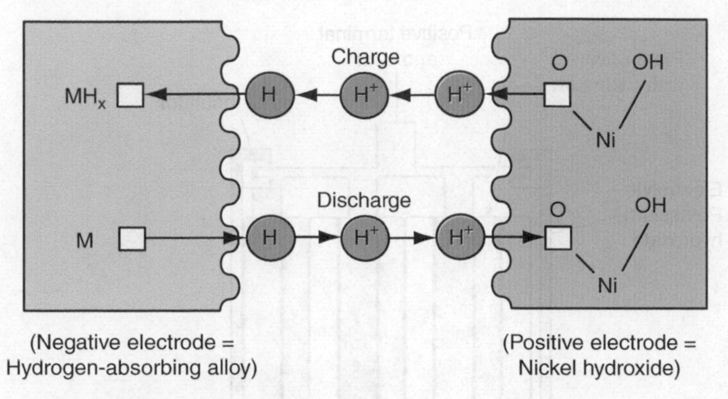

Figure 5-27 The chemical action of a NiMH cell.

Under load, the cell discharges and the hydrogen moves from the anode to the cathode electrode **(Figure 5-27)**. Hydrogen moves from the cathode to the anode electrode when recharging the cell. Since the electrolyte supports only the ion movement from one electrode to the other, it has no active role in the chemical reaction. This means that the electrolyte level does not change because of the chemical reaction.

A 300-volt battery has 240 cells that produce 1.2 volts each. Six cells form a module. Each module is actually a self-contained 7.2-volt battery. Connecting the modules in series to creates the total voltage (**Figure 5-28**).

A service disconnect disables the HV system if repairs or service to any part of the system are to be performed (**Figure 5-29**). Two functions of the service connector separate the HV battery pack into two separate batteries, with approximately 150 volts each. First, lifting the service disconnect's handle opens a high-voltage interlock loop (HVIL); then, with the service disconnect fully removed, it opens the high-voltage connector. When the HVIL is open, **contactors** should open. Contactors are heavy-duty relays connected to the positive and negative sides of the HV battery. The contactors are normally open and require a 12-volt supply to keep them closed. Lifting the service disconnect

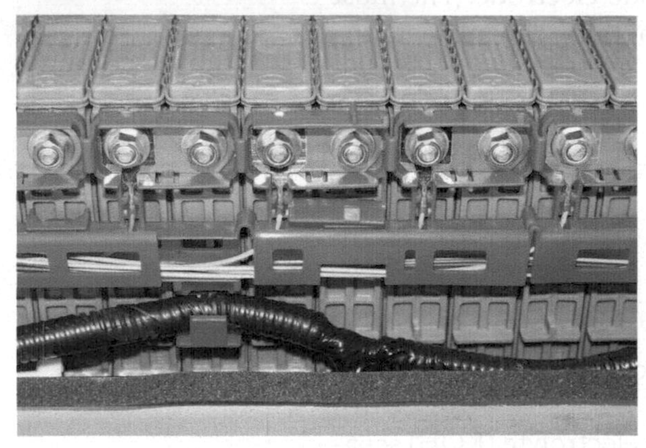

Figure 5-28 Cell module connections.

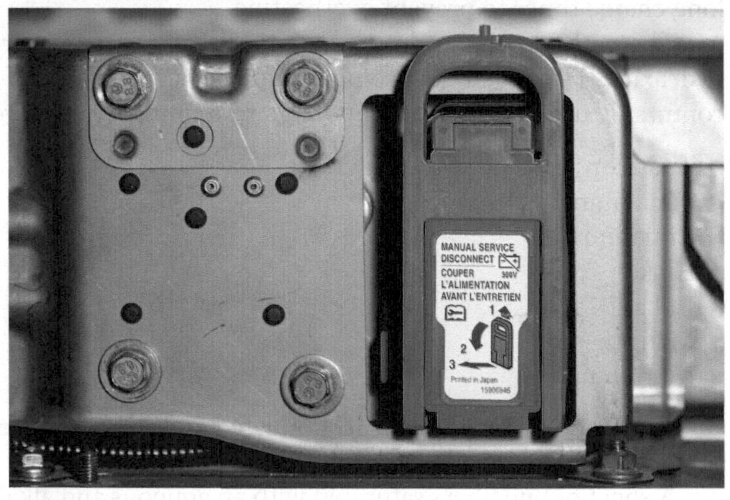

Figure 5-29 Service disconnect plug.

opens the HVIL and interrupts the voltage supply to the contactors, and the contactors should open. However, the contactors' contacts may weld together because of arcing and prevent the circuit from opening. This will result in a DTC being set.

Lithium-Ion (Li-Ion) Batteries

Rechargeable lithium-based batteries are very similar in construction to nickel-based batteries. Positives of using lithium batteries include high energy density, limited memory effect, and their environment-friendly property. The negatives are that lithium is an alkali metal and oxidizes very rapidly in air and water, which makes lithium highly flammable and slightly explosive when exposed to air and water. Lithium metal is also corrosive.

AUTHOR'S NOTE Lithium is the lightest metal and provides the highest energy density of all known metals.

The anode electrode of the Li-ion battery is made of graphite (a form of carbon). The cathode is composed mainly of graphite and lithium alloy oxide. Due to the safety issues associated with lithium metal, the Li-ion battery uses a variety of lithium compounds. The development of a manganese Li-ion battery for use in hybrid vehicles has the potential to last twice as long as the NiMH battery.

The electrolyte is a lithium salt mixed in a liquid. Polyethylene membranes separate the plates inside the cells and, in effect, separate the ions from the electrons. The membranes have extremely small pores that allow the ions to move within the cell.

As with most other rechargeable cells, ions move from the anode to the cathode when the cell is providing electrical energy. During recharging, the ions move back from the cathode to the anode **(Figure 5-30)**.

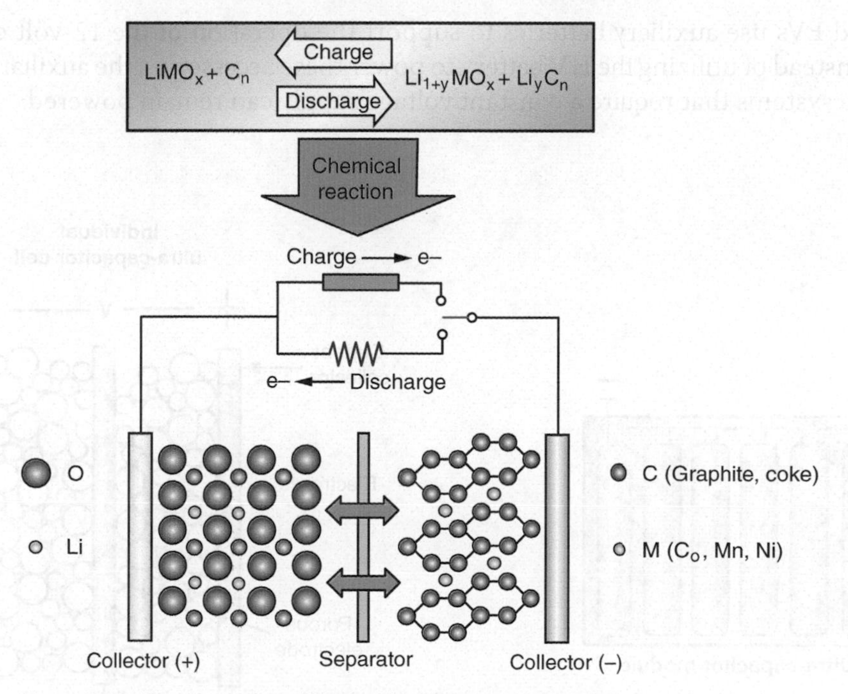

Figure 5-30 Chemical action within the Li-ion battery at it is cycled.

Lithium-Polymer (Li-Poly) Batteries

The lithium-polymer battery is nearly identical to the Li-ion battery and shares the same electrode construction. The difference is in the lithium salt electrolyte. The Li-Poly cell holds the electrolyte in a thin solid, polymer composite (polyacrylonitrile) instead of as a liquid. The solid polymer electrolyte is not flammable.

The dry polymer electrolyte does not conduct electricity. Instead, it allows ions to move between the anode and cathode. The polymer electrolyte also serves as the separator between the plates. Since the dry electrode has very high resistance, it's unable to provide bursts of current for heavy loads. Increasing the cell temperature above 140° **F** (60° C) increases its efficiency. The voltage of the Li-Poly cell is about 4.23 volts when fully charged.

Ultra-Capacitors

Ultra-capacitors are capacitors constructed with a large electrode surface area and a minimal distance between the electrodes. Unlike conventional capacitors that use a dielectric, ultra-capacitors use an electrolyte (**Figure 5-31**). It also stores electrical energy at the boundary between the electrodes and the electrolyte. Although an ultra-capacitor is an electrochemical device, no chemical reactions are involved in storing electrical energy. This means that the ultra-capacitor remains an electrostatic device. The design of the ultra-capacitor increases its capacitance capabilities to as much as 5,000 farads.

Many present-day HEVs, and some experimental fuel cell EVs, use ultra-capacitors because of their ability to quickly discharge high voltages and then be quickly recharged. This makes them ideal for increasing boost to electrical motors during acceleration or heavy loads. Ultra-capacitors are also very good at absorbing the energy from **regenerative braking**.

Auxiliary Batteries

HEVs and EVs use auxiliary batteries to support the operation of the 12-volt electrical system. Instead of utilizing the HV battery to power these accessories, the auxiliary battery is used so systems that require a constant voltage supply can remain powered.

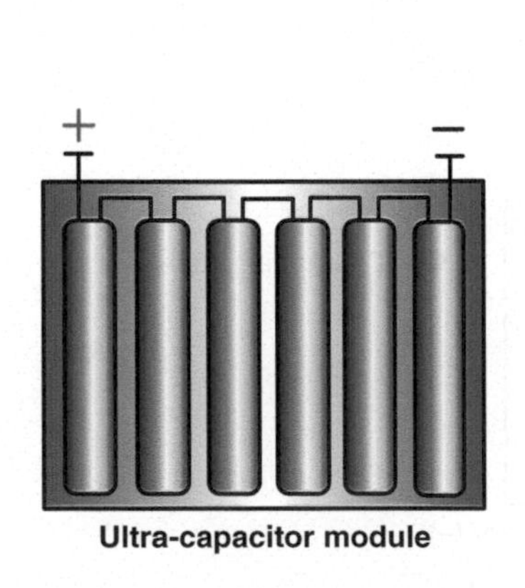

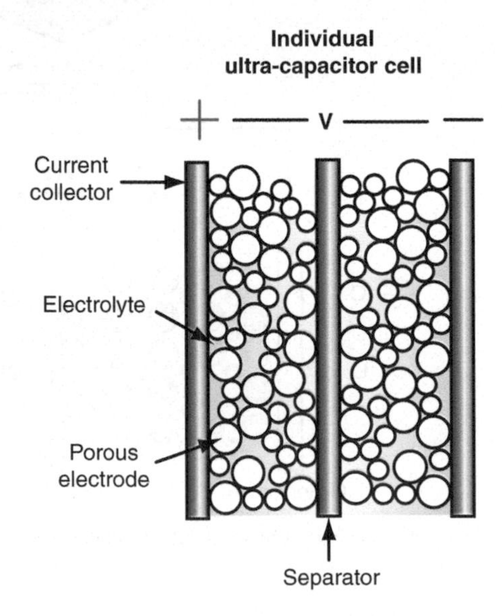

Figure 5-31 Ultra-capacitor cell construction.

The big difference between HEV and EV auxiliary battery systems from that of the conventional vehicle is the method used to recharge the battery. HEVs and EVs do not use an alternator to charge the auxiliary battery; instead, they are recharged by the HV battery using the inverter/converter.

Although there are exceptions, many HEVs do not use the 12-volt auxiliary battery for starting the internal combustion engine (ICE). Generally, the auxiliary battery is used to support all 12-volt electrical systems on the vehicle. The exceptions are the air conditioning and heating systems. In most cases, the auxiliary battery supplies power to:

- Accessory systems
- Headlights
- Audio systems
- Computer controls

HEVs and EVs are not the only vehicles that use an auxiliary battery. An auxiliary battery can be used as a safety backup to support the main battery when required or to provide constant voltage for specific vehicle systems. Many vehicles with Start/Stop and ADAS systems may also utilize an auxiliary battery alongside the main vehicle starter battery.

Summary

- An automotive battery is an electrochemical device capable of storing chemical energy that can be converted to electrical energy. The chemical reaction that occurs between two dissimilar plates that are immersed in an electrolyte solution produces electrical energy in the battery.
- Electrical loads placed on the battery when the ignition switch is OFF are called key-off or parasitic loads.
- The amount of electrical energy that a battery is capable of producing depends on the size, weight, and active area of the plates and the specific gravity of the electrolyte solution.
- The conventional battery has seven basic components:

  1. Positive plates.
  2. Negative plates.
  3. Separators.
  4. Case.
  5. Plate straps.
  6. Electrolyte.
  7. Terminals.

- Electrolyte solution used in automotive batteries consists of 64% water and 36% sulfuric acid by weight.
- The electrolyte solution causes chemical actions between the lead dioxide of the positive plates and the sponge lead of the negative plates. The electrolyte is also the carrier that moves electric current between the positive and negative plates through the separators.
- The automotive battery has a fully charged specific gravity of 1.265 corrected to 80° F.
- Grid growth is the formation of little metallic fingers that extend from the grid through the separators and short out the plates.
- Deep cycling is discharging the battery almost completely before recharging it.
- In a conventional battery, lead dioxide covers the positive plate and sponge lead covers the negative plate.
- In maintenance-free batteries, the cell plates contain calcium, cadmium, or strontium to reduce gassing and self-discharge.
- The grid construction of the hybrid battery consists of approximately 2.75% antimony alloy on the positive plates and a calcium alloy on the negative plates.
- The recombination battery uses separators that hold a gel-type material in place of liquid electrolyte.
- Absorbed glass mat (AGM) batteries hold their electrolyte in a moistened fiberglass matting that sandwiched between the battery's high-purity lead plates. Acid-permeated vitreous separators that act as sponges to absorb acid separate the plates.

- Within the VRLA battery, the negative plate absorbs the oxygen produced on the positive plate, causing a decrease in the amount of hydrogen produced at the negative plate. The small amount of hydrogen combines with the oxygen to produce water and returns to the electrolyte.
- The three most common types of battery terminals are:
  1. Post or top terminals.
  2. Side terminals.
  3. L terminals.
- The most common methods of battery rating are cold cranking, cranking amps, reserve capacity, and ampere-hour.
- Cell construction of the NiCad battery consists of the cathode (positive) electrode made of fiber mesh covered with nickel hydroxide, while the anode (negative) electrode is a fiber mesh that is covered with cadmium. The electrolyte is aqueous potassium hydroxide (KOH).
- The cathode electrode of the NiMH battery is a fiber mesh that contains nickel hydroxide. The anode electrode is made of hydrogen-absorbing metal alloys. A sheet of fine fibers saturated with an aqueous and alkaline electrolyte (KOH) separate the cathode and anode electrodes.
- A 300-volt NiMH battery has 240 cells that produce 1.2 volts each. Each module has six cells. Connecting the modules in series produces the total voltage.
- A service disconnect in the HV battery disables the HV system if repairs or service to any part of the system is being performed.
- Contactors are heavy-duty relays connected to the positive and negative sides of the HV battery.
- Ultra-capacitors are capacitors constructed to have a large electrode surface area and a very small distance between the electrodes.
- HEVs use ultra-capacitors because of their ability to quickly discharge high voltages and then be quickly recharged.

Review Questions

Short-Answer Essays

1. Explain the purposes of the battery.
2. Describe how you can determine the correct battery to install into a vehicle.
3. Describe the methods used to rate batteries.
4. What is the purpose of the ultra-capacitor?
5. Explain the effects that temperature has on battery performance.
6. Describe the different loads or demands placed on a battery during different operating conditions.
7. List and describe the seven main elements of the conventional battery.
8. What is the purpose of the service disconnect on a HV battery?
9. Describe the process a battery undergoes during charging.
10. Describe the difference in construction of the hybrid battery as compared to the conventional battery.

Fill in the Blanks

1. An automotive battery is an ________ device capable of storing ________ energy that can be converted to electrical energy.
2. When discharging the battery, it changes ________ energy into ________ energy.
3. The assembly of the positive plates, negative plates, and separators is called the ________ ________.
4. The electrolyte solution used in automotive batteries consists of ________% water and ________% sulfuric acid.
5. A fully charged automotive battery has a specific gravity of ________ corrected to 80° F (26.7° C).
6. ________ ________ is a condition where the grid grows metallic fingers that extend through the separators and short out the plates.
7. The ________ ________ rating indicates the battery's ability to deliver a specified amount of current to start an engine at low ambient temperatures.

8. The electrolyte solution causes the chemical actions to take place between the lead dioxide of the ________plates and the sponge lead of the ________plates.

9. Some of the aspects that determine the battery rating required for a vehicle include engine ________, engine________, ________ conditions, and vehicle ________.

10. Electrical loads that are still present when the ignition switch is in the OFF position are called ________loads.

Multiple Choice

1. *Technician A* says the battery provides electricity by releasing free electrons.
   Technician B says the battery stores energy in chemical form.
   Who is correct?
   A. A only
   B. B only
   C. Both A and B
   D. Neither A nor B

2. *Technician A* says the largest demand on the battery is when it must supply current to operate the starter motor.
   Technician B says the current requirements of a starter motor may be over 100 amps.
   Who is correct?
   A. A only
   B. B only
   C. Both A and B
   D. Neither A nor B

3. Which of the following statements about NiMH cells is NOT true?
   A. When the NiMH cell discharges, hydrogen moves from the anode to the cathode electrode.
   B. Nickel-metal hydride batteries have an anode electrode that contains nickel hydroxide.
   C. The alkaline electrolyte has no active role in the chemical reaction.
   D. A sheet of fine fibers saturated with an aqueous and alkaline electrolyte separate the plates.

4. The current capacity rating of the battery is being discussed.
   Technician A says the amount of electrical energy that a battery is capable of producing depends on the size, weight, and active area of the plates.
   Technician B says the current capacity rating of the battery depends on the types of materials used in the construction of the battery.
   Who is correct?
   A. A only
   B. B only
   C. Both A and B
   D. Neither A nor B

5. The construction of the battery is being discussed.
   Technician A says the 12-volt battery consists of positive and negative plates connected in parallel.
   Technician B says the 12-volt battery consists of six cells wired in series.
   Who is correct?
   A. A only
   B. B only
   C. Both A and B
   D. Neither A nor B

6. Which of the following statements about battery ratings is true?
   A. The ampere-hour rating is the amount of steady current that a fully charged battery can supply for 1 hour at 80° F (26.7° C) without the cell voltage falling below a predetermined voltage.
   B. The cold cranking amps rating represents the number of amps that a fully charged battery can deliver at 0° F (-17.7° C) for 30 seconds while maintaining a voltage above 9.6 volts for a 12-volt battery.
   C. The cranking amp rating expresses the number of amperes a battery can deliver at 32° F (0° C) for 30 seconds and maintain at least 1.2 volts per cell.
   D. The reserve-capacity rating expresses the number of amperes a fully charged battery at 80° F can supply before the battery's voltage falls below 10.5 volts.

7. Battery terminology is being discussed.
   Technician A says grid growth is a condition where the grid grows little metallic fingers that extend through the separators and short out the plates.
   Technician B says deep cycling is discharging the battery almost completely before recharging it.
   Who is correct?
   A. A only C. Both A and B
   B. B only D. Neither A nor B

8. Battery rating methods are being discussed.
   Technician A says the ampere-hour is determined by the load in amperes a battery is able to deliver for 30 seconds at 0° F (-17.7° C) without terminal voltage falling below 7.2 volts for a 12-volt battery.
   Technician B says the cold cranking rating is the amount of steady current that a fully charged battery can supply for 20 hours at 80° F (26.7° C) without battery voltage falling below 10.5 volts.
   Who is correct?
   A. A only C. Both A and B
   B. B only D. Neither A nor B

9. The hybrid battery is being discussed.
   Technician A says the hybrid battery can withstand at least six deep cycles and still retain 100% of its original reserve capacity.
   Technician B says the grid construction of the hybrid battery consists of approximately 2.75% antimony alloy on the positive plates and a calcium alloy on the negative plates.
   Who is correct?
   A. A only C. Both A and B
   B. B only D. Neither A nor B

10. *Technician A* says battery polarity must be observed when connecting the battery cables.
    Technician B says the battery must be secured in the vehicle to prevent internal damage and the possibility of shorting across the terminals if it tips.
    Who is correct?
    A. A only C. Both A and B
    B. B only D. Neither A nor B

CHAPTER 6

STARTING SYSTEMS AND MOTOR DESIGNS

Upon completion and review of this chapter, you should be able to:

- Explain the purpose of the starting system.
- List the components of the starting system.
- Explain the principle of operation of the DC motor.
- Describe the purpose and operation of the armature.
- Explain the purpose and operation of the field coil.
- Describe the differences between the types of magnetic switches used.
- Describe the differences between starter drive mechanisms.
- Explain the differences between the positive-engagement and solenoid shift starter.
- Explain the operation and features of the permanent magnet starter.
- Explain the principles of operation of the three-phase AC motor.
- Explain the purpose of the inverter module.
- Describe the operating principles of integrated starter generator (ISG) systems.
- Explain the operation of the Stop/Start system components.

Terms to Know

Amortisseur winding
Armature
Belt alternator starter (BAS)
Brushes
Commutator
Compound motor
Counter electromotive force (CEMF)
Double-start override
Drive coil
Eddy currents
Enhanced starter
Field coils
Hold-in windings
Induction motor
Integrated starter generator (ISG)
Laminated construction
Overrunning clutch
Permanent magnet gear reduction (PMGR)
Pole shoes
Pull-in windings
Pullout torque
Ratio
Rotating magnetic field
Sentry key
Shunt
Slip
Start clutch interlock switch
Starter drive
Static neutral point
Stop/Start
Synchronous motor
Synchronous speed

Introduction

The internal combustion engine (ICE) must be rotated before it will run under its own power. The starting system is a combination of mechanical and electrical parts that work together to start the engine. The starting system converts the energy stored in the battery into mechanical energy. To accomplish this conversion, a starter or cranking motor is used. The starting system includes the following components:

Shop Manual
Chapter 6, page 268

1. Battery.
2. Cable and wires.

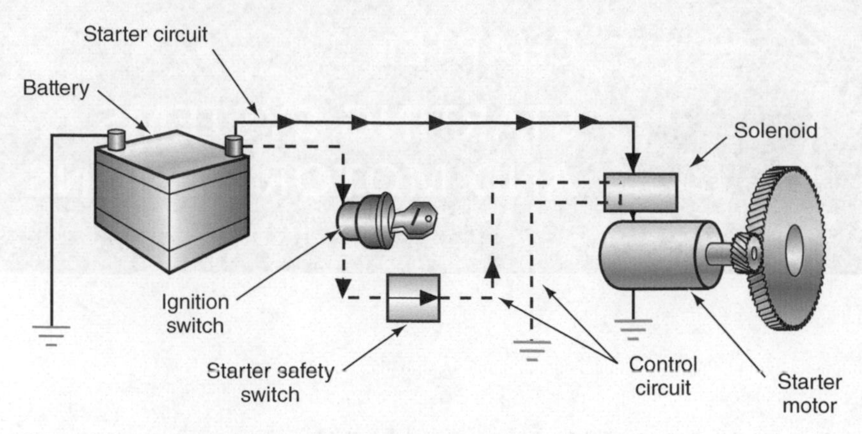

Figure 6-1 Major components of the starting system. The solid line represents the starting (cranking) circuit and the dashed line indicates the starter control circuit.

3. Ignition switch.
4. Starter solenoid or relay.
5. Starter motor.
6. Starter drive and flywheel ring gear.
7. Starter safety switch.

Figure 6-1 illustrates the components of a simplified cranking system circuit. This chapter examines both this circuit and the fundamentals of electric motor operation.

Direct-Current Motor Principles

Shop Manual
Chapter 6, page 273

DC motors use the interaction of magnetic fields to convert electrical energy into mechanical energy. Magnetic lines of force flow from the north pole to the south pole of a magnet (**Figure 6-2**). Placing a current-carrying conductor within the magnetic field results in two fields (**Figure 6-3**). On the left side of the conductor, the lines of force are in the same direction. This will concentrate the flux density of the lines of force on the left side. This will produce a strong magnetic field because the two fields will reinforce each other. The lines of force oppose each other on the right side of the conductor. This results in a weaker magnetic field. The conductor will tend to move from the strong field to the weak field (**Figure 6-4**). This principle of electromagnetism converts electrical energy into mechanical energy in a starter motor.

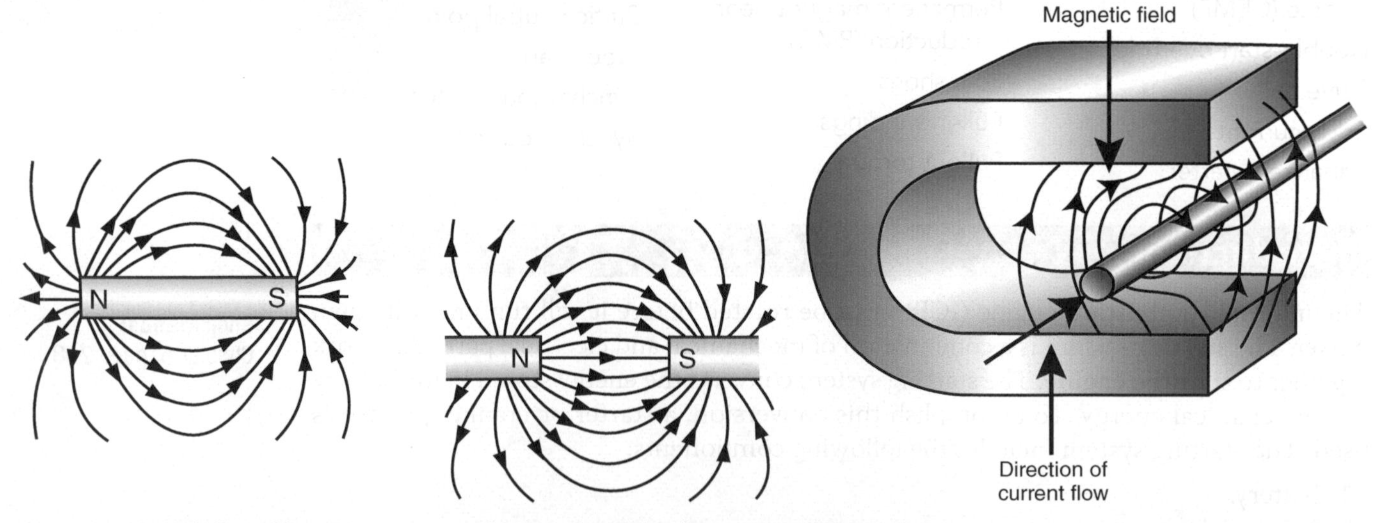

Figure 6-2 Magnetic lines of force flow from the north pole to the south pole.

Figure 6-3 Interaction of two magnetic fields.

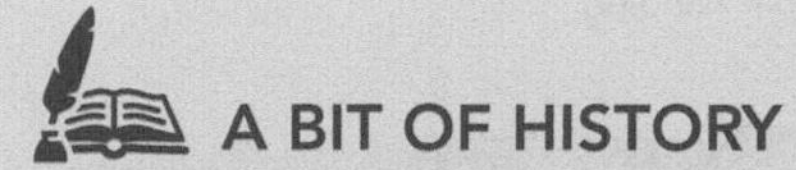

A BIT OF HISTORY

In the early days of the automobile, vehicles did not have a starter motor. The operator had to use a starting crank to turn the engine by hand. Charles F. Kettering invented the first electric self-starter, which was developed and built by the Delco Electrical Plant. The self-starter first appeared on the 1912 Cadillac and was actually a combination starter and generator.

Figure 6-5 illustrates a simple electromagnet-style starter motor. The **armature** is the inside windings and is the moveable component of the motor. The armature consists of a conductor wound around a laminated iron core. It's used to create a magnetic field. The armature rotates within the stationary outside windings, called the **field coils**, which has windings coiled around **pole shoes** (**Figure 6-6**). Field coils are heavy copper wire wrapped around an iron core to form an electromagnet. Pole shoes are made of high–magnetic permeability material to help concentrate and direct the lines of force in the field assembly.

Shop Manual Chapter 6, page 272

Shop Manual Chapter 6, page 288

Applying current to the field coils and the armature results in both producing magnetic flux lines (**Figure 6-7**). The direction of the windings will place the left pole at a south polarity and the right side at a north polarity. The lines of force move from north to south in the field. In the armature, the flux lines circle in one direction on one side of the loop and in the opposite direction on the other side. Current will now set up a

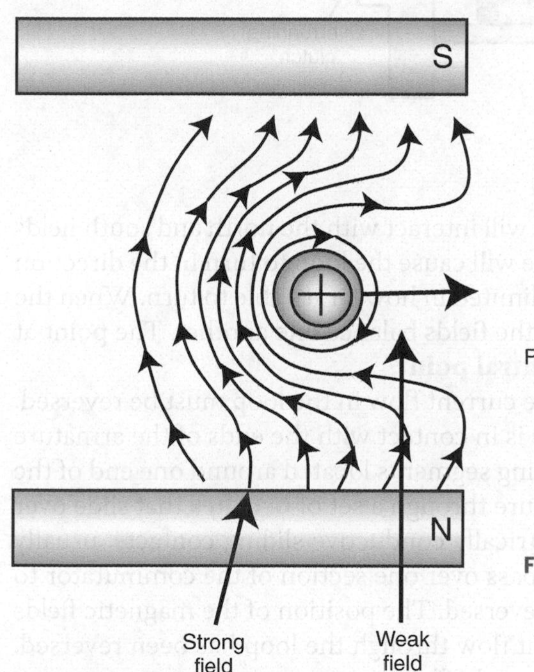

Figure 6-4 Conductor movement in a magnetic field.

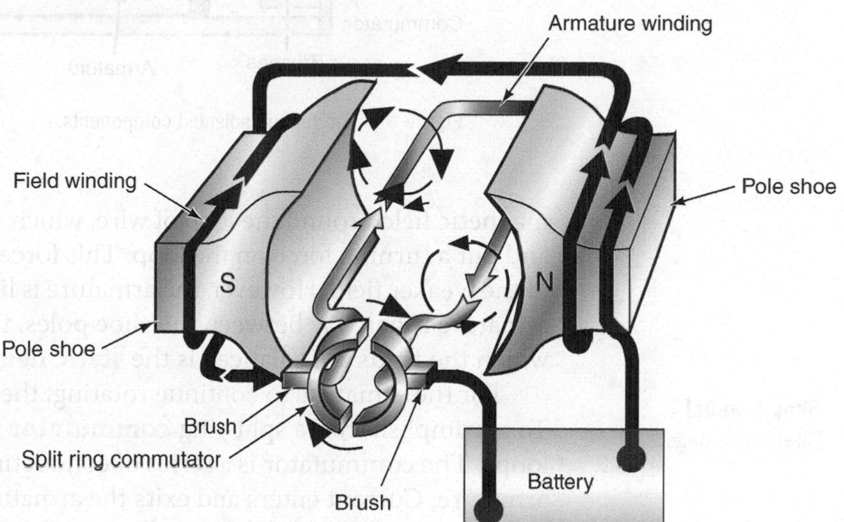

Figure 6-5 Simple electromagnetic motor.

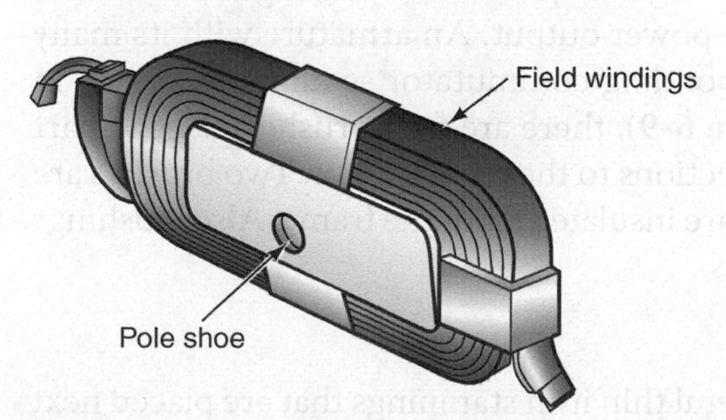

Figure 6-6 Field coil wound around a pole shoe.

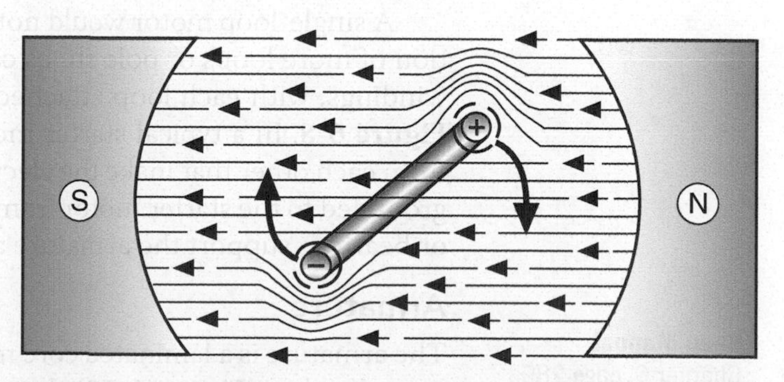

Figure 6-7 Rotation of the conductor is in the direction of the weaker field.

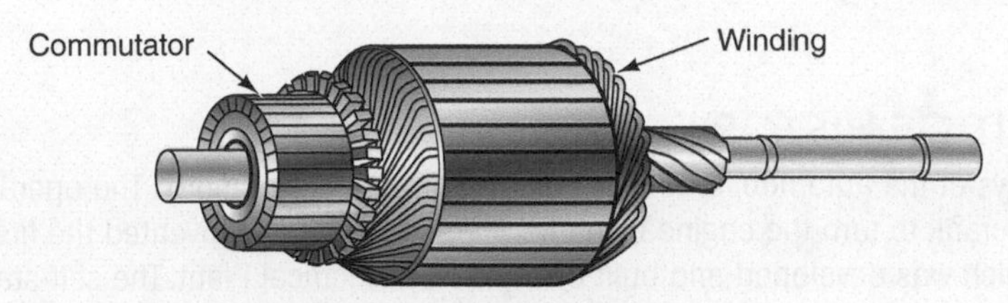

Figure 6-8 Starter armature.

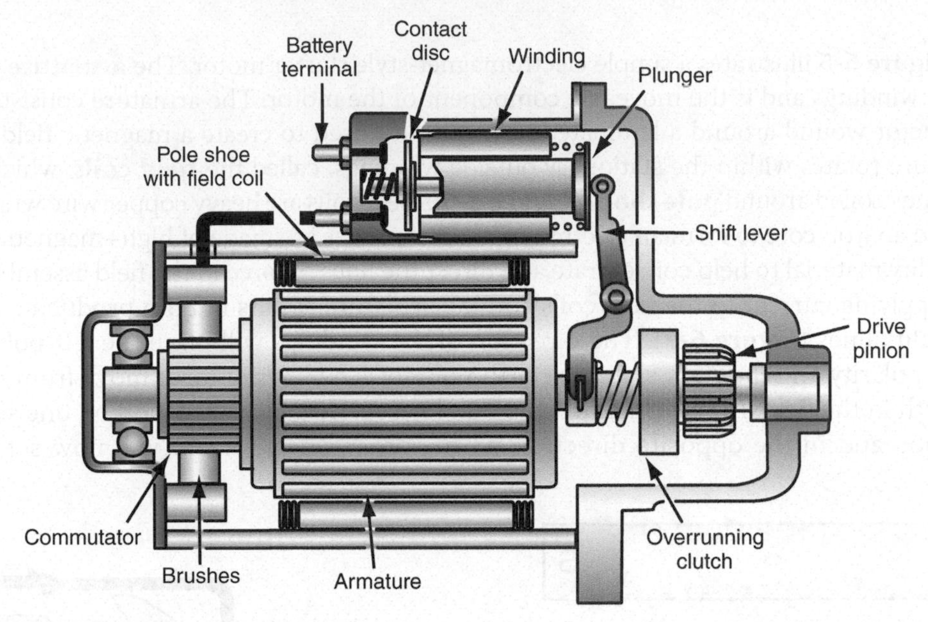

Figure 6-9 Starter and solenoid components.

magnetic field around the loop of wire, which will interact with the north and south fields and put a turning force on the loop. This force will cause the loop to turn in the direction of the weaker field. However, the armature is limited in how far it's able to turn. When the armature is halfway between the shoe poles, the fields balance one another. The point at which the fields are balanced is the **static neutral point**.

Shop Manual
Chapter 6, page 289

For the armature to continue rotating, the current flow in the loop must be reversed. To accomplish this, a split-ring **commutator** is in contact with the ends of the armature loops. The commutator is a series of conducting segments located around one end of the armature. Current enters and exits the armature through a set of **brushes** that slide over the commutator's sections. Brushes are electrically conductive sliding contacts, usually made of copper and carbon. As the brushes pass over one section of the commutator to another, the current flow in the armature is reversed. The position of the magnetic fields is the same. However, the direction of current flow through the loop has been reversed. This will continue until the current flow is turned off.

Shop Manual
Chapter 6, page 289

A single-loop motor would not produce enough torque to rotate an engine. The addition of more loops or pole shoes can increase power output. An armature with its many windings, with each loop attached to corresponding commutator sections, is shown in **Figure 6-8**. In a typical starter motor (**Figure 6-9**), there are four brushes placed apart from each other that make the electrical connections to the commutator. Two brushes are grounded to the starter motor frame and two are insulated from the frame. Also, bushings or bearings support the armature at both ends.

Armature

Shop Manual
Chapter 6, page 289

The armature is a laminated core made of several thin iron stampings that are placed next to each other (**Figure 6-10**). **Laminated construction** is used because, in a solid iron

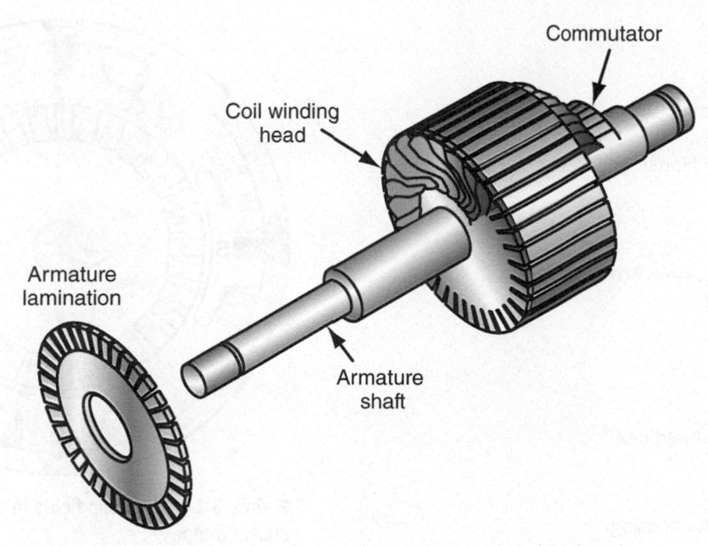

Figure 6-10 Lamination construction of a typical motor armature.

core, the magnetic fields would generate **eddy currents**. These are counter voltages induced in a core. They cause heat to build up in the core and waste energy. By using laminated construction, eddy currents in the core are minimized.

Eddy currents are closed loops of electrical current that is created within a conductor by a changing magnetic field in the conductor. Eddy currents flow in planes that are perpendicular to the magnetic field, creating a magnetic field that opposes the magnetic field that created it. The eddy current reacts back onto the source of the magnetic field.

The slots on the outside diameter of the laminations hold the armature windings. The windings loop around the core and connect to the commutator. Each commutator segment is insulated from the adjacent segments. A typical armature can have more than 30 commutator segments.

A steel shaft is fitted into the center hole of the core laminations. The commutator is insulated from the shaft.

The armature uses two basic winding patterns: lap winding and wave winding. The lap winding connects the two ends of the winding to adjacent commutator segments (**Figure 6-11**). In this pattern, the wires passing under a pole field have their current flowing in the same direction. In the wave-winding pattern, each end of the winding connects to commutator segments that are 90° or 80° apart (**Figure 6-12**). In this pattern design, some windings will have no current flow at certain positions of armature rotation. This occurs because the segment ends of the winding loop are in contact with brushes that have the same polarity. The wave-winding pattern is the most commonly used due to its lower resistance.

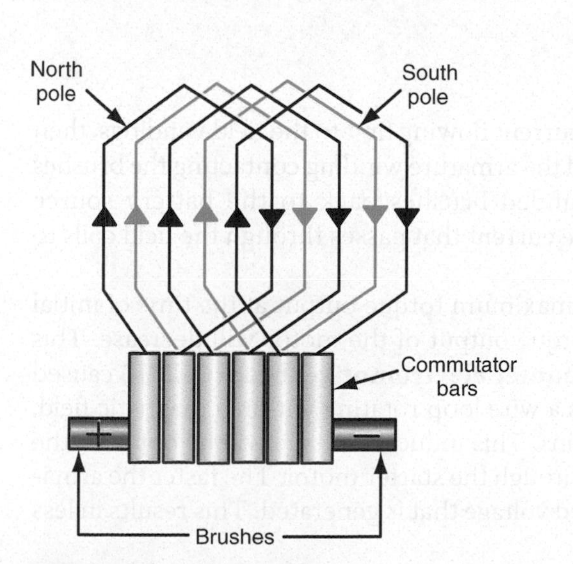

Figure 6-11 Lap winding diagram.

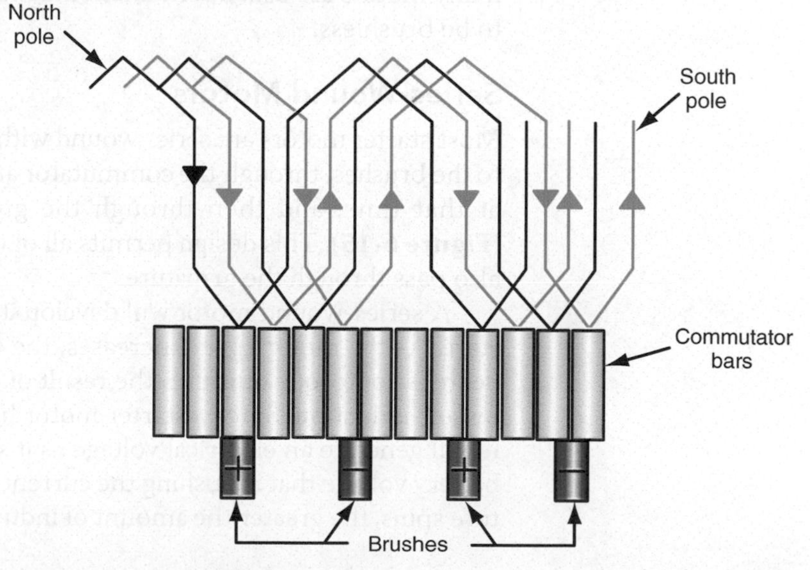

Figure 6-12 Wave-wound armature.

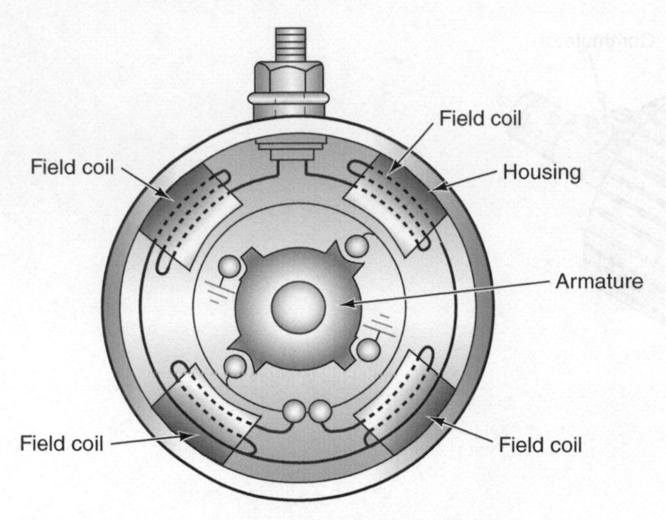

Figure 6-13 Field coils mounted to the inside of starter housing.

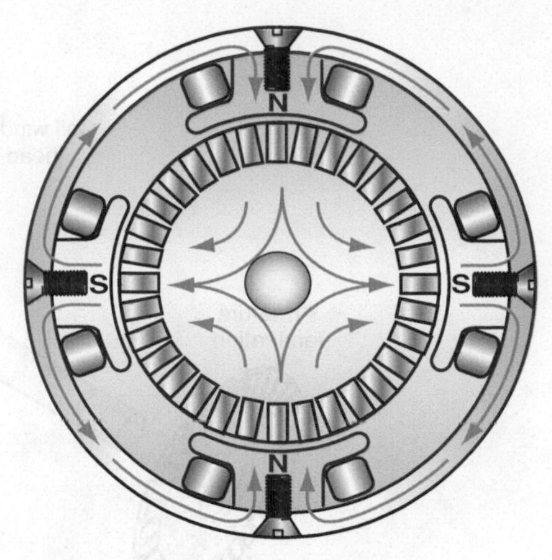

Figure 6-14 Magnetic fields in a four-pole starter motor.

Field Coils

Shop Manual
Chapter 6, page 288

The field coils are electromagnets constructed of wire ribbons or coils wound around a pole shoe. The pole shoes are constructed of heavy iron and are attached to the inside of the starter housing (**Figure 6-13**). Most starter motors use four field coils. The iron pole shoes and the iron starter housing work together to increase and concentrate the field strength of the field coils (**Figure 6-14**).

Current flow through the field coils creates strong stationary electromagnetic fields. The fields have a north and south magnetic polarity based on the direction the windings are wound around the pole shoes. The polarity of the field coils alternates to produce opposing magnetic fields.

In any DC motor, there are three methods of connecting the field coils to the armature: series, parallel (shunt), and a compound connection that uses both series and shunt coils.

DC Motor Field Winding Designs

The field windings and armature of the DC motor can be wired in various ways. The motor design refers to the method these two components are wired together. In addition, many motors are using permanent magnet fields. Also, many newer motors are designed to be brushless.

Series-Wound Motors

Most starter motors are series wound with current flowing first to the field windings, then to the brushes, through the commutator and the armature winding contacting the brushes at that time, and then through the grounded brushes back to the battery source (**Figure 6-15**). This design permits all of the current that passes through the field coils to also pass through the armature.

A series-wound motor will develop its maximum torque output at the time of initial start. As the motor speed increases, the torque output of the motor will decrease. This decrease of torque output is the result of **counter electromotive force (CEMF)** caused by self-induction. Since a starter motor has a wire loop rotating within a magnetic field, it will generate an electrical voltage as it spins. This induced voltage will be opposite the battery voltage that is pushing the current through the starter motor. The faster the armature spins, the greater the amount of induced voltage that is generated. This results in less

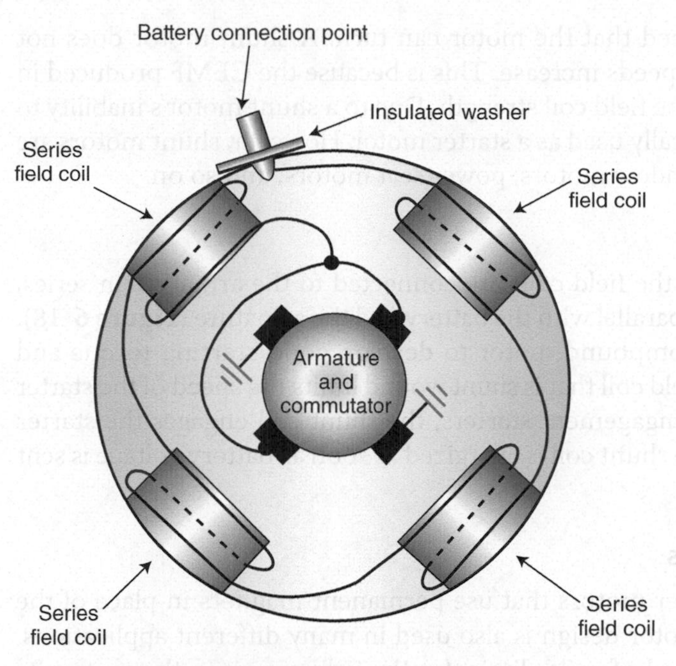

Figure 6-15 A series-wound starter motor.

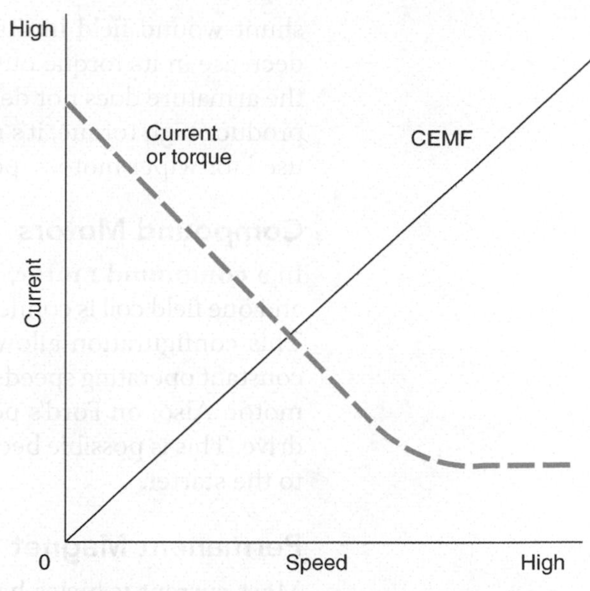

Figure 6-16 Graph illustrating the relationship between CEMF, starter motor speed, and current draw. As speed increases, so does CEMF, reducing current draw and torque.

current flow through the starter from the battery as the armature spins faster. **Figure 6-16** shows the relationship between starter motor speed and CEMF. Notice that at 0 (zero) rpm, CEMF is also at 0 (zero). At this time, maximum current flow from the battery through the starter motor will be possible. As the motor spins faster, CEMF increases and current decreases. Since current decreases, the amount of rotating force (torque) also decreases.

Shunt-Wound Motors

Electric motors, or **shunt** motors, have the field windings wired in parallel across the armature (**Figure 6-17**). Shunt means there is more than one path for current to flow. A

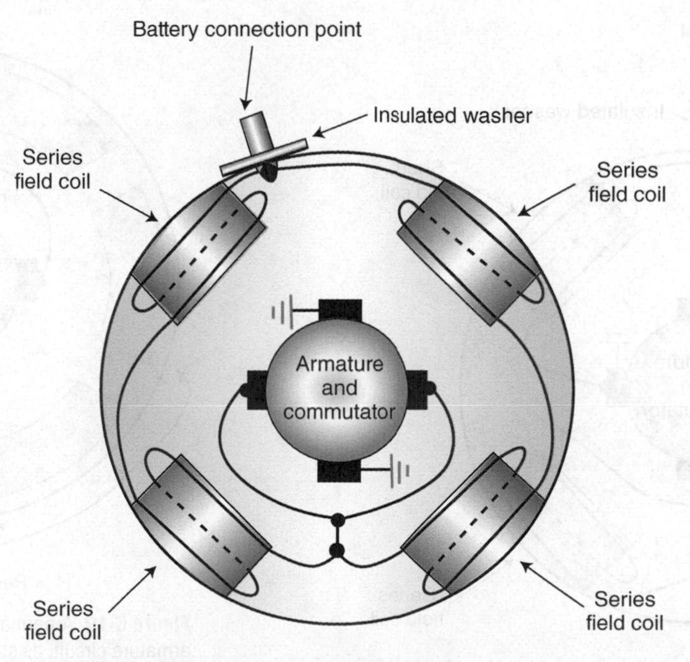

Figure 6-17 A shunt-wound (parallel) starter motor.

shunt-wound field limits the speed that the motor can turn. A shunt motor does not decrease in its torque output as speeds increase. This is because the CEMF produced in the armature does not decrease the field coil strength. Due to a shunt motor's inability to produce high torque, it's not typically used as a starter motor. However, shunt motors are used for wiper motors, power window motors, power seat motors, and so on.

Compound Motors

In a **compound motor**, most of the field coils are connected to the armature in series, and one field coil is connected in parallel with the battery and the armature (**Figure 6-18**). This configuration allows the compound motor to develop good starting torque and constant operating speeds. The field coil that is shunt wound limits the speed of the starter motor. Also, on Ford's positive-engagement starters, the shunt coil engages the starter drive. This is possible because the shunt coil is energized as soon as battery voltage is sent to the starter.

Permanent Magnet Motors

Most current vehicles have starter motors that use permanent magnets in place of the field coils (**Figure 6-19**). This motor design is also used in many different applications. Using a permanent magnet instead of coils eliminates the field circuit in the motor. By eliminating this circuit, potential electrical problems are also eliminated, such as field-to-housing shorts. Another advantage to using permanent magnets is a weight reduction of 50% compared to a typical starter motor. Most permanent magnet starters are gear-reduction–type starters.

Multiple permanent magnets are positioned in the housing around the armature. These permanent magnets are an alloy of boron, neodymium, and iron. The field strength of these magnets is much greater than typical permanent magnets. The operation of these motors is the same as other electric motors, except there is no field circuit or windings.

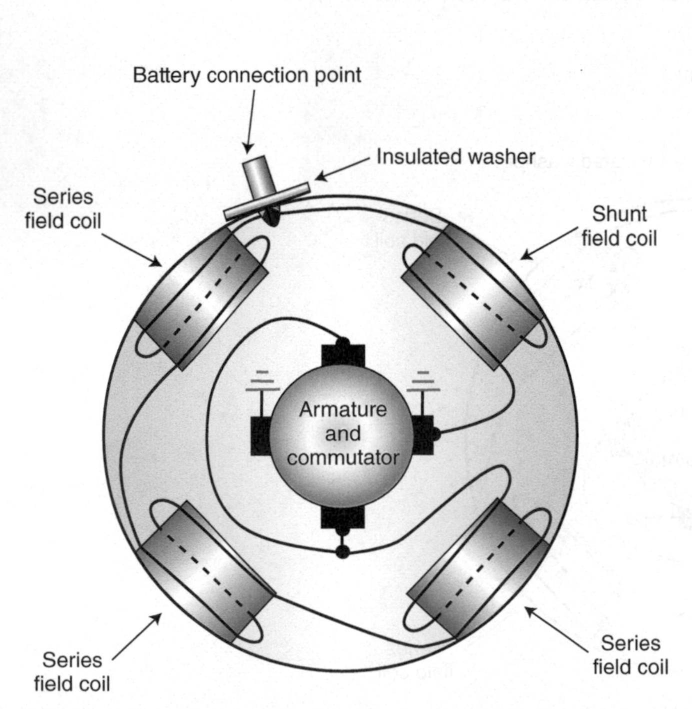

Figure 6-18 A compound motor uses both series and shunt coils.

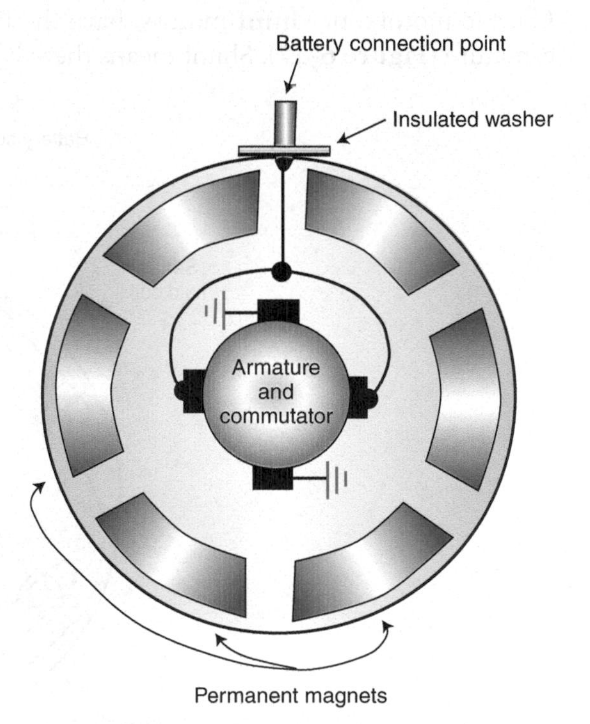

Figure 6-19 A permanent magnet motor has only an armature circuit, as strong permanent magnets create the field.

Brushless Motors

The brushless motor uses a permanent magnet rotor and electromagnet field windings (**Figure 6-20**). Since the motor design is brushless, the potential for arcing is decreased and longer service life is expected. In addition, arcing can cause electromagnetic interference that can adversely affect electronic systems. High-output brushless DC motors are used in some HEVs (**Figure 6-21**).

The field windings of a brushless motor are also called the stator.

Control of the stator is by an electronic circuit that switches the current flow as needed to keep the rotor turning. Power transistors wired as "H" gates reverse current

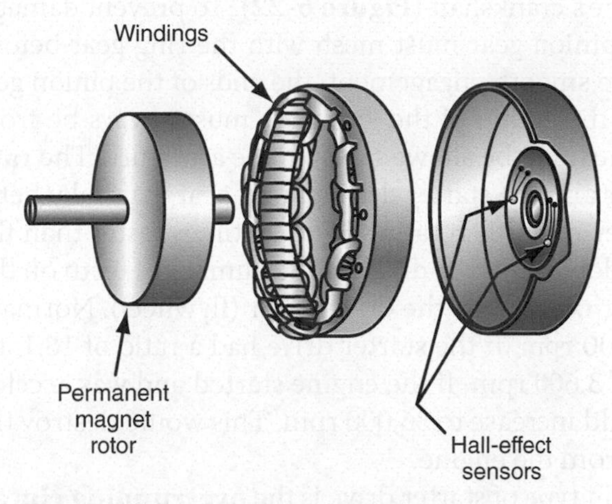

Figure 6-20 Components of a brushless DC motor. The Hall-effect sensor is used to determine rotor position.

Figure 6-21 Brushless motor used by Honda in some of its HEVs.

flow according to the position of the rotor. Pulse-width modulation (PWM) of the driver circuits control the motor's speed. Rotor position is usually monitored by the use of Hall-effect sensors. However, rotor position can also be determined by monitoring the CEMF that is present in stator windings that are not energized.

Starter Drives

The **starter drive** is the part of the starter motor that engages the armature to the engine flywheel ring gear. A starter drive includes a pinion gear set that meshes with the flywheel ring gear on the engine's crankshaft (**Figure 6-22**). To prevent damage to the pinion gear or the ring gear, the pinion gear must mesh with the ring gear before the starter motor rotates. To help assure smooth engagement, the ends of the pinion gear teeth are tapered (**Figure 6-23**). Also, the action of the armature must always be from the motor to the engine. The engine must not be allowed to spin the armature. The **ratio** of the number of teeth on the ring gear and the starter drive pinion gear is usually between 15:1 and 20:1. This means the starter motor is rotating 15 to 20 times faster than the engine. The ratio of the starter drive is determined by dividing the number of teeth on the drive gear (pinion gear) into the number of teeth on the driven gear (flywheel). Normal cranking speed for the engine is about 200 rpm. If the starter drive had a ratio of 18:1, the starter would be rotating at a speed of 3,600 rpm. If the engine started and was accelerated to 2,000 rpm, the starter speed would increase to 36,000 rpm. This would destroy the starter motor if it was not disengaged from the engine.

Shop Manual
Chapter 6, page 291

The most common type of starter drive is the **overrunning clutch**. The overrunning clutch is a roller-type clutch that transmits torque in one direction and freewheels in the other direction. This allows the starter motor to transmit torque to the ring gear but prevents the ring gear from transferring torque to the starter motor.

In a typical overrunning-type clutch (**Figure 6-24**), the clutch housing is internally splined to the starter armature shaft. The drive pinion turns freely on the armature shaft within the clutch housing. When torque is transmitted through the armature to the clutch housing, the spring-loaded rollers are forced into the small ends of their tapered slots (**Figure 6-25**). They are then wedged tightly against the pinion barrel. The pinion barrel and clutch housing are now locked together; torque is transferred through the starter motor to the ring gear and engine.

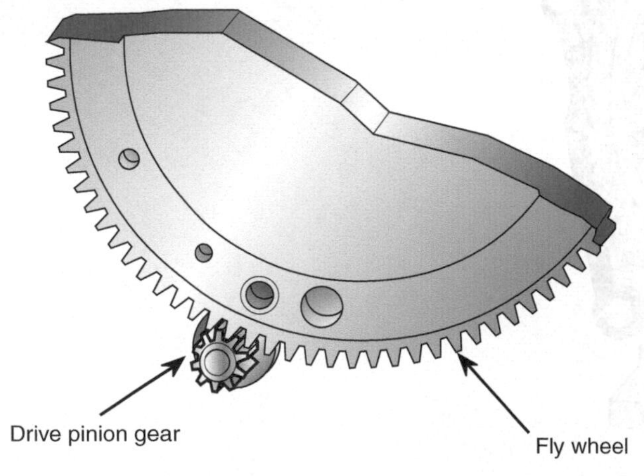

Figure 6-22 Starter drive pinion gear is used to turn the engine's flywheel.

Figure 6-23 The pinion gear teeth are tapered to allow for smooth engagement.

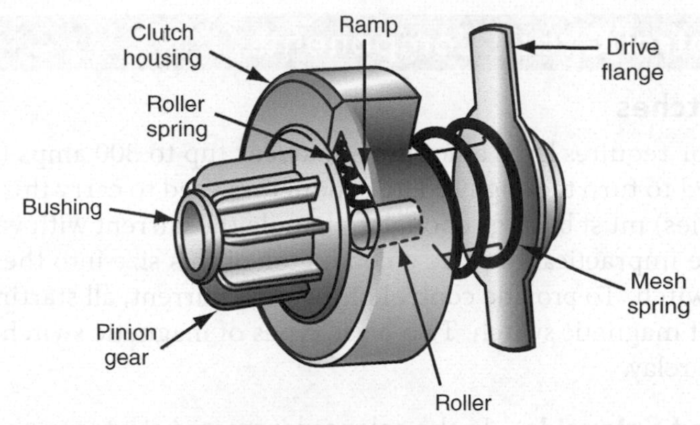

Figure 6-24 Overrunning clutch starter drive.

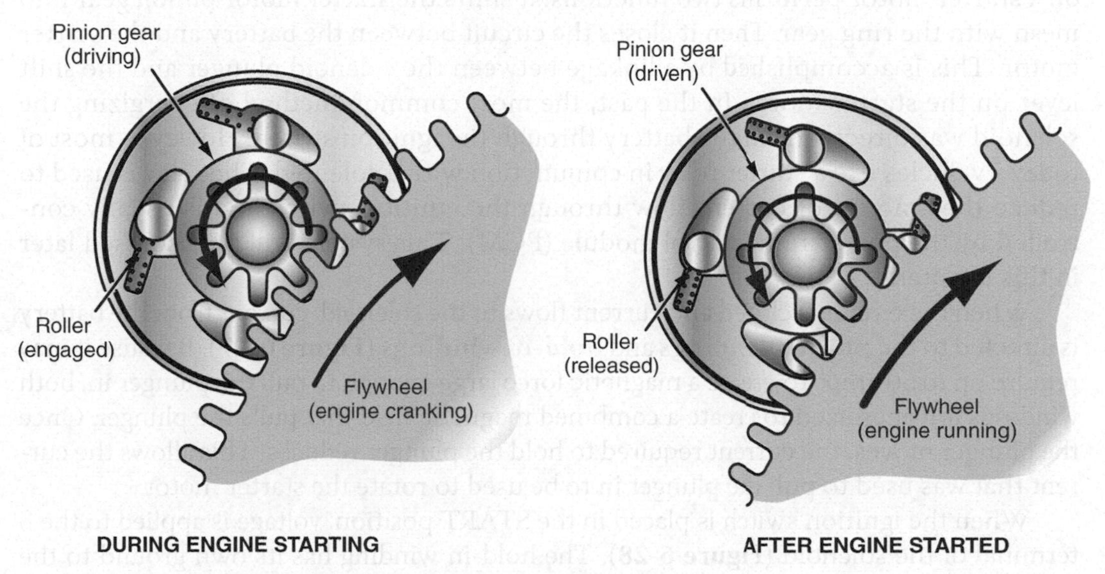

Figure 6-25 When the armature turns, it locks the rollers into the tapered notch.

When the engine starts, and is running under its own power, the ring gear attempts to drive the pinion gear faster than the starter motor. This unloads the clutch rollers and releases the pinion gear to rotate freely around the armature shaft.

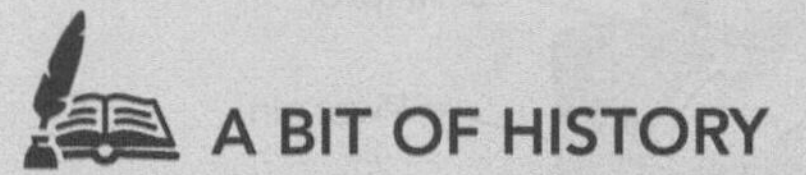

A BIT OF HISTORY

The integrated key starter switch was introduced in 1949 by Chrysler. Before this, the key turned the system on and the driver pushed a starter button.

Cranking Motor Circuits

The starting system of the vehicle consists of two circuits: the starter control circuit and the motor feed circuit. These circuits are separate but related. The control circuit consists of the starting portion of the ignition switch, the starting safety switch (if applicable), and the wire conductor to connect these components to the relay or solenoid. The motor feed circuit consists of heavy battery cables from the battery to the relay and the starter or directly to the solenoid if the starter is so equipped.

Starter Control Circuit Components

Magnetic Switches

Shop Manual
Chapter 6,
pages 279, 280

The starter motor requires large amounts of current [up to 300 amps (A)] to generate the torque needed to turn the engine. The conductors used to carry this amount of current (battery cables) must be large enough to handle the current with very little voltage drop. It would be impractical to place a conductor of this size into the wiring harness to the ignition switch. To provide control of the high current, all starting systems contain some type of magnetic switch. Two basic types of magnetic switches are used: the solenoid and the relay.

Starter-Mounted Solenoids. In the solenoid-actuated starter system, the solenoid is mounted directly on top of the starter motor (**Figure 6-26**). The solenoid switch on a starter motor performs two functions: it shifts the starter motor pinion gear into mesh with the ring gear. Then it closes the circuit between the battery and the starter motor. This is accomplished by a linkage between the solenoid plunger and the shift lever on the starter motor. In the past, the most common method of energizing the solenoid was directly from the battery through the ignition switch. However, most of today's vehicles use a starter relay in conjunction with a solenoid. The relay is used to reduce the amount of current flow through the ignition switch and is usually controlled by the powertrain control module (PCM). This system will be discussed later in this chapter.

The two windings of the solenoid are called the **pull-in windings** and the **hold-in windings**. Their names explain their functions.

When the circuit is closed and current flows to the solenoid, current from the battery is directed to the **pull-in windings** and **hold-in windings** (**Figure 6-27**). Because it may require up to 50 amps to create a magnetic force large enough to pull the plunger in, both windings are energized to create a combined magnetic field that pulls the plunger. Once the plunger moves, the current required to hold the plunger reduces. This allows the current that was used to pull the plunger in to be used to rotate the starter motor.

When the ignition switch is placed in the START position, voltage is applied to the S terminal of the solenoid (**Figure 6-28**). The hold-in winding has its own ground to the

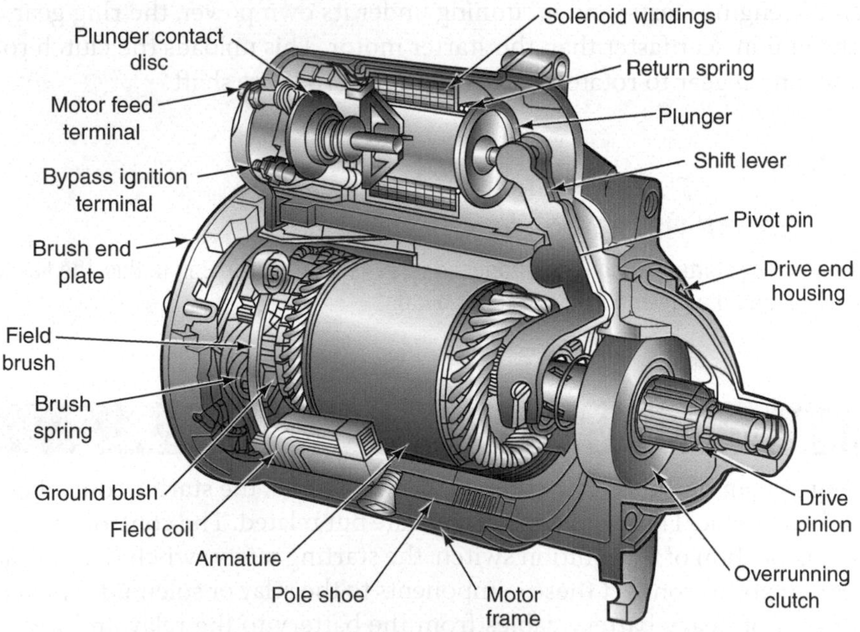

Figure 6-26 Solenoid-operated starter has the solenoid mounted directly on top of the motor.

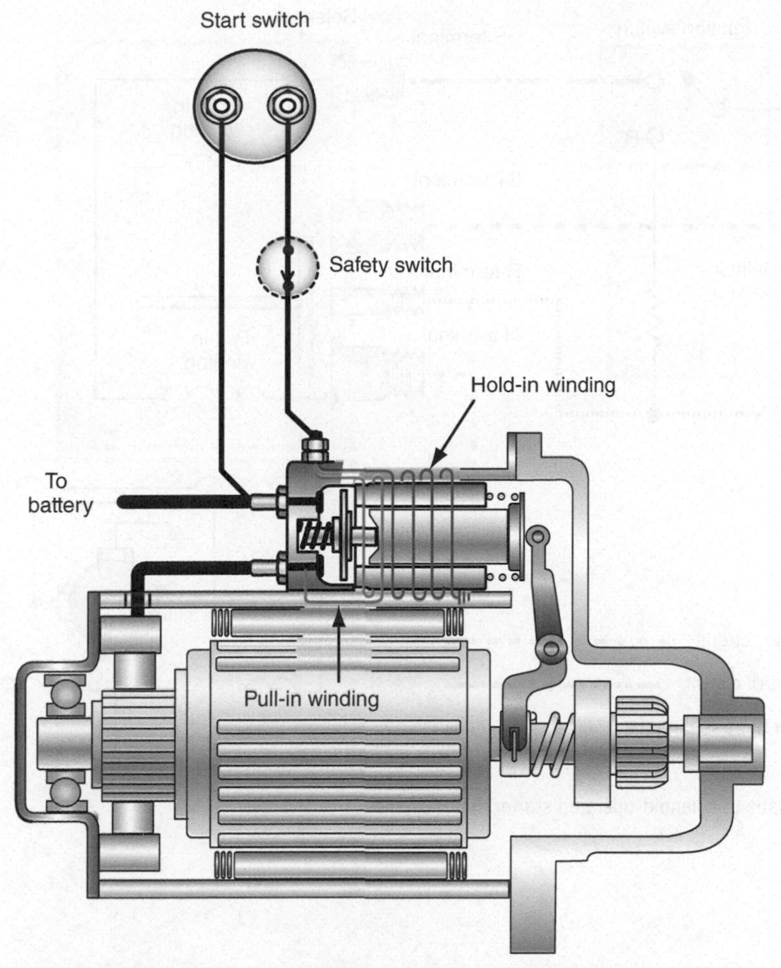

Figure 6-27 The solenoid uses two windings. Both are energized to draw the plunger; then only the hold-in winding is used to hold the plunger in position.

case of the solenoid. The pull-in winding's ground is through the starter motor. Current will flow through both windings to produce a strong magnetic field. When the plunger is moved into contact with the main battery and motor terminals, the pull-in winding is de-energized. The pull-in winding is not energized because the contact places battery voltage on both sides of the coil (**Figure 6-29**). The current that was directed through the pull-in winding is now sent to the motor.

Because the contact disc does not close the circuit from the battery to the starter motor until the plunger has moved the shift lever, the pinion gear is in full mesh with the flywheel before the armature starts to rotate.

After the engine is started, releasing the key to the RUN position opens the control circuit. Voltage no longer is supplied to the hold-in windings, and the return spring causes the plunger to return to its neutral position.

In Figures 6-28 and 6-29, an R terminal is illustrated. This terminal provides current to the ignition bypass circuit that is used to provide full battery voltage to the ignition coil while the engine is cranking. This circuit bypasses the ballast resistor. The bypass circuit is not used on most ignition systems today.

A common problem with the control circuit is that low system voltage or an open in the hold-in windings will cause an oscillating action to occur. The combination of the pull-in winding and the hold-in winding is sufficient to move the plunger. However, once the contacts are closed, there is insufficient magnetic force to hold the plunger in place.

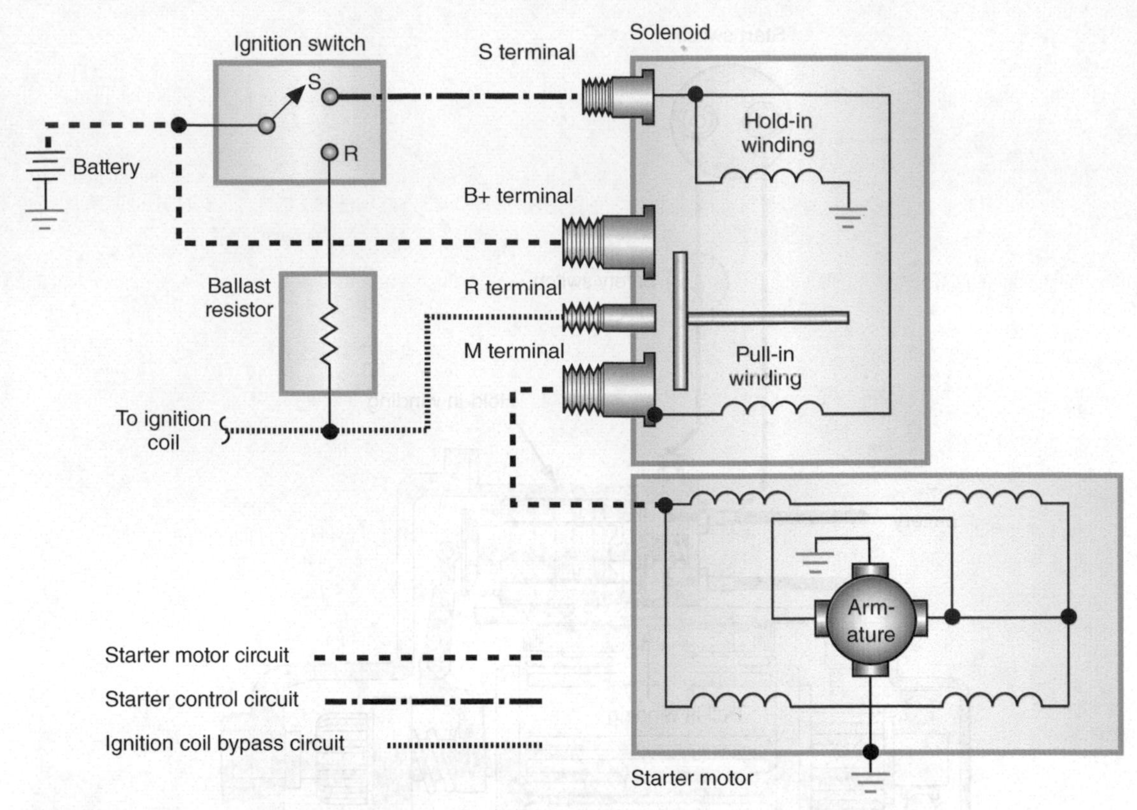

Figure 6-28 Schematic of solenoid-operated starter motor circuit.

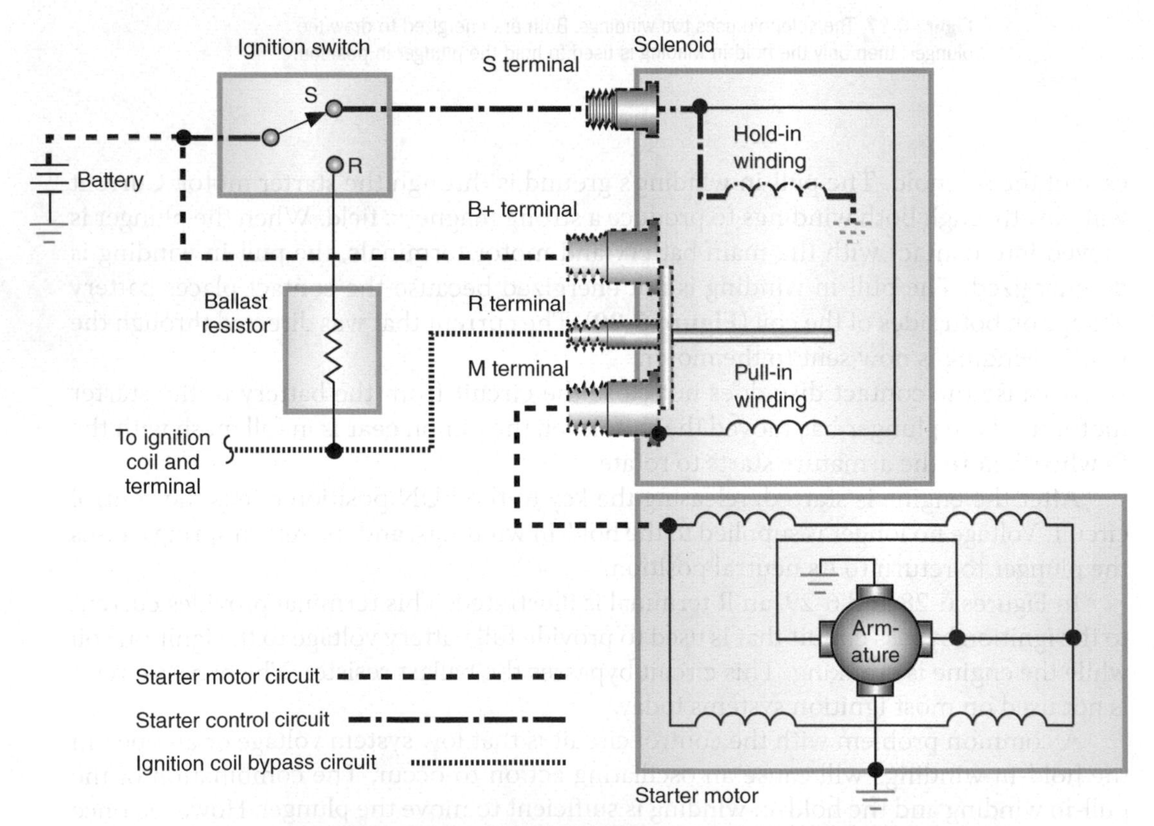

Figure 6-29 Once the contact disc closes the terminals, the hold-in winding is the only one that is energized.

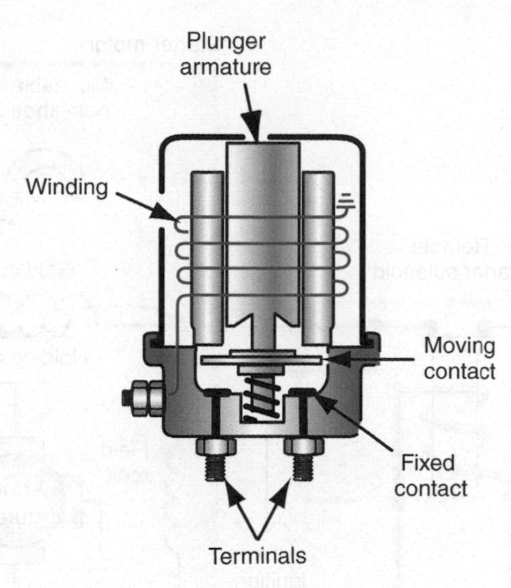

Figure 6-30 A remote starter solenoid, often referred to as the starter relay.

This condition is recognizable by a series of clicks when the ignition switch is turned to the START position. Before replacing the solenoid, check the battery condition; a low battery charge will cause the same symptom.

AUTHOR'S NOTE Some manufacturers use a starter relay in conjunction with a solenoid relay. The relay is used to reduce the amount of current flow through the ignition switch.

Remote Solenoids. Some manufacturers use a starter solenoid that is mounted near the battery on the fender well or radiator support (**Figure 6-30**). Unlike the starter-mounted solenoid, the remote solenoid does not move the pinion gear into mesh with the flywheel ring gear.

Shop Manual
Chapter 6, page 280

When the ignition switch is turned to the START position, current is supplied through the switch to the solenoid windings. The windings produce a magnetic field that pulls the moveable core into contact with the internal contacts of the battery and starter terminals (**Figure 6-31**). With the contacts closed, full battery current is supplied to the starter motor.

Many manufacturers call the remote solenoid the starter relay.

A secondary function of the starter relay is to provide for an alternate path for current to the ignition coil during cranking. This is done by an internal connection that is energized by the relay core when it completes the circuit between the battery and the starter motor.

Starter Relay Controls

Most modern vehicles will use a starter relay in conjunction with a starter motor–mounted solenoid to control starter motor operation. The relay can be controlled through the ignition switch or by the PCM.

In a system that uses the ignition switch to control the relay, the switch will usually be installed on the insulated side of the relay control circuit (**Figure 6-32**). When the ignition switch is turned to the START position, battery voltage is applied to the coil of the relay.

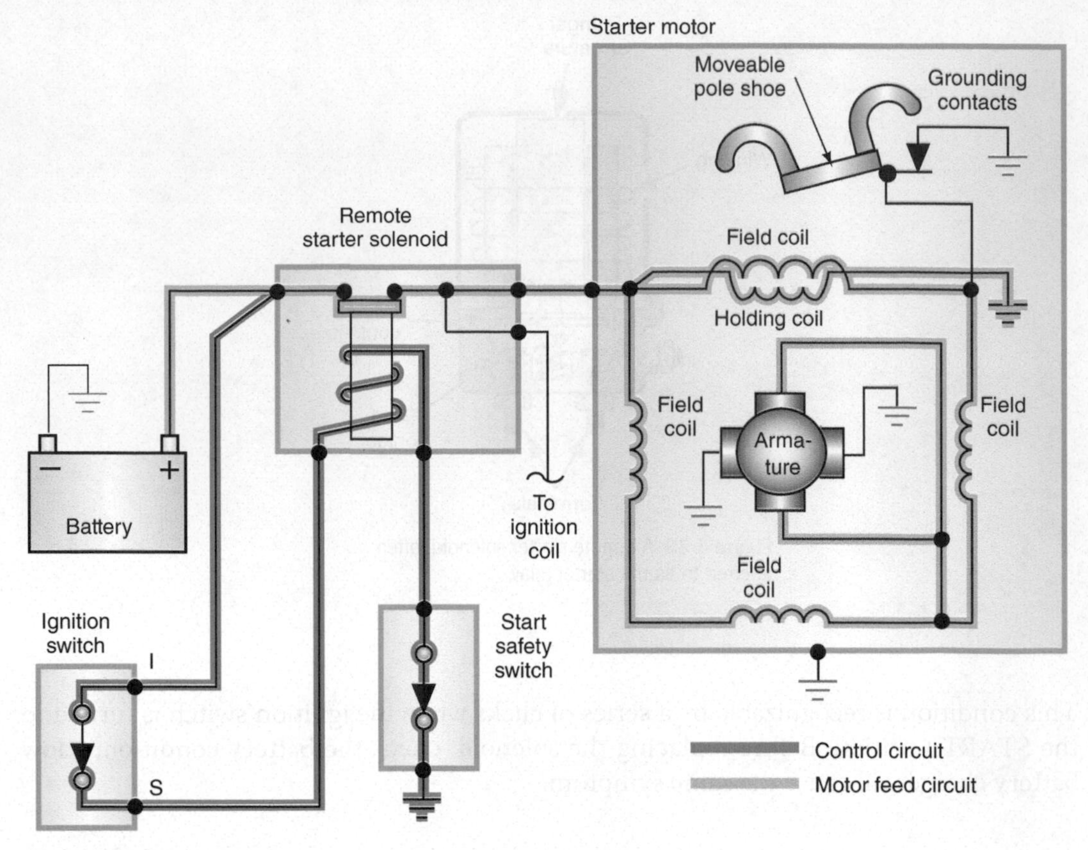

Figure 6-31 Current flow when the remote starter solenoid is energized.

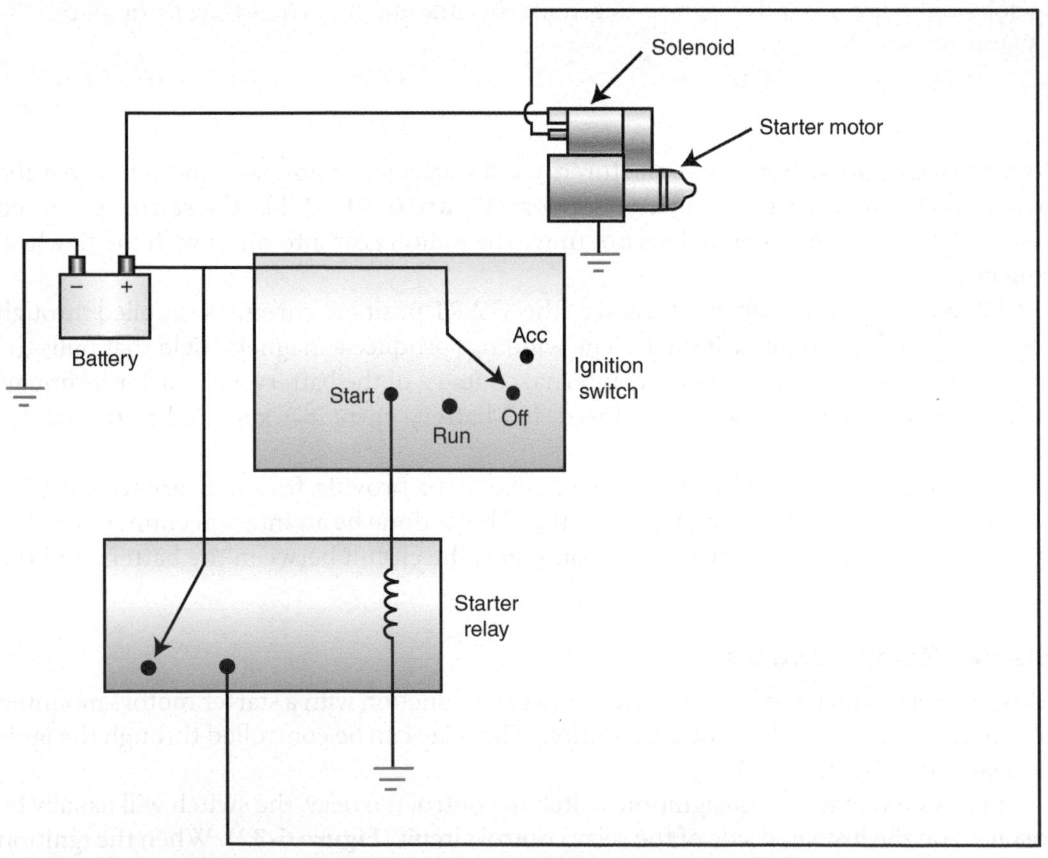

Figure 6-32 Starter control circuit using an insulated side relay to control current to the starter solenoid.

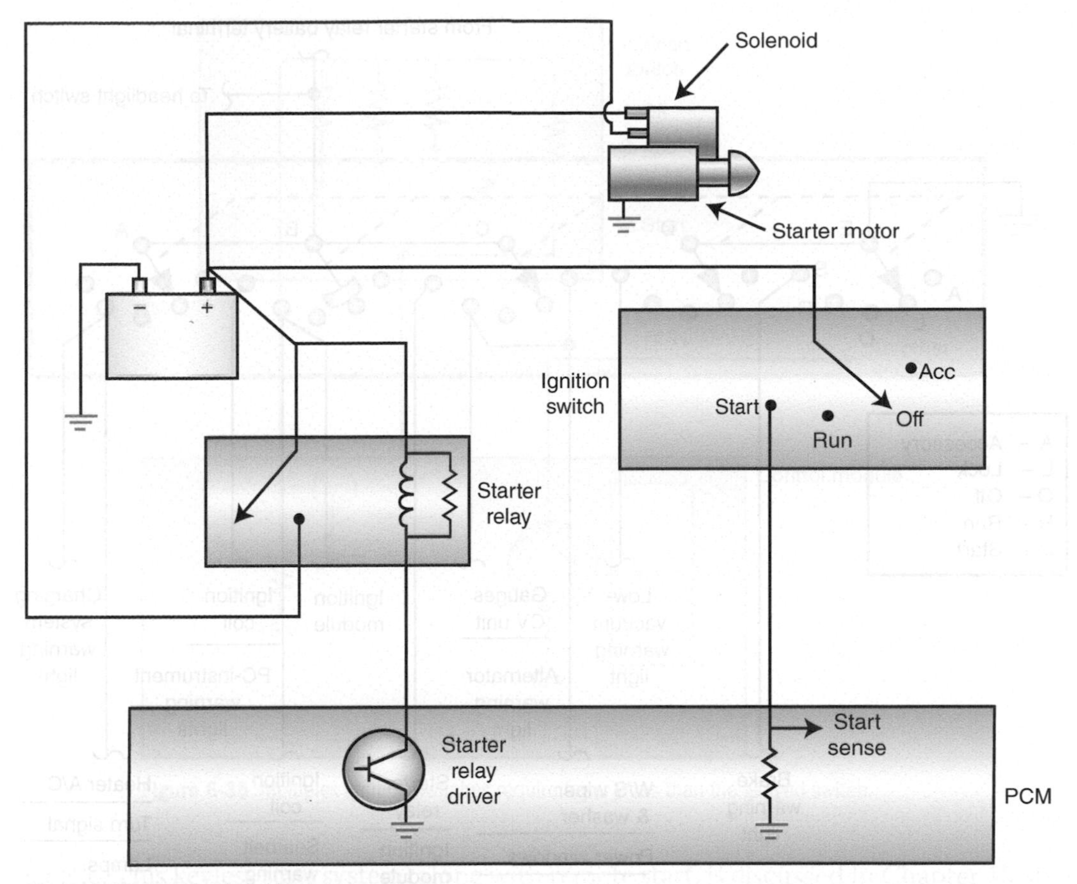

Figure 6-33 Typical PCM starter control circuit.

Since the relay coil is grounded, the coil is energized and pulls the contacts closed. With the contacts closed, battery voltage is applied to the control side of the starter solenoid. The solenoid operates in the same manner as discussed previously.

In this type of system, a very small wire can be used as the ignition switch. This reduces the size of the wiring harness.

In a PCM-controlled system, the PCM will monitor the ignition switch position to determine if the starter motor should be energized. System operation differs among manufacturers. However, in most systems, the PCM will control the starter relay coil ground circuit (**Figure 6-33**). Control by the PCM allows the manufacturer to install software commands such as **double-start override**, which prevents the starter motor from being energized if the engine is already running, and **sentry key** within the PCM.

Sentry key is one of the terms used to describe a sophisticated antitheft system that prevents the engine from starting unless a special key is used.

Ignition Switch

The ignition switch is the power distribution point for most of the vehicle's primary electrical systems (**Figure 6-34**). Most ignition switches have five positions:

Shop Manual
Chapter 6, page 282

1. ACCESSORIES: Supplies current to the vehicle's electrical accessory circuits. It will not supply current to the engine control circuits, starter control circuit, or the ignition system.
2. LOCK: Mechanically locks the steering wheel and transmission gear selector. All electrical contacts in the ignition switch are open. Most ignition switches must be in this position to insert or remove the key from the cylinder.
3. OFF: All circuits controlled by the ignition switch are opened. The steering wheel and transmission gear selector are unlocked.

the switch is closed, allowing current to flow to the starter circuit. If the transmission is in a gear position, the switch is opened and current cannot flow to the starter circuit.

The neutral safety switch feature can also be a function of the transmission range switch (**Figure 6-37**). The range switch is used by the transmission control module to determine transmission range position. This information is then broadcasted on the network bus to the PCM. If the PCM receives the PARK or NEUTRAL position input, it will allow the starter relay to be energized.

Many vehicles equipped with manual transmissions use a similar type of safety switch. The **start clutch interlock switch** is usually operated by movement of the clutch pedal (**Figure 6-38**). When the clutch pedal is pushed downward, the switch closes and current can flow through the starter circuit. If the clutch pedal is left up, the switch is open and current cannot flow.

Some vehicles use a mechanical linkage that blocks movement of the ignition switch cylinder unless the transmission is in PARK or NEUTRAL (**Figure 6-39**).

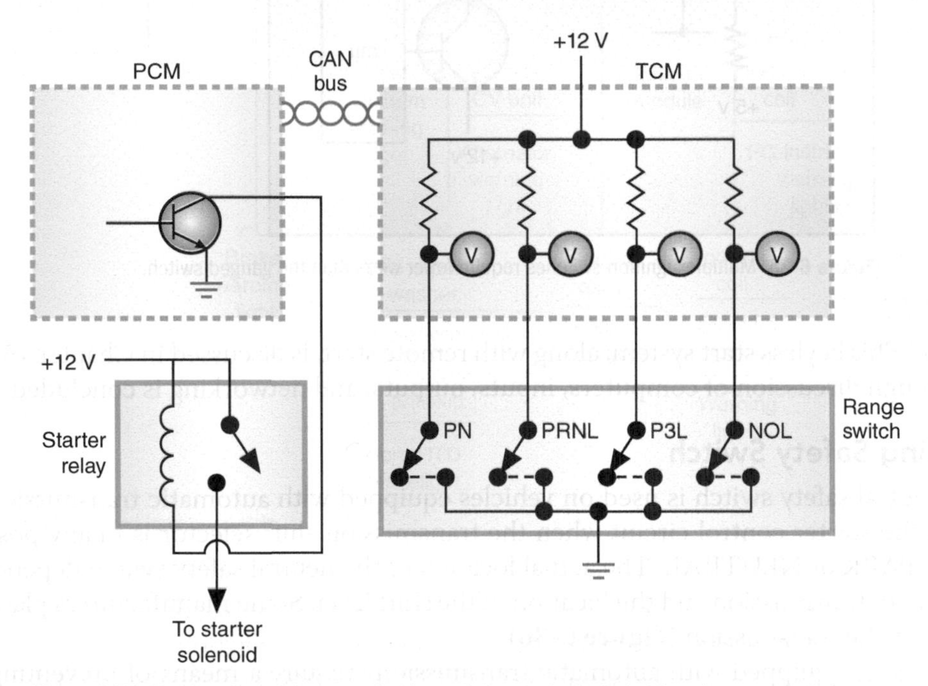

Figure 6-37 The function of the neutral safety switch may be included into the range switch.

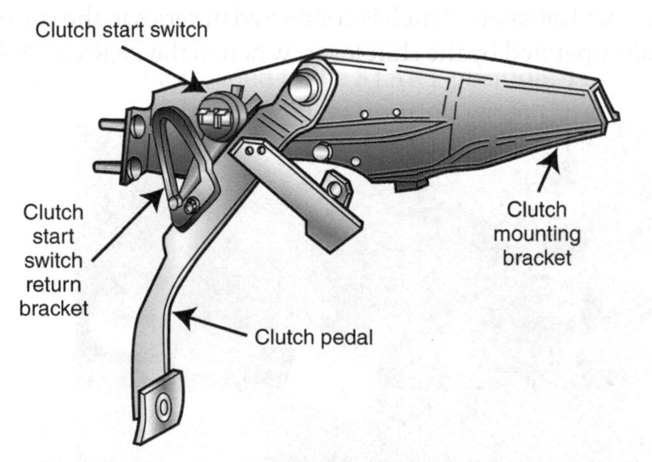

Figure 6-38 Most vehicles with a manual transmission use a clutch start switch to prevent the engine from starting unless the clutch pedal is pressed.

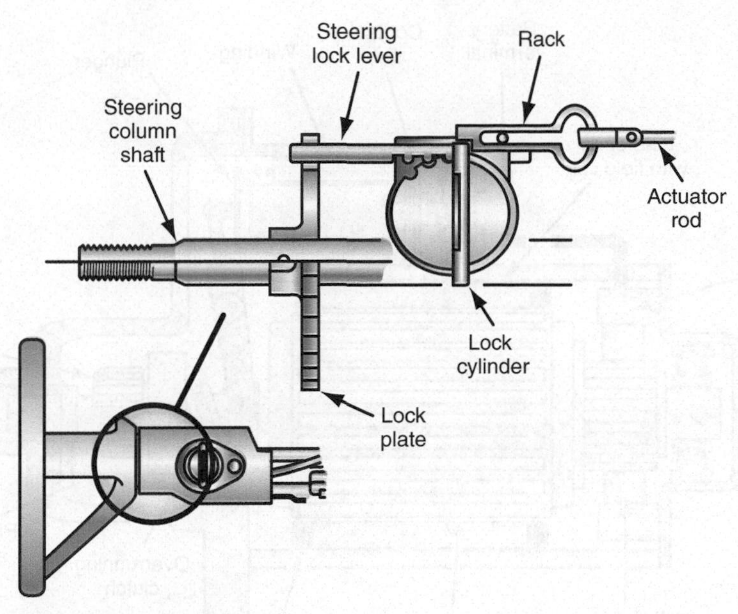

Figure 6-39 Mechanical linkage used to prevent starting the engine while the transmission is in gear.

AUTHOR'S NOTE One-touch and remote starting systems will be discussed in Chapter 14.

Cranking Motor Designs

The most common type of starter motor used today incorporates the overrunning clutch starter drive instead of the old inertia-engagement Bendix drive. There are four basic groups of starter motors:

Shop Manual
Chapter 6, page 285

1. Direct drive.
2. Gear reduction.
3. Positive-engagement (moveable pole).
4. Permanent magnet.

Direct Drive Starters

A common type of starter motor is the solenoid-operated direct drive unit (**Figure 6-40**). Although there are construction differences between applications, the operating principles are the same for all solenoid-shifted starter motors.

The direct drive starter motor can be either series wound or compound motors.

When the ignition switch is placed in the START position, the control circuit energizes the pull-in and hold-in windings of the solenoid. The solenoid plunger moves and pivots the shift lever, which in turn locates the drive pinion gear into mesh with the engine flywheel.

When the solenoid plunger is moved all the way, the contact disc closes the circuit from the battery to the starter motor. Current now flows through the field coils and the armature. This develops the magnetic fields that cause the armature to rotate, thus turning the engine.

Gear Reduction Starters

Some manufacturers use a gear reduction starter to provide increased torque (**Figure 6-41**). The gear-reduction starter differs from most other designs, in that the armature does not drive the pinion gear directly. In this design, the armature drives a small gear that is in

Some gear reduction starter motors are compound motors.

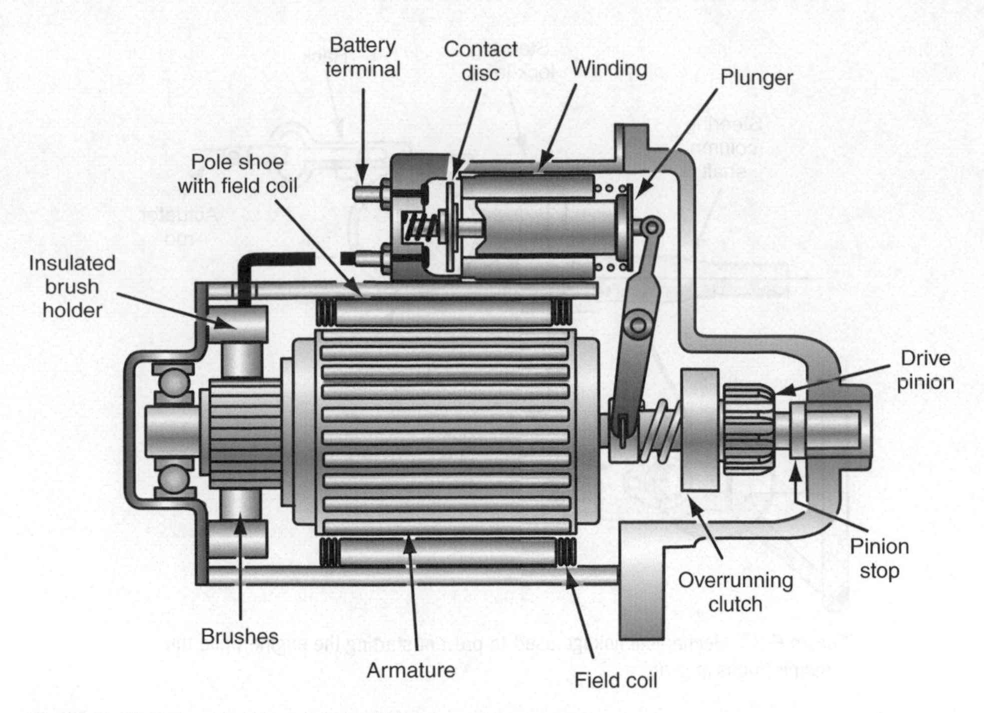

Figure 6-40 Solenoid-operated Delco MT series starter motor.

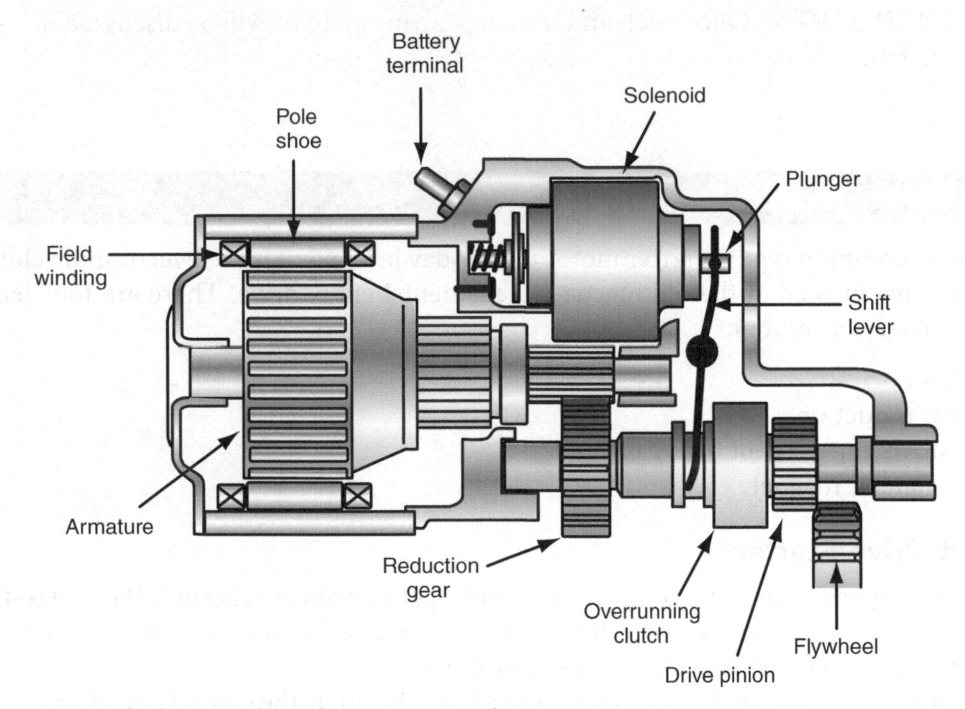

Figure 6-41 Gear reduction starter motor construction.

Many gear reduction starters have the commutator and brushes located in the center of the motor.

constant mesh with a larger gear. Depending on the application, the ratio between these two gears is between 2:1 and 3.5:1. The additional reduction allows for a small motor to turn at a greater torque with less current draw.

The solenoid operation is similar to that of the solenoid-shifted direct drive starter in that the solenoid moves the plunger, which engages the starter drive.

Positive-Engagement Starters

Shop Manual
Chapter 6, page 285

A commonly used starter on Ford applications in the past was the positive-engagement starter (**Figure 6-42**). Positive-engagement starters use the shunt coil windings of the starter motor to engage the starter drive. The high starting current is controlled by a starter

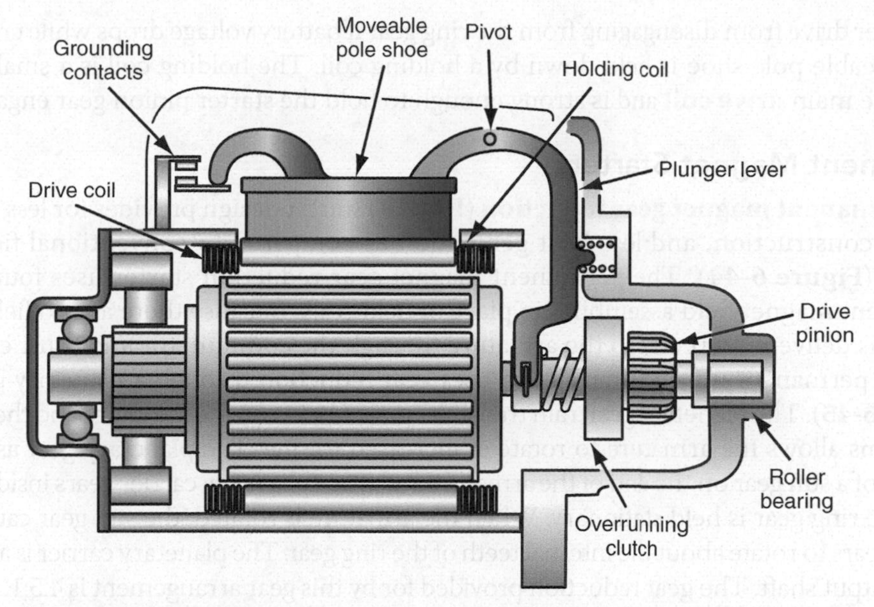

Figure 6-42 Positive-engagement starters use a moveable pole shoe.

solenoid mounted close to the battery. When the solenoid contacts are closed, current flows through a drive coil. The drive coil creates an electromagnetic field that attracts a moveable pole shoe. The moveable pole shoe is attached to the starter drive through the plunger lever. When the moveable pole shoe moves, the drive gear engages the engine flywheel.

Positive-engagement starters are also called moveable-pole shoe starters.

As soon as the starter drive pinion gear contacts the ring gear, a contact arm on the pole shoe opens a set of normally closed grounding contacts (**Figure 6-43**). With the return to ground circuit opened, all the starter current flows through the remaining three field coils and through the brushes to the armature. The starter motor then begins to rotate. To prevent

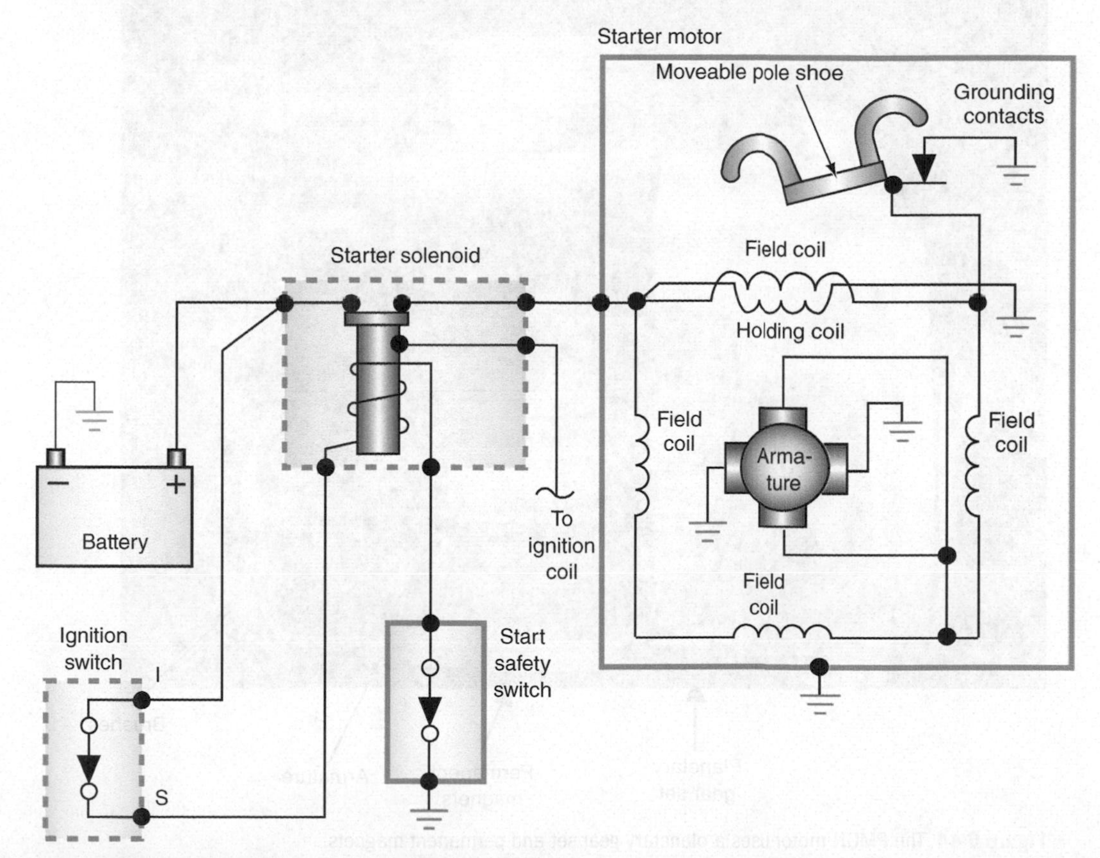

Figure 6-43 Schematic of positive-engagement starter.

The **drive coil** is a hollowed field coil that is used to attract the moveable pole shoe.

the starter drive from disengaging from the ring gear if battery voltage drops while cranking, the moveable pole shoe is held down by a holding coil. The holding coil is a smaller coil inside the main **drive coil** and is strong enough to hold the starter pinion gear engaged.

Permanent Magnet Starters

Shop Manual
Chapter 6, page 285

The **permanent magnet gear reduction (PMGR)** starter design provides for less weight, simpler construction, and less heat generation as compared to conventional field coil starters (**Figure 6-44**). The permanent magnet gear reduction starter uses four or six permanent magnet field assemblies in place of field coils. Because there are no field coils, current is delivered directly to the armature through the commutator and brushes.

The permanent magnet starter also uses gear reduction through a planetary gear set (**Figure 6-45**). The planetary geartrain transmits power between the armature and the pinion shaft. This allows the armature to rotate at increased torque. The planetary gear assembly consists of a sun gear on the end of the armature and three planetary carrier gears inside a ring gear. The ring gear is held stationary. When the armature is rotated, the sun gear causes the carrier gears to rotate about the internal teeth of the ring gear. The planetary carrier is attached to the output shaft. The gear reduction provided for by this gear arrangement is 4.5:1. By providing for this additional gear reduction, the demand for high current is lessened.

The electrical operation between the conventional field coil and PMGR starters remains basically the same (**Figure 6-46**).

AUTHOR'S NOTE The greatest amount of gear reduction from a planetary gear set is accomplished by holding the ring gear, inputting the sun gear, and outputting the carrier.

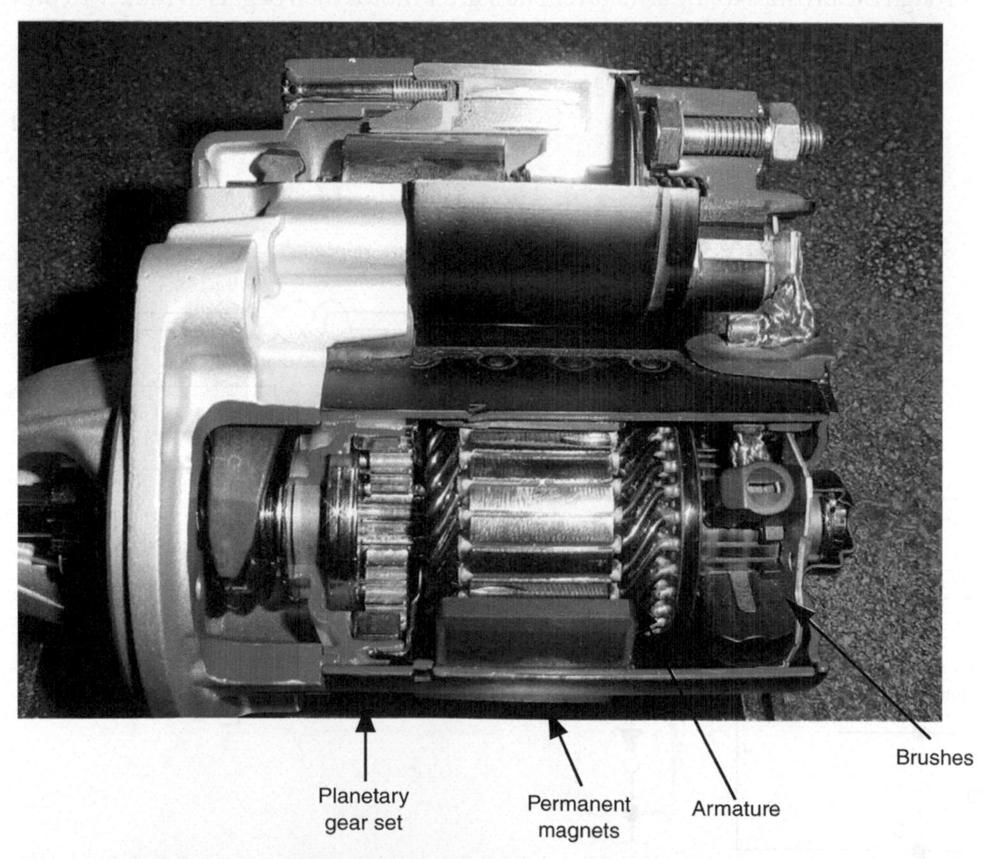

Figure 6-44 The PMGR motor uses a planetary gear set and permanent magnets.

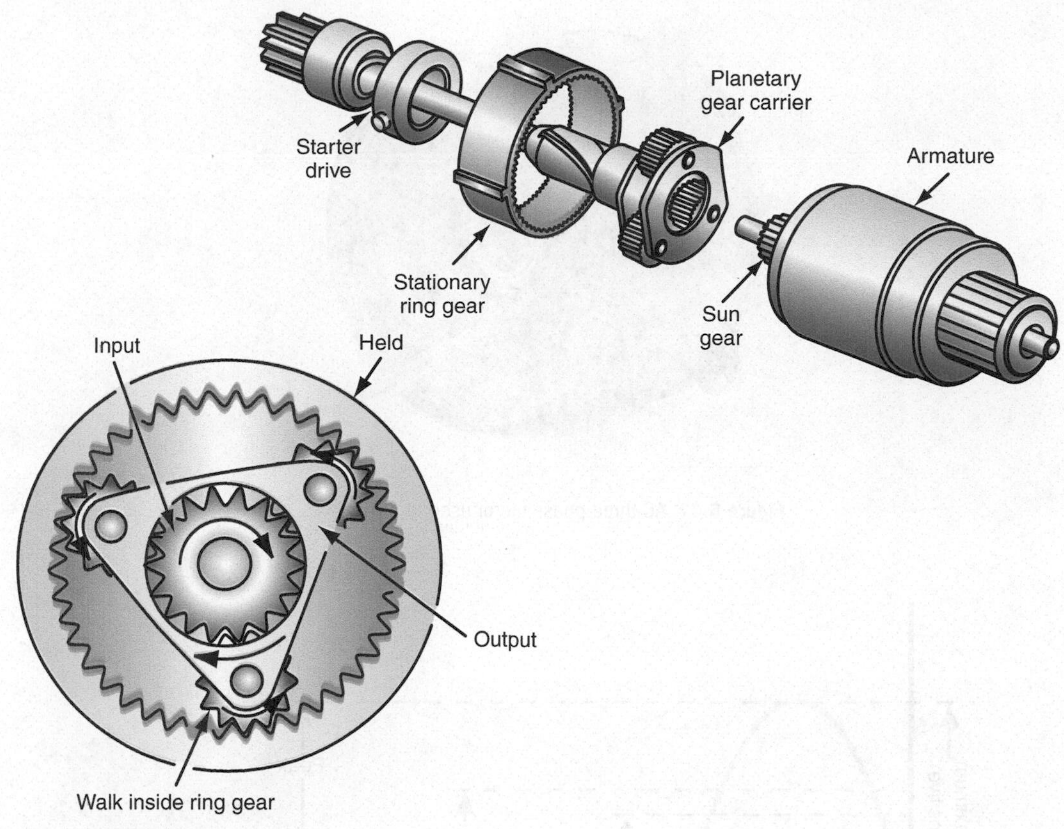

Figure 6-45 Planetary gear set.

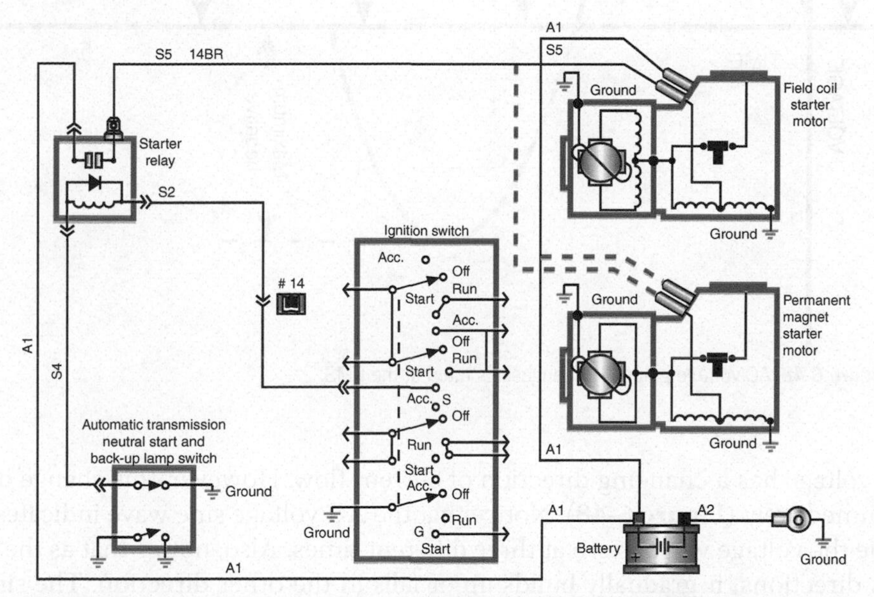

Figure 6-46 Comparison of the electrical circuits used in field coil and PMGR starters.

AC Motor Principles

A few years ago, the automotive technician did not need to be concerned much about the operating principles of the AC motor. With the increased focus on HEVs and EVs (electric vehicles), this is no longer an option since most of these vehicles use AC motors (**Figure 6-47**).

Figure 6-47 AC three-phase motor used in a HEV.

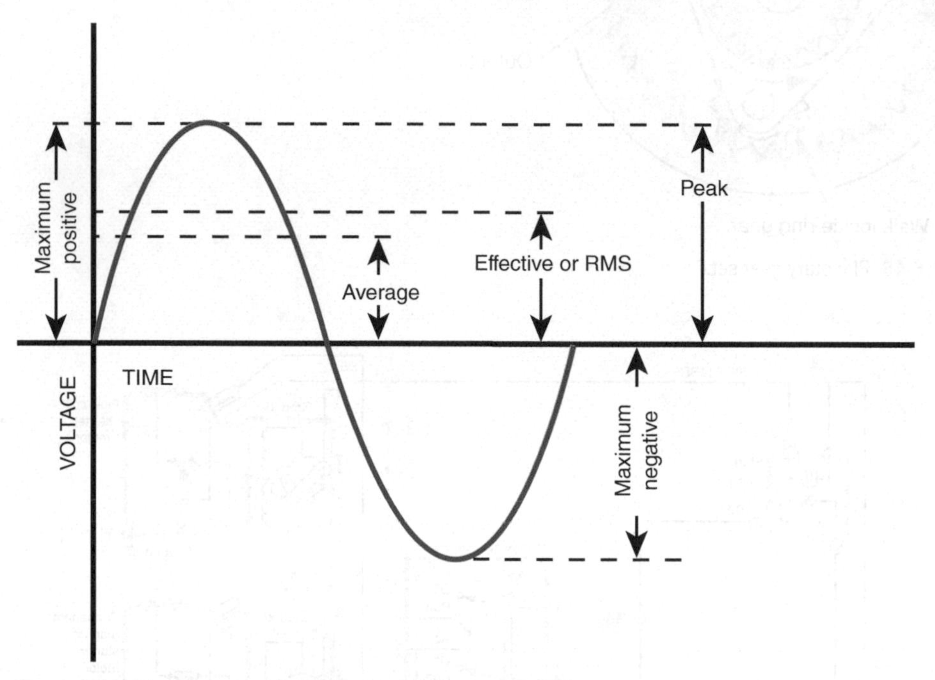

Figure 6-48 AC voltage gradually changes. It's rated at the RMS.

AC voltage has a changing direction of current flow. However, this change does not occur immediately (**Figure 6-48**). Notice that the AC voltage sine wave indicates that in one cycle the voltage will be zero at three different times. Also, notice that as the current changes directions, it gradually builds up or falls in the other direction. The sine wave illustrates that the amount of current in an AC circuit always varies. The current rating is based on the average referred to as a root mean square (RMS) value.

AC Motor Construction

A **synchronous motor** operates at a constant speed regardless of load. It generates its own rotor current as the rotor cuts through the magnetic flux lines of the stator field.

Like the DC motor, the AC motor uses a stator (field winding) and a rotor. Common types of AC motors are the **synchronous motor** and the **induction motor**. In both motor types, the stator comprises individual electromagnets that are either electrically connected to each other or connected in groups. The difference is in the rotor designs. AC motors can use either single-phase or three-phase AC current. Since the three phase is the most common motor used in HEV and EV, we will focus our discussion on these.

AUTHOR'S NOTE Three-phase AC voltage is commonly used in motors because it provides a smoother and more constant supply of power. Three-phase AC voltage is like having three independent AC power sources, which have the same amplitude and frequency but are out of phase with each other.

As in a DC motor, the movement of the rotor is the result of the repulsion and attraction of the magnetic poles. However, the way this works in an AC motor is very different. Because the current is alternating, the polarity in the windings constantly changes. The principle of operation for all three-phase motors is the **rotating magnetic field**. The rotor turns because it's pulled along by a rotating magnetic field in the stator. The stator is stationary and does not physically move. However, the magnetic field does move from pole to pole. There are three factors that cause the magnetic field to rotate (Figure 6-48). The first is the fact that the voltages in a three-phase system are 120° out of phase with each other. The second is the fact that the three voltages change polarity at regular intervals. Finally, the third factor is the arrangement of the stator windings around the inside of the motor.

In **Figure 6-49** the stator is a two-pole, three-phase motor. Two pole means that there are two poles per phase. The motor is wired with three leads: L_1, L_2, and L_3.

AUTHOR'S NOTE The stator of AC motors does not have actual pole pieces, as shown in Figure 6-49. Pole pieces are illustrated to help understand how the rotating magnetic field is created in a three-phase motor.

Referring to **Figure 6-49**, note that the pole pieces labeled as 1A and 1B are opposite each other. This is true for the poles labeled 2A and 2B, as well as for poles 3A and 3B. Each of the poles is wound in such a manner that when current flows through the winding they develop opposite magnetic polarities. All three windings are joined to form a wye connection for the stator. Since each phase reaches its peak at successively later times (**Figure 6-50**), the strongest point of the magnetic field in each winding is also in succession. This succession of the magnetic fields is what creates the effect of the magnetic field continually moving around the stator.

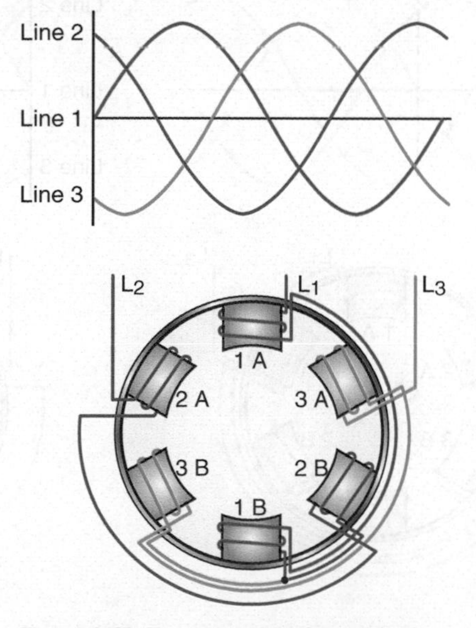

Figure 6-49 The motor stator is energized with the three-phase AC voltage.

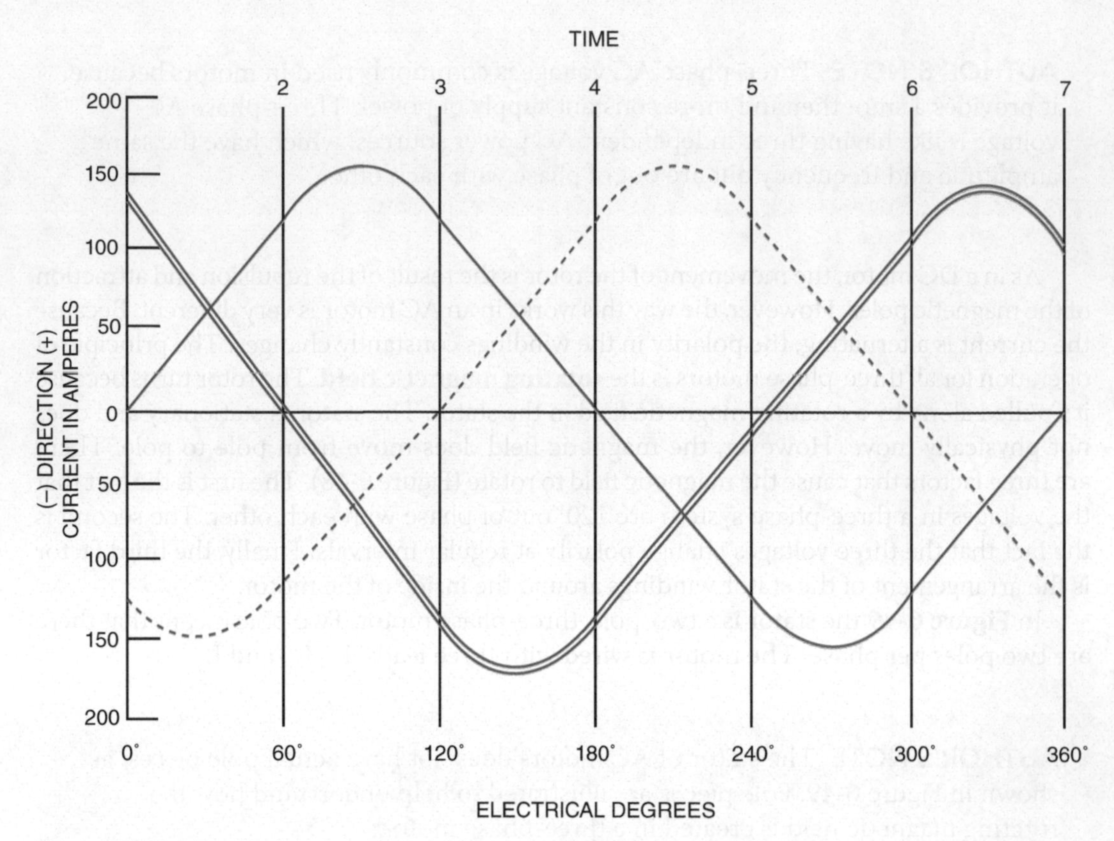

Figure 6-50 The three AC sine waves are apart. At any one time, there are two voltages at the same polarity.

To understand the concept of the rotating magnetic field around the inside of the stator, refer to **Figure 6-51**. The dashed lines that intersect the sine waves illustrate the voltage values of the three lines at a point in time. In addition, the arrows inside the motor illustrate the greatest concentration of magnetic lines of flux at this point in time.

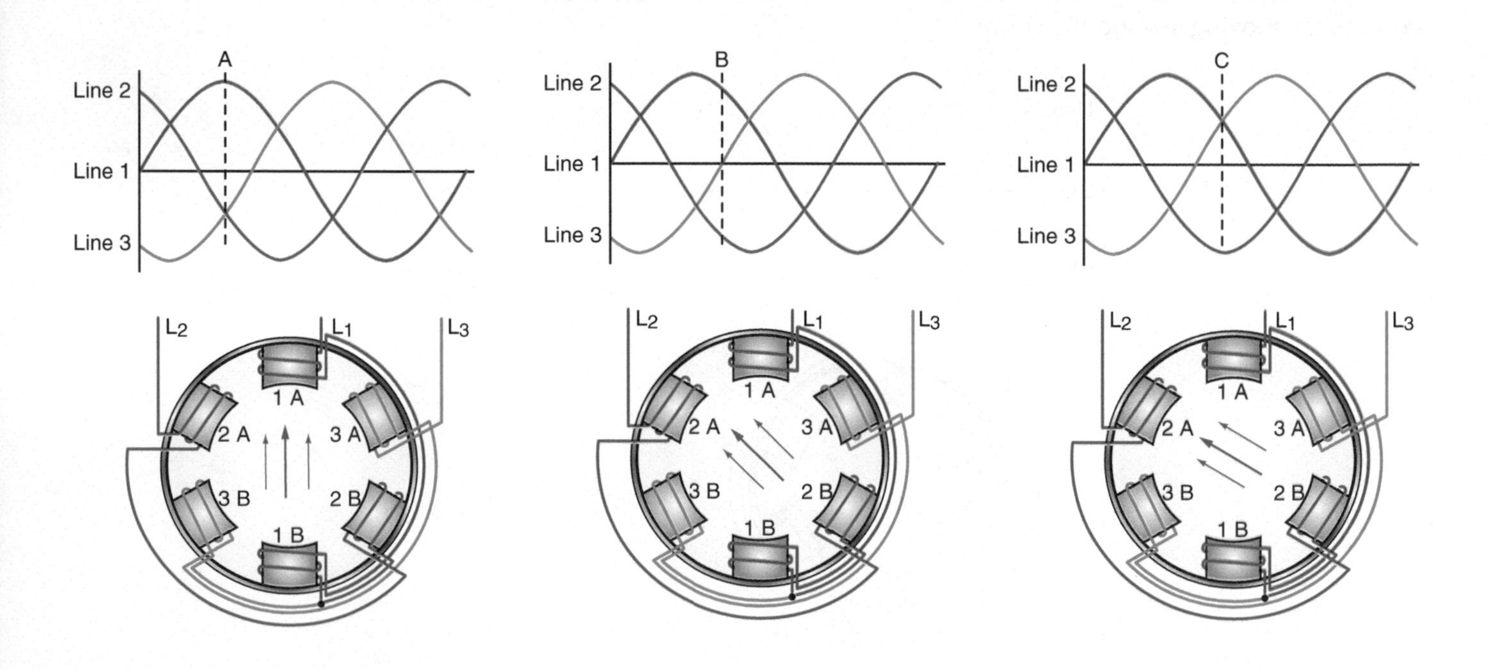

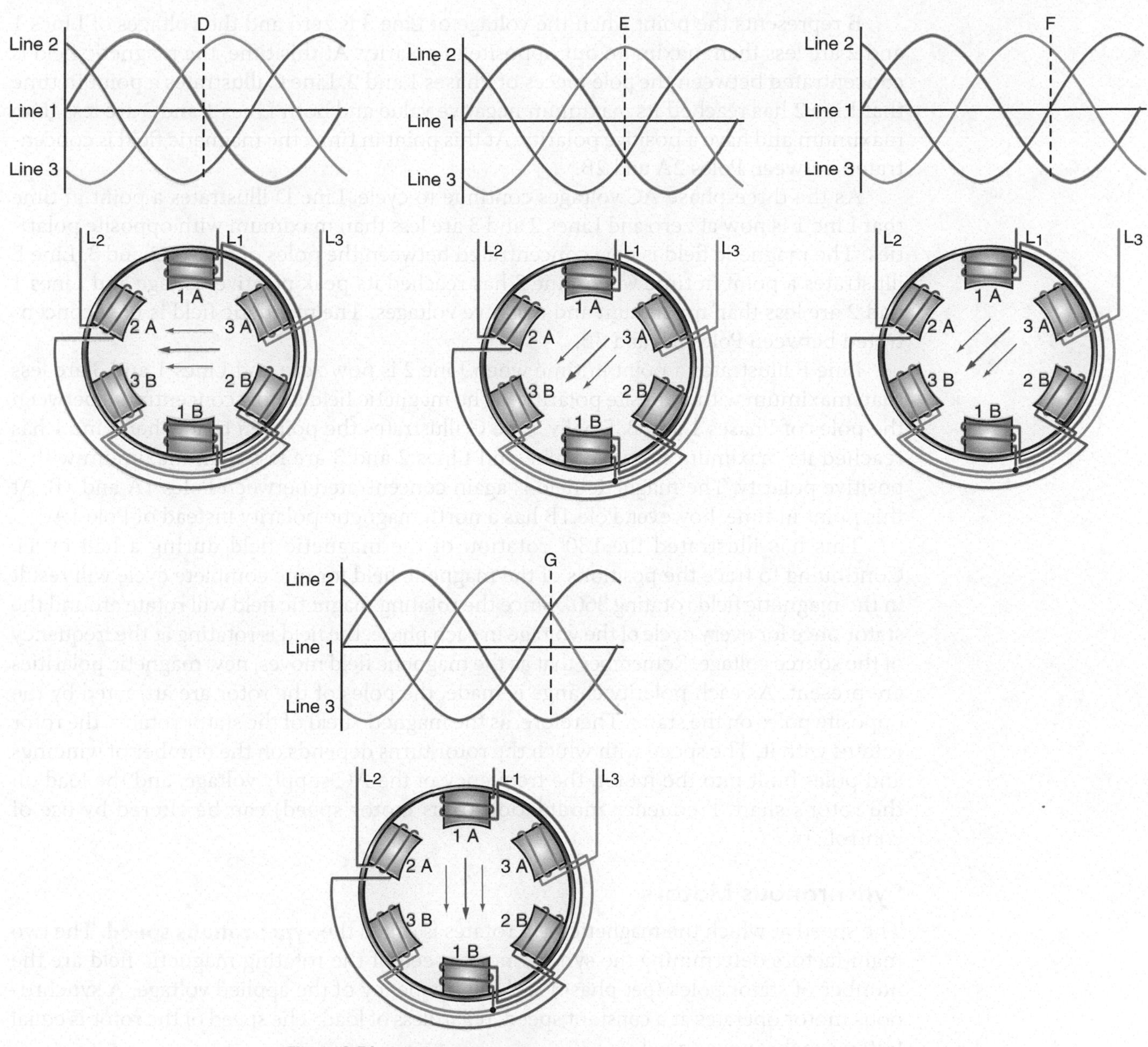

Figure 6-51 Illustration of the concentration of the magnetic field at any point in time.

Assuming that the arrows are pointing toward the north magnetic polarity, Line 1 has reached its maximum peak positive voltage, while both Lines 2 and 3 have negative voltages. This is represented by the point in time labeled A. At this time, the magnetic field is concentrated between poles 1A and 1B.

AUTHOR'S NOTE The three phases are identified by the position of the magnetic field, not by its polarity. When the magnetic field is concentrated between poles 1A and 1B, this position is referred to as Phase 1. When the magnetic field is concentrated between poles 2A and 2B, this position is referred to as Phase 2. Phase 3 is when the magnetic field is concentrated between poles 3A and 3B. Phase 1 is reached again when the magnetic field is concentrated between poles 1A and 1B, but this time Pole 1B has a north magnetic polarity instead of Pole 1A.

B represents the point when the voltage of Line 3 is zero and the voltages of Lines 1 and 2 are less than maximum but opposite in polarity. At this time, the magnetic field is concentrated between the pole pieces of Phases 1 and 2. Line C illustrates a point in time that Line 2 has reached its maximum negative value and both Lines 1 and 3 are less than maximum and have a positive polarity. At this point in time, the magnetic field is concentrated between Poles 2A and 2B.

As the three-phase AC voltages continue to cycle, Line D illustrates a point in time that Line 1 is now at zero and Lines 2 and 3 are less than maximum with opposite polarities. The magnetic field is now concentrated between the poles of Phases 2 and 3. Line E illustrates a point in time when Line 3 has reached its peak positive voltage and Lines 1 and 2 are less than maxi mum and negative voltages. The magnetic field is now concentrated between Poles 3A and 3B.

Line F illustrates a point in time when Line 2 is now zero and Lines 1 and 3 are less than maximum with opposite polarities. The magnetic field is now concentrated between the poles of Phases 1 and 3. Finally, Line G illustrates the point in time when Line 1 has reached its maximum negative value and Lines 2 and 3 are less than maximum with a positive polarity. The magnetic field is again concentrated between Poles 1A and 1B. At this point in time, however, Pole 1B has a north magnetic polarity instead of Pole 1A.

This has illustrated the 180° rotation of the magnetic field during a half cycle. Continuing to trace the positions of the magnetic field for one complete cycle will result in the magnetic field rotating 360°. Since the rotating magnetic field will rotate around the stator once for every cycle of the voltage in each phase, the field is rotating at the frequency of the source voltage. Remember that as the magnetic field moves, new magnetic polarities are present. As each polarity change is made, the poles of the rotor are attracted by the opposite poles on the stator. Therefore, as the magnetic field of the stator rotates, the rotor rotates with it. The speed with which the rotor turns depends on the number of windings and poles built into the motor, the frequency of the AC supply voltage, and the load on the rotor's shaft. Frequency modulation (thus motor speed) can be altered by use of controllers.

Synchronous Motors

The speed at which the magnetic field rotates is called the **synchronous speed**. The two main factors determining the synchronous speed of the rotating magnetic field are the number of stator poles (per phase) and the frequency of the applied voltage. A synchronous motor operates at a constant speed, regardless of load. The speed of the rotor is equal to the synchronous speed.

The synchronous motor does not depend on induced current in the rotor to produce torque. The strength of the magnetic field determines the torque output of the rotor, while the speed of the rotor is determined by the frequency of the AC input to the stator.

Synchronous motors cannot be started by applying three-phase AC power to the stator. This is because when the AC voltage is applied to the stator windings, a high-speed rotating magnetic field is present immediately. The rotating magnetic field will pass the rotor so quickly that the rotor does not have time to start turning.

In order to start the motor, the rotor contains a squirrel-cage-type winding made of heavy copper bars connected by copper rings **(Figure 6-52)**. The squirrel cage is known as the **amortisseur winding**. When voltage is first applied to the stator windings, the resulting rotating magnetic field cuts through the squirrel-cage bars. The cutting action of the field induces a current into the squirrel cage. Since the squirrel cage is shorted, the low voltage that is induced into the squirrel-cage windings results in a relatively large current flow in the cage. This current flow produces a magnetic field within the rotor that is attracted to the rotating magnetic field of the stator. The result is the rotor begins to turn in the direction of rotation of the stator field.

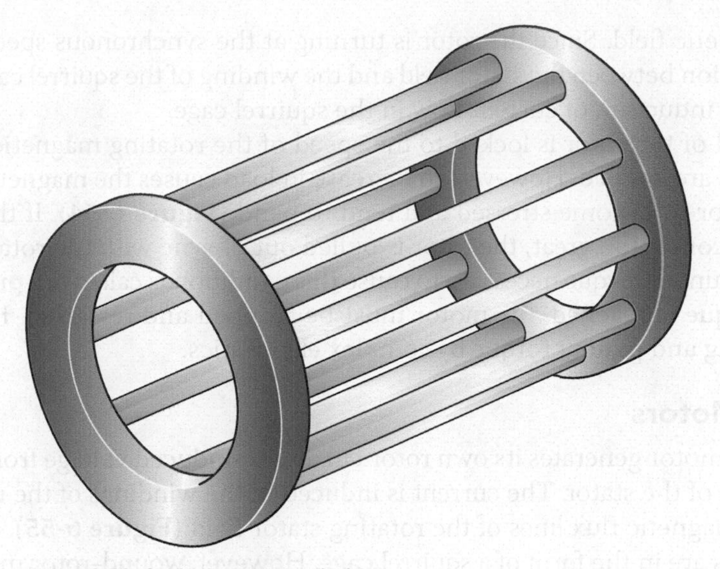

Figure 6-52 Squirrel-cage rotor design.

During this start-up time, the synchronous motor actually behaves as an induction motor since it is using induced voltage and current to get the rotor turning. However, it is impossible for the rotor to rotate at synchronous speed using the principle of induction. This is because if the rotor and the rotating field are at the same speed, there would be no relative motion between them. As a result, no lines of force would be cut by the rotor's conductors, and there would be no induced voltage in the rotor.

The construction the rotor of a synchronous motor includes wound pole pieces that become electromagnets when DC voltage is applied to them **(Figure 6-53)**. The excitation current can be applied to the rotor through slip rings or by a brushless exciter.

As the rotor is accelerated to a speed of 95% of the speed of the rotating magnetic field, DC voltage is connected to the rotor through the slip rings on the rotor shaft or by a brushless exciter. The application of DC voltage to the rotor windings results in the creation of electromagnets. The electromagnetic field of the rotor is locked in step with the rotating magnetic field of the stator. The rotor will now turn at the same speed as the

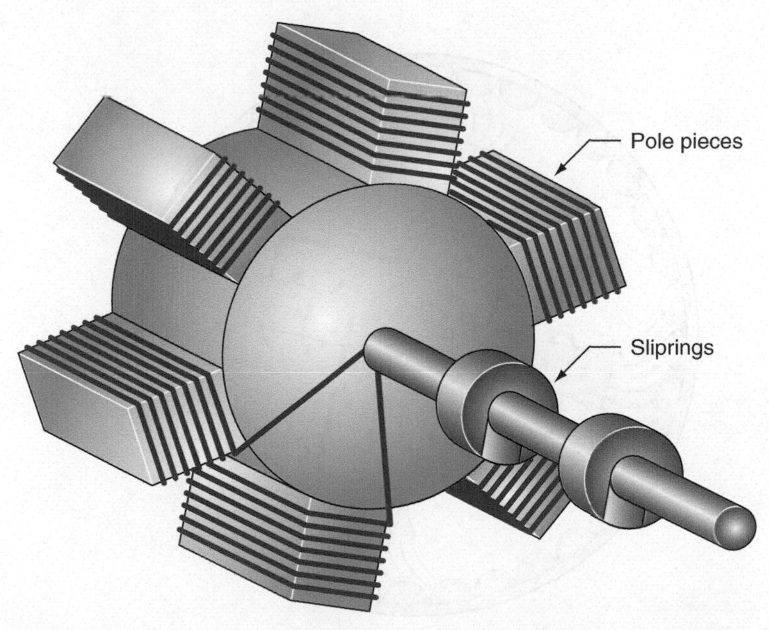

Figure 6-53 The synchronous motor with pole pieces that become electromagnets.

rotating magnetic field. Since the rotor is turning at the synchronous speed of the field, the cutting action between the stator field and the winding of the squirrel cage has ceased. This stops the induction of current flow in the squirrel cage.

The speed of the rotor is locked to the speed of the rotating magnetic field even as different loads are applied. However, an increase in load causes the magnetic fields of the rotor and stator to become stressed and tend to bend **(Figure 6-54)**. If the load on the rotor shaft becomes too great, the rotor is pulled out of sync with the rotating magnetic field. The amount of torque necessary to cause this condition is called the **pullout torque.** If pullout torque is reached, the motor must be stopped and restarted. HEVs and EVs control starting and pullout torque by complex electronics.

Induction Motors

The induction motor is also referred to as an asynchronous motor.

An induction motor generates its own rotor current by induced voltage from the rotating magnetic field of the stator. The current is induced in the windings of the rotor as it cuts through the magnetic flux lines of the rotating stator field (**Figure 6-55**). Generally, the rotor windings are in the form of a squirrel cage. However, wound-rotor motors are constructed by winding three separate coils on the rotor 120° apart. The rotor will contain as many poles per phase as the stator winding. These coils are connected to three slip rings located on the rotor shaft so rushes can provide an external connection to the rotor.

When voltage is first applied to the stator windings, the rotor does not turn. To start the squirrel-cage induction motor, the magnetic field of the stator cuts the rotor bars that induce a voltage into the cage bars. This induced voltage is of the same frequency as the voltage applied to the stator. Since the rotor is stationary, maximum voltage is induced into the squirrel cage and causes current to flow through the cage's bars. The current flow results in the production of a magnetic field around each bar.

The magnetic field of the rotor is attracted to the rotating magnetic field of the stator. The rotor begins to turn in the same direction as the rotating magnetic field. As the rotor increases in speed, the rotating magnetic field cuts the cage bars at a slower rate, resulting in less voltage being induced into the rotor. This also results in a reduction of rotor current. With the decrease in rotor current, the stator current also decreases. If the motor is operating without a load, the rotor continues to accelerate until it reaches a speed close to that of the rotating magnetic field. This means that when a squirrel-cage induction motor is first started, it has a current draw several times greater than its normal running current.

Figure 6-54 As a load is placed on the rotor, the magnetic field becomes stressed.

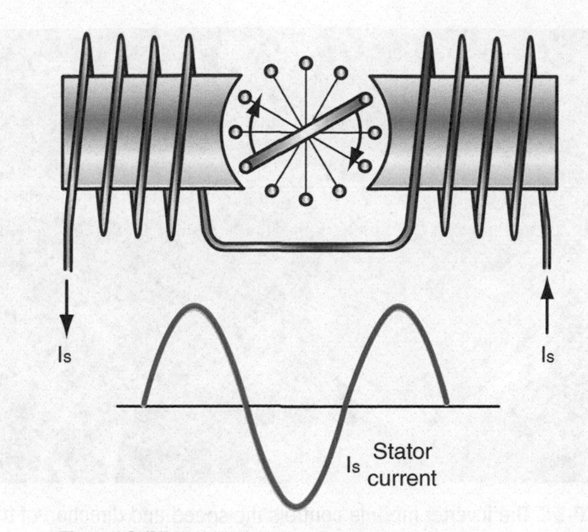

Figure 6-55 Concept of the induction motor.

The amount of torque produced by an AC induction motor is determined by three factors:

- The strength of the magnetic field of the stator.
- The strength of the magnetic field of the rotor.
- The phase angle difference between rotor and stator fields.

If the rotor were to turn at the same speed as the rotating magnetic field, there would be no induced voltage in the rotor and, consequently, no rotor current. This means that an induction motor can never reach synchronous speed. If the motor is operated with no load, the rotor will accelerate until the torque developed is proportional to friction losses. As loads are applied to the motor, additional torque is required to overcome the load. The increase in load causes a reduction in rotor speed. This results in the rotating magnetic field cutting the cage bars at a faster rate. This in turn increases the induced voltage and current in the cage and produces a stronger magnetic field in the rotor; thus, more torque is produced. The increased current flow in the rotor causes increased current flow in the stator. This is why motor current increases as load is added.

The difference between the synchronous speed and actual rotor speed is called **slip**. Slip is directly proportional to the load on the motor. When loads are on the rotor's shaft, the rotor tends to slow and slip increases. The slip then induces more current in the rotor and the rotor turns with more torque, but at a slower speed and therefore produces less CEMF.

In HEVs and EVs, the direction of motor rotation will need to change to meet certain operating requirements. In a three-phase AC motor, the direction of rotation can be changed by simply reversing any two of its stator leads. This causes the direction of the rotating magnetic field to reverse.

An electronic controller is used to manage the flow of electricity from the high-voltage (HV) battery pack to control the speed and direction of rotation of the electric motor(s). The intent of the driver is relayed to the controller by use of an accelerator position sensor. The controller monitors this signal plus other inputs regarding the operating conditions of the vehicle. Based on these inputs, the controller provides a duty cycle control of the voltage levels to the motor(s).

If the HEV or EV uses AC motors, an inverter module is used to convert the DC voltage from the HV battery to a three-phase AC voltage for the motor (**Figure 6-56**). This conversion is done by using sets of power transistors. The transistors modulate the voltage using pulse width while reversing polarity at a fixed frequency (**Figure 6-57**). The inverter module is usually a slave module to the hybrid control processor. Often the inverter module is called the motor control processor since it provides for not only current modification but also motor control.

Figure 6-56 The inverter module controls the speed and direction of the AC motor.

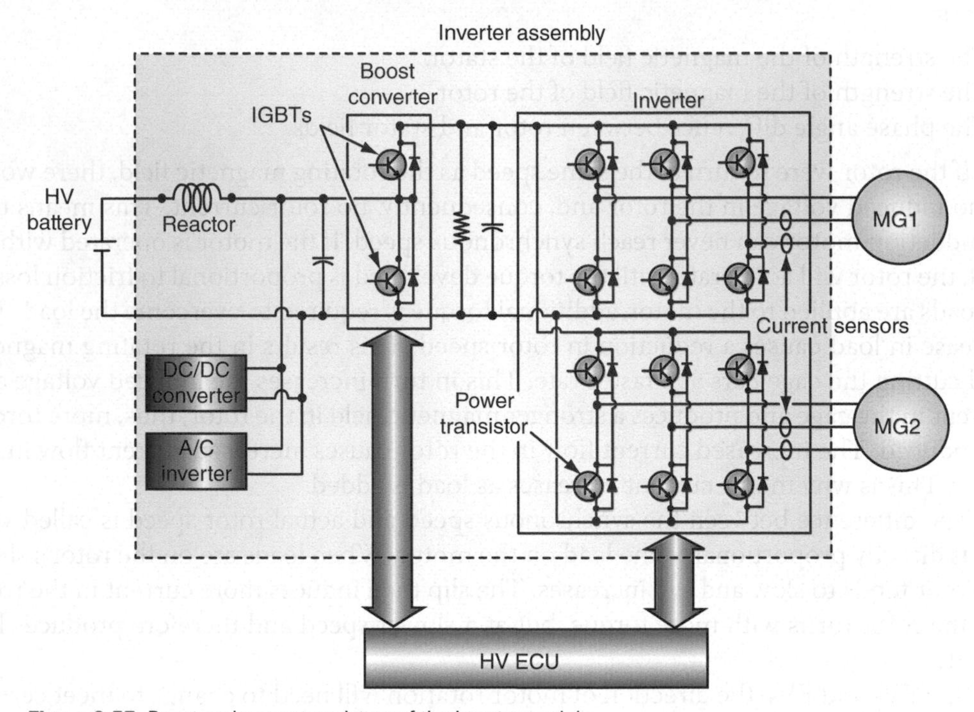

Figure 6-57 Boost and power transistors of the inverter module.

Another possible function of the inverter is to convert the AC voltage generated during regenerative braking or by the generator to DC to charge the HV and auxiliary batteries. Built into the inverter is a DC-to-DC converter that drops some of the high DC voltage to the low voltage required to recharge the auxiliary 12-volt battery.

The inverter may also include an air conditioning inverter. The purpose of this inverter is to change the high–DC battery voltage of the HV battery to a low-voltage AC for use by the motor that operates the air conditioning compressor

The motor may also use a permanent magnet rotor. This makes the starting of the motor an easier task. In order to operate the power transistors, the inverter needs to know the direction and speed of the rotor. If the position of the rotor is not known, the current to the stator windings cannot be timed accurately. This could result in the rotor being turned in the wrong direction. A resolver sensor is integrated into the motor assembly.

Integrated Starter Generator

One of the newest technologies to emerge is the **integrated starter generator (ISG)**. Although this system can be used in conventional engine-powered vehicles, one of the key contributors to the hybrid's fuel efficiency is its ability to automatically stop and restart the engine under different operating conditions. A typical hybrid vehicle uses a 14-kW electric induction motor or ISG between the ICE and the transmission. The ISG performs many functions such as fast, quiet starting, automatic engine Stop/Start to conserve fuel, recharging the vehicle batteries, smoothing driveline surges, and providing regenerative braking.

Hybrid vehicles utilize the automatic Stop/Start feature to shut off the ICE when the vehicle is not moving or when power from the ICE is not required. Usually, this feature is activated when the vehicle is stopped, no engine power is needed, and the driver's foot is on the brake pedal. On manual transmission—equipped vehicles, this feature may be activated when the vehicle is stopped, no ICE power is needed, the transmission is in neutral, and the clutch pedal is released. Once the driver's foot is removed from the brake pedal (or the clutch is engaged), the starter automatically restarts the ICE in less than one-tenth of a second. To further save fuel and reduce emissions, the engine is accelerated to idle speed by the starter/generator prior to the start of the combustion process and the injection of fuel.

The ISG is a three-phase AC motor. At low vehicle speeds, the ISG provides power and torque to the vehicle. It also supports the engine when the driver demands more power. During vehicle deceleration, ISG regenerates the power that is used to charge the traction batteries.

The ISG can also convert kinetic energy from AC to DC voltage. When the vehicle is traveling downhill and there is zero load on the engine, the wheels can transfer energy through the transmission and engine to the ISG. The ISG then sends this energy to the HV battery for storage.

An ISG can be mounted externally to the engine and connected to the crankshaft with a drive belt (**Figure 6-58**). This design is called a **belt alternator starter (BAS)**. In these applications, the unit can function as the engine's starter motor as well as a generator driven by the engine.

Both the BAS and the ISG use the same principle to start the engine. Current flows through the stator windings, which generates magnetic fields in the rotor. This will cause the rotor to turn, thus turning the crankshaft and starting the engine. In addition, this same principle is used to assist the engine as needed when the engine is running.

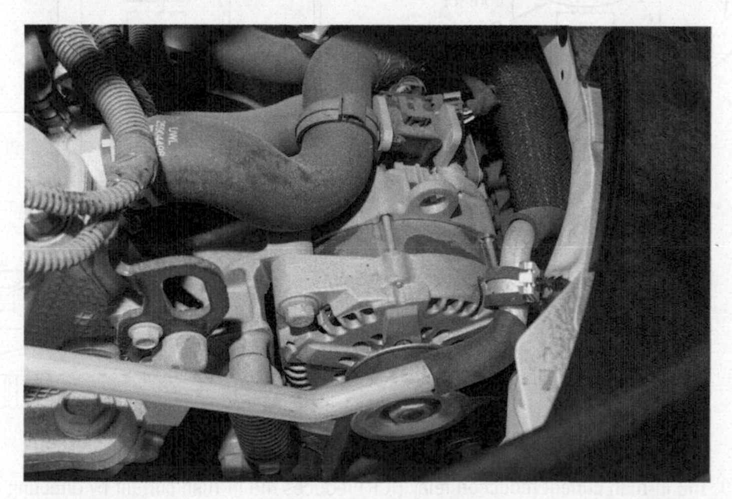

Figure 6-58 A BAS mounted external to the engine.

Stop/Start Systems

Shop Manual
Chapter 6, page 293

Hybrid vehicles utilize the automatic **Stop/Start** feature to shut off the internal combustion engine (ICE) whenever the vehicle is not moving or when power from the ICE is not required. Conventional vehicles can also benefit from Stop/Start. Usually, this feature is activated when the vehicle is stopped, no engine power is required, and the driver's foot is on the brake pedal. On manual transmission—equipped vehicles, this feature may be activated when the vehicle is stopped, no ICE power is required, and the clutch pedal is released with the transmission in neutral. Once the driver's foot is removed from the brake pedal (or the clutch pedal is pressed), the starter automatically restarts the ICE in less than one-tenth of a second.

The enhanced starter is also called an *advanced engagement starter.*

Because conventional vehicles do not have motor generators like an HEV, they may utilize a BAS or an **enhanced starter**. The enhanced starter is basically a conventional starter motor that has been modified to meet the requirement of multiple restarts. Modifications include dual layer brushes and a unique pinion spring mechanism. This is the least expensive method for adding the Stop/Start feature.

Stop/Start systems that use an enhanced starter may incorporate an in-rush current reduction relay (ICR). When the starter motor is initially energized, high current is required to start the rotation. Because voltage drop increases as the amount of current flow increases, during this time there is an increase in the volt drop to the starter motor. The ICR is dual path relay that adds resistance to the starter circuit during initial cranking to reduce the current draw (**Figure 6-59**). Prior to closing of the starter motor solenoid contact, the ICR is energized to open the shorting bar contacts. All current is routed to the starter motor through a resistor bar (approximately 10 ohms). Since the resistance bar is in series with the circuit, the increased resistance reduces the initial in-rush current spike and reduces the voltage drop. The ICR is energized for a very short time (about 185 msec), and it's de-energized. This closes the shorting bar contacts and the current bypasses the resistor bar, allowing full electrical power to the starter motor. Although there is a rebound of the current surge at this time the subsequent voltage drop is not as great as it would be without the ICR.

The tandem solenoid starter uses a co-axial dual solenoid that provides independent control of the starter's pinion gear engagement fork and starter motor rotation (**Figure 6-60**).

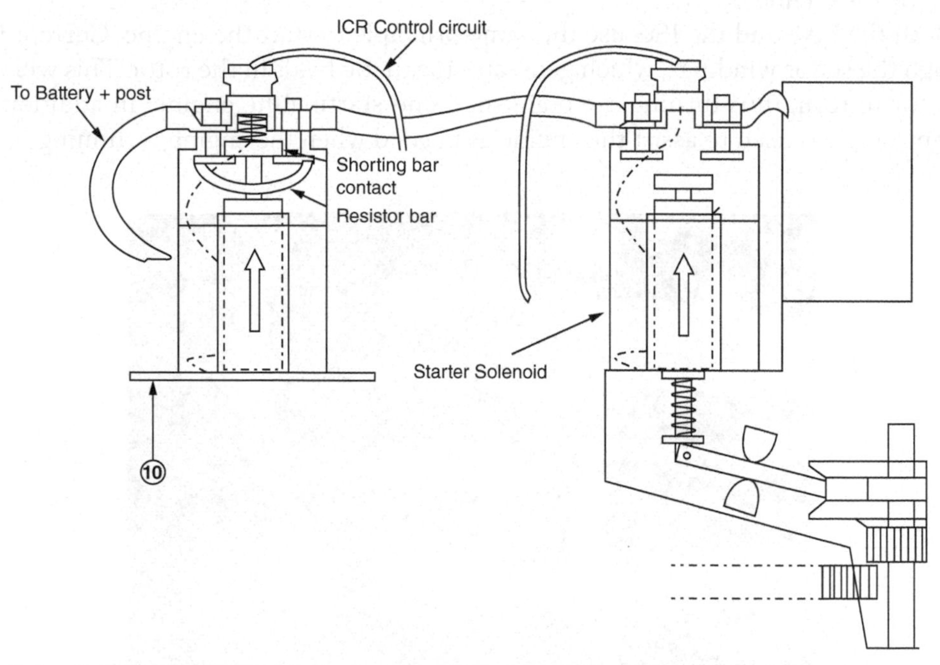

Figure 6-59 The in-rush current reduction relay (ICR) reduces the in-rush current by directing the starter current through a resistor during initial engine cranking.

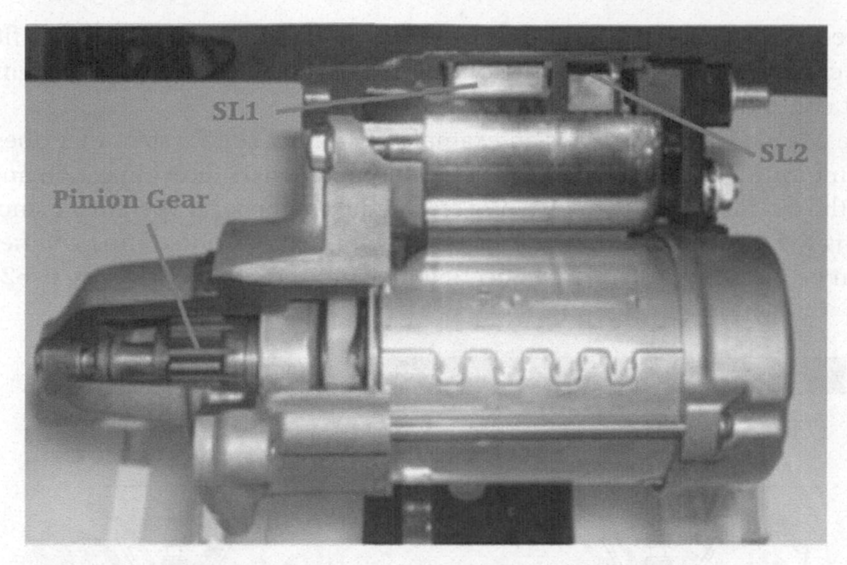

Figure 6-60 The tandem solenoid starter is capable of energizing the starter before engaging the pinion gear.

Solenoid SL1 is used to engage the pinion gear with the flywheel ring gear. Solenoid SL2 is used to energize the starter motor. The tandem solenoid allows for engine starting even if the engine has not yet come to a complete stop. This scenario can come into play if the driver has a "change of mind" after the engine has been shut down. When the engine is shutdown at 600 rpms, it takes between 0.5 and 1.5 seconds for the crankshaft to stop rotating. The use of a starter motor that does not allow for engagement to a spinning flywheel requires the engine to come to a complete stop before the starter can be engaged. The tandem solenoid system is capable of first spinning the starter motor, and then engaging the pinion gear. Software controls the timing and synchronization aspects for pinion gear shifting into the spinning flywheel. At higher flywheel speeds the motor is energized first to increase the speed of the pinion gear, and then the pinion gear is shifted forward when the rotation speed of the ring gear and pinion gear match. If the flywheel rpm is slow enough to allow the pinion gear to engage the flywheel, the pinion gear is first moved forward and then the motor is energized.

Another starter design is the permanent engaged (PE) starter. The pinion gear engagement fork is eliminated in the PE starter since the starter motor is mounted to be permanently engaged to the flywheel (**Figure 6-61**). This eliminates the issue of engaging the pinion gear into a rotating flywheel. When the engine is restarted, the motor is simply

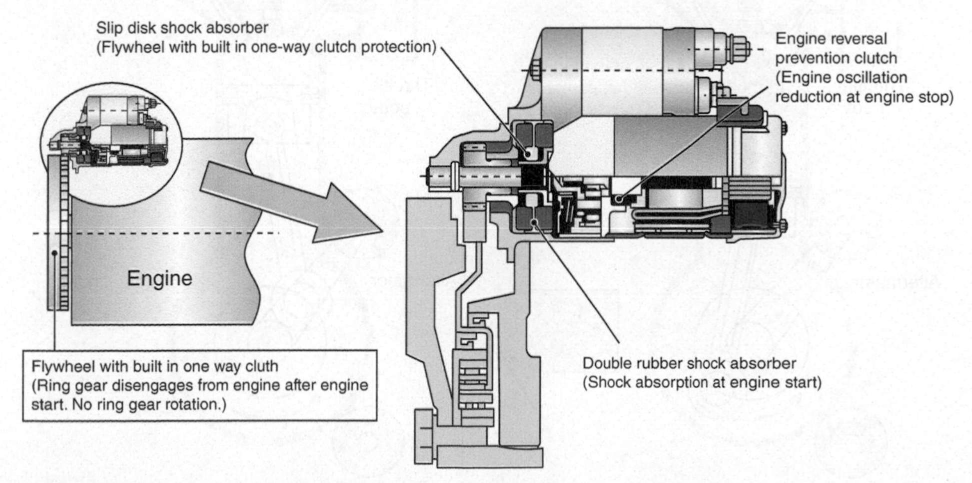

Figure 6-61 The PE starter is always engaged to the flywheel.

energized and immediately begins to rotate the crankshaft. The flywheel is fitted with a special clutching mechanism to disconnect it from the engine after the engine starts to prevent continued rotation of the starter motor.

The Mazda i-Stop system is an example of a direct start system that does not use a starter motor for the Stop/Start function. This system uses direct injection and combustion of the air/fuel mixture to instantly restart the engine. The operating principle of this system is the placement of the pistons into an optimal position during engine shutdown so it can be instantly restarted by injecting fuel into the cylinder (**Figure 6-62**).

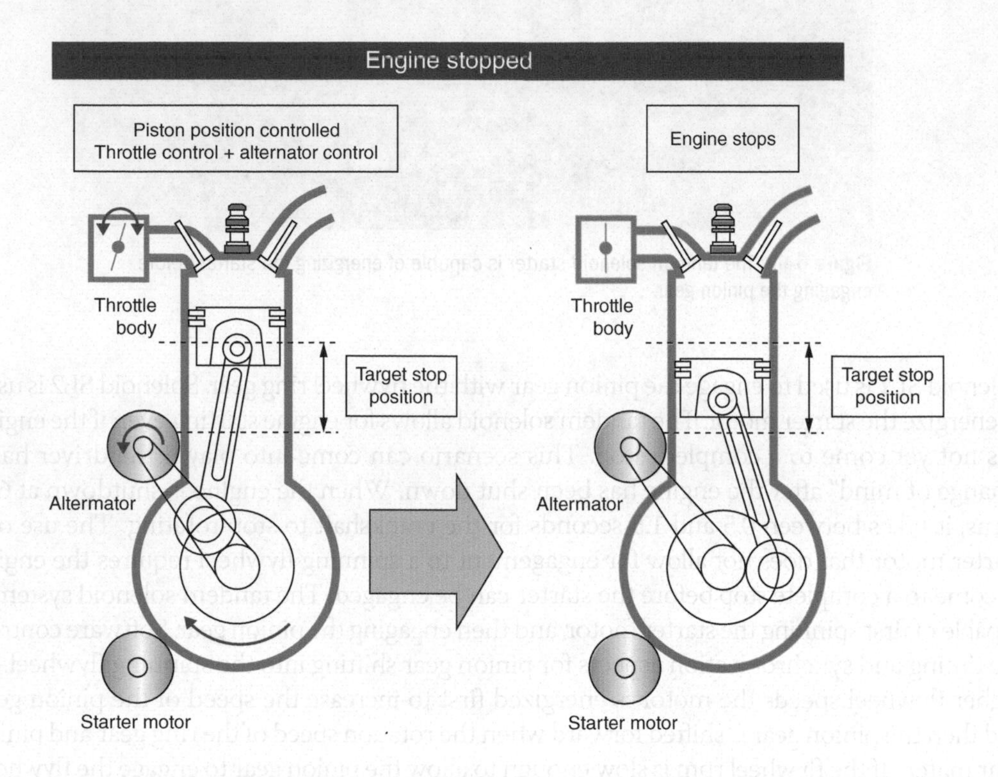

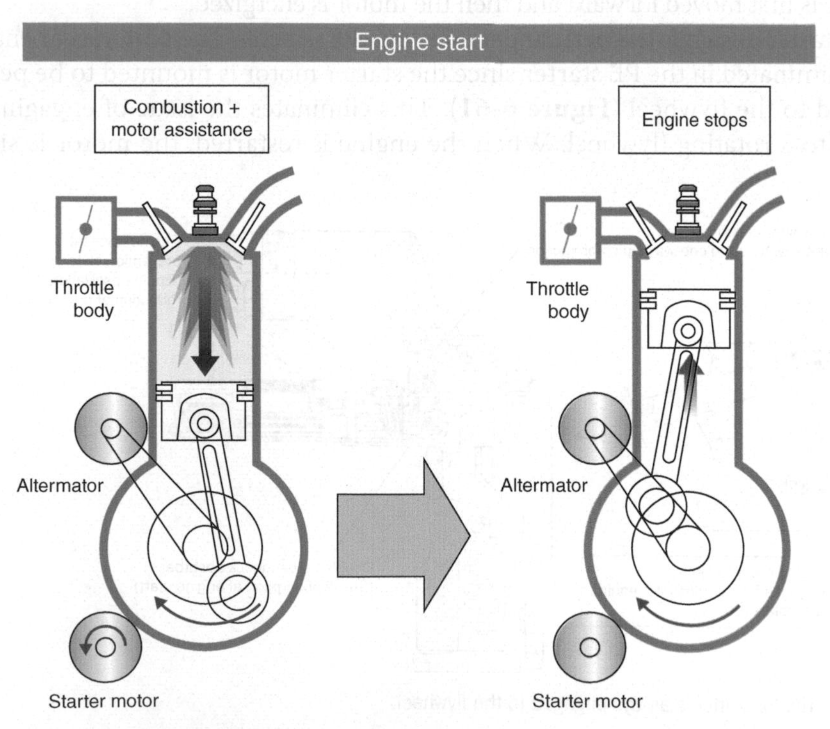

Figure 6-62 Mazda's i-Stop system.

The engine is restarted by directly injecting the fuel into the cylinder and then igniting it to create downward piston force. The control module is responsible for identifying and providing precise control over the piston position during engine shutdown. Stopping the pistons when all of them are level with each other provides the correct balance of air volumes in the cylinders and is key to quick restarts.

As the vehicle is coming to a stop, the control module will allow the engine to "pulse" until the cylinder air volumes are balanced. When the engine is stopped, one of the cylinders will be in the combustion stroke. The control module identifies this cylinder and injects fuel directly into it. The atomized fuel is then ignited to allow combustion to take place and forcing the piston to move downward, rotating the crankshaft. At the same time, the starter motor applies a small amount of additional momentum to the crankshaft. As engine speed increases, the cylinders are continuously selected for ignition until the engine reaches its idle speed.

Summary

- The starting system is a combination of mechanical and electrical parts that work together to start the engine.
- The starting system components include the battery, cable and wires, the ignition switch, the starter solenoid or relay, the starter motor, the starter drive and flywheel ring gear, and the starting safety switch.
- The armature is the moveable component of the motor that consists of a conductor wound around a laminated iron core. It's used to create a magnetic field.
- Pole shoes are made of high-magnetic permeability material to help concentrate and direct the lines of force in the field assembly.
- The magnetic forces will cause the armature to turn in the direction of the weaker field.
- Within an electromagnetic style of starter motor, the inside windings are called the armature. The armature rotates within the stationary outside windings, called the field, which has windings coiled around pole shoes.
- The commutator is a series of conducting segments located around one end of the armature.
- A split-ring commutator is in contact with the ends of the armature loops. So, as the brushes pass over one section of the commutator to another, the current flow in the armature is reversed.
- Two basic winding patterns are used in the armature: lap winding and wave winding.
- The field coils are electromagnets constructed of wire coils wound around a pole shoe.
- When current flows through the field coils, strong stationary electromagnetic fields are created.
- In any DC motor, there are three methods of connecting the field coils to the armature: series, parallel (shunt), and a compound connection that uses both series and shunt coils.
- A starter drive includes a pinion gear set that meshes with the engine flywheel ring gear on the engine.
- To prevent damage to the pinion gear or the ring gear, the pinion gear must mesh with the ring gear before the starter motor rotates.
- The Bendix drive depends on inertia to provide meshing of the drive pinion with the ring gear.
- The most common type of starter drive is the overrunning clutch. This is a roller-type clutch that transmits torque in one direction and freewheels in the other direction.
- The starting system consists of two circuits called the starter control circuit and the motor feed circuit.
- The components of the control circuit include the starting portion of the ignition switch, the starting safety switch (if applicable), and the wire conductor to connect these components to the relay or solenoid.
- The motor feed circuit consists of heavy battery cables from the battery to the relay and the starter or directly to the solenoid if the starter is so equipped.
- There are four basic groups of starter motors: direct drive, gear reduction, positive engagement (moveable pole), and permanent magnet.
- A synchronous motor operates at a constant speed regardless of load.
- An induction motor generates its own rotor current as the rotor cuts through the magnetic flux lines of the stator field.
- The principle of operation for all three-phase motors is the rotating magnetic field.

- In order to start the synchronous motor, the rotor contains a squirrel-type winding to act as an induction motor.
- Induction motor rotor windings can be in the form of a squirrel cage or constructed by winding three separate coils on the rotor apart.
- The ISG can also convert kinetic energy to storable electric energy. When the vehicle is traveling downhill and there is zero load on the engine, the wheels can transfer energy through the transmission and engine to the ISG. The ISG then sends this energy to the battery for storage and use by the electrical components of the vehicle.
- The belt alternator starter (BAS) is about the same size as a conventional generator and is mounted in the same way.
- The ISG is a three-phase AC motor. At low vehicle speeds, the ISG provides power and torque to the vehicle. It also supports the engine when the driver demands more power.
- Both the BAS and the ISG use the same principle to start the engine. Current flows through the stator windings, which generates magnetic fields in the rotor. This will cause the rotor to turn, thus turning the crankshaft and starting the engine.
- Usually, the Stop/Start feature is activated when the vehicle is stopped, no engine power is needed, and the driver's foot is on the brake pedal. Once the driver's foot is removed from the brake pedal, the starter automatically restarts the ICE.
- The enhanced starter is basically a conventional starter motor that has been modified to meet the requirement of multiple restarts. Modifications include dual layer brushes and a unique pinion spring mechanism.
- Stop/Start systems that use an enhanced starter may incorporate an in-rush current reduction relay (ICR).
- The tandem solenoid starter uses a co-axial dual solenoid that provides independent control of the starter's pinion gear engagement fork and starter motor rotation to allow for engine starting even if the engine has not yet come to a complete stop.
- The permanent engaged (PE) starter eliminates the pinion gear engagement fork since the starter motor is mounted to be permanently engaged to the flywheel.
- The Mazda i-Stop system is a direct start system that does not use a starter motor for the Stop/Start function. This system uses direct injection and combustion of the air/fuel mixture to instantly restart the engine.

Review Questions

Short-Answer Essays

1. What is the purpose of the starting system?
2. List and describe the purpose of the major components of the starting system.
3. Explain the principle of operation of the DC motor.
4. Describe the types of magnetic switches used in starting systems.
5. Describe the operation of the overrunning clutch drive.
6. Describe the differences between the positive engagement and solenoid shift starter.
7. Explain the operating principles of the permanent magnet starter.
8. Describe the purpose and operation of the armature.
9. Describe the purpose and operation of the field coil.
10. Describe how the rotor turns in a three-phase AC motor.

Fill in the Blanks

1. DC motors use the interaction of magnetic fields to convert ______________ energy into ______________ energy.
2. The ______________ is the moveable component of the motor, which consists of a conductor wound around a ______________ iron core and is used to create a ______________ field.
3. Pole shoes are made of high-magnetic ______________ material to help concentrate and direct the ______________ in the field assembly.

4. The starter motor electrical connection that permits all of the current that passes through the field coils to also pass through the armature is called the ______________ motor.
5. ______________ ______________ is voltage produced in the starter motor itself. This current acts against the supply voltage from the battery.
6. A starter motor that uses the characteristics of a series motor and a shunt motor is called a ______________ motor.
7. The ______________ ______________ is the part of the starter motor that engages the armature to the engine flywheel ring gear.
8. The ______________ ______________ is a roller-type clutch that transmits torque in one direction and freewheels in the other direction.
9. The two circuits of the starting system are called the ______________ ______________ circuit and the ______________ ______________ circuit.
10. There are two basic types of magnetic switches used in starter systems: the ______________ and the ______________.

Multiple Choice

1. The armature:
   A. Is the stationary component of the starter that creates a magnetic field.
   B. Is the rotating component of the starter that creates a magnetic field.
   C. Carries electrical current to the commutator.
   D. Prevents the starter from engaging if the transmission is in gear.
2. What is the purpose of the commutator?
   A. To prevent the field windings from contacting the armature.
   B. To maintain constant electrical contact with the field windings.
   C. To reverse current flow through the armature.
   D. All of these choices.
3. The field coils:
   A. Are made of wire wound around a nonmagnetic pole shoe.
   B. Are always shunt wound to the armature.
   C. Are always series wound with the armature.
   D. None of these choices.
4. Which of the following describes the operation of the starter solenoid?
   A. An electromagnetic device that uses movement of a plunger to exert a pulling or holding force.
   B. Both the pull-in and hold-in windings are energized to engage the starter drive.
   C. When the starter drive plunger is moved, the pull-in winding is de-energized.
   D. All of the above.
5. In the ISG, how does current flow to make the system perform as a starter?
   A. Through the rotor to create an electromagnetic field that excites the stator, which causes the rotor to spin.
   B. Through the rotor coils, which cause the magnetic field to collapse around the stator and rotate the crankshaft.
   C. Through the stator windings, which generate magnetic fields in the rotor, causing the rotor to turn the crankshaft.
   D. From the start generator control module to the rotor coils that are connected to the delta wound stator.
6. Permanent magnet starters are being discussed.
   Technician A says the permanent magnet starter uses four or six permanent magnet field assemblies in place of field coils.
   Technician B says the permanent magnet starter uses a planetary gear set.
   Who is correct?
   A. A only
   B. B only
   C. Both A and B
   D. Neither A nor B
7. Typical components of the control circuit of the starting system include:
   A. Ring gear.
   B. Magnetic switch.
   C. Pinion gear.
   D. All of these choices.
8. The characteristic of the series-wound motor is:
   A. Current flows from the armature to the brushes, and then to the field windings.
   B. Current flows from the field windings, to the brushes, and to the armature.
   C. Current flows through shunts to the field windings and the armature.
   D. All of these choices.

9. The gear reduction starter uses:
   A. A starter drive that is connected directly to the armature.
   B. A larger gear to drive a smaller gear that is attached to the starter drive.
   C. A smaller gear to drive a larger gear that is attached to the starter drive.
   D. A starter drive that is attached to the commutator ring.

10. A characteristic of permanent magnet starters is:
    A. The use of planetary gears.
    B. Current flows from the field windings to the brushes and to the armature.
    C. Connection directly to the armature.
    D. All of these choices.

CHAPTER 7
CHARGING SYSTEMS

Upon completion and review of this chapter, you should be able to:

- Explain the purpose of the charging system.
- Identify the major components of the charging system.
- Describe the function of the major components of the AC generator.
- Identify the two styles of stators.
- Explain the process of rectifying AC to DC in the AC generator.
- Describe the three principal circuits used in the AC generator.
- Explain the relationship between regulator resistance and field current.
- Explain the relationship between field current and AC generator output.
- Identify the differences between A circuit, B circuit, and isolated circuit regulation.
- Describe the operation of computer-controlled regulation.
- Explain the operating principles of the smart charging system.
- Describe the operation of charge indicators, including lamps, electronic voltage monitors, ammeters, and voltmeters.
- Explain the use of the ISG and AC motors in an HEV to recharge the HV battery.
- Explain the process of recharging the HV battery using regenerative braking.
- Explain the function of the DC/DC converter for charging the HEV auxiliary battery.

Terms to Know

Delta connection	Heat sink	Rectification
Diode rectifier bridge	Inductive reactance	Rotor
Diode trio	Interfaced generator	Sensing voltage
Electronic regulator	Load shedding	Slip rings
Full-wave rectification	Pulse-width modulation (PWM)	Stator
Half-wave rectification		Wye-wound connection

Introduction

The automotive storage battery is not capable of supplying the demands of the electrical system for an extended period. Every vehicle must have a means of replacing the current drawn from the battery. The charging system restores the electrical power the battery lost during engine starting. In addition, the charging system must react quickly to high-load demands required by the electrical system. It's the charging system that generates the current to operate all of the electrical accessories while the engine is running.

In an attempt to standardize terminology in the industry, the term **"generator"** has replaced **"alternator."** Often an alternator is referred to as an AC generator.

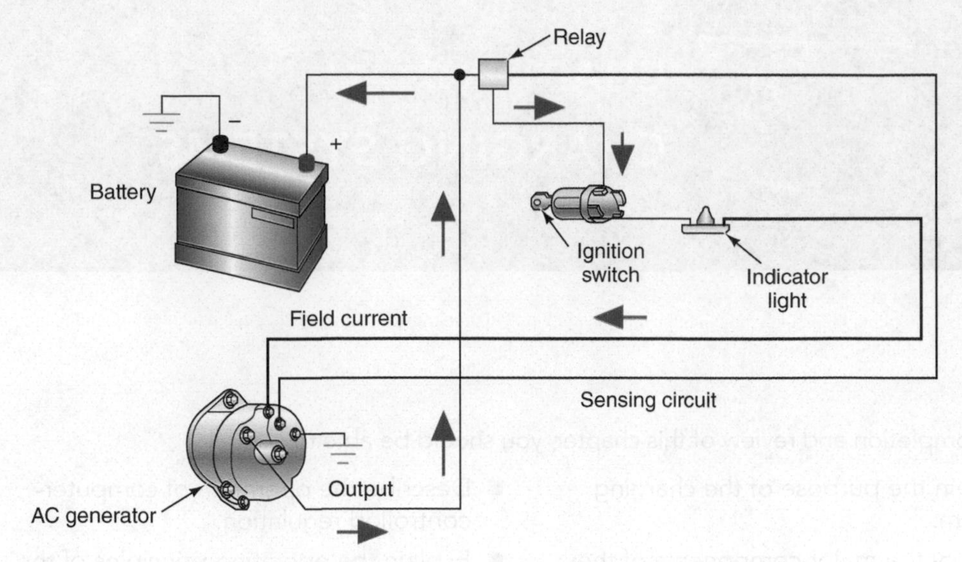

Figure 7-1 Current flow through a basic charging system when recharging the battery.

Two basic types of charging systems have been used. The first was a DC (direct current) generator, which was discontinued in the 1960s. Since then, the AC (alternating current) generator has been the predominant charging device. The DC generator and the AC generator use similar operating principles.

The purpose of the conventional charging system is to convert the engine's mechanical energy into electrical energy to recharge the battery and run the electrical accessories. The battery supplies all the current required by the starting and ignition systems during engine starting.

As battery drain continues, and engine speed increases, the charging system is able to produce more voltage than the battery can deliver. When this occurs, the electrons from the charging device flow in a reverse direction through the battery's positive terminal. The charging device now supplies the electrical system's load requirements; and the reserve electrons build up and recharge the battery.

If there is an increase in the electrical demand, and a drop in the charging system's output equal to the voltage of the battery, the battery and charging system work together to supply the required current.

The entire conventional charging system consists of the following components (**Figure 7-1**):

1. Battery.
2. Generator.
3. Drive belt.
4. Voltage regulator.
5. Charge indicator (lamp or gauge).
6. Ignition switch.
7. Cables and wiring harness.
8. Starter relay (some systems).
9. Fusible link (some systems).

The ignition switch is considered a part of the charging system if it has a circuit that stimulates the field coil.

This chapter also covers the charging systems used on EVs and HEVs. HEVs can recharge the high-voltage (HV) battery by running the engine and using the ISG or AC motors as generators. They can also use regenerative braking. They may use a DC/DC converter to charge the auxiliary battery.

Principle of Operation

All charging systems use the principle of electromagnetic induction to generate electrical power (**Figure 7-2**). The electromagnetic principle states that motion between a conductor and a magnetic field produces a voltage. The amount of voltage produced is affected by:

1. The speed at which the conductor passes through the magnetic field.
2. The strength of the magnetic field.
3. The number of conductors passing through the magnetic field.

Shop Manual
Chapter 7, pages 313, 316

Figure 7-3 illustrates how electromagnetic induction produces an AC voltage by rotating a magnetic field inside a fixed conductor (stator). None of the flux lines cut the conductor when the conductor is parallel to the magnetic field (**Figure 7-3A**). No voltage or current is produced at this point in the revolution. As the magnetic field rotates 90°,

The sine wave produced by a single conductor during one revolution is called single-phase voltage.

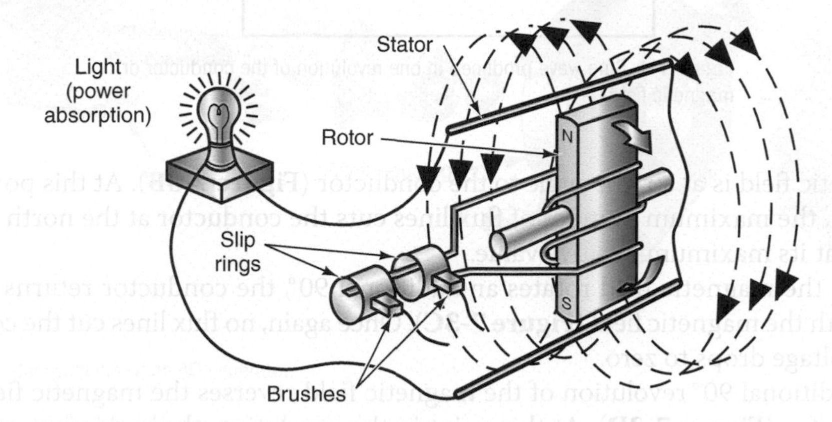

Figure 7-2 Simplified AC generator indicating electromagnetic induction.

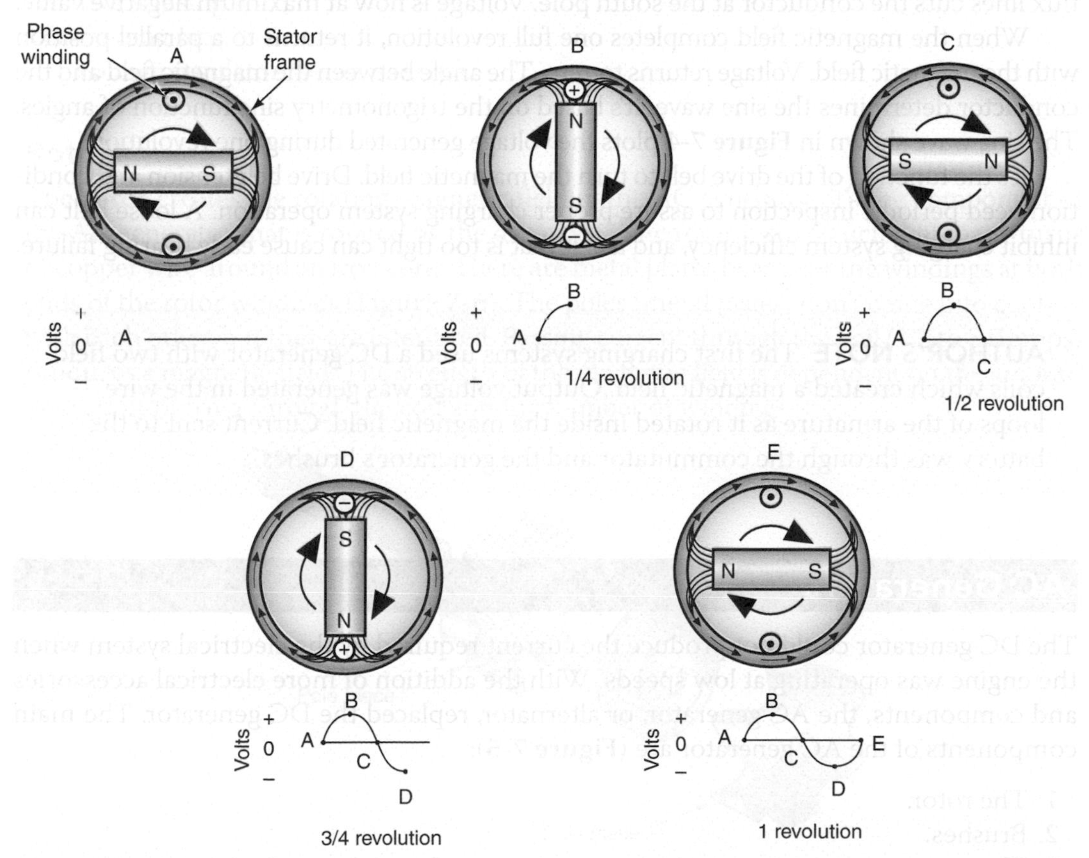

Figure 7-3 Rotation of the magnetic field produces an alternating current.

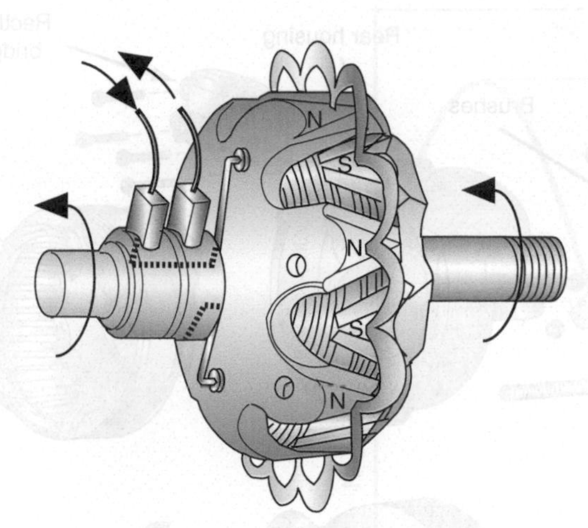

Figure 7-7 The north and south poles of a rotor's field alternate.

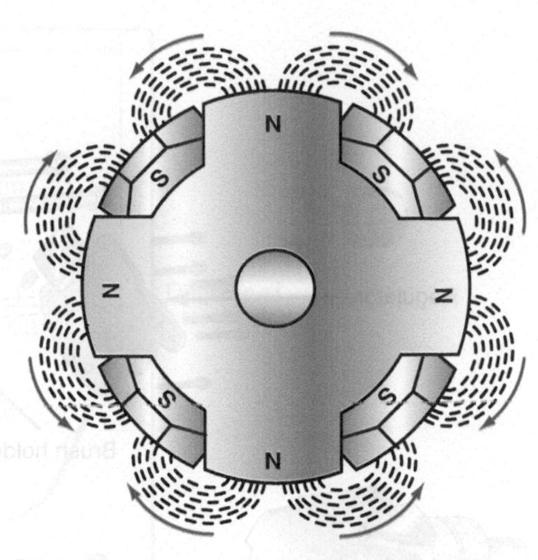

Figure 7-8 Magnetic flux lines move in opposite directions between the rotor poles.

Most rotors have 12 to 14 poles.

The poles will take on the polarity (north or south) of the side of the coil they touch. The right-hand rule will show whether it creates a north or south pole magnet. The assembled rotor has poles that alternate from north to south around the rotor (**Figure 7-7**). The alternating arrangement of poles moves the magnetic flux lines in opposite directions between adjacent poles (**Figure 7-8**). This arrangement allows several alternating magnetic fields to intersect the stator as the rotor turns. These individual magnetic fields produce a voltage by induction in the stationary stator windings.

The wires from the rotor coil are attached to two **slip rings** that are insulated from the rotor shaft. The slip rings function much like the armature commutator in the starter motor, except they are smooth. The insulated stationary carbon brush passes field current into a slip ring, then through the field coil, and back to the other slip ring. Current then passes through a grounded stationary brush (**Figure 7-9**) or to a voltage regulator.

Brushes

Shop Manual
Chapter 7, page 345

The field winding of the rotor receives current through a pair of brushes that ride against the slip rings. The brushes and slip rings provide a means of maintaining electrical continuity between stationary and rotating components. The brushes (**Figure 7-10**) ride the surface of the slip rings on the rotor and are held tight against the slip rings by spring tension provided by the brush holders. The brushes conduct only the field current (2 to 5 amps). The low current that the brushes must carry contributes to their longer life.

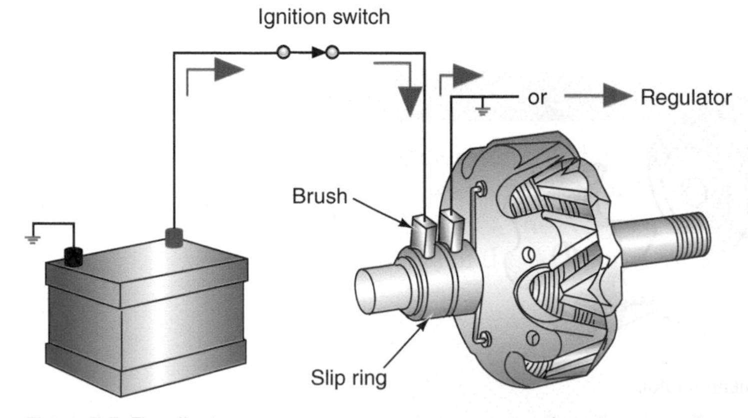

Figure 7-9 The slip rings and brushes provide a current path to the rotor coil.

Figure 7-10 Brushes are the stationary electrical contact to the rotor's slip rings.

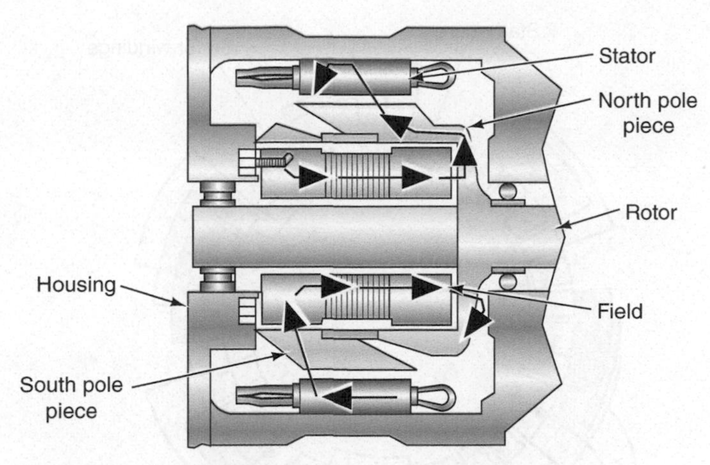

Figure 7-11 Brushless AC generator with stationary field and stator windings.

Direct current from the battery flows to the rotating field through the field terminal and the insulated brush. The second brush may be the ground brush, which is attached to the AC generator housing or to a voltage regulator.

Not all AC generators require the use of brushes or slip rings. Brushless AC generators hold the field and stator windings stationary (**Figure 7-11**). A screw terminal makes the electrical connection. The rotor contains the pole pieces and fits between the field and stator windings.

Applying current to the field winding produces the magnetic field. The air gaps in the magnetic path contain a nonmetallic ring to divert the lines of force into the stator winding.

The pole pieces on the rotor concentrate the magnetic field into alternating north and south poles. When the rotor is spinning, the north and south poles alternate as they pass the stator winding. The moving magnetic field produces an electrical current in the stator winding. The alternating current is rectified in the same manner as in conventional AC generators.

Stators

The **stator** contains three main sets of windings wrapped in slots around a laminated, circular iron frame (**Figure 7-12**). The production of electricity is within the stationary stator coil. Each of the three windings has the same number of coils as the rotor has pairs

Shop Manual
Chapter 7, page 346

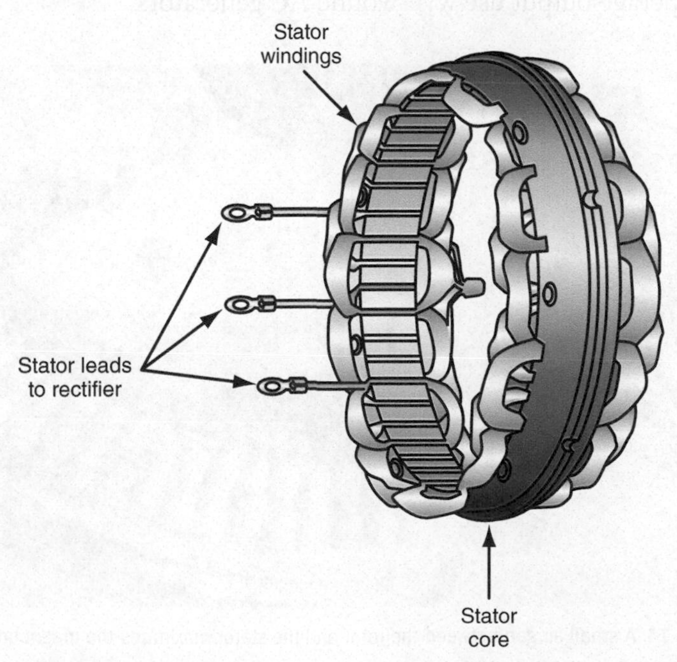

Figure 7-12 Components of a typical stator.

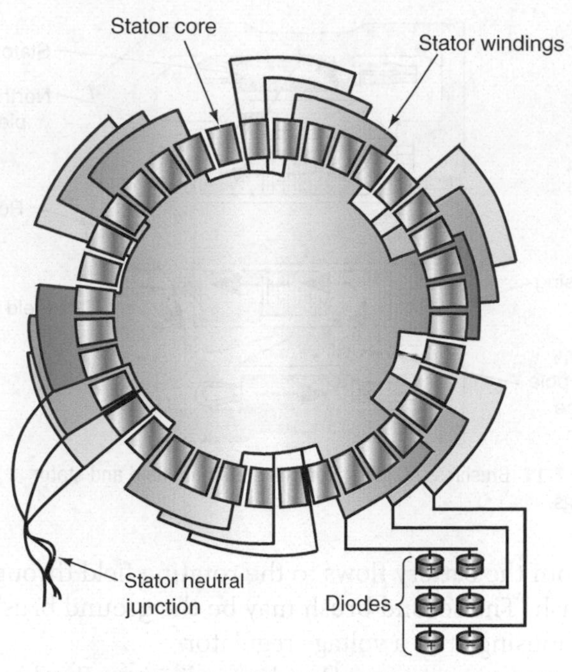

Figure 7-13 Overlapping stator windings produce the required phase angles.

of north and south poles. The coils of each winding are evenly spaced around the core. The three sets of windings alternate and overlap as they pass through the core (**Figure 7-13**). The overlapping is needed to produce the required phase angles.

The rotor is fitted inside the stator (**Figure 7-14**). A small air gap (approximately 0.015 inch or 0.381 mm) between the rotor and the stator allows the rotor's magnetic field to energize all of the windings of the stator at the same time and to maximize the magnetic force.

The common junction in the wye-connected winding is called the stator neutral junction.

Each group of windings has two leads. The first lead is for the current entering the winding. The second lead is for the current leaving. There are two basic means of connecting the leads. The first method is the **wye-wound connection** (**Figure 7-15**). In the wye connection, one lead from each winding connects to one common junction. From this junction, the other leads branch out in a Y pattern. Generally, applications that don't require high-amperage output use wye-wound AC generators.

Figure 7-14 A small air gap between the rotor and the stator maximizes the magnetic force.

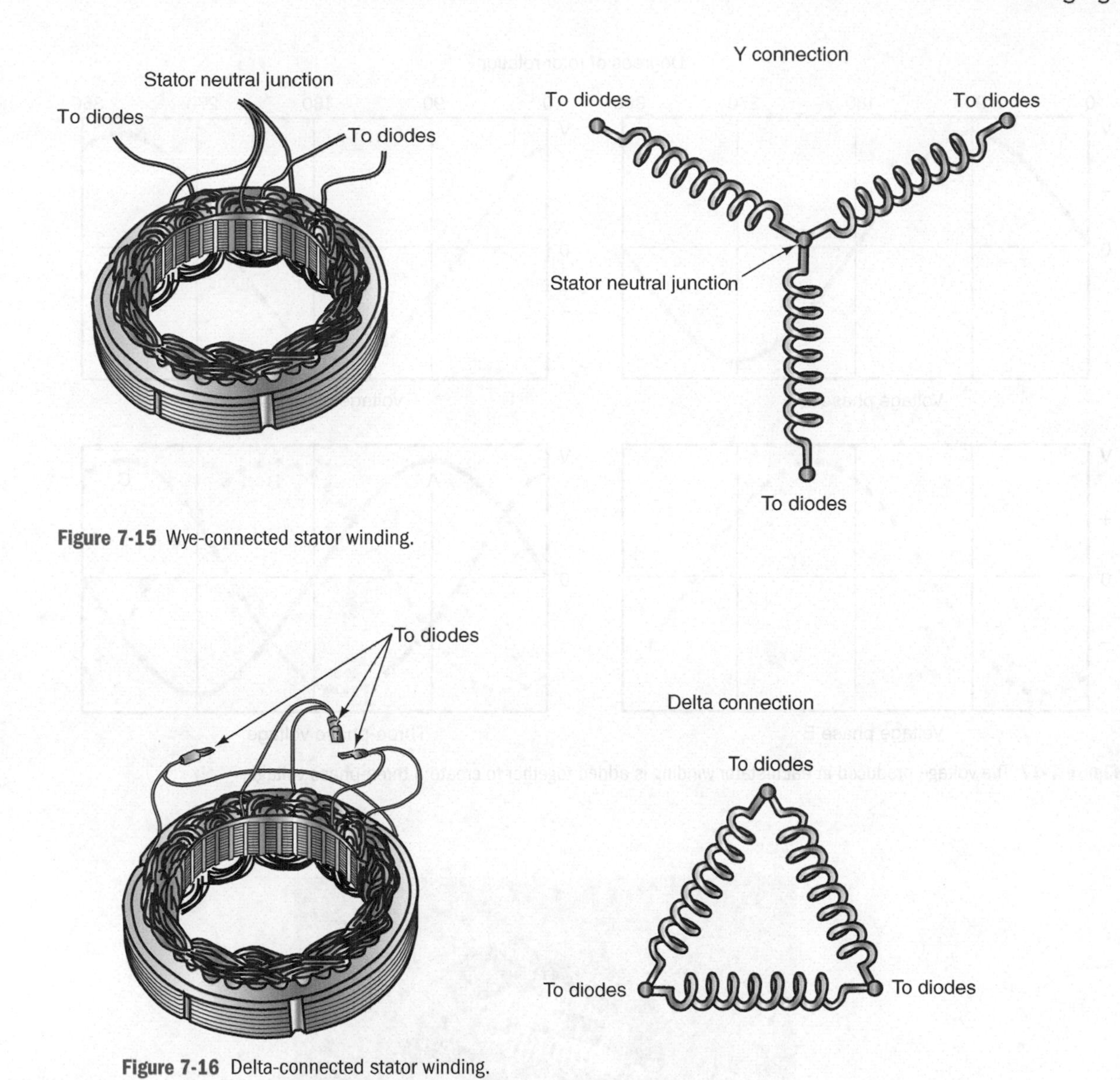

Figure 7-15 Wye-connected stator winding.

Figure 7-16 Delta-connected stator winding.

The second method of connecting the windings is called the **delta connection** (**Figure 7-16**). The delta connection attaches the lead of one end of the winding to the lead at the other end of the next winding. Applications requiring high amperage output commonly use the delta connection.

In a wye or delta-wound stator winding, each group of windings occupies one-third of the stator, or of the circle. Rotating the rotor in the stator produces a voltage in each loop of the stator at different phase angles. **Figure 7-17** illustrates the resulting overlap of sine waves. Each of the sine waves is at a different phase of its cycle at any given time. As a result, the output from the stator is divided into three phases.

Diode Rectifier Bridge

Shop Manual
Chapter 7, pages 324, 347

The battery and the electrical system cannot accept or store AC voltage. The alternating current requires conversion to direct current for the vehicle's electrical system to use the voltage and current generated in the AC generator. This process is called **rectification**. A **diode rectifier bridge** converts the current in an AC generator (**Figure 7-18**). Acting as a one-way check valve, the diodes switch the current flow back and forth so it flows from the AC generator in only one direction.

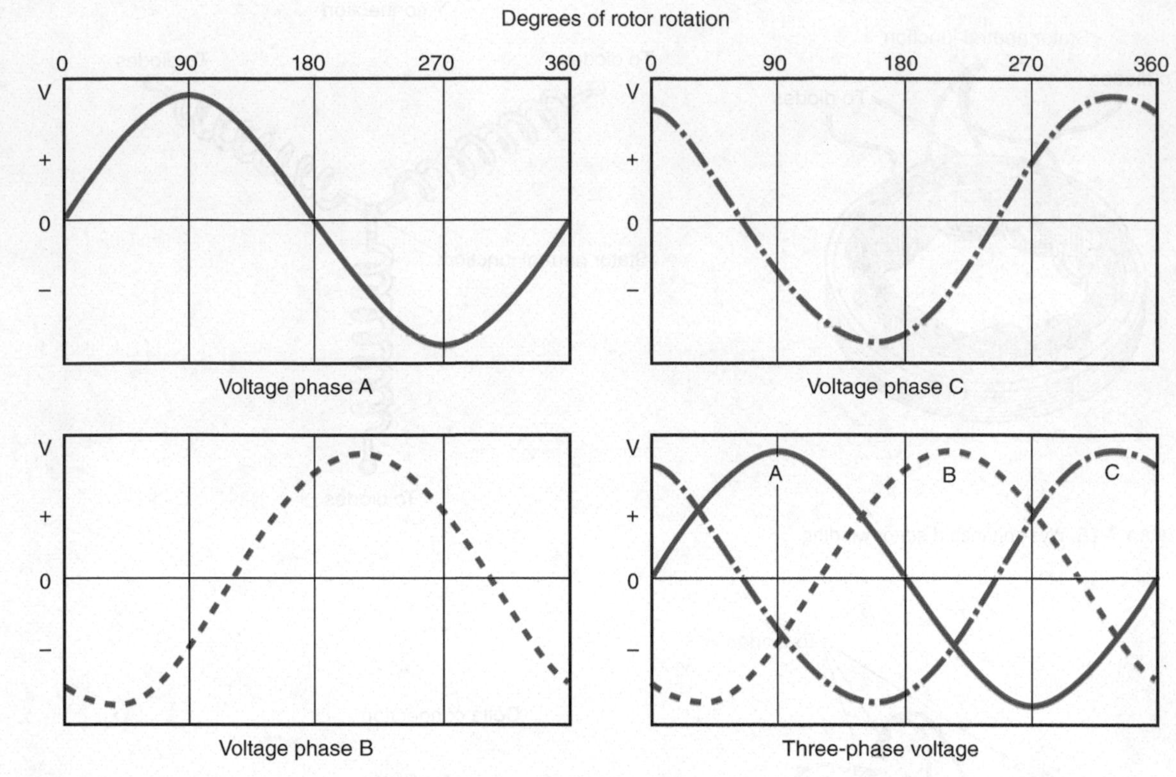

Figure 7-17 The voltage produced in each stator winding is added together to create a three-phase voltage.

Figure 7-18 General Motors' rectifier bridge.

The rectifier bridge is also known as a rectifier stack.

When the alternating current reverses itself, the diode blocks, and no current flows. If alternating current passes through a positively biased diode, the diode will block off the negative pulse. This changes the alternating current to a pulsing direct current **(Figure 7-19)**. This process is called **half-wave rectification**.

An AC generator usually uses a pair of diodes for each stator winding, for a total of six diodes (**Figure 7-20**). Three of the diodes are positive biased and are mounted in a **heat sink** to dissipate the heat (**Figure 7-21**). The three remaining diodes are negative biased and are attached directly to the frame of the AC generator (**Figure 7-22**). By using a pair of diodes that are reverse-biased to each other, rectification of both sides of the AC sine wave is achieved (**Figure 7-23**). The process of converting both sides of the sine wave to a DC voltage is called **full-wave rectification**. The negative-biased diodes allow the conducting current from the negative side of the AC sine wave to be utilized by the circuit. Diode rectification changes the negative current into positive output.

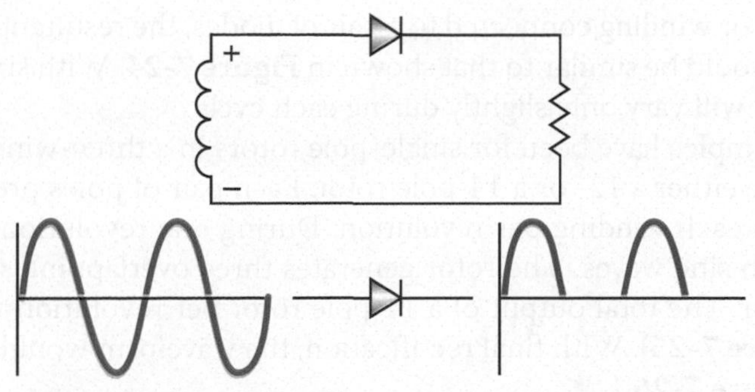

Figure 7-19 Alternating current rectified to a pulsating direct current after passing through a positive-biased diode. This is called half-wave rectification.

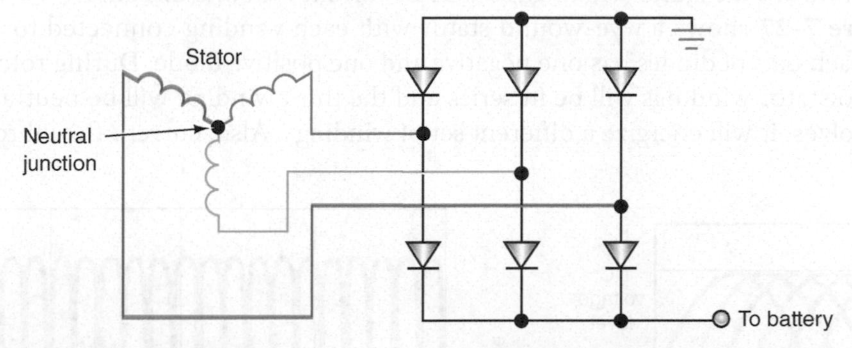

Figure 7-20 A simplified schematic of the AC generator windings connected to the diode rectifier bridge.

Figure 7-21 The positive-biased diodes are mounted into a heat sink to provide protection.

Figure 7-22 Negative-biased diodes pressed into the AC generator housing.

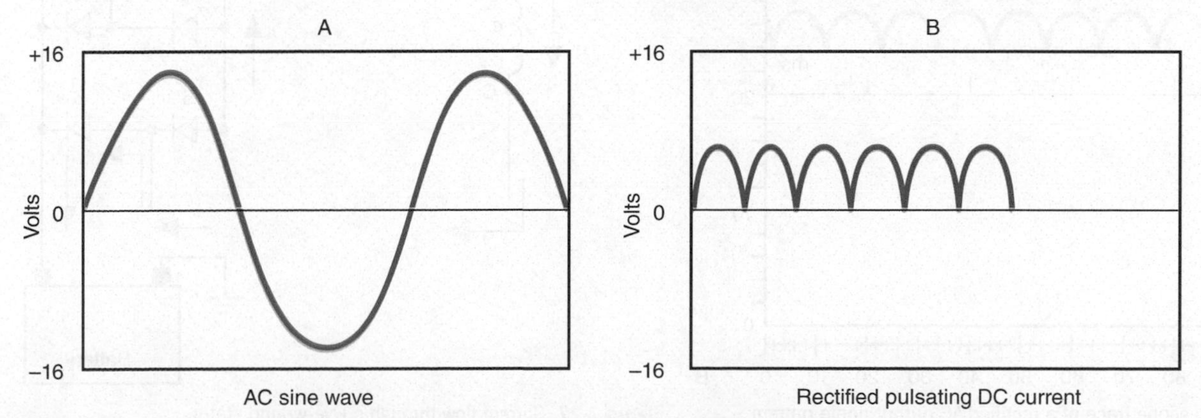

Figure 7-23 Full-wave rectification uses both sides of the AC sine wave to create a pulsating direct current.

With each stator winding connected to a pair of diodes, the resultant waveform of the rectified voltage would be similar to that shown in **Figure 7-24**. With six peaks per revolution, the voltage will vary only slightly during each cycle.

So far, the examples have been for single-pole rotors in a three-winding stator. Most AC generators use either a 12- or a 14-pole rotor. Each pair of poles produces one complete sine wave in each winding per revolution. During one revolution, a 14-pole rotor will produce seven sine waves. The rotor generates three overlapping sine wave voltage cycles in the stator. The total output of a 14-pole rotor per revolution would be 21 sine wave cycles (**Figure 7-25**). With final rectification, the waveform would be similar to the one shown in **Figure 7-26**.

Half-wave rectification is inefficient since it wastes the other half of the alternating current. Full-wave rectification of the stator output uses the total potential by redirecting the current from the stator windings so that all current is in one direction.

Figure 7-27 shows a wye-wound stator with each winding connected to a pair of diodes. Each pair of diodes has one negative and one positive diode. During rotor movement, two stator windings will be in series and the third winding will be neutral. As the rotor revolves, it will energize a different set of windings. Also, current flow through the

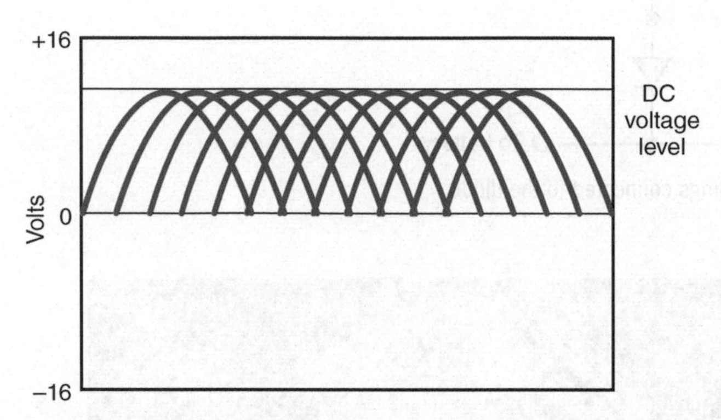

Figure 7-24 With three-phase rectification, the DC voltage level is uniform.

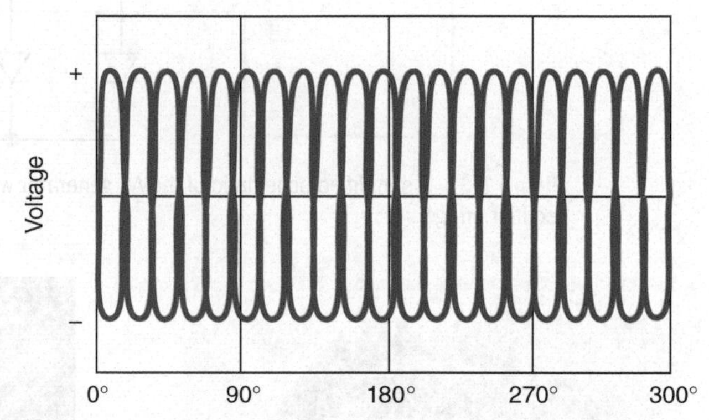

Figure 7-25 Sine wave cycle of a 14-pole rotor and three-phase stator.

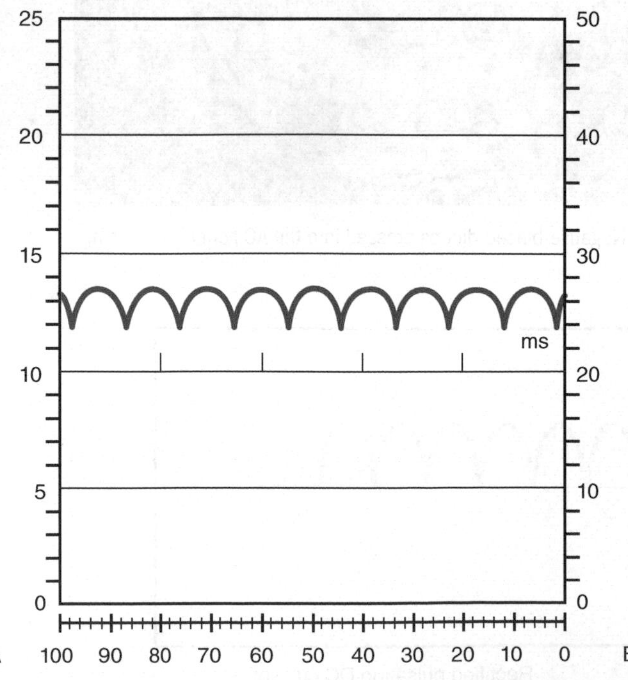

Figure 7-26 Lab scope trace of a rectified AC output ripple pattern.

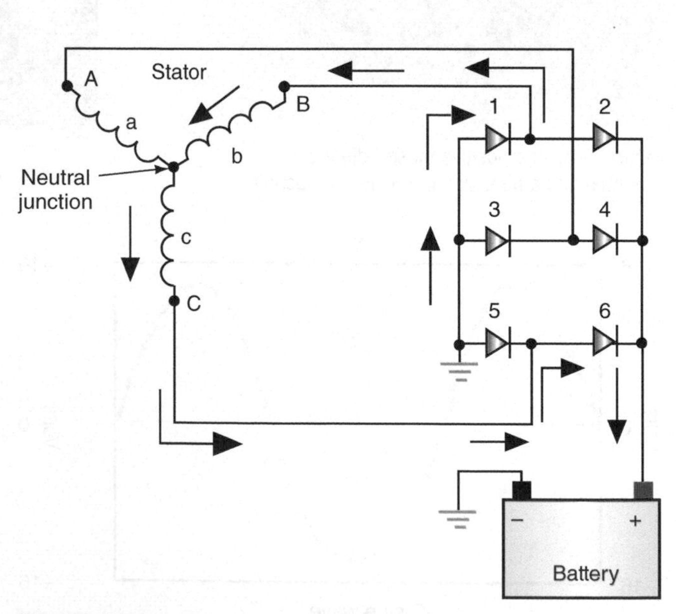

Figure 7-27 Current flow through a wye-wound stator.

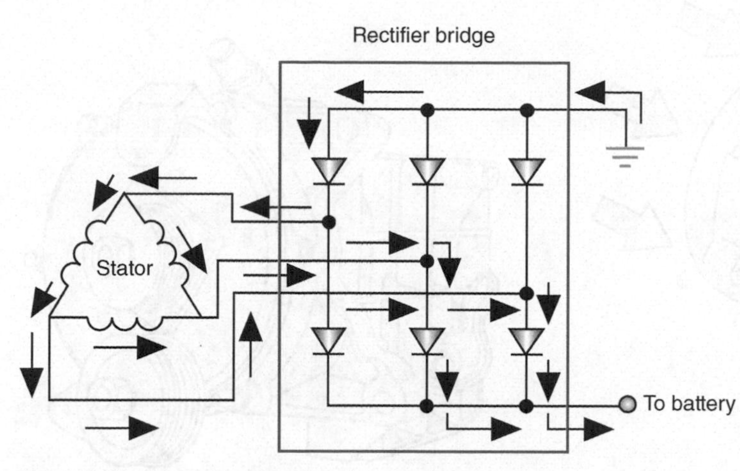

Figure 7-28 Current flow through a delta-wound stator.

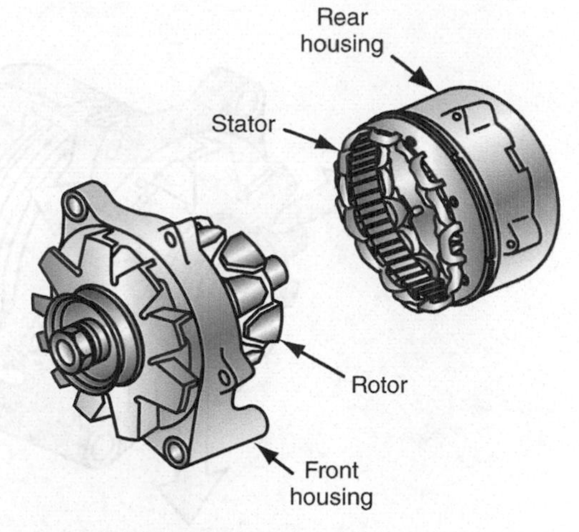

Figure 7-29 Typical two-piece AC generator housing.

windings is reversed as the rotor passes. Current in any direction through two windings in series will produce direct current.

Figure 7-28 illustrates the current flow through the delta-wound stator. Instead of two windings in series, the three windings of the delta stator are in parallel. The parallel arrangement makes more current available since the parallel paths allow more current to flow through the diodes. Since the three outputs of the delta winding are in parallel, current flows from each winding continuously.

AUTHOR'S NOTE Not only do the diodes rectify stator output, but they also block battery drain back when the engine is not running.

AC Generator Housing and Cooling Fan

Most AC generator housings are a two-piece construction, made from cast aluminum (**Figure 7-29**). The two end frames provide support for the rotor and the stator. In addition, the end frames contain the diodes, regulator, heat sinks, terminals, and other components of the AC generator.

Shop Manual
Chapter 7, page 345

The drive end housing holds a bearing that supports the front of the rotor shaft. The rotor shaft extends through the drive end housing. The drive pulley and cooling fan are installed onto the end of the shaft.

The slip ring end housing holds a bearing that supports the rear rotor shaft. In addition, it contains the brushes and has all of the electrical terminals. The slip ring end housing holds the integral regulator (if applicable).

The wire coils that make up the rotor and stator can produce heat during the conversion process. Heat increases resistance and decreases the AC generator's capabilities. A cooling fan installed behind the pulley keeps the generator's components cool. The spinning fan draws air in through openings in the slip ring end housing, through the generator, and out openings behind the cooling fan in the drive end housing (**Figure 7-30**).

Liquid-Cooled Generators

High-output generators tend to have higher internal temperatures that can shorten the life of the diodes. To help reduce diode temperatures, some manufacturers are using a liquid-cooled generator (**Figure 7-31**). The water-cooled generator has water jackets cast into its housing and is connected to the engine's cooling system by hoses.

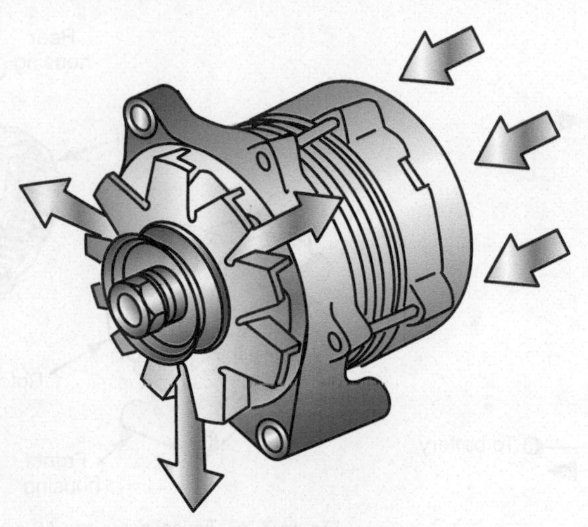

Figure 7-30 The cooling fan draws air in from the rear of the AC generator to keep the diodes cool.

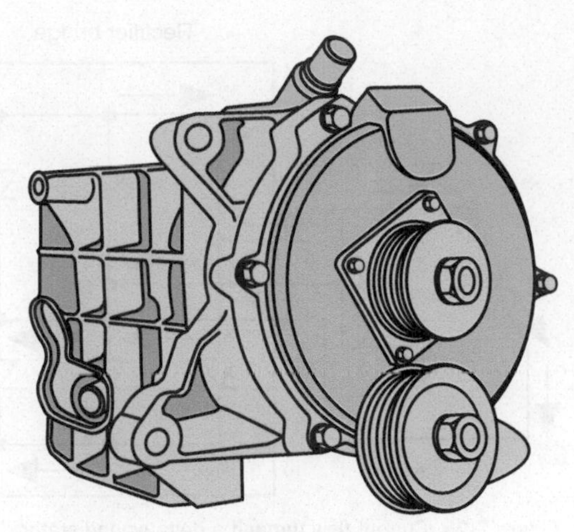

Figure 7-31 Water-cooled generator.

AC Generator Circuits

There are three principal circuits used in an AC generator:

1. The charging circuit: Consists of the stator windings and rectifier circuits.
2. The excitation circuit: Consists of the rotor field coil and the electrical connections to the coil.
3. The preexcitation circuit: Supplies the initial current for the field coil that starts the buildup of the magnetic field.

Shop Manual
Chapter 7, page 317

The field coil must develop a magnetic field for the AC generator to produce current. The AC generator creates its own field current in addition to its output current.

Shop Manual
Chapter 7, page 347

Excitation of the field occurs when the voltage induced in the stator rises to the point that it overcomes the forward voltage drop of at least two of the rectifier diodes. Excitation of the field can be accomplished in various ways, depending on generator design. Many generators with an internal regulator use a **diode-trio** to rectify current from the stator to create the magnetic field in the rotor's field coil. Before the diode trio can supply field current, the anode side of the diode must be at least 0.6 volt (V) more positive than the cathode side (**Figure 7-32**). The early General Motors' SI generator excited the field coil by sending current through the ignition switch in the RUN and START positions, through the warning lamp, and onto the field (**Figure 7-33**). This small magnetizing current pre-excites the field, reducing the speed required to start its own supply of field current.

With the engine running, the stator's output flows through the diode trio and back to the regulator and becomes the field current source. At this time, the alternator is self-sustaining.

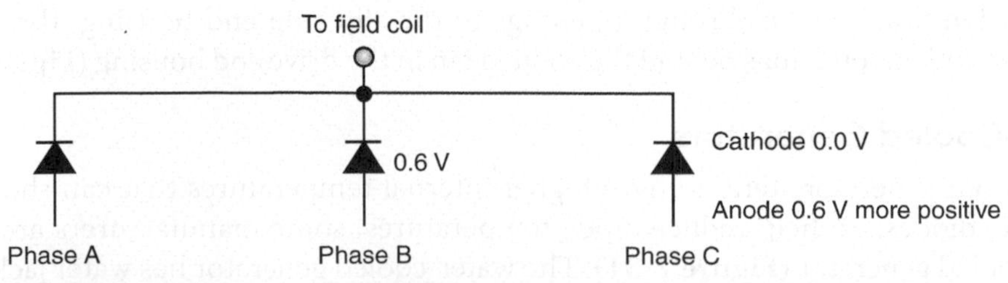

Figure 7-32 The diode trio connects the phase windings to the field. To conduct, there must be 0.6 volt more positive on the anode side of the diodes.

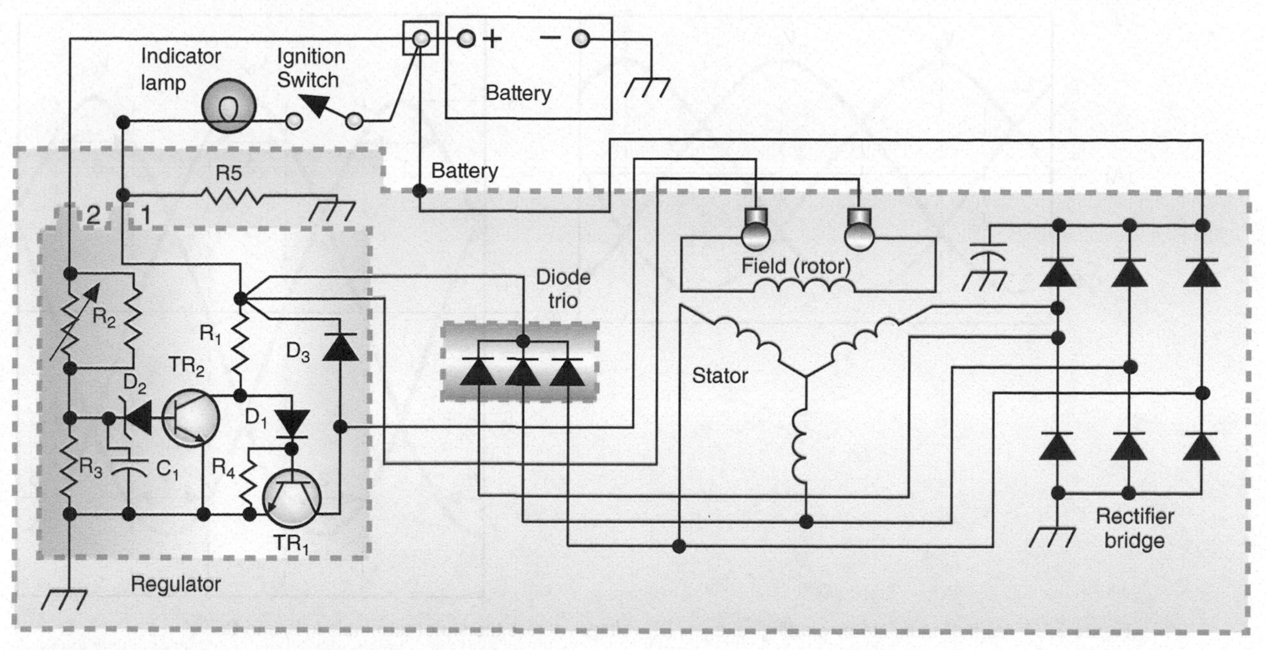

Figure 7-33 Schematic of a charging system with a diode trio.

Modern, computer-controlled charging systems excite the field coil using the control module to supply the current t the field coil when the ignition is in the RUN and START position. Computer-controlled charging systems are discussed later in this chapter.

AUTHOR'S NOTE If the battery is completely discharged, the vehicle cannot be push started because there is no excitation of the field coil.

AC Generator Operation Overview

Shop Manual
Chapter 7, page 313

Shop Manual
Chapter 7, Page 319

When the engine is running, the drive belt spins the rotor inside the stator windings. The magnetic field inside the rotor generates a voltage in the stator's windings. Field current flowing through the slip rings to the rotor creates alternating north and south poles on the rotor.

The induced voltage in the stator is an alternating voltage because the magnetic fields are alternating. As the magnetic field begins to induce voltage in the stator's windings, the induced voltage increases. The amount of voltage will peak when the magnetic field is the strongest. As the magnetic field begins to move away from the stator windings, the voltage will decrease. Each of the three windings of the stator generates voltage, so the three combine to form a three-phase voltage output.

In the wye connection (refer to Figure 7-27), output terminals (A, B, and C) apply voltage to the rectifier. Because only two stator windings apply voltage (the third winding always connects to diodes that are reverse-biased), the voltages come from points A to B, B to C, and C to A.

To determine the amount of voltage produced in the two stator windings, find the difference between the two points. For example, to find the voltage applied from points A and B, subtract the voltage at point B from the voltage at point A. Suppose the voltage at point A is 8 volts positive and the voltage at point B is 8 volts negative, the difference is 16 volts. This procedure can be performed for each pair of stator windings at any point in time to get the sine wave patterns (**Figure 7-34**). The designations for voltages in the windings are V_a, V_b, and

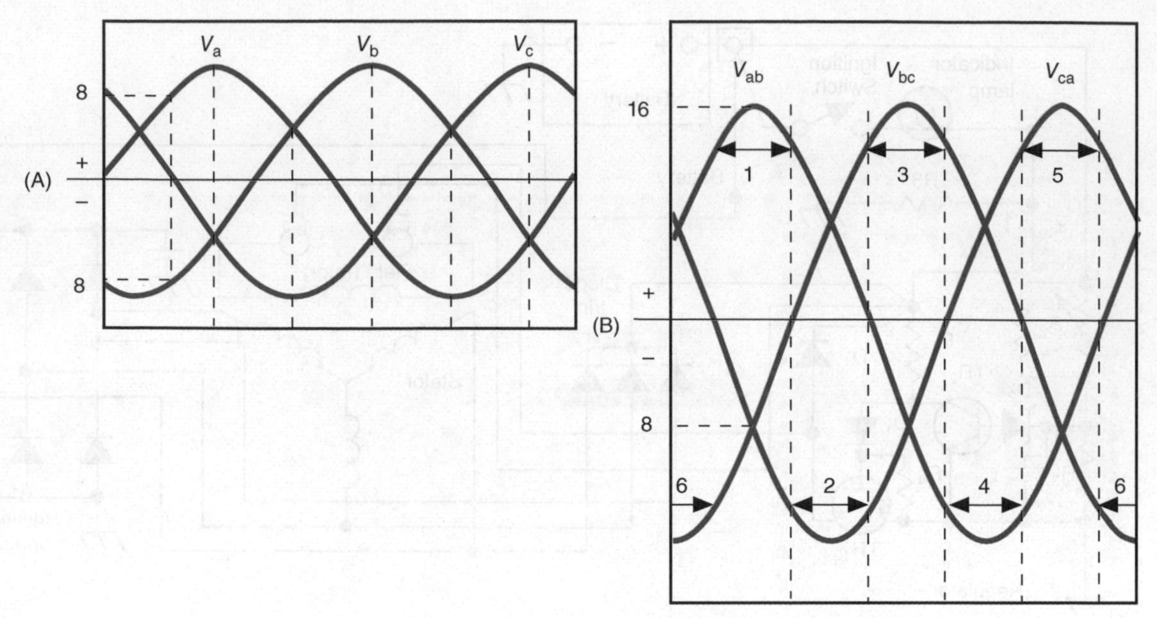

Figure 7-34 (A) Individual stator winding voltages; (B) voltages across the stator terminals A, B, and C.

V_c. Designations V_{ab}, V_{bc}, and V_{ca} refer to the voltage difference in the two stator windings. In addition, the numbers refer to the diodes used for the voltages generated in each winding pair.

AUTHOR'S NOTE Alternating current is constantly changing, so this formula would have to be performed at several different times.

The current induced in the stator passes through the diode rectifier bridge, consisting of three positive and three negative diodes. At this point, there are six possible paths for the current to follow. The path taken depends on the stator terminal voltages. If the voltage from points A and B is positive (point A is positive in respect to point B), current is supplied to the battery's positive terminal from terminal A through diode 2 (**Figure 7-35**). The negative return path is through diode 3 to terminal B.

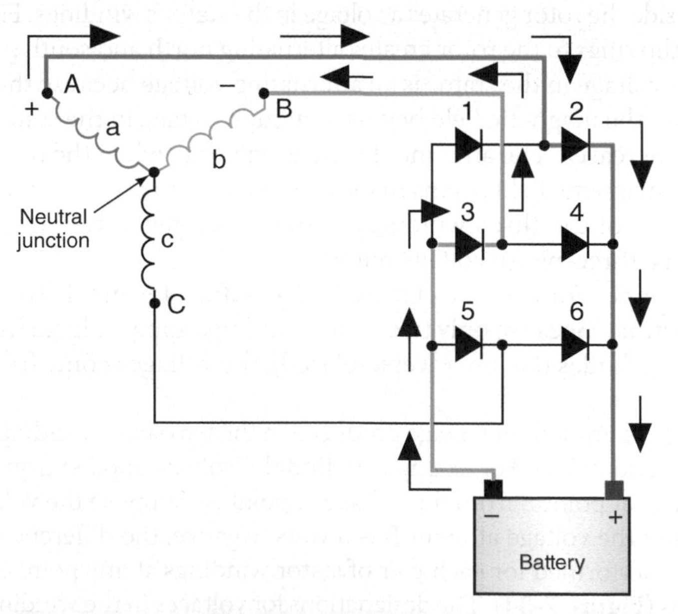

Figure 7-35 Current flow when terminals A and B are positive.

Both diodes 2 and 3 are forward-biased. The stator winding labeled C does not produce current because it's connected to reverse-biased diodes. Rectified stator current charges the battery and supplies current to the vehicle's electrical system.

When the voltage from terminals C and A is negative (point C is negative in respect to point A), current flow to the battery positive terminal is from terminal A through diode 2 (**Figure 7-36**). The negative return path is through diode 5 to terminal C.

This procedure is repeated through the four other current paths (**Figures 7-37** through **7-40**).

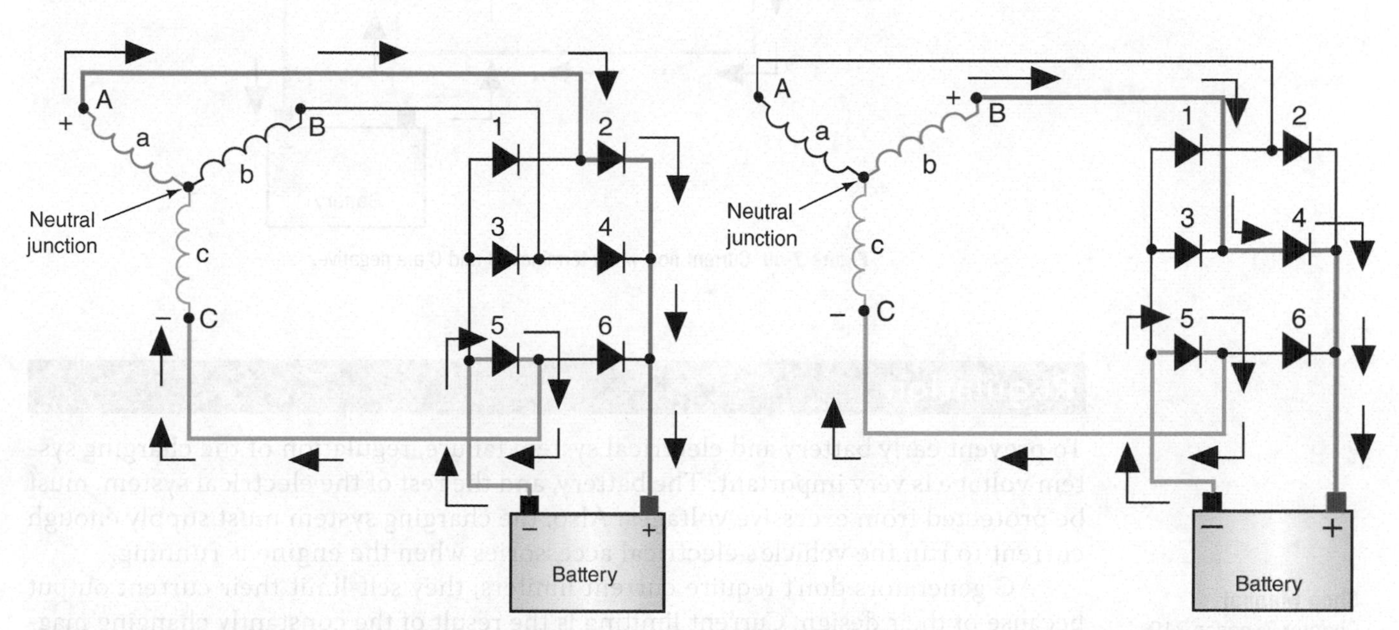

Figure 7-36 Current flow when terminals A and C are negative.

Figure 7-37 Current flow when terminals B and C are positive.

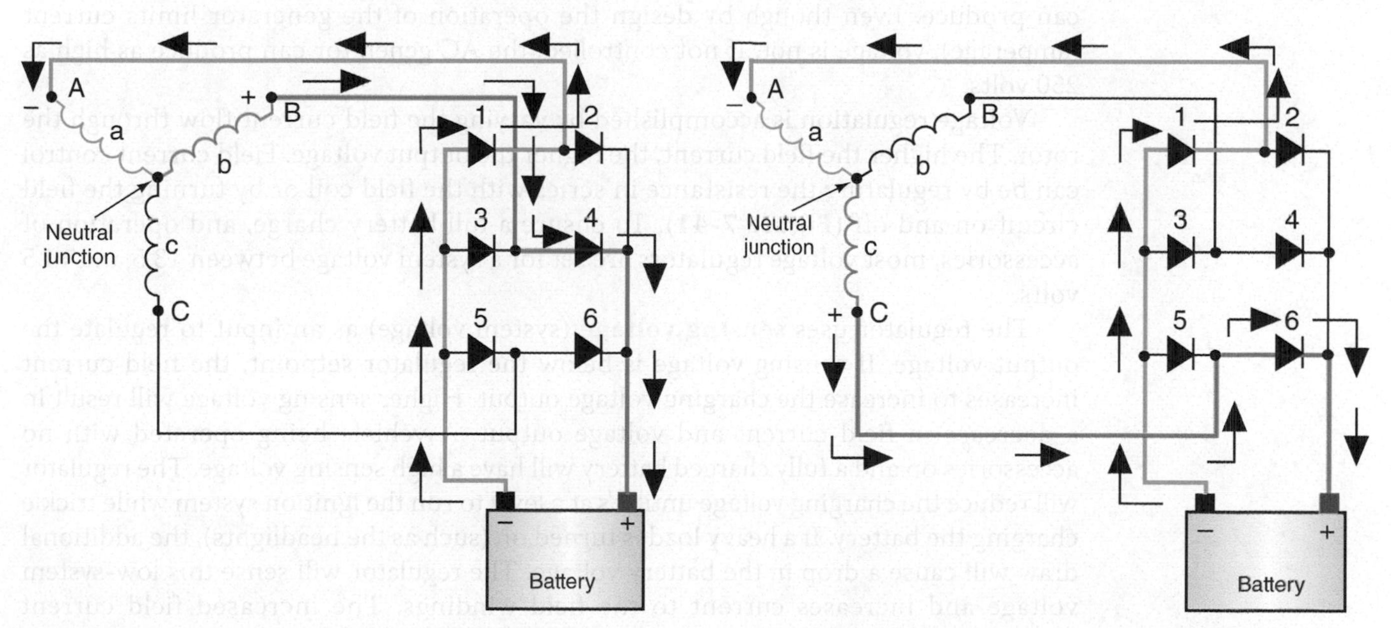

Figure 7-38 Current flow when terminals A and B are negative.

Figure 7-39 Current flow when terminals A and C are positive.

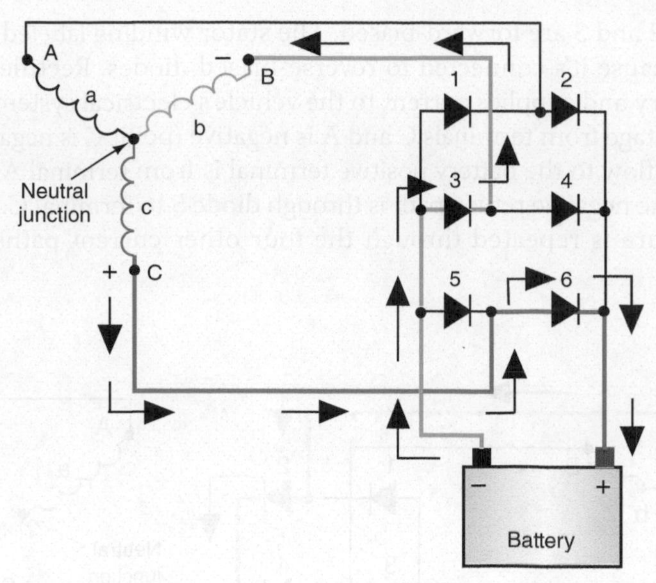

Figure 7-40 Current flow when terminals B and C are negative.

Regulation

To prevent early battery and electrical system failure, regulation of the charging system voltage is very important. The battery, and the rest of the electrical system, must be protected from excessive voltages. Also, the charging system must supply enough current to run the vehicle's electrical accessories when the engine is running.

Shop Manual
Chapter 7, pages 319, 324

AC generators don't require current limiters; they self-limit their current output because of their design. Current limiting is the result of the constantly changing magnetic field. As the magnetic field changes, it induces an opposing current in the stator windings. This **inductive reactance** limits the maximum current that the AC generator can produce. Even though by design the operation of the generator limits current (amperage), voltage is not. If not controlled, the AC generator can produce as high as 250 volts.

Voltage regulation is accomplished by varying the field current flow through the rotor. The higher the field current, the higher the output voltage. Field current control can be by regulating the resistance in series with the field coil or by turning the field circuit on and off (**Figure 7-41**). To ensure a full battery charge, and operation of accessories, most voltage regulators are set for a system voltage between 13.5 and 14.5 volts.

The regulator uses **sensing voltage** (system voltage) as an input to regulate the output voltage. If sensing voltage is below the regulator setpoint, the field current increases to increase the charging voltage output. Higher sensing voltage will result in a decrease in field current and voltage output. A vehicle being operated with no accessories on and a fully charged battery will have a high sensing voltage. The regulator will reduce the charging voltage until it's at a level to run the ignition system while trickle charging the battery. If a heavy load is turned on (such as the headlights), the additional draw will cause a drop in the battery voltage. The regulator will sense this low-system voltage and increases current to the field windings. The increased field current strengthens the magnetic field, and AC generator voltage output increases. Turning off the load causes a rise in system voltage that the regulator senses and cuts back the amount of field current and ultimately AC generator voltage output.

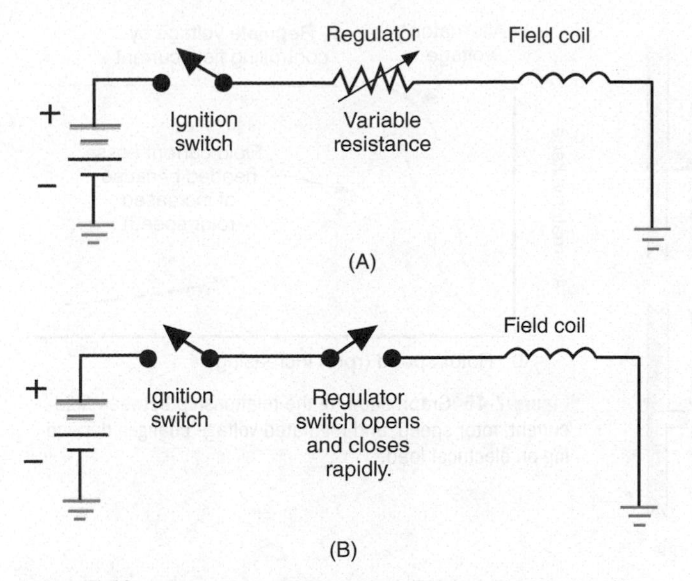

Figure 7-41 The regulator can control the field current by (A) controlling the resistance in series with the coil or (B) by switching the field on and off.

Temperature	Volts	
	Minimum	Maximum
20 °F	14.3	15.3
80 °F	13.8	14.4
140 °F	13.3	14.0
Over 140	Less than 13.3	–

Figure 7-42 Chart indicating the relationship between temperature and charge rate.

Another input that affects regulation is temperature. Because ambient temperatures influence the rate of charge that a battery can accept, regulators are temperature compensated (**Figure 7-42**). Temperature compensation is required because the battery is more reluctant to accept a charge at lower ambient temperatures. The regulator will increase the system voltage until it's at a higher level so the battery will accept it.

Field Circuits

Shop Manual
Chapter 7, page 317

There are three basic types of field circuits automakers use. The location of the regulator in relation to the field winding determines the circuit type. The first type is called the A circuit. It has the regulator on the ground side of the field coil. Battery voltage for the field coil is picked up from inside the AC generator (**Figure 7-43**). By placing the regulator on the ground side of the field coil, the regulator will allow the control of the field current by varying the current flow to ground.

The A circuit is called an external grounded field circuit.

The second type of field circuit is called the B circuit. In this case, the voltage regulator controls the power side of the field circuit. Also, the field coil is grounded from inside the AC generator.

Usually, the B circuit regulator is mounted externally to the AC generator. The B circuit is an internally grounded circuit.

AUTHOR'S NOTE To remember these circuits: Think of "A" for "After" the field and "B" for "Before" the field.

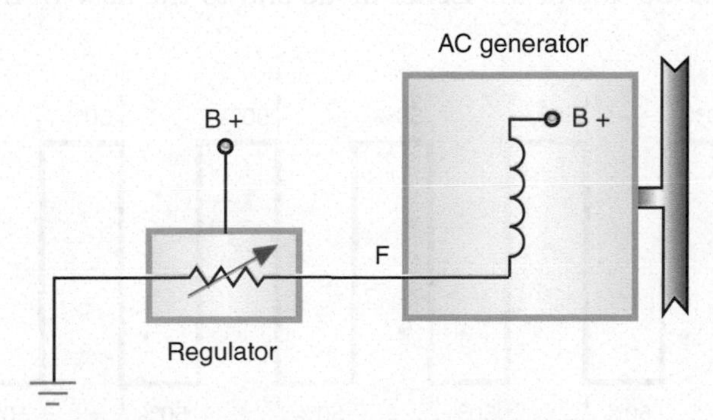

Figure 7-43 Simplified diagram of an A circuit field.

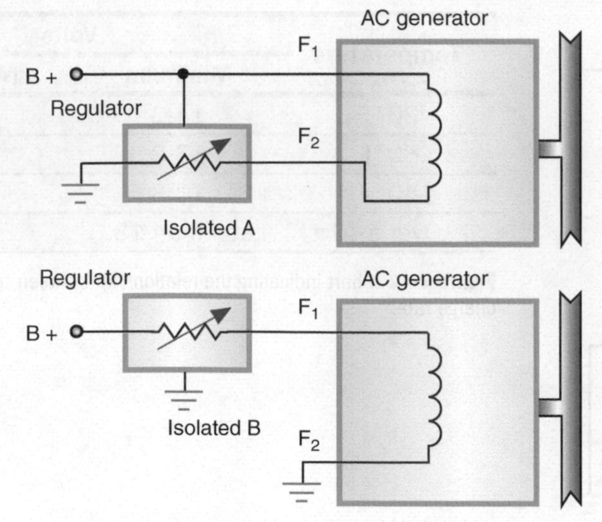

Figure 7-44 In the isolated circuit field AC generator, the regulator can be installed on either side of the field.

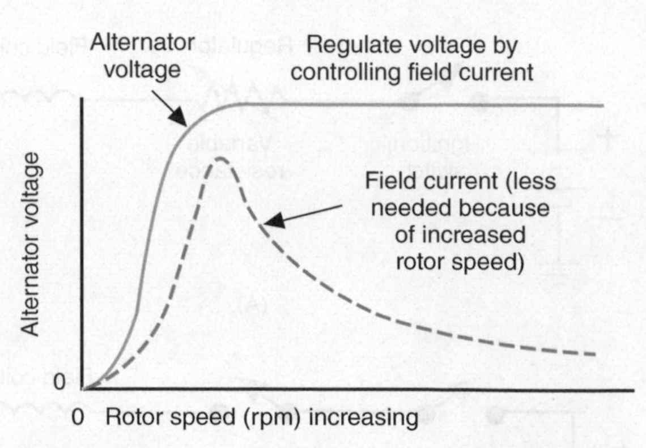

Figure 7-45 Graph showing the relationship between field current, rotor speed, and regulated voltage changes depending on electrical load.

The third type of field circuit is called the isolated field. The AC generator has two field wires attached to the outside of the case. The voltage regulator can be located either on the ground (A circuit) or on the B+ (B circuit) side (**Figure 7-44**). Isolated field AC generators pick up B+ and ground externally.

Regardless of which type of field circuit used, generator voltage output is regulated by controlling the amount of current through the field windings. As rotor speed increases, field current must decrease to maintain regulated voltage. **Figure 7-45** illustrates the relationship between the field current, rotor speed, and regulated voltage.

Electronic Regulation

Shop Manual
Chapter 7, page 324

The **electronic regulator** uses solid-state circuitry to perform the regulatory functions. Electronic regulators do not have any moving parts, so they can cycle between 10 and 7,000 times per second. This rapid cycling provides more accurate field current control through the rotor.

Pulse-width modulation (PWM) controls AC generator output by varying the amount of time the field coil is energized. For example, assume that a vehicle equipped with a 100-amp generator has an electrical demand of 50 amps; the regulator would energize the field coil for 50% of the time (**Figure 7-46**). If the electrical system's demand increased to 75 amps, the regulator would energize the field coil 75% of the cycle time.

The electronic regulator uses a Zener diode that blocks current flow until a specific voltage is reached, at which point it allows the current to flow (**Figure 7-47**). Battery voltage is applied to the cathode side of the Zener diode and to the base of transistor number 1.

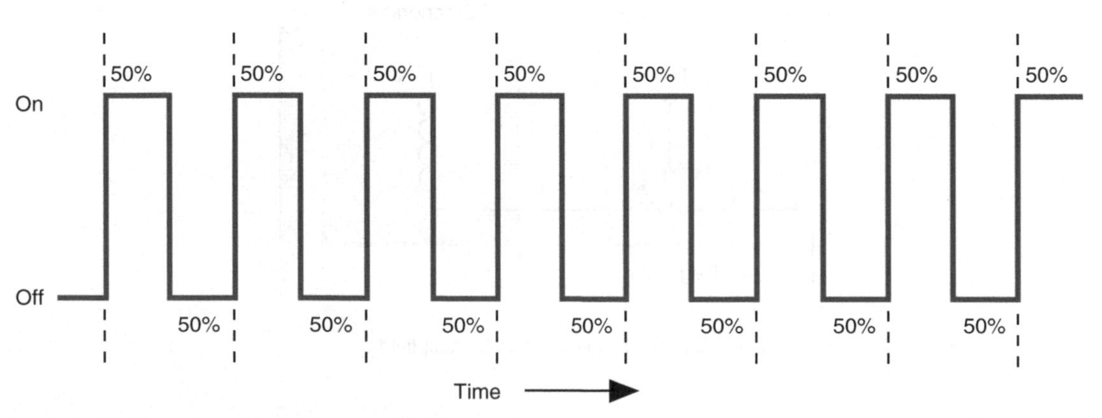

Figure 7-46 Pulse-width modulation with 50% on time.

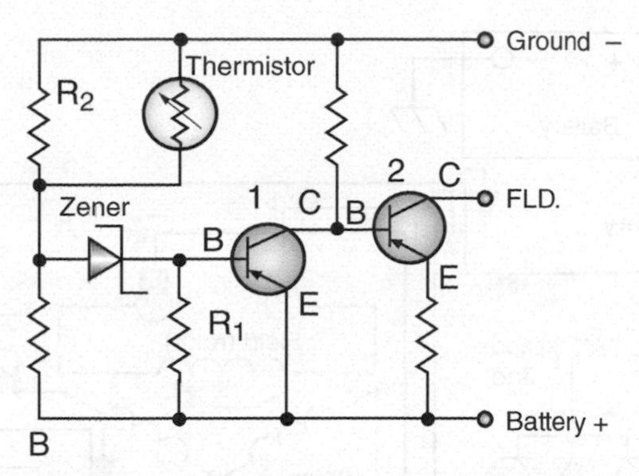

Figure 7-47 A simplified circuit diagram of an electronic regulator utilizing a Zener diode.

No current will flow through the Zener diode, since the voltage is too low to push through the Zener. However, as the AC generator produces voltage, the voltage at the cathode will increase until it reaches the upper limit (14.5 volts) and is able to push through the Zener diode. Current now flows from the battery, through the resistor (R_1), through the Zener diode, through the resistor (R_2) and the thermistor in parallel, and to ground. Since current is flowing, each resistance in the circuit will drop voltage. As a result, voltage to the base of transistor number 1 will be less than the voltage applied to the emitter. Since transistor number 1 is a PNP transistor and the base voltage is less than the emitter voltage, transistor number 1 turns on. The base of transistor number 2 will now have battery voltage applied to it. Since the voltage applied to the base of transistor number 2 is greater than that applied to its emitter, transistor number 2 turns off. Transistor number 2 is in control of the field current and generator output.

The thermistor changes circuit resistance according to temperature. This provides the temperature-related voltage change necessary to keep the battery charged in cold-weather conditions.

The following illustrates regulation control under different operating conditions. With the engine off and the ignition in the RUN position, battery voltage is applied to the field coil through the common point above R_1 **(Figure 7-48)**. TR_1 conducts the field current

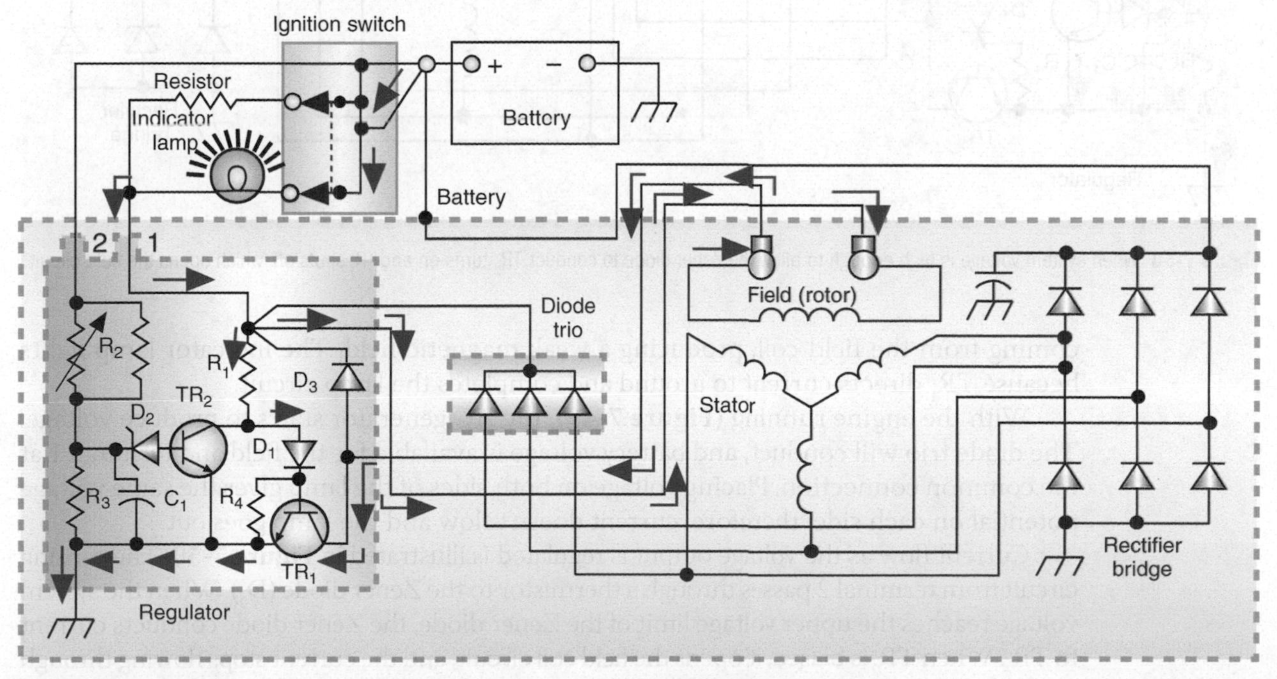

Figure 7-48 Current flow to the rotor with the ignition switch in the RUN position and the engine OFF.

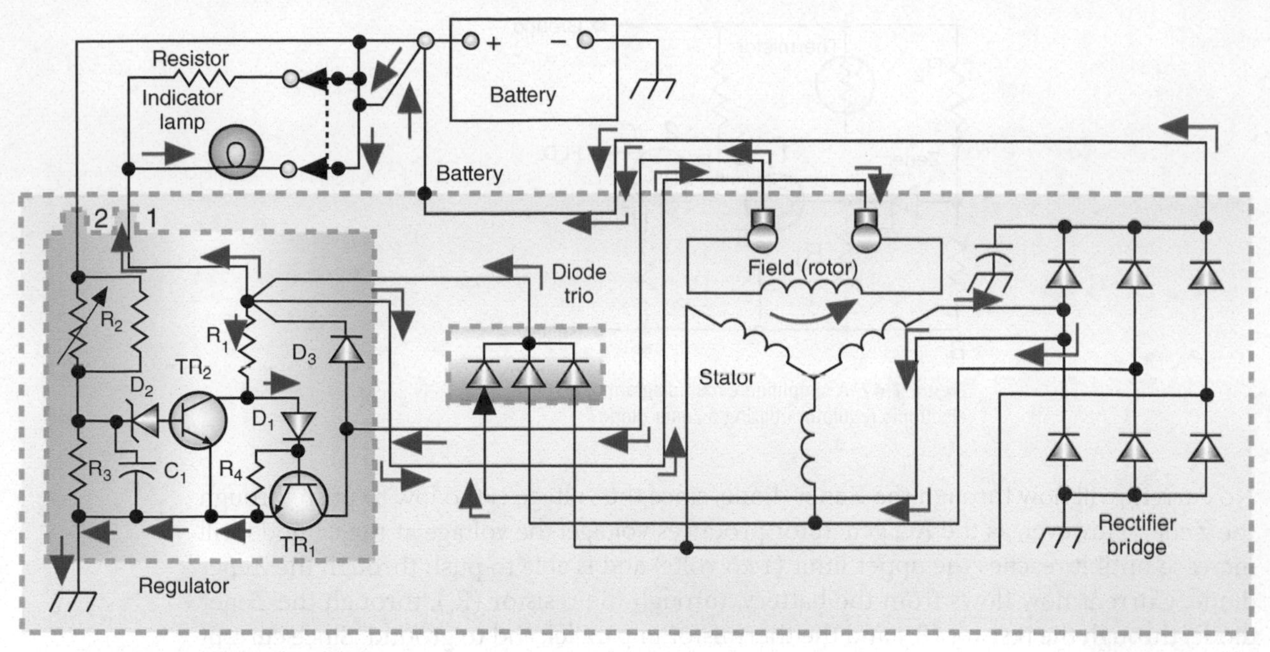

Figure 7-49 Current flow with the engine running and AC generator producing voltage.

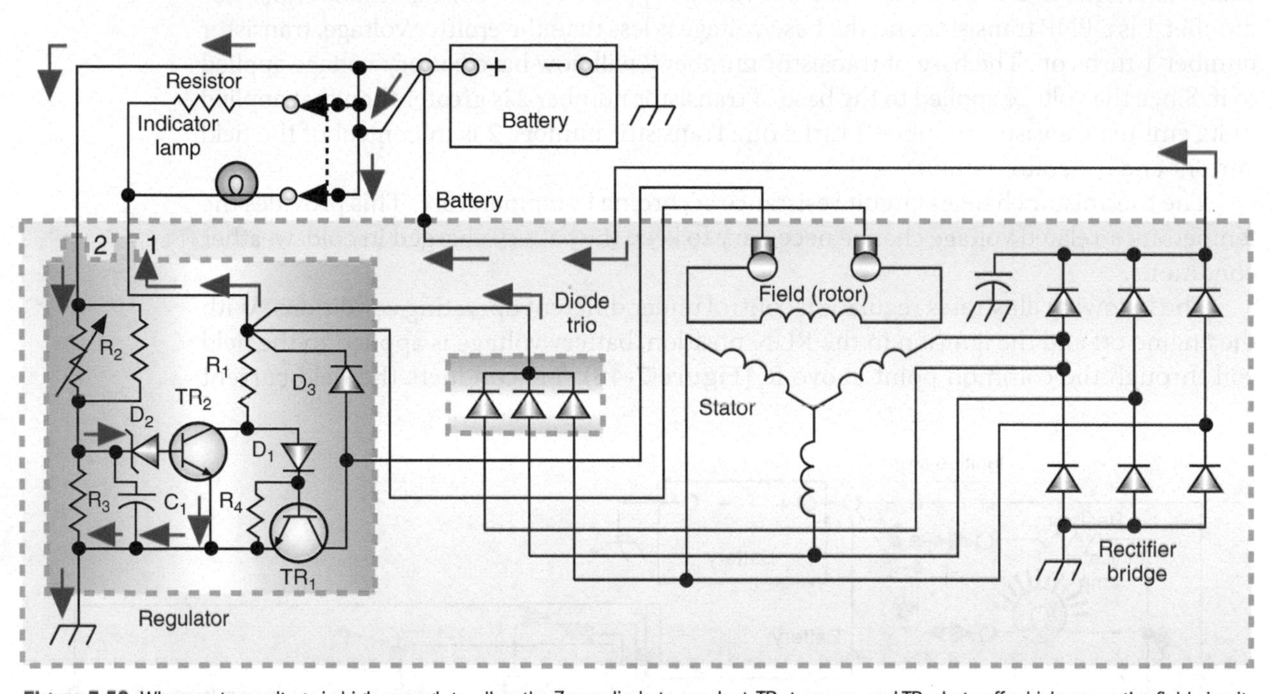

Figure 7-50 When system voltage is high enough to allow the Zener diode to conduct, TR_2 turns on and TR_1 shuts off, which opens the field circuit.

coming from the field coil, producing a weak magnetic field. The indicator lamp lights because TR_1 directs current to ground and completes the lamp circuit.

With the engine running **(Figure 7-49),** the AC generator starts to produce voltage. The diode trio will conduct, and battery voltage is available for the field and terminal 1 at the common connection. Placing voltage on both sides of the lamp gives the same voltage potential on each side; therefore, current doesn't flow and the lamp goes out.

Current flow as the voltage output is regulated is illustrated in **Figure 7-50**. The sensing circuit from terminal 2 passes through a thermistor to the Zener diode (D_2). When the system voltage reaches the upper voltage limit of the Zener diode, the Zener diode conducts current to TR_2. When TR_2 is biased, it opens the field coil circuit, and the current stops flowing through it. Regulation of this switching on and off is based on the sensing voltage received through terminal 2. With the circuit to the field coil opened, the sensing voltage decreases, and the Zener diode stops conducting. TR_2 is turned off, and the circuit for the field coil is closed.

Computer-Controlled Regulation

On many vehicles after the mid-1980s, voltage regulation became a function of the powertrain control module (PCM) (**Figure 7-51**). The computer-controlled regulation system precisely maintains and controls the changing voltage rate according to the

Shop Manual
Chapter 7, pages 321, 330

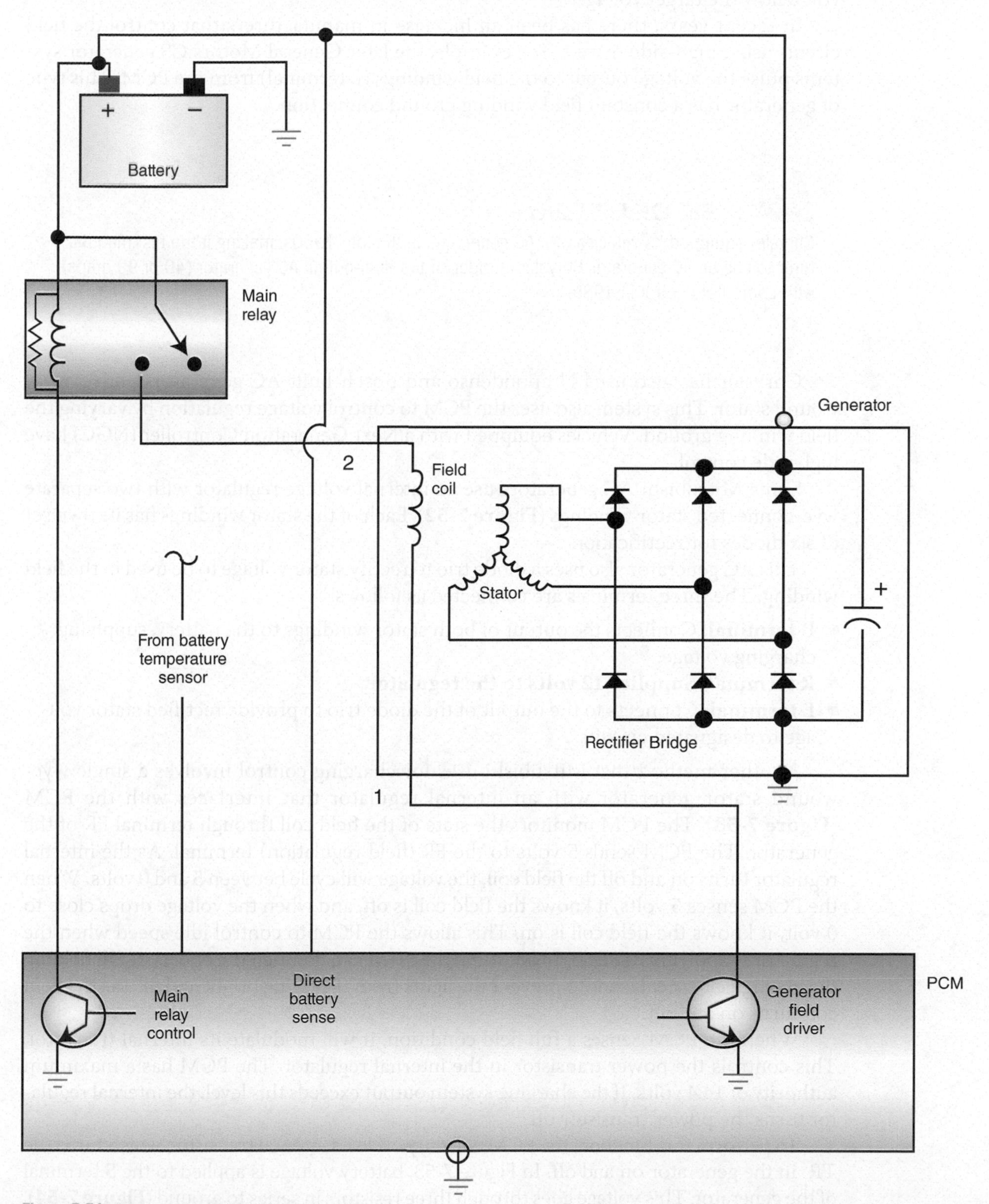

Figure 7-51 Computer-controlled voltage regulator circuit.

electrical requirements and battery (or ambient) temperature. Regulation of the field circuit is through the ground (A circuit). The PCM will determine the SOC of the battery as inputted by the battery sense circuit. This information and ambient temperature or battery temperature sensor inputs determine a target battery voltage. The PCM controls the excitation of the field coil to maintain the target voltage. Typically, field coil activation continues until the voltage on the sense circuit reaches about 0.5 volt above the target voltage, then turns off. The field coil reactivates when the sense battery voltages falls 0.5 volt below the target voltage.

In recent years, there has been an increase in manufacturers that control the field circuit using high-side drivers. For example, the later General Motors CS generator systems pulse the voltage output to the field windings (L terminal) from the PCM. This type of generator has a constant field winding ground connection.

A BIT OF HISTORY

Chrysler equipped its vehicles with AC generators in the late 1950s, making it the first manufacturer to use an AC generator. Chrysler introduced the dual-output AC generator (40 or 90 amps) with computer control in 1985.

Chrysler has also used Nippondenso and Bosch-built AC generators with a wye-wound stator. This system also uses the PCM to control voltage regulation by varying the field winding ground. Vehicles equipped with a Next Generation Controller (NGC) have high-side control.

Some Mitsubishi AC generators use an internal voltage regulator with two separate wye-connected stator windings (**Figure 7-52**). Each of the stator windings has its own set of six diodes for rectification.

This AC generator also uses a diode trio to rectify stator voltage to be used in the field winding. The three terminals are connected as follows:

- **B terminal:** Connects the output of both stator windings to the battery, supplying charging voltage.
- **R terminal: Supplies 12 volts to the regulator.**
- **L terminal:** Connects to the output of the diode trio to provide rectified stator voltage to designated circuits.

Another method that Mitsubishi uses for charging control involves a single wye-wound stator generator with an internal regulator that interfaces with the PCM (**Figure 7-53**). The PCM monitors the state of the field coil through terminal FR of the generator. The PCM sends 5 volts to the FR (field regulation) terminal. As the internal regulator turns on and off the field coil, the voltage will cycle between 5 and 0 volts. When the PCM senses 5 volts, it knows the field coil is off, and when the voltage drops close to 0 volt, it knows the field coil is on. This allows the PCM to control idle speed when the regulator is applying full field. In addition, the PCM can dampen the effects of full fielding during high electrical loads to prevent the lights from flickering bright and dim as the field coil turns on and off.

When the PCM senses a full field condition, it will modulate its internal transistor. This controls the power transistor in the internal regulator. The PCM has a maximum authority of 14.4 volts. If the charging system output exceeds this level, the internal regulator turns the power transistor off.

To perform this function, the PCM will duty cycle its internal transistor, which turns the TR_1 in the generator on and off. In Figure 7-53, battery voltage is applied to the S terminal of the generator. This voltage goes through three resistors in series to ground (**Figure 7-54**).

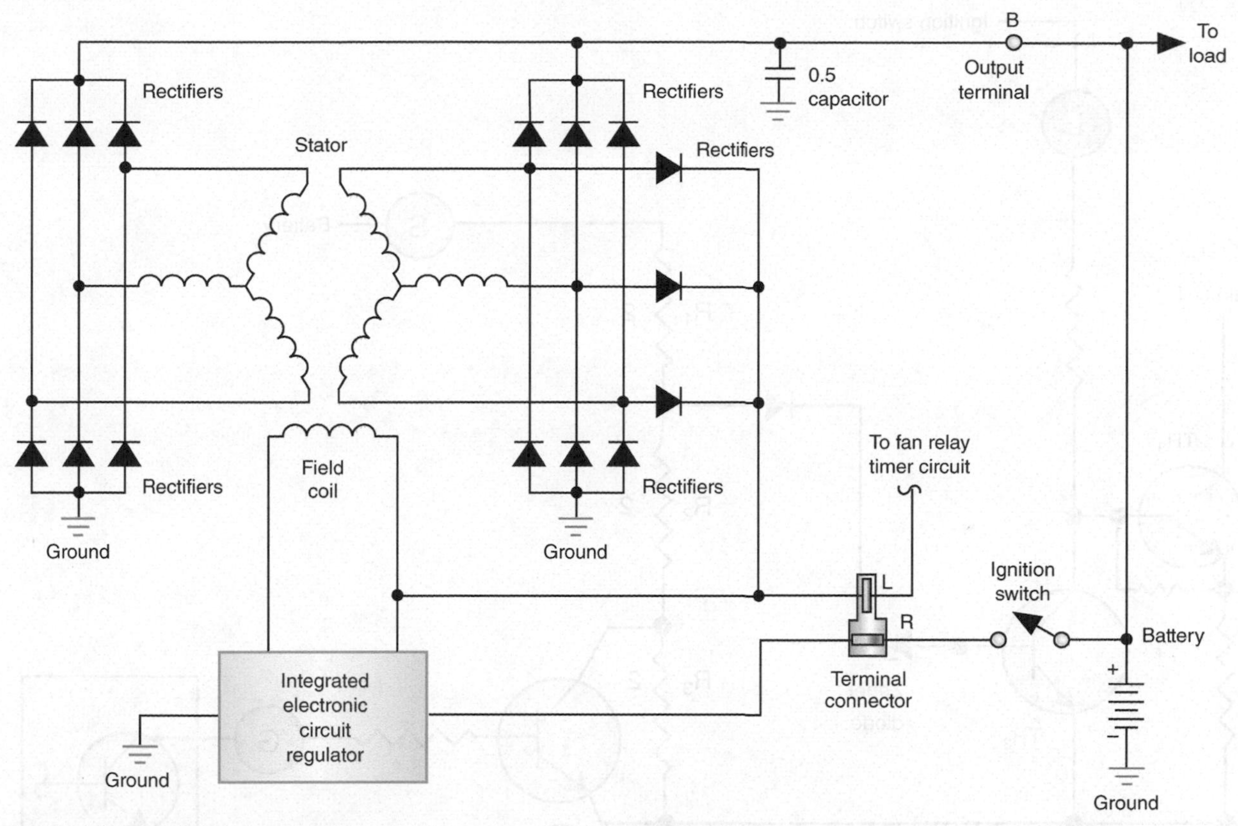

Figure 7-52 The Mitsubishi AC generator uses two separate stator windings and a total of 15 diodes.

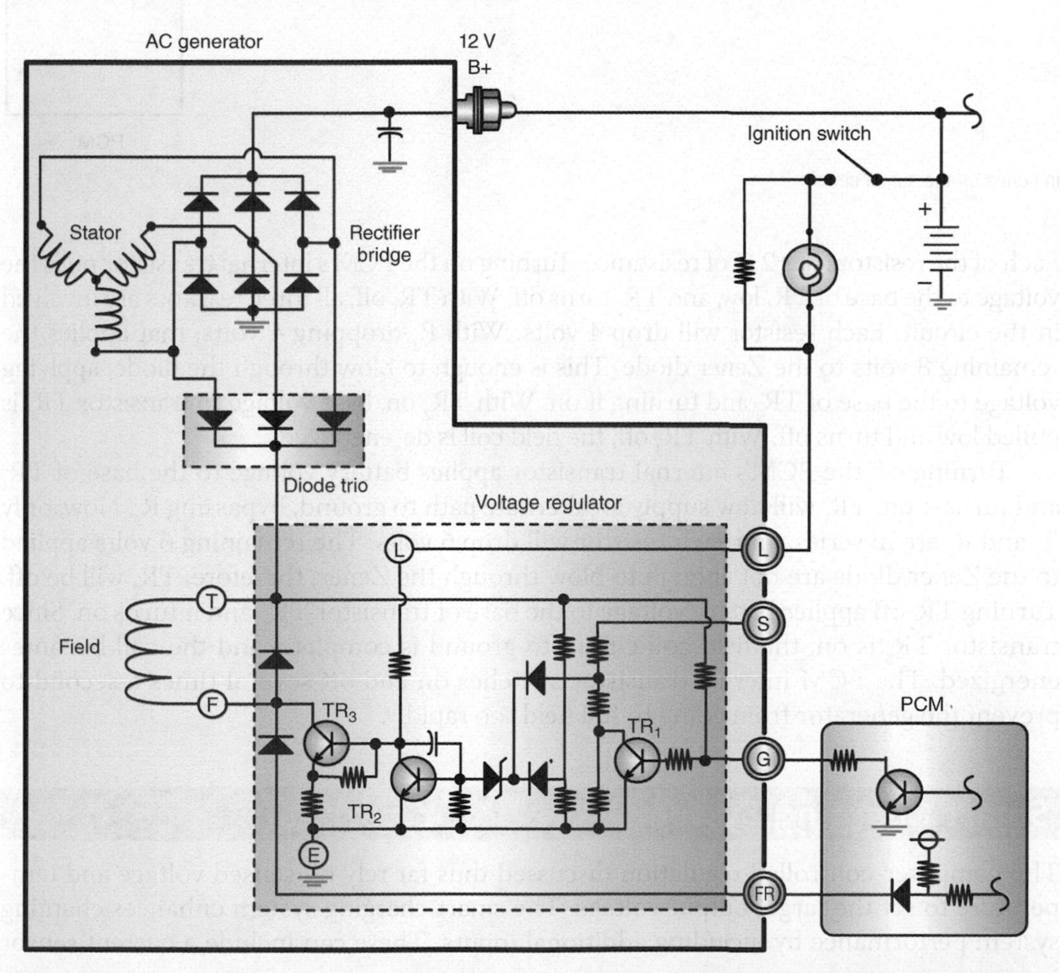

Figure 7-53 Schematic of Mitsubishi charging system using both an internal voltage regulator and a PCM to control output.

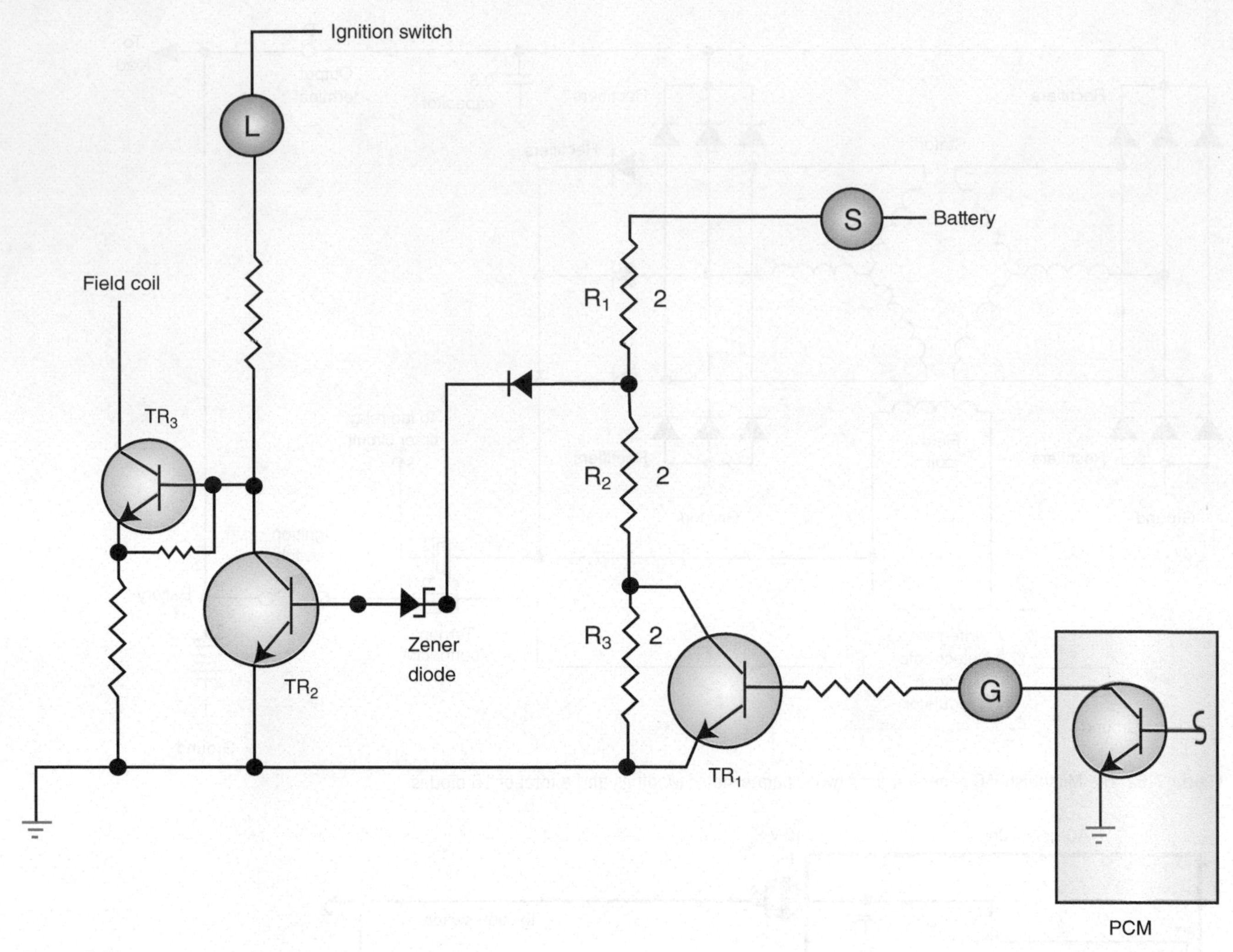

Figure 7-54 Three resistors in series on the sense circuit.

Each of the resistors has 2 Ω of resistance. Turning on the PCM's internal transistor pulls the voltage to the base of TR_1 low, and TR_1 turns off. With TR_1 off, all three resistors are involved in the circuit. Each resistor will drop 4 volts. With R_1 dropping 4 volts, that applies the remaining 8 volts to the Zener diode. This is enough to blow through the diode, applying voltage to the base of TR_2 and turning it on. With TR_2 on, base voltage to transistor TR_3 is pulled low and turns off. With TR_3 off, the field coil is de-energized.

Turning off the PCM's internal transistor applies battery voltage to the base of TR_1 and turns it on. TR_1 will now supply an alternate path to ground, bypassing R_3. Now, only R_1 and R_2 are in series, and each resistor will drop 6 volts. The remaining 6 volts applied to the Zener diode are not enough to blow through the Zener; therefore, TR_2 will be off. Turning TR_2 off applies battery voltage to the base of transistor TR_3, and it turns on. Since transistor TR_3 is on, the field coil circuit to ground is complete, and the coil becomes energized. The PCM internal transistor switches on and off several times a second to prevent the generator from going to full field too rapidly.

Smart Charging Systems

Shop Manual
Chapter 7, page 330

The computer-controlled regulation discussed thus far rely on sensed voltage and temperature to set the target output voltage. The smart charging system enhances charging system performance by including additional inputs. These can include a current sensor

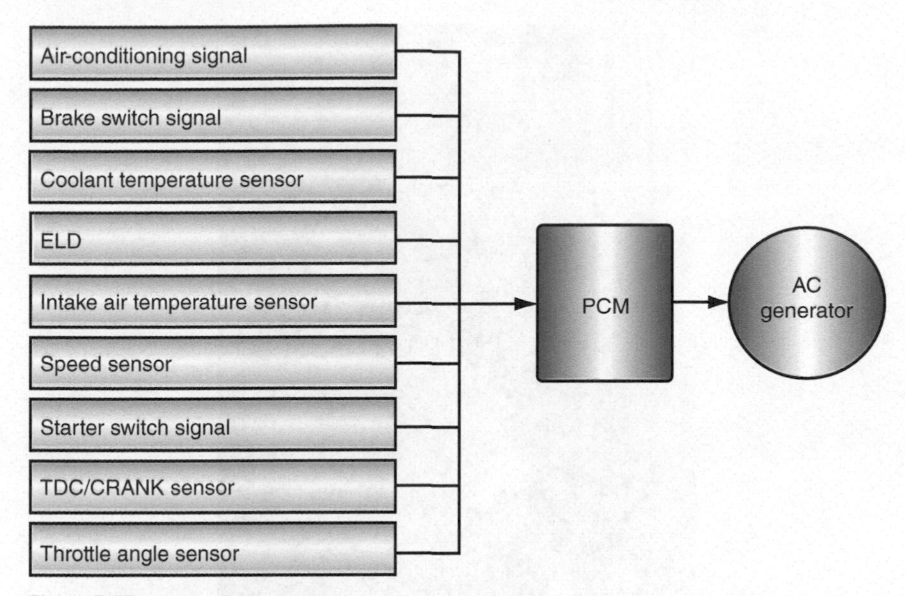

Figure 7-55 The smart charging system uses various inputs to regulate AC generator output.

and the shared inputs from the accelerator pedal position (APP) sensor, manifold absolute pressure (MAP) sensor, and vehicle speed sensors to determine the necessary charging rate **(Figure 7-55)**.

Also, the smart charging system may use more than one module. Usually, one of the modules is the PCM. The other module can be the body control module (BCM) and, in some cases, the generator itself.

The **interfaced generator (Figure 7-56)** has an internal eight-bit microprocessor that is a slave module on the local interconnect network (LIN) bus. The PCM communicates current and voltage requirements over the LIN bus to the microprocessor. Electronic driver stages control the generator output.

When the generator is a slave module on a bus network, it's often called an **interfaced generator.**

The smart charging system uses an intelligent battery sensor (IBS) **(Figure 7-57)** to determine the current flow out of and into the battery, and to monitor the battery's SOC.

Depending on the manufacturer, the IBS communicates either to the PCM or the BCM. **Figure 7-58** shows an example of the IBS communicating the battery status to the BCM. Here, the PCM regulates the charging system's output by PWM of the HSD circuit to the field winding. The PCM receives a voltage input from the generator and the power

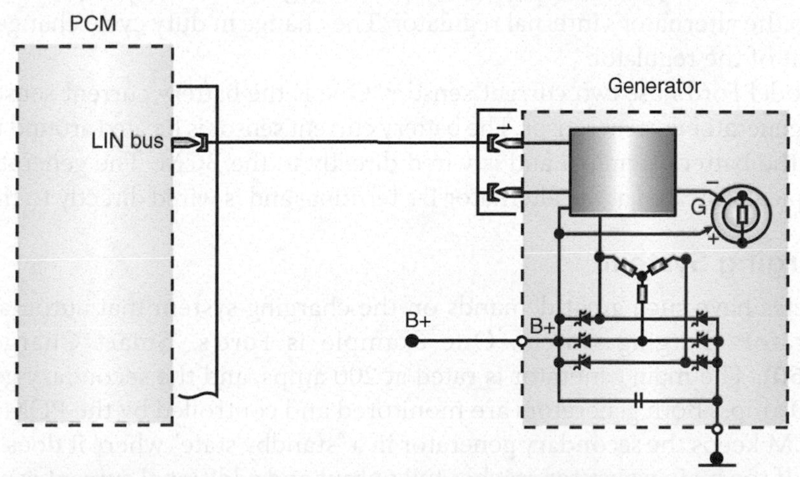

Figure 7-56 A generator with an internal microprocessor using a bus network for communications to the PCM.

Figure 7-57 The IBS may be connected to the battery's negative or the positive terminal to monitor current flow.

distribution center (PDC). These voltage levels are compared to the desired voltage levels programmed in the software.

Terminal 2 is internally connected to the B+ terminal to provide a voltage reading to the PCM on the sense circuit. The internal connection allows for the monitoring of the B+ voltage. If the B+ terminal stud is loose or disconnected, the PCM will shut down the generator field.

During engine cruise and acceleration, the charging system delivers enough voltage to provide for battery stabilization. During deceleration, the charging system outputs the maximum regulated voltage to replenish the battery.

GM's integrated regulated voltage control (RVC) system is a dual-module charging system using the ECM and the BCM **(Figure 7-59)**. The BCM receives battery current sensor inputs and determines the required alternator output. The BCM sends this data over the GMLAN bus to the ECM that commands the output of the generator.

The ECM sends a duty cycled 5-volt signal to terminal L of the alternator. When an increase in voltage output is necessary, the ECM changes the duty cycle. Terminal L feeds the signal to the alternator's internal regulator. The change in duty cycle changes the voltage set point of the regulator.

Some vehicles, such as those with the 2.7 turbocharged engine, mount the generator current sensor to the battery's negative post.

Late-model Ford's use two current sensors. One is the battery current sensor, and the other is the generator current sensor. The battery current sensor is located around the ground cable from the battery terminal and is wired directly to the BCM. The generator current sensor is typically located at the alternator B+ terminal and is wired directly to the PCM.

Dual Charging System

Some vehicles have such great demands on the charging system that automakers use a dual generator charging system. One example is Ford's "Smart Charge" system **(Figure 7-60)**. The main generator is rated at 200 amps, and the secondary generator is rated at 160 amps. Both generators are monitored and controlled by the PCM.

The PCM keeps the secondary generator in a "standby state" where it does not generate voltage. If the main generator reaches full output and additional current is needed, the PCM activates the secondary generator.

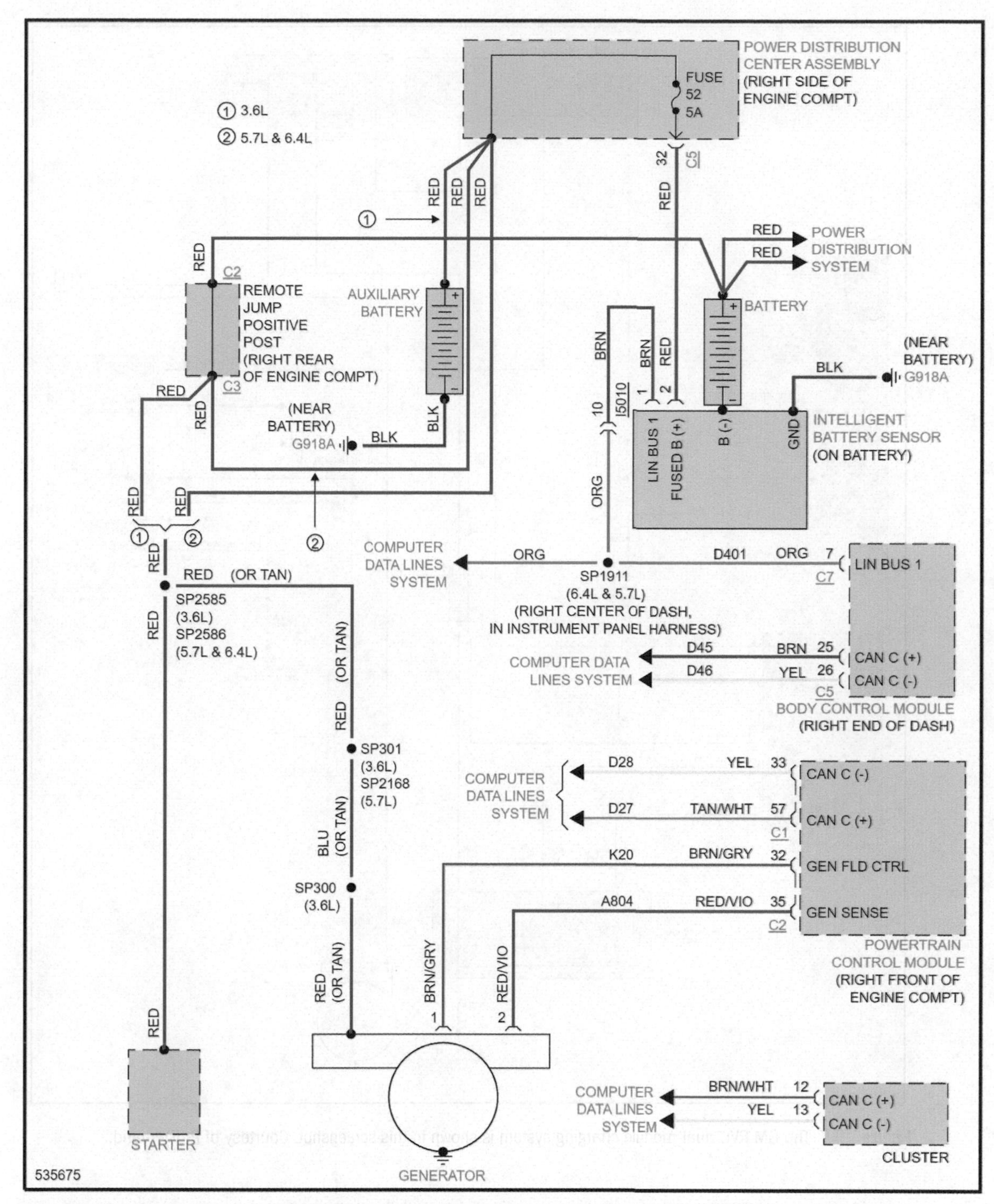

Figure 7-58 A screenshot shows that the IBS can communicate with modules on the LIN bus. Courtesy of ProDemand.

The circuits of the Ford Smart Charge system include:

- B+- Alternator output to the batteries and the vehicle electrical system.
- RC- Sends the duty cycle commands (ranging from 3% to 98%) from the PCM circuit to the generator's voltage regulator to control voltage output. This circuit is called GENCOM.
- LI- Provides feedback from the voltage regulator to the PCM by way of the GENMON circuit.
- AS- Senses battery voltage.

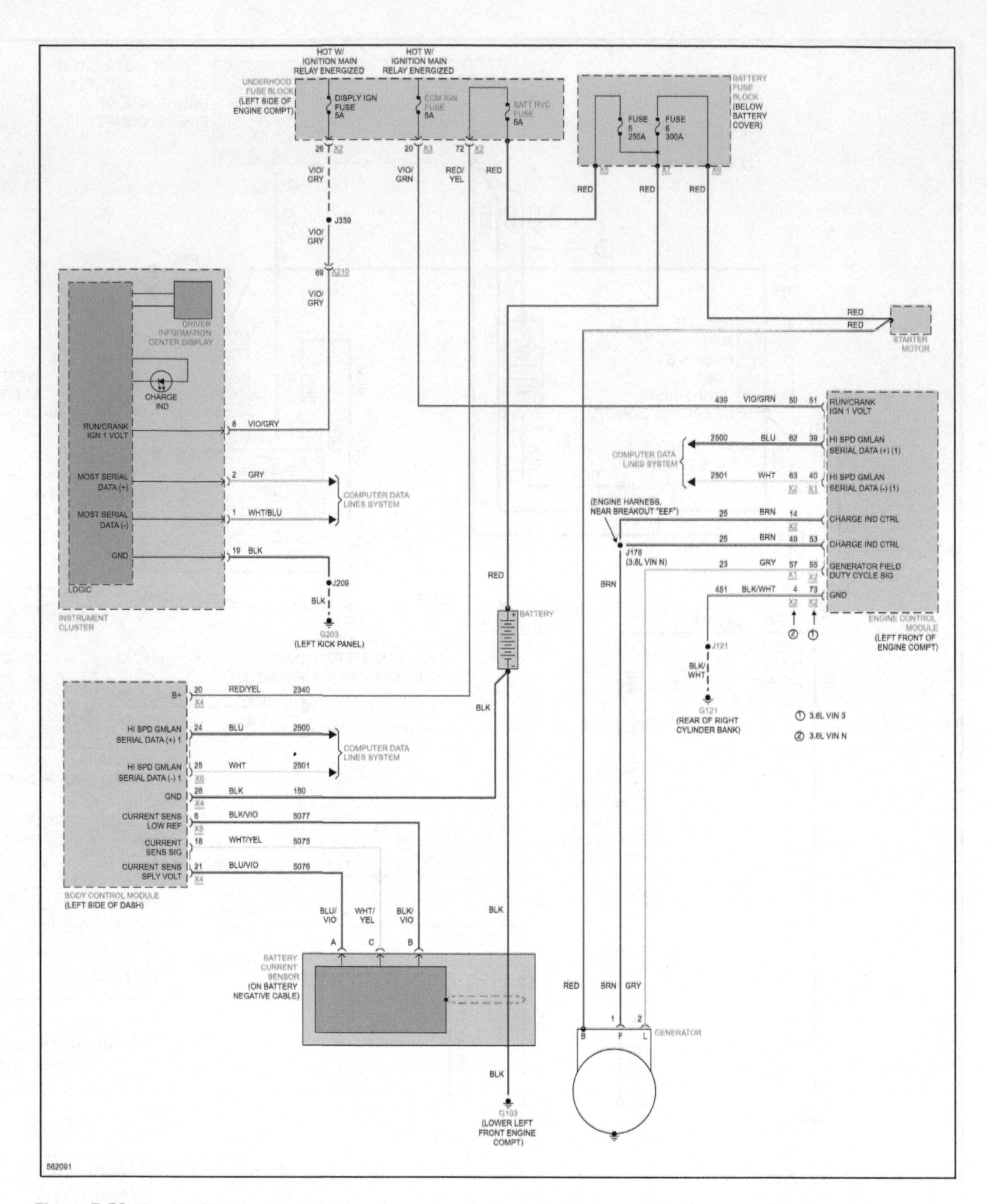

Figure 7-59 The GM RVC dual-module charging system is shown in this screenshot. Courtesy of ProDemand.

If the GENCOM circuit fails to initiate charging, the generator may still be able to charge the battery and keep the engine running. If the engine momentarily operates at more than 2000 RPM, the generator will "self-excite" and enters default mode. Output voltage during the default condition is approximately 13.5 volts. The charging system warning lamp will illuminate, or the message center will display a "Check Charging System" message.

Load Shedding

Shop Manual
Chapter 7, page 332

If the IBS detects the charging system is failing, or the vehicle battery SOC is too low, electrical **load shedding** actions are initiated. Load shedding reduces electrical power consumption by turning off non-essential electrical loads to extend the driving time and distance. The aggressiveness of the load shedding depends on the battery's SOC **(Figure 7-61)**.

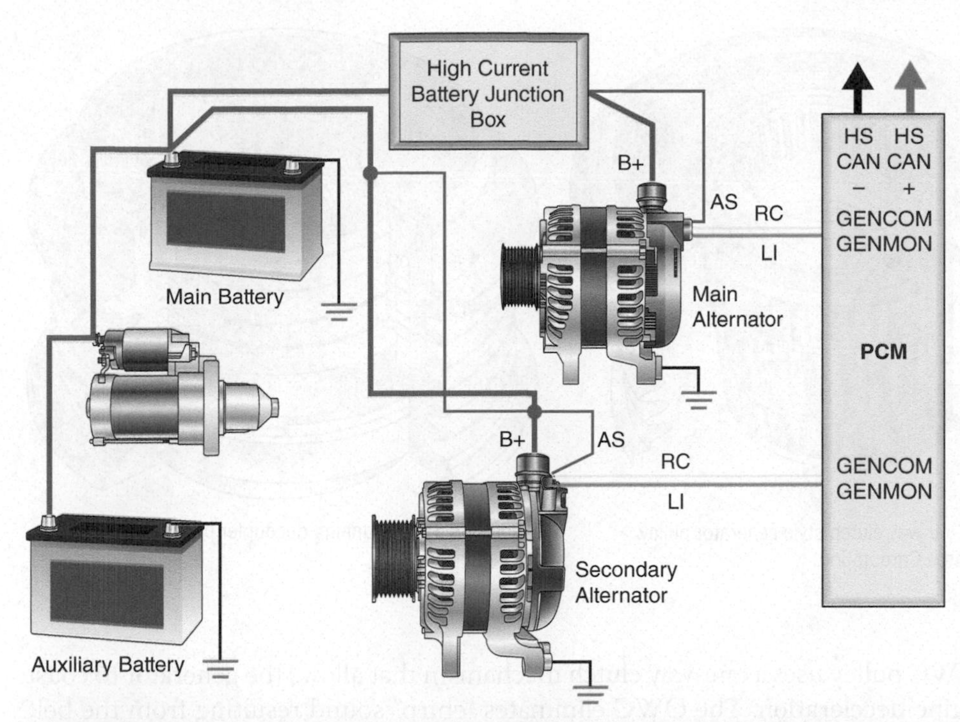

Figure 7-60 Ford's Smart Charge system uses two generators if demand requires additional current.

Level	System	Action
0	None	None
	Heated Outside Mirrors	"L" Cycled at 80% Duty Cycle. Off 4 seconds out of every 20 seconds. Indicator not affected.
	Message Center	No messages or indicators displayed. Data stored to indicate that Load Shed was entered.
2	Heated Outside Mirrors, Rear Defroster, Heated Seats	"L" Cycled at 50% Duty Cycle. Off 10 seconds out of every 20 seconds.
	Message Center	"Battery Save Action" message displayed or flashing indicators.
3	Heated Outside Mirrors, Rear Defroster	System turned OFF, Indicators not affected.
	Message Center	"Battery Save Action" message displayed and/or "Battery Charging System Failure" icon is illuminated. Chime is activated until Load Shed 3 is exited. flashing indicators.

Figure 7-61 Load shedding levels progressively turn off non-essential accessories.

Load shedding can occur anytime generator output does not meet the current demands of the system. Engine idling with heavy electrical accessory loads is one condition that may initiate load shedding. In this case, the battery current sensor detects a discharge of current from the battery, and load shedding may be commenced along with idle boost mode to improve charging system performance.

Alternator Pulley Types

In 1979 accessory belt drive systems began the transition from multiple V-belt drives to serpentine drives with automatic belt tensioners. The most common alternator pulley for these drives has been the standard "solid" design. 1997 saw the introduction of the one-way clutch (OWC) pulley **(Figure 7-62)**. The first overrunning alternator decoupler (OAD) pulley debuted in 1999 **(Figure 7-63)**.

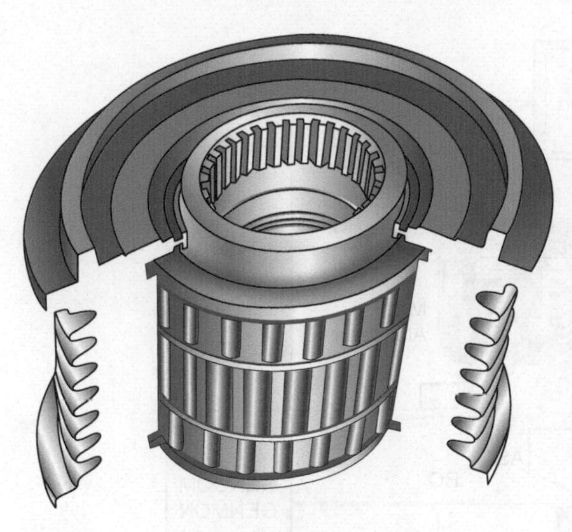

Figure 7-62 One-way clutch style generator pulley. Courtesy of Gates Corporation.

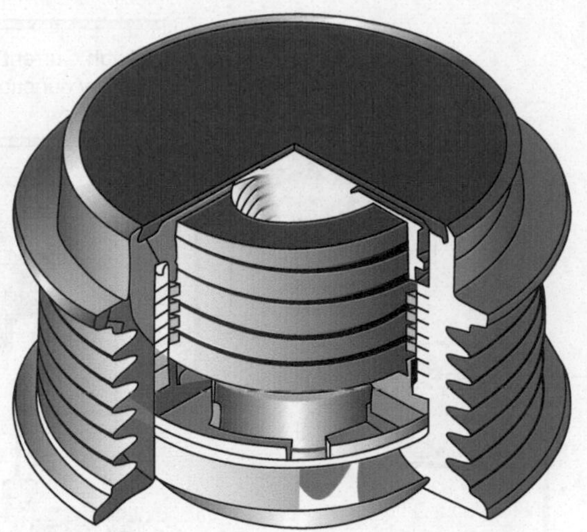

Figure 7-63 Overrunning decoupler pulley. Courtesy of Gates Corporation.

The OWC pulley uses a one-way clutch mechanism that allows the generator to coast during engine deceleration. The OWC eliminates "chirp" sound resulting from the belt slipping when engine speed quickly decelerates during shut down or transmission shifting. The OWC allows automakers more tolerance to decrease the drive belt tension, which increases belt-driven component life.

The OAD pulley also uses a one-way clutch mechanism that allows alternators to coast during deceleration. In addition, OAD pulleys incorporate torsion springs that absorb energy that creates vibrations in the accessory belt drive system. The springs are tuned for "engine specific" operation to absorb cylinder firing pulses before they reach the accessory drive system.

AC Generator Output Ratings

Shop Manual
Chapter 7, page 328

The output rating of an AC generator is usually expressed in the amount of amperage that the generator can produce at a specific rotational speed. The International Standards Organization and the Society of Automate Engineers standards (ISO 8854 and SAE J 56) indicate that the rated output is at a rotor shaft speed of 6,000 rpm.

AUTHOR'S NOTE There is a difference between the AC generator's amperage rating and the amount of current it can produce at idle speed. As discussed in the Shop Manual, it's possible to measure the actual output of an alternator under a simulated load. This will allow you to determine what the AC generator can actually produce.

Both ISO and SAE require a format of AC generator ratings that identifies the amperage output at the idle speed (ISO identifies this as the actual engine idle speed and ASE identifies this as 1,500 rpm), the rated amperage output, and the test voltage. This rating is typically stamped or labeled on the housing. For example, a label with "80/200A 13.5V" would indicate that the rated output is 80 amps at idle and 200 amps at 6,000 rpm (shaft speed) at a test voltage of 13.5 volts.

Several factors determine the current output of the alternator. These include the rotational speed of the rotor, the size of the poles, the size of the stator core, the thickness of the lamination, the number of turns in the coil, and the wire size used for the windings. In addition, high output AC generators use a higher amperage rated diode rectifier bridge. Standard duty rectifiers use 55 amp diodes, while the high output rectifier typically uses 70 amp diodes.

AC Generator Output Supply and Demand

The amperage rating on an AC generator is the amount of current the generator is capable of producing, not the amount it always produces. The output of the AC generator depends on the demands placed on it by the vehicle's electrical system. An AC generator will not produce more current than these demands require. For example, an AC generator rated at 180 amps will output only 40 amps if that is the total draw of the electrical system.

An AC generator that is under rated for the vehicle can cause problems since it cannot meet the requirements of the electrical system. However, an AC generator with a higher output than required represents wasted potential.

Since electrical demand dictates current output, an oversized AC generator will not damage the battery or electrical system.

Some conventional vehicles with stop/start use a high output (220 amp) AC generator due to the frequent demands placed on the battery. Since shutting the engine off stops the AC generator from producing any output, a voltage stabilization module (VSM) is used to allow vehicle accessories to continue to function.

AUTHOR'S NOTE With aftermarket high wattage amplifiers, and other high current components, it's common to install high output AC generators in place of the OEM units. Even if the demands of the added accessories require 150 amps, and the AC generator produces this amount, the rest of the vehicle's electrical system will not have to deal with this high amperage. Since resistance determines the current flow in a circuit, any given electrical component will only draw as much amperage as required to operate. While an aftermarket amplifier may require 150 amps, you do not need to worry about 150 amps going to the headlights and blowing them out.

If you install a high output alternator, replace the ground straps and the cable between the AC generator to the battery with heavier gauge cables.

Charging Indicators

There are two basic methods of informing the driver of the charging system's condition: indicator lamp and voltmeter.

AUTHOR'S NOTE Chapter 13 covers the operation of indicator lamps and gauges, including computer-controlled systems.

Indicator Light Operation

Earlier charging systems often used opposing voltages to operate the indicator lamp. If the AC generator output is less than battery voltage, there is an electrical potential difference in the lamp circuit, and the lamp will light. If there is no stator output through the diode trio, then the lamp circuit is completed to ground through the rotor field and TR_1 (**Figure 7-64**).

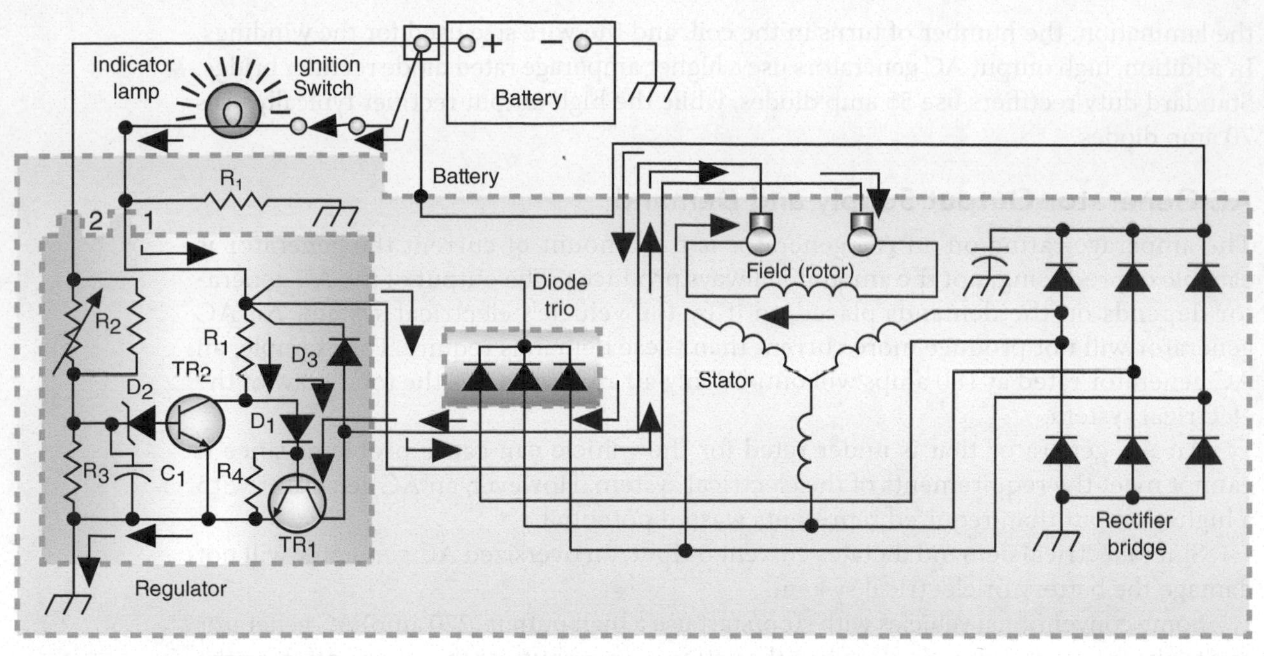

Figure 7-64 Electronic regulator with an indicator light on due to no AC generator output.

The warning lamp will be "proofed" when the ignition switch is in the RUN position before the engine starts. This indicates that the bulb and indicator circuit are operating properly. When proofing the bulb, ground is through the field coil since there is no stator output without the rotor turning.

Today's computer controlled charging systems control the indicator lamp by requests between modules. For example, suppose the PCM monitors the charging system and detects a failure or low battery SOC. In that case, it can send a signal to the instrument cluster module requesting illumination of the warning lamp. Many automakers have replaced the warning lamp with message displays.

Voltmeter Operation

Early systems connected a simple analog voltmeter between a voltage output circuit of the ignition switch and ground (**Figure 7-65**). The premises is the same as using a voltmeter to measure the voltage across the battery terminals.

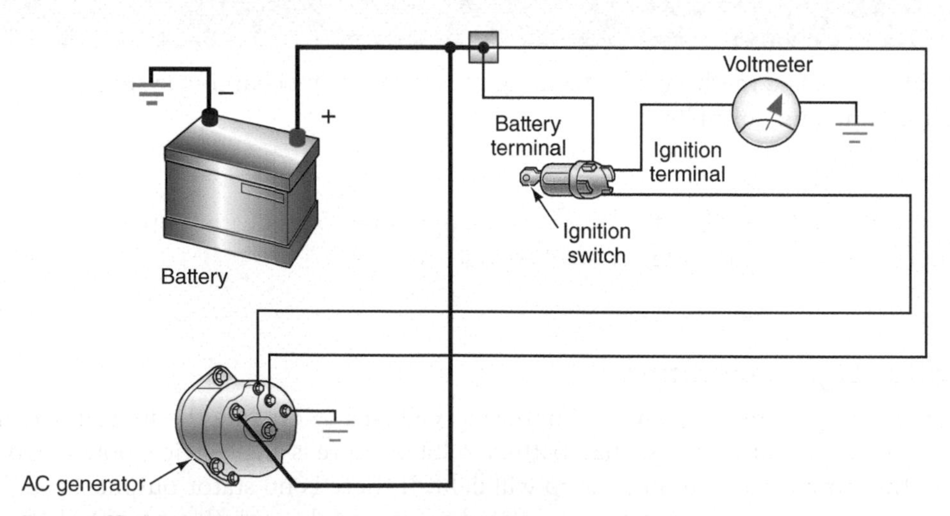

Figure 7-65 Voltmeter connected to the charging circuit to monitor operation.

When the engine starts, it is normal for the voltmeter to indicate a reading between 13.2 and 15.2 volts. If the voltmeter indicates a voltage level below 13.2, it may mean that the battery is discharging. If the voltmeter indicates a voltage reading above 15.2 volts, the charging system is overcharging the battery. Higher-than-normal charging system voltage output can damage the battery and electrical circuits.

Most modern systems control the voltmeter either directly by the PCM or by information sent to the instrument cluster from the PCM. A dedicated circuit from the battery to the PCM allows the PCM to monitor the battery voltage constantly.

HEV Charging Systems

HEVs utilize the automatic stop/start feature to shut off the engine whenever power from the engine is not required. A starter/generator unit starts the engine when needed. The starter/generator also recharges the HV battery. The difference between a motor and a generator is that the motor uses two opposing magnetic fields, and the generator uses one magnetic field with rotating conductors. Electronics control the direction of current flow, allowing the unit to function as both a motor and a generator.

There are two basic designs of the starter/generator. One design uses a belt alternator starter (BAS) that is about the same size as a conventional generator and mounts in the same way (**Figure 7-66**). The second design mounts an integrated starter generator (ISG) at either end of the crankshaft. Most designs have the ISG mounted at the rear of the crankshaft between the engine and transmission (**Figure 7-66**).

The ISG is a three-phase AC motor that can provide power and torque to the vehicle. It also supports the engine when the driver demands more power. The ISG includes a rotor and stator located inside the transmission bell housing **(Figure 7-67)**. The stator is attached to the engine block and is comprises of two separate lamination stacks. The rotor is bolted to the engine crankshaft and has both wire-wound and permanent magnet sections. Traditional diodes rectify the voltage and current. The advantage of this rotor and stator design is that the output at engine idle speed is up to 240 amps. Maximum output can exceed 300 amps.

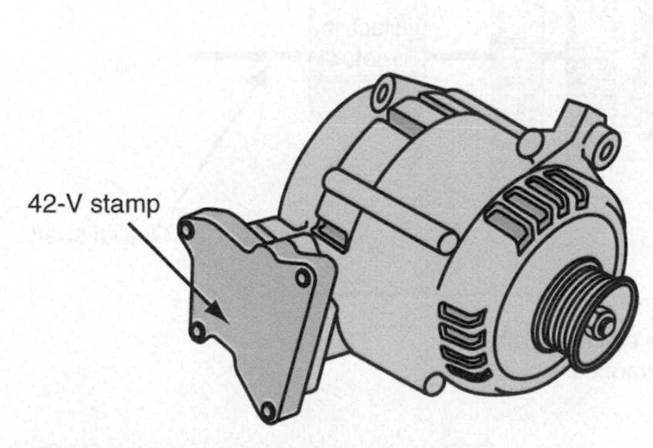

Figure 7-66 A BAS is used on some HEVs.

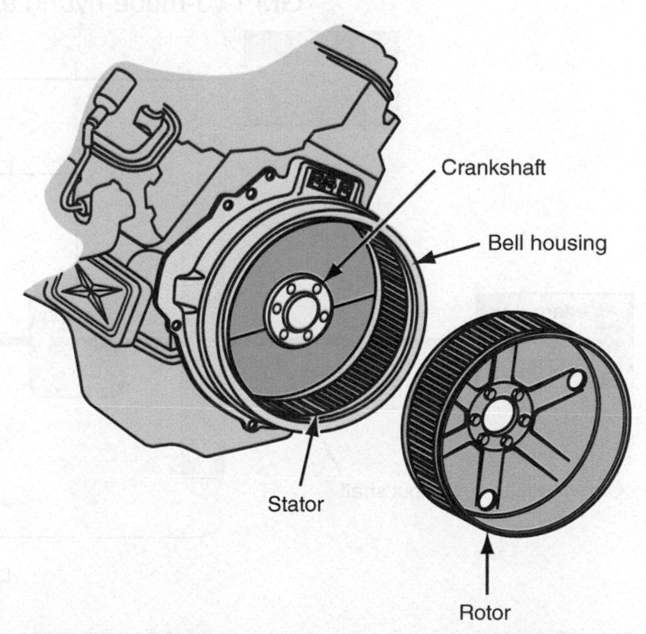

Figure 7-67 The ISG is usually located at the rear of the crankshaft in the bell housing.

Generation occurs anytime the engine is running. With the rotor connected to the crankshaft, it turns at the speed of the engine. Since the speed at which the magnetic poles move influences the amount of current output, the output will be the lowest at idle. During times of low engine speeds, current demand is likely to be at its highest. To increase output at lower speeds, the hybrid rotor has permanent magnets located between the pole pieces of the rotor. The magnetic flux from these permanent magnets goes into the pole piece, through the rotor shaft, and then back through the pole piece on the opposite side of the magnet. The permanent magnet fills the gap between the pole pieces, forcing more of the flux from the rotor into the stator windings. Maximum output is achieved when engine speed is increased.

Regulation uses a technique referred to as "boost-buck." At low speed and high electrical loads, the wire-wound section is fully energized, and the extra magnetic flux boosts the output of the permanent magnet section. The field current is off when the engine is operated at a medium speed and with a medium electrical load. During this time, only the permanent magnet section is producing the output. During high-speed, low-electrical-load conditions, the field current is reversed. This bucks the permanent magnet's field and maintains a constant output voltage.

Full hybrid vehicles that are capable of propelling the vehicle in an electric-only mode require HV batteries to power the three-phase AC motors. These batteries may have a capacity of over 300 volts. Many full hybrid vehicles have at least two AC motors located in the transmission or transaxle assembly to operate the planetary gear sets that constantly provide variable gear ratios (**Figure 7-68**). These motors can also function as generators. If the HV SOC becomes too low, the engine starts, and the crankshaft drives motor "A" to generate high-voltage alternating current. The current is rectified to DC voltage and sent to recharge the HV battery.

Voltage generation can also occur whenever one of the motors slips. In most cases, one of the motors slips at all times. The slipping causes a cutting of the magnetic field and results in an alternating current to supply electrical energy to the other motor (**Figure 7-69**).

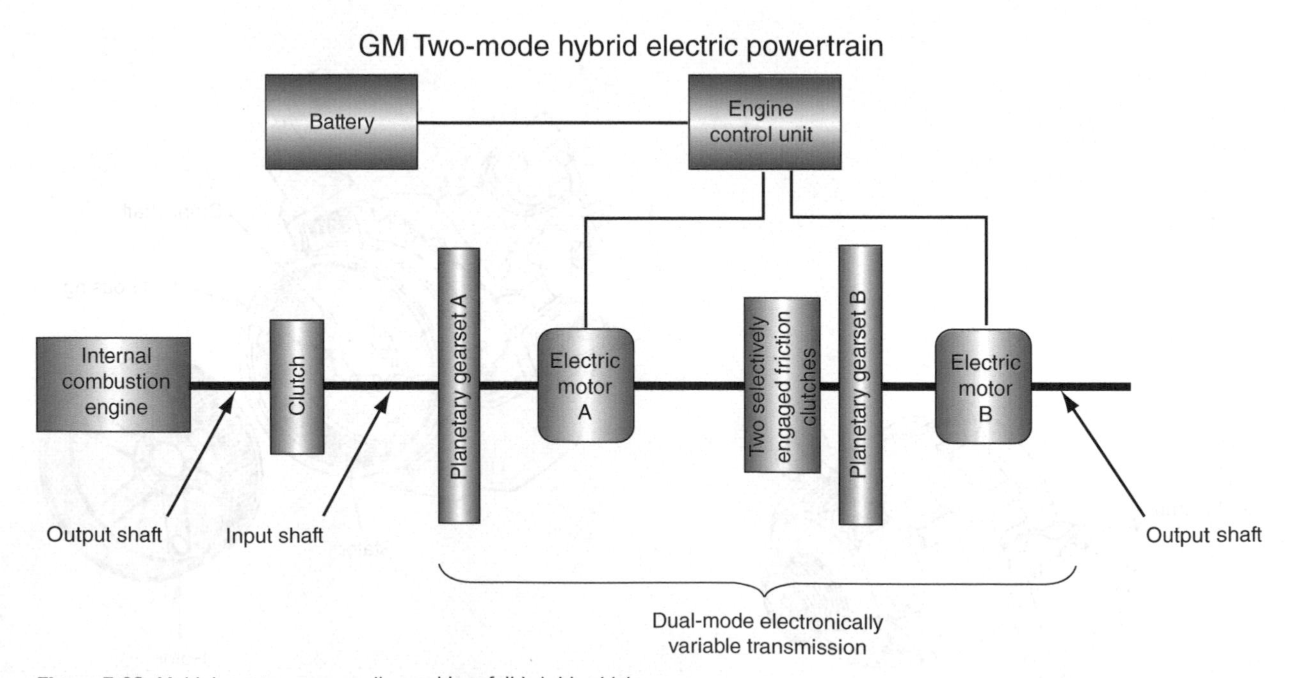

Figure 7-68 Multiple motors are usually used in a full hybrid vehicle.

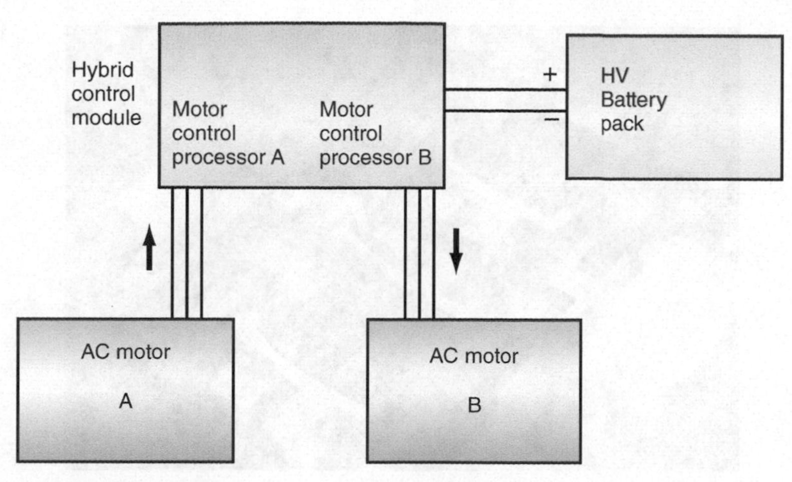

Figure 7-69 Current generated in one motor can be used to power the other.

Regenerative Braking

During vehicle deceleration, the ISG regenerates the energy used to slow the vehicle to recharge the HV and auxiliary batteries. About 30% of the kinetic energy lost during braking is in heat. During deceleration (decreasing acceleration), regenerative braking minimizes energy loss by recovering the energy used to slow the vehicle by converting rotational energy into electrical energy through the ISG or AC motors. Regenerative braking undertakes some of the stopping duties of the conventional friction brakes and uses the electric motor to help slow the vehicle.

Regenerative braking is mainly a function of light brake pedal application, using inputs from sensors such as the pedal angle sensor, the vacuum sensor, and the accelerator pedal position sensor. The hybrid control module will initiate regenerative braking as soon as the accelerator pedal is released. As the kinetic energy continues to move the vehicle forward, the wheels continue to turn the electric motors. Since the rotor is turning within the stator windings, the motor acts as a generator to recharge the HV battery. If additional vehicle deceleration is required, the hybrid controller can increase the force required to rotate the electric motors and increase the rotational resistance of the wheels to slow the vehicle further. The antilock brake system (ABS) activates the hydraulic brakes if additional braking power is needed.

DC/DC Converter

Shop Manual
Chapter 7, Page 348

The inverter module rectifies the high AC voltage from the motors during regenerative or charging modes into high DC voltage to recharge the HV battery. The high DC voltage can't be used to recharge the auxiliary battery. An additional function of the inverter is that of a DC/DC converter. The converter is a bidirectional, solid-state, DC conversion device that charges the 12-volt system from the 300-volt direct current system. The DC/DC converter changes the DC voltage level between the high-voltage and the low-voltage (LV) systems. The converter replaces the function of the engine-driven generator while maintaining isolation of the HV system.

High voltage to low voltage conversion is by magnetic fields within sets of coils instead of physical wired connections (**Figure 7-70**). The coils operate as step-down transformers to reduce the high DC voltage to the low DC voltage. Smooth low voltage output is maintained by sequentially inducing and collapsing the coils' magnetic field.

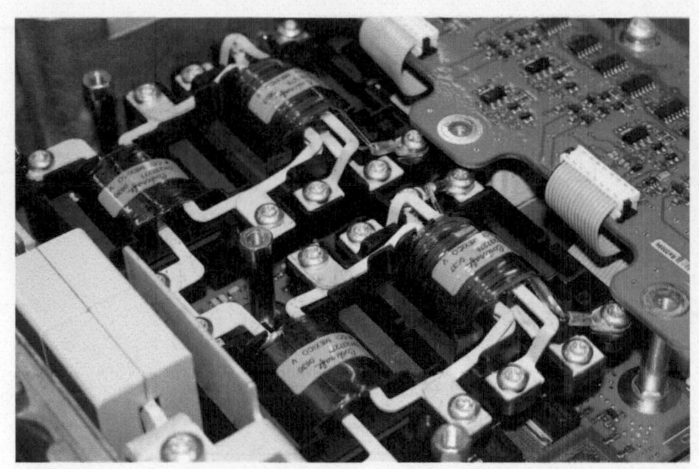

Figure 7-70 DC/DC converter coils are used as transformers to reduce the 300 volts DC to about 14 volts DC.

Using the jump assist mode requires the DC/DC converter to charge the 300-volt system using the 12-volt battery. In this mode, the coils operate as step-up transformers to increase the voltage.

AUTHOR'S NOTE A high current conversion rate within the inverter/converter module produces a significant amount of heat. Many manufacturers will use a separate cooling system to run coolant through these modules to prevent damage from the heat.

The inverter is PWM controlled by the hybrid control module or by bus communication from the control module. The hybrid control module requests the required amount of voltage the LV system needs. This will generally be a commanded PWM to the converter between 33% and 90%. In this range, the converter increases or decreases the charging voltage between 12.5 and 15.5 volts. A loss or corruption of the signal from the control module activates a fail-safe default low-voltage setting of 13.8 volts.

Summary

- The most common method of stator connection is called the wye connection, where one lead from each winding connects to one common junction. From this junction, the other leads branch out in a Y pattern.
- The delta connection connects the lead of one end of the winding to the lead at the other end of the next winding.
- The diode rectifier bridge provides reasonably constant DC voltage to the vehicle's electrical system and battery. The diode rectifier bridge changes the current in an AC generator.
- The conversion of AC to DC is called rectification.
- The three principal circuits used in the AC generator are the charging circuit, which consists of the stator windings and rectifier circuits; the excitation circuit, which consists of the rotor field coil and the electrical connections to the coil; and the preexcitation circuit, which supplies the initial current for the field coil that starts the buildup of the magnetic field.
- The voltage regulator controls the output voltage of the AC generator based on charging system demands by controlling the field current. The higher the field current, the higher the output voltage.

- The regulator must have system voltage as an input to regulate the output voltage. The input voltage to the regulator is called sensing voltage.
- Because ambient temperatures influence the rate of charge that a battery can accept, regulators are temperature compensated.
- The A circuit is an external grounded field circuit. The B+ for the field coil is picked up from inside the AC generator.
- The B circuit is an internally grounded circuit. In the B circuit, the voltage regulator controls the power side of the field circuit. Usually, the B circuit regulator is mounted externally to the AC generator.
- Isolated field AC generators pick up B+ and ground externally. The AC generator has two field wires attached to the outside of the case. The voltage regulator can be located either on the ground (A circuit) or on the B+ (B circuit) side.
- An electronic regulator uses solid-state circuitry to perform the regulatory functions using a Zener diode that blocks current flow up to a specific voltage; then, it allows current flow.
- The regulator function has been incorporated into the vehicle's engine computer on most modern vehicles. Regulation of the field circuit is through the ground (A circuit).
- The smart charging system adds a current sensor and the shared inputs from the accelerator pedal position (APP) sensor, manifold absolute pressure (MAP) sensor, and vehicle speed sensors to determine the necessary charging rate.
- Load shedding reduces electrical power consumption by turning off non-essential electrical loads to extend the driving time and distance.
- There are two basic methods of informing the driver of the charging system's condition: indicator lamps and voltmeter.
- Early indicator lamps operate on the basis of voltage drop. If the charging system output is less than battery voltage, there is an electrical potential difference in the lamp circuit, and the lamp will light.
- Most modern vehicles monitor charging system operation and will use computer-controlled indicator lights or voltmeters to alert the driver of any issues.
- The hybrid AC generator design consists of a rotor assembly with both wire-wound and permanent magnet sections. The permanent magnets are located between the pole pieces of the rotor.
- The ISG is a three-phase AC motor that can provide power and torque to the vehicle and generate voltage whenever the rotor is turning.
- During vehicle deceleration, the ISG regenerates the energy used to slow the vehicle to recharge the HV and auxiliary batteries.
- The AC motors used in full hybrid vehicles can be used as generators whenever the engine is running and during vehicle deceleration.
- While the vehicle is coasting, the electric motors are turned by the wheels and act as generators to recharge the HV battery. If additional vehicle deceleration is required, the hybrid controller can increase the force required to rotate the electric motors.
- The inverter module rectifies the AC voltage from the motors during regenerative or charging modes. This module converts the high AC voltage into high DC voltage to recharge the HV battery.
- The DC/DC converter allows the conversion of electrical power between the HV system and the LV system. It replaces the function of the engine-driven generator while maintaining isolation of the HV system.

Review Questions

Short-Answer Essays

1. List the major components of the charging system.
2. List and explain the function of the major components of the AC generator.
3. What is the relationship between field current and AC generator output?
4. Identify the differences between A, B, and isolated circuits.
5. Explain the purpose of the battery current sensor and its influence on load shedding.
6. Describe the two styles of stators.
7. What is the difference between half and full-wave rectifications?
8. Describe how AC voltage is rectified to DC voltage in the AC generator.
9. What is the purpose of the charging system?
10. Explain the meaning of regenerative braking.

Fill in the Blanks

1. The charging system converts the ____________ energy of the engine into ____________ energy to recharge the battery and run the electrical accessories.
2. All charging systems use the principle of ____________ to generate electrical power.
3. The ____________ creates the rotating magnetic field of the AC generator.
4. ____________ are electrically conductive sliding contacts, usually made of copper and carbon.
5. In the ____________ connection stator, one lead from each winding is connected to one common junction.
6. The ____________ controls the output voltage of the AC generator, based on charging system demands, by controlling ____________ current.
7. In an electronic regulator, ____________ controls AC generator output by varying the amount of time the field coil is energized.
8. Full-wave rectification in the AC generator requires ____________ pair of diodes.
9. The ____________ is the stationary coil that produces current in the AC generator.
10. ____________ recovers the heat energy used to brake by converting rotational energy into ____________ energy through a system of electric motors and generators.

Multiple Choice

1. The magnetic field current of the AC generator is carried in the:
   A. Rotor.
   B. Diode trio.
   C. Rectifier bridge.
   D. Stator.
2. The voltage induced in one conductor by one revolution of the rotor is called:
   A. Three phase.
   B. Half wave.
   C. Single phase.
   D. Full wave.
3. Rectification is being discussed.
   Technician A says the AC generator uses a segmented commutator to rectify alternating current.
   Technician B says the DC generator uses a pair of diodes to rectify alternating current.
   Who is correct?
   A. A only
   B. B only
   C. Both A and B
   D. Neither A nor B
4. Rotor construction is being discussed.
   Technician A says the poles will take on the polarity of the side of the coil that they touch.
   Technician B says the magnetic flux lines will move in opposite directions between adjacent poles.
   Who is correct?
   A. A only
   B. B only
   C. Both A and B
   D. Neither A nor B
5. The amount of voltage output of the AC generator is related to:
   A. Field strength.
   B. Stator speed.
   C. Number of rotor segments.
   D. All of these choices.
6. The delta-wound stator:
   A. Shares a common connection point.
   B. Has each winding connected in series.
   C. Doesn't require rectification.
   D. None of these choices.
7. Non-computer–controlled indicator lamp operation is being discussed.
   Technician A says in a system with an electronic regulator, the lamp will light if there is no stator output through the diode trio.
   Technician B says when there is stator output, the lamp circuit has voltage applied to both sides, and the lamp will not light.
   Who is correct?
   A. A only
   B. B only
   C. Both A and B
   D. Neither A nor B

8. *Technician A* says only two stator windings apply voltage because the third winding is always connected to diodes that are reverse-biased.
   Technician B says AC generators that use half-wave rectification are the most efficient.
   Who is correct?
   A. A only C. Both A and B
   B. B only D. Neither A nor B

9. Charging system regulation is being discussed.
   Technician A says the voltage regulation is done by varying the amount of field current flowing through the rotor.
   Technician B says the field current can be controlled either by regulating the resistance in series with the field coil or by turning the field circuit on and off.
   Who is correct?
   A. A only C. Both A and B
   B. B only D. Neither A nor B

10. Which component is responsible for recharging the 12-volt battery from the 300-volt DC system in a full hybrid HEV?
    A. The battery control module.
    B. The belt-driven generator.
    C. The DC/DC converter.
    D. The LV control module.

CHAPTER 8

INTRODUCTION TO THE BODY COMPUTER

Upon completion and review of this chapter, you should be able to:

- Describe the basic functions of the computer.
- Explain the principle of analog and digital voltage signals.
- Explain the principle of computer communications.
- Explain the basics of logic gate operation.
- Describe the basic function of the microprocessor.
- Describe the basic process by which the microprocessor makes determinations.
- Explain the differences in memory types.
- Explain the operation of low and high-side drivers.
- Explain the operation of common output actuators.

Terms to Know

Actuators
Adaptive memory
Adaptive strategy
Binary code
Bit
Clock circuit
High-side drivers
Interface
Logic gates
Low-side drivers
Microprocessor
Nonvolatile
Output driver
Output signal
Program
Sequential logic circuits
Sequential sampling
Servomotor
Stepper motor
Volatile

Introduction

Shop Manual
Chapter 8, page 369

A computer is an electronic device that stores and processes data. It's also capable of controlling other devices. Computer usage in automobiles has expanded to include control and operation of several functions, including climate control, lighting circuits, cruise control, antilock braking, electronic suspension systems, and electronic shift transmissions. Some of these are functions of what is known as a body computer module (BCM). Some body computer-controlled systems include direction lights, rear window defoggers, illuminated entry, intermittent wipers, and other systems once thought of as basic. This chapter introduces the basic theory and operation of the digital computer used to control most of the vehicle's electrical accessories.

AUTHOR'S NOTE When first installed on the automobile, the aura of mystery surrounding computer controls was so great that some technicians were afraid to work on them. Most technicians coming into the field today have grown up around computers and do not experience this anxiety. Regardless of your comfort level with computers, knowledge is the key to understanding their function. Although it's not necessary to understand all the concepts of computer operation to service the systems they control, knowledge of the digital computer will help you feel more comfortable working on these systems.

Computer Functions

Shop Manual
Chapter 8, page 370

A computer processes the physical conditions that represent information (data). The operation of the computer is divided into four basic functions:

1. *Input:* A voltage signal sent from an input device. This device can be a sensor, or a switch activated by the driver or technician.
2. *Processing:* The computer uses the input information and compares it to programmed instructions. The logic circuits process the input signals into output demands.
3. *Storage:* An electronic memory stores the program instructions. The memory also stores some input signal data for later processing.
4. *Output:* After the computer has processed the inputs and checked its programmed instructions, it will put out control commands to various output devices. These output devices may be the instrument panel display or a system actuator. The output of one computer may be an input to another computer.

Understanding these four functions will help today's technician organize the troubleshooting process. When diagnosing an issue in a system, you will be attempting to isolate the problem to one of these functions.

A BIT OF HISTORY

"Computer" was originally a job title, so the first computers were actual people. The job title of computer described those people who performed the repetitive calculations required to compute navigational tables, tide charts, and planetary positions. Electronic computers were given this name because they performed the work previously assigned to people. People who had the job title of computer relied upon a tool known as the abacus to aid their memory while performing the calculations. Since people who worked as computers would often suffer from boredom that resulted in carelessness and mistakes, inventors started searching for ways to mechanize this task.

During World War II, the United States had battleships capable of firing shells weighing over a ton at targets up to 25 miles away. To determine the trajectory of the shells, physicists provided the equations but solving these equations using human computers proved to be labor intensive. To solve its problems, the U.S. military looked for means of automating these computations. An early success was a computer built from a partnership between Harvard and IBM in 1944 known as the Harvard Mark I. The Harvard Mark I was the first programmable digital computer made in the United States. However, the Harvard Mark I was not exclusively electronic and was constructed out of switches, relays, rotating shafts, and clutches. The computer was 8 feet tall and 51 feet long and weighed 5 tons. It used 500 miles of wire and required a 50 feet rotating shaft that ran its length. A 5-horsepower electric motor turned the shaft. The Harvard Mark I contained three-quarters of a million components, yet it could store only 72 numbers. Today's home computers can store 30 million numbers in RAM and over a trillion numbers on their hard disk. The Harvard Mark I ran nonstop for 15 years. Grace Hopper was one of the primary programmers for the Harvard Mark I and is said to be the founder of the first computer "bug." A dead moth had gotten into the Harvard Mark I, and its wings blocked the reading of the holes in the paper tape.

Analog and Digital Principles

Remembering the basics of electricity, voltage does not flow through a conductor; current flows and voltage is the pressure that "pushes" the current. However, voltage can be used as a signal—for example, the difference in voltage levels, frequency of change, or switching from positive to negative values can be used as a signal. The computer uses these voltage signals to perform its functions.

A **program** is a set of instructions the computer must follow to achieve desired results. The program used by the computer is "burned" into integrated circuit (IC) chips using a series of numbers. These numbers represent various combinations of voltages that the computer can understand. The voltage signals to the computer can be either analog or digital. Many of the inputs from the sensors are analog variables. For example, ambient temperature sensors do not change abruptly. The temperature varies in infinite steps from low to high. The same is true for several other inputs such as engine speed, vehicle speed, and fuel flow etc.

Compared to an analog voltage representation, digital voltage patterns are square-shaped because the transition from one voltage level to another is abrupt (**Figure 8-1**). Switching an on/off or high/low voltage produces a digital signal. The simplest digital signal generator is a switch (**Figure 8-2**). If the circuit has 5 volts applied to it, the voltage sensor will read 5 volts (a high-voltage value) when the switch is open. Closing the switch will result in the voltage sensor reading close to 0 volt. This measuring of voltage drops sends a digital signal to the computer. A series of digits represents the voltage values, which create a **binary code**. Binary code is represented by the numbers 1 and 0. Any number and word can be translated into a combination of binary 1s and 0s.

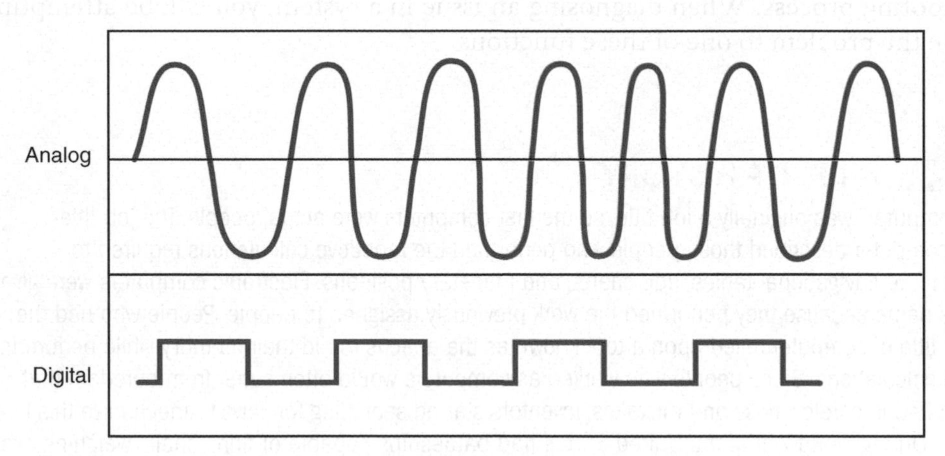

Figure 8-1 Analog voltage signals are constantly variable. Digital voltage patterns are either on or off. Digital signals are referred to as a square sine wave.

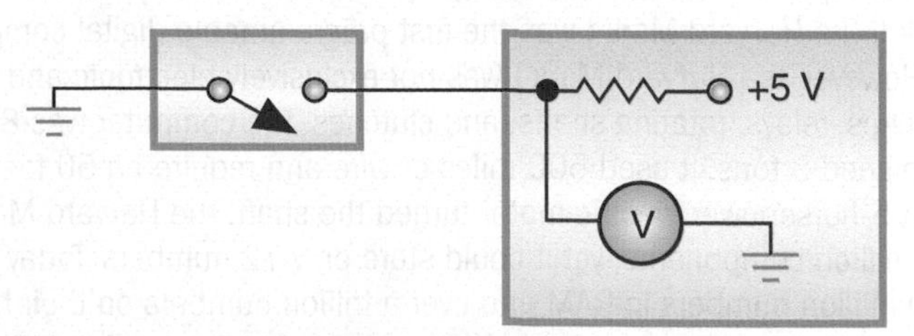

Figure 8-2 Simplified voltage sensing circuit that indicates if the switch is opened or closed.

Binary Numbers

A transistor that operates as a relay is the basis of the digital computer. As the input signal switches from off to on, the transistor output switches from cutoff to saturation. The on and off output signals represent the binary digits 1 and 0.

The computer converts the digital signal into binary code by translating voltages above a given value to 1 and voltages below a given value to 0. Each 1 or 0 represents one **bit** of information. As shown in **Figure 8-3**, when the switch is open, the sensed 5 volts translates into a 1 (high voltage). When the switch is closed, the lower voltage value is translated into a 0.

A bit is a 0 or a 1. Eight bits is called a byte.

In the binary system, whole numbers are grouped from right to left. Because the system uses only two digits, the first portion must equal a 1 or a 0. To write the value of 2 requires the use of the second position. In binary, the value of 2 would be represented by 10 (one two and zero one). To continue, a 3 would be represented by 11 (one two and one one). **Figure 8-4** illustrates the conversion of binary numbers to digital

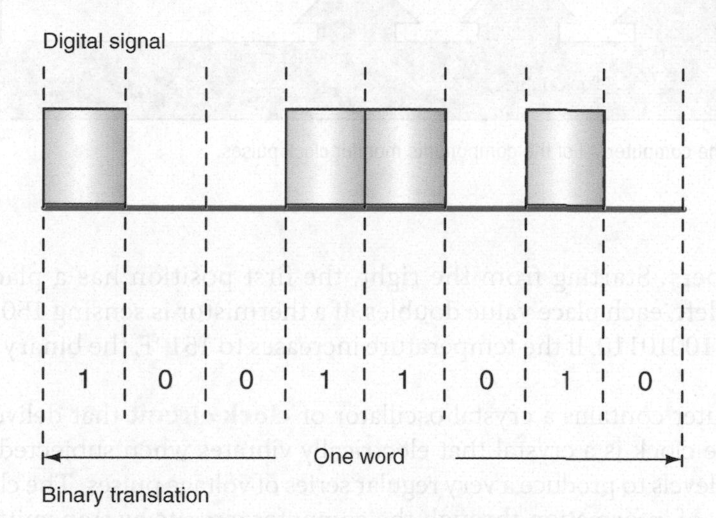

Figure 8-3 Each binary 1 and 0 is one bit of information. Eight bits equal one byte.

Decimal number	Binary number code 8 4 2 1	Binary to decimal conversion
0	0000	= 0 + 0 = 0
1	0001	= 0 + 1 = 1
2	0010	= 2 + 0 = 2
3	0011	= 2 + 1 = 3
4	0100	= 4 + 0 = 4
5	0101	= 4 + 1 = 5
6	0110	= 4 + 2 = 6
7	0111	= 4 + 2 + 1 = 7
8	1000	= 8 + 0 = 8
9	1001	= 8 + 1 = 9

Figure 8-4 Binary number code conversion to base 10 numbers.

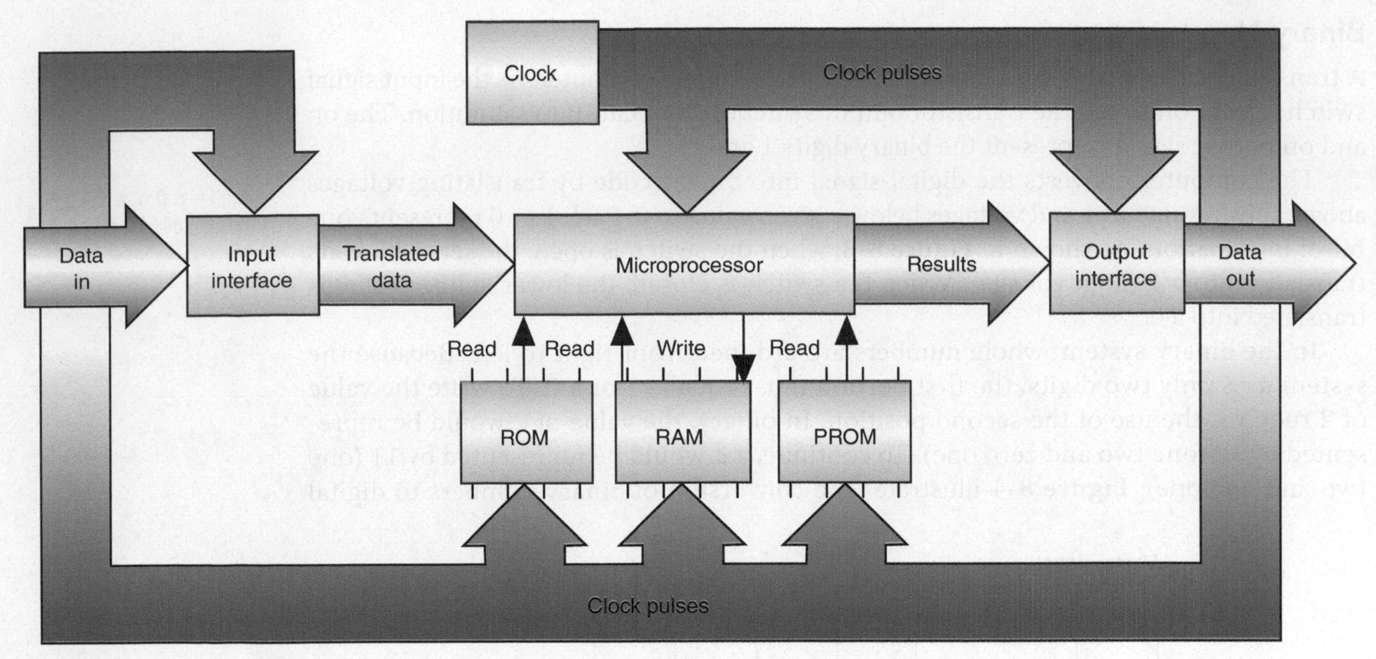

Figure 8-5 Interaction of the main components of the computer. All of the components monitor clock pulses.

base ten numbers. Starting from the right, the first position has a place value of 1. Moving to the left, each place value doubles. If a thermistor is sensing 150 °F, the binary code would be 10010110. If the temperature increases to 151 °F, the binary code changes to 10010111.

The computer contains a crystal oscillator or **clock circuit** that delivers a constant time pulse. The clock is a crystal that electrically vibrates when subjected to current at certain voltage levels to produce a very regular series of voltage pulses. The clock maintains an orderly flow of information through the computer circuits by transmitting one bit of binary code for each pulse (**Figure 8-5**). In this manner, the computer can distinguish between the binary codes such as 101 and 1001.

Signal Conditioning and Conversion

The input and/or output signals may require conditioning to be used. This conditioning may include amplification and/or signal conversion.

Some input sensors produce a very low-voltage signal of less than 1 volt. This signal has an extremely low-current flow and requires amplification (increased) before sending it to the microprocessor. An amplification circuit in the input conditioning chip performs this function (**Figure 8-6**).

For the computer to receive information from the sensor and give commands to actuators, it requires an **interface**. The computer will have two interface circuits: input and output. Interfaces protect the computer from excessive voltage levels and translate input and output signals. The digital computer cannot accept analog signals from the sensors and requires an input interface to convert the analog signal to a digital signal. The analog-to-digital (A/D) converter continually scans the analog input signals at regular intervals. For example, if the A/D converter scans the TPS signal and finds the signal at 5 volts, the A/D converter assigns a numeric value to this specific voltage. Then the A/D converter changes this numeric value to a binary code (**Figure 8-7**).

Also, some of the controlled actuators may require an analog signal. In this instance, an output digital to analog (D/A) converter is used.

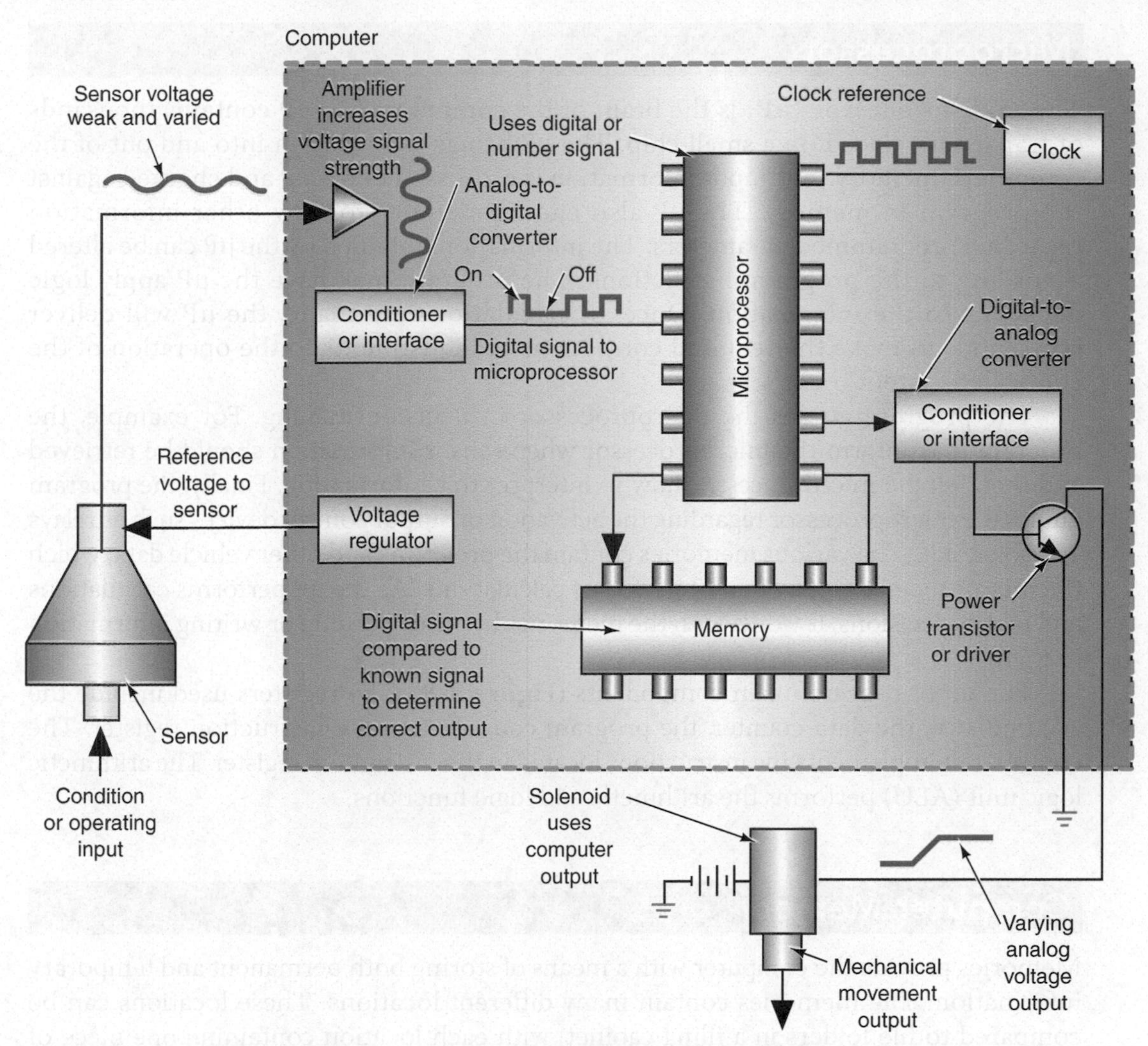

Figure 8-6 Amplification and interface circuits in the computer. The amplification circuit boosts the voltage and conditions it. The interface converts analog inputs into digital signals. The digital-to-analog converter changes the output voltage from digital to analog.

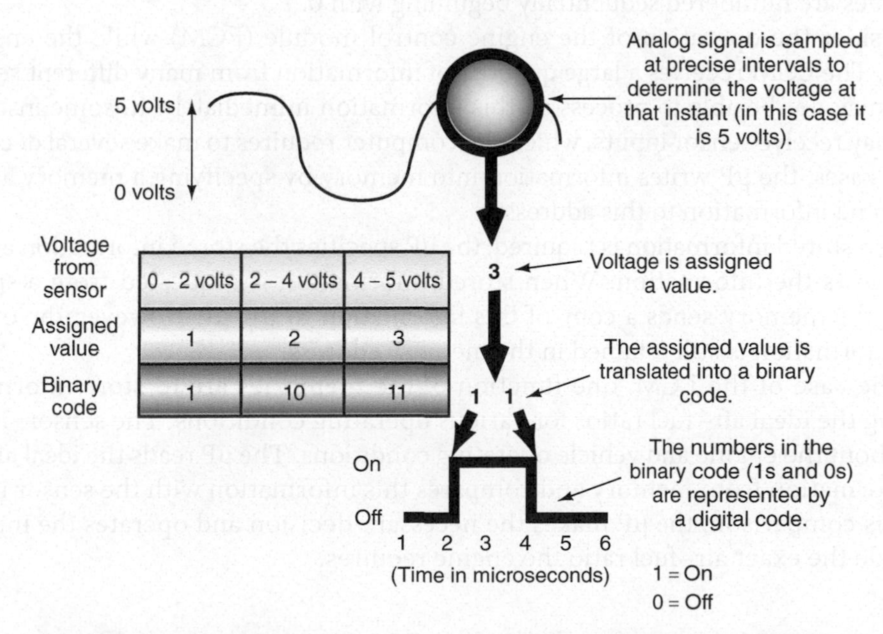

Figure 8-7 The A/D converter assigns a numeric value to input voltages and changes this numeric value to a binary code.

Microprocessor

The terms *microprocessor* and *central processing unit* are basically interchangeable.

The **microprocessor (μP)** is the brain of the computer. The μP contains thousands of transistors placed on a small chip. The μP brings information into and out of the computer's memory. The input information is processed in the μP and checked against the program in memory. The μP also checks memory for any other information regarding programmed parameters. The information obtained by the μP can be altered according to the program instructions. The program may have the μP apply logic decisions to the information. Once all calculations are made, the μP will deliver commands to make the required corrections or adjustments to the operation of the controlled system.

The program guides the microprocessor in decision-making. For example, the program may inform the microprocessor when sensor information should be retrieved and then tell the microprocessor how to interpret this information. Finally, the program guides the microprocessor regarding the activation of output control devices such as relays and solenoids. The various memories contain the programs and other vehicle data, which the microprocessor refers to as it performs calculations. As the μP performs calculations and makes decisions, it works with the memories by either reading or writing information to them.

The μP has several main components (**Figure 8-8**). The registers used include the accumulator, the data counter, the program counter, and the instruction register. The control unit implements the instructions located in the instruction register. The arithmetic logic unit (ALU) performs the arithmetic and logic functions.

Computer Memory

Memories provide the computer with a means of storing both permanent and temporary information. The memories contain many different locations. These locations can be compared to file folders in a filing cabinet, with each location containing one piece of information. Each memory location is assigned an address (similar to the lettering or numbering arrangement on file folders). Each address is written in a binary code, and these codes are numbered sequentially beginning with 0.

Consider the operation of the engine control module (ECM) while the engine is running. The ECM receives a large quantity of information from many different sensors. The μP may not be able to process all this information immediately. In some instances, the μP may receive sensor inputs, which the computer requires to make several decisions. In these cases, the μP writes information into memory by specifying a memory address and sending information to this address.

When stored information is required, the μP specifies the stored information address and requests the information. When stored information is requested from a specific address, the memory sends a copy of this information to the μP. However, the original stored information is still retained in the memory address.

In the case of the ECM, one function of the memories are to store information regarding the ideal air–fuel ratios for various operating conditions. The sensors inform the μP about the engine and vehicle operating conditions. The μP reads the ideal air–fuel ratio information from memory and compares this information with the sensor inputs. After this comparison, the μP makes the necessary decision and operates the injectors to provide the exact air–fuel ratio the engine requires.

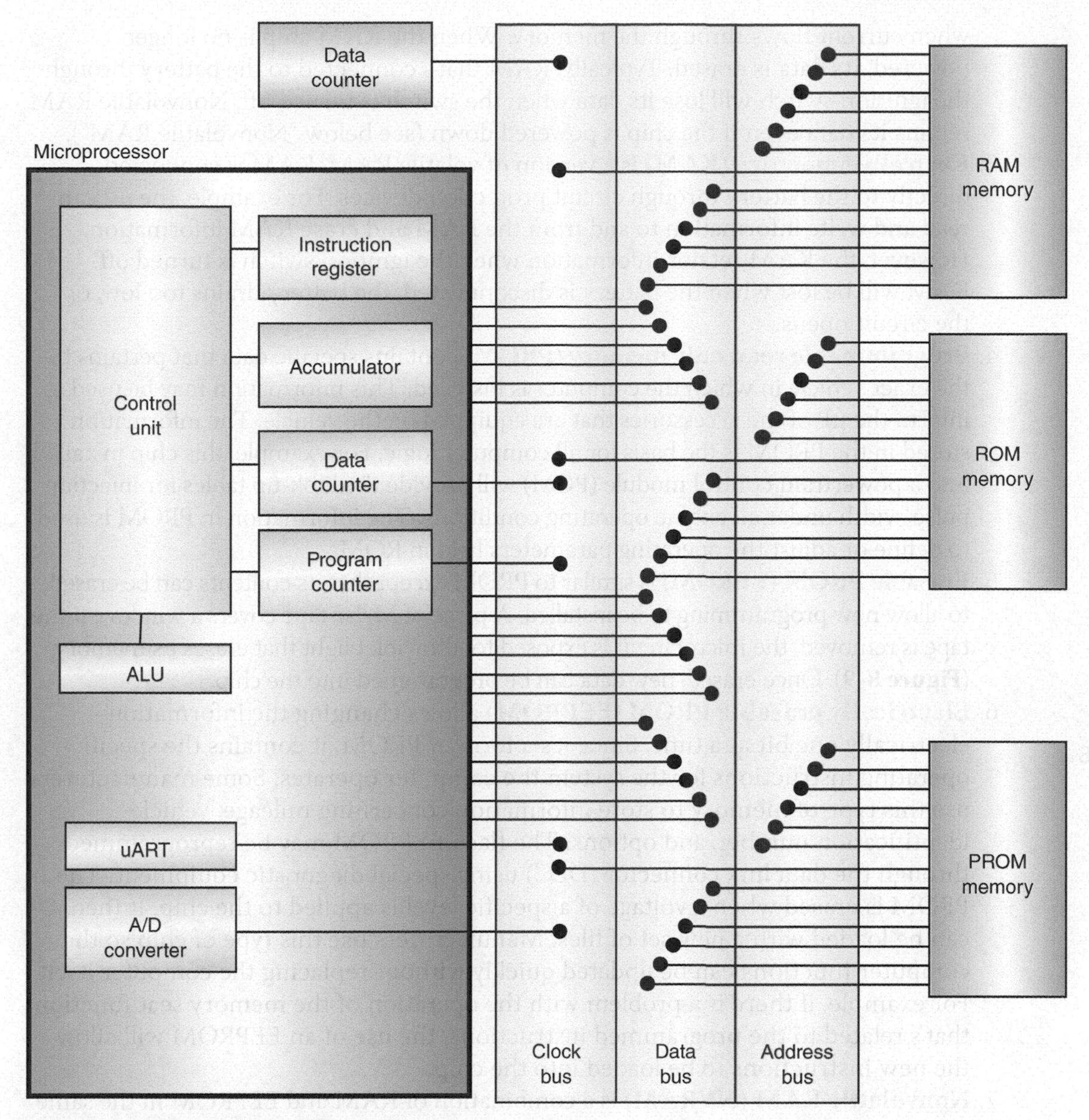

Figure 8-8 Main components of the µP.

Several types of memory chips may be used in the computer:

1. **Read-only memory (ROM)** contains a fixed pattern of 1s and 0s that represent permanent stored information. This information is used to instruct the µP on what to do in response to input data. The µP reads the information contained in ROM but it cannot write to it or change it. ROM is permanent memory that's programmed in. This memory is not lost when power to the computer is removed. ROM contains formulas and calibrations that identifies the basic purpose and functions of the computer, along with identification data.
2. **Random access memory (RAM)** is constructed from flip-flop circuits formed into the chip. The RAM will store temporary information that can be read from or written to by the µP. RAM stores information that's waiting to be acted upon and the output signals that are waiting to be sent to an output device. RAM can be designed as **volatile** or **nonvolatile**. In volatile RAM, the data will be retained only

The terms *ROM, RAM,* and *PROM* are used fairly consistently in the computer industry. However, the names vary between automobile manufacturers.

when current flows through the memory. When the RAM chip is no longer powered, its data is erased. Typically, RAM that's connected to the battery through the ignition switch will lose its data when the switch is turned off. Nonvolatile RAM retains its data even if the chip is powered down (see below "Nonvolatile RAM").

Shop Manual
Chapter 8, page 372

3. **Keep alive memory (KAM)** is a version of volatile RAM. KAM is connected directly to the battery through circuit protection devices. For example, the μP can read and write information to and from the KAM and erase KAM information. However, the KAM retains information when the ignition switch is turned off. KAM will be lost when the battery is disconnected, the battery drains too low, or the circuit opens.
4. **Programmable read only memory (PROM)** contains specific data that pertains to the exact vehicle in which the computer is installed. This information may be used to inform the μP of the accessories that are equipped on the vehicle. The information stored in the PROM is the basis for all computer logic. For example, this chip installed into a powertrain control module (PCM) will provide the look-up tables for injection pulse-width under all engine operating conditions. The information in PROM is used to define or adjust the operating parameters held in ROM.
5. **Erasable PROM (EPROM)** is similar to PROM except that its contents can be erased to allow new programming to be installed. A piece of Mylar tape covers a window. If the tape is removed, the microcircuit is exposed to ultraviolet light that erases its memory (**Figure 8-9**). Once erased, new data can be programmed into the chip.

Shop Manual
Chapter 8, page 388

6. **Electrically erasable PROM (EEPROM)** allows changing the information electrically one bit at a time. Since it's a form of PROM, it contains the specific operating instructions for the system the computer operates. Some manufacturers use this type of memory to store information concerning mileage, vehicle identification number, and options. The flash EEPROM may be reprogrammed through the data link connector (DLC) using special diagnostic equipment. The PROM is erased when a voltage of a specific level is applied to the chip. It then can be loaded with a new set of files. Manufacturers use this type of chip so the computer functions can be updated quickly without replacing the computer itself. For example, if there is a problem with the operation of the memory seat function that's related to the programmed instructions, the use of an EEPROM will allow the new instructions to be loaded into the chip.
7. **Nonvolatile RAM (NVRAM)** is a combination of RAM and EEPROM in the same chip. During normal operation, data is written to and read from the RAM portion of the chip. If the power is removed from the chip, or at programmed timed intervals, the data is transferred from RAM to the EEPROM portion of the chip. When the power is restored to the chip, the EEPROM will write the data back to the RAM. One example of this type of memory would be vehicle mileage tracking. As vehicle mileage is accumulated, the current value is stored in RAM. At set intervals (or during key off) the current vehicle mileage is load into a specified block of the EEPROM chip by first erasing the stored value and then loading the current value.

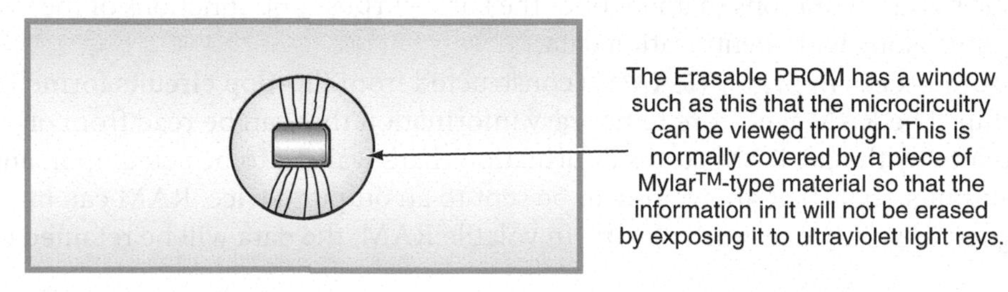

The Erasable PROM has a window such as this that the microcircuitry can be viewed through. This is normally covered by a piece of Mylar™-type material so that the information in it will not be erased by exposing it to ultraviolet light rays.

Figure 8-9 EPROM memory is erased when the ultraviolet rays contact the microcircuitry.

Adaptive Strategy and Memory

If a computer has **adaptive strategy** capabilities, it can actually learn from past experience. For example, the normal voltage input range from an ambient temperature sensor may be 0.6 to 4.5 volts. If the sensor sends a 0.4-volt signal to the computer, the μP interprets this signal as an indication of component wear and stores this altered calibration in the RAM. The μP now refers to this new calibration during calculations, and normal system performance is maintained. If a sensor output is erratic or considerably out of range, the computer may ignore this input. When a computer has adaptive strategy, a short learning period may be necessary under the following conditions:

1. After the battery has been disconnected.
2. When a computer system component has been replaced or disconnected.
3. A new vehicle.
4. Replacement computer is installed.

Adaptive memory is the ability of the computer system to store changing values and to correct operating characteristics. For example, a transmission control module may monitor the transmission's input and output shaft speeds to determine gear ratio. If the input speed sensor indicates a speed of 1,000 rpm and the output speed sensor indicates a speed of 333 rpm, then the controller determines that the ratio is 3:1 (first gear). When the controller determines that it will make the shift to second gear (2:1 ratio), it monitors the sensors to see how long it takes to achieve the ratio change from 3:1 to 2:1. The length of time required represents the amount of fluid needed to stroke the clutch piston and lock up the clutch element. This value is learned so the timing of the shifts can be altered as the clutch elements wear, yet the quality of the shifts will not deteriorate over the life of the transmission.

Information Processing

The air charge temperature (ACT) sensor input will be used as an example of how the computer processes information. If the air temperature is low, the air is denser and contains more oxygen per cubic foot. Warmer air is less dense and therefore contains less oxygen per cubic foot. The cold, dense air requires more fuel compared to the warmer air that's less dense. The μP must supply the correct amount of fuel in relation to air temperature and density.

An ACT sensor is positioned in the intake manifold where it senses air temperature. This sensor contains a resistive element that has an increased resistance when the sensor is cold. Conversely, the ACT sensor resistance decreases as the sensor temperature increases. When the ACT sensor is cold, it sends a high-analog voltage signal to the computer, and the A/D converter changes this signal to a digital signal.

When the μP receives this ACT signal, it addresses the tables in the ROM. The look-up tables list air density for every air temperature. When the ACT sensor voltage signal is very high, the look-up table indicates very dense air. This dense air information is relayed to the μP, and the μP operates the output drivers and injectors to supply the exact amount of fuel the engine requires (**Figure 8-10**).

Logic Gates

Logic gates are the thousands of field effect transistors (FETs) incorporated into the computer circuitry. These circuits are called logic gates because they act as gates to output voltage signals depending on different combinations of input signals. The FETs use the

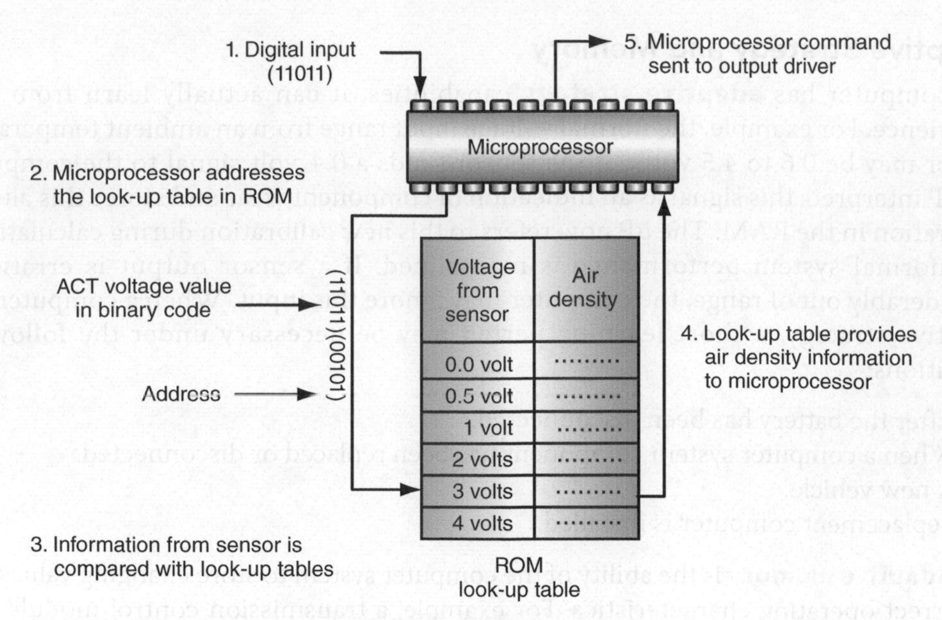

Figure 8-10 The microprocessor addresses the look-up tables in the ROM, retrieves air density information, and issues commands to the output devices.

incoming voltage patterns to determine the pattern of pulses leaving the gate. The following are some of the most common logic gates and their operations. The symbols represent functions and not electronic construction:

1. **NOT gate:** A NOT gate simply reverses binary 1s to 0s and vice versa (**Figure 8-11**). A high input results in a low output and a low input results in a high output.
2. **AND gate:** The AND gate will have at least two inputs and one output. The operation of the AND gate is similar to two switches in series to a load (**Figure 8-12**). The only way the light will turn on is if switches A *and* B are closed. The output of the gate will be high only if both inputs are high. Before current can be present at the output of the gate, current must be present at the base of both transistors (**Figure 8-13**).

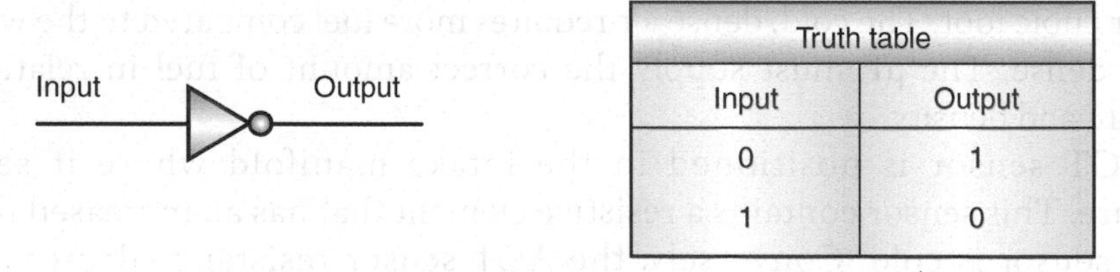

Figure 8-11 The NOT gate symbol and truth table. The NOT gate inverts the input signal.

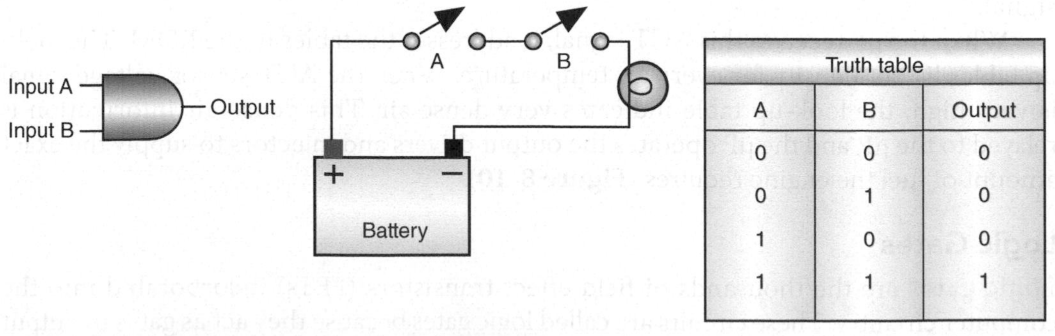

Figure 8-12 The AND gate symbol and truth table. The AND gate operates similar to switches in series.

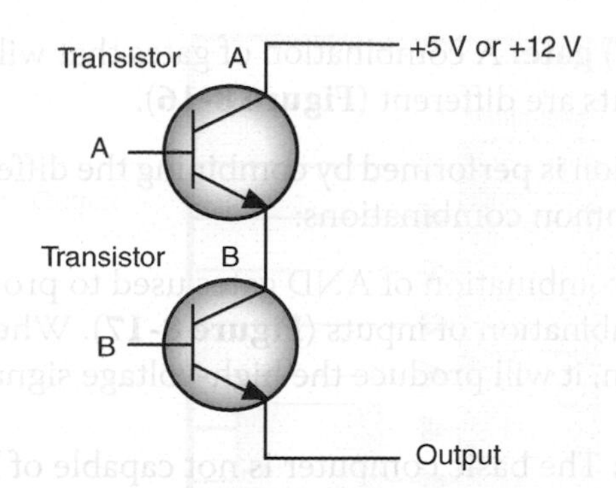

Figure 8-13 The AND gate circuit.

3. **OR gate:** The OR gate operates similarly to two switches that are wired in parallel to a light (**Figure 8-14**). If switch A *or* B is closed, the light will turn on. A high signal to either input will result in a high output.
4. **NAND and NOR gates:** A NOT gate placed behind an OR or AND gate inverts the output signal (**Figure 8-15**).

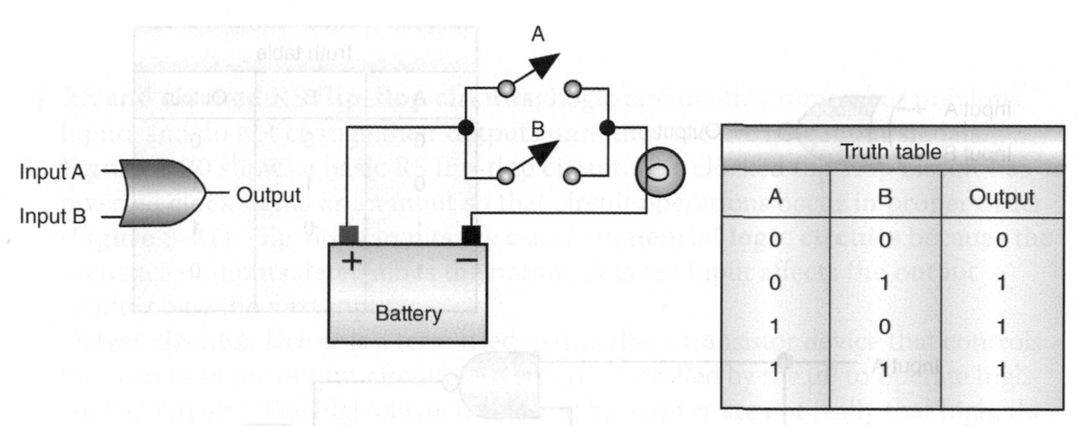

Truth table

A	B	Output
0	0	0
0	1	1
1	0	1
1	1	1

Figure 8-14 OR gate symbol and truth table. The OR gate is similar to parallel switches.

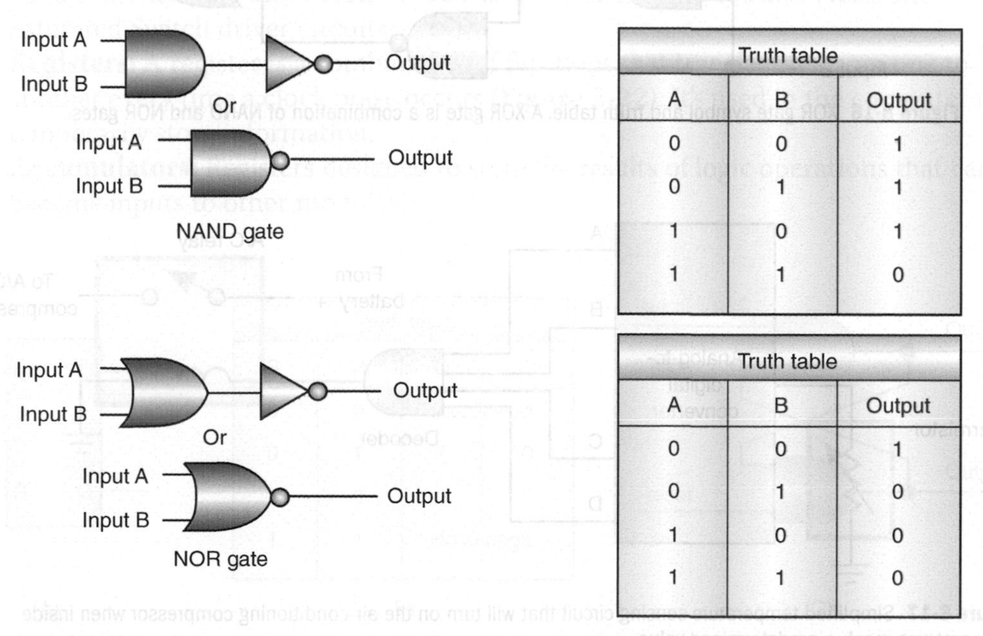

Truth table

A	B	Output
0	0	1
0	1	1
1	0	1
1	1	0

Truth table

A	B	Output
0	0	1
0	1	0
1	0	0
1	1	0

Figure 8-15 Symbols and truth tables for NAND and NOR gates. The small circle represents an inverted output on any logic gate symbol.

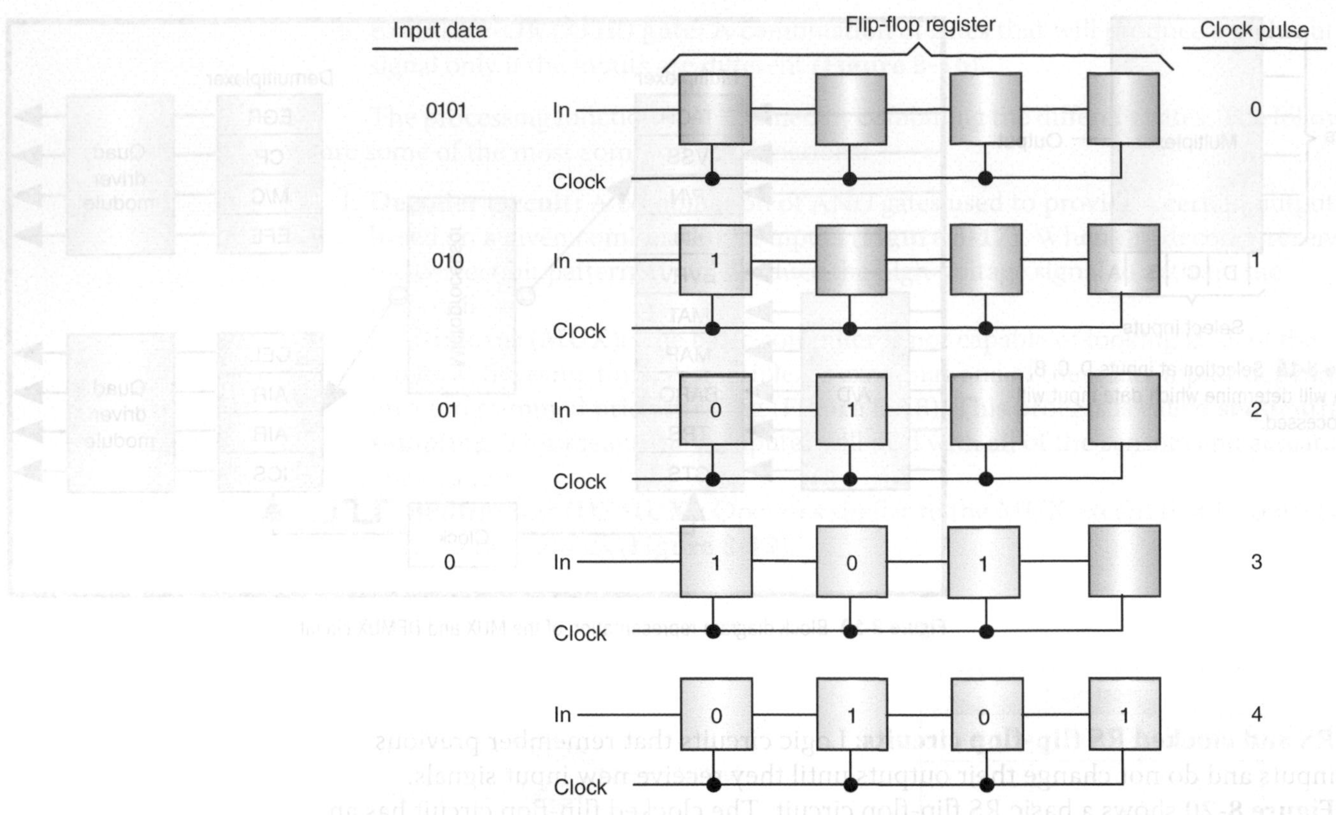

Figure 8-22 It takes four clock pulses to load 4 bits into the register.

High-Side and Low-Side Drivers

Usually, the computer will control an actuator using **low-side drivers**. These drivers will complete the path to ground through an FET transistor to control the output device. The computer may monitor the voltage on this circuit to determine if the actuator operates when commanded (**Figure 8-23**). The system can be monitored either by measuring voltage on the circuit or by measuring the current draw of the circuit.

Many manufacturers use **high-side drivers to** control the output device by varying the positive (12-volt) side. High-side drivers consist of a Metal Oxide Field Effect Transistor (MOSFET) controlled by a bipolar transistor. The bipolar transistor is controlled by the microprocessor. The advantage of the high-side driver is that it may provide quick-response self-diagnostics for shorts, opens, and thermal conditions. It also reduces vehicle wiring.

High-side driver diagnostic capabilities may include the ability to determine a short- or open-circuit condition. In this case, the high-side driver will take the place of a fuse. In the event of a short-circuit condition, it senses the high-current condition and will turn off the power flow. It may then store a diagnostic trouble code (DTC) in memory. The driver will automatically reset once the short-circuit condition is removed. In addition, the high-side driver monitors its internal temperature. The driver reports the junction temperature to the μP. If a slow-acting resistive short occurs in the circuit, the temperature will begin to climb. Once the temperature reaches 300℉ (150℃), the driver will turn off and set a DTC.

The high-side driver is also capable of detecting an open circuit, even with the system off. When the driver is off, a feedback voltage is read by the μP. For example, a 5-volt, 50 μ*A* current can be fed through the circuit, which also has a resistor wired in parallel. Low voltage (less than 2.25 volts) will indicate a normal circuit. If the voltage is high (above 2.25 volts), it indicates the circuit has high resistance or an open. The detection of an open circuit sets a DTC.

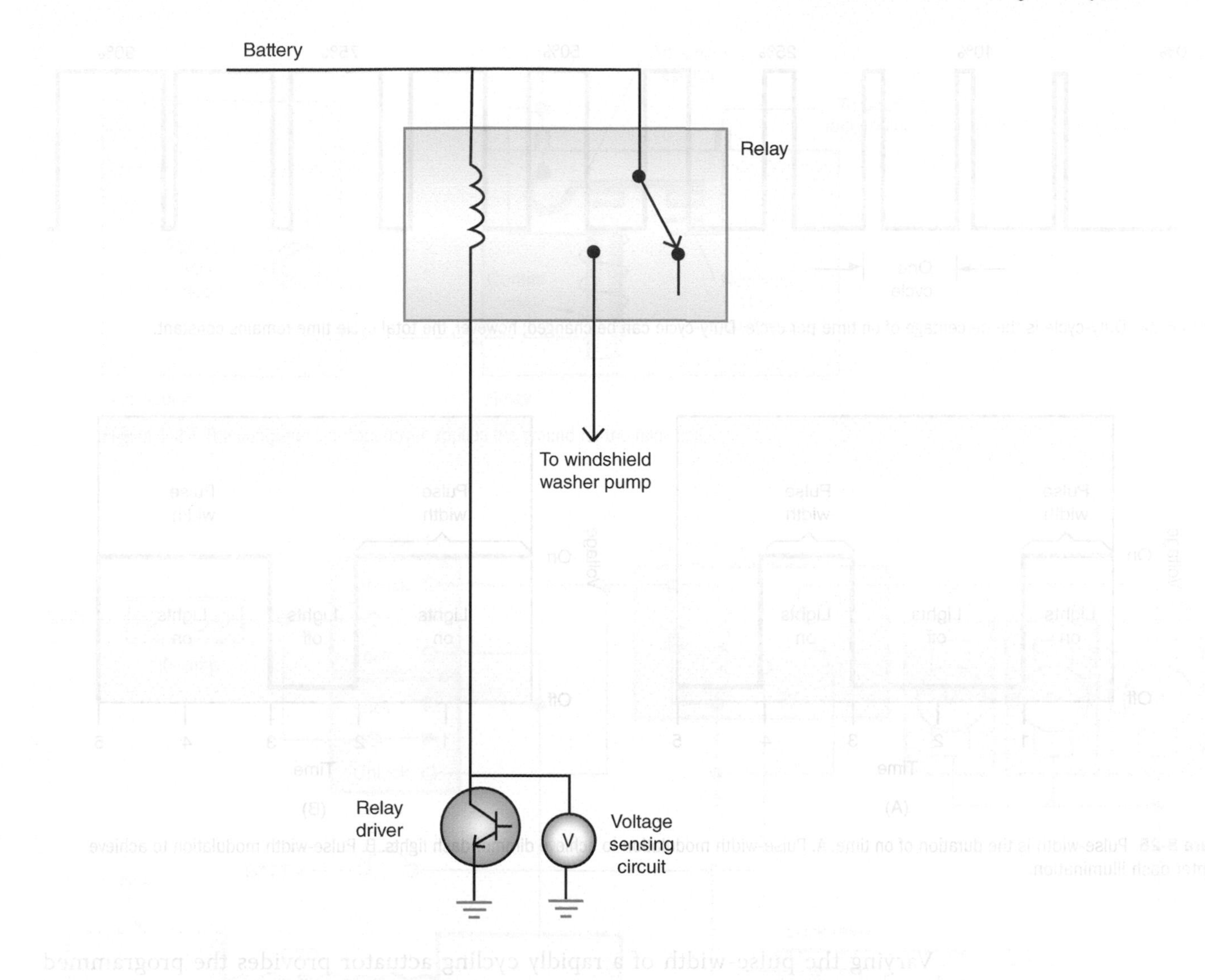

Figure 8-23 Computers using low-side drivers may be able to monitor the circuit for proper operation. When the relay is off (coil de-energized), the sense circuit should see a high voltage (12 V). The voltage should go low (0 V) when the coil is energized to turn on the relay.

Outputs

When the computer's programming determines that a correction, adjustment, or activation within the controlled system is necessary, an **output driver** sends an **output signal** to an actuator. This involves translating the electronic signals into mechanical motion.

The output driver usually controls the ground circuit of the actuator. The ground can be applied steadily if the actuator requires activation for a specific amount of time. For example, if the BCM inputs indicate that the automatic door locks are to be activated, the actuator is energized steadily until the locks latch. Then the ground is removed.

Other systems require turning the actuator on and off very rapidly or for a set number of cycles per second. It is duty-cycled if it's turned on and off a set number of cycles per second. To complete a cycle, it must go from off to on to off again. If the cycle rate is ten times per second, one actuator cycle is one-tenth of a second. If the actuator is on for 30% of each tenth of a second and off for 70%, it's referred to as a 30% duty-cycle (**Figure 8-24**).

Solenoids. A solenoid is commonly used as an actuator because it operates well under duty-cycling conditions. The solenoid can be controlled by a high-side or a low-side output driver.

An example of using a solenoid is to control vacuum to other components. Automatic climate control systems may use vacuum motors to move the blend doors. The computer can control the operation of the doors by controlling the solenoid.

Motors. Many computer-controlled systems use a **stepper motor** to move the controlled device to the desired position. A stepper motor contains a permanent magnet armature with two, four, or more field coils (**Figure 8-28**). By applying voltage pulses to selected coils of the motor, the armature will turn a specific number of degrees. Applying the same voltage pulses to the opposite coils causes the armature to rotate the same number of degrees in the opposite direction.

Some applications require using a permanent magnet field **servomotor** (**Figure 8-29**). A servomotor produces rotation of less than a full turn. A feedback mechanism is used to position itself to the exact degree of rotation required. The polarity of the voltage applied to the armature windings determines the direction the motor rotates. The computer can apply a continuous voltage to the armature until the desired result is obtained.

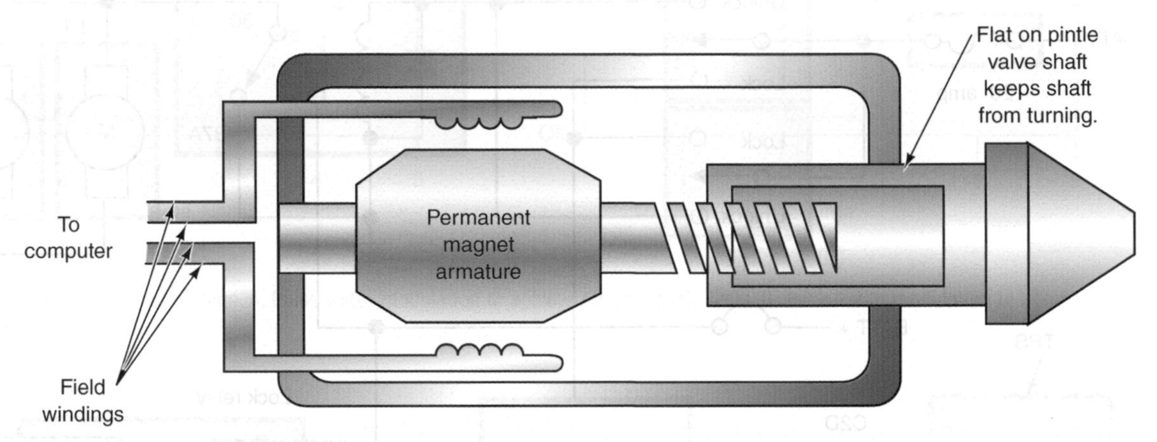

Figure 8-28 Typical stepper motor.

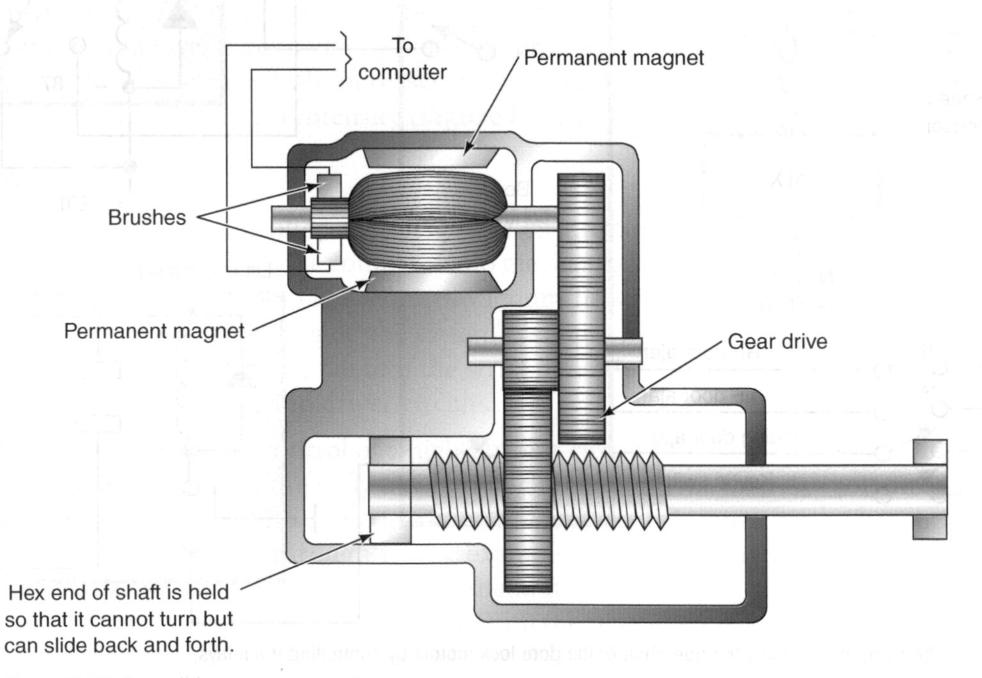

Figure 8-29 Reversible permanent magnet motor.

Summary

- A computer is an electronic device that stores and processes data and can operate other devices.
- The operation of the computer is divided into four basic functions: input, processing, storage, and output.
- Binary numbers are represented by the numbers 1 and 0. A transistor that operates as a relay is the basis of the digital computer. As the input signal switches from off to on, the transistor output switches from cutoff to saturation. The on and off output signals represent the binary digits 1 and 0.
- Logic gates are the thousands of field-effect transistors incorporated into the computer circuitry. The FETs use the incoming voltage patterns to determine the pattern of pulses that leave the gate. The most common logic gates are NOT, AND, OR, NAND, NOR, and XOR gates.
- The body computer uses several types of memory chips; ROM, RAM, and PROM are the most common types.
- ROM (read-only memory) contains a fixed pattern of 1s and 0s representing permanent stored information used to instruct the computer on what to do in response to input data.
- RAM (random access memory) will store temporary information that can be read from or written to by the μP.
- PROM (programmable read-only memory) contains specific data that pertains to the exact vehicle in which the computer is installed.
- EPROM (erasable PROM) is similar to PROM, except its contents can be erased to allow installing new data.
- EEPROM (electrically erasable PROM) allows changing the information electrically one bit at a time.
- NVRAM (nonvolatile RAM) combines RAM and EEPROM into the same chip.
- Actuators are devices that perform the actual work commanded by the computer. They can be in the form of a motor, relay, switch, or solenoid.
- A servomotor produces rotation of less than a full turn. A feedback mechanism is used to position itself to the exact degree of rotation required.
- A stepper motor contains a permanent magnet armature with two, four, or more field coils. It's used to move the controlled device to whatever location is desired by applying voltage pulses to selected coils of the motor.

Review Questions

Short-Answer Essays

1. What is binary code?
2. Describe the basics of NOT, AND, and OR logic gate operation.
3. List and describe the four basic functions of the microprocessor.
4. What is the difference between ROM, RAM, and PROM?
5. Explain the differences between analog and digital signals.
6. What is adaptive strategy?
7. Describe the basic function of a stepper motor.
8. Explain the function of a high-side driver.
9. What is the difference between duty-cycle and pulse-width?
10. What are the purposes of the interface?

Fill in the Blanks

1. In the binary code, number 4 is represented by ______________.
2. The ______________ is a crystal that electrically vibrates when subjected to current at certain voltage levels.
3. ______________ are registers designed to store the results of logic operations.
4. The ______________ is the brain of the computer.
5. ______________ contains specific data that pertains to the exact vehicle in which the computer is installed.
6. The ______________ gate reverses binary code.
7. ______________ drivers complete the actuator control circuit to ground.
8. If a control circuit to an actuator is turned on and off a set number of cycles per second, it's called ______________.

9. The ______________ function of the computer holds the program instructions.

10. The input ______________ converts analog signals to digital signals.

Multiple Choice

1. *Technician A* says during the processing function, the computer uses input information and compares it to programmed instructions.
   Technician B says during the output function, the computer will put out control commands to various output devices.
   Who is correct?
   A. A only C. Both A and B
   B. B only D. Neither A nor B

2. Which of the following is correct?
   A. Analog signals are either high/low, on/off, or yes/no.
   B. Digital signals are infinitely variable within a defined range.
   C. All of these choices.
   D. None of these choices.

3. Logic gates are being discussed.
   Technician A says NOT gate operation is similar to two switches in series to a load.
   Technician B says an AND gate simply reverses binary 1s to 0s and vice versa.
   Who is correct?
   A. A only C. Both A and B
   B. B only D. Neither A nor B

4. All of the following statements about computer memory are true, EXCEPT:
   A. RAM stores temporary information that can be written to and read by the CPU.
   B. ROM can be read only by the CPU.
   C. All PROM memory is flashable.
   D. Volatile memory is erased when voltage is removed.

5. Nonvolatile memory is retained if removed from its power source.
   A. True B. False

6. *Technician A* says the EEPROM can be reprogrammed with new files.
   Technician B says electrostatic discharge can destroy the memory chip.
   Who is correct?
   A. A only C. Both A and B
   B. B only D. Neither A nor B

7. *Technician A* says high-side drivers control the ground side of the circuit.
   Technician B says high-side drivers may be capable of determining circuit faults.
   Who is correct?
   A. A only C. Both A and B
   B. B only D. Neither A nor B

8. Which of the following would represent the number "255" in binary code?
   A. 00000000 C. 00001111
   B. 11111111 D. 11110000

9. Which of the following is responsible for sequential sampling?
   A. DEMUX C. MUX
   B. Driver D. Register

10. *Technician A* says the microprocessor commands actuators by output drivers.
    Technician B says that outputs are never controlled by supplying voltage to the actuator.
    Who is correct?
    A. A only C. Both A and B
    B. B only D. Neither A nor B

CHAPTER 9
COMPUTER INPUTS

Upon completion and review of this chapter, you should be able to:

- Describe the function of input devices.
- Explain the purpose of the thermistor and how it's used in a circuit.
- Explain the difference between NTC and PTC thermistors.
- Explain the operation and purpose of the Wheatstone bridge.
- Describe the operation and purpose of piezoelectric devices.
- Describe the operation and purpose of piezoresistive devices.
- Explain the function of the potentiometer and how it's used.
- Explain the purpose and operation of magnetic pulse generators.
- Explain the purpose and operation of Hall-effect sensors.
- Describe the function of accelerometers.
- Describe the function of photocell components such as the photodiode, photoresistor, and phototransistor.
- Explain the function of the photoelectric sensor.
- Explain the function of the pull-down sense circuit.
- Explain the function of the pull-up sense circuit.
- Explain the purpose and operation of feedback systems.

Terms to Know

Accelerometers
Capacitance discharge sensor
Dual-range circuit
Feedback
Floating
G force
Hall-effect sensor
Infrared temperature sensor
Linearity
Magnetic pulse generators
Magnetically coupled linear sensors
Magnetoresistive effect
Magnetoresistive (MR) sensors
Negative temperature coefficient (NTC) thermistors
Photocell
Photoconductive mode
Photoresistor
Photovoltaic mode
Pickup coil
Piezoelectric device
Piezoresistive device
Positive temperature coefficient (PTC) thermistors
Potentiometer
Potentiometric pressure sensors
Pull-down circuit
Pull-down resistor
Pull-up circuit
Pull-up resistor
Schmitt trigger
Sensors
Shutter wheel
Strain gauge
Thermistor
Timing disc
Wheatstone bridge

Introduction

Shop Manual
Chapter 9, page 407

Several different types of input devices are used to gather information for the computer to use in determining the desired output. Many input devices are also used as feedback signals to confirm the proper positioning of the actuator. Depending on the input, the computer will control the actuator(s) until the programmed results are obtained (**Figure 9-1**). The inputs can come from other computers, the vehicle operator, the technician, or through a variety of sensors.

Typically, momentary contact switches provide driver input signals by momentarily applying a ground through a switch. The computer receives this signal and performs the desired function. For example, if the driver wishes to reset the trip odometer on a digital instrument panel, they would push the reset switch. The switch provides a momentary ground that the computer receives as an input and sets the trip odometer to zero.

Switches can be used as an input for any operation that requires only a yes/no, or on/off condition. Other inputs include those supplied by a sensor, and those signals returned to the computer in the form of feedback.

Linearity refers to the sensor signal being as constantly proportional to the measured value as possible. It's an expression of the sensor's accuracy.

This chapter discusses the many different designs of **sensors** and inputs. Sensors convert some measurement of vehicle operation into an electrical signal. Some sensors are nothing more than a switch that completes the circuit. Others are complex chemical reaction devices that generate their own voltage under different conditions. Repeatability, accuracy, operating range, and **linearity** are all requirements of a sensor.

Sensors discussed in this chapter include common forms of electrical and electronic devices used in body and chassis systems. These include the following:

- **Thermistor**—a solid-state variable resistor made from a semiconductor material that changes resistance in relation to temperature changes.
- **Wheatstone bridge**—A series-parallel arrangement of resistors between an input terminal and ground. The sensing circuit will receive a voltage reading that is proportional to the amount of resistance change.

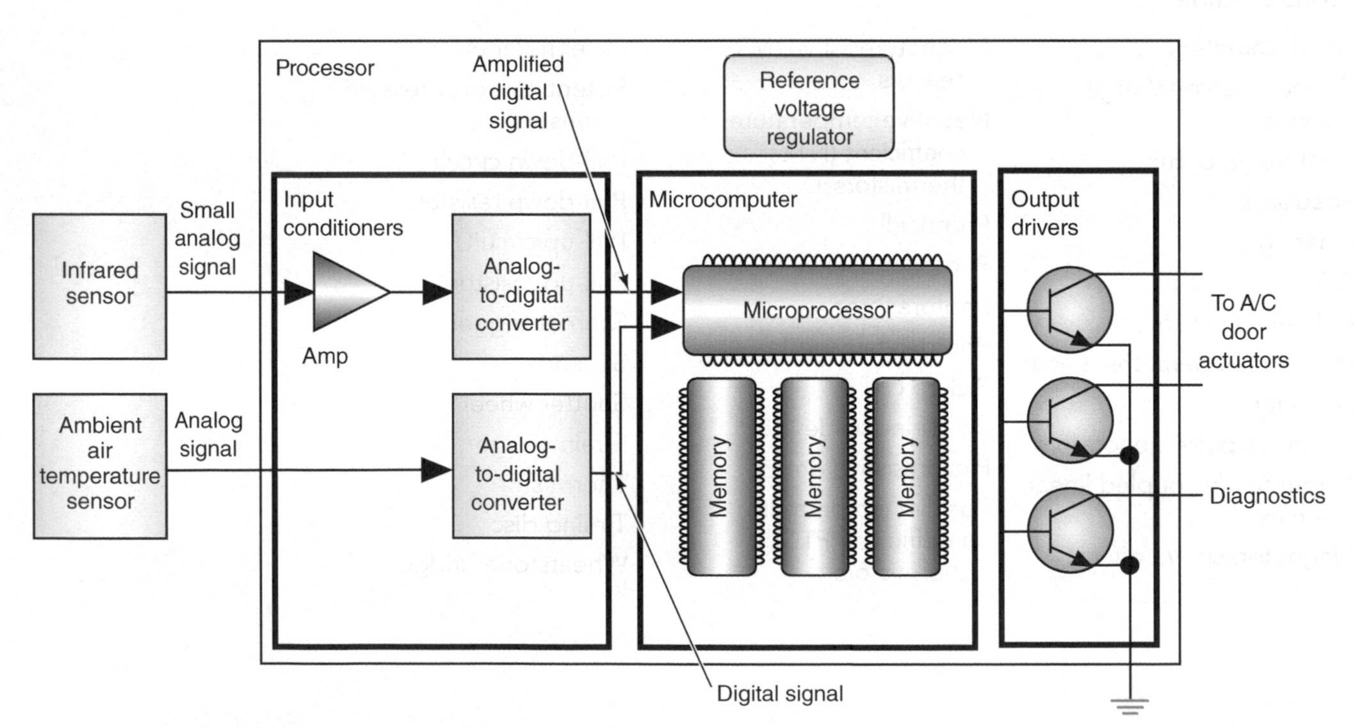

Figure 9-1 The microprocessor processes the input signals and then directs the output drivers to activate actuators as instructed by the program.

- **Piezoelectric device**—A voltage generator with a resistor connected in series. This sensor can measure fluid and air pressures.
- **Piezoresistive device**—Similar to a piezoelectric device, except they operate like a variable resistor. Its resistance value changes as the pressure applied to the crystal changes.
- **Potentiometer**—A voltage divider that provides a variable direct current (DC) voltage reading to the computer. The potentiometer usually consists of a wire-wound resistor with a movable center wiper.
- **Magnetic pulse generators**—Commonly used to send voltage signals to the computer about the speed of the monitored component. They use the principle of magnetic induction to produce an alternating current (AC) voltage signal that is conditioned by an analog to digital (A/D) converter.
- **Hall-effect sensor**—A sensor that operates on the principle current flowing through a thin conducting material exposed to a magnetic field produces another voltage. The sensor contains a permanent magnet, a thin semiconductor layer made of gallium arsenate crystal (Hall layer), and a trigger wheel.

AUTHOR'S NOTE This chapter discusses sensor types that are used in multiple systems. For example, the potentiometer can be used as a throttle position sensor, accelerator position sensor, an A/C mode door position sensor, a seat track position sensor, and so on. Sensors with specific functions (such as radar, laser, yaw rate, etc.) will be discussed as we explore the system that uses them. Engine management systems use special sensors. *Today's Technician: Automotive Engine Performance* covers these types of sensors. For detailed explanation of sensor construction and operation, refer to *Today's Technician: Advanced Automotive Electronic Systems.*

Thermistors

Thermistors measure the temperature of liquids and air. A thermistor is a solid-state variable resistor made from a semiconductor material, such as metal oxides, possessing very reproducible resistance versus temperature properties.

Shop Manual
Chapter 9, page 409

The computer can observe very small temperature changes by monitoring the thermistor's resistance value. The computer sends a reference voltage to the thermistor (usually 5 volts) through a fixed resistor. As the current flows through the thermistor resistance to ground, a voltage-sensing circuit measures the voltage after the fixed resistor (**Figure 9-2**). The voltage dropped over the fixed resistor will change as the resistance of the thermistor changes. Using its programmed values, the computer is able to translate the voltage drop into a temperature value.

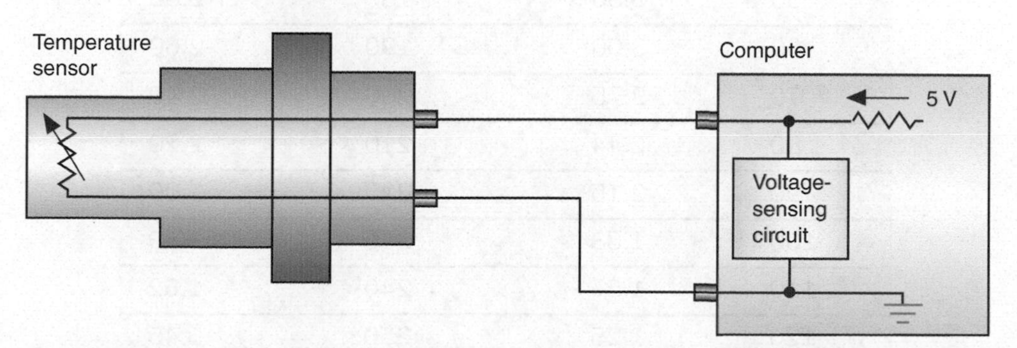

Figure 9-2 Thermistors measure temperature. The sensing unit measures the resistance change and translates the data into temperature values.

There are two types of thermistors: **negative temperature coefficient (NTC) thermistors** and **positive temperature coefficient (PTC) thermistors**. NTC thermistors reduce their resistance as the temperature increases, while PTC thermistors increase their resistance as the temperature increases.

Using the circuit shown in Figure 9-2, if the value of the fixed resistor is 10K ohms (Ω) and the value of the thermistor is also 10K ohms, the voltage-sensing circuit will read a voltage value of 2.5 volts (V). If the thermistor is an NTC, its resistance decreases as the ambient temperature increases. If the resistance of the NTC is now 8K ohms, the voltage reading by the voltage-sensing circuit will now be 2.22 volts. As ambient temperature increases and the NTC value continues to decrease, the voltage-sensing circuit will measure a voltage decrease (**Figure 9-3**). If the thermistor was a PTC, the opposite would be true, and the voltage-sensing circuit would measure an increase in voltage as the ambient temperature increases.

Dual-range circuits are also referred to as dual ramping circuits.

Shop Manual
Chapter 9, page 410

Some temperature-sensing circuits are **dual-range circuits** (**Figure 9-4**). This circuit provides for switching the resistance values to allow the microprocessor to measure temperatures more accurately. When the voltage-sensing circuit records a calibrated voltage value (1.25 volts, for example), the microprocessor turns on the transistor, which places the 1K resistor in parallel with the 10K resistor. The circuit will operate as described until the voltage reaches 1.25 volts. With the 1K resistor now involved in the circuit, the resistance of the fixed resistor portion of the circuit is now 909 ohms. At this occurrence, the voltage-sensing circuit will record the sharp voltage increase and use the second range of values. Voltages can represent two different temperatures depending on which side of the switch the voltage is (**Figure 9-5**).

Temperature Sensor			
Voltages versus Temperature Values			
Cold Temperature		Hot Temperature	
Degrees F	Volts	Degrees F	Volts
−20	4.70	110	4.20
−10	4.57	120	4.10
0	4.45	130	4.00
10	4.30	140	3.60
20	4.10	150	3.40
30	3.90	160	3.20
40	3.60	170	3.02
50	3.30	180	2.80
60	3.00	190	2.60
70	2.75	200	2.40
80	2.44	210	2.20
90	2.15	220	2.00
100	1.83	230	1.80
110	1.57	240	1.62
120	1.25	250	1.45

Figure 9-3 Chart of temperature and voltage correlation.

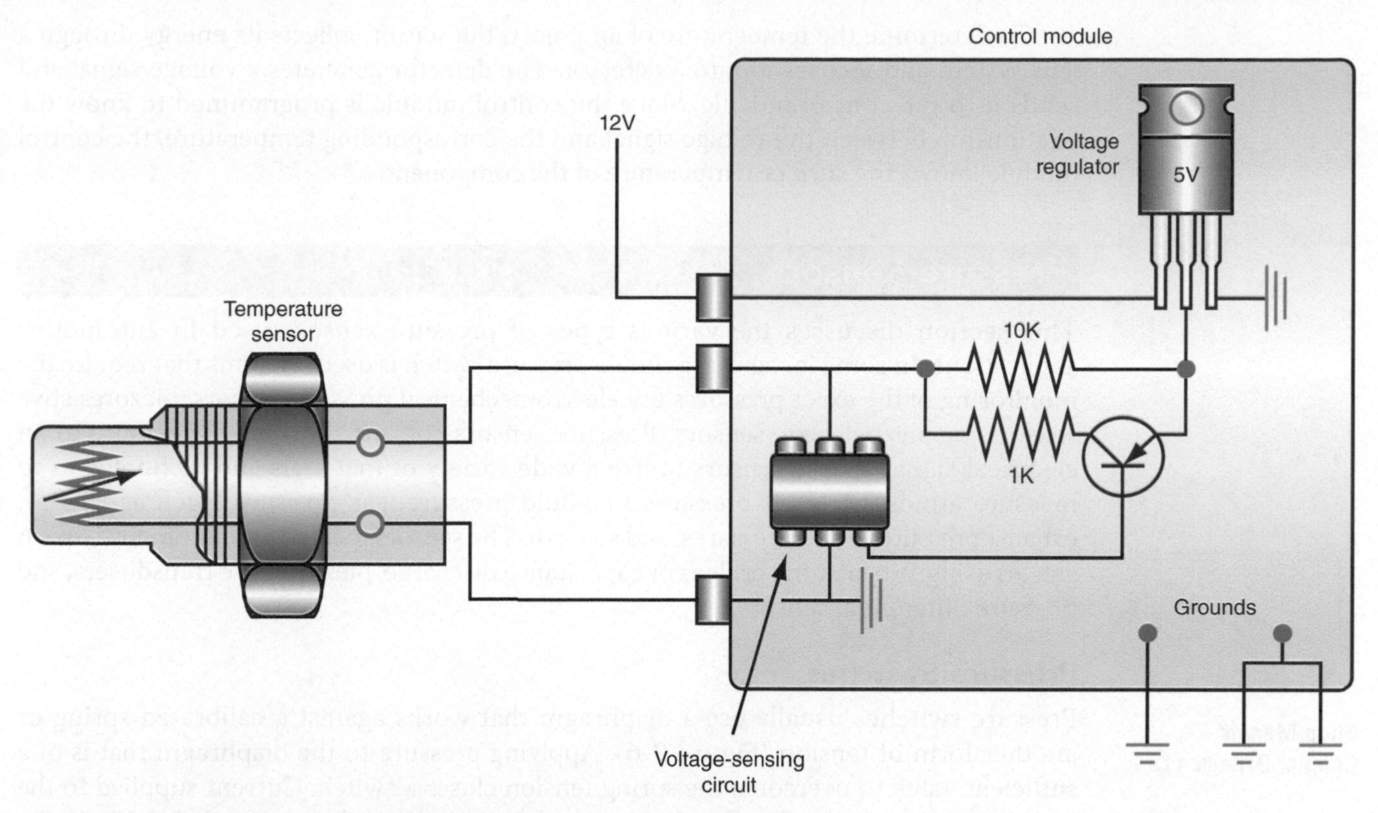

Figure 9-4 Dual-range temperature sensor circuit.

COLD		HOT	
10K-ohm resistor		909-ohm resistor	
−20°F	4.7 V	110°F	4.2 V
0°F	4.4 V	130°F	3.7 V
20°F	4.1 V	150°F	3.4 V
40°F	3.6 V	170°F	3.0 V
60°F	3.0 V	180°F	2.8 V
80°F	2.4 V	200°F	2.4 V
100°F	1.8 V	220°F	2.0 V
120°F	1.2 V	240°F	1.6 V

Figure 9-5 The same voltage value can represent different temperatures.

Infrared Temperature Sensors

It's becoming more common for manufacturers to use an **infrared temperature sensor** instead of in-vehicle temperature sensors. The infrared sensor measures surface temperatures without physically touching the surface. Infrared temperature sensors use the principle that all objects emit energy and as the temperature of an object rises, so does the amount of energy it emits. The infrared sensor receives the heat energy radiated from an object and converts the heat to an electrical potential. In climate control systems, this information assists in determining the best control to maintain the desired temperature set by the vehicle occupants.

To determine the temperature of an object, the sensor collects its energy through a lens system and focuses it onto a detector. The detector generates a voltage signal and sends it to the control module. Since the control module is programmed to know the relationship between the voltage signal and the corresponding temperature, the control module knows the surface temperature of the component.

Pressure Sensors

This section discusses the various types of pressure sensors used in automotive applications. In some instances, a simple pressure switch is used. Systems that require the monitoring of the exact pressures use electromechanical pressure sensors, piezoresistive sensors, or piezoelectric sensors. Pressure sensors convert the applied pressure to an electrical signal. These sensors utilize a wide variety of materials and technologies to measure atmospheric air pressure, manifold pressure, gas pressure (such as R134a), exhaust pressures, fluid pressures, and so forth. The sensors include potentiometric, strain gauges using Wheatstone bridges or capacitance discharge, piezoelectric transducers, and pressure differential sensors.

Pressure Switches

Shop Manual
Chapter 9, page 413

Pressure switches usually use a diaphragm that works against a calibrated spring or another form of tension (**Figure 9-6**). Applying pressure to the diaphragm that is of a sufficient value to overcome the spring tension closes a switch. Current supplied to the switch now has a completed path to ground. In a very simple warning light circuit, the closed pressure switch completes the circuit for the bulb and alerts the driver to an unacceptable condition. For example, a simple oil pressure warning lamp circuit will use a pressure switch.

Pressure switches monitor the presence of pressure above or below a set point; they don't indicate the exact amount of applied pressure. Computer-monitored pressure switch circuits use the change in voltage as an indication of pressure. When the pressure changes (from either low to high or high to low), it changes the state of the switch, and the computer interprets the voltage change (**Figure 9-7**).

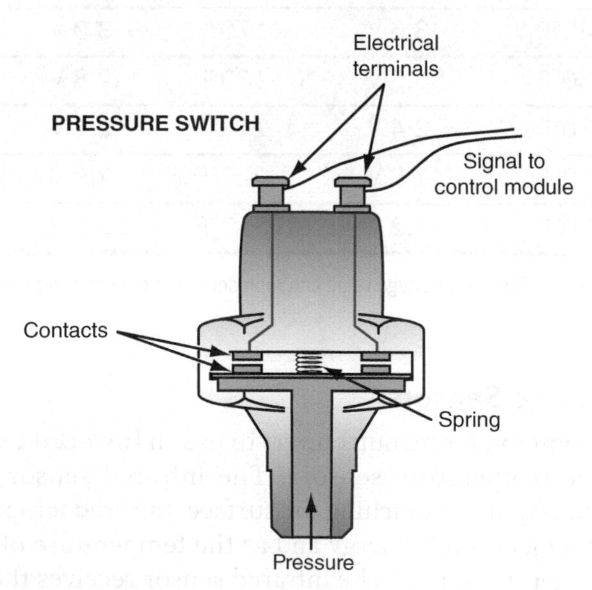

Figure 9-6 Simple pressure switch uses contacts to complete the electrical circuit.

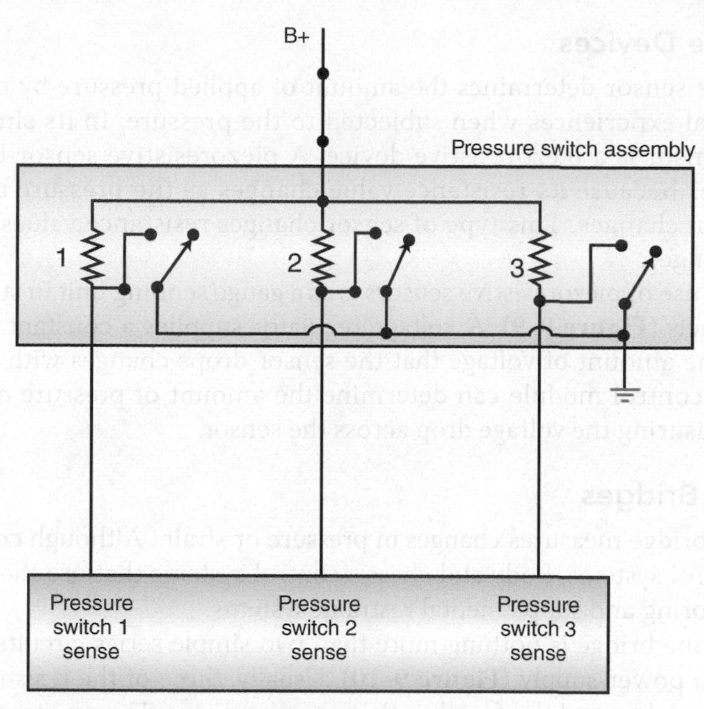

Figure 9-7 Computer-monitored pressure switch circuit.

Potentiometric Pressure Sensor

One of the basic types of pressure sensor is the **potentiometric pressure sensor**. The potentiometric pressure sensors use a Bourdon tube, a capsule, or bellows to move a wiper arm on a resistive element (**Figure 9-8**). Using the principle of variable resistance, the movement of the wiper across the resistive element will record a different voltage reading to the computer. Although this type of sensor can be used as a computer input, a computer is not always involved. Some early analog instrument panels used this sensor unit with an air-coil gauge to display engine oil pressure.

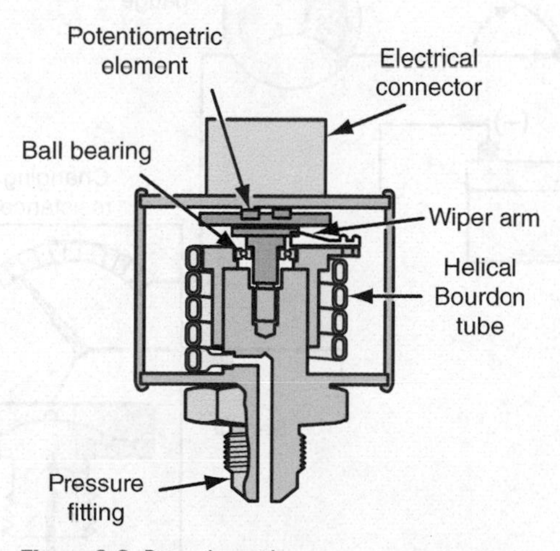

Figure 9-8 Potentiometric pressure sensors use a Bourdon tube, capsule, or bellows to drive a wiper arm on a resistive element.

Piezoresistive Devices

Shop Manual
Chapter 9, page 415

Strain gauges are also called stress gauges.

The term *piezo* refers to pressure and is derived from the Greek to mean "to be pressed."

A **strain gauge** sensor determines the amount of applied pressure by measuring the strain a material experiences when subjected to the pressure. In its simplest form, a strain gauge sensor is a piezoresistive device. A piezoresistive sensor behaves like a variable resistor because its resistance value changes as the pressure applied to the sensing material changes. This type of sensor changes resistance values as a function of pressure changes.

A common use of piezoresistive sensors is as a gauge sending unit in standard analog instrument panels (**Figure 9-9**). A voltage regulator supplies a constant voltage to the sensor. Since the amount of voltage that the sensor drops changes with the change of resistance, the control module can determine the amount of pressure on the sensing material by measuring the voltage drop across the sensor.

Wheatstone Bridges

Shop Manual
Chapter 9, page 415

Wheatstone bridges can have different configurations, such as using four variable resistors instead of one sensing resistor and three fixed resistors.

A Wheatstone bridge measures changes in pressure or strain. Although commonly used for engine control systems, body and chassis control systems that use them include tire pressure monitoring and supplemental restraint systems.

A Wheatstone bridge is nothing more than two simple series circuits connected in parallel across a power supply (**Figure 9-10**). Usually, three of the resistors are kept at exactly the same value, and the fourth is the sensing resistor. The resistors are placed on a silicon chip that flexes. As the chip flexes, the resistance of the sensing resistor changes. When all four resistors have the same value, the bridge is balanced, and the voltage sensor will indicate a value of 0 volt. The output from the amplifier acts as a voltmeter. Remember, since a voltmeter measures electrical pressure between two points, it will display this value. For example, if the reference voltage is 5 volts and the resistors have the same value, then the voltage drop over each resistor is 2.5 volts. Since the voltmeter is measuring the potential on the line between R_s and R_1; and between R_2 and R_3, it will read 0 volts because both of these points have 2.5 volts on them. If there is a change in the resistance value of the sensing resistor, a change will occur in the circuit's balance. The sensing circuit will receive a voltage reading that is proportional to the amount of resistance change. The sensing resistor is a variable resistor that changes resistance as the silicon chip flexes.

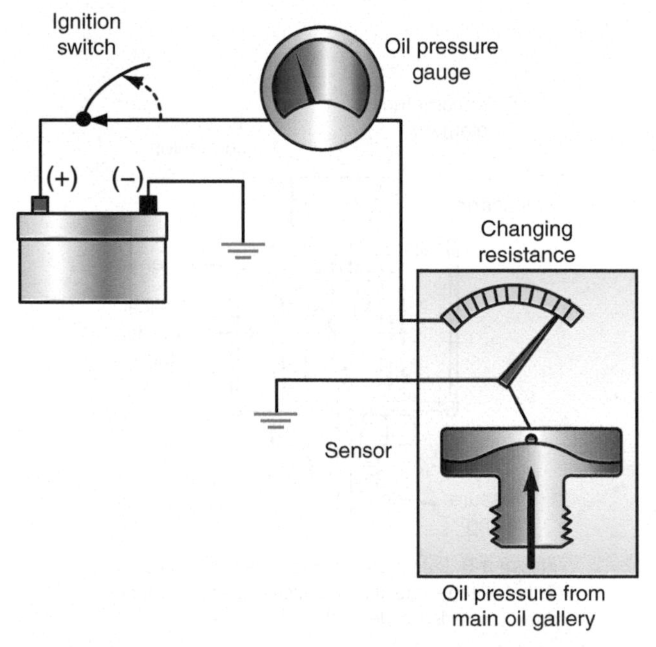

Figure 9-9 Oil pressure sensor used in gauge indicator circuit.

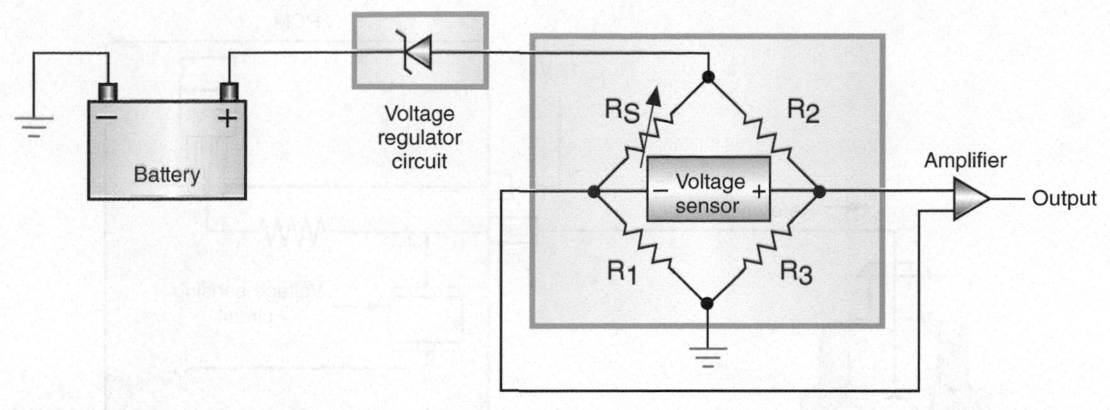

Figure 9-10 Wheatstone bridge.

Piezoelectric Devices

Shop Manual
Chapter 9, page 420

Piezoelectric devices measure pressures by the generation of voltage. Piezoelectric sensors are constructed from alumina ceramics, metalized quartz, single crystals, or ultrasonic transducer materials that make up a bidirectional transducer capable of converting stress into an electric potential (**Figure 9-11**). The piezoelectric materials consist of polarized ions within the crystal. Applying pressure on the piezoelectric material creates some mechanical deformation in the polarized crystal, which produces a proportional output charge due to the displacement in the ions. In the automotive industry, uses for this type of sensor include piezoelectric accelerometers, piezoelectric force sensors, and piezoelectric pressure sensors.

The sensor is a voltage generator with a resistor connected in series (**Figure 9-12**). The resistor protects the sensor from excessive current flow in case the circuit becomes shorted. The voltage generator is a thin ceramic disc attached to a metal diaphragm. Pressure on the diaphragm transfer pressure onto the piezoelectric crystals in the ceramic disc. The disc generates a voltage that is proportional to the amount of pressure. The voltage generated ranges from zero to one or more volts.

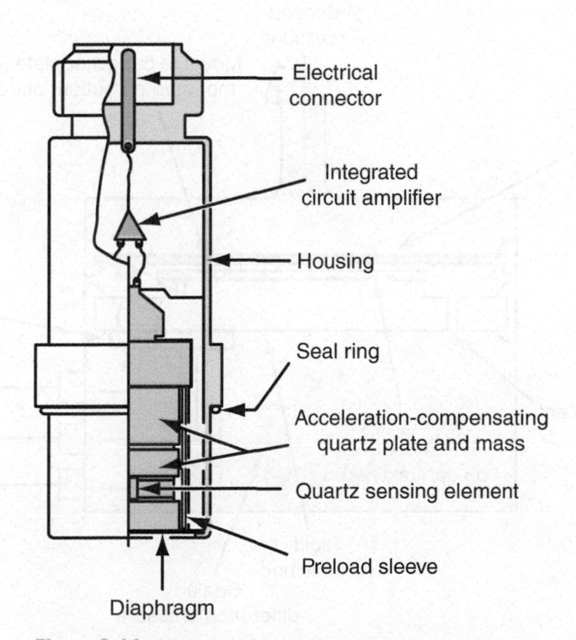

Figure 9-11 Piezoelectric sensors convert stress into an electric potential and vice versa. Sensors based on this technology are used to measure varying pressures.

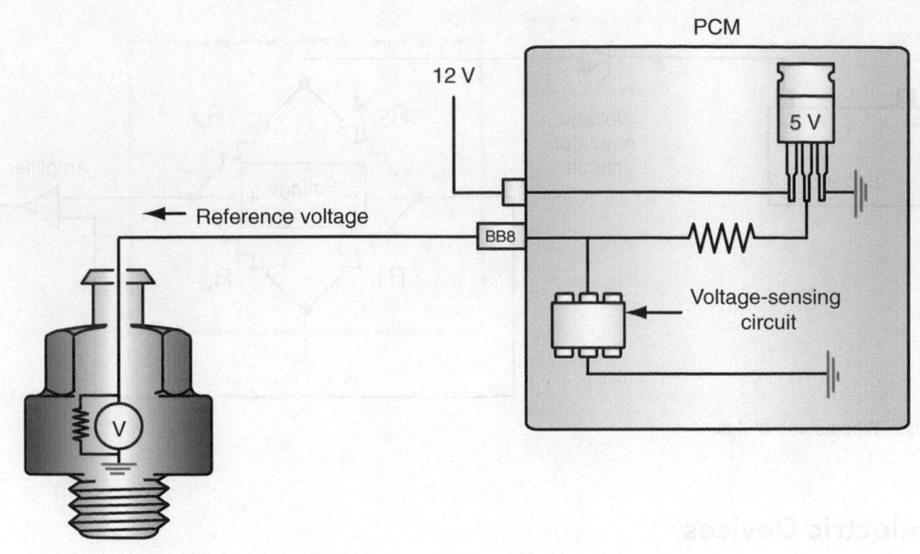

Figure 9-12 Piezoelectric sensor circuit.

Shop Manual
Chapter 9, page 418

Capacitance Discharge Sensors

Another variation of the piezosensor uses capacitance discharge. Instead of using a silicon diaphragm, the **capacitance discharge sensor** uses a variable capacitor. In the capacitor capsule–type sensor, two flexible alumina plates are separated by an insulating washer (**Figure 9-13**). A film electrode is deposited on each plate's inside surface, and a connecting lead extends out for external connections. The result is a parallel plate capacitor with a vacuum between the plates. This capsule is placed inside a sealed housing that connects to the sensed pressure. If constructed to measure vacuum, as the pressure increases (goes toward atmospheric), the alumina plates deflect inward, decreasing the distance between the electrodes.

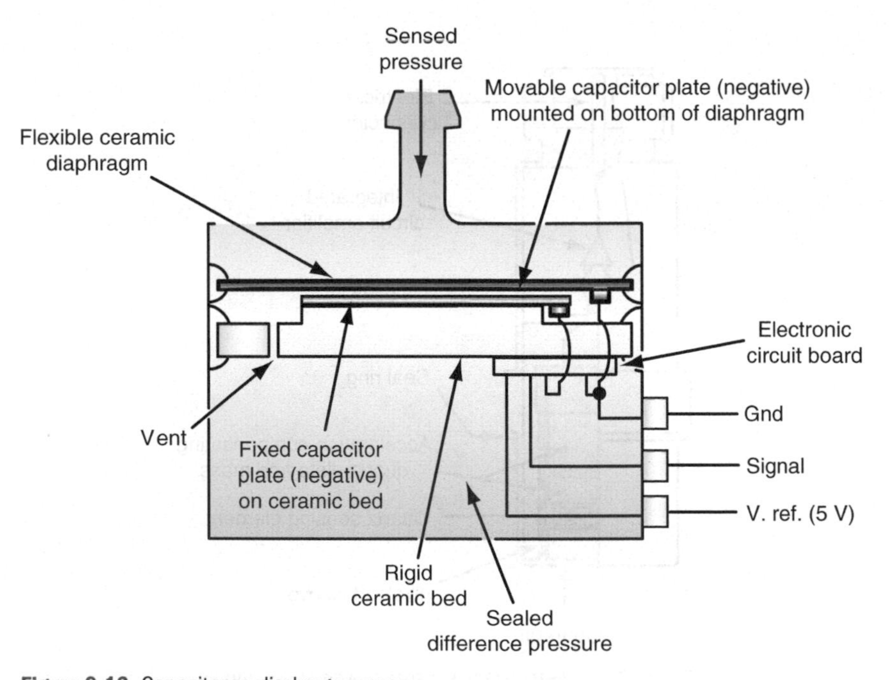

Figure 9-13 Capacitance discharge sensor.

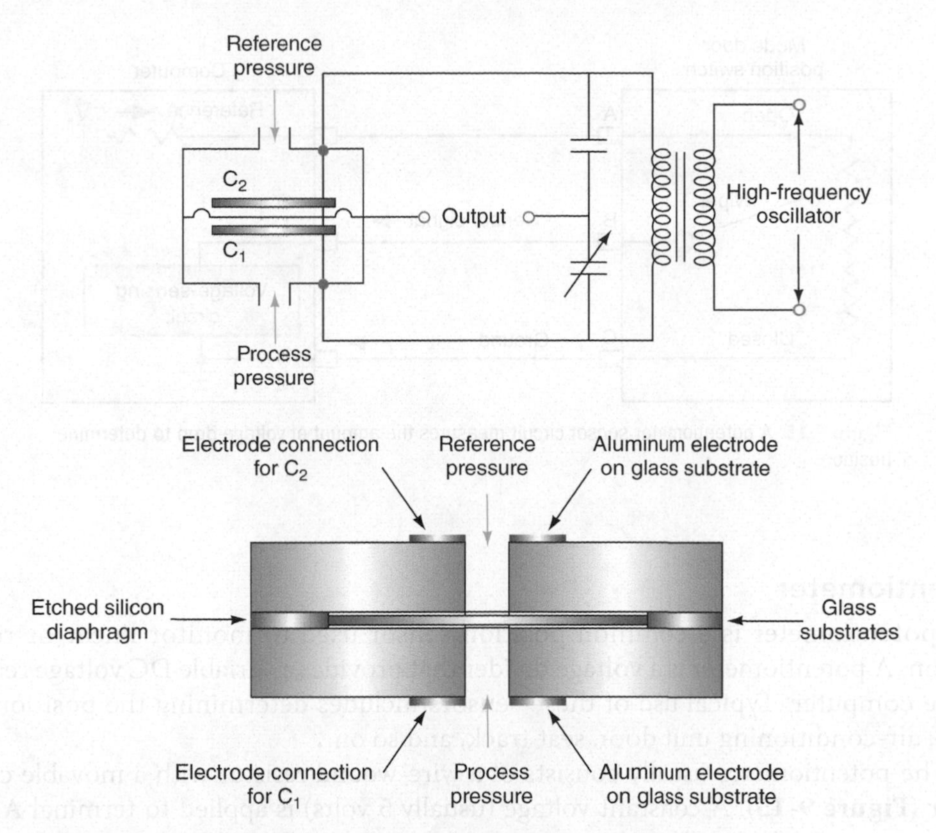

Figure 9-14 Capacitance pressure transducer construction and schematic.

As the distance between the electrodes changes, so does the capacity of the capacitor. A measure of capacitance constitutes a measurement of pressure that is detected by a bridge circuit. The output from the bridge circuit can be either an analog DC voltage or applied to a chip that produces a frequency modulated digital signal.

Capacitance Pressure Transducers. Capacitance pressure transducers measure pressure by changes in capacitance as the result of the movement of a diaphragm element (**Figure 9-14**). The diaphragm must physically travel a distance that is only a few microns. One side of the diaphragm is exposed to the measured pressure and the other side to the reference pressure. The change in capacitance may control the frequency of an oscillator or to vary the coupling of a voltage signal. Depending on sensor construction and the type of reference pressure, the capacitive transducer can be either an absolute, gauge, or differential pressure transducer.

Position and Motion Detection Sensors

Many electronic systems require input data concerning position, motion, and speed. Most motion and speed sensors use a magnet as the sensing element or sensed target to detect rotational or linear speed. The types of magnetic speed sensors include magnetoresistive (MR), inductive, variable reluctance (VR), and Hall-effect. In addition, potentiometers and commutator pulse counting also are used to monitor position.

Some systems require photoelectric sensors that use light-sensitive elements to detect the movement of an object. In addition, solid-state accelerometers, axis rotation sensors, yaw sensors, and roll sensors are becoming common components on many systems. This chapter will explore the operation of common position and motion detection sensors.

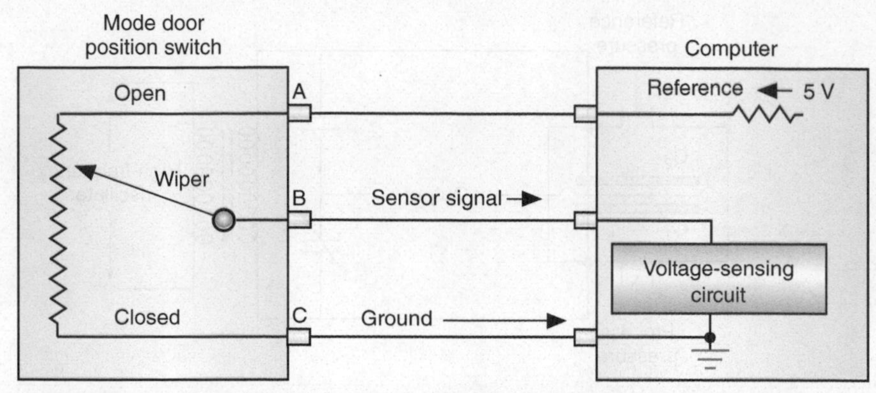

Figure 9-15 A potentiometer sensor circuit measures the amount of voltage drop to determine position.

Potentiometer

Shop Manual
Chapter 9, page 422

The potentiometer is a common position sensor used to monitor linear or rotary motion. A potentiometer is a voltage divider that provides a variable DC voltage reading to the computer. Typical use of these sensors includes determining the position of a valve, air-conditioning unit door, seat track, and so on.

The potentiometer usually consists of a wire-wound resistor with a movable center wiper (**Figure 9-15**). A constant voltage (usually 5 volts) is applied to terminal A. The computer tracks the unit's position as the wiper moves across the resistor. If the wiper (which is connected to the shaft or movable component of the unit that is being monitored) is located close to terminal A, there will be a low voltage drop represented by a high-voltage signal back to the computer through terminal B. As the wiper moves toward terminal C, the sensor signal voltage to terminal B decreases. The computer interprets the different voltage values into different shaft positions. The potentiometer can measure linear or rotary movement.

Since applied voltage must flow through the entire resistance, temperature and other factors don't create false or inaccurate sensor signals to the computer. A rheostat is not as accurate and has limited use in computer systems.

Magnetic Pulse Generator

Shop Manual
Chapter 9, page 425

An example of the use of magnetic pulse generators is to determine vehicle and individual wheel speed. The signals from the speed sensors are inputs for computer-driven instrumentation, cruise control, antilock braking, speed-sensitive steering, and automatic ride control systems.

The components of the pulse generator are as follows (**Figure 9-16**):

The **timing disc** is known as an armature, reluctor, trigger wheel, pulse wheel, toothed wheel, or timing core. It conducts the lines of magnetic force.

The **pickup coil** is also known as a stator, sensor, or pole piece.

1. A **timing disc** that is attached to the rotating shaft or cable. The manufacturer determines the number of teeth on the timing disc based on application. The teeth will cause a voltage generation that is constant per revolution of the shaft. For example, a vehicle speed sensor may deliver 4,000 pulses per mile. The number of pulses per mile remains constant regardless of speed. The computer calculates how fast the vehicle is going based on the signal's frequency.
2. A **pickup coil** consists of a permanent magnet wound around by fine wire.

An air gap is maintained between the timing disc and the pickup coil. As the timing disc rotates in front of the pickup coil, the generator sends an A/C signal (**Figure 9-17**). As a tooth on the timing disc aligns with the core of the pickup coil, it repels the

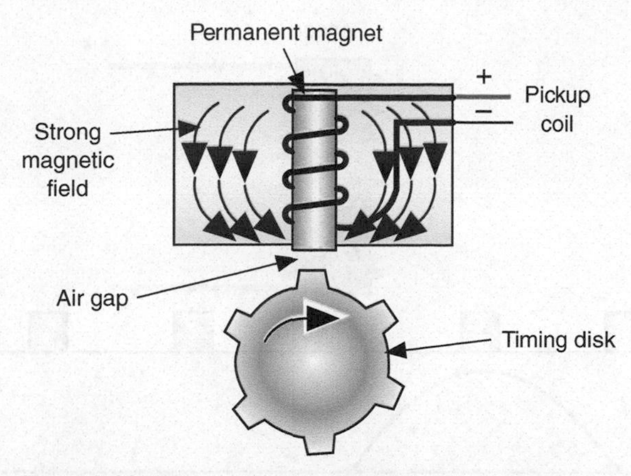

Figure 9-16 Components of the magnetic pulse generator. The pickup coil produces a strong magnetic field as the teeth align with the core.

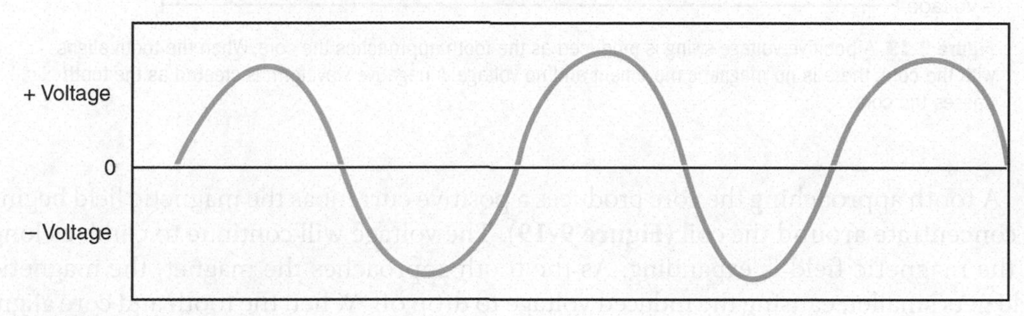

Figure 9-17 Pulse signal sine wave.

magnetic field. The magnetic field is forced to flow through the coil and pickup core (refer to Figure 9-16). Since the magnetic field is not expanding, no voltage is induced in the pickup coil. As the tooth passes the core, the magnetic field begins to expand (**Figure 9-18**). The expanding magnetic field cuts across the windings of the pickup coil. This movement of the magnetic field induces a voltage in the windings. This action repeats every time a tooth passes the core. The moving lines of magnetic force cut across the coil windings and induce a voltage signal.

Magnetic pulse generators are also called magnetic induction sensors.

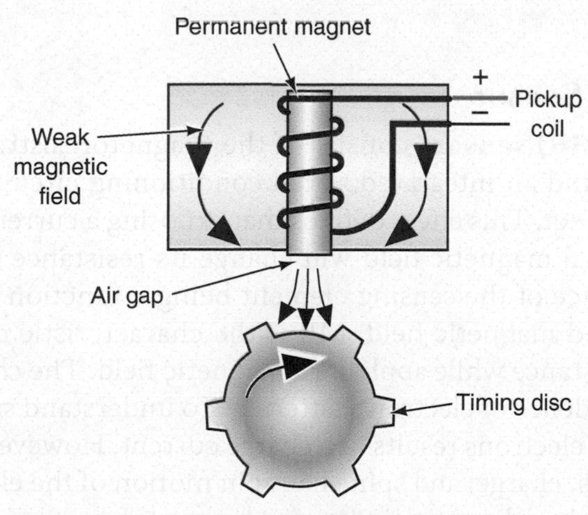

Figure 9-18 The magnetic field expands as the teeth pass the core.

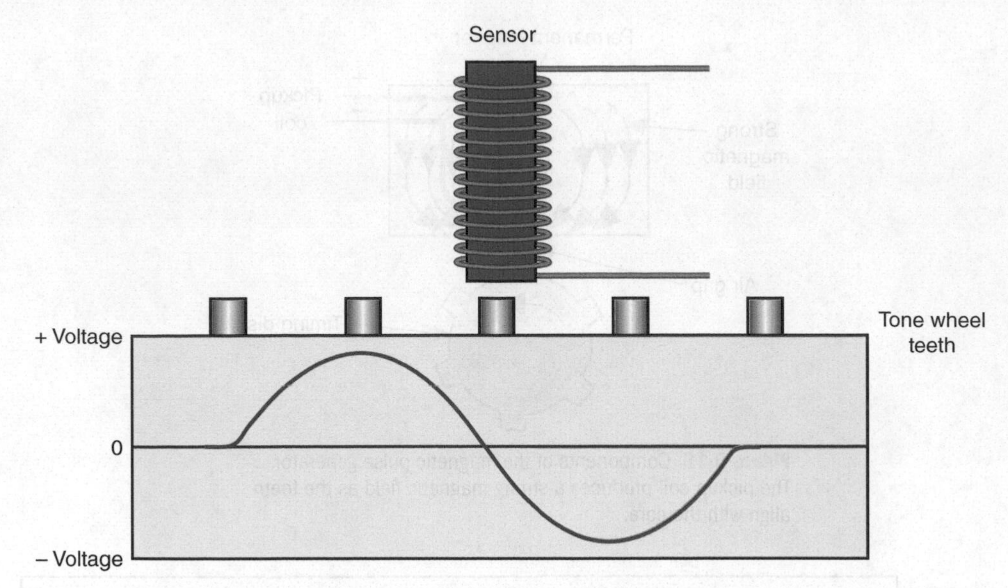

Figure 9-19 A positive voltage swing is produced as the tooth approaches the core. When the tooth aligns with the core, there is no magnetic movement and no voltage. A negative waveform is created as the tooth passes the core.

A tooth approaching the core produces a positive current as the magnetic field begins to concentrate around the coil (**Figure 9-19**). The voltage will continue to climb as long as the magnetic field is expanding. As the tooth approaches the magnet, the magnetic field gets smaller, causing the induced voltage to drop off. When the tooth and core align, there is no more expansion or contraction of the magnetic field (thus no movement), and the voltage drops to zero. When the tooth passes the core, the magnetic field expands, producing a negative current. The resulting pulse signal is digitalized and sent to the microprocessor.

AUTHOR'S NOTE The magnetic pulse (PM) generator operates on basic magnetic principles. Remember that a voltage is induced only when a magnetic field moves across a conductor. The pickup unit provides the magnetic field, and the rotating timing disc provides the movement of the magnetic field needed to induce voltage.

Magnetoresistive Sensor

Magnetoresistive sensors can determine the direction of rotation based on the north and south pole influences of the magnets in the reluctor ring.

Shop Manual
Chapter 9, page 428

Magnetoresistive (MR) sensors consist of the magnetoresistive sensor element, a permanent magnet, and an integrated signal conditioning circuit to make use of the **magnetoresistive effect**. This effect defines that exposing a current-carrying magnetic material to an external magnetic field will change its resistance characteristics. This results in the resistance of the sensing element being a function of the direction and intensity of an applied magnetic field. MR is the characteristic of some materials to change electrical resistance while applying a magnetic field. The change in resistance is due to the spin dependence of electron scattering. To understand spin current, consider that the movement of electrons results in a charge current. However, electrons also have the properties of mass, charge, and spin. The spin motion of the electron creates a spin current. Within a semiconductor, spin is a random process.

A thin spacer layer closely separates two magnetic layers in the magnetic multilayered structure. The first magnetic layer will allow electrons in only one spin channel to pass through with little resistance (shown in **Figure 9-20** as the spin-up channel).

The spin-down channel has resistance as it passes through the first magnetic layer. Alignment of the second magnetic layer results in the spin-up channel having low resistance as it passes through the structure. Both channels have high resistance if the second magnetic layer is misaligned, as shown in **Figure 9-21**. The MR sensor measures the difference in angle between the two magnetic layers. Small angles give a low resistance, while large angles give a higher resistance.

> **AUTHOR'S NOTE** The magnetoresistive principle provides rotational speed measurements down to zero. For this reason, they are sometimes called "zero speed sensors."

Automotive applications include the antilock brake system (ABS) where MR wheel-speed sensors provide an extremely accurate output, even at very low speeds. Also, navigational systems, lane change detection systems, and proximity sensors use MR sensors to provide more precise readings. MR sensors cannot generate a signal voltage on their own and must have an external power source. The magnetoresistive bridge changes resistance due to the relationship between the tone wheel and the magnetic field surrounding the sensor.

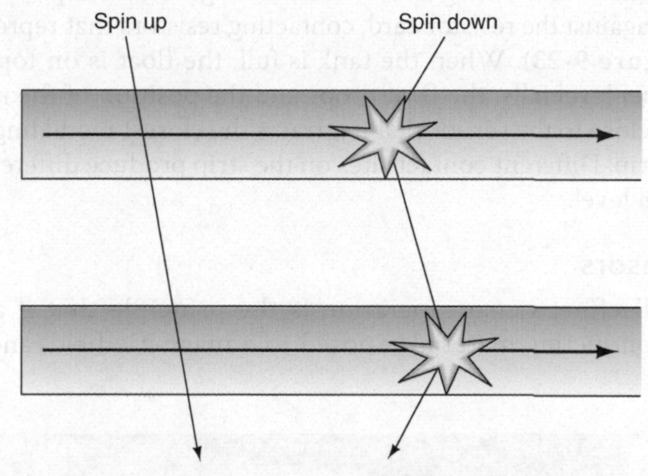

Figure 9-20 The spin-up channel is illustrated as allowing electrons to flow through the magnetic layer with little resistance (i.e., if the second magnetic layer is aligned).

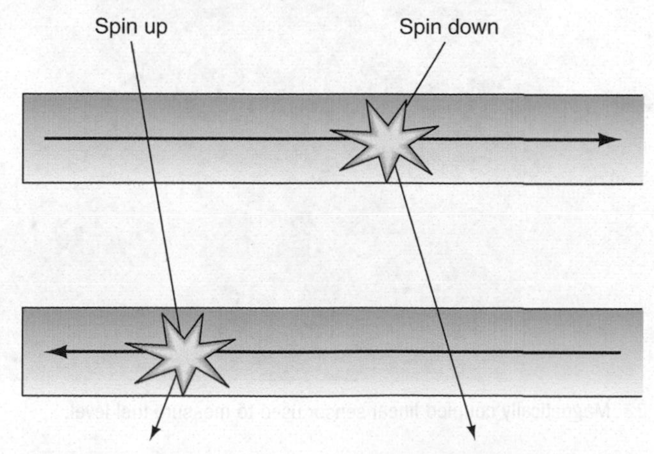

Figure 9-21 If the second magnetic layer is misaligned, then both the spin-up and spin-down channels have high resistance.

A circuit provided by the computer provides the integrated circuit (IC) in the sensor with its 12-volt power. The IC supplies a constant 7 mA power to the computer. The relationship of the tooth on the tone wheel to the permanent magnet in the sensor signals the IC to enable a second 7-mA power supply. The output of the sensor, sent to the computer, is a DC voltage signal with changing voltage and current levels. When a valley of the tone wheel aligns with the sensor, the voltage signal is approximately 0.8 volt and a constant 7 mA current is sent to the computer. As the tone wheel rotates, the tooth shifts the magnetic field and the IC enables a second 7-mA current source. The computer senses a voltage signal of approximately 1.6 volts and 14 mA. The computer measures the amperage of the digital signal and interrupts the signal as component speed.

The alternating magnetic poles allow the computer to determine the direction of rotation.

Magnetically Coupled Linear Sensors

Linear sensors are used for such functions as fuel level sending units. The most common type of fuel level sensor is a rheostat style with a wire-wound resistor and a movable wiper. The wiper is in constant contact with the winding and may eventually rub through the wire. Many manufacturers are now using **magnetically coupled linear sensors** that are not prone to this type of wear (**Figure 9-22**).

Magnetically coupled linear sensors used for fuel level sensing have a magnet attached to the end of the float arm. Also, a resistor card and a magnetically sensitive comb are located next to the magnet. When the magnetic field passes the comb, the fingers are pulled against the resistor card, contacting resistors that represent the various levels of fuel (**Figure 9-23**). When the tank is full, the float is on top along with the magnet. As the fuel level falls, the float drops and the position of the magnet changes. The magnet is so close to the sensor that it attracts the closest metal fingers. The fingers contact a metal strip. Different contact sites on the strip produce different resistances to determine the fuel level.

Hall-Effect Sensors

Shop Manual
Chapter 9, page 431

The basis of Hall-effect sensor operation is the principle that if a current flows through a thin conducting material exposed to a magnetic field, another voltage is

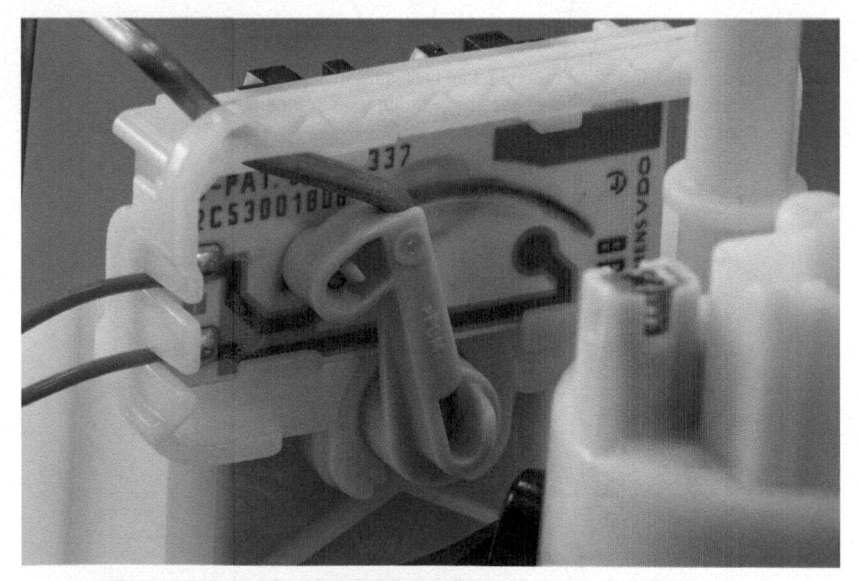

Figure 9-22 Magnetically coupled linear sensor used to measure fuel level.

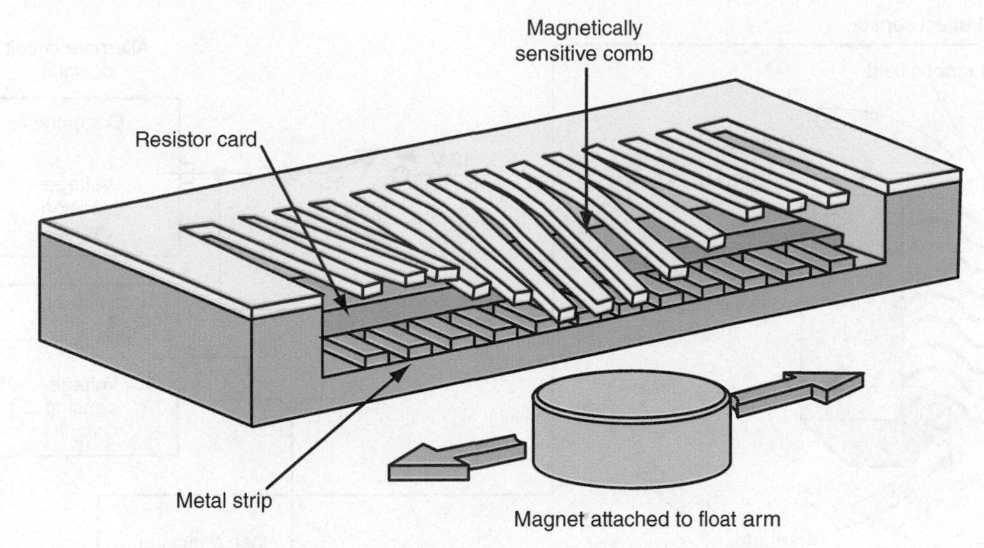

Figure 9-23 Comb design magnetic sensor pulls the fingers against the resistor card.

produced (**Figure 9-24**). The sensor contains a permanent magnet, a thin semiconductor layer made of gallium arsenate crystal (Hall layer), and a **shutter wheel** (**Figure 9-25**). The Hall layer has a negative and a positive terminal connected to it. Two additional terminals located on either side of the Hall layer provide the output circuit. The shutter wheel consists of a series of alternating windows and vanes. It creates a magnetic shunt that changes the strength of the magnetic field from the permanent magnet.

The permanent magnet is located directly across from the Hall layer so that its lines of flux will bisect at right angles to the current flow. The permanent magnet is mounted so that a small air gap is between it and the Hall layer. The crystal of the Hall layer has a steady current applied to it and produces a signal voltage perpendicular to the direction of current flow and magnetic flux. The signal voltage produced results from the magnetic field's effect on the electrons. When the magnetic field bisects the supply current flow, the electrons are deflected toward the Hall layer negative terminal (**Figure 9-26**). This result produces a weak voltage potential in the Hall sensor.

As the shutter wheel (attached to a rotational component) rotates, the shutters (vanes) pass in this air gap. When a shutter vane enters the gap, it intercepts the magnetic field and shields the Hall layer from its lines of force. The electrons in the supply current are no longer disrupted and return to a normal state. This results in low voltage potential in the signal circuit of the Hall sensor.

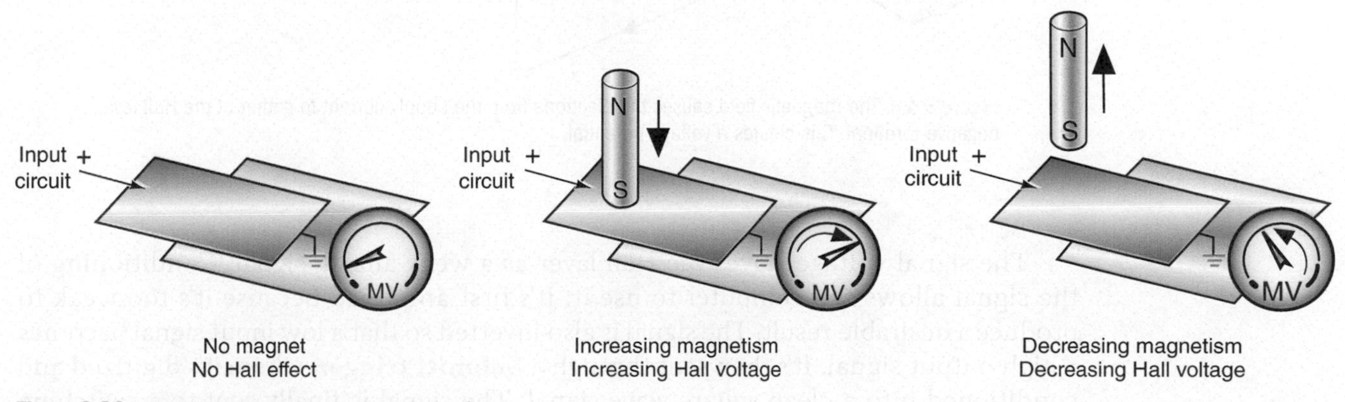

Figure 9-24 Hall-effect principles of voltage induction.

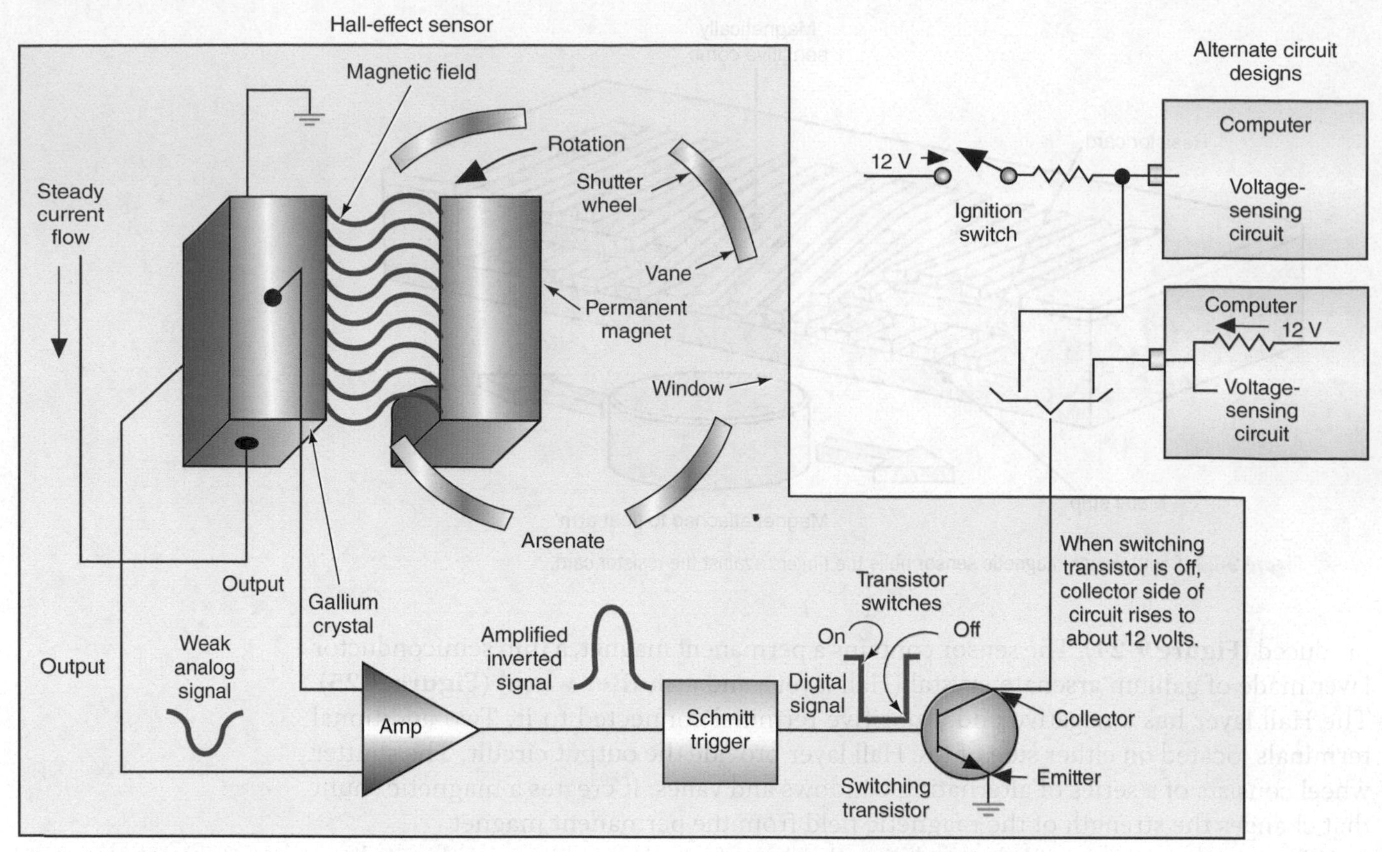

Figure 9-25 Typical circuit of a Hall-effect sensor.

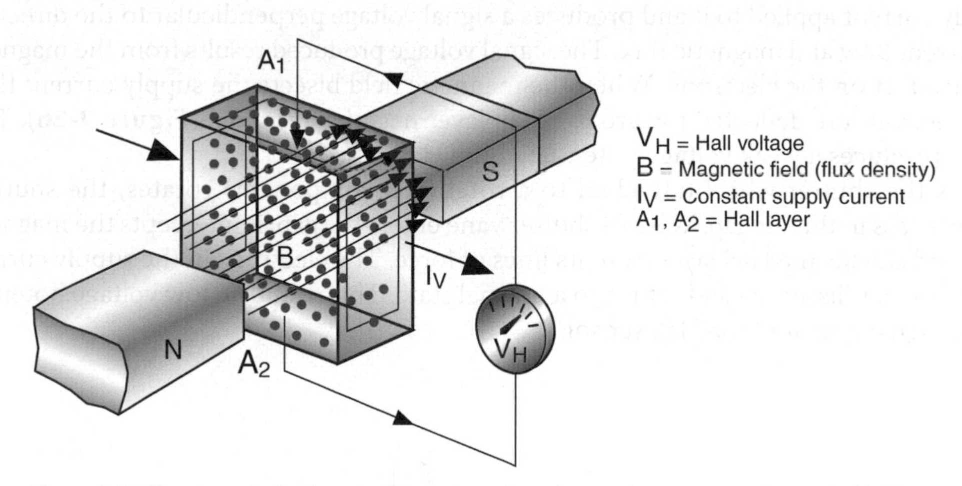

Figure 9-26 The magnetic field causes the electrons from the supply current to gather at the Hall layer negative terminal. This creates a voltage potential.

The signal voltage leaves the Hall layer as a weak analog signal. Conditioning of the signal allows the computer to use it. It's first amplified because it's too weak to produce a desirable result. The signal is also inverted so that a low input signal becomes a high output signal. It's then sent through a **Schmitt trigger,** where it's digitized and conditioned into a clean square wave signal. The signal is finally sent to a switching transistor. The computer senses the turning on and off of the switching transistor to determine the frequency of the signals and calculates the speed.

The Hall effect discussed describes its usage as a switch to provide a digital signal. It can also be designed as an analog (or linear) sensor that produces an output voltage that is proportional to the applied magnetic field. This makes them useful for determining the position of a component instead of just rotation. For example, this type of sensor can monitor fuel level or track seat positions in memory seat systems.

A Hall-effect sensor can measure fuel level by attaching a magnet to the float assembly (**Figure 9-27**). As the float moves up and down with the fuel level, the gap between the magnet and the Hall element changes. The change in gap changes the Hall effect, and thus the output voltage.

Typical Hall-effect sensors use three wires; however, linear Hall-effect sensors can use two-wire circuits (**Figure 9-28**). This is common on systems that use a DC motor drive. The reference voltage to the sensor passes through a pull-up resistor. Typically, this reference voltage will be 12 volts. The reference voltage is applied when the motor is operating and remains for a short time after the motor is turned off.

Internal to the motor assembly is a typical three-terminal Hall sensor. The reference voltage is supplied to terminal 1 of the Hall sensor. A pull-up resistor also connects the reference voltage to terminal 3 of the Hall sensor. This becomes the signal circuit. The two pull-up resistors will be of equal value. Terminal 2 of the Hall sensor is connected to the sensor return circuit. A magnet is attached to the motor armature to provide a changing magnetic field once per motor revolution.

When the Hall sensor is off, the voltage supplied to the Hall sensor will be close to that of the source voltage. Since this is an open circuit condition in the Hall sensor at terminal 3, the voltage drop over the signal circuit pull-up resistor will be 0.

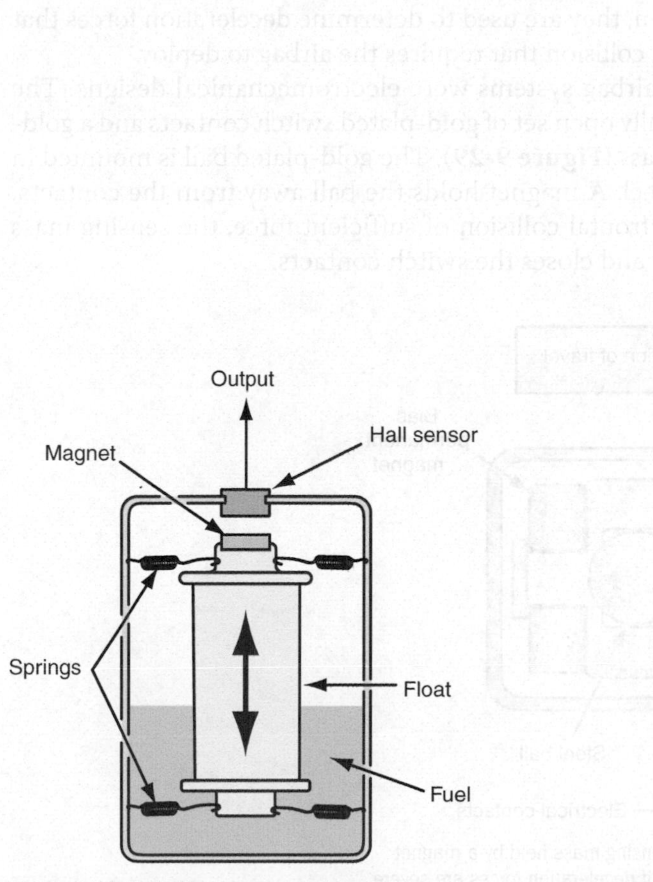

Figure 9-27 Hall-effect sensor used for fuel level indication.

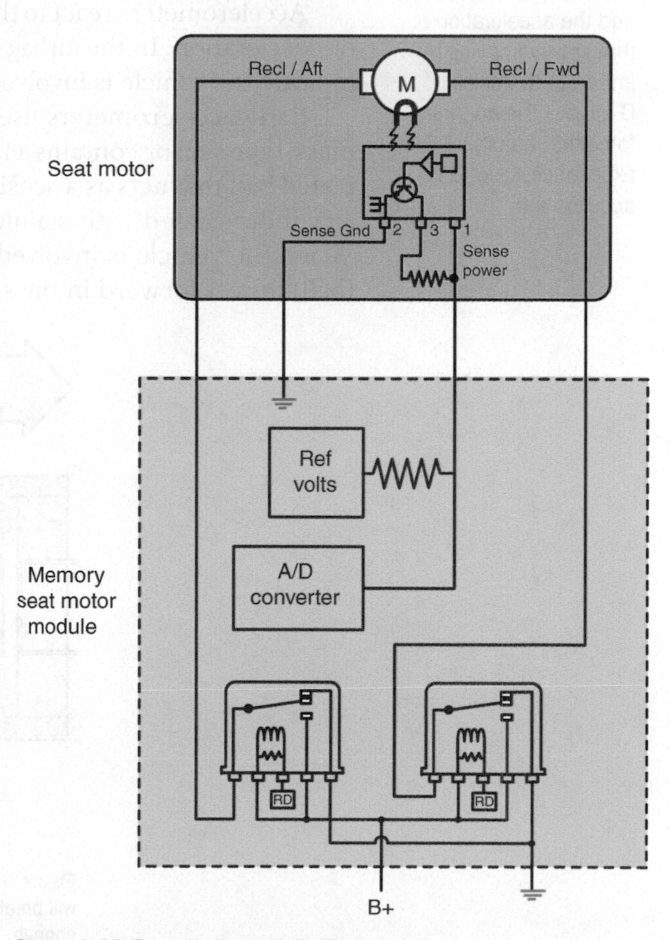

Figure 9-28 Two-wire linear Hall-effect sensor.

A BIT OF HISTORY

The American physicist, Edwin Herbert Hall (1855–1938), discovered the principle of the Hall-effect sensor in 1879.

When the motor rotates, and the influence of the magnetic field turns on the Hall sensor, the signal terminal 3 is connected to ground within the sensor. This pulls the signal voltage low and results in the formation of a series circuit from the reference supply to terminal 3. Since each of the pull-up resistors is equal, the voltage drop is split between the two. Approximately half the voltage will be dropped across the pull-up resistor in the computer and the other half over the pull-up resistor in the motor assembly. The Hall-effect sensor will remain powered since the reference voltage to terminal 1 is connected between the two resistors, and the 6 volts on the circuit is sufficient to operate the sensor.

Accelerometers

Accelerometers are sensors designed to measure the rate of acceleration or deceleration. Common sensors include mass-type, roller-type, and solid-state accelerometers. The first extensive use of the accelerometer was in the airbag system. The use of accelerometers has expanded greatly in today's vehicles. Vehicle stability, roll-over mitigation, hill-hold control, electronic steering, and navigational systems can use these sensors. Accelerometers may perform specific functions other than forward acceleration and deceleration forces. For example, they can operate as a gyro to determine direction change and rotation.

G force describes the measurement of the net effect of the acceleration that an object experiences and the acceleration that gravity is trying to impart to it. Basically, G force is the apparent force that an object experiences due to acceleration.

Accelerometers react to the amount of **G force** associated with the rate of acceleration or deceleration. In the airbag system, they are used to determine deceleration forces that indicate the vehicle is involved in a collision that requires the airbag to deploy.

Early accelerometers used in airbag systems were electromechanical designs. The mass-type sensor contains a normally open set of gold-plated switch contacts and a gold-plated ball that acts as a sensing mass (**Figure 9-29**). The gold-plated ball is mounted in a cylinder coated with stainless steel. A magnet holds the ball away from the contacts. When the vehicle is involved in a frontal collision of sufficient force, the sensing mass (ball) moves forward in the sensor and closes the switch contacts.

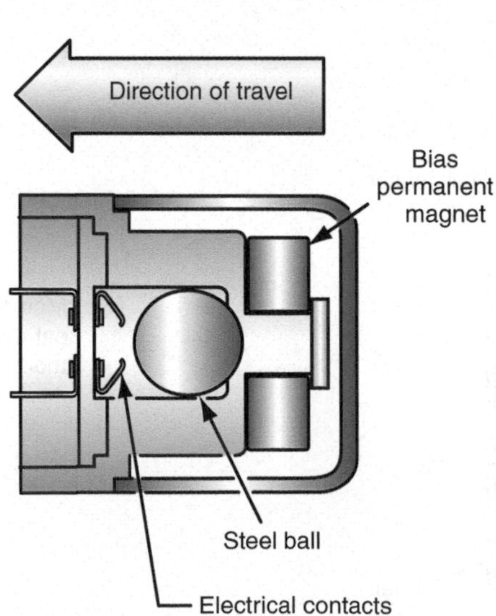

Figure 9-29 Sensing mass held by a magnet will break loose if deceleration forces are severe enough.

In many airbag systems, solid-state accelerometers sense deceleration forces. The accelerometer contains a piezoelectric element that is distorted during a high G force condition and generates an analog voltage proportional to the G force. The analog voltage from the piezoelectric element is sent to a collision-judging circuit in the airbag computer. The computer deploys the airbag if the collision impact is great enough.

Accelerometers can also be piezoresistive sensors using a silicon mass suspended from four deflection beams. The deflection beams are the strain sensing elements. The four strain elements are in a Wheatstone bridge circuit. The beam's strain elements generate a signal proportional to the G forces. An internal chip interprets the resistance changes over the bridge and then communicates the status to the control module using a frequency modulated digital pulse.

Photocells

AUTHOR'S NOTE Diagnostics of the photocell is covered in Chapter 11 when these components are introduced as an input for automatic lighting systems.

Shop Manual
Chapter 11, page 533

A **photocell** describes any component that is capable of measuring or determining light. It's used where settings need to be adjusted to ambient light conditions, such as automatic temperature control, automatic headlight operation, night vision assistance, and other convenience and safety systems. Uses also include automatically changing display intensity of displays, backlighting of instrument panels and button controls, and adjusting daytime running lights to full power when ambient light levels are low.

Several different types of semiconductor devices can measure light intensity. The following are some of the most common.

Photodiode

A photodiode is a light-receiving device containing a semiconductor PN junction, enclosed in a case with a convex lens (**Figure 9-30**). The lens allows ambient light to enter the case and strike the photodiode.

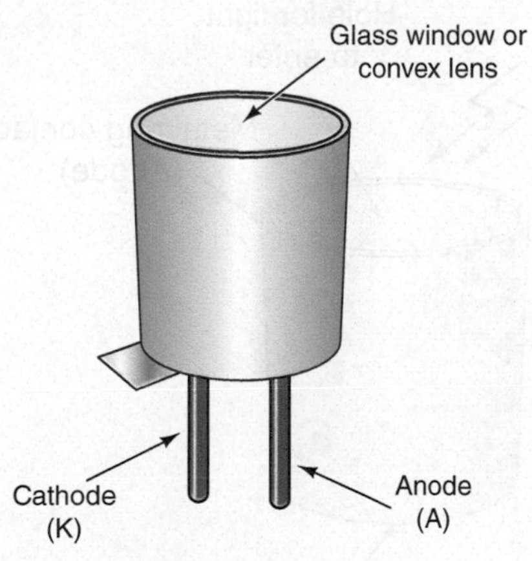

Figure 9-30 Photodiode

Photodiodes can operate in two modes: **photovoltaic mode** or **photoconductive mode**. In the photovoltaic mode, the photodiode generates a voltage in response to light. When light strikes the photodiode, it creates electron-hole pairs. The negatively charged electrons generated in the depletion region are attracted to the positively charged ions in the N-type material. The holes are attracted to the negatively charged ions in the P-type material. The result is a separation of charges and the development of a small voltage drop of about 450 mV across the diode. Connecting a load resistor across the voltage source will result in a small current flow from the cathode to the anode.

AUTHOR'S NOTE Operating the photodiode in photovoltaic mode is the principle of the solar cell.

In the photoconductive mode, the conductance of the diode changes when light is applied. In this mode, the photodiode is reverse biased when no light is applied. Being reverse biased results in a very wide depletion region and a high resistance across the diode. In this state, there will be only a small reverse current through the diode.

Applying light to the photodiode generates electron-hole pairs. The electrons are attracted to the positive bias voltage, and the holes are attracted to the negative bias voltage. This movement of electrons and holes causes an increase in the reverse current flow. When light is applied, the resistance of the photodiode is very low and decreases as the intensity of light increases. As the resistance continues to decrease, current flow increases.

Photoresistor

A **photoresistor** is a passive light-detecting device composed of a semiconductor material that changes resistance when its surface is exposed to light (**Figure 9-31**). The semiconductor material is shaped into a zigzag strip, and the ends are attached to the external terminals. Common materials are either cadmium sulfide (CdS) or cadmium selenide (CdSe). A transparent cover allows the ambient light to pass through. The semiconductor material is light sensitive; thus, light energy creates free electrons

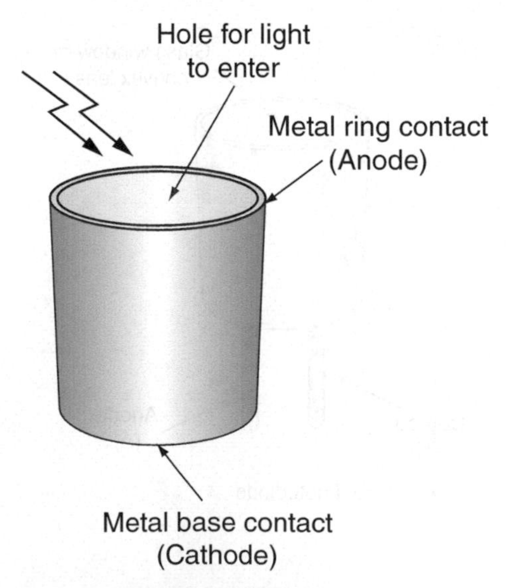

Figure 9-31 Photoresistor construction.

instead of heat energy as in conventional semiconductor devices. The creation of the free electrons causes the resistance of the semiconductor material to decrease. As the applied light intensity increases, so does the number of free electrons. Thus, as light increases the resistance decreases. This, in turn, means the sense voltage drops as light intensity increases.

Phototransistor

A phototransistor is a light-detecting transistor that uses the application of light to generate carriers to supply the base leg current. The intensity of the light controls the collector current of the transistor. Unlike a normal transistor, the phototransistor has only two terminals (**Figure 9-32**). Like a normal transistor, the output current is amplified.

Photoelectric Sensors

Photoelectric sensors use light-sensitive elements to detect rotation (**Figure 9-33**). Photoelectric sensors consist of an emitter (light source) and a receiver. There are three basic forms of photoelectric sensors:

1. *Direct reflection*—has the emitter and receiver in the same module and uses the light reflected directly off the monitored object for detection.
2. *Reflection with reflector*—also has the emitter and receiver in the same module but requires a reflector. The motion of an object is detected by its interruptions of the light beam between the sensor and the reflector.
3. *Thru beam*—separates the emitter and receiver and detects the motion of an object when it interrupts the light beam between the emitter and the receiver.

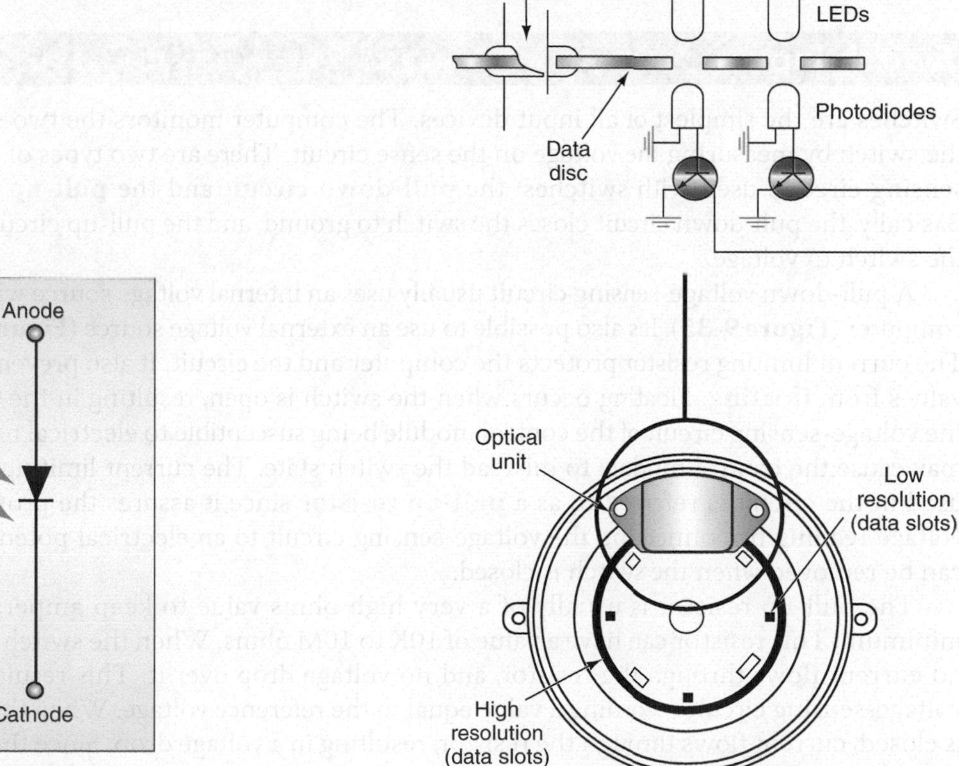

Figure 9-33 A photoelectric-type position sensor.

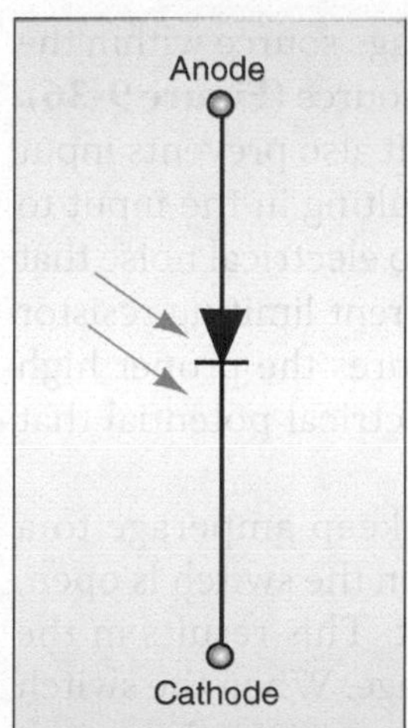

Figure 9-32 Symbol for phototransistor.

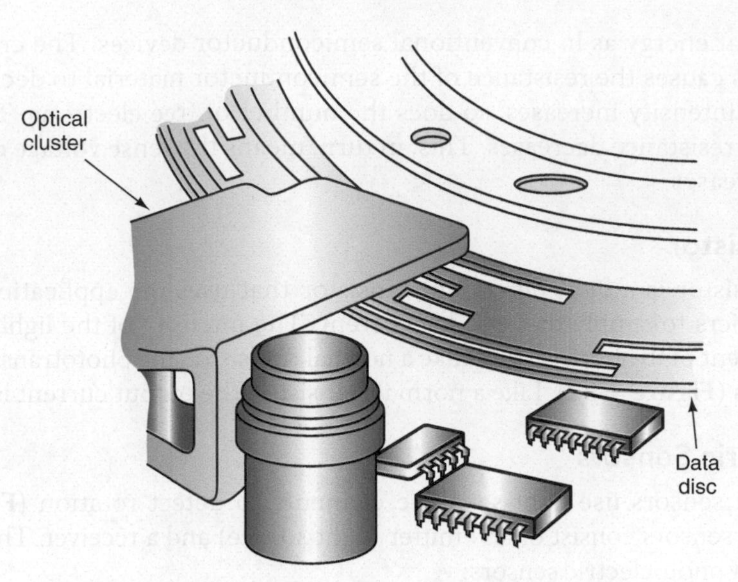

Figure 9-34 The data disc interrupts the light beams from the optical cluster to detect motion.

A thru beam–type photoelectric sensor has a series of LEDs situated across from the same number of photocells. If used as a steering wheel sensor, a code disc (**Figure 9-34**) is attached to the steering shaft and interrupts the infrared light beams from the optics cluster. The optics cluster has three rows of four light detectors that provide a bit code that determines steering wheel position and rotation based on the order of light beam interruption. Based on this input from the photoelectric sensor, the ECU can determine the direction and rate the steering wheel is being turned.

Switch Inputs

Shop Manual
Chapter 9, page 407

Switches are the simplest of all input devices. The computer monitors the two states of the switch by measuring the voltage on the sense circuit. There are two types of voltage-sensing circuits used with switches: the **pull-down circuit** and the **pull-up circuit**. Basically, the pull-down circuit closes the switch to ground, and the pull-up circuit closes the switch to voltage.

A pull-down voltage-sensing circuit usually uses an internal voltage source within the computer (**Figure 9-35**). It's also possible to use an external voltage source (**Figure 9-36**). The current limiting resistor protects the computer and the circuit. It also prevents input values from **floating**. Floating occurs when the switch is open, resulting in the input to the voltage-sensing circuit of the control module being susceptible to electrical noise that may cause the control module to misread the switch state. The current limiting resistor used in the circuit is referred to as a **pull-up resistor** since it assures the proper high voltage reading by connecting the voltage-sensing circuit to an electrical potential that can be removed when the switch is closed.

The pull-up resistor is usually of a very high ohms value to keep amperage to a minimum. This resistor can have a value of 10K to 10M ohms. When the switch is open, no current flows through the resistor, and no voltage drop over it. This results in the voltage-sensing circuit recording a value equal to the reference voltage. When the switch is closed, current flows through the resistor, resulting in a voltage drop. Since the switch should provide a clean contact to ground, the voltage-sensing circuit should read a value close to 0 volt.

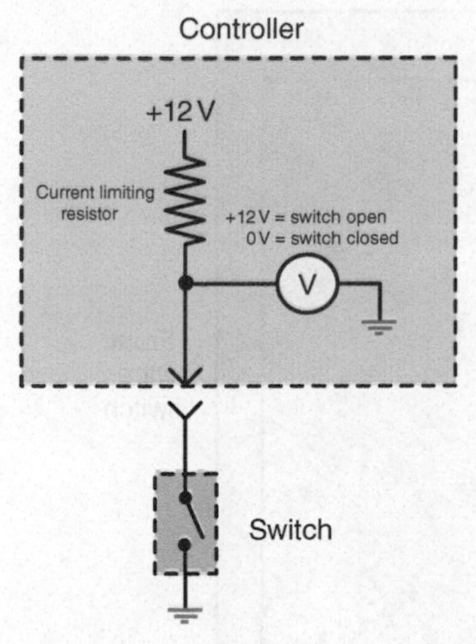

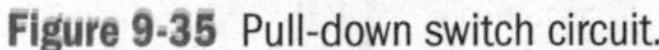

Figure 9-35 Pull-down switch circuit.

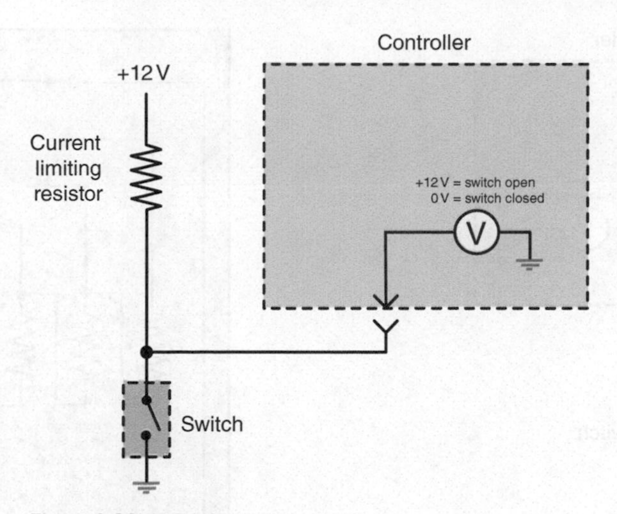

Figure 9-36 Pull-down switch circuit with an external voltage source.

The pull-up circuit will have a reference voltage through the switch (**Figure 9-37**). Usually, the battery or the ignition switch provides the reference voltage. The current limiting resistor performs the same function in this circuit as in the pull-down circuit. This resistor is referred to as a **pull-down resistor** since it assures a proper low-voltage reading by preventing float when the switch is open. With the switch in the open position, the voltage-sensing circuit will read 0 volt. With the switch closed, the sense circuit should read close to the reference voltage.

Both of these circuits have limited ability to determine circuit faults. Since there are only two states for the switch, there are two voltage values that the computer expects to see. An open or short to ground will not produce an unexpected voltage value but will result in improper system operation. However, the computer may be capable of determining a functionality problem with the input circuit if the seen voltage is implausible for the conditions. For example, if the switch is an operator-activated switch that requests A/C operation and the voltage indicates that the switch may be stuck, the computer can set a stuck switch diagnostic trouble code (DTC) and ignore the input.

To provide continuity diagnostics, the circuit may have a diagnostic resistor wired parallel to the switch (**Figure 9-38**). The computer will be able to recognize three different voltage values. In the example, the current limiting resistor has a value of 10K ohms, while the diagnostic resistor has a value of 2K ohms. With the switch in the open state, the voltage-sensing circuit would read 2 volts. With the switch closed, the voltage reading will be close to 0 volt. A reading of 12 volts would indicate an open in the circuit.

Another typical switch is the resistive multiplex switch. This switch provides multiple inputs from a single switch using one circuit (**Figure 9-39**). The control module sends a

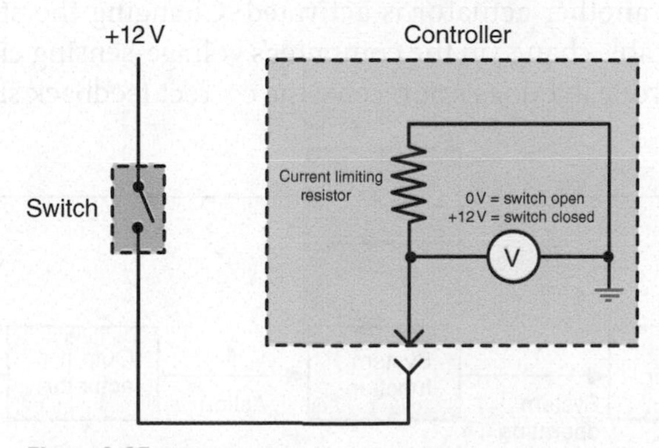

Figure 9-37 Pull-up switch circuit.

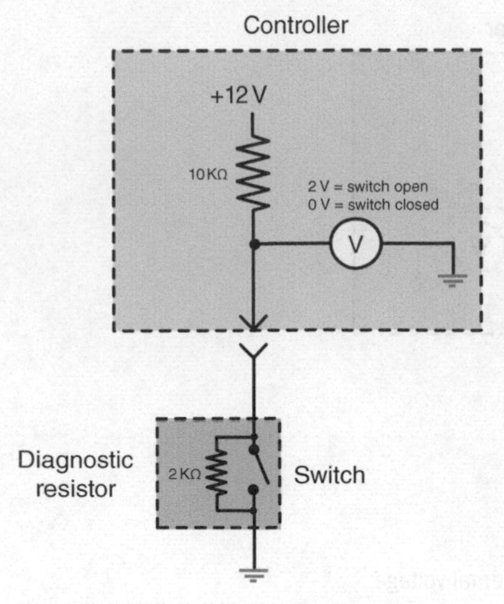

Figure 9-38 Pull-down circuit with a diagnostic resistor.

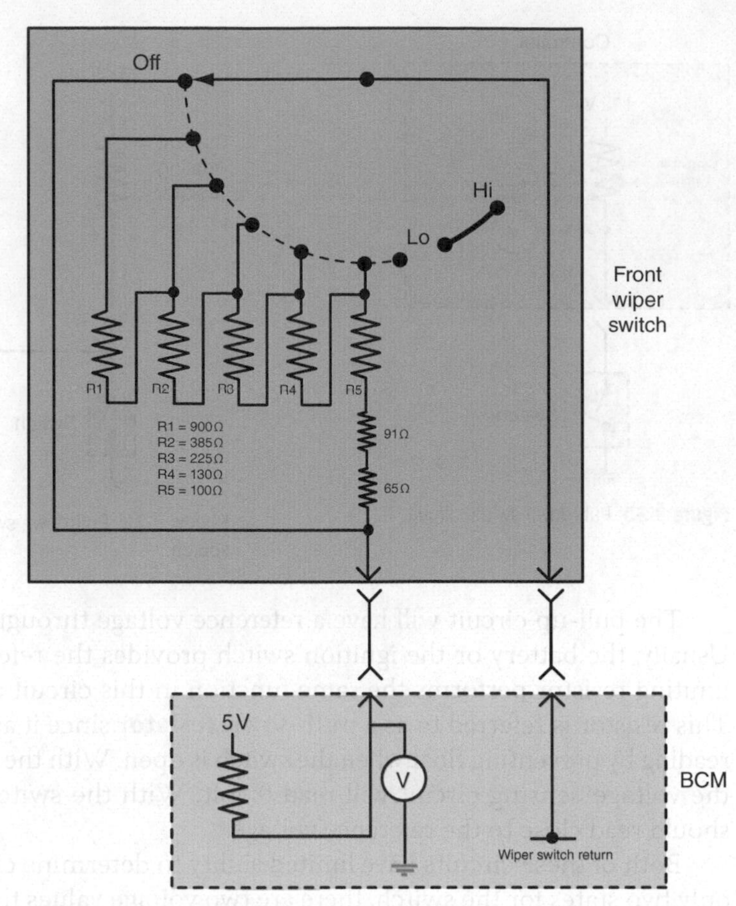

Figure 9-39 Resistive multiplex switch.

signal voltage to the switch through a fixed resistor. Each switch position has a unique resistance value placed in series with the resistor in the control module. Selecting different switch positions results in different amounts of voltage dropped over the fixed resistor in the control module, changing the sensed voltage level. Based on the sensed voltage value, the control module interprets the operation the driver requests.

Feedback Signals

If the computer sends a command signal to open a blend door in an automatic climate control system, a **feedback** signal from the actuator may inform the computer that the task was performed. The feedback signal will confirm both the door position and actuator operation (**Figure 9-40**). Another form of feedback involves the computer monitoring voltage as a switch, relay, or another actuator is activated. Changing the states of the actuator will result in a predictable change in the computer's voltage-sensing circuit. The computer may set a diagnostic code if it does not receive the correct feedback signal.

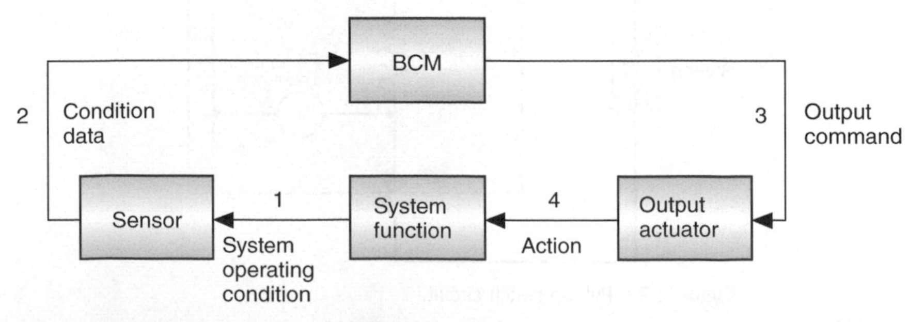

Figure 9-40 Principle of feedback signals.

Summary

- Inputs provide the computer with system operation information or driver requests.
- Switches can be used as an input for any operation that only requires a yes/no, or on/off condition.
- Sensors convert some measurement of vehicle operation into an electrical signal.
- A thermistor is a solid-state variable resistor made from a semiconductor material that changes resistance in relation to temperature changes.
- Negative temperature coefficient (NTC) thermistors reduce their resistance as the temperature increases.
- Positive temperature coefficient (PTC) thermistors increase their resistance as the temperature increases.
- Some temperature-sensing circuits are dual-range circuits to provide more accurate temperature measurements.
- Pressure switches usually use a diaphragm that works against a calibrated spring or another form of tension. A switch is closed when pressure is applied to the diaphragm that is of a sufficient value to overcome the spring tension.
- A strain gauge sensor determines the amount of applied pressure by measuring the strain a material experiences when subjected to the pressure.
- Piezoresistive devices change resistance values as the pressure applied to the crystal changes.
- The Wheatstone bridge is a series-parallel arrangement of resistors between an input terminal and ground.
- Piezoelectric devices are voltage generators with a resistor connected in series that measure fluid and air pressures.
- The capacitance discharge sensor uses a variable capacitor constructed of two flexible alumina plates separated by an insulating washer. As the distance between the electrodes changes, so does the capacity of the capacitor.
- Capacitance pressure transducers may measure pressure. The change in capacitance results from the movement of a diaphragm element.
- A potentiometer is a variable resistor that usually consists of a wire-wound resistor with a movable center wiper.
- Magnetic pulse generators use the principle of magnetic induction to produce a voltage signal and are commonly used to send data concerning the speed of the monitored component to the computer.
- Magnetoresistive (MR) sensors consist of the magnetoresistive sensor element, a permanent magnet, and an integrated signal conditioning circuit to change resistance due to the relationship of the tone wheel and the magnetic field surrounding the sensor.
- Magnetically coupled linear sensors use a movable magnet attached to the measured element, a resistor card, and a magnetically sensitive comb. When the magnetic field passes the comb, the fingers are pulled against the resistor card contacting resistors that represent the various positions of the measured element.
- Hall-effect sensors operate on the principle that if a current flows through a thin conducting material exposed to a magnetic field, another voltage is produced.
- Accelerometers are sensors designed to measure the rate of acceleration or deceleration. Accelerometers react to the amount of G force associated with the rate of acceleration or deceleration.
- The piezoelectric accelerometer generates an analog voltage proportional to a G force.
- A photocell describes any component that is capable of measuring or determining light.
- A photodiode is a light-receiving device containing a semiconductor PN junction, enclosed in a case with a convex lens that allows ambient light to enter the case and strike the photodiode.
- Photodiodes can operate in two modes: photovoltaic mode or photoconductive mode.
- A photoresistor is a passive light-detecting device composed of a semiconductor material that changes resistance when its surface is exposed to light.
- A phototransistor is a light-detecting transistor that uses the application of light to generate carriers to supply the base leg current.
- Photoelectric sensors use light-sensitive elements to detect rotation using an emitter (light source) and a receiver.
- The pull-down circuit will close the switch to ground.
- The pull-up circuit will close the switch to voltage.
- The pull-up resistor assures proper high voltage reading by connecting the voltage-sensing circuit to an electrical potential that can be removed when the switch is closed.
- The pull-down resistor assures a proper low-voltage reading by preventing float when the switch is open.
- The resistive multiplex switch provides multiple inputs from a single switch using one circuit.
- Feedback signals confirm the position and operation of an actuator.

Review Questions

Short-Answer Essays

1. What are the functions of input devices?
2. Explain the purpose of the thermistor and how it's used in a circuit.
3. Describe the operation and purpose of the Wheatstone bridge.
4. Explain the operation and purpose of piezoelectric devices.
5. What is the difference between NTC and PTC thermistors?
6. How does the Hall-effect sensor generate a voltage signal?
7. What is the purpose of the multiplex switch?
8. What is meant by feedback as it relates to computer control?
9. Describe the operation of the pull-down sense circuit.
10. Describe the operation of the pull-up sense circuit.

Fill in the Blanks

1. The piezoelectric accelerometer generates an analog voltage proportional to a _______________.
2. The _______________ resistor assures the proper high-voltage reading by connecting the voltage-sensing circuit to an electrical potential that can be removed when the switch is closed.
3. The resistive multiplex switch is used to provide multiple inputs from a single switch using _______________ circuit.
4. Magnetoresistive (MR) sensors consist of the magnetoresistive sensor element, a permanent magnet, and an integrated signal conditioning circuit to change _______________ due to the relationship of the tone wheel and magnetic field surrounding the sensor.
5. The capacitance discharge sensor changes its capacitance by the difference in _______________ between the electrodes.
6. _______________ convert some measurement of vehicle operation into an electrical signal.
7. Negative temperature coefficient (NTC) thermistors _______________ their resistance as the temperature increases.
8. _______________ sensors operate on the principle that if a current is allowed to flow through a thin conducting material exposed to a magnetic field, another voltage is produced.
9. Magnetic pulse generators use the principle of _______________ to produce a voltage signal.
10. _______________ means that data concerning the effects of the computer's commands are fed back to the computer as an input signal.

Multiple Choice

1. All of the following can measure movement or position, EXCEPT:
   A. Potentiometer
   B. Magnetic pulse generator
   C. Piezoelectric device
   D. Hall-effect sensor
2. *Technician A* says the piezoresistive sensor changes resistance as a function of temperature. *Technician B* says the piezoresistive sensor outputs current based on the pressure it's exposed to. Who is correct?
   A. A only
   B. B only
   C. Both A and B
   D. Neither A nor B
3. The Wheatstone bridge is:
   A. A pressure sensing device that uses a variable capacitor.
   B. A pressure sensing device that uses varying resistances in a series-parallel circuit design.
   C. Used to measure motion by use of magnetic inductance.
   D. None of these choices.
4. The piezoelectric sensor operates by:
   A. Altering the resistance values of the bridge circuit located on a ceramic disc.
   B. Dropping voltage over a fixed resistor as pressure is applied to the switching transistor.
   C. Generating a voltage within a thin ceramic disc voltage generator attached to a diaphragm, which stresses the crystals in the disc.
   D. None of these choices.

5. Capacitance discharge sensors are being discussed.
   Technician A says the size of the electrodes alters as the sensing element is exposed to different pressures.
   Technician B says the distance between the electrodes alters as the sensing element is exposed to different pressures.
   Who is correct?
   A. A only
   B. B only
   C. Both A and B
   D. Neither A nor B

6. *Technician A* says a potentiometer is a voltage divider circuit used to measure the movement of a component.
   Technician B says a magnetoresistive sensor alters current flow through the sense circuit when influenced by the magnetic field.
   Who is correct?
   A. A only
   B. B only
   C. Both A and B
   D. Neither A nor B

7. *Technician A* says negative temperature coefficient thermistors reduce their resistance as the temperature decreases.
   Technician B says positive temperature coefficient thermistors increase their resistance as the temperature increases.
   Who is *correct*?
   A. A only
   B. B only
   C. Both A and B
   D. Neither A nor B

8. *Technician A* says magnetic pulse generators send voltage signals to the computer concerning the speed of the monitored component.
   Technician B says an on/off switch sends a digital signal to the computer.
   Who is correct?
   A. A only
   B. B only
   C. Both A and B
   D. Neither A nor B

9. Speed *sensors* are being discussed.
   Technician A says the timing disc is stationary, and the pickup coil rotates in front of it.
   Technician B says the number of pulses produced per mile increases as rotational speed increases.
   Who is *correct*?
   A. A only
   B. B only
   C. Both A and B
   D. Neither A nor B

10. *Technician A* says a Hall-effect sensor uses a steady supply current to generate a signal.
    Technician B says a Hall-effect sensor consists of a permanent magnet wound with a wire coil.
    Who is correct?
    A. A only
    B. B only
    C. Both A and B
    D. Neither A nor B

CHAPTER 10
VEHICLE COMMUNICATION NETWORKS

Upon completion and review of this chapter, you should be able to:

- Explain the principle of multiplexing.
- Describe the different OBD II multiplexing communication protocols.
- Describe the different classes of communications.
- Explain the operation of a class A multiplexing system.
- Explain the operation of a class B multiplexing system.
- Explain the operation of the Controller Area Network (CAN) bus system.
- Detail the purpose and operation of different supplemental data bus networks.
- Explain the operation of the Local Interconnect Network (LIN) data bus system.
- Describe the operation of the Media-Oriented System Transport (MOST) data bus using fiber optics.
- Explain the operation of wireless networks using Bluetooth technology.
- Describe the purpose of the Security Gateway Module.

Terms to Know

Asynchronous
Baud rate
Bluetooth
Bus (–)
Bus (+)
Central gateway (CGW)
Chrysler Collision Detection (CCD)
Controller Area Network (CAN)
Fiber optics
ISO 14230-4
ISO 9141-2
ISO-K
J1850
K-line
Linear network
L-line
Local Interconnect Network (LIN)
Master module
Media-Oriented System Transport (MOST)
Multiplexing (MUX)
Network architecture
Node
Programmable Communication Interface (PCI)
Protocol
Ring network
Security Gateway
Slave modules
Smart sensors
Stub network
Supplemental bus networks
Termination resistors
Total reflection
Wireless networks

Introduction

In the past, adding an accessory to the vehicle that required sensor input information involved additional sensors or splicing the new accessory into an existing sensor circuit. Either way, the cost of production was increased due to added components and wiring. For example, some early vehicles had three separate engine coolant temperature sensors. The powertrain control module (PCM) used one sensor for fuel and ignition strategies, the cooling fan module used the other to operate the radiator fans at the correct speed based on temperature, and the instrument cluster used the third for temperature gauge operation.

Today vehicle manufacturers use **multiplexing (MUX)** systems to enable different control modules to share information. Multiplexing provides the ability to use a single circuit to distribute and share data between several control modules throughout the vehicle. Transmitting the data through a single circuit eliminates bulky wiring harnesses. A MUX wiring system uses bus data links that connect each module and allow for the transporting of data from one module to another. Each module can transmit and receive digital codes over the bus data links. Each computer connected to the data bus is called a **node**. The sensor's signal can be sent to any module and then shared with other modules. Before multiplexing, if several controllers needed information from the same sensing device, a wire from each controller needed to be connected in parallel to that sensor. If the sensor signal is analog, the controllers need an analog to digital (A/D) converter to read the sensor information. Multiplexing eliminates the need for separate conductors from the sensor to each module and reduces the number of drivers in the controllers.

Additionally, multiplexing systems have increased system reliability by reducing circuits and improving efficiency by less power demands and more accurate control. Multiplexing has increased diagnostic capabilities dramatically by allowing the technician to access diagnostic trouble codes (DTCs), see live data streams, and use bidirectional controls to actuate different components. Government regulations have dictated the use of certain multiplexing systems on the vehicle. Other benefits of the multiplexing system include improved emissions, safety, and fuel economy.

Communication messages (both internally and with other controllers) use binary code. Each digital message is preceded by an identification code that establishes its priority. If two modules attempt to send a message simultaneously, the message with the higher priority code is transmitted first. A chip prevents the digital codes from overlapping by allowing the transmission of only one code at a time.

The major difference between a multiplexed and a nonmultiplexed system is how data is gathered and processed. Nonmultiplexed systems send an analog signal from a sensor through a dedicated wire to the computer or computers. The computer converts the signal from an analog signal to a digital signal. Because each sensor requires its own dedicated signal wire, the number of wires required to feed data from all of the sensors and transmit control signals to all of the output devices is great.

In a MUX system, the signal is sent to a computer, where it's converted from analog to digital if needed. Since the computer or control module of any system can process only one input at a time, it calls for input signals as it needs them. Timing data transmission from the sensors to the control module allows using a single data circuit. Between each data transmission to the control module, the sensor is electronically disconnected from the control module.

Multiplexing Communication Protocols

A **protocol** is a language computers use to communicate with one another over the data bus. Protocols may differ in baud rate and the method of delivery. For example, some protocols use pulse-width modulation while others use variable pulse width. In addition, there may be differences in the voltage levels that equal a 1 or a 0 bit.

The Society of Automotive Engineers (SAE) has defined different classes of protocols according to their **baud rate** (speed of communication):

- Class A—An **asynchronous** low-speed protocol that has a baud rate of up to 10 kb/s (10,000 bits per second). *Asynchronous protocol* means communication between nodes only occurs when needed instead of continuously.
- Class B—Medium-speed protocol has a baud rate between 10 kb/s and 125 kb/s.

Multiplexing Systems

The following examples show how data bus messages are transmitted based on the different classes and protocols. Although protocols are in place, manufacturers have some freedom to design the system they wish to use. The following examples will explain the common methods that are employed.

Class A Data Bus Network

The **Chrysler Collision Detection (CCD)** system is also referred to as (C squared D).

Shop Manual
Chapter 10, page 474

One of the earliest multiplexing systems was developed by Chrysler in 1988 and used through the 2003 model year. This system is called **Chrysler Collision Detection (CCD)**. The term *collision* refers to the collision of data occurring simultaneously.

The CCD system uses a twisted pair of wires to transmit the data in digital form. One of the wires is called the **bus (+),** and the other is the **bus (−)**. Negative voltages are not used. The (+) and (−) indicate that one wire is more positive than the other when the bus is sending the dominant bit "0." All modules that connect to the CCD bus system have a special CCD chip (**Figure 10-4**). In most vehicles (but not all), the body control module (BCM) provides the bias voltage to power the bus circuits. Since the BCM powers the system, its internal components are illustrated (**Figure 10-5**). The other modules will operate the same as the BCM to send messages.

The bias voltage on the bus (+) and bus (−) circuits is approximately 2.50 volts when the system is idle (no data transmission). Biasing is accomplished through a regulated 5-volt circuit and a series of resistors. The regulated 5 volts sends current through a 13-kΩ resistor to the bus (−) circuit (**Figure 10-6**). The current flows through the two 120-Ω resistors wired in parallel and to the bus (+) circuit. Finally, the current flows to ground through a second 13-kΩ resistor. **Figure 10-7** is a simplified schematic of the biasing circuit with the normal voltage drops resulting from the resistors. The two 120-Ω resistors are referred to as **termination resistors**. Termination resistors control induced voltages.

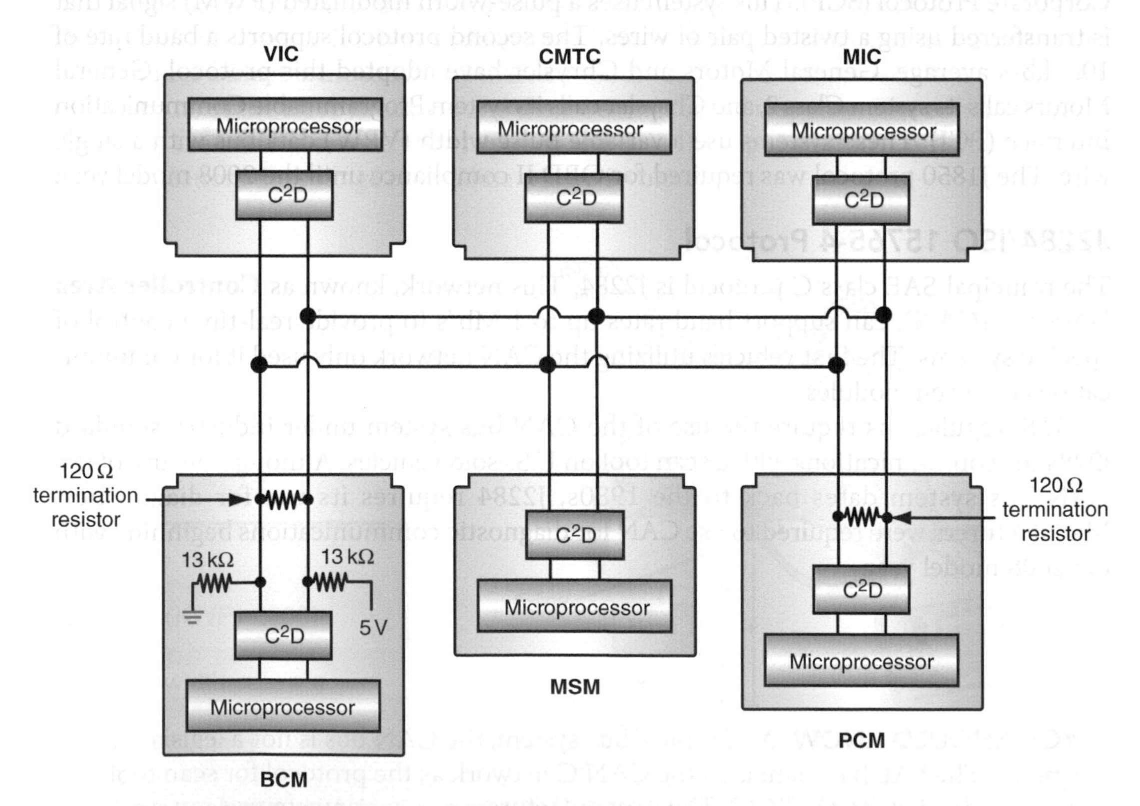

Figure 10-4 Each module on the CCD bus system has a CCD (C^2D) chip.

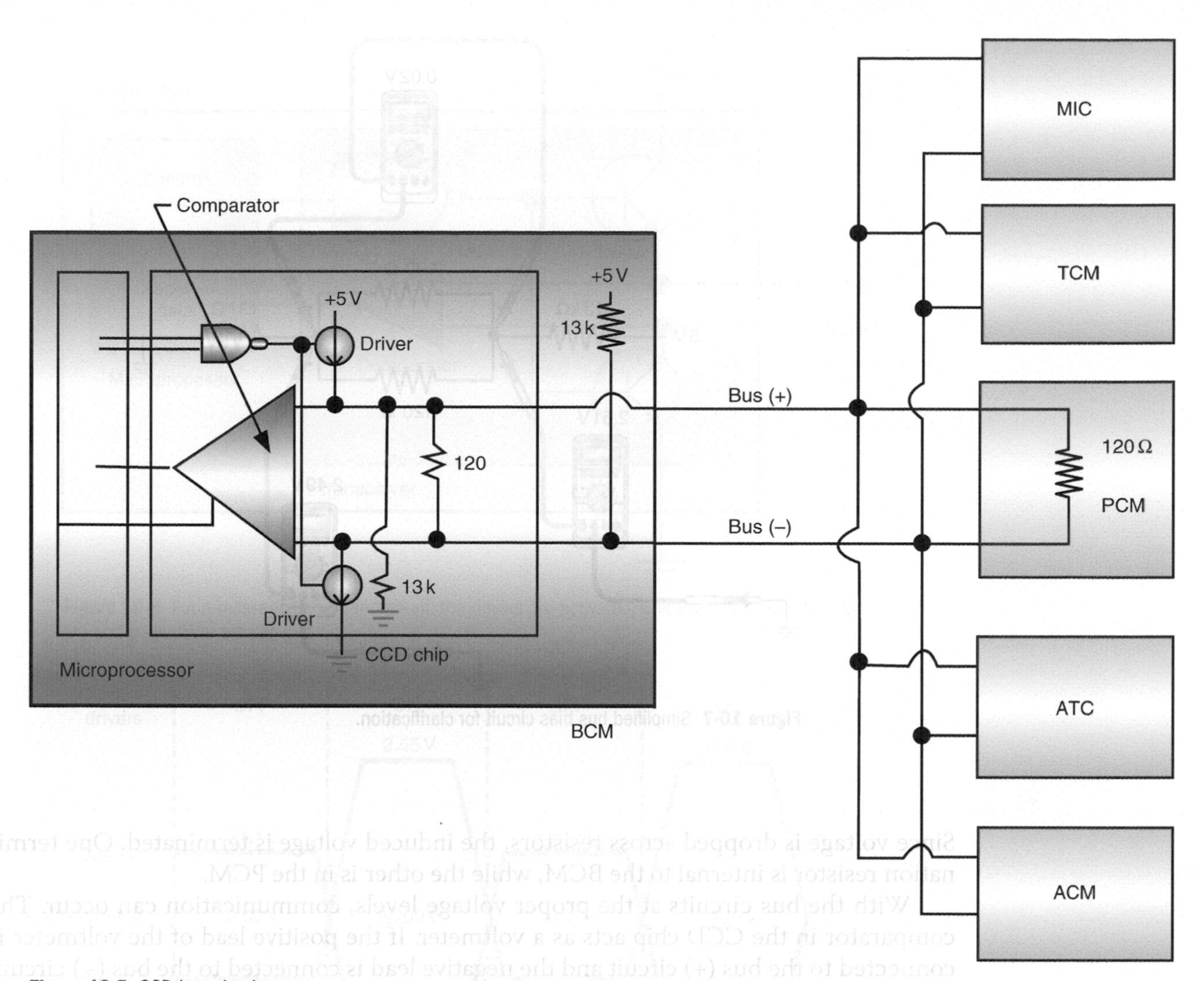

Figure 10-5 CCD bus circuit.

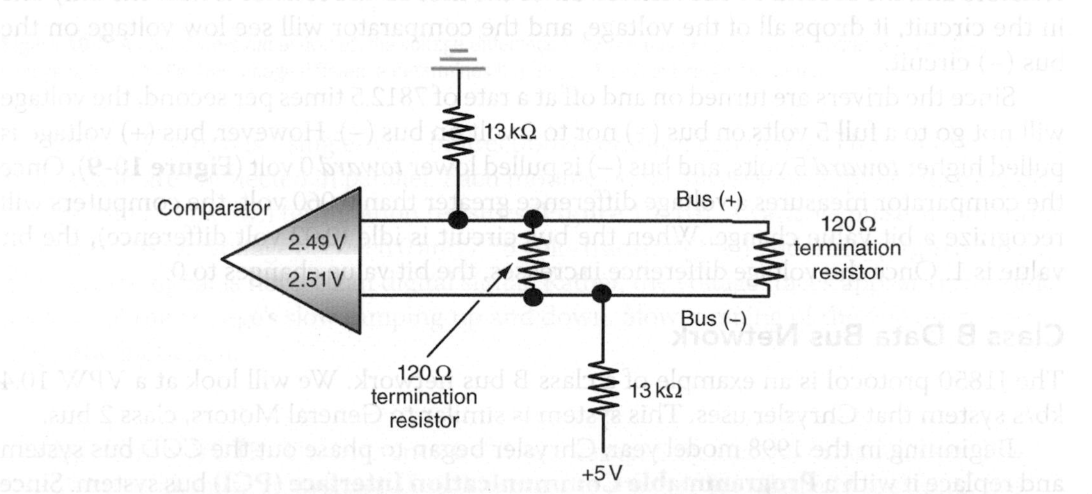

Figure 10-6 The bus is supplied 2.5 volts using pull-up and pull-down resistors.

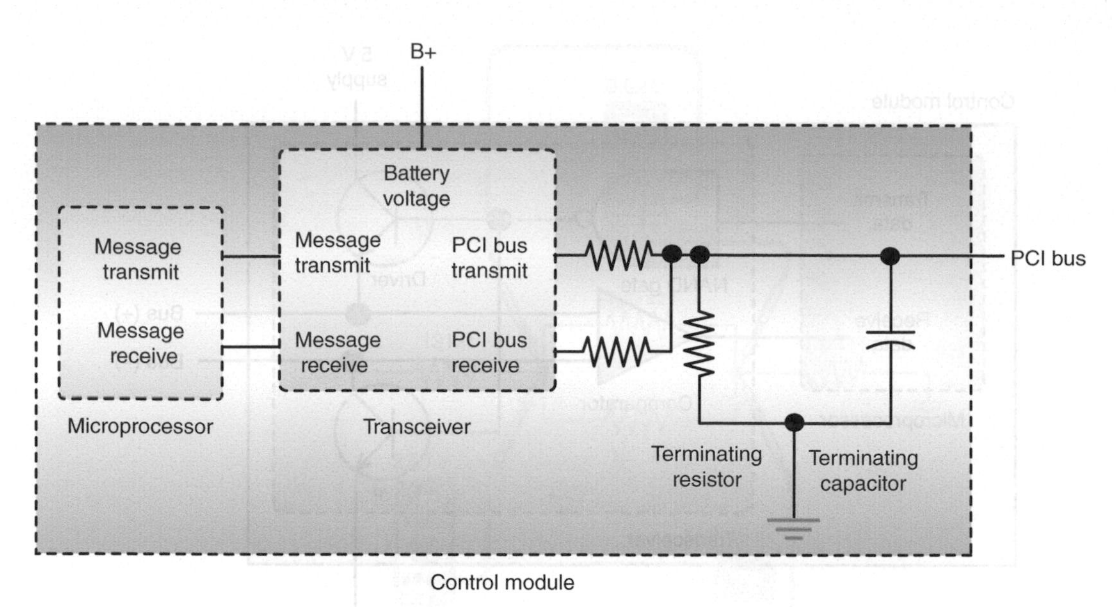

Figure 10-10 Each module supplies bias and termination on the PCI bus system.

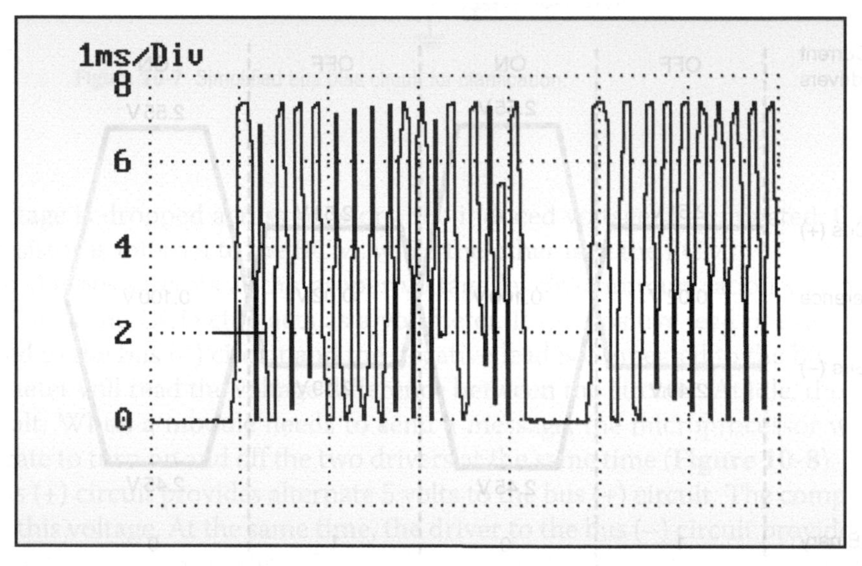

Figure 10-11 Lab scope trace of PCI bus voltages.

The length of time the voltage is high or low determines if the bit value is 1 or 0 (**Figure 10-12**). The typical PCI bus message will have the following elements (**Figure 10-13**):

- SOF—Start of frame pulse used to notify other modules that a message is going to be transmitted.
- Header—One to three bytes of information concerning the type, length, priority, target module, and sending module.
- Data byte(s)—The message that is being sent. This can be up to 8 bytes in length.
- Cyclic redundancy check (CRC) byte—Detects if the message has been corrupted or if there are any other errors.
- In-Frame Response (IFR) byte(s)—If the sending module requires an acknowledgment or an immediate response from the target module, this request is included with the message. The IFR is the target module sending the requested information to the original sending module.

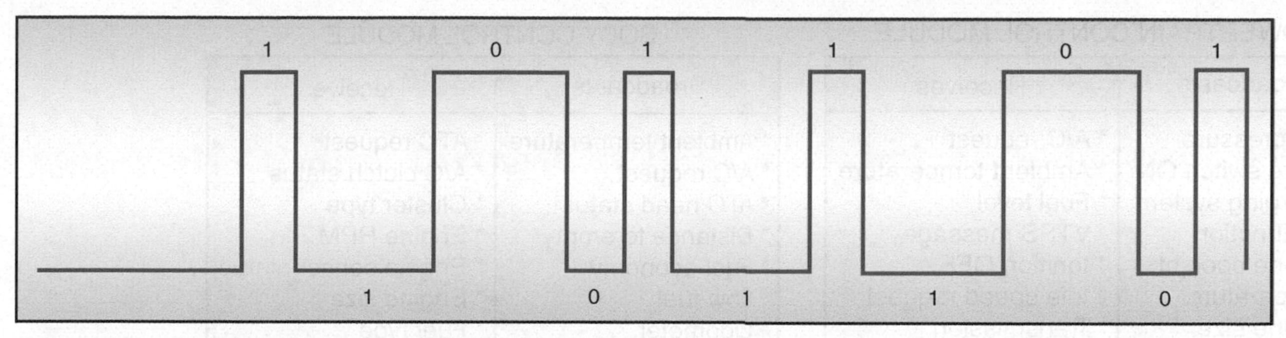

Figure 10-12 The VPWM determines the bit value.

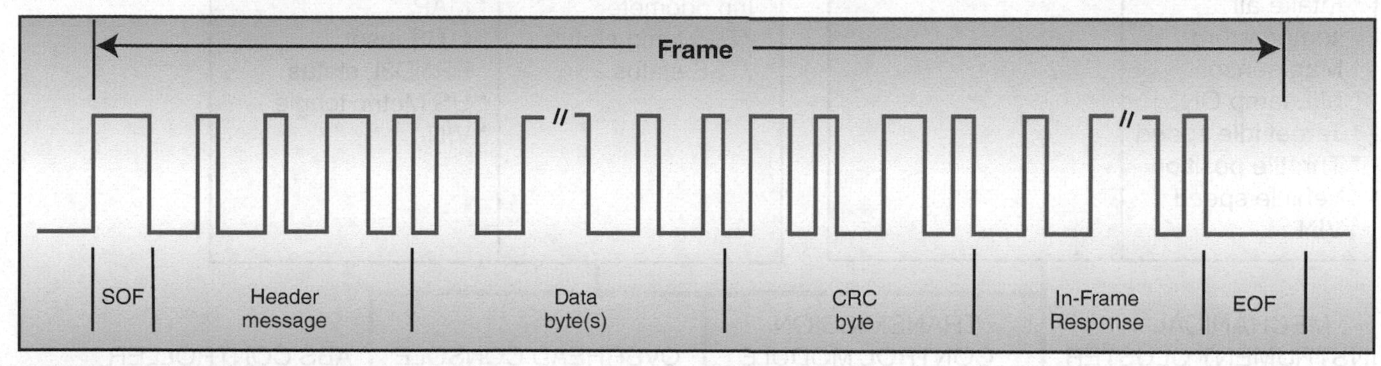

Figure 10-13 Components of a typical PCI bus message.

- EOF—A period of time with no voltage that identifies the module is finished communicating its message.

Figure 10-14 illustrates the type of information sent over the PCI bus system.

Controller Area Network

AUTHOR'S NOTE The following is an example of the CAN bus system. Understanding this system should enable you to grasp any system design. Again, baud rates and design vary between manufacturers. The following discussion covers a common method used.

Most vehicles that follow the J2284 protocol integrate CAN bus networks that operate at different baud rates. A high-speed bus supports real-time functions, such as engine management and antilock brake operation. There may be more than one high-speed bus. For example, powertrain functions may use one high-speed bus network, the driver assistance systems may use a second, and diagnostics use a third. Vehicle body functions such as seat, window, radio, and instrumentation typically use the lower-speed bus.

Shop Manual
Chapter 10, page 479

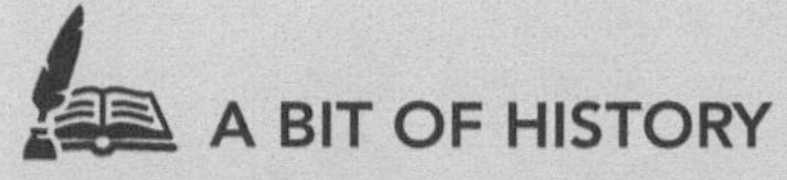

A BIT OF HISTORY

The first production car to use a CAN network was the 1991 Mercedes S-Class.

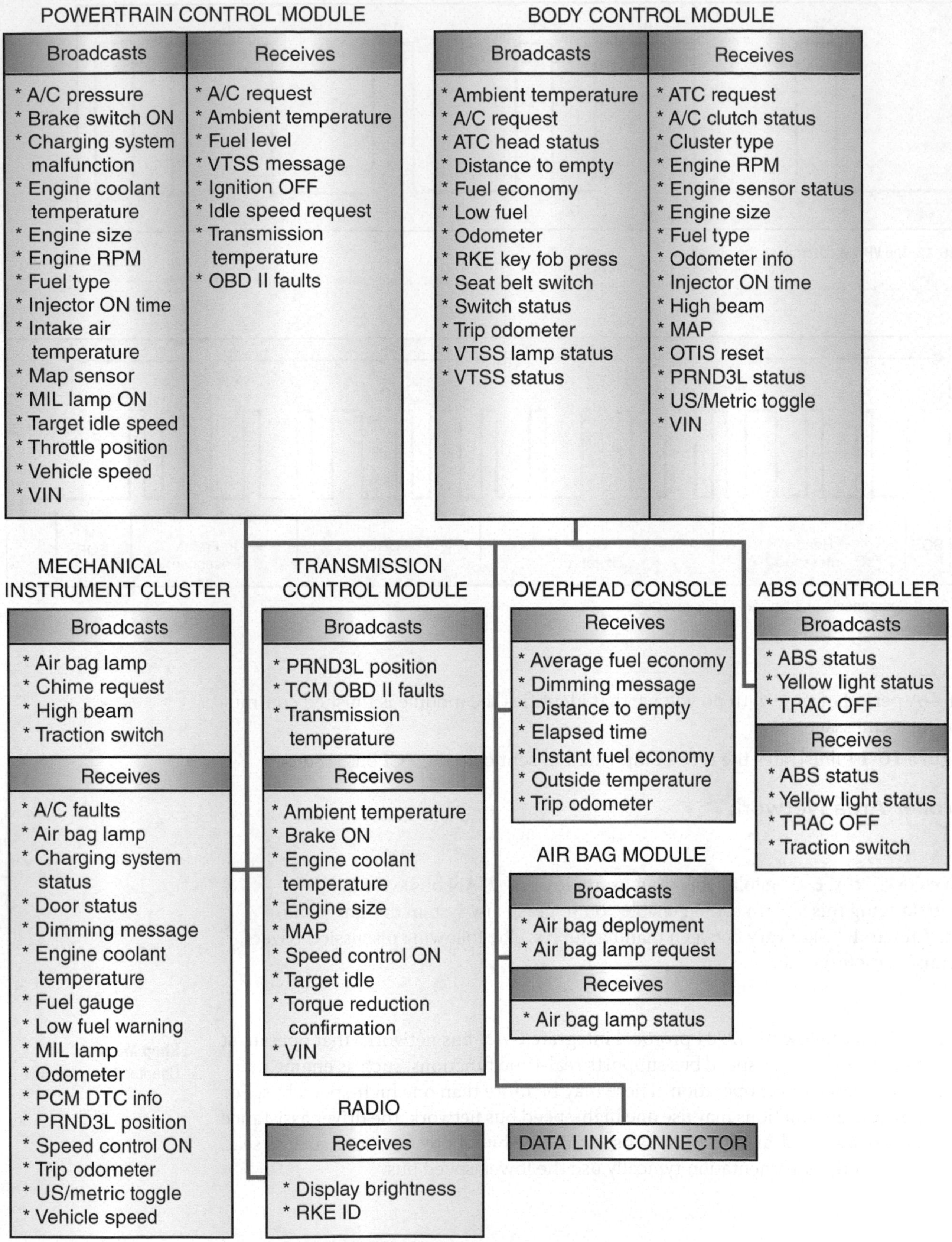

POWERTRAIN CONTROL MODULE
Broadcasts
* A/C pressure
* Brake switch ON
* Charging system malfunction
* Engine coolant temperature
* Engine size
* Engine RPM
* Fuel type
* Injector ON time
* Intake air temperature
* Map sensor
* MIL lamp ON
* Target idle speed
* Throttle position
* Vehicle speed
* VIN
Receives
* A/C request
* Ambient temperature
* Fuel level
* VTSS message
* Ignition OFF
* Idle speed request
* Transmission temperature
* OBD II faults
BODY CONTROL MODULE
Broadcasts
* Ambient temperature
* A/C request
* ATC head status
* Distance to empty
* Fuel economy
* Low fuel
* Odometer
* RKE key fob press
* Seat belt switch
* Switch status
* Trip odometer
* VTSS lamp status
* VTSS status
Receives
* ATC request
* A/C clutch status
* Cluster type
* Engine RPM
* Engine sensor status
* Engine size
* Fuel type
* Odometer info
* Injector ON time
* High beam
* MAP
* OTIS reset
* PRND3L status
* US/Metric toggle
* VIN
MECHANICAL INSTRUMENT CLUSTER
Broadcasts
* Air bag lamp
* Chime request
* High beam
* Traction switch
Receives
* A/C faults
* Air bag lamp
* Charging system status
* Door status
* Dimming message
* Engine coolant temperature
* Fuel gauge
* Low fuel warning
* MIL lamp
* Odometer
* PCM DTC info
* PRND3L position
* Speed control ON
* Trip odometer
* US/metric toggle
* Vehicle speed
TRANSMISSION CONTROL MODULE
Broadcasts
* PRND3L position
* TCM OBD II faults
* Transmission temperature
Receives
* Ambient temperature
* Brake ON
* Engine coolant temperature
* Engine size
* MAP
* Speed control ON
* Target idle
* Torque reduction confirmation
* VIN
RADIO
Receives
* Display brightness
* RKE ID
OVERHEAD CONSOLE
Receives
* Average fuel economy
* Dimming message
* Distance to empty
* Elapsed time
* Instant fuel economy
* Outside temperature
* Trip odometer
AIR BAG MODULE
Broadcasts
* Air bag deployment
* Air bag lamp request
Receives
* Air bag lamp status
DATA LINK CONNECTOR
ABS CONTROLLER
Broadcasts
* ABS status
* Yellow light status
* TRAC OFF
Receives
* ABS status
* Yellow light status
* TRAC OFF
* Traction switch

Figure 10-14 Chart of messages received and broadcasted by each module on the PCI bus.

The CAN bus system uses terminology such as CAN B and CAN C. The letters *B* and *C* distinguish the speed of the bus. CAN B is a medium-speed bus with a speed of up to 125,000 bits per second. The CAN C bus has a speed of 500,000 bits per second. The CAN B bus is also referred to as CANLS (CAN low-speed) and the CAN C as CANHS (CAN high-speed).

The CAN bus circuit consists of a pair of twisted wires. Twisting the wires for the CAN bus system reduces the effects of electromagnetic interference. The transfer of digital data is done by simultaneously pulling the voltage on one circuit high and pulling the voltage on the other circuit low. This requires 33 to 50 twists per meter. To maintain the twist, the bus wire pair are in adjacent cavities at connectors. In addition, careful routing of the bus circuits avoids parallel paths with high-current sources, such as ignition coil drivers, motors, and high-current PWM circuits.

On a CAN bus system, each module provides its own bias. Because of this, communication between groups of modules is still possible if an open occurs in the bus circuit. The CAN bus transceiver has drivers internal to the transceiver chip to supply the voltage and ground to the bus circuit.

Each CAN bus system has its advantages and limitations. For example, the high-speed CAN C bus may be functional only when the ignition is on. On the other hand, the CAN B bus can remain active with the ignition turned off if a module requires it to be active. The requirements of the module determine which bus system it's connected to. Using more than one bus network on the same vehicle gives the manufacturer the optimum features of each system.

AUTHOR'S NOTE Some CAN C bus networks become active based on an event on the CAN B bus. Also, if the CAN C bus is used for the vehicle's interior systems, it's event driven.

When the CAN C bus becomes active, the bus is biased to approximately 2.5 volts. When both CAN C (+) and CAN C (–) are equal, the bus is recessive, and the bit "1" is transmitted. When CAN C (+) is pulled high and CAN C (–) is pulled low, the bit "0" is transmitted. When the bit "0" is transmitted, the bus is considered dominant (**Figure 10-15**). To be dominant, the voltage difference between CAN C (+) and CAN C (–) must be at least 1.5 volts and not more than 3.0 volts. To be recessive, the voltage difference between the two circuits must not be more than 50 mV.

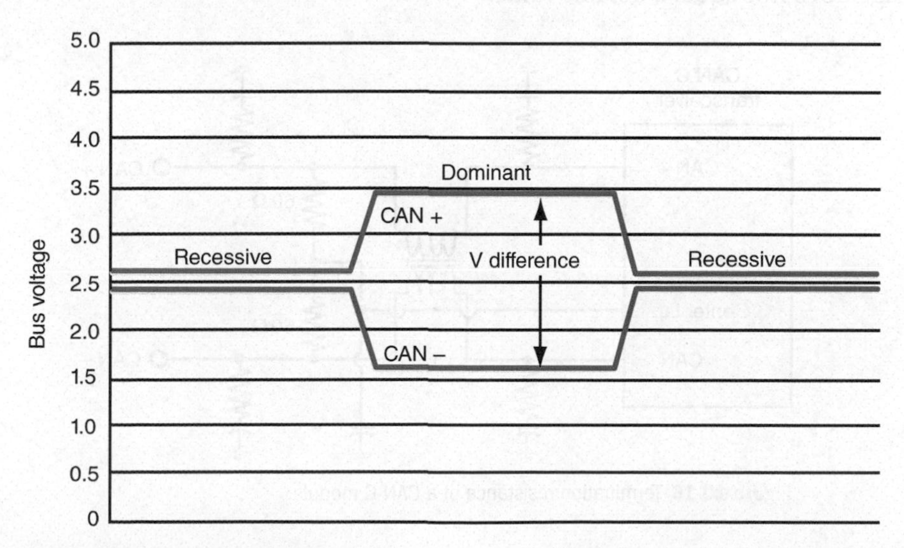

Figure 10-15 Voltages on the CAN C bus.

The optimum CAN C bus termination is 60 ohms. Two CAN C modules each provide 120 ohms of termination. Since the modules are in parallel, the total resistance is 60 ohms. The two modules providing termination are typically located the farthest apart from each other. The terminating modules have two 60-ohm resistors connected in series to equal the 120 ohms. Common to both resistors are the connections to the center tap and ultimately through a capacitor to ground. This center tap may also connect to the transceiver (**Figure 10-16**). The other ends of the resistors are connected to CAN (+) and CAN (–).

AUTHOR'S NOTE Some CAN bus networks will have additional termination resistance in other modules. Due to this, the total resistance of the bus may be lower than 60 ohms.

The bus is idle, or recessive, when CAN B (+) is approximately 0 to 0.2 volt and CAN B (–) is 4.8 to 5 volts. In this state, the logic is "1." When CAN B (+) is pulled between 3.6 and 5 volts and CAN B (–) is pulled low between 1.4 and 0 volts, the bus is considered dominant, and the logic is "0" (**Figure 10-17**). The bus is asleep when CAN B (+) is approximately 0 volt and CAN B (–) is near battery voltage.

Each module on the CAN B bus supplies its own termination resistance. Total bus termination resistance depends on the number of modules installed on the vehicle. Internal to CAN B modules are two termination resistors. The resistors connect CAN B (+) and CAN B (–) to their respective transceiver termination pins (**Figure 10-18**). To provide termination and bias, the transceiver internally connects the CAN B (+) resistor to ground and the CAN B (–) resistor to a 5-volt source. When the CAN B bus goes into sleep mode, the termination pin connected to CAN B (–) switches from 5 volts to battery voltage by the transceiver.

The following defines some of the network messages:

- Cyclic: A message launched on a periodic schedule. An example is the ignition on status broadcasts on the CAN B bus every 100 milliseconds.
- Spontaneous: An application-driven message.
- Cyclic and change: A message launched on a periodic schedule as long as the signal D is not changing. The message is relaunched whenever the signal changes.
- By active function: A message that is transmitted only at a specific rate when the message does not equal a default value.

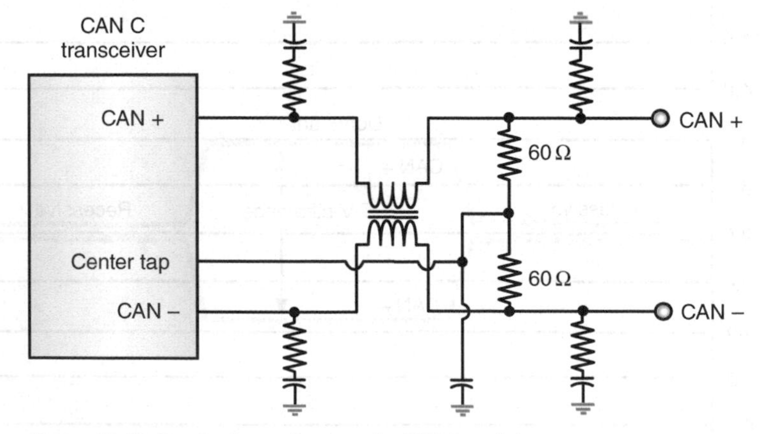

Figure 10-16 Termination resistance of a CAN C module.

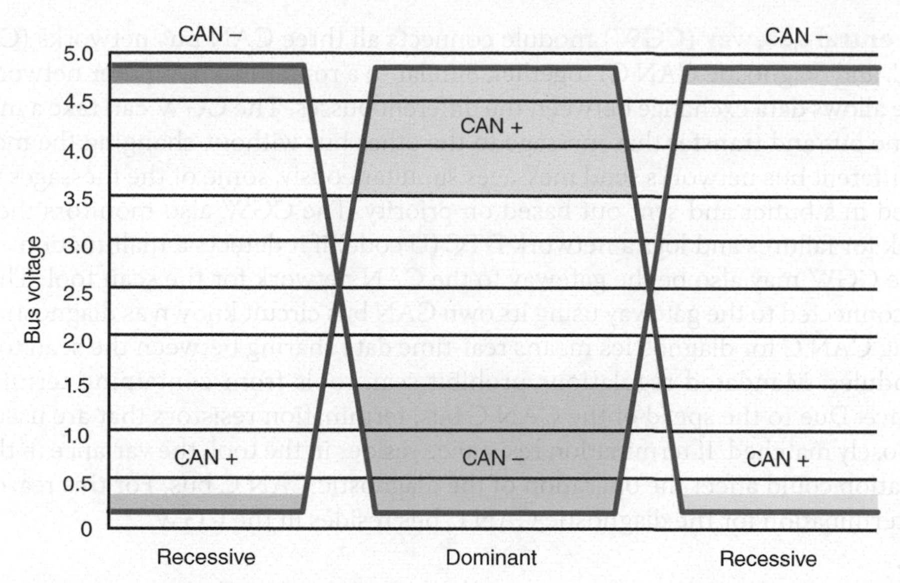

Figure 10-17 Typical CAN B bus voltages.

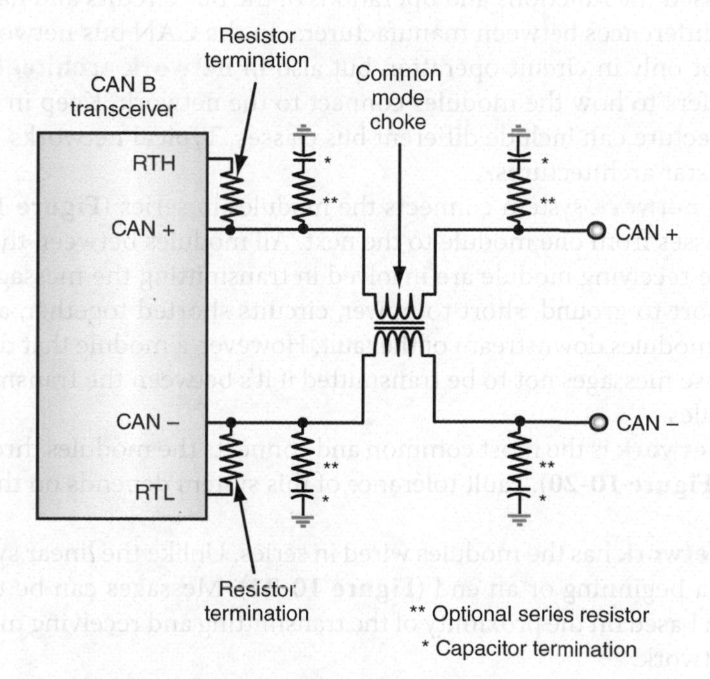

Figure 10-18 CAN B bus module termination resistance.

Most CAN B bus systems are fault tolerant and can operate with one of their conductors shorted to ground or with both of their conductors shorted together. Provided there is an electrical potential between one of the CAN B circuits and chassis ground; communication may still be possible. Due to its high speed, CAN C is not fault tolerant.

YOU SHOULD KNOW Since the design of the CAN B bus allows for a baud rate between 10.4 kb/s and 125 kb/s, fault tolerance diminishes as the baud rate increases.

A **central gateway (CGW)** module connects all three CAN bus networks (CAN B, CAN C, and diagnostic CAN C) together. Similar to a router in a computer network, this module allows data exchange between the different busses. The CGW can take a message from one bus and transfer that message to the other bus without changing the message. If the different bus networks send messages simultaneously, some of the messages will be captured in a buffer and sent out based on priority. The CGW also monitors the CAN network for failures and logs a network DTC (U code) if it detects a malfunction.

The CGW may also be the gateway to the CAN network for the scan tool. The scan tool is connected to the gateway using its own CAN bus circuit known as diagnostic CAN C. Using CAN C for diagnostics means real-time data sharing between the scan tool and the modules. Mandated regulations prohibit scan tools from containing termination resistance. Due to the speed of the CAN C bus, termination resistors that are used need to be closely matched. If termination resistance resides in the tool, the variance in the tool termination could affect the operation of the diagnostic CAN C bus. For this reason, the entire termination for the diagnostic CAN C bus resides in the CGW.

Network Architecture

Network Architecture is also called *topology*.

We have discussed the functions and operations of the bus circuits and have mentioned that there are differences between manufacturers. In the CAN bus networks, these differences are not only in circuit operation but also in **network architecture**. *Network architecture* refers to how the modules connect to the network. Keep in mind that the network architecture can include different bus classes. Typical networks include linear, stub, ring, and star architectures.

The linear network is also known as a daisy chain.

The **linear network** system connects the modules in series (**Figure 10-19**). In this system, data passes from one module to the next. All modules between the transmitting module and the receiving module are involved in transmitting the message. In this type of system, a short to ground, short to power, circuits shorted together, and opens will affect only the modules downstream of the fault. However, a module that does not power on will also cause messages not to be transmitted if it's between the transmitting and the receiving modules.

The stub network is also called a backbone network since it uses a single main line to connect the modules. It can also be referred to as simply "bus network" because it was the first type used.

The **stub network** is the most common and connects the modules through a parallel circuit layout (**Figure 10-20**). Fault tolerance of this system depends on the speed of the network.

The **ring network** has the modules wired in series. Unlike the linear system, the ring does not have a beginning or an end (**Figure 10-21**). Messages can be transmitted in either direction based on the proximity of the transmitting and receiving modules to each other in the network.

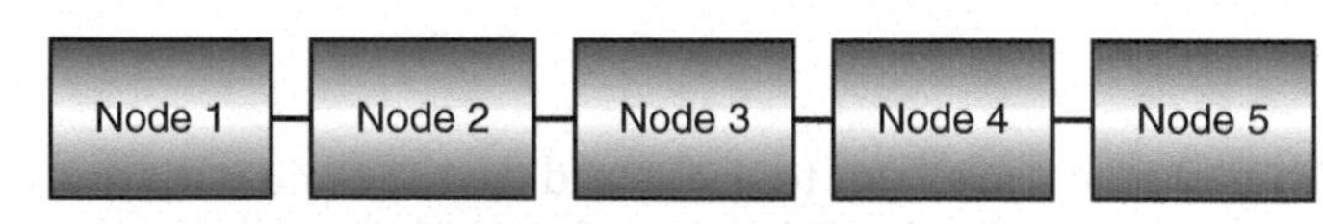

Figure 10-19 Linear network architecture configuration.

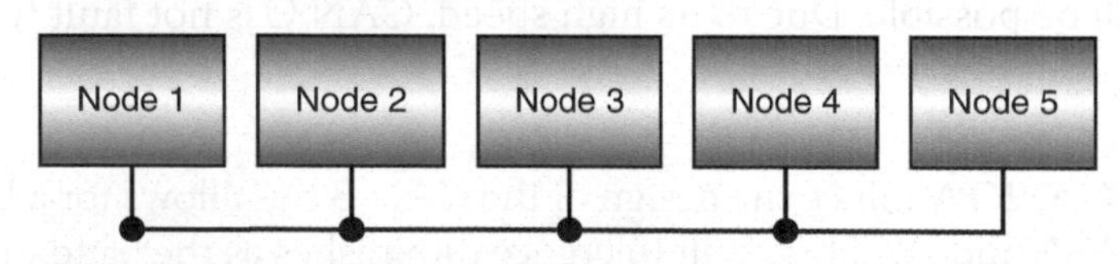

Figure 10-20 The stub network architecture is a common configuration.

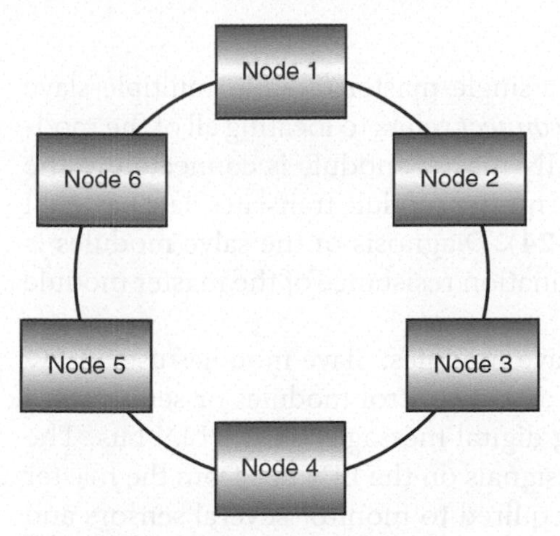

Figure 10-21 The ring network architecture configuration.

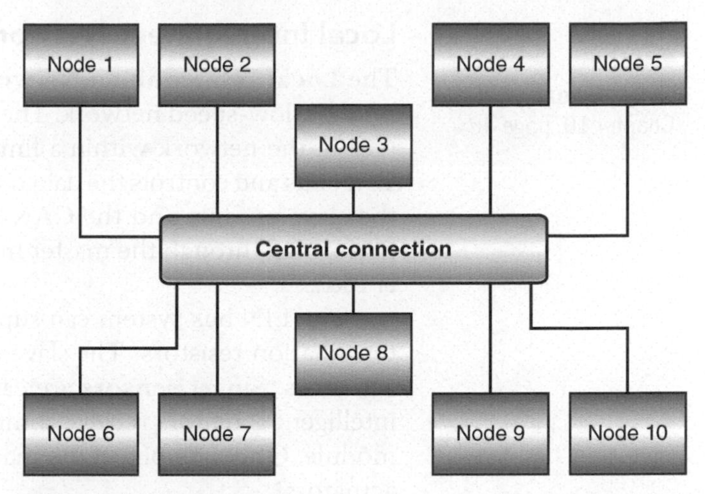

Figure 10-22 The star network architecture uses a common connection point.

Figure 10-23 The star connector is the central location point and may have the termination resistors residing within them.

The **star** (or hub) **network** connects all modules to a central point in a parallel circuit (**Figure 10-22**). In this system, termination may resistors reside in the star connector(s) instead of the module(s) (**Figure 10-23**).

Shop Manual
Chapter 10, page 483

Supplemental Data Bus Networks

Since a single data bus cannot handle all of the requirements of computer-controlled operations on today's vehicles, **supplemental bus networks** are also used. These are bus networks that are on the vehicle in addition to the main bus network. For example, a vehicle may have the CAN bus network and supplemental bus network to handle specified requirements. This section discusses the common bus networks that may also be on the vehicle.

Local Interconnect Network Data Bus

Shop Manual
Chapter 10, page 484

The **Local Interconnect Network (LIN)** bus is a single-master module, multiple-slave module, low-speed network. The term *local interconnect* refers to locating all of the modules in the network within a limited area. The LIN master module is connected to the CAN bus and controls the data transfer speed. The master module translates data between the slave module and the CAN bus (**Figure 10-24**). Diagnosis of the salve modules is performed through the master module. The termination resistance of the master module is 1 kΩ.

The LIN bus system can support up to 15 slave modules. Slave modules use 30-kΩ termination resistors. The slave modules can be actual control modules or sensors and actuators. **Smart sensors** are capable of sending digital messages on the LIN bus. The intelligent actuators receive commands in digital signals on the LIN bus from the master module. Only one pin of the master module is required to monitor several sensors and actuators.

The data transmission is variable between 1 kb/s and 20 kb/s over a single wire. The specific baud rate is programmed into each module. The master module or the slave module can send messages. A steady 12 volts on the circuit indicates no messages are being transmitted, and the recessive bit is sent (bit 1). To transmit a dominant bit (bit "0"), the circuit is pulled low by a transceiver in the module that is transmitting the message (**Figure 10-25**).

Media-Oriented System Transport Data Bus

The standards of the **Media-Oriented System Transport (MOST)** cooperation data bus system are the result of cooperative efforts between automobile manufacturers, suppliers, and software programmers. The MOST data system is specifically designed to transmit audio and video data. The MOST bus uses **fiber optics** to transmit data at a speed of up

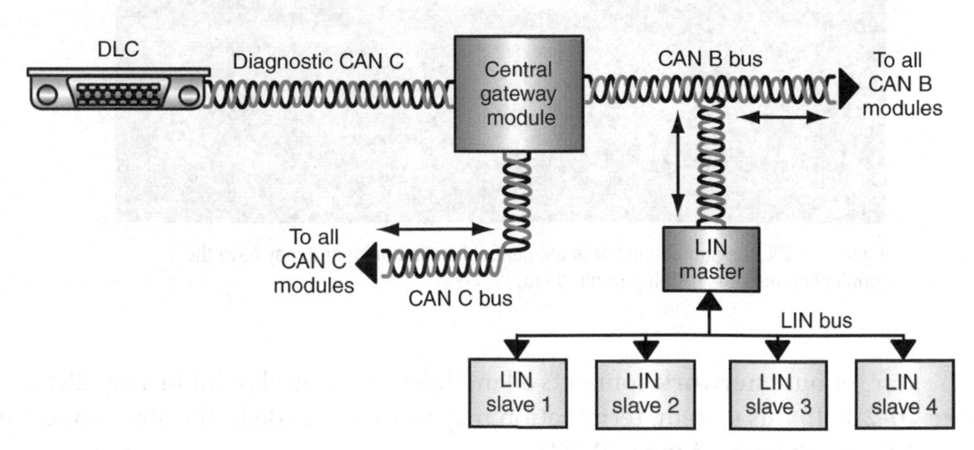

Figure 10-24 The LIN master communicates messages from the slaves onto the CAN bus.

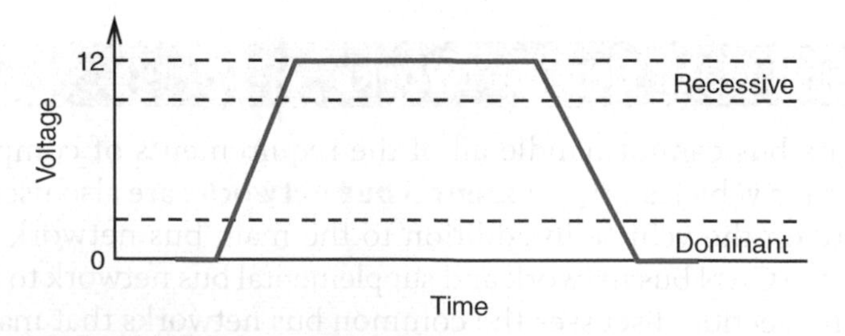

Figure 10-25 Voltages of the LIN bus.

to 25 Mb/s. Fiber optics use light waves to transmit the data without the effects of electromagnetic or radio frequency interference. In the past, video and audio transmission signals were analog. The MOST system uses a fiber optics data bus to transmit digital video and audio communications.

Modules on the MOST data bus use an LED (light-emitting diode), photodiode, and a MOST transceiver to communicate with light signals (**Figure 10-26**). The LED and photodiode are part of the fiber-optic transceiver. The photodiode changes light signals into voltage signals and sends them to the MOST transceiver. The LED converts the voltage signals from the MOST transceiver into light signals.

The conversion of light to voltage signals in the photodiode is by subjecting the PN junction of the photodiode to light. When light penetrates the junction, the energy converts to free electrons and holes. The electrons and holes pass through the junction in direct proportion to the amount of light. The photodiode is connected in series with a resistor (**Figure 10-27**). As the voltage through the photodiode increases, the voltage drop across the resistor also increases. Since the voltage drop changes with light intensity, the light signals change to voltage signals.

The microprocessor commands the MOST transceiver to send messages to the fiber-optic transceiver as voltage signals. Also, the MOST transceiver sends voltage signals from the fiber-optic transceiver to the microprocessor.

Fiber-optic cable connects the modules in a ring fashion (**Figure 10-28**). Messages are sent in one direction only. The master module usually starts the message, but not always. The master module sends the message onto the data bus with a duty cycle frequency of 44.1 kHz. This frequency corresponds to the frequency of digital audio and video equipment. A module sends its message to the next module in the ring. That module then sends it to the next module, and this continues until the originating module receives its own message. At this time, the ring is closed, and the message is no longer transmitted. If a module receives a message that it does not need, the message is sent through the MOST transceiver and back to the fiber-optic transceiver without being transmitted to the microprocessor. If the MOST bus is powered down (asleep), turn on the ignition, or an

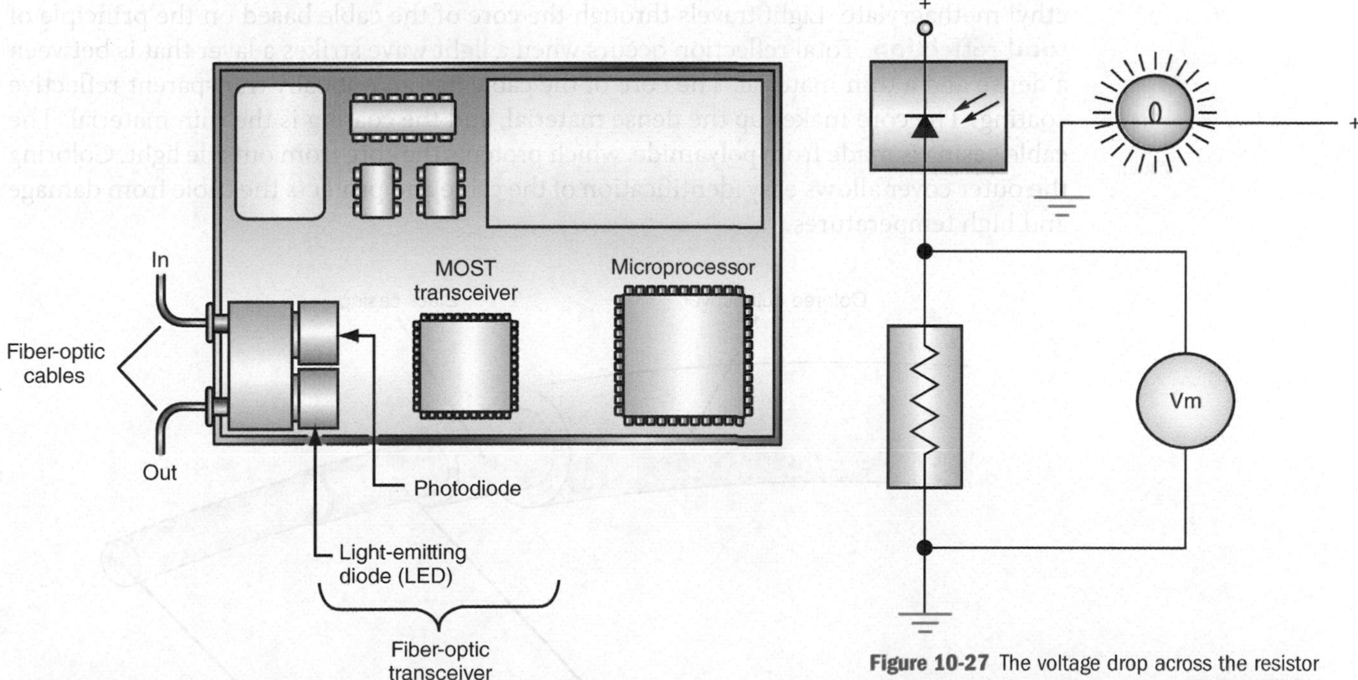

Figure 10-26 Typical MOST data system controller components.

Figure 10-27 The voltage drop across the resistor changes in relation to the amount of light applied to the photodiode.

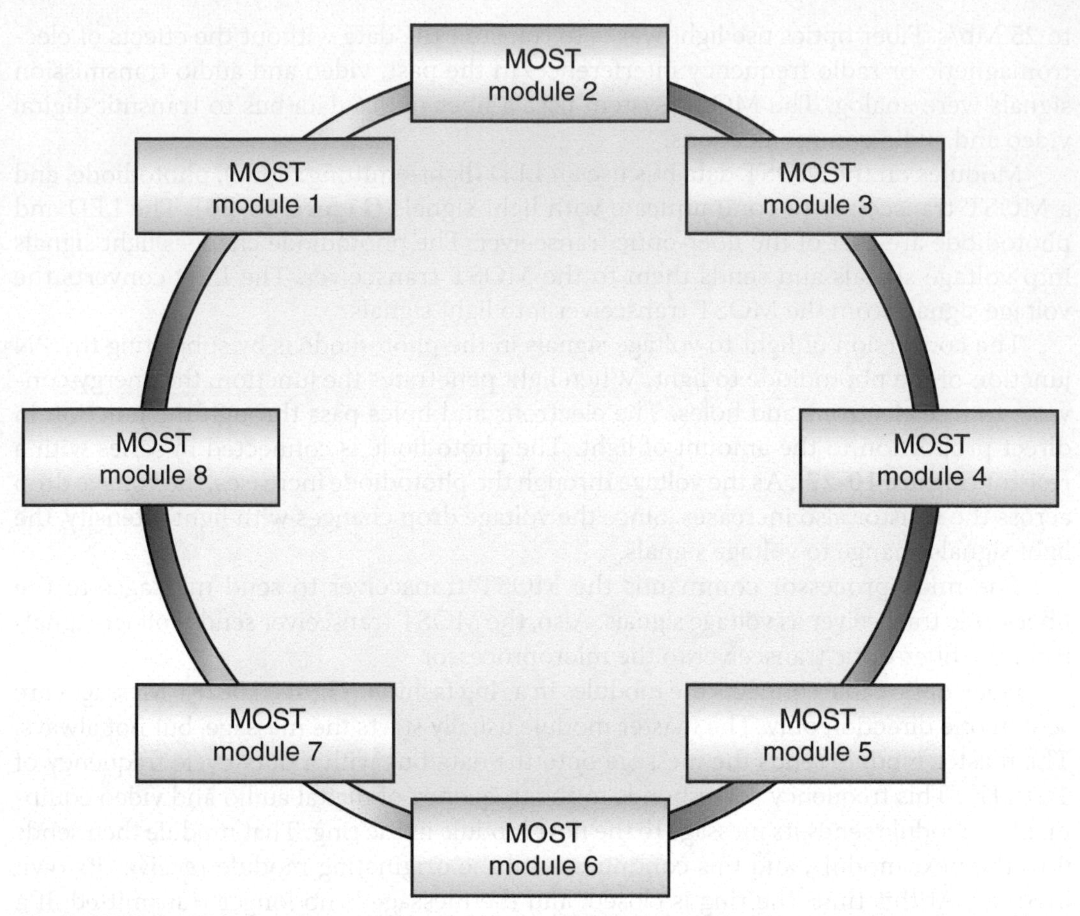

Figure 10-28 The MOST data system transfers data in a single direction through a ring configuration.

input from a module can awaken it. When an input is received, the master module will send a wake-up message to all of the modules in the ring.

The fiber-optic cable has several layers (**Figure 10-29**). The core consists of polymethyl methacrylate. Light travels through the core of the cable based on the principle of **total reflection**. Total reflection occurs when a light wave strikes a layer that is between a dense and a thin material. The core of the cable has an optically transparent reflective coating. The core makes up the dense material, and the coating is the thin material. The cable casing is made from polyamide, which protects the core from outside light. Coloring the outer cover allows easy identification of the cable and protects the cable from damage and high temperatures.

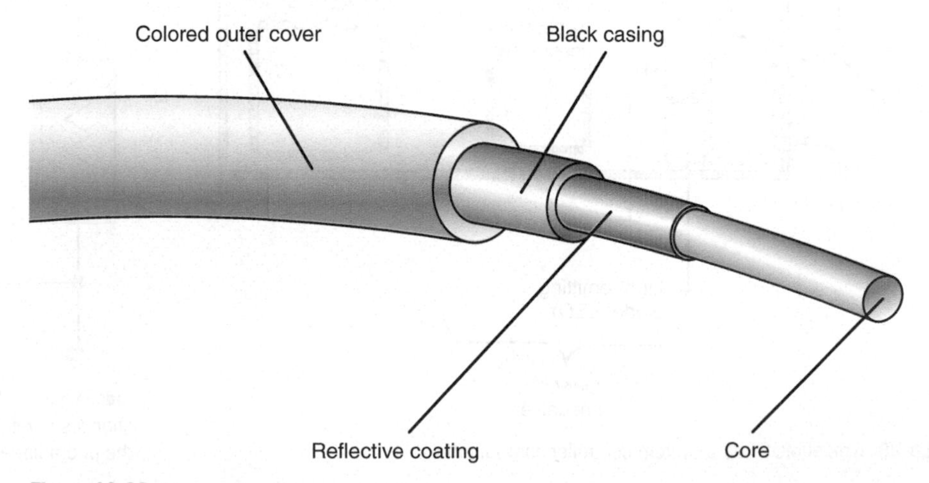

Figure 10-29 Fiber-optic cable construction.

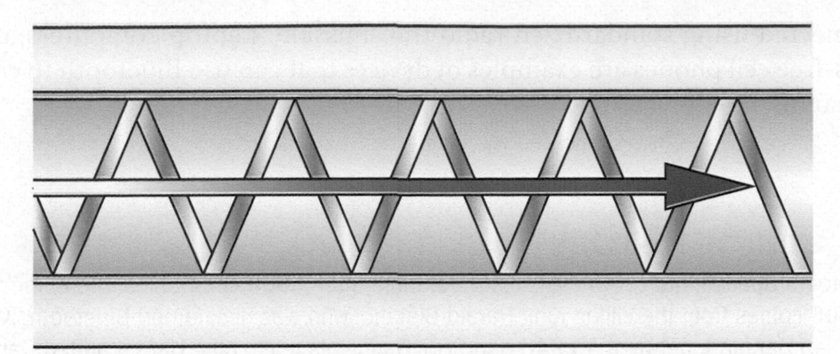

Figure 10-30 Light waves traveling through a straight section of the fiber-optic cable.

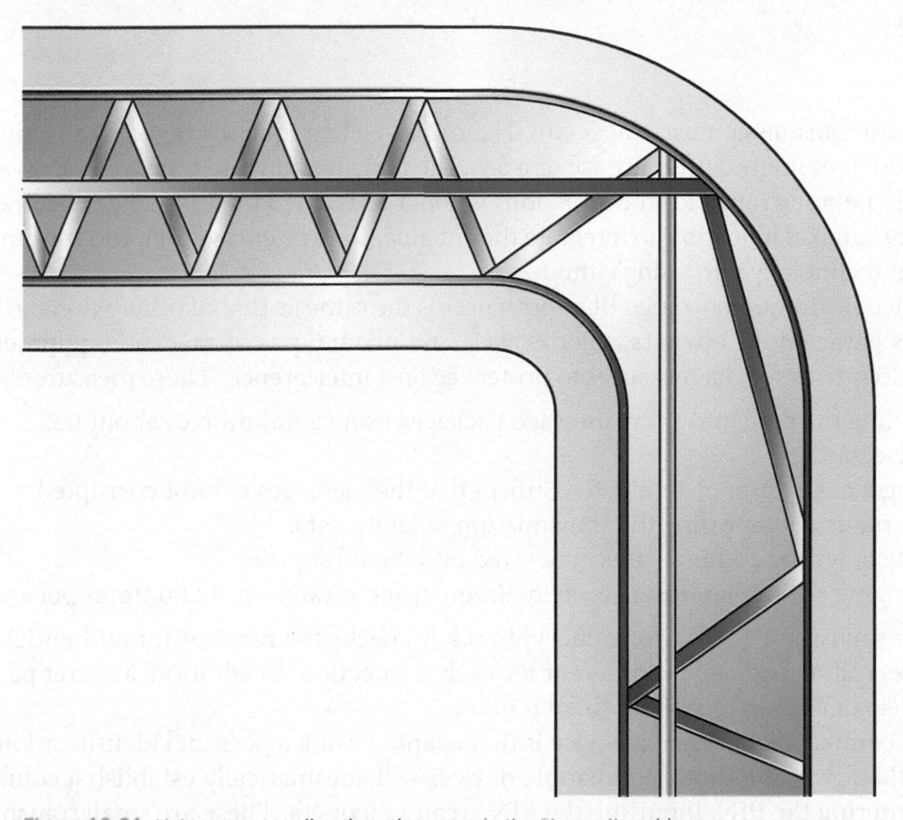

Figure 10-31 Light waves travelling through a curve in the fiber-optic cable.

Laying the cable out straight results in some of the light waves traveling through the core in a straight line. However, most of the light waves travel in a zigzag pattern (**Figure 10-30**). The zigzag pattern is a result of the total reflection. If the fiber-optic cable is bent, total reflection reflects the light waves at the borderline of the core coating and guides the waves through the bend (**Figure 10-31**).

Wireless Bus Networks

Wireless networks can connect modules together to transmit information without using physical connections by wires. For example, tire pressure information can be transmitted from a sensor in the tire to a module on the vehicle without wires. Although there are different technologies used for wireless communications, a popular one is called **Bluetooth**. Bluetooth technology allows several modules from different manufacturers to be connected using standardized radio transmission. Laptop computers, notepads, and hands-free cell phones are examples of devices that can use Bluetooth to connect to vehicle systems.

A BIT OF HISTORY

The Bluetooth Special Interest Group consists of more than 2,000 companies. The name "Bluetooth" comes from the Viking King Harald Blåtand, who was nicknamed Bluetooth. King Harald ruled Denmark between AD 940 and 985. During his reign, King Harald united Denmark and Norway. "Bluetooth" was adopted for this wireless communications technology because it shares the same philosophy as the king: global unity. Bluetooth unifies multinational companies. Bluetooth was initially a code name for the project; however, it has now become the trademark name.

Radio transmission uses the 2.40-GHz to 2.48-GHz frequency range. Transmitting on this band does not require a license or a fee. The transmission rate is up to 1 Mb/s, and the normal operating range for transmissions is about 33 feet (10 meters). The short transmitting range makes it possible to integrate the antenna, control, encryption, and transmission/receiver technology into a single module.

Since the frequency range Bluetooth uses is the same as that of other wireless devices such as garage door openers, microwaves, and many types of medical equipment, the technology uses special measures to protect against interference. These measures include:

- Dividing the data into short message packages using a duration of about 625 milliseconds.
- Using a check sum of 16 bits to confirm that the messages are not corrupted.
- Automatically repeating the transmission of faulty data.
- Using language coding that is converted into digital signals.
- Changing the transmitting/receiving frequencies at random, 1,600 times per second.

For security, a 128-bit long encryption key checks the receiver for authenticity. The key uses rolling code and is different for each connection. In addition, a secret password is used so devices can connect to each other.

To connect devices, each device is first adapted using a personal identification number (PIN). Two Bluetooth-compatible devices will automatically establish a connection after entering the PIN. Inputting the PIN creates piconets. These are small transmission cells that assist in the organization of the data. Each piconet will allow for up to eight devices to operate at the same time. One device in each piconet is assigned the role of the master. The master is responsible for establishing the connection and synchronizing the other devices.

AUTHOR'S NOTE Each device has an address that is 48 bits long and is unique worldwide. This means over 281 trillion devices can be identified worldwide.

Wi-Fi has gained use as a means of vehicle communications. Scan tools, internet, local infrastructure, homes, other vehicles, and so forth can connect to the vehicle through Wi-Fi. Chapter 14 discusses using Wi-Fi for vehicle connectivity

Vehicle Network Security

Vehicle connectivity to the outside world is a reality. Using Wi-Fi and/or Bluetooth has opened up a new world of possibilities as the vehicle becomes an extension to cell phone services and the internet. Today most scan tools use wireless "Dongles" that plug into the DLC and communicate with a separate display device. Through Wi-Fi, the scan tool connects to the internet and accesses servers for acquiring service information, shop, forms, and even customer information. In addition, vehicle owners may connect devices to the DLC that communicate over the internet for vehicle tracking, driving evaluation, and so forth. All of these devices and actions leave the vehicle open to hacks or malware. A security breach could allow a hacker to take control of different vehicle systems. This may be as "harmless" as changing radio stations but can also be as serious as taking control of the throttle and steering.

Any system using an antenna is susceptible to hacking, including the tire pressure monitoring system. However, the most common hacking uses the radio as an unwitting accomplice.

At the request of Congress, SAE formed a working group to consider threats posed to the OBD II port and Cyber Security. Manufacturers are developing different methods for mitigating these threats. As an example of this implementation, we will look at Fiat Chrysler America's (FCA) approach to vehicle network security.

Beginning in the 2018 model year, Chrysler started adding a new ECU called the **Security Gateway** (SGW). The SGW splits the vehicle's communication networks into two categories: the less secure "public" and a very secure "private" network. The SGW divides the vehicle's network into three sections. Two sections are public, and one is private **(Figure 10-32).**

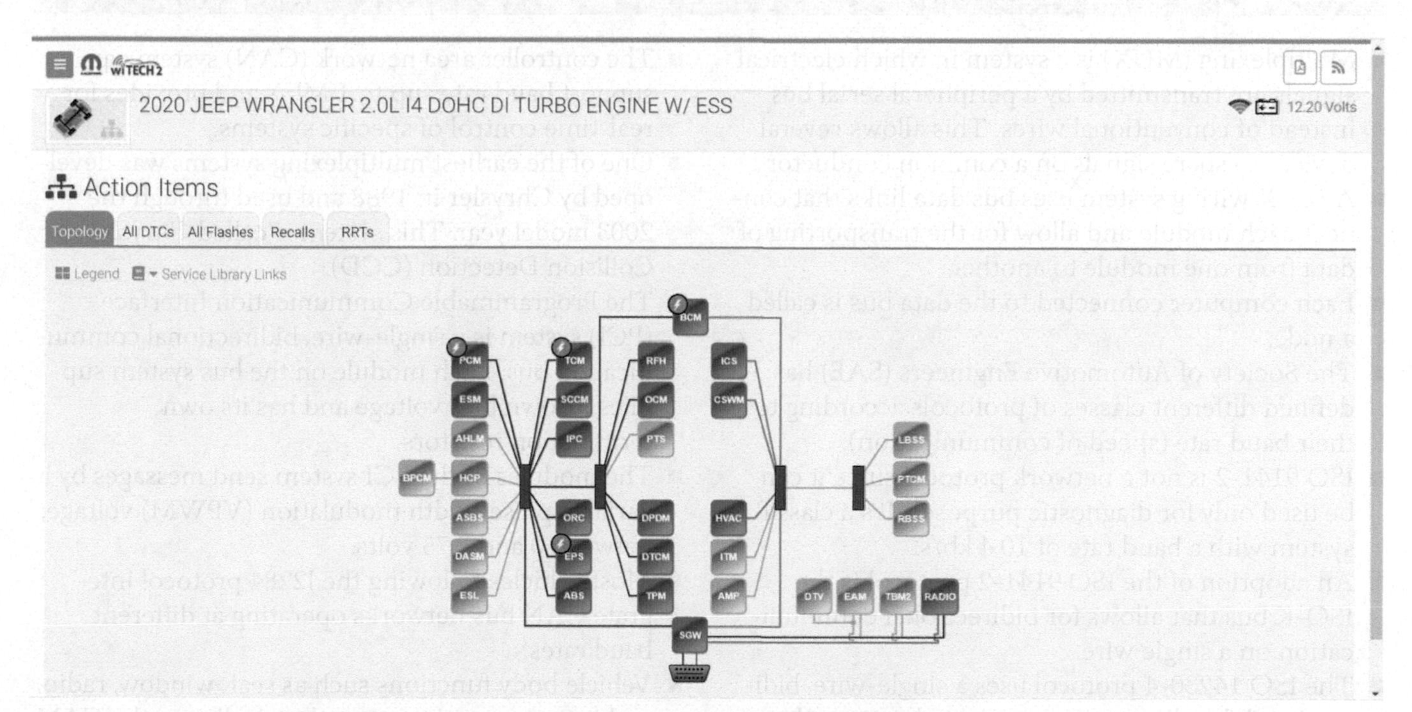

Figure 10-32 This wiTECH2® screen shot shows that the SGW splits the communication network into three sections.

The direct connection between the DLC and the SGW is one of the public sections. The other public section is the radio. The radio is independently wired to the SGW and is isolated from the DLC. All of the other modules are on the "private" network.

This configuration makes the SGW the gateway between three independent sections. However, the SGW has no drivers and cannot directly operate or control any vehicle components. The SGW acts as a firewall as it monitors the information passing between the three sections. The SGW blocks any inappropriate messages transmitted on any of the public networks from accessing the secure private network(s).

Device Authentication

Shop Manual
Chapter 10, page 488

The SGW also authenticates any external device connected to the DLC, including scan tools. Authentication is by passing code between the device and the SGW. The SGW sends a "code" to the device, and the device must send a correct but different code back to the SGW. Through this process, the SGW verifies that the device is safe.

Vehicle manufacturers' scan tools may have a verification tool that seamlessly performs code sharing in the background during the vehicle selection process. Aftermarket scan tools will also have to go through an authorization process. Not only does the scan tool need to meet the manufacturer's security requirements, but the technician using the tool must be an authenticated user. If the tool or the technician is not authenticated, the scan tool will have limited "read-only" access. For example, you will be able to read DTCs, but the SGW will block all other unapproved communications. This includes such functions as bi-directional control, module reprogramming, module initialization, and so forth. Authenticated tools and users are granted access to these advanced functions.

Summary

- Multiplexing (MUX) is a system in which electrical signals are transmitted by a peripheral serial bus instead of conventional wires. This allows several devices to share signals on a common conductor.
- A MUX wiring system uses bus data links that connect each module and allow for the transporting of data from one module to another.
- Each computer connected to the data bus is called a node.
- The Society of Automotive Engineers (SAE) has defined different classes of protocols according to their baud rate (speed of communication).
- ISO 9141-2 is not a network protocol since it can be used only for diagnostic purposes. It's a class B system with a baud rate of 10.4 kb/s.
- An adoption of the ISO 9141-2 protocol is the ISO-K bus that allows for bidirectional communication on a single wire.
- The ISO 14230-4 protocol uses a single-wire, bidirectional data line to communicate between the scan tool and the nodes.
- J1850 protocol is the class B standard for OBD II. The J1850 standard allows for two different versions based on baud rate.
- The controller area network (CAN) system can support baud rates up to 1 Mb/s and provides for real-time control of specific systems.
- One of the earliest multiplexing systems was developed by Chrysler in 1988 and used through the 2003 model year. This system is called Chrysler Collision Detection (CCD).
- The Programmable Communication Interface (PCI) system is a single-wire, bidirectional communication bus. Each module on the bus system supplies its own bias voltage and has its own termination resistors.
- The modules of the PCI system send messages by a variable pulse-width modulation (VPWM) voltage; between 0 and 7.75 volts.
- Most vehicles following the J2284 protocol integrate CAN bus networks operating at different baud rates.
- Vehicle body functions such as seat, window, radio, and instrumentation control typically use the CAN B (LS) bus. Functions that require real-time data transmission use the CAN C (HS) bus.

- A central gateway (CGW) module connects the CAN bus networks together. Similar to a router in a computer network, this module allows data exchange between the different busses.
- The local interconnect network (LIN) is a supplement network to the CAN bus system. The term *local interconnect* refers to all of the modules in the LIN network being located within a limited area.
- The LIN bus is a single-master, multiple-slave, low-speed network.
- The media-oriented system transport (MOST) cooperation data bus system is based on standards established by a cooperative effort between automobile manufacturers, suppliers, and software programmers that resulted in a data system specifically designed for the data transmission of media-oriented data. MOST uses fiber optics to transmit data up to 25 Mb/s.
- Modules on the MOST data bus use an LED, a photodiode, and a MOST transceiver to communicate with light signals.
- Wireless networks can connect modules together to transmit information without using a physical connection by wires.
- Bluetooth technology allows several modules from different manufacturers to be connected using a standardized radio transmission.
- The SGW splits the vehicle's communication networks into two categories: the less secure "public" and a very secure "private" network.
- The SGW is the gateway between three independent sections and acts as a firewall by monitoring the information passing between networks. It also authenticates any external device connected to the DLC, including scan tools.

Review Questions

Short-Answer Essays

1. Explain the principle of multiplexing.
2. Briefly describe the different multiplexing communication protocols.
3. Explain the different classes of communications.
4. Briefly explain the principle of operation of the CCD bus system as an example of class A multiplexing.
5. Briefly explain the principle of operation of the PCI bus system as an example of class B multiplexing.
6. Briefly explain the principle of operation of the controller area network (CAN) bus system.
7. Describe the purpose of the supplemental data bus networks.
8. Explain the operation of the local interconnect network (LIN) data bus system.
9. Describe the operation of the media-oriented system transport data bus using fiber optics.
10. Explain the operation of wireless networks using Bluetooth technology.

Fill in the Blanks

1. Multiplexing provides the ability to use a ______________ circuit to distribute and share data between several control modules.
2. Each computer connected to the data bus is called a ______________.
3. The ______________ protocol is the class B standard for OBD II.
4. The CCD system uses a twisted pair of wires to transmit the data in ______________ form.
5. ______________ are used to control induced voltages.
6. The PCI system uses a ______________ pulse-width modulation voltage between 0 and 7.75 volts to represent the 1 and 0 bits.
7. A ______________ module is used where all three CAN bus networks (CAN B, CAN C, and diagnostic CAN C) connect together.
8. In the MOST data bus system, the ______________ changes light signals into voltage that is then transmitted to the MOST transceiver.

9. The modules of the MOST data bus system are connected in a ________ fashion by fiber-optic cable.

10. ________ technology allows several modules from different manufacturers to be connected using a standardized radio transmission.

Multiple Choice

1. In the CAN bus system:
   A. CAN B is high-speed and not fault tolerant.
   B. CAN C is fault tolerant.
   C. Can C is low speed and used for many body control functions.
   D. None of these choices.

2. All of the following statements concerning multiplexing are true **EXCEPT:**
   A. Reduction of hardwiring requirements and vehicle weight.
   B. A single computer controls all of the vehicle functions.
   C. Enhanced diagnostics are possible.
   D. Reduces driver requirements in the computer.

3. The purpose of the central gateway module in a CAN bus system is:
   A. To provide a means for the modules on the different CAN bus networks to communicate with each other.
   B. To provide a method for the scan tool to communicate with the modules.
   C. Both A and B.
   D. Neither A nor B.

4. *Technician A* says some multiplexing systems use a data bus consisting of a twisted pair of wires. *Technician B* says some multiplexing systems use a data bus that consists of a single wire. Who is correct?
   A. A only
   B. B only
   C. Both A and B
   D. Neither A nor B

5. On a *data* bus that uses two wires, why are the wires twisted?
   A. To increase the physical strength of the wires.
   B. For identification of the circuit.
   C. To reduce the effect of the high-current flow through the bus wires on other circuits.
   D. To minimize the effects of an induced voltage on the data bus.

6. In the local interconnect network (LIN) bus:
   A. The master controller is connected to the CAN bus and controls the data transfer speed.
   B. The master controller translates data between the slave modules and the CAN bus.
   C. Supporting up to 15 slave modules is possible.
   D. All of these choices.

7. Protocol is defined as:
   A. A common communication method.
   B. A method of reducing electromagnetic interference.
   C. Data transmission through a single circuit.
   D. All of these choices.

8. *Technician A* says multiplexed circuits communicate multiple messages over a single circuit. *Technician B* says multiplexed circuits communicate by transmitting serial data. Who is correct?
   A. A only
   B. B only
   C. Both A and B
   D. Neither A nor B

9. In a single-wire bus network, electromagnetic interference is controlled by:
   A. Using a shielded wire.
   B. Slowly ramping up and down the voltage levels.
   C. Locating the wire outside of the normal wiring harness.
   D. None of these choices.

10. A computer that can communicate on a data bus is known as:
   A. Node.
   B. Byte.
   C. Protocol.
   D. Transceiver.

CHAPTER 11
LIGHTING CIRCUITS

Upon completion and review of this chapter, you should be able to:

- Describe the operation and construction of automotive lamps.
- Describe the operation and construction of various headlights.
- Explain the function of the headlight system, including the computer-controlled headlight system.
- Explain the function of the automatic headlight on/off and time delay features.
- Describe the operation of the most common types of automatic headlight dimming systems.
- Explain the operation of the SmartBeam headlight system as an example of today's sophisticated headlight systems.
- Describe the function of automatic headlight leveling systems.
- Explain how adaptive headlight systems operate.
- Explain the operation of the night vision system.
- Explain the purpose and function of daytime running lamps.
- Describe how various exterior light systems operate, including parking, tail, brake, turn, side, clearance, and hazard warning lights.
- Describe the operating principles of the turn signal and hazard light flashers.
- Explain how adaptive brake light systems work.
- Explain the operation of the various interior light systems, including illuminated entry and instrument panel lights.
- Describe the use and function of fiber optics.
- Explain the purpose and operation of lamp outage indicators.

Terms to Know

Adaptive brake light
Adaptive headlight system (AHS)
Automatic headlight dimming
Automatic on/off with time delay
Ballast
Bi-xenon headlamps
Composite headlight
Courtesy lights
Daytime running lamps (DRL)
Dimmer switch
Double-filament lamp
Flasher
Halogen
Headlight leveling system (HLS)
High-intensity discharge (HID)
Ignitor
Illuminated entry systems
Incandescence
Lamp
Lamp outage module
LUX
Night vision
Prisms
Sealed-beam headlight
Sensitivity control
Timer control
Vaporized aluminum
Wake-up signal

Introduction

Today's technician must understand the operation and purpose of the various lighting circuits on the vehicle. If a lighting circuit is not operating properly, the safety of the driver, passengers, people in other vehicles, and pedestrians is in jeopardy. When today's technician performs repairs on the lighting systems, the repairs must meet at least two requirements: they must assure vehicle safety and meet all applicable laws.

The lighting circuits of today's vehicles can consist of more than 50 light bulbs and hundreds of feet of wiring. Incorporated within these circuits are circuit protectors, relays, switches, lamps, and connectors. In addition, most of today's vehicles have very sophisticated lighting systems that use computers and sensors. The lighting circuits consist of an array of interior and exterior lights, including headlights, taillights, parking lights, stop lights, marker lights, dash instrument lights, and courtesy lights.

With the addition of solid-state circuitry in the automobile, manufacturers have been able to incorporate several different lighting circuits or modify the existing ones. Some of these systems use sophisticated body computer-controlled circuitry and fiber optics. Some refinements to the lighting system include automatic headlight washers, automatic headlight dimming, automatic on/off with timed-delay headlights, and illuminated entry systems.

Some manufacturers have included such basic circuits as turn signals into their body computer to provide for pulse-width dimming in place of a flasher unit. The body computer can also control instrument panel lighting based on inputs, including if the side marker lights are on or off. Utilizing the body computer to control many of the lighting circuits has reduced the amount of wiring necessary. In addition, computer control has provided a means of self-diagnosis in some applications.

This chapter discusses the types of lamps used, describes the headlight circuit, and explores the various exterior and interior light circuits individually.

Lamps

Shop Manual
Chapter 11, page 524

A **lamp** generates light through a process of changing energy forms called **incandescence**. The lamp produces light as a result of current flow through a filament. The filament is enclosed within a glass envelope and is a type of resistance wire generally made from tungsten (**Figure 11-1**). As current flows through the tungsten filament, it gets very hot. The conversion of electrical energy to heat energy in the resistive wire filament is so intense that the filament starts to glow and emits light. The lamp must have a vacuum surrounding the filament to prevent it from burning so hot that the filament burns in two. The glass envelope that encloses the filament maintains the presence of a vacuum. During manufacturing, all the air is removed, and the glass envelope seals out the air. If air is allowed to enter the lamp, the oxygen will cause the filament to oxidize and burn.

Many lamps can execute more than one function. A **double-filament lamp** has two filaments, allowing the bulb to perform more than one function (**Figure 11-2**). It can be combined in the stop light, taillight, and turn signal circuits.

Burned-out lamps require replacement with the correct lamp. The technician can determine what lamp to use by checking the lamp's standard trade number (**Table 11-1**).

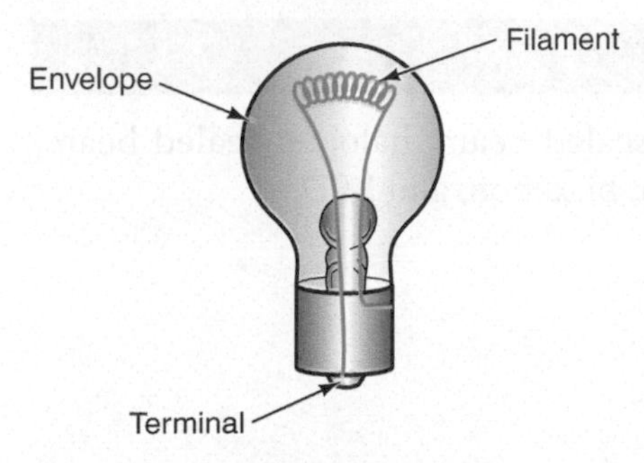

Figure 11-1 A single-filament bulb.

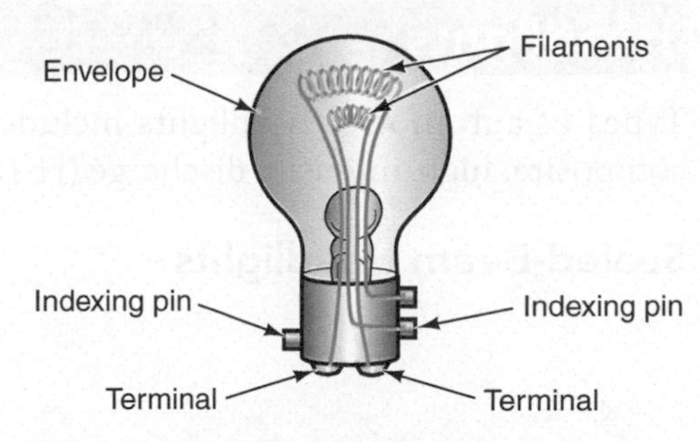

Figure 11-2 A double-filament lamp.

Table 11-1 A table of some typical automotive light bulb examples

Typical Automotive Light Bulbs			
Trade Number	**Design Volts**	**Design Amperes**	**Watts: P = A × V**
168	14.0	0.35	4.9
192	13.0	0.33	4.3
194	14.0	0.27	3.8
194E-1	14.0	0.27	3.8
194NA	14.0	0.27	3.8
912	12.8	1.00	12.8
921	12.8	1.40	17.92
1141	12.8	1.44	18.4
1142	12.8	1.44	18.4
1156	12.8	2.10	26.9
1157	12.8	2.10/0.59	26.9/7.6
1157A	12.8	2.10/0.59	26.9/7.6
1157NA	12.8	2.10/0.59	26.9/7.6
2057	12.8	2.10/0.48	26.9/6.1
2057NA	12.8	2.10/0.48	26.9/6.1
3057	12.8–14.0	2.10/0.48	26.9/6.72
3156	12.8	2.10	26.9
3157	12.8–14.0	2.10/0.59	26.9/8.26
3457	12.8–14.0	2.23/0.59	28.5/8.26
4157	12.8–14.0	2.23/0.59	28.5/8.26
6411	12.0	0.833	10.0
6418	12.0	0.417	5.0
7440	12.0	1.75	21.0
7443	12.0	1.75/0.417	21.0/5.0
7507	12.0	1.75	21.0

Headlights

Types of automotive headlights include standard sealed beam, halogen sealed beam, composite, high-intensity discharge (HID), projector, bi-xenon, and LED.

Sealed-Beam Headlights

A BIT OF HISTORY

In 1940, Federal Motor Vehicle Safety Standard 108 required all cars sold in the United States use two 7-inch round sealed-beam headlights. The headlight was technically known as the Par 56 for parabolic aluminized reflector that is 56 eights of an inch (7-inches) in diameter. The headlight incorporated both the low and high beams. In 1957, updated rules allowed two pairs of 6-inch headlamps, one low beam and one high beam on each side. In 1975, rectangular sealed-beam headlights were approved. The last vehicles to use round 7-inch headlights were the Jeep Wrangler, Toyota FJ Cruiser, and the Mercedes-Benz G-Wagon. The Jeep Wrangler used sealed-beam headlights until 2006.

Shop Manual
Chapter 11, page 525

The **sealed-beam headlight** is a self-contained glass unit composed of a filament, an inner reflector, and an outer glass lens (**Figure 11-3**). The standard sealed-beam headlight does not surround the filament with its own glass envelope (bulb). The glass lens is fused to the parabolic reflector, which is sprayed with **vaporized aluminum** that gives a reflecting surface that is comparable to silver. The reflector intensifies the light produced by the filaments, and the lens directs the light to form the required light beam pattern. The inside of the lamp is filled with argon gas. All oxygen is removed from the standard sealed-beam headlight during manufacturing to prevent the filament from oxidizing.

The lens produces a broad, flat beam. The light from the reflector passes through concave **prisms** in the glass lens (**Figure 11-4**). Lens prisms redirect the light beam and create a broad, flat beam. **Figure 11-5** shows the light beam's horizontal spreading and vertical control to prevent upward glaring.

By placing the filament in different locations on the reflector, the direction of the light beam is controlled (**Figure 11-6**). In a dual-filament lamp, the lower filament is for the high beam, and the upper filament is for the low beam.

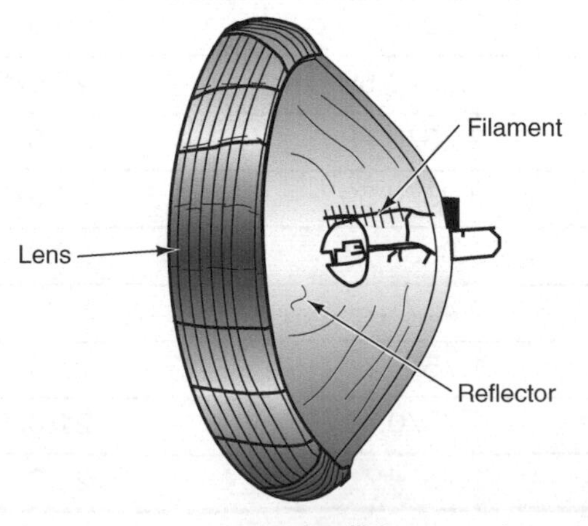

Figure 11-3 Sealed-beam headlight construction.

Figure 11-4 The lens uses prisms to redirect the light.

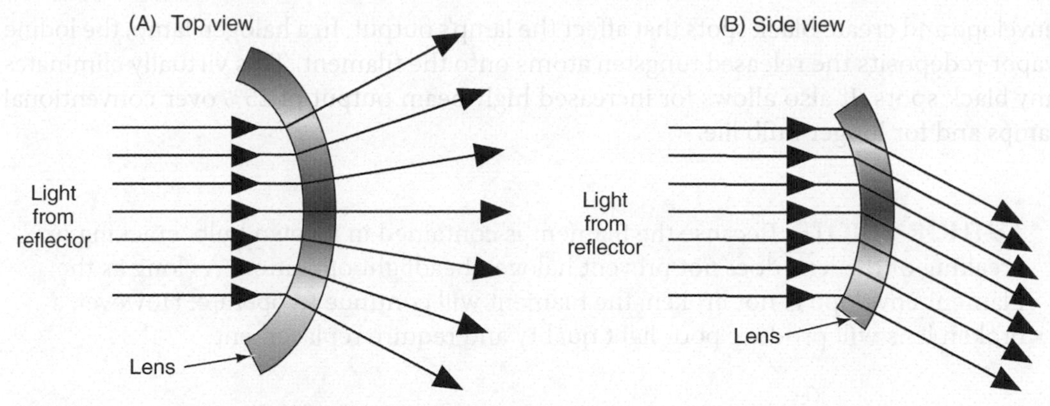

Figure 11-5 The prism directs the beam into (A) a flat horizontal pattern and (B) downward.

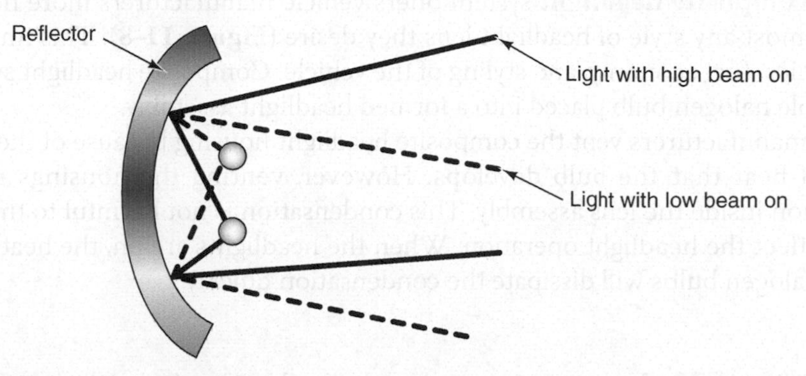

Figure 11-6 Filament placement controls the projection of the light beam.

Halogen Headlights

The sealed-beam **halogen** lamp used in automotive applications consists of a small bulb filled with iodine vapor. The inner bulb has a high-temperature-resistant quartz bulb that surrounds a tungsten filament. The inner bulb resides in a sealed glass housing (**Figure 11-7**). Adding halogen to the bulb allows the tungsten filament to withstand temperatures higher than conventional sealed-beam lamps. The higher temperatures mean the bulb can burn brighter.

Halogen is the term used to identify a group of chemically related nonmetallic elements. These elements include chlorine, fluorine, and iodine.

In a conventional sealed-beam headlight, heating the filament causes the release of tungsten atoms from the surface of the filament. These atoms deposit on the glass

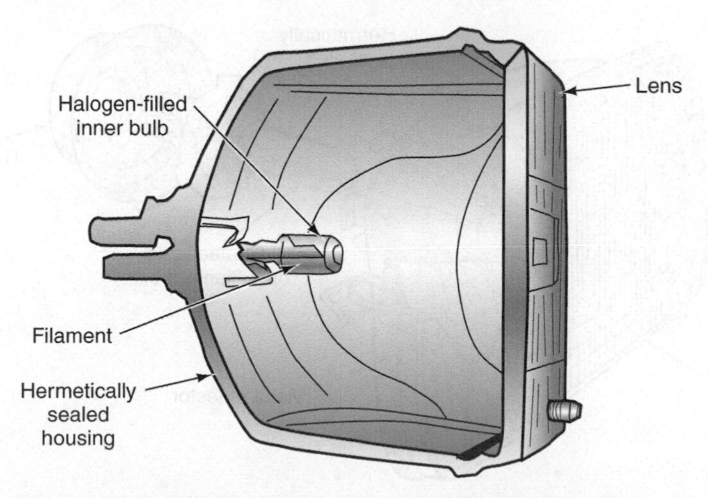

Figure 11-7 A sealed-beam halogen headlight with iodine vapor bulb.

envelope and create black spots that affect the lamp's output. In a halogen lamp, the iodine vapor redeposits the released tungsten atoms onto the filament. This virtually eliminates any black spots. It also allows for increased high-beam output of 25% over conventional lamps and for longer bulb life.

AUTHOR'S NOTE Because the filament is contained in its own bulb, cracking or breaking of the lens does not prevent halogen headlight operation. As long as the filament envelope is not broken, the filament will continue to operate. However, a broken lens will produce poor light quality and require replacement.

Composite Headlights

Shop Manual
Chapter 11, page 526

Composite headlights are units that have a lens and housing assembly with a replaceable headlight bulb. The bulb can be halogen or HID.

Using the **composite headlight** system offers vehicle manufacturers more flexibility to produce almost any style of headlight lens they desire (**Figure 11-8**). This improves the aerodynamics, fuel economy, and styling of the vehicle. Composite headlight systems use a replaceable halogen bulb placed into a formed headlight assembly.

Many manufacturers vent the composite headlight housing because of the increased amount of heat that the bulb develops. However, venting the housings may cause condensation inside the lens assembly. This condensation is not harmful to the bulb and does not affect the headlight operation. When the headlights are on, the heat generated from the halogen bulbs will dissipate the condensation quickly.

AUTHOR'S NOTE Ford uses integrated non-vented composite headlights. On these vehicles, condensation is not considered normal and requires assembly replacement.

HID Headlights

Shop Manual
Chapter 11, page 527

High-intensity discharge (HID) headlamps use an inert gas to amplify the light produced by arcing across two electrodes. HID headlamps (**Figure 11-9**) put out three times more light and twice the light spread on the road than conventional halogen headlamps (**Figure 11-10**). They also use about two-thirds less power to operate and will last two to three times longer. HID lamps produce light in both ultraviolet and visible wavelengths. This advantage allows highway signs and other reflective materials to glow.

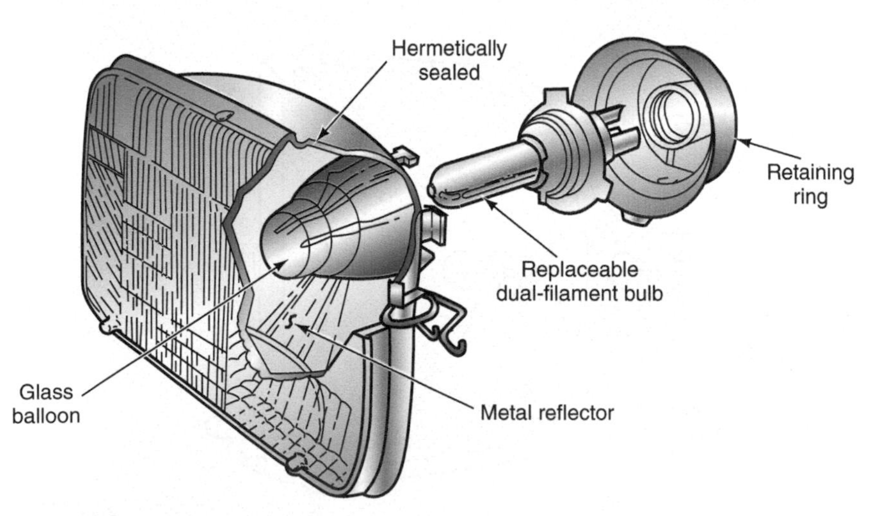

Figure 11-8 A composite headlight system with a replaceable halogen bulb.

Figure 11-9 HID headlamps; note the reduced size of the headlamp assemblies within the lens.

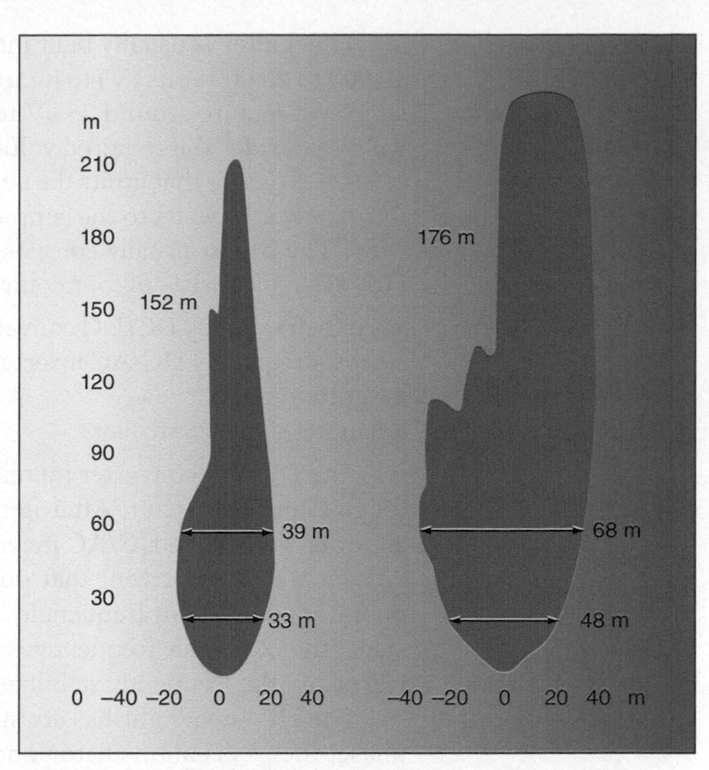

Figure 11-10 Comparison between light intensity and pattern. The halogen lamp is on the left, and the xenon (HID) lamp is on the right.

The HID lamp (**Figure 11-11**) consists of an outer bulb made of cerium-doped quartz that houses the inner bulb (arc tube). The inner bulb is made of fused quartz and contains two tungsten electrodes. It also is filled with xenon gas, mercury, and metal halides (salts).

The HID lamp does not rely on a glowing filament for light. Instead, it uses a high-voltage arcing bridge across the air gap between the electrodes. The xenon gas amplifies the light intensity given off by the arcing. The HID system requires the use of an **ignitor** and a **ballast** to provide the electrical energy required to arc the electrodes (**Figure 11-12**).

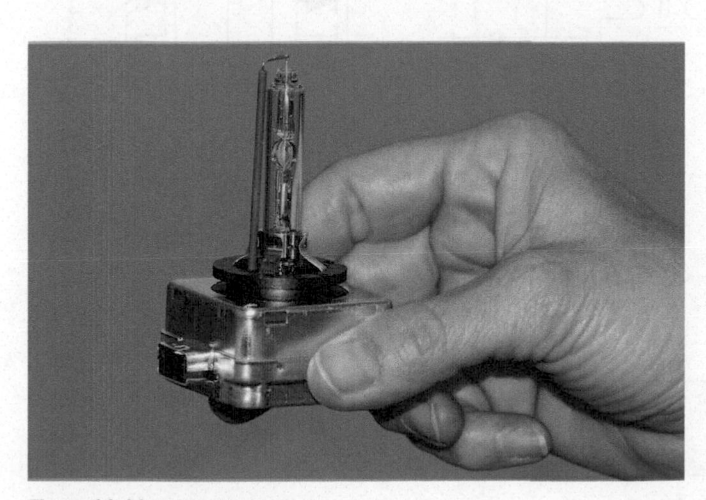

Figure 11-11 HID bulb element.

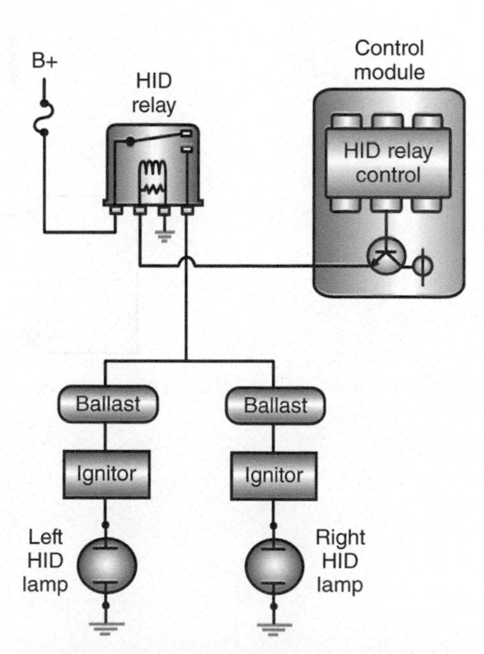

Figure 11-12 HID headlight schematic showing the use of a ballast and an ignitor.

The ignitor is usually built into the base of the HID bulb and will provide the initial 10,000 to 25,000 volts (V) to jump the gap. A cold lamp requires about 10 kV, while a hot lamp will require around 25 kV to ignite. With the gap jumped, the gas warms, and the ballast provides the required voltage to maintain the current flow across the gap. *Ballast* refers to a device that limits the current delivered through an electrical circuit. The ballast must deliver 35 watts to the lamp when the voltage across the lamp is between 70 and 110 volts. The ballast usually consists of a transformer and a capacitor (**Figure 11-13**). The HID system consists of four components:

- High-frequency DC/DC converter.
- Low-frequency DC/AC inverter.
- Ignition circuit.
- Digital signal controller.

The DC/DC converter increases the 12 volts from the battery to the level required for the ignition circuit. After ignition, the output voltage from the converter drops to about 85 volts. The DC/AC inverter converts the voltage output of the converter to an AC square wave current that puts an equal charge on the electrodes. The converter operates on different frequencies depending on the stage the lamp is in. For the turn-on stage, the converter frequency is 1 kHz and increases to 20 Hz during warm-up. After warm-up, the frequency stabilizes at 200 Hz.

The HID headlight has six stages of start-up (**Figure 11-14**). Prior to the turn-on phase, the gas is an insulator, and the resistance is infinite. During the turn-on phase,

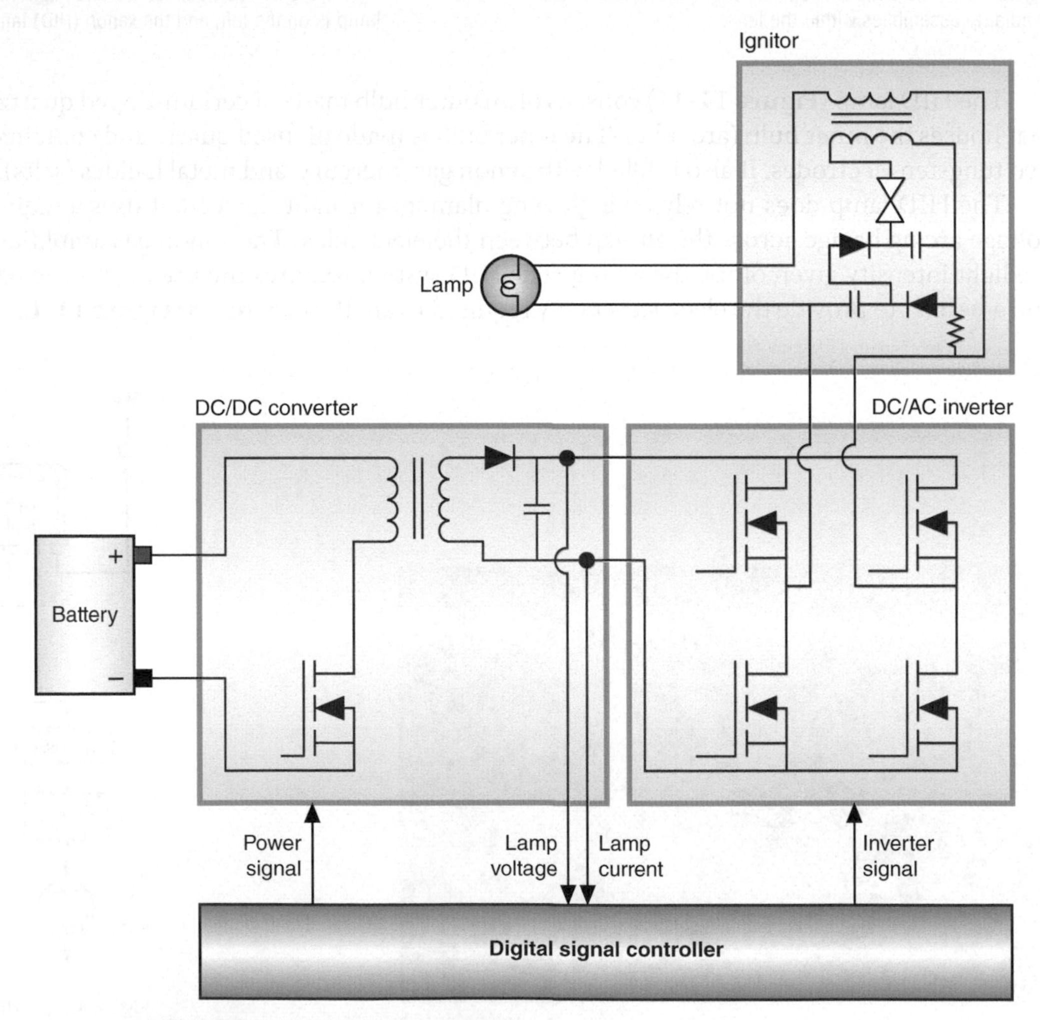

Figure 11-13 The components of the ballast assembly.

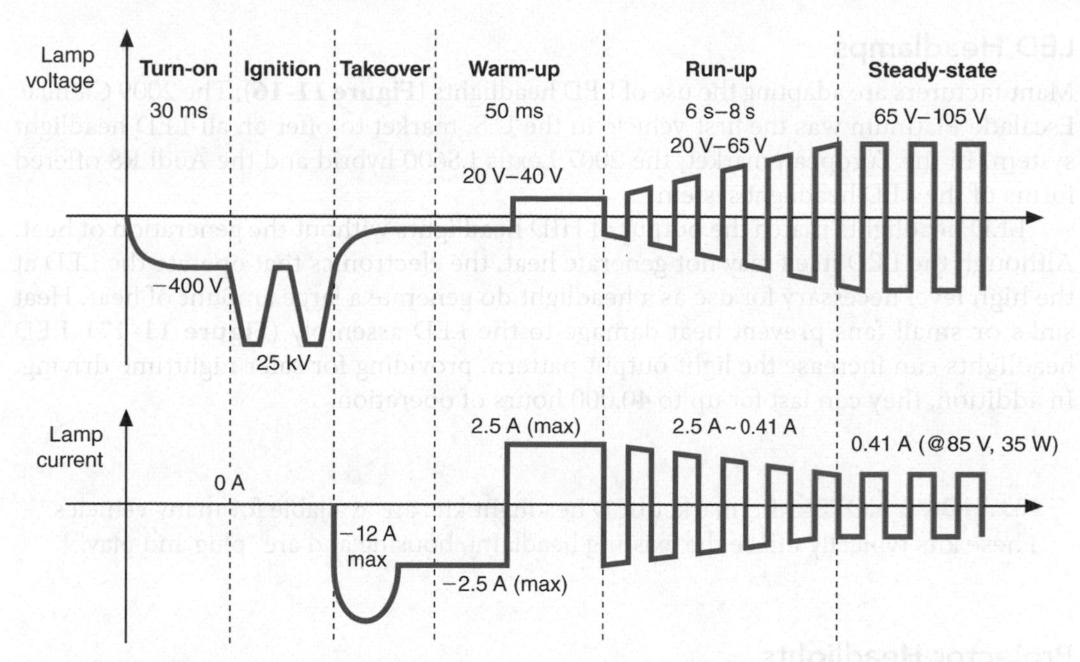

Figure 11-14 Graph showing the lamp voltage and current during the six stages of lighting the HID headlight.

the voltage generated in the ballast is sent to the ignitor. The ignitor then produces a high-voltage pulse and sends it to the HID lamp electrodes. Under the presence of the high voltage, the gas that was an insulator becomes conductive. An arc jumps across the electrodes, producing light. Once the gas has ignited, a large amount of current is required to maintain the arc. The ballast's capacitor initially provides this takeover high current. Next, the converter provides the current required to warm the gas. The warm-up of the gas takes about 50 milliseconds; then, the run-up stage starts. During this stage, the voltage increases while the current decreases. Once a steady state is reached, the voltage output is held to about 85 volts while the current is held at 410 mA.

The greater light output of these lamps allows the headlamp assembly to be smaller and lighter. These advantages allow designers more flexibility in body designs as they attempt to make their vehicles more aerodynamic and efficient. For example, the Infiniti Q45 models have a seven-lens HID system (**Figure 11-15**) to provide stylish looks and high lamp output.

Figure 11-15 Seven-lens HID headlamp.

LED Headlamps

Manufacturers are adapting the use of LED headlights (**Figure 11-16**). The 2009 Cadillac Escalade Platinum was the first vehicle in the U.S. market to offer an all-LED headlight system. In the European market, the 2007 Lexus LS600 hybrid and the Audi R8 offered forms of the LED headlight system.

LED headlights match the output of HID headlights without the generation of heat. Although the LED itself may not generate heat, the electronics that operate the LED at the high level necessary for use as a headlight do generate a large amount of heat. Heat sinks or small fans prevent heat damage to the LED assembly (**Figure 11-17**). LED headlights can increase the light output pattern, providing for safer nighttime driving. In addition, they can last for up to 40,000 hours of operation.

> **AUTHOR'S NOTE** Aftermarket LED headlight kits are available for many vehicles. These kits typically utilize the existing headlamp housing and are "plug and play."

Projector Headlights

The projector is a type of headlight; it's not a bulb type. It can use an HID, halogen, bi-xenon, or LED bulb. Like a light bulb used in a projector, the light is focused in front of the vehicle.

Projector headlights (**Figure 11-18**) provide very accurate light dispersion with less scatter. They offer a crisp line of light that is focused on the ground and illuminates most everything under the line. A projector headlight produces a stronger light beam in a more focused area.

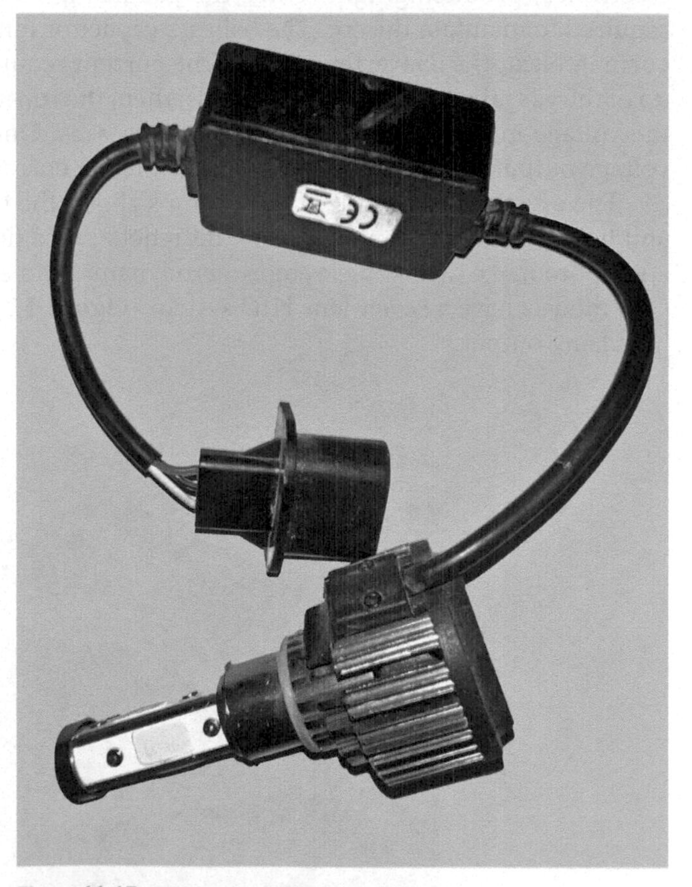

Figure 11-17 The LED headlight unit may use heat sinks or fans to remove heat.

Figure 11-16 LED headlight.

Figure 11-18 Projector headlight.

The bulb is positioned at the rear of an elliptical reflector, and a condenser lens is located in front of the bulb (**Figure 11-19**). A shield placed between the bulb and the lens provides cutoff of the light beam. This configuration harnesses and focuses the light from the reflector, and light from the reflector is not lost.

Bi-Xenon Headlamps

Due to the increased amount of light produced by the HID and the projector type headlight, they are not used as high-beam lamps in a quad projector headlight system. Instead, they are used only as low-beam lamps, and a halogen bulb provides the high beam. Because the quad headlamp system uses all four bulbs for high-beam operation, using a quad lamp system with HID or projector lamps would blind any oncoming drivers due to the excessive light output. Also, it is not possible to reduce the light intensity of an HID projector headlight by pulse-width modulation of the current to the element.

To overcome these issues, many manufacturers use **bi-xenon headlamps** in their dual headlamp systems. *Bi-xenon* refers to using a single xenon lamp to provide both the high-beam and the low-beam operations. The full light output produces the high beam. Low beam is formed by either moving the xenon bulb within the lens or moving a shutter between the bulb and the lens.

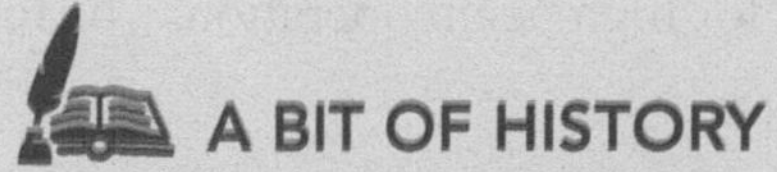

A BIT OF HISTORY

The Model T Ford used a headlight system that had a replaceable bulb. The owner's manual warned against touching the reflector except with a soft cloth.

Systems that use the shutter will have a motor within the headlamp assembly that raises and lowers the shutter (**Figure 11-20**). The position of the shutter dictates the amount of projected light and its pattern.

AUTHOR'S NOTE The actual direction the shutter travels to block or unblock varies based on application.

Some systems use a motor to change the position of the bulb. The bulb is physically raised in the reflector housing to produce the high-beam output. Lowering the bulb in the reflector housing places the headlight in low-beam mode. The amount of reflection dictates the light intensity and pattern.

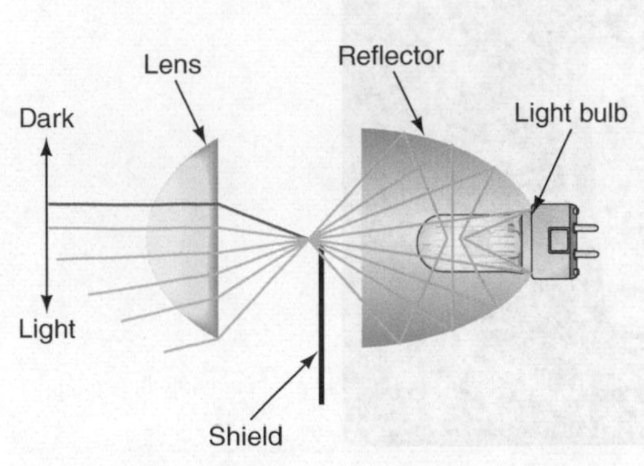

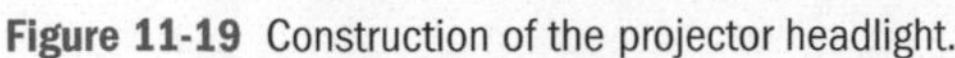

Figure 11-19 Construction of the projector headlight.

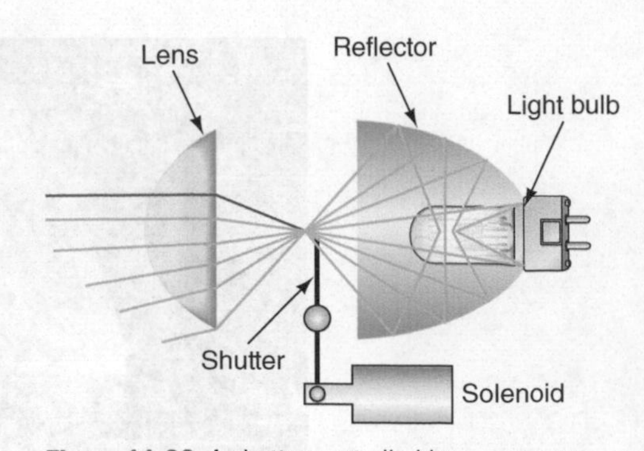

Figure 11-20 A shutter controlled by an actuator motor controls the amount of light projected ahead of the vehicle.

Using the same lamp for both low- and high-beam operations permits both modes to have the same light color. This produces less visual contrast when switching between modes and is less stressful to the eyes of the driver.

Laser Headlamps

Laser headlights (**Figure 11-21**) are the latest entry in the technology race to improve automotive lighting. The laser headlight is not a laser beam; instead, it is a defused light pattern generated by lasers.

The laser headlight assembly consists of one or more solid state indium gallium nitride diode lasers positioned at the rear of the assembly (**Figure 7-22**). The laser diodes project a blue light onto a set of mirrors located at the front of the laser channel. The mirrors focus the laser energy into a lens filled with yellow phosphorus. When struck by the blue laser energy the yellow phosphorus becomes excited and creates a bright, highly intense white light. The intense white light is directed backward onto a special reflector to diffuse it and reduce its intensity. The reflector then projects the light outward through the headlight assembly's outermost lens. The result is a light 10 times brighter than that of an LED headlight.

The illumination of the laser generated light reaches out to 2,000 feet (600 meters) ahead of the vehicle. This is twice the distance that an LED high beam can reach. Because of the intense output of these lights, they are only used for high beam operation. The low beams use conventional LED headlamps.

Figure 11-21 BMW was one of the first manufacturers to use laser headlights.

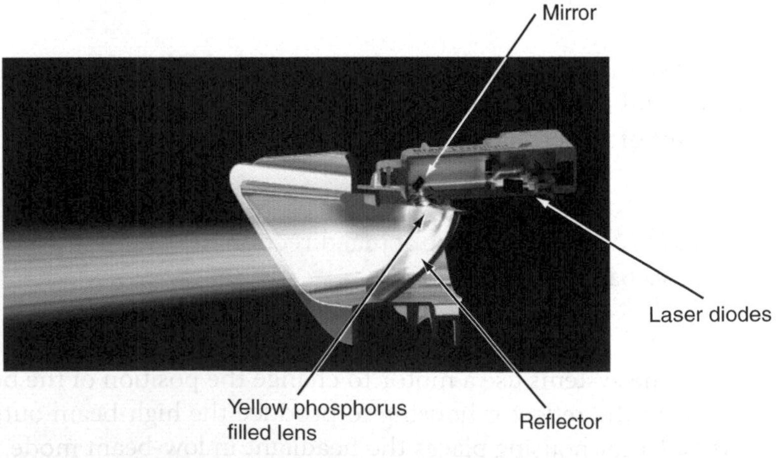

Figure 11-22 The laser headlight assembly. Courtesy of BMW.

Although used for high beam operation, the laser circuit is always on when the headlights are on. During low beam operation, a shutter closes to block the light. The shutter opens to project the high beams.

Headlight Switches

Shop Manual
Chapter 11, page 554

The headlight switch may be located either on the dash by the instrument panel or on the steering column multifunction switch (**Figure 11-23**). Depending on the application, the headlight switch is one of two types. The first is a switch that carries current to the lamps directly or to a control device, such as a relay. The second type is a switch that is only an input to the BCM and does not directly activate the lights. Either switch will control a majority of the vehicle's lighting systems.

The most common style of current-carrying headlight switch is the three-position type with OFF, PARK, and HEADLIGHT positions. The headlight switch will generally receive direct battery voltage to two switch terminals. This allows the operation of the light circuits without having the ignition in the RUN or ACC (accessory) position.

When the headlight switch is in the OFF position, the open contacts prevent battery voltage from continuing to the lamps. When the switch is in the PARK (parking lights) position, the closed contacts apply the battery voltage at terminal 5 to the side marker, taillight, license plate, and instrument cluster lights (**Figure 11-24**). These circuits are usually protected by a 15- to 20-amp fuse that is separate from the headlight circuit.

Shop Manual
Chapter 11, page 552

When the switch is in the HEADLIGHT position, the closed contacts apply battery voltage that is present at terminal 1 through the circuit breaker to light the headlights. The circuit breaker prevents temporary overloads to the system from totally disabling the headlights. Battery voltage from terminal 5 continues to illuminate the lights that were on in the PARK position (**Figure 11-25**)

The headlight circuits just discussed are designed with insulated side switches and grounded bulbs. In this system, battery voltage is applied to the headlight switch. The switch must be closed for current to flow through the filaments and to ground. The circuit is complete because the headlights are grounded to the vehicle body or chassis. Many manufacturers use a system design that has insulated bulbs and ground side switches. In this system, locating the headlight switch in the HEADLIGHT position closes the contacts to complete the circuit path to ground. Battery voltage is applied directly to the headlights when the relays are closed. But the headlights will not light until the switch completes the ground side of the relay circuits. In this system, both the headlight and dimmer switches complete the circuits to ground.

Figure 11-23 (A) Instrument panel-mounted headlight switch. (B) Steering column-mounted headlight switch.

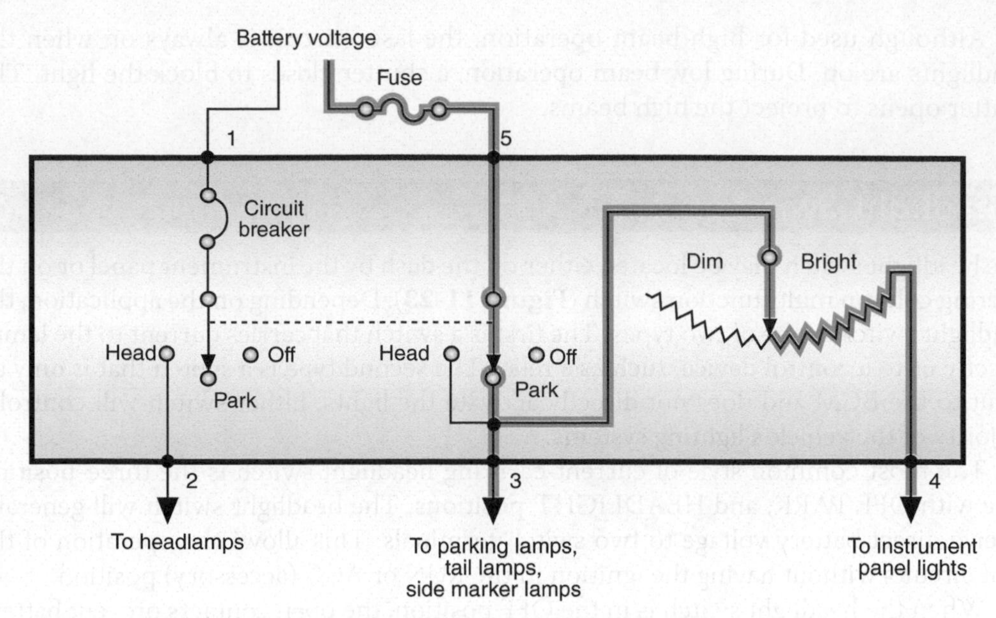

Figure 11-24 Operation with the headlight switch in the PARK position.

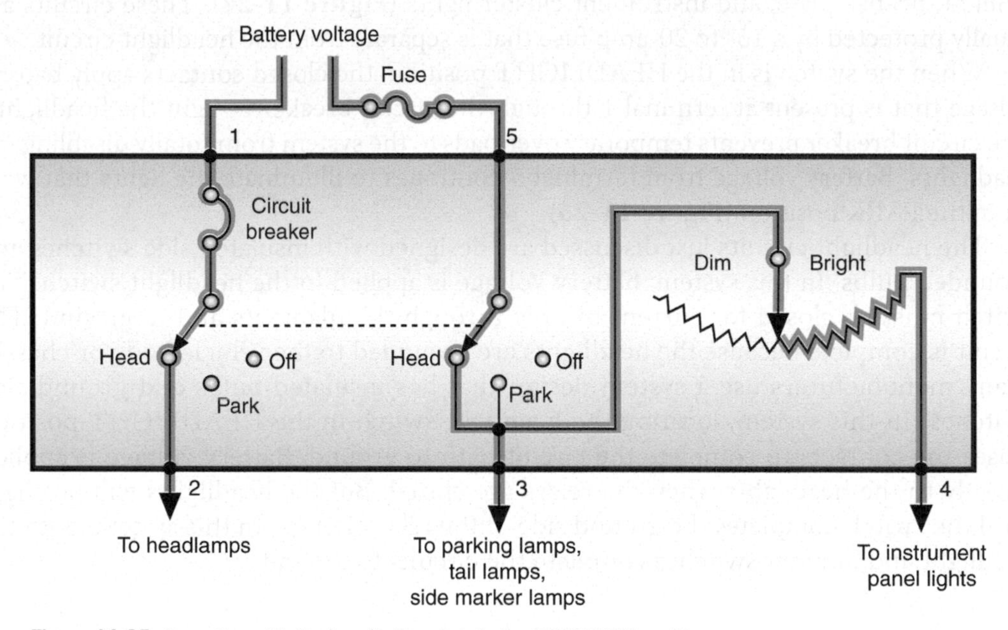

Figure 11-25 Operation with the headlight switch in the HEADLIGHT position.

The headlight switch rheostat is a variable resistor the driver uses to control the brightness of the instrument cluster illumination lamp. As the driver turns the light switch knob, the resistance in the rheostat changes. The greater the resistance, the dimmer the instrument panel illumination lights glow. In vehicles with the headlight switch in the steering column, the rheostat may be a separate unit located on the dash near the instrument panel.

Shop Manual
Chapter 11, page 529

In computer-controlled headlight systems, the headlight switch is usually a resistive multiplex switch that provides multiple inputs over a single circuit.

AUTHOR'S NOTE The resistive multiplex switch was discussed in Chapter 9.

The BCM will control the relays associated with the exterior lighting. If input from the switch indicates that the headlights are requested, the BCM will energize the low-beam headlight relay (**Figure 11-26**). With this relay energized, current flows to the low-beam lamps. Also, voltage the multifunction switch (dimmer switch function) receives battery voltage. Placing the multifunction switch in the high-beam position completes the path for current flow to the coil of the high-beam headlight relay. Since this relay is connected to ground, the coil is energized and the contacts move to supply current to the high-beam lamps. The low-beam relay must be energized to turn on the high beams unless using the flash-to-pass function. The multifunction switch's flash-to-pass function bypasses the circuit's BCM control. The flash-to-pass circuit illuminates the high-beam headlights even with the headlight switch in the OFF or PARK position (**Figure 11-27**). In this illustration, battery voltage is supplied to terminal B1 of the headlight switch and to the dimmer switch. Battery voltage is available to the dimmer switch through this wire in both the OFF and PARK positions of the headlight switch. When the driver activates the flash-to-pass feature, the contacts in the dimmer switch complete the circuit to the high-beam filaments.

The **dimmer switch** provides the means for the driver to select high- or low-beam operation and to switch between the two. In systems that do not utilize computer control, the dimmer switch is connected in series within the headlight circuit and controls the

Shop Manual
Chapter 11, page 556

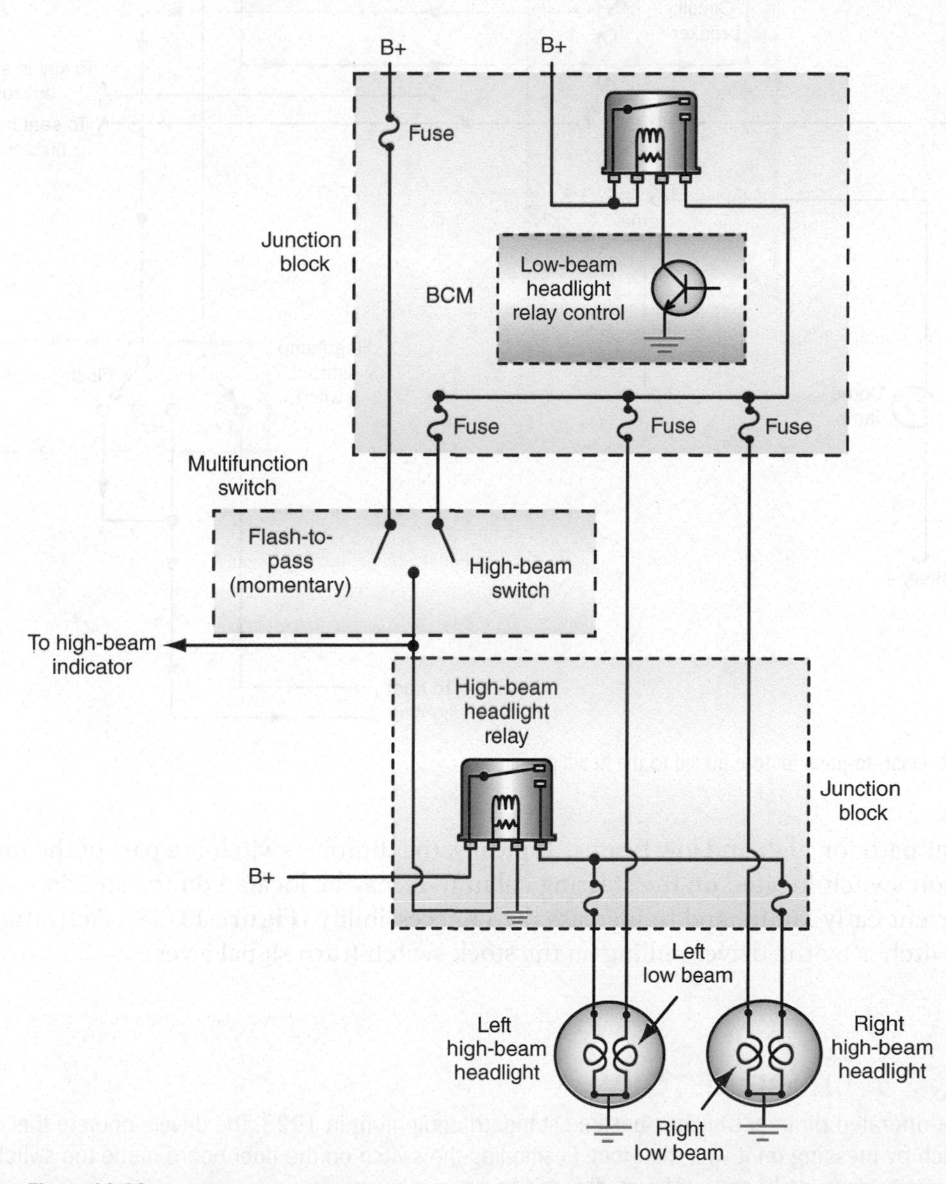

Figure 11-26 The body computer controls the relay of the headlight system.

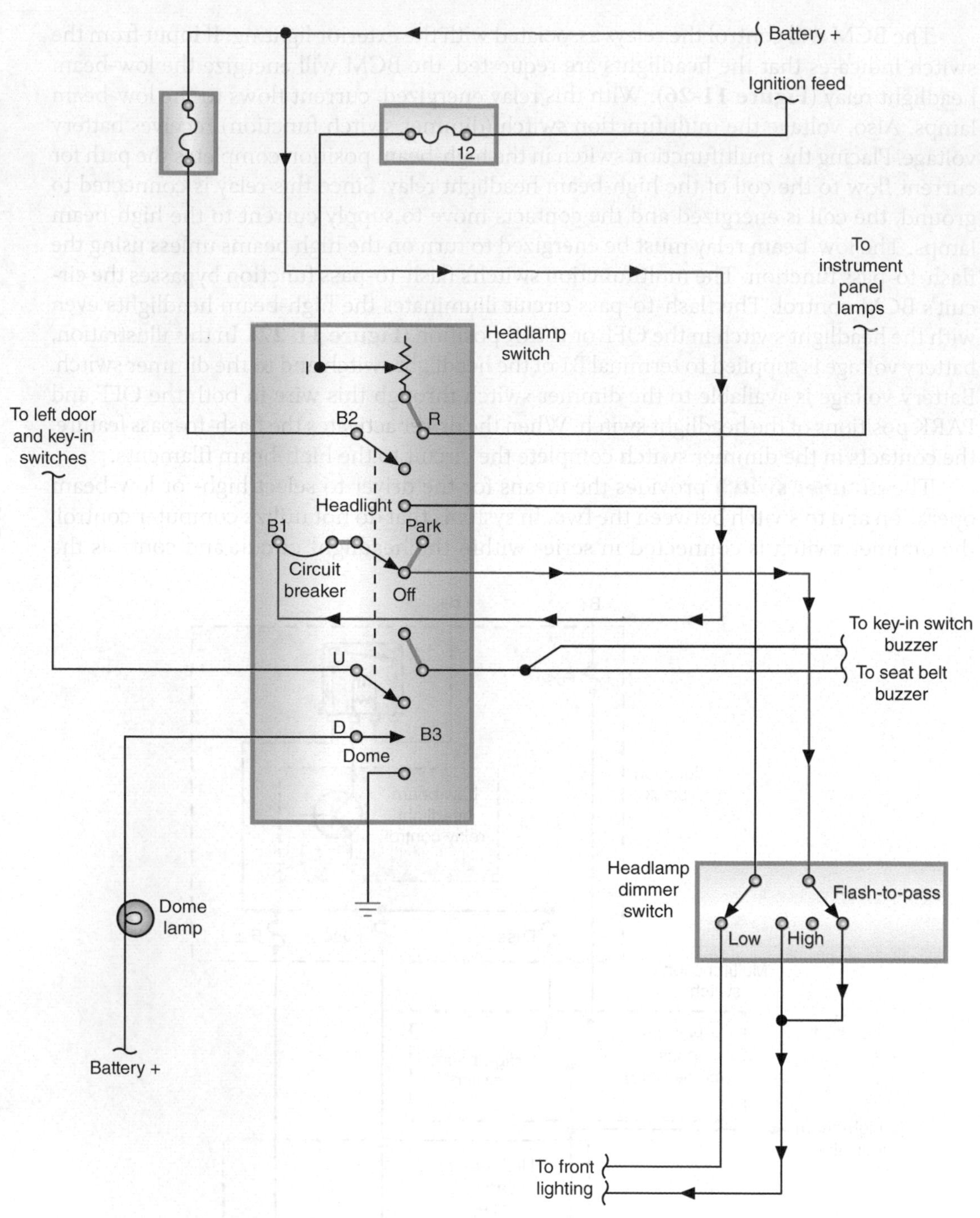

Figure 11-27 Flash-to-pass feature added to the headlight circuit.

current path for high and low beams. Typically, the dimmer switch is a part of the multi-function switch located on the steering column. It may be located on the steering wheel to prevent early failure and to increase driver accessibility (**Figure 11-28**). Activation of this switch is by the driver pulling on the stock switch (turn signal lever).

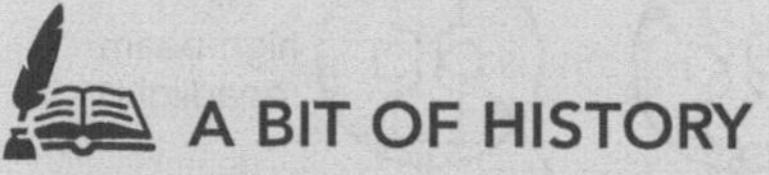

A BIT OF HISTORY

Foot-operated dimmer switches became standard equipment in 1923. The drivers operate this switch by pressing on it with their foot. Positioning the switch on the floor board made the switch subject to damage because of rust, dirt, and so on.

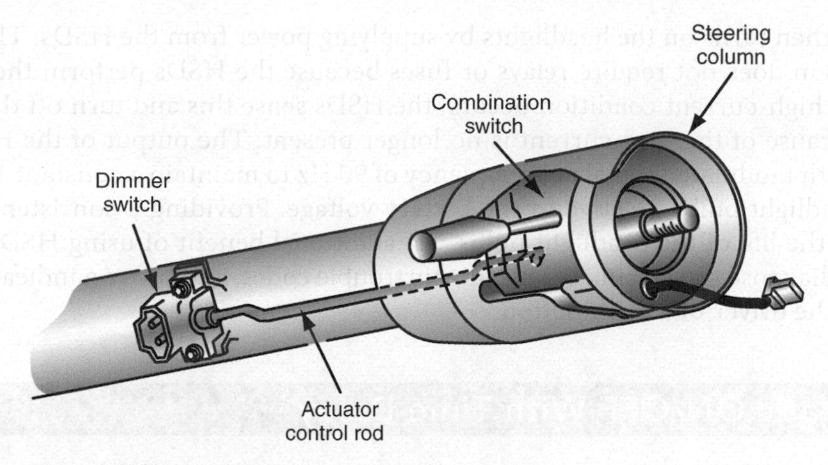

Figure 11-28 A steering column–mounted dimmer switch.

In most modern systems, the dimmer switch is an input to the BCM. In this case, the BCM will activate the high-beam relay. The BCM may also control the flash-to-pass feature.

Shop Manual
Chapter 11, page 531

Many modern computer-controlled headlight systems use high-side drivers (HSD) (**Figure 11-29**). Low-side drivers (relay drivers) control the park and fog lamp relays in the example shown. The difference is that the BCM receives the headlight switch input and sends the request to the Integrated Power Module (IPM). The IPM then turns on and off the relays as needed. The operation of the headlights is unique. Based on the input from the headlight switch, the BCM sends the requested state to the IPM.

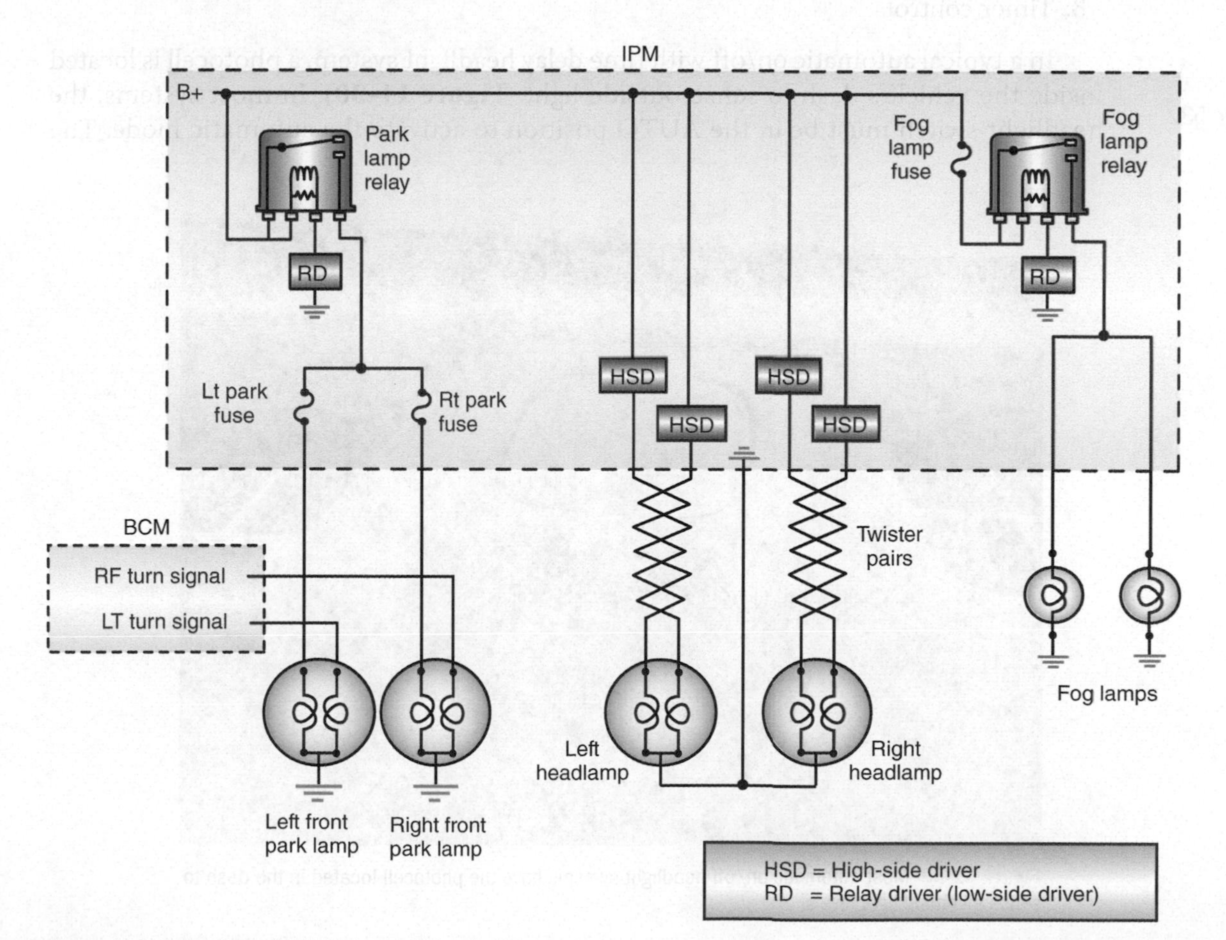

Figure 11-29 Headlight circuit that uses high-side drivers.

The IPM then turns on the headlights by supplying power from the HSDs. This headlamp system does not require relays or fuses because the HSDs perform these functions. If a high-current condition occurs, the HSDs sense this and turn off the circuit until the cause of the high current is no longer present. The output of the HSDs is a pulse-width modulated signal at a frequency of 90 Hz to maintain a constant 13.5 volts to the headlight bulbs, relative to the battery voltage. Providing a consistent voltage increases the life of the headlight bulbs. An additional benefit of using HSDs is their ability to diagnose the system, set diagnostic trouble codes, and turn on indicator lights to notify the driver of a malfunction.

Automatic ON/OFF With Time Delay

Shop Manual
Chapter 11, page 532

AUTHOR'S NOTE Various manufacturers give this system several different names. Some common names include Twilight Sentinel, Auto-lamp/Delayed Exit, and Safeguard Sentinel.

The photocell is a variable resistor that uses light to change resistance.

The **automatic on/off with time delay** has two functions: to turn on the headlights automatically when ambient light decreases to a predetermined level and to allow the headlights to remain on for a certain amount of time after turning the vehicle off. The common components of the automatic on/off with time delay include the following:

1. Photocell and amplifier.
2. Power relay.
3. Timer control.

Shop Manual
Chapter 11, page 532

In a typical automatic on/off with time delay headlight system, a photocell is located inside the vehicle's dash to sense outside light (**Figure 11-30**). In most systems, the headlight switch must be in the AUTO position to activate the automatic mode. The

Figure 11-30 Most automatic on/off headlight systems have the photocell located in the dash to sense incoming light levels.

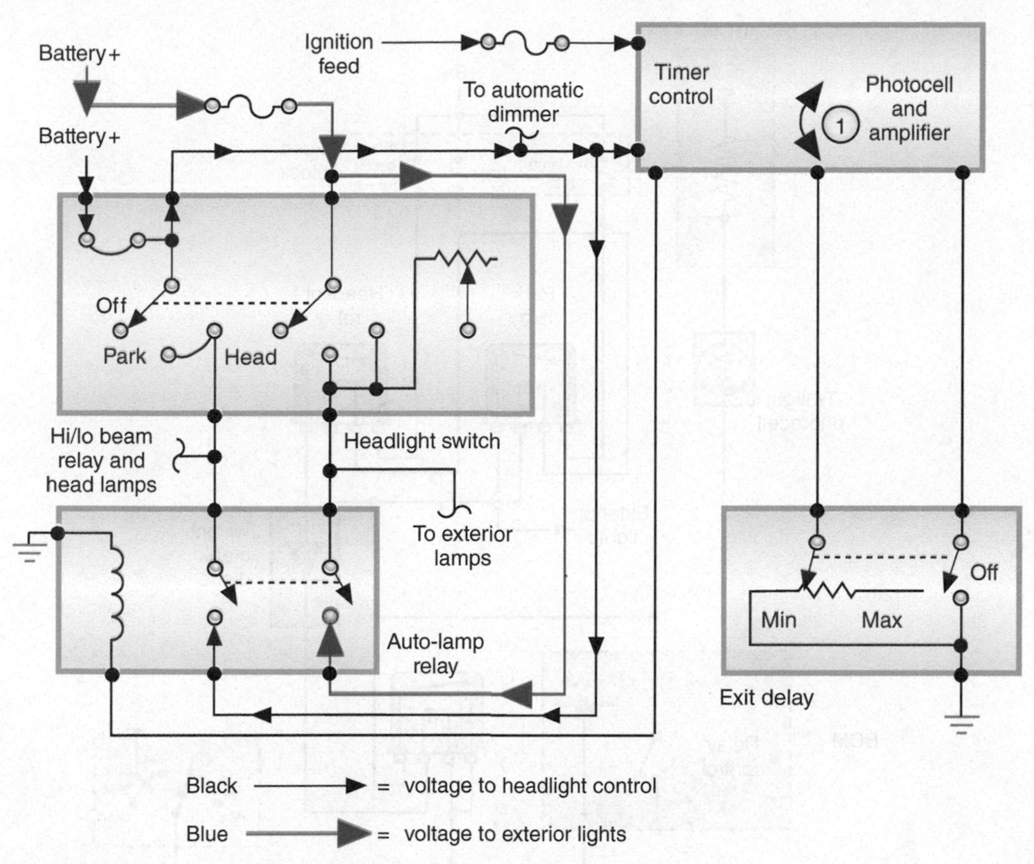

Figure 11-31 Schematic of an automatic headlight on/off with time delay system.

driver can override the automatic on/off feature by placing the headlight switch in the HEADLIGHT or OFF position.

Shop Manual
Chapter 11, page 534

Early systems utilize a **timer control** that works along with a potentiometer in the headlight switch (**Figure 11-31**). The timer control unit controls the automatic operation of the system and the length of time the headlights stay on after turning off the ignition. The timer control signals the sensor-amplifier module to energize the relay for the requested length of time.

To activate the automatic on/off feature, the photocell and amplifier receives voltage from the ignition switch. As the ambient light level decreases, the internal resistance of the photocell increases. When the resistance value reaches a predetermined level, the photocell and amplifier trigger the sensor-amplifier module. The sensor-amplifier module energizes the relay, turning on the headlights and exterior parking lights.

General Motors's (GM) Twilight Sentinel System is an example of early use of the BCM to control system operation (**Figure 11-32**). The BCM senses the voltage drop across the photocell and the delay control switch. If the ambient light level drops below a specific value, the BCM grounds the headlamp and park lamp relay coils. The BCM also keeps the headlights on for a specific length of time after turning the ignition off.

Most current systems continue to use the same principle. The difference is that the time control is a function of the module (usually the BCM). An interface module, such as the instrument panel or overhead console, provides a means for the driver to select the desired amount of time delay after the ignition is turned off. The BCM keeps the headlight relay energized for the length of time programmed. Turning off the ignition de-energizes all other relays controlling exterior lighting, except the headlight relay.

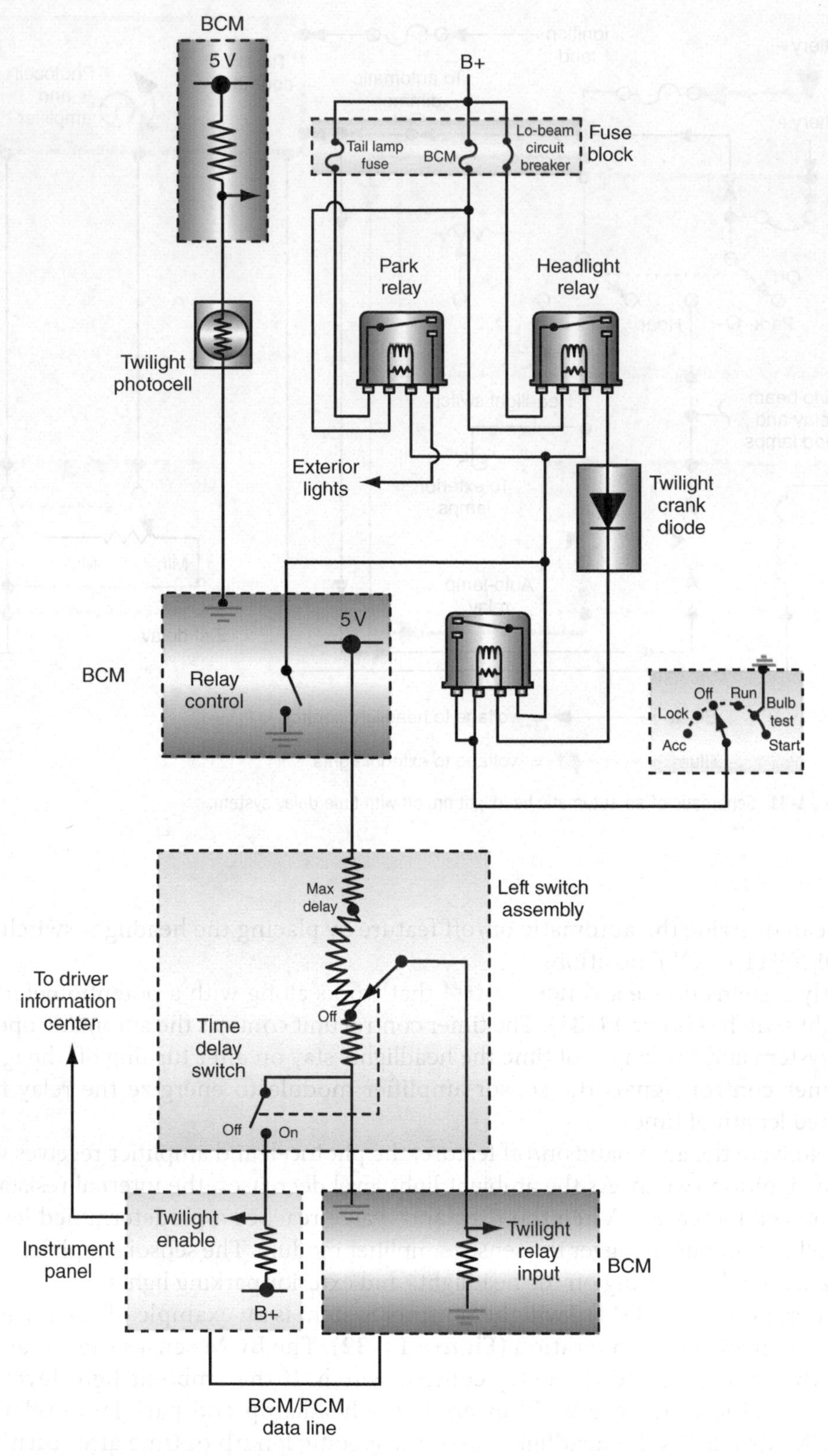

Figure 11-32 Some systems use the BCM to sense inputs from the photocell and delay control switch.

AUTHOR'S NOTE Some systems that do not have the automatic headlight on/off feature can activate the time delay feature when the headlight switch is moved from the HEADLIGHT position to OFF after the ignition is turned off.

Automatic Headlight Dimming

Modern **automatic headlight dimming** systems use solid-state circuitry and electromagnetic relays or HSDs to control beam switching. Automatic headlight dimming automatically switches the headlights from high beams to low beams under two different conditions: when light from oncoming vehicles strikes the photocell–amplifier or light from the taillights of a passing vehicle strikes the photocell–amplifier.

Most systems consist of the following major components:

1. Light-sensitive photocell and amplifier unit.
2. High–low beam relay.
3. Sensitivity control.
4. Dimmer switch.
5. Flash-to-pass relay.
6. Wiring harness.

The photocell–amplifier is usually mounted behind the front grill, but ahead of the radiator. Some systems use a **sensitivity control** that the driver uses to set the intensity level at which the photocell–amplifier will energize. The control is located next to, or is a part of, the headlight switch assembly (**Figure 11-33**). Increasing the sensitivity level will make the headlights switch to the low beams sooner (approaching vehicle is farther away). A decrease in the sensitivity level will switch the headlights to low beams when the approaching vehicle is closer. Rotating the knob to the full counterclockwise position places the system into manual override.

AUTHOR'S NOTE Many vehicle manufacturers install the sensor-amplifier in the rearview mirror support.

The high–low relay is a single-pole, double-throw unit that provides the switching of the headlight beams. The relay also contains a clamping diode for electrical transient damping to protect the photocell and amplifier assembly.

The dimmer switch is usually a flash-to-pass design. Pulling the turn signal lever partway up energizes the flash-to-pass relay. The high beams stay on as long as the lever is held in this position, even if the headlights are off. In addition, the driver can select either low beams or automatic operation through the dimmer switch.

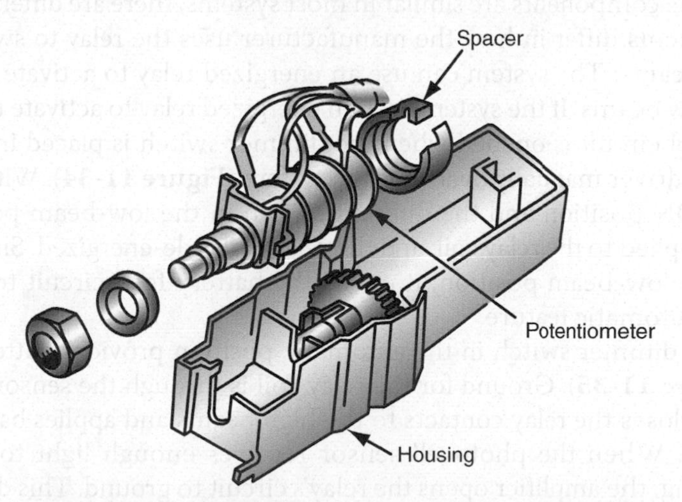

Figure 11-33 The driver sets the sensitivity of the automatic headlight dimmer system by rotating the potentiometer to change resistance values.

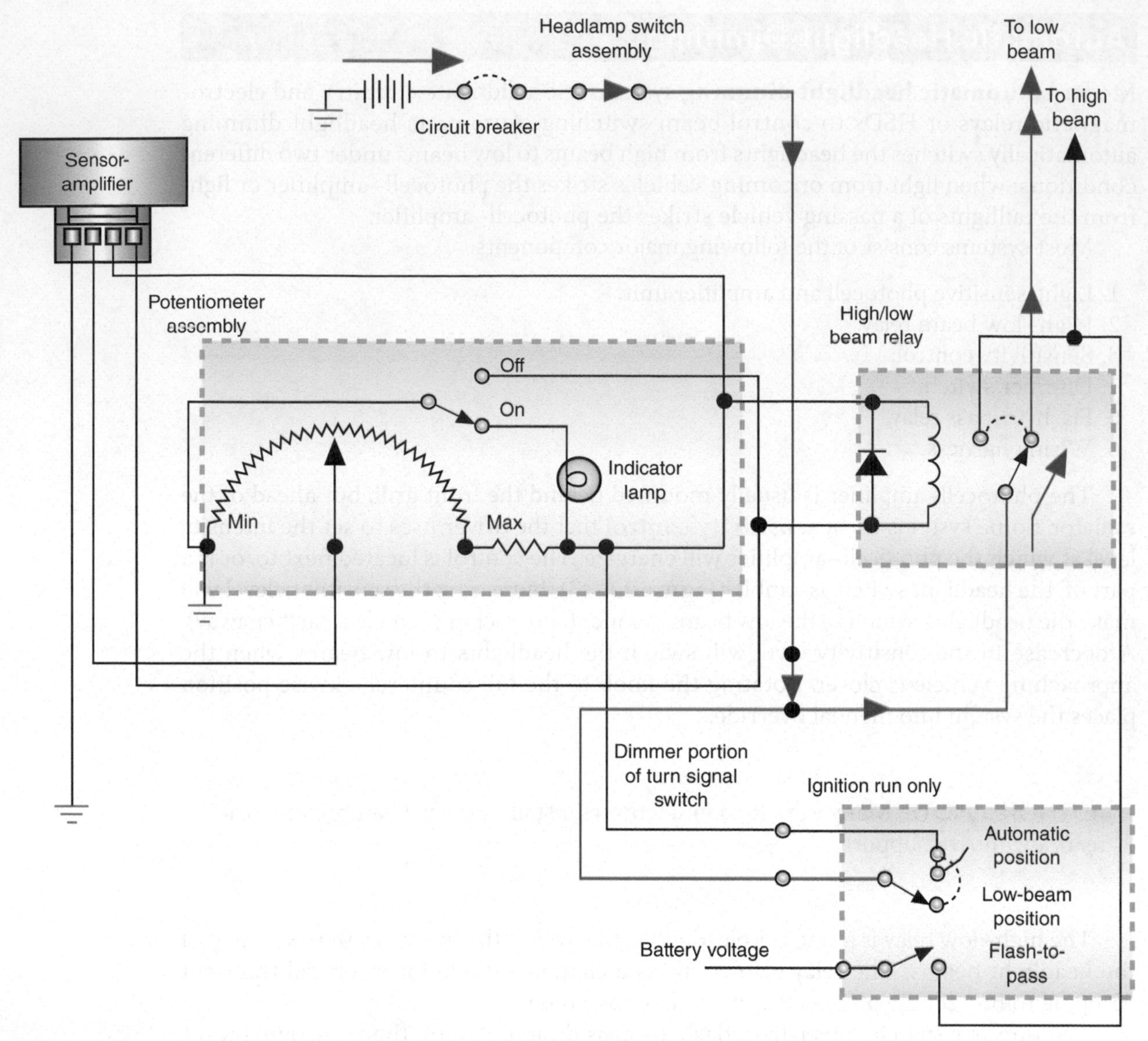

Figure 11-34 Automatic headlight dimming circuit with the dimmer switch located in the LOW-BEAM position.

Although the components are similar in most systems, there are differences in system operations. Systems differ in how the manufacturer uses the relay to switch from high beams to low beams. The system can use an energized relay to activate either the high beams or the low beams. If the system uses an energized relay to activate the high beams, the relay control circuit is opened when the dimmer switch is placed in the low-beam position, or the driver manually overrides the system (**Figure 11-34**). With the headlight switch in the ON position and the dimmer switch in the low-beam position, battery voltage is not applied to the relay coil, and the relay coil is de-energized. Since the dimmer switch is in the low-beam position, it opens the battery feed circuit to the relay coil, bypassing the automatic feature.

Placing the dimmer switch in the automatic position provides battery feed for the relay coil (**Figure 11-35**). Ground for the relay coil is through the sensor-amplifier. The energized coil closes the relay contacts to the high beams and applies battery voltage to the headlamps. When the photocell sensor receives enough light to overcome the sensitivity setting, the amplifier opens the relay's circuit to ground. This de-energizes the relay coil and switches battery voltage from the high-beam to the low-beam position.

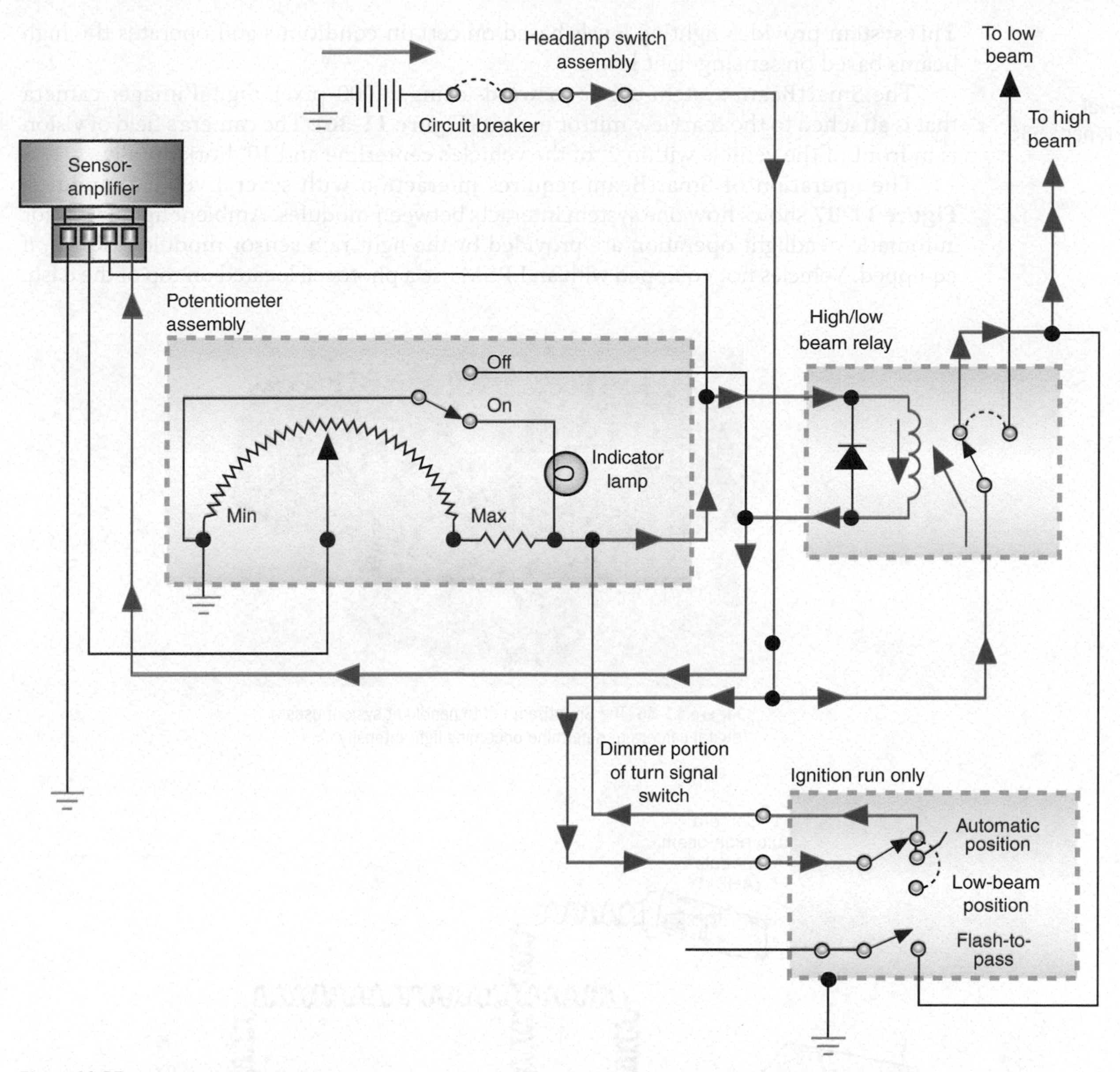

Figure 11-35 Automatic headlight dimming circuit with the dimmer switch located in the automatic position and no oncoming light sensed.

If the system uses an energized relay to switch to low beams, placing the dimmer switch in the low-beam position will energize the relay. With the headlights turned on and the dimmer switch in the automatic position, battery voltage is applied to the photocell–amplifier, one terminal of the high–low control, and through the relay contacts to the high beams. The voltage drop through the high–low control is an input to the photocell–amplifier. When enough light strikes the photocell–amplifier to overcome the sensitivity setting, the amplifier allows battery current to flow through the high–low relay, closing the contact points to the low beams. Once the light has passed, the photocell–amplifier opens battery voltage to the relay coil and the contacts close to the high beams.

When flash-to-pass is activated, the switch closes to ground. This bypasses the sensitivity control and de-energizes the relay to switch from low beams to high beams.

Today the headlight system can be very sophisticated. The following is an example of how the SmartBeam system performs auto headlamp and auto high-beam operation.

This system provides lighting levels based on certain conditions and operates the high beams based on sensing light levels.

Shop Manual
Chapter 11, page 536

The SmartBeam system uses a forward-facing, 5,000-pixel, digital imager camera that is attached to the rearview mirror mount (**Figure 11-36**). The camera's field of vision is in front of the vehicle within 2° of the vehicle's centerline and 10° horizontally.

The operation of SmartBeam requires interaction with several vehicle modules. **Figure 11-37** shows how one system interacts between modules. Ambient light levels for automatic headlight operation are provided by the light rain sensor module (LRSM), if equipped. Vehicles not equipped with an LRSM use a photocell located on top of the dash.

Figure 11-36 The SmartBeam auto headlight system uses a digital camera to determine oncoming light intensity.

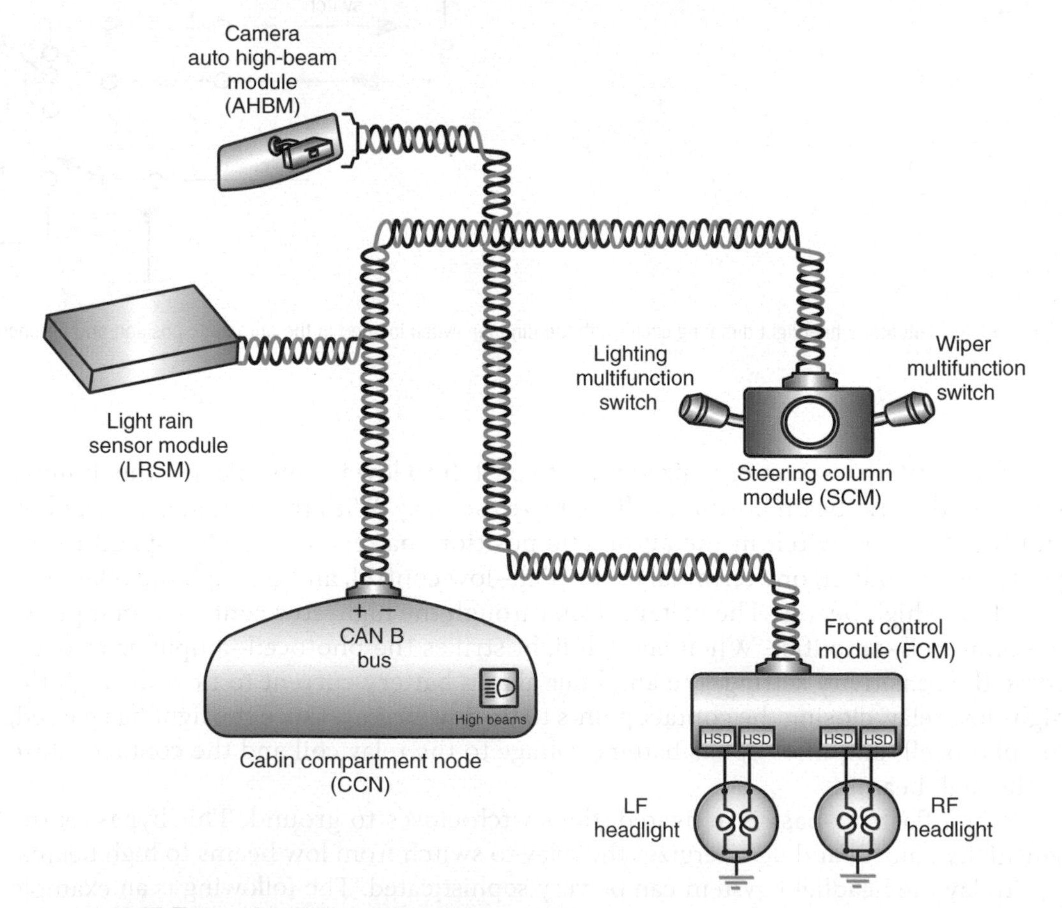

Figure 11-37 The auto headlight system integrates several modules.

The lighting multifunction switch signals the headlamp switch position to the steering column module (SCM). The SCM sends switch position status over the data bus. The front control module (FCM) uses HSDs to provide power to both the low- and high-beam bulbs. The cabin compartment node (CCN) controls the operation of the high-beam indicator.

The decisions made for headlight intensity are based on the sensed intensity of light, the light's location, and the light's movement. The system is capable of distinguishing between light types, such as mercury vapor used for street lighting. In addition, it can distinguish colors, so it is possible to identify the red lights used on tail lamps and sign colors. Distinguishing light types and colors is done by identifying the wavelength of the light source. If the approaching light source is determined to be another vehicle, a data bus message is sent from the auto high-beam module (AHBM) to the module that controls headlight operation (FCM in this case) to deactivate the high beams.

For this system to be operational, the headlight switch must be in the AUTO headlamp position, and the "LOW/HIGH BEAM" option must be selected from the configurable display (**Figure 11-38**). Also, the engine must be running, and the vehicle speed must be over 20 mph.

The auto headlight function will use either a photocell in the dash or a part of the mirror assembly to sense ambient light intensity. When the engine is running, and the ambient light levels are less than 1,000 **LUX**, the auto headlamp low beam becomes operational. LUX is the International System unit of measurement of the intensity of light. It is equal to the illumination of a surface 1 meter away from a single candle (one lumen per square meter).

Once vehicle speed exceeds 20 mph, if the ambient light level sensed at the SmartBeam camera is 5 LUX or less, a PWM voltage is applied to the high-beam circuit by the controlling module (FCM). Within 2½ to 5 seconds, the high beams will be at full intensity. By using PWM, the high beam intensity ramps up and down; this eliminates the usual flash that occurs as the high beams are turned on and off. Drivers in oncoming vehicles do not see any indication of the beam change since it is gradual and based on distance.

When another vehicle approaches, the camera determines the light intensity from its headlights. Once the light intensity reaches a predetermined level, a bus message is sent to the control module (FCM) to deactivate the high beams. PWM decreases the voltage to the high beams until they turn off.

Figure 11-38 For the auto headlight system to operate, the drive must activate it.

AUTHOR'S NOTE Federal law specifies that the high-beam indicator turns on at the initial start of the ramping up to high-beam operation and remain on during all high-beam operation. High beam operation includes the ramp-down phase; thus, the indicator light stays on until the high beams turn off.

If the driver uses the high-beam switch to manually turn on the high beams, SmartBeam operation is defeated. Also, the system is momentarily defeated if the driver uses the flash-to-pass function. Placing the headlamp switch in any other position other than AUTO defeats both the auto headlamp and the SmartBeam functions.

Shop Manual
Chapter 11, page 538

Initial camera calibration and verification are performed at the factory as the vehicle is near completion. During this time, the camera is precisely aligned. Once the camera is properly aligned, logic used by the AHBM will make adjustments to fine-tune the alignment based on sensed lighting inputs during vehicle operation. These adjustments occur as the processing logic looks for a light source that represents oncoming headlights. This would include light sources that start as low intensity and then gradually increase in brightness and have movement, indicating a gradual rate of approach. In addition, the light source must be coming from just to the left of center. These conditions indicate a vehicle is approaching from a distance. The computer logic of the AHBM will apply a weighting factor to its calibration to correct the aim. If the system is out of calibration, the light-emitting diode (LED) in the mirror will flash.

As in the other systems discussed, a photocell or camera determines the light intensity of the headlights from oncoming vehicles. Headlight systems that utilize a single high-intensity discharge (HID) bulb can use a shutter to control light intensity between low- and high-beam operation. A motor or solenoid alters the position of the shutter (**Figure 11-39**). Usually, the shutter is spring loaded to the low-beam position. In this position, the shutter blocks some of the HID's light from being directed at the lens. When the high beams are selected, the shutter moves so the HID is completely uncovered, and its full light intensity is directed through the lens. On solenoid- controlled systems, the solenoid is de-energized to allow the shutter to return to its default position of low-beam operation. Energizing the solenoid opens the shutter for high-beam operation. When the oncoming vehicle is close enough that the headlights must change from high to low beams, the solenoid is de-energized until the oncoming light has passed.

Figure 11-39 Single HID headlight systems can use a shutter blade operated by a solenoid or motor to control high- and low-beam operations.

Headlight Leveling System

Shop Manual
Chapter 11, page 543

The **headlight leveling system (HLS)** uses front lighting assemblies with a leveling actuator motor. Some systems use a leveling switch that the driver controls. The switch allows adjusting the headlights into different vertical positions (usually four) so the driver can compensate for headlight position changes that can occur when the vehicle is loaded.

Electric motors use a pushrod assembly to change the position of the headlight reflector (**Figure 11-40**). When different voltage levels are input to leveling motors from the multiplexed switch, the motors move the reflector to the selected position.

The automatic headlight level system (AHLS) places the headlight reflectors in the correct position based on vehicle height and wheel speed sensor inputs. This system will maintain a constant level without input from the driver and can continuously alter the level to different driving conditions. The AHLS can be a stand-alone system with its own control module, or be part of the adaptive headlight system. Regardless, the operation is very similar.

The height sensor is attached at the rear of the vehicle to measure the distance between the rear suspension and the vehicle body (**Figure 11-41**). This information is monitored whenever the ignition switch is in the RUN position (**Figure 11-42**). The antilock brake system (ABS) module shares speed sensor information over the bus network. The inputs from the height and speed sensors indicate the vehicle's posture.

When the ignition is in the RUN position, the ECU will command the leveling actuators to lower the beams to their lowest travel and then command them to their proper position. This provides an initial set point. The actuators use a stepper motor, and the ECU counts the steps to determine the position of the reflectors. If the vehicle posture changes due to additional loads or different driving conditions, the ECU commands the actuator to move the reflector up or down as needed to match the vehicle's posture.

Figure 11-40 Beam level actuator with a push rod connection to the headlight reflector.

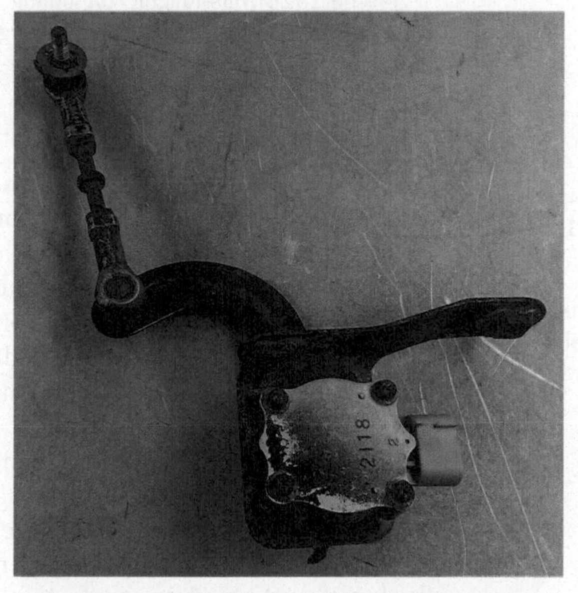

Figure 11-41 Automatic leveling headlight system height sensor.

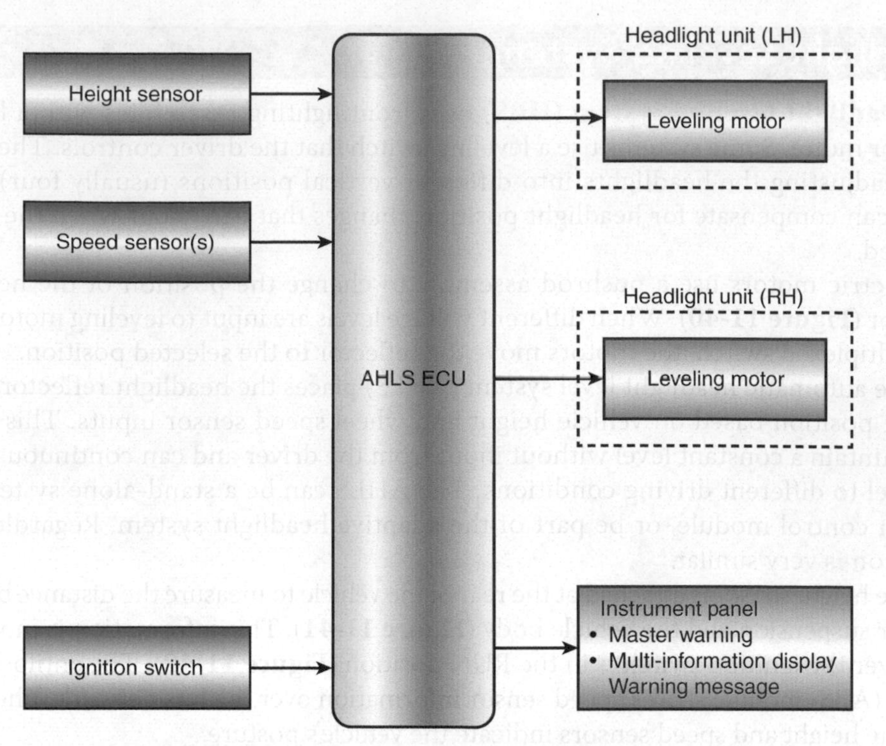

Figure 11-42 Schematic of an automatic headlight leveling system.

Adaptive Headlights

A BIT OF HISTORY

The 1928 Wills-Knight 70A Touring car had a third headlight mounted above the front bumper on the vehicle's centerline. This headlight was a directional light attached to the steering system so it would rotate left and right as the steering wheel was turned.

Shop Manual
Chapter 11, page 546

Adaptive headlights are also referred to as active headlights.

The **adaptive headlight system (AHS)** enhances nighttime safety by turning the headlight beams to follow the direction of the road as the vehicle enters a turn (**Figure 11-43**). This provides the driver additional reaction time if approaching any road hazards. Conventional fixed headlights illuminate the road straight in front of the vehicle. On turns, the light beam can shine into oncoming traffic or leave the road beyond the turn into darkness. The AHS uses headlight assemblies that swivel into the direction the vehicle is steering. This provides a 90% improvement in the illumination of the road, so obstacles become visible sooner.

The AHS control module monitors vehicle speed, steering angle, and yaw (degree of rotation around the vertical axis) (**Figure 11-44**). The input from the steering angle sensor indicates how much the headlight beams need to rotate. As these sensor inputs indicate the vehicle is entering a turn, electric motors turn the headlights in the direction of the turn. As the headlight beams rotate, they follow the road and

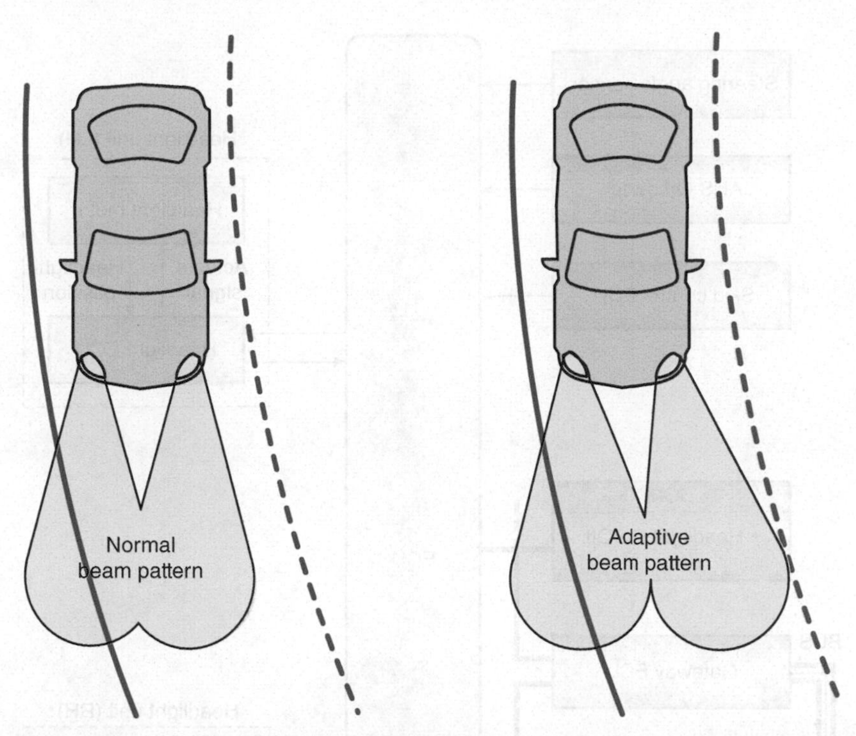

Figure 11-43 The adaptive headlight system will illuminate the road beyond a turn for added safety.

guide the driver into the turn (**Figure 11-45**). The headlights can follow the turns of the road through an arc of 15°. The headlight on the side of the vehicle opposite the turn direction will rotate about 7°.

AUTHOR'S NOTE The AHS is activated whenever the vehicle is traveling forward. If the vehicle is in reverse, or the steering wheel is turned when the vehicle is not moving, then the AHS is deactivated.

Some systems are capable of predicting when the vehicle will be entering a turn. These systems use steering angle, vehicle speed, and yaw rate inputs along with GPS-fed road data. They adjust not only the rotation of the headlamp motors but also the pattern of illumination of the bi-xenon lamps.

Night Vision

AUTHOR'S NOTE While only 28% of driving is at night, more than 62% of pedestrian fatalities occur at night. When driving at 60 mph (96 kph), the standard automobile headlights allow only 3.5 seconds of reaction time.

Night vision systems use military-style thermal imaging to allow drivers to see things that they may not be able to see with the naked eye (**Figure 11-46**). The system allows the driver to see up to five times more of the road than with just headlights.

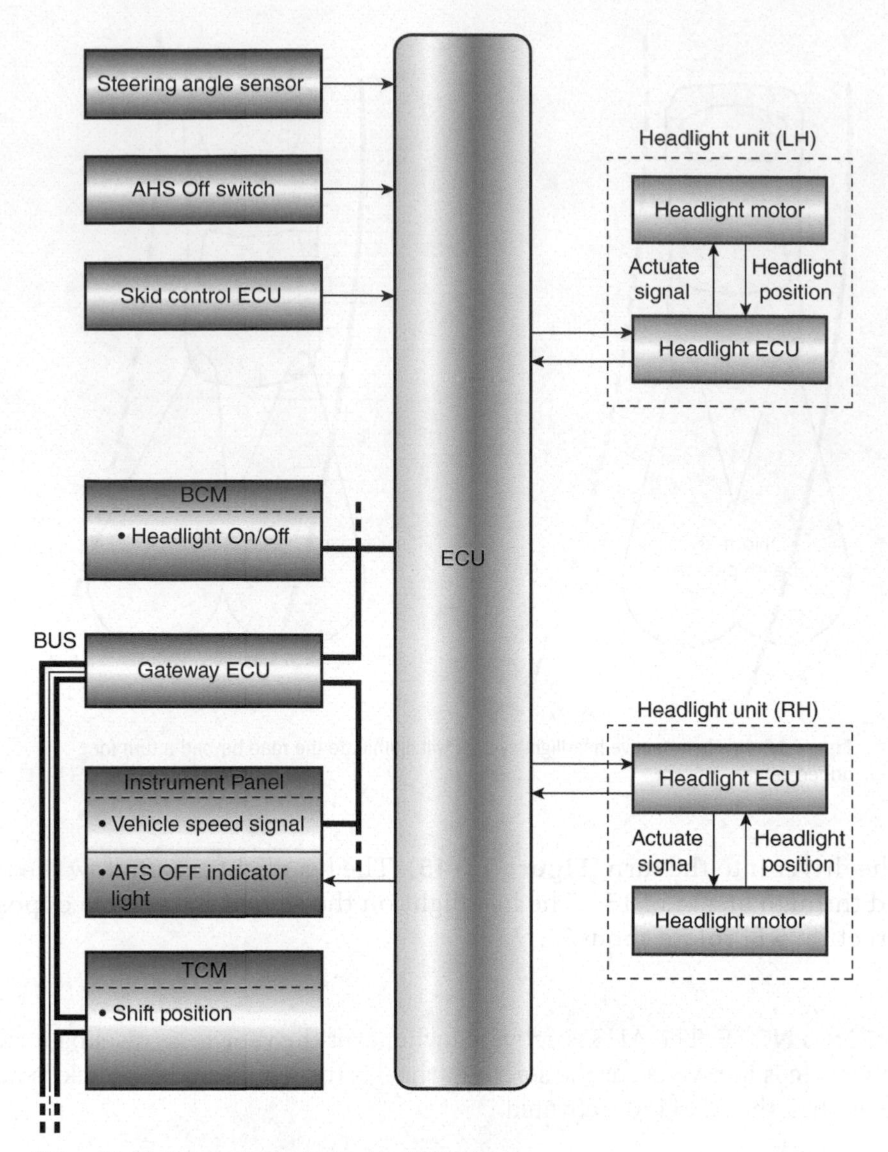

Figure 11-44 Schematic of adaptive headlight system.

Figure 11-45 The adaptive headlight assembly uses motors to rotate the beam.

A fixed-lens, high-resolution, "far infrared" camera located behind the front grill generates an electronically processed video image. The head-up display (HUD) or a thin-film transistor (TFT) monitor in the instrument panel displays the real-time image.

Figure 11-46 Night vision uses an infrared camera to project objects onto a screen. The objects may be out of the range of the headlights.

The camera's lens is an infrared sensor that operates at room temperature. The lens has its own heating and cooling system to maintain this temperature. The special camera creates a temperature pattern (thermogram) that is refreshed 30 times a second. The heat from a person or an animal is greater than the heat radiation from the surroundings of the camera.

The system works by registering small differences in temperature and displaying them in 16 different shades of gray on the HUD screen. A signal processor translates the thermogram data and sends it to the display. Using a wavelength of 6 to 12 micrometers, the camera's lens detects the infrared heat emitted from the vehicle's surroundings and displays it as a negative image. The display color of people, animals, or other heat-generating objects depends on the object's temperature. Cold objects appear as dark images, while warmer objects appear as white or a light color.

Another variation of the night vision system uses near-infrared. This system has two infrared emitters integrated into the headlight assemblies. A camera behind the rearview mirror captures the infrared reflection from the objects in front of the vehicle. The camera converts the reflection into a digital signal by a charge-coupled device (CCD). The signal from the CCD goes to an image processor that changes the signal into a format that can be viewed on the display.

Daytime Running Lamps

All late-model Canadian and most newer domestic vehicles have **daytime running lamps (DRL)**. The basic idea behind DRLs is to dimly light the headlamps during the day so other drivers and pedestrians to see the vehicle from a distance. Manufacturers have taken many different approaches to achieve this lighting. Most have a control module or relay (**Figure 11-47**) that turns the lights on when the engine is running and allows normal headlamp operation when the driver turns on the headlights. Daytime running lamps generally use the high-beam or low-beam headlight system at a reduced intensity.

Shop Manual
Chapter 11, page 546

The dimmer headlights can result from headlight current being passed through a resistor when the daytime running lamps are activated. The resistor reduces the voltage available, and the current flowing through the circuit, to the headlights. Voltage bypasses the resistor during normal headlamp operation.

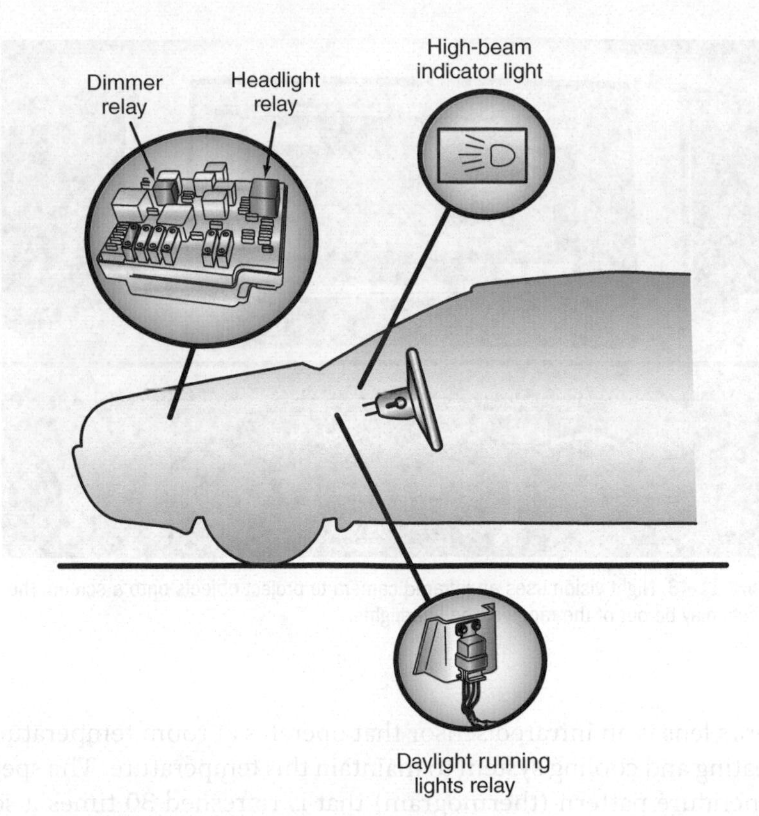

Figure 11-47 A daytime running light relay.

Other systems use a control module that sends a duty-cycled output to either the high-beam or the low-beam headlights (depending on the manufacturer). The duty cycle reduces the output of the headlights by 50% to 75%. Most systems that use the high-beam headlights have a method of turning off the high-beam indicator lamp in the instrument cluster if DRLs are activated. However, on some systems, it is normal for the high-beam indicator to illuminate dimly during DRL operation.

Some GM's DRL systems include a solid-state control module assembly, a relay, and an ambient light sensor assembly. During daylight, the system lights the low-beam headlights at a reduced intensity when the ignition switch is in the RUN position. The DRL system illuminates the low-beam headlamps at full intensity when low-light conditions exist.

As the intensity of the light reaching the ambient light sensor increases, the electrical resistance of the sensor assembly decreases. When the DRL control module assembly senses the low resistance, the module allows voltage to be applied to the DRL diode assembly and then to the low-beam headlamps. Because of the voltage drop across the diode assembly, the low-beam headlamps are on with a low intensity.

As the intensity of the light reaching the ambient light sensor decreases, the electrical resistance of the sensors increases. When the DRL module assembly senses high resistance in the sensor, the module closes an internal relay, which allows the low-beam headlamps to illuminate with full intensity.

If PWM controls the normal headlight operation from a high-side driver (refer to Figure 11-29), the same drivers control the DRLs. The headlight switch inputs to the computer if the switch position is in the OFF or AUTO modes of operation. In these modes, PWM reduces the voltage to the lamps to activate DRL. The low PWM results in a low-luminous output of the lamps.

AUTHOR'S NOTE Some manufacturers use a DRL system integrated into the corona rings of the HID headlights.

Most DRL systems also use the parking brake switch as an input. Applying the parking brake while the engine is running, turns off the daytime running light feature.

Exterior Lights

The headlight switch controls parking lights and taillights. The headlights do not need to be on for these lights to illuminate. Usually, the parking lights turn on when the headlight switch is in its first detent position. Placing the headlight switch in the PARK or HEADLIGHT position turns on the front parking lights, taillights, side marker lights, and rear license plate light(s). The front and rear parking lights may use dual-filament bulbs. The other filament is for the turn signals and hazard lights.

The parking lights can be turned on with the ignition switch in the OFF position.

Taillight Assemblies

Most taillight assemblies include the brake, parking, rear turn signal, and rear hazard lights. The center high-mounted stop light (CHMSL), back-up lights, and license plate lights can be a part of the taillight circuit design. Depending on the manufacturer, the taillight assembly can use single-filament or dual-filament bulbs. When using single-filament bulbs, the taillight assembly is wired as a three-bulb circuit. A three-bulb circuit uses one bulb each for the tail, brake, and turn signal lights on each side of the vehicle. When using dual-filament bulbs, the system is wired as a two-bulb circuit. Each bulb can perform more than one function.

Shop Manual
Chapter 11, page 561

In a three-bulb taillight system, the brake lights are controlled directly by the brake light switch. The brake light switch is attached to the brake pedal in most applications. When the brakes are applied, the pedal moves down, and the switch plunger closes the contact points and lights the brake lights (**Figure 11-48**). Some vehicles use a pressure-sensitive brake light switch located in the brake master cylinder. When the brakes are applied, the pressure developed in the master cylinder closes the switch to light the lamps.

Shop Manual
Chapter 11, page 561

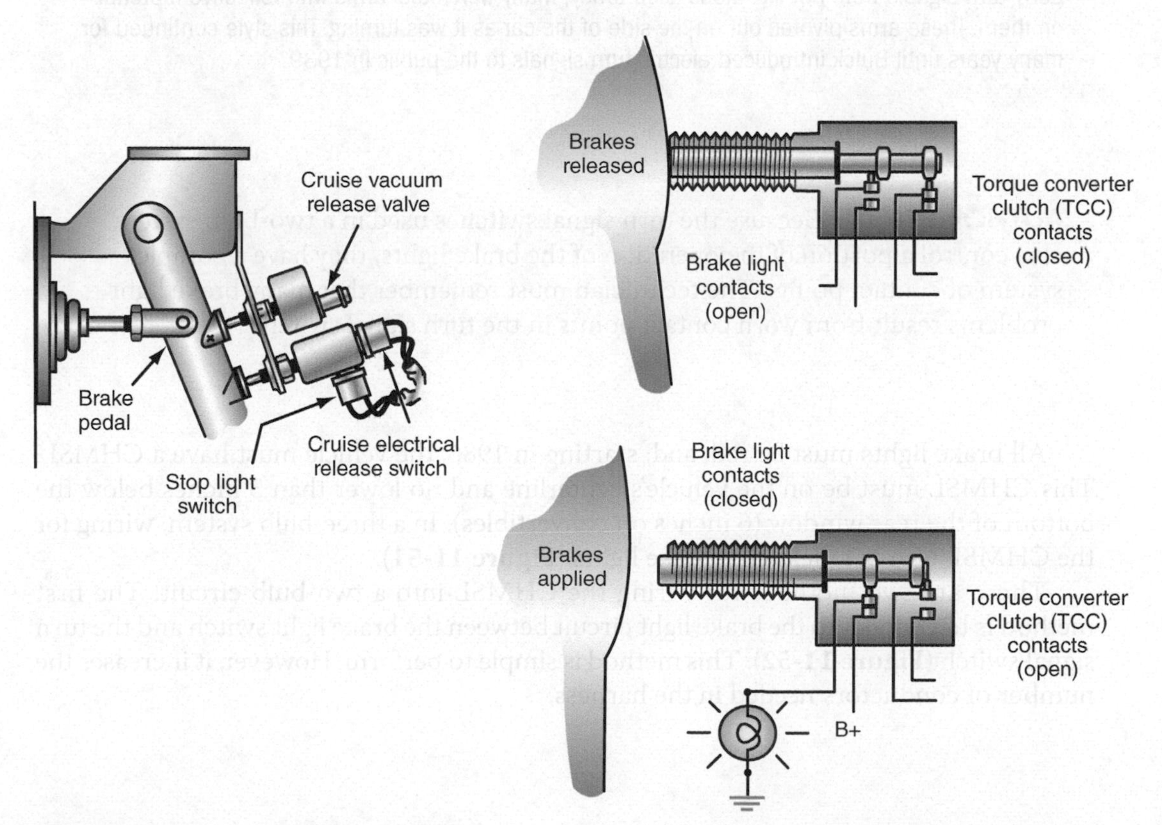

Figure 11-48 Operation of a brake light switch.

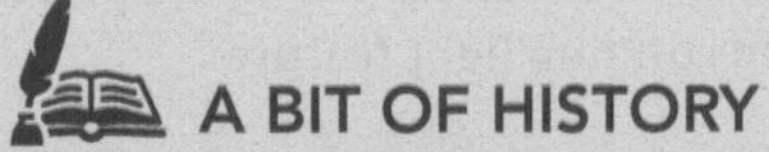

A BIT OF HISTORY

Taillights on both sides of the car didn't appear until 1929.

The brake light switch receives direct battery voltage through a fuse, which allows the brake lights to operate when the ignition switch is in the OFF position. Once the brake light switch is closed, voltage is applied to the brake lights. The brake lights on both sides of the vehicle are wired in parallel. The bulb is grounded to complete the circuit.

Many brake light systems use dual-filament bulbs that perform multiple functions. Usually, the filament of the dual-filament bulb used for the brake lights (the high-intensity filament) is the same filament used by the turn signal and hazard lights. In this type of circuit, the brake lights are wired through the turn signal and hazard switches. If neither turn signal is on, the current is sent to both brake lights (**Figure 11-49**). If the left turn signal is on, the current for the right brake light flows through turn signal switch terminal 5. The left brake light does not receive any voltage from the brake switch because the turn signal switch opens that circuit (**Figure 11-50**). The left-rear lamp will flash as the turn signal flasher provides pulsed voltage into switch terminal 3 and out terminal 8.

A BIT OF HISTORY

In 1921, turn signals were made standard equipment by Leland Lincoln. This marquee later joined Ford Motor Company. Leland Lincolns were built by Henry Leland, the originator of Cadillac. Early turn signals were not like those used today; many were steel arms with reflective material on them. These arms pivoted out on the side of the car as it was turning. This style continued for many years until Buick introduced electric turn signals to the public in 1939.

AUTHOR'S NOTE Because the turn signal switches used in a two-bulb system also control a portion of the operation of the brake lights, they have a complex system of contact points. The technician must remember that many brake light problems result from worn contact points in the turn signal switch.

All brake lights must be red, and, starting in 1986, the vehicle must have a CHMSL. This CHMSL must be on the vehicle's centerline and no lower than 3 inches below the bottom of the rear window (6 inches on convertibles). In a three-bulb system, wiring for the CHMSL is in parallel to the brake lights (**Figure 11-51**).

These are two methods for wiring the CHMSL into a two-bulb circuit. The first method is to connect to the brake light circuit between the brake light switch and the turn signal switch (**Figure 11-52**). This method is simple to perform. However, it increases the number of conductors needed in the harness.

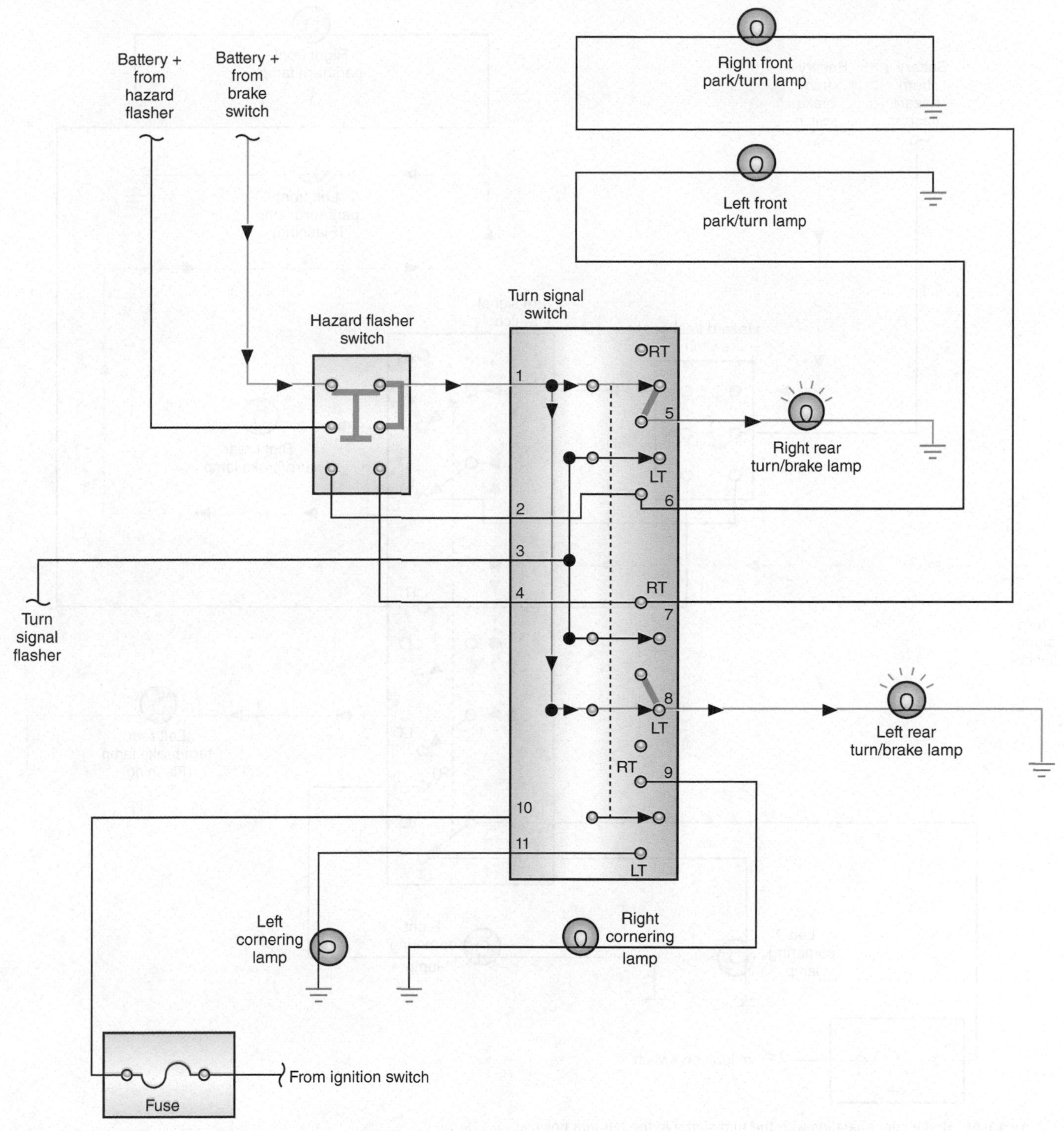

Figure 11-49 Brake light operation with the turn signals in the neutral position.

A common method manufacturers use is to install diodes in the conductors connected between the left- and right-side bulbs (**Figure 11-53**). When applying the brakes with the turn signal switch in the neutral position, the diodes will allow voltage to flow to the CHMSL. If the turn signal switch is in the left-turn position, the left light

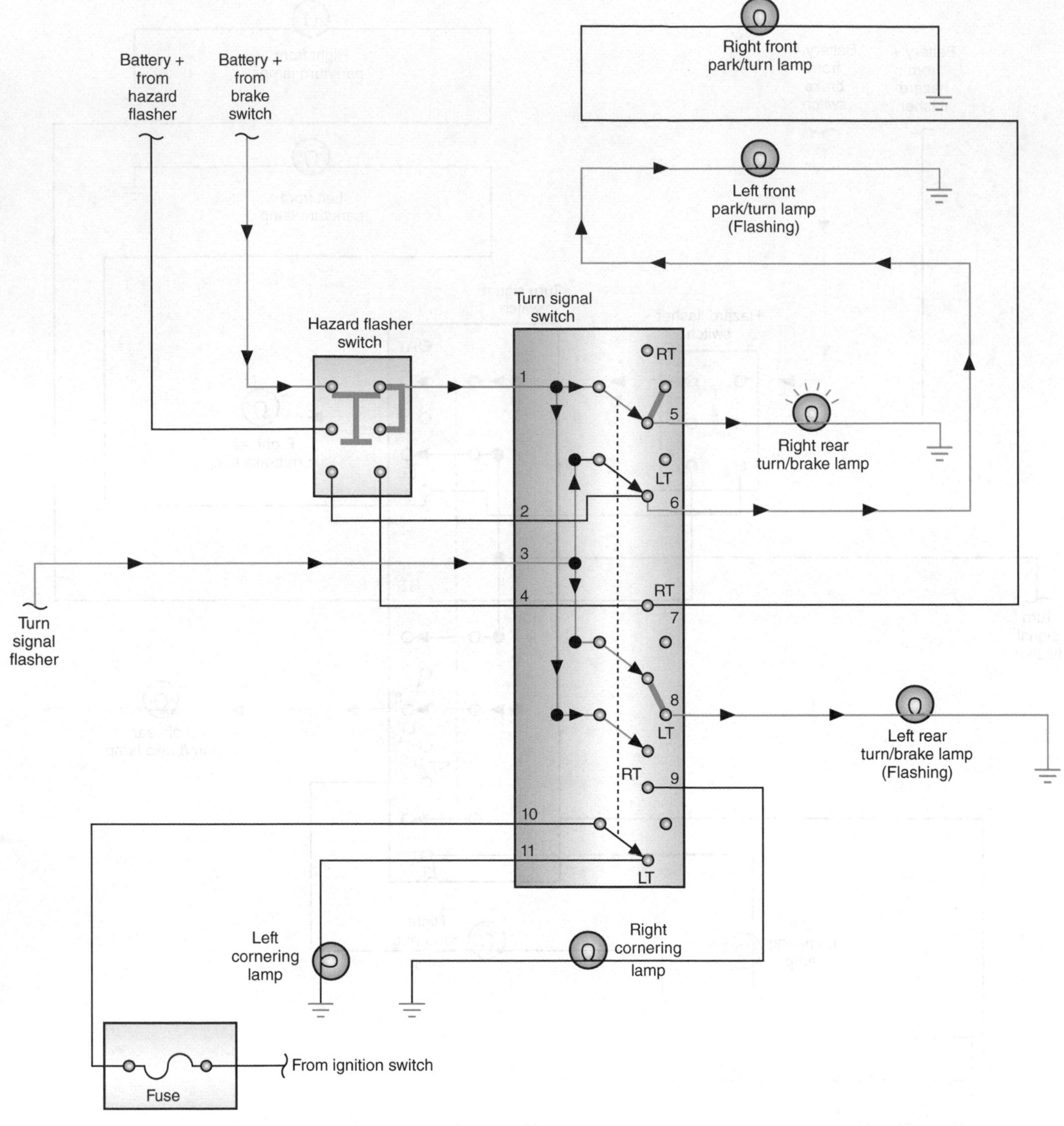

Figure 11-50 Brake light operation with the turn signal in the left-turn position.

must receive a pulsating voltage from the flasher. However, the steady voltage applied to the right brake light would cause the left light to burn steady without using the diode. Diode 1 will block the voltage from the right lamp, preventing it from reaching the left light. Diode 2 will allow voltage from the right brake light circuit to be applied to the CHMSL.

Computer control of the parking light and taillight systems has greatly simplified the operation of these systems. Because the switches are inputs, this eliminates the complexity and contacts associated with the steering wheel switch. However, not all

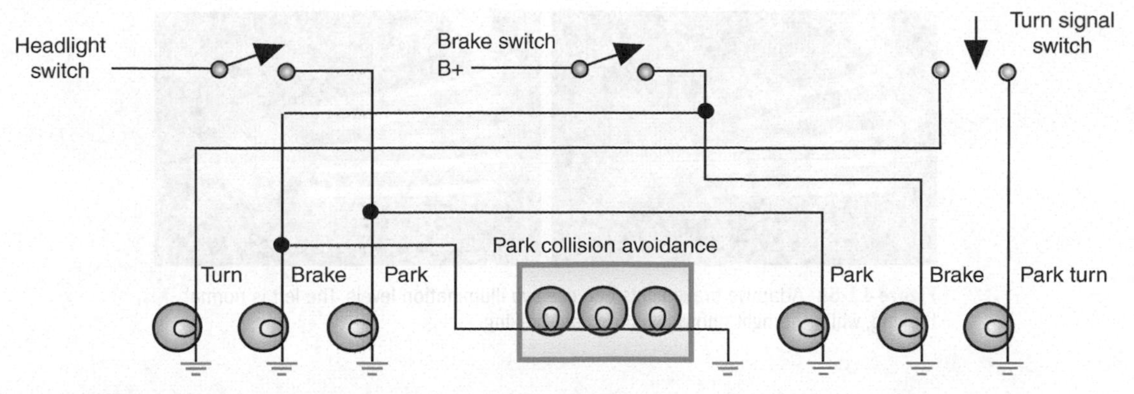

Figure 11-51 Wiring of a CHMSL in a three-bulb circuit.

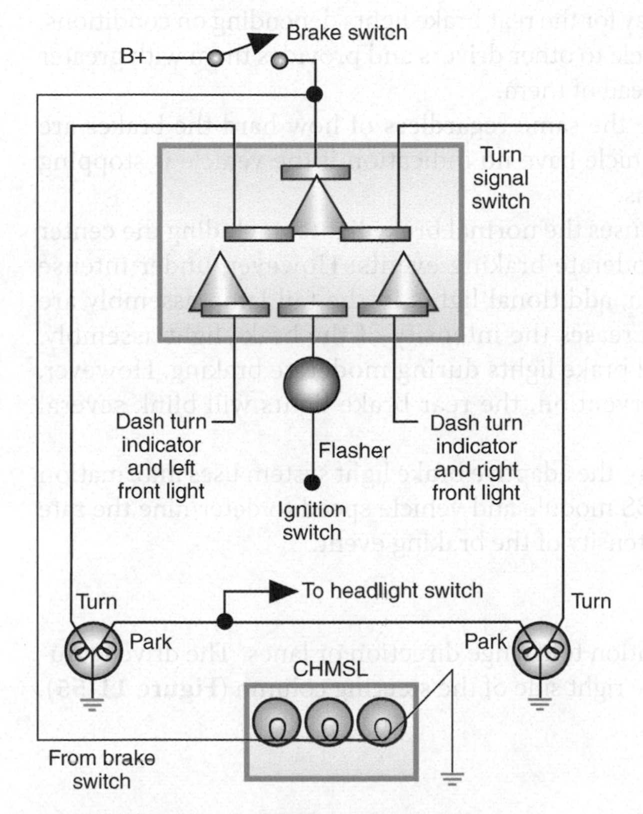

Figure 11-52 Wiring the CHMSL into the two-bulb circuit between the brake light switch and the turn signal switch.

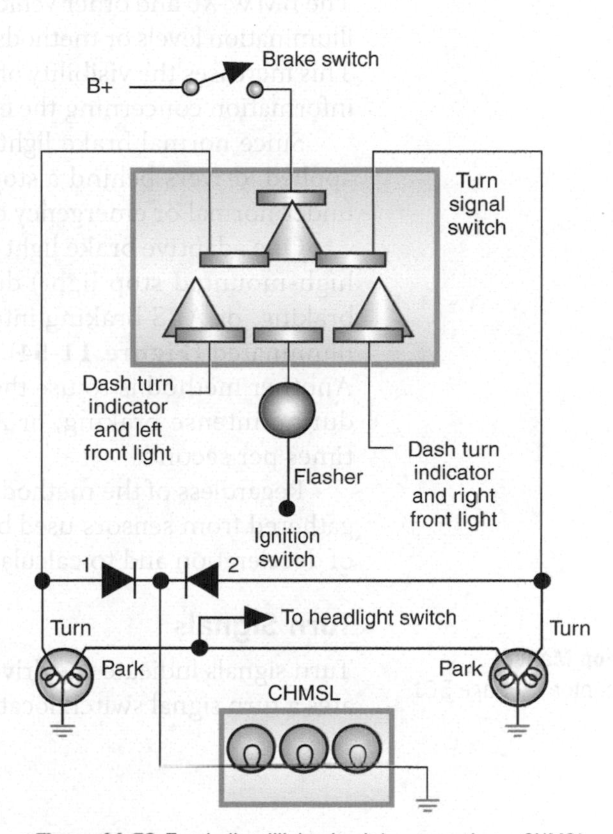

Figure 11-53 Two-bulb taillight circuit incorporating a CHMSL into the brake light system.

inputs go directly to the BCM. For example, the brake light switch input may be an input to the ABS module. This module sends a data bus message to the BCM to illuminate the brake lights. In addition, the operation of the exterior lighting system may involve several bus networks.

Computer-controlled parking light relays are controlled by the BCM in the same manner as discussed for headlight control. Because the BCM controls the operation of the relays, it can turn off the exterior lights if the driver forgets to do so. Turning off the ignition starts a timer. After a programmed length of time, the BCM will turn off all the relays associated with the exterior lighting system, even if the switch is still in the headlight position.

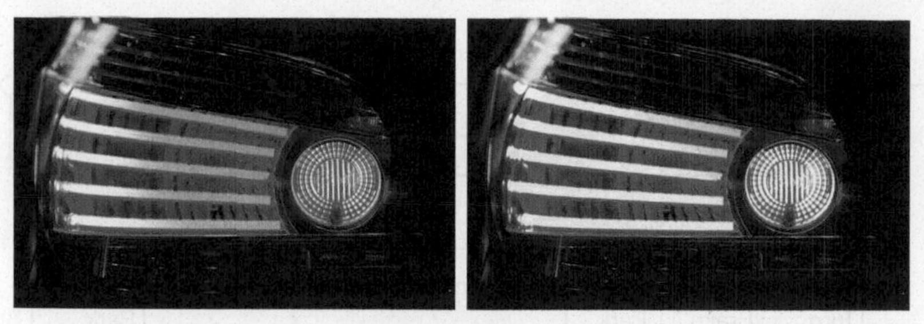

Figure 11-54 Adaptive brake lights can use two illumination levels. The left is normal braking, while the right photo is during hard braking.

Adaptive Brake Lights

The BMW X6 and other vehicles use an **adaptive brake light** system that selects different illumination levels or methods of display for the rear brake lights depending on conditions. This increases the visibility of the vehicle to other drivers and provides them with greater information concerning the events ahead of them.

Since normal brake lights behave the same regardless of how hard the brakes are applied, drivers behind a stopping vehicle have no indication if the vehicle is stopping under normal or emergency conditions.

One adaptive brake light method uses the normal brake lights (including the center high-mounted stop light) during moderate braking events. However, under intense braking, or ABS braking intervention, additional lights in the tail lamp assembly are illuminated (**Figure 11-54**). This increases the intensity of the brake light assembly. Another method is to use the normal brake lights during moderate braking. However, during intense braking, or ABS intervention, the rear brake lights will blink several times per second.

Regardless of the method of display, the adaptive brake light system uses information gathered from sensors used by the ABS module and vehicle speed to determine the rate of deceleration and to calculate the intensity of the braking event.

Turn Signals

Shop Manual
Chapter 11, page 561

Turn signals indicate the driver's intention to change direction or lanes. The driver actuates a turn signal switch located on the right side of the steering column (**Figure 11-55**).

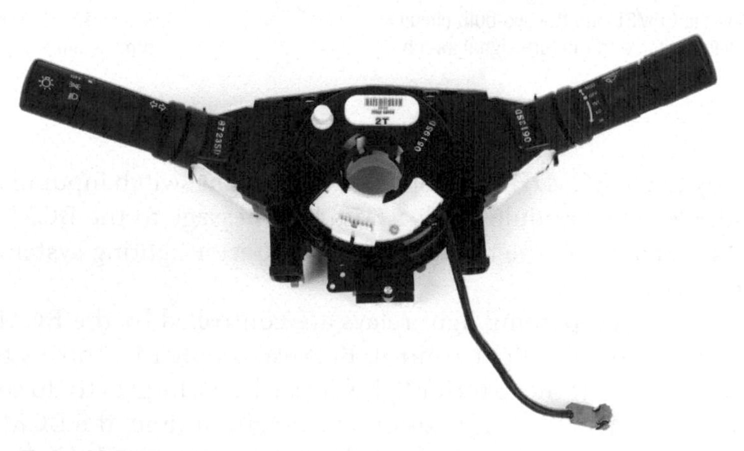

Figure 11-55 Typical turn signal switch.

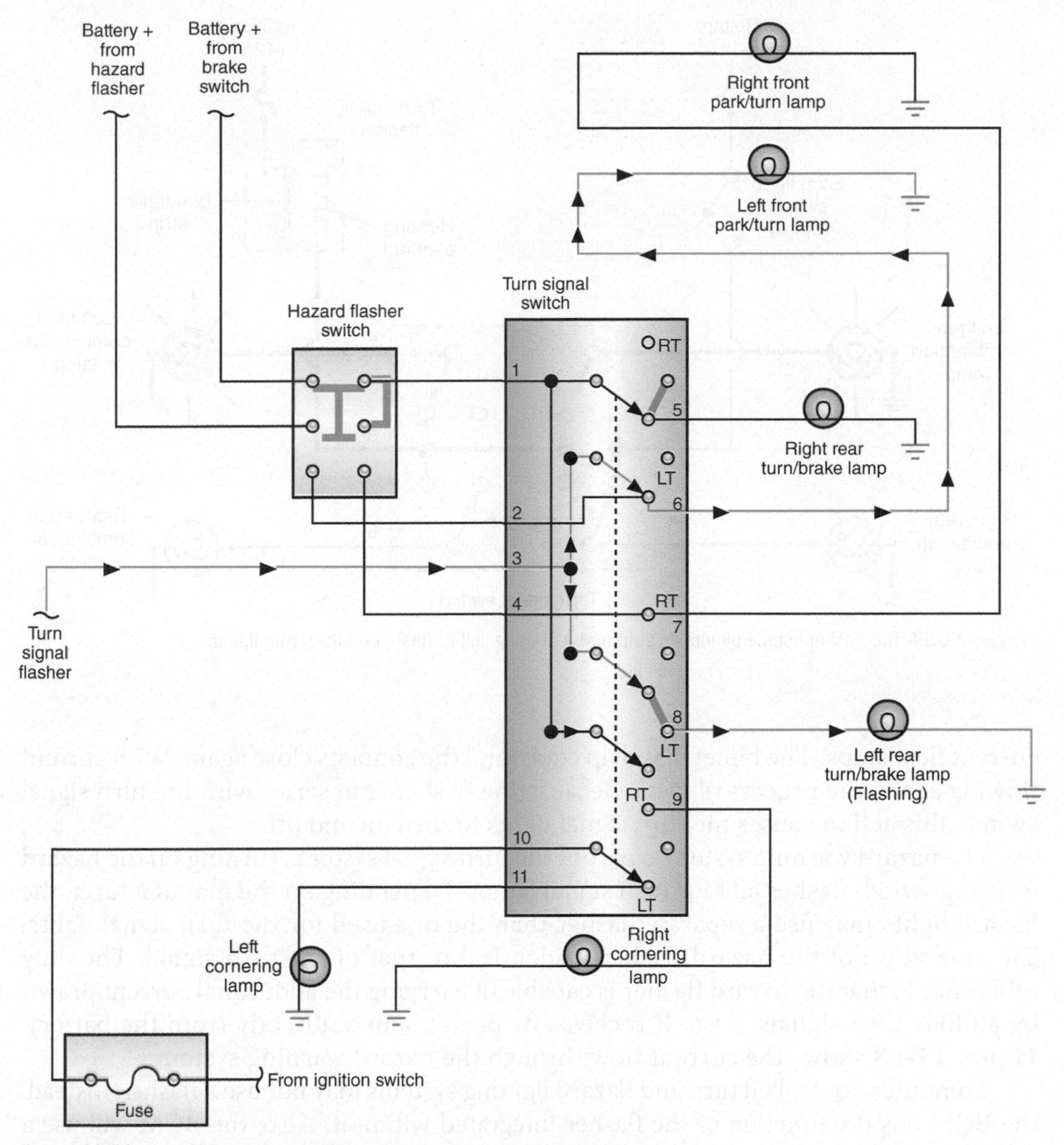

Figure 11-56 Turn signal circuit with the switch in the left-turn position.

In the neutral position, the contacts open, preventing current flow. When the driver moves the turn signal lever to indicate a left turn, the turn signal switch closes the contacts to direct voltage to the front and rear lights on the left side of the vehicle (**Figure 11-56**). When moved to indicate a right turn, the turn signal switch contacts move to direct voltage to the front and rear turn signal lights on the right side of the vehicle.

Most of today's vehicles use the turn signal switch as an input. The signal is sent to a control module that can send the voltage to the lights using PWM drivers. The drivers can be either HSD or LSD. Computer-controlled turn signal circuits may offer enhanced diagnostic capabilities.

Shop Manual
Chapter 11, page 565

A **flasher** opens and closes the turn signal circuit at a set rate. With the contacts closed, power flows from the flasher through the turn signal switch to the lamps. The flasher consists of a set of normally closed contacts, a bimetallic strip, and a coil heating element (**Figure 11-57**). These three components are wired in series. As current flows through the heater element, it increases in temperature, which heats the bimetallic strip. The strip then bends and opens the contact points. Once the points are open,

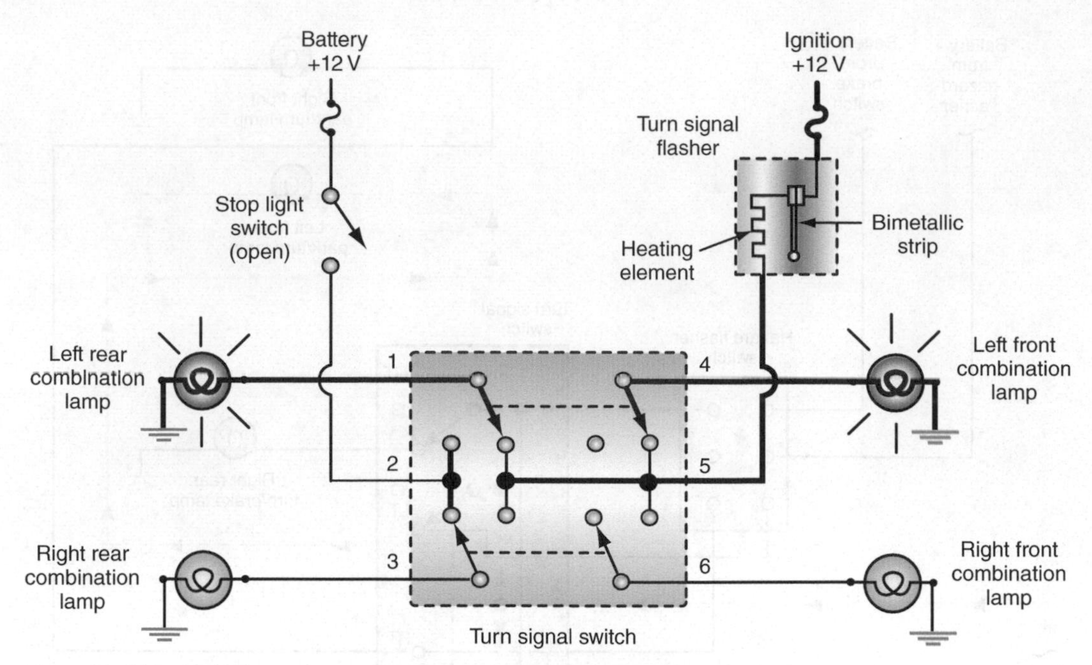

Figure 11-57 The flasher uses a bimetallic strip and a heating coil to flash the turn signal lights.

current flow stops. The bimetallic strip cools, and the contacts close again. With current flowing again, the process repeats. Because the flasher is in series with the turn signal switch, this action causes the turn signal lights to turn on and off.

The hazard warning system is part of the turn signal system. Turning on the hazard warning switch flashes all four turn signal lamps. Depending on the manufacturer, the hazard lights may use a separate flasher than the one used for the turn signal lights. The operation of the hazard flasher is identical to that of the turn signal. The only difference is that the hazard flasher is capable of carrying the additional current drawn by all four turn signals. Also, it receives its power source directly from the battery. **Figure 11-58** shows the current flow through the hazard warning system.

Computer-controlled turn and hazard lighting systems may not use a flasher. Instead, the BCM has the function of the flasher integrated within it. Also, the BCM will use a noise generator to make the audible clicking sound to indicate that the turn signals or hazards are activated.

Fog Lights

Some vehicles are equipped with fog lights for increased safety when driving in snow, sleet, heavy rain, and heavy fog. Fog lights emit a specialized beam to penetrate through the snow, rain, or fog, providing the driver with a better and safer field of vision. Headlights will reflect off of heavy, dense fog and cause a white haze that can reduce visibility.

Fog lights are installed on each side of the vehicle, generally low on the front fascia. Due to their mounting location, fog lights illuminate below the normal line of sight. This minimizes the amount of reflected light to help the driver see better.

Common fog light circuits use a relay (**Figure 11-59**). Also, the wiring of most fog light circuits allows the fog lights to only operate when the headlight switch is in the PARK or low-beam position.

Back-Up Lights

All vehicles sold in North America after 1971 must have back-up lights. Back-up lights illuminate the road behind the vehicle and warn other drivers and pedestrians of the

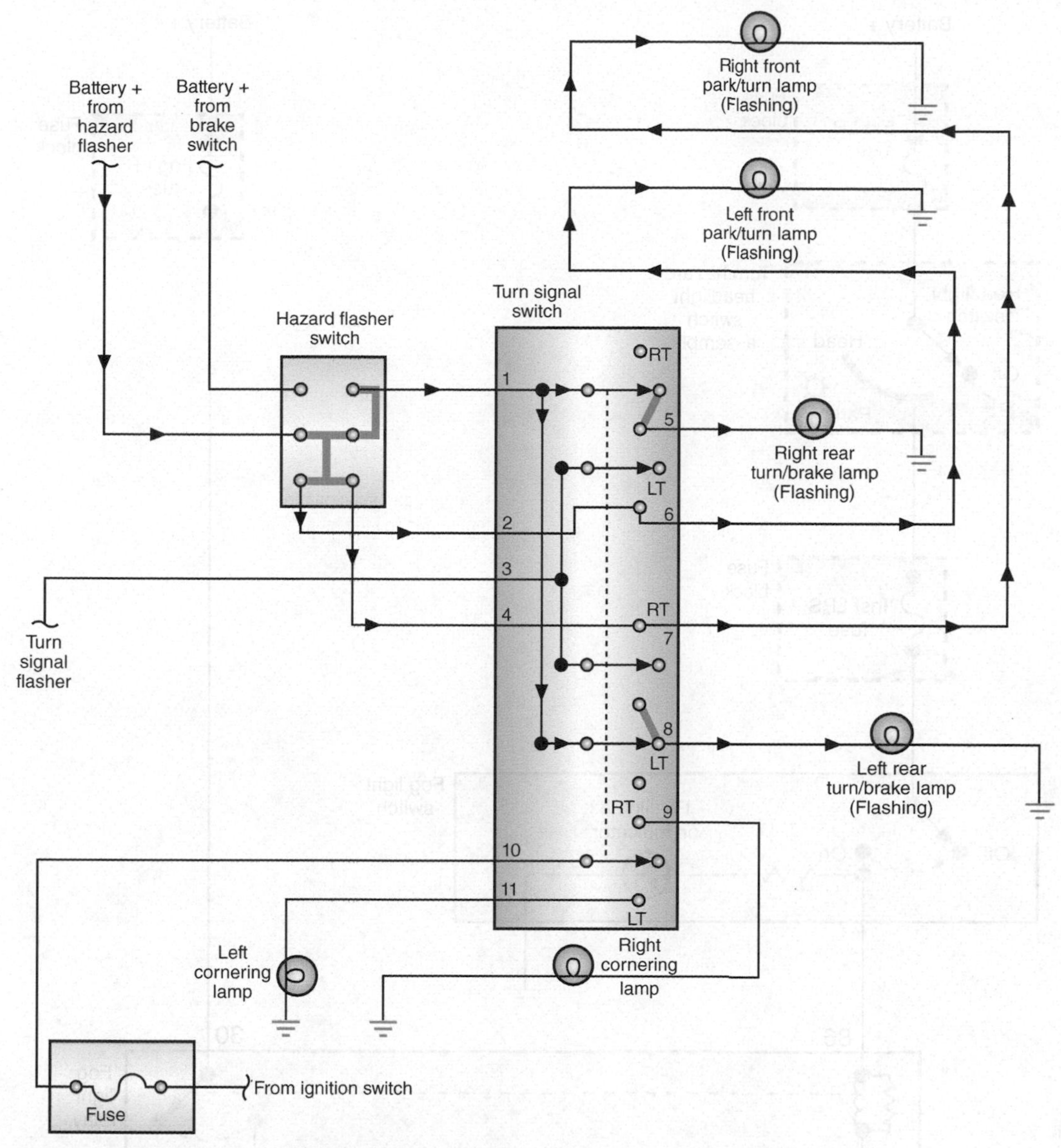

Figure 11-58 Current flow when the hazard warning system is activated.

driver's intention to back up. **Figure 11-60** illustrates a back-up light circuit. Power is supplied through the ignition switch when it is in the RUN position. When the driver shifts the transmission into reverse, the back-up light switch contacts are closed, completing the circuit.

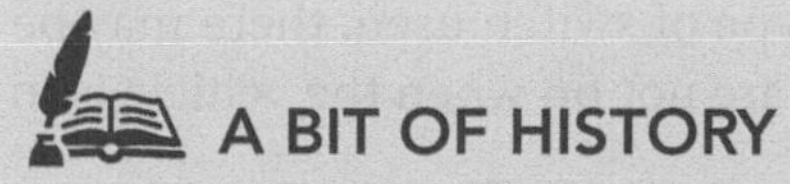

The 1921 Wills-St. Claire was the first car to feature a back-up lamp.

Many vehicles with automatic transmissions incorporate the back-up light switch into the neutral safety switch. Most manual transmissions have a separate back up switch. Either style of switch can be located on the steering column, on the floor console, or on

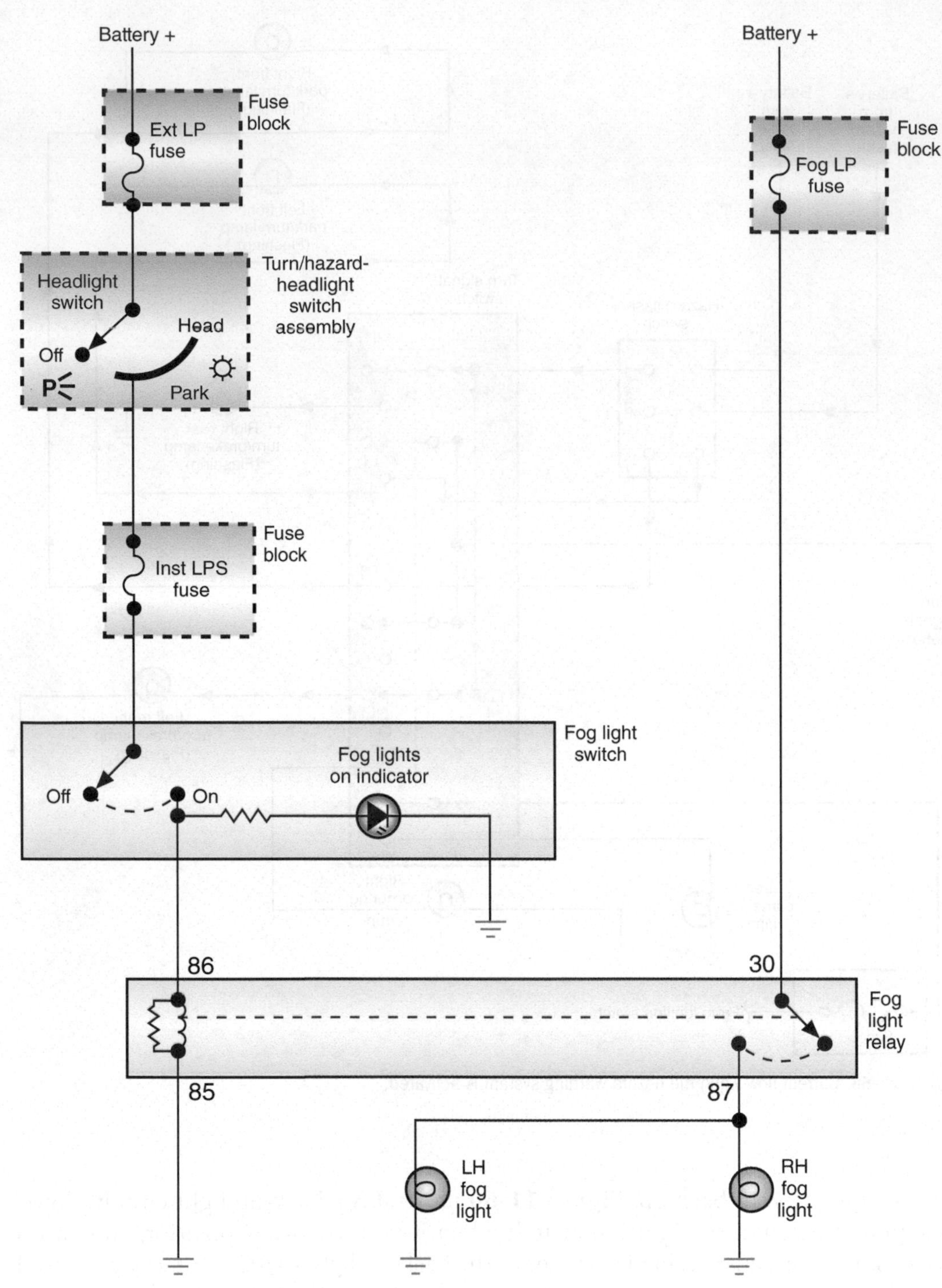

Figure 11-59 Typical fog light circuit.

the transmission (**Figure 11-61**). Depending on the type of switch used, there may be a means of adjusting the switch to assure that the lights are not on when the vehicle is in a forward gear selection.

Side Marker Lights

Side marker lights permit the vehicle to be seen from the side when entering a roadway. This also provides a means for other drivers to determine vehicle length. The front side marker light lens must be amber, and the rear lens must be red. Vehicles that use a wrap-around headlight and taillight assemblies use this lens for the side marker

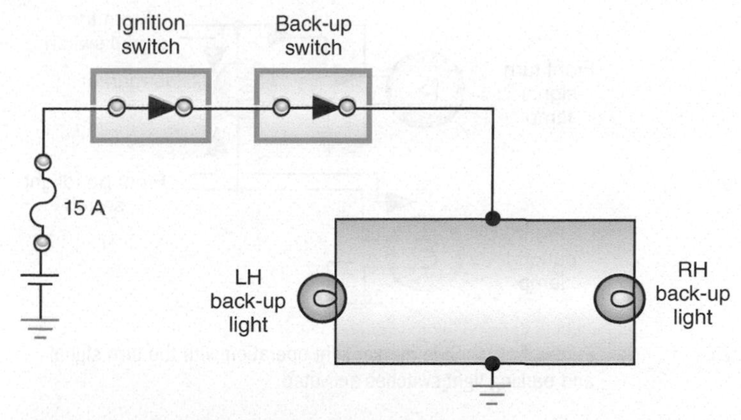

Figure 11-60 Back-up light circuit.

Figure 11-61 A combination back-up and neutral safety switch installed on an automatic transmission.

Figure 11-62 Wrap-around headlights serve as side marker lights.

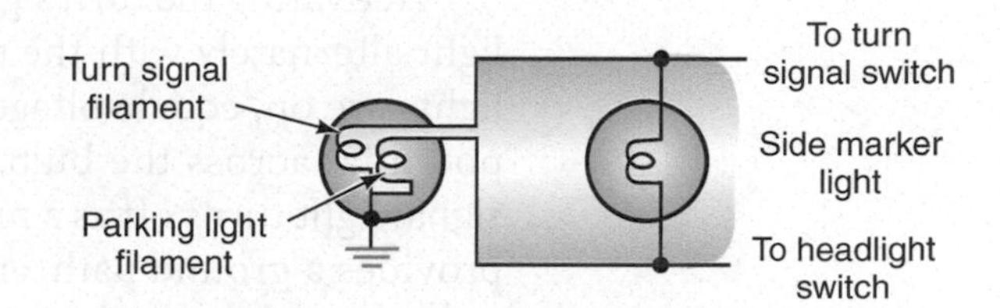

Figure 11-63 A side marker light wired across two circuits.

lights (**Figure 11-62**). Vehicles that surpass certain length and height limits are also required to have clearance lights that face both the front and rear of the vehicle.

The common method of wiring the side marker lights is parallel to the parking lights. Wired in this manner, the side marker lights will illuminate only when the headlight switch is in the PARK or HEADLIGHT position.

Many vehicle manufacturers use a method of wiring the side marker lights to flash when the turn signals are activated. The side marker light is wired across the parking light and turn signal light (**Figure 11-63**). If the parking lights are on, voltage is applied to the side marker light from the parking light circuit. Ground for the side marker light is through the turn signal filament. Because of the large voltage drop across the side marker lamp, the turn signal bulb will barely illuminate. In this circumstance, the side marker light stays on constantly (**Figure 11-64**).

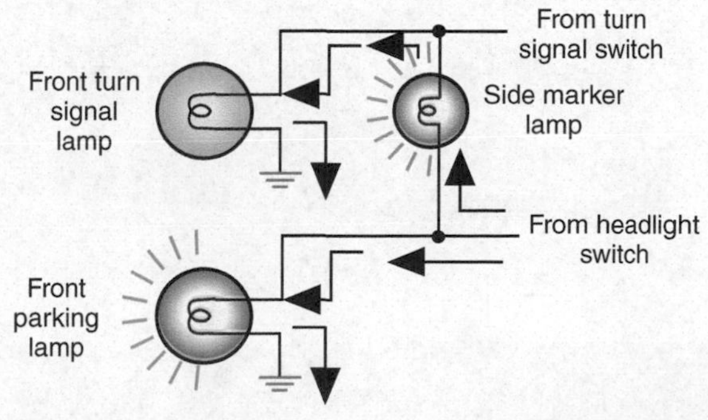

Figure 11-64 Current flow to the side marker light with the parking light on.

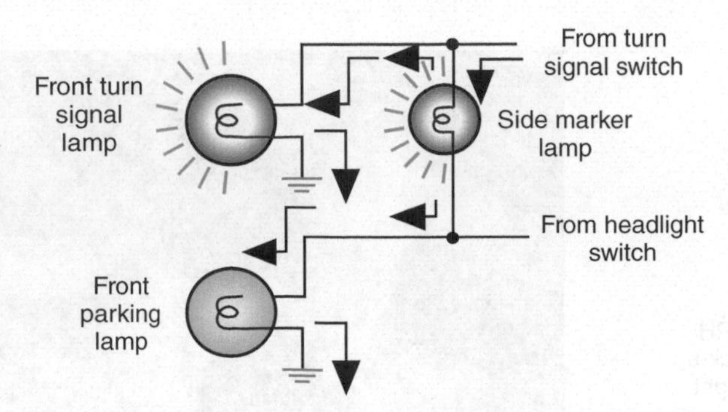

Figure 11-65 Side marker operation, with the turn signal switch activated.

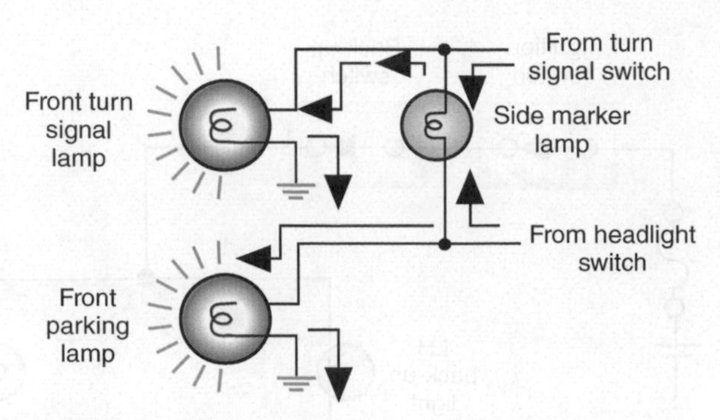

Figure 11-66 Side marker light operation with the turn signal and parking light switches activated.

If the parking lights are off and the turn signal is activated, the side marker light receives its voltage source from the turn signal circuit. Ground for the side marker light is through the parking light filament. The voltage drop over the side marker light is so high that the parking light will not illuminate. The side marker light will flash with the turn signal light (**Figure 11-65**).

Activating the turn signal with the parking lights illuminated flashes the side marker light alternately with the turn signal light. When both the turn signal and the parking lights are on, equal voltage is on both sides of the side marker light. There is no voltage potential across the bulb, so the light does not illuminate (**Figure 11-66**). The turn signal light turns off as a result of the flasher opening. Then the turn signal light filament provides a ground path, and the side marker light comes on. The side marker light will stay on until the flasher contacts close, turning on the turn signal light again.

LED Exterior Lighting

Many car manufacturers use LED lighting technology for exterior lighting functions, including the CHMSL, taillight assemblies (**Figure 11-67**), side marker lights, and turn-indicating outside mirrors. Using LEDs in rear-lighting applications (especially the CHMSL) provides one means of increasing traffic safety. The reaction time of the following driver in response to brake light illumination is shorter for CHMSLs equipped with LEDs than for those equipped with conventional incandescent bulbs. This is due to the shorter

Figure 11-67 LED taillight assembly.

LED illumination time of less than 100 milliseconds compared to 300 milliseconds of an incandescent bulb. At 75 mph (120.7 km/hour), the faster illumination time means an additional 21 feet (6.4 meters) of braking distance for the following drivers. Another advantage of LED lighting is the extended life compared to a bulb.

An example of the use of LED technology in rear-lighting systems is the Cadillac STS. Each tail lamp assembly has 30 points of illumination using two vertical boards, each consisting of 15 LEDs. The CHMSL is approximately 1/2 inch (12 mm) thick and has 78 points of illumination.

When using LEDs, the assemblies use a larger number of points of illumination since LEDs don't have the same light intensity as an incandescent bulb, and the light emitted by an LED is in a narrow angle of view. The increased number of LEDs assures that the light will be viewed from several angles.

Figure 11-68 illustrates the diagram for an LED taillight assembly. The brightness of the assembly is dependent upon the resistance in the circuit. Because resistance controls the amount of current in the circuit, changing circuit resistance will also change circuit current. When the park light switch is closed, the emitter for TR_1 is pulled low through the 33-ohm (Ω) resistance of R_2. This turns on TR_1 and TR_2 and illuminates the LEDs. When the brake light switch is closed, the emitter for TR_1 is pulled low through the 15 ohms of resistance of R_3. Now the circuit has less resistance than when the park switch was closed. The current increases, resulting in an increase in the LEDs' illumination level.

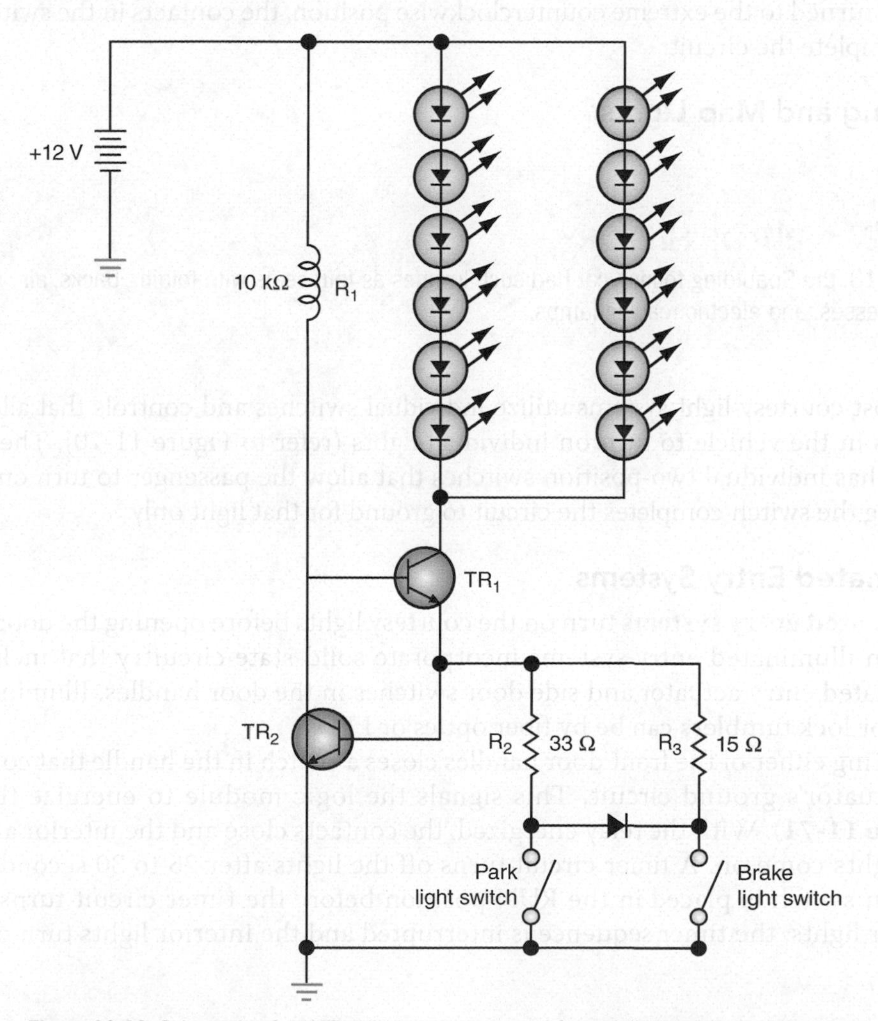

Figure 11-68 Schematic of an LED taillight assembly used for park light and brake light circuits.

Red, green, blue (RGB) LEDs can emit any color light from the same LED unit. A unit can consist of three different LEDs that emit the red, green, and blue colors that make up the color spectrum. A single LED can be doped with three different substances to create the colors. With either system, varying the voltage to the LEDs produces different colors. Mixing the red, green, and blue when they are emitting equal brightness produces a white light. Varying the voltages will vary the brightness of the LED, and mixing the red, green, and blue will result in different colors and shades. This allows using a single LED strip for multiple functions, such as taillight assemblies.

Interior Lights

Interior lighting includes courtesy lights, map lights, and instrument panel lights.

Courtesy Lights

Shop Manual
Chapter 11, page 566

Courtesy lights illuminate the vehicle's interior when the doors are open. Courtesy lights operate from the headlight and door switches and receive their power source directly from a fused battery connection. The switches can be either ground switch circuit (**Figure 11-69**) or insulated switch circuit design (**Figure 11-70**). In the insulated switch circuit, the switch is the power relay to the lights. In the grounded switch circuit, the switch controls the grounding portion of the circuit for the lights.

The headlight switch may also activate the courtesy lights. When the headlight switch knob is turned to the extreme counterclockwise position, the contacts in the switch close and complete the circuit.

Reading and Map Lights

A BIT OF HISTORY

In 1913, the Spaulding touring car had such luxuries as four seats with folding backs, air mattresses, and electric reading lamps.

Most courtesy light systems utilize individual switches and controls that allow passengers in the vehicle to turn on individual lights (refer to Figure 11-70). The system shown has individual two-position switches that allow the passenger to turn on a light. Pressing the switch completes the circuit to ground for that light only.

Illuminated Entry Systems

Shop Manual
Chapter 11, page 569

Illuminated entry systems turn on the courtesy lights before opening the doors. Most modern illuminated entry systems incorporate solid-state circuitry that includes an illuminated entry actuator and side door switches in the door handles. Illumination of the door lock tumblers can be by fiber optics or LEDs.

Lifting either of the front door handles closes a switch in the handle that completes the actuator's ground circuit. This signals the logic module to energize the relay (**Figure 11-71**). With the relay energized, the contacts close and the interior and door lock lights come on. A timer circuit turns off the lights after 25 to 30 seconds. If the ignition switch is placed in the RUN position before the timer circuit turns off the interior lights, the timer sequence is interrupted and the interior lights turn off.

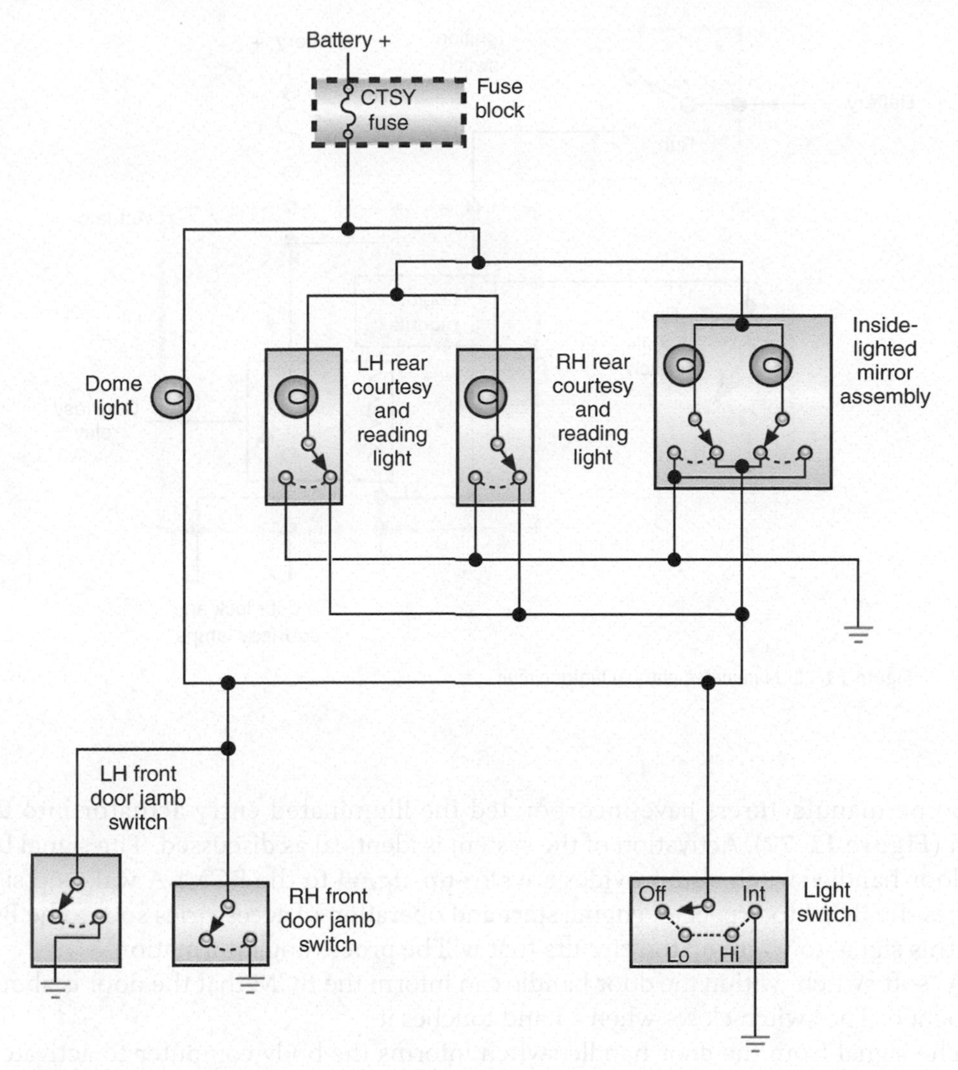

Figure 11-69 Courtesy lights using ground side switches.

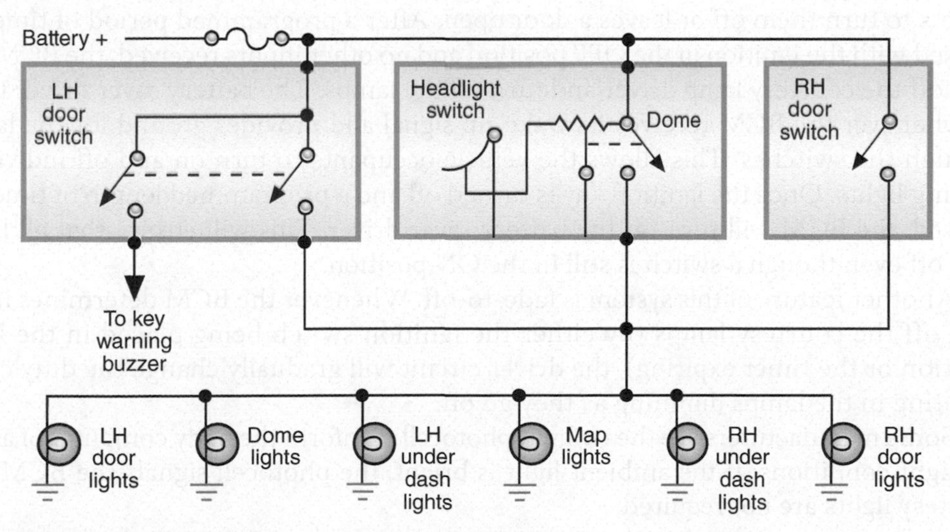

Figure 11-70 Courtesy lights using insulated side switches.

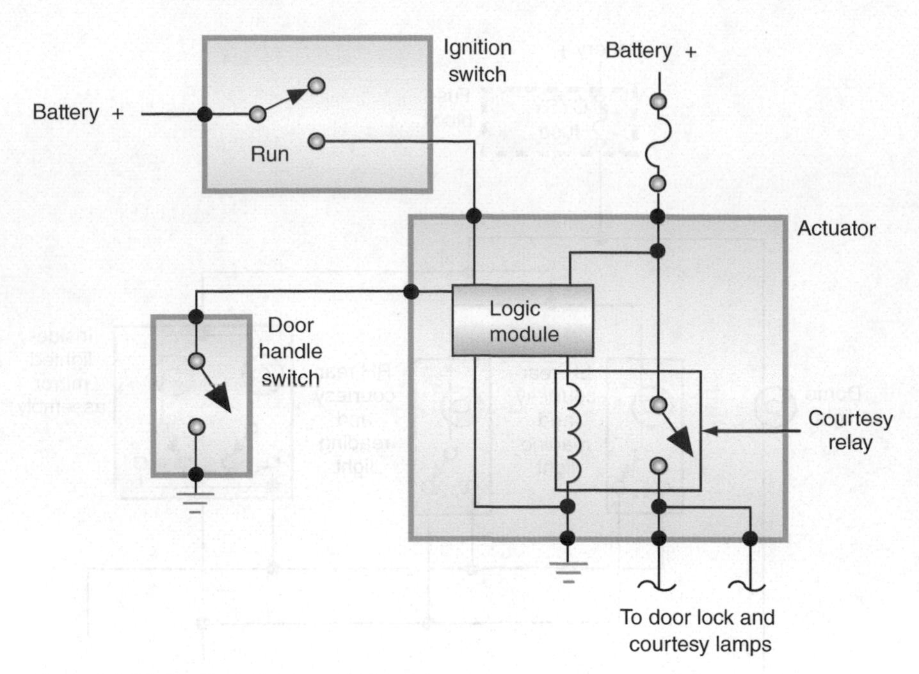

Figure 11-71 Illuminated entry actuator circuit.

Some manufacturers have incorporated the illuminated entry actuator into their BCM (**Figure 11-72**). Activation of the system is identical as discussed. The signal from the door handle switch also provides a **wake-up signal** to the BCM. A wake-up signal notifies the BCM to expect an engine start and operation of accessories soon. The BCM uses this signal to warm up the circuits that will be processing information.

A "soft switch" within the door handle can inform the BCM that the door is about to be opened. The switch closes when a hand touches it.

The signal from the door handle switch informs the body computer to activate the courtesy light driver. Some systems use a pair of door jamb switches that signal the body computer to keep the courtesy lights on when the door is open. The lights turn off when the door is closed and the ignition is in the RUN position.

Referring to Figure 11-72, this system can turn off the interior lights if the driver forgets to turn them off or leaves a door open. After a programmed period of time has elapsed with the ignition in the OFF position and no other inputs received, the BCM will turn off the courtesy lamp driver and turn off the lamps. The battery saver driver turns on whenever the BCM receives its wake-up signal and provides ground for the lamps through the switches. This allows the vehicle occupants to turn on and off individual reading lights. Once the ignition key is turned off and a programmed length of time has elapsed, the BCM will turn off the battery saver driver. This will ensure that all lights turn off even though a switch is still in the ON position.

Another feature of this system is fade-to-off. Whenever the BCM determines it will turn off the courtesy lamps (by either the ignition switch being placed in the RUN position or the timer expiring), the driver circuit will gradually change the duty cycle, resulting in the lamps dimming as they go off.

Some manufacturers use the twilight photocell to inform the body computer of ambient light conditions. If the ambient light is bright, the photocell signals the BCM that courtesy lights are not required.

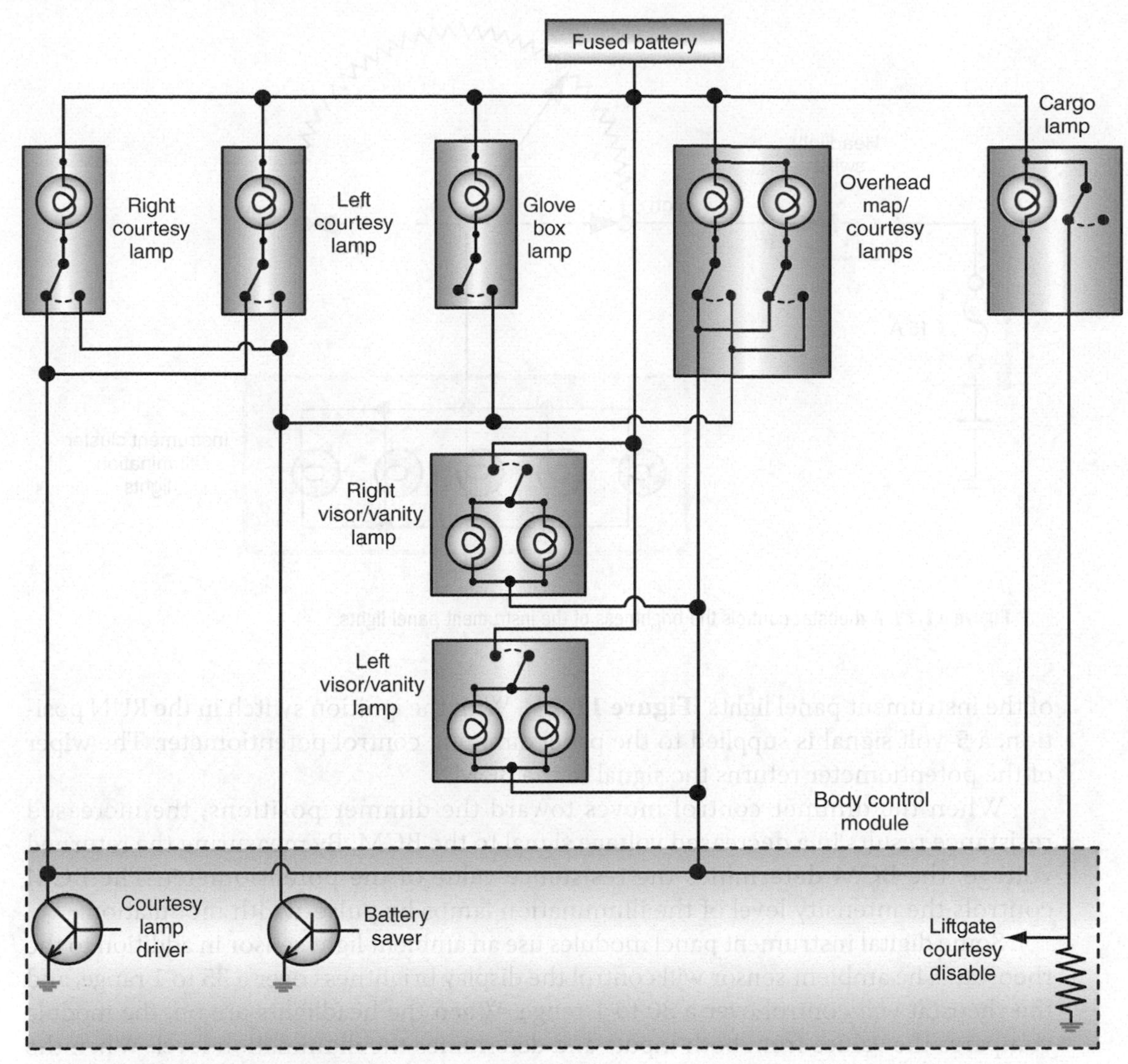

Figure 11-72 Body computer control of the illuminated entry system.

Instrument Cluster and Panel Lamps

Shop Manual
Chapter 11, page 573

Consider the following three types of lighting circuits within the instrument cluster:

1. *Warning lights* alert the driver to potentially dangerous conditions such as brake failure or low oil pressure.
2. *Indicator lights* include turn signal indicators.
3. *Illumination lights* provide indirect lighting to illuminate the instrument gauges, speedometer, heater controls, clock, ashtray, radio, and other controls.

In a conventional system, the headlight switch provides the power source for the instrument panel lights. The contacts are closed when the headlight switch is in the PARK or HEADLIGHT position. The current must flow through a variable resistor (rheostat) that is either a part of the headlight switch or a separate dial on the dash. Turning the knob varies the resistance of the rheostat and changes the current flow to the lamps, controlling the brightness of the lights (**Figure 11-73**).

Instrument panel dimming can be a function of the BCM. Using inputs from the panel dimming control switch and a photocell, the BCM determines the illumination level

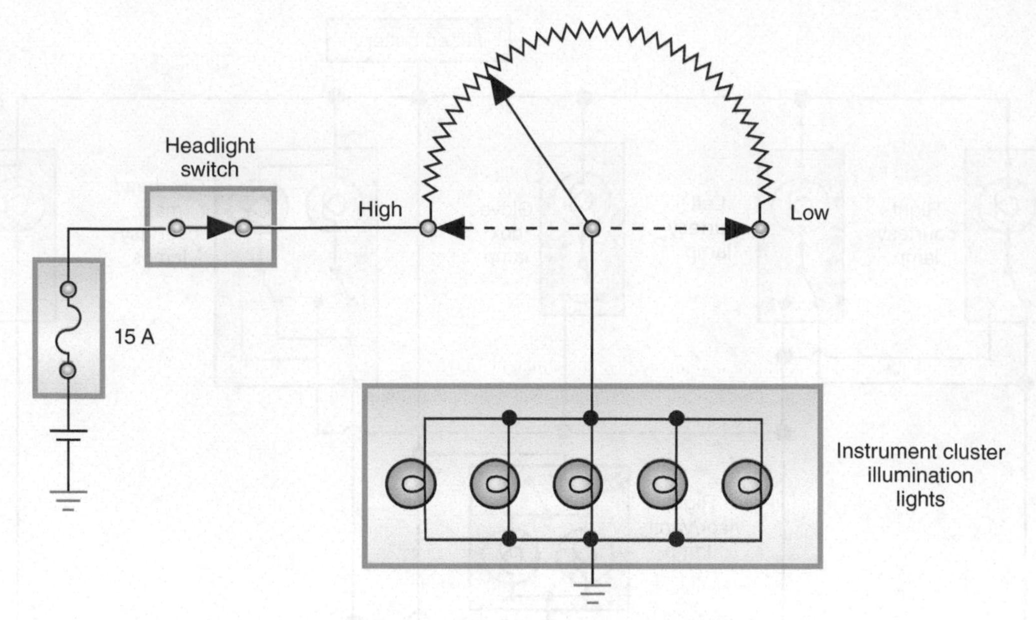

Figure 11-73 A rheostat controls the brightness of the instrument panel lights.

of the instrument panel lights (**Figure 11-74**). With the ignition switch in the RUN position, a 5-volt signal is supplied to the panel dimming control potentiometer. The wiper of the potentiometer returns the signal to the BCM.

When the dimmer control moves toward the dimmer positions, the increased resistance results in a decreased voltage signal to the BCM. By measuring the returned voltage, the BCM determines the resistance value of the potentiometer. The BCM controls the intensity level of the illumination lamps by pulse-width modulation.

Some digital instrument panel modules use an ambient light sensor in addition to the rheostat. The ambient sensor will control the display brightness over a 35 to 1 range, and the rheostat will control over a 30 to 1 range. When the headlights are on, the module compares the values from both inputs and determines the illumination level. When the headlights are off, the module uses only the ambient light sensor for its input.

A variation of this operation is that the BCM will receive the light intensity level request from the headlight rheostat as it did before, but the requested level is then sent to the instrument cluster by the multiplexing circuit. The instrument cluster uses this input information to control lamp intensity through its own microprocessor.

Fiber Optics

Shop Manual
Chapter 11, page 578

Fiber optics is the transmission of light through polymethyl methacrylate plastic that keeps the light rays parallel even if extreme bends are in the plastic. Fiber optics enable the illumination of several objects using a single light source (**Figure 11-75**). Plastic fiber-optic strands transmit light from the source to the object to be illuminated. The plastic strands are sheathed by a polymer that insulates the light rays as they travel within the strands. The light rays travel through the strands by means of internal reflections.

Some manufacturers use fiber optics to provide illumination of the lock cylinder halo during illuminated entry operation (**Figure 11-76**). The light collector provides

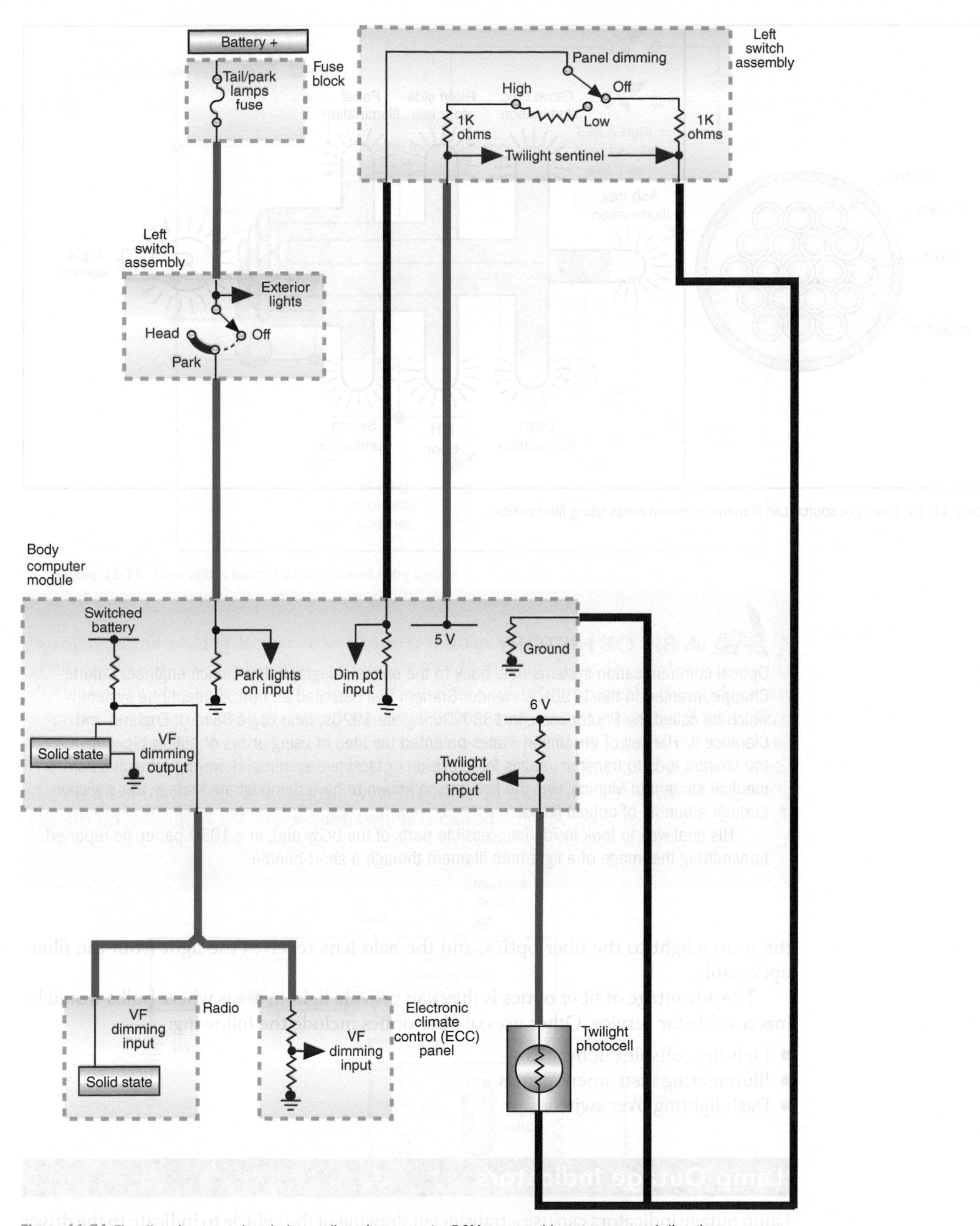

Figure 11-74 The dimming control and photocell are inputs to the BCM to control instrument panel dimming.

AUTHOR'S NOTE Winding the coils in opposite directions creates the opposing magnetic fields.

Some manufacturers use a **lamp outage module** either as a stand-alone module or in conjunction with the BCM. A lamp outage module is a current-measuring sensor that contains a set of resistors wired in series with the power supply to the headlights, taillights, and stop lights. If the module is a stand-alone unit, it will operate the warning light directly. The module monitors the voltage drop of the resistors. Proper operation of the circuits is indicated by a 500-mV input signal to the module. If one of the monitored bulbs burns out, the voltage input signal drops to about 250 mV. The module completes the ground circuit to the warning light to alert the driver that a bulb has burned out. The module is capable of monitoring several different light circuits.

Many vehicles today use a computer-driven information center to inform the driver of the condition of monitored circuits (**Figure 11-78**). The vehicle information center usually receives its signals from the BCM (**Figure 11-79**). In this system, the lamp outage module sends signals to the BCM. The BCM will illuminate a warning light, give a digital message, or activate a voice warning device that alerts the driver of the burned out light bulb.

A burned-out light bulb is identified when there is a loss of current flow in one of the resistors of the lamp outage module. A monitoring chip in the module compares the voltage drop across the resistor. If there is no voltage drop across the resistor, there is an open in the circuit (burned-out light bulb). When the chip measures no voltage drop across the resistor, it signals the BCM, which then gives the necessary message to the vehicle information center.

AUTHOR'S NOTE In these systems the bulbs are monitored only when current is supplied to them.

Figure 11-78 The computer-driven vehicle information center keeps the driver aware of the condition of monitored systems.

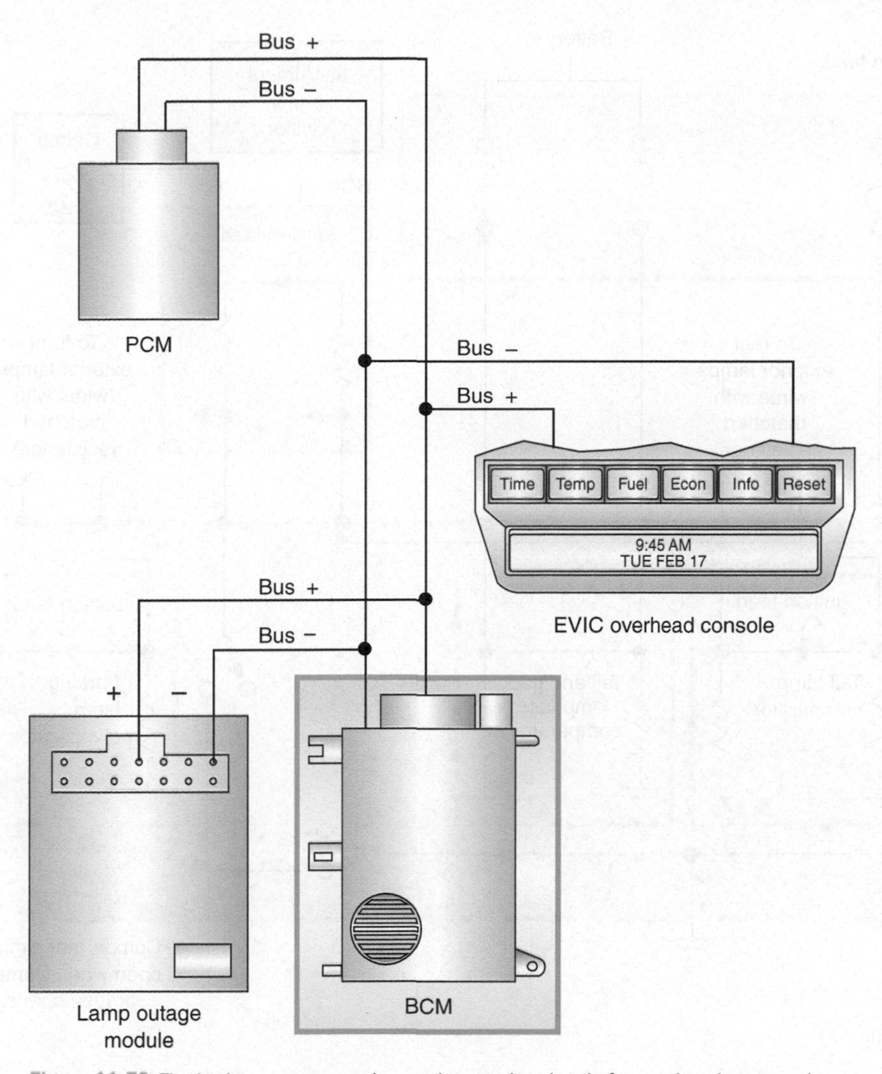

Figure 11-79 The body computer can be used to receive signals from various inputs and to give signals to control the information center.

General Motors uses the lamp monitor module to connect the light circuits to ground (**Figure 11-80**). When the circuits are operating properly, the ground connection in the module causes a low-circuit voltage. Input from the lamp circuits is through two equal-resistance wires. The module output to the bulbs is from the same module terminals as the inputs.

If a bulb burns out, the voltage at the lamp monitor module terminal will increase. The module will open the appropriate circuit from the BCM, signaling the BCM to send a communication to the instrument panel cluster computer, which displays the message in the information center.

Shop Manual
Chapter 11, page 531

Vehicles that use HSDs to control lamp illumination can detect lamp outages. Often these systems can determine if a lamp circuit is open without the need to turn on the lamp system by sending a small diagnostic current through the circuit. If the circuit is complete, an expected voltage drop will occur. If the circuit is open, the voltage reading will remain high and the module will request the lamp outage indicator to come on.

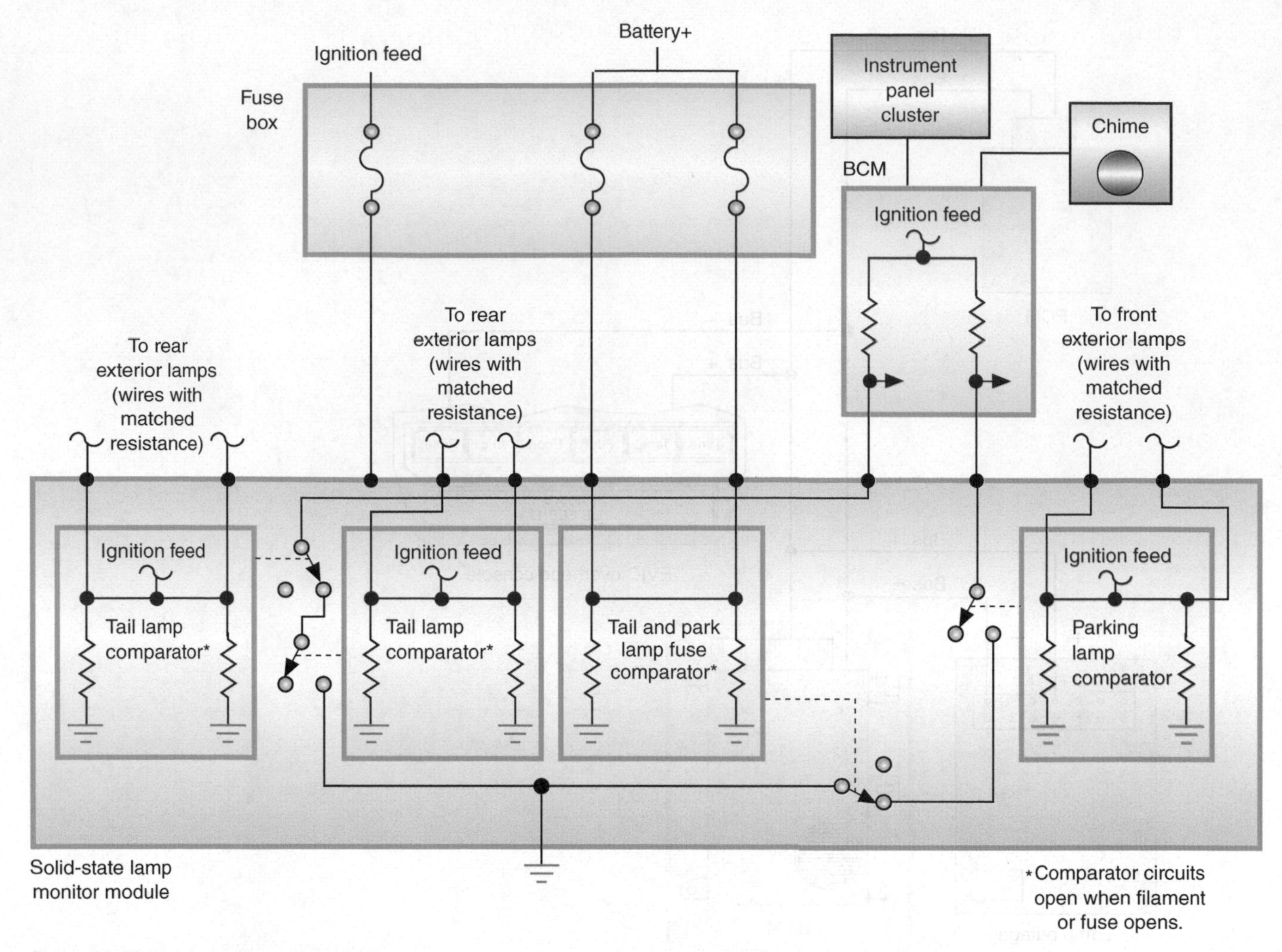

Figure 11-80 Lamp monitor module circuit.

Summary

- Different types of lamps are used to provide illumination for the systems. The lamp may be either a single-filament bulb that performs a single function or a double-filament bulb that performs several functions.
- The headlight lamps can be one of four designs: standard sealed beam, halogen sealed beam, composite, or high-intensity discharge (HID).
- The headlight filament is located on a reflector that intensifies the light, which is then directed through the lens. The lens is designed to change the circular light pattern into a broad, flat light beam. Placement of the filament in the reflector provides for low- and high-beam light patterns.
- In addition to the headlight system, the lighting systems include the following:

  Stop lights.
  Turn signals.
  Hazard lights.
  Parking lights.
  Taillights.
  Back-up lights.
  Side marker lights.
  Courtesy lights.
  Instrument panel lights.
- The headlight switch can control many of the lighting systems.

- A rheostat used with the headlight switch controls the brightness of the instrument panel illumination lights.
- Many computer-controlled headlight systems use resistive multiple headlight switches as an input.
- Computer-controlled headlight systems can use relays operated by the control module or high-side drivers (HSDs) to illuminate the lamps.
- The automatic on/off with time delay has two functions: to turn on the headlights automatically when ambient light decreases to a predetermined level and to allow the headlights to remain on for a certain amount of time after turning off the vehicle.
- Most automatic headlight dimming systems consist of a light-sensitive photocell and amplifier unit, high–low beam relay, sensitivity control, dimmer switch, flash-to-pass relay, and a wiring harness.
- The SmartBeam system uses a forward-facing, 5,000-pixel, digital imager camera.
- The operation of SmartBeam requires interaction with several vehicle modules, including the light rain sensor module (LRSM), the steering column module (SCM), the front control module (FCM), and the cabin compartment node (CCN).
- Decisions about headlight intensity are based on the sensed intensity of light, the light's location, and the light's movement.
- A headlight leveling system (HLS) uses front lighting assemblies with a leveling actuator motor.
- The adaptive headlight system (AHS) enhances nighttime safety by turning the headlight beams to follow the direction of the road as the vehicle enters a turn.
- The AHS uses sensors that measure vehicle speed, steering angle, and yaw (degree of rotation around the vertical axis). Based on this information, small electric motors turn the headlights, so the beam falls on the road ahead, guiding the driver into the turn.
- The night vision system uses an infrared camera to transfer images to a display panel, enabling the driver to identify and react to obstacles outside the headlight range.
- Daytime running lamps can use a relay or module to illuminate the low- or high-beam lamps at a reduced output.
- The adaptive brake light system selects different illumination levels or display methods for the rear brake lights depending on conditions.
- The illuminated entry system turns on the interior lights before opening the door. The system may also be capable of turning off the lights if the driver fails to shut a door when exiting.
- Instrument panel dimming is usually done by the BCM providing a pulse-width modulation to the illumination lamps or LEDs.
- Fiber optics is the transmission of light through polymethyl methacrylate plastic that keeps the light rays parallel even if there are extreme bends in the plastic.
- The lamp outage indicator alerts the driver, through an information center on the dash or console, that a light bulb has burned out.

Review Questions

Short-Answer Essays

1. List the common components of the automatic headlight dimming system.
2. Explain the operation of body computer–controlled instrument panel illumination dimming.
3. Explain the operation of the SmartBeam headlight system.
4. Describe the function of automatic headlight leveling systems.
5. Describe the purpose and function of daytime running lamps.
6. Explain the use and function of fiber optics.
7. What is meant by pulse-width dimming?
8. What three lighting circuits are incorporated within the instrument cluster?
9. Describe the operation of HID headlights.
10. Describe the purpose of bi-xenon headlights.

Fill in the Blanks

1. The photocell will have ____________ resistance as the ambient light level increases.
2. In some illuminated entry systems, the ____________ signals the body computer that the courtesy lights are not required.

3. The body computer uses inputs from the _____________ and _____________ to determine the illumination level of the instrument panel lights.
4. The body computer dims the illumination lamps by using a _____________ signal to the panel lights.
5. Most computer-controlled headlight systems use a _____________ switch for an input.
6. Fiber optics are commonly used as _____________ lights.
7. Lamp outage modules detect _____________ in a normally operating circuit.
8. HSDs supply _____________ to the lamps.
9. When today's technician performs repairs on the lighting systems, the repairs must meet at least two requirements: they must assure vehicle _____________ and meet all _____________.
10. A _____________ is a device that produces light as a result of current flow through a _____________.

Multiple Choice

1. All of the following statements about the SmartBeam system are true, **EXCEPT:**
   A. The system uses a digital camera to determine light intensity.
   B. The system is capable of detecting the movement of oncoming light.
   C. The system is capable of distinguishing colors.
   D. The AHBM turns off the high-beam relay when the oncoming light intensity is 10 LUX or more.

2. Computer-controlled instrument panel dimming is being discussed.
   Technician A says the body computer dims the illumination lamps by varying resistance through a rheostat wired in series to the lights.
   Technician B says the body computer can use inputs from the panel dimming control and photocell to determine the illumination level of the instrument panel lights on certain systems.
   Who is correct?
   A. A only C. Both A and B
   B. B only D. Neither A nor B

3. Which statement about fiber optics is correct?
   A. Fiber optics is the transmission of light through several plastic strands sheathed by a polymer.
   B. Fiber optics is used only for exterior lighting.
   C. Fiber optics can be used only in applications where the conduit can be laid straight.
   D. All of these choices.

4. The purpose of the headlight leveling system is to:
   A. Reduce the need to align the light beams.
   B. To allow the driver to raise or lower the light beams as vehicle loads change.
   C. To allow the driver to raise or lower the light beams as the vehicle ascends and descends hills.
   D. All of these choices.

5. *Technician A* says computer-controlled headlight systems can use relays that the BCM operates.
   Technician B says computer-controlled headlight systems can use high-side drivers to operate the lamps.
   Who is correct?
   A. A only C. Both A and B
   B. B only D. Neither A nor B

6. In the SmartBeam system, the headlight intensity is based on which of the following:
   A. Movement of the light.
   B. Intensity of the light.
   C. Location of the light.
   D. All of these choices.

7. Which statement is the most correct?
   A. Lamp outage modules can use voltage drop to determine circuit operation.
   B. HSDs can only detect opens in activated circuits.
   C. Low-side, driver-controlled lamps cannot determine lamp outage conditions.
   D. All of these choices.

8. In a composite headlight:
   A. The bulb is replaceable.
   B. A cracked lens will prevent lamp operation.
   C. Both A and B.
   D. None of these choices.

9. The CHMSL circuit is being discussed.
   Technician A says the diodes are used to assure proper turn signal operation.
   Technician B says the diodes are used to prevent radio static when the brake light is activated.
   Who is correct?
   A. A only
   B. B only
   C. Both A and B
   D. Neither A nor B

10. Which of the following best describes the function of the bi-xenon headlight system?
    A. Uses double-filament headlights to provide the high-beam output.
    B. Uses a chamber filled with multiple xenon gases that are ignited at different temperatures to provide high-beam and low-beam operations.
    C. Uses multiple chambers of xenon gas that are ignited based on dimmer switch position. One chamber is for low beam, and both chambers are ignited for high beam.
    D. None of these choices.

CHAPTER 12
INSTRUMENTATION AND WARNING LAMPS

Upon completion and review of this chapter, you should be able to:

- Describe the operation of electromagnetic gauges, including d'Arsonval, three-coil, two-coil, and air-core.
- Describe the operation of electronic fuel, temperature, oil, and voltmeter gauges.
- Describe the operation of quartz analog instrumentation.
- Explain the function and operation of the various gauge sending units, including thermistors, piezoresistive, and mechanical variable resistors.
- Describe the purpose of speedometers and odometers.
- Describe the purpose of the tachometer.
- Describe the operating principles of the digital speedometer.
- Explain the operation of integrated circuit (IC) chip and stepper motor odometers.
- Explain the operation of various warning lamp circuits.
- Explain the operation of various audible warning systems.

Terms to Know

Air-coil gauge
Bucking coil
Buffer circuit
d'Arsonval gauge
Digital instrument clusters
Electromagnetic gauges
Electromechanical gauge
Gauge
Head-up display (HUD)
H-gate
High-reading coil
International Standards Organization (ISO)
Low-reading coil
Odometer
Oscillator
Pinion factor
Prove-out circuit
Sending unit
Three-coil gauge
Two-coil gauge
Warning lamp
Watchdog circuit

Introduction

Instrument gauges and indicator lights provide information to the driver concerning the current operation of various vehicle systems (**Figure 12-1**). Warning devices also provide information to the driver; however, they are commonly associated with an audible signal. Some vehicles use a voice module to alert the driver to certain conditions.

Early instrument cluster gauges were analog or swing needle type. Although many modern vehicles still use the analog gauge, they are now computer-driven. These instruments provide far more accurate readings than their conventional analog counterparts. This chapter introduces you to the most commonly used computer-driven instrumentation systems. These systems include the speedometer, odometer, fuel, oil, tachometer, and temperature gauges.

The computer-driven instrument panel uses a microprocessor (μP) to process information from various sensors and to control the gauge display. Depending on the manufacturer, the microprocessor can be a separate computer that receives direct information from the sensors and makes the calculations, or can use the body control module (BCM) to perform all functions.

In addition, there are many types of information systems used today. These systems inform the driver of various monitored conditions, including vehicle maintenance, trip information, and navigation.

Electromechanical Gauges

Shop Manual
Chapter 12, pages 622, 631

A **gauge** is a device that displays the measurement of a monitored system using a needle or pointer that moves along a calibrated scale. The **electromechanical gauge** operates electrically, but its movement is mechanical. The gauge acts as an ammeter since its reading changes with variations in resistance. There are two basic types of electromechanical gauges: the bimetallic gauge and the electromagnetic gauge. Conventional analog instrument clusters that used these types of gauges had a direct connection to the sending unit. The resistance of the sending unit determined the location of the needle on the gauge face. A short study of the different types of gauges is offered as a foundation for studying computer-driven gauges.

Electromagnetic gauges produce needle movement by magnetic forces instead of heat. There are four types of electromagnetic gauges: the d'Arsonval, the three-coil, the two-coil, and the air-core.

The **d'Arsonval gauge** uses the interaction of a permanent magnet and an electromagnet, and the total field effect to cause needle movement. The d'Arsonval gauge consists of a permanent horseshoe–type magnet that surrounds a movable electromagnet (armature) attached to a needle (**Figure 12-2**). When current flows through the armature, it becomes an electromagnet and is repelled by the permanent magnet. When current flow through the armature is low, the electromagnet's strength is weak, and needle movement is small. Increasing the current flow increases the magnetic field created in the armature, and needle movement is greater. The armature has a small spring attached to it to return the needle to zero when no current flows to the armature.

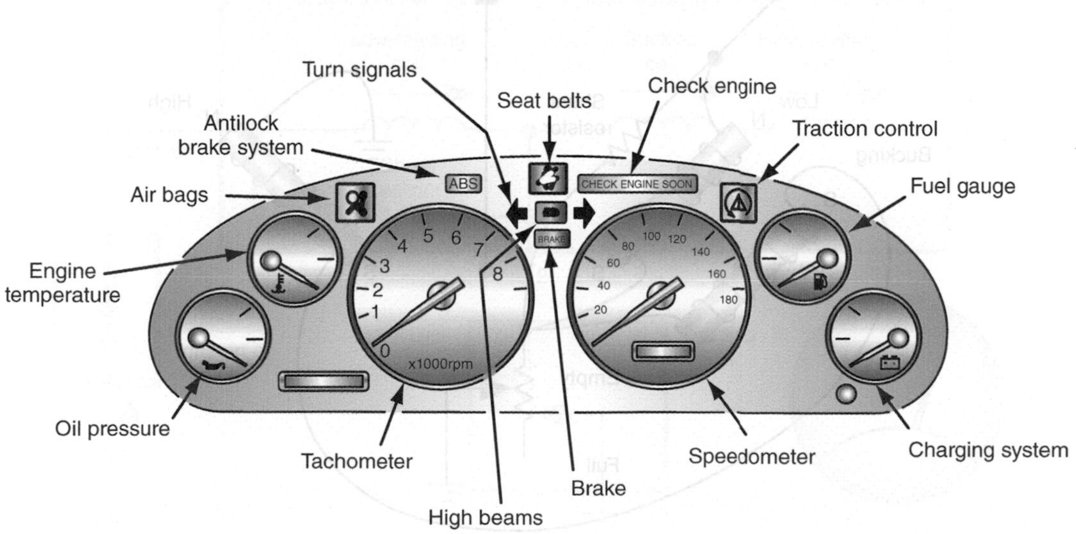

Figure 12-1 Typical gauge and warning indicator layout of an instrument panel.

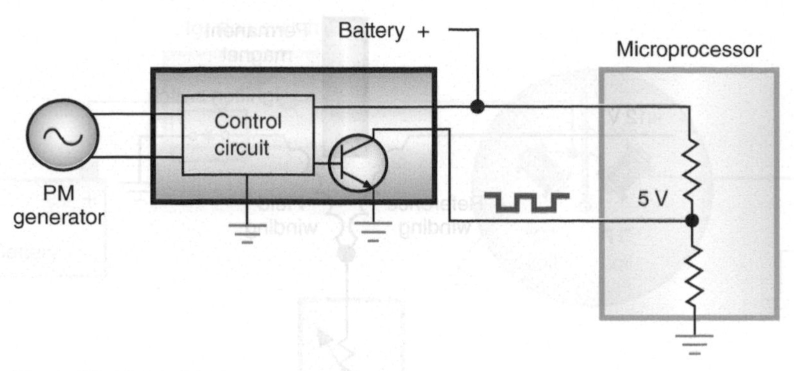

Figure 12-10 A buffer circuit.

the speedometer assembly. This system relied on a rotating permanent magnet that produced a rotating magnetic field around a drum. The rotating magnetic field generated circulating eddy currents in the drum that produced a small magnetic field that interacted with the field of the rotating magnet. This interaction of the two magnetic fields pulled the drum and needle around with the rotating magnet. It's not hard to see that this system would not be extremely accurate. Today's vehicles use sensors and computer logic to display vehicle speed.

Shop Manual
Chapter 12, page 628

Many quartz analog speedometer gauge systems place a permanent magnet (PM) generator sensor in the transaxle, transmission, or differential. As the PM generator rotates, it induces a small alternating current (AC) in its coil. A **buffer circuit** changes the AC voltage from the PM generator into a digitalized signal (**Figure 12-10**) and sends the signal to the processing unit (**Figure 12-11**). The signal passes through a quartz clock circuit, a gain selector circuit, and a driver circuit. The driver circuit sends voltage pulses to the coils of the gauge; the coils operate like conventional air-coil gauges to move the needle.

A BIT OF HISTORY

One of the early styles of speedometers used a regulated amount of air pressure to turn a speed dial. The air pressure was generated in a chamber containing two intermeshing gears. A flexible shaft connected to a front wheel or the driveshaft drove the gears. The air was applied against a vane inside the speed dial. The amount of air applied was proportional to the speed of the vehicle.

Often the sensor used to determine vehicle speed has multiple purposes. As a result, the signal may not accurately represent vehicle speed because it comes before the final drive unit. The control module may need to do additional calculations to make the speedometer accurate. For example, Chrysler vehicles equipped with the 41TE or 42LE electronic shift transaxles use an output speed sensor that generates an AC signal from a 24-tooth tone wheel on the rear planetary unit. The transmission control module (TCM) receives this signal and applies **pinion factor** to the hertz count to calculate vehicle speed. Pinion factor is a calculation using the final drive ratio and the tire circumference to obtain accurate vehicle speed signals. The TCM then transmits this information to the powertrain control module (PCM) at a set rate of 8,000 pulses per mile by pulsing the dedicated circuit. The PCM then sends the vehicle speed signal over the data bus circuit to all modules that require it.

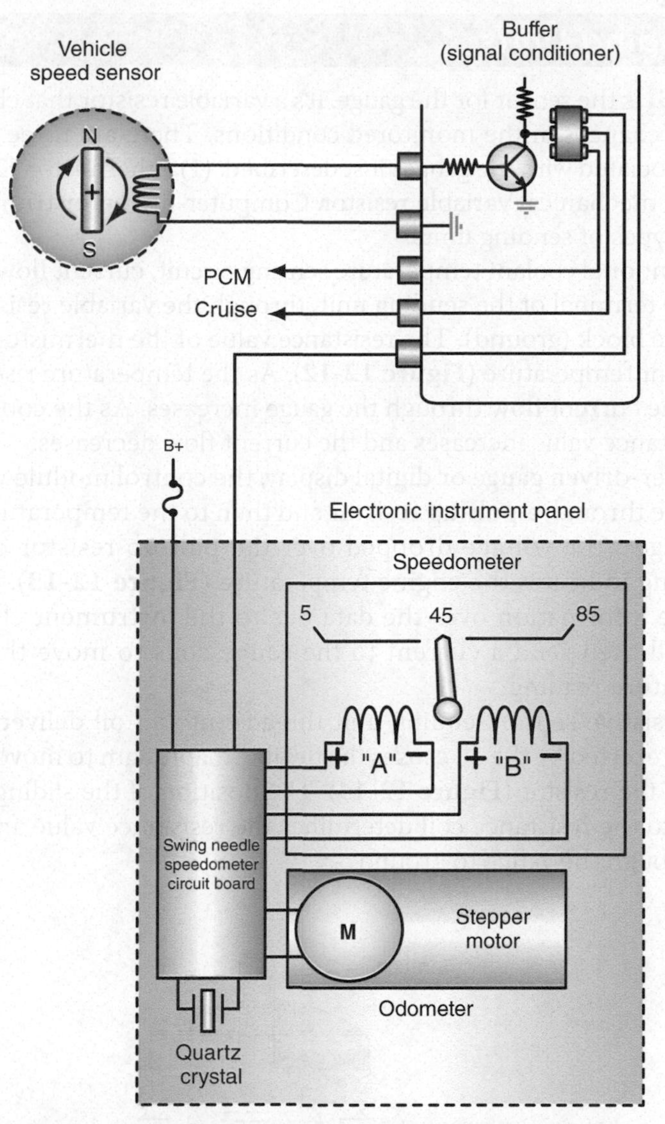

Figure 12-11 Quartz swing needle speedometer schematic. The "A" coil is connected to system voltage, and the "B" coil receives a voltage that is proportional to input frequency. The magnetic armature reacts to the changing magnetic fields.

Pinion factor information is programmed into the TCM at the factory. Replacing the TCM in the field requires a scan tool to program the tire size. In some systems, the gear ratio also has to be programmed. If the pinion factor is not programmed into the TCM, the speedometer and cruise control systems will not function.

Some manufacturers will use the wheel speed sensors from the antilock brake system (ABS) to determine vehicle speed. Usually, the two front or the two rear sensor inputs are averaged. The ABS module then determines vehicle speed and broadcasts the information on the data bus.

The other air-coil gauges (temperature, fuel level, etc.) work as described earlier with conventional instrument clusters. The difference is that the sending unit input goes to a module. Based on the sending unit resistance, the module controls current flow through the gauge coils.

Gauge Sending Units

The **sending unit** is the sensor for the gauge. It's a variable resistor that changes resistance values based on changes in the monitored conditions. There are three types of sending units that are associated with the gauges just described: (1) a thermistor, (2) a piezoresistive sensor, and (3) a mechanical variable resistor. Computer-driven instrument clusters also use these same types of sending units.

Shop Manual
Chapter 12, page 635

In the conventional coolant temperature sensing circuit, current flows from the gauge unit into the top terminal of the sending unit, through the variable resistor (thermistor), and to the engine block (ground). The resistance value of the thermistor changes in proportion to coolant temperature (**Figure 12-12**). As the temperature rises, the resistance decreases and the current flow through the gauge increases. As the coolant temperature lowers, the resistance value increases and the current flow decreases.

In a computer-driven gauge or digital display, the control module will send a 5-volt reference voltage through a pull-up resistor and then to the temperature sensor. As the resistance changes, the voltage dropped over the pull-up resistor changes and the voltmeter reading indicates the engine temperature (**Figure 12-13**). The PCM sends the temperature information over the data bus to the instrument cluster (or BCM). The module will then send a current to the gauge coils to move the pointer to the correct temperature reading.

The piezoresistive sensor sending unit threads into an oil delivery passage of the engine. Pressure exerted by the oil causes the flexible diaphragm to move, and the contact arm slides along the resistor (**Figure 12-14**). The position of the sliding contacts on the arm in relation to the resistance coil determines the resistance value and the amount of current flow through the gauge to ground.

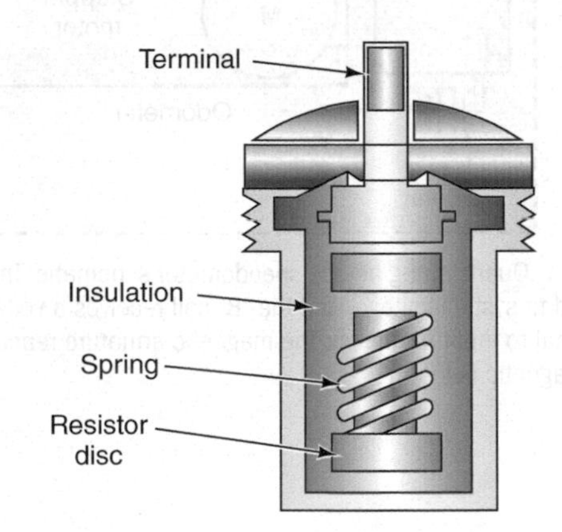

Figure 12-12 A thermistor used to sense engine temperature.

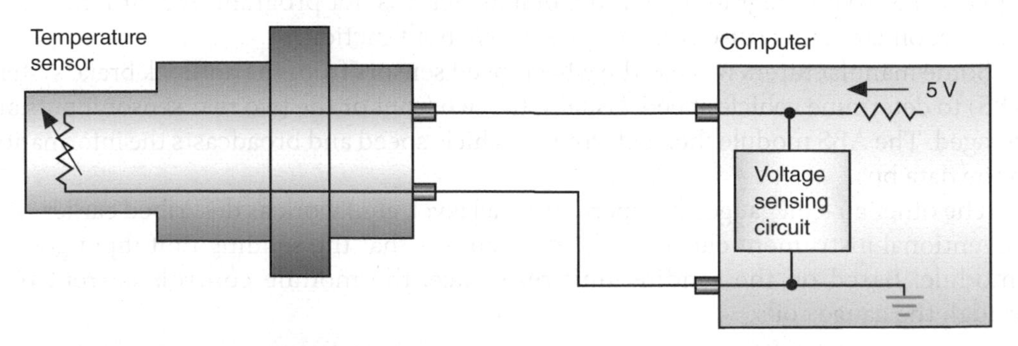

Figure 12-13 A typical temperature sensing circuit.

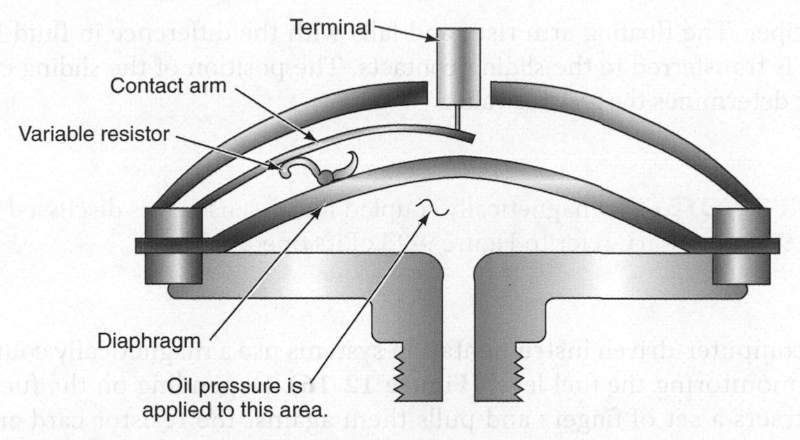

Figure 12-14 Piezoresistive sensor used for measuring engine oil pressure.

Another style is a transducer that operates like a Wheatstone bridge MAP sensor. The function of the gauge is the same as that discussed for the computer-driven temperature gauge, based on data bus messages from the PCM.

Some computer-driven instrument clusters have an oil gauge but don't use a sensor. These systems use an oil pressure switch. When oil pressure exceeds 6 psi (41 kPa), the switch opens the sense circuit. This pulls the sense voltage high. The gauge displays an oil pressure based on a calculated value determined from engine run time, engine temperature, engine load value, and ambient temperature. The gauge will indicate normal oil pressure as long as the switch is open. If the oil pressure drops below 6 psi (41 kPa), the switch closes the circuit to ground, pulling the sense voltage low. The instrument cluster gauge now reads 0.

A fuel level sending unit is an example of a mechanical variable resistor (**Figure 12-15**). The sending unit is located in the fuel tank and has a float connected to a variable

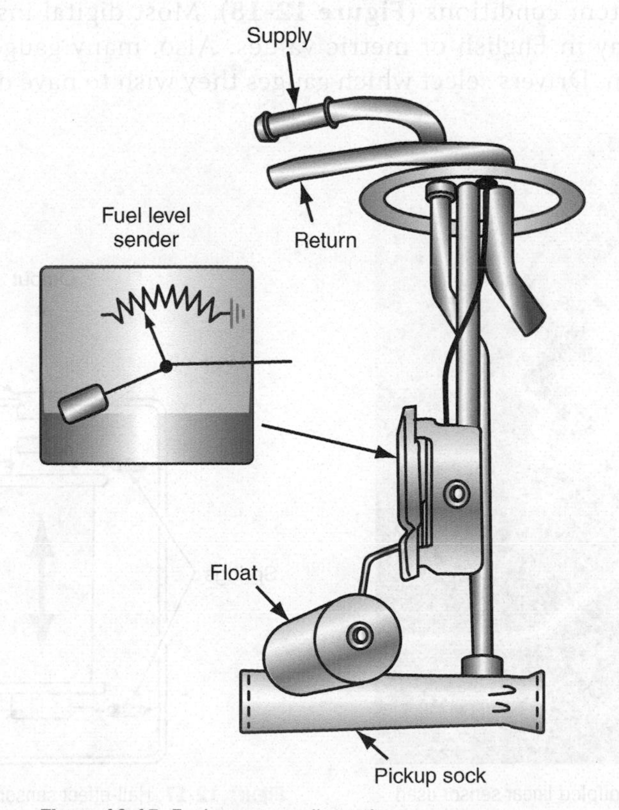

Figure 12-15 Fuel gauge sending unit.

resistor's wiper. The floating arm rises and falls with the difference in fluid level. Float movement is transferred to the sliding contacts. The position of the sliding contacts on the resistor determines the resistor value.

AUTHOR'S NOTE The magnetically coupled linear sensor was discussed in Chapter 9. If necessary, refer to Figure 9-23 of its operation.

Some computer-driven instrumentation systems use a magnetically coupled linear sensor for monitoring the fuel level (**Figure 12-16**). Depending on the fuel level, the magnet attracts a set of fingers and pulls them against the resistor card and a metal strip. Each segment of the resistor card produces different resistances to represent the current fuel level.

Another sensor used to determine the fuel level is the linear Hall-effect sensor (**Figure 12-17**). These sensors produce an output voltage that is proportional to the applied magnetic field. When used as a fuel level sending unit, a magnet is attached to the float assembly. As the float moves up and down with the fuel level, the gap between the magnet and the Hall element changes. The gap changes the Hall-effect and thus the output voltage.

Digital Instrumentation

Shop Manual
Chapter 12, page 634

Digital instrumentation is far more precise than conventional analog gauges. Analog gauges display an average of the readings received from the sensor; a digital display will present exact readings. In some systems, the information to the gauge is updated as often as 16 times per second.

Digital instrument clusters use digital and linear displays to notify the driver of monitored system conditions (**Figure 12-18**). Most digital instrument clusters provide for display in English or metric values. Also, many gauges are a part of a multigauge system. Drivers select which gauges they wish to have displayed. Most of

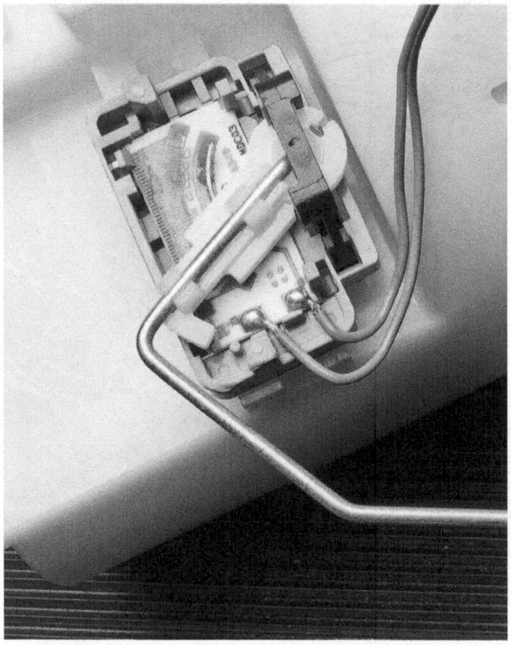

Figure 12-16 Magnetically coupled linear sensor used to measure fuel level.

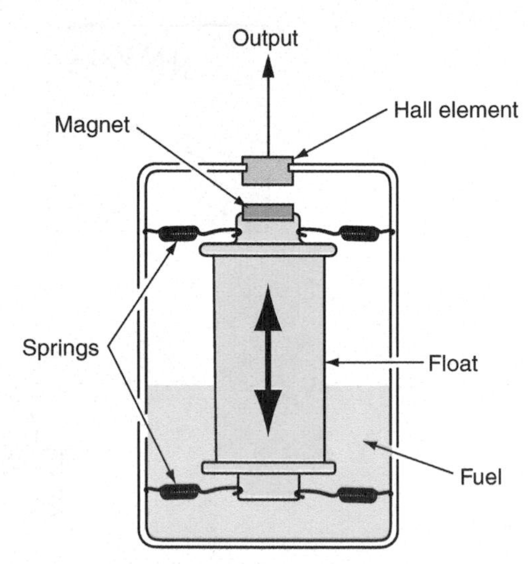

Figure 12-17 Hall-effect sensor used for fuel level indication.

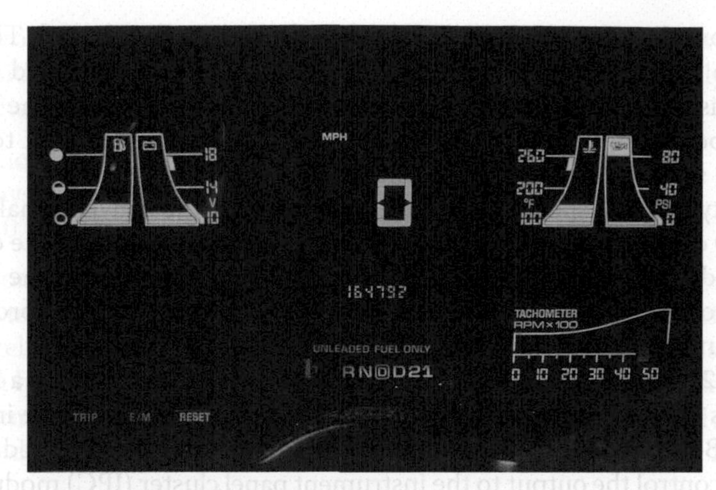

Figure 12-18 Digital instrument cluster.

these systems will automatically switch gauges to display the gauge indicating a potentially dangerous situation. For example, suppose the driver has chosen to display the oil pressure gauge, and the engine temperature increases above set limits. In that case, the temperature gauge will automatically replace the oil pressure gauge to warn the driver of the overheating condition. A warning light and/or a chime will also activate to get the driver's attention.

Shop Manual
Chapter 12, page 620

Most electronic instrument panels have self-diagnostic capabilities. The self-diagnostic test may be initiated using a scan tool or by pushing selected buttons on the instrument panel. The instrument panel cluster also initiates a self-test every time the ignition is placed in ACC or RUN. Usually, the entire dash illuminates and every display segment is lighted. **International Standards Organization (ISO)** symbols represent the gauge function (**Figure 12-19**). These symbols may flash during the test. After completion of the test, all gauges display the current readings. If the test detects a fault, the gauge may display a code to alert the driver.

Speedometers

Shop Manual
Chapter 12, page 629

Ford, General Motors (GM), and Toyota have used optical vehicle speed sensors (VSSs). The Ford and Toyota optical sensors operated from the conventional speedometer cable that rotates a slotted wheel between a light-emitting diode (LED) and a phototransistor (**Figure 12-20**). As the slots in the wheel break the light, the transistor conducts an electronic pulse signal to the speedometer. An integrated circuit rectifies the analog

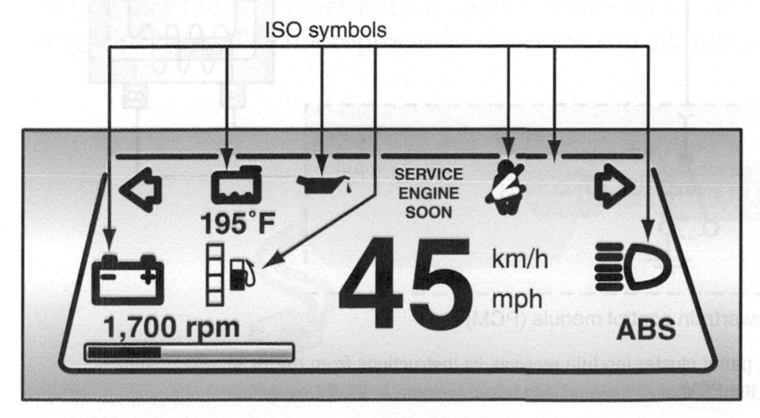

Figure 12-19 A few of the ISO symbols used to identify the gauge.

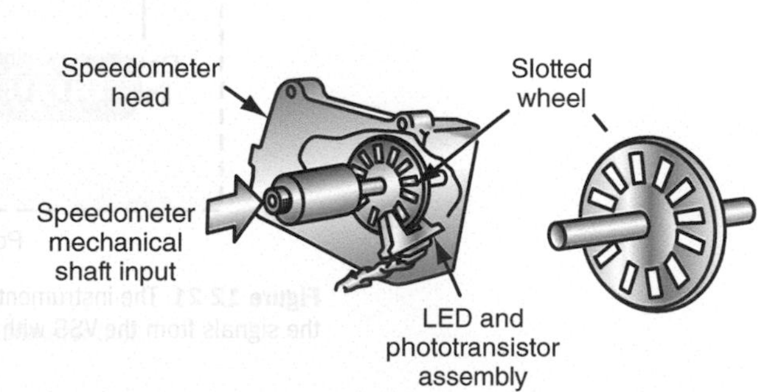

Figure 12-20 Optical speed sensor.

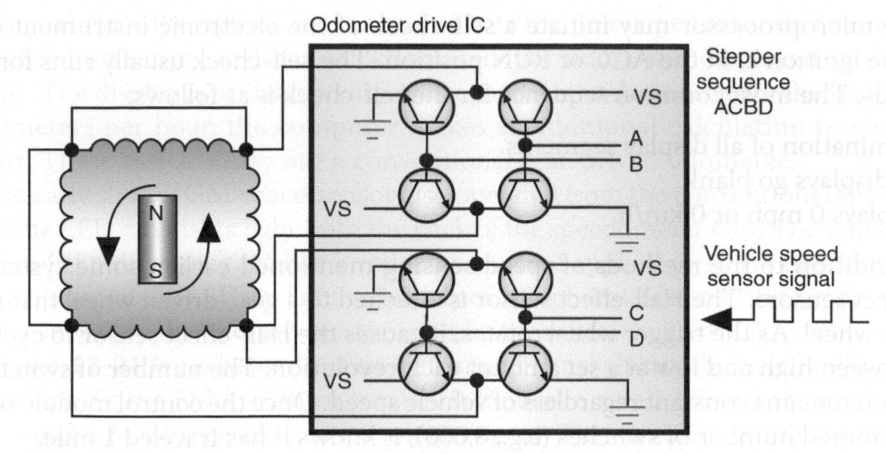

Figure 12-23 The H-gate energizes two coils at a time and constantly reverses system polarity.

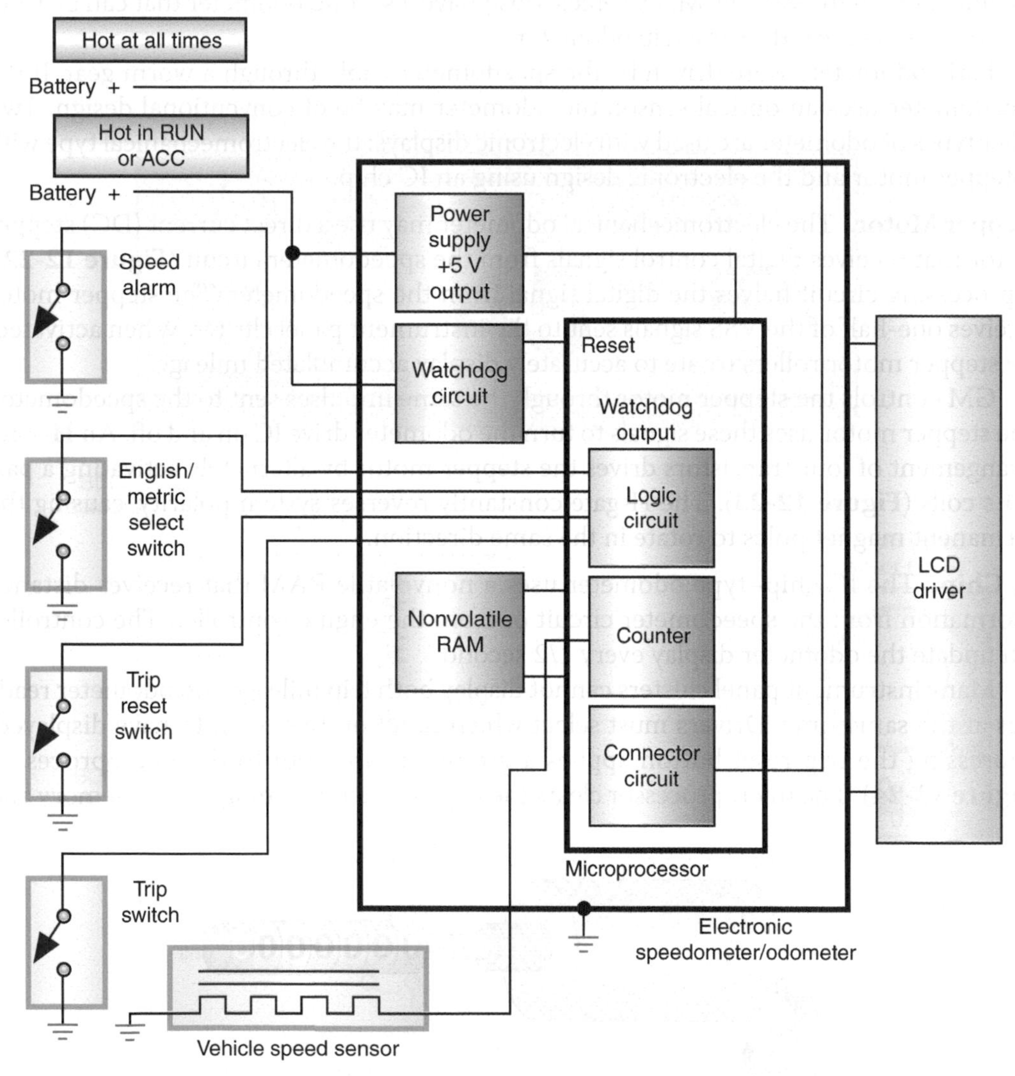

Figure 12-24 The trip reset button provides a ground signal to the logic circuit, which then erases the trip odometer memory while retaining the total accumulated mileage in the odometer.

returns the display to zero. The trip odometer continues to store trip mileage even if this function is not selected for display.

If the IC chip fails, some manufacturers provide for the replacement of the chip. Depending on the manufacturer, the new chip may be programmed to display the last odometer reading. Most replacement chips display an X, S, or * to indicate the odometer chip is not the original. If the odometer IC chip cannot be programmed, a door sticker is fixed to the vehicle to indicate that the odometer has been replaced.

AUTHOR'S NOTE Federal Motor Vehicle Safety Standards require the odometer be capable of storing up to 500,000 miles in nonvolatile memory. Most odometer readouts are up to 199,999.9 miles.

If an error occurs in the odometer circuit, the display changes to notify the driver. The form of error message differs among manufacturers. Some systems display "ERROR," while others may use dashed lines.

Federal and state laws prohibit tampering with the correct mileage as indicated on the odometer. If the odometer requires replacement, follow all applicable laws.

Tachometers

Shop Manual
Chapter 12, page 630

The electric tachometer receives voltage pulses from the ignition system, usually the ignition coil (**Figure 12-25**). The negative (−) side of the coil provides the tachometer signal as the switching unit opens the primary circuit. Each voltage pulse represents the generation of one spark at the spark plug. The rate of spark plug firing is in direct relationship to the speed of the engine. A circuit within the tachometer converts the ignition pulse signal into a varying voltage that rotates the needle.

The digital tachometer can be a continuously displayed gauge, or be a part of a multigauge display. The digital tachometer receives its voltage signals from the ignition module or PCM via the bus network and displays the readout in a bar graph (**Figure 12-26**) or quartz analog gauge. The PCM or ignition module typically receives the engine speed data from the Crankshaft Position (CKP) sensor.

The multigauge system has a built-in power supply that provides a 5-volt reference signal to the other monitored systems for the gauge. Also, the gauge has a **watchdog circuit** incorporated into it. The power on/off watchdog circuit supplies a reset voltage to the microprocessor in the event that pulsating output signals from the microprocessor are interrupted.

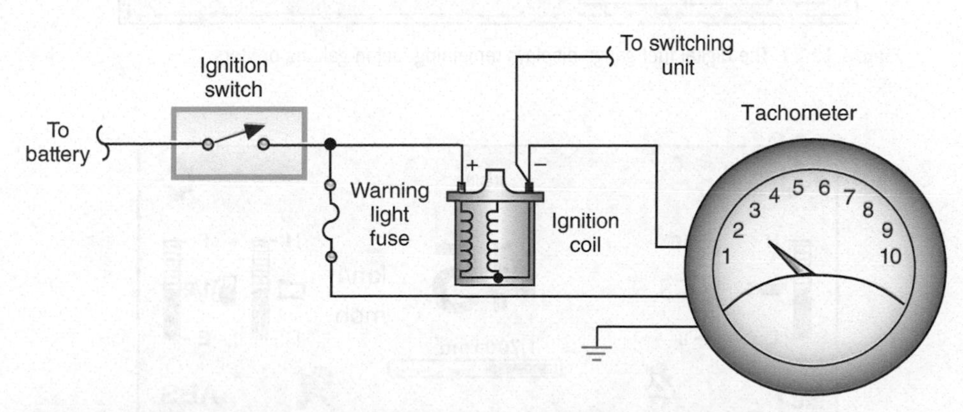

Figure 12-25 Electrical tachometer wired into the ignition system.

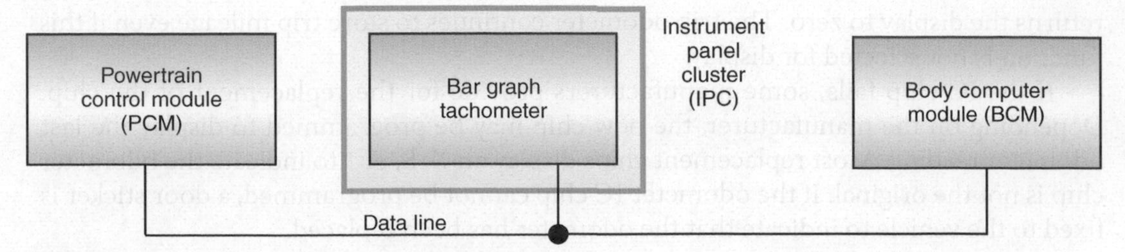

Figure 12-26 The IPC "listens in" on the communications between the PCM and the BCM to gather information on engine speed.

Electronic Fuel Gauges

Shop Manual
Chapter 12, page 635

Most digital fuel gauges use a fuel level sender that decreases resistance value as the fuel level decreases. The microprocessor converts the resistance values to voltage values. A voltage-controlled **oscillator** changes the signal into a frequency signal. The microprocessor counts the cycles and sends the appropriate signal to operate the digital display (**Figure 12-27**).

The display shows an "F" when the tank is full, and an "E" when less than 1 gallon is in the tank. Other warning signals include incandescent lamps, a symbol on the dash, or flashing of the fuel ISO symbol. If a bulb displays the warning, usually a switch is located in the sending unit that closes the circuit. The microprocessor usually controls flashing digital displays.

The bar graph–style gauge uses segments that represent the amount of fuel remaining in the tank (**Figure 12-28**). The segments divide the tank into equal levels. The display also includes the F, 1/2, and E symbols along with the ISO fuel symbol. A "Low Fuel" warning appears when only one bar remains lit. The gauge also alerts the driver to problems in the circuit. A common method of indicating an open or short is to flash the F, 1/2, and E symbols while the gauge reads empty.

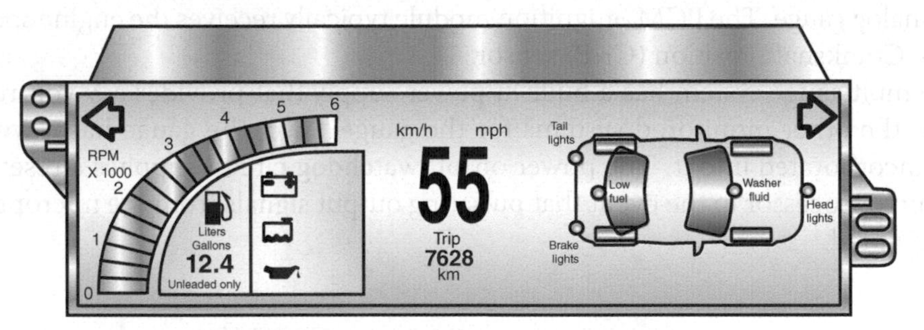

Figure 12-27 The digital fuel gauge displays remaining fuel in gallons or liters.

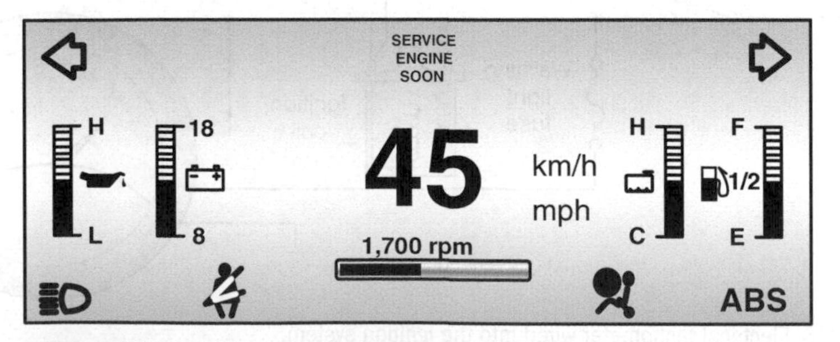

Figure 12-28 Bar graph style of electronic instrumentation. Each segment represents a different value.

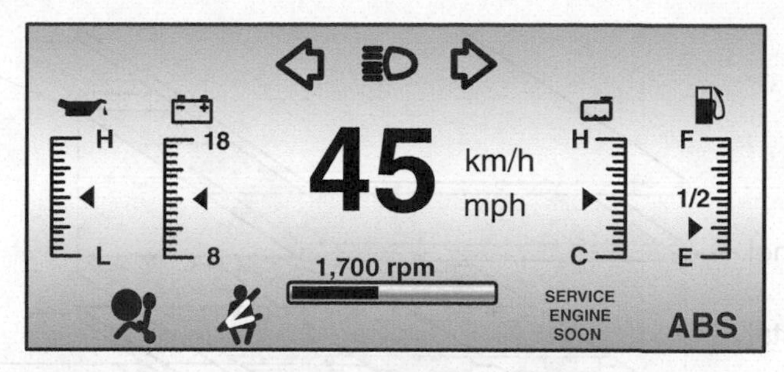

Figure 12-29 Floating pointer indicates the value received from the sensor.

Other Digital Gauges

Most of the gauges used to display temperature, oil pressure, and charging voltage are of bar graph design. Another popular method is to use a floating pointer (**Figure 12-29**).

The temperature gauge usually receives its input from a negative temperature coefficient (NTC) thermistor. When the engine is cold, the resistance value of the thermistor is high, resulting in a high-voltage input to the microprocessor. The input signal is translated into a low-temperature reading on the gauge. As the engine coolant warms, the resistance value drops. At a predetermined resistance level, the microprocessor will activate an alert function to warn the driver of excessive engine temperature.

The voltmeter calculates charging voltage by comparing the voltage supplied to the instrument panel module to a reference voltage signal. The oil pressure gauge uses a piezoresistive sensor that operates like those used for conventional analog gauges.

Digital gauges perform self-tests. The detection of a fault results in the display of a warning message to the driver. A "CO" indicates the circuit is open, and a "CS" indicates the circuit is shorted. The gauge continues to display these messages until the problem is corrected.

LCD Monitors

Thin-film liquid crystal display (LCD) monitors are popular for displaying instrument panel gauge readings (**Figure 12-30**). In addition, it's commonly used for radio, audio, and navigational information. LCD monitors provide a high-resolution graphic display.

The thin-film transistor (TFT) is the most common switching device. TFT is an arrangement of tiny transistors and capacitors in a matrix on the glass of the display. TFT LCD monitors have a sandwich-type structure with liquid crystals filled between two glass plates (**Figure 12-31**). The TFT glass has as many TFTs as the number of pixels displayed.

Figure 12-30 Monitor display of instrument panel gauges.

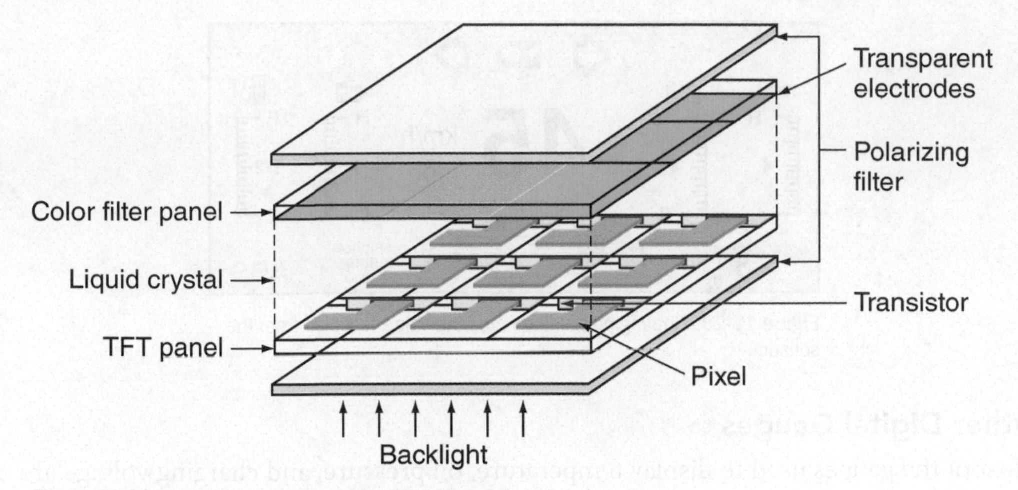

Figure 12-31 Sandwich structure of the TFT LCD monitor.

A TFT LCD module consists of a TFT panel, driving-circuit unit, backlight system, and assembly unit (**Figure 12-32**). The monitor is divided into millions of pixel units that are formed by liquid crystal cells. The cells change the polarization direction of light passing through them in response to an electrical voltage. As the polarization direction changes, the amount of light allowed to pass through a polarizing layer changes. The LCD monitor uses the matrix driving method that displays characters and pictures in sets of dots.

The TFT substrate contains the TFTs, storage capacitors, pixel electrodes, and circuit wiring (**Figure 12-33**). The color filter generates the colors of the display. The movement

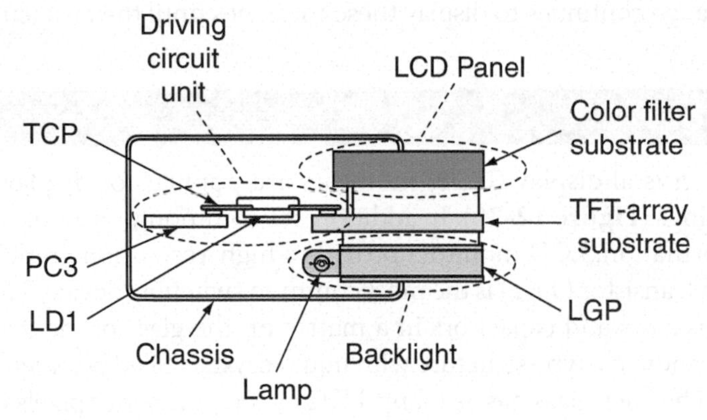

Figure 12-32 Color TFT LCD panel.

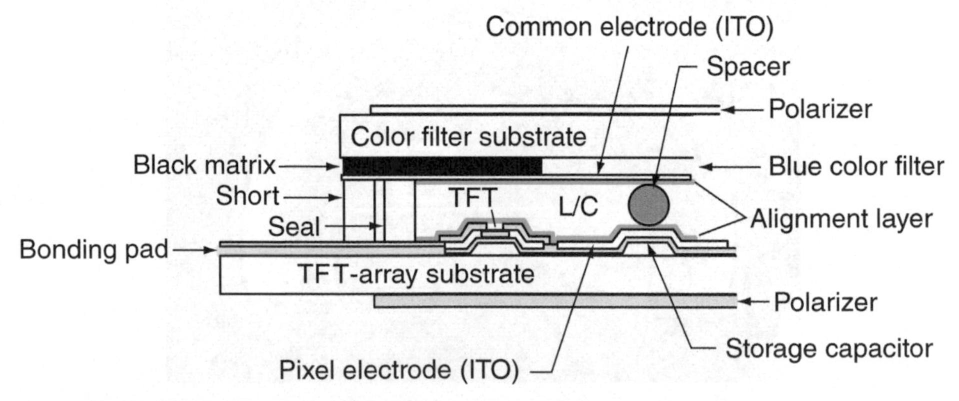

Figure 12-33 Cross section of the TFT substrate.

of the liquid crystals is based on the difference in voltage potential between the TFT glass and the color filter glass. The amount of backlight supplied is determined by the amount of movement of the liquid crystals in such a way as to generate color. The color filter contains a black matrix and a resin film. The resin film includes the three primary color pigments of red, green, and blue (RGB).

Head-Up Display

Some manufacturers have equipped select models with a **head-up display (HUD)** feature. This system displays visual images onto the inside of the windshield in the driver's field of vision (**Figure 12-34**). The images project onto the windshield from a vacuum fluorescent light source, much like a movie projector. With the display located in this area, drivers don't need to remove their eyes from the road to check the instrument panel.

Shop Manual
Chapter 12, page 640

The head-up control module contains a computer and an optical system that projects images to a holographic combiner integrated into the windshield above the module. The holographic combiner projects the images in the driver's view just above the front end of the hood. The HUD may contain the following displays and warnings:

1. Speedometer reading with US/metric indicator.
2. Turn signal indicators.
3. High-beam indicator.
4. Low-fuel indicator.
5. Check gauges indicator.

The head-up control switch contains a head-up display on/off switch, USC/metric switch, and a head-up dimming switch. The head-up dimming switch is a rheostat that sends an input signal to the head-up module. A switch in the control switch assembly provides for changing the vertical position of the head-up display. The switch is connected to the head-up module through a mechanical cable-drive system. Moving the vertical position switch moves the position of the head-up module.

The PCM sends the VSS signal information to the head-up module for the speedometer display. The instrument cluster sends the check gauges and low-fuel signals to the head-up module. The dimmer switch sends a high-beam indicator input signal to the head-up module. This module also receives inputs from the signal light switch to operate the signal light indicators.

Figure 12-34 The HUD displays various information inside the windshield.

Travel Information Systems

Shop Manual
Chapter 12, page 638

The travel information system can be a simple calculator that computes fuel economy, distance to empty, and remaining fuel (**Figure 12-35**). Other systems provide a much larger range of functions.

Fuel data centers display the amount of fuel remaining in the tank and provide additional information for the driver (**Figure 12-36**). By depressing the RANGE button, the BCM calculates the distance until the tank is empty by using the amount of fuel remaining and the average fuel economy. When the INST button is depressed, the fuel data center displays instantaneous fuel economy. The BCM makes the necessary calculations and updates the display every 1/2 second.

Depressing the AVG button displays average fuel economy for the total distance traveled since the last press of the reset button. FUEL USED displays the amount of fuel used since the last time this function was reset. The RESET button resets the average fuel economy and fuel-used calculations. The function displayed in the fuel data center is the one that resets.

Deluxe systems may incorporate additional features such as outside temperature, compass, elapsed time, estimated time of arrival, distance to destination, day of the week, time, and average speed. **Figure 12-37** shows the inputs used to determine many of these functions. The sensors shown in **Figure 12-38** determine fuel system calculations. Injector on time and vehicle speed pulses determine the amount of fuel flow. Some manufacturers use a fuel flow sensor that provides pulse information to the microprocessor concerning fuel consumption (**Figure 12-39**).

Figure 12-35 Fuel data display panel.

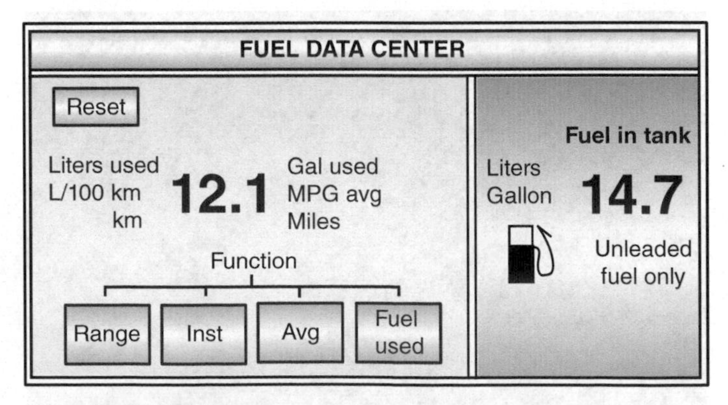

Figure 12-36 Fuel data center.

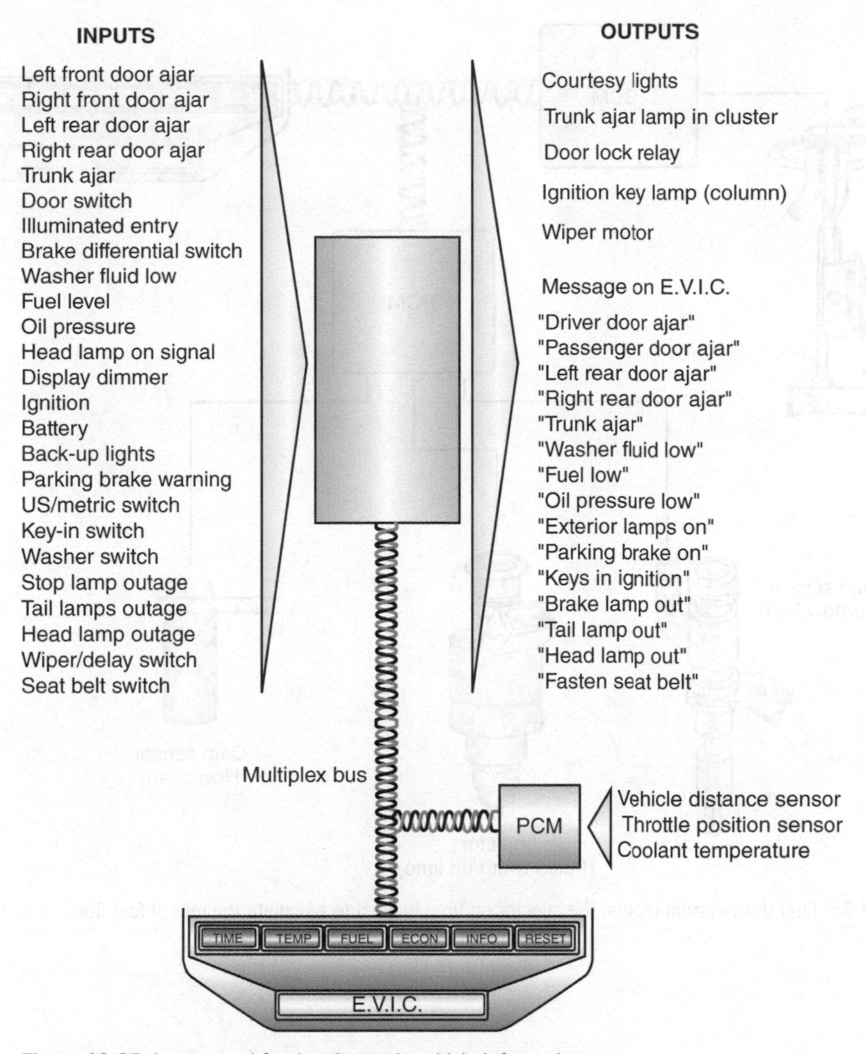

Figure 12-37 Inputs used for the electronic vehicle information center.

Warning Lamps

A **warning lamp** illuminates to warn the driver of a possible problem or hazardous condition. Uses of a warning lamp include the warning of low oil pressure, high coolant temperature, defective charging system, or a brake failure. Two methods can operate a warning lamp: a sending unit circuit hardwired to the instrument cluster or computer-controlled lamp drivers.

Shop Manual
Chapter 12, page 636

Sending Unit–Controlled Lamps

Unlike gauge sending units, the sending unit for a warning lamp is nothing more than a simple switch. The type of switch can be either normally open or normally closed, depending on the monitored system.

Shop Manual
Chapter 12, page 637

Most sending unit-controlled oil pressure warning circuits use a normally closed switch (**Figure 12-40**). The oil pressure works upon the sending unit's diaphragm. The movement of the diaphragm controls the switch contacts. With no oil pressure working against the diaphragm, the contacts remain closed, and the circuit is complete to ground. The oil warning lamp turns on when the ignition is in the RUN position without the engine running. This proves out the bulb. Starting the engine causes the oil pressure to build, and the diaphragm moves the contacts apart. This opens the circuit, and the

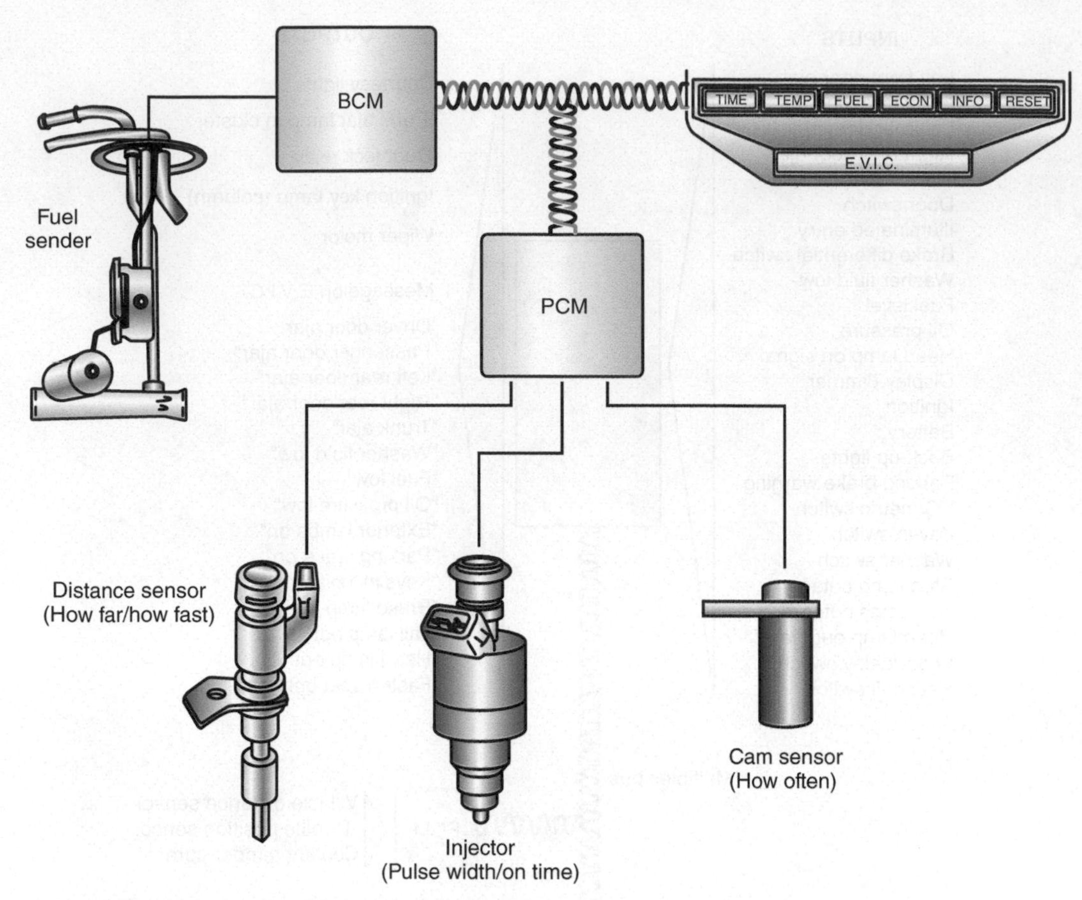

Figure 12-38 Fuel data system inputs. The injector on time is used to calculate the rate of fuel flow.

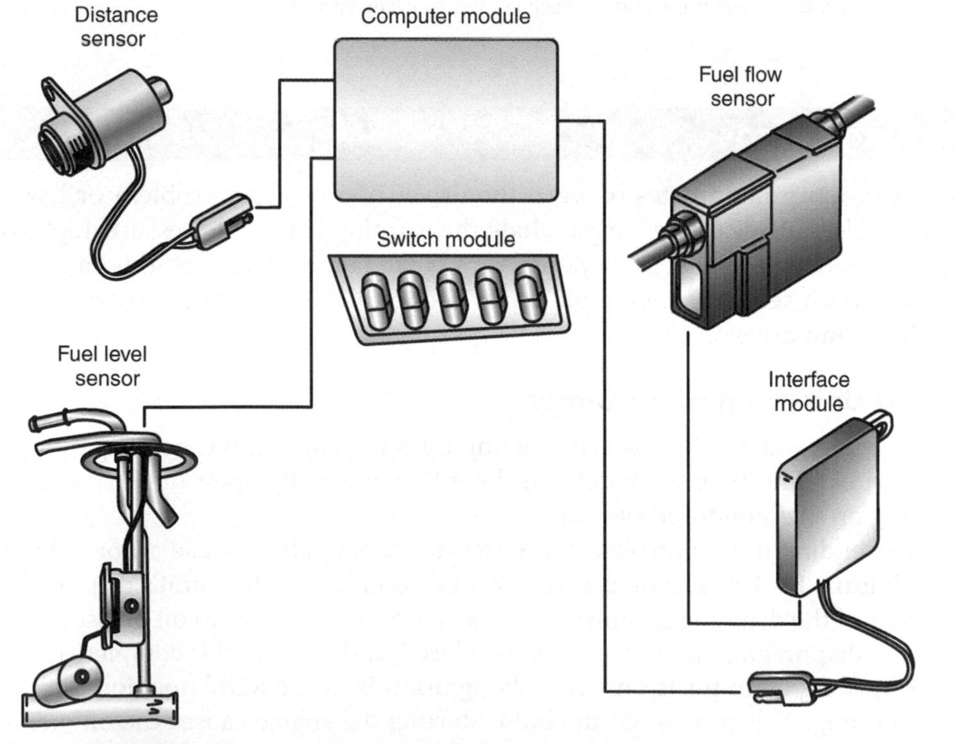

Figure 12-39 Some information centers use a fuel flow sensor.

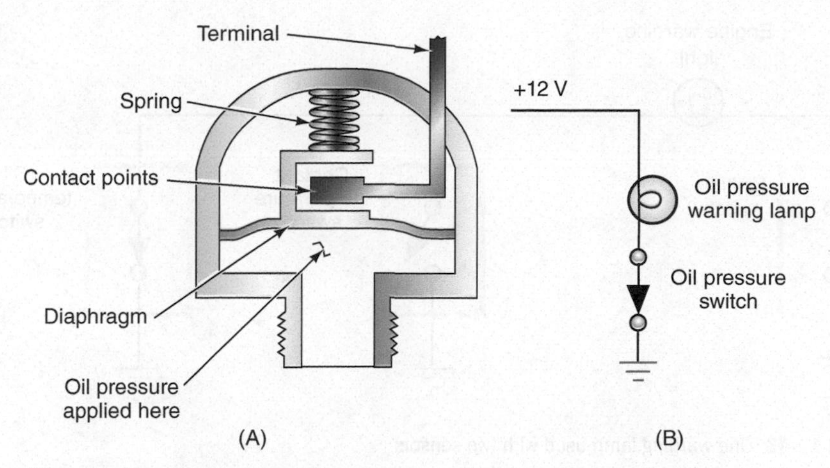

Figure 12-40 (A) Oil pressure light sending unit. (B) Oil pressure warning lamp circuit.

warning lamp goes off. The amount of oil pressure required to move the diaphragm is about 3 psi (21 kpa). Illumination of the oil pressure warning lamp with the engine running indicates that the oil pressure has dropped below the 3 psi (21 kpa)limit.

Most coolant temperature warning lamp circuits use a normally open switch (**Figure 12-41**). The temperature sending unit consists of a fixed contact and a contact on a bimetallic strip. As the coolant temperature increases, the bimetallic strip bends. As the strip bends, the contacts move closer to each other. The contacts close when the temperature exceeds the predetermined level, completing the circuit to ground. The completed circuit illuminates the warning lamp.

Shop Manual
Chapter 12, page 637

The bulb check of a system using normally open–type switches (where the contacts are not closed when the ignition is ON) uses a **prove-out circuit** (**Figure 12-42**). A prove-out circuit completes the warning light circuit to ground through the ignition switch when it's in the START position. The warning light will be on during engine cranking to indicate to the driver that the bulb is working properly.

The prove-out function is also known as "Bulb Test" or "Bulb Check" position.

It's possible to have more than one sending unit connected to a single bulb. The illustration (**Figure 12-43**) shows a wiring circuit of a dual-purpose warning lamp. The lamp comes on whenever oil pressure is low, or the coolant temperature is too high.

A warning lamp may communicate braking system failures to the driver. **Figure 12-44** shows a brake system combination valve. The center portion of the valve senses differences in the hydraulic pressures between the two outlets of the valve. With the differential

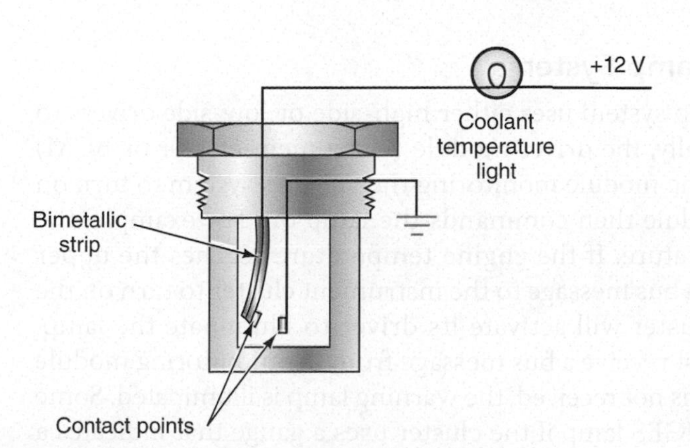

Figure 12-41 Temperature indicator light circuit.

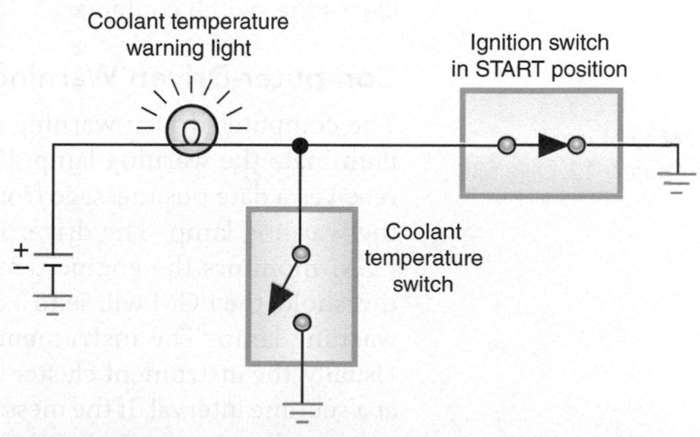

Figure 12-42 A prove-out circuit included in a normally open (NO) coolant temperature light system.

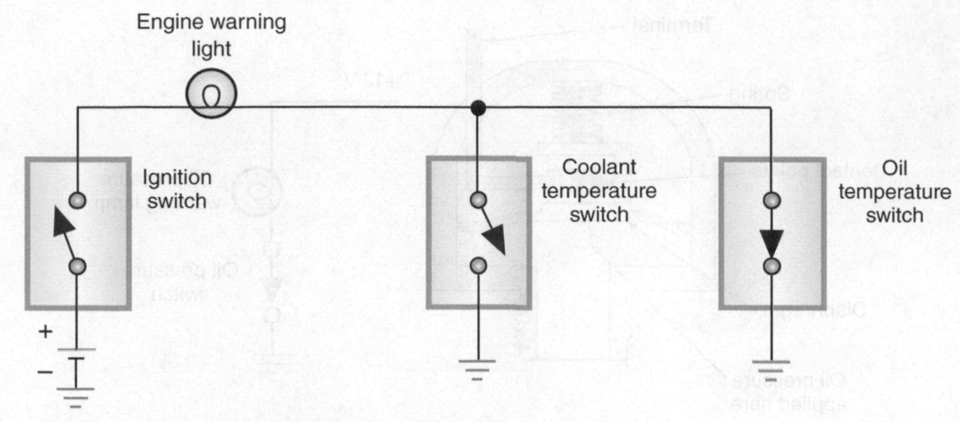

Figure 12-43 One warning lamp used with two sensors.

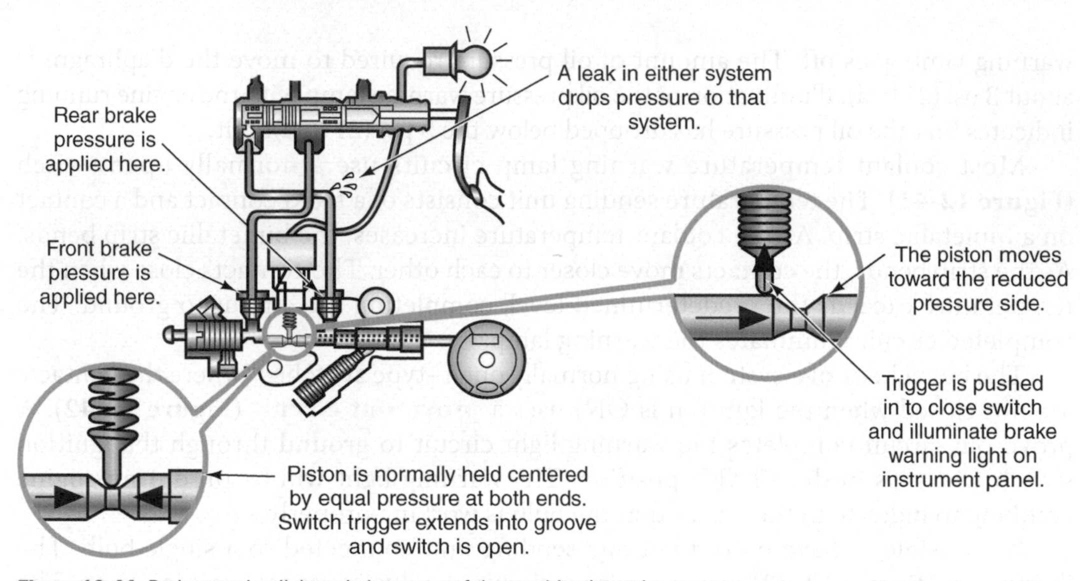

Figure 12-44 Brake warning light switch as part of the combination valve.

valve centered, the plunger on the warning lamp switch is in the recessed area of the valve. A pressure drop in either side of the brake system forces the differential valve to move by hydraulic pressure. Movement of the differential valve pushes the switch plunger up, and closes the switch contacts.

Computer-Driven Warning Lamp Systems

Shop Manual
Chapter 12, page 638

The computer-driven warning lamp system uses either high-side or low-side drivers to illuminate the warning lamp. Usually, the driver module (instrument cluster or BCM) receives a data bus message from the module monitoring the affected system to turn on the warning lamp. The driver module then commands the lamp on. For example, the PCM monitors the engine temperature. If the engine temperature reaches the upper threshold, the PCM will send a data bus message to the instrument cluster to turn on the warning lamp. The instrument cluster will activate its driver to illuminate the lamp. Usually, the instrument cluster must receive a bus message from the monitoring module at a set time interval. If the message is not received, the warning lamp is illuminated. Some systems illuminate a CHECK GAUGES lamp if the cluster uses a gauge that indicates a condition requiring the driver's attention.

Summary

- Through the use of gauges and indicator lights, the driver is capable of monitoring several engine and vehicle operating systems.
- The gauges include a speedometer, odometer, tachometer, oil pressure, charging indicator, fuel level, and coolant temperature.
- The most common types of electromechanical gauges are the d'Arsonval, three-coil, two-coil, and air-core.
- Computer-driven quartz swing needle displays are similar in design to the air-core electromagnetic gauges used in conventional analog instrument panels.
- All gauges require the use of a variable resistance sending unit. Styles of sending units include thermistors, piezoresistive sensors, and mechanical variable resistors.
- Digital instrument clusters use digital and linear displays to notify the driver of monitored system conditions.
- The most common types of displays used on electronic instrument panels are: light-emitting diodes (LEDs), liquid crystal displays (LCDs), vacuum fluorescent displays (VFDs), and a phosphorescent screen that is the anode.
- A head-up display system displays visual images onto the inside of the windshield in the driver's field of vision.
- In the absence of gauges, warning lamps indicate alters about important engine and vehicle functions.

Review Questions

Short-Answer Essays

1. What are the most common types of electromagnetic gauges?
2. Describe the operation of the piezoresistive sensor.
3. What is a thermistor used for?
4. What is meant by *electromechanical?*
5. Describe the operation of the air-coil gauge.
6. What is the basic difference between conventional analog and computer-driven analog instrument clusters?
7. Describe the operating principles of the digital speedometer.
8. Explain the operation of IC chip–type odometers.
9. Describe the operation of the electronic fuel gauge.
10. Describe the operation of quartz analog speedometers.

Fill in the Blanks

1. The purpose of the tachometer is to indicate ______________.
2. A piezoresistive sensor monitors ______________ changes.
3. The most common style of fuel level sending unit is ______________ variable resistor.
4. The combination valve activates the brake warning light by ______________ pressure in the brake hydraulic system.
5. In a three-coil gauge, the ______________ ______________ produces a magnetic field that bucks or opposes the low-reading coil. The ______________ ______________ coil and the bucking coil are wound together, but in opposite directions. The ______________ ______________ coil is positioned at 90° angle to the low-reading and bucking coils.
6. A ______________ ______________ circuit completes the warning light circuit to ground through the ignition switch when it's in the START position.
7. Digital instrument clusters use ______________ and ______________ displays to notify the driver of monitored system conditions.
8. ______________ ______________ is a calculation using the final drive ratio and the tire circumference to obtain accurate vehicle speed signals.

9. Most digital fuel gauges use a fuel level sender that _______________ resistance value as the fuel level decreases.

10. Computer-driven quartz swing needle displays are similar in design to the _______________ _______________ electromagnetic gauges used in conventional analog instrument panels.

Multiple Choice

1. Odometer replacement is being discussed.
   Technician A says it's permissible to turn back the odometer reading as long as the customer is notified.
   Technician B says that all federal and state laws must be followed.
   Who is correct?
   A. A only C. Both A and B
   B. B only D. Neither A nor B

2. Electromagnetic gauges are being discussed.
   Technician A says that the d'Arsonval gauge uses the interaction of a permanent magnet and an electromagnet, and the total field effect to cause needle movement.
   Technician B says that the three-coil gauge uses the interaction of three electromagnets and the total field effect upon a permanent magnet to cause needle movement.
   Who is correct?
   A. A only C. Both A and B
   B. B only D. Neither A nor B

3. The three-coil gauge is being discussed. *Technician A* says the gauge uses the principle that the majority of current seeks the path of least resistance.
   Technician B says that the three coils are the low-reading coil, the bucking coil, and the high-reading coil.
   Who is correct?
   A. A only C. Both A and B
   B. B only D. Neither A nor B

4. Warning light circuits are being discussed.
   Technician A says most oil pressure warning circuits use a normally closed switch.
   Technician B says most conventional coolant temperature warning light circuits use a normally open switch.
   Who is correct?
   A. A only C. Both A and B
   B. B only D. Neither A nor B

5. The brake failure warning system is being discussed.
   Technician A says if the pressure drops in either side of the brake system, the switch plunger is pushed up, and the switch contacts close.
   Technician B says if the pressure is equal on both sides of the brake system, the warning light comes on.
   Who is correct?
   A. A only C. Both A and B
   B. B only D. Neither A nor B

6. The IC chip odometer is being discussed.
   Technician A says if the chip fails, some manufacturers provide for the replacement of the chip.
   Technician B says depending on the manufacturer, the new chip may be programmed to display the last odometer reading.
   Who is correct?
   A. A only C. Both A and B
   B. B only D. Neither A nor B

7. Computer-driven quartz swing needle displays are being discussed.
   Technician A says the A coil is connected to system voltage, and the B coil receives a voltage that is proportional to input frequency.
   Technician B says the quartz swing needle display is similar to air-core electromagnetic gauges.
   Who is correct?
   A. A only C. Both A and B
   B. B only D. Neither A nor B

8. *Technician A* says digital instrumentation displays an average of the readings received from the sensor.
   Technician B says conventional analog instrumentation gives more accurate readings but is not as decorative.
   Who is correct?
   A. A only
   B. B only
   C. Both A and B
   D. Neither A nor B

9. The microprocessor-initiated self-check of the electrical instrument cluster is being discussed.
   Technician A says during the first portion of the self-test all segments of the speedometer display are lit.
   Technician B says the display should not go blank during any part of the self-test.
   Who is correct?
   A. A only
   B. B only
   C. Both A and B
   D. Neither A nor B

10. *Technician A* says bar graph–style gauges don't provide for self-tests.
    Technician B says the digital instrument panel may display "CO" to indicate that the circuit is shorted.
    Who is correct?
    A. A only
    B. B only
    C. Both A and B
    D. Neither A nor B

CHAPTER 13 ACCESSORIES

Upon completion and review of this chapter, you should be able to:

- Explain the operation and function of the horn circuit.
- Explain the operation of standard two- and three-speed wiper motors, both permanent magnet and electromagnetic field designs.
- Describe the operation of intermittent wipers.
- Explain the operation of depressed-park wipers.
- Describe the operation of the intelligent windshield wiper system.
- Explain the operation of windshield washer pump systems.
- Explain the operation and methods used to control blower fan motor speeds.
- Explain the operation of electric defoggers.
- Explain the operational principles of power mirrors.
- Explain the operation of power windows, power seats, and power locks.
- Explain the operation of the memory seat feature.
- Explain the operation of climate-controlled seats.
- Describe the operation of automatic door lock systems.
- Explain the operation of the keyless entry system.
- Explain the operation of common antitheft systems.
- Explain the operating principles of the electronic cruise control system.
- Explain how the adaptive cruise control system functions.
- Describe the concepts of electronically controlled sunroofs.

Terms to Know

Adaptive cruise control (ACC)
Armed
Automatic door locks (ADL)
Child safety latch
Climate-controlled seats
Clockspring
Cruise control
Depressed-park
Diaphragm
Electrochromic mirror
Express down
Express up
Grid
Keyless entry
Negative logic
Park contacts
Peltier element
Resistor block
Sector gear
Trimotor
Window regulator

Introduction

Electrical accessories provide additional safety and comfort. This chapter explains the principles of operation for some of the most common electrical accessories. Systems not discussed here are similar in concept.

This chapter explores the operation of safety accessories, such as the horn, windshield wipers, and windshield washers. Comfort accessories explored in this chapter include the blower motor, electric defoggers, power mirrors, power windows, power seats, and power door locks.

In today's automobile, no system is immune to computer control. The vehicle may have computer-controlled wipers, transmissions, locking differentials, brakes, suspensions, all-wheel drive systems, and so on. It would be impossible to cover the various operations of all these systems. It's important for today's technician to have an understanding of how electronics work and a basic knowledge of the control system. Regardless if you are working on a domestic or foreign-built automobile, electricity and electronics work the same. Always refer to proper service information to gain an understanding of a system that may be new to you.

In addition to electrical accessories, this chapter discusses several of the electronic systems found in today's automobiles. Some of these systems are covered in greater detail in other *Today's Technician* series books. Refer to these textbooks for more information.

In this chapter, you will learn the operation of cruise control systems and the many electrical accessory systems with electronic controls added to them for added features and enhancement. These accessories include memory seats, electronic sunroofs, antitheft systems, and automatic door locks.

The comfort and safety of the driver and/or passengers depend on the technician's proper diagnosis and repair of these systems. As with all electrical systems, the technician must have a basic understanding of the operation of these systems before attempting to perform any service.

Horns

The automotive electrical horn operates on an electromagnetic principle that vibrates a **diaphragm** to produce a warning signal. The diaphragm is a thin, flexible plate held around its outer edge by the horn housing, allowing the middle to flex. Most electrical horns consist of an electromagnet, a movable armature, a diaphragm, and a set of normally closed contact points (**Figure 13-1**). The contact points are wired in series with the field coil. One of the points attaches to the armature. A magnetic field develops when current flows through the field coil, attracting the movable armature. The diaphragm attaches to, and moves with, the armature. Movement of the armature results in the contact points opening, which breaks the circuit. The diaphragm is released and returns to its normal position. The contact points close again, and the cycle repeats. Vibrating the diaphragm happens several times per second. As the diaphragm vibrates, it causes a column of air in the horn to vibrate and produces sound.

Shop Manual
Chapter 13, page 666

A horn is a device that produces an audible warning signal.

In two horn systems, one is the low-note horn and the other is the high-note horn.

Most vehicles have two horns wired in parallel to each other and in series with the horn switch or relay. One of the horns will have a slightly lower pitch than the other.

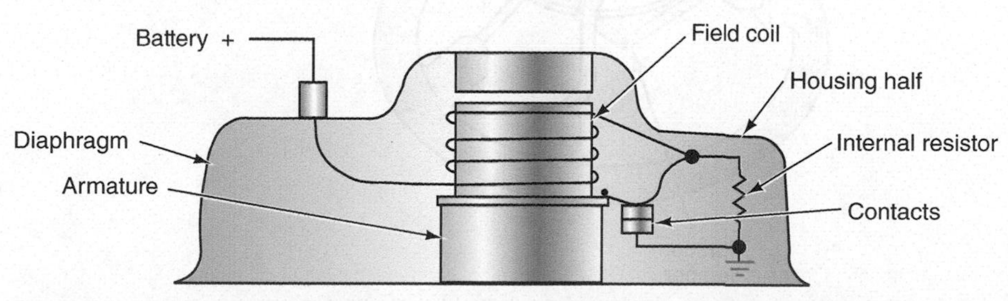

Figure 13-1 Horn construction. The internal resistor allows a weak magnetic field to remain after the contacts open, reducing the time required to rebuild the field when the contacts close again.

Figure 13-2 Horn design determines sound quality.

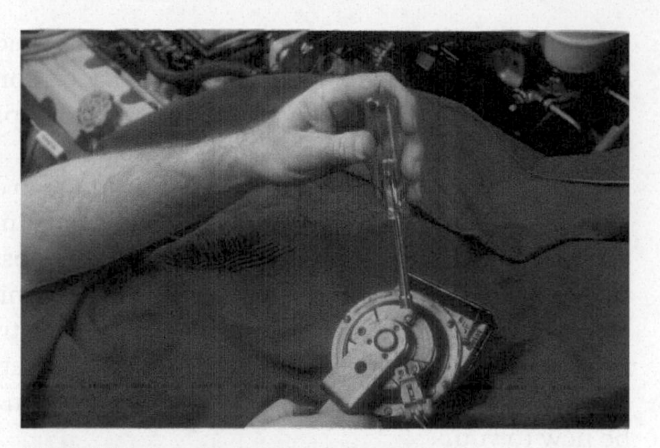

Figure 13-3 Horn pitch adjustment screw.

The horn's design and shape determine the sound's frequency and tone (**Figure 13-2**). The number of times the diaphragm vibrates per second determines the horn's pitch. The faster the vibration, the higher the pitch. On some horns, the pitch can be adjusted by changing the spring tension of the armature. This alters the magnetic pull on the armature and changes the rate of vibrations. An adjustment screw on the outside of the horn alters the pitch (**Figure 13-3**).

Horn Switches

Shop Manual
Chapter 13, pages 666, 669

Horn switches are either in the center of the steering wheel or as a part of the multifunction switch. Most horn switches are normally open switches.

The steering wheel–mounted horn switch can be a single button in the middle of the steering wheel. Another design is to have multiple buttons in the horn pad. Switches mounted on the steering wheel require a slip ring (**Figure 13-4**). The slip ring has contacts that provide continuity for the horn control in all steering wheel positions.

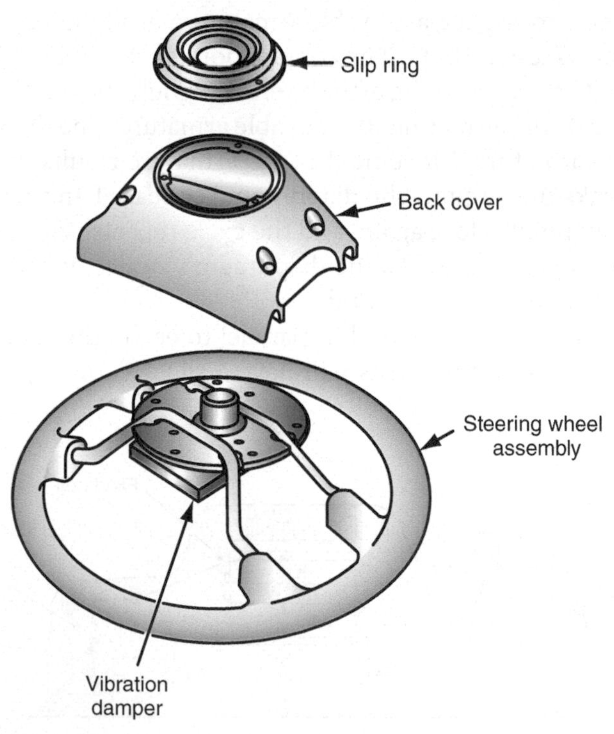

Figure 13-4 Slip ring contact provides horn continuity in all steering wheel positions.

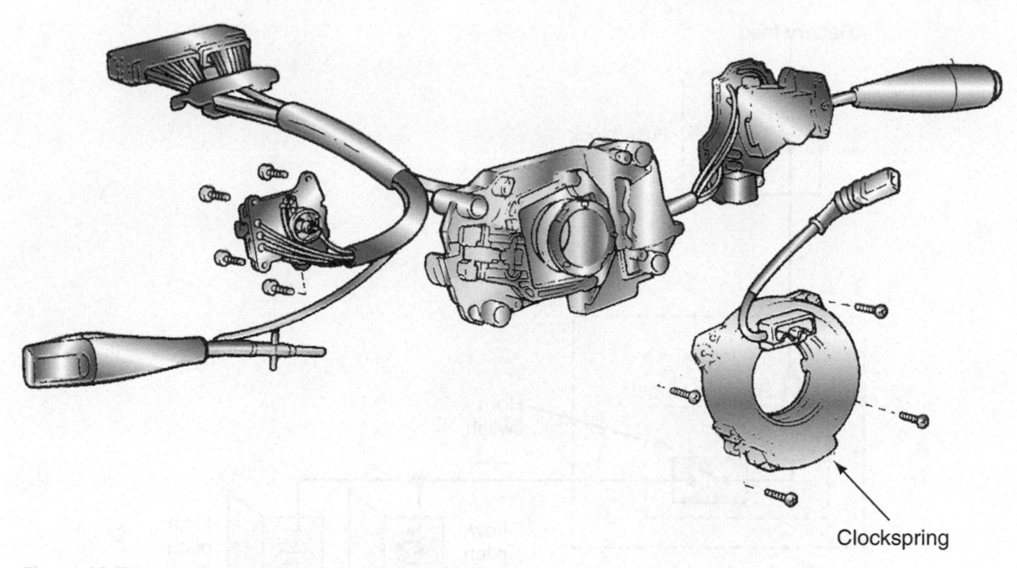

Figure 13-5 Most vehicle manufacturers now use a clockspring instead of sliding contacts.

The contacts consist of a circular contact in the steering wheel that slides against a spring-loaded contact in the steering column. Vehicles equipped with a steering wheel–mounted air bag use a **clockspring** (**Figure 13-5**) to provide continuity between the steering wheel components—horn switch, cruise control switches, air bag, and so on—and the steering column wiring harness. A clockspring is a winding of electrical conducting tape enclosed within a plastic housing. The clockspring maintains continuity between the steering wheel switches, the air bag, and the wiring harness in all steering wheel positions. The clockspring provides a more reliable connection than the sliding contacts.

Horn switches that are a part of the multifunction switch usually operate by a push button on the end of the lever.

Either switch type can be an input to the BCM. The BCM monitors the switch for a change of state and then commands the horn.

Horn Circuits

Shop Manual
Chapter 13, pages 666, 669

There are two methods of circuit control: with or without a relay. If the horn circuit does not use a relay, the horns must be of low-current design because the horn switch carries the total current. Depressing the horn switch completes the circuit from the battery to the horns (**Figure 13-6**).

The most common type of circuit control is to use a relay (**Figure 13-7**). Most circuits have battery voltage applied to the lower contact plate of the horn switch. When the switch is depressed, the contacts close and complete the circuit to ground. The horn switch does not have to carry the heavy current requirements of the horns since the relay coil only requires a low current. Pressing the horn button closes the switch and energizes the relay core. The core attracts the relay armature, which closes the contacts and completes the horn circuit. Current flows from the battery to the grounded horns.

Shop Manual
Chapter 13, page 670

Many vehicle manufacturers use a module (or series of modules) to operate the horn. This reduces wiring and utilizes modules that can perform multiple functions. As a benefit of using modules, many can perform diagnostic routines and set diagnostic trouble codes (DTCs) to assist in diagnostics. Consider the system illustrated in **Figure 13-8**. In this system, the steering wheel switches control the radio and navigational systems. The left steering switch is hard wired to the right switch. However, the right switch is more than just a switch. It's a microprocessor that puts inputs it receives onto the LIN bus to the

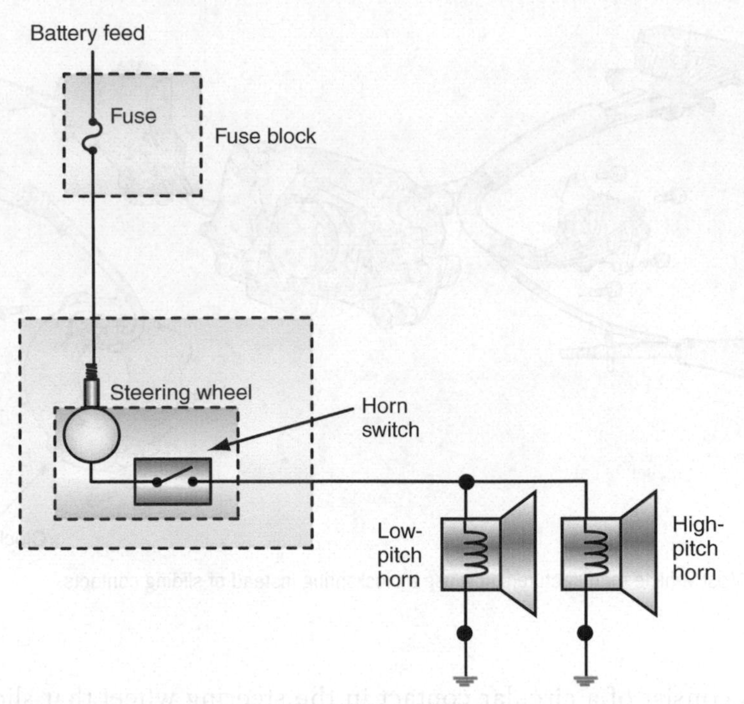

Figure 13-6 Insulated side switch without a relay.

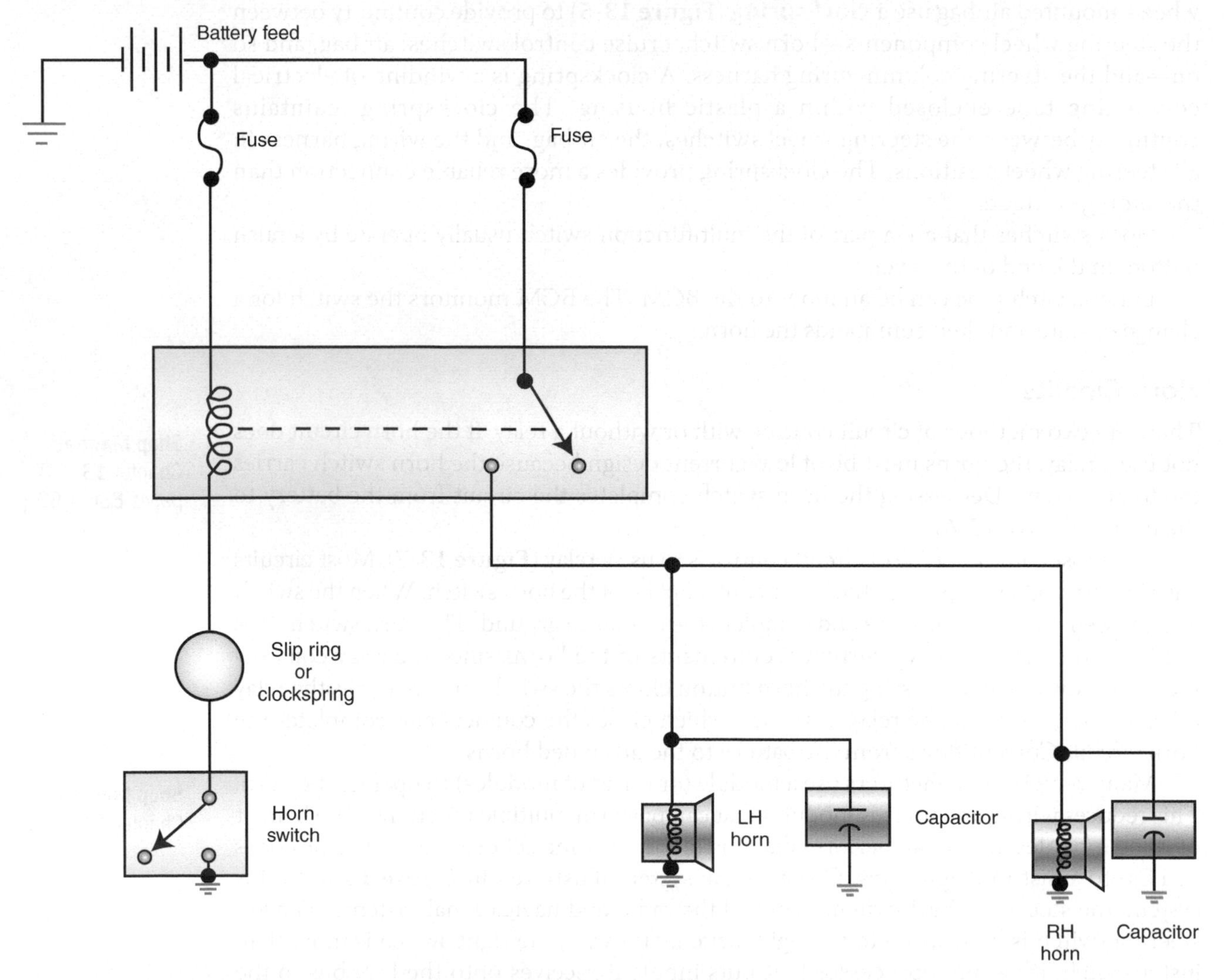

Figure 13-7 Relay controlled horn circuit. The horn button completes the relay coil circuit.

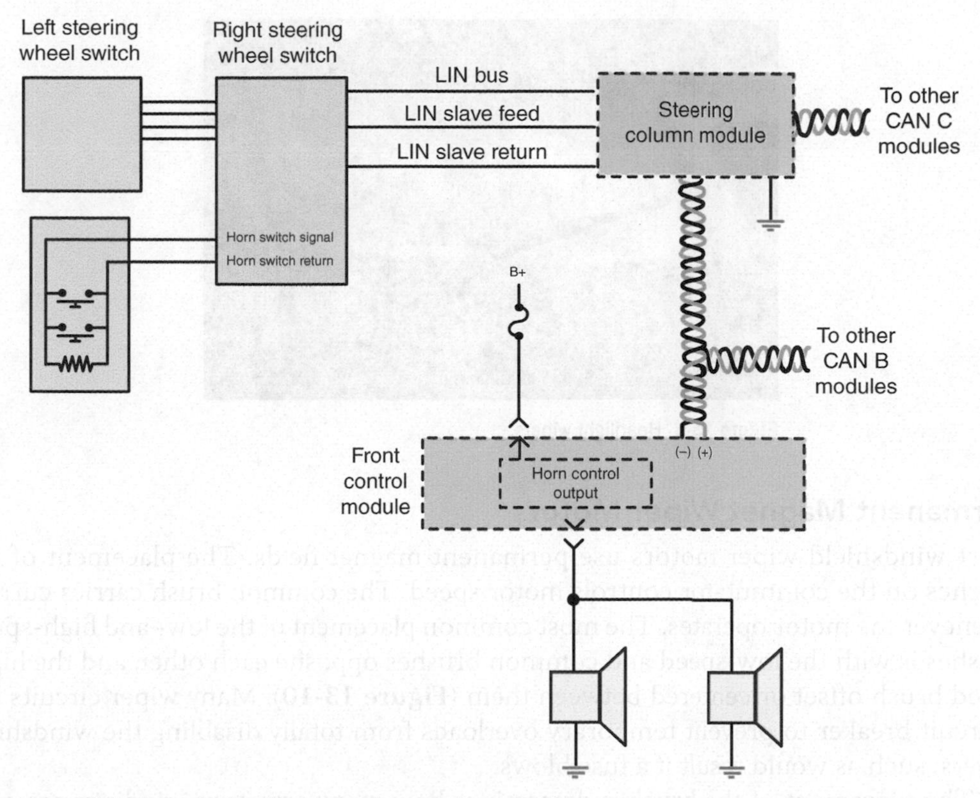

Figure 13-8 Horn circuit utilizing computers and bus networks.

master LIN module (steering column module). The steering column module then puts the message onto the CAN B bus network so other modules receive the input. The right steering wheel switch (LIN module) monitors the horn switch input. If the driver presses the horn switch, the right steering wheel switch sees a change in voltage on the signal circuit and broadcasts the horn request signal to the steering column module. The steering column module in turn broadcasts the request on the CAN B bus network to the front control module. The front control module then activates a high-side driver (HSD) to supply current to the horns. As you can see from this example, this system uses three microprocessors and two bus networks to operate the horns.

Windshield Wipers

The windshield wiper system typically provides a two- or three-speed wiper system with an intermittent wipe feature. The wiper system may also provide a **depressed-park** feature. In systems equipped with depressed park, the wiper blades drop down below the lower windshield molding to hide them.

Shop Manual
Chapter 13, page 670

Many vehicles use a single-speed rear window wiper and washer. In addition, many luxury vehicles have headlight wipers that operate in union with the windshield wipers (**Figure 13-9**). The operation of these accessories is the same as the windshield wipers.

A BIT OF HISTORY

Windshield wipers were introduced at the 1916 National Auto Show by several manufacturers. These wipers were hand operated. Vacuum-operated wipers were standard equipment by 1923. Electric wipers became common in the 1950s.

Figure 13-9 Headlight wipers.

Permanent Magnet Wiper Motors

Shop Manual
Chapter 13,
pages 670, 672

Most windshield wiper motors use permanent magnet fields. The placement of the brushes on the commutator controls motor speed. The common brush carries current whenever the motor operates. The most common placement of the low- and high-speed brushes is with the low-speed and common brushes opposite each other, and the high-speed brush offset or centered between them (**Figure 13-10**). Many wiper circuits use a circuit breaker to prevent temporary overloads from totally disabling the windshield wipers, such as would result if a fuse blows.

The placement of the brushes determines how many armature windings are connected in the circuit. When the wiper control switch is in the HIGH-SPEED position, wiper 1 passes battery voltage to the high-speed brush (**Figure 13-11**). Wiper 2 moves with wiper 1 but does not complete any circuits. Current flows through the high-speed brush, through the armature, to the common brush, and to ground. There are fewer armature windings connected between the common and high-speed brushes.

Applying battery voltage to fewer windings creates less magnetism in the armature and a lower counter electromotive force (CEMF). With less CEMF in the armature, there is greater armature current. The greater armature current results in higher speeds. Because the ground connection is after the park switch, the park switch position does not affect motor operation.

Placing the switch in the LOW-SPEED position supplies battery voltage through wiper 1 to the low-speed brush (**Figure 13-12**). Wiper 2 also moves to the LOW position, but does not complete any circuits. Current flows through the low-speed brush, through the armature, and through the common brush to ground. More armature windings are connected in the circuit between the common and low-speed brushes, increasing the magnetic field in the armature. This results in greater CEMF, reduces the amount of current in the armature, and slows the speed of the motor. Park switch position does not affect motor operation.

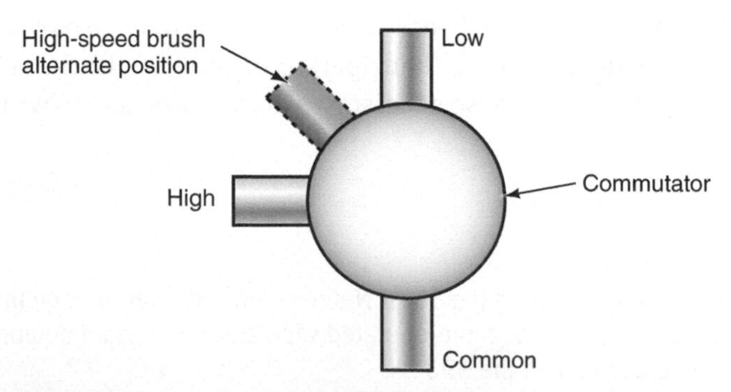

Figure 13-10 The most common brush arrangement is to place the low-speed brush opposite the common brush.

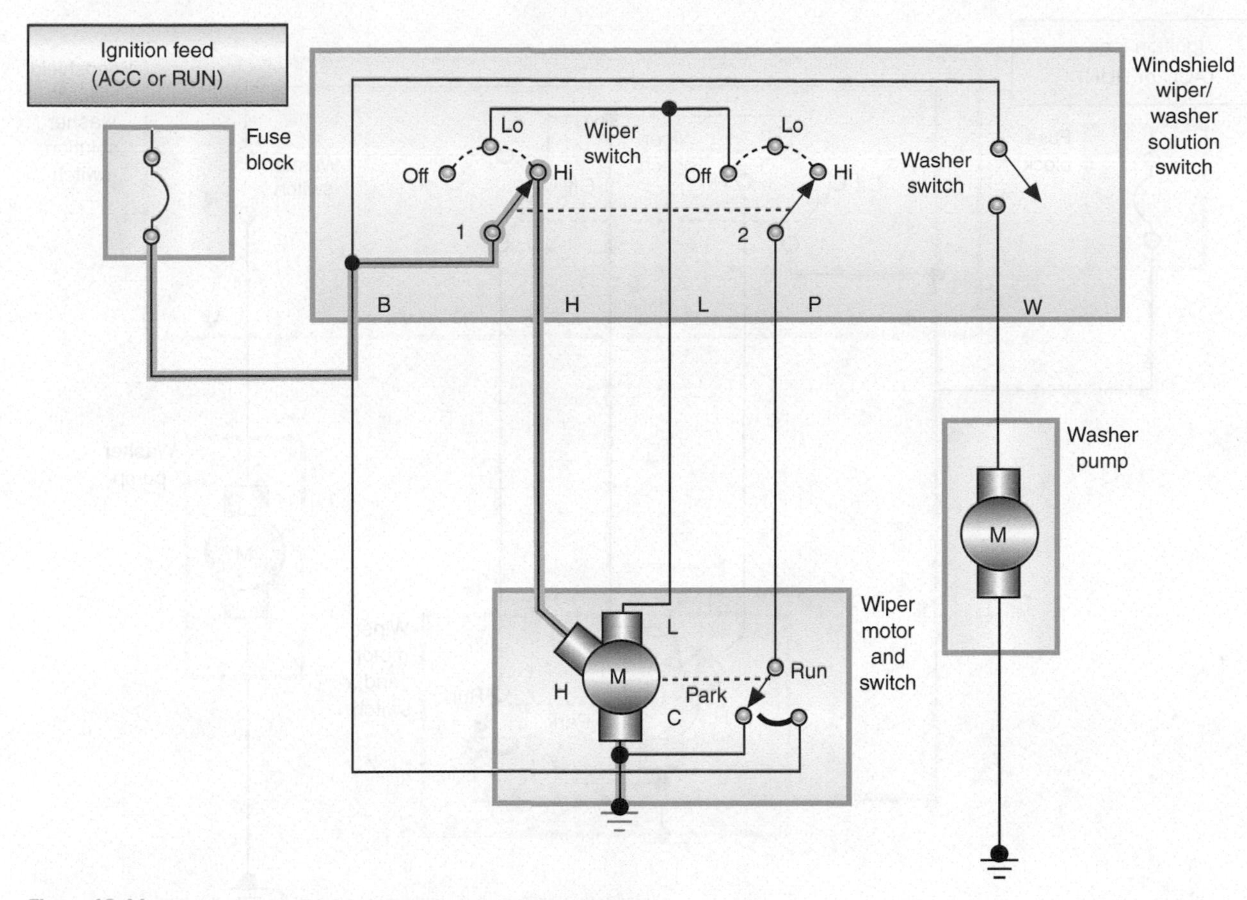

Figure 13-11 Current flow in HIGH, using a permanent magnet motor.

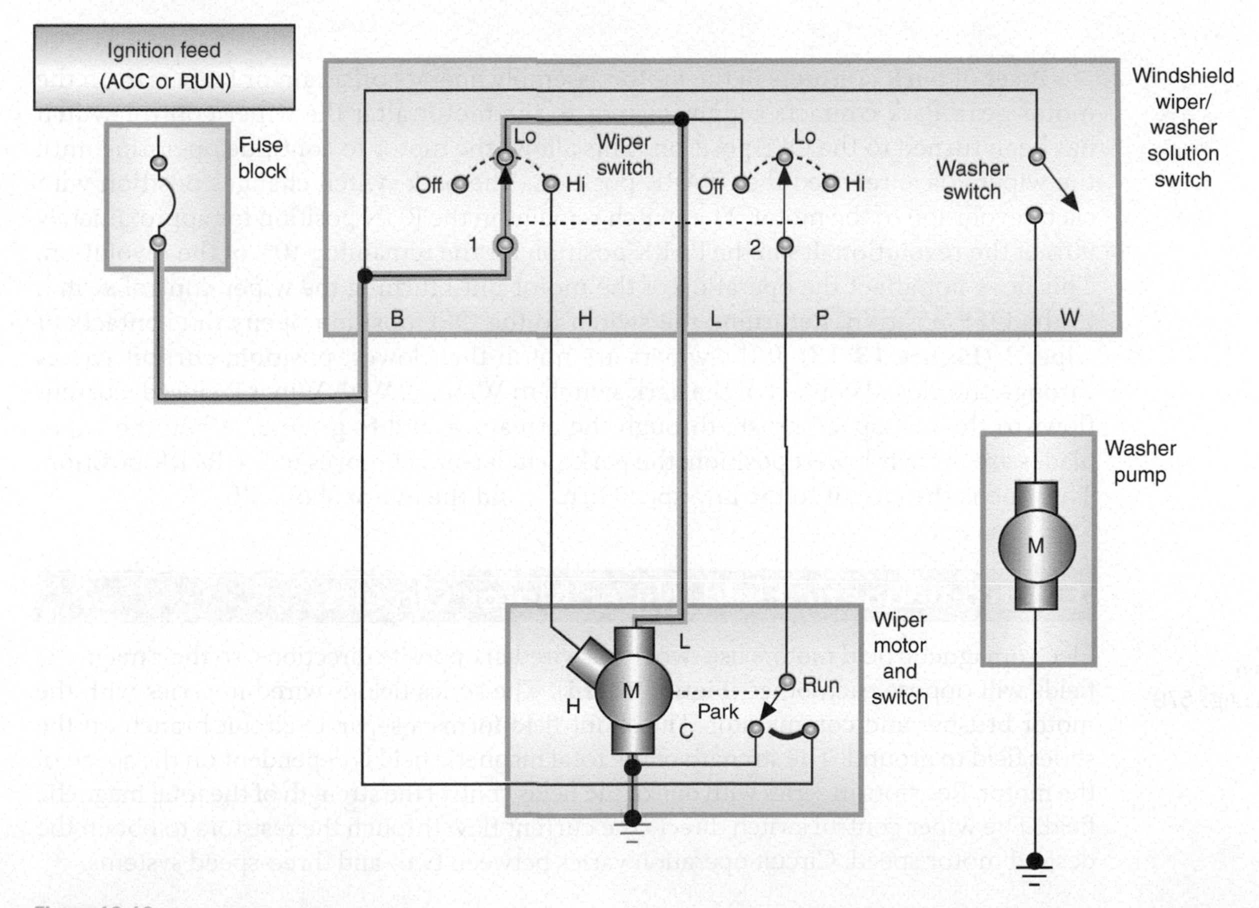

Figure 13-12 Current flow in LOW position.

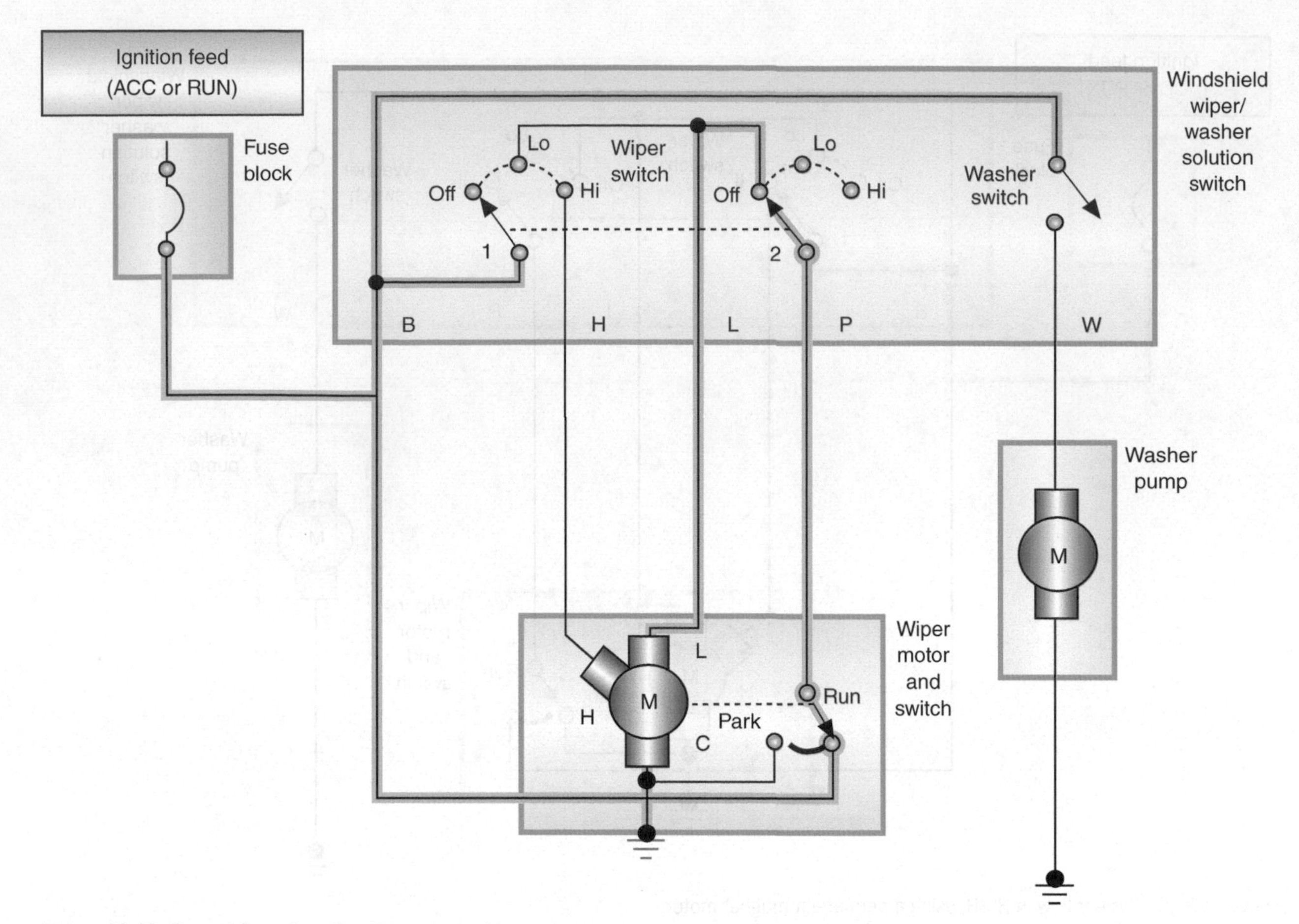

Figure 13-13 Current flow when the wipers are parking.

Shop Manual
Chapter 13, page 673

A set of **park contacts** in the motor assembly operate off a cam or latch arm on the motor gear. Park contacts supply current to the motor after the wiper control switch has been turned to the OFF position. This allows the motor to continue operating until the wipers have reached their PARK position. The park switch changes position with each revolution of the motor. The switch remains in the RUN position for approximately 90% of the revolution. It's in the PARK position for the remaining 10% of the revolution. This does not affect the operation of the motor until turning the wiper control switch to the OFF position. Returning the switch to the OFF position opens the contacts of wiper 1 (**Figure 13-13**). If the wipers are not in their lowest position, current passes through the closed contact of the park switch to Wiper 2. With Wiper 2 closed, current flows to the low-speed brush, through the armature, and to ground. When the wiper blades are in their lowest position, the park switch contact moves to the PARK position. This opens the circuit to the low-speed brush, and the motor shuts off.

Electromagnetic Field Wiper Motor Circuits

Shop Manual
Chapter 13, page 670

Electromagnetic field motors use two fields wired in opposite directions, so their magnetic fields will oppose each other (**Figure 13-14**). The series field is wired in series with the motor brushes and commutator. The shunt field forms a separate circuit branch off the series field to ground. The strength of the total magnetic field is dependent on the speed of the motor. Resistors in series with one of the fields control the strength of the total magnetic field. The wiper control switch directs the current flow through the resistors to obtain the desired motor speed. Circuit operation varies between two- and three-speed systems.

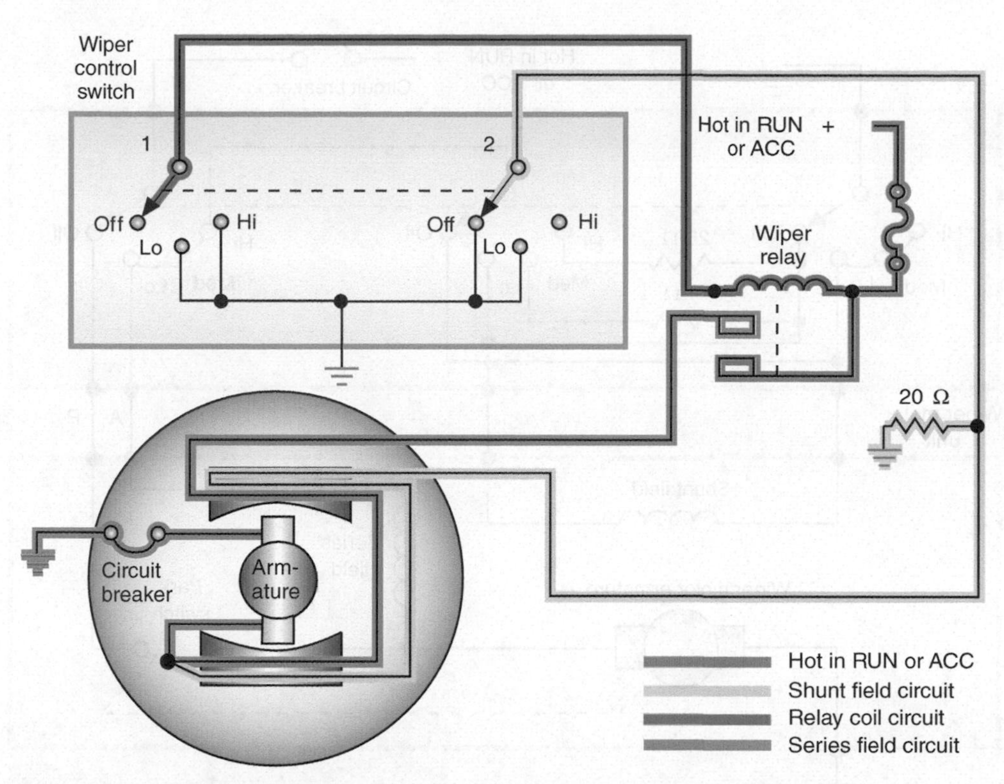

Figure 13-14 Simplified diagram of an electromagnetic field, two-speed wiper system.

Shop Manual
Chapter 13, page 671

Two-Speed Motors. The ground side switch will determine the current path. One path is directly to ground after the field coil; the other is through a 20-ohm (Ω) resistor (refer to Figure 13-14).

With the switch in the OFF position, switch wiper 1 breaks the circuit through the relay to ground. With the relay de-energized, no current flows to the motor.

With the switch in the LOW-SPEED position, wiper 1 completes the relay coil circuit to ground. The energized relay closes the contacts and applies current to the motor, through the series field and shunt field coils. Wiper 2 provides the ground path for the shunt field coil. Because there is less resistance to ground through wiper 2, the 20-ohm resistor is bypassed. With no resistance in the shunt field coil, the shunt field is very strong and bucks the magnetic field of the series field. The result is slow motor operation.

With the switch in the HIGH-SPEED position, wiper 1 completes the relay coil circuit to ground. This closes the relay contacts to the series field and shunt field coils. Wiper 2 opens the circuit to ground, and current must now pass through the 20-ohm resistor to ground. The resistor reduces the current flow and strength of the shunt field. With less resistance from the shunt field, the series field is able to turn the motor at an increased speed.

Returning the switch to the OFF position opens the contact of wiper 1 and turns off the relay. The relay is de-energized, but the park switch manually holds the contact points to the series and shunt field coils closed until the wipers are in their lowest position. Wiper 2 closes the circuit path to ground. As long as the park switch is closed, current flows through the series field, shunt field, and wiper 2 to ground. Once the wipers are at their lowest point of travel, the park switch opens, and the motor turns off.

Shop Manual
Chapter 13, page 671

Three-Speed Motors. The three-speed motor system offers a low-, medium-, and high-speed selection. The wiper control switch position determines what resistors, if any, are connected to the circuit of one of the fields (**Figure 13-15**).

With the wiper control switch in the LOW-SPEED position, both field coils receive equal current, so the total magnetic field is weak and the motor speed is slow.

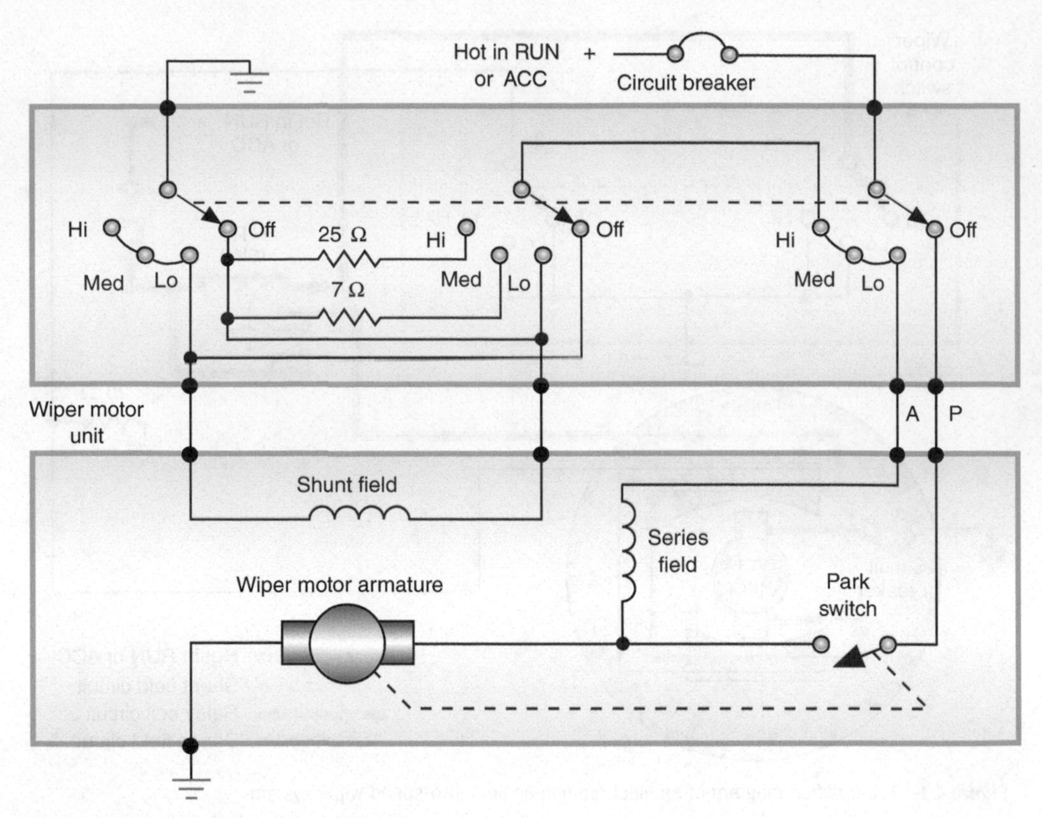

Figure 13-15 Three-speed wiper motor schematic.

With the switch in the MEDIUM-SPEED position, the current flows through a resistor before flowing to the shunt field (**Figure 13-16**). The resistor weakens the shunt coil's strength but strengthens the motor's total magnetic field. The speed increases over that of the LOW-SPEED position.

Placing the switch in the HIGH-SPEED position, connects a resistor of greater value to the shunt field circuit. The resistor weakens the magnetic field of the shunt coil, allowing a stronger total field to rotate the motor at a high speed.

Relay-Controlled, Two-Speed Wiper Systems

To reduce the size of the wires through the steering column harness, a wiper system may use relays to supply current to the wiper motor (**Figure 13-17**). Placing the ignition in the RUN or ACC position supplies battery voltage to the wiper motor and the coils of the ON/OFF and HI/LO relays. Placing the wiper switch in the LOW-SPEED position energizes the coil of the ON/OFF relay. The contacts of the ON/OFF relay now provide ground for the wiper motor. With the HI/LO relay coil de-energized, current flows through the motor's low-speed brush to ground.

With the wiper switch in the HIGH-SPEED position, the switch provides ground for both the wiper ON/OFF and the wiper HI/LO relay coils. The wiper HI/LO relay is a circuit diverter and now provides ground for the wiper motor's high-speed brush through the contacts of the wiper ON/OFF relay (which remains energized).

Intermittent Wipers

Shop Manual
Chapter 13, page 676

Most wiper systems offer an intermittent mode that provides a variable interval between wiper sweeps. Many of these systems use a module located in the steering column. Initiating the intermittent wiper mode while the wipers are in their parked position, the park switch is in the ground position. Current flows to the solid-state module's "timer

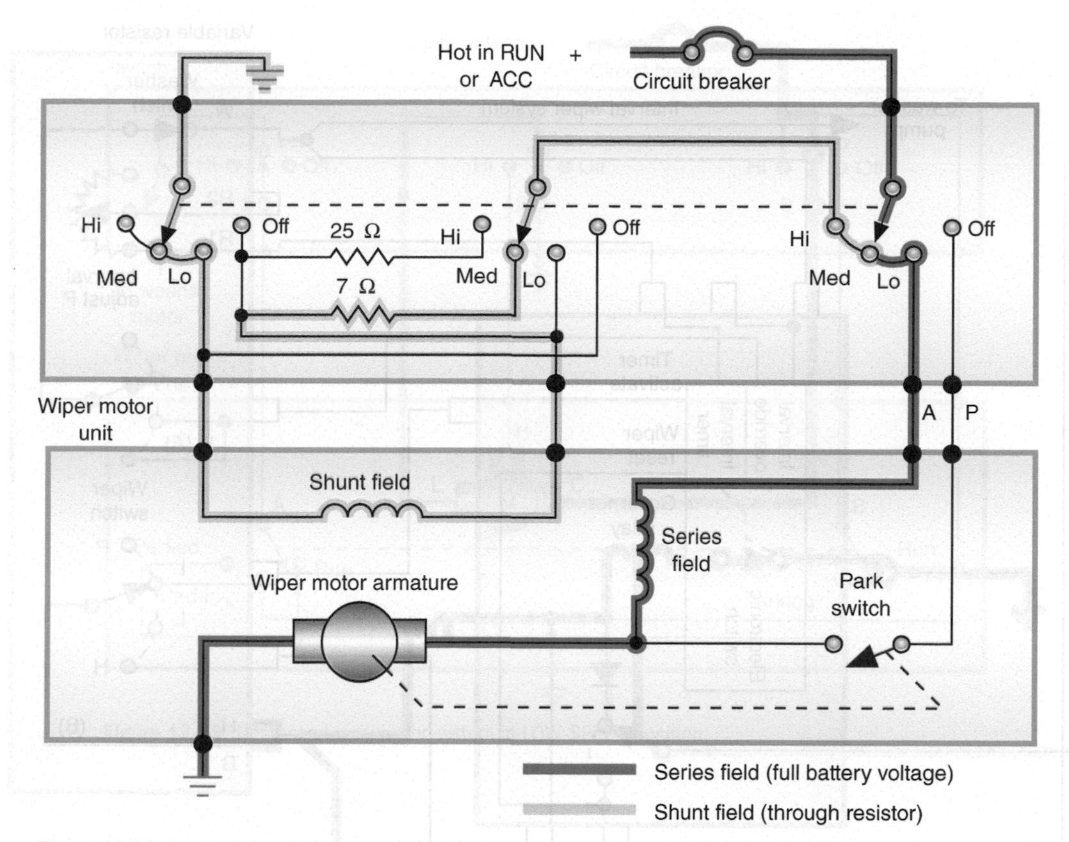

Figure 13-16 Current flow in MEDIUM position.

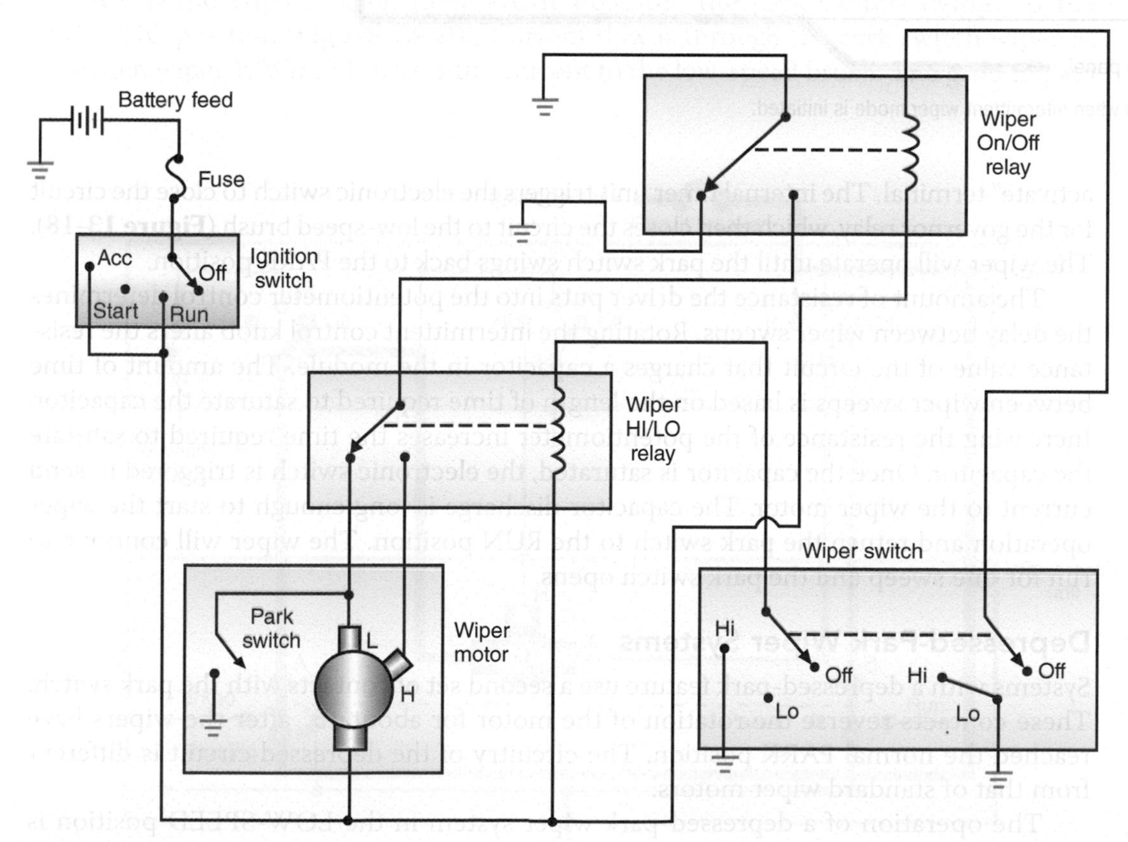

Figure 13-17 Relay-controlled wiper system.

is through the common brush, switch wiper 3, and park switch wiper B. This reversed current flow continues until the wipers reach the depressed-park position, where park switch wiper A swings to the PARKED position.

AUTHOR'S NOTE Note the difference in operation between the depressed-park wiper system and the one shown in Figure 13-11.

Computer-Operated Wipers

Shop Manual
Chapter 13, page 678

For much the same reasons as going to computer control of the horn system, manufacturers are also using computers to operate the wiper system (**Figure 13-21**). In this system, the multifunction switch provides the driver's request for front wiper operation using a resistive multiplexed signal. The steering column module then broadcasts the request over the CAN B bus network to the front control module. This front control module operates the relays by low-side drivers.

The wiper motor is usually a permanent magnet motor. The relays control current flow to the appropriate brush of the motor. When selecting the LOW-SPEED position of the multifunction switch, the steering column module broadcasts a wiper switch LOW message to the BCM. The BCM uses an HSD to energize the wiper ON/OFF relay.

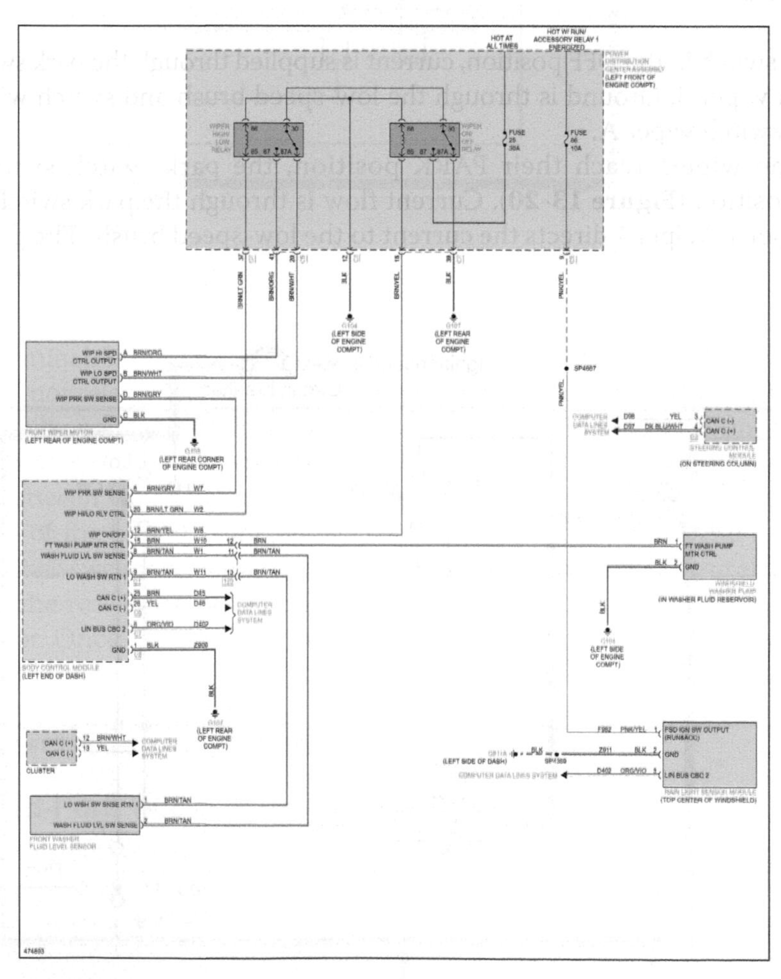

Figure 13-21 A computer-operated wiper system is shown in this screenshot.
Courtesy of ProDemand.

This directs battery current through the closed contacts of the energized wiper ON/OFF relay and the normally closed contacts of the de-energized wiper HI/LO relay to the low-speed brush of the wiper motor, causing the wipers to cycle at low speed.

When placed in the HIGH-SPEED position, the steering column module broadcasts a wiper switch HIGH message to the BCM. The BCM uses HSDs to energize both the wiper ON/OFF relay and the wiper HI/LO relay. This directs battery current through the closed contacts of the energized wiper ON/OFF relay and the closed contacts of the energized wiper HI/LO relay to the high-speed brush of the wiper motor, causing the wipers to cycle at high speed.

When the multifunction switch is rotated to the OFF position, the steering column module sends an electronic wiper switch OFF message to the BCM. The park switch does not operate the motor as described earlier; instead it's used as an input to the BCM. If the wipers are not in the PARK position on the windshield when the OFF request is broadcast, the BCM keeps the wiper ON/OFF relay energized, until the wiper blades are in the PARK position as indicated by the park switch input. If the OFF request is broadcasted while the wiper motor is at high speed, the BCM de-energizes the wiper HI/LO relay, causing the wiper motor to return to low-speed operation before parking the wipers.

The computer-controlled wiper system also provides for intermittent wiper operation. When the multifunction switch is in one of the intermittent interval positions, the steering column module broadcasts the delay message to the BCM. The BCM uses an intermittent wipe logic circuit that calculates the correct length of time between wiper sweeps based on the selected delay interval input. The BCM monitors the state of the wiper motor park switch to determine the proper intervals at which to energize and de-energize the wiper ON/OFF relay to operate the wiper motor intermittently for one low-speed cycle at a time.

With computer-controlled wiper operation, the driver can select to have the headlights turn on automatically whenever the wipers complete a minimum of five automatic wipe cycles within about 60 seconds. This meets the legal requirements in some states, which stipulate that the headlights must be on whenever the wipers are in use. The headlights will also turn off automatically after the wipers are turned off, and four minutes elapse without any wipe cycles.

Intelligent Windshield Wipers

Shop Manual
Chapter 13, page 681

Manufacturers have developed intelligent wiper systems to avoid making the driver having to adjust the correct speed of the windshield wipers according to the amount of rain. We will explore the operation of two intelligent wiper systems. One senses the amount of rainfall, and the other adjusts wiper speed according to vehicle speed.

The automatic wiper system selects the wiper speed needed to keep the windshield clear by sensing the presence and amount of rain on the windshield. The system relies on a series of LEDs that shine at an angle onto the inside of the windshield glass and an equal number of light collectors (**Figure 13-22**). The outer surface of a dry windshield will reflect the lights from the LEDs back into a series of collectors. The presence of water on the windshield will refract some of the light away from the collectors (**Figure 13-23**). When this happens, the wipers turn on. If one complete travel of the wipers fails to clear the water, the wipers operate again. The amount of water sensed on the windshield determines the frequency and speed of wiper operation.

Speed-sensitive wipers don't require additional components for this function since most use the BCM. Speed-sensitive wipers compensate for extra moisture that normally accumulates on the windshield at higher speeds in the rain. At higher speeds, the delay between wipers shortens when the wipers are operating in the interval mode.

Figure 13-22 The rain sensor is mounted to the rearview mirror.

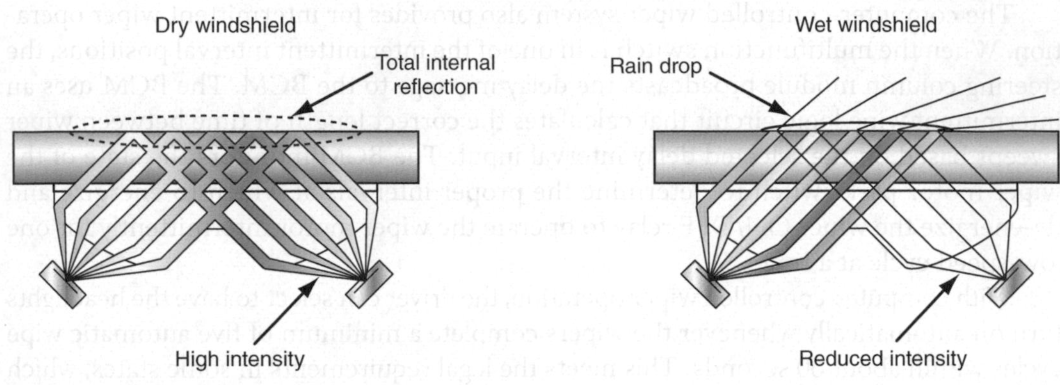

Figure 13-23 Water on the windshield deflects the light beams.

This delay automatically adjusts at speeds between 10 and 65 mph (16 and 105 kph). Basically, this system functions according to the input the computer receives about vehicle speed.

Washer Pumps

Shop Manual
Chapter 13, page 683

Windshield washers spray a washer fluid solution onto the windshield and work in conjunction with the wiper blades to clean the windshield of dirt. Some vehicles with composite headlights incorporate a headlight washing system along with the windshield washer (**Figure 13-24**). Most systems have the washer pump motor installed into the reservoir (**Figure 13-25**). General Motors uses a pulse-type washer pump that operates off the wiper motor (**Figure 13-26**).

The washer system includes plastic or rubber hoses to direct fluid flow to the nozzles and produce the spray pattern.

Pressing the washer switch activates the windshield washer system (**Figure 13-27**). If the wiper/washer system also has an intermittent control module, the module receives the washer switch's activation signal (**Figure 13-28**). An override circuit in the module operates the wipers on low speed for a programmed length of time. Depending on system design, the wipers will either return to the parked position or operate in intermittent mode.

Figure 13-24 Headlight washer system may operate with the windshield washer or have a separate switch.

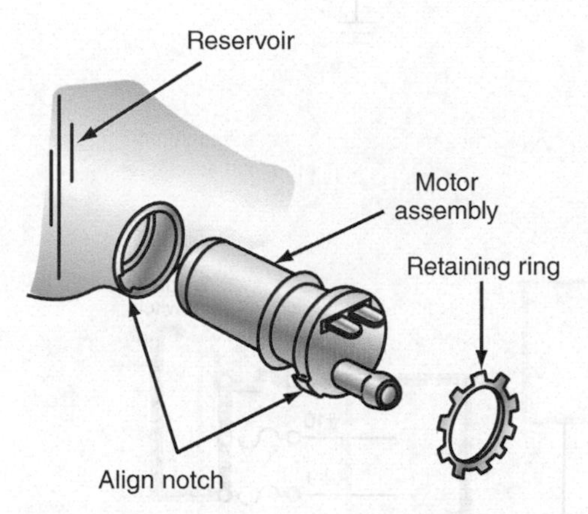

Figure 13-25 Washer motor installed into the reservoir.

Figure 13-26 General Motors' pulse-type washer system incorporates the washer motor into the wiper motor.

Computer-Operated Washer Systems

Referring to Figure 13-21, with the WASH button on the control stalk of the multifunction switch depressed, the steering column module broadcasts a WASH request message to the BCM over the CAN C bus. The BCM then uses an HSD circuit to supply current to the washer pump. At the same time, the wipers turn on and operate for about three wipes.

AUTHOR'S NOTE Some systems equipped with a rear window wiper system may use an H-gate driver circuit to operate the washer pump. This system uses only one washer pump. The H-gate controls the direction of current flow to the pump that dictates which direction the pump rotates. Based on the direction of rotation, the washer fluid flows to either the front windshield or the rear window.

If wash is requested while the wipers are already turned on and operating in one of the intermittent interval positions, the washer pump operation is the same. However, during

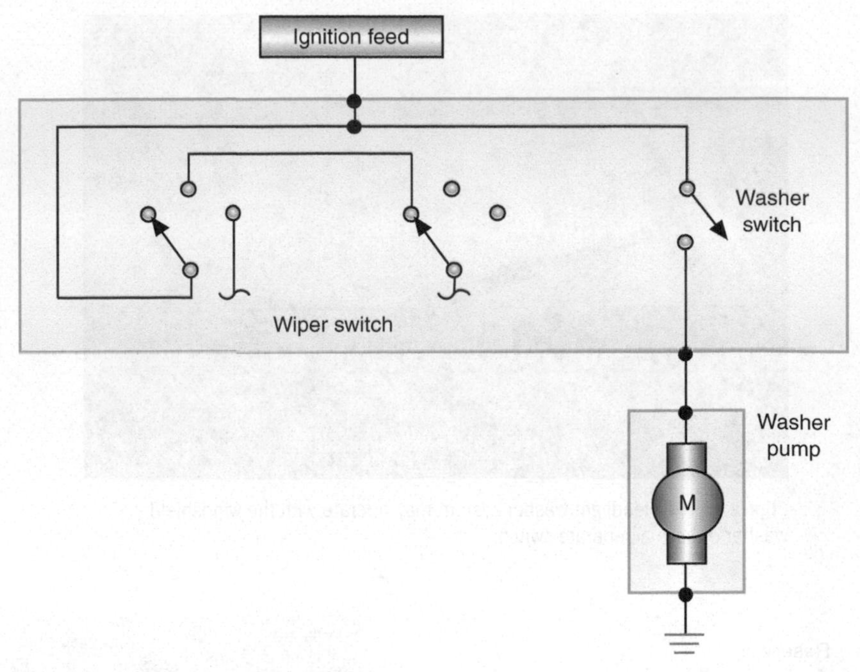

Figure 13-27 Windshield washer motor circuit.

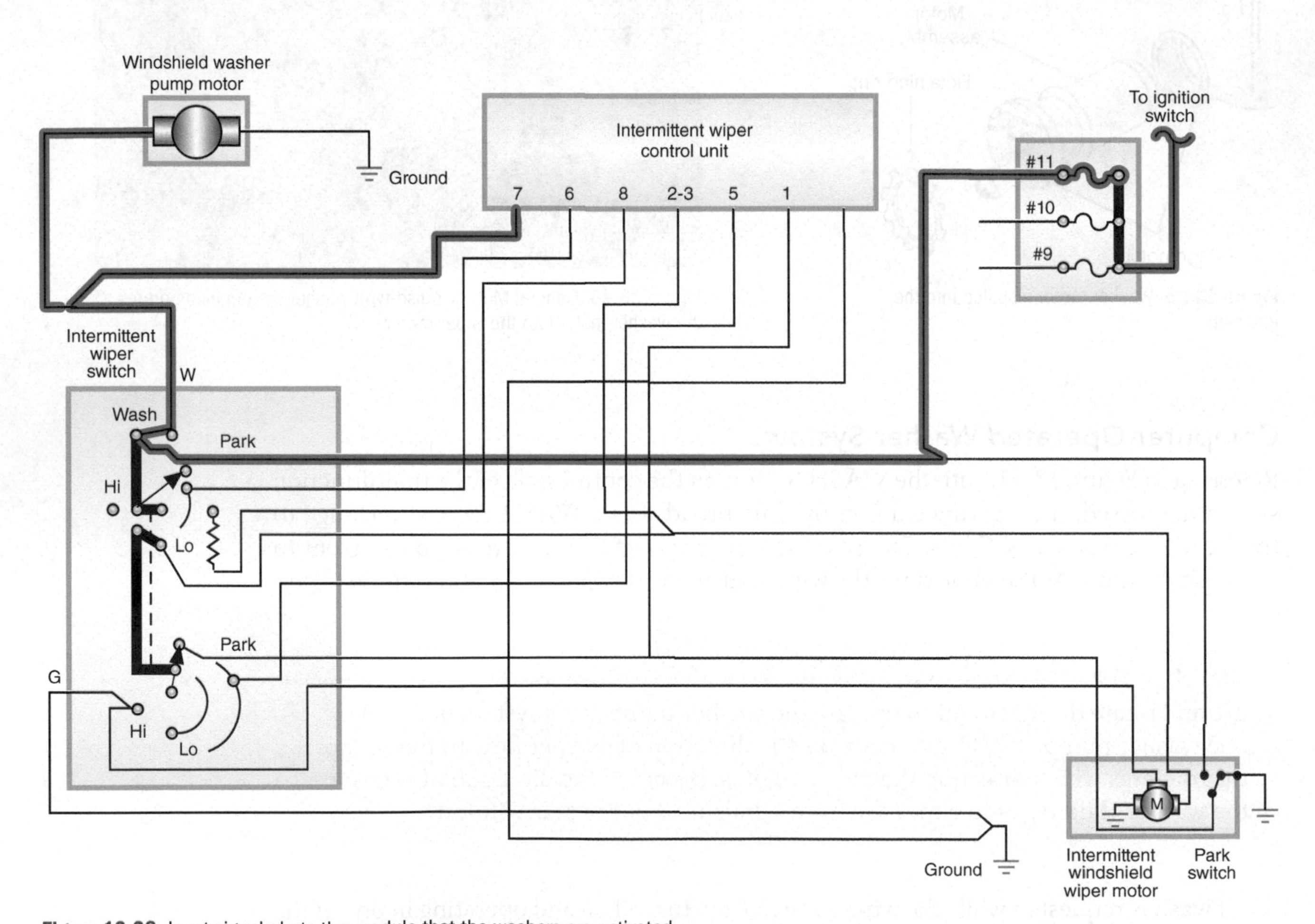

Figure 13-28 Input signal alerts the module that the washers are activated.

this time, the BCM will abort the delay feature and will energize the wiper ON/OFF relay to operate the wiper motor in a continuous low-speed mode for as long as the WASH switch is closed. When the WASH request is no longer present, the BCM will resume the selected delay mode interval.

The headlamp washer system uses a separate high-pressure pump that operates when the headlamps are on and the windshield washer switch is closed. The high-pressure pump will direct two-timed, high-pressure sprays onto the headlamp lens.

Blower Motor Circuits

Shop Manual
Chapter 13, page 686

The blower motor moves air inside the vehicle for air conditioning, heating, defrosting, and ventilation. The motor is usually a permanent magnet, single-speed motor located in the heater housing assembly (**Figure 13-29**). A blower motor switch, mounted in the HVAC panel, controls the fan speed. The switch position directs current flow to a **resistor block** wired in series between the switch and the motor (**Figure 13-30**) and consists of two or three helically wound wire resistors wired in series.

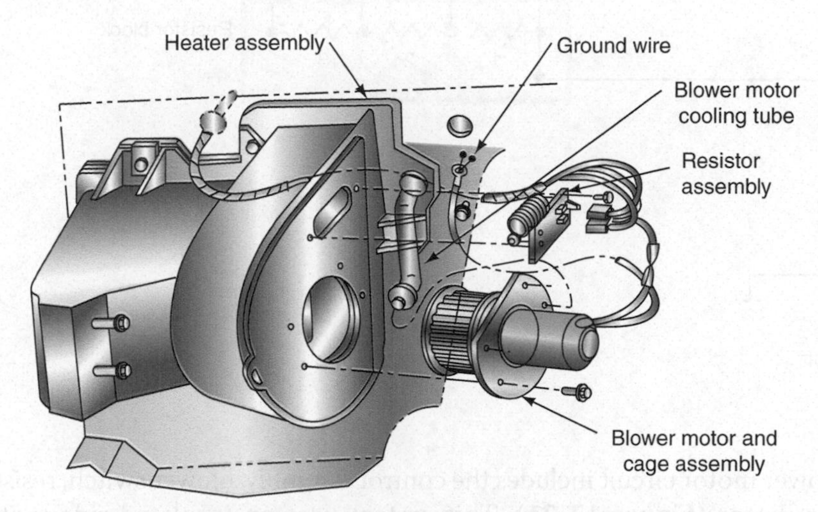

Figure 13-29 The blower motor is usually installed in the heater assembly. Mode doors control if vent, heater, or A/C-cooled air is blown by the motor cage.

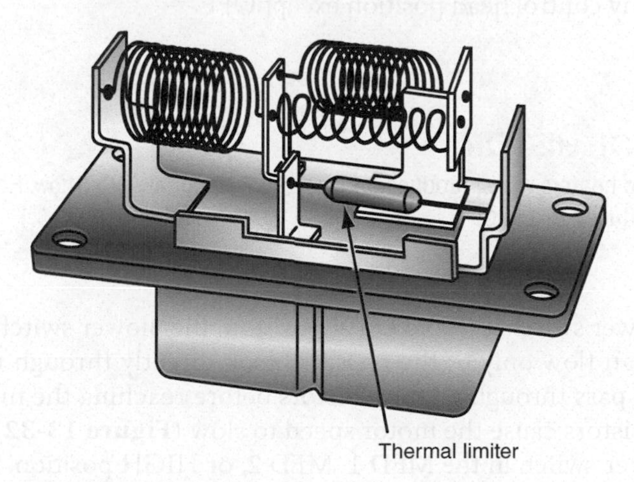

Figure 13-30 Fan motor resistor block.

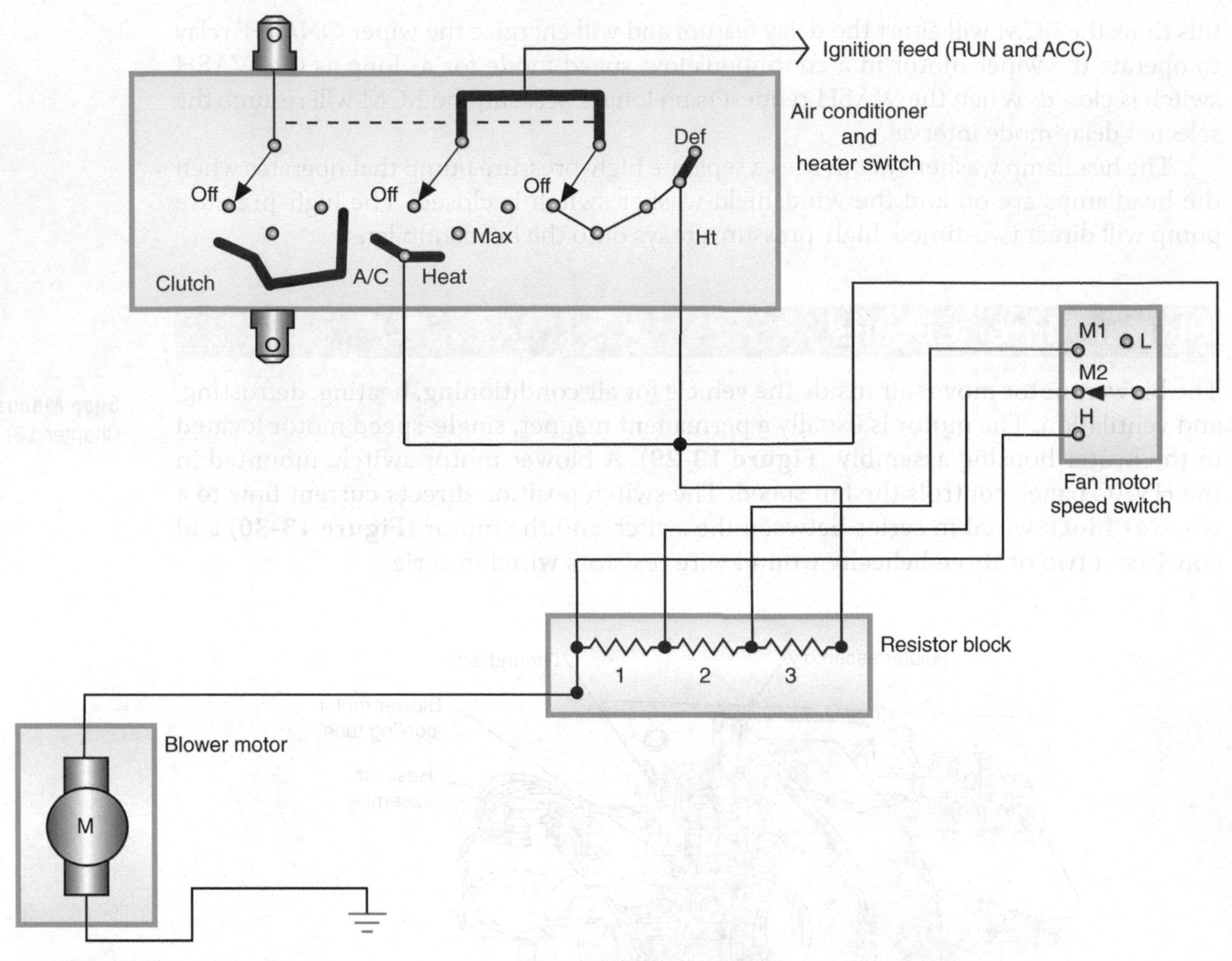

Figure 13-31 Blower motor circuit.

The blower motor circuit includes the control assembly, blower switch, resistor block, and blower motor (**Figure 13-31**). This system uses an insulated side switch and a grounded motor. With the ignition in the RUN or ACC position, the control head receives battery voltage. The current can flow from the control head to the blower switch and resistor block in any control head position except OFF.

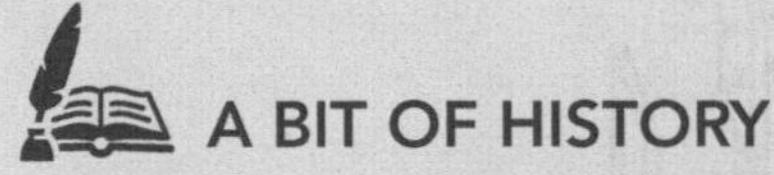

A BIT OF HISTORY

Automotive electric heaters were introduced at the 1917 National Auto Show. Hot water in-car heaters were introduced in 1926.

When the blower switch is in the LOW position, the blower switch wiper opens the circuit. Current can flow only to the resistor block directly through the control head. The current must pass through all the resistors before reaching the motor. The voltage drops over the resistors cause the motor speed to slow (**Figure 13-32**).

With the blower switch in the MED 1, MED 2, or HIGH position, the current flows through the blower switch to the resistor block. Depending on the speed selection, the current must pass through one, two, or none of the resistors. As the amount of resistance decreases, the more voltage applied to the motor and fan speed increases.

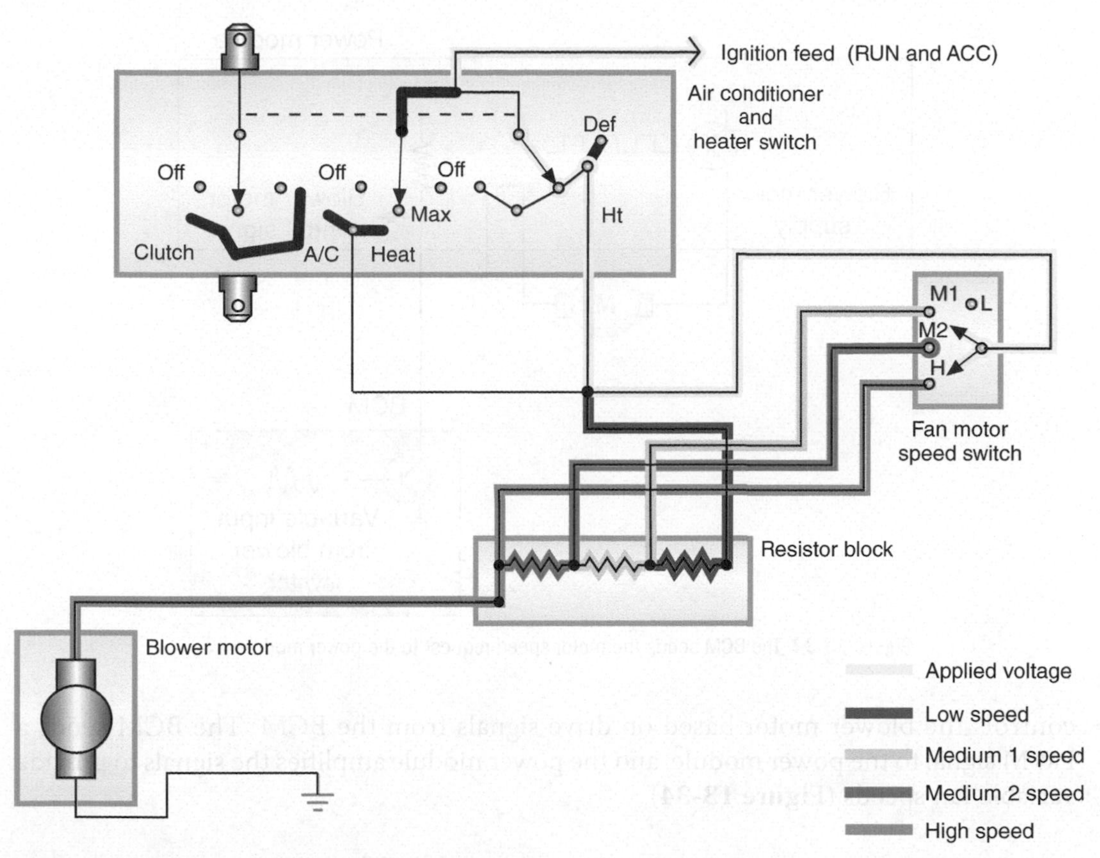

Figure 13-32 Current flow in the different speed selections.

Some manufacturers use ground side switching with an insulated motor. The switch completes the circuit to ground. The operating principles are identical to that of the insulated switch already discussed.

Many of today's vehicles use the BCM to control fan speed by pulse width modulation (PWM). Typically the BCM does not directly operate the motor. This is because of the high-amperage requirements of the motor. Instead, a power module (**Figure 13-33**)

Shop Manual
Chapter 13, page 688

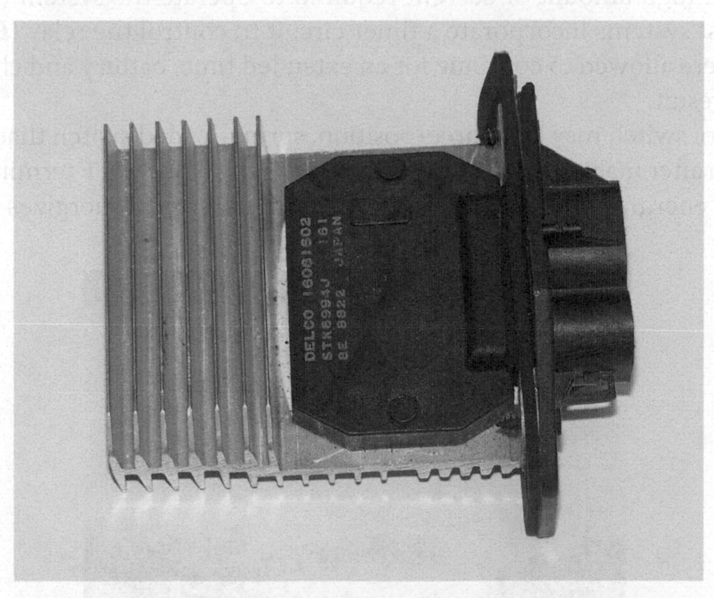

Figure 13-33 The power module amplifies the BCM signal to control motor speed.

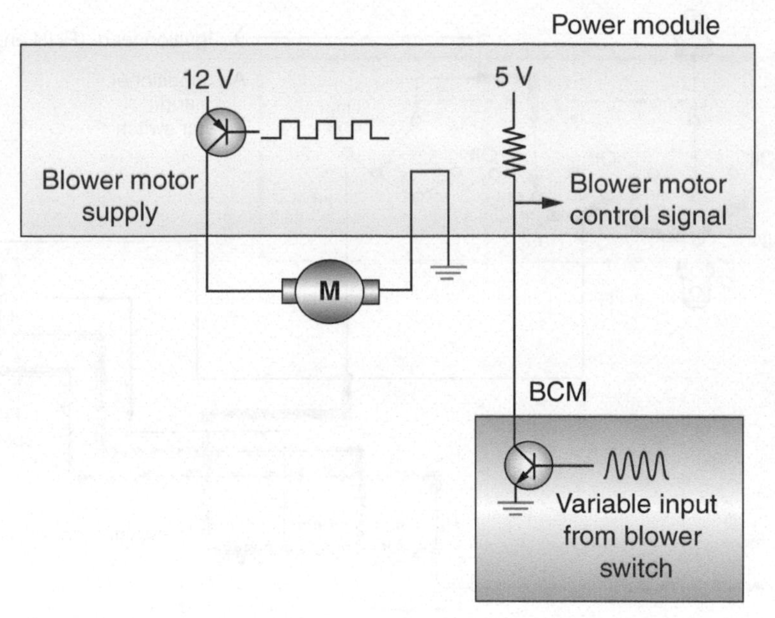

Figure 13-34 The BCM sends the motor speed request to the power module.

controls the blower motor based on drive signals from the BCM. The BCM sends a PWM signal to the power module, and the power module amplifies the signals to provide variable fan speeds (**Figure 13-34**).

Electric Defoggers

Shop Manual
Chapter 13, page 689

Many manufacturers refer to the electric rear window defogger as an electric backlight.

The ON indicator can be either a bulb or a light-emitting diode (LED).

Electric defoggers heat the rear window to remove ice and/or condensation. Some vehicles use the same circuit to heat the outside side mirrors. The principle behind electric defoggers is that forcing electrons to flow through a resistance generates heat. Electric defoggers use controlled resistance to heat the glass. The resistance is through a **grid** baked on the inside of the glass (**Figure 13-35**). The rear window defogger grid is a series of horizontal, ceramic silver–compounded lines. The terminals are soldered to the vertical bus bars. One terminal supplies the current from the switch; the other provides the ground (**Figure 13-36**).

Due to the high amount of current required to operate the system (approximately 30 amps), most systems incorporate a timer circuit to control the relay (**Figure 13-37**). If this drain were allowed to continue for an extended time, battery and charging system failure could result.

The control switch may be a three-position, spring-loaded switch that returns to the center position after making momentary contact with the ON or OFF terminals. Activation of the switch energizes the electronic timing circuit, which energizes the relay coil.

Figure 13-35 Rear window defogger grid.

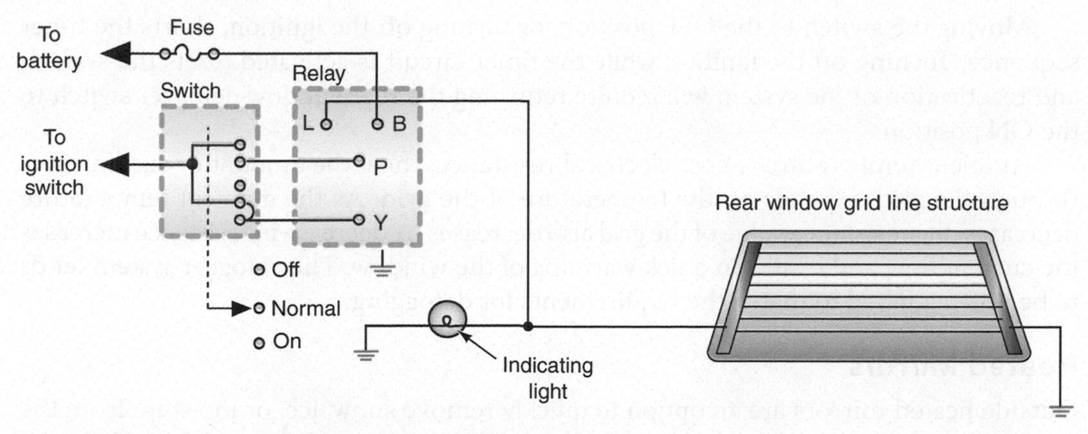

Figure 13-36 Rear window defogger circuit schematic.

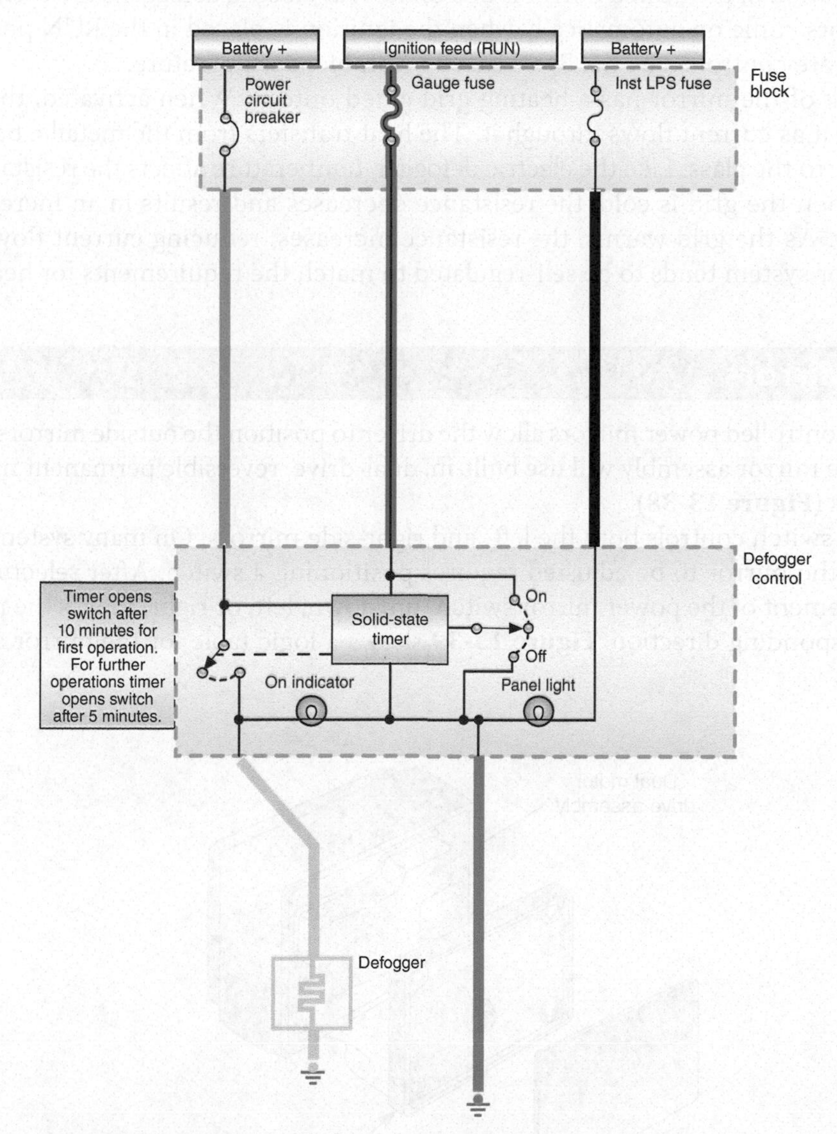

Figure 13-37 Defogger circuit using a solid-state timer.

With the relay contacts closed, the heater grid receives direct battery voltage. At the same time, the ON indicator receives voltage. The timer is activated for 10 minutes. At the completion of the timed cycle, the relay is de-energized and the circuits to the grid and indicator light open. A second activation of the switch results in the timer energizing the relay for 5 minutes.

Moving the switch to the OFF position, or turning off the ignition, aborts the timer sequence. Turning off the ignition while the timer circuit is activated resets the system, and reactivation of the system will require returning the rear window defogger switch to the ON position.

Ambient temperatures affect electrical resistance; thus, the amount of current flow through the grid depends on the temperature of the grid. As the ambient temperature decreases, the resistance value of the grid also decreases. A decrease in resistance increases the current flow and results in quick warming of the window. The defogger system tends to be self-regulated to match the requirements for defogging.

Heated Mirrors

Outside heated mirrors are an option to quickly remove snow, ice, or moisture from the mirror glass. Heated mirrors operate on the same principle as the rear electric defogger. Usually, activation of the heated mirror is tied to the rear electric defogger. However, some heated mirrors come on automatically when the ignition is placed in the RUN position, while others are controlled by the BCM based on ambient temperature.

The back of the mirror has a heating grid glued onto it. When activated, the grid generates heat as current flows through it. The heat transfers from the metallic backing of the mirror to the glass. Like the electric defogger, temperature affects the resistance of the grid. When the grid is cold, the resistance decreases and results in an increase in current flow. As the grid warms, the resistance increases, reducing current flow. The heated mirror system tends to be self-regulated to match the requirements for heating.

Power Mirrors

Electrically controlled power mirrors allow the driver to position the outside mirrors using a switch. The mirror assembly will use built-in, dual-drive, reversible permanent magnet (PM) motors (**Figure 13-38**).

A single switch controls both the left- and right-side mirrors. On many systems, the selection of the mirror to be adjusted requires positioning a switch. After selecting the mirror, movement of the power mirror switch (up, down, left, or right) moves the mirror in the corresponding direction. **Figure 13-39** shows a logic table for the mirror switch and motors.

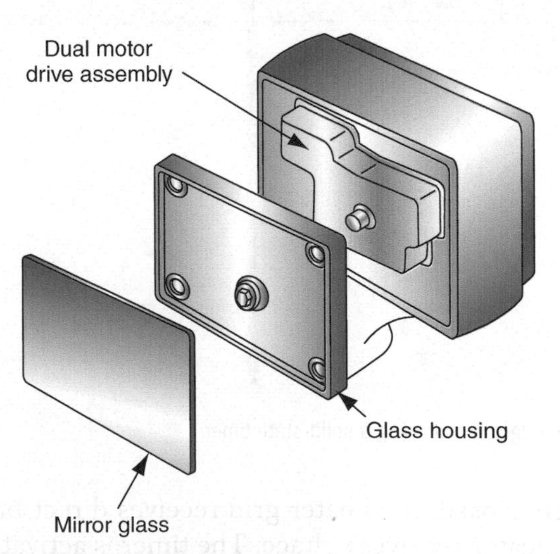

Figure 13-38 Power mirror motor.

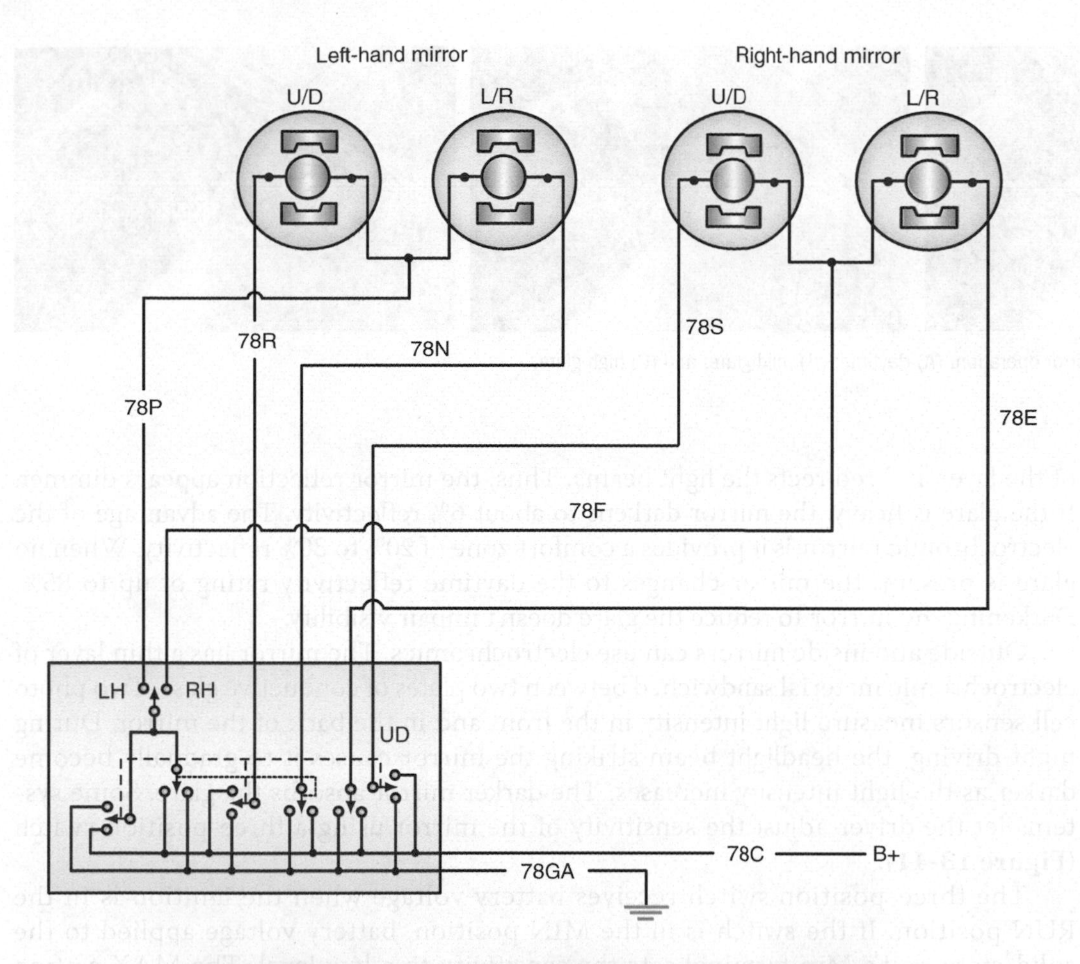

Switch function	Circuit function	
	Left mirror	Right mirror
Left	(78P+78C)(78N+78GA)	(78F+78C)(78E+78GA)
Right	(78N+78C)(78P+78GA)	(78E+78C) (78F+78GA)
Up	(78R+78C)(78P+78GA)	(78S+78C)(78F+78GA)
Down	(78P+78C)(78R+78GA)	(78F+78C)(78S+78GA)

Figure 13-39 Power mirror logic table.

Automatic Rearview Mirror

Some manufacturers have developed interior rearview mirrors that automatically tilt when the intensity of light that strikes the mirror is sufficient enough to cause discomfort to the driver.

The system has two photo cells mounted in the mirror housing. One of the photo cells measures the intensity of light inside the vehicle. The second measures the intensity of light the mirror receives. When the intensity of the light striking the mirror is greater than that of ambient light, by a predetermined amount, a solenoid is energized to tilt the mirror.

Electrochromic Mirrors

Electrochromic mirrors automatically adjust the amount of reflectance based on the intensity of glare (**Figure 13-40**). The electrochromic mirror uses forward- and rearward-facing photo sensors and a solid-state chip. The chip applies a small voltage to the silicon layer based on light intensity differences. Applying voltage rotates the molecules

Figure 13-40 Electrochromic mirror operation: (A) daytime; (B) mild glare; and (C) high glare.

of the layer and redirects the light beams. Thus, the mirror reflection appears dimmer. If the glare is heavy, the mirror darkens to about 6% reflectivity. The advantage of the electrochromic mirror is it provides a comfort zone of 20% to 30% reflectivity. When no glare is present, the mirror changes to the daytime reflectivity rating of up to 85%. Darkening the mirror to reduce the glare doesn't impair visibility.

Outside and inside mirrors can use electrochromics. The mirror has a thin layer of electrochromic material sandwiched between two plates of conductive glass. Two photo cell sensors measure light intensity in the front and in the back of the mirror. During night driving, the headlight beam striking the mirror causes it to gradually become darker as the light intensity increases. The darker mirror absorbs the glare. Some systems let the driver adjust the sensitivity of the mirror using a three-position switch (**Figure 13-41**).

The three-position switch receives battery voltage when the ignition is in the RUN position. If the switch is in the MIN position, battery voltage applied to the solid-state unit's Min terminal sets the sensitivity to a low level. The MAX setting causes the mirror to darken more at a lower glare level. Placing the transmission in reverse activates the reset circuit and returns the mirror to the daytime setting for clearer viewing to back up.

Power Windows

Shop Manual
Chapter 13, page 692

Most vehicles no longer use a hand crank to raise and lower the side windows. Reversible PM or two-field winding electric motors have replaced the mechanical crank. In addition, most sport utility models use electric motors to operate the rear tailgate window.

The power window system usually consists of the following components:

1. Master control switch.
2. Individual control switches.
3. Individual window drive motors.
4. Lock-out or disable switch.

Another design is to use rack-and-pinion gears. The rack is a flexible strip of gear teeth with one end attached to the window (**Figure 13-42**).

A **window regulator** converts the motor's rotary motion into the vertical movement of the window. The motor operates the window regulator either directly or through a cable. On direct-drive motors, the motor pinion gear meshes with gear teeth on the regulator called the **sector gear** (**Figure 13-43**). Lowering the window winds up the spiral spring. Raising the window unwinds the spring to assist the motor. This reduces the current the motor would need to raise the window by itself.

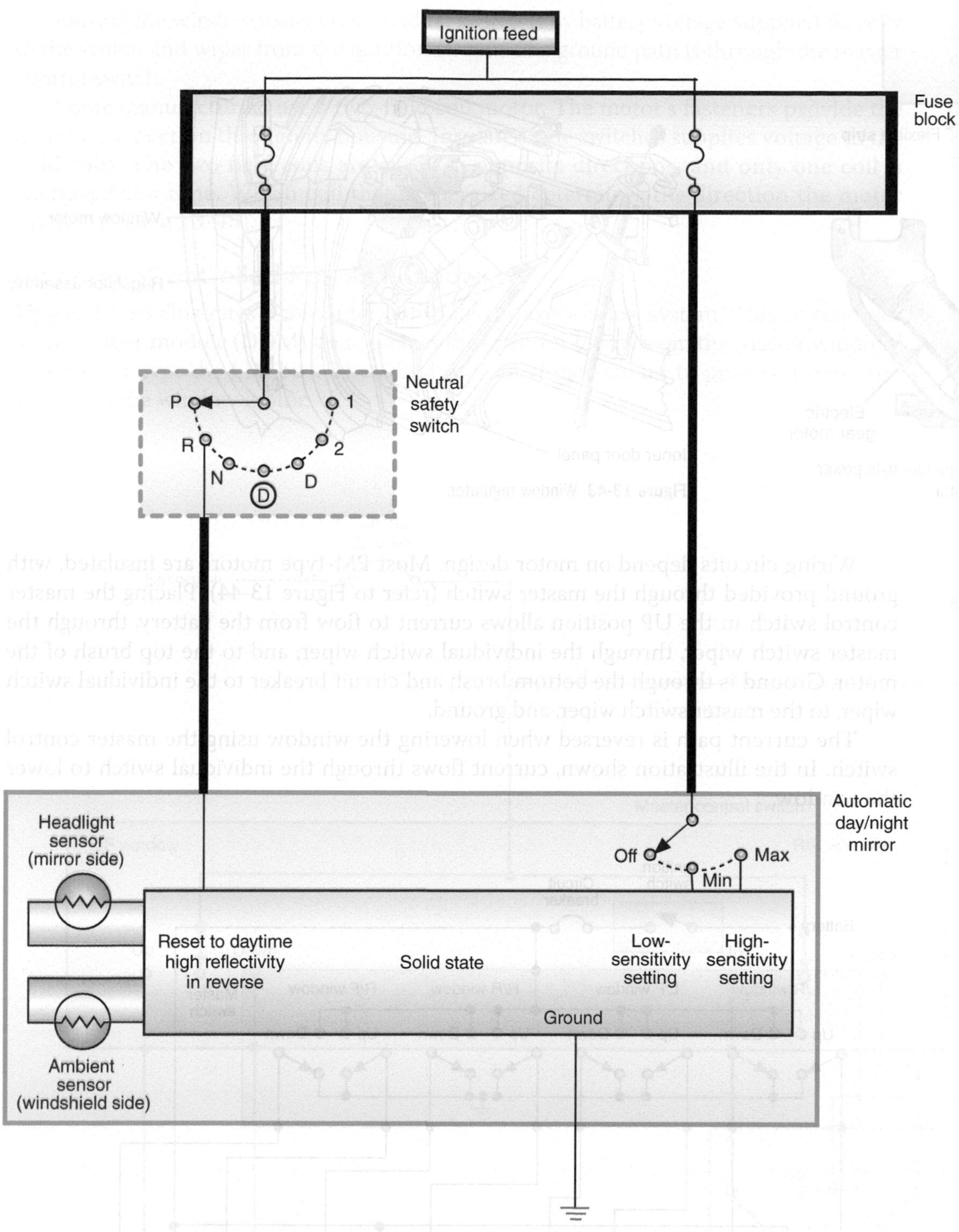

Figure 13-41 Automatic electrochromic day/night mirror diagram.

The master control switch provides the overall control of the system (**Figure 13-44**). The master switch provides power to the individual switches. The master switch may also have a safety lock switch that prevents the operation of the windows using the individual switches. When the safety switch is activated, it opens the circuit to the other switches, and control is only by the master switch. As an additional safety feature, some systems prevent the operation of the individual switches unless the ignition is in the RUN or ACC position.

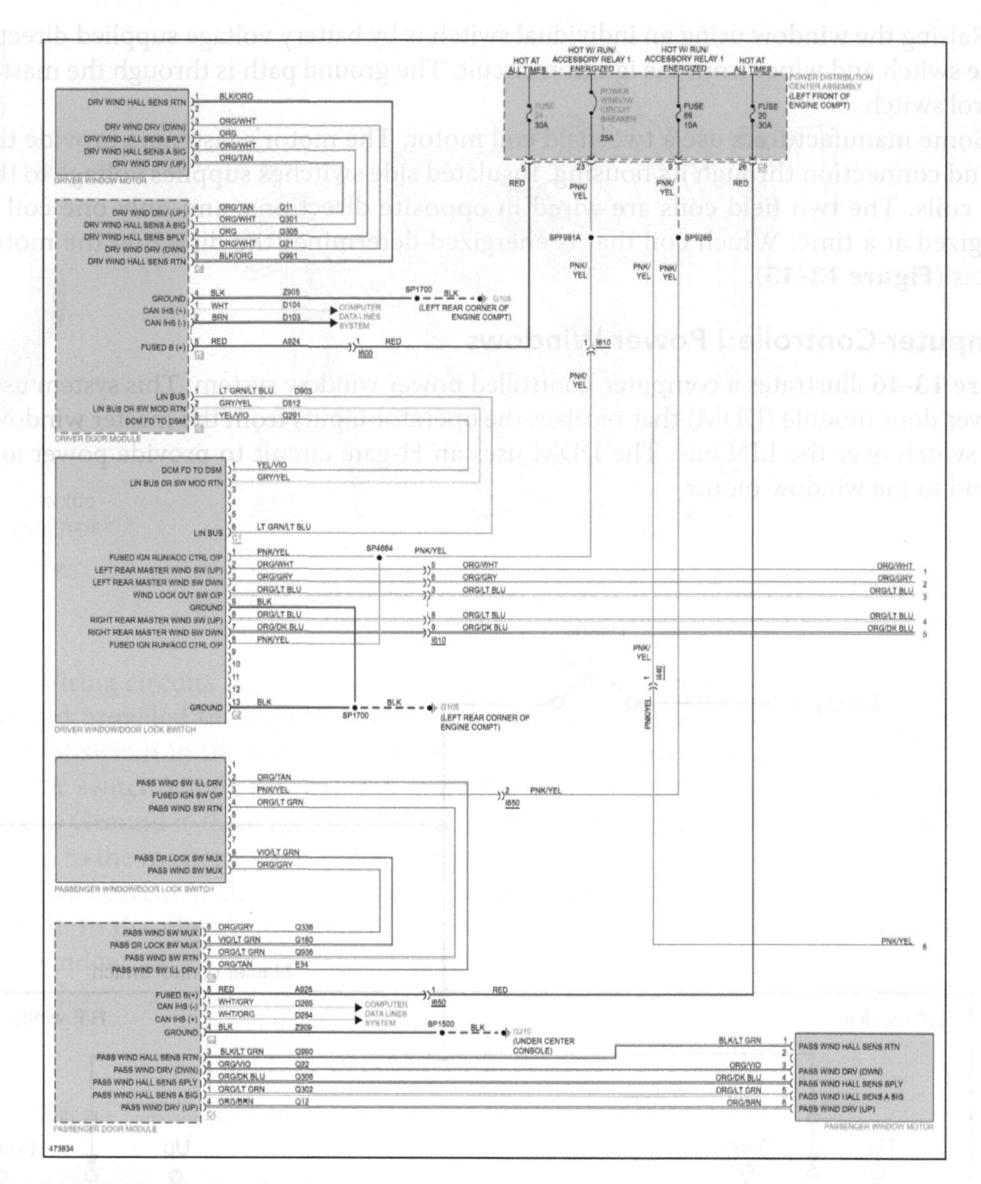

Figure 13-46 A power window system operated by a door module is shown in this screenshot. Courtesy of ProDemand.

Amperage or commutator pulses can determine if the window has reached full travel. The control module will shut off motor operation regardless of the switch position.

In this example, the master window/lock switch controls the inputs for all motor operation except the front passenger motor. The PDM controls the front passenger window based on MUX inputs from the passenger window/door switch. The DDM and the PDM communicate over the CAN IHS bus. When the driver operates the master window/lock switch to open or close the front passenger window, the DDM sends the request to the PDM over the CAN IHS bus network to operate that window motor.

The rear passenger door power window switches receive their battery feed from the master window/lock switch. Placing the lockout switch in the lock position interrupts the battery feed for the rear passenger door power window switches. The DDM sends a "Lock Request" bus message to the PDM to prevent the operation of the front passenger window.

The system can offer a front window **express down** feature. Express down allows a front window to lower to its full travel without having to hold the switch. Express down is activated by momentarily placing the switch into the second switch detent.

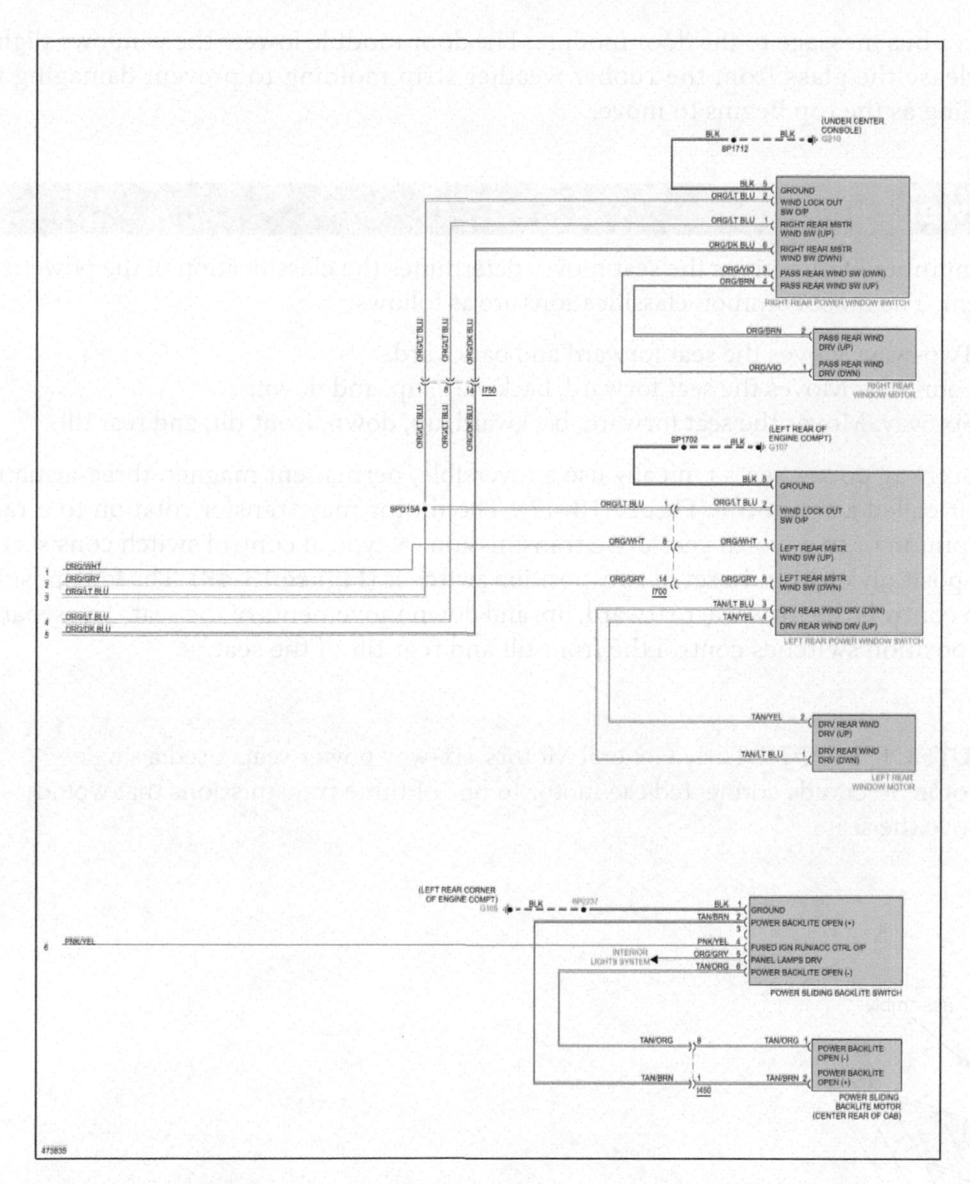

Figure 13-46 Continued

The corresponding door module operates the power window to the fully open position. Monitoring motor current and a Hall-effect switch determine glass travel distance and speed. An increase in current, or if the counts from the Hall-effect switch slow, indicate a possible obstruction, and the door module will stop operation. When the Hall-effect counts match the stored full up or full down value, the door module stops motor operation.

The front window **express up** feature allows raising a front window glass to the full up position without having to hold the switch. Monitoring of motor current and the Hall-effect switch are like in the express down feature. However, if an obstruction is detected, the window reverses direction.

Momentarily placing the switch in either the up or down position aborts the express down and express up operations.

Vehicles with convertible tops may use an input from the top control switch to operate the windows. If the top is in the closed position moving the switch to lower it sends a bus message to the door module. The door module lowers the windows slightly to release the glass from the rubber weather strip molding to prevent damaging the molding as the top begins to move.

Power Seats

Shop Manual
Chapter 13, page 695

The number of directions the seat moves determines the classification of the power seat system. The most common classifications are as follows:

1. Two-way: Moves the seat forward and backward.
2. Four-way: Moves the seat forward, backward, up, and down.
3. Six-way: Moves the seat forward, backward, up, down, front tilt, and rear tilt.

Some seat back latches use a solenoid to lock the seat unless the door is open. The door jamb switch controls the solenoid.

Six-way power seats typically use a reversible, permanent magnet, three-armature motor called a **trimotor** (**Figure 13-47**). The motor may transfer rotation to a rack-and-pinion or to a worm gear drive transmission. A typical control switch consists of a four-position knob and a set of two-position switches (**Figure 13-48**). The four-position knob controls the forward, rearward, up, and down movements of the seat. The separate two-position switches control the front tilt and rear tilt of the seat.

AUTHOR'S NOTE Early General Motors' six-way power seats used a single motor. Solenoids connected the motor to one of three transmissions that would move the seat.

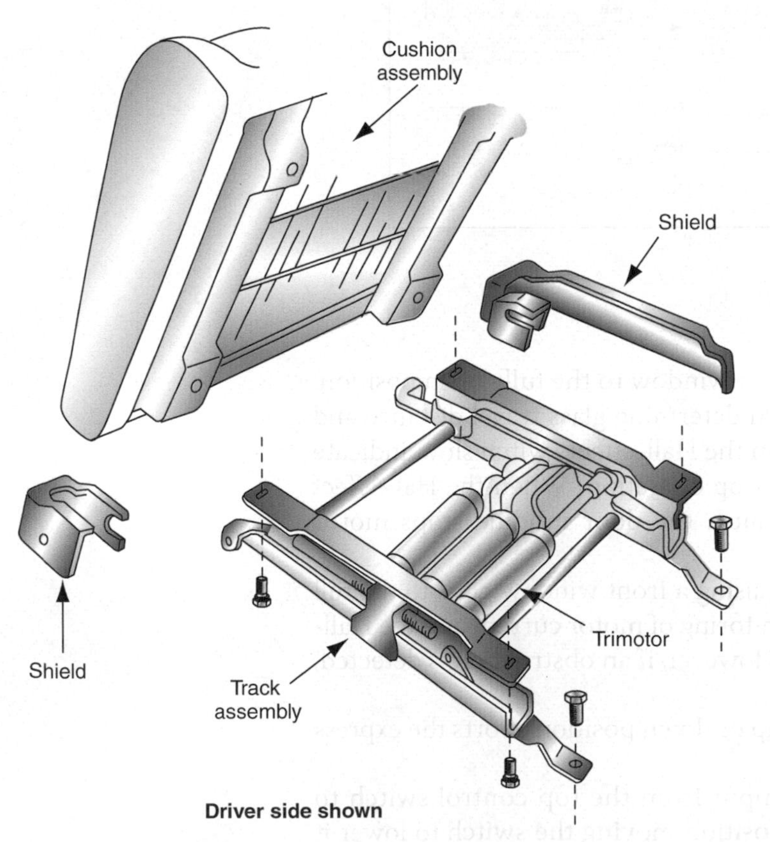

Figure 13-47 Trimotor power seat installation.

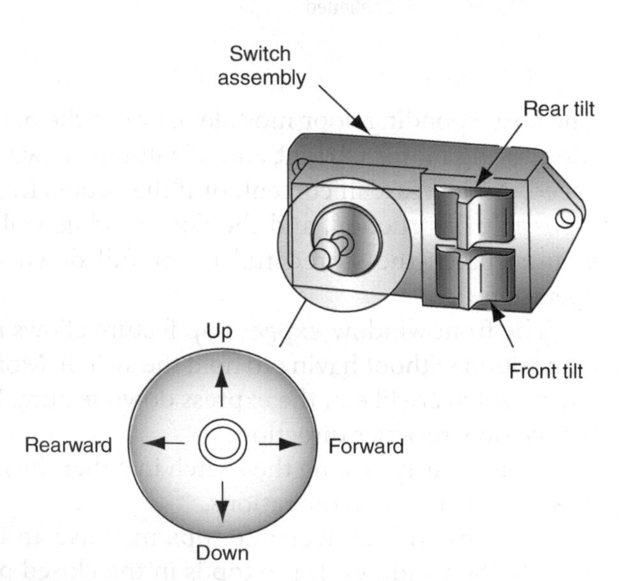

Figure 13-48 Power seat control switch.

The switch wipers control the current flow direction, thus the motor's rotation direction. If the driver pushes the four-way switch into the down position, the entire seat lowers (**Figure 13-49**). Switch wipers 3 and 4 move to the left, and current flows through wiper 4 to wipers 6 and 8. These wipers direct the current to the front and rear height motors. The ground circuit is through wipers 5 and 7, to wiper 3 and ground.

Some manufacturers equip their seats with adjustable support mats that shape the seat to fit the driver (**Figure 13-50**). The lumbar support mat provides additional comfort by supporting the back curvature. Some systems pump air into the mats; others use a motor to roll the lumbar support.

This power seat system uses the circuit breakers in the motor assemblies as limit switches. When the end of travel is reached, the higher current draw opens the circuit breaker and stops motor operation.

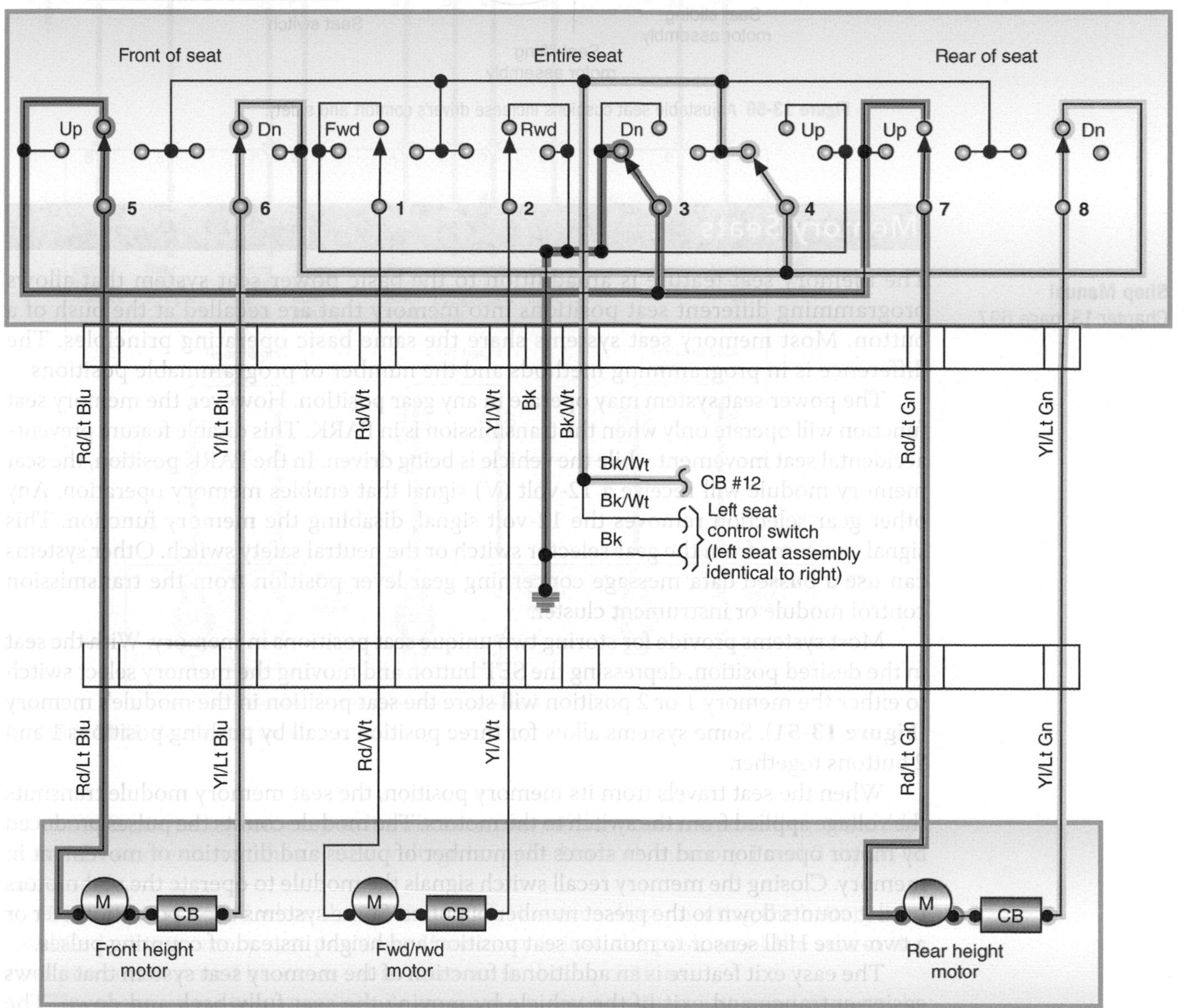

Figure 13-49 Current flow in the seat LOWER position.

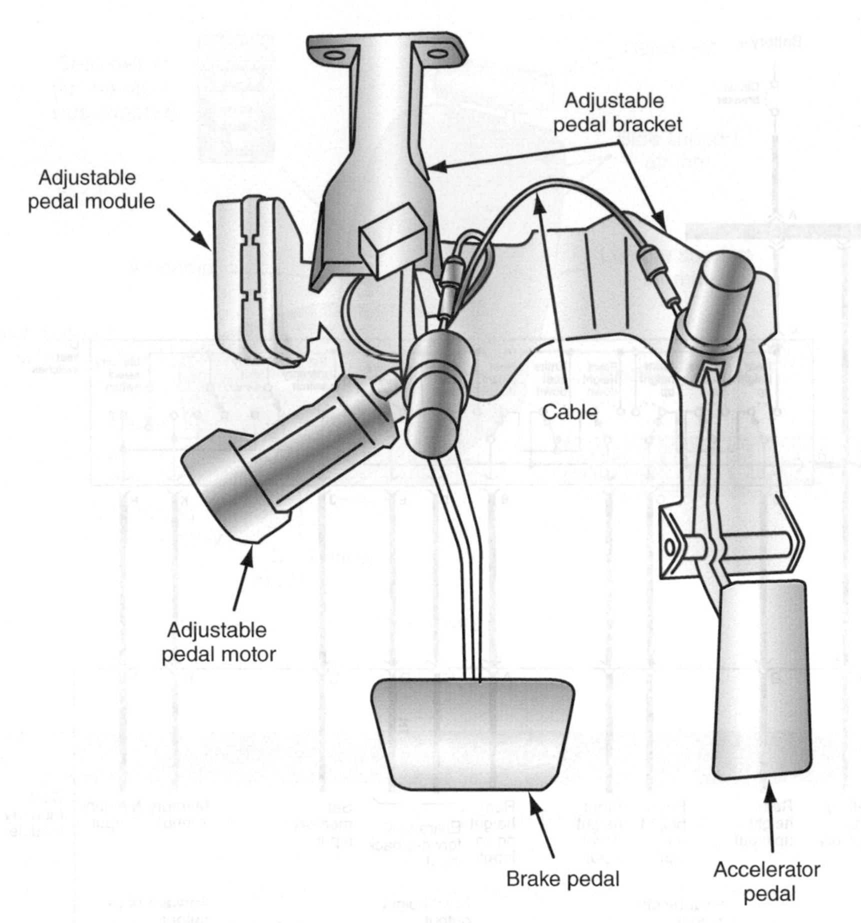

Figure 13-52 Electrically adjustable pedal assembly.

Climate-Controlled Seats

Shop Manual
Chapter 13, page 699

Climate-controlled seats provide additional comfort by heating and/or cooling the seat cushion and seat back. **Figure 13-53** is a schematic of a heated seat system that uses four heating element grids. Two elements are integral to each seat: one in the seat back and the other in the seat cushion. The heated seat module (HSM) contains the control logic and software for the heated seat system. Two heated seat switches control each heated seat individually. The cabin compartment node (CCN) is part of the instrument cluster on this vehicle and is the link between the heated seat switches and the HSM.

The heated seat system operates on battery current received through a fused ignition feed output. Using ignition feed limits operating the heated seat system to only when the ignition is in RUN. When either of the heated seat switches is depressed, the CCN receives a multiplexed resistance signal. The CCN broadcasts the requested heating level to the HSM over the data bus. Based on the heat level requested, the HSM controls the 12-volt output to the heating elements using HSDs. The HSDs use PWM to control the current flow through the seat elements. The carbon fiber–heated seat elements consist of multiple heating circuits wired in parallel. The heated seat elements are located between the leather trim cover and the seat cushion. As electrical current passes through the heated seat element, the resistance of the wire used in the element converts electrical energy into heat energy. The heat radiates through the seat cushion and seat back trim covers. The element grids are wired in parallel, so an open in one or more of the carbon fiber circuits will not prevent the other circuits from operating.

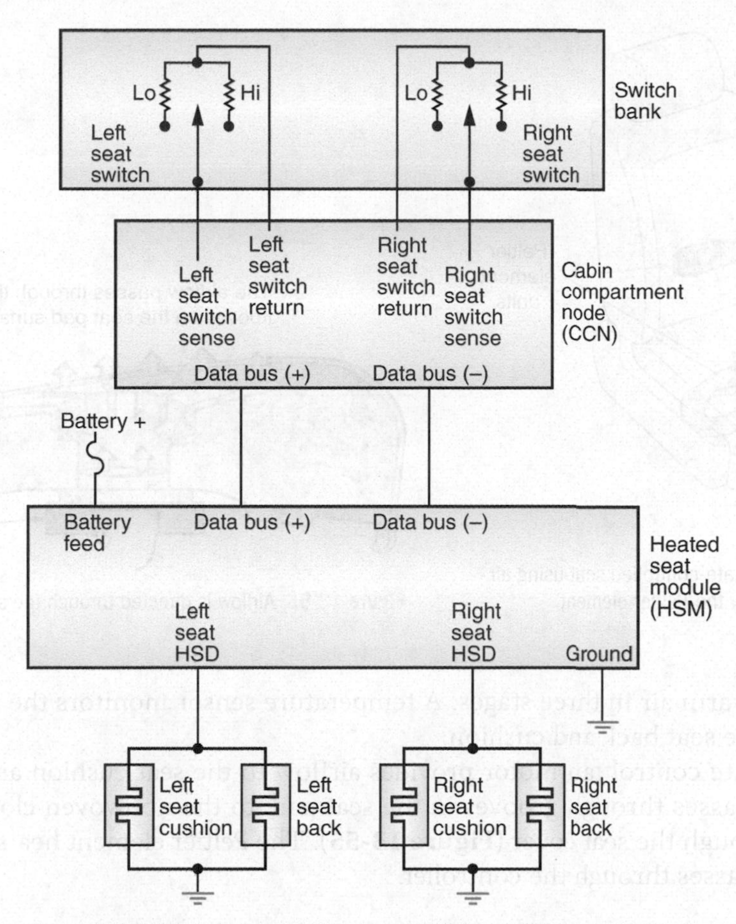

Figure 13-53 Heated seat system using grids in the cushion and seat back.

When the switch selection indicates a high-temperature heating request, the HSM will provide a boosted heat level using a 95% duty cycle during the first 4 minutes of operation. The heating output then drops to the normal high-temperature maintain level of a 30% duty cycle. This will maintain a temperature of about 107.6°F (42°C). The HSM will automatically switch to the low level after 2 hours of continuous operation. The low heat level uses a 15% duty cycle to hold the seat temperature to about 100.4°F (38°C). The HSM will turn off the low heat level operation after 2 hours.

The HSDs in the HSM monitor the operation of the heater element circuits. The HSM will turn off the heating elements if it detects an open or a short in the heating element circuit.

Some systems use a thermistor to measure the seat temperature. This system will energize the seat elements at a 100% duty cycle until it reaches the desired heat level. Once reached, the HSM will stop current flow. If the seat temperature drops a programmed amount, the HSM will re-energize the seat elements. This process repeats throughout the heated seat operation.

Some manufacturers use airflow to warm or cool the seat surface. A **Peltier element** operates similar to a bimetal switch. The element consists of two different types of metals, which are joined together. The joint area generates or absorbs heat when an electric current is applied to the element at a specified temperature. The Peltier element is integral to the climate controller (**Figure 13-54**). The climate controller cools or warms the airflow from the climate control fan motor based on the climate control electronic control unit (ECU) activation. The setting selection of the control switch determines the ECU output. The control switches provide for cool air in three stages,

Shop Manual
Chapter 13, page 701

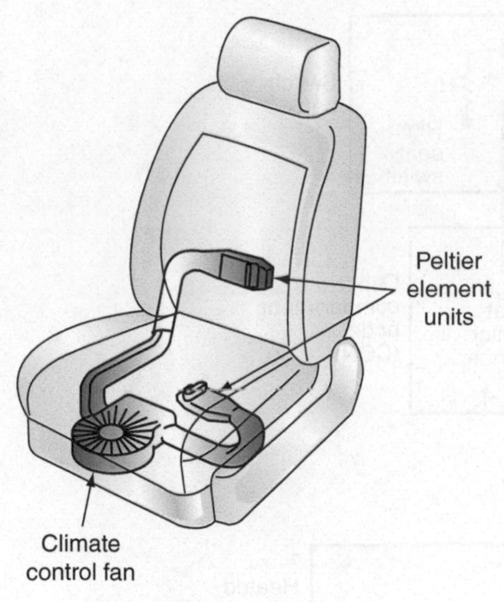

Figure 13-54 Climate-controlled seat using air heated or cooled by the Peltier element.

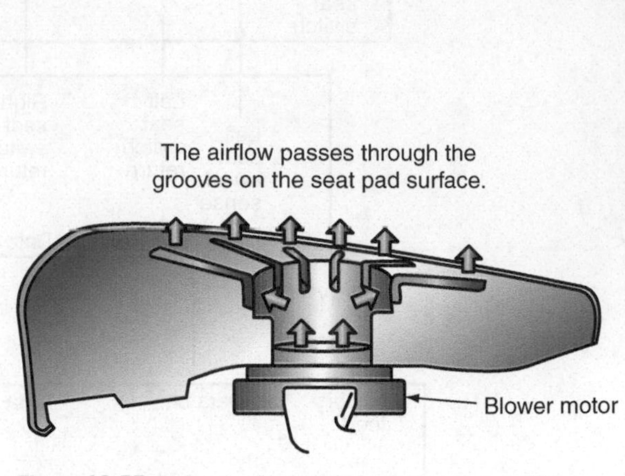

Figure 13-55 Airflow is directed through the seat cushion.

airflow, and warm air in three stages. A temperature sensor monitors the surface temperature of the seat back and cushion.

The climate control fan motor provides airflow to the seat cushion and seat back. The airflow passes through grooves in the seat pad, to the nonwoven cloth layer, and dissipates through the seat cover (**Figure 13-55**). The Peltier element heats or cools the airflow as it passes through the controller.

Power Door Locks

Shop Manual
Chapter 13, page 703

Electric power locks use either a solenoid or a permanent magnet reversible motor. Due to the high-current demands of solenoids, most modern vehicles use PM motors (**Figure 13-56**). Depending on circuit design, the system may incorporate a relay (**Figure 13-57**). The relay has two coils and two sets of contacts to control current direction. In this system, the door lock switch energizes one of the door lock relay coils to send battery voltage to the motor. **Figure 13-58** illustrates the current flow with the door lock switch in the LOCK position. **Figure 13-59** shows current flow with the door lock switch placed in the UNLOCK position.

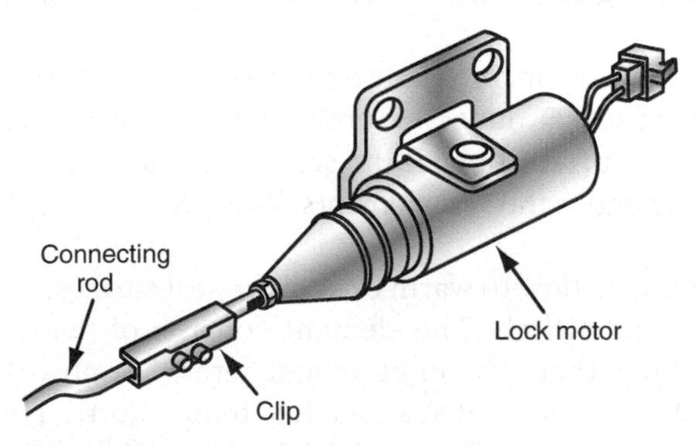

Figure 13-56 PM power door lock motor.

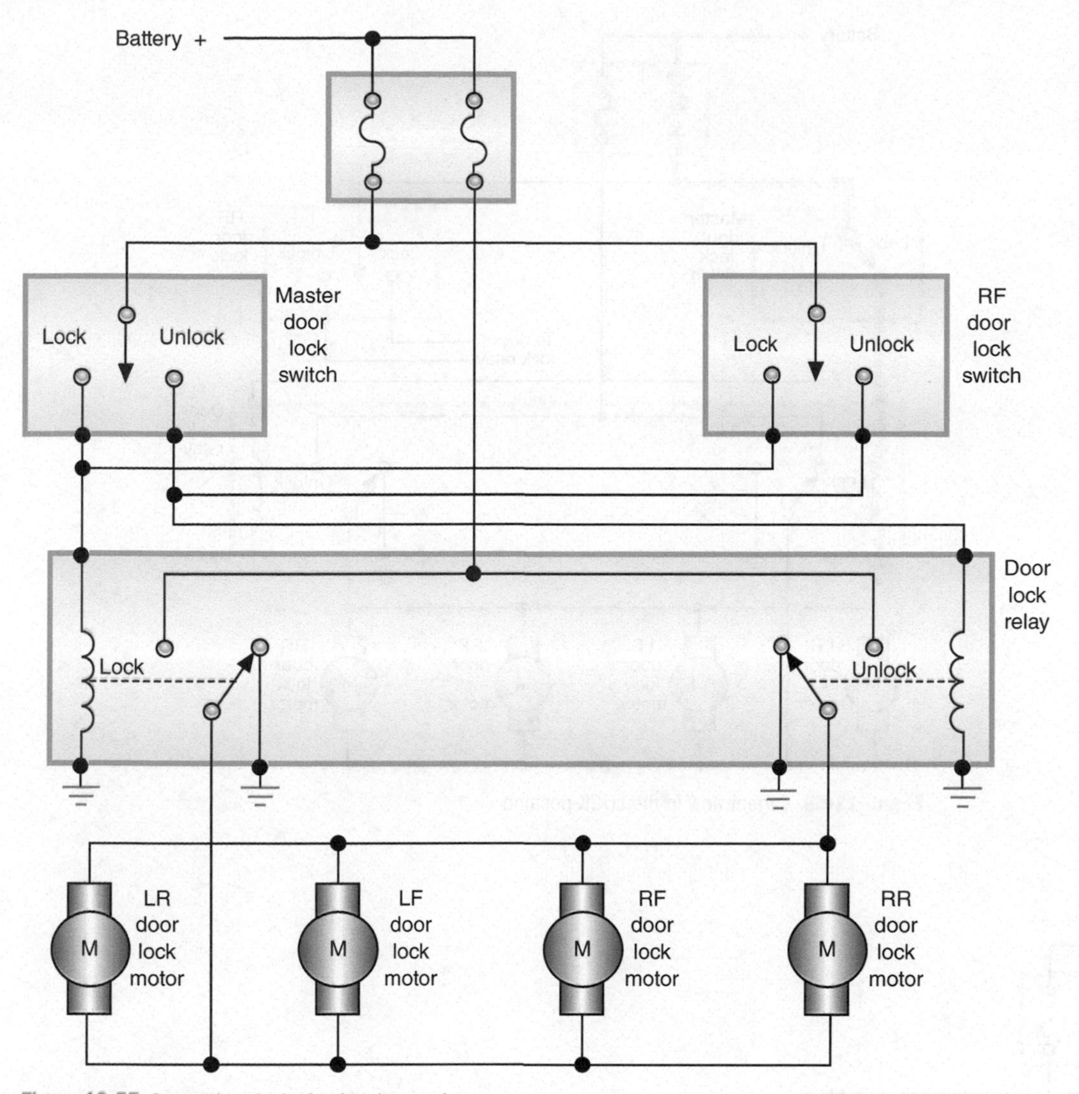

Figure 13-57 Power door lock circuit using a relay.

Systems that don't use relays use the switch to provide control of current flow in the same manner as the power seat or power window systems.

A **child safety latch** in the door lock system prevents opening the door from the inside, regardless of the position of the door lock knob. The child safety latch activates from a switch designed into the latch bellcrank (**Figure 13-60**). The door operates as normal when the latch is in the deactivated mode.

AUTHOR'S NOTE Like the power windows, the BCM or door modules can operate the power locks. In this case, the switch is nothing more than an input to the door module, and the door module directly controls the relays or the solenoid.

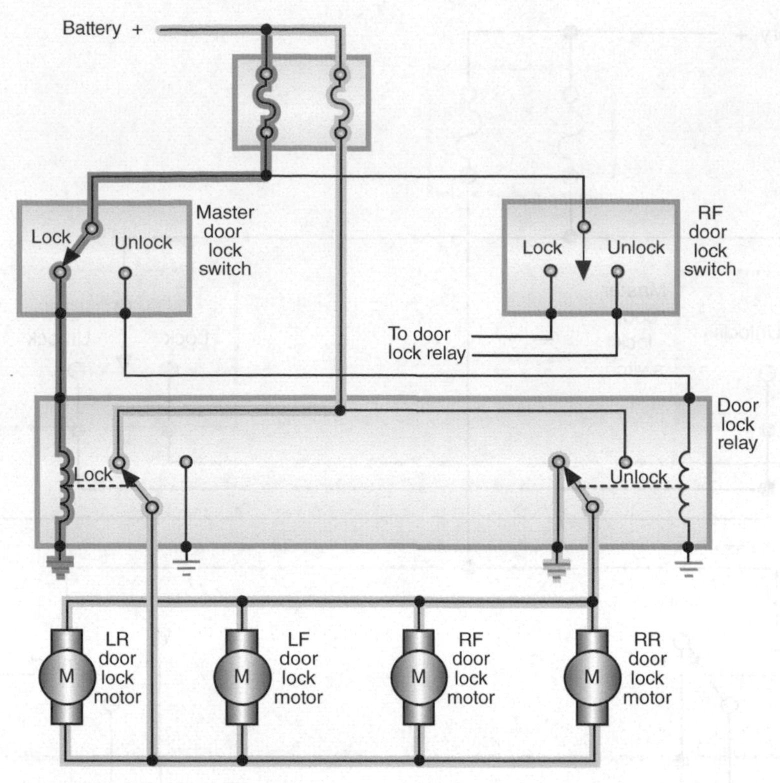

Figure 13-58 Current flow in the LOCK position.

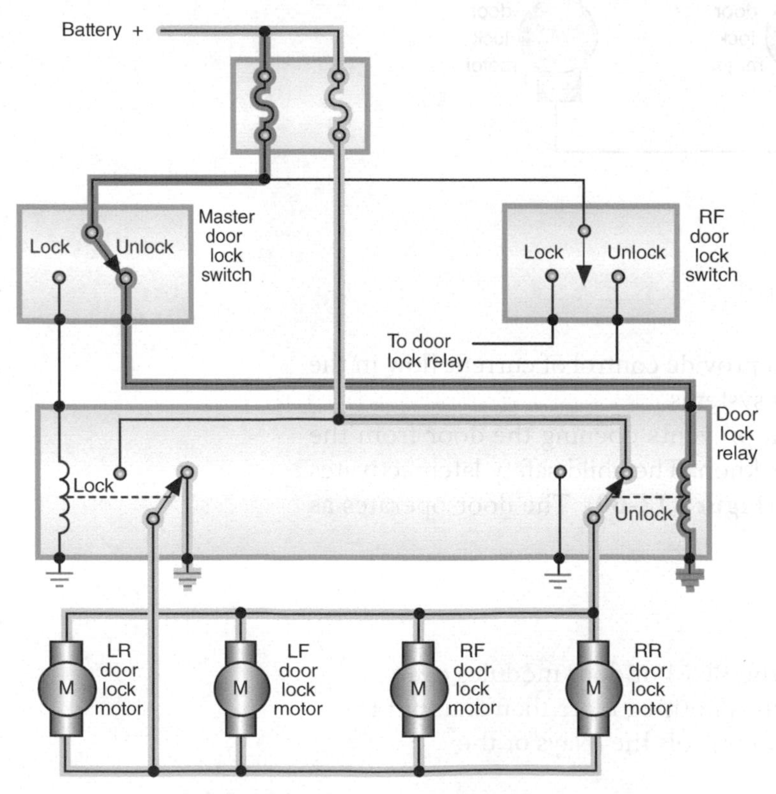

Figure 13-59 Current flow in the UNLOCK position.

Figure 13-60 Child safety latch.

Automatic Door Locks

Automatic door locks (ADL) is a passive system that locks all doors when the required conditions are met. The ADL system is an additional safety and convenience system that uses the existing power door function. Most systems lock the doors after shutting all doors, the ignition is in RUN, and the gear selector is in drive. Some systems lock the doors when the gear shift selector passes through the reverse position; others don't lock the doors unless vehicle speed is 8 mph (12.9 km/h) or faster.

Shop Manual
Chapter 13, page 703

The system may use the BCM to control the door lock relays (**Figure 13-61**) or a separate controller. The controller (or body computer) takes the place of the door lock switches for automatic operation.

When all of the door jamb switches are open (doors closed), the ground is removed from the door jamb input circuit to the controller (**Figure 13-62**). This signals the controller to enable the lock circuit. When the gear selection moves from the PARK position, the neutral safety switch removes the power signal from the controller. The controller sends voltage through the LH seat switch to the lock relay coil. Current flows through the motors to lock all doors.

Returning the gear selector to the PARK position applies voltage to the controller through the neutral safety switch. The controller then sends power to the unlock relay coil to reverse current flow through the motors. Inputs concerning the gear position can be bussed instead of hardwired.

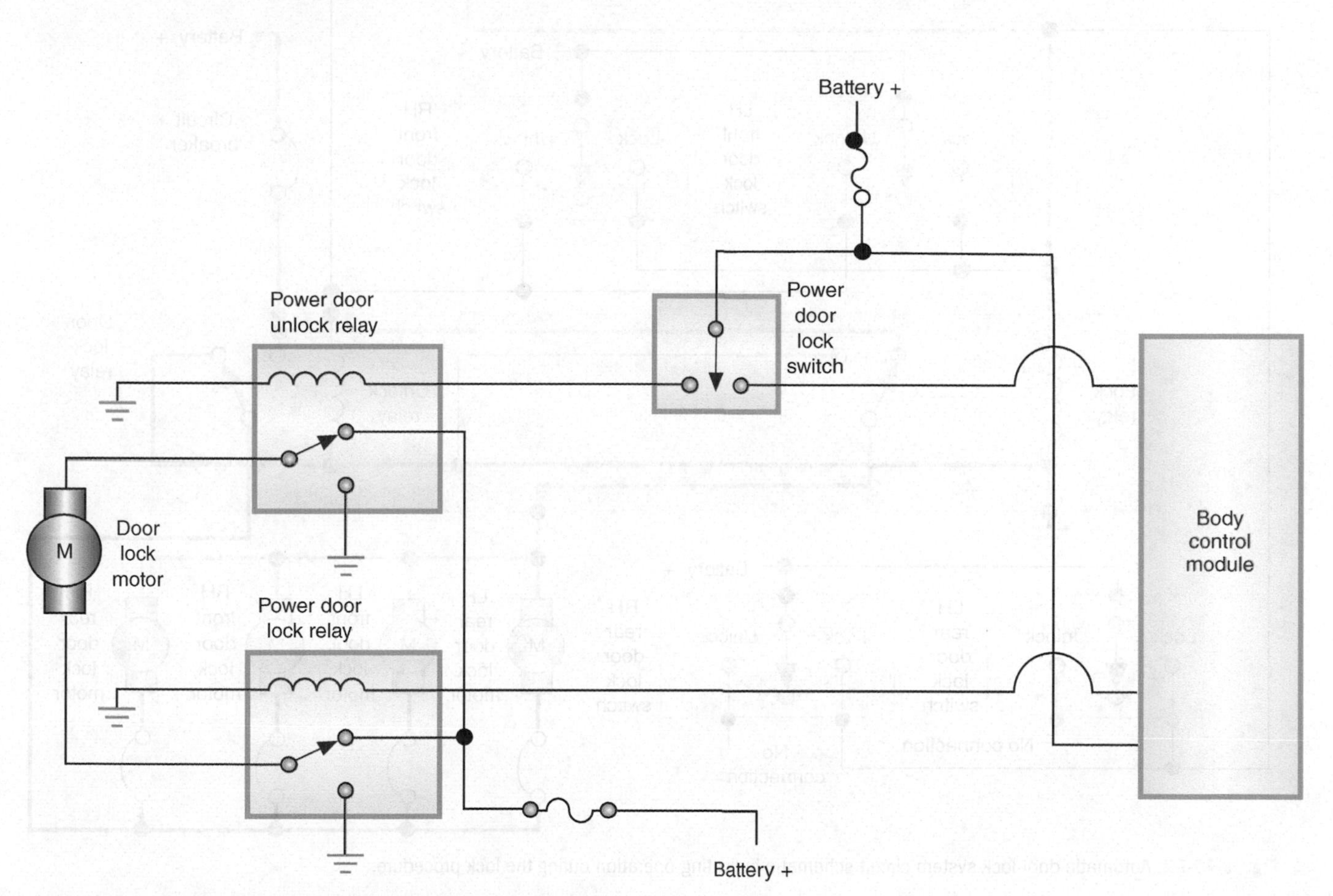

Figure 13-61 Automatic door lock system utilizing the body computer.

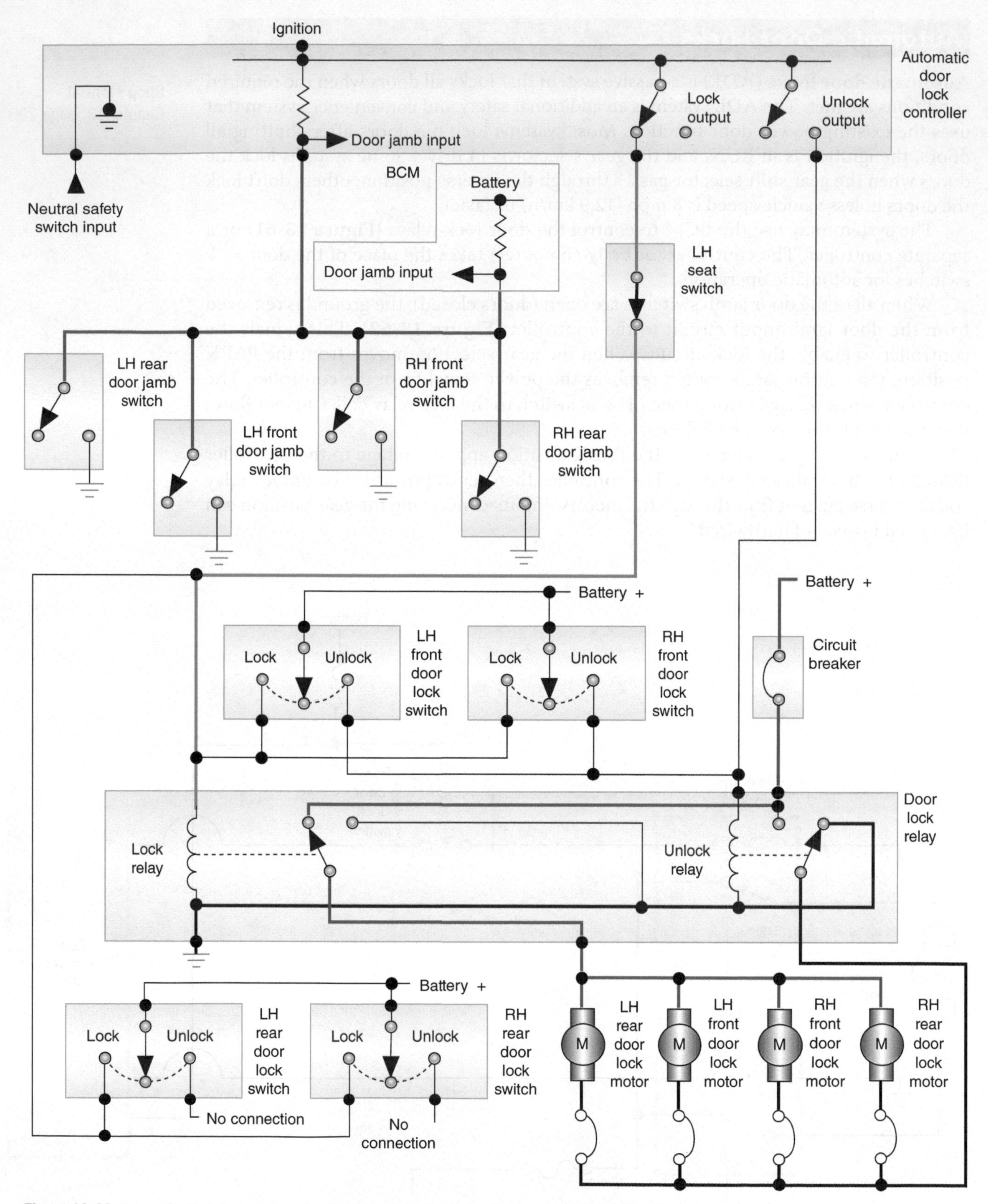

Figure 13-62 Automatic door lock system circuit schematic indicating operation during the lock procedure.

Keyless Entry

The **keyless entry** system provides a means for the driver to unlock the doors or the deck lid (trunk) from outside the vehicle without using a key. The main components of the keyless entry system are the control module, a coded-button keypad located on the driver's door, and the door lock motors.

The keypad consists of five normally open, single-pole, single-throw switches. Each switch represents two numbers: 1-2, 3-4, 5-6, 7-8, and 9-0 (**Figure 13-63**).

The keypad provides input to the control module. The control module is programmed to lock the doors when the 7-8 and 9-0 switches are closed at the same time. Using the keypad to enter a five-digit code unlocks the driver's door. The controller is programmed with the unlock code at the factory. However, the driver may enter a second code. Either code will operate the system.

In addition to the aforementioned functions, the keyless entry system also:

1. Unlocks all doors by pressing the 3–4 button within 5 seconds of entering the five-digit code.
2. Releases the deck lid lock if the 5–6 button is pressed within 5 seconds of code entry.
3. Activates the illuminated entry system if any of the buttons is pressed.
4. Operates in conjunction with the automatic door lock system and may share the same control module.

See the schematic (**Figure 13-64**) of the keyless entry system used by Ford. With the 7–8 and 9–0 buttons on the keypad pressed, the controller applies battery voltage to all motors through the lock switch.

Entering the correct five-digit code triggers the controller to close the driver's switch to apply voltage in the opposite direction to the driver's door motor. If the driver presses the 3–4 button, the controller applies reverse voltage to all motors to unlock the rest of the doors.

Antitheft Systems

A vehicle theft occurs in the United States every 26 seconds. In response to this problem, vehicle manufacturers offer various levels of antitheft systems. These systems are deterrents designed to scare off would-be thieves by sounding alarms and/or disabling

Shop Manual
Chapter 13, page 704

Figure 13-63 Keyless entry system keypad.

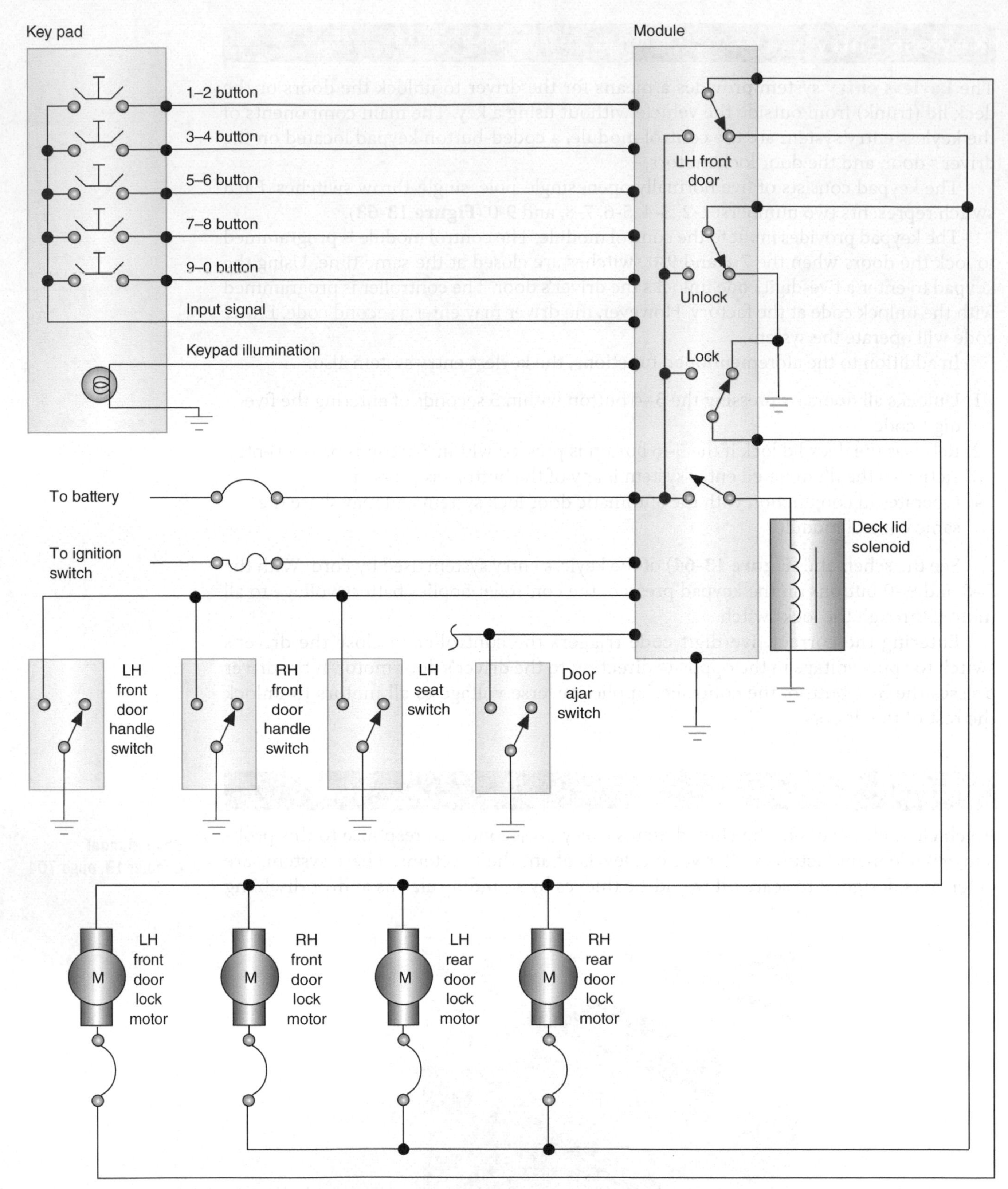

Figure 13-64 Simplified keyless entry system schematic.

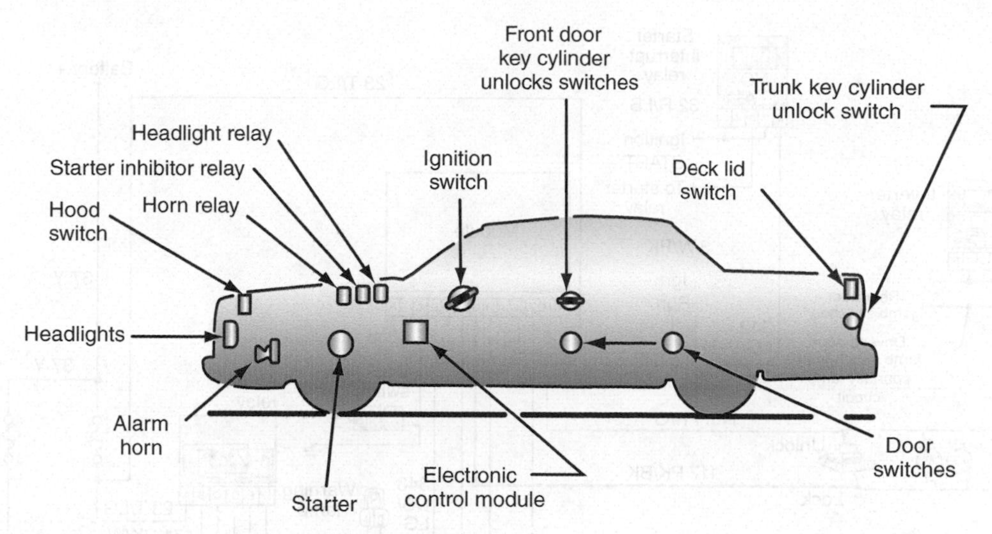

Figure 13-65 Typical components of an antitheft system.

the ignition system. **Figure 13-65** shows many of the common components used in an antitheft system. These components include the following:

1. An electronic control module.
2. Door switches at all doors.
3. Trunk key cylinder switch.
4. Hood switch.
5. Starter inhibitor relay.
6. Horn relay.
7. Alarm.

In addition, many systems incorporate the exterior lights into the system. Triggering the system flashes the lights.

For the system to operate, it must first be **armed**. Most systems arm when the ignition is off and the doors are locked using the electric door lock feature. When the driver's door is shut, a security light will illuminate for approximately 30 seconds to indicate that the system is armed and ready to function. If any other door is open, the system will not arm until it's closed. Once armed, the system is ready to detect an illegal entry.

The control module monitors the switches and activates the alarms if they indicate opening of the doors or trunk, or the rotating of the key cylinders. The control module will sound the alarm and flash the lights until the timer circuit has counted down. At the end of the timer function, the system will automatically rearm itself.

Some systems use ultrasonic sensors that signal the control module if someone attempts to enter the vehicle through the door or window. The sensors can monitor the parameter of the vehicle and sound the alarm if someone enters within the protected parameter distance.

The system can also use current-sensitive sensors that will activate the alarm if there is a change in the vehicle's electrical system. The change can occur if the courtesy lights come on or if there is an attempt to start the engine.

The following systems are presented as sample types of antitheft systems used. **Figure 13-66** illustrates an antitheft system that uses a separate control module and an inverter relay. When triggered, the system will sound the horn. It also flashes the low-beam headlights, the taillights, and the parking lamps. The system may also disable the ignition system.

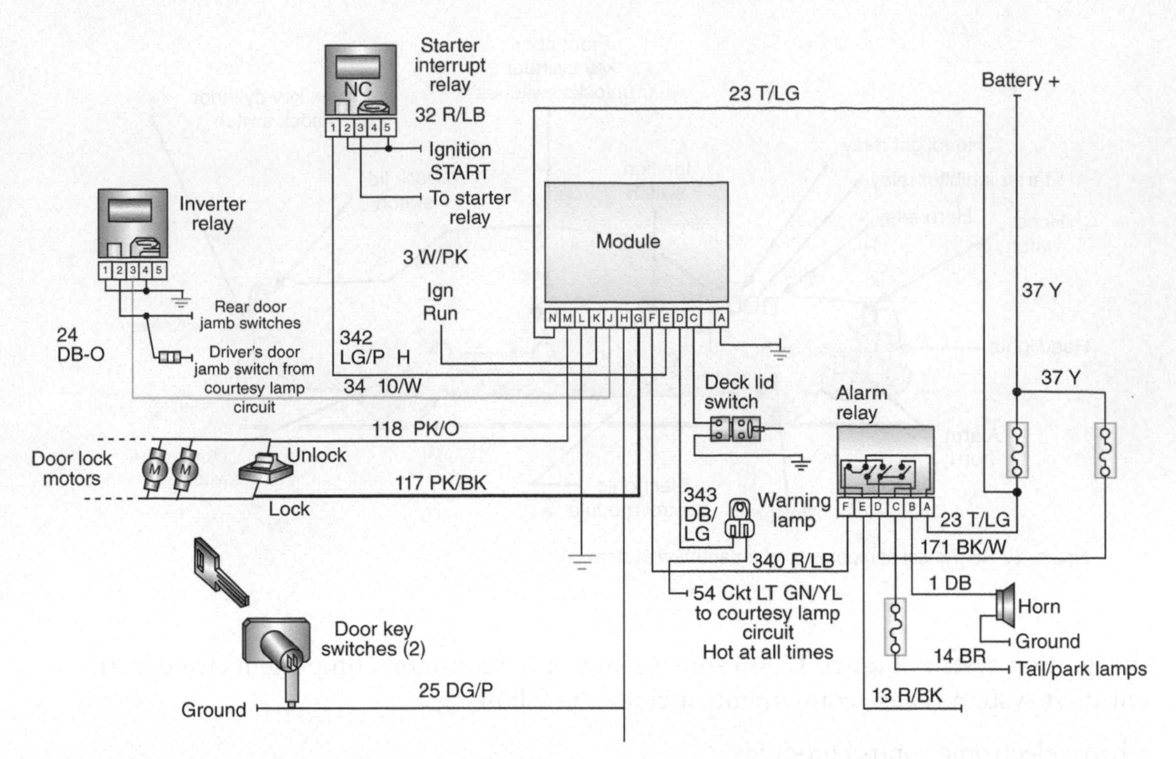

Figure 13-66 Circuit schematic of Ford's antitheft system.

Turning off the ignition starts the arming process by removing voltage to terminal K of the module. Opening the door applies voltage to the courtesy lamp circuit through the closed switch to terminal 2 of the inverter relay. This voltage energizes the inverter relay and provides a ground for module terminal J. The control module uses this signal to provide an alternating ground at terminal D, causing the indicator lamp to blink to alert the driver that the system is not armed. Placing the door lock switch in the LOCK position applies battery voltage to terminal G of the module. The module uses this signal to apply a steady ground at terminal D, causing the indicator light to stay on continuously. Closing the door electrically opens the door switch and de-energizes the inverter relay coil. Terminal J no longer has a ground and the indicator light goes out after a couple of seconds.

With the system armed, the control module triggers the alarm if terminals J and C receive a ground signal. Terminal C is grounded if the trunk tamper switch contacts close. Terminal J is grounded when the inverter relay contacts are closed. The door jamb switches control the inverter relay. Opening a door closes the switch and energizes the inverter relay's coil. The contacts close and provide ground for terminal J.

When the alarm is triggered, terminal F of the module has a pulsating ground. This pulsating ground energizes and de-energizes the alarm relay. As the relay contacts open and close, the horns and exterior lights receive a pulsating voltage.

At the same time, the start interrupt circuit is activated. The start interrupt relay receives battery voltage from the ignition switch when it's in the START position. When triggered, the module provides ground through terminal E, energizing the relay coil. The energized relay opens the circuit to the starter system, preventing starter operation.

Disarming the system requires unlocking the door(s) using a key, inputting the correct code into the keyless entry keypad at one of the front doors, or pressing the unlock button on the FOB. Unlocking the door closes the lock cylinder switch and provides ground for terminal H of the module.

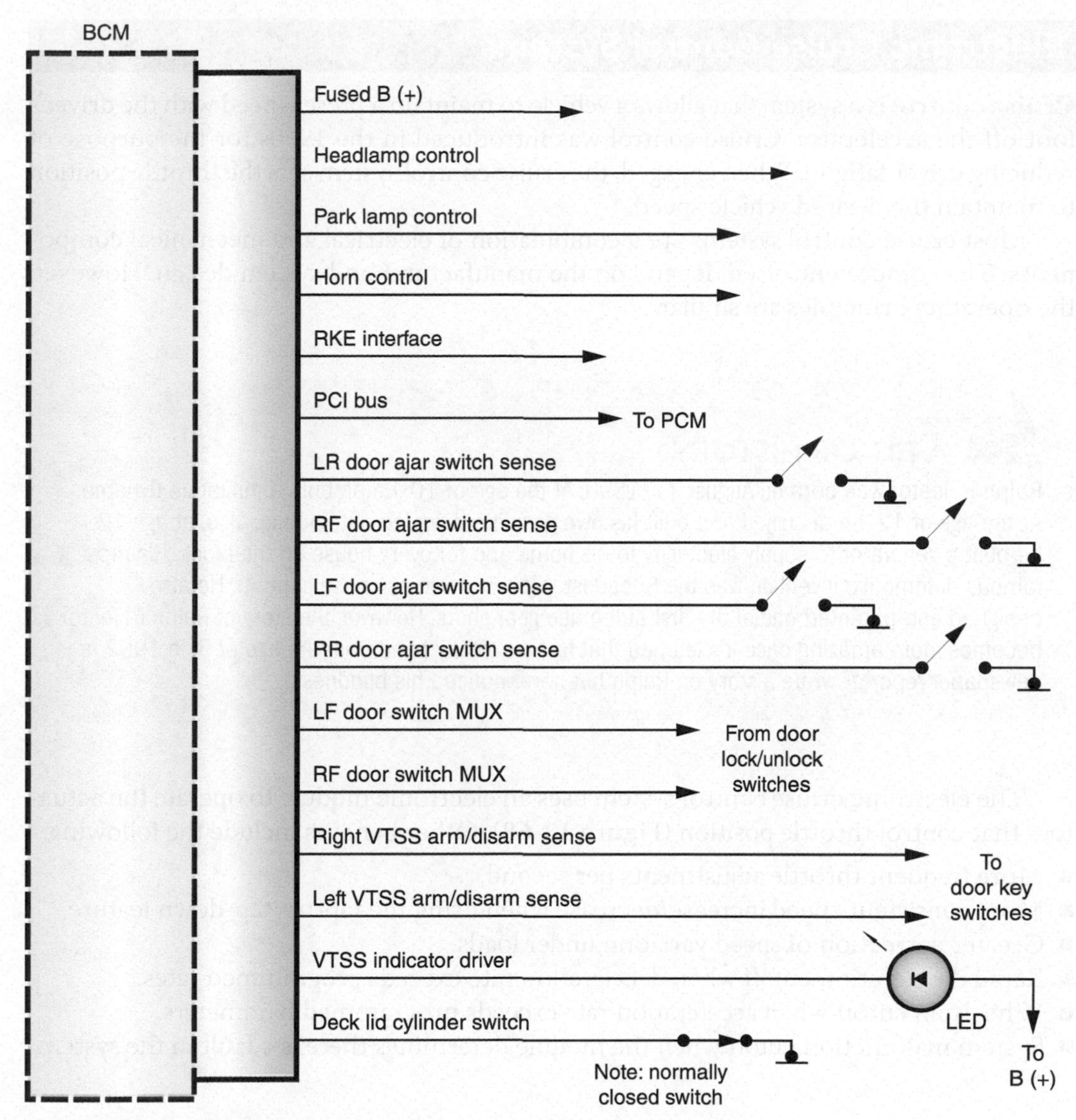

Figure 13-67 BCM-controlled antitheft system.

Often the BCM controls the functions of the antitheft system. In this case, the BCM receives the inputs and outputs (**Figure 13-67**). The BCM monitors the arming process and then triggers the alarm if there is an unauthorized entry. At this time, the BCM controls the exterior lamps to cause them to flash and cycles the horn on and off. The BCM also sends a data bus message to the powertrain control module (PCM) not to start the engine. The PCM programming requires that it must receive a data bus message that it's alright to start. If the data bus circuit should fail, the engine may not start.

Systems incorporating an intrusion monitor detect the movement of a person or an object inside the passenger compartment. The intrusion monitor uses a single sensor that transmits 40-kHz ultrasonic sound waves. The sensor also receives the ultrasonic sound waves. Movement of an object within the sensor's coverage area distorts the received ultrasonic signals. The module detects and processes this distortion, which then triggers the antitheft system.

AUTHOR'S NOTE An enhancement of the antitheft system includes the immobilizer system. This system is discussed in Chapter 14 since it uses radio frequencies.

Shop Manual
Chapter 13, page 708

Electromechanical cruise control receives its name from two subsystems: the electrical and the mechanical portion.

Electronic Cruise Control Systems

Cruise control is a system that allows a vehicle to maintain a preset speed with the driver's foot off the accelerator. Cruise control was introduced in the 1960s for the purpose of reducing driver fatigue. When engaged, the cruise control system sets the throttle position to maintain the desired vehicle speed.

Most cruise control systems are a combination of electrical and mechanical components. The components used depend on the manufacturer and system design. However, the operating principles are similar.

A BIT OF HISTORY

Ralph R. Teetor was born on August 17, 1890. At the age of 10, Ralph built a miniature dynamo. At the age of 12, he designed and built his own gasoline-powered automobile. Also, at age 12, he built a generator to supply electricity to his home and for every house on the block. His most famous automotive invention was the Speedostat (now known as cruise control). He also designed and patented one of the first automatic gear shifts. However, the story of Ralph R. Teetor becomes more amazing once it's learned that he was totally blind from the age of 5. In 1902, a newspaper reporter wrote a story on Ralph but never noticed his blindness.

The electronic cruise control system uses an electronic module to operate the actuators that control throttle position (**Figure 13-68**). Other benefits include the following:

Some manufacturers combine the transducer and servo into one unit. They usually refer to this unit as a servomotor.

- More frequent throttle adjustments per second.
- More consistent speed increase/decrease when using the tap-up/tap-down feature.
- Greater correction of speed variation under loads.
- Rapid deceleration cutoff when deceleration rate exceeds programmed rates.
- Wheelspin cutoff when acceleration rate exceeds programmed parameters.
- System malfunction cutoff when the module determines there is a fault in the system.

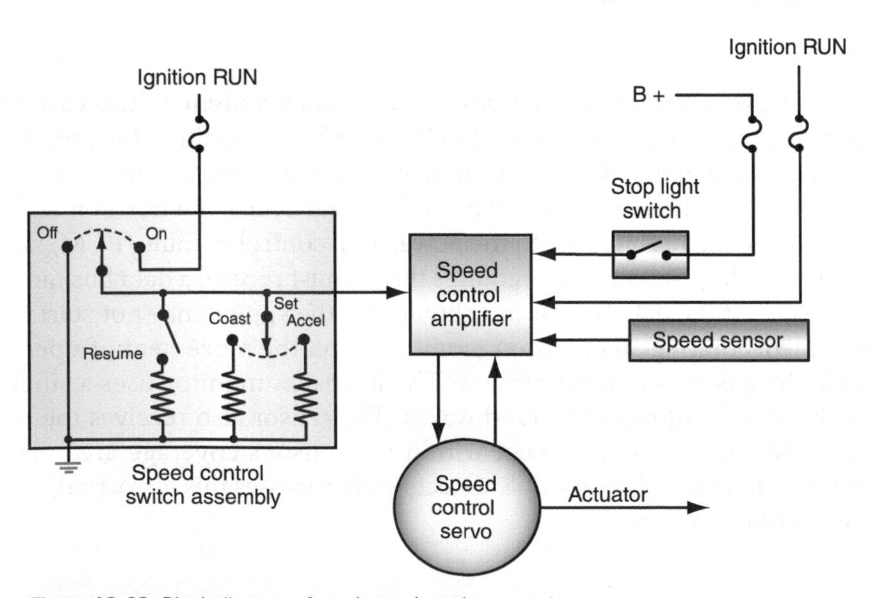

Figure 13-68 Block diagram of an electronic cruise control system.

Common Components

Common components of the electronic cruise control system include the following:

Shop Manual
Chapter 13, page 711

1. The control module: The module can be a separate cruise control module, the PCM, or the BCM. The operation of the systems is similar regardless of the module used.
2. The control switch (**Figure 13-69**): Depending on system design, the control switch contacts apply the ground circuit through resistors. Because each resistor has a different value, the control module receives a different voltage level for each button. In some systems, the control switch will send a 12-volt signal to different control module terminals.
3. The brake or clutch switch.
4. Vacuum release switch.
5. Servo: The servo operates on a vacuum controlled by supply and vent valves. These operate from controller signals to solenoids.

Depending on system design, the sensors used as inputs to the control module include the vehicle speed sensor, servo position sensor, and throttle position sensor. Other inputs include the brake switch, instrument panel switch, control switch, and the park-neutral switch.

The control module receives signals from the speed sensor and the control switch. When the vehicle's speed is fast enough to allow cruise control operation, and the driver pushes the SET button on the control switch, an electrical signal is sent to the controller, and the voltage level is set in its memory. This signal is the base value for the creation of two additional signals with values set at 1/4 mph (0.4 km/h) above and below the set speed. The module uses the comparator values to change vacuum levels at the servo to maintain the set vehicle speed.

The control module employs three safety modes:

1. Rapid deceleration cutoff: If the module determines that the deceleration rate is greater than programmed values, it disengages the cruise control system and returns operation back over to the driver.
2. Wheelspin cutoff: If the control module determines that the acceleration rate exceeds programmed values, it will disengage the system.
3. System malfunction cutoff: The module checks the operation of the switches and circuits. If it determines there is a fault, it will disable the system.

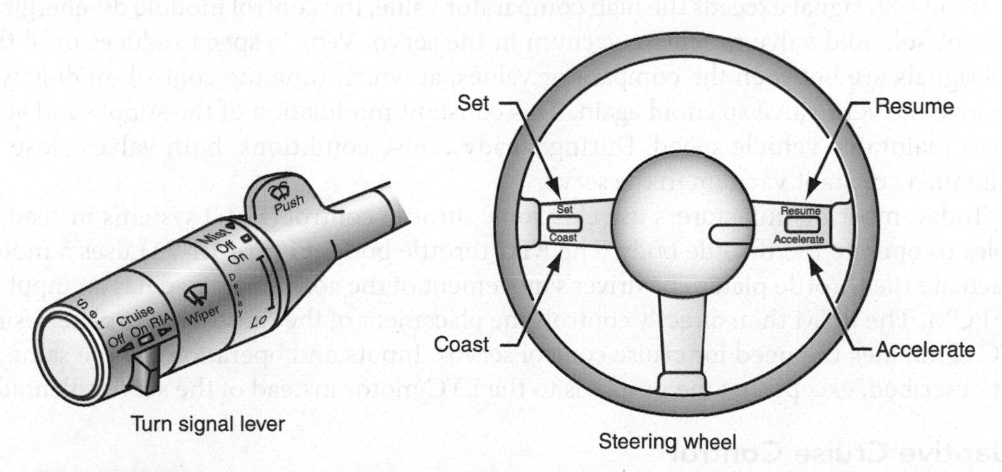

Figure 13-69 The control switch(s) can be mounted on the turn signal stock or into the steering wheel. The switch provides driver inputs for the system.

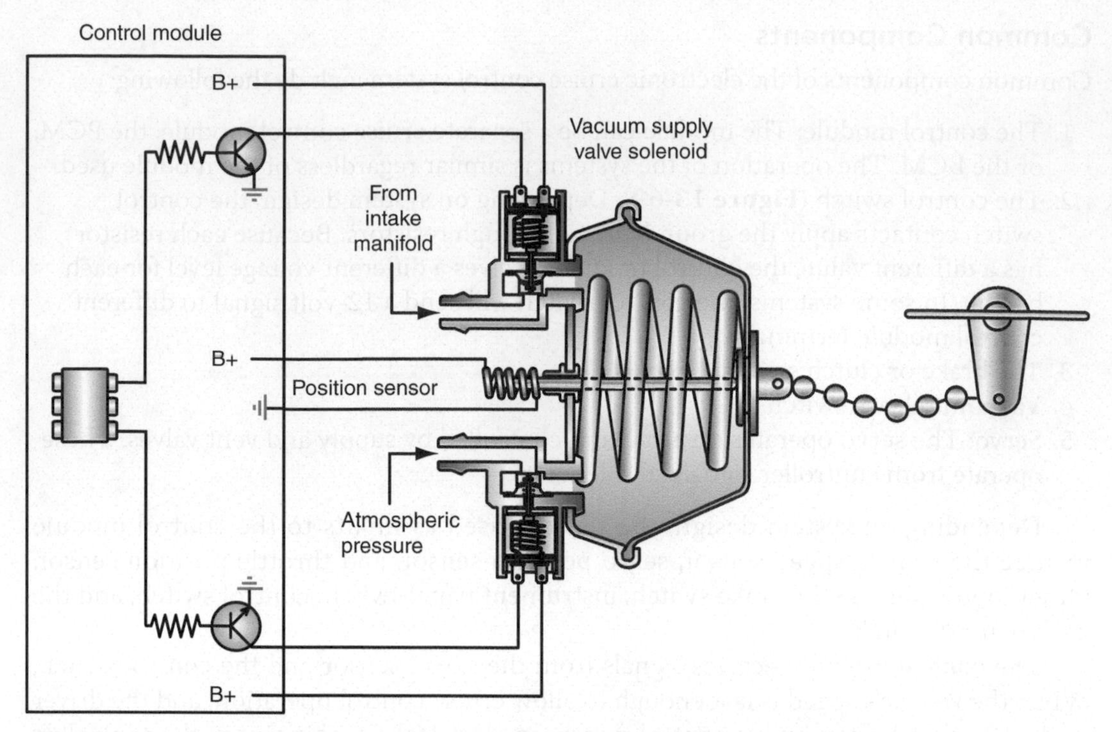

Figure 13-70 Servo valve operation in the electronic control system. The servo position sensor informs the controller of servo operation and position.

The vacuum-modulated servo is the primary actuator. A supply and a vent solenoid control vacuum to the servo. The vent valve is normally open, and the supply valve is normally closed (**Figure 13-70**). The controller energizes the supply and vent valves to allow manifold vacuum or atmospheric pressure to enter the servo. The servo uses the vacuum and pressure to move the throttle to maintain the set speed. Balancing the vacuum in the servo maintains the set vehicle speed. The vacuum used to move the servo may be an engine vacuum or may come from a vacuum pump.

If the voltage signal from the VSS drops below the low comparator value, the control module energizes the supply valve solenoid to allow more vacuum into the servo and increases the throttle opening. When the VSS signal returns to a value within the comparator levels, the supply valve solenoid is de-energized.

If the VSS signal exceeds the high comparator value, the control module de-energizes the vent solenoid valve to release vacuum in the servo. Vehicle speed reduces until the VSS signals are between the comparator values, at which time the control module will energize the vent valve solenoid again. This constant modulation of the supply and vent valves maintains vehicle speed. During steady cruise conditions, both valves close to maintain a constant vacuum in the servo.

Today, most manufacturers use electronic throttle control (ETC) systems instead of cables to operate the throttle body. The ETC throttle body (**Figure 13-71**) uses a motor to actuate the throttle plate. The driver's movement of the accelerator pedal is an input to the PCM. The PCM then directly controls the placement of the throttle body plate. Using ETC eliminates the need for cruise control servos. Inputs and operation are the same as just described, except that the output is to the ETC motor instead of the servo solenoids.

Adaptive Cruise Control

Shop Manual
Chapter 13, page 713

The **adaptive cruise control (ACC)** system developed by TRW Automotive is similar to conventional cruise control systems, with the added function of adjusting and maintaining appropriate following distances between vehicles. This system uses laser or radar

Figure 13-71 An electronic throttle control throttle body.

sensors to determine the vehicle-to-vehicle distances and relational speeds. Like other cruise control systems, the adaptive cruise control system maintains a fixed speed that the driver has set. However, the system also allows the driver to set a distance between their vehicle and any vehicle in front of them in the same lane. If the distance to the vehicle ahead is closing, the ETC will close the throttle plate to decelerate the vehicle. If additional deceleration is necessary, the transmission shifts into a lower gear. If further deceleration is still necessary, the system controls the brake actuator to apply the brakes.

The system maintains the set distance between vehicles until the vehicle in front is no longer there (due to lane change). At this time the vehicle accelerates slowly to regain the set vehicle speed, and then the system maintains the fixed speed.

A steering wheel–mounted switch provides the driver with a selection of three following distances. The first button push sets the system to long, which is approximately 245 feet (75 meters) between vehicles. The second button push selects a middle range of approximately 165 feet (50 meters). A third button push sets the distance range to short, which is approximately 100 feet (30 meters).

Laser is an acronym for light amplification by stimulated emission of radiation.

The laser or radar sensor are typically mounted behind the front grill (**Figure 13-72**). Radar is a system that uses electromagnetic waves to identify an object's range, direction, or speed. Radar emits a radio wave reference signal and measures the time it takes to

Figure 13-72 The laser or radar sensor is usually located behind the grill or in the front fascia.

echo back. When the waves hit an object, they reflect back to the receiver. Consider that radio waves travel at the speed of light (186,000 miles per second or 300,000,000 meters per second) and that the waves must travel to the object and echo back to the receiver; distance is calculated by measuring the length of the pulse multiplied by the speed of light, divided by two.

A laser is a device that controls the way that energized atoms release photons (light). Laser light has the following properties:

- Monochromatic—The light is only one color since it contains one specific wavelength of light.
- Coherent—The waves have the same wavelength and a fixed-phase relationship.
- Directional—The laser light produces a very tight beam.

The laser light uses stimulated emission to accomplish these properties. Stimulated emission results in the laser light atoms releasing photons in a very organized state. A typical laser emits light in a narrow, low-divergence beam with a well-defined wavelength.

Laser distance sensors function on the same basic principle as the ultrasonic and radar distance sensors. They use a "time-of-flight" measurement that compares the outgoing and returning light waveform signals to determine the distance to an object. An internal clock measures the time the laser light waveform takes to be transmitted and then returned.

The laser sensor has three parts (**Figure 13-73**): the laser emitter that radiates laser rays forward, the laser-receiving portion that receives the laser beams as they are reflected back by the vehicle that is ahead, and the processing unit that determines the length of time it takes for the reflected beams to return to the sensor; the unit calculates the distance to the vehicle ahead and the relative speed. This data is then transmitted to the distance control ECU (**Figure 13-74**).

Regardless of the sensor type, a control module (either a separate distance ECU or the PCM) receives the distance and relative speed information. The control module determines which vehicle to follow based on the information provided by the sensor. The control module also calculates the target acceleration signals for following the vehicle. The control module sends acceleration or deceleration requests to maintain the set distance. If necessary, it will also request a transmission downshift and brake application.

If brake control is necessary, the distance control module will determine a target deceleration rate. The maximum rate will be set at 0.30 G. Current vehicle speed, the distance to the vehicle in front, and the relative speed determine the target deceleration rate. The antilock brake system (ABS) control module receives the brake apply request

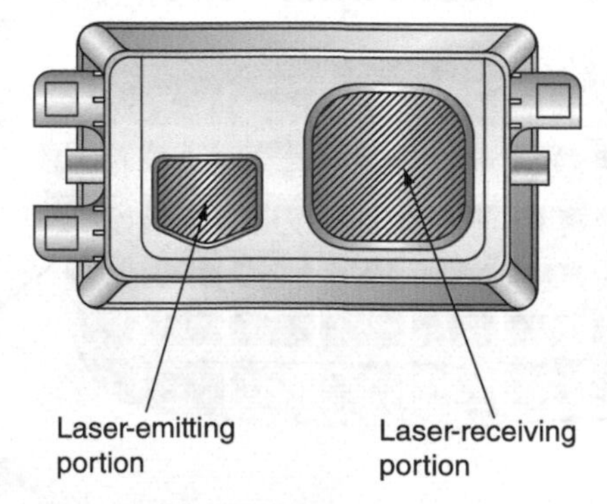

Figure 13-73 Components of the laser sensor.

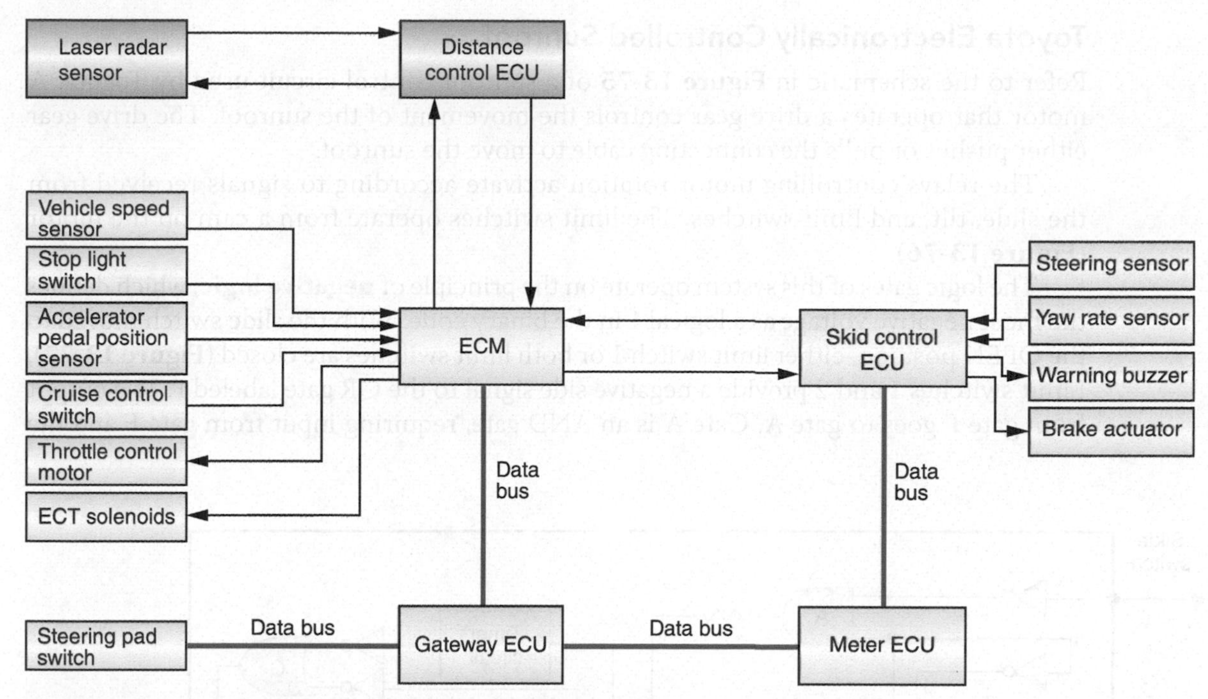

Figure 13-74 Block diagram of the adaptive cruise control system.

signal and controls the brake actuator to apply the brakes. If the vehicle does not decelerate at a rate fast enough to avoid a collision, a warning buzzer sounds to prompt the driver to apply the brakes.

If a slower-moving vehicle is in front, the ACC system enters follow mode, matches the speed of the vehicle ahead, and maintains the set distance. While in follow mode, if either vehicle changes lanes the system requests an acceleration mode. To regain the set speed, the throttle plate opens to gradually increase vehicle speed. At this time, the system enters fixed speed mode and maintains the set speed.

Speed Sign Recognition

Using the ACC system fitted with a camera, the speed sign recognition system reads the speed limit signs as the vehicle passes them. The posted speed limit is communicated to the ACC control module. If the current set speed is greater than the posted speed limit, the ACC system reduces the speed of the vehicle to match the posted speed limit. If the speed limit increases again to the previously set speed of the ACC, the system will resume the new speed, but not greater than the new speed limit.

When the vehicle is slowing down due to a slower speed limit, a speed limit icon is illuminated in the instrument cluster to notify the driver. The icon also displays the current speed limit.

The driver can turn the sign recognition feature on or off as they desire. In addition, they can set a speed to maintain that is above or below the posted speed limit.

Electronic Sunroof Concepts

Many manufacturers have introduced electronic control of their electric sunroofs. These systems incorporate a pair of relay circuits and a timer function into the control module. Although there are variations between manufacturers, the systems discussed here provide a study of the two basic types of systems.

Shop Manual
Chapter 13, page 719

Toyota Electronically Controlled Sunroof

Refer to the schematic in **Figure 13-75** of a sunroof control circuit used by Toyota. A motor that operates a drive gear controls the movement of the sunroof. The drive gear either pushes or pulls the connecting cable to move the sunroof.

The relays controlling motor rotation activate according to signals received from the slide, tilt, and limit switches. The limit switches operate from a cam on the motor (**Figure 13-76**).

The logic gates of this system operate on the principle of **negative logic**, which defines the most negative voltage as a logical 1 in the binary code. With the slide switch moved to the OPEN position, either limit switch 1 or both limit switches are closed (**Figure 13-77**). Limit switches 1 and 2 provide a negative side signal to the OR gate labeled F. The output from gate F goes to gate A. Gate A is an AND gate, requiring input from gate F and the

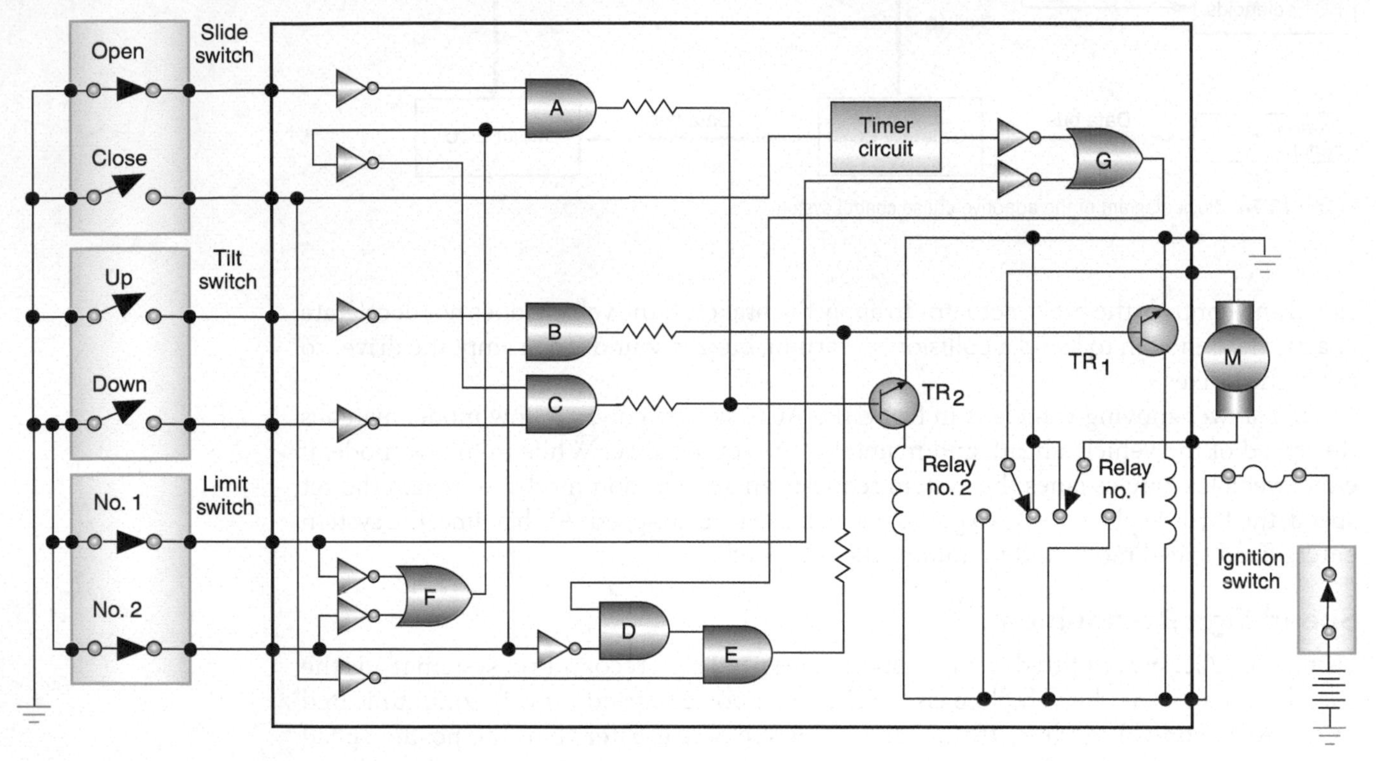

Figure 13-75 Toyota sunroof circuit using electronic controls.

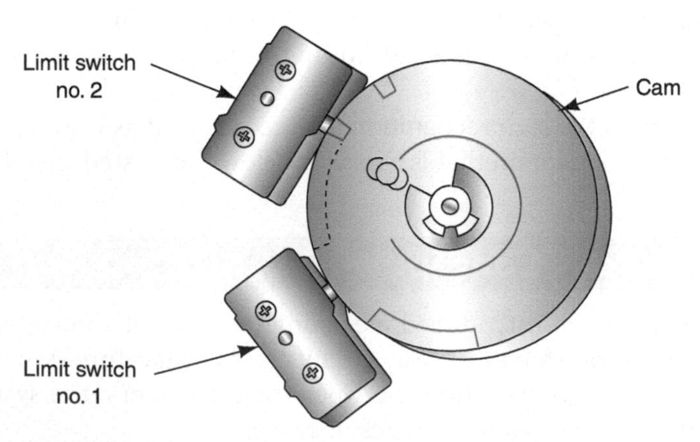

Figure 13-76 The limit switches operate off a cam on the motor.

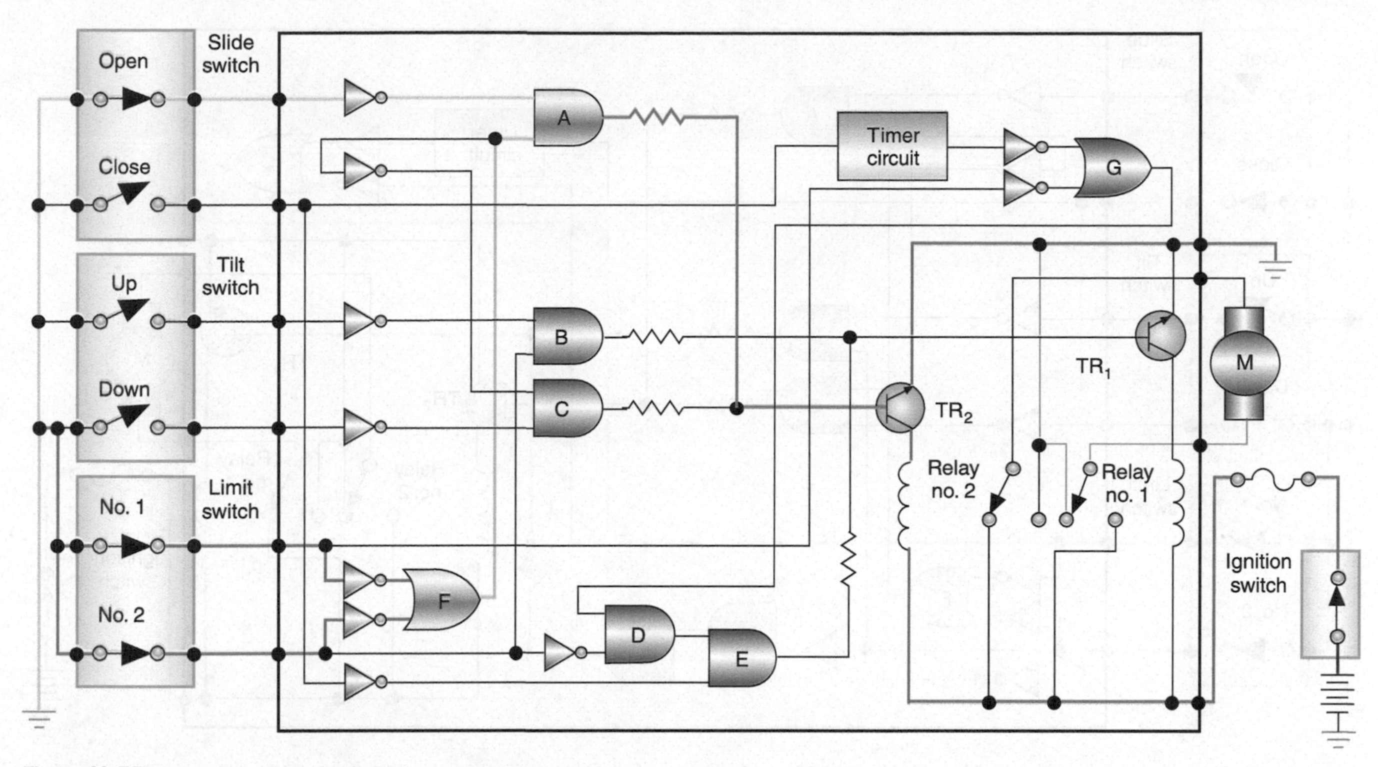

Figure 13-77 Circuit operation with the switch in the OPEN position.

open slide switch. The output signal from gate A turns on TR_2, providing a ground path for the coil in relay 2. Relay 2 applies battery voltage to the motor; the ground path is through the de-energized relay 1. Current is sent to the motor as long as the OPEN switch is depressed. Holding the OPEN switch in this position for too long disengages a clutch in the motor that unlocks the motor from the drive gear.

AUTHOR'S NOTE The schematics used to explain the operation of the Toyota sunroof use logic gates. If needed, refer to Chapter 11 of this manual to review the operation of the gates.

How the system operates during closing depends on how far the sunroof is open. Placing the slide contact to the CLOSE position with the sunroof open more than 7.5 inches (19 cm) sends an input signal to gate E (**Figure 13-78**). The limit switches supply the other input signals required at gate E. The limit switch 1 signal passes through the OR gate G to the AND gate D. Limit switch 2 provides the second signal required by gate D. The output signal from D is the second input signal required by gate E. The output signal from E turns on TR_1. This energizes relay 1 and reverses the current flow through the motor. The motor will operate until the slide switch or limit switch 2 opens.

If the sunroof is open less than 7.5 inches (19 cm) when placing the slide switch in the CLOSE position, the timer circuit is activated (**Figure 13-79**). The CLOSE switch signals the timer and provides an input signal to gate E. Limit switch 1 is open when the sunroof is open less than 7.5 inches. The timer provides the second input signal required by gate D. The timer is activated for 0.5 second. This turns on TR_1 and operates the motor for 0.5 second, or long enough for rotation of the motor to close limit switch 1. When limit switch 1 is closed, the operation is the same as described when the sunroof is closed after it's more than 7.5 inches (19 cm) open.

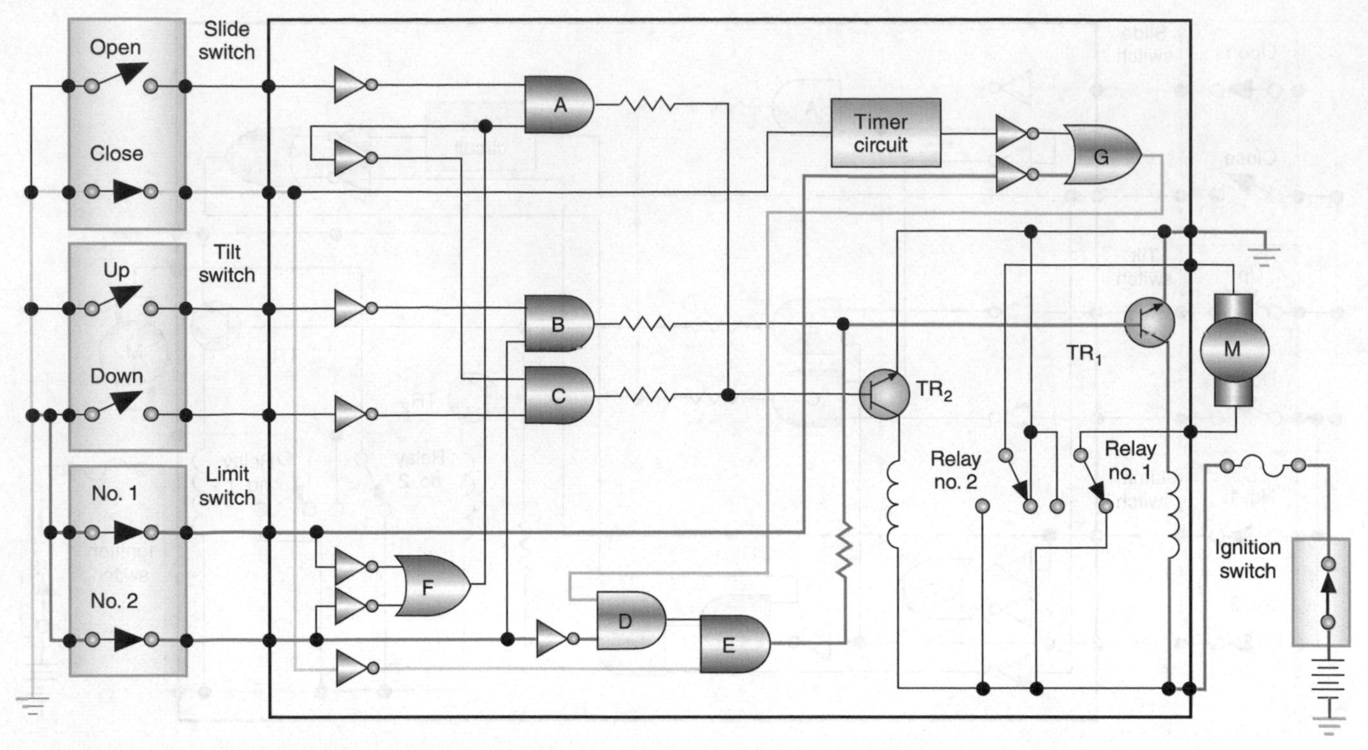

Figure 13-78 Circuit operation when the switch is in the CLOSE position, and the sunroof is open more than 7.5 inches (19 cm).

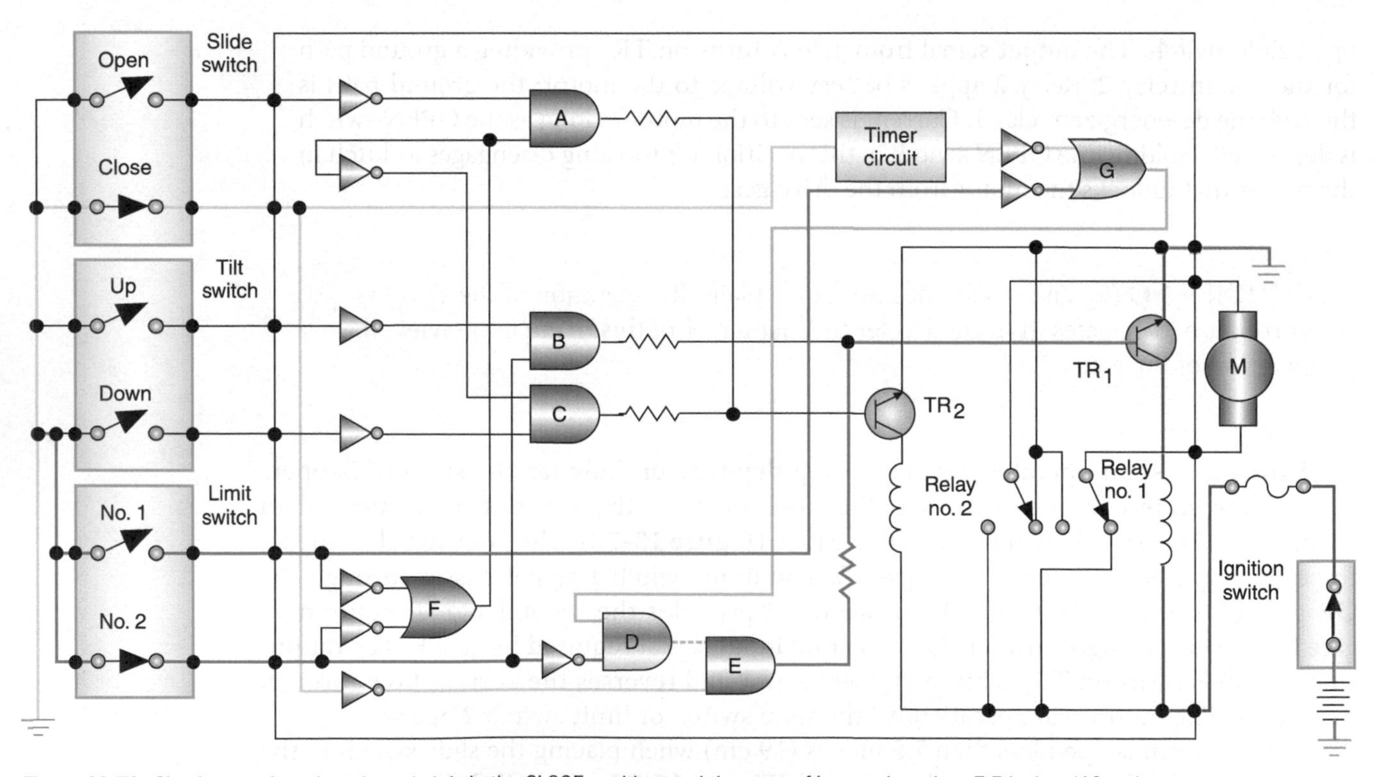

Figure 13-79 Circuit operation when the switch is in the CLOSE position, and the sunroof is open less than 7.5 inches (19 cm).

Placing the tilt switch in the UP position imposes a signal on gate B (**Figure 13-80**). The NOT gate inverts this signal and equals the value received from the opened number 2 limit switch. The output signal from gate B turns on TR$_1$, which energizes relay 1 to turn on the motor. Holding the switch in the closed position for longer than needed disengages the motor clutch.

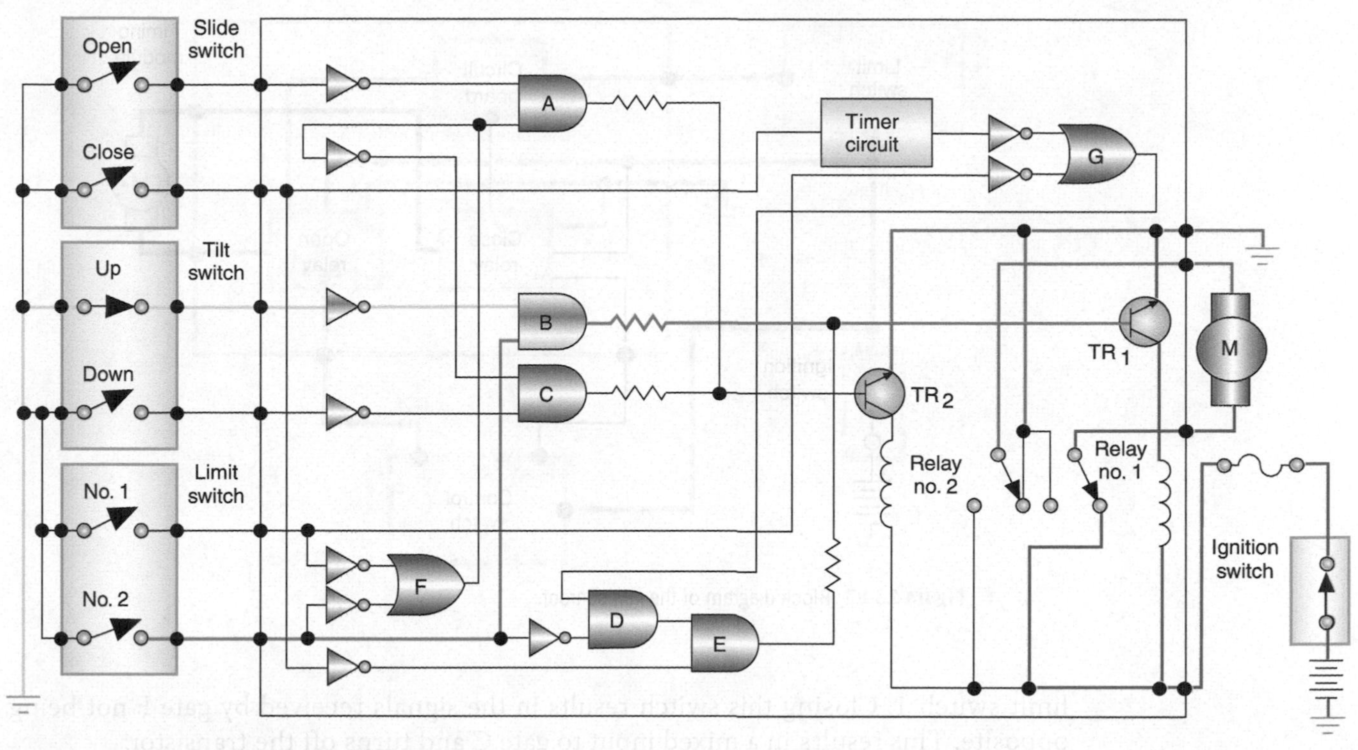

Figure 13-80 Circuit operation in the TILT UP position.

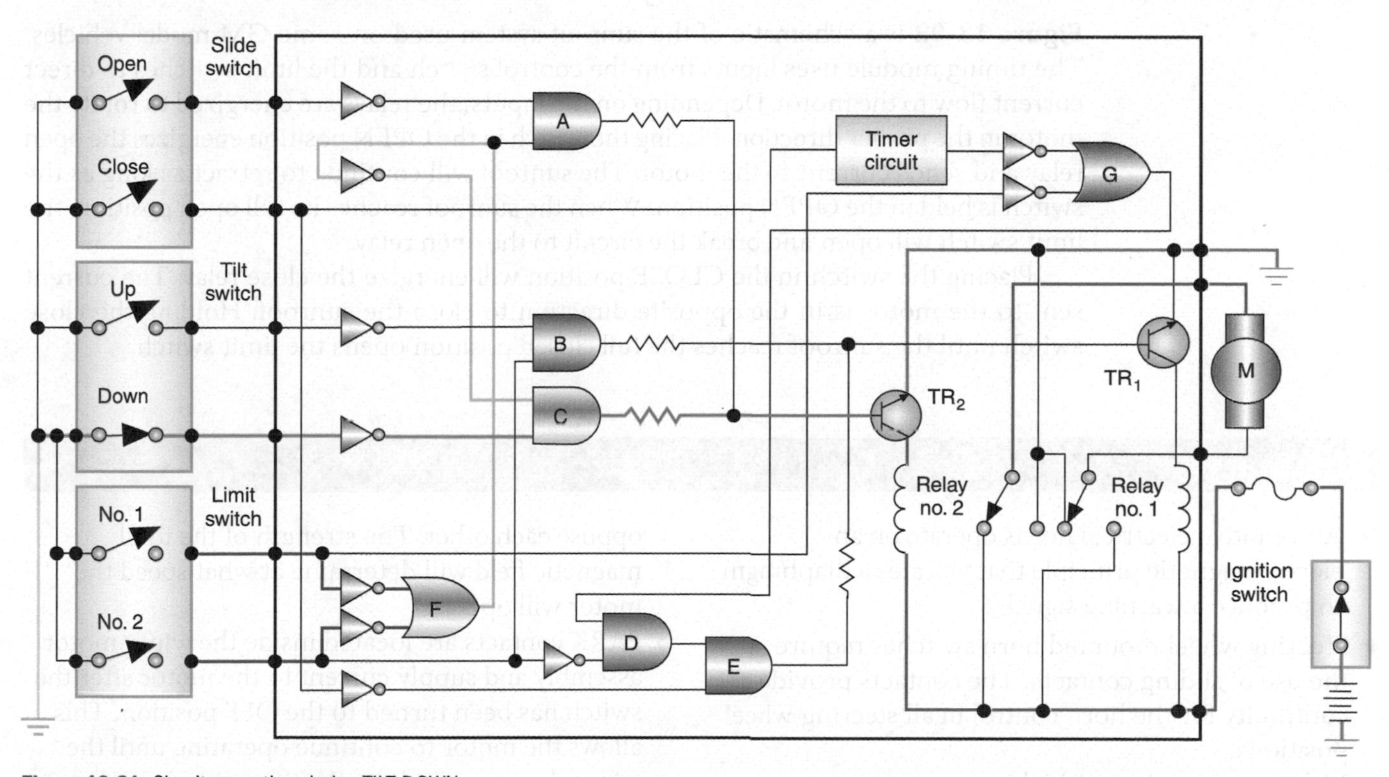

Figure 13-81 Circuit operation during TILT DOWN.

Placing the tilt switch in the DOWN position imposes a signal on gate C (**Figure 13-81**). The second signal to gate C comes from the limit switches (both are open) through gate F. The NOT gate inverts the signal from gate F and is equal to that from the DOWN switch. The output signal from gate C turns on TR_2 and energizes relay 2 to lower the sunroof. Holding down the DOWN switch for longer than necessary closes

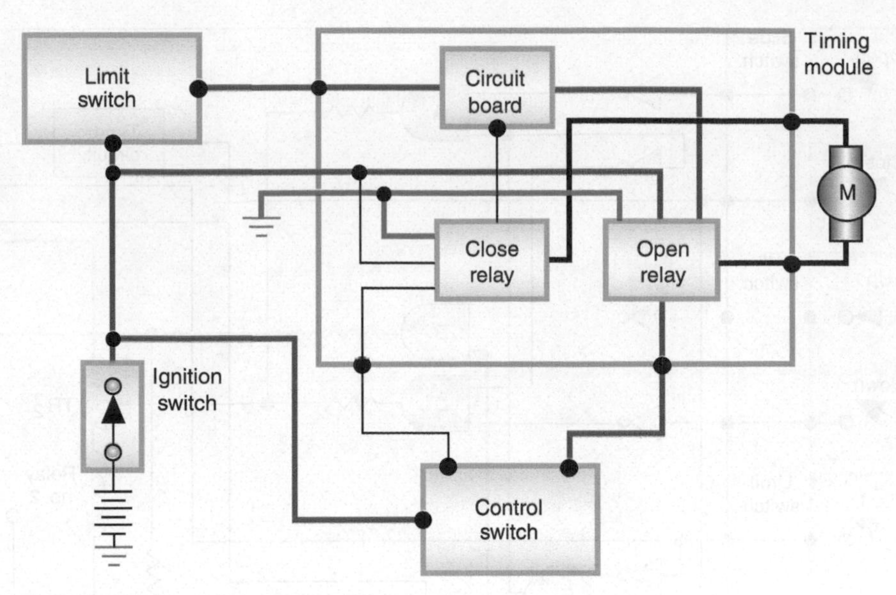

Figure 13-82 Block diagram of the GM sunroof.

limit switch 1. Closing this switch results in the signals received by gate F not being opposite. This results in a mixed input to gate C and turns off the transistor.

General Motors Electronically Controlled Sunroof

Figure 13-82 is a schematic of the sunroof system used on some GM model vehicles. The timing module uses inputs from the control switch and the limit switches to direct current flow to the motor. Depending on the inputs, the relays are energized to rotate the motor in the proper direction. Placing the switch in the OPEN position energizes the open relay and sends current to the motor. The sunroof will continue to retract as long as the switch is held in the OPEN position. When the sunroof reaches its full open position, the limit switch will open and break the circuit to the open relay.

Placing the switch in the CLOSE position will energize the close relay. The current sent to the motor is in the opposite direction to close the sunroof. Holding the close switch until the sunroof reaches the full closed position opens the limit switch.

Summary

- Automotive electrical horns operate on an electromagnetic principle that vibrates a diaphragm to produce a warning signal.
- Steering wheel mounted horn switches require the use of sliding contacts. The contacts provide continuity for the horn control in all steering wheel positions.
- Most two-speed windshield wiper motors use permanent magnet fields and the placement of the brushes on the commutator control motor speed.
- Some two-speed and all three-speed wiper motors use two electromagnetic field windings: series field and shunt field. The two field coils are wound in opposite directions, so their magnetic fields will oppose each other. The strength of the total magnetic field will determine at what speed the motor will operate.
- PARK contacts are located inside the wiper motor assembly and supply current to the motor after the switch has been turned to the OFF position. This allows the motor to continue operating until the wipers have reached the PARK position.
- Intermittent wiper mode uses a solid-state module to provide a variable interval between wiper sweeps.
- Systems with a depressed-park feature use a second set of contacts with the park switch, which reverses the rotation of the motor for about 15° after the wipers have reached the normal PARK position.

- Blower fan motors control fan speed using a resistor block that consists of two or three helically wound wire resistors connected in series.
- The blower motor circuit includes the control assembly, blower switch, resistor block, and blower motor.
- Some blower motor systems use BCM control of a power module.
- Electric defoggers heat the rear window using a resistor grid.
- Electric defoggers may incorporate a timer circuit to prevent the high current required to operate the system from damaging the battery or charging system.
- The electrically controlled mirror allows the driver to position the outside mirrors by use of a switch that controls dual-drive, reversible PM motors.
- Power windows, seats, and door locks usually use reversible PM motors, whereby the direction of current flow determines motor rotational direction through the switch wipers.
- Cruise control is a system that allows the vehicle to maintain a preset speed with the driver's foot off the accelerator.
- The cruise control module energizes the supply and vent valves to allow manifold vacuum to enter the servo. The servo moves the throttle to maintain the set speed. Balancing the vacuum in the servo maintains the set vehicle speed.
- Adaptive cruise control uses laser or radar sensors to maintain a set distance between vehicles. They can also slow or stop the vehicle if the distance closes.
- The memory seat feature is an addition to the basic power seat system that allows programming different seat positions into memory that are recalled at the push of a button.
- The easy exit feature is an additional function of the memory seat that provides for easier entrance and exit of the vehicle by moving the seat all the way back and down.
- Seats can be climate controlled using heater grids and fans.
- Antitheft systems are deterrent systems designed to scare off would-be thieves by sounding alarms and/or disabling the ignition system.
- The control module monitors the switches and activates the alarms if they indicate opening of the doors or trunk, or the rotating of the key cylinders.
- Automatic door locks is a passive system that locks all doors when the required conditions are met. Many automobile manufacturers are incorporating the system as an additional safety and convenience feature.
- The keyless entry system allows the driver to unlock the doors or the deck lid from outside the vehicle without using a key.

Review Questions

Short-Answer Essays

1. Describe the function of the adaptive cruise control system.
2. Explain the basic operating principles of the electronic cruise control system.
3. Describe the operation of a relay-controlled horn circuit.
4. Explain how brush placement determines the speed of a two-speed, permanent magnet motor.
5. How do wiper motor systems that use a three-speed, electromagnetic motor control wiper speed?
6. List and describe the safety modes incorporated into electronic cruise control systems.
7. List the main components of common antitheft systems.
8. Explain two methods the memory seat control module uses to determine seat position.
9. Describe the operation of electric defoggers.
10. Briefly explain the principles of operation for power windows.

Fill in the Blanks

1. Electrical accessories provide for additional ______________ and ______________
2. The ______________ is a thin, flexible, circular plate that is held around its outer edge by the horn housing, allowing the middle to flex.
3. Horn switches that are mounted on the steering wheel require the use of ______________ ______________ to provide continuity in all steering wheel positions.

4. The ______________ feature is an additional function of the memory seat that provides for easier entrance and exit of the vehicle.

5. Adaptive cruise control uses ______________ or ______________ sensors to maintain a set distance between vehicles.

6. ______________ is a system that uses electromagnetic waves to identify the range, direction, or speed of an object.

7. A ______________ is a device that controls the way that energized atoms release photons (light).

8. Most blower motor fan speeds are controlled through a ______________ that is wired in series to the fan motor.

9. The principle behind electric defoggers is that forcing electrons to flow through a ______________ generates heat.

10. Climate-controlled seats may use a ______________ element to cool the seat cushion.

Multiple Choice

1. The generation of sound from the horn is by:
   A. Heat causes the diaphragm to vibrate.
   B. Vibrating a column of air.
   C. Pulse width modulation of the horn relay.
   D. All of these choices.

2. The horn circuit is being discussed.
   Technician A says if the circuit does not use a relay, the horn switch carries the total current requirements of the horns.
   Technician B says most systems that use a relay have battery voltage applied to the lower contact plate of the horn switch and the switch closes the path for the relay coil.
   Who is correct?
   A. A only
   B. B only
   C. Both A and B
   D. Neither A nor B

3. Electronic cruise control systems are being discussed.
   Technician A says if the voltage signal from the VSS drops below the low comparator value, the control module energizes the vent valve solenoid.
   Technician B says the control module energizes the supply solenoid valve if the VSS signal is greater than the high comparator value.
   Who is correct?
   A. A only
   B. B only
   C. Both A and B
   D. Neither A nor B

4. Electronic cruise control is being discussed.
   Technician A says the electronic cruise control system offers more precise speed control than the electromechanical system.
   Technician B says the throttle position sensor provides smooth throttle changes while the cruise control is engaged.
   Who is correct?
   A. A only
   B. B only
   C. Both A and B
   D. Neither A nor B

5. Memory seats are being discussed.
   Technician A says the power seat and memory seat functions only operated when the transmission is in the PARK position.
   Technician B says as the seat moves from its memory position, the module stores the number of pulses and direction of movement in memory.
   Who is correct?
   A. A only
   B. B only
   C. Both A and B
   D. Neither A nor B

6. The operation of Toyota's electronically controlled sunroof is being discussed.
   Technician A says it's not necessary to understand logic gate operation to understand the control of the sunroof.
   Technician B says a motor that operates a drive gear controls the movement of the sunroof.
   Who is correct?
   A. A only
   B. B only
   C. Both A and B
   D. Neither A nor B

7. The keyless entry system is being discussed.

   Technician A says an additional function of the system is releasing the deck lid lock by pressing the 5-6 button.

   Technician B says a second code can be entered into the system.

   Who is correct?

   A. A only C. Both A and B
   B. B only D. Neither A nor B

8. The adaptive cruise control system is being discussed.

   Technician A says if the vehicle is approaching another vehicle in the same lane, ACC may activate the brake system to prevent a rear-end collision.

   Technician B says the radar or laser sensor determines the distance between the vehicle and any other vehicle ahead of it.

   Who is correct?

   A. A only C. Both A and B
   B. B only D. Neither A nor B

9. Which statement about permanent magnet wiper motors is true?

   A. The more armature windings between the high-speed and common brushes result in less magnetism in the armature and lower CEMF.

   B. The lower the CEMF in the armature, the greater the armature current.

   C. The fewer windings between the low-speed and common brushes result in increased magnetism and higher CEMF.

   D. All of these choices.

10. *Technician A* says the master control switch for power windows provides the overall control of the system.

    Technician B says current direction through the power seat motor determines the rotation direction of the motor.

    Who is correct?

    A. A only C. Both A and B
    B. B only D. Neither A nor B

CHAPTER 14

RADIO FREQUENCY, INFOTAINMENT, AND CONNECTED VEHICLE TECHNOLOGY

Upon completion and review of this chapter, you should be able to:

- Explain the fundamentals of radio wave generation.
- Explain the construction and operation of the antenna.
- Describe the three common modulations used for radio waves.
- Detail how radio waves are received.
- Explain the operation of the remote keyless entry system.
- Explain the operation of the keyless start system.
- Describe the operation of the Tire Pressure Monitoring (TPM) system.
- Explain the function of immobilizer security systems.
- Describe the operating characteristics of speakers.
- Explain how a radio tuner operates.
- Detail the use of radio frequency in vehicle audio entertainment systems.
- Describe the operation of satellite radio systems.
- Explain the operation of the telematics system, including Wi-Fi connectivity.
- Explain the operation of hands-free cell phone systems.
- Describe the operation of the navigational system.
- Explain the functionality of various vehicle connectivity systems.

Terms to Know

Angular momentum
Antenna
Cellular
Dead reckoning
Dedicated Short-Range Communications (DSRC)
Detector
Gyroscope
Immobilizer system
Infotainment
Keyless start system
Micro-Electro-Mechanical System (MEMS)
Multiplexer
Navigational system
Radio choke
Reader
Receiver
Rolling code
Satellite radio
Speakers
Telematics
Tire Pressure Monitoring (TPM) system
Transmitter
Tweeters
Wi-Fi
Woofers

Introduction

Radio wave transmitting and receiving devices are commonplace in the automobile. The first of these was the amplitude modulation (AM) radio system. Today, many vehicle systems use radio frequency. These include remote keyless entry, keyless start, immobilizer, and so forth.

Infotainment is a type of media system that combines information and entertainment systems. For example, the infotainment system can provide GPS directions and road conditions along with satellite radio, satellite TV, and Wi-Fi connectivity.

This chapter will explore the process of generating and receiving radio frequency waves. In addition, we will explore the operation of navigational systems and satellite receivers.

Today we are in the first phases of converging automotive and information technologies. This is possible thanks to the availability of high-speed broadband internet, the development of sophisticated sensors (both in vehicles and infrastructure), and advancements in communication technologies. This chapter highlights the verging technology and its implementation into different vehicle connectivity technologies.

Radio Frequency Generation

Any system using radio frequency to broadcast data must have a **transmitter** and a **receiver**. The transmitter generates the radio frequency by taking the data and encoding it onto a sine wave. Radio wave then transmit the data. The receiver receives the radio waves and decodes the message from the sine wave.

Both components require an **antenna** to radiate and capture the radio wave signal. Antennas convert radio frequency electrical currents into electromagnetic waves and vice versa. The antenna is simply a wire or a metal stick that increases the amount of metal the transmitter's waves can interact with. The transmitting antenna will use the oscillating current that enters the antenna to create a magnetic field around the antenna. The magnetic field will induce an electric field (voltage and current) in space. This electric field, in turn, induces another magnetic field in space, which induces yet another electric field. This process repeats as these electric and magnetic fields (electromagnetic fields) induce each other in space at the speed of light outward away from the antenna.

Shop Manual
Chapter 14, page 782

Radio wave transmission is actually a simple technology based on the principle of induction. Radio waves are electromagnetic (EM) radiation. Remember that induced voltage occurs only when the magnetic field is in motion.

The transmitter induces a continuously varying AC electrical current (**Figure 14-1**). **Figure 14-2** illustrates a simple transmitter that can produce this type of sine wave using

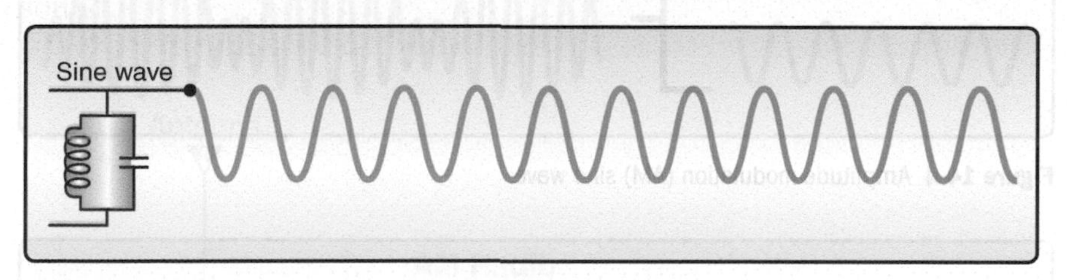

Figure 14-1 The transmitter produces an AC sine wave.

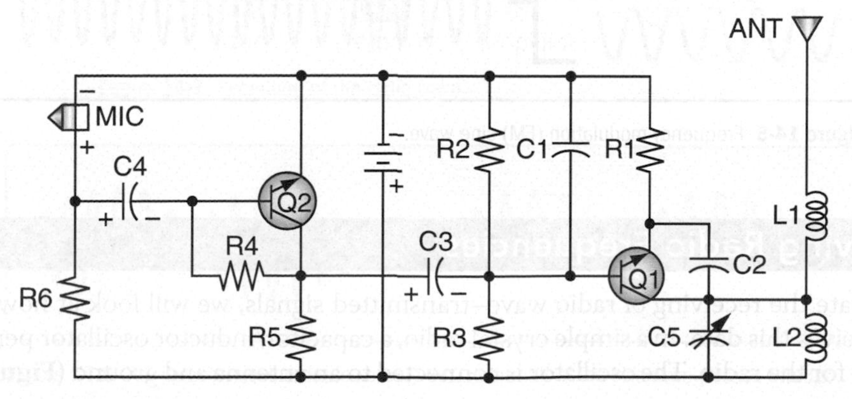

Figure 14-2 A simple transmitter using a capacitor and an inducer.

Remote Keyless Entry

Shop Manual
Chapter 14, page 765

Remote keyless entry differs from the keyless entry system discussed in Chapter 13 since this system can perform the keyless entry functions from a distance.

Many vehicles utilize a remote keyless entry system to lock and unlock the doors, turn on the interior lights, and release the trunk latch. A small receiver module is installed inside the vehicle. The transmitter assembly is a handheld item referred to as a FOB (**Figure 14-9**). Pressing a button on the FOB allows the operation of the system from a distance of 25 to 50 feet (7.6 to 15.2 m). Pressing the UNLOCK button unlocks the driver's door and illuminates the interior lights. If the vehicle has a theft-deterrent system, it's also disarmed by pressing the unlock button. A driver exiting the vehicle can activate the door locks and arm the security system by pressing the lock button. Many transmitters also have a third button for opening the deck lid. Some systems will also include a PANIC button that, when pressed, signals the control module to honk the horns and flash the lights.

A BIT OF HISTORY

The acronym "FOB" originally was the word "fob" referring to a small pocket in a waistcoat or pants used to hold a watch and chain. "FOB" now means a finger-operated button.

The system operates at a fixed radio frequency. If the unit does not work from a normal distance, check for two conditions: weak batteries in the remote transmitter or a stronger radio transmitter close by (radio station, airport transmitter, etc.).

Many keyless entry systems incorporate the ability to start the engine without entering the vehicle. A separate button on the FOB initiates this function. The engine start button usually requires multiple presses or depressing it for a couple of seconds to activate the

Figure 14-9 Remote keyless entry system transmitter.

remote start function. When the remote keyless entry module receives the transmission of an engine start request, it will send bus messages to various modules to gather status information and to request activation.

The FOB has an RF modulator circuit that will generate the identification signal. A small lithium coin cell battery powers this circuit. The identification code is either a 32- or 64-bit **rolling code**. Rolling code uses a counting system that requires a different signal every transmission for improved security. The RF generator modulates the carrier code creating an output for the RKE receiver.

An antenna in the ECM receives the input from the transmitter's RF modulator. The receiver in the ECM detects the transmission signal called the message authentication code (MAC). The code will include the transmitter ID, rolling code, and command code. The receiver validates the MAC by comparing its stored rolling code with the FOB's ID rolling code. The transmitter ID rolling code is stored in the receiver's nonvolatile memory. To validate the rolling code, the transmitted rolling code count should be greater than or equal to the stored count in the receiver. To be a match, the count must be within a range. If the rolling counts do not match, the receiver will not initiate the requested command. If the rolling counts match, the ECM generates a command to initiate the requested function.

The electronics of the RKE system are responsible for translating coded input from the FOB into commands for driver outputs (Figure 9-20). Code generation in the FOB is detected, processed, and then verified through these electronics. Upon acceptance of the code, the ECM will generate a bus message command and sends it to the PCM to start the engine.

Shop Manual
Chapter 14, page 767

Upon receiving the transmission from the FOB, the parking lights will turn on (or flash), and the horn will chirp. The next command is to lock all doors. Some systems will also raise the windows. With the vehicle secure, network bus messages confirm proper gear selection, proper parking brake position, and so forth. The PCM will start the engine if all are within the correct parameters.

Most systems will only allow the engine to run for 5 to 10 minutes, and then shuts off the engine. Typically, the remote start feature only allows the engine to start a couple of times without physically placing the ignition in the RUN position. If the remote start function is used too many times in a row, the control module denies the FOB request and the horn will double chirp to indicate rejection of the engine start request. Reactivating the remote start function requires placing the ignition in the RUN position.

Other inputs to the remote start system include door ajar and hood switches. Open doors or an unlatched hood can disable the remote start function.

The remote FOB can also shut the engine off (if started remotely). Pressing the remote start button once sends the message to shut off the engine. The horn may chirp, the engine shuts down, and the parking lights turn off. The doors will remain locked.

The remote start feature may be interconnected with the automatic temperature control (ATC) system. Upon receiving the start request, the ECM broadcasts a bus message to the ATC controller. The ATC controller analyzes the in-car temperature sensor to determine the in-car heat load. Based on this input, the ATC system activates A/C or heater/defrost modes. It's also possible that the climate-controlled seats and steering wheel functions interact with the remote start feature.

To prevent the theft of the vehicle while it's running, the control module will monitor different inputs and the sequence they are received. For example, the engine will shut off if the brake pedal is depressed prior to placing the ignition in the RUN position. The clutch pedal switch provides this input on vehicles with manual transmissions. In addition, opening the hood without the ignition in the RUN position shuts the engine off. If the vehicle has an antitheft system, anything that triggers the system will cause the engine to shut down.

Shop Manual
Chapter 14, page 767

Aftermarket Kits. Vehicles that are factory equipped with power door locks can be upgraded to remote keyless entry and remote start by installing aftermarket kits. The kits will include a control module, antenna, transmitters, and wiring harnesses. Some kits also include relays, diodes, and a program switch. Depending on the system, add-on features may be installed or programmed with the remote keyless entry feature. Some of these add-on systems may require additional control units.

Typically, the aftermarket remote keyless system operates very similar to the factory system. However, adding remote start to the system can become very complex (**Figure 14-10**). The kit will usually require the installation of additional relays. The relays bypass the operation of factory system, allowing for the original operation of the system without interruption from the add-on kit. For many systems, such as the horn, this is a better practice than attempting to wire the system using the existing relays. Keep in mind that these systems may now have two relays.

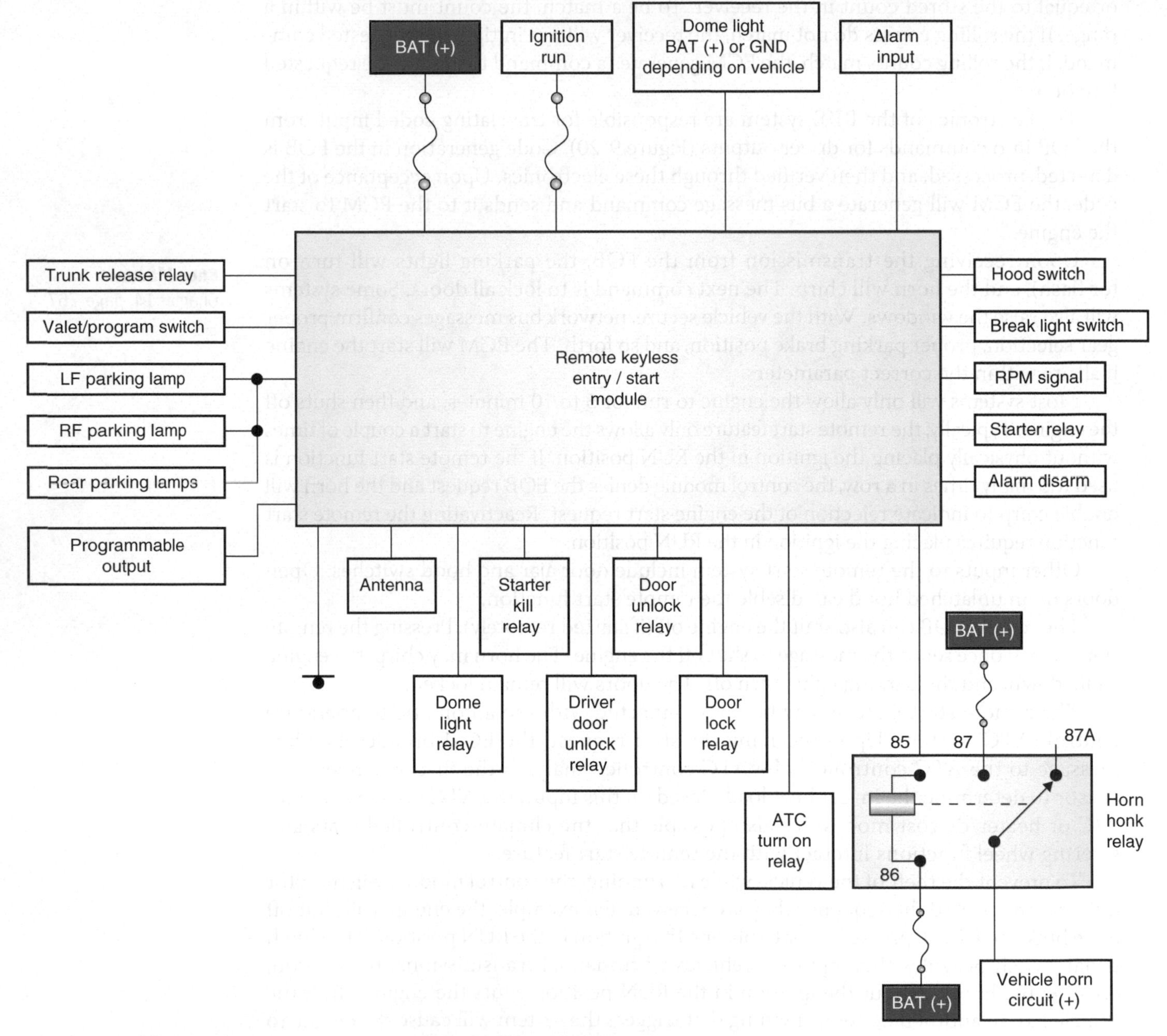

Figure 14-10 A typical aftermarket remote keyless entry system with remote start.

There are exceptions; for example, the factory door lock and unlock relays can be used by splicing the wires from the control module to the vehicle's wiring harness for these circuits. If the vehicle does not have relays, these same circuits from the ECU are wired to provide a push/pull (H-gate) function of the actuators (if the manufacturer states that the circuits can handle the current). Some systems only have a 200-mA output to the door lock/unlock system; thus, the installation of relays will be necessary.

There are different methods for the aftermarket system to bypass the factory immobilizer system. One method is to connect one of the programmable output circuits of the ECU to control a bypass module. The bypass module allows for the add-on system to be installed but maintains the integrity of the manufacturer's system. Vehicles equipped with the GM VATS system will have a resistor that matches the resistance of the key installed in a bypass circuit from the ignition switch to the VATS module (**Figure 14-11**). The VATS module will see the correct resistance when the relay is energized and allow the engine to start.

Some of the circuits from the control module will be wired to battery voltage or ground, depending on the vehicle. For example, the dome light input connects to battery voltage if the door pin circuit is positive and to ground if the circuit is negative. The control module may be configurable to operate the parking lamps by either positive feed or ground. Changing the control of the lamps is by placement of a jumper harness or may require a Bitwriter program.

Keyless Start

An enhancement to the remote keyless entry system is the **keyless start system**. This system allows starting the vehicle without an ignition key. Early versions of these systems use the key FOB to unlock the vehicle doors and disarm the antitheft system. At the same

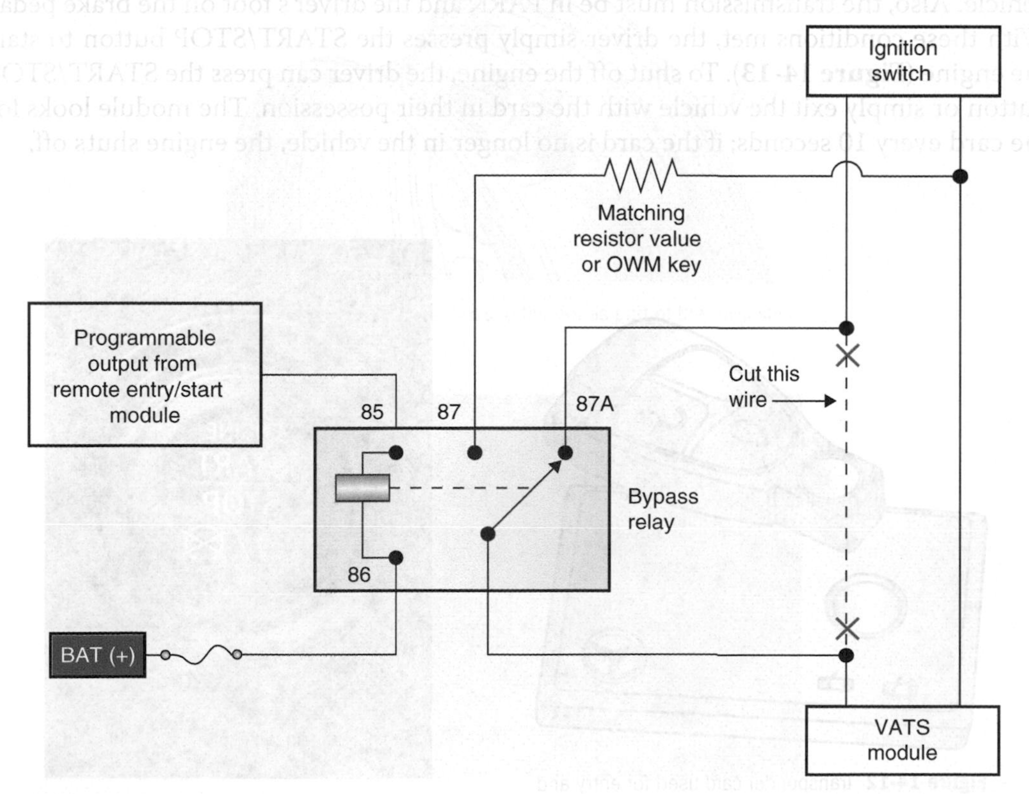

Figure 14-11 VATS bypass circuit.

The TPS will send pressure and temperature data to the receiver module when any of the following occurs:

- If the vehicle is parked for more than 15 minutes, the TPS monitors the pressure every minute but only sends tire pressure data once every 13 hours, unless there is a change in pressure.
- While the vehicle is in motion, the TPS sends data once every 15 seconds for the first 30 data blocks and once per minute after that.
- If a drop in pressure greater than 1 psi (6.9 kPa) occurs when the vehicle is stationary, the TPS will send data once per minute.
- If a drop in pressure greater than 1 psi (6.9 kPa) occurs when the vehicle is in motion, the TPS will send data once every 5 seconds.

Logic in the receiving module prevents false warning displays to the driver. For example, since the signal transmission frequency depends on wheel speed, if a signal from one of the TPSs isn't received as often as the other sensors, the receiving module will assume that the tire has been relocated and the spare is in use. Also, rapid tire pressure changes can occur due to temperature changes. The TPMS may indicate a low-tire warning if the vehicle leaves a heated garage on a cold day. To prevent this, the system may use temperature information and pressure data to filter the data and compensate for rapid pressure changes resulting from temperature changes.

The receiver module can set a DTC when one of the transmitters fails to produce a signal. Learning the ID of replacement transmitters occurs as the vehicle is driven. When the receiver module receives an unrecognized signal at the same transmission intervals as the recognized transmitters, the receiver module stores the ID and begins monitoring that transmitter. The receiver module will relearn the TPS IDs every time the vehicle is driven after being stopped for at least 15 minutes.

AUTHOR'S NOTE It may be necessary to drive the vehicle up to 10 miles (16 kilometers) before the replacement transmitter's ID is learned.

Premium systems use the same basic TPSs as just described, but use additional components or programming to identify which tire is out of limits. The premium system records TPS ID locations, so this system is capable of displaying the actual tire pressures to the driver (**Figure 14-16**).

Figure 14-16 Premium systems display the pressures of each tire.

Some systems require programming the ID location into the receiving module. On some systems, this is done by entering training mode and following the specified order of tire location to place a training magnet around each tire's valve stem. Placing the magnet around the valve stem pulls a reed switch closed and causes the TPS to send its data. After learning the location, the ID is locked to that position. The learning procedure needs to be performed anytime the tires are rotated.

Some systems don't require ID location learning. These use a transponder (**Figure 14-17**) located behind the wheel splash shields at three locations. A fourth transponder is unnecessary since the system infers the location of the fourth tire. The receiving module will infer the fourth TPS ID it's receiving while the vehicle is moving is in this position. The tire position location learning process repeats each time the vehicle is in motion.

The transponders use two ground terminals. One of the ground terminals is common for all transponders. Which one of the other three ground terminals used identifies the transponder location. This arrangement allows the receiving module to determine the location of the transponder.

When the vehicle is in motion, all four TPSs send periodic signals to the receiving module. The receiving module then sends a signal to one of the transponders. This causes the transponder to emit a 125-kHz signal in the area surrounding the wheel. This signal excites the TPS at that location, causing it to send a constant signal to the receiving module. The receiving module will then identify the TPS and its location. The other two transponder locations go through the same process. Upon completion of the identification process for the three wheel locations, the fourth location can be inferred and identified. This learning process occurs when the vehicle is driven after being stationary for at least 15 minutes.

AUTHOR'S NOTE Because of this procedure, training magnets are not required to learn TPS locations.

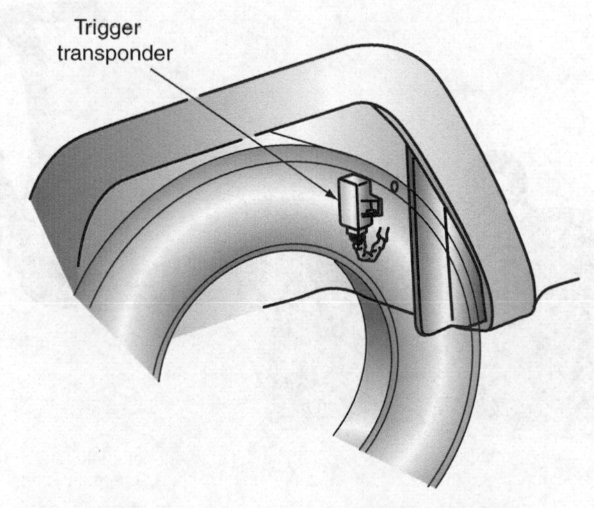

Figure 14-17 Transponder used to identify sensor location on the vehicle.

Another method of tire location ID places the tire pressure monitor module in a specific location, such as in the left front fender well. The TPS broadcasts their pressures as described, but also broadcasts the direction of tire rotation. The module determines tire location by the strength of the signal from the TPS, and its rotation direction.

A fourth method is to use the antilock brake systems' (ABS) wheel speed sensors. Since the tire's diameter dictates the number of revolutions per mile, tire speed can determine if a tire has low pressure. As the tire pressure drops, the tire's diameter becomes smaller. This results in more revolutions per mile. Compared to the other wheel speed sensors, the system determines a low tire when it sees a higher wheel speed sensor reading than the others. Since the ABS identifies the location of the wheel speed sensors, that data also identifies the location of the low tire.

Immobilizer Systems

Shop Manual
Chapter 14, page 768

The **immobilizer system** protects against unauthorized vehicle use by disabling engine starting. Using an invalid key, or attempting to hot-wire the ignition system disables the engine from starting. Manufacturers have several different names for this system. The following are examples of system operation.

The primary components of the system include the immobilizer module, ignition key or FOB with transponder chip, the powertrain control module (PCM), and an indicator lamp. The immobilizer module is composed of a magnetic coil and a charge/detect module (transceiver). The immobilizer module is either mounted to the steering column or in the instrument panel. It may include an integrated antenna that surrounds the ignition switch lock cylinder (**Figure 14-18**). Modules mounted in a remote location are connected to an antenna by a cable. The signal communication that occurs between the transponder and the charge/detect circuit of the ECM is the heart of system operation (**Figure 14-19**)

The system includes special keys with a transponder chip under the covering (**Figure 14-20**). Some systems are capable of recognizing up to eight different keys. To use any additional keys they must be programmed to the vehicle's immobilizer module. Most systems provide a procedure that allows the customer to program additional keys if they already have two valid immobilizer keys programmed to the vehicle. Keyless entry systems have the transponder chip in the FOB.

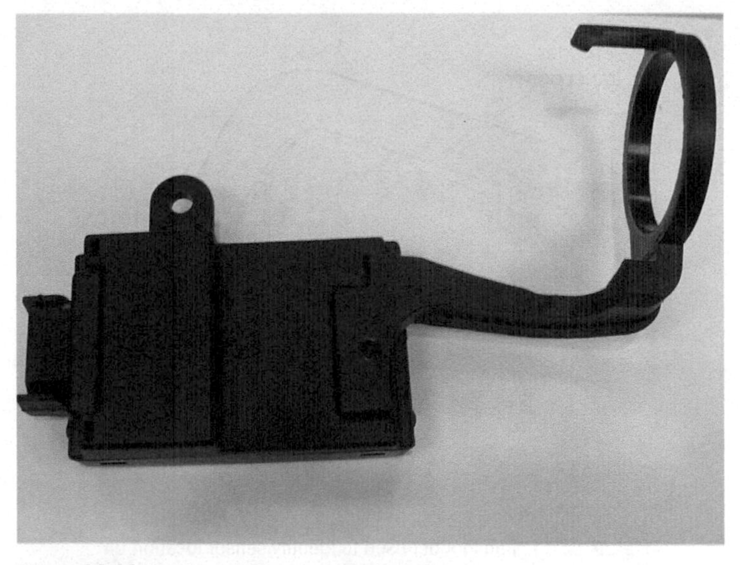

Figure 14-18 The immobilizer module with halo antenna.

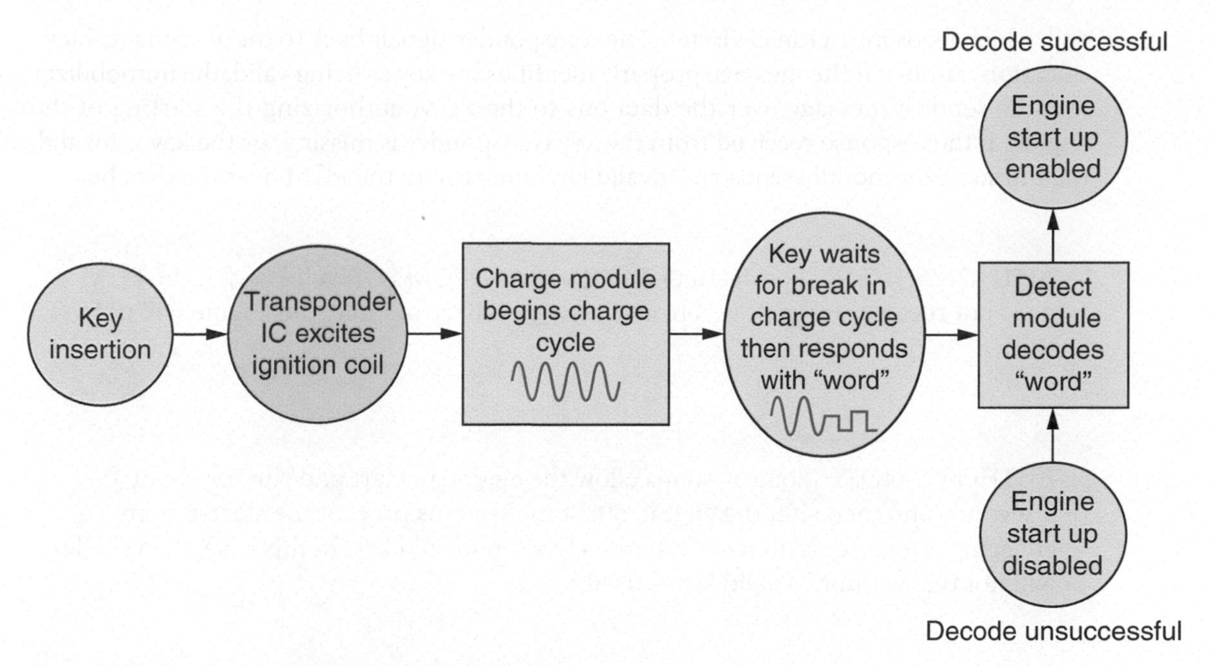

Figure 14-19 Flow chart of immobilizer system functionality.

Figure 14-20 The immobilizer key has an internal transponder chip. The key cover was removed for this illustration.

System Operation

Inserting the ignition key, equipped with the radio frequency identification (RFID) transponder, into the ignition switch cylinder and turning it to the RUN position signals the immobilizer module to radiate an RF signal through its antenna to the key. The transponder's integrated circuit induces a current in the magnetic coil and excites the

The Keyless GO system is incorporated within the immobilizer system.

coil, which feeds into a tuned circuit. The transponder signals back to the antenna its identification number. If the message properly identifies the key as being valid, the immobilizer module sends a message over the data bus to the PCM authorizing the starting of the engine. If the response received from the key transponder is missing, or the key is invalid, the immobilizer module sends an "invalid key" message to the PCM over the data bus.

AUTHOR'S NOTE The default condition in the PCM is "invalid key." If the PCM does not receive any messages from the immobilizer module, the engine will not start.

AUTHOR'S NOTE Some systems allow the engine to start and run for about 2 seconds and then shut the engine off. Some systems prevent the starter from engaging after a set number of attempts with an invalid key. In this case, the vehicle will not recover until a valid key is used.

The **reader** system serves as an interface between the transponder and the microprocessor.

Within the immobilizer module is a **reader** that allows the module to read and process the identifier code from a transponder (**Figure 14-21**). The interface operates in two directions. In one direction, energy transfers from the reader to the transponder as the

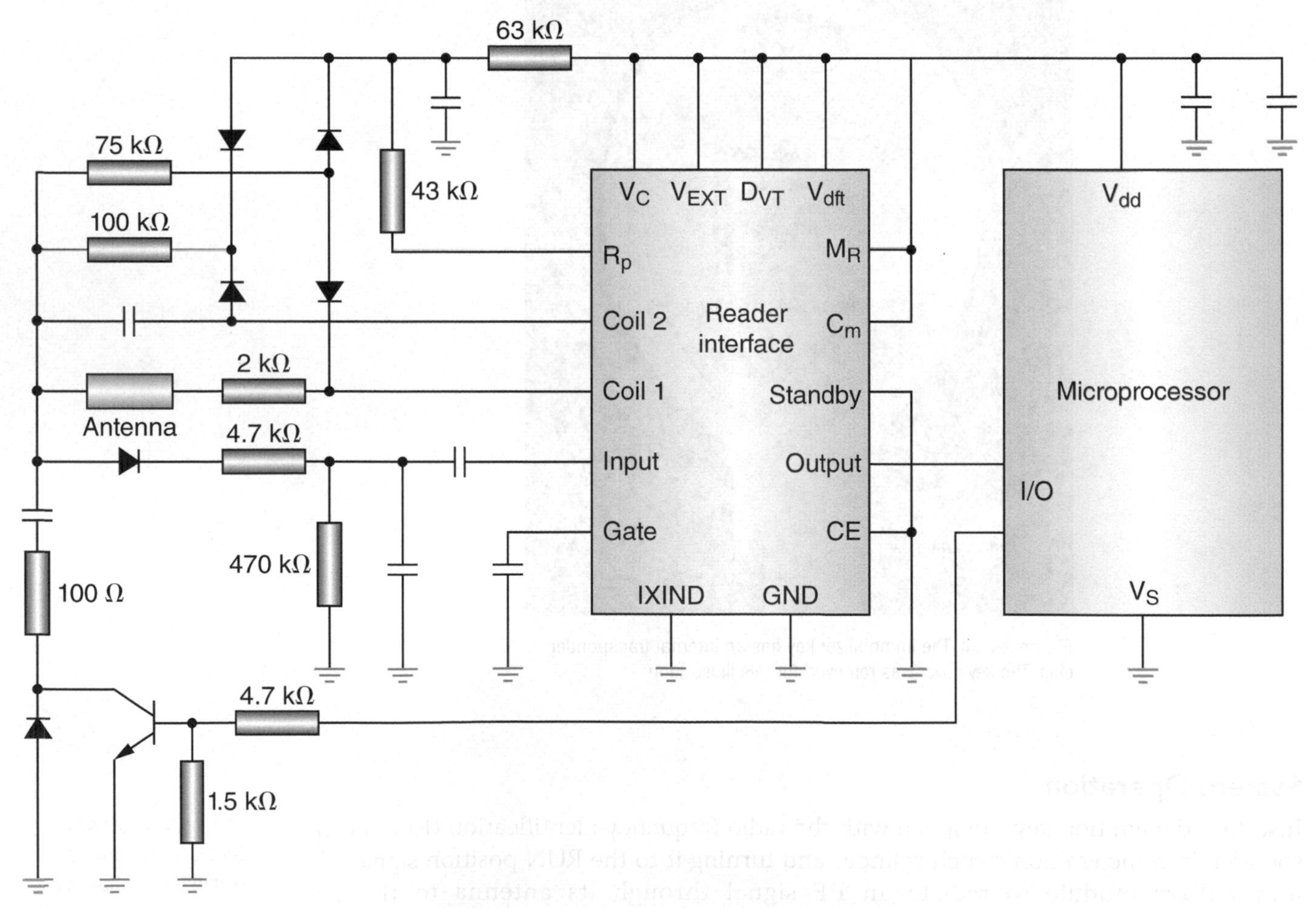

Figure 14-21 Reader interface circuit.

reader creates a magnetic field through the antenna. Energizing the antenna is by using series resonance. The magnetic field induces a voltage in the transponder key's resonant circuit. This voltage powers the transponder IC. The current in the transponder coil generates a magnetic field superimposed to the reader's field. Provided the supply voltage to the transponder's IC is high enough, the transponder will transmit by damping the resonant circuit in accordance with the data signal.

In the other direction of interface operation, data transfer is from the transponder to the microprocessor. The voltage signal from the transponder is smaller than that of the reader voltage. This results in a slight voltage modulation at the reader coil. Demodulation of the incoming signal is accomplished by a rectifier and decoupling capacitor (**Figure 14-22**). The signal is then amplified and conditioned into the appropriate digital output data.

A unique secret key code is programmed into the immobilizer module. The module also retains the unique ID number of all keys programmed to the system in its memory. The transponder stores the secret key code and transmits it to the keys during the programming function. In addition, the PCM records the secret key code. Gaining secured access to the immobilizer module for service requires a personal identification number (PIN). The immobilizer module also records the vehicle identification number (VIN). Scrambling all messages transmitted by the immobilizer module reduces the possibility of unauthorized immobilizer module access or disabling.

The immobilizer module also sends indicator lamp status messages. The indicator lamp will normally illuminate for 3 seconds for a bulb check after placing the ignition in the RUN position. After the bulb check is complete, the lamp should go out. If the lamp remains on after the bulb check, this indicates that the immobilizer module has detected a system malfunction or that the system has become inoperative. If the lamp flashes after the completion of the bulb check this indicates the detection of an invalid key, or a key-related fault is present.

Each key has a unique transponder identification code permanently programmed into it. Programming a key into the immobilizer module stores the transponder identification code in the immobilizer's memory. In addition, the key learns the secret key code from the immobilizer module and programs it into its transponder memory. For the engine to start and run, all of the following must be in place:

- Each key's transponder ID must be programmed into the immobilizer module.
- The immobilizer's secret key code must be programmed into each key.
- The VIN programmed in the immobilizer module must match the VIN in the PCM.
- The data bus network must be intact to allow communications between the immobilizer module and the PCM.
- The immobilizer module and the PCM must have properly functioning power and ground circuits.

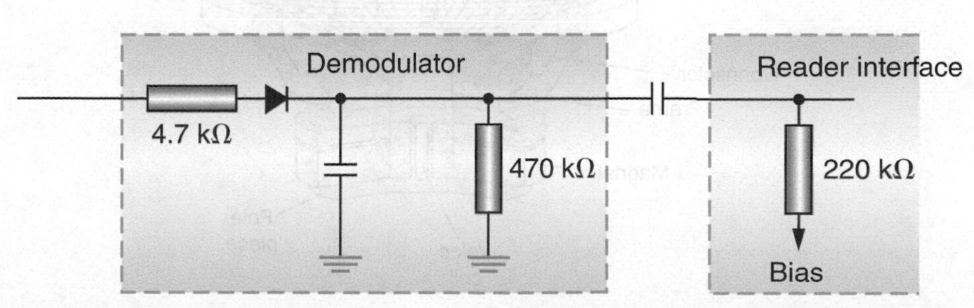

Figure 14-22 The signal from the transponder passes through a demodulation circuit to the reader.

AUTHOR'S NOTE In 1996, GM introduced an immobilizer system that does not use a chip in the key. This system has a Hall-effect sensor in the key cylinder that measures the key's magnetic properties. The cut pattern of every key has its own magnetic identity. Inserting the wrong key into the lock cylinder prevents the engine from starting, even if the lock cylinder turns.

Speakers

Shop Manual
Chapter 14, page 785

Speakers turn the electrical energy from the radio receiver amplifier into acoustical energy. The acoustical energy moves air to produce sound. A speaker moves the air using a permanent magnet and electromagnet (**Figure 14-23**). Current from the amplifier to the voice coil energizes the electromagnet. The coil forms magnetic poles that cause the voice coil and speaker cone to move in relation to the permanent magnet. The current to the speaker is a rapidly changing alternating current (AC) that results in the speaker cone moving in and out in time with the electrical signal. When the cone moves forward, the air immediately in front of it's compressed. This will cause a slight increase in air pressure in front of the cone. The cone then moves back past its rest position, resulting in a reduction in the air pressure (rarefaction). This process continues so that a wave of alternating high- and low-pressure waves radiates away from the speaker cone at the speed of sound. These changes in air pressure are actually sound.

Since one speaker cannot reproduce the entire hearing frequency range of approximately 20 Hz to 20 kHz, speakers reproduce only parts of the desired frequency. Large speakers, called **woofers**, produce the low frequencies of midrange and bass better than small speakers. Smaller speakers, called **tweeters**, produce the high frequencies of treble better than large speakers. Coaxial speakers have two separate speakers combined in one speaker frame and cover a broader frequency range than a single-cone speaker. Coupling sub-woofer speakers with the coaxial speaker (or a separate tweeter and mid-range speaker) increase coverage of the hearing range and maximizes sound quality.

Infotainment Systems

Shop Manual
Chapter 14, page 779

The first radios introduced to the automotive market produced little more than tinny noise and static. Today's audio sound systems produce music and sound that rivals the best that home sound systems can produce, and with nearly the same or even greater amounts of volume or sound power!

The most common sound system configuration is the all-in-one unit called a receiver. Housed in this unit are the radio tuner, amplifier, tone controls, and unit controls for all

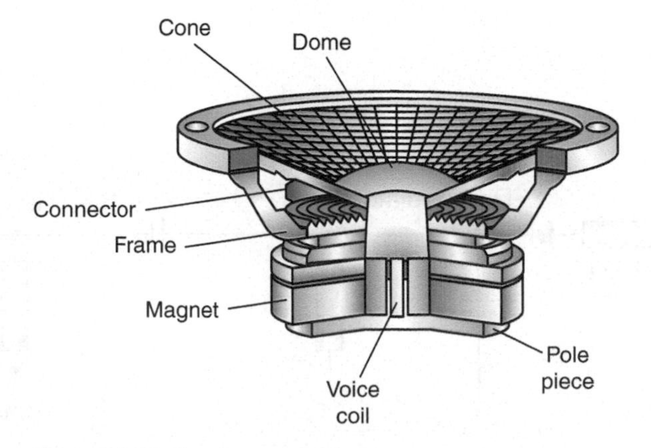

Figure 14-23 Speaker construction.

functions. These units may also include internal capabilities such as compact disc players, DVD players, digital audiotape players, and/or graphic equalizers. Most will be electronically tuned with a display showing all accessed or controlled functions and digital clock functions (**Figure 14-24**).

A BIT OF HISTORY

Recently manufacturers have taken individual functions (tape, disc, equalizer, control head, tuner, amplifier, etc.) and put them in individual boxes and call them components. This allows owners greater flexibility in selecting options to suit their needs and tastes (**Figure 14-25**). Componentizing has allowed greater dash design flexibility. Some components, such as multiple-CD changers, can be remotely mounted in a trunk area for greater security.

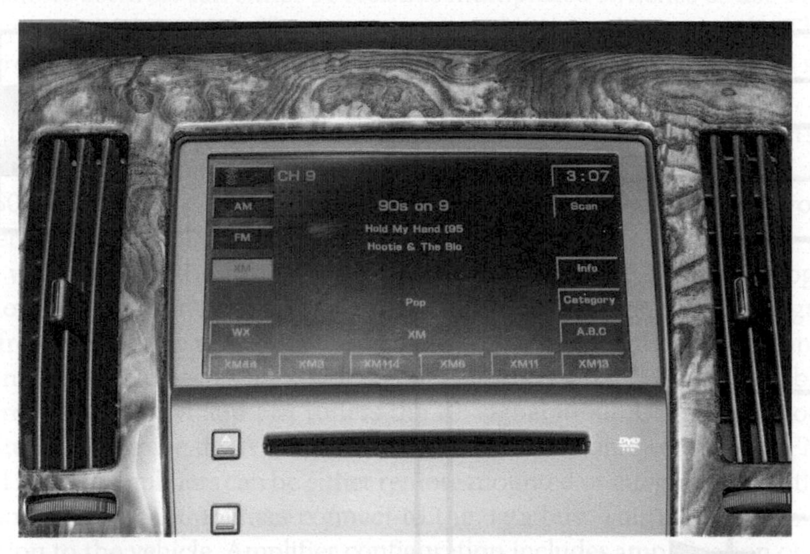

Figure 14-24 Radios are actually receivers. They contain the basic elements of a tuner, an amplifier, and a control assembly in one housing and can also contain a tape or CD player.

Figure 14-25 Components, like those shown here, allow for more sound system options.

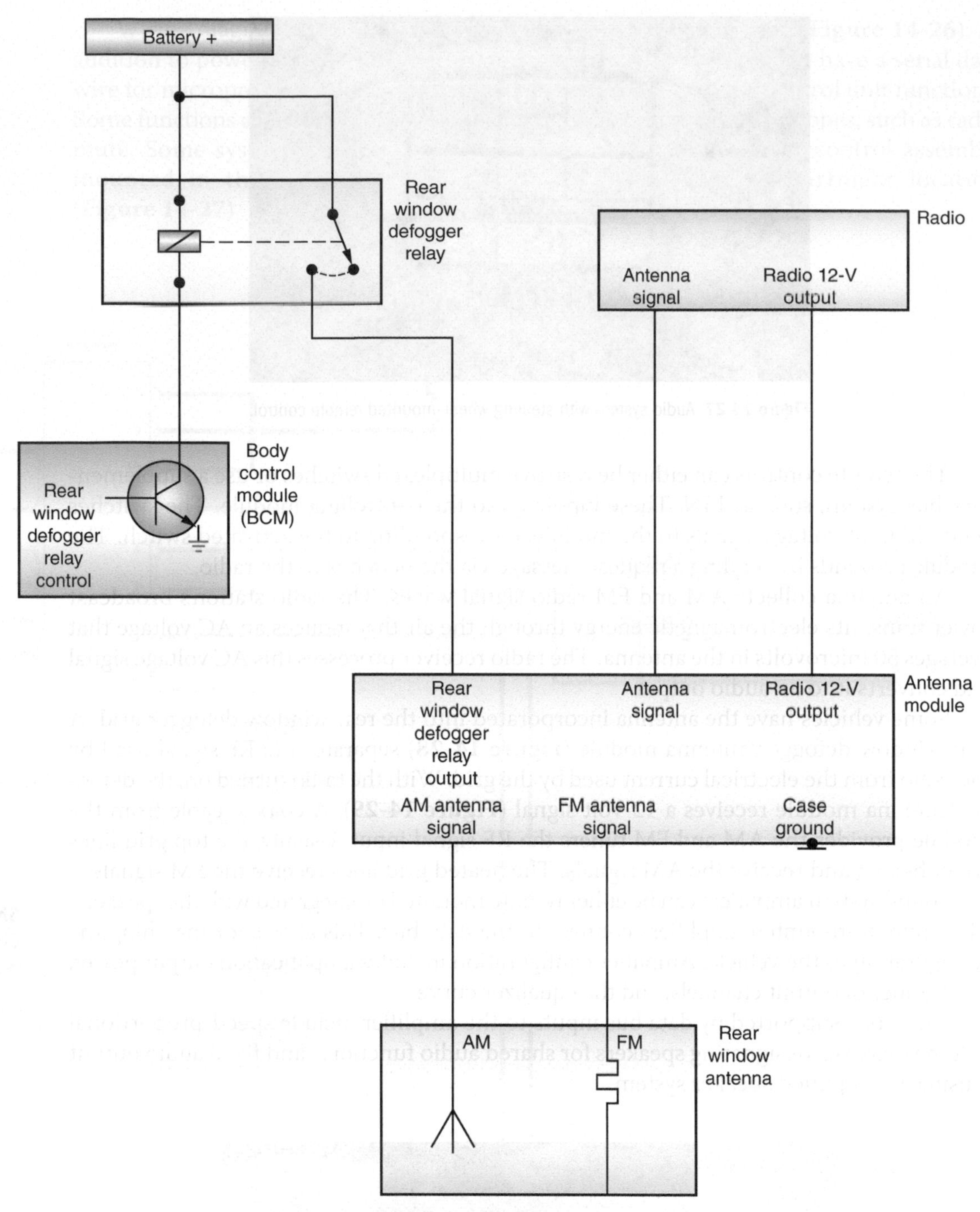

Figure 14-29 Rear window defogger/antenna circuit.

Integrated amplifiers (mounted in the speakers) receive power through the amplifier relay. Turning the radio on supplies 12 volts (V) to energize the relay.

Shop Manual
Chapter 14, page 783

Reception to the radio receiver from the antenna can be interfered with by noise resulting from radio frequency interference (RFI) and electromagnetic interference (EMI). This is especially true of the AM band. The radio receiver picks up the noises and amplifies it through the audio circuits. The FM band is susceptible to RFI and EMI, but usually isn't as noticeable compared to AM. Proper radio antenna base, radio receiver, engine-to-body, and heater core grounds are necessary to control RFI and EMI noise. In addition, resistor-type spark plugs and radio suppression secondary ignition wiring reduce this noise. The radio will also have internal suppression devices, such as capacitors. The capacitors shunt

AC noise to ground and slows sudden voltage changes in a circuit. Some systems will use a **radio choke**. The choke is a winding of wire. In a direct current (DC) circuit the choke acts like a short, but in an AC circuit it represents high resistance. The choke blocks the noisy alternating current but allows the direct current to pass normally.

Satellite Radio

Satellite radio (**Figure 14-30**) provides several music and talk show channels using orbiting satellites to provide a digital signal. An in-vehicle receiver receives the digital signal, which travels to the conventional vehicle radio as an auxiliary audio input and plays through the normal speaker system. In this case, the satellite radio function becomes an additional mode of the radio. Satellite radio operation is available only when the owner purchases the subscription service.

Shop Manual
Chapter 14, page 781

AUTHOR'S NOTE If all satellite signals are lost, the radio becomes totally silent, unlike the loss of an AM or FM signal where some static or hiss may be heard.

The satellite digital audio receiver (SDAR) is an additional audio receiver installed separately from the vehicle's radio. The SDAR processes the satellite signal, which provides an input signal to the conventional radio over dedicated circuits for right and left channels. The usual radio controls for mode selection and tuning are bussed to the SDAR. This allows the radio controls to operate the SDAR. The radio screen displays any data bussed text information (such as channel numbers, track, and artist) that the stations transmits.

The SDAR signals the satellite antenna located on the centerline of the roof (**Figure 14-31**), and the tuning of the antenna calibration uses the ground plane of the vehicle. Signal reception is possible only when the antenna and the satellite are in a direct line. Obstacles such as buildings, overpasses, and tunnels may temporarily disrupt the signal. Adding a buffer prevents brief interruptions when the vehicle passes under bridges.

Some radios have only one audio input. In order to add multiple audio sources such as SDAR, CD/DVD, and hands-free cell phone may require using a **multiplexer**. The multiplexer acts as an electronic switch to change between the different audio sources. Depending on which source requires the use of the system, the multiplexer will make the switch and send the output to the radio.

Figure 14-30 Satellite radio system.

Figure 14-31 Satellite radio antenna

DVD Systems

Shop Manual
Chapter 14, page 781

Several digital video disc/video entertainment systems (DVD/VES) are available on today's vehicles. Most DVD systems display the video on a flip-down, roof-mounted monitor (**Figure 14-32**) or on a monitor attached to the back of the front seats (**Figure 14-33**). Most will play the audio through the vehicle's regular audio system. If a DVD is playing, the audio system automatically switches to a "surround sound" mode that is biased toward the rear of the vehicle. In addition, the system may include remote control and auxiliary input jacks that permit the display of video cameras and video games.

If the system supports using headphones, the speakers may play audio from one source while the headphones can play audio from another source. If the headphones and the radio are playing in the same mode, the radio is the master of the system. Headphones use either a 900-MHz signal or infrared. The headphones enable rear-seat occupants to listen to audio from one source while front-seat occupants listen to another source on the front speakers. The headphones receive their signals from the radio or DVD changer. Two separate channels are used to minimize interference. If another audio/video system is within range, the headphones automatically switch to the strongest channel.

Figure 14-32 Flip-down, roof-mounted DVD monitor.

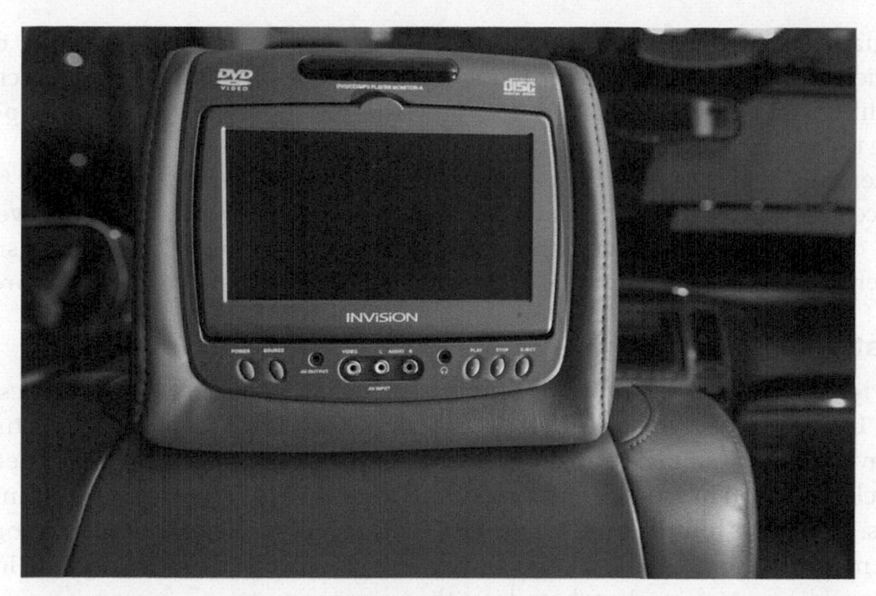

Figure 14-33 DVD monitor located in the seat back.

Telematics

Shop Manual
Chapter 14, page 785

Telematics refers to the ability to connect the vehicle to the outside world. The vehicle telematics system can receive one-way transmissions from GPS satellites, XM Radio stations, HD Radio stations, and other sources. In addition, it can perform two-way communications to exchange data with cell phone networks by using the customer's smartphone. The telematics system may include the navigational system, smartphones, music players, emergency services, satellite and HD radio, and Wi-Fi connectivity. The vehicle's telematics system acts as the central hub where all the services connect.

Hands-Free Cell Phone System

Shop Manual
Chapter 14, page 785

The hands-free cell phone system can use different wireless technologies. Bluetooth™ wireless technology is the most common. This technology allows for communication between modules. In this case, the communication is between a compatible cell phone and the vehicle's on-board receiver. The system communicates with a cell phone that is anywhere within the vehicle. The system can recognize several cell phones. Each cell phone user provides an identification number, name, and priority during the setup process. The assigning of this information pairs the cell phone to the system. The pairing process stores the cell phone's IP address in the hands-free module (HFM) and the HFM's IP address into the cell phone.

Bluetooth is a wireless connection technology that uses the 2.4-GHz frequency band to connect a Bluetooth capable phone to the telematics system. This allows for hands-free function of the cellular phone.

The system uses voice recognition technology to control operation. When a programmed cell phone is within range of the HFM, communication between the two components is established. During a cell phone call, the radio speakers broadcast the incoming voice. If any other audio device is using the speakers at that time, the hands-free system automatically overrides it.

When the HFM broadcasts a data bus message announcing the initiation of hands-free operation, the vehicle radio stores its current volume level and mode. The radio then switches to the stored hands-free volume level. The amplifier fades all of the speakers and then transmits the hands-free audio through the speakers. When the HFM broadcasts a data bus message signaling the termination of the cell phone call, the radio stores the HFM volume level and returns to the previous radio mode and volume level. At this time, the amplifier returns all speakers to their previous audio level.

A dual-element microphone module (DEMM) (mounted in the dash, center console, or rearview mirror) consists of microphone elements and electrical circuitry, including a preamplifier network. The DEMM can tune the microphone frequency response to improve the voice recognition function.

Systems incorporating the hands-free feature into the telematics system will typically use the cell phone to route audio and data to the telematics control head. However, some systems can access the local cell network directly to allow the vehicle to transmit and receive emergency voice communication and data even when no cell phone is present.

Navigational Systems

Shop Manual
Chapter 14, page 788

The **navigational system is** one of the fundamental components of the telematics system (**Figure 14-34**). These systems use satellites to direct drivers to their desired destinations. Most navigational systems use a global positioning system (GPS) antenna to determine the vehicle's location by latitude and longitude coordinates. A gyroscope determines vehicle turns. Usually, a thin-film transistor (TFT) or liquid crystal display (LCD) color screen displays map data (provided on a DVD or over-the-air download) and navigation information. The audio system speakers broadcast the voice prompts.

Most systems will provide at least some of the following features:

- Full-screen map display.
- Vehicle location and route guidance.
- Turn-by-turn distance in feet or meters.
- Points of interest.
- The storage of favorite routes and locations.
- The storage of recent routes.

During most conditions, the gyroscope, data bus information, and the map data locate the vehicle on the displayed map. The system uses GPS data if the vehicle travels into an unmapped area.

A vehicle-tracking system may incorporate the GPS system. This provides a method of using satellites to track a stolen vehicle. By tying the navigational system to the hands-free cell phone system, the driver can get on-road assistance. For example, if the driver locks the keys in the vehicle, they can call the assistance line and the representative can send a signal to unlock the doors. If the airbags deploy, the system can automatically send a signal of this event. A representative from the tracking subscription company will

Figure 14-34 Navigational system monitors and controls.

attempt to get in touch with the vehicle occupants to see if assistance is required. They can dispatch emergency personnel if needed since the satellite system informs the representative of the vehicle's exact location.

Some navigational systems utilize local highway department data concerning construction zones, traffic congestion, and accidents to display real-time driving conditions. Based on this information, the navigational system will calculate an alternate route and guide the driver around the congested traffic.

Gyro Sensors. Vehicle control, rollover mitigation, and navigational systems use different types and names of X-Y-Z axis sensors. They detect yaw, lateral, and roll movements of the vehicle. Each of these measurements involves individual sensors; however, the sensors reside in a single unit.

The X-Y-Z coordinates represent positions in space. These coordinates influence the forces acting upon a vehicle in motion. The navigational system can use the yaw and lateral sensors to determine vehicle acceleration rate and turning.

A **gyroscope** is a device used to measure orientation in space (**Figure 14-35**), using the principle of conservation of **angular momentum**. This is similar to inertia in that the angular momentum is the measure of the extent that an object rotating around a reference point will continue to rotate unless acted upon by an external torque. This "gyroscopic inertia" resists changes in orientation and makes the instrument a very good indicator of orientation. In fact, high-quality compasses use gyroscopes as their element instead of typical magnets. This is because a magnetic compass will point to magnetic north, which is actually some distance from true north. This variance needs to be compensated for when using a magnetic compass. The gyroscope possesses a much more accurate directional indicator.

Since gyroscopes of this type are not practical in an automotive application, it was not until the development of **Micro-Electro-Mechanical System (MEMS)** gyros that allowed the installation of more advanced safety and navigational systems onto the vehicle. The vibrating structure gyroscope incorporates two proof masses that are induced to vibrate at a frequency along a plane within these small units. The Coriolis effect induces an acceleration on these masses. Measuring lateral motion is due to the force of the Coriolis effect. This produces a signal that corresponds to the rate of rotation.

Micro-Electro-Mechanical System (MEMS) integrates mechanical elements, sensors, actuators, and electronics onto a common silicon substrate using a process called micro-fabrication technology. Fabrication of the electronic portion of the MEMS uses the integrated circuit (IC) process. Fabrication of the micromechanical components uses a process that etches away parts of the silicon wafer and then adds new layers to the structure that will form the desired mechanical and electromechanical device.

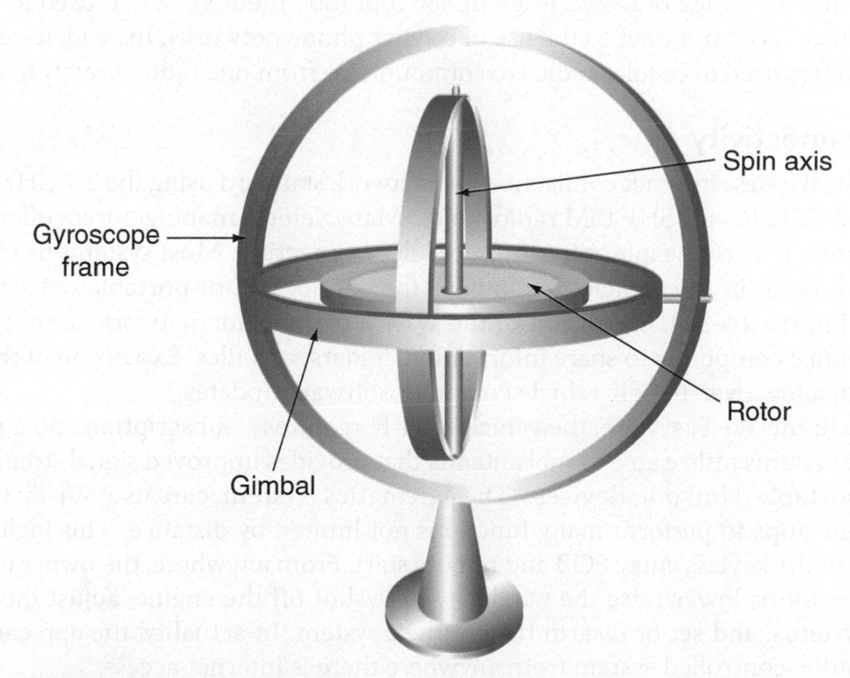

Figure 14-35 Gyroscope construction.

Navigational systems that rely on satellites and GPS to determine the vehicle's position typically use MEMS gyro sensors. The gyro informs the system of the speed and direction of the vehicle. If the vehicle is making a turn, the direction displayed on the monitor will indicate this change.

The gyro sensor measures the rotation of the vehicle around the Z-axis. This input, along with a lateral acceleration sensor and steering angle sensor, indicates the vehicle's path of travel.

Dead Reckoning. In the event that GPS satellite signals are lost, the navigation system will invert to **dead reckoning**. Dead reckoning calculates the vehicle's position using the previously determined location and then uses vehicle speed, time, and gyro inputs to estimate the current location. Converting to dead reckoning is sometimes evidenced on the monitor as showing the vehicle in a location that it really isn't. For example, showing the vehicle driving down railroad tracks instead of on the highway. Signal loss can occur when driving through a tunnel, in a canyon, or in parking garages. Using the last known position and direction of travel, the navigational system will use vehicle speed and gyro sensor inputs to calculate the approximate position of the vehicle. Many systems will display a message such as, "Satellite Signal Lost" to alert the driver that the system converted to dead reckoning and the displayed vehicle position may not be accurate.

Vehicle Connectivity

The concept of a "connected vehicle" has been a goal of the automotive industry for many years. Other industries have already adopted the technologies to fulfill this goal. Currently, seven vehicle connectivity types fall under the category of Vehicle to Everything (V2X) communications. The communication technology used may differ for each type. Today, the three most common communication protocols are **Dedicated Short-Range Communications (DSRC), cellular**, and Bluetooth.

DSRC is a variant of Wi-Fi wireless communications that enables vehicles to communicate with each other without relying on cellular or other infrastructure. A vehicle with DSRC broadcasts its location, heading, and speed to other DSRC equipped vehicles. The transmission range of DSRC is about 980 foot (300 meters). When used to identify V2X, "cellular" does not refer to the use of cellular phone networks. Instead, it references the technology used in cellular radios to communicate from one radio directly to another.

Wi-Fi Connectivity

Shop Manual
Chapter 14, page 788

Wi-Fi is the wireless internet connection and network standard using the 2.4 GHz (12 cm) UHF and 5 GHz (6 cm) SHF ISM radio bands. Many vehicle manufacturers offer vehicle Wi-Fi connectivity to the internet using a cellular connection. Most systems use Wi-Fi to create a Hotspot in the vehicle to allow for the connection of portable devices to the internet (**Figure 14-36**). Expansion of the system can include network connectivity to home or office computers to share information, folders, and files. Expansion of the Wi-Fi system can allow over-the-air vehicle computer software updates.

Like a home Wi-Fi system, the vehicle's Wi-Fi requires a subscription and a modem. In-vehicle systems utilize an external antenna that provides improved signal strength over that of portable Hotspot devices. The telematics system can use Wi-Fi to allow smartphone apps to perform many functions not limited by distance. This includes the functions of the keyless entry FOB and remote start. From anywhere, the owner can lock/unlock the doors, lower/raise the windows, start/shut off the engine, adjust the climate control systems, and set or disarm the antitheft system. In actuality, the app can access any computer-controlled system from anywhere there is internet access.

Figure 14-36 The telematics system may offer Wi-Fi connectivity.

Vehicle to Device Connectivity

Vehicle to cell phone connectivity is the most common form of Vehicle to Device (V2D) connectivity. Typically, V2D uses Bluetooth protocol to exchange information between the vehicle smart devices, such as tablets or wearables. The exchange may utilize the vehicle's infotainment system. Once connected, the device can share certain file formats using Apple's CarPlay® or Google's Android Auto®.

Vehicle to Vehicle Connectivity

Many of today's cars already come equipped with advanced safety systems that use radar, lidar, ultrasonic sensors, and cameras to detect vehicles, pedestrians, and other objects. Vehicle to Vehicle (V2V) connectivity allows moving vehicles to communicate with each other and share the information gathered by these sensors. Line of sight between the vehicles is unnecessary for sharing of information (**Figure 14-37**).

Vehicles with V2V technology become nodes of a mesh network since they gather, send, and retransmit signals. The communication of shared data between vehicles is in real time. This gives the V2V equipped vehicle a virtual 360-degree view of its surroundings. The data can hop from vehicle to vehicle, extending the transmission range from the initial 980 foot (300 meters) to well over a mile (1.6 kilometers).

The shared information can alert drivers about an imminent hazard that they may be unaware of. For example, suppose that a line of vehicles has five cars. The lead and the last vehicles have V2V, but the vehicles between them don't. The driver of the last vehicle may not have a clear view of the lead vehicle. If the lead vehicle driver needs to stop their vehicle quickly, the driver at the end of the line may not see it. The driver second in line may see what is happening and be able to serve around the lead vehicle. This may lead to the other drivers applying their brakes or doing evasive maneuvers. By the time the driver of the last vehicle can react, it may be too late. But with V2V, a "Hard Brake Application" message is broadcasted as soon as the lead driver applies their brakes. The vehicle at the end of the line receives the message and delivers a warning that a driver ahead of them is applying their brakes hard. The received message also alerts the collision avoidance system, so it is "readied."

By monitoring almost all vehicle systems, hazardous conditions can be identified and broadcasted. For example, broadcasting the activation of the electronic stability program indicating a slick road alerts other drivers of the slippery road conditions. Fast steering wheel movements that may indicate an object is on the road are transmitted to alert other drivers. Even transmitting that the wipers have been on for an extended period indicates rain.

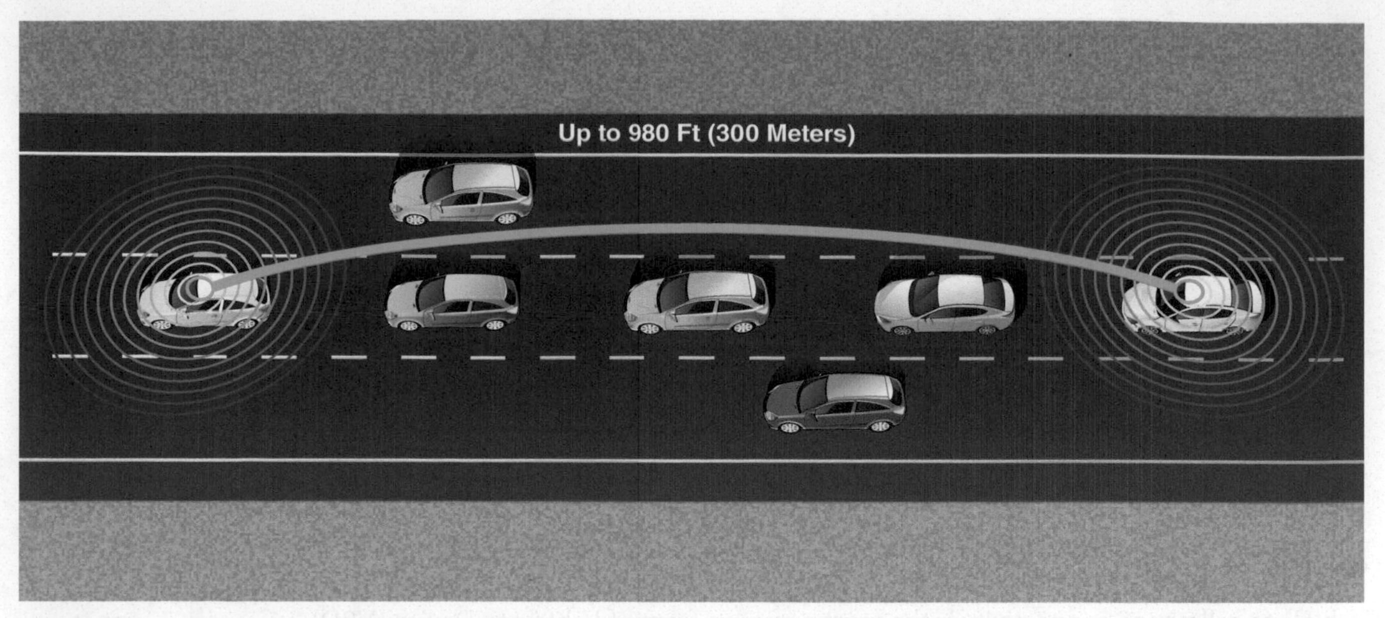

Figure 14-37 V2V provides an additional layer of safety since it can gather information from other vehicles to determine a possible hazard without requiring a line of sight.

Vehicle to Infrastructure Connectivity

Vehicle to Infrastructure (V2I) connectivity has been used in a limited form for several years. Many navigational systems use data acquired from the Intelligent Transportation System (ITS) or local authorities to display posted speed limits, traffic hazards, and construction zones.

The V2I uses the ITS to gather information from road sensors (such as cameras, traffic lights, streetlights, lane markers, road signs, etc.) and vehicle-generated traffic data from V2X equipped vehicles to monitor road and traffic conditions. V2I connects the vehicle to its surrounding infrastructure and provides real-time, bidirectional information exchange between the two. If there is a hazard ahead, the driver is alerted in advance.

V2I equipped vehicles can receive real-time data from traffic controllers concerning signal phasing and timing. This information can alert the driver that the traffic light ahead is about to turn red. In this case, the driver can begin to slow down before the light turns yellow. In addition, the V2I system can alert drivers that a vehicle is potentially going to run a red light.

Vehicle to Pedestrian Connectivity

Vehicle to Pedestrian (V2P) has the potential to save lives. However, currently there are many challenges to implementing this system. Implementing smart sensors into wheelchairs, bicycles, and strollers so the V2P system can be aware of these items is easy. However, implanting smart sensors to identify pedestrians, children playing, and pets is not practical. Some current safety systems use radar lidar, 360-degree cameras, and so forth to detect pedestrians. These devices work well for line-of-sight but offer limited advance notice if the pedestrian is behind something.

Vehicle to Grid Connectivity

Vehicle to Grid (V2G) communication exchanges data between plug-in hybrid and electric vehicles with the electric grid. Data gathered allows the electric grid to balance loads and reduce electrical utility costs efficiently.

Vehicle to Network Connectivity

Vehicle to Network (V2N) is the connection that ties all connectivity types together and is the backbone to V2X. V2N uses DSRC or cellular communications to create an ad hoc Wi-Fi hotspot to interact with other vehicles, the road infrastructure, data centers, laptops, and other devices connected to the internet.

V2N can control devices of the smart home. With V2N, the driver can control their home's HVAC system, turn on or off the oven, lights, coffee makers, entertainment systems, and so forth.

Vehicle to Cloud Connectivity

Vehicle to Cloud (V2C) communication uses the V2N connection to provide data exchange with the cloud. Some applications of this technology include:

- Over-the-air software updates to the vehicle's on-board computers
- Remote vehicle diagnostics
- Redundancy to DSRC communication
- Bidirectional communication with smart home appliances that connect to the cloud (such as security and camera systems)
- Storing and retrieving drivers' preferences for seat position, mirrors, radio stations, etc.

Summary

- Any system using radio frequency to broadcast data must have a transmitter and a receiver.
- The transmitter generates the radio frequency by taking the data and encoding it onto a sine wave. A receiver receives and decodes the sine wave.
- The transmitter and the receiver require an antenna to radiate and capture the radio wave signal. Antennas convert radio frequency electrical currents into electromagnetic waves and vice versa.
- There are three common ways to modulate a sine wave: pulse modulation, amplitude modulation, and frequency modulation.
- In a simple crystal radio, a capacitor/inductor oscillator performs as the tuner for the radio that resonates and amplifies one frequency and ignores all the others.
- A detector extracts the data (voices and music) from the sine wave.
- The electronics of the remote keyless entry (RKE) system are responsible for translating coded input from the key FOB into commands for driver outputs.
- Rolling code uses a counting system that requires a different signal every transmission to improve security. The RF generator modulates the carrier code, creating an output for the RKE receiver.
- The keyless start system is an enhancement to the remote keyless entry system that allows starting the vehicle without using of an ignition key.
- The keyless start system uses a transmitter card or a FOB, signal acquisition and actuation modules (SAMs) installed throughout the vehicle, an electronic ignition switch (EIS), and antennas. The electromagnetic field of the antennas causes the transmitter card to send its code by RF signal and determines if the transmitter card is located within the vehicle or outside of the vehicle.
- Keyless start systems will only perform the engine-starting function if the antenna determines that the transmitter is inside the vehicle.
- The tire pressure monitoring (TPM) system is a safety system that notifies the driver if a tire is under- or overinflated.
- Most TPM systems use a pressure sensor transmitter in each wheel. The TPS's internal transmitter uses a dedicated radio frequency signal to broadcast tire pressure and temperature information. In addition, it transmits its transmitter ID.
- Premium TPM systems use additional components or programming to identify which tire is out of limits. These systems can use a transponder located behind the wheel splash shields, signal strength and rotation, or the ABS wheel speed sensors to determine tire location.

- The immobilizer system protects against unauthorized vehicle use by disabling the engine if an invalid key is used to start the vehicle or an attempt to hot-wire the ignition system is made.
- The primary components of the immobilizer system include the immobilizer module, ignition key with transponder chip, the PCM, and an indicator lamp.
- Inserting an ignition key equipped with the radio frequency identification (RFID) transponder into the ignition switch cylinder and turning it to the RUN position, the immobilizer module radiates an RF signal through its antenna to the key. The transponder signals back to the antenna its identification number. If the message properly identifies the key as being valid, the immobilizer module sends a message over the data bus to the PCM authorizing engine starting.
- The immobilizer module is programmed with a unique secret key code and retains in memory the unique ID number of all keys programmed to the system. The secret key code is transmitted to the keys during the programming function and is stored in the transponder.
- To gain secured access to the immobilizer module for service requires a PIN code.
- Infotainment refers to a type of media system that provides a combination of information and entertainment.
- Speakers turn the electrical energy from the radio receiver amplifier into acoustical energy. The acoustical energy moves air to produce sound.
- Vehicle audio entertainment systems are generally only a single component (receiver). Recently, manufacturers have been developing multiple components (tuner, amplifier, control head, etc.) to allow for greater flexibility.
- Some audio component systems utilize a serial data line to provide for communication and control of functions between components.
- The infotainment system remote controls can either be resistive multiplexed switches or use a supplementary bus system such as LIN.
- Sound system amplifiers can be either remote mounted or integrated with the speakers.
- Satellite radio provides several music and talk show channels using orbiting satellites to provide a digital signal. An in-vehicle receiver receives the digital signal, which travels to the conventional vehicle radio as an auxiliary audio input and plays through the normal speaker system.
- The satellite digital audio receiver (SDAR) is an additional audio receiver, separate from the vehicle's radio. The SDAR processes the satellite's signal and provides an input signal to the conventional radio over dedicated circuits for right and left channels.
- The multiplexer acts as an electronic switch to switch between the different audio sources. Depending on which source requires use of the system, the multiplexer makes the switch and sends the output to the radio.
- Most DVD systems display the video on a flip-down, roof-mounted monitor or on a monitor attached to the back of the front seats, and most play the audio through the vehicle's regular audio system.
- Hands-free cell phone system uses wireless technologies such as Bluetooth™, which allows for communication between a compatible cell phone and the vehicle's on-board receiver.
- Navigational systems use satellites to direct drivers to their desired destinations using a GPS antenna, to determine the vehicle's location by latitude and longitude coordinates, and a gyroscope to determine vehicle turns.
- Incorporating the navigational system into a vehicle-tracking system enables tracking the vehicle, using the satellite, if it is stolen.
- Dedicated Short-Range Communications (DSRC), cellular, and Bluetooth are the most common V2X protocols.
- V2V connectivity allows vehicles that are within range to communicate with each other and share data to improve road safety by alerting drivers of potential hazards in advance.
- V2V can gather information from almost all vehicle systems to determine the current state of vehicle handling (such as driving on an icy road) and warn other drivers of road hazards.
- V2I connectivity allows the vehicle to communicate with traffic signals, road sensors, highway cameras, and other ITS systems.

Review Questions

Short-Answer Essays

1. Explain the operation of the immobilizer system.
2. Explain how radio waves are generated.
3. Explain the operation of the transmitting antenna.
4. How does the keyless start system determine if the transmitter is in the vehicle?
5. Explain the operating characteristics of speakers.
6. What methods are used to reduce RFI and EMI in the radio?
7. What three methods can be used by the tire pressure monitoring system to determine the location of the tire on the vehicle?
8. Describe how the audio system changes outputs when a hands-free cell phone call is initiated.
9. Explain the principle of the MEMS gyro sensor.
10. How does the navigational system track vehicle movement using dead reckoning?

Fill in the Blanks

1. The _______________ acts as an electronic switch to alternate between the different audio sources that travel to the conventional vehicle radio as an auxiliary audio input and play through the normal speaker system.
2. The tire pressure monitoring system uses a(n) _______________ _______________ signal to broadcast tire pressure and temperature information.
3. Satellite radio systems use an in-vehicle receiver that receives the _______________ signal.
4. The _______________ resonates and amplifies one frequency and ignores all the others.
5. The transmitter generates the radio frequency by taking the data and encoding it onto a(n) _______________ wave.
6. _______________ convert radio frequency electrical currents into electromagnetic waves.
7. Speakers turn the electrical energy from the radio receiver amplifier into _______________ energy.
8. A tire pressure monitoring system can excite the tire pressure sensors to cause them to transmit a constant data stream by using a _______________ Hz frequency.
9. The keyless start system will use _______________ to determine the location of the transponder.
10. Any system that is using radio frequency to broadcast data must have a _______________ and a _______________.

Multiple Choice

1. The tire pressure monitoring system can identify the location of the tire on the vehicle by:
   A. Training magnets.
   B. Electronic transponders.
   C. Both A and B.
   D. None of these choices.
2. Transmitting antennas work by:
   A. Oscillating current to create a magnetic field.
   B. Magnetic field induction of an electric field.
   C. Both A and B.
   D. Neither A nor B.
3. What is the function of the radio's detector?
   A. Determines the frequency to amplify.
   B. Extracts the data from the sine wave.
   C. Receives the sine waves from space.
   D. All of these choices.
4. If a radio system has only one audio input but multiple audio sources, it may use a(n) _______________ to switch between the sources.
   A. Decoder
   B. Translator
   C. Multiplexer
   D. Grid/antenna module

5. The speaker produces sound by:
   A. Creating a noise from a vibrating diaphragm.
   B. Changing air pressures in front of the speaker cone.
   C. Rubbing two dissimilar materials together at the speed of sound.
   D. Changing resistance of the permanent magnet.
6. Each key of the immobilizer system has a unique:
   A. Secret key code.
   B. Transponder ID code.
   C. Cover interface.
   D. TFT overlay.
7. The purpose of transmitting temperature information along with the tire pressure readings is to:
   A. Alert the driver of tire over temperature conditions.
   B. Prevent the tire pressure monitoring system from using the data sent by the tire pressure sensors when the tires are cold.
   C. To prevent displaying false low tire pressure warning messages.
   D. To alert the driver that ice may be on the road.
8. The satellite radio signal is processed by:
   A. The antenna.
   B. The radio mode selection transformer.
   C. Data bus interface chip.
   D. The satellite digital audio receiver.
9. Technician A says over-the-air updates to the vehicle's on-board computers is an example of V2C connectivity. Technician B says V2X groups may interconnect with almost all vehicle systems. Who is correct?
   A. Technician A.
   B. Technician B.
   C. Both Technician A and Technician B.
   D. Neither Technician A nor Technician B.
10. The vibrating structure gyroscope can measure lateral motion by:
    A. The force of the Coriolis effect on two elements.
    B. Stress on the Wheatstone bridge.
    C. Capacitance discharge between vibrating plates.
    D. All of these choices.

CHAPTER 15

PASSIVE RESTRAINT AND OCCUPANT SAFETY SYSTEMS

Upon completion and review of this chapter, you should be able to:

- Explain the purpose of passive restraint systems.
- Describe the basic operation of passive seat belt systems.
- Describe the common components of an air bag system.
- List the components and explain the function of the air bag module.
- Describe the function of the clockspring.
- Explain the functions of the ORC used in air bag systems.
- List and describe the operation of the different types of air bag system sensors.
- List the sequence of events that occur during air bag deployment.
- Describe the normal operation of the air bag system warning light.
- Describe the operation of a hybrid inflator module and explain the advantages of this type of module.
- Explain the function of multistage air bags.
- Explain the function of the side-impact air bags and describe the locations of the modules and sensors.
- Describe the operation and purpose of factory-installed air bag on/off switches.
- Describe the purpose and operation of seat belt pretensioners.
- Describe the function of occupant classification systems (OCS).

Terms to Know

Air bag
Belt tension sensor (BTS)
Carriers
Crash sensors
Driver-side inflatable knee blocker (IKB)
Fifth percentile female
Fuel pump inertia switch
G force
Hybrid air bag
Igniter
Inertia lock retractors
Micro-Electro-Mechanical System (MEMS)
Multistage air bags
Passive restraint systems
Piezoelectric accelerometers
Pretensioners
Safing sensor
Seat track position sensor (STPS)
Squib

Introduction

Federal regulations have mandated the use of automatic **passive restraint systems** in all vehicles sold in the United States after 1990. Passive restraint systems operate automatically, with no action required on the part of the driver or occupant. Two- or three-point automatic seat belt and **air bag** systems meet this requirement. Expansion of Federal law requires equipping all passenger vehicles with front passenger-side air bags, along with side-impact and head protection.

In a two-point system, the occupant must manually lock the lap belt.

In this chapter, you will learn the operation of the automatic passive restraint and air bag systems. The safety of the driver and/or passengers depends on the technician properly diagnosing and repairing these systems. As with all electrical systems, the technician must have a basic understanding of the operation of the restraint system before attempting to perform any service.

There are many safety cautions associated with working on air bag systems. Safe service procedures are accomplished through proper use of the service manual and by understanding the operating principles of these systems.

In addition, Federal Motor Vehicle Safety Standard 201, known as the Interior Trim and Safety Refinements Acts, regulate the requirements for occupant protection during accidents. In particular, they address head injuries that may occur during an accident. This act resulted in the development of honeycombed plastic structures for use as trim and door materials in place of steel. The act also covers other structural components such as the instrument panel, the windshield, the rear window, and all side glass, along with all of the pillars, headers, roof, and side rails. It also includes the stitching of the seat belt loops. This means the technician must use all fasteners and proper service procedures whenever they perform any service that requires the removal of any of these components.

Passive Seat Belt Systems

Shop Manual
Chapter 15, page 823

The passive seat belt system automatically puts the shoulder and/or lap belt around the driver or occupant (**Figure 15-1**). The automatic seat belt system utilizes direct current (DC) motors that move the belts by **carriers** on tracks (**Figure 15-2**). The carriers are attached to the shoulder anchor to move or carry the anchor from one end of the track to the other.

One end of the seat belt is attached to the carrier; the other end attaches to the **inertia lock retractors** (**Figure 15-3**). Inertia lock retractors use a pendulum mechanism to lock the belt tightly during sudden movement. Opening the door moves the outer end of the shoulder harness forward (to the A-pillar), allowing for easy entry or exit (**Figure 15-4**). When the door is closed and the ignition placed in the RUN position, the motor moves the outer end of the harness to the locked position in the B-pillar (**Figure 15-5**).

Figure 15-1 Passive automatic seat belt system operation.

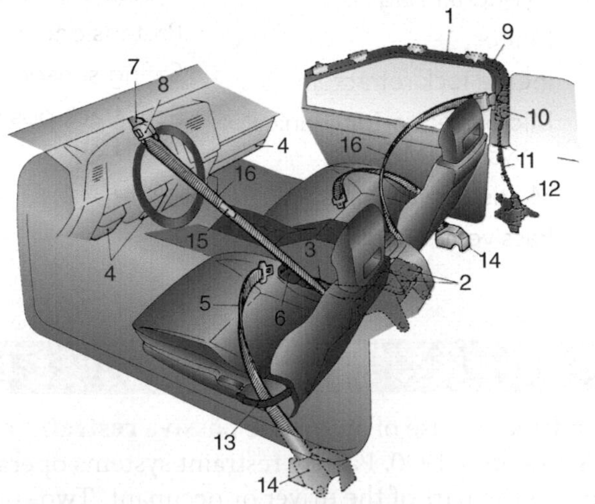

1 Rail and motor assembly
2 Emergency locking retractor assembly
3 Belt guide
4 Knee panel
5 Outer bell assembly (manual tap belt)
6 Inner belt assembly (manual lap belt)
7 Shoulder anchor
6 Emergency release buckle
9 Rail
10 Locking device
11 Tube
12 Motor
13 Belt holder
14 Emergency locking retractor assembly (manual lap belt)
15 Caution label
16 Shoulder belt

Figure 15-2 Passive seat belt restraint system uses a motor to put the shoulder harness around the occupant.

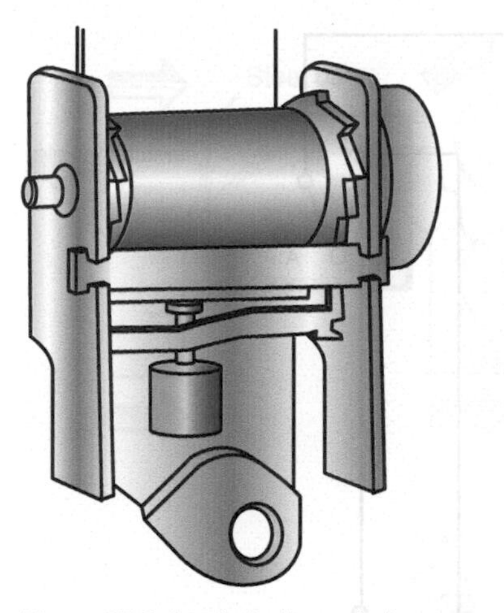

Figure 15-3 Inertia lock seat belt retractor.

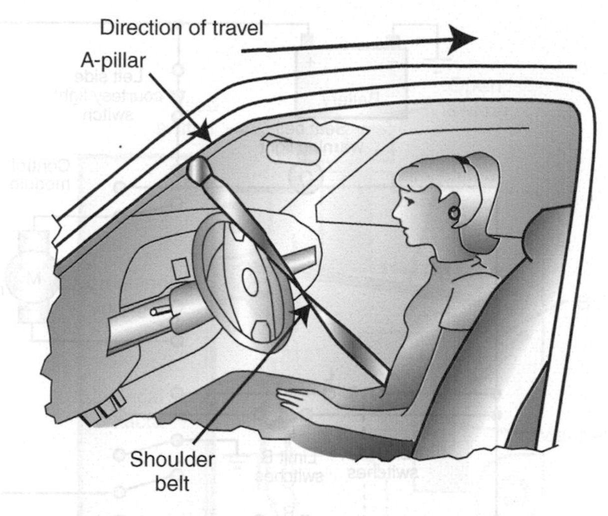

Figure 15-4 When the door is opened, the motor pulls the harness to the A-pillar.

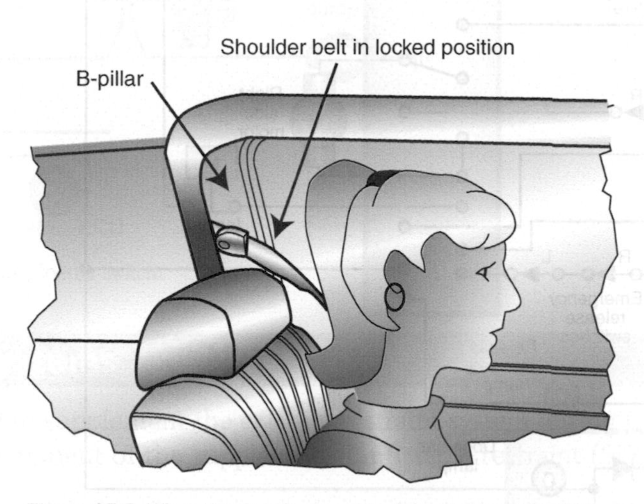

Figure 15-5 When the door is closed and the ignition is in RUN, the motor draws the harness to its lock position.

The automatic seat belt system uses a control module to monitor operation (**Figure 15-6**). The monitor receives inputs from door ajar switches, limit switches, and the emergency release switch.

A BIT OF HISTORY

In the 1920s, there was a movement by some American physicians to have the automobile manufacturers install padded dashboards and seat belts in cars. However, it took until the 1950s before vehicle manufacturers began incorporating these recommendations. Volvo and Saab were the first manufacturers to install padded dashboards and seat belts.

The door ajar switches signal the position of the door to the module. The switch is open when the door is closed. The control module uses this signal to activate the motor and move the harness to the lock point behind the occupant's shoulders. If the module receives a signal that the door is open, regardless of the ignition switch position, it will activate the motor to move the harness to the FORWARD position.

A BIT OF HISTORY

Although there were several early attempts at developing air bags, it was not until the mid-1980s that many manufacturers made air bags available as an option. In 1988, Chrysler was the first automotive manufacturer to offer the driver-side air bag as standard equipment.

The air bag system contains an inflatable air bag module that attaches to the steering wheel. If the vehicle is involved in a frontal collision, the air bag inflates rapidly to keep the driver's body from flying ahead and hitting the steering wheel or windshield. The frontal impact must be within 30° of the vehicle centerline to deploy the front air bag. The air bag system helps to prevent head and chest injuries during a collision. The air bag system may be referred to as a passive restraint because it doesn't require active participation by the driver.

Common Components

A typical air bag system consists of sensors, an occupant restraint controller (ORC), a clockspring, and air bag modules. **Figure 15-9** illustrates typical locations of the common components of the SRS system.

AUTHOR'S NOTE Many of the components used for driver-side air bags are similar to those used in passenger-side air bags. The basic operation of the two systems is the same.

Air Bag Module

Shop Manual
Chapter 15, page 831

The air bag module contains the air bag and inflator assembly packaged into a single module. **Figure 15-10** illustrates a driver's side air bag module mounted in the center of the steering wheel.

The purpose of the air bag module is to inflate the air bag in a few milliseconds when the vehicle is involved in a frontal collision. A typical fully inflated driver-side air bag has a volume of 2.3 cubic feet (65 L).

The air bag module uses pyrotechnology (explosives) to inflate the air bag. The **igniter** (**Figure 15-11**) is a combustible device that converts electrical energy into thermal energy to ignite the inflator propellant. The igniter starts the chemical reaction to inflate the air bag.

Shop Manual
Chapter 15, page 828

At the center of the igniter assembly is the **squib**, which contains zirconic potassium percolate (ZPP). Squib is a pyrotechnic term used for a firecracker that burns but doesn't explode. The squib is similar to a blasting cap. The squib starts the process of air bag deployment. The air bag deploys when as little as 400 mA flows through the squib.

The air bag module is non-serviceable. Replace as a unit if it has been deployed or is defective.

An explosion requires fuel, oxygen, and heat. The squib and the igniter charge of barium potassium nitrate (a very fast-reacting explosive) provide the heat necessary for inflator module explosion. In early air bag modules, the fuel is a generant containing sodium azide and cupric oxide. The sodium azide provides hydrogen, and the cupric oxide provides oxygen. The exploding of the chemicals in the inflator module produces large quantities of hot, expanding nitrogen gas very quickly. The expanding nitrogen gas flows through the igniter assembly diffuser to be filtered and cooled before inflating the air bag. Four layers of screen are positioned on each side of the ceramic in the filter. The filter traps

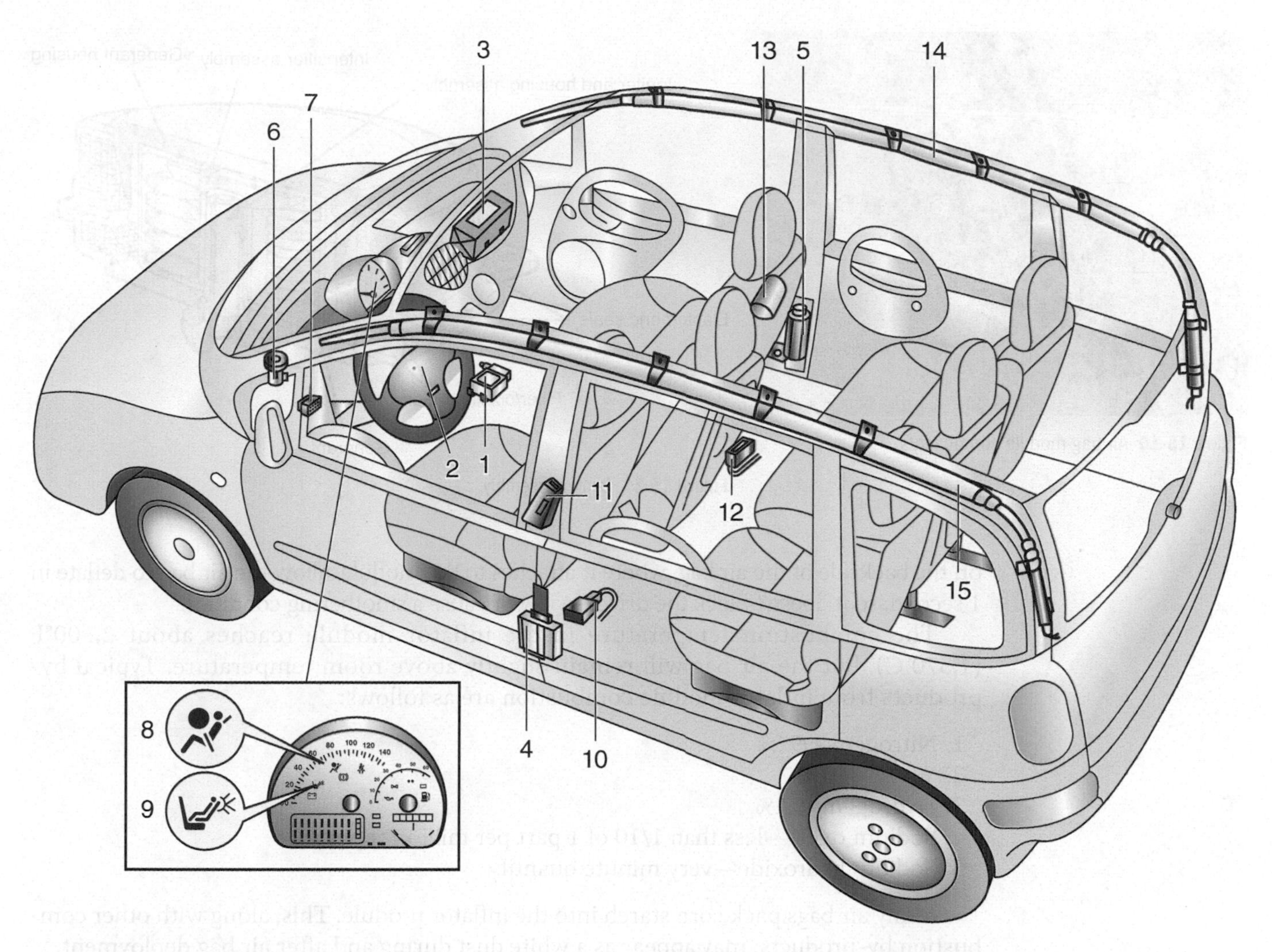

Figure 15-9 Typical location of components of the air bag system. Courtesy of 4CarData.info.

sodium oxide dust. Sodium hydroxide is an irritating caustic. Therefore, automotive technicians must adhere to warnings concerning wearing safety goggles and protective gloves when servicing deployed air bags. Within seconds after air bag deployment, the sodium hydroxide changes to sodium carbonate.

Tear seams in the steering wheel cover and in the instrument panel cover above the passenger-side air bag split easily and allow the air bag to exit from the module. Large openings

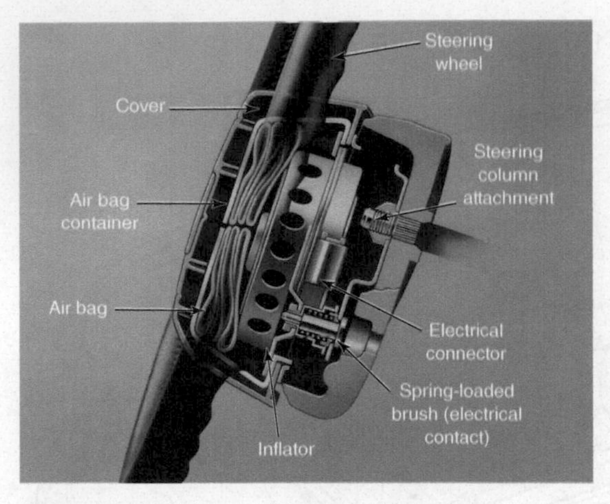

Figure 15-10 Air bag module components.

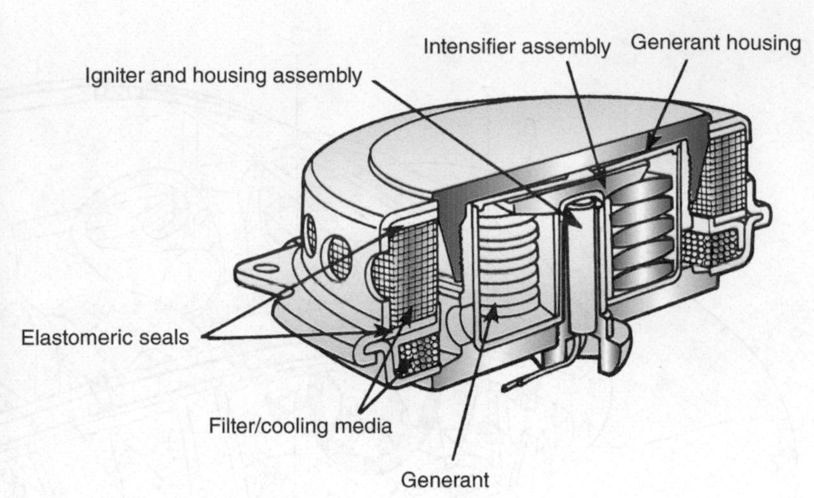

Figure 15-11 Igniter assembly.

on the backside of the air bag, where it attaches to the module, allow the air bag to deflate in 1 second, so it doesn't block the driver's view or cause a smothering condition.

The combustion temperature in the inflator module reaches about 2,500°F (1,370°C), but the air bag will remain slightly above room temperature. Typical by-products from inflator module combustion are as follows:

1. Nitrogen—99.2%.
2. Water—0.6%.
3. Hydrogen—0.1%.
4. Sodium oxide—less than 1/10 of 1 part per million (ppm).
5. Sodium hydroxide—very minute quantity.

Many air bags pack corn starch into the inflator module. This, along with other combustion by-products, may appear as a white dust during and after air bag deployment.

Hybrid Air Bags

Most air bag systems have transitioned from sodium azide to **hybrid air bag** modules. The hybrid air bag module uses compressed gas to fill the air bag. There are three common types of hybrid air bag modules.

Solid Fuel with Argon Gas. The hybrid inflator module contains an initiator similar to the squib in other inflator modules. However, the hybrid inflator module also has a container of pressurized argon gas (**Figure 15-12**). Energizing the initiator in the hybrid inflator uses the same method as conventional systems. Upon energizing the initiator, the surrounding propellant explodes and pushes out the burst disc. As the pressurized argon escapes through the exhaust holes and fills the air bag, the burning propellant heats the argon gas (**Figure 15-13**). Heating the gas makes it expand quickly to fill the air bag.

Early versions of the hybrid system used a pressure sensor mounted at the end of the argon gas chamber opposite the initiator and propellant. This sensor monitors the pressure in the chamber, and if the gas pressure decreases below a preset value, the ORC illuminates the air bag warning light.

Liquid Fuel with Argon Gas. The liquid-fueled air bag module uses a small quantity of ethanol alcohol to blow out the burst disc (**Figure 15-14**). The fuel blows the burst disc, and as the fuel continues to burn, the heat expands the argon gas. The expanding gas pushes through an orifice cup, over the diffuser, and into the air bag. The benefits of using alcohol are it ignites at a lower temperature and doesn't leave a harmful residue.

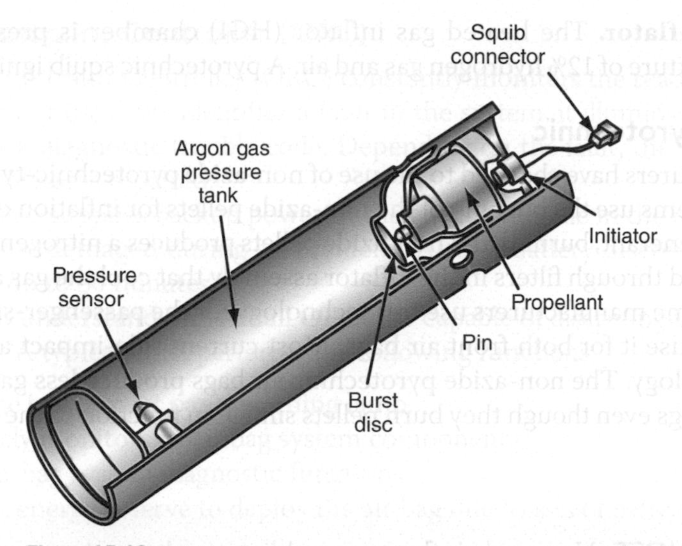

Figure 15-12 Hybrid inflator module with argon gas pressure chamber.

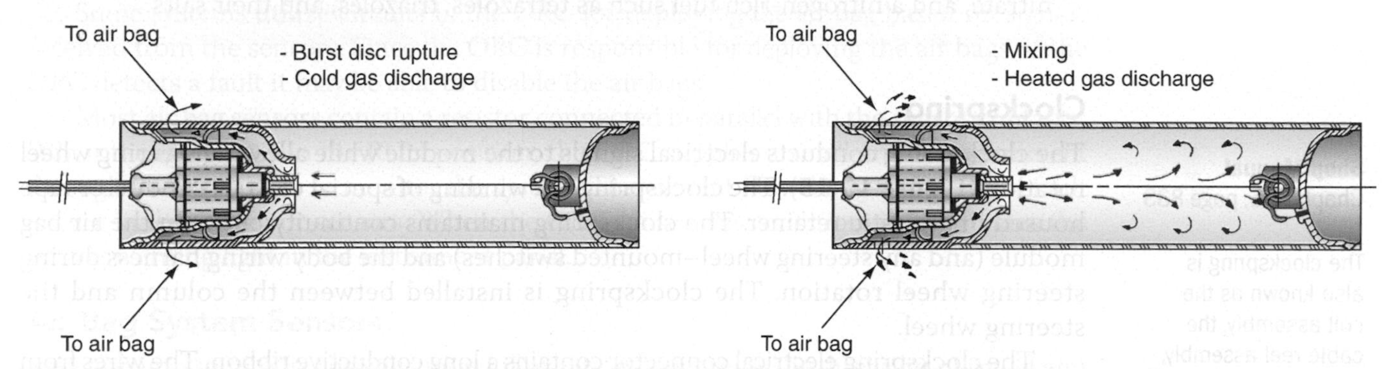

Figure 15-13 Energizing the initiator causes the propellant to explode and puncture the container, allowing pressurized argon gas to fill the air bag.

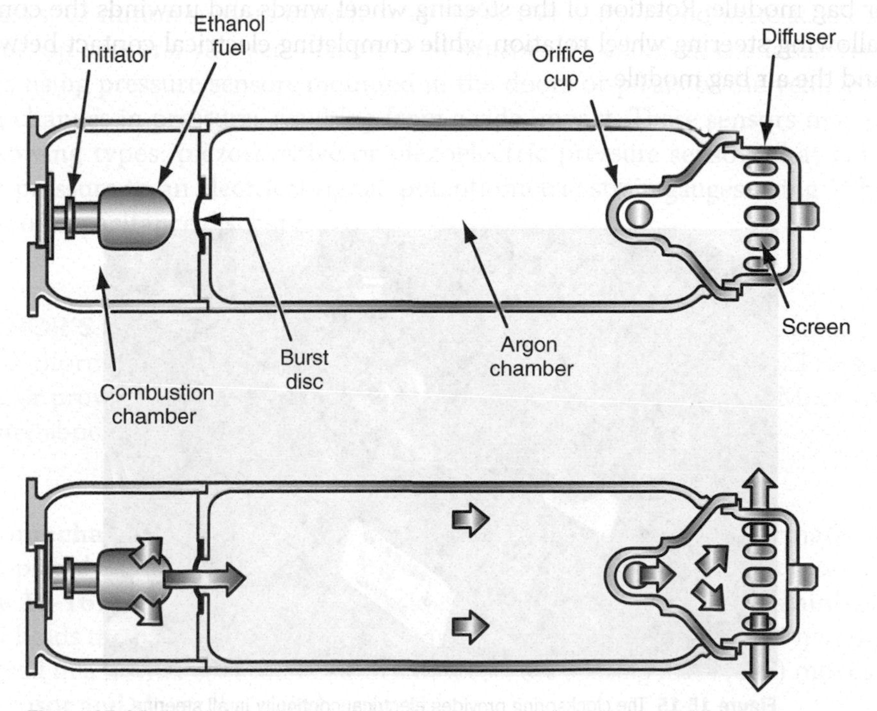

Figure 15-14 Liquid-fueled hybrid inflator operation.

Most sensors are directional. A forward marking (usually an arrow) on the sensor faces toward the front of the vehicle. In addition, only mount the sensor in the original position designed by the manufacturer. Replace any sensor brackets that are bent or distorted.

Some mass-type air bag sensors contain a pivoted weight connected to a moving contact. When the vehicle is involved in a frontal collision with sufficient impact to deploy the air bag, the sensor weight moves in a circular path until the moving contact touches a fixed contact (**Figure 15-17**).

The roller-type sensor has a roller mass mounted on a ramp (**Figure 15-18**). One sensor terminal is connected to the ramp. The second sensor terminal is connected to a spring contact extending through an opening in the ramp without contacting the ramp. A 10,000-ohm (Ω) resistor is connected in parallel to the sensor contacts. Small retractable springs on each side hold the roller against a stop. These springs are similar to a retractable tape measure. If the vehicle is involved in a frontal collision at a high enough deceleration rate to deploy the air bag, the roller moves up the ramp and strikes the spring contact. In this position, the roller completes the circuit between the ramp and the spring contact.

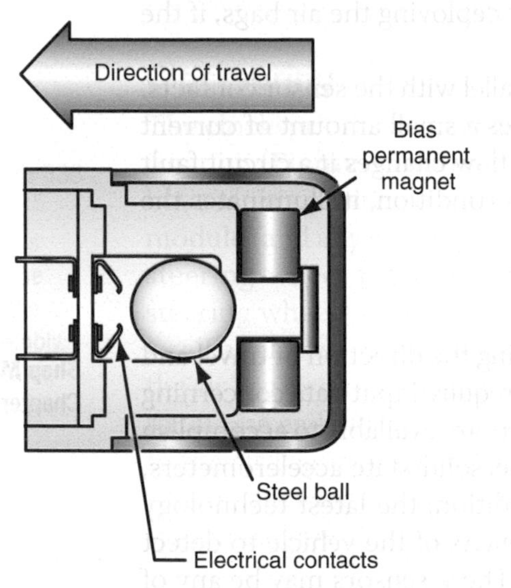

Figure 15-16 Some crash sensors hold the sensing mass by magnetic force. If the impact is severe enough to free the ball, it travels forward and closes the electrical contacts.

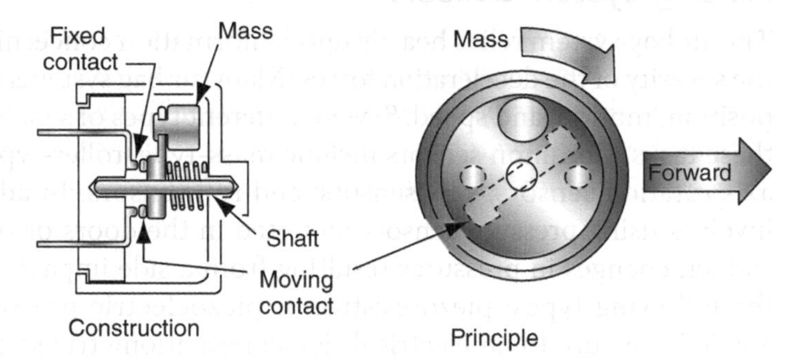

Figure 15-17 Mass-type air bag sensor with pivoted weight connected to a moving contact.

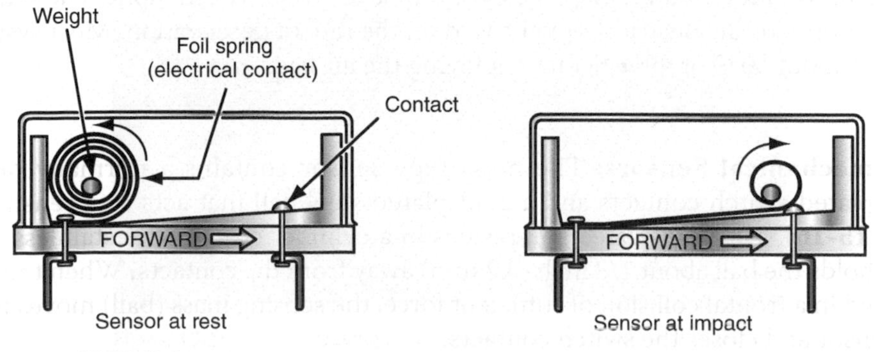

Figure 15-18 Roller-type air bag sensor.

Solid State Accelerometers. **Micro-Electro-Mechanical System (MEMS)** integrates mechanical elements, sensors, actuators, and electronics onto a common silicon substrate using a process called micro-fabrication technology. Fabrication of the electronic portion of the MEMS uses the integrated circuit (IC) process. Fabrication of the micromechanical components utilizes a process that etches away parts of the silicon wafer and then adds new layers to the structure that will form the desired mechanical and electromechanical device. This technology provides expanded capabilities for the use of microsensors and microactuators.

Shop Manual
Chapter 15, page 838

As a sensor, MEMS can gather information about the monitored system by measuring mechanical, thermal, chemical, optical, and magnetic events.

Current systems use MEMS **piezoelectric accelerometers** as the primary impact detection sensors. The accelerometer senses deceleration forces and the direction of impact (**Figure 15-19**). The accelerometer contains a piezoelectric element constructed of zinc oxide (ZnO) that is distorted during a high **G force** condition and generates an analog voltage directly related to the amount of G force. A collision-judging circuit in the ORC receives the analog voltage. The computer deploys the air bag if the collision impact is great enough.

The **piezoelectric accelerometer** generates an analog voltage proportional to a G force.

Many MEMS-type piezoelectric sensors also contain a communication chip. This sensor will condition the analog voltage signal and send a pulsed digital signal to the ORC. The frequency of the pulses is proportional to the G force applied to the sensor. The MEMS sensor offers the advantages of being smaller than the piezoelectric sensor they replaced, and because these sensors have their own IC chip, they can perform self-diagnostic routines.

G force describes the measurement of the net effect of the acceleration that an object experiences and the acceleration that gravity is trying to impart to it.

Another type of MEMS sensor is a piezoresistive sensor that uses a silicon mass suspended from four deflection beams. The deflection beams are the strain-sensing elements. The four strain elements are in a Wheatstone bridge circuit. The beam's strain elements generate a signal proportional to the G forces. An internal chip interprets the changes in resistance over the bridge, then communicates the status to the control module using a frequency-modulated digital pulse.

Both single-point and distributed systems can use accelerometers. The single-point systems use only one accelerometer located within the ORC (**Figure 15-20**). Distributed systems use an accelerometer in the ORC along with remotely mounted high-G sensors at other locations on the vehicle (**Figure 15-21**). Side-impact air bag systems, as well as rollover protection systems, also use accelerometers.

Shorting Bars

The SRS electrical system is a dedicated system and isn't interconnected with other electrical systems on the vehicle. All wiring harness connectors in the system are the same

Shop Manual
Chapter 15, page 826

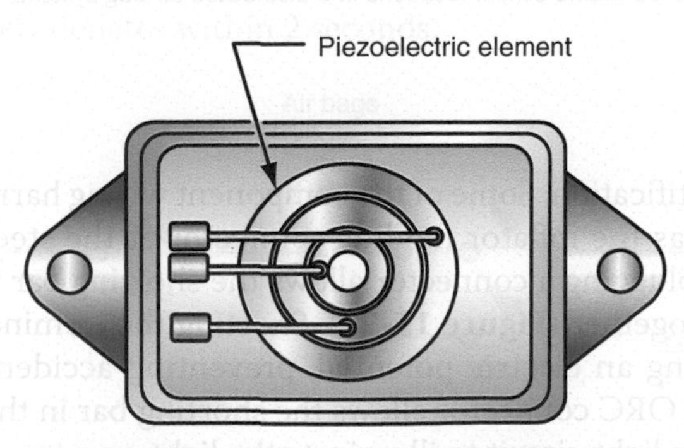

Figure 15-19 Accelerometer air bag sensor with a piezoelectric element.

Passenger-Side Air Bags

Federal law expanded to require that all passenger vehicles produced after 1995 provide front passenger-side air bags (**Figure 15-24**). In most systems, the passenger-side air bag only deploys if the driver-side air bag deploys. Since there is a greater distance between the passenger and the instrument panel compared to the distance between the driver and the steering wheel, the passenger-side air bag is much larger. A typical passenger-side air bag has a fully inflated volume of 7 cubic feet (198 L).

AUTHOR'S NOTE Reference here to passenger-side air bags means the front-seat passenger air bag deployed from the instrument panel. This system isn't to be confused with side-impact air bags, which are discussed later.

Air Bag Warning Lamp

The air bag system warning lamp indicates the system condition to the driver. The ORC operates the warning lamp. The ORC receives ignition on and crank signals. When the ignition is in the RUN position, the air bag warning lamp should illuminate for a bulb check. In some systems, the lamp will flash seven to nine times and then remain steadily illuminated while the engine is cranking. Once the engine starts, the air bag warning lamp should turn off. Any of the following warning lamp conditions may indicate an air bag system failure:

1. With the ignition on, the lamp remains on but doesn't flash.
2. With the ignition on, the lamp flashes seven to nine times and then remains on.
3. The lamp comes on when the engine is running.
4. The lamp doesn't come on at any time.
5. The lamp doesn't come on steadily while the engine is cranking.

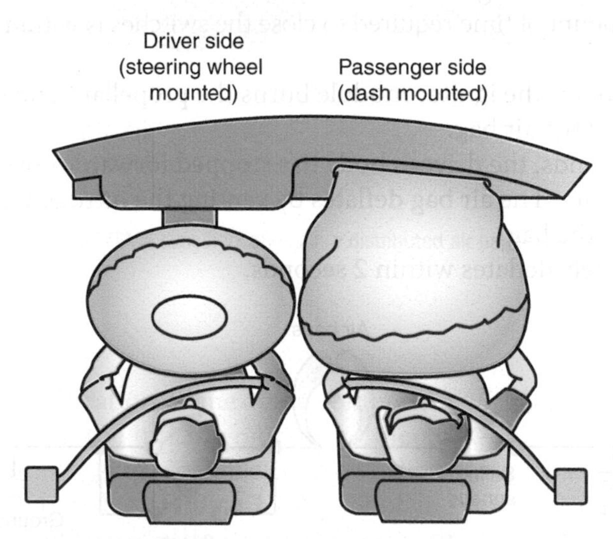

Figure 15-24 Components of typical driver- and passenger-side air bag system.

Multistage Air Bag Deployment

Other than the azide and liquid fuel hybrid air bags, all others module types discussed can be designed for multistage deployment. **Multistage air bags** are hybrid or non-azide air bags that use two squibs to control the rate of inflation (**Figure 15-25**). In the hybrid systems, these modules apply the principle of using heat to expand the argon gas to fill the air bag. The bag fills faster as the heat increases.

The firing of one squib begins the air bag deployment. The firing of the second squib generates additional heat so the air bag fills faster (**Figure 15-26**). The length of time between the firing of the two squibs determines the rate of air bag deployment.

Figure 15-25 The multistage air bag module uses two squibs.

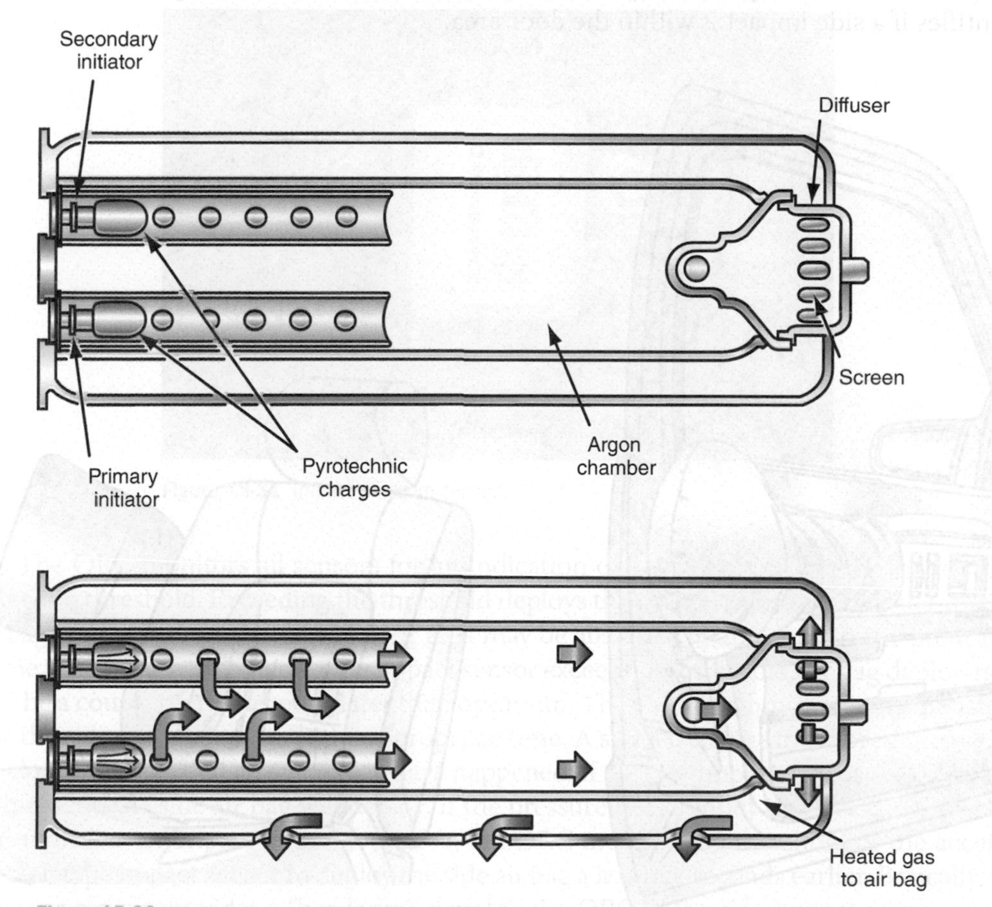

Figure 15-26 Sequence of events within the multistage inflator.

Air Bag On/Off Switches

For the air bags to perform safely, there should be at least 10 inches (25 cm) between the air bag module and the occupant. Also, children should not sit in the front seat with an air bag. NEVER use a rear-facing infant seat (RFIS) in the front seat with an air bag. Some vehicles with limited rear seating may be factory equipped to turn off the passenger air bag if deployment would not be safe.

Earlier systems relied on the driver or passenger to turn a switch (**Figure 15-32**). Placing the switch in the off position disconnects the passenger side air bag from the ORC and cannot deploy. However, placing the switch in the off position connects a resistor to the squib circuit (**Figure 15-33**). The resistor tricks the ORC into believing the circuit is still intact so it will not set false fault codes.

Figure 15-32 Passenger-side air bag on/off switch.

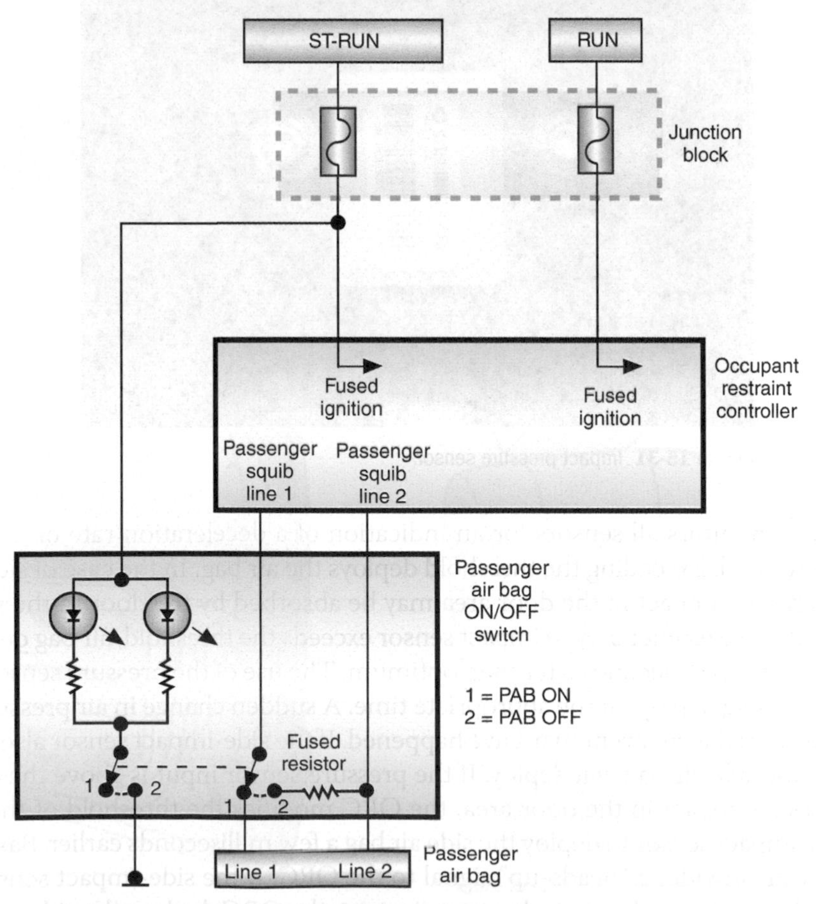

Figure 15-33 Wiring diagram of a hardwired passenger-side air bag on/off switch.

Newer systems use a MUX signal from an on/off module (**Figure 15-34**). The ORC provides a pulsed signal to the on/off module at a frequency of 10 Hz with a 3% duty cycle. The ORC monitors the voltage drop across the switch. With the switch in the ON position, the monitoring circuit sees 4 volts (V). In the OFF position, the ORC sees about 10 volts. A reading of 20 volts indicates an open circuit, while a reading of zero volts sets a short-circuit fault. If the switch is in the OFF position, the ORC will deactivate the passenger air bag internally. This system doesn't interrupt the circuit as earlier systems did.

Seat Belt Pretensioners

To assure that the occupants stay in position during an accident, some vehicle manufacturers have added seat belt **pretensioners**. Pretensioners tighten the seat belt and shoulder harness around the occupant during an accident severe enough to deploy the air bag. The control module deploys the pretensioner and the front air bags at the same time.

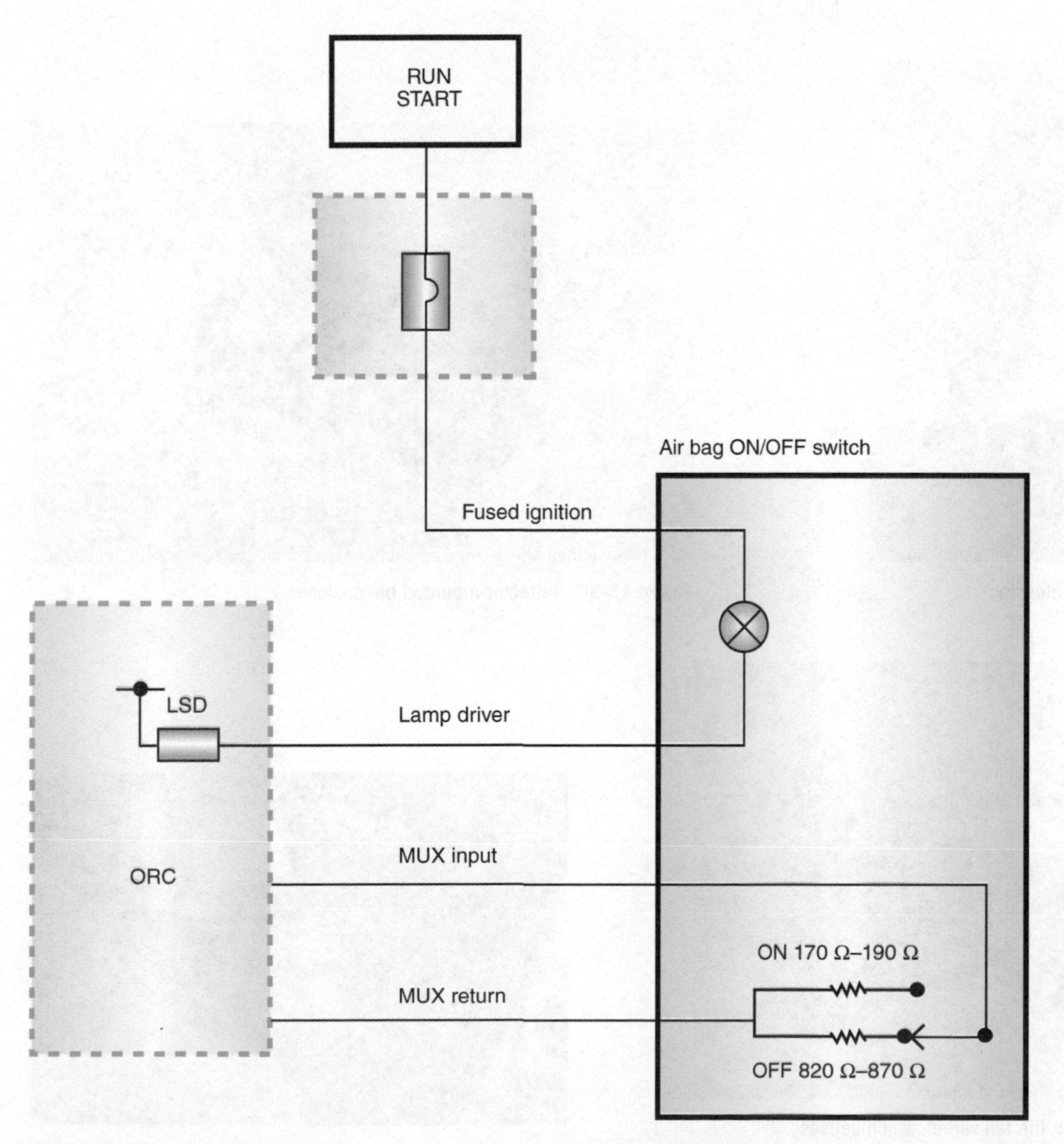

Figure 15-34 Wiring diagram of MUX circuit for passenger-side air bag on/off switch.

All of the vehicle's seat belt assemblies can have pretensioners. The pretensioner can be integrated with the buckle side of the seat belt (**Figure 15-35**). A small piston is attached to a cable connected to the buckle. There is a pyrotechnic charge below the piston. Firing the pretensioner expands the gas, which forces the piston to travel up the cylinder and pulls the buckle tight by the cable.

The pretensioner can also be integral to the retractor (**Figure 15-36)** The system type shown has a retractor assembly with a fan wheel–type unit attached to one end (**Figure 15-37**). Firing the pretensioner's pyrotechnic charge shoots out a series of balls from the channel that strike the fan wheel. As the balls hit the fan wheel, the retractor rotates and pulls the seat belt tight. The last ball is a little bigger than the others and lodges into the fan wheel, causing the seat belt to lock.

The seat belts used with the pretensioner systems usually use a special process of looping the end of the web called an energy management loop. The lower part of the seat belt webbing is looped and stitched with a special rip stitching (**Figure 15-38**) that will release at a predetermined load. The seat belt doesn't come apart, but the ripping of the stitching reduces the load being transmitted through the belt to the occupant by giving the belt some additional "give." Replace the seat belt anytime the vehicle is involved in an accident where the pretensioners have deployed.

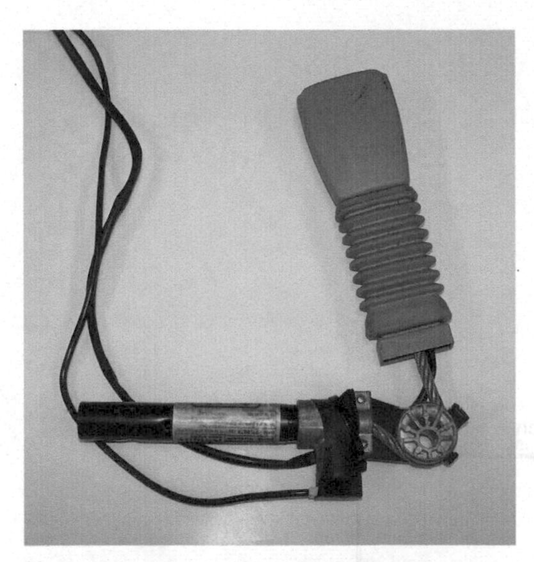

Figure 15-35 Buckle-mounted pretensioner.

Figure 15-36 Retractor-mounted pretensioner.

Figure 15-37 The balls are shot at the fan wheel, which causes the retractor to wind up the seat belt tight against the occupant.

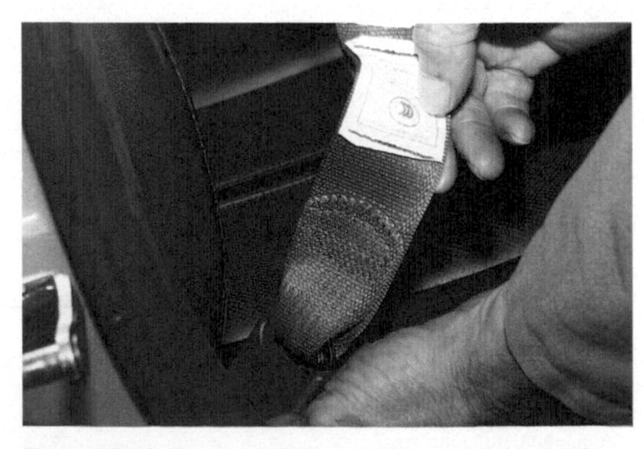

Figure 15-38 The seat belt loop uses a special rip stitching.

Inflatable Knee Blockers

A **driver-side inflatable knee blocker (IKB)** is located beneath the instrument panel cover and attaches to the instrument panel reinforcement. (**Figure 15-39**). The IKB deploys simultaneously with the driver-side air bag to provide upper-leg protection and to maintain the driver's position. Deployment of the IKB pushes a tethered plate against the driver's knees. This keeps the driver in the correct upright position during a collision, making the air bag more efficient.

Occupant Classification Systems

Shop Manual
Chapter 15, page 838

As the result of an amendment to the Federal Motor Vehicle Safety Standard 208, manufacturers have designed and installed air bag systems that reduce the risk of injuries resulting from air bag deployment. The goal is to reduce injuries suffered by children and small adults in the **fifth percentile female** weight classification. The fifth percentile female classifies those individuals who weigh less than 100 pounds (45 kg). The amendment mandates suppressing the passenger-side air bag when an infant in an RFIS occupies the front passenger seat. The amendment also mandated notification to indicate if the passenger air bag is active or suppressed (**Figure 15-40**).

AUTHOR'S NOTE The fifth percentile female is determined by averaging all potential occupants by size and then plotting the results on a graph. The middle of the bell curve on the graph would indicate the majority of occupants. At the far right of the bell curve would be those occupants who are very large, while on the left side of the curve would be occupants who are very small. At the fifth percentile range on the left of the curve, the majority of occupants would be female.

To meet these new requirements, manufacturers have taken different approaches. This section presents the Delphi and TRW systems as examples of two different methods of meeting this regulation. The Delphi system uses a bladder to determine occupant weight, and the TRW system uses strain gauges. Both systems use an occupant

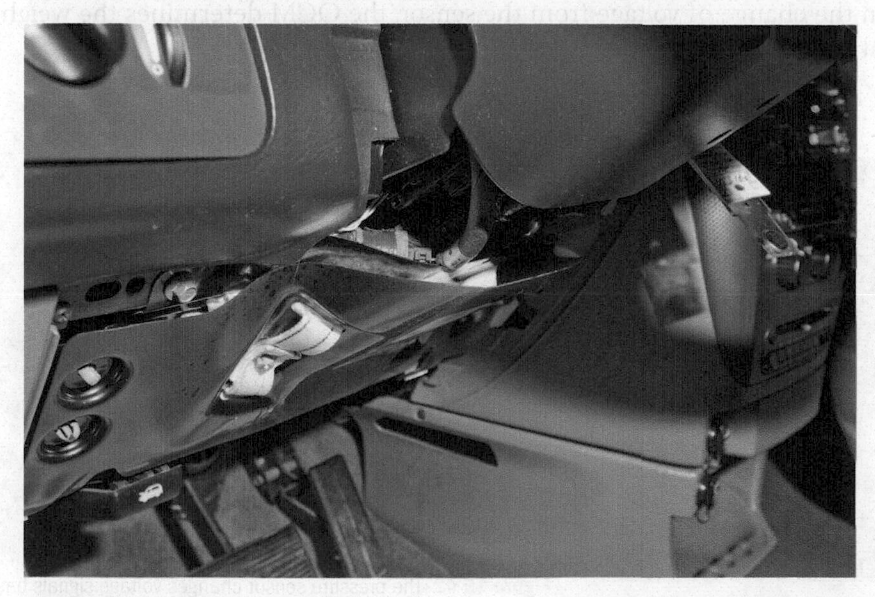

Figure 15-39 Inflatable knee blocker.

Figure 15-40 The PADL illuminates if the air bag is suppressed.

classification module (OCM) that determines the weight classification of the front passenger and sends this information to the ORC. Also, both systems use multistage passenger-side air bags.

Delphi Bladder System

The bladder system uses a silicone-filled bladder positioned between the seat foam and the seat support (**Figure 15-41**). A hose connects a pressure sensor to the bladder (**Figure 15-42**). When someone occupies the seat, pressure applied to the bladder disperses the silicone, and the pressure sensor reads the increase in pressure. The OCM receives this data (**Figure 15-43**).

Shop Manual
Chapter 15, page 838, 846

Since pressure determines seat occupation, and ultimately occupant classification, the system corrects for changes in atmospheric pressures. In addition, the OCM learns the natural aging of the seat foam by monitoring gradual changes. The OCM stores seat aging and calibration information in the ORC. After installing a new OCM, it retrieves this information from the ORC, so the system continues functioning properly.

When the seat is empty, the OCM compares the sensor voltage with the value stored in memory. Adding weight to the seat changes the voltage values measured by the sensor. Based on the change of voltage from the sensor, the OCM determines the weight of the occupant.

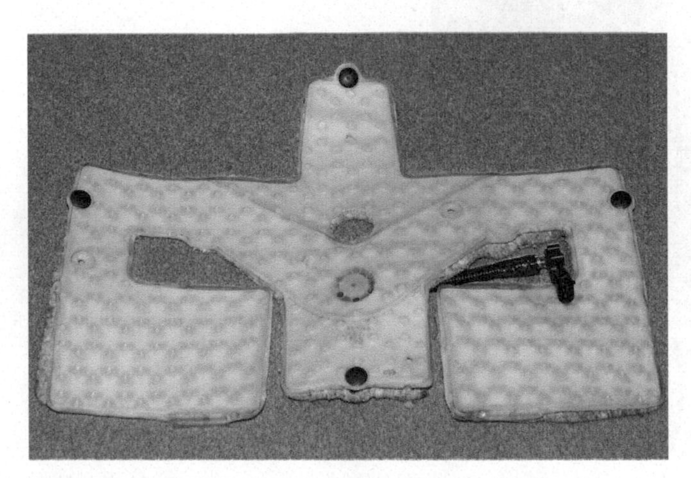
Figure 15-41 Bladder used to determine occupant classification.

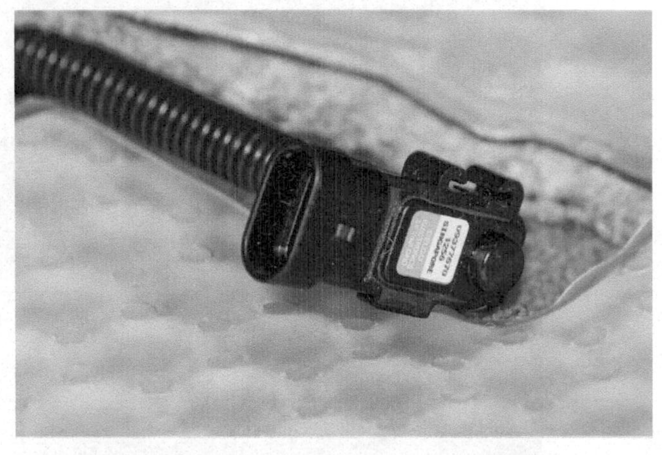
Figure 15-42 The pressure sensor changes voltage signals based on the weight placed on the bladder.

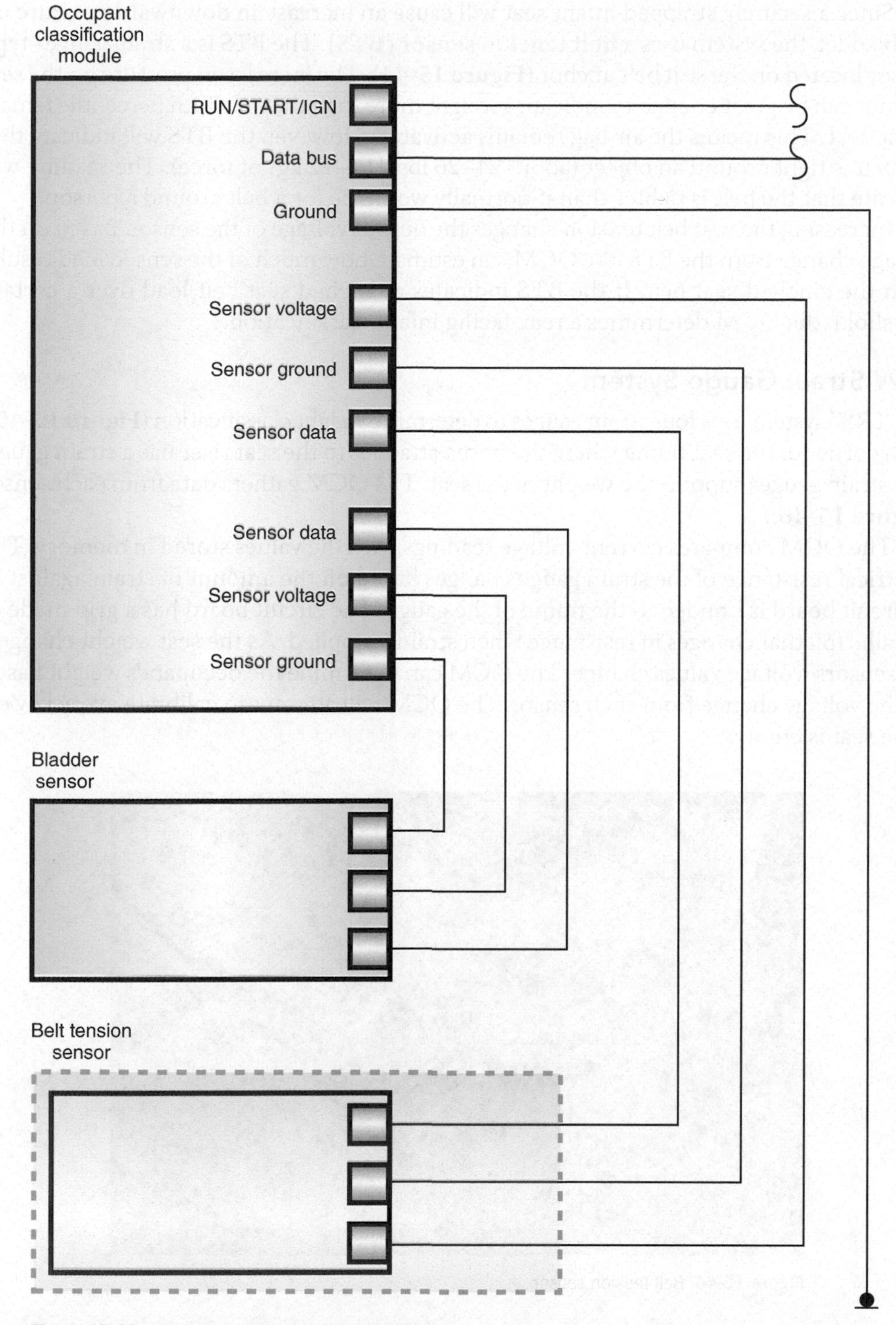

Figure 15-43 Schematic of bladder occupant classification system.

Based on the weight information that the OCM sends to the ORC, the following are determined:

- An empty seat. The PADL is off, and the ORC suppresses the air bag.
- Weight equivalent to or less than that of a six-year-old child. The PADL illuminates, and the ORC suppresses the air bag.
- Weight equivalent to or greater than that of a fifth percentile female. The PADL light is off, and the front passenger air bag is enabled. Deployment rate depends on the severity of the impact.

Shop Manual
Chapter 15, page 839

Since a securely strapped infant seat will cause an increase in downward pressure on the bladder, the system uses a **belt tension sensor (BTS)**. The BTS is a strain gauge–type sensor located on the seat belt anchor (**Figure 15-44**). The increase in pressure on the seat bladder can be great enough to indicate a weight more than that of a fifth percentile female in the seat. This means the air bag remains activated. However, the BTS will indicate that the belt is tight around an object (about 24–26 lbs. [11–12 kg] of force). The reading will indicate that the belt is tighter than it normally would be for a belt around a person.

Increasing the seat belt tension changes the output voltage of the sensor. Based on the voltage change from the BTS, the OCM can estimate how much of the sensed load results from the cinched seat belt. If the BTS indicates a cinched seat belt load over a certain threshold, the OCM determines a rear-facing infant classification.

TRW Strain Gauge System

Shop Manual
Chapter 15, page 847

The TRW system uses four strain gauges to determine weight classification (**Figure 15-45**). Each corner of the seat frame where the frame attaches to the seat riser has a strain gauge. The strain gauges support the weight of the seat. The OCM gathers data from each sensor (**Figure 15-46**).

The OCM compares current voltage readings with the values stored in memory. The electrical resistance of the strain gauge changes based on the amount of strain against it. A circuit board is bonded to the frame of the gauge. The circuit board has a grid made of metallic foil that changes in resistance when strain is applied. As the seat weight changes, the sensors' voltage values change. The OCM can determine the occupant's weight based on the voltage change from each sensor. The OCM will attempt to calibrate every key off if the seat is empty.

Figure 15-44 Belt tension sensor.

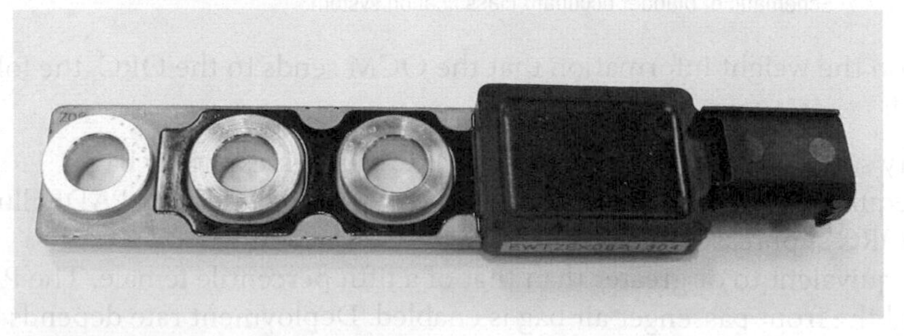

Figure 15-45 Strain gauge.

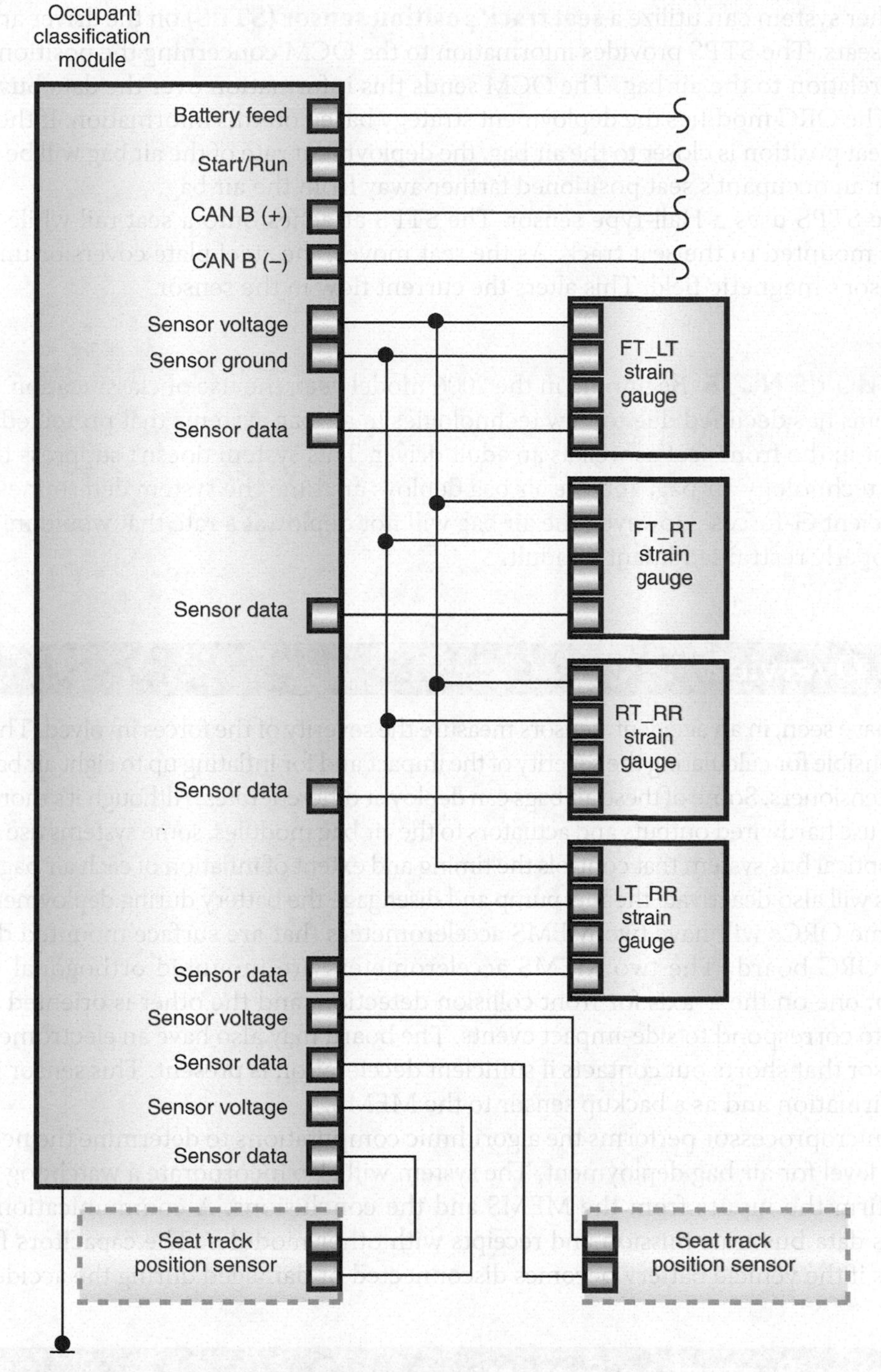

Figure 15-46 Schematic of strain gauge occupant classification system.

Based on the weight information that the OCM sends to the ORC, the following are determined:

- An empty seat. The PADL is off, and the air bag is suppressed.
- Weight equivalent to or less than that of an RFIS. The PADL is illuminated, and the air bag is suppressed.
- Weight equivalent to a child. The PADL is off, and the air bag is enabled. Deployment is at the low-risk deployment level.
- Weight equivalent to or greater than that of a fifth percentile female. The PADL is off, and the front passenger air bag is enabled. Deployment rate depends on the severity of the impact.

Either system can utilize a **seat track position sensor (STPS)** on the driver and passenger seats. The STPS provides information to the OCM concerning the position of the seat in relation to the air bag. The OCM sends this information over the data bus to the ORC. The ORC modifies the deployment strategy based on this information. If the occupant's seat position is closer to the air bag, the deployment rate of the air bag will be slower than for an occupant's seat positioned farther away from the air bag.

The STPS uses a Hall-type sensor. The STPS attaches onto a seat rail while a steel plate is mounted to the seat track. As the seat moves, the steel plate covers or uncovers the sensor's magnetic field. This alters the current flow in the sensor.

AUTHOR'S NOTE Beginning in the 2006 model year, the use of classification systems has declined due to new technologies in air bag systems that protected an infant in the front seat as well as an adult driver. This system doesn't suppress the new technology air bag, and the air bag deploys anytime the system determines sufficient G-forces. However, the air bag will not deploy at a rate that would injure a properly restrained infant or adult.

ORC Overview

As we have seen, in an accident, sensors measure the severity of the forces involved. The ORC is responsible for calculating the severity of the impact and for inflating up to eight air bags and the pretensioners. Some of these air bags can deploy at different rates. Although it's more common to use hardwired outputs and actuators to the air bag modules, some systems use a high-speed optical bus system that controls the timing and extent of inflation of each air bag. Some systems will also deactivate the fuel pump and disengage the battery during deployment.

Some ORCs will have two MEMS accelerometers that are surface mounted directly to the ORC board. The two MEMS accelerometers are mounted orthogonal to one another, one on the Y-axis for front collision detection, and the other is oriented on the X-axis to correspond to side-impact events. The board may also have an electromechanical sensor that shorts out contacts if sufficient deceleration is present. This sensor is used as confirmation and as a backup sensor to the MEMS.

A microprocessor performs the algorithmic computations to determine the need and proper level for air bag deployment. The system will also incorporate a watchdog circuit to confirm the inputs from the MEMS and the conclusions. A communications chip handles data bus transmission and receipts with other modules. The capacitors fire the air bags if the vehicle battery becomes disconnected or damaged during the accident.

Rollover Protection Systems

Shop Manual
Chapter 15, page 848

Rollover detection is an additional function of the air bag system deploys side air bag curtains and the seat belt pretensioners upon detection of a rollover event. This is a feature of the rollover protection systems (ROPS). The MEMS accelerometer, an additional low-G acceleration sensor, and a gyroscopic sensor detect if the driver made too quick of a turn or if the vehicle has swerved to avoid an obstacle. Some systems use a MEMS inclinometer to make these determinations. The inclinometer measures lateral and vertical acceleration (vehicle speed and roll rate) to predict an impending rollover. The sensors may be located internal to the ORC. The inputs from the sensors will determine the rotation angle and rate around the X-axis (**Figure 15-47**). The ORC uses the information gathered to evaluate the potential of a rollover. If a rollover event is imminent, the ORC deploys the side-curtain air bags and the seat belt pretensioners to help protect the occupant from contact with the side of the vehicle's interior and to help prevent occupant ejection. The ORC will determine the appropriate time to deploy the specific air bags or belt pretensioners. In addition, during a potential rollover event, enhanced traction or stability control can cut the throttle and pulse the brakes to correct the vehicle's trajectory.

The ORC compares the tilt angle and rate-of-roll to determine if the vehicle is going to recover, or whether a roll is inevitable. If the vehicle is going to roll, the system deploys the side air bag curtain at a matched deployment speed to the event. A high-speed curb launch will trigger a faster deployment than a corkscrew roll on an embankment.

The side air bag curtains are a new style "safety canopy" curtain hidden above the headliner. The canopy extends from the A-pillar to the C-pillar or D-pillar. To prevent occupant ejection from the vehicle, once the curtain has deployed tethers located on the bottom corners of the bag lock it in place. The controlled deployment speed may happen much faster than with other air bag curtains. The deployment zone is from the roof rail to the bottom of the window, the full length of the vehicle, and the bag is about 5 inches (127 mm) thick.

The curtain uses low-porous materials, so it retains its volume longer. The curtain can remain inflated for up to 6 seconds. The curtain is folded inside the module using "roll fold technology." Rollover protection air bag curtains actually roll down between the window glass and the occupant. The rolling effect allows these curtains to deploy even if the occupant is out of position. In fact, the bag will actually set the occupant upright in the seat.

Convertibles that cannot use side air bag curtains use a ROPS system utilizing deployable roll bars that flip up or pop up (**Figure 15-48**). The flip-up ROPS roll bar remains concealed until activated; then, it flips up in less than three-tenths of a second. Usually, the flip-up roll bar deploys hydraulically and then locks into position.

The pop-up system hides the roll bars within the rear seats (or directly behind them). When deployed, they pop straight up. These systems can be either hydraulically deployed or spring-loaded and operated by an actuator latch device. The spring-loaded system uses an electromagnetic actuator system that releases the rollover protection system in fractions of a second in the event of an accident (**Figure 15-49**).

AUTHOR'S NOTE The Volvo C70 uses safety canopy air bags that deploy upward from the window sill and remain rigid to protect occupants in case the vehicle flips multiple times.

The roll-bar systems use an inclinometer to sense vehicle inclination and lateral acceleration and a MEMS accelerometer to detect vehicle weightlessness in the event the

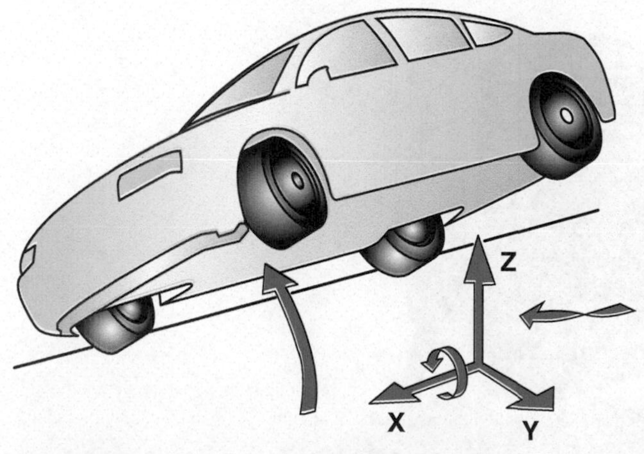

Figure 15-47 Sensors used for rollover protection measure vehicle movement on the X-Y-Z axis.

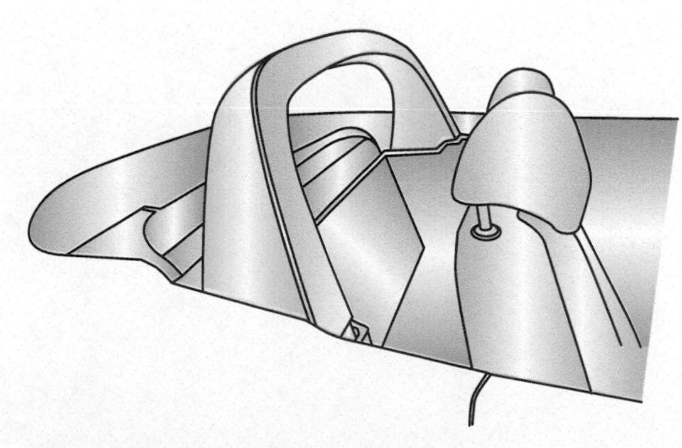

Figure 15-48 ROPS using pop-up roll bars.

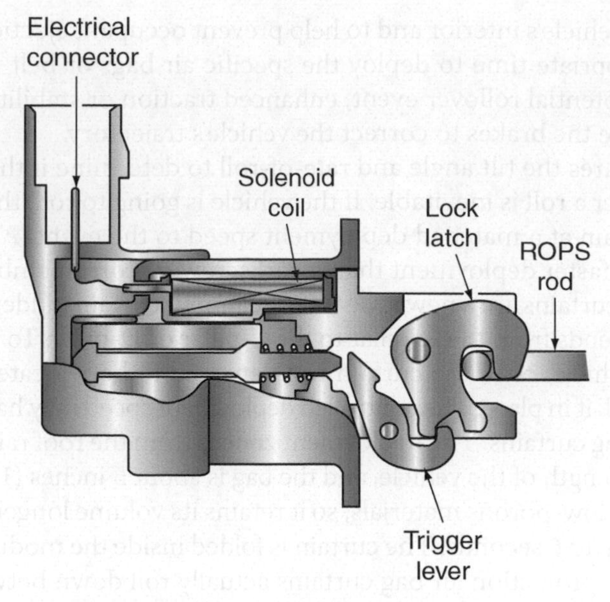

Figure 15-49 Spring-loaded ROPS may use a solenoid latch that releases the roll bar.

vehicle becomes airborne. Based on this information, the control module determines when to deploy the roll bars. Most systems only deploy the roll bars in the following scenarios:

- When the vehicle experiences a lateral acceleration of approximately 3 Gs.
- When the vehicle approaches a lateral angle of 62°.
- When the vehicle approaches its longitudinal angle of about 72°.
- When a combination of longitudinal acceleration and longitudinal angle would cause the vehicle to roll over in the forward direction.
- When the vehicle becomes airborne and for at least 80 milliseconds.

Active Headrests

One of the largest factors contributing to injury during an accident results from the sudden, jerking movement of the head and neck. Active headrests incorporated into the seats support the head during an impact to prevent these types of injuries (**Figure 15-50**). Within 20 milliseconds of detecting an impact, the active headrests are pushed forward, reducing sudden head movement. The reduction of head movement helps to prevent neck injuries.

Figure 15-50 Active headrests don't appear much different than standard headrests.

Summary

- Passive restraints operate automatically with no action required on the part of the driver or occupant.
- The automatic seat belt system uses a control module to monitor operation by receiving inputs from door ajar switches, limit switches, and the emergency release switch.
- The air bag is a supplemental restraint. The seat belt is the primary restraint system.
- The air bag module is composed of the air bag and inflator assembly. It's packaged in a single module and mounted in the center of the steering wheel.
- The ORC constantly monitors the readiness of the SRS electrical system. If the battery or cables are damaged during an accident, it supplies backup power to the air bag module.
- The igniter is a combustible device that converts electrical energy into thermal energy to ignite the inflator propellant.
- Air bags will deploy if the vehicle is involved in a frontal collision of sufficient impact and the collision force is within 30° on either side of the vehicle centerline.
- The total air bag deployment time from the instant of impact until the air bag full inflation is less than 160 milliseconds.
- Most current air bag systems use MEMS piezoelectric accelerometers to sense deceleration forces and direction.
- The clockspring electrical connector maintains electrical contact between the inflator module and the air bag electrical system.
- The air bag warning light indicates an inoperative air bag system.
- A hybrid inflator module contains a pressurized argon gas cylinder, which is punctured by the exploding propellant to inflate the air bag.
- Shorting bars connect air bag system squib terminals together when the connector is disconnected to prevent accidental air bag deployment.
- Federal law expanded to require equipping all passenger vehicles produced after 1995 with front passenger-side air bags.
- The air bag module of the multistage system uses two squibs. Firing one of the squibs begins the air bag deployment process. The second squib fires to generate more heat to fill the air bag faster.
- Side-impact air bags can come out of the door panel, from the seat back, between the A-pillar and the headliner, or from a roof-mounted curtain in the headliner that protects both the front- and rear-seat occupants.
- To meet Federal Motor Vehicle Safety Standards (FMVSS) for vehicle-to-pole side impacts, some manufacturers utilize a pressure-sensitive sensor in the doors.
- Some vehicles with limited rear seating may be factory equipped to turn off the passenger air bag if it would not be safe to have it deploy.
- To assure the driver and/or passenger stay in position during an accident, some vehicles have seat belt pretensioners.
- There are two ways of mounting the pretensioner—on the buckle side or retractor side of the seat belt.
- The inflatable knee blocker (IKB) deploys simultaneously with the driver-side air bag to provide upper-leg protection and positioning of the driver.
- Occupant classification systems are a mandatory requirement designed to reduce the risk of injuries resulting from air bag deployment.
- The Delphi system uses a bladder to determine weight, and the TRW system uses strain gauges. Both systems use an occupant classification module (OCM) that determines the weight classification of the front passenger and sends this information to the ORC. Also, both systems use multistage passenger-side air bags.
- The belt tension sensor (BTS) is a strain gauge–type sensor located on the seat belt anchor to indicate if an infant seat is cinched into the passenger-side front seat.
- The seat track position sensor (STPS) provides information to the OCM concerning the position of the seat in relation to the air bag.
- Some ORCs will have two MEMS accelerometers that are surface mounted directly to the ORC board, one on the Y-axis for front collision detection and the other oriented on the X-axis to correspond to side-impact events.
- A microprocessor performs the algorithmic computations to determine the need and level for air bag deployment.
- The rollover detection system provides an additional function of the air bag system that deploys side air bag curtains and/or seat belt pretensioners upon detection of a rollover event.
- The inclinometer measures lateral and vertical acceleration to predict an impending rollover.
- Convertibles that cannot use side air bag curtains will use a ROPS system utilizing deployable roll bars that flip up or pop up.

Review Questions

Short-Answer Essays

1. Define the term *passive restraint.*
2. Describe the basic operation of automatic seat belts.
3. List and describe the design and operation of three different types of air bag system sensors.
4. List and explain two functions of the OCM used in air bag systems.
5. List the sequence of events that occur during air bag deployment.
6. What is the purpose of the clockspring?
7. Describe how the non-azide multistage air bag system deploys.
8. What is the purpose of multistage air bags?
9. Where is the side-impact air bag control module or sensor usually located?
10. List the common mounting locations of the seat belt pretensioner.

Fill in the Blanks

1. The ______________ conducts electrical signals to the air bag module while permitting steering wheel rotation.
2. In the automatic seat belt system, the ______________ switches inform the module of the position of the harness.
3. The ______________ is a combustible device that converts electrical energy into thermal energy to ignite the inflator propellant.
4. The ORC supplies ______________ to the air bag ______________ in the event that the battery or cables are damaged during an accident.
5. To prevent accidental deployment of the air bag, most electromechanical systems require that at least ______________ sensor switches be closed to deploy the air bag.
6. The frontal impact must be within ______________ degrees of the vehicle centerline to deploy the air bag.
7. An accelerometer-type air bag sensor produces an analog voltage in relation to ______________ ______________.
8. The current flow through the squib required to deploy the air bag is approximately ______________ amperes.
9. A hybrid inflator module contains a cylinder filled with compressed ______________ gas.
10. The OCM system that uses a bladder may also use a ______________ ______________ ______________ to determine if an infant seat is securely strapped into the seat.

Multiple Choice

1. A pressure sensor mounted in the door as an input to the ORC is used to:

   A. Determine if a frontal impact has occurred.

   B. Determine if a side impact at the C-pillar has occurred.

   C. Determine if a side impact in the door area has occurred.

   D. Confirm that a door is properly closed.

2. All of the following are characteristics of the occupant classification system **EXCEPT:**

   A. The OCM is responsible for the deployment of the passenger-side air bag.

   B. The system determines if a rear-facing infant seat is in the front passenger seat.

   C. The PADL illuminates if the passenger-side air bag is suppressed.

   D. Weight classification of fifth percentile female and greater allows air bag deployment.

3. Air bag components are being discussed.
   Technician A says the igniter is a combustible device that converts electrical energy into thermal energy.
   Technician B says the inflation of the air bag uses an explosive release of compressed air.
   Who is correct?
   A. A only
   B. B only
   C. Both A and B
   D. Neither A nor B

4. The air bag system is being discussed.
   Technician A says the clockspring is located at the bottom of the steering column.
   Technician B says the clockspring conducts electrical signals to the module while permitting steering wheel rotation.
   Who is correct?
   A. A only
   B. B only
   C. Both A and B
   D. Neither A nor B

5. The air bag system components are being discussed.
   Technician A says the ORC constantly monitors the readiness of the air SRS electrical system.
   Technician B says a crash sensor may be composed of a gold-plated ball held in place by a magnet.
   Who is correct?
   A. A only
   B. B only
   C. Both A and B
   D. Neither A nor B

6. Air bag sensors are being discussed.
   Technician A says the arrow on each sensor must face toward the rear of the vehicle.
   Technician B says air bag sensor brackets must not be bent or distorted.
   Who is correct?
   A. A only
   B. B only
   C. Both A and B
   D. Neither A nor B

7. Accelerometer-type air bag system sensors are being discussed.
   Technician A says an accelerometer senses collision force and direction.
   Technician B says an accelerometer produces a digital voltage.
   Who is correct?
   A. A only
   B. B only
   C. Both A and B
   D. Neither A nor B

8. Which of the following is a characteristic of the strain gauge–type occupant classification system?
   A. If the seat is empty, the PADL is illuminated.
   B. The system uses a belt tension sensor to determine the presence of an infant seat.
   C. The air bag will deploy if a child is determined to be sitting in the seat.
   D. None of these choices.

9. The air bag deployment loop is being discussed.
   Technician A says if the arming sensor contacts close, this sensor completes the circuit from the inflator module to ground.
   Technician B says the air bag deploys if the contacts close in two crash sensors.
   Who is correct?
   A. A only
   B. B only
   C. Both A and B
   D. Neither A nor B

10. Hybrid inflator modules are being discussed.
    Technician A says a pressure sensor in the argon gas cylinder sends a signal to the ORC in relation to gas pressure in the cylinder.
    Technician B says when the initiator is energized, the propellant explodes and pierces the propellant container, allowing the pressurized argon gas to escape into the air bag.
    Who is correct?
    A. A only
    B. B only
    C. Both A and B
    D. Neither A nor B

CHAPTER 16

ADVANCED DRIVER ASSISTANCE SYSTEMS

Upon completion and review of this chapter, you should be able to:

- Describe the use of proximity sensors in ADAS and safety systems.
- Explain the operating principles of the ultrasonic sensor.
- Describe the operation of visual imaging systems.
- Describe how the infrared sensor determines objects and detects obstructions.
- Explain the purpose of using an infrared sensor in conjunction with a camera.
- Detail the process of distance and speed detection using radar sensors.
- Explain the principle of stimulated emission in the production of laser light.
- Describe how a laser sensor determines the distances between objects.
- Detail the properties of laser light.
- Explain the function of the laser diode.
- Define the levels of autonomy.
- Explain the functional characteristics of the park assist system.
- Explain the operating fundamentals of the self-parking vehicle.
- Explain the function of Lane Departure Warning systems.
- Describe the operation of the Lane Keep Assist and Lane Centering systems.
- Explain the operation of the Collision Avoidance System.
- Describe the operation of Rollover Mitigation Systems.

Terms to Know

Autonomous
Complementary metal oxide semiconductor (CMOS)
Doppler shift
Electronic Roll Mitigation (ERM)
Gain medium
Haptic
Injection Laser Diode (ILD)
Lane Departure Warning (LDW)
Laser
Laser diodes
Light Detecting and Ranging (LIDAR)
Low Voltage Differential Signaling (LVDS)
Millimeter-wave radar sensors
Odometry
Optical cavity
Park assist system
Population inversion
Proximity sensor
Pumped
Radar
Steering Angle Sensor (SAS)
Stimulated emission
Ultrasonic sensors
Ultrasound

Introduction

Since the invention of the automobile, there have been continuous changes in the technologies added to it. Owners were thrilled at the introduction of the electric starter motor since it meant they no longer had to crank the engine by hand to start it. They also celebrated that the new lighting system did not require oil and a flame. Anyone who has followed the evolution of the automobile has witnessed the rapid growth in technologies added to the vehicle. We have explored many of these in the preceding chapters. Safety is the primary source driving these new innovations.

This chapter discusses some of the more advanced safety and driver assistance systems. These systems have become what is known as Advanced Driver Assistance Systems (ADAS). As we advance through the various technologies used by ADASs it should be a simple step to understanding **autonomous** self-driving technology that enables a vehicle with the ability to sense its environment and operate without human involvement.

The future is here, and you will be called upon to service these new technologies. It's important to realize that different manufacturers have different approaches to these systems. For this reason, gathering the proper service information for the vehicle you are servicing is critical.

Vision Systems

ADAS and many of the newest safety systems use vision imaging. The automotive vision system covers a broad range of different subsystems that include such automotive safety technologies as:

- *Adaptive cruise control with collision mitigation system*—uses sensors (along with laser or radar technologies) and cameras to determine the control of the throttle and brakes. The purpose is to maintain a safe following distance of the vehicle in front even under changes in traffic speeds or if another vehicle pulls into the lane. If the system determines that an accident is possible, it will automatically apply the brakes, and some systems will tighten the seat belts.
- *Blind-spot detection system*—designed to alert the driver about vehicles or other objects that are in the "blind spots" of the driver. Usually, system activation is by turning on the turn signals. Detection of an object activates a warning light in the mirror to flash, along with an audio warning tone. In addition, some manufacturers will cause the seat or steering wheel to vibrate to provide a warning indication.
- *Lane-departure warning*—similar to blind-spot detection, this system determines an approaching vehicle's speed and distance to warn the driver of the potential danger if they should change lanes. This system will also provide a warning if it determines that the vehicle is wandering out of the lane.
- *Rearview camera*—designed to protect both people and property, this system provides the driver with a visual view of any objects directly behind the vehicle. This system provides a view of the areas most mirrors cannot.
- *Adaptive headlights system*—can be as simple as providing automatic headlight activation in low ambient light conditions. More advanced systems also provide automatic dimming based on speed and other conditions and may also have headlights that follow the direction of the vehicle while it navigates around corners.
- *Night vision assist systems*—use infrared headlamps or thermal imaging cameras to allow the driver a better view of what is farther down the road during low ambient light conditions. An image of objects on the road is displayed for the driver to see.

Shop Manual
Chapter 16, page 897

Proximity Sensors

The ADAS and safety systems can use various sensor types to detect proximity, motion, **odometry**, and light intensity. The system can also use imaging cameras.

Odometry uses data from GPS, XYZ, accelerometers, and other motion detection sensors to estimate the change in position over time.

Most systems rely on a **proximity sensor** that detects objects without physical contact. These sensors use several different technologies such as ultrasonic, radar, laser, camera, and infrared.

This section covers the operation of the various sensors used by the systems we will be discussing. Some of these sensors were introduced in earlier chapters; here, we will dive deeper into their operation.

Ultrasonic Sensors

Shop Manual
Chapter 16, pages 883, 887, 897

Ultrasonic sensors evaluate the attributes of a target by interpreting the echoes from sound waves. The field of robotics use these sensors for obstacle avoidance and as range finding for cameras. The automotive industry employs this technology for basically the same purpose. Today, ultrasonic sensors on the back of the vehicle detect if an object is in the path while the driver is backing up (**Figure 16-1**). Some systems also include these sensors in the front of the vehicle to assist in parallel parking maneuvers (**Figure 16-2**).

Ultrasound is a cycling sound pressure wave that is at a frequency greater than the upper limit of human hearing (about 20 kHz).

Ultrasonic sensors emit a short burst of **ultrasound** waves that reflect or "echo" from a target. Most sensors operate in the range of about 40 kHz. After the waves strike an object, they echo back (**Figure 16-3**) and are evaluated. Calculating the distance the vehicle is from an object is based on the time interval between sending the signal and receiving the echo and dividing that time value by the speed of sound.

Either the module or the sensor contains a transmitter that generates the ultrasonic sound waves by turning electrical energy into sound. The transducer circuit uses

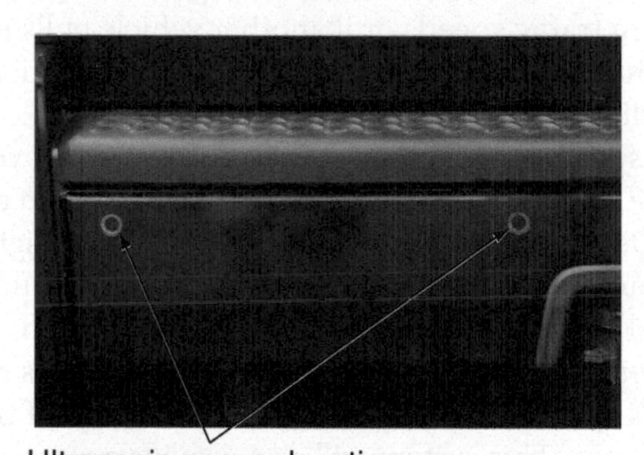

Figure 16-1 Ultrasonic sensors used for backing up obstacle detection.

Figure 16-2 Front ultrasonic sensors used in park assist systems.

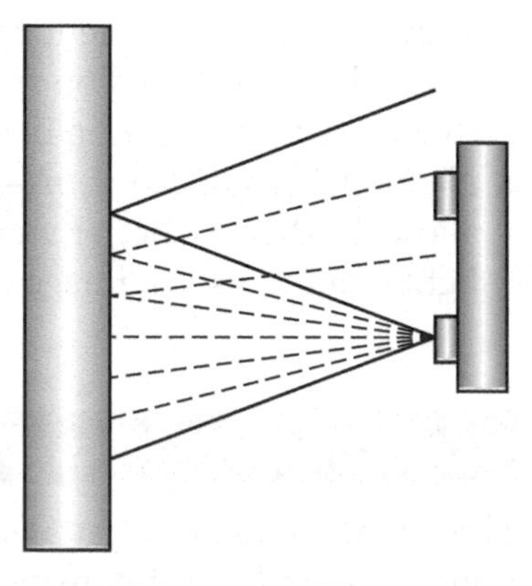

Figure 16-3 The echo effect of ultrasonic sound as it hits an object.

piezoelectric crystals that change size when a voltage is applied. This creates an alternating voltage across the crystals that causes them to oscillate at very high frequencies. This, in turn, produces high-frequency sound waves.

The detector is a piezoelectric crystal that generates a voltage by the sound wave pressure forces applied to it. The detector circuit of the ultrasonic sensor receives the echo, which converts the sound waves into electrical energy. Usually, a single piezoelectric transceiver combines the transmitter and detector.

Radar Sensors

Radar is a system that uses electromagnetic waves to identify an object's range, direction, or speed. After emitting a radio wave reference signal, the time it takes for the receiver to detect the echo wave reflected off the object is measured. Consider that radio waves travel at the speed of light (186,000 mi/s or 300,000,000 km/s) and that the waves must travel to the object and echo back to the receiver; the calculation of distance is by measuring the length of the pulse multiplied by the speed of light, divided by two.

Shop Manual
Chapter 16, pages 887, 897

A duplexer switches the radar sensor from being a transmitter to a receiver at a set rate. In most cases, the receiver does not detect the return while transmitting the signal. Based on the switching rate of the duplexer, the sensor operates as either a long-range or a short-range sensor. Long-range detection systems use long pulses with long delays between them, while short-range detection systems use short pulses with less time delays between them.

Doppler shift provides another method of calculating the speed of an object. Doppler shift occurs when radio waves reflect from a moving object but use the frequency of the waves instead of time of flight. Since Doppler shift occurs with sound waves as well, we will use them to explain this concept. You may have experienced the sound change in a train's horn as it passes by you. If you are in front of the train and it's headed toward you with the horn honking, you will hear one tone or note from the horn. However, the tone (note) changes to a lower note when the train passes you. Even though the horn's note does not change, the note you hear does change due to the Doppler shift. To further illustrate, consider that sound travels at approximately 600 mph (965.6 km/h). If the train is stationary at a distance of 1 mile (1.6 km) from you, and it sounds its horn for sixty seconds, the sound waves from the horn will propagate from the train toward you at a rate of 600 mph (965.6 km/h). There will be a six-second delay before you hear the horn; then, you will hear sixty seconds of the sound. The tone you hear, and anyone else in any direction, and the train's engineer will be the same.

Now assume the train is moving toward you at 60 mph (96.6 km/h) and it starts to blow its horn when it's 1 mile (1.6 km) away from you and continues to blow the horn for sixty seconds. There will be a six-second delay from when the horn starts to sound until you hear it, just as before. However, you will only hear the horn for fifty-four seconds. Since the train will be next to you in one minute, the sound at the end of the minute gets to you instantaneously. The engineer in the train will hear the horn for the full minute; you hear the horn for only fifty-four seconds since the sixty seconds worth of sound is compressed into fifty-four seconds. What happens is the same number of sound waves are compressed into a shorter time frame. The compacting of the waves results in a higher frequency. This makes the horn tone sound higher to you than to the engineer who is moving at the same speed as the train.

The process reverses if the horn is still sounding as the train passes you. Now the sound waves expand to fill more time and result in a lower frequency (**Figure 16-4**).

In the radar system, Doppler shift determines speed by sending out a radio wave toward a vehicle and determining the compression of the returning wave. For example, if the vehicle is coming toward you, the echoes become compressed and have a higher frequency. The same would be true if you are driving a vehicle and coming up to a slower-moving vehicle in front of you. If the vehicle is moving away from you, the echoes would have a

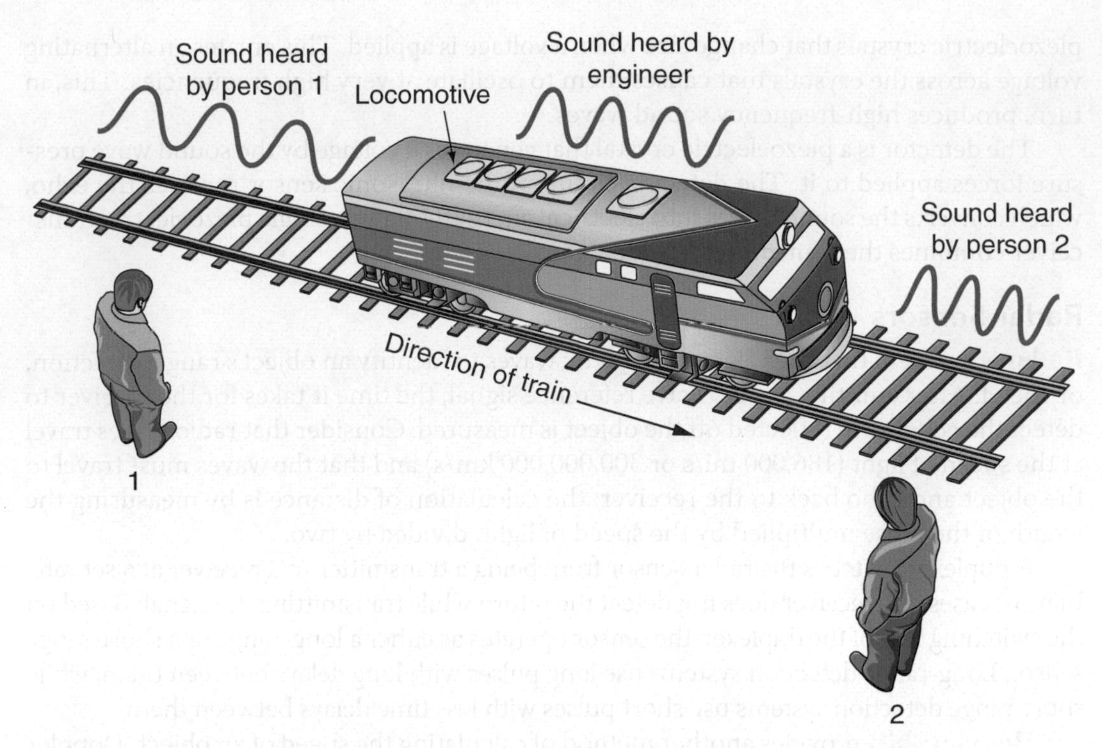

Figure 16-4 When in front of the approaching train, person 2 hears a higher tone than the engineer because the train is approaching them. When the train passes, person 1 hears a lower tone than the engineer because the train is moving away from them.

lower frequency. The time it takes for the echo to arrive determines distance; speed is determined by the frequency of the echo.

Blind spot and lane departure radar sensors are **millimeter-wave radar sensors** that can provide stable detection of targets even under inclement weather conditions. These sensors use a higher frequency (77 GHz) signal than conventional radar. Millimeter-wave radar sensors use Frequency Modulated-Continuous Wave (FM-CW) radar technology. The sensor can measure the distance from a target and its velocity simultaneously by using a principle known as "beat signal." This signal results from mixing the transmitted frequency modulated millimeter-wave signal and the reflected signal from the target. The beat signals result from the pair of frequency-modulated waves having different time variations. A time delay results from the distance from the radar to the target and a Doppler shift due to the relative velocity of the target.

Laser Sensors

Shop Manual
Chapter 16, page 897

A **laser** is a device that controls the way that energized atoms release photons (light). Laser light has the following properties:

- Monochromatic—The light is only one color since it contains one specific wavelength of light.
- Coherent—The waves have the same wavelength and a fixed-phase relationship.
- Directional—The laser light produces a very tight beam.

Laser is an acronym for *l*ight *a*mplification by *s*timulated *e*mission of *r*adiation.

The laser light uses **stimulated emission** to accomplish these properties. In standard incandescent light, the atoms release photons in a large solid angle and over a wide spectrum of wavelengths. For example, a flashlight releases light in many directions. This results in a weak and diffused light. Stimulated emission results in the laser light atoms releasing photons in a very organized state. A typical laser emits light in a narrow, low-divergence beam with a well-defined wavelength.

A laser contains the following components:

- A gain medium
- An optical cavity
- A means of supplying energy to the gain medium

The **gain medium** is a material (such as gas, liquid, solid, or free electrons) that possess the appropriate optical properties. The gain medium transfers external energy into the laser beam and is inside the **optical cavity**. The cavity consists of two mirrors, one at each end of the cavity (**Figure 16-5**). This arrangement forces the light to bounce back and forth. As the light of a specific wavelength passes through the gain medium, it's amplified. Each time the light bounces it must again pass through the gain medium. The mirrors ensure that a majority of the light must make several passes through the gain medium. The mirror mounted to the cavity's output side is partially transparent. Part of the light in the cavity will pass through the partially transparent mirror and appears as a beam of laser light.

The gain medium is **pumped** (energized) to get the atoms into an excited state. Pumping is the process of supplying the energy required for amplification. The energy supplied is either by an electrical current or a flash of light at a different wavelength. The pump energy is absorbed in the medium and creates a large collection of atoms that possess higher-energy electrons. The higher-energy electrons migrate to a higher orbit around the nucleus. The atoms are excited to a level two or three times greater than their relaxed state. This increases the degree of **population inversion.** Population inversion refers to the number of atoms in the excited state being greater than the number in the relaxed state.

Just as the electrons absorb energy to reach this excited level, they will also release this energy. Turning off the pump energy relaxes the electrons, and they shed their energy as they fall to a lower energy level. The released energy is in the form of photons that have a specific wavelength. The amount of energy released when the electron drops to a lower energy level determines the wavelength of light (**Figure 16-6**).

Two identical atoms with electrons in the same states release photons with identical wavelengths. If a photon encounters another atom with an electron in the same excited state, stimulated emission occurs, and amplifies the light. Stimulated emission results from the first photon stimulating a second atom to emit a photon that vibrates with the same frequency and direction as the first photon.

The mirrors on each end of the optical cavity cause the photons to travel back and forth through the gain medium. During this process, they will stimulate other electrons to drop to a lower energy level. The photon emitted from the second atom is in the same direction as the photon passing by it. This ultimately emits more photons of the same wavelength and phase and causes a chain reaction of events to occur. Much like a snowball rolling downhill, this effect causes the propagation of several photons of the same wavelength and phase. Finally, the outlet mirror will emit some of the photons out of the optical cavity. **Figure 16-7** illustrates this process using a laser pumped with a flash of light.

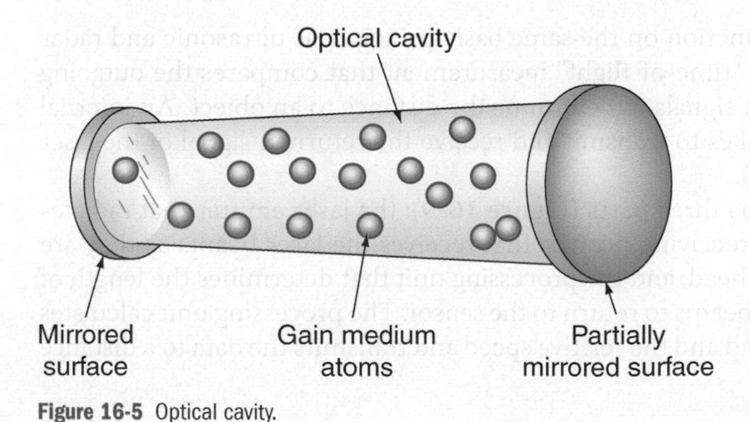

Figure 16-5 Optical cavity.

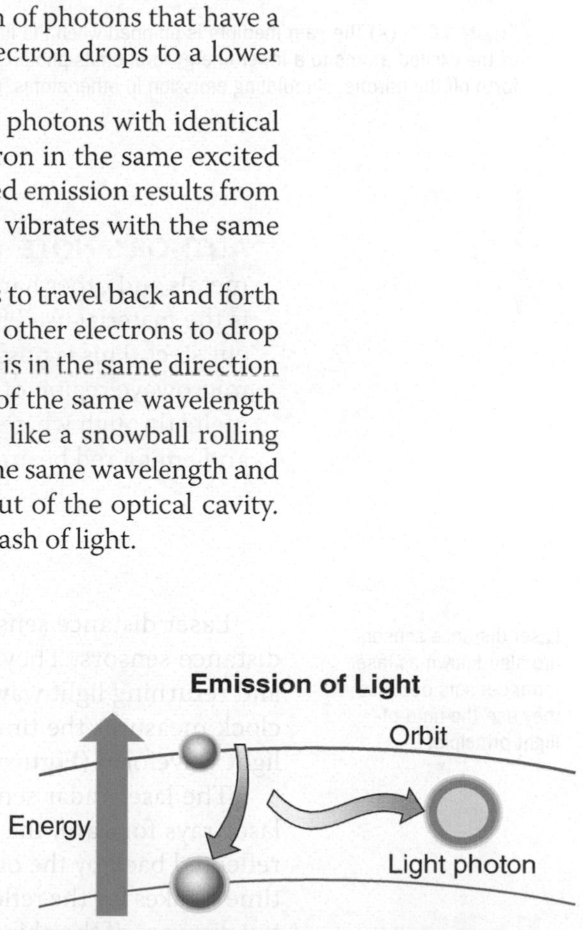

Emission of Light
Orbit
Energy
Light photon

Figure 16-6 The wavelength of light is determined by the amount of energy released when the electron drops to a lower energy level.

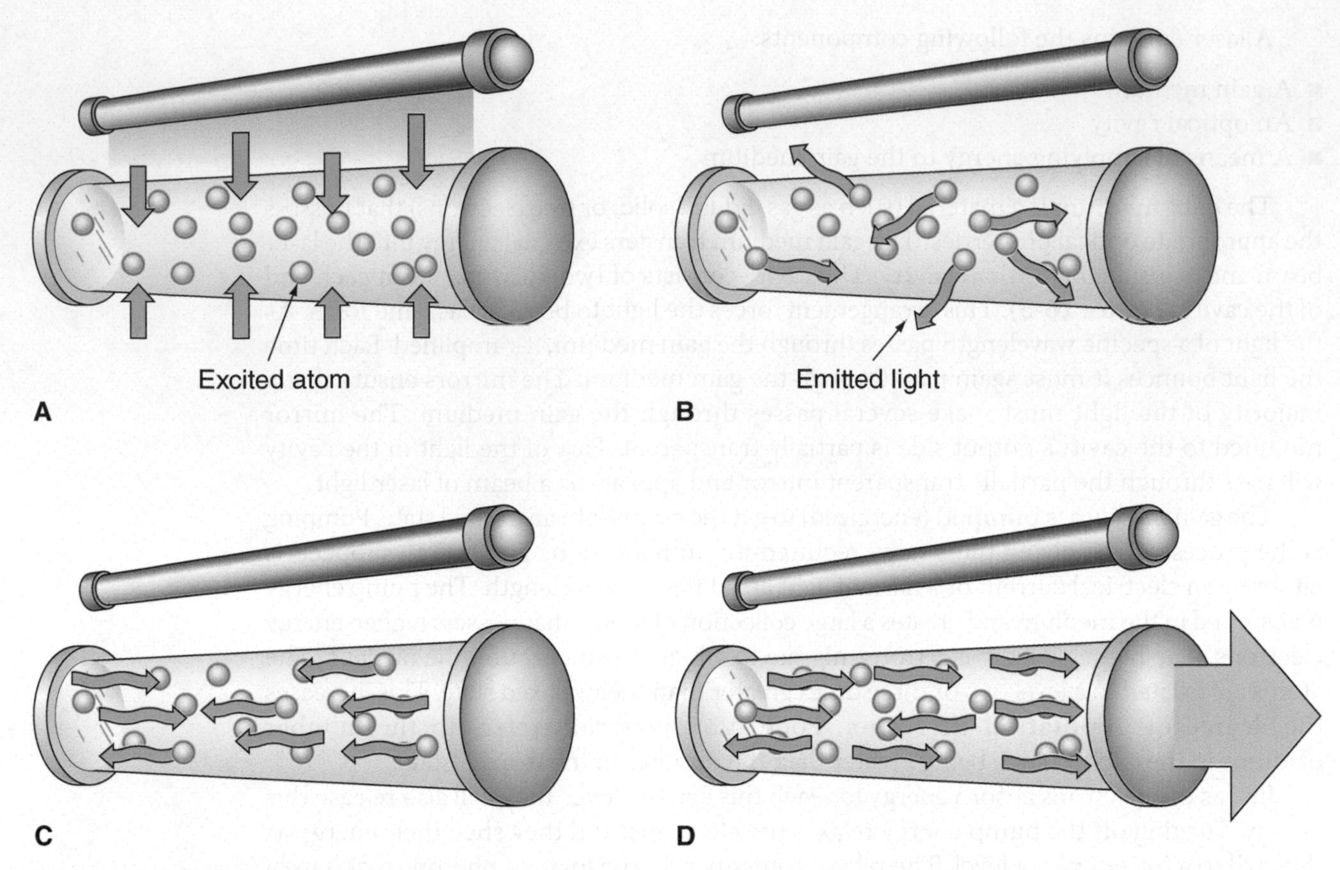

Figure 16-7 (A) The gain medium is pumped when the flash tube injects light into the optical cavity, exciting the atoms. (B) Relaxing the electrons of the excited atoms to a lower energy level emits photons. (C) The photons travel inside the cavity through the gain medium and bounce back and forth off the mirrors, stimulating emission in other atoms. (D) Some photons exit the optical cavity's outlet side as laser light.

AUTHOR'S NOTE Often, lasers are associated with light beams that cut through metals and other hard substances. One factor determining the strength of the laser is the material used for the gain medium. The CO_2 laser is very powerful and can cut steel. This is possible because the CO_2 laser emits laser light in the infrared and microwave region of the spectrum. Infrared radiation is heat, and this laser basically melts through whatever it's focused upon. On the other hand, diode lasers are weak and emit a red beam of light that has a wavelength between 630 and 680 nm.

Laser distance sensors are also known as laser radar sensors because they use the time-of-flight principle.

Laser distance sensors function on the same basic principle as ultrasonic and radar distance sensors. They use a "time-of-flight" measurement that compares the outgoing and returning light waveform signals to determine the distance to an object. An internal clock measures the time it takes to transmit and receive the returned signal of the laser light waveform (**Figure 16-8**).

The laser radar sensor has three parts (**Figure 16-9**): the laser emitter that radiates laser rays forward, the laser-receiving portion that receives the laser beams as they are reflected back by the object ahead, and the processing unit that determines the length of time it takes for the reflected beams to return to the sensor. The processing unit calculates the distance of the object ahead and the relative speed and transmits the data to a distance control ECU (**Figure 16-10**).

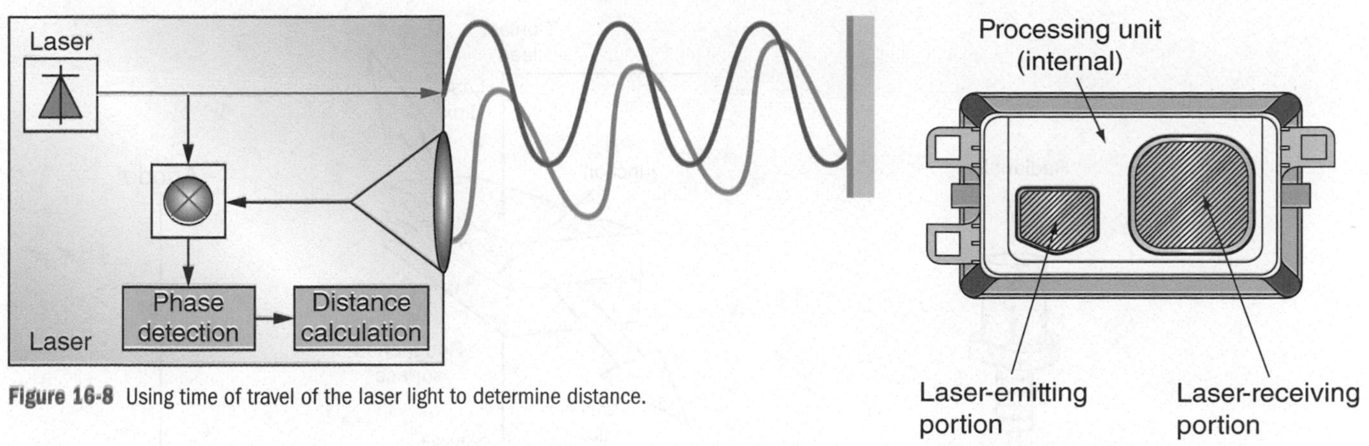

Figure 16-8 Using time of travel of the laser light to determine distance.

Figure 16-9 The laser radar sensor components.

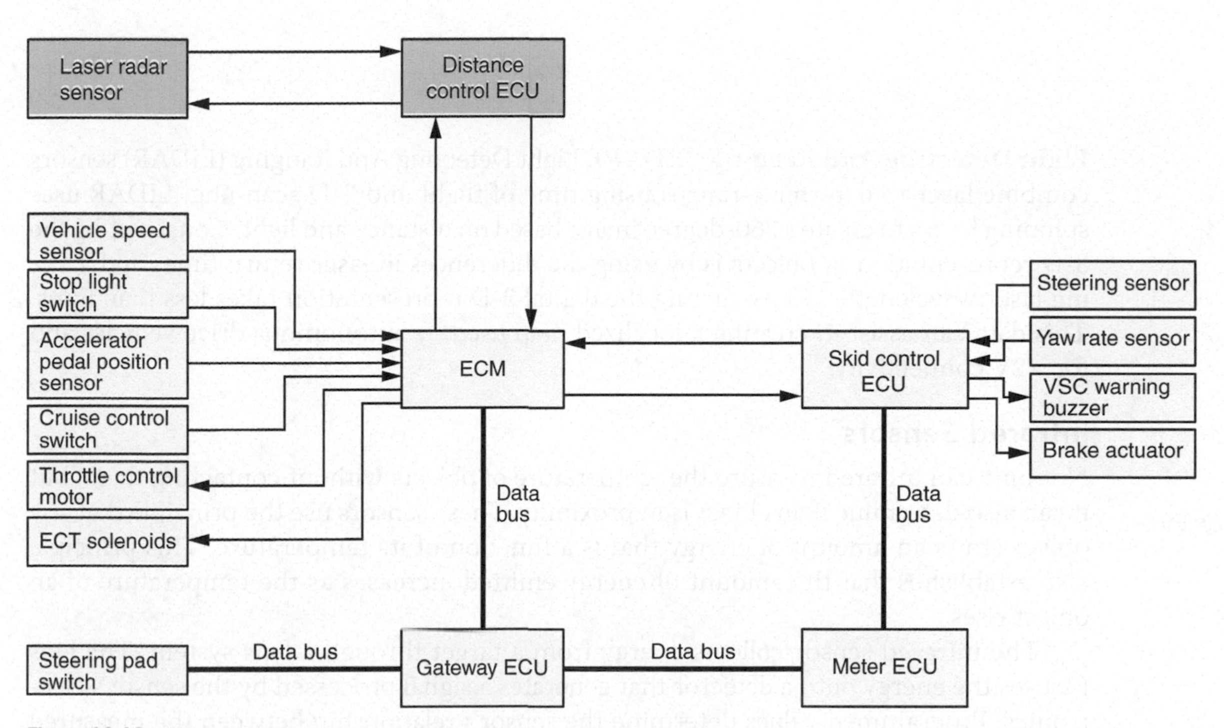

Figure 16-10 Schematic of laser sensor used in collision avoidance system.

Laser Diodes. Laser diodes are light-emitting devices capable of emitting laser beams (**Figure 16-11**). A common laser diode is known as the **Injection Laser Diode (ILD)**. ILDs are lasers that use a forward-biased semiconductor junction as the gain medium.

Laser diodes work the same as a regular light-emitting diode (LED). **Figure 16-12** illustrates the construction of the laser diode with the P-type and N-type materials constructed of aluminum gallium arsenide (AlGaAs). Forward biasing the diode emits light at the PN junction. The difference between the laser diode and the regular LED is the method of emitting the light. The laser diode emits its light from the end faces. One of the faces will have a coating of reflective material to focus the emitted light in one direction only. This results in a coherent and monochromatic light.

The schematic symbol for the laser diode uses zigzagged light emission arrows (**Figure 16-13**) to differentiate it from the standard LED that are straight.

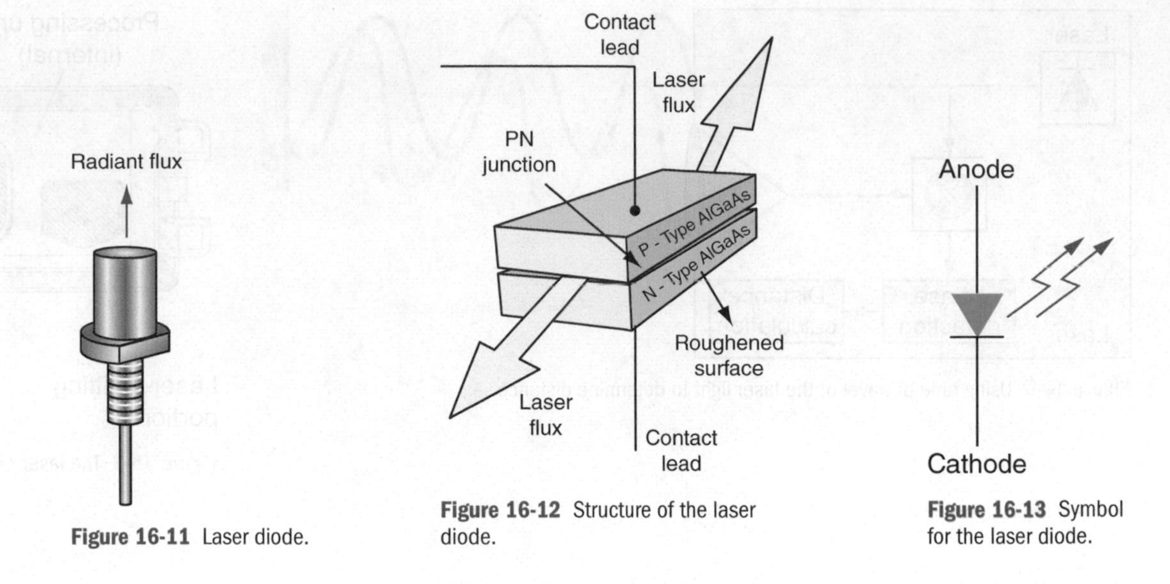

Figure 16-11 Laser diode.

Figure 16-12 Structure of the laser diode.

Figure 16-13 Symbol for the laser diode.

Light Detecting And Ranging (LIDAR). Light Detecting And Ranging (LIDAR) sensors combine laser to determine ranges using time of flight and 3-D scanning. LIDAR uses spinning lasers to create a 360-degree image based on distance and light. Constructing the 3-D representation of objects is by using the differences in laser return times and varying laser wavelengths. Constructing the digital 3-D representation takes less than 5 ms. This data can assist in creating a localized map used by autonomous drive vehicles and for V2V connectivity.

Infrared Sensors

Shop Manual
Chapter 16, page 897

Not only can infrared measure the temperature of objects without contacting them, but it can also determine if an object is in proximity. These sensors use the principle that any object emits an amount of energy that is a function of its temperature. This principle also establishes that the amount of energy emitted increases as the temperature of an object rises.

The infrared sensor collects energy from a target through a lens system. The lens focuses the energy onto a detector that generates a signal processed by the sensor's electronics. Programmed values determine the sensor's relationship between the measured signal and the corresponding temperature. When used as an obstruction sensor, the infrared light beams are monitored for distortions.

Cameras

Shop Manual
Chapter 16, pages 884, 897

Many safety systems utilize a camera to provide visual assistance for backing up the vehicle. However, cameras can determine light intensity and speed of oncoming vehicles for such systems as automatic headlight dimming, as discussed in Chapter 11.

Also, cameras can enhance the functionality of a system. For example, a camera can provide a system with the ability to recognize and learn patterns. Since radar sensors cannot determine if the object is a pedestrian, a vehicle, or a road sign, camera-based pattern recognition provides the capability of recognizing these objects. In addition, night vision assistance systems utilize cameras. The use of the digital camera in public life has expanded greatly in the past decade; it's this technology that the camera systems on today's vehicles employ. However, when used in the confines of the automobile, efficient processing of the captured image and displaying it's an obstacle. This obstacle is overcome by intelligent image capture systems.

Intelligent image capture systems perform local image processing to correct for lens "fisheye" or for full-object recognition. The video controller receives the information over the bus network to. **Figure 16-14** illustrates a typical microprocessor-based system that controls and processes the video data. Processing the data conditions it for the type of media receiving it. The processing functions are performed before bussing the video data to the display.

The captured parallel video data image is converted into a serial stream and transmitted over a **Low Voltage Differential Signaling (LVDS)** interface (**Figure 16-15**). The display processes the data back to its original form. **Figure 16-16** provides an example of a system using an LCD monitor.

Low Voltage Differential Signaling (LVDS) is a high-speed data bus used in point-to-point configurations for visual displays.

When used in conjunction with a camera, infrared sensors assist the camera in determining the shape and size of an object under low-light conditions. This enhances the ability to see objects that the eye may not and is the technology used in some night vision systems. When used as a camera system to determine the identity of an object, the sensor uses the emitted energy to determine the object's shape.

The demands of ADAS require extracting depth information from camera imagery accurately. Stereo cameras can extract this information by measuring objects with two camera sensors. This offers a parallax view that facilitates the ability to measure distances with great precision. With stereo cameras, it becomes possible to generate stationary objects like a traffic sign or a white line and distance information of a moving object.

360-degree cameras provide a view of the area surrounding the vehicle. This system uses several cameras located around the vehicle. Advanced systems use software to interpret each camera's images and stitch them into a single image. In most cases, the displayed image is a top-down view; but some systems have the software to provide the means to adjust the image to provide different perspectives.

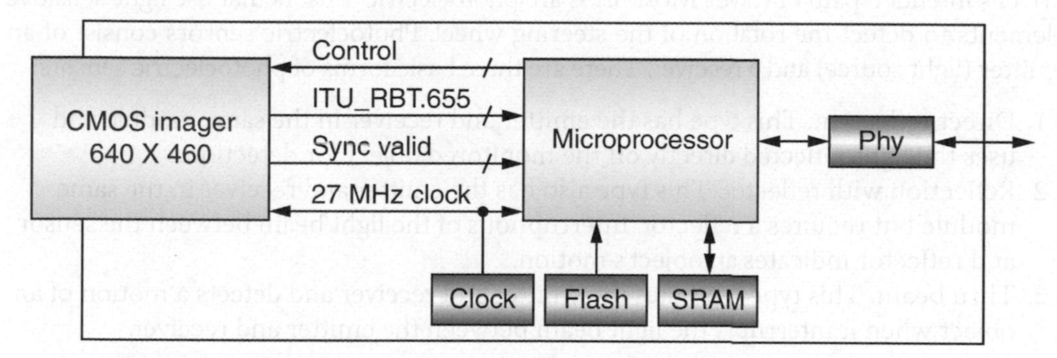

Figure 16-14 Intelligent image capture system.

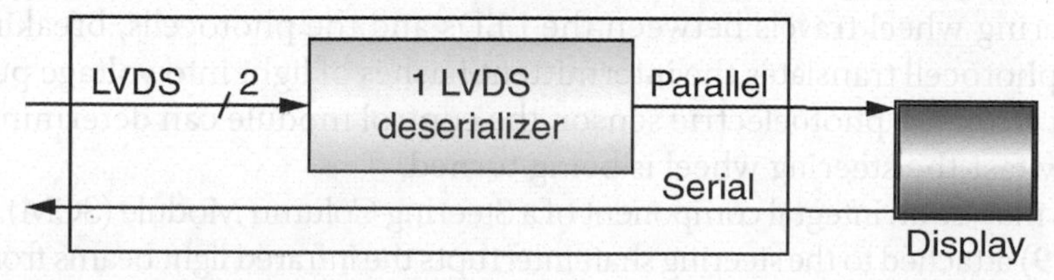

Figure 16-15 Transmission of captured video.

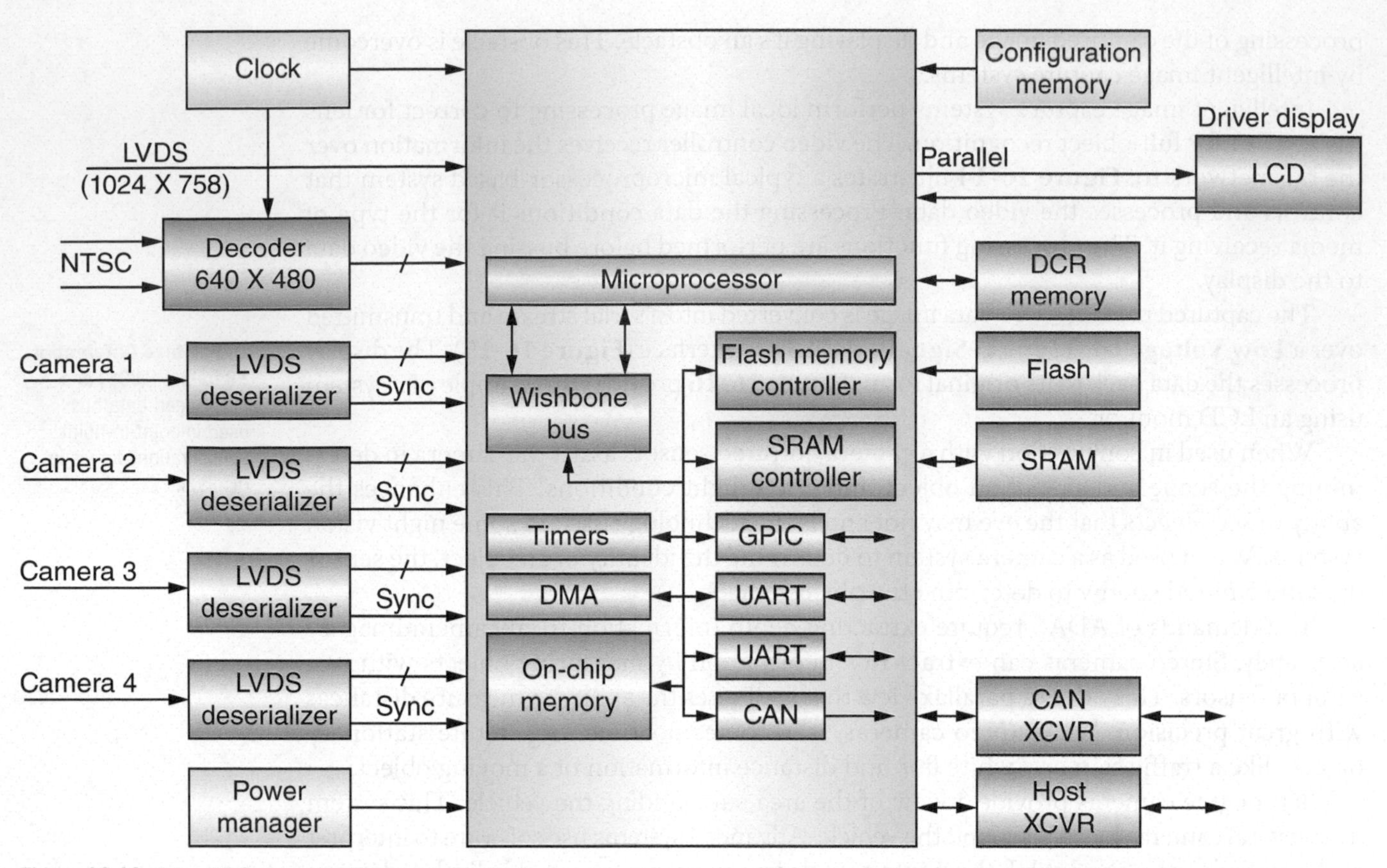

Figure 16-16 Using LVDS to transmit images to the display.

Steering Angle Sensors

Shop Manual
Chapter 16, page 881

The **Steering Angle Sensor (SAS)** monitors actual steering wheel position to determine the driver's intended path of travel. Most SASs are photoelectric sensors that use light-sensitive elements to detect the rotation of the steering wheel. Photoelectric sensors consist of an emitter (light source) and a receiver. There are three basic forms of photoelectric sensors:

1. Direct reflection. This type has the emitter and receiver in the same module and uses the light reflected directly off the monitored object for detection.
2. Reflection with reflector. This type also has the emitter and receiver in the same module but requires a reflector. Interruptions of the light beam between the sensor and reflector indicates an object's motion.
3. Thru beam. This type separates the emitter and receiver and detects a motion of an object when it interrupts the light beam between the emitter and receiver.

Regardless of the type, photoelectric sensors operate about the same. We will look at the thru beam type as an example of this operating principle (**Figure 16-17**).

An example would be one that has a series of LEDs situated across from the same number of photocells (**Figure 16-18**). A shutter wheel (light beam interrupter) that rotates with the steering wheel travels between the LEDs and the photocells, breaking the light beams. The photocell translates the intermittent flashes of light into voltage pulses. Based on this input from the photoelectric sensor, the control module can determine the direction and how fast the steering wheel is being turned.

The SAS may be an integral component of a Steering Column Module (SCM). A code disc (**Figure 16-19**) attached to the steering shaft interrupts the infrared light beams from the optics cluster. The optics cluster has three rows of four light detectors that provide a bit code that determines steering wheel position based on the order of light beam interruption. Underneath the data disc are two Hall-effect switches that track the number of revolutions the data

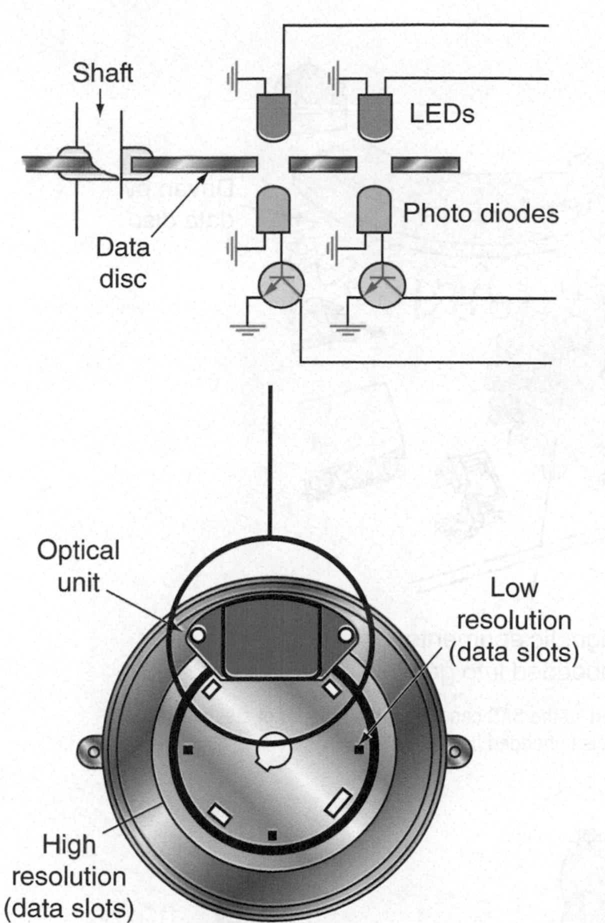

Figure 16-17 A thru beam photoelectric-type SAS.

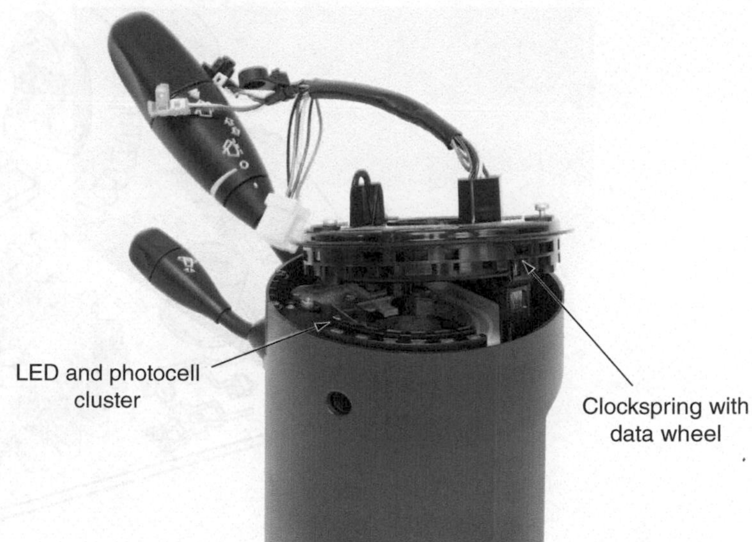

Figure 16-18 SAS using a roll of LEDs and photocells to detect steering wheel motion and position.

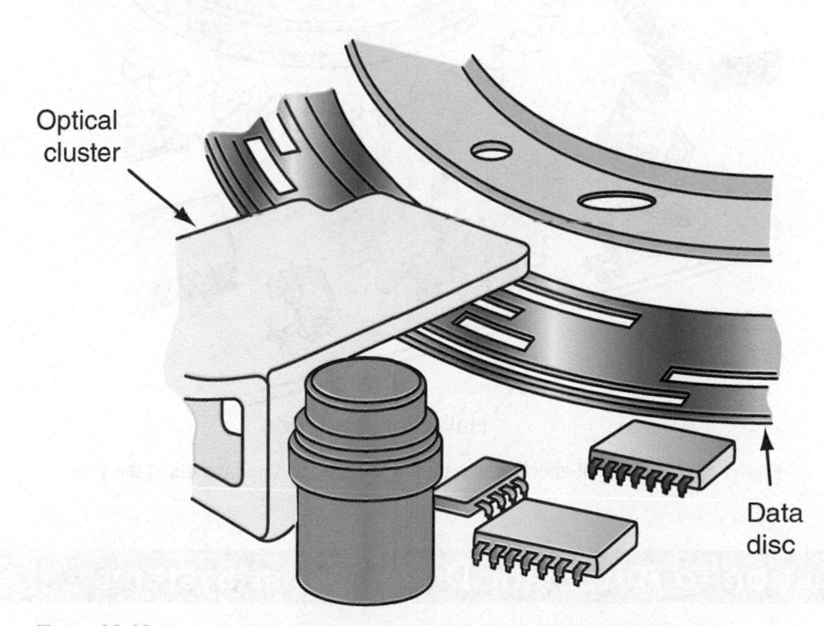

Figure 16-19 The data disc interrupts the light beams from the optical cluster to detect motion.

disc has made. The trigger for the Hall-effect switches is driven by a fifty-one-tooth gear connected to the data disc (**Figure 16-20**). This gear drives an eighteen-tooth gear that contains magnet segments. The Hall-effect sensors are located under the smaller gear (**Figure 16-21**). As the eighteen-tooth gear rotates, the magnetic segments influence the Hall-effect switches.

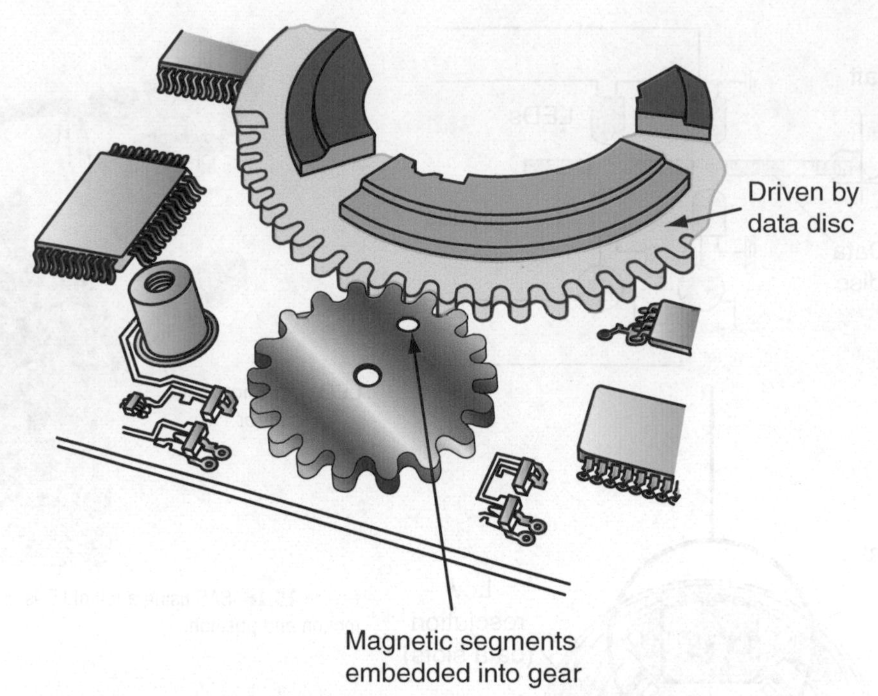

Figure 16-20 A series of gears are used so the SAS can detect the number of steering wheel revolutions. The smaller gear has magnets imbedded into it.

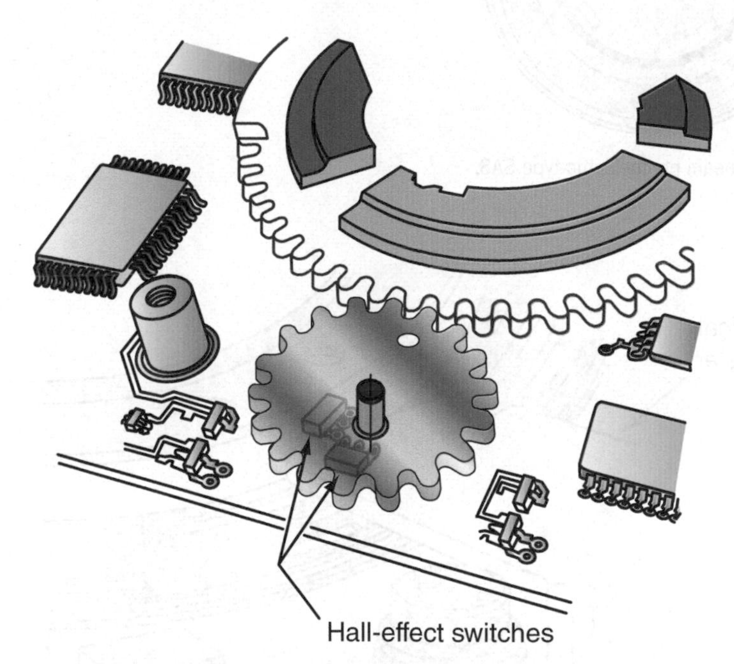

Figure 16-21 The Hall-effect switches are influenced by the magnets in the gear.

Introduction to Automatic Driver Assist Systems

Shop Manual
Chapter 16, page 878

The National Highway Traffic Safety Administration (NHTSA) has adopted the Society of Automotive Engineers (SAE) six levels of autonomy (0 – 5):

- Level Zero: No Automation—The driver must always maintain complete control of the vehicle. Only the driver can control acceleration, braking, and steering. The use of warning tones or safety intervention systems does not classify the vehicle at a higher level.
- Level One: Driver Assistance—Hands on the wheel. The car can either take control of the steering wheel or the pedals in certain driving modes. Adaptive cruise control (ACC)

and park assist systems are examples of level one automation. The computer is never in control of both steering and acceleration/braking.

- Level Two: Partial Automation—Hands off the wheel, eyes on the road. A level two vehicle has certain modes in which the car can take over both the pedals AND the wheel, but only under certain conditions. The driver must maintain ultimate control over the vehicle. Self-parking systems are an example of level 2 autonomy.
- Level Three: Conditional Automation—Under certain conditions, a level 3 vehicle will fully take over the driving responsibilities; however, the driver must retake control when necessary. A level 3 vehicle can decide when to change lanes and how to respond to dynamic incidents on the road but uses the human driver as the backup system.
- Level Four: High Automation—A level four vehicle does not need a driver. Under the right circumstances, the vehicle can drive itself full time.
- Level Five: Full Automation—A level 5 vehicle would not require a steering wheel since the vehicle is completely self-driving under all conditions. A level 5 vehicle does not require a human to be on board.

The ADAS uses many sophisticated electronic components. These include the various sensor types and cameras we have discussed. Outputs can include motors, solenoids, relays, and a vast array of other electromechanical or electronic devices.

Most ADASs interconnect with other vehicle systems. Not many ADASs are stand-alone systems and will rely on inputs, outputs, and data from other systems. Also, the different ADASs share these elements with each other. Each level of autonomy requires different involvement of various Electronic Control Modules (ECMs). Most of these are networked with the Electronic Steering Control (ESC), Automatic Braking System (ABS), and Electronic Throttle Control (ETC).

Some of the shared inputs used by the ADAS include:

- Wheel speed
- Steering/torque
- Yaw rate
- Accelerometer
- GPS

Interconnectivity involves using data bus networks, including different CAN bus networks, dedicated LIN buses, MOST, and manufacturer-specific bus network protocols. It's common for one ADAS to use several different bus networks.

Park Assist Obstacle Detection

A **park assist system** is a parking aid that alerts the driver to obstacles located in the path immediately behind the vehicle. The system can use visual indicators such as a warning message in the instrument cluster, a vehicle information monitor, or an LED display. The system will also use warning chimes to alert the driver of an object's presence.

Shop Manual
Chapter 16, page 883

The park assist system may include the following major components:

- *Instrument cluster*—used to display textual warnings and error messages related to the current operating status of the park assist system.
- *Park assist display*—mounted at the rear of the headliner just above the back glass, provides a visual indication of the presence of an obstacle by use of LEDs (**Figure 16-22**). Vehicles with front park assist use a second display unit mounted in the center of the instrument panel's top pad near the windshield.
- *Park assist module*—the central component of the system that supplies voltage to the park assist sensors and the park assist displays; processes the data from the sensors, calculates the display information, and performs system diagnostics.

Figure 16-22 Park assist system rear display.

- *Park assist sensors*—ultrasonic sensors located behind the rear bumper fascia. In addition, vehicles equipped with the front park assist system have ultrasonic sensors behind the front bumper fascia.
- *Park assist switch*—allows the driver to manually disable or enable the park assist feature. This allows the driver to disable the system if towing a trailer.

The park assist system is active anytime the ignition switch is in the RUN position, the parking brake is not applied, and the vehicle speed is less than 10 mph (16 Km/h). Rear park assist is active only with the transmission gear selector in the REVERSE position. If the vehicle also has a front park assist system, it's active with the transmission gear selector lever in the DRIVE or REVERSE position (automatic transmissions).

Ultrasonic transceiver park assist sensors in the bumpers locate and identify the proximity of obstacles in the path of the vehicle. The module triggers the sensors to generate ultrasonic sound pulses. The sensor then signals the module upon receiving the echo of the reflected sound pulses.

Each sensor receives battery voltage and ground from the module. The voltage supply and ground circuits are wired in parallel from the module. However, each sensor has a dedicated serial bus communication circuit to the module (**Figure 16-23**). The module alternately oscillates and then quits the membrane of each sensor at the same time. When the sensor membrane is oscillating, it emits an ultrasonic signal. The ultrasonic signal echoes from objects in the path of the vehicle.

When the membrane is quieted, each membrane will receive the echoes of the ultrasonic signals. These echoes may be from the sensor's own signals or from another sensor. The control module receives the echo data over the serial bus. The module uses the intervals between the ultrasonic transmission and reception data from the sensors to calculate the distance the vehicle is from any obstacles that the sensors may have identified.

The sensor area of detection from the vehicle is between 11.8 inches (0.3 meters) and 59.1 inches (1.5 meters), and the height from the ground is between 7.8 inches (0.2 meters), and 31 inches (0.8 meters). The detection area extends around the corners of the vehicle.

The module uses preprogrammed algorithms and calibrations to determine the appropriate outputs based on the data from the sensors. If needed, the module sends a warning message to the display unit(s) over a dedicated serial bus.

Figure 16-24 illustrates a display that contains two sets of LED indicators with eight on each side. Also, the display units house a chime tone transducer. While the park assist

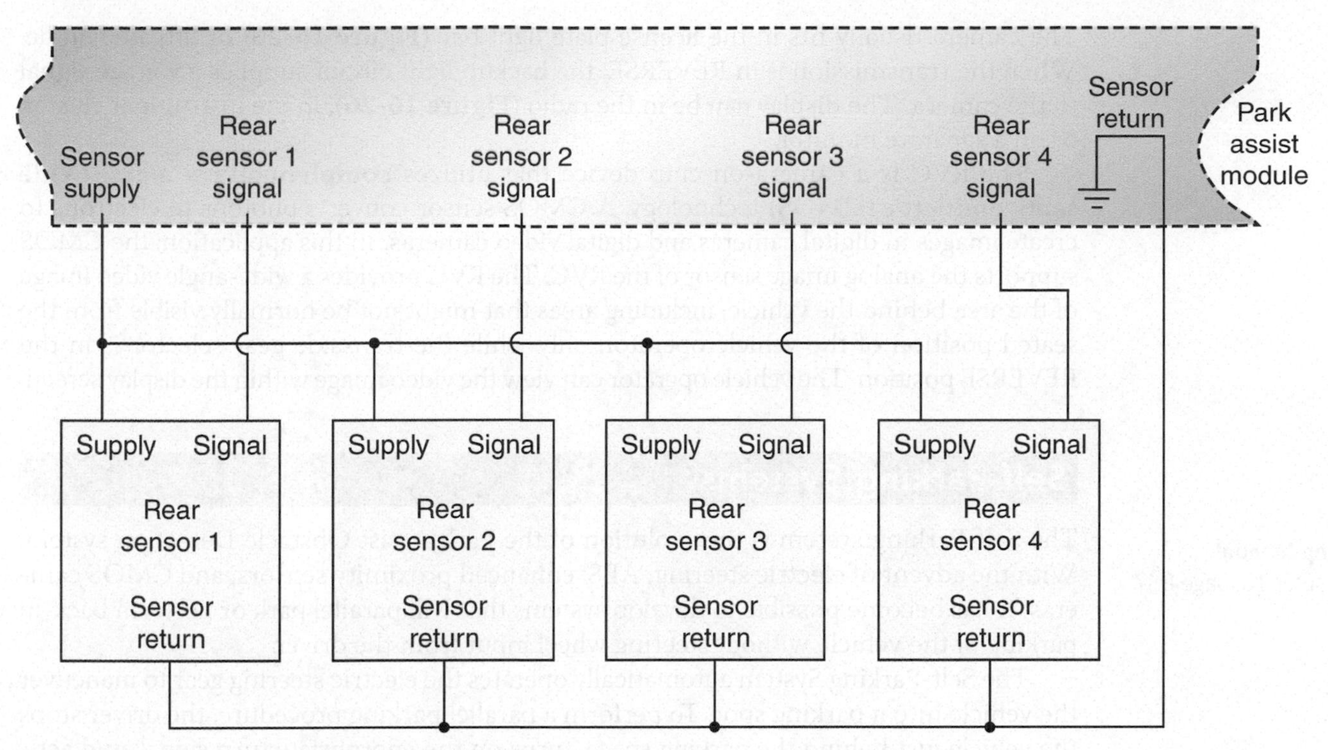

Figure 16-23 Wiring of the park assist system transponders.

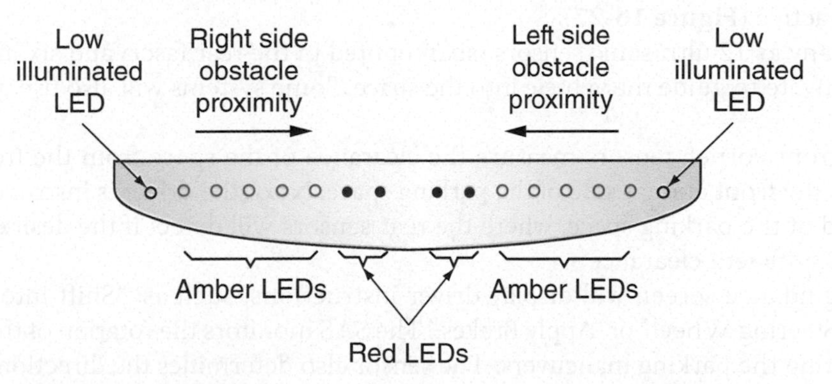

Figure 16-24 The display uses LEDs and audio tones to illustrate the proximity of an object.

system is active, the number, position, and color of the illuminated LEDs and the frequency of the audible signal indicate the distance of obstacles from the vehicle. The visual indication of an object is by illuminating one or more LED indicators. The outer amber units will light first and move toward the display's center the closer the vehicle gets to the obstacle. If the vehicle is closer than 31 inches (40 centimeters) of the obstacle, one of the red LED units illuminates. Also, the display will emit a series of short, intermittent, audible beeps. If the vehicle comes within 12 inches (30.5 centimeters) of the obstacle, the second red LED illuminates and the audible warning changes to a continuous tone.

When the park assist system is active, and there are no obstacles detected, the two outermost amber LED units illuminate at a reduced intensity. This provides a visual confirmation that the system is operating.

Beginning in the 2018 model year, all vehicles sold in the United States must have a rearview camera (RVC). The RVC can work in conjunction with the park assist system.

Shop Manual
Chapter 16, page 884

The camera usually fits in the license plate light bar (**Figure 16-25**) or liftgate handle. When the transmission is in REVERSE, the backup light circuit supplies a voltage signal to the camera. The display can be in the radio (**Figure 16-26**), in the instrument cluster, or on a separate monitor.

The RVC is a camera-on-chip device that utilizes **complementary metal oxide semiconductor (CMOS)** technology. A CMOS sensor converts photons to electrons to create images in digital cameras and digital video cameras. In this application, the CMOS supports the analog image sensor of the RVC. The RVC provides a wide-angle video image of the area behind the vehicle, including areas that might not be normally visible from the seated position of the vehicle operator, only while the transaxle gear selector is in the REVERSE position. The vehicle operator can view the video image within the display screen.

Self-Parking Systems

Shop Manual
Chapter 16, page 887

The Self-Parking system is the evolution of the Park Assist Obstacle Detection system. With the advent of electric steering, ABS, enhanced proximity sensors, and CMOS cameras, it has become possible to develop systems that will parallel park or perform back-in parking of the vehicle without steering wheel input from the driver.

The Self-Parking System automatically operates the electric steering gear to maneuver the vehicle into a parking spot. To perform a parallel parking procedure, the driver stops the vehicle just behind the parking space, turns on the appropriate turn signal, and activates the system with an input button. The turn signal indicates to the control module which side of the vehicle the parking space is. The driver receives a message that the system is active (**Figure 16-27**).

As many as 12 ultrasonic sensors (six mounted in the rear fascia and six in the front fascia) activate to guide the vehicle into the space. Some systems will also use up to eight cameras.

The front corner sensors measure the clearance of the space from the front of the vehicle to the front and far side of the parking space. Next, the driver is instructed to pull just ahead of the parking space, where the rear sensors will detect if the desired parking space has sufficient clearance

The guidance screen will display driver instructions, such as: "Shift into Reverse," "Release Steering Wheel," or "Apply Brakes." The SAS monitors the rotation of the steering wheel during the parking maneuvers. The sensor also determines the direction the front wheels are pointing to determine if there will be enough steering gear rotation to

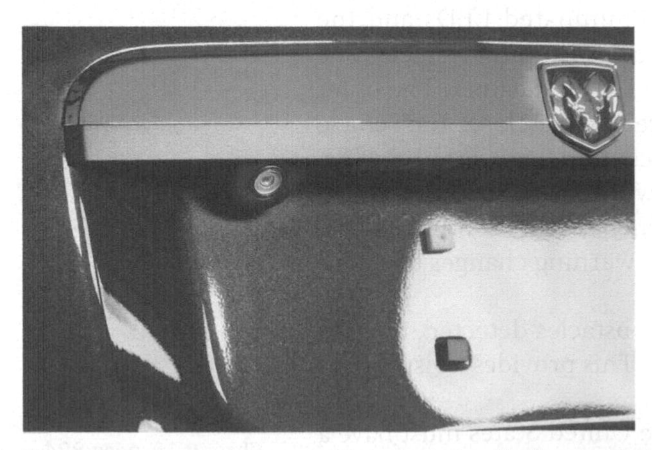

Figure 16-25 Rearview camera used for park assist and back-up monitoring.

Figure 16-26 Camera display shown on the radio screen.

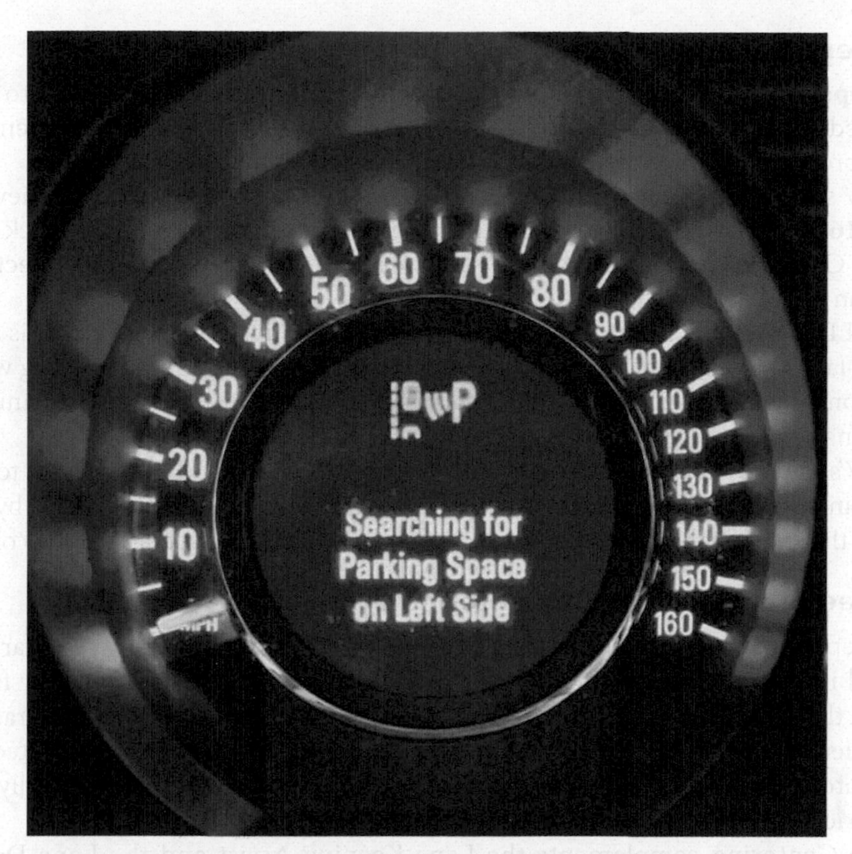

Figure 16-27 The driver will receive notifications of system activations along with instructional messages.

maneuver the vehicle into the space. In addition, the sensor may determine if the driver has removed their hands from the steering wheel by monitoring for extra resistance to turning that would result from the driver influencing the steering wheel input.

Using the ultrasonic sensors, the system will control the steering gear to maintain sufficient clearance within the parking space. During the parking maneuver, the ABS activates to apply the brakes if the sensors detect an obstacle that could result in a collision.

To park in a space on the left side of the vehicle, the driver turns on the automatic parking system and activates the left turn signal. This action informs the system that the desired parking space is on the left side of the vehicle. The left side ultrasonic sensors activate to determine the size of the parking space. The remainder of the process is the same as for parking in a right-side parking space.

For back-in parking, the driver stops the vehicle perpendicular to the center of the target parking spot. The driver activates the parking system, and the message center will display instructions. First, the system determines if there is sufficient space for the vehicle to enter. The system then controls the steering gear to back the vehicle into the space. The system can use multiturn maneuvering to back the vehicle into the parking spot if necessary.

Lane Departure Warning, Lane Keep Assist, and Lane Centering Systems

Arising from the Blind Spot Warning system's technologies that alert the driver if a vehicle is coming alongside them that they may not see in the rearview mirror are the Lane Departure Warning, Lane Keep Assist, and Lane Centering systems.

Shop Manual
Chapter 16, page 897

Lane Departure Warning

Shop Manual
Chapter 16, pages 880, 897

Lane Departure Warning (LDW) systems determine if the vehicle is about to leave its designated lane and then alert the inattentive or drowsy driver. Advanced systems maintain the proper lane until the driver takes control and manually changes lanes.

LDW systems use a forward-facing camera, often located behind the rearview mirror (**Figure 16-28**), to detect white lines, yellow lines, or dots in the road and to track the road features. Cameras used in the latest systems can operate even if it only detects a lane marker on one side of the vehicle.

Haptic is from the Greek word meaning "pertaining to the sense of touch."

The LDW system can inform the driver of the potential that their vehicle is about to leave the lane by sounding warning buzzers, by **haptic** feedback in the steering wheel, by tugging on the seat belt. Usually, the displaying of a visual warning accompanies these indications as well (**Figure 16-29**).

TRW's LDW system uses the electric power–assisted steering (EPAS) system to provide feedback in the steering wheel. The system simulates the rumble strip sensation felt by the cuts placed in the shoulder of the road. It's also configurable to provide actual steering correction.

Lane Keep Assist and Lane Centering Systems

Shop Manual
Chapter 16, pages 880, 897

Lane Keep Assist and Lane Centering systems can use millimeter-wave radar sensors mounted in the exterior mirrors or the outside corners of the front and rear fascia. In addition, these active systems use the Lane Departure Warning system's camera.

An alert sounds if the system determines the vehicle will leave its designated lane. If the operator fails to take corrective action, the control module will automatically operate the electric steering gear to maneuver the vehicle back into the proper lane.

Lane Centering complements the Lane Keeping Assist and the Lane Departure Warning systems. Lane Centering uses the sensors and cameras of the LDW and Lane Keep Assist systems to keep the vehicle centered in the lane. In addition, the system may provide steering assistance to guide the vehicle through gentle turns at highway speeds.

Collision Avoidance Systems

Shop Manual
Chapter 16, pages 880, 897

Collision Avoidance Systems are an extension of the ABS and ACC systems and will automatically apply the brakes to prevent an accident. Most early systems relied solely on millimeter-wave radar to detect an object's presence in the vehicle's path. Upon detection

Figure 16-28 The forward-facing lane departure camera can detect lines and dots in the road paint to track the vehicle's path of travel.

Figure 16-29 The driver will receive audible and visual warnings that the vehicle is drifting from its lane.

of an object, the Collision Avoidance System sounds an alarm and may display a warning message. If the driver does not respond, and the sensors indicate the likelihood of a collision, the ABS activates to turn on the pump and pulse the brakes to stop the vehicle. If the probability of an accident is unavoidable, the ABS activates to mitigate the collision force. If the vehicle has an active suspension system, it's activated to prevent excessive weight transfer to the front of the vehicle. In addition, the automatic seatbelt retractors may pull the seat belt tighter against the driver and passenger to securely retain them.

The Collision Avoidance system also becomes active due to sudden braking or the vehicle skidding: The ABS control module monitors brake pressure information and other inputs to determine if the driver has suddenly applied the brakes or lost control of the vehicle. If so, the ABS control module sends a signal to the seat belt control module over the CAN C bus to activate the automatic seatbelt retractors.

Advanced systems utilize an object recognition camera along with a driver monitor camera. The object recognition camera determines if a vehicle, pedestrian, animal, or other object is in a range of view in front of the vehicle. The camera determines the distance of the object and the closing speed before impact. If a collision may occur, an alert sounds, and vehicle speed slows by the closing of the Electronic Throttle Control (ETC). If more aggressive actions are required, the system operates as described for systems using radar sensors.

The driver monitor camera detects if the driver is not facing forward, if the driver's eyes are not looking forward, or if their eyelids are closed. The system may also monitor the SAS to identify if the steering wheel has not moved for an extended period. In addition, pressure sensors in the steering wheel determine if the driver is firmly gripping the steering wheel. If it's determined that the drive is not attentive, a warning buzzer sounds.

Vehicles with self-driving or self-parking capabilities that use several cameras and sensors to monitor the area around the vehicle can also control the electric steering system. If the proximity cameras and sensor indicate an empty lane or other "escape route," the vehicle can self-steer into that area to avoid a collision.

Shop Manual
Chapter 16, pages 887, 897

Rollover Mitigation System

The purpose of **Electronic Roll Mitigation (ERM)** is to attempt to prevent a vehicle rollover situation due to extreme lateral forces experienced during cornering or hard evasive maneuvers. This program is similar to electronic stability control (ESP) but is more aggressive in engine torque management. SUVs and other high-profile vehicles are the usual beneficiaries of Rollover mitigation systems.

Electronic Roll Mitigation (ERM) is also called proactive roll avoidance (PRA), roll stability control (RSC), active roll control (ARC), along with several other names.

About 5% of rollovers occur due to high-speed lane-change maneuvers and overwhelming sideways momentum (**Figure 16-30**). Most vehicle rollovers are the result of the vehicle being "tripped." Tripping occurs when a tire hits uneven ground, disrupting the vehicle's balance. This can happen if the tire runs up on a soft shoulder or road obstruction. Usually, the ESP system is capable of correcting these types of events. However, the vehicle can also lose its balance as a result of excessive cornering forces. The rollover mitigation feature uses additional sensors and actuators within the ESP system to prevent rollovers caused by these events.

Shop Manual
Chapter 16, page 888

Some systems function by attempting to reduce the friction with the road by applying individual brake calipers to induce a skid. This allows the vehicle to follow in the current direction that its momentum is taking it. By not attempting to change the vehicle's path and allowing the tires to skid on the road surface, it's less likely that the vehicle will overturn.

If equipped, activation of the vehicle's electronic suspension system adds further stability by controlling body roll and weight transfer. For example, the TRW system uses hydraulic actuators to actively alter the stiffness of the stabilizer bars by changing the length of the link to reduce body roll (**Figure 16-31**). Other systems change spring or shock absorber rates to prevent body roll.

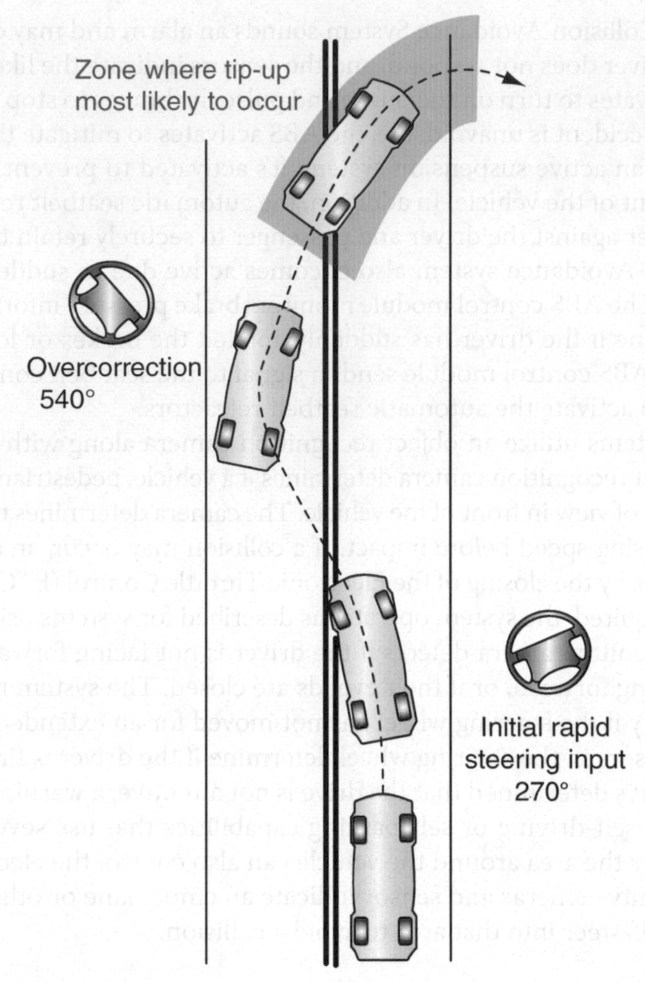

Figure 16-30 High-speed maneuvers can cause the vehicle to rollover.

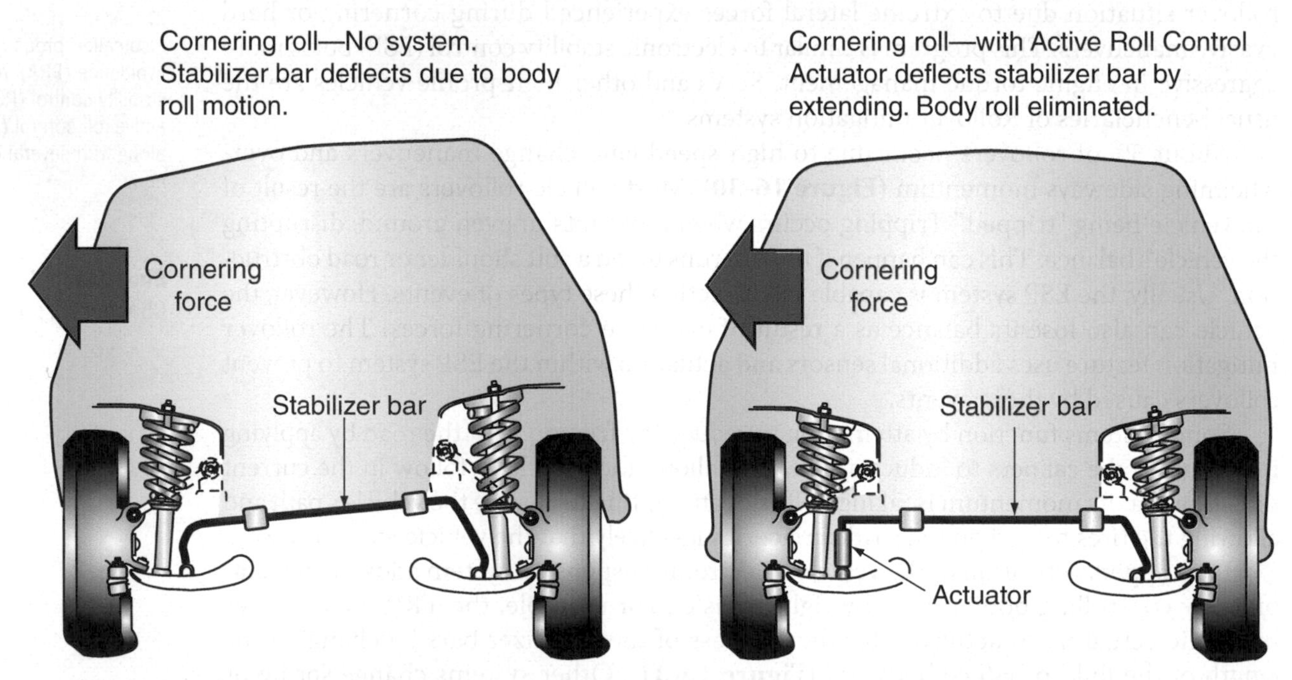

Figure 16-31 The actuator changes the stabilizer bar deflection to prevent body roll.

Autonomous Vehicles

If you consider all the technologies, sensors, and systems discussed thus far; it's not hard to see that self-driving autonomous vehicles were inevitable. The autonomous vehicle utilizes all of the available techniques discussed to detect its surroundings and to identify appropriate navigation paths. They also detect obstacles and relevant signage.

Some systems (such as Tesla Motors' Autopilot) have a "summon" feature. This feature allows the "driver" to leave the vehicle and send it to search out a parking space and self-park. When the "driver" needs the car to pick them up, they can use their smartphone to summon the vehicle.

Some self-drive systems use the information gathered from sensors and systems to generate a localization map that pinpoints the vehicle's location on an HD map. This enables the vehicle to establish road and lane structures so it can detect upcoming forks in the road, passing lanes, and lanes that are merging. Not all manufacturers use HD maps.

Summary

- Ultrasonic sensors evaluate the attributes of a target by interpreting the echoes from sound waves.
- Radar is a system that uses electromagnetic waves to identify an object's range, direction, or speed.
- The frequencies of Doppler shift provide measurement of an objects distance and speed.
- A laser is a device that controls the way that energized atoms release photons (light) using stimulated emission to cause the laser light atoms to release photons in a very organized state.
- The laser radar sensor has three parts: the laser emitter that radiates laser rays forward, the laser-receiving portion that receives the laser beams as they are reflected back by the object that is ahead, and the processing unit that determines the length of time it takes for the reflected beams to return to the sensor; the unit calculates the distance to the object ahead and the relative speed.
- Laser diodes are light-emitting devices that are capable of emitting laser beams.
- The infrared sensor can determine if an object is in proximity and detect obstructions by using the principle that any object emits an amount of energy that is a function of its temperature.
- When used as an obstruction sensor, the infrared light beams are monitored for distortions.
- Infrared sensors, used in conjunction with a camera, assist the camera in determining the shape and size of an object under low-light conditions.
- A park assist system is a parking aid using ultrasonic transceiver sensors to alert the driver of obstacles located in the path immediately behind the vehicle. Upon detection of an object, the system uses an LED display and warning chimes to provide the driver with visual and audible warnings of the object's presence.
- Lane departure warning (LDW) systems use a forward-facing camera to detect lane markers in the road and to track the road features. The system determines if the vehicle is about to leave its designated lane and then alerts the inattentive or drowsy driver.
- Lane Keep Assist will automatically operate the electric steering gear to maneuver the vehicle back into the proper lane.
- Lane Centering uses the sensors and cameras of the LDW and Lane Keep Assist systems to keep the vehicle centered in the lane. In addition, the system may provide steering assistance to guide the vehicle through gentle turns at highway speeds.
- Collision avoidance systems are an extension of the ABS and ACC systems and will automatically apply the brakes in an attempt to prevent an accident.
- Advanced collision avoidance systems utilize an object recognition camera along with a driver monitor camera. The driver monitor camera can detect if the driver is not facing forward, or if the driver's eyes are not looking forward or are closed.
- The purpose of electronic roll mitigation (ERM) is to attempt to prevent a vehicle rollover situation due to extreme lateral forces experienced during cornering or hard evasive maneuvers.

Review Questions

Short-Answer Essays

1. Explain the process of speed and distance detection using radar sensors
2. Explain the function of the laser diode.
3. Explain the principle of stimulated emission in the production of laser light.
4. Describe the operation of ultrasonic sensors.
5. Describe how a laser sensor can determine distances between objects.
6. What is the function of the Lane Keep Assist and the Lane Centering systems?
7. Describe how the infrared sensor can be used in conjunction with a camera to determine an object's shape and size under low-light conditions.
8. Explain the operating fundamentals of the self-parking vehicle.
9. Explain the operation of the Collision Avoidance System.
10. Explain how the Electronic Roll Mitigation (ERM) can use the brakes to prevent a rollover.

Fill in the Blanks

1. Radar is a system that uses _______________ waves to identify the range, direction, or speed of an object.
2. In the radar sensor system, the speed of an object is determined by the _______________ of the echo.
3. A laser is a device that controls the way that energized atoms release _______________.
4. A CMOS sensor converts _______________ to _______________ to create images in digital cameras and digital video cameras.
5. _______________ sensors can also be used in conjunction with a camera to assist the camera in determining the shape and size of an object under low-light conditions.
6. The Self-Parking System automatically operates the _______________ _______________ _______________ to maneuver the vehicle into a parking spot.
7. LIDAR uses spinning _______________ to create a 360-degree image based upon distance and light.
8. Level _______________ automation allows the car to take control of the steering wheel and/or the pedals in certain driving modes.
9. The _______________ _______________ camera can detect if the driver is not facing forward, or if the driver's eyes are not looking forward or are closed.
10. When used in a camera system, infrared sensors use the _______________ _______________ to determine the object's shape.

Multiple Choice

1. In the ultrasonic sensor, the transmitter generates the ultrasonic sound waves by:
   A. Using a piezoelectric crystal with an alternating voltage across it that causes them to oscillate at very high frequencies.
   B. Using a piezoresistive bridge that varies the current flow through a sound generator.
   C. Converting the sound waves into electrical energy.
   D. Using a capacitive discharge to oscillate a thin-film transistor sheet at a high frequency.
2. Ultrasonic sensors evaluate the attributes of a target by interpreting the echoes from:
   A. radio waves.
   B. frequency distortion.
   C. sound waves.
   D. None of these choices.
3. Radar sensors:
   A. use electromagnetic waves.
   B. emit a reference signal and measure the time it takes to echo back.
   C. use Doppler shift to determine speed.
   D. All of these choices.
4. Stimulated emission results from:
   A. two atoms rubbing together.
   B. two atoms forced to stay apart.
   C. a photon stimulating a second atom to emit a photon.
   D. two photons that vibrate at differing frequencies.

5. When electrons fall to a lower energy level, they release the energy in the form of:
   A. heat.
   B. photons.
   C. kinetic energy.
   D. sound.

6. The self-parking system is being discussed. *Technician A* says the system only performs parallel parking maneuvers if the parking space is on the right hand side of the vehicle.
   Technician B says when maneuvering into a back-in parking space, the system can use multiturn maneuvering if necessary.
   Who is correct?
   A. A only
   B. B only
   C. Both A and B
   D. Neither A nor B

7. *Technician A* says Lane Keep Assist will automatically operate the electric steering gear to maneuver the vehicle back into the proper lane.
   Technician B says Lane Centering may provide steering assistance to guide the vehicle through gentle turns at highway speeds.
   Who is correct?
   A. A only
   B. B only
   C. Both A and B
   D. Neither A nor B

8. Self-parking systems are being discussed.
   Technician A says the ultrasonic sensors determine if the parking space is of sufficient size.
   Technician B says the ultrasonic sensors detect obstacles.
   Who is correct?
   A. Technician A
   B. Technician B
   C. Both Technician A and Technician B
   D. Neither Technician A nor Technician B

9. *Technician A* says the Collision Avoidance System may use the Steering Angle Sensor to determine if the driver has removed their hands from the steering wheel.
   Technician B says each ADAS must be a stand-alone system with its own inputs and outputs.
   Who is correct?
   A. A only
   B. B only
   C. Both A and B
   D. Neither A nor B

10. *Technician A* says the park assist obstruction detection system is an example of Level Two automation.
    Technician B says Lane Keep Assist is an example of Level Four automation.
    Who is correct?
    A. A only
    B. B only
    C. Both A and B
    D. Neither A nor B

CHAPTER 17

HEV, EV, AND ALTERNATIVE POWER SOURCES

Upon completion and review of this chapter, you should be able to:

- Describe the typical operation of a hybrid vehicle.
- Explain how regenerative braking recharges the HV battery.
- Explain the difference between parallel, series, and series/parallel hybrids.
- Describe the power flow through Mild and Strong HEVs.
- Explain the basic operation of electric vehicles.
- Describe the methods used for thermal management of the HV systems and battery.
- Explain the 42-volt system used for Stop/Start.
- Explain the operating principles of the Battery Management System (BMS).
- Describe the differences between HV battery charging levels.
- Explain the necessity for the HV battery system to perform cell balancing.
- Describe how a proton exchange membrane produces electricity in a fuel cell system.
- List and describe the different fuels used in a fuel cell system.
- Describe the purpose of the reformer.
- Explain how different types of reformers operate.
- Explain the differences in charging levels.

Terms to Know

Battery balancing
Battery Management System (BMS)
Blended brake system
Direct Current Fast Chargers (DCFCs)
Electric Vehicle (EV)
Electric Vehicle Supply Equipment (EVSE)
Electrolysis
Extended Range Electric Vehicle (EREV)
Fuel cell
Full parallel hybrid
High-voltage ECU (HV ECU)
Hybrid electric vehicle (HEV)
Lean burn technology
Mild parallel hybrid
Parallel hybrid
Plug-in Hybrid Electric Vehicle (PHEV)
Proton Exchange Membrane (PEM)
Reformer
Regenerative braking
Series hybrid
Series/Parallel hybrid
Starter Generator Control Module (SGCM)
Ultra capacitor

Introduction

Expanded emission regulations and the public's desire to become less dependent on carbon-based fuels have led most major automotive manufacturers to develop alternative-powered vehicles. The California Air Resources Board (CARB) established a low-emission vehicles/clean fuel program to reduce mobile source emissions in California during the late 1990s. This program established emission standards for five vehicle types (**Figure 17-1**):

	CV	TLEV	LEV	ULEV	ZEV
NMOG	0.25*	0.125	0.075	0.040	0.0
CO	3.4	3.4	3.4	1.7	0.0
NOx	0.4	0.4	0.2	0.2	0.0

(*) Emission standards of NIMHC

Figure 17-1 California tailpipe emission standards in grams per mile at 50,000 miles.

Conventional Vehicle (CV), Transitional Low-Emission Vehicle (TLEV), Low-Emission Vehicle (LEV), Ultra-Low-Emission Vehicle (ULEV), and Zero-Emission Vehicle (ZEV).

This chapter explores common Hybrid Electric Vehicle (HEV) and Electric Vehicle (EV) power systems. In addition, this chapter covers fuel cell theories and some of the methods manufacturers use to approach this alternative power source.

Hybrid Vehicles

A BIT OF HISTORY

The 1896 Armstrong is the world's first known gasoline/electric hybrid automobile. Designed by Harry E. Dey, it was built in Bridgeport, Connecticut by the American Horseless Carriage Company. There is only one known in existence, and was purchased in 2016 for $483,400 by Dutch collector Evert Louwman, who has it on display in his museum in The Hague, Netherlands. The vehicle has a 6,500 cc opposed, twin-cylinder gasoline engine with a dynamo wound flywheel. This design allowed the engine to charge the storage batteries for use by the ignition and lighting systems but could also rotate the engine for starting. Solenoids installed into the intake valve housings would release compression while the engine was rotated electrically. The size of the flywheel dynamo allowed for propelling the vehicle using only electric power. Other innovations included an automatic spark advance, an electric clutch, rear-wheel brakes with a regenerative electric motor assist, and a three-speed (with reverse) constant mesh semi-automatic transmission. A sliding key system engaged the transmission. Half of the gears were cut from rawhide to reduce noise.

The fully battery-powered vehicle was an early attempt at an alternate powered vehicle. However, the battery has a limited energy supply and restricts the traveling distance which is typically much shorter than that provided by a fuel. This limitation was a major stumbling block for many consumers. One method to improve the range was the addition of an on-board power generator assisted by an Internal Combustion Engine (ICE). The result was the **Hybrid Electric Vehicle (HEV)**. An HEV has two power sources: an ICE and an electric motor. The addition of the ICE means the HEV does not meet the classification of a ZEV. However, it does reduce emission levels and increases fuel economy significantly.

A typical hybrid vehicle uses a 14-kW or larger Motor/Generator (MG) or Integrated Starter Generator (ISG) between the ICE and the transmission (**Figure 17-2**). These units provide fast, quiet-starting, automatic engine stop/start, recharging the vehicle batteries, and smoothing driveline surges.

Shop Manual
Chapter 17, page 924

The components of a typical hybrid vehicle include the following:

- *Batteries.* The types of batteries used or experimented with include lead-acid, nickel-metal hydride, and lithium-ion batteries. Also, passenger safety is a major concern. The batteries reside in sealed containers to ensure complete protection (**Figure 17-3**).

Shop Manual
Chapter 17, page 928

Figure 17-2 The MGs (or ISG) is usually located between the engine and the transmission in the bell housing.

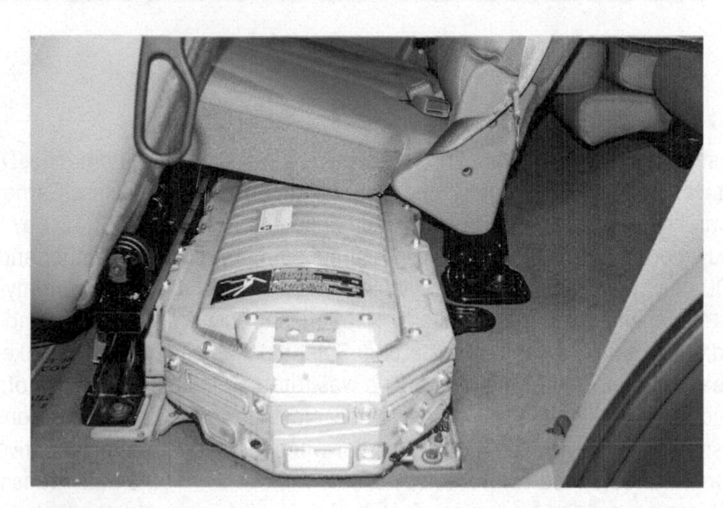

Figure 17-3 Battery pack.

- *Electric motors*. The electric motor converts electrical energy to mechanical energy. The mechanical energy drives the wheels of the vehicle. The electric motor design allows for maximum torque at low revolutions per minute (rpm). This gives the electric motor the advantage of having better acceleration than the conventional ICE.
- *Ultra capacitors*. The **ultra capacitor** stores energy as an electrostatic charge. It's the primary device in the power supply during hill climbing, acceleration, and braking energy recovery. To create a larger storage capacity for the ultra capacitors requires increasing the surface area, and in turn, the voltage is increased. Additional electronics are required to maintain a constant voltage because the voltage drops as energy is discharged.
- ***Starter Generator Control Module (SGCM)***. Controls the flow of torque and electrical energy into or out of the MG or ISG. The function of the SGCM is to control the engine cranking, torque, speed, and active damping functions.

Shop Manual
Chapter 17, page 938

The SGCM is also called a **High-Voltage ECU (HV ECU).**

Regenerative Braking

Shop Manual
Chapter 17,
page 936, 938

The MG or ISG can also convert kinetic energy to Direct Current (DC) voltage through **regenerative braking**. Regenerative braking turns the energy used to slow the vehicle back into electricity. Regenerative braking occurs under two conditions; when the vehicle is traveling downhill with no load on the ICE and when applying the brakes to slow or stop the vehicle. Regenerative braking will only occur if the HV battery can store the energy.

When the vehicle travels downhill, the wheels transfer energy through the transmission and ICE to the MG or ISG. The HV battery receives this energy for storage and use by the vehicle's electrical components.

During deceleration, the vehicle slows as the SGCM allows the transfer of a certain amount of tire rotation to the MG or ISG. As the vehicle decelerates several inputs calculate the reduction in speed. Common inputs include the accelerator pedal position sensor, brake pedal travel sensor, and a longitude accelerometer that measures vehicle inertia.

The service brakes operate differently than most conventional vehicles. Most HEVs and EVs use a **blended brake system** that uses conventional ABS and regenerative braking. About 30% of the kinetic energy lost during braking is in heat. When decreasing acceleration, regenerative braking helps minimize energy loss by recovering the energy used for braking.

AUTHOR'S NOTE Some blended braking systems have a direct hydraulic connection between the master cylinder and the rear brakes.

Under normal operating conditions, the master cylinder generates a hydraulic "signal" to the ABS module to indicate the driver's demand for braking. The master cylinder doesn't directly apply the brake calipers. Regenerative braking assumes some of the stopping duties from the conventional friction brakes and uses the electric motor to help stop the vehicle. To do this, the electric motor operates as a generator when the brakes are applied, recovering some of the kinetic energy and converting it into electrical energy.

If regenerative braking alone fails to decelerate the vehicle at a sufficient rate, the ABS module will apply the service brakes hydraulically using the Hydraulic Control Unit (HCU). If regenerative braking is turned off due to the battery having a high SOC, the system reverts to hydraulic braking.

Since the master cylinder only provides an input to the ABS control module, the driver may notice a different brake pedal feel compared to a conventional vehicle. A brake pedal stroke simulator sensor provides feedback to the driver's foot that simulates the expected feel of braking on a conventional vehicle to overcome this difference.

Plug-In HEV

A **Plug-In Hybrid Electric Vehicle (PHEV)** offers increased all-electric mode operation without needing to engage the ICE within a certain range and speed. The PHEV typically has a larger battery size than the HEV. The biggest difference is the batteries of a PHEV can be charged using a home electric outlet or a public charging station.

Since the main power source for PHEVs is electricity, recharging the HV battery from the electricity grid reduces the need to start the ICE. This increases the vehicle's range while further reducing emissions. However, when the HV battery SOC drops to a specified level, the ICE starts to drive the generator, replenish the battery, and propel the vehicle.

HEV Propulsion

Basically, the HEV relies on power from the electric motor, the engine, or both (**Figure 17-4**). When the vehicle moves from a stop and has a light load, the electric motor moves the vehicle. Power for the electric motor comes from stored electricity in the battery pack. During normal driving conditions, the engine is the main power source. Engine power also rotates a generator that recharges the storage batteries (**Figure 17-5**). The output from the generator can also power the electric motor to provide additional power

Figure 17-4 Hybrid power system.

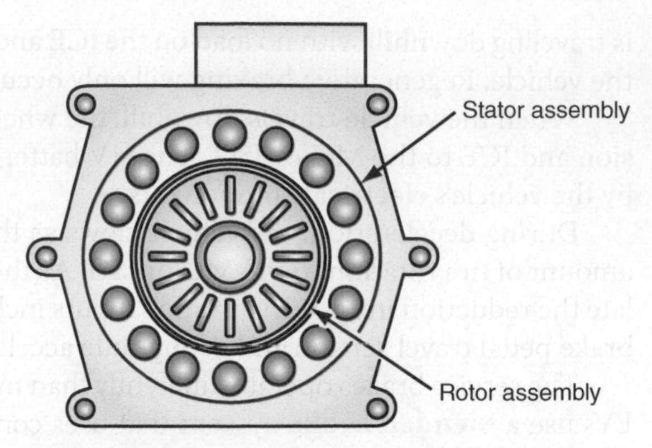

Figure 17-5 Engine power also rotates a generator that recharges the storage batteries and drives the vehicle. The rotor assembly is a very powerful magnet that, when rotating, induces voltage into the stator windings.

to the powertrain (**Figure 17-6**). A computer controls the electric motor's operation depending on the vehicle's power needs. During full throttle, or heavy load operation, the battery provides additional energy to the motor to increase the output of the powertrain.

There are three typical ways to arrange power flow in an HEV: parallel, series, and series/parallel. The **parallel hybrid** uses the ICE as well as the MG or ISG to propel the vehicle (**Figure 17-7**). The parallel hybrid configuration has a direct mechanical connection between the ICE and the wheels. Both the engine and the electric motor can turn the transmission at the same time.

There is a further distinction between a **mild parallel hybrid** and a **full parallel hybrid** vehicle. A mild parallel hybrid vehicle has an electric motor that is large enough to provide regenerative braking, instant engine start-up, and a boost to the ICE. However, the electric motor cannot propel the vehicle on its own.

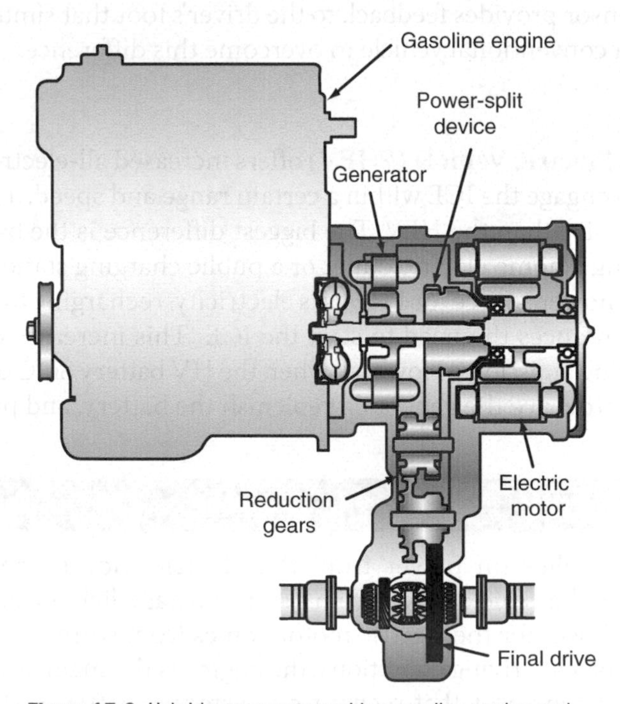

Figure 17-6 Hybrid power system with a gasoline engine and electric propulsion motor.

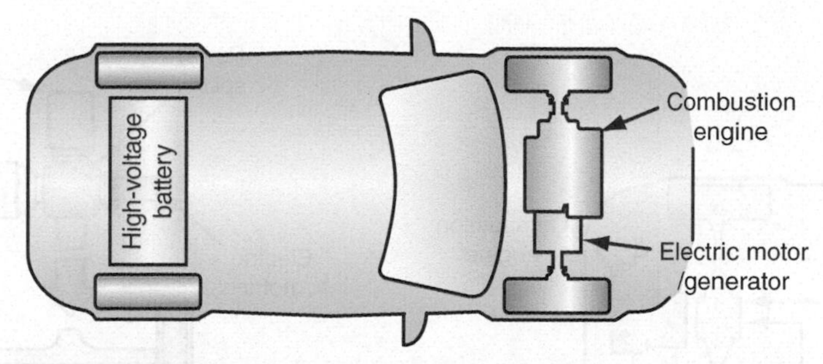

Figure 17-7 Parallel hybrid configuration.

A full (strong) parallel hybrid vehicle uses an electric motor that is powerful enough to propel the vehicle on its own. The electric motor(s) propel the vehicle during most city driving conditions and at road speeds below 30 mph (50 km/h). When needed, the ICE and electric motors combine to provide a boost in power. For example, the ICE is used for periods of long highway driving, while short, low-intensity drives around town uses the electric motor. The electric motor also provides the vehicle with added acceleration. Typically, the boost in acceleration is only sustained until the vehicle reaches a certain speed. At this speed, the ICE starts and replaces the electric motor. The parallel hybrid combines the alternator, starter, and wheels to create a system that starts the ICE, electronically balances it, takes power from it and turns it into electricity, and provides extra power to the driveline power-assist is needed for hill climbing or quick acceleration.

Other parallel hybrid vehicle configurations include using an ICE to power one axle and an electric motor to power the other (**Figure 17-8**). Another concept is to use a combination where the ICE (coupled with the electric motor) powers one axle and another electric motor powers the other axle (**Figure 17-9**).

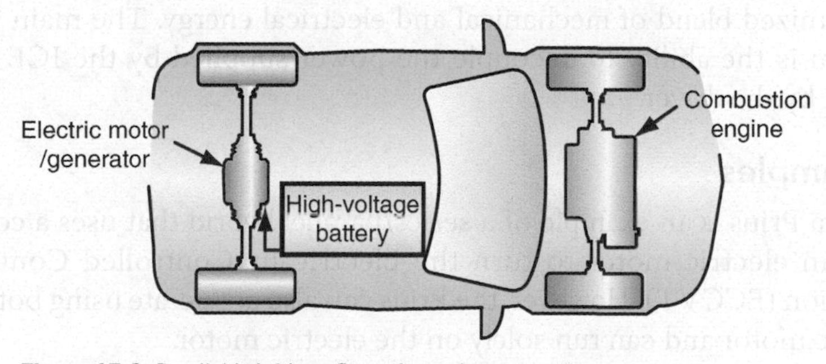

Figure 17-8 Parallel hybrid configuration using an engine to power one axle and an electric motor to power the other axle.

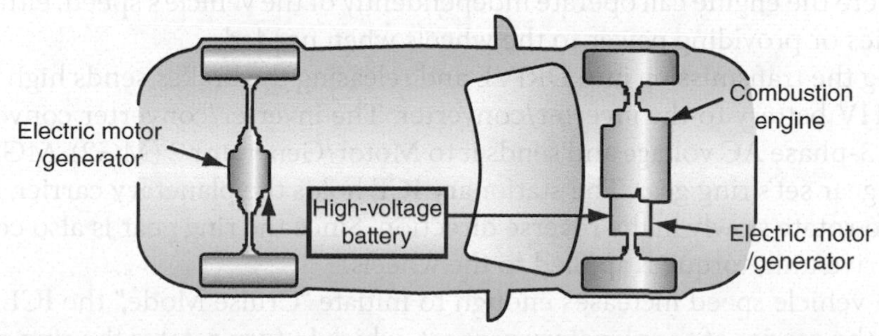

Figure 17-9 Parallel hybrid configuration using a combination where the engine coupled with an electric motor powers one axle and another electric motor powers the other axle.

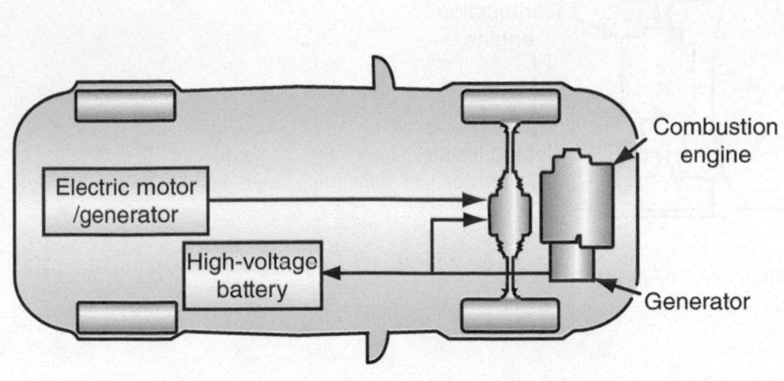

Figure 17-10 Series hybrid configuration.

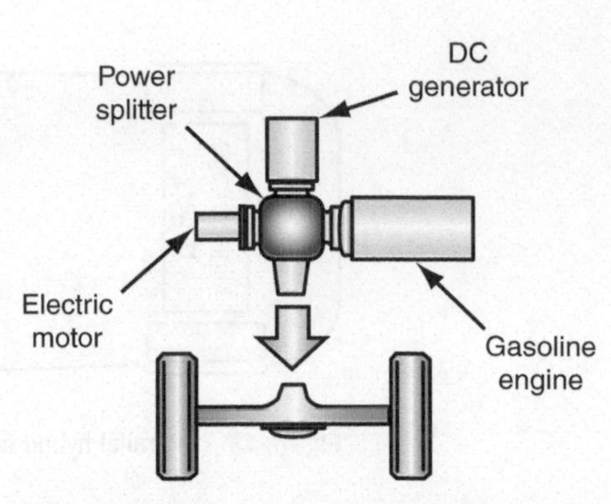

Figure 17-11 The power splitter allows for acceleration using both the engine and the electric motor and can run solely on the electric motor.

The **series hybrid** vehicle doesn't have a mechanical connection between the ICE and the wheels. The electric motor(s) are the only means of propelling the vehicle. The ICE turns a generator, and the generator either charges the batteries or powers the electric motor, which in turn drives the transmission. Therefore, the ICE never directly powers the automobile (**Figure 17-10**).

The power used to give the vehicle motion transforms from chemical energy into mechanical energy, then into electrical energy, and finally back to mechanical energy to drive the wheels. This configuration is efficient in that it never idles. The automobile turns off completely at rest, such as at a stop sign or traffic light. This feature greatly reduces emissions. There are various options in the configuration and mounting of all the components. Some series hybrid vehicles do not use a transmission.

The **series/parallel hybrid** is a combination of the two drive types. This system can utilize a smaller and highly efficient ICE and incorporates a power-split device (**Figure 17-11**). The power-split allows a power path from the ICE to the wheels that can be an optimized blend of mechanical and electrical energy. The main principle behind this system is the ability to decouple the power supplied by the ICE from the power demanded by the driver.

HEV Examples

The Toyota Prius is an example of a series/parallel hybrid that uses a combination of an ICE and an electric motor to turn the Electrically Controlled Continuous Variable Transmission (ECCVT). However, the Prius can also accelerate using both the engine and the electric motor and can run solely on the electric motor.

Operating the engine independently of the vehicle speed means that even though the vehicle is traveling at highway speeds, the engine can be close to idle speed since it only acts as a generator.

Using a set of planetary gears (**Figure 17-12**), the vehicle can operate like a parallel vehicle in that either the electric motor or the gasoline engine powers the vehicle. If needed, both power the vehicle together. However, the vehicle can also operate as a series hybrid where the engine can operate independently of the vehicle's speed, either charging the batteries or providing power to the wheels when needed.

Placing the transmission into DRIVE and releasing the brakes sends high DC voltage from the HV battery to the inverter/converter. The inverter/converter converts the DC voltage to 3-phase AC voltage and sends it to Motor/Generator 2 (MG2). MG2 drives the planetary gear set's ring gear. The stationary ICE holds the planetary carrier, causing the sun gear to rotate slowly in the reverse direction. Since the ring gear is also connected to the final drive unit, torque is applied to the wheels.

Shop Manual
Chapter 17, page 936

When vehicle speed increases enough to initiate "Cruise Mode," the ICE starts. The ICE turns the carrier of the planetary gear set, which in turn rotates the ring gear and the drive wheels. The sun gear also drives MG1. The electrical energy produced by MG1 flows

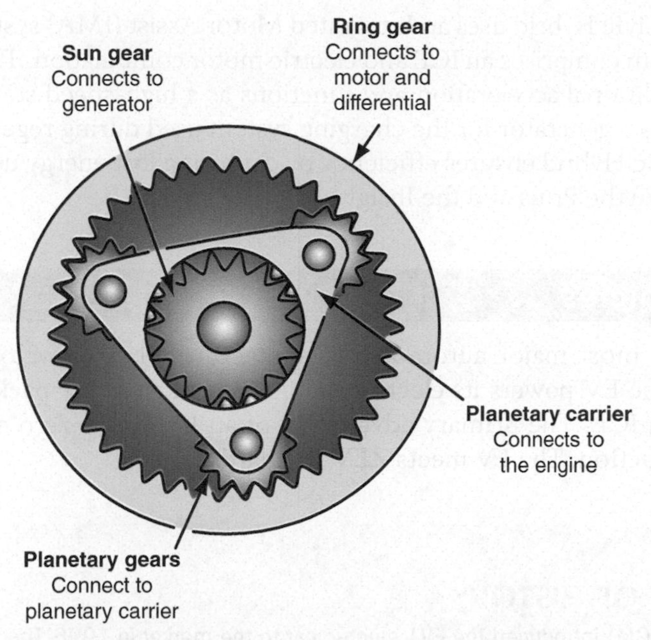

Figure 17-12 Planetary gears transfer power to the drive wheels.

to the inverter/converter, which directs it to MG2. Since MG2 is also connected to the ring gear, it rotates the drive wheels. So, when started, the ICE has a mechanical connection to the drive wheels, and at the same time, MG1 powers MG2, which assists the ICE. The HV battery doesn't provide any power during this time. The amount of assist by the ICE can reduce to zero as the load decreases.

Under heavy load conditions, both the ICE and MG2 propel the vehicle. This is similar to cruise mode, except now the inverter/converter directs energy from the HV battery to MG2. The ICE produces full power, so MG1 also produces full power. This power transfers to MG2, which also receives energy from the HV battery. Thus, MG2 is operating at full power also.

If the HV battery state of charge drops too low, the ICE starts to recharge it. The inverter/converter shuts down power flow from the HV battery, and MG2 turns off. The ICE propels the vehicle through the carrier to ring gear connection of the planetary gear set. The rotating sun gear drives MG1 with acts as a generator to produce AC voltage that the inverter/converter changes to DC voltage and directs it to the HV battery.

During deceleration, the ICE shuts off. The drive wheels turn MG2 by the ring gear of the planetary gear set. The stationary ICE holds the carrier. The sun gear rotates slowly, and MG1 doesn't produce any energy. Since the drive wheel/ring gear combination drives MG2, it becomes a generator. The AC voltage from MG2 goes to the inverter/converter, which changes it into DC voltage and directs it to the HV battery.

The Prius uses a "drive-by-wire" accelerator. The driver inputs their acceleration request to the motor management system. The management system then decides if the necessary power should come from the engine, the battery, or both. The same accounts for the brake-by-wire system: the driver calls for the appropriate amount of retardation, and the motor management coordinates this between the wheel brakes and the regenerative braking system.

The Honda Insight is also a parallel hybrid vehicle. In this system, the ICE provides the majority of the power. The electric motor assists the ICE by providing additional power during acceleration. The Insight uses regenerative braking technology to capture energy lost during braking. The Insight also has a lightweight engine that uses **lean burn technology** to maximize efficiency. Lean burn technology, developed in the 1960s, uses high air-fuel ratios to increase fuel efficiency.

The Honda Civic Hybrid uses an Integrated Motor Assist (IMA) system to power the vehicle. The system comprises an ICE and electric motor combination. The electric motor is a source of additional acceleration and functions as a high-speed starter. The electric motor also acts as a generator for the charging system used during regenerative braking. This way the Civic Hybrid ensures efficiency by capturing lost energy using regenerative braking, much like the Prius and the Insight.

Electric Vehicles

Since the 1990s, most major automobile manufacturers have developed an **Electric Vehicle (EV)**. The EV powers its electric motor(s) from a battery pack (**Figure 17-13**) without using an ICE. The primary advantages of an EV are its zero emissions and its drastic noise reduction. The EV meets ZEV standards.

A BIT OF HISTORY

General Motors (GM) introduced the EV1 electric car to the market in 1996. The original battery pack in this car contained 26 12-volt batteries that delivered electrical energy to a 3-phase 102-kilowatt (kW) AC electric motor.

Historically, the driving range of EVs has been their biggest disadvantage. Other disadvantages include the cost of replacement batteries and the danger associated with the high voltage and high frequency of the motors. Much research continues on extending the range and decreasing the required recharging times. Currently, the use of lithium-ion, nickel/metal/hydride, or lead-acid batteries and permanent magnet motors have extended the operating range. Other improvements, such as **regenerative braking** and highly efficient accessories (e.g., a heat pump for passenger heating and cooling), are also improving the EV range.

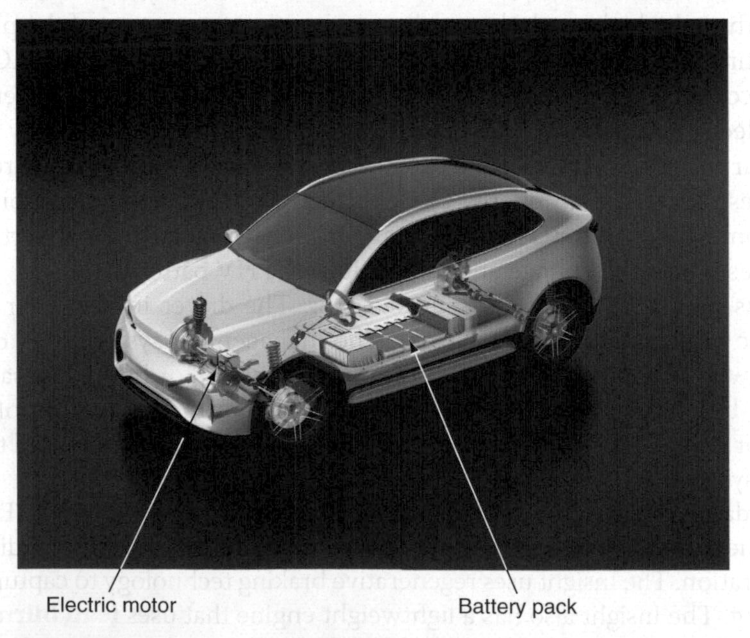

Figure 17-13 An electric motor that receives its energy from a battery pack powers the electric vehicle.

Extended Range Electric Vehicles

The **Extended Range Electric Vehicle (EREV)** is much like a series PHEV in that it has an ICE that drives a generator to recharge the HV batteries and extend the vehicle's range. Unlike the PHEV, the electric motors are the only means of driving the vehicle's wheels. The ICE never propels the vehicle and starts only when the vehicle's HV battery is nearly depleted.

High-Voltage System Thermal Management

Shop Manual
Chapter 17, page 928

In the development of the high-voltage electronics and battery, thermal management must be a consideration. Consequently, all HV vehicles need to control the environment that the HV system and battery reside. High voltage, high amperage circuits generate heat. In particular, the inverter/converter and electric motors generate enough heat to require cooling.

HEVs and PHEVs have the additional challenge of heating and maintaining the temperature of an engine that doesn't run all the time. It's common for HEVs and PHEVs to have at least two individual and separate cooling systems; one for the ICE and the other for the HV electronics (**Figure 17-14**). Some systems will also have a third separate loop used to control the temperature of the HV batteries.

The inverter/converter has its own cooling system, complete with a reservoir tank, hoses, bleeders, radiator, and pump. The 12-volt electric pump circulates coolant to the inverter/converter's cooling housing (**Figure 17-15**). Coolant can also be pumped to the MG(s) or ISG. The pump turns on anytime the vehicle is in READY mode.

A temperature sensor monitors the temperature of the inverter/converter. The ECU uses the input to the control module to control the operation of the pump. The coolant in the system flows in a loop from the pump to the DC/DC converter, to the inlet port of the motor electronics radiator, out of the radiator, into the transaxle, and back to the pump. The cooling system has a degassing system built into the loop that bleeds air/gases into the coolant reservoir.

HV batteries have an ideal operating temperature window between 65° and 105°F (18° and 40°C). Service life decreases at operating temperatures higher than 105°F (40°C). Efficiency drops, and output decreases at temperatures below 14°F (−10°C). Also, cooler

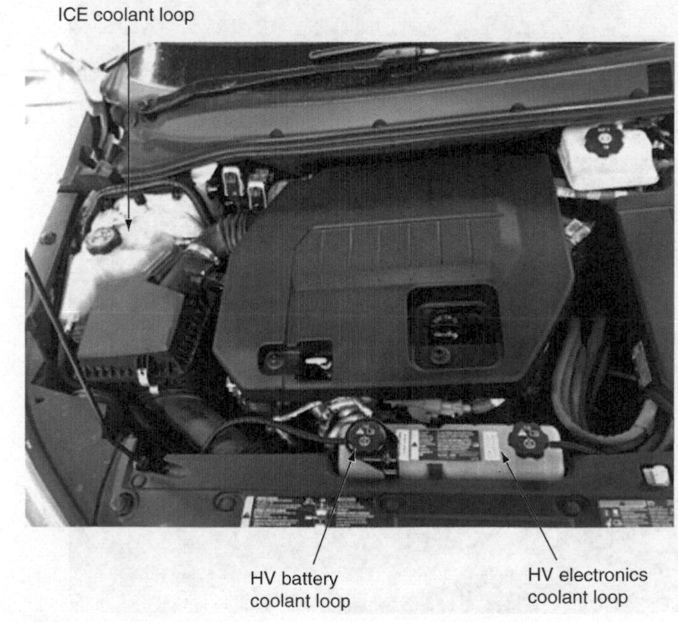

Figure 17-14 The ICE and HV systems have separate cooling systems.

Figure 17-15 Coolant pumped through the inverter/converter's housing controls the temperature of the unit. © iStockPhoto.com/supersmario

ambient temperatures increase the battery's resistance to charging. Furthermore, maintaining the temperature differential between individual battery cells within a defined range optimizes battery health.

Battery temperature can vary from module to module. An imbalance in temperature affects the power and capacity of the battery, the charge acceptance during regenerative braking, and vehicle operation.

A significant increase in the temperature of the cells occurs during peak loads with high current flows that can result from regeneration and boosting. Also, high ambient temperatures can cause HV battery temperature to reach the critical 105°F (40 °C) level quickly.

There are three common methods of cooling the HV battery; air cooling, liquid cooling, and refrigerant cooling. Some manufacturers will combine methods.

The air-cooled system can be passive and rely solely on the convection of the surrounding air. Active systems use cooling fans that direct cabin air into the battery (**Figure 17-16**). Temperature sensors placed in different locations around the battery cells monitor the battery's temperature. The ECU uses PWM to control the speed of the blower fan motor.

Another method is to plumb cabin air from the HVAC system to the battery. The major disadvantage of active air cooling is its inefficiency since the cooling fans require large amounts of energy.

Liquid cooling offers greater cooling potential than air cooling since liquid coolants have higher thermal conductivities. Coolant can either submerge the batteries or be pumped through passages in the battery assembly (**Figure 17-17**). Liquid cooling typically involves a dedicated cooling system that includes a:

- Coolant radiator
- Pump
- Surge tank
- Flow control valve
- Air separator

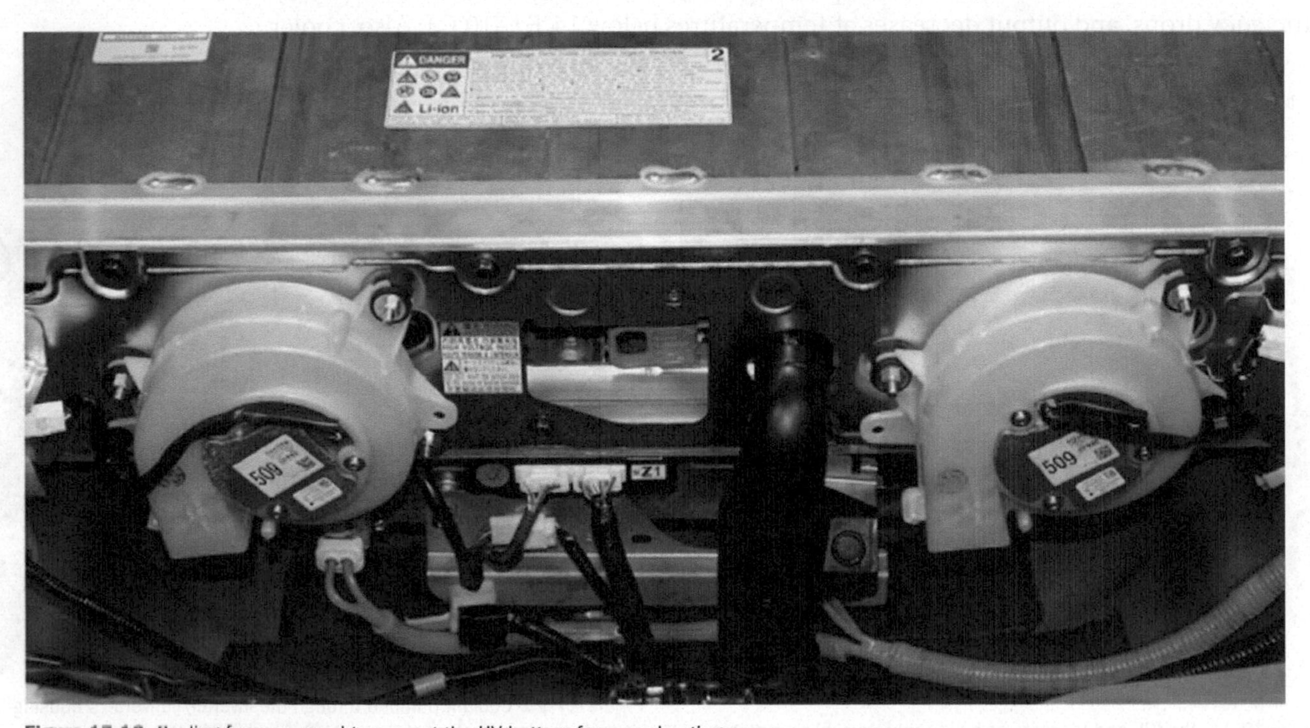

Figure 17-16 Cooling fans are used to prevent the HV battery from overheating.

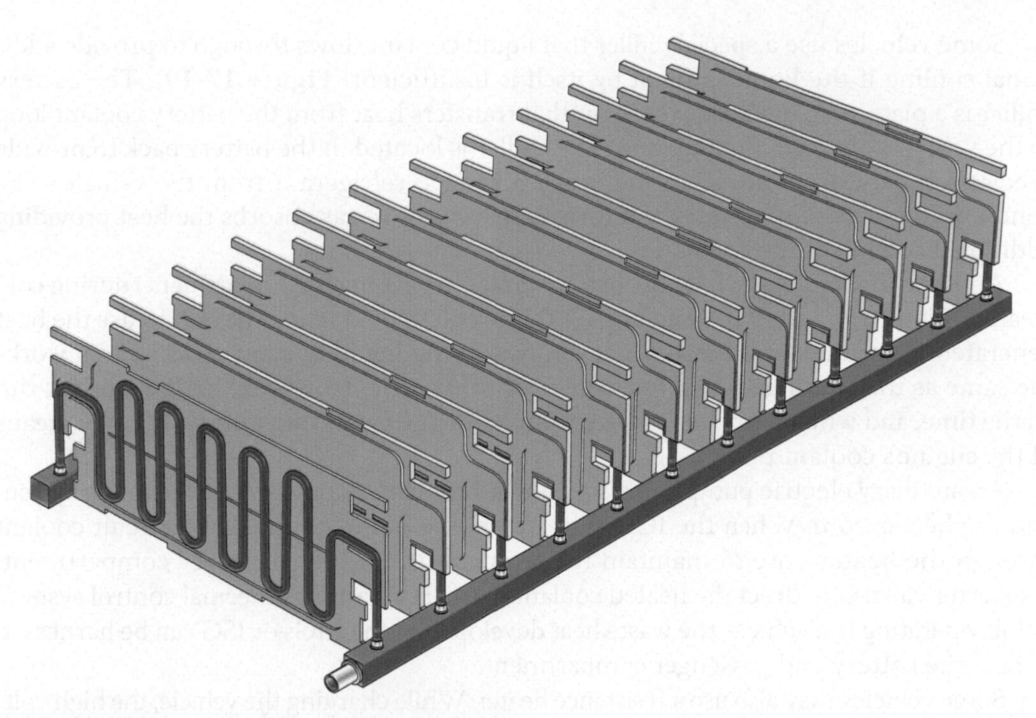

Figure 17-17 Coolant flow through an HV battery pack.

An electric pump circulates coolant when the system is in "Ready" mode. PHEVs and EVs may also turn the pump on during HV battery charging using AC power. PWM controls pump speed, and a speed sensor provides feedback concerning the pump's actual speed.

Another method of cooling the HV battery is to route refrigerant from the vehicle's air conditioning system through a high-voltage battery coolant chiller circuit that removes heat from the battery coolant. This system has its own TXV and coolant to refrigerant heat exchanger (chiller) that functions as the evaporator.

Figure 17-18 is an example of a system that controls HV battery temperature using refrigerant only. Note the two blowers, two thermal expansion valves, two evaporators, and the drive battery cooling solenoid. During operation, the PHEV-ECU monitors battery temperature sent by the Battery Management Unit (BMU). The BMU operates the battery blower based on signals from the PHEV-ECU. If the temperature continues to rise, the PHEV-ECU signals the Climatc Control head to engage the A/C compressor. The PHEV-ECU operates the A/C condenser fans.

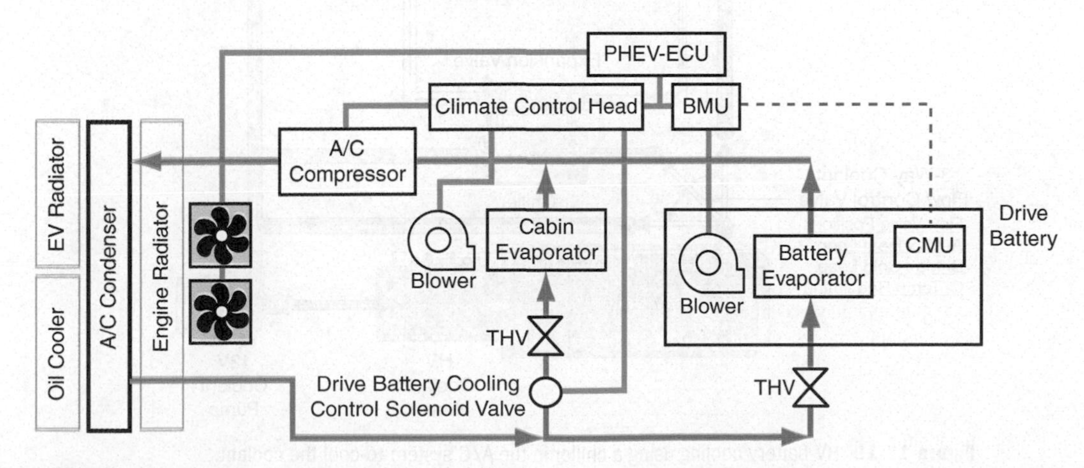

Figure 17-18 HV battery cooling using the vehicle's A/C system to cool the battery.

Some vehicles use a special chiller that liquid coolant flows through to provide additional cooling if the liquid coolant by itself is insufficient (**Figure 17-19**). The battery chiller is a plate-to-plate heat exchanger that transfers heat from the battery coolant loop to the vehicle's air conditioning loop. The chiller is located in the battery pack to provide precise temperature control. In the heat exchanger, refrigerant from the vehicle's air-conditioning system evaporates and turns into a gas. The gas absorbs the heat providing additional cooling of the coolant.

Another challenge is to heat the battery (and the passenger compartment) during cold weather. The HEV and PHEV can answer this problem by starting the ICE to use the heat generated from the combustion process to warm the liquid coolant. This system works the same as the heating system in a conventional vehicle. However, the ICE doesn't run all the time, and when it's not running coolant doesn't flow through the systems by means of the engine's coolant pump.

An auxiliary electric pump can also circulate coolant through the ICE cooling system and the heater core. When the ICE shuts down, the pump continues to circuit coolant through the heater core to maintain the temperature in the passenger compartment. A diverter valve can direct the heated coolant to the HV battery's thermal control system. While operating the vehicle, the waste heat developed in the MGs or ISG can be harnessed to heat the battery and passenger compartment.

Some vehicles may also use a resistance heater. While charging the vehicle, the high voltage supplied by the charging station preheats the battery while the vehicle. This prepares the vehicle for driving. The heater turns on and off as needed during vehicle operation. During this time, the heater uses power from the HV battery through the inverter/converter.

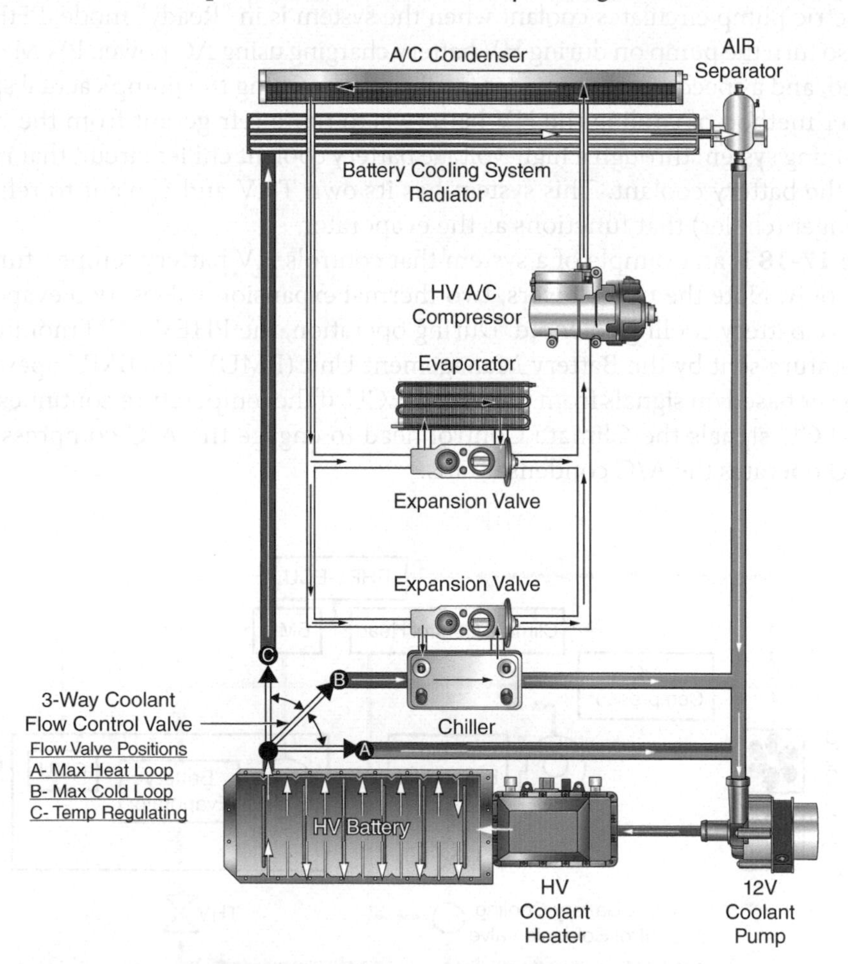

Figure 17-19 HV battery cooling using a chiller in the A/C system to cool the coolant. Courtesy of gm-volt.com.

A newer technology is the PTC film heater core (**Figure 17-20**). The PTC element heats as electric current flows through it. The PTC is a self-regulating resistor. Initially, the film requires high power to start the heating process. As the film heats, the resistance increases, and the current demand decreases. Once the film's temperature reaches its threshold, it doesn't require much energy to maintain the temperature. A fan behind the film blows the heated air into the passenger compartment and/or the HV battery case.

Another method is to use a heat pump. Like the air conditioning system, the heat pump comprises of a compressor, evaporator, and condenser (**Figure 17-21**). However, the heat pump works in reverse to air conditioning. The advantage of the heat pump is it doesn't have the energy depletion attributes of resistive heating during vehicle operation. The heat pump uses the difference in temperature between a refrigerant and the outside air (**Figure 17-22**). The heat pump also scavenges waste heat from the drive motors, on-board chargers, inverter/converter, and the HV battery.

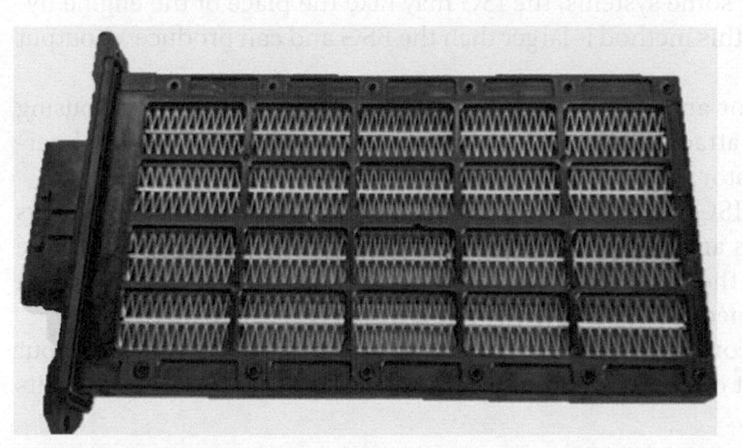

Figure 17-20 A PTC heater used to warm the passenger compartment.

Figure 17-21 The heat pump is not a heater. The heat pump transfers heat to different locations.

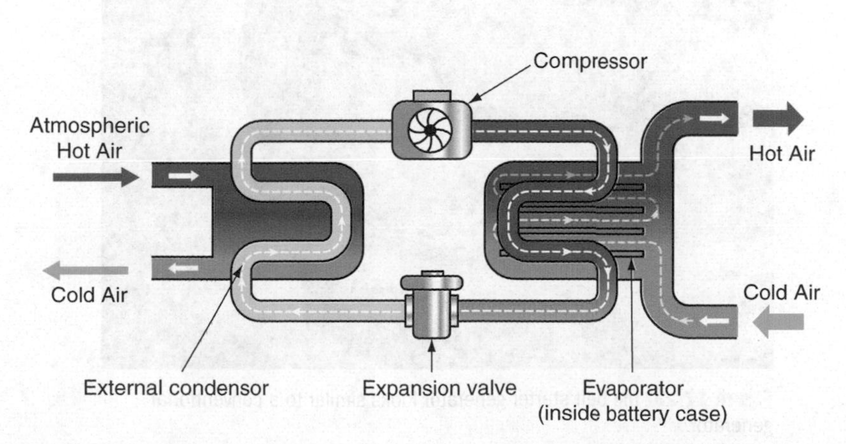

Figure 17-22 The heat pump works in reverse of the air conditioning system.

Reversible heat pumps can both warm and cool the battery. When cooling, the excess heat from the battery can be directed to the cabin heater.

When in use, heat from the surrounding air vaporizes the liquid refrigerant in the condenser into a gas. The compressor compresses the gas and discharges it into the evaporator as a high-pressure gas. Here the gas converts back into a liquid as it gives off its heat.

Since the heat pump will eventually reach a limit in which it's no longer capable of bringing heat in from outside. A PTC film may provide secondary heating to increase the temperature of the compartment and/or the battery.

42-Volt Stop/Start Systems

Some HEVs use a 42-volt ISG system to operate the automatic stop/start system since the power requirements are higher than what the 12-volt system can provide. Currently, there are two main system designs.

The first design uses a Belt Alternator Starter (BAS) that is about the same size as a conventional generator and mounts in the same way (**Figure 17-23**). BASs have a maximum power output of around 5 kW. BAS designs include permanent magnet and induction types.

The second design is to mount the ISG at the end of the crankshaft between the engine and transmission. In some systems, the ISG may take the place of the engine flywheel. The ISG mounted in this method is larger than the BSG and can produce an output of 6 to 15 kW.

The ISG includes a rotor and a stator located inside the transmission bell housing (**Figure 17-24**). The stator attaches to the engine block and uses coils formed by laser-welding copper bars. The rotor bolts to the engine crankshaft.

Both the BAS and the ISG use the same principle to start the ICE. Current flows through the stator windings and generates magnetic fields in the rotor. This causes the rotor to turn, thus turning the crankshaft and starting the ICE. In addition, this same principle assists the ICE as needed when it is running.

The DC/DC converter configuration provides a 14-volt output from the 42-volt input (**Figure 17-25**). The 14-volt output can supply electrical energy to the electrical circuits that do not require 42 volts.

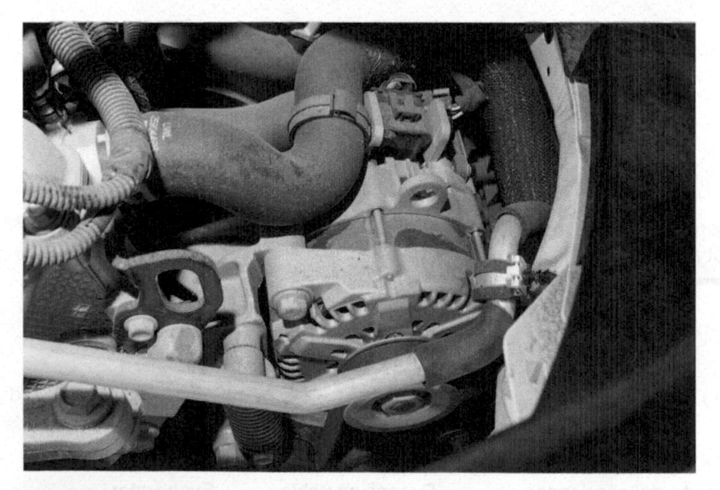

Figure 17-23 The belt starter generator looks similar to a conventional generator.

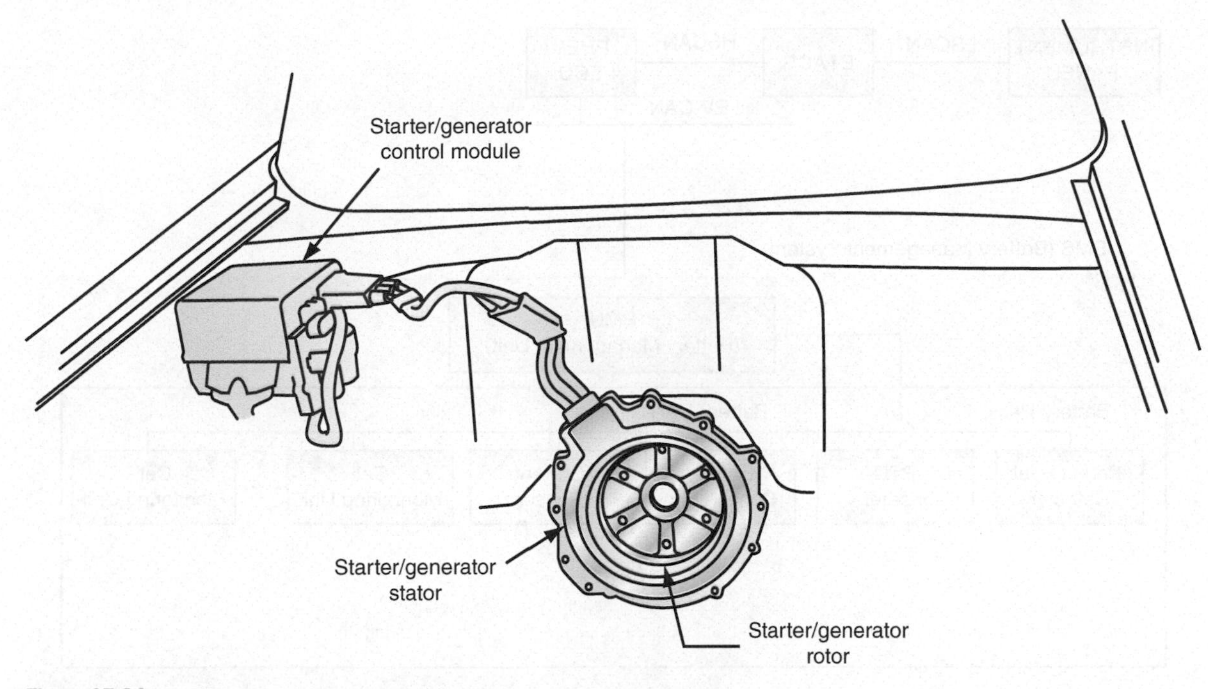

Figure 17-24 The ISF stator and rotor assembly.

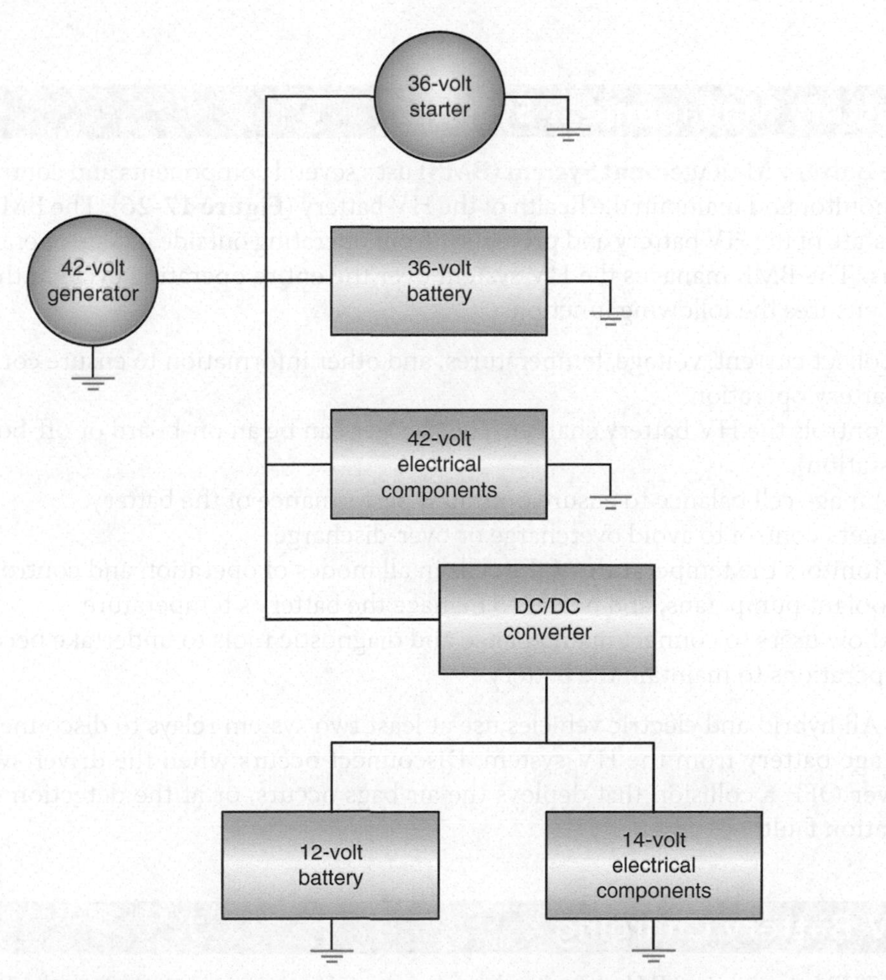

Figure 17-25 The DC/DC converter uses the 42-volt input and converts it to a 14-volt output.

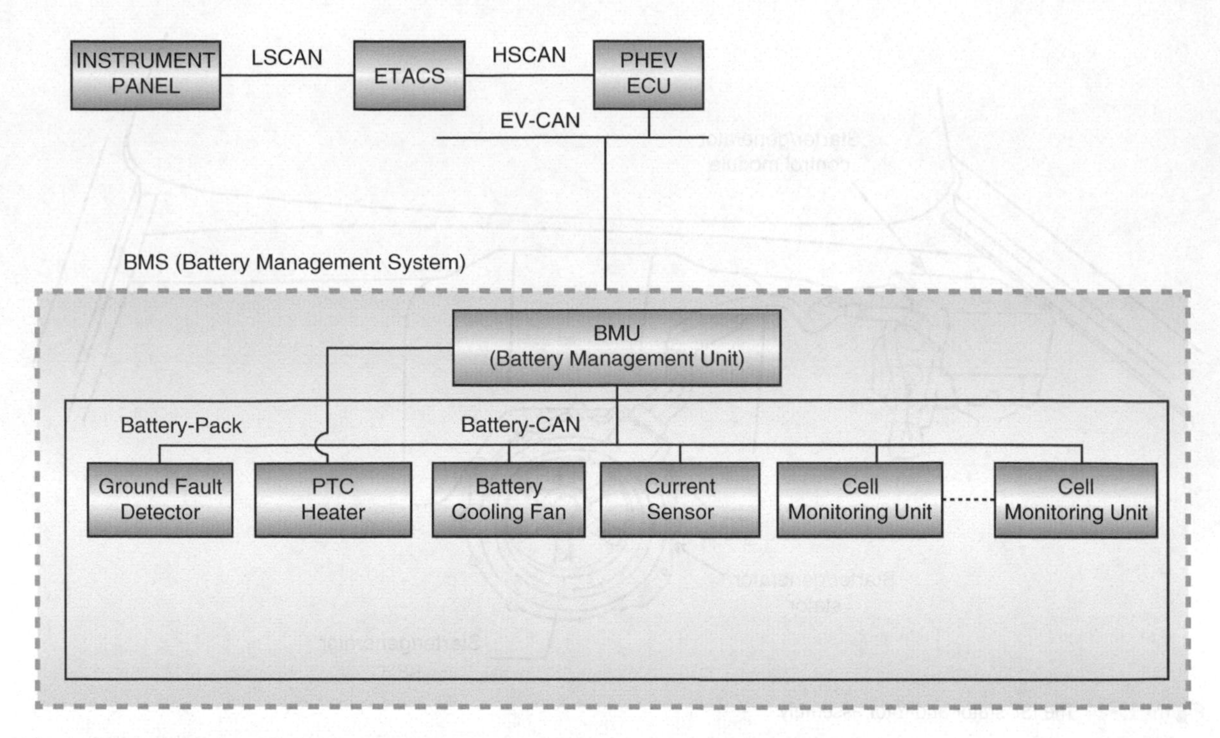

Figure 17-26 The Battery Management System.

Battery Management System

Shop Manual
Chapter 17, page 928

The **Battery Management System (BMS)** uses several components and control modules to monitor and maintain the health of the HV battery (**Figure 17-26**). The BMS monitors the state of the HV battery and protects it from operating outside its safe operating conditions. The BMS manages the HV system over the entire operating cycle of the EV/HEV and ensures the following functions:

- Collect current, voltage, temperatures, and other information to ensure correct battery operation.
- Controls the HV battery charger. The charger can be an on-board or off-board (station).
- Manage cell balance to ensure optimum performance of the battery.
- Safety control to avoid overcharge or over-discharge.
- Monitors the temperature of the cells in all modes of operation and controls the coolant pump, fans, and heater to manage the battery's temperature.
- Allow users to connect maintenance and diagnostic tools to undertake necessary operations to maintain the battery.

Shop Manual
Chapter 17, page 940

All hybrid and electric vehicles use at least two system relays to disconnect the high voltage battery from the HV system. Disconnect occurs when the driver switches the power OFF, a collision that deploys the air bags occurs, or at the detection of a loss of isolation fault.

HV Battery Charging

Shop Manual
Chapter 17,
pages 928, 930

The HEV charges its HV battery by driving one of the electric motors with the ICE to operate the motor like a generator. This charging function occurs when the ICE is running while the vehicle is in Park. This is the preferred method for charging the HV battery of an HEV.

EVs and PHEVs require a means of recharging the HV battery (**Figure 17-27**). This may mean installing a charging unit at the owner's home, the availability of charging stations at public locations, and even a means for the service shop to recharge the batteries.

Charging Standards and Rates

The PHEV and EV have an on-board charger to recharge the HV batteries. This charger may take over 15 hours to fully recharge the HV batteries. Other faster chargers are available. HV battery charging times depend on variables such as:

- The size of the battery.
- The charging station's maximum power capacity.
- The vehicle's on-board charger power capacity.
- Very cold or very hot battery temperatures can limit the maximum power intake of the battery and increase charging time.

Currently, there are three levels of charging stations as determined by the station's input voltage and charging load.

Level 1 charging stations can plug into any 120 VAC electrical service using a standard 3-prong outlet (**Figure 17-28**). Vehicle manufacturers usually include Level 1 chargers with the purchase of the vehicle.

Figure 17-27 Recharging the EV.

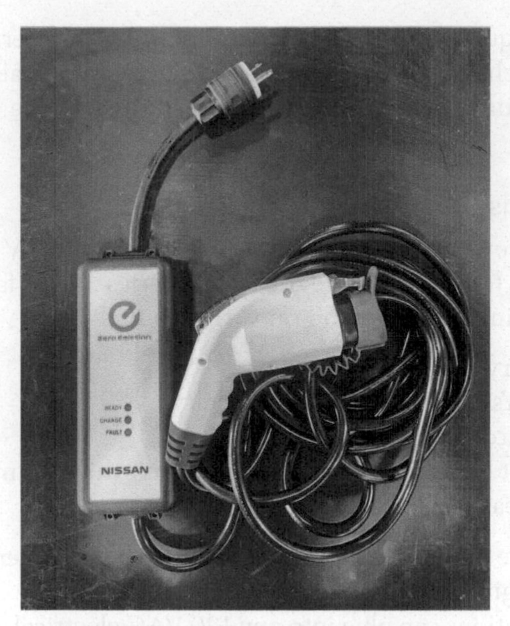

Figure 17-28 The Level 1 charger plugs into a standard 120-volt socket.

Shop Manual
Chapter 17, page 932

Level 1 charging is the slowest, supplying an average AC power output of 1.4 kW to 1.9 kW. This charge rate adds only 3 to 5 miles (5 to 8 km) of range per hour of charging. This means that plugging in the charger for eight hours only increases the vehicle's range by 24 to 40 miles (19 to 64 km). Although used to charge EVs, Level 1 charging is more suitable for PHEVs that can use the ICE to complete the battery charge.

Level 2 chargers are setup at the vehicle owner's home (**Figure 17-29**) or at public charging station locations. Level 2 chargers use 240 VAC and have an AC power output between 2.5 and 19 kW. The Level 2 charger provides between 80 and 160 miles (128 and 256 km) of charge per hour, making it good for overnight charging.

Level 3 chargers are called **Direct Current Fast Chargers (DCFCs)**. The output of the DCFC is a DC voltage instead of the AC voltage used with Level 1 and Level 2 chargers. The DCFC is the fastest of the charger types. With a maximum output of 350 kW DC, DCFCs can charge an HV battery to 80% SOC in 20–30 minutes and 100% in 60–90 minutes. To use the DCFC station, the vehicle must have a DCFC port. Most DCFC stations provide multiple connector options to allow using them with most EVs and PHEVs.

Figure 17-29 Level 2 charging station.

DCFC charging stations are expensive and require a 3-phase 480 VAC input voltage. For these reasons, the primary uses of these chargers are as public or commercial charging stations (**Figure 17-30**).

Level 1 and 2 charging stations are more of a "smart" extension cord than an actual charger. The Level 1 and 2 charging systems connect to the vehicle's on-board charger module, not to the HV battery. The charger provides AC voltage to the charging module, which converts the AC voltage to DC voltage and sends it to the battery. Since Level 3 chargers supply DC voltage, they plug directly into the battery, bypassing the charger module.

Electric Vehicle Supply Equipment (EVSE)

Electric Vehicle Supply Equipment (EVSE) is a safe charging protocol developed for the EV and PHEV. Early charging stations required the operator to input information. This left open the opportunity for mistakes that could present a safety hazard. EVSE eliminated the need for inputting data by the user by using a communication link between the vehicle and the station.

As a component of the BMS, an EV Communication Controller (EVCC) provides two-way communication between the charger and the vehicle. The correct output charging current is set based on the maximum current the charger can provide and the maximum current the charging module can receive.

The EVCC communicates over a single-line pilot wire. The pilot signal interface requires the transmission of a 1-kHz, ±12-volt PWM signal to the vehicle. The duty cycle communicates the current limit the charging station can supply to the vehicle. The vehicle can use up to this amount of current for its charging circuitry. The vehicle returns the battery's current state by placing a load on the line, which causes a voltage drop.

A safety line uses a series of resistors to identify the proper insertion and locking of the connector into the plug. If not inserted, the lock-out feature prevents current from flowing through the charge cord. EVSE can also detect hardware faults and disconnect power to prevent damage to the battery, electrical shorts, or fire.

Connector Types

Each charger level requires its own connector type (**Figure 17-31**). Level 1 chargers use a Type 1 plug, Level 2 uses a Type 2, and Level 3 uses Type 3. In the US, the plug and connector must meet J1772 standards. All EVs in the US and Canada can use this plug for charging, including Tesla cars with an adapter.

Figure 17-30 Level 3 charging stations are typically used at public locations. Fees may apply for charging vehicles at these stations.

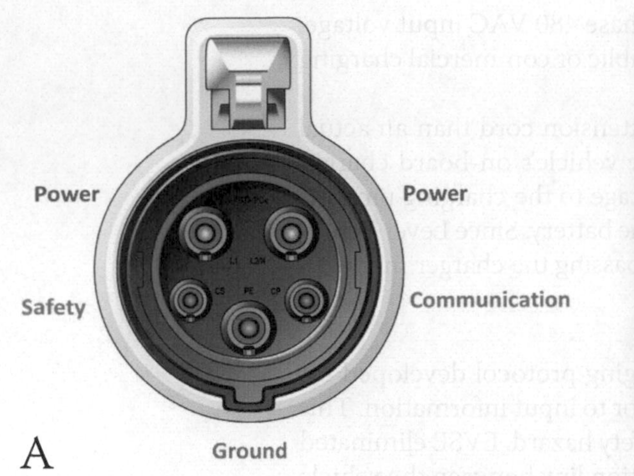

Figure 17-31 (A) Type 1 connector. (B) Type 2 connector and plug. (C) For fast charging, the CCS is the most common connector.

DCFC is vehicle-specific and not available for all EVs and PHEVs. There are three types of fast charging ports: CHAdeMO, CCS, and Tesla supercharge. CHAdeMO is a Japanese standard used by many Asian vehicle manufacturers. Currently, only Nissan and Mitsubishi have used CHAdeMO in the U.S.

The Combo Charging System (CCS) is a combination of Type 2 (with the four AC pins for neutral and 3-phase removed) and DCFC plugs. The CCS connector allows for both AC and DC charging.

Tesla uses a proprietary connector. It's the same connector for Level 1, Level 2, and DCFC. Only Tesla vehicles can use Tesla's DC fast charger stations, called Superchargers. Tesla installs and maintains the Superchargers, which are for the exclusive use of Tesla customers. Even with an adapter cable, charging a non-Tesla EV at a Supercharger station is impossible. When connecting a vehicle to the station, an authentication process must identify it as a Tesla before granting access. Tesla offers adapters that allow charging stations other than Superchargers to charge Tesla vehicles.

HV Battery Balancing

Shop Manual
Chapter 17, page 930

Battery balancing refers to the battery's cells being within a specific voltage difference of each other. Each manufacturer has a different specification, but this difference can be as low as approximately 0.01 volt. The weakest cell limits the total performance of the multi-cell battery. Balancing the cells equalizes the voltage so all cells recharge to the same level. Battery related DTCs set if the imbalance condition is outside allowable limits.

The BMS controls cell balancing according to a predetermined strategy or algorithm. The BMS can balance the battery passively or actively. Passive balancing bleeds high-voltage cells through a resistor during charging in the 70–80% SOC curve. Active balancing uses the DC/DC converter to shuttle the extra charge from higher-voltage cells during discharge to those with a lower voltage.

Fuel Cells

A **fuel cell** produces current from hydrogen and aerial oxygen. They combine the reach of conventional internal combustion engines with high efficiency, low fuel consumption, and minimal or no pollutant emissions. At the same time, they are extremely quiet. Because they work with regenerative fuels such as hydrogen, they reduce the dependence on crude oil and other fossil fuels.

A fuel cell–powered vehicle (**Figure 17-32**) is basically an EV. Like the EV, it uses an electric motor to supply torque to the drive wheels. The difference is that the fuel cell produces and supplies electric power to the electric motor instead of the batteries. Most vehicle manufacturers and several independent laboratories are involved in fuel cell research and development programs. Manufacturers have produced many prototype fuel cell vehicles, placing them in fleets in North America and Europe.

Fuel cells electrochemically combine oxygen from the air with hydrogen to produce electricity. The oxygen and hydrogen provide "fuel" to the fuel cell for the electrochemical reaction. There are different types of fuel cells, but the most common type is the **Proton Exchange Membrane (PEM)**. Normally, hydrogen and oxygen bond with a loud bang, but in fuel cells a special PEM impedes the oxyhydrogen gas reaction by ensuring that only protons (H+) and not elemental hydrogen molecules react with the oxygen.

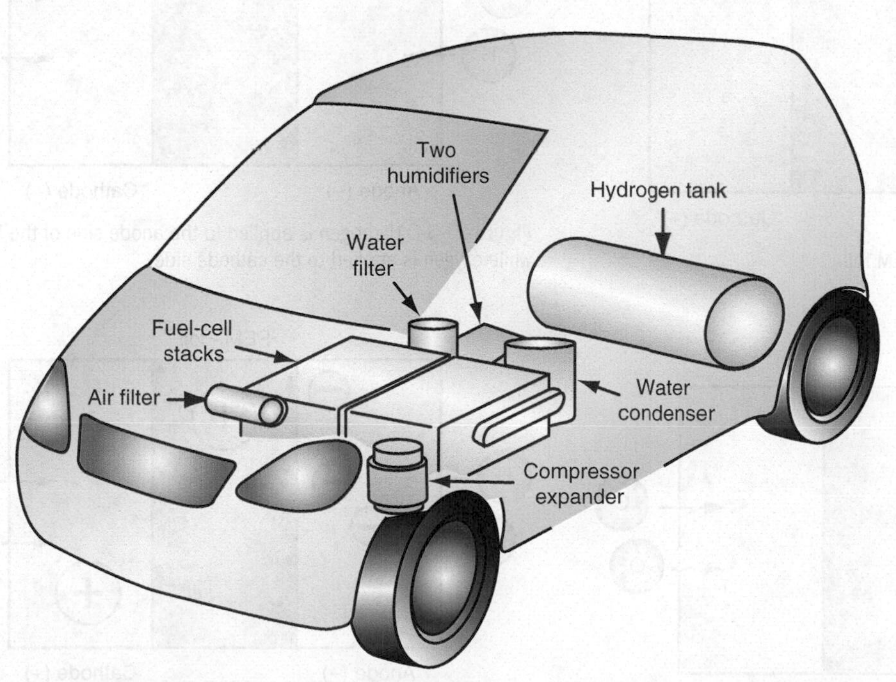

Figure 17-32 Fuel cell vehicle components. Technology has allowed engineers to design fuel cell vehicles without losing passenger and cargo space.

How the Fuel Cell Works

The construction of the PEM fuel cell is a sandwich structure (**Figure 17-33**). The electrolyte is between two electrodes of gas-permeable graphite paper. The electrolyte is a polymer membrane. Hydrogen is applied to the anode side of the PEM, and ambient oxygen is applied to the cathode side (**Figure 17-34**). The membrane keeps the distance between the two gases and provides a controlled chemical reaction.

The anode is the negative post of the fuel cell. It conducts the freed electrons from the hydrogen molecules for use in an external circuit. It has channels etched into it that disperse the hydrogen gas evenly over the catalyst's surface. The cathode is the positive post of the fuel cell. It also has channels etched into it that distribute the oxygen to the catalyst's surface. It also conducts the electrons back from the external circuit to the catalyst where they can recombine with the hydrogen ions and oxygen to form water.

A fine coating of platinum applied to the foil acts as a catalyst. The catalyst accelerates the decomposition of the hydrogen atoms into electrons and protons (**Figure 17-35**). The platinum-coated side of the catalyst faces the PEM. The platinum is rough and porous to expose the hydrogen or oxygen to the maximum surface area.

The PEM is the electrolyte. This specially treated material (which looks similar to ordinary kitchen plastic wrap) allows only the protons to move across from the anode to the cathode (**Figure 17-36**). As a result, the anode has a surplus of electrons and the cathode has a surplus of protons. Connecting the anode and cathode outside the cell causes the conductor to flow current (**Figure 17-37**). The electrons move through the conductor to the cathode and then recombine with the protons and oxygen to produce water.

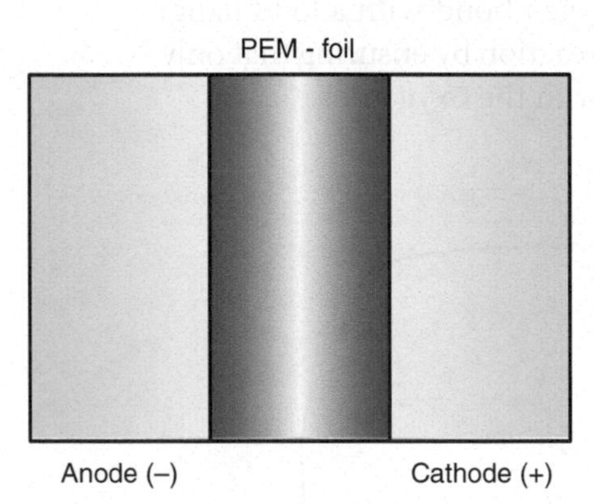

Figure 17-33 The PEM foil.

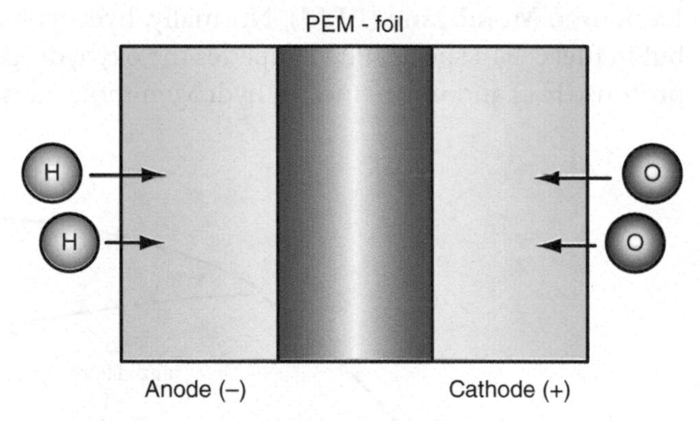

Figure 17-34 Hydrogen is applied to the anode side of the PEM, while oxygen is applied to the cathode side.

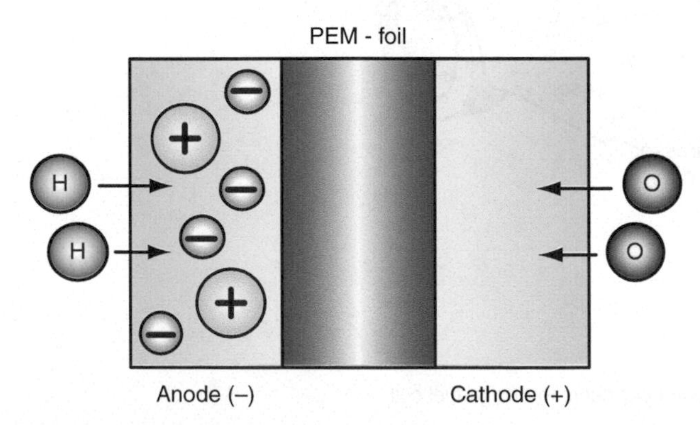

Figure 17-35 The catalyst breaks down into protons and electrons.

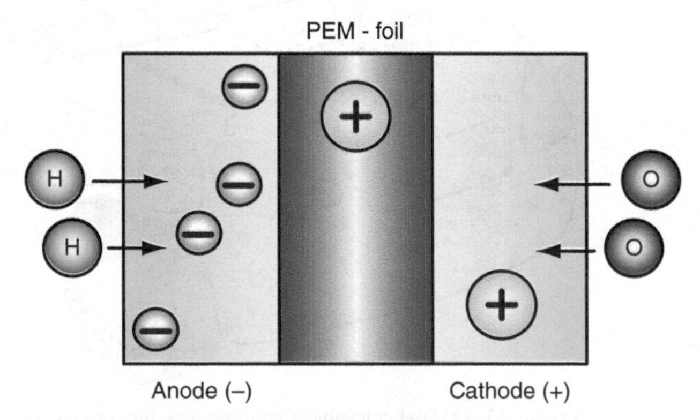

Figure 17-36 The PEM foil only allows protons to migrate to the cathode, leaving the electrons on the anode.

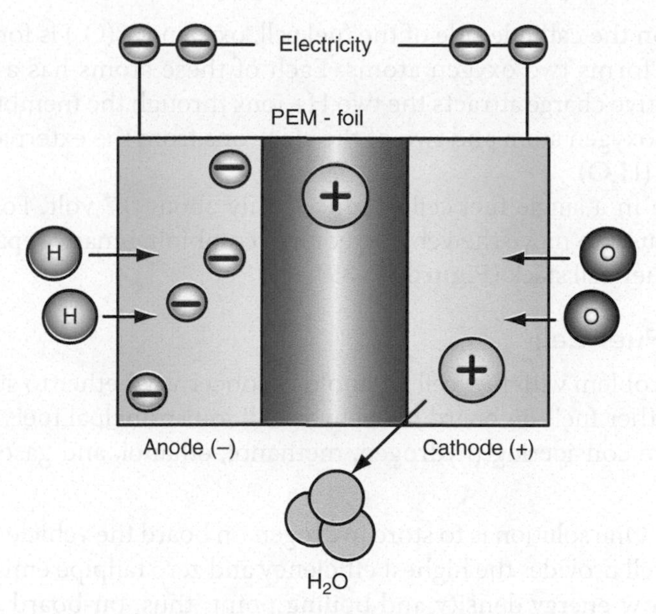

Figure 17-37 With an excess amount of electrons on the anode and an excess amount of protons on the cathode, current flows through an external conductor. The electrons then react with the protons and oxygen to produce water.

Figure 17-38 illustrates the entire process. Pressurized hydrogen gas (H_2) enters the fuel cell on the anode side. The pressure forces the hydrogen gas through the catalyst. When an H_2 molecule contacts the platinum on the catalyst, it splits into two H+ ions and two electrons (e–). The electrons conduct through the anode, where they make their way through the external circuit (doing useful work such as turning a motor) and return to the cathode side of the fuel cell.

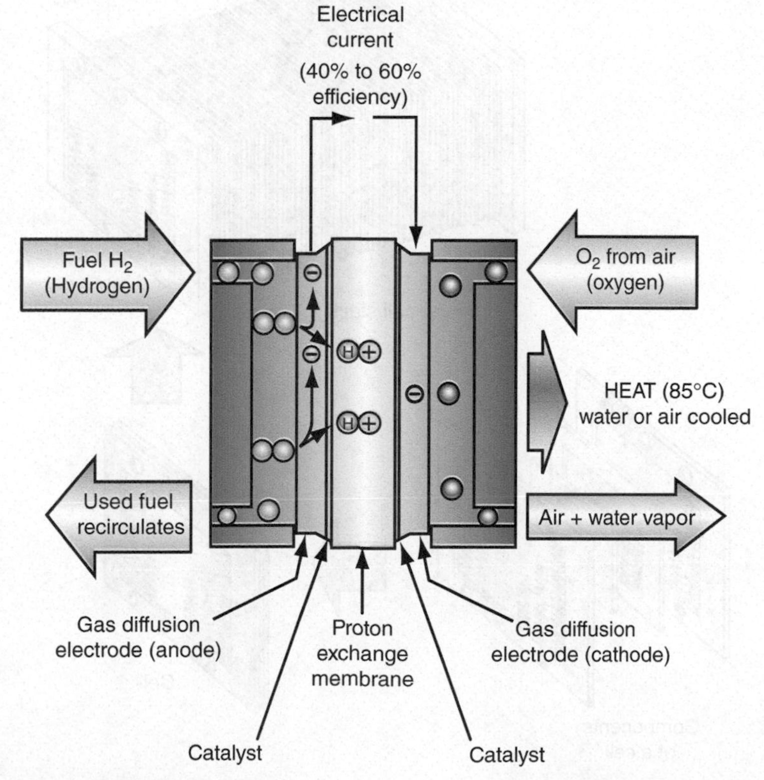

Figure 17-38 PEM fuel cell.

Meanwhile, on the cathode side of the fuel cell, oxygen gas(O_2) is forced through the catalyst where it forms two oxygen atoms. Each of these atoms has a strong negative charge. This negative charge attracts the two H+ ions through the membrane, where they combine with an oxygen atom and two of the electrons from the external circuit to form a water molecule (H_2O).

This reaction in a single fuel cell produces only about 0.7 volt. For this voltage to become high enough to move the vehicle requires combining many separate fuel cells in series to form a fuel-cell stack (**Figure 17-39**).

Fuels for the Fuel Cell

A fundamental problem with fuel cell technology concerns whether to store hydrogen or convert it from other fuels on board the vehicle. All four principal fuels that automotive manufacturers are considering (hydrogen, methanol, ethanol, and gasoline) pose some challenges.

Hydrogen Fuel. One solution is to store hydrogen on board the vehicle. Using hydrogen directly in a fuel cell provides the highest efficiency and zero tailpipe emissions. However, hydrogen has a low energy density and boiling point; thus, on-board storage requires large, heavy tanks. Three types of hydrogen storage methods under development include compressed hydrogen, liquefied hydrogen, and binding hydrogenate to solids in metal hydrides or carbon compounds.

Compressed hydrogen offers the least expensive method for on-board storage. However, at normal CNG-operating pressures of 3,500 psi (241 bar), reasonably sized, commercially available pressure tanks provide limited range for a fuel cell vehicle (about 120 miles or 193 km). Daimler and Hyundai are now using pressure tanks capable of 5,000 psi (345 bar). Quantum is researching high-performance hydrogen storage systems,

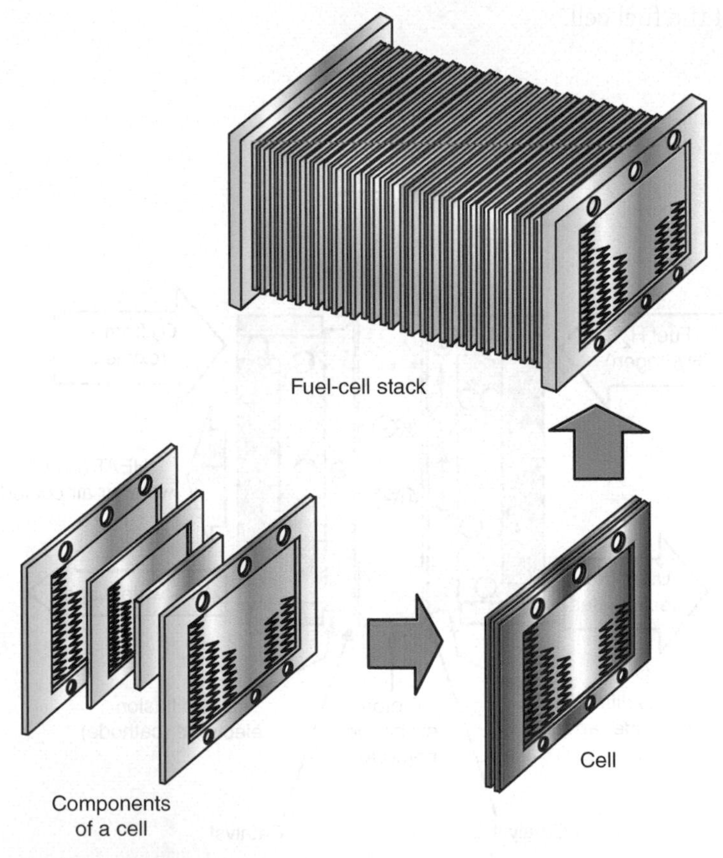

Figure 17-39 PEM fuel-cell stack.

looking at pressure tanks capable of up to 10,000 psi (689 bar). This capability would permit a 400-mile (644-km) driving range.

Liquefied hydrogen can be stored in large cylinders containing a hydride material (something like steel wool). Liquefied hydrogen doesn't require the high storage capacity of compressed hydrogen for the same amount of driving range. However, the very low boiling point of hydrogen requires that the tanks have excellent insulation. Maintaining the extremely cold temperature of −423°F (−253°C) during refueling and storage is difficult. Estimates indicate that up to 25% of the liquid hydrogen may boil off during the refueling process. In addition, the on-board storage can lose about 1% per day. Storing liquid hydrogen on a vehicle also involves some safety concerns. As the fuel tank warms, the pressure increases and may activate the pressure relief valve. This action discharges flammable hydrogen into the atmosphere, creating a source of danger and pollution.

Methanol. Several automotive manufacturers are using methanol to power their fuel cells. Methanol fuel cells may bridge the gap over the next few decades while a hydrogen distribution infrastructure is built. Using "methanolized" hydrogen as a fuel has the advantage that storing it in the vehicle is similar to storing gasoline. For the reaction in the fuel cell, a **reformer** on board the vehicle produces hydrogen from the methanol fuel. A reformer is a high-temperature device that converts hydrocarbon fuels to carbon monoxide (CO) and hydrogen. Mixing the methanol fuel with water produces hydrogen. When it evaporates, it decomposes into hydrogen and carbon dioxide (CO_2). Additional steps purify the hydrogen and CO_2 prior to sending it to the fuel cell.

Methanolized hydrogen contains more hydrogen atoms and has an energy density greater than liquid hydrogen. Like hydrogen, methanolized hydrogen is also independent of mineral oil. Compared with hydrogen, vehicles driven by methanol are not completely emissions-free, but they produce very little pollutants and much CO_2 less than internal combustion engines.

A special type of PEM fuel cell called the Direct Methanol Air Fuel Cell (DMAFC) utilizes methanol combined with water directly as a fuel and ambient air for oxygen. This technology enables the use of a liquid fuel without needing an on-board reformer while still providing a zero-emissions system. However, current research has demonstrated that the power density is lower than for other PEM fuel cells.

Ethanol. Ethanol is less toxic than either gasoline or methanol. An ethanol system requires adding a reformer to the vehicle, similar to a methanol system. The fuel cell could use E100, E95, or E85.

Gasoline. A special, more pure gasoline can drive the fuel cell. Using reformers for on-board hydrogen extraction from gasoline is one approach to commercializing fuel cell vehicles since the gasoline infrastructure is already in place. However, producing hydrogen from gasoline in a vehicle system is much more difficult than producing hydrogen from methanol or ethanol. Gasoline reforming requires higher temperatures and more complex systems than methanol or ethanol reforming. The reformation reactions occur at 1,562°F to 1,823°F (850°C to 995°C), making the devices slow to start and the chemistry temperamental. Thus, the drive would work less efficiently and produce more emissions. Moreover, the capabilities for cold starts would be restricted. The size of the reformer is also an issue, making it difficult to fit under the hood of a standard-sized vehicle. Furthermore, there is concern about the sulfur levels in current gasoline and CO in the reformer poisoning the fuel cell.

Reformer. As mentioned, some fuel cell systems may require an on-board reformer to extract hydrogen from liquid fuels such as gasoline, methanol, or ethanol. On-board reformation of a hydrocarbon fuel into hydrogen allows the use of more established infrastructures but adds additional weight and cost and reduces vehicle efficiency. In addition, the reformer does create some emissions.

PEM fuel cell reformers combine fuel and water to produce additional H_2 and convert the CO to CO_2. The CO_2 is then released into the atmosphere. Reformer technologies include Steam Reforming (SR), partial oxidation, and high-temperature electrolytes reforming.

SR uses a catalyst to convert fuel and steam to H_2, CO, and CO_2. Further reforming of the CO with steam to form more H_2 and CO_2. Next, a purification step removes CO, CO_2, and any impurities to achieve a high hydrogen purity level (97% to 99.9%). SR of methanol is the most developed and least expensive method to produce hydrogen from a hydrocarbon fuel on a vehicle, resulting in 45% to 70% conversion efficiency.

Partial Oxidation (POX) reforming is similar to stream reforming since both technologies combine fuel and steam, but this process adds oxygen in an additional step. The process is less efficient than SR, but the heat-releasing nature of the reaction makes it more responsive than SR to variable load. POX can use heavier hydrocarbon (HC), but it has lower carbon-to-hydrogen ratios, which limits hydrogen production.

The process of **electrolysis** can also obtain hydrogen from water. Electrolysis is the splitting of water into hydrogen and oxygen. The drawback to this process is that it requires a great deal of electrical energy. Recently, the development of high-temperature electrolytes that can operate at temperatures in excess of 212°F (100°C) has shown some positive results. The benefits of high-temperature electrolytes include the following:

- *Improved CO tolerance.* This allows the manufacturer to reduce or remove the need for an oxidation reactor and for air bleed. Reducing these requirements can increase system efficiency by 5% to 10%. There is also a considerable reduction in start-up time. The remaining CO is combusted in a catalytic tail-gas burner to prevent emissions of CO.
- *Facilitated stack cooling.* This reduces the size of the radiator and reduces the fuel-cell stack cooling plate requirements.
- *Humidity-independent operation.* Generally, high-temperature membranes require humidifiers and water recovery, whereas this system doesn't.

Solid Oxide Fuel Cells. Planar Solid Oxide Fuel Cells (SOFCs) operate at high temperatures of 932°F to 1,472°F (500°C to 800°C) and can use CO and H_2 fuel. SOFCs have a good tolerance to fuel impurities and use ceramic as an electrolyte. Currently, SOFCs use gasoline fuel and require a reformer.

Summary

- The HEV relies on power from the electric motor, the engine, or both.
- Regenerative braking turns the energy used to slow the vehicle back into electricity. Regenerative braking occurs under two conditions; when the vehicle is traveling downhill with no load on the ICE, and during brake apply to slow or stop the vehicle.
- Most HEVs and EVs use a blended brake system in which the master cylinder may not directly apply the service brakes.
- Configurations of the hybrid vehicle include parallel, series, and series/parallel.
- A mild parallel hybrid vehicle has an electric motor that is large enough to provide regenerative braking, instant engine start-up, and a boost to the ICE. However, the electric motor cannot propel the vehicle on its own.
- A full (strong) parallel hybrid vehicle uses an electric motor that is powerful enough to propel the vehicle on its own.

- EVs powered by an electric motor run from a battery pack.
- It's common for HEVs and PHEVs to have at least two individual and separate cooling systems; one for the ICE and the other for the HV electronics. Some systems will also have a third separate loop used to control the temperature of the HV batteries.
- The 12-volt electric pump circulates coolant to the inverter/converter's cooling housing. Coolant can also be pumped to the MG(s) or ISG.
- HV batteries have an ideal operating temperature window between 65° and 105°F (18° and 40°C). Service life decreases at operating temperatures higher than 105°F (40°C). Efficiency drops, and output decreases at temperatures below 14°F (−10°C).
- A significant increase in the temperature of the cells occurs during peak loads with high current flows that can result from regeneration and boosting. Also, high ambient temperatures can cause HV battery temperature to reach the critical 105°F (40°C) level quickly.
- There are three common methods of cooling the HV battery; air cooling, liquid cooling, and refrigerant cooling. Some manufacturers will combine methods.
- The air-cooled system can be passive and rely solely on the convection of the surrounding air. Active systems use cooling fans that direct cabin air into the battery.
- Some vehicles use a special heat exchanger (chiller) that the coolant flows through if the cooling by the battery cooler is insufficient.
- Some vehicles use a resistance heater to warm the battery.
- A newer technology is the PTC film heater core The PTC is a self-regulating resistor.
- Another method to manage HV battery temperature is to use a heat pump.
- The Battery Management System (BMS) monitors the state of the HV battery and protects it from operating outside its safe operating conditions. The BMS manages the HV system over the entire operating cycle of the EV/HEV.
- Both the BAS and the ISG use the same principle to start the engine. Current flows through the stator windings and generates magnetic fields in the rotor. This causes the rotor to turn, thus turning the crankshaft and starting the engine.
- The PHEV and EV have an on-board charger to recharge the HV batteries.
- Currently, there are three levels of charging stations as determined by the station's input voltage and charging load.
- Level 1 charging stations can plug into any 120 VAC electrical service using a standard 3-prong outlet and supply an average AC power output of 1.4 kW to 1.9 kW.
- Level 2 chargers use 240 VAC and have an AC power output between 2.5 and 19 kW.
- Level 3 chargers output DC voltage with a maximum output of 350 kW DC.
- The Level 1 and 2 charging systems connect to the vehicle's on-board charger module, not to the HV battery.
- An EV Communication Controller (EVCC) provides two-way communication between the charger and the vehicle over a single line pilot line. The correct output charging current is set based on the maximum current the charger can provide and the maximum current the charging module can receive.
- A safety line uses a series of resistors to identify the proper insertion and locking of the connector into the plug. If not inserted, the lock-out feature prevents current from flowing through the charge cord.
- Battery balancing refers to the battery's cells being within a specific voltage difference of each other. The weakest cell limits the total performance of the multi-cell battery. Balancing the cells equalizes the voltage so all cells recharge to the same level.
- A fuel cell–powered vehicle is basically an EV except the fuel cell produces and supplies electric power to the electric motor instead of batteries.
- Fuel cells electrochemically combine oxygen from the air with hydrogen to produce electricity.
- The most common type of fuel cell is the PEM.
- Possible fuels for the fuel cell include hydrogen, methanol, ethanol, and gasoline.
- Most fuel cell systems require the use of a reformer to extract hydrogen from liquid fuels such as gasoline, methanol, or ethanol.

Review Questions

Short-Answer Essays

1. Explain the meaning of regenerative braking.
2. Describe the basic operation of a typical hybrid vehicle.
3. Explain the difference between parallel and series hybrids.
4. Briefly describe how the Proton Exchange Membrane (PEM) fuel cell produces electrical energy.
5. What is the purpose of the reformer?
6. Explain the power flow of a series hybrid.
7. Describe the different levels of PHEV/EV charging stations.
8. Explain the importance of HV battery cell balancing.
9. What methods of HV electronics and battery cooling are used?
10. Explain how a heat pump can be used to warm an HV battery.

Fill in the Blanks

1. ______________ ______________ recovers the energy used to brake by converting rotational energy into ______________ energy through a system of electric motors and generators.
2. Fuel cells ______________ combine oxygen from the air with hydrogen to produce electricity.
3. Most fuel cell systems require the use of a ______________ to extract hydrogen from liquid fuels such as gasoline, methanol, or ethanol.
4. In a ______________ hybrid configuration, there is a direct mechanical connection between the engine and the wheels.
5. In the ______________ hybrid vehicle, there is no mechanical connection between the engine and the wheels.
6. The ______________ is the negative post of the fuel cell.
7. In the ISG system, the ______________ is attached to the engine block and coils that are formed by laser-welding copper bars.
8. A significant increase in the temperature of the cells occurs during ______________ ______________ with high current flows that can result from ______________ and ______________.
9. A ______________ is used to control the flow of torque and electrical energy.
10. In a hybrid system, the ______________ is a device that stores energy as electrostatic charge and is the primary device in the power supply during hill climbing, acceleration, and the recovery of braking energy.

Multiple Choice

1. *Technician A* says regenerative braking recovers the energy used to brake by converting rotational energy into electrical energy through a system of electric motors and generators.
   Technician B says when the brakes are applied, the MG or ISG becomes a generator by using the kinetic energy of the vehicle to store power in the battery for later use.
   Who is correct?
   A. A only
   B. B only
   C. Both A and B
   D. Neither A nor B
2. Electric vehicles power the motor by:
   A. A generator.
   B. A battery pack.
   C. An engine.
   D. None of these choices.
3. *Technician A* says in a parallel hybrid vehicle, propulsion comes directly from the electric motor.
   Technician B says in a series hybrid vehicle, both the engine and the electric motor can turn the transmission at the same time.
   Who is correct?
   A. A only
   B. B only
   C. Both A and B
   D. Neither A nor B

4. *Technician A* says a fuel cell produces electrical energy by breaking down H_2 atoms into electrons and protons.
   Technician B says the compressed hydrogen system requires a reformer to cool the fuel cell.
   Who is correct?
   A. A only
   B. B only
   C. Both A and B
   D. Neither A nor B
5. All of the following statements about Direct Current Fast Chargers is true, **EXCEPT**:
   A. The output is a DC voltage.
   B. The charger connects to the vehicle's battery charger.
   C. It can charge an HV battery to 80% SOC in 20–30 minutes.
   D. The vehicle must have a DCFC port.
6. The splitting of water into hydrogen and oxygen is an example of:
   A. Steam reforming.
   B. Hydrocarbon reforming.
   C. Partial oxidation reforming.
   D. Electrolysis.
7. What is a characteristic of the series/parallel hybrid?
   A. Uses the ICE as a generator only.
   B. Doesn't have the capability to utilize regenerative braking.
   C. The power-split allows a power path from the ICE to the wheels that can be an optimized blend of mechanical and electrical energy and has the ability to decouple the power supplied by the ICE from the power demanded by the driver.
   D. The electric motor cannot propel the vehicle on its own.
8. On which side of a PEM fuel cell does the pressurized hydrogen gas enter?
   A. Anode.
   B. Cathode.
   C. Drain.
   D. Gate.
9. All of the following methods can be used for thermal control of the HV battery, **EXCEPT**:
   A. Submersion in coolant.
   B. Heat pump.
   C. Refrigerant air conditioning system.
   D. Ultra capacitor heat dissipation.
10. The Level 1 charging station:
    A. Can plug into any 120 VAC electrical service using a standard 3-prong outlet.
    B. Supplies AC power to the vehicle's battery charger.
    C. Is more suitable for PHEVs that can use the ICE to complete the battery charge.
    D. All of these choices.

Glossary

Note: **Terms are highlighted in bold,** followed by **Spanish translation in color.**

Absorbed glass mat (AGM) battery A variation of the recombination batteries that hold their electrolyte in a moistened fiberglass matting instead of using a gel. The plates are made of high-purity lead and are tightly compressed into six cells.

Batería de malla de fibra de vidrio absorbente (AGM, por su sigla en inglés) Una variante de las baterías de recombinación que contienen sus electrolitos en una malla de fibra de vidrio humedecido en lugar de utilizar gel. Las placas están fabricadas con plomo de alta pureza y están bien comprimidas en seis celdas.

Accelerometer Generates an analog voltage in relation to the severity of deceleration forces. The accelerometer also senses the direction of impact force.

Acelerómetro Genera un voltaje análogo en relación a la severidad de las fuerzas de deceleración. El acelerómetro tambien detecta la dirección de la fuerza de un impacto.

A circuit A generator regulator circuit that uses an external grounded field circuit. In the A circuit, the regulator is on the ground side of the field coil.

Circuito A Circuito regulador del generador que utiliza un circuito inductor externo puesto a tierra. En el circuito A, el regulador se encuentra en el lado a tierra de la bobina inductora.

Actuators Devices that perform the actual work commanded by the computer. They can be in the form of a motor, relay, switch, or solenoid.

Accionadores Dispositivos que realizan el trabajo efectivo que ordena la computadora. Dichos dispositivos pueden ser un motor, un relé, un conmutador o un solenoide.

Adaptive brake light A brake light system that selects different illumination levels or methods of display for the rear brake lights depending on conditions.

Luz de freno adaptable Sistema de luces de freno que selecciona diferentes niveles de iluminación o métodos de visualización para las luces del freno traseras, de acuerdo con las condiciones.

Adaptive cruise control (ACC) A cruise control system using laser radar or infrared transceiver sensors to adjust and maintain appropriate following distances between vehicles. The sensor determines vehicle-to-vehicle distances and relational speeds.

Control de crucero adaptable (CCA) Sistema de control de crucero que utiliza un radar láser o sensores de transceptor infrarrojo para ajustar y mantener la distancia de seguimiento adecuada entre los vehículos. El sensor determina la distancia de vehículo a vehículo y la velocidad de relación.

Adaptive headlight system (AHS) Headlight system that is designed to enhance night-time safety by turning the headlight beams to follow the direction of the road as the vehicle enters a turn.

Sistema de faros delanteros adaptables (AHS, por su sigla en inglés) Sistema de faros diseñado para mejorar la seguridad durante la noche, ya que las luces de los faros giran para seguir la dirección del camino al tomar una curva.

Adaptive memory The ability of the computer system to store changing values in order to correct operating characteristics.

Memoria adaptable La capacidad del sistema de computación de almacenar valores cambiantes para corregir las características operativas.

Adaptive strategy The capacity of the computer to make corrections to the normal strategy based on input signals.

Estrategia adaptable La capacidad de una computadora de realizar correcciones a la estrategia normal sobre la base de las señales recibidas.

Air bag Inflates in a few milliseconds when the vehicle is involved in a frontal collision. A typical fully inflated air bag has a volume of 2.3 cubic feet.

Bolsa de aire Infla en unos milisegundos cuando un vehículo se ha involucrado en una colisión delantera. Una bolsa de aire típica tiene un volúmen de 2.3 pies cúbicos al estar completamente inflada.

Air bag module Composed of the air bag and inflator assembly, which is packaged into a single module.

Unidad del Airbag Formada por el conjunto del Airbag y el inflador. Este conjunto se empaqueta en una sola unidad.

Air bag system Designed as a supplemental restraint that, in the case of an accident, will deploy a bag out of the steering wheel or passenger-side dash panel to provide additional protection against head and face injuries.

Sistema de bolsa de aire Diseñada como una restricción suplemental que, en el caso de un accidente, desplegará una bolsa del volante de dirección o del tablero lateral del pasajero para proveer la protección adicional contra los daños a la cabeza y a la cara.

Air-coil gauge Gauge design that uses the interaction of two electromagnets and the total field effect upon a permanent magnet to cause needle movement.

Calibrador de núcleo de aire Calibrador diseñado para utilizar la interacción de dos electroimanes y el efecto inductor total sobre un imán permanente para generar el movimiento de la aguja.

Alternating current (AC) Electrical current that changes direction between positive and negative.

Corriente alterna Corriente eléctrica que recorre un circuito ya sea en dirección positiva o negativa.

Ambient temperature The temperature of the outside air.

Temperatura ambiente Temperatura del aire ambiente.

Ambient temperature sensor A thermistor used to determine the ambient temperature.

Sensor de la temperatura ambiente Termostato que se usa para determinar la temperatura ambiente.

American wire gauge (AWG) System used to determine wire sizes based on the cross-sectional area of the conductor.

Calibrador americano de alambres Sistema utilizado para determinar el tamaño de los alambres, basado en el área transversal del conductor.

Ammeter A test meter used to measure current draw.

Amperímetro Instrumento de prueba utilizado para medir la intensidad de una corriente.

Amortisseur winding The name given to the bars of a squirrel cage placed around the AC motor's rotor.

Amortisseur que enrolla El nombre dado a las barras de una jaula de ardilla colocó alrededor del rotor del motor de C.A.

Ampere-hour rating Indicates the amount of steady current a battery can supply for 20 hours.

Límite de amperio-hora Indica la cantidad de corriente fijo que un puede proveer una batería durante 20 horas.

Ampere (A) See current.

Amperios *Véase* corriente.

Analog A voltage signal that is infinitely variable or that can be changed within a given range.

Señal analógica Señal continua y variable que debe traducirse a valores numéricos discontinuos para poder ser trataba por una computadora.

Angular momentum The rotational analog of linear momentum. that remains constant unless acted on by an external torque.

Momento angular El análogo de rotación del momento lineal que permanece constante a menos que sea actuado por un esfuerzo de torsión externo.

Anode The positive charge electrode in a voltage cell.

Ánodo Electrodo de carga positiva de un generador de electricidad.

Antenna A wire or a metal stick that increases the amount of metal the transmitter's waves can interact with to convert radio frequency electrical currents into electromagnetic waves and vice versa.

Antena Un cable o varilla de metal que aumenta la cantidad de metal con la que pueden interactuar las ondas del transmisor para convertir las corrientes eléctricas de radio frecuencia en ondas electromagnéticas y viceversa.

Antilock brakes (ABS) A brake system that automatically pulsates the brakes to prevent wheel lockup under panic stop and poor traction conditions.

Frenos antibloqueo Sistema de frenos que pulsa los frenos automáticamente para impedir el bloqueo de las ruedas en casos de emergencia y de tracción pobre.

Antitheft system A system that prevents illegal entry or driving of a vehicle. Most are designed to deter entry.

Dispositivo a prueba de hurto Un dispositivo o sistema que previene la entrada o conducción ilícita de un vehículo. La mayoría se diseñan para detener la entrada.

Armature The movable component of an electric motor, which consists of a conductor wound around a laminated iron core and is used to create a magnetic field.

Armadura Pieza móvil de un motor eléctrico, compuesta de un conductor devanado sobre un núcleo de hierro laminado y que se utiliza para producir un campo magnético.

Armed The process of activating the antitheft system by shutting all doors and locking them.

Armado El proceso de activar el sistema antirrobo al cerrar todas las puertas y trabarlas con seguro.

Arming sensor A device that places an alarm system into "ready" to detect an illegal entry.

Sensor de armado Un dispositivo que pone "listo" un sistema de alarma para detectar una entrada ilícita.

Aspirator Tubular device that uses a venturi effect to draw air from the passenger compartment over the in-car sensor. Some manufacturers use a suction motor to draw the air over the sensor.

Aspirador Dispositivo tubular que utiliza un efecto Venturi para extraer aire del compartimiento del pasajero sobre el sensor dentro del vehículo. Algunos fabricantes utilizan un motor de succión para extraer el aire sobre el sensor.

Asynchronous Data that is sent on the bus network intermittently (as needed) rather than continuously.

Asíncronos Datos que se envían en el bus múltiple de modo intermitente (como sea necesario) en lugar de modo continuo.

Atom The smallest part of a chemical element that still has all the characteristics of that element.

Átomo Partícula más pequeña de un elemento químico que conserva las cualidades íntegras del mismo.

Audio system The sound system for a vehicle; can include radio, cassette player, CD player, amplifier, and speakers.

Sistema de audio El sistema de sonido de un vehículo; puede incluir el radio, el tocacaset, el toca discos compactos, el amplificador, y las bocinas.

Automatic door locks (ADL) A system that automatically locks all doors by activating one switch.

Cerrojos de compuertas automatizados Cerrojos de compuertas automatizados eléctricamente que utilizan o un solenoide o un motor reversible de imán permanente para cerrar y abrir las puertas.

Automatic headlight dimming An electronic feature that automatically switches the headlights from high beam to low beam under two different conditions: headlights from oncoming vehicles strike the Photocell-amplifier, or from the taillights of a vehicle being passed strikes the Photocell-amplifier.

Reducción automática de intensidad luminosa de los faros delanteros Característica electrónica que conmuta los faros delan teros automáticamente de luz larga a luz corta dadas las siguientes circunstancias: la luz de los vehículos que se aproximan alcanza el amplificador de fotocélula, o la luz de los faros traseros de un vehículo que se ha rebasado alcanza el amplificador de fotocélula.

Automatic on/off with time delay Turns on the headlights automatically when ambient light decreases to a predetermined level; also allows the headlights to remain on for a certain amount of time after the vehicle has been turned off. This system can be used in combination with automatic dimming systems.

Prendido/apagado automático con temporización Prende los faros automáticamente cuando la luz ambiental se oscurece a un nivel predeterminado. Tambien permite que los faros quedan prendidos por un tiempo determinado después de que se ha apagado el vehículo. Este sistema se puede utilizar en combinación con los sistemas de regulación de intensidad luminosas automáticos.

Automatic temperature control (ATC) A passenger comfort system that is capable of maintaining a preset temperature level as selected by the operator. Sensors are used to determine the present temperatures, and the system can adjust the level of heating or cooling as required using actuators to open and close air-blend doors to achieve the desired in-vehicle temperature.

Control automático de la temperatura (CAT) Un sistema de comodidad para el pasajero es capaz de mantener un nivel de temperatura previamente fijo tal como lo selecciona el operador. Los sensores se utilizan para determinar las temperaturas actuales, y el sistema puede ajustar el nivel de calentamiento o enfriamiento como se requiera al usar actuadores para abrir y cerrar las compuertas de recirculación para lograr la temperatura deseada dentro del vehículo.

Automatic traction control A system that prevents slippage of one of the drive wheels. This is done by applying the brake at that wheel and/or decreasing the engine's power output.

Control Automático de Tracción Un sistema que previene el patinaje de una de las ruedas de mando. Esto se efectúa aplicando el freno en esa rueda y/o disminuyendo la salida de potencia del motor.

Autonomous When referenced to an automobile, means it is able to self-drive.

Autónomo Aplicado a un automóvil, que funciona solo, sin que lo conduzca una persona.

Avalanche diodes Diodes that conduct in the reverse direction when a reverse-bias voltage of about 6.2 volts or higher is applied. Causes the avalanche effect to occur when the reverse electric field moves across the PN junction, causing a wave of ionization that leads to a large current.

Diodo Zener Diodos que hacen conducción en dirección contraria cuando se aplica velocidad invertida de transmisión de baudios de alrededor de 6.2 voltios o más. Produce el efecto Zener o de avalancha cuando el campo eléctrico invertido se mueve a cruzar la unión PN, y causa así una onda de ionización que provoca una gran corriente.

B circuit A generator regulator circuit that is internally grounded. In the B circuit, the voltage regulator controls the power side of the field circuit.

Circuito B Circuito regulador del generador puesto internamente a tierra. En el circuito B, el regulador de tensión controla el lado de potencia del circuito inductor.

Balanced Refers to an atom that has the same number of electrons as the number of protons.

Balanceado Se refiere a un átomo que tiene el mismo número de electrones y de protones.

Ballast An electrical device that limits the current delivered through a circuit.

Balastra Un dispositivo eléctrico que limita la corriente que se distribuye a través de un circuito.

Base The center layer of a bipolar transistor.

Base Capa central de un transistor bipolar.

BAT The terminal identifier for the conductor from the generator to the battery positive terminal.

BAT El terminal que identifica el conductor del generador al terminal positivo de la batería.

Battery balancing A cycling process that brings the cells of an HV battery within a specific voltage difference of each other. Balancing the cells equalizes the voltage so all cells recharge to the same level.

Balanceo de la batería Proceso cíclico que coloca las celdas de una batería de AV dentro de una diferencia de voltaje específica entre una y otra. El balanceo de las celdas ecualiza el voltaje de modo que todas las celdas se recarguen al mismo nivel.

Battery cables High-current conductors that connect the battery to the vehicle's electrical system.

Cables de batería Conductores de alta corriente que conectan la batería al sistema eléctrico del vehículo.

Battery cell The active unit of a battery.

Acumulador de batería Componente activo de una batería.

Battery holddowns Brackets that secure the battery to the chassis of the vehicle.

Portabatería Los sostenes que fijan la batería al chasis del vehículo.

Battery Management System (BMS) HEV, PHEV, EV battery management that uses several components and control modules to monitor and maintain the health of the HV battery and protects it from operating outside its safe operating conditions.

Sistema de gestión de baterías (BMS) Sistema para gestionar baterías de VEH, VEHE y VE que usa varios componentes y módulos de control para monitorear y mantener la salud de la batería AV, y protegerla para que no funcione fuera de su área de condiciones operativas seguras.

Battery terminals Terminals of the battery to which the positive and the negative battery cables are connected. The terminals may be posts or threaded inserts.

Bornes de la batería Los bornes en la batería a los cuales se conectan los cables positivos y negativos. Los terminales pueden ser postes o piezas roscadas.

Baud rate The measure of computer data transmission speed in bits per second.

Razón de baúd Medida de la velocidad de la transmisión de datos de una computadora en bits por segundo.

Belt alternator starter (BAS) A high-voltage starter/alternator combination that uses current flow through the stator windings to generate magnetic fields in the rotor, causing the rotor to turn, thus turning the crankshaft and starting the engine magnetic fields in the rotor, and starting the engine.

Arrancador del alternador por faja (AAF) Una combinación de arrancador/alternador de alto voltaje que utiliza el flujo de corriente a través del devanado estatórico para generar campos magnéticos en el rotor, lo que provoca que el rotor dé vueltas, haciendo girar así el cigüeñal y arrancando el motor.

Belt tension sensor (BTS) A strain gauge-type sensor located on the seat belt anchor of the passenger-side front seat, used to determine if an infant seat is being used.

Sensor de la tensión de la faja (STF) Sensor de tipo medidor de tensiones localizado en el ancla del cinturón del asiento del lado del asiento frontal del pasajero y que se usa para determinar si se usa un asiento para bebés.

Bendix drive A type of starter drive that uses the inertia of the spinning starter motor armature to engage the drive gear to the gears of the flywheel. This type of starter drive was used on early models of vehicles and is rarely seen today.

Acoplamiento Bendix Un tipo del acoplamiento del motor de arranque que usa la inercia de la armadura del motor de arranque giratorio para endentar el engrenaje de mando con los engrenajes del volante. Este tipo de acoplamiento del motor de arranque se usaba en los modelos vehículos antiguos y se ven raramente.

Bias voltage Voltage applied across a diode.

Tensión polarizadora Tensión aplicada a través de un diodo.

Bimetallic strip A metal contact wiper consisting of two different types of metals. One strip will react quicker to heat than the other, causing the strip to flex in proportion to the amount of current flow.

Banda bimetálica Contacto deslizante de metal compuesto de dos tipos de metales distintos. Una banda reaccionará más rápido al calor que la otra, haciendo que la banda se doble en proporción con la cantidad de flujo de corriente.

Binary code A series of numbers represented by 1s and 0s. Any number and word can be translated into a combination of binary 1s and 0s.

Código binario Serie de números representados por unos y ceros. Cualquier número y palabra puede traducirse en una combinación de unos y ceros binarios.

Bipolar The name used for transistors because current flows through the materials of both polarities.

Bipolar Nombre aplicado a los transistores porque la corriente fluye por conducto de materiales de ambas polaridades.

Bit A binary digit.

Bit Dígito binario.

Bi-xenon headlamps HID headlamps that use a single xenon lamp to provide both the high-beam and the low-beam operations. The full light output is used to produce the high beam, while the low beam is formed by moving a shutter between the bulb and the lens.

Faros delanteros de doble xenón Faros de descarga de alta intensidad (HID, por su sigla en inglés) que utilizan una sola bombilla de xenón para cubrir las funciones de luces largas y luces cortas. La potencia de luz completa se utiliza para las luces largas, mientras que las luces bajas se forman moviendo un obturador entre la bombilla y el lente.

Blend-air door actuator An electric motor that controls the position of the blend-air door, in order to supply the in-vehicle temperature the driver selected.

Actuador de puertas por aire mezclado Un motor eléctrico que controla la posición de las puertas por aire mezclado para proporcionar la temperatura dentro del vehículo que seleccione el conductor.

Blended brake system EV, HEV, and PHEV braking system that uses conventional ABS and regenerative braking.

Sistema de freno combinado Sistema de frenos de VE, VEH y HEVE, que usa el sistema convencional de frenos antibloqueo y regenerativo.

Bluetooth Technology that allows several modules from different manufacturers to be connected using a standardized radio transmission.

Bluetooth Tecnología que permite que se conecten varios módulos de diferente manufactura por medio del uso de transmisión de radio estándar.

Brushes Electrically conductive sliding contacts, usually made of copper and carbon.

Escobillas Contactos deslizantes de conduccion eléctrica, por lo general hechos de cobre y de carbono.

Bucking coil One of the coils in a three-coil gauge. It produces a magnetic field that bucks or opposes the low-reading coil.

Bobina compensadora Una de las bobinas de un calibre de tres bobinas. Produce un campo magnético que es contrario o en oposición a la bobina de baja lectura.

Buffer A buffer cleans up a voltage signal. Buffers are used with PM generator sensors to change AC voltage to a digitalized signal.

Separador Un separador aguza una señal del tensión. Estos se usan con los sensores generadores PM para cambiar la tensión de corriente alterna a una señal digitalizado.

Buffer circuit Changes AC voltage from the PM generator into a digitalized signal.

Circuito separador Cambia el voltaje de corriente alterna del generador PM a una señal digital.

Bulkhead connector A large connector that is used when many wires pass through the bulkhead or firewall.

Conectador del tabique Un conectador que se usa al pasar muchos alambres por el tabique o mamparo de encendios.

Bus Used in reference to data transmission since data is being transported from one place to another. The multiplex circuit is often referred to as the bus circuit.

Colectiva Se usa en referir a la transmisión de datos que se estan transportando de un lugar a otro. El circuito multiplex suele referirse como el circuito colectivo.

Bus bar A common electrical connection to which all of the fuses in the fuse box are attached. The bus bar is connected to battery voltage.

Barra colectora Conexión eléctrica común a la que se conectan todos los fusibles de la caja de fusibles. La barra colectora se conecta a la tensión de la batería.

Bus (−) The bus circuit that is most negative when the dominant bit is being transmitted.

Bus negativo (−) Circuito del bus que es más negativo cuando se transmite el bit predominante.

Bus (+) The bus circuit that is most positive when the dominant bit is being transmitted.

Bus positivo (+) Circuito del bus que es más positivo cuando se transmite el bit predominante.

Buzzer An audible warning device that is used to warn the driver of possible safety hazards.

Zumbador Dispositivo audible de advertencia utilizado para prevenir al conductor de posibles riesgos a la seguridad.

Capacitance The ability of two conducting surfaces to store voltage.

Capacitancia Propiedad que permite el almacenamiento de electricidad entre dos conductores aislados entre sí.

Capacitance discharge sensor A pressure sensor that uses a variable capacitor. As the distance between the electrodes changes, so does the capacity of the capacitor.

Sensor de descarga de capacitancia Un sensor de presión que utiliza un capacitor variable. A medida que cambia la distancia entre los electrodos, también lo hace la capacidad del capacitor.

Capacitor An electrical device made by wrapping two conductor strips around an insulating strip used to temporarily store electrical energy.

Capacitor Un dispositivo eléctrico que se fabrica al envolver dos tiras conductoras alrededor de una tira de aislante y que se utiliza para almacenar energía eléctrica en forma temporal.

Carbon monoxide An odorless, colorless, and toxic gas that is produced as a result of combustion.

Monóxido de carbono Gas inodoro, incoloro y tóxico producido como resultado de la combustión.

Carriers Attached to the shoulder anchor to move or carry the anchor from one end of the track to the other.

Portadoras Conectados al reborde de anclaje para mudar o transportar el anclaje de una extremidad del carril a la otra.

Cartridge fuses *See* maxi-fuse.

Fusibles cartucho *Véase* maxifusible.

Cathode Negatively charged electrode of a voltage cell.

Cátodo Electrodo de carga negativa de un generador de electricidad.

Cathode ray tube Similar to a television picture tube. It contains a cathode that emits electrons and an anode that attracts them. The screen of the tube will glow at the points that are hit by the electrons.

Tubo de rayos catódicos Parecidos a un tubo de pantalla de televisor. Contiene un cátodo que emite los electrones y un ánodo que los atrae. La pantalla del tubo iluminará en los puntos en donde pegan los electrones.

Caustic A material that has the ability to destroy or eat through something. Caustic materials are considered extremely corrosive.

Caustico Una materia que tiene la habilidad de destruir o carcomer algo. Las materias causticas se consideran extremamente corrosivas.

Cell element The assembly of a positive and a negative plate in a battery.

Elemento de pila La asamblea de una placa positiva y negativa en una bateria.

Cellular When used to identify V2X references the technology used in cellular radios to communicate from one radio directly to another.

Celular Cuando se usa para identificar V2X, hace referencia a la tecnología empleada en radios celulares para comunicar una radio directamente con otra.

Central gateway (CGW) A module on the CAN bus network that is the hub between the different networks.

Puerta central Un módulo de la red de bus CAN que está en el núcleo entre las diferentes redes.

Charging system Converts mechanical energy of the engine into electrical energy to recharge the battery and run the electrical accessories.

Sistema de carga Convierte la energía mecánica del motor en energía eléctrica para recargar la batería y hacer trabajar los accesorios eléctricos.

Child safety latch A switch that is designed into the door lock system that prevents the door from being opened from the inside, regardless of the position of the door lock knob.

Cerrojo de seguridad para niños Un interruptor diseñado en el sistema de bloqueo de las puertas para evitar que se abra la puerta desde el interior, sin importar la posición del botón de bloqueo de la puerta.

CHMSL The acronym for center high-mounted stop light, often referred to as the third brake light.

CHMSL El acrónimo de luz de freno central montada en la parte superior; se conoce comúnmente como la tercera luz de freno.

Choke An inductor in series with a circuit.

Reactancia Un inductor en serie con un circuit.

Choke coil Fine wire wound into a coil used to absorb oscillations in a switched circuit.

Bobina de inducción Alambre fino devanado en una bobina, utilizado para absorber oscilaciones en un circuito conmutado.

Chrysler Collision Detection (CCD) Chrysler's data bus network first used in 1988; uses a twisted pair of wires to transmit data.

Detección de colisión de Chrysler Red del bus de datos de Chrysler que se usó primero en 1988. Utilizó un par de hilos torcidos para transmitir datos.

Circuit The path of electron flow consisting of the voltage source, conductors, load component, and return path to the voltage source.

Circuito Trayectoria del flujo de electrones, compuesto de la fuente de tensión, los conductores, el componente de carga y la trayectoria de regreso a la fuente de tensión.

Circuit breaker A mechanical fuse that opens the circuit when amperage is excessive. In most cases, the circuit breaker resets when overload is removed.

Interruptor Un fusible mecánico que abre el circuito cuando la intensidad de amperaje es excesiva. En la mayoría de los casos, el interruptor se reengancha al eliminarse la sobrecarga.

Clamping diode A diode that is connected in parallel with a coil to prevent voltage spikes from the coil reaching other components in the circuit.

Diodo de bloqueo Un diodo que se conecta en paralelo con una bobina para prevenir que los impulsos de tensión lleguen a otros componentes en el circuito.

Climate-controlled seats Vehicle seats that provide additional comfort by heating and/or cooling the seat cushion and seat back.

Asientos climatizados Asientos del vehículo que brindan comodidad adicional al calentar y/o enfriar el cojín y el respaldo del asiento.

Clock circuit A crystal that electrically vibrates when subjected to current at certain voltage levels. As a result, the chip produces very regular series of voltage pulses.

Circuito de reloj Cristal que vibra electrónicamente cuando está sujeto a una corriente a ciertos niveles de tensión. Como resultado, el fragmento produce una serie sumamente regular de impulsos de tensión.

Clockspring A winding of electrical conducting tape enclosed within a plastic housing. The clockspring maintains continuity between the steering wheel, switches, the air bag, and the wiring harness in all steering wheel positions.

Muelle espiral Una bobina de cinta conductiva eléctrica encerrada en una caja de plástico. El muelle espiral mantiene la corriente continua entre el volante de dirección, los interruptores, la bolsa de aire, y el mazo de alambres en cualquier posición del volante de dirección.

Closed circuit A circuit that has no breaks in the path and allows current to flow.

Circuito cerrado Circuito de trayectoria ininterrumpida que permite un flujo continuo de corriente.

Coil Assembly An Assembly of coils that contains two coils or more.

Asamblea de bobina Una asamblea de bobinas que contiene dos bobinas o más.

Cold cranking amps The battery rating that indicates the battery's ability to deliver a specified amount of current to start an engine at low-ambient temperatures.

Amperios de arranque en frío Tasa indicativa de la capacidad de la batería para producir una cantidad específica de corriente para arrancar un motor a bajas temperaturas ambiente.

Cold engine lock-out switch Prevents blower motor operation until the air entering the passenger compartment reaches a specified temperature.

Interruptor de cierre en motor inactivo Previene la operación del motor del ventilador hasta que el aire que entra en el compartimiento del pasajero alcanza su temperatura específica.

Collector The portion of a bipolar transistor that receives the majority of current carriers.

Dispositivo de toma de corriente Parte del transistor bipolar que recibe la mayoría de los portadores de corriente.

Collision Avoidance System An extension of the ABS and ACC systems that automatically applies the brakes to prevent an accident.

Sistema anticolisión Extensión de los sistemas ABS y ACC, que aplica los frenos de manera automática para evitar un accidente.

Common connection Splices in the wiring that are used to share a source of power or a common ground with different systems.

Conexión común Empalmes en el cableado que se utilizan para compartir una fuente de alimentación o una puesta a tierra común con distintos sistemas.

Common connector A connector that is shared by more than one circuit and/or component.

Conector común Un conector que se comparte entre más de un circuito y/o componente.

Commutator A series of conducting segments located around one end of the armature.

Conmutador Serie de segmentos conductores ubicados alrededor de un extremo de la armadura.

Complementary metal oxide semiconductor (CMOS) A technology for constructing integrated circuits using a metal gate electrode physical structure on top of an oxide insulator, that is placed on top of a semiconductor material.

Semiconductor de óxido de metal complementario (SOMC) Tecnología para construir circuitos integrados usando una estructura física de electrodo de compuerta de metal sobre un aislante de óxido, que se coloca sobre un material semiconductor.

Component locator A service manual that lists and describes the exact location of components on a vehicle.

Localizador de componentes Un manual de servicio que cataloga y describe la posición exacta de los componentes en un vehículo.

Composite bulb A headlight assembly that has a replaceable bulb in its housing.

Bombilla compuesta Una asamblea de faros cuyo cárter tiene una bombilla reemplazable.

Composite headlight A halogen headlight system that uses a replacement bulb.

Faro compuesto Sistema de faros halógenos que usa un foco de recambio.

Compound motor A motor that has the characteristics of a series-wound and a shunt-wound motor.

Motor compuesta Un motor que tiene las características de un motor exitado en serie y uno en derivación.

Computer An electronic device that stores and processes data and is capable of operating other devices.

Computadora Dispositivo electrónico que almacena y procesa datos y que es capaz de ordenar a otros dispositivos.

Concealed headlight System used to help improve fuel economy and styling of the vehicle.

Faros ocultos Un sistema que se usa para mejorar el rendimiento del combustible y el estilo del vehículo.

Condenser A capacitor made from two sheets of metal foil separated by an insulator.

Condensador Capacitador hecho de dos láminas de metal separadas por un medio aislante.

Conduction Bias voltage difference between the base and the emitter has increased to the point that the transistor is switched on. In this condition, the transistor is conducting. Output current is proportional to that of the base current.

Conducción La diferencia de la tensión polarizadora entre la base y el emisor ha aumentado hasta el punto que el transistor es conectado. En estas circunstancias, el transistor está conduciendo. La corriente de salida está en proporción con la de la corriente conducida en la base.

Conductor A material in which electrons flow or move easily.

Conductor Una material en la cual los electrones circulen o se mueven fácilmente.

Contactors Heavy-duty relays that are connected to the positive and negative sides of the HV battery.

Contactos Relés para servicio pesado conectados al positivo y al negativo de la batería HV.

Continuity Refers to the circuit being continuous with no opens.

Continuidad Se refiere al circuito ininterrumpido, sin aberturas.

Control panel assembly Provides for driver input into the automatic temperature control microprocessor.

Asamblea de controles Permite la entrada del conductor al microprocesador del control de temperatura automático. La asamblea de control tambien se refiere como el tablero de instrumentos.

Controller area network (CAN) A two-wire bus network that allows the transfer of data between control modules.

CAN (Red del área del controlador) Red de bus de dos hilos que permite la transferencia de datos entre los módulos de control.

Conventional theory Electrical theory which states that current flows from a positive point to a more negative point.

Teoría convencional Teoría de electricidad la cual enuncia que el corriente fluye desde un punto positivo a un punto más negativo.

Cornering lights Lamps that illuminate when turn signals are activated. They burn steady when the turn signal switch is in a turn position, to provide additional illumination to the road in the direction of the turn.

Faros de viraje Los faros que iluminen cuando se prenden los indicadores de virajes. Quedan prendidos mientras que el indicador de viraje esta en una posición de viraje para proveer mayor ilumincaión del camino en la dirección del viraje.

Counter electromotive force (CEMF) An induced voltage that opposes the source voltage.

Fuerza cóntraelectromotriz Tensión inducida en oposición a la tensión fuente.

Courtesy lights Lamps that illuminate the vehicle's interior when the doors are open.

Luces interiores Lámparas que iluminan el interior del vehículo cuando las puertas están abiertas.

Covalent bonding The bonding where atoms share valence electrons with other atoms.

Enlace covalente Cuando los átomos comparten electrones de valencia con otros átomos.

Cranking amps A battery rating system that indicates the battery's ability to provide a cranking amperage at 32° F (0° C).

Amperios de arranque Un sistema de clasificación de baterías que indica la capacidad de una batería de suministrar un amperaje de arranque a 32° F (0° C).

Crash sensors Normally open electrical switches designed to close when subjected to a predetermined amount of jolting or impact.

Sensores de impacto Conmutador normalmente abierto diseñado para cerrarse al ser sacudido por una fuerza predeterminada o al recibir un impacto.

Cruise control A system that allows a vehicle to maintain a preset speed with the driver's foot off the accelerator.

Control crucero Un sistema que permite que el vehículo mantenga una velocidad predeterminada sin que el pie del conductor dceprime al accelerador.

Crystal A term used to describe a material that has a definite atom structure.

Cristal Término utilizado para describir un material que tiene una estructura atómica definida.

Current The aggregate flow of electrons through a wire. One ampere represents the movement of 6.25 billion billion electrons (or one coulomb) past one point in a conductor in one second.

Corriente Flujo combinado de electrones a través de un alambre. Un amperio representa el movimiento de 6,25 mil millones de mil millones de electrones (o un colombio) que sobrepasa un punto en un conductor en un segundo.

Cutoff The state where reverse-bias voltage is applied to the base leg of a transistor. In this condition, the transistor does not conduct and no current flows.

Corte Cuando se aplica tensión polarizadora inversa a la base del transistor. En estas circunstancias, el transistor no está conduciendo y no fluirá ninguna corriente.

Cycle Completed when the voltage has gone positive, returned to zero, gone negative, and returned to zero.

Ciclo Completado cuando el voltaje ha sido positivo, regresado al cero, ha sido negativo y regresado al cero.

Darlington pair An arrangement of transistors that amplifies current by one transistor acting as a preamplifier that creates a larger base current to the second transistor.

Par Darlington Conjunto de transistores que amplifica la corriente. Un transistor actúa como preamplificador y produce una corriente base más ámplia para el segundo transistor.

D'Arsonval gauge A gauge design that uses the interaction of a permanent magnet and an electromagnet and the total field effect to cause needle movement.

Calibrador d'Arsonval Calibrador diseñado para utilizar la interacción de un imán permanente y de un electroimán, y el efecto inductor total para generar el movimiento de la aguja.

Daytime running lamps (DRL) Generally use a high-beam or low-beam headlight system at a reduced intensity to provide additional visibility of the vehicle for other drivers and pedestrians.

Faros diurnos Generalmente usan el sistema de faros de alta y baja intensidad en una intensidad disminuada para proporcionar el vehículo más visibilidad para los otros conductores y los peatones.

DC/DC converter The DC/DC converter is configured to provide a 14-volt output from the high-voltage input. The 14-volt output can be used to supply electrical energy to those components that do not require the high voltage.

Convertidor de CC/CC El convertidor CC/CC está configurado para proporcionar una salida de 14 V de una entrada de alto voltaje. La salida de 14 V puede usarse para suministrar energía eléctrica a los componentes que no requieren alto voltaje.

Dead reckoning The process of calculating current position using previously determined positions, and advancing the current position based upon speeds over elapsed time and course.

Cuenta muerta El proceso de calcular la posición actual usando posiciones previamente determinadas, y avanzar de la posición actual basada en velocidades sobre el tiempo transcurrido y curso.

Dedicated Short-Range Communications (DSRC) A variant of Wi-Fi wireless communications that enables vehicles to communicate with each other without relying on cellular or other infrastructure. A vehicle with DSRC broadcasts its location, heading, and speed to other DSRC equipped vehicles.

Comunicaciones dedicadas de corto alcance (DSRC) Variante de las comunicaciones inalámbricas wifi que permite a los vehículos comunicarse entre sí sin depender de infraestructura celular o similar. Un vehículo con DSRC transmite su ubicación, rumbo y velocidad a otros DSRC vehículos equipados.

Deep cycling Discharging the battery to a very low state of charge before recharging it.

Ciclo profundo Proceso por el que la batería se descarga hasta alcanzar un nivel muy bajo de carga antes de volver a cargarse.

Delta connection A connection that receives its name from its resemblance to the Greek letter delta (Δ).

Conexión delta Una conexión que recibe su nombre a causa de su aparencia parecida a la letra delta griega.

Delta stator A three-winding AC generator stator, with the ends of each winding connected to each other.

Estátor delta Estátor generador de corriente alterna de devanado triple, con los extremos de cada devanado conectados entre sí.

Depletion-type FET Cuts off current flow.

FET tipo agotamiento Corta el flujo del corriente.

Depressed-park A system in which the blades drop down below the lower windshield molding to hide them.

Limpiaparabrisas guardadas Un sistema en el cual los brazos se guardan abajo del borde inferior de la parabrisa para asi esconderlos.

Detector Extracts data (voices and music) from the sine wave.

Detector Extrae datos (voces y música) de la onda senoidal.

Diagnostic module Part of an electronic control system that provides self-diagnostics and/or a testing interface.

Módulo de diagnóstico Parte de un sistema controlado electronicamente que provee autodiagnóstico y/o una interfase de pruebas.

Diaphragm A thin, flexible, circular plate that is held around its outer edge by the horn housing, allowing the middle to flex.

Diagragma Una placa redonda flexible y delgada sostenido en el cárter del claxon por medio de su borde exterior, así permitiendo flexionar la parte central.

Dielectric An insulator material.

Dieléctrico Material aislante.

Digital A voltage signal is either on/off, yes/no, or high/low.

Digital Una señal de tensión está Encendida-Apagada, es Sí-No o Alta-Baja.

Digital instrument clusters Use digital and linear displays to notify the driver of monitored system conditions.

Grupo de instrumentos digitales Usan los indicadores digitales y lineares para notificar al conductor de las condiciones de los sistemas regulados.

Dimmer switch A switch in the headlight circuit that provides the means for the driver to select either high-beam or low-beam operation and to switch between the two. The dimmer switch is connected in series within the headlight circuit and controls the current path for high beam and low beam.

Conmutador reductor Conmutador en el circuito para faros delanteros que le permite al conductor que elegir la luz larga o la luz corta, y conmutar entre las dos. El conmutador reductor se conecta en serie dentro del circuito para faros delanteros y controla la trayectoria de la corriente para la luz larga y la luz corta.

DIN The abbreviation for Deutsche Institut füer Normung (German Institute for Standardization) and the recommended standard for European manufacturers to follow.

DIN La abreviatura de Deutsche Institut füer Normung (Normas del Instituto Aleman) y se recomienda que los fabricantes europeos siguen estas normas.

Diode An electrical one-way check valve that will allow current to flow only in one direction.

Diodo Válvula eléctrica de retención, de una vía, que permite que la corriente fluya en una sola dirección.

Diode rectifier bridge A series of diodes that are used to provide a reasonably constant DC voltage to the vehicle's electrical system and battery.

Puente rectificador de diodo Serie de diodos utilizados para proveerles una tensión de corriente continua bastante constante al sistema eléctrico y a la batería del vehículo.

Diode trio Used by some manufacturers to rectify the stator voltage of an AC generator current so the voltage can be used to create the magnetic field in the field coil of the rotor.

Trío de diodos Utilizado por algunos fabricantes para rectificar el estátor de la corriente de un generador de corriente alterna y poder así utilizarlo para crear el campo magnético en la bobina inductora del rotor.

Direct current (DC) Electric current that flows in one direction.

Corriente continua Corriente eléctrica que fluye en una dirección.

Direct Current Fast Chargers (DCFCs) A Level 3 charger for charging the HV battery of a EV or PHEV that outputs a DC voltage to the battery. DCFC is the fastest of the charger types with a maximum output of 350 kW DC.

Cargador rápido de corriente continua (CRCC) Cargador de nivel 3 para cargar la batería de AV de un VE o VEHE, que proporciona voltaje CC a la batería. Con una capacidad máxima de salida de 350 kW CC, es el tipo de cargador más rápido que existe.

Direct drive A situation where the drive power is the same as the power exerted by the device that is driven.

Transmisión directa Una situación en la cual el poder de mando es lo mismo que la potencia empleada por el dispositivo arrastrado.

Discrete devices Electrical components that are made separately and have wire leads for connections to an integrated circuit.

Dispositivos discretos Componentes eléctricos hechos uno a uno; tienen conductores de alambre para hacer conexiones a un circuito integrado.

Discriminating sensors Part of the air bag circuitry; these sensors are calibrated to close with speed changes that are great enough to warrant air bag deployment. These sensors are also referred to as crash sensors.

Sensores discriminadores Una parte del conjunto de circuitos de Airbag; estos sensores se calibran para cerrar con los cambios de la velocidad que son bastante severas para justificar el despliegue del Airbag. Estos sensores también se llaman los sensores de impacto.

Doping The addition of another element with three or five valence electrons to a pure semiconductor.

Impurificación La adición de otro elemento con tres o cinco electrones de valencia a un semiconductor puro.

Doppler shift The frequency that occurs when radio waves reflect from a moving object.

Efecto Doppler Frequencia producida por dos ondas de radio que se generan a partir de un objeto en movimiento que las refleja.

Double-filament lamp A lamp designed to execute more than one function. It can be used in the stop light circuit, taillight circuit, and the turn signal circuit combined.

Lámpara con filamento doble Lámpara diseñada para llevar a cabo más de una función. Puede utilizarse en una combinación de los circuitos de faros de freno, de faros traseros y de luces indicadoras para virajes.

Double-start override Prevents the starter motor from being energized if the engine is already running.

Sobremarcha de doble marcha Previene que se excita el motor del encendido si ya esta en marcha el motor.

Drain The portion of a field-effect transistor that receives the holes or electrons.

Drenador Parte de un transistor de efecto de campo que recibe los agujeros o electrones.

Drive coil A hollowed field coil used in a positive-engagement starter to attract the movable pole shoe of the starter.

Bobina de excitación Una bobina inductora hueca empleada en un encendedor de acoplamiento directo para atraer la pieza polar móvil del encendedor.

Dual-climate control Provides separate temperature settings for the driver and the front-seat passenger. This system is similar to previous systems except two blend doors are used to control separate temperature settings.

Control de clima doble Provea la regulación de temperatura individual para el conductor ye el pasajero del asiento delantero. Este sistema es parecido a los sistemas anteriores menos que los dos compuertas de mezcla se usan para controlar la regulación individual de la temperatura.

Dual-range circuit A circuit that provides for a switch in the resistance values to allow the microprocessor to measure temperatures more accurately.

Circuitos de doble rango Circuito que proporciona un interruptor en los valores de resistencia para permitir que el microprocesador mida las temperaturas de manera mas precisa.

Duty cycle The percentage of on time to total cycle time.

Ciclo de trabajo Porcentaje del trabajo efectivo a tiempo total del ciclo.

Easy exit An additional function of the memory seat that provides for easier entrance and exit of the vehicle by moving the seat all the way back and down. Some systems also move the steering wheel up and to full retract.

Salida fácil Una función adicional de la memoria del asiento que provee una entrada y salida más fácil del vehículo al mover el asiento hasta su posición más extrema hacia atrás y abajo. Algunos sistemas tambien muevan el volante de dirección hacia arriba y a su posición más alejada.

Eddy currents Small induced currents.

Corriente de Foucault Pequeñas corrientes inducidas.

Electric defoggers Heat the rear window to remove ice and/or condensation. Some vehicles use the same circuit to heat the outside driver's-side mirror.

Desneblador eléctrica Calientan la ventanilla trasxera para remover el hielo y/o la condensación. Algunos vehículos usan el mismo circuito para calentar el espejo lateral del conductor.

Electric vehicle (EV) A vehicle that powers its motor off of a battery pack.

Vehículo eléctrico (VE) Vehículo que apaga su motor por medio de un paquete de baterías.

Electric Vehicle Supply Equipment (EVSE) A protocol developed for safe charging of the EV and PHEV by using a communication link between the vehicle and the charging station.

Equipo de carga de un vehículo eléctrico (ECVE) Protocolo desarrollado para cargar un VE y VEHE de manera segura, usando un enlace de comunicación entre el vehículo y la estación de carga.

Electrical accessories Electrical systems or components that provide for additional safety and comfort, including safety accessories such as the horn, windshield wipers, and windshield washers. Comfort accessories include the blower motor, electric defoggers, power mirrors, power windows, power seats, and power door locks.

Accesorios eléctricos Sistemas o componentes eléctricos que proporcionan seguridad y comodidad adicionales, y que incluyen accesorios de seguridad tales como el claxon, limpia brisas y parabrisas. Los accesorios de comodidad incluyen un motor de aire, desnubilizador eléctrico, espejos mecánicos, asientos mecánicos y cierre mecánico de puertas.

Electrical load The working device of the circuit.

Carga eléctrica Dispositivo de trabajo del circuito.

Electrical symbols Used to represent components in the wiring diagram.

Símbolos electrónicos Se usan para representar los componentes en uyn esquema de conexiones.

Electrically erasable PROM (EEPROM) Memory chip that allows for electrically changing the information one bit at a time.

Capacidad de borrado electrónico PROM Fragmento de memoria que permite el cambio eléctrico de la información un bit a la vez.

Electrochemical The chemical action of two dissimilar materials in a chemical solution.

Electroquímico Acción química de dos materiales distintos en una solución química.

Electrochromic mirror Automatically adjusts to light using forward- and rearward-facing photo sensors and a solid-state chip. Based on light intensity differences, the chip applies a small voltage to the silicon layer. As voltage is applied, the molecules of the layer rotate and redirect the light beams. Thus, the mirror reflection appears dimmer.

Espejo electrocrómico Se ajusta automáticamente a la luz usando los sensores orientados hacia afrente y atrás juntos con un chip de estado sólido.

Electrolysis The production of chemical changes by passing electrical current through an electrolyte; the splitting of water into hydrogen and oxygen.

Electrólisis La producción de los cambios químicos al pasar un corriente eléctrico por un electrolito.

Electrolyte A solution of 64% water and 36% sulfuric acid.

Electrolito Solución de un 64% de agua y un 36% de ácido sulfúrico.

Electromagnetic gauges Gauges that produces needle movement by magnetic forces.

Calibradores electromagnéticos Calibradores que generan el movimiento de la aguja mediante fuerzas magnéticas.

Electromagnetic induction (EMI) The production of voltage and current within a conductor as a result of relative motion within a magnetic field.

Inducción electromagnética Producción de tensión y de corriente dentro de un conductor como resultado del movimiento relativo dentro de un campo magnético.

Electromagnetic interference (EMI) An undesirable creation of electromagnetism whenever current is switched on and off.

Interferencia electromagnética Fenómeno de electromagnetismo no deseable que resulta cuando se conecta y se desconecta la corriente.

Electromagnetism A form of magnetism that occurs when current flows through a conductor.

Electromagnetismo Forma de magnetismo que ocurre cuando la corriente fluye a través de un conductor.

Electromechanical A device that uses electricity and magnetism to cause a mechanical action.

Electromecánico Un dispositivo que causa una acción mecánica por medio de la electricidad y el magnetismo.

Electromechanical gauge A gauge that operates electrically, but its movement is mechanical. The gauge acts as an ammeter since its reading changes with variations in resistance.

Calibrador electromecánico Calibrador que funciona en forma eléctrica pero cuyo movimiento es mecánico. El calibrador actúa como amperímetro porque su lectura cambia según las variaciones en la resistencia.

Electromotive force (EMF) *See* voltage.

Fuerza electromotriz *Véase* tensión.

Electron Negatively charged particles of an atom.

Electrón Partículas de carga negativa de un átomo.

Electronic roll mitigation (ERM) A vehicle safety system that helps to prevent a vehicle rollover situation due to extreme lateral forces experienced during cornering or hard evasive maneuvers using brake and engine torque management.

Mitigación de rollos electrónicos (ERM) Un sistema de seguridad para vehículos que ayuda a prevenir una situación de vuelco del vehículo debido a las fuerzas laterales extremas experimentadas durante las curvas o fuertes maniobras de evasión utilizando el freno y el esfuerzo de torción del motor.

Electron theory Defines electrical movement from negative to positive.

Teoría del electrón Define el movimiento eléctrico como el movimiento de lo negativo a lo positivo.

Electronic regulator Uses solid-state circuitry to perform regulatory functions.

Regulador electrónico Usa los circuitos de estado sólido para llevar a cabo los funciones de regulación.

Electronic stability control An additional function of the antilock brake and traction control system that uses additional sensors and inputs to determine if the vehicle is actually moving in the direction intended by the driver, as indicated by steering wheel position sensors, yaw sensors, and lateral sensors. If the actual path is not the intended path, the module will apply the appropriate brake to bring the vehicle back onto the correct path.

Mando electrónico de estabilidad Función adicional de un sistema de antibloqueo de frenos y de control de la tracción que utiliza sensores adicionales y entradas para determinar si efectivamente se mueve el vehículo en la dirección que desea el conductor, tales como indican los sensores de posición del volante, los sensores de guiñada y los sensores laterales. Si el recorrido real no es el corrido deseado, el módulo aplicará el freno apropiado para regresar el vehículo al recorrido correcto.

Electrostatic field The field that is between two oppositely charged plates.

Campo electrostático Campo que se encuentra entre las placas de carga opuesta.

Emitter The outer layer of a transistor, which supplies the majority of current carriers.

Emisor Capa exterior del transistor que suministra la mayor parte de los portadores de corriente.

Energy density The amount of energy that is available for a given amount of space.

Densidad de la energía La cantidad de energía disponible en un espacio dado.

Engine vacuum Formed during the intake stroke of the cylinder. Engine vacuum is any pressure lower than atmospheric pressure.

Vacío del motor Formado durante la carrera de entrada de un cilindro. El vacío de motor es cualquier presión más baja de la presión atmosférica.

Enhancement-type FET Improves current flow.

FET tipo de acrecentamiento Mejor ael flujo del corriente.

Enhanced starter A conventional starter motor that has been modified to meet the requirement of multiple restarts by using dual layer brushes and a unique pinion spring mechanism.

Motor de arranque reforzado Motor de arranque convencional modificado para satisfacer la exigencia de múltiples arranques usando escobillas de capa dual y un sistema de piñón y resorte único.

Equivalent series load (equivalent resistance) The total resistance of a parallel circuit, which is equivalent to the resistance of a single load in series with the voltage source.

Carga en serie equivalente (resistencia equivalente) Resistencia total de un circuito en paralelo, equivalente a la resistencia de una sola carga en serie con la fuente de tensión.

Erasable PROM (EPROM) Similar to PROM except that its contents can be erased to allow for new data to be installed. A piece of Mylar tape covers a window. If the tape is removed, the microcircuit is exposed to ultraviolet light and erases its memory.

Memoria de solo lectura programable borrable (MSLPB) Memoria similar a la memoria de solo lectura programable cuyo contenido puede borrarse para permitir la instalación de nuevos datos. Un trozo de cinta Mylar cubre una ventana; si se remueve la cinta, el microcircuito queda expuesto a la luz ultravioleta y borra la memoria.

Excitation current Current that magnetically excites the field circuit of the AC generator.

Corriente de excitación Corriente que excita magnéticamente al circuito inductor del generador de corriente alterna.

Express down A power window feature that allows the operator to lower the window with a single press of the switch instead of having to hold the switch during window operation.

Bajada rápida Una función de los levantavidrios automáticos que permite que el operador baje la ventanilla oprimiendo un interruptor una sola vez en lugar de mantener el interruptor oprimido durante el funcionamiento de la ventanilla.

Express up A power window feature that allows the operator to raise the window with a single press of the switch instead of having to hold the switch during window operation.

Subida rápida Una función de los levantavidrios automáticos que permite que el operador suba la ventanilla oprimiendo un interruptor una sola vez en lugar de mantener el interruptor oprimido durante el funcionamiento de la ventanilla.

Extended Range Electric Vehicle (EREV) An EV that has an ICE that drives a generator to recharge the HV batteries and extend the vehicle's range. The ICE never propels the vehicle and starts only when the vehicle's HV battery is nearly depleted.

Vehículo eléctrico de rango extendido (VERE) Vehículo eléctrico con un motor de combustión interno que impulsa a un generador para recargar las baterías de AV y extender el rango del vehículo. El motor de combustión interno nunca impulsa al vehículo y se enciendo solo cuando la batería de AV del vehículo está casi agotada.

Face shield A clear plastic shield that protects the entire face.

Máscara protectora Una máscara de plástico transparente que proteje la cara entera.

Feedback 1. Data concerning the effects of the computer's commands is fed back to the computer as an input signal; used to determine if the desired result has been achieved. 2. A condition that can occur when electricity seeks a path of lower resistance, but the alternate path operates a component other than that intended. Feedback can be classified as a short.

Realimentación 1. Datos referentes a los efectos de las órdenes de la computadora se suministran a la misma como señal de entrada. La realimentación se utiliza para determinar si se ha logrado el resultado deseado. 2. Condición que puede ocurrir cuando la electricidad busca una trayectoria de menos resistencia, pero la trayectoria alterna opera otro componente que aquel deseado. La realimentación puede clasificarse como un cortocircuito.

Fiber optics A medium of transmitting light through polymethyl methacrylate plastic that keeps the light rays parallel even if there are extreme bends in the plastic.

Transmisión por fibra óptica Técnica de transmisión de luz por medio de un plástico de polimetacrilato de metilo que mantiene los rayos de luz paralelos aunque el plástico esté sumamente torcido.

Field coils Heavy copper wire wrapped around an iron core to form an electromagnet.

Bobina del campo El alambre grueso de cobre envuelta alrededor de un núcleo de hierro para formar un electroimán.

Field current The current going to the field windings of a motor or generator.

Corriente inductora El corriente que va a los devanados inductores de un motor o generador.

Field-effect transistor (FET) A unipolar transistor in which current flow is controlled by voltage in a capacitance field.

Transistor de efecto de campo (TEC) Transistor unipolar en el cual la tensión en un campo de capacitancia controla el flujo de corriente.

Field relay The relay that controls the amount of current going to the field windings of a generator. This is the main output control unit for a charging system.

Relé inductor El relé que controla la cantidad del corriente a los devanados inductores de un generador. Es la unedad principal de potencia de salida de un sistema de carga.

Fifth percentile female The fifth percentile female is determined to be those who weigh less than 100 pounds (45 kg). The fifth percentile female is determined by averaging all potential occupants by size and then plotting the results on a graph.

Quinto percentil femenino La determinan aquellos que pesan menos de 45 kg (100 lbs. Se determina al hacer el promedio de todos los posibles ocupantes por su tamaño, y luego al trazar los resultados en una gráfica.

Fixed resistors Have a set resistance value and are used to limit the amount of current flow in a circuit.

Resistores fijos Tienen un valor de resistencia fijo y se usan para limitar la cantidad de flujo del corriente en un circuito.

Flammable A substance that supports combustion.

Inflamable Una substancia que ampara la combustión.

Flasher Used to open and close the turn signal circuit at a set rate.

Pulsador Se usa para abrir y cerrar el circuito del indicador de vueltas en una velocidad predeterminada.

Floating The movement of voltage levels when a switch is open.

Flotante El movimiento de los niveles de voltaje cuando un interruptor se encuentra abierto.

Flooded battery A battery that completely submerges the plates with liquid electrolyte.

Batería inundada Batería que sumerge las placas completamente en un electrolito líquido.

Floor jack A portable hydraulic tool used to raise and lower a vehicle.

Gato de pie Herramienta hidráulica portátil utilizada para levantar y bajar un vehículo.

Flux density The number of flux lines per square centimeter.

Densidad de flujo Número de líneas de flujo por centímetro cuadrado.

Flux lines Magnetic lines of force.

Líneas de flujo Líneas de fuerza magnética.

Forward-biased A positive voltage that is applied to the P-type material and negative voltage to the N-type material of a semiconductor.

Polarización directa Tensión positiva aplicada al material P y tensión negativa aplicada al material N de un semiconductor.

Fuel cell A battery-like component that produces current from hydrogen and aerial oxygen.

Celda de combustible Componente tipo batería que produce corriente del hidrógeno y del oxígeno en el aire.

Fuel pump inertia switch An NC switch that opens if the vehicle is involved in an impact at speeds over 5 mph or if it rolls over. When the switch opens, it turns off power to the fuel pump. This is a safety feature to prevent fuel from being pumped onto the ground or hot engine compartments if the engine dies. The switch has to be manually reset if it is triggered.

Interruptor inercia de la bomba de combustible Un interruptor NC que se abre si el vehículo se involucra en un choque en una velocidad que exceda 5 millas por hora o si se invierte de arriba abajo. Cuando el interruptor se abre, corta la corriente a la bomba del combustible. Este es una precaución de seguridad para prevenir que la bomba vierte el combustibel en el suelo o sobre un compartimento caliente del motor si se muere el motor. El interruptor se tiene que reenganchar a mano si se acciona.

Full field Maximum AC generator output.

Campo completo Salida máxima de un generador de corriente alterna.

Full parallel hybrid Uses an electric motor that is powerful enough to propel the vehicle on its own.

Híbrido en paralelo completo Utiliza un motor eléctrico que es lo suficientemente potente para que el vehículo se impulse por sí mismo.

Full-wave rectification The conversion of a complete AC voltage signal to a DC voltage signal.

Rectificación de onda plena La conversión de una señal completa de tensión de corriente alterna a una señal de tensión de corriente continua.

Fuse A replaceable circuit protection device that melts should the current passing through it exceed its rating.

Fusible Dispositivo reemplazable de protección del circuito que se fundirá si la corriente que fluye por el mismo excede su valor determinado.

Fuse block The term used to indicate the central location of the fuses contained in a single holding fixture.

Bloque de fusibles El término que su usa para indicar la ubicación central de los fusibles contenidos en una fijación central.

Fuse box A term used to indicate the central location of the fuses contained in a single holding fixture.

Caja de fusibles Término utilizado para indicar la ubicación central de los fusibles contenidos en un solo elemento permanente.

Fusible link A wire made of meltable material with a special heat-resistant insulation. When there is an overload in the circuit, the link melts and opens the circuit.

Cartucho de fusible Alambre hecho de material fusible con aislamiento especial resistente al calor. Cuando ocurre una sobrecarga en el circuito, el cartucho se funde y abre el circuito.

G force Term used to describe the measurement of the net effect of the acceleration that an object experiences and the acceleration that gravity is trying to impart to it. Basically G force is the apparent force that an object experiences due to acceleration.

Fuerza G Término utilizado para describir la medida del efecto neto de la aceleración que experimenta un objeto y la aceleración que la gravedad está intentando impartirle. Básicamente, la fuerza G es la fuerza aparente que un objeto experimenta debido a la aceleración.

Gain The ratio of amplification in an electronic device.

Ganancia Razón de amplificación en un dispositivo electrónico.

Gain medium A material (such as gas, liquid, solid, or free electrons) inside the optical cavity that possess the appropriate optical properties to transfer external energy into the laser beam.

Medio activo Material (como un gas, un líquido, un sólido o electrones libres) dentro de la cavidad óptica que cuenta con las propiedades ópticas adecuadas para transferir energía externa a un rayo láser.

Ganged switch Refers to a type of switch in which all wipers of the switch move together.

Acoplado en tándem Se refiere a un tipo de conmutador en el cual todos los contactos deslizantes del mismo se mueven juntos.

Gassing The conversion of a battery's electrolyte into hydrogen and oxygen gas.

Burbujeo La conversión del electrolito de una bateria al gas de hidrógeno y oxígeno.

Gate The portion of a field-effect transistor that controls the capacitive field and current flow.

Compuerta Parte de un transistor de efecto de campo que controla el campo capacitivo y el flujo de corriente.

Gauge 1. A device that displays the measurement of a monitored system by the use of a needle or pointer that moves along a calibrated scale. 2. The number that is assigned to a wire to indicate its size. The larger the number, the smaller the diameter of the conductor.

Calibrador 1. Dispositivo que muestra la medida de un sistema regulado por medio de una aguja o indicador que se mueve a través de una escala calibrada. 2. El número asignado a un alambre indica su tamaño. Mientras mayor sea el número, más pequeño será el diámetro del conductor.

Gear reduction Occurs when two different-sized gears are in mesh and the driven gear rotates at a lower speed than the drive gear but with greater torque.

Desmultiplicación Ocurre cuando dos engranajes de distinctos tamaños se endentan y el engrenaje arrastrado gira con una velocidad más baja que el engrenaje de mando pero con más par.

Grid 1. The frame structure of a battery that normally has connector tabs at the top. It is generally made of lead alloys. 2. The rear window defogger grid is a series of horizontal, ceramic silver–compounded lines that are baked into the surface of the window.

Rejilla 1. Estructura de una batería que normalmente tiene puntos de conexión en la parte superior. En general, está hecha de aleaciones de plomo. 2. La rejilla para desempañar de la ventana trasera consiste en una serie de líneas horizontales compuestas de plata y cerámica que se integran a la superficie de la ventana.

Grid growth A condition where the grid grows little metallic fingers that extend through the separators and short out the plates.

Expansión de la rejilla Una condición en la cual la rejilla produce protrusiones metálicas que se extienden por los separadores y causan cortocircuitos en las placas.

Ground The common negative connection of the electrical system. It is the point of lowest voltage.

Tierra Conexión negativa común del sistema eléctrico. Es el punto de tensión más baja.

Ground side The portion of the circuit that is from the load component to the negative side of the source.

Lado a tierra Parte del circuito que va del componente de carga al lado negativo de la fuente.

Grounded circuit An electrical defect that allows current to return to ground before it has reached the intended load component.

Circuito puesto a tierra Falla eléctrica que permite el regreso de corriente a tierra antes de alcanzar el componente de carga deseado.

Ground straps Electrical conductors that complete the return path to the battery from components that are insulated from the vehicle's chassis. In addition, ground straps help suppress EMI conduction and radiation by providing a low-resistance circuit ground path.

Conectores de puesta a tierra Conductores eléctricos que completan el circuito de regreso hacia la batería de los componentes que están aislados desde el chasis del vehículo. Además, los conectores de puesta a tierra ayudan a suprimir la conducción IEM y la radiación al proporcionar un recorrido de tierra del circuito de resistencia.

Gyroscope A spinning wheel or disc that axis of rotation is capable assuming any orientation by itself.

Giroscopio Una rueda giratoria o disco cuyo eje de rotación es capaz de asumir cualquier orientación por sí mismo.

Half-field current The current going to the field windings of a motor or generator after it has passed through a resistor in series with the circuit.

Corriente de medio campo El corriente que va a los devanados inductores de un motor o a un generador después de que haya pasado por un resistor conectado en serie con el circuito.

Half-wave rectification Rectification of one half of an AC voltage.

Rectificación de media onda Rectificación en la que la corriente fluye únicamente durante semiciclos alternados.

Hall-effect sensor A sensor that operates on the principle that if a current is allowed to flow through a thin conducting material being exposed to a magnetic field, another voltage is produced.

Sensor de efecto Hall Sensor que funciona basado en el principio de que si se permite el flujo de corriente a través de un material conductor delgado que ha sido expuesto a un campo magnético, se produce otra tensión.

Halogen The term used to identify a group of chemically related nonmetallic elements. These elements include chlorine, fluorine, and iodine.

Halógeno Término utilizado para identificar un grupo de elementos no metálicos relacionados químicamente. Dichos elementos incluyen el cloro, el flúor y el yodo.

Hand tools Tools that use only the force generated from the body to operate. They multiply the force received through leverage to accomplish the work.

Herramientas manuales Herramientas que para funcionar sólo necesitan la fuerza generada por el cuerpo. Para llevar a cabo el trabajo, las herramientas multiplican la fuerza que reciben por medio de la palancada.

Haptic The use of the sense of touch.

Háptico Uso del sentido del tacto.

Hazardous material Materials that can cause illness, injury, or death; or pollute water, air, or land.

Material peligroso Las materias que puedan causar la enfermedad, los daños, la muerte o que puedan contaminar el agua, el aire o la tierra.

Headlight leveling system (HLS) Uses front lighting assemblies with a leveling actuator motor to allow the headlights to be adjusted into different vertical positions to compensate for headlight position that can occur when the vehicle is loaded.

Sistema de nivelación de los faros (SNF) Utiliza el ensamblaje de los faros frontales con un motor accionador de nivelación para permitir el ajuste de los faros en diferentes posiciones verticales para compensar la posición que el faro pudiera tomar si se carga el vehículo.

Head-up display (HUD) Displays images onto the inside of the windshield so the driver can see them without having to take his eyes off the road.

Presentación en pantalla (PEP) Proyecta las imágenes en la parte inferior del parabrisas para que el conductor las pueda ver sin tener que quitar la vista de la pista.

Heat sink An object that absorbs and dissipates heat to another object.

Dispersador térmico Objeto que absorbe y disipa el calor de otro objeto.

Heated windshield system A specially designed windshield that allows current flow through the glass without interfering with the driver's vision; it is capable of melting ice and frost from the windshield three to five times faster than conventional defroster systems.

Sistema de parabrisas térmico Parabrisas especialmente diseñado para permitir el flujo de la corriente a través del vidrio sin interferir con la visión del conductor; está capacitado para derretir el hielo y la escarcha que haya en el parabrisas de 3 a 5 veces más rápido que los sistemas convencionales anticongelantes.

Heater core flow valve Shuts off the coolant flow through the heater core when the A/C system is in the max air mode.

Válvula del flujo térmico del núcleo Cierra el flujo del enfriador a través del núcleo del calentador cuando el sistema AC está en el mando de aire máximo.

H-gate A set of four transistors that can reverse current.

Compuerta H Juego de cuatro transistores que pueden invertir la corriente.

HID High-intensity discharge; a lighting system that uses an arc across electrodes instead of a filament.

HID Descarga de Alta Intensidad; un sistema de iluminación que utiliza un arco por dos electrodos en vez de un filamento.

High-intensity discharge (HID) Uses an inert gas to amplify the light produced by arcing across two electrodes.

Descarga de alta intensidad (HID) Usa un gas inerte para amplificar la luz producida al conectar dos electrodos con una arca.

High-reading coil Gauge coil that is positioned at an angle to the low-reading and bucking coils.

Bobina de lectura de alta tensión Bobina que se posiciona en ángulo con respecto a las bobinas de lectura de baja tensión y las bobinas compensadoras.

High-side drivers Control the output device by varying the positive (12-volt) side.

Impulsores del lado de alto potencial Controlan el dispositivo de salida en variar el lado positivo (12 voltíos).

High-voltage ECU (HV ECU) *See* starter generator control module (SGCM).

UCE de alto voltaje (UCE AV) Vea el módulo de control del generador de arranque (MCGA).

Hoist A lift that is used to raise the entire vehicle.

Elevador Montacargas utilizado para elevar el vehículo en su totalidad.

Holddowns Secure the battery to reduce vibration and to prevent tipping.

Portador Aseguran la batería para disminuir la vibración y prevenir que se vierte.

Hold-in winding A winding that holds the plunger of a solenoid in place after it moves to engage the starter drive.

Devanado de retención Un devanado que posiciona el núcleo móvil de un solenoide después de que mueva para accionar el acoplamiento del motor de arranque.

Hole The absence of an electron in an element's atom. These holes are said to be positively charged since they have a tendency to attract free electrons into the hole.

Agujero Ausencia de un electrón en el átomo de un elemento. Se dice que dichos agujeros tienen una carga positiva puesto que tienden a atraer electrones libres hacia el agujero.

Horn A device that produces an audible warning signal.

Claxon Un dispositivo que produce una señal de advertencia audible.

Hybrid air bag Modules use compressed gas to fill the air bag instead of burning a chemical to produce gas.

Bolsa de aire híbrido Los modulos usan el gas comprimido para llenar la bolsa de aire en vez de quemar una química para producir un gas.

Hybrid battery A battery that combines the advantages of low-maintenance and maintenance-free batteries.

Batería híbrida Una batería que combina las ventajas de las baterías de bajo mantenimiento y de no mantenimiento.

Hybrid electric vehicle (HEV) System that has two different power sources. In most HEV, the power sources are a small displacement gasoline or diesel engine and an electric motor.

Vehículo eléctrico híbrido (VEH) Sistema con dos diferentes fuentes de potencia. En la mayoría de los VEH, las fuentes de potencia son un motor de diesel o de gasolina de desplazamiento menor y un motor eléctrico.

Hydrometer A test instrument used to check the specific gravity of the electrolyte to determine the battery's state of charge.

Hidrómetro Instrumento de prueba utilizado para verificar la gravedad específica del electrolito y así determinar el estado de la carga de la batería.

Igniter 1. A combustible device that converts electric energy into thermal energy to ignite the inflator propellant in an air bag system. 2. An electronic device used in an HID headlight to provide the initial 10,000 to 25,000 volts (V) to jump the gap of the electrodes.

Encendedor 1. Dispositivo combustible que convierte la energía eléctrica en energía térmica para encender el propelente inflador del sistema de bolsas de aire. 2. Dispositivo electrónico usado en un faro DAI para proporcionar los 10,000 a 25,000 V iniciales requeridos para saltar la separación de los electrodos.

Ignition coil A step-up transformer that builds up the low-battery voltage of approximately 12.6 volts to a voltage that is high enough to jump across the spark plug gap and ignite the air–fuel mixture.

Bobina de encendido Transformador multiplicador que sube el bajo voltaje de la batería de aproximadamente 12.6 voltios a uno lo suficientemente alto para brincar sobre el huelgo de la bujía y encender la mezcla de aire y combustible.

Ignition system Responsible for delivering properly timed high-voltage surges to the spark plugs.

Sistema de encendido Es responsable de llevar subidas de alto voltaje reguladas apropiadamente a las bujías.

Ignitor Component of the HID headlight that is used to create a high-voltage arc across the lamp electrodes to ignite the gas.

Encendedor Componente del faro DAI que se utiliza para crear un arco de alto voltaje a través de los electrodos de la lámpara para encender el gas.

Injection laser diode (ILD) Lasers that use a forward-biased semiconductor junction as the gain medium.

Diodo láser (DL) Láser que usa una unión de polarización directa del semiconductor como medio activo.

Illuminated entry systems Turns on the courtesy lights before the doors are opened.

Sistemas de entrada iluminada Enciendan las luces de cortesía antes de que se abren las puertas.

Immobilizer system Designed to provide protection against unauthorized vehicle use by disabling engine starting if an invalid key is used to start the vehicle or if an attempt to hot-wire the ignition system is made.

Sistema de inmovilización Sistema diseñado para proporcionar protección contra el uso no autorizado del vehículo que desactiva el motor si una llave no válida se utiliza para encender el motor o para hacer un puente con la intención de encender el motor sin la llave correspondiente.

Incandescence The process of changing energy forms to produce light.

Incandescencia Proceso a través del cual se cambian las formas de energía para producir luz.

Induced voltage Voltage that is produced in a conductor as a result of relative motion within magnetic flux lines.

Tensión inducida Tensión producida en un conductor como resultado del movimiento relativo dentro de líneas de flujo magnético.

Induction The magnetic process of producing a current flow in a wire without any actual contact to the wire. To induce 1 volt, 100 million magnetic lines of force must be cut per second.

Inducción Proceso magnético a través del cual se produce un flujo de corriente en un alambre sin contacto real alguno con el alambre. Para inducir 1 voltio, deben producirse 100 millones de líneas de fuerza magnética por segundo.

Induction motor An AC motor that generates its own rotor current as the rotor cuts the magnetic flux lines of the stator field.

Motor de inducción Un motor de CA que genera su propia corriente de rotor a medida que el rotor corta las líneas de flujo magnético del campo del estator.

Inductive reactance The result of current flowing through a conductor and the resultant magnetic field around the conductor that opposes the normal flow of current.

Reactancia inductiva El resultado de un corriente que circule por un conductor y que resulta en un campo magnético alrededor del conductor que opone el flujo normal del corriente.

Inductive reluctance A statement of a material's ability to strengthen the magnetic field around it.

Reluctancia a la inducción Una indicación de la habilidad de una materia en reenforzar el campo que la rodea.

Inertia The tendency of an object that is at rest and an object that is in motion to stay in motion.

Inercia La tendencia de un objeto que esta en descanso quedarse en descanso y un objeto en movimiento de quedarseen movimiento.

Inertia engagement A type of starter motor that uses rotating inertia to engage the drive pinion with the engine flywheel.

Conexión por inercia Tipo de motor de arranque que utiliza inercia giratoria para engranar el piñon de mando con el volante de la máquina.

Inertia lock retractors Use a pendulum mechanism to lock the seat belt tightly during sudden movement.

Retractores de cierre tipo inercia Usan un mecanismo de péndulo para enclavar fuertemente la cinta durante un movimiento repentino.

Inflatable knee blocker (IKB) A small air bag that deploys simultaneously with the driver's-side airbag to provide upper-leg protection and positioning of the driver.

Bloqueante inflable de la rodilla (BIR) Bolsa de aire pequeña que se despliega simultáneamente con la bolsa de aire al lado del conductor para darle protección en la parte superior de la pierna y posicionar al conductor.

Infotainment A type of media system that provides a combination of information and entertainment.

Infotenimiento Un tipo de sistema de medios que brinda una combinación de información y entretenimiento.

Infrared temperature sensor A senor that measures the surface temperature of an object or person by measuring the intensity of the energy given off by an object.

Sensor infrarrojo para la temperatura Sensor que mide la temperatura de la superficie de un objeto o persona al medir la intensidad de la energía que desprende un objeto.

Esquemas de instalación Proveen una duplicación más precisa de donde se encuentran el cableado preformado, los conectores, y los componentes en el vehículo.

Instrument voltage regulator (IVR) Provides a constant voltage to the gauge, regardless of the voltage output of the charging system.

Instrumento regulador de tensión (IRT) Le provee tensión constante al calibrador, sin importar cual sea la salida de tensión del sistema de carga.

Insulated side The portion of the circuit from the positive side of the source to the load component.

Lado aislado Parte del circuito que va del lado positivo de la fuente al componente de carga.

Insulator A material that does not allow electrons to flow easily through it.

Aislador Una material que no permite circular fácilmente los electrones.

Integrated circuit (IC) A complex circuit of thousands of transistors, diodes, resistors, capacitors, and other electronic devices that are formed into a small silicon chip. As many as 30,000 transistors can be placed on a chip that is 1/4 inch (6.35 mm) square.

Circuito integrado (CI) Circuito complejo de miles de transistores, diodos, resistores, condensadores, y otros dispositivos electrónicos formados en un fragmento pequeño de silicio. En un fragmento de 1/4 de pulgada (6,35 mm) cuadrada, pueden colocarse hasta 30.000 transistores.

Integrated starter generator (ISG) A combination starter generator in one unit that attaches directly to the crankshaft to allow for the automatic stop/start function of an HEV. It can also convert kinetic energy to DC voltage when the vehicle is traveling downhill and there is zero load.

Generador de arranque integrado (GAI) Combinación de generador de arranque en una unidad que se adhiere directamente al cigüeñal para permitir la función automática de encendido y apagado de un VEH. También puede convertir la energía cinemática a voltaje de corriente continua cuando el vehículo va de bajada y no lleva carga.

Intelligent windshield wipers A wiper system that uses a monitoring system to detect if water is present on the windshield and that automatically turns on the wiper system.

Limpiaparabrisas inteligente Sistema de limpiadores que utiliza un sistema de monitoreo para detectar si hay agua en el parabrisas, y esto hace que automáticamente se encienda el sistema de los limpiadores.

Interface Used to protect the computer from excessive voltage levels and to translate input and output signals.

Interfase Utilizada para proteger la computadora de niveles excesivos de tensión y traducir señales de entrada y salida.

Interfaced generator A generator that has an internal eight-bit microprocessor that is a slave module on the local interconnect network (LIN) bus.

Generador con interfases Generador con un microprocesador interno de ocho bits que actúa como módulo esclavo en un bus de red de interconexión local o LIN.

International Standards Organization (ISO) As relating to gauges, identifies standardized symbols used to represent the gauge function.

Organización Internacional de Normalización (ISO, según su sigla en inglés) En lo que respecta a calibradores, identifica los símbolos normalizados usados para representar el funcionamiento de los calibradores.

In-vehicle sensor The in-vehicle sensor contains a temperature-sensing NTC thermistor to measure the average temperature inside the vehicle.

Sensor en el vehículo El sensor dentro del vehículo contiene un termostato de coeficiente de temperatura negativo (CTN) para percibir la temperatura que mide la temperatura promedio dentro del vehículo.

Ion An atom or group of atoms that has an electrical charge.

Ion Átomo o grupo de átomos que poseen una carga eléctrica.

Ionize To electrically charge.

Ionizar Cargar eléctricamente.

ISO An abbreviation for International Standards Organizations.

ISO Una abreviación de las Organizaciones de Normas Internacionales.

ISO 14230-4 A bus data protocol that uses a single-wire bidirectional data line to communicate between the scan tool and the nodes. This data bus is used only for diagnostics and maintains the ISO 9141 protocol with a baud rate of 10.4 kb/s.

ISO 14230-4 Protocolo de un bus de datos que utiliza una línea de datos en dos direcciones en un hilo sencillo para comunicarse entre el instrumento de exploración y los nodos. Este bus de datos se utiliza solamente para diagnósticos y mantiene el protocolo de ISO 9141 con una velocidad de transmisión de baudios de 10.4 kb/s.

ISO 9141-2 A class B system with a baud rate of 10.4 kb/s used only for diagnostic purposes between the nodes on the data bus and an OBD II standardized scan tool.

ISO 9141-2 Sistema B de clase A con una velocidad de transmisión de baudios de 10.4 kb/s que se utiliza sólo con un propósito de diagnóstico entre los nodos en el bus de datos y un instrumento de exploración estandarizado del sistema de diagnóstico a bordo II o DAB II.

ISO-K An adoption of the ISO 9141-2 protocol that allows for bidirectional communication on a single wire. Vehicles that use the ISO-K bus require that the scan tool provide the bias voltage to power up the system.

ISO-K La adopción del protocolo del ISO 9141-2 es que permite la comunicación en dos direcciones en un hilo sencillo. Los vehículos que utilizan el bus de ISO-K requieren que el instrumento de exploración proporcione la tensión de polarización para hacer funcionar el sistema.

ISO relays Conform to the specifications of the International Standards Organization (ISO) for common size and terminal patterns.

Relés ISO Conforman a las especificaciones de la Organización Internacional de Normas (ISO) en tamaño normal y conformidades de terminales.

J1850 The bus system that is the class B standard for OBD II. The J1850 standard allows for two different versions based on baud rate. The first supports a baud rate of 41.6 kb/s, which is transmitted by a pulse-width–modulated (PWM) signal over a twisted pair of wires. The second protocol supports a baud rate of 10.4 kb/s average, which is transmitted by a variable pulse width (VPW) data bus over a single wire.

J1850 El sistema de bus que es el estándar de clase B para el sistema de diagnóstico a bordo II o DAB II. El estándar J1850 permite dos diferentes versiones que se basan en la velocidad de transmisión de baudios. El primero respalda una velocidad de transmisión de baudios de 41.6 kb/s que transmite una señal de modulación de duración de impulsos (MDI o PWM) mediante un par torcido de hilos. El segundo protocolo respalda una velocidad de transmisión de baudios de 10.4 kb/s promedio que transmite un bus de datos de anchura variada entre impulsos (AVI o VPW) mediante un hilo sencillo.

Keyless entry system A lock system that allows for locking and unlocking of a vehicle with a touch keypad instead of a key.

Entrada sin llave Un sistema de cerradura que permite cerrar y abrir un vehículo por medio de un teclado en vez de utilizar una llave.

Keyless start system System that allows an engine to start without using an ignition key.

Sistema de arranque sin llave Sistema que permite arrancar un motor sin utilizar una llave de encendido.

Kirchhoff's current law Electrical law which states that the total flow of current in a parallel circuit will be the sum of the currents through all the circuit branches.

Ley de corriente de Kirchhoff Ley eléctrica que afirma que el flujo total de corriente en un circuito paralelo será la suma de las corrientes a través de todas las ramas del circuito.

Kirchhoff's voltage law Electrical law which states that the sum of the voltage drops in a series circuit must equal the source voltage.

Ley de voltaje de Kirchhoff Ley eléctrica que afirma que la suma de las caídas de voltaje en un circuito en serie debe ser igual al voltaje fuente.

K-line One circuit of the ISO 9141-2 data bus that is used for transmitting data from the module to the scan tool. The scan tool provides the bias voltage onto this circuit and the module pulls the voltage low to transmit its data.

Línea K Un circuito del bus de datos ISO 9141-2 que se utiliza para transmitir datos de un módulo a un instrumento de exploración. El instrumento de exploración proporciona la tensión de polarización sobre este circuito, y el módulo baja el voltaje para transmitir sus datos.

Laminated construction Construction of the armature from individual stampings.

Construcción laminada La armadura esta construida de un matrizado individual.

Lamination The process of constructing something with layers of materials that are firmly connected.

Laminación El proceso de construir algo de capas de materiales unidas con mucha fuerza.

Lamp A device that produces light as a result of current flow through a filament. The filament is enclosed within a glass envelope and is a type of resistance wire that is generally made from tungsten.

Lámpara Dispositivo que produce luz como resultado del flujo de corriente a través de un filamento. El filamento es un tipo de alambre de resistencia hecho por lo general de tungsteno, que es encerrado dentro de una bombilla.

Lamp outage module A current-measuring sensor that contains a set of resistors, wired in series with the power supply to the headlights, taillights, and stop lights. If the sensor indicates that a lamp is burned out, the module will alert the driver.

Unidad de avería de la lámpara Sensor para medir corriente que incluye un juego de resistores, alambrado en serie con la fuente de alimentación a los faros delanteros, traseros y a las luces de freno. Si el sensor indica que se ha apagado una lámpara, la unidad le avisará al conductor.

Lane departure warning (LDW) A vehicle safety system that uses a forward-facing camera, warning buzzers, haptic feedback in the steering wheel, and/or tugging on the seatbelt if the vehicle is about to leave its designated lane.

Advertencia de salida de carril (LDW) Un sistema de seguridad del vehículo que utiliza una cámara orientada hacia delante, zumbadores de advertencia, retroalimentación háptica en el volante y/o tirón en el cinturón de seguridad si el vehículo está a punto de abandonar su carril designado.

Laser Acronym for light amplification by stimulated emission of radiation. A device that controls the way that energized atoms release photons (light).

Láser Acrónimo que en inglés significa amplificación de la luz por emisión estimulada de radiación. Dispositivo que controla la manera en que los átomos energizados liberan fotones (luz).

Laser diodes Light-emitting devices capable of emitting laser beams.

Diodos láser Dispositivos que emiten luz y son capaces de emitir rayos láser.

Leading edge The edges of the rotating blade that enter the switch in a Hall-effect switch.

Borde anterior Los bordes de la ala giratorio que entran al interruptor en un terruptor efecto Hall.

Lean burn technology Uses lean air–fuel ratios to increase fuel efficiency.

Tecnología de quema limpia Determina las relaciones aire-combustible limpios para aumentar la eficacia del combustible.

Light-emitting diode (LED) A gallium-arsenide diode that converts the energy developed when holes and electrons collide during normal diode operation into light.

Diodo emisor de luz (o LED, por su sigla en inglés) Diodo semiconductor de galio y arseniuro que convierte en luz la energía producida por la colisión de agujeros y electrones durante el funcionamiento normal del diodo.

Light Detecting and Ranging (LIDAR) Determines ranges by time of flight and 3-D scanning using spinning lasers to create a 360-degree image based on distance and light.

Dispositivo de detección de luz y determinación de rangos (DLDR) Dispositivo que determina rangos por tiempo de vuelo y escaneo 3-D con láser rotatorio para crear una imagen de 360 grados basada en la distancia y la luz.

Lighting system Electrical system that consists of all of the lights used on the vehicle, including headlights, front and rear park lights, front and rear turn signals, side marker lights, daytime running lights, cornering lights, brake lights, back-up lights, instrument cluster backlighting, and interior lighting.

Sistema de iluminación Sistema eléctrico que consta de todas las luces que usa el vehículo, incluyendo los faros, las luces frontales y traseras de estacionamiento, las luces intermitentes frontales y traseras, luces de posición, luces de marcha diurna, luces de esquina?, luces de freno, luces de marcha atrás, iluminación trasera de tablero de controles e iluminación interior.

Limit switch A switch used to open a circuit when a predetermined value is reached. Limit switches are normally responsive to a mechanical movement or temperature changes.

Disyuntor de seguridad Un conmutador que se emplea para abrir un circuito al alcanzar un valor predeterminado. Los disyuntores de seguridad suelen ser responsivos a un movimiento mecánico o a los cambios de temperatura.

Linearity Refers to the sensor signal being as constantly proportional to the measured value as possible. It is an expression of the sensor's accuracy.

Linealidad Significa que la variación del valor de una magnitud es lo más proporcional posible a la variación del valor de otra magnitud. Expresa la precisión del sensor.

Linear network Bus communication network that connects the nodes in series so all modules between the sending and the receiving module are involved in the data transmission.

Red lineal Red de comunicación en bus que conecta los nodos en serie de manera que todos los módulos entre el módulo emisor y receptor intervienen en la transmisión de datos.

Liquid crystal display (LCD) A display that sandwiches electrodes and polarized fluid between layers of glass. When voltage is applied to the electrodes, the light slots of the fluid are rearranged to allow light to pass through.

Visualizador de cristal líquido (LCD, según su sigla en inglés) Visualizador digital que consta de dos láminas de vidrio selladas, entre las cuales se encuentran los electrodos y el fluido polarizado. Cuando se aplica tensión a los electrodos, se rompe la disposición de las moléculas para permitir la formación de caracteres visibles.

L-line One circuit of the ISO 9141-2 data bus that is used by the module to receive data from the scan tool. The module provides the bias onto this circuit and the scan tool pulls the voltage low to communicate.

Línea L Un circuito del bus de datos del ISO 9141-2 que utiliza un módulo que recibe datos de un instrumento de exploración. El módulo proporciona la tensión sobre este circuito y el instrumento de exploración baja el voltaje para comunicarse.

Load device The component that performs some form of work.

Dispositivo de carga El componente que lleva a cabo algun forma de trabajo.

Load shedding The process of reducing electrical power consumption by turning off non-essential electrical loads.

Desconexión de cargas Proceso por el que se reduce el consumo de energía eléctrica desconectando cualquier carga eléctrica que no sea esencial.

Local interconnect network (LIN) A bus network that was developed to supplement the CAN bus system. The term *local interconnect* refers to all of the modules in the LIN network being located within a limited area.

Red local de interconexiones (LIN, según su sigla en inglés) Una red de bus que se desarrolló para complementar el sistema de bus CAN. El término "interconexión local" se refiere a todos los módulos en la red de LIN que se encuentran dentro de un área.

Logic gates Electronic circuits that act as gates to output voltage signals depending on different combinations of input signals.

Compuertas lógicas Circuitos electrónicos que gobiernan señales de tensión de salida, dependiendo de las diferentes combinaciones de señales de entrada.

Look-up tables The part of a microprocessor's memory that indicates how a system should perform in the form of calibrations and specifications.

Tablas de referencia La parte de la memoria de una microprocesora que indica como debe ejecutar las calibraciones y las especificaciones la sistema.

Low-reading coil A coil of the three-coil gauge that is wound together with the bucking coil but in the opposite direction.

Bobina de lectura de baja tensión Bobina del calibrador de tres bobinas enrollada con la bobina compensadora pero en la dirección opuesta.

Low-side drivers Transistor used to complete the path to ground to turn on an actuator.

Controlador del lado a tierra Transistor usado para completar el circuito a tierra y activar un actuador.

Low voltage differential signaling (LVDS) A signaling method used for high-speed bus transmission of binary data that is used in point- to-point configurations for visual displays.

Señalización diferencial de baja tensión (SDBT) Método de señalización utilizado para la transmisión de datos binarios de buses a alta velocidad que se utiliza en configuraciones punto a punto para pantallas visuales.

LUX The International System unit of measurement of the intensity of light. It is equal to the illumination of a surface 1 meter away from a single candle (1 lumen per square meter).

LUX (lumen por metro cuadrado) Unidad del sistema internacional de medida de la intensidad de la luz. Es igual a la iluminación de una superficie a un metro de distancia de una vela sencilla (un lumen por metro cuadrado).

Magnetic field The area surrounding a magnet where energy is exerted due to the atoms aligning in the material.

Campo magnético Espacio que rodea un imán donde se emplea la energía debido a la alineación de los átomos en el material.

Magnetic flux density The concentration of the magnetic lines of force.

Densidad de flujo magnético Número de líneas de fuerza magnética.

Magnetic pulse generators Sensors that uses the principle of magnetic induction to produce a voltage signal. Commonly used to send data to the computer concerning the speed of the monitored component.

Generadoress de impulsos magnéticos Sensores que funcionan según el principio de inducción magnética para producir una señal de tensión. Los generadores de impulsos magnéticos se utilizan comúnmente para transmitir a la computadora datos relacionados con la velocidad del componente regulado.

Magnetically coupled linear sensors Sensors that can be used to measure movement by use of a moveable magnet, a resistor card, and a magnetically sensitive comb. Changes in the location of the magnet on the card provide a variable output.

Sensores lineales acoplados magnéticamente Sensor que se puede utilizar para medir el movimiento mediante el uso de un imán móvil, una tarjeta de resistor y un peine sensible al magnetismo. Los cambios en la ubicación del imán en la tarjeta proporcionan información variable.

Magnetism An energy form resulting from atoms aligning within certain materials, giving the materials the ability to attract other metals.

Magnetismo Forma de energía que resulta de la alineación de átomos dentro de ciertos materiales y que le da a éstos la capacidad de atraer otros metales.

Magnetoresistive effect The increase in resistance of a current-carrying magnetic substance when an external magnetic field is applied perpendicular to its surface.

Efecto magneto resistivo El aumento en la resistencia de una substancia magnética transportadora de corriente cuando se aplica un campo magnético externo, perpendicular a su superficie.

Magnetoresistive (MR) sensors Speed detection sensor consisting of a permanent magnet and an integrated signal conditioning circuit to change resistance due to the relationship of the tone wheel and magnetic field surrounding the sensor. The change in resistance results in a digital reading of current levels by the control module.

Sensores de magneto resistivos (MR) Sensor de detección de velocidad que está compuesto por un imán permanente y un circuito de acondicionamiento de señales integradas para cambiar la resistencia debido a la relación entre la rueda de virado y el campo magnético que rodea al sensor. El cambio en la resistencia tiene como resultado una lectura digital de los niveles actuales por parte del módulo de control.

Maintenance-free battery A battery that has no provision for the addition of water to the cells. The battery is sealed.

Sin mantención Que no tiene provisión para añadir el agua a la células. Es una batería sellada.

Master module Controller on the network that translates messages between different network systems.

Instancia maestra Controlador o entidad única, dentro de una red distribuida, que traduce los mensajes entre los diferentes sistemas de la red.

Material expanders Fillers that can be used in place of the active materials in a battery. They are used to keep the cost of manufacturing low.

Expansores de materias Los rellenos que se pueden usar en vez de las materiales activas de una bateria. Se emplean para mantener bajos los costos de la fabricación.

Matrix A rectangular array of grids.

Matriz Red lógica en una rejilla de forma rectangular.

Maxi-fuse A circuit protection device that looks similar to a blade-type fuse except that it is larger and has a higher amperage capacity. Maxi-fuses are used because they are less likely to cause an under-hood fire when there is an overload in the circuit. If the fusible link burned in two, it is possible that the hot side of the fuse could come into contact with the vehicle frame and the wire could catch fire.

Maxifusible Dispositivo de protección del circuito parecido a un fusible de tipo de cuchilla, pero más grande y con mayor capacidad de amperaje. Se utilizan maxifusibles porque existen menos probabilidades de que ocasionen un incendio debajo de la capota cuando ocurra una sobrecarga en el circuito. Si el cartucho de fusible se quemase en dos partes, es posible que el lado "cargado" del fusible entre en contacto con el armazón del vehículo y que el alambre se encienda.

Media-oriented system transport (MOST) A data bus system based on standards established by a cooperative effort between automobile manufacturers, suppliers, and software programmers that resulted in a data system specifically designed for the data transmission of media-oriented data. MOST uses fiber optics to transmit data at a rate up to 25 Mb/s.

MOST Sistema de bus de datos basado en estándares establecidos por un esfuerzo cooperativo entre los fabricantes de vehículos, los proveedores y los programadores de software que resultó en un sistema de datos específicamente diseñado para la transmisión de datos informativos. MOST utiliza fibras ópticas para transmitir datos a una velocidad de 25 megabitios por segundo (25Mb/s).

Memory effect The battery failing to fully charge because it remembers its previous charge level. This results in a low-battery charge due to a battery that is not completely discharged before it is recharged.

Efecto memoria Fallas en la carga completa de la batería debido a que "recuerda" su nivel de carga previo. En consecuencia, se produce una baja carga de la batería ya que ésta no está completamente descargada antes de la recarga.

Memory seat Power seats that can be programmed to return or adjust to a point designated by the driver.

Asientos con memoria Los asientos automáticos que se pueden programar a regresar o ajustarse a un punto indicado por el conductor.

Metri-pack connector Special wire connectors used in some computer circuits. They seal the wire terminals from the atmosphere, thereby preventing corrosion and other damage.

Conector metri-pack Los conectores de alambres especiales que se emplean en algunos circuitos de computadoras. Impermealizan los bornes de los alambres, así previniendo la corrosión y otros daños.

Microprocessor The brains of the computer where most calculations take place.

Microprocesador El cerebro de la computadora en donde se realizan la mayoría de los cálculos.

Micro-Electro-Mechanical System (MEMS) A technology that combines miniaturized mechanical and electro-mechanical elements using the techniques of microfabrication.

Sistema microelectromecánico (SMEM) Una tecnología que combina elementos mecánicos y electromecánicos miniaturizados utilizando las técnicas de microfabricación.

Mild parallel hybrid Uses an electric motor that is large enough to provide regenerative braking, instant engine start-up, and a boost to the combustion engine.

Híbrido de medio paralelo Utiliza un motor eléctrico que es lo suficientemente grande para proveer freno regenerativo, encendido instantáneo del motor y un aumento a la combustión del motor.

Millimeter-wave radar sensors Provide stable detection of targets even under inclement weather conditions using 77 GHz Frequency Modulated-Continuous Wave (FM-CW) radar technology. The sensor can measure the distance from a target and its velocity simultaneously by mixing the transmitted frequency modulated millimeter-wave signal and the reflected signal from the target.

Sensores de radar de onda milimétrica Ofrecen la detección estable de objetivos incluso en condiciones meteorológicas desfavorables usando tecnología de radar de onda continua modulada en frecuencia. El sensor mide la distancia desde un objetivo y su velocidad combinando simultáneamente la señal de onda milimétrica modulada en frecuencia transmitida con la señal reflejada por el objetivo.

Mode door actuator An electric motor that is linked to the mode door to supply air flow to the floor ducts, A/C panel ducts, or defrost ducts.

Actuador de mando puerta Motor eléctrico que está unido al mando puerta para suministrar flujo de aire a los conductos del piso, los conductos del panel de corriente alterna o a los conductos de descongelamiento.

Momentary contact A switch type that operates only when held in position.

Contacto momentáneo Tipo de conmutador que funciona solamente cuando se mantiene en su posición.

Multiplexer An electronic switch that switches between the different audio sources.

Multiplexor Interruptor electrónico que se usa para hacer cambios entre las diferentes fuentes de audio.

Multiplexing (MUX) A means of transmitting information between computers. It is a system in which electrical signals are transmitted by a peripheral serial bus instead of conventional wires, allowing several devices to share signals on a common conductor.

Multiplexaje Medio de transmitir información entre computadoras. Es un sistema en el cual las señales eléctricas son transmitidas por una colectora periférica en serie en vez de por líneas convencionales. Esto permite que varios dispositivos compartan señales en un conductor común.

Multistage air bags Hybrid air bags that use two squibs to control the rate of inflation.

Bolsas de aire de etapas múltiples Las bolsas de aire híbridas que usan dos petardos para controlar la velocidad de la inflación.

Mutual induction An induction of voltage in an adjacent coil by changing current in a primary coil.

Inducción mutua Una inducción de la tensión en una bobina adyacente que se efectúa al cambiar la tensión en una bobina primaria.

MUX Common acronym for multiplexing.

MUX Una sigla común del proceso de multiplex.

Navigational system Use satellites to direct drivers to desired destinations.

Sistema de navegación Usa los satélites para dirigir el conductor a la destinación deseada.

Negative logic Defines the most negative voltage as a logical 1 in the binary code.

Lógica negativa Define la tensión más negativa como un 1 lógico en el código binario.

Negative temperature coefficient (NTC) thermistors Thermistors that reduce their resistance as the temperature increases.

Termistores con coeficiente negativo de temperatura Termistores que disminuyen su resistencia según aumenta la temperatura.

Nematic Describes a fluid that is a liquid crystal with a threadlike form. It has light slots that can be rearranged by applying small amounts of voltage.

Nemático Describe un flúido que es un cristal líquido con una forma de filamento. Tiene aberturas de luz que se pueden reorganizar por medio de la aplicación de pequeñas cantidades de voltaje.

Neon lights A light that contains a colorless, odorless inert gas called neon. These lamps are discharge lamps.

Luces de neón Una luz que contiene un gas inerto sin color, inodoro llamado neón. Estas lámparas son lámparas de descarga.

Network Incorporating the vehicle's electrical systems together through computers so information gathered by one system can be used by another.

En red Incorporar los sistemas eléctricos del vehículo mediante el uso de computadoras para que la información que obtenga un sistema pueda usarla otro sistema.

Network architecture Describes how modules are connected on a network. Common types include linear, stub, ring, and star.

Arquitectura de red Describe la forma en que se conectan los módulos en una red. Los tipos comunes son: lineal, aislada, anillo y estrella.

Neutral atom *See* balanced atom.

Átomo neutro *Véase* átomo equilibrado.

Neutral junction The center connection to which the common ends of a Y-type stator winding are connected.

Empalme neutro Conexión central a la cual se conectan los extremos comunes de un devanado del estátor de tipo Y.

Neutral safety switch A switch used to prevent the starting of an engine unless the transmission is in PARK or NEUTRAL.

Disyuntor de seguridad en neutral Un conmutador que se emplea para prevenir que arranque un motor al menos de que la transmisión esté en posición PARK o Neutral.

Neutrons Particles of an atom that have no charge.

Neutrones Partículas de un átomo desprovistas de carga.

Night vision Describes the ability to see objects at night by use of infrared cameras.

Visión nocturna Describe la capacidad de ver objetos en la noche mediante el uso de cámaras infrarrojas.

Node A computer that is connected to a data bus network and capable of sending or receiving messages.

Nodo Computadora conectada a una red de bus de datos y con capacidad de mandar o recibir mensajes.

Nonvolatile RAM memory that will retain its memory if battery voltage is disconnected. NVRAM is a combination of RAM and EEPROM into the same chip. During normal operation, data is written to and read from the RAM portion of the chip. If the power is removed from the chip, or at programmed timed intervals, the data is transferred from RAM to the EEPROM portion of the chip. When the power is restored to the chip, the EEPROM will write the data back to the RAM.

Memoria de acceso aleatorio no volátil (NV RAM) Memoria de acceso aleatorio (RAM) que retiene su memoria si se desconecta la carga de la batería. La NV RAM es una combinación de RAM y EEPROM en el mismo fragmento. Durante el funcionamiento normal, los datos se escriben en y se leen de la parte RAM del fragmento. Si se remueve la alimentación del fragmento, o si se remueve ésta a intervalos programados, se transfieren los datos de la RAM a la parte del EEPROM del fragmento. Cuando se restaura la alimentación en el fragmento, el EEPROM volverá a escribir los datos en la RAM.

Normally closed (NC) switch A switch designation denoting that the contacts are closed until acted upon by an outside force.

Conmutador normalmente cerrado Nombre aplicado a un conmutador cuyos contactos permanecerán cerrados hasta que sean accionados por una fuerza exterior.

Normally open (NO) switch A switch designation denoting that the contacts are open until acted upon by an outside force.

Conmutador normalmente abierto Nombre aplicado a un conmutador cuyos contactos permanecerán abiertos hasta que sean accionados por una fuerza exterior.

N-type material When there are free electrons, the material is called an N-type material. The N means negative and indicates that it is the negative side of the circuit that pushes electrons through the semiconductor and the positive side that attracts the free electrons.

Material tipo N Al material se le llama material tipo N cuando hay electrones libres. La N significa negativo e indica que el lado negativo del circuito empuja los electrones a través del semiconductor y el lado positivo atrae los electrones libres.

Nucleus The core of an atom that contains protons and neutrons.

Núcleo Parte central de un átomo que contiene los protones y los neutrones.

Occupant classification systems A mandated requirement to reduce the risk of injuries resulting from air bag deployment by determining the weight classification of the front-seat passenger.

Sistemas de clasificación de los ocupantes Mandato para reducir el riesgo de daños que resulten del desarrollo de la bolsa de aire al determinar la clasificación del peso del pasajero del asiento de enfrente.

Occupational safety glasses Eye protection that is designed with special high-impact lens and frames and provides for side protection.

Gafas de protección para el trabajo Gafas diseñadas con cristales y monturas especiales resistentes y provistas de protección lateral.

OCS service kit Special kit that consists of the seat foam, the bladder, the pressure sensor, the occupant classification module (OCM), and the wiring.

Kit de servicio SCO Kit especial que consiste en hule-espuma del asiento, el depósito, el sensor de presión, el módulo de clasificación del ocupante (MCO) y el alambrado.

OCS validation test A test that confirms that the system can properly classify the occupant.

Prueba de revalidación del SCO Prueba que confirma que el sistema puede clasificar apropiadamente al ocupante.

Odometer A mechanical or digital counter in the speedometer unit indicating total miles accumulated on the vehicle.

Odómetro Contador mecánico o digital del velocímetro que indica el total de millas recorridas por el vehículo.

Odometry Uses data from GPS, XYZ, accelerometers, and other motion detection sensors to estimate the change in position over time.

Odometría Usa datos de GPS, XYZ, acelerómetros y otros sensores de detección del movimiento para estimar el cambio de posición con el tiempo.

Ohms Unit of measure for resistance. One ohm is the resistance of a conductor such that a constant current of 1 amp in it produces 1 volt between its ends.

Ohmio Unidad de resistencia eléctrica. Un ohmio es la resistencia de un conductor si una corriente constante de 1 amperio en el conductor produce una tensión de 1 voltio entre los dos extremos.

Ohmmeter A test meter used to measure resistance and continuity in a circuit.

Ohmiómetro Instrumento de prueba utilizado para medir la resistencia y la continuidad en un circuito.

Ohm's law Defines the relationship between current, voltage, and resistance.

Ley de Ohm Define la relación entre la corriente, la tensión y la resistencia.

Open Term used to describe a break in the circuit. The break can be from a switch in the off position or a physical break in the wire.

Abierto Término utilizado para describir una interrupción en el circuito. La interrupción puede ser debido a un interruptor en la posición de apagado o a una rotura física en el cable.

Open circuit A term used to indicate that current flow is stopped. By opening the circuit, the path for electron flow is broken.

Circuito abierto Interrupción en el circuito eléctrico que causa que pare el flujo de corriente.

Optical cavity The cavity of the laser consisting of two mirrors, one at each end of the cavity. That forces the light to bounce back and forth. The mirrors ensure that a majority of the light must make several passes through the gain medium.

Cavidad óptica Cavidad del láser que consta de dos espejos ubicados respectivamente a cada extremo de la cavidad. Esto hace que la luz rebote hacia atrás y hacia adelante. Los espejos aseguran que la mayor parte de la luz atraviese varias veces el medio activo.

Optical horn A name Chrysler uses to describe their flash-to-pass headlamp system.

Claxón óptico Un nombre que usa Chrysler para describir su sistema de faros "relampaguea para rebasar."

Oscillator Creates a rapid back-and-forth movement of voltage.

Oscilador Crea un movimiento de oscilación rápido de voltaje.

Output driver A transistor circuit that controls the operation of an actuator. It can send a voltage to the actuator or provide the ground for the actuator.

Controlador de salida Un circuito de transistores que controla la operación de un actuador. Puede enviar un voltaje al actuador o proporcionar la puesta a tierra para el actuador.

Output signal A command signal sent from the computer to an actuator.

Señal de salida Una señal de comando que se envía desde la computadora hacia un actuador.

Overload Excess current flow in a circuit.

Sobrecarga Flujo de corriente superior a la que tiene asignada un circuito.

Overrunning clutch A starter drive that uses a roller clutch to transmit torque in one direction and freewheels in the other direction.

Embrague de sobremarcha Una asamblea de embrague en un acoplamiento del motor de arranque que se emplea para prevenir que el volante del motor dé vueltas al armazón del motor de arranque.

Oversteer The tendency of the back of the vehicle to turn on the vehicle's center of gravity and come around the front of the vehicle.

Tener la dirección muy sensible Tendencia de la parte trasera de un vehículo de dar vuelta en el centro de gravedad del vehículo y de doblar al frente del vehículo.

Oxygen sensor A voltage-generating sensor that measures the amount of oxygen present in an engine's exhaust.

Sensor de oxígeno Un sensor generador de tensión que mide la cantidad del oxígeno presente en el gas de escape de un motor.

Parallel circuit A circuit that provides two or more paths for electricity to flow.

Circuito en paralelo Circuito que provee dos o más trayectorias para que circule la electricidad.

Parallel hybrid A hybrid vehicle configuration that has a direct mechanical connection between the engine and the wheels. Both the engine and the electric motor can turn the transmission at the same time.

Híbrido en paralelo completo Utiliza un motor eléctrico que es lo suficientemente potente para que el vehículo se impulse por sí mismo.

Parasitic loads Electrical loads that are still present when the ignition switch is in the OFF position.

Cargas parásitas Cargas eléctricas que todavía se encuentran presente cuando el botón conmutador de encendido está en la posición OFF.

Park contacts Located inside the motor assembly and supply current to the motor after the wiper control switch has been turned to the PARK position. This allows the motor to continue operating until the wipers have reached their PARK position.

Contactos de Park Ubicado dentro de la asamblea del motor y proveen el corriente al motor después de que el interruptor de control de la limpiaparabrisa se ha puesto en la posición de estacionamiento. Esto permite que el motor continua operando hasta que los brazos de la limpiaparabrisas hayan llegado a su posición de estacionamiento.

Park switch Contact points located inside the wiper motor assembly that supply current to the motor after the wiper control switch has been turned to the PARK position. This allows the motor to continue operating until the wipers have reached their PARK position.

Conmutador PARK Puntos de contacto ubicados dentro del conjunto del motor del frotador que le suministran corriente al motor después de que el conmutador para el control de los frotadores haya sido colocado en la posición PARK. Esto permite que el motor continue su funcionamiento hasta que los frotadores hayan alcanzado la posición original.

Park assist system A parking aid that alerts the driver to obstacles located in the path immediately behind or in front of the vehicle.

Sistema de ayuda al aparcamiento Ayuda al aparcamiento que avisa al conductor de los obstáculos que se encuentran en el camino inmediatamente detrás o delante del vehículo.

Pass key A specially designed vehicle key with a coded resistance value. The term *pass* is derived from personal automotive security system.

Llave maestra Una llave vehícular de diseño especial que tiene un valor de resistencia codificado. El termino pass se derive de las palabras Personal Automotive Security System (sistema personal de seguridad automotriz).

Passive restraints systems A passenger restraint system that automatically operates to confine the movement of a vehicle's passengers.

Correas passivas Un sistema de resguardo del pasajero que opera automaticamente para limitar el movimiento de los pasajeros en el vehículo.

Passive suspension systems Use fixed spring rates and shock valving.

Sistemas pasivos de suspensión Utilizan elasticidad de muelle constante y dotación con válvulas amortigadoras.

Peak reverse voltage (PRV) Indicates the maximum reverse-bias voltage that may be applied to a diode without causing junction breakdown.

Voltaje inverso pico (PRV, por su sigla en inglés) Indica el voltaje de polarización inversa máximo que se puede aplicar a un diodo sin causar una avería a un empalme.

Peltier element An element consisting of two different types of metals that are joined together. The area of the joint will generate or absorb heat when an electric current is applied to the element at a specified temperature.

Elemento Peltier Un elemento compuesto por dos tipos diferentes de metales unidos. El área de la unión generará o absorberá calor cuando se aplique una corriente eléctrica al elemento a una temperatura específica.

Permanent magnet gear reduction (PMGR) A starter that uses four or six permanent magnet field assemblies in place of field coils.

Reducción de engranaje de imán permanente (PMGR) Un arrancador que usa cuatro o seis asambleas permanentes de campo magnético en vez de las bobinas de campo.

Permeability Term used to indicate the magnetic conductivity of a substance compared with the conductivity of air. The greater the permeability, the greater the magnetic conductivity and the easier a substance can be magnetized.

Permeabilidad Término utilizado para indicar la aptitud de una sustancia en relación con la del aire, de dar paso a las líneas de fuerza magnética. Mientras mayor sea la permeabilidad, mayor será la conductividad magnética y más fácilmente se comunicará a un cuerpo propiedades magnéticas.

Piezoelectric accelerometers A piezoelectric element constructed of zinc oxide that is distorted during a high G force condition and generates an analog voltage in direct relationship to the amount of G force. Used to sense deceleration forces.

Acelerómetros piezoeléctricos Un elemento piezoeléctrico construido de óxido de zinc que se distorsiona durante una condición de alta fuerza G y genera un voltaje análogo en relación directa con la cantidad de fuerza G. Se utiliza para detectar fuerzas de desaceleración.

Photocell A variable resistor that uses light to change resistance.

Fotocélula Resistor variable que utiliza luz para cambiar la resistencia.

Photoconductive mode Operating mode of the photodiode when an external reverse-bias is applied that results in a small reverse current through the diode.

Modo fotoconductor Modo de operación del fotodiodo cuando se aplica una polarización inversa externa, la cual produce una pequeña corriente inversa a través del diodo.

Photodiode Allows current to flow in the opposite direction of a standard diode when it receives a specific amount of light.

Fotodiodo Permite que fluye el corriente en la dirección opuesta de él de un diodo normal al recibir una cantidad específica de luz.

Photoresistor A passive light-detecting device composed of a semiconductor material that changes resistance when its surfaced is exposed to light.

Fotorresistor Un dispositivo detector de luz pasivo, compuesto por un material semiconductor que cambia su resistencia cuando su superficie se expone a la luz.

Phototransistor A light-detecting transistor that uses the application of light to generate carriers to supply the base leg current.

Fototransistor Un transistor detector de luz que utiliza la aplicación de luz para generar portadores que suministren corriente a la pata base.

Photovoltaic diodes Diodes capable of producing a voltage when exposed to radiant energy.

Diodos fotovoltaicos Diodos capaces de generar una tensión cuando se encuentran expuestos a la energía de radiación.

Photovoltaic mode An operating mode of the photodiode with no bias applied where the generated current or voltage is in the forward direction.

Modo fotovoltaico Un modo de operación del fotodiodo sin aplicar polarización en donde la corriente o voltaje que se genera es en dirección positiva.

Photovoltaics (PV) One of the forces that can be used to generate electrical current by converting solar radiation through the use of solar cells.

Fotovoltaica (PV) Una de las fuerzas que puede utilizarse para generar corriente eléctrica al convertir la radiación solar a través del uso de celdas solares.

Pickup coil The stationary component of the magnetic pulse generator consisting of a weak permanent magnet that has fine wire wound around it. As the timing disc rotates in front of it, the changes in magnetic lines of force generate a small voltage signal in the coil.

Bobina captadora Componente fijo del generador de impulsos magnéticos compuesta de un imán permanente débil devanado con alambre fino. Mientras gira el disco sincronizador enfrente de él, los cambios de las líneas de fuerza magnética generan una pequeña señal de tensión en la bobina.

Piconets Small transmission cells that assist in the organization of data.

Picoredes Pequeñas células de transmisión que ayudan a organizar los datos.

Piezoelectric device A voltage generator with a resistor connected in series that is used to measure fluid and air pressures.

Dispositivo piezoeléctrico Un generador de voltaje con un resistor conectado en series que se utiliza para medir las presiones de fluido y aire.

Piezoelectricity Voltage produced by the application of pressure to certain crystals.

Piezoelectricidad Generación de polarización eléctrica en ciertos cristales a consecuencia de la aplicación de tensiones mecánicas.

Piezoresistive device Similar to a piezoelectric device except that it operates like a variable resistor. Its resistance value changes as the pressure applied to the crystal changes.

Dispositivo piezoresistivo Similar a uno piezoeléctrico excepto porque operan como un resistor variable. Su valor de resistencia cambia a medida que lo hace la presión que se aplica al cristal.

Piezoresistive sensor A sensor that is sensitive to pressure changes.

Sensor piezoresistivo Sensor susceptible a los cambios de presión.

Ping (or denotation) A knocking sound that occurs as two flame fronts collide.

Picado (o detonación) Ruido del impacto que se produce cuando colisionan dos frentes de llama.

Pinion factor A calculation using the final drive ratio and the tire circumference to obtain accurate vehicle speed signals.

Factor de piñón Una calculación que usa la relación de impulso final y la circunferencia de la llanta para obtener unas señales precisad de la velocidad del vehículo.

Pinion gear A small gear; typically refers to the drive gear of a starter drive assembly or the small drive gear in a differential assembly.

Engranaje de piñon Un engranaje pequeño; tipicamente se refiere al engranaje de arranque de una asamblea de motor de arranque o al engranaje de mando pequeño de la asamblea del diferencial.

Plate straps Metal connectors used to connect the positive or negative plates in a battery.

Abrazaderas de la placa Los conectores metálicos que sirven para conectar las placas positivas o negativas de una batería.

Plates The basic structure of a battery cell; each cell has at least one positive plate and one negative plate.

Placas La estructura básica de una celula de batería; cada celula tiene al menos una placa positiva y una placa negativa.

Plug-in hybrid electric vehicle (PHEV) An HEV that increases all-electric mode operation without needing to engage the ICE within a certain range and speed by charging the batteries using a home electric outlet or a public charging station.

Vehículo eléctrico híbrido enchufable (VEHE) VEH que incrementa la operación en modo eléctrico sin necesidad de que el motor de combustión interna funcione en un cierto rango o a una determinada velocidad cargando las baterías mediante un tomacorriente doméstico o una estación de carga pública.

P-material Silicon or germanium that is doped with boron or gallium to create a shortage of electrons.

Material-P Boro o galio añadidos al silicio o al germanio para crear una insuficiencia de electrones.

PMGR An abbreviation for permanent magnet gear reduction.

PMGR Una abreviación de desmultiplicación del engrenaje del imán permanente.

Pneumatic tools Power tools that are powered by compressed air.

Herrimientas neumáticas Herramientas mecánicas accionadas por aire comprimido.

PN junction The point at which two opposite kinds of semiconductor materials are joined together.

Unión pn Zona de unión en la que se conectan dos tipos opuestos de materiales semiconductores.

Polarizers Glass sheets that make light waves vibrate in only one direction. This converts light into polarized light.

Polarizadores Las láminas de vidrio que hacen vibrar las ondas de luz en un sólo sentido. Esto convierte la luz en luz polarizada.

Polarizing The process of light polarization or of setting one end of a field as a positive or negative point.

Polarizadora El proceso de polarización de la luz o de establecer un lado de un campo como un punto positivo o negativo.

Pole The number of input circuits.

Poste El número de los circuitos de entrada.

Pole shoes The components of an electric motor that are made of high-magnetic permeability material to help concentrate and direct the lines of force in the field assembly.

Expansión polar Componentes de un motor eléctrico hechos de material magnético de gran permeabilidad para ayudar a concentrar y dirigir las líneas de fuerza en el conjunto inductor.

Population inversion In a laser, refers to the number of atoms in the excited state being greater than the number in the relaxed state.

Inversión de población En un láser, indica que el número de átomos en estado de excitación es mayor que el número de átomos en estado de relajación.

Positive-engagement starter A type of starter that uses the magnetic field strength of a field winding to engage the starter drive into the flywheel.

Acoplamiento de arranque positivo Un tipo de arrancador que utilisa la fuerza del campo magnético del devanado inductor para accionar el acoplamiento del arrancador en el volante.

Positive plate The plate connected to the positive battery terminal.

Placa positiva La placa conectada al terminal positivo de la batería.

Positive temperature coefficient (PTC) thermistors Thermistors that increase their resistance as temperature increases.

Termistores con coeficiente positivo de temperatura Termistores que aumentan su resistencia según aumenta la temperatura.

Potential The ability to do something; typically voltage is referred to as the potential. If you have voltage, you have the potential for electricity.

Potencial La capacidad de efectuar el trabajo; típicamente se refiere a la tensión como el potencial. Si tiene tensión, tiene la potencial para la electricidad.

Potentiometer A variable resistor that acts as a circuit divider to provide accurate voltage drop readings proportional to movement.

Potenciómetro Resistor variable que actúa como un divisor de circuito para obtener lecturas de perdidas de tensión precisas en proporcion con el movimiento.

Potentiometric pressure sensor Sensor used to measure pressure by use of a Bourdon tube, a capsule, or bellows to move a wiper arm on a resistive element. The movement of the wiper across the resistive element records a different voltage reading.

Sensor de presión potenciométrica Sensor que se utiliza para medir la presión mediante un tubo Bourdon, una cápsula o un tubo flexible ondulado para mover el brazo de un contacto deslizante en un elemento resistivo. El movimiento del contacto deslizante contra el elemento resistivo registrará una lectura de voltaje diferente.

Power (P) The rate of doing electrical work.

Potencia La tasa de habilidad de hacer el trabajo eléctrico.

Power formula A formula used to calculate the amount of electrical power a component uses. The formula is, where P stands for power (measured in watts), I stands for current, and E stands for voltage.

Formula de potencia Una formula que se emplea para calcular la cantidad de potencia eléctrica utilizada por un componente. La formula es, en el que el P quiere decir potencia (medida en wats), I representa el corriente y el E representa la tensión.

Power tools Tools that use forces other than those generated from the body. They can use compressed air, electricity, or hydraulic pressure to generate and multiply force.

Herramientas mecánicas Herramientas que utilizan fuerzas distintas a las generadas por el cuerpo. Dichas fuerzas pueden ser el aire comprimido, la electricidad, o la presión hidráulica para generar y multiplicar la fuerza.

Pressure control solenoid A solenoid used to control the pressure of a fluid, commonly found in electronically controlled transmissions.

Solenoide de control de la presión Un solenoide que controla la presión de un fluido, suele encontrarse en las transmisiones controladas electronicamente.

Pretensioners Used to tighten the seat belt and shoulder harness around the occupant during an accident severe enough to deploy the air bag. Pretensioners can be used on all seat belt assemblies in the vehicle.

Pretensadores Se usan para apretar la cinta de seguridad y el arnés del cuerpo alrededor del ocupante durante un accidente bastante severo como para activar la bolsa de aire. Los pretensadores se pueden usarse en cualquier asamblea de cinta de seguridad del vehículo.

Primary circuit All the components that carry low voltage through the system.

Circuito pirmario Todos los componentes que llevan un voltaje bajo dentro del sistema.

Primary coil winding The second set of winding in the ignition coil. The primary winds will have about 200 turns to create a magnetic field to induce voltage into the secondary winding.

Devanado primario El grupo segundo de devanados en la bobina del encendido. Los devandos primarios tendrán unos 200 vueltas para crear un camp magnético para inducir el voltaje al devanado secundario.

Primary wiring Conductors that carry low voltage and current. The insulation of primary wires is usually thin.

Hilos primarios Hilos conductores de tensión y corriente bajas. El aislamiento de hilos primarios es normalmente delgado.

Printed circuit boards Made of thin phenolic or fiberglass board with copper deposited on it to create current paths. These are used to simplify the wiring of circuits.

Circuito impreso Un circuito hecho de un tablero de fenólico delgado o de fibra de vidrio el cual tiene depósitos del cobre para crear los trayectorios para el corriente. Estos se emplean para simplificar el cableado de los circuitos.

Prism lens A light lens designed with crystal-like patterns, which distort, slant, direct, or color the light that passes through it.

Lente prismático Un lente de luz con diseños cristalinos que distorcionan, inclinan, dirigen o coloran la luz que lo atraviesa.

Prisms Redirect the light beam and create a broad, flat beam.

Prismas Dirigen un rayo de luz y crean un rayo ancho y plano.

Program A set of instructions that the computer must follow to achieve desired results.

Programa Conjunto de instrucciones que la computadora debe seguir para lograr los resultados deseados.

Program number Represents the amount of heating or cooling required to obtain the temperature set by the driver.

Número de programa Representa la cantidad de calefacción o enfriamiento requerido para obtener la temperatura indicado por el conductor.

Programmable Communication Interface (PCI) A single-wire, bidirectional communication bus where each module supplies its own bias voltage and has its own termination resistors. As a message is sent, a variable pulse width modulation (VPWM) voltage between 0 and 7.75 volts is used to represent the 1 and 0 bits.

Interfaz de comunicación programable (PCI) Bus de comunicación en dos direcciones en un hilo sencillo en donde cada módulo proporciona su propia velocidad de transmisión de baudios y tiene sus propias resistencias de unión. Mientras se envía un mensaje, un voltaje de anchura variada entre impulsos entre 0 y 7.75 voltios se utiliza para representar los bits 1 y 0.

Programmer Controls the blower speed, air mix doors, and vacuum motors of the SATC system. Depending on manufacturer, they are also called servo assemblies.

Programador Controla la velocidad del ventilador, las puertas de mezcla de aire y los motores de vacío de un sistema SATC. Según el fabricante, tambien se llaman asambleas servo.

Projector headlight A headlight system using a HID or halgen bulb to produce a very focused light beam.

Faro proyector Un sistema de faro que utiliza un HID o bulbo de halógeno para producir un haz de luz muy concentrado.

PROM (programmable read only memory) Memory chip that contains specific data which pertains to the exact vehicle in which the computer is installed. This information may be used to inform the CPU of the accessories that are equipped on the vehicle.

PROM (memoria de sólo lectura programable) Fragmento de memoria que contiene datos específicos referentes al vehículo particular en el que se instala la computadora. Esta información puede utilizarse para informar a la UCP sobre los accesorios de los cuales el vehículo está dotado.

Protection device Circuit protector that is designed to turn off the system that it protects. This is done by creating an open to prevent a complete circuit.

Dispositivo de protección Protector de circuito diseñado para "desconectar" el sistema al que provee protección. Esto se hace abriendo el circuito para impedir un circuito completo.

Protocol A language used by computers to communicate with each other over a data bus.

Protocolo Lenguaje que se utiliza en computadoras para comunicarse entre sí sobre un mando de bus.

Proton Positively charged particles contained in the nucleus of an atom.

Protón Partículas con carga positiva que se encuentran en el núcleo de todo átomo.

Proton exchange membrane (PEM) Impedes the oxyhydrogen gas reaction in a fuel cell by ensuring that only protons (H+), and not elemental hydrogen molecules, react with the oxygen.

Membrana de intercambio de protones (MIP) Impide la reacción de gas de oxigeno-hidrógeno en una célula de combustible al asegurar que sólo los protones reaccionen con el oxígeno y no las moléculas de hidrógeno elemental.

Prove-out circuit A function of a circuit that completes the warning light circuit to ground when the ignition switch is in the START position. The warning light will be on during engine cranking to indicate to the driver that the bulb is working properly.

Circuito de prueba Función de un circuito que completa el circuito de la luz de advertencia a tierra cuando el interruptor de encendido se encuentra en la posición ENCENDIDO. La luz de advertencia se encenderá durante el arranque del motor para avisar al conductor que la bombilla funciona correctamente.

Proximity sensor A sensor able to detect the presence of nearby objects without any physical contact.

Sensor de proximidad Un sensor capaz de detectar la presencia de objetos cercanos sin ningún contacto físico.

Pull-down circuit Closes the switch to ground.

Circuito de bajada Cierra el interruptor a tierra.

Pull-down resistor A current-limiting resistor used to assure a proper low-voltage reading by preventing float when the switch is open.

Resistor de bajada Un resistor que limita la corriente y que se utiliza para garantizar una lectura de voltaje bajo adecuada al evitar la flotación cuando el interruptor se encuentra abierto.

Pull-in windings An electrical coil internal to a solenoid that is energized to create a magnetic field used to move the solenoid plunger to the engaged position.

Devanados de puesta en trabajo Una bobina eléctrica que es íntegra a un solenoide que se excita para crear un campo magnético que sirve para mover el relé de solenoide a la posición de engranaje.

Pullout torque The amount of torque required to cause the rotor of an AC motor to be pulled out of sync with the rotating magnetic field.

Par de desenganche Cantidad de torque requerida para que el rotor de un motor de CA pierda el sincronismo con el campo magnético giratorio.

Pull-up circuit Closes the switch to voltage.

Circuito de subida Cierra el interruptor al voltaje.

Pull-up resistor A current-limiting resistor used to assure proper high-voltage reading of a voltage sense circuit by connecting the voltage sense circuit to an electrical potential that can be removed when the switch is closed.

Resistor de subida Un resistor que limita la corriente y que se utiliza para garantizar una lectura de voltaje alto adecuada de un circuito de sensores de voltaje al conectar el circuito de sensores de voltaje a un potencial eléctrico que se puede quitar cuando el interruptor se encuentra cerrado.

Pulse width The length of time in milliseconds that an actuator is energized.

Duración de impulsos Espacio de tiempo en milisegundos en el que se excita un accionador.

Pulse-width modulation (PWM) On/off cycling of a component. The period of time for each cycle does not change; only the amount of on time in each cycle changes.

Modulación de duración de impulsos Modulación de impulsos de un componente. El espacio de tiempo de cada ciclo no varía; lo que varía es la cantidad de trabajo efectivo de cada ciclo.

Pumped Energizing the gain medium to get the atoms into an excited state to supply the energy required for amplification.

Bombeado Energizar el medio de ganancia para que los átomos entren en un estado excitado estado para suministrar la energía necesaria para la amplificación.

Radar A system that uses electromagnetic waves to identify an object's range, direction, or speed.

Radar Sistema que usa ondas electromagnéticas para identificar el rango, la dirección o la velocidad de un objeto.

Radial grid A type of battery grid with vertical branches that extend from a common center.

Rejilla radial Un tipo de rejilla de bateria cuyos diseños extienden de un centro común.

Radio choke Absorbs voltage spikes and prevents static in the vehicle's radio.

Inductor de radio Absorbe los picos de tensión y previene la estática de la radio de un vehículo.

Radio frequency interference (RFI) Radio and television interference caused by electromagnetic energy.

Interferencia de frecuencia radioeléctrica Interferencia en la radio y en la televisión producida por energía electromagnética.

RAM (random access memory) Stores temporary information that can be read from or written to by the CPU. RAM can be designed as volatile or nonvolatile.

RAM (memoria de acceso aleatorio) Almacena datos temporales que la UCP puede leer o escribir. La RAM puede ser volátil o no volátil.

Ratio A mathematical relationship between two or more things.

Razón Una relación matemática entre dos cosas o más.

Reactivity A statement of how easily a substance can cause or be a part of a chemical.

Reactividad Una indicación de cuan fácil una sustancia puede causar o ser parte de una química.

Reader An interface between the transponder and the microprocessor that allows the module to read and process the identifier code from a transponder.

Lector Una interfaz entre el transpondedor y el microprocesador que permite al módulo leer y procesar el código identificador de un transpondedor.

Receiver Receives radio waves from the transmitter and decodes the message from the sine wave.

Receptor El receptor recibe las ondas de radio del transmisor y decodifica el mensaje de la onda senoidal.

Recirc/air inlet door actuator An electric motor that is linked to the recirculation door to provide either outside air or in-vehicle air into the A/C heater case.

Actuador de la compuerta de entrada de recirculación del aire Motor eléctrico que está unido a una compuerta de recirculación para proporcionar ya sea aire de fuera o que provenga del vehículo, dentro de la caja del calentón de corriente alterna.

Recombination battery A type of battery that is sometimes called a dry-cell battery because it does not use a liquid electrolyte solution.

Batería de recombinación Un tipo de batería que a veces se llama una pila seca porque no requiere una solución líquida de electrolita.

Rectification The conversion of AC current to DC current.

Rectificación Proceso a través del cual la corriente alterna es transformada en una corriente continua.

Reflectors A device whose surface reflects or radiates light.

Reflectores Un dispositivo cuyo superficie refleja o irradia la luz.

Reformer A high-temperature device that converts hydrocarbon fuels to CO and hydrogen.

Reformador Dispositivo de alta temperatura que convierte los combustibles de hidrocarburo a monóxido de carbono CO e hidrógeno.

Regenerative braking Braking energy that is converted into electricity instead of heat.

Frenado regenerativo La energía de frenado se convierte nuevamente en electricidad en lugar de calor.

Relay A device that uses low current to control a high-current circuit. Low current is used to energize the electromagnetic coil, while high current is able to pass over the relay contacts.

Relé Dispositivo que utiliza corriente baja para controlar un circuito de corriente alta. La corriente baja se utiliza para excitar la bobina electromagnética, mientras que la corriente alta puede transmitirse a través de los contactos del relé.

Reluctance A term used to indicate a material's resistance to the passage of flux lines.

Reluctancia Término utilizado para señalar la resistencia ofrecida por un circuito al paso del flujo magnético.

Reserve capacity The amount of time a battery can be discharged at a certain current rate until the voltage drops below a specified value.

Capacidad de reserva La cantidad de tiempo que puede descargarse una batería a una cierta proporción de corriente, hasta que el voltaje caiga debajo de un valor especificado.

Reserve-capacity rating An indicator, in minutes, of how long a vehicle can be driven, with the headlights on, if the charging system should fail. The reserve-capacity rating is determined by the length of time, in minutes, that a fully charged battery can be discharged at 25 amps before battery cell voltage drops below 1.75 volts per cell.

Clasificación de capacidad en reserva Indicación, en minutos, de cuánto tiempo un vehículo puede continuar siendo conducido, con los faros delanteros encendidos, en caso de que ocurriese una falla en el sistema de carga. La clasificación de capacidad en reserva se determina por el espacio de tiempo, en minutos, en el que una batería completamente cargada puede descargarse a 25 amperios antes de que la tensión del acumulador de la batería disminuya a un nivel inferior de 1,75 amperios por acumulador.

Resistance Opposition to current flow.

Resistencia Oposición que presenta un conductor al paso de la corriente eléctrica.

Resistor block Consists of two or three helically wound wire resistors wired in series.

Bloque de resistencia Consiste de dos o tres cables helicoilades rostáticos intercalados en serie.

Resistive multiplex switch Provides multiple inputs over a single circuit. Since each switch position has a different resistance value, the voltage drop is different. This means a switch can have one power supply wire and one ground wire instead of a separate wire for each switch position.

Interruptor multiplex resistente Provee múltiples entradas de energía sobre un circuito simple. Debido a que la posición de cada interruptor tiene un valor diferente de resistencia, el voltaje de la caída de tensión es diferente. Esto significa que un interruptor puede tener un cable de abastecimiento de energía y un cable conductor a tierra, sin que sea necesario un cable individual para cada posición del interruptor.

Reverse-biased A positive voltage is applied to the N-type material and a negative voltage is applied to the P-type material of a semiconductor.

Polarización inversa Tensión positiva aplicada al material N y tensión negativa aplicada al material P de un semiconductor.

Rheostat A two-terminal variable resistor used to regulate the strength of an electrical current.

Reóstato Resistor variable de dos bornes utilizado para regular la resistencia de una corriente eléctrica.

Right-hand rule Identifies the direction of the lines of force of an electromagnet.

Regla de la mano derecha Identifica la dirección de las líneas de fuerza de un electroimán.

Ring network A bus communication network that has the modules wired in series, but the ring does not have a beginning or an end. Messages can be transmitted in either direction based on the proximity of the transmitting and receiving modules.

Red en anillo Red de comunicación en bus con módulos conectados en serie en la que el anillo no tiene principio ni fin. Los mensajes pueden transmitirse en cualquier dirección según la proximidad que exista entre los módulos de transmisión y de recepción.

Rolling code A counting system that requires a different signal every transmission to improve security.

Código evolutivo Sistema de conteo que requiere una señal diferente en cada transmisión para mejorar la seguridad.

ROM (read only memory) Memory chip that stores permanent information. This information is used to instruct the computer on what to do in response to input data. The CPU reads the information contained in ROM, but it cannot write to it or change it.

ROM (memoria de sólo lectura) Fragmento de memoria que almacena datos en forma permanente. Dichos datos se utilizan para darle instrucciones a la computadora sobre cómo dirigir la ejecución de una operación de entrada. La UCP lee los datos que contiene la ROM, pero no puede escribir en ella o puede cambiarla.

Rotating magnetic field The magnetic field of the stator windings in an AC motor. The stator is stationary, but the field rotates from pole to pole.

Campo magnético rotativo El campo magnético de los giros de un estator en un motor de CA. El estator permanece fijo, pero el campo rota de polo en polo.

Rotor The component of the AC generator that is rotated by the drive belt and creates the rotating magnetic field of the AC generator.

Rotor Parte rotativa del generador de corriente alterna accionada por la correa de transmisión y que produce el campo magnético rotativo del generador de corriente alterna.

Safety goggles Eye protection device that fits against the face and forehead to seal off the eyes from outside elements.

Gafas de seguridad Dispositivo protector que se coloca delante de los ojos para preservarlos de elementos extraños.

Safety stands Support devices used to hold the vehicle off the floor after it has been raised by the floor jack.

Soportes de seguridad Dispositivos de soporte utilizados para sostener el vehículo sobre el suelo después de haber sido levantado con el gato de pie.

Safing sensor Determines if the collision is severe enough to inflate the air bag.

Monitor de seguridad (safing sensor) Determina si el impacto del choque es lo suficientemente grave para inflar las bolsas de aire.

Satellite radio Provide several music and talk show channels using orbiting satellites to provide a digital signal.

Radio por satélite Sistema que ofrece diversos canales de música y de entrevistas usando satélites en órbita para proporcionar una señal digital.

Saturation 1. The point at which the magnetic strength eventually levels off, and where an additional increase of the magnetizing force current no longer increases the magnetic field strength. 2. The point where forward-bias voltage to the base leg is at a maximum. With bias voltage at the high limits, output current is also at its maximum.

Saturación 1. Máxima potencia posible de un campo magnético, donde un aumento adicional de la corriente de fuerza magnética no logra aumentar la potencia del campo magnético. 2. La tensión de polarización directa a la base está en su máximo. Ya que polarización directa ha alcanzado su límite máximo, la corriente de salida también alcanza éste.

Schematic An electrical diagram that shows how circuits are connected, but not details such as color codes.

Esquemático Diagrama eléctrico que muestra cómo se conectan los circuitos, pero no los detalles tales como las claves por colores.

Schmitt trigger An electronic circuit used to convert analog signals to digital signals or vice versa.

Disparador de Schmitt Un circuito electrónico que se emplea para convertir las señales análogas en señales digitales o vice versa.

Sealed-beam headlight A self-contained glass unit that consists of a filament, an inner reflector, and an outer glass lens.

Faro delantero sellado Unidad de vidrio que contiene un filamento, un reflector interior y una lente exterior de vidrio.

Seat track position sensor (STPS) A sensor that provides information to the occupant classification module (OCM) concerning the position of the seat in relation to the air bag.

Sensor de la posición del carril del asiento (SPCA) Sensor que proporciona información al módulo de control de salida (MCS) referida a la posición del asiento en relación con la bolsa de aire.

Secondary wiring Conductors, such as battery cables and ignition spark plug wires, that are used to carry high voltage or high current. Secondary wires have extra thick insulation.

Hilos secundarios Conductores, tales como cables de batería e hilos de bujías del encendido, utilizados para transmitir tensión o corriente alta. Los hilos secundarios poseen un aislamiento sumamente grueso.

Sector gear The section of gear teeth on the regulator.

Engranaje de cables La sección de los dientes de engranaje en el regulador.

Security Gateway Splits the vehicle's communication networks into two categories: the less secure "public" and a very secure "private" network.

Puerta de seguridad Separa las redes de comunicación de un vehículo en dos categorías: las menos seguras o "públicas" y las muy seguras o "privadas"

Self-induction The generation of an electromotive force by a changing current in the same circuit.

Autoinducción Este es el generamiento de la fuerza electromotriz cuando la corriente cambia en el mismo circuito.

Semiconductor An element that is neither a conductor nor an insulator. Semiconductors are materials that conduct electric current under certain conditions, yet will not conduct under other conditions.

Semiconductor Elemento que no es ni conductor ni aislante. Los semiconductores son materiales que transmiten corriente eléctrica bajo ciertas circunstancias, pero no la transmiten bajo otras.

Sending unit The sensor for the gauge. It is a variable resistor that changes resistance values with changing monitored conditions.

Unidad emisora Sensor para el calibrador. Es un resistor variable que cambia los valores de resistencia según cambian las condiciones reguladas.

Sensing voltage Input voltage to the AC generator.

Detección del voltaje Determina la tensión de entrada de energía al generador de Corriente Alterna AC.

Sensitivity control A potentiometer that allows the driver to adjust the sensitivity of the automatic dimmer system to surrounding ambient light conditions.

Controles de sensibilidad Un potenciómetro que permite que el conductor ajusta la sensibilidad del sistema de intensidad de iluminación automático a las condiciones de luz ambientales.

Sensors Any device that provides an input to the computer.

Sensors Cualquier dispositivo que le transmite información a la computadora.

Sentry key Describes a sophisticated antitheft system that prevents the engine from starting unless a special key is used.

Llave guardiante centinela Describe un sofisticado sistema de guardia anti-robo el que evita prender el motor, a menos que se haga con una llave específicamente creada.

Separators Normally constructed of glass with a resin coating. These battery plates offer low resistance to electrical flow but high resistance to chemical contamination.

Separadores Normalmente se construyen del vidrio con una capa de resina. Estas placas de la batería ofrecen baja resistencia al flujo de la electricidad pero alta resistencia a la contaminación química.

Sequential logic circuits Flip-flop circuits in which the output is determined by the sequence of inputs. A given input affects the output produced by the next input.

Circuitos de lógica secuenciales Cambia los circuitos en los cuales la salida de energía es determinada según la secuencia de las entradas de corriente. Una entrada de energía afecta la salida de corriente que va ser producida en una próxima entrada de energía.

Sequential sampling The process that the MUX and DEMUX operate on. This means the computer will deal with all of the sensors and actuators one at a time.

Muestreo secuencial La forma en que funcionan los sistemas de las abreviaturas (MUX y DEMUX). Esto significa que la computadora se encargará de que todos los monitores y actuadores funcionen uno por uno.

Series circuit A circuit that provides a single path for current flow from the electrical source through all the circuit's components and back to the source.

Circuito en serie Circuito que provee una trayectoria única para el flujo de corriente de la fuente eléctrica a través de todos los componentes del circuito, y de nuevo hacia la fuente.

Series hybrid Hybrid configuration where propulsion comes directly from the electric motor.

Híbrido en serie Configuración del híbrido en donde la propulsión llega directamente del motor eléctrico.

Series-parallel circuit A circuit that has some loads in series and some in parallel.

Circuito en series paralelas Circuito que tiene unas cargas en serie y otras en paralelo.

Series/parallel hybrid A combination of the parallel and the series HEV drive types that incorporates a power-split device that allows a power path from the ICE to the wheels that can be an optimized blend of mechanical and electrical energy.

Híbrido en serie/paralelo Combinación de los tipos de VEH en serie y en paralelo que incorpora un dispositivo de potencia dividida para posibilitar el paso de energía del motor de combustión interna a las ruedas y optimizar la fusión de energía mecánica y eléctrica.

Series-wound motor A type of motor that has its field windings connected in series with the armature. This type of motor develops its maximum torque output at the time of initial start. Torque decreases as motor speed increases.

Motor con devanados en serie Un tipo de motor cuyos devanados inductores se conectan en serie con la armadura. Este tipo de motor desarrolla la salida máxima de par de torsión en el momento inicial de ponerse en marcha. El par de torsión disminuye al aumentar la velocidad del motor.

Servomotor An electrical motor that produces rotation of less than a full turn. A feedback mechanism is used to position itself to the exact degree of rotation required.

Servomotor Motor eléctrico que genera rotación de menos de una revolución completa. Utiliza un mecanismo de realimentación para ubicarse al grado exacto de la rotación requerida.

Shell The electron orbit around the nucleus of an atom.

Corteza Órbita de electrones alrededor del núcleo del átomo.

Short An unwanted electrical path; sometimes this path goes directly to ground.

Corto Una trayectoria eléctrica no deseable; a veces este trayectoria viaja directamente a tierra.

Shorted circuit A circuit that allows current to bypass part of the normal path.

Circuito corto Este circuito permite que la corriente pase por una parte del recorrido normal.

Shunt More than one path for current to flow.

Desviación Más de una derivación para que la corriente pueda fluir.

Shunt circuits The branches of the parallel circuit.

Circuitos en derivación Las ramas del circuito en paralelo.

Shunt-wound motor A type of motor whose field windings are wired in parallel to the armature. This type of motor does not decrease its torque as speed increases.

Motor con devanados en derivación Un tipo de motor cuyos devanados inductores se cablean paralelos a la armadura. Este tipo de motor no disminuya su par de torsión al aumentar la velocidad.

Shutter wheel A metal wheel consisting of a series of alternating windows and vanes. It creates a magnetic shunt that changes the strength of the magnetic field from the permanent magnet of the Hall-effect switch or magnetic pulse generator.

Rueda obturadora Rueda metálica compuesta de una serie de ventanas y aspas alternas. Genera una derivación magnética que cambia la potencia del campo magnético, del imán permanente del conmutador de efecto Hall o del generador de impulsos magnéticos.

Sine wave A waveform that shows voltage changing polarity.

Onda senoidal Una forma de onda que muestra un cambio de polaridad en la tensión.

Single 42-volt system An electrical system that uses 42 volts for all circuit operations including starter, lighting, and accessories.

Sistema único de 42 voltios Un sistema eléctrico que utiliza 42 voltios para todas las operaciones del circuito, incluyendo el arrancador, la iluminación y los accesorios.

Single-phase voltage The sine wave voltage induced in one conductor of the stator during one revolution of the rotor.

Tensión monofásica La tensión en forma de onda senoidal inducida en un conductor del estator durante una revolución del rotor.

Slave modules Controllers on the network that must communicate through a master controller.

Módulos esclavos Controladores de la red que deben comunicarse por medio de un control maestro.

Slip The difference between the synchronous speed and actual rotor that is directly proportional to the load on the motor.

Slip La diferencia entre la velocidad síncrona y rotor real es directamente proporcional a la carga en el motor.

Slip rings Rings that function much like the armature commutator in the starter motor or generator; however, they are smooth.

Anillos colectores Anillos que funcionan de manera muy similar al conmutador inducido en el arranque del motor o generador, con la diferencia de que estos anillos son lisos.

Smart sensors Sensors that are capable of sending digital messages on the data bus.

Sensor inteligente Sensores capaces de enviar mensajes digitales en el bus de datos.

Solenoid An electromagnetic device that uses movement of a plunger to exert a pulling or holding force.

Solenoide Dispositivo electromagnético que utiliza el movimiento de un pulsador para ejercer una fuerza de arrastre o de retención.

Sound generator *See* buzzer.

Generador de sonido Vea "timbre."

Source The portion of a field-effect transistor that supplies the current-carrying holes or electrons.

Fuente Terminal de un transistor de efecto de campo que provee los agujeros o electrones portadores de corriente.

Spark plug Electrodes provide gaps inside each combustion chamber across which the secondary current flows to ignite the air–fuel mixture in the combustion chambers.

Enchufe de chispa Electrodos para proveer intervalos dentro de cada cámara de combustión a través de la corriente secundaria que fluye para encender la mezcla de aire/combustible en las cámaras de combustión.

Speakers Convert electrical energy provided by the radio receiver amplifier into acoustical energy using a permanent magnet and an electromagnet to move air to produce sound.

Altavoces Convierten la energía eléctrica que suministra el amplificador receptor de radio en energía acústica mediante el uso de un imán permanente y un electroimán, los cuales mueven el aire para producir sonido.

Specific gravity The weight of a given volume of a liquid divided by the weight of an equal volume of water.

Gravedad específica El peso de un volumen dado de líquido dividido por el peso de un volumen igual de agua.

Splice A common connection point for the electrical circuit; used to eliminate the need for multiple connection points.

Empalme Un punto de conexión común para el circuito eléctrico; se utiliza para eliminar la necesidad de varios puntos de conexión.

Squib A pyrotechnic term used for a firecracker that burns but does not explode. The squib starts the process of air bag deployment.

Mecha Un término pirotécnico usado para prender una pólvora que se quema pero no explota. La mecha inicia el proceso de la salida de la bolsa de aire.

Star network Bus communication network that connects all modules to a central point in a parallel circuit.

Red en estrella Red de comunicación en bus que conecta todos los módulos con un punto central en un circuito paralelo.

Start/clutch interlock switch Used on vehicles equipped with manual transmissions that only allows the starter to be engaged if the clutch is disengaged.

Interruptor de seguridad de embrague Usado en los vehículos equipados con transmisiones manuales que permite que el arranque se accione si el embrague no está accionado.

Starter drive The part of the starter motor that engages the armature to the engine flywheel ring gear.

Transmisión de arranque Parte del motor de arranque que engrana la armadura a la corona del volante de la máquina.

Starter generator control module (SGCM) Also called a high-voltage ECU (HV ECU). Used to control the flow of torque and electrical energy into and from the motor generator of the HEV.

Módulo de control del generador de encendido (MCGE) También se le llama UCE de alto voltaje (UCE AV). Se utiliza para controlar el flujo del par motor y la energía eléctrica dentro y fuera del generador del motor del VHE.

Starting system A combination of mechanical and electrical parts that work together to start the engine by changing the electrical energy that is being supplied by the battery into mechanical energy by use of a starter or cranking motor.

Sistema de encendido Combinación de partes mecánicas y eléctricas que trabajan unidas para encender el motor al cargar la energía eléctrica que proporciona la batería, en energía mecánica mediante el uso de un encendedor o motor de arranque.

State of charge The condition of a battery's electrolyte and plate materials at any given time.

Estado de carga Condición del electrolito y de los materiales de la placa de una batería en cualquier momento dado.

Static electricity Electricity that is not in motion.

Electricidad estática Electricidad que no está en movimiento.

Static neutral point The point at which the fields of a motor are in balance.

Punto neutral estático El punto en que los campos de un motor estan equilibrados.

Stator The stationary coil of the AC generator where current is produced.

Estátor Bobina fija del generador de corriente alterna donde se genera corriente.

Stator neutral junction The common junction of Wye stator windings.

Unión de estátor neutral La unión común de los devanados de un estátor Y.

Stepped resistor A resistor that has two or more fixed resistor values.

Resistor de secciones escalonadas Resistor que tiene dos o más valores de resistencia fija.

Stepper motor An electrical motor that contains a permanent magnet armature with two or four field coils; can be used to move the controlled device to the desired location. By applying voltage pulses to selected coils of the motor, the armature will turn a specific number of degrees. When the same voltage pulses are applied to the opposite coils, the armature will rotate the same number of degrees in the opposite direction.

Motor de pasos Contiene una armadura magnética permanente con dos, cuatro o más bobinas del campo.

Steering Angle Sensor (SAS) Monitors actual steering wheel position to determine the driver's intended path of travel.

Sensor del ángulo de dirección (SAD) Controla la posición real del volante de dirección para determinar el camino que pretende hacer el conductor.

Stimulated emission Develops the properties of the laser light that results in the laser light atoms releasing photons in a very organized state.

Emisión estimulada Desarrolla las propiedades de la luz láser de manera tal que los átomos de la luz láser liberan fotones en un estado muy organizado.

Stop/Start Feature that shuts off the engine when the vehicle is not moving or when power from the engine is not required. Once the driver's foot is removed from the brake pedal (or the clutch is engaged), the starter automatically restarts the engine.

Detención/Arranque Función que detiene el motor cuando el vehículo no está en movimiento o cuando no se requiere la potencia del motor. Cuando el conductor levanta el pie del pedal de freno (o cuando se acciona el embrague) el arranque reinicia automáticamente el motor.

Strain gauge A sensor that determines the amount of applied pressure by measuring the strain a material experiences when subjected to pressure.

Galga extensiométrica Un sensor que determina la cantidad de presión aplicada medir la tensión que experimenta un material cuando se somete a presión.

Extensómetro de resistencia eléctrica Un sensor que determina la cantidad de presión aplicada al medir la resistencia eléctrica que experimenta un material cuando se lo somete a presión.

Stranded wire A conductor comprised of many small solid wires twisted together. This type conductor is used to allow the wire to flex without breaking.

Cable trenzado Un conductor que comprende muchos cables sólidos pequeños trenzados. Este tipo de conductor se emplea para permitir que el cable se tuerza sin quebrar.

Stub network Bus communication network that connects the modules through a parallel circuit layout.

Red aislada Red de comunicación en bus que conecta los módulos mediante una configuración de circuito paralelo.

Sulfation A condition in a battery that reduces its output. The sulfate in the battery that is not converted tends to harden on the plates, resulting in permanent damage to the battery.

Sulfatación Una condición en una batería que disminuya su potencia de salida. El sulfato en la batería que no se convierte suele endurecerse en las placas y resulta en daños permanentes en la batería.

Supplemental bus networks Bus networks that are on the vehicle in addition to the main bus network.

Redes de bus complementarios Redes de bus que hay en el vehículo aparte de la red del bus principal.

Synchronous motor Type of AC motor that operates at a constant speed regardless of load. The rotor always rotates at the speed of the rotating magnetic field.

Motor sincrónico Tipo de motor de CA que opera a velocidad constante sin importar la carga. El rotor siempre gira a la velocidad del campo magnético rotativo.

Synchronous speed The speed at which the magnetic field of the stator rotates in an AC motor.

Velocidad sincrónica La velocidad a la cual el campo magnético del estator gira en un motor de CA.

Telematics A system that encompasses telecommunications, instrumentation, wireless communications, road transportation, multimedia, and internet connectivity.

Telemática Sistema que abarca telecomunicaciones, instrumentación, comunicaciones inalámbricas, transporte por carretera, multimedia y conectividad a internet.

Termination resistors Used to control induced voltages. Since voltage is dropped over resistors, the induced voltage is terminated.

Resistores de terminación Usados para controlar la conducción de los voltajes. Como el voltaje cae sobre los resistores, es así como se determina el voltaje.

Thermistor A solid-state variable resistor made from a semiconductor material that changes resistance in relation to temperature changes.

Termistor Resistor variable de estado sólido hecho de un material semiconductor que cambia su resistencia en relación con los cambios de temperatura.

Three-coil gauge A gauge design that uses the interaction of three electromagnets and the total field effect upon a permanent magnet to cause needle movement.

Calibrador de tres bobinas Calibrador diseñado para utilizar la interacción de tres electroimanes y el efecto inductor total sobre un imán permanente para producir el movimiento de la aguja.

Throw Term used in reference to electrical switches or relays referring to the number of output circuits from the switch.

Posición activa Término utilizado para conmutadores o relés eléctricos en relación con el número de circuitos de salida del conmutador.

Thyristor A semiconductor switching device composed of alternating N and P layers. It can also be used to rectify current from AC to DC.

Tiristor Dispositivo de conmutación del semiconductor compuesto de capas alternas de N y P. Puede utilizarse también para rectificar la corriente de corriente alterna a corriente continua.

Timer circuit Uses a bimetallic strip that opens as a result of the heat being generated by the current flow.

Circuito Sincronizador Consisite de una cinta bimetálica la que se abre, debido al calor generado por el flujo de corriente.

Timer control A potentiometer that is part of the headlight switch in some systems. It controls the amount of time the headlights stay on after the ignition switch is turned off.

Control temporizador Un potenciómetro que es parte del conmutador de los faros en algunos sistemas. Controla la cantidad del tiempo que quedan prendidos los faros después de apagarse la llave del encendido.

Timing disc Known as armature, reluctor, trigger wheel, pulse wheel, or timing core. It is used to conduct lines of magnetic force.

Disco medidor de tiempo Se conoce como una armadura de inducción, rueda disparadora, rueda de pulsación o un núcleo de tiempo. Este disco es usado para conducir líneas de fuerza magnética.

Tire pressure monitoring (TPM) system A safety system that notifies the driver if one or more tires are underinflated or overinflated.

Sistema de monitoreo de presión de las llantas (MPLL) Sistema de seguridad que le hace saber al conductor si una llanta o más llantas están desinfladas o sobre infladas.

Torque converter A hydraulic device found on automatic transmissions. It is responsible for controlling the power flow from the engine to the transmission; works like a clutch to engage and disengage the engine's power to the drive line.

Convertidor de par Un dispositivo hidraúlico en las transmisiones automáticas. Se encarga de controlar el flujo de la potencia del motor a la transmisión; funciona como un embrague para embragar y desembragar la potencia del motor con la flecha motríz.

Total reflection A phenomenon wherein a light wave reflects off the surface 100% when the light wave advances from a medium of high index of refraction to a medium of low index of refraction.

Reflexión total Fenómeno en el que una onda de luz se refleja 100% de una superficie cuando la onda de luz avanza de un índice medio de refracción de uno alto a uno índice medio de refracción de uno bajo.

Tracer A thin or dashed line of a different color than the base color of the insulation.

Traza líneas Una línea delgada o instrumento de color diferente al color básico de la insulación.

Trailing edge In a Hall-effect switch, the edges of the rotating blade that exit the switch.

Borde de salida Un interruptor de efecto Hall indica los bordes de la paleta giratoria que sale del interruptor.

Transducer A device that changes energy from one form into another.

Transductor Dispositivo que cambia la energía de una forma a otra.

Transistor A three-layer semiconductor used as a very fast switching device.

Transistor Semiconductor de tres capas utilizado como dispositivo de conmutación sumamente rápido.

Transmitter Generates a radio frequency by taking data and encoding it onto a sine wave that is transmitted to a receiver.

Transmisor Genera una radiofrecuencia al tomar los datos y codificarlos en una onda senoidal que se transmite a un receptor.

Trimotor A three-armature motor.

Trimotor Motor de tres armaduras.

Turn-on voltage The voltage required to jump the PN junction and allow current to flow.

Voltaje de conección El voltaje requerido para hacer funcionar el cable de empalme PN y permitir que la corriente fluya.

TVRS An abbreviation for television-radio-suppression cable.

TVRS Una abreviación del cable de supresíon del televisión y radio.

Tweeters Smaller speakers that produce the high frequencies of treble.

Tweeters Pequeñas bocinas que producen altas frecuencias agudas.

Two-coil gauge A gauge design that uses the interaction of two electromagnets and the total field effect upon an armature to cause needle movement.

Calibrador de dos bobinas Calibrador diseñado para utilizar la interacción de dos electroimanes y el efecto inductor total sobre una armadura para generar el movimiento de la aguja.

Ultra-capacitor A device that stores energy as electrostatic charge. It is the primary device in the power supply during hill climbing, acceleration, and the recovery of braking energy.

Ultra capacitor Dispositivo que guarda energía como carga electrostática. Es el dispositivo primario en la fuente de energía durante una subida, una aceleración y el recobro de la energía de frenado.

Ultrasonic sensors A proximity detector that measures the distances to nearby objects using acoustic pulses and a control unit that measures the return interval of each reflected signal and calculates the object's distance.

Sensores ultrasónicos Detector de proximidad que mide las distancias a objetos cercanos usando pulsos acústicos y una unidad de control que mide el intervalo de retorno de cada señal reflejada y calcula la distancia del objeto.

Ultrasound A cycling sound pressure wave that is at a frequency greater than the upper limit of human hearing (about 20 kHz).

Ultrasonido Una onda de presión de sonido cíclico que está en una frecuencia mayor que el límite superior de la audición humana (aproximadamente 20kHz).

Vacuum distribution valve A valve used in vacuum-controlled concealed headlight systems. It controls the direction of vacuum to various vacuum motors or to vent.

Válvula de distribución al vacío Válvula utilizada en el sistema de faros delanteros ocultos controlado al vacío. Regula la dirección del vacío a varios motores al vacío o sirve para dar salida del sistema.

Vacuum fluorescent display (VFD) A display type that uses anode segments coated with phosphor and bombarded with tungsten electrons to cause the segments to glow.

Visualización de fluorescencia al vacío Tipo de visualización que utiliza segmentos ánodos cubiertos de fósforo y bombardeados de electrones de tungsteno para producir la luminiscencia de los segmentos.

Valence ring The outermost orbit of the atom.

Anillo de valencia Órbita más exterior del átomo.

Valve body A unit that consists of many valves and hydraulic circuits. This unit is the central control point for gear shifting in an automatic transmission.

Cuerpo de la válvula Una unedad que consiste de muchas válvulas y circuitos hidráulicos. Esta unedad es el punto central de mando para los cambios de velocidad en una transmisión automática.

Valve-Regulated Lead–Acid (VRLA) battery Another variation of the recombination battery that uses lead–acid. The oxygen produced on the positive plates is absorbed by the negative plate causing a decrease in the amount of hydrogen produced at the negative plate which is then combined with the oxygen to produce water.

Batería de plomo-ácido de válvula regulada (VRLA, por su sigla en inglés) Otra variante de la batería de recombinación que utiliza plomo-ácido. El oxígeno que se produce en las placas positivas es absorbido por la placa negativa, lo que causa una disminución en la cantidad de hidrógeno que se produce en la placa negativa y que luego se combina con el oxígeno para formar agua.

Vaporized aluminum The process of applying a very thin layer of aluminum onto a substrate and then protected by a coating of silica to give the headlight a reflecting surface that is comparable to silver.

Aluminio vaporizado Proceso por el cual se aplica una capa muy delgada de aluminio a un sustrato y luego se lo protege con una capa de sílice para dar al faro una superficie brillante comparable con la plata.

Variable resistor A resistor that provides for an infinite number of resistance values within a range.

Resistor variable Resistor que provee un número infinito de valores de resistencia dentro de un margen.

Vehicle instrumentation systems A system that monitors the various vehicle operating systems and provides information to the driver of their correct operation.

Sistemas de instrumentación del vehículo Sistema que monitorea varios sistemas de operación del vehículo y proporciona información de su operación correcta al conductor.

Volatile Easily vaporizes or explodes.

Volátil Vaporiza o explota fácilmente.

Volatile RAM memory that is erased when it is disconnected from its power source. Also known as Keep Alive Memory.

Volátil Memoria RAM cuyos datos se perderán cuando se la desconecta de la fuente de alimentación. Conocida también como memoria de entretenimiento.

Volt The unit used to measure the amount of electrical force.

Voltio Unidad práctica de tensión para medir la cantidad de fuerza eléctrica.

Voltage The difference or potential that indicates an excess of electrons at the end of the circuit the farthest from the electromotive force. It is the electrical pressure that causes electrons to move through a circuit. One volt is the amount of pressure required to move 1 amp of current through 1 ohm of resistance.

Tensión Diferencia o potencial que indica un exceso de electrones al punto del circuito que se encuentra más alejado de la fuerza electromotriz. La presión eléctrica genera el movimiento de electrones a través de un circuito. Un voltio equivale a la cantidad de presión requerida para mover un amperio de corriente a través de un ohmio de resistencia.

Voltage drop A resistance in the circuit that reduces the electrical pressure available after the resistance. The resistance can be the load component, conductors, any connections, or unwanted resistance.

Caída de tensión Resistencia en el circuito que disminuye la presión eléctrica disponible después de la resistencia. La resistencia puede ser el componente de carga, los conductores, cualquier conexión o resistencia no deseada.

Voltage limiter Connected through the resistor network of a voltage regulator. It determines whether the field will receive high, low, or no voltage. It controls the field voltage for the required amount of charging.

Limitador de tensión Conectado por el red de resistores de un regulador de tensión. Determina si el campo recibirá alta, baja o ninguna tensión. Controla la tensión de campo durante el tiempo indicado de carga.

Voltage regulator Used to control the output voltage of the AC generator, based on charging system demands, by controlling field current.

Regulador de tensión Dispositivo cuya función es mantener la tensión de salida del generador de corriente alterna, de acuerdo a las variaciones en la corriente de carga, controlando la corriente inductora.

Voltmeter A test meter used to read the pressure behind the flow of electrons.

Voltímetro Instrumento de prueba utilizado para medir la presión del flujo de electrones.

Wake-up signal An input signal used to notify the body computer that an engine start and operation of accessories is going to be initiated soon. This signal is used to warm up the circuits that will be processing information.

Señal despertadora Señal de entrada para avisarle a la computadora del vehículo que el arranque del motor y el funcionamiento de los accesorios se iniciarán dentro de poco. Dicha señal se utiliza para calentar los circuitos que procesarán los datos.

Warning lamp A lamp that is illuminated to warn the driver of a possible problem or hazardous condition.

Luz de aviso Lámpara que se enciende para avisarle al conductor sobre posibles problemas o condiciones peligrosas.

Watchdog circuit Supplies a reset voltage to the microprocessor in the event that pulsating output voltages from the microprocessor are interrupted.

Circuito de vigilancia Proporciona un voltaje de reposición para el microprocesador en caso de que se interrumpan los voltajes de potencia útil de pulsaciones del microprocesador.

Watt The unit of measure of electrical power, which is the equivalent of horsepower. One horsepower is equal to 746 watts.

Watio Unidad de potencia eléctrica, equivalente a un caballo de vapor. 746 watios equivalen a un caballo de vapor (CV).

Wattage A measure of the total electrical work being performed per unit of time.

Vataje Medida del trabajo eléctrico total realizado por unidad de tiempo.

Watt-hour rating Equals the battery voltage times ampere-hour rating.

Contador de voltaje-hora nominal Su función es igualar el tiempo del voltaje de la batería con el voltaje-hora nominal.

Weather-pack connector A type of connector that seals the terminal's ends. This type connector is used in electronic circuits.

Conectador impermeable Un tipo de conectador que sella las extremidades de los terminales. Este tipo de conectador se emplea en los circuitos electrónicos.

Wheatstone bridge A series–parallel arrangement of resistors between an input terminal and ground. Flexing of the disc on which the resistors are laid changes their value.

Puente de Wheatstone Conjunto de resistores en series paralelas entre un borne de entrada y la conexión a tierra. La flexión del disco sobre el que se apoyan los resistores cambia su valor.

Wi-Fi A wireless local area networking technology based on the IEEE 802.11 standards.

Wi-Fi Tecnología de red local inalámbrica basada en los estándares IEEE 802.11.

Window regulator Converts rotary motion of the motor into the vertical movement of the window.

Regulador del vidrio parabrisas Convierte la acción rotatoria del motor en un movimiento vertical sobre el vidrio parabrisas.

Wiper Moveable contact of a variable resistor.

Limpiador Consiste de un movimiento deslizable de resistores variables.

Wireless networks Connection of modules together to transmit information without the use of physical connection by wires.

Redes inalámbricas Conexión entre los módulos para transmitir información sin el uso de conexiones físicas por medio de alambres.

Wiring diagram An electrical schematic that shows a representation of actual electrical or electronic components and the wiring of the vehicle's electrical systems.

Esquema de conexiones Esquema en el que se muestran las conexiones internas de los componentes eléctricos o electrónicos reales y las de los sistemas eléctricos del vehículo.

Wiring harness A group of wires enclosed in a conduit and routed to specific areas of the vehicle.

Cableado preformado Conjunto de alambres envueltos en un conducto y dirigidos hacia áreas específicas del vehículo.

Woofers Large speakers that produce low frequencies of midrange and bass.

Woofers Bocinas grandes que producen bajas frecuencias de media distancia y bajo.

Worm gear A type of gear whose teeth wrap around the shaft. The action of the gear is much like that of a threaded bolt or screw.

Engranaje de tornillo sin fin Un tipo de engranaje cuyos dientes se envuelven alrededor del vástago. El movimiento del engranaje es muy parecido a un perno enroscado o una tuerca.

Wye-wound connection A type of stator winding in which one end of the individual windings is connected at a common point. The structure resembles the letter Y.

Conexión Y Un tipo de devanado estátor en el cual una extremidad de los devanados individuales se conectan en un punto común. La estructura parece la letra "Y."

Yaw The tendency of a vehicle to rotate around its center of gravity.

Giro longitudinal Tendencia de un vehículo a girar sobre su propio eje de gravedad.

Y-type stator A three-winding AC generator that has one end of each winding connected at the neutral junction.

Estátor de tipo Y Generador de corriente alterna de devanado triple; un extremo de cada devanado se conecta al empalme neutro.

Zener diode A diode that allows reverse current to flow above a set voltage limit.

Diodo Zener Diodo que permite que el flujo de corriente en dirección inversa sobrepase el límite de tensión determinado.

Zener voltage The voltage that is reached when a diode conducts in reverse direction.

Tensión de Zener Tensión alcanzada cuando un diodo conduce en una dirección inversa.

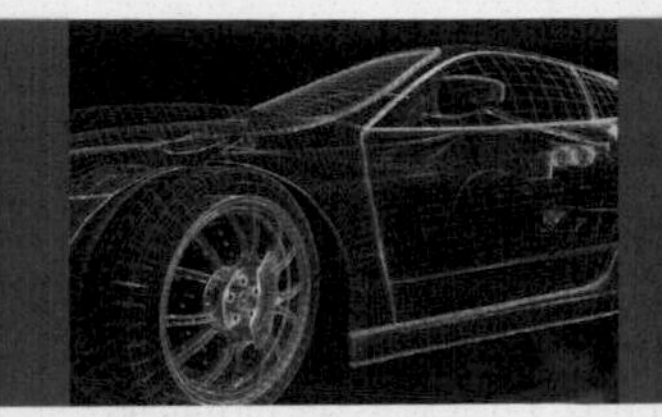

Index

A

B

C

D

J

K

L

O

P

T